Kelker/Hatz

Handbook of Liquid Crystals

Hans Kelker/Rolf Hatz

Handbook of Liquid Crystals

With a Contribution by

Christian Schumann

Verlag Chemie
Weinheim · Deerfield Beach, Florida · Basel · 1980

Prof. Dr. Hans Kelker
Dr. Rolf Hatz
Dr. Christian Schumann
Hoechst AG
Postfach 800320
D-6230 Frankfurt/M. 80
Germany

Copy Editing: Dr. Gerd Giesler

This book contains 438 figures and 48 tables.

CIP-Kurztitelaufnahme der Deutschen Bibliothek

Kelker, Hans
Handbook of liquid crystals / Hans Kelker; Rolf Hatz. – Weinheim, Deerfield: Verlag Chemie, 1980.
 ISBN 3-527-25481-1 (Weinheim)
 ISBN 0-89573-008-1 (Deerfield)
NE: Hatz, Rolf

Printer: Zechnersche Buchdruckerei, D-6720 Speyer. Bookbinder: Klambt-Druck GmbH, D-6720 Speyer
Printed in West Germany

This book is dedicated to

Professor Dr. phil. nat. Wilhelm Kast

who has performed so much vital research
in the field of "liquid crystals"

Preface

Twelve years ago we were requested by Verlag Chemie to write a short monograph on liquid crystals. Encouraged by the research director of Hoechst AG, where we are employed as analytical chemists, we hoped to realize this project in the form of a two hundred page volume. Some time after Weygand's monograph and Gray's book, there appeared to be a real need for a revised text, and the first international liquid crystal conference (1965) in Kent, Ohio, had forecast the first possibilities of technical applications.

We were encouraged by the memory of Wilhelm Maier and our admiration for Wilhelm Kast and Horst Sackmann, who continued traditional work after the second world war in Germany at Freiburg and Halle. We derived enthusiasm from the subject matter itself, from its fruitful investigations in the past, from enticing and interesting questions touching upon an extensive area of science, and from the possibilities of technical applications. This book aimed to be both a contribution to the revival of liquid crystal research at that time and a valuable enrichment of our own analytical activities especially concerning the development of separation methods, spectroscopic techniques, and structure determination.

Our intention was to recapitulate all the results and merits of the previous generation once more but within the framework of a single book, avoiding all attempts at indoctrination. We soon found that we had to cope with an exponentially increasing flood of knowledge, facts, theories, and suggested practical applications. Computer-assisted literature searches in this field began in the mid-seventies and a number of books appeared which elucidated various aspects of the liquid-crystalline state. Considerable efforts had to be invested in the documentation of the literature. Two important factors in our assessment of the situation were the lack of a book giving a comprehensive survey of the available literature and the foreseeable impossibility of writing such a book in a few years time due to the high rate of progress in this field.

This progress is evident in the multitude of publications issued yearly and an annually increasing production of millions of liquid crystal display devices. Other possibilities of perhaps greater importance make use of liquid crystals for the storage of visual information, a flat television screen, and thermographic detectors for medicinal diagnostics and nondestructive testing of materials. It is impossible to present the scientific importance of liquid crystals in a few words and we only mention the difficulties in defining this intermediary state of matter between true liquids and solid crystals. There may be continuous transitions, and the field of liquid crystals is also closely related to living systems. Much is still under discussion and some of the problems could even be extended to philosophical considerations made already by O. Lehmann, who earned the violent criticism of his contemporaries. The ambition to pursue all ramifications of the field of liquid crystals would lead to boundless deviation.

Nevertheless, we persevered trying to arrive at a reasonable compromise with respect to what was feasible and what was really desirable. "Handbook of Liquid Crystals" is the result of our endeavors.

Because particular emphasis has been placed on the bibliography, the references include the titles of the publications cited (often in abbreviated form) and the reference to Chemical Abstracts. The transliterations of names in Russian Cyrillic and other characters are also taken from Chemical Abstracts, as are the English translations of all publications which did not appear in English, French, or German. The bibliography also contains references, of which only the titles are known and which do not provide sufficient information for assignment to a particular part of the text. Moreover, the volume of literature is so overwhelming and widely scattered that detailed examination and evaluation of all references turned out to be impossible. For these reasons, this work will not stand up to close scrutiny with regard to absolute comprehensiveness, and it may also fail to satisfy some other expectations; nevertheless, it will have fulfilled its main purpose if it succeeds as a convenient introduction to the field and as a guide to the literature.

This volume, however, endeavors to document completely the literature up to 1976, and it includes most of the 1977 publications. It also cites older work awaiting reinvestigation and could therefore stimulate further research.

A minute subdivision of the material should assist in providing a comprehensive survey and facilitate searches for detailed information. In addition, many cross-references and a detailed subject index provide ready access to information contained in other sections and chapters. Cross-references are cited in one chapter but itemized in the list of references belonging to another, and are given by a combination of two numbers separated by a colon. The first number refers to the chapter; for example, [8:168] means reference 168 in chapter 8. References are otherwise given as a single number and will be found in the list of references belonging to the chapter in question.

We gratefully acknowledge the extensive support of the research directors of Hoechst AG during the preparation of this book. In particular, we thank Dr. Ch. Schumann, Hoechst AG, who wrote chapter 10 on NMR and ESR spectroscopy. We are also indebted to Dr. E. Paulus, Dr. E. Perplies, and Dr. G. Baur for valuable suggestions and contributions, to Mrs. A. Geisler for typing the manuscript, and to our colleagues in the literature department of Hoechst AG for their unstinting cooperation. Thanks also go to J. T. Burridge for translating the German text, W. J. Steele and especially R. E. Goozner for correcting the manuscript, W. Girbig for drawing many figures, and Dr. G. Giesler of Verlag Chemie for his editorial support.

Contents – a Summary

Contents

Microscopic Investigations Including Basic Concepts, Phenomenology, and Morphology

1

1 Microscopic Investigations Including Basic Concepts, Phenomenology, and Morphology

1.1 Introduction

The terms liquid crystals, crystalline liquid, mesophase, and mesomorphous state are used synonymously to describe a state of aggregation that exhibits a molecular order in a size range similar to that of a crystal but acts more or less as a viscous liquid. The traditional use of these admittedly contradictory terms are continued here because none better have been suggested, although there have been heated discussions on the topic in the past. A substance that forms a mesophase after melting or dissolving is called a mesogen. When the type of mesophase formed is known, a more precise terminology can be applied, e.g. nematogen, smectogen, or cholesteric nematogen.

A careful optical investigation of morphology and of typical structural characteristics is required for a complete description of the liquid-crystalline state. The phenomena resulting from the anisotropy of liquid crystals are so diverse as to warrant a chapter to themselves, in which we will discuss the observation of liquid crystals under the polarizing microscope.

The basic information in structure investigations is the description of the texture, the term being defined as the sum of topological elements which are large enough to be observed under the polarizing microscope. For clarity, we limit this concept to optical phenomena (e. g. homogeneous, homeotropic and focal-conic). All textures depend on typical molecular short range order that is in turn dependent on molecular structure. Thus, texture is the liquid-crystal analog to "morphology" in solid crystals. Whereas different short range order corresponds to thermodynamically different phases (whose transitions can be of different order), one and the same phase can exist in many different textures depending on geometrical, mechanical, magnetic, electric and any other boundary conditions. Certain textures, called type-textures, are characteristic of a given microstructure. There are other textures that do not correspond to a specific microstructure. We cannot overemphasize the dependency of liquid-crystal morphology on external influences such as boundary surfaces, sample thickness, purity, the effects of electric or magnetic fields, and, last but not least, the thermal and temporal pretreatment.

As introduction to polarizing microscopy, four of the most useful and best known monographs are cited: the classic standard work of F. Pockels [281] and, for more advanced study, the writing of C. Burri [25], N. H. Hartshorne et al. [103, 105], F. Rinne and M. Berek [302], and, in addition, W. Voigt's "Kristall-Physik" is the standard volume on crystal physics other than optics [381]. We also refer to some books dealing with subjects of related interest [16, 79, 120, 124, 299, 363].

The historical development began with morphology. The honor of discovery and for the pioneering work goes principally to O. Lehmann [128–262], F. Reinitzer [300, 301], L. Gattermann [70–72], D. Vorländer [384–390, 394], and R. Schenck [340–349]. There was a long and fruitless argument concerning the right of discovery, but both O. Lehmann and F. Reinitzer contributed considerable amounts of knowledge deserving recognition. O. Lehmann (1855–1922) was the first to recognize the importance of polarizing microscopy, beginning in 1877 with his dissertation "Über Physikalische Isomerie", in which he thoroughly described his observations on "plastic crystals", especially those on silver iodide [128]. Lehmann has constructed special polarizing microscopes with a hot stage and projection apparatus. He called this type of instrument a "crystallization microscope". His first important monograph appeared in 1904 [145]. The botanist F. Reinitzer discovered in 1888 new phenomena exhibited by cholesteryl acetate and, more pronounced, by cholesteryl benzoate [300]. These substances exhibited iridescent colors in certain temperature ranges which are characteristic of a state that was to become later known as liquid-crystalline. Reinitzer's description, which was correct in all details, included the melting point (146.6 °C) and clearing point (180.6 °C) for the benzoate. These values are remarkable in that they vary by only about 1 °C from the values accepted today. Similar color phenomena which may have been based on liquid-crystal observations had been mentioned even earlier, but it is difficult to separate reality from fiction. For instance in chapter 19 of his fictional work "The Narrative of Arthur Gordon Pym" (1838) Edgar Allen Poe describes a miraculous water that could be a liquid crystal [282].

R. Virchow [380] and C. Mettenheimer [275] were the first to report a birefringent body material, myelin. F. Wallerant performed the first French research on liquid crystals, those of cholesteryl esters [391] and p-azoxyanisole [392], P. Gaubert investigated ergosteryl esters [73]. Thanks to the work of O. Lehmann [145], F. Reinitzer [300], R. Schenck [340–348] (and especially [349]), and L. Gattermann [70–72], the concept of liquid crystals was, at least in outline form, clearly defined by the turn of the century. Soon thereafter, D. Vorländer recognized the principles of molecular structure leading to mesogens [389], see chapter 2.

The accepted classification and terminology of liquid crystals is based almost entirely on G. Friedel's observations with the polarizing microscope [54–66]. His pioneering work "Les Etats Mésomorphes de la Matière" contains 32 typical microphotographs and additionally introduces and defines the terms used today to describe the liquid-crystalline state: smectic, nematic, and nematic-cholesteric [62]. More recently H. Sackmann et al. thoroughly investigated liquid-crystal textures and phase diagrams, especially with regard to the iso- and polymorphism of smectic phases [305–337]. H. Sackmann's monograph is of special interest for its compilation of data characterizing smectic phases [321]. There is also an index of publications by the Halle-team [298] and a study on the effect of deuteration on the mesomorphism of carboxylic acids [41].

G. W. Gray's book (1962), dealing especially with the relationships between the mesomorphous state and chemical structure has been indispensable [90]. It continues the work begun by D. Vorländer [389] and C. Weygand [393]. S. Chandrasekhar's book [25a] gives the most recent treatment of the physics of liquid crystals. This theme has already been represented by P. G. de Gennes in an outstanding monograph [3:176]. The reviews by G. H. Brown and W. Shaw [21] and by V. A. Usol'tseva and I. G. Chistyakov [30] summarize a large amount of physical data. P. Châtelain (Montpellier) has produced an instructive film on the optical behavior of liquid crystals [29], as already O. Lehmann has done [258]. Recently Y. Bouligand presented a further film on this subject [20]. For introduction the reader can be referred to the reviews given in refs. [36, 92, 96, 119, 339]. The most comprehensive treatises which represent a synopsis of detailed lectures written by authorities in the different fields were edited by G. W. Gray and P. A. Winsor [93], G. H. Brown [22], and E. B. Priestley et al. [291].

A microfilm bibliography compiling the literature on liquid crystals up to September 1975 is available from Eastman Kodak [50]. Finally we mention the related group of plastic crystals [373], cf. chapter 8.

The required degree of assurance in ascribing the diversity of textures involved is attained only from repeated personal observations and comparison to a great number of standards. To this end, a collection of mesogens showing type-textures is recommended in the list at the end of this chapter (section 1.7, page 30). In order to reduce the technical equipment and make physical and chemical investigations more easily, there is a constant demand for the synthesis of liquid crystals combining a very low melting point and a high clearing point [10], cf. chapter 2.

A polarizing microscope with a continuously variable adjustable hot stage is indispensable for observation of liquid crystals. For such work, we use the Zeiss Standard GFL polarizing microscope with 2.5/6.3/16/40/100 objectives and 8 X eyepiece. For microphotography, we added a Zeiss camera with light meter, beam-splitting system (Grundkörper II), and Bertrand lense. In order to obtain a general view, the simple hot stage described by W. Stürmer is satisfactory [369]. It can be built into the stage of the microscope, and is basically a glass plate coated with SnO_2, mounted in a plastic frame, and heated with a continuously variable current. The temperature can be estimated from precalibrated values. However, more precise measurements require a hot stage with built-in temperature control such as the stages offered by the Reichert and Mettler companies. H. Sackmann [305] uses a Boetius hot stage made by F. Küstner Nachfolger, Dresden, which is similar to the apparatus of L. and A. Kofler [123]. For lecture purposes, a projection apparatus developed by G. W. Gray [88, 89] is suitable, as are similar arrangements suggested by G. van Iterson [378] and others [102, 104, 106, 278, 287, 351, 352, 370]. Proven models of Mettler, Leitz, Zeiss, Zeiss-Jena, Olympus, and other equipment are also available for more sophisticated work. To investigate the optical behavior of liquid crystals in electric or magnetic fields, G. Meier and H. Gruler use a hot stage and objectives having very long focal length, e.g. Zeiss UD 40/0, 65C or UD 16/0.17 [273]. Interesting possibilities may be offered by several methods of investigation that have been suggested by several authors, but which have been used only rarely, if at all, for observations on liquid crystals. These include G. Nomarski's method [280], phase contrast microscopy [52], and reflected-light techniques. Ultramicroscope experiments were made by P. P. von Veimarn so early as 1909 [383].

A short introduction to the phase transitions of liquid crystals will facilitate the later descriptions of textures observed with the polarizing microscope. For a single-component system, the thermodynamically stable mesomorphous range is limited by the melting point and the clearing point, the latter being higher than the former in a stable, enantiotropic mesophase. Many compounds are known whose clearing points are lower than their melting points and in which monotropic mesophases appear, but this is a supercooled state that is thermodynamically unstable with respect to the crystalline solid phase. Below the clearing point, depending on conditions, enantiotropic or monotropic transitions can occur in which liquid-crystal modifications are in equilibrium with one another. Unfortunately there is a certain ambiguity in the use of the terms enantiotropic and monotropic, but misunderstandings can be avoided by specifying first a known phase to which the phase under discussion bears either a monotropic or an enantiotropic relationship. Thus, these terms refer mainly to the thermodynamic stability of a given mesophase in comparison with the solid crystal. However, if transitions within the liquid-crystal range are considered, there can often exist enantiotropic transitions between mesophases that are both monotropic with regard to a third solid phase. I. G. Chistyakov et al. have studied some vitrified liquid crystal films under the polarizing microscope [32].

1.2   Polymorphism

Since the work of G. Friedel, two basic types of mesophase, smectic and nematic, must be distinguished, with subtypes of each [62]. Smectic phases consistently exhibit a higher (two- or three-dimensional) degree of order thus more resembling to solid crystals. They always represent the low-temperature modifications in systems, where both nematic and smectic phases coexist. The polymesomorphism was reviewed by A. Loesche [263a].

As will also be detailed later, it now appears that the nematic phase exhibits two modifications differing from each other very slightly in energy content. A. de Vries suggested that these be called homeotropic-nematic and cybotactic-nematic [45], although both of these terms had already been adopted for similar but different phenomena (cf. G. W. Stewart [357–362]). The former designation applies to the low-temperature modification that shows, as the name implies, a pronounced tendency toward the formation of a homeotropic texture. The typical patterns shown in polarized light, with their threads, streaks and distinctly visible interference colors are attributed to the cybotactic-nematic phase. However, the relationships among nematic phases are still a matter of discussion. So in most present cases, we can speak only of a single nematic phase, and treat the various subtextures as more incidental phenomena rather than as phase characteristics. A microstructure model was developed by L. Pohl and R. Stein-strässer [283].

Nematic phases (and, according to recent experiments, certain smectic phases as well) can acquire an optically active "superstructure". This is a frequently observed phenomenon known as the Mauguin texture [267]. It forms in preparations on strongly adsorbing boundary surfaces when a twisting torque is applied to molecules that were originally oriented parallel.

The characteristic texture called nematic-cholesteric, or simply cholesteric is related to the Mauguin texture and derives its optical activity from either the addition of chiral solutes or the introduction of chiral centers. There are dextro- and levorotary systems. In a two-phase system, a cholesteric phase can coexist only with a smectic phase, but it cannot with a nematic phase. In multicomponent cholesteric systems it is possible to find a temperature and composition such that the individual dextro- and levorotary cholesteric components cancel one another. Such an optically inactive system exhibits nematic properties and is called a compensated cholesteric mixture (E. Sackmann, H. Stegemeyer, cf. chapter 7 and ref. [402]). Cholesteric material can also be obtained from the separation of nematic racemates [11, 13]. Since structure bears no obvious relation to the sense of optical rotation, it is necessary to define dextrorotary and levorotary systems for cholesteric textures and to differentiate between dextrorotary (+) and levorotary (−) activity in a certain wavelength region (see chapter 7). For the present, an example of each type will suffice: cholesteryl chloride forms a dextrorotary system, and cholesteryl benzoate a levorotary one. Almost all cholesteryl esters presently known form levorotary systems. A very simple method to discern the sense of optical rotation has been described by G. Heppke and F. Oestreicher [107]. There are still other suggestions [108, 123a]. The observation of two distinctly different cholesteric textures goes back to G. Friedel [63]. Upon cooling, an isotropic melt of a cholesteric substance yields the undisturbed focal-conic texture that is transformed under slight pressure into the disturbed form known as the Grandjean planar texture [80–87]. The important basic investigations on this phenomenon were made by D. Vorländer [385], E. Dorn [396], and F. Stumpf [365–368]. Further characteristics of the cholesteric texture include the so called cross-hatching and commas that have been described by G. Friedel in detail [62].

Polymorphism is much more obvious in smectic phases, and has been thoroughly investigated in the United States [2–5, 372], in the Soviet Union [31, 35], and by H. Sackmann in Halle, Germany [305–337]. Although the final number of smectic phases must be left undecided, the existence of the following phases seems assured:

smectic A

    smectic D

        smectic C

            smectic B

                smectic E

                    smectics F, G, H ...

In this scheme (A, D, C, B, E, ...), the lower-temperature modifications (i.e., the ones showing higher short range order symmetry) stand farther to the right. H. Sackmann has developed this classification as an aid to avoid future confusion, since more smectic modifications exist. The alphabetic order merely indicates the chronological order of discovery.

The various phases and their range of existence can be advantageously indicated by a uniform line notation [407] based on the work of H. Kelker [117], G. W. Gray [91], and L. Verbit [379]. This notation incorporates the following symbols that will be generally used in the following text and seem superior to capital letters

k	crystalline solid
a	anisotropic liquid ≡ liquid crystal ≡ mesophase
i	isotropic liquid
s	smectic mesophase

n   nematic mesophase
c   cholesteric mesophase
c(R)   dextrorotary cholesteric mesophase ⎫
c(L)   levorotary cholesteric mesophase   ⎬   (as defined by geometry of structure)
p   polymesomorphous liquid crystal   ⎭
d   thermal decomposition
( )   monotropic transition.

In this scheme (illustrated below) the melting point appears to the right of k, and the clearing point to the left of i (but not between k and i). The clearing point of a monotropic phase is enclosed in parentheses. This disposition brings the transition temperatures rising from left to right, as the following examples will illustrate:

Case 1. Both transition temperatures and structures unknown:

k a
        enantiotropic liquid crystal
k i
        no liquid crystal
k i(a)
        monotropic liquid crystal
k a p
        enantiotropic polymesomorphous substance.

Case 2. Transition temperatures known, structures unknown.

k 134 a 178 i
        the crystalline solid melts at 134°C to a liquid crystal having a clearing point of 178°C.
k 259 i
        the crystalline solid melts at 259°C to an isotropic liquid (no liquid crystal).
k 184 i (156 a)
        the crystalline solid melts at 184°C to an isotropic liquid and forms a liquid crystal monotropically at 156°C.

Case 3. Both transition temperatures and structures known:

k 136 s 141 s_C 148 s_A 196 n 210 i (s 128)
        the crystalline solid melts at 136°C to a smectic phase, transforms at 141°C into a smectic C phase, at 148°C into a smectic A phase, at 196°C into a nematic phase, and at 210°C (clearing point) into an isotropic liquid; at 128°C another smectic phase forms monotropically.
k 117 s 124 c 196 id
        the crystalline solid melts at 117°C to a smectic phase, transforms at 124°C into a cholesteric phase, and decomposes at the clearing point of 196°C.
k 300 n 330 d
        the crystalline solid melts at 300°C to a nematic phase and decomposes at 330°C without having reached the clearing point.

Table 1.1 summarizes these introductory remarks. The higher the entry is in the table, the higher the relative temperature zone in which the phase exists. In other words: the higher entries in the table correspond to the lower state of order in the mesophase. Transition phenomena are also tabulated.

Table 1.1. Key-word characterization of mesophases.

i	isotropic liquid	
clearing point		
n	**Nematic phase**	
	probably dimorphous:	
	n_c: nematic-cybotactic	
	n_h: nematic-homeotropic	
optically inactive:	optically active:	
optically uniaxial, in	Mauguin texture	cholesteric texture
homeotropic texture		(induced or inherent)
optically positive		optically negative in the
		"Grandjean texture" and in
		focal-conic or "undisturbed"
		microstructure domains
	dextrorotary type: reflection of dextro-circularly polarized light, at shorter wavelengths in the shortwave range of the dispersion zone	levorotary type: reflection of levo-circularly polarized light, at shorter wavelengths in the shortwave range of the dispersion zone
transition point		
	supercooling of nematic and nematic-cholesteric phase	
	Smectic phase	
polymorphism:	s_A	
	s_D	
	s_C	
	s_B	
	s_E	
	$s_{F,G,H}$	
Optically positive in homeotropic texture and microstructure, optically uni- or biaxial [371], occasionally pseudomonoclinic, s_C and s_H can be optically active; s_D is optically isotropic		
melting point		
k	**Crystalline phase**	
		supercooling of nematic and smectic phases, glass-type solidification [32]

Still one more term remains to be defined, a term that has been unfortunately used ambiguously. The so called lyotropic phases or lyophases are considered in a separate group within multicomponent liquid-crystal systems, but these contain a solvent (e. g.: water, chloroform, dimethylsulfoxide) that neither is mesomorphous nor in any way related to mesomorphous substances. However, since liquid-crystal mixtures with benzene, aniline, or nitrobenzene are considered to be thermotropic, the division into lyotropic and thermotropic systems is meaningless unless the type of solvent is also indicated (e. g.: water-like). A prime example of a lyotropic liquid crystal system is sodium palmitate/water which has textures that are in many ways similar to that of smectic phases. But up to the present, no unambiguous relationship of lyotropic phases to the thermotropic-smectic phases s_A to s_H has been established. As will be discussed in chapter 11, phases are either described by the still-used soap classifications "neat", "subneat", "superneat", "waxy", "middle", etc., or they are given an entirely new classification based on their X-ray diffraction patterns. Furthermore, there is no sharp delineation between lyophases and birefringent colloidal "solutions", especially in the area of dyestuffs. Here we mention the substances called tactosols that were discovered by H. Zocher [405] and more recently studied by J. F. Dreyer [46]. There are also lyotropic phases that are analogous to the cholesteric textures of thermotropic systems, represented by the optically active solutions of certain polymers, of which the prototype is poly-γ-benzyl-L-glutamate (see section 12.1).

It is difficult to define clearly the boundary between apparent and actual birefringence for liquid-crystal phases and textures. Liquid crystals appear to the naked eye as opalescent, cloudy liquids that are reminiscent of soap solutions due to their frequently white to yellowish color. In thin films, however, they may become transparent. Nematic phases have low viscosities near that of water, but all smectic and most cholesteric phases are highly viscous and in some cases so extremely viscous that it is difficult to recognize them as "liquids". The experimenter finds himself frequently scratching free droplets with needles or moving the cover glasses on closed preparations to convince himself that the sample is truly fluid. In this instance, it is often found that this action transforms a supercooled substance into a crystalline one. Figs. 1.1, 1.2, and 1.3 illustrate respectively a nematic, a nematic-cholesteric, and a smectic A phase as normally seen under the polarizing microscope with crossed polarizers and without special preparations of support and cover slide. Fig. 1.4 shows a lyotropic-smectic phase and fig. 1.5 the lyotropic-cholesteric phase of poly-γ-benzyl-L-glutamate.

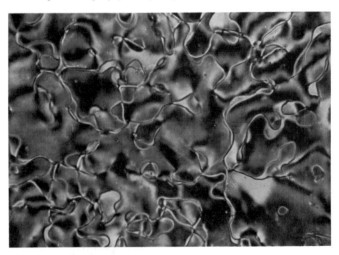

Fig. 1.1. Nematic Schlieren texture.

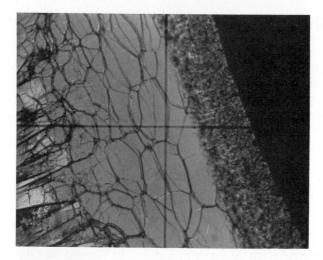

Fig. 1.2. Disturbed and undisturbed cholesteric texture between the crystalline phase (left) and the isotropic liquid (right).

Fig. 1.3. Focal-conic texture and bâtonnets of the smectic A phase.

Fig. 1.4. Lyotropic-smectic phase. Fanlike texture of the middle phase of an aqueous solution of potassium laurate.

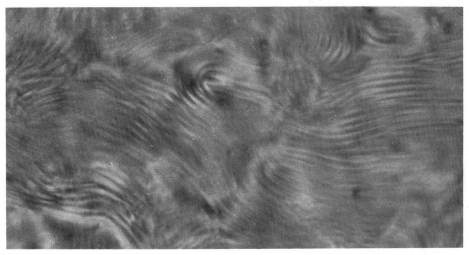

Fig. 1.5. Texture of a lyotropic-cholesteric phase (PBLG in dioxane).

1.3   Solid Surface Boundary Action

Since most liquid crystal device applications and physical investigations require uniformly oriented layers, this field has been thoroughly studied in the last few years. As pointed out later, the textures differ drastically whether the anchoring conditions are tangential or homeotropic. Thus surfaces favoring homeotropic alignment may untwist cholesteric phases of low twisting power and a continuous change of the anchoring angle between 0 and 90° is possible [356]. The roles of surface topography and surface chemistry were distinguished by L. T. Creagh and A. R. Kmetz [38]. Substrates with surface energy lower than the liquid crystal surface tension cause homeotropic alignment; otherwise homogeneous alignment parallel or inclined to the surface is formed. However, observations by I. Haller cast doubt upon the claims that predict the tendency of a nematic liquid crystal to align transverse to a solid surface using the difference between the surface free energies of the nematic liquid and of the substrate [100]. Some adhesion free energies of liquid crystals have been measured [297b]. The solid/liquid-crystal interface energy crucial for the tilt angle has been measured in other works [49, 286, 296, 297], but theoretical considerations scarcely exist [40]. Methods for the measurement of tilt angles are described in the literature [6, 39, 276b, 401]. On grooved surfaces the long axes of the molecules may align parallel or skewed with respect to the grooves [9].

Uniform alignment which is superior to that achieved by rubbing or liquid coating emerges from oblique deposition of thin films, e.g. by a 7 nm thick film of Au or SiO [112, 376]. This technique generates microgrooves in one direction of the surface that can be visualized by electron microscopy [126, 400]. Alignment additives and their actions are discussed in chapter 14 and in refs. [113, 114, 264, 292–295, 375].

Liquid crystal orientation has been studied on crystal surfaces [353], synthetic high polymers [48, 49, 354a], silane treated glass [375a], metal oxides [354a], and metal films [44b]. A method described by G. Ryschenkow allows one to measure the angular variation of the director at the level of superficial twisted loops more accurately than by observation between polarizer and analyzer [304]. In addition we mention that adsorption layers of nonmesogenic substances were found to resemble to nematic liquid crystals [285].

It has already been mentioned that it is difficult to observe liquid crystals free of the influences of boundary surfaces on their orientation. The surface forces can often penetrate several tenths of a millimeter into the sample corresponding to a specific "coherence length" [8, 26–28, 33, 75–78, 115, 276, 355, 390].

In this way, heterogeneities of the substrate (e. g. scratches, defects [401 a], and ferroelectric domain boundaries [69]) are indicated as disturbancies of a uniform texture or by the nucleation occurring at the defects during cooling through the isotropic/nematic transition [122]. In mesophases, with their extremely low elasticity, the tendency to minimize internal energy is of very considerable influence on the surface physics (cf. chapter 3). K. L. Wolf wrote in his highly recommended book "Tropfen, Blasen und Lamellen" on such boundary and deformation phenomena in general [399]. Because of these surface phenomena, microscopic studies should be repeated under three different sets of conditions:

1. with a cover slide;
2. with an open extended layer;
3. with a small droplet of predominantly uniform texture, possibly on a heavier immiscible liquid.

This method will normally permit the identification of textures and phases when temperature gradients in both directions are applied. It is advisable to first determine the number of phase transitions with a preliminary experiment in which the sample is rapidly heated on a heating stage having a low thermal capacity. After a change in temperature, however, the appearance of a new texture may take up to several hours to reach equilibrium especially in the case of the highly viscous smectic phases. For more detailed investigations, "contact preparations" and the method of H. Sackmann are advised [311].

As early as 1911 H. Stolzenberg [364] had published a paper on the determination of the melting point and J. Billard [14] and R. D. Ennulat [51] surveyed identification aids. Differential thermoanalysis and differential calorimetry are now gaining favor in the investigations of liquid crystal phase-transitions (cf. chapter 8).

The so called homeotropic layers are pseudoisotropic and technically important. They are easily distinguished from isotropic melts by using conoscopic observation. They can also be identified readily by their tendency to become translucent when touched lightly. Uniaxial phases with an oblique orientation are also found frequently, especially in cases where the boundary surface more strongly interacts with the liquid-crystal phase. The border-line case of a planar orientation parallel to the boundary surface is the so called homogeneous texture. It should be noted that completely uniform homogeneous textures showing only one direction of the optical axis are certainly not common. Most preparations show a tendency either to partial pseudoisotropy or to an analog with twin plates having symmetrically oblique axes inclined toward the center. This can be easily seen in convergent light (D. Vorländer and H. Hauswaldt [384]). The various mesophases are now described according to their appearance under the polarizing microscope. Only the most typical identifying characteristics are noted here. Some of them will be detailed in section 3.3.

All those studying liquid crystal morphology find an indispensable aid in D. Demus' and L. Richter's book: Textures of Liquid Crystals [44 a].

1.4   Textures in Nematic Phases

1.4.1 Homogeneous Texture

The homogeneous texture occurs in the nematic and only in very special cases of smectic phases, and thus it can serve as a rather typical texture. The term homogeneous texture can easily be misunderstood. It means a special kind of homogeneity. The optical axis (resp. the bisectrix) lies parallel to the surface boundary of the layer. Sometimes the term planar texture is used in the same meaning.

In order to form a homogeneous texture having a uniform direction of extinction throughout the entire preparation, the glass surfaces are rubbed in the direction of the desired orientation of the molecular axes. It is helpful to add polyethyleneglycol or other substances, e.g. crown ethers and related compounds [284] or chromium complexes [265]. The so called mosaic texture which is a subform of the homogeneous texture is typically encountered in the smectic B phases. It is recognizable by its curved grain boundaries and very different interference colors. In contrast to this, another variant of nematic homogeneous texture, called marbled texture exhibits sharp, straight bordered areas that impart a rock-like appearance to the preparation (H. Sackmann [321], fig. 1.6).

Fig. 1.6. Marbled texture of a nematic phase showing transitions to the Schlieren texture.

Order within homogeneous mesophases is so intensely influenced by external forces that samples can occur in which surfaces are delineated with varying degrees of sharpness. Each of these surfaces has a different orientation but exhibits a regular direction of extinction. This results in marbled texture.

Occasionally the weak optical activity of the Mauguin texture can be observed. This indicates that the phase under investigation is nematic. In such preparations a new type of disclination lines appears which has a coreless asymmetric alignment of molecules around the line that delineates regions of opposite chiralities [397].

The orienting influences of boundary surfaces similarly affect the texture of liquid crystals obtained by melting crystalline aggregates. These surfaces are depicted by the orientation of the mesophase and are maintained even at temperatures far above the transition point to the isotropic phase. The effect corresponds to the well known phenomenon of pseudomorphism* in the solid crystalline state. Many investigations of this phenomenon have been published most notably those of F. Grandjean [80–87], J. F. Dreyer [47], and E. T. Wherry [395].

H. Zocher speaks of "tribo-epitaxy" as a typical epitaxial effect that is exerted by the cleavage or growth planes of solid crystals [406]. This has been investigated most thoroughly by F. Grandjean [80–87], G. Friedel [65], H. Zocher [403, 404], and others [263].

The homogeneous texture is exceptionally stable due to adsorption on the solid boundary surface. Its outward appearance consequently remains virtually unchanged even after repeated crossing of the nematic/isotropic transition temperature [121]. The constant and strong Brownian movement indicated by enclosed dust particles is surprising when this apparent stability is considered. These particles even cross boundaries between areas of different axial orientation. This is direct proof that the orienting influences of boundary surfaces are superimposed on fluctuation phenomena occurring in the wavelength range of visible light. Such Brownian movements also cause the typical cloudiness always observed in thick (>1 mm) nematic layers as well as the nematic sparking phenomenon (R. Fürth and K. Sitte [67, 68], H. Tropper [374], J. T. Hartman [101], cf. chapter 3.

Chemical treatment in conjunction with mechanical manipulations (e.g., with acids, alkalis, detergents, emulsifiers) also affects the orienting influences of boundary surfaces.

Incomplete homogeneous preparations in thin layers often exhibit streaks, nuclei, or convergent lines. Thicker layers show so called disclination lines, dark threads from which the nematic phase derives its name (νεμ̟ατος, thread). These are discussed in section 3.3, see also [125].

1.4.2 Homeotropic Texture

The homeotropic texture is of great interest as a special case of uniform texture and even from the standpoint of practical technology. This texture occurs when the long axes of the molecules are at right angle to the boundary surface in what D. Vorländer calls uniaxial order [390]. Methods for investigating such preparations have been developed principally by C. Mauguin [266–271], E. Sommerfeldt [354], and H. Hauswaldt [384]. These preparations are analogous to optically uniaxial crystal plates that have been cut at right angles to the optical axis. With such liquid "single crystals" the field of view using crossed polarizers and orthoscopic illumination remains uniformly dark as the preparation is turned ("pseudoisotropy"). The conoscopic interference pattern corresponds to that of a uniaxial system showing the optically positive character of nematic and smectic A and B phases with a half wave plate being used.

To build up a homeotropic texture, the slide and cover glass are coated with a very thin film of lecithin or a similar amphiphile. This is done either by rubbing or by allowing a dilute solution to evaporate from the glass surfaces. Then a pinhead-sized drop of a nematic liquid such as MBBA (4-N-**m**ethoxybenzylidene-*p*-*n*-**b**utyla**n**iline, cf. section 1.7, p. 30) is placed on the slide and the cover glass pressed lightly over it. In preparations over ca. 30 μm thick it takes several minutes to form a uniform homeotropic texture (figs. 1.7, 1.8, 1.9 show a partially formed homeotropic texture). If the cover glass is now touched, the originally dark

---

* or paramorphism

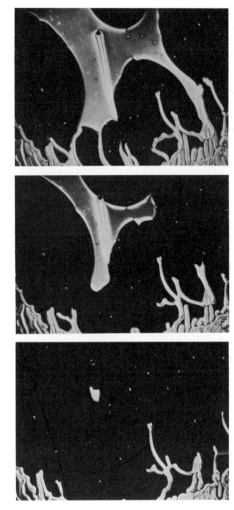

Figs. 1.7 to 1.9. Formation of a homeotropic nematic texture.

field of view brightens instantly (analogous to the "sparking phenomenon") and, using convergent light, a view of a biaxial interference figure appears for a short time in which the orientation is such that the bisectrix, usually showing a small axial angle, lies perpendicular to the plane of preparation. The original uniaxial view reappears after the reestablishment of a mechanically stable state. Like nematic phases, smectic A phases easily form homeotropic texture.

A uniaxially ordered layer can usually be obtained only under carefully controlled conditions. Even then, it will not occur with all liquid crystals. Homeotropic textures form fastest in very thin films that are influenced by capillary forces. Even cholesteric mixtures can order themselves homeotropically and optically positive provided that their helical twisting power is low. They lose their structural activity due to the weak attachment to the boundary layers.

D. Vorländer made an unsuccessful search for a correlation between constitution and the tendency to homeotropy [390]. His work gives one the impression that molecules terminally substituted in different or similar ways will all show practically the same behavior. The relationship apparently depends more on the direction and the magnitude of the molecular dipole moment, but no reliable generalization has been made until now. When a nematic-homeotropic preparation is heated, the transformation to the nematic-cybotactic phase (or, more carefully stated, to the normal nematic centered texture) is often accompanied by decomposition into droplets (fig. 1.10).

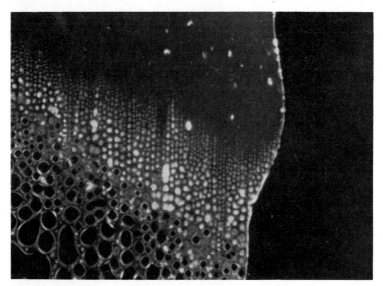

Fig. 1.10. Transition of a nematic-homeotropic phase into the isotropic liquid. Formation of droplets (cybotactic interphase?).

1.4.3 Schlieren Texture (Centered Texture)

Schlieren texture appears in the smectic C and B phase as well as in the nematic phase. It also occurs in other less well-investigated smectic phases. It is unmistakably characterized by its singularities with 2 or 4 "brushes" (fig. 1.11) with only the latter appearing in smectic C phases.

The English expression centered texture corresponds to Friedel's plages à noyaux and to Lehmann's Schlierentextur. Recently Lehmann's expression is also used in English literature while D. Vorländer and O. Lehmann speak of nuclei and convergence points ("Kernpunkte", "Konvergenzpunkte"). To create this texture, a few small crystals of a nematogen are melted between two glass plates that have been neither chemically nor mechanically pretreated. The melting is carried out so that the substance (under the influence of capillary forces) forms a thin film that does not completely fill the interstitial space.

The orthoscopic examination of an ideal sample of the Schlieren texture reveals a centered, symmetrical orientation of the planes of extinction. These are almost always double or quadrupole formations with nuclei or convergence points in the center. The quadrupole formations are reminiscent of windmill arms, and the convergence points are joined across the arms. In thicker preparations, the points develop spatially into disclination lines that are identical with the well-known nematic threads (fig. 1.12).

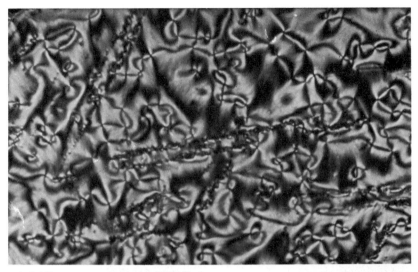

Fig. 1.11. Smectic Schlieren texture (s_C phase of TBBA).

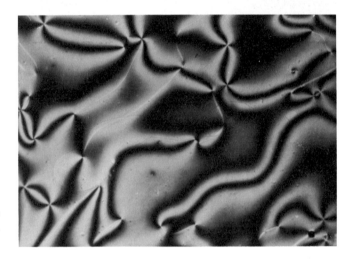

Fig. 1.12. Different types of nematic threads and singularities.

Double brushes can be additionally connected by a "disclination line of second order". These lines are easily made visible using a half wave plate. There are also morphologically different threads (called by O. Lehmann, D. Vorländer, G. Friedel, and others "threads with or without hollow centers") corresponding to the different numbers of nuclei and convergence points. Fig. 1.13 shows such an inversion wall and its topological relationship to the Schlieren texture. In the nematic-cholesteric phase, the inversion walls and disclination lines also represent the geometric loci of the extreme direction changes of the optical axes. Figs. 1.14 and 1.15 illustrate their relationship to the patterns called oily streaks by F. Reinitzer.

When a) the polarizer or b) the object stage is rotated, the windmill-like arms turn a) in the opposite or b) in the same direction. These are then designated "minus" singularities (cf. section 3.3). There is another type of singularity not affected by such rotation. This is

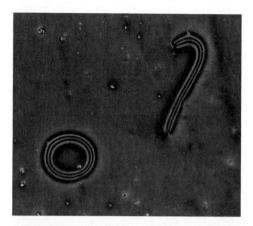

Fig. 1.13. Inversion walls in the nematic Schlieren texture.

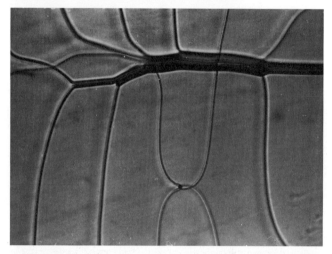

Fig. 1.14. Disturbed cholesteric texture with oily streaks.

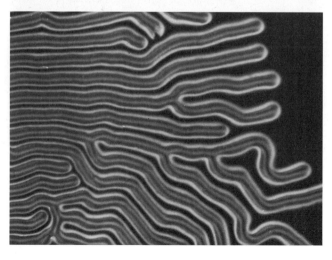

Fig. 1.15. Disturbed cholesteric texture (perpendicular view to the helical axis, "fingerprint texture").

called a noyau non tournant and reflects a simple spherical configuration of the type known in German literature as "Tropfen in 1. Hauptlage". The optical character of the sphere can be determined orthoscopically by the use of the half wave plate. Singularities are not mutually independent but may interact mutually attempting to extinguis one another, i.e., there is a tendency toward a minimum-tension state of the phase. A connected series of singularities is called an inversion wall. This describes the situation in which the optical axis of the spatial element undergoes a sharp change (e.g., 180°) upon crossing the wall so that it turns back upon itself on the other side of the wall.

1.4.4 Nematic Droplets

Nematic droplets characterize a type-texture of the nematic phase since they occur nowhere else (fig. 1.16).

Upon cooling an isotropic melt, the nematic phase begins to separate at the clearing point in the form of typical free-form droplets with cylindrical symmetry called C_∞ point-symmetry by Schönfliess. Fig. 1.17 shows this for a drop lying on one side. The direction of vibration n_γ can be determined orthoscopically by use of a half wave plate. Similarly in more complex textures the vectors of the vibration having the higher and the lower index can be designated. The direction of the higher index always coincides with that of the long molecular axis.

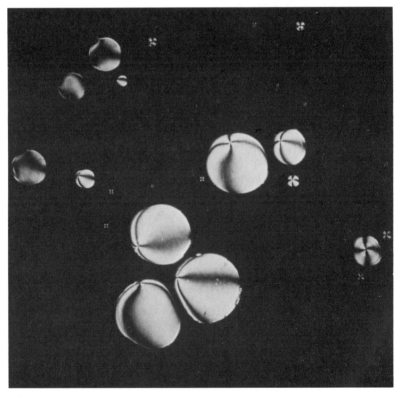

Fig. 1.16. Nematic droplets.

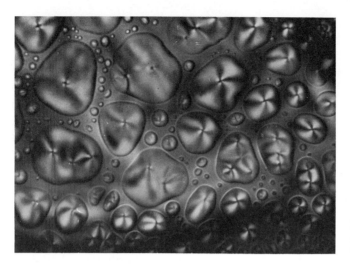

Fig. 1.17. Nematic droplets observed with a quarter wave plate.

O. Lehmann and R. Schenck [349] have studied the phenomenon of nematic droplets and observed the Airy spiral on rotating drops. D. Vorländer and H. Hauswaldt obtained this interference figure which is typical of two oppositely rotating layers in an optically active nematic-cholesteric drop [384]. R. Schenck found that mechanical rotation can create the optically active form from the normal nematic phase (cf. C. Mauguin). Computed arrangements with predicted and observed optical patterns that occur in liquid-crystal droplets were recently described in the literature [288]. R. Schenck also discovered the dichroism of free nematic drops [349]. He attributed this phenomenon as well to optical activity induced by torsional forces within the drop. D. Vorländer described similar phenomena as diagonal and "quadrant colorations" of homeotropic nematic-cholesteric phases. F. Stumpf [365–368], W. Voigt [382], G. Friedel [62], and D. Vorländer himself [385, 386] contributed to these discussions of interference coloration phenomena.

Schenck's experiments with free nematic drops in a magnetic field showed clearly that the symmetry axis of the liquid "monocrystal" (a free moving drop of p,p'-azoxyphenetole) tries to orient itself at right angles to the lines of the magnetic field. This again proves that there is, within a drop, a cylindrical rotational symmetry of the elementary structure. Later investigations showed that the longitudinal molecular axes, as axes of minimum diamagnetic susceptibility, are always parallel to the magnetic field [53, see however 2:191a].

Liquid-crystal elementary structures near air bubbles often orient themselves in a characteristic way. It can be determined from this whether the structures orient radially or tangentially with respect to the indicatrix of an ordered group of molecules. These clusters of molecules formerly were called a swarm [75]. A new decoration mode was found when MBBA was spread on the surface of a thick drop of diethylene glycol [37].

In a glass capillary (diameter ca. 0.1 mm) with a hydrophobic surface, a nematic phase can arrange itself so that the optical axes of the spatial elements are at right angles to the inner surface of the capillary. Under crossed polarizers one can see distinctly the dark axis of the liquid cylinder which corresponds to the center of a nematic spherite (fig. 1.18).

Other observation on nematic [273a], smectic and cholesteric samples [290] enclosed in capillaries are reported in the literature.

Fig. 1.18. Nematic phase in a
glass capillary with homeotropic boundary conditions.

1.4.5 *Mauguin Texture and Nematic-Cholesteric Phases*

The close relationship between nematic and cholesteric phases was first recognized by G. Friedel and is supported by the following experiments. R. Schenck has demonstrated on rotating drops that the nematic structure becomes optically active by the application of torsion. Similarly, C. Mauguin was able to show that a weak optical activity develops in a homogeneously oriented nematic phase between two plane-parallel plates when these boundary surfaces are twisted [267]. In this way the sterically determined intermolecular forces impose a new (helical) axis of symmetry at right angles to the boundary surfaces with their homogeneously oriented adhesion layers. Since the torsion angle is never greater than $\pi/2$, there is only weak optical activity.

Much stronger optical activity is obtained by mixing an optically active substance with the nematic phase to induce structural activity. Another possibility is offered by introducing chirality into the molecule of a nematogen. Note that the chiral solute must not possess the characteristics of a liquid crystal, but it must be chemically "compatible" with the nematogen. O. Lehmann and D. Vorländer mainly used colophony (abietic acid) for this purpose [387, 388]. H. Stegemeyer investigated camphene and chiral molecules having no asymmetric carbon atoms. At first glance, it may seem curious that the optical activity of a nematic-cholesteric phase can reach higher values under certain conditions when the concentration of the chiral solute is reduced. This will be elucidated through the dependence of the rotational dispersion on the concentration (cf. chapter 7). For certain wavelengths of light the cholesteric texture exhibits an extraordinarily strong dispersion of optical activity having specific rotations up to several thousand degrees per millimeter of film thickness. It seems possible that the influence of the chiral solute is similar to that of the twisted boundary layers in Mauguin's experiment. It seems also that in principle two molecules will suffice to replace the twisted boundary layers and to induce a helical arrangement.

Many nematic-cholesteric phases show a characteristic color effect that is usually, but not always, temperature-sensitive. The appearance is reminiscent of the coloration of butterfly wings. This is not always observed in such substances since the effect can manifest itself at frequencies outside the visible spectrum. A nematic-cholesteric preparation that was stroked with a brush exhibits distinct longitudinal channels. Disregarding any optical activity, it is easy to see that the waves subjected to the larger index are oriented largely in the direction of the strokes. Turning the stage 90° when coupled with the half wave plate (red 1) causes a change in the interference color from blue (addition) to yellow (subtraction). Likewise in the undisturbed focal-conic texture of a nematic-cholesteric phase (figs. 1.19 and 1.20), use of the half wave plate reveals the principal optical axes of the elementary domains (blue/yellow) arranged at right angles where ω (the larger index) vibrates tangentially ("negative spheroid").

Fig. 1.19. Undisturbed cholesteric texture.

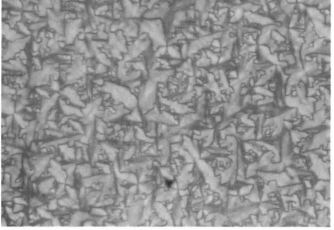

Fig. 1.20. The same preparation as in fig. 1.19 observed with a half wave plate (red I).

The relationship of the elementary domains of the cholesteric phase to those of the nematic can be understood when considered first from the standpoint of the rod-like shape of the molecules. We can then consider the formation of "smallest units" (e. g., double molecules or astatic pairs with quadrupolar characteristics), and final aggregates with and without inherent twist angles. The smallest unit Δv can always be assigned an ω and an ε ($\varepsilon > \omega$, i.e., in every case positive birefringence).

Only in the case of a planar helical arrangement of larger spatial elements do elementary structural domains $\Delta v'$ exhibit D_∞ symmetry. These have a new axis of symmetry that, of course, corresponds to the smallest index of the elementary domain, Δv. This explains the negative birefringence of the cholesteric phase.

In addition, if the degree of freedom of the translational movement $\rightarrow F$ is considered, it can be seen that both arrangements behave similarly in contrast to the positions of the refraction indices. Translation in the direction of the negative optical axis is thus "blocked" in the cholesteric phase. This results in a quasismectic behavior with focal-conic texture and its Dupin cyclide areas (cf. below).

A mixture of dextro- and levorotary cholesteric substances in proper proportions becomes purely nematic at a certain well-defined inversion temperature at which point it loses its optical activity. The optical rotation above this temperature is in the direction opposite to that exhibited below the inversion temperature. Correspondingly, for a given temperature there exist one or more compositions in which the dextro- and levorotary components compensate each other. Such systems, whose compensation characteristics have been described fully by O. Lehmann [262] and G. Friedel [62], are of great interest in spectroscopy (cf. chapter 6).

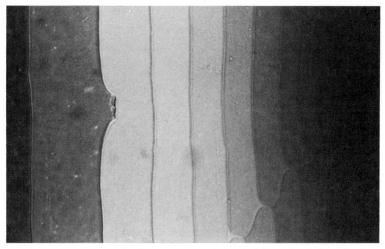

Fig. 1.21. Diffusion zone in a contact preparation between a nematic (at the left) and a cholesteric phase.

Fig. 1.21 illustrates a simple demonstration to show the dependence of nematic-cholesteric characteristics and of texture on concentration. This experiment uses a contact preparation consisting of a nematic phase (e.g., MBBA) and a cholesteric phase [118, see also 276a]. Each of the distinctly recognizable "steps" corresponds to a definite texture and to a certain number of rotations of the helical axis i.e. to a certain number of helical pitches. Each step has a different chemical composition (diffusion zone). The number and length (p) of the pitches changes step by step until $p/2$ finally becomes identical with the thickness of the phase in direct contact with the nematic component. p is constant within a step, and the texture exhibits a single interference color. Such experiments additionally permit diffusion coefficient determinations [99].

Use of a wedge-shaped slit and a cholesteric substance of constant composition leads to the formation of the Grandjean steps [87, 272], (fig. 1.22) that look alike but differ in stability since no concentration gradient is applied.

The optical activity, or more correctly the structural activity of a planar cholesteric layer, the so called Grandjean texture, is easily recognized by the asymmetrical color changes that appear upon rotating the analyzer. The extent of the optical rotation can be estimated in this way according to the earliest investigation of D. Vorländer and F. Stumpf [365–367]. One can thus immediately determine whether the system under investigation is dextro- or levorotary. In polarization microscopy, a preparation is conventionally designated as dextrorotary when a small clockwise rotation of the analyzer, as seen from above, shifts the color to longer wavelengths.

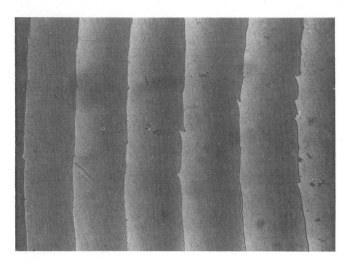

Fig. 1.22. Grandjean steps in a wedge shaped preparation.

For a long time the optical behavior of cholesteric phases caused confusion and could not be explained until it was recognized that the dispersion of optical activity showed a range of anomalous dispersion analogous to the index of refraction. The selective reflection of light occurs in the center of this wavelength region (cf. chapter 7). The temperature dependence of the colors reflected from cholesteric phases is applied in medicine and technology to measure surface or skin temperatures (e.g., thermography or thermal mapping, cf. section 14.3). For any given composition under constant angles of illumination and observation there is an unequivocal color temperature characteristic of cholesteric phases if they are not exposed to pressure and shear. It must be emphasized that the colored reflection does not spontaneously appear when the preparation is inclosed between glass plates. If the isotropic melt of a cholesteric substance such as cholesteryl nonanoate is cooled below its clearing point one obtains the texture shown in fig. 1.19 with almost no color effect. This phenomenon is easily reproducible, almost independent of the nature of the boundary surface, and it leads to the "undisturbed texture". It consequently calls attention to its similarity with the focal-conic texture of smectic A and C phases. Only after a brush stroke to the uncovered preparation or light pressure on the cover glass the reflective coloration appears. The "oily streaks", bands of disclination walls, simultaneously emerge indicating an incomplete Grandjean texture. In a convergent beam one can also see the typical axial figures in the areas enclosed by the oily streaks. After some time "recrystallization" to the undisturbed focal-conic texture occurs. This thermo-optical behavior is described in many earlier works and in refs. [69a, 126a].

Many cholesteric substances exhibit a strongly birefringent smectic phase upon further cooling that in most cases were identified as a smectic A modification [94, 321]. In the case of cholesteryl nonanoate, the modification appears monotropic and finally solidifies to form spherulites [1, 289, see also 7, 127], fig. 1.23.

When the spherulite is remelted one sees the solid state give way to a pseudomorph of the nematic-cholesteric phase (fig. 1.24) which exhibits the Grandjean texture with oily streaks only after the cover glass is disturbed. The morphology of nematic-cholesteric phases is discussed in the papers of B. Williamson Irvin [398], I. G. Chistyakov [34], K. Jabarim [109–111], W. G. Merritt [274], and A. Adamczyk [1a].

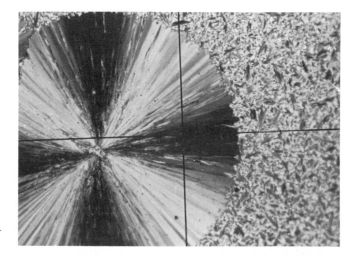

Fig. 1.23. Spherulite of cholesteryl nonanoate.

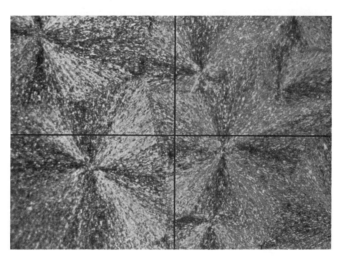

Fig. 1.24. Undisturbed cholesteric texture showing pseudomorphism to spherulitic crystals.

Our knowledge of cholesteric textures was conclusively enriched by Y. Bouligand [18, 19]. Apart from classical focal curves there also exist arrangements in a set of domains which are not exactly focal and differ concerning their rules of association from those classically known for smectics. Rotation dislocations and edge dislocations occur in certain polygonal fields and at their boundaries. In some cases rotation dislocations are also focal segments and introduce strong variations of the cholesteric pitch. Screw dislocations are occasionally superimposed on a focal segment. The distribution of the cholesteric axis along straight segments stretched between focal curves, which are not focal conics, is selenoidal and leads to screw dislocations. The polygon sides are split into several kinds. By these observations evidence was found that the envelopes of cholesteric axis are not exactly straight segments but strongly curved in the vicinity of the focal lines and the edge dislocations. Furthermore, local variations of the helical pitch are involved. An additional variety of cholesteric patterns is circumscribed by the term fan shaped textures [19]. Upon cooling from an isotropic melt cholesteryl myristate exhibits a transition to the cholesteric phase which ordinarily occurs in two steps. The first is described as a rapid transformation to a "turbid blue nonbirefringent state" while the second is the common transition to a focal conic texture which is formed much slower [110, 111]. For cholesteric textures shown by cholesteryl esters of dicarboxylic acids see refs. [23, 24].

1.5   Textures in Smectic Phases

Smectic A and C̲ phases, hitherto the most frequently observed of the presently known smectic phases, are recognized by the typical fan shaped or focal-conic texture. The latter term refers to the ellipses and hyperbolae possessing a common focus and which thus appear as lines of discontinuity in a molecular arrangement following Dupin cyclide surfaces (Lehmann's konische Texturen or Vorländer's Pocken). Smectic C phases show this fan texture less distinctly. H. Sackmann called them "broken focal-conic texture". A similar modification is formed in free drops. It should be noted that smectic C phases and smectic B phases can also exhibit Schlieren textures [44]. A. de Vries recently found an interesting relationship between the smectic C phase and the nematic-cybotactic texture (or phase) [45]. Like nematic phases smectic C phases adopt a twisted structure on addition of an optically active compound, e.g. cholesteryl chloride or camphor [15]. Smectic C phases represent a new type of symmetry in so far as they show biaxial character of the volume elements (symmetry group C_{2h}). This phenomenon is not very easily observed because of the mostly dominant focal conic texture. It can only be observed with a quasi-homeotropically ordered phase giving a conoscopic picture as these are well known for rhombic or monoclinic crystal plates. It also happens that the angle between the two optical axis is very small, and the conoscopic interference picture looks like that of an uniaxial system whose optical axis forms an angle with respect to the optical axis of the microscope. This angle is generally strongly temperature-dependent.

G. Friedel recognized the formation mechanism underlying the focal-conic texture. It leads to a polygonal structure that is the result of the maximum occupation of the available space. J. P. Meunier and J. Billard have recently succeeded in measuring accurately the typical parameters of a Dupin cyclide on a macroscopic preparation of diethylene monolaurate thus proving the formation laws quantitatively [12].

D. Coates and G. W. Gray have recently written a detailed review on microscopic observation of smectic textures [97, see also 280a].

The so called bâtonnets are closely related to the focal-conic texture. Bâtonnets are rod-shaped with cylindrically symmetrical protrusions and are formed when an isotropic melt of a smectic phase cools just before separation of the smectic phase as a whole. Occasionally they are also observed upon the first separation of nematic-cholesteric phases where they correspond to the focal-conic fans. Bâtonnets correspond to nematic droplets, with regard to formation conditions. They exhibit cylindrical rather than centered symmetry. The formation of bâtonnets out of an isotropic melt indicates the existence of a smectic phase, fig. 1.25 illustrates this for the smectic A phase of ethyl *p*-azoxybenzoate. The bâtonnets join together to form larger structures from which the compact focal-conic texture finally forms.

Bâtonnets also form in lyotropic systems. They are, together with nematic droplets, the true "liquid crystals" as defined by O. Lehmann, i.e., liquid monocrystals. W. Voigt discusses Lehmann's definition and its limitations [381], (see also [338]). One of his old research papers on liquid monocrystals free from influences of boundary surfaces has recently found renewed interest (whiskers, crystal growth from solutions).

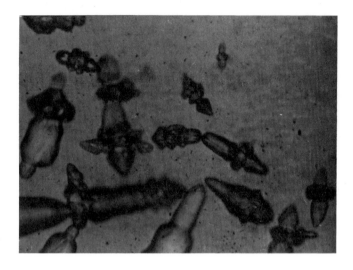

Fig. 1.25. Bâtonnets of a smectic A phase.

The smectic B phase and some other smectic phases rarely observed exhibit the unmistakable mosaic texture. Numerous pictures and special features are represented in the literature [42, 277, 350]. The remaining smectic phases, auch as the cubic D phase, the E phase, and others, are still too little known to be able to give diagnostic criteria for their identification [331, 335], and it may happen that problems appear as was recently observed with smectic biphenyl derivatives [95]. The cubic smectic D phase occupies a key position in the correlation to both the plastic crystals and lyotropic mesomorphism [98].

Note that particularly in the case of the smectic A phase a mechanically less stable texture (e.g. homeotropic) can appear instead of the equilibrium focal-conic texture. Pseudomorphs of smectic phases are often observed either as epitaxial growth [263] or depicting crystalline solid or other low-temperature phases of higher order. Transition points in such cases are sometimes difficult to detect. Conoscopic observations indicate the tilt angle and are thus an important aspect of the identification of the different phases [279].

From the observance of the oily streaks that form in both cases and from other common features (with the exception of birefringence) cholesteric and smectic phases repetedly were misconceived to be closely related which is only true where their textures are concerned. Smectic A and nematic phases resemble in the tendency to form homeotropic textures. The homeotropy in an open preparation may persist through a phase transition thus concealing it from microscopic observation. In an open smectic A preparation, "genuine" stepped (or "terrassed") drops can be observed with the microscope. They are not to be confused with the so called stepped drops of the cholesteric phase, which are more properly termed Grandjean steps. Using electron microscopy, microsteps have been detected in stepped drops which directly depict the layer structure [377]. The smectic layer thickness can also be evaluated from the optical observation of a free film deposited on water [297a]. According to publications of H. Sackmann et al. the textures and phase transitions observed up to now are patterned in the following scheme.

	Textures							Phase transitions										
	1	2	3	4	5	6	7	i	n	n_c	n_h	c	s_A	s_D	s_C	s_B	s_E	s_F
n	+	+	+					+	⌐									
n_c								+		⌐								
n_h								+		+	⌐							
c	+	+		+				+				⌐						
s_A	+	+		+		+		+	+		+	+	⌐					
s_D				+		+		+				+		⌐				
s_C	+		+	+				+	+	+		+	+	+	⌐			
s_B	+	+	+		+	+		+	+			+			⌐			
s_E		+			+			+				+			+	⌐		
s_F			+	+											+			⌐
s_G				+				+				+			+	+	+	+

1: homogeneous texture
2: homeotropic texture
3: Schlieren texture
4: focal-conic, broken focal-conic and fan-like textures

5: mosaic texture
6: terrassed drops
7: isotropic texture

1.6 Lyotropic Mesophases

The focal-conic texture also appears in lyophases where special smectic phase modifications appear which are difficult to differentiate with the polarizing microscope. F. B. Rosevear gives an instructive summary of the forms appearing in both natural soaps and synthetic detergents [303], see section 11.2. The phase and X-ray diagrams which are prerequisite to a consideration of such systems will be discussed together in chapter 11. Here we should mention the so called myelin structure (fig. 1.26) that also appears in organic material, and that structure was described by R. Virchow even before Reinitzer's discovery of thermotropic mesophases [380].

In his early writings, O. Lehmann often stressed the importance of mesophases in biochemical processes [158, 165, 183]. This was underscored by the title of his book "Flüssige Kristalle und die Theorien des Lebens" [165, 183]. The fact that he could evaluate these matters so well, despite the primitive experimental facilities and techniques of the time, may possibly have been due to the fact that liquid-crystal structure is exhibited by a large number

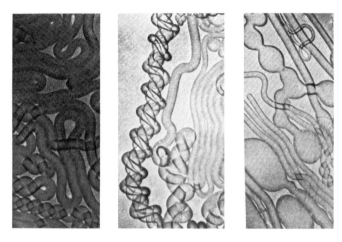

Fig. 1.26. Myelin forms after a schematic drawing by O. Lehmann.

of biological materials (e.g., the tobacco mosaic virus and the erythrocyte membrane, see chapter 12). Fig. 1.27 is an example from a more recent study on the structure of a chromosome.

Cholesteric phases form a special class of lyotropic systems in the biochemical field. C. Robinson (cf. section 12.1) devoted a thorough study to these phases using poly-γ-benzyl-L-glutamate (fig. 1.5, p. 11). P. Gaubert showed that incident light is selectively reflected in the same way by beetle wing-covers as by disturbed cholesteric phases [74]. In fact, the cuticula of the insects' bodies form from cholesteric lyophases that persist as pseudomorphs after solidification. This was recently shown by Y. Bouligand [17], cf. section 12.2.

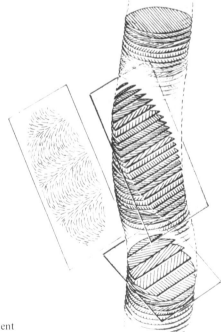

Fig. 1.27. Schematic drawing of the twisted arrangement of the DNA filaments in a chromosome (from 17).

1.7   Selected Mesogens

1   Nematic Phases

p,p'-azoxyanisole

PAA   CH₃O—◯—N=N(O)—◯—OCH₃   k 117 n 135 i

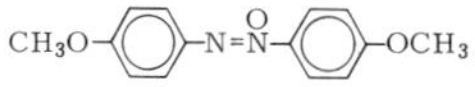

p,p'-azoxyphenetole

PAP   C₂H₅O—◯—N=N(O)—◯—OC₂H₅   k 137 n 168 i

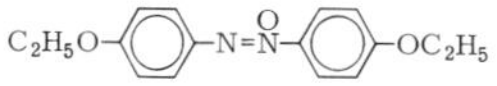

p-methoxybenzylidene-*p*-*n*-butylaniline

MBBA   CH₃O—◯—CH=N—◯—C₄H₉   k 21 n 45 i

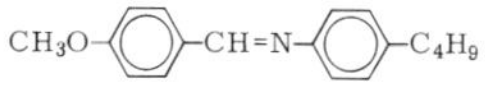

p-ethoxybenzylidene-*p*-*n*-butylaniline

EBBA   C₂H₅O—◯—CH=N—◯—C₄H₉   k 36 n 80 i

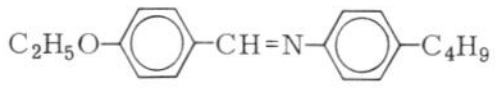

p-methoxy-*p'*-*n*-butylazoxybenzene (isomeric mixture)

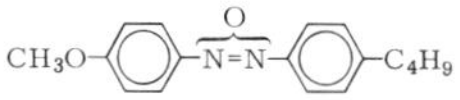

CH₃O—◯—N=N(O)—◯—C₄H₉   k 20 n 74 i

anisylidene-*p*-aminophenylacetate

APAPA   CH₃O—◯—CH=N—◯—OCCH₃   k 83 n 110 i

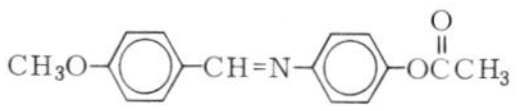

p-methoxybenzylidene-*p*-oxybutyrylaniline

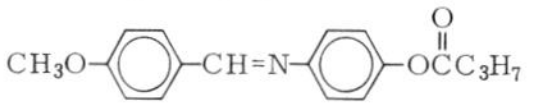

CH₃O—◯—CH=N—◯—OCC₃H₇   k 50 n 112 i

2   Smectic Phases

The following substances can serve well for the study of smectic phases. Most were investigated by H. Sackmann and are easily obtainable, or of exceptional interest.

ethyl *p*-azoxybenzoate

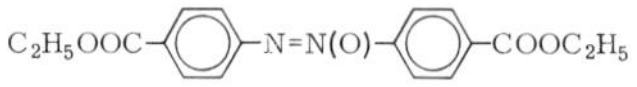

C₂H₅OOC—◯—N=N(O)—◯—COOC₂H₅   k 113.7 s$_A$ 122.5 i

ethyl *p*-azoxycinnamate

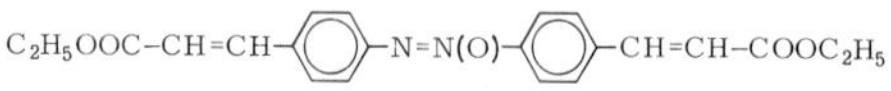

C₂H₅OOC–CH=CH—◯—N=N(O)—◯—CH=CH–COOC₂H₅   k 135.7 s$_A$ 267 i
                                                 (100.4 s$_C$?)

p-cyanobenzylidene-*p'*-*n*-octyloxyaniline

CBOOA   NC—◯—CH=N—◯—OC₈H₁₇   k 73 s$_A$ 82.6 n 108 i

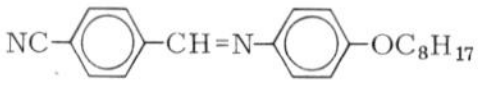

p,p'-di-*n*-dodecyloxyazoxybenzene

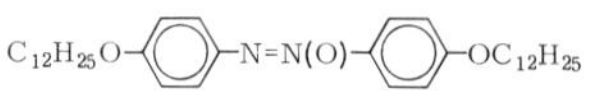

C₁₂H₂₅O—◯—N=N(O)—◯—OC₁₂H₂₅   k 80.8 s$_C$ 121.9 i

hexyl *p*-azoxycinnamate

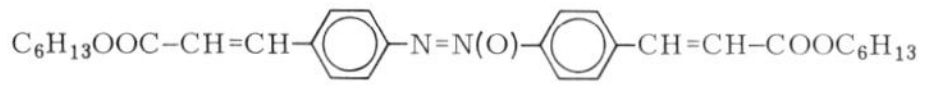

$$C_6H_{13}OOC-CH=CH-\langle O\rangle-N=N(O)-\langle O\rangle-CH=CH-COOC_6H_{13}$$ k 95.1 s$_C$ 156.1 s$_A$ 188.7 i

ethyl *p*-ethoxybenzylidene-*p*-aminocinnamate

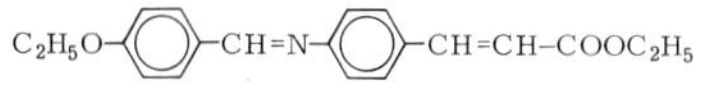

$$C_2H_5O-\langle O\rangle-CH=N-\langle O\rangle-CH=CH-COOC_2H_5$$ k 81.4 s$_B$ 119.3 s$_A$ 157.3 n 160.2 i

4-*n*-pentylbenzylidene-4'-*n*-hexylaniline

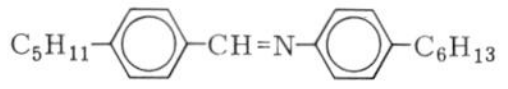

$$C_5H_{11}-\langle O\rangle-CH=N-\langle O\rangle-C_6H_{13}$$ k -2 s$_B$ 36.2 n 38.2 i

Numerous other compounds of this type also exhibit s$_B$ phases showing mosaic texture which is depicted in the literature [277].

p-*n*-alkoxybenzylidene-*p*-aminocinnamic acid *n*-amyl esters are easily obtainable, thermally stable smectic trimorphs, e.g.:

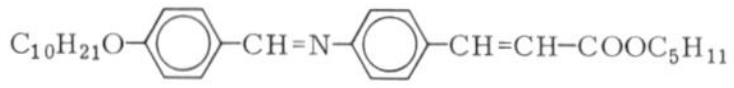

$$C_{10}H_{21}O-\langle O\rangle-CH=N-\langle O\rangle-CH=CH-COOC_5H_{11}$$ k 73.8 s$_B$ 97.0 s$_C$ 105.6 s$_A$ 136.8 i

The s$_A$ phase has fan texture, the s$_C$ phase broken fan texture or Schlieren texture, and the s$_B$ phase mosaic texture. The s$_A$ and s$_B$ phases can also be homeotropic.

4'-*n*-alkoxy-3'-nitrobiphenyl-4-carboxylic acids exhibit another smectic trimorphism having different phases, e.g.:

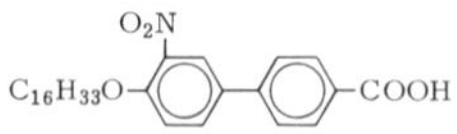

$$C_{16}H_{33}O-\langle O\rangle-\langle O\rangle-COOH$$ k 126.8 s$_C$ 171.0 s$_D$ 197.2 s$_A$ 201.9 i

(with O$_2$N substituent)

The s$_D$ phase appears in a homeotropic or mosaic-like texture.

terephthalidene-bis-(4-*n*-butylaniline)

TBBA $$C_4H_9-\langle O\rangle-N=CH-\langle O\rangle-CH=N-\langle O\rangle-C_4H_9$$

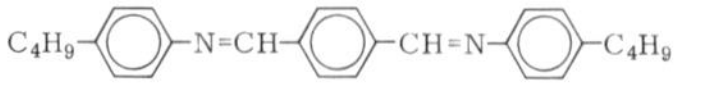

k 113 s$_H$ 144.5 s$_C$ 172.5 s$_A$ 199.6 n 236.5 i (s$_G$? 68 s$_5$ 52)

2-(4-*n*-pentylphenyl)-5-(4-*n*-pentyloxyphenyl)-pyrimidine, a smectic tetramorph, is an excellent example showing that the multiplicity of smectic phases is not yet exhausted since it suggests the existence of new phases: s$_F$ and s$_G$

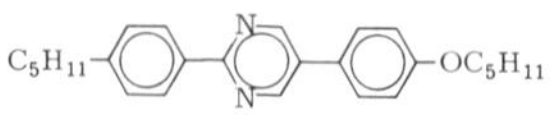

$$C_5H_{11}-\langle O\rangle-\langle O\rangle-\langle O\rangle-OC_5H_{11}$$ k 79 s$_G$ 102.7 s$_F$ 113.8 s$_C$ 144 s$_A$ 210 i

The s$_A$ phase has focal-conic fan texture, the s$_C$ phase broken fan texture or Schlieren texture, the s$_F$ phase a striped fan texture and a Schlieren texture that are very similar to that of the s$_C$ phase, and s$_G$ phase two still unnamed but distinctly different textures. The Schlieren texture of the s$_C$ and s$_F$ phases appear remarkably like the nematic Schlieren texture.

3   *Salt-like Compounds*

H. Sackmann et al. adopted the terminology normally used for lyotropic systems in their descriptions of the little-known thermotropic phases of salt-like compounds. Except for neat and middle phases, none of the named phases possesses a specific texture. Neat phases form a focal-conic texture, normally a fan texture, like that found in the s_A phases of nonionic substances, e.g.: sodium phenyloxyacetate (sodium salt of racemic mandelic acid)

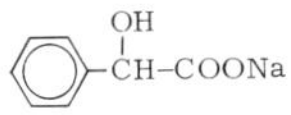

$$\underset{\displaystyle}{\bigcirc}-\overset{\text{OH}}{\underset{|}{\text{CH}}}-\text{COONa} \qquad\qquad\qquad\qquad \text{k 289 neat 343 i}$$

Middle phases exhibit a texture, illustrated but without name, having no hint of focal-conic sections, e.g.:

sodium diphenylacetate   k 249 middle 316 i
sodium dibenzylacetate   k 266 middle 311 i

Since their textures are unspecific, transitions of the mesomorphous low-temperature phases are recognized only by sudden changes in light intensity. In very rare instances they may escape from microscopic examination and must be ascertained by thermal analysis, e.g.:

Na stearate   k 116 subwaxy 135 waxy 165 superwaxy 208 subneat 255 neat 283 i
K stearate   k 170 waxy 267 neat 348 i

The best presently available data on mesogenic substances have been tabulated by W. Kast [116] and quite recently and complete by H. Demus and H. Zaschke [43]. D. Demus and L. Richter have written another indispensable book on the textures of liquid crystals [44a].

Chemical Constitution

2

2 Chemical Constitution

2.1 Structural Features

Our conception of the correlations between chemical constitution and liquid-crystalline properties is largely based on the work of D. Vorländer [1: 389, 782–834], C. Weygand [1: 393, 845–858], C. Wiegand [865–871], and G. W. Gray [1: 90, 254–305]. These investigators discovered four general criteria that indicate a molecule's predisposition to forming a liquid crystal:

1. Long, narrow, rod-shaped molecules have the most suitable geometry, especially in cases where rigid groups are implied. Such molecules must have a minimum length of 1.3 to 1.4 nm. This characteristic led D. Vorländer to coin the term "dominant unidimensionality" [825].

2. Permanent dipolar groups are present in almost all molecules that form mesophases. The size and direction of both the total moment and the group moments must be considered [149 b, 305 a, 469, 604, 892].

3. Almost all mesomorphically arranging molecules have a high anisotropy of polarizability. These strongly anisotropic directional forces result from dispersion forces and are supported by permanent dipoles and induced dipoles. A newer theory considers the quadrupole interactions to play the dominant role in intermolecular forces [658]. However, this theme cannot be treated here, and the reader is referred to the literature [98, 217, 218, 306, 307, 390, 595] and section 3.4. Repulsive forces have also been considered [399].

4. In addition, the melting point must not be too high, lest only supercooled metastable mesophases be formed monotropically.

Numerous compounds illustrate this phenomenon. Other ways of investigating the tendency of substances to form mesophases are by their behavior in mixtures (cf. section 8.8.1) and their depolarized light scattering [160a], cf. chapter 6, p. 267.

This, then, is the central theme of the following discussion of the structural features of liquid crystals. The common phenomena will be explained and quantitatively evaluated where possible. While the text must of necessity be limited to a selective explanation, the table aspires to completeness (section 2.13, following Kast's example [1:116]). The vast majority of liquid-crystalline substances are based on the following structure:

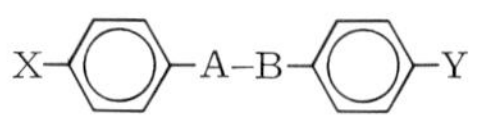

They possess:
1. Two or more aromatic (or, more rarely, heteroaromatic and/or cycloaliphatic) rings, usually benzene rings, as shown:
2. One or more bridging groups, A—B, that bind the rings together;
3. Two terminal groups, X and Y, usually on the long axis of the molecule.

Table 2.1. Aromatic and heteroaromatic rings in mesogens.

standard unit

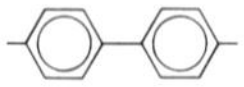

next most frequent unit

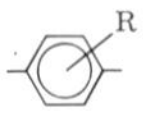

R: Cl, Br, I, OH, OAlk, OOCAlk, CH₃, CN, NO₂

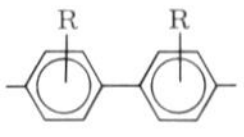

R: F, Cl, (Br), OCH₃, CH₃, NO₂, NH₂

rare units

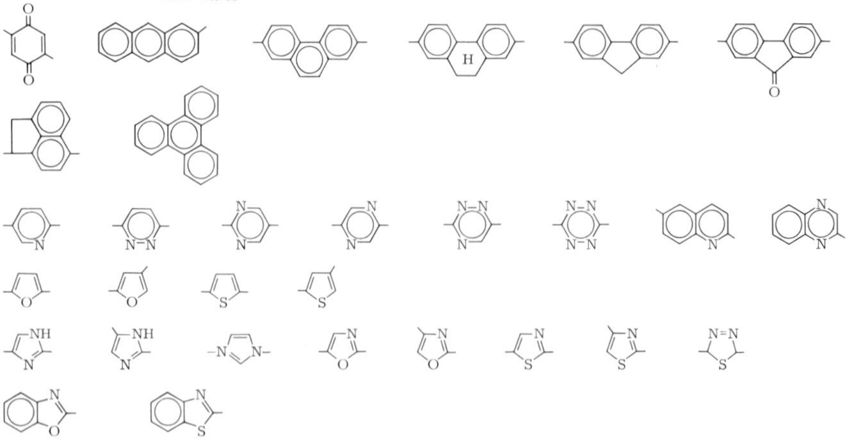

Table 2.2. Bridging groups in mesogens.

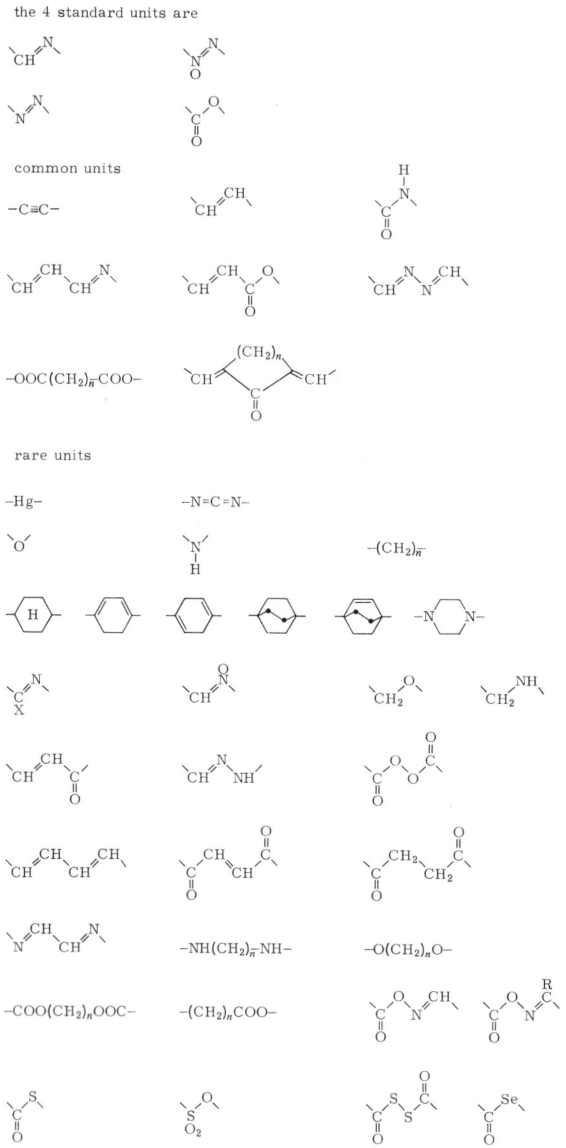

the 4 standard units are
common units
−C≡C−
−OOC(CH2)n−COO−
rare units
−Hg−
−N=C=N−
−(CH2)n−
−NH(CH2)n−NH−
−O(CH2)nO−
−COO(CH2)nOOC−
−(CH2)nCOO−

Representatives of a condensed structure without bridging groups can also be considered as belonging to this family of structures:

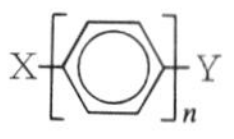

Table 2.3. Terminal groups in mesogens (R = *n*-alkyl, R′ = branched or unsaturated alkyl).

standard units

—OR	—R	—COOR
—OOCR	—OOCOR	—CN

common units

—Cl	—NO$_2$	—COR

$\overset{CH}{\underset{CH}{\diagdown}}\diagup\underset{COOR}{}$

rare units

—H	—F	—Br	—I	—Ñ=C̄
—N=C=O	—N=C=S	—N$_3$		—R′
—OH		—OR′		—OCOR′
—COR′		—COOR′		—CX=CY—COOR
—NH$_2$		—NHR		—NR$_2$
—NHCHO		—NHCOR		
—SR		—COSR		—OCOSR
—HgCl		—HgOCOCH$_3$		—OCF$_3$
—R$_{Si}$		—O(CH$_2$)$_n$OR		—CH$_2$CH$_2$OH

Following these structural principles first formulated by D. Vorländer, the tables 2.1 to 2.3 list the known molecular building blocks [784, 798]. The great number of infrequently used groups is conspicuous. Most of these occur only in exceptional cases, and then in combination with the more usual groups found in liquid crystals. Concerning reviews, the reader is referred to the literature cited on p. 4 and [123, 281, 289, 304, 430, 449, 574, 591, 759].

2.2 Aromatic and Heteroaromatic Rings

2.2.1 Linear Molecules

The 1,4-disubstituted benzene ring forms the standard building-block from which the majority of liquid-crystal molecular structures are derived. It is highly polarizable and determines the basic rod-shaped structure. This is a distinct contrast to the few mesomorphous materials derived from 1,2- or 1,3-disubstituted benzenes whose molecules are bent at the aromatic ring. The following examples serve as a comparison between the *para* disubstituted molecules on one hand with the *ortho* and *meta* structures on the other:

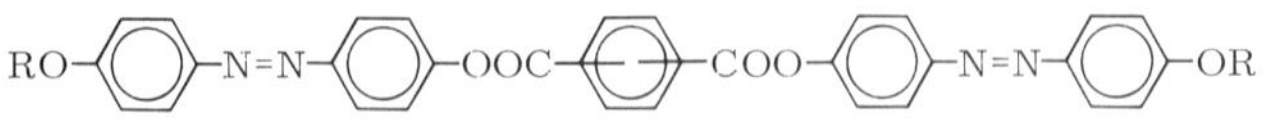

RO—⬡—N=N—⬡—OOC—⬡—COO—⬡—N=N—⬡—OR   [632, 811]

R: CH$_3$, *p-* k 252–255 a >300 d
       m- k 255–257 a 263–266 i
       o- k 200 i
R: C$_2$H$_5$, *p-* k 240–242 a >300 d
       m- k 247–249 a 252–254 i
       o- k 214–215 i

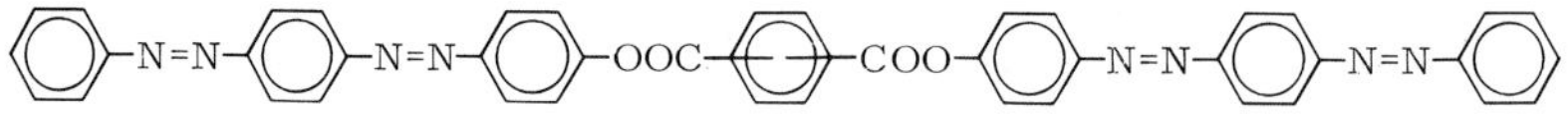

[632, 811]

p-	k d
m-	k 286 a 292 i
o-	k 257 i

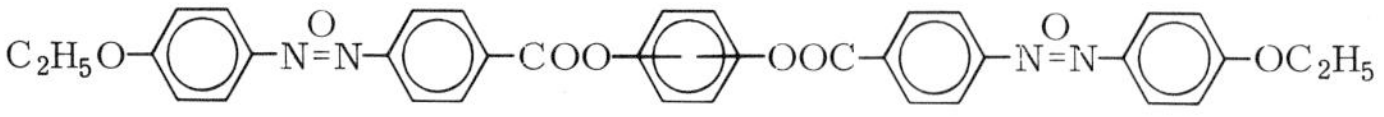

[818], (see note on p. 45)

p-	k 231 ad
m-	k 184 a 218 i
o-	k 164 a 213 i

(In fact, the last example was for a long time the only known liquid crystal among the more acutely angled ortho compounds.) This small tendency of angular molecules to enter the mesomorphous state also holds true for cases where the bending is caused by an angular bridging group, as well as for naphthalene derivatives. Molecules with two bends on aromatic rings form no liquid crystals. Contrast the following examples:

[321]

p-	k 141 a 246 i
m-	k 78.5–79 i
o-	k 102–103 i

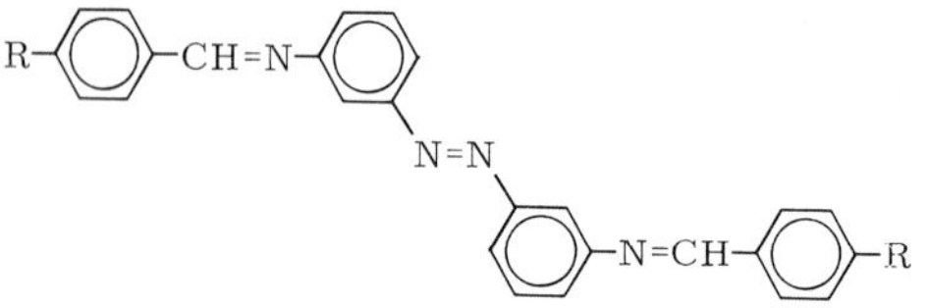

with R: H, CH$_3$, OCH$_3$,
OC$_2$H$_5$, NO$_2$,
N(CH$_3$)$_2$

[579]

None of the double bent molecules exhibit mesophases, nor do their azoxy analogs.

Naphthalene offers three possibilities for a disubstitution by which mostly linear molecules are formed. In order of decreasing tendency to form mesomorphs, the three combinations are 2,6-, 1,5-, and 1,4-disubstituted naphthalenes. Only these configurations yield liquid crystals as an example by C. Wiegand illustrates [869]. Mesogens are also known from 1,4,5,8-tetrasubstituted naphthalenes [468].

CH_3O—⟨O⟩—CH=N—⟨OO⟩—N=CH—⟨O⟩—OCH_3

2,6- k 188.5–189.5 a 354–356 i
1,5- k 196 a 282.5 i
1,4- k 182.5–184 a 262–263 i
1,7- k 142–143 i
2,7- k 204–206 i
2,5- k 133–134 i

This data aided in assigning the structure of a naphthalene derivative for which two possible structures had been suggested [393]:

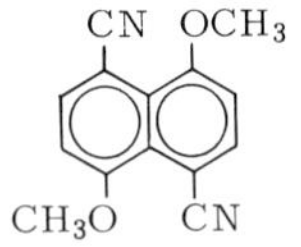

I 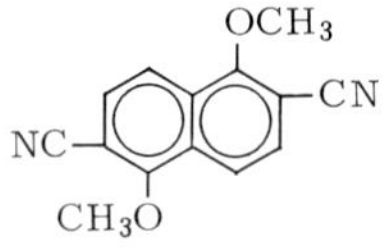   II

Compound I was transformed into I', and II' was synthesized independently. Since II' had the higher clearing point, it was demonstrated that the questionable naphthalene derivative had structure I.

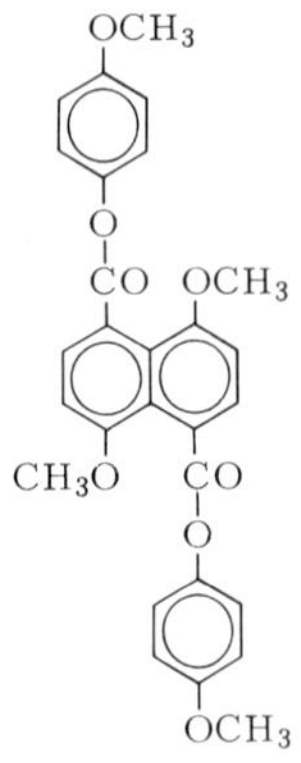

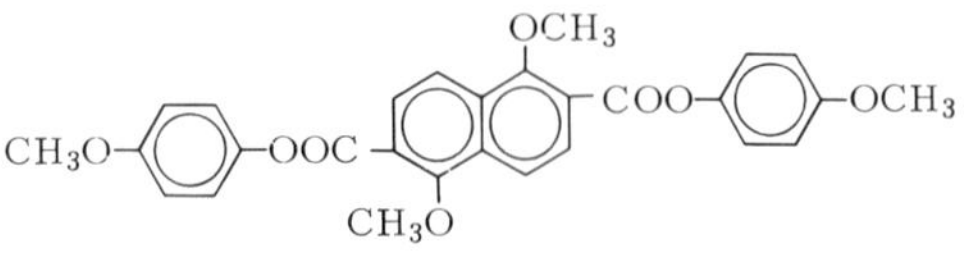

II'   k 218.5 a 286–287 i

I'   k 235–236 a 275 i

Similarly, a much older example of the use of mesophases in the determination of molecular structure is worthy of note. Two structures, open-chain and cyclic, had been considered for the oxidized reaction products of α-dicarbonyl compounds (e. g., Glyoxal) with phenylhydrazines:

R—⟨O⟩—N=N–CH=CH–N=N—⟨O⟩—R   and

CH=N–N—⟨O⟩—R
|
CH=N–N—⟨O⟩—R

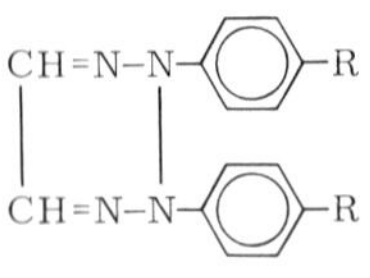

D. Vorländer proved that the linear configuration was the correct structure since many compounds in that series formed liquid crystals [804].

The first mesomorphous quinones have just recently been synthesized [660], while mesomorphous fulvenes have not yet come to light.

2.2.2 Sheet Molecules (Platelets, Disc-Shaped Molecules)

Early inquiries were made as to what types of molecules other than rod-shaped ones might form liquid crystals [799]. A few unproven examples seem to show mesomorphism in sheet-shaped molecules. On the other hand, a vast number of these molecules are nonmesomorphous, indicating that this structure has little tendency to form liquid crystals [799, 873]. However, the structures of the following consistently monotropic-mesomorphous salicylaldehyde derivates seem certain:

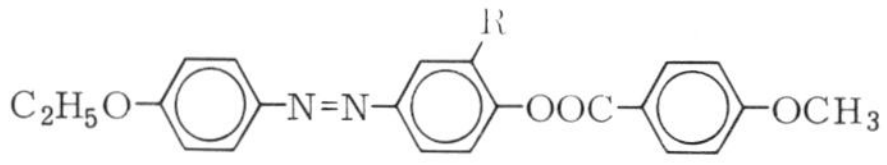

C_2H_5O —◯— N=N —◯— OOC —◯— OCH_3

R: —CH=N—◯—OC_2H_5 k 145 i (59a)

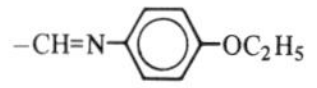

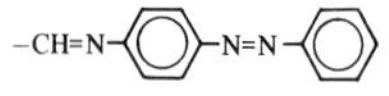

—CH=N—◯—N=N—◯   k 159 i (101a)

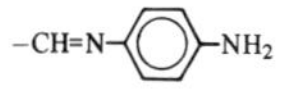

—CH=N—◯—NH_2 k 262 i (a)

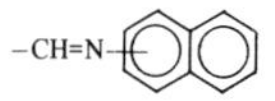

—CH=N—◯◯   k 160 i (107a)

A similar structure is also possible for the mesomorphous benzoate and anisate of *p*-phenetolazosalol [427]. However, one might doubt the mesomorphous character of the following molecules since they are not planar:

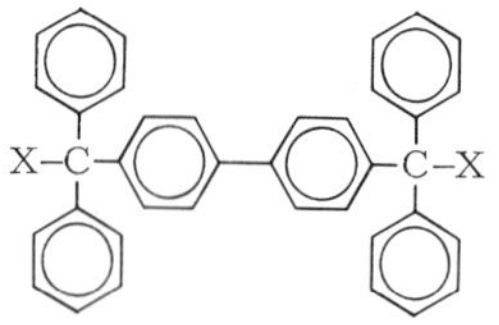

X—C—◯—◯—C—X   X: —OH   k 160 a 186 i
                                —Cl   k 219 a 223 i
                               [601]

The prototype of sheet-shaped molecules' mesomorphism can be seen in the carbonaceous mesophases and the synthesis of mesomorphous hexasubstituted benzenes [50a, 98a] and triphenylene derivatives [183c] has removed all doubt about its existence. Such compounds may form unaxial mesophases with an optically negative sign, the socalled discotic phases.

2.2.3 Heteroaromatic Rings

H. Schubert and coworkers have provided a wealth of data on mesomorphous nitrogen-heteroaromatic compounds [636, 639–643, 646, 652, 654–656]. Also D. Vorländer and his students [453, 830, 843] and C. Weygand [848] directed attention to liquid crystals of this kind and synthesized derivatives of both dehydrothiotoluidine and pyridazine. The reader is referred to further literature on this subject [95, 213, 391, 483, 537, 881] and the table of mesogens. If the heteroatom introduced is nitrogen, as it is in the majority of cases investigated, the lowering of the clearing point is slight, and that of the melting point is sometimes considerable. In addition, such substitution greatly favors the smectic state. The first systematic studies on a limited number of heterocyclic mesogens suggest that the dominant effect of the heteroatom is to produce changes in conjugative interactions within the molecule, thus affecting factors such as polarizability and dipolarity [287]. Fig. 2.1 gives an impression on the magnitude of effects.

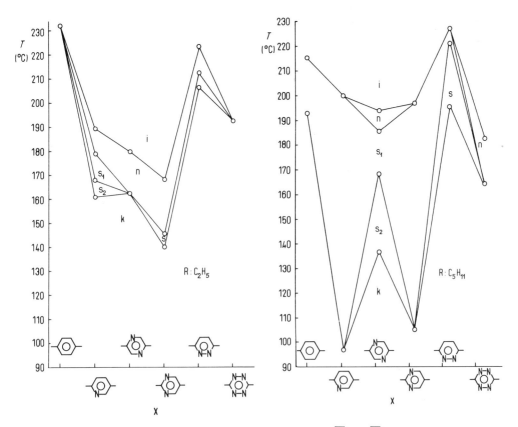

Fig. 2.1. Transition temperatures of two 4,4″-dialkyl-*p*-terphenyls ʀ—⬡—x—⬡—ʀ and their heterocyclic analogs (from [287]).

2.3   *Bridging Groups*

The most reasonable classification of bridging groups is based on their stereochemistry, determining three types of structures:
1. Linear bridging groups have a structural angle of 180°. However, only the tolanes are of any significance. In addition, liquid-crystalline derivatives of diphenyl mercury [566, 799] and diphenylcarbodiimide are known [590, 810].

2. Angular bridging groups with parallel configuration are far more frequently encountered and exhibit far greater thermal stability than any other group. This structure forces the 1,4-axes of the benzene rings to lie parallel, but only when there are an even number of atoms (usually two) in the bridging group as in case of the standard types: azomethines, azo compounds,

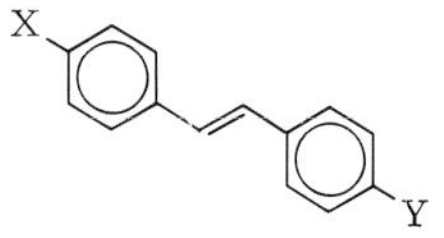

azoxy compounds, and carboxylic acid esters. Less frequently, there may be four or more atoms in the chain, as in azomethines of cinnamic aldehyde and cinnamic acid esters. Obviously only *trans*-double bonds give the molecule the necessary stretched configuration. Other examples are stilbenes, nitrones, carboxylic acid amides, benzyl phenyl ethers, benzyl phenyl amines azines, and glyoxal-bis-azomethines.

3. Angular bridging groups with nonparallel configuration are, on the contrary, much less likely to form liquid crystalline compounds. They have an odd number of atoms in the chain causing a W-shaped central configuration. In this type of molecule, e.g., hydrazones or distyryl ketones, the 1,4-axes of the aromatic groups are nonparallel:

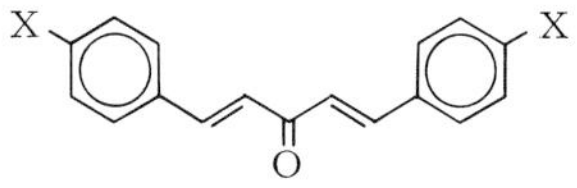

Angular, monatomic bridging groups (such as O , S , NH , CH_2 , and C) are to be avoided in the synthesis of liquid crystals. However, despite its angular nonparallel structure, the carbonic acid ester

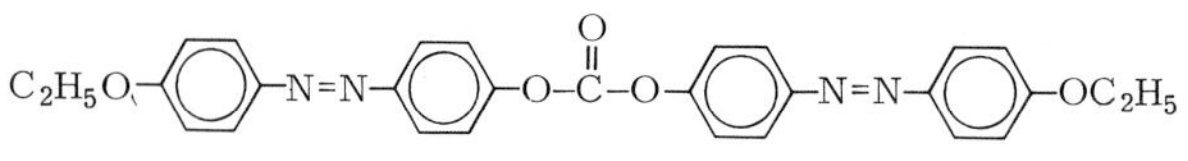

(F: 183 °C)

is noteworthy for its high clearing point of 305 °C. Preferred configurations include a *trans*-double bond in the chain of the central group. In addition to the desired geometry of the molecule, the bond supplies a certain rigidity that supports the parallel alignment in the liquid-crystalline state.

This is not to be overemphasized, however, because many mesomorphous carboxylic acid esters lack such rigidity.

The homologous series of dicarboxylic acid esters shown in fig. 2.2 offers an informative example. The distinctly alternating clearing temperatures show relative maxima for angular parallel structures ($n = 0$, 2, 4, 6, 8) and relative minima for angular nonparallel structures ($n = 1$, 3, 5, 7). The differences become markedly less with increasing chain length [805, 816, 837].

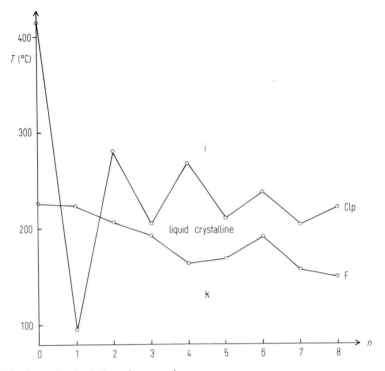

Fig. 2.2. Melting and clearing points in the homologous series

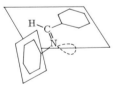

2.3.1 *Azomethines*

Aside from the rod-shaped molecules, a plane-shaped molecule was long considered to have a significant tendency toward the formation of the mesomorphous state. However, this is no longer true since the nonplanar structure of many liquid-crystalline aromatic azomethines has now been confirmed:

Several physical and chemical parameters prove the existence of this nonplanar configuration of aromatic azomethines, e. g., the lack of stable *cis*-isomers, the reduced intensity of long-wave absorption by the $\pi_1 \rightarrow \pi_2$ transition [193, 490], the lack of luminescence, and the NMR and IR spectra [64a, 721].

As the X-ray analysis indicates in the crystalline state, the two benzene rings are twisted with respect to the plane of the azomethine group [41, 84, 113]. CBOOA [754a] and bisazometh-

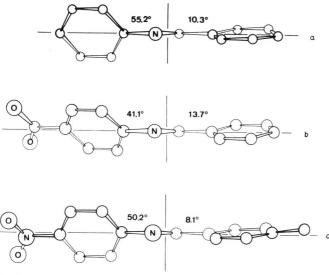

Fig. 2.3. Molecular structure of azomethines a) benzylideneaniline, b) benzylidene-*p*-aminobenzoic acid, c) *p*-methylbenzylidene-*p*-nitraniline (from [84]).

ines have also been studied [448]. The striking influence on the molecular structure by the chelating of salicylaldehyde azomethines is of special interest [729]. Even significantly planar azines such as anisaldehyde azine are not completely so [43, 44, 229, 230], fig. 2.4. The —CH=N—N=CH— group lies in a plane, but the two planes containing the benzene rings are twisted about 7° 30′ with respect to each other. As with numerous other anisole derivatives, the angles at carbons 4 and 4′ differ markedly from 120°, and the C—O bonds are also shortened. Like the transesterification of esters, the azomethines show equilibrium properties in mixtures [683].

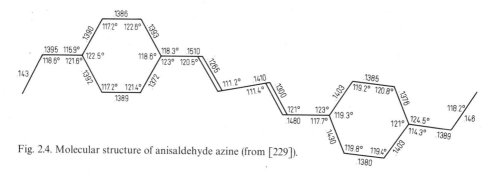

Fig. 2.4. Molecular structure of anisaldehyde azine (from [229]).

2.3.2 Azo Compounds

Azobenzene and its *p,p′*-disubstituted derivatives have been described as completely or essentially planar [6, 74, 75, 230–232, 335]. As an example, fig. 2.5 illustrates the bond lengths and angles of trans-*p,p′*-dichloroazobenzene [335]. The crystal structure of *p*-methoxybenzylidene-*p*-phenylazoaniline is described in the literature [770].

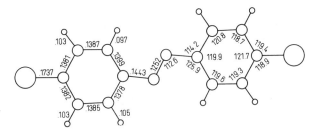

Fig. 2.5. Molecular structure of *p,p'*-dichloroazobenzene (from [335])

2.3.3 Azoxy Compounds*

The aplanar structure of azoxy compounds can be quite distinct, although less extreme than that of the azomethines. *p*-Azoxyanisole (PAA), certainly the most thoroughly investigated of the nematics, has a precisely known structure [40, 423, 875, fig. 2.6]. The —N=N— bonds deviate by about 4.2° from the planar *trans*-form, and the normals for the benzene rings are at angles of about 22.6°.

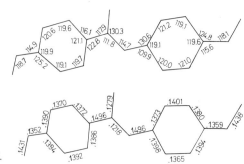

Fig. 2.6. Molecular structure of PAA (from [423]).

In contrast to PAA, both polymorphs of crystalline ethyl *p*-azoxybenzoate are formed by nearly planar molecules whereas other parameters differ markedly, e.g., the O—N—N angle averages 124° or 134.7° as compared with 130.8° for solid PAA [424].

2.3.4 Biphenyl Compounds

There exists extensive literature on the structural analyses of biphenyl bodies [31–33, 36]. Biphenyl has a planar molecule in the crystalline state, but the benzene rings in the vapor phase are twisted about 40° with respect to each other. As indicated in table 2.4, twisting of the rings, elongation of the 1-1′ bond, and changing the angles at the 2,2′ and 6,6′ positions are the result of steric hindrance.

---

* Unsymmetrical azoxy compounds are indicated by R—⬡—N=N—⬡—R' when it is not clear to which nitrogen the oxygen belongs. The notation is not to be considered as implying a ring structure.

Table 2.4. Molecular structure of crystalline biphenyls.

	angle between benzene rings	length of 1-1′ bond, nm	references
biphenyl	0°	0.1506 ± 0.0017	[322, 738]
4,4′-dinitrobiphenyl	33°	0.150 ± 0.0016	[66]
3,3′-dichlorobenzidine	21°	0.1515 ± 0.0024	[100]
4-acetyl-2′-fluorobiphenyl	50.5°	0.1479 ± 0.0010	[880]
4-acetyl-2′-chlorobiphenyl	49.2°	0.1490	[716]
2,2′-dichlorobenzidine	72°	0.153	[678]
decafluorobiphenyl	≈ 60°	0.149 (?)	[533]
p-terphenyl	≈ 0°	0.1494	[147]

From 2,2′-dichlorobenzidine, which is the most strongly twisted of the examples in the above table, liquid-crystalline Schiff bases are derived. This is also the case for 2,2′,6-trichlorobenzidine, although very much less pronounced. However, the limit has been reached, since no such liquid-crystalline Schiff bases are known to have been derived from 2,2′,6,6′-tetrachlorobenzidine [275].

For description and comparison of the molecular structures of compounds from which mesogens are derived, we also mention: stilbene [592], tolane and derivatives [115, 593], 9,10-dihydrophenanthrene derivatives [114], dibenzyl [73], diphenyl oxide [735], and diphenyl sulfide [734].

2.4   Terminal Groups and Homologous Series

D. Vorländer had called attention very early to the mesomorphous behavior in homologous series. The —CH_2— group can be introduced either in a bridging or in a terminal group. As an example of the former case, we can present the homologous series shown in fig. 2.2 [805, 816, 837]:

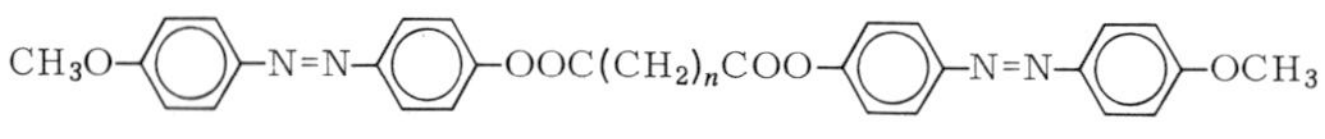

$$CH_3O-\bigcirc-N{=}N-\bigcirc-OOC(CH_2)_nCOO-\bigcirc-N{=}N-\bigcirc-OCH_3$$

In the following two series, D. Vorländer found an alternation of clearing points as mentioned earlier, although lower in magnitude [805]:

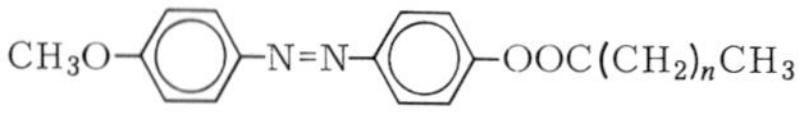

$$CH_3O-\bigcirc-N{=}N-\bigcirc-OOC(CH_2)_nCH_3$$

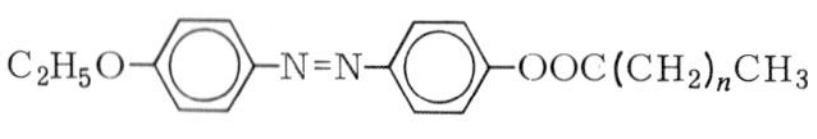

$$C_2H_5O-\bigcirc-N{=}N-\bigcirc-OOC(CH_2)_nCH_3$$

C. Weygand [853] and G. W. Gray [276] later payed attention to these phenomena. In homologous series containing —R, —OR, or —COOR (where R = n-alkyl) as a terminal group, the following regularities become apparent:

1. The clearing points of nematics decrease with increasing chain length.
2. The clearing points of nematics decrease with a fairly distinct alternation, the amplitude of the alternation also being reduced with increasing chain length. If the first members of a homologous series are already relatively long, the alternation is less apparent.
3. Smectics are rarer with shorter chains, but become most common with longer chains.
4. The plot of the clearing points of smectics and their transition points to nematics usually reaches a maximum at a moderate chain length and decreases gradually with further elongation of the chain.
5. This plot normally shows no distinct odd-even effect (cf. p. 50).
6. In the case of very long alkyl radicals, the clearing points approach the approximate range of 80–120°C. If the clearing points of the first members of the series are already in or below this temperature range, the points alternate only within this limit, or they rise alternately until this range is reached. In such cases, the points 1, 2, and 4 naturally do not apply (fig. 2.8).

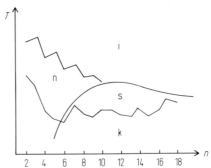

Fig. 2.7. Typical transition point curve for a homologous series (temperature, T, vs. number of atoms, n, in the chain of the terminal group).

Fig. 2.7 illustrates these points by showing the normal characteristics of a clearing point curve that is valid for many homologous series.

In all cases of polymesomorphous single compounds studied at normal pressure, the smectic modification represents the lower-temperature form. The curve of the n/s transition temperatures approaches that of the nematic clearing temperature from below as the number of atoms in the terminal groups increase and the two curves coincide. Rules for the behavior of melting points are given in the literature [8:346, 8:347; 401, 532, 743].

The behavior of clearing points in homologous series is still under discussion, and the key to the solution of the problem is being sought in conformational properties. Longer terminal groups prevent the molecules from sliding and thus favor the formation of smectics at the expense of nematics. On the other hand, these groups tend to coil with increasing length, thus generally hindering the parallel alignment. It thus follows that the clearing temperatures of smectics in a homologous series generally rise initially with increasing chain length, reach a maximum at a moderate chain length, then decrease slightly, and finally acquire a nearly constant value in very long chains. There are no indications of a stepwise change in lateral forces of orientation with a lengthening of the chain. This is supported by the fact that the clearing points of smectics normally do not alternate in any temperature range.

The main remaining problem is an explanation of the alternation of clearing temperatures in nematics. This always occurs such that the clearing points are relatively higher when the chains of the terminal groups have odd numbers of atoms. G. W. Gray attributed this to

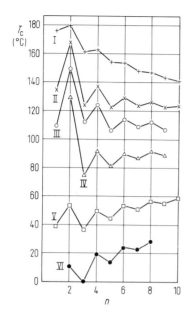

Fig. 2.8. Clearing temperatures, T_c, as a function of the number n in the following homologous series

I CH_3O—◯—$CH{=}CH$—◯—OC_nH_{2n+1}

II $C_nH_{2n+1}O$—◯—$N{=}N(O)$—◯—OC_nH_{2n+1}

III $C_nH_{2n+1}O$—◯—$N{=}N$—◯—OC_nH_{2n+1}

IV $C_nH_{2n+1}O$—◯—$C(O){-}O$—◯—OC_nH_{2n+1}

V C_6H_{13}—◯—$C(O){-}o$—◯—OC_nH_{2n+1}

VI $C_nH_{2n+1}CH_2$—◯—$C(O){-}o$—◯—$CH_2C_nH_{2n+1}$

(from [151]).

the regularly recurring properties of conformation derived from the thermodynamically stable zigzag structure of the alkyl chain [276].

Lengthening the chain from 2 to 3, from 4 to 5, or from 6 to 7 creates an extra bond parallel to the axis of the *p*-substituted benzene ring; the polarizability vector in the direction of the molecular axis is incremented twice as much as that at the right angles. On the other hand, if the chain is lengthened from 3 to 4, from 5 to 6, or from 7 to 8, the extra bond is more in the direction of an *m*-axis. In this case, the increments in the polarizability vectors are about the same for the direction of the molecular axis and for the perpendicular [149, 462]. This peculiar behavior has fascinated many authors and prompted them to look for a quantitative explanation by mathematical treatment [3, 149, 471, 472, 554, 555]. We also mention the calculation of interaction energy between isolated, parallel oriented, and saturated aliphatic hydrocarbon chains by E. Shapiro et al. [661]. The reproduction must be confined to the works of S. Marcelja [471, 472].

Starting from the works of Flory on statistical mechanics of chain molecules and from the works of Maier-Saupe on molecular interaction energy, S. Marcelja derived the expression for the internal energy in the nematic phase as

$$E = -\tfrac{1}{2}C_a(N)\,V_{aa}\eta_a^2 - \tfrac{1}{2}C_c(N)\,V_{ac}\eta_c^0 - NC_a(N)\,V_{ca}\eta_a(\eta_c - \eta_c^0)$$
$$-\tfrac{1}{2}NC_c(N)\,V_{cc}[\eta_c^2 - (\eta_c^0)^2] + E_{cnf}(X_c) - E_{cnf}(O),$$

with the free energy difference given by $F = E - TS$

The parameters are:

$C_a(N)$, $C_c(N)$:   Volume fraction of the aromatic resp. chain ($\equiv$ aliphatic) part

V_{aa}, V_{ac}, V_{ca}, V_{cc}: Four coupling constants

$\eta_a \eta_c$:   Order parameters, related to the Maier-Saupe form by

$$\eta_c = \left\langle (1.88/N) \sum_{i=1}^{N} \left(\frac{3}{2} \cos^2 \vartheta_i - \frac{1}{2} \right) \right\rangle$$

E_{cnf}:   Average conformational energy of an alkyl chain

$$X_c = C_c(N) \eta_c V_{cc} + C_a(N) \eta_a V_{ca}$$

The calculations of n/i transition entropies and their comparison with experimental data resulted in calculated values which were too high, but still reproduced well the odd-even effect and the increase with alkyl chain lengths (cf. section 8.4.2). On the other hand, in homologous series with high transition temperatures where the anisotropic interaction between the rigid structures of the molecule is rather strong, the elongation of chains decreases the average anisotropic interaction. In the opposite case with low transition temperatures, the elongation of chains increases the average anisotropic interaction between the molecules. W. H. de Jeu and J. van der Veen have also succeeded in an evaluation of n/i transition temperatures from Maier-Saupe theory [149a].

The occurrence of the different smectics has been investigated to date only in a few homologous series, and the disappearance of the s_E, s_B, and s_C phases with increasing chain length was quite striking ([295, 300], fig. 2.9). The first attempts to correlate the chemical structure with the type of a smectogen have been reported [149b, 305a, 309, see also 469].

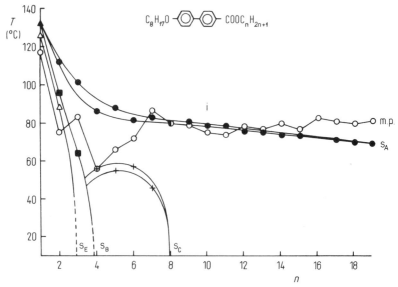

Fig. 2.9. Transition temperatures as a function of the number of carbon atoms, n, in the n-alkyl chain of the alkyl 4'-n-octyloxybiphenyl-4-carboxylates.
Key: ▲ s_{AB}/i; ● s_A/i; ■ s_B/s_A; △ s_E/s_B; + s_C/s_A; ○ $k/s_E, s_A$ or i; some of the transitions are "virtual" (from [300]).

Supplementary to point 5 (p. 47), two examples must be mentioned that exhibit unusual alternation of smectic clearing points. In the series of ω-phenylalkyl 4-(4'-phenylbenzylidene-amino)cinnamates:

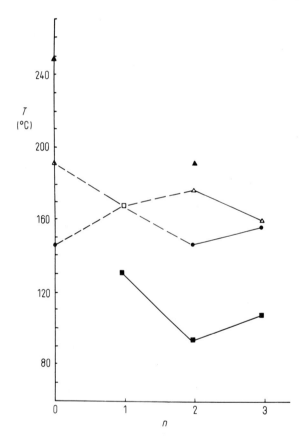

nematic phases only exist with an even number of *n*. Also with the smectic phases of this series an alternating effect was observed. However, s_A thermal stabilities culminated with even values of *n* while s_B and s_E thermal stabilities culminated with odd values of *n* [291, fig. 2.10]. The alternation of smectic clearing points appears only slightly within the homologous series of 2,5-bis-(4'-*n*-alkylbiphenylyl-(4))-thiophenes [650].

Fig. 2.10. Transition temperatures as a function of the number of methylene units, *n*, in the alkylene chain of 3''-methyl-ω-phenylalkyl 4-(4'-phenyl-benzylideneamino)cinnamates (from [291]).
Key: ▲, n/i; △, s_A/i or n; □, s_{AB}/i; ●, s_B/s_A; ■, s_E/s_B.

A method for a first estimation of clearing temperature derived from molecular parameters has been suggested [149, 382], and even a computer program for this purpose has been written [764]. A comparison of actual and calculated clearing points for azomethines supports the possibility of such an estimation method, but it also indicates its limitations [405]. With small substituents, a linear relationship between the clearing temperature and the anisotropy in the polarizability for the different C_{ar}—X bonds correlates rather accurately [753].

General statements concerning the influence of the terminal group on the thermal stability of a mesophase should not be given undue weight for specific cases. Thus, the series CN, C_2H_5O, CH_3O, NO_2, $N(CH_3)_2$, Cl, Br, CH_3, F, H is in approximate order of decreasing stability of nematics. G. W. Gray also points at deficiencies in the prediction of trends in homologous series [296]. There are frequent displacements, however, even though the highest clearing temperatures are to be expected with CN, C_2H_5O, CH_3O or NO_2, and practically all substituents raise the stabilities above that for H [276, 277, 627]. Introduction of the vinyl cyanide group is not advisable because this group undergoes *trans-cis* isomerization upon exposure to light [293].

D. Vorländer found that the introduction of a branched alkyl group, —R', such as —OR', —OOCR', or —CH=CH—COOR', strongly depressed the clearing points as compared to the analogous *n*-alkyl isomers. This is attributed to the broadening of the molecules [525, 786]. In practice, such compounds of interest have either asymmetric carbon atoms (especially the esters of amyl alcohol, $HOCH_2CH^*(CH_3)C_2H_5$) or bear those groups, R', which cause low melting points. Numerous chiral azomethines [93, 285, 343, 608], esters [403], and other compounds are reported and listed separately in the table of mesogens (section 2.13). As a rule, the clearing temperatures and melting points decrease when progressing from cinnamic acid to α-methylcinnamic acid and β-methylcinnamic acid derivatives:

	R: H		R: CH₃	
	F [°C]	$T_{n/i}$ [°C]	F [°C]	$T_{n/i}$ [°C]
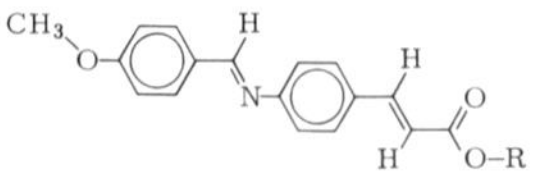	208	265	161	183
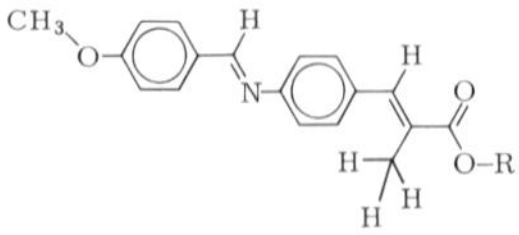	180	265	133	123
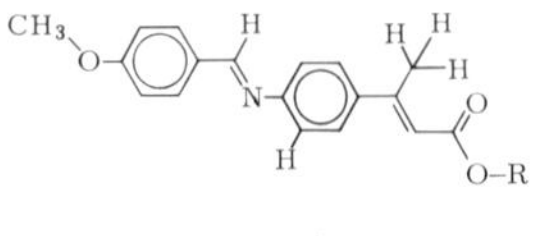	170	240	114	91

Since the ether function plays an outstanding role as a terminal group, ethers containing more than one oxygen atom have been thoroughly investigated, e. g., in the following compounds [51, 855].

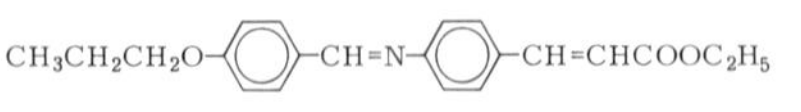 $CH_3CH_2CH_2O$—⬡—CH=N—⬡—CH=CHCOOC₂H₅	k	64	s	159	i
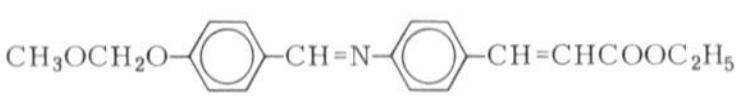 CH_3OCH_2O—⬡—CH=N—⬡—CH=CHCOOC₂H₅	k	54	s	79	i
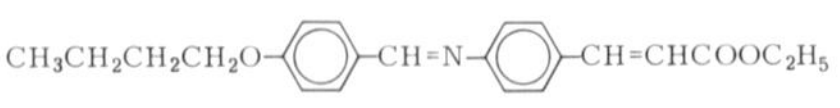 $CH_3CH_2CH_2CH_2O$—⬡—CH=N—⬡—CH=CHCOOC₂H₅	k	66	s	162	i
 $CH_3OCH_2CH_2O$—⬡—CH=N—⬡—CH=CHCOOC₂H₅	k	65	s	154	i
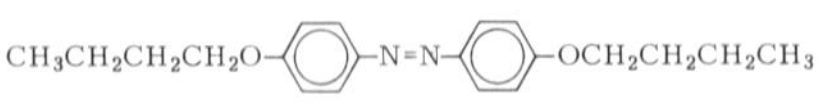 $CH_3CH_2CH_2CH_2O$—⬡—N=N—⬡—OCH₂CH₂CH₂CH₃	k	135	n	124	i
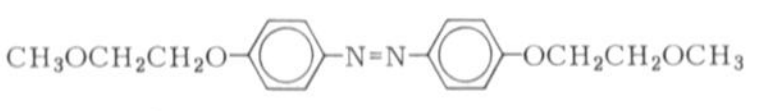 $CH_3OCH_2CH_2O$—⬡—N=N—⬡—OCH₂CH₂OCH₃	k	109	n	116	i

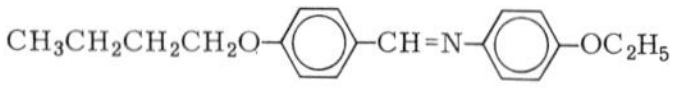

$$CH_3CH_2CH_2CH_2O-\!\!\left<\!\!\bigcirc\!\!\right>\!\!-CH\!=\!N-\!\!\left<\!\!\bigcirc\!\!\right>\!\!-OC_2H_5 \qquad\qquad k\quad105.5\quad n\quad129.5\quad i$$

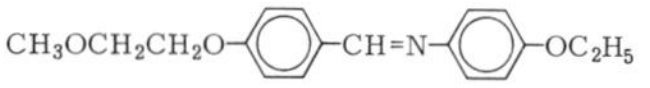

$$CH_3OCH_2CH_2O-\!\!\left<\!\!\bigcirc\!\!\right>\!\!-CH\!=\!N-\!\!\left<\!\!\bigcirc\!\!\right>\!\!-OC_2H_5 \qquad\qquad k\quad109\quad n\quad116\quad i$$

All terminal groups with a second oxygen atom lower the clearing points of smectics and nematics. The depression strongly depends on the oxygen's position in the chain [319]:

$$\left.\begin{array}{l}CH_3CH_2CH_2O-\\CH_3OCH_2O-\end{array}\right\}\Delta Clp:\ \approx 50\,°C \qquad\qquad \left.\begin{array}{l}CH_3CH_2CH_2CH_2O-\\CH_3OCH_2CH_2O-\end{array}\right\}\Delta Clp:\ \approx 10\,°C$$

The data is interpreted in this way so that the dispersion forces dominate over the permanent dipole interactions in determining the mesophase stability [319]. In this connection, J. P. Schroeder et al. recall the insolubility of polyformaldehyde in water, whereas polyethylene oxide is readily soluble [627].

Investigations to date indicate that lower clearing temperatures are to be expected if oxygen is replaced by sulfur [90, 542]. Only one study has been made on the mesomorphism of compounds with silicon in the side chain [884]. The larger atomic volumes and frequently different bond angles appear to be unprofitable and are not offset by the gain in polarizability. Terminal groups which enter into intermolecular hydrogen bondings such as OH or NH_2, are unfavourable but nondeterrent to mesomorphism [629, 630].

A comparative study by G. H. Brown and W. G. Shaw, reproduced here in a somewhat modified form, comes much closer to describing the real difficulties [1:21]. Fig. 2.11 gives an idea of the extent that clearing points can change when the bridging and terminal groups are varied simultaneously.

The comparison of isomeric compounds gave rise to speculation about how mesomerism might influence the thermal stability of mesophases. Even though this seems to be a simple explanation of the behavior of the two isomeric esters

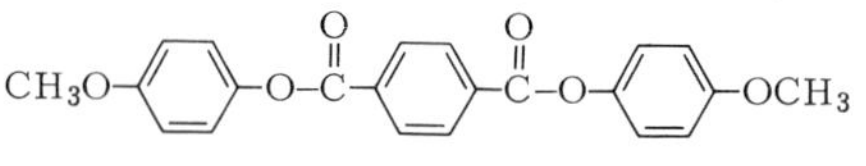

$$CH_3O-\!\!\left<\!\!\bigcirc\!\!\right>\!\!-O-\overset{O}{\overset{\|}{C}}-\!\!\left<\!\!\bigcirc\!\!\right>\!\!-\overset{O}{\overset{\|}{C}}-O-\!\!\left<\!\!\bigcirc\!\!\right>\!\!-OCH_3 \qquad\qquad k\ 205\ n\ 277\ i$$

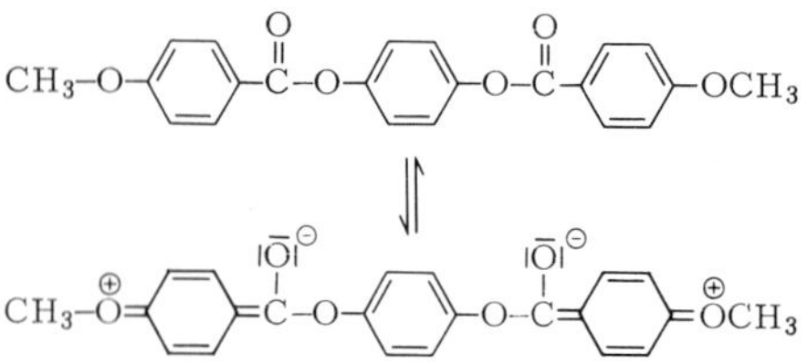

$$CH_3-O-\!\!\left<\!\!\bigcirc\!\!\right>\!\!-\overset{O}{\overset{\|}{C}}-O-\!\!\left<\!\!\bigcirc\!\!\right>\!\!-O-\overset{O}{\overset{\|}{C}}-\!\!\left<\!\!\bigcirc\!\!\right>\!\!-OCH_3 \qquad\qquad k\ 213\ n\ 297\ i$$

$$\Updownarrow$$

$$CH_3-\overset{\oplus}{O}=\!\!\left<\!\!\bigcirc\!\!\right>\!\!=\overset{\overset{|\overline{O}|^{\ominus}}{|}}{C}-O-\!\!\left<\!\!\bigcirc\!\!\right>\!\!-O-\overset{\overset{|\overline{O}|^{\ominus}}{|}}{C}=\!\!\left<\!\!\bigcirc\!\!\right>\!\!=\overset{\oplus}{O}CH_3$$

the authors themselves caution against drawing conclusions based on it [162, 246, see also 168, 169]. For di-*p*-methoxyphenyl terephthalate, which is largely symmetrically substituted on all benzene rings, there is no possibility of mesomerism. However, the proposed mesomeric form of hydroquinone-bis-[*p*-methoxybenzoate] is quite possible. This form makes the molecule

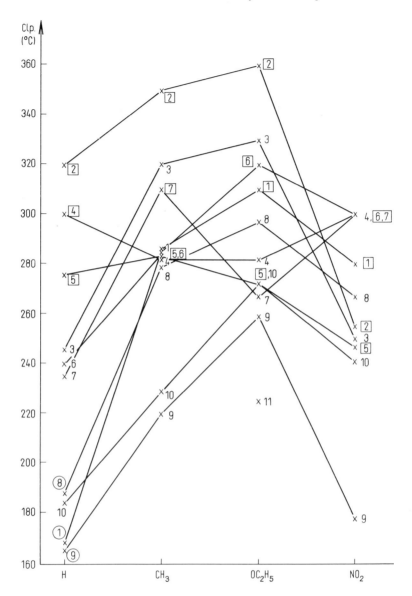

Fig. 2.11. Clearing temperatures in the system

$$R—\bigcirc—CH=N—\bigcirc—XY—\bigcirc—N=CH—\bigcirc—R$$

Key: 1, $-CH_2-CH_2-$ 2, $-CH=CH-$ 3, $-C\equiv C-$ 4, $-N=N-$
         5, $-N(O)=N-$ 6, $-CO-O-$ 7, $-CO-NH-$ 8, $-CH_2-O-$
         9, $-CH_2-NH-$ 10, $-Hg-$ 11, $-NH-$

         1 , the transition is below this temperature
         2 , the transition is above this temperature or it occurs under decomposition

markedly more polar at the ends of the chains, and it in fact exhibits a 20°C higher clearing temperature. The third isomer

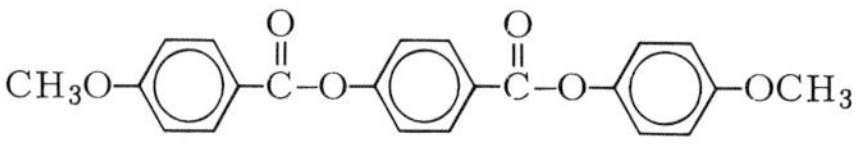

$$CH_3O-\!\!\bigcirc\!\!-\overset{\overset{\displaystyle O}{\|}}{C}-O-\!\!\bigcirc\!\!-\overset{\overset{\displaystyle O}{\|}}{C}-O-\!\!\bigcirc\!\!-OCH_3 \qquad\qquad k\ 145\ n\ 284\ i$$

has an intermediate thermal stability [758].

Most isomers differ much less with regard to their clearing points, and the following list of azomethines illustrates the difficulties of generalizing:

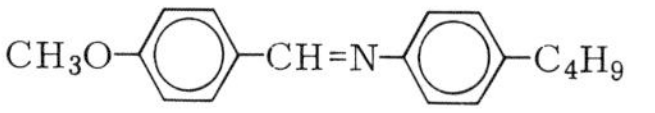

$$CH_3O-\!\!\bigcirc\!\!-CH\!=\!N-\!\!\bigcirc\!\!-C_4H_9 \qquad\qquad k\ 21\ n\ 47\ i$$

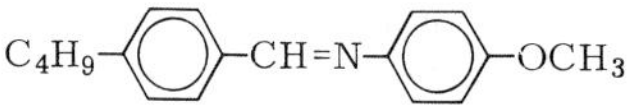

$$C_4H_9-\!\!\bigcirc\!\!-CH\!=\!N-\!\!\bigcirc\!\!-OCH_3 \qquad\qquad k\ 40\ n\ 43\ i$$

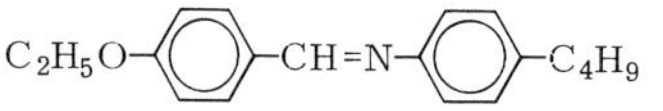

$$C_2H_5O-\!\!\bigcirc\!\!-CH\!=\!N-\!\!\bigcirc\!\!-C_4H_9 \qquad\qquad k\ 34\ n\ 68\ i$$

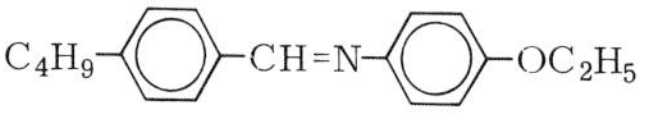

$$C_4H_9-\!\!\bigcirc\!\!-CH\!=\!N-\!\!\bigcirc\!\!-OC_2H_5 \qquad\qquad k\ 60\ n\ 79\ i$$

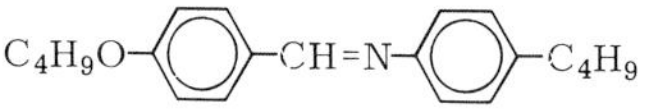

$$C_4H_9O-\!\!\bigcirc\!\!-CH\!=\!N-\!\!\bigcirc\!\!-C_4H_9 \qquad\qquad k\ 32\ n\ 55\ i$$

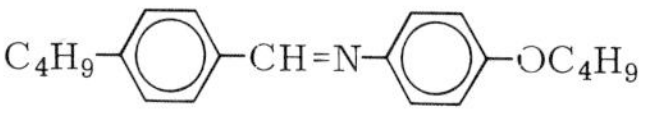

$$C_4H_9-\!\!\bigcirc\!\!-CH\!=\!N-\!\!\bigcirc\!\!-OC_4H_9 \qquad\qquad k\ 49\ n\ 74\ i$$

Deuterated azomethines [342] and other mesogens are described in the literature [305c].

Replacing aromatic units by cycloaliphatic rings has substantially reduced the stability of mesophases in some cases, even where the molecular form is hardly modified, e. g., the rigid linear 1,4-bicyclo-[2.2.2]-octyl grouping. This is due to the loss of polarizability. The stability is even more reduced by the 1,4-cyclohexyl group [163, 651, see also 380].

In other cases, the introduction of a *trans*-1,4-cyclohexyl group instead of a benzene ring hardly affected the clearing points of nematics [154, 644], and the following compounds deserve special note:

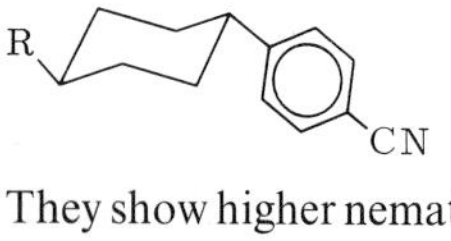

$$R\!-\!\langle\text{cyclohexyl}\rangle\!-\!\bigcirc\!-CN$$

They show higher nematic clearing points and a wider persistence range than the analogous biphenyl compounds [189–191], and the mesomorphism of the corresponding bicyclohexyl compound was quite surprising [191a].

A comparison of the following esters also shows that, with respect to the nematic phase thermal stability, the polarizability of the ends of the molecule is more significant than that of the center of the molecule [163, 587].

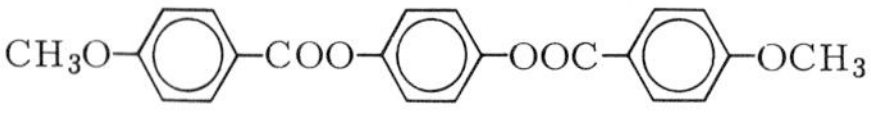

$$CH_3O-\!\!\bigcirc\!\!-COO-\!\!\bigcirc\!\!-OOC-\!\!\bigcirc\!\!-OCH_3 \qquad\qquad k\ 213\ n\ 297\ i$$

$$CH_3O-\!\!\bigcirc\!\!-COO-\!\!\langle\!\cdot\!\rangle\!-OOC-\!\!\bigcirc\!\!-OCH_3 \qquad\qquad k\ 185\ n\ 269\ i$$

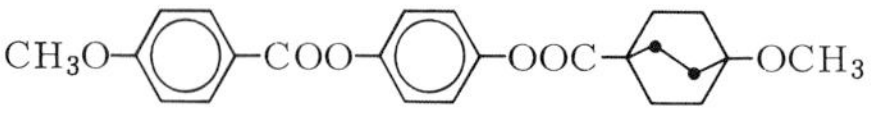

$$CH_3O-\!\!\bigcirc\!\!-COO-\!\!\bigcirc\!\!-OOC-\!\!\langle\!\cdot\!\rangle\!-OCH_3 \qquad\qquad k\ 189\ n\ 221\ i$$

A few examples studied by M. J. S. Dewar et al. seem to suggest that the replacement of a *p*-phenylene group by 1,4-bicyclo[2.2.2]octenylene, —⬡—, does not change the n/i transition temperatures while greatly lowering the melting points. In some cases the thio ester linkage led to more thermally stable mesophases than those observed with esters [164–166, 421a, 586, 587]. A few nematic selenolesters are known [327a].

2.5 *Broadening of Molecules*

From the very beginning, molecular breadth was considered an important feature in the investigations of the correlations between thermal stability and chemical constitution of mesophases. We have already discussed the influence of branched terminal groups on p. 51, but the increase in molecular breadth due to substitutions on the aromatic rings remains to be considered. This would normally reduce the anisotropy of molecular shape, thus depressing the stability of both nematics and smectics (the exceptional case of fluorine will be covered on p. 58). The experience with the biphenyl structures points out that twisting often accompanies substitution in these and other molecules, and this, like the increase in molecular diameter, lowers the clearing temperature. The following examples give the clearing points of some nematics:

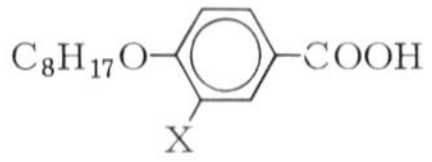

$C_8H_{17}O$—⬡—COOH   X: H 147°C   F   120.5°C [260]
        X 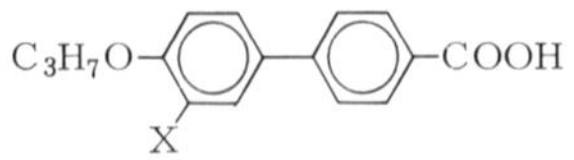

C_3H_7O—⬡—⬡—COOH   X: H 287°C   Cl   248.5°C
        X   Br 239°C   NO$_2$ 224°C [262, 267]

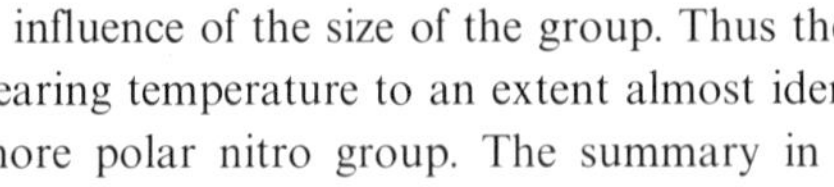

Cl—⬡—CH=N—⬡—N=CH—⬡—Cl X: H 290°C   CH$_3$ 228°C   Cl 216.5°C [14]
   X   X

Gray's quantitative studies of the R—⬡—⬡—COOH system yielded a good approximation of the linear correlation for nematics (shown in fig. 2.12 A) [276]. Smectics, shown in fig. 2.12 B, exhibit unity because of the influence of side-chain polarity in addition to the influence of the size of the group. Thus the smaller low-polarity methyl group depresses the clearing temperature to an extent almost identical with that shown by the distinctly larger but more polar nitro group. The summary in table 2.5 is taken from G. W. Gray [276].

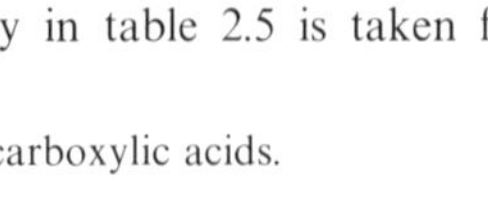

Table 2.5. 3'-Substituted 4'-*n*-alkoxybiphenyl-4-carboxylic acids.

	F	Cl	Br	I	NO$_2$	Me
decrease in smectic thermal stability (°C)	0.5	30	41	62.5	41	42
decrease in nematic thermal stability (°C)	9	31.5	40.5	50.5	50.5	27.5

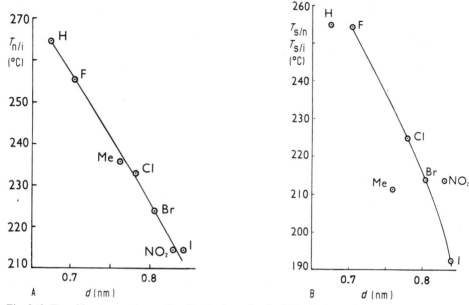

Fig. 2.12. Transition temperatures, $T_{n/i}$, (fig. A, the point for NO$_2$ in fig. A indicates the $T_{s/i}$), $T_{s/n}$ or $T_{s/i}$ (fig. B), as a function of molecular diameter, d, for different 3'-substituted 4'-n-alkoxybiphenyl-4-carboxylic acids (from [276]).

Comparison of the two homologous series RO—⬡—COO—⬡—OOC—⬡—OR
(where R'=—H and —CH$_3$) shows that the introduction of the methyl group depresses nematic clearing points by about 35–40°C and the melting points by about 20–50°C [16]. The s_C phase is confined to a much narrower range of existence. When R'=—CH$_3$, it first emerges enantiotropically with R=—C$_{12}$H$_{25}$, whereas it already exists with R=—C$_8$H$_{17}$ when R'=H. With increasing size of the lateral substituent (see table 2.6, where it is expressed by the van der Waals radius $r_{(X)}$ in nm, a steady decrease in the n/i transition temperature is observed. This is not due to a decrease in the heat of transition (which actually increases with increasing size of the substituent) but to a steady increase in the entropy of transition [167].

Table 2.6. Effect of molecular broadening on the thermodynamic parameters of the compounds.

CH$_3$O—⬡—COO—⬡—OOC—⬡—OCH$_3$

(ΔH: Heat of transition in kJ/mol, ΔS: entropy of transition in J/mol·K).

X	$r_{(x)}$	$T_{(n/i)}$	$\Delta H_{(n/i)}$	$\Delta S_{(n/i)}$
H	0.12	301	1.71	2.98
F	0.135	278.5	1.69	3.07
Cl	0.18	252.4	1.94	3.69
Br	0.195	241.1	2.04	3.97
I	0.215	222.9	2.07	4.16
CH$_3$	0.20	252.1	2.22	4.23

The literature gives many more examples of the influence of molecular broadening on the mesomorphous state [71, 161, 265, 484, 868]. Biphenyl bodies comprise one of the most interesting research areas in the study of relations between chemical constitution and mesomorphous properties. G. W. Gray has performed the most exhaustive investigations on these molecules, e. g., in the series

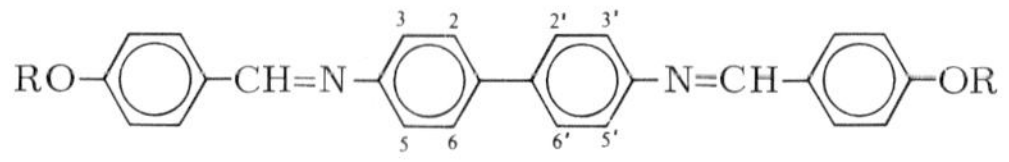

with R=—C_7H_{15} to —$C_{10}H_{21}$ and with —Cl, —Br, or —CH_3 on the biphenyl structure [273–275]. The substituents —Cl and —CH_3 cause similar depressions in clearing temperature, but —Br causes a distinctly greater one. The average depressions for —Cl compounds are given as a quantitative guide; in certain cases the additivity is obvious:

$T_{n/i}$ with Cl in   3 > 2 > 3,3′ > 2,3′ > 2,5 > 2,6 > 2,2′ > 2,2′,6 ≫ 2,2′,6,6′
ΔClp_n 48.4   84.1   94.2   126.4 134.6 152.9 160.2 222   no mesogen

The mesomorphous properties gradually disappear with increased twisting of the biphenyl, and this continues until mesomorphism disappears completely with 2,2′,6,6′-tetra substitution (exception: fluorine compounds discussed below). If, however, a substitution on the aromatic causes little or no increase in molecular diameter or twisting, then the increased polarity or polarizability will induce a higher clearing point. We can cite here only a few examples of this phenomenon, the most striking being that given by G. W. Gray. Among the *n*-alkoxynaphthalene carboxylic acids thus far investigated, only those substituted in the 2,6 positions have proven to be liquid-crystalline [258]. In these cases, the clearing temperatures can be raised by proper substitution in position 5 [255, 261]:

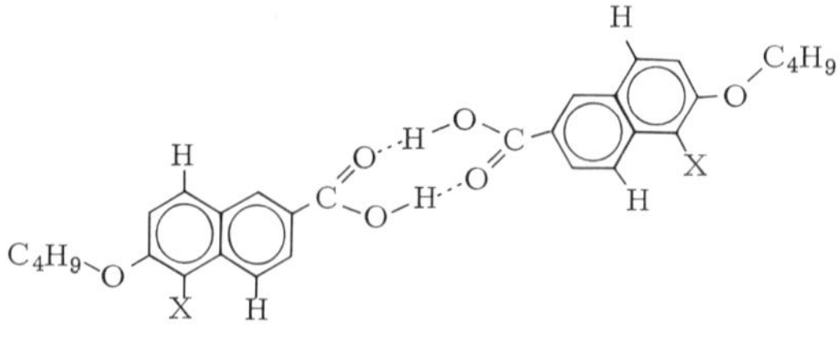

X	$T_{n/i}$ [°C]
H	208.5
Cl	216.5
Br	(213)
I	< 208.5

Other liquid-crystalline naphthalene derivatives can be expected to exhibit this effect too, e. g.:

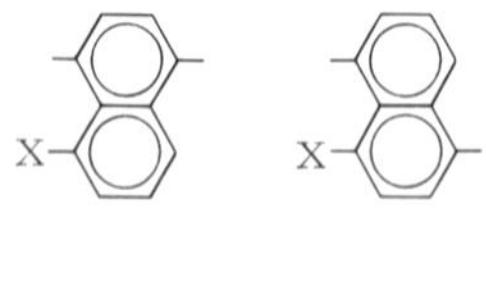

In the special case of the following compounds where intramolecular hydrogen bonds are involved, the clearing points are raised by replacement of H by OH [433, 434, 436, 495, 729]:

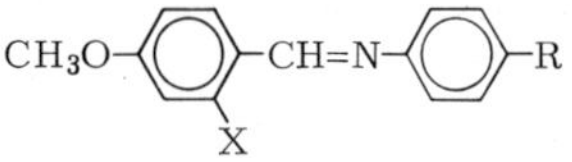

$$CH_3O-\langle\bigcirc\rangle-CH=N-\langle\bigcirc\rangle-R$$
$$\underset{X}{}$$

	X: H		X: OH	
	F	$T_{(n/i)}$	F	$T_{(n/i)}$
R: C_4H_9	21	46	44	64.5
$CH=CH_2$	97	111	111	124
C_6H_5	161	177	165	197
CN	106	117	141	147

The high clearing point of $C_3H_7-\langle\bigcirc\rangle-\underset{O\ H}{\overset{}{\langle\bigcirc\rangle}}-\langle\bigcirc\rangle-C_3H_7$, which exceeds that of the corresponding terphenyl compound by about 100°C, is also attributed to hydrogen bonding [151]. Because of low k/n transition temperatures, a variety of compounds were synthesized that derive from the formula

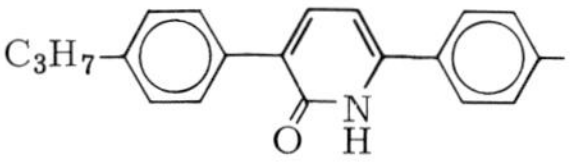

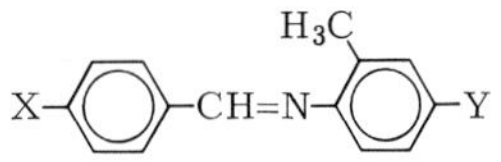

$$X-\langle\bigcirc\rangle-CH=N-\overset{H_3C}{\langle\bigcirc\rangle}-Y$$

[521, 680, 682].

Fluorine compounds occupy a special position. Unfortunately, the literature is rather sparse, but the following cases will serve to illustrate the problem [247, 274, 277, 527a]:

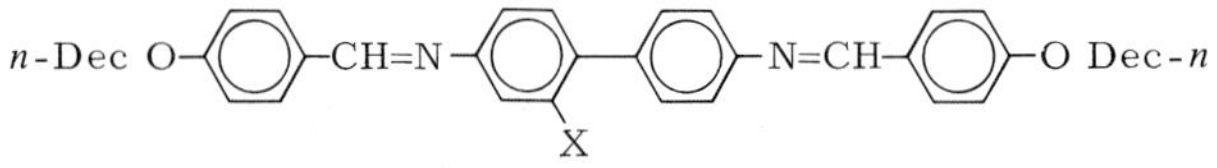

$$n\text{-Dec }O-\langle\bigcirc\rangle-CH=N-\underset{X}{\langle\bigcirc\rangle}-\langle\bigcirc\rangle-N=CH-\langle\bigcirc\rangle-O\text{ Dec-}n$$

X: H   k 202.5 s 311.5 n 324 i
X: F   k 123 s   253 n   296 i

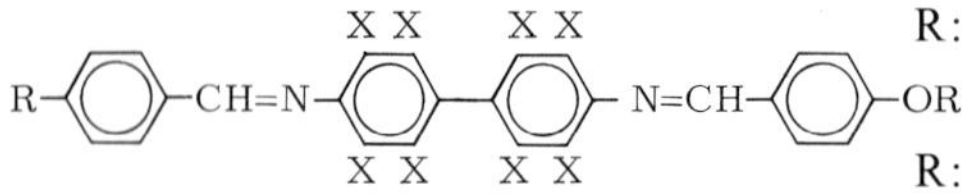

$$R-\langle\bigcirc\rangle-CH=N-\underset{X\ X}{\overset{X\ X}{\langle\bigcirc\rangle}}-\underset{X\ X}{\overset{X\ X}{\langle\bigcirc\rangle}}-N=CH-\langle\bigcirc\rangle-OR$$

R: CH_3 X: H   k 266 n >390 i
             X: F   k 219 n 370 i
R: nC_3H_7 X: H   k 255.5 n >390 i
             F: k 205 s 210 n 315 i

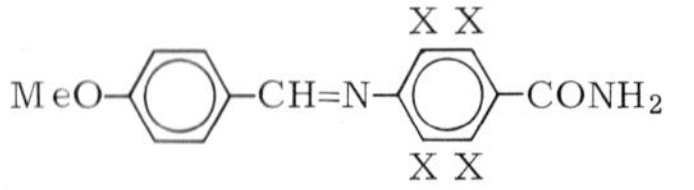

$$MeO-\langle\bigcirc\rangle-CH=N-\underset{X\ X}{\overset{X\ X}{\langle\bigcirc\rangle}}-CONH_2$$

X: H   k 176 s 195 i
X: F   k 238 s 268 i

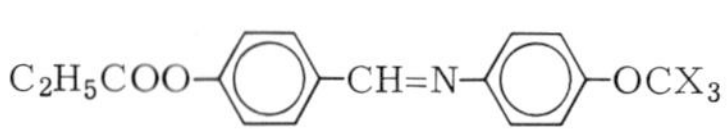

$$C_2H_5COO-\langle\bigcirc\rangle-CH=N-\langle\bigcirc\rangle-OCX_3$$

X: H   k 86 n 118 i
X: F   k 90 s 136 i

The first two examples show that the clearing temperatures are lowered when the introduction of fluorine causes a broadening or twisting of the molecule. The final two examples clearly indicate that a fluorine substitution can raise the thermal stability of a smectic phase. As a rule, such substitutions seem to favor the smectic over the nematic state. G. W. Gray has found the CF_3 group to be less efficient in inducing nematic mesomorphism than the CN and CH_3 groups [297]. The literature on low temperature mesomorphism is summarily cited here because of its technological importance [7, 8, 34, 47–50, 56–62, 64, 72, 93, 95–97, 112, 116, 118–120, 122, 124, 141, 144, 152, 154, 155, 157, 158, 170–174, 176, 179–183, 189–191, 204, 214, 223, 224, 226, 241, 242, 245, 277, 281, 283, 287, 288, 290, 293–296, 298, 300–302, 305, 312, 313, 317, 320, 337–339, 343, 344, 350, 353, 354, 356–359, 364, 369–372, 374–377, 380, 386–389, 391, 392, 394, 395, 400, 403, 415, 416, 420, 421, 432–434, 436–438, 442, 445–447, 456–458, 460, 465, 466, 482, 483, 485, 489, 491–493, 495, 496, 500–527, 534, 536, 540, 541, 546, 548, 552, 553, 578, 586, 599, 606, 608, 609, 624, 628, 644, 655, 656, 664, 665, 680–682, 685–690, 692–701, 710, 712, 717–720, 727–729, 731, 740, 747, 748, 750, 751, 756–758, 760, 761, 840, 842, 878, 879, 883, 886–891, 893, 894, 1:10, 1:12].

p-Methoxybenzylidene-*p'*-*n*-butylaniline

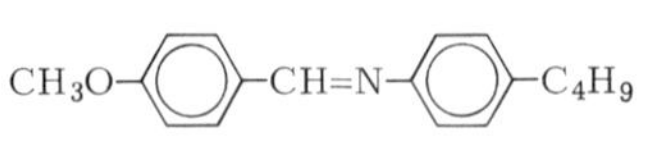

(MBBA, k 21 n 46 i)

was the first substance of sufficient stability which exhibited the nematic state at room temperature [392].

2.6 *Aromatic Carboxylic Acids*

With over 250 representatives, the aromatic carboxylic acids form a significant and well-investigated group of thermotropic liquid-crystalline compounds. The most numerous are the *p*-substituted benzoic and cinnamic acids, biphenyl-4-carboxylic acids, and 6-substituted naphthalene-2-carboxylic acids. In addition, mesomorphous 7-alkoxyfluorene-2-carboxylic acids are known, but the analogous 7-alkoxyfluorenone-2-carboxylic acids are not mesomorphous. We have already often mentioned carboxylic acids because they lend themselves to convenient classification in the structural system of liquid-crystalline substances having aromatic as well as bridging and terminal groups, if we consider the carboxyl dimer

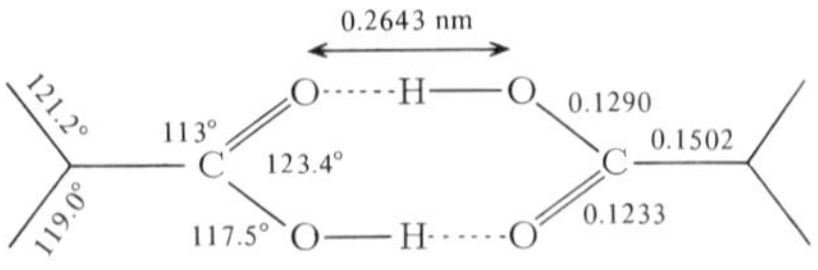

in the role of the bridging group. However, these acids occupy a special position in the system due to their dissociation. In homologous series and with side-chain substitutions, carboxylic acids behave as previously indicated. Liquid-crystalline dicarboxylic acids derived from *p*-terphenyl deserve to be mentioned [146]. Primary aromatic carboxylic acid amides, which dimerize like carboxylic acids, form liquid crystals, but they have been infrequently studied to date [247, 277, 722].

The molecular structure of carboxylic acids is well known from X-ray investigations [78, 80–82, 736, 737]. The above data applies to anisic acid [80]. A detailed X-ray study of *p-n*-butoxybenzoic acid revealed two crystallographically independent molecules [81]. In one of them, the two carboxy C—O distances are not significantly different, whereas there is the expected distinction between them in the other. The different patterns of distortions in both molecules confirm that the deviations from planarity have their origin in intermolecular rather than in intramolecular interactions.

2.7   *Carboxylic Acid Salts and Ammonium Salts*

The consistently smectic phases of these compounds are derived from aromatic and aliphatic structures, such as the following (see section 2.13):

1. [RCOO]$_n$M, where R = *n*-alkyl, *i*-alkyl, or alkenyl, and M = Na, K, Tl, or, more rarely, Rb, Cs, Mg, Ca, Cd, Al, Cu, or Pb. The clearing temperatures in homologous series of the salts of *n*-alkane carboxylic acids obey the curve expected for smectic phases.

2. 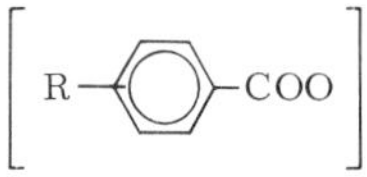 $\left[R - \bigcirc - COO \right]$ M, in which M = Li, Na, K, Rb, Cs, and Tl. Salts of numerous,

usually *p*-substituted benzoic acids form liquid crystals as do salts of cinnamic acids, phenylacetic acid, mandelic acid, naphthalene-1- (and -2-) carboxylic acids and, some other acids.

There are two ways in which a long-chain radical can be introduced in ammonium salts, and neither method has been thoroughly investigated to date. For instance, it is stated that tetramethylammonium stearate is liquid-crystalline, but no other details are given. Somewhat more data are available on pyridinium salts having a long *n*-alkyl group attached to the nitrogen in the α- or β-position, and the highest clearing temperature is found with the stearyl group [413]. Recently mesomorphism has been found with *p-n*-alkoxybenzamidine hydrochlorides [642, 643] and *p-n*-alkoxyaniline hydrochlorides [631]. In contrast to almost all other types of liquid-crystalline substances, the aliphatic representatives of this class lack any rigid component in the molecule. This feature is also found in the smectic alkyl stearates [425] and the liquid crystals of long chain alkylamines [536] and carbonyl bis(amino acid esters) [452] (cf. section 2.9).

All carbon atoms in the B form of crystalline sodium palmitate assume the form of a stretched zigzag chain and lie in one plane. A 16.4° rotation of the carboxyl group about the first C—C bond in the molecule causes one oxygen atom to lie on each side of the plane [184]. The length of the C—C bond in the paraffin chain averages 0.1521 nm, and the C—C—C angle is 114.4°. The C—O bonds of the carboxylate group are much shorter at 0.121 and 0.127 nm, and the O—C—O angle is 123°. The stretched hydrocarbon chain and the stratified lattice both contain features of the smectic order. For information on X-ray diffraction analyses of the salts of other carboxylic acids, we refer the reader to the literature [79, 450, 668, 669, 779]. Since D. Vorländer's pioneering work in 1910 [791], only H. Sackmann has thoroughly investigated the behavior of organic salts upon melting [1 : 322–1 : 324]. Why these substances are declared as a novel class of liquids in refs. [745, 746] is not evident.

The salts of sulfonic acids also include some mesogens, the sole examples being the sulfanilates of the structure 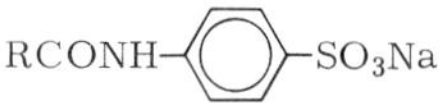 RCONH–$\bigcirc$–SO$_3$Na [417].

2.8   Steroids

Several hundred mesogens derive from steroids, mainly cholesterol:

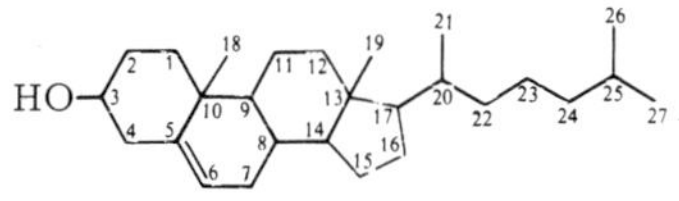

The optical activity of these derivatives of naturally occurring substances imparts the nematic-cholesteric texture to their mesophases (see chapters 1 and 7). The existence of a mesophase again requires a rod-shaped molecular geometry, polarity, and polarizability. While cholesterol derivatives are liquid-crystalline, those of epicholesterol are not because the axial position of the hydroxyl group causes a bent molecule. Neither coprostanol nor epicoprostanol derivatives are liquid-crystalline because the cis-connection between the A and B rings also imparts an angular shape to the molecules.

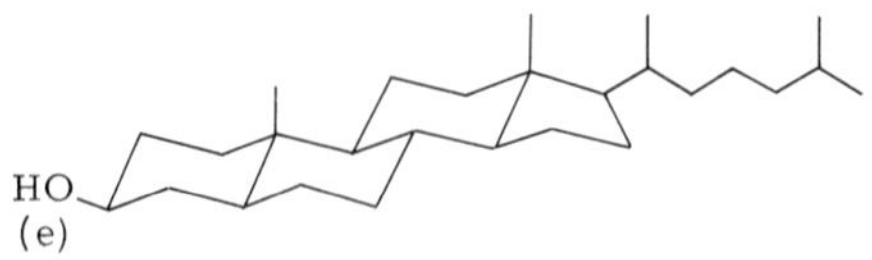

cholestanol

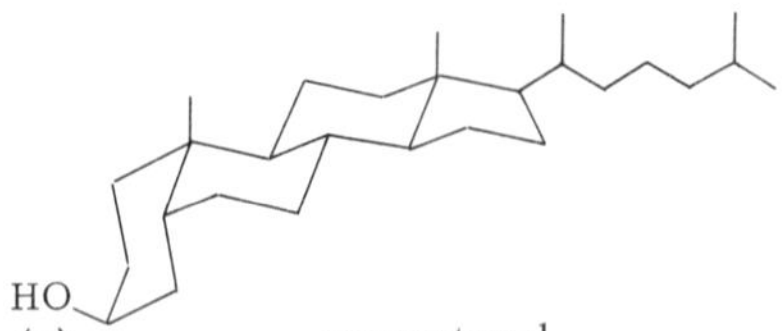

coprostanol

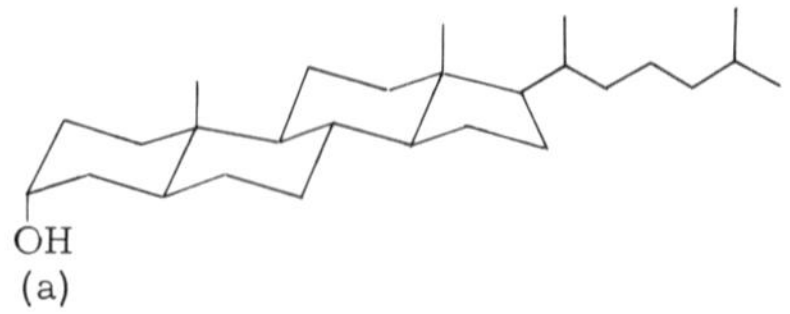

epicholestanol

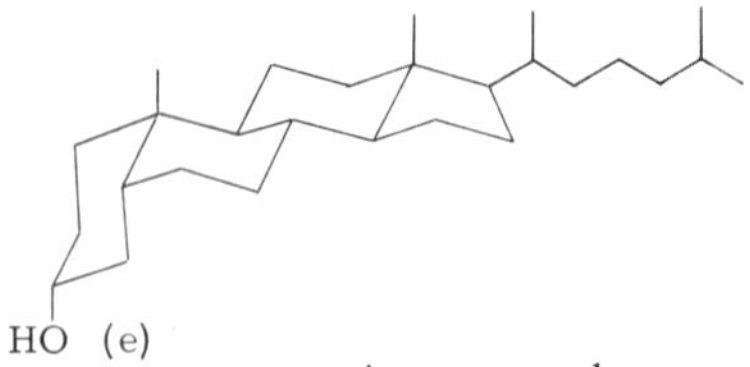

epicoprostanol

By abstracting C. Wiegand's results, one can conclude that steroid mesomorphism requires a stretched configuration of the ring system [866]. This means that there must be a *trans*-bonding of all rings and an equatorial orientation of substituents with respect to the long axis of the molecule, i.e., at the 3β- and 17β-positions.

The n-alkanoates have been most thoroughly investigated of the aliphatic cholesteryl esters. In addition, there are many known mesomorphous esters of branched and unsaturated carboxylic acids, ether carboxylic acids, and the homologous series of cholesteryl-n-alkyl carbonates of cholesteryl-5-n-alkylthio carbonates (RSCOOCh), thiocholesteryl alkanoates (RCOSCh), and S-cholesteryl-O-n-alkyl thiocarbonates (ROCOSCh) (see section 2.13).

The position, number, and length of branches in cholesteryl alkanoates profoundly affect their mesomorphous behavior [241, 242]. Physical and chemical properties of cholesteryl esters [556, 663] as well as reviews [7:106, 203, 731] are in the literature. With the homologous series compared in fig. 2.13, increased thermal stability of smectic phases is observed by the replacement of oxygen by sulfur [202].

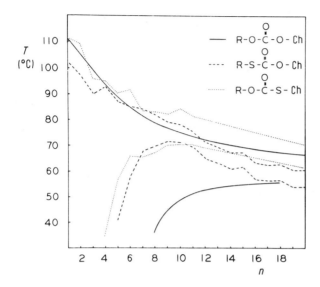

Fig. 2.13. Transition temperatures, $T_{c/i}$ and $T_{s/c}$, of cholesteryl alkyl-carbonates, cholesteryl S-alkyl thiocarbonates, and S-cholesteryl alkyl thiocarbonates as a function of the number of carbon atoms, n, in the alkyl chain (from [202]).

The cholesteryl esters of succinic, adipic, and sebacic acids are also mesomorphous as are the inorganic cholesterol derivatives, e. g., cholesteryl chloride and carbonate. Numerous aromatic cholesteryl esters form mesophases, particularly those derived from *p*-substituted benzoic and cinnamic acids [616, 780, 781]. The mesophases of cholesteryl-6-alkoxy-2-naphthoates [136, 139] and cholesteryl *p*-alkoxybenzylidene-*p'*-aminobenzoates [142]

are significant due to their high thermal stability.

The latter compound may be regarded as a hybrid between steroids and the standard type of mesogens, but such combinations are not yet frequently employed. Other connecting links distantly related to steroids are the following derivatives of estrone [333] and *trans*-hexahydrochrysene [725]:

Mesomorphous esters of estradiol- [3, 17] were recently reported [648] and the mesomorphism of the hydrocarbons of the structure

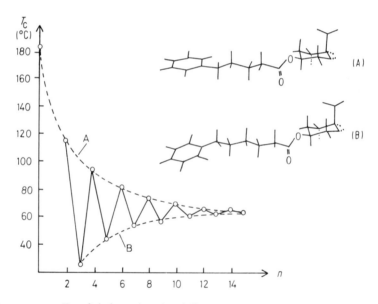

deserves special note [470a].

In contrast to the scarcely alternating clearing temperatures of the homologous *n*-alkanoates of steroids [530], those of the ω-phenylalkanoates alternate very strongly as is shown in fig. 2.14 [199, 564].

Fig. 2.14. Transition temperatures, $T_{c/i}$, of cholesteryl ω-phenylalkanoates as a function of the number of methylene units, *n*, in the alkylene chain (from [199]).

Recently the thio-analogs derived from 3β-mercaptocholesterol have been described [201].

Variations of structure include:

a) the acid that esterifies the 3β-hydroxyl,
b) the 17β-side chain, and
c) the steroid skeleton.

Systematic studies show that the 17β-chain of steroids can be modified and shortened without the compound losing its mesomorphous properties. This is possible right down to hydrogen, methyl, ethyl, isopropyl, and the following two structures [198, 467, 562, 563]:

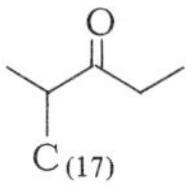

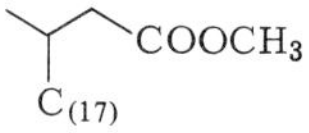

Increased branching of the 17β-chain, as in campesteryl [198] and β-sitosteryl esters [428, 560]

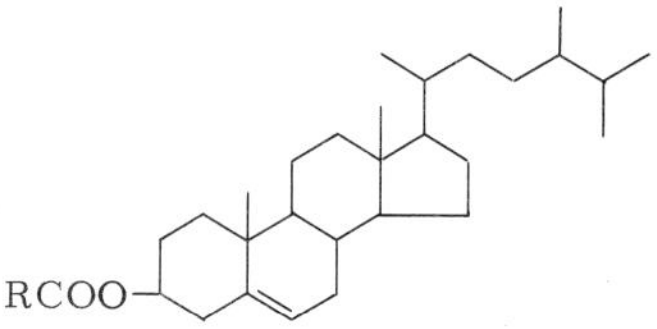

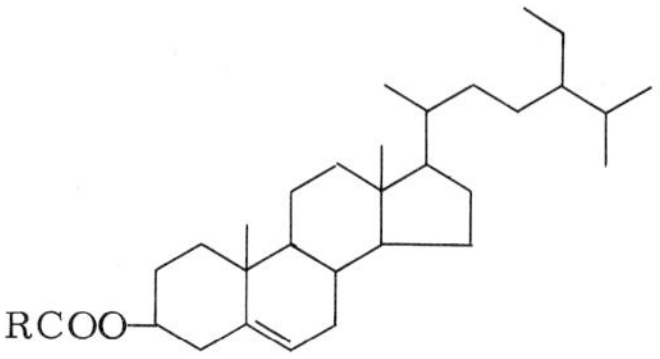

impairs the formation of mesophases. This also applies to the extra introduced double bonds, as in stigmasteryl esters [104, 107, 428, 559].

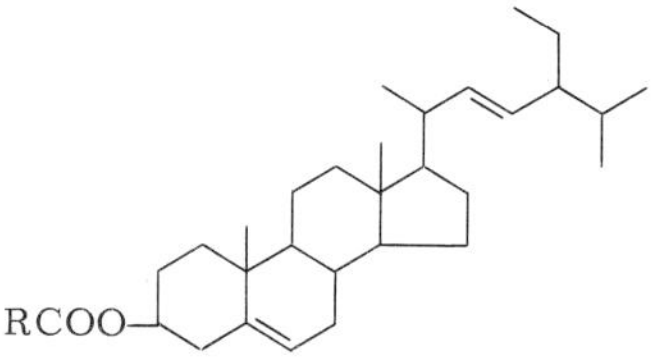

5α-cholestanyl esters have lower clearing points than the analogous cholesteryl esters [195, 199, 561, 616, 677, 866].

Several quite stable but exclusively smectic mesophases are known for the homologous series of ergosteryl *n*-alkanoates [233, 410, 428, 616]:

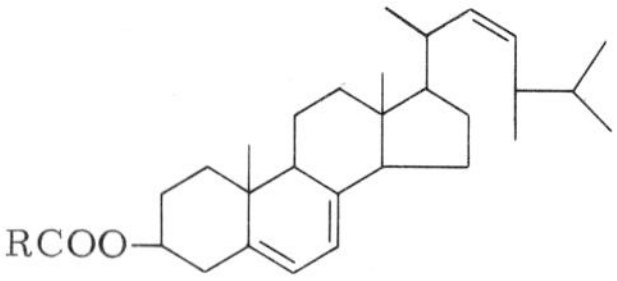

Only a slight difference in melting and clearing points is seen from the examples below, differing only in positions and number of double bonds in ring B [866]:

	$T_{k/c}$ [°C]	$T_{c/i}$ [°C]
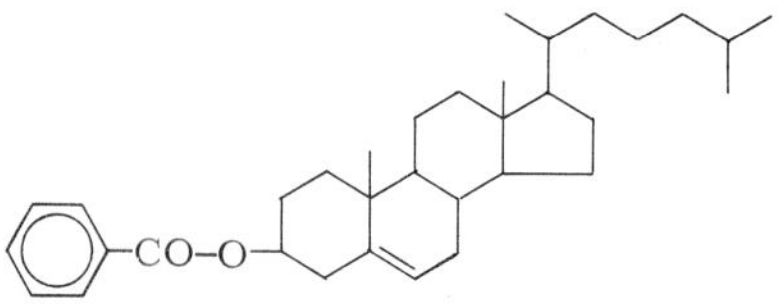	150	178

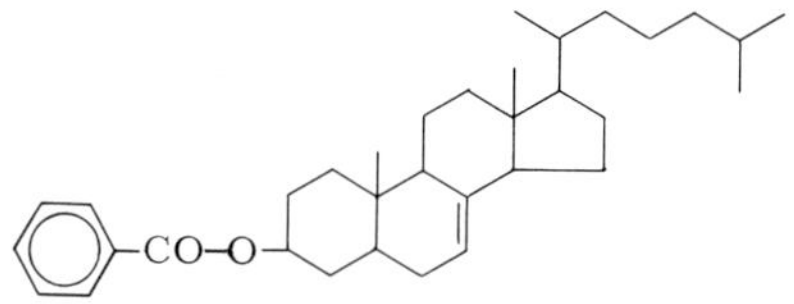

158	176	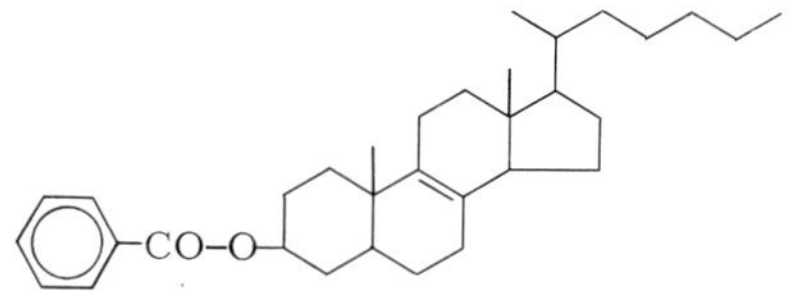
147	174	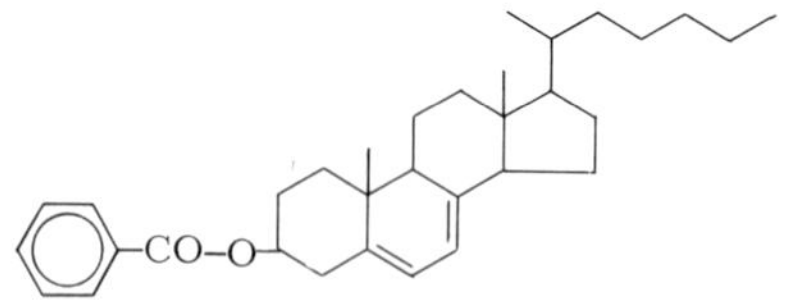
143	188	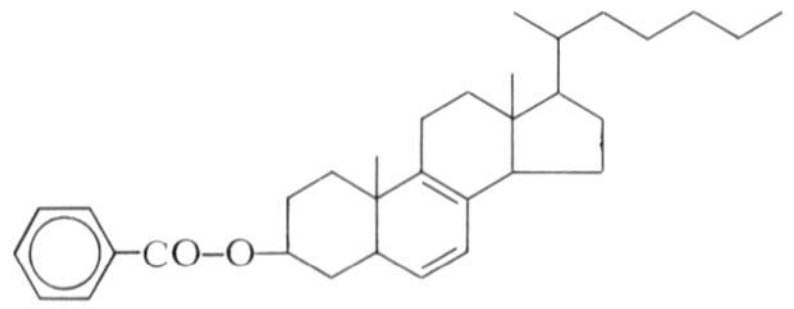
146	180	

Examples having double bonds in other positions are too rare to draw general conclusions. Three cases are known where a 14,15 double bond (in ring D) destroyed the mesomorphous properties [866]. A number of right-rotatory cholesterics were found among the $\Delta^{8(14)}$-cholestanol esters (doristeryl esters) [109, 110, 445–447]. Similar results occur with a double bond in ring A, e. g., in esters of 5α-cholest-1-en-3β-ol and cholest-4-en-3β-ol [562]. The literature also mentions liquid crystals among the triterpenes, such as the cycloartenyl alkanoates [407–409, 411]:

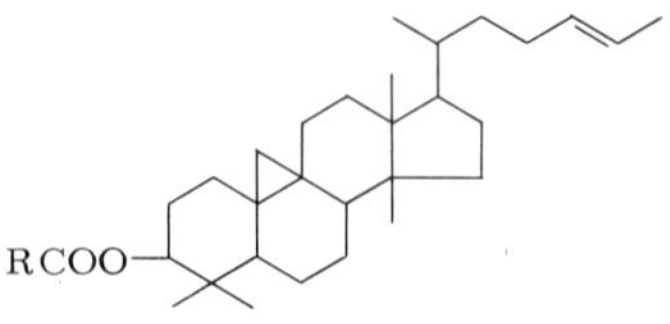

A detailed description of the structure of steroids would be beyond the scope of this book, but one may get a good impression of the extended nature of the steroid molecule from the "dimensions": 0.75 nm broad (C_7 to C_{11}), 0.43 nm thick (C_{19} over C_6), and 2.14 nm long (in cholesteryl iodide from I to C_{27}) [86, 89].

For two typical cholesteric mesogens the molecular shape shown in fig. 2.15 is assumed [844].

Other crystallographic studies of these compounds are reported [26, 121, 184a, 428a].

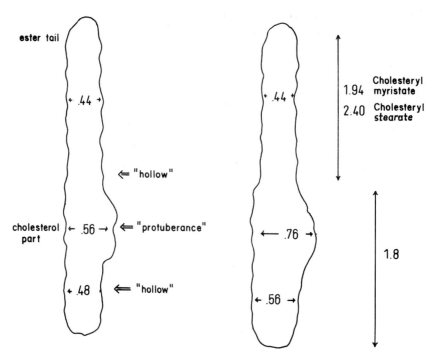

Fig. 2.15. Molecular shapes and dimensions (in nm) of cholesteryl myristate and cholesteryl stearate deduced from space filling models (from [844]).

2.9   Aliphatic Compounds

Except for the salts of aliphatic carboxylic acids and the aliphatic esters of steroids, only a few aliphatic compounds form mesophases. The mesomorphism of 1-alkadiene-2,4 acids proved that a high enough anisotropy of molecular polarizability is sufficient to cause the formation of a mesophase [463, 854]. But this is not a condition sine qua non, because *trans*-4-*n*-alkylcyclohexane-1-carboxylic acids are also mesogens [644]. We should also mention alkadiene-(2,4)-al-(1)-azines [463], vinyloleate with the data k −45 s −18 i [277], esters with proven mesomorphism, e. g., the *n*-alkylstearates [46, 425, 581], aminoacid derivatives of the structure

$$CH_2\!\!=\!\!C(CH_3)CONH(CH_2)_4CH\genfrac{}{}{0pt}{}{COOH}{NHCO(CH_2)_nCH_3}\ [664],$$

carbonyl-bis-amino acid esters [452, 9:112, 9:113], and *trans,trans*-4'-alkyl-4-cyanobicyclo-hexyls [191 a].

2.10 Inorganic and Miscellaneous Compounds

In the melts of some metals ordered states are met that are reminiscent of mesophases, but which cannot really be classified as such [177, 215, 340, 341, 528, 623]. Nor can any of the following be called liquid crystals: ammonium oleate, which exhibits only lyotropic mesomorphism [494, 835]; silver and thallium halogenides, which separate as plastic crystals [707] as is the case with carborane- [404] and ferrocene derivatives [253, 769, 873a], but not for those described in ref. [470].

The postulated mesophases of polychlorinated metallocenes [163] are probably also plastic crystals. All in all, mesomorphous structures are found only very rarely if at all among inorganic compounds, e. g. silicates [205, 535], organosilicon compounds [70, 187], organonickel compounds [244a], and organopalladium compounds [84a, 597].

Liquid-crystalline substances whose chemical nature is largely unknown, but whose existence has been fully confirmed, are found among the products of the liquid-phase pyrolysis of aromatic mixtures, e. g. coal tar pitches [334, 347–349, 402, 451a, 451b, 451c, 454a, 463a, 610, 742, 859, 896]. At about 400°C, the isotropic phase transforms into a spherulitic mesophase that suddenly and completely coagulates to a bulk mesophase at approximately 465°C. This first solidification phase remains to about 525°C, above which a second solidification to semicoke sets in. The literature cited contains detailed descriptions of these mesophases.

A number of recent works confirm that the processes of graphitization and carbonization imply a mesophase transition state under certain conditions [24, 178, 244, 250, 351, 351a, 355, 373, 451, 473–481, 481a, 538, 539, 573, 585, 611–615, 662, 666, 667, 668a, 723, 741, 860–864, 876, 877].

The first mesogenic free radicals to be synthesized were described [185, 186].

2.11 Conclusion

Ninety years of liquid crystal research have revealed thousands of mesogens that belong to very different classes of chemical compounds. This gives evidence that the prime requirements for long-range self-ordering in a liquid system come from molecular geometry. On the other hand, certain chemical structural units provide optimal molecular shape together with other dispositions for mesomorphism, and this is the reason for its very frequent occurrence in certain classes of compounds. This can be seen from the following table which should also allow to discern missing links and thus assist in the further expansion of this field. Homologous series are described in the boldfaced references. We also refer to the sections 5.7 (Bibliography of X-ray studies on liquid crystals), 10.2.5 (Bibliography of NMR studies on liquid crystals), and 10.3.5 (Table of nematic solvents for NMR spectroscopy).

2.12 Classification of Mesogens

1 Aromatic compounds without bridging group

2 Heteroaromatic compounds without bridging group

3 Aromatic compounds with one bridging group

3.1 Azomethines
3.2 Azo compounds
3.3 Azoxy compounds
3.4 Carboxylic acid esters

11 Steroids

12 Aliphatic compounds

13 Miscellaneous

2.13 Table of Mesogens

1   Aromatic compounds without bridging group

1.1   hexasubstituted benzenes X = OR [50a] X = OCOR [99]

1.2   4,4'-disubstituted biphenyls

X, Y = R, —OR, —OCOR, —OCOOR, —COR, —CN, —NO$_2$, —CH$_2$COOR, —CO(CH$_2$)$_n$-COOR, and other groups [62, 67, **152, 182, 183**, 224, 228, **262**, 281, 283, 288, 290, 293, **295, 296, 300**, 353, 354a, 363, 376, 542a, 548, 601, **624, 645**, 748, 754, 805]
X = R (chiralic): [282, 294, 301, 305d, 740]

3,4,4'-trisubstituted biphenyls [267]

1.3   4,4''-disubstituted p-terphenyls

X, Y = R, —OR, —COR, —COOR, —(CH$_2$)$_n$COOCH$_3$, —CO(CH$_2$)$_n$COOCH$_3$, —CN, —NO$_2$, Cl, Br, NH$_2$, and other groups [1:331, **146**, 272, 283, 286, 483a, 484, 607, **637, 645**, 805, 830]
X = R (chiralic): [290, 301]

3,4,4''-trisubstituted p-terphenyls [272]

1.4   X—⬡—⬡—⬡—⬡—Y   4,4‴-disubstituted quaterphenyls [212, 222, 643, 645, 805]

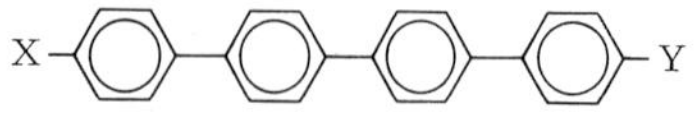

1.5   X—(⬡)—Y   quinquiphenyls and sexiphenyls [805]
          5,6

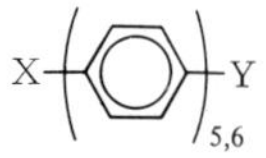

1.6   ⬡⬡   2,6-disubstituted naphthalenes [379]

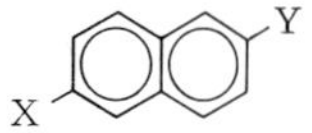

1.7   X—⬡⬡—Y   2,7-disubstituted fluorenes [47, 50, **268**, 290]

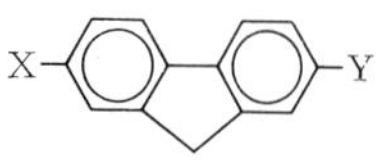

1.8   X—⬡⬡—Y   2,7-disubstituted fluorenones [268]
          O

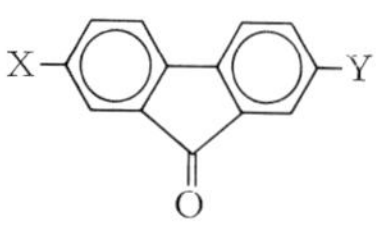

1.9   X—⬡⬡—Y   2,7-disubstituted 9,10-dihydrophenanthrenes [**2**, 47, 48, 114a]

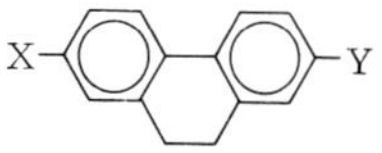

1.10   (triphenylene structure with OR groups)   2,3,6,7,10,11-hexaalkoxytriphenylenes [183c]

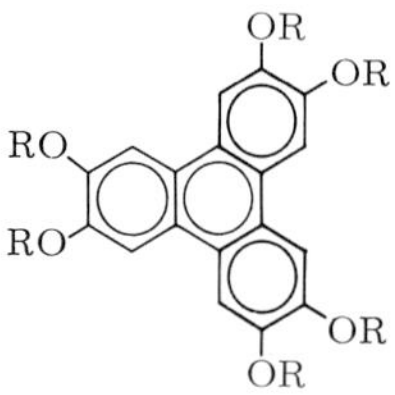

2   *Heteroaromatic compounds without bridging group*

2.1   X—⬡—⬡N   2-hydroxy-4-phenylpyridines [643]
              OH

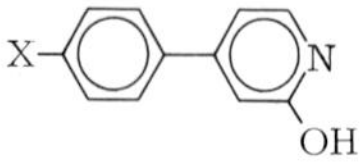

2.2   X—⬡—⬡—OH   2-hydroxy-5-phenylpyrazines [643]

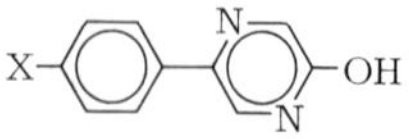

2.3   X—⬡—⬡—Y   2-phenylpyrimidines [61, 63a, 63b, 655, **893, 894**]

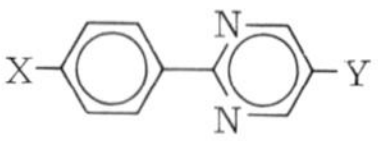

2.4   X—⬡—⬡—Y   5-phenylpyrimidines [63b, 656c, 894a]

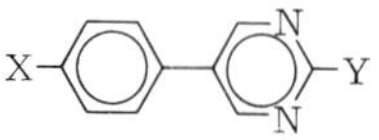

2.5	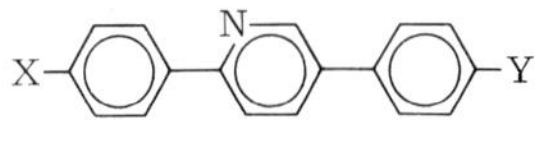	2,5-diphenylpyridines [151, **643**]
		2,5-diphenyl-6-hydroxypyridines [151, **643**]
2.6	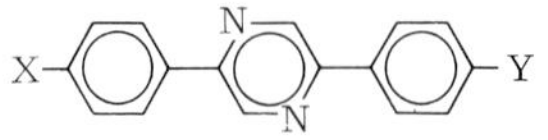	2,5-diphenylpyrazines [1:328, 1:330, 151, **641,** 643]
		2,5-diphenyl-3-hydroxypyrazines [**636, 639, 640,** 643]
2.7	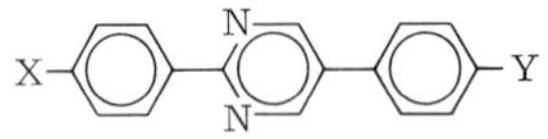	2,5-diphenylpyrimidines [1:325, 63a, 151, **642,** 643, 646]
		2,5-diphenyl-4-hydroxypyrimidines [**642**]
2.8	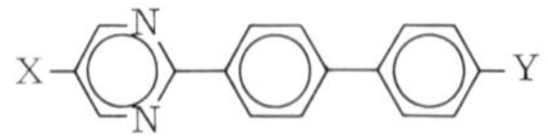	2-biphenylylpyrimidines [63a]
2.9	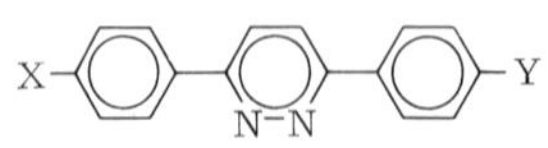	3,6-diphenylpyridazines [151, **643, 848**]
2.10	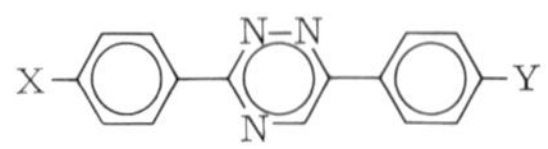	3,6-diphenyl-1,2,4-triazines [656a]
2.11	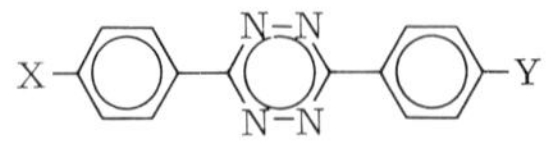	3,6-diphenyl-s-tetrazines [151, **643**]
2.12	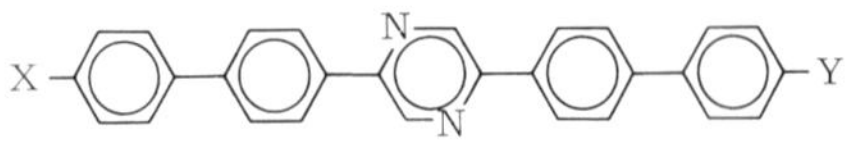	2,5-di-(*p*-biphenyl)-pyrazines [**643**]
2.13	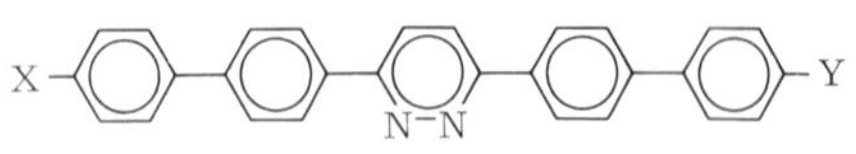	3,6-di-(*p*-biphenyl)-pyridazines [643]
2.14	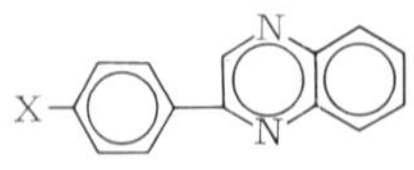	2-phenylquinoxalines [652] X = radical: [186]
2.15	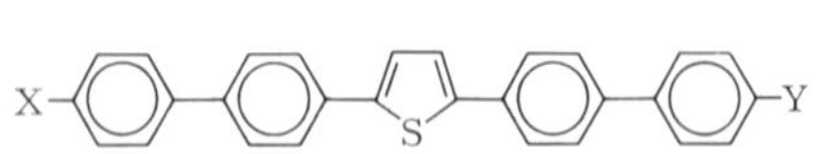	2,5-bis-(biphenylyl-(4))-thiophenes [650]

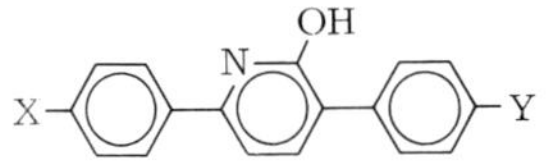

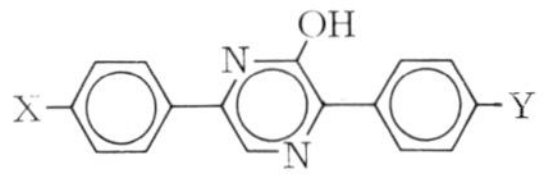

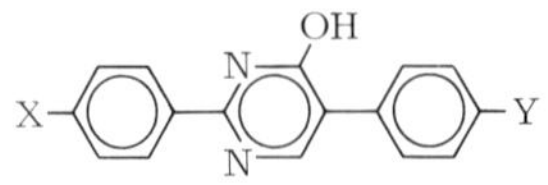

2.16   X—⬡—[thiazole N, S]—⬡—Y   2,5-diphenylthiazoles [654]

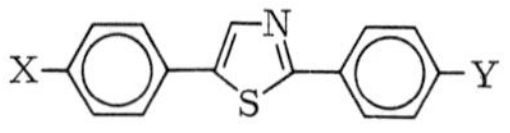

2.17   X—⬡—[N=N, S thiadiazole]—⬡—Y   2,5-diphenyl-1,3,4-thiadiazoles [654, 656]

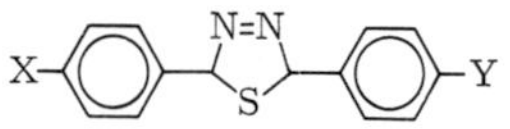

2.18   X—⬡—⬡—C(=O)—[imidazole N, N-H]—⬡—⬡—Y   2,4-disubstituted imidazoles [**643**, 654]

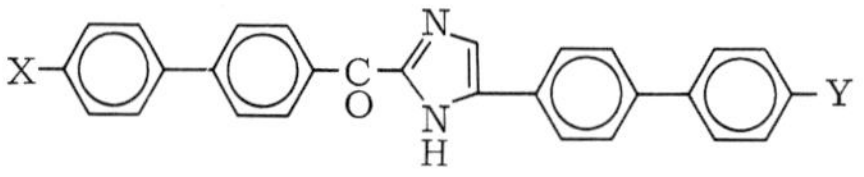

2.19   X—[benzoxazole N, O]—⬡—Y   2-phenylbenzoxazoles [550]

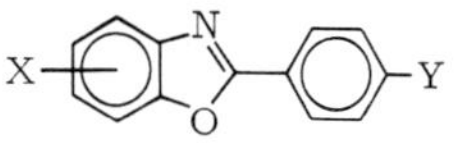

2.20   ⬡—[benzoxazole N, O]—⬡⬡—[benzoxazole N, O]—⬡   benzoxazole derivatives [391]

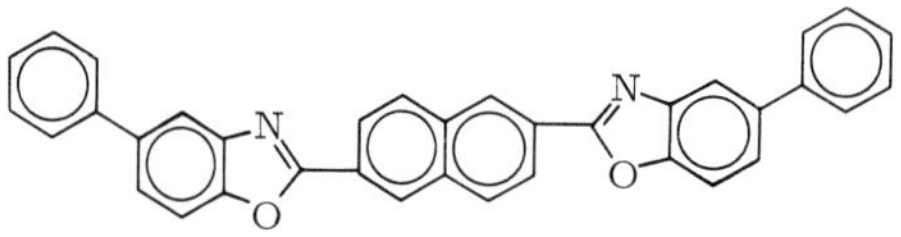

2.21   X—[benzthiazole N, S]—⬡—Y   2-phenylbenzthiazoles [550]

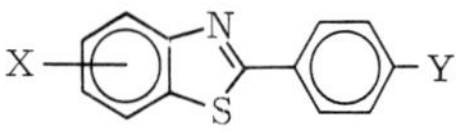

2.22   other heterocyclic compounds [417a]

3   *Aromatic compounds with one bridging group:*

—CH=N—,   —N=N—,   —N(O)=N—,   or —COO—

3.1   *Azomethines*

subtypes:

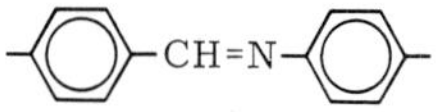

⬡—CH=N—⬡—   benzylidene anilines 3.1.1–26

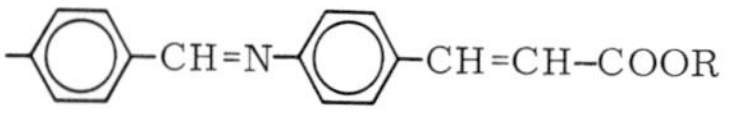

⬡—CH=N—⬡—CH=CH—COOR   benzylidene *p*-aminocinnamates 3.1.28–33

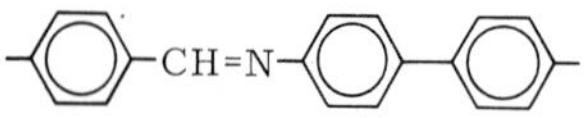

⬡—CH=N—⬡—⬡—   benzylidene *p*-aminobiphenyls 3.1.34

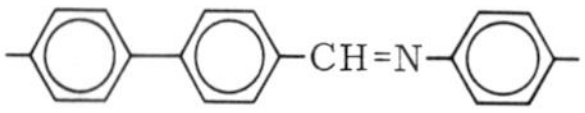

⬡—⬡—CH=N—⬡—   biphenylidene anilines 3.1.35

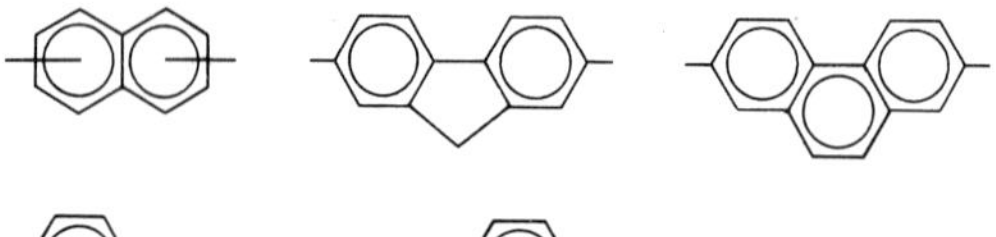

[naphthalene]   [fluorene]   [phenanthrene]   azomethines with naphthalene, fluorene or phenanthrene moiety 3.1.38–43

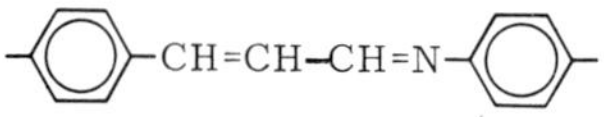

⬡—CH=CH—CH=N—⬡—   cinnamylidene anilines 3.1.44

3.1.1	X—⬡—CH=N—⬡—X	p,p′-symmetrically substituted benzylidene-anilines [1:277, 51, 90, 122, 212, 214, 222, **227**, 316, 321, 381, 487, 541, 541a, 787, 817, 830, 833, 834]
3.1.2	X—⬡—CH=N—⬡—Hal	p-X-benzylidene-p-halogenoanilines [45, 49, 180, 222]
3.1.3	Hal—⬡—CH=N—⬡—X	p-halogenobenzylidene-p-X-anilines [49, 180, 222]
3.1.4	X—⬡—CH=N—⬡—CN	p-X-benzylidene-p-cyanoanilines [49, **56, 60**, 90, 180, 222, 223, 344, 416a, 437, **458**, 482, 540, 733] X = R (chiralic): [93, 432, 505, 537a]
3.1.5	NC—⬡—CH=N—⬡—X	p-cyanobenzylidene-p-X-anilines [49, **60**, 90, 180, **338**, 346, 350, 482, 492] X = R (chiralic): [514]
3.1.6	X—⬡—CH=N—⬡—NO₂	p-X-benzylidene-p-nitranilines [212, 493, **496**, 839a]
3.1.7	O₂N—⬡—CH=N—⬡—X	p-nitrobenzylidene-p-X-anilines [784]
3.1.8	X—⬡—CH=N—⬡—NRR′	p-X-benzylidene-p-phenylenediamines [426]
3.1.9	RR′N—⬡—CH=N—⬡—X	p-aminobenzylidene-p-X-anilines [90, 784]
3.1.10	RO—⬡—CH=N—⬡—OOCR′	p-alkoxybenzylidene-p-aminophenol alkanoates [38, 90, 94, 148, 155, 173, 249, 321, 391, 456, 577, 599, 685, **692, 697**, 749] R′ (chiralic): [93, 192]
3.1.11	RO—⬡—CH=N—⬡—COOR′	p-n-alkoxybenzylidene-p-aminobenzoic acid esters [1:309, 45, 90, **129**, 213, 635]
3.1.12	X—⬡—CH=N—⬡—COR	p-X-benzylidene-p-n-alkanoylanilines R = CH₃: [15, 90, **148**, 316, **319**, 321, 730, 787, 816, 837] R ≠ CH₃: [13, 122, **213**, 319, 542]
3.1.13	X—⬡—CH=N—⬡—OR′	p-X-benzylidene-p-alkoxyanilines [122, **148**, 598, 685, 710]
3.1.14	RO—⬡—CH=N—⬡—OR′	p-alkoxybenzylidene-p-alkoxyanilines [45, 90, **127, 128**, 148, **227**, 316, 321, 489, 523, 524, 685, 834, **845, 847, 851, 853**]
3.1.15	RO(CH₂)ₙO—⬡—CH=N—⬡—OR′	p-ω-alkoxyalkoxybenzylidene-p-alkoxyanilines [51]

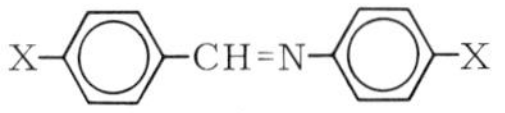

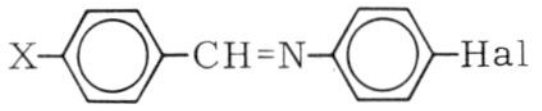

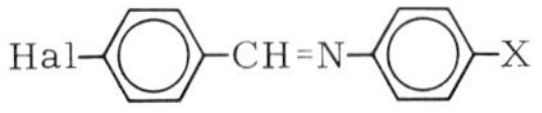

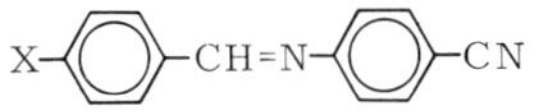

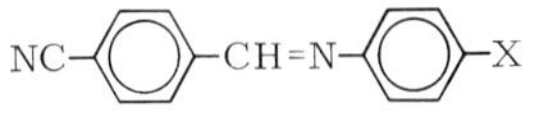

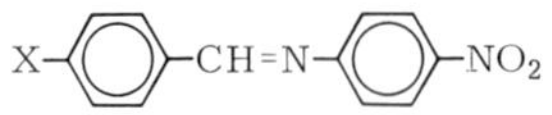

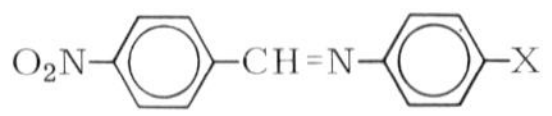

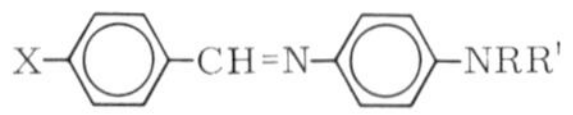

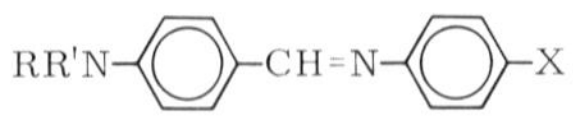

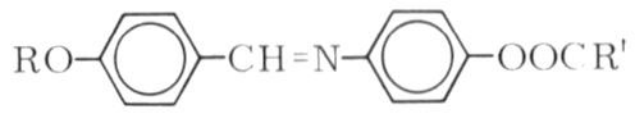

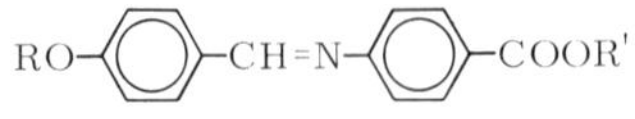

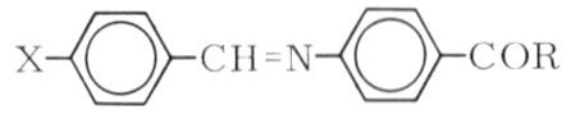

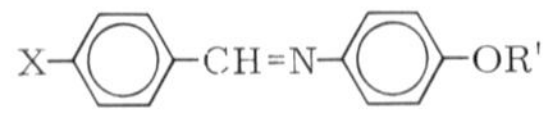

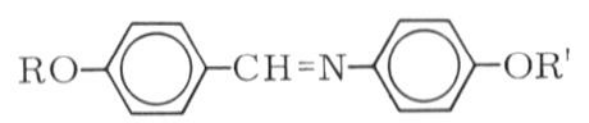

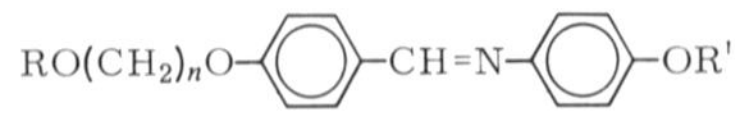

3.1.16 RCOO—⟨benzene⟩—CH=N—⟨benzene⟩—X

p-n-alkoyloxybenzylidene-*p*-X-anilines
R (saturated): [90, 94, 122, 126, 213, 321, 421, 685, 688, 689]
R (unsaturated): [344, 551, 711, 712, 713]
R,X (unsaturated), X (chiralic): [714]

3.1.17 RO—⟨benzene⟩—CH=N—⟨benzene⟩—R'

p-n-alkoxybenzylidene-*p*-alkylanilines
R' = CH$_3$ [1:319, 45, 103, 148, 227, 392, 457, 598, 850, 851]
R' = C$_2$H$_5$ [1:336, 148, 227, 392, **523**, 524, 546, 598, 850, 851]
R' = *n*-C$_3$H$_7$ [148, 227, 339, 392, 416, **523**, 524, 598, 710, 850, 851, 857]
R' = *n*-C$_4$H$_9$ [30, 38, 94a, **148,** 159, 160, **170, 214,** 339, 356, 391, 392, 394, 395, 398, 456, 491, 523, 524, 546, 578, 598, 681, 685, 710]
R' = *n*-C$_5$H$_{11}$ [148, 339, 420, 598, 681, 710]
R' = *n*-C$_6$H$_{13}$ [339, 416, 420, 609, 681, 710]
R' = *n*-C$_7$H$_{15}$ [339, 681, 710]
R' = *n*-C$_8$H$_{17}$ [421, 681, 710]
R' = chiralic alkyl   [176, **343,** 504]
R' = —CH=CH$_2$ [344, 544]
R' = —(CH$_2$)$_n$—OH [**148,** 248]
R' = *n*-C$_4$H$_9$, R = chiralic alkyl: [608]
R' = *n*-C$_4$H$_9$, R = variety of substituents: [170, 172, 685]

3.1.18 X—⟨benzene⟩—CH=N—⟨benzene⟩—SR

p-X-benzylidene-*p*-thioalkylanilines [90]

3.1.19 RS—⟨benzene⟩—CH=N—⟨benzene⟩—X

p-thioalkylbenzylidene-*p*-X-anilines [90]

3.1.20 R$_{Si}$—⟨benzene⟩—CH=N—⟨benzene⟩—X

p-Si-substituted benzylidene-*p*-X-anilines [884]

3.1.21 X—⟨benzene⟩—CH=N—⟨benzene⟩—HgY

p-X-benzylidene-*p*-mercuriated anilines [799]

3.1.22 fluorinated benzylidene anilines [247, 277, 297, 527a]

3.1.23 deuterated azomethines [342]

3.1.24 X—⟨benzene⟩—CH=N—⟨benzene⟩—Y

p,p'-unsymmetrically substituted benzylidene anilines other than those already included in 3.1.2 to 23 [17, 122, 126, 170, 171, 212, 214, 345, 414, **415,** 421, 485, 487, 541, 542, 679, 685, 687, 732, 795, 855, 872]

3.1.25 X—⟨benzene Y⟩—CH=N—⟨benzene⟩—Z

disubstituted benzylidene-*p*-substituted anilines [433, 434, 436, 495, 729, 784]

3.1.26 X—⟨benzene⟩—CH=N—⟨benzene Z⟩—Y

p-substituted benzylidene-disubstituted anilines [148, 521, **522,** 680, 682, 749, 888]

3.1.27   X—⟨○⟩—C=N—⟨○⟩—Y   (Z = Cl, CN) imidoyl chlorides and cyanides [752]
$\quad\quad\quad\quad\quad\quad\quad$ |
$\quad\quad\quad\quad\quad\quad\quad$ Z

3.1.28   RO—⟨○⟩—CH=N—⟨○⟩—CH=CH–COOR'

n-alkyl 4-(*p*-alkoxybenzylideneamino)-cinnamates
R: CH$_3$ [1:308, 1:309, 28, 40, 108, 221, **278,** 316, 381, 385, 422, 441, 730, 783, 787, 789, 799,
$\quad\quad$ 832, 833, 872]
R: C$_2$H$_5$ [1:308, 1:309, 1:318, 28, 316, 385, 441, 789, 799, 830, 833, 872]

RO—⟨○⟩—CH=N—⟨○⟩—CH=CH–COOC$_2$H$_5$

ethyl 4-(*p-n*-alkoxybenzylideneamino)-cinnamates [221, **227,** 237, 789, 815, 830, 836]

3.1.29   X—⟨○⟩—CH=N—⟨○⟩—CH=CH–COOC$_5$H$_{11}$ act

(act.)amyl 4-(*p*-X-benzylideneamino)-cinnamates
X: OR [**277,** 285, 396, 441, 706, 789]
X: other substituent [1:13, 1:386, 88, 237, 277, 285, 316, 346, 441, 706, 714, 789, 816]

3.1.30   X—⟨○⟩—CH=N—⟨○⟩—CH=CH–COOR

alkyl 4-(*p*-X-benzylideneamino)-cinnamates
X: —CN [28, 278, 279, 346, 441, 833]
X: —NO$_2$ [**278,** 833]

other 4-(*p*-X-benzylideneaminocinnamates [1:309, 1:319, 90, 278, 279, 316, 345, 422, 542, 582,
713, 789, 803, 830, 833, 855, 872]
R (chiralic): [397, 714]

3.1.31   X—⟨○⟩—CH=N—⟨○⟩—CH=CH–COO–(CH$_2$)$_n$—⟨○⟩—Y

ω-phenylalkyl 4-(*p*-X-benzylideneamino)-cinnamates [279, 290]

3.1.32   X—⟨○⟩—CH=N—⟨○⟩—CH=C–COOR
$\quad\quad\quad\quad\quad\quad\quad\quad\quad\quad\quad$ |
$\quad\quad\quad\quad\quad\quad\quad\quad\quad\quad\quad$ Y

4-(*p*-X-benzylideneamino)-α-Y-cinnamates
Y = CH$_3$ $\quad\quad$ R = *n*-alkyl $\quad\quad$ X = OCH$_3$ $\quad\quad$ [28, 316, 330, 385, 441, 789, 799, 803]
Y = CH$_3$ $\quad\quad$ R = *n*-alkyl $\quad\quad$ X = OC$_2$H$_5$ $\quad\quad$ [28, 316, 385, 715, 730, 789, 799, 815, 830]
Y = CH$_3$ $\quad\quad$ R = active amyl $\quad$ [1:385, 1:387, 316, 346, 396, 789]
Y = CH$_3$ $\quad\quad$ R = other alkyl $\quad$ [1:309, 28, 316, 441, 542, 789, 830]
Y = C$_2$H$_5$ $\quad\quad$ [28, 316, 385, 441, 789]
Y = other substituents [42, 789]
R (chiralic): [396]

3.1.33   X—⟨○⟩—CH=N—⟨○⟩—C=CH–COOR
$\quad\quad\quad\quad\quad\quad\quad\quad\quad\quad\quad$ |
$\quad\quad\quad\quad\quad\quad\quad\quad\quad\quad\quad$ Y

4-(*p*-X-benzylideneamino)-β-Y-cinnamates
Y = CH$_3$ $\quad\quad$ R = CH$_3$ $\quad\quad$ [316, 486, 799, 830]
Y = CH$_3$ $\quad\quad$ R = C$_2$H$_5$ $\quad$ [1:309, 316, 486, 799]

3.1.34   X—⬡—CH=N—⬡—⬡—Y

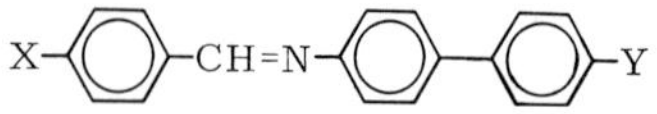

p-X-benzylidene-*p'*-Y-*p*-diphenylamines
Y=H   X=OR   [145, 222, 228, **264**, 816]
Y=OR   X=OR'   [276, 884]
Y=OR   X=other substituents   [276, 300]
Y=NRR'   X=other substituents   [67, 332, 802]
Y and X=other substituents [1:309, 1:318, 67, 212, 228, 332, 381, 422, 691, 817, 830]

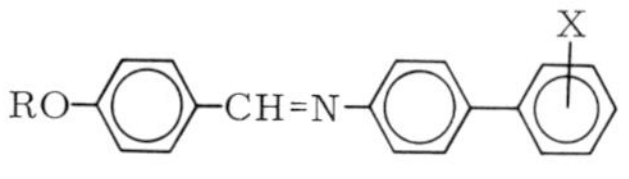

RO—⬡—CH=N—⬡—⬡(X)

p-(*n*-alkoxybenzylidene)-2' (or 3')-X-*p*-diphenylamines [273, 277]

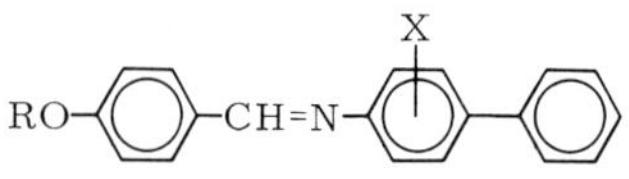

RO—⬡—CH=N—⬡(X)—⬡

p-(*n*-alkoxybenzylidene)-2 (or 3)-X-*p*-diphenylamines [273, 277]

X—⬡(Y)—CH=N—⬡—⬡—Z

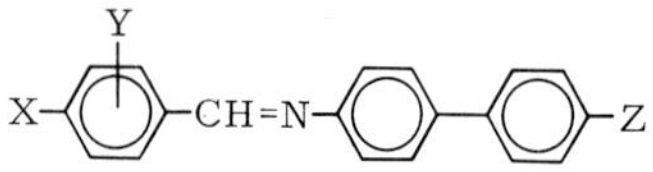

p-X-2 (or 3)-Y-benzylidene-*p'*-Z-*p*-diphenylamines [212, 729]

3.1.35   ⬡—⬡—CH=N—⬡—X   *p*-phenylbenzylidene-*p*-X-anilines [1:309, 45, 90, 790, 834]

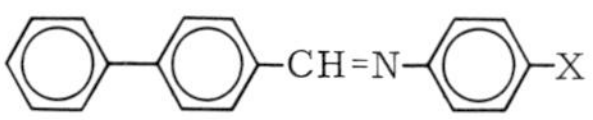

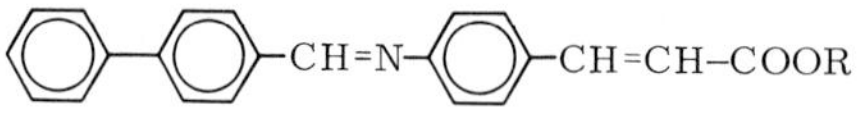

⬡—⬡—CH=N—⬡—CH=CH—COOR

p-phenylbenzylidene-*p*-aminocinnamates
R=*n*-alkyl [1:318, 40, 45, 278, **285**, 316, 381, 715, 730, 803, 816]
R=*i*-alkyl [278, 279, 706]

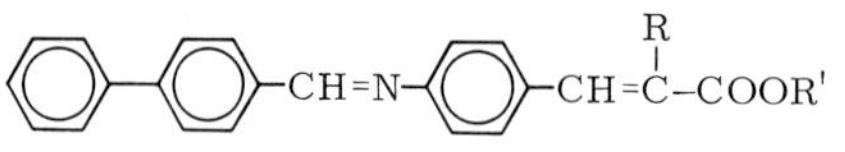

⬡—⬡—CH=N—⬡—CH=C(R)—COOR'

p-phenylbenzylidene-*p*-amino-α-alkylcinnamates [1:309]

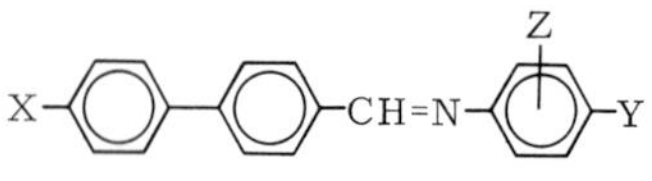

X—⬡—⬡—CH=N—⬡(Z)—Y

p'-X-*p*-phenylbenzylidene-*p*-Y-2 (or 3)-Z-anilines [45, 90]

3.1.36 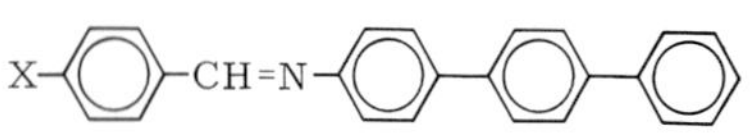   X—⬡—CH=N—⬡—⬡—⬡   *p*-X-benzylidene-*p*-amino-*p*-terphenyls [484]

3.1.37 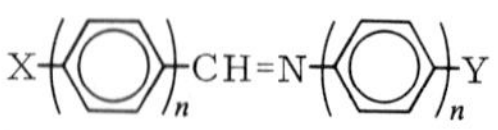   X—⟮⬡⟯ₙ—CH=N—⟮⬡⟯ₙ—Y   *p*-X-*p'*-phenylbenzylidene-*p*-Y-*p'*-phenyl anilines other than those already included in 3.1.34 to 36 [833]

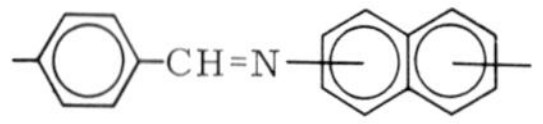

⬡—CH=N—⬡⬡   ⬡⬡—CH=N—⬡

benzylidene naphthylamines 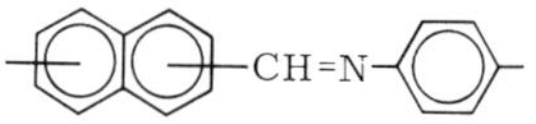   naphthylidene anilines

3.1.38

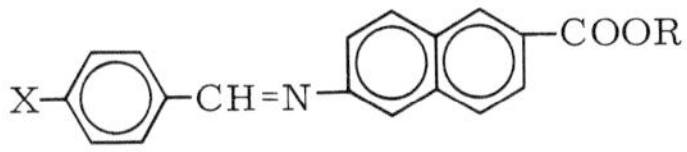

p-X-benzylidene-2-aminonaphthalene-6-carboxylates [391]

3.1.39 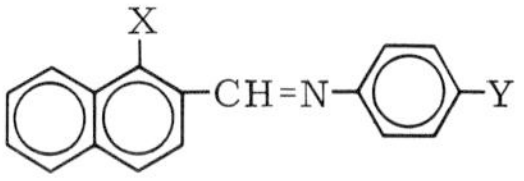   1-X-2-naphthylidene-*p*-Y-anilines [784, 799]

3.1.40 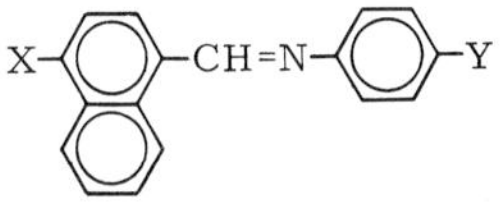   4-X-1-naphthylidene-*p*-Y-anilines [145a]

3.1.41 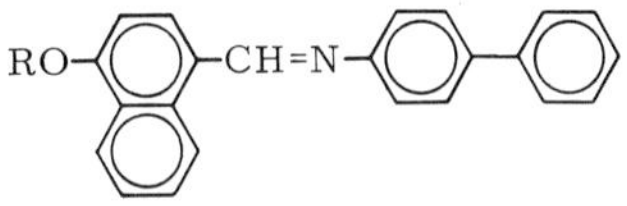   4-alkoxy-1-naphthylidene-*p*-phenylanilines [145]

3.1.42    *p*-(*n*-alkoxybenzylidene)-2-aminofluorenes [**264**]

3.1.43 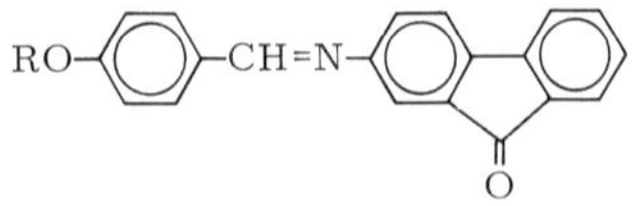   *p*-(*n*-alkoxybenzylidene)-2-aminofluorenones [**264**]

3.1.44    *p*-(*n*-alkoxybenzylidene)-2-aminoanthracenes [609a]

3.1.45 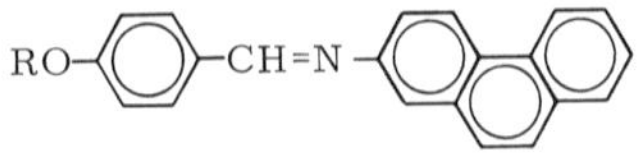   *p*-(*n*-alkoxybenzylidene)-2-aminophenanthrenes [**269**]

3.1.46 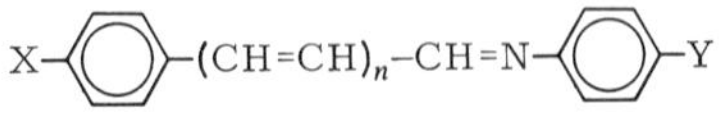   azomethines of ω-phenylpolyene aldehydes (cinnamylidene aniline type) [67, 212, 243, 332, 459, 486, 487, 691, 784, 801, 808, 812, 833]

[484, 830]

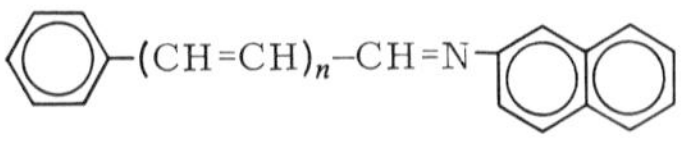

[243, 812]

3.2   *Azo compounds*

3.2.1 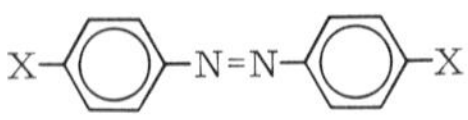   *p,p'*-symmetrically substituted azobenzenes
X = *n*-alkyl [**148, 750**]
X = chiralic alkyl [502, 503]
X = OR [5, 52, 53, **148**, 787, 831, 834, 837, **845, 851,** 872]
X = other substituents [321, 488, 542, 837, 855]

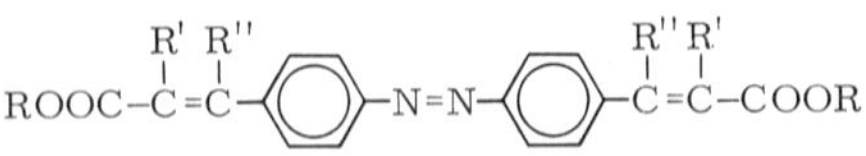

alkyl *p*-azocinnamates [1:309, 321, 385, 486, 784, 789, 831]

3.2.2   X—⬡—N=N—⬡—CN

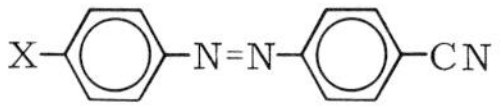

p-X-*p'*-cyanoazobenzenes [39, 222, 437, 438, 510, 515, 732]
X (chiralic): [512]

3.2.3   RO—⬡—N=N—⬡—OOCR'   *p*-alkoxybenzene-azophenol alkanoates

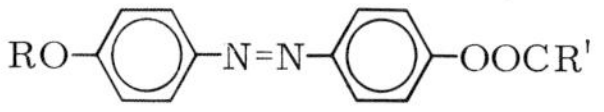

R=CH₃   R'=*n*-alkyl   **[35, 92,** 693, **805,** 872]
R=C₂H₅   R'=*n*-alkyl   **[35,** 38, **92, 148,** 531, 693, **805,** 806, 853]
R,R'=other alkyl groups [35, **92,** 328, 442, 706, 805, 872]

3.2.4   RO—⬡—N=N—⬡—X

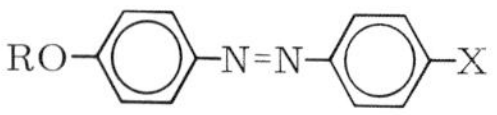

p'-X-*p*-alkoxyazobenzenes other than those already included
in 3.2.3

R=CH₃ [1:309, 1:318, **148,** 395, 438, 590, 621, 622, 693, 698, 699, 706, 830, 831, 837, 872, 879]
R=C₂H₅ [1:309, 395, 438, 590, 621, 622, 693, 698, 699, 706, 763, 810, 830, 831, 834, 879]
R=other alkyl groups [526, 527, 693, 698, 699, 879]
X (chiralic): [506, 508, 509, 518, 520]

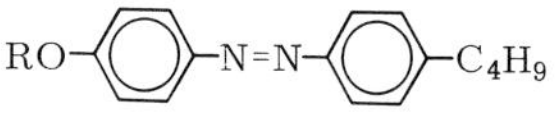

RO—⬡—N=N—⬡—C₄H₉   *p*-alkoxy-*p'*-*n*-butylazobenzenes [**148, 693, 698**]

3.2.5   RCOO—⬡—N=N—⬡—X

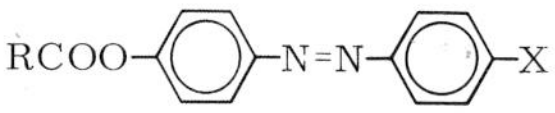

p'-X-benzeneazophenol alkanoates [1:309, 1:318, 693, 698,
699, 706, 730, 783, 831, 872, 878]

3.2.6   X—⬡—N=N—⬡—Y

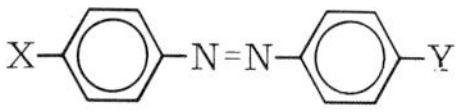

p,p'-disubstituted azobenzenes other than those already
included in 3.2.2 to 5 [1:309, 1:318, 39, 222, 316, 460, 542,
590, 622, 690, 732, 784, 830, 831, 834, 872]
X (chiralic): [519]

3.2.7   X—⬡—N=N—⬡(Z)—Y

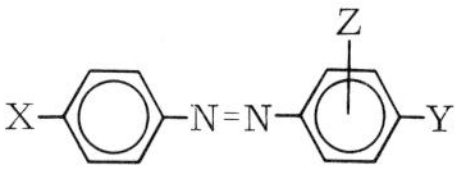

p-X-*p'*-Y-2' (or 3')-Z-azobenzenes [51a, 427, 717, **751**]

3.2.8   X—⬡—N=N—⬡—⬡—Y

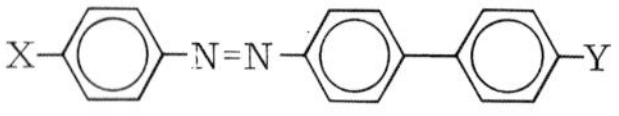

p-X-benzeneazo-*p'*-Y-*p*-biphenyls [67, 222, 802]

3.2.9   X—⬡—⬡—N=N—⬡—⬡—Y   *p'*-X-*p'*-Y-*p*-azobiphenyls [228, 802, 819]

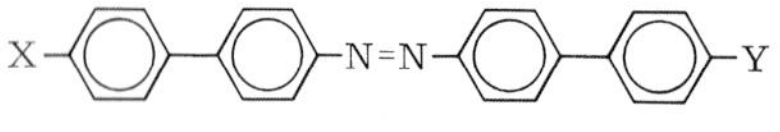

3.2.10   X—⬡—N=N—⬡⬡—Y

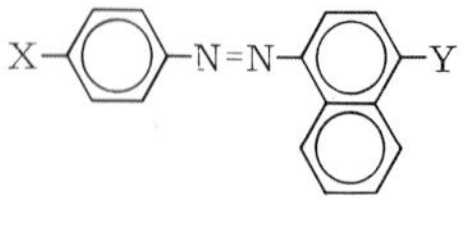

p-X-benzeneazo-4-Y-1-naphthalenes [633, 799]

3.2.11   X—⬡—N=N—C(Y)=C(Z)—N=N—⬡—X   bis-(benzeneazo)-ethylenes [804]

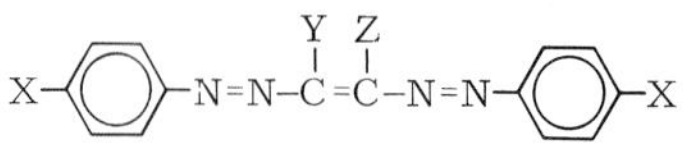

3.3 Azoxy compounds

3.3.1 X—◯—N=N—◯—X (with O above N=N)

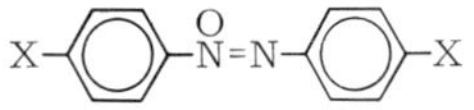

p,p′-symmetrically substituted azoxybenzenes
X = *n*-alkyl [87, **148, 552, 750**]
X = chiralic alkyl [500, 501]
X = OR [1:70, 1:131, 1:306, 1:308, 1:309, 5, 38, 40, 52, 53, 54, 69, 83, 87, **148**, 150, 188, **227**,
    237, 315, 316, 422, **461**, 575, 589, 594, 784, 787, 827, 837, **845, 851,** 853]
X = O(CH$_2$)$_n$OR [51, 855]
X = COOR [1:308, 1:309, 83, 188, 381, 422, 488, 621, 722, 730, 782, 827, 829, 837]
X = CH=CHCOOR
    R = *n*-alkyl [1:152, 1:308, 1:309, 1:317, 83, **227, 385,** 486, 782, 789, 799, 805, 853]
    R ≠ *n*-alkyl [1:309, 321, 385, 706, 782, 799]
X = CH=C(CH$_3$)COOR [1:309, 385, 486, 789, 799]
X = CR=CRCOOR [1:309, 316, 486, 783]
X = other substituents [1:309, 87, 321, 381, 542, 621, 782, 787, 872]

3.3.2 X—◯—N=N—◯—CN (with O above N=N)   *p*-X-*p′*-cyanoazoxybenzenes [364, 437, 438, 513, 516, 732]
                                                        X (chiralic): [511]

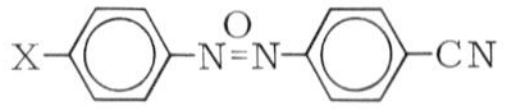

3.3.3 RO—◯—N=N—◯—X (with O above N=N)   *p*-alkoxy-*p′*-X-azobenzenes

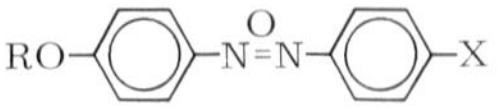

R = CH$_3$ [1:131, 92, 148, 331, 395, 694, 720, 839]
R = C$_2$H$_5$ [**92**, 331, 381, 694, 720, 730, 827, 839]
R = other alkyl groups [**148**, 331, 526, 527, 694, 698, 720]
X (chiralic): [507, 517, 520]

3.3.4 X—◯—N=N—◯—Y (with O above N=N)   *p,p′*-unsymmetrically substituted azobenzenes other than
                                                        those already included in 3.3.2 to 3 [**92**, 174, 460, 621,
                                                        705, 719]

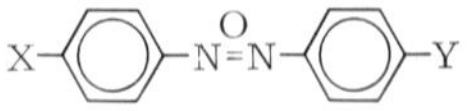

3.3.5 X—◯—N=N—◯(Z)—Y (with O above N=N)   *p*-X-*p′*-Y-2′ (or 3′)-Z-azoxybenzenes [64, 317, 751]

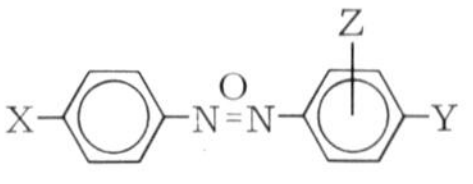

3.3.6 X—◯—◯—N=N—◯—◯—Y (with O above N=N)   *p,p′*-disubstituted azoxybiphenyls [228, 802, 819]

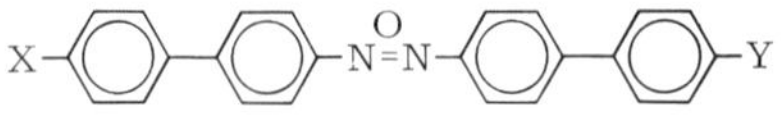

3.4 Carboxylic acid esters

3.4.1 X—◯—COO—◯—Y   *p,p′*-disubstituted phenyl benzoates

X = R	Y = R′	[34, 695, 727, 757]
X = R	Y = OR′	[8, 155, 245, 337, 387, 534, 695, 700, 727, 757]
X = OR	Y = R′	[**141**, 337, 387, 534, 695, 700, 757]
X = OR	Y = OR′	[7, 155, 245, 633, 695, 726, 755, 757]
X = OR	Y = OCOR′	[695, 700, 726]
X = OCOR	Y = OR′	[387, 665, 700]
X = R	Y = OCOR′	[695, 700]
X = OCOR	Y = R′	[386, 700]

X or Y (chiralic): [354, 403]

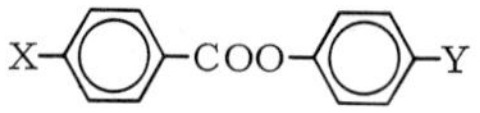

X—◯—COO—◯—Y   (continued)

X = other substituent   Y = COR [**144**, 418, 726]
X = OCOOR   Y = other substituent [34, 38, **59**, **91**, 245, 337, 386, 388, 389, 633]
X = other substituent   Y = OCOOR [34]
X = CN   Y = other substituent [374, 718]
X = other substituent   Y = CN [57, **59**, 226b, 354, 375, 377, 389, 714b, 732]
X or Y = other substituent than listed before [58, 158, 245, 665, 722, 732, 754, 799]

3.4.2   X—◯(Z)—COO—◯(Z')—Y   *p,p'*-disubstituted phenyl benzoates + further substituent(s)
[1, 183b, 312, 313, 337, 388, 389, 747]

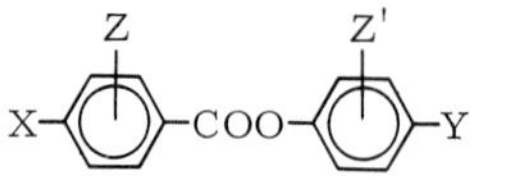

3.4.3 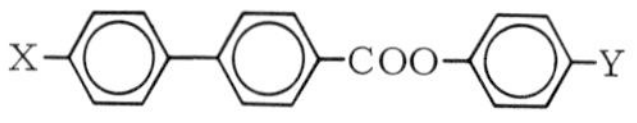   X—◯—◯—COO—◯—Y   4-Y-phenyl 4″-X-biphenyl-4′-carboxylates [240, 292, **300**, 303, 305a, 305b, 326, 352, 353, 491a, 702, 703]
Y, X = R (chiralic): [305a]

3.4.4   X—◯(Z)—COO—◯—◯—Y   4″-Y-4′-biphenylyl 4-X-2 (or 3)-Y-benzoates [303, 605, 702, 703]

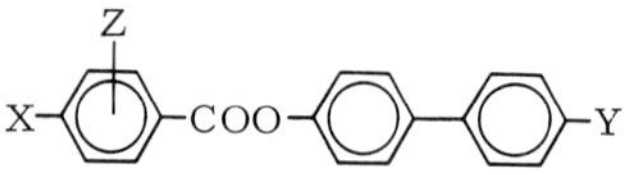

3.4.5 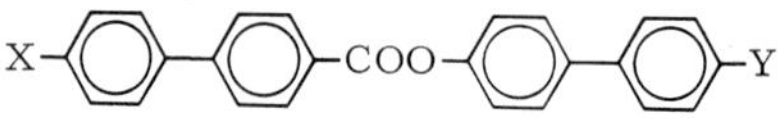   X—◯—◯—COO—◯—◯—Y   4-Y-4′-biphenylyl 4‴-X-biphenyl-4″-carboxylates [303]

3.4.6 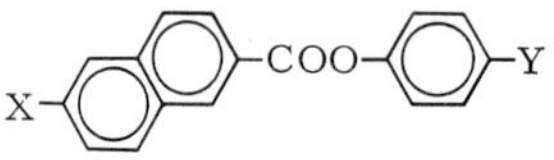   X—◯◯—COO—◯—Y   4′-Y-phenyl 6-X-2-naphthoates [303]

3.4.7   X—◯(Z)—COO—◯◯—Y   6-Y-2-naphthyl 4′-X-2′,3′-Z-benzoates [303, 379, 600]

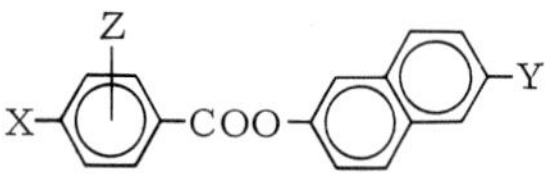

3.4.8 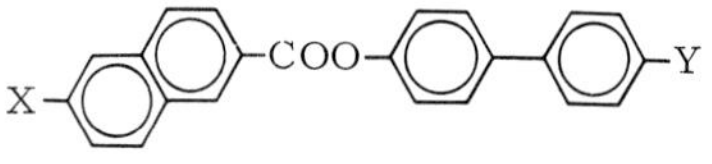   X—◯◯—COO—◯—◯—Y   4-Y-4′-biphenylyl 6″-X-2″-naphthoates [303]

3.4.9 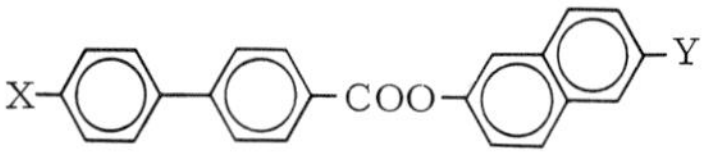   X—◯—◯—COO—◯◯—Y   6-Y-2-naphthyl 4″-X-biphenyl-4′-carboxylates [303]

3.4.10 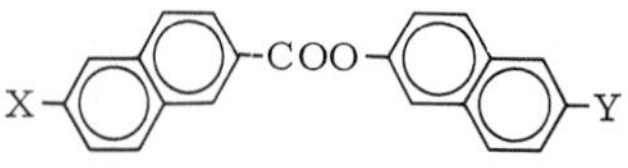   X—◯◯—COO—◯◯—Y   6-Y-2-naphthyl 6′-X-2′-naphthoates [303]

3.4.11   X—◯(Z)—CH=CH—COO—◯(Z')—Y   phenyl cinnamates [**63**, 153, 365, 369, 370, 371, 372, 422, 549, 628, 633, 732, 837]

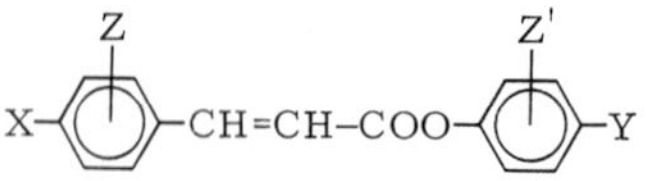

3.4.12 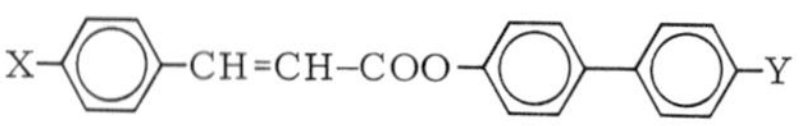   X—◯—CH=CH—COO—◯—◯—Y   biphenylyl cinnamates [143]

3.4.13 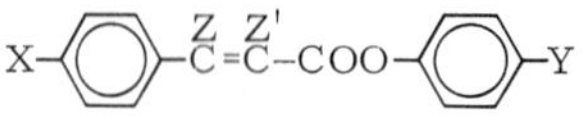   X—◯—C(Z)=C(Z')—COO—◯—Y   α- or β-substituted phenyl cinnamates [153, **606**]

3.4.14 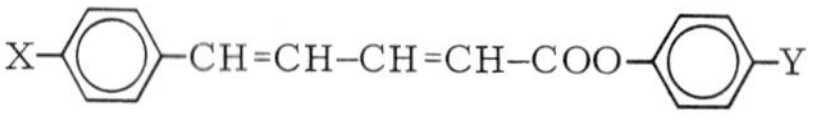   X—◯—CH=CH—CH=CH—COO—◯—Y   phenyl cinnamyledeneacetates [243, 813]

4   *Aromatic compounds with multiple identical bridging groups:*

—CH=N—,   —N=N—,   —N(O)=N—,   or —COO—

4.1   *Bisazomethines*

main types:

Ar–N=CH–⟨◯⟩–CH=N–Ar   terephthalbisazomethines 4.1.1 to 4

Ar–CH=N–⟨◯⟩–N=CH–Ar   p-phenylenediaminebisazomethines 4.1.5 to 8

Ar–CH=N–⟨◯⟩–⟨◯⟩–N=CH–Ar   benzidine bisazomethines 4.1.10 to 17

4.1.1   ROOC–C(R')=CH–⟨◯⟩–N=CH–⟨◯⟩–CH=N–⟨◯⟩–CH=C(R')–COOR

terephthalbis-*p*-aminocinnamates [1:319, 40, 346, 381, 817, 830]
R (chiralic): [396]

4.1.2   X–⟨◯⟩–N=CH–⟨◯⟩–CH=N–⟨◯⟩–Y   terephthalbis-anilines

X=Y: [1:309, 1:318, 1:336, 11, 17, 30, 42, 222, 297, 346, **534a**, 542, 609]
X=Y (chiralic): [285, 299]
X ≠ Y: [30]

4.1.3   X–⟨◯⟩–⟨◯⟩–N=CH–⟨◯⟩–CH=N–⟨◯⟩–⟨◯⟩–X

terephthalbis-*p*-aminobiphenyls [212]

4.1.4   X–⟨◯◯⟩–N=CH–⟨◯⟩–CH=N–⟨◯◯⟩–X   terephthalbis-α-naphthylamines [346]

4.1.5   X–⟨◯(Y)⟩–CH=N–⟨◯(Z)⟩–N=CH–⟨◯(Y)⟩–X   bisbenzylidene-*p*-phenylenediamines

Y=Z=H [**12, 14**, 17, **213, 264**, 316, 321, 484, 551, 787, 799]
Y=H, Z ≠ H [14, 18, **112**, 208, 728]
        X (chiralic): [93, 324]
Z=H, Y ≠ H [484]

4.1.6   X–⟨◯⟩–⟨◯⟩–CH=N–⟨◯⟩–N=CH–⟨◯⟩–⟨◯⟩–X

bis-(*p*-phenylbenzylidene)-*p*-phenylenediamines [45]

4.1.7   RO—CH=N—N=CH—OR

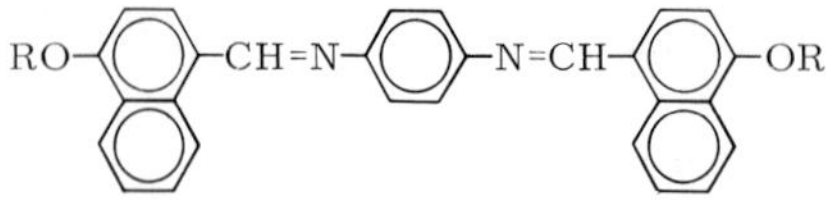

bis-(4-alkoxy-1-naphthylidene)-*p*-phenylenediamines [**133**]

4.1.8   X—CH=CH–CH=N—N=CH–CH=CH—X

biscinnamylidene-*p*-phenylenediamines [799]

4.1.9    X—CH=N—CH=N—Y   benzylidene-*p*-aminobenzylidene anilines [484, 830]

4.1.10   X—CH=N—N=CH—X

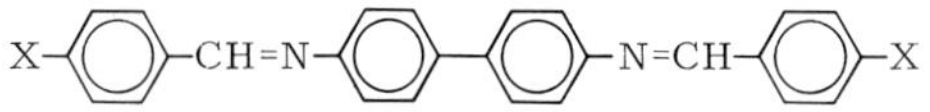

bis-(*p*-X-benzylidene)-benzidines
X = OR: [1:72, **264**, 270, 484, 799, 805, 829]
X other substituent: [1:71, 1:138, 40, 69, 106, 222, 484, 551, 798, 799, 805, 829]

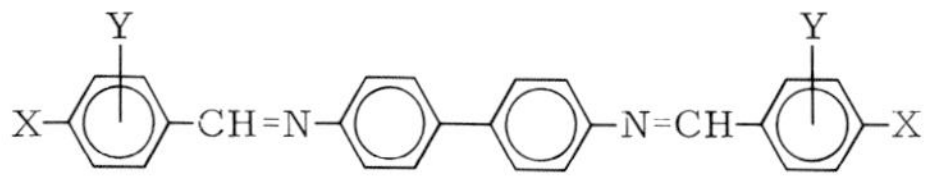

X—CH=N—N=CH—X (with Y substituents)

bis-(*p*-X-2- (or 3)-Y-benzylidene)-benzidines [222, 484]

4.1.11   RO—CH=N—N=CH—OR

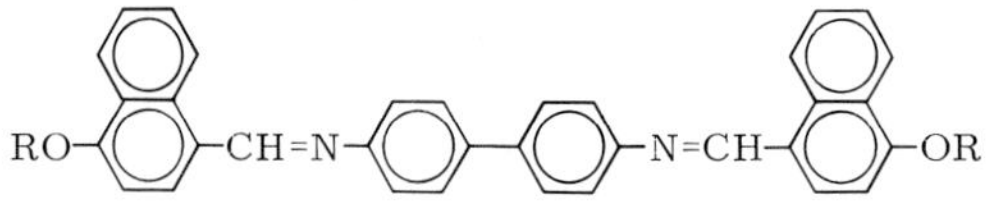

bis-(*p-n*-alkoxy-1-naphthylidene)-benzidines [**131**]

4.1.12   RO—CH=N—N=CH—OR (with X substituent)

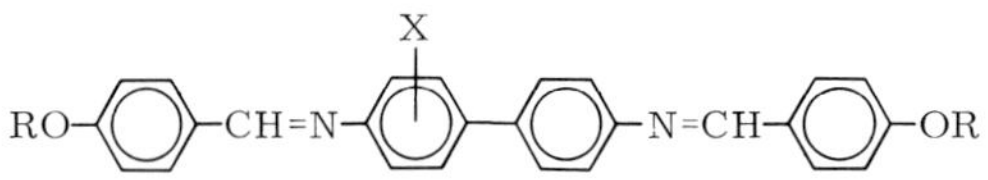

bis-(*p-n*-alkoxybenzylidene) monosubstituted benzidines
3-substitution: [274]
2-substitution: [274, 277]

4.1.13   RO—CH=N—N=CH—OR (with X, X substituents)

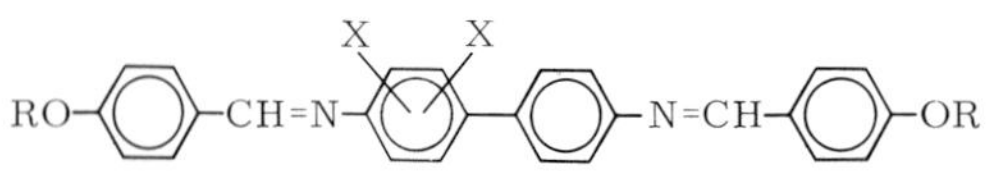

bis-(*p-n*-alkoxybenzylidene) disubstituted benzidines
2,5-substitution: [275]
2,6-substitution: [275]

4.1.14   RO—CH=N—N=CH—OR (with X, X substituents)

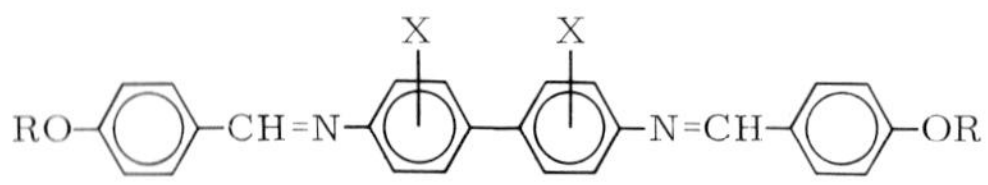

bis-(*p-n*-alkoxybenzylidene) disubstituted benzidines
2,2'-substitution: [275, 868]
2,3'-substitution: [275]
3,3'-substitution (also with other terminal groups than OR): [71, 275, 868]

4.1.15 RO—⟨◯⟩—CH=N—⟨◯⟩—⟨◯⟩—N=CH—⟨◯⟩—OR

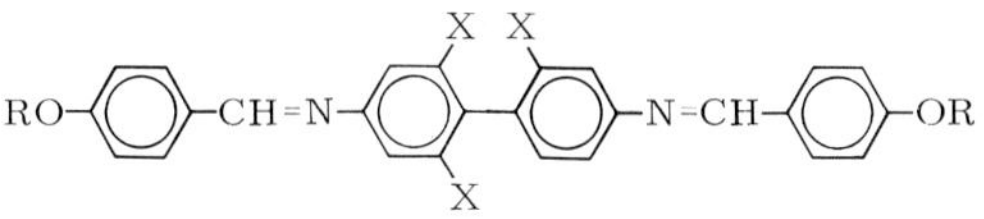

bis-(*p-n*-alkoxybenzylidene)-2,2',6-trisubstituted benzidines [275]

4.1.16 RO—⟨◯⟩—CH=N—⟨◯⟩—⟨◯⟩—N=CH—⟨◯⟩—OR

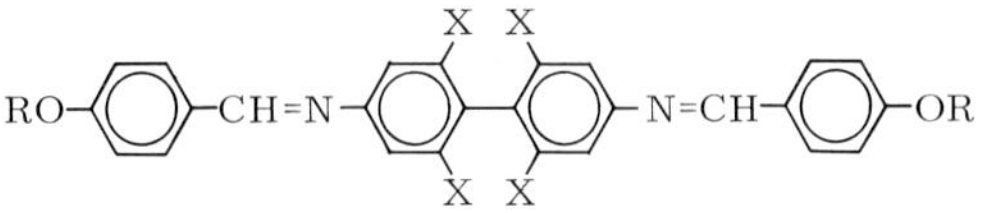

bis-(*p-n*-alkoxybenzylidene)-2,2',6,6'-tetrasubstituted benzidines [275, 277, 868]

4.1.17 RO—⟨◯⟩—CH=N—⟨◯⟩—⟨◯⟩—N=CH—⟨◯⟩—OR

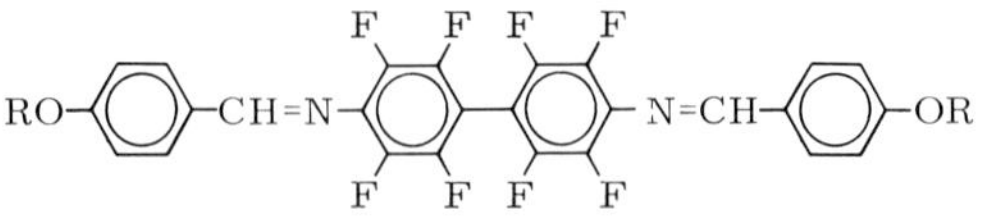

bis-(*p-n*-alkoxybenzylidene)-octafluorobenzidines [247, 277]

4.1.18 X—⟨◯⟩—CH=N—[⟨◯⟩—N=CH]$_n$—⟨◯⟩—X

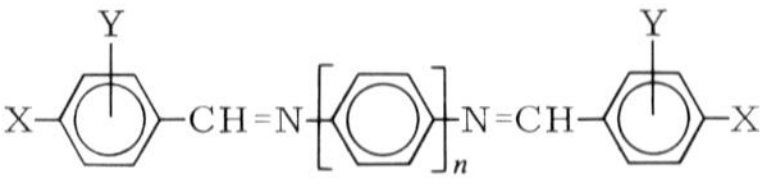

bisbenzylidene-*p,p*-diamino-*p*-phenyls [212, 484]

4.1.19 X—⟨◯⟩—CH=N—⟨◯⟩—⟨◯⟩—⟨◯⟩—N=CH—⟨◯⟩—X

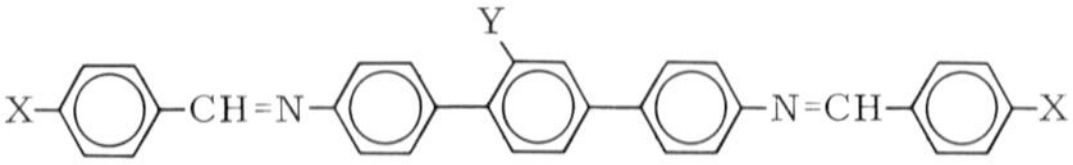

bisbenzylidene-2'-Y-*p,p''*-diamino-*p*-terphenyls [484]

4.1.20 CH$_3$O—⟨◯⟩—CH=N—⟨◯◯⟩—N=CH—⟨◯⟩—OCH$_3$

bisanisylidene-naphthylenediamines [40, 869]

4.1.21 RO—⟨◯⟩—CH=N—⟨◯◯⟩—N=CH—⟨◯⟩—OR

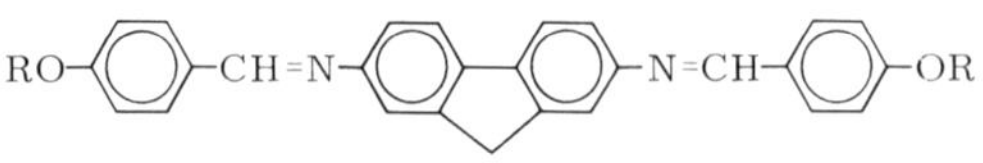

bis-(*p-n*-alkoxybenzylidene)-2,7-diaminofluorenes [**264**]

4.1.22 RO—⟨◯⟩—CH=N—⟨◯◯⟩—N=CH—⟨◯⟩—OR

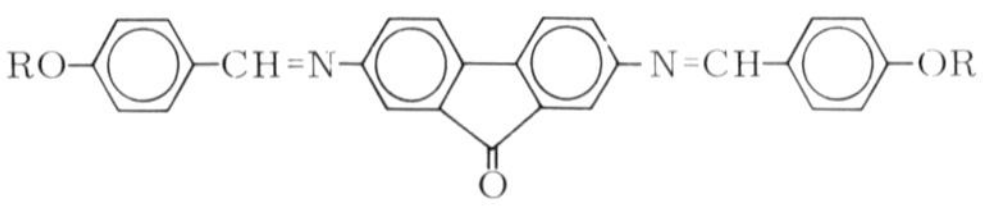

bis-(*p-n*-alkoxybenzylidene)-2,7-diaminofluorenones [**264**]

4.1.23 X—⟨◯⟩—CH=CH–CH=N—⟨◯⟩—CH=N—⟨◯⟩—OC$_2$H$_5$

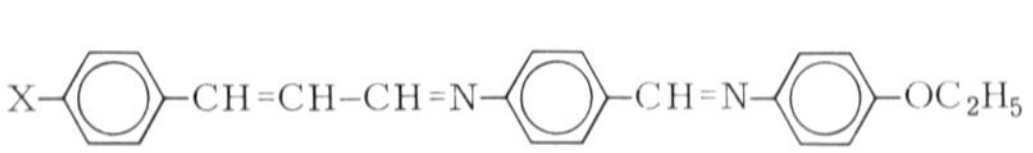

cinnamylidene-*p*-aminobenzylidene anilines [414]

4.1.24   X—◯—CH=CH–CH=N—[—◯—]$_n$—N=CH–CH=CH—◯—X

bis-cinnamylidene-*p,p*-diamino-*p*-phenyls [484]

4.1.25   Y—◯—CH=CH–CH=N—◯—◯—N=CH–CH=CH—◯—Y

(X substituents above the central rings)

bis-cinnamylidene-3,3′-disubstituted benzidenes [71]

4.2   Bisazo compounds

4.2.1   X—◯—N=N—◯—N=N—◯—X   phenyl-*p*-bisazobenzenes [590]

4.2.2   X—◯—N=N—◯—◯—N=N—◯—X   biphenyl-*p*-bisazobenzenes [71, 830]

4.2.3   X—◯—N=N—◯—◯—N=N—◯—X   3,3′-disubstituted   biphenyl-*p*-bisazoben-
(Y substituents above the central rings)   zenes [71]

4.3   Bisazoxy compounds

(*Note:* Liquid-crystalline bisazoxy compounds without additional bridging groups have not been described up to now, although there is no doubt as to their existence.)

4.4   Bis and oligo esters

4.4.1   X—◯—COO—◯—OOC—◯—Y   hydroquinone bisbenzoates

X = Y = OR: [**16**, 161, **162, 163**, 245, 633]
X = Y = COOR: [161, 633]
X, Y = other substituents: [162, 626, 627, 629, 630, 726]

X—◯—COO—◯—OOC—◯—Y   hydroquinone bisbenzoates [1c, 16, **17**, 157, 489,
(Z, Z′, Z″ substituents above the rings)   628, 714a, 842]
                                                             X (chiralic): [403]

4.4.2   X—◯—COO—◯—COO—◯—OOC—◯—OOC—◯—X

bis-(*p*-oxybenzoylbenzoyl)-hydroquinones [633]

4.4.3   X—◯—OOC—◯—COO—◯—X   diphenyl terephthalates
(Y substituent above the central ring)

X = OR, Y = H: [**162**, 245, **391**]
X, Y other substituents: [162, 756]

4.4.4   X—⟨⟩—COO—⟨⟩—COO—⟨⟩—Y   phenyl *p*-benzoyloxybenzoates

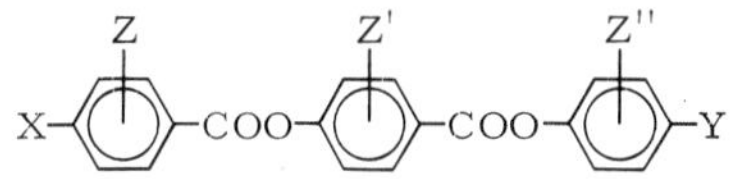

Z = Z′ = Z″ = H: [377c, 633, 700, 722, 755, 758, 799, 837]
Z, Z′, Z″ other substituents: [10, 183b, 226, 377a, 377b, 701, 755, 760, 761, **888,** 889, 891]
X (chiralic): [226a, 344a, 403]

4.4.5   X—⟨⟩—COO—⟨⟩—COO—⟨⟩—COO—⟨⟩—Y

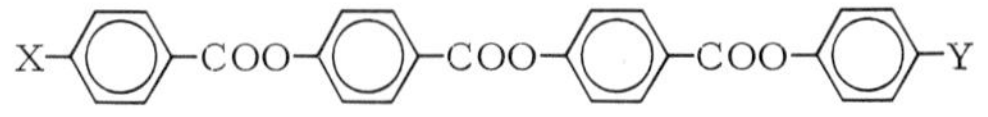

p-benzoyloxy-*p*-benzoyloxy-*p*-benzoyloxy benzenes [700, 722, 799, 816]

4.4.6   X—⟨⟩—COO—⟨⟩—⟨⟩—OOC—⟨⟩—X

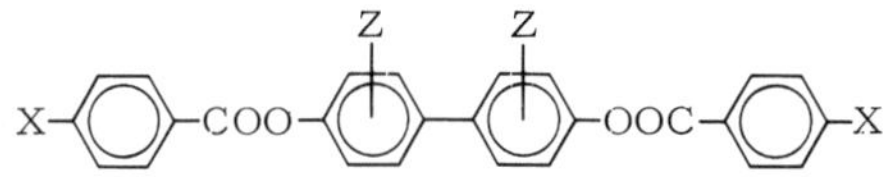

4,4′-dihydroxybiphenyl-bisbenzoates [71, 161, 228, 378, 799]

4.4.7   X—⟨⟩—OOC—⟨⟩—⟨⟩—COO—⟨⟩—X

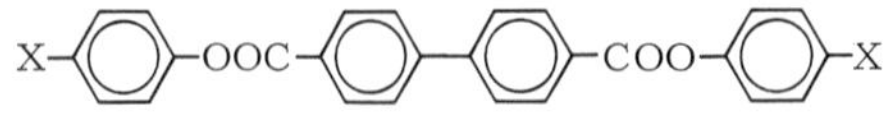

biphenyl-4,4′-dicarboxylates [326, 799]

4.4.8   X—⟨⟩—COO—⟨⟩—⟨⟩—⟨⟩—OOC—⟨⟩—X

p,p″-dihydroxy-*p*-terphenyl-bisbenzoates [484]

4.4.9   X—⟨⟩—OOC—⟨⟩⟨⟩—COO—⟨⟩—X

diphenyl naphthalenedicarboxylates [393]

4.4.10 X—⟨⟩—COO—⟨⟩⟨⟩—OOC—⟨⟩—X   naphthalenediol bisbenzoates [138, 545]

4.4.11 X—⟨⟩—COO   OOC—⟨⟩—X
        X—⟨⟩—COO—⟨⟩⟨⟩—OOC—⟨⟩—X   1,4,5,8-naphthalenetetrol tetrabenzoates [468]

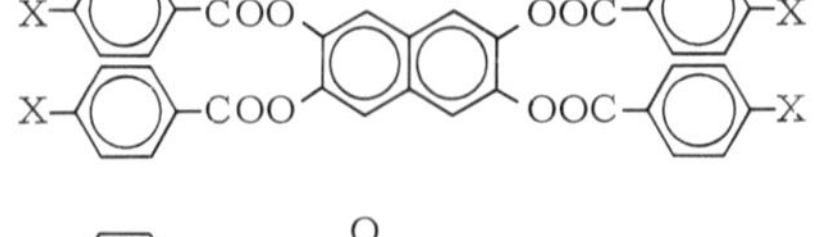

4.4.12 X—⟨⟩—COO   Y   2,5-dihydroxybenzoquinone bisbenzoates [660]
        Y   OOC—⟨⟩—X

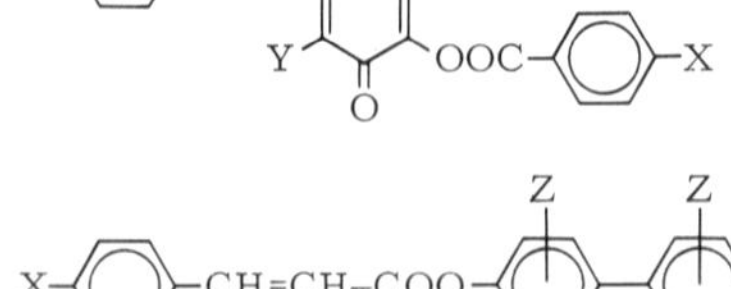

4.4.13 X—⟨⟩—CH=CH—COO—⟨⟩—⟨⟩—OOC—CH=CH—⟨⟩—X

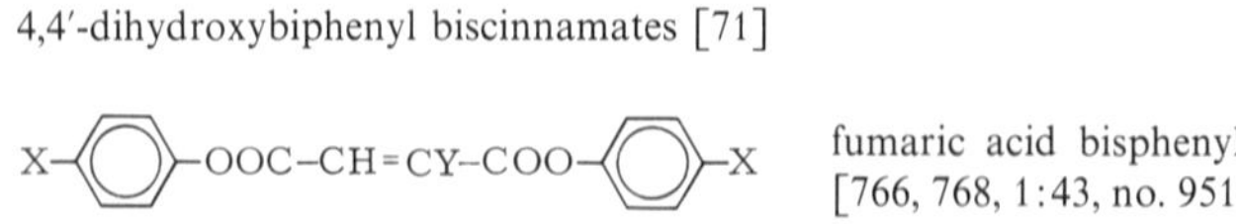

4,4′-dihydroxybiphenyl biscinnamates [71]

4.4.14 X—⟨⟩—OOC—CH=CY—COO—⟨⟩—X   fumaric acid bisphenyl esters (Y = H or alkyl) [766, 768, 1:43, no. 951–960]

5   *Aromatic compounds with multiple different bridging groups:*

—CH=N—,   —N=N—,   —N(O)=N—,   or —COO—

5.1   *Azomethine-azo compounds*

5.1.1   X—◯—CH=N—◯—N=N—◯—Y

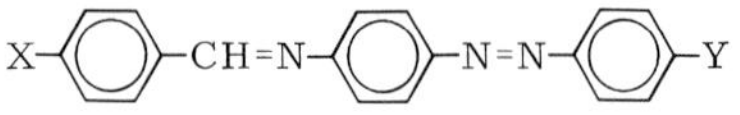

benzylidene-p-aminoazobenzenes [1:318, 1:326, **17,** 106, **213,** 227, 237, 316, 542, 590, 711, 715, 830]

5.1.2   X—◯—◯—CH=N—◯—N=N—◯—Y

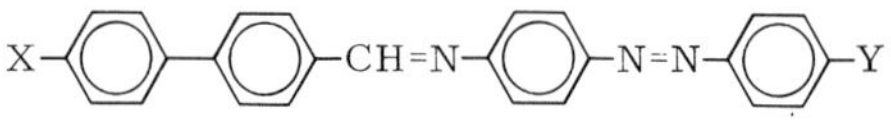

p-phenylbenzylidene-p-aminoazobenzenes [45]

5.1.3   X—◯—CH=N—◯—◯—N=N—◯—Y

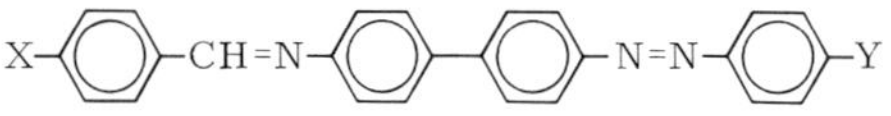

benzylidene-p-aminobiphenylazobenzenes [332, 802]

5.1.4   X—◯—CH=N—◯—N=N—◯—N=CH—◯—X

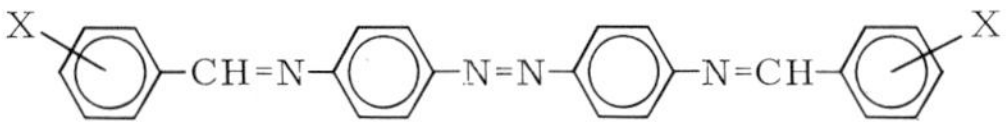

4,4′-bis-(benzylideneamino)-azobenzenes [579, 798]

5.1.5   X—◯—N=CH—◯—N=N—◯—CH=N—◯—X

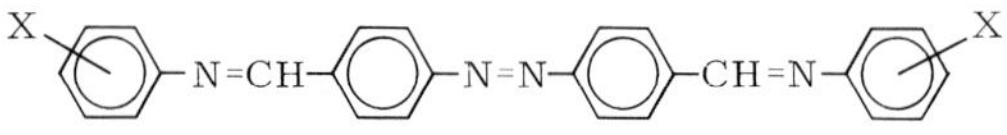

4,4′-azodibenzal dianilines [1:309, 209, 381, 453, 830]

5.1.6   X—◯—CH=N—◯—◯—N=N—◯—◯—N=CH—◯—X

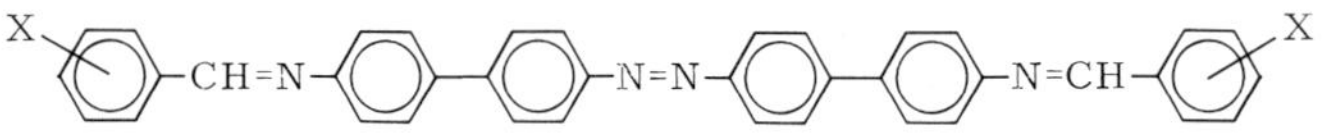

benzylidene-p,p′-diaminoazobiphenyls [212, 830]

5.1.7   X—◯—N=N—◯—N=CH—◯—CH=N—◯—N=N—◯—X

terephthal-bis-p-aminoazobenzenes [830]

5.1.8   X—◯—N=N—◯—N=CH—◯—N=N—◯—CH=N—◯—N=N—◯—X

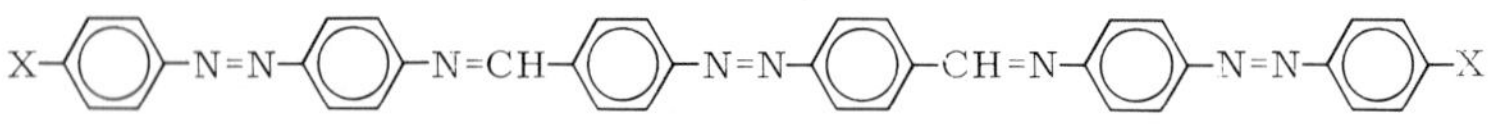

4,4′-azodibenzal-di-p-aminoazobenzenes [453]

5.1.9   X—◯—CH=N—◯—N=N—◯—Y

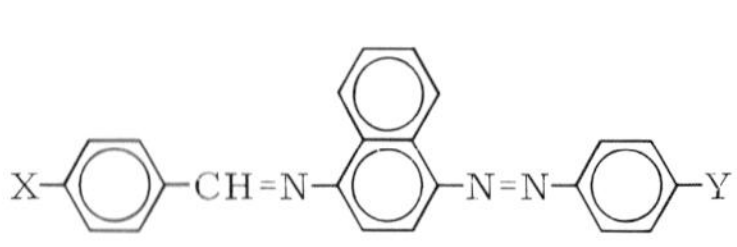

4-benzylideneamino-1-naphthaleneazobenzenes
X=OR; Y=H: [105, 227, 316, 422, 814, 822, 849, **853**]
X, Y other substituents: [25, 227, 427, 588, 633, 822, 855]

5.1.10 RO—CH=N—N=N—

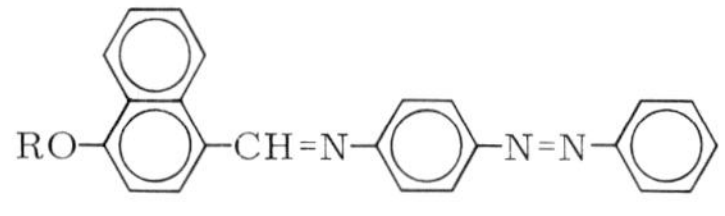

4-n-alkoxy-1-naphthylidene-4'-aminoazobenzenes [140]

5.1.11 RO—CH=N—N=N—N=CH—OR

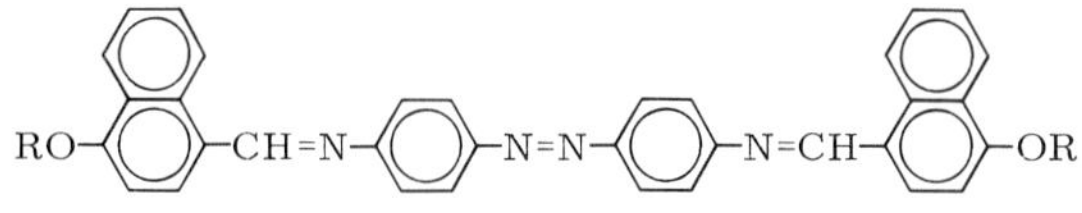

bis(4-n-alkoxy-1-naphthylidene)-p-azoanilines [140]

5.1.12 X—CH=CH–CH=N—N=N—Y

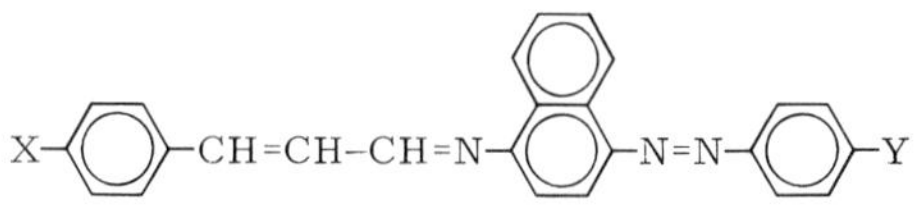

4-cinnamylideneamino-1-naphthaleneazobenzenes [25]

5.1.13 X—(CH=CH)$_n$CH=N—N=N—Y

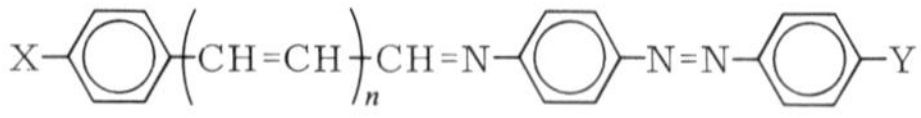

ω-phenylpolyenylidene-4-aminoazobenzenes [243, 590, 812]

5.1.14 X—CH=CH–CH=N(—)$_n$N=N(—)$_n$N=CH–CH=CH—X

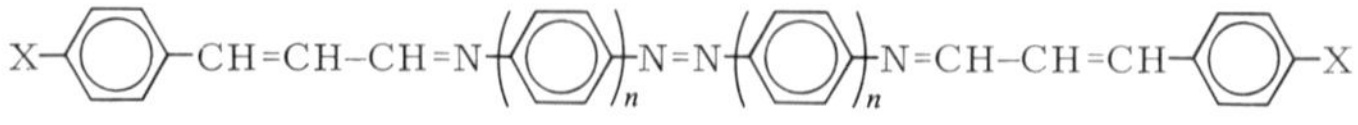

4,4'-bis-(cinnamylidene)-aminoazopolyphenyls [579]

5.2 Azomethine-azoxy compounds

5.2.1 X—CH=N—N=$\overset{O}{N}$—Y

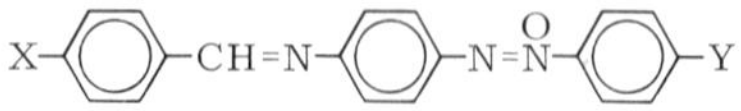

a-benzylidene-p-aminoazoxybenzenes [826]

5.2.2 X—CH=N—$\overset{O}{N}$=N—Y

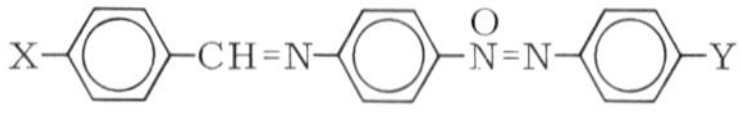

b-benzylidene-p-aminoazoxybenzenes [826]

5.2.3 X—CH=N—N=$\overset{O}{N}$—Y

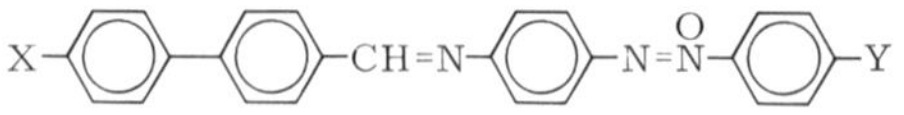

a-p-phenylbenzylidene-p-aminoazoxybenzenes [826]

5.2.4 X—CH=N—$\overset{O}{N}$=N—Y

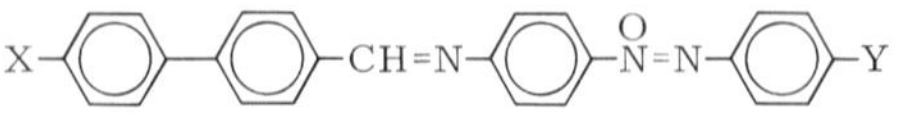

b-p-phenylbenzylidene-p-aminoazoxybenzenes [826]

5.2.5 X—CH=N—$\overset{O}{N}$=N—N=CH—X

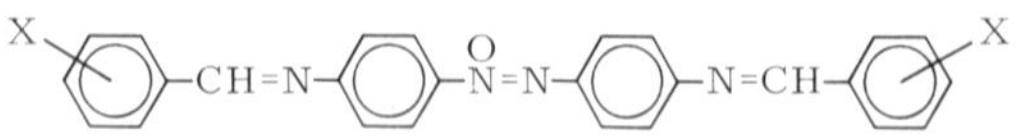

4,4'-azoxydibenzylideneaminoazoxybenzenes [579]

5.2.6

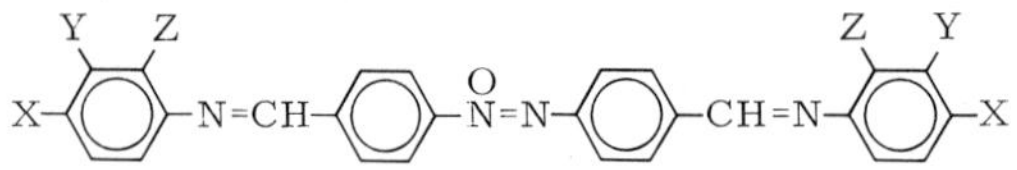

4,4'-azoxydibenzylidene anilines
Y = Z = H: [1:309, 1:319, 209, 453, 542, 830]
Y, Z other substituents: [1:309, 1:317, 209, 316, 453, 830]

5.2.7

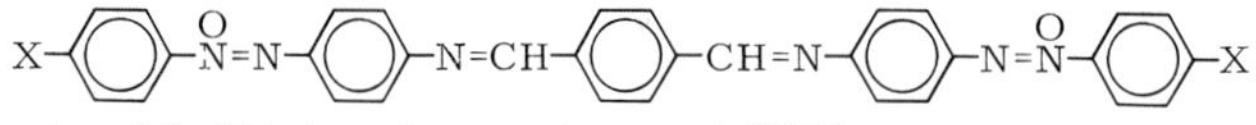

a-terephthal-bis-(*p*-aminoazoxybenzenes) [826]

5.2.8

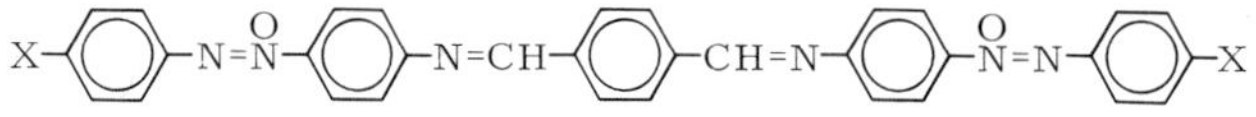

b-terephthal-bis-(*p*-aminoazoxybenzenes) [826]

5.2.9

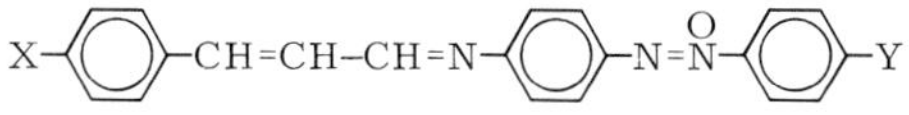

a-cinnamylidene-*p*-aminoazoxybenzenes [826]

5.2.10

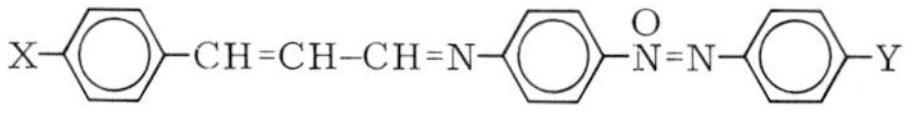

b-cinnamylidene-*p*-aminoazoxybenzenes [826]

5.2.11

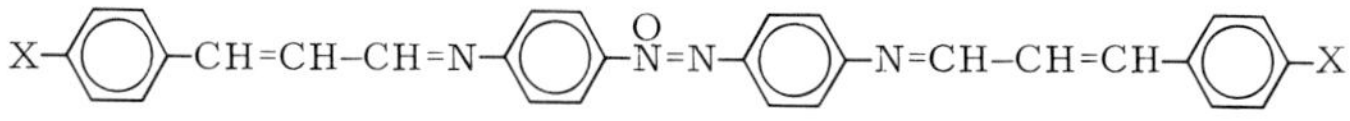

4,4'-dicinnamylidenaminoazoxybenzenes [579]

5.3   *Azomethine carboxylic acid esters*

5.3.1

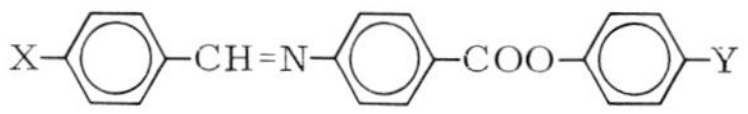

phenyl benzylidene-p-aminobenzoates [1:309, 706, 799, 837]

5.3.2

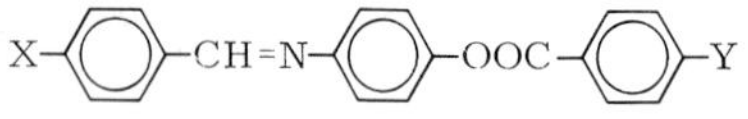

benzylidene-*p*-aminophenolbenzoates [422, 691]

5.3.3

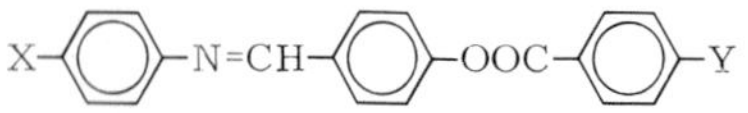

benzoyl-*p*-hydroxybenzylidene anilines [1:318, **145b,** 470, 830, 833, 872]

5.3.4

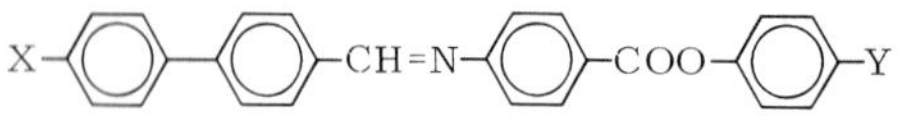

phenyl *p*-phenylbenzylidene-*p*-aminobenzoates [706]

5.3.5

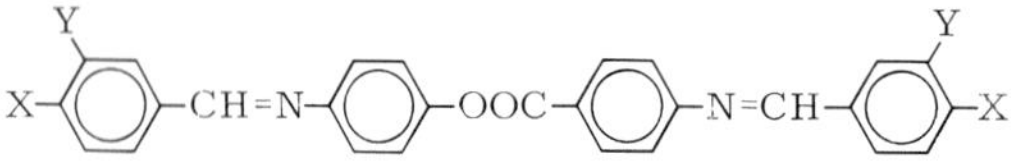

4,4'-bisbenzylidenaminophenylbenzoates [419]

5.3.6 X—◯—CH=CH–CH=N—◯—COO—◯—Y

phenyl cinnamylidene-*p*-aminobenzoates [830]

5.3.7 X—◯—CH=CH–CH=N—◯—OOC—◯—N=CH–CH=CH—◯—X

4,4'-bis-(cinnamylidenamino) phenyl benzoates [419]

5.3.8 X—◯—CH=N—◯—CH=CH–COO—◯—Y

phenyl benzylidene-*p*-aminocinnamates [1:309, 278, 279, 706]

5.3.9 X—◯—◯—CH=N—◯—CH=CH–COO—◯—Y

phenyl *p*-phenylbenzylidene-*p*-aminocinnamates [1:309, 278, 279, 706]

5.3.10 X—◯—CH=CH–CH=N—◯—CH=CH–COO—◯—Y

phenyl cinnamylidene-*p*-aminocinnamates [706]

5.4 *Azo-compound carboxylic acid esters*

5.4.1 X—◯—N=N—◯—OOC—◯—Y   *p*-benzoyloxyazobenzenes

Y = H: [35, 39, 222, 542, 706, 763, 872]
X = H: [17, 39, 837]
X = OR: [39, 211, **649,** 805, 834]
Y = H, X = CR = CR'COOR'': [1:309, 1:318, 486, 830, 831]
X, Y other substituents: [1:309, 39, **649,** 659, 830]

X—◯—N=N—◯(Z)—OOC—◯—Y   *p*-benzoyloxyazobenzenes [633, 840]

5.4.2 X—◯—N=N—◯—COO—◯—Y   phenyl benzeneazobenzoates [621]

5.4.3 X—◯—◯—N=N—◯—OOC—◯—Y   biphenylazophenolbenzoates [67, 222]

5.4.4 X—◯—N=N—◯—OOC—◯—◯—Y   benzeneazophenol-*p*-biphenylcarboxylates [1:309, 326]

5.4.5 X—◯—COO—◯—N=N—◯—OOC—◯—X   *p*-azophenol-bisbenzoates [715, 784, 831]

5.4.6 X—◯—OOC—◯—N=N—◯—COO—◯—X   diphenyl *p*-azobenzoates [621]

5.4.7   RO—⬡—N=N—⬡—OOC—⬡—COO—⬡—N=N—⬡—OR

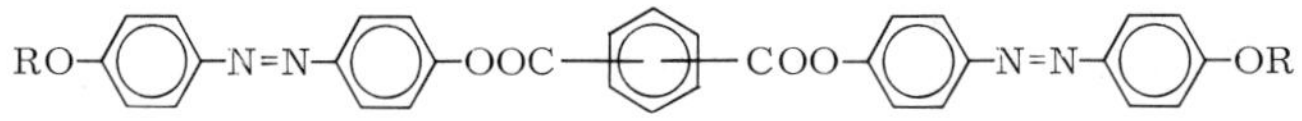

benzenedicarboxylic acid bis-(*p*-alkoxyazophenol esters) [632, 811, 829]

5.4.8   X—⬡—N=N—⬡—N=N—⬡—OOC—⬡—COO—⬡—N=N—⬡—N=N—⬡—X

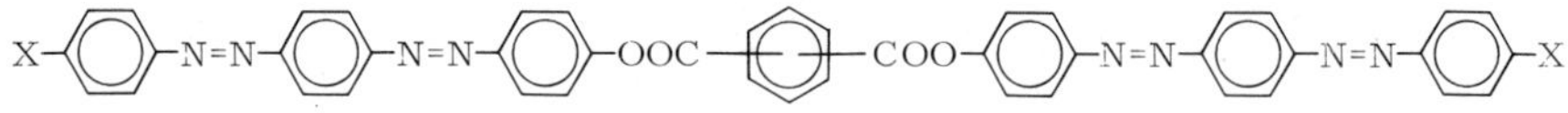

benzenedicarboxylic acid bis-(azophenolazobenzene esters) [632, 811]

5.4.9   X—⬡—COO—⬡(Y)—N=N—⬡(Z)—⬡(Z)—N=N—⬡(Y)—OOC—⬡—X

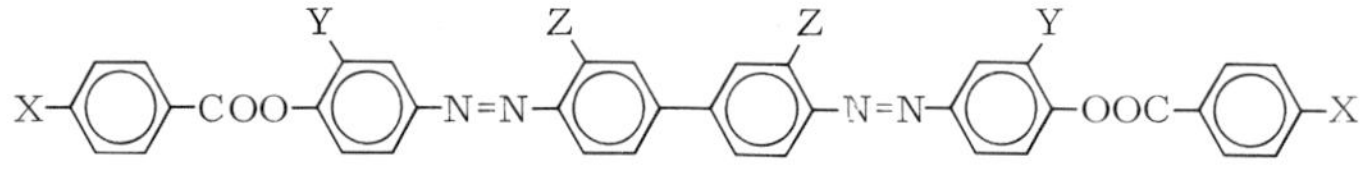

biphenylbisazophenol-bis-benzoates [71]

5.4.10   X—⬡—N=N—⬡(naphthyl)—OOC—⬡—Y

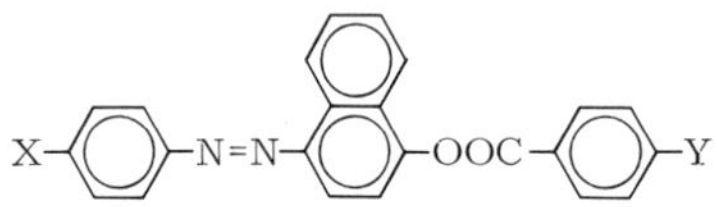

benzeneazo-α-naphth-4-ol benzoates [427, 633]

5.4.11   X—⬡—N=N—⬡—OOC—C(R)=C(R')—⬡—Y

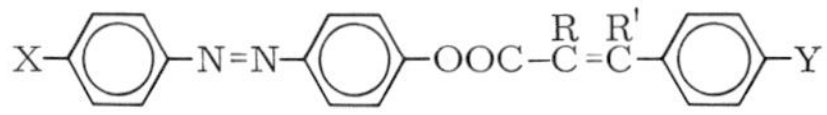

benzeneazophenolcinnamates [327, 805, 874]

5.4.12   X—⬡—N=N—⬡—OOC–CH=CH–CH=CH—⬡—Y

benzeneazophenol cinnamyleneacrylates [35, 805]

5.5   *Azo azoxy compounds*

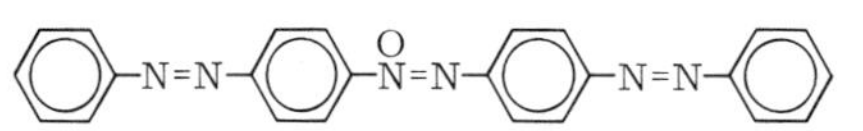

⬡—N=N—⬡—N(→O)=N—⬡—N=N—⬡   4,4'-bisazobenzenazoxybenzenes [829]

(*Note:* To date no additional liquid crystals have been described that contain these two bridging groups and no others.)

5.6   *Azoxy-compound carboxylic acid esters*

5.6.1   X—⬡—N(→O)=N—⬡—OOC—⬡—Y   benzeneazoxyphenolbenzoates [659, 839]

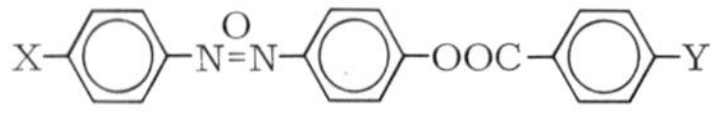

5.6.2   X—⬡—N(→O)=N—⬡—COO—⬡(Z)(Z')—Y   phenyl azoxybenzene-*p*-carboxylates [621]

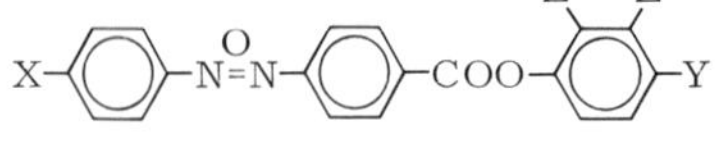

5.6.3   X—⬡—COO—⬡—N(→O)=N—⬡—OOC—⬡—X   *p*-azoxyphenol-bis benzoates [321]

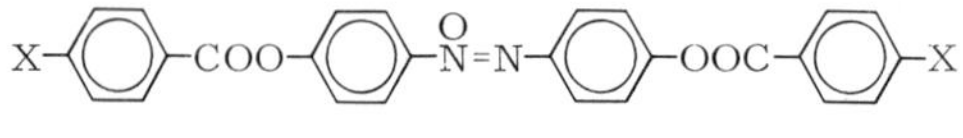

5.6.4 X—⟨○⟩—OOC—⟨○⟩—N=N—⟨○⟩—COO—⟨○⟩—X
                               O

diphenyl *p*-azoxybenzoates [1:309, 381, 706, 722]

5.6.5 C₂H₅O—⟨○⟩—N=N—⟨○⟩—COO—⟨○⟩—OOC—⟨○⟩
C_2H_5O—⟨○⟩—N=N—⟨○⟩—COO—⟨○⟩—OOC—⟨○⟩

p-phenetolazoxybenzoyl dihydroxybenzene benzoate [818]

5.6.6 C_2H_5O—⟨○⟩—N=N—⟨○⟩—COO—⟨○⟩—OOC—⟨○⟩—N=N—⟨○⟩—OC_2H_5

bis-(*p*-phenetolazoxybenzoyl)-dihydroxybenzene [818]

5.6.7 X—⟨○⟩—OOC–CH=CH—⟨○⟩—N=N—⟨○⟩—CH=CH–COO—⟨○⟩—X

diphenyl *p*-azoxycinnamates [1:309, 706]

5.7 Compounds with three different bridging groups

5.7.1   azomethine-azo-azoxy compounds [453]

6 Tolanes, stilbenes, carboxylic acid amides, chalcones, hydrazine and glyoxal derivatives (sometimes with other bridging groups)

6.1 Tolanes

6.1.1   X—⟨○⟩—C≡C—⟨○⟩—Y   *p,p'*-disubstituted tolanes

       X = R   Y = OR'   [357, 358, 465, 466]
       X = OR   Y = OR'   [179, **357, 358**, **465, 466**]
       X = OR   Y = OCOR'   [116, **179**, 181]
       X = OR   Y = OCOOR'   [116, 179, 181]
       X = OCOR   Y = OCOR'   [181, 711, 787]
       X = CN   Y = other substituent [72, **118, 120, 302**]
       X, Y = other substituents [466]

6.1.2   X—⟨○⟩—C≡C—⟨○⟩—Y   *p,p'*-disubstituted tolanes and further substituents [359]
            Z   Z'

6.1.3   X—⟨○⟩—⟨○⟩—C≡C—⟨○⟩—Y   *p*-phenyltolanes [302]

6.1.4   X—⟨○⟩—CH=N—⟨○⟩—C≡C—⟨○⟩—N=CH—⟨○⟩—X
          Y   Z   Z   Y

bis-(benzylidene-*p*-amino)-tolanes [363, 799]

6.1.5   X—⟨○⟩—CH=CH–CH=N—⟨○⟩—C≡C—⟨○⟩—N=CH–CH=CH—⟨○⟩—X

bis-(cinnamylidene-*p*-amino)-tolanes [363]

6.1.6 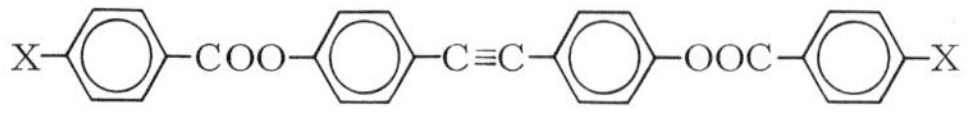   X—⟨O⟩—COO—⟨O⟩—C≡C—⟨O⟩—OOC—⟨O⟩—X   bis-(*p*-benzoyloxy)-tolanes [787]

6.1.7 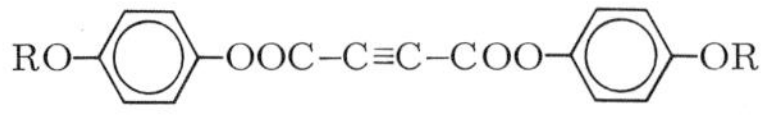   RO—⟨O⟩—OOC—C≡C—COO—⟨O⟩—OR   di-*p-n*-alkoxyphenyl acetylenedicarboxylates [765]

6.2   Stilbenes

6.2.1 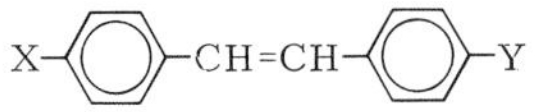   X—⟨O⟩—CH=CH—⟨O⟩—Y   4,4'-disubstituted stilbenes

    X = R   Y = OR' [695, 887, 890]
    X = OR   Y = OR' [117, 228, 787, **886**, 887, 890]
    X = OCOR   Y = OCOR' [787, 790]
    X = CN   Y = other substituent [**119**, 183a, 590, 723a]
    X, Y = other substituents [183a, 429, 686]

6.2.2   X—⟨O⟩(Z)—CH=CH—⟨O⟩(Z')—Y   4,4'-disubstituted stilbenes and further substituents [117, 887]

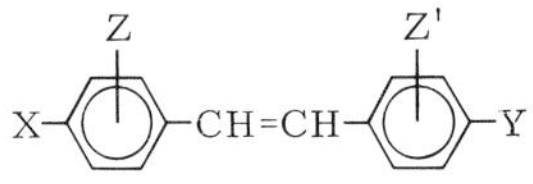

6.2.3   X—⟨O⟩—C(Z)=C(Z')—⟨O⟩—Y   4,4'-disubstituted stilbenes substituted on the double bond

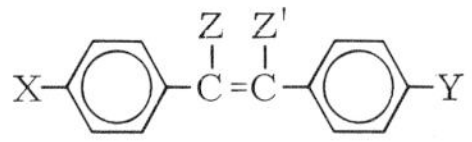

    Z or Z' = Cl, CH$_3$ [117, 148, 320, 499, 725, 883, 887, 890]
    Z or Z' = CN [124, 553]

6.2.4    X—⟨O⟩—CH=CH—⟨O⟩—CH=CH—⟨O⟩—Y   bis-phenyl-1,4-divinylbenzenes [H. Kelker, R. Hatz, unpublished]

6.2.5 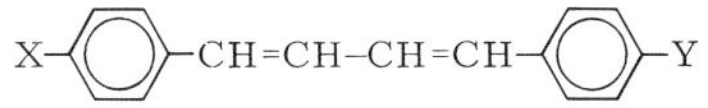   X—⟨O⟩—CH=CH–CH=CH—⟨O⟩—Y   1,4-diphenylbutadienes [567, 601]

6.2.6 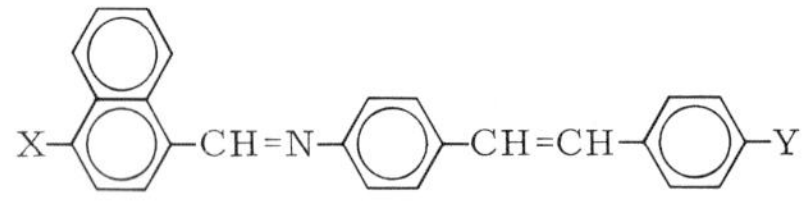   X—⟨naphthyl⟩—CH=N—⟨O⟩—CH=CH—⟨O⟩—Y   1-naphthylidene-*p*-aminostilbenes [227]

6.2.7   X—⟨O⟩(Y Z)—CH=N—⟨O⟩—CH=CH—⟨O⟩—N=CH—⟨O⟩(Z Y)—X

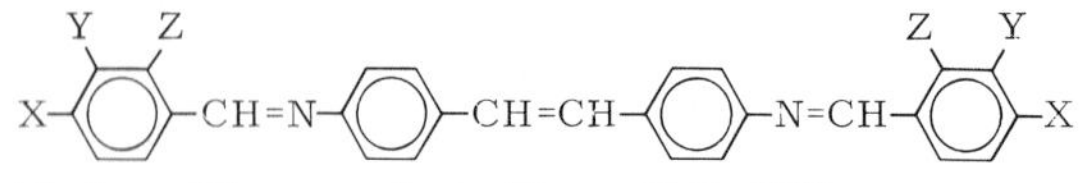

    bis-(benzylidene-*p*-amino)-stilbenes [363, 799]

6.2.8   X—⟨O⟩—CH=CH–CH=N—⟨O⟩—CH=CH—⟨O⟩—N=CH–CH=CH—⟨O⟩—X

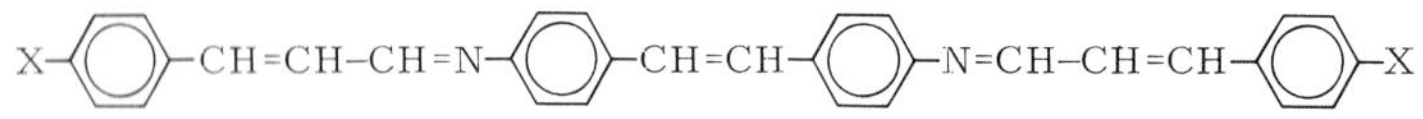

    bis-(cinnamylidene-*p*-amino)-stilbenes [363]

6.2.9   X—⟨O⟩—COO—⟨O⟩—CH=CH—⟨O⟩—OOC—⟨O⟩—X

    p,p'-dihydroxystilbene dibenzoates [787]

6.3   *Carboxylic acid amides*

6.3.1   X—◯—CO—NH—◯—Y   *p,p′*-disubstituted benzoyl anilines [67, 529, 722]

6.3.2   X—◯—◯—CO—NH—◯—Y   biphenyl-4-carboxylic acid anilides [326]

6.3.3   X—◯—CO—NH—◯—◯—Y   benzoyl-4-aminobiphenyls [67]

6.3.4   X—◯—NH—CO—(◯)ₙ—CO—NH—◯—X   *p*-phenyl-*p,p′*-dicarboxylic acid anilides [326, 799]

6.3.5   X—◯—CO—NH—(◯)ₙ—NH—CO—◯—X   bis-(benzoyl-*p*-amino)-*p*-phenyls [67]

6.3.6   X—◯—CO—NH—CO—◯—X   dibenzoyl imides [722]

6.3.7   X—◯—(CH=CH)ₙ—CO—NH—◯—Y   ω-phenylpolyenic acid anilides [243, 874]

6.3.8   X—◯—CH=CH—CO—HN—◯—◯—NH—CO—CH=CH—◯—X

dicinnamoyl benzidines [67]

6.3.9   X—◯—CH=N—◯—CO—NH—◯—COOR

benzylidene-*p*-aminobenzoyl-*p*-aminobenzoates [1:309, 529, 799]

6.3.10   X—◯—CH=N—◯—◯—NH—CO—◯—Y

benzylidene-monobenzoyl benzidines [67]

6.3.11   X—◯—CH=N—◯—CO—NH—◯—N=CH—◯—X

bis-(benzylidene-*p*-amino)-benzoyl anilines [314]

6.3.12   X—◯—CH=N—◯—CONH—◯—◯—NHCO—◯—N=CH—◯—X

benzidine-bis-(*p*-benzylidenaminobenzoic amide) [799]

6.3.13   X—◯—CH=CH—CH=N—◯—CO—NH—◯—Y

cinnamylidene-*p*-aminobenzoyl anilines [67, 529, 813]

6.3.14   X—◯—CH=CH—CH=N—◯—◯—NH—CO—◯—Y

cinnamylidene-monobenzoyl benzidines [67]

6.3.15   X—⟨○⟩—CH=CH–CH=N—⟨○⟩—CO–NH—⟨○⟩—N=CH–CH=CH—⟨○⟩—X

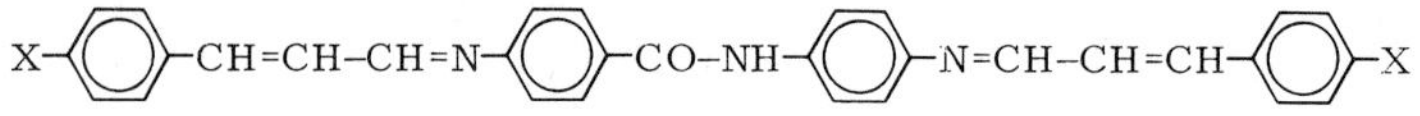

bis-(cinnamylidene-*p*-amino)-benzoyl anilines [314]

6.3.16   X—⟨○⟩—CO–NH—⟨○⟩—N=N—⟨○⟩—Y

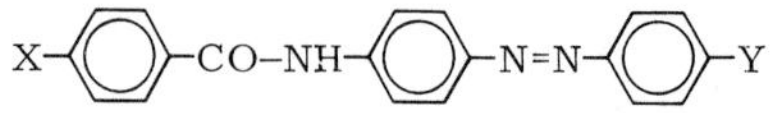

benzoyl-*p*-aminoazobenzenes [590, 810]

6.3.17   X—⟨○⟩—⟨○⟩—CO–NH—⟨○⟩—N=N—⟨○⟩—Y

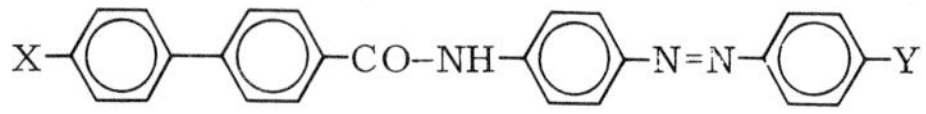

p-phenylbenzoyl-*p*-aminoazobenzenes [326]

6.3.18   X—⟨○⟩—COO—⟨○⟩—CO–NH—⟨○⟩—Y

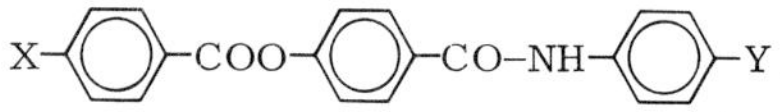

p-benzoyloxybenzoic acid anilides [722]

6.3.19   X—⟨○⟩—(CH=CH)–CH=N—⟨○⟩—NH–CO–NH—⟨○⟩—N=CH–(CH=CH)—⟨○⟩—X

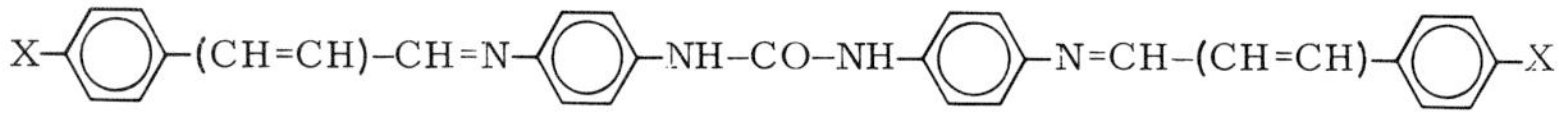

derivatives of diphenylurea [1:43, no. 1196–1201]

6.4   *Chalcones* (characteristic group —CH=CH—CO—)

subtypes:
compounds with open-chain chalcone group 6.4.1–6
compounds with cyclic chalcone group 6.4.7–10
chalcones with additional bridging groups 6.4.11–18

6.4.1 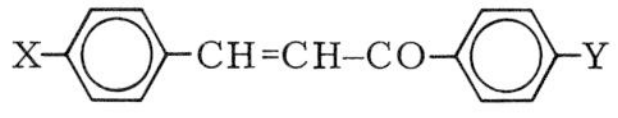   X—⟨○⟩—CH=CH–CO—⟨○⟩—Y   benzylidene-acetophenones [45, 633, 800]

6.4.2   X—⟨○⟩—CH=CH–CO—⟨○⟩—⟨○⟩—CO–CH=CH—⟨○⟩—X

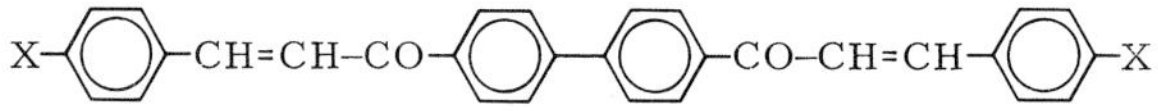

bis-(benzylidene-*p*-acetyl)-biphenyls [326]

6.4.3   X—⟨○⟩—(CH=CH)$_n$–CO–(CH=CH)$_n$—⟨○⟩—X

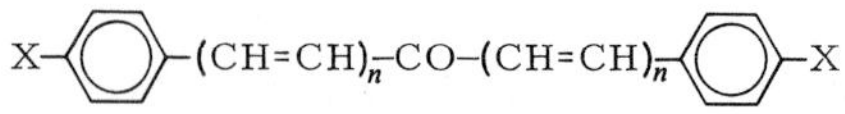

ω-phenylalkylidene acetones [45, 243, 801, 808]

6.4.4   X—⟨○⟩—CO–CH=CH–CO—⟨○⟩—X

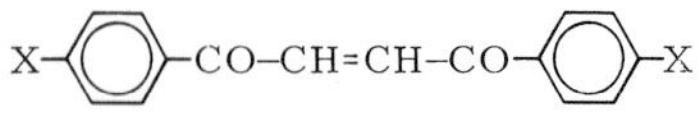

1,2-bis-(benzoyl)-ethylenes [**638, 846**]

6.4.5   X—⟨○⟩—⟨○⟩—CO–CH=CH–CO—⟨○⟩—⟨○⟩—X

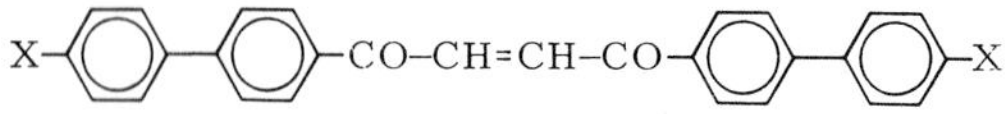

1,2-bis-(*p*-phenylbenzoyl)-ethylenes [**643**]

6.4.6   X—⟨○⟩—CH=CH–CH=CH–CO—⟨○⟩—Y

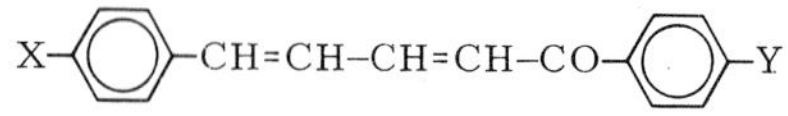

cinnamylidene acetophenones [243]

6.4.7   X—⬡—(CH=CH)ₙ—CH=⬠=CH—(CH=CH)ₙ—⬡—X

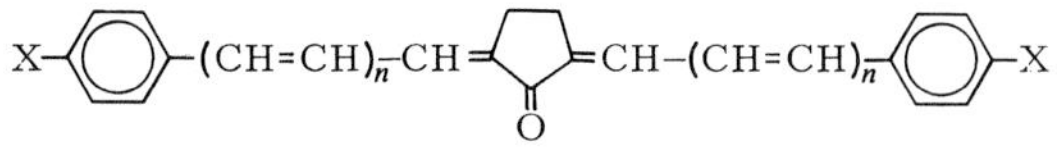

bis-(ω-phenylalkenylidene)-cyclopentanones [45, 243, 316, 625, 797]

6.4.8   X—⬡—CH=⬡=CH—⬡—X

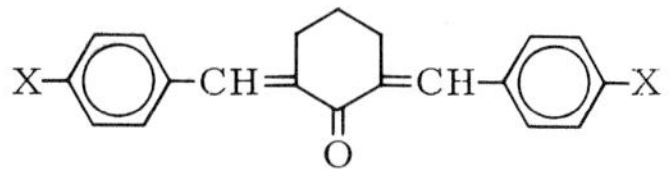

bisbenzylidene cyclohexanones [45, 156, 346, 464, 498, 542, 625, **656d**, 787, 797, 800]

6.4.9   X—⬡—(CH=CH)ₙ—CH=⬡=CH—(CH=CH)ₙ—⬡—X

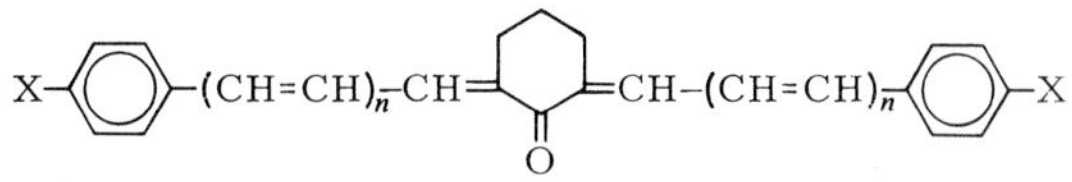

bis-(ω-phenylalkenylidene)-cyclohexanones [243, 801, 808]

6.4.10   X—⬡—(CH=CH)ₙ—CH=⬡=CH—(CH=CH)ₙ—⬡—X

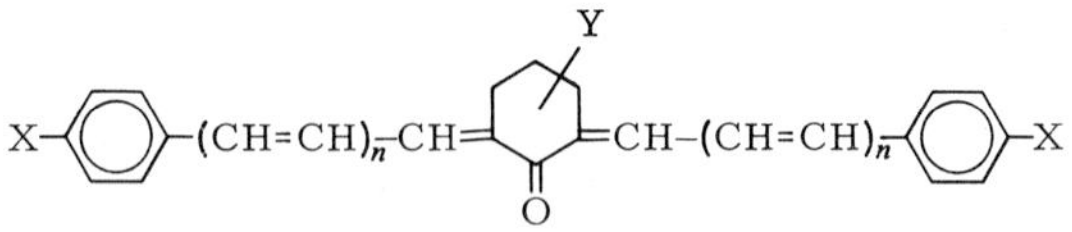

bis-(ω-phenylalkenylidene)-cyclohexanones [243, 346, 625, 800, 822]

6.4.11   X—⬡—CH=⬡=CH—⬡—X

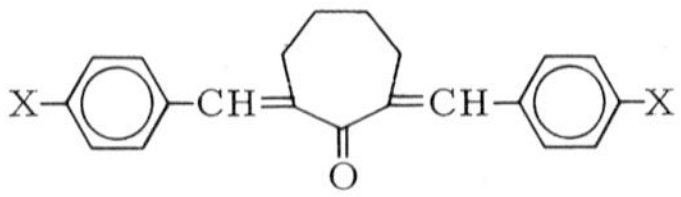

bisbenzylidene cycloheptanones [1:43, no. 170–178]

6.4.12   X—⬡—CH=N—⬡—CO–CH=CH—⬡—X

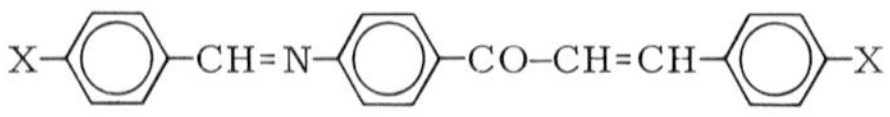

bis-(benzylidene)-p-aminoacetophenones [872]

6.4.13   X—⬡—CH=N—⬡—CH=CH—CO—⬡—Y

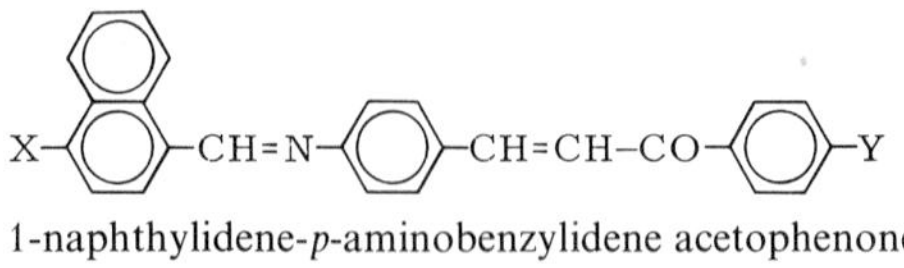

1-naphthylidene-p-aminobenzylidene acetophenones [227]

6.4.14   X—⬡—CH=CH–CH=N—⬡—CO–CH=CH–CH=CH—⬡—X

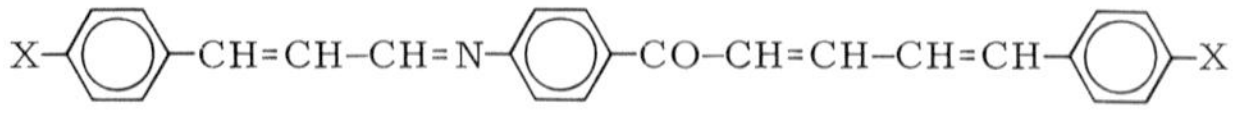

bis-(cinnamylidene)-p-aminoacetophenones [872]

6.4.15   X—⬡—CH=CH–CO—⬡—N=N—⬡—CO–CH=CH—⬡—X

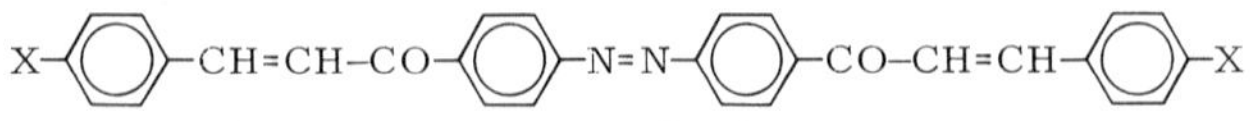

bis-(benzylidene)-azoacetophenones [209]

6.4.16   X—⬡—CH=CH–CO—⬡—N=N—⬡—CO–CH=CH—⬡—X

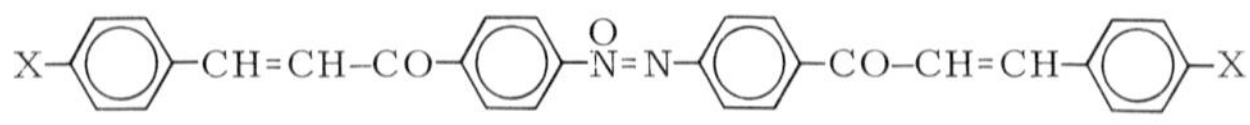

bis-(benzylidene)-azoxyacetophenones [111, 782, 872]

6.4.17   X—⬡—CO–CH=CH—⬡—N=N—⬡—CH=CH–CO—⬡—X

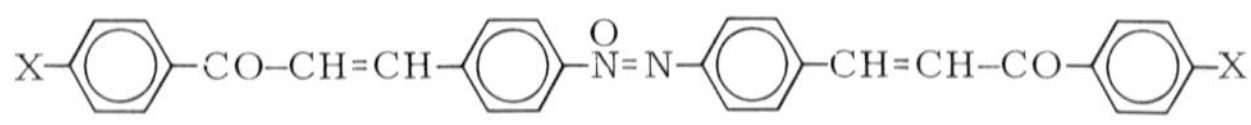

azoxybenzal bisacetophenones [209]

6.4.18   X—⟨○⟩—[CH=CH]ₙ—CO—⟨○⟩—OOC—⟨○⟩—Y

ω-phenylalkylidene-*p*-oxybenzoyl acetophenones [243, 633]

6.4.19   X—⟨○⟩—COO—⟨○⟩—CH=CH—CO—⟨○⟩—OOC—⟨○⟩—X

bis-(benzoyl)-*p*-oxybenzylidene-*p*-oxyacetophenones [243, 633]

6.4.20   other chalcones [1:43, nos. 160, 161, 169]

6.5   *Hydrazine derivatives*

(this series only includes compounds with N—N single bond)
subtypes:
azines (characteristic group —CH=N—N=CH—)
hydrazones (characteristic group —CH=N—NH—)

6.5.1   X—⟨○⟩—CH=N—N=CH—⟨○⟩—X   benzylideneazines [1:138, 45, 69, **76,** 83, 345, 542, **603,** 711, 782, 787, 837]

6.5.2   X—⟨○⟩—C(R)=N—N=C(R)—⟨○⟩—X   phenylketazines [345, 602, 787]

6.5.3   X—⟨○⟩—⟨○⟩—C(R)=N—N=C(R)—⟨○⟩—⟨○⟩—X   *p*-phenylbenzylideneazines [45, 691]

6.5.4   X—⟨○⟩—CH=CH–CH=N—N=CH–CH=CH—⟨○⟩—X   cinnamylideneazines [601]

6.5.5   X—⟨○⟩(Y,Z)—CH=N—⟨○⟩—CH=N—N=CH—⟨○⟩—N=CH—⟨○⟩(Z,Y)—X

bis-(benzylidene-*p*-amino)-benzylideneazines [414]

6.5.6   X—⟨○⟩—CH=CH–CH=N—⟨○⟩—CH=N—N=CH—⟨○⟩—N=CH–CH=CH—⟨○⟩—X

bis-(cinnamylidene-*p*-amino)-benzylideneazines [414]

6.5.7   X—⟨○⟩—COO—⟨○⟩—CH=N—N=CH—⟨○⟩—OOC—⟨○⟩—X

bis-(*p*-benzoyloxy)-benzylideneazine [345, 782]

6.5.8   X—⟨○⟩—CH=CH–COO—⟨○⟩—CH=N—N=CH—⟨○⟩—OOC–CH=CH—⟨○⟩—X

bis-(*p*-cinnamoyloxy)-benzylideneazine [345]

6.5.9   X—⟨○⟩—CH=N–NH—⟨○⟩

benzylidenephenylhydrazones [77, 601]

6.5.10 X—⬡—CH=CH–CH=CH–CH=N–NH—⬡

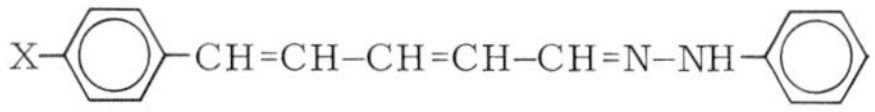

5-phenylpentadienal-phenylhydrazones [243, 812]

6.5.11 X—⬡—NH–N=CH—⬡—N=N—⬡—CH=N–NH—⬡—X

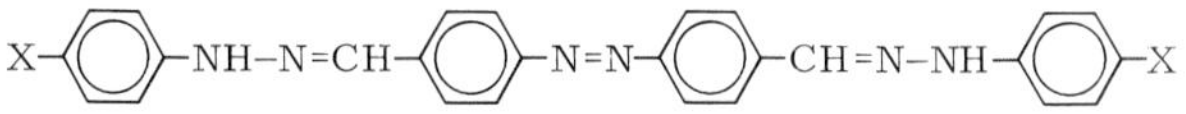

p,p'-azodibenzal-bis-phenylhydrazones [453]

6.5.12 X—⬡—NH–N=CH—⬡—$\overset{O}{N}$=N—⬡—CH=N–NH—⬡—X

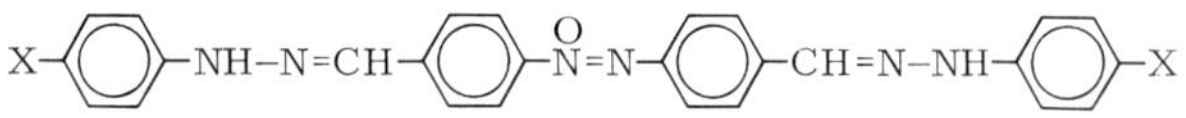

p,p'-azoxydibenzal-bis-phenylhydrazones [453]

6.5.13 alkadienalazines cf. 12.6

6.6 *Glyoxal derivatives*

6.6.1 X—⬡—N=$\overset{Y}{C}$–$\overset{Z}{C}$=N—⬡—X   glyoxalanilides [804, 895]

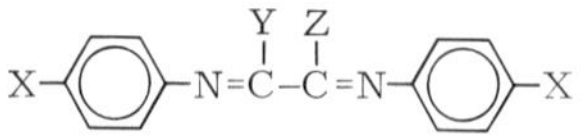

7   *Aromatic compounds with rare bridging groups,* mostly used in conjunction with —CH=N—, —N=N—, —N(O)=N—, or —COO—

7.1 *Linear bridging groups*

7.1.1 —⬡—Hg—⬡—   diphenylmercury derivatives [566, 799]

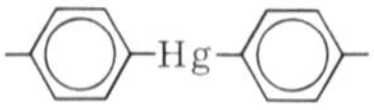

7.1.2 —⬡—N=C=N—⬡—   diphenylcarbodiimide derivatives [590, 810]

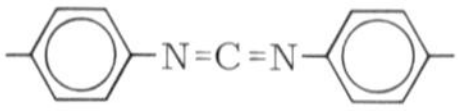

7.1.3 tolanes cf. 6.1

7.2 *Monatomic bent bridging groups*

7.2.1 —⬡—O—⬡—   diphenyl ether derivatives [798, 799, 841]

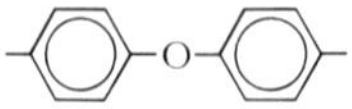

7.2.2 —⬡—S—⬡—   diphenylsulfide derivatives [798, 799, 841]

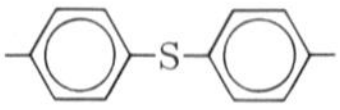

7.2.3 —⬡—NH—⬡—   diphenylamine derivatives [222, 798, 799, 841]

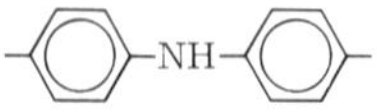

7.2.4 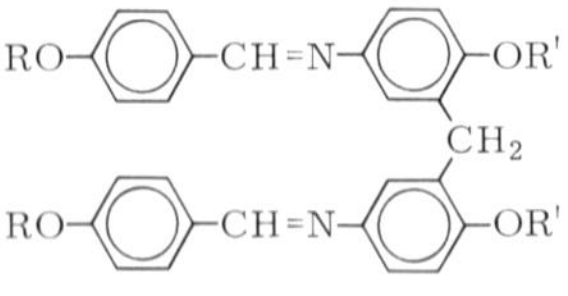 RO—⬡—CH=N—⬡—OR'   2,2'-dialkoxy-5,5'-di-N-(*p*-alkoxybenzylideneamino)-
                            $\overset{|}{CH_2}$ diphenylmethanes [308]
        RO—⬡—CH=N—⬡—OR'

7.3   Aliphatic, alicyclic, and heterocyclic bridging groups
(heteroaromatic compounds see 2)

7.3.1   α,ω-diphenylalkanes [68, 298, 305, 364, 596, 798, 799, 805]

7.3.2   1,4-disubstituted cyclohexanes [1:43, 154, 163, 166, 189, 190, 191, 240a, 380, 644, 645, 651, 766, 768]

7.3.3   4,4′-disubstituted *trans*-cyclohexylcyclohexanes [378]

7.3.4   1,4-disubstituted dihydroresorcines [633]

7.3.5   d-camphoric acid esters [327, 839]

7.3.6   1,4-disubstituted bicyclo-(2.2.2)-octanes [**163,** 166, 168]

7.3.7   1,4-disubstituted cyclohexa-1,3-dienes [166]

7.3.8   1,4-disubstituted cyclohexa-1,4-dienes [166]

7.3.9   1,4-disubstituted bicyclo-(2.2.2)-octenes [166]

7.3.10   N,N′-disubstituted piperazines [151, 426, **643, 651**]

7.3.11   heterocyclic chalcones

Z = O: derivatives of γ-pyrone [1:43, no. 3938–3947]
Z = S: derivatives of γ-thiopyrone [1:43, no. 3948–3957]
Z = NCH$_3$: derivatives of piperidone-4 [1:43, no. 3958–3968]

7.3.12   hexahydrochrysene derivatives [725]

7.4   Ethers

7.4.1   benzylphenyl ethers [722, 798, 799]

7.4.2   p-phenylbenzylphenyl ethers [305]

7.4.3 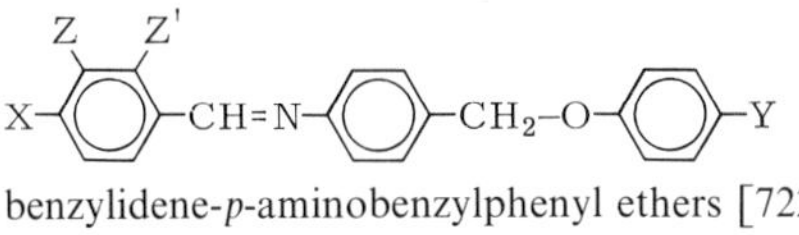   benzylidene-*p*-aminobenzylphenyl ethers [722, 799, 830]

7.4.4 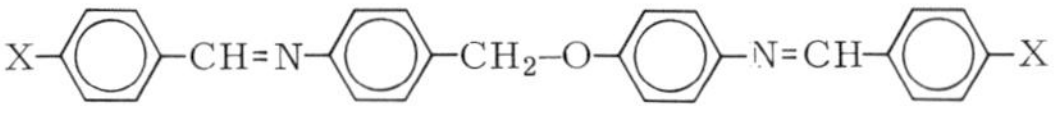   4,4′-bis-(benzylideneamino)-phenylbenzyl ethers [596, 799]

7.4.5 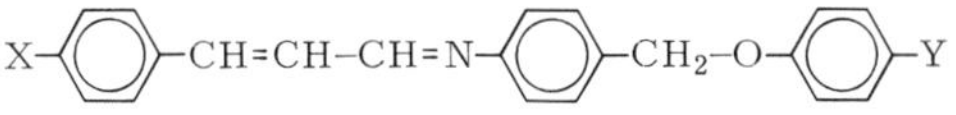   cinnamylidene-*p*-aminobenzylphenyl ethers [722]

7.4.6 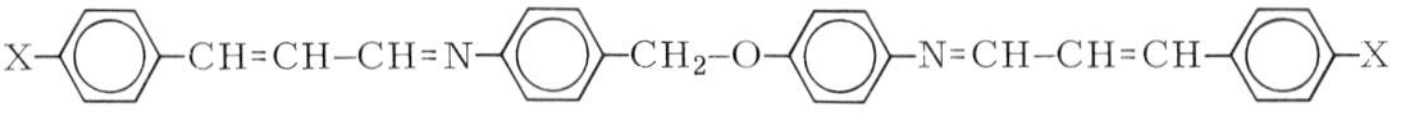   4,4′-bis-(cinnamylideneamino)-phenylbenzyl ethers [596]

7.4.7 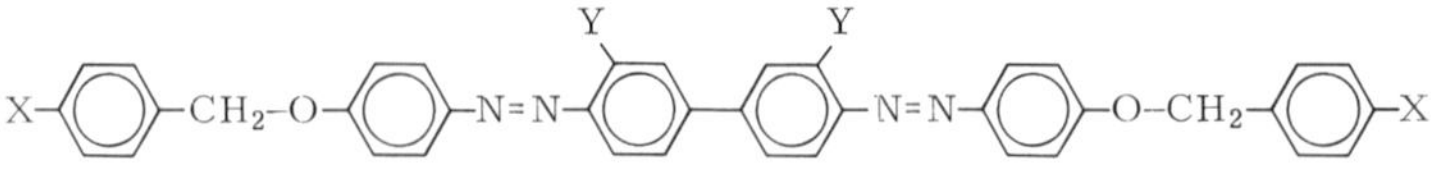   biphenylbis-(azophenoldibenzyl ethers) [71]

7.4.8 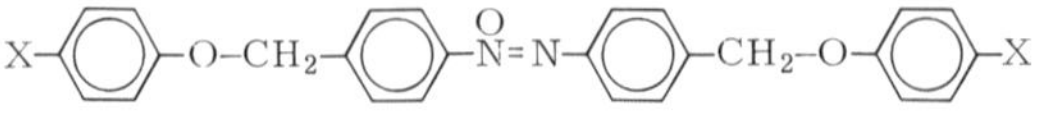   *p*-azoxybenzylbis-(phenyl ethers) [1:309, 722]

7.4.9 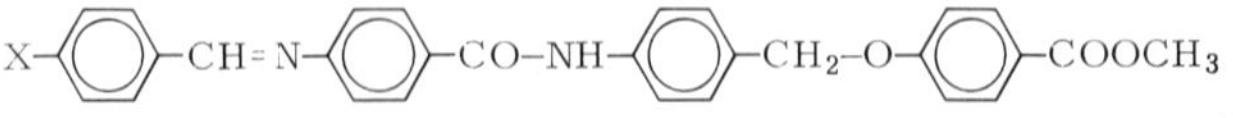   methyl benzylidene-*p*-aminobenzoyl-*p*-aminobenzyl-*p*-oxybenzoates [722]

7.4.10 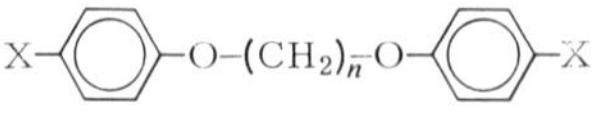   *n*-methylenebis-(phenol ethers) (*n* = 2 or 3) [798, 799]

7.4.11 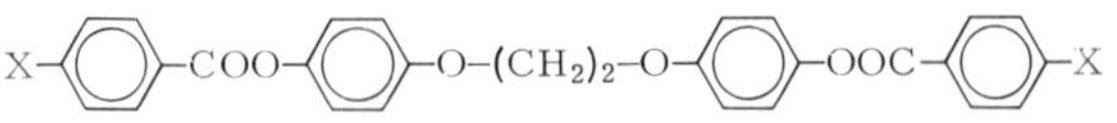   ethylenebis-(hydroquinone dibenzoates) [837]

7.5   Amines

7.5.1 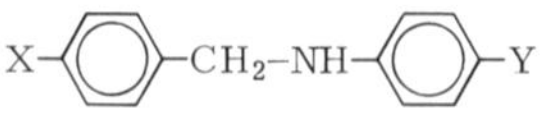   benzylanilines [798, 799]

7.5.2 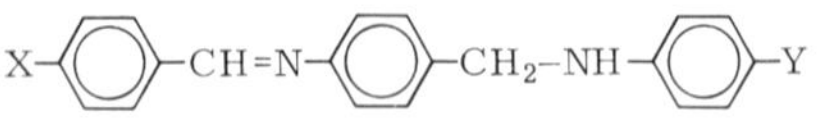   benzylidene-*p*-aminobenzylanilines [314, 414, 529]

7.5.3 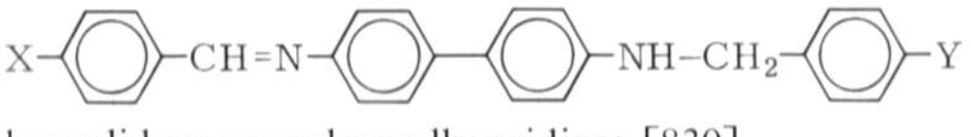   benzylidene-monobenzylbenzidines [830]

7.5.4   X—[ring Y]—CH=N—[ring]—CH₂—NH—[ring]—N=CH—[ring Y]—X

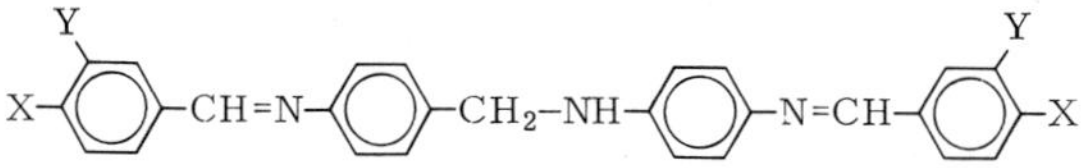

bis-(p-benzylideneamino)-benzylanilines [314, 799]

7.5.5   X—[ring]—CH=N—[ring]—CH₂—NH—[ring]—[ring]—N=CH—[ring]—X

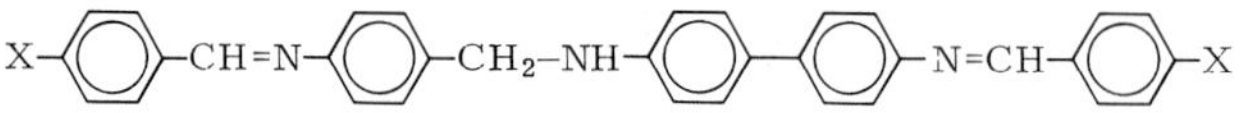

bisbenzylidene-mono-p-aminobenzylbenzidines [830]

7.5.6   X—[ring]—CH=N—[ring]—CH₂—NH—[ring]—CH₂—NH—[ring]—Y

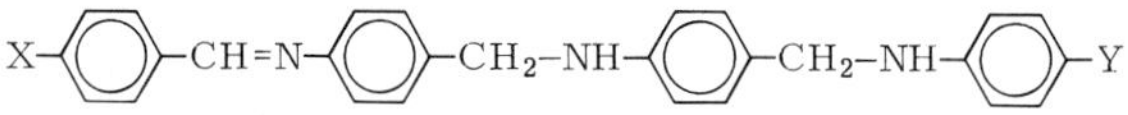

benzylidenebis-(p-aminobenzyl)-anilines [414]

7.5.7   X—[ring]—CH=N—[ring]—CH₂—NH—[ring]—NH—CH₂—[ring]—N=CH—[ring]—X

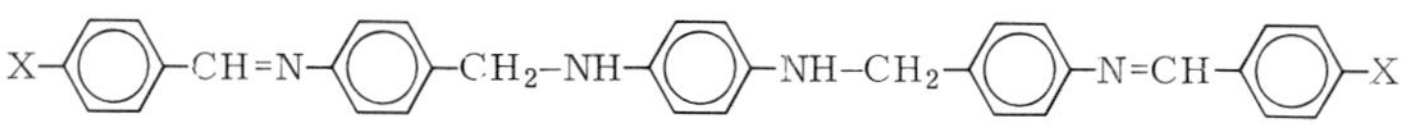

bis-(p-benzylideneaminobenzyl)-p-phenylenediamines [1:43, no. 1169, 1170]

7.5.8   X—[ring]—CH=N—[ring]—CH₂—NH—[ring]—[ring]—NH—CH₂—[ring]—N=CH—[ring]—X

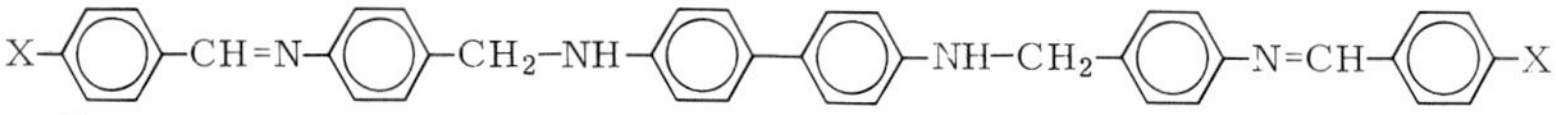

bis-(p-benzylideneaminobenzyl)-benzidines [799]

7.5.9   X—[ring]—CH=CH—CH=N—[ring]—CH₂—NH—[ring]—Y

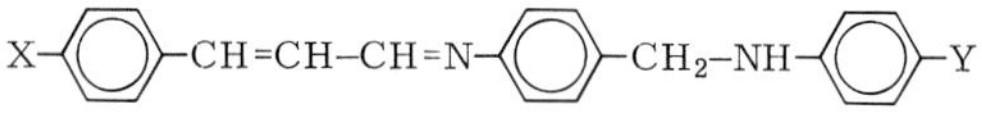

p-cinnamylideneaminobenzylanilines [529]

7.5.10   X—[ring]—CH=CH—CH=N—[ring]—CH₂—NH—[ring]—N=CH—CH=CH—[ring]—X

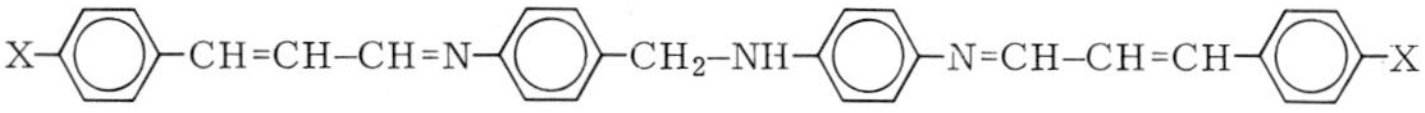

bis-(p-cinnamylideneamino)-benzylanilines [314]

7.5.11   X—[ring]—CH=CH—CH=N—[ring]—CH₂—NH—[ring]—CH₂—NH—[ring]—Y

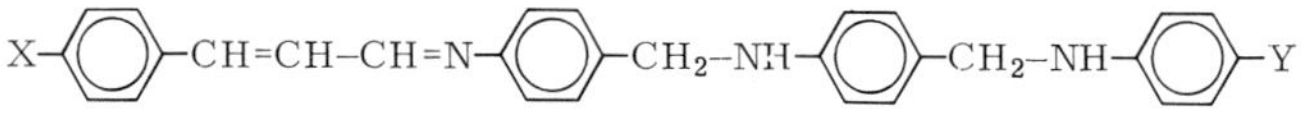

cinnamylidenebis-(p-aminobenzyl)-anilines [414]

7.5.12   X—[ring]—CH=N—[ring]—NH—(CH₂)ₙ—NH—[ring]—N=CH—[ring]—X

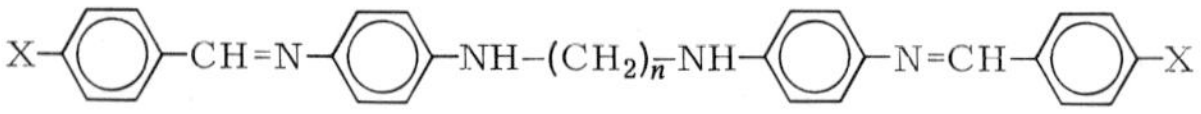

bis-(p-benzylideneamino)-n-methylene-α,ω-dianilines (n = 2 or 3) [426, 798, 799]

7.6   Carboxylic acid esters with aliphatic carbon atoms in the bridging group

subtypes and characteristic groups:

glycol esters	—COO(CH₂)OOC—
alkanedicarboxylic acid esters	—OOC(CH₂)COO—
ω-phenylalkanecarboxylic acid esters	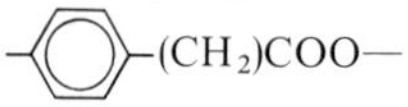 [ring]—(CH₂)COO—
ω-phenylalkyl esters	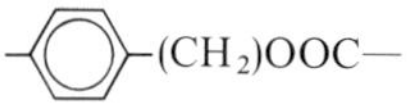 [ring]—(CH₂)OOC—

7.6.1   X—[ring]—COO—(CH₂)ₙ—OOC—[ring]—X   .   alkandiol dibenzoates [635]

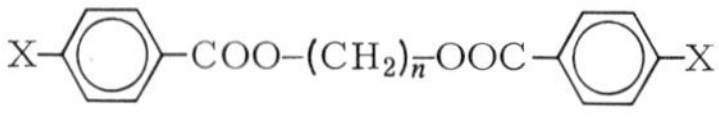

7.6.2 X—⬡—CH=N—⬡—COO—(CH₂)ₙ—OOC—⬡—N=CH—⬡—X

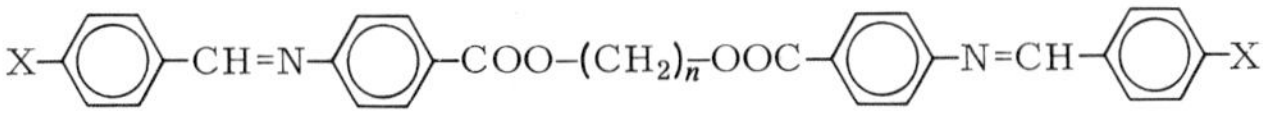

alkandiol bis(p-benzylideneaminobenzoates) [1:309, 635]

7.6.3 X—⬡—N=N—⬡—COO—(CH₂)ₙ—OOC—⬡—N=N—⬡—X

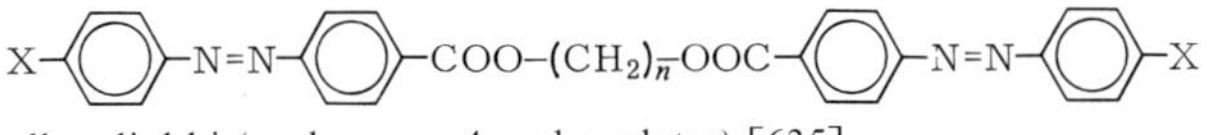

alkandiol bis(azobenzene-4-carboxylates) [635]

7.6.4 X—⬡—COO—⬡—N=N—⬡—COO—(CH₂)ₙ—OOC—⬡—N=N—⬡—OOC—⬡—X

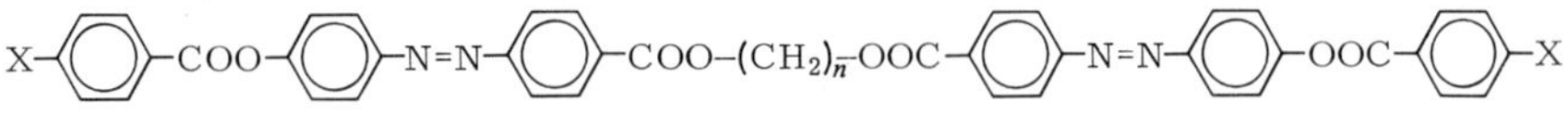

alkandiol bis(4-benzoyloxyazobenzene-4′-carboxylates) [635]

7.6.5 X—⬡—OOC—(CH₂)ₙ—COO—⬡—X   diphenyl alkandicarboxylates [897]
with n = 0: diphenyl oxalates [1:43, no. 921–928]

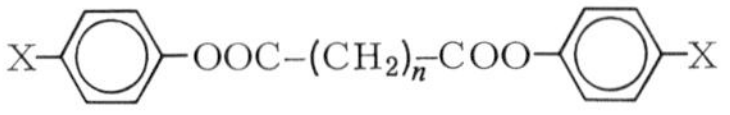

7.6.6 X—⬡—CH=N—⬡—OOC—(CH₂)ₙ—COO—⬡—N=CH—⬡—X

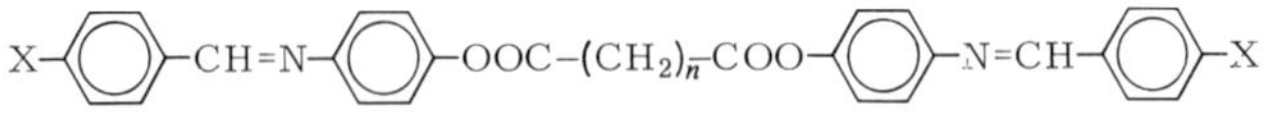

bis(p-benzylidenaminophenyl)succinates and -adipates [419]

7.6.7 X—⬡—CH=CH–CH=N—⬡—OOC—(CH₂)ₙ—COO—⬡—N=CH–CH=CH—⬡—X

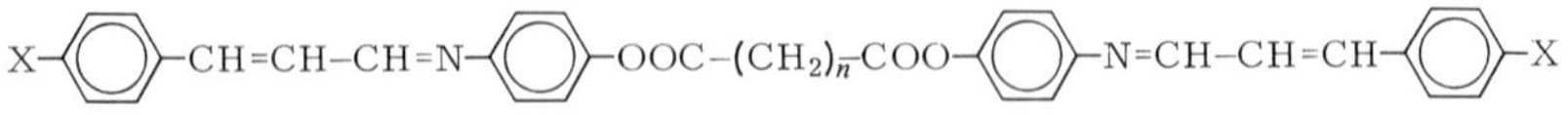

bis(p-cinnamylidenaminophenyl)succinates and -adipates [419]

7.6.8 RO—⬡—N=N—⬡—OOC—(CH₂)ₙ—COO—⬡—N=N—⬡—OR

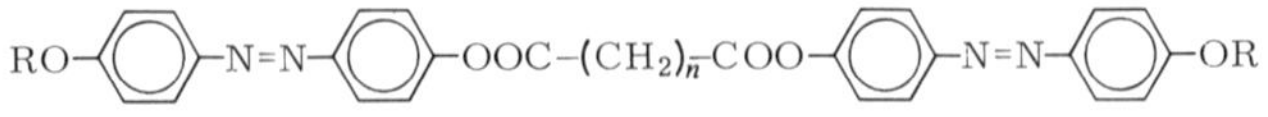

bis(p-azophenol)alkandicarboxylates [805, 816, **837**]

7.6.9 X—⬡—N=N—⬡—OOC—CH₂CH(Cl)—COO—⬡—N=N—⬡—X

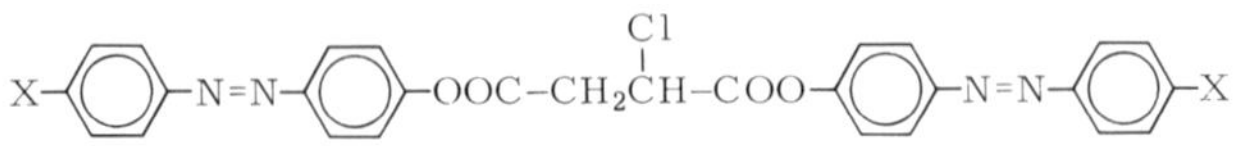

bis-(benzenazophenol)-chlorosuccinates [327]

7.6.10 X—⬡—N=N—⬡—OOC—CH₂CH(CH₃)–CH₂–CH₂–COO—⬡—N=N—⬡—X

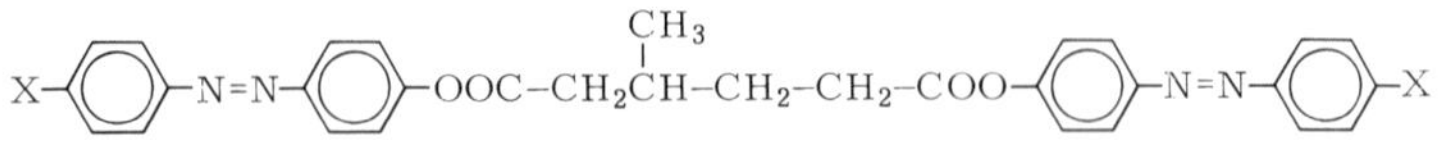

bis-(benzenazophenol)-β-methyladipates [839]

7.6.11 X—⬡—N=N(O)—⬡—OOC—CH₂CH(CH₃)–CH₂–CH₂–COO—⬡—N=N(O)—⬡—X

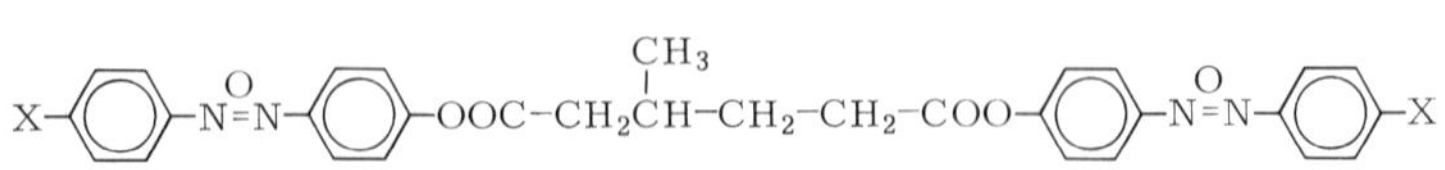

bis-(benzenazoxyphenol)-β-methyladipates [839]

7.6.12 X—⬡—N=N—⬡—OOC—(CH₂)ₙ—⬡

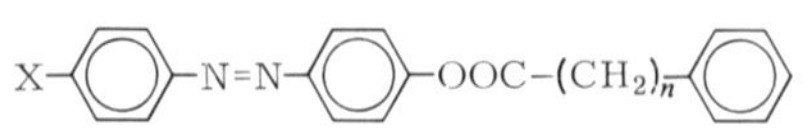

benzenazophenol-ω-phenylalkanoates [35, 633, 805]

7.6.13 X—⬡—CH₂–COO—⬡—N=N—⬡—⬡—N=N—⬡—OOC–CH₂—⬡—X

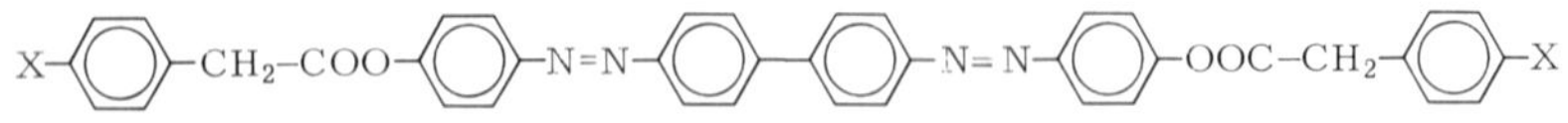

biphenyl-4,4′-azophenol bis(phenylacetates) [71, 830]

7.6.14　RO—⬡—N=N—⬡—OOC–CH–CH₂–⬡—OR'

with X on the CH

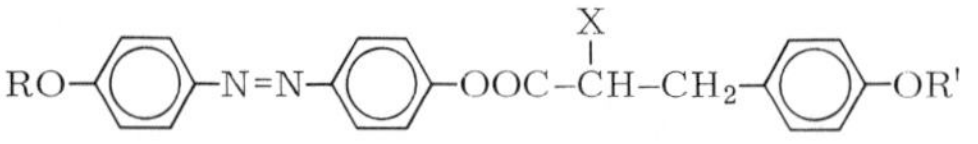

alkoxyphenolazophenol-*p*-alkoxydihydrocinnamates [327]

7.6.15　X—⬡—CH=N—⬡—CH=CH–COO–(CH₂)ₙ—⬡

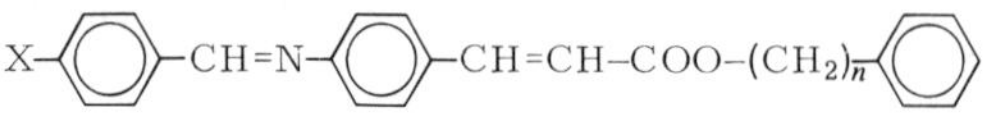

ω-phenylalkyl *p*-benzylideneaminocinnamates [278]

7.6.16　⬡—⬡—CH=N—⬡—CH=CH–COO–(CH₂)ₙ—⬡—X

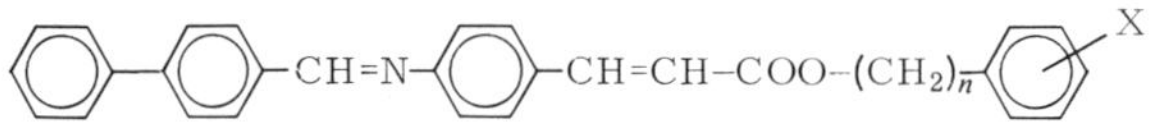

ω-phenylalkyl *p*-biphenylideneaminocinnamates [291, 295]

7.7　*Nitrones*

7.7.1　RO—⬡—CH=N—⬡—OR　　　N-(*p*-alkoxyphenyl)-*p*-alkoxyphenyl nitrones [784, **882, 885**]

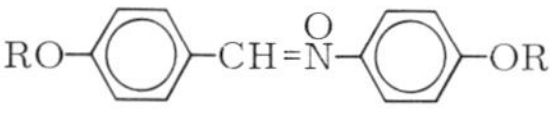

7.7.2　other nitrones [4, 584, 882]

7.8　*Compounds with bridging groups containing sulfur or selenium*

7.8.1　—⬡—S–S—⬡—　　　　　　　diphenyldisulfides [542]

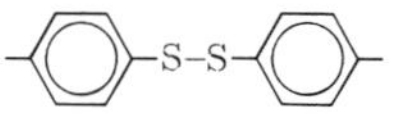

7.8.2　X—⬡—CO–S—⬡—Y　　　　phenylthiolbenzoates [**400, 421a, 586**]

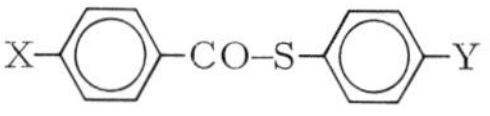

7.8.3　X—⬡—COS—⬡—SOC—⬡—X　thiohydroquinone bis benzoates [166]

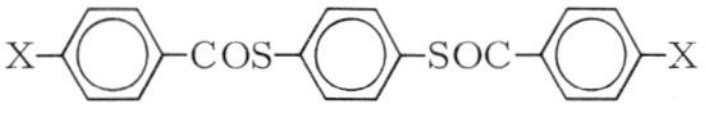

7.8.4　X—⬡—SOC—⬡—COS—⬡—X　terephthalic acid bis(thiophenyl esters) [166]

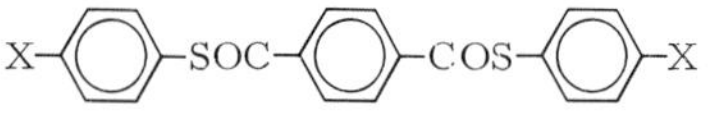

7.8.5　X—⬡—CO–S–S–CO—⬡—X　dibenzoyldisulfides [1:116, no. 1354]

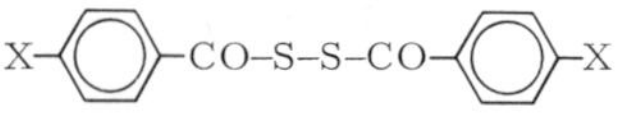

7.8.6　X—⬡—CO–Se—⬡—Y　　　phenylselenolbenzoates [327a]

No mesophases have been found up to now among phenylbenzylsulfides, phenyl benzenesulfonates, benzenesulfonic acid anilides [219], and diphenylthioureas. Diphenylsulfides see 7.2.2.

7.9　*1-Benzylidene-6-benzylidenaminoacenaphthenes*

7.9　X—⬡—CH=⬡⬡—N=CH—⬡—X　　[H. Kelker, R. Hatz, unpublished]

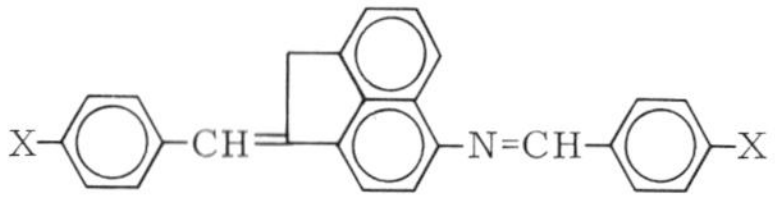

7.10 α,ω-*Bis-benzoyl-alkanes*

7.10 R—⬡—⬡—CO–CH$_2$–CH$_2$–CO—⬡—⬡—R

1,4-bis-(4′-n-alkyl-biphenylyl-(4))-butane-1,4-diones [650]
bis-benzoyl-methane derivatives [1:43, no. 221]

7.11 *Bis-(p-benzylideneamino)-desoxybenzoines*

7.11 X—⬡—CH=N—⬡—CH$_2$–CO—⬡—N=CH—⬡—X   [363]

7.12 *Aldoxime esters*

7.12 X—⬡—CH=N–O–CO—⬡—Y   [653, 1:43, no. 2078–2175]

7.13 *Ketoxime esters*

7.13 X—⬡—C=N–O–CO—⬡—Y   [656b]
R group on C

7.14 *Benzoylperoxides*

7.14 X—⬡—CO–O–O–CO—⬡—X   [1:43, no. 464–478]

8 *Heteroaromatic compounds with bridging groups*

8.1 *Azomethines*

8.1 X—⬡—N=CH—furyl—Y   N-furfurylideneanilines [535a]

X—⬡—CH=N—pyridyl—Y   2-benzylideneaminopyridines [17]

X—⬡—CH=N—pyridyl—Y   3-benzylideneaminopyridines [17, 95, 96, **97**, **213**, 537]

X—pyridyl—N=CH—⬡—CH=N—pyridyl—X   terephthal bis(2-aminopyridines) [17]

X—pyridyl—N=CH—⬡—CH=N—pyridyl—X   terephthal bis(3-aminopyridines) [17] benzthiazole-azomethines [391, 453, 550, 830, 843]

benzoxazole-azomethines [391]

other heterocyclic azomethines

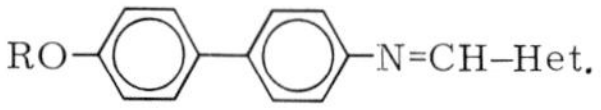

RO—⟨○⟩—⟨○⟩—N=CH–Het.

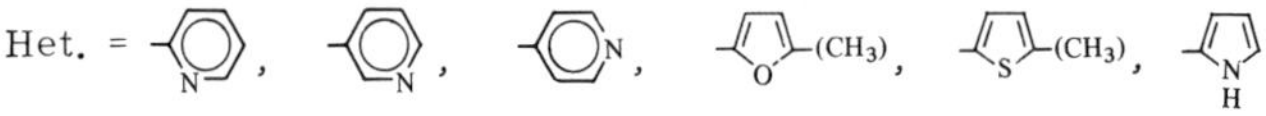

Het. = ⟨○⟩ , ⟨○⟩ , ⟨○⟩N , ⟨○⟩–(CH₃) , ⟨○⟩–(CH₃) , ⟨○⟩
[287, 881]

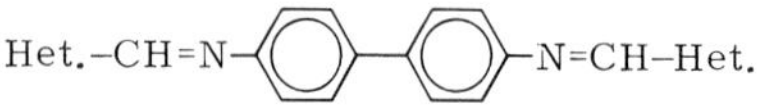

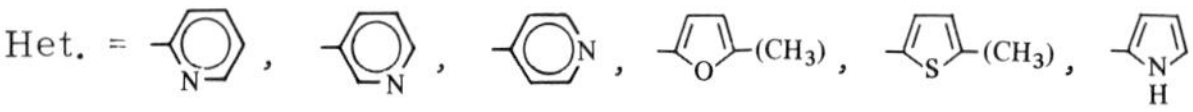

Het.–CH=N—⟨○⟩—⟨○⟩—N=CH–Het.

Het. = ⟨○⟩ , ⟨○⟩ , ⟨○⟩N , ⟨○⟩–(CH₃) , ⟨○⟩–(CH₃) , ⟨○⟩
[287]

8.2   *Azo compounds*

phenylazobenzthiazoles [550]

8.3   *Carboxylic acid esters*

8.3 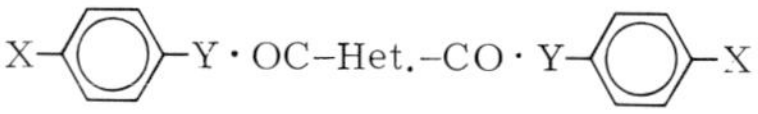   X—⟨○⟩—Y · OC–Het.–CO · Y—⟨○⟩—X

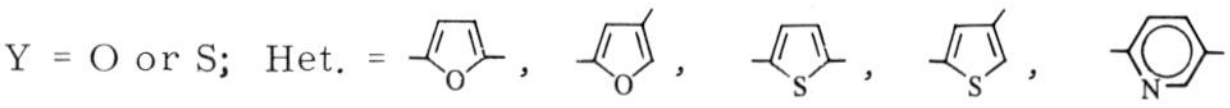

Y = O or S; Het. = ⟨○⟩ , ⟨○⟩ , ⟨○⟩ , ⟨○⟩ , ⟨○⟩N—
[166]

phenyl benzthiazole-2-carboxylates [550]

8.4   *Styryl compounds*

stryrylbenzthiazoles [550]

styrylbenzoxazoles [550]

8.5   *Heterocyclics with different bridging groups*

8.5 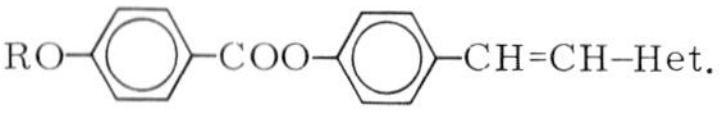   RO—⟨○⟩—COO—⟨○⟩—CH=CH–Het.

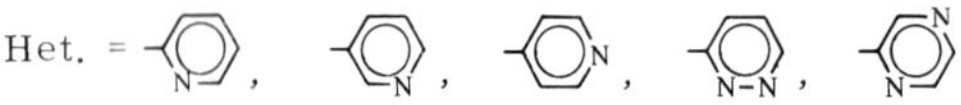

Het. = ⟨○⟩ , ⟨○⟩ , ⟨○⟩N , ⟨○⟩N=N , ⟨○⟩
[287]

9 Aromatic carboxylic acids

9.1 Benzoic acids

9.1   X—⟨◯⟩—COOH   *p*-substituted benzoic acids

X = OR: [**1:41**, 1:309, **130**, 228, **256**, 366, **367**, **368**, 431, 799, 837, **852**, **853**]
X = other substituent: [228, 785, 799, 853]

RO—⟨◯⟩—COOD   *p-n*-alkoxydeuterobenzoic acids [**1:41**]

RO—⟨◯⟩—COOH   (F(Cl))   *p-n*--alkoxy-*m*-fluoro (chloro) benzoic acids [**260**, 277]

9.2 Cinnamic acids, ω-phenylpolyene carboxylic acids

9.2   X—⟨◯⟩—CH=CH—COOH   *p*-substituted-cinnamic acids

X = OR: [150, 188, **259**, **368**, 575, 601, 708, 709, 762, 783, 787, 789, 799, 837, 839]
X = other substituent: [45, 105, 345, 547]

RO—⟨◯⟩—CH=CH—COOH   (Cl(Br))   *p-n*-alkoxy-*m*-chloro (bromo) cinnamic acids [**265**]

X—⟨◯⟩—CH=C(Y)—COOH   *p*-substituted-α-substituted-cinnamic acids [228, 345, 783, 789, 837]

X—⟨◯⟩—C(Y)=CH—COOH   *p*-substituted-β-substituted-cinnamic acids [634]

X—⟨◯⟩—(CH=CH)$_{\overline{n}}$COOH   ω-phenylpolyene carboxylic acids [243, 805, 809, 813, 837]

9.3 Biphenylcarboxylic acids and p-phenylcinnamic acids

9.3   X—⟨◯⟩—⟨◯⟩—COOH   *p'*-substituted-biphenyl-*p*-carboxylic acids

X = OR: [1:11, 1:13, 1:333, **262**]
X = other substituent: [45, 277, 787, 828]

RO—⟨◯⟩(Y)—⟨◯⟩—COOH   *p'-n*-alkoxy-*m'*-substituted biphenyl-*p*-carboxylic acids

Y = Cl: [**267**, 271]
Y = Br: [1:333, **267**, 271]
Y = NO$_2$: [1:320, 1:333, **267**, 271]
Y = other substituents: [271, 277]

X—⟨◯⟩—⟨◯⟩—CH=CH—COOH   *p*-phenylcinnamic acids [787, 828]

9.4   *p-Terphenylcarboxylic and -dicarboxylic acids*

9.4 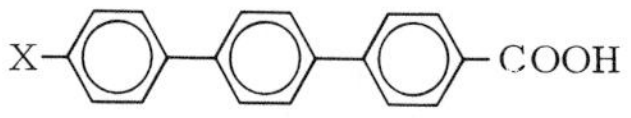   X—◯—◯—◯—COOH   *p*-terphenyl-4-carboxylic acids [276]

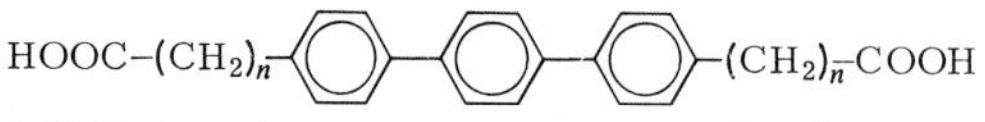

HOOC–$(CH_2)_{\overline{n}}$—◯—◯—◯—$(CH_2)_{\overline{n}}$COOH

4,4″-bis-(ω-carboxy-*n*-alkyl)-*p*-terphenyls [**146**]

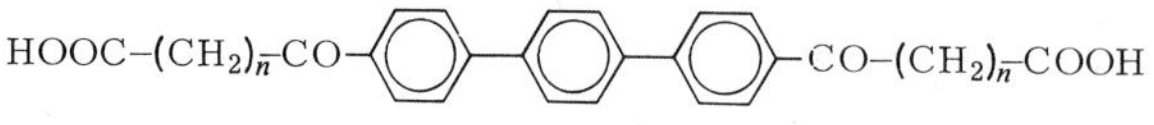

HOOC–$(CH_2)_{\overline{n}}$–CO—◯—◯—◯—CO–$(CH_2)_{\overline{n}}$COOH

4,4″-bis-(ω-carboxy-*n*-alkyryl)-*p*-terphenyls [146]

9.5   *Naphthalene-, fluorene-, and fluorenonecarboxylic acids*

9.5 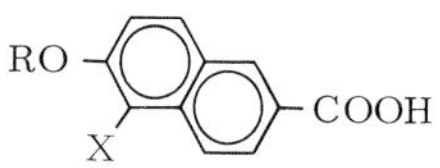   RO—◯◯—COOH   6-*n*-alkoxy-5-substituted-naphthalene-2-carboxylic acids
     X   [254, **255**, **258**, **261**, 277]

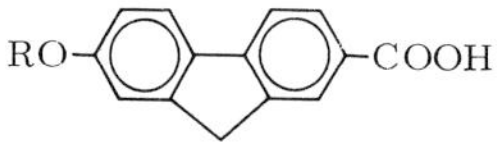

RO—◯◯◯—COOH   7-*n*-alkoxyfluorene-2-carboxylic acids [263, **268**]

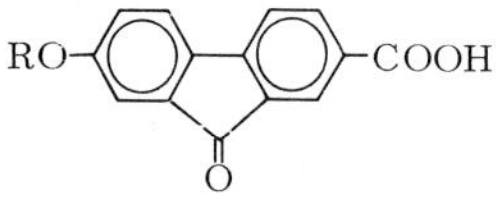

RO—◯◯◯—COOH   7-*n*-alkoxyfluorenone-2-carboxylic acids [263, 268]
    O

9.6   *Heteroaromatic carboxylic acids*

9.6 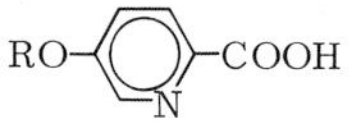   RO—◯—COOH   5-alkoxypicolinic acids [550] 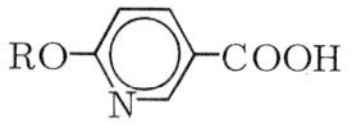
    N

RO—◯—COOH   6-alkoxynicotinic acids [550]
    N

9.7   *Aromatic carboxylic acids with bridging groups*

azomethinecarboxylic acids

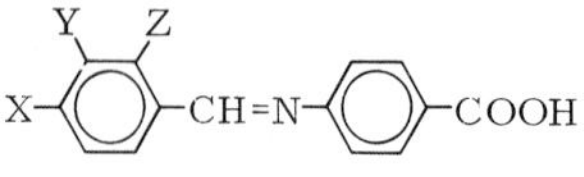

    Y   Z

9.7   X—◯—CH=N—◯—COOH   benzylidene-*p*-aminobenzoic acids [**129**, 222, 321, 346,
                              487, 582, 711, 787, 830]

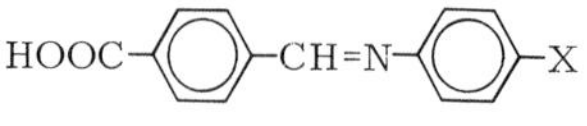

HOOC—◯—CH=N—◯—X   *p*-carboxybenzylidene anilines [90, 830]

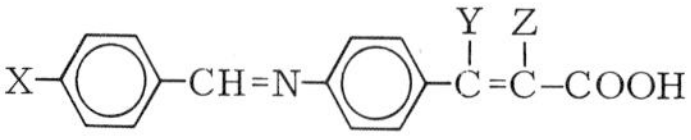

                         Y   Z

X—◯—CH=N—◯—C=C–COOH   benzylidene-*p*-aminocinnamic acids [321, 486,
                              679, 708, 799, 833]

X─⟨◯⟩─CH=CH─CH=N─⟨◯⟩─$\overset{Y}{\underset{|}{C}}$=$\overset{Z}{\underset{|}{C}}$─COOH   cinnamylidene-*p*-aminocinnamic acids
[486]

RO─⟨◯◯⟩─CH=N─⟨◯⟩─COOH   4-alkoxy-1-naphthylidene-*p*-aminobenzoic acids
[134]

carboxylic acids with azo linkage [25, 582, 621, 834]

carboxylic acids with azoxy linkage [621, 821, 827, 831]

carboxylic acids with ester linkage [228, 345, 722, 785, 799, 829, 888]

carboxylic acids with multiple different bridging groups [529]

10   *Ammonium and carboxylic-acid salts*

10.1   *Salts of aliphatic carboxylic acids*

RCOOM   *n*-alkanoates of monovalent metals
M = Li: [1:322, 37, 776, 779]
M = Na: [1:322, 37, 572, 670, 684, 772, 773, 774, 775, 776, 777, **791**]
M = K: [1:322, 37, 672, 774, 776, **791**]
M = Tl: [1:322, 1:324, 320a, 485a, 791, 838]

RCOORb(Cs)
Rubidium and Cesium alkanoates (R = *n*-, *i*-, and unsaturated) [37, 676, 776, 791]

R'COOM   *i*-alkanoates of monovalent metals
M = Na: [744, 791]
M = K: [791]
M = Tl: [791, 838]

R''COOM   alkenoates of monovalent metals
M = Na: [774, 775, 791]
M = K: [791]
M = Tl: [838]

(RCOO)$_n$M   alkanoates of multivalent metals
[1:322, 1a, 1b, 220, 454, 671, 673, 674, 675, 778]

10.2   *Salts of aromatic carboxylic acids*

10.2 $\overset{X}{\underset{|}{}}$⟨◯⟩─COOM   benzoates [1:323, 791]

$\left(\text{X}─⟨◯⟩─\overset{Y}{\underset{|}{C}}=\overset{Z}{\underset{|}{C}}─COO\right)_n$M   cinnamates [1:323, 791]

⟨◯⟩─$\overset{X}{\underset{|}{C}}$H─COOM   phenylacetates and salts of mandelic acid [1:323, 791, 838]

salts of other aromatic carboxylic acids [1:323, 791]

10.3 *Salts of nitrogen bases*

aliphatic ammonium salts [1:10, 791]

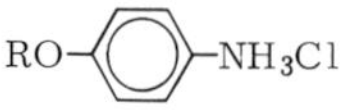

RO—⟨O⟩—NH₃Cl   *p*-alkoxyanilinium chlorides [631]

pyridinium salts [225, 325, 413]

amidine salts

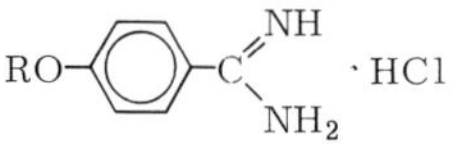

RO—⟨O⟩—C⟨=NH / NH₂⟩ · HCl   *p*-*n*-alkoxybenzamidinium chlorides [**642**, 643]

11   *Steroids* (including related compounds)

11.1 *Aliphatic cholesteryl esters*

11.1   RCOO—   cholesteryl alkanoates

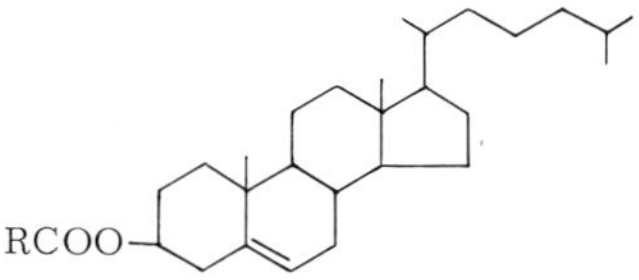

R = *n*-alkyl [1:163, 1:274, 1:300, 65, 69, 101, 102, **206**, 241, 242, **266**, **310**, **360**, 361, **362**, 439, 440, 530, 543, 557, 568, **569**, 570, **571**, 575, 619, 620, 706]

R = *i*-alkyl or halogenealkyl [214, 316, 318, 361, 575]

R = chiralic [242, 435]

R = unsaturated [27, 175, 570, 706, 724]

R = alkoxyalkyl (R'O—(CH₂)ₙ—) [**310**, 618]

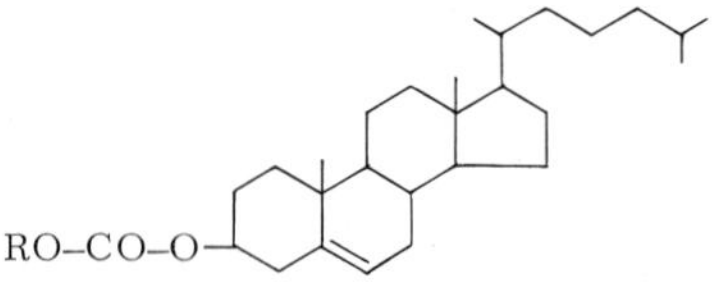

RO—CO—O—   cholesteryl alkylcarbonates

R = *n*-alkyl [125, **194**, **200**, 316, 730, 799]

R = *i*-alkyl, unsaturated or alkoxyalkyl [125, 204, 236, 391, 443, 455, 619, 620]

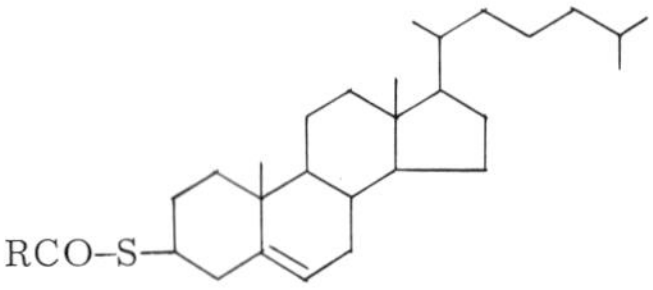

RCO—S—   thiocholesteryl-*n*-alkanoates

[**197**, **206**]

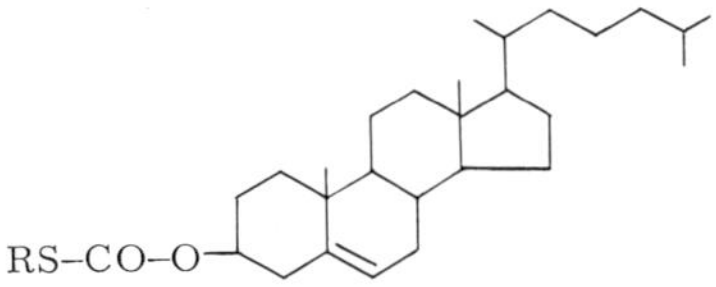

cholesteryl-S-*n*-alkylthiocarbonates [**195, 196**]
R = cholesteryl or H: [251]

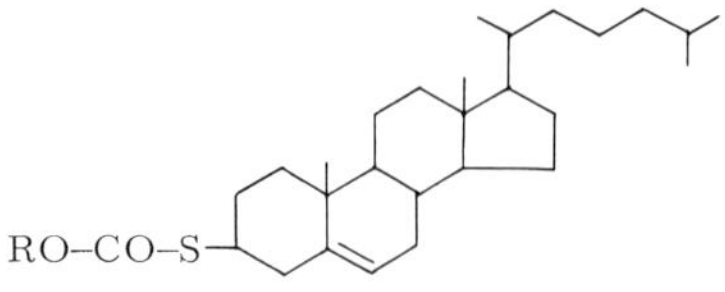

RS–CO–O

S-cholesteryl-O-*n*-alkylthiocarbonates [**195**, 202]

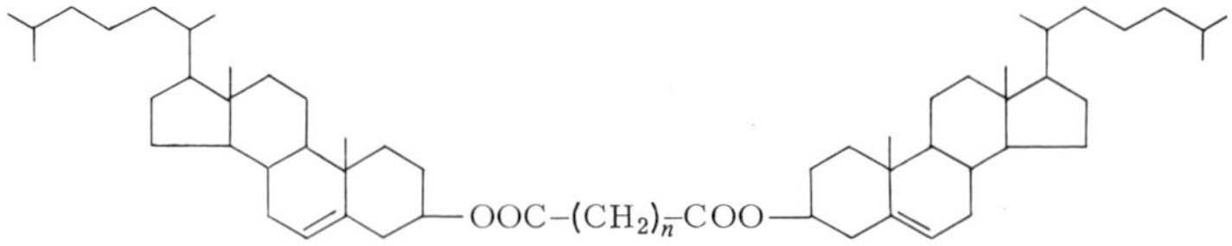

RO–CO–S

OOC–$(CH_2)_{\overline{n}}$–COO

bis-cholesteryl-alkanedicarboxylates [**1:24**, 27, 85, 236, **311**, 543, 580]

bis-cholesteryl *trans*-1,4-cyclohexanedicarboxylate [768]

11.2 *Aromatic cholesteryl esters*

11.2

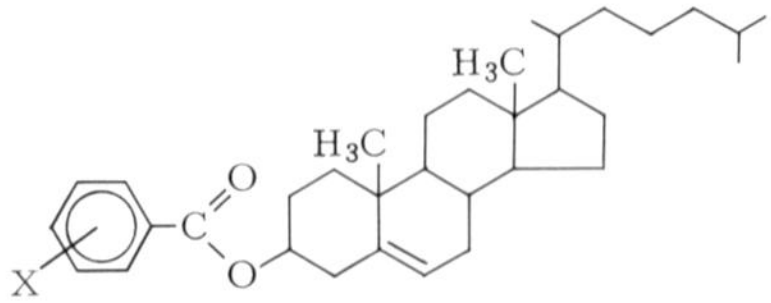

cholesteryl benzoates

[1:129, 1:274, 1:300, 27, 29, 65, 101, **130,** 135, 235, 360, 361, 391, 568, 575, 616, 633, 706, 739, 866]

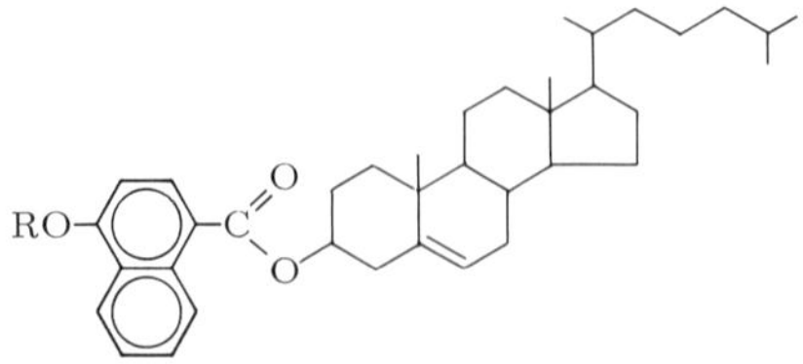

cholesteryl 4-*n*-alkoxy-1-naphthoates [**139**]

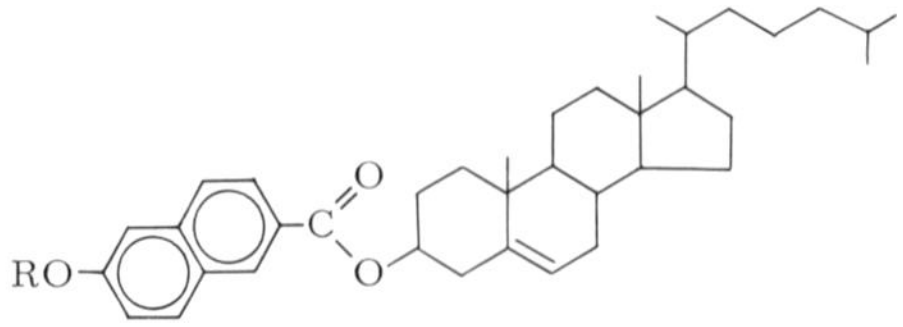

RO

cholesteryl 6-*n*-alkoxy-2-naphthoates [**136, 139**]

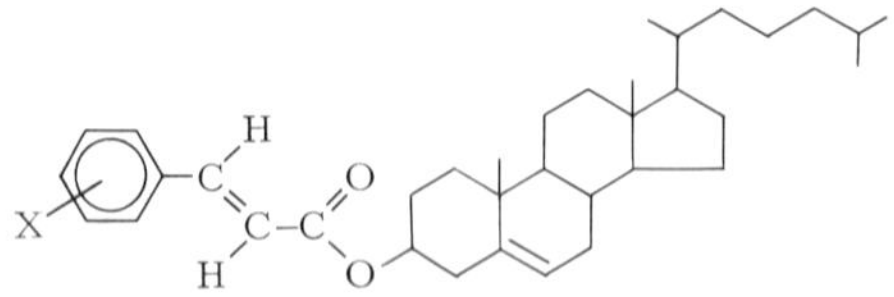

RO

cholesteryl cinnamates

[1:274, 27, 29, 101, **132,** 235, 360, 361, 616, 617, 706]

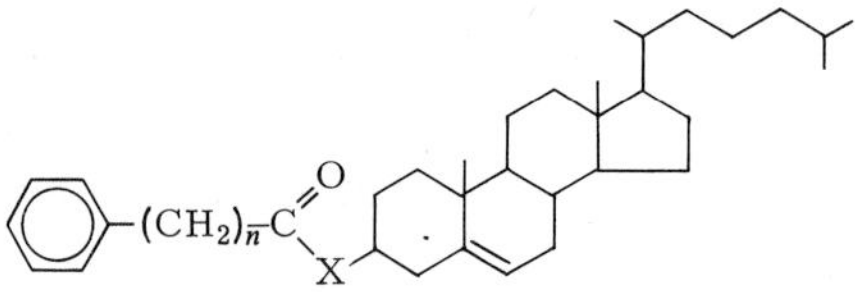

X = O: cholesteryl-ω-phenylalkanoates [**206, 207,** 210, **564,** 706]

X = S: thiocholesteryl-ω-phenylalkanates [**197, 199, 201, 206**]

cholesteryl esters with aromatic moieties comprising other bridging groups

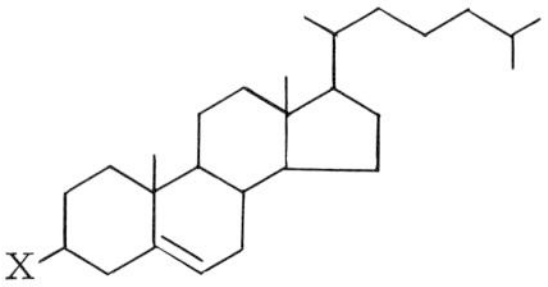

X = 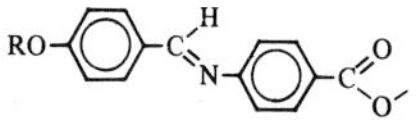   [**142**]

X = 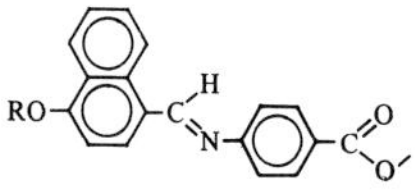   [**137,** 145c]

X = 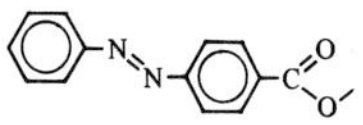   [27]

other aromatic cholesteryl esters [27, 235, 767]

11.3 *Inorganic cholesteryl esters*

[1:385, 40, 125, 252, 316, 444, 619, 620, 647, 730]

11.4 *5α-Cholestanyl esters*

11.4

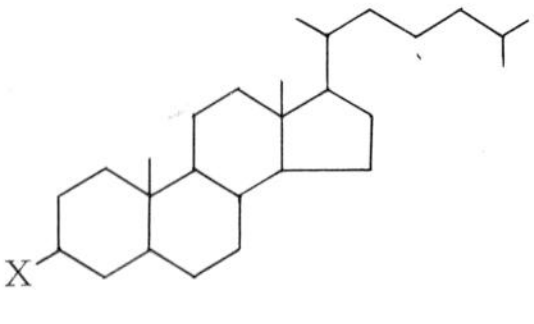

X = RCOO–   5α-cholestanyl alkanoates [362, 677]

X = ROCOO–   5α-cholestanyl-n-alkylcarbonates [**195, 561**]

X = RSCOO–   5α-cholestanyl-n-alkylthiocarbonates [**195**]

X = Y—⟨ ⟩—COO–   5α-cholestanyl benzoates [616, 866]

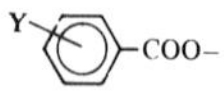

X = ⟨ ⟩–(CH₂)ₙ–COO–   5α-cholestanyl-ω-phenylalkanoates [**197, 199, 565**]

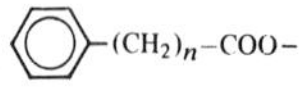

X = other substituents [329, 561]

11.5 5α-Cholest-8(14)-en-3β-yl esters (doristeryl esters)

11.5

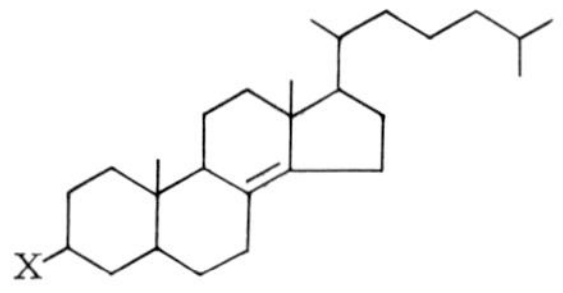

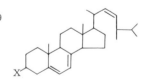

[**109, 110,** 445, 558]

11.6 Campesteryl esters

11.6

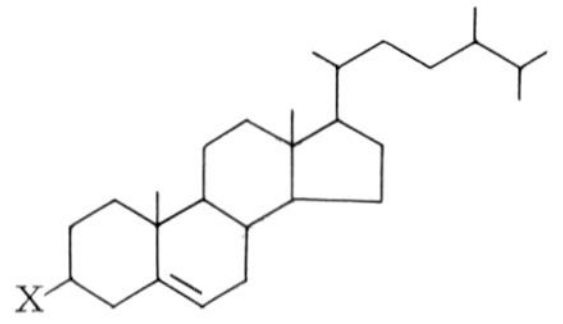

[198, 616]

11.7 β-Sitosteryl esters

11.7

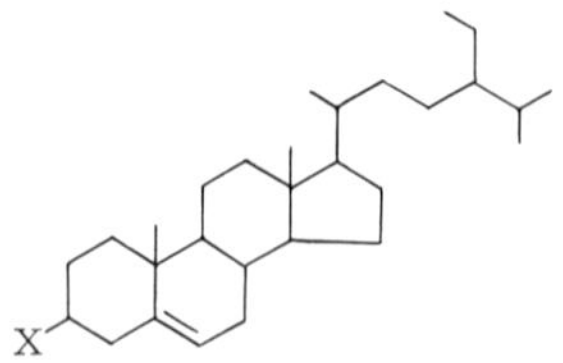

[428, 560]

11.8 Stigmasteryl esters

11.8

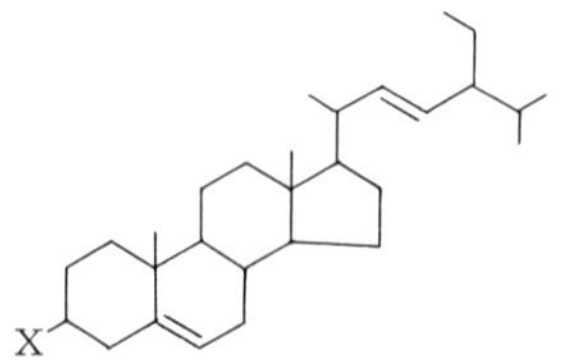

[104, 107, 428, **559,** 616]

11.9 Ergosteryl esters

11.9

[233, 236, **410,** 412, 428, 616]

11.10 24-Methylenecycloartanyl-n-alkanoates

11.10   [406, **407**, 408, 411]

without double bond [407, 408, 411]

11.11 Cycloartenyl alkanoates

11.11   [407, 408, 409]

without double bond [407]

11.12 Azasteroids [198]

11.13 Isolated examples of other cholesteryl and steroid esters and distant steroid analogs

[19, 20, 21, 22, 23, 198, 234, 333, 408, 412, 444, 446, 447, 467, 470a, 497, 562, 563, 616, 648, 739, 866]

12   Aliphatic compounds

12.1 Aliphatic carboxylic acids [1:43, 644, 651]

The mesomorphism of agaricinic acid is still uncertain [238]

12.2 Aliphatic esters [1:10, 1:12, 1:43, 46, 277, 425, 581]

12.3 Aliphatic amines [536]

12.4 Aliphatic ethers [1:43]

12.5 Aminoacid derivatives

ROOC—CHR′—NHCONH—CHR′—COOR
carbonyl bis (amino acid esters)
[9:112, 9:113, 9:114, 452]
other aminoacid derivatives [664]

12.6 Alkadiene derivatives

R—CH=CH—CH=CH—COOH
n-alkadiene-(2,4)-carboxylic acids-1 [463, 854]
R—CH=CH—CH=CH—CH=N—N=CH—CH=CH—CH=CH—R
n-alkadiene-(2,4)-al-(1)-azines [463]

12.7 *trans-trans-4'-Alkyl-4-cyanobicyclohexyls*

12.7
R ⬡—⬡ CN   [191a] (cf. [1:43, no. 16–24])

see also:
10.1 salts of aliphatic carboxylic acids
10.3 aliphatic ammonium salts
11 steroids; the major part is of aliphatic character

13   Miscellaneous

13.1 Primary carboxylic acid amides [277, 722]

13.2 Mannich bases [771]

13.3 Alkali phenolates [791]

13.4 Sulfanilates

13.4   RCONH—⬡—SO₃Na   [417]

13.5 Inorganic or metalorganic compounds

(Hg-compounds see section 7.1.1, 3.1.21)
[70, 84a, 187, 205, 244a, 253, 323, 535, 597, 769]

The liquid crystals of $Ca_3(PO_4)_2$ seem to be lyotropic [239]

Theory of the Liquid-Crystalline State

3

3 Theory of the Liquid-Crystalline State

3.1 Introduction

There have been two basic theories to describe and to explain the liquid-crystalline state, the continuum theory and the swarm theory. These are founded on completely different concepts: on one hand the continuum theory conceives the liquid crystal as an anisotropic elastic medium with its own symmetry, viscosity, and elasticity parameters [2, 4, 22, 148, 189, 218, 223, 231–233, 452, 462, 644, 794]; on the other hand the swarm theory interprets the state as the result of intermolecular interactions and the resulting statistical and thermodynamic equilibrium. In the first case, a molecular interpretation of the macroscopically interpreted parameters is more or less relinguished but by no means suppressed (C. W. Oseen), while in the second the molecular interaction forms the basis of the theory. The mathematical models of C. W. Oseen [598–623], H. Zocher [4:962, 857–862], and F. C. Frank [253, 254] paved the way for the development of the theory. The molecular statistical aspect was supported by the works of E. Bose [55–62], M. Born [54], L. S. Ornstein and W. Kast [586–591], G. W. Stewart [768], Weiss' magnetism theory (cf. p. 133), and finally by the experiments of A. Smekal [759], V. N. Tsvetkov [802], W. Maier and A. Saupe [506–510]. The contrary interpretations of isolated phenomena, especially concerning the influence of external fields, were often discussed in the thirties and developed into a confrontation between proponents of the swarm theory and supporters of the continuum theory [1:67, 1:68]. The theories and models of today can be considered as a unified picture based on Oseen's concept, joining together different ideas in a successful synthesis. The most prominent among recent works are those of P. G. de Gennes [154–179], F. M. Leslie [460–464, 464a], and J. L. Ericksen [205–226, 226a, 226b].

Some reviews deal with theoretical aspects of liquid crystals [170, 252, 425, 502, 696, 728, 737], of nematics [172, 173, 204, 402, 690, 832, 833, 862], and smectics [171, 834].

Above all, P. G. de Gennes' monograph "The Physics of Liquid Crystals" imparts a comprehensive view [176]. There are also suggestions for computer simulation [440, 697, 495a].

3.2   The Continuum Theory

We begin with Oseen's molecular-statistical model, which had sometimes been paradoxically interpreted as the prototype for a continuum theory, especially after F. C. Frank's 1958 revision [254] that completely avoided any molecular-statistical basis. C. W. Oseen [609] started with molecular pairs (1) and (2), and he described the most probable condition of the internal energy of the system by using the molecular-statistical method developed by Boltzmann and Planck. He confined himself to the parameters r; $L_{(1)} \cdot \frac{r}{r}$; $[L_{(1)} - L_{(2)}] \cdot \frac{r}{r}$; $1 - L_{(1)} \cdot L_{(2)}$; $[L_{(1)} \times L_{(2)}] \cdot \frac{r}{r}$ with r as the vector of the distance between the molecular centers of gravity of (1) and (2). The potential energy, E, of the relatively limited area (that later became known as a swarm) and within which the direction of the molecules, $(L)^*$, is a continuous function of the coordinates of their centers of gravity is

$$E = \frac{1}{2m^2} \int \int \rho_1 \rho_2 F(\xi_1, \xi_2, A_1, A_2) \partial \omega_1 \partial \omega_2 \tag{3.1}$$

$$(\partial \omega_i = \partial x_i \partial y_i \partial z_i)$$

where m is the mass of a molecule, ρ is the density of the liquid, and A and ξ are defined as follows:

$$r = \xi_1; \quad \frac{r}{r} \cdot L_{(1)} = \xi_2; \quad [L_{(1)} - L_{(2)}] \frac{r}{r} = A_1;$$

$$1 - L_{(1)} \cdot L_{(2)} = A_2; \quad [L_{(1)} \times L_{(2)}] \frac{r}{r} = A_3$$

The development of eq. 3.1 into an exponential series leads to the following expression for the potential energy:

$$E = \frac{1}{2m^2} \int \int \rho_1 \rho_2 Q(\xi_1, \xi_2) d\omega_1 d\omega_2 + \frac{1}{2m^2} \int \rho^2 \{ K_1 L \cdot \mathrm{rot} L$$

$$+ K_{11}(L \, \mathrm{rot} \, L)^2 + K_{22}(\mathrm{div} \, L)^2 + K_{33}((LV) L)^2 + 2 K_{12} \mathrm{div} \, L \cdot L \, \mathrm{rot} \, L \} \, d\omega \tag{3.2}$$

If the potential energy of a pair of molecules is dependent only on the A-terms, the terms K_1 and K_{12} are zero. Eq. 3.2 is often written in operator notation as

$$E = \frac{1}{2m^2} \int \int \rho_1 \rho_2 Q(\xi_1, \xi_2) d\omega_1 d\omega_2 + \frac{1}{2m^2} \int \rho^2 \{ K_1 L \nabla \times L$$

$$+ K_{11}(L \nabla \times L)^2 + K_{22}(\nabla L)^2 + K_{33}((L \nabla) L)^2 + 2 K_{12}(\nabla L)(L \nabla \times L) \} \, d\omega \tag{3.3}$$

---

* L appears in later papers as the so called director d.

In eqs. 3.2 and 3.3, the second integral is that of the energy component dependent on the orientation, i.e., it is a function of twist, splay, and bend. These individual terms are described by the eqs. 3.4, 3.5, and 3.6.

Twist: $L \cdot \mathrm{rot}\, L$ $\left(\dfrac{\partial L_y}{\partial L_x} - \dfrac{\partial L_x}{\partial L_y} \right)$ (3.4)

Splay: $\mathrm{div}\, L$ $\left(\dfrac{\partial L_x}{\partial x} + \dfrac{\partial L_y}{\partial y} \right)$ (3.5)

Bend: $[(L\nabla) \cdot L]^2$ $\left(\dfrac{\partial L_x}{\partial z} \right)^2 - \left(\dfrac{\partial L_y}{\partial z} \right)^2$ (3.6)

The connection between these differential expressions and the energy, E, of the system is made by the modules of eq. 3.2, such as K_{11}, K_{22}, and K_{33}. While Oseen has confined himself to a formal introduction of the coefficients K_1, K_{11}, etc. into the development of his series and merely mentioned their temperature dependence, F. C. Frank has started from the concept of "curvature elasticity" by defining the corresponding elasticity modules and applying Oseen's ideas [254]. So

$$G = \int g \, d\tau \tag{3.7}$$

with the energy density, g, being considered as a quadratic function of deformation according to

$$g = k_i \cdot a_i + \tfrac{1}{2} k_{ij} \cdot a_i a_j \ast \quad i, j = 1 \ldots 6, \ k_{ij} = k_{ji} \tag{3.8}$$

where the k-terms are the elasticity modules, and the a-terms characterize the various possibilities of the deformation as

$$
\begin{array}{lll}
a_3 = \partial L_x / \partial z & a_4 = \partial L_y / \partial x & a_2, a_4: \text{twist} \\
a_1 = \partial L_x / \partial x & a_2 = \partial L_x / \partial y & a_1, a_5: \text{splay} \\
a_5 = \partial L_y / \partial y & a_6 = \partial L_y / \partial z & a_3, a_6: \text{bend}
\end{array}
\qquad
\begin{array}{l}
(3.4\,\mathrm{a}) \\
(3.5\,\mathrm{a}) \\
(3.6\,\mathrm{a})
\end{array}
$$

The correlation with Oseen's notation is evident when it is considered that

$$
\begin{aligned}
L \cdot \nabla \times L &= a_4 - a_2 \quad \text{(twist)} \\
\nabla \cdot L &= a_1 + a_5 \quad \text{(splay)} \\
[(L \cdot \nabla) L]^2 &= a_3^2 + a_6^2 \quad \text{(bend)}
\end{aligned}
\qquad
\begin{array}{l}
(3.4\,\mathrm{b}) \\
(3.5\,\mathrm{b}) \\
(3.6\,\mathrm{b})
\end{array}
$$

Frank finally derived eq. 3.9 expressing the energy density:

$$
\begin{aligned}
g = {} & [k_1 (\nabla \cdot L)] - k_2 (L \nabla \times L) \\
& + \frac{1}{2} k_{11} (\nabla \cdot L)^2 + \frac{1}{2} k_{22} (L \cdot \nabla \times L) + \frac{1}{2} k_{33} ((L \cdot \nabla) L)^2 \\
& - k_{12} (\nabla \cdot L)(L \cdot \nabla \times L) \left[-(k_{22} + k_{24}) \left(\frac{\partial L_x}{\partial x} \cdot \frac{\partial L_y}{\partial y} + \frac{\partial L_y}{\partial x} \cdot \frac{\partial L_x}{\partial y} \right) \right]
\end{aligned}
\tag{3.9}
$$

---

* given in Einstein's summation rule

The square brackets enclose terms missing in Oseen's equation. Table 3.1 contrasts the various notations, with A. Saupe and J. Nehring using Frank's notation [733, 734].

Table 3.1. Notation correlations.

Oseen	Frank, Saupe, Nehring	Zocher	Deformation
$-\dfrac{\rho^2}{2m^2}\cdot K_1$	k_2	—	
$-\dfrac{\rho^2}{2m^2}\cdot K_{12}$	k_{12}	—	
$\dfrac{\rho^2}{m^2}\cdot K_{11}$	k_{22}	k_t	twist
$\dfrac{\rho^2}{m^2}\cdot K_{22}$	k_{11}	k_1	splay
$\dfrac{\rho^2}{m^2}\cdot K_{33}$	k_{33}	k_2	bend

The three important modules are illustrated in fig. 3.1 showing their corresponding changes in texture.

Fig. 3.1. The three simple deformations of liquid crystals.

SPLAY k_{11} TWIST k_{22} BEND k_{33}

Frank evaluated the prerequisites that reasonably allow to ignore the "square-bracket" terms. The coefficient K_1 is zero because the condition $L = -L$ (nonpolar structure) is always here fulfilled. Modules with mixed indices are required only for nonplanar textures. Here it may be sufficient to refer to H. Zocher, who limited himself to planar textures right from the start and thus found three constants sufficient (cf. table 3.1). In their critical comparison of the concepts of Frank and Oseen, A. Saupe and J. Nehring returned to a formulation much like that originally used by Oseen himself*:

$$\begin{aligned}
2g = &-2k_2(L\nabla \times L) + k'_{11}(\nabla\cdot L)^2 \\
&+ k_{22}(L\nabla \times L)^2 + k'_{33}(L\nabla L)^2 \\
&+ 2k^{(2)}_{13}(\nabla\cdot(\nabla\cdot L)L)
\end{aligned}$$

(3.10)

---

* Compare the notations of Ericksen and Leslie, p. 134. Newer presentations also include flexoelectric terms [316, 778], see p. 176 and section 4.4.7. Photoelastic phenomena were also observed in liquid crystals [366].

where

$$k'_{11} = k_{11} - 2k^{(2)}_{13} \qquad\qquad (3.11\,a)$$
$$k'_{33} = k_{33} + 2k^{(2)}_{13} \qquad\qquad (3.11\,b)$$

Frank did not use the term $2k^{(2)}_{13}(\nabla\cdot(\nabla\cdot L)\cdot L)$, but the original work of Oseen had contained the corresponding terms with the coefficients k_{12} and $(k_{22}+k_{24})$. The first term, $2k_2(L\nabla\cdot L)$, applies to cholesteric behavior and corresponds to Oseen's term 2. For purely nematic behavior, it disappears because k_2 is zero. Much effort was expended before it was possible to obtain an idea on the magnitude and relationships of the elastic constants from experimental evidence and theoretical calculations. J. Nehring and A. Saupe finally came to the conclusion that the experimentally observable elastic constants are not k_{11} and k_{33} proper [734], but rather the parameters k'_{11} and k'_{33} (eq. 3.11). However, experimentally determined values showing the ratio $k'_{11}:k_{22}:k'_{33} = 1.6:1:3.2$ deviate from the theoretical prediction that requires $k'_{11}:k_{22}:k'_{33} = 5:11:5$. J. P. Straley comments on such calculations [777]. It can be stated for nematics that the twist constant is the smallest one. Therefore twist deformations happen most easily.

3.3   *Explanation of Some Texture Characteristics*

In summary, it can be said that the texture characteristics of nematics, cholesterics, and smectics are correlated by the following sets of nonzero modules (using Frank's notation for the modules):

A. Nematic Phase: No polar axes in spatial element, and no certain indication of enantiomorphous forms. The values k_{11}, k_{22}, and k_{33} are sufficient for a description, except that k_{12} must be added for nonplanar textures. A. Saupe and J. Nehring have discussed the meaning of k_{12}, which Frank sets equal to zero [733].

Early workers like O. Lehmann [1:250], V. N. Tsvetkov [798, 802], V. Zolina [863], A. Saupe [730, 731], T. C. Lubensky [482], J. Prost [691], G. Meier [540], R. G. Priest [682], P. E. Cladis [124], and others considered the experimental possibilities for the determination of the elastic constants. More recently, the modules have been determined for several nematics [48a, 296, 304, 382a, 389, 455, 659, 734, 749, 824]. J. Wahl and F. Fischer used a new method involving the observation of the change in the interference pattern of a homeotropic layer between counter-rotating glass plates as a function of rotation speed [242, 243]. Another possibility is the measurement of threshold fields [188, 541, 548, 751, see chapter 4]. W. H. de Jeu has critically evaluated this method and underlines the importance of strong anchoring power and strict orthogonality between the magnetic field and the nematic director [180]. The elastic constants depend on the molecular length and the mean field energy in a complicated way; figs. 3.2 and 3.3 illustrate the behavior in the homologous series of *p-n*-alkoxyazoxybenzenes [541]. H. Gruler has considered intrinsic splay, twist, and bend deformation due to the molecular shape of nematogens [297]. This work also submits an evaluation of the frequency dependence of the elastic constants.

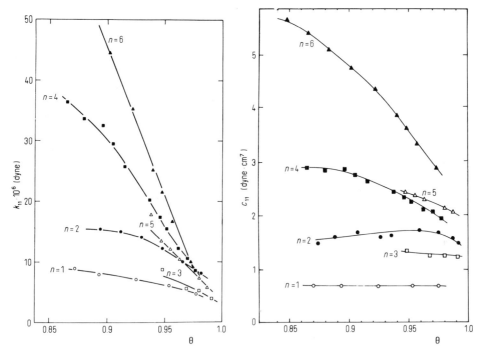

Fig. 3.2. Splay elastic constant, k_{11}, and reduced splay elastic constant, c_{11}, of p-alkoxyazoxybenzenes (n is the number of carbon atoms in the alkoxy group) as a function of reduced temperature, $\theta = (T/T_c)(V/V_c)^2$ (from [541]).

In fig. 3.2, the reduced splay elastic constant defined by A. Saupe as

$$c_{11} = k_{11}\frac{V^{7/3}}{S^2}$$

is nearly temperature independent for the lower homologs ($n=1$, 2, and 3). However for the higher homologs, c_{11} is no longer temperature independent and shows that the changes with molar volume, V, and the degree of order, S, can no longer be exclusively applied to explain the temperature dependence of the splay elastic deformation [541]. The influence of short range order indicated by this behavior can be expressed by a power law, $(T-T^*)^{-\gamma}$, which derives from de Gennes' theory and has been proven for the bend elastic constants in the n/s_A pretransition temperature range [126, 128, 456, see also 76]. Fig. 3.3 contrasts the three elastic constants and their reduced values showing the stronger influence of temperature on the bend deformation [298].

B. *Cholesteric phase:* In addition to the elastic structure considered for the nematic phase, the cholesteric phase involves spontaneous molecular twist of the spatial element. Thus, the constant k_2 (Oseen's k_1) is nonzero, and the pitch is determined by the quotient $\pi/(k_2/k_{22})$ if the elementary cell of the texture exhibits no polarity. This has been accepted by most authors (symmetry D_∞), but different opinions exist, too [398].

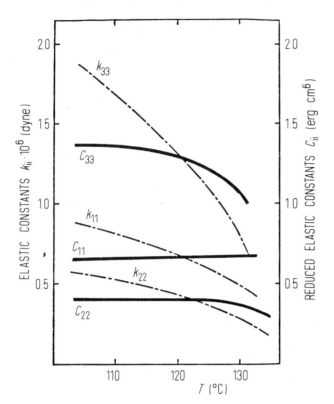

Fig. 3.3. Elastic and reduced elastic constants of PAA as a function of temperature (from [298]).

C. Smectic phase: The theory has thus far been unable to provide an unequivocal correspondence of molecules to the various polymorphous modifications (cf. [2:309]). Smectic A, the phase with parallel bent molecular layers, whose minimum-energy state is the focal-conic texture (fig. 1.3), obeys the relationships $k_{11} \cong 0$, $k_{22} \neq 0$ and $k_{33} \neq 0$. It is possible to ignore k_{11} because no tension occurs in splay, this being a characteristic property of the layer arrangement. The layers lack splay in the planes perpendicular to L, and they glide smoothly over one another.

In the observation of textures, the defects (i.e., dislocations, disclinations, and singularities) are of primary interest and are already described in reviews [415, 416] and basic discussions [421b, 437, 709].

It must be required that the continuum theory explain the quite different directional distribution of L, which is visible in the polarizing microscope, and its dependence on the elastic constants and on surface influences. It should be noted that L is simultaneously the direction of the optical axis of the spatial element having this same L as its average orientation. We must obtain from the general relationship between energy and space coordinates those solutions in which the corresponding field vector L is really oriented. This is a variation problem that has been thoroughly investigated in the last years [574, 735, 736]. The general relation is given as:

$$\delta \int g \, d\omega = 0 \qquad\qquad (3.12)$$

where g is defined as in eq. 3.9. A few examples should suffice to illustrate the principle behind the further calculations. Φ, α, x_3 are cylindrical coordinates which describe the orientation of L as demonstrated in fig. 3.4. A. Saupe and J. Nehring have shown that the following solutions result from the Euler-Lagrange equations:

$$k\left[\frac{\partial^2\Phi}{\partial\alpha^2}+\varepsilon\left(\frac{\partial^2\Phi}{\partial\alpha^2}\cos 2\left(\frac{\partial\Phi}{\partial\alpha}-\alpha\right)\right)+\frac{\partial\Phi}{\partial\alpha}\left(2-\frac{\partial\Phi}{\partial\alpha}\right)\sin 2(\Phi-\alpha)\right]=0 \tag{3.13}$$

$$\delta\cdot\left[\frac{\partial\Phi}{\partial x_3}\cdot\frac{\partial\Phi}{\partial\alpha}\cos(\Phi-\alpha)+2\frac{\partial^2\Phi}{\partial\alpha\,\partial x_3}\sin(\Phi-\alpha)\right]=0 \tag{3.14}$$

$$\eta\cdot\frac{\partial^2\Phi}{\partial x_3^2}=0 \tag{3.15}$$

where

$$k=\tfrac{1}{2}(k'_{11}+k_{22}\cos^2\theta+k'_{33}\sin^2\theta)\cdot\sin^2\theta \tag{3.16a}$$
$$\varepsilon k=\tfrac{1}{2}(k'_{11}-k_{22}\cos^2\theta-k'_{33}\sin^2\theta)\cdot\sin^2\theta \tag{3.16b}$$
$$\delta=\tfrac{1}{2}(k_{22}-k'_{33})\cdot\sin 2\theta\cdot\sin^2\theta \tag{3.16c}$$
$$\eta=[k(1-\varepsilon)-2\delta\cdot\mathrm{ctg}\,2\theta]\sin^2\theta \tag{3.16d}$$

and k, ε, δ, and η are constants containing the elastic modules, while θ is the angle that gives the deviation of L from the x_3 direction. From eqs. 3.16a–d follows Zocher's well known solution:

$$\Phi(\alpha,x_3)=q\cdot x_3+\varphi \tag{3.17}$$

where q and φ are functions of α that can be determined with the aid of eqs. 3.13 and 3.14.

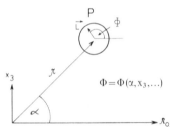

$$\Phi=\Phi(\alpha,x_3,\ldots)$$

Fig. 3.4. Coordinate system for the description of disclinations.

Now a few special cases of "singularity" will be discussed. We first consider the case where $q=0$, i.e., the nematic phase without twist. When $q=0$, eq. 3.17 says that $\varphi=\Phi(\alpha,x_3)$, and eq. 3.14 disappears because a prerequisite to a planar texture is

$$\frac{\partial\Phi}{\partial x_3}=0$$

One obtains φ as the solution of eq. 3.13, noting that according to eq. 3.16b and using a justifiable approximation

$$k'_{11}\cong k_{22}\cong k'_{33};\qquad \varepsilon\cdot k\cong 0$$

it follows that

$$k \cdot \frac{\partial^2 \Phi}{\partial \alpha^2} = 0$$

$$\Phi(\alpha, x_3) = 0$$

$$\varphi = s\alpha + c_0 \qquad\qquad (3.18)$$

with: $s = \pm 1, 2, 3, \ldots$ for $\theta \neq \dfrac{\pi}{2}$

and: $s = \pm \dfrac{1}{2}, 1, \dfrac{3}{2}, 2 \ldots$ for $\theta = \dfrac{\pi}{2}$

and finally

and $c_0 = $ const. in the ranges

$$\Phi(\alpha, x_3) = s\alpha + c_0 \qquad (3.17\,\text{a})*$$

$0 \le c_0 \le 2\pi$ for $\theta \neq \dfrac{\pi}{2}$ and

$0 \le c_0 \le \pi$ for $\theta = \dfrac{\pi}{2}$

J. Nehring and A. Saupe [735] made the analogous calculations for smectics C, W. L. McMillan [500] and J. A. Geurst [281] did so for smectics A, and P. G. de Gennes mainly considered the phase transitions [168], see chapter 8.

Eq. 3.17a is a fundamental relationship for describing the socalled Schlieren texture (fig. 1.12). For the director L lying parallel to the boundary surface, singularities with two and four brushes are frequently encountered corresponding to $s = \pm\frac{1}{2}$ and $s = \pm 1$. Nehring and Saupe were able to show that s_C phases only form singularities having an integer, i.e. those with four brushes. A precise description of the behavior of nematic disclinations can be found in work so early as that of G. Friedel [1:62, see also 1:33]. H. Zocher also discussed the general topology, while Frank wrote an instructive summary (fig. 3.5, in which the older designation and that of Saupe-Nehring have been added). J. F. Dreyer has studied the molecular alignment in nematics using dichroitic dyes [191]. The most important features of the Schlieren texture are:
1. A series of interlocking streaks similar to those illustrated in fig. 3.6. It has the tendency to reduce the number of centers with time so that everything finally reduces to zero and all singularities have vanished; this is written $\sum_i s_i \to 0$ (see also H. Imura et al. [334, 335]).

2. The singularities with two brushes (but never those with four brushes) cause the formation of so called second-order disclination walls, which appear as light "threads" showing a constant tendency to shorten until they disappear (II in fig. 3.6). Two-brush-singularities without adhering second order walls can also exist, as it has been proved by own observations.

3. A streaked pattern known as a disclination wall (fig. 3.7) that stretches across an elongated area. Disclination walls separate areas that have different but uniform directional vectors, L. Disclination walls can also form closed figures (fig. 3.8) or appear as circular streaks without

---

* This is given by Frank as

$$\Phi = \tfrac{1}{2} n\, \Psi + \Phi_0$$

$n = $ integer and in Cartesian coordinates.

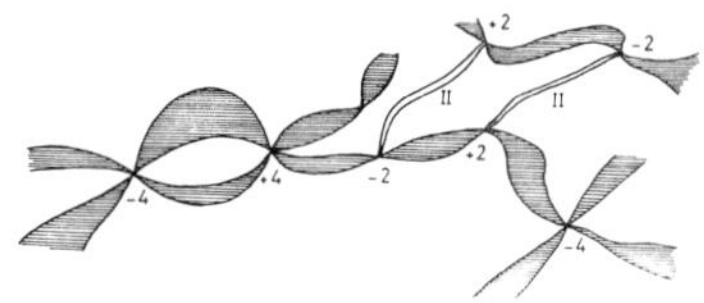

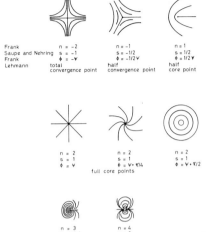

Frank	n = -2	n = -1	n = 1
Saupe and Nehring	s = -1	s = -1/2	s = 1/2
Frank	φ = -ψ	φ = -1/2 ψ	φ = 1/2 ψ
Lehmann	total convergence point	half convergence point	half core point

	n = 2	n = 2	n = 2
	s = 1	s = 1	s = 1
	φ = ψ	φ = ψ + π/4	φ = ψ + π/2
		full core points	

	n = 3	n = 4
	s = 3/2	s = 2
	φ = 3/2 ψ	φ = 2 ψ

Fig. 3.6. Arrangement of singularities in the nematic Schlieren texture.

Fig. 3.5. Singularities of the Schlieren texture (from [735]).

convergence points. These are called "zero streaks" and are apparently the results of the spatial changable directional field developing only in the *z*-direction (fig. 3.9).

Fig. 3.7. Disclination wall in the Schlieren texture.

Fig. 3.8. Closed zone and zero streaks.

The streaks can also be of "higher order", arising from multiple transition of identity periods. Compare this with other views [594] and the calculations of the free energy of the disclination lines according to the molecular field models by C. P. Fan [237, 238] and R. B. Meyer [544]. In cholesterics with Grandjean texture, areas of different pitches are separated from one another by a type of disclination line peculiar to these systems [293, 563, 594, 736]. The singularities and the second-order streaks are of special note. The symmetry properties of a singularity with four brushes and the typical interference pattern indicate that the molecules lift up in the *z*-direction and change the angle θ away from zero.

Fig. 3.9. Molecular arrangement in a zero streak.

The singularities are apparently due to disturbances in the boundary surfaces, i.e., the local homeotropy of a singular "N-texture" as classified by H. Zocher. A combination of singularities with two brushes of opposite sign has the arrangement shown by fig. 3.10 in the *z*-direction.

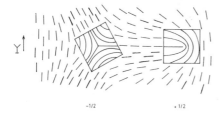

Fig. 3.10. Correlation between singularities with two brushes.

Let us regard a complex system of directly related 2-fold brushes (see fig. 3.11): Between the centers (area A, A′, B, B′) and far away from the cores the L-directions change steadily, but in the very vicinity of each core a quite different topology and symmetry must be mentioned: The left ones in fig. 3.10 and 3.11 (A, A′ region) are closely surrounded by a steadily changing direction field, as it is also the case for a 4-fold brush, for example typ $n = \pm 2$ ($s = \pm 1$, $\Phi = \pm \Psi$, see fig. 3.5). The contrary is true for the two-fold system (B, B′ in fig. 3.11), where the direction field is severly interrupted at the core itself. As in the case $n = 2$ ($s = 1$, $\Phi = \Psi + \pi/2$), a discontinuity of a higher degree exists, because a direction change of formally $\pi/2$ between neighbouring molecules has taken place. The most interesting question considers the character of the zone connecting centers of equal sign (A, A′ = $-\frac{1}{2}$; B, B′ = $+\frac{1}{2}$). It has been stated experimentally that two cores of the same sign are connected by a faint line (crossed nicols) which can bee observed better by use of a half wave plate, where it becomes bright and white, see fig. 1.12. Such lines are well known and called "lines of second order" (Saupe, Nehring, de Gennes). They are observed mostly between brushes of rank $\frac{1}{2}$, in some rare cases of rank 1, but their topology and the condition of their existence seem to be not quite clear. They look like twist walls being created between Mauguin textures of different screw sense, but their topology is different. They never form boundaries around closed areas but always end in core singularities. We suppose a rather abrupt change in tilt angle may be the reason for such a disclination "line". Let the molecules be tilted in a constant angle with respect to the boundary. Going from A to A′ the tilt direction vector will change with respect to a fixed coordinate system, and a zone with a strong change in tilt angles will be created, ressembling a Mauguin texture viewed perpendicularly to the tilt axis. In fig. 3.11 b and c the possible formation of such a wall is demonstrated.

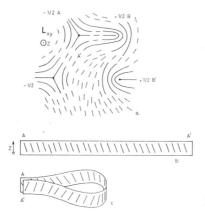

Fig. 3.11. Correlation between singularities with two brushes.

Regarding the fact that intermediate orientations (with one fixed symmetry plane) between the extreme cases of homogeneous and homeotropic texture are often found ("weak anchoring" at the surface) it would be feasible to have a simple term for this texture. Let us call it simply "quasi-homogeneous" if the optical axis is fixed. De Gennes' conical or continuously degenerated corresponds to this case, but with variable local axis.

A "broken form" where the domains show quasi-homogeneous texture with different symmetry planes is known and described as the marbled texture.

An experimental pursuit of the topology of singularities can also start from the observation that the Schlieren texture preferably forms where a homeotropic texture is converted into the isotropic liquid by heating or by the action of an electric field. In such a case ($24\,\mu m$ thick MBBA layer, undisturbed homeotropic texture; $25\,V$) the streaks appear simultaneously with the interference colors showing exclusively singularities with four brushes. The development of uniform interference colors that are tunable by the applied voltage is of technological interest (cf. chapters 4 and 14) and is due to a texture called "NPN" by H. Zocher. A closely related phenomenon is observed on heating a $10-20\,\mu m$ thick layer of a-4-methoxy-4'-n-butyl-azoxybenzene exhibiting homogeneous and Schlieren texture. The preparation assumes a pseudo-isotropic appearance slightly below the clearing point, but on further heating it brightens again a few degrees before it transforms into the isotropic liquid.

A small surface irregularity can occasionally be observed at the singularity, having (as already mentioned) four brushes and an optically positive character which indicates a radial arrangement of n_γ. Twist disclinations have found special interest since the discovery of the Schadt-Helfrich mode [464, 763, 806–808, see also 318, 650]. Other details are reported in the literature [16, 67, 117, 125, 133, 165, 183, 329, 409, 417, 419, 544, 574, 666a, 709a, 760, 792]. Disclination centers at the boundary nematic/isotropic liquid were studied by R. B. Meyer [543]. Observations in specially treated capillaries indicate nonsingular $S = +1$ screw disclination lines and other disclinations [38, 39, 68, 127, 825, 826]. Various views exist concerning the molecular arrangement in nematic droplets [677]. In all probability, accurate observations have led to false generalizations, e.g., in [572].

The interesting optically active textures of cholesterics, especially of those having large pitch, must be studied in three dimensions. R. Cano [86], J. Friedel et al. [268, 269], J. Rault [714, 715, 717], M. Brunet-Germain [80], and Y. Bouligand [63, 69] have made valuable contributions to the systematic description of these textures (cf. p. 21). Space limitation confines us to a reference list on the morphological observations concerning dislocations in cholesterics [1:18, 1:19, 7, 8, 65–67, 105, 108, 123, 270, 553a, 679, 680, 716, 736]. Likewise, we merely cite the literature more closely connected with theoretical aspects of the subject: W. J. A. Goossens [286, 287], J. T. Jenkins [345, 346], P. N. Keating [395], H. Kimura [401], T. C. Lubensky [484], C. J. Gerritsma [276, 277], other authors [7:20, 1, 31, 32, 192], and theoretical considerations on long-known changes in texture caused by shearing action [542, 660–664].

Details on the textures of smectics, especially on the topology and formation mechanism of focal-conics, cf. [199], can be found in the older writings of W. Bragg [70, 71], R. Gibrat [283, 284], D. G. Kim [399], and the more recent papers by W. L. McMillan [500, 501], J. A. Geurst [278, 280, 281], P. G. de Gennes [166, 167, 178], G. Meier et al. [539], J. P. Straley [774, 775], N. A. Clark [131, 550b], S. Sakagami et al. [729], M. Delaye et al. [181], the Orsay group [596], M. Kléman et al. [413, 414, 418, 421, 421a], C. E. Williams et al. [827–830], P. S. Pershan [646, 647], F. P. Price [680], and P. E. Cladis [127, 130]. The paper by Y.

Bouligand provides an especially illustrative presentation that includes the polygonal texture [64]. Here, we also mention the undulation instability of smectics A due to a dilation normal to the layers [40a, 721] and the instability of smectics A induced by a stress [186, 546].

3.4   The Maier-Saupe Theory (Mean-Field Approximations)

We have seen how Oseen's molecular-statistical approach has led Oseen himself to the continuum theory. The much-debated swarm theory goes back to L. S. Ornstein's interpretation of W. Kast's experiment on the orientation effect caused by an electric field [587, see also 1:53, 800].

H. Zocher, the most important advocate of the continuum theory, was able to show that the field effects investigated by W. Kast, M. Jezewski, V. Frederiks, and others could also be explained by the continuum theory. This explanation is now generally recognized as fact and expressed in the equation for inner energy:

$$g = \int (g_{\text{elast}} + g_F) \, d\omega \tag{3.19}$$

now expanded to include

$$g_F = -\tfrac{1}{2} \boldsymbol{F} \, \hat{\chi} \boldsymbol{F}' \tag{3.20}$$

In the field-dependent energy term g_F, $\boldsymbol{F}'$ is the electric (or magnetic) field strength and $\hat{\chi}$ the dielectric constant (or diamagnetic susceptibility) tensor. The molecular interactions remained unexplained at this point. V. N. Tsvetkov's work brought a revival in this area, and A. Saupe pointed out that these new studies are based largely on P. Weiss' theory of ferromagnetism [822a] and the experiments by W. Gorsky, and Bragg and Williams.

A molecular-statistical theory purporting to be a complete theory of the liquid-crystalline phase has to start from molecular properties to account for the interacting forces, and it has finally to pass from the pair to the total interaction by allowing a quantitative estimate of macroscopic thermodynamic parameters, e.g. stability regions for various phases, and the nature of the phase transitions [731]. However, since this approach has not been successful even for isotropic liquids (cf. Kohler's monograph [427]), it is understandable that molecular-statistical theories on the liquid-crystalline state are questionable even today, as are ab-initio model-based calculations of the characteristic parameters. The works of L. K. Runnels and C. Colvin appear to be of interest here [726, 727]. De Gennes' monograph has thrown much light upon the whole field [176]. The well-known method of the internal field is an important aid upon which the quantitative theories of the liquid-crystalline state are based. An interpretation of this concept is given by C. F. J. Böttcher who also explains the compensation of the molecular dipole moments within the short range order [4:104]. This has been often discussed, especially concerning the influences of electric fields. The historical development of a model of the short range order is highlighted by the studies of M. Born [54], P. Debye [152, 153], H. A. Stuart [779], W. Kast and W. Maier [390], and W. Maier and A. Saupe [506–509, 732].

It is immediately apparent that there must be direct dependence of the elastic constants on the intermolecular forces. V. N. Tsvetkov used this correlation to obtain an approach to the mean field [802]. Purely thermodynamic considerations could overlook this correlation. A. Saupe and W. Maier went a step further and employed the dispersion forces, i.e., second-order interference factors of a Coulomb interaction, as intermolecular forces. Only the dipole-dipole term of the dispersion interaction was implied here. Higher-order terms were first introduced by later investigators, e.g., W. J. A. Goossens in his theory of the cholesteric phase [287]. The final results of the Maier-Saupe calculation of the interaction energy is the equation

$$\tilde{u}_1 = F\{A(\boldsymbol{R}_{1k})/R_{1k}^6\} \cdot (1 - \tfrac{3}{2}\sin^2\theta_1)(1 - \tfrac{3}{2}\overline{\sin^2\theta}) \cdot g(C(\boldsymbol{R}_{1k})/R_{1k}^6) \tag{3.21}$$

It might be sufficient only to explain the symbols in eq. 3.21 at this point. $\tilde{u}_1$ is the average interaction energy of a molecule ("l") within the nematic phase; the function F, which incorporates a summation, $\sum_{lk}$, of all interaction potentials, lk, contains the distance vectors, $\boldsymbol{R}_{1k}$, and the distances, R_{1k}. F is independent of orientation and applies only to the molecular centers of gravity. For present purposes, the second term is the most important because it reflects the dependence of the interaction energy of molecule "l" on its "internal field strength", determined by the degree of order in its environment:

$$S = \langle 1 - \tfrac{3}{2}\overline{\sin^2\theta}\rangle \tag{3.22}$$

The angle θ is one of the three "Euler angles", namely the one formed by the longitudinal axis, ξ, of the molecule itself and the spatially constant z axis. As one can best learn from V. N. Tsvetkov or A. Saupe, averaging of the Euler angles leads to the recurring central expression of eq. 3.22. The concept of the degree of order as it was first used by V. N. Tsvetkov developed from this and not from some arbitrary definition. The second variable, determining the momentary interaction energy, is the inclination of the "central molecule" (index l) itself, i.e., the quantity $1 - \tfrac{3}{2}\sin^2\theta_1$. The g-factor in the second term corresponds to the first sum F, and it contains the summation of all paired interactions of l with the environment. For the portion of $\tilde{u}_1$ depending on the arrangement of the molecule, Saupe introduced the concept of "energy of order":

$$D_1 = -\frac{A}{V^2} \cdot S \cdot \left(1 - \frac{3}{2}\sin^2\theta_1\right) \tag{3.23}$$

where A/V^2 (corresponding to the coefficients, z·B, used by J. C. Raich et al. [699] for the Hamiltonian of intermolecular interaction) explicitly stands for the following expression of the function G in eq. 3.24:

$$\frac{A}{V^2} = g(C(\boldsymbol{R})/R^6) = \sum_{\mu\nu}{}' \frac{\delta o_\mu \delta o_\nu}{E_{\mu\nu} - E_\infty} \sum C(\boldsymbol{R}_{1k})/R_{1k}^6 \tag{3.24}$$

Note the remarkable simplification obtained by an identification of the double summation with a simple "lattice constant", A. The temperature dependence is mainly included in the term R_{1k}^6, i.e., the molar volume, V. The first product sum is also called the anisotropy factor. For further considerations especially, the quantity D_1 and its dependence on S is important. This additional calculation proceeds, after insertion of the angle-dependent part of the energy of order D_1, with the aid of a Boltzmann calculation to form the average value for $\sin^2\theta_1$

and to express this as a function of S (or of $\overline{\sin^2\theta}$, which amounts to the same thing). Note the different definitions of the "force constants" in the Boltzmann calculation used by Maier-Saupe and that used by Weber: $\alpha_{(M,S)}$ corresponds to the $\alpha/kT=\varkappa$ of Weber; otherwise the two calculations are largely analogous.

This now yields

$$\overline{\sin^2\theta_1} = \frac{\int\limits_0^{\pi/2} \sin^3\theta_1 \cdot e^{D_1/kT} \cdot d\theta_1}{\int\limits_0^{\pi/2} \sin\theta_1 \cdot e^{D_1/kT} \cdot d\theta_1}$$

$$= \frac{\int\limits_0^{\pi/2} \sin^3\theta_1 \cdot e^{\frac{A\cdot S}{k\,TV^2}\cdot\left(1-\frac{3}{2}\sin^2\theta\right)} \cdot d\theta_1}{\int\limits_0^{\pi/2} \sin\theta_1 \cdot e^{\frac{A\cdot S}{k\,TV^2}\cdot\left(1-\frac{3}{2}\sin^2\theta\right)} \cdot d\theta_1} \tag{3.25}$$

One can now regard eq. 3.25 as an implicit function for the calculation of AS/V^2 or vice versa; the function can be calculated when AS/V^2 is given (fig. 3.13). If the system is in equilibrium, then

$$\overline{\sin^2\theta_1} = \overline{\sin^2\theta} \tag{3.26}$$

This corresponds to the 45° line in the illustration.

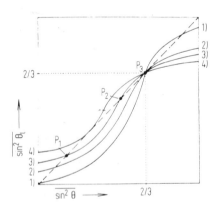

Fig. 3.12. $\overline{\sin^2\theta_1}$ as a function of $\overline{\sin^2\theta}$ following eq. 3.25, using as a parameter $A/kTV^2>5$ (line 1), $=5$ (line 2), $=4.4876$ to 5 (line 3), and $=4.4876$ (line 4) (from [506]).

As can be seen from this figure, there are basically four different areas delineated by definite values of the parameter A/V^2. The additional stability criterion is

$$\partial\overline{\sin^2\theta_1}/\partial\overline{\sin^2\theta}<1$$

This is a plausible relationship that is fulfilled, for example, at point P_1 in fig. 3.12. Junction P_3 characterizes the isotropic case. The following limiting cases characterize conditions that restrict the existence of isotropic and nematic phases. The value $A/kTV^2=5$ sets a lower limit for the isotropic phase, while 4.4876 sets an upper limit for the nematic phase. Line 1 ($A/k\,TV^2 > 5$) corresponds to solutions of eq. 3.25 with $\overline{\sin^2\theta} > \frac{2}{3}$. Here the molecular axes are preferentially perpendicular to the optical axis. An attempt to create this state by superheating of a nematic phase would result in a reversal to the isotropic state. However, there still exists the possibility of a correlation with the polymesomorphous forms of the nematic phase, especially with the "cybotactic" structure [1:45, 1:283]. A. Saupe has calculated the corresponding values for $\sin^2\theta_1$ and S, as well as for $A/k\,TV^2$ given in table 3.2.

Table 3.2. Equilibrium values characterizing the nematic order (from A. Saupe's dissertation, "Eine einfache Theorie nematischer Flüssigkeiten und ihre Anwendung auf PAA", Freiburg, 1958). Maximum errors for $\overline{\sin^2\theta_1}$ are 3 for S and $A/k\,TV^2$ 4 units in the last decimal place.

a	$\overline{\sin^2\theta_1}$	S	$A/k\,TV^2$
2.89	0.383 558	0.424 663	4.53 693
2.9241	0.380 514	0.429 229	4.54 163
2.9584	0.377 476	0.433 786	4.54 664
2.9929	0.374 434	0.438 349	4.55 178
3.0276	0.371 402	0.442 897	4.55 727
3.0625	0.368 370	0.447 445	4.56 294
3.0976	0.365 343	0.451 986	4.56 887
3.1329	0.362 319	0.456 522	4.57 503
3.1684	0.359 302	0.461 047	4.58 146
3.2041	0.356 290	0.465 565	4.58 812
3.24	0.353 284	0.470 074	4.59 502
3.61	0.323 710	0.514 435	4.67 827
4	0.295 373	0.556 941	4.78 806
4.5	0.263 295	0.605 058	4.95 820
5	0.235 734	0.646 399	5.15 677
5.5	0.212 227	0.681 660	5.37 903
6	0.192 292	0.711 562	5.62 144
7	0.162 916	0.758 626	6.15 147
8	0.137 931	0.793 104	6.72 463
9	0.120 649	0.819 026	7.32 577
10	0.107 272	0.839 092	7.54 510

The practical value of this tabulation is that for a given or experimentally determined $A/k\,TV^2$, the corresponding degree of order, S, can be found.

The next and most important step is to find a correlation to easily obtained experimental data, e.g., the volume change at the clearing point and the molar volume at that point. This correlation, the derivation of which is detailed in Saupe's work, is given by

$$\frac{\Delta V_{cl}}{V_{n,cl}} = \frac{k\,T_{cl}\,V_{n,cl}^2}{A\,S_{cl}^2}\left(2\int_0^1 e^{a_{cl}x^2}\,dx - \frac{A}{k\,T_{cl}\,V_{n,cl}^2}\,S_{cl}(S_{cl}+1)\right) \qquad (3.27)$$

For this equation as well, Saupe supplied numerical values, e.g., if $\Delta V_{cl}/V_{n,cl} = 0.00336$, then $A/k\,T_{cl}\,V_{cl}^2 = 4.5573$ and $S_{cl} = 0.4429$. Two examples illustrate the validity of Saupe's data (table 3.3).

Table 3.3. Comparison of some material constants at the clearing point (from [508]).

	$\Delta V_{cl}/V_{n,cl}$	S_{cl}		$A/k\,TV^2$	T_{cl}	$V_{n,cl}$	A
		theor.	det.		[K]	[ml]	[erg·cm^6]
PAA	0.0035	0.443	0.42	4.558	408	225	13
2,4-nonadienic acid	0.0047	0.448		4.564	326	126	5.4

Experiments revealed three common features:

1. There is good agreement in the degree of order ($S = 0.44$) at the clearing point.
2. The values of $A/k\,T_{cl}\,V_{cl}^2$ are also in agreement (4.55).
3. A plot of S vs. $\dfrac{V^2}{T_{cl} \cdot V_{cl}^2}$ gives an identical curve for nematics.

Thus one might speak of a "corresponding state" at clearing point conditions. In fact, measurements of the activity coefficients of solutes near the clearing point support the concept of coincident behavior (cf. section 9.1.2.5). Thus, the experience of a nematic phase forming contrary to the molecular movement is a mere external manifestation of the agreement between the important parameters of intermolecular interaction. It should be stressed that many newer works on the molecular theory of nematics and on the n/i phase transition are comments expanding Saupe's view (e.g. [51, 52, 248, 275, 282, 344, 350, 405, 473, 474a, 488, 505a, 741, 750, 756, 757, 819, 835, 850, 852]; for the n/s$_A$ transition, see [550a]). The Maier-Saupe theory can also be applied to liquid-crystalline mixtures, e.g. to determine the activity coefficients of solutes in gas chromatography using a stationary liquid crystal [487], see also [534, 535] and chapter 9. Similar behavior is found with superconductors where the discontinuity at a temperature showing a corresponding state comparable to that of liquid crystals at the clearing point [164, 305].

It seems problematic to dispense with a treatment of the influence of fields in discussing the theory of liquid crystals, since such field experiments have just yielded decisive information on the nature of this aggregate state. A great abundance of supplementary data will be considered, and other chapters return to the fundamentals discussed here. Many theoretical works extending the statements of the "classical" continuum (elasticity) theory and the molecular statistical theory deserve special mention, especially those of S. Chandrasekhar et al. [92–103], A. Wulf [836–840, 845, 845a], G. Lashner [449–451], J. Vieillard-Baron [815–817], K. Kobayashi [423, 424, 426], R. L. Humphries and G. R. Luckhurst [487–495, 495a, 495b], M. A. Cotter [135–139, 138a], J. G. J. Ypma and G. Vertogen [851, 852, 852a, 852b, 852c] and others [19, 53, 104, 134, 183a, 264, 275a, 349a, 403, 404, 453a, 502, 541a, 571, 575, 676a, 681, 683, 684, 687, 740a, 745a, 749a, 754, 764, 764a, 774–776, 784, 804, 820, 821, 851]. The statistical theories of nematics are reviewed in M. S. Rapport's dissertation [721b] and J. C. Raich et al. consider the molecular crystals of mesogens [700].

R. L. Humphries and G. R. Luckhurst have developed the most comprehensive concept, and fig. 3.13 shows the temperature dependence of the orientational order parameter introduced by them [490].

Y. Poggy et al. elaborated a comparison between the order parameter variation and the temperature dependence of the macroscopic physical coefficients of nematics [665, 666].

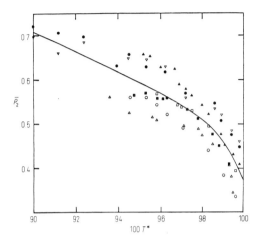

Fig. 3.13. Orientational order parameter, $\bar{P}_2$, for PAA as a function of the reduced temperature, $T^* = T/T_c$, and the theoretical curve indicated by the solid line; data compiled by R. L. Humphries et al. (from [490]).

Orientational order measurements are the topic of T. J. McKee's dissertation [499]. Other chapters continue this text describing the methods for order parameter measurements (see also [303, 436]). A somewhat difficult point in considering the model of Maier-Saupe comprises the "statistical microgroups", small aggregations of only a few molecules allowing "free" rotation about all group axes. This calls for a correlation between the degree of order as a mean quantity and the short range orientational order, cf. [45, 46, 505, 752]; this raises the question of whether the fluctuation effect about the angle θ (as described by the degree of order) actually applies to individual molecules or to such microgroups. Of special note in this regard is a paper by P. de Gennes, who correctly interprets the theory unequivocally in the sense of two levels of fluctuations. The parameter $\xi(\tau)$, called coherence length, has been introduced to describe the characteristic distance of persisting local order. This comes very close to the concept of "cybotactic" groups. G. W. Stewart and R. M. Morrow coined the term "cybotaxis" in their explanation of X-ray diffraction phenomena in liquids [1:357]. Cybotactic groups are the smallest groups of molecules having similar (constant) orientation. Vibrations of the individual molecules (or of local groups) with respect to the "director" comprise the thermal energy component. Fluctuations of larger molecular aggregates, and thus those of the "director" itself, are superimposed on the former. Quantitative considerations of both relaxation mechanisms can be found in the papers of P. A. Pincus [10:859], J. W. Doane [10:329, 10:331], C. C. Sung [781], the Orsay group [593], H. Imura [332], and in C. S. Shih's dissertation [755]. Many works show how closely we approach the models of defects in solids, "paracrystalline" textures, and the "zone" concept of the theory of ferromagnetism [10, 11, 59, 79, 151–153, 165, 175, 229–233, 236a, 269, 292, 336, 411, 419, 477, 561a, 566, 740, 753, 758, 847]. In isotropic liquids, the ordered swarms are too small to be recognizable by birefringence. R. Alben considers the existence of "negative nematics" where planar (i.e. disc-shaped) molecules are arranged with their shortest axis parallel [12].

Finally we summarize the literature concerning the molecular theories of cholesterics [286, 287, 471, 472a, 474, 486, 685, 791, 812, 813, 841–843], smectics A [84, 281, 500, 686, 688, 689, 738], smectics C [84, 501, 549, 686, 688, 844], and other smectics [162, 549, 550].

3.5   Hydrodynamics, Equilibrium Theories

There is no lack of stimulus for further expansion of the continuum theory beyond Oseen and Frank; J. L. Ericksen [205–226, 226a, 226b] and F. M. Leslie [460–464, 464a] have already been mentioned. Ericksen proceeds from the concept of an existing (unified) vector field in a liquid crystal

$$l \cdot n = 1 \tag{3.28}$$

where l is a unit vector which describes the direction of a single "rigid rod" molecular unit underlying a temperature dependent distribution. This distribution function of the smallest units which is described by the well known ("molecular") degree of order, S, gives rise to a macroscopic uniaxial order within a certain domain. The direction of the "macroscopic" local domain is described by another unit vector n, the director. l and n are interrelated by the equation

$$\langle l_i l_j \rangle = S n_i n_j + \tfrac{1}{3}(1 - S)\delta_{ij} \tag{3.28a}$$
$$\text{where}\ \ S = \tfrac{1}{2}\langle 3(l \cdot n)^2 - 1 \rangle$$

The basic problem of hydrodynamic theory is the coupling of order and flow, the former is described on the basis of the Oseen-Frank theory and the latter is given by a vector function which describes the velocity field. In the meantime excellent reviews have been written so that is seems useless to reproduce papers covering this difficult matter. It shall be sufficient to sketch the general trend and to refer to an excellent survey given by P. G. de Gennes in his monograph [176].

The basic equation with the four modules, $K_{11} \dots K_{12}$, see p. 117, has been interpreted by Ericksen and made plausible by L. Davidson who applied the general laws of elasticity physics [148]. This made the meaning of the individual modules clearer. Flügge's "Tensor Analysis" [247] and Leipholz' "Einführung in die Elastizitätstheorie" [458] are recommended as introductions to the mathematical bases for these ideas. The molecular-statistical and elasto-mechanical models consider only the (stable or metastable) static condition and its disturbance, especially by external forces emanating from the boundary surfaces. Since a mathematical treatment of the dynamic flow requires considerable expansion and generalization of the theoretical dispositions, the description of a moving liquid crystal proved quite difficult. C. P. Fan [238], E. Guyon et al. [301], P. G. de Gennes [165, 175], S. Blaha [50] and other authors [204a, 818a] wrote on the analogies between the nematic and superfluid state. The influences of sound or ultrasound waves, being closely related to flow properties, have been studied recently with increasing interest and form a complex and technologically promising subject. The inclusion of hydrodynamic principles (excellently summarized by M. J. Stephen [765] in the continuum theory of media at rest) led to the largely comparable results of J. L. Ericksen and F. M. Leslie. This theory is based on the more general concept of multipolar continuum mechanics, which seemed applicable to anisotropic media as well. The previously mentioned central idea of a "director" as well as the force- and force-pair-induced displacements are the main bases for the kinematic-dynamic theory of the anisotropic continuum and its flow characteristics.* In principle, Ericksen and Leslie go back to the works of A. E. Green

---

* A different approach to the problem is thoroughly discussed by P. G. de Gennes [176].

and R. S. Rivlin [294, 295]. It is assumed that stretched (or even cylindrically symmetrical) molecules predetermine the preferred direction of the director field. In the kinematic treatment of flow characteristics, there is given the location of a particle at time t as a function of a past time period $\tau(-\infty<\tau\leq t)$:

$$\xi_i(\tau)=\xi_i(x_1,x_2,x_3,t)$$
$$x_i=\xi_i(t) \tag{3.29}$$

Analogously, the director $d_i(\tau)$ is introduced as a function of such coordinates. For each position and for each directional vector, a velocity as a function of $\tau:v(\tau)$ and $w(\tau)$, respectively, is defined. These two variables are connected to each other by a vector, (N_i), and a tensor, (N_{ij}).

The flow characteristics due to additional mechanical stresses can be considered as disturbances of the basic characteristics. The subsequent thermodynamic treatment of the problem strikes the energy balance in which the position velocity, v, and the directional velocity, w, now appear as the characteristic independent variables.

$$\frac{D}{Dt}\int_V \rho\left(\frac{1}{2}v_iv_i+\frac{1}{2}w_iw_i+U\right)dV$$

$$=\int_V \rho(r+F_iv_i+L_iw_i)dV+\int_A (t_iv_i+p_iw_i-h)dA \tag{3.30}$$

where V is a material volume bounded by a surface A. ρ is the density, U the internal energy per unit mass, r the heat supply function per unit mass per unit time, and h the flux of heat per unit area of surface A per unit time. F_i is a body force per unit mass, L_i a director body force per unit mass, t_i the surface force per unit area of A, and p_i the director surface force per unit area of A.

The entropy generation is given by

$$\int_V \rho\dot{S}dV-\int_V \rho\frac{r}{T}dV+\int_A \frac{h}{T}dVA\geq 0 \tag{3.31}$$

If there is a motion superimposed upon this basic behavior, e.g., a translation or the rotation of a Couette viscosimeter body, the equation is considerably simplified:

$$\rho r-\rho\dot{U}-q_{i,i}+\sigma_{ki}v_{i,k}+\pi_{ki}w_{i,k}+\Gamma_iw_i=0, \tag{3.32}$$
$$(\Gamma_i=\rho(L_i-\dot{w}_i)+\pi_{ki,k})$$

where σ_{ki} are the surface tension components, π_{ki} are the component of the director surface tension, and q the heat flow.

For further calculations, it is now necessary to derive "constitutional equations" for the thermodynamic function U (or A or S) and for σ_{ij}, π_{ij}, Γ_i, and q_i. Thus, these quantities must be expressed as functions of the variables ρ, T, d, w, v, and the temperature gradient. Under the simplified conditions, e.g., for linear constitutional relationships, and assuming certain conditions of symmetry $(-d_i = d_i)$, the equation becomes even simpler for incompressible fluids as

$$\rho r - \rho \dot{U} - q_{ii} + (\sigma_{ik} + \rho \delta_{ik}) A_{ik} + (\Gamma_i - \gamma d_i) N_i = 0 \tag{3.33}$$

and the entropy expression simplifies to

$$(\sigma_{ik} + p \delta_{ik}) A_{ik} + (\Gamma_i - \gamma d_i) N_i - q_i T_i / T \geq 0 \tag{3.34}$$

The constitutional equations for σ_{ij}, Γ_i, and q_i can also be simplified under restrictive conditions until finally one obtains

$$\begin{aligned}
&v_{i,i} = 0, \\
&\rho \dot{v}_i = \rho F_i + \sigma_{ki,k}, \\
&\rho_1 \dot{w}_i + \gamma d_i + \gamma_1 N_i + \gamma_2 d_j A_{ji} = 0, \\
&\sigma_{ij} = -p \delta_{ij} + \alpha_1 d_k d_p A_{kp} d_i d_j + \alpha_2 d_i N_j + \alpha_3 d_j N_i \\
&\quad + \alpha_4 A_{ij} + \alpha_5 d_i d_k A_{kj} + \alpha_6 d_j d_k A_{ki},
\end{aligned} \tag{3.35}$$

where $\gamma_1 = \alpha_3 - \alpha_2$, $\quad \gamma_2 = \alpha_6 - \alpha_5$, $\quad$ the α's are constants, and σ_{ij} is the stress tensor.

O. Parodi showed that eq. 3.36 expresses the relations among the "viscosity" coefficients [144, 634]:

$$\alpha_2 + \alpha_3 = \alpha_6 - \alpha_5 \tag{3.36}$$

The literature and section 3.8 report measurements of these coefficients [78, 129, 592, 855]. The solutions of eq. 3.35 can describe simple shear flow, Poiseuille flow, and the Couette flow of anisotropic fluids specified in the literature [320–323, 460]. Always the six constants, α_1 to α_6, appear, five of them being mutually independent, as stated above. This is the most important statement of the theory up to now.

Under the special conditions that

$$v_x = k y \qquad v_y = 0 \qquad v_z = 0 \qquad d_x = \cos \Phi \qquad d_y = \sin \Phi \qquad d_z = 0$$

where k is a constant, and Φ is a function of y, simple shear flow can be described by

$$\begin{aligned}
&2\gamma \cos \Phi - k(\gamma_1 - \gamma_2) \sin \Phi = 0 \\
&2\gamma \sin \Phi + k(\gamma_1 + \gamma_2) \cos \Phi = 0
\end{aligned} \tag{3.37}$$

If $\quad k \neq 0$ and $|\gamma_1| \leq |\gamma_2|$, it is obtained as

$$\begin{aligned}
&\gamma = -\gamma_2 k \sin \Phi \cos \Phi \\
&\cos^2 \Phi = (\gamma_2 - \gamma_1)/2\gamma_1 \gamma_2 \\
&\sin^2 \Phi = (\gamma_1 + \gamma_2)/2\gamma_2
\end{aligned} \tag{3.38}$$

The components of the stress tensor are:

$$\sigma_{xx} = -p + A k; \quad \sigma_{yy} = -p + B k; \quad \sigma_{zz} = -p; \quad \sigma_{xy} = \sigma_{yx} = C k \tag{3.39}$$

where

$$A = f(\alpha_1 \ldots \alpha_6, \Phi) \qquad B = f(\alpha_1 \ldots \alpha_6, \Phi) \qquad C = g(\alpha_1 \ldots \alpha_6, \Phi)$$

A second solution is obtained for the case where the director is always parallel to the y-axis:

$$\sigma_{xx} = \sigma_{yy} = \sigma_{zz} = -p$$
$$\sigma_{xy} = \sigma_{yx} = \tfrac{1}{2}\alpha_4 \cdot k \tag{3.40}$$

Similar considerations lead to the solutions for the cases of Poiseuille and Couette flow [129, 145, 197b, 240, 406]. Twist wave propagation has also been treated by using the Leslie-Ericksen theory [746, 748].

Unfortunately, a correlation to the static theory of Oseen and Frank has not yet been performed [472]. P. Martinoty's dissertation [514], W. Helfrich [317], and Ericksen himself [226] provide a good overview. A comparison between the Leslie-Ericksen theory and the Eringen-Lee theory [227–236] is found in the literature [232, 748a].

T. C. Lubensky [485], A. C. Eringen and J. D. Lee [233], and other authors [441a, 575a] have deduced equations describing the hydrodynamics of cholesterics. E. Klein et al. [410] and S. Chandrasekhar et al. [407] have studied the shear-flow behavior of cholesterics in relation to their scattering properties.

In a moving medium which is simultaneously anisotropically viscous and elastic, the molecular interactions overlap with elastic-hydrodynamic forces and external forces such as pressure and surface phenomena. The kinetics and dynamics of the systems are interrelated in such a way that fluid motion (velocity gradient) and molecular arrangement have a mutual influence. This is mathematically expressed as

$$\boldsymbol{n} \times \boldsymbol{h} - \boldsymbol{\Gamma} = J \cdot \frac{\mathrm{d}\boldsymbol{\Omega}}{\mathrm{d}t}{}^{\dagger} \tag{3.41}$$

where $\boldsymbol{n} \times \boldsymbol{h}$ is a force couple that develops from an elastic coupling of a molecule with its neighbors, $\boldsymbol{h}$ is the functional derivative of the elastic energy $-\partial F/\partial n_i$, $\boldsymbol{\Gamma}$ is a force couple created by the effect of hydrodynamic flow on the molecule, and $J \cdot \partial\boldsymbol{\Omega}/\partial t$ is the force couple created by molecular rotation where

$$\boldsymbol{\Omega} = \boldsymbol{n} \times \frac{\mathrm{d}\boldsymbol{n}}{\mathrm{d}t} \quad (\cong 0, \text{ i.e. } 10^{-14}\,\mathrm{g\,cm^{-1}} = \rho \cdot a^2) \tag{3.42}$$

with ρ being the density, and a the molecular dimension. The motion of the system is described by

$$\rho \cdot \frac{\mathrm{d}V_i}{\mathrm{d}t} = \frac{\partial}{\partial x_j} \cdot \sigma_{ji}^* \tag{3.43}$$

Here, σ_{ji}^* is the tensor implying terms for pressure dependence: energy dissipation, σ_{ij}, and elasticity, σ_{ij}^0. For small deviations about the average position $\boldsymbol{d}$, σ_{ij}^0 is a quadratic function

---

† $\boldsymbol{n}$ is identical with $\boldsymbol{d}$, the director, being introduced to avoid "dd".

of the displacement from d and can therefore be neglected. For the dissipation term, it holds that

$$\sigma_{ij}=\alpha_1\, n_k n_p A_{kp} n_i n_j+\alpha_2\, n_i N_j+\alpha_3\, n_j N_i+\alpha_4 A_{ij}+\alpha_5\, n_i n_k A_{kj}+\alpha_6\, n_j n_k A_{ki} \tag{3.44}$$

where $A_{ij}=\dfrac{1}{2}\left[\dfrac{\partial V_i}{\partial x_j}+\dfrac{\partial V_j}{\partial x_i}\right]$ $\qquad N=\dfrac{\partial n}{\partial t}-\omega\times n$ $\qquad \omega=\dfrac{1}{2}\,\mathrm{rot}\,V$

where ω is the angular velocity and N describes the rotational velocity of the director in relation to the fluid. The α-coefficients have the dimensions of viscosity and are called Leslie friction coefficients. The unsymmetrical tensor σ_{ij} obeys the relation

$$\sigma_{ij}-\sigma_{ij}=\Gamma_k\cdot\varepsilon_{ijk} \tag{3.45}$$

where ε_{ijk} is 0 for two even indices, 1 for an even permutation of the indices, or -1 for an odd permutation of the indices. Considering the expression for σ_{ij}, it follows that

$$\Gamma=n\times\left[\gamma_1\cdot N+\gamma_2\bar{A}\cdot n_2\right] \tag{3.46}$$

where $\gamma_1=\alpha_3-\alpha_2$ $\qquad \gamma_2=\alpha_6-\alpha_5$ $\tag{3.47}$

The number of components is reduced to five due to an Onsager relationship, see eq. 3.36 [144, 634].

If the problem is simplified by assuming incompressibility,
a) $\mathrm{div}\,V=0$,
b) setting the elastic constants equal to each other, $h=K\nabla h$,
c) neglecting the expression $J\cdot\partial\Omega/\partial t$, then the following relatively simple relationships evolve:

$$\rho\cdot V_i=\frac{\partial}{\partial x_j}\sigma_{ij}-\frac{\partial p}{\partial x_j}\cdot\delta_{ij} \tag{3.48}$$

$$K\,n\times\nabla n=\Gamma \tag{3.49}$$
$$\alpha_2+\alpha_3=\alpha_6-\alpha_5=\gamma_2 \tag{3.50}$$
$$\gamma_1=\alpha_3-\alpha_2$$

These equations allow not only to interpret correctly the viscosity measurements as well as Rayleigh and Brillouin scattering, but they also predict results. Nothing might be more desirable in proving a theory of the liquid-crystalline state to be comprehensive. Further hydrodynamic considerations are reported in the literature [5, 17, 185, 198, 235, 241, 251, 257a, 459, 692, 747].

3.6   The de Gennes Theory

Special attention is called to another phenomenological concept for the description of liquid-crystalline properties that depends on elastomechanical characteristics and the degree of order. This concept was developed by P. G. de Gennes and the Orsay group [156, 159].

De Gennes characterizes the macroscopically effective arrangement from the magnetic susceptibility, χ, by using a symmetrical tensor ("tensor order parameter") cf. Sullivan's "anisotropy tensor" [782]:

$$A_{ij} = \langle \cos\theta_i \cos\theta_j \rangle - \tfrac{1}{3}\delta_{ij}$$
$$Q_{\alpha\beta} = \chi_{\alpha\beta} - \tfrac{1}{3}\chi_{\gamma\gamma} \cdot \delta_{\alpha\beta} \qquad\qquad\qquad (3.51)$$

Assuming rigid molecules, this quantity correlates with the degree of order (as usually defined) according to

$$Q_{\alpha\beta} = n S_{ij}^{\alpha\beta} \chi^{ij} \qquad\qquad\qquad (3.52)$$

where $S_{ij}^{\alpha\beta} = \langle \tfrac{1}{2}(3\,i_\alpha j_p - \delta_{ij} \cdot \delta_{\alpha\beta}) \rangle$ $\qquad\qquad\qquad (3.53)$

    α, β, γ: working coordinates
    i, j, k: molecular coordinates
    n: molecular density (numerical)

The valid relationship for uniaxial nematics is

$$\chi_\parallel - \chi_\perp = Q_{zz} - Q_{xx} = \tfrac{3}{2} S_{ij} \chi^{ij} \qquad\qquad\qquad (3.54)$$

The theory is also applicable to biaxial nematic domains [265, 266].

The purpose of this statement is to use a certain property (which is theoretically well defined, measurable and free from intermolecular influences) in a definition of the degree of order and to join it together with molecular characteristics by a transforming relationship. The next step would then be to show the dependence of the free energy on the state of order, Q, where stability conditions require $F(Q,H)$ to be minimized in analogy with Saupe's theory. However, de Gennes purposely restricted himself to the vicinity of the clearing point, cf. [793a]. An extrapolation far into the nematic range was challenged because it is not certain what role the higher Q-terms might play, being neglected at present.

The theory predicts a plausible dependence on field strength and temperature for magnetic birefringence. The considerations assume a uniform arrangement of the mean nematic axis. Generalizing the function $F = F(Q)$ permits a local change in Q (whereby, however, only two elastic modules are required) as long as $T \cong T_{\rm clp}$. An especially important parameter typical of the de Gennes theory, ξ, describes the spatial change in the degree of order occurring when approaching the coexistent isotropic phase from a (parallel or perpendicularly) ordered phase. These socalled coherence lengths depend on the temperature and the elastic constants:

$$\xi^2(T) = \left(L_1 + \frac{1}{6} L_2 \right) \cdot \frac{1}{A(T)} \qquad\qquad\qquad (3.55)$$

where $A(T) = a \cdot (T - T^*)$ and $T^* \cong T_{\rm clp}$

ξ is very large near the clearing point. Can coherence lengths be experimentally determined? Intensity measurements of scattered light do not help in this case because the intensity is independent of the wavenumber indicating that $q\xi \ll 1$, i.e., the coherence length is small compared to the wavelength [475, 476]. The scattered-light intensity itself is inversely proportional to the clearing point difference A, i.e., it increases sharply when approached from higher temperatures [772], see section 6.2.

Light reflectivity measurements at the n/i interface enabled determination of the interface tension and an estimation of the coherence length to be about 12 nm in MBBA at the clearing point [446, see also 443, 447, 773]. It is also significant that there are precisely as many different coherence lengths as there are different Q gradients near the boundary surface, e. g., normal, tangential, and conical (fig. 3.14).

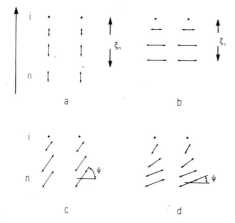

Fig. 3.14. Possible structures for the nematic/isotropic interface: (a) normal case, (b) tangential case, (c) conical case with constant angle in the transition layer, (d) conical case with variable angle (from [159]).

Recall in this regard the remarkable aberrations in the behavior of homeotropic phases of various molecular polarities just before reaching the clearing point, wherein the homeotropy reverses at a specific temperature before the final temperature increase that finally causes the transition to the isotropic phase.

The time-dependence of state-of-order fluctuations, which is naturally of special importance in streaming systems, can be definitely attributed to a viscous position change with only a minor quasi-elastic factor [6:218, 74, 593, 595]. The relaxation time, τ_g, is given by

$$1/\tau_g = \frac{K \cdot g^2}{\eta_{\text{eff.}}} \tag{3.56}$$

where K is an "average" elastic module, and η an "average" viscosity number. The isotropic melt, having the lower order, is a stronger dissipative medium (friction losses) than the nematic phase. Relaxation effects also play an important role in the ultrasound absorption characteristics of isotropic media (see section 3.10). For the "quasicrystalline" liquid crystals, A. Gierer and K. Wirtz developed a relaxation theory that can stimulate interesting comparisons with the theory developed for anisotropic systems [285]. It was obvious to express these phenomena in the terminology of Onsager's "Irreversible Thermodynamics". For the creation of entropy, de Gennes derived the function

$$T \cdot S = \Phi_{\alpha\beta} \cdot R_{\alpha\beta} + \tfrac{1}{2}\sigma_{\alpha\beta} \cdot e_{\alpha\beta} \tag{3.57}$$

with the fluxes defined as follows (using Onsager's terminology):

$$R_{\alpha\beta} = \frac{\partial Q_{\alpha\beta}}{\partial t} \qquad \text{(change of order with time)} \tag{3.58}$$

$$e_{\alpha\beta} = \partial_\alpha v_\beta + \partial_\beta v_\alpha \quad \text{(shear gradient of the streaming system)} \tag{3.59}$$

and forces as follows:

$$\Phi_{\alpha\beta} = -\frac{\partial F}{\partial Q_{\alpha\beta}} \qquad \text{(change of free energy with the energy of order)} \tag{3.60}$$

where $\sigma_{\alpha\beta}$ is the stress tensor of the viscous fluid.

The enormous simplification when compared to the Leslie-Ericksen calculations is obvious. However, the elimination of all but one elasticity and two viscosity constants naturally leaves open the question of how far this theory can be applied to the liquid-crystalline state remote from the clearing temperature. In spite of this possible restriction, the de Gennes theory nevertheless permits a clear explanation for the streaming birefringence, the Rayleigh scattering, and the damping of shear waves.

3.7 Pretransition Phenomena

Another model attempting to explain the temperature-dependent degree of order has already been mentioned. This adopts the concept of equilibrium formation from chemistry, and it is therefore known as the quasichemical theory based on the theory of intermolecular interactions. Frenkel had an especially important role in its development. In simple terms, this theory states that with increasing temperature the ordered domains reversibly disintegrate to an "unordered" state. The heterophase fluctuations thus replace the concept of degree of order, and they characterize the pretransition phenomenon occurring near the clearing point.

Independently of the heterophase fluctuation model, the pretransition can be divided into a translation component ("position premelting") and a rotation and a vibration component [809]. In this way, quite plausible models were developed not only for the melting process in solid crystals but also for the transition processes of liquid crystals. For this development, we cite the works of R. Alben [9], I. G. Chistyakov [112, 114], J. R. McColl et al. [497], M. A. Cotter and D. E. Martire [135, 530–533], C. P. Fan and M. J. Stephen [237], C. F. Frank [253], M. J. Freiser [266], F. K. Gorskii et al. [288, 289], A. P. Kapustin et al. [355, 356, 360, 361], N. I. Koshkin [430], V. N. Tsvetkov et al. [803], W. A. Hoyer et al. [325, 326], R. B. Meyer et al. [545, 547], H. Imura et al. [331], and others [106, 432, 673, 848]. In chapter 8, we return to this subject in connection with the thermodynamics of phase transitions. The literature on PAA is summarized in a review [706]. Pretransition phenomena are always observed when the short-range order exerts a measurable influence, and they have been studied in detail for most of the physical properties to be discussed in the following chapters.

3.8   Viscosity and Flow Studies

Due to the interdependence between flow characteristics and other transport properties (e. g., diffusion, heat conductivity), the subject is described here altogether. The works of J. A. Fisher [246, 259, 260], I. G. Chistyakov [116], and R. S. Porter and J. F. Johnson's series "Order and flow of liquid crystals" [668–676] impart a good overview, and the last contains abundant experimental data whose theoretical explanation by kinematic-dynamic modes is still pending. A general introduction into the problems of rheology is given by S. Peter [649]. Investigations in this area began with the experiments of R. Schenck [1:342], E. Eichwald [2:188], E. Bose [55–60], F. Dickenschied [184], R. Schachenmeier [739], F. Krüger [438], A. H. Krummacher [439], and E. C. Bingham [48]. Wo. Ostwald has worked out analogies to colloid chemistry which formed an interesting contribution to the old argument about "emulsion theory" [624, 625], and Ostwald's work was an important contribution to Frenkel's equilibrium theory at the same time [267]. An earlier work points out to the correlation between molar volume and the degrees of freedom to translation and rotation [319]. Almost simultaneously with E. Bose, H. Pick noted parallels between hardness, flow pressure of solid (plastic) bodies, and the viscosity of liquid crystals which had already been implied in O. Lehmann's earliest works [651]. One result that is still important was Pick's discovery that PAP, having the higher clearing point than PAA, also possesses greater internal friction than the latter. Viscosity and clearing point curves for the binary mixtures run symbathically with the composition.

M. Miesowicz introduced three viscosity coefficients according to the geometry shown in fig. 3.15 and obtained the following values for magnetically oriented PAP at 144.4 °C: $\eta_1 = 0.013 \pm 0.0005$, $\eta_2 = 0.083 \pm 0.004$, $\eta_3 = 0.025 \pm 0.003 \, \mathrm{g \, cm^{-1} \, s^{-1}}$ [551–553].

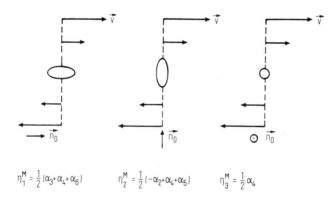

$$\eta_1^M = \frac{1}{2}(\alpha_3 + \alpha_4 + \alpha_6) \qquad \eta_2^M = \frac{1}{2}(-\alpha_2 + \alpha_4 + \alpha_5) \qquad \eta_3^M = \frac{1}{2}\alpha_4$$

Fig. 3.15. The definition of the Miesowicz viscosities (from [342]).

This geometry allows an understanding of the relatively high differences observed between η_1 and η_3 when compared to η_2. The experimental apparatus consisted of two plates that could be moved in opposite directions and a perpendicular magnetic field; this corresponds to an arrangement used earlier by G. H. Quincke [698]. Similarly, C. K. Yun has evaluated Tsvetkov's earlier measurements in a rotating magnetic field [854].

Many measurements have indicated that the viscosity is very sensitive to changes in molecular order in the pretransition range (fig. 3.16).

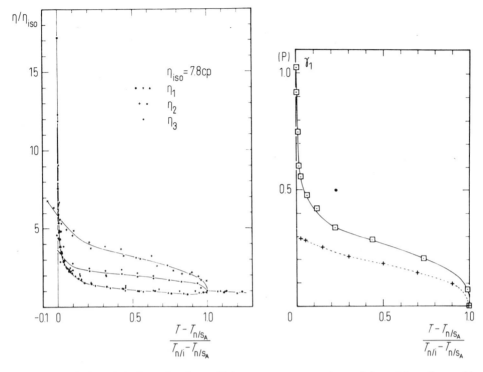

Fig. 3.16. Variation of the four viscosity coefficients, η_1, η_2, η_3, and γ_1, of 4-*n*-octyloxy-4′-cyanobiphenyl (k 54.5 s$_A$ 66.6 n 79.3 i) versus the reduced temperature (from [457]; measurements of the twist viscosity, γ_1, are also reported in refs. [182, 694]).

The pronounced increase in η_1 observed with the temperature approaching the n/s$_A$ transition temperature obeys the equation

$$\frac{\eta_1}{\eta_{\text{iso}}} = A\,e^{+W/kT} + B(T - T_{\text{ns}_A})^{-\nu} \tag{3.61}$$

with

$$
\begin{aligned}
A &= 0.651 \cdot 10^{-6} \\
W &= 1.788 \cdot 10^{-14} \text{ erg} \\
B &= 2.106 \\
T_{\text{ns}_A} &= 66.661\,°C \\
\nu &= -0.364
\end{aligned}
$$

η_{iso} is the measured viscosity in the isotropic phase close to $T_{\text{n/i}}$ [457].

Here, we point at different definitions of η_1 and η_2 used by Miesowicz (M) and Helfrich (H) and give the correlations to the Leslie-Ericksen viscosities together with the values (in centipoise) obtained for MBBA at room temperature [272, 274, see also 783]:

η_1^M	η_2^H	$\frac{1}{2}(\alpha_3+\alpha_4+\alpha_6)^*$	23.8 ± 0.3
η_2^M	η_1^H	$\frac{1}{2}(\alpha_4+\alpha_5-\alpha_2)$	103.5 ± 1.5
η_3^M	η_3^H	$\frac{1}{2}\alpha_4$	41.6 ± 0.7
	η_{12}	α_1	6.5 ± 4
	χ_1	$-\alpha_2$	77.5 ± 1.6
	χ_2	α_3	1.2 ± 0.1

* due to eq. 3.36, this can also be expressed by $\alpha_3+\frac{1}{2}(\alpha_2+\alpha_5+\alpha_6)$

The relationship for the twist viscosity, γ_1, is $\gamma_1=\alpha_3-\alpha_2$ [457].

In the search for correlations between Oseen and Franck's elastic theory and the parameters of structural viscosity, the works of M. Miesowicz [551–553], J. L. Ericksen [207], T. C. Lubensky [249, 482, 483], the Orsay group [592, 593], J. A. Geurst [279], H. W. Huang [327, 328], F. Jähnig and H. Schmidt [338, 339], P. C. Martin et al. [644], and others [814] are important.

Other experimental data on the viscosities of nematics are reported in the literature [2:38, 2:63b, 2:191, 2:330, 2:622, 2:832, 3, 42, 48a, 49, 81, 121, 130b, 145a, 260, 274, 330, 342, 391, 400, 441, 529, 529a, 536, 538, 648, 667–669, 694, 795, 796, 801, 856]. Some papers pay special attention to the behavior at the n/s$_A$ transition [76, 182, 306, 307], see also fig. 3.17 and n/i transition [393, 394, 527, 630].

Without going into details, we cite the literature concerning the viscosity of cholesterics [73, 290, 291, 667, 670, 671, 674–676, 693, 846], smectics [2:622, 2:832, 400, 457, 538, 538a, 671, 674], and anhydrous soap mesophases [2:572].

According to C. Gähwiller, the flow alignment angle in the nematic phase reaches a maximum value at the clearing point and apparently can drop to zero within the nematic range [273] (fig. 3.17).

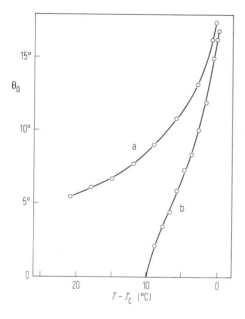

Fig. 3.17. Flow alignment angle, θ_0, of MBBA (curve a, $T_c=43\,°C$) and *p-n*-hexyloxybenzylidene-*p'*-cyanoaniline (curve b, $T_c=101.8\,°C$) as a function of temperature (from [273]).

Remarks on the flow properties of nematics are found on p. 136, dealing with hydrodynamics. The most important fact to be recognized is that both flow and diffusion parameters are anisotropic and exhibit minimum resistance to material transport in nematics parallel to the optical axis, i.e. in the direction of the longitudinal molecular axis [111, 113]. In smectics, the path of least resistance is tangential to the molecular layers [195, 597]. A polarization due to dipole orientation was observed on shearing ferroelectric smectics C parallel to the smectic layers [654]. The literature also reports flow properties of cholesterics [18, 513, 520, 664] and the mesophases of molten salts [790, 810]. The chief work was done in the flow alignment [107, 116, 163, 250, 273, 315, 408, 537] and the flow deformation and instabilities of nematics [1:259, 130a, 179, 197a, 244, 245, 302, 464a, 652, 655–658, 658a, 743].

The flow alignment was also pursued across the n/i transition [132], and other flow properties of nematics have been studied [82, 324, 470, 633, 653]. W. Helfrich's studies of nematics, smectics, and cholesterics revealed that the Poiseuille law is not valid for the flow through capillaries [309–313]. The motion resembles that of a plug rather than being parabolic, cf. [383, 789], and it depends on external fields [576, 797, 799]. Two types of instabilities have been considered in smectics A under stress deformation, a molecular tilt inside the layers and an undulation of the layers [723].

3.9 Diffusion and Heat Conductivity

As an effect of the correlation between external and intermolecular forces, the anisotropy of viscosity has an analog in the anisotropy of diffusion. Here, the concentration differences compensate anisotropically, and this anisotropy is especially pronounced in smectics. Heat conductivity and electric conductivity behave analogously. The latter is of special significance for electro-optical phenomena and is considered in detail in chapter 4. T. Svedberg in 1918 was the first to study the dependence of diffusion on the orientation of a nematic phase [785]. Using m-nitrophenol in an PAA/PAP mixture, he compared the equilibrium in a magnetic field oriented parallel or perpendicular to the direction of diffusion, and he found a ratio of 1.41 in favor of longitudinal diffusion. For a time, no further work was undertaken except by A. Perrier [643] until J. F. Sullivan [782], C. K. Yun [261–263, 853], and I. Teucher, H. Baessler and M. M. Labes [33] reported other observations, while A. F. Martins studied the behavior in the isotropic liquid of nematogens [528]. W. Franklin [255–257] worked out the diffusion theory for anisotropic liquids. Some nonconforming values of diffusion coefficients are found in the literature [119]; fig. 3.18 represents the values measured for PAA, for which W. Franklin reports $D_{\parallel} = 4.25 \cdot 10^{-6}$ and $D_{\perp} = 2.87 \cdot 10^{-6}$ cm^2 s^{-1} at 122°C and gives the ratio $D_{\parallel}/D_{\perp} = 1.48$ [255].

F. Rondelez has studied the diffusion of a dye in MBBA [725]. Much experimental data results from neutron scattering experiments (see section 5.3) and NMR investigations (see chapter 10).

The opposite case where $D_{\perp}$ is much larger than $D_{\parallel}$ is found in smectics; a value of $D_{\perp}/D_{\parallel} \sim 10$ has been measured and calculated in a s$_A$ phase [120]. The diffusion in cholesterics can be studied by using an optical method [2:318].

Aside from the translational diffusion considered previously, the rotational diffusion has also been studied [118, 255, 257b, 512], see sections 4.4.1, 5.3, and 6.5.

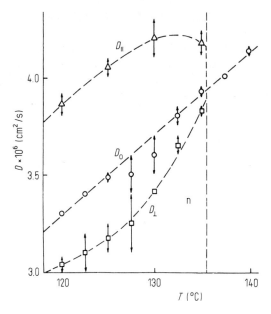

Fig. 3.18. Self-diffusion coefficients of PAA as a function of temperature. D_0 is the value of a disoriented liquid; $D_\parallel$ and $D_\perp$ were determined in a magnetic field of 3500 Gauss in the directions parallel and perpendicular to the mass flow, respectively (from [261]).

With the exception of the early works of G. W. Stewart et al. who concentrated on the orientation effects [1:362, 766, 767, 769–771], all studies of thermal conductivity date from recent years and mainly consider nematics [8:264, 3, 258, 259, 498, 640, 641, 701–705, 818], see also chapter 4, p. 164, but scarcely smectics [572a]. Other papers try to build a bridge to hydrodynamic theories [149, 150, 236, 805]. Thermoconvective instabilities have occupied much attention because their low threshold of temperature gradient is 2 to 3 orders of magnitude lower than in isotropic liquids [20, 21, 39a, 87, 142, 193, 195–197, 234, 299, 300, 302a, 459a, 554, 707, 708]. E. Dubois-Violette et al. provide an especially detailed description containing pictures of the textures due to the convective motion [197]. Thermoconvective instabilities are also known in cholesterics [177, 194, 463, 638] and smectics [722, 738a]. A molecular rotation in cholesterics subject to a time dependent temperature gradient along the helical axis has already been observed by O. Lehmann [748b]. Such studies allow to draw a parallel to electrohydrodynamic instabilities and we should recall the remarkable disturbances in W. Maier's first measurements of dielectric constants [4:592].

3.10 Ultrasonic Studies

S. Candau and P. Martinoty in Strasbourg [514–527, 527a, 527b] and several other authors, have investigated the sound propagation in nematics. Intense study began with P. Martinoty's dissertation [514] and the nearly simultaneously published results of J. D. Litster et al. [475]. Previously, A. P. Kapustin had begun his extensive investigations which included all types of liquid crystals and the behavior in homologous series [351–365, 367–376]. All these experiments contributed to a coherent picture based on the hydrodynamic theory and de Gennes' theory [141, 227, 228, 333, 340, 585]. The subject is summarized in some reviews [368, 370, 378, 527a, 583, 849], and potential applications have been outlined [147, 380]. Of particular value are the successful measurements of the individual viscosity coefficients and relaxation times. Most of the works employ a magnetic field to measure the anisotropy in oriented samples. P. Martinoty and S. Candau used a measuring apparatus in which a transverse ultrasonic wave is reflected from the boundary surface of a glass body as shown in fig. 3.19. The geometry of the equipment can be selected for either a homogeneous or a homeotropic boundary layer, according to the pretreatment of the plate surface. In other words, the sound wave is reflected from and/or passed across the boundary either parallel or perpendicular to the director, as the two cases in the figure illustrate.

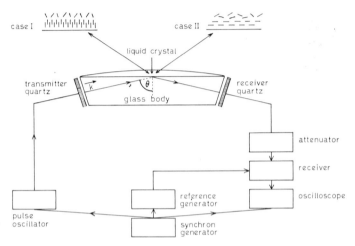

Fig. 3.19. Martinoty and Candau's experimental set-up (from [514]).

Note that the sound wave has low penetration (ca. 1 μm). The solution of a dispersion equation derived from the Leslie-Ericksen expression contains a characteristic parameter, μ, given by

$$\mu = \frac{K \cdot \rho}{\bar{\eta}^2} \tag{3.62}$$

The constant μ, having a magnitude $\ll 1$, contains the elasticity K ($\simeq 10^{-6}$ dyne), the density, ρ, and the average viscosity number, $\bar{\eta}$. This leads to the following solutions of the dispersion equation for the three wave numbers $k_0^\perp$, $k_1^\perp$, and $k_2^\perp$ for the case where $K \perp n_0$ or $K \| n_0$:

$$k_0^\perp = (1+i)\left(\frac{\omega M \gamma_1}{2LK}\right)^{1/2} \qquad L = \frac{1}{2}(\alpha_3 + \alpha_4 + \alpha_6)$$

$$k_1^\perp = (1+i)\left(\frac{\omega \rho}{2M}\right)^{1/2} \qquad M = L - \frac{\alpha_3}{2}\left(1 + \frac{\gamma_2}{\gamma_1}\right)$$

$$k_2^\perp = (1+i)\left(\frac{\omega \rho}{2N}\right)^{1/2} \qquad N = \frac{1}{4}\alpha_4$$

$$k_0^\| = (1+i)\left(\frac{\omega P \gamma_1}{2QK}\right)^{1/2} \qquad Q = \frac{1}{2}(\alpha_4 + \alpha_5 - \alpha_2)$$

$$k_1^\| = (1+i)\left(\frac{\omega \rho}{2P}\right)^{1/2} \qquad P = Q + \frac{\alpha_2}{2}\left(1 - \frac{\gamma_2}{\gamma_1}\right)$$

$$[M = P] \tag{3.63}$$

The correlation between attenuation and the Leslie viscosity coefficients is evident. $k_1^\perp$ and $k_2^\perp$ are dependent on the viscosity as in normal liquids, while $k_0^\perp$ is strongly orientation dependent and, like $k_0^\parallel$, strongly damped. Quantitative statements have been derived concerning the "acoustic impedance" based on conditions determined by the experimental parameters and molecular arrangement. The results are summarized as follows:

$$\sigma = ZV$$

where

$$
Z = \begin{vmatrix} (i-1)\left(\dfrac{\omega\rho M}{2}\right)^{1/2} & [=Z_{11}] & 0 & [=Z_{21}] \\ 0 & [=Z_{12}] & (i-1)\left(\dfrac{\omega\rho M}{2}\right)^{1/2} & [=Z_{22}] \end{vmatrix}
\tag{3.64}
$$

and σ is the tensor defining the experimental conditions; V is the velocity vector. The ratio, r, of the velocity of the reflected wave to that of the incident wave, the socalled reflection coefficient, is obtained from the matrix elements, Z_{11} or Z_{22}, and the acoustic impedance of the wave conductor, Z_s, according to:

$$r_1 = \frac{V_1^r}{V_1^i} = \frac{Z_s - Z_{11}}{Z_s + Z_{11}}$$

$$r_2 = \frac{V_2^r}{V_2^i} = \frac{Z_s - Z_{22}}{Z_s + Z_{22}}
\tag{3.65}$$

These simple relationships are valid only for sound waves incident either parallel or perpendicular to the director of the boundary layer, n_0.

Reflection at oblique incidence is accompanied by a rotation of the plane of polarization due to different damping of the sound waves in the major vibration directions. It should also be considered that a directional change of n_0 plays an important part in the sound-wave impact zone. Here the uniform arrangement of the absorption zone is gradually lost, so three cases can be differentiated:

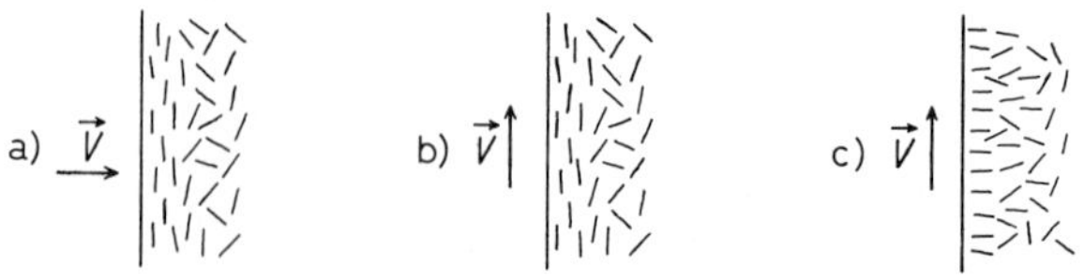

In case a there is no coupling between the director and the propagation direction:

$$\eta_a = \frac{\alpha_4}{2}
\tag{3.66}$$

The sound wave in cases b and c induces a local change in n. This means that the effective viscosity, η, in each case contains an additional viscosity-dependent parameter:

case b): $\eta_b = \dfrac{\alpha_3+\alpha_4+\alpha_6}{2} - \dfrac{\alpha_3}{2}\left(1+\dfrac{\gamma_2}{\gamma_1}\right)$ (3.67)

case c): $\eta_c = \dfrac{\alpha_4+\alpha_5-\alpha_2}{2} + \dfrac{\alpha_2}{2}\left(1-\dfrac{\gamma_2}{\gamma_1}\right)$ (3.68)

Candau and Martinoty have shown that the most precise results are obtained for oblique incidence, and the same effective viscosity is found in all three geometries (a, b, and c) for both oblique and perpendicular incidence. For this experiment, eq. 3.69 states that

$$ r_\theta = \frac{Z_s \cdot \cos\theta - Z_{11}}{Z_s \cdot \cos\theta + Z_{11}} \tag{3.69}$$

where θ is the angle of incidence and Z is defined according to eq. 3.64. Very interesting results were also obtained from capillaries.

F. Brochard performed analogous investigations on cholesterics with the result that the penetration length can be very much smaller or larger than the pitch, depending on the ratio between the pitch and the wavelength [73]. These observations are strongly reminiscent of optical reflection and polarization phenomena. Thus, two waves having an opposite sense of rotation are obtained in the case of a large penetration length compared with a relatively small pitch [711]. No definite value has been obtained on the anisotropy of sound velocity in nematics which seems to be very small and of the order of 0.1 % or less [453, 478, 573]. In the n/i transition region of MBBA, the sound velocity decreases sharply by about 9 % [573]. The attenuation which shows a distinct anisotropy and dispersion deserves the major interest [469, 823], fig. 3.20.

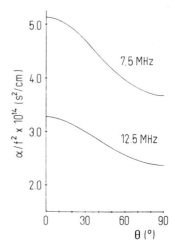

Fig. 3.20. Anisotropy of the attenuation coefficient at 7.5 and 12.5 MHz in MBBA at 25 °C (from [469]).

At temperatures far from a mesophase transition, the attenuation and velocity dispersion can be described by a single relaxation process [15, 202], but this is not valid close to a transition, as fig. 3.21 illustrates [524].

Relaxation mechanisms are discussed in the literature [6, 341]. Other ultrasonic studies have concentrated on the n/i [16a, 83, 392, 431, 525, 536a, 562, 573], n/s$_A$ [26, 28, 40, 83],

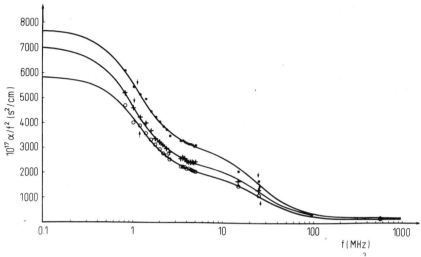

Fig. 3.21. Frequency dispersion of the attenuation coefficient in CBOOA (k 73 s_A 82.6 n 108 i) at 82.7 °C. The magnetic field and the direction of the sound propagation are parallel (●), perpendicular (○), or at 45° (+). The arrows indicate the relaxation frequencies (from [524]).

c/s_A [27, 627], and c/i pretransition phenomena [627, 710]. As already mentioned on p. 148 an acoustic field exerts an aligning effect on nematics [185a, 185b, 308, 314, 371, 379]. The results for MBBA in fig. 3.22 were obtained from numerous papers reporting viscosity data derived from ultrasonic studies.

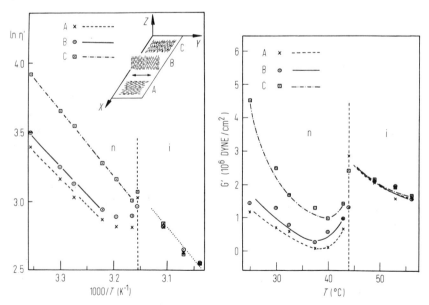

Fig. 3.22. The dynamic viscosities, η', and shear storage moduli, G', of MBBA at 10 MHz for three orientations; ↔ denotes the reflecting bar surface (from [78]).

    With sufficient incident acoustic energy, a scattering phenomenon is observed which resembles the dynamic scattering mode (see section 4.4.5.3) and possesses both threshold and gray-scale properties [47, 85, 308a, 382, 397, 511, 558a, 567, 744, 745]. Acoustic surface waves may lead to similar patterns [146, 377, 558, 584, 811]. A. P. Kapustin et al. were the first to observe the acoustohydrodynamic instability of a nematic liquid crystal [351, 352]. Plasma-acoustic cavitation phenomena (sonoluminescence) have also been studied [285a].

    There are a great many other papers on the acoustic properties of nematics. However, since these largely consider the problems already discussed, we limit ourselves to a reference list [24, 25, 34, 35, 37, 43, 44a, 143, 190, 203, 271, 307a, 326, 381, 387, 388, 396, 422, 430a, 454, 466–468, 481, 558a, 560, 561, 568–570, 581, 582, 585a, 626, 650a, 719, 720, 761, 762, 786]. Ultrasonic second harmonic generation was observed in MBBA [88, 89].

    A typical example of a cholesteric mesophase also showing pretransitional behavior is given in fig. 3.23 [429].

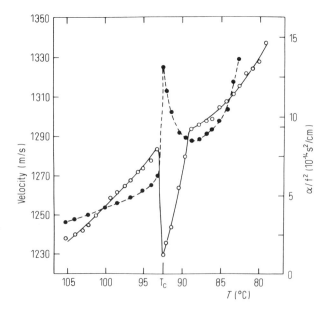

Fig. 3.23. Ultrasonic velocity (○) and attenuation coefficient (●) of cholesteryl nonanoate at 2 MHz as a function of temperature (from [429]).

    Concerning other ultrasonic studies of cholesterics we refer to the literature [23, 36, 73, 200, 201, 203, 248, 271, 326, 353, 367, 372, 376, 384–386, 428, 479, 556, 557, 565, 582a, 627, 632, 678, 864]. The ultrasonic attenuation in smectics is significantly more anisotropic than it is in nematics, and even the velocity has a measurable anisotropy of about 5% [480]. In addition to the normal longitudinal wave propagation, called the first sound, the socalled second sound which corresponds to an overdamped transverse wave has been considered [29, 75, 77, 343]. Details on s_A, s_B, s_C, and s_E phases can be drawn from the literature [30, 44, 47a, 354, 355, 363, 465, 522, 527b, 555–557, 712a]. The velocity and absorption of hypersound are accessible from Brillouin-scattering experiments in liquid crystals [115, 645], see sections 6.7 and 7.8. Yu. S. Alekhin describes the methods for studying acoustical properties of liquid crystals at high frequencies [13, 14].

3.11 Surface and Interface Properties

Aside from the normal decrease of surface tension, γ, with increasing temperature, T, anomalies have been observed in the $\gamma(T)$ characteristic where a positive slope or discontinuity occurs as indicated in fig. 3.24. As is also evident from fig. 3.24 (curve c), the anomalous behavior must not be related to a phase transition, and it even can take place in the isotropic liquid. C. A. Croxton and S. Chandrasekhar offer an explanation that the slope, $\partial\gamma/\partial T$, be determined by the competing influences of orientational order and spatial delocalization within the surface [102]. The continuity seen in the cases b and c of fig. 3.24 reveals that the breakdown of the macroscopic bulk orientational order can take place without dramatic change in surface orientational order.

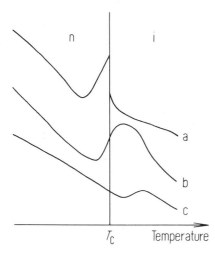

Fig. 3.24. Variation of surface tension with temperature shown schematically for *p*-anisalazine (a, $T_c = 182.2\,°C$), PAA (b, $T_c = 135.2\,°C$), and PAP (c, $T_c = 166.6\,°C$) (from [434]).

Numerous measurements have confirmed the behavior delineated above [109, 110, 239, 433–435, 580, 642, 742, 788, 822]. F. M. Jäger's contribution is of historical interest [337] and S. K. Ray has used these data in search of a correlation between parachor and constitution of nematogens [718]. Measurements have also been performed with cholesterics [122, 577–579, 822], but only little data exists on the surface tension of smectics [435, 742, 822], an example is shown in fig. 3.25. The subject is also summarized in reviews [559, 790a]. Thin nematic droplets subject to thermal surface tension gradients show a streaming analogous to the Marangoni effect in isotropic fluids [725a].

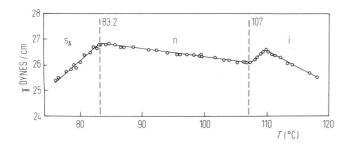

Fig. 3.25. Surface tension, γ, of CBOOA vs. temperature, T (from [435]).

M. Papoular et al. [628, 629, 631] and D. Langevin et al. [442–445, 447, 448] have shown that irregularities of about 1 nm exist on the surface of nematics such as PAA or MBBA. These irregularities are of thermal origin, and their influence on the scattering and polarization of electromagnetic waves (138–509 cm^{-1} band) permits the calculation of both the surface tension and the viscosity coefficients. This technique has also been applied to PAP [503], smectics A [504, 635, 712], and cholesterics [635, 637, 712]. A special wave described as a propagating disclination surface might be able to propagate in nematics without damping [140, 348].

There are close correlations between surface tension and the mechanics of quasi-two-dimensional nematic systems [72, 90, 91, 161, 420, 787]. This area is linked to biochemical problems; see section 12.7. H. Sackmann et al. have studied the compressibility of azoxy-α-methyl-cinnamate surface films on aqueous substrates [187]. Other studies on thin film structures found a strong dependence of molecular alignment on the thickness [245a, 695]. The interesting question of molecular orientation on the free surface has been considered for nematics [636] and cholesterics [713]. Molecular theories of surface tension in nematics are discussed in the literature [564, 639]. C. Robinson followed this line of study so early as 1958 in connection with biochemical problems [724]. The literature reports other aspects of related interset [41, 412, 496].

Two independent determinations of the interfacial tension between nematic and isotropic liquid of MBBA agree within the accuracy of measurement. A value of $2.3 \cdot 10^{-2}$ erg/cm^2 from light scattering [446] and $1.6 \cdot 10^{-2}$ erg/cm^2 from the sessile drop method [831] have been obtained. An earlier estimation seems to be too low [349]. J. T. Jenkins and P. J. Barratt have attempted to calculate the anisotropic n/i interface energy [347], see also p. 140. Problems associated with the solid/mesomorphous interface are considered in chapter 1 and 14, see also [225, 642, 780].

Behavior in Magnetic and Electric Fields

4

4   Behavior in Magnetic and Electric Fields

From the very beginnings of research on mesomorphism, the influence of magnetic and electric fields on this state of matter has been of great importance. Merely the goals have changed with time. The discussion about the emulsion theory stimulated much work in the pioneering days e.g., the unsuccessful attempts to separate the purportedly multiphase system by electrophoresis [1:138, 1:139, 1:215, 1:270, 2:69, 2:715, 2:763, 2:787, 619, 777, 927]. The various principles by which the influence of fields were interpreted were of exceptional theoretical importance; these include the orientation effects of fields in the sense of the Ornstein-Kast swarm theory, or, as Zocher demonstrated a bit later, in a continuous model adapted from Oseen's hypothesis. Today's accent on technological applications has necessitated a refinement of these theories, but the older models are still directly applicable. The main interest now is in the electro-optical effects that open new possibilities for display and storage of visual information, e.g., alphanumeric displays (cf. chapter 14).

A sufficiently strong magnetic field (> 3000 Gauss) is undoubtedly the best instrument to impose the character of a single crystal on a liquid-crystalline phase. The molecules in such a field are oriented such that only thermal vibrations occur, and they are directed along one principal axis as determined by the direction of the field. Similar changes are caused by electric fields, but the conditions are much more complex due to such factors as uncontrollable ion transfer, convection currents, and field inhomogeneities. The early research, on which later developments are based was done by T. Svedberg [862–866], Y. Björnstahl [88–90],

H. Zocher [962–964], W. Kast [498–509], and V. Fréedericksz [291–305]*. Field effects are covered in several reviews and reports [94, 171, 435, 439, 601, 651, 731, 775, 14:741].

4.1   Nematic Phases in Magnetic Fields

If a nematic phase, preferably several tenths of a millimeter thick, is observed under a microscope, then it will be noticed that the typical Schlieren texture with its threads and disclination centers disappear after the application of a magnetic field [910, 914, 920]. The preparation also becomes transparent in the macroscopic sense, even if it is not isotropic with regard to the polarization and intensity distribution of light scattering. T. Svedberg described this phenomenon as follows [864]: "The anisotropy (in the sense of alignment) caused by a magnetic field increases, slowly at first, with field strength. This is followed by a period of rapidly increasing alignment, and, after a time of smaller increase, the anisotropy approaches a nearly constant saturation value" (fig. 4.1). It is of great interest to note that while A. Cotton and H. Mouton obtained quite similar curves during their optical examination of certain iron oxide hydrates, these same investigators found curves of the form shown in fig. 4.1b for molecular-disperse systems. These results indicate that the degree of dispersion in liquid crystals is quite low.

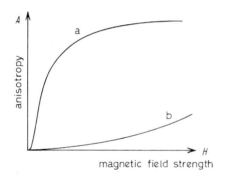

Fig. 4.1. Behavior in a magnetic field (anisotropy, A, vs. field strength, H) (from [864]). a) nematic liquid crystal, b) isotropic liquid.

One may expect that the complex orientation effects will manifest themselves on boundary surfaces under the influence of the magnetic field in competition with surface forces. For the free surface of a nematic liquid crystal, P. G. de Gennes predicted that domains would form when the field exceeded a threshold strength of 3000 Gauss [198]. The following example shows the extent to which these interactions may depend on the nature of the material being observed. Nematic droplets of MBBA in glycerol observed under the polarizing microscope appear much different than those of the analogous azoxy compound in poly-trichlorofluoroethylene, since the former molecules arrange themselves perpendicular to the droplet surface under the influence of the magnetic field, while the latter molecules do so tangentially [120, 575]. The arrangement of the molecules near the boundary is evidently determined by the polar interactions between the nematic phase and the neighboring medium. Calculations have been carried out [741]. This is generally supported by measurements made on the reflection of polarized light from a free MBBA surface in a magnetic field that runs parallel to the surface [105], see also [513] and section 3.11. While the PAA molecules remain lying in the plane,

---

* We use the simple transcription "Frederiks".

the MBBA molecules are diagonal at an angle of about 75° and remain so at the surface, even under the influence of the field, cf. [664].

4.1.1 Magnetic Susceptibility

The anisotropy of the molecular diamagnetic susceptibility will now be considered as the driving force for the orientation effect of an external magnetic field. J. P. Dias gives a theoretical treatment [246a]. In this discussion, we will follow this standard magnetic alignment theory first proposed by G. Foëx and L. Royer and which went unchallenged for a long time except by J. O. Kessler [524]. The first systematic investigation of the diamagnetism of nematics began with G. Foëx [286, 289] and, a short time later, W. Kast [505]. A review of the measurement apparatus [659] and the description of a NMR method [776] are found in the literature. From the newer work we use as examples the magnetic properties of PAA [219, 314] and MBBA [51, 219, 314, 776] shown in figs. 4.2 and 4.3. Both compounds have also been studied in the glassy nematic state [738].

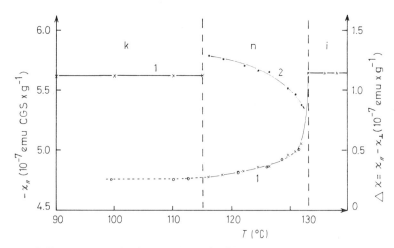

Fig. 4.2. Magnetic susceptibility (curve 1) and anisotropy (curve 2) of PAA vs. temperature, *T*, measured with increasing (×) and decreasing temperature (o) (from [314]).

MBBA samples of different degrees of purity show little difference in magnetic anisotropy [835].

The following equation relates the degree of order, *S*, and the magnetic susceptibility, χ [314].

$$S = \frac{\chi_{\parallel} - \chi_{\perp}}{\chi_a - \chi_b} = \frac{\Delta\chi}{\chi_1 - \frac{1}{2}(\chi_3 + \chi_2)}$$

where $\chi_{\parallel}$ is the susceptibility parallel to the magnetic field, $\chi_{\perp}$ the corresponding value perpendicular to the field, χ_a the susceptibility parallel to the molecular axis, χ_b the corresponding value perpendicular to the axis; and χ_1, χ_2, and χ_3 the principal susceptibilities of the molecules as derived from single crystals. It could be advantageous to relate $\chi_{\parallel}$ and $\chi_{\perp}$ to the optical axis.

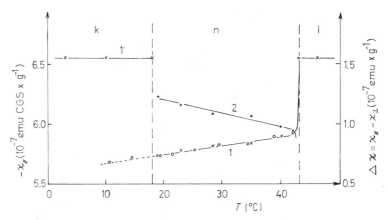

Fig. 4.3. Magnetic susceptibility (curve 1) and anisotropy (curve 2) of MBBA vs. temperature, T, measured with increasing ($\times$) and decreasing temperature ($\circ$) (from [314]), see also fig. 4.6.

For PAA, this method gives the value of $S = 0.53$ at $116\,°C$ that reduces to 0.37 at $131\,°C$ [3:509, 314], for other nematics, see [2:320, 3:303, 3:666, 737, 739]. It has already been implied in fig. 4.1 that as long as the magnetic field is too weak to cause complete orientation in the direction of the field, the measured magnetic susceptibilities depend on the field strength [613, 614, 736]. Complete orientation for PAA is approximated at 2550 Oe.

The diamagnetic anisotropy exhibits an alternating effect within homologous series at the n/i and s_A/n transition temperatures [373a]. It is always positive with aromatic mesogens but it was found to be negative for trans-trans-4'-alkyl-4-cyanobicyclohexyls [2:191a].

Magnetic effects in liquid crystals normally require relatively high field strengths (ca. 10^4 Oe) due to the small anisotropy of the diamagnetic susceptibility ($\Delta\chi = $ca. 10^{-7}). This led F. Brochard and P. G. de Gennes to the idea that suspending small ferromagnetic grains in nematics could form "ferronematics" or "ferrocholesterics" [200]. Experiments of J. Rault et al. [756] and C. F. Hayes [377] showed the practicality of this suggestion. When fine γ-Fe_2O_3 (0.35 by 0.04 µm particles) is suspended in MBBA and magnetized, there is observed a strong remnant magnetism that disappears when the nematic phase is heated above its clearing point.

Together with calorimetric measurements, those of magnetic susceptibility can serve for classification of phase transitions which, for instance, is very weakly first order in the s_A/n transition of CBOOA [321] and quasi-second order in certain mixtures [322]. An unusual decrease in χ with decreasing temperature was observed during the n/s_B transition of p-hexyloxy-benzylidene-p-toluidine [52a].

4.1.2 Elastic Deformation (Frederiks Deformation)

The elastic deformation of a nematic liquid crystal in a magnetic field was first recognized and described by V. Frederiks [294]. For concave-plane layers and a given magnetic field, he found a definite thickness which would limit the action of the field. This threshold thickness separates the thin layers not deformed by the field from the thicker layers that do show a deformation, one which increases with thickness. V. Frederiks also calculated the deformation profile (fig. 4.4), the consequently modified birefringence [294], and, in a later paper, two elastic constants of PAA [296, 300]. The elastic constants of MBBA were also determined in this way [315]; a theoretical interpretation of Frederiks' experiment is given in the literature [760].

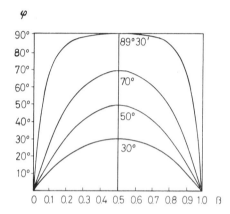

Fig. 4.4. Change in orientation, φ, of a homeotropic layer with the thickness of the layer, β, for various angles φ_0 between the magnetic field and the preferred axis (from [294]).

When applied to a nematic layer (of constant thickness, d, and homogeneous planar texture) and a magnetic field normal to it, these observations signify that the deformation first appears above a threshold field, H_c. This is related to the anisotropy of magnetic susceptibility, $\Delta\chi$, by [246, 770]

$$H_c = \frac{\pi}{d}\sqrt{\frac{k_{11}}{\Delta\chi}}$$

For a homeotropic layer and a magnetic field parallel to the nematic director, the threshold field H'_c is

$$H'_c = \frac{\pi}{d}\sqrt{\frac{k_{33}}{\Delta\chi}}$$

Measurements of the threshold fields allow determinations of the elastic constants of splay (k_{11}) and bend (k_{33}) [3:296, 3:304, 3:382a, 3:541, 3:731, 246], provided anchoring conditions are really parallel and perpendicular [3:180, 219]. The orientational kinetics of MBBA in a magnetic field was pursued by a conoscopic [109, 719], and an electrochemical technique [407] by means of light-transparency [507] or electric conductivity measurements [411, 412, 415] (this work also considers metastable elastic deformations). An important prerequisite for undisturbed observation is a uniform orientation of the liquid crystal on the boundary surface, as stressed by V. Frederiks [293]. Nevertheless, during the deformation there often appear domains that deviate at the same angles but in various directions from the magnetic field and are separated from each other by disclination walls. L. Leger [573] and E. F. Carr

[150] investigated the static and dynamic behavior of these domains. As expected, the arrangement in a magnetic field is not connected with any change in volume of the anisotropic phase [73]. Further papers deal with the magnetic Frederiks deformation [71, 110, 112, 282a, 360, 536, 571, 574, 604, 652, 750, 751, 786].

As Frederiks himself discovered, an electric field of suitable frequency and strength can create a molecular arrangement that is not optically distinguishable from that which is created by a magnetic field. Since this variant is more important today, it will be considered further in section 4.4.3.

Other recent studies consider the elastic deformation of Mauguin-texture nematics (cf. section 4.4.4) and the action of oblique magnetic fields [240].

A comparison of the capacitance and the optical transmission between parallel polarizers in a twisted MBBA layer under the influence of a magnetic field shows that the "optical threshold" occurs at a higher value of the field strength than the "capacitive threshold" [325, 911], see fig. 4.5. The critical field strength for the untwisting, $H_{c,\text{ntw}}$, is related to the thickness, d, by

$$(\chi_\parallel - \chi_\perp) H_{c,\text{ntw}}^2 d^2 = k_{11}\pi^2 + (k_{33} - 2k_{22})(2\Phi_0)^2$$

where k_{11}, k_{22}, and k_{33} are the elastic modules and Φ_0 is the twist angle [325, 492a, 831]. For MBBA, $H_{c,\text{ntw}}d$ is 6.8 to 7.4 Oe cm [831].

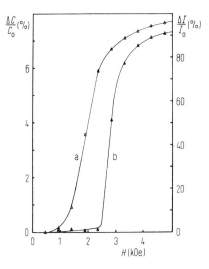

Fig. 4.5. Relative changes in the capacitance, $\Delta C/C_0$, (plot a) and in the intensity of transmitted polarized light, $\Delta I/I_0$, (plot b) vs. the magnetic field strength, H, in a 54 µm MBBA layer (from [325]).

A method of observing the Frederiks deformation in a direction inclined at a large angle to the twist axis allows a determination of the twist elastic constant, k_{22} [158, 219]. The equilibrium state between the magnetic torque and the anchoring energy of twist on the surface allows a determination of this interphase energy [834]. Distorted nematic layers can be composed of oppositely twisted regions separated by disclination lines, the motions of which were studied in a magnetic field by C. J. Gerritsma et al. [327, 328, 332]. The relaxation process by which a field-aligned nematic state returns to the helicoidal structure was also studied [112, 496]. J. L. Ericksen examined the opposite effect, twisting created by applying uniform magnetic fields [264]. However, the technological interest is now more directed at the analogous untwisting of nematic layers in electric field.

Quite early it was questioned whether or not electric surface charges might be induced by the elastic deformation and the associated molecular orientation when a magnetic field is applied to a liquid crystal [499, 500, 502, 867, 868]. Previous observations had indicated that no nematic substance possessed a spontaneous magnetic or electric polarization. In contrast to ferromagnetics, the axis of the elementary domain is nonpolar and thus nonvariable with respect to a reversal of the field-strength vector. However, in a magnetic field, polar zones can be expected on the surfaces or on inversion walls [250, 391, 677]. The thin inversion walls between uniformly oriented domains loosely resemble the Bloch and Néel walls in ferro-magnetics; W. Helfrich has considered these alignment inversion walls theoretically [391].

4.1.3 Action of a Rotating Magnetic Field (Tsvetkov Effect)

The movement of an anisotropic liquid in a rotating magnetic field was used for measurements of the viscosity and diamagnetic anisotropy by V. N. Tsvetkov [888–890, 903]. MBBA, having a homogeneous texture, displays a periodic texture change in a uniformly rotating magnetic field [742]. The static twisting of a nematic liquid crystal of the type long known as the Mauguin texture cannot be greater than $\pi/2$ in the case of the classical experimental condition. A greater twist can be created with the aid of a rotating magnetic field, whereby a series of easily observed boundaries appears. These are twisted 180° and have a migration speed in agreement with the theoretical expectations [283, 572]. In a rotating magnetic field, it should be possible to twist a nematic slab with one free surface to such a degree that "cholesteric" reflection colors appear [203].

H. Gasparoux and J. Prost suggested a new interpretation for the behavior of nematics in a rotating magnetic field. Their values for the diamagnetic anisotropy of PAA vary by about 20% from those of V. N. Tsvetkov [316]. Their data for MBBA are shown in figs. 4.6 and 4.7. Mainly theoretical aspects are considered in some other papers [113, 114, 320, 429b, 553, 684, 730].

The twist viscosity coefficients and their temperature dependence have been measured for several nematics [284]. G. Heppke and F. Schneider used a modified method for the determination of the twist viscosity in a rotating magnetic field [406, 410]. Defined deformation profiles can also be generated by torsional shear flow in electric or magnetic fields as has been demonstrated by photographs of the interference patterns [281], see section 3.8.

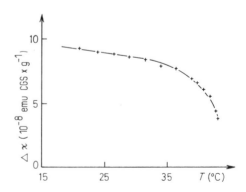

Fig. 4.6. Diamagnetic anisotropy, $\Delta\chi$, of MBBA (from [316]), see also fig. 4.3.

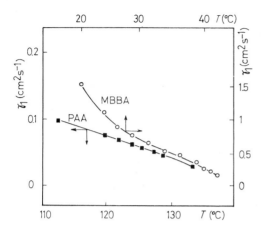

Fig. 4.7. Twist viscosity coefficients, γ_1, of MBBA and PAA (from [316], see also [3:691]).

4.1.4 Thermal Effects of a Magnetic Field (Moll-Ornstein Effect)

W. J. H. Moll and L. S. Ornstein performed the first thermal measurements on liquid crystals in magnetic fields and concluded that this "effect probably has its origin in the elastic change of form of liquid-crystalline particles" [688, 689], fig. 4.8.

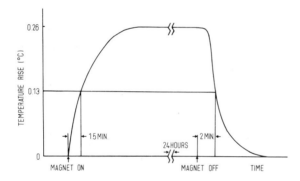

Fig. 4.8. A typical time course of the temperature rise in *p-n*-decyloxy-benzoic acid at 126°C and 3940 Gauss (from [959]).

M. Miesowicz and M. Jezewski described the magnetothermal effect as follows [455]: "Turning on a 2300 Gauss magnetic field whose lines of force ran perpendicular to the plate causes a relatively slow rise in temperature, which for PAA at 121.2°C amounted to 1.05°C. The temperature increase is reversible. After the immediate removal of the magnetic field, the temperature returned to its earlier value. However, if after applying the magnetic field, one waits for a thermal equilibrium and then removes the field, then a reduction in temperature is observed. The effect diminishes with increasing temperature and disappears after the transition to the isotropic phase. A magnetic field parallel to the plate had no effect."

During their study of magnetic deformation, P. Pieranski et al. measured the anisotropy of the thermal conductivity and obtained the following data on MBBA [111], see also section 3.9.

$$k_{\parallel} = (21 \pm 1.1) \cdot 10^{-4} \frac{J}{cm \cdot s \cdot deg} \qquad k_{\perp} = (12.8 \pm 1.1) \cdot 10^{-4} \frac{J}{cm \cdot s \cdot deg}$$

$$\frac{(k_{\parallel} - k_{\perp})}{k_{\perp}} = 0.64 \pm 0.04$$

This anisotropy of thermal conductivity could be the reason for the observation that the temperature rise is permanent rather than transient [583, 959], see fig. 4.8. Other studies are reported in the literature [716].

4.1.5 Optical Rotatory Power and Linear Electro-Optical Effect

If a nematic phase that has been oriented by a magnetic field perpendicular to the optical axis is observed through crossed polarizers, four extinctions will occur when they are rotated 360° in the same direction. These always appear when one of the polarizers is parallel to the magnetic field, and the entire preparation appears uniformly dark at these positions. Under precisely vertical illumination and in an orthoscopic light path, however, only the birefringence is detectable. In order to observe the simultaneous rotatory power in a magnetic field, a stereomicroscope can be used [937, 938]. The preparation is thusly observed at a somewhat oblique angle, e.g., at $\pm 7°$ out of vertical. The preparation is then rotated to an extinction point so that the entire field of view has a uniformly medium tone. Upon turning one of the crossed polarizers a certain amount ($\approx 10°$), some facets of the sample become light and others dark. Turning through the same angle in the opposite direction makes the formerly light areas dark, and vice versa. Thus one can speak of *levo-* and *dextro*-rotary domains. If one observes at an angle of $\pm 7°$ deviation, then the rotation changes sign, but not absolute value. Many other factors influence the observed optical rotation besides the direction of the magnetic field and that of observation. The preparation comprises nearly equal proportions of *levo-* and *dextro*-rotary domains having sharp boundaries that change very little during observation times of ten minutes or more. Such changes that do occur are generally those where the smaller domains dissolve into the larger. In a transition to either the crystalline or the isotropic phase followed by a reformation of the liquid crystal, the *levo-* and *dextro*-rotary domains recur with new boundaries, but always in nearly equal proportions. If nematic droplets, regardless of size, are allowed to form out of an isotropic melt under the influence of a magnetic field, then each will contain *levo-* and *dextro*-rotary domains of nearly equal size. The rotation increases linearly with the thickness of the preparation, and it decreases approximately 10% within the entire nematic range of PAA from 117 to 135°C. The dispersion of the rotation shows a normal increase with decreasing wavelength. The rotation is independent of field strength from 500 to 2400 Gauss, and the aforementioned phenomena do not appear with only the common cloudiness observed in fields below 500 Gauss.

Levo- and *dextro*-rotary domains created by a magnetic field appear equally bright between crossed polarizers observed at an angle of $\pm 7°$ inclined towards the perpendicular of the magnetic field. As soon as an electric field is applied at right angles to the magnetic field, all domains of one rotational sense become lighter, and the others become darker. Reversing the electric field makes the previously lighter domains darker and vice versa. This electro-optical effect is linear to a certain electric field strength where the *levo-* and *dextro*-rotary domains both lighten, and a poor reproducibility of intensity measurements signals the unstable conditions that foreshadow a complete molecular reorientation. The time of response for this electro-optical effect is about 50 ms. The effect does not appear in parallel magnetic and electric fields, but still another phenomenon is apparent. As described above, PAA under the influence of a magnetic field contains nearly equal proportions of *levo-* and *dextro-*rotary domains that may be recognized as light and dark facets. A sufficiently strong electric field parallel to the magnetic field distinctly displaces the equilibrium, favoring one sense of rotation. Reversing the electric field favors the opposite sense. This probably general effect infers different free energies of *levo-* and *dextro-*rotary domains in an electric field due to a polar contribution to the structure.

When in excess of 500 Gauss, the strength of the magnetic field and its direction do not influence either the optical rotation of the electro-optical effect, but merely act to orient the molecules. The basic element of symmetry for both of these effects is a vertical mirror-image plane in the direction of the magnetic field. Incident light parallel to this plane of symmetry is not affected by either effect. For incident light at other angles, however, they differ only by the sign and not by absolute value when a light path inclined to this plane is substituted by its mirror-image. In fig. 4.9 it is attempted to correlate the observations to the molecular structure of PAA which has a plane of mirror-image symmetry, but no other element of symmetry if a planar structure is assumed (C_s), cf. [3:401] and p. 45.

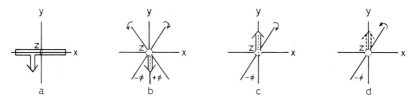

Fig. 4.9. a) The PAA molecule is in the plane of the paper, the plane of symmetry in the x,y-plane. b) Molecule A turned 90° on the y-axis, the plane of symmetry is in the y,z-plane. c) Molecule B turned 180° on the z-axis, the plane of symmetry is in the y,z-plane. d) Molecule C turned 180° on the y-axis, the plane of symmetry is in the y,z-plane.

Thus PAA is among those molecules that cause a rotation of linearly polarized light, provided that the light falls neither parallel nor perpendicular to the molecular plane of symmetry. The same absolute value of rotation is therefore found for the angles of incidence $+\Phi$ and $-\Phi$, but with opposite sign (see fig. 4.9). Since the optical rotation is not influenced by an opposite observation direction, a 180° rotation of the molecule on its longitudinal axis is not detectable. Identical rotational values are hence in both the b and c cases for any given angle $-\Phi$. However, a 180° rotation of the molecule on its short axis has quite a different result: a light ray striking such a molecule at angle $-\Phi$ (fig. 4.9d) is analogous to one striking the molecule of fig. 4.9b) at an angle of $+\Phi$. Such a rotation reverses the optical rotation of the molecule. In an isotropic phase this would not be detectable because the statistical distribution of molecular axes cancel to yield a total rotation of zero. The molecules of PAA arrange themselves in a magnetic field with their longitudinal axes parallel to H; in fig. 4.9b), c), and d), H coincides with the z-axis perpendicular to the plane of the paper. If the two optically equivalent molecular orientations noted in fig. 4.9b) and c) are rotated 180° on the y-axis (fig. 4.9d shows one of them), then the opposite optical positions are obtained. The optical activity of larger domains indicates a polar arrangement parallel or antiparallel to the magnetic field. Experiments confirm that both orientations are energetically equivalent from the standpoint of the magnetic field. Polarity arises only from the dipole orientation that occurs randomly parallel or antiparallel, but in such a way that it is uniform within larger domains.

If an electric field is now applied perpendicularly to the magnetic field, then the molecules are deflected from their parallel or antiparallel orientation to the magnetic field thus altering the optical activity and birefringence. A reversal of the electric field also reverses the displacement and the sign of the electro-optical effect, while a reversal of the magnetic field does not change the situation.

On the other hand, an electric field applied parallel to the magnetic partially and reversibly inverts the polarity, wherein the molecules turn 180° around the axis perpendicular to the longitudinal axis of the molecules [938].

Quantitative measurements have been performed in the nematic and isotropic phase of MBBA [811, 812, see also 458], fig. 4.10.

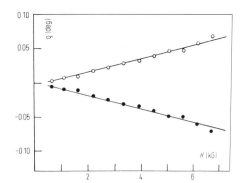

Fig. 4.10. Optical rotation of MBBA, ρ, (at 22°C and 546 nm) vs. the magnetic field strength, H; ○ first and ● reversed field direction (from [811]).

4.1.6 Other Investigations in Magnetic Fields

The rotation of the microwave polarization plane in PAA at 129°C can be considered analogous to the Faraday effect [608, 609].

J. L. Ericksen [263] and F. M. Leslie [577] have considered a nematic phase subject to a shear flow in a static magnetic field (see section 3.8).

Other magnetic studies on liquid crystals are reported in the literature [119, 909a].

4.1.7 Cotton-Mouton Effect

The birefringence of PAA measured in a magnetic field above the clearing point is positive and much stronger than in normal liquids [879, 891, 960]. The temperature coefficient is large, and it becomes even larger as the clearing point is approached. For MBBA, the following relation has been experimentally confirmed [852]:

$$\frac{\Delta n}{H^2} = \frac{(\varepsilon_{\|}-\varepsilon_{\perp})(\chi_{\|}-\chi_{\perp})}{6\left[\frac{1}{3}(2\varepsilon_{\perp}+\varepsilon_{\|})\right]^{1/2} a(T-T_c^*)}$$

Both the magnetic and electric birefringence of p-n-hexyl-p'-cyanobiphenyl show the same dependence on $(T-T_c^*)^{-1}$ as shown in fig. 4.11 (see p. 206).

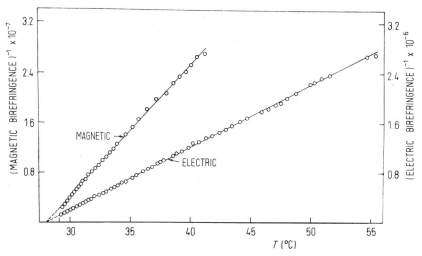

Fig. 4.11. The reciprocal of the magnetic and electric birefringence of *p-n*-hexyl-*p'*-cyanobiphenyl plotted vs. temperature both give the same $T_c^* = 28\,°\text{C}$; $T_c - T_c^* = 1.1\,°\text{C}$ (from [754]).

Measurements of magnetic and electric birefringence offer an excellent tool for studying pretransitional phenomena and checking de Gennes' theory [157, 159, 274, 277, 280, see also 6:248, 8:364]. For information concerning the light absorption of nematics in magnetic fields, we refer to the literature [685–687, 693, 694, 887] and chapter 6.

4.2   Cholesteric Phases in Magnetic Fields

All meaningful contributions in this area are relatively recent because, without the support of de Vries' theory (1951), earlier observations had remained inexplicable [578, 918, 919]. The theoretical aspects that evolve from the representations in chapter 3 cannot be detailed here and the reader is referred to the literature [440, 587]. Cholesteric droplets (having a pitch much smaller than the diameter) display a spiral optical pattern with or without a radial disclination. This pattern becomes oval in a strong magnetic field [120].

4.2.1 Helical Unwinding

P. G. de Gennes' prediction on the untwisting of a cholesteric helix has found remarkable proof [3:158, 196, 318, 636, 697], figs. 4.12 and 4.13. He derived the equation

$$H_{cn}\,p_0 = \pi^2 \left(\frac{k_{22}}{\Delta\chi}\right)^{1/2}$$

Above a critical field strength ($H_{cn} = $ ca. $15\cdot10^3$ Oe) the structure is purely nematic, and the pitch, p, diverges at H_{cn}.

Measurements of the diamagnetic anisotropy, $\Delta\chi$, the twist elastic constant, k_{22} [3:124, 3:693, 3:824, 78, 317, 821] and theoretical considerations are to be mentioned here

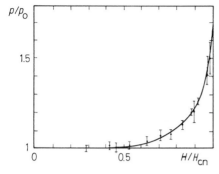

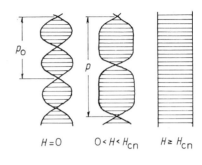

Fig. 4.12. Dependence of the reduced helical pitch, p/p_0, on the reduced magnetic field strength, H/H_{cn}, in PAA/cholesteryl acetate; theoretical curve after P. G. de Gennes [196]. On the right, the helical unwinding and the increase in periodicity are shown schematically (from [636]).

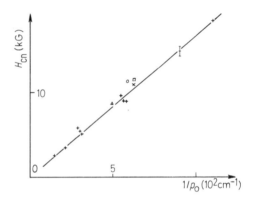

Fig. 4.13. Linear variation of H_{cn} vs. $1/p_0$ for various PAA cholesteric mixtures (+ cholesteryl chloride, ○ acetate, □ decanoate, × nonanoate, △ palmitate). The low-concentration regime is close to the origin (from [253]).

[576, 635]. If the cholesteric helix restores itself in a diminishing magnetic field, then, below H_{cn}, a buckling is to be expected at the nucleation point [197].

Fig. 4.14 contrast the optical behavior of a cholesteric substance with and without a magnetic field. The four polarizations are $\binom{1}{0}$ and $\binom{0}{1}$ linearly polarized in the x and y directions, respectively, and $\binom{1}{i}$ and $\binom{1}{-i}$ right and left circularly polarized.

Besides the shift of the transmission minimum from $\lambda/p_0 = 1$ to $\lambda/p_0 \cong 1.1$ for right-circularly polarized light, both the complete opacity of right-circularly and the complete transparency of left-circularly polarized light fade away in a magnetic field. A region of partial reflection centered at half the wavelength of the primary reflection peak occurs in the significantly disturbed helical structure for linearly and circularly (both left and right) polarized light [638]. R. Dreher has discussed conditions of the possible formation of higher order reflection bands without a displacement of the main reflection band [248].

4.2.2 Periodic Deformations

Still other phenomena of cholesterics have been observed in magnetic fields below the critical intensity, H_{cn}. There has been a unanimously verified report of a periodic deformation

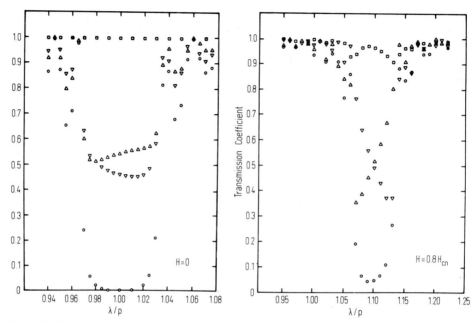

Fig. 4.14. Transmission coefficients about the reflection region for four polarizations of incident light, E_{inc}: △, $\binom{1}{0}$; ▽, $\binom{0}{1}$; ○, $\binom{1}{i}$; □, $\binom{1}{-i}$, when $H=0$ and $H=0.8\,H_{cn}$ (from [638]).

in the area $H_H < H < H_B$, and it has been variously described as a grid-like deformation [809] or a lattice of dislocation lines analogous to that observed in the Grandjean wedge geometry [424, 758]. J. Rault summarizes the shapes and causes of periodic distortions in cholesterics [759]. J. Marignan et al. explain the dynamics by means of static and hydrodynamic theories [607], and J. M. Delrieu discusses critical field strengths and the influence of a tilted magnetic field [221].

Slightly higher field strengths, $H_B < H < H_{cn}$, bring about the "fingerprint texture" [770]. Above the critical field strength H_B, the helical axes turn 90° perpendicular to the magnetic field starting from a parallel orientation of field and helical axis (unlike the situation in fig. 4.12). Fig. 4.15 illustrates such complex relationships in the example of a 250 μm thick MBBA-cholesteryl nonanoate preparation at 22°C and having a pitch of 16.5 μm.

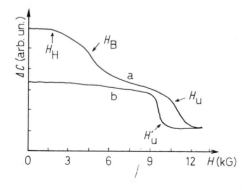

Fig. 4.15. Variation of the capacitance, ΔC, with increasing (curve a) and decreasing fields, H (curve b). H_H, H_B, and H_u denote respectively the critical fields for the onset of Helfrich's periodic deformations, the appearance of the "fingerprint texture", and the unwinding of the helix. H'_u corresponds to the reappearance of the "fingerprint texture". The sample is a mixture of MBBA and cholesteryl nonanoate with the pitch $p_0 = 16.5$ μm at 22°C; thickness of the layer: 250 μm (from [770]).

Supplementary to chapter 1, we insert the observations of J. Rault and P. E. Cladis on fingerprint texture [757]. Near the clearing point and under the influence of only very weak anchoring forces of the glass surface, the helical axis lies parallel to the plates and also parallel to the c/i phase boundary. With crossed polarizers, a nearly parallel or symbathic striping is observed for which the term "fingerprint texture" is quite appropriate. Similar patterns appear in magnetic fields, and a variant "herringbone texture" has also been described.

In a Cano wedge geometry, there appears a series of "virgules" that grow as the temperature approaches T_c. A cholesteric phase placed in rapidly increasing thickness wedges exhibits a new type of disclination line with torsion jumps twice as large as usual [3:594, 695]. In region II of fig. 4.16, thicker lines are observed that buckle in a zigzag when a critical field strength of $H_{zz} \sim 0.52 H_{cn}$ is attained, while the pattern of normal disclination lines (in region I) does not yet change. The motion of the first disclination lines in a magnetic field is characterized by a long relaxation time [602, 603].

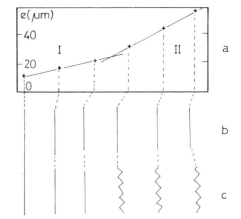

Fig. 4.16. Partial pattern of disclination lines viewed under a microscope in conical geometry: (a) plot of cell thickness, e, vs. line number with the slope II is twice I; (b) $H < H_{zz}$; (c) $H > H_{zz}$. The sample is a 3% solution of cholesteryl chloride in PAA with $p_0 = 11 \, \mu m$ (from [695]).

4.2.3 Elastic Deformation

The discussion up to now has assumed that the cholesteric layer is thicker than the pitch of the helix. The reverse has been investigated in a mixture of MBBA and cholesteryl nonanoate (fig. 4.17).

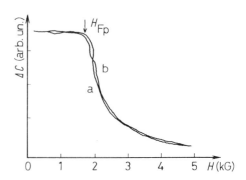

Fig. 4.17. Variation of the capacitance, ΔC, with increasing (curve a) and decreasing fields, H, (curve b) in a 50 μm cholesteric layer with $p_0 = 120 \, \mu m$ (from [770]).

As in a nematic layer, a deformation analogous to a Frederiks transition occurs above a threshold field H_{Fp}. At this point we compile the equations that have been proposed to describe the critical fields of the different deformations.

Frederiks deformation in a nematic layer:

$$H_c = \sqrt{\frac{k_{ii}}{\Delta \chi}} \frac{\pi}{d} \quad \text{homogeneous texture: } k_{ii} = k_{11}; \quad \text{homeotropic texture } k_{ii} = k_{33}.$$

Elastic deformation in a twisted nematic layer:

$$H_{cntw} = \sqrt{\frac{k_{11} \pi^2 + (k_{33} - 2 k_{22})(2 \Phi_0)^2}{\Delta \chi}} \frac{1}{d}$$

Elastic deformation in a cholesteric layer:

$$H_{Fp} = \sqrt{\frac{k_{11}}{\Delta \chi} \left(1 + \frac{k_{33}}{k_{11}} \cdot \frac{4 d^2}{p^2} \right)} \frac{\pi}{d}$$

Untwisting of a cholesteric layer:

$$H_{cn} = \sqrt{\frac{k_{22}}{\Delta \chi}} \frac{\pi^2}{p_0}$$

Helfrich deformation in a cholesteric layer [436]:

$$H_H^2 = \sqrt{\frac{6 k_{22} k_{33}}{p_0 d}} \frac{2 \pi^2}{\Delta \chi}$$

Each of these critical values indicates the beginning or the saturation of a certain deformation of a liquid crystal in a magnetic field, and is the result of an internal rigidity of the molecular arrangement. Therefore, the three elastic constants discussed in chapter 3 enter the equations as determining values. These critical values do not rigorously delineate since there is a certain oscillation around the average position of the molecules. In addition, a threshold need not be at the same field-strength for increasing and decreasing fields when dynamic effects are significant, because the three elastic modules are defined for the rest state.

ESR spectroscopy permits investigations of the structure of the helix itself and its change in a magnetic field [10:702]. The pitch of a cholesteric substance can be easily determined by using the diffraction patterns of a He-Ne laser (632.8 nm) with or without a magnetic field [780], cf. chapter 7.

4.2.4 Storage Effect

The fingerprint and the focal-conic textures are restored below H_{cn}, and they may persist if $0 \leq H < H_B$; they transform within one minute to several days into the Grandjean texture [434]. Both of the former textures seem to increase in stability with increasing ratio of sample thickness to pitch. This storage effect as well as the other changes in texture can also be observed in an electric field. The modes underlying light scattering of cholesterics in a static magnetic field have been discussed [706], cf. chapter 7.

4.3 Smectic Phases in Magnetic Fields

Only a few aspects of the problem have been examined since the systematic investigations by G. Foëx [287, 288, 290], leaving a void in our knowledge. According to G. Foëx, the diamagnetism of smectic A ethyl *p*-azoxybenzoate is quite similar to that of nematics.

All transitions in *p*-octyloxybenzylidene-*p*-toluidine [52] and in *p*-butoxybenzylidene-*p'*-octylaniline [321], fig. 4.18, are associated with a discontinuity in the $\chi_\parallel$ and $\Delta\chi$ curve.

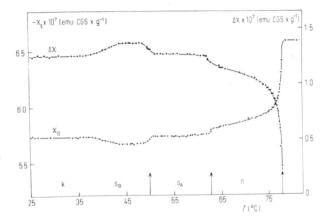

Fig. 4.18. Magnetic susceptibility, $\chi_\parallel$, and anisotropy, $\Delta\chi$, of *p*-butyl-oxybenzylidene-*p'*-octylaniline (from [321]).

No measurable change in magnetic properties was observed in the quasi-second order s_A/n transition of CBOOA having a transition enthalpy of 0.25 J/g [319, 321, 373]. This was also true of the s_A/s_C and s_4/s_B transitions (s_4 is probably s_H, i.e. tilted s_B) observed in *p*-heptyloxybenzylidene-*p'*-pentylaniline [321, fig. 4.19].

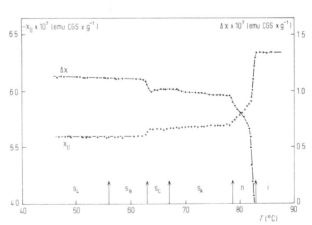

Fig. 4.19. Magnetic susceptibility, $\chi_\parallel$, and anisotropy, $\Delta\chi$, of *p*-heptyl-oxybenzylidene-*p'*-pentylaniline (from [321]).

Still other smectics have been studied [319, 321, 373, 373a]. The n/s_A transition has been observed in a magnetic field under a polarized laser beam [53]. A. Rapini has calculated the magnetic threshold field for the Frederiks deformation in a dozen possible s_C geometries [752]. In addition, magnetic-field effects have been considered in smectics A [258, 346, 705], at the free surface of smectics A [568–570], in smectics C [141, 285, 429, 730, see also 8:215c], and in the largely unknown mesophase of coal tar pitch [425].

4.4   Nematic Phases in Electric Fields

An electric field, like a magnetic field, can cause an elastic deformation of a nematic layer with the dielectric susceptibility replacing the diamagnetic susceptibility. Thus, the preferred molecular arrangement is determined by the dielectric constants. In addition to these orientation effects, generally called Frederiks deformation or Frederiks transition, numerous phenomena emerge in the "conduction regime" that are governed by both the anisotropy of the dielectric constant and of the electric conductivity. The term "conduction regime" refers to a voltage and frequency range characterized by electrohydrodynamic phenomena, particularly by the so called Williams domains and the dynamic scattering mode.

The term "dielectric regime", on the other hand, refers to that region of an ac electric field above a certain "cutoff" frequency in which only elastic deformations and oscillations occur. The characteristic form is the chevron pattern of the "fast turn-off mode". The literature contains brief descriptions of this [174, 359, 400, 530, 631, 675, 732, 816, 941], a consideration of two dimensional nematics [795], and a comparison with the electro-optical characteristics of dipole suspensions [610].

4.4.1 Dielectric Susceptibility; Dielectric Constants

The dielectric anisotropy, dispersion, and relaxation deserve practical interest since technical applications of the electro-optical effects have been realized. It is still uncertain whether substances of positive or negative dielectric anisotropy are to be more important. Sufficient care was not taken in all the sources to assure that only measurements of unidirectionally oriented samples were obtained, but it is impossible here to analyse all the publications dealing with this subject. From the numerous contributions of L. S. Ornstein [3:586–3:589, 690–692], W. Kast [498, 501, 503, 504, 506, 508, 509], M. Jezewski [449–454], W. Maier [2:461, 592–600], A. Axmann [37–40], E. F. Carr [121–129, 131, 132, 139, 145], W. H. de Jeu [206–220], and other authors [3:799, 1, 69, 85, 106, 117, 259, 265, 305, 307, 526, 579, 832, 846], only selected papers can be discussed together with the recent results of several teams. For a sophisticated study we recommend the books of C. P. Smyth [841], C. J. F. Böttcher [104], V. I. Minkin et al. [646], W. E. Vaughan et al. [916], the treatment of O. Fuchs and K. L. Wolf [306], and the reviews of G. Meier [628] and W. H. de Jeu [220]. The optical axis holds the most convenient and reliable reference axis in the discussion of anisotropic phenomena. Its unidirectional orientation is usually effected by a magnetic field of sufficient strength such that the optical axis and the field vector coincide. The axis of preferred molecular orientation, the nematic director, is also used as a reference, but it is not known whether this axis can vary significantly from the longitudinal axis of simple molecules. This axis can be defined by connecting the centers of the benzene rings. These axes need not be considered identical, even if no difference can be discerned in case of simple molecules.

W. Maier and G. Meier derived the equations that correlate dielectric properties to molecular parameters by using the Onsager theory as [597, see also 252, 551, 552, 656].

$$\frac{\varepsilon_{is}-1}{4\pi} = N_L \cdot \frac{\rho_{is}}{M} \cdot h_{is} \cdot F_{is} \left\{ \bar{\alpha} + F \frac{\mu^2}{3kT} \right\}$$

$$\frac{\varepsilon_{\parallel}-1}{4\pi} = N_L \cdot \frac{\rho_n}{M} \cdot h_n \cdot F_n \left\{ \bar{\alpha} + \frac{2}{3}\Delta\alpha S + F \frac{\mu^2}{3kT}[1-(1-3\cos^2\beta)S] \right\}$$

$$\frac{\varepsilon_{\perp}-1}{4\pi} = N_L \cdot \frac{\rho_n}{M} \cdot h_n \cdot F_n \left\{ \bar{\alpha} - \frac{1}{3}\Delta\alpha S + F \frac{\mu^2}{3kT}\left[1 + \frac{1}{2}(1-3\cos^2\beta)S\right] \right\}$$

$$\frac{\Delta\varepsilon}{4\pi} = N_L \cdot \frac{\rho_n}{M} \cdot h_n \cdot F_n \left\{ \Delta\alpha - F \cdot \frac{\mu^2}{2kT}(1-3\cos^2\beta) \right\} S$$

where

$$h_n = \frac{3\bar{\varepsilon}}{2\bar{\varepsilon}+1} \qquad \bar{\varepsilon} = \frac{1}{3}(\varepsilon_{\parallel}+2\varepsilon_{\perp})$$

$$F_n = \frac{1}{1-\bar{\alpha}f} \qquad f = \frac{2\bar{\varepsilon}-2}{2\bar{\varepsilon}+1}\frac{4\pi}{3}N_L \cdot \frac{\rho}{M} \quad \text{and} \quad \bar{\alpha} = \frac{1}{3}(\alpha_{\parallel}+2\alpha_{\perp})$$

for h_n: $\bar{\varepsilon}=\varepsilon$; and for F_{is}: $\bar{\varepsilon}=\varepsilon$ and $\bar{\alpha}=\alpha$

From the equation for $\Delta\varepsilon$, its experimental value, and the dipole moment ($\mu=2.30\,\mathrm{D}$), the angle β between the vector of the dipole moment and the preferred axis of PAA has been calculated as 64°, with the partial moment of the azoxy group forming an angle of 54° [598]. W. Maier and G. Meier, who assumed a spherical cavity, obtained elevated theoretical values from their calculations, but corrections have been suggested by V. N. Tsvetkov [898–901], and A. I. Derzhanski and A. G. Petrov [224, 225, 718].

V. N. Tsvetkov introduced two new parameters (X_1 and X_2) that stand for the hindered rotation about the minor and major molecular axes, respectively [898], table 4.1, see also [951].

Table 4.1. Static dielectric constant, dielectric constant at $4\cdot10^{10}\,\mathrm{Hz}$, Onsager factors (PQ) for the internal field, molar dielectric susceptibilities (σ), degree of nematic order (S), and parameters of the hindrance of molecular rotation (X_1 and X_2) for PAA at various temperatures (from [898]).

t [°C]	$\varepsilon_{\parallel}$	$\varepsilon_{\perp}$	$\varepsilon_{\parallel\infty}$	$\varepsilon_{\perp\infty}$	$(PQ)_{\parallel}$	$(PQ)_{\perp}$	$\left(\frac{\sigma}{PQ}\right)_{\parallel}$	$\left(\frac{\sigma}{PQ}\right)_{\perp}$	S	X_1	X_2
113	5.630	5.865	3.57	2.79	2.11	1.93	38.7	44.3	0.685	−0.402	0.137
125	5.607	5.780	3.50	2.85	2.10	1.95	38.9	43.4	0.600	−0.405	0.094
135	5.620	5.700	3.42	2.92	2.07	1.98	39.9	42.3	0.400	−0.395	0.033
146	5.647		3.013		1.975		42.3		0	0	0

The different signs of X_1 and X_2 indicate that the rotation about the minor molecular axis is more strongly hindered in the nematic phase and that the rotation about the major axis is less than in the isotropic phase. The hindrance of the rotation about the minor axis in the nematic phase is almost independent of temperature, but the facilitation of rotation around the major axis increases with decreasing temperature. Somewhat at variance with W. Maier and G. Meier, V. N. Tsvetkov found the angle between the dipole and major axes of PAA to be $\beta = 59.3°$ [898].

A further improvement was made by A. I. Derzhanski and A. G. Petrov who proposed an ellipsoidal cavity to describe the dielectric anisotropy of nematics [224, 225, 718]. This yielded a value of $\beta = 54° 16'$ for the angle under discussion [718]. They also considered possible correlations between dielectric and piezoelectric parameters [225]. Also starting from an ellipsoidal cavity, G. R. Luckhurst and C. Zannoni developed a theory comprising the frequency dependence [588, 589]. New impetus to the consideration of possible polar order will arise from the observation of dielectric hysteresis loops [923]. The hysteresis area is proportional to the voltage applied in a 10 μm layer of a mixture of PAA and PAP. The remaining polarization is bound to dielectric domains, which are created by the field and remain fixed in the liquid crystal after the field is switched off; photomicrographs of these domains are shown. There is a close connection with the electret effect that is signified by a depolarization current when solid PAA that is obtained from an electrically oriented nematic phase is heated above its melting point. The total quenched charge in the crystal depends linearly on the polarizing field up to a saturation strength of 600 V/cm [488]. It must still be proven that this thermoelectric effect does not imply electrochemical processes [489, 869]. "Frozen" space charge has also been studied in supercooled nematics [177b].

Even the early measurements of M. Jezewski with and without a magnetic field indicated negative dielectric anisotropy of PAA and PAP [451]. W. Maier and G. Meier have obtained more recent values [595, 622, see also 178, 549, 950], fig. 4.20.

The dielectric anisotropy is remarkably low for azo compounds of the type RO⟨◯⟩-N=N-⟨◯⟩-OR where the strong positive anisotropy of the displacement-polarization

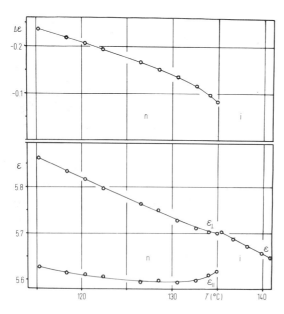

Fig. 4.20. Dielectric anisotropy, $\Delta\varepsilon = \varepsilon_\| - \varepsilon_\perp$, and dielectric constant, ε, of PAA (from [622]).

conteracts the negative anisotropy of the orientation-polarization of the ether dipole while the strong azoxy moment is lacking. The values given in table 4.2 are those of W. Maier and G. Meier [600]. They found the angle between the dipole moment vector and the axis of preferred molecular orientation to be 80°, a considerably larger angle than that of the analogous azoxy compounds.

Table 4.2. Susceptibility data on 4,4′-di-*n*-alkoxyazobenzenes. Nematic phase with $S = 0.676$. a = displacement and μ = orientation components (from [600]).

-alkyl-	$T \cdot C$	DK-axis	σ_α	σ_μ	σ_{cal}	σ_{exp}
-hexyl-	105.7	‖	64.86	8.73	73.59	71.69
		⊥	43.67	28.53	72.20	70.31
-heptyl-	100.6	‖	67.62	8.37	75.99	74.06
		⊥	47.42	27.37	74.79	72.86
-octyl-	104.8	‖	70.43	8.06	78.49	76.41
		⊥	50.82	26.35	77.17	75.08

It must be noted in the case of di-*n*-octyloxyazoxybenzene that the anisotropy of the dielectric constant passes through a maximum in the nematic phase [600, 622], fig. 4.21. It is very important to consider the frequency dependence of the dielectric constant in the quantitative evaluation of such data (see p. 179).

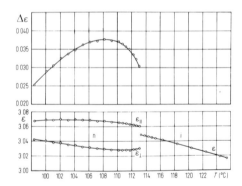

Fig. 4.21. Dielectric anisotropy, $\Delta\varepsilon = \varepsilon_\| - \varepsilon_\perp$, and dielectric constant, ε, of *p,p′*-di-*n*-octyloxyazobenzene (from [622]).

MBBA shows a negative dielectric anisotropy which is approximately three times that of PAA (see fig. 4.22). A 10 kG magnetic field can provide a nearly complete orientation of the MBBA molecules in the direction of the field. Fewer than 0.1 % of the molecules are misoriented at the boundary surfaces as determined from the saturation values of the dielectric constant [255]. On the other hand, the orientation by means of an applied d.c. voltage is less certain [191], and other measurements also seem to be of questionable accuracy [796]. The influence of external fields on the dielectric constants of MBBA is also evident from other studies [18, 54, 437]. The dielectric properties of solid MBBA were also measured [657].

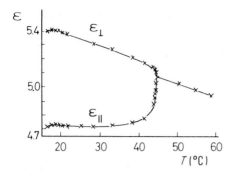

Fig. 4.22. Dielectric constant, ε, of MBBA (from [255], see also [3:665]).

Comparison of the data in fig. 4.23 shows that each added polar group raises the average dielectric constant, $\bar{\varepsilon} = \frac{1}{3}(\varepsilon_{\parallel} + 2\varepsilon_{\perp})$. The dipole moment of the compound C_7H_{15}–⬡–OOC–⬡–C_4H_9 is probably due to the ester group, and a weakly positive dielectric anisotropy is the result. This anisotropy changes sign upon introduction of an ether group, which contributes more the $\varepsilon_{\perp}$ than to $\varepsilon_{\parallel}$ due to the 72° angle between the dipole and p,p' axes. The compound with three rings has higher values of $\bar{\varepsilon}$ and $\Delta\varepsilon$. An additional point of special note is a discontinuity between $\bar{\varepsilon}$ and ε_{iso} at the clearing point of the compound C_2H_5O–⬡–OOC–⬡–O–$COOC_4H_9$ [212], see fig. 4.23.

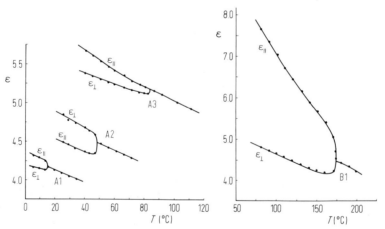

Fig. 4.23. Dielectric constants, ε, of some phenylbenzoates in the nematic and isotropic states. The structures are:

A_1 C_7H_{15}–⬡–OOC–⬡–C_4H_9

A_2 $C_7H_{15}O$–⬡–OOC–⬡–C_4H_9

A_3 C_2H_5O–⬡–OOC–⬡–$OCOOC_4H_9$

B_1 C_4H_9–⬡–OOC–⬡–OOC–⬡–C_4H_9

(from [212]).

Mesomorphous nitriles have an exceptionally high anisotropy of dielectric constant as shown in fig. 4.24 and in the compound C_5H_{11}—⟨○⟩—⟨○⟩—CN with $\varepsilon_\parallel = 17$ and $\varepsilon_\perp = 6$ [26, 54, 192, 755].

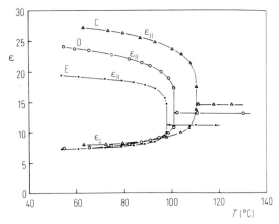

Fig. 4.24. Dielectric constants, ε, of some nitrile compounds in the nematic and isotropic states. The structures are:

C C_4H_9O—⟨○⟩—CH=N—⟨○⟩—CN

D $C_6H_{13}O$—⟨○⟩—CH=N—⟨○⟩—CN

E $C_7H_{15}COO$—⟨○⟩—CH=N—⟨○⟩—CN

(from [803]).

These examples of a very high positive dielectric anisotropy due to the terminal nitrile group are complemented by an example (fig. 4.25) of a very highly negative dielectric anisotropy caused by a nitrile group standing out at an angle from the longitudinal axis of the molecule.

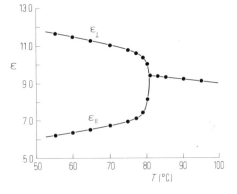

Fig. 4.25. Dielectric constant, ε, of p-ethoxy-p′-hexyl-oxy-α-cyano-trans-stilbene:

C_2H_5O—⟨○⟩—C⟨$^H_{}$⟩=C⟨$^{}_{CN}$⟩—⟨○⟩—OC_6H_{13}

(from [211]).

The literature includes additional data on the dielectric constants of nematics [2:49, 2:181, 2:183, 2:191, 2:281, 2:302, 2:303, 2:380, 2:466, 2:732, 2:750, 2:752, 2:63b, 3:665, 36, 178, 214–218, 220, 261, 340, 491, 533, 545, 546, 678, 747, 749, 778, 913, 546c, 755a, 920a].

Up to this point, the figures and text have applied to the static and quasistatic dielectric constants. We now consider the frequency-dependence of the dielectric constants. The dispersion has been well investigated only up to about 10^9 Hz; it results from the relaxation of the orientation-polarization from the static case up to about 10^{12} Hz. It affects only $\varepsilon_\parallel$ at first, and it is followed by $\varepsilon_\perp$. The dielectric relaxation time of $\varepsilon_\parallel$ depends strongly on the nematic potential parameter, and it can be several orders of magnitude larger in the nematic phase than in the isotropic liquid if the polarization is parallel to the axis of symmetry [192, 623]. The activation energy for the rotational diffusion around the short molecular axis can be derived from the relaxation frequency of $\varepsilon_\parallel$ and its temperature dependence. The influence of molecular order on the dipole relaxation in a nematic phase is obvious. For example, there is a distinct rise in activation energy in the n/s pretransition region [647, 675]. Depending on the frequency of the electric field, the dielectric anisotropy is determined by the permanent and induced dipole moments, and, on the short-wave side of the dispersion steps (first $\varepsilon_\parallel$, then $\varepsilon_\perp$), only by the induced dipole moment ("displacement" or "atomic" polarization). W. Maier and G. Meier discuss this dispersion curve (fig. 4.26); however, it has not been completely confirmed by measurements [599, 622].

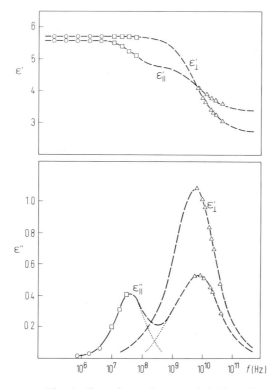

Fig. 4.26. Expected dispersion of the real and imaginary parts of the dielectric constants of PAA (from [623]).

The $\varepsilon'_\perp$-dispersion and a second $\varepsilon'_\parallel$-dispersion step lead to the limiting values of $\varepsilon'_{\parallel\infty} = n_1^2$ and $\varepsilon'_{\perp\infty} = n_2^2$, respectively, which are obtained from the indices of diffraction when the values found in the visible wavelength region are extrapolated to infinitely long waves. A compensation that amounts to 10 % of the electron polarizability is added to account for the IR dispersion step [599, see also 61, 779].

The barrier to rotation around the long axis of PAA in the nematic phase was calculated from dielectric relaxation times $(2 \cdot 10^{-11}\,\text{s})$ to be 11.3 kJ/mol [444–447, see also 68]. 4,4'-dimethoxybenzalazine is similar to PAA, and it has $\varepsilon_\parallel < \varepsilon_\perp$ in the low frequency range and $\varepsilon_\parallel > \varepsilon_\perp$ at microwavelengths [123]. More examples of such sign changes are undoubtedly awaiting discovery. This should be consistent for negative dielectric anisotropy since nematics are always optically positive. For dielectrically positive materials, the expected shape of the dispersion curve permits the anisotropy to change sign twice. G. Baur, A. Stieb, and G. Meier seem to have discovered such a mixture, but only one frequency has been verified to date under which $\Delta\varepsilon$ is positive and above which it is negative [629]. Figs. 4.27 and 4.28 present examples of $\varepsilon_\parallel$-dispersion.

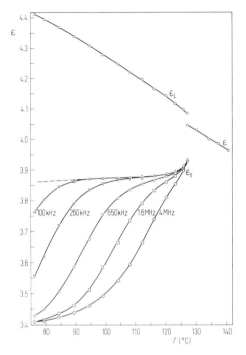

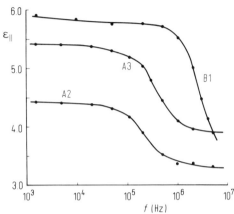

Fig. 4.28. Relaxation of $\varepsilon_\parallel$ at a reduced temperature of 0.93 $T_{n/i}$, corresponding to S=0.65. Structures, see fig. 4.23 (from [212]).

◀ Fig. 4.27. Dielectric constant, ε, of p,p'-di-n-hexyloxyazoxybenzene for different frequencies (from [599] and [622]), concerning the homologous series, see [647]).

The $\varepsilon_\parallel$-relaxation of these esters can be described using a single relaxation time τ_R derived from the Cole-Cole equation [185]:

$$\varepsilon - \varepsilon_\infty = \frac{\varepsilon_0 - \varepsilon_\infty}{1 + (i\omega\tau_R)^{1-\alpha}}$$

where ε_0 and ε_∞ are the static and infinite frequency dielectric constants, ω is the angular frequency, τ_R is the effective relaxation time, and α is a measure of the deviation from a single relaxation time. If $\varepsilon_\parallel$ is apportioned into a real and an imaginary component, then the concept requires that the measured points lie on a semicircle, the Cole-Cole plot (fig. 4.29).

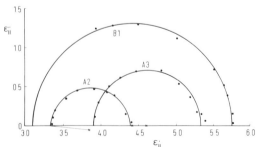

Fig. 4.29. Cole-Cole plots of $\varepsilon_\parallel$ at 0.93 $T_{n/i}$. Structures, see fig. 4.23 (from [212]).

Dielectric relaxation phenomena were also investigated in MBBA [20, 60, 256, 447a, 497, 589, 620, 774], nitriles [194, 249, 310, 492b, 546a, 755b, 803], other nematogens [66–68, 103, 170, 220, 491, 543, 544, 546b, 704, 747, 908, 909, 950a], and are reviewed in the literature [676, 740a, 6:53]. A pulse generator has been developed toward this end [358]. Recently it has been demonstrated that a sole change in temperature may be sufficient to change the sign of dielectric anisotropy of certain nematics. The temperature at which the dielectric anisotropy vanishes is not frequency dependent from 10 to 10^5 Hz, and it is not accompanied by a thermal phase transition or a sign reversal of conductivity anisotropy [534].

4.4.2 Electric Conductivity

Accurate measurement of the electric conductivity requires a high-purity liquid-crystalline material. For this reason, the usual technique of producing a homeotropic texture by coating with an amphiphilic substance such as egg lecithin is normally to be avoided. Purification of nematics (e.g. of MBBA by electrodialysis or vacuum degassing) may increase the resistivity by a factor which varies from 2 to 10 [63, 63a, 311, 510, 668a, 670, 876]. The uniform orientation by a magnetic field, already noted by T. Svedberg in the early days, best surveys these phenomena [862–866, see also 874]. After T. Svedberg's pioneer work, there followed other investigations by W. Voigt [925], M. Jezewski [449], W. Kast [498], L. S. Ornstein [690], V. Frederiks [297, 305], and Y. Björnstahl [89]. Nematics are known with both positive and negative anisotropy of electric conductivity. In the following discussion, $\sigma_\parallel$ and $\sigma_\perp$ are the components of the conductivity tensor parallel and perpendicular to the optical axis, respectively. Experience was especially gained with MBBA (fig. 4.30), where the conductivity has been investigated both in the pure substance [82, 228, 255, 266] and with ionic additives [2:159, 164, 165, 408, 409, 606, 793, 855].

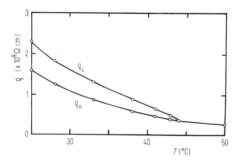

Fig. 4.30. Resistivity, ρ, vs. temperature, T, for MBBA (from [256]).

The current, I, that is based on the self-conductivity of MBBA obeys the equation [32]:

$$I = I_0 \exp(\alpha d) \exp\left[-\frac{U_0 - \beta\sqrt{v/d}}{kT} \right]$$

This means that the temperature dependence illustrated in fig. 4.30 can be linearized as shown in fig. 4.31. In MBBA and its azoxy analogs, the conductivity can vary widely with temperature and the concentration of the added electrolyte (tetraalkylammonium picrates and perchlorates), while the quotient $\sigma_\parallel/\sigma_\perp$ is fairly independent of the concentration of the additive (fig. 4.32), and the quotient $\dfrac{\sigma_\parallel - \sigma_\perp}{\sigma_{iso}}$ is similarly independent of the temperature [404, 408], fig. 4.33.

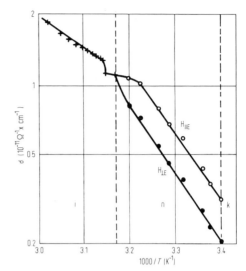

Fig. 4.31. Conductivity, σ, of MBBA vs. the reciprocal temperature, $1000/T$ (from [311]).

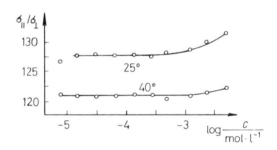

Fig. 4.32. Ratio of conductivities, $\sigma_\parallel/\sigma_\perp$, vs. concentration of tetrapropylammonium picrate in MBBA (from [404]).

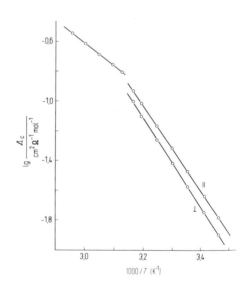

Fig. 4.33. Equivalent conductivity, Λ_c, at a concentration of 1.4×10^{-4} mol tetrapropylammonium picrate/l MBBA vs. the reciprocal temperature, $1000/T$ (from [404]).

Onsager's equation for electrolytic conductivity can be applied to doped MBBA samples [849]. The temperature dependence shown in fig. 4.33 inspired G. Heppke and F. Schneider to propose the following equation relating conductivity to the degree of order [404, 405].

$$S = 3 \frac{\sigma_{iso}}{\sigma_\parallel - \sigma_\perp} \cdot \frac{\sigma_\parallel / \sigma_\perp - 1}{\sigma_\parallel / \sigma_\perp + 2}$$

A pronounced frequency dependence has been found which can change the sign of the conduction anisotropy, $\sigma_\parallel - \sigma_\perp$. The two relaxation times, $\tau_\parallel$, measured by dielectric and conductivity dispersion are in reasonable agreement thus proving that the conductivity relaxation is due to the dispersion of the orientation polarization [806]. Conductivity measurements made at angles to the orienting magnetic field show the expected values according to [404]

$$\sigma = \sigma_\perp + (\sigma_\parallel - \sigma_\perp) \cos^2 \theta$$

and were used for a determination of the bend and splay elastic constants of MBBA [418, 820]. Depending on the angle between the electric and magnetic fields, a cross voltage can be observed due to surface charges generated by the anisotropic conductance [405], fig. 4.34.

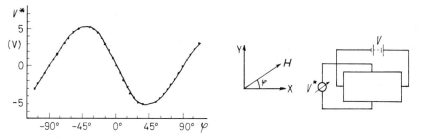

Fig. 4.34. Cross voltage, V^*, vs. the angle, φ, between the axis of alignment and the cell axis. On the right a schematic diagram of the cell and the axes of the coordinate system is shown. The magnetic field vector points in the direction of orientation axis (from [405]).

Estimation of the U_0 constant supports neither the earlier-assumed Schottky emission nor the field induced dissociation. Progressive voltage measurements on a 5-mm thick MBBA layer show that a 10 to 30 μm thick sheath exists on the cathode at all d.c. voltages. The electric field becomes uniform at distances greater than 100 μm from the cathode (fig. 4.35).

It can be deduced from this that some form of negative ion is ejected from the cathode into the dielectric fluid. The literature contains considerations on the nature of the charge carrier and its relation to the properties of MBBA [108]. The anionic radical (MBBA)$^-$ has been shown to be a product of the electrochemical reduction of MBBA [222], while butylaniline is considered responsible for the degradation of MBBA because it is an easily oxidized product of hydrolysis [2:159, 2:160], see also p. 193. The literature contains other investigations on the conductivity of nematics [26, 43, 107, 178, 270, 311, 343, 379, 421, 422, 490, 491, 534, 550, 554, 648, 665, 679, 797, 847, 848, 877, 953, 956, 957].

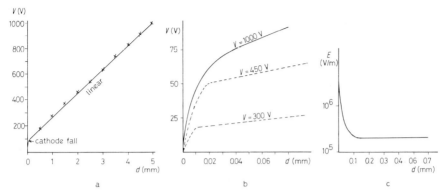

Fig. 4.35. Electric potential distributions (voltage, *V*, and field strength, *E*, respectively, vs. the distance from the cathode, *d*). a) Distribution across the total cell; $V = 1000$ V. b) Distribution near the cathode for different applied voltages in a 5 mm thick cell. c) Distribution near the cathode for 1000 V applied to a 5 mm thick cell (from [32], see also [846a]).

As already mentioned, a d.c. voltage (e. g. 4V) applied to a liquid crystal such as MBBA or APAPA shows a low, steady current after an initial surge peak. If the capacitor electrodes are short-circuited after the potential has been applied for some time, then a gradually decreasing reverse current flows that can be explained from the voltage curve [226].

Pretransitional effects sometimes lead to unexpected results, e. g., those found with the conductivities of *p*-octyloxyazoxybenzene (fig. 4.36). This peculiarity can be interpreted in terms of a smectic short range order in the nematic mesophase near the s_C/n transition [647, 771]. The anisotropy ratio, $\sigma_\parallel/\sigma_\perp$, can either be >1 (curves 1 and 5 in fig. 4.37), $\gtrsim 1$ (curve 8), or <1 (curves 6, 7, 9, and 10).

Other studies using an addition of tetrabutylammonium picrate confirmed this behavior, but they already found the ratio of anisotropy $\sigma_\parallel/\sigma \gtrsim 1$ with *p*-dihexyloxyazoxybenzene [417].

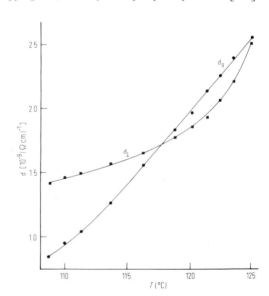

Fig. 4.36. Conductivity, *σ*, of *p,p'*-di-*n*-octyl-oxyazoxybenzene vs. temperature, *T* (from [647]).

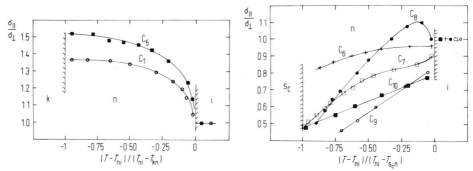

Fig. 4.37. Ratio of conductivities, $\sigma_\parallel/\sigma_\perp$, versus the reduced temperature in the homologous di-*p*-alkoxy-azoxybenzenes. The curves are labeled by the number of carbon atoms in the alkoxy group (from [647]).

In the nematic phase, just after the melting of the crystal, *p-n*-nonyloxybenzoic acid exhibits a positive anisotropy of conductivity that becomes negative after a few hours [144, 145, 147]. E. F. Carr considers the dependence of the conductivity anisotropy on molecular alignment [142], and the mechanisms of the electric conductivity have also been discussed [675]. A differential conductivity analysis which plots $d\log\sigma/dT$ against temperature has been suggested as an analog to the thermal methods (DTA and DSC), and it allows the detection of phase transitions [2:320a, 419, 819a].

4.4.3 *Elastic Deformation of Ordinary Nematics (Frederiks Deformation)*

Although they were not the first to publish on this subject, V. Frederiks and V. Tsvetkov determined the direction of future investigations [299, 302, 303]. In their work, the orientation effects were attributed to the dielectric anisotropy, and the frequency dependence was used to separate hydrodynamic effects from pure orientation phenomena. Earlier workers concentrated more on the phenomena themselves rather than their interpretation [88, 350], and a dissertation considers the influence of smectic-like ordering in nematics [77]. The literature stresses the similarities between the Frederiks deformation in electric and magnetic fields [625], (fig. 4.38).

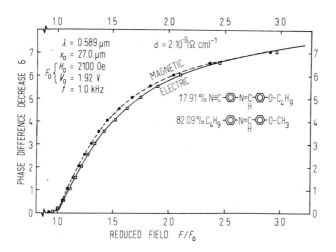

Fig. 4.38. Comparison between Frederiks deformations induced by electric or magnetic fields in a mixed nematic on a reduced field scale (from [625]).

The term "Frederiks deformation" is synonymous with H. Zocher's normal deformation. These terms refer to the simplest deformation with reference only to the normal of the plane (z-axis). For research purposes, a magnetic field is usually preferred to ensure uniform and complete orientation at a given sufficient field strength, while still higher field strengths do not disrupt this arrangement. Of course, this does not say anything about a dependence of the degree of order on the field strength, see p. 205; the decisive factors are the temperature and the constitution of the sample.

Only the Frederiks deformation in electric field seems to hold promise for technical applications. For this purpose, the liquid crystal is sandwiched as a uniformly thin layer between two glass plates that bear a transparent electrode material such as SnO_2 or In_2O_3. Unless otherwise specified, sandwich cells of this type with a liquid crystal layer about 10 to 25 μm thick are a standard in the following descriptions. Occasionally other electrode dispositions are mentioned [843, 845]. A uniform orientation with and without a field as well as a suppression of hydrodynamic effects must be ensured because higher electric field strengths, in contrast to the case of magnetic fields, usually destroys the orientation. While a rather strong magnetic field (over 2000 Gauss) is required to force a magnetic Frederiks deformation to saturation, only a low voltage is sufficient to saturate an electric Frederiks deformation (fig. 4.39). This is usually studied by observing the optical phase differences, δ, of light that is polarized in the plane of the sandwich-cell plates. The experimental values agree well with the theoretical values obtained from the following calculations [360, 624, 625], in which n_o is the ordinary and n_e the extraordinary index of refraction.

$$\delta = d_{(E)} - d_{(0)} = \frac{x_0 n_e n_o}{\lambda n_1} \left\{ (1 + \sin^2 \varphi_R)^{1/2} = \frac{2E}{\pi E_c} \int_{\Phi_R}^{\Phi_M} \frac{(1 + K \sin^2 \Phi)^{1/2} d\Phi}{(1 + v \sin^2 \Phi)^{1/2} (\sin^2 \Phi_M - \sin^2 \Phi)^{1/2}} \right\}$$

for $\varepsilon_\| > \varepsilon_\perp$: $n_1 = n_o$ $K = \dfrac{k_{33} - k_{11}}{k_{11}}$ $v = \dfrac{n_e^2 - n_o^2}{n_o^2}$ $\Phi = \varphi + \varphi_R$

for $\varepsilon_\| < \varepsilon_\perp$: $n_1 = n_e$ $K = \dfrac{k_{11} - k_{33}}{k_{33}}$ $v = \dfrac{n_o^2 - n_e^2}{n_e^2}$ $\Phi = \varphi + \varphi_R \pm \dfrac{\pi}{2}$

Here φ_R is the angle between the boundary surface and the undisturbed optical axis of the liquid crystal, φ is the angle between the director vector and the undisturbed optical axis, and φ_M is the minimum value of φ in the center of the layer $(d/2)$. For high $\Delta\varepsilon$, where the electric field can no longer be considered as uniform, φ can still be approximated by the same calculations [624].

H. Gruler and G. Meier derived the deformation profile from the measured phase differences [624], figs. 4.40 and 4.41.

A threshold voltage, V_c, exists for

homogeneous alignment: $V_{c(\varphi_R = 0)} = E_c \cdot x_0 = \pi \left(\dfrac{k_{11}}{\varepsilon_0 (\varepsilon_\| - \varepsilon_\perp)} \right)^{1/2}$

and homeotropic alignment: $V_{c(\varphi_R = \pi/2)} = E_c \cdot x_0 = \pi \left(\dfrac{k_{33}}{\varepsilon_0 (\varepsilon_\perp - \varepsilon_\|)} \right)^{1/2}$

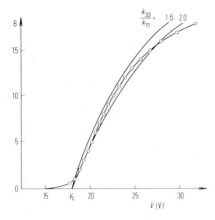

Fig. 4.39. Measured and calculated phase differences, δ, for different voltages, V. Cell thickness: 100 μm, temperature: 105 °C, frequency: 1 kHz, wavelength: 589 nm, compound: di-*p*-*n*-heptyloxyazoxybenzene (from [624]).

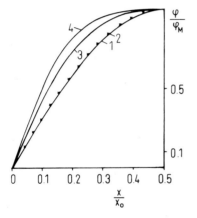

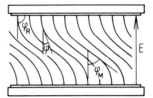

Fig. 4.40. Relative deformation, φ/φ_M, vs. location in the sample, x/x_0, where 1, 2, 3, and 4 correspond to V/V_c values of 1.0, 1.2, 2.1, and 2.6. On the right side, the deformation profile and the definition of the angles are sketched (from [624]).

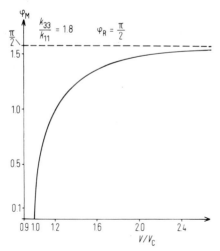

Fig. 4.41. Deformation angle in the center of the sample, φ_M, vs. reduced voltage, V/V_c (from [624]).

Close to this voltage, $V_{c(\varphi_R = \pi/2)}$, interference rings of laser light have been observed [786a, 786b].

In order to avoid the formation of disclinations, pretilted nematic layers have been examined, and the disappearance of the threshold voltage was found to be already at a pretilt angle of 7° [365, 819]. The elastic and dielectric constants can be calculated from the birefringence-voltage curve [25, 168, 621, 643, 764, 839]. Corresponding to the frequency dependence of the local electric field strength, the Frederiks deformation can also be frequency dependent [362, 548, 817, 818]. Both the phase difference and the threshold voltage strongly depend on the temperature [33]. Further details on the Frederiks deformation have been published, especially on its kinetics [8, 34, 65, 75, 133, 138, 140, 176, 184, 204, 232, 353, 369, 433, 462, 482, 486, 495, 564, 565, 626, 629, 644, 674, 680, 681, 762–764, 842, 885, 897]. Work in depth has also been done on its theoretical aspects [72, 235, 236, 403, 426a] and the formation of disclinations [630, 753].

Solutes in nematics orient themselves similarly. Such a guest-host alignment involving a dichroitic dye is called color switching and holds promise for practical applications [382, 386, 653, 674a], cf. section 14.4.2.

The possible applications of the Frederiks deformation of pure nematics were especially well investigated and expanded by M. Schiekel et al. [813, 814, 817–819]. The intensity, I, of transmitted light parallel to the axis through an optically uniaxial substance (e. g., a nematic liquid crystal) and under crossed polarizers can be calculated from the initial intensity, I_0, and the phase difference, δ, between ordinary and extraordinary ray according to

$$I = I_0 \sin^2 \frac{\delta}{2}, \quad \text{where} \quad \delta = 2\pi \cdot d \cdot \Delta n \cdot \frac{1}{\lambda}$$

$$I = 0, \quad \text{if} \quad \delta = 2\pi m \quad \text{and} \quad m = 0, 1, 2, \ldots$$

Thus, the wavelength, λ, the layer thickness, d, and the difference, Δn, $(\Delta n = n_{e, \text{var}} = n_o)$ determine the transmission. Decisively, the variable index of refraction, $n_{e, \text{var}}$, and therefore the optical transmission of the nematic layer is an unequivocal function of the Frederiks deformation, which in turn is controlled by the voltage applied. This possibility of tunable birefringence is illustrated in figs. 4.42 and 4.43 which show measurements made on the same substance, but by different authors and under different conditions [591, 763, 814].

Fig. 4.43 shows five sharp interference maxima and minima ranging from 5 to 35 V between ordinary and extraordinary light rays. This means that the intensity of monochromatic

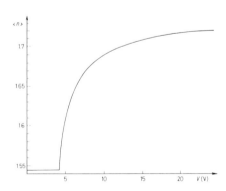

Fig. 4.42. Average index of refraction, $\langle n \rangle$, vs. applied voltage, V, for MBBA (from [763]).

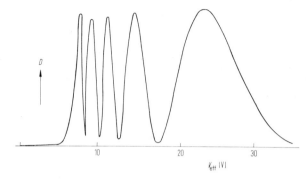

Fig. 4.43. Optical transmission for monochromatic light, D, ($\lambda = 632.8$ nm) vs. applied voltage, V_{eff}, ($f = 1$ kHz) for a 20 µm thick MBBA cell with homeotropic alignment at 25 °C (from [814]).

light can thus be very easily controlled. However, the creation of interference colors from white light is of much greater practical interest. White-light transmission attains a wide plateau under the proper experimental conditions (fig. 4.44). The transmitted light is colored (e.g., blue) when yellow is extinguished by a phase difference of 600 nm. A liquid crystal cell can thus be used to create all of the interference colors known from crystal optics. Each color appears at least once within a voltage range of 5 to 10 V (fig. 4.45).

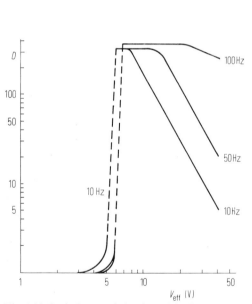

Fig. 4.44. Optical transmission for "white light", D, vs. applied voltage, V_{eff}, for a 20 µm thick MBBA/EBBA cell with homeotropic alignment at 25 °C (from [814]).

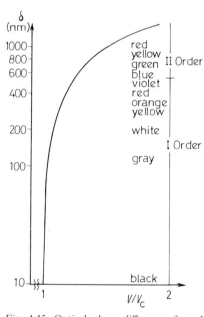

Fig. 4.45. Optical phase difference, δ, and interference colors vs. reduced voltage, V/V_c, in the Frederiks deformation of a homeotropically aligned phase (Schiekel's DAP effect (from [814]).

The definite threshold voltage of the Frederiks deformation is obvious from the last four illustrations. R. A. Soref et al. report results in agreement with those of M. Schiekel [844]. Switching the voltage on and off alters the transparency in a complex manner until after a few seconds when a constant value is attained. In nematics with a frequency-dependent sign reversal of the dielectric anisotropy, rise and decay times of the Frederiks deformation can be controlled by switching the frequency of the electric field [633]. A number of optical waveguide structures have been designed by utilizing the tunable birefringence [166, 167, 431, 432, 829, 830]. Potential applications in integrated optics include index tuning for nonlinear optics, frequency tuning of distributed feedback lasers, deflection, and switching.

In addition to $\varepsilon_{\parallel}$ and $\varepsilon_{\perp}$, ε-values at angles to the orienting magnetic field have also been measured [949]. In order to study the relative effectiveness of electric and magnetic fields, E. F. Carr exposed nematic anisal-p-aminoazobenzene [$\Delta\varepsilon > 0$; $\Delta\chi > 0$] to the simultaneous cross-action of both fields [130, 131]. At a value of dielectric loss that corresponds to a random orientation, the two fields (H_0 in Oe and E_0 in V/cm) were found to be related to one another by the equation [139]:

$$\frac{H_0}{E_0} = 300 \left[\frac{\varepsilon_{\parallel} - \varepsilon_{\perp}}{\chi_{\parallel} - \chi_{\perp}} \right]^{1/2}$$

The $\Delta\chi/\Delta\varepsilon$ ratio of anisal-p-aminoazobenzene and MBBA appears to be independent of temperature, in contrast to PAA [139]. The ratio E_0/H_0 was found to be fairly constant for the random value of the dielectric constant of MBBA (see fig. 4.46), and this indicates the absence of other major effects than that due to the dielectric anisotropy. As shown in fig. 4.47, the threshold fields in the dielectric regime do not coincide [748].

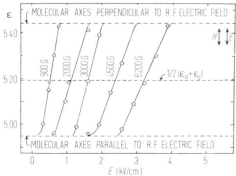

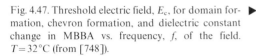

Fig. 4.46. Dielectric constant, ε, of MBBA at 1 MHz measured for various values of a static magnetic field applied parallel to the external electric field, E, ($f = 200$ Hz) and r.f. measuring field. $T = 32\,°C$ (from [748]).

Fig. 4.47. Threshold electric field, E_c, for domain formation, chevron formation, and dielectric constant change in MBBA vs. frequency, f, of the field. $T = 32\,°C$ (from [748]).

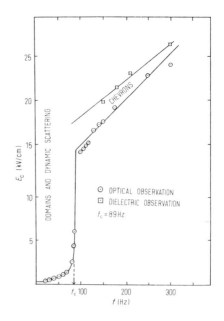

Other investigations [146, 239, 241, 904, 917] and NMR studies [139] (cf. chapter 10) for the comparison of the relative effectiveness of electric and magnetic fields for producing molecular alignment in nematics are reported in the literature.

4.4.4 Elastic Deformation of Twisted Nematics (Schadt-Helfrich Mode, TN Devices)

M. Schadt and W. Helfrich were the first to recognize the advantages of the elastic deformation of a nematic phase in the Mauguin texture [802]. A cell was employed to which a twist of 90° was imposed by rubbing the electrodes such that the preferred orientations are at a right angle. The twisted layer appears on melting a solid nematogen, e. g., a nitrile with a high positive dielectric anisotropy between these "crossed" electrodes. The pitch of the helix in a 10 μm layer is 40 μm and thus much greater than the wavelength of the visible light that can follow the molecular twist. The plane of polarization is such that the rotation amounts to the expected 90°. A potential of only one volt can be sufficient to reversibly convert the Mauguin texture into a homeotropic one, with only the narrowest boundary layer retaining its original order. The electric field reduces the rotation practically to zero, and the cell returns to its original state after the voltage is switched off. Thus, the effect exhibits no hysteresis and is independent of frequency in the range of 0.1 to 80 kHz.

Like Schiekel's DAP cell, the Schadt-Helfrich cell can be driven using crossed or parallel linear polarizers. The big advantage of the latter cell lies in the possibility of attaining very low threshold voltages, ($V_{c,ntw}$), when $(k_{33} - 2k_{22}) < 0$ in the equation

$$\frac{\varepsilon_{\parallel} - \varepsilon_{\perp}}{4\pi} V_{c,ntw}^2 = k_{11} \left(\frac{\pi}{2}\right)^2 + (k_{33} - 2k_{22})\varphi_0^2$$

where φ_0 is the twist angle, which can maximally be $\pi/2$. The low values of $V_{c\,ntw}$ (e. g., 0.9 V) are mainly due to the size of the elastic modules [805]. G. Meier corrected this equation and obtained the following expression for the threshold voltage, V_c, when $\varphi_0 = \pi/2$ * [626].

$$V_{c,ntw} = \pi \sqrt{\frac{k_{11} + \dfrac{k_{33} - 2k_{22}}{4}}{\varepsilon_0 \Delta \varepsilon}}$$

H. J. Deuling has derived equations to describe the deformation of twisted nematics in an electric field [237].

Details concerning the kinetics [163, 330, 563, 766, 905, 912], disclinations [632], comparisons of electric and optical threshold voltages [517, 518, 561], electric field-induced twisting of nematics [678a], and other observations [463, 645, 807, 906] are reported in the literature (cf. section 14.4.3). Twisted nematics and cholesterics display similar electro-optical behavior if the dielectric anisotropy is positive [765].

4.4.5 Electrohydrodynamic Instabilities

Both V. Frederiks et al. [295, 298, 301] and V. Naggiar [663] almost simultaneously observed that "in a low-frequency electric field, the (anisotropic) liquid enters a state of intensive agitation that grows with an increase in field strength" [301]. The phenomena of orientation

---

* We reproduce the authors' original notations that have sometimes been derived using the "absolute" system with a dielectric constant of the dimension 1 ("dimensionless") in the fundamental relation between field strength and dielectric displacement: $D = \varepsilon E$. Other authors have preferred in their derivations the basic relation $D = \varepsilon' \varepsilon_0 E$ where ε_0 has the dimension $I \cdot t / V \cdot L$.

and motion were still earlier observed by O. Lehmann with a nematic-cholesteric phase in an inhomogeneous electric field. The further details of V. Frederiks apply more to the phenomenon that was later to become known by the name "Williams domains", since he speaks of a "series of intertwined dark threads" that are visible in both natural and polarized light. Only higher voltages could be involved with the state of rapid turbulence and intensive light scattering that has technical importance today. Thus, the hydrodynamic deformation of a nematic liquid crystal can take the form of either regular or completely irregular domains. These observations were later taken up by V. Tsvetkov [886], A. P. Kapustin [3:351, 475, 476, 478], R. Williams [933, 934], G. Elliot and J. G. Gibson [260], and G. H. Heilmeier [378]. This matter is also reviewed in the literature [537, 642, see also 35] and studied in dissertations [285a, 702a].

4.4.5.1 Electrochemical Behavior

The electrohydrodynamic instability is attributed to the motion of negative ions, which induce a shear flow within the liquid crystal [30]. At a certain threshold voltage, the movement of charges causes the destabilizing torque to exceed the stabilizing forces. The unstable state extends to a cutoff frequency up to which the charge carriers can follow the ac electric field. Conductivity measurements confirm that the space charge is responsible for the electrohydrodynamic instability [615]. There is as yet no clear picture of the nature and source of the charge carriers. G. Heilmeier et al. discussed the influence of water [389]. MBBA, studied preferably, shows no instability when the number of charge carriers is reduced to fewer than 10^{10} to 10^{12} per cubic centimeter by extremely careful cleaning under vacuum [670, see also 63, 311, 510, 876]. This contrasts with reports by other authors about the unipolar injection of charge carriers into MBBA and on the migration of space-charge fronts [269, 566]. The first polarographic reduction of MBBA occurs at -2.06 V (vs. saturated calomel) and is nearly reversible at scan rates above 5 V/s (fig. 4.48). A second reduction wave at -2.60 V shows no reversal current even at a high scan rate. The radical anion MBBA$^-$ formed at -2.06 V decomposes with a half-life of about four seconds. Fig. 4.48 also shows the two-step oxidation of MBBA (1.45 and 2.0 V vs. saturated calomel). Neither wave yielded a reversal current at scan rates up to 20 V/s. Other electrode potentials are reported in the literature [13]. As this means a degradation of the nematic compound, the addition of dopants showing reversible redox reactions at low potentials is recommended in composites for the dynamic scattering mode [59]. Redox dopants seem to be superior to salt dopants [580, 581a].

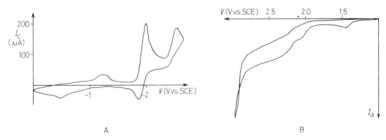

Fig. 4.48. Cyclic voltammograms at a Pt disc electrode: (A) 6.65 mM MBBA in N,N-dimethylformamide, 0.1 M tetra-*n*-butylammonium perchlorate. The scan rate was 2 V/s. (B) 2.24 mM MBBA in acetonitrile, 0.1 M tetra-*n*-butylammonium perchlorate. The scan rate was 0.2 V/s. (SCE: aqueous saturated calomel electrode) (from [582]).

MBBA is electrochemically negative with respect to a n-type CdS photoconductor inducing a negative charge on the CdS surface [531, see also 184a, 604a]. Other nematics have been studied in the literature [223, 348, 364, 926], and the electrochemistry of nematics is also treated in a review [860].

S. Lu and D. Jones used a special probe to study the potential drop in MBBA [584]. At a potential of 2170 V across electrodes 1 cm apart, they found an almost constant internal field of about 200 V/cm that increased to ca. 10^4 V/cm on approaching the cathode and to ca. 10^5 V/cm near the anode. This uneven increase in the electric field on the two electrodes is caused by a piling up of external charges (see fig. 4.35 and p. 184). Such measurements support the observation that the dynamic scattering at a certain low voltage (see below) only occurs at the anode, the side of the highest field strength. However, a direct correlation between electro-chemical and electrohydrodynamic behavior is still missing.

The carrier mobility in PAA is estimated to be either $3.1 \cdot 10^{-4}$ cm^2/Vs [427] or in the range of 10^{-6} to 10^{-5} cm^2/Vs [199, see also 567, 948]. In the case of direct current, the threshold voltage is markedly reduced by the addition of charge-transfer acceptors to nematics, but this does not hold for alternating current [313, 555].

An overview of the behavior of nematics in electric fields can be much more easily obtained from studies in ac fields than in dc fields. This is because, as soon as frequency exceeds more than a few Hz, the charge transfer on the electrodes is suppressed such that only the phase selfconductivity remains apparent. This is also evident from the shorter lifetime in a dc field as compared to ac excitation [857]. This charge transfer is directly visible in an isotropic phase in the form of a fixed convection flow that, however, does not emerge in ac fields [199, see also 357]. This was confirmed by insulating (by means of thin plates) the electrodes with no effect on the threshold voltage, thus indicating that the ac effects are due to the selfconductivity.

4.4.5.2 Periodic Deformations, Type Williams Domains

Between the two extreme phenomena:
a) the Williams domains that consists of a regular and stationary array of convection flows, and
b) the dynamic scattering mode created by turbulent flow
all intermediate state are possible [464, 467, 472a, 745]. The dynamic scattering mode corresponds to the higher excitation. Some publications give the impression that dynamic scattering can also occur immediately adjacent to the threshold voltage without the observation of Williams domains. The appearance of stationary parallel domains lead to the synonymous term "Williams striations". Instructive illustrations are depicted by R. Williams himself [936, and in refs. 426b, 472c]. In place of striped domains, round domains can also occur, e. g., in butyl-*p*-anisyli-dene-*p'*-aminocinnamate [380, 540]. Evidence of a spontaneous polarization is drawn from the hysteresis loops [380]. A network of striations occasionally appears. The network spacing depends on the thickness of the sample and is visible only in light polarized parallel to the direction in which the molecules were originally aligned. Small particles suspended in the sample describe circles within each cell of the network.

Another network pattern that consists of much more closely spaced lines appears at higher frequencies [201]. Still other patterns are known [100, 102, 179, 466, 541a, 854].

The instability just above the threshold voltage leads to regular disturbances that yield He-Ne laser diffraction patterns with a two-fold symmetry axis [31]. The same diffraction pattern was observed with dc and ac excitation; the periodic organization persists even in the presence of the dynamic scattering mode. This is supported by several authors who report well defined diffraction patterns from electrically excited nematics [151, 428, 585]. A regular pattern similar to Williams domains has also been observed perpendicular to a static electric field [79]. For those studies, experimental designs other than the sandwich cell have only a peripheral interest [939], e. g., films with one or two free surfaces [940].

According to the intensive studies of P. A. Penz, Williams domains represent a rotational state of high order occurring in selfclosed tubular vortices [707–714, see also 527]. The liquid crystal appears microscopically as a striation pattern of cylindrical lenses forming real images of the microscope lamp above the plane of the sample, and similar virtual images below that are shifted laterally by half a stripe distance (fig. 4.49). The quotient of the stripe distance and the sample thickness is independent of the latter and of the voltage up to about 3 V above the threshold, which is itself independent of the thickness of the layer (experimental conditions for PAA: 130°C, applied voltage 7.8 V, sample thickness 38 µm, spacing 31 µm). Fine enclosed dust particles describe a nearly circular motion that is perpendicular to the glass plates, and oppositely directed in adjacent domains [709, see also 148, 467]. P. A. Penz has presented a dynamical analysis finding the turn-on rates of Williams domains being bounded by the space charge relaxation time [714].

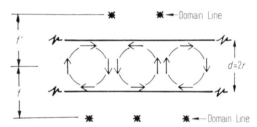

Fig. 4.49. Schematic drawing of a sandwich cell with Williams domains in cross section. The stream lines inside the cell represent the fluid vortex pattern observed by tracer particle motion. The upper domain lines appear between the vortices, the lower domain lines below the centers of the vortices (from [709]).

The relative focal length, f'/r or f/r, is reduced with increasing voltage. It is more or less different for top and bottom lines, but it probably approaches a common limiting value. Other domain patterns described above are also due to a steady-state flow [254, 782], fig. 4.50.

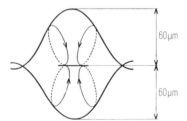

Fig. 4.50. Schematic drawing of hydrodynamic motion observed in a cellular domain pattern of MBBA (from [254]).

Both optical and dielectric measurements on MBBA gave identical threshold voltages for electrohydrodynamic instabilities [137]. This boundary, although a characteristic of the medium, is only slightly dependent on the electric conductivity [390]. The domain structure itself depends on the amplitude and shape of the electric field [789]. Further investigations are listed [22, 156, 175, 177a, 180, 182, 205, 333, 367, 441, 472, 481, 484, 487, 492, 529, 649, 671, 729, 740, 787, 794a, 871, 921, 922, 929, 930]. Nematic droplets [492] suspended films [266], and nematics with some smectic ordering [323a] have also been studied. An additional magnetic field raises the threshold voltage [875, 884, see also 323] and the flow cell width [149]. If the domain-creating voltage is modulated with another voltage of proper amplitude and frequency, the threshold for the combined signal is higher than that for the former signal alone [932].

Very thin nematic layers (3–6 µm in place of the usual 10–30 µm) no longer exhibit any dynamic scattering effect, but rather a parallel domain texture whose spacing decreases linearly from ca. 10 µm to about 1 µm with increasing voltage. This is known as the variable grating mode [740]. With such a tunable phase grid and an additional space filter, it is possible to create spectral colors by scattering white collimated light [354, 355].

Unusually thick samples (800–1200 µm) have also been studied [703, 767]. Williams domains and chevron patterns are also known in twisted nematics [945]; other types of electrohydrodynamic instabilities are reported in tilted nematic layers [726, 727].

4.4.5.3 Dynamic Scattering Mode (DSM)

The characteristic properties of this effect are [672, 768]:

a) A prerequisite of a nematic substance with a negative dielectric anisotropy and sufficient conductivity (ca. 10^{-8}–$10^{-10} \Omega^{-1}$ cm^{-1}). The literature reports data on the current density [472b].

b) Intensity and angular distribution of the scattered light are independent of the wavelength and polarization of the incident light, and the scattering is directed strongly forward.

c) Voltage and frequency play an important role. Instability begins just above a threshold voltage, V_c. Williams domains often appear first, and they gradually switch over to the dynamic scattering mode with increasing voltage. This reaches maximum scattering (saturation) at potentials of about 5 to 10 times V_c. The threshold voltage is practically independent of the thickness of the liquid-crystal layer and hardly varies from substance to substance (5 to 10 volts). The dynamic scattering mode disappears at a cutoff frequency above which the fast turn-off mode occurs, showing a domain pattern that scatters light less strongly and is called the "chevron" pattern.

d) The original orientation of the molecules relative to the electric field and to the boundary surfaces does not play a crucial role, but it does have importance for technical applications, particularly regarding the reproducibility of the contrast.

e) The response times are influenced by many parameters, e.g., viscosity, sample thickness, conductivity, voltage, and others.

The turbulent flow creates optical dishomogeneities that act as individual scattering centers [80, 193]. A measurement of the velocity of electrohydrodynamic flow has been performed [83]. It is remarkable that even a state of random appearance yields diffraction patterns indicating that all scattering centers lie in one plane and are (in a cell 6 µm thick) 13.1 µm apart in a square or nearly square rectangular pattern [27]. Without distortion, the width of the first diffraction maximum should be 3′, but 1° 15′, corresponding to a broadening factor of 23.8, is actually measured. The scattering arises from a more or less continuous but periodic change in the dielectric constant. Thus, the scattering centers are not confined per se but can be located within spheres of a radius of ca. 3 μm, which is in reasonable accordance with the so called lattice constant of 13.1 μm.

In doped nematics, the intervortex spacing decreases monotonically with increasing applied voltage [161].

Laser beams with limited spatial coherence can be used to determine the characteristic times of the turbulent motion [24]. From the observations on the microstructure of MBBA in an electric field, R. Chang derived possible mechanisms explaining the formation of Williams domains and the dynamic scattering mode [162]. J. Nehring et al. described the formation of threads in the dynamic scattering mode of nematics [673]. Patchy areas may emerge after prolonged excitation [23]. Many features of the dynamic scattering mode have been studied with regard to technical applications (cf. section 14.4.4). This especially applies to rise and decay times; the time was found to be directly proportional to the viscosity and inversely proportional to the square of the electric field strength [538, 539, see also 335, 701, 702, 815]. R. Williams has suggested that the reorientation of nematics (disordered by electric currents) are initiated at the boundary surface and propagates at a diffusion-controlled rate. The temperature dependence of the molecular alignment and realignment points to a mechanism governed by Brownian motion [163]. A nematic phase near the critical frequency when switched to the off-state on returning to the original homeotropic texture from the dynamic scattering mode may form a birefringent layer [859]. Many technical studies are concerned with the contrast ratio, e.g., its improvement by spatial filtering.

In the dynamic scattering mode, a nematic liquid-crystal layer destroys the coherence of scattered laser light. Technical applications of this are possible in holography and for the sparklefree projection of pictures using coherent light [525]. The coherence of laser light scattered by a liquid crystal cell can be adjusted depending on the scattering angle, the applied voltage, and the dimensions of the illuminated spot [823]. The forward-scattering contrast ratio of a DSM cell can be increased about one order of magnitude by fitting the cell with polarizer and analyzer. C. Deutsch and P. N. Keating report on the dynamic scattering of coherently polarized light, e.g., the angular distribution of the scattering intensity versus the positions of the polarizers [242], fig. 4.51.

Thus the relative energy, I, scattered into the polarization opposite to the incident polarization increases markedly as the voltage increases. I is less than 8 % at 10 V with little depolarization, whereas I is almost 70 % at 60 V; i.e. more than half of the incident energy

Fig. 4.51. Scattering intensity of light, I, ($\lambda = 632.8$ nm) vs. detector angle, φ, for a 25 μm thick APAPA cell under an applied voltage of 20 V. The scattered light with the same polarization as that of incident light was observed for two different cases: ○, polarizer and analyzer parallel to the optical axis; □, polarizer and analyzer perpendicular to the optical axis (from [242, see also 790–792]).

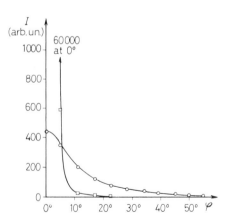

is scattered into the other polarization. Intensity polarization and angular distribution have been subjects of further studies [25a, 630, 650, 650a, 836, 851a]. Measurements of the second order correlation function of the scattered light have indicated a deviation from Gaussian statistics (for sufficiently small spot size) and the presence of different kinds of scattering centers [81, 84, 465, 822, 965, see also 150a, 423, 947]. Phase fluctuations in the emergent wavefront seem to be dominant, but amplitude fluctuations are not entirely negligible [744]. The Orsay group very cautiously interprets such measurements [840]. Further studies in the non-Gaussian regime (illuminated region of sample comparable in size to the scatterer structure) are necessary.

High-contrast, two-color displays can be made as follows: colored, polarized light is passed through an oriented liquid-crystal layer, a following delay foil, and a polarizer. This produces a mixture of the original color and the interference color due to the phase displacement. In the dynamic scattering state, the liquid crystal depolarizes the light so that the delay foil is ineffective, and the light is transmitted in its original color [355].

Our knowledge of the dynamic scattering mode is enlarged by a great number of other publications that can be cited here only altogether [2:96, 64, 115, 172, 261, 262, 266, 285a, 312, 334, 351, 352, 381, 384, 448, 457, 468, 470, 474, 520, 528, 532, 547, 560, 581, 605, 606, 617, 627, 645b, 658, 683, 708, 733, 788, 789, 794, 858, 881–883].

4.4.5.4 Theory of Electrohydrodynamic Instabilities

Table 4.3 summarizes the previous discussion using the terminology of the Orsay liquid crystal group [698].

Table 4.3. Electric instabilities of nematics with negative dielectric anisotropy.

dc, or *ac* field where $f < f_c$	*ac* field where $f > f_c$
Williams domains (striations or cellular patterns) and dynamic scattering mode, associated with hydro-dynamic flow	Fast-turnoff mode (striations or chevrons with a few μm-period, associated with bend oscillations

The key to the understanding of these effects lies in their frequency-dependence and can be explicated using the typical example of MBBA (fig. 4.52).

R. Williams suggested a static model with a ferroelectric molecular order, while A. Saupe considered these phenomena to be associated with material flow. E. Carr and W. Helfrich succeeded in showing that the anisotropy of the conductivity responsibly participates. The following experiment helps discern the formation of Williams domains: By rubbing with a cotton cloth one prepares a homogeneous, planar texture in a cell having SnO_2-coated glass plates 25 μm apart. The direction of rubbing determines the direction of the optical axis (both are parallel) and, in a field, also that of the Williams domains that are at right angles to the direction of rubbing [393]. This suggests a periodically distorted molecular order whose wave vector runs parallel to the optical axis in the field-free state. Many later investigations are based on W. Helfrich's work [392]. He calculated the torques per unit volume in some simple distorted orientation patterns:

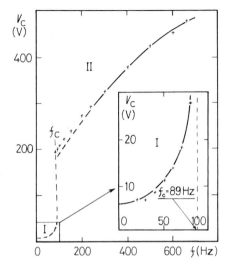

Fig. 4.52. Voltage threshold, V_c, of instabilities in MBBA ($T=25\,°C$, $d=100\,\mu m$) versus the frequency, f, of the applied field (I=conduction regime with Williams domains and dynamic scattering, II=dielectric regime: fast-turnoff mode with chevrons, solid line = theory [202] (from [696, see also 96, 749]).

Torque produced by the shear:

$$m_{s,y}=(k_1\cos^2\theta+k_2\sin^2\theta)S\,.$$

Insertion of the Leslie-Ericksen equations yields:

$$m_{s,y}=-\frac{K_1\cos^2\theta+K_2\sin^2\theta}{\eta_1\cos^2\theta+\eta_{12}\cos^2\theta\sin^2\theta+\eta_2\sin^2\theta}\left(\Delta\sigma\,\frac{\varepsilon_{\|}\cos^2\theta+\varepsilon_{\perp}\sin^2\theta}{\sigma_{\|}\cos^2\theta+\sigma_{\perp}\sin^2\theta}-\Delta\varepsilon\right)\frac{\cos\theta\sin\theta\,E_z^2}{4\pi}$$

Torque due to dielectric polarizations: $m_{p,y}$

$$m_{p,y}=-\Delta\varepsilon\cos\theta\sin\theta\,\frac{E_z^2-E_x^2}{4\pi}+\Delta\varepsilon(\sin^2\theta-\cos^2\theta)\frac{E_zE_x}{4\pi}$$

Torque due to distortion of the orientation pattern: $m_{d,y}$

$$m_{d,y}=-(k_{33}\cos^2\theta+k_{11}\sin^2\theta)\frac{d^2\theta}{dx^2}-(k_{11}-k_{33})\sin\theta\cos\theta\left(\frac{d\theta}{dx}\right)^2$$

In the presence of a magnetic field H_x, one has a fourth torque per unit volume: [875, 917]

$$m_{H,y}=\Delta\chi\cos\theta\sin\theta\,H_x^2$$

Starting with these torques, which he modifies for infinitesimally distorted orientation patterns, W. Helfrich considered the stability of a nematic phase in a dc electric field [392]. The calculation of the threshold voltage, V_c, from the material constants of the nematic liquid crystal constituted an especially remarkable conformation of his theory.

$$V_c=\pi\left(\frac{k_{33}}{\dfrac{K_1\,\varepsilon_{\|}\varepsilon_0}{\eta_1}\left(\dfrac{\varepsilon_{\perp}}{\varepsilon_{\|}}-\dfrac{\sigma_{\perp}}{\sigma_{\|}}\right)+\Delta\varepsilon\varepsilon_0\dfrac{\sigma_{\perp}}{\sigma_{\|}}}\right)^{1/2}$$

E. Dubois-Violette, P. G. de Gennes, and O. Parodi expanded Helfrich's theory to ac electric fields [202, 840]. The threshold field, E_c, depends only on two parameters characterizing the nematic phase [202, 696]; the first one is the relaxation time for charges, τ, and the second one is the dimensionless coefficient, ζ^2.

$$E_c^2 = E_0 \cdot \frac{1 + \omega^2 \tau^2}{\zeta^2 - (1 + \omega^2 \tau^2)}$$

$$\zeta^2 = \left[1 - \frac{\sigma_\perp}{\sigma_\parallel} \frac{\varepsilon_\parallel}{\varepsilon_\perp} \right] \left[1 - \frac{\varepsilon_\parallel}{\Delta\varepsilon} \frac{2\gamma_1^2}{(\gamma_1 - \gamma_2)(\gamma_1 + \eta_0)} \right]$$

The correlation to the Helfrich parameter, θ_H, is $\theta_H = \Delta\varepsilon(\zeta^2 - 1)$.

For increasing frequency, the equation for E_c^2 leads to the cut off frequency:

$$\omega_c = \frac{(\zeta^2 - 1)^{1/2}}{\tau}$$

In the dielectric regime ($\omega > \omega_c$) the charges do not have enough time to flow during one cycle; the charge density q at the threshold becomes time independent. The expression for the threshold field

$$E_c^2 = \frac{X_m \zeta^2}{\lambda} \omega$$

is an asymptotic approach for high ω. It can also be adopted for the frequency range up to ω_c by addition of a constant. The literature also concerns the possibility of hydrodynamic instabilities within dielectrically positive nematics [202]. Synchronous recording of the electric field (sine wave) exciting the fast-turnoff mode and of the scattered light intensity in the first Bragg beam shows the anharmonic property of the light intensity, which peaks when the field goes to zero [698], in this paper "chevron" patterns are depicted. Considerations on the works of W. Helfrich and E. Dubois-Violette et al. are stated by E. W. Aslaksen [28, 29], T. O. Carroll [152], P. A. Penz and G. W. Ford [711, 712], S. A. Pikin [55, 56, 58, 722–725], and others [22a, 57, 62a, 72, 118, 168a, 189, 361, 426, 654, 655, 761, 961]. A nonlinear theory of electrohydrodynamics has recently been submitted [655a]. The most comprehensive treatment is given by the Orsay group, who delineate the different modes as illustrated in fig. 4.53 in terms of field and material characteristics [251, 761, 840]. Based

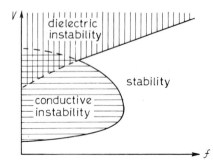

Fig. 4.53. Regions of stability and instability in electric fields depending on the voltage and the frequency. A nematic layer of "normal" thickness is considered (after [761, 840, see also 211]).

on a general phenomenological model, E. Guyon and P. Pieransky explain the similarities between the three phenomena: electrohydrodynamic instabilities, hydrodynamic shear-flow instabilities, and thermoconvective instabilities [366, see also 62, 459, 469], sections 3.8 and 3.9.

Stability charts showing stability and instability ranges for nematics can be calculated as a function of the frequency of an applied electric field, the material constants, and the wavelength of the disturbance [801]. With reduced quantities for field and frequency the curves depend on two parameters; the first parameter depends only on the material constants, and the second parameter is equal to the ratio between the decay time for the bend mode and the dielectric relaxation time [801, 824].

A "phase diagram" of MBBA has been published showing the frequency and voltage dependences of the dissipative structures [471].

4.4.5.5 *Special Observations*

Further investigations must determine if the discrepancies observed in the threshold voltages and domain spacings of 10 μm MBBA cells with homeotropic alignment are to be considered a special case or a frequent occurrence [856]. One case has been mentioned where the domain patterns could be observed only several degrees below the n/i transition temperature [784]; the characteristics of the dynamic scattering mode also depend on temperature [177c, 258a, 535], as does the cut-off frequency [616]. A laser induced orientation was observed in a nematic liquid crystal at voltages just below the threshold voltage of Williams domain formation [746].

Some investigations unanimously accentuate the influence of the applied ac voltage (i. e., sine-wave or square-wave) on the behavior of nematics in an electric field [207, 699, see also 170a, 430], fig. 4.54. Inhomogeneous fields have also been considered [233].

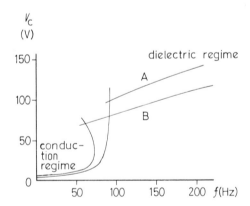

Fig. 4.54. Voltage-frequency stability diagram for the electrohydrodynamic instability of a 50-μm MBBA layer under a sinusoidal (A) or square (B) waveform excitation (from [699]).

A dielectric anisotropy of about −0.2 is desirable for DSM display devices [441a]. Substances with particularly high negative dielectric anisotropies ($\Delta\varepsilon \approx -5$) deviate from the defined normal behavior of substances where $-0.5 < \Delta\varepsilon < 0$ in two ways [211]. First, the conduction regime has not only a lower threshold voltage at which the hydrodynamic flow begins, but also an upper limiting voltage at which this streaming again disappears. This supports the observations that the dielectric anisotropy for the dynamic scattering mode must be negative, but only just slightly less than zero. The upper boundary for such substances is to be expected at much higher (perhaps unattainable) voltages. Second, the chevron pattern, typical of the bend oscillations in the dielectric regime, can also be observed in the frequency range of hydrodynamic effects. Thus ω_c is no longer a frequency separating two regimes, see also [840], fig. 4.53.

The predicted instability of nematics having positive dielectric anisotropy [202, 713] has been repeatedly observed in the form of domains, but barely as a dynamic scattering effect [183, 206, 208–210, 213, 345, 347, 420, 495, 522, 872, 961a]. Bis-*p*-octyloxyazoxybenzene and *p*-butoxybenzylidene-*p'*-octylaniline are interesting study objects because they are characterized by a strongly temperature dependent conductivity anisotropy that changes sign at a specific temperature [345, 347, 420]. Anomalous observations are reported for bis-*p*-heptyloxy-azoxybenzene [590] and CBOOA [721]. Phenomena similar to the dynamic scattering mode have been observed in *p*-butoxybenzoic acid (which exhibits positive dielectric and conductivity anisotropies [143]), in *p,p'*-dibutylazoxybenzene [557], and twisted nematics [861].

The ester mixture of fig. 4.55 provides a productive study material because its dielectric anisotropy changes sign at a certain frequency, f_0.

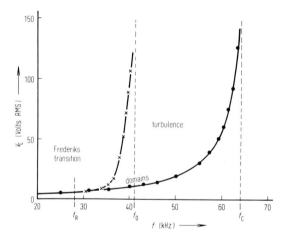

Fig. 4.55. Threshold voltage for deformation, V_c, (plotted as ●) of a planar layer of the mixture

C$_4$H$_9$—〇〇—OOC—〇〇—OOC—〇〇—R

(with R = C$_4$H$_9$ and OCH$_3$)

around the frequency of dielectric isotropy, f_0; see text for the meaning of × ($d = 50\,\mu$m, $\sigma = 6 \cdot 10^{-9}\,\Omega^{-1}\,$cm^{-1}, $T = 70\,°$C) (from [213]).

A Frederiks transition is observed at low frequencies ($f < 30\,$kHz, see fig. 4.55), starting at a threshold voltage, V_c, that is from 3 to 6 V. The correlation of V_c with $\Delta\varepsilon$ according to

$$V_c = \pi \sqrt{4\pi \frac{k_{11}}{\Delta\varepsilon}}$$

can be subject to an elegant proof using the frequency dependence (see fig. 4.56).

Williams domains have been observed in the region of small positive $\Delta\varepsilon$-values between 30 kHz and f_0. The domains appear at a relatively low threshold voltage and continue to a higher voltage as an upper limit of existence (curve xxx in fig. 4.55), above which only a Frederiks deformation is found. In the frequency range with negative dielectric anisotropy, the ester mixture behaves in the manner detailed earlier [213, 336]. The existence of an oscillatory mode which can propagate in a nematic liquid crystal under the influence of an electric field has been demonstrated [715].

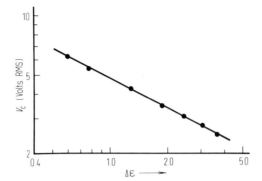

Fig. 4.56. Threshold voltage, V_c, vs. dielectric anisotropy, $\Delta\varepsilon$, for frequencies up to 30 kHz in a planar layer of the mixture described in fig. 4.55 (from [213]).

4.4.6 Fast Turn-off Mode

As detailed above, hydrodynamic effects are limited by a critical frequency above which the socalled dielectric regime extends to where the liquid crystal only responds to the ac electric field by orientational oscillations [388], figs. 4.52 and 4.54. This phenomenon is also tied to surpassing a threshold voltage [202]:

$$E_c^2 = \text{const} + \frac{X_m \zeta^2}{\lambda}\,\omega$$

Lower voltages can only induce Frederiks deformation. The term "fast turn-off mode" expresses the abrupt decay of less than 3 ms when the exciting voltage is switched off. A blurred domain texture ("chevrons") sometimes reminiscent of palm fronds can be observed microscopically [101, 116, 493]. This scatters the light far more strongly than Williams domains, but not so strongly as the dynamic scattering mode. For a correct interpretation, it is important to note that the light passing through a liquid crystal that is excited to the fast turn-off mode is modulated at twice the frequency of the exciting electric field [479]. The cutoff frequency, which increases with the temperature and electric conductivity of the sample, can be very low, e. g., less than 5 Hz in ultrapure MBBA [493]. Such MBBA is thrice recrystallized from absolute ethanol, distilled under vacuum, and stored on a molecular sieve; its clearing point is 47 °C, and ρ is $5 \cdot 10^{11}\ \Omega \cdot \text{cm}$. The elastic constant, k_{33}, can be derived from the spatial periodicity of the chevrons stabilized under the influence of an additional ac field; for MBBA, $k_{33} = 7.3 \pm 1.5$ dynes [308]. It has been theoretically predicted that the spatial frequency of threshold striations increases with the increased frequency of excitation, and this is analogous to the creation of standing waves in an elastic medium [189, 309]. A new type of domain structure was observed at frequencies above the disappearance of the chevron pattern [717].

4.4.7 Curvature Electricity (Piezo- or Flexo-electricity)

R. B. Meyer predicted that liquid crystals should exhibit a special type of piezoelectricity linked with curvature of the orientation pattern [637]. W. Helfrich then discussed this phenomenon assuming wedge- or banana-shaped molecules with a permanent electric dipole moment of a specific magnitude and direction [397, 399, 401], fig. 4.57.

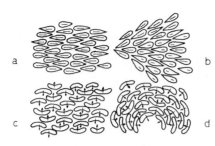

Fig. 4.57. R. B. Meyer's model of curvature electricity. The two liquid crystals consisting of dipolar molecules are nonpolar if not deformed (a and c) but polar under splay (b) or bend (d) (from [637]).

If the normal parallel order is disturbed so that a weak divergence results, then such molecules will preferentially orient themselves in a polar alignment. R. B. Meyer and W. Helfrich lead us to expect a curvature electricity in the range of 10^8 elementary charges per cm^2, but this has not been measured successfully [399, 637]. The piezoelectric energy density in alignment fluctuations should be negligible when compared to the corresponding elastic energy density. For this reason, it will be difficult to find a liquid crystal whose piezoelectricity is measurable through its effect on alignment fluctuations [399]. The literature discusses the possible influences of flexoelectric effects on the elastic deformation [234a, 236, 237, 402], the hydrodynamic behavior [268, 618], and the possibilities to measure flexoelectric coefficients [234b, 234c, 238]. A correlation with piezoelectricity is more apparent for other phenomena. Thus, an electric field that is parallel to the glass plates and far below the threshold voltage of Frederiks deformation causes a small but reproducible phase difference in MBBA. W. Helfrich connected this socalled bending mode with a flexoelectric deformation of the orientation pattern, and he was able to calculate the flexoelectric coefficient for bend as $e_{33} = 3.7 \cdot 10^{-5}$ dyne$^{1/2}$ [401, see also 230, 231, 234]. There has also been mention of a splay wave distortion of a nematic liquid crystal, and this can hardly be explained without postulating a flexoelectric cause [639, 640]. The value of splay curvature electric constant seems surprisingly high: $e_{11} \sim 10^{-3}$ dyne$^{1/2}$. A. Derzhanski and A. G. Petrov suggested a "piezoelectric" correction of the dielectric constants of liquid crystals [225, 227, 229, 230], and S. G. Dmitriev defined the conditions for the formation of piezoelectric domains [247]. Aside from these studies, piezoelectric effects in liquid crystals are still a matter of discussion, and a detailed treatment does not seem advisable here [3:778, 743].

4.4.8 Ferroelectric Effects

Definite proof of the existence of ferroelectric phenomena in nematics has not been found as yet [cf. 8:153], and p. 218. Remanent polarization (hysteresis, e. g. observable as a thermocurrent at the crystal to liquid crystal transition) and the nonlinearity of the low frequency impedance have sometimes been considered evidence of such effects [87, 935], see p. 176, but other authors favored an electrochemical explanation [523], see p. 193. At present, an explanation of Williams domains without assuming ferroelectric effects appears more reasonable [477], see p. 198.

4.4.9 Phase Transitions in External Fields

The typical anisotropic interaction in a mesophase causes the influences of fields to propagate easily in the liquid-crystal order [512]. The orientation also seizes solute molecules or suspended particles with a strongly anchoring surface. Experiments using a field-oriented "liquid single crystal" as a matrix for other substances are discussed in other chapters (NMR, IR, UV spectroscopy). The Rayleigh scattering (cf. chapter 6) in liquid crystals results from thermally induced fluctuations. It is actuated without an external field threshold and is clearly to be distinguished from dynamic scattering due to electrohydrodynamic fluctuations [443]. Only little is known on the influence of external fields on the thermally induced fluctuations of the nematic director, i.e., on the degree of order [739a, 931]. "Light beating" spectroscopy allows a quantitative pursuit of the scattering intensity and damping time of these thermally excited angular fluctuations [257]. The Frank elastic constant and the bend viscosity can be determined in this way.

Furthermore, the influence of an electric field on the n/i transition temperature is an interesting example of a purely scalar property. W. Helfrich's calculation starts from the equation [395]:

$$\frac{\partial \Delta g}{\partial T} = -\frac{q}{T}$$

with the approximation:

$$\Delta g = -\frac{q}{T_{cl}} \Delta T \quad \text{(valid only near } T = T_{cl})$$

where Δg is the difference between the Gibbs free energies per unit mass of the phases ($\Delta g = g_2 - g_1$) and q is the heat of transition per unit mass. In an electric field, E (cgs units), the difference of the polarization energies is

$$\Delta g = -\frac{q}{T_{cl}} \Delta T + \frac{\varepsilon_2 - \varepsilon_1}{8 \pi \rho} E^2$$

where ε_1 and ε_2 are the dielectric constants below and above the transition, and ρ is the density. Since Δg is zero at phase equilibrium, the shift of the clearing point, ΔT, as a function of the electric field strength, E, at constant pressure is

$$\Delta T = \frac{T_{cl}}{q} \frac{\varepsilon_2 - \varepsilon_1}{8 \pi \rho} E^2$$

According to this equation, an electric field of $\approx 1.2 \cdot 10^5 \, \text{V cm}^{-1}$ is required for a 1 °C rise in the clearing point of p-ethoxybenzylidene-p'-cyanoaniline, which has an especially high dielectric anisotropy. The experimental results agree with these calculations and prove the linear correlation between $\log \Delta T$ and E. As expected, transition temperatures associated with an extremely small change in enthalpy (i.e. of second or quasisecond order) are more sensitive to external fields [342, 785, 943]. C. P. Fan and M. J. Stephen have considered the n/i transition with and without applied fields from the standpoint of the molecular-field theory [267]. There is also a discussion of the s_C/s_A transition in a magnetic field [429a] and the distortion of a uniformly polarized s_C phase in an electric field [645a]. Volta potential measurements exhibit a discontinuity at the clearing point of MBBA [99]. However, a more interesting phenomenon seems to be the photovoltaic effect of a nematic liquid crystal, the delivery of a constant current during the illumination of the cell with light of constant intensity [473]. Additionally, we mention pyroelectric detectors using nematics [363] and the generation of induced radiation [438].

4.4.10 Kerr Effect

Pretransition phenomena are always observed in the isotropic phase of a mesogen if the short range order exerts a measurable influence. Regarding the electrically induced Frederiks deformation as a special case of the Kerr effect with an extremely high Kerr constant, it was logical to study the behavior of the isotropic phase near the phase-transition as V. N. Tsvetkov first did [892–896, 907]. The behavior exhibited in fig. 4.58 is typical of many mesogens.

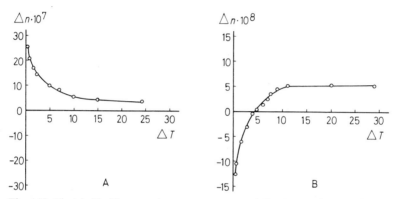

Fig. 4.58. Electric birefringence, Δn, vs. temperature, $\Delta T = T - T_{n/i}$, for A, anisal-p-aminoazobenzene, E: 30 cgs units; B, PAA, E: 40 cgs units (from [894, see also 160]).

All measurements at constant temperature confirm the dependence of the electric birefringence, $\Delta n \equiv (n_e - n_o)$, on the field strength, E, and the light wavelength, λ, as

$$\Delta n = K \lambda E^2$$

where K is the Kerr constant. For a nitrile mixture with a strongly positive dielectric anisotropy, M. Schadt and W. Helfrich have measured [804].

$$K = (3.5 \pm 0.1) \cdot 10^{-10} \, \mathrm{mV}^{-2} \qquad (\lambda = 540 \, \mathrm{nm} \ \text{and} \ T = T_{cl} + 0.1 \, ^\circ\mathrm{C})$$

This value is about two orders of magnitude greater than that of the nitrobenzene standard. The Kerr constants exhibit an alternating effect in homologous series [947a]. As a function of electric field strength and of temperature, Δn can be expressed as

$$\Delta n \propto (T - T^*)^{-\gamma} E^2$$

This equation contains a temperature, T^*, of a second-order phase transition that is apparently cut off by the first-order transition at T_{cl} [2:98, 754, 804]. As fig. 4.58 and newer studies show, the temperature range of pretransitional phenomena in the isotropic phase is about 15°C [280]. Additional details have been published [70a, 87a, 159, 160, 187, 188, 271–276, 278, 279, 370–372, 456, 880, 944, 6:136a, 8:305], including the dispersion of the Kerr constant [808, 808a]. Fig. 4.11 contrasts electric and magnetic birefringence, see also [274, 276]. The linear relationship (between the inverse molar Kerr constant and the temperature) proves de Gennes' theory and it exists even for solutions of MBBA in carbon tetrachloride where the clearing temperature has been depressed by about 30°C relative to the undiluted liquid crystal [70, see also 186].

In two cases reported there are strong indications of a conduction induced alignment [398].

4.5 Cholesteric Phases in Electric Fields

Attempts to utilize the electro-optical properties of cholesterics have not as yet yielded technical success. Given necessary further improvements, the storage effect could find use in technology.

4.5.1 Dielectric Susceptibility; Dielectric Constants

Reliable measurements that take into account the texture of the cholesteric phase are rare. A mixture comprising 1 ppw cholesteryl chloride and 1.75 ppw cholesteryl myristate which

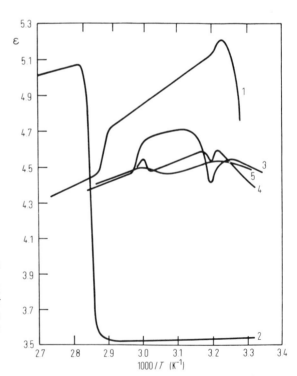

Fig. 4.59. Dielectric constant, ε, vs. the reciprocal temperature, $1000/T$: $1=$ cholesteryl myristate; $2=$ cholesteryl chloride; 3 to $5=1.75/1.00$ mixture of chloride/myristate at sample thicknesses of 10, 25.4, and 127 μm, respectively (from [45]).

was introduced by E. Sackmann was thoroughly investigated [44–46]. This mixture is nematic at 43 °C and is characterized by a strongly temperature-dependent pitch at temperatures above and below that point.

Fig. 4.59 shows the static dielectric constants of the mixture and each of its components in the Grandjean texture that exhibit discontinuities at the i/c transition (the data was obtained during cooling). The striking difference is that $\varepsilon_c > \varepsilon_i$ for cholesteryl myristate, but for the chloride $\varepsilon_c < \varepsilon_i$. This can be explained by the different angles between the dipole moments and the longitudinal molecular axes, which are estimated to be 60° and < 10°, respectively [45]. However, a theoretical value is calculated from the composition of the mixture that considerably deviates from the value actually measured. This is explained by assuming an angle δ_0 between the axis of the helix and the longitudinal molecular axis that is not $\pi/2$ but varies from $\delta_0 = 25°$ at 60 °C to $\delta_0 = 20°$ at 25 °C. Note that there is no reason to assume an angle not equal to $\pi/2$ for the two pure substances, so the ε-values illustrated in fig. 4.59 can be considered $\varepsilon_\perp$ values. A remarkable dip occurs near $T_{nem} = 43$ °C in thin films where, due to the large pitch, undisturbed helices cannot be formed. In thicker layers, this dip disappears, as does the discontinuity at the clearing point. This is because without an applied field, the helical axes are no longer uniformly oriented.

Unlike its components, whose dielectric constants are independent of dc fields up to 10^2 kV/cm, the molecular order in the mixture is sensitive to electric fields (fig. 4.60).

Two possibilities must be considered: a turning of the entire helix with the angle between the helical axis and the molecular axis remaining constant, and a turning of the molecule

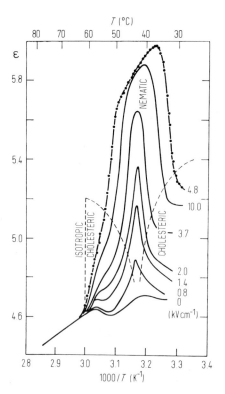

Fig. 4.60. Dielectric constant, ε, of a 1.75/1.00 mixture of cholesteryl chloride/cholesteryl myristate vs. the reciprocal temperature, $1000/T$, at various d.c. field strengths on a 127 μm thick sample (from [45]).

with the helical axis remaining perpendicular to the boundary surface and parallel to *E*. The first is impeded by additional splay, and thus increasing the angle δ_0 by an angle δ in an electric field seems more probable [45].

Two threshold fields were observed. The helical structure is first affected above a lower field strength, E_1, and is destroyed when an upper field strength, E_u, is exceeded. From these two threshold values, the elastic moduli of bend and twist can be determined by using the following correlations of R. B. Meyer [635].

$$E_1 = E_c \left(\frac{k_{33}}{k_{22}}\right)^{1/2} \qquad E_u = E_c \left(\frac{k_{22}}{k_{33}}\right)^{1/2}$$

$$\dot{E}_c = \frac{2\pi}{p_0}\left(\frac{k_{22}}{\Delta\chi}\right)^{1/2}$$

where p_0 is the pitch without field and $\qquad \Delta\chi = \dfrac{\Delta\varepsilon}{4\pi}$

Experiments with the same mixture of cholesteryl chloride and myristate in magnetic fields indicate a magnetic susceptibility of about 10^{-9}. The helical axis orients itself parallel to the magnetic field, and simultaneously the angle δ_0 is reduced by the angle δ [47]. Fig. 4.61 again shows the great influence of the magnetic field near the nematic temperature.

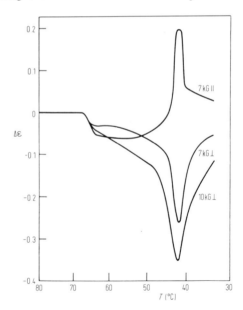

Fig. 4.61. Change in the dielectric constant, $\Delta\varepsilon$, vs. temperature, *T*, for magnetic fields applied perpendicular and parallel to the plane of a 127 μm thick cholesteryl chloride/myristate sample (from [47]).

The stepwise decrease of the dielectric constants in a.c. electric fields with decreasing temperature indicates the presence of two relaxation processes [49, 668], fig. 4.62.

In the literature, one finds a number of other works on the dielectric properties of cholesterics [2:282, 7:109, 76, 154, 155, 170, 337, 341, 666, 667, 678, 825, 828, 853, 878], but interpretation is often difficult because of the confusing influence of the texture. This influence manifests itself by yielding different values in a heating cycle starting from the crystalline phase or in a cooling cycle starting from the isotropic liquid [15–17, 50, 338, 339, 850].

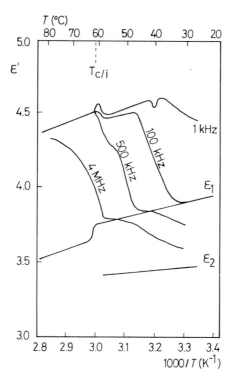

Fig. 4.62. Real part of the dielectric constant, ε', at various frequencies vs. the reciprocal temperature, $1000/T$. Sample: cholesteryl myristate (from [49]).

4.5.2 *Electric Conductivity*

Certain basic principles have been elucidated, but the state of our knowledge in this area is still incomplete at present [825–827]. It may be that the exceptionally small discontinuities in conductivity at the phase transitions are partially due to very high sample thicknesses. Other papers report a stepwise change [43a, 88, 344, 554, 833, 952, 954, see also 74, 98, 956, 957] and show that the focal-conic and Grandjean textures (fig. 4.63) are stable at sufficiently low

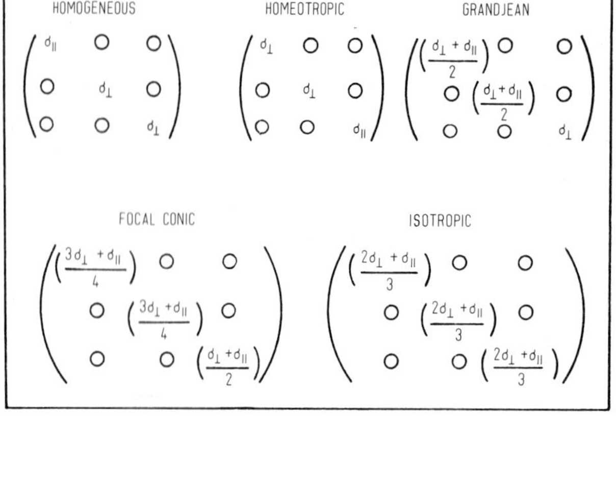

$$
\text{HOMOGENEOUS} \quad \begin{pmatrix} d_\| & 0 & 0 \\ 0 & d_\perp & 0 \\ 0 & 0 & d_\perp \end{pmatrix} \qquad \text{HOMEOTROPIC} \quad \begin{pmatrix} d_\perp & 0 & 0 \\ 0 & d_\perp & 0 \\ 0 & 0 & d_\| \end{pmatrix} \qquad \text{GRANDJEAN} \quad \begin{pmatrix} \left(\dfrac{d_\perp + d_\|}{2}\right) & 0 & 0 \\ 0 & \left(\dfrac{d_\perp + d_\|}{2}\right) & 0 \\ 0 & 0 & d_\perp \end{pmatrix}
$$

$$
\text{FOCAL CONIC} \quad \begin{pmatrix} \left(\dfrac{3d_\perp + d_\|}{4}\right) & 0 & 0 \\ 0 & \left(\dfrac{3d_\perp + d_\|}{4}\right) & 0 \\ 0 & 0 & \left(\dfrac{d_\perp + d_\|}{2}\right) \end{pmatrix} \qquad \text{ISOTROPIC} \quad \begin{pmatrix} \left(\dfrac{2d_\perp + d_\|}{3}\right) & 0 & 0 \\ 0 & \left(\dfrac{2d_\perp + d_\|}{3}\right) & 0 \\ 0 & 0 & \left(\dfrac{2d_\perp + d_\|}{3}\right) \end{pmatrix}
$$

Fig. 4.63. Characterization of the different textures by the components $\sigma_\|$ and $\sigma_\perp$ of the conductivity tensor (from [11]).

currents [11]. The conformation of Walden's rule between mobility and viscosity and the mobility values (of the order of 10^{-5} to 10^{-7} cm^2/Vs) indicate that ionic transport may be the dominant conduction mechanism in the cholesteric phase of cholesteryl esters [952, 954, see also 41, 42]. The influence of texture was also investigated [11, 850], see fig. 4.63.

4.5.3 Electro-Optical Effects

To draw an analogy with the phenomena in nematics may seem somewhat strained, but it is a valuable aid for obtaining an overview of these rather complex phenomena:
Hydrodynamic effects:
a) Excitation of storage mode via turbulent flow analogous to dynamic scattering mode ($\Delta\varepsilon < 0$).
b) Periodic two-dimensional (grid-like) deformation analogous to Williams domains ($E\|$ helical axis, $\Delta\varepsilon \gtrless 0$).
Elastic deformations:
a) Erasing of the storage mode analogous to Frederiks deformation ($E\|$ helical axis, $\Delta\varepsilon < 0$).
b) Periodic deformation leading to the fingerprint texture ($E\|$ helical axis, $\Delta\varepsilon > 0$).
c) Unwinding of the helix analogous to Frederiks deformation ($E\perp$ or $\|$ helical axis, $\Delta\varepsilon > 0$).
d) Conical deformation, blue shift mode ($E\|$ helical axis, $\Delta\varepsilon > 0$).

The behavior of cholesterics in electric fields is generally far less researched than nematics [71a, 95a]. The first intensive studies, those of J. H. Muller [660–662] and W. J. Harper [376], are only about ten years old, and many questions remain unanswered despite the greater efforts of the last decade. Again, behavior in ac electric fields is surveyed more easily than in dc fields [928].

The light-reflection of cholesterics is very sensitive to disturbances of the helix (chapter 7). R. Dreher considered the following deformations [7:96], fig. 4.64.

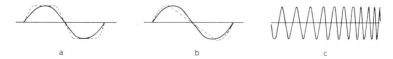

Fig. 4.64. a) Nonpolar deformation symmetrical to $n\frac{p}{4}$, where $n = 0, \pm 1, \pm 2, \ldots$ b) Polar deformation symmetrical to $n\frac{p}{2}$, where $n = 0, \pm 1, \pm 2, \ldots$ c) Aperiodic deformation (from [7:96]).

Case a) If the reflection spectrum is compared with that of an undisturbed helix, bands of higher orders and greater intensities are found along with the weak first-order reflection. These can be observed in a magnetic field.

Case b) In the reflection spectrum, the first-order band appears at twice the wavelength due to the doubling of the periodicity from $p/2$ to p, but this has not yet been observed due to the very low intensity. The second-order band thus becomes dominant, and the third-order band now also appears at the doubled wavelength. Such structures are probably created in electric fields [7:96].

Case c) The principal band is displaced in proportion to the degree of disturbance, and it loses intensity to the benefit of other bands. The reason for this could be a temperature or concentration gradient.

4.5.3.1 Storage Modes

The interest in this area is aroused more by the possibilities of technical applications than by theoretical considerations. If an ionic migration is actuated by a short dc or low-frequency ac pulse, the uniform Grandjean texture of a cholesteric sample is completely disintegrated into a focal-conic texture. The helical structure in this state is still present with the same pitch on a microscopic level. Macroscopically, the liquid crystal is broken up into a myriad of randomly oriented domains. The size of these domains is on the order of a few µm, and consequently this texture strongly scatters visible light of all wave lengths. This texture is said to be stable for up to several days, during which time it gradually reverts to the Grandjean texture. Given a negative dielectric anisotropy, the Grandjean texture can be restored at any time by the application of a high-frequency (>700 Hz) ac pulse [521]. Nematic/cholesteric mixtures also may show typical electrohydrodynamic instability [645c].

It must be remembered that the Grandjean texture can also be transformed into the focal-conic texture by thermal treatment, e. g. with an IR laser beam, where the temperature of transition to the isotropic phase is exceeded momentarily. A strongly scattering focal-conic texture may also be formed when the helical structure is restored from a field-induced nematic state of a cholesteric material after the field has been switched off or diminished. Therefore one can see the common feature of cholesteric storage modes in a stable focal-conic texture that is accessible in different ways.

Since both Grandjean and focal-conic textures are fairly stable, either one can be chosen as the memory state [10]. G. H. Heilmeier and J. E. Goldmacher were the first to describe the storage effect [383]. Their data for the contrast ratio (ca. 1:7) as well as for rise and decay times form a margin that has not as yet been decisively improved (cf. fig. 4.67). As mentioned, the stored information slowly decays, but it persists in certain cholesterics for several days after the excitation is removed [634]. Thickness [12, 586] and boundary conditions [494, 496] were found to play a crucial role in the memory properties of the Grandjean focal-conic texture transitions. Other texture transitions occurring in dielectrically positive nematic/cholesteric mixtures were also suggested as a storage mode [356]. The literature also reports studies in a small angle wedge geometry [515]. Other investigations envisage possible technological applications, especially by studying dynamic properties [3, 6, 169, 349, 485, 519, 521, 799, 837, 838, 850], see also chapter 14. The optical storage mode can also be induced by magnetic fields [10].

4.5.3.2 Periodic Deformations

An electric field applied parallel to the helical axis can cause a grid pattern deformation of the planar texture [3:277, 769]. This can be observed both with positive dielectric anisotropy (e. g., di-p-n-butylazoxybenzene containing 0.5, 1.0, or 2.5 % by weight of cholesteryl nonanoate, the corresponding pitches being 17, 9, and 4 µm) and negative dielectric anisotropy (e. g., MBBA containing 0.1 to 5 % by weight cholesteryl nonanoate, the corresponding pitches varying from 112 to 4 µm) [772]. In the case of larger dielectric anisotropy ($-5 \leq \varepsilon \leq -1$), the square grid deformation of a Grandjean texture is not immediately followed by space charge induced turbulence but by a new planar texture in which the number of pitches across the layer has increased by one [243, 245]. This anomalous effect can be observed several times in succession at higher voltages, but the planar texture usually becomes less stable, which finally leads to turbulence [244].

Only two-dimensional periodic patterns have as yet been observed. However, with an electric field applied perpendicular to the helical axis, a static array of stripes may occur [14, 21, 424, 559, 669]. Picturesque patterns may arise in the case of gridlike deformations, sometimes remaining periodic even at high voltages both in the conduction and the dielectric regime [773]. Analogous to the storage mode, various processes display a similar manifestation. As with nematics, a threshold voltage, V_c, must be surpassed obeying the equation [400, 436, 772, 773]:

$$V_c^2 = 4\pi^2 \left(\frac{\varepsilon_\parallel + \varepsilon_\perp}{2\varepsilon_0 \varepsilon_\perp}\right) \left(\frac{1 + \omega^2 \tau^2}{\theta_H + \Delta\varepsilon \omega^2 \tau^2}\right) \left(\frac{3}{2} k_{22} k_{33}\right)^{1/2} \frac{l}{p_0}$$

where τ is the space-charge limited dielectric relaxation time, ω is the frequency of the applied electric field, and θ_H is a positive "dimensionless" constant. With positive dielectric anisotropy, V_c reaches a saturation value for $\omega\tau \gg 1$. With negative dielectric anisotropy, a periodic deformation is possible in the conduction regime only below a critical frequency, f_c. The analogy with nematics can be carried even further and is also apparent in the dielectric regime above f_c in the form of a square grid of focals having a far shorter period than in the conduction regime [773], cf. Williams domains vs. chevrons. In the dielectric regime, there is a field threshold, E_c, that varies with the excitation frequency according to a parabolic law, $E_c \propto f^{1/2}$. The threshold voltage of the deformation of the Grandjean texture in the conduction regime is greater than that of the nematic state by a factor of $\sim$(thickness/pitch)$^{1/2}$ [396]. W. Helfrich believes that the threshold voltage of a static periodic deformation is smaller by a factor of $\sim$(pitch/thickness)$^{1/2}$ than that of the complete helical unwinding [394]. J. P. Hurault [436] and J. Rault [759] discuss details of the static Helfrich distortion.

In large pitch cholesteric mixtures, a spherulitic texture was observed having a domain size which can be changed by small dc or ac voltages and is stable in the absence of an electric field [368]. Still other domain shapes [86, 514], periodic deformations [95, 542], and textural transitions [607a] are reported in the literature.

4.5.3.3 Helical Unwinding

Two mechanisms have been considered for the disappearence of the helical structure in an electric field: helical unwinding and conical deformation [46], see also p. 208.

The field is perpendicular to the helical axis in the former case. Under certain conditions, the helical axis can be turned $\pi/2$ from an original position parallel to the field. The helix is gradually unwound by increasing the pitch until at the critical field strength, E_{cn}, the helix has vanished (figs. 4.65 and 4.66). This process, which requires a positive dielectric anisotropy, finally leads to a homeotropic nematic phase. The field acts parallel to the helical axis in a conical deformation. A positive dielectric anisotropy is again prerequisite to the process. The helix is extinguished in this case by a reduction of the angle between its axis and that of the molecules (normally $\pi/2$) until the helix completely disappears at E_c', but his process is still being disputed.

J. J. Wysocki, J. Adams, and W. Haas were the first to transform a cholesteric phase into a nematic one with the aid of an electric field [2]. However, in spite of positive dielectric anisotropies, numerous cholesterics do not undergo such a transformation because they would be irreversibly destroyed by an electric field of the required strength, E_{cn}, given by the equation:

$$E_{cn} = \frac{\pi^2}{p_0} \left(\frac{k_{22}}{\varepsilon_0 \Delta \varepsilon} \right)^{1/2}$$

E. Sackmann et al. have found that the onset of the grid pattern deformation at E_c^I and the fingerprint texture at E_c^{II} (starting from the Grandjean texture) is clearly visible in the optical rotation, and the helical pitch remains nearly constant [781].

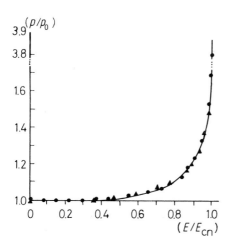

Fig. 4.65. Normalized pitch, p/p_0, vs. normalized electric field, E/E_{cn}, normal to the helical axis for a mixture of 37.5% cholesteryl chloride, 25% cholesteryl nonanoate, and 37.5% cholesteryl oleyl carbonate, $p = 340$ nm (circles); and a mixture of 48.05% chloride, 20.9% nonanoate, and 31.05% oleyl carbonate (triangles). Theoretical curve after de Gennes [196]. The critical field is that for c/n transition (from [461]).

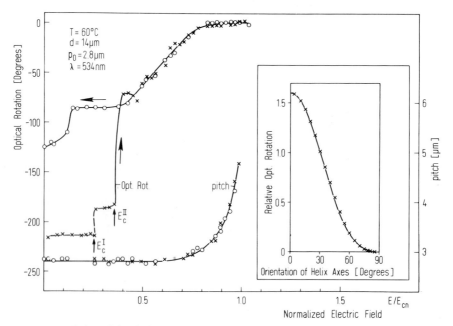

Fig. 4.66. Variation of the pitch and the optical rotation with the electric field strength in a mixture of cholesteryl chloride : cholesteryl laurate $= 1.72 : 1$ by weight. The inset shows the optical rotation as a function of the angle between the helical axis and the direction of light propagation (from [781]).

In mixtures with cholesteryl halide, long-chain cholesteryl esters lead to significantly higher threshold voltages [4, see also 190], while a mixture of the chloride and oleylcarbonate exhibits a minimum of E_{cn} near the 1:1 composition [9]. In mixtures with high positive dielectric anisotropy, E_{cn} can be as low as $2.5 \cdot 10^4$ V/cm [387]. Other studies concern mainly the kinetics of the c/n transition [170b, 195, 516a, 682, 781, 915, 946], the frequency dependence of E_{cn} [48, 173], and applications [5, 385, 442, 511, 800]. When the field is switched off, the nematic phase reverts to a strongly light-scattering cholesteric structure [7], see also p. 212. The contrast ratio obtained by the field-induced c/n transition reaches only between one-third and one-half that of dynamic scattering mode (fig. 4.67).

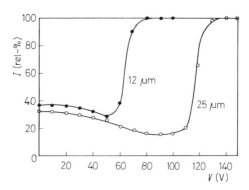

Fig. 4.67. Transmitted-light intensity, I, vs. applied 60 Hz a.c. voltage, V, for a mixture of 55% *p-n*-butoxybenzylidene-*p'*-cyanoaniline, 34% *p*-methylbenzylidene-*p'-n*-butylaniline, and 11% cholesteryl oleyl carbonate observed at 20°C with 521 nm plane-polarized light without an analyzer (from [682, see also 658a]).

4.5.3.4 *Conical Deformation, Blue Shift Mode*

While the expansion of the helix is accompanied by a shift of the maximum reflection toward longer wavelengths, the reverse shift toward shorter wavelengths may be also undoubtedly recognized. Here also, there appears to be a threshold field strength, E'_c, but that is about all that can be stated with any degree of certainty. One suggested equation

$$E'_c = \frac{2\pi}{p_0} \left(\frac{k_{33}}{\varepsilon_0 \Delta \varepsilon} \right)^{1/2}$$

is unsatisfactory while that of T. J. Scheffer [810]

$$V'_c = \pi \left[\frac{k_{11} + 4k_{33} \left(\frac{l}{p_0} \right)^2}{\varepsilon_0 \Delta \varepsilon} \right]^{1/2}$$

is not yet proven.

H. Baessler and M. M. Labes have given special attention to these phenomena (see p. 208), and the response and relaxation times also were measured [556, 558]. Similar study objects are comprised of cholesteryl chloride with laurate, myristate, or palmitate [798].

In place of the conical deformation, which is derived from a pitch contraction, C. J. Gerritsma and P. van Zanten submitted a different explanation of the blue shift [324]. Fig. 4.68 illustrates the measurements on a mixture containing 20% cholesteryl chloride and 80% cholesteryl oleyl carbonate that do not agree well with expectations because the pitch

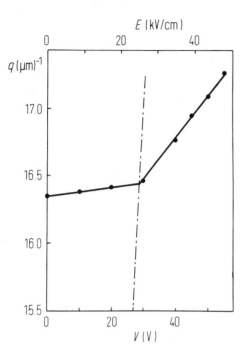

Fig. 4.68. Helix wave vector, $q = \pi/p$, vs. applied voltage, V, and field, E, respectively. ———— experimental curve; —·—·— theoretical curve drawn through E_c and the origin (from [324]).

undergoes some slight change even below the threshold field, and the slope of the curve is considerably smaller above it. According to this explanation, the equilibrium pitch is the average of a distribution of pitches. Domains with $p > p_0$ reflect light of wavelengths longer than those reflected from equilibrium regions. The long pitch domains in an electric field first undergo a grid deformation reducing the intensity of their selective reflection [327, 329]. The observations suggesting a periodic distortion instead of a pitch concentration are discussed these papers.

4.5.3.5 Special Observations

In a magnetic field applied perpendicularly to an electric field, the helical axis of a cholesteric mixture having a positive dielectric anisotropy should prefer a direction perpendicular to both field directions [134]. As with the hydrodynamic effects already discussed, valuable information has been obtained by using a mixture where the sign of $\Delta\varepsilon$ changes as a consequence of dipole relaxation of $\varepsilon_{\parallel}$ at frequencies where $\varepsilon_{\perp}$ still remains constant [326]. The data illustrated in fig. 4.69 confirms that the c/n transition occurs only at frequencies $< f_0$ where $\Delta\varepsilon > 0$. This proves that the helical unwinding is a purely dielectric effect, and it contrasts to the instability of the planar texture (curve 2) which occurs both for $\varepsilon > 0$ and $\varepsilon < 0$.

F. Fischer discusses the influence of external fields on the planar cholesteric texture with homeotropic boundary conditions [282]. Concerning the influence of an electric field on a cholesteric phase under constant shearing conditions, the reader is referred to the literature [460, 734]. Other observations comprise bubble domains [86a, 516] and surface potentials [14a]. In conclusion, a work on the Kerr constants of cholesteryl esters is cited [902]. This reveals experiences similar to those found with nematogens.

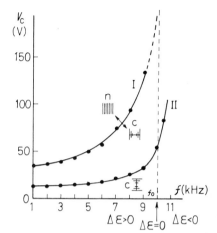

Fig. 4.69. Threshold voltage, V_c, for the c/n transition (curve I, "fingerprint" texture $\leftrightarrow$ homeotropic nematic) and for the instability of a planar texture (curve II, square grid deformations) in a cholesteric mixture as a function of frequency, f. Sample 27 µm thick, $T = 25.0\,°C$, $p_0 = 3.4$ µm, and $f_0 = 10$ kHz (from [326]).

4.6 Smectic Phases in Electric Fields

Systematic investigations in this field are lacking because no technological applications of smectic phases are known. Thus, only a fragmentary picture can be obtained at present from the few existing isolated investigations.

4.6.1 Dielectric Constants

Usually, measurements of dielectric constants reveal a discontinuity at phase transition temperatures as shown in fig. 4.70* [19, 124, 135, 145, 249, 546a, 656, 678, 942].

The anomalous decrease of $\varepsilon_{\parallel}$ and increase of $\varepsilon_{\perp}$ observed in the n/s_A transition range of p,p'-di-n-alkylazoxybenzenes (see fig. 4.71*) is explained by using the Kirkwood-Fröhlich theory of dielectrics to describe the dipole-dipole interaction, and this becomes especially important for neighboring molecules in the same smectic layer [214, 217].

4.6.2 Electric Conductivity

If the electric conductivity is plotted as a function of temperature, phase transitions stand out as peaks or kinks in the curve (see fig. 4.72*).

A striking difference in the electric conductivities $\sigma_{\parallel}$ and $\sigma_{\perp}$ (respectively parallel and perpendicular to the long axis of the molecules) was found in smectics A (fig. 4.72*).

This behavior similarly recurring with diffusion coefficients can be correlated to X-ray results which indicate a bimolecular association within the layers of the cyanobiphenyl compound [648].

Other smectics show a similarly pronounced anisotropy of electric conductivity [413, 416]. The literature reports further studies of electric conductivity and carrier mobility [42, 97, 414, 422, 554, 870, 953, 955–958, 958a], including salts of nitrogen bases [364a].

---

* Figs. 4.70 to 4.72 are added at the end of this text on p. 219 and 220.

4.6.3 *Electric Deformations*

V. Frederiks and A. Repiewa stated that bâtonnets of ethyl *p*-azoxybenzoate and *p*-azoxy-cinnamate orient themselves with their longitudinal axes parallel to the electric field [304]. Compared to nematics, the much more rigid smectics require considerably higher voltages for their deformations [195a, 480, 483]. On the other hand, the study of electro-optical storage modes seems promising. Texture transitions strongly suggestive of those known from cholesterics involve planar, focal-conic, or homeotropic textures of smectics A [177, 374, 375, 851]. Without the existence of a threshold electric field, a uniaxial s_A phase can be transformed to a birefringent scattering texture [735]. J. A. Geurst and W. J. A. Goossens have published predictions on electrically induced hydrodynamic instabilities in smectics A [331]. E. F. Carr has demonstrated on the s_A phase of ethyl *p*-(*p*-methoxybenzylidene)amino cinnamate that a magnetic field is preferred to an electric field in the orientation of smectics [136].

Still to be mentioned are the storage effects in the otherwise not described smectic phase of *p*-cyanobenzylidene-*p*-*n*-octylaniline [873] and the rotation of domains in the fan-shaped s_A phase of *p*-octyloxybenzoic acid at a certain threshold voltage [181], for *p*-nonyloxybenzoic acid, see [177a]. A report that an electric field can initiate the s_B/n transition must be evaluated as a provisional conclusion [153]. Compare this with the early experiments that indicated field-induced orientation when the field is applied to the isotropic state and maintained during the transition to the smectic phase.

Chiral s_C and s_H phases deserve special interest since their possible ferroelectricity has been discussed [2:396, 2:397, 91–93, 541, 562, 611, 612, 641, 641a, 700, 720, 721a, 728, 958b], see section 7.12 and fig. 4.73*.

Electro-optical effects were also studied in smectics C [783, 835a, 924]. A s_A phase formed by a chiral compound that is also a chiral C smectogen shows an electroclinic effect, i.e., a direct coupling of molecular tilt to the applied field [641b].

---

* Fig. 4.73 is added on p. 220.

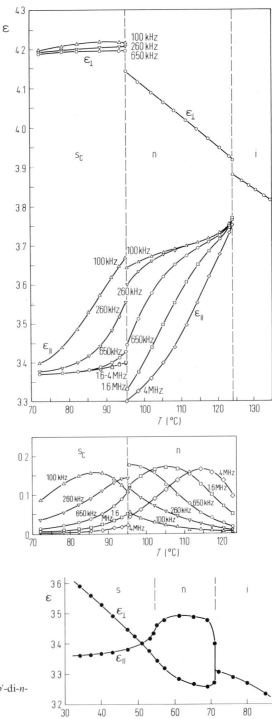

Fig. 4.70. Dielectric constant, ε, and dielectric loss, $\varepsilon'_{\parallel}$, of *p,p'*-di-*n*-heptyloxy-azoxybenzene (from [599, see also 447a]).

Fig. 4.71. Dielectric constant, ε, of *p,p'*-di-*n*-heptylazoxybenzene (from [217]).

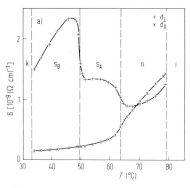

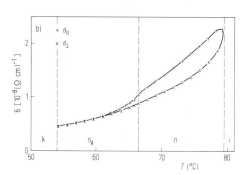

Fig. 4.72. Electric conductivities, σ, of a) *p*-butyloxybenzylidene-*p'*-octylaniline and b) *p*-cyano-*p'*-octyl-oxybiphenyl vs. temperature, T (from [648]).

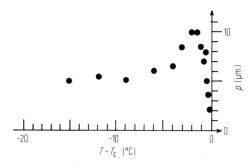

Fig. 4.73. Helical pitch, p, in an uniform texture domain of 2-methylbutyl *p*-tetradecyloxy-benzylidene-*p'*-amino-α-cyanocinnamate vs. the temperature below the s_A/s_C transition point, $T - T_c$ (from [611]).

5 X-Ray and Neutron Scattering, Positron Annihilation, Mössbauer Investigations, and Cherenkov Radiation

5

5 X-Ray and Neutron Scattering, Positron Annihilation, Mössbauer Investigations, and Cherenkov Radiation

5.1 X-Ray Scattering

The first investigations into the structure of liquid crystals by X-ray diffraction were made in the twenties by E. Hückel [157], J. S. van der Lingen [265, 266], C. V. Raman [226], M. de Broglie [69], and mostly by G. Friedel [1:62, 113, 114]. Friedel's studies established the criteria for differentiating between smectic and nematic bodies as well as the assignment of cholesteric phases to the nematic. Until now more than one hundred papers dealt with this subject (see section 5.7) and many of these have given detailed insights into the molecular order in liquid crystals; for reviews see ref. [57, 92, 112]. Still earlier, in 1907, the existence of a three-dimensional lattice in mesophases was considered, but this was rejected [115]. The first experiments were not conclusive due to the primitive state of the art. However, even the pioneering X-ray diagrams of W. Kast [169–173] and the work of K. Herrmann and A. H. Krummacher [128–136] on nematic PAA showed, in addition to the diffuse diffraction patterns typical of liquids, a sharpening of the interference figure in a magnetic field as is not the case in isotropic phases. The interference patterns resembled that of a fibrous structure, i.e., it showed sickle-shaped interference patterns that correspond to a parallel orientation of the molecules. K. Herrmann and A. H. Krummacher confirmed this molecular alignment with alkyl phenetolazoxybenzoates. G. W. Stewart detected a distinct, if weak, anisotropy of the nematic phase by X-ray diffraction [1:358–1:361, 245]. He considered this a confirmation of the swarm theory; applicable here because his measurements involved relatively large volumes.

In the case of anisotropic molecular shape, X-ray diffraction figures of liquids reveal a cybotactic structure similar to that of the mesomorphous state except that the dimensions of the ordered groups are much smaller [70, 176, 207].

J. D. Bernal and D. Crowfoot made a structural analysis of solid crystalline PAA and were able to determine the parameter of the elementary cell by use of the crystallographic data [2:40].

The elementary cell of the monoclinic crystal ($a=1.10$ nm, $b=0.81$ nm, $c=1.495$ nm, $\beta=107°\,30'$) comprises four molecules. The indices of refraction are $n_\gamma=2.198$, $n_\beta=1.573$, and $n_\alpha=1.564$, of which only the first shows a strong absorption. The molecules are oriented almost parallel to the c-axis, and the crystal structure is very close to tetragonal (uniaxial). Furthermore, the X-ray patterns indicate that a certain translational freedom exists in the c-direction according to the blurring of the interference figures as is characteristic of fibrous structures such as asbestos and serpentine. A remarkable similarity can be seen in the symmetry of the nematic spatial unit. It is also noteworthy that the molecules are already detached in the direction of their long axes with the ether groups opposite the azoxy groups (fig. 5.1).

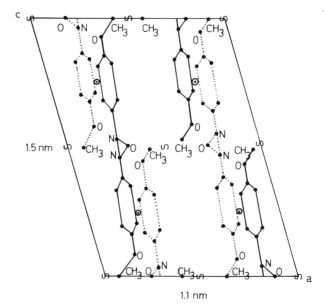

Fig. 5.1. Structure of PAA projected on the ac-plane: ●———● molecule y=0, ●·······● molecule y=½. ʊ screw axes, ⊙ centers of symmetry y=¼ or ¾. Direction of glide plane at y=0, ½. The planes of the rings are about 40° out of the (010) planes (from [2:40]).

A similar structure is observed in PAP.

Ethyl anisylidene-p-aminocinnamate revealed, on the other hand, a totally different structure. In contrast to PAA and PAP, the molecules here lie in double layers with a glide plane and their ends lying parallel. This explains very simply how the smectic phase is created upon melting. Bernal supplemented the X-ray diffraction studies with optical investigations concentrating especially on the metastable (white) form of PAA that appears upon supercooling below 84 °C. This metastable modification (e. g., with respect to its epitaxis on mica) exhibits extraordinary similarities with the nematic state. They even have nearly identical indices of refraction. Thus, the characteristics of the nematic phase can be in large part quite well understood from those of the solid phase (or, as in the present example, from a metastable phase). An important conclusion leading from this is that there exist structural analogies between liquid crystals and their crystalline equilibrium forms. Bernal called such a transformation having substantial retention of the molecular orientation, be it between solid or liquid-crystal phases, a homeomorphic transformation. This contrasts with a morphotropic transformation, i. e. one in which the molecular orientation and their freedom of motion is considerably altered. The relations appear best summarized by Bernal's scheme for liquid-crystalline phases (fig. 5.2).

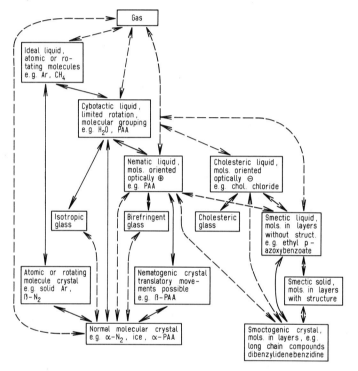

Fig. 5.2. Bernal's transformation scheme (from [2:40]).
⟵⟶ indicate homeomorphous transformation
⟵--⟶ indicate morphotropic transformation

One can therefore safely assume that a lattice-like order exists in liquid crystals. However, while it is possible to rotate a true crystal and mirror its smallest units to show their mutual identity; this is not readily possible in the cases of partially or wholly amorphous bodies. The crystalline spatial element always has three translations. It is the rotation and mirror manipulations that define the symmetry classes of the crystal. C. Hermann developed his "theory of statistical translations" or symmetry operations in order to use an analogous terminology for describing bodies having indistinctly localized molecules [127]. This theory still — or more aptly stated, again — occupies center stage in all X-ray diffraction studies. A. J. Mabis recently published an excellent interpretation of Hermann's work [204], cf. fig. 5.3 showing the principle used in illustrating point-groups in a reciprocal lattice.

Hermann applied the term "statistical translation group" to those groups in which the point density, N, of a volume element, dV, and their deviations, $\sqrt{NdV}$, have a constant value. A statistical translation, S, denotes the parallel displacement of a volume element in such a way that two volume elements, dV, of equal statistical translations coincide. In the case of crystals, this operation obviously leads to translations of the crystal lattice. However, it also describes the progress from one to any other given molecule of an (isotropic) gas. A reciprocal translation, R, is defined such that the geometrical location of this translation vector is a bundle of equidistant parallel planes. A direct translation, D, leads to equivalent structural elements that lie equidistant along parallel straight lines. Finally, the pseudodirect translations, P, indicate those special D translations that are directionally restricted: for a D translation with two degrees of freedom called P_2, for one with a single degree of freedom P_1, and for one with no degree of freedom P_0 ($P_3 = D$). The types of translations thus defined are combined so that three are used for any complete description. An amorphous body is described by the translation symbol SSS, and the crystal for which three independent linear and three subordinate reciprocal translations exist by (RD)(RD)(RD). Hermann investigated the number of possible translations, particularly with respect to their interdependencies, and found that a triple D or triple R requires the supplementary reciprocal translations. Plausible physical considerations furthermore require that any structure with two D or with two P translations have an R translation in the third direction. On the other hand two linearly independent R translations call for a D translation in the third direction. Disregarding the pseudotransformations for the moment, we have seen seven types of translation that can be assigned to liquid crystals:

1. SSS (amorphous)
2. SSR
3. SSD
4. SS(RD)
5. RDS
6. RRD
7. DDR
8. RD(RD)
9. (RD)(RD)(RD) (crystalline).

If the direct translations in each of these seven combinations are replaced by pseudotranslations with the permissible degree-of-freedom index for the chain direction, we obtain the additional combinations:

3. a) SSP_2
3. b) SSP_1
3. c) SSP_0
4. a) $SS(RP_1)$
4. b) $SS(RP_0)$
5. a) RP_1S
5. b) RP_0S
7. a) DP_0R
7. b) P_1P_0R
7. c) P_0P_0R
8. a) $RD(RP_0)$

Details concerning implicit limiting conditions are discussed extensively in Hermann's original paper [127]. 18 models, including the pseudotranslation combinations are designed to describe the liquid-crystalline state.

Two examples, SSR and SSD (fig. 5.3), will be treated in detail using the very instructive Mabis model [204].

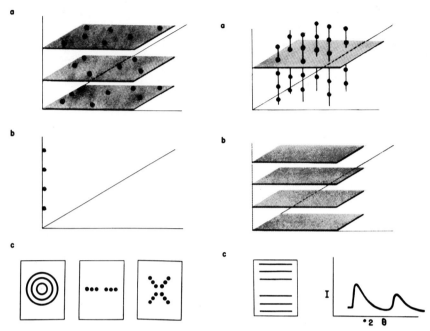

Fig. 5.3. Mabis' model of SSR and SSD translations (from [204]).

In these cases, statistical molecular distributions are restricted to planes. R requires that there be equidistant parallel planes, $E_1 \ldots E_n$, with corresponding reciprocal lattice having points on lines perpendicular to the E planes. A Debye-Scherrer picture of such a structure shows equidistant concentric circles, while a single-crystal picture indicates one or more rows of equidistant spots depending on whether the axis of rotation lies in the plane or at some angle to it. This model introduces the molecular distribution of a smectic phase with this example being the most general case. Fig. 5.3 also has the translation type SSD presented standing for the nematic phase. The molecules form chains that randomly penetrate any plane of perpendicular orientation. The chains are arbitrarily movable with respect to each other and only the intermolecular distances within the chains being fixed. Thus the reciprocal lattice comprises equidistant planes and the rotating crystal picture shows parallel lines, while the powder diagram exhibits uniformly symmetrical lines that are sharp on the side facing the smaller angle and broader on the other side. This shape of the lines is characteristic of a plane arrangement in the reciprocal lattice.

An additional benefit of the Hermann's model is that it can serve as a single base for both the structures of thermotropic and lyotropic phases. In this state, where polar and apolar groups are bound together to form amphiphilic molecules, double layers and complex lattices are observed that form several of Hermann's types (cf. chapter 11). For a recent discussion see ref. [121].

The next step was to consistently develop a theory of partially deformed lattices that would connect the principles of crystalline translations with Hermann's statistical (i. e., completely unordered) translations. Proof of such translations since then, which F. Rinne called paracrystals, has become unequivocal, and this has been shown especially by the works of R. Hosemann

[144–156]. The field of paracrystals is more comprehensive and totally includes that of liquid crystals. The Fourier space regarding the scattering of coherent X-rays by an ideal paracrystal can be divided into six embracing spheres that gradually merge from one to another.

1. Central spot $B_{0,r}$. This zone of small angle scattering is always a crystal reflection and thus enables one to determine the crystallite size.

2. Zone $B_{1,r}$. Within this zone are the undisturbed (crystal) reflections that exhibit symbathic changes of relative intensity, from which the shape of the paracrystal and the statistical shape distribution can be calculated.

3. Interzone $B_{2,r}$. This is the zone of reflections which exhibits varying degrees of disturbance. The integral line widths increase with decreasing crystallite size and with increasing type-2 lattice disturbances.

4. Zone $B_{3,r}$. This is the zone of strongly disturbed reflections, but the nodes of the lattice factor still have maximum intensities many times higher than the diffuse background.

5. Zone $B_{4,r}$. This zone exhibits the socalled liquid interference with the nodes of the lattice factor having maximum intensities only about 0.3 higher than that of the diffuse background.

6. Reflectionless zone $B_{5,r}$. The scattering phenomenon in this zone is only a superposition of the single diffraction figures of the individual lattice components. The paracrystal lattice has completely vanished in this zone insofar as the interference action is concerned.

Hosemann's new conception of paracrystallinity thus accounts for the vast diversity of aberrations from crystalline order [144–156]. In the above model, he used the distance-distribution function $\widehat{H_k(X)}$ of the cell-edge vector $\bar{a}_k$ ($k = 1, 2, 3$) that indicates the degree of deviation from an ideal crystal. The mean distance vector is given by

$$\bar{a}_k = \int X \, \widehat{H_k(X)} \, dV_X$$

where $\widehat{H_k(X)}$ is the distance-distribution function of $\bar{a}_k$. The standard deviation of $\bar{a}_k$ is defined as

$$\Delta X_{ki} = \left[\frac{1}{\bar{a}_i^2} \int (X - \bar{a}_k \cdot a_i)^2 \, H_k(X) \, dV_X \right]^{1/2}$$

Thus nine statistical parameters are introduced into the three-dimensional point lattice as shown in fig. 5.4.

The standard deviation ΔX_{ki} relative to the vector component $\bar{a}_i$ is defined by the G^{ki} components of the fluctuation tensor G_{ki} according to

$$G_{ki} = \frac{\Delta X_{ki}}{\bar{a}_i}$$

The ideal lattice is given when $G^{ki} = 0$. If fluctuations around a_i occur, then the reflections become more and more blurred with increasing scattering angle. If n_{ik} is the number of reflections in direction i, then

$$n_{ik} = \frac{0.3}{G_{ki}}$$

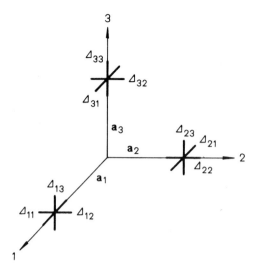

Fig. 5.4. The statistical fluctuations, Δ_{ik}, of three distance vectors, $\mathbf{a}_i$, in an orthorhombic paracrystalline lattice (from [154]).

Hosemann distinguishes between different degrees of lattice distortion, discussed above, according to the value of G_{ki}. Paracrystals appear in the range of the smallest recurring period as well as in medium and large elementary cells (see table 5.1 and fig. 5.5). Such periods as long as 10 to 50 nm have been observed in linear polyethylene [152–154], cf. chapter 13. Since units of even much larger dimensions were found in tensides, lipids, and as well in cholesteric phases, it seems reasonable to adopt the older but proven classification of the elementary cells based on their degrees of distortion [150, 155].

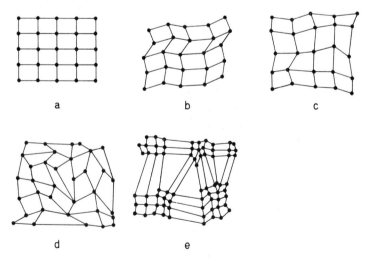

Fig. 5.5. Different types of two-dimensional lattices and superlattices: a) crystal, b) ideal paracrystal, c) real paracrystal, d) amorphous, e) microparacrystallites as knots of a paracrystalline superstructure with a three-dimensional network of tie molecules (from [154]).

Table 5.1. Degrees of distortion g_{ik} (from [150]).

g_{ik}	
None	Crystal
$<2\%$	Atomic or ionic lattices in high polymers, alloys, spinels, etc.
$2\text{--}10\%$	Macrolattices in natural and synthetic high polymers
$10\text{--}20\%$	Liquids
$>20\%$	Some glasses, and liquids at higher temperatures

Lattice constants a_k

> $0.1\text{--}1\,\text{nm}$ atomic and ionic lattices;
> $1\text{--}100\,\text{nm}$ molecular lattices, colloids;
> $>100\,\text{nm}$ optically visible particles.

R. Hosemann distinguishes the Hermann types according to their g'_{ik} $1\perp2$ values. We again limit our discussion to the combination SSR and SSD:

	g_{11} *	g_{12}	g_{31}	g_{23}	g_{22}	
SSR	$>20\%$	0	$>20\%$	$>20\%$	0	smectic
SSD	$>20\%$	$>20\%$	$>20\%$	0	0	nematic

* Steric identities limit the g_{ik} to these five cases, since $g_{11}=g_{33}$, $g_{12}=g_{32}$, $g_{13}=g_{31}$, and $g_{21}=g_{23}$.

Although Hosemann has not as yet investigated liquid crystals — their elementary super-structure cells attaining the dimensions of light-wave lengths — it was possible experimentally to obtain diffraction pictures corresponding to Hermann's g_{ik}-modified structural types on the basis of two-dimensional point lattices. These lattices were either calculated or drawn

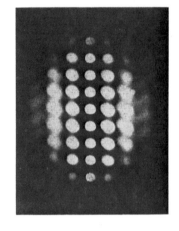 

Fig. 5.6. Model and optically obtained diffraction picture of a paracrystal showing totally crystalline, liquid, and amorphous X-ray reflections. Example: β-keratin from a feather (from [147]).

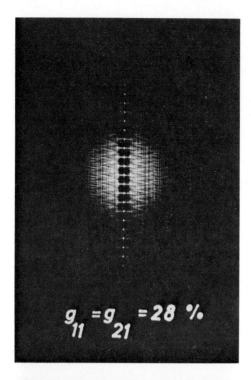

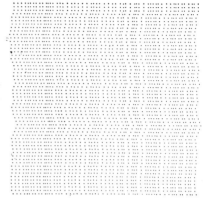

Fig. 5.7. Two-dimensional model of the meso-phase SSR, if the horizontal axis has the same features as the axis orthogonal to the drawing plane (from [150]).

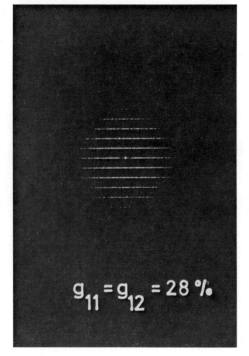

Fig. 5.8. Two-dimensional model of the meso-phase SSD under the same assumption of fig. 5.7 (from [150]).

using a computer and plotter and then reduced photographically [147, 150]. The diffraction patterns of the point models (point-diameter 85 μm) were taken with the Fraunhofer camera using a He-Ne laser beam (figs. 5.6, 5.7, 5.8).

The reader is referred to the literature for further information on optical models used in the study of liquid-crystalline structures [37, 39, 50, 55, 158–162, 180] as well as a discussion of the possibility of estimating the paracrystalline distortion from the broadening of the Debye-Scherrer lines [151]. Computers can be applied for the normalization and absorption correction of arbitrary X-ray scattering intensities from paracrystals [104, see also 2]. Real paracrystals, which are not liquid crystals, were frequently described. Some examples are biological materials (cf. chapter 12), synthetic polymers [149, 152–154], cf. chapter 13, carbon black [3], molten metals, alloys, and other inorganic materials [125, 148, 149, 156, 187, see also 9]. The applicability of the paracrystalline model has been discussed [14a].

Space limitations require the following subjects be mentioned here only as literature citations: details of X-ray diffraction techniques [23, 29, 35, 58, 66, 102, 129, 143, 178, 197a, 199, 206a, 234, 274], X-ray studies of supercooled liquid crystals [22, 24, 83], preparations oriented under magnetic fields [1:358, 20, 21, 26–29, 33–35, 45, 46, 56, 73, 105, 109, 129, 131, 133, 169, 200], those oriented under electric fields [26, 27, 30, 32, 34, 36, 38, 42–44, 123, 132, 133, 170, 171, 179, 206], streaming [3:116, 105], liquid crystals oriented by the influence of boundary surfaces [108], and calculations of the molecular distribution derived from the intensities of scattered X-rays [17, 75, 105, 106].

The standard technique in the case with the most examples includes the alignment by an external field. Some results will now be discussed in detail.

5.2   X-Ray Studies of Special Mesophases

Fourier analysis of X-ray diffraction data from liquid crystals cannot yield a definite atomic location. One method of investigation is to construct models and to compare their scattering behavior with the distribution of intensities observed experimentally. Another method is to begin with the model of a solid crystal as was first done by J. D. Bernal, more recently by J. Falgueirettes et al. for PAA [111], and for PAP [2:231].

Usually one can identify one or more "inner" rings ($2\theta \approx 5°$) and one more "outer" ring ($2\theta \approx 20°$). Cholesteryl compounds can be distinguished in all phases by a much narrower outer ring ($2\theta \approx 16°$) [88]. The diffuse rings of unoriented nematics grow sharper in aligned samples. Smectics exhibit sharp inner rings and differ markedly by their outer rings. The inner ring yields for a smectic phase the layer thickness and for an isotropic phase the molecular length. In all phases the outer ring yields the average distance between neighbouring parallel molecules.

5.2.1 Nematic Phases

Using the model of an aggregation of parallel molecules with cylindrical spatial requirements, E. Buchwald was able to explain Herrmann's observation of X-ray diffraction in nematic PAA as an intermolecular interference [16]. In addition to the intensive sickle-shaped intermolecular interference at 20°, J. Falgueirettes observed weak sickles caused by intramolecular interference perpendicular to this at 12°, 28°, and 44° 40′ [107]. Most of the newer investigations aim at the determination of the cylindrical distribution of atoms or the molecular axes [51, 74, 110, 111]. Using PAP (k 136 n 167 i) as an example, the most probable distance between neighboring molecules is 0.46 nm. Four coordination cylinders can be recognized around each molecule (fig. 5.9).

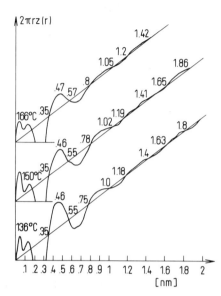

Fig. 5.9. Cylindrical distribution functions for the atoms of PAP (from [76]).

I. G. Christyakov et al. made an instructive comparison of crystalline, liquid, and mesomorphous phases (fig. 5.10) extending their X-ray studies to include the orienting influences of a.c. and d.c. electric fields [30, 32, 34].

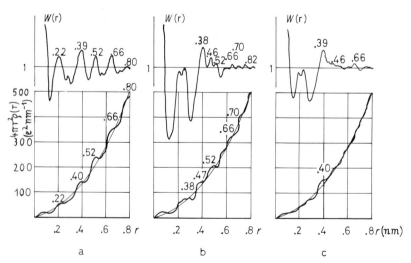

Fig. 5.10. Radial-distribution curves $4\pi r^2\rho(r)$ and probability functions $W(r)$ for PAA in the solid (a), nematic (b), and isotropic-liquid (c) states (from [30]).

In addition to the classical study objects, PAA and PAP, the entire homologous series has been thoroughly investigated [33]. I. G. Christyakov et al. compared the mean space between centers of molecules along the axis of texture (d) with the molecular length (l) calculated from the model, and they derived from this the angle, $\beta = \arccos d/l$, that is in agreement with the experimentally determined angle β (fig. 5.11). If the degree of order, $S = \langle 1 - \frac{3}{2}\sin^2\beta \rangle$, is calculated, then this plot shows an analogous alternation to that of the clearing points when the X-ray diagrams are taken at temperatures 5°C above the transition point from either solid crystal to nematic liquid crystal or from smectic to nematic mesophases. This is shown in fig. 5.12 [33, 44–46, see also 73, 180].

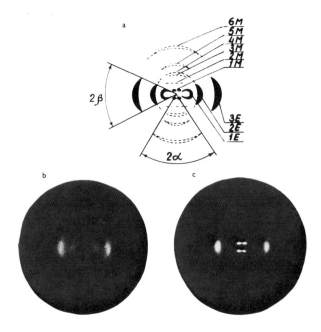

Fig. 5.11. The X-ray diagrams (CuK$_\alpha$) and their scheme (CuK$_\alpha$, MoK$_\alpha$) for nematic phases in the series   RO—◯—N(O)=N—◯—OR   where a is the schematic of b; b shows 1,3 members, and c shows 2,4–10 members (from [33]).

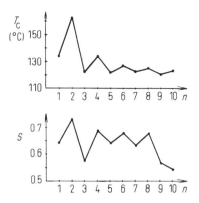

Fig. 5.12. The clearing temperature, T_c, and degree of order, S, vs. number of alkyl-chain carbon atoms, n, in the series

RO—◯—N(O)=N—◯—OR   (from [33]).

Some critical comments on cylindrical distribution functions were stated by A. de Vries [85, 86]; on the other hand, P. Delord et al. underline their validity and give a calculation [77, 78, see also 71]. The radial distribution curve for nematic anisalazine shows peaks at $r = 0.131$, 0.246, 0.403, 0.476, 0.566, and 0.654 nm. While the first two result from intramolecular atomic spacing, the four latter peaks are intermolecular interference maxima. Their nearly equal sizes indicate the lack of a certain preferred intermolecular distance [234].

C. C. Gravatt and G. W. Brady used the small-angle X-ray technique to pursue quantitatively the enlargement of cybotactic groups in the isotropic phase upon approaching the clearing temperature. These phenomena depend strongly on the purity of the material [122, 123]. W. L. McMillan found a strong pretransition scattering in the nematic phase upon approaching the n/s phase transition [200, 201, see also 252]. A high-resolution X-ray study of the critical fluctuations in nematic CBOOA near the n/s_A transition was made to elucidate the nature of this quasi second order transition [1c]. In the pretransition range of the nematic phase adjacent to a n/s_C transition, the tilt angle between the plane normal and the nematic director emerges [202].

Secondary nematic structures were found by A. de Vries during the X-ray investigation of bis-(4'-*n*-alkoxybenzylidene)-2-chloro-1,4-phenylenediamines [82]. In addition to the outer ring, which has an angle of diffraction of about 20° corresponding to an intermolecular distance of 0.5 nm there appears an inner ring with an angle of diffraction that decreases as the molecular length increases. This phenomenon is incompatible with the classical concept of the nematic phase and suggests an additional order of molecules within cybotactic groups for which a probable structure is proposed.

5.2.2 Cholesteric Phases

X-ray analysis to date has not furnished criteria for differentiation between nematics and cholesterics. It is only mentioned that the helical twist causes a broadening of the peaks on the cylindrical distribution curve due to reduced parallelism (see fig. 5.13 and 5.14).

A small but normally detectable discontinuity accompanies the c/i phase transition in contrast to the very distinct fashion exhibited by the s_A/c transition point (fig. 5.15).

By X-ray scattering the presence of smectic clusters in the pretransitional cholesteric phase was proven and the number of molecules was estimated to be to the order of 400 [231]. A layer structure is also suggested in ref. [272]. Other X-ray studies of cholesterics are reported in the literature [103, 104a, 112a, 206a].

Recently, H. Stegemeyer* et al. have published systematic studies on the "blue phase" that has been shown by optical and calorimetric measurements to be a polymorphic variant of the nematic-cholesteric state. It exhibits the properties of ordinary cholesterics, the selective reflection of circularly polarized light within a narrow wavelength range and a strong angular dependence of this reflection. Evidence has been found for the existence of two "blue phases" of cholesteryl myristate; one is observed between 84.0 and 84.45 °C, and the other between 84.45 and 84.6 °C. A characteristic feature is their lack of birefringence which H. Stegemeyer explains by assuming an angle θ between the helical axis and the molecular long axes; at a critical angle $\theta = 54.74°$ the negative indicatrix of cholesteryl myristate becomes spherical.

---

* H. Stegemeyer et al., Z. Naturforsch. *34a*, 251, 253, 1031 (1979).

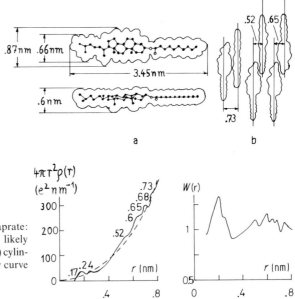

Fig. 5.13. Structure of cholesteryl caprate: a) shape of the molecule, b) most likely arrangement in the cholesteric phase, c) cylindrical distribution curve, d) probability curve (from [23]).

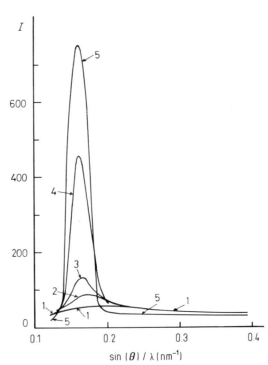

Fig. 5.14. Measured X-ray scattered intensity, I (arbitrary units), per unit solid angle for cholesteryl myristate. Key: 1, 95.4°C (i); 2, 89.8°C (c); 3, 76.2°C (c); 4, 71.0°C (s); 5, 56.8°C (s) (from [199]).

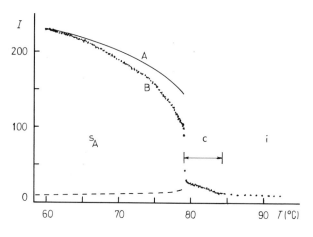

Fig. 5.15. Measured X-ray scattering intensity, I, vs. temperature, T, at the Bragg angle for cholesteryl myristate. The points are data taken while cooling from the isotropic liquid, and these reproduced well on heating. The dashed line is the calculated diffuse-scattering and fluctuation-scattering contribution. Solid lines A and B show Bragg scattering plus diffuse- and fluctuation-scattering calculated from the microscopic theory for model potentials A and B respectively. The theoretical intensity has been fit to the experimental intensity at the lowest temperature (from [199]).

5.2.3 Smectic Phases

A comparison of the X-ray patterns indicates three main classes of smectics [89]:
1. with a diffuse outer ring: s_A, s_C, s_F, s_D
2. with a single sharp outer ring: s_B
3. with several sharp outer rings: s_E, s_G, s_H.

s_B phases are divided in three subgroups [90, see also 190]. s_E phases exhibit three sharp outer rings, while s_G and s_H phases have several sharp outer rings [88]. Neighbouring parallel molecules are usually separated by an intermolecular distance of 0.5 ± 0.015 nm, and the exact values in the different phases are compiled in ref. [88].

That the variety of smectic polymorphism is not yet exhausted has been demonstrated recently [79 b]. The X-ray diffraction patterns of two biphenyl compounds are unlike any other previously reported; they contain a sharp inner ring and two diffuse outer rings [196]. Surprisingly this new smectic phase is completely miscible with a conventional s_A phase, it is called bilayered smectic A.

5.2.3.1 Smectic A Phases

Thorough X-ray investigations have uncovered the following data relating to ethyl p-ethoxybenzylidine-p-aminobenzoate [83]. The intermolecular distance is smallest in the smectic phase at 0.4894 nm and rises only slightly (0.09 pm/°C) with increasing temperature. The intermolecular distance is 0.4932–0.4950 nm in the nematic phase and 0.495–0.5182 nm in the isotropic liquid, but the temperature dependence is much stronger here (fig. 5.16).

Like the intermolecular distance, the layer thickness in the smectic phase increases only slightly (0.07 pm/°C) with temperature (fig. 5.17). It is difficult to explain why the average 1.994 nm thickness of the smectic layers is about 7% smaller than the molecular length of 2.14 nm measured in the isotropic liquid. Since in the s_A phase the angle between the major molecular axes and the layer planes is very close to 90°, and since a non-stretched arrangement appears improbable, an interpenetration of the layers has been suggested as a possible explanation.

The observations delineated above recur with 4-alkoxybenzylidene-4'-ethylanilines [87], fig. 5.17. However, new insight was gained from the fact that dD/dT in the nematic phase is greater than in the s_A, and this is greater than in the isotropic phase.

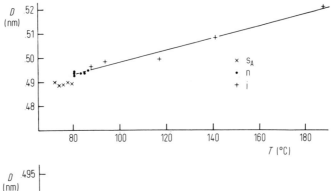

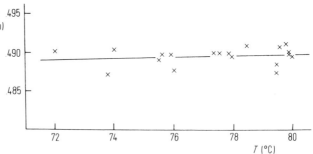

Fig. 5.16. Intermolecular distance, D, in ethyl p-ethoxybenzylidene-p'-aminobenzoate: k 93.4 i (87.6 n 80.4 s_A) as a function of temperature (from [83]).

W. R. Krigbaum, J. C. Poirier, and M. J. Costello offer an explanation of the X-ray scattering that contradicts the previously accepted distinction between s_A and s_C phases [66, 181]. They suggest that molecules, dimers, or molecular clusters can be inclined to the smectic

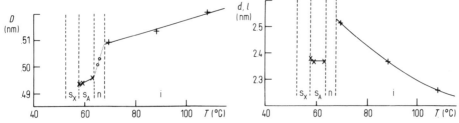

Fig. 5.17. Plots of intermolecular distance, D, and smectic layer thickness, d, or apparent length, l, of molecules in the isotropic phase versus temperature. The dashed lines mark the transition temperatures. s_X, s_A, n, and i denote the phases of 4-n-heptyloxybenzylidene-4'-ethyl aniline (from [87]).

plane even in the A phase, e.g. in ethyl-p-azoxybenzoate and thallium stearate (s_A?). Some examples of socalled bilayer s_A phases are also known where the smectic layer thickness is about twice the molecular length [1:95, 124a].

X-ray studies of s_A phases have been made that agree with McMillan's theory [246].

5.2.3.2 Smectic C Phases

The polymorphous substance n-amyl-4-(4-n-dodecyloxybenzylideneamino) cinnamate (k 73.9 s_B 95 s_C 106.7 s_A 134.3 i) provides a valuable study material. H. Sackmann et al. observed nearly the same distribution function with maxima at 0.46 and 0.9 nm for each of the three smectic modifications s_A, s_C, and s_B [232]. Twisted smectics C exhibit diffraction patterns similar to normal smectics C [21 a].

5.2.3.3 Smectic B, Smectic G, and Smectic H Phases

Both vertical and tilted alignment were found in s_B phases, but tilted s_B phases are now designated as s_H phases. Examples are ethyl *p*-ethoxybenzylidene-*p*-aminocinnamate (EBAC) and TBBA introduced in fig. 5.18 and table 5.2 [98, 188].

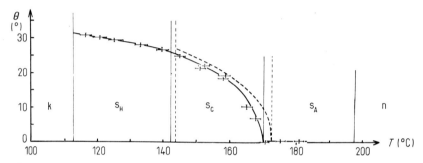

Fig. 5.18. Tilt angle variation of three smectic phases of TBBA (from [98]).

Table 5.2. Structural data of s_B and s_H phases (from [188]).

	EBAC (s_B)	TBBA (s_H)
Smectic layer thickness c (nm)	2.15	2.3
Apparent molecular length (nm)	2.15	2.63
Hexagonal lattice parameter in the plane of the layers (nm)	0.485	0.52
Mean angle of the molecular axis with respect to the smectic planes	90°	58°
Mean fluctuation of this angle	6°	2°

An example of a s_G phase is *p*-butoxybenzylidene-*p'*-ethylaniline (BBEA) which was first designated as s_H [84]. Evidently the structure is that illustrated in figs. 5.19 and 5.20. s_H phases occupy special interest because in these phases the two dimensional lattice of the

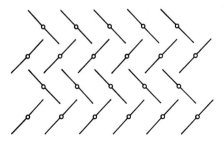

Fig. 5.19. Schematic representation of a herringbone-type packing as seen along the long axes of the molecules. The lines represent the directions of the planes of the molecules, and the circles indicate the positions of the long axes of the molecules, which occupy the lattice points of a hexagonal lattice (from [84]).

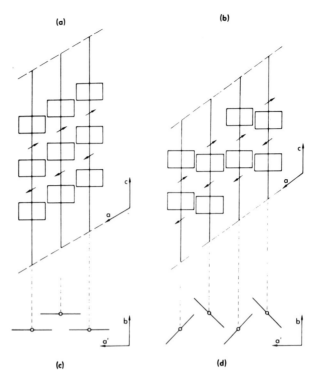

Fig. 5.20. In (a) and (b), schematic drawings are given of the arrangement of the molecules in a layer seen along the b axis of the monoclinic lattice. (a): TBBA, s_H and (b): BBEA, s_G. The long lines represent the molecular axes, the rectangles the phenylene groups (C_6H_4; the corners of the rectangles give the positions of the hydrogens), and the short arrows the dipoles. In (c) and (d) are shown the corresponding arrangements of the molecular planes as seen long the c axis (from [91]).

molecules within the smectic layers is extended by a special order between the layers to form a three dimensional (e. g., monoclinic) lattice [72a, 80, 91, 101a]. Two models of packing are reproduced in fig. 5.20.

5.2.3.4 Smectic E Phases

Three s_E phases showed a molecular alignment perpendicular to the layers [95, 100, 101, 233]. Within each layer the lattice is orthorhombic and was possibly induced by a small deformation of the hexagonal lattice of the higher-temperature s_B phase (see fig. 5.21).

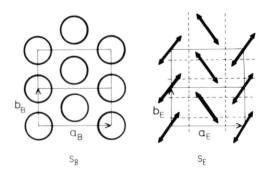

Fig. 5.21. Positions of the molecules in s_B and s_E phase (from [100]).

5.2.3.5 Smectic D Phases

The X-ray diagrams of s_D phases show a diffuse reflection at large angles around $s \simeq (0.45\,nm)^{-1}$ and six reflections in the small angle part [233, 247]. Their spacing ratios are indicative of a body centered cubic lattice, and the parameter of the unit cell was determined to be 10.2 nm. The s_D modification thus combines a high level of long range order in three dimensions with a high disorder at the atomic level. In this respect they are closely related to cubic lyophases, but they are completely different from plastic crystals.

5.2.4 Mesophases of Carboxylic Acid Salts

We recall the similarity of these mesophases to the smectic phases of non-saltlike compounds. It is greatest in the two high-temperature modifications, neat and s_A, although there is no continuous miscibility [1:322]. A. Skoulious deserves special note for his X-ray investigations [2:670–2:676, 236–242]. The structural parameters of all mesophases are known for some sodium salts (table 5.3).

Table 5.3. Structural parameters of sodium-salt mesophases; temperatures in °C, distances in nm (from [2:670]).

Phase	Structure	Sodium laurate	Sodium myristate	Sodium palmitate	Sodium stearate
waxy	two-dimensional rectangular	166 a = 7.52 b = 3.09	155 a = 8.00 b = 3.45	152 a = 8.66 b = 3.83	140 a = 8.00 b = 4.03
superwaxy	two-dimensional rectangular	195 a = 6.85 b = 3.03	193 a = 6.90 b = 3.36	195 a = 7.42 b = 3.60	198 a = 7.25 b = 3.79
new phase	three-dimensional orthorhombic	205 a = 5.55 b = 2.83 c = 3.27			
subneat	two-dimensional rectangular	234 a = 4.98 b = 2.70	233 a = 5.38 b = 2.91	236 a = 5.62 b = 3.06	238 a = 6.24 b = 3.44
neat	lamellar	290 d = 2.53	271 d = 2.75	278 d = 2.91	285 d = 3.06

The following sections discuss the main structural features of the mesophases of saltlike compounds. Section 5.7 of this chapter contains a complete list of literature references since space limits us here to a few examples.

5.2.4.1 Lamellar Structure

The lamellar structure is built up by parallel, equidistant layers of metal ions. No order has been recognized in the distances within the layer or in the conformation of the paraffin chains. This structural type is seen in the neat phases, and temperature-dependent layer distances are significant. Small clusters of the lamellar arrangement persist a few degrees above the melting point.

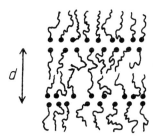

Fig. 5.22. Schematic drawing of the lamellar structure in the neat phase of anhydrous soaps; d is the layer distance.

5.2.4.2 Two-Dimensional Centered Rectangular Structure

This structural type is found in waxy, subwaxy, superwaxy, and subneat phases. It comprises bands arranged as shown in fig. 5.23.

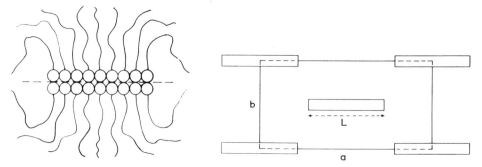

Fig. 5. 23. Schematic drawing of soap molecules in a band, and of bands in the mesomorphous state (from [239]).

The parameters a and b of the two-dimensional centered rectangular structure and the breadth, L, of the ribbons are in nm:

Soap	T (°C)	a	b	L
Li palmitate	200	8.76	3.7	3.6
Li palmitate	220	8.01	3.55	3.1

With potassium soaps a two-dimensional oblique arrangement of ribbons was found [237, 240]. In a certain sense, this can be considered a broken lamellar structure since the length of the bands is unlimited.

5.2.4.3 Structures with Disc-Shaped Groups

Disc-shaped structural elements with unordered paraffin chains can take the place of bands [2:675, 2:676, 240]. This type resembles a new phase of sodium laurate in which weakly asymmetrical discs form a three-dimensional body-centered orthorhombic structure [2:670], see fig. 5.24.

5.2.4.4 Structures with Cylindrical Groups

Soaps with bivalent metals form structures comprising a two dimensional hexagonal arrangement of cylinders [2:671, 2:673–2:675] or a body-centered cubic arrangement [193].

Fig. 5.24. Model of the structure containing disc-shaped groups (from [2:675]).

5.3   Neutron Scattering

The quasi-elastic portion in the scattering spectrum of low-energy neutrons through a liquid crystal contains information on stochastic phenomena as the molecular translational, rotational, and vibrational movements that occur within approximately 10^{-11} s (fig. 5.25). If the nonelastic scattering and a weakly elastic component caused by the sample holder are eliminated the quasi-elastic scattering peaks are obtained (fig. 5.26). Interesting details on molecular dynamics can be drawn from these experiments although their interpretation is rather complicated and cannot be elucidated here [5, 7a, 7b, 67, 93, 183, 215, 216, 230, 247a, 269a].

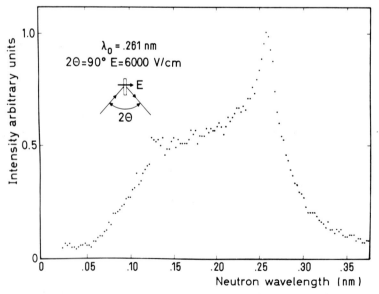

Fig. 5.25. A recorded time-of-flight spectrum in nematic PAA at 125°C plotted as a function of the neutron wavelength. The scattering geometry is denoted by the direction of the incoming and outgoing neutron beams and by the position of the flat sample holder. A vector E indicates the direction of the applied electric field (from [165]).

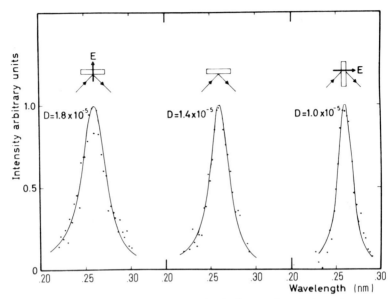

Fig. 5.26. Quasi-elastic neutron scattering in PAA at 125°C. A sketch above each peak shows the scattering geometry used. The direction of the electric field, when applicable, is indicated by a vector. The smooth curves represent a convolution of the instrumental Gaussian function with a Lorentzian giving the best fit to be experimental data (from [165]).

It has been shown that for very small momentum transfer the translational diffusion dominates quasi-elastic scattering. If the momentum transfer is large three peaks superimpose arising from anharmonic inelastic, translational, and rotational effects which are now the most important [230]. The information is discerned from the broadening of the neutron "line" (i.e. energy distribution) after scattering from a liquid crystal.

For $Q^2 \to 0$, the following equation is valid [12, 165]:

$$\Delta E = 2\hbar D' Q^2$$

where ΔE is the quasi-elastic line width and Q is equal to $\frac{4\pi}{\lambda_0}\sin\delta/2$, with δ: scattering angle and λ_0: de Broglie neutron wave length. D' is an apparent diffusion coefficient that is the sum of the true self-diffusion coefficient D, a fictitious coefficient D_{real} (that describes the motion of the average proton relative to its molecular center of gravity), and a small mixture term D_{mix} (between the relative motion and the center-of-gravity motion). Thus:

$$D' = D + D_{\mathrm{real}} + D_{\mathrm{mix}}$$

The dependence of the peak broadening ΔE and the diffusion coefficient D' upon temperature is illustrated in figs. 5.27 and 5.28.

The apparent diffusion coefficient, D', is temperature-independent in nematic PAA, anisalazine, and 4,4'-diheptyloxyazoxy benzene [11–13]. Equal values of D' have been measured in both nematic and smectic phases of 4,4'-diheptyloxyazoxy benzene. Neutron scattering on oriented nematic PAA (fig. 5.26) at 125°C gave the expected values of D'. $D'_{\parallel} = 1.8 \cdot 10^{-5}$,

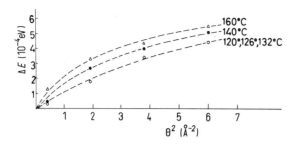

Fig. 5.27. Quasi-elastic broadening of the incoherently scattered neutron line in PAA as a function of the square of the momentum transfer (from [12]).

$D'_\perp = 1.0 \cdot 10^{-5}$, and $D'_{av} = 1.4 \cdot 10^{-5}$ cm²/s [165]. Compared to the ratio of the viscosity coefficients $\eta_{av}/\eta_{\parallel} = 2$, the ratio $D'_{\parallel}/D'_{av} = 1.3$ is markedly smaller. Since the Einstein equation, $D = kT/6\pi \cdot r \cdot \eta$ is applicable only to spherical molecules, it is invalid for molecules of this type. PAA was also reported with the values $D = 3.4 \cdot 10^{-6}$ and $D_{av} = 4.1 \cdot 10^{-6}$ cm²/s from measurements performed at relatively small scattering vectors [253]. Other studies on PAA [4, 6, 167], for solid PAA, see [142a], other nematics [7, 7a, 7b, 186, 186a] and a review [227a] are reported in the literature. Long before these experiments, which were all performed in the seventies, J. A. Janik et al. had studied the motion and orientation of PAA molecules in the nematic phase by use of neutron scattering [163, 164]. The incoherent cross-section for neutron quasi-elastic scattering in a nematic liquid crystal was calculated by K. Rosciezewski [228, 229]. An additional inelastic scattering peak was found among the quasi-elastic scattered neutrons in various liquid crystals [96].

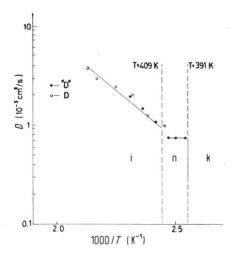

Fig. 5.28. Temperature dependence of the diffusion coefficients in PAA, D' is the apparent diffusion coefficient obtained from the neutron scattering data as $K \to 0$, and D is the self-diffusion coefficient determined by NMR (from [12]).

In MBBA evidence of rotational proton jumps was found whose correlation times are of the order of 10^{-12} s [166]. With the same substance large differences in the values of the diffusion coefficients are reported if NMR and neutron data are compared [14], see also section 3.9.

The protons play the dominant role in incoherent neutron scattering. Their scattering cross section being about one order of magnitude greater than that of other nuclei. In order to eliminate the incoherent proton scattering, perdeuterated liquid crystals have been studied.

Inelastic and elastic coherent neutron scattering can be observed in such mesophases. Data from these observations provide information on harmonic motions which always have a collective character. A concept of quasi-phonons has been derived from the quantum model of crystal lattice vibrations [14b]. Based on this concept, the nematic order parameter and pretransition phenomena have been studied in d_{14}-PAA [68, 221–225, 227]. N. Niimura found 5 peaks in the molecular structure factor of d_{14}-PAA within the momentum transfer range from 5 to 150 nm^{-1} [211, 213], fig. 5.29.

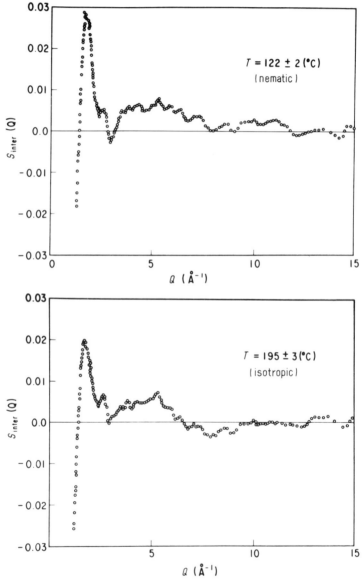

Fig. 5.29. The structure factor of d_{14}-PAA depending on the intermolecular correlations (from [213]).

Two are attributed to intramolecular distances of 0.114 and 0.203 nm. The most probable intermolecular distance both for the nematic and the isotropic liquid is 0.35 nm. The intensities as shown in fig. 5.29 and the range between 60 and 80 nm^{-1} in $S_{inter}(Q)$ differ clearly.

The dispersion of coherent inelastic neutron scattering in nematic d_{14}-PAA has also been studied [65, 65a, 217a]. Neutron scattering experiments were also applied or suggested to study the critical temperature gradients for the onset of convection and turbulence in nematics [10] or electrohydrodynamic instabilities [212]. Neutron-scattering studies of smectics have mainly concerned TBBA (s_A, s_C, s_H, $s_{VI}=s_G$, and S_{VII}). Solid-crystalline TBBA is discussed in ref. [267]. The literature reports details on orientational order [94b, 138, 140, 141, 269a], pretransition phenomena [268], tilt angle measurements [139, 142], translational motions [94, 94a], rotational motions [137, 167, 269], and an undulation mode contribution [94a] in TBBA. Other neutron scattering experiments were performed with smectics A [217], e. g. diethyl *p*-azoxydibenzoate [18, 19], ethyl 4-(4'-acetoxybenzylidene)aminocinnamate [186, 186a], CBOOA [65b] smectics B, e. g. 4-*n*-hexyloxybenzylidene-4'-propylaniline [235] and smectics E [205a], e. g. d_9-butyl 4(4'-phenylbenzylidene)aminocinnamate [185].

5.4   *Positron Annihilation*

Valuable informations about the irradiated material can be obtained from the fact that the average lifetime of positrons depends on their environment. Details on this subject are discussed elsewhere [1]. This method applied to liquid crystals shows that only those phase transitions are distinctly observable which are connected with a sudden change in the intermolecular dipole-dipole interaction: k/n, k/c, k/s, s/n, s/c, and naturally also k/i (figs. 5.30 and 5.31). Experience indicates that the higher degree of order corresponds to the shorter positron life. The lifetime spectra are composed of a short-lived component (lifetime τ_1 and intensity I_1) which is attributed to annihilation of free positrons and of parapositronium, whereas the long-lived component (τ_2 and I_2) is considered to be the result of orthopositronium annihilation [210].

The lifetime of positrons is but little affected by the phase transitions n/i, c/i, and s/i. It may be assumed from this that positrons are destroyed in similar environments slightly

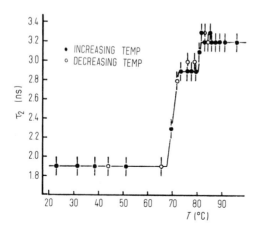

Fig. 5.30. Position lifetime, τ_2, vs. temperature, T, ("lifetime spectrum") in cholesteryl myristate: k 71 s_A 81 c 86.5 i (from [63]).

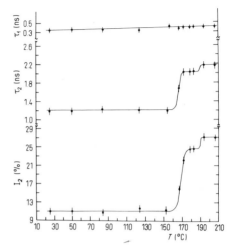

Fig. 5.31. Positron lifetimes τ_1 and τ_2 and intensity I_2 of the long-lived component as a function of temperature in p-methoxycinnamic acid: k 175.5 n 189 i (from [210]).

above and below the clearing point. Thus, the cybotactic groups in the liquid and the liquid-crystal swarms must possess similar structures. An annihilation scheme including both conversion and pickoff annihilation has been suggested [210]:

$$\text{annihilation} \xleftarrow{\lambda t} \text{oPs} \underset{\gamma}{\overset{3\gamma}{\rightleftarrows}} \text{pPs} \xrightarrow{\lambda s} \text{annihilation}.$$

More information can be deduced from the angular distribution of the annihilation quanta, but only a few papers deal with the dependence on molecular order in nematics [15b, 203, 275]. Cholesteryl esters were the liquid-crystalline substances first and most intensively investigated using this method [61–64, 205]. Positron lifetimes were also studied in plastic crystals and compared with those of liquid crystals [182]. Fig. 5.32 presents an illustrative example of a polymesomorphous compound [208, 270].

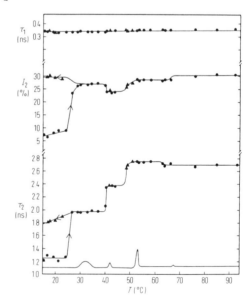

Fig. 5.32. Positron lifetimes, τ_1 and τ_2, and intensity I_2 of the long-lived component as a function of temperature, T, in 4-n-butyloxybenzylidene-4'-ethylaniline: k 28.5 x 41.5 s 51.5 n 66.5 i (from [270]).

5.5   *Mössbauer Investigations*

By their nature, liquid crystals do not seem likely to exhibit recoil-free emission and absorption of γ-rays when it is recalled that the Mössbauer effect of solute molecules can be observed in liquids only under very special circumstances, e. g. in glycerol whose hydrogen bonds give a sufficiently rigid matrix. Experience has shown that the smectic state also made a recoil-free absorption possible. One of the first observations [255] had to be reinterpreted because the iron bearing compound was not dissolved but rather incorporated as a suspension [261]. However, it has been interesting to find that the crystals were ordered by the s_C phase in such a way as to feign a solution. Aside from the study of the liquid crystalline state proper, Mössbauer spectroscopy can also be applied to the study of the solute material analogous to the NMR-technique of A. Saupe and G. Englert (figs. 5.33 and 5.34, cf. section 10.3).

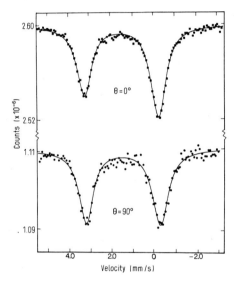

Fig. 5.33. Mössbauer spectrum of a solution of 3.5% (by weight) of triethyltinpalmitate in the supercooled s_H phase of 4-*n*-hexyloxybenzylidene-4'-*n*-propylaniline at 77 K for two angles, θ, between the direction of an external magnetic field and the γ-ray beam from the $BaSnO_3$ source (from [258]).

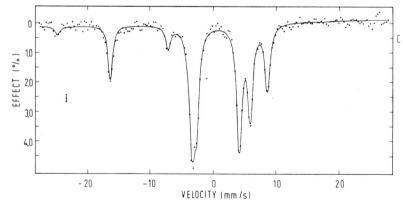

Fig. 5.34. Mössbauer spectrum of 6-heptyloxy-5-iodo-2-naphthoic acid samples arranged in a nematic order by a 4 kG field, quenched to −90°C and measured at 55 K (from [218]).

The smectic matrix may serve to create a "monocrystal" of the solute that can be oriented by a magnetic field. The literature reports some tilt angle and order parameter measurements. In some s_B glasses a remarkable low degree of order was measured for different solutes [1a, 9a, 264b, 264c]. At low temperatures, the Mössbauer effect can also be observed with other quenched mesophases, with nematics (see fig. 5.34), and with cholesterics [49, 53, 54]. In many cases it is uncertain whether supercooled or glassy mesophases have been studied, but a glassy state seems to be much more probable (see section 8.5).

A comparison of the recoil-free absorption of 1,1'-diacetylferrocene in the crystalline and in the glassy s_H phase of 4-*n*-hexyloxybenzylidene-4'-*n*-propylaniline shows that the probe is much more rigidly anchored in the solid [257, 259]. The tilt angle and the lattice contribution to the vibrational anisotropy in a s_H liquid crystal are evaluated from the angular dependence of the Mössbauer recoil-free fraction [260]. Mössbauer studies have indicated the onset of diffusion occurring at the glassy s_H/supercooled s_H transition temperature, T_g, e. g., $T_g = 233$ K for *p*-hexyloxybenzylidene-*p*'-*n*-propylaniline [264a]. Probability calculations for the Mössbauer effect in mesophases were published in the literature [116, 168]. J. M. Wilson and D. L. Ulrich considered the theoretical aspects of the intensity asymmetry of ^{57}Fe quadrupole split lines in smectic or frozen nematic solutions [256]. Anomalies arising from the atoms bonding to the free surface of smectics are expected and could possibly be seen in Mössbauer or neutron experiments [72]. Other Mössbauer investigations of liquid crystals are described in the literature [6:53, 81, 117–119, 219, 220, 262–264, 276].

5.6   Cherenkov Structure Radiation

If a charged high-velocity particle passes through a crystal, it emits a coherent radiation due to the spatial periodicity of the electron density. The observation of a similar phenomenon in the Grandjean texture of cholesterics can be seen as a valuable aid to the understanding of the structures of these phases [8, 174]. In these cases, the radiation is not due to a periodicity of density as it is in crystals but to the regular change in molecular orientation within a helical arrangement. The wavelengths of this radiation, which ranges from visible light to soft X-rays, changes with the pitch of the helix and with the anisotropy of the molecular polarizability.

An important difference is found in the polarization of the radiation. In a crystal this is linear and independent of the direction of emission, but in a cholesteric liquid crystal it is elliptical and dependent on the direction of emission. For a particle moving along the optical axis, the polarization varies from circular (for emission in this direction) to linear (for emission perpendicular to the path of the particle).

About 10^{-3} photon per particle is emitted when the thickness of the layer is $\approx 100\,\mu$m, the pitch ≈ 100 nm, the anisotropy parameter $\approx 10^{-2}$, and the particle velocity ≈ 0.9 of the phase velocity of light.

Only short notices have been published on the detection of ionizing particles with liquid crystals [120, 209], see also p. 610. A γ-ray dosimeter using cholesterics is mentioned [1b], but little is known on the radiation stability and radiolysis of liquid crystals [227b, 233a], see also p. 610.

5.7   Bibliography of X-Ray Studies on Liquid Crystals

1   Derivatives of azoxybenzene

PAA   [1:358–1:361, 16, 20, 26, 28, 30, 32, 34, 36, 39, 42, 55, 73–78, 105, 107, 109–111, 122–124, 129, 132, 133, 157, 169–173, 175, 179, 191, 226, 245, 265]
PAP   [76, 108, 157, 175]
p-alkoxyazoxybenzenes (others than PAA and PAP)   [33, 42, 44–47, 52, 202, 232]
ethyl p-azoxybenzoate   [66, 113, 134, 136, 181]
ethyl p-azoxycinnamate   [113]
n-hexyl p-azoxycinnamate   [232]
allyl p-azoxy-α-methylcinnamate   [232]
allyl phenethylazoxybenzoate   [130, 132–134, 136]

2   Derivatives of benzylideneaniline

p-n-alkoxybenzylidene-p-toluidine   [23, 27, 201]
p-n-alkoxybenzylidene-p-ethylaniline   [84, 87, 88, 91]
p-n-alkoxybenzylidene-p-n-propylaniline   [79, 79a, 79b]
MBBA   [184, 197, 248, 250, 251, 277]
EBBA   [184, 277]
p-phenylbenzylidene-p-n-butylaniline   [192]
other p-n-alkoxybenzylidene-p-n-alkylanilines   [72a, 101a]
p-methoxybenzylidene-p-cyanoaniline   [15, 126, 184]
CBOOA   [1c, 200]
ethyl p(p-ethoxybenzylideneamino)benzoate   [83, 88]
methyl p(p-ethoxybenzylideneamino)cinnamate   [132, 134, 136]
ethyl p(p-methoxybenzylideneamino)cinnamate   [1:35,21,38,40–42,48,55,136,160,180,206,232]
ethyl p(p-ethoxybenzylideneamino)cinnamate   [88, 136, 188, 232, 48 (here also higher homologs)]
ethyl p(p-ethoxybenzylideneamino)-α-methylcinnamate   [136]
amyl p(p-alkoxybenzylideneamino)cinnamate   [88, 232]
ethyl p(p-methylbenzylideneamino)cinnamate   [232]
alkyl p(p-phenylbenzylideneamino)cinnamate   [100, 134, 136]
TBBA   [8:312, 15a, 88, 91, 97–99, 188, 189, 191, 192, 244]
homologs of TBBA   [15a]
terephthal-bis-(ethyl p-aminocinnamate)   [134, 136]
bis-(p-n-alkoxybenzylidene)-2-chloro-1,4-phenylenediamine   [82, 88]
1,5-bis(p-methoxybenzylideneamino)naphthalene   [131]
bis-(p-n-alkoxybenzylideneamino)-4,4′-biphenyl   [243, 244, 244a, 244b]

3   Azomethine-azo compounds

p-anisylideneaminoazobenzene   [3:116, 42]
p-(p-n-nonyloxybenzylideneamino)azobenzene   [232]
1-phenylazo-4(p-alkoxybenzylideneamino)naphthalene   [22, 31]
N-(p-dimethylaminobenzylidene)-4(p-carbethoxy-phenylazo)-1-naphthylamine   [24, 25]
1-phenylazo-4(p-alkylbenzylideneamino)naphthalene   [55]

4   Azo compounds

ethyl *p*-acetoxybenzene-*p*-azocinnamate   [88, 136]

5   Steroids (cholesteryl esters)

chol.-formate   [1:274]
chol.-acetate   [273]
chol.-propionate   [157, 226]
chol.-butyrate   [1:274]
chol.-caprate   [23, 27, 112a]
chol.-pelargonate   [199, 250, 251, 272]
chol.-myristate   [2:121, 199, 272]
chol.-stearate   [195, 272]
chol.-esters of saturated fatty acids with various chain lengths   [271]
chol.-ester mixtures   [21a, 24, 177, 231]
chol.-esters of dicarboxylic acids   [103]
chol.-benzoate   [1:274, 157]
chol.-cinnamate   [1:274]

6   Carboxylic acid salts

Na butyrate   [254]
Na isovalerate   [254]
Na laurate   [2:670, 236]
Na myristate   [2:670, 236]
Na palmitate   [2:670, 214, 236]
Na stearate   [2:670, 198, 236, 237]
Li soaps   [2:779, 237, 239]
K soaps   [2:672, 237, 240]
Rb soaps   [241]
Cs soaps   [2:676]
Tl soaps   [66, 135, 181]
Mg soaps   [2:673]
Ca soaps   [2:454, 2:671, 2:675, 2:778, 193, 194]
Sr soaps   [2:454, 193, 238]
Ba soaps   [193]
Cd soaps   [2:674]
di-soaps   [242]

7   Miscellaneous

saturated aliphatic esters   [2:425]
p-nitrophenyl *p'-n*-decyloxybenzoate   [124a]
benzene-hexa-n-alkanoates   [2:98a]
p,p'-alkoxybenzalazine   [132, 157, 234]
1-(*p-n*-pentyloxybiphenyl)-4'-glyoxal-2-phenylhydrazine   [233]

p-nonyloxybenzoic acid   [1:35, 43, 55, 59, 160]

4'-*n*-alkoxy-3'-nitro-biphenyl-4-carboxylic acid   [232, 233, 247]

n-alkyl-4'-*n*-alkoxybiphenyl-4-carboxylates   [2:300, 249, 251]

p-*n*-alkoxyphenyl 4'-*n*-alkoxybiphenyl-4-carboxylates   [2:300]

bisalkyl-*p*-terphenyl-4,4''-dicarboxylate   [15a, 60, 95, 233]

bis-(*p*-methoxyphenyl) trans-1,4-cyclohexanedicarboxylate   [15a]

bis-(*p*-*n*-hexyloxyphenyl)mercury   [66, 181]

2-(*p*-*n*-pentylphenyl)-5-(*p*-*n*-pentyloxyphenyl)-pyrimidine   [1:325]

2-(*p*-*n*-decyloxybiphenyl)-quinoxalline   [233]

4'-*n*-alkyl-4-cyanobiphenyl and 4'-*n*-alkoxy-4-cyanobiphenyl   [1:95, 184, 196]

p-*n*-hexyloxybenzylidene-5-amino-2-butoxypyridine   [88]

Optical Properties of Optically Inactive Mesophases

6

6 Optical Properties of Optically Inactive Mesophases

Continuing the discussion begun in chapter 1, this chapter first covers birefringence which manifests itself as the visible expression of the anisotropic molecular order in the previously described textures. The thermal fluctuations of these ordered domains are the cause of the light scattering properties of liquid crystals. On the other hand the orientation of the individual molecule and its vibration (especially the intermolecular correlation of the vibration phenomena) determine the light absorption properties of liquid crystals and solute molecules. This offers an excellent method for the measurement of linear dichroism. Analogously, measurements of fluorescence polarization allow the assignment of the direction of transition moments in molecules. Of particular interest are the low-frequency Raman scattering bands which correspond to certain types of external vibrations of the lattice. The basic problem in the investigation and discussion of the light absorption of mesophases is the distinction between molecular and phase characteristics which is, in other words, the typical difficulty of differentiating between actual and form birefringence.

6.1 Birefringence

From a vast amount of research material D. Vorländer [11, 319] and P. Gaubert [115, 117, 118] have found that all liquid crystals showing coloration (i.e. all cholesterics having Grandjean texture) are optically negative, while all others (i. e., the achiral) are optically positive. O. Wiener [323] and H. Zocher [1:404, 331] have explained the positive character of the birefringence of rod-shaped molecules; the work of H. Zocher, who, being engaged in the manufacture of dichroitic layers, has been especially important. P. Gaubert also investigated the birefringence of liquid-crystalline mixtures [116, 120] and the pleochroism of optically negative and positive liquid crystals colored with iodophenol [119]. As P. Châtelain has stressed, special attention must be given to the orientation of the sample as is always the case in measuring anisotropic properties [61, 72]. For this reason, early studies must be very critically reevaluated [1:368, 26, 88, 189, 214].

6.1.1 Experimental Results

The first reliable results on the temperature dependence of the birefringence were reported by E. Dorn [89], (fig. 6.1). W. Harz, using the same material, later studied the dispersion

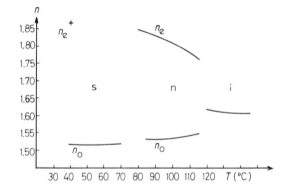

Fig. 6.1. Refractive indices, n, vs. temperature, T, for ethyl p-(p-ethoxybenzyl-ideneamino)-α-methylcinnamate   (from [89]).

of the indices [137], and C. Mauguin showed that the index n_i (isotropic) at the clearing point corresponds to the median index for the liquid crystal at the clearing point which is calculated as equal to $(2n_o + n_e)/3$ [207].

Here n_e is the index of the extraordinary ray (electric vector parallel to the optical axis), and n_o is the index of the ordinary ray (electric vector normal to the optical axis). The birefringence of MBBA has been measured with special accuracy [3:665, 8:222, 9, 30, 58, 133, 185], fig. 6.2. We refer the reader to the literature for information concerning the ex-

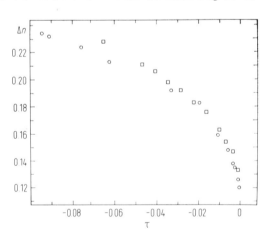

Fig. 6.2. Birefringence of MBBA, Δn, vs. reduced temperature, $\tau = (T - T_{ni})/T_{ni}$ at 589.3 nm (squares) and 632.8 nm (circles) (from [133]).

perimental techniques [1:148, 9, 54, 124, 125, 146, 153, 175, 175a, 185, 188]; we also mention a table of values of index measurements [205].

In the visible wavelength region, where the absorption of MBBA is negligibly small, n_e decreases while n_o increases with increasing wave length [55, 56, see also 175]. The indices of PAA have been measured many times at various temperatures and wavelengths [5:173, 29, 59, 60, 156]. According to P. Châtelain the following relationship is valid [60]:

$$\frac{1}{3}\left[\frac{n_e^2-1}{n_e^2+2}+2\frac{n_o^2-1}{n_o^2+2}\right]\frac{1}{d}=\frac{n_i^2-1}{n_i^2+2}\frac{1}{d'}=\text{const.}$$

where d and d' are the densities.

The indices of mixtures of PAA and PAP are additive of those of their components if n_e and n_o for the mixtures and pure substances are measured at temperatures equidistant from the corresponding clearing points and if the values of the pure components are multiplied by the mole fraction [29, 71]. This additivity is also known from one example of s_A phases [222].

G. Pelzl and H. Sackmann are noted especially for the determination of refractive indices of smectics [1:324, 1:327]. n_e decreases steadily with rising temperature within a phase and discontinuously at a phase transition, but n_o can go through a very flat minimum (figs. 6.3 and 6.4, see also fig. 8.18). Within homologous series, G. Pelzl and H. Sackmann observed an alternating decrease in n_e, n_o, and (n_e-n_o) as indicated in fig. 6.5 [1:327].

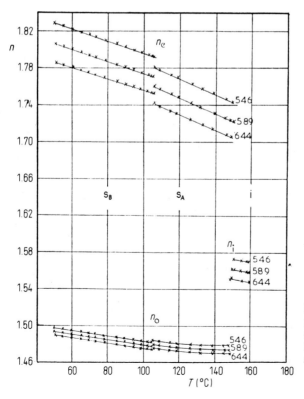

Fig. 6.3. Refractive indices, n, vs. temperature, T, for *n*-propyl *p*-(*p*-*n*-octyloxybenzylideneamino)cinnamate: k 66.4 s_B 105.4 s_A 154.2 i at 546, 589, and 644 nm (from [1:327]).

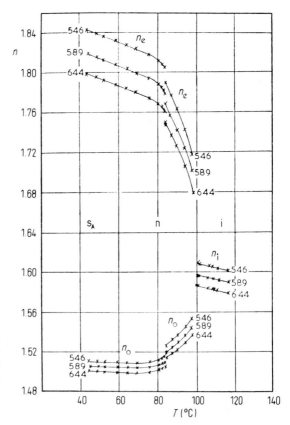

Fig. 6.4. Refractive indices, *n*, vs. temperature, *T*, for *n*-amyl *p*-(*p*-ethoxybenzylideneamino)-α-methylcinnamate: k 65.4 s$_A$ 84.2 n 100.5 i at 546, 589, and 644 nm (from [1:327]).

This phenomenon corresponds to the clearing point alternation that is due to different contributions of the dispersion interactions parallel and perpendicular to the molecular axis, thus having the same basis as does the anisotropy of the refractive index. Details are characterized as follows: "With an increasing number of carbon atoms in the side chain, the refractive indices n_e, n_o, and n_i decrease. When proceeding from odd to even numbers of C-atoms in the alkyl chain, n_e decreases by a larger amount than when proceeding from even to odd numbers. The alternating decrease of n_o is inverse to the change of n_e. The birefringence of homologous compounds shows the same alternation as n_e. In the homologous series the dispersion of n_e, n_i, and, for the lower homologs, n_o, decrease with increasing chain length".

The s$_D$ modification is optically isotropic, and thus behaves optically like a cubic crystal. Measurements on s$_C$ phases are complicated by the problem of orienting them in one direction. Only recently has it been possible, after determining the angle of the optical axes (or main axis of symmetry) to the normal of the layer, to measure the axial angle of a (weakly) biaxial s$_C$ phase [114a]. An earlier study showed that Δn decreases with increasing temperature. Δn of the homogeneously oriented phase proceeds continuously into the nematic domain. While in a preparation perpendicular to this, Δn passes to zero at the transition to the nematic phase. This must be due to surface phenomena and measurement technique [92], see fig. 6.6.

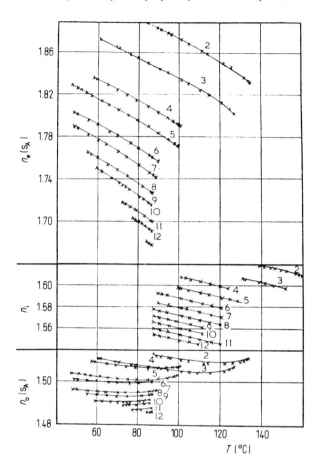

Fig. 6.5. Refractive indices, *n*, at 589 nm, vs. temperature, *T*, for the homologous di-*n*-alkyl *p,p'*-azoxy-α-methylcinnamates, where the numbers indicate the number of alkyl carbon atoms (from [1:327]).

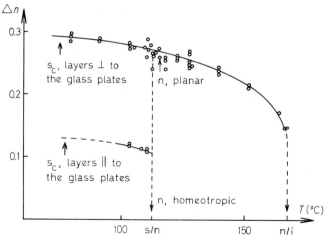

Fig. 6.6. Birefringence, Δn, vs. temperature, *T*, for N,N'-bis(4-*n*-decyloxybenzylidene)-2-chloro-1,4-phenylenediamine in different textures and mesophases (from [92]).

Index measurements are also reported of the homologous p,p'-dialkoxyazoxybenzenes [136, 228, 302], p-butoxybenzoic acid [103], p,p'-disubstituted tolanes [31, 32, 177], and other liquid crystals [2:176, 2:343, 3:389, 3:665, 4:170, 4:194, 4:778, 7:233, 146, 174, 223, 225, 290, 303].

The intensity of a laser beam passing through a liquid crystal layer between crossed polarizers shows a close series of maxima and minima upon changing temperature because the liquid crystal acts as a "wave plate". Transmission then disappears altogether at the clearing point [8].

G. Pelzl and H. Sackmann have found characteristic differences between the birefringence of the neat phases of anhydrous soaps and that of smectic phases of aromatic compounds [1:324]. As fig. 6.7 shows, the lower homologs exhibit a negative birefringence which decreases

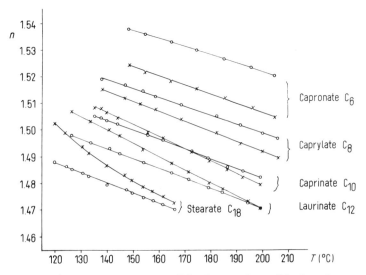

Fig. 6.7. Refractive indices, $n_o = o$, and $n_e = \times$, vs. temperature, T, for the neat phase of five homologous thallium soaps at 589 nm (from [1:324, see also 2:838]).

with increasing chain-length and decreasing temperature. The middle members of homologous series change the sign of the birefringence depending on the temperature. The higher members exhibit a positive birefringence that increases with increasing chain-length and with decreasing temperature.

6.1.2 Theoretical Considerations

The birefringence of liquid crystals is the visible manifestation of their long-range order and it is defined only for a uniformly ordered domain. Its value is determined by the degree of order, S, and by the principal polarizabilities, α_e and α_o, for uniaxial systems [3:509, 204]. These are interrelated by eq. 6.1

$$S = \frac{(\alpha_e - \alpha_o)\,\text{nemat.}}{(\alpha_e - \alpha_o)S = 1} \tag{6.1}$$

where the indices are introduced by the following equations that A. Saupe and W. Maier obtained from Neugebauer [3:509].

$$\frac{n_e^2 - 1}{n_e^2 + 2 - 2a(n_e^2 - 1)} = \frac{4\pi}{3} N\alpha_e \quad \text{and} \tag{6.2}$$

$$\frac{n_o^2 - 1}{n_o^2 + 2 + a(n_o^2 - 1)} = \frac{4\pi}{3} N\alpha_o \tag{6.3}$$

Here a is a constant by which the Clausius-Mosotti relationship can be adapted to optically uniaxial crystals as shown in eqs. (6.2), (6.3), (6.4), and (6.5).

$$\frac{n_e^2 - 1}{n_e^2 + 2} = \frac{4\pi}{3} \cdot \frac{N}{A_1 + \dfrac{1}{\alpha_1}} \tag{6.4}$$

$$\frac{n_o^2 - 1}{n_o^2 + 2} = \frac{4\pi}{3} \cdot \frac{N}{A_2 + \dfrac{1}{\alpha_2}} \tag{6.5}$$

where

$$A_1 + 2A_2 = 0 \quad \text{and} \quad a = \frac{3A_1}{8\pi N}$$

In order to calculate the indices from the main polarizabilities of the molecule, the following equations are compared with one another [289].

For the liquid phase, the Lorenz-Lorentz relation is valid

$$(\alpha_{\parallel} + 2\alpha_{\perp}) = \frac{9}{4\pi N_1} \left(\frac{n^2 - 1}{n^2 + 2} \right) \tag{6.6}$$

The Neugebauer relation for the crystal has the form

$$\frac{1}{\alpha_x} + \frac{1}{\alpha_y} + \frac{1}{\alpha_z} = \frac{4\pi N_c}{3} \left[\frac{n_x^2 + 2}{n_x^2 - 1} + \frac{n_y^2 + 2}{n_y^2 - 1} + \frac{n_z^2 + 2}{n_z^2 - 1} \right] \tag{6.7}$$

and for the nematic liquid crystal

$$\frac{1}{\alpha_e} + \frac{2}{\alpha_o} = \frac{4\pi N}{3} \left[\frac{n_e^2 + 2}{n_e^2 - 1} + \frac{2(n_o^2 + 2)}{n_o^2 - 1} \right] \tag{6.8}$$

The indices e and o stand for extraordinary and ordinary, respectively, and the symbols $\parallel$ and $\perp$ label the principal axis of the molecule.

The internal field constants, γ_e and γ_o, and the degree of order, S, can be calculated from the equations [289, see also 67, 70]:

$$n_e^2 - 1 = 4\pi N \alpha_e (1 - N \alpha_e \gamma_e)^{-1} \tag{6.9}$$
$$n_o^2 - 1 = 4\pi N \alpha_o (1 - N \alpha_o \gamma_o)^{-1} \tag{6.10}$$

$$S = \frac{\alpha_e - \alpha_o}{\alpha_{\parallel} - \alpha_{\perp}} \tag{6.1}$$

The calculated γ-factor is always less than $4\pi/3$ in the case when the electric vector is parallel to the optical axis, and in the other case, when the electric vector is perpendicular to the optical axis, it is larger than $4\pi/3$. Some effective polarizabilities and internal field constants have been calculated [2:320, 3:303, 4:778, 136a, 174, 275, 290, 297a]. The degree of order in nematic phases has been determined from birefringence measurements using Vuks formula [3:96, 3:389, 57, 136], which was not preferred to the Neugebauer relationship by some [3:509, 228, 291, 292, see also 3:303, 52, 134, 146, 275], and other authors have developed a new theory [10]. S. Chandrasekhar suggested the most comprehensive theory starting with the total interaction energy, U, of molecule i with all of its neighbors, where

$$U_i = -(u_0 + u_1 \cos\theta_i + u_2 \cos^2\theta_i + u_4 \cos^4\theta_i + \cdots) \tag{6.11}$$

In order to explicitly express the dependence of U_i on temperature, T, and volume, V, it may be written

$$u_0 = \left(\frac{G_1}{V^2} + \frac{R_a}{V^4}\right) + \frac{G_2}{kTV^4} + \frac{G_3}{k^2T^2V^4},$$

$$u_1 = \frac{G_4}{kTV^2} + \frac{G_5}{k^2T^2V^4} + \frac{G_6}{k^3T^3V^4},$$

$$u_2 = \left(\frac{G_7}{V^2} + \frac{R_b}{V^4}\right) + \frac{G_8}{kTV^4} + \frac{G_9}{k^2T^2V^4},$$

$$u_4 = \frac{R_c}{V^4}, \text{ etc.} \tag{6.12}$$

where G_1, G_2, G_7, G_8 are dispersion terms; G_4, G_6 are dipole terms; G_3, G_5, G_9 are dipole-dispersion cross terms and R's are repulsion terms. From the Lorenz-Lorentz relationship and P. Châtelain's suggestion of considering an ellipsoidal cavity for effective polarizabilities instead of a spherical one as in the Lorenz-Lorentz case, S. Chandrasekhar et al. deduced the equation [52]:

$$n_e^2 - n_o^2 = 4\pi N(\alpha'_{\parallel} - \alpha'_{\perp}) \left[\frac{1}{kTV^4}\left(\frac{2R_b}{15} + \frac{4R_c}{35} + \cdots\right) + \frac{2G_7}{15kTV^2} \right.$$

$$\left. + \frac{1}{k^2T^2V^4}\left(\frac{2G_8}{15} + \frac{4G_7^2}{315}\right) + \frac{2G_9}{15k^3T^3V^4} + \frac{G_4^2}{15k^4T^4V^4} \right] \tag{6.13}$$

However, insufficient reliable information is available to calculate the birefringence of a nematic phase from this theory. From the theory of temperature dependence of birefringence it can be concluded that the birefringence is mainly determined by the dispersion and the repulsion forces and not by the dipole-dipole forces as suggested in an earlier work [51]. The theory corresponds with the experimentally observed results that the temperature coefficient of the extraordinary index is large and negative, whereas that of the ordinary index is small and positive.

Calculations of the permittivity tensor for an unlimited number of parallel planes with the molecular axes being tilted on these planes have permitted a deduction of the biaxial character of the s_C phase and of the "herringbone structure" from structural models [81, 147].

6.1.3 Special Refractive Modes

Some publications are concerned with the potentials offered by nematic liquid crystals waveguides [121, 121a, 144, 145]. Their usefulness, however, seems to be limited, e. g. due to absorption loss (which was found to be highly dichroitic), thermal self-focusing, and substantial scattering losses (depending on the polarization and molecular alignment). Devices employing nematic waveguides usually include electrically tunable birefringence (see section 4.4.3).

The focusing and defocusing of laser light observed in a thin nematic layer of MBBA were attributed to a thermal lens effect [318]. The limiting diameter of the self-focused beam can be due to induced stimulated Brillouin scattering [136b], see section 6.7. Self-focusing is also known in isotropic liquids close to the nematic transition temperature, where the effect is enhanced in the pretransition range as compared to normal Kerr-liquids [112, 136a, 235, 236, 238, 324–327]. For the steady state case, the self-focusing threshold power obeys a linear dependence on $(T-T^*)$ that can be extrapolated to the second order phase transition temperature, T^*. The reader is referred to the literature for a discussion of the underlying intensity dependence of the index of refraction.

For a comparison of optical and magnetic birefringence see L. Royer [248] and section 4.4.10. Studies of the Kerr effect and the Rayleigh scattering offer a good possibility for observing both the slow fluctuations due to a collective response of strongly correlated molecules and the fast fluctuations due to individual molecular movement [178–180, 180a]. The literature submits an estimation of the correlation times and the amplitude ratio of both modes. The fast fluctuations do not exhibit any critical behavior vs. temperature near the i/n transition.

P. E. Cladis et al. describe a device containing a smectic liquid crystal that can split a beam of unpolarized light into three distinctly polarized beams [77].

There is a prediction of evanescent modes in liquid crystals [210b].

6.2 Rayleigh Scattering

6.2.1 Nematic Liquid Crystals

The thermally excited angular fluctuations of the director in nematics cause an intensive and strongly depolarized scattering of light. A striking angular dissymmetry in the angle dependent intensities of the scattered light is observed in all nematics which is analogous to that found in colloidal solutions. When dealing with performance and interpretation of scattering experiments, especially with respect to the polarization characteristics of incident and scattered radiation, the following parameters and conditions are of fundamental importance [173]:

1. The sample to be investigated is about 0.2 mm thick and oriented either homogeneously or homeotropically (see fig. 6.8).

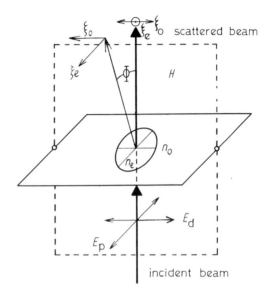

Fig. 6.8. Scattering geometry and scattering parameters.

2. The "plane of scattering" is defined as that plane which contains the incident light beam and the scattering beams under measurement. The plane of scattering (H in fig. 6.8) is a plane of symmetry, and the components, I, of incident and scattered light vectors, E, are usually given with respect to this plane (parallel or perpendicular). The plane of scattering contains the (variable) angle Φ that is read on the goniometer circle, under which scattered radiation beams are observed (fig. 6.8).

3. A homogeneously ordered sample can be oriented in two limiting and most important directions: The optical axis lies in the plane H or it is turned perpendicular to it (figs. 6.8 and 6.9).

4. The direction of the incident light electric vector can be chosen parallel (d) or perpendicular (p) with respect to H (as marked in fig. 6.8, E_d or E_p resp.).

The vector components of the scattered radiation, I, are functions of conditions 1 to 4. Following the nomenclature of P. Châtelain, we have I_e, the intensity of the extraordinary beam, vibrating perpendicularly to H and I_o, the intensity of the ordinary beam vibrating within H (case H_d). I_e and I_o are indexed as I_{ed} and I_{od}. If the incident light vibrates perpendicularly to H, then the scattered components are written as I_{ep} and I_{op}. (In the original papers: $\xi_{e'p}$, ξ_{op}, ξ_{ed}, and ξ_{od}).

The most important result of measurements is the interdependence between the I's and Φ in the two different cases of H_p and H_d (fig. 6.9). The surprising result of P. Châtelain's experiments was the fact that the most intensive component of the polarized radiation is

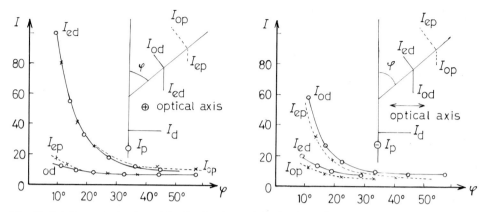

Fig. 6.9. Relative scattering intensities, I_{op}, I_{od}, I_{ep}, and I_{ed}, for the ordinary and the extraordinary ray vs. scattering angle, φ, measured in PAA. The optical axis is perpendicular to the scattering plane (left fig., case H_p) or in this plane (right fig., case H_d). The incident vibration is perpendicular (p) to the scattering plane or is in this plane (d) (from [63]).

always that one which lies perpendicular to the vibration direction of the incident light (62, 63, 65). In the H_p case, the intensity I_{ep} varies strongly with Φ.

The curves of fig. 6.9 obey the relation

$$I' = k \cdot (\sin\varphi)^{-X} \tag{6.14}$$

By the observation of identity or great similarly, i.e. $I_{ed} \simeq I_{op}$, P. Châtelain succeeded in stating the principle of reprocity with an anisotropic liquid, as is expected for a medium possessing an axis of revolution. P. Châtelain found no temperature influence on the scattering [62, 63, 65, see also 3:97]. He compared the scattering of nematics to that of particles (~ 0.2 to 0.3 µm diameter) attributing it to the thermal fluctuations of swarms or domains of this size [64, 66, 68, 69, see also 4:887]. As a measure of their orientation he introduced the expression

$$\Phi_{(t)} = 1 - \tfrac{3}{2}\sin^2\theta \tag{6.15}$$

where $\sin^2\theta$ is derived from either n_e or n_0 [70].

With a light-beat laser in the experimental setup of P. Châtelain, the Orsay group observed two low-frequency purely dissipative modes in nematic PAA arising from thermal fluctuations of the director, the local mean molecular orientation [3:592, 91, 217]. Such interpretation requires the absence of static defects and disclinations which can be critical in smectics, see [83]. These fluctuations, δn, can be decomposed into two uncoupled components: (a) δn_1 in the wave vector-n_o plane, a combination of bend and splay deformations; and (b) δn_2 perpendicular to this plane, a combination of bend and twist deformations. For the usual nematic materials these modes are purely relaxational with Lorentzian spectral densities of half-width U_{S1} and U_{S2}. It is possible to observe δn_1 and δn_2 separately by using the geometry sketched in fig. 6.10.

The incident laser beam of wave vector k is normal to n_o; the scattered wave vector $k = k + q$ (wave vector of the fluctuations) is in the (k, n_o) plane, and $\varphi = (k, k)$ is the scattering

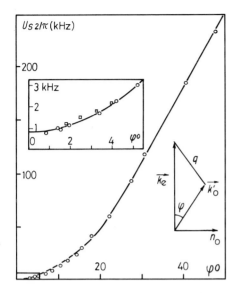

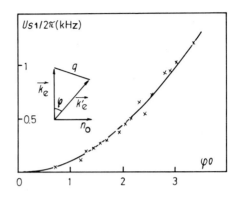

Fig. 6.10. Angular dependence of the half-width of the Lorentzian spectral density for the two dissipative modes in PAA (from [91]).

angle. The polarizations of incoming and scattered beams are purely ordinary (o) or extraordinary (e). Both scattering modes were observed also in PAP [197] but different interpreting equations are suggested [91, 197].

The intense scattering of light by liquid crystals can be measured as an absorption coefficient which is equal to the total scattering cross section. D. Langevin and M. A. Bouchiat present a computation of the cross section, σ (only the equation can be reproduced here), and an evaluation of the three elastic constants of MBBA derived from the scattering experiments shown in fig. 6.11 [183].

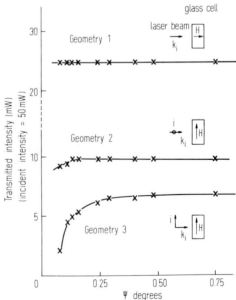

Fig. 6.11. Transmitted intensities by a MBBA cell 1.3 mm thick at 23°C, as a function of the half aperture, ψ, of the cone used as solid angle of detection (from [183]).

$$\sigma=\left(\frac{\omega^2}{4\pi c^2}\right)^2\frac{1}{n_i\cos\delta_i}\sum_{j=1,2}\int d\Omega_j^k\frac{n_j}{\cos^2\delta_j}\times\sum_{\alpha=1,2}\varepsilon_a^2kT\frac{[(e_\alpha\cdot i)(f_j\cdot v_0)+(e_\alpha\cdot f_j)(i\cdot v_0)]^2}{K_{33}q_{z_j}^2+K_{xx}q_{\perp j}^2+\chi_a H^2} \tag{6.16}$$

The literature reports other measurements of the three Frank elastic constants [3:48a, 3:749, 4:257, 98, 104, 105, 182], the viscosity coefficients [3:48a, 3:182, 3:595, 4:257, 105] and of the degree of order [15] by Rayleigh scattering studies.

According to the investigations of I. Haller et al., the normalized light intensity scattered by fluctuations in the nematic phase of MBBA (kT devided by the intensity of the scattered light) is only slightly dependent on temperature, but the linewidth of the Lorentz spectrum is more sensitive to temperature changes [131, 132].

A scattering experiment sees the average situation over the time of measurement since the situation in a mesophase is a dynamic state with the domain boundaries and orientation directions continually changing. This limitation is imposed as long as a sufficiently small volume of sample (containing only a few domains) cannot be observed for a sufficiently short time (less than the domain orientation relaxation time). For liquid crystal systems essentially all of the light scattered arises from orientational fluctuations having correlation distances on the order of several hundred nm [3:154, 3:158, 288]. Thin, ordered liquid-crystal layers generally exhibit characteristics similar to those of a lone, supposedly resting swarm. However, the different boundary and adsorption influences (tribo-epitaxy) must be considered in any comparison. By the application of an external field, a steady-state orientation distribution will result which will give rise to a steady-state Ω dependence of $I(\alpha,\Phi)$ which is independent of the relaxation time (see fig. 6.12).

The intensity of scattered light can be expressed by the relationship

$$I(\theta,\Omega)=K_L'\langle[\eta(r_1)]^2\rangle_{av}\iiint\gamma(r_{12})\cos[k(r_{12}S)]r_{12}^2\sin\alpha_{12}dr_{12}d\alpha_{12}d\Phi_{12} \tag{6.17}$$

where $r_{12}=r_1-r_2$ is the vector separation of scattering elements 1 and 2 located by vectors r_1 and r_2; α_{12} and Φ_{12} are the angular coordinates of r_{12}. The light scattering correlation function is

$$\gamma(r_{12})=\frac{\langle\eta(r_1)\eta(r_2)\rangle r_{12}}{\langle[\eta(r_1)]^2\rangle_{av}} \tag{6.18}$$

The symbol $\langle\ \rangle r_{12}$ designates a time and space average over all pairs of volume elements separated by r_{12} whereas $\langle\ \rangle_{av}$ designates a time and space average over all volume elements. $\eta(r)$ is proportional to the refractive index fluctuation

$$\eta(r)=c[n(r)-n_0] \tag{6.19}$$

where $n(r)$ is the refractive index at r and n_0 is the average refractive index. (r_{12}) may be obtained from Fourier inversion of the light scattering data in a method similar to X-ray scattering. The spherical symmetry reversion is expressed by

$$I(\theta)=4\pi K_L'\langle[\eta(r_1)]^2\rangle_{av}\int\gamma(r_{12})\frac{\sin(hr_{12})}{hr_{12}}r_{12}^2dr_{12} \tag{6.20}$$

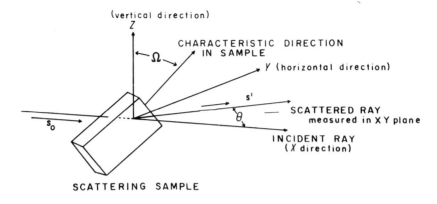

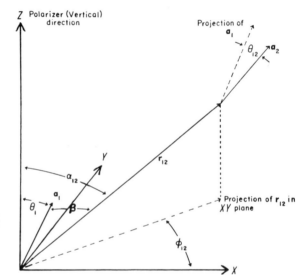

Fig. 6.12. The coordinates defining the scattering direction and a pair of scattering elements with their principal polarizability directions (from [288]).

This equation can be applied to systems without spherical symmetry using an ellipsoidal correlation function of the type

$$\gamma(r_{12}) = \exp\left[-\left(\frac{x^2}{a_2^2} + \frac{y^2}{b_2^2} + \frac{z^2}{c_2^2}\right)\right] \tag{6.21}$$

where a_2, b_2, and c_2 are the correlation distances in the direction of X, Y, and Z axes which are assumed to be the principal axes of symmetry of correlation. The components of the vector r_{12} along these axes are x, y, and z. The concept of non-randomness of orientation correlations goes back in great degree to the work of R. D. Stein [288].

T. D. Gierke developed a concept of static orientational pair correlations which allows an estimation of a nonmesogenic molecule's tendency toward the nematic state from the depolarized Rayleigh scattering [2, 122, 123, see also 2:160a, 220a]. J. J. Nemec considers three correlation coefficients that correspond to splay, twist and bend deformations, and he pursues their temperature dependence [213].

The possible influence of piezoelectric charges on the light scattering of nematics has also been discussed [218]. By assuming that the molecules are arranged parallel to one another within spherical volume elements, D. Krishnamurti and H. J. Subramhanyam developed a theory that, when corrected by an empirical form factor, fits the observed phenomena [173].

There is a series of other papers dealing with theoretical aspects of the Rayleigh scattering in nematics [3:476, 3:593, 8:364, 101, 101a, 110, 173, 213, 218–220, 266, 328 and polymer films, 137a]. Photographs of the polarized light scattering patterns for nematic, cholesteric, and smectic (A, B, and C) phases have been published [259]. We also refer here to other papers on light scattering in liquid crystals, but these only enrich the text above by giving further details [1d, 159, 206, 230, 231, 242, 243, 261, 262, 277, 301, 314, and, for a review, 109].

6.2.2 Smectic Liquid Crystals

The angular fluctuations in smectics are as strongly coupled to light waves as they are in nematics, but the layer structure forbids the thermally excited angular fluctuations of the director to have a large amplitude (as in nematics) except for an undulation of the layers [3:157, 94, 239]. The thermal average of the angular fluctuations, θ, can be expressed by the equation

$$\langle\theta\rangle = \frac{kT}{K(q_\perp^2 + l^2 q_c^4/q_\perp^2)} \sim I \tag{6.22}$$

where k is the Boltzman constant, K is the Frank splay elastic constant of the smectic material, q is the wave vector of the undulation mode with the critical value q_c, and l is an integer from 1 to ∞ [96, see also 100].

The maximum appears at $q_\perp = q_c$ for the fundamental mode ($l=1$). G. Durand observed a linear dependence of q_c^2 on the inverse thickness, d^{-1}, of a s_A slab. This was predicted by the equation $q_c^2 \lambda d = \pi$, where λ, the penetration length of de Gennes, is 2.2 ± 0.3 nm [96], see fig. 6.13; another work reports $\lambda = 1.4$ nm [245].

Such experiments must try to exclude static wall induced layer undulations. Therefore, in order to minimize the influence of the inevitable surface defects in the homeotropic geometry,

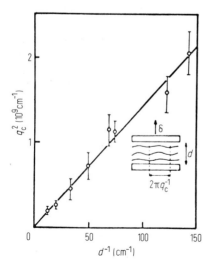

Fig. 6.13. Linear dependence of q_c^2 on the inverse thickness d^{-1} of a homeotropic slab of CBOOA at 78°C (from [96]).

further studies preferred the planar texture [244]. Again using CBOOA, it was shown that only splay distortions can be observed in smectics A while twist and bend distortions decay within molecular distances. The penetration length diverges according to $(T-T^*)^{-1/6}$ at the s_A/n transition. Other studies of the static and dynamic behavior of smectics A by Rayleigh scattering are reported in the literature [3:131, 75, 84, 95, 97, 246]. In *p*-heptyloxybenzylidene-*p'*-heptylaniline with a s_H/s_B transition at 55°C, the light scattering by undulation instabilities indicates that the layers are highly decoupled and are free so slip on each other [240].

One of the two modes of quasielastic Rayleigh scattering (mode 1, where the director oscillates in the scattering plane, cf. p. 264) is not observed in a s_C liquid crystal [93]. However, with the configurations $\{K_e, K'_0\}$ and $\{K_0, K'_e\}$, a strong signal is observed which is probably due to twisting of the s_C planes.

6.2.3 Pretransition Phenomena

In general, the depolarization of scattered light offers a method for the study of pretransitional effects and the establishment of thermal equilibria [3:477, 43]. Figs. 6.14 and 6.15 show the results of measurements with the incoming polarization perpendicular and parallel to the plane formed by the incident and scattered waves. Usually there exist fairly constant values of the depolarization factors except in a region of temperature near the phase transition. The zone of scattering fluctuations coincides with the range of latent heat absorption of the n/i transition, but it does not include the maximum, see fig. 6.16.

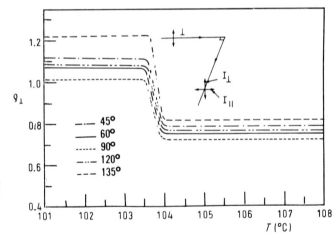

Fig. 6.14. Perpendicular depolarization factor, $\varrho_\perp = I_\parallel/I_\perp$, for APAPA from 101 to 108 °C (from [43]).

In a few experiments, light scattering has been used like polarization microscopy or differential thermal analysis for determining phase transitions and microstructures [167, 267]. We also mention the experiment of W. Kast and L. S. Ornstein, who used measurements of the transparency of PAA in an attempt to choose between the swarm and continuum theories [3:586, 3:588, 4:685–4:687, 4:693, 4:694, 155, see also 4:90].

The main activities in the last several years have been concerned with the scattering in the isotropic-liquid phase of mesogens close to the mesophase transition temperature. Such experiments give valuable information on the size of ordered groups and their temperature

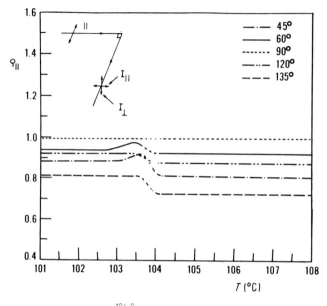

Fig. 6.15. Parallel depolarization factor, $\varrho_\parallel = I_\perp / I_\parallel$, for APAPA from 101 to 108 °C (from [43]).

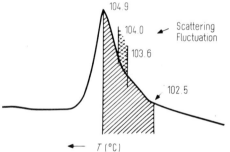

Fig. 6.16. Regions of latent heat absorption and scattering fluctuation for the n/i transition of APAPA (from [43]).

dependence as predicted by de Gennes [3:159]. Most of the papers submit an estimation of the coherence lengths [3:772, 6, 73, 74, 76, 80, 80a, 166, 286].

From de Gennes' phenomenological theory of short-range orientational order, the Rayleigh ratio, R, is predicted to depend on the temperature according to $R \sim T/(T - T^*)$. As fig. 6.17 demonstrates, the data measured for p-dioctyloxyazoxybenzene agree with a linear prediction, while that of p-diundecycloxyazoxybenzene (having a direct i/s$_C$ transition) distinctly deviate [286].

E. Gulari and B. Chu analyse their Rayleigh intensity data obtained in the isotropic phase of MBBA to calculate the coherence length, ξ, which was found to obey the equation

$$\xi = (0.55 \pm 0.02) \left(\frac{T - T^*}{T^*} \right)^{-0.50 \pm 0.01} \tag{6.23}$$

This is represented as the straight line in fig. 6.18.

An extension of the Maier-Saupe theory also gives a good agreement with experimental data of light scattering intensities measured in the pretransition range [3:757, 3:852]. From

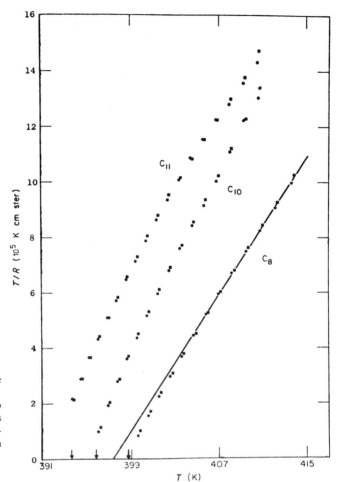

Fig. 6.17. Intensities for three
p-di-*n*-alkoxyazoxybenzenes
divided by the Rayleigh ratio
vs. temperature. The arrows
indicate the respective transition temperatures (from
[286]).

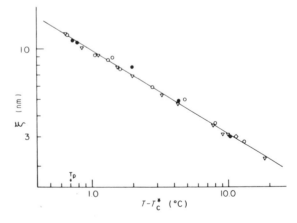

Fig. 6.18. Temperature dependence of
the coherence length, ξ, in the isotropic
phase of MBBA; ○ from depolarized
scattered light, ● from polarized scattered light, and ▽ from unpolarized
(total) scattered light. The angular anisotropy was measured to a precision
of 0.1% corresponding to a coherence
length resolution of better than 1 nm
(from [76]).

the spectrum of scattered light, the relaxation time of the order parameter fluctuations has been determined [286, see also 3:475, 178, 179]. In the isotropic liquid, a critical slowing down of the fluctuations is observed as the transition to an ordered phase is approached.

6.3   UV- and Visible Absorption Spectroscopy

The anisotropy of the molecules, which is most often present, remains undetected when the standard techniques of spectroscopy are applied. Thus, the spectroscopy of ordered molecules deserves greater interest due to the increased information that can be obtained from it. There are detailed works recommended on this subject. One is by S. Chandrasekhar and N. V. Madhusudana [53], and the other by E. Sackmann [14:741, 254]; other reviews have appeared [212, 269].

In addition to the methods by which the molecules are incorporated into suitable host single-crystals or stretched polymer foils (e.g. [27, 181, 294]), anisotropic solvents are more frequently employed. A uniform orientation of these solvents can easily be obtained by a surface treatment or the application of electric or magnetic fields. The solute molecules arrange themselves according to their shapes in the liquid-crystal matrix. If they are elongated, the degree of order is roughly the same as that of the liquid crystal itself. J. Fischer was the first to show that the light absorption of the liquid crystals is much closer to that of the amorphous melt than to that of the crystals [107, 108]. All spectroscopic data provide strong evidence for the generalization that solute molecules with a long axis tend to align with that axis parallel to the long axis of the liquid crystal molecules. Even without the application of external fields and polarizers, information on the orientation of transition moments can be obtained if the liquid crystal solvent can be uniformly oriented by surface treatment [284]. Fig. 6.19 illustrates the UV spectrum of PAA.

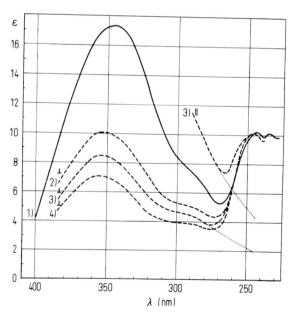

Fig. 6.19. UV spectra of PAA in the isotropic liquid (1); and in the nematic phase at 130°C (2), 109°C (3), and 86°C (4). The electric vector of the lineary polarized light is normal or parallel to the optical axis (from [198]).

In order to study molecular interactions and the spacial distribution of longitudinal molecular axes W. Maier and A. Saupe have begun to investigate the UV-dichroism in nematics [198]. The authors assign the UV bands of PAA as follows [199]: ($\pi_1 \rightarrow \pi_1^*$ transition at 348 nm; $\pi_0 \rightarrow \pi^*$ (or $\pi_2 \rightarrow \pi_1^*$) transition at about 290 nm; and $\Phi_1 \rightarrow \pi^*$ transition at 243, 231, and 223 nm. The n$\rightarrow \pi^*$ transition is hidden by the $\pi_1 \rightarrow \pi_1^*$ transition. Fig. 6.20 indicates that dichrism can be observed in smectics as well as in nematics. Consider the example in fig. 6.21 before the dichroism of solute molecules is discussed in detail.

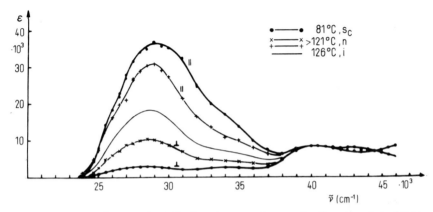

Fig. 6.20. Molar extinction coefficients of *p,p'*-di-*n*-heptyloxyazoxybenzene for linearly polarized light, parallel (‖) and perpendicular (⊥) to the optical axis (from [269]).

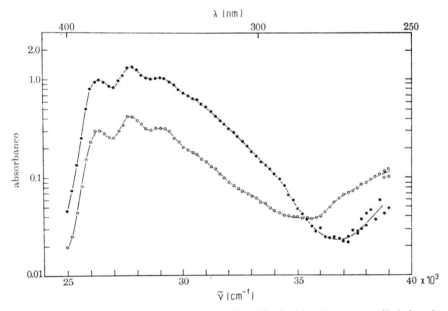

Fig. 6.21. Absorbance of all trans-1,6-diphenyl-1,3,5-hexatriene dissolved in a "compensated" cholesteric mixture. The absorbance are measured parallel (●) and perpendicular (○) to the electric field direction (from [47]).

The transition moments are observed as being directed along the longitudinal molecular axis in the longwave UV band and in a direction perpendicular to the molecular axis in the shortwave band [47].

E. Sackmann has especially studied spectroscopy in anisotropic solvents and has provided valuable information on the polarization of optical transitions [249–256]. From the examples shown in fig. 6.22, it can be seen that the transition corresponding to the visible band of β-carotene is polarized in the direction of the long molecular axis, and the 425 nm transition of octa-*t*-butyldiphenochinone is polarized in the short molecular axis [249, see also 129].

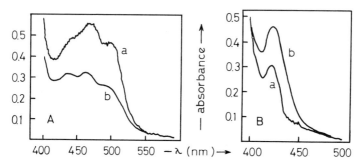

Fig. 6.22. Spectra of β-carotene (A) and octa-*t*-butyldiphenoquinone (B) in "compensated" mixtures aligned for 12 hours at 35 °C and 20 kOe. In the spectra (a) the electric vector of the light is perpendicular, and in (b) it is parallel to the magnetic field (from [249]).

Compensated cholesteric mixtures have proven especially useful for practical work because they absorb only below 250 nm and can be quenched to a glassy state below −50 °C. Such mixtures are obtained from cholesterol derivatives having short substituents at C_3 forming right-handed helices combined with those having long substituents at C_3 forming left-handed helices [255, 256], fig. 6.23. Studies employing compensated mixtures should ensure that the system after the addition of the solute is still in the compensated nematic state, because cetyl alcohol dissolved in such a mixture transforms it into the cholesteric state [211a].

By using anthracene, E. Sackmann and H. Moehwald demonstrated the applicability of this method to localize weak hidden bands [255], figs. 6.24 and 6.25.

Fig. 6.24 shows that the ^1L_a transition with the 0—0 vibrational band at 380 nm is preferentially polarized along the short inplane molecular axis (order parameter $S_{xx} \approx -0.08$), while the 0—0 band at 250 nm, corresponding to the ^1B_b transition, is polarized parallel to the long axis of anthracene (order parameter $S_{zz} \approx +0.33$). A comparison of fig. 6.24a and 6.24b clearly shows that the sign of the circular dichroism reverses when the degree of absorption polarization, $N \equiv$ optical density $\|E/$ optical density $\perp E$, changes from a value of $N > 1$ to a value of $N < 1$. According to fig. 6.25, the degree of fluorescence polarization, $N = I_{\|E}/I_{\perp E}$, is smaller than the one in the whole wavelength region. This difference in the absorption and the fluorescence polarization spectrum clearly indicates the appearance of the weak ^1L_b transition which is buried below the considerably stronger ^1L_a band. The fact that the circular dichroism also changes sign at $29000\,cm^{-1}$ provides strong evidence that the 0—0 band of the ^1L_b transition is situated at about this wavenumber [255].

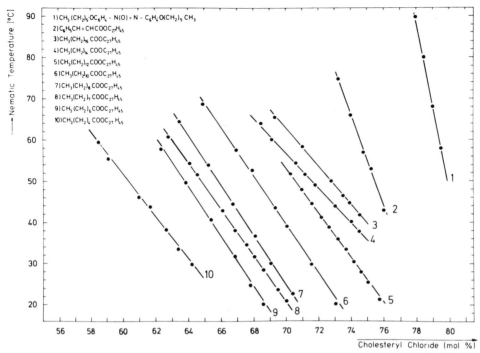

Fig. 6.23. Nematic temperature of several "compensated" mixtures of cholesteryl chloride with different cholesteryl esters and with *n*-hexyloxyazoxybenzene (curve 1) (from [256]).

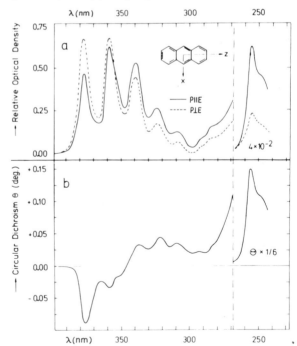

Fig. 6.24. Absorption polarization spectrum (a) of anthracene in a compensated mixture and circular dichroism spectrum (b) of anthracene in a cholesteric solvent with a pitch of 800 nm (from [255]).

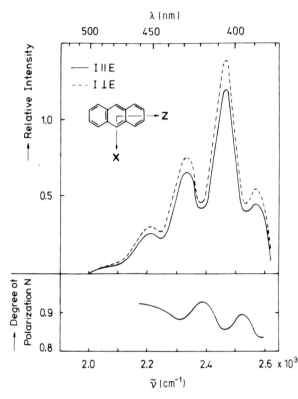

Fig. 6.25. Fluorescence polarization spectrum of anthracene in a compensated mixture. The intensities were measured with the polarizer oriented parallel or perpendicular to the electric field (from [255]).

Other studies concern dyestuffs [17–22, 22a, 22b, 297, 321, 322] e.g. in chlorophyll a, where the direction of electronic transitions relative to the orientation in the nematic mesophase could be localized [111a, 140–142]. Since the investigation of dyestuff absorption in a liquid crystal matrix — whose orientation can be controlled by the voltage applied — aim at a technological application, one typical spectrum is recorded in fig. 6.26.

The field-induced helical unwinding of a cholesteric solution to a homeotropic nematic solution was proposed by F. D. Saeva et al. and illustrated in the following way:

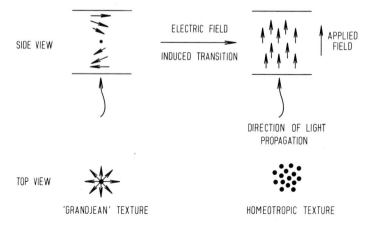

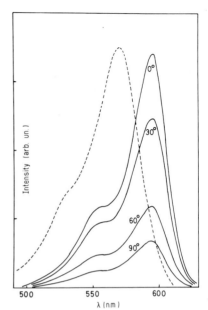

Fig. 6.26. Absorption spectrum 3,3′,9-triethyl-6,6′-di-methoxythiacarbocyanine:

in MBBA solution (———) with various angles between the electric field and the incident light and in alcoholic solution (·······) (from [322]).

It allows a convenient measurement of circular dichroism and linear dichroism in one solution [257].

Most spectra of oriented solutes contain vibrational bands which are polarized purely parallel to an inplane symmetry axis and which do not overlap appreciably with other vibrational bands. E. Sackmann has provided a wealth of data (see figs. 6.22, 6.24, 6.25) and shall be cited in the following text. "The so called 0—0-transitions between the vibrational ground levels of the electronic states normally represent such "purely" polarized bands in the spectrum. All three order matrix elements of molecules with C_{2v}- or D_{2h}-symmetry can be determined if two perpendicularly polarized 0—0-bands appear in the spectrum. In favorable cases it is also possible to determine the average orientation of one molecular axis of asymmetric molecules" [253, 256].

Both a direct and a local field corrected calculation of the degree of order, S, have been used in the literature, e.g.

$$S = \frac{D_{\parallel}^* - D_{\perp}^*}{D_{\parallel}^* + 2D_{\perp}^*} = \frac{N^* - 1}{N^* + 2} = \frac{gN - 1}{gN + 2} \tag{6.24}$$

In this form, the asterics denote that the optical densities, D^*, and the dichroitic ratio, $N^* = D_{\parallel}^*/D_{\perp}^*$, have been corrected by taking local field anisotropy into account according to [19]

$$g = \frac{N^*}{N} = \frac{n_{\parallel}}{n_{\perp}} \left(\frac{n_{\perp}^2 + 2}{n_{\parallel}^2 + 2}\right)^2 < 1 \tag{6.25}$$

Figs. 6.27 and 6.28 illustrate the relations between the degree of order and molecular parameters found by E. Sackmann [256, see also 22a, 22b].

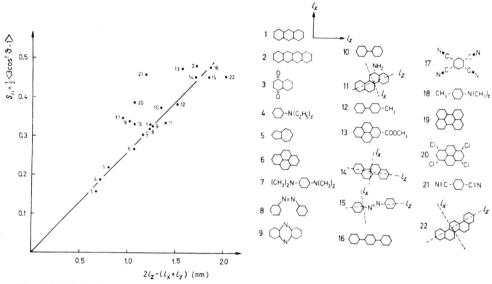

Fig. 6.27. Relation between the degree of order, S_{zz}, of the long molecular axis, l_z, of elongated aromatic molecules and the difference, $2l_z - l_x - l_y$, of the molecular dimensions in the direction of the x-, y-, and z-axis, respectively (from [256]).

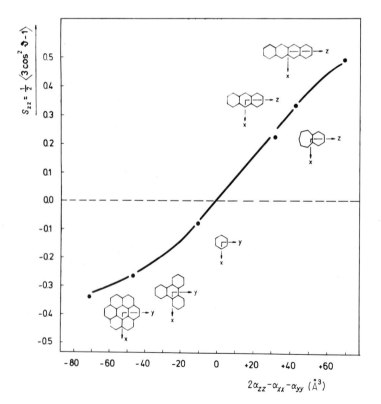

Fig. 6.28. Relation between the degree of order, S_{zz}, of several aromatic molecules in a compensated cholesteric mixture and the anisotropy of polarizabilities (from [256]).

Finally we mention the studies of other compounds not described in the text [14, 14a, 22, 45, 128–130, 221, 226, 284, 285], a proposed correction [82], the older work of R. Riwlin [241], a study of the transparency of anhydrous mesomorphous soaps [13], photoelastic phenomena [169], and photoacoustic observations [247].

6.4   *Fluorescence Spectroscopy*

Continuing the discussion begun in the previous section, it is advisable to record both absorption and fluorescence spectra in order to compare the degree of polarization. However, only one of these two spectra is required in order to determine the orientation of transition moments in the molecule (fig. 6.29).

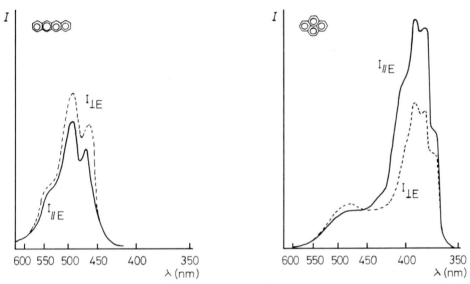

Fig. 6.29. Fluorescence polarization spectra (relative intensity, I, vs. wavelength, λ) of tetracene and pyrene in "compensated" mixtures (from [251]).

The "guest molecules" are assumed to adopt the orientation of their longitudinal axes parallel to the nematic matrix molecules that are oriented in an electric field, E. For the measurement unpolarized light can be applied. Due to the anisotropy of the various spectral transitions in the ordered matrix, the fluorescent light is polarized. After passing through the analyzer, which may be set parallel or perpendicular to the nematic director, the receiver is supplied either with an intensity $I_{\|E}$ or $I_{\perp E}$. Again the cases where $I_{\|E} > I_{\perp E}$ and $I_{\|E} < I_{\perp E}$ have been found. Assuming that the orientation of the guest molecules can actually be determined, the direction of polarization belonging to the transition moments of the bands under discussion can be ascertained. If the transition moment in question lies in the longitudinal axis of the molecule, then the difference between the intensity in the direction of the electric vector and that direction which is perpendicular to it ($I_{\|E} - I_{\perp E}$) will be greater than $I_{\perp E} - I_{\|E}$ in other orientations of transition moment, especially when the transition moment is perpendicular to the longitudinal axis (fig. 6.29). It must be stressed that the monomer fluorescence (380 nm, polarized parallel to the longitudinal molecular axis) and the excimer fluorescence (470 nm, perpendicular to the longitudinal molecular axis) are oppositely polarized with pyrene, as shown in fig. 6.29 [251].

The monomer fluorescence (structured, short-wave) and the hetero-excimer fluorescence (broad, long-wave) in the fluorescence spectrum of a pyrene/diethylaniline mixture (fig. 6.30,

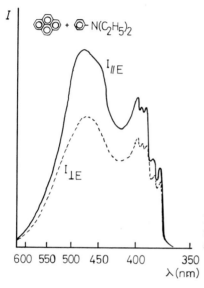

Fig. 6.30. Fluorescence polarization spectrum (relative intensity, I, vs. wavelength, λ) of pyrene and diethylaniline in a "compensated" mixture (from [252]).

a socalled hetero-excimer in the excited state) are polarized parallel to the electric field [252]; this is in contrast to the spectrum of pyrene shown in fig. 6.29.

Fluorescence polarization is very sensitive to phase transitions as is expected [208, 209, 258, 260], fig. 6.31. In fig. 6.32 a dramatic change of the quenching rate appears at the n/i transition.

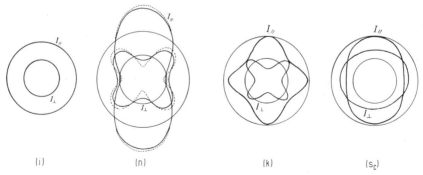

Fig. 6.31. Angular distributions of the polarized components of fluorescence observed in the isotropic state, the aligned nematic state, the s_C state, and the crystalline state of *p-n-octyloxybenzoic* acid. The analyzer is rotated so that the direction of the electric vector of the fluorescent light lies either parallel (I) or perpendicular (I) to that of the exciting light (from [258]).

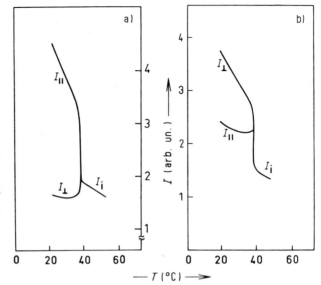

Fig. 6.32. Polarized fluorescence emission (intensity, *I*, parallel and perpendicular to the optical axis vs. temperature, *T*) in the case of unpolarized excitation: a) *p*-dimethylamino-*p*-nitrostilbene in MBBA at 645 nm, b) tetracene in MBBA at 486 nm (from [209]).

In addition, fluorescence polarization measurements allow a study of rotational diffusion. In an isotropic solvent it may occur within about 1 ns, while in a nematic solution no rotation is observed during the fluorescence decay time [48]. Other fluorescence studies in nematic solutions are reported [49, 172, 260, 284, 285, 304] as are polarized luminescence spectra [87]. From the fluorescence intensities under unpolarized as well as polarized excitation, the degree of order, *S*, can be estimated. However, in some cases there can be substantial differences in the results when compared to those obtained by other methods [12, 208, 209, 223a]. The potentials of fluorescence switching by means of liquid crystals have been tested by R. D. Larrabee [184]. The liquid crystal must satisfy the condition that it be relatively transparent in the near-UV in order to allow the exciting radiation to reach the dissolved fluorescent dye. In addition the liquid crystal must provide a suitable molecular environment for the dissolved dichroitic dye.

6.5 *Infrared Spectroscopy*

Most works in the field follow one of two courses. They either measure the dichroism of solute molecules in a liquid-crystalline matrix, or they are direct investigations into the structure of the mesophase itself [5:265, 41, 42, 212, 269].

6.5.1 *Nematic Phases*

W. Maier and G. Englert were the first to use IR-dichroism for the assignment of the IR bands in a liquid crystal [200, 201]. They derived a relationship between the dichroitic ratio, δ, and the degree of order, $S_{zz} = (1 - 3/2 \sin^2 \theta)$ [203], cf. chapter 3:

$$\delta \equiv \frac{\varepsilon_{\parallel}}{\varepsilon_{\perp}} = \frac{2(1 - \sin^2 \alpha) - (2 - 3\sin^2 \alpha)\overline{\sin^2 \theta}}{\sin^2 \alpha + (1 + \frac{3}{2}\sin^2 \alpha)\overline{\sin^2 \theta}} \tag{6.26}$$

where $\varepsilon_{\parallel}$ and $\varepsilon_{\perp}$ are the extinction coefficients of the liquid-crystal layer for linearly polarized light whose electric vector oscillates parallel or perpendicular, respectively, to the optical axis. Eq. (6.26) is valid for bands whose vibrational moment lies in the molecular plane and forms an angle, α, with the longitudinal axis. In order to operate with eq. (6.26), it is necessary to find bands whose vibrational moments lie, as mentioned, in the molecular plane and in addition are either parallel or perpendicular to the major axis of the molecule, respectively the so called A_1 and B_1 bands. For B_2 bands (whose moments are perpendicular to the molecular plane), the relationship simplifies to

$$\delta = \frac{2\overline{\sin^2\theta}}{2-\sin^2\theta} \tag{6.27}$$

Errors in and corrections of this method have been discussed elsewhere [203]. There are occasionally variations up to 20% when compared with other methods [3:507, 3:508, 135, 204]. It has been shown, contrary to theoretical predictions, that higher order parameters may be found by taking aromatic core absorption bands as a basis rather than using bands of hydrocarbon chains of the same compound [114]. If the nonplanar configuration of many liquid-crystal molecules (e.g. PAA, MBBA) is considered (cf. chapter 2) together with the fact that measurements are possible only on rotation-symmetrical systems (averaging the x, y transitions), the limits of the method readily become apparent. The values suggested in ref. [28] seem remarkably low, while that of CBOOA are noteworthy for their extension to the s_A phase [106, 307].

The degree of order, S, can be determined from the IR spectra of nematic substances in homogeneous and homeotropic alignment, even without the use of polarized light [38]. An alternative to measuring dichroism without the aid of polarized light consists of comparing the spectrum of a homeotropic nematic layer to that of the same layer in the isotropic state [157, 158], fig. 6.33. The relationship between the dichroitic ratio and the degree of order is expressed by an equation containing δ when linearly polarized light is used or δ' when unpolarized light is used. For parallel-polarized (A_1) bands these are ($\varepsilon_{\odot}$ is the extinction coefficient of the homeotropic sample, ε_i refers to the isotropic one)

$$\delta \equiv \frac{\varepsilon_{\parallel}}{\varepsilon_{\perp}} = \frac{1+2S}{1-S} \quad \text{and} \quad \delta' \equiv \frac{\varepsilon_{\odot}}{\varepsilon_i} = 1-S \tag{6.28}$$

and for perpendicularly polarized (B_1 or B_2) bands these are

$$\delta \equiv \frac{\varepsilon_{\parallel}}{\varepsilon_{\perp}} = \frac{2(1-S)}{2+S} \quad \text{and} \quad \delta' \equiv \frac{\varepsilon_{\odot}}{\varepsilon_i} = 1 + \frac{1}{2}S \tag{6.29}$$

The "orthogonality" in the IR spectrum of MBBA is greatest in the $1513\,\text{cm}^{-1}$ band with $\delta' = 0.4$ and in the $831\,\text{cm}^{-1}$ band with $\delta' = 1.4$, which may be classified as a deformation vibration and a γ-CH vibration of the benzene ring, respectively. The band at $1030\,\text{cm}^{-1}$, which can be termed the ether-CO band, has a dichroitic ratio of approximately one, and this might be expected from the angular position of the transition moment. The CH valence vibrations in MBBA exhibit various dichroisms. Below $3000\,\text{cm}^{-1}$ and above $3020\,\text{cm}^{-1}$ it is type B, but two A bands are in betweeen them. This method, circumventing the use of linearly polarized light, can be extended to measure the dichroism of nonmesomorphous substances dissolved in a liquid-crystalline matrix [158].

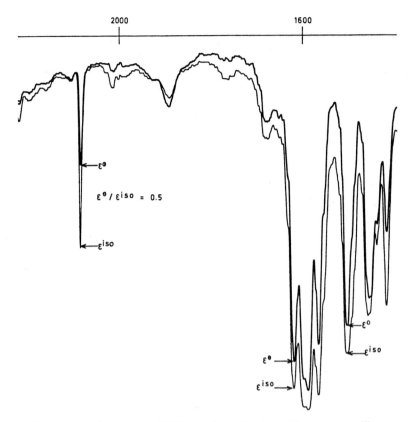

Fig. 6.33. Part of the IR spectrum of a mixture of MBBA and *p*-butoxybenzylidene-*p*-cyanoaniline: $\varepsilon_\ominus$, homeotropic nematic phase; ε_i, same layer in the isotropic state (from [158]).

Independently, force constant calculations and IR polarization studies have enabled assignment of the bands of $Mn_2(CO)_{10}$ and $Re_2(CO)_{10}$ in the following way [46, 190, 191]: There are three IR-allowed normal modes which correspond to $C≡O$ stretching in $M_2(CO)_{10}$ molecules of D_{4d} symmetry, two of B_2 symmetry, and one of E_1 symmetry. The B_2 modes should be parallel polarized $(\varepsilon_z > \varepsilon_{x,y})$, whereas E_1 should be perpendicular polarized $(\varepsilon_{xy} > \varepsilon_z)$. Since symmetry requires $\varepsilon_z - \varepsilon_{x,y}$ to be positive for two of the three $C≡O$ stretching fundamentals, the fact that $\varepsilon_\parallel - \varepsilon_\perp$ is found to be positive for two observed bands establishes them as B_2 transitions and determines that S_{zz} is positive. Furthermore, the perpendicularly polarized band must be assigned as the E_1 fundamental.

The activation energies for rotational diffusion, U_{or}, can be determined from the temperature dependence of the half width of IR-bands polarized parallel or perpendicular to the long molecular axis [162, 163, 280, 282]. These values are obtained when one considers the half width, d, of a band, i, at temperature, T, to be composed of a zero half width, d_{i0}, and an increment, Δr, due to the Brownian rotating movement. Then

$$d_{i(T)} = d_{i0} + \Delta r(T)$$

The temperature-dependent band broadening (Δr) is related to the average lifetime, τ_0, and to a potential barrier, U_{or}, by the equations

$$\Delta r = \frac{1}{\pi \cdot c \cdot \tau_0} \quad \text{and} \quad \tau_0 = \tau'_{\exp}(U_{or}/kT)$$

where τ' is the period of the reversive rocking of the molecules. Evaluation of diagrams can reveal the values of U_{or} without determination of the zero width. Some typical values (given in kJ/mol) of the activation energies for rotational diffusion, $U_{or}^{\parallel}$ and $U_{or}^{\perp}$, are [282]

	nematic phase		isotropic liquid	
	$U_{or}^{\parallel}$	$U_{or}^{\perp}$	$U_{or}^{\parallel}$	$U_{or}^{\perp}$
MBBA (293–318 K)	26.4	84	20.5	54
EBBA (308–350 K)	31.4	105	22.2	67
APAPA (358–380 K)	43.1	146	24.3	67
PAA (390–407 K)	38.9	>167	26.4	75

Further the literature reports an estimation of the average angular jumps of rotation and of the rotational correlation times of benzene rings in the s, n, and i phases [192, 194], and, in addition, quantitative measurements of the integral intensity of IR bands [165].

Only a few papers report finding indentical IR spectra in both mesomorphous and iso-tropic phases [160, 195, 202], and differences were frequently observed [33, 34, 161, 211, 295, 296, 317]. Thus, several IR bands of crystalline p-alkoxyazoxybenzenes disappear upon transition to the nematic phase, while only the relative intensities of certain bands change upon transition from the nematic phase to the isotropic liquid [33, 34]. Such changes arise consequently for dichroitic bands when, on average, the radiation meets domains in a preferred direction. However, other authors report a shift of IR-bands at the n/i transition [211]. The sharp change in the $100 \, cm^{-1}$ absorption band of PAP at the clearing point is noteworthy [196]. The far-IR spectrum of MBBA is dominated by a strong band near $130 \, cm^{-1}$ which may be shown to arise from the librational mode about the long axis of the molecule [37, 102, 149, 150, 313]. Figs. 6.34 and 6.35 give an impression of the differences in spectra taken in various phases and textures.

The far-IR spectra of PAA [40, 127, 279, 306], p-cyano-p'-n-heptylbiphenyl [102a], and other nematics [1b] are discussed in the literature, and a force constant calculation in the lattice vibration region (10 to $100 \, cm^{-1}$) is also reported. An absorption band in the far IR spectrum of p-butoxybenzylidene-p'-cyanoaniline is observed at $24 \, cm^{-1}$, and it disappears abruptly at the n/i transition and can be used for a calculation of the degree of order [1, 1c].

IR spectra taken from Mauguin textured nematics give evidence of an IR optical activity in this state [39, fig. 6.36]. The difference intensity of IR radiation transmitted by nematic MBBA can be decreased markedly by applying a potential that exceeds the threshold voltage of the dynamic scattering mode [216, 315]. Liquid crystals were also studied by using IR multiply perturbed total internal reflection [332, see also 114]. The literature contains further studies on IR dichroism of nematics [7, 38, 143, 151, 154, 164, 170, 176, 227, 312, 316, 7:99] and suggests a relation between IR spectra and mesogenic behavior [126a, 126b, 211b]. No details are known on ref. [7a].

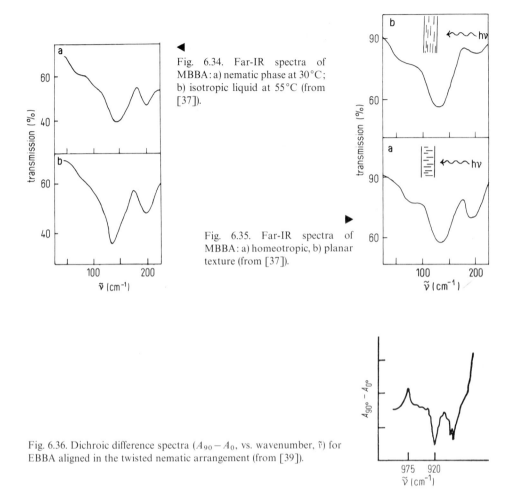

Fig. 6.34. Far-IR spectra of MBBA: a) nematic phase at 30°C; b) isotropic liquid at 55°C (from [37]).

Fig. 6.35. Far-IR spectra of MBBA: a) homeotropic, b) planar texture (from [37]).

Fig. 6.36. Dichroic difference spectra ($A_{90} - A_0$, vs. wavenumber, $\tilde{\nu}$) for EBBA aligned in the twisted nematic arrangement (from [39]).

6.5.2 Smectic and Other Mesophases

The IR dichroism of smectics has been much less researched than that of nematics. Most publications only describe the changes in absorption observed at phase transitions [1b, 7, 154, 161, 170, 171, 307a, 8:151]. P. Simova et al. have extended their studies on rotational diffusion to smectic mesophases [283], see also p. 283. Comparing the IR spectra of methyl stearate, the spectrum of the smectic phase is more like that of the crystalline solid than that of the isotropic liquid (fig. 6.37). These features become increasingly pronounced as the temperature of the liquid crystal is lowered [50]. The CH_2-rocking vibrations in the IR spectra of solid hydrocarbons manifest themselves as well defined peaks in the region $720–1\,000\,cm^{-1}$ and the CH_2-wagging and twisting vibrations in the region $1\,180–1\,360\,cm^{-1}$. In each region the peaks are approximately equally spaced, but they diminish rapidly in intensity with increasing wavenumber. These observations indicate the predominance of linear zig-zag chains in the smectic phase which is abolished upon entering the isotropic liquid.

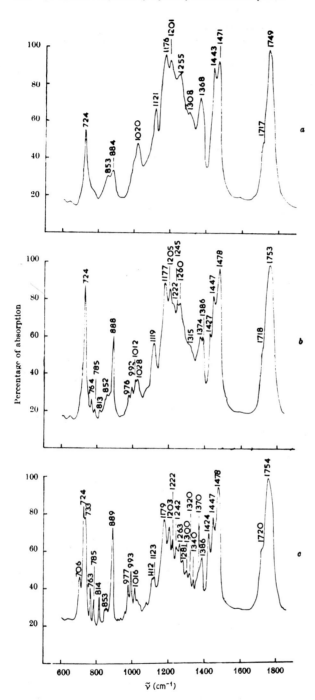

Fig. 6.37. Vibrational spectrum of methyl stearate recorded in the isotropic liquid (a), smectic (b), and crystalline state (c) (from [50]).

The IR spectra of alkali stearates change markedly upon entering mesomorphous states [90, 298, 299], especially in the ranges 400–500 cm^{-1}. (C—C—C deformation and C—C torsion), 750–900 cm^{-1} (CH$_2$-rocking), 950–1070 cm^{-1} (C—C-valence vibration), 1180–1350 cm^{-1} (CH$_2$-wagging or twisting), and at 540, 580, and 698 cm^{-1} (symmetrical and asymmetrical out-of-plane and in-plane deformations of the ionized carboxyl group).

6.6 Raman Scattering

In the expection of finding information on intermolecular forces, the low frequency region corresponding to certain types of external vibrations of the lattice has been studied preferably [42a, 212].

6.6.1 Nematic Phases

Similar to PAA, crystalline MBBA exhibits distinct crystal-lattice bands at low frequencies that are less obvious in the nematic state and lack the isotropic liquid, provided, of course, that the temperature is not close to the clearing point [16, 25, 138, 187]. G. Vergoten has found a pronounced pretransitional ordering in the isotropic phase of MBBA [308], fig. 6.38. The total Raman spectrum of MBBA is discussed in the literature [85, 310].

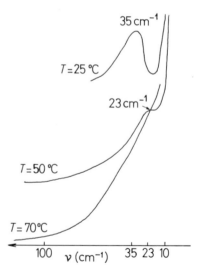

Fig. 6.38. Low frequency Raman spectra of MBBA in the nematic ($T=25\,°C$) and isotropic liquid ($T=50$ and $70\,°C$) phases (from [308]).

PAA, which is more often studied than MBBA, was not chosen as example here because the literature contains some contradictions, and it is therefore cited altogether [3, 4, 35, 40, 113, 127, 148, 261, 262, 264, 265, 270–272, 279, 329, 330]. The Raman spectra of higher *p*-alkoxyazoxybenzenes (not that of PAA) exhibit a band which is attributed to a longitudinal, "accordion-like" mode, and it is supposed to be characteristic of the side-chain backbone [271]. Since its intensity decreases discontinuously with rising temperature near the phase transitions, it seems probable that different alkoxy tail conformations exist in all phases. It is further conspicuous that the 80 cm^{-1} band of nematic PAA is much stronger with the higher homologs, and this indicates that the length-to-breadth ratio of the molecules has an important effect on lateral intermolecular associations [272]. The solution of PAA in cholesteric solvents shows an ordering of the PAA which can be understood by comparison with Raman spectra of the single crystal [36].

Concurringly, the total intensity curve of Raman scattering, $I_{tot}(T)$, and the bend/splay ratio curve, $k_{33}/k_{11}(T)$, show four different regions in the homologous p-alkoxyazoxybenzenes suggesting a correlation between both physical quantities [193].

The Raman spectra have also been investigated in the nematic state of EBBA [215], APAPA [281, 292a], 4-cyano-4'-pentylbiphenyl [126], anisaldazine [329, 330], N-anisylidene-4-(phenylazo)-1-naphthylamine [23, 24, 300, in the vitrified nematic state], other compounds [1c, 86, 278, 292a, 311, 317], and in the dynamic scattering state of nematics [4:701, 4:702].

The depolarization ratio reflects the change of short-range order near the phase transition far more than does the total scattering intensity. However, only some experiments have been published to date on the polarized and depolarized Raman spectra of liquid crystals [168, 320]. As with other optical methods, one can derive useful data for the assignment of Raman bands from comparison of intensities in ordered samples. This is because energy can only be absorbed when the vibration has a component perpendicular to the direction of irradiation. Distinct minima of the depolarization ratio, I_V^H/I_V^V, occur in the Raman spectrum of p-heptyloxyazoxybenzene near the phase transitions [186], fig. 6.39. Raman polarization studies have also been performed with MBBA and other nematics [139, 210a, 223b, 229, 309].

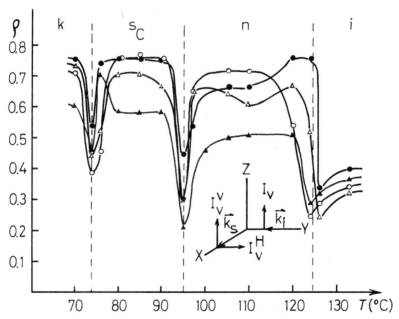

Fig. 6.39. Depolarization ratio, ϱ, vs. temperature, T, in the different phases of p-heptyloxyazoxybenzene for the lines: $\triangle$, 1267 cm^{-1}; $\bullet$, 1459 cm^{-1}; $\circ$, 1489 cm^{-1}; $\blacktriangle$, 2925 cm^{-1}. The insert shows the scattering geometry, and the Z-axis is parallel to the magnetic field (from [186]).

Defining two depolarization ratios $\rho_1 = I_{zx}/I_{zz}$ and $\rho_2 = I_{xz}/I_{xx}$, the equation has been derived

$$\langle \cos^2\theta \rangle = \frac{3\rho_2(2\rho_1+1)}{8\rho_1 + 3\rho_2 + 12\rho_1\rho_2}$$

allowing the determination of the degree of order [229].

Values obtained in this way are reported [139, 152a, 210, 210a, 223b, 229], but since marked discrepancies have been observed, S. Jen et al. have used Raman scattering to obtain a new quantitative measure of orientational statistics of individual molecules in a nematic liquid crystal [152].

6.6.2 Smectic and Other Mesophases

Analogous to nematics, the two esters diethyl azoxybenzoate (DEAB) and diethyl azoxy-cinnamate each show a low-frequency intermolecular Raman mode in the s_A state. This band has a far lower intensity and is shifted to lower frequencies when compared to that of the crystalline state [5, 148, 329]. The Raman spectra of crystalline and s_H TBBA are remarkably similar below $200\,\text{cm}^{-1}$, while each higher temperature mesophase exhibits a less distinct phonon band structure as fig. 6.40 shows [273].

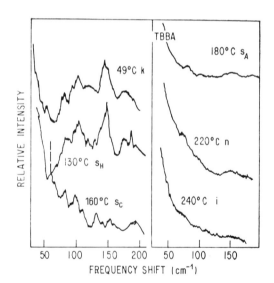

Fig. 6.40. Raman spectra in the 50 to $200\,\text{cm}^{-1}$ region showing all phases of TBBA. Spectral resolution is $1\,\text{cm}^{-1}$ (from [273]).

In addition, the solid phase exhibits a pronounced peak at $22\,\text{cm}^{-1}$ which is also present, although somewhat broadened, in the s_H phase; this peak disappears at the s_H/s_C transition point [273, see also 99, 274, 293, 307a].

The literature reports Raman spectra of other smectic mesophases [5:79a, 263, 264, 276, 293a, 311] and polarized Raman spectra of TBBA [111] and other smectics [152a].

6.7   Brillouin Scattering

The threshold power of Raman scattering (the most intense line of MBBA is shifted by $1610\,cm^{-1}$) and of Brillouin scattering (shifted in the backward direction by $0.25\,cm^{-1}$) strongly depend on temperature [232], fig. 6.41.

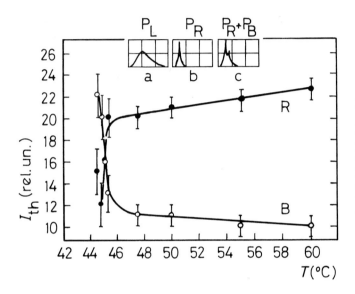

Fig. 6.41. Threshold power, I_{th}, for back-scattered stimulated Raman ($\bullet$, R) and Brillouin ($\circ$, B) emission vs. temperature in the isotropic liquid of MBBA. Insert: Oscilloscope pictures, a) laser pulse, b) and c) back-scattered pulse for sample temperature 44.8 and 45.1 °C. Horizontal scale 20 ns per division (from [232]).

An intensive and sharp Raman spike is observed in MBBA. As can be seen from the inserts in fig. 6.41, the width of this spike is less and its intensity is higher than instantaneous laser power [232–234]. A 10% conversion of pump laser to the back-scattered Raman Stokes power appears to be a conservative estimate. The forward Stokes power is an order of magnitude lower than the back-scattered Stokes power. Increasing the sample temperature to 45.1 °C (and above) the Raman Stokes spike is followed by Brillouin light. The Brillouin treshold power decreases strongly near the phase transition and more weakly at higher temperatures. This is apparently due to linear absorption by the sample. The complicated curve of the Raman threshold power is seen in connection with the self-focusing of the laser beam. It has been shown that molecular ordering in the isotropic phase of MBBA can be induced by an intense laser field [324], see also p. 262. In the isotropic phase of MBBA no dispersion of the Brillouin shifts was observed in the pretransition range; fig. 6.42 represents a typical spectrum [79].

From the Brillouin splitting and width, the adiabatic sound velocity and the sound attenuation coefficient in the GHz spectral range have been determined [3:645, 1a, 1e, 27a, 27b, 78, 136c, 305]. Compared to measurements at low frequencies, a dispersion of the sound velocity was observed indicating that substantial parts of the bulk and shear viscosities have relaxed out at 6 GHz. The local field dielectric anisotropy has been evaluated by analyzing the ratio of the peak Brillouin intensity to the depolarized Rayleigh tail intensity [78]. Other notes on Brillouin scattering in the isotropic liquids of mesogens be mentioned [3:115, 1a, 44, 136c, 224, 237, 287].

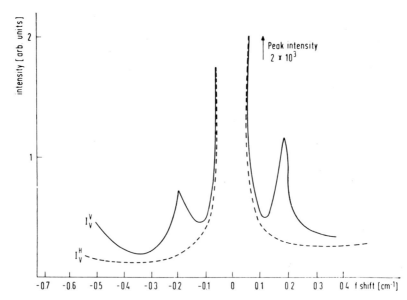

Fig. 6.42. Polarized (VV) and depolarized (VH) Brillouin spectrum in the isotropic liquid of MBBA at $T - T_c = 0.8 \pm 0.1\,^\circ C$ (from [79]).

Optical Properties of Optically Active Mesophases

7

7 Optical Properties of Optically Active Mesophases

Analogous to nematic-cholesteric phases, a few "smectic-cholesteric" phases are known. However, this term is not commonly used, the name twisted smectic phases being preferred (and used in this text). The latter designation prevents confusion when the short form "cholesteric" is used for nematic-cholesteric, as is common practice [37, 258].

7.1 Selective Reflection and Optical Activity

Cholesteric bodies are particularly noteworthy for their unique optical properties, which can be characterized as follows:
1. The optical rotation dispersion shows extremely high values.
2. The wavelength regions, having opposite signs of rotatory power, are separated by a region of selective reflection of nearly circularly polarized light.
3. The electric vectors of the reflected and the transmitted light rotate in opposite directions, but the reflection does not change the sense of the circularly polarized incident light.
4. The wavelength, λ_{max}, of the reflection band depends upon the angle of incidence.
5. In a wedge-shaped geometry or in a contact preparation with a nematic or a second cholesteric phase, the Grandjean zones appear as a typical texture, see p. 21 and 335.

As early as 1910, F. Giesel considered the appearance of reflection colors such "that cholesteryl ester layers must be able to be used as circular polarizers for ordinary light" [128].

Among the older papers, [1:149, 1:164, 1:201, 2:706, 114, 121, 166, 227, 228], those of F. Stumpf [1:365–1:368] and D. Vorländer [1:385–1:388] deserve special interest as well as the observation of P. Gaubert that "the sense of rotation does not change simultaneously but for each color under more or less different conditions, exactly those where one of the two circular waves is reflected, and there is consequently no rotatory power" [120]. Thorough measurements of the rotational dispersion and ellipticity performed by J. P. Mathieu showed that the reflected light is more nearly circularly polarized than is the transmitted light [176]. All of these characteristics of the cholesteric state rest upon its optical properties, but this state cannot be considered an individual thermodynamic phase apart from the nematic. The similarity to a multilayer interference filter is limited to optical phenomena since no molecular layers are observed by X-ray diffraction. This indicates that the most probable structure is that of a spontaneously twisted nematic phase and is thus in close analogy to the Mauguin texture. The optical properties of cholesterics are reviewed in some papers [6:53, 48, 63, 106, 147, 197, 217, 270, 273, 300, 301].

7.1.1 Helical Structures

In principle, a helical structure is possible with five of the eighteen Hermann translation types [5:127]: SSR, SSD, SS(RD), SP_1R, and P_1P_0R. The following discussion is taken from a paper by R. Dreher [95].

"In type SSR, the elements of symmetry are randomly arranged in equidistant planes (R translation). Two cases can be distinguished: a) There is a continuous distribution of mass along the helical axis, and thus there is a true translation in this direction. The molecular centers of gravity are disturbed randomly (two S translations perpendicular to the helical axis). However, an order exists in the orientation of molecules. In planes perpendicular to the axis of the helix, the molecules tend to a preferred orientation that changes uniformly from plane to plane; in this way, identical orientations are found in planes at intervals equal to the height of one turn of the helix (R translation). b) Instead of being continuous, a discrete distribution of mass on different planes is assumed. In general this means an improper translation since the height of one turn of the helix is not an even multiple of the interplanar distance, i.e. it is complete at an interstitial point.

In contrast to the SSR type, the symmetry elements of the SSD type are aligned in parallel straight chains. These can be shifted arbitrarily in their longitudinal direction and cut any given plane at randomly distributed points. The lateral spacing of chains is random. The individual molecules are held in groups by lateral forces, thus producing a helical structure.

Type SS(RD) can be derived from translation type SSR with a discrete mass distribution by the periodic recurrence of the R planes (D translation). The D translation need not be perpendicular to the R planes but can be inclined to them at any angle.

In type SP_1R, the elements of symmetry lie in equidistant planes (R translation). Each of these planes contains a swarm of parallel chains of molecules (P_1 translation) that are themselves statistically distributed within the plane. The chains in two adjacent planes are twisted to one another.

Finally, the translation type P_1P_0R has a helical structure and also exhibits the highest order. While the chains of molecules in type SP_1R are randomly distributed within a plane, those in type P_1P_0R are equidistant and immobile with respect to one another. The elements of symmetry are distributed netlike on each plane in the manner that adjacent nets have a constant angle of twist with respect to each other."

R. Dreher prefers type SSR as the most probable model of the cholesteric phase [95]. A. Wulf discusses a model system [3:843] and the possible but very small biaxiality of cholesterics [3:841, 3:842]. The rod-like molecules are stacked in such a way that the preferred direction (that of the longitudinal molecular axes) changes continuously in the direction perpendicular to it, thus forming a helical structure. As with nematics, the molecular centers of gravity are randomly distributed. Since there is no polarity the translation period is equal to half the length of one helical turn. The helix is usually described by its pitch, p, defined as the distance along the axis of the helix required to cover a 360° turn in the helix. Conversely, the description is often made in terms of "helicity", which is p^{-1}.

By means of scanning electron microscopy, the helical structure of cholesterics has been visualized using the freeze-etching preparation [290]. A. G. Khachaturyan considers a ferroelectric nematic liquid crystal which he finds to be unstable transforming into a helical cholesteric structure [146]. F. Fischer presents a calculation of the planar cholesteric texture with homeotropic boundary conditions [4:282, see also 141a].

7.1.2 Basic Observations

It was a long time until the unique properties of the cholesteric phase could be explained by a twofold chirality (i.e., a releasing chirality of the individual molecule and a secondary chirality in the molecular alignment within the phase). By the synthesis of racemic and optically active 1-butyl-1-d 4-(p-cyanobenzylideneamino) cinnamate, it has been demostrated that H-D asymmetry is sufficient to produce a (large pitch) cholesteric mesophase [136].

Long before C. W. Oseen [3:621] and H. de Vries [93] established the helical structure of the cholesteric phase, E. Reusch had observed a rotation of the polarization plane of linearly polarized light on a stack of twisted mica lamellae, cf. [42]. This assembly imitated with crystals and in step a cholesteric phase. Using a setup of twisted metal wires, H. J. Gerritsen and R. T. Yamaguchi demonstrated in an analogous experiment the rotation of the polarization plane of the microwave vector [126]. A. Rapini has considered the acoustical rotation power of cholesterics [3:711]. The observations correlated qualitatively with optical findings on cholesterics (fig. 7.1). Within a certain wavelength region, unpolarized or linearly polarized light is split into largely left- and right-circularly polarized components. Depending on the sense of the helix, one of these components is transmitted, while the other is reflected without a change in phase. The zones of anomalous rotation dispersion and of circular dichroism

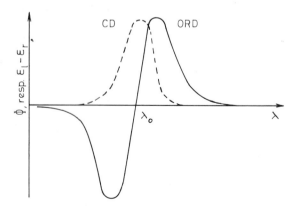

Fig. 7.1. Optical rotatory dispersion (ORD) and circular dichroism (CD) of cholesterics (from [257]).

nearly coincide analogous to the molecular Cotton effect, and it causes the λ_{max} of the dichroism and the λ_0 of the dispersion to be almost identical.

The natural optical rotation arises from a dissymmetric refractive medium such as a dissymmetric molecular arrangement (e.g., quartz crystal or cholesteric mesophase) or a solution of inherently dissymmetric molecules [49]. The electric field vectors of a linearly polarized light beam are established in a helix pattern. The medium is converted into its nonsuperimposable mirror image by a symmetry operation of space inversion, thus the chirality of the helical pattern of the electric field vectors in the medium is reversed. This corresponds to a reversal of the sense of optical rotation because the reversed direction of light propagation does not affect the sense of the optical rotation. Natural optical rotation and Faraday rotation both conserve parity and are invariant under time reversal [28].

The two structures, molecular and helical, are interdependent because the cholesteryl skeleton induces a right-handed helix that converts to a left-handed by increasing the distance, d, of substituents from the 3β-carbon atom [6, 163]. An approach is given by

$$\frac{1000}{\lambda_0} = 75d - 15.6$$

A rule applicable to non-sterol cholesterics appears in outlines [43a, 137a]. As a first approximation and according to the Bragg equation, the reflection follows a cosine function (fig. 7.2).

Other quantitative measurements on a mixture of equal parts of cholesteryl benzoate, acetate, and palmitate revealed the optical behavior shown in fig. 7.3.

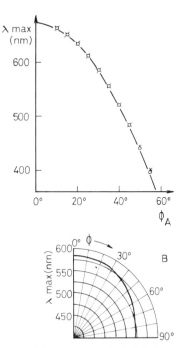

Fig. 7.2. Wavelength of the reflection maximum, λ_{max}, as a function of the angle of observation, ϕ_A, for cholesteryl ethoxyethoxypropionate. Conditions of measurement: temperature 26 °C; sample thickness 20 µm; angle of incidence = angle of observation; ×, measurement in reflection; o, measurement in transmission (from [46]).

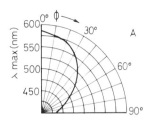

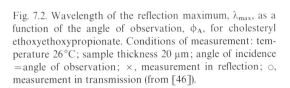

Fig. 7.3. The wavelength of the reflection maximum, λ_{max}, of a mixture containing equal parts of cholesteryl benzoate, acetate, and palmitate measured: a) if the angle of incidence = angle of observation, and b) if the angle of observation varies and the film is illuminated normal to the surface (from [112]).

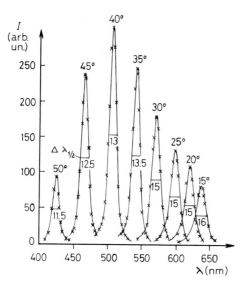

Fig. 7.4. Reflection spectra (rel. intensity, I, vs. wavelength, λ) of cholesteryl ethoxyethoxypropionate (sample thickness: 100 µm) for different angles of incidence at 27 °C (from [46]).

Perpendicular incidence and perpendicular observation generally yield the greatest wavelength of the reflection maxima [112]. The theoretical aspects will be discussed later. If the reflection band is designated by its spectral half-width, $\Delta\lambda$ (fig. 7.4), a multilayer interference filter obeys the equation

$$\Delta\lambda = \lambda / z\, n_{bi}$$

where z is the order of interference (normally 1 for cholesterics) and n_{bi} is the number of beams interfering.

If one considers n (and thus the quality of the resonator) as independent of wavelength, then the half-width should increase linearly with the wavelength (fig. 7.5). However, the observed values distinctly exceed those calculated from

$$\Delta\lambda = \frac{n_e^2 - n_o^2}{n_o^2} \cdot \lambda_{max}$$

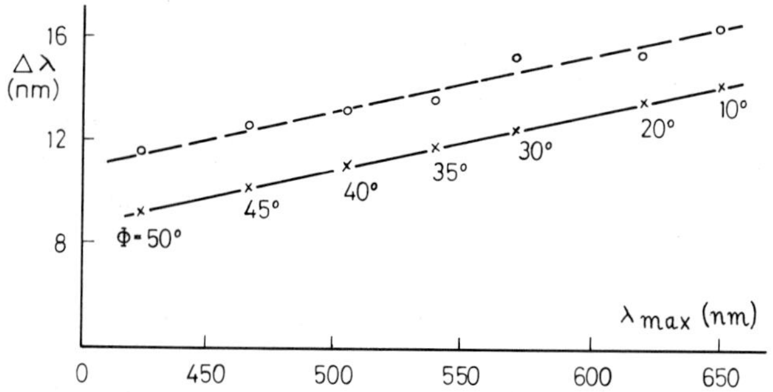

Fig. 7.5. Measured (○) and calculated (×) half-widths, $\Delta\lambda$, of the reflection bands (see fig. 7.4) as a function of the wavelength of maximum reflection, λ_{max}, for different angles of incidence, ϕ_i, on cholesteryl ethoxyethoxypropionate. Conditions of measurement: temperature 27 °C; sample thickness 100 µm; angle of incidence = angle of observation (from [46]).

The high thickness and the observation of almost twice the half-widths in transmission when compared to those measured in reflection (due to the longer light path) indicate that the falsification is mainly by disorientation. Other measurements on cholesterics have proven the linear relationship between $\Delta\lambda$ and λ_{max} [164].

The exceptional properties of the cholesteric state are not completely limited to the Grandjean texture (which is also called the disturbed or planar texture). However, reliable interpretation of quantitative measurements depends on the greatest possible proportion of the helical axes being aligned perpendicularly to the boundary surfaces. This restriction can imply considerable experimental difficulties. Reflection colors can be seen in both disturbed and undisturbed layers, but the uniform domains are much smaller in the undisturbed layer, even though the pitch appears to be about the same [4]. Microscopically, the undisturbed regions excel by their high birefringence but low optical activity [2].

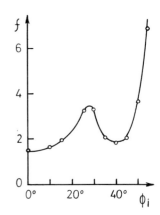

Fig. 7.6. Ellipticity of transmitted light, f, vs. angle of incidence, ϕ_i, on cholesteryl ethoxyethoxypropionate with the incident light linearly polarized perpendicular to the plane of incidence; temperature 28 °C; sample thickness 20 μm (from [46]).

Measurements of the ellipticity, f, of light transmitted (fig. 7.6) qualitatively confirm the relationship

$$f \equiv \frac{E^2_{max}}{E^2_{min}} \sim \frac{1}{\cos^2 \Phi_E}$$

This relationship holds except for angles between approximately 25° and 45°, where a distinct minimum is at 40° and one of the two circularly polarized partial waves is completely reflected (Brewster case, f is theoretically one). Normally, the oblique incidence of linearly polarized light results in two elliptically polarized components (positive and negative), whereby the partial waves migrate with different phase velocities. After leaving the cholesteric layer, superimposition of the two components yields an elliptical wave. The major axis of the wave ellipsoid is twisted against the incident light polarization plane.

Comparisons concerning the pitch of various cholesterics have been published [283]. The pitch of cholesteryl 2-(2-ethoxyethoxy)-ethyl carbonate is exceptionally large, but quite diverse values have been given for it. On the other hand, dicholesteryl compounds show very low values, e.g., 102 nm for dicholesteryl carbonate. The pitch varies with the temperature in different ways. It changes considerably in most cases as indicated by the optical data presented in fig. 7.7.

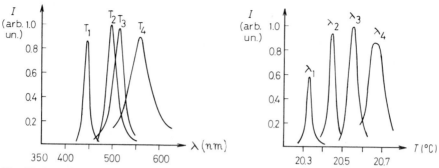

Fig. 7.7. Intensity of selectively reflected light, I, as a function of: a, wavelength, λ; T_1, 20.68°C; T_2, 20.55°C; T_3, 20.50°C; T_4, 20.44°C; and b, temperature, T; λ_1, 650 nm; λ_2, 550 nm; λ_3, 500 nm; λ_4, 450 nm; measured on cholesteryl oleyl carbonate (from [107]).

Among the substances investigated, the largest temperature coefficient has been found with cholesteryl oleyl carbonate [107] and S-cholesteryl esters [108]. Such materials are capable of indicating temperature differences of less than a millidegree directly to the human eye (cf. p. 606).

A mixture of two parts of cholesteryl capronate and one part of cholesteryl acetate exhibits a wide temperature-color range, it being blue at 87°C, green at 46°C, and red at 20°C [144]. Certain mixtures of nematics and cholesterics (which will be discussed in detail later) maintain a constant color within a large temperature range, and the angular dependency still remains [198]. The analogy to the Curie-Weiss law can be supposed from fig. 7.8, and the reasons for the aberration from a hyperbolic dependence have been discussed [107].

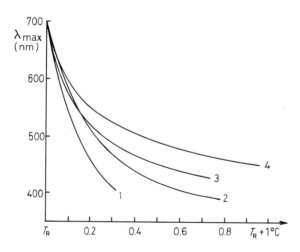

Fig. 7.8. Wavelength of maximum selective reflection, λ_{max}, as a function of temperature for 1, cholesteryl erucyl carbonate $T_R = 36.3$°C; 2, cholesteryl oleyl carbonate $T_R = 20.3$°C; 3, 1:1 mixture of 1 and 2 $T_R = 23.5$°C; 4, cholesteryl nonanoate $T_R = 76$°C (from [107]).

For perpendicular incidence, the required symmetry between reflected and transmitted light has been observed (fig. 7.9). The reflected signal is diffuse because of actual variation in pitch and a deviation from flatness at the surface. If a defined surface quality can be attained, the inhomogeneity of the sample seems to play the dominant role [4].

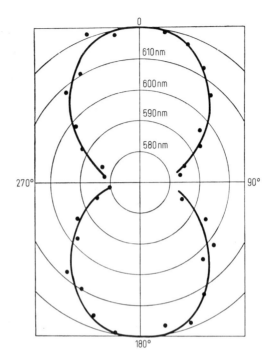

Fig. 7.9. Symmetry in reflection and transmission measurements for normal incidence; wavelength, λ_{max}, vs. angle of observation (from [4]).

7.1.3 Theory of Light Propagation

After the pioneering works of C. W. Oseen [3:606, 3:608, 3:612, 3:617, 3:621] and H. de Vries [93], many authors have investigated the theory of light propagation in cholesterics such as: G. H. Conners [88], S. Chandrasekhar [67–74], P. Châtelain [80], D. Taupin [281], E. T. Kats [145, 145a], A. S. Marathay [170], M. Aihara [16–19], S. V. Subramanyam [278], R. Dreher [95, 96, 182, 183], D. W. Berreman and T. J. Scheffer [31–36], R. M. A. Azzam [22–24], B. Böttcher [44, 46], C. Elachi [105], J. S. Prasad [212, 213], R. Nityananda [192, 193], and others [30, 118, 181, 220, 277a, 284]. Reviews have also appeared [30b, 166a].

H. de Vries derived the following relationships from Maxwell's equations [93]. The expressions $\partial\Psi/\partial z$ stand for optical activity and f for the ellipticity in a helicoidal propagation medium. Additionally normal incidence and no relative rotation between the electric vector of the linearly polarized light and the molecular arrangement are assumed.

$$f = \frac{1 - \alpha - m'^2 - \lambda'^2}{2m'\lambda'}$$

$$\frac{\partial\Psi}{\partial z} = -\frac{2\pi}{p}\cdot\frac{\alpha^2}{8\lambda'^2(1-\lambda'^2)}$$

where p is the pitch, λ is the wavelength, ε_1 and ε_2 are the dielectric constants and

$$\alpha = \frac{\varepsilon_2 - \varepsilon_1}{\varepsilon_1 + \varepsilon_2} \qquad \lambda' = \frac{\lambda}{p\sqrt{\varepsilon}} \qquad \varepsilon = \frac{\varepsilon_1 + \varepsilon_2}{2} \qquad m' = \frac{m}{\sqrt{\varepsilon}}$$

m plays the role of a refractive index.

The de Vries equation is usually given in the form

$$\frac{\partial \Psi}{\partial z} = 2\pi p \left[\frac{\alpha^2 \varepsilon}{8\lambda^2 \left(1 - \frac{\lambda^2}{p^2 \varepsilon}\right)} \right]$$

Substituting $\varepsilon = \frac{1}{2}(n_1^2 + n_2^2)$ in the equation simplifies it if the following special cases are considered:
1. when $p \gg \lambda$ and n_1 and n_2 are not too different

$$\frac{\partial \Psi}{\partial z} = -\pi p \frac{(n_2 - n_1)^2}{8\lambda^2}$$

This relationship has been repeatedly substantiated experimentally [78, 79].

2. $\left(1 - \frac{\lambda^2}{p^2 \varepsilon}\right)$ becomes zero when $\lambda_0 = p \cdot \sqrt{\frac{1}{2} n_1^2 + n_2^2}$

This last equation, which correlates the wavelength of maximum reflection with the pitch and index of refraction, has proven the most important relationship in the experimental proofs of this theory [81, 171].

S. Chandrasekhar suggested another idea which can be considered a variant of the de Vries theory [67]. Analogous to the scattering factor of a single atom used in the X-ray case, he introduced a reflection coefficient of a single molecular layer to develop a dynamic theory of cholesteric reflection. This approach points out the analogies to X-ray diffraction, and it can be expanded to compensated cholesteric mixtures and to the absorbing case (fig. 7.10).

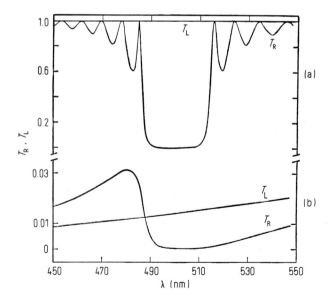

Fig. 7.10. Transmission coefficients, T_R and T_L, for right and left circular waves for a film of thickness 25 p (a) non-absorbing, (b) absorbing. The enhanced transmission for the right circular component in (b) is the analog of the Borrmann effect (from [72]).

With a model that regards the liquid crystal as consisting of a set of parallel planes the interference of multiply reflected waves with one another and with the primary light can be calculated directly [69]. The same papers also extend this treatment to finitely thick specimens, and they suggest a refinement that considers the dependence of the reflection coefficient and birefringence on each other and on the wavelength. The intensity of the reflection peak increases with increasing film thickness to finally give a single flat-topped maximum of decreasing width.

Without detailing the derivation, M. Aihara and H. Inaba suggest the following for the reflection coefficient [16]. They predict the existence of minor side lobes adjacent to the reflection maximum as do other authors.

$$R_r = \frac{|\sin\frac{1}{2}(\zeta_1 - \zeta_2)|^2}{\left|\frac{1+\zeta_1}{1+\zeta_2}r_2 e^{-i\zeta_1 kd} - \frac{1+\zeta_2}{1+\zeta_1}r_1 e^{-i\zeta_2 kd}\right|^2}$$

where

$$\beta = \lambda/p, \quad k = (\omega/c)\varepsilon^{1/2}, \quad \zeta_{1,2} = \beta \pm [\beta^2 + 1 - 2\{\beta^2 - (\tfrac{1}{2}\delta)^2\}^{1/2}]^{1/2}, \quad r_{1,2} = (\zeta_{1,2}^2 - 1)/\delta$$

δ: dielectric anisotropy constant

The equation $\dfrac{d\chi}{dz} = -n_{12}\chi^2 + (n_{22} - n_{11})\chi + n_{21}$ deduced by R. M. A. Azzam and N. M. Bashara is a first-order differential equation whose solution, $\chi(z, \chi_0)$, gives the evolution of the ellipse of light polarization as it propagates through an anisotropic medium. It starts from the initial polarization state χ_0 at $z = 0$ [22]. For a homogeneous anisotropic medium, the N matrix is independent of position along the direction of propagation. The helicity of the cholesteric structure can be represented in the form

$$N = g_0 \begin{pmatrix} 0 & -ie^{-i2az} \\ -ie^{i2az} & 0 \end{pmatrix}$$

where g_0 is given by $g_0 = \frac{1}{2}(n_y - n_x)$, and n_x and n_y are the principal propagation constants. The solution of the differential equation is given explicitly by the equation

$$\chi(z, \chi_0) = \frac{[\beta - \frac{1}{2}(n_{11} - n_{22})\tan\beta z]\chi_0 + [n_{21}\tan\beta z]}{[n_{12}\tan\beta z]\chi_0 + [\beta + \frac{1}{2}(n_{11} - n_{22})\tan\beta z]}$$

where $\beta = [-\frac{1}{4}(n_{11} - n_{22})^2 - n_{12}n_{21}]^{1/2}$.

For propagation along the positive direction of the helical axis the general solution is given by the equation:

$$\chi(z, \chi_0) = \left[\frac{(\beta - i\alpha\tan\beta z)\chi_0 + (-ig_0\tan\beta z)}{(-ig_0\tan\beta z)\chi_0 + (\beta + i\alpha\tan\beta z)}\right]e^{i2az}$$

where $\beta = g_0[(d/g_0)^2 + 1]^{1/2}$ and $\alpha = 2\pi/p$.

It shows that, starting from an arbitrary polarization state χ_0 at $z=0$, this state never repeats itself along the helical axis. Although both factors of this equation are periodic functions of z, the entire function $\chi(z,\chi_0)$ is not periodic, because the periods of the two factors are different. R. M. A. Azzam et al. discuss examples of trajectories that describe the evolution of the ellipse of light polarization along the helical axis of a cholesteric liquid crystal [23, 24]. Such ellipses are shown to be distorted hypo- or epicycloids appearing in the form of open nonrepetitive multilobed or multibranched trajectories depending on the initial state, χ_0, and the properties of the liquid crystal.

For the conditions described below, D. W. Berreman and T. J. Scheffer reduced Maxwell's equations to the matrix form [31, 32]:

$$\frac{\partial}{\partial z}\begin{bmatrix} E_x \\ iH_y \\ E_y \\ -iH_x \end{bmatrix} = \frac{\omega}{c}\begin{bmatrix} \left(-i\dfrac{kc\varepsilon_{xz}}{\omega\varepsilon_{zz}}\right) & \left[1-\dfrac{1}{\varepsilon_{zz}}\left(\dfrac{kc}{\omega}\right)^2\right] & \left(-i\dfrac{kc\,\varepsilon_{yz}}{\omega\,\varepsilon_{zz}}\right) & 0 \\ \left(-\varepsilon_{xx}+\dfrac{\varepsilon_{xz}^2}{\varepsilon_{zz}}\right) & \left(-i\dfrac{kc\,\varepsilon_{xz}}{\omega\,\varepsilon_{zz}}\right) & \left(\dfrac{\varepsilon_{xz}\varepsilon_{yz}}{\varepsilon_{zz}}-\varepsilon_{xy}\right) & 0 \\ 0 & 0 & 0 & 1 \\ \left(\dfrac{\varepsilon_{xz}\varepsilon_{yz}}{\varepsilon_{zz}}-\varepsilon_{xy}\right) & \left(-i\dfrac{kc\,\varepsilon_{yz}}{\omega\,\varepsilon_{zz}}\right) & \left[\dfrac{\varepsilon_{yz}^2}{\varepsilon_{zz}}-\varepsilon_{yy}+\left(\dfrac{kc}{\omega}\right)^2\right] & 0 \end{bmatrix}\begin{bmatrix} E_x \\ iH_y \\ E_y \\ -iH_x \end{bmatrix}$$

Here, the dielectric properties of the medium are represented locally by a symmetric dielectric tensor, ε (fig. 7.11), and the medium is nonmagnetic so that the permeability is equal to one in Gaussian units.

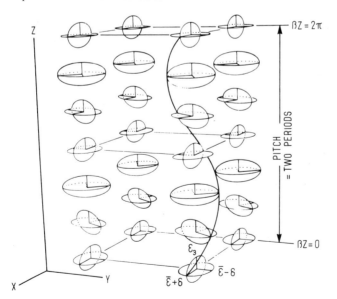

Fig. 7.11. Spiraling dielectric ellipsoids in Oseen's optical model of a cholesteric liquid crystal (from [32]).

The optical model of perfectly ordered cholesteric liquid crystal used by C. W. Oseen [3:621] and H. de Vries [93] was modified to consider the oblique incidence of light. The dielectric tensor with this generalization is given by the following:

$$\varepsilon = \begin{bmatrix} \bar{\varepsilon}+\delta\cos 2\beta z & \delta\sin 2\beta z & 0 \\ \delta\sin 2\beta z & \bar{\varepsilon}-\delta\cos 2\beta z & 0 \\ 0 & 0 & \varepsilon_3 \end{bmatrix}$$

As a satisfying approximation for the local propagation matrix, it is proposed that

$$\mathbf{P}(z,h) \approx \mathbf{P}_0(h) + (h\omega/c)D_2(z)$$

where

$$\mathbf{P}_0(h) = \begin{bmatrix} \cos(abh\omega/c) & (a/b)\sin(abh\omega/c) & 0 & 0 \\ (-b/a)\sin(abh\omega/c) & \cos(abh\omega/c) & 0 & 0 \\ 0 & 0 & \cos(vh\omega/c) & (1/v)\sin(vh\omega/c) \\ 0 & 0 & (-v)\sin(vh\omega/c) & \cos(vh\omega/c) \end{bmatrix}$$

$$a = [1-(kc/\omega)^2/\varepsilon_3]^{1/2}$$
$$b = (\bar{\varepsilon})^{1/2}$$
$$v = [\bar{\varepsilon}-(kc/\omega)^2]^{1/2}$$

and

$$D_2(z) = \begin{bmatrix} 0 & 0 & 0 & 0 \\ -\delta\cos 2\beta z & 0 & -\delta\sin 2\beta z & 0 \\ 0 & 0 & 0 & 0 \\ -\delta\sin 2\beta z & 0 & +\delta\cos 2\beta z & 0 \end{bmatrix}$$

With obliquely incident light, higher order reflections may be observed and calculated from the formula above (s. fig. 7.12); a 4×4 matrix technique applied by D. W. Berreman proved to be particularly well suited for a computer [35].

Matrix operations are commonly used to describe the migration of polarized light in a helicoidal propagation medium. This might be done in terms of four 2-by-2 matrix operators for propagation along the helical axis [170].

R. Dreher's calculations show that in a helical structure amplitude-modulated waves can propagate and that, for certain waves, forbidden frequency bands exist [95, 96, 182, 183]. This method of calculation differs from others in that the partial considerations of intensity and polarization are completely separated from the problem of reflectivity. Given that the dielectric is nonmagnetic, a system of differential equations whose solutions describe amplitude-modulated waves can be derived from the Maxwell equations. Two functions are determined by ordinary fourth-order differential equations with periodic coefficients, while the third is obtained essentially by differentiation of one of the first two:

$$\frac{d^2 E_1}{d\alpha^2} + (a_1 + a_2\cos 2\alpha)E_1 = -a_2 E_2 \sin 2\alpha$$

$$\frac{d^2 E_2}{d\alpha^2} + (b_1 - b_2\cos 2\alpha)E_2 = -b_2 E_1 \sin 2\alpha$$

$$E_3 = -i\frac{c}{\omega}\frac{m}{m^2-\varepsilon_3}\frac{dE_1}{dz}$$

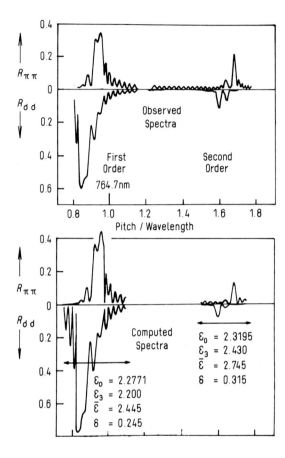

Fig. 7.12. First- and second-order reflectance spectra of a cholesteric liquid-crystal film 15 pitch lengths or 11.47 μm thick confined between two glass prisms of optical dielectric constant ε_0. The light beam is incident at 45°. Polarizer and analyzer were parallel to the plane of reflection for R_π and normal to it for R_σ measurements (from [32]).

The constants a_1, a_2, b_1, and b_2 are determined by the structure parameters ε_1, ε_2, and p; the frequency, ω, of the waves and the quantity m which is in turn determined largely by the incident angle of the light wave. The first two equations form a system of coupled Mathieu equations that may also be used to describe vibrational systems. Considering that we have a locally uniaxial dielectric, i.e., two of the three principal dielectric constants are equal to one another, we obtain

$$\frac{\mathrm{d}^2\varphi}{\mathrm{d}\alpha^2} - 2\frac{\mathrm{d}\chi}{\mathrm{d}\alpha} + a_1\varphi + a_2\varphi\cos 2\alpha = 0$$

$$\frac{\mathrm{d}^2\chi}{\mathrm{d}\alpha^2} + 2\frac{\mathrm{d}\varphi}{\mathrm{d}\alpha} + b_1\chi - a_2\varphi\sin 2\alpha = 0$$

or, if one passes to a fourth-order differential equation

$$\frac{\mathrm{d}^4\varphi}{\mathrm{d}\alpha^4} + (a + a_2\cos 2\alpha)\frac{\mathrm{d}^2\varphi}{\mathrm{d}\alpha^2} - (6a_2\sin 2\alpha)\frac{\mathrm{d}\varphi}{\mathrm{d}\alpha} + (b + \mathrm{d}\,a_2\cos 2\alpha)\varphi = 0$$

which uses the abbreviated notations

$$a = 4 + a_1 + b_1 \qquad b = a_1 b_1 \qquad d = b_1 - 8$$

The above fourth-order differential equation usually has a fundamental system of solutions of the form

$$\varphi_j(\alpha) = e^{i\mu j\alpha} P_j(\alpha) \qquad (j = 1, 2, 3, 4)$$

Periodic structures generally act as bandpass filters for the propagation of waves of certain polarizations. R. Dreher compared the waves in a helicoidal structure with the Bloch waves in solids [95]. Analogous to the energy band schematics for solids, he traced out stability maps that show under which conditions stable (progressive) and instable (damped) waves appear (fig. 7.13). One characteristic curve is given by the equation

$$\lambda = p \sqrt{\varepsilon_2 - (n \cdot \sin \Psi)^2}$$

where Ψ is angle of incidence.

In the fundamental system

$$j(\alpha) = e^{i\mu j\alpha} P_j(\alpha) \qquad (j = 1, 2, 3, 4)$$

the functions $P_j(d)$ are periodic, steady, and thus also limited, so the entire character of the stability is given by the typical coefficient, μ. R. Dreher's discussion [95, 183] cannot

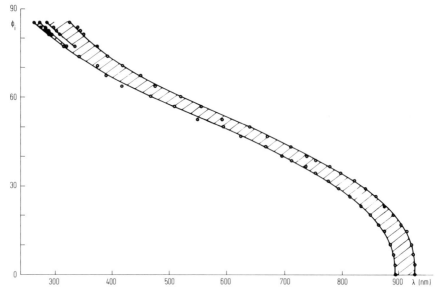

Fig. 7.13. Stability chart (wavelength, λ, vs. angle of incidence, ϕ_i) obtained experimentally at room temperature for a mixture of cholesteryl oleate, pelargonate and chloride in the weight ratio $11:40:29$ (from [95]).

be repeated in full here, and we are limited to the following abstract taken from his work: "The value of the typical exponent depends on the magnitude of the constants ε_1, ε_2, and p; and on the frequency (or wavelength) as well as on the chosen value of the parameter m. With a given structure (ε_1, ε_2, and p fixed), the stability or instability of the solutions depends only on λ and m. In a λ, m plane, the behavior can thus be illustrated by sketching the areas of stability and instability."

Fig. 7.13 indicates that with normal incidence ($\Psi = 0$) the reflection occurs at the longest wavelengths and shifts only slightly at small angles of incidence, but it changes markedly at medium angles of incidence and finally splits into two branches of different widths at strongly inclined incidence. For any direction of propagation other than those parallel and perpendicular to the helical axis, there is an infinite number of further reflection bands that correspond to higher harmonics. Higher-order reflections occur in addition to the basic band when the helical structure is distorted. This phenomenon can occur even at normal incidence [96]. Measurements of pitch by the standard method of maximum reflectivity can lead to erroneous results on aperiodically perturbed samples such as those having thermal or concentration gradients. R. Dreher discusses polar, nonpolar, and aperiodic deformations of the helical structure [96], cf. section 4.5.3.

The literature also considers special cases of cholesterics and their optical properties, e.g., distorted and inhomogeneous structures [24, 30c, 178–180, 252], absorbing cholesterics [30a, 71, 92a, 279] and the influence of external fields [29], see chapter 4. The measured and calculated reflection spectra agree as seen in fig. 7.14. The characteristics of circular dichroism, $D = (I_l - I_r)/(I_l + I_r)$, are clearly related to the selective reflection, R, of cholesterics. At normal incidence and for wave propagation along the helical axis, the relation was shown to

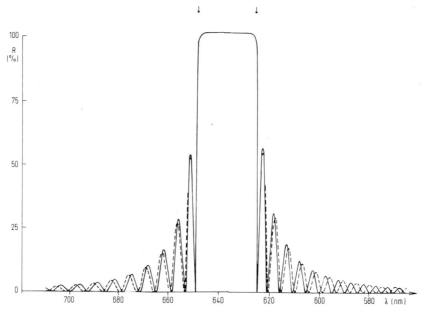

Fig. 7.14. Reflection spectra at normal incidence: dashed curve, computed spectrum on a 21.0 μm layer with pitch 427.3 nm; solid curve, experimental spectrum (from [95]).

be $D = R/(2 - R)$ [215]. The CD spectra measured therefore closely resembles those of the reflectance [165, 167, 214, 215, 253].

J. Adams and W. E. L. Haas used the relationship

$$\lambda = 2np\cos\tfrac{1}{2}[\cos^{-1}\{1/n^2\,(\sqrt{(n^2 - \sin^2\theta_i)(n^2 - \sin^2\theta_s)} - \cos\varphi_s\sin\theta_s\sin\theta_i)\}]$$

where p corresponds only to a 180° turn, n is the refractive index, φ_i is the angle of incidence, and φ_s is the angle of reflection [3, 5].

The dielectric constants may be calculated from the angles of total reflection. The dielectric constants then serve in the calculation of the reflection spectrum illustrated in fig. 7.14 [182]. The Beer-Lambert law is not applicable to "cholesteric" absorption bands. Dreher's calculations were confirmed by C. Elachi and C. Yeh [105]. According to these works, the stop-band characteristic for wave propagation in cholesterics can split into two or three stop-bands. This splitting depends on the dielectric constant and the angle of incidence. J. S. Prasad et al. [212, 213] reported experiments that support the theoretical considerations of R. Dreher, S. Chandrasekhar, and M. Aihara, and refute those of S. V. Subramanyam [278]. Optical transmission through a single-domain cholesteric film has not yet been thoroughly studied [117, 297]; light reflection by a cholesteric prism was considered theoretically [200].

Raising the temperature increases the libration of the molecules and decreases the eccentricity of the dielectric tensor [36], fig. 7.15. The measured changes in the reflection spectra roughly agree with the prediction of the Maier-Saupe theory applied to cholesterics, see also [284b]. Defining the order parameter as $S = \tfrac{1}{2}\langle 3\cos^2\theta - 1\rangle$ and assuming negligible correlation for values of θ among neighbouring molecules, the dielectric tensor is smeared out to a more nearly isotropic form according to the formulas:

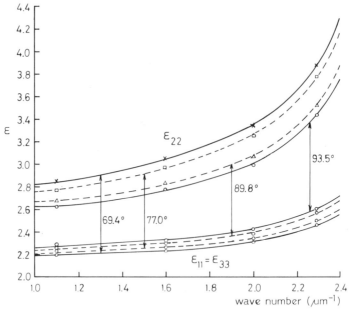

Fig. 7.15. Optical dielectric dispersion curves computed from the equations below to give the best fit to the observed reflectance spectra at four different temperatures (from [36]).

$$\varepsilon_{11} = \varepsilon_{33} \tfrac{1}{3}[(2+S)\varepsilon_{11}^0 + (1-S)\varepsilon_{22}^0]$$
$$\varepsilon_{22} = \tfrac{1}{3}[(2-2S)\varepsilon_{11}^0 + (1+2S)\varepsilon_{22}^0]$$

Setting the order parameter at unity each principal axis of the optical dielectric tensor can be expressed by

$$\varepsilon_{11}^0 \approx \varepsilon_i + \tfrac{1}{3}[\alpha(T_c - T)(\varepsilon_i^2 + \varepsilon_i - 2)]$$

The dispersion of the dielectric tensor components (fig. 7.15) can be represented by the following classical harmonic oscillator dispersion function, which contains a far-infrared and a near-ultraviolet resonance term.

$$\varepsilon_i = -(A_{1i}/v)^2 + A_{2i} + \{A_{3i}/[1-(v/A_{4i})^2]\}$$

7.1.4 Influence of Temperature and Pressure

An improvement of the relationship $\lambda_{max} \sim 1/T$ which describes the temperature dependence of the wavelength of maximum reflection has been elaborated by P. N. Keating [3:395, see also 3:812, 3:813, 108] as

$$\lambda_{max} = \frac{A}{T}\left(1 + \frac{B}{T-T_0}\right)^2$$

where A and B are constants, and T_0 is the transition temperature to the crystalline or smectic phase. B. Böttcher has modified the Keating theory without assuming neighboring planes of molecules [47]. The pitch of cholesterics is especially sensitive to temperature changes or chemical additives in the pretransition range. Thus, a mixture of 23% cholesteryl chloride and 77% cholesteryl nonanoate, which at temperatures below 30°C is in a pretransitional state, behaves contrarily to a mixture of 45% chloride and 55% nonanoate in the normal cholesteric state above 20°C [232]. An increase in the number and size of smectic clusters causes an increase in pitch [252a], and the pitch vs. composition curve has a maximum for those mixtures in which the appearance of a smectic phase is approached [156a]. Based on the Frenkel theory of heterophase fluctuations, E. Sackmann predicted the pitch to increases exponentially with decreasing temperature [232], fig. 7.16. S-cholesteryl 14-phenyltetradecane-thioate provides an example that shows an extremely high temperature coefficient of selective reflectance [108].

Upon approaching the transition to the s_A phase, the coherence length, ξ, of the smectic clusters diverges exponentially toward an apparent transition temperature, T^*, according to the equation

$$\xi = \xi_0(T-T^*)^{-\gamma}$$

which manifests itself in a corresponding increase in the twist and bend elastic constants, and pitch such that

$$p = p_0 + c(T-T^*)^{-\gamma}$$

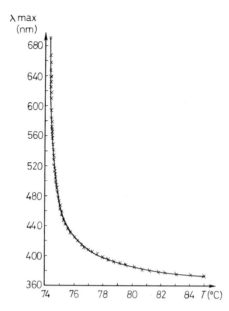

Fig. 7.16. Temperature dependence of the wavelength of maximum reflection in cholesteryl nonanoate, s_A 74 c 91 i (from [206]).

where p_0 is the pitch in the absence of smectic ordering and

$$\gamma = 0.675 \pm 0.025 \qquad ([206, \text{ see also } 20])$$

Numerous other authors have studied the temperature dependence of the reflection and optical rotation of cholesterics [2:203, 2:210, 2:294, 2:301, 2:435, 3:31, 89, 113, 125, 140, 146a, 158, 165, 167, 174, 175, 184, 253, 254, 274]. Special attention has been paid to the divergence of the helical pitch near c/s_A phase transitions [50, 143, 206–208, 295].

To describe the critical enhancement of optical rotatory power in the pretransitional state of the isotropic to the cholesteric liquid, R. B. Meyer et al. report a similar equation

$$\varphi = \varphi_0 + \text{const.} \cdot T(T - T^*)^{-\gamma}$$

where φ is the molecular optical activity [3:106, 185, 186], fig. 7.17.

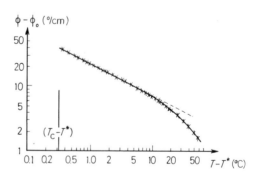

Fig. 7.17. Optical rotation, $\phi - \phi_0$, as a function of the relative temperature, $T - T^*$, showing the -0.50 exponent in the pretransition region of $(+)$-p-ethoxybenzylidene-p'-(β-methylbutyl)aniline (from [185]).

The rotatory power varies significantly over a large temperature range becoming singular as $T = T^*$, but the singular behavior at the second-order transition temperature, T^*, is cut off at the temperature T_c by the first order transition. The coherence length, ξ, of the local spiral conformation in the isotropic phase is estimated from de Gennes' phenomenological theory to be ≈ 25 nm at T_c. Pretransition phenomena have been studied in other cholesterics [229]. When cholesterics are supercooled, the iridescent state remains similar to that of the fluid cholesteric in intensity, angular dependence, and reflective circular dichroism, but not in pressure- or temperature-dependence [169].

Cholesterics undergo a progressive untwisting of the helix under pressure. This occurs up to a pressure, p_∞, where λ_0 approaches infinity (fig. 7.18).

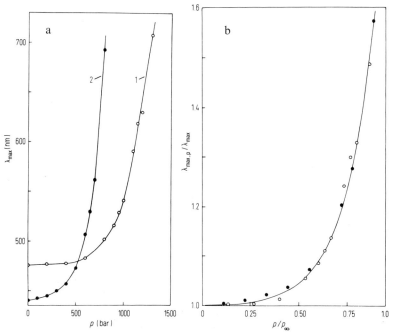

Fig. 7.18. Pressure dependence of the wavelength of maximum reflection of two cholesteric mixtures of cholesteryl oleyl carbonate (COC) and cholesteryl chloride (CC) (mixture 1: COC:CC = 74.8 : 25.2 mol-%; mixture 2: COC:CC = 80.1 : 19.9 mol-%): a, maximum reflection wavelength λ_{max} at room temperature; b, relative maximum reflection wavelength $\lambda_{max,p}/\lambda_{max}$ at reduced pressure p/p_∞ with mixture 1 (○) at 476 nm and 1500 bar p_∞ and mixture 2 (●) at 440 nm and 900 bar p_∞ (from [267]).

Experimental details are described in the literature [210] but it is still too sparse to make generalizations [75, 269].

7.1.5 Special Observations

Observations of D. W. Berreman and T. J. Scheffer throw considerable doubt on the acceptance of a uniform pitch in the boundary proximity and the middle region of a cholesteric liquid-crystal film. Some measurements indicate an apparent average pitch change which is considerably less than what the model of a Grandjean discontinuity predicts to be a change of one half turn within the film thickness [33, see also 91]. Other measurements agree with de Gennes' model of a Grandjean discontinuity, while still other observations indicate that the wavelength dependence of the rotatory power varies greatly with film thickness [62, 147a]. The intensities of both the first and the second order reflection bands depend strongly on the experimental conditions such as the molecular orientation, the polarization, the film thickness, and the angles of incidence and reflection [34, 68, 91, 92]. Therefore one always must be aware of the possibility of finding misleading results. Experimental details for studying the color response have been published in the literature [86, 94, 284a], as has been the influence of the refractive index of the glass [287].

The color changes of cholesterics, which can be used as temperature indicators, can be disturbed by thermal hysteresis [15a, 94], fig. 7.19.

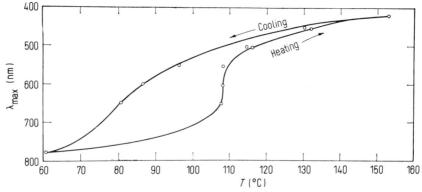

Fig. 7.19. Wavelength of maximum reflection, λ_{max}, vs. temperature, T, showing the hysteresis of a 45:55 mixture of cholesteryl cinnamate and benzoate (from [94]).

In order to determine the pitch of cholesterics, one can start from one of the following experiments by using the distance of the Grandjean lines in a wedge geometry, the wavelength of maximum reflection, light transmittance (especially in the infrared), and the angular dependence of the Bragg reflection. H. Baessler and M. M. Labes were able to observe distinctly the optical absorption of the helical structure free from absorption due to molecular vibrations for wavelengths under 6 µm. Fig. 7.20 shows the data obtained from a 1.75:1.00 mixture of cholesteryl chloride and cholesteryl myristate. The helicity (pitch^{-1}), which is strongly temperature dependent, disappears at 40.0°C. E. H. Korte recently reinvestigated this system [152].

On shearing, cholesteric liquid-crystal films lose intensity of reflected light, but the wavelength of the reflection maximum does not change appreciably [87, 3:663], fig. 7.21. The original intensity, after the cessation of shearing, is restored proportional to the logarithm of time. The decrease in wavelength of maximum intensity as a function of shear was observed to be smaller for thicker films.

A strain, reducing the thickness of a cholesteric layer, induces a temporary blue shift until the original pitch is restored. This occurs within a few seconds [11].

Right-handed cholesterics which are formed only by a few cholesteryl esters are more common with doristeryl ester [2:445].

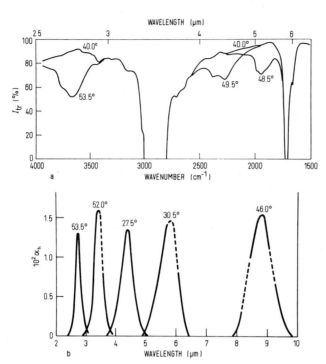

Fig. 7.20. Transmission and absorption spectra in a 1.75 : 1.00 mixture of cholesteryl chloride and cholesteryl myristate: a, IR transmittance, I_t, at various temperatures; b, helical absorption coefficient, α_h, at various temperatures with the dashed regions indicating wavelengths where there is masking by strong vibrational absorption (from [25]).

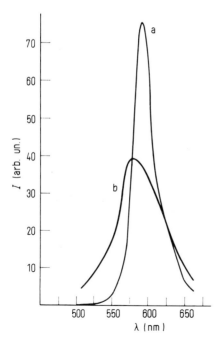

Fig. 7.21. Intensity of selectively reflected light, I, vs. wavelength, λ, a, at rest and b, under shear ($V = 1$ cm/min) for a 25 μm cholesteric film (from [87]).

7.1.6 Behavior of Mixtures

Some results obtained from the mixtures of two cholesterics have already been stated insofar as these properties are common with pure cholesterics. Fig. 7.22 illustrates an example of ternary compound-color relationship.

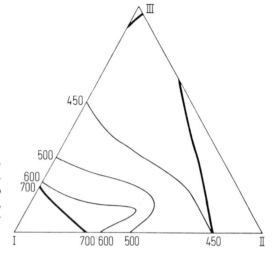

Fig. 7.22. Isochromes (λ_{max} in nm) of a ternary system (in weight-%) prepared by spontaneous cooling from the melt after cover slip displacement. *I*, cholesteryl cinnamate; II, cholesteryl hydrogenphthalate; III, cholesteryl hydrogenisophthalate (from [169]).

The light reflection of cholesterics is very sensitive to both solute molecules and impurities [195, 230, 296]. An analytical application of this effect has been proposed [6:42, 230]. In a cholesteric solution, the photochemical cis-trans isomerization of azobenzene or stilbene induces visible color changes as do systems employing a photolabile chiral compound [187]. Large shifts to lower wavelengths are favored by dipolar effects, small molecular size, and hydrogen bonding. Figs. 7.23 and 7.24 show that reflectance and gas chromatographic studies (cf. section 9.1) together reveal much information about thermodynamic characteristics of cholesterics.

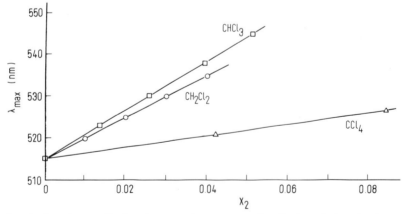

Fig. 7.23. Wavelength of maximum reflection, λ_{max}, vs. solute concentration, X_2, in a cholesteric mixture of 24% cholesteryl chloride, 45.6% cholesteryl oleyl carbonate and 30.4% cholesteryl nonanoate at 18°C (from [296]).

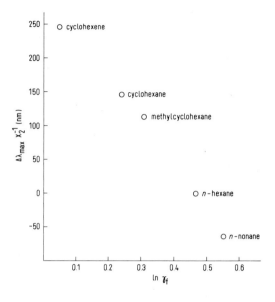

Fig. 7.24. Shift in the wavelength of maximum reflection per mole fraction of solute, $\Delta\lambda_{max} X_2^{-1}$, vs. natural logarithm of solute activity coefficients, $\ln\gamma_f$, in the mixture of fig. 7.23 at 18 °C (from [296]).

UV radiation that can decompose a substance, such as cholesteryl iodide, changes the reflection color of a cholesteric liquid crystal analogously and can thus be visualized [1].

For mixtures of nematics and cholesterics, the measurements allow for a verification of the theories of de Vries and Maier-Saupe [51, 52, 60, 61, 77, 172]. Fig. 7.25 shows experimental confirmation of the linear correlation between rotatory power and pitch as derived from the de Vries equation when $\lambda^2/p^2 N^2$ is small compared with unity [60, 61]. With regard to the enormously high rotational values, it must be mentioned that the rotatory power can be expected to revert to normal values when the reflection is negligible irrespective of whether λ is larger or smaller than λ_0 [69]. The pitch-concentration dependence in mixed nematic-cholesterics may show an inverse proportionality in a narrow concentration region of the chiral compound [59, 77, 84, 85, 199, 286], fig. 7.26.

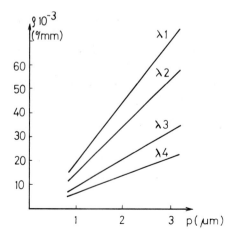

Fig. 7.25. Optical rotatory power, ϱ, vs. pitch, p, in a mixture of cholesteryl benzoate and PAP at 140 °C. λ_1, 480; λ_2, 508.6; λ_3, 579.1; λ_4, 690.8 nm (from [60]).

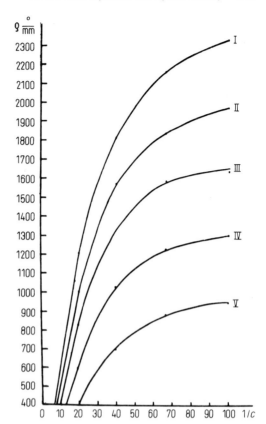

Fig. 7.26. Optical rotatory power, ϱ, vs. the inverse concentration, $1/c$ of cholesteryl propionate in PAA at 100°C for various wavelengths. I, 471; II, 489; III, 513; IV, 555; V, 637 nm (from [85]).

The area of cholesterics is enormously enlarged by the fact that both the mixtures of cholesterics and nematics as well as the mixtures of nematic and optically active materials exhibit the behavior of pure cholesterics as G. Friedel first predicted [116]. Since the addition of an optically active substance to a nematic liquid crystal induces a helical structure whose rotatory power may be larger by a factor of ca. 10^3 than that displayed in an isotropic solvent, very small amounts of optically active substances can be detected [162a, 257], and the nature of conglomerates can be recognized in this way [40, 43]. Nematics, which are valuable aids in the study of oriented solutions (cf. sections 6.3, 6.4, 6.5, and 6.6), are suited to the study of optically active solute molecules only when the spontaneously appearing cholesteric phase is untwisted by applying electric or magnetic fields [56].

As the concentration of the chiral compound increases, three different types of behavior occur [199]:

a) the pitch slowly saturates as the concentration approaches to the pure cholesteric compound.

b) the pitch decreases to a minimum value and then increases again to that of the pure cholesteric compound.

c) the pitch decreases to a minimum value, increases to infinity, and then decreases again to that of the pure cholesteric compound [138a].

Examples of the second case are found in mixtures of cholesteryl propionate with APAPA or MBBA where the helical twisting power of certain mixtures is larger than that of the pure cholesteric compound [190, 191]. Ternary mixtures consisting of 1 cholesteric and 2 nematics have also been studied [156b]. Additional peculiarities emerge when mixtures of two cholesterics are considered [265, 268], fig. 7.27.

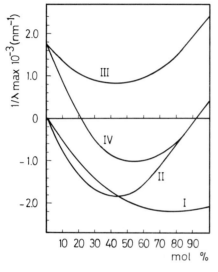

Fig. 7.27. Dependence of the reciprocal wavelength of maximum reflection, $1/\lambda_{max} = (\bar{n}p)^{-1}$, on the mole fraction of various cholesteric mixtures described in the text (from [268]).

Mixture I. Cholesteryl propionate (*l*-helix) and MBBA: case b, no inversion, minimum of $1/\lambda_0$ at x_{min}: 0.74.

Mixture II. Cholesteryl 2-(2-ethoxyethoxy)ethyl carbonate (*d*-helix) and EBBA: case c, helical inversion at x_n: 0.89, minimum of $1/\lambda_0$ at x_{min}: 0.43.

Mixture III. Cholesteryl chloride (*d*-helix) and act. amyl-*p*-(*p*-cyanobenzylidene)-aminocinnamate (*d*-helix): no inversion, minimum of $1/\lambda_0$ at x_{min}: 0.42.

Mixture IV. Cholesteryl 2-(2-ethoxyethoxy)ethyl carbonate (*d*-helix) and act. amyl *p*-(*p*-cyanobenzylidene)-aminocinnamate (*d*-helix): helical inversions at x_n: 0.20 and 0.91, minimum of $1/\lambda_0$ at x_{min}: 0.54.

H. Finkelmann and H. Stegemeyer have described this system in detail [266]. From the molecular statistical theory of Goossens extended to binary cholesteric mixtures, H. Stegemeyer et al. derive the relationship $X_{min} = \frac{1}{2}(x_{n,1} + x_{n,2})$ [113, 268, 271]. The behavior of mixture IV is quite surprising (fig. 7.27). It indicates that identical or opposite helical twist of the pure components cannot be predicted from the existence or absence of a nematic phase in a contact preparation of two cholesterics. This refutes an earlier supposition [41, cf. 1:108].

In many binary and ternary systems, helicity (inverse pitch) changes linearly with concentration when there is no influence by pretransition phenomena [3, 7, 127], fig. 7.29, but large deviations are also known [127], e.g. in the ternary system consisting of MBBA, cholesteryl chloride, and cholesteryl nonanoate [9]. For a series of mixtures, an additive law

$$p = \left| \frac{100}{\sum\limits_{i} x_i \theta_i} \right|$$

is valid which relates the pitch, p, of a cholesteric mixture to the percentages, α_i, and the effective rotatory power, θ_i, of the constituents [8]. S. Chandrasekhar et al. proposed a theory dealing with the optical properties of a mixture of right- and left-handed cholesterics [70]. Defining a specific rotation, θ, as $\theta = (2np)^{-1}$, J. Adams and W. Haas extend the linear additive law for pitch dependence in binary cholesteric mixtures to include first order interactive effects. The simple equation

$$\theta = \alpha\theta_A + (1-\alpha)\theta_B + \alpha(1-\alpha)k(A,B)$$

reproduces the different cases illustrated in fig. 7.28 for which examples are mentioned in the literature [10]. The literature contains other experimental and theoretical studies of the pitch-concentration relationship in cholesteric mixtures [2:443, 2:444, 4:800, 26, 27, 82a, 86a, 113, 138, 153–156, 156a, 175, 177, 209, 271, 288, 298].

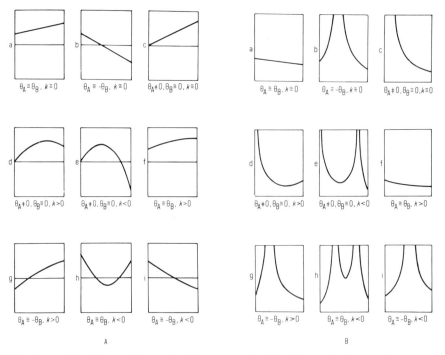

Fig. 7.28. Specific rotation (fig. A, plotted vertically) and pitch (fig. B, plotted vertically) as a function of the weight fraction varying from 0 to 1 (from [10]).

If the concentration dependence of the helical pitch is known, the diffusion coefficient can be determined by visual observation of a contact preparation [161, see also 1:118].

The socalled compensated cholesteric mixtures are especially notable. Here the pitch approaches infinity, and the helicity becomes zero (figs. 7.29 and 7.30). Such mixtures behave as pure nematics and can be prepared for a wide range of temperature from pure cholesterics of varying helicity [3:32, 8:280], fig. 7.30, cf. p. 274.

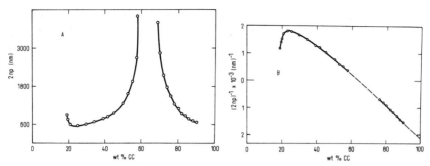

Fig. 7.29. Pitch (A) and helicity (B) vs. composition in mixtures of cholesteryl chloride (CC) and cholesteryl nonanoate (from [7]).

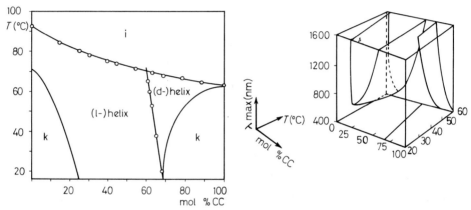

Fig. 7.30. System cholesteryl chloride (CC)/cholesteryl nonanoate: A, phase diagram; B, wavelength of reflection maximum vs. temperature and composition (from [260]).

Aberrations from optical-rotation symmetry around T_{nem} of a compensated cholesteric system are due to the temperature dependence of the birefringence [282].

G. Durand has applied the Debye-Scherer X-ray experiment using laser light on cholesterics. He could thus show that around the temperature where cholesterics behave as nematics the pitch becomes infinite, and the sense of the helicoidal structure changes sign [101], fig. 7.31.

Specialities appear among mixture components with small powers of optical rotation, such as cholesteryl iodide, which can induce either chirality, and the sense of rotation depends on the environment. From the chirality of cholesterics produced from a nematic solvent (MBBA or butyl *p*-(*p*-ethoxy phenoxy carbonyl) phenyl carbonate) and the following chiral solutes:

Chiral solute	Induced cholesteric mesophase chirality
I (S)-*sec* amyl-*p*-aminocinnamate	R
II (R)-2-octyl-*p*-aminobenzoate	R
III (S)-2-octyl-*p*-aminocinnamate	L
IV (S)-*sec* amyl-*p*-aminobenzoate	R
V (R)-2-octyl-ethylcarbonate	R
VI (S)-2-octyl-ethylcarbonate	L
VII *l*-menthol	R

the conclusion can be drawn that the cholesteric mesophase chirality cannot be predicted from the chirality of the active solute alone, even though the mesophases produced appear always to be enantiomeric [237, see also 133, 134].

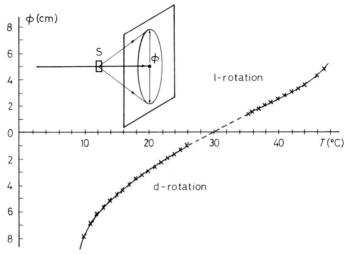

Fig. 7.31. Scattering of laser light, $\lambda = 632.8$ nm, by cholesteryl ethoxyethoxypropionate in the cholesteric state (from [101]).

As might be expected, cholesteric characteristics can be induced in nematics by molecules having an asymmetric carbon atom as well as by those with a general chiral structure [256, 257]. Thus, solutions of *d*- or *l*-6,7-diphenyl-5,8-diaza-(dinaphtho-2′,1′;2;1″,2″;3,4)-cyclooctatetraene (I) in MBBA exhibit the optical characteristics of a cholesteric liquid crystal (fig. 7.32).

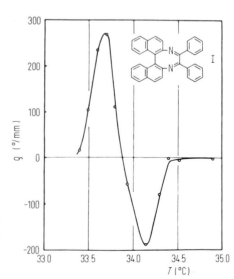

Fig. 7.32. Rotatory power, ρ, vs. temperature measured for a solution of 1.82 mole-% d–I in MBBA at 546 nm (from [256]).

In addition to the induced cholesteric behavior, we mention some other observations such as that of mobile rings and droplets in PAA with phlorizine [123], and cholesterol or cholesteryl-ester additives [124]. A weak interaction between the nematic solvent and the chiral solute (e.g., MBBA and *l*-menthol) causes a large bandwidth of circular dichroism of about 100 nm which is large enough to prevent the perception of reflection colors by the eye [255, 262]. In agreement with the molecular theory of Goossens, the induced twist of this system increases sharply with concentration causing a maximum prior to the breakdown of the liquid-crystal structure.

Other studies on cholesteric mixtures are reported in the literature [162] and in section 12.1, where the cholesteric lyophases of polypeptides are described; such phases are probably also derived from amino acid derivatives [244].

7.2   UV and Visible Absorption Spectroscopy

A Cotton effect is induced within the absorption bands of achiral molecules that are dissolved in a cholesteric solvent. This effect is superimposed upon the anomalous rotatory dispersion and has the same sign [259, 261]. For instance, the optical rotatory dispersion of rubrene shows a positive Cotton effect at longer wavelengths which is similar to that of the cholesteric solvent used. In the short-wave curve an additional positive Cotton effect with a small amplitude is observed in the range of absorption bands of rubrene (fig. 7.33). Even circular dichroism bands, the pitch band as well as the liquid-crystal induced band, are of the same sign [237]. A helical arrangement of the solute molecules in the cholesteric

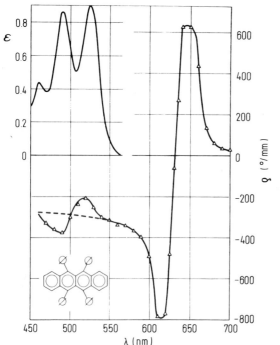

Fig. 7.33. ORD spectrum of rubrene in cholesteric solution and absorption spectrum of rubrene in isooctane (from [259]).

matrix appears most likely to be responsible for the appearance of the induced Cotton effect [259].

According to F. D. Saeva et al., the sign of the liquid-crystal induced circular dichroism (LCICD) depends on the wavelength of the absorption band, λ_{ab}, the reflective wavelength of the cholesteric pitch band, λ_{max}, and the cholesteric matrix in the following way [238, 240]:

Helix sense	$\lambda_{max}/\lambda_{ab}$	LCICD sign*
right-handed	>1	−
	<1	+
left-handed	>1	+
	<1	−

* max. change with a variation in the cholesteric matrix

The molecular ellipticity is independent of the solute concentration over a limited concentration range. This indicates that the primary cause of the LCICD is the chiral organization of the solute within the cholesteric mesophase rather than the mere exposure of the solute molecules to a helical arrangement of solvent molecules [238, 259]. The LCICD spectrum of pyrene has been thoroughly investigated, and the 0—0 bands are assigned thusly: the ^1L_b transition band at 372 nm, the ^1L_a transition band at 339 nm, and the ^1B_b transition band at 277 nm [238, 240], fig. 7.34. The 0—0 bands of the ^1L_b and ^1B_b transitions, which are known to be transversely polarized, show positive CD in the right-handed cholesteric matrix of fig. 7.34 ($\varepsilon_L > \varepsilon_R$).

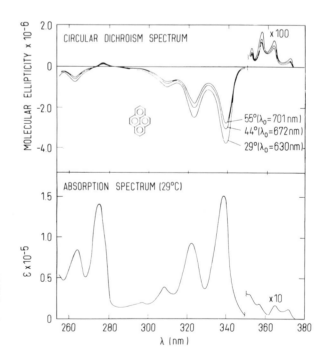

Fig. 7.34. CD and absorption spectrum of pyrene in 70:30 (wt%) cholesteryl nonanoate/ cholesteryl chloride (from [238]).

All the vibrational bands within the ^1L_b transition appear to be of the same polarization. This contrasts to the ^1B_b transition which appears to be of mixed polarization or may contain overlapping transitions which are of opposite CD signs as indicated by the lack of band matching between the absorption and LCICD spectra. The ^1L_a transition, on the other hand, is longitudinally polarized and shows negative circular dichroism [238].

In a series of disubstituted benzene compounds, the *o*- and *m*-isomers exhibit negative LCICD of the ^1L_b transition, while the *p*-isomers show bands with a positive sign. However, this does not apply to the substitution combinations hydroxy/methyl or chloro/cyano [239, 240, 242]. This *p*-effect is believed to be a spectroscopic effect and the result of conformational variations between the isomers in a cholesteric mesophase of single chirality (fig. 7.35).

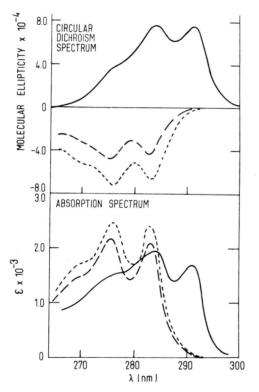

Fig. 7.35. LCICD- and absorption spectra of *o*-, *m*-, and *p*-methoxy-chlorobenzene (from [239]).

Independent of whether the solvent or solute is chiral, the LCICD shows common features that are influenced by concentration, helix chirality, pitch, temperature, and texture [234, 240].

The LCICD offers the possibility of determining the chirality of large pitch materials where the cholesteric pitch band may be inaccessible and the Grandjean texture unstable [235]. The sense of the preferred absorption indicates the sense of the helix, i.e., preferred absorption of left-circularly polarized light, indicates the existence of a left-handed cholesteric mesophase, and vice versa (fig. 7.36).

An instructive comparison is drawn in fig. 7.37. The absorption spectra show a slight Ham effect, while a marked increase in positive molecular circular dichroism is observed

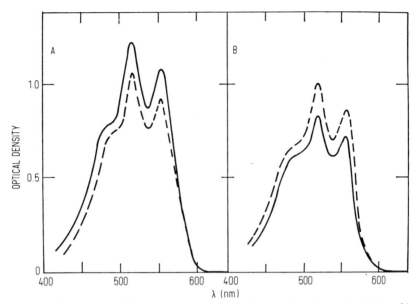

Fig. 7.36. Absorption spectra of (A) 2% 1-amino-2-phenoxy-4-hydroxyanthraquinone in 51/49 wt% cholesteryl nonanoate (CN)/cholesteryl chloride (CC) right handed helix, and (B) in 64/36 wt% CC/CN left handed helix using right (————) and left (—————) circularly polarized light (from [235]).

Fig. 7.37. Electronic spectra of anthracene: absorption (A) in ethylenedibromide, and (B) in cyclohexane; MCD (C) in ethylenedibromide and (D) in cyclohexane; (E) LD spectrum in stretched polyethylene film; and (F) CD spectrum in a uniaxial cholesteric mesophase with a 50 μm path length and consisting of a 1:1 mixture by weight of cholesteryl chloride and cholesteryl nonanoate with a pitch band (left circular light reflected) near 1 μm. The ^1L_a origin lies at 26.60, 26.23, 26.42, and 26.39 (10^3 cm^{-1}) in cyclohexane, ethylenedibromide, polyethylene, and the mesophase, respectively. The frequency scale refers to the cyclohexane solution spectra and the spectra in other media are each blue-shifted to bring the ^1L_a origins into coincidence (from [173]).

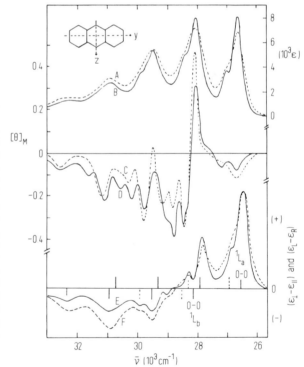

on changing the solvent from cyclohexane to ethylenedibromide. The LCICD produces essentially the same information as the stretched-film linear dichroism [6:257, 173, 240].

For the three heterocyclic compounds: N-ethylcarbazole, dibenzofuran, and dibenzothiophene (1.0 wt%) in cholesteryl chloride/cholesteryl nonanoate (60/40 wt%, right handed helix), a positive Cotton effect was found for the first electronic transition ($^1L_b \leftarrow {^1}A$) and a negative Cotton effect for all other bands [236]. Other studies on this subject concern the LCICD of methyl 2-pyrenecarboxylate [231, 246], β-carotene [219], dipyrromethene derivatives [110], and other compounds [91, 99, 216]. Only one short note mentions the optical activity of a solute measured in a field-aligned compensated mixture [160].

Spectroscopic observations of mesophases from deoxycholic acid and π-systems are reported [64]. Reviews have also been published [6:53, 241].

7.3   Fluorescence Spectroscopy

The temperature dependence of the circular fluorescence polarization in cholesteric solutions has been studied for *trans-p*-dimethylamino-*p'*-nitrobenzene [264, 274a], pyrene [196, 268, 285], phenanthrene [196], and other aromatics [6:260, 272, 274a].

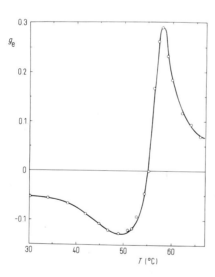

Fig. 7.38. Temperature dependence of the circular fluorescence polarization, $g_e = (I_L - I_R)/\frac{1}{2}(I_L + I_R)$, of *trans-p*-dimethylamino-*p'*-nitrostilbene in a 65:35 mol-% mixture of cholesteryl chloride and cholesteryl nonanoate (from [264]).

Fig. 7.38 illustrates a system in which no detectable circular fluorescence polarization exists above the clearing point. In the compensated mixture at T_n, the dissymmetry factor, g_e, of the emission at maximum fluorescence changes its sign.

The dissymmetry factor, g_e, depends on the sense of the helical pitch and the components of the fluorescence transition moments, $|M|$, parallel and perpendicular to the longitudinal molecular axis in the following way [272]:

	$\lvert M_\parallel \rvert > \lvert M_\perp \rvert$	$\lvert M_\parallel \rvert < \lvert M_\perp \rvert$
d-helix	$g_e > 0$	$g_e < 0$
l-helix	$g_e < 0$	$g_e > 0$

Regardless of large changes in the crystalline range below a temperature of 125°C, the fluorescence decay time of pyrene in cholesteryl benzoate is fairly uniform above this point. It shows a most remarkable step decrease in the vicinity of 145°C, which is the k/c transition point. At this temperature, interactions between the medium and the fluorescent solute lead to radiationless deactivation of the solute, and the decay time is considerably shortened [285]. The interesting case where the wavelength of emitted fluorescence coincides with the wavelength of the reflection maximum seems to be characterized by a markedly lengthened fluorescence decay time [97].

7.4   IR Spectroscopy

Only a very small IR rotation dispersion is to be expected for molecular vibrations. The amplitude is by several orders of magnitude smaller than with electronic transitions, as could be observed until now only in the shortwave IR range. However, distinct anomalies appear when the chiral substance is dissolved in a nematic liquid crystal. The frequencies of anomalous IR rotation dispersions agree with those of the IR absorption bands of the nematic solvent, while the vibrations of the chiral solute do not emerge as anomalies in the dispersion curve [247], figs. 7.39 and 7.40.

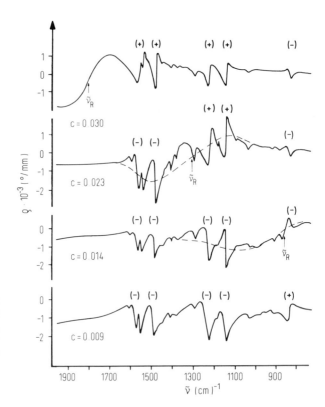

Fig. 7.39. IR-ORD of different solutions of cholesteryl chloride in a MBBA/EBBA mixture. $\tilde{\nu}_R$ designates the center of the "reflection Cotton effect" (from [54, 150]).

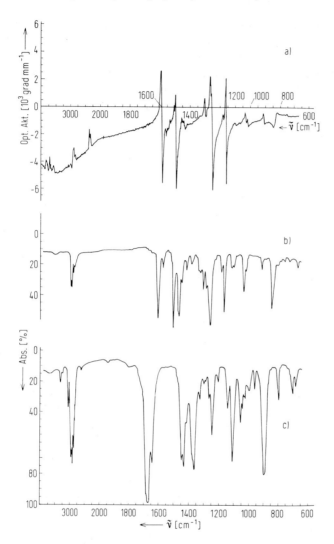

Fig. 7.40. IR optical properties of cholesterics: a) IR-ORD of 2 mol-% Carvone in *p*-methoxy-*p'*-*n*-butylazoxybenzene; b) absorption spectrum of the solvent; c) absorption spectrum of the solute (from [247]).

In MBBA containing 2% (−)-menthol, the vibrational circular dichroism observed is predominantly positive with minor negative bands at high and low frequencies. However, no reflection band was discerned in this system [98, see also 255, 262]. The infrared circular dichroism (IRCD) spectra of solutions of a *l*- or *d*-enantiomer in a nematic solvent are mirror images of each other [65, 152a]. The sign of the IRCD-peaks depends upon the absolute configuration of the chiral solute and the position of the reflection band with respect to the absorption band, and the sign is inverted in going from 1% to 2% to 4% samples of cholesteryl 2-(2-ethoxyethoxy)ethyl carbonate in *p*-methoxy-*p'*-*n*-butylazoxybenzene with the sense of the helix being unchanged.

IRCD bands only occur within the region of the cholesteric reflection band, thus 13% and 27% samples of the latter system showed no observable circular dichroism between 1000 and 1650 cm^{-1}. The sharp IRCD bands are superimposed on a broad background of the reflection band. As expected from the spectra shown in fig. 7.40, the positions of the IRCD bands are essentially independent of the particular optically active guest molecule. A number of small frequency shifts in the IRCD spectra of different samples perhaps can be seen in relation to different intermolecular interactions. The zero point between the positive and negative CD peaks is at the center of the absorption band. These types of bands possibly result from the coupling of an external radiation field to two polarization directions of the host molecule vibration with the directions being non-equivalent in the chiral cholesteric matrix. The IRCD bands without change in sign are apparently due to the excitation of vibrational modes of highly anisotropic host molecules in a cholesteric helical array. By using the measured IR linear dichroism data to define the optical properties of the individual layers of the cholesteric structure; the signs, amplitudes, positions and shapes of the IRCD were calculated by means of the Oseen-de Vries model for cholesterics [6:143]. By addition of a frequency-dependent, complex contribution to the spiralling dielectric tensor, the theory of light propagation in nonabsorbing cholesterics can be extended to the absorbing case. The experimentally observed features of the CD and ORD spectra can be explained with the estimated values of the absorption coefficient and the background linear birefringence [142].

To measure the very small Cotton effects of the rotatory dispersion in the IR region, special instruments have been constructed [66, 149, 151]. The literature reports other measurements of IR rotatory dispersion in cholesteric liquid-crystal solutions [90, 134, 151].

There have been only some investigations as yet of the IR spectra of cholesterics consisting only of one chemical species [6:41, 6:195, 6:283, 55, 109, 132a, 132b, 188, 189].

7.5 Raman Scattering

In the Raman spectra of cholesteryl chloride, acetate, and propionate, the Raman lines of the C—H stretching region change position and intensities during phase transitions [57], fig. 7.41.

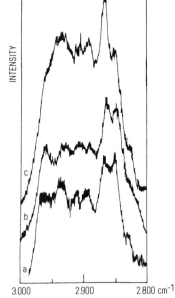

Fig. 7.41. Raman spectra of cholesteryl chloride in the C—H stretching region: (a) solid at 25°C, (b) cholesteric at 57°C, and (c) isotropic liquid at 130°C (from [57]).

Since absorptions due to intermolecular interactions lie in the far IR spectral region, they are more conveniently observed in Raman scattering as frequency shifts in the visible region. R. Chang [76], G. Vergoten et al. [289] and other authors [6:292a] have mainly studied the crystalline state.

Only in one case, with the band of cholesteryl chloride at $2860\,\text{cm}^{-1}$, does the intensity not suddenly change at the phase transition, but it changes continuously as the temperature is raised in the cholesteric phase thus increasing the probability of pitch-dependence. However, many other systems indicate that generally the Raman spectrum is not particularly sensitive to pitch [57]. B. J. Bulkin et al. observed differences in the intensity of Raman lines of solutes in cholesterics and in isotropic liquids [6:36]. They conclude that "the optical activity and refractive index discontinuities of the mesophases give rise to a polarization scrambling which affects the intensity of polarized and depolarized solute bands unequally. One may expect a variation in solute relative intensities as a rule, particularly in high symmetry solutes where complete polarization is possible".

7.6   Birefringence

As has been repeatedly accentuated, nematics and cholesterics differ in the character of their birefringence if observed in the homeotropic or planar textures respectively [6:11, 6:115, 6:118, 6:319, 119, 159], see p. 22.

According to the arrangement of the rod-like molecules within the mesophases nematics exhibit positive birefringence ($n_{e,n} > n_{o,n}$) whereas that of the cholesterics is negative ($n_{e,c} < n_{o,c}$). It has been interesting to note how the birefringence might indicate the appearance of a compensated mixture. The conversion of the helical twist sense in the range of a compensated mixture, as shown in fig. 7.42, is indicated by a strong increase of birefringence and an exchange of the refractive indices of the ordinary and extraordinary rays [263].

By surface treatment with lecithin the molecules are turned over to form the homeotropic nematic phase within a range of $\pm 3°C$ around T_n. The maxima of $n_{e,n}(T)$ of all mixtures coincide with the compensation temperature, T_n, where the optical rotatory power has vanished. Cholesteryl nonanoate shows birefringence typical of that of cholesteric esters (fig. 7.43).

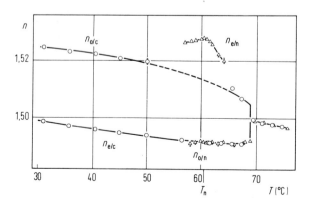

Fig. 7.42. Refractive indices, n, vs. temperature, T, of a cholesteric mixture (cholesteryl chloride/cholesteryl nonanoate = 62:38 mol-%) at 589 nm (from [263]).

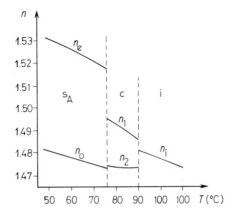

Fig. 7.43. Refractive indices, n, vs. temperature, T, of cholesteryl nonanoate (k 79.8 c 90.3 i (76.0 s$_A$)) at 589 nm (from [233]).

In the homeotropic s$_A$ phase as well as in the homeotropic nematic phase, the optical axis, the preferred direction of the longitudinal axes of the molecules, and the direction of maximum polarizability are all coincident, thus $n_e > n_o$. Fig. 7.44 shows the normal dispersion of the indices of refraction.

Using the method of R. Dreher et al. [182] which is based on the angles of total reflection, γ_1 and γ_2, from a homogeneously oriented cholesteric phase, one can calculate approximate values for the indices of refraction of the corresponding untwisted nematic structure. γ_1 and γ_2 are calculated according to

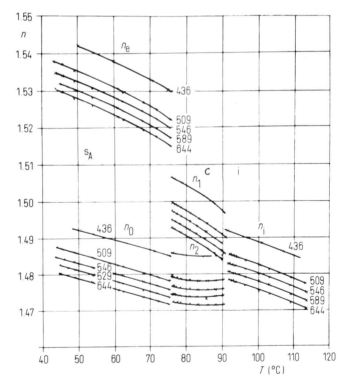

Fig. 7.44. Refractive indices, n, vs. temperature, T, of cholesteryl nonanoate at different wave lengths (from [233]).

$$\sin^2\gamma_2 = \frac{n_o^2}{n^2} \quad \text{and} \quad \sin^2\gamma_1 = \frac{n_e^2 + n_o^2}{2n^2}$$

where n is the index of refraction of the prism used [233]. Between the refractive indices of a cholesteric structure, $n_{o,c}$ and $n_{e,c}$, and the indices of the corresponding untwisted nematic structure, $n_{o,n}$ and $n_{e,n}$, the following relationships have been derived [263]:

$$n_{o,n} = n_{e,c} \qquad n_{e,n} = [2n_{o,c}^2 - n_{e,c}^2]^{1/2}$$

P. Adamski and A. Dylik-Gromiec used the equation

$$S = \frac{(\alpha_e - \alpha_o)_{\text{liq.cryst.}}}{(\alpha_e - \alpha_o)_{S=1}}$$

which has already been mentioned in section 6.1.2 together with the Clausius-Mossotti formula to determine the polarizabilities, α, and the degree of order, S, in cholesterics [15].

The literature describes the indices of other cholesterics [2:176, 2:310, 4:170, 6:89, 12–14, 45, 88a, 109, 148, 203a, 224, 280, 282, 294].

7.7   Rayleigh Scattering

Compared to the unique scattering due to the helical structure, the Rayleigh scattering exhibits no special features and can be described by de Gennes' phenomenological theory [205, 218a, 299], see also section 6.2. C. C. Yang studied the Rayleigh scattering of cholesteryl 2-(2-ethoxyethoxy)ethyl carbonate in the isotropic liquid and in the pretransition region to the cholesteric phase [299]. Other authors found contradictory results with cholesteryl oleyl carbonate [139, 168]. The four intensities, H_h, H_v, V_h, and V_v, of scattered light were measured in cholesteryl myristate [291].

R. S. Stein et al. have compared the light-scattering patterns of cholesteryl esters with those of spherulitic polymers, and they have drawn conclusions from the polarization of the scattered light (and its angular dependence) about the shape and arrangement of the scattering regions in the various phases of some cholesteryl esters [275–277]. In the future, such structure investigations of liquid crystals by light scattering deserve increased attention because they give insight into nucleation phenomena, growing of the new phase, and the disappearance of the previous phase [3:131, 6:43, 83, 157, 211]. All transitions occurring in cholesteryl nonanoate and other cholesteryl esters manifest themselves in the light transmission. Furthermore, there occur non-reproducible peaks of an as yet uncertain source [211]. The literature gives data on cholesteryl myristate [1:109]. Two modes which should be readily observable by light scattering of cholesterics have been predicted [111]: the first mode is a twisting and untwisting of the helical structure and the second mode is a combination viscous-splay mode.

7.8   Brillouin Scattering

In cholesterics such as cholesteryl 2-(2-ethoxyethoxy)ethyl carbonate, Doppler-shifted components of the scattering spectrum were observed [100, 104, 194, 221, 226, see also 6:136c,

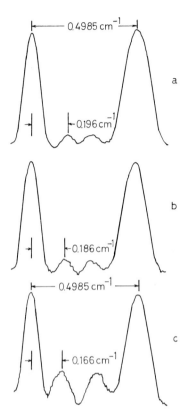

Fig. 7.45. Scattering of laser light (intensity vs. frequency) by cholesteryl 2-(2-ethoxyethoxy)ethyl carbonate at three significant temperatures: a, 23.5°C; b, 35°C; c, 62°C. Only one of the Brillouin peaks associated with each of the larger Rayleigh peaks is shown (from [194]).

141]. The spectrum shown in fig. 7.45 was recorded normally to the He-Ne laser light with a laser output polarized in a plane perpendicular to the scattering plane.

The Brillouin component of the scattered light is polarized almost entirely in a plane perpendicular to the scattering plane. With increasing temperature, the Brillouin peaks become larger and shift toward their related Rayleigh peak. They are shifted on either side of the exciting frequency, ω_0, by the frequency of the hypersonic excitation according to

$$\Omega = (2\omega_0 n v/c)\sin(\theta/2)$$

where n is the refractive index, v the acoustic velocity, and θ the angle between the directions of incident and scattering radiation [226]. J. D. Parsons and C. F. Hayes give a theoretical treatment [202, 203]. It is usually difficult to measure the inelastic scattering intensity alongside the elastic scattering intensity, which can under certain conditions be several orders of magnitude larger. This could be a reason for the contradictory observations made at the c/i phase transition [104, 226, 292, 293], the phase transition temperature of cholesteryl palmitate at 146°C seems to be a mistake. All measurements of the stimulated Brillouin scattering threshold power below 48°C (showing the higher gain) and above 52°C (showing the lower gain) can be given by a single gain curve which is almost independent of temperature. In cholesteryl 2-(2-ethoxy-ethoxy)ethyl carbonate the threshold power increases gradually around the c/i transition [221].

7.9 Nonlinear Optics

From de Vries theory it can be deduced that the phase-matching conditions of optical third-harmonic generation (THG) can be fulfilled in a cholesteric liquid crystal by the rotational dispersion [218, 221]. The THG signals shown in fig. 7.46 are observed for fundamental pump waves left- and right-circularly polarized at two temperatures where the pitch is of the correct value.

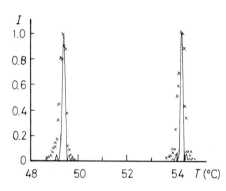

Fig. 7.46. Normalized third-harmonic intensity, I, versus temperature, T, near the phase-matching temperatures for a mixture of 1.75 parts cholesteryl chloride and 1 part cholesteryl myristate (by weight) in a cell 130 μm thick. The peak at the lower temperature corresponding to left helical structure) is generated by left-circularly polarized fundamental waves and the one at the higher temperature by right-circularly polarized fundamental waves. The solid line is the theoretical phase-matching curve and the dots are experimental data points. The uncertainty in the experimental third-harmonic intensity is about 20% (from [248]).

Analogous with the interaction of electrons propagating in a periodic lattice, the phase-matched THG processes in cholesterics are considered coherent normal and coherent optical umklapp processes, or nonlinear Bragg reflection [249–251]. J. W. Shelton and Y. R. Shen discuss many different colinear phase-matching conditions [250].

These results and the observation of nonlinear optical properties in many materials have encouraged the investigation of the second-harmonic generation of light (SHG). From this effect, complementary to the X-ray and linear optical experiments, structural information could be found that would prove the medium lacks a center of inversion (to be piezoelectric) on the scale of an optical wavelength [103]. However, in the cholesteric phase of cholesteryl 2-(2-ethoxyethoxy)ethyl carbonate that has been carefully purified of any crystalline particles, there appears to be no signal corresponding to the SHG [102, 103]. Positive results reported in the literature [115] were due to crystalline impurities of the investigated samples [102].

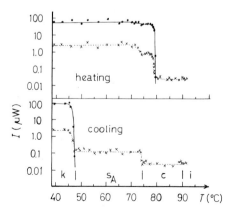

Fig. 7.47. Intensity, I, of second- (———) and third-harmonic (– – – –) radiation vs. temperature, T, in 25 μm thick cholesteryl nonanoate (from [129]).

Confirming this, cholesteryl nonanoate in the crystalline state yields a strong SHG signal whose magnitude is of the same order as the SHG produced in a comparison (control) sample of finely ground quartz powder [129]. Heating through the k/c transition near 79°C results in a sharp decrease in SHG of more than four orders of magnitude and going below the limit of detectability. Correspondingly upon cooling cholesteryl nonanoate from the isotropic liquid through the cholesteric and smectic mesophases, the SHG signal is recovered precisely at the point where recrystallization is observed to occur (fig. 7.47).

Other interesting studies in harmonic generation of light in liquid crystals are described in the literature [19a, 130, 251a, 251b].

7.10 Other Optical Observations

A special optical phenomenon which sometimes accompanies the c/i phase transition is the appearance of socalled platelets. They seem to constitute a phase in which a process of building up (on cooling) or breaking down (on heating) of the cholesteric order occurs [135]. Since the color of the platelets does not change upon rotation of the stage, their optical axis must lie along the line of sight, but rotation of the analyzer alters the color intensity of the platelets indicating that the platelets rotate the plane of polarization. The platelet texture purportedly comprises conglomerations of small, independent regions of differing and changing thicknesses and having structures of the classical cholesteric type but of varying pitch. Whenever the platelets are observed microscopically, a "blue phase" also exists over exactly the same temperature range [1:110, 1:111, 21, 135, 137]. These and the following observations lead to the supposition that certain details of the cholesteric structure (cf. section 1.4.5, p. 127 and 234) remain to be discovered. Thus, P. Châtelain et al. [82] found three different types of Grandjean steps [1:118, 1:272, 3:80, 3:86, 3:105, 3:155, 3:276, 122] and J. Rault observed a periodic array of "commas" [223] and helical dislocation lines that have the same chirality and pitch as the cholesteric phase and wind around straight dislocation lines [222]. T. Sarada's dissertation concerns the influence of linear polymers on the optical properties of the cholesteric mesophase [245].

7.11 Twisted Nematic Phases

A few considerations concerning Mauguin textures need to be mentioned. For twisted nematic layers of very small thickness (ca. 0.3 µm), J. Billard has measured and calculated a difference of a few degrees between the ellipticities of the two privileged electromagnetic waves [38, 39, 42]. Depending on the wavelength and on the material's parameters, a twisted nematic layer may produce elliptically polarized light from incident plane polarized light [131].

Other experimental studies [132] and theoretical considerations of light propagation in a twisted nematic structure have been published [4:765, 37a, 225]. The analogy to cholesterics manifests itself in a weak circular dichroism observable in twisted nematics [58, 243]. Optical rotatory effects that occur in nematic phases under the influence of magnetic fields are discussed in section 4.1.5. Other observations are mentioned in sections 1.4.5, 4.4.4, and 14.4.3.

7.12 Twisted Smectic Phases

Twisted smectics were mentioned at the beginning of this chapter. Examples of these have only recently been exposed [2:324]. In certain cases, the nematic-cholesteric phase transforms directly into a twisted s_C phase (labeled s_C^*) instead of the normal s_A. The s_A/s_C^* transition is also known [2:396, 204], where other properties are also described. There exists a single Bragg reflection band at normal incidence for either type of liquid crystal [2:301]. However, D. W. Berreman predicts that with obliquely incident light the s_C^* phases would show additional Bragg reflection bands at optical frequencies intermediate between the bands that would appear in both samples [37]. O. Parodi has found the de Vries' theory applicable to twisted smectics C [201].

H. Stegemeyer et al. discovered two coexistent phases in the system cholesteryl myristate/MBBA, both of which selectively reflect light [258], fig. 7.48. While the properties of phase II identify it definitely as cholesteric, those of phase I indicate a s_C^* structure.

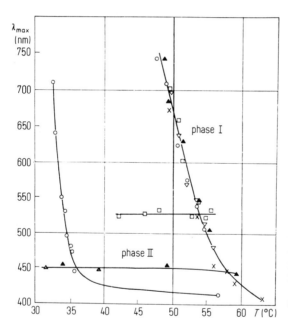

Fig. 7.48. Temperature dependence of the wavelength of maximum circular dichroism in the mixtures of cholesteryl myristate/MBBA in the following mole ratios of myristate to MBBA: × — × 0.9; ▽—▽ 0.7; ○—○ 0.5; ▲—▲ 0.4; □—□ 0.3 (from [258]).

It is interesting to note that a mixture of bis (4-*n*-decyloxybenzylidene) 2-chloro-1,4-phenylenediamine and 5% of cholesteryl cinnamate forms a s_C^* phase [53]. This example suggests that other C-smectogens with chiral solutes will show induced twist [1:15, 5:21a, 21a]. Examples of chiral s_H phases (labeled s_H^*) have recently come to light [2:299, 204], fig. 7.49.

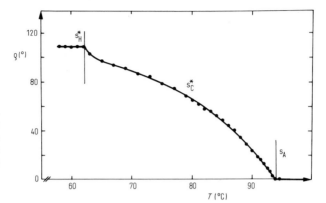

Fig. 7.49. Rotatory power, ρ, vs. temperature, *T*, in the mesophases of *p*-decycloxybenzylidene *p*-amino 2-methylbutyl cinnamate: k 76 s$_C^*$ 95 s$_A$ 117 i (63 s$_H^*$) (from [204]).

Thermodynamic Properties

8

8 Thermodynamic Properties

In the controversy concerning the existence of liquid crystals the small but distinctly measurable changes in heat content accompanying the transition from the mesomorphous to the isotropic liquid have been an important point favoring the consideration of the liquid crystalline phases as a separate state of aggregation.

8.1 Classification of Phase Transitions

We shall begin with a short overview. Certain transitions where no sudden change in energy content or volume is encountered, but rather only a jump in their temperature-dependence (e.g., specific heat, coefficient of expansion), have led P. Ehrenfest to classify them in various orders [107, 114], figs. 8.1 and 8.2. A discussion of free energy diagrams can

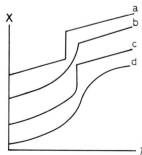

Fig. 8.1. Schematic presentation of a property, X, e.g. energy content, volume, whose behavior is a) first order; b) second order; c) first order with pretransition; d) infinite order.

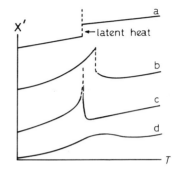

Fig. 8.2. Schematic presentation of a property, X′, whose behavior is a) first order; b) second order, e.g. specific heat, coefficient of expansion; c) first order with pretransition; d) infinite order.

aid to illustrate the different types of phase transitions. Fig. 8.3 shows the chemical potential with respect to temperature for a system with one second order and two first order transitions.

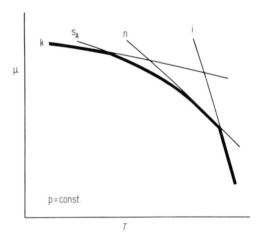

Fig. 8.3. Idealized phase diagram (chemical potential, μ, vs. temperature, T, at constant pressure) illustrating a suspected second order s_A/n transition (from [349]).

However, there was considerable doubt as to the existence of genuine second-order phase transitions [159, 160]. It seemed more likely that there was a higher-order transition where only a higher differential quotient of the free energy changed discontinuously. This means that the specific heat and the coefficient of expansion do not change discontinuously but merely in an exceptionally strong fashion in the transition zone. It is evident, that in the case of a convergent overlapping of curves a and b (c in fig. 8.1) a separation will be nearly impossible. Since the second order model is also thermodynamically doubtful in certain respects, it seems better to consider gradually smeared transitions and differentiate them on a purely phenomenological basis. The theoretical background has been cleared much by the works of A. Eucken and E. Bartholomé [109], E. F. Lype [201], F. C. Nix and W. Shockley [226]. Less important papers are [136, 266]. The basic and extensive investigations of W. L. Bragg and E. J. Williams are worthy of a more detailed account [51]. Detailed reviews [306, 318] and brief reviews [3, 170, 205, 270] have appeared laying stress upon the theory of mesophase transitions, see also [48, 89 214].

The type of transition illustrated in fig. 8.1 b and 8.2 b has been designated the ammonium chloride type or, using Ehrenfest's nomenclature, the Λ-type. On the other hand, it even seems questionable whether a strictly first-order transition is possible in molecular crystals.

There is still another problem: The purity of the substance is of the utmost importance if a homogeneous pretransformation is to be analyzed, and for a multicomponent system it will be difficult to separate the homogeneous and the heterogeneous part of the overall melting behavior. While a first-order phase transition is described thermodynamically by the Clausius-Clapeyron equation as

$$\frac{dp}{dT} = \frac{S_1 - S_2}{V_1 - V_2} \tag{8.1}$$

the Keesom equation is valid fot those of second order

$$\frac{dp}{dT} = \frac{\dfrac{dS_1}{dT} - \dfrac{dS_2}{dT}}{\dfrac{dV_1}{dT} - \dfrac{dV_2}{dT}} = -\frac{\dfrac{dV_1}{dT} - \dfrac{dV_2}{dT}}{\dfrac{dV_1}{dp} - \dfrac{dV_2}{dp}} \tag{8.2}$$

Figs. 8.1d and 8.2d illustrate a completely continuous transition between two states with different characteristics, as with a homogeneous gas or solution equilibrium, whereas transitions of the types shown by figs. 8.1a, b and 8.2a, b occur only if condensed phases are involved. The mesophase/isotropic phase transitions are of the first order, but the differences in the free energy are generally very small ("weakly of first order"). These transitions are generally accompanied by a "pretransition" period, which can be interpreted as second or even higher order transition. But such a superposition of transitions of different order shows quite clearly that the basic processes are not independent from one another, otherwise two different temperatures T_{tr} would exist. All phase transitions are also characterized by the occurrence of new parameters which describe the alteration in symmetry.

Quite another model of a phase transition is given by the concept of the coexistence of two "inner" phases. This was used by J. Frenkel [3:267] in his theory of heterophase fluctuations and explicated at full length by E. Donth [103]. This model suggests that microscopic domains differing in at least one property, are at equilibrium with one another across a wide range of pressures and temperatures. Naturally, this is only possible in the sense of the Gibbs phase rule when the composition — whatever one means by that in a "one-component-system" — also changes continuously, perhaps in such respects as molecular association or configuration in both "phases".

Landau considers the free energy, $F = F(p, T)$, or the free enthalpy, G (in Landau's notation Φ), as a function of pressure, temperature, and the "degree of order", ξ, and develops it as an exponential series of ξ (eq. 8.4). p, T, and ξ are considered independent variables determining the free energy, and such a model must also include nonequilibrium states so as not to be overdetermined. The "degree of order", ξ, and the entropy, S, are basically corresponding expressions, and for a given F, the entropy is determined by the fundamental expression:

$$S = -(dF/dT)_V \tag{8.3}$$

In studying the original documents, we must remember that the two most important papers by Landau use different symbols and expressions, and only a purely phenomenological description of the higher order phase transitions was given at first [182, 338]. Symmetry factors are explicitly handled in the later works. J. Frenkel's book deals much more exhaustively with the simple model pointing up the relations to earlier theories based on the model of Bragg and Williams [51]. Landau goes much farther in his later publications, but unfortunately, the text is partly not easy to understand, and a specific difficulty is the lack of convergence criteria in the power series expansions.

In this respect, Ubbelohde's monograph "Melting and Crystal Structure" [347] is recommended, especially in connection with the terms "pretransition and premelting". But curiously, there is no reference to Landau, and even Frenkel's "Kinetic Theory of Fluids" [3:267] is hardly considered. Of great aid in understanding Landau's theory are the corresponding parts in H. E. Stanley's monograph [327], some books [128, 259, 337], and the textbook on theoretical physics by Landau and Lifschitz [185].

At the root of all considerations is the fact that there exist very different types ("orders") of phase transitions whose characteristics can be described somewhat as follows:
a) Continuous liquid/gas transitions "detouring" the critical point, analogous to "critical melting" [347].
b) Normal, discontinuous, first-order phase transitions such as melting or boiling, with typical discontinuities of enthalpy, entropy, specific volume, etc.
c) Continuous transitions such as those of the socalled Λ-type, or second-order whose prototype is the magnetic orientation below the Curie temperature [107]. The term "Curie transition" is thus used generally for this entire class.
d) Combinations of the above, and the pretransition effects that are classified as positional, orientational, vibrational, or surface-melting effects. This group is characteristic of liquid crystals, especially for polymesomorphous transition and those to the isotropic state [122].

8.2   Theory of Phase Transitions

L. D. Landau considered a model substance (without at first explicitly mentioning liquid crystals) that will undergo transitions depending on F as a function of the degree of order as well as on temperature and pressure [182–184].

Starting step is a general expression relating F with the degree of order which must be adapted to boundary conditions given by the molecular species.

$$F = F_0 + \alpha \xi + \beta \xi^2 + \tfrac{1}{3}\gamma \xi^3 + 0(\xi^4) \tag{8.4}$$

Higher terms are supposed to vanish with higher order. α, β, γ are functions of pressure and temperature. ξ has been conceived in the sense of an occupational probability of lattice positions, such that when it is equal to zero, then there is a completely random distribution (see note), corresponding to the entropy maximum. ξ should not be simply identified with the degree of order, as used by Maier and Saupe (see section 3.4). In a more sophisticated way and in direct relation to the nematic order, de Gennes replaced ξ by the "tensor order parameter", Q_{ik}, which gives a much more translucent picture, because Q is well defined:

$$F = F_0 + \tfrac{1}{2}A(T)Q_{\alpha\beta}Q_{\beta\alpha} + \tfrac{1}{3}B(T)Q_{\alpha\beta}Q_{\beta\gamma}Q_{\beta\gamma} + 0(Q^4) \tag{8.5}$$

Nevertheless, it may be useful to follow at first the original concept of Landau, that was not at all concerned with mesophases but with phase transitions of higher order. Main object of Landau's considerations was the Curie-temperature and the temperature dependence of the specific heat and other thermodynamic functions in its vicinity.

To fit eq. (8.4) with experimental results the coefficients must be choiced adequately, regarding their temperature dependence and, as the most essential problem, stability criteria have to be introduced. Special cases of stability conditions are demonstrated in fig. 8.4, the curves being selected quite arbitrarily. Thermodynamically stable and metastable states are represented by the minima $\partial F/\partial \xi = 0$, the isotropic state is defined by $\xi = 0$, and curve C for example represents a stable equilibrium between two phases of finite and of zero degree of order, respectively. The most important feature of Landau's theory is the introduction of a "critical condition" which is defined by

$$\xi = 0; \quad \frac{\partial F}{\partial \xi} = 0; \quad \frac{\partial^2 F}{\partial \xi^2} = 0; \quad \frac{\partial^3 F}{\partial \xi^3} > 0 \tag{8.6}$$

The corresponding temperature is T_{cr} and the critical pressure is p_{cr} (fig. 8.4, curve a). From eqs. 8.4 and 8.6 it follows, that a critical point is necessarily characterized by the conditions $\alpha = 0$ and $\beta = 0$, whilst $\gamma \neq 0$. No ordered phase is stable if p, T exceed p_{cr} or T_{cr}. Whether or not such a condition can be realized by experiment has been discussed elsewhere (Ubbelohde). If one of the critical parameters is chosen arbitrarily (e. g. $p < p_{cr}$) a certain temperature

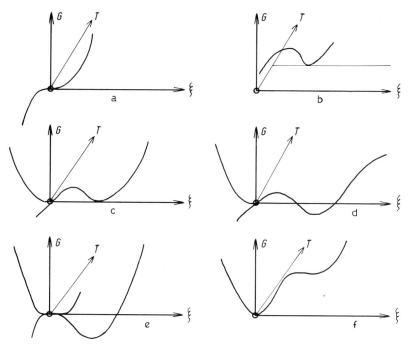

Fig. 8.4. Some arbitrarily chosen curves showing the free enthalpy, G, as a function of the temperature, T, and the "degree of order", ξ.

$T^0 \sim T_{cr}$ will exist, where the degree of order vanishes for $T \to T^0$ ($T \lesssim T^0$). Such a transition may be of the first or of higher order. A first order transition is represented by a curve like a and c in fig. 8.4 and a second order transition can be described by adjusting the term α so that it vanishes for $T \to T^0$:

$$\alpha = a(T - T_0) = a\Delta T \qquad (8.7)$$

Then we get

$$F = F_0 + a\Delta T\, \xi + \beta\, \xi^2 + \tfrac{1}{3}\gamma\, \xi^3 \qquad (8.8)$$

which becomes for $T \to T_0$:

$$F = F_0 + \beta\, \xi^2 + \tfrac{1}{3}\xi^3 \qquad (8.9)$$

with a real solution $\xi = 0$, as demanded.

On the other hand, the state $\xi = 0$, $T \to T^0$ is approached steadily, but $\partial^2 F / \partial \xi^2 \neq 0$ under the stability conditions

$$F = F_{min}; \quad \partial F / \partial \xi = 0 \quad \text{for} \quad T \neq T^0$$

$$\frac{\partial F}{\partial \xi} = \alpha\Delta T + 2\beta\, \xi + \gamma\, \xi^2 = 0$$

$$\frac{\partial^2 F}{\partial \xi^2} = 2(\beta + \gamma\, \xi) > 0$$

$$\xi_{min} = \frac{-\beta + \sqrt{\beta^2 - \alpha\gamma\,\Delta T}}{\gamma} \qquad (8.10)$$

The entropy is obtained from eqs. 8.3 and 8.8 neglecting the temperature dependence of β and γ as

$$S = -\frac{\partial F}{\partial T} = -a\xi \qquad (8.11)$$

This relationship shows clearly the direct relation between the entropy and the "degree of order".

The specific heat is derived as

$$c = \frac{T \cdot dS}{dT} = \frac{a^2 T}{2\sqrt{\beta^2 - a\gamma\Delta T}} \simeq \frac{a^2 \cdot T^0}{2\sqrt{\beta^2 - a\gamma\Delta T}} \qquad (8.12)$$

Introducing

$$\frac{\beta^2}{a\gamma} - \Delta T = \tau \qquad (8.12a)$$

one obtains

$$c = \frac{A}{\sqrt{T}} \quad \text{where} \quad A = \frac{1}{2} T^0 \cdot a \cdot \sqrt{\frac{a}{\gamma}} \tag{8.13}$$

and (from eq. 8.10)

$$\xi = -\frac{\beta}{\gamma} + \sqrt{\frac{a}{\gamma}} \cdot \sqrt{\tau} \tag{8.14}$$

If $\Delta T = 0$, the specific heat is derived as

$$c = \frac{T^0 \cdot a^2}{2\beta} \tag{8.15}$$

It has been deduced analogously for a first order transition

$$c = \frac{T^0 \cdot a^2}{|\beta|} \tag{8.16}$$

The relation between ξ and the entropy, S, has been elucidated in the following way. Using the Bragg-Williams model, Frenkel derived the following equation which holds for small values of

$$S = kN \ln 2 - \frac{kN}{2} \left(\xi + \frac{1}{6} \xi^2 + \frac{1}{15} \xi^3 \right) \tag{8.17}$$

The inner energy is given as

$$E = \frac{1}{4} N W (1 - \xi) \tag{8.18}$$

where W is the increase in inner energy due to lattice rearrangement and N is the total number of molecules.

From eqs. 8.17 and 8.18, and with $F = E - TS$, we obtain an equation as formulated generally by Landau (eq. 8.4), but with coefficients being expressed explicitly:

$$F = F_0 + \alpha \xi + \beta \xi^2 + \frac{1}{3} \gamma \xi^3 + 0(\xi^4)$$

with

$$F_0 = \frac{1}{4} N W \quad \alpha = \left(-\frac{W}{4} + \frac{kT}{2} \right) N \quad \beta = \frac{1}{12} N k T \quad \gamma = \frac{N}{30} k T \tag{8.19}$$

The relation between F, S and ξ is especially distinct when we set

$$F = -kT \ln Z \simeq -TS \tag{8.20}$$

for high temperatures where the entropy term is dominant. Here the state sum, Z, is an expression of the quantity of possible combinations of translational space distributions in the lattice model, and these are assumed to correspond to a certain degree of order. Introducing ξ, Landau obtains:

$$F = N k T \{ \ln \tfrac{1}{2} + \tfrac{1}{2}\xi + \tfrac{1}{12}\xi^2 + \tfrac{1}{30}\xi^3 \} \tag{8.21}$$

In the following only the long range order is regarded which is representative for the lattice symmetry of any crystal. If the position of a molecule (1) is given, the probability that at a distance r_{12} a molecule (2) will be found is a function of the distance vector r_{12}. The probability density, ρ_{12}, is thus a function of r_{12}, and the symmetry characteristics of ρ_{12} determine the symmetry of the system. If ρ_{12} is periodic in the sense of a long range order translation lattice, the 230 Schönfliess spatial groups are obtained. An isotropic liquid (or gas) is characterized by $\rho = \text{const.}$, which is also the case for anisotropic real liquids, i.e. for nematics. But in the latter case there is an additional anisotropic correlation function $\rho_{12}(r_{12})$, whilst the long range translational symmetry has vanished. The only symmetry elements which rest are those of the point groups, not being limited to the 32 classes of solid crystals (C_∞, $C_{\infty h}$, $C_{\infty v}$, D_∞, $D_{\infty h}$, or cyclic groups with $n = 2, 3, 4, 5$ (!) etc.).

Landau correlates the density function with the degree of order, a certain state being represented by a symmetry group $\mathfrak{G}°$. A transition from a lower ordered state to a higher-ordered state (i.e. one with lower symmetry) is indicated by a change of the density function $\rho_0 \to \rho_0 + \delta\rho = \rho$. The density function is expressed as an exponential series analogous to eq. (8.4). Landau develops ρ and $\delta\rho$ according to

$$\rho = \sum_n \sum_i C_i^{(n)} \varphi_i^{(n)}$$

$$\delta\rho = {\sum_n}' \sum_i C_i^{(n)} \varphi_i^{(n)} \tag{8.22}$$

which is by groups each characterized by the number of irreducible representations n and comprising i functions φ that all transform in the same way.

In ρ one of the functions $\varphi^{(n)}$ is invariant with respect to all transformations of the group $\rho_0(\mathfrak{G}°)$. We regard this function to belong to $\mathfrak{G}°$ only. In other words: the "1-representation" of the group $\mathfrak{G}°$ is excluded from the sum (symbolized by $\sum'$) representing $\partial\rho$.

The free energy depends on ρ, T and p as already stated, but now G, the Gibbs free energy, is preferred as the adequate function. $G(\rho_0 + \delta\rho)$ is developed in terms of the degree of order according to the symmetry terms, i.e. as exponentials $C_i^{(n)}$ and $\varphi_i^{(n)}$. It can be shown that all first-order terms are always zero, and the sum of all second-order terms for a given $\delta\rho$ also disappears at the transition point. G is a function of the C_i only, and terms of the second order have the form

$$A \cdot \sum_i C_i^2$$

The behavior of the third-order terms is highly characteristic of the mode of phase transition:

a) If these are identical to zero, there remains but one condition for the phase transition, namely

$$A(p,T)=0$$

This yields a curve in the p/T plane describing the behavior of the Curie point.

The term $\sum_i C_i^2$ is used to define a new order parameter, η:

$$\sum_i C_i^2 = \eta^2$$

The new symbol η is introduced instead of the former ξ, and an order dependent parameter, γ_i, being related to single coefficients C_i, is defined as

$$C_i/\eta = \gamma_i$$

With these new parameters one obtains:

$$G = G_0 + A\eta^2 + B(\gamma_i)\cdot\eta^4 + \ldots \tag{8.23}$$

b) If the third-order term is not identical zero, then the following applies as a condition for continuous phase transitions. The second- and third-order terms disappear when $A(p,T)$ is zero on one hand (see above) or when the third-order terms obtain a coefficient $B(p,T)$ analogous to the A and is pressure- and temperature-dependent as

$$B(p,T)\cdot b(\gamma)\cdot\eta^3 \tag{8.24}$$

One now obtains

$$\Phi = \Phi_0 + A(p,T)\eta^2 + B(p,T)b(\gamma)\eta^3 + C(p,T,\gamma)\eta^4 \tag{8.25}$$

Now both A and B are zero at a continuous transition point, and in this case there is an isolated point in place of a Curie line (fig. 8.5).

Now this theory can be applied to liquid crystals. Landau viewed these as "crystals" obtained from an isotropic phase by continuous transitions of the above described type. This

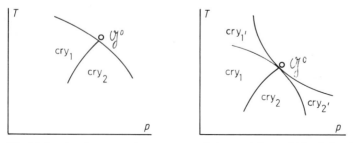

Fig. 8.5. Landau-theory coexistence curves of phases, a) for vanishing 3rd order term, b) for the special case of 4th order terms (from [183]).

is justified insofar as the transition is accompanied by only small first-order effects, and it undoubtedly contains higher-order elements. A continuous transition of one of the two above types is assumed according to whether the liquid crystal possesses centers of symmetry or not. Thus there is a Curie-point transition, $f(p, T)$, for systems lacking a center of symmetry, whereas there are only singular transition points in systems having such centers (see fig. 8.5).

In this regard, it must be recalled that one of Landau's early examples, "liquid-crystalline helium", has only just recently again become the subject of a series of experimental and theoretical investigations. In fact, one of the two superfluid phases of helium can be considered a liquid crystal order [202].

Since the writings of Landau have a very formal character, we will repeat the calculations by using a practical example, and the connection with the de Gennes formulae will become clearer in this way. Consider the transition from an isotropic liquid to a crystal corresponding to a transition from ρ_0 to $\rho = \rho_0 + \delta\rho$, and the latter corresponds to the symmetry of the crystal. Landau develops $\delta\rho$ as

$$\delta\rho = \sum_f a_f e^{i(f \cdot r)} \tag{8.26}$$

where f are the vectors of the reciprocal lattice, and r is that of the distance ($\delta\rho$ being real means $a_f = a^*_f$). For the free energy of the crystal, we have:

$$\varphi = \varphi(\delta\rho) = \varphi\left(\sum_f a_f e^{i(f \cdot r)}\right) \tag{8.27}$$

Near the transition point, φ is developed as an exponential series with respect to a_f. Three product terms appear of the following kind

$$a_{f_1} \cdot a_{f_2} \cdot a_{f_3} \tag{8.28}$$

On the other hand only those terms can enter exponential series for which

$$f_1 + f_2 + f_3 = 0 \tag{8.29}$$

One can already see the analogy to the de Gennes definition of "tensor degree of order", $Q_{\alpha\beta}$, where

$$Q_{\alpha\beta} = \begin{bmatrix} Q_1 & 0 & 0 \\ 0 & Q_2 & 0 \\ 0 & 0 & -(Q_1 + Q_2) \end{bmatrix} ; \quad (\text{Spur } Q_{\alpha\beta} = 0) \tag{8.30}$$

Since $\sum f_i$ is zero, the first-order terms in the developed series disappear. The second-order terms contain products of the form $|a_f|^2$, and the series becomes

$$\varphi = \varphi_0 + \sum_f A_f |a_f|^2 + \dots \tag{8.31}$$

Above the transition point, φ exhibits a minimum for all values of A_f; thus A_f is always positive. At the transition point, all second-order terms disappear when $\delta\rho$ is not zero. It can be seen from this that one A_f becomes zero at the transition point, and thus the $A_f(f)$ curve touches the f-axis. That leads to a $\delta\rho$ that corresponds to plane waves with a definite wavelength. This wavelength is defined by those values of f that are derived from the disappearing A_f. All A_f belonging to other values of f are zero, and $A_f \equiv A$. Then

$$\varphi = \varphi_0 + A_f \sum_f |a_f|^2 = \varphi_0 + A \sum |a_f|^2 \tag{8.32}$$

The third-order terms are of the form

$$\sum_{f_1 f_2 f_3} B_{f_1 f_2 f_3} \quad a_{f_1} a_{f_2} a_{f_3} \tag{8.33}$$

where again

$$f_1 + f_2 + f_3 = 0$$

As stated previously, point lattice arrangements showing periodicities of equal size appear at the transition point.

The third-order terms therefore contain only those f_1, f_2, and f_3 that differ only in direction, and thus form identical equilateral triangles that vary only in orientation. In addition, due to the isotropy of the fluid phase, the $B_{f_1 f_2 f_3}$ can depend only on the size of these vectors but not upon their orientation, and the $B_{f_1 f_2 f_3}$ are thus all of identical magnitude. So $B_{f_1 f_2 f_3} \equiv B$, and it follows that

$$\varphi = \varphi_0 + A(p, T) \sum_f |a_f|^2 + B(p, T) \sum_f a_{f_1} a_{f_2} a_{f_3} \tag{8.34}$$

A comparison with de Gennes' formula gives eq. 8.5.

Other theoretical approaches describing mesophase transitions by modifications of the theories discussed are reported in the literature [7b, 132, 138, 153, 216, 217, 272, 282, 284, 286–290, 309, 342a], as are applications to the following phase transitions:

n/i   [122, 239, 307, 354, 364],
c/i   [122],
n/s_A [1, 62, 204, 309],
n/s_C [62, 205a, 333a],
k/s_A [188],
k/s_B [134],
s_A/s_C [114a, 205a]
s_A/s_C^* [215, 366, 215a, 215b, and considering a Lifshitz point: 215c],
s_A/s_B [147],
s_C^* (uniformly polarized)/s_C^* (distorted) [4:645a].

Numerous other considerations of mesophase transitions are based on molecular statistical calculations described in section 3.4, especially the works of S. Chandrasekhar et al. [3:94, 3:98, 3:100], M. A. Cotter and D. E. Martire [3:137, 3:530, 3:532, 3:533], and others [3:84, 3:666, 3:852].

A comparison between experimental data and theoretically predicted thermodynamic properties has been performed for PAA [100], and we also mention a study on the solid state polymorphism of this compound [41].

8.3 *Pretransition Phenomena*

We shall now consider the details of pretransition phenomena and the various attempts to use them as the basis for a theory on liquid-crystalline/isotropic-liquid phase transitions. Placed in front, fig. 8.6 depicts two differential thermograms that are typical of mesomorphism (a) or plastic crystallinity (b) as it is displayed by dodecamethylcyclotetrasilazane [356, 357, see also 2:70 and J. N. Sherwood, Edt. The Plastically Crystalline State (Wiley: Chichester). 1979. 383 pp.].

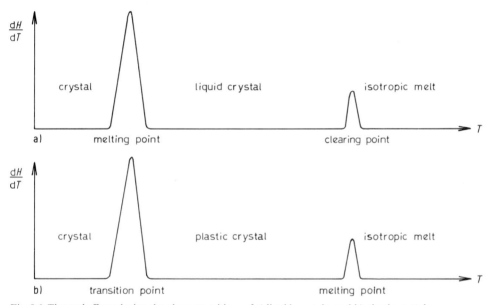

Fig. 8.6. Thermal effects during the phase transitions of a) liquid crystals; and b) plastic crystals.

The phenomenon of a measurable characteristic first increasingly and then asymptotically approaching the value corresponding to the other phase is known as a pretransition effect occurring before the temperature of a recognizable first-order transition is reached. Various reasons are to be considered, e.g., structure-dependent translational factors or the excitation of rotation or vibration. The pretransition due to flaws and the "configurational pretransition" localized in individual molecules must also be mentioned along with the socalled surface melting [179a]. However, one factor must be considered in the evaluation of pretransition phenomena, namely, a socalled heterogeneous premelting (caused by impurities) which is as frequent as it is disturbing. The paraffins exemplify this behavior by their specific heats which increase strongly and in a temperature-dependent manner before reaching the melting point [110, see also 179]. Ubbelohde's monograph cites other examples [347]. Although the premelting of solid crystals to isotropic melts and the analogous reverse phenomenon of prefreezing have been thoroughly investigated, only a few examples can be found that implicate liquid crystals in this type of behavior.

Density measurements indicating a hysteresis are particularly suited to the study of pretransition effects (see section 8.6). E. Bauer and J. Bernamont were the first to measure the density and expansion coefficient of PAP in a dilatometer, discovering a discontinuity following a distinct pretransition. W. Maier and A. Saupe's measurements confirm similar behavior with PAA, and the most important result of this experiment is that the density and density-change entered as significant parameters in the general theory of the nematic state (cf. section 3.4). Valuable information on pretransition effects is also gained by measurements of dielectric and Kerr constants [305], cf. section 4.4.10. The behavior of the viscosity in the pretransition range should also be mentioned here, although due to its dynamic nature, it seems less suited to be compared to the behavior of a static system. Solvents have also been investigated in the pretransition range, and the limiting activity coefficient of the solute was found to be an order-dependent parameter with a pronounced pretransition effect (see section 9.1.2.2).

In a streaming system, the change in birefringence exhibits a pretransition, as does the sound velocity (see section 3.10). Thorough investigations on the temperature-dependence of polarized IR absorption and the bandwidth of NMR signals could also give support to the conclusion that the pretransition range must be considered a region of a more rapidly changing degree of order. One can even state that the concept of pretransition is really nothing more than the qualitative description of the molecular-statistically dependent degree of order. NMR and other studies also permit predictions about the state of order above the transition point. In the transition from the isotropic to the liquid-crystalline melt, a line-broadening that corresponds to a decay in the spin/lattice relaxation time has been observed rather early.

The theory of J. A. Pople and F. E. Karasz [241], which is a progressive treatment of the two-lattice model of J. E. Lennard-Jones and A. F. Devonshire [189], satisfactorily explains many observations on the n/i phase transition, e.g., the variation in degree of order with temperature, the temperature-dependence of the molar volume and of the expansion coefficient, and the pressure-dependence of the transition temperature [53, 54].

Based on this theory S. Chandrasekhar could show that the n/i transition is always of the first order, and there always remains a short-range order in the isotropic phase in spite of the sudden disappearance of any long-range order at the clearing point. This correlates well with the results of earlier experimental work, such as that on light scattering (cf. section 6.2).

If the melting process is considered a transition from higher to lower order, it can be separated into changes of translational position and orientation, the latter being also deemed a special case of rotational freedom [52–55]. The theory predicts two transitions for minor orientational barriers. With rising temperature, a crystal is first formed having lattice particles in free rotation, and this passes into the isotropic melt by a second transition [62a]. If the forces determining positions and orientations are within a certain range, then the two transitions coincide accompanied by a corresponding increase in the entropy of fusion. Major orientational barriers cause the positional melting to preceed the orientational melting. Thus a mesophase is formed which is free in diffusion but limited in rotation. In the extreme case, this is a nematic phase with only orientational order but without positional order. A more precise examination must take into account that even the positional order does not have unlimited freedom, but it is still anisotropic with respect to the anisotropy of diffusion. The characteristic parameter, v, is introduced as a measure of the relative barriers for the rotation of a molecule and its diffusion to an interstitial site. The approximate ranges of v given in table 8.1, define the different types of phase transitions.

Table 8.1. Typical phase-transition phenomena characterized by ranges of the parameter v (from [53]).

$v < 0.298$	Two transitions, solid/solid preceding melting
$v < 0.264$	Second-order solid/solid transition
$0.264 < v < 0.298$	First-order solid/solid transition
$0.298 < v < 0.975$	Single transition
$v > 0.975$	Two transitions, positional melting preceding rotational melting
$0.975 < v < 1.047$	Second-order mesomorphous/isotropic transition
$v > 1.047$	First-order mesomorphous/isotropic transition

Fig. 8.7 shows the behavior of the transition entropy for the various values of v obtained by S. Chandrasekhar (curve a) and J. A. Pople and F. E. Karasz (curve b).

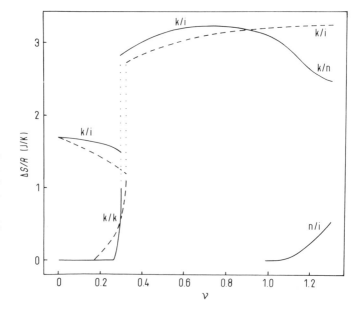

Fig. 8.7. Changes of transition entropy as a function of v according to Chandrasekhar (solid lines) and Pople and Karasz (dashed lines): k/k, solid/solid; k/i, solid/isotropic; k/n, solid/nematic; n/i, nematic/isotropic transitions (from [53]).

From the Chandrasekhar theory as well as from Pople and Karasz' model, it can be seen that pre- and posttransition effects cannot be separated from the transition per se, and these effects are qualitatively included within the theory (the behavior of the specific heat will be considered below in detail). At this time there still exist small discrepancies in the theory with regard to the coefficient of expansion and the isothermal compressibility. It seems necessary and justified to treat the concepts of degree-of-order and pre- and posttransition effects in the sense of action and reaction as directly connected to one another.

A. R. Ubbelohde has thoroughly studied the phenomena of pretransition effects [2:743, 343–347, see also 173, 328] which are also closely related to the cybotactic structure of crystalline liquids [362, 363], see also section 3.7.

D. W. McClure has tried to calculate the change in lattice energy accompanying the rotational phase transition in paraffin crystals [203]. Models similar to those described here were developed in a modified double-lattice theory [231] and in a "expandable" lattice theory [137, see also 103a].

As an alternative to Chandrasekhar, who proceeds from a lattice or discrete model of the liquid state, K. K. Kobayashi has suggested a theory of translational and orientational melting that proceeds from a continuum model of the liquid state [169–171]. Data calculated according to this theory agrees well with the experimental results [187].

In the thermodynamic sense, there is no cholesteric/nematic phase transition. There are only mixtures that exhibit a change in texture (similar to this kind of phenomena) that is dependent on pressure and temperature. The specific heat and optical activity of such mixtures change steadily with temperature, and the optical rotation is zero at a certain temperature. Above and below this point the substance acts as a cholesteric phase with an opposite sense of rotation [280], cf. section 7.1.6. The amount by which the s_A/n transition temperature, $T_{s_A/n}$, is lowered by additional forces producing a helicoidal cholesteric structure are estimated to be of the order of $10^{-3} \, T_{s_A/n}$ [198].

A first-order transition is discussed in isotropic liquid cholesteryl stearate just above the normally reported transition temperature [357a].

The s_A/s_C phase transition is compared with the Λ-transition of helium by P. G. de Gennes [87], while A. de Vries considers two kinds of this transition [98a]. It is impossible to comprehend here all observations of pretransitional behavior because it is found with most properties, and this will be discussed in other chapters. In addition we refer to the following literature [72, 192, 237, 252, 305, 353].

8.4   Calorimetric Measurements

Along with observations with the polarizing microscope, calorimetric measurements are best suited for the recognition of phase transitions and the determination of their temperatures.

D. Vorländer et al. were able to prove unequivocally the polymorphism of the smectic state by using this method [2:730, 2:815, 2:816]. Calorimetric phenomena found with liquid crystals have been mainly studied by H. Arnold [10–18] and by E. M. Barrall II, R. S. Porter, and J. F. Johnson [27–40]. We refer here to literature concerning experimental details on the construction and standardization of an adiabatic precision calorimeter [10, 28, 42, 42a, 164, 178] and the application of DTA in the determination of mesophase transition temperatures [1:3–1:5, 1:35, 1:274, 2:15, 2:103, 2:214, 2:330, 2:778, 2:84a, 2:320a, 2:485a, 2:609a, 4:364a, 7:55, 65, 113a, 158, 167a, 177, 223, 225, 252, 285, 301, 322, 336, 360, 367, 368].

We additionally mention that, like DTA, measurements of the depolarizing current can detect the crystalline/liquid-crystalline phase transition [342]. The literature also contains discussions of calorimetric measurements of phase transitions in two-dimensional liquid crystals [155, 181, 200, 255, 307, 361].

8.4.1 Molar Heat

The behavior of the molar heat, c_p, as a function of temperature exhibits the following features. In the solid state, c_p first increases linearly with temperature and then rises at an increasingly faster rate upon approaching the melting point. Similarly, c_p generally rises with increasing temperature, and this rise is linear in some ranges. The pretransition ranges are characterized by a sharp increase in c_p below, and a sharp decrease in c_p above the transition temperature when observed in the direction of increasing temperature. After the transition to the isotropic liquid, c_p first decreases sharply, passes through a minimum, and again rises slightly and linearly with the temperature [11–13]. This means that in every case, a flat minimum whose two branches rise very differently appears a few degrees above the transitions (figs. 8.8, 8.9, and 8.10). If the nearly linear behavior of c_p is extrapolated from the isotropic region into the nematic and vice versa, a usually markedly larger specific heat is found in the liquid crystal, e.g., as Arnold has stated for the p-alkoxyazoxybenzenes [12, see also 296]. However, this experience cannot be generalized or applied to all smectic phases (fig. 8.10).

A few degrees below the transition temperature it seems that a single order parameter can self-concistently describe the transition anomalies in the specific heat, c_p, the isobaric thermal expansion coefficient, α, and the isothermal compressibility, β [4], fig. 8.11.

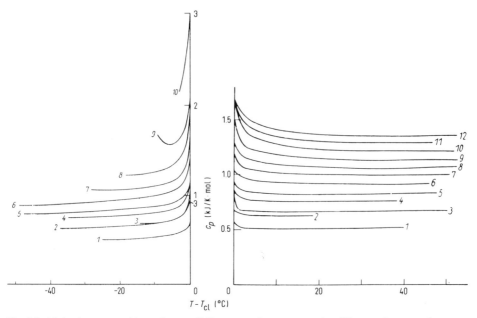

Fig. 8.8. Molar heats, c_p, of homologous dialkoxyazoxybenzenes vs. the difference between the measurement and clearing temperature, $T - T_{cl}$, for: a) the nematic phase, b) the isotropic liquid (from [12]).

H. Arnold has tried to bridge the gap to the Maier-Saupe molecular statistical theory, i.e., to use the Maier-Saupe parameters to express the difference between the specific heats of the nematic and isotropic phases as a function of the degree of order. However, this did not succeed convincingly, and it is unclear whether this was due to the theory being still incomplete or to systematic measurement errors. H. Arnold also demonstrated characteristic

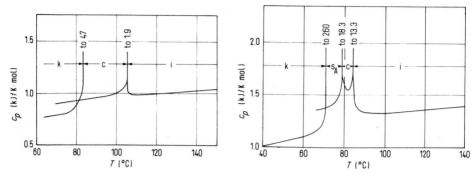

Fig. 8.9. Molar heats, c_p, vs. temperature, T, for a) cholesteryl ethyl carbonate, b) cholesteryl myristate (from [16]).

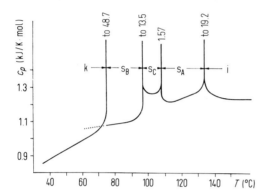

Fig. 8.10. Molar heats, c_p, vs. temperature, T, for *n*-amyl 4-*n*-dodecyloxybenzylidene-4-aminocinnamate (from [17]).

pretransition effects in the behavior of molar heats in smectics. Deduced from the theories of de Gennes [152] or Maier-Saupe [124], the following equations have been proposed to express c_p as a function of temperature:

$$c_{p(T)} = c_0 T^2 (T - T_c)^{1/2}$$

$$c_p^n \propto ((T_c - T)/T)^{-\beta}$$

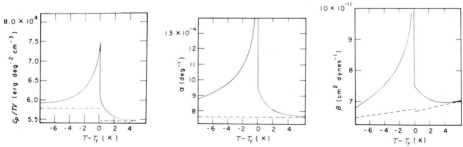

Fig. 8.11. Experimental values of c_p, α, and β in PAA; the temperature scales were set by taking T_r as the point of maximum effect. Dashed lines show the estimated non-transition background (from [4]).

The order parameter, η, is a function of the volume V and the temperature T according to $\eta = \eta^{V_0 T_0}(V^n T)$, where $n = V \cdot \dfrac{\partial n}{\partial V}\bigg| \dfrac{\partial n}{\partial T}$. For PAA, n is 4.3 1 K below the n/i transition. The constancy of the parameter of order is derived for the coexistence curve [4].

Other measurements of specific heats mainly concerned MBBA [4:447a, 6:151, 7a, 140, 162, 163, 229, 308] and other mesogens [4:321, 4:447a, 7, 7a, 22, 151, 161, 212, 314a, 324, 370, 371].

8.4.2 Transition Enthalpy and Entropy

As the measurements of H. Arnold [11] and J. van der Veen et al. [348] show, the n/i transition enthalpies and entropies alternate in the same way as the transition temperatures (fig. 8.12).

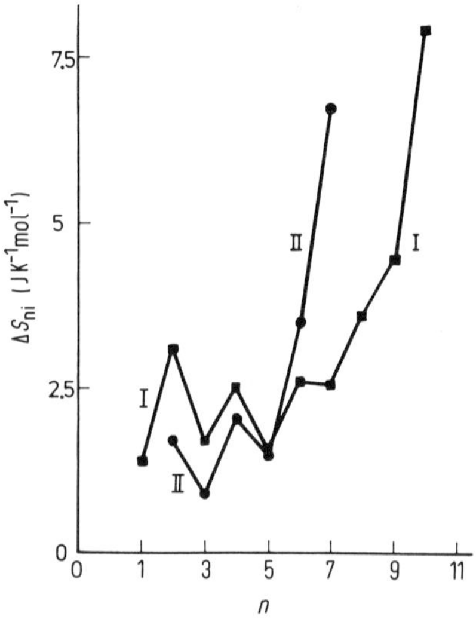

Fig. 8.12. n/i transition entropies, ΔS_{ni}, as a function of the number of the carbon atoms, n, in R (from [11] and [348]).

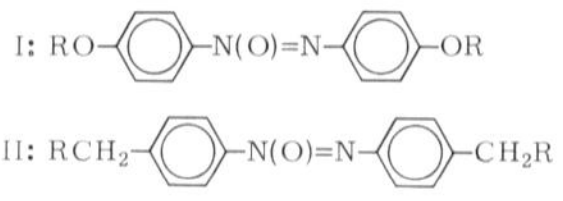

I: RO—⬡—N(O)=N—⬡—OR

II: RCH₂—⬡—N(O)=N—⬡—CH₂R

No quantitative explanation for the alternation of the clearing entropy can be deduced from Maier-Saupe theory at this time [348].

An odd-even variation of the n/i transition entropy with chain length is also observed in 4-ethoxy-4'-n-alkanoyloxyazobenzenes that occurs in the same manner as with the transition temperatures, but it shows a markedly higher alternation [139].

While the n/i and c/i transitions do not differ calorimetrically, the s/i transitions are associated with significantly higher values of ΔH and ΔS [16, 35]. Thus the different molecular orders in nematics and cholesterics cause no marked variance in the enthalpy and entropy differences. As a rule, the mesophases are thermodynamically more closely related to the liquids than to the crystals due to their normally distinctly lower transition heats from the liquid-crystalline to the isotropic-liquid states as compared to those of liquid-crystalline to crystalline-solid phases. However, the special case of p-butoxybenzylidene-p-ethylaniline should be mentioned; the largest entropy change is found in the s/n transition rather than in the k/s one [350].

Due to polymorphism in the crystalline state, crystals with different pretreatments can exhibit different heats of transition when passing into the mesomorphous state [2:531, 6:151, 41, 213, 226a]; there is also an influence on the phase diagrams of mixtures [219a]. The melting enthalpies of nematogens show no alternating effect [190a].

A very weak alternation in the values of the c/i transition entropies for aliphatic cholesteryl esters can be discerned from the values shown in fig. 8.13. However, this is no longer the case for the s/c transitions within the same homologous series of substances, as shown in fig. 8.14 [2:569, 2:571].

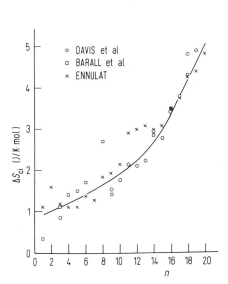

Fig. 8.13. c/i transition entropies, ΔS_{ci}, of cholesteryl n-alkanoates vs. the number, n, of carbons in the acid (from [2:571]).

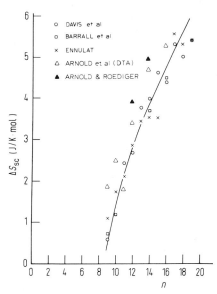

Fig. 8.14. s/c transition entropies, ΔS_{sc}, of cholesteryl n-alkanoates vs. the number, n, of carbons in the acid (from [2:571]).

Among the higher members of homologous series, the mesomorphous isotropic transition entropy increases linearly with molecular weight [34, 208]. In the series of aliphatic cholesteryl esters, there is additionally observed a sharp discontinuity in the crystalline/mesomorphous transition entropies between the octanoate and the nonanoate, corresponding to the first appearance of the smectic phase [2:571, 33, 34, 36].

The obviously wide range of values, shown in figs. 8.13 and 8.14, can probably be attributed to impurities in the investigated materials. Fig. 8.15 indicates the very different DSC thermograms obtained from 97.4% and 99.70% cholesteryl heptadecanoate.

Attention must also be given to traces of solvent and to the decomposition tendency of the cholesteryl esters [37]. For purification, recrystallization from n-pentanol (not from ethanol!) [2:568] chromatography [367], or zone refining [7, 225, 302] are recommended. The very distinct alternation of the c/i clearing points in the series of cholesteryl-ω-phenylalkanoates corresponds to an odd-even effect where the odd chain length branch has the higher values of transition enthalpies and entropies that are shown in fig. 8.16 (cf. chapter 2). However,

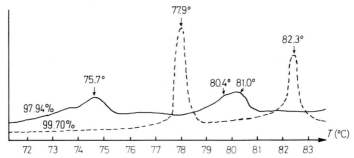

Fig. 8.15. DSC curves for 97.4% (————) and 99.70% (-----) pure cholesteryl heptadecanoate (from [32, 38]).

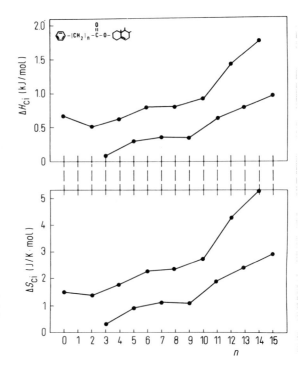

Fig. 8.16. c/i transition enthalpies, ΔH_{ci}, and entropies, ΔS_{ci}, of cholesteryl-ω-phenylalkanoates vs. the number, n, of methylene units (from [2:207]).

the even chain length branch in the s/c transitions has the higher values (fig. 8.17), while the transition temperatures do not alternate [2:207].

    The literature contains additional data on the k/c, s/c, and c/i transition heats [2:27, 2:29, 2:110, 2:169, 2:196, 2:200–2:202, 2:204, 2:206, 2:285, 2:343, 2:565, 2:570, 2:677, 29, 30, 39, 108, 116, 186, 210, 245, 274]. R. S. Porter discusses possible generalizations including numerous steroid moieties [246]. More information on k/n, s/n, and n/i transition heats has been published in the literature [1:277, 2:30, 2:92, 2:120, 2:164, 2:168, 2:169, 2:191, 2:279, 2:283, 2:285, 2:288, 2:328, 2:344, 2:400, 2:466, 2:537, 2:754, 2:884–2:886, 2:888, 2:149b, 2:183b, 5:79, 5:79a, 5:79b, 7, 27, 29, 95, 112, 135, 174, 175, 186, 207, 274, 278, 315, 370, 371].

The values measured for MBBA are 12.85 kJ/mol for the k/n and 285 J/mol for the n/i transition [308, see also 6:151, p. 400; concerning EBBA see ref. 324]. Measurements of the transition heats of the three homologous *p*-di-*n*-alkylazoxybenzenes (see table 8.2) indicating a practically second order, a weakly first order, and a clearly first order s_A/n transition have

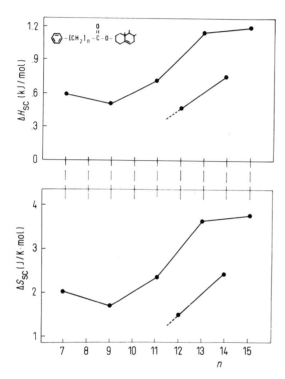

Fig. 8.17. s/c transition enthalpies, ΔH_{sc}, and entropies, ΔS_{sc}, of cholesteryl-ω-phenylalkanoates vs. the number, *n*, of methylene units (from [2:207]).

found elegant confirmation by measurements of the birefringence, where no change of birefringence could be discerned at the second order s/n transition of *p*-di-*n*-hexylazoxybenzene [88], fig. 8.18.

Table 8.2. Transition data of *p*-di-*n*-alkylazoxybenzenes (from [88]).

Alkyl		s/n	n/i
C_6H_{13}	k 24 n 54.5 i (17 s_A)		
	H (kJ/mol)	0.02	0.57
	S (J/K mol)	0.07	1.73
C_7H_{15}	k 34 s_A 54.5 n 71 i		
	H (kJ/mol)	0.16	1.12
	S (J/K mol)	0.50	3.26
C_8H_{17}	k 39 s_A 64.5 n 67 i		
	H (kJ/mol)	2.3	2.3
	S (J/K mol)	6.8	6.75

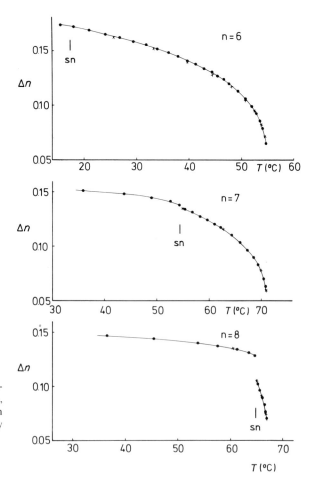

Fig. 8.18. Birefringence, Δn, vs. temperature, T, of di-n-hexyl-, heptyl-, and octylazoxybenzene at 632.8 nm and different thicknesses marked by circles and crosses (from [88]).

The following data obtained for the di-n-dodecyl-4,4′-azoxymethylcinnamate transition enthalpies provides an interesting comparison [14]:

k 79 s_c: 75.2 kJ/mol
s_c 82.5 s_A: 0.1 kJ/mol
s_A 87.5 i: 8.8 kJ/mol

Other substances also show an exceptionally small s_C/s_A transition heat, being only about 5% of that between the B and A modifications [15, 17].

This indicates a close structural relationship between the s_C and s_A phases (see section 5.2.3). Measurements of the heats of transition between smectic modifications can also serve to proove the relationships between them that were derived from the rules of isomorphism [18].

The ranges of existence of the mesomorphous sodium soaps and their transition enthalpies indicate a complex picture (figs. 8.19 and 8.20) due to the large number of phases involved [2:775, 2:776, 230, 271; for lead(II) carboxylates see 2:1a].

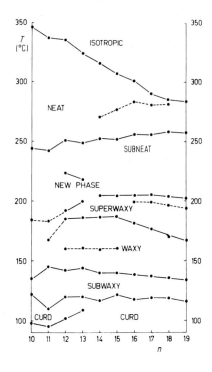

Fig. 8.19. Transition temperatures, T, of normal sodium soaps as a function of the number of carbon atoms, n (from [230]).

Table 8.3. Enthalpies of various transitions (from [95]).

Transition	Number of cases	Transition enthalpy (kJ/mol)
n/i	202	0.84 – 9.6
c/i	79	0.84 – 3.8
s_A/i	93	2.9 – 12.6
s_C/i	21	10 – 42.7
s_D/i	1	10.5
s_A/n	65	0.21 – 4.6
s_A/c	26	0.42 – 1.9
s_B/n	1	8.8
s_C/n	17	0.67 – 9.6
s_C/c	2	2.1 – 4.6
s_B/s_A	55	0.42 – 4.6
s_B/s_C	12	1.84 – 10.5
s_C/s_A	59	<0.04 – 2.8
s_C/s_D	2	2.85 – 4.2
s_D/s_A	1	6.7
s_E/s_A	2	6.2 – 7.9
s_E/s_B	28	0.5 – 1.84
s_F/s_C	3	0.17 – 0.5
s_G/s_C	1	2.34
k/i (melting)	391	7.1 – 117

Summarizing a great number of measurements, table 8.3, published by D. Marzotko and D. Demus, gives an impression of the wide range of enthalpy changes observed with the transitions of mesophases [95]; a wealth of transition data is found in the review of E. M. Barrall, III and J. F. Johnson [40].

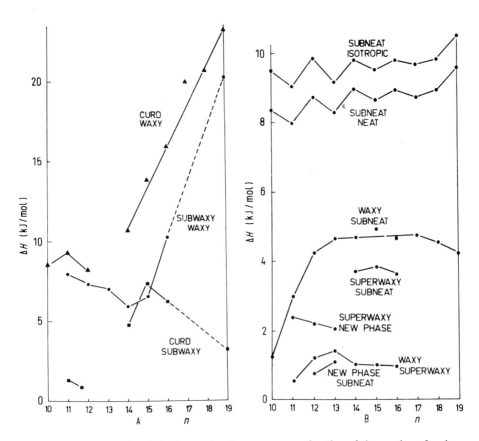

Fig. 8.20. Transition enthalpies, ΔH, of normal sodium soaps as a function of the number of carbon atoms, n (from [230]). A: from 90 to 140°C. B: from 140 to 350°C.

In lower-temperature transitions (curd/waxy and subwaxy/waxy), the transition enthalpies increase markedly with chain length. The subneat/neat transition contains evidence of an alternating effect of the enthalpy that is a function of chain length. The neat/isotropic transition enthalpy (not illustrated) is small, being about 0.84 kJ/mol. This is consistent with the model of an ordered liquid in the isotropic state which is similar to the neat phase.

From the difference between the s_C/i transition heats of the homologous p-alkoxyazoxybenzenes with $C_{18}H_{37}$- and $C_{12}H_{25}$-side chains, an increment for each contribution of a CH_2-group can be calculated to be $\Delta(\Delta H) = 1.4$ kJ/mol. Analogously, the increment for the transition entropy is $\Delta(\Delta S) = 3.8$ J/K mol $= 0.46$ R [17]. The literature also reports on the calorimetric study of ethyl *p*-azoxybenzoate [325], a smectic trimorphous substance [17], the smectic tetramorphous material 2-(4-*n*-pentylphenyl)-5-(4-*n*-pentyloxyphenyl)-pyrimidine [1:325], and other smectics [1:277, 2:47, 2:92, 2:110, 2:279, 2:283, 2:285, 2:343, 2:344, 2:452, 2:466, 2:537, 2:602, 2:603, 2:652, 2:754, 2:884–2:886, 2:149b, 2:183b, 5:79, 5:79a, 5:79b, 6:273, 95, 99, 112, 133, 150, 315, 319]. Transition entropies have been measured for the n/s_A transitions in the homologous series of 4-*n*-alkoxybenzylidene-4'-phenylazoanilines having two to ten carbon atoms in the side chain [105]. No correlation has been found between the change in the optical behavior and the magnitude of the enthalpy change with the smectic modifications of bis-(4'-*n*-octyloxybenzylidene)-1,4-phenylenediamine [112].

The melting behavior of s_H phases was found to be consistent for distinguishing different types of them [98].

8.5   *Kinetics of Phase Transitions, Glassy State*

The supercooling at the i/n transition is usually less than 5°C and always far less than at the n/k transition [31]. High viscosity and branched chains in substances of the type

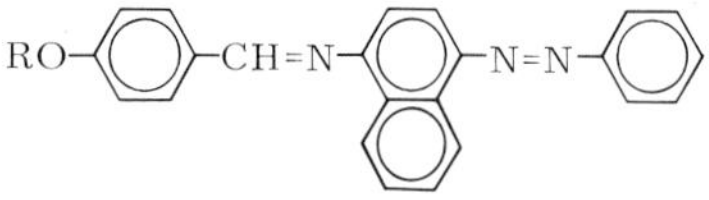

$$\text{RO}-\underset{}{\bigcirc}-\text{CH}=\text{N}-\underset{}{\bigcirc}-\text{N}=\text{N}-\bigcirc$$

lead to the possibility of a significant supercooling of the isotropic melt with respect to the nematic phase, although no other examples are known of such exceptions [2:227, 2:849]. The kinetics of the i/n transition can be pursued by the microscopic observation of the growth of nematic droplets as the sample is cooled at a constant rate on the heating stage. Under these conditions, the phase-transition rate of PAA is determined by the transport of the heat of transition from the phase boundary to the slide rather than by the transition reaction itself [228]. Fig. 8.21 shows that nucleation of the nematic phase is already beginning with very slight supercooling ($\Delta T \approx 1 \cdot 10^{-2}$ K).

Unlike PAA, its homolog, PAP, shows anomalously fast crystallization if the surface is about 3 to 6°C supercooled [206, see also 277]. See section 3.9 for data on the heat conductivity of mesophases [264].

The nucleation rate for the formation of the cholesteric phase of cholesteryl nonanoate shows a typical maximum because the nucleation is transport-controlled at lower temperatures, and it is governed by the thermodynamic driving force of supercooling at higher temperatures [238, see also 275]. Crystallization in mixtures can take place forming a pattern of concentric rings [190]. J. Rault suggests a mechanism for the nucleation of the focal conic texture from the planar texture [265].

All three transitions (k/s/c/i) of cholesteryl myristate can be described kinetically by the Avrami equation

$$F_{(t)} = e^{-Kt^n}$$

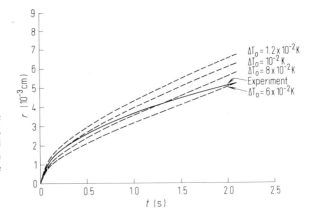

Fig. 8.21. The radius, r, of a nematic droplet of PAA as a function of time, t: comparison of experiment and theory, with initial supercooling given by parameter T_0 and a cooling rate of 0.2 K per minute (from [228]).

where $F_{(t)}$ is the volume fraction not transformed, t is time, n is a constant dependent on the modes of nucleation and growth of the transforming regions, and K is a constant dependent on the mode of nucleus injection and the shape of the transforming regions. Glasses of liquid crystals, which have been but sparingly studied to date, have the order characteristics of the nematic, cholesteric, or smectic mesophase from which each material was quenched [2]. N-(o-hydroxy-p-methoxybenzylidene)-p-butylaniline, whose DTA curve and molar heats are reproduced in figs. 8.22 and 8.23, has been recommended for the study of glass transitions in liquid crystals.

Evidence of the existence of a glassy state was also found in MBBA [322], cholesteryl hydrogen phthalate [2:739], some smectics [322], and in twisted nematics [165].

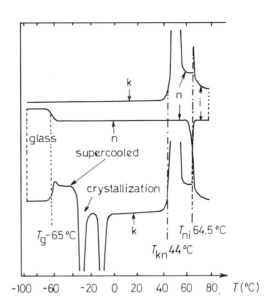

Fig. 8.22. DTA curves for N-(o-hydroxy-p-methoxybenzylidene)-p-butylaniline (from [321]).

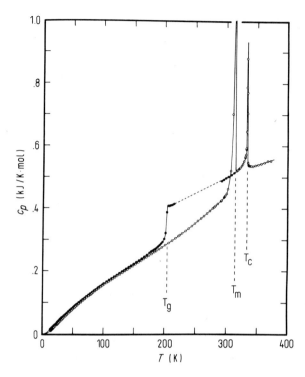

Fig. 8.23. Molar heats, c_p, of N-(*o*-hydroxy-*p*-methoxybenzylidene)-*p*-butylaniline vs. temperature, *T*. ○: crystal, nematic, and isotropic liquid; ●: glassy and supercooled liquid crystal (from [323, 333]).

The glassy states of mesophases were only identified by thermodynamic observations and up to now no auxiliary evidence has been obtained that supports the distinction between the glassy and the supercooled mesomorphous state (see section 5.5).

8.6   Temperature-Dependence of Density

From a series of very precise measurements, the generalization can be made that the density shows a pronounced pretransition behavior only toward the state of lower order on the low-temperature side of each transition (figs. 8.24 to 8.27). On the higher-temperature side, e. g., in the isotropic liquid, the pretransition behavior can be detected only by high-sensitivity measurements [256a]. The volume changes in the n/i transition alternate like the heats of transition. A comparison with theoretical values indicates a satisfactory correspondence between ΔH and ΔV, ΔH and the degree of order, etc. One result of such a comparison that seems especially important is the possibility of estimating the short range order parameter, i.e., the number of molecules in those groups that for steric reasons form the smallest units in the nematic melt; such groups are the basis for all further theoretical considerations. PAA shows coefficients of expansion for the crystalline, nematic, and isotropic phases at $4.1 \cdot 10^{-4}$, $9.4 \cdot 10^{-4}$, and $8.4 \cdot 10^{-4}$ per degree C., respectively [251], fig. 8.24.

The coefficient of expansion in the cholesteric phase of cholesteryl myristate also exceeds that of the isotropic liquid [248], fig. 8.25. However, the thermal expansion coefficient of the s_A phase of diethyl *p*-azoxybenzoate is smaller than that of the isotropic liquid [130].

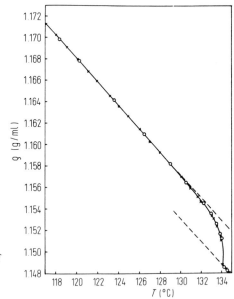

Fig. 8.24. Density-temperature plot (ρ vs. T) of nematic PAA: ×, heating; ○, cooling (from [251]).

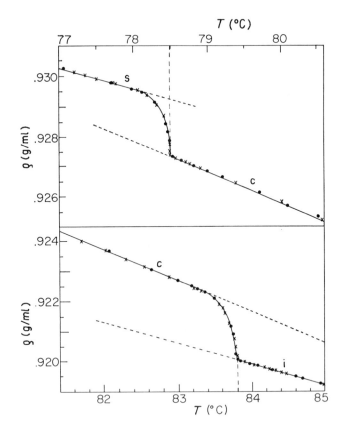

Fig. 8.25. Density-temperature plots (ρ vs. T) of cholesteryl myristate measured in heating and cooling cycles (from [248]).

The ΔV is 6.9% for the k/n transition and 0.34% for the n/i transition of PAA [251], another work reports 7.74% and 0.35%, respectively [24, see also 120, 256c]. A much higher density change, e.g. 2% for diethyl *p*-azoxydibenzoate, is observed at the s_A/i transition [130].

In the pretransition region slightly below the clearing point of MBBA, its volume-temperature relationship can be described by the five-parameter nonlinear equation

$$V = a + bT + c(T_{cl} - T)^n$$

[60], density measurements of MBBA are also reported in refs. [3:130b, 59, 125, 253, 256c]; for EBBA see refs. [6:291, 23, and CBOOA 3:130b]. R. Pynn has suggested an equation correlating the density and temperature of a fluid at the n/i transition in terms of molecular interactions [262]. Usually, as shown in fig. 8.26, the density within the temperature intervals

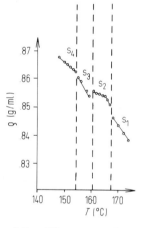

Fig. 8.26. Density-temperature plots (ρ vs. T) for bis-(4-*n*-heptyloxy-benzylidene)-1,4-phenylenediamine (from [92]).

of the different smectic modifications continuously decreases, but it curiously increases at the transition discontinuity of smectic 3 to smectic 2 [92]. Transitions associated with a large change in density also exhibit large enthalpies of transition, indicating basic structural

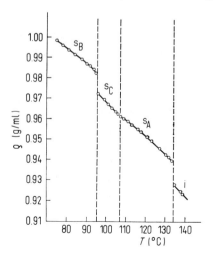

Fig. 8.27. Density-temperature plots (ρ vs. T) for *n*-amyl 4-(4-*n*-dodecyloxybenzylideneamino)cinnamate (from [93]).

changes. According to this, the transition between the closely related s_A and s_C phases stands out for its extremely small change in density (fig. 8.27).

When the thermal expansion coefficient, defined as

$$\alpha = \frac{V_{T_1} - V_{T_2}}{\frac{1}{2}(V_{T_1} + V_{T_2})(T_1 - T_2)}$$

is plotted against temperature, the phase transitions become more apparent than in plots of density changes [93], fig. 8.28, cf. fig. 8.11, p. 356.

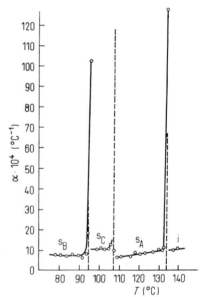

Fig. 8.28. Thermal expansion coefficient, α, of *n*-amyl 4-(4-*n*-dodecyloxybenzylideneamino)cinnamate vs. temperature, T (from [93]).

Results similar to those found for cholesteryl myristate (fig. 8.25) have also been obtained for cholesteryl acetate [249], cholesteryl nonanoate [250, 254], other cholesteryl esters [131], and their mixtures with nematics [219]. Dilatometric studies have indicated a change from first to second order of the c/s_A transition in a mixture of 62 mol% cholesteryl oleyl carbonate and 38 mol% cholesteryl chloride [329, 330]. The influence of cholesterol was also studied [331]. We additionally mention that a reversible texture change was observed in the cholesteric phase of cholesteryl myristate at a well-defined temperature, while there were two such changes in the nonanoate [248, 250].

We refer here to the literature for a densitometric study of the smectic tetramorphous substance 2-(4-*n*-pentylphenyl)-5-(4-*n*-pentyloxyphenyl)-pyrimidine [1:325], other smectics [311–313, 185a], and anhydrous soaps [2:37, 2:684, 2:772, 2:773, 2:776]. The discontinuity in the volume/temperature curve at the clearing point was unequivocally demonstrated by F. Conrat as early as 1909 [73]. Such a volume discontinuity could not be observed at the quasi-second order s_A/n transition of CBOOA, and it is believed to be less than 10^{-5} ml/g [256]. The literature describes other investigations on the density and expansion coefficients of mesogens, especially in the transition regions [2:188, 2:320, 2:622, 3:34, 3:184, 3:579, 3:627, 3:130b, 5:244, 21, 43–45, 94, 154, 197, 211, 212, 218, 247, 276, 310, 341, 3a, 185a, 256b], and it also contains a review [25].

8.7   High-Pressure Investigations

Many authors have studied the pressure-dependence of the transition temperatures of mesophases. However, the first data published [117, 118, 232, 261, 268, 279, 291, 339] varies within a wide range (as explicated by B. Deloche et al. [90]), therefore only recent papers have been selected here. The melting point, T_m, and the clearing point, T_c, of PAA rise with temperature in a similar way with $T_m = 116\,°C + 0.245$ (± 0.01)$°C/kPa$ and $T_c = 133\,°C + 0.27$ (± 0.01)$°C/kPa$ [90]. The phase diagram of MBBA shown in fig. 8.29 is typical of many other nematogens [49, 50, 56, 167, 168, 190c, 292, 294, 326, 332, 358]. Some of these papers provide thermodynamic data that was measured under high pressures.

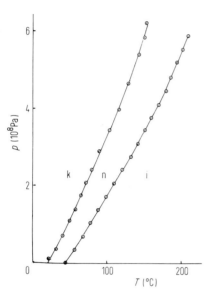

Fig. 8.29. Phase diagram (pressure, p, vs. temperature, T) of MBBA (from [167]).

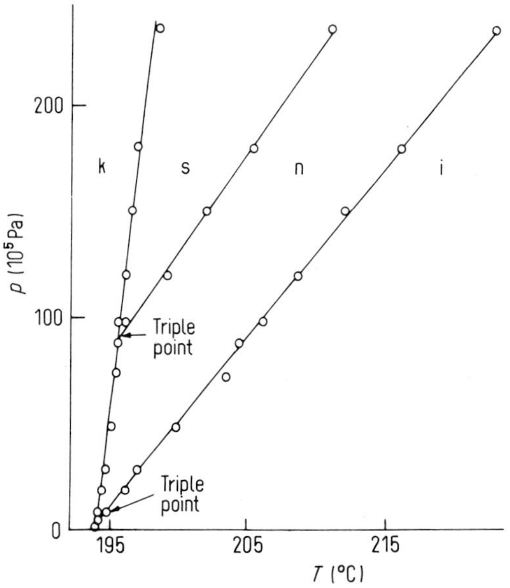

Fig. 8.30. Phase diagram (pressure, p, vs. temperature, T) of p-methoxybenzoic acid (from [57]).

Examples are known of the pressure-induced appearance (fig. 8.30) and disappearance (fig. 8.31) of mesophases [56, 57, 167, 302a].

Above a critical comprehensive stress, a tilt is imposed on a s_A phase transforming it to a s_C phase [267, see also 4:177]. Some cholesterics have also been investigated [57, 167, 240, 293, 295, 297–299, 302a, 340], fig. 8.32.

The DTA curves of the $s_{A/C}$ transition in cholesteryl oleyl carbonate suggest a tricritical point at about $2.67 \cdot 10^8\,Pa$ where the first order character of the transition has diminished to become practically second order [57, 166, 240a]. Such tricritical points were also observed in mixtures and without the application of pressure [4:645a, 5, 113, 133b, 157, 233, 240b, 257]. Earlier works have found the pressure dependence of the transition temperature to

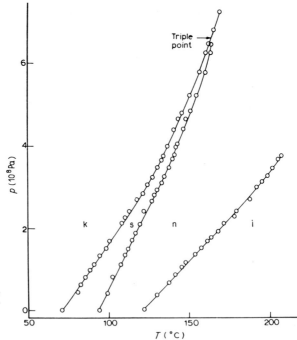

Fig. 8.31. Phase diagram (pressure, p, vs. temperature, T) of p,p'-bis(heptyloxy)azoxybenzene (from [57]).

vary only slightly within homologous series [12], but newer works observed a distinct alternation [180]. The Kraut-Kennedy equation

$$\frac{p_m}{a} = \left(\frac{T_m}{T_0}\right)^c - 1$$

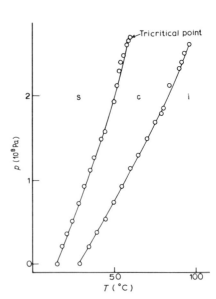

Fig. 8.32. Phase diagram (pressure, p, vs. temperature, T) of cholesteryl oleyl carbonate (from [57]).

gives a useful approximation describing the pressure dependence of melting and clearing points [180].

Using the method of Mattis and Schultz, J. C. Chin and V. D. Neff derived equations describing the isobaric thermal expansivity [63, 64]. The molar compressibility or Wada constants showed an abrupt fall in the vicinity of the n/i transition temperature [26, see also 3:626, 44, 45, 132a, 167b, 365], fig. 8.11. Measurements have also been performed with some cholesteryl esters [3:385, 3:386, 3:627, 224]. Experimental details for the visual observation [58] and DTA studies [266a] of mesophase transitions in a high-pressure cell and piezothermograms of liquid crystals [260, 338a] are discussed in the literature.

In a series of publications, V. K. Semenchenko et al. have studied the changes in specific volumes and the phase transitions of several liquid crystals [292–300, 300a, 300b]. High-pressure studies of CBOOA have provided the first example of a liquid crystal phase transition in which the volume discontinuity is negative upon heating [191, see also 256, 258]. The $p-T$ phase diagrams of other smectics are reported in the literature [106]. Bilayered smectics A formed by certain cyano compounds exhibit a reentrant nematic phase at high pressure [71a, 190b].

The pressure-dependence of the degree of order obtained from NMR studies of PAA was found to be $S_{(Tn/k)}: 0.55 \pm 0.015$ and $S_{(Tn/i)}: 0.40 \pm 0.015$ [90].

8.8   *Mixed Systems with Mesophases*

The early study of mesogenic behavior in mixed systems was mainly due to two reasons. One was an attempt to obtain a mesomorphous state as close to room temperature as possible, and the other was to probe the possibility of "cryoscopy" by lowering the clearing point. Furthermore, the tendency of nonmesogenic substances to form mesophases can be determined from a study of phase diagrams (A. Bogojawlenski and N. Winogradow [2:52, 2:53]). The most important result of investigations of mixed systems was H. Sackmann's classification of the smectic modifications. An extensive scheme describing phase diagrams of mesophases has been given by M. Domon [101, 102, see also 273]. The Schroeder-van Laar equation was found inapplicable for calculations of composition, eutectic-mixture melting points, and k/n transition heats for a few mesomorphous mixtures [156], but it has proven applicable in numerous other mixtures of nematogens [61, 143, 145]. Other authors go back to the Flory-Huggins theory [176], cf. p. 403 and to the Hildebrand theory [188a]. Concerning the general thermodynamics of mixed phases, we refer to R. Haase's book [128]. Reviews concerning mixed mesomorphism have also appeared [86, 146].

8.8.1 *Mixtures of an Enantiotropic Mesogenic Substance with a Monotropic or Nonmesogenic Substance*

Chronologically, the works that should be mentioned first are those of R. Schenck [1:341–1:344, 1:349], K. Auwers [19, 20], and others [104, 148, 281, 320]. Around the turn of the century, some of them had already tried to determine molecular weights by "cryoscopy" in liquid crystals.

From the van't Hoff equation

$$X = \frac{\Delta H \cdot \Delta T}{R T_m^2}$$

it can be seen that ΔT is inversely proportional to ΔH, and as ΔH becomes smaller the melting point depression, ΔT must become larger. However this expectation is not applicable to liquid-crystalline solvents since its validity would require a substance to be soluble in the isotropic liquid but to be insoluble in the mesophase, which is seldom the case. For this reason, all of these experiments must end in frustration due to the inconstancy of the "cryoscopic constant". Nevertheless, investigations of this type are valuable in allowing a quantitative estimation of the tendency of a substance to form a mesomorphous phase. From the behavior of the n/i transition curve, the unmeasurable "transition temperature" of a nonnematogenic substance can be extrapolated [2:52, 2:53, 2:129, 76, 77, 83, 101, 263]. Figs. 8.33 and 8.34 show typical examples of this type of mixture [1:133, 74].

As indicated in table 8.4 and figs. 8.33 and 8.34, a lower slope of the n/i transition curve indicates a greater tendency to form a liquid crystal [75].

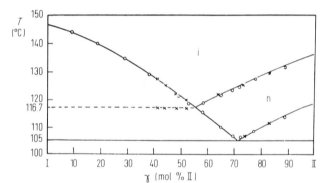

Fig. 8.33. Phase diagram (temperature, T, vs. composition γ) of the system anisylidene-p-anisidine (I)/PAA (II) (from [74]).

Table 8.4. Transition-curve slopes for compounds of the type p-$AC_6H_4BC_6H_4C$-p mixed with PAA (in °C per 10% change in molar composition) (from [75]).

A	B	C	Slope	A	B	C	Slope
H	CH:N	H	29.5	OMe	CH:N	Br	11.0
H	N:N	H	28.0	Cl	N:NO	Cl	10.0
Me	CH:N	H	26.0	OMe	CH:N	Cl	9.5
H	N:NO	H	25.0	Cl	CH:N	OMe	9.0
H	CH:N	NMe$_2$	23.0	Me	CH:N	OMe	9.0
H	CH:N	OMe	22.0	OMe	CH:N	Me	8.5
Me	CH:N	Br	16.0	NO$_2$	CH:N	Me	8.0
Me	CH:N	Me	14.5	OMe	CH:N	NMe$_2$	7.5
Cl	CH:N	Cl	14.5	OMe	CH:N	OMe	4.0
Cl	CH:N	Me	14.5	NO$_2$	CH:N	OMe	2.5
Me	CH:N	NMe$_2$	13.0				

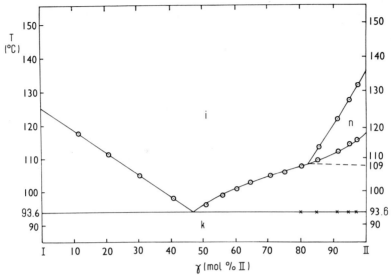

Fig. 8.34. Phase diagram (temperature, *T*, vs. composition, *γ*) of the system *p*-chlorobenzylidene-*p*-bromoaniline (I)/*p*-acetoxybenzylidene-*p*-phenetidine (II) (from [2:129]).

In mixed systems, the influence of terminal groups on the mesomorphous state can thus be quantitatively measured [81], cf. chapter 2. However, reliable results require both components to be similar in shape and structure [79, 196]. Fig. 8.35 shows this limitation.

Other examples have confirmed the observations described above [9, 85, 126, 193, 195, 351].

As expected, C_4 to C_{16} *n*-alkanols dissolved in MBBA become less effective in their disturbance of the nematic phase with increasing chain length [70]. Other authors have studied the systems MBBA/benzene [314, 359], MBBA/biphenyl [303, 304], mixtures of MBBA with spherical aliphatic compounds [209], and similar systems [206a, 227]. J. S. Dave et al. studied the mixed systems of ethyl *p*-azoxybenzoate with nonmesogenic substances and discovered two types of phase diagrams [80]:

a) The usual curve, where the transition line s_A/i is depressed regularly (fig. 8.36).

b) A rising curve exhibiting a maximum that indicates an enhancement in the smectic thermal stability (fig. 8.37). This seems to be particularly true for nitro compounds [283].

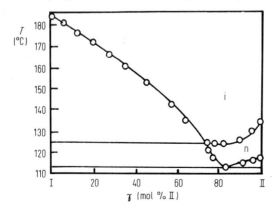

Fig. 8.35. Phase diagram (temperature, *T*, vs. composition, *γ*) of the system *p*-methoxybenzoic acid (I)/PAA (II) (from [79]).

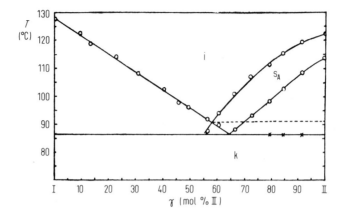

Fig. 8.36. Phase diagram (temperature, T, vs. composition, γ) of the system p-chlorobenzylidene-p-toluidine (I)/ethyl p-azoxybenzoate (II) (from [80]).

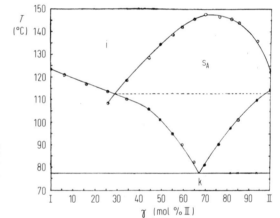

Fig. 8.37. Phase diagram (temperature, T, vs. composition, γ) of the system p-nitrobenzylidene-p-phenetidine (I) / ethyl p-azoxybenzoate (II) (from [80]).

It has been long known that two individual nonenantiotropic or nonmesogenic substances can yield an enantiotropic nematic mixture, as the example from Dave et al. in fig. 8.38 indicates [83, 194, 195, see also 123].

It should also be mentioned here that vapor-pressure measurements of a volatile component dissolved in a nematic liquid crystal allow the detection of a solute induced n/i transition

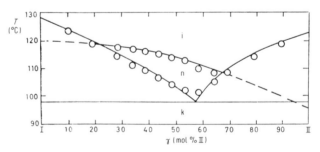

Fig. 8.38. Phase diagram (temperature, T, vs. composition, γ) of the system anisal-p-phenetidine (I)/p-nitrobenzylidene-p-phenetidine (II) (from [76]).

[236]. Since these phenomena play a role in gas chromatography, they will be discussed in detail in section 9.1.

8.8.2 *Mixtures of Two Enantiotropic Mesogenic Substances*

The diagrams of figs. 8.39 and 8.40 are typical for mixtures of two nematics [142].

Other examples of ideal behavior in mixed liquid crystals with a straight n/i transition line have been found [78, 96, 143, 144]. However, the curves often deviate from the linear due to difficulties in packing dissimilar molecules, for which a statistical theory has been

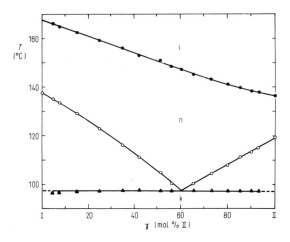

Fig. 8.39. Phase diagram (temperature, T, vs. composition, γ) of the system PAP (I)/ PAA (II) (from [142, see also 172, 317]).

suggested [199]. As fig. 8.40 shows, this deviation is expressed at most by a flattened minimum [91, 96, 141, 143, 144]. Since the mixtures also usually show a pronounced eutectic, the mesomorphous range of suitable mixtures is notably larger than that of the pure components. This fact was recognized early [1:146, 1:156, 1:205, 1:242, 2:575, 2:576, 269] and is of importance today with respect to technical applications [2:91, 2:117, 2:162, 2:163, 2:298, 2:350, 2:370, 2:436, 2:692, 2:694, 2:695, 2:837, 46, 67, 68, 111, 149, 219a, 220, 300c], see section 14.4.

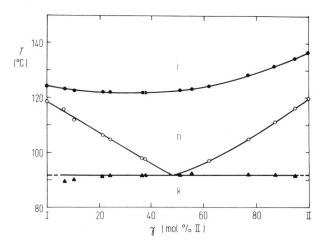

Fig. 8.40. Phase diagram (temperature, T, vs. composition, γ) of the system p,p'-n-dipropoxyazoxybenzene (I)/PAA (II) (from [142, see also 352]).

The MBBA/EBBA system deserves special interest [316]. Due to charge transfer complexing, the binary mixture MBBA(donor) / 4-cyano-4′-pentylbiphenyl(acceptor) exhibits maxima of transition temperatures at a 1:1 molar mixture, two eutectics, and other nonlinear properties [234], see fig. 8.41. Even a nonmesogenic solute may increase the n/i transition temperature

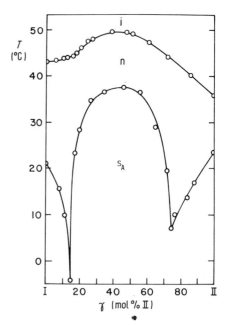

Fig. 8.41. Phase diagram (temperature, T, vs. composition, γ) of the system MBBA (I)/p-cyano-p′-pentylbiphenyl (II) (from [234]).

when the solute and solvent enter into a donor-acceptor interaction [235]. Still other reasons have been discussed and smectic mesomorphism can be induced by mixing "incompatible" nematics [69, 69a, 134a, 226b]. Such anomalous behavior seems to be the rule with mixtures consisting of cyano compounds and nonpolar or slightly polar nematogens if a certain chain length is exceeded [134a].

Nematic mixtures with more than two components are often used, and phase diagrams of such systems are described in the literature [2:91, 221, 334, 335, see also 129]. An apparent paradox has been repeatedly discussed, i.e. the question of why it should be thermodynamically possible that a two-component mixture of practically any desired composition at a constant temperature (i.e., the clearing point of the mixture) can instantly pass into a liquid-crystalline phase of exactly the same composition without forming any two-phase (solid/liquid) range. The regular appearance of a two-phase range, first postulated by A. C. de Kock, could not be experimentally reproduced [2:150]. J. S. Dave's explanation, with which H. Sackmann agreed, seemed plausible: the composition of coexisting fluid phases, x_i and x_a, obeys the relation

$$x_i - x_a = \frac{dT/dx}{T \cdot (\partial \mu_1/\partial x - \partial \mu_2/\partial x)} \cdot (H_i - H_a)$$

Thus, except for the trivial case where $dT/dx = 0$, the equality $x_i = x_a$ is dependent on the heat of transition, $H_i - H_a$. Since this is an order of magnitude smaller than that of the solid, the effects are usually small. In many mixtures, J. S. Dave et al. could find no indication of two fluid phases, isotropic and the other anisotropic, coexisting in a series of mixed crystals corresponding to solid/liquid "lenses" [74, 75]. The behavior of transition heats has also been studied in mixed liquid crystals [1, 142]. The approach from a first to a second order of the n/s_A transition can be observed by using appropriate mixtures (see p. 370).

Binary mixtures containing n, s_A, and s_C phases have found special interest in search of polycritical points, second-order triple points and Lifshitz points [62, 133c, 157a, 205a].

As fig. 8.42 illustrates, there are no new concepts set forth in the phase diagrams of systems with cholesteric components. Cholesteryl myristate and cholesteryl nonanoate form ideal "co-mesophases" because the cholesteric and smectic mesophases coexist through the total concentration range [243]. Note in fig. 8.42 that the solid phase comprises a solid solution rather than pure myristate in the range of 0–60% myristate [242].

The cholesteryl oleate/linolenate system yielded a continuous series of solid solutions over the entire composition interval and a continuous linear behavior of the mesophase transitions [244]. In other cases, solid solutions are rarely observed in the phase diagrams of mesogens. The literature describes other mixtures containing cholesteryl esters [115, 116, 119, 121, 127, 220, 243, 314b, 369, see also 2:93, 2:241, 7:125].

Fig. 8.43 illustrates two phase diagrams for mesogenic carboxylic acid salts.

Fig. 8.43A shows ideal miscibility since each mesophase in one substance has an equivalent in the other, but this is no longer completely the case in fig. 8.43B.

The literature contains phase diagrams of systems containing other carboxylic acid salts [1:322, 1:323, 2:1b, 355].

An important enrichment of our knowledge of smectic phases is contributed by the work of H. Arnold and H. Sackmann, who in 1959 began to systematically investigate the miscibility of liquid crystals as a criterion of isomorphism [1:305–1:337]. Fig. 8.44 shows

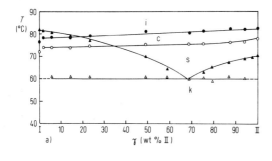

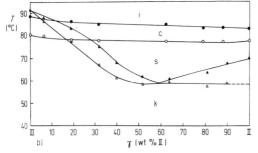

Fig. 8.42. Phase diagrams (temperature, T, vs. composition, γ) of the systems. a) cholesteryl stearate (I)/cholesteryl myristate (II); b) cholesteryl undecanoate (III)/cholesteryl myristate (II) (from [242]).

an example of a system with continuous miscibility in the isotropic-liquid, nematic, smectic, and crystalline-solid states [1:306]. Fig. 8.45 shows a system that illustrates how the addition

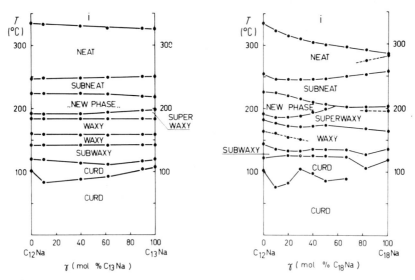

Fig. 8.43. Phase diagrams (temperature, T, vs. composition, γ) of the systems. a) sodium laurate ($C_{12}Na$)/ sodium tridecanoate ($C_{13}Na$); b) sodium laurate ($C_{12}Na$)/sodium stearate ($C_{18}Na$) (from [230]).

Fig. 8.44. Phase diagram (temperature, T, vs. composition, γ) of the system p,p'-n-dinonyloxyazoxybenzene (I)/p,p'-n-dioctyloxyazoxybenzene (II) from [1:306]).

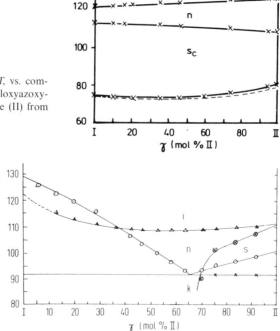

Fig. 8.45. Phase diagram (temperature, T, vs. composition, γ) of the system anisal-p-phenetidine (I)/p-n-dodecyloxybenzylidene-p-n-butoxyaniline (II) (from [82]).

of a nematic component to a smectic one reduces the range of existence of the latter [82, 84].

Numerous later investigations have supported Arnold's thesis that smectic phases are selective in their miscibility. They have also proven true the widely-held conclusion that all continuously miscible smectic modifications can be grouped into one phase type. Great amounts of data have been gathered toward this end, and we can illustrate here only a few of the most interesting diagrams. Unlimited and limited miscibility of smectic modifications are illustrated in fig. 8.46. In addition, the diagrams in figs. 8.47 and 8.48 show that phases can appear in mixtures that are not to be found with the pure substances.

Mixtures of *p*-(*p*-hexyloxybenzylidene)-aminobenzonitrile and *p*-cyanobenzylidene-*p'*-*n*-octyloxyaniline are noteworthy for the occurance of a nematic phase at both a higher and a lower temperature than the smectic phase (socalled reentrant nematics) [71].

Figs. 8.49 to 8.52 illustrate the introduction of additional smectic phases and simultaneously document the arrangement of the smectic phases according to decreasing energy content in the order A-D-C-B-E.

Fig. 8.52 is an example with numerous smectic phases, and also illustrates that two substance possessing only different smectic modifications may show limited miscibility only in the mesomorphous state.

Phase diagrams of mixtures containing s_F and s_G phases have recently been reported [47], and the conjecture of a complete miscibility between s_B and s_H phases has been disproved [97], fig. 8.53.

As already mentioned, the works of H. Sackmann, usually in collaboration with H. Arnold or D. Demus, would alone fill a book. For this reason we can refer only to the original papers concerning other mixed smectic systems [1:305–1:337, 2:152, see also 2:49, 2:182, 2:291, 2:295, 2:300, 2:537, 2:609, 2:652, 2:834, 2:183b, 5:124a, 66, 94].

Mesophase transitions in the supercritical immiscible region of binary systems have been discussed in the literature [8]. It should be noted that the classification of smectic phases based on the thermodynamic criterion of miscibility has been supported by many X-ray structural studies (cf. section 5.2.3). Mixtures of rodlike and platelike mesogens have been already considered theoretically [6] and studied experimentally [133a].

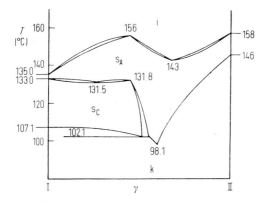

Fig. 8.46. Phase diagram (temperature, *T*, vs. composition, *γ*) of the system di-*n*-hexadecyl *p,p'*-azoxycinnamate (I)/methyl *p*-ethylmer-captobenzylidene-*p*-aminocinnamate   (II) (from [1:312]).

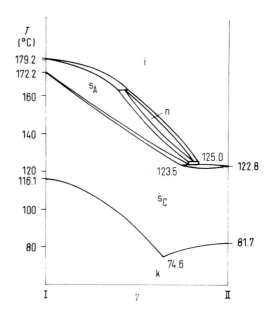

Fig. 8.47. Phase diagram (temperature, T, vs. composition, γ) of the system 2,5-bis-(4-n-hexylphenyl)pyrazine (I)/4,4′-n-didodecyloxy-azoxybenzene (II) (from [1:328]).

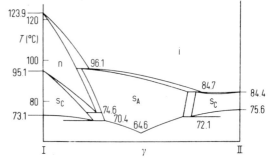

Fig. 8.48. Phase diagram (temperature, T, vs. composition, γ) of the system p,p'-n-diheptyloxyazoxybenzene (I)/di-n-hexadecyl p,p'-azoxy-α-methylcinnamate (II) (from [1:311]).

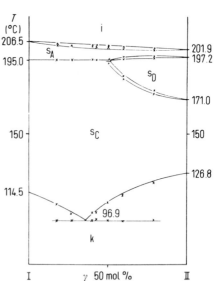

Fig. 8.49. Phase diagram (temperature, T, vs. composition, γ) of the system 4′-n-tetradecycloxy-3′-nitro-biphenyl-4-carboxylic acid (I) / 4′-hexadecyloxy-3′-nitrobiphenyl-4-carboxylic acid (II) proving the sequence A–D–C (from [1:320]).

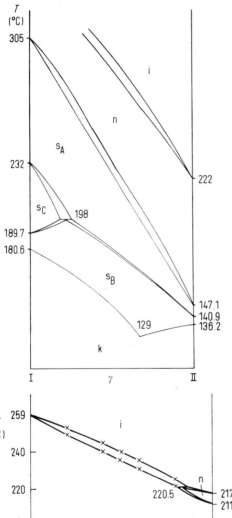

Fig. 8.50. Phase diagram (temperature, T, vs. composition, γ) of the system terephthal-bis-(ethyl-p-aminocinnamate) (I)/terephthal-bis-(p-methoxymethyleneoxyaniline) (II) proving the sequence C–B (from [1:319]).

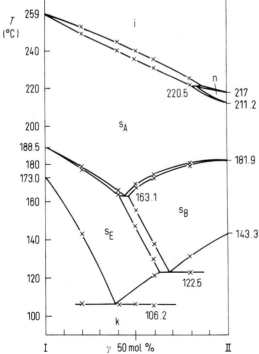

Fig. 8.51. Phase diagram (temperature, T, vs. composition, γ) of the system diethyl p-terphenyl-4,4″-dicarboxylate (I)/ethyl 4-(4-phenylbenzylidene)aminocinnamate (II) proving the sequence B–E (from [1:331]).

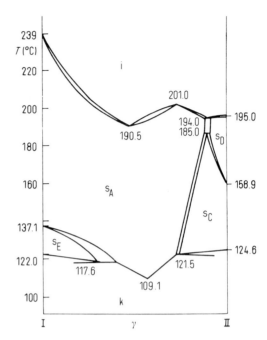

Fig. 8.52. Phase diagram (temperature, *T*, vs. composition, *γ*) of the system *n*-dipropyl *p*-terphenyl-4,4''-dicarboxylate (I) / 4'-*n*-octadecyloxy-3'-nitrobiphenyl-4-carboxylic acid (II) (from [1:331]).

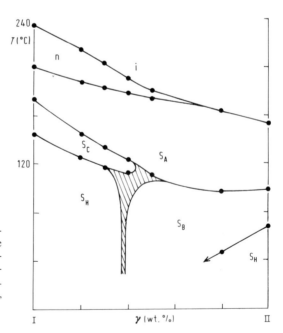

Fig. 8.53. Phase diagram (temperature, *T*, vs. composition, *γ*) of the system TBBA (I) / 4-*n*-octylphenylester of 4'-*n*-octyloxybiphenyl-4-carboxylic acid (II) (from J. W. Goodby, G. W. Gray, J. Phys. (Paris), C-3, *40*, 363 (1979)).

Liquid Crystals as Solvents in Gas Chromatography and in Chemical Reactions

9

9 Liquid Crystals as Solvents in Gas Chromatography and in Chemical Reactions

9.1 Liquid Crystals in Gas Chromatography

Work in gas chromatography using liquid crystals as stationary phases began in 1960. The first practical problem was the analytical separation of m- and p-xylene, a system typical of substituted aromatic compounds. The situation is illustrated by the boiling and melting points of the components to be separated (table 9.1). Note the extremely different melting points and the almost identical boiling points of the m- and p-isomers. The relatively high melting point of p-xylene is indicative of a high lattice energy. It is well known that their separation and purification can easily be performed using cryogenic techniques, but distillation is extremely difficult when the separation is based only upon the volatility ratio of pure m- to p-xylene. Thus, the problem was finding a solvent system in which this ratio could be altered by specific solute/solvent interaction.

Table 9.1. Melting and boiling points of isomeric xylenes.

Isomeric xylenes	m. p. (°C)	b. p. (°C)
o-xylene	-25.3	144
m-xylene	-53.5	139
p-xylene	$+13.2$	138

9.1.1 Experimental Results

There are adsorption systems, such as modified Bentone [133], which show a typical affinity for these two different isomers, but it was an interesting question as to how liquid-crystalline phases would perform in this respect. The most useful effect for analytical purposes is the enhancement of solvent affinity for p-isomers compared with m-isomers. This effect is demonstrated in fig. 9.1 which shows the separation of the xylenes using nematic PAP.

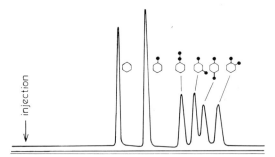

Fig. 9.1. Separation of aromatic hydrocarbons by gas chromatography with nematic PAP at 140 °C (from [89]).

The isotropic liquids and adsorbents (table 9.2) used to separate m- and p-xylenes retain the m-isomer more strongly than the p-isomer. With a nematic solvent, e. g. PAP, the reverse is observed. An important property to the analyst is the much higher absolute value of the activity coefficient in the nematic phase. This results in a shorter analysis time.

Table 9.2. Affinity of m- and p-xylene for different stationary phases.

Stationary phase	$\dfrac{V_g \, (meta)}{V_g \, (para)}$	$f_2^{\prime} \; meta$	$f_2^{\prime} \; para$
a) phenanthrene (110 °C)	1.055	1.439	1.476
b) 7,8-benzoquinoline (100 °C)	1.061	1.608	1.667
c) "bentone 34"/silicon oil (70 °C)	1.13	—	—
(90 °C)	1.05	—	—
d) PAP (140 °C)	0.918	2.761	2.51

Preference has been given to the separation of position isomers of aromatic compounds. Some examples of aliphatic substances have additionally been found that were not easily separated using isotropic phases, but which have been split into their components successfully by using mesophases. Examples are the system oleic acid/elaidic acid, and the homologs:

a) H_3C-N ⟨ ⟩ $\overset{\displaystyle O}{\overset{\|}{C}}-O-CH \overset{CH_3}{\underset{CH_3}{}}$, C_6H_5
b) H_3C-N ⟨ ⟩ $\overset{\displaystyle O}{\overset{\|}{C}}-O-CH_2-CH_3$, C_6H_5

having been separated with a relative retention of $r_{a,b} = 1.15$ using 4,4′-bis-benzylidene-benzidine as solvent. Enantiomorphic amino acids have been successfully resolved into their optical antipodes [112, 114], and the separation of α- from β-naphthol is a typical example in the analysis of substituted naphthalenes. Another stationary phase is the bis-phenetidyl-terephthalic aldehyde:

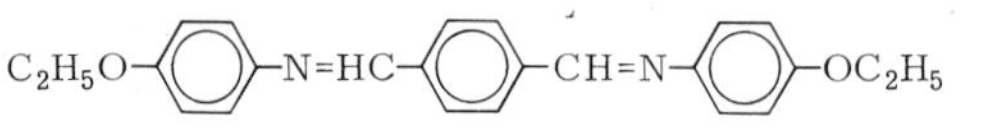

$$C_2H_5O-\langle\bigcirc\rangle-N=HC-\langle\bigcirc\rangle-CH=N-\langle\bigcirc\rangle-OC_2H_5 \qquad k\ 198\ n\ 323\ i$$

which has an optimal working temperature of ca. 210 °C and is recommended for numerous solute systems. These systems include intermediates for dyestuffs and other industrial products. Table 9.3 shows the results obtained with this compound and with the two solvents mentioned above. Other suitable substances with closely related structures have been synthesized and tested, e.g. phenylcarboxylic acid esters by J. P. Schroeder [44, 45] and, quite recently, others by L. E. Cook and R. C. Spangelo [37].

Table 9.3. Relative, V_r, and specific retention volumes, $V_g^{(T)}$, from [92]. The relative values are referred to dimethyl *o*-phthalate.

Solvent and temperature	1) bis(phenetidyl)terephthalaldehyde *T*: 206 °C					
	2) 4,4′-bis(benzylidene)benzidine *T*: 240 °C					
	3) bis(*p*-methoxybenzylidene-*trans*-4,4′-diaminostilbene *T*: 270 °C					
Solute	1) V_r	1) $V_g^{(T)}$ (ml/g)	2) V_r	2) $V_g^{(T)}$ (ml/g)	3) V_r	3) $V_g^{(T)}$ (ml/g)
n-hexadecane	0.21	33	0.22	30	0.13	4.4
methyl laurate	0.33_5	53	0.35	48	0.19	6.5
dimethyl sebacate	1.0_1	161	0.98	133	0.72	25
o-xylene	0.054	9	0.06	8	0.06	2
		m: 7				
		p: 8				
o-cresol	0.22_5	36	0.213	29	0.19_4	6.6
m-cresol	0.28	$44._5$	0.26	$35._4$	0.36	12
p-cresol	0.29	46	0.25_4	$34._5$	0.36	12
o-dimethoxy benzene	0.20	32	0.22_8	31	0.25	8.5
m-dimethoxy benzene	0.25	40	0.26_7	$36._3$	0.31	10.5
p-dimethoxy benzene	0.30	48	0.27_5	$37._4$	0.37	12.5
o-dimethyl phthalate	1.00	159	1.00	136	1.00	34
m-dimethyl phthalate	1.4_2	226	1.2_4	169	1.2_5	$42._5$
p-dimethyl phthalate	1.8_4	293	1.2_7	173	1.5_3	52
o-dibromo benzene	0.41	65	0.43	58	0.45	15
m-dibromo benzene	0.35	56	0.35	48	0.38	13
p-dibromo benzene	0.41	65	0.37	50	0.37	$12._5$
o-phenyl phenol	1.7_1	272	1.57	214	1.62	55
p-phenyl phenol	7.8	1 240	—	—	—	—
2,3-dichloro toluene	0.22	35	0.25	34	0.30	10

Table 9.3. (Continued.)

Solvent and temperature	1) bis(phenetidyl)terephthalaldehyde *T*: 206 °C
	2) 4,4′-bis(benzylidene)benzidine *T*: 240 °C
	3) bis(*p*-methoxybenzylidene-*trans*-4,4′-diaminostilbene *T*: 270 °C

Solute	1) V_r	1) $V_g^{(T)}$ (ml/g)	2) V_r	2) $V_g^{(T)}$ (ml/g)	3) V_r	3) $V_g^{(T)}$ (ml/g)
2,4-dichloro toluene	0.18	29	0.20	27	0.22	7.5
2,5-dichloro toluene	0.19	30	0.21	28.$_5$	0.22	7.5
2,6-dichloro toluene	0.18	29	0.20	27	0.23	8
3,4-dichloro toluene	0.23	37	0.26	35	0.29	10
3,5-dichloro toluene	0.16	25	0.19$_5$	26.5	0.21	7
1,2-dimethyl-3-nitrobenzene	0.56	89	0.60	82	0.65	22
1,2-dimethyl-4-nitrobenzene	0.93	143	0.91	124	1.03	35
1,3-dimethyl-2-nitrobenzene	0.59	94	0.60	82	0.67	23
1,3-dimethyl-4-nitrobenzene	0.27	43	0.32	44.$_5$	0.29	10
1,4-dimethyl-2-nitrobenzene	0.50	79.$_5$	0.52	71	0.55	19
1-nitro-2-ethylbenzene	0.34	54	0.39	53	0.40	13.$_5$
naphthalene	0.37	59	0.38	52	0.45	15
α-naphthol	3.0$_4$	484	—	—	—	—
β-naphthol	3.9$_3$	625	—	—	—	—
1-chloronaphthalene	0.83	132	0.86	117	1.0$_4$	35.$_5$
2-chloronaphthalene	0.96	153	0.87	118	1.0$_7$	36.$_5$
diphenyl	0.74	118	0.70	95	0.76	26
diphenylmethane	0.60	96	0.69	94	0.61	21
trans-stilbene	4.0	640	2.85	388	2.75	93.$_5$
acenaphthene	1.31	208	1.28	174	1.37	46.$_5$
phenanthrene	7.0	1 100	5.32	724	5.6$_8$	193
anthracene	9.6	1 530	6.00	816	6.7$_4$	229
pyridine	0.15	24	0.05	7	0.07	2.$_5$
quinoline	0.61	97	0.60	82	0.80	27
α-benzylpyridine	0.90	143	1.00	136	1.01	34
γ-benzylpyridine	1.26	200	1.5$_2$	207	1.5$_8$	54
carbazole	16.4	2550	9.8	1 330	11.6	390
4-methyl-acetidinone-2	0.31	49	0.22	30	0.38	13
o-nitrobenzaldehyde	0.94	149				
m-nitrobenzaldehyde	1.47	234				
p-nitrobenzaldehyde	1.62	258				
o-nitrobenzotrifluoride	0.26	41				
m-nitrobenzotrifluoride	0.20	32				
p-nitrobenzotrifluoride	0.23	36.5				
o-trifluoromethylphenylisocyanate	0.16	25				
m-trifluoromethylphenylisocyanate	0.14	22				
p-trifluoromethylphenylisocyanate	0.14	22				

We shall now discuss the gas-liquid partition equilibrium of anisotropic phases and its relationship to the structural attributes of nematogens. The most important relationship which follows from gas chromatography experiments is well-known, and it allows the calculation of the Raoult-Henry activity coefficient, f_2^∞, from GC-experiments [121]:

$$f_2^\infty = \frac{RT}{V_g^{(T)} M_1 p_2^0} = \frac{273\,R}{V_g^{(N)} M_1 p_2^0} \tag{9.1}$$

The meaning of the symbols is given in table 9.4. From the temperature dependence of f^∞, the molar excess enthalpy and the molar excess entropy of mixing, h^E resp. s^E, may be calculated.

f^∞ is the activity coefficient for an infinitely dilute solution, but different notations are frequently used. Thus we often find γ_2^0 instead of f_2^∞, but both symbols have the same meaning [59, 142, 143]. However, we prefer f_2^∞ because γ is also used with respect to a different standard state, cf. E. A. Guggenheim [59]. The other symbols, definitions, and relationships used in this discussion are summarized in table 9.4.

Table 9.4. Symbols and relationships.

Temperature dependence of the retention volume:

$$\frac{\partial \ln V_g^N}{\partial 1/T} = \frac{\Delta_v H}{R} = \frac{\Delta_{ev} H - h^E}{R} \tag{9.2}$$

Temperature dependence of the activity coefficient:

$$\frac{\partial \ln f_2^\infty}{\partial 1/T} = \frac{h^E}{R} \tag{9.3}$$

Relationship between enthalpy of volatilization, $\Delta_v H$, evaporation, $\Delta_{ev} H$, and mixing, h^E:

$$H = \Delta_{ev} H - h^E \tag{9.4}$$

Relationships between excess free enthalpy, entropy, and activity coefficients:

$$\left(\frac{\partial G_{(mix)}^E}{\partial X_2}\right) = \mu_2^E = RT \ln f_2^\infty = h^E - T s^E \tag{9.5}$$

with

$$G_{(mix)}^E = \Delta_{(mix)} G - \Delta_{(id\,mix)} G \tag{9.6}$$

$$\mu_2^E = G_{(mix)}^E + (1 - X_2)\,\frac{\partial G_{(mix)}^E}{\partial X_2} \tag{9.7}$$

$$\left(X_2 \to 0 \ldots G_{(mix)}^E \to 0, \text{ but } \frac{\partial G_{(mix)}^E}{\partial X_2} \neq 0\right)$$

Table 9.4. (Continued.)

where

$G^E_{(mix)}$ is the excess free enthalpy of mixing and
$\Delta_{(mix)} G$ is the free enthalpy change after mixing.

V^T_g	specific retention volume (T)
V^N_g	specific retention ("reduced", 273 K)
M_1	molar mass of solvent ("1")
P^0_2	vapor pressure of solute ("2")
f^∞_2	activity coeff. of solute ($\equiv \gamma^0_2$)
K	partition coefficient (Ostwald)
C^{fl}_2	concentration (molar basis), liquid phase
C^g_2	concentration (molar basis), gas phase
$\rho(T)$	density
$\Delta_v H$	molar enthalpy of "volatilization" (from solution)
$\Delta_{ev} H$	molar enthalpy of evaporation (pure component)
h^E	molar enthalpy of mixing (identical with excess enthalpy of mixing)
s^E	molar excess entropy of mixing
μ^E, G^E	molar excess free enthalpy, as defined above
δ_i	solubility parameter of a component i (Hildebrand): $\delta_i = \left(\dfrac{E_i}{V_i}\right)^2$
E_i	evaporation energy of a component i
V_i	molar volume of a component i
S	degree of order, def. as $S_i = 1 - 3/2\,\overline{\sin^2\theta}$ or: $\overline{P^{(i)}_2}$ *
E_K	"Configuration energy" by Saupe $E_K = -\dfrac{N}{m} \cdot \dfrac{A}{2} \cdot \dfrac{S^2}{V^2}$, where m is the number of the next neighbors in a "statistical short range order" group. A is a general constant.

* Legendre-Polynomial of 2nd degree

It is necessary to explain the true meaning of the activity coefficient, f^∞_2. Compare an ideal solution of the solute (2) with a real one. Their partial pressures, p_2, are equal to $x_2 p^0_2$, or, using the Henry constant, k, we have

$$x_2 f^\infty_2 p^0_2 = k_2 x_2$$

Thus, it is easy to see that

$$f^\infty_2 = k_2/p^0_2$$

and μ^E is the molar free-energy change when going from p^0_2 to $f^\infty_2 p^0_2$, the "maximum work" being

$$\mu^E = R T \ln p^0_2 f^\infty_2 /p^0_2 = R T \ln f^\infty_2$$

Now the question is raised as to what kinds of general behavior can be considered as typical of mesomorphous phases. The previously mentioned "selectivity" for *m-/p-*isomers is certainly the most striking difference when compared with that of an isotropic phase, but this is only a secondary effect. The basic phenomena will be described immediately below, and the theoretical aspects that lead to an explanation of the effects will be discussed later. Reviews on the application of liquid crystals in gas chromatography have been published in the literature [31, 81, 156, 183a]. All the phenomena being typical for the use of liquid crystals in GC can be summarized by the following statements.

9.1.1.1 First Statement

Experimental data thus far available leads to the conclusion that liquid-crystalline phases, when used as solvents, always exert a reduced affinity for the solute when compared with the isotropic phases under the same conditions.

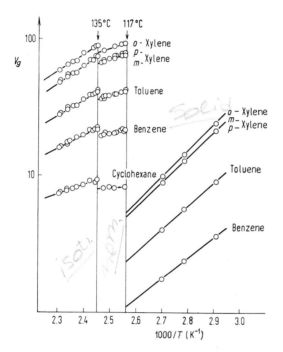

Fig. 9.2. Specific retention volumes, V_g, vs. the reciprocal temperature, $1000/T$, in the solid, nematic, and isotropic phases of PAP (from [88]).

Fig. 9.2 shows the transition from the crystalline solid to the nematic mesophase, and the transition from the nematic mesophase to the isotropic melt as reflected by gas-chromatographic measurements of solute activity, i.e., the "specific retention volume" as a function of $1/T$. The melting point is characterized by a strong change in affinity, and this marks the change from an adsorption process to a partition one. To a lesser extent, the same is observed at the n/i transition where two different partition coefficients interchange to give relatively sharp maxima in the curves. Maxima like these have also been observed in nonmesomorphous systems, e.g., in the change of modifications of solids into other modifications [1, 49, 65, 66]. Guillet et al. [60–64] and others [110, 118] have recently reported additional similar observations very closely related to liquid-crystal phase transitions (fig. 9.3).

The general behavior illustrated in fig. 9.2 has been frequently confirmed, for example by D. E. Martire et al. [120] and E. M. Barrall II et al. [18] on cholesteryl esters, especially

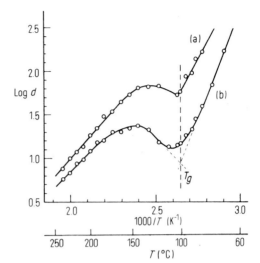

Fig. 9.3. Retention times for *n*-hexadecane on a polymethylmethacrylate column (0.357 g on Chromosorb W, two plotting methods), time in arbitrary units (from [61]).

those exhibiting more than one mesophase. Further examples are PAA [88], PAP [88, 89], and their eutectic mixture [154, 155]; the system PAP/cholesteryl benzoate [91]; *p,p'*-dihexyloxy-azoxybenzene [24c]; 4-methoxybenzylidene-4'-cyanoaniline; 4-methoxybenzylidene-4'-acetoxy-aniline; and 4-methoxybenzylidene-4'-butyryloxyaniline [92]. The last example is illustrated in fig. 9.4. Thus the solubility behavior as measured by the gas chromatographic retention offers a general method for detecting structural changes in a solvent.

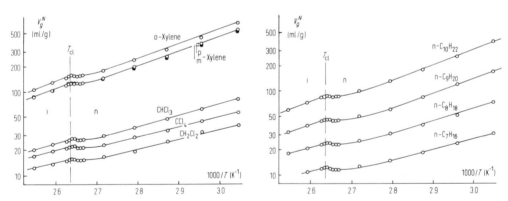

Fig. 9.4. Specific retention volumes, V_g^N, vs. the reciprocal temperature, $1000/T$, for methylene chloride, chloroform, carbon tetrachloride, xylene isomers (left fig.), and some *n*-alkanes (right fig.) on 4-methoxy-benzylidene-4'-butyryloxyaniline (from [93]).

Smectics A exhibit a different behavior. There is no maximum but merely a change in slope (fig. 9.5). Eight transitions have been detected in solid and mesomorphous sodium stearate taking the log V_g^T vs. $1/T$ plots [181].

The above differences in partition behavior have been observed in a random assortment of substances such as toluene and xylene. Now even more interesting questions are raised.

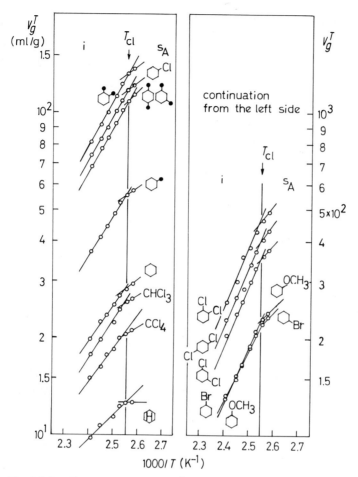

Fig. 9.5. Specific retention volumes, V_g^T, vs. the reciprocal temperature, $1000/T$, of different solutes in diethyl *p*-azoxybenzoate (from [89]).

Do differences exist among different classes of solutes? Are these identical in isotropic and anisotropic solvents? The "*m-p*-selectivity" is one difference that has already been mentioned. However, if this effect is neglected as being largely limited to the vicinity of the melting point and diminishing with rising temperature until it vanishes at the transition temperature, T_c, then the behavior discussed below is observed.

9.1.1.2 Second Statement

The general behavior of the solute activity, the form of temperature dependence (especially in the vicinity of the clearing point), and the relative enhancement of the partition coefficient at the n/i transition temperature are all characteristics of the solvent alone (figs. 9.4 and 9.5), and they are thus largely independent of the solute species [93]. The smaller the slope of the curve, the lower the absolute value of the partition coefficient. As will be seen later, this rule governing the relationship between entropy and energy is quite general. Note that the extrapolated absolute values of V_g (or partition coefficients) are smaller in the anisotropic phase, and the slope is steeper in the isotropic phase. This means that there is a higher (stronger endothermic) heat of mixing in the mesophase than in the isotropic phase. The heats of mixing in mixed mesophases are generally lower than is the case with pure components.

9.1.1.3 Third Statement

The linear relationship between $\log V_g^N$ and $1/T$, which holds well in the lower temperature ranges up to a few degrees below the transition, is no longer valid upon approaching the clearing point. A typical pretransition period exists in which V_g^N (i.e., the affinity vs. the solvent) rises with rising temperature. This leads, as mentioned previously, to the maximum at T_c. An increase of the retention volume with temperature is indeed a curious phenomenon. Such pretransition effects have been observed in numerous other physical qualities that are dependent on the degree of order, e.g., specific volume, viscosity, diffusion, conductivity, and absorption of electromagnetic energy (UV, IR, NMR, and dielectric constant).

Going back to our systems, we can now state that the activity coefficient reflects the temperature-dependent degree of order. A few degrees below the n/i transition point there is observed an analogous increase in the heat of mixing, h^E, which becomes larger and larger and even changes sign ($h^E > \Delta^V H$) while approaching the transition point. The process of mixing solute and solvent obviously needs more and more energy (latent heat) to the point where the amount of energy exceeds even the evaporation enthalpy of the pure solute! This shows clearly the disordering of the solvent which is obviously an entropic effect.

In the search for a simple explanation of such an effect, a direct relationship to the specific volume seems obvious. The specific volume is indeed closely related to the interaction energy, the socalled configuration energy of the mesophase. It is clear that a higher density corresponds to a higher resistance to the penetration of the solvent by solute molecules. Sharp discontinuities at the freezing point of isotropic solvents commonly used in gas/liquid chromatography are also well known [36]. Details of theoretical considerations and models will be discussed later.

9.1.1.4 Summary

In summary, the following are the most important properties of a mesogenic solvent:
a) The "anisotropic" activity coefficient, ^af, of any solute is higher than the "isotropic" one when extrapolated to the same temperature. The ratio $(^if/^af)_{T_c}$ is in the first approximation independent of the solute.
b) The activity coefficient of a given p-isomer is always smaller than those of the corresponding m- and o-isomers.
c) The heat of mixing, $\dfrac{\partial \log f_2^c}{\partial 1/T}$, is higher (more endothermic) in the liquid-crystal phase system than in the isotropic solution. The larger the h^E, the larger the free-energy term $RT \log f_2^c$, and also the larger the s^E.
d) The n/i transition begins some degrees below T_c, this "pretransition period" being reflected by the specific retention volume.
Table 9.5 reviews substances and systems under GC-investigation with respect to any mesomorphous behavior.

Table 9.5. Solvents, solutes, and analytical problems being investigated by gas chromatography using anisotropic liquids.

Stationary phase	Object, problem, and literature
PAA	*m*-, *p*-xylene [88, 89]; *m*-, *p*-dialkylbenzenes [135]; vapor pressure measurements [162]
PAP	mono-, dimethylnaphthalenes [174]; retention on solid PAP [173]; and eutectic mixtures [175, 176]; alkenes and cycloalkenes [98]
eutectic mixture of PAA/PAP	general questions [91]; *m*-, *p*-xylene [55a, 177]; aromatic hydrocarbon mixtures [119, 179]
comparison of an isotropic phase (Apiezon L) with PAA	aromatic compounds of the type *o*-, *m*-, *p*-xylene [90]
p,p'-dihexyloxyazoxybenzene	divinylbenzenes, thermodynamics [123]
p,p'-dialkoxyazoxybenzenes and their mixtures	xylenes, general investigations, phase diagrams [130, 155]
p,p'-diheptyloxyazoxybenzene and *p*-(*p*-ethoxyphenylazo)-phenylundecylenate	*n*-alkanes, cyclohexane, benzene, xylenes [28]
bis-(phenetidyl)-terephthalaldehyde and other Schiff bases	see table 9.3 [92]
N,N'-bis(*p*-methoxybenzylidene)-α,α'-bi-*p*-toluidine	separation of 2–6 ring polycyclic aromatic hydrocarbons including carcinogenic compounds [82, 83]; of isomeric alkylnaphthalenes [180]; of androstanol and cholestanol epimers [184]; of isomeric benoxaprofens

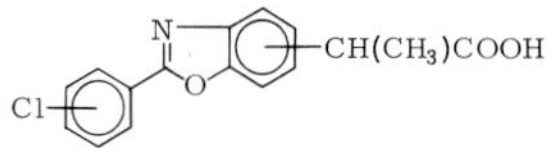

[67]; and azaheterocyclics [136]

N,N'-bis(*p*-phenylbenzylidene)-α,α'-bi-*p*-toluidine and N,N'-bis(*p*-hexyloxybenzylidene)-α,α'-bi-*p*-toluidine	polycyclic aromatic hydrocarbons including carcinogenic compounds and cigarette smoke [83a, 84]

esters of the type

aromatics, homologous series [2:391]

esters of the type

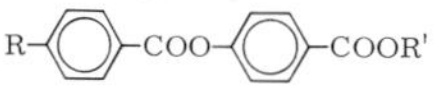

aromatics, basic investigations [44–46, 154]

PAA, PAP, and 4,4'-biphenylene bis-(*p-n*-heptyloxybenzoate):

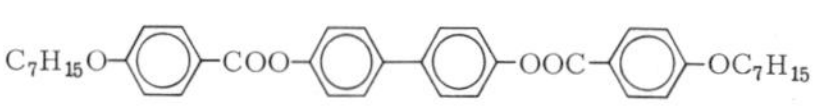

o-, *m*-, *p*-isomers of: xylene, diethylbenzene; cresol, its methyl ethers and acetates; fluorochlorobenzene; fluorobromobenzene; fluoroiodobenzene; fluoroanisole; fluorobenzylchloride; fluorobenzaldehyde; fluoronitrobenzene; dichlorobenzene; chlorotoluene; chloroaniline; chloromethylbenzoate; bromoanisole; trifluoromethylnitrobenzene; dimethoxybenzene; diethoxybenzene; methyl methylbenzoate; methyl chloromethylbenzoate; methyl nitrobenzoate; methylaniline; *m*-, *p*-di-*i*-propylbenzene [144]

Table 9.5. (Continued.)

Stationary phase	Object, problem, and literature
di(*p*-methoxyphenyl)-*trans*-cyclohexane-1,4-dicarboxylate	thermodynamic investigations, transition phenomena [148]
2,6-naphthalene-bis-(*p-n*-hexyloxybenzoate): 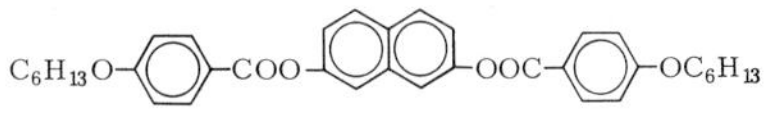 $C_6H_{13}O$—◯—COO—◯◯—OOC—◯—OC_6H_{13}	*o*-, *m*-, *p*-isomers of dichlorobenzene and dimethoxybenzene; *m*-, *p*-isomers or chlorotoluene, cresol methyl ethers, methyltoluate, methylacetophenone [32]; 1- and 2-substituted naphthalenes [34]
1,4-naphthalene-bis-(*p-n*-hexyloxybenzoate): 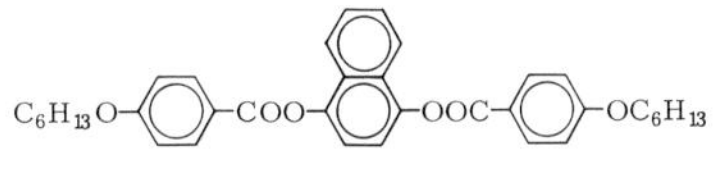 $C_6H_{13}O$—◯—COO—◯—OOC—◯—OC_6H_{13}	solutes similar to those investigated in [32, 33]
p-hexyloxycinnamic acid	fatty acids, thermodynamic relationships [94]
derivatives of 2,5-diphenylpyrimidines	methylanisoles; dimethoxybenzenes; chlorotoluenes, bromonitrobenzenes; *o*-, *p*-iodoanisole; β-, γ-picoline [99]
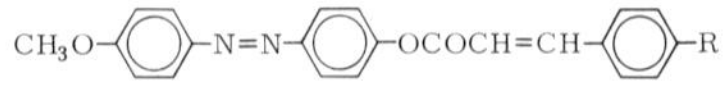 CH_3O—◯—N=N—◯—OCOCH=CH—◯—R	butyrylacetic acid ester, chlorobenzaldehydes, chlorobenzotrichlorides, chlorothiophenolmethyl ethers [94]
(*p*-ethoxyphenylazo)phenyl crotonate	benefin, trifluralin [67a]
(*p*-ethoxyphenylazo)phenyl heptanoate	thermodynamic investigations [85]
(*p*-ethoxyphenylazo)phenyl undecylenate	column efficiency using glass beads as support material [57, 58]
4-hydroxy-4′-methoxyazobenzene 4-methoxycinnamate	separation of oleic, elaidic, and stearic acid; traces of 2-hydroxynaphthalene in 1-hydroxynaphthalene; *o*-, *m*-, and *p*-methylphthalate; 1- and 2-ethylnaphthalene
4-hydroxy-4′-methoxyazoxybenzene 4-(ethoxyethoxy-carbethoxyoxy)cinnamate	phenanthrene, anthracene [140]
smectic carbonyl-bis-(D-leucine isopropyl ester)	enantiomers [112]
smectic carbonyl-bis-(L-valine esters)	enantiomers [113, 114]
p-(*p*-methoxybenzylidene)-aminophenylacetate, *p*-(*p*-ethoxyphenylazo)-phenylundecylenate, *p*-(*p*-ethoxyphenylazo)-phenylheptanoate, *p*-(*p*-methoxybenzylidene)-amino phenylbenzoate, *p*-(*p*-methoxybenzylidene)-amino phenylbenzoate	monosubstituted phenols [37]
p-phenylene-bis-4*n*-heptyloxybenzoate, *p*-(*p*-ethoxyphenylazo)-phenylcrotonate, cholesteryl cinnamate	cymenes; cresols; methylacetophenones; carvestyrene; dipentene; chlorophenols; carvacrols; thymols; ethylnaphthalenes; pyrazines; isoprene-isopropylidene pairs; trienes; epoxides; cyclohexenes (Diels-Alder adducts); nerol; geraniol; α-, β-, γ-*n*- and isomethylionones [141]
"monolayers" of some liquid crystals	*m*-, *p*-xylene [172]
cholesteryl palmitate	thermodynamic quantities of different solutes in the smectic, cholesteric and isotropic phase [86]

Table 9.5. (Continued.)

Stationary phase	Object, problem, and literature
cholesteryl esters and their mixtures	surface effects [122, 125] thermodynamic properties [129, 183]
1.743:1.0 (weight:weight) mixture of cholesteryl chloride and cholesteryl myristate	thermodynamic study of the influence of helical inversion and "nematic point" on solubility properties [129]
cholesteryl benzoate, 5α-cholestan-3β-yl benzoate, cholesteryl p-phenylbenzoate	5α-, 5β-androstane; 5α-, 5β-pregnane; 5α-, 5β-cholestane; Δ⁴-, Δ⁵-androstene; Δ⁴-, Δ⁵-cholestene [96]
PAA	GLC in capillary columns, supercooling, hysteresis effects in transport and distribution coefficients [56]
MBBA and EBBA in glass capillary columns	o-, m-, p-isomers of xylene, chlorotoluene, bromotoluene, methylanisole, dichlorobenzene [100]
4-n-pentyl-acetophenone-O-(4-n-alkoxybenzoyl-oximes) in glass capillary columns	partial molar enthalpies and entropies of solution in a homologous series [101]
p-methoxybenzylidene-p'-n-propylaniline	solute activity coefficients as a function of solute mole fraction and temperature [128]
cinnamic acid esters, cholesteryl myristate, and other cholesteryl esters	influence of electric fields [19, 167, 182]
diverse mesogens	influence of the support material [172a]
standard gas-chromatography columns	liquid-crystal GC-IR cell proposed to fractionate samples, from carrier gas [105]
collagen	behavior in liquid chromatography [150]
7,8-benzoquinoline in the vicinity of the melting point	comparison with mesogens and thermodynamic studies [87, 103]

The literature contains other studies of thermodynamic properties of dilute solutions in MBBA [38, 111] and a review considering orientational order in solution thermodynamics [129a, 166]. No details could be obtained on refs. [12, 178].

9.1.2 Models and Theory

9.1.2.1 Theoretical Background

The use of anisotropic solvents and the theoretical explanations of their behavior are based mainly on the following considerations. These include the fundamental laws and principles cited in the literature [3:267, 8:347, 53, 59, 77, 78, 121, 142, 143].

Similar structures enhance the affinity between solute and solvent. Since nematogens are characterized by a repeated 4,4'-substitution, it was expected that a solute with a similar constitution (e.g., the p-isomers) should approach more closely the state of ideal solution than would be the case with either an o- or a m-isomer. Thus

$$^i\mu_{(p)}^E < {}^i\mu_{(o,m)}^E \tag{9.8}$$

$$^a\mu_{(o,m,p)}^E > {}^i\mu_{(o,m,p)}^E \tag{9.9}$$

If a solvent phase is compressed to a smaller molar volume by "inner pressure", it is expected that the activity will generally be enhanced analogously to well-known thermodynamic relationships such as the influence of pressure on the activity coefficient. Nematic phases are in a denser (and more highly ordered) condition than the coexisting isotropic state, and the activity coefficient can be expected to become larger if the solute passes from the isotropic to the anisotropic state. One could therefore conclude that nematogens allow the change from normal pressure to a high (inner) pressure only through phase transitions. The small but necessary temperature change can be corrected by extrapolation. The same is obviously true when a liquid is compared with its solid crystal phase, but here equilibrium studies (partition!) are not possible because the diffusion is inhibited in the crystal. It would be interesting to discover whether or not any liquid-crystal phases exist that show an "anomalous" volume change such as the case with water: ΔV^{trans}, i.e., $^aV - {}^iV > 0$.

It must be emphasized that f cannot easily be measured as a function of x, although this is possible if static conditions are provided. The tendency can only be observed from the peak shape, i.e., from "tailing" or "fronting". A typical experiment revealed a strong concentration dependence and fronting tendency with the retention volume being larger with higher concentration (fig. 9.6).

The result is easily understood because higher solute concentration means lower meso-phase stability and degree of order [35]. This behavior can be demonstrated by lowering

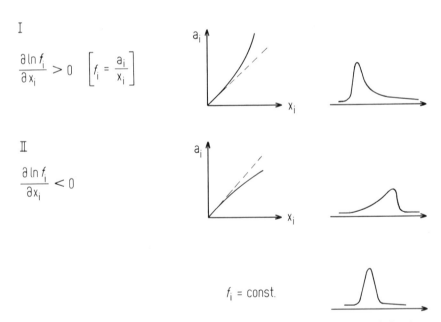

I

$$\frac{\partial \ln f_i}{\partial x_i} > 0 \quad \left[f_i = \frac{a_i}{x_i} \right]$$

II

$$\frac{\partial \ln f_i}{\partial x_i} < 0$$

$f_i = $ const.

Fig. 9.6. Distribution isotherms and peak shapes showing: I, positive; and II, negative deviation from the Raoult-Henry law.

the n/i transition temperature, which is strongly dependent on the low transition energy of nematics as the data of PAA and MBBA show (cf. p. 360).

	k/n		n/i	
	T, °C	ΔH, kJ/mol	T, °C	ΔH, kJ/mol
PAA	137	26.8	168	1.37
MBBA	20	13.9	44	0.34

9.1.2.2 The Regular-Solution Concept

Discussions on the interaction energy of solvent/solute systems usually start with the relationship (symbols, cf. table 9.4):

$$\Delta E = E_{11} - 2E_{21} + E_{22} \tag{9.10}$$

We can determine from Saupe's work [3:506–3:509] that the "energy of order" is

$$E_{11} = -kS^2/V^2 \tag{9.11}$$

The basic difficulty with the "regular solution" concept arises from neglecting the entropy terms and setting μ^E nearly equal to h^E. In other words, the free energy is set equal to the energy of mixing. To check the validity of making such an approximation, let us use experimental values derived from the system xylene/PAP:

$$^a\mu^E = 2.1 \text{ kJ/mol} \qquad ^ah = 25 \text{ kJ/mol}$$
$$^i\mu^E = 1.3 \text{ kJ/mol} \qquad ^ih = 12.6 \text{ kJ/mol}$$

It so happens that the only really valid relationship to come from these data is $\text{sign } \mu^E = \text{sign } h^E$. On the other hand, it is easily seen that

$$T^a S^E = 25 \ \ -2.1 \approx 23 \text{ kJ/mol} = {}^a h^{(E)}$$
$$T^i S^E = 12.6 - 1.3 \approx 11 \text{ kJ/mol} = {}^i h^{(E)}$$

Thus, we can state that the excess molar entropy of the solute is positive, and the latent-heat term, TS^E, comes closer to $h^{(E)}$ than it does to vanishing.

A system that nearly reaches the ideal-solution condition after having passed the n/i transition point has been found among the fatty-acid series in a mesogenic acid solvent, hexyloxycinnamic acid. The typical differences in μ^E, h, and S^E can be easily seen in this case (fig. 9.7).

The regular solution approximation was also applied to binary *p*-alkoxyazoxybenzene systems [8:61].

Ignoring the argument that the enthalpy term cannot be used to derive a quantitative expression for the excess free enthalpy, we can try explaining qualitatively the typical effects in terms of the regular solution theory.

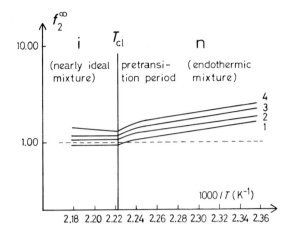

Fig. 9.7. Limiting activity coefficients, f_2^∞, vs. the reciprocal temperature, $1000/T$, for various fatty acids (1, propionic; 2, butyric; 3, valeric; 4, caproic acid) in p-hexyloxy-cinnamic acid as stationary phase (from [94]).

Starting from the well-known equation

$$\ln f_2 = \frac{V_2}{RT}\left\{\left(\frac{E_2}{V_2}\right)^{1/2} - \left(\frac{E_1}{V_1}\right)^{1/2}\right\}^2 \tag{9.12}$$

which was used extensively by H. Hildebrand [77, 78], a very simple relationship is obtained for comparing the anisotropic and isotropic solvents. As we know, the molar volume is reduced in the anisotropic phase, so we have $^aV_1 < {}^iV_1$. With respect to E_1, we should be able to set $^aE_1 \gtrless {}^iE_1$. There exists only one measurement of the temperature dependence of the vapor pressure of a mesophase, $^aE_1 > {}^iE_1$. Thus we obtain

$$^a\left(\frac{E_1}{V_1}\right) > {}^i\left(\frac{E_1}{V_1}\right) \quad \text{or} \quad {}^i\delta_1 > {}^a\delta_1 \tag{9.13}$$

On this basis, we should at least be able to predict a typical enhancement of $\ln f_2$, which has been reported, when the isotropic phase changes into the anisotropic one. We have already seen that S, V, and $\ln f$ show a somehow monotonous dependence on T (fig. 9.8). The only result that goes a step further is shown in fig. 9.9 where the function μ^E is plotted against S^2/V^2, the temperature-dependent part of the configuration energy. This might already represent a universal ("reduced") interdependence between the free energy and the degree of order, but note that S and V are not independent of each other.

9.1.2.3 *The Entropic Part of f*

It has been often suggested that the activity coefficient be split into two parts: the "energy" portion and the "entropy" portion as shown below

$$f = f_h + f_s \qquad \ln f_h = \frac{h}{RT} \qquad \ln f_s = \frac{-s^E}{R} \tag{9.14}$$

where $\ln f_h = 1$ defines the athermal and $\ln f_s = 1$ the regular solution. Let us first consider the f_s of an anisotropic solution. Since a discussion of f_h is fruitless when f_s has not been

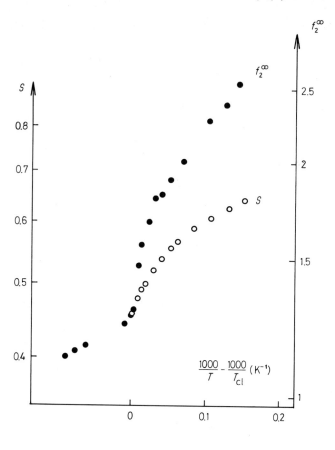

Fig. 9.8. Limiting activity co-efficient, f_2^∞, and degree of order, S, vs. the reciprocal temperature, $1000/T - 1000/T_{cl}$, measured for PAP (from [93]).

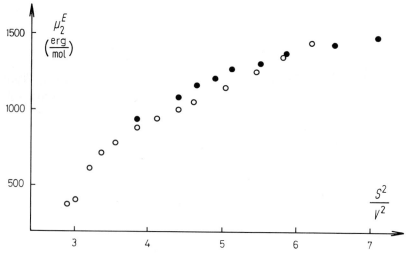

Fig. 9.9. Excess free energy of solute, $\mu_2^E = RT \ln f_2^\infty$, vs. the configuration term, S^2/V^2, of PAA (●) and PAP (○) (from [93]).

taken into account, M. L. Huggins [79, 80] and P. J. Flory [50, 51] have independently found that

$$\ln f_{2,s} = 1 + \ln \frac{1}{r} - \frac{1}{r} \qquad r = \frac{V_1}{V_2} \tag{9.15}$$

The molar volume, V_1, of the anisotropic phase is smaller than that of the isotropic, V_2, and it thus follows that

$$\ln {}^a f_2 > \ln {}^i f_2$$

as long as the condition $V_2/V_1 < 1$ holds. However, it should be noted that this condition leads to a value of f_s smaller than one, contrary to the regular solution model where f is always greater than 1 *.

9.1.2.4 The Free Energy Concept

The following concept, developed by R. L. Humphries and G. R. Luckhurst [3:487–3:495], is basically an application of the Maier-Saupe model. The theory, which started with a consideration of mixtures of rodlike nematogenic molecules, has been extended to describe the interaction of nonmesogenic molecules (2) with a nematic solvent (1).

The molar internal orientation energy is given as the sum over $N_L/2$ pairs of molecules**:

$$^a\bar{E}^m = -\tfrac{1}{2} N_L \{ x^2 \varepsilon_{11} \bar{P}^2_{(1)} + x(1-x)\varepsilon_{12} \bar{P}_{(1)} \bar{P}_{(2)} + (1-x)^2 \varepsilon_{22} \bar{P}^2_{(2)} \} \tag{9.16}$$

The products containing ε and P are identical with the "energy of order" introduced by A. Saupe (see eq. (9.11)). The corresponding entropy term $^aS^m$ is derived as follows:
With Z_1 resp. Z_2 (Humphries et al.) being defined as

$$Z_{1,(2)} = \int \exp -(E_{1,(2)}/kT)\sin\vartheta \, d\vartheta \tag{9.17}$$

and the orientational entropy of the mixture as

$$^aS^m = x \cdot S_{(1)} + (1-x) S_{(2)} \tag{9.18}$$

---

* We feel that a more convincing definition is that of E. A. Guggenheim, using the term "simple solution" and being defined by

$$G^E = x(1-x)w \qquad w = w(T, p)$$

$$\lim_{x \to 0} \frac{\partial G^E}{\partial x} = RT \ln \gamma_2 = w$$

This definition is in agreement with Henry's law, giving no details with respect to "energy-" and "entropy-" effects (see also [52]).

** 1) For sake of agreement with the original text we set x for the solvent and $1-x$ for the solute which is contrary to general use. 2) Instead of writing $S \equiv$ degree of order, Humphries et al. introduced the term "2nd order Legendre Polynomial, P_2" with exactly the same meaning (see table 9.4). So we can use the letter S for the entropy, as usual.

From $S_i = R \ln Z_i$; $^aF^m = {}^a\bar{E}_m - T^a S^m$ it follows that

$$^aF^m = {}^a\bar{E}_m - R T\{x \ln Z_1 + (1-x) \ln Z_2\} \tag{9.19}$$

This equation will be used later in deriving the activity coefficient of such a system.

It is a fruitful and theoretically sound procedure to consider the relationship between h^E and s^E, see [16a, 19a]. D. E. Martire has shown that the relationship

$$T\Delta S = \alpha \Delta H + \beta \tag{9.20}$$

which has been derived earlier for small molecules dissolved in a solvent of heigh molecular weight also exists for liquid crystal solution systems. Only one example will be given here. For a paraffin solute and cholesteryl myristate solvent (cholesteric stationary phase), we have

$$\Delta S_2^{(sol)} = 2.843 \cdot 10^{-3} \Delta H_2^{(sol)} - 2.1, J/Kmol \tag{9.21}$$

It should be noted that Martire's definition of the term ΔS is

$$\Delta S_2^{(sol)} = S_2^E - \Delta S_2^{(vap)} = S_2^E - \frac{\Delta H_2^{(evap)}}{T} \tag{9.22}$$

Let us use this as a starting point for a theoretical consideration by Humphries, Luckhurst, and James [3:487, 3:489] of the principal differences in free energy between the isotropic and liquid-crystalline states. For an example explained by E. A. Guggenheim [59], it can be written

$$R T \ln f_2^{\infty} = \left(\frac{\partial G_{mix}^E}{\partial x_2}\right)_{x_2 \to 0} = -\left(\frac{\partial G_{mix}^E}{\partial x_1}\right)_{x_1 \to 1} \tag{9.23}$$

where $G_{mix}^E = \Delta_{mix} G - \Delta_{mix} G_{(id\,mix)}$ (see eq. (9.2) and the following relations). R. L. Humphries et al. here now set $x_1 = x$, where x_1 is the mole-fraction of the solvent ($x \to 1$, see note** on p. 403). We retrain also the original nomenclature as such as the mixing terms are concerned (Δ^m means Δ_{mix}).

An orientation dependent activity coefficient f_{or} is introduced by the relation:

$$\ln f_{2,or}^{\infty} \equiv \ln {}^a f_2^{\infty} - \ln {}^i f_2^{\infty} = -\left(\frac{\partial \Delta^m F_{or}}{\partial x}\right)_{x \to 1} \tag{9.24}$$

where $\Delta^m F$ means the gain in Helmholtz free energy (but only the orientation dependent part) after mixing x mol of solvent with $(1-x)$ mol of solute*.

$$\Delta^m F = F_m - x F_m^{(1)} - (1-x) F_m^{(2)} \tag{9.25}$$

---

* It should be kept in mind that the term $\Delta^m F$ thus defined does not correspond to the excess function G_m^E, but this is defined as the surplus ΔG over that of the ideal mixture, $\Delta G_{id\,mix}$. The $\Delta^m F_{or}$ used here is defined in terms of the free energy of both phases by:

$$\Delta^m F_{or} = \Delta^{ma} F - \Delta^{mi} F = [({}^aF - {}^iF)_m - x({}^aF^{(1)} - {}^iF^{(1)}) - (1-x)({}^aF^{(2)} - {}^iF^{(2)})$$

F_m can be approximately written as a function of interaction energy ε_{ik}, orientation P, and the distribution functions Z_1 and Z_2 as follows:

$$F_m = -N\{x^2\varepsilon_{11}\,\overline{P_2^{(1)}}^2 + 2x(1-x)\varepsilon_{12}\,\overline{P_2^{(1)}}\,\overline{P_2^{(2)}} + (1-x)^2\varepsilon_{22}\,\overline{P^{(2)}}^2\}/2$$
$$- RT\{x\ln Z_1 + (1-x)\ln Z_2\} \tag{9.26}$$

where Z_i is defined as $Z_i = \int \exp[-U^{(i)}/kT]\sin\theta\,d\theta$, and the $P_2^{(i)}$ are the 2nd Legendre polynomials which characterize the "degree of order".

Differentiation of eq. (9.25) and eq. (9.26) yields

$$\frac{\partial\Delta^m F}{\partial x} = N(\varepsilon_{11}\,\overline{P_2^{(1)}}^2 - \varepsilon_{22}\,\overline{P_2^{(2)}}^2)/2 + RT\ln(Z_2^+/Z_2) \tag{9.27}$$

where

$$Z_2^+ = \int \exp\{-\varepsilon_{12}\,\overline{P_2^{(1)}}\,\overline{P_2^{(2)}}/kT\}\sin\theta\,d\theta \tag{9.28}$$

In eq. (9.27), Z_2^+ is the only magnitude depending on the solute-solvent interaction. The activity coefficient contribution, $\ln f_{2,\mathrm{or}}^{\infty}$, is obtained directly as

$$\ln f_{2,\mathrm{or}}^{\infty} = -(\varepsilon_{11}\,\overline{P_2^{(1)}}^2 - \varepsilon_{22}\,\overline{P_2^{(2)}}^2)2kT - \ln(Z_2^+/Z_2) \tag{9.29}$$

Since the orientation order of the pure solute is equal to zero, i.e., $P^{(2)}=0$ and $Z_2=1$, it follows that

$$\ln f_{2,\mathrm{or}}^{\infty} = -\varepsilon_{11}\,\overline{P_2^{(1)}}^2/2kT - \ln Z_2^+ \tag{9.30}$$

Experimental data shows that the activity coefficient in the anisotropic phase is larger [126, 127]. ε_{11} is negative in the nematic phase, therefore the first term is positive and greater than $|\ln Z_2^+|$. Z_2^+ vanishes in the vicinity of the clearing point analogous to the disappearance of ε_{12}. The maximum value of $\ln f_{2,\mathrm{or}}$ when $Z_2^+ \to 0$ is

$$\ln f_{2,\mathrm{or}}^{\infty} = -\varepsilon_{11}\,\overline{P_2^{(1)}}^2/2kT_{\mathrm{cl}}^{(1)} \tag{9.31}$$

It has been shown by G. R. Luckhurst that the n/i temperature can be obtained by a numerical evaluation of the Helmholtz function. This yields

$$T_{\mathrm{cl}}^{(1)} = -\frac{\varepsilon_{11}}{4.542k} \tag{9.32}$$

On the other hand, W. Maier and A. Saupe have found that an orientational degree of order of 0.4292 is typical of the n/i transition point. Introducing this value gives

$$\ln f_{2,\mathrm{or}}' = 0.418 \qquad f_{2,\mathrm{or}}' = 1.52$$

A nearly constant value of $^a f^{\infty}/^i f^{\infty} = 1.1_4$ has been found experimentally for the hydrocarbons benzene, toluene, the three xylenes and cyclohexane [88].

The pretransition range does not allow for a direct reading of the discontinuity of $\ln f^\infty$, and $\Delta \ln f^\infty$ was evaluated by extrapolation from the values of $\ln f^\infty$ in the nematic phase (PAP). D. E. Martire et al. report other studies on infinite dilution activity coefficients and their interpretation [8:208, 8:209].

9.1.2.5 The Liquid-crystalline/Isotropic Transition as a Corresponding-State Condition [95]

The following discussion is based on the traditional notation in the thermodynamics of mixtures where free energies, enthalpy, and other quantities are almost always given on a molar basis. For example, the Gibbs free energy of an ideal mixture is defined as

$$\Delta G^m = R T \sum_i n_i \ln \frac{n_i}{\sum_i n_i} = R T \sum_i n_i \ln x_i \tag{9.33}$$

For a two-component system, there is the question of how to represent in the most rational way the affinity of a solute with respect to a liquid-crystalline solvent. We begin with the Raoult-Henry law, expressed in a molar basis as

$$p_2 = p_2^0 f_2 x_2 \tag{9.34}$$

Experimental results have shown that values of the activity coefficient, f_2^∞, determined for different liquid-crystalline solvents in a single solute (e.g., *o*-xylene) vary widely (fig. 9.10).

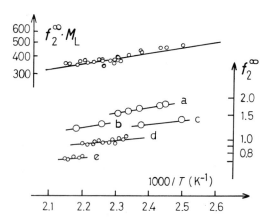

Fig. 9.10. Limiting activity coefficients, f_2^∞, and the molar expression, $f_2^\infty \cdot M_L$, vs. the reciprocal temperature, $1000/T$, for *o*-xylene in the isotropic range of the stationary phases a, PAA; b, PAP; c, diethyl *p*-azoxybenzoate; d, mixture of cholesteryl benzoate and PAP; e, cholesteryl benzoate (from [91]).

The question was whether another definition of affinity could reflect the special conditions of interaction in liquid-crystalline phases, particularly at their transition points to the isotropic liquids. This transition could be compared with the normal boiling point or critical temperature. In fact, the interpretation of the transition point as a special case of "corresponding states" is already included in the Maier-Saupe theory of the n/i transition. In our search for general rules governing the partial molar quantities of solutes under the influence of an anisotropic solvent we found that the following relationship exists. If the Raoult-Henry law is defined on a *weight*-fraction basis, then nearly equal activity coefficients result. This is true almost over the whole temperature range of the anisotropic phase. In fig. 9.11, the curve $f_2^{\infty} M_1$ shows this where M_1 is the molar mass ("molecular weight") of the solvent and $_w f_2^{\infty}$ defined as

$$p_2 = p_2^0 f_2^{\infty} \underset{x_2 \to 0}{x_2} \cong p_2^0 f_2^{\infty} \frac{m_2/M_2}{m_1/M_1} \tag{9.35}$$

With $_w f_2^{\infty} = f_2^{\infty} \cdot \dfrac{M_1}{M_2}$ it follows that

$$p_2 = {_w f_2^{\infty}}\, p_2^0 \frac{m_2}{m_1} \tag{9.36}$$

If only one solute is considered, then the constant factor M_2 can be dropped as in fig. 9.11. What is the reason for such a simple relationship, which means that in different anisotropic solvents, the activity of a solute is always proportional in the same degree to the vapor pressure of the pure solute p_2^0?

We can be sure that the activity of the solute molecules, at least as far as the polarization-dependent portion is concerned, is determined by a predominating dispersion-force interaction among the mesogenic molecules themselves. Thus, we know that the benzene nucleus interacts with the solvent molecules by a space-dependent rather than by a number-dependent mechanism. Such an interaction has recently been indicated as acting in the similar case of polymer homologs [149]. In other words, if we assume that two systems (fig. 9.11) have the same solute activity and that this activity depends on the probability of molecules 1 and 2 contacting one another, then we can easily understand that this probability does not depend on the number ratio but rather on the ratio of contacting sites being equal in the two cases. This is independent of the number of individual solvent molecules.

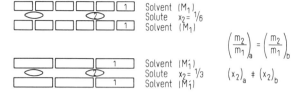

Fig. 9.11. Solute-solvent interaction depending on the mass (or volume) ratio rather than on the molar ratio (from [95]).

Coming back to the experimental results, we obtain the general function:

$$\ln\{f_2^{\infty} M_1\} = \frac{A}{T} + \text{const} \tag{9.37}$$

Introduction of M_2, the molar mass ("molecular weight") of the solute, which is constant in the following calculations, gives

$$\ln\left\{f_2^{\infty} \frac{M_1}{M_2}\right\} = \ln {_w f_2^{\infty}} = \frac{A}{T} + \text{const}' \tag{9.38}$$

From this a simple derivation yields

$$\ln\left\{f_2^{\infty}\frac{M_1}{M_2}\right\} = \frac{\mu^E}{RT} + \ln\frac{M_1}{M_2} \tag{9.39}$$

setting

$$\mu^E = \frac{h}{RT} - \frac{s^E}{R}$$

it follows that

$$\ln{}_wf_2^{\infty} = \frac{h}{RT} - \frac{s^E}{R} + \ln\frac{M_1}{M_2} \tag{9.40}$$

In a first approximation, the sum of the last two terms is constant even with a large variation in M_1. From this we see that s^E is strongly dependent on the molar mass ratio, M_1/M_2 (or, if M_2 is held constant throughout the experiment, on M_1):

$$s^E = R\ln\frac{M_1}{M_2} + \text{const} \tag{9.41}$$

Both h and s^E are temperature-dependent functions, but this dependence is relatively small. It may be stated that the general and uniform behavior has led to a special expression concerning the excess entropy, s^E, which is defined in its original sense, i.e. on a molar basis.

The molar mass enters explicitly, and the value of the constant is obtainable by experiment and can be regarded as characteristic of mesophases. In the same way one could define a new μ^E on a mass fraction basis as has been previously done. This is written

$$RT\ln{}_wf_2^{\infty} = {}_w\mu^E \tag{9.42}$$

So one obtains:

$${}_w\mu^E = h - T\left[s^E - R\ln\frac{M_1}{M_2}\right] \tag{9.43}$$

with

$$s^E - R\ln\frac{M_1}{M_2} \equiv {}_ws^E \tag{9.44}$$

One consequence of the newly defined activity coefficient, ${}_wf_2^{\infty}$, is worthy of special mention. If the specific retention volume of a solute is measured just above or just below the clearing point of the mesophase, then we get a straight line if the different values of V_g^T are plotted against $1/T_{cl}$, where T_{cl} is the temperature of the clearing point. To a good approximation, the same linear dependence (fig. 9.12) is also valid for the partition coefficients, K, because $K = V_g^{(T)}\rho$ (cf. table 9.3).

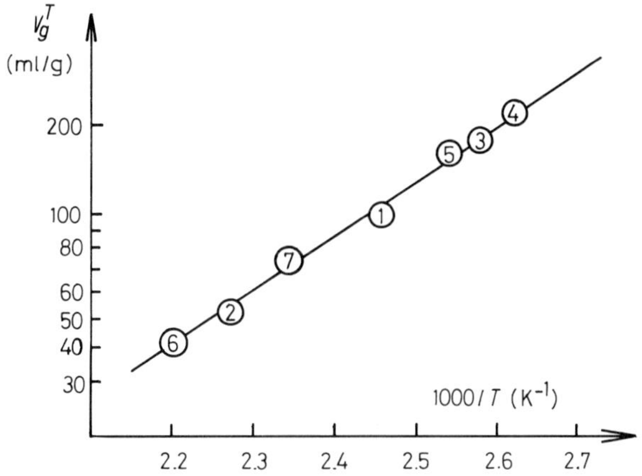

Fig. 9.12. Specific retention volumes of o-xylene, V_g^T, vs. the reciprocal temperature, $1000/T$, for different mesogens just above the clearing point: 1, PAA; 2, PAP; 3, p-methoxybenzylidene-p'-cyanoaniline; 4, p-methoxybenzylidene-p'-carbomethoxyaniline; 5, diethyl p-azoxybenzoate; 6, cholesteryl benzoate; 7, mixture of 2 and 6 in the ratio 1:1 mol (from [1:117]).

The explanation is quite simple. Consider a fixed solute, denoted by the subscript 2, that is kept constant while M_j varies:

$$V_g^{(T)}(j) = K\rho = \frac{273\,R}{p_2^0\,M_j\,f_2^c} \tag{9.45}$$

and

$$\ln V_g^{(T)}(j) = -\ln p_2^0 - \ln M_j f_2^c + \ln 273\,R = \left[\frac{-\Delta H^v}{T} + \text{const}\right] - \left[\frac{A}{T} + \text{const}'\right] + \text{const}'' \tag{9.46}$$

In eq. (9.46) the first term is independent of the solvent, and the second is, as a first approximation, a universal function of temperature. Hence the linear relationship

$$\ln V_g^{(T)} = \frac{-\Delta H^v - A}{T} + \text{const} \tag{9.47}$$

is completely analogous to eq. (9.38). In eq. (9.47), T can naturally be chosen as T_c, the clearing point of any nematic melt.

9.1.2.6 Pretransition Phenomena

Since the degree of order, S, is a complex function of the temperature, and cooperative phenomena are involved in the phase transition, any physical function of a liquid crystal depends on the degree of order. Even in the case of "normal" melting, the relationship between the activity coefficient and the temperature must also be related to the degree of order. One example of this is the specific retention volume where the pretransition effect has been qualitatively verified (fig. 9.13).

A quantitative treatment is difficult because even minute amounts of impurities (including those introduced by a partial decomposition of the substance being analyzed) leads to a less-pronounced "heterogeneous premelting" effect.

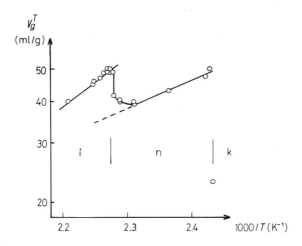

Fig. 9.13. Specific retention volumes, V_g^T, of *o*-xylene on PAP vs. the reciprocal temperature, $1000/T$, in the vicinity of the clearing point (H. Kelker, unpublished data, courtesy of Hoechst AG; see also [1:117]).

It is certainly impossible to explain pretransition phenomena using only the one temperature-dependent parameter, S, the (molecular) degree of order. In Saupe's theory, the degree of order is very clearly defined as the degree of fluctuation of single molecules or of short-range order statistical groups. However, the classical pretransition theories use the concept of a partial phase change (heterophase fluctuation) where different modes of motion such as rotation, vibration, or translation come into action, and these modes are fully developed when the higher-energy phase is attained. Using the model of J. T. Frenkel [3:267], A. R. Ubbelohde [8:347], and G. W. Stewart's concept of "cybotactic groups" [1:357–1:362], we can accept the picture of quasi-heterophase fluctuations between regions of higher order, a, and those that represent an already-transformed "isotropic" state, i. In accordance with the work of H. C. Longuet-Higgins and G. R. Luckhurst, it has been shown that pretransition phenomena, i.e., a temperature-dependent equilibrium between these two "phases", may also explain the temperature-dependent activity of a solute distributed in solution between the two phases [93]. The distribution in the solvent system is governed by the temperature-dependent coefficient

$$K = {}^aC_2/{}^iC_2$$

with the phase ratio being temperature-dependent as well. For a homogeneous phase, it is found that

$$ {}^1V_R = m_1 \, {}^1V_g^{(T)} \tag{9.48}$$

where 1 represents the solvent as before. It consequently follows that for a heterophase solvent system with one anisotropic phase, a, and one isotropic one, i, we have

$$V_{R\,(\mathrm{mix})}^{(T)} = {}^am_1 \cdot {}^aV_g^{(T)} + {}^im_1 \cdot {}^iV_g^{(T)} \tag{9.49}$$

Assuming a temperature-dependent equilibrium between the masses of "phase" a and "phase" i (this model including the existence of mixed crystals), and introducing mass fractions of the solvent (2) as defined by

$$^aX_1 + {}^iX_1 = \frac{^am_1}{^am_1 + {}^im_1} + \frac{^im_1}{^am_1 + {}^im_1} = 1 \tag{9.50}$$

We can now define a specific "mixed retention volume" by dividing eq. (9.49) by $(^am_1 + {}^im_1)$ with this being the proportion of stationary phase:

$$V_{g\,(\mathrm{mix})}^{(T)} = {}^aX_1 \, {}^aV_g^{(T)} + (1 - {}^aX_1) \, {}^iV_g^{(T)} \tag{9.51}$$

Note that x_1 is an extremely temperature dependent parameter. Introducing the activity coefficients, and with $^aM_1 = {}^iM_1$ we obtain

$$V_{g\,(\mathrm{mix})}^{(T)} = \frac{RT}{M_1 \, p_2^0} \left\{ \frac{^ax_1}{^af_2^\infty} + \frac{1 - {}^ax_1}{^if_2^\infty} \right\} \tag{9.52}$$

From here H. C. Longuet-Higgins continues as follows: Assuming also that M_1 is identical above and below the clearing point, division by $^iV_g^{(T)}$ yields:

$$\frac{V_{g\,(\mathrm{mix})}^{(T)}}{^iV_g^{(T)}} = {}^if_2^\infty \left\{ \frac{^ax_1}{^af_2^\infty} + \frac{1 - {}^ax_1}{^if_2^\infty} \right\} \tag{9.53}$$

The quotient on the left side can easily be determined experimentally. An extrapolated value should be used due to its strong temperature dependence. Then one obtains:

$$\frac{V_{g\,(\mathrm{mix})}^{(T)}(2)}{^iV_g^{(T)}(2)} = \frac{^aK\rho(T)}{^iK\rho(T)} = {}^if_2^\infty \left\{ \frac{^ax_1}{^af_2^\infty} + \frac{(1 - {}^ax_1)}{^if_2^\infty} \right\} = \frac{^if_2^\infty}{^af_2^\infty} \left\{ ^ax_1 + \frac{^af_2^\infty}{^if_2^\infty}(1 - {}^ax_1) \right\} \tag{9.54}$$

("2" being the solute)

The partition coefficient $K(i/a)$ as defined for the solvent system can be identified in dilute solution as

$$K = {}^if_2^\infty / {}^af_2^\infty \tag{9.55}$$

To obtain $^if_2^\infty / {}^af_2^\infty$ from gas chromatographic measurements we need one more value, namely the ratio $^ax_1/{}^ix_1$ which is called the "phase ratio". On the other hand, a quite independent method to determine K is proposed by H. C. Longuet-Higgins and G. R. Luckhurst [10:677] in which K is defined as

$$K = \frac{^aC_2}{^iC_2} = \frac{^an_2}{^am_1} \Big/ \frac{^in_2}{^im_1} = \frac{^an_2}{^ax} \, \frac{(1 - {}^ax)}{^in_2} \tag{9.56}$$

The hyperfine splitting, ^aA, of the solute being completely ordered can be calculated by a method given by A. Carrington and G. R. Luckhurst [10:676]. The hyperfine splitting, ^iA, in the isotropic phase is measured easily, and the value $\langle a \rangle$ of hyperfine splitting really measured in the mesophase is defined as the mean value between ^aA and ^iA as being

$$\langle a \rangle = \frac{^an_2\,{}^aA + {}^in_2\,{}^iA}{^an_2 + {}^in_2} = \frac{(^an_2/{}^in_2)^aA + {}^iA}{(^an_2/{}^in_2) + 1} \tag{9.57}$$

The mass ratio of partition, ${}^{a}n_2/{}^{i}n_2$, is obtained directly from this equation, but the situation is the same as above; the partition coefficient K can only be calculated from eq. 9.56 if the phase ratio $({}^{a}x_1/(1 - {}^{a}x_1))$ is available.

E. McLaughlin et al. made an independent determination of ${}^{a}V_1/{}^{i}V_1$ (which can be identified with the mass ratio) of 0.18 at the melting point and 0.12 at the clearing point [8:345]. A comparison of these values with gas chromatographic data led H. C. Longuet-Higgins and G. R. Luckhurst to the assumption that McLaughlin's phase ratio values were too small. Corrected values for the melting and the clearing point of about 0.90 and 0.60, respectively, agree with some otherwise contradictory results. Based on this "two-phase" model, L. C. Chow and D. E. Martire have calculated the degree of order, S, from gas-liquid chromatographic measurements [124].

The dependence of activity coefficients on texture changes and higher-order phase transitions, especially in cholesterics, have scarcely been investigated at all [129, 153, 183]. Our own experiments to date lead to the conclusion that at least certain cholesterics yield irreproducible results because the partition coefficient is dependent on the thermal history of the preparation*. Our attempts to separate enantiomers on optically active mesophases were unsuccessful. C. H. Lochmüller and R. W. Souter recently reported a separation of enantiomers using an optically active smectic phase [112–114]. H-bond complexing apparently favors separation of optical isomers. This is analogous to the findings of E. Gil Av et al. with ordinary isotropic liquids [139, see also 132].

The general effects of the dependence of transport phenomena (diffusion) on phase anisotropy have been well known for about fifty years, beginning with the work of T. Svedberg. In gas chromatography, there has been no observation of a significant change in diffusion coefficient in a nematic solvent.

Recently there have been some indications that the influence of an electric field would change both diffusion and partition behavior. The dependence of partition coefficients on external-field influences have been reported. Tailor et al. have interpreted their experiments in such a way that f_2 should change when an electric field is applied [19]. However, further confirmation of these highly interesting results will be necessary. We have attempted a similar experiment using a magnetic field of up to 10,000 Gauss, but with no effect.

To date, only a few smectic phases have been used in gas chromatography, and no systematic observations concerning diffusion are available. Although observable effects with highly viscous substances are to be expected, their practical use promises little. Reminiscent of chromatography, W. Helfrich expects different velocities of different molecules permeating a smectic layer structure or migrating along the helical axis of a cholesteric structure [76].

9.2   Chemical Reactions in Anisotropic Liquids

9.2.1 Thermotropic Mesophases

Only recently, several decades after T. Svedberg's pioneering work, have his ideas again come into favor and been used as the basis of new experiments. In four reports that carry

---

* E. Grushka and J. F. Solsky have used capillary columns to show that either HTP values or distribution coefficients indicate strong hysteresis effects under the special conditions of a uniform liquid layer as used in the capillary. A texture influence may be possible. Curiously enough, these differences are observed even for the isotropic phase [56].

the same title as this section, T. Svedberg himself investigated the system PAP and picric acid or pyrogallol [163]. He measured the reaction rate using the increase in electric conductivity with time at various temperatures. Although the chemical reactions taking place in these early experiments never came to light, the following qualitative predictions can be considered valid. The reaction rate, which rises but slightly with increasing temperature in the nematic phase, jumps at the clearing point and continues steeply upward in the isotropic liquid. In addition, the reaction rate exhibits a distinct change when the nematic phase is under the ordering influence of a magnetic field. These observations cannot be used as generally valid principles, because even now not enough data can be extracted from the meager experimental material. Examples are known that exhibit each case: The first, where the reactions in the isotropic and the nematic fluid differ distinctly; and the second, where no noteworthy differences could be observed. Reviews have already appeared [15, 165].

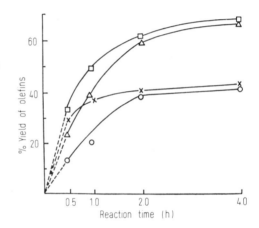

Fig. 9.14. Xanthate pyrolysis of (I) at 160°C in various solvents (from [17]).

o: decalin

x: MeO—◯—CH=N—◯ (Cl)

□: MeO—◯—O-CO—(H)—CO-O—◯—OMe

△: MeO—◯—CH=N—◯(Cl)—◯(Cl)—N=CH—◯—OMe

A nematic solvent favors the following xanthate pyrolysis where the olefin yield can be raised from 40% to 65% [15, 17], fig. 9.14:

I R: $C_{16}H_{33}$

In contrast, the almost equal activation parameters and first-order rate constants show a nearly identical progression of the Claisen rearrangement in both nematic and isotropic-liquid solvents [15, 25, 26, 47]. The following is considered the most likely intramolecular reaction mechanism in both media.

where R: CH_3, Cl, CN, NO_2
with $k_{(NO_2)} < k_{(CN)} < k_{(Cl)} < k_{(CH_3)}$

However, the Claisen rearrangement is strongly inhibited in a channel type clathrate of cinnamyl phenyl ether [47].

Two examples of stereospecific reactions in cholesteric solvents are known. F. D. Saeva et al. report a stereospecific reaction transmitting the chirality of a cholesteric solvent to the reaction product of the following Claisen rearrangement [152]:

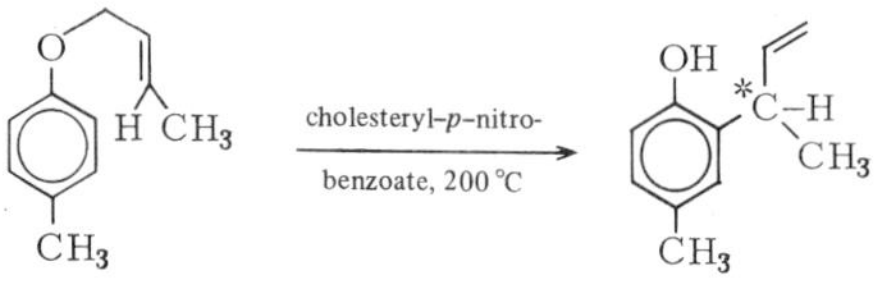

L. Verbit et al. have found an optical purity of 18% in the following decarboxylation of ethylphenylmalonic acid [171]:

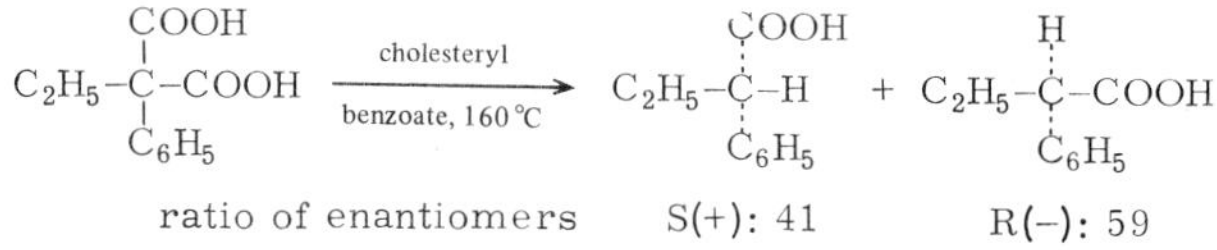

$$\text{ratio of enantiomers} \qquad \text{S}(+)\text{: }41 \qquad \text{R}(-)\text{: }59$$

Racemic solutes consisting of interconvertible enantiomers were shown to assume a nonracemic composition in cholesteric solvents [140a].

Another first-order reaction, the thermal isomerization of 2,4,6-trimethoxy-s-triazine, proved to be independent of the anisotropy or isotropy of the reaction medium [27]. On the other hand, the thermal isomerization of *cis*-stilbene to *trans*-stilbene is reported to show a higher rate constant in a nematic mesophase than in an isotropic solution [168]. A slight solvent influence can be expected for other unimolecular reactions, while for higher-order reactions the influence will be greater that is in reactions where a collision step is the rate determining step, but this is subject to further confirmation. This is obvious from the isomerization of α-benzyloxystyrene that is assumed to proceed by a second-order reaction according to the following free radical chain mechanism [15].

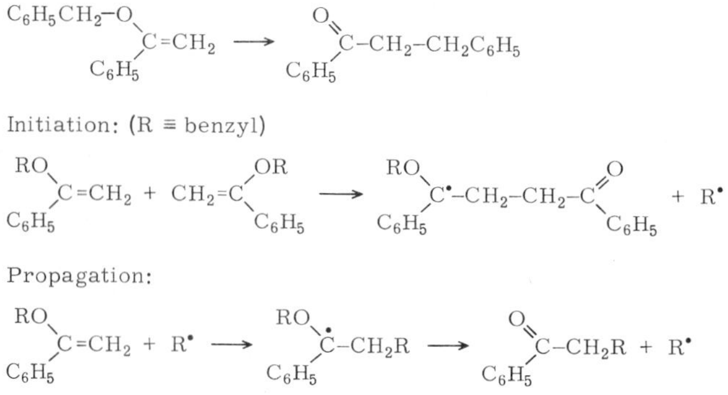

Initiation: (R ≡ benzyl)

Propagation:

The rate constants are determined by the initiation step and are distinctly higher in a nematic solvent (di(*p*-methoxyphenyl)*trans*-cyclohexane-1,4-dicarboxylate) than in diphenyl ether. In addition, a different slope of the Arrhenius plots was observed indicating the rate of the reaction and the energy of activation to be higher in the nematic solvent than in the diphenyl ether.

A typical solid state controlled reaction is seen in the photodimerization of tetraphenylbutatriene:

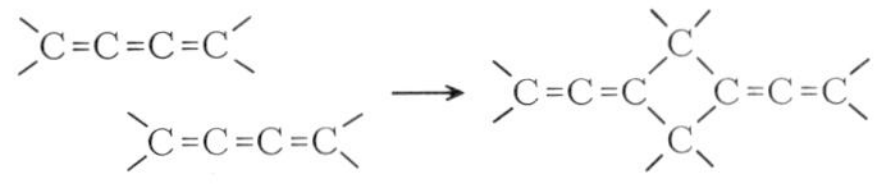

This reaction was also observed in a polymer film and in a nematic mesophase as reaction matrices. However, no dimeric product could be detected when the solution was irradiated at a temperature above the clearing point, and the spectrum of the isotropic solution remained unchanged [14].

Two examples are finally mentioned where the wavelength of maximum reflection has been measured to follow the kinetics of a reaction in a cholesteric solvent [102, 153a]. No details could be obtained on refs. [81a, 138a].

9.2.2 Polymerization Reactions in Thermotropic Mesophases

In view of technical interest, most work has been directed to polymerizations in liquid-crystalline phases. However, summarizing the results, this technique cannot compete with the standard methods of stereospecific polymerization. A series of examples may show this, e.g. the data given in table 9.6.

Table 9.6. Tacticity of poly-(p-methacryloyloxy)benzoic acid expressed as average stereosequence length, μ, of isotropic and syndiotactic sequences (from [20]).

Preparation condition	Conversion in %	μ isotactic	μ syndiotactic
Bulk	55	2.0	2.8
Isotropic solution	50	2.0	2.8
Nematic solution	55	4.8	4.3
Smectic solution	55	5.3	4.7

Polymerization of vinyloleates in the crystalline or smectic state probably yields stereoregular polymers (m. p. 34–38 °C), while the supposed atactic polymer formed in isotropic liquids cannot be crystallized [2, 4].

In the case of N-(p-methoxy-o-hydroxybenzylidene)-p-aminostyrene no significant difference was detected in the polymerization of the nematic and isotropic phases. Differences were found neither in the polymerization rate nor in the nature of the polymer. Even a magnetic field had no influence [137, 138].

Similar results were also obtained with cholesteryl acrylate [40]. The experimental results of H. Saeki et al. confirm that there is no significant difference in tacticity between poly-(cholesteryl methacrylate) obtained in the mesophase and that obtained in the isotropic liquid phase [151]. A. C. de Visser et al. concluded in a later paper that the rate of bulk thermal polymerization of cholesteryl acrylate increases with a decreasing order of the monomer

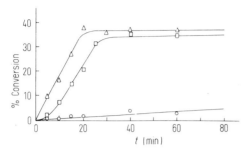

Fig. 9.15. Conversion-time plots for the bulk polymerization of cholesteryl acrylate: (○) in the solid phase at 120 °C; (□) in the cholesteric phase at 123 °C; (△) in the isotropic liquid at 126 °C (from [41]).

system [41], fig. 9.15, and the rather low conversions result from the formation of inhibiting byproducts during the polymerization.

None of the polymers obtained from cholesteryl acrylate, methacrylate or cholestanyl acrylate, and methacrylate showed any liquid crystallinity when swollen or dissolved [42].

A. A. Baturin et al. postulated that the acrylate having the structure

$CH_2=C(CH_3)COO$—⟨○⟩—OOC—⟨○⟩—$OC_{16}H_{33}$ in some solvents forms liquid crystals aligned in such a way that the overall polymerization rate is increased [9]. Radical initiation gives a polymer of *p*-hexyloxyphenyl *p*′-acryloyloxybenzoate showing a layer structure independent of the state of the monomer [115]. Starting from the isotropic liquid, the polymer formed is optically isotropic and the ordered polymer domains become visible only if the monomer has been in the nematic state. A nematic sample oriented by a magnetic field or by surface forces yields an oriented polymer showing a uniform orientation of the normal layer [115, see also 147].

The system $CH_2=C(CH_3)COO$—⟨○⟩—$COOH/C_{16}H_{33}O$—⟨○⟩—$COOH$ exhibits a higher rate of polymerization and conversion, and it leads to a higher average molecular weight upon polymerization in the smectic state than in the isotropic liquid [3, 5, see also 21, 22]. However, the same authors confirm that the polymerization rate of cholesteryl methacrylate is lower in the cholesteric phase than in the isotropic liquid as stated earlier [7]. One must thus consider both an acceleration and a retardation in chemical reactions due to the liquid-crystalline state. Investigations of this type will certainly add to our knowledge of the molecular mechanism of chemical reactions. The accordance or discrepancy between the steric requirements for propagation and the mesomorphous order may be decisive. Some reviews discuss the role of anisotropic states in polymerization processes [10, 24b, 134a]; V. P. Shibaev's comprehensive review will be published in English [157b].

E. Perplies, H. Ringsdorf, and J. H. Wendorff have found the overall polymerization rate of *p*-(acryloyloxybenzylidene)-*p*′-methoxyaniline to be lower and less temperature dependent in the nematic phase than in the isotropic liquid [2:551, 145], see fig. 9.16.

The same authors have shown that the nematic structure of *p*-(methacryloyloxybenzylidene)-*p*′-ethoxyaniline was transformed to a smectic structure during the polymerization [2:551, 146, 147a]; another example of this is known, but monomer and polymer can also form mesophases of the same type [24]. A quite similar polymer was obtained by anionic polymerization in solution at −78 °C and subsequent tempering, while in solution at 20 °C only an isotropic polymer resulted. The X-ray patterns indicate a long-range order normal to the direction of the layer with a periodicity of 3.3 nm and a short-range order perpendicular to it. Instead of this "broken" s_A-order, however, a typical fiber structure emerges if the

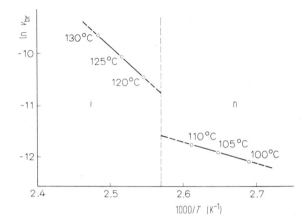

Fig. 9.16. Overall polymerization rate, v_{br}, of N-(p-acryloyloxybenzylidene)-p'-methoxyaniline vs. the reciprocal temperature, $1000/T$ (from [145]).

polymer is prepared in the nematic phase that has been oriented by a magnetic field of 70 kGauss [2:551, 147]. The fiber axis, the direction of the layer normal, and that of the side chains are parallel to the direction of the magnetic field. As fig. 9.17 illustrates, a magnetic field increases the polymerization rate in the nematic phase but not in the isotropic liquid.

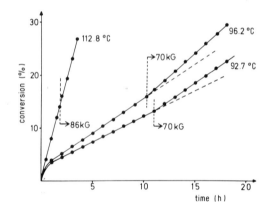

Fig. 9.17. Polymerization of N-(p-methacryl-oyloxybenzylidene)-p'-ethoxyaniline at different temperatures in the nematic phase (92.7 and 96.2 °C) and in the isotropic melt (112.8 °C) with and without application of a magnetic field (from [147]).

Phenylacetylene provides an interesting study material forming cyclic and linear oligomers as well as high polymers in a second order reaction for which W. E. Bacon has found remarkable differences. These are reproduced in table 9.7 [15, 16].

Table 9.7a. Rate constants for the thermal polymerization of phenylacetylene at 150 °C (from [16]).

Solvents	$k \cdot 10^6 \, \mathrm{l} \cdot \mathrm{mol}^{-1} \, \mathrm{s}^{-1}$	Relative rate
p-xylene	2.4 ± 0.3	1
Isotropic liquid of nematogen[a]	8.2 ± 0.4	3
Cholesteric[b]	44.9 ± 0.5	18
Nematic[a]	87.2 ± 0.5	36

Table 9.7b. Polymerization of phenylacetylene at 150°C for 120 hours (from [16]).

Solvents	Monomer conc. in M	1,3,5-TPB[c] in mg	1,2,4-TPB[c] in mg	1,2,4-1,3,5	Percent aromatic	Percent polymer	Molar mass
Nematic phase[a]	0.77	43.7	53.9	1.2	11.2	62	917
Cholesteric phase[b]	0.75	41.8	54.1	1.3	11.1	65	810
Isotropic phase[d]	1.39	22.5	67.5	3.0	6.0	56	607
p-xylene	0.84	7.3	37.6	5.2	4.9	53	508

[a] Di(p-methoxyphenyl)-trans-cyclohexane-1,4-dicarboxylate
[b] Prepared by the addition of 4.4 mol% of cholestanyl p-methylbenzoate to the nematic
[c] TPB is triphenylbenzene
[d] Isotropic liquid of the nematogen

The presence of an even more-ordered mesomorphous phase similar to the liquid-crystal-line state has been suggested for a monomer polymerized near the melting point at a maximum rate [68, fig. 9.18].

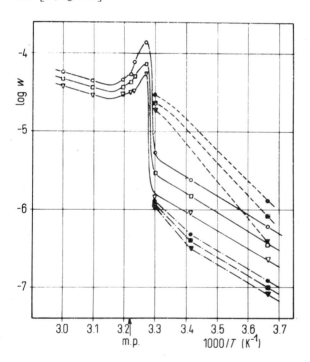

Fig. 9.18. Temperature dependence of the polymerization rate, W, $[W] = mol^{-1} s^{-1}$, of dimethyl itaconate in the liquid, solid, and supercooled liquid (dotted line) phase and of the rate of solid-state post-polymerization at (○) 117.0; (□) 71.0; and (▽) 48.5 r/h dose rates (from [68]).

In the case of 1-vinyl-o-carborane, the mesophase assumed is probably a plastic crystal and is thought to have a role in polymerization [97].

The specialities which might accompany the copolymerizations in liquid crystalline phases have been discussed by P. G. de Gennes [39]. After crosslinking has occurred in a nematic phase, the residue on washing out the solvent will be able to relax without any major obstacle into the isotropic conformation. In a smectic solvent, the chains of the polymer can be expected to be contained between the layers. This would result, upon removal of the solvent, in a piling up of these plates of the copolymer, and the polymer would be strongly crosslinked within the plate but only very weakly between the plates. Crosslinking in a cholesteric mesophase will cause links within the spiral and between the spirals. While the links between the spirals can relax, those within a spiral will not be able to do so. However, experimental proof is still required before we can say definitely that the helical structure can be fixed by copolymerization so that optically active copolymers can be created from optically inactive reagents in an optically active solvent. Copolymers exhibiting a blocked nematic or cholesteric phase morphology were obtained from a variety of mesomorphous monomers under the conditions of a cross-linking polymerization [109]. In solid-state polymerization, the rate is usually more dependent on crystal-lattice parameters than on the reactivity of the monomer, but in the liquid-crystalline state a wide range of organized mobile monomer arrays can markedly favor polymerization or copolymerization. An example is the copolymeriza-

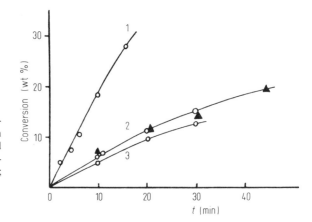

Fig. 9.19. Copolymerization of *p*-methacryloyloxybenzoic acid with styrene at a 1:1 initial monomer feed in different solvents: 1, *p*-cetyloxybenzoic acid; 2, dioxane; 3, DMF; △, acetic acid (from [8]).

tion of p-methacryloyloxybenzoic acid with styrene (fig. 9.19). Of the systems shown, only that with p-cetyloxybenzoic acid is liquid-crystalline (probably smectic) [8]. The copolymer formed in the mesophase exhibits short blocks, while the alternate copolymer predominates in isotropic liquids. Based on the local composition, a homogeneous liquid-crystalline mixture cannot be considered a priori to be a continuum throughout the whole reaction medium. Other studies of mesophase copolymerizations are reported in the literature [73].

Table 9.8 gives an overview of the investigations to date on polymerization in anisotropic liquids.

Table 9.8. Polymerizations in thermotropic mesophases.

Monomer	Object, Problem, and Literature
vinyl oleate	radiation-induced bulk polymerization in the smectic state from -32 to $-18\,^\circ$C [2, 4, 72]
phenylacetylene	thermal polymerization in nematic and cholesteric solvents [15, 16]
2,4-nonadien-1-oic acid	photochemical polymerization in nematic MBBA [104]

Table 9.8. (Continued.)

Monomer	Object, Problem, and Literature
methyl methacrylate	photochemical polymerization in nematic MBBA [104]
alkyl methacrylates	radical polymerization in nematic EBBA and Na oleate [152a]
p-methacryloyloxybenzoic acid CH₂=C(CH₃)COO—◯—COOH	thermal polymerization at 120 or 130°C, probably two different smectic phases; solvent: p-nonyloxybenzoic acid or p-hexadecyloxybenzoic acid [6] radical polymerization with benzoyl peroxide or t-butyl peroxide in the smectic phase, 83–143°C solvent: p-hexadecyloxybenzoic acid [3, 5] radical polymerization with benzoyl peroxide or UV-induced polymerization in the nematic and smectic phases; solvent: p-heptyloxybenzoic acid or p-hexadecyloxybenzoic acid [20–22]
p-(2-vinyloxyethoxy)benzoic acid CH₂=CHOCH₂CH₂O—◯—COOH	radical bulk polymerization in the nematic and smectic phase [165a]
p-hexyloxyphenyl p'-acryloyloxybenzoate CH₂=CHCOO—◯—COO—◯—OC₆H₁₃	radical polymerization in the nematic phase, influence of a magnetic field, oriented polymers [115]
p-[p-(alkoxy)benzoyloxy]phenyl methacrylate CH₂=C(CH₃)COO—◯—OOC—◯—OR	polymerization in bulk and in various solvents, including liquid-crystalline solutions [9, 11]
p-R-phenyl p-(ω-methacryloyloxy) alkoxybenzoates CH₂=C(CH₃)COO(CH₂)ₙO—◯—COO—◯—R	radical polymerization in solution. The polymers exhibit enantiotropic nematic and smectic phases [147h]
p-(ω-methacryloylamino)alkanoyloxyphenyl p-alkoxybenzoates CH₂=C(CH₃)CONH(CH₂)ₙCOO—◯—OOC—◯—OR	no details known to the authors [157a]
N-(p-acryloyloxy)benzylidene-p-aminobenzoic acid CH₂=CHCOO—◯—CH=N—◯—COOH	thermal bulk polymerization in the smectic state [2:711]
N-(p-methacryloyloxy)benzylidene p-aminobenzoic acid CH₂=C(CH₃)COO—◯—CH=N—◯—COOH	thermal bulk polymerization in the smectic and nematic phase [23]
N-(p-acryloyloxy)benzylidene p'-alkoxyanilines CH₂=CH—COO—◯—CH=N—◯—OR	thermal bulk polymerization forming a cross-linked polymer [2:551, 145, 146, 147a]

Table 9.8. (Continued.)

Monomer	Object, Problem, and Literature
N-(p-methacryloyloxy) benzylidene-p'-ethoxyaniline 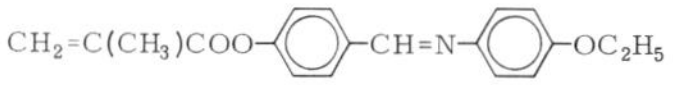 $CH_2=C(CH_3)COO-$⬡$-CH=N-$⬡$-OC_2H_5$	thermal bulk polymerization forming a smectic polymer [2:551, 146, 147a] influence of a magnetic field [2:551, 147, 147a]
alkyl N-(p-acryloyloxybenzylidene)-p-aminocinnamates 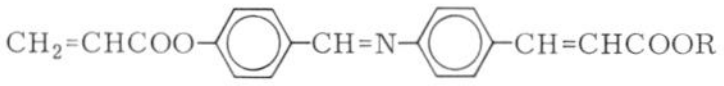 $CH_2=CHCOO-$⬡$-CH=N-$⬡$-CH=CHCOOR$	photochemical polymerization in the s_A phase [2:713]
p-acryloyloxy-p'-methylazoxybenzene 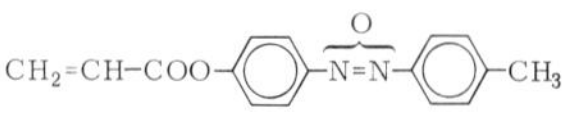 $CH_2=CH-COO-$⬡$-\overset{O}{N=N}-$⬡$-CH_3$	investigation of the monomer/polymer equilibrium in the nematic phase [74]
N-(p-butoxybenzylidene)-p-aminostyrene 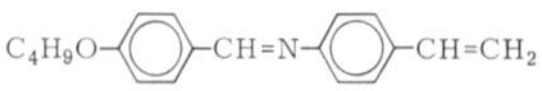 C_4H_9O-⬡$-CH=N-$⬡$-CH=CH_2$	thermal bulk polymerization in the nematic phase [23]
N-(p-methoxy-o-hydroxybenzylidene)-p-aminostyrene 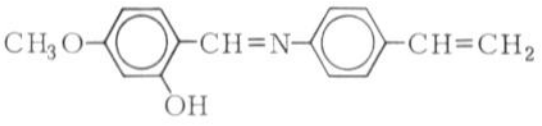 CH_3O-⬡$-CH=N-$⬡$-CH=CH_2$ $\overset{|}{OH}$	radical polymerization with benzoyl peroxide in the nematic phase [137, 138]
N-(p-cyanobenzylidene)-p-aminostyrene 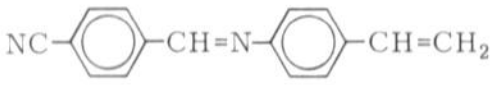 $NC-$⬡$-CH=N-$⬡$-CH=CH_2$	free-radical polymerization in the nematic phase [24a]
N,N'-bis-[p-(acryloyloxy)-benzylidene]-p-diaminobenzene 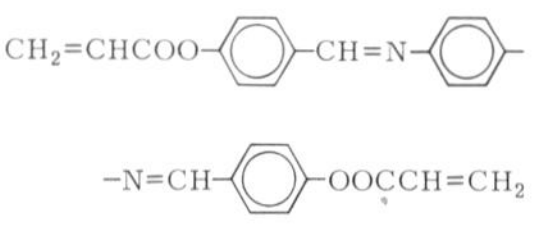 $CH_2=CHCOO-$⬡$-CH=N-$⬡$-$ $-N=CH-$⬡$-OOCCH=CH_2$	polymerization in the nematic and smectic phases, and thermal bulk copolymerization, e.g., with N-p-(acryloyloxy)-benzylidene-p-cyanoaniline in a magnetic field to form a birefringent copolymer film [24, 106–108]
di(N-p-acryloyloxybenzylidene)hydrazine 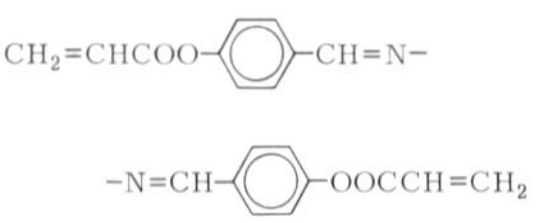 $CH_2=CHCOO-$⬡$-CH=N-$ $-N=CH-$⬡$-OOCCH=CH_2$	thermal bulk polymerization [24]
p-phenylene bis(N-methylene)-p-aminostyrene  $CH_2=CH-$⬡$-N=CH-$⬡$-CH=N-$⬡$-CH=CH_2$	thermal bulk polymerization [24]
cholesteryl acrylate	thermal bulk polymerization [40–42, 169, 170], radical and radiation-induced polymerization in solution [4, 69, 70], and structure of the polymer [71]

Table 9.8. (Continued.)

Monomer	Object, Problem, and Literature
cholesteryl methacrylate	thermal bulk polymerization [42, 151, 164]
cholesteryl N-methacryloyl-aminoalkanoates	no details known to the authors [157, 157b]
cholesteryl vinyl succinate	thermal bulk polymerization [75]
cholesteryl 11-methacryloyloxyundecanoate	radical polymerization [131]
cholesteryl ω-(p-methacryloyloxyphenyl)alkanoates	radical polymerization and copolymerization in solution. The polymers are smectic and cholesteric [147i]
cholestanyl acrylate	thermal bulk polymerization [42]
cholestanyl methacrylate	thermal bulk polymerization [42, 43]

9.2.3 Lyotropic Mesophases

Because of their biological importance (cf. chapter 12), reactions in lyophases deserve far more interest than do chemical reactions in thermotropic mesophases. The little work done to date on this subject leaves large areas of investigation open to future research.

Micelle formation has a very marked influence on the reaction rate. This has been confirmed by numerous experiments, one of which will be discussed here. Chapter 11 will introduce more information on the chemistry of lyophases. The nucleophilic attack of hydroxide or fluoride ions in water at 25°C on p-nitrophenyl diphenyl phosphate is strongly catalyzed by cationic micelles of cetyltrimethylammonium bromide (fig. 9.20), mildly inhibited by salts, and strongly inhibited by anionic micelles of sodium lauryl sulfate or by uncharged micelles of Igepal® [29, see also 30].

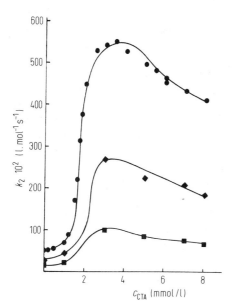

Fig. 9.20. Reaction rates, k_2, vs. the concentration, c, for the hydrolysis of p-nitrophenyl phosphate with 0.01 M sodium hydroxide catalyzed by cetyltrimethylammonium bromide (●) at 25°C; (◆) at 15.4°C; and at (■) 5°C (from [29]).

These problems of colloid chemistry can only be mentioned here, since our main interest is how the formation of a lyophase manifests itself in the reaction process. The reader interested will find a wealth of information in J. Fendler's and E. Fendler's book: "Catalysis in Micellar and Macromolecular Systems" [48]. Supplementary to his studies on liquid-crystalline polymers, H. Ringsdorf has also concerned himself with polymerization reactions in monolayer [147b, 147g], multilayer [147d, 147f], and micellar system [147c, 147e]. For the hydrolysis of procaine in an alkaline medium:

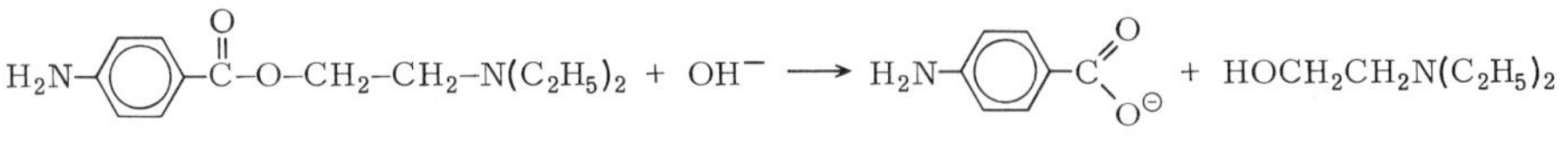

$$H_2N-\!\!\!\bigcirc\!\!\!-\overset{\overset{O}{\|}}{C}-O-CH_2-CH_2-N(C_2H_5)_2 \;+\; OH^- \;\longrightarrow\; H_2N-\!\!\!\bigcirc\!\!\!-C\overset{\nearrow O}{\underset{\searrow O^\ominus}{}} \;+\; HOCH_2CH_2N(C_2H_5)_2$$

markedly lower first-order rate constants were found in the neat phase (comprising 55% polyoxyethylene tridecyl ether in water) than in the isotropic aqueous solutions [134]. Nevertheless, conclusions are difficult to make because the hydrolysis proceeds slowly even in nonmesomorphous solutions of polyoxyethylene. The hydrolysis in the neat phase is characterized by a low Arrhenius activation energy and a large negative entropy.

Reaction rates of the same order of magnitude are found in the hydrolysis of *p*-nitrophenyl laurate in the various phases of the system water/hexadecyltrimethylammonium-bromide/hexanol: isotropic liquid, neat phase, and middle phase [54, 55]. In the lamellar mesophase, the rate constant increases in the water content range from 50 to 70% w/w [55].

The Fischer indole cyclization proceeds at somewhat slower intrinsic rates in a lamellar lyophase as compared to an isotropic liquid of similar chemical composition [148a].

In ternary mixtures of soap, water, and monomer (e. g., styrene, isoprene, dimethylbutadiene), the mesophase structure is always destroyed during polymerization. In all investigations to date the initial organization is lost with the expulsion of the polymer [117]. Under certain circumstances, binary liquid-crystalline mixtures of soap and monomer can polymerize such that the mesomorphous structure is retained. However, in these cases we are not dealing with an organized polymer because after removal of the soap there remains a polymer that no longer possesses any of the original structural features. Organized polymers do form in most cases where the characteristics of both soap and monomer are united in a single substance, as in the case of the vinyl soaps. Monomers and polymers exhibit the same optical anisotropy and yield identical X-ray diagrams (figs. 9.21 and 9.22).

In order to preserve the mesophase structure, such as that of sodium ω-styrylundecanoate in the neat or middle phase, addition of a crosslinking agent, e. g., divinylbenzene, is recommended [116, 160]. One may go one step further by synthesizing organized copolymers from mesomorphous solutions of a polymer in a monomeric solvent. This yields a solid copolymer that retains the structure of a mesomorphous gel (cf. chapter 11). Organized polymers of various structures may be obtained starting with an AB-type block copolymer containing polystyrene and poly-(oxyethylene) dissolved in acrylic acid or methyl methacrylate [158]. A polymerizable solvent is not necessarily a requirement if the mesomorphous polymer solution can be fixed by a crosslinking reaction [159, 161]. This was also observed with a cholesteric PBLG solution with 1,6-hexanediamine as a crosslinking agent [13], cf. section 12.1. Polymers obtained from N,N-dimethacrylamide containing 20% dissolved PBLG exhibit cholesteric reflection colors [170a].

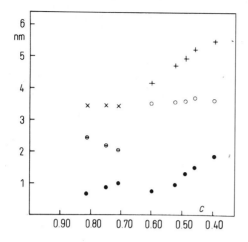

Fig. 9.21. Structure parameters of the mesophases of the system sodium 11-styrylundecanoate/water at 117°C with different concentrations.
Middle phase: +, distance between cylinder axes; o, cylinder diameter.
Neat phase: ×, distance between layers; o, thickness of the soap layer.
Middle at neat phases: ●, thickness of the water layer (from [117]).

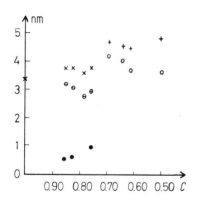

Fig. 9.22. Structure parameters of poly(sodium 11-styrylundecanoate) prepared in the mesomorphous state at different concentrations. The polymers preserve the structure shown in fig. 9.21, the symbols have equivalent meaning (from [117]).

Nuclear Magnetic Resonance and Electron Spin Resonance Studies

10

10 Nuclear Magnetic Resonance and Electron Spin Resonance Studies

(by Ch. Schumann)

10.1 Introduction

This chapter is subdivided into three sections, NMR of mesophases, NMR of solutes oriented in a mesophase matrix, and ESR of mesomorphous systems. The author is aware of the somewhat arbitrary character of the boundaries between these sections. Section 10.2 includes references that are mainly concerned with the study of the mesophase itself, even when dissolved non-mesomorphous molecules are used as probes. Section 10.3 includes papers that are mainly interested in the properties of the dissolved molecules; some of them give results on the properties of the mesophase as well. Section 10.4 consists of studies employing ESR including some papers which report the use of both NMR and ESR.

Because of the wealth of the literature, it was impossible to give a full discussion. Only a concise one was attempted in sections 10.2 and 10.3, and the discussion was omitted in section 10.4. However, a complete registration of the literature was attempted in the tables. For a more detailed discussion, the reader is referred to the review articles. These can be grouped according to the following general subjects:

a) NMR and ESR of thermotropic mesophases: R. Blinc [62, 66, 75], N. Boden [79, 80], R. Ewing [431], J. Jonas [563], H. Kamei [564], A. Lösche [662], G. R. Luckhurst [703, 706], and A. Saupe [6:269, 935].

b) NMR of lyophases: M. Bloom [78], C. L. Khetrapal [583], K. D. Lawson [610], B. Lindman [629, 639, 648], G. J. T. Tiddy [1039, 1042].

c) NMR and ESR of lipids and membranes: M. P. Klein [591], Y. K. Levine [623], J. C. Metcalfe [814].

d) NMR of lipids and membranes: H. Akutsu [1], N. J. M. Birdsall [58], S. I. Chan [154], G. Klose [593], L. W. Reeves [903].

e) ESR of lipids and membranes: F. S. Axel [15], O. H. Griffith [518], D. E. Holmes [549], A. D. Keith [572], J. Seelig [969].

f) NMR of mesophase-oriented solutes: R. A. Bernheim [41], P. M. Borodin [88], J. Bulthius [117], B. P. Dailey [230], P. Diehl [290, 295, 298, 305, 316], J. W. Emsley [417], G. R. Luckhurst [684], L. Lunazzi [721], C. MacLean [768], S. Meiboom [806, 807], L. W. Reeves [889], A. Saupe [935].

g) ESR of mesomorphous systems: R. Lenk [618], G. R. Luckhurst [691], K. Moebius [818], M. Schara [953], J. Seelig [981].

The literature also contains numerous theoretical studies on the three main sections of this chapter. Only an enumeration can be given here.

a) Theory of NMR in mesophases: V. A. Andreev [11], A. Caillé [134], S. I. Chan [153], J. W. Doane [329–331, 342], R. Lenk [619], G. Lindblom [635], A. Loesche [656], T. C. Lubensky [3:483], A. F. Martins [792], P. L. Nordio [841, 842, 846], P. A. Pincus [858–860], C. C. Sung [3:781], C. E. Tarr [1027, 1028], M. Vilfan [1074], C. G. Wade [1079, 1082, 1086], K. H. Weber [1096, 1097], H. Wennerstrom [1100], A. Wulf [1107, 1108], S. Zumer [1117].

b) Theory of NMR of mesophase-oriented solutes: J. M. Anderson [10], M. Barfield [28], A. D. Buckingham [107, 116], D. Canet [144], B. P. Dailey [215], P. Diehl [269, 275, 280, 299, 309, 320], J. W. Emsley [423], D. F. R. Gilson [485], C. W. Haigh [525], J. P. Jacobsen [557], H. Kato [569–571], C. L. Khetrapal [587], N. J. D. Lucas [673–675], C. MacLean [774], S. Meiboom [809], J. I. Musher [829, 830], P. T. Narasimhan [831], A. Saupe [927, 943], L. C. Snyder [992, 993].

c) Theory of ESR in mesophases: H. R. Falle [434], J. H. Freed [445, 449], G. C. Fryburg [457], J. Israelachvili [556], J. I. Kaplan [567], G. R. Luckhurst [3:490, 677, 681, 692, 693, 714], P. L. Nordio [839, 840], T. Pietrzak [857], E. Sackmann [923].

The reader interested in special details may be referred to the following dissertations:

a) NMR studies in mesophases: [30, 38, 151, 192, 197, 198, 392, 435, 460, 473, 546, 551, 594, 606, 671, 788, 797, 816, 821, 828, 832, 833, 852, 853, 916, 961, 1029, 1045, 1075, 1101, 1102, 1110, 1116].

b) ESR studies in mesophases: [135, 795, 872, 982, 1043].

10.2 NMR Studies of Mesophases

10.2.1 Basic Principles

In NMR spectroscopy, the observer is encountered with four phenomena: chemical shifts, spin-spin couplings, relaxation times, and interactions of the nuclear quadrupole moments with electric field gradients. Chemical shifts and indirect, electron-transmitted spin-spin couplings can usually be obtained only in liquids of low viscosity, where the dipolar and quadrupolar couplings are averaged out because of their directional dependence. Conversely, the spectra in solids are obscured by the large intra- and intermolecular dipolar and quadrupolar interactions, and the inherent information is lost in the linewidth which is increased by 3 or 4 orders of magnitude over the liquid spectra. As mesophases hold an intermediate position, it is expected that their NMR spectra present some of the information which is lost in the liquid spectra because of motional averaging, and in the solid spectra because of line broadening. Thus, NMR should be a particularly valuable tool to study these phases and they attracted the attention of the early NMR spectroscopists [995–997]. However, the line widths observed in mesophase spectra are still much larger than in isotropic liquids because of the large number of magnetically active nuclei constituting the mesogen.

Complementary information may be obtained from the measurement of relaxation times, and this proved to be a valuable tool to study the molecular dynamics of the mesophases [837]. Quadrupolar interactions may be studied, for example, after replacement of protons by deuterons and observation of the deuterium resonance [855]. Also, many other nuclei such as ^{7}Li, ^{23}Na, ^{35}Cl, and ^{79}Br, which otherwise are not very useful because of short relaxation times, are open to this approach because their spectra in mesophases are dominated by large quadrupole splittings.

10.2.2 Available Information

10.2.2.1 Orientation

Information about the orientation of mesophases in the presence of a magnetic field or in the presence of both a magnetic and an electric field can be obtained in favorable cases from dipolar splittings, if the distance of the coupling nuclei is known, or from the quadrupolar splittings, if the electric field gradient around the quadrupolar nucleus is known. The technique of measuring the signals from small probe molecules has also been applied (see section 10.3), but it has been restricted mainly to ESR spectroscopy because of its much higher sensitivity.

10.2.2.2 Phase Structure

Information on phase structure — that is the arrangement of molecules within the mesophase — can be drawn from line shape studies. This is particularly applicable to lyophases with their wide variety of phase structures.

10.2.2.3 Phase Transitions

Phase transitions can be determined from discontinuities in the line width parameters or the relaxation times during a change in temperature or composition. Also, the behavior of these parameters close to the transition points can be studied to yield information about the nature of the transition.

10.2.2.4 Molecular Dynamics

Molecular dynamics within a mesophase can be studied with the help of the line width parameters, but the measurement of relaxation times is more informative here. The equilibrium magnetization of an ensemble of nuclear spins is restored after a disturbing process via interaction of the nuclear spins with the motions of the surrounding molecules. This interaction is possible if the Fourier transform of these motions contains a frequency which corresponds to the Larmor frequency of the precessing nucleus. Therefore conclusions concerning the nature and frequency of these motions can be drawn from the relaxation times.

10.2.3 Experimental Methods

10.2.3.1 Line Shape Analysis

Line shape analysis has been used in liquid crystal NMR work since the first investigations by R. D. Spence [995–997] and H. Lippmann and K. H. Weber [654, 655, 1094–1098]. Line shapes can be analyzed in terms of half-widths [376] or second moments [378]. In more favorable cases, line separations can additionally be used to describe the spectra [743].

10.2.3.2 Relaxation Times and Multiple Pulse Techniques*

The first instance of relaxation measurements in mesogens was reported by A. W. Nolle, who used pulse techniques to determine t_1 and t_2 and their temperature dependence in the isotropic liquid phase close to the i/n transition point [837]. The investigation could not be carried to the nematic phase because its relaxation times were too short. More advanced methods later permitted the measurements to be performed in the liquid crystalline state itself [607], allowing the frequency and temperature dependence of the spin-lattice relaxation times to be obtained [60]. P. A. Pincus predicted a $\omega^{1/2}$-dependence of t_1 [859], causing many experimental investigations (R. Blinc [60, 64], B. Cabane [128, 131], J. W. Doane [328], R. Y. Dong [349], C. S. Johnson, Jr. [562], A. F. Martins [791], C. G. Wade [1081, 1082], M. M. Pintar [865]).

With the increasing availability of pulse techniques, an increasing number of papers have appeared concerned with the measurement of t_1, t_2, and the relaxation in the rotating frame, $t_{1\rho}$ [167, 350, 352, 863]. Nuclei other than protons have become accessible to relaxation measurements [130, 131, 178, 183, 624, 630, 631, 643, 644, 1035]. Multiple-pulse techniques have also been used to determine the anisotropy of relaxation times [331, 346, 348, 562, 1017, 1018] and the self diffusion coefficients [60, 61, 63, 65, 334, 476, 540, 1004, 1006].

10.2.3.3 Sample Orientation at the Magic Angle

When the angle, θ, between an internuclear connection vector and the magnetic field is about 55° ($3\cos^2\theta - 1 = 0$), the dipolar coupling between these nuclei is zero. This effect was used by H. J. C. Berendsen who recorded a sharp signal from K-oleate solutions which were oriented between glass plates with their normal being inclined 55° relative to the magnetic field [31]. Spin echoes were measured in the same system by C. G. Wade [1084]. D. Chapman used high-frequency spinning at the magic angle to average out dipolar interactions in liquid-crystalline phospholipids [175].

10.2.3.4 Influence of an Electric Field

W. D. Phillips investigated the influence of a d.c. electric field on the NMR spectra of nematics, observing an increase in ordering parallel to the direction of the applied electric field [856]. R. Y. Dong [346] and C. E. Tarr [1017] measured the anisotropy of t_1 by changing the orientation of the nematic phase by a d.c. electric field. C. E. Tarr found, in the nematic phase of p-methoxybenzilidene-p'-cyano-aniline, an isotropic t_1 for the $=$CH- and CH$_3$-group

---

* Note: Using t for the relaxation times instead of T, we break common practice. This change, however, is in accordance with the recommendations of SI-units and German legislation. T stands for the temperature throughout this book.

and an anisotropic t_1 for the benzene ring protons [1017]. He extended these measurements to MBBA by using low and high frequency a.c. electric fields [1018, 1019]. Here it was found that high frequency fields applied parallel to the magnetic field can change the orientation of the director of the nematic phase relative to the magnetic field from a parallel configuration to a perpendicular one, whereas low frequency electric fields cause turbulences due to the competing influence of the anisotropies of conductivity and of dielectricity. Before this, it had been observed by E. F. Carr that an a.c. electric field causes an orientation which is determined by the anisotropy of the dielectric constant [148]. Therefore, when the dielectric constant is greater perpendicular to the long axis of the molecules than parallel to it, they will orient perpendicular to the magnetic field when the electric field is parallel to it. The angular dependence of the dipolar splitting agreed with the calculated curve when the two fields are at an angle other than 90° [4:139]. t_1 and $t_{1\rho}$ in APAPA as a function of molecular alignment in a d.c. field were investigated by K. R. K. Easwaran [394].

10.2.3.5 Influence of Pressure

The influence of pressure on the nematic order parameter was studied by J. R. McColl [723]. He found the density dependence of intermolecular interactions closer to that predicted for the hard rod model than to the dispersion force model.

10.2.3.6 Magnetic Resonance of Quadrupolar Nuclei

The first study of deuterium resonance in liquid crystals was reported by W. D. Phillips, who used the observed quadrupolar splittings to obtain information on structure and orientation of the thermotropic nematic phase investigated [855, 856]. The deuteron relaxation times of p-hexyloxybenzoic acid were determined by B. Cabane [131] and J. W. Doane [332]. More extensive use of deuterium resonance was made in lyophases [178, 183, 560, 561, 612, 615, 758]. Perdeuterated alkyl chains were used by D. Chapman to study the chain mobility in biological membrane systems [172, 173]. Selectively deuterated lipids were used by J. Seelig analogously to the spin label method but with the advantage that no structural changes are caused by the introduction of deuterium [976]. A number of other quadrupolar nuclei were studied in lyophases, notably by the group of G. Lindblom and B. Lindman. These include ^{7}Li [632], ^{14}N [129, 130, 132], ^{23}Na [560, 625, 628, 632, 643, 781, 885], ^{35}Cl and ^{37}Cl [626, 631], ^{39}K [632], ^{81}Br [624, 627, 630, 631, 643], ^{85}Rb and ^{87}Rb [632, 644], and ^{133}Cs [632].

10.2.3.7 Use of Small Probe Molecules

In this section, work is described which was mainly conducted in order to study the properties of the solvent mesophase by observation of the spectra of solute molecules. A lyophase of a special kind, namely PBLG/methylene chloride was studied by observing the doublet splitting of the methylene chloride component [994]. It was found that in a magnetic field, a slow c/n transition occurs [1077]. The same phase was studied by spin echo experiments [1076]. The structure of the dimethyl tin ion was determined by L. W. Reeves with the intention to obtain information about its participation in the electric double layers in a soap mesophase [893]. Monofluorostearic acids were used as probe molecules in lecithin vesicles for observing the fluorine resonance linewidths [49]. S. Meiboom used methylene chloride dissolved in several smectogens to study their behavior as a function of temperature and under sample rotation relative to the magnetic field [810, 812]. J. W. Doane [333] and H. Spiesecke [1004, 1006] measured diffusion coefficients and their anisotropies of TMS and other small molecules dissolved in nematics.

10.2.4 Results

10.2.4.1 Orientation

It was pointed out by S. K. Ghosh that the measurement of the order parameter from line positions introduces an uncertainty which can be avoided by use of the complete analysis of the involved *p*-disubstituted aromatic spin system [475]. Order parameter measurements of six nematics as a function of temperature were reported by M. N. Avadhanlu, and it was concluded that the O—C-bond rotation is hindered in the acetoxy compound [12, 13]. The orientation parameter of MBBA was measured by J. P. Le Pesant [620], and its temperature dependence could be fitted to an expression given by P. A. Pincus. Nematic APAPA was investigated by H. A. Moses [825] and, more recently, by K. R. K. Easwaran [394]. The orientation parameter was deduced from a doublet splitting due to adjacent benzene ring protons and its temperature dependence was determined. A continuously decreasing orientation was found with increasing temperature. Additionally, t_1 and $t_{1\rho}$ and their dependence on a d.c. electric field were studied [394]. The orientational behavior of nematic PAA on rotation in the magnetic field was studied by H. Lippmann [654]. The non-spinning sample is aligned parallel to the magnetic field. When the sample is rotated about an axis perpendicular to the field, the orientational director is twisted away from the direction of the magnetic field. The angle between the magnetic field and the orientational director increases until a critical spinning speed is reached. The behavior of the nematic director above the critical spinning speed is not yet fully understood [695].

S. Meiboom's results confirmed earlier work which had shown that orientation of smectics is possible by cooling an i or n phase in the magnetic field [810, 812]. The use of small probe molecules permitted a more detailed analysis than the use of line shapes and second moments [1094]. An oriented smectic phase retains its director with respect to the sample tube when rotated in the magnetic field [810], see also section 10.2.4.2.

K. D. Lawson and T. J. Flautt discovered a nematic lyophase which is oriented macroscopically by a magnetic field [611]. The orientation is exhibited by the doublet splitting in the deuteron resonance of the heavy water solvent. It is retained on spinning the sample, indicating that the director is perpendicular to the magnetic field. Most lyophases cannot be oriented macroscopically by a magnetic field. R. Blinc found powder type deuteron resonance spectra for the smectic, lamellar neat phase as well as for the hexagonal middle phase of the sodium palmitate/heavy water system [61]. The quadrupole splitting is small in both cases, indicating a small microscopic anisotropy in the water motion with an orientation parameter of some 10^{-3}. It increases first with increasing temperature, the maximum being positioned in the middle of the mesophase temperature region. A. Johansson also observed powder spectra for the lamellar mesophase of H_2O, *n*-octylamine, and *n*-octylamine hydrochloride [559]. He was able to deduce theoretical lineshapes for the non-spinning and the spinning case by using the assumption that the lineshape arises from magnetic susceptibility anisotropies experienced by the water molecules. Additionally, he found that it is possible to achieve a macroscopic orientation of the lamellae by a very slow cooling of an isotropic solution in the magnetic field. The orientation is such that the lamellae are perpendicular to the magnetic field and the long axes of the surfactant molecules parallel to the magnetic field, as evidenced by measured and calculated line shapes of the water resonance. M. P. McDonald achieved orientation of a smectic lyophase by stacking a mixture of 1-mono-octanoin between glass plates [746]. Two doublet splittings were found, a small one for the hydroxyl protons and a large one for the alkyl chain protons. Both showed the same $(3\cos^2\theta - 1)$ dependence on rotating the glass plates with respect to the magnetic field. This angular dependence identifies the splittings as the result of dipole-dipole couplings, and their magnitude shows that the anisotropy of the motion of the water molecules is much smaller than that of the lipid molecules.

10.2.4.2 Phase Structure

S. Meiboom obtained macroscopically oriented samples of smectics A, C, and B [810], see also section 10.2.4.1. He showed that the line shapes of the mesophases and of small probe molecules dissolved in the mesophases are in agreement with a layer structure, the long axes of the molecules being arranged parallel to the layer normals for the s_A phase. Going to the s_C phase, the molecules retain their orientation, but a tilt angle is introduced between the long axes of the molecules and the layer normals. This leaves the layer orientation undefined, and the sample breaks up into domains with different layer orientations. The layer normals of the different domains are distributed evenly on a cone with a vertex angle of twice the tilt angle. When the sample is rotated in the field, the smectic layers retain their orientation with respect to the sample tube, but the molecules are free to reorient, keeping the tilt angle constant but minimizing the angle between their long axes and the magnetic field. Using this model and the tilt angle as adjustable parameter, it was possible to account for the experimental line shapes and splittings observed for the s_C phases. Very similar results were obtained by J. W. Doane [337]. S. Meiboom observed a pronounced difference in the NMR-spectra of smectics B_A and B_C [810]. Lines are considerably wider in both s_B phases than in the s_A and s_C phases; there exists a two-dimensional order within the layers, and intermolecular dipolar interactions are not averaged to zero. In the smectic B_A-phase, there is a pronounced $(3\cos^2\theta - 1)$ dependence of the line spacings on sample rotation indicating a uniaxial structure, whereas there remain wide lines for all orientations in the smectic B_C because there is again a tilt angle between the layer normal and the molecular long axes. There is no angle at which all dipolar interactions vanish simultaneously (note: smectic B_C is now s_H).

From measurements of spin-spin-relaxation times in MBBA and HOAB (*p-n*-heptyloxy-azobenzene), R. Y. Dong concluded that a short range smectic order exists even 10° above the s_C/n transition temperature in HOAB [355]. A maximum of t_2 was found on rotating the sample 75° from its original position. This maximum did not change with temperature, indicating a constant tilt angle for this particular s_C phase.

Lyophases exhibit a great variety of structures and have been studied extensively by NMR methods (A. Johannson, B. Lindman). For instance, aqueous solutions of sodium caprylate, N-(trimethylamino)-dodecanimide, and Aerosol® OT were studied by NMR and other methods; fibrous, lamellar, and statistically isotropic structures were found [11:420]. Cubic isotropic phases, which give high resolution NMR spectra, were also found in aqueous dodecyl trimethyl-ammoniumchloride solutions.

Biological membranes have structures similar to those found in many lyotropic mesophases, and consequently have received much attention by NMR spectroscopists. L. D. Bergelson succeeded in differentiating the interior and exterior surfaces of vesicles produced by treating egg lecithin dispersions in water with ultrasound [35, 36]. On the addition of paramagnetic ions such as Mn^{2+} or Eu^{3+}, two signals appear for the $\overset{+}{N}(CH_3)_3$-groups which were ascribed to molecules at the inside and the outside of the membranes. D. Chapman found a lamellar arrangement of myelin membranes. After sonication of these membranes, it could be seen by high resolution NMR that cholesterol influences the lipid chain mobility and is itself prevented from moving isotropically [163]. E. G. Finer investigated the lecithin/cholesterol interactions in more detail [440]. He found that lecithin and cholesterol form an equimolar complex with a lifetime of greater than 30 ms. In this complex, the mobility of the first ten methylene groups of each hydrocarbon chain is restricted severely, while the chain ends are almost unaffected. The chain-melting transition is removed by the influence of cholesterol, resulting in a material which is not sensitive to temperature changes over a range of 50°. The influence of Ca^{2+} and local anesthetics on membrane structure was investigated by D. O. Shah using a model membrane of hexadecane, water, potassium oleate, and hexanol [985]. A destructive influence of these substances on the liquid crystal structure was reported. Apparently the anesthetic effect is correlated to the influence of these molecules on the membrane structure.

J. M. Steim used high-resolution techniques to study the structures of lipoproteins [1011]. The results were consistent with a micellar structure rather than a hydrophobic binding, which was expected to cause extensive line broadening.

10.2.4.3 Phase Transitions

Phase transitions can be studied in more detail by NMR spectroscopy than is possible by other methods. This was demonstrated by J. W. Doane [335]. Here a homologous series of thermotropic liquid crystals and their n/s_A transitions were investigated by measurement of the order parameter. It was found that the transition approaches a second-order characteristic with decreasing alkyl chain length. Other workers used relaxation time measurements to detect phase transitions, e.g. two different mesophases were found in vinyl oleate at low temperatures [789]. An increase in proton spin-lattice relaxation time in a small temperature range close to the n/k transition in MBBA indicated the existence of a biaxial nematic phase [354]. The c/n transition in mixtures of PAA and cholesteryl-chloride was studied by means of t_1 and $t_{1\rho}$ measurements. Above a certain temperature, the magnetic field is strong enough to unwind the cholesteric and a nematic phase results. This is indicated by a change in the temperature dependence of the relaxation times, which increase with temperature in the cholesteric but remain constant in the nematic. A phase change within the s_A phase was discovered by means of proton relaxation measurements [359].

Phase transitions in lyophases have been determined since the original work of K. D. Lawson [609]. Examples are found in the work of C. A. Gilchrist [11:420], M. P. McDonald [744], and D. Chapman [159, 161, 164].

10.2.4.4 Molecular Dynamics

NMR spectroscopy has contributed a wealth of information about the dynamic properties of liquid crystals. Since the notable theoretical work of P. A. Pincus [859] and J. W. Doane [329, 330], many papers have appeared concerning spin relaxation measurements which arrived at conclusions regarding the molecular dynamics. Only a few of the more recent ones can be mentioned here. R. Blinc found that two different mechanisms are responsible for proton spin-lattice relaxation in nematics [64]; in MBBA it is controlled by molecular diffusion, but it is controlled by collective order fluctuations in PAA. R. Y. Dong showed that the extended Pincus theory of long-range order fluctuations and short-range molecular reorientations cannot account for the observed relaxation data of several nematics [353]. C. G. Wade performed a relaxation study of PAA and methyl-deuterated PAA in fully deuterated PAA, in order to separate intermolecular and intramolecular contributions [1083]. It was found that both mechanisms make substantial contributions to the observed spin-lattice relaxation times in PAA and that the interaction of methyl protons with phenyl protons is essentially intermolecular [1086]. These results seem to be in contradiction to the observed $\omega^{1/2}$-dependence of t_1^{-1} [64]; however, the influence of the end chains is not fully understood and may obscure the results. It has been shown in an earlier study that in the presence of long end chains the relaxation rates are greater, indicating relaxation by internal motions of the chains [1081].

The Pincus theory of long range order fluctuations seems to be completely inapplicable for lyophases. L. A. McLachlan [757] investigated the lyotropic system described by K. D. Lawson [614] with spin-lattice relaxation measurements and obtained a smooth decrease of relaxation rates versus temperature, regardless of the intervening phase transitions from viscous isotropic to nematic and from nematic to fluid isotropic. It was possible to fit the curves to two motional correlation times: one intramolecular, describing rotation about the long axis of the alkyl chains, and one intermolecular, describing translational diffusion between micelles or over their surface. The broad lines which were found in the nematic phase despite the comparatively long spin-lattice relaxation times were explained by incomplete averaging of dipolar interactions, the molecules spending most of their time incorporated in the anisotropic micelles. Similarly, J. Charvolin concluded from t_D, t_1, and $t_{1\rho}$-measurements in the potassium laurate/D_2O system that there is rapid translational diffusion in all phases encountered [185]. Chain conformation jumps give an additional mechanism for spin-lattice relaxation. It was suggested that the two types of motion might be not completely independent. In the sodium caprylate/decanol/water system, G. J. T. Tiddy found similar t_1 and t_2-values for the alkyl chain protons in three different phases and concluded that the average rotational correlation time does not change from phase to phase [1033]. In phases with high water content, the water t_1 values show that there is little restriction on water mobility. There is a sharp increase in the relaxation rate when the caprylate/water molar ratio exceeds 0.05 or the average water spacings decrease below 2 nm. This is suggested to be due to the restriction of water motion by counter-ion binding. The water t_2-values are dominated by the contribution from the decanol-OH proton due to OH proton exchange. The conclusion was drawn that molecular mobilities are similar in three different lamellar phases, the main parameter being the average spacing of water molecules. Similar results were obtained for mesophases containing fluorocarbon surfactants [1032, 1034, 1035]. The fluorine t_1-values indicate that the rotation of CF_2-groups about the molecular axis is slower than in hydrocarbon chains; proton t_1-values of hydrocarbon alcohols contained in the same phase showed that chain motions are mainly determined by the chain length [1032]. There is no noticeable distribution of motional correlation times within the same chain, even in the 12,12,13,13,14,14,14-heptafluoromyristate chain where part of the protons are replaced by fluorines [1035].

The study of molecular dynamics by NMR techniques also seems to be promising for biological liquid crystal and membrane systems. Notably, the work of D. Chapman should be mentioned here. It was shown that natural abundance ^{13}CMR is a valuable tool to study liquid crystal systems of biological significance [169]. 220 MHz proton high resolution spectra and relaxation measurements were used to study the mobility of hydrocarbon chains and head groups in lipid/water systems [171]. 220 MHz studies were performed on the effect of cholesterol on cerebrosides and sphingomyelins [174]. It was found that cholesterol removes the phase transition in both cases, and this produces a liquid crystalline state where the hydrocarbon chains have a mobility which is an intermediate between the two phases formed without cholesterol and is not very sensitive to temperature changes. It was concluded that the function of cholesterol is the regulation of fluidity and permeability of membranes in mammalian systems. E. G. Finer studied the frequency dependence of high resolution spectra obtained by sonication of lecithin dispersions and attributed this dependence to the different extent of averaging out dipolar interactions by particle tumbling [438]. S. I. Chan performed relaxation measurements on similar unsonicated systems and analyzed the t_1 and t_2 data in terms of mobility of hydrocarbon chains [153]. He suggested that the hydrophobic bilayer core is more appropriately described as a soft solid than as a mobile fluid.

10.2.5 Table of Thermotropic Mesophases

In this table, the mesophases are arranged according to their sum formula. The columns "methods" and "results" correspond to the same headings of the previous sections. If no nucleus is specified, only proton resonance spectra are studied. The following abbreviations are used:

Methods:

LS	Line shape analysis (line separations, line width, 2nd moment)
$t_1, t_{1\rho}, t_2$:	Relaxation times
Δt_1:	Anisotropy of t_1
SpE:	Spin echo technique
MPT:	Multiple pulse technique
MA, MAR:	Studies of phases inclined at an angle of 55° (magic angle) with respect to the magnetic field or spinning about an axis inclined at 55°.
E:	Application of an electric field
p:	Application of pressure
PrMol:	Use of small probe molecules
DNP:	Dynamic nuclear polarization

Results:

S	Orientation
Str	Phase structure
Tr	Phase transitions
Dy	Molecular dynamics
$\Delta\sigma$	Anisotropy of chemical shifts

Formula	Structure	Author and ref.	Methods	Results
$C_4H_{14}B_{10}$	HC——C–CH=CH$_2$ / $B_{10}H_{10}$	Klingen [592]	LS	Str
$C_6H_{11}KO_2$	$C_5H_{11}COOK$	Dunell [381]	LS	Tr, Dy
$C_8H_{15}KO_2$	$C_7H_{15}COOK$	Dunell [381]	LS	Tr, Dy
$C_{10}H_{12}O_3$	 C_3H_7O—⬡—COOH	Phillips [856]	LS (^{1}H, ^{2}H)	S, Str
$C_{10}H_{19}KO_2$	$C_9H_{19}COOK$	Dunell [379]	LS	Tr
$C_{11}H_{14}O_3$	C_4H_9O—⬡—COOH	Breternitz [94] \ Loesche [661]	LS, SpE \ t_1	Dy
$C_{11}H_{21}NaO_2$	$C_{10}H_{21}COONa$	van Putte [1046, 1047]	t_1	Dy
$C_{12}H_{16}O_3$	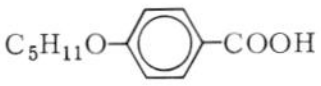 $C_5H_{11}O$—⬡—COOH	Breternitz [94]	LS, SpE	Dy
$C_{12}H_{23}KO_2$	$C_{11}H_{23}COOK$	Dunell [379]	LS	Tr
$C_{12}H_{23}NaO_2$	$C_{11}H_{23}COONa$	Lawson [608] \ van Putte [1046, 1047]	LS \ t_1	Tr, Str \ Dy
$C_{13}H_{18}O_3$	$C_6H_{13}O$—⬡—COOH	Breternitz [94] \ Cabane [131] \ Phillips [856] \ Phillips [855]	LS, SpE \ t_1 \ ^{1}H, ^{2}H, LS \ ^{2}H, LS	Dy \ S, Dy \ S, Str \ S, Str
$C_{13}H_{25}NaO_2$	$C_{12}H_{25}COONa$	van Putte [1046, 1047]	t_1	Dy
$C_{14}H_{14}N_2O_3$	MeO—⬡—N=N—⬡—OMe / O \ (PAA)	Allison [6] \ Blinc [60] \ Blinc [63] \ Blinc [64] \ Blinc [69] \ Cabane [128] \ Cabane [129] \ Cabane [130] \ Diehl [312] \ Doane [328] \ Doane [343] \ Dong [349] \ Dong [350] \ \ \ \ Dong [352, 361] \ Dong [363]	^{13}C \ t_1, SpE \ SpE \ t_1 \ \ t_1 \ ^{14}N, LS \ ^{14}N, t_1, t_2 \ LS (^{2}H, deut.) \ t_1 \ LS (^{1}H, deut.) \ t_1, $t_{1\rho}$ \ t_1, $t_{1\rho}$ \ \ \ \ t_1 \ SpE (deut.)	S \ Dy \ Dy \ Dy \ \ Tr, Dy \ Tr, Dy \ Dy \ S, Dy \ S, Dy \ Dy \ Dy \ Tr, Dy \ (+chol- \ esteryl \ chloride) \ Dy \ Dy

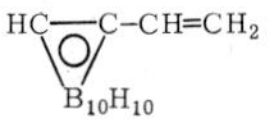

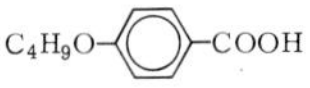

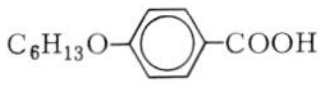

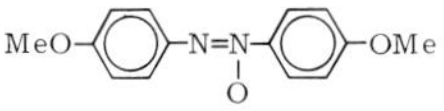

Formula	Structure	Author and ref.	Methods	Results
		Dong [356]	t_1 (^{2}H, deut.)	Dy
		Dong [365]	LS (^{1}H, ^{2}H, deut.)	S, Dy
		Dong [368]	LS (^{14}N)	S, Dy
		Dong [369]	t_1 (^{1}H, ^{2}H, deut.)	S. Dy
		Dong [370]	LS (^{2}H, deut.)	S, Str
		Dong [371]	t_1 (^{1}H, deut.)	Dy
		Dong [372]	t_1, $t_{1\rho}$	S, Tr
		Ghosh [476]	SpE	Dy
		Ghosh [477]	LS	S
		Ghosh [478]	t_1, t_2	Dy
		Goren [514]	t_1	S
		Grant [515]	^{13}C, t_1	Dy
		Hayamizu [537]	t_1 (^{1}H, deut.)	S, Tr, Dy
		Johnson [562]	t_1	Dy
		Lippmann [654]	LS	S
		Lippmann [655]	LS	Str, Dy
		McColl [723]	LS, P	S, Str
		McColl [724]	LS, PrMol	S, Tr
		McColl [725]	LS	S
		Martins [790, 793]	t_1 (deut.)	Dy
		Noack [835]	t_1	Dy
		Nolle [837]	t_1, t_2	Dy
		Nordio [844]	t_1 (^{1}H, ^{2}H, ^{13}C)	S, Dy
		Nordio [845]	t_1	Dy
		Pines [861]	LS (^{13}C)	S, Tr, Dy
		Phillips [855]	^{2}H, LS	S, Str
		Phillips [856]	^{1}H, ^{2}H, LS	S, Str
		Pintar [863]	$t_{1\rho}$	Dy
		Pintar [865, 866]	t_1	Dy
		Pintar [868, 869]	t_1, $t_{1\rho}$	Dy
		Pocsik [871]	LS	Dy
		Spence [995, 999]	LS	S
		Spence [996, 997]	LS	S, Str
		Wade [1081, 1083]	t_1	Dy
		Wade [1089]	t_1 (^{2}H, deut.)	Dy
		Wade [1090]	t_1 (^{1}H, deut.)	Dy
		Weber [1094]	LS	S, Str
	(deut.)	Spence [998]	LS	S, Str
$C_{14}H_{27}KO_2$	$C_{13}H_{27}COOK$	Dunell [379]	LS	Tr
$C_{14}H_{27}LiO_2$	$C_{13}H_{27}COOLi$	van Putte [1046, 1047]	t_1	Dy
$C_{14}H_{27}NaO_2$	$C_{13}H_{27}COONa$	Lawson [608]	LS	Tr, Str
		van Putte [1046, 1047]	t_1	Dy
$C_{15}H_{12}N_2O$	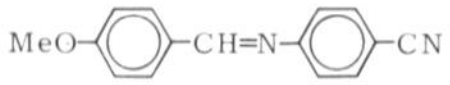 MeO—⬡—CH=N—⬡—CN	Dong [348]	t_1	Dy
		Dong [365]	LS	S, Dy
		Moses [827]	LS	S
		Tarr [1017]	t_1, E	Dy

Formula	Structure	Author and ref.	Methods	Results
$C_{15}H_{22}O_3$	$C_8H_{17}O$—⬡—COOH	Meiboom [810]	LS, PrMol	S, Tr, Str
		Pintar [868]	$t_1, t_{1\rho}$	Dy
$C_{15}H_{29}LiO_2$	$C_{14}H_{29}$COOLi	van Putte [1046, 1047]	t_1	Dy
$C_{15}H_{29}NaO_2$	$C_{14}H_{29}$COONa	van Putte [1046, 1047]	t_1	Dy
$C_{16}H_{14}N_2O$	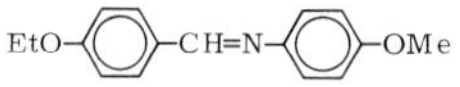 EtO—⬡—CH=N—⬡—OMe	Moses [827]	LS	S
$C_{16}H_{15}NO_2$	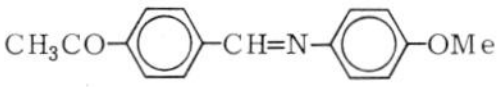 CH_3CO—⬡—CH=N—⬡—OMe	Avadhanlu [13]	LS	S, Dy
$C_{16}H_{15}NO_3$	MeO—⬡—CH=N—⬡—OCOCH$_3$ (APAPA)	Carr [148]	E	
		Derzhanski [264]	LS	
		Dong [365]	LS	S, Dy
		Easwaran [394]	LS, $t_1, t_{1\rho}$	S, Dy
		Goren [512, 514]	t_1	S
		Moses [825]	LS	
$C_{16}H_{16}N_2O_2$	MeO—⬡—CH=N—N=CH—⬡—OMe	Blinc [59]	t_1	Tr, Dy
		Blinc [63]	SpE	Dy
		Dong [365]	LS	S, Dy
		Spence [997]	LS	S, Str
$C_{16}H_{16}N_2O_3$	EtO—⬡—N=N—⬡—OCOCH$_3$	Tarr [1023]	LS, E	S
$C_{16}H_{18}N_2O_3$	EtO—⬡—N=N—⬡—OEt, O (PAP)	Blinc [60]	t_1	Dy
		Blinc [63]	SpE	Dy
		Dong [365]	LS	S, Dy
		Phillips [855]	^{2}H, LS	S, Str
		Phillips [856]	^{1}H, ^{2}H, LS	S, Str
		Spence [997, 998]	LS	S, Str
		Spence [999]	LS	S
		Wade [1081]	t_1	Dy
		Weber [1094]	LS	S, Str
$C_{16}H_{24}O_3$	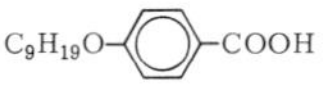 $C_9H_{19}O$—⬡—COOH	Loesche [661]	t_1	
$C_{16}H_{31}KO_2$	$C_{15}H_{31}$COOK	Dunell [379]	LS	Tr
$C_{16}H_{31}LiO_2$	$C_{15}H_{31}$COOLi	van Putte [1046, 1047]	t_1	Dy
$C_{16}H_{31}NaO_2$	$C_{15}H_{31}$COONa	Lawson [608]	LS	Str, Tr
		van Putte [1046, 1047]	t_1	Dy

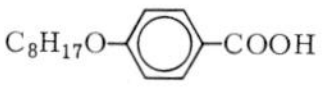

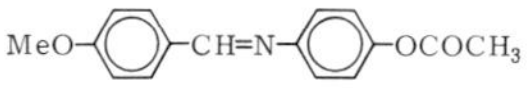

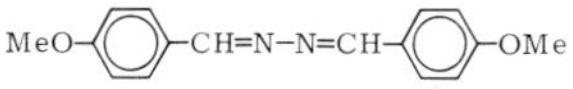

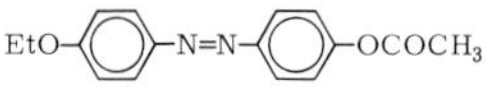

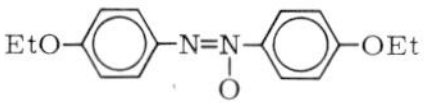

Formula	Structure	Author and ref.	Methods	Results
$C_{17}H_{17}NO_3$	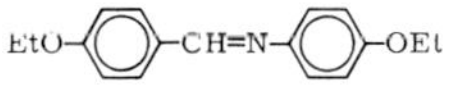 EtO—⬡—CH=N—⬡—OEt	Moses [827]	LS	S
	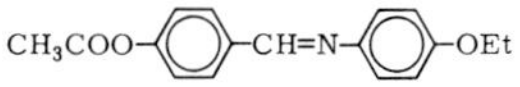 CH_3COO—⬡—CH=N—⬡—OEt	Avadhanlu [12]	LS	S, Dy
$C_{17}H_{20}N_2O_2$	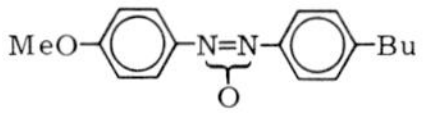 MeO—⬡—N=N—⬡—Bu (O)	Spiesecke [1006]	SpE, PrMol	Dy
$C_{17}H_{33}LiO_2$	$C_{16}H_{33}COOLi$	van Putte [1046, 1047]	t_1	Dy
		van Putte [1048]	t_1 (^{2}H, deut.)	Dy
$C_{17}H_{33}NaO_2$	$C_{16}H_{33}COONa$	van Putte [1046, 1047]	t_1	Dy
$C_{18}H_{18}N_2O_5$	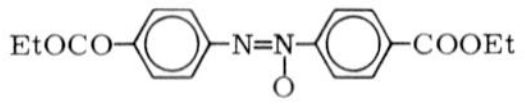 EtOCO—⬡—N=N—⬡—COOEt (O)	Allison [6]	^{13}C	S
		Dong [358]	t_1	Dy
		Dong [368]	LS (^{14}N)	S, Dy
		Meiboom [810]	LS, PrMol	S, Tr, Str, Dy
		Moses [826]	LS	Tr, Str
		Weber [1098]	LS	Tr, Str, Dy
$C_{18}H_{19}N$	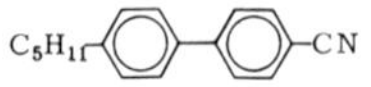 C_5H_{11}—⬡—⬡—CN	Luckhurst [711]	LS (^{2}H, deut.)	S, Dy
$C_{18}H_{19}NO$	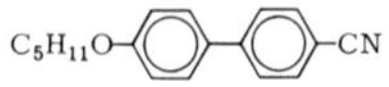 $C_5H_{11}O$—⬡—⬡—CN	Kamezawa [566]	LS, PrMol	S, Dy
$C_{18}H_{20}N_2O_3$	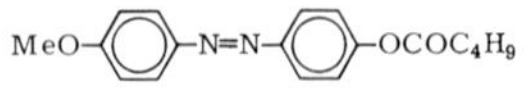 MeO—⬡—N=N—⬡—OCOC$_4$H$_9$	Spiesecke [1004]	t_1, SpE, PrMol	Dy
$C_{18}H_{21}NO$	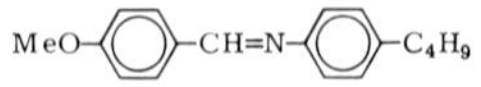 MeO—⬡—CH=N—⬡—C$_4$H$_9$ (MBBA)	Avadhanlu [14]	LS	S, Dy
		Blinc [64]	t_1	Dy
		Blinc [65, 68, 71]	MPT (deut.)	Dy
		Blinc [67]	t_1, LS (deut.)	Tr, Dy
		Blinc [69]		
		Blinc [73]	t_1	S, Dy
		Blinc [76]	LS, t_1, $t_{1\rho}$	S, Dy
		Bossard [92]	SpE, PrMol	Dy
		Clin [195]	LS (^{13}C)	
		Diehl [312]	LS (^{2}H, deut.)	S, Dy
		Doane [332]	$t_{1\rho}$, t_1, PrMol	Dy
		Doane [333]	PrMol	Dy
		Doane [339]	$t_{1\rho}$	Dy
		Dong [351, 361]	t_1	S, Dy
		Dong [354]	t_1	Tr, Dy
		Dong [355]	t_2	Str, Tr, Dy
		Dong [357, 364]	t_1	Dy
		Dong [360]	LS, t_1	Dy
		Dong [363]	SpE	Dy
		Dong [365]	LS (^{1}H, ^{2}H, deut.)	S, Dy
		Dong [366]	t_1 (^{1}H, deut.)	S, Dy

Formula	Structure	Author and ref.	Methods	Results
		Dong [369]	t_1 (^{2}H, deut.)	S, Dy
		Dong [370]	LS (^{2}H, deut.)	S, Str
		Dong [371]	t_1 (^{1}H, deut.)	Dy
		Ghosh [474]	t_1	Dy
		Ghosh [476]	SpE	Dy
		Hayward [540]	t_1, t_2, SpE	Dy
		Johnson [562]	t_1, LS, E	S, Dy
		Lee [617]	LS (^{1}H, ^{2}H, deut.)	S, Str
		Le Pesant [620]	LS	S
		Le Pesant [621]	DNP	Dy
		Loesche [660]	LS	S
		Noack [836]	t_1	Dy
		Nordio [845]	t_1	Dy
		Pines [862]	^{13}C, LS	S, Tr
		Pintar [870]	t_1, $t_{1\rho}$	S, Dy
		Rose [914]		
		Spiesecke [1006]	SpE, PrMol	Dy
		Tarr [1018]	LS, t, E	S, Dy
		Tarr [1019]	t_1, E	S, Dy
$C_{18}H_{22}N_2O_3$	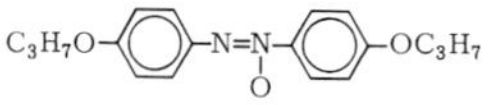 C_3H_7O—⬡—N=N—⬡—OC_3H_7 ‖ O	Moses [826]	LS	Tr, Str
		Wade [1081]	t_1	Dy
		Weber [1095]	LS	Tr, S, Str
$C_{18}H_{33}NaO_2$	$C_{17}H_{33}COONa$	Dunell [385]	LS, t_1	Dy
		Spence [997]	LS	
		van Putte [1047]	t_1	Dy
$C_{18}H_{35}CsO_2$	$C_{17}H_{35}COOCs$	Dunell [380]	LS	Tr
$C_{18}H_{35}KO_2$	$C_{17}H_{35}COOK$	Dunell [377]	LS	Tr, Dy
$C_{18}H_{35}LiO_2$	$C_{17}H_{35}COOLi$	Dunell [383]	LS	Str, Tr
		Dunell [384]	LS	Tr
$C_{18}H_{35}NaO_2$	$C_{17}H_{35}COONa$	Dunell [376]	LS	Tr
		Dunell [378, 382]	LS	Tr, Dy
		Dunell [385]	LS, t_1	Dy
		Lawson [608]	LS	Tr, Str
		Spence [997]	LS	
$C_{18}H_{35}O_2Rb$	$C_{17}H_{35}COORb$	Dunell [380]	LS	Tr
$C_{18}H_{36}O_2$	$C_{17}H_{35}COOH$	Dunell [383]	LS	Tr
$C_{19}H_{19}NO_3$	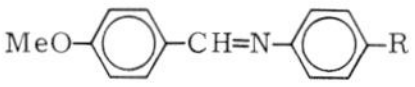 MeO—⬡—CH=N—⬡—R R: CH=CH—COOEt	Dong [359]	t_1	Tr, Dy
		Meiboom [810]	LS, PrMol	S, Tr, Str
		Tarr [1020]	t_1	Dy
		Tarr [1021]	LS, t_1	S, Dy
		Tarr [1022]	MPT	Dy

Formula	Structure	Author and ref.	Methods	Results
$C_{19}H_{23}NO$	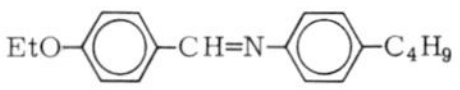 EtO—⬡—CH=N—⬡—C_4H_9 (EBBA)	Avadhanlu [14]	LS	S, Dy
		Blinc [76]	LS, t_1, $t_{1\rho}$	S, Dy
		Dong [353]	t_1	Dy
		Goren [510]	t_1	Dy
		Hayward [540]	t_1	Dy
		Loesche [660]	LS	S
		Loesche [661]	t_1	
		Prasad [873]	LS (^{1}H,^{2}H, deut.)	S, Str
$C_{19}H_{23}NO_2$	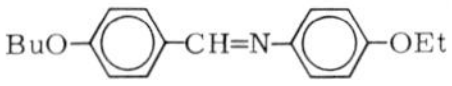 BuO—⬡—CH=N—⬡—OEt	Avadhanlu [12, 14]	LS	S, Dy
$C_{19}H_{37}NaO_2$	$C_{18}H_{37}COONa$	van Putte [1047]	t_1	Dy
$C_{20}H_{14}F_2N_2$	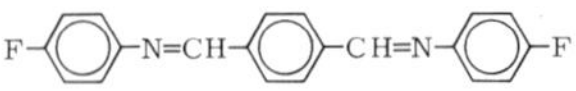 F—⬡—N=CH—⬡—CH=N—⬡—F	Doane [332]	t_1, $t_{1\rho}$	Dy
$C_{20}H_{17}N_3O$	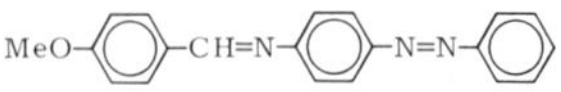 MeO—⬡—CH=N—⬡—N=N—⬡	Avadhanlu [13]	LS	S, Dy
		Loesche [661]	t_1	
$C_{20}H_{21}NO_3$	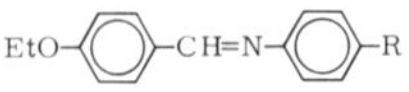 EtO—⬡—CH=N—⬡—R R: CH=CH–COOEt	Spiesecke [1009]	t_1, SpE	Dy
$C_{20}H_{22}O_6$	 EtO—⬡—OOC—⬡—OCOOBu	Fung [466]	PrMol	Dy
		Johnson [562]	t_1, LS	S, Dy
		Spiesecke [1004]	t_1, SpE, PrMol	Dy
$C_{20}H_{23}N$	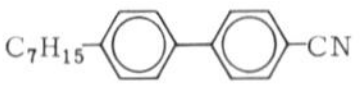 C_7H_{15}—⬡—⬡—CN	Kamezawa [566]	LS, PrMol	S, Dy
$C_{20}H_{23}NO$	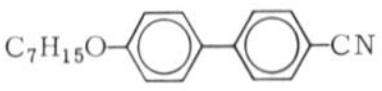 $C_7H_{15}O$—⬡—⬡—CN	Kamezawa [566]	LS, PrMol	S, Dy
$C_{20}H_{24}N_2O_3$	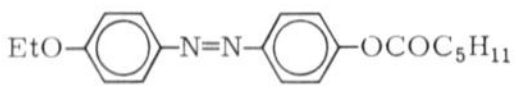 EtO—⬡—N=N—⬡—$OCOC_5H_{11}$	Johnson [562]	LS, t_1	S, Dy
$C_{20}H_{24}O_4$	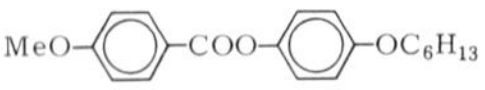 MeO—⬡—COO—⬡—OC_6H_{13}	Loesche [663]	LS	Tr, Dy
		Schulze [963]	LS	S
$C_{20}H_{26}N_2O_3$	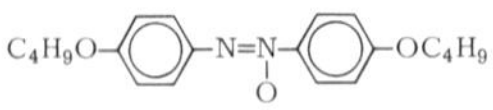 C_4H_9O—⬡—N=N—⬡—OC_4H_9 　　　　　　O	Moses [826]	LS	Tr, Str
		Wade [1081]	t_1	Dy
		Weber [1094]	LS	S, Str
$C_{21}H_{17}N_3O$	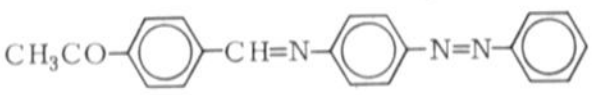 CH_3CO—⬡—CH=N—⬡—N=N—⬡	Avadhanlu [13]	LS	S, Dy

Formula	Structure	Author and ref.	Methods	Results
$C_{21}H_{19}N_3O$	EtO–⟨⟩–CH=N–⟨⟩–N=N–⟨⟩	Avadhanlu [12]	LS	S, Dy
		Doane [335]	LS	S, Tr
$C_{21}H_{23}NO_3$	MeO–⟨⟩–CH=N–⟨⟩–R R: $CH=C-COOC_3H_7$ 　　　$|$ 　　　CH_3	Dong [346]	t_1, E	S, Dy
		Dong [347]	t_1, $t_{1\rho}$	Tr, S, Dy
$C_{21}H_{25}N$	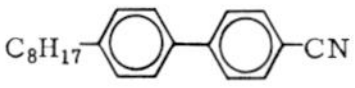 C_8H_{17}–⟨⟩–⟨⟩–CN	Emsley [424]	t_1 (^{2}H, deut.)	S, Dy
$C_{21}H_{25}NO$	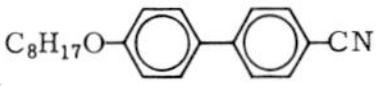 $C_8H_{17}O$–⟨⟩–⟨⟩–CN	Kamezawa [566]	LS, PrMol	S, Dy
$C_{21}H_{25}NO_5$	$C_8H_{17}O$–⟨⟩–COO–⟨⟩–NO_2	Loesche [659]	^{13}C	Dy
		Loesche [657]	LS	S, Dy
$C_{21}H_{26}N_2O_3$	EtO–⟨⟩–N=N–⟨⟩–$OCOC_6H_{13}$	Dong [353]	t_1	Dy
		Johnson [562]	LS, t_1	S, Str, Dy
$C_{21}H_{28}N_2O_2$	MeO–⟨⟩–N=N–⟨⟩–C_8H_{17} 　　　　$|$ 　　　　O	Kamezawa [565]	LS, PrMol	S, Dy
$C_{22}H_{21}N_3O$	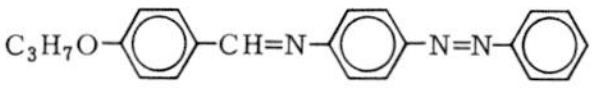 C_3H_7O–⟨⟩–CH=N–⟨⟩–N=N–⟨⟩	Doane [335]	LS	S, Tr
$C_{22}H_{22}N_2O_5$	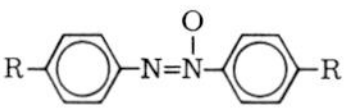 $\overset{O}{\underset{\|}{}}$ R–⟨⟩–N=N–⟨⟩–R R: CH=CHCOOEt	Meiboom [810]	LS, PrMol	S, Tr, Str, Dy
$C_{22}H_{26}N_2O$	NC–⟨⟩–CH=N–⟨⟩–OC_8H_{17} (CBOOA)	Charvolin [190]	LS (^{2}H, deut.)	S, Dy
		Doane [340]	MPT	Dy
		Dong [362]	t_1	Tr, Dy
		Dong [367]	t_1, LS	S, Dy
		Tarr [1024]	$t_{1\rho}$	Dy
		Tarr [1026]	$t_{1\rho}$, E	S, Tr, Dy
$C_{22}H_{29}NO$	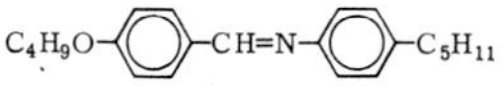 C_4H_9O–⟨⟩–CH=N–⟨⟩–C_5H_{11}	Köhler [595]	LS	Str
$C_{22}H_{30}N_2O_3$	$\overset{O}{\underset{\|}{}}$ RO–⟨⟩–N=N–⟨⟩–OR R: nC_5H_{11}	Prasad [874]	LS	S
		Wade [1081]	t_1	Dy
		Weber [1094]	LS	S, Str
	R: $CH_2-CH-C_2H_5$ 　　　　$|$ 　　　　CH_3	Meiboom [800]	LS	Tr

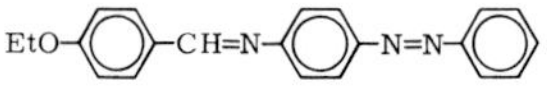

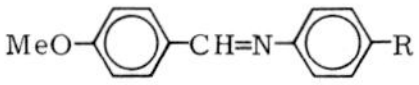

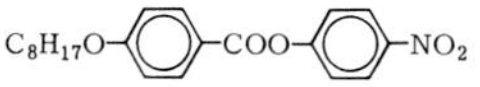

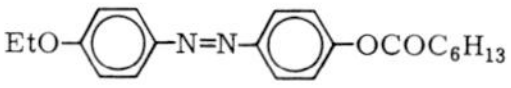

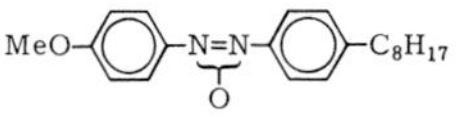

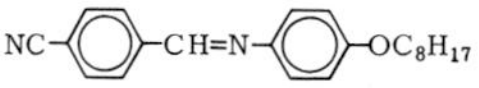

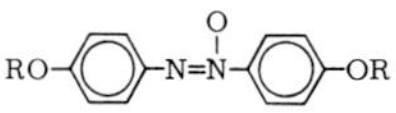

Formula	Structure	Author and ref.	Methods	Results
$C_{23}H_{28}N_2$	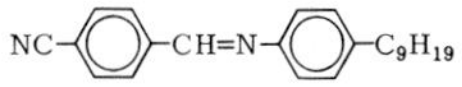 NC—〇—CH=N—〇—C_9H_{19}	McColl [726]	LS	S
$C_{23}H_{30}O_3$	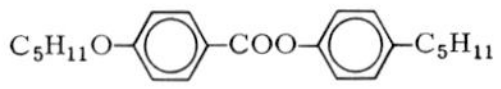 $C_5H_{11}O$—〇—COO—〇—C_5H_{11}	Loesche [659]	^{13}C	Dy
		Pintar [869]	$t_1, t_{1\rho}$	Dy
$C_{24}H_{33}NO$	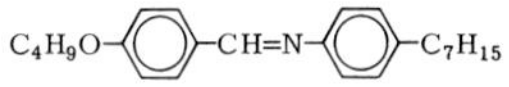 C_4H_9O—〇—CH=N—〇—C_7H_{15}	Doane [340]	MPT	Dy
$C_{24}H_{34}N_2O_3$	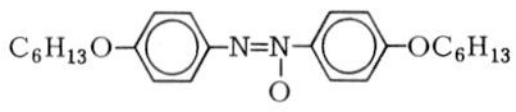 $C_6H_{13}O$—〇—N=N—〇—OC_6H_{13} (O)	Boden [81]	SpE	S, Dy
		Dong [361]	t_1	Dy
		Dong [365]	LS	S, Dy
		Meiboom [800]	LS	Tr
		Phillips [856]	$^1H, ^2H$, LS	S, Str
		Wade [1081]	t_1	Dy
		Weber [1094]	LS	S, Str
$C_{25}H_{25}N_3O_2$	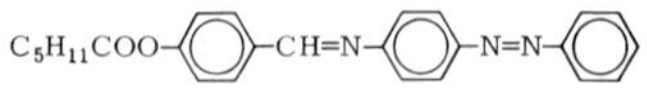 $C_5H_{11}COO$—〇—CH=N—〇—N=N—〇	Spiesecke [1010]	t_1	S, Dy
$C_{25}H_{33}N_2O_3$	 EtO—〇—N=N—〇—$OCOC_{10}H_{19}$	Johnson [562]	LS, t_1	S, Str, Dy
$C_{25}H_{35}NO$	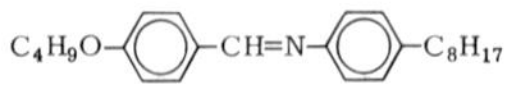 C_4H_9O—〇—CH=N—〇—C_8H_{17}	Blinc [65]	MPT	Dy
		Charvolin [187, 190]	LS, ^2H, (deut.)	S, Dy
		Doane [334]	t_2	Dy
		Doane [336]	PrMol	Dy
$C_{26}H_{29}N_3O$	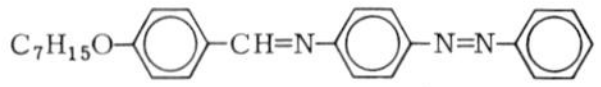 $C_7H_{15}O$—〇—CH=N—〇—N=N—〇	Doane [335]	LS	S, Tr
$C_{26}H_{33}NO$	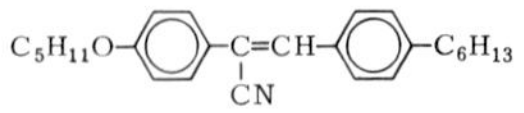 $C_5H_{11}O$—〇—C=CH—〇—C_6H_{13} (CN)	Borodin [91]	LS	Str
$C_{26}H_{36}O_2S$	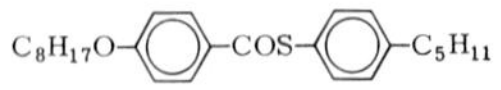 $C_8H_{17}O$—〇—COS—〇—C_5H_{11}	Doane [341]	$t_1 (^2H)$	S, Dy
$C_{26}H_{36}O_4$	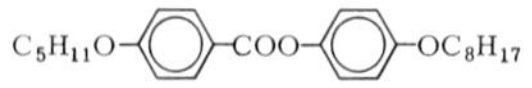 $C_5H_{11}O$—〇—COO—〇—OC_8H_{17}	Loesche [659]	^{13}C	Dy
		Loesche [663]	LS	Tr, Dy
	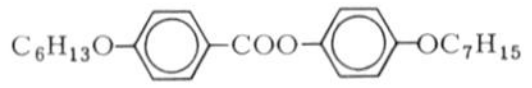 $C_6H_{13}O$—〇—COO—〇—OC_7H_{15}	Loesche [658]	t_1	Dy
$C_{26}H_{38}N_2O_2$	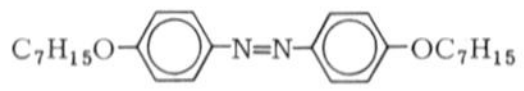 $C_7H_{15}O$—〇—N=N—〇—OC_7H_{15}	Weber [1094]	LS	S, Str

Formula	Structure	Author and ref.	Methods	Results
$C_{26}H_{38}N_2O_3$	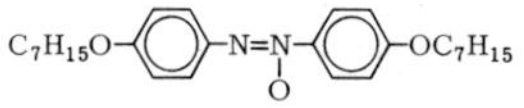 $C_7H_{15}O$–⟨⟩–N=N($\downarrow$O)–⟨⟩–OC_7H_{15}	Boden [82]	SpE	S, Dy
		Doane [328]	t_1	S, Dy
		Doane [337]	LS	S, Str, Dy
		Doane [338]	MPT, PrMol	S, Tr, Dy
		Doane [344]	LS (^{2}H, deut.)	S
		Dong [355]	t_2	Str, Tr, Dy
		Dong [361]	t_1	Dy
		Dong [365]	LS	S, Dy
		Meiboom [810]	LS, PrMol	S, Str, Tr
		Wade [1089]	t_1 (^{2}H, deut.)	Dy
		Weber [1094]	LS	S, Str
		Weber [1098]	LS	S, Str, Tr
$C_{27}H_{38}O_3$	 $C_5H_{11}O$–⟨⟩–COO–⟨⟩–C_9H_{19}	Loesche [661]	t_1	
$C_{27}H_{38}O_4$	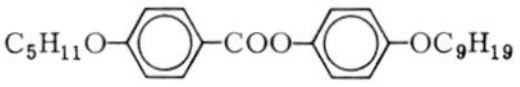 $C_5H_{11}O$–⟨⟩–COO–⟨⟩–OC_9H_{19}	Loesche [663]	LS	Tr, Dy
$C_{27}H_{45}Cl$	cholesteryl chloride	McColl [725]	LS	S
$C_{28}H_{31}N_3O_2$	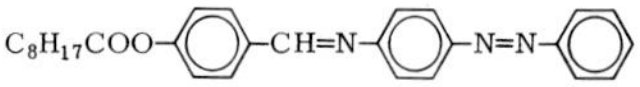 $C_8H_{17}COO$–⟨⟩–CH=N–⟨⟩–N=N–⟨⟩	Spiesecke [1010]	t_1	S, Dy
$C_{28}H_{32}N_2$	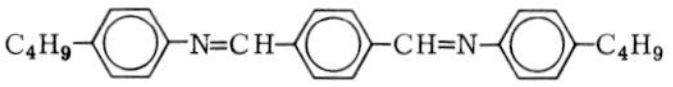 C_4H_9–⟨⟩–N=CH–⟨⟩–CH=N–⟨⟩–C_4H_9 (TBBA)	Blinc [70]	MPT	Dy
		Blinc [72]	t_1	Tr, S, Dy
		Blinc [74]	t_1	S, Dy
		Charvolin [187]	LS (^{2}H, deut.)	S, Dy
		Charvolin [189]	LS (^{2}H, deut.)	Dy
		Doane [337]	LS	S, Str, Dy
		Krueger [599]	SpE	Dy
		Meiboom [810, 812]	LS, PrMol	S, Str, Tr, Dy
		Meiboom [813]	LS (^{2}H, deut.)	S, Str, Dy
		Spiesecke [1009]	t_1, SpE	Dy
$C_{28}H_{35}N_2O_5$	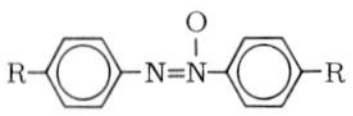 R–⟨⟩–N=N($\downarrow$O)–⟨⟩–R R: $CH=CH-COOC_5H_{11}$	Meiboom [810]	LS, PrMol	S, Tr, Str, Dy
$C_{28}H_{42}N_2O_3$	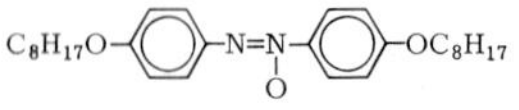 $C_8H_{17}O$–⟨⟩–N=N($\downarrow$O)–⟨⟩–OC_8H_{17}	Boden [82]	SpE	S, Dy
$C_{29}H_{33}N_3O_2$	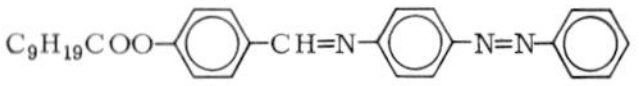 $C_9H_{19}COO$–⟨⟩–CH=N–⟨⟩–N=N–⟨⟩	Spiesecke [1010]	t_1	S, Dy
$C_{29}H_{35}N_3O$	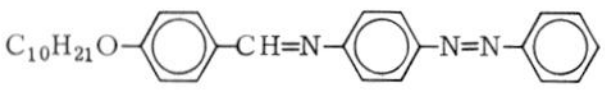 $C_{10}H_{21}O$–⟨⟩–CH=N–⟨⟩–N=N–⟨⟩	Doane [335]	LS	S, Tr

Formula	Structure	Author and ref.	Methods	Results
$C_{30}H_{50}O_2$	cholesteryl propionate	Nolle [837]	t_1, t_2	Dy
		Wade [1080]	t_1	Tr, Dy
$C_{31}H_{37}N_3O_2$	$C_{11}H_{23}COO$—◯—$CH=N$—◯—$N=N$—◯	Spiesecke [1007]	t_1, PrMol	Dy
		Spiesecke [1008]	SpE	Dy
		Spiesecke [1010]	t_1	S, Dy
$C_{31}H_{52}O_2$	cholesteryl butyrate	Cutler [210]	t_1	Dy
		Nolle [837]	t_1, t_2	Dy
$C_{33}H_{43}N_3O$	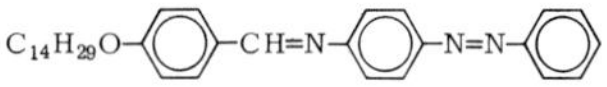 $C_{14}H_{29}O$—◯—$CH=N$—◯—$N=N$—◯	Doane [335]	LS	S, Tr
$C_{34}H_{49}FO_2$	cholesteryl fluorobenzoate	McColl [727]	LS (^{19}F)	S
$C_{34}H_{50}O_2$	cholesteryl benzoate	Nolle [837]	t_1, t_2	Dy
		Spence [997]	LS	S, Str
$C_{36}H_{62}O_2$	cholesteryl nonanoate	Cutler [210]	t_1	Dy
		Kamezawa [566]	LS, PrMol	S, Dy
		Tarr [1024]	$t_{1\rho}$	Dy
$C_{37}H_{64}O_2$	cholesteryl decanoate	Tarr [1026]	$t_{1\rho}$, E	S, Dy, Tr
		Wade [1080]	t_1	Tr, Dy
$C_{41}H_{72}O_2$	cholesteryl myristate	Wade [1080]	t_1	Tr, Dy
$C_{45}H_{76}O_2$	cholesteryl linoleate	Wade [1088]	t_1	
$C_{45}H_{78}O_2$	cholesteryl oleate	Wade [1080]	t_1	Tr, Dy
		Wade [1088]	t_1	
$C_{45}H_{80}O_2$	cholesteryl stearate	Wade [1088]	t_1	
$C_{46}H_{78}O_3$	cholesteryl oleyl carbonate	Tarr [1025]	MPT, LS, E	Dy
$C_{57}H_{110}O_6$	$CH_2OCOC_{17}H_{35}$ $CHOCOC_{17}H_{35}$ $CH_2OCOC_{17}H_{35}$	Barrall [29]	LS	Str

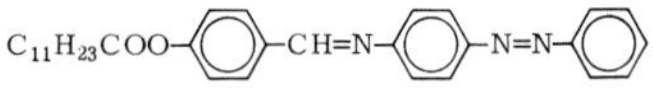

10.2.6 Table of Synthetic Lyophases

The arrangement of this table is similar to that of table 2.5. As lyophases are composed of several constituents, there appears an additional column "Other Components". Water and inorganic counter ions are not mentioned explicitly. However, in many cases, it can be seen from the kind of nucleus studied which counterion was used or if H_2O or D_2O was the solvent.

Formula	Structure	Other Components	Author and ref.	Methods	Results
$C_4H_{10}O$	C_4H_9OH	Na octanoate	Seelig [976]	^{2}H, LS	S, Str, Dy

Formula	Structure	Other Components	Author and ref.	Methods	Results
$C_6H_{11}O_2^-$	$C_5H_{11}COO^-$		Lindman [646]	^{13}C, LS	Str
$C_6H_{14}O$	$C_6H_{13}OH$	cetyltrimethyl-ammonium bromide	Johansson [560]		
		Na octanoate	Seelig [976]	^2H, LS	S, Str, Dy
		hexadecane, K oleate	Shah [985]	LS	S, Str
$C_8F_{15}O_2^-$	$C_7F_{15}COO^-$		Jasinski [558]	MPT, ^{19}F	S, $\Delta\sigma_F$
			Tiddy [1032]	^1H, ^{19}F, t_1, t_2	Str, Dy
			Tiddy [1034]	^1H, ^2H, t_1, t_2	Str, Dy
			Tiddy [1036, 1040]	MPT	Dy
		octanol	Tiddy [1041]	LS, ^2H	S, Str
$C_8H_{15}O_2^-$	$C_7H_{15}COO^-$		Gilchrist [11:420]	LS	S, Str
			Johansson [560]	^2H, ^{23}Na, LS	Str, Dy
			Johansson [561]	^1H, ^2H, LS	Str, Dy
			Lindblom [628]	^{23}Na, LS	S, Str
			Lindman [644]	^{85}Rb, t_1	Dy
			Lindman [646]	^{13}C, LS	Str
			Lindman [647]	^2H, LS	S, Dy
			Lindman [653]	LS (^2H, ^{14}N)	S, Str
			McDonald [743]	LS	Str, Dy
		CCl$_4$	Henriksson [543]	^2H, LS (deut.)	S
		$(CH_3)_4N^+$	Lindman [653]	LS (^2H, ^{14}N)	S, Str
		butanol	Seelig [976]	^2H, LS	S, Str, Dy
		hexanol	Seelig [976]	^2H, LS	S, Str, Dy
		octanol	Boden [84]	SpE	Str, Dy
			Seelig [976]	^2H, LS	S, Str, Dy
		decanol	Ekwall [395]	^1H, ^{23}Na, LS	Str, Dy
			Johansson [560]		
			Lindblom [632]	^7Li, ^{23}Na, ^{39}K, ^{87}Rb, ^{135}Cs, LS	Str, Dy
			Lindblom [636]	LS, t_1, t_2 (^2H, ^{14}N, ^{23}Na)	S, Str, Dy
			Lindblom [637, 641]	LS (^{23}Na)	S
			Lindman [643]	^{23}Na, t_1	Dy
			Lindman [645]	^2H, LS	Str, Dy
			Lindman [651]	LS (^2H)	S, Str
			Lindman [652]	LS (^2H)	Str
			Seelig [976]	^2H, LS	S, Str, Dy
			Tiddy [1031]	t_2	
			Tiddy [1033]	t_1, t_2	Str, Dy
		octyl sulfate, decanol	Lindman [650]	LS (^2H)	S, Str

Formula	Structure	Other Components	Author and ref.	Methods	Results
$C_8H_{17}O_3S^-$	$C_8H_{17}SO_3^-$	decanol	Johansson [560]	^{2}H, ^{23}Na, LS	Str, Dy
			Lindblom [632]	^{23}Na, LS	Str, Dy
			Lindblom [636]	LS, t_1, t_2 (^{2}H, ^{23}Na)	S, Str, Dy
			Lindman [651]	LS (^{2}H)	S, Str
$C_8H_{17}O_4S^-$	$C_8H_{17}OSO_3^-$		Johansson [560]	^{2}H, ^{23}Na, LS	Str, Dy
			Lindman [653]	LS (^{2}H)	S, Str
		octanol	Lawson [611]	^{1}H, ^{2}H, LS	S
		decanol	Lindblom [632]	^{23}Na, LS	Str, Dy
			Lindblom [636]	LS, t_1, t_2 (^{2}H, ^{23}Na)	S, Str, Dy
			Lindman [651]	LS (^{2}H)	S, Str
		decanol, octanoate	Lindman [650]	LS (^{2}H)	S, Str
$C_8H_{18}O$	$C_8H_{17}OH$	Na octanoate	Boden [84]	SpE	Str, Dy
			Seelig [976]	^{2}H, LS	S, Str, Dy
		Li perfluoro-octanoate	Tiddy [1041]	LS, ^{2}H	S, Str
		Na octyl sulfate	Lawson [611]	^{1}H, ^{2}H, LS, PrMol	S
		octylamine + Na dodecyl sulfate	McDonald [743]	LS	Str
$C_8H_{19}N$	$C_8H_{17}NH_2$		Johansson [559]	LS	Str
			Johansson [561]	^{1}H, ^{2}H, LS	Str, Dy
			McDonald [747]	LS, ^{1}H	S, Str
		HCl	Berendsen [34]	LS, MA	S, Str, Dy
			Lindblom [631]	^{35}Cl, LS	Str, Dy
		octanol + Na dodecyl sulfate	McDonald [743]	LS	Str
		decanol + HCl	Lindblom [626]	^{35}Cl, ^{37}Cl, LS	Str
$C_9H_{22}N^+$	$C_9H_{19}NH_3^+$		Lindblom [631]	^{35}Cl, ^{81}Br, LS	Str, Dy
$C_{10}H_{19}O_2^-$	$C_9H_{19}COO^-$	decanol	Seelig [973, 976]	^{2}H, LS	S, Str, Dy
$C_{10}H_{21}O_4S^-$	$C_{10}H_{21}OSO_3^-$	decanol	Diehl [315]	LS (^{2}H)	S, Str
			Reeves [902]	^{2}H, ^{7}Li, ^{23}Na, ^{133}Cs	Str
			Reeves [904, 905]	LS (^{2}H, deut.)	Str, Dy
			Reeves [910]	^{2}H, ^{7}Li, ^{23}Na, ^{133}Cs	S, Str

Formula	Structure		Other Components	Author and ref.	Methods	Results
			decanol + Na_2SO_4	Lawson [611]	^1H, ^2H, LS, PrMol	S
				McLachlan [757]	t_1	Dy
				Reeves [885]	^{23}Na, LS	S, Str, Dy
				Reeves [896]	^2H, LS	
				Reeves [898]	^2H, ^7Li, ^{23}Na, LS	Str, Dy
$C_{10}H_{22}O$	$C_{10}H_{21}OH$			Johansson [561]	^1H, ^2H, LS	Str, Dy
			Na octanoate	Ekwall [395]	^1H, ^{23}Na, LS	Str, Dy
				Johansson [560]	^2H, ^{23}Na, LS	Str, Dy
				Lindblom [632]	^7Li, ^{23}Na, ^{39}K, ^{87}Rb, ^{133}Cs, LS	Str, Dy
				Lindblom [636]	LS, t_1, t_2 (^2H, ^{23}Na)	S, Str, Dy
				Lindblom [637, 641]	LS (^{23}Na)	S
				Lindman [643]	^{23}Na, t_2	Dy
				Lindman [645]	^2H, LS	Str, Dy
				Lindman [651]	LS (^2H)	S, Str
				Lindman [652]	LS (^2H)	Str
			NH_4 octanoate	Lindblom [636]	LS, t_1, t_2 (^2H, ^{14}N)	S, Str, Dy
			octanoate, octyl sulfate	Lindman [650]	LS (^2H)	S, Str
			octylammonium chloride	Lindblom [626]	^{35}Cl, ^{37}Cl, LS	Str
			Na octyl sulfate	Johansson [560]		
				Lindblom [632]	^{23}Na, LS	Str, Dy
				Lindblom [636]	LS, t_1, t_2 (^2H, ^{23}Na)	S, Str, Dy
			octyl sulfate	Lindman [651]	LS (^2H)	S, Str
			Na octyl sulfonate	Johansson [560]		
				Lindblom [636]	LS, t_1, t_2 (^2H, ^{23}Na)	S, Str, Dy
			octyl sulfonate	Lindman [651]	LS (^2H)	S, Str
			Na decanoate	Seelig [973, 976]	^2H, LS	S, Str, Dy
			Na decyl sulfate	Diehl [315]	LS (^2H)	S, Str
			decyl sulfates	Reeves [898]	^2H, ^7Li, ^{23}Na, LS	Str, Dy
				Reeves [902]	^2H, ^7Li, ^{23}Na, ^{133}Cs	Str
				Reeves [904, 905]	LS (^2H, deut.)	Str, Dy
				Reeves [910]	^2H, ^7Li, ^{23}Na, ^{133}Cs	S, Str

Formula	Structure	Other Components	Author and ref.	Methods	Results
		Na decyl sulfate $+ Na_2SO_4$	Lawson [611]		
			McLachlan [757]	t_1	Dy
			Reeves [885]	^{23}Na, LS	S, Str, Dy
			Reeves [896]	^2H, LS	
		K laurate	Schaumburg [958]	LS (^2H), t_1	Dy
		K laurate, KCl	Goldstein [504]	LS (^2H)	Str
$C_{10}H_{23}NO$	$C_8H_{17}N(CH_3)_2$ $\mid$ O		Lawson [607]	t_1	Str, Dy
$C_{10}H_{24}N^+$	$C_{10}H_{21}NH_3^+$		Lindblom [631]	^{35}Cl, ^{81}Br, LS	Str, Dy
			Reeves [905]	LS (^2H, deut.)	Str, Dy
		CH_3COO^-	Reeves [899]	^1H, ^2H, LS	Str, Dy
		cholesterol	Reeves [909]	LS, ^2H	S
$C_{11}H_{26}N^+$	$C_8H_{17}\overset{+}{N}(CH_3)_3$		Lindblom [631]	^{35}Cl, LS	Str, Dy
		Na dodecyl sulfate	Tiddy [1037]	LS, SpE	Str, Dy
$C_{12}H_{23}O_2^-$	$C_{11}H_{23}COO^-$		Charvolin [178]	^2H, LS	Str
			Charvolin [179, 180]	t_1	Dy
			Charvolin [181]	LS, t_1, t_2	Dy
			Charvolin [183]	^2H, t_1	Dy
			Charvolin [184]	^2H, LS	S, Dy
			Charvolin [185]	t_1, $t_{1\rho}$, t_D	Dy
			Charvolin [186]	LS (^2H)	Str, Dy
			Charvolin [188]	LS (^2H)	S, Str, Dy
			Lawson [609]	LS	Str, Dy
			Roberts [913]	SpE	Dy
		decanol	Schaumburg [958]	LS (^2H), t_1	Dy
		decanol, KCl	Goldstein [504]	LS (^2H)	Str
$C_{12}H_{25}O_4S^-$	$C_{12}H_{25}OSO_3^-$		Roberts [912]	^{13}C, t_1	Dy
		octylamine + octanol	McDonald [743]	LS	Str
		octyltrimethylam- monium bromide	Tiddy [1037]	LS, SpE	Str, Dy
		palmitic acid	McDonald [748]	LS	Str
	$\begin{array}{c}CH_3\\ \mid +\\ C_{16}H_{33}N(CH_2)_3SO_3^-\\ \mid\\ CH_3\end{array}$		Tiddy [1038]	LS	Str

Formula	Structure	Other Components	Author and ref.	Methods	Results
$C_{12}H_{28}N^+$	$C_9H_{19}\overset{+}{N}(CH_3)_3$		Lindblom [631]	^{35}Cl, ^{81}Br, LS	Str, Dy
$C_{13}H_{30}N^+$	$C_{10}H_{21}\overset{+}{N}(CH_3)_3$		Lindblom [631] Schulman [962]	^{35}Cl, ^{81}Br, LS LS	Str, Dy Str, Dy
$C_{14}H_{20}F_7O_2^-$	$CF_3(CF_2)_2(CH_2)_{10}COO^-$		Tiddy [1035]	^1H, ^{19}F, t_1, t_2	
$C_{14}H_{24}N^+$	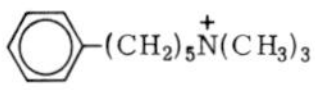 ⬡$-(CH_2)_5\overset{+}{N}(CH_3)_3$		Inoue [554]	LS	Str, Dy
$C_{14}H_{27}O_2^-$	$C_{13}H_{27}COO^-$		Lawson [609]	LS	Tr, Str
$C_{14}H_{31}NO$	$C_{12}H_{25}\underset{O}{\overset{\|}{N}}(CH_3)_2$		Lawson [609] Lawson [612] Lawson [615]	LS ^1H, ^2H, LS t_1, t_2	Tr, Str S, Tr, Str Dy
$C_{15}H_{26}N^+$	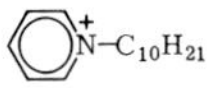 ⬡$\overset{+}{N}-C_{10}H_{21}$		Lindblom [631]	^{81}Br, LS	Str, Dy
$C_{15}H_{32}N_2O$	$(CH_3)_3\overset{+}{N}-\overset{-}{C}OC_{11}H_{23}$		Gilchrist [11:420]	LS	S, Str
$C_{15}H_{34}N^+$	$(CH_3)_3\overset{+}{N}C_{12}H_{25}$		Goodman [11:10] Lindman [649]	LS MPT, t_2	Str Str, Dy
$C_{16}H_{31}O_2^-$	$C_{15}H_{31}COO^-$		Blinc [61] Blinc [63] Blinc [67] Boden [83] Burnell [127] Hayashi [538] Hayashi [539] Lawson [609]	^2H, t_1, SpE SpE LS (^2H) SpE LS, SpE LS LS LS	S, Dy Dy Dy Str, Dy Dy Tr Str, Tr Str, Tr
$C_{16}H_{32}O_2$	$C_{15}H_{31}COOH$	Na dodecyl sulfate	McDonald [748]	LS	Str
$C_{16}H_{34}$	$C_{16}H_{34}$	K oleate + hexa- decanol	Shah [985]	LS	S, Str
$C_{17}H_{30}N^+$	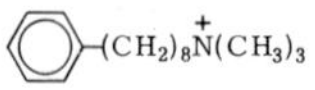 ⬡$-(CH_2)_8\overset{+}{N}(CH_3)_3$		Inoue [554]	LS	Str, Dy
$C_{17}H_{38}N^+$	$C_{14}H_{29}\overset{+}{N}(CH_3)_3$		Lindblom [631]	^{35}Cl, ^{81}Br, LS	Str, Dy
$C_{18}H_{33}O_2^-$	$C_{17}H_{33}COO^-$ (oleate)		Berendsen [31] Berendsen [34] Lawson [609] Wade [1084] Wade [1087] Wade [1091]	LS, MA LS, MA LS LS, MA SpE, t_1, $t_{1\rho}$, t_2 MPT	Dy, Str S, Str, Dy Str, Tr Dy Dy Dy
		hexadecane + hexanol	Shah [985]	LS	S, Str

Formula	Structure	Other Components	Author and ref.	Methods	Results
	$C_{17}H_{33}COO^-$ (elaidate)		Lawson [609]	LS	Str, Tr
$C_{18}H_{35}O_2$	$C_{17}H_{35}COO^-$		Blinc [60]	t_1	Dy
			Charvolin [188]	LS (^2H)	S, Str, Dy
			Charvolin [191]	LS (^2H)	Str, Dy
			Lawson [609]	LS	Str, Tr
$C_{19}H_{42}N^+$	$C_{16}H_{33}\overset{+}{N}(CH_3)_3$		Johansson [561]	^1H, ^2H, LS	S, Str, Dy
			Lindblom [630]	^{81}Br, Ls, PrMol	Str, Dy
			Lindblom [631]	^{35}Cl, ^{81}Br, LS	Str, Dy
		hexanol	Johansson [560]	^2H, LS	Str, Dy
			Lindblom [624]	^{81}Br, LS	Str, Dy
			Lindblom [627]	^{81}Br, LS	Str
$C_{20}H_{36}O_2$	$C_{17}H_{33}COOCH=CH_2$ (vinyl oleate)		Martins [789]	t_1, t_2	Tr
$C_{20}H_{37}O_7S$ (Aero-sol®-OT)	ROOCCH$_2$CHCOOR $\mid$ SO$_3^-$ R: CH$_2$CHC$_4$H$_9$ $\mid$ C$_2$H$_5$		Lindblom [636]	LS, t_1, t_2 (^2H, ^{23}Na)	S, Str, Dy
			Lindman [650]	LS (^2H)	S, Str
			Lindman [651, 653]	LS (^2H)	S, Str
$C_{20}H_{42}O_6$	$C_{10}H_{21}(OCH_2CH_2)_5OH$		Clemett [194]	t_1	Str, Dy
$C_{21}H_{45}NO_3S$	$(CH_3)_2\overset{+}{N}(CH_2)_3SO_3^-$ $\mid$ C$_{16}$H$_{33}$	Na dodecyl sulfate	Tiddy [1038]	LS	Str
$C_{24}H_{50}O_7$	$C_{12}H_{25}(OCH_2CH_2)_6OH$		Goodman [509]	LS, t_1	Str
$C_{27}H_{48}O_7$	C$_9$H$_{19}$—◯—(OCH$_2$CH$_2$)$_6$OH		Johansson [561]	^1H, ^2H, LS	S, Str, Dy
$C_{35}H_{64}O_{11}$	C$_9$H$_{19}$—◯—(OCH$_2$CH$_2$)$_{10}$OH		Johansson [561]	^1H, ^2H, LS	S, Str, Dy
PBLG	poly(γ-benzyl-L-glutamate)		Sobajima [994]	LS, PrMol	S, Str
			Vold [1076]	LS	S
			Vold [1077]	LS	S, Str
			Wade [1085]	t_1	Str, Dy

10.2.7 Table of Biological Lyophases

Monoglycerides. M. P. McDonald [744–747].

Diglycerides. E. G. Finer [441].

Lysolecithins. J. M. Steim [1011].

Lecithins (phosphatidylcholines). L. D. Bergel'son [35–37], N. J. M. Birdsall [49, 50, 53–57], S. I. Chan [152, 153, 155–157], D. Chapman [158, 159, 161, 164–172, 175], J. Charvolin [182], H.

Dreeskamp [375], E. G. Finer [436–444], M. P. Klein [590], G. Lindblom [625, 633, 634, 638], J. A. Magnuson [784], J. C. Metcalfe [815], J. Seelig [974, 975, 979, 980], B. Sheard [987], J. S. Waugh [1093].

Phosphatidylethanolamines. E. G. Finer [441, 443].

Phosphatidylserines. E. G. Finer [443], J. Seelig [978].

Sphingomyelins. D. Chapman [174].

Cerebrosides. D. Chapman [174, 176].

Lipoproteins. J. M. Steim [1011].

Myelins. D. Chapman [163], H. Lecar [616].

Erythrocyte menbranes. N. J. M. Birdsall [50], D. Chapman [160, 162, 165], J. A. Magnuson [781], J. M. Steim [1012].

Sarcoplasmic reticulum membranes. N. J. M. Birdsall [51].

A coleplasma membranes. N. J. M. Birdsall [52], D. Chapman [173].

Miscellaneous membranes and materials. R. E. Block [77], D. Chapman [177], J. A. Magnuson [782, 783], F. Millett [817], H. Monoi [820], M. M. Pintar [864, 867], G. Vass [1049].

10.3 NMR Studies of Mesophase-Oriented Solutes

10.3.1 Basic Principles

In 1963, A. Saupe discovered that it is possible to obtain well-resolved NMR spectra of small molecules dissolved in nematics [926]. Since then, NMR spectroscopy using liquid crystalline solvents has become a valuable tool for obtaining information about properties of solute molecules. Its unique importance is that this is the only method of obtaining direct geometrical information from the liquid phase.

When a nematic is subjected to a magnetic field, it is oriented to give a liquid "single crystal" instead of a liquid "crystalline powder". This means that the molecular motions are macroscopically anisotropic with a unique direction of preference for the whole sample volume. Molecules dissolved in such a phase are forced to move anisotropically, (similar to the solvent) even if they do not alone tend to form a liquid crystalline order.

The anisotropic motion makes spin-spin couplings due to magnetic dipole-dipole interactions observable, which in isotropic liquids are averaged to zero, but to a finite value in anisotropic liquids. The reason for this is the angular dependence within the expression for the magnetic dipolar interaction energy in a magnetic field. The resulting dipolar coupling constant is defined by

$$D_{ij} = -\frac{h \cdot \gamma_i \cdot \gamma_j}{4\pi^2} \cdot \frac{1}{2} \cdot \left\langle \frac{3\cos^2(H, r_{ij}) - 1}{r_{ij}^3} \right\rangle \tag{10.1}$$

where γ_i and γ_j are the magnetogyric ratios of the magnetic nuclei i and j, r_{ij} is the distance between the nuclei, and (H, r_{ij}) is the angle between the magnetic field and the internuclear connection vector. Averaging over all possible distances and orientations has to be taken into account. In solids, there is not sufficient molecular motion to average the angular term to zero as in isotropic liquids. However, a large number of strong intra- and intermolecular couplings prevents the observation of narrow NMR signals. In liquid crystals, on the other hand, there is rapid but anisotropic motion which averages out intermolecular couplings completely, but intramolecular couplings only partially. As liquid crystalline solvent molecules contain a large number (≥ 14) of magnetically active nuclei, the large number of energy levels caused by strong couplings all over the molecule results in many overlapping transitions and still prevents the observation of individual lines. When a smaller molecule is dissolved in such a solvent, however, well-resolved spectra, which contain a wealth of information can be obtained.

Prominent features of these spectra are large couplings up to several thousand Hertz which is much more than any coupling usually observed in isotropic liquids. In most cases, they are in the same order of magnitude or larger than the chemical shifts involved. Therefore strongly coupled spin systems arise which can only be analysed with the help of computer programs. According to P. Diehl, the usual programs for the analysis of spectra from isotropic liquids are easily modified for liquid crystal spectra [271]. For an extensive treatment of spectral analysis and discussion of many simple spin systems, the reader is referred to the review of P. Diehl [290]. Another striking feature of liquid crystal spectra may be pointed out here: Even if there is chemical and magnetic equivalence of coupling nuclei, the anisotropic coupling causes a line splitting. For example, the spectrum of benzene, which is a case of chemical equivalence and exhibits only a single line in the isotropic phase, is composed of more than 50 lines in the nematic phase [290].

10.3.2 Available Information

10.3.2.1 Dipolar Coupling Constants, Orientation and Geometry

The angular term in equation (10.1) may be averaged separately, implying that orientational and vibrational motions are independent, and the degree of orientation is defined as

$$S_{ij} = \tfrac{1}{2}\langle 3\cos^2(H, r_{ij}) - 1\rangle \tag{10.2}$$

for a pair of nuclei i and j. In a rigid molecule, the degrees of orientation of different internuclear connection vectors are related to each other by the equation:

$$S_{ij} = \sum_{p,q} \cos(p, r_{ij}) \cdot \cos(q, r_{ij}) \cdot S_{pq} \tag{10.3}$$

where p and q are the x, y, and z coordinates of a molecule-fixed system and S_{pq} the elements of an order matrix, which is defined by

$$S_{pq} = \tfrac{1}{2}\langle 3\cos(H, p)\cos(H, q) - \delta_{pq}\rangle \tag{10.4}$$

Since it is symmetric and traceless, it contains a maximum of five independent elements for a completely asymmetric, rigid molecule.

From equation (10.1) it can be seen that only ratios of the degree of order and the cube of the internuclear distances can be obtained, so the determination of the absolute values of both parameters requires the knowledge of a scaling factor which has to be introduced from some other method. When more coupling constants are observable than the number of order matrix elements necessary for the description of the orientation, ratios of distances can be calculated. In fact, much precise geometrical data has been derived from H-H-couplings. However, several assumptions must be made before an evaluation of equation (10.1) with respect to distance ratios is possible:

1) D_{ij} is purely dipolar
2) r_{ij} is constant, i. e. the molecule is rigid.

These assumptions seem to yield useful approximations for H-H-couplings, but there is considerable discussion about their feasibility when nuclei other than protons are involved (see sections 10.3.3.2 and 10.3.3.3).

10.3.2.2 Intramolecular Motions

Information about intramolecular motions such as vibrations and rotations can be drawn from the NMR spectra of oriented molecules using equation (10.1), which shows that the direct couplings are dependent on the average of the internuclear distances. The computational handling of averaging is much easier for the case of rotations than for vibrations. In addition, other methods of structural analysis, such as electron diffraction and microwave spectroscopy, yield a different type of vibrational average and operate in the gas phase, so it becomes difficult to compare the results of various methods. This is the reason why vibrational averaging has been discussed in most cases as a small disturbance of the accuracy of the method, which may lead to errors of a few percent in the calculated bond lengths. No generally applicable method of correcting for these errors has been found yet, whereas some valuable information concerning intramolecular rotations has been collected from nematic phase NMR spectra.

10.3.2.3 Anisotropy of Indirect Coupling Constants

As shown by N. F. Ramsey the indirect electron transmitted nuclear spin-spin coupling (J) is a second-rank tensor which, in the most general case of an asymmetric molecule, has nine components [880]. Only the average of its diagonal term, one third of its trace, is observable in isotropic liquids. In anisotropic liquids, however, the anisotropy of the tensor has to be taken into account, and there is no way to separate the total observed anisotropic coupling into its direct dipole-dipole and indirect electron-transmitted components by spectral analysis only.

Since the anisotropy of the indirect coupling constant is, in most cases, small compared to the dipole-dipole coupling, it has been neglected completely in most investigations, and it manifests itself, if at all, in small deviations of the molecular geometry calculated on the basis of dipolar couplings, from the expected geometry on grounds of symmetry considerations, or from other experimental methods (electron diffraction, microwave and X-ray studies). It is difficult, however, to interpret these deviations in terms of anisotropic indirect coupling because the influence of vibrations is also not exactly known.

If it is possible to prepare nematic solutions of molecules with symmetry lower than C_{3v} in such a way that essentially different ratios of orientation parameters are found, then information about the J-anisotropy can be obtained because of the different orientational dependence of direct and indirect coupling. This method postulates, of course, that molecular geometry is independent of solvent effects and orientation.

10.3.2.4 Anisotropy of Chemical Shifts

Like the indirect spin-spin coupling constants, the chemical shift is a tensorial quantity, and only an average value is observable in isotropic liquids [880]. Information about the individual tensor components is of considerable interest with respect to the architecture of the chemical bonds. In principle, all six components of the symmetric tensor can be obtained from measurements of single crystals: the magnitudes in the direction of the principal axes and the orientation of the principal axes with respect to the crystal lattice. Only the diagonal elements can be obtained from crystal powder measurements [1092]. However, there is a strong dipolar line broadening in solids, and the accuracy is low in most cases. The partial ordering offered by liquid crystals is therefore of considerable advantage, although only a linear combination of the diagonal elements is available from liquid crystal measurements.

10.3.2.5 Quadrupole Coupling Constants

Quadrupole coupling constants may be obtained from the line splittings in the spectra of quadrupolar nuclei, e.g. deuterium, and the orientation parameter in the nematic phase. Since they depend only on ground state electronic properties and the nuclear electric quadrupole moment, they offer direct information on the electronic structure around the nucleus studied. On the other hand, quadrupole coupling constants from solid state measurements may be used to determine orientation and geometric parameters from the nematic phase quadrupolar splittings.

10.3.2.6 Signs of Spectral Parameters

Only the relative signs of the parameters can be obtained from the analysis of NMR spectra. Therefore, additional information has to be used in order to obtain absolute signs. Since direct coupling, indirect coupling, chemical shift anisotropy, and orientation parameters are linked to each other in the nematic phase NMR spectra, all signs can be related. In most cases, the sign of the orientation parameter is used as a starting point because the orientational behavior is known to be determined mainly by dispersion forces [932], so reasonable assumptions about its orientation can be deduced from the shape of the molecule in many cases. Also, there have been experimental [97] and theoretical [524] studies of the indirect ^{13}C-H-coupling constant across one chemical bond, which shows a positive sign. In isotropic phase NMR work, the signs of many other couplings have been related to this one. So the indirect coupling constant may, in many cases, be also used for the deduction of the absolute signs. Similary, if there is theoretical or experimental evidence concerning the sign of the chemical shift anisotropy, then the signs of the other parameters may be fixed absolutely [234].

10.3.3 Experimental Methods

10.3.3.1 Nematic Solvents

In the first years of NMR spectroscopy with nematic solvents, elevated temperatures of at least 50–80°C had to be used to obtain the nematic phase. As a consequence, the spectral resolution was limited by temperature and concentration inhomogeneities, which in the probes of commercial NMR spectrometers are much greater at elevated than at normal temperatures. Therefore, low-melting eutectic mixtures were developed [1000], and some room temperature nematics, e.g., MBBA. It is now possible to record nematic phase spectra down to $-35°$ [811]. It is advantageous to use a mixture that melts above room temperature and further lower its melting point by addition of a solute. In this way, a maximum solute concentration is achieved [964]. Recently, *trans-trans*-4′-alkyl-4-cyanobicyclohexyls having negative diamagnetic anisotropy have been recommended as nematic solvents allowing sample rotation to obtain high resolution NMR spectra [2:191a].

10.3.3.2 Lyotropic Solvents

In 1969 it was discovered that it is possible to obtain well-resolved spectra of molecules oriented in a lyophase [613, 614]. The optical axis of the liquid crystal used by K. D. Lawson is perpendicular to the magnetic field [614]; therefore, it is possible to improve the resolution in conventional magnets by spinning the sample tube without disturbing the orientation of the nematic phase.

Lyophases offer a quite interesting alternative to the thermotropic mesophases as solvents for NMR studies because the orientational behavior of solutes is so different. Recently, an increasing amount of work has been directed into this field, notably by J. H. Goldstein and coworkers [495–499]. In addition to the original lyophase used by K. D. Lawson [614], another suitable solvent phase with a different pH-range has been found by R. C. Long, Jr. [672].

10.3.3.3 Slow Spinning Technique

Several workers have investigated the behavior of liquid crystals on slow rotation about an axis perpendicular to the applied magnetic field [267, 1094, 1095]. It is understood now that rotation, if at least 20 % lower than a critical speed, does not appreciably disturb the uniform orientation, but it twists the optical axis of the nematic phase away from the magnetic field direction [398]. This seems to be the only simple experimental method for changing the degree of orientation without changing temperature, and it is of considerable interest in the determination of chemical shift anisotropies [1115] and the separation of isotropic and anisotropic spin-spin couplings [402, 406].

10.3.3.4 Spectrometers with Superconducting Solenoids

When superconducting solenoids are applied in NMR spectroscopy, the magnetic field is parallel to the axis of rotation of the sample tube, so the usual thermotropic solvents can be spun rapidly without disturbing the orientation. This causes a considerable increase in resolution, which is now the same as in slightly viscous isotropic liquids. Additionally, the high magnetic fields are advantageous because of the increased chemical shifts and sensitivity, so more complex spin-systems can be analyzed. The limit at this time seems to be 10 spins using a 63 kG magnet [142, 296], but it will probably be further extended by the application of still higher magnetic fields.

10.3.3.5 Fourier Transform Techniques

The technique of Fourier transform spectroscopy has been applied for the detection of various nuclei pertaining to solutes in nematic liquid crystals [232, 304, 944, 945]. Since the whole spectrum is excited at the same time, the measurements of line positions are free of systematic errors caused by temperature drifts and inhomogeneities in concentration or temperature. Any one of these experimental faults causes only a line broadening, but not a shift of the intensity maximum. Computer listing of line positions considerably speeds up evaluation of the spectra. Proton spectra can be recorded in a few minutes [965], and even natural abundance ^{13}C resonances are accessible to this method [964].

10.3.3.6 Double Resonance Techniques

Double resonance techniques have been used to aid in spectral analysis [140, 260, 463] and to remove deuterium couplings [808]. With the large splittings in the deuteron spectra because of quadrupole interactions, it is necessary to induce double quantum transitions to obtain an effective decoupling [808]. When heavy deuterium substitution is used and the deuterons are distributed statistically, most of the remaining protons are contained in singly protonated molecular species, and only a few are found in doubly protonated species. Dipolar couplings can be taken, while decoupling the deuterons, from simple two-spin systems. Therefore, this is a promising approach toward the analysis of more complicated molecules by means of nematic phase NMR spectroscopy [808].

Irradiation of the deuteron resonance while observing the proton resonance in selectively deuterated molecules was used to obtain the relative signs of dipolar couplings and the magnitude and signs of the quadrupole interactions [400, 401, 409, 410]. Conversely, proton decoupling while observing deuteron resonance was used to determine deuteron-deuteron interactions [304]. ^{1}H-$\{^{14}$N$\}$-double resonance yielded the nitrogen chemical shift anisotropy and quadrupole coupling constant [1114]. ^{13}C, ^{15}N, and ^{199}Hg were irradiated while observing protons in order to determine the chemical shift anisotropy of these nuclei [749, 750].

10.3.3.7 Nuclei Studied

^{1}**H and** ^{19}**F.** The most work has been done on proton and fluorine resonance (see table 10.3.6) because these nuclei have the largest magnetic moment and 100% natural abundance. There have been reservations about fluorine anisotropic couplings, however, because they may not be purely dipolar and are therefore difficult to interpret (see 10.3.4.4).

^{2}**H.** Deuteron resonance has been applied for the detection of quadrupole splittings since 1964 [855]. The lines are rather wide, and dipolar splittings by interaction with protons cannot be resolved in most cases. Therefore, information about orientation has to be taken from additional measurements. Deuterium relaxation of C_6D_6 has been measured by A. Loewenstein to obtain information about the motion of solutes in the nematic phase [664].

^{13}**C.** The first instance of ^{13}C resonance has, because of its low sensitivity, only been reported in 1970 with samples of CH_3I and CH_3CN enriched to 60% of ^{13}C [822]. Since then, it has been applied to a number of compounds enriched with ^{13}C mainly to determine the anisotropy of ^{13}C chemical shifts [227, 229, 234, 238, 374, 427, 428]. It has also been shown in a favorable case that it can be applied to natural abundance samples by using the present NMR equipment [964].

^{15}**N.** ^{15}N shielding anisotropies were determined by B. P. Dailey in N_2O [226].

^{31}**P.** The first paper reporting a ^{31}P spectrum of an oriented solute appeared in 1973 [228]. Since then, further reports have been published [231–233, 236, 237]. All of them applied FT-techniques because of the low sensitivity of ^{31}P nuclei.

10.3.3.8 Isotopic Substitution

^{2}**H.** While complete deuterium substitution is useful only for measuring quadrupole splittings in the deuteron resonance, partial substitution can be used to simplify proton spectra. The H-D-couplings are about one-sixth of the corresponding H-H-couplings. Spectra are converted from strongly coupled to weakly coupled types because of the large shift between proton and deuteron resonances [400, 667], and double resonance techniques can be applied [303, 400, 409, 410], see section 10.3.2.5. S. Meiboom used a high deuterium substitution for obtaining the direct dipolar couplings in cyclohexane [808], see section 10.3.2.5. This technique will probably be applied more frequently for simplification of the analysis of complicated spin systems.

^{13}C. ^{13}C-substitution was used initially to obtain, from the proton resonance, the direct carbon-proton couplings which give information about the position of carbon atoms in a molecule [931]. It has been utilized for structural investigations in several substances [44, 93, 112–114, 234, 293, 374, 499, 895, 937, 1001, 1002], and the question of anisotropy in the indirect carbon-proton coupling has also been raised. ^{13}C magnetic resonance has been applied to enriched compounds in order to obtain the ^{13}C chemical shift anisotropy.

^{15}N. ^{15}N substitution has been used to study the ^{15}N-H anisotropic couplings in the proton resonance, to examinate their use for structural analysis, and to eliminate line broadening caused by ^{14}N quadrupole interactions [882, 891, 895, 937, 965].

10.3.3.9 Additional Application of an Electric Field

It was suggested by P. Diehl that a strong electric field perpendicular to the magnetic field can be used to rotate the optical axis of the nematic solvent by 90° [277]. This offers a means of changing the solute orientation without changing either the solvent or temperature, and it can be used to study the signs of indirect couplings and chemical shift anisotropies.

10.3.4 Results

10.3.4.1 Dipolar Coupling Constants, Orientation and Geometry

The analytical procedures appropriate for obtaining coupling constants and chemical shifts from NMR spectra of oriented solute molecules are well known [927, 992]. For details, the reader is referred to the review by P. Diehl [290]. Some peculiarities are discussed in the original papers of P. Diehl [269, 275, 280], J. I. Musher [829, 830], and A. Saupe [934]. The bulk of results on molecular geometry shall not be discussed here in detail, but it is presented in table 10.3.6. Only a few aspects of structural analysis will be treated in the following sections.

From the anisotropic coupling constants, information about the orientation of solutes can be obtained. Systematic studies of the dependence of orientation on molecular properties were presented by A. Saupe showing that there is practically no influence of permanent electric dipole moments, but the orientational behavior can be satisfactorily described by the influence of dispersion forces [932, 938]. Only small changes, which may be due to specific interactions, are found in different nematic solvents. This was confirmed by J. M. Anderson [9] and D. F. R. Gilson [482], who derived a description of orientation from the molecular inertia tensor and from molecular dimensions. An interesting complementation came from J. H. Goldstein who found, for some solutes, a quite different orientational behavior for lyotropic as opposed to thermotropic nematic solvents, and he concluded that specific interactions dominate orientation for certain polar solute-solvent combinations [498, 499].

A surprising result was published by S. Meiboom, who found a dipolar splitting in the spectra of tetrahedral molecules like tetramethyl-silane and neopentane, although these should not be oriented by the nematic matrix because of their high symmetry [798]. More recently, this phenomenon was investigated in considerable detail by A. Loewenstein [665, 668] and by L. W. Reeves [894, 897, 900], who found splittings in the tetrahedral molecules CH_4, NH_4^+, BF_4^-, SiH_4, GeH_4, $N(CH_3)_4^+$ or their deuterated analogs. An explanation was given which accounts for the splittings in that there exist small distortions of the molecules of the order of 10^{-2} degrees [116]. This means that vibrational and orientational averaging may not be treated independently, and this may be of importance for structure determinations of molecules with small degrees of orientations.

K. A. McLauchlan described a method of determining the absolute sign of the orientation parameters of polar solutes in nematic solvents from measurements of the anisotropy of the dielectric constant of the solution [760]. This may provide the additional information needed to proceed from the relative signs of parameters furnished by NMR spectroscopy to absolute signs (see section 10.3.4.6).

10.3.4.2 Intramolecular Motions

The problem of correcting the anisotropic coupling constants for molecular vibrations occurs mainly when coupling constants between directly bonded carbons and protons are considered, yielding C—H-distances several percents too large. Vibrational corrections were considered in general by S. Meiboom [807] but C. MacLean did not succeed in computing improved molecular structures for the methyl halides probably because of insufficient knowledge of the molecular force field, notably the anharmonicity of the potential functions [763]. N. J. D. Lucas found an improved structure for cyclopropane [674] but not for methylfluoride [673], and this was ascribed to a possible interaction of vibrational and reorientational motions [675], see also section 10.3.4.3. P. Diehl was able to reconcile the results on ^{13}C-benzene [293] with electron diffraction and Raman results by regarding only harmonic corrections [299].

The problem of ring puckering vibrations in four-membered rings was considered by S. Meiboom for the analysis of the NMR spectrum of oriented cyclobutane [805]. A satisfactory interpretation of the spectrum was achieved by assuming rapid interconversion of two bent conformers. Later, the data was reinterpreted by D. F. R. Gilson with the introduction of a square-quartic puckering potential and a better fit of calculated and measured spectra was obtained [485]. A similar potential exists in trimethylene oxide, but the energy barrier between the bent forms is lower than the vibrational ground state, so the planar form is the equilibrium form. These findings from microwave and IR spectroscopy were confirmed by NMR-spectroscopy [483, 488, 947], where an averaging of two forms with a small tilt angle of about 10° is needed to explain the measured dipolar couplings. Trimethylene sulphide [488, 718, 947, 1015] is bent in the ground state, and a tilt angle of 13° explains the NMR spectrum [947]. On the other hand, cyclobutanone [718, 1015] is interpreted as rigidly planar [718]. The same was found for propiolactone [888] and tetrafluoro-1,3-dithietane [495]. In five- and six-membered rings the situation is more complicated, and specific interpretation of spectral data in terms of puckering frequencies or bond angles is not possible. L. W. Reeves merely arrived at the conclusion that in ethylene carbonate and ethylene trithiocarbonate, ring puckering is more pronounced in the lyotropic than in the thermotropic nematic phase [886, 887, 890]. Only two conformationally mobile six-ring compounds have been studied, s-trioxane [196] and cyclohexane [808], but no energy barriers of frequencies for puckering vibrations were derived.

A considerable amount of research has been devoted to the study of intra-molecular rotations, notably by P. Diehl [282, 283, 285, 288, 289, 291, 296, 297, 300] and L. Lunazzi and C. A. Veracini [717, 1051–1053, 1055, 1057, 1059, 1060, 1062]. The first instance where an approximate height and position of the methyl group rotation barrier could be determined was in *o*-chlorotoluene [288]. Similar values were reported for *o*-bromo- and *o*-iodo-toluene [289]. Predominance of "geared" rotation in *o*-xylene could be excluded from NMR-data [296]. For *p,p'*-dichlorobiphenyl, unambiguous results concerning the rotational potential could be obtained only with the help of electron diffraction data [300]. A. A. Bothner-By postulated two unevenly populated rotamers from his results for propene, but could not exclude a free rotation [93]. I. J. Gazzard found the *gg*-conformation of 1,2,2,3-tetrachloropropane to be most stable [467]. There are a number of cases where the spectra indicated a hindered methyl rotation but, because of a lack of accuracy, no conclusion could be reached about the small energy potentials. These include the following: ethane and 1,1,1-trifluoroethane [217], methyl silane [666, 667], methyl germane [666], the methylammonium ion [895], γ-picoline [944], and 2,5-dichloro-*p*-xylene [139]. Energy barriers to rotation are much more pronounced in the heterocyclic compounds studied by L. Lunazzi and C. A. Veracini. Therefore, most of these spectra could be interpreted in terms of two rapidly interconverting rotamers: furane-2-aldehyde [1057], thiophene-2,5-dialdehyde [1055], 2,2'-dithienyl [1062] and its 5,5'-disubstituted derivatives [1059] and 3,3'- and 5,5'-bis-isoxazol [1051, 1053]. The SO-cis conformer was exclusively found for thiophene-2-aldehyde [1060].

A. D. Buckingham performed a study of organometallic carbonyl compounds substituted with ^{13}C in the carbonyl part in order to obtain information about internal molecular rotations [112–114]. He could show that in the case of π-cyclobutadienyl iron tricarbonyl [114] and π-cyclopentadienyl manganese tricarbonyl [112], free rotation of the organic ring with respect to the rest of the molecule takes place.

10.3.4.3 Anisotropy of Indirect Coupling Constants

Until 1969, no systematic research was directed toward the problem of anisotropic indirect coupling constants, probably because of the difficulties mentioned in section 10.3.2.3 and the miniscularity of the effects. From H-H-coupling constants, reasonable molecular geometries were obtained which compared well with the ones from microwave spectroscopy and electron diffraction work. This was expected from the theoretical foundation laid by N. F. Ramsey, who showed that in the case of proton couplings the Fermi contact term, which is isotropic, predominates [881].

A significantly different situation may arise for couplings involving fluorine, where spin-dipolar and orbital terms could be much larger. A. D. Buckingham could not reach a concise conclusion about the anisotropic contributions to J_{HF} and J_{FF} in *cis*-1,2-difluoroethene and 1,1-difluoroethene [104, 108], whereas A. Saupe concluded there is a strong anisotropy in J_{FF} in 1,1-difluoroethene and tetrafluoroethene [940]. C. MacLean was able to evaluate the individual *J*-tensor components of J_{FF} in 1,1-difluoroethene using essentially different orientations and found them to be much larger than their isotropic average [764]. R. A. Bernheim investigated ^{13}CH$_3$F and was unable to interpret the anisotropic couplings in terms of purely dipolar interactions, concluding therefore that there exist large anisotropies in J_{CH} or J_{CF} or both [42, 44]. These results prompted a large amount of theoretical work both on the influence of vibrations and on the possibility of large *J*-anisotropies. C. MacLean [763] and N. J.

D. Lucas [673] could not find suitable vibrational corrections to explain Bernheim's results, but semiempirical calculations by M. Barfield [28], A. D. Buckingham [107], and H. Kato [569–571] showed that the anisotropies in the indirect couplings could not be as large as claimed by R. A. Bernheim. L. C. Snyder performed an ab initio calculation on the same topic and reached similar conclusions, in that the anisotropy in H-H- and C-H-couplings is completely negligible, and in C-F- and H-F-couplings still small enough as not to affect geometry calculations seriously [993]. He pointed out, that the anisotropies calculated according to Bernheim depend heavily on the choice of geometry for methyl fluoride. This finding was confirmed by a more recent experimental work of B. P. Dailey, who measured the NMR spectra of methyl fluoride in three different nematic solvents and found the geometry slightly solvent dependent [234]. B. P. Dailey argued that this could not be due to solvent dependence of indirect couplings as in isotropic media. These are known to be virtually solvent independent. E. E. Burnell was able to interpret his experimental results on the same molecule in terms of a rapid exchange between two sites with orientation parameters of opposite signs [121]. The same interpretation was successfully applied to the spectra of ^{13}C-enriched acetylene by P. Diehl [303]. It therefore seems that the large J-anisotropies claimed by R. A. Bernheim [44] are disproved thoroughly. In ^{13}C-enriched mercury dimethyl, H. Dreeskamp found anisotropies in J_{CC} and J_{HgC} which are about one half of the isotropic value [374].

Considerable effort was devoted by C. MacLean to the investigation of anisotropic indirect couplings between two fluorines. In *o*-difluorobenzene all indirect couplings except J_{FF} were considered isotropic, and the corresponding direct couplings were used to calculate the molecular geometry [766]. The position of the fluorines found was used to calculate the direct F-F dipole-dipole coupling constant, which agreed with the observed coupling. Thus the observed anisotropic coupling could be explained from dipole-dipole interaction only. A small anisotropy was indeed found for the F-F indirect coupling in *m*-difluorobenzene by employing the method of essentially different orientations through the inclusion of several thermotropic solvents and a lyotropic nematic solvent [772]. A study of *p*-difluorobenzene in a lyotropic medium was performed to obtain an orientation different from previous work, and all these experiments were interpreted with a single geometry [480, 761]. This also resulted in a small anisotropy in J_{FF} [775]. C. MacLean's results on F-F-couplings seem to confirm the calculations of couplings including fluorine by C. W. Haigh [525], who found that most types of F-F-couplings are appreciably anisotropic while most H-F-couplings are not.

At this point, one can conclude that for geometry calculations, H-H, H-F, H-C, and probably C-F anisotropic couplings can be used with the assumption that the indirect contribution is negligible. In the case of H-C-couplings, care must be taken of vibrational effects; anisotropic indirect contributions may arise from F-F-interactions. However, there are results which seek a detailed explanation concerning anisotropic indirect couplings, e. g., the spectra of tetrafluoro-1,3-dithietane [495], pentafluoropyridine [397], and phosphoryl fluoride [252]. More experimental and theoretical work remains to be done.

10.3.4.4 Anisotropy of Chemical Shifts

In ^{1}HMR chemical shifts and also their anisotropies are small. Therefore, the corrections due to referencing, temperature changes, and phase changes may be larger than the measured effects [98, 100, 109, 533–536, 819, 1114], and all earlier results based on the phase subtraction method should be regarded with utmost care. More reliable results are obtained by plotting the observed chemical shift versus orientation; from the equation:

$$\delta_{obs} = \delta_{iso} + \tfrac{1}{2} S_{zz} (2 \delta_{zz} - \delta_{xx} - \delta_{yy})$$

for molecules with three-fold or higher symmetry, it follows that the isotropic part can be obtained as the intercept and the anisotropic part as the slope, if a linear plot is obtained. The orientation parameter may be changed by temperature; this again introduces uncertainties due to a possible temperature dependence of the chemical shifts. C. S. Yannoni suggested changing δ by slow rotation [1114]; this possibly is the only reliable way to obtain ^{1}H chemical shift anisotropies.

The resonances of heavier nuclei such as ^{13}C, ^{14}N, ^{19}F and ^{31}P exhibit much larger chemical shifts, and the phase subtraction method may be applied here without introducing excessive uncertainties. However, few systematic studies have been done. ^{19}F shift anisotropies were studied in fluorobenzenes by D. F. R. Gilson [481] and A. Saupe [939], but both authors were unable to obtain a satisfactory theoretical interpretation. ^{19}F shift anisotropies in some organic and inorganic compounds were determined by B. P. Dailey [220, 228, 234, 235, 237]. More recently, with the increased sensitivity of spectrometers, ^{13}C chemical shifts were investigated by I. Morishima [822] in methyl iodide and acetonitrile and by B. P. Dailey in HCN [221], methanol [225], the methyl halides [229, 234], and in chloroform [227]. A trend to more positive values was observed for the ^{13}C shift anisotropy in the methyl halides with increasing halogen electronegativity [234]. G. Englert studied the ^{13}C shift anisotropy in benzene [427] and in acetylene [428]. Recently, B. P. Dailey measured ^{31}P shift anisotropies in PH$_3$ [236], PF$_3$ [228], several halogenophosphazenes [233], POF$_3$ [237] and P$_4$S$_3$ [231], and the ^{15}N shift anisotropies in N$_2$O [226]. W. McFarlane obtained the ^{199}Hg chemical shift anisotropy in CH$_3$HgBr and the ^{13}C and ^{15}N shift anisotropies in CH$_3$CN by double resonance methods [749, 750]. Similarly, the ^{14}N shift anisotropy in CH$_3$NC had been obtained earlier by C. S. Yannoni [1114].

10.3.4.5 Quadrupole Coupling Constants

Most of the work about quadrupole coupling constants has been concerned with deuteron couplings. The early work of W. D. Phillips used quadrupole coupling constants from crystal powder spectra and deuteron quadrupole splittings from nematic phase spectra to evaluate the degree of orientation and bond angles in deuterated benzoic acid, benzene, toluene, naphthalene and stilbene [855, 856]; all these substances were oriented in nematic solvents.

Most of the quadrupole coupling constants measured only from nematic phase spectra are from B. P. Dailey [216, 224, 232] and B. M. Fung [461, 462, 464]. B. P. Dailey used a mixture of protonated and deuterated molecular species [224]. He obtained the degree of orientation from the proton spectrum and thus was able to evaluate the quadrupole coupling constant from the deuteron spectrum. An increase of the quadrupole coupling constant and thus the electric field gradient around the nucleus was obtained with increasing C-D-force constant for hydrogen cyanide, acetylene, methylacetylene, ethane, methyl iodide, methyl bromide, acetonitrile, benzene, and cyclopropane. Similar results were obtained by B. M. Fung, using the same method. He investigated phosphine, thiophenole, phenyl phosphine, phenyl silane [461], and nitro benzene, m-dinitrobenzene, and 1,3,5-trinitrobenzene [462]. P. Diehl pointed out that deuterium substitution causes a noticeable change in orientation, thus introducing a systematic error in B. P. Dailey's method [276]. The deuterium quadrupole coupling constant of monodeutero-benzene was given, taking in account the isotope effects on molecular geometry and orientation [274]. A. Loewenstein found quadrupolar splittings for fully deuterated neopentane, confirming S. Meiboom's results [798] for CDH_3, CD_2H_2, CD_3H, CD_4 [665], and for the analogous silicon compounds [668]. He concluded that even for small molecules, distortion mechanisms are operative in the nematic phase to change their effective symmetry. The deuterium quadrupolar couplings in methyl silane and methyl germane were measured by the same author [666, 667]. For acetone and dimethylsulfoxide, quadrupolar couplings were obtained by A. Azman [17]. J. W. Emsley obtained the magnitudes and relative signs of dipolar and quadrupolar interactions from ^{1}H-$\{^2$H$\}$-double resonance experiments in CH_3CD_2OH, CD_3CH_2OH, CD_3COCH_3, 4-D-pyridine, and CH_3CDO [400, 401, 409, 410]. For the CD_2-nuclei in ethanol, the individual tensor components of the quadrupole coupling constant were obtained by using five different concentrations.

C. S. Yannoni performed ^{1}H-$\{^{14}$N$\}$-double resonance experiments in methyl isocyanide to obtain the ^{14}N-quadrupole coupling constant, which is unusually small in this molecule [1114]. H. J. C. Berendsen obtained the quadrupole tensor elements for all three kinds of deuterons in dimethyl formamide, and the nitrogen quadrupolar interaction was also measured from ^{14}N-resonance [32]. The nematic lyophase of poly-L-glutamic acid/water/dimethyl formamide was used for these studies.

10.3.5 Table of Nematic Solvents

I Benzylidene-aniline derivatives R—⟨O⟩—CH=N—⟨O⟩—R′

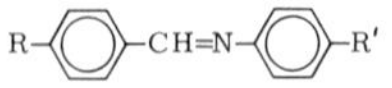

No.	Name	R	R′	Nematic range of p (°C)	Used pure (p), mixture with No.	Recent literature example
1	MBBA	OCH_3	C_4H_9	20–45	p, 2	112
2	EBBA	OC_2H_5	C_4H_9	37–80	p, 1, 24	303
3	APAPA	OCH_3	$OCOCH_3$	82–111	24	526
4	*p′*-Methoxybenzylidene-*p*-aminophenylbutyrate	OCH_3	$OCOC_3H_7$	53–112	23	965
5	*p′*-Methoxybenzylidene-*p*-amino-α-methyl-*n*-propyl-cinnamate	OCH_3	$CH{=}C{-}COOC_3H_7$ $\quad\mid$ $\quad CH_3$	54–89	p, 6, 6+17	576

II Azobenzene derivatives R—⟨O⟩—N=N—⟨O⟩—R′

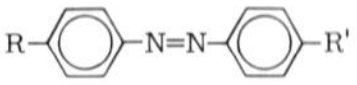

No.	Name	R	R′	Nematic range of p (°C)	Used pure (p), mixture with No.	Recent literature example
6	*p*-Methoxy-*p′*-caproyl-oxyazobenzene	OCH_3	$OCOC_5H_{11}$	66–106	p, 5, 5+17	576
7	*p*-Ethoxy-*p′*-valeroyl-oxyazobenzene	OC_2H_5	$OCOC_4H_9$	79–125	8, 10+24	249
8	*p*-Ethoxy-*p′*-caproyl-oxyazobenzene	OC_2H_5	$OCOC_5H_{11}$	70–126	7, 24	249
9	*p*-Ethoxy-*p′*-heptanoyl-oxyazobenzene	OC_2H_5	$OCOC_6H_{13}$	68–118	10, 10+24	771

No.	Name	R	R'	Nematic range of p (°C)	Used pure (p), mixture with No.	Recent literature example
10	p-Ethoxy-p'-undecenoyl-oxyazobenzene	OC_2H_5	$OCO(CH_2)_8CH=CH_2$	62–106	9, 7+24	771
11	p-Ethoxy-p'-crotyloxy-azobenzene	OC_2H_5	$OCOCH=CH-CH_3$	112–196	p	811
12	p,p'-Di-n-hexyloxy-azobenzene	OC_6H_{13}	OC_6H_{13}	102–114	17	932

III Azoxybenzene derivatives*   R—⟨◯⟩—N=N—⟨◯⟩—R'
                                 O

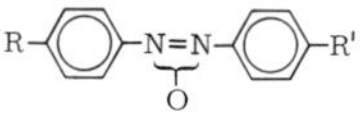

No.	Name	R	R'	Nematic range of p (°C)	Used pure (p), mixture with No.	Recent literature example
13	PAA	OCH_3	OCH_3	118–136	14	932
14	PAP	OC_2H_5	OC_2H_5	138–168	13	932
15	p,p'-Di-n-butoxyazoxy-benzene	OC_4H_9	OC_4H_9	107–134	16+17	772
16	p,p'-Di-n-pentyloxy-azoxybenzene	OC_5H_{11}	OC_5H_{11}	82–119	15+17	772
17	p,p'-Di-n-hexyloxy-azoxybenzene	OC_6H_{13}	OC_6H_{13}	81–127	p, 12, 18, 15+16	1059
18	p,p'-Di-n-heptyloxy-azoxybenzene	OC_7H_{15}	OC_7H_{15}	74–122	17	811
19	p-Methoxy-p'-n-butyl-azoxybenzene	OCH_3	C_4H_9	20–75	p, 20	719
20	p-Ethoxy-p'-n-butyl-azoxybenzene	OC_2H_5	C_4H_9	55–99	19	811
21	p-Ethoxy-p'-caproyloxy-azoxybenzene	OC_2H_5	$OCOC_5H_{11}$	81–138	24	819
22	p-Butoxy-p'-heptyloxy-azoxybenzene	OC_4H_9	OC_7H_{15}		24	526

IV Phenyl benzoate derivatives   R—⟨◯⟩—O—C—⟨◯⟩—R'
                                        O

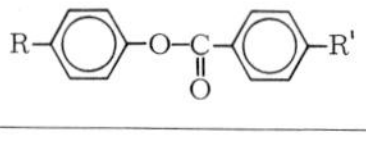

No.	Name	R	R'	Nematic range of p (°C)	Used pure (p), mixture with No.	Recent literature example
23	p-Ethoxyphenyl-p-caproyl-oxybenzoate	OC_2H_5	$OCOC_5H_{11}$		4	965
24	p-Ethoxyphenyl-p-butoxy-carbonyloxybenzoate	OC_2H_5	$OCOOC_4H_9$	54–67	2, 3, 8, 21, 22, 7+10, 9+10	150

V Benzoic acid derivatives   R—⟨◯⟩—COOH

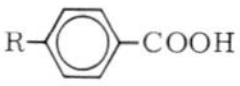

No.	Name	R		Nematic range of p (°C)	Used pure (p), mixture with No.	Recent literature example
25	p-n-Butoxybenzoic acid	OC_4H_9		147–160	26	251
26	p-n-Octyloxybenzoic acid	OC_8H_{17}		108–147	p, 25	251

VI Other compounds

No.	Name			Nematic range of p (°C)	Used pure (p), mixture with No.	Recent literature example
27	p,p'-Di-n-hexyloxybenzalazine 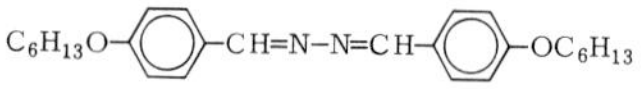			127–150	p	932

$C_6H_{13}O$—⟨◯⟩—CH=N–N=CH—⟨◯⟩—OC_6H_{13}

No.	Name	R	R'	Nematic range of p (°C)	Used pure (p), mixture with No.	Recent literature example
28	TBBA 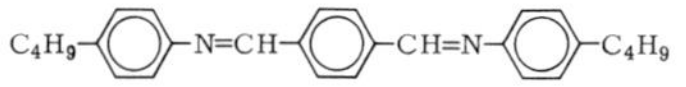			200–236	p	811
	C_4H_9—⬡—N=CH—⬡—CH=N—⬡—C_4H_9					
29	*p*-Bis-(*p'*-*n*-heptyloxybenzoyloxy)benzene 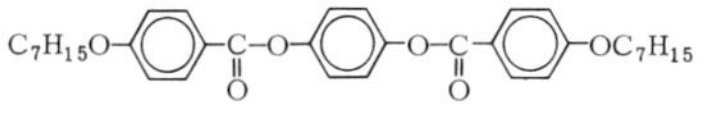			122–199	p	1113
	$C_7H_{15}O$—⬡—C‖O—O—⬡—O—C‖O—⬡—OC_7H_{15}					
30	6-*n*-Hexyloxy-2-naphthoic acid 			147–198	p	930
	$C_6H_{13}O$—⬡⬡—COOH					
31	*trans-trans*-4'-alkyl-4-cyanobicyclohexyls 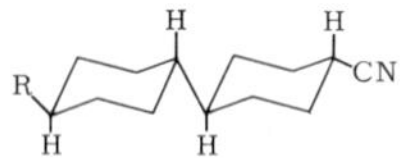			≈60–80	p	2:191a
	R—⬡⬡—CN					

* see note on p. 45

10.3.6 Table of Solutes

In this table, the solutes are arranged according to their sum formula. The columns "methods" and "results" correspond to the same headings of the previous sections. If no nucleus is specified, only proton resonance spectra are studied. In the column "results" the results are indicated which are of primary interest in the paper in consideration. The following abbreviations are used:

Methods

T	Thermotropic nematic solvent
Sm	Thermotropic smectic solvent
L	Lyotropic nematic solvent
Sp	Slow-spinning techniques
^{1}H-{^{19}F}	Double resonance technique: observing ^{1}H while irradiating ^{19}F
^{1}H (^{13}C-Sat.)	Study of ^{13}C-satellites of natural abundance in proton resonance spectra
E	Application of an electric field
La	Application of a lanthanide shift reagent
t	Relaxation times

Results

S	Orientation
G	Geometry
I	Intramolecular motions
J	Indirect coupling constants
sign (*J*)	Sign of indirect coupling constants
Δ*J*	Anisotropy of indirect coupling constants
Δσ	Anisotropy of chemical shifts
QCC	Quadrupole coupling constants

Formula	Structure	Author and ref.	Methods	Results
BF_4^-		Buckingham [116]	L, ^{11}B, ^{19}F	S
		Reeves [897]	L, ^{11}B, ^{19}F	S
$Cl_6N_3P_3$	$(NPCl_2)_3$	Dailey [233]	T, ^{31}P	$\Delta\sigma_P$
$Cl_8N_4P_4$	$(NPCl_2)_4$	Dailey [233]	T, ^{31}P	$\Delta\sigma_P$
F_3OP	$O{=}PF_3$	Dailey [237]	3T, ^{19}F, ^{31}P	ΔJ_{PF}, $\Delta\sigma_F$, $\Delta\sigma_P$
		de Lange [252]	T, ^{19}F	ΔJ_{PF}
F_3P		Dailey [228]	T, ^{19}F, ^{31}P	$\Delta\sigma_F$, $\Delta\sigma_P$, sign (J)
		Dailey [241]	Sm, ^{31}P	$\Delta\sigma_P$
F_3PS	$S{=}PF_3$	Dailey [241]	Sm, ^{19}F, ^{31}P	G, $\Delta\sigma_F$, $\Delta\sigma_P$, sign (J), ΔJ
$F_6N_3P_3$	$(NPF_2)_3$	Dailey [231]	T, $^{31}P\text{-}\{^{19}F\}$	$\Delta\sigma_P$
F_6S		Buckingham [100]	3T, ^{19}F	$\Delta\sigma_F$
GeH_4		Loewenstein [668]	T, ^1H, ^2H (GeH_3D)	S
H_2		Buckingham [99]	T, ^1H	$\Delta\sigma_H$
H_2O		Niederberger [834]	L, ^{17}O ($D_2^{17}O$)	S
H_3P		Dailey [232]	T, ^2H, ^{31}P (PD_3)	QCC
		Dailey [236]	T, ^1H, ^{31}P	$\Delta\sigma_H$, $\Delta\sigma_P$
		Fung [461]	T, ^1H, ^2H (PD_3)	QCC
		Spiesecke [1003]	T	G
H_4N^+		Buckingham [116]	L, ^1H, ^2H (NH_4^+, ND_4^+)	S
		Chen [193]	L, ^1H, ^2H, ^{14}N (ND_4^+)	G, I, QCC
		Loewenstein [670]	3L, ^2H, ^{14}N (ND_4^+)	S, G
		Reeves [894]	L, ^1H, ^2H (NH_4^+, ND_4, ND_3H)	S, QCC
		Saupe [949]	L	S, G
H_4Si		Loewenstein [668]	T, ^1H, ^2H (SiH_nD_{4-n})	S, QCC
N_2O		Dailey [226]	T, ^{15}N ($^{15}N_2O$)	$\Delta\sigma_N$, sign (J)
$CBrF_3$		Dailey [247]	Sm, ^{19}F	$\Delta\sigma_F$
$CClF_3$		Dailey [247]	Sm, ^{19}F	$\Delta\sigma_F$
CCl_3F		Dailey [235]	T, ^{19}F (^{13}C-Sat), Sp	$\Delta\sigma_F$
		Dailey [244]	T, ^{13}C	$\Delta\sigma_C$

Formula	Structure	Author and ref.	Methods	Results
CF_3I		Dailey [220]	T, ^{19}F	$\Delta\sigma_F$
		Dailey [247]	Sm, ^{19}F	$\Delta\sigma_F$
CF_4		Buckingham [100]	T, ^{19}F	$\Delta\sigma_F$
$CHCl_3$		Courtieu [209]	T, t_1 ($^{13}CHCl_3$)	
		Dailey [227]	T, ^1H, ^{13}C ($^{13}CHCl_3$)	$\Delta\sigma_H$, $\Delta\sigma_C$
		Dailey [244]	T, ^{13}C	$\Delta\sigma_C$
		Vold [1078]	T, t_1	
CHF_3		Bernheim [43]	T, ^1H, ^{19}F	$\Delta\sigma_H$, $\Delta\sigma_F$, sign (J_{HF})
		Dailey [240]	Sm	S, $\Delta\sigma_H$
		Dailey [246]	Sm, ^1H, ^{19}F	S, $\Delta\sigma_F$, sign (J)
		de Lange [254]	T, ^1H (^{13}C-Sat)	G, ΔJ, sign (J)
CHN	H—C≡N	Dailey [221]	T, ^1H, ^{13}C ($H^{13}CN$)	$\Delta\sigma_H$, $\Delta\sigma_C$
		Dailey [224]	T, ^1H, ^2H (DCN)	QCC
		Spiesecke [1000]	T, ^1H ($H^{13}CN$)	
CH_2Cl_2		Borodin [87]	T	$\Delta\sigma_H$
		Courtieu [209]	T, t_1	
		Fung [465]	L, ^1H, ^{35}Cl	
		Morishima [823]	T	S
CH_2F_2		Dailey [246]	Sm, ^1H, ^{19}F	S, $\Delta\sigma_F$, sign (J)
		de Lange [254]	T, ^1H (^{13}C-Sat)	G, ΔJ, sign (J)
CH_2I_2		Courtieu [208]	^{13}C, ^{13}C-{^1H} ($^{13}CH_2I_2$)	
		Morishima [823]	T	S
CH_3Br		Dailey [214]	T	$\Delta\sigma_H$
		Dailey [216]	T, ^1H, ^2H, (CD_3Br)	QCC
		Dailey [229]	T, ^1H, ^{13}C ($^{13}CH_3Br$)	$\Delta\sigma_C$, sign (J)
		Dailey [242]	Sm	$\Delta\sigma_H$
		Diehl [310]	^2H (CD_3Br)	QCC
		Diehl [313]	2T (^2H, CD_3Br)	S, QCC
		Hayamizu [536]	L	$\Delta\sigma_H$
		Morishima [823]	T	S
CH_3BrHg	CH_3HgBr	McFarlane [749]	T, ^1H-{^{199}Hg}	$\Delta\sigma_{Hg}$
		McFarlane [752]	2T, ^1H-{^{199}Hg}	$\Delta\sigma_{Hg}$
		Saupe [946]	T	G, $\Delta\sigma_H$
CH_3Cl		Dailey [214]	T	$\Delta\sigma_H$
		Dailey [229]	T, ^1H, ^{13}C ($^{13}CH_3Cl$)	$\Delta\sigma_C$
		Dailey [242]	Sm	$\Delta\sigma_H$
		Hayamizu [536]	L	$\Delta\sigma_H$

Formula	Structure	Author and ref.	Methods	Results
CH_3ClHg	CH_3HgCl	McFarlane [752]	2T, ^1H-$\{^{199}Hg\}$	$\Delta\sigma_{Hg}$
		McLauchlan [759]	T	G
		Saupe [946]	T	G, $\Delta\sigma_H$
CH_3Cl_3Si	CH_3SiCl_3	Dailey [243]	Sm	S, $\Delta\sigma_H$
$CH_3Cl_4NP_2$	$CH_3N(PCl_2)_2$	McFarlane [754]	T, ^1H-$\{^{13}C\}$ $^{13}CH_3N(PCl_2)_2$ ^1H-$\{^{15}N\}$ $CH_3{}^{15}N(PCl_2)_2$	G, ΔJ, $\Delta\sigma_C$, $\Delta\sigma_N$, $\Delta\sigma_P$
CH_3F		Bernheim [39]	T, ^1H ($^{13}CH_3F$)	sign (J_{CH}), sign (J_{CF})
		Bernheim [40]	T, ^1H, ^{19}F	$\Delta\sigma_F$
		Bernheim [42]	T, ^1H ($^{13}CH_3F$)	ΔJ_{CH}, ΔJ_{CF}
		Bernheim [43]	T, ^1H, ^{19}F	sign (J_{HF}), $\Delta\sigma_H$, $\Delta\sigma_F$
		Buckingham [98]	T, ^{19}F	$\Delta\sigma_F$
		Burnell [121]	T, Sp	G, S
		Dailey [214]	T	$\Delta\sigma_H$
		Dailey [234]	3T, ^1H, ^{13}C, ^{19}F ($^{13}CH_3F$)	$\Delta\sigma_C$, $\Delta\sigma_F$, ΔJ
		Dailey [239]	T, ^1H, ^2H (CD_3F)	QCC
		Dailey [242]	Sm	$\Delta\sigma_H$
		Dailey [246]	Sm, ^1H, ^{19}F	S, $\Delta\sigma_F$, sign (J)
		Hayamizu [536]	L	$\Delta\sigma_H$
		Kato [571]	(theory)	
		Lucas [673]	(theory)	
		MacLean [763]	(theory)	
		Snyder [993]	(theory)	
CH_3Hg^+		Reeves [908]	L, ^1H (^{199}Hg-Sat)	G
CH_3HgI		McFarlane [752]	2T, ^1H-$\{^{199}Hg\}$	$\Delta\sigma_{Hg}$
		Saupe [946]	T	G, $\Delta\sigma_H$
CH_3I		Azman [21]	T, t_1	
		Dailey [214]	T	$\Delta\sigma_H$
		Dailey [216]	T, ^1H, ^2H (CD_3I)	QCC
		Dailey [229]	T, ^1H, ^{13}C ($^{13}CH_3I$)	$\Delta\sigma_C$, $\Delta\sigma_H$
		Dailey [242]	Sm	$\Delta\sigma_H$
		Diehl [310]	^2H (CD_3I)	QCC
		Diehl [313]	2T (^2H, CD_3I)	S, QCC
		Hayamizu [536]	L	$\Delta\sigma_H$
		Morishima [822]	T, ^{13}C ($^{13}CH_3I$)	$\Delta\sigma_C$
		Morishima [823]	T	S
CH_3NO	NH_2CHO	Reeves [883]	L	S
		Reeves [891]	L, ^1H ($^{15}NH_2CHO$)	
CH_3NO_2		Diehl [310]	^2H (CD_3NO_2)	QCC
		Diehl [313]	T, ^2H (CD_3NO_2)	S, QCC

Formula	Structure	Author and ref.	Methods	Results
$CH_3O_3P^{2-}$	$CH_3-\overset{\overset{O}{\|\|}}{\underset{\underset{O^-}{\|}}{P}}-O^-$	Goldstein [499]	L, ^1H, ^{13}C ($^{13}CH_3PO_3^{2-}$)	sign (J)
		Goldstein [501]	L, ^1H, ^{13}C ($^{13}CH_3PO_3^{2-}$)	S, G, I
CH_4		Buckingham [100]	3T, ^1H	$\Delta\sigma_H$
		Loewenstein [665, 668]	T, ^1H, ^2H ($CH_{4-n}Dn$)	S, QCC
		Loewenstein [669]	T, T_1, ^1H, ^2H (CD_4)	
		Yannoni [1112]	T, ^1H (CH_3D)	S
CH_4O	CH_3OH	Dailey [225]	T, ^{13}C ($^{13}CH_3OH$)	G, $\Delta\sigma_C$
		Diehl [267]	T, Sp	S
		Diehl [313]	2L, 2T, ^1H ($^{13}CH_3OH$)	S, G
		Goldstein [499]	L	
		Goren [513]	T	S, $\Delta\sigma_H$
		Khetrapal [580]	T, La	
		Reeves [911]	L, ^2H (CD_3OH)	S
		Saupe [931]	T, ^1H ($^{13}CH_3OH$)	
		Saupe [933]	T	S
		Schaumburg [960]	L, ^1H, ^2H (CD_3OD)	QCC
		Wooten [1106]	T, ^2H (^{13}C-Sat) (CDH_2OH)	QCC
CH_5N	CH_3NH_2	Schaumburg [960]	L, ^1H, ^2H (CD_3ND_2)	QCC
CH_6Ge	CH_3GeH_3	Loewenstein [666]	T, ^1H, ^2H (CH_3GeD_3, CD_3GeH_3)	G, I, QCC
CH_6N^+	$CH_3NH_3^+$	Reeves [894]	L, ^1H, ^2H ($CH_3ND_3^+$)	S, QCC
		Reeves [895]	L, ^1H ($^{13}CH_3^{15}NH_3^+$)	G, I
$CH_6N_3^+$	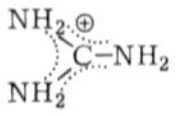 $\underset{NH_2^+}{\overset{NH_{2\,\oplus}}{C-NH_2}}$	Diehl [308]	T	G, I
CH_6Si	CH_3SiH_3	Loewenstein [666]	T, ^1H, ^2H (CH_3SiD_3, CD_3SiH_3)	G, I, QCC
		Loewenstein [667]	T, ^1H, ^2H (CH_3SiD_3)	I, QCC
$C_2Cl_3F_3$	CF_3-CCl_3	Dailey [220]	T, ^{19}F	$\Delta\sigma_F$
		Dailey [244]	T, ^{13}C	$\Delta\sigma_C$
		Dailey [247]	Sm, ^{19}F	$\Delta\sigma_F$
		Spiesecke [1005]	T, ^{19}F	
		Yannoni [1111]	T, ^{19}F	S
		Yannoni [1115]	T, Sp	$\Delta\sigma_F$

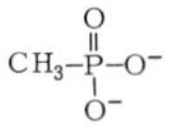

Formula	Structure	Author and ref.	Methods	Results
C_2F_4	$CF_2{=}CF_2$	Saupe [940]	T, ^{19}F	ΔJ_{FF}
$C_2F_4S_2$	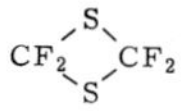	Goldstein [495]	L, ^{19}F	$\Delta\sigma_F$, ΔJ_{FF}, sign (J)
C_2HF_3	$CF_2{=}CHF$	MacLean [776]	T, L	S, G, sign (J), ΔJ
$C_2HF_3O_2$	CF_3COOH	Dunn [386]	T, ^{19}F	$\Delta\sigma_F$
		Emsley [402]	T, ^{19}F, Sp	J, sign (J)
C_2H_2	$CH{\equiv}CH$	Dailey [224]	T, ^{1}H, ^{2}H (C_2D_2)	QCC
		Diehl [303]	2T, ^{1}H ($H^{13}C{\equiv}CH$, $H^{13}C{\equiv}^{13}CH$)	G, S
		Englert [428]	T, ^{1}H, ^{13}C ($H^{13}C{\equiv}CH$)	$\Delta\sigma_C$
		Mohanty [819]	T, L	$\Delta\sigma_H$
		Saupe [936]	T	$\Delta\sigma_H$, S
		Spiesecke [1002]	T, ^{1}H ($H^{13}C{\equiv}CH$, $H^{13}C{\equiv}^{13}CH$)	G, S
$C_2H_2Cl_2$	$CH_2{=}CCl_2$	Diehl [267]	T, Sp	S
		Kato [568]	T	S
	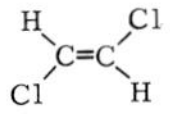	Diehl [267]	T, Sp	S
		Kato [568]	T	S
	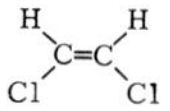	Diehl [267]	T, Sp	S
		Diehl [277]	T, E	S
		Kato [568]	T	S
$C_2H_2F_2$	$CH_2{=}CF_2$	Buckingham [104]	T, ^{1}H, ^{19}F	ΔJ_{FF}
		Haigh [525]	(theory)	
		MacLean [764, 765]	2T, ^{1}H, ^{19}F	ΔJ_{FF}
		MacLean [770]	T, ^{19}F	$\Delta\sigma_F$
		Saupe [940]	T, ^{19}F	ΔJ_{FF}
	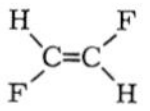	MacLean [773]	T, L, ^{1}H, ^{19}F	ΔJ_{FF}, sign (J)
		MacLean [774]	T, L, ^{1}H, ^{19}F	ΔJ_{FF}
		MacLean [777]	T, ^{1}H, ^{19}F (^{13}C-Sat)	G, I, ΔJ, sign (J)
	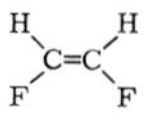	Buckingham [108]	T, ^{1}H, ^{19}F	S
		Emsley [406]	T, ^{1}H, ^{19}F, Sp	G, S, ΔJ_{FF}, I, sign (J)
		MacLean [777]	T, ^{1}H, ^{19}F (^{13}C-Sat)	G, I, ΔJ, sign (J)

Formula	Structure	Author and ref.	Methods	Results
$C_2H_2N_2Se$		Veracini [1073]	2L, T, ^1H (^{13}C-Sat, ^{77}Se-Sat)	G, I
$C_2H_3Cl_3$	CH_3-CCl_3	Dailey [243]	Sm	S, $\Delta\sigma_H$
		Yannoni [1115]	T, Sp	$\Delta\sigma_H$
C_2H_3F	$CH_2=CHF$	Buckingham [108]	T, ^1H, ^{19}F	G, S
$C_2H_3F_3$	CH_3-CF_3	Dailey [217]	T, ^1H, ^{19}F	$\Delta\sigma_H$, $\Delta\sigma_F$, sign(J_{HF})
$C_2H_3F_3O$	CF_3CH_2OH	Azman [18]	T, ^1H, ^2H, ^{19}F (CF_3CD_2OH, CF_3CD_2OD)	G, S, QCC
C_2H_3N	$CH_3C\equiv N$	Azman [19]	T, T_1	
		Dailey [216]	T, ^1H, ^2H (CD_3CN)	QCC
		Dailey [218]	T, ^1H	$\Delta\sigma_H$
		Dailey [238]	T, ^{13}C ($CH_3{}^{13}CN$)	$\Delta\sigma_C$
		Diehl [267]	T, Sp	S
		Diehl [276]	T, ^1H, ^2H ($CH_{3-n}D_nCN$)	QCC
		Diehl [310]	^2H (CD_3CN)	QCC
		Diehl [313]	2T, ^2H (CD_3CN)	S, QCC
		Emsley [398]	T, Sp	S
		McFarlane [750]	T, $^1H\text{-}\{^{13}C\}$, $^1H\text{-}\{^{15}N\}$ ($CH_3{}^{13}C^{15}N$)	$\Delta\sigma_C$, $\Delta\sigma_N$
		Morishima [822]	T, ^{13}C ($^{13}CH_3CN$)	$\Delta\sigma_C$
		Morishima [823]	T	S
		Saupe [931]	2T, ^1H ($^{13}CH_3CN$)	sign (J_{CH})
		Saupe [937]	3T, ^1H ($^{13}CH_3CN$, $Me^{13}CN$, $MeC^{15}N$, $^{13}CH_3C^{15}N$)	G, sign ($^1J_{CH}$, $^2J_{CH}$, $^3J_{CH}$)
		Wooten [1106]	T, ^2H (^{13}C-Sat) (CDH_2CN)	QCC
	$CH_3N\equiv C$	Dailey [218]	T	$\Delta\sigma_H$
		Spiesecke [1001]	T, ^1H ($^{13}CH_3NC$)	sign (J_{CH})
		Yannoni [1114]	T, $^1H\text{-}\{^{14}N\}$	$\Delta\sigma_N$, QCC, ΔJ_{NH}
$C_2H_3O_2^-$	CH_3COO^-	Reeves [899]	L, ^1H, ^2H (CD_3COO^-, $^{13}CH_3COO^-$, $CH_3{}^{13}COO^-$)	G, S, QCC
		Schaumburg [960]	L, ^1H, ^2H (CD_3COO^-)	QCC

Formula	Structure	Author and ref.	Methods	Results
C_2H_4	$CH_2\!=\!CH_2$	Diehl [311]	T, ^1H ($H_2{}^{13}C\!=\!CH_2$, $H_2{}^{13}C\!=\!{}^{13}CH_2$)	G, I
		Lindblom [642]	T, ^2H ($CH_2\!=\!CHD$)	S, QCC
		MacLean [762]	T	G
		Spiesecke [1000]	T	G
$C_2H_4Br_2$	$CH_2Br\!-\!CH_2Br$	Swinton [1013]	T	sign (J)
$C_2H_4Cl_2$	$CH_2Cl\!-\!CH_2Cl$	Diehl [267]	T	S
		Fung [465]	L, ^1H, ^{35}Cl	
		Swinton [1013]	T	sign (J)
$C_2H_4F_2$	$CH_2F\!-\!CH_2F$	MacLean [771]	2T, ^1H, ^{19}F	G, I
$C_2H_4I_2$	$CH_2I\!-\!CH_2I$	Swinton [1013]	T	sign (J)
C_2H_4O		Canet [145]	T, ^1H (^{13}C-Sat)	G, S, sign (J)
		Gazzard [468]	T	G, S
	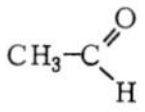	Borodin [89]	T	$\Delta\sigma_H$
		Emsley [410]	T, ^1H, ^{13}C ($^{13}CH_3{}^{13}CHO$, CH_3CDO	G, I, QCC
$C_2H_4O_2$	CH_3COOH	Reeves [911]	L, ^2H (CD_3COOH)	S
$C_2H_4O_3S$	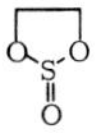	Reeves [884]	T	G
		Spiesecke [1005]	T	G
C_2H_4S		Canet [145]	T, ^1H (^{13}C-Sat)	G, S, sign (J)
		Gazzard [468]	T	S, G
C_2H_5Br	$CH_3\!-\!CH_2Br$	Emsley [414]	T, ^1H-$\{^2H\}$ (CH_2DCH_2Br, CH_3CD_2Br)	I, QCC
C_2H_5ClO	$ClCH_2\!-\!CH_2OH$	Azman [20]	T	G, I
C_2H_5F	$CH_3\!-\!CH_2F$	Buckingham [103]	T, ^1H, ^{19}F	$\Delta\sigma_F$, sign (J)
C_2H_5I	$CH_3\!-\!CH_2I$	Woodman [1105]	T	
C_2H_5N		Gazzard [469]	L	S, G

Formula	Structure	Author and ref.	Methods	Results
C_2H_5NO	CH_3NHCHO	Khetrapal [578]	T	G, I
		Khetrapal [585]	T, ^1H	G
			$(^{13}CH_3^{15}NH^{13}CHO)$	
C_2H_6	$CH_3—CH_3$	Dailey [217]	T	I, $\Delta\sigma_H$, sign (J)
		Dailey [224]	T, ^1H, ^2H	QCC
			(CH_3CD_3)	
C_2H_6Cd	CH_3CdCH_3	McFarlane [755]	T, ^1H-$\{^{111/113}Cd\}$	$\Delta\sigma_{Cd}$, G
C_2H_6Hg	CH_3HgCH_3	Dreeskamp [374]	T, ^1H, ^{13}C	ΔJ_{HgC}, ΔJ_{CC}
			$(^{13}CH_3HgCH_3,$	
			$^{13}CH_3Hg^{13}CH_3)$	
		Englert [426]	T	QCC
		McFarlane [752]	2T, ^1H-$\{^{199}Hg\}$	$\Delta\sigma_{Hg}$
C_2H_6O	$CH_3—CH_2OH$	Emsley [396]	T, ^1H-$\{^2H\}$	
			(CH_3CD_2OH)	
		Emsley [400]	T, ^1H-$\{^2H\}$	QCC, I
			$(CD_3CH_2OH,$	
			$CH_3CD_2OH)$	
		Lawson [614]	L	
		Reeves [911]	L, ^2H (deut.)	S
C_2H_6OS	CH_3SOCH_3	Azman [17]	T, ^1H, ^2H	QCC
			(CD_3SOCD_3)	
		Dailey [222]	L	S
		Narasimhan [831]	T, ^1H (theory)	S, $^4J_{HH}$
C_2H_6Se	CH_3SeCH_3	Diehl [326]	T, ^1H (^{13}C-Sat,	G
			^{77}Se-Sat)	
$C_2H_6Sn^{2+}$	$CH_3Sn^{2+}CH_3$	Reeves [892]	L	I
		Reeves [893]	L	G, I
C_2H_6Te	CH_3TeCH_3	Diehl [326]	T, ^1H (^{13}C-Sat,	G
			^{123}Te-Sat,	
			^{125}Te-Sat)	
$C_2H_6Tl^+$	$CH_3Tl^+CH_3$	Reeves [901]	2L	S, G
$C_2H_8N^+$	$(CH_3)_2NH_2^+$	Reeves [895]	L, ^1H	I
			$(^{13}CH_3CH_3NH_2^+,$	
			$(CH_3)_2^{15}NH_2^+)$	
C_3F_6	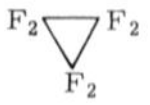 $F_2 \triangledown F_2$ F_2	Emsley [416]	T, ^{19}F	ΔJ, sign (J)
C_3HF_3	$CF_3C{\equiv}CH$	Buckingham [101]	T, ^1H, ^{19}F	$\Delta\sigma_H$, $\Delta\sigma_F$,
				sign (J_{HF})
C_3H_3Br	$CH_2BrC{\equiv}CH$	Saupe [931]	T	$\Delta\sigma_H$

Formula	Structure	Author and ref.	Methods	Results
C_3H_3Cl	$CH_2ClC\equiv CH$	Saupe [931]	T	$\Delta\sigma_H$
$C_3H_3Cl_3O$	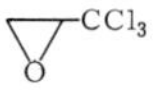	Meiboom [802]	T	S
C_3H_3FO	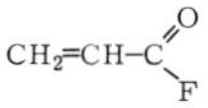	Courtieu [206]	T	G, I
C_3H_3NO		Fung [463]	T, ^1H-$\{^1H\}$	
$C_3H_3N_3$		Canet [147]	T, ^1H-$\{^{13}C\}$, ^1H-$\{^{15}N\}$	G
C_3H_4	$CH_2=C=CH_2$	Sackmann [918] Spiesecke [1000]	T, ^1H (^{13}C-Sat) T	G, I
	$CH_3-C\equiv CH$	Canet [143]	T, ^1H (^{13}C-Sat)	G
		Dailey [224]	T, ^1H, ^2H ($CH_3C\equiv CD$)	QCC
		Saupe [931]	T	G, S, $\Delta\sigma_H$
		Saupe [936]	T	G, S, $\Delta\sigma_H$, sign (J)
$C_3H_4Cl_2$	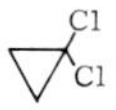	Gilson [487]	T	G, I
$C_3H_4Cl_4$	$CH_2ClCCl_2CH_2Cl$	Gazzard [467]	T	I
C_3H_4O	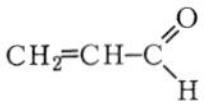	Courtieu [206]	T	G, I
$C_3H_4O_2$		Reeves [888]	2T	G
$C_3H_4O_2S$	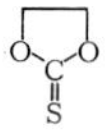	Swinton [1014]	T	G
	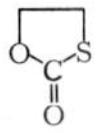	Reeves [887] Reeves [890]	T L	G, I G, I
$C_3H_4O_3$	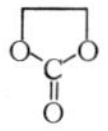	Reeves [887] Reeves [890] Swinton [1014]	T L T	G, I G, I G
$C_3H_4S_3$	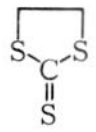	Reeves [886]	T	G, I

Formula	Structure	Author and ref.	Methods	Results
C_3H_5Br	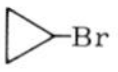 ▷—Br	Gilson [491]	2T	S, G, I
C_3H_5BrO	CH_3COCH_2Br	Courtieu [205]	T	I
C_3H_5Cl	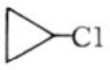 ▷—Cl	Gilson [491]	2T	S, G, I
C_3H_5ClO	CH_3COCH_2Cl	Courtieu [205]	T	I
C_3H_5FO	CH_3COCH_2F	Courtieu [205]	2T	G, I
$C_3H_5O_2^-$	$CH_3CH_2COO^-$	Schaumburg [960]	L, ^{1}H, ^{2}H ($C_2D_5COO^-$)	QCC
C_3H_6	$CH_3—CH{=}CH_2$	Bothner-By [93]	T, ^{1}H ($CH_3{}^{13}CH{=}CH_2$)	G, S, I
	▽	Dailey [216] Lucas [674] Meiboom [799] Meiboom [801]	T, ^{1}H, ^{2}H (C_3D_6) (theory) T T, ^{1}H (^{13}C-Sat)	QCC G G, sign (J)
C_3H_6O	CH_3COCH_3	Azman [17] Borodin [90] Courtieu [203] Dailey [222] Emsley [401] Narasimhan [831] Shcherbakov [986]	T, ^{1}H, ^{2}H (CD_3COCD_3) 2T T L T, ^{1}H-$\{^2H\}$ (CH_3COCD_3) T, ^{1}H (theory) T	QCC S S S QCC S, $^4J_{HH}$
	⬠O (epoxide)	Gilson [483, 488] Saupe [947]	T T	G, I G, I
$C_3H_6O_2$	CH_3CH_2COOH	Reeves [911]	L, ^{2}H (CH_3CD_2COOH)	S
	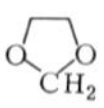 (dioxolane CH_2)	de Lange [255]	T	G, I
$C_3H_6O_3$	 (trioxane)	Cocivera [196]	T, ^{1}H (^{13}C-Sat)	G, I, sign (J_{CH})
C_3H_6S	(thietane)	Gilson [488] Lunazzi [718] Saupe [947] Swinton [1015]	T T T T	G, I G, I G, I G, S
C_3H_7NO	$(CH_3)_2NCHO$	Berendsen [32] Bopp [86] Phillips [12:299]	L, ^{1}H, ^{2}H, ^{14}N ($(CD_3)_2NCDO$) T L	QCC_D, QCC_N I S

Formula	Structure	Author and ref.	Methods	Results
	$CH_3CONHCH_3$	Khetrapal [584]	T	S, G
C_3H_8O	$CH_3CH_2CH_2OH$	Diehl [304]	T, ^2H-$\{^1H\}$ ($CH_3CH_2CD_2OH$)	G, S, QCC
		Reeves [911]	L, ^2H (deut.)	S
C_3H_9OP	$(CH_3)_3PO$	Albrand [4]	T, ^{13}C, ^{31}P (($^{13}CH_3)_3PO$)	G, $\Delta\sigma_C$, $\Delta\sigma_P$, ΔJ
		Dailey [245]	3Sm, ^{31}P	$\Delta\sigma_P$, sign (J)
		McFarlane [751]	T, ^1H-$\{^{31}P\}$	$\Delta\sigma_P$
C_3H_9P	$(CH_3)_3P$	Albrand [4]	T, ^{13}C, ^{31}P (($^{13}CH_3)_3P$)	G, $\Delta\sigma_C$, $\Delta\sigma_P$, ΔJ
		Dailey [245]	3Sm, ^{31}P	$\Delta\sigma_P$, sign (J)
		McFarlane [751]	T, ^1H-$\{^{31}P\}$	$\Delta\sigma_P$
C_3H_9PS	$(CH_3)_3PS$	Albrand [4]	T, ^{13}C, ^{31}P (($^{13}CH_3)_3PS$)	G, $\Delta\sigma_C$, $\Delta\sigma_P$, ΔJ
		Dailey [245]	Sm, ^{31}P	$\Delta\sigma_P$, sign (J)
		McFarlane [751]	T, ^1H-$\{^{31}P\}$	$\Delta\sigma_P$
C_3H_9PSe	$(CH_3)_3PSe$	Albrand [4]	T, ^{13}C, ^{31}P (($^{13}CH_3)_3PSe$)	G, $\Delta\sigma_C$, $\Delta\sigma_P$, ΔJ
		Dailey [245]	Sm, ^{31}P	sign (J)
$C_3H_9Pb^+$	$(CH_3)_3Pb^+$	Reeves [908]	L, ^1H (^{207}Pb-Sat)	G
$C_3H_9Sn^+$	$(CH_3)_3Sn^+$	Reeves [908]	L, ^1H (^{117}Sn-Sat, ^{119}Sn-Sat)	G
$C_3H_{10}N^+$	$(CH_3)_3NH^+$	Reeves [900]	L	
$C_4Cl_2F_4$	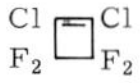 Cl⬜Cl F_2 F_2	Harris [526]	T, ^{19}F	$^2J_{FF}$
$C_4Cl_4F_4$	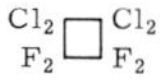 Cl_2⬜Cl_2 F_2 F_2	Harris [526]	T, ^{19}F	$^2J_{FF}$
C_4F_6	$CF_3-C\equiv C-CF_3$	Buckingham [102]	T, ^{19}F	$\Delta\sigma_F$, sign (J_{FF})
$C_4F_{12}P_4$	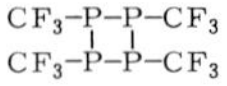 $CF_3-P-P-CF_3$ $CF_3-P-P-CF_3$	Albrand [3] Albrand [5]	T, ^1H, ^{19}F, ^{31}P T, ^{31}P	G, $\Delta\sigma$ $\Delta\sigma_P$, G, I, J
$C_4H_2Cl_2N_2$	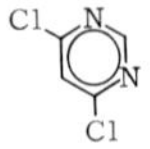 (structure)	Diehl [290]	T	
$C_4H_2N_2$	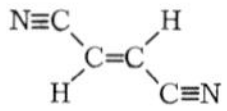 (structure)	Kato [568]	T	S
C_4H_3N	$CH_3-C\equiv C-C\equiv N$	Canet [141]	T, ^1H (^{13}C-Sat)	G
$C_4H_4Br_2$	$CH_2Br-C\equiv C-CH_2Br$	Borodin [87]	T	$\Delta\sigma_H$

Formula	Structure	Author and ref.	Methods	Results
$C_4H_4Cl_2$	$CH_2Cl-C\equiv C-CH_2Cl$	Saupe [936]	T	S
$C_4H_4N_2$		de Lange [249]	T	G, S
		Goldstein [498]	L	G, S
		Khetrapal [577]	L	S, G
		Diehl [284]	T	G, S
		Goldstein [498]	L	G, S
		Khetrapal [577]	L	S, G
		Diehl [268]	T	$\Delta\sigma_H$
		Diehl [318]	T, ^{1}H (^{13}C-Sat)	G
		Goldstein [498]	L	G, S
		Khetrapal [577]	L	S, G, sign (J)
$C_4H_4NiS_4$	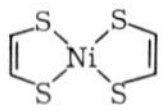	Bailey [23]	T	G, S, I
C_4H_4O		Burnell [122]	T, ^{1}H (^{13}C-Sat)	G
		Diehl [266]	T	G, S, sign (J)
		Diehl [277]	T, E	S
		Diehl [317]	T, ^{1}H (^{13}C-Sat)	G, I
		Goldstein [497]	L	G, S, sign (J)
$C_4H_4O_2$		Russell [915]	T	G
C_4H_4S		Dereppe [257]	L	G
		Diehl [265]	T	G
		Diehl [277]	T, E	S
		Diehl [317]	T, ^{1}H (^{13}C-Sat)	G, I
		Goldstein [497]	L	G, S, sign (J)
$C_4H_4S_2$		Goldstein [496]	L	G, S
		Russell [915]	T	G
C_4H_4Se		Dahlquist [212]	L	G
C_4H_4Te		Lunazzi [719]	T	G, S
C_4H_5N		Emsley [422]	T, ^{1}H ($C_4H_5^{15}N$)	G, I
		Randall [882]	T, ^{1}H ($C_4H_5^{15}N$)	G, S, ΔJ_{NH}
	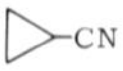 $\triangleright$-CN	Gilson [491]	2T	S, G, I
C_4H_6	$CH_3-C\equiv C-CH_3$	Buckingham [102]	T	G, sign (J_{HH})
	$CH_2=CH-CH=CH_2$	Castellano [150]	T, ^{1}H	G, S, I

Formula	Structure	Author and ref.	Methods	Results
	□	Günther [522]	T	G, S
	▨	Meiboom [803]	T	G
		Meiboom [804]	T, ^{1}H-{^{1}H}	
C_4H_6O		Gilson [489]	T	G, I
		Lunazzi [718]	T	G, S, I
		Swinton [1015]	T	G, S
		Courtieu [204]	T	G, S
		de Kowalewski [248]	T	G
		Gilson [484]	T	G, I
		Swinton [1016]	T	G, S
$C_4H_6O_2$	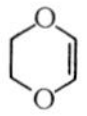	de Lange [253]	T	G, S, I
		Goldstein [500]	2L, ^{1}H, ^{13}C (^{13}C-Sat)	S, G
$C_4H_6O_2S$		Khetrapal [586]	L	G
		Saupe [942]	T	G, S
C_4H_8	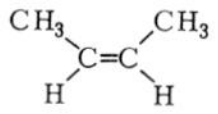 $\begin{smallmatrix}CH_3\\ \\H\end{smallmatrix}C{=}C\begin{smallmatrix}CH_3\\ \\H\end{smallmatrix}$	Diehl [301]	T	G, I
	□	Gilson [485]	(theory)	I
		Meiboom [799]	T	G
		Meiboom [805]	T, ^{1}H (^{13}C-Sat)	G, I
$C_4H_8O_2$	n-C_3H_7COOH	Reeves [911]	L, ^{2}H ($C_2H_5CD_2COOH$)	S
C_4H_9NO	$CH_3CON(CH_3)_2$	Anderson [7]	L, ^{1}H ($CD_3CON(CH_3)_2$)	S, I
$C_4H_{10}O$	n-C_4H_9OH	Reeves [911]	L, ^{2}H (deut.)	S
$C_4H_{12}N^+$	$(CH_3)_4N^+$	Loewenstein [670]	3L, ^{2}H, ^{14}N ($N(CD_3)_4^+$)	S, G
		Reeves [900]	L	S
$C_4H_{12}Si$	$(CH_3)_4Si$	Dailey [243]	Sm, ^{1}H (^{13}C-Sat)	S
		Meiboom [798]	T	S
C_5F_5N		Emsley [397]	T, ^{19}F	ΔJ_{FF}, $\Delta \sigma_F$, sign (J)
		Emsley [421]	T, ^{19}F	G, I
$C_5H_2F_3N$		Emsley [421]	T, ^{1}H, ^{19}F, Sp	G, I, ΔJ, sign (J)

Formula	Structure	Author and ref.	Methods	Results
$C_5H_3Br_2N$		Diehl [269]	T	S
$C_5H_3Cl_2N$		Orrell [850]	T	S, sign (J)
		Azman [16]	T	S
		Orrell [850]	T	S, sign (J)
$C_5H_3F_2N$		Emsley [421]	T, ^{1}H, ^{19}F, Sp	G, I, ΔJ, sign (J)
		Orrell [850]	T	G, S, sign (J)
C_5H_4	$CH_3-C\equiv C-C\equiv CH$	Saupe [931]	T	G, S
		Saupe [936]	T	$\Delta\sigma_H$
C_5H_4OS	CHO	Veracini [1060]	T	G, I
		Veracini [1069]	L	G, I
$C_5H_4O_2$	CHO	Veracini [1057]	T	G, I
		Goldstein [502]	2L	S, G
C_5H_5As		Wong [1104]	T	G
$C_5H_5Br_2N$	NBr$_2$	Veracini [1058]	T	S
$C_5H_5I_2N$	NI$_2$	Veracini [1066]	2T	S, G
C_5H_5N		Burnell [120]	T, La	
		de Lange [249]	T	S
		Diehl [271]	T	G, S
		Emsley [409]	T, ^{1}H-$\{^2$H$\}$	G, S, I, QCC
		Goldstein [507]	L	S, G
		Khetrapal [579]	L	G, S
		Schumann [965]	T, ^{1}H ($C_5H_5^{15}$N)	G
C_5H_5NNiO		Emsley [403]	T	G, I
C_5H_5NO	OH	Goldstein [506]	L	S, G
		Khetrapal [582]	L	G
		Khetrapal [579]	L	G, S

Formula	Structure	Author and ref.	Methods	Results
C_5H_5NS		Goldstein [506]	L	S, G
C_5H_5P	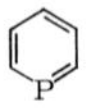	Wong [1104]	T	$\Delta\sigma_P$, sign (J), G
C_5H_6		Emsley [405]	T	G, S, I
		Günther [523]	T	G, S
		Veracini [1061]	T	G
$C_5H_6N_2$		Goldstein [506]	L	S, G
C_5H_6O		Gilson [486]	T	G
$C_5H_6O_3$	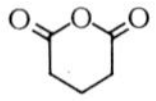	Bulthius [118]	T	S, G, I
C_5H_8		Buckingham [106]	T, ^1H	G, S
	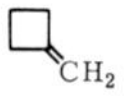	Gilson [489]	T	G, I
$C_5H_{10}O_2$	n-C_4H_9COOH	Reeves [911]	L, ^2H (n-$C_3H_7CD_2COOH$)	S
	$(CH_3)_3CCOOH$	de Lange [251]	T	G, S
		Courtieu [202]	T, ^1H (($CD_3)_2$)	S
C_5H_{12}	$(CH_3)_4C$	Dailey [243]	Sm	S
		Loewenstein [665]	T, ^1H, ^2H (($CD_3)_4C$)	S, QCC
		Meiboom [798]	T	S
		Wei [1099]	T, ^1H, ^2H (($CD_3)_4C$, $CH_3C(CD_3)_3$)	S, G, I, QCC
$C_5H_{12}O$	n-$C_5H_{11}OH$	Reeves [911]	L, ^2H (deut.)	S
$C_6Br_2F_4$	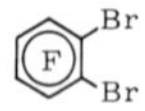	MacLean [767]	2T, ^{19}F	ΔJ_{FF}
	 Br-F-Br	MacLean [767]	2T, ^{19}F	ΔJ_{FF}
$C_6Br_3F_3$	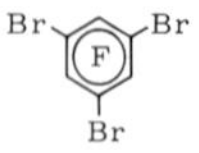	Dailey [220]	T, ^{19}F	$\Delta\sigma_F$

Formula	Structure	Author and ref.	Methods	Results
$C_6Cl_3F_3$	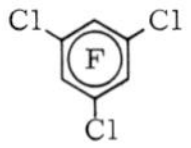	Emsley [419]	T, ^{19}F (^{13}C-Sat)	G, I, ΔJ
$C_6F_4I_2$		MacLean [767]	2T, ^{19}F	ΔJ_{FF}
C_6F_6		Goldstein [505]	L, ^{19}F	S, G, ΔJ
		MacLean [769]	2T, ^{19}F	ΔJ_{FF}
		MacLean [778]	T, ^{19}F	G, ΔJ
		Saupe [938]	2T, ^{19}F	S
		Saupe [939]	T, ^{19}F	$\Delta\sigma_F$
		Snyder [991]	T, ^{19}F	S, $\Delta\sigma_F$
C_6HCl_4F	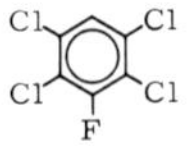	Saupe [928]	T	S
C_6HF_5		Goldstein [505]	L, ^1H, ^{19}F	S, G, ΔJ
		MacLean [778]	T, ^1H, ^{19}F	G, ΔJ
$C_6H_2BrClF_2$	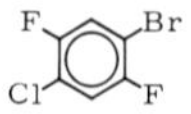	Anderson [9]	T	S
$C_6H_2Cl_2F_2$	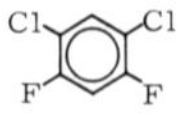	Anderson [9]	T	S
	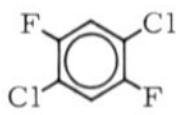	Anderson [9]	T	S
$C_6H_2Cl_3NO_2$	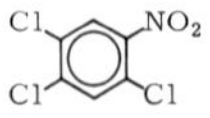	Diehl [280]	T	S
$C_6H_2Cl_4$	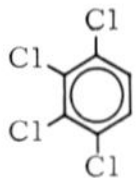	Goren [511]	T, Sp	S
	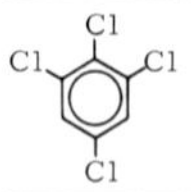	Saupe [928]	T	S
	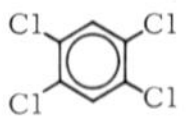	Saupe [928]	T	S
$C_6H_2F_3NO_2$	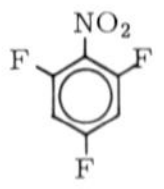	MacLean [780]	T, ^{19}F	$\Delta\sigma_F$
$C_6H_2F_4$	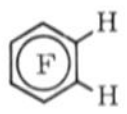	Goldstein [505]	L, ^1H, ^{19}F	S, G, ΔJ
		MacLean [767]	T, ^1H, ^{19}F	ΔJ_{FF}

Formula	Structure	Author and ref.	Methods	Results
		Goldstein [505]	L, ^{1}H, ^{19}F	S, G, ΔJ
		MacLean [778]	T, ^{1}H, ^{19}F	G, ΔJ
		Saupe [938]	2T, ^{1}H, ^{19}F	S
		Goldstein [505]	L, ^{1}H, ^{19}F	S, G, ΔJ
		MacLean [761]	T, ^{1}H, ^{19}F	ΔJ_{FF}, sign (J)
		MacLean [767]	4T, ^{1}H, ^{19}F	ΔJ_{FF}
		Saupe [938]	2T, ^{1}H, ^{19}F	S
		Saupe [939]	T, ^{19}F	$\Delta \sigma_F$
$C_6H_3Br_3$	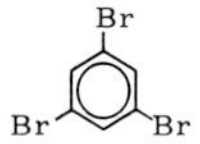	Courtieu [201]	T	S
		Hayamizu [533]	T	$\Delta \sigma_H$
		Hayamizu [534]	T, L	$\Delta \sigma_H$
$C_6H_3Cl_3$	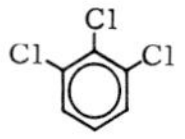	Saupe [932]	2T	S
	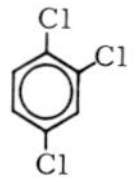	Diehl [269]	T	
	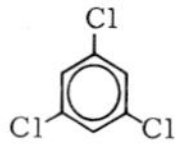	Courtieu [201]	T	S
		Emsley [419]	T	G, I
		Hayamizu [533]	T	$\Delta \sigma_H$
		Saupe [928]	T	$\Delta \sigma_H$
		Saupe [932]	T	S
		Schumann [964]	T, ^{1}H, ^{13}C (^{13}C-Sat)	G, ΔJ_{CH}
		Yannoni [1115]	T, Sp	$\Delta \sigma_H$
$C_6H_3F_2I$	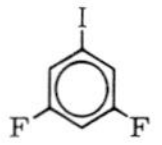	Gilson [481]	T, ^{19}F	$\Delta \sigma_F$
$C_6H_3F_3$	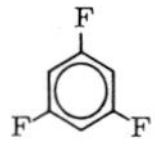	Dailey [239]	T, ^{1}H, ^{2}H ($C_6D_3F_3$)	QCC
		Gilson [479]	T, ^{1}H, ^{19}F	$\Delta \sigma_H$, $\Delta \sigma_F$, sign (J)
		Goldstein [505]	L, ^{1}H, ^{19}F	S, G, ΔJ
		Saupe [938]	2T, ^{1}H, ^{19}F	S
		Saupe [939]	T, ^{19}F	$\Delta \sigma_F$
$C_6H_3N_3O_6$	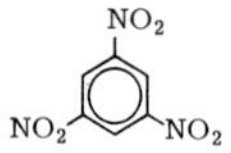	Courtieu [201]	T	S
		Fung [462]	T, ^{1}H, ^{2}H ($C_6D_3N_3O_6$)	QCC
C_6H_3P	$P(C{\equiv}CH)_3$	McFarlane [753]	T	G
C_6H_4BrCl	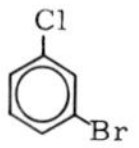	Khetrapal [574]	T	

Formula	Structure	Author and ref.	Methods	Results
C_6H_4BrF		Diehl [280]	T	
		Gilson [481]	T, ^{19}F	$\Delta\sigma_F$
$C_6H_4Br_2$	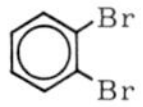	Diehl [270]	T	G, S
		Diehl [321]	T, ^1H (^{13}C-Sat)	G
	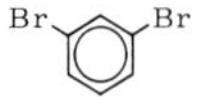	Diehl [272]	T	S, G
		Burnell [119]	3T, ^1H (^{13}C-Sat)	G, S, J_{CH}
		MacLean [761]	T	sign (J)
C_6H_4ClF		Diehl [280]	T	
		Gilson [481]	T, ^{19}F	$\Delta\sigma_F$
$C_6H_4Cl_2$	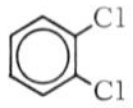	Dereppe [262]	T, ^1H-$\{^1H\}$	
		Diehl [270]	T	G, S
		Diehl [321]	T, ^1H (^{13}C-Sat)	G
	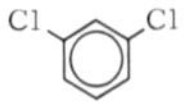	Diehl [272]	T	G, S
		Diehl [323]	T, ^1H (^{13}C-Sat)	G
		Diehl [302]	T, ^1H (^{13}C-Sat)	G, J_{CH}
		MacLean [761]	T	sign (J)
		Saupe [932]	T	S
$C_6H_4FNO_2$		Gilson [481]	T, ^{19}F	$\Delta\sigma_F$
$C_6H_4F_2$	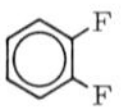	Gilson [480]	T, ^1H, ^{19}F	$\Delta\sigma_F$
		Goldstein [505]	L, ^1H, ^{19}F	S, G, ΔJ
		MacLean [765, 766]	4T, ^1H, ^{19}F	ΔJ_{FF}
		Saupe [938]	2T, ^1H, ^{19}F	S
		Saupe [939]	T, ^{19}F	$\Delta\sigma_F$
	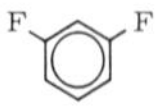	Gilson [480]	T, ^1H, ^{19}F	$\Delta\sigma_F$
		Goldstein [505]	L, ^1H, ^{19}F	S, G, ΔJ
		MacLean [772]	3T, L, ^1H, ^{19}F	ΔJ_{FF}
		Saupe [938]	2T, ^1H, ^{19}F	S
		Saupe [939]	T, ^{19}F	$\Delta\sigma_F$
		Gilson [480]	T, ^1H, ^{19}F	$\Delta\sigma_F$
		Goldstein [505]	L, ^1H, ^{19}F	S, G, ΔJ
		MacLean [761]	T, ^1H, ^{19}F	sign (J)
		MacLean [775]	L, ^1H, ^{19}F	ΔJ_{FF}
		Saupe [938]	2T, ^1H, ^{19}F	S
		Saupe [939]	T, ^{19}F	$\Delta\sigma_F$

Formula	Structure	Author and ref.	Methods	Results
$C_6H_4I_2$	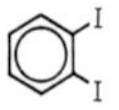	Diehl [321]	T, ^{1}H (^{13}C-Sat)	G
		MacLean [761]	T	sign (*J*)
$C_6H_4NO_2^-$	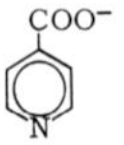	Reeves [906]	L	G
$C_6H_4N_2$		Goldstein [507]	L	S, G
$C_6H_4N_2O_2$	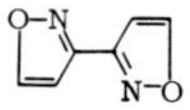	Veracini [1051]	T	S, I
	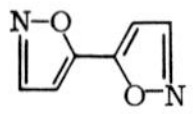	Veracini [1053]	T	S, I
	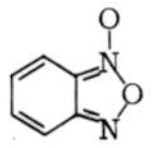	Diehl [273]	T	G, S
$C_6H_4N_2O_4$	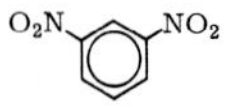	Fung [462]	T, ^{1}H, ^{2}H (2,4,6-D$_3$)	QCC
		Veracini [1050, 1056]	T	G
$C_6H_4N_2S$	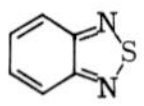	Khetrapal [576]	T	G, S
$C_6H_4O_2$	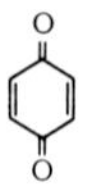	Burnell [125] Diehl [267] Diehl [268]	2T, ^{1}H (^{13}C-Sat) T, Sp T	G S G, S, $\Delta\sigma_H$
$C_6H_4O_2S$	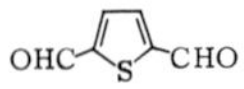	Huckerby [552] Veracini [1055] Veracini [1069]	T T L	I I G, I
$C_6H_4O_3$	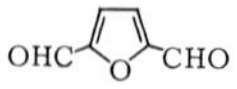	Huckerby [552] Veracini [1052]	T T	I I
$C_6H_4S_2$	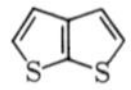	Lunazzi [716]	T	G, S, sign (*J*)

Formula	Structure	Author and ref.	Methods	Results
$C_6H_4S_4$		Burnell [123]	T, ^1H (^{13}C-Sat)	S, G
		Burnell [126]	T, ^1H (^{13}C-Sat)	G, I
C_6H_5Br	Br	Canet [138]	T	S
		Schaumburg [959]	T, ^1H, ^2H (C_6D_5Br)	G, QCC
		Tracey [1044]	2L	S
C_6H_5Cl	Cl	Canet [138]	T	S
		Canet [140]	T, ^1H-$\{^1H\}$	G, S
		Schaumburg [959]	T, ^1H, ^2H (C_6D_5Cl)	G, QCC
		Tracey [1044]	2L	S
$C_6H_5Cl_3Si$	 $SiCl_3$	Canet [138]	T	S
C_6H_5F	F	Canet [138]	T	S
		Goldstein [505]	L, ^1H, ^{19}F	S, G, ΔJ
		Hillenbrand [547]	T	S
		MacLean [769]	2T, ^1H, ^{19}F	ΔJ_{HF}
		Snyder [992]	T, ^1H, ^{19}F	G, S, $\Delta\sigma_F$
		Tracey [1044]	2L	S, G
C_6H_5FO	 F ... OH	Gilson [481]	T, ^{19}F (FC_6H_4OD)	$\Delta\sigma_F$
C_6H_5I	I	Canet [138]	T	S
		Tracey [1044]	2L	S
C_6H_5NO	 CHO ... N	Orrell [851]	T	S, G, I
	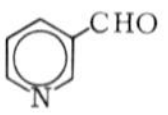 CHO ... N	Veracini [1072]	T	G, I
	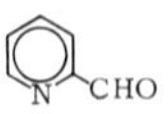 N ... CHO	Veracini [1064]	T	G, I
	 NO	Canet [138]	T	S
$C_6H_5NO_2$	NO_2	Canet [138]	T	S
		Fung [462]	T, ^1H, ^2H (2,4,6-D_3)	QCC

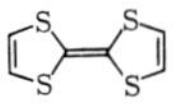

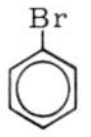

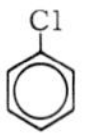

Formula	Structure	Author and ref.	Methods	Results
C_6H_6		Dailey [216]	T, ^{1}H, ^{2}H (C_6D_6)	QCC
		Dailey [219]	T	$\Delta\sigma_H$
		Diehl [274]	T, ^{1}H, ^{2}H (C_6H_5D)	QCC
		Diehl [277]	T, E	S
		Diehl [293]	T, ^{1}H (1-^{13}C)	G, I
		Diehl [299]	(theory)	
		Englert [427]	T, ^{13}C (1-^{13}C)	$\Delta\sigma_C$
		Goldstein [505]	L	S, G
		Henriksson [544]	L, ^{2}H (C_6D_6)	S
		Lawson [613]	L	S, sign (J)
		Loewenstein [664]	T, ^{2}H (C_6D_6)	relaxation
		Saupe [928, 929]	T	$\Delta\sigma_H$, sign (J)
		Saupe [938]	T	S
		Snyder [990]	T, ^{1}H	sign (J)
		Tracey [1044]	2L	S
	$CH_3-C{\equiv}C-C{\equiv}C-CH_3$	Saupe [931]	T	S
		Saupe [936]	T	S, G, sign (J)
	$CH_2{=}C{=}CH-CH{=}C{=}CH_2$	Hopf [550]	T	G
C_6H_6FN		Gilson [481]	T, ^{19}F	$\Delta\sigma_F$
$C_6H_6N_2O$	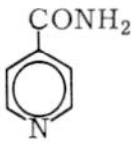	Goldstein [506]	L	S, G
$C_6H_6N_2S$	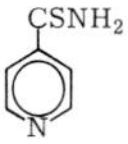	Goldstein [506]	L	S, G
C_6H_6O	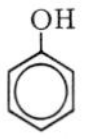	Diehl [292]	T	S, G, I
C_6H_6S		Fung [461]	T, ^{1}H, ^{2}H (C_6H_5SD)	QCC
C_6H_7N		Goldstein [503]	L	S, G, I
		Goldstein [506]	L	S, G
		Saupe [944]	T	G, I
$C_6H_7N_3O$	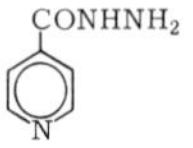	Goldstein [506]	L	S, G
C_6H_7P	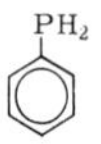	Fung [461]	T, ^{1}H, ^{2}H ($C_6H_5PD_2$)	QCC

Formula	Structure	Author and ref.	Methods	Results
C_6H_8		Buckingham [105]	T	G, I, sign (J)
$C_6H_8N^+$	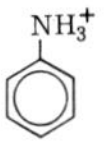 NH_3^+	Diehl [307]	L, ^1H, ^2H ($C_5D_5NH_3^+$)	S, G, QCC
C_6H_8Si	 SiH_3	Fung [461]	T, ^1H, ^2H ($C_6H_5SiD_3$)	QCC
C_6H_{12}		Henriksson [544] Meiboom [808]	L, ^2H (C_6D_{12}) T, ^1H-$\{^2H\}$ ($C_6H_2D_{10}$)	S G, S
$C_6H_{12}N_4$		Frey [451] Veracini [1065]	T L, 2T, ^1H (^{13}C-Sat)	S, G S, G
$C_6H_{12}O_2$	n-$C_5H_{11}COOH$	Reeves [911]	L, ^2H (n-$C_4H_9CD_2COOH$)	S
$C_6H_{14}O$	n-$C_6H_{13}OH$	Reeves [911]	L, ^2H (deut.)	S
C_7HF_5O	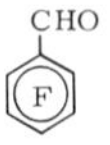 CHO F	Emsley [411]	T	G, I
$C_7HF_5O_2$	 COOH F	MacLean [779]	L, 2T, ^{19}F	S, G, ΔJ
$C_7H_4Cl_2O$	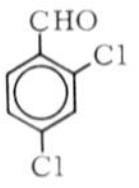 CHO Cl Cl	Diehl [320]	T	G, I
$C_7H_4Cl_2O_2$	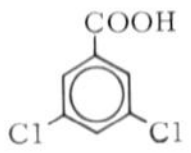 COOH Cl Cl	Saupe [930]	T	S
$C_7H_4Cl_4O$	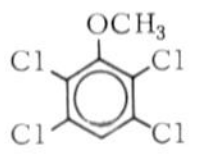 OCH$_3$ Cl Cl Cl Cl	Saupe [930]	T	S
	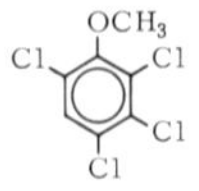 OCH$_3$ Cl Cl Cl Cl	Saupe [930]	T	S
C_7H_4FN	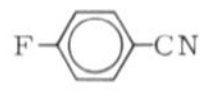 F—CN	Gilson [481]	T, ^{19}F	$\Delta\sigma_F$
$C_7H_4F_2$	 F$_2$	Günther [521]	T, ^1H, ^{19}F	G, S

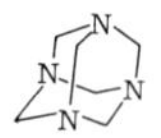

Formula	Structure	Author and ref.	Methods	Results
$C_7H_4F_2O_2$	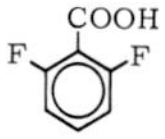 COOH, F, F	MacLean [779]	L, 2T, ^1H, ^{19}F	S, G, ΔJ
$C_7H_4F_4O$	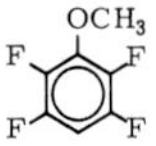 OCH_3, F, F, F, F	Emsley [413]	T, ^1H, ^{19}F	G, I
$C_7H_4FeO_3$	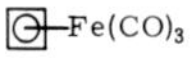 $Fe(CO)_3$	Buckingham [114] Dailey [213] Emsley [407]	T, ^1H (^{13}CO) T (theory)	G, I G
$C_7H_4NO_4^-$	 COO^-, NO_2	Reeves [907]	2L	S
$C_7H_4N_2O_2$	 NO_2, CN	Veracini [1056]	T	G
C_7H_5BrO	 CHO, Br	Wong [1103]	T	S, G, I
C_7H_5ClO	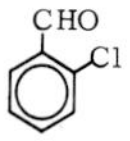 CHO, Cl	Diehl [320]	T	G, I
	 CHO, Cl	Wong [1103]	T	S, G, I
$C_7H_5ClO_2$	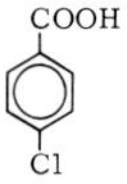 COOH, Cl	Saupe [928]	T	S
$C_7H_5Cl_2F$	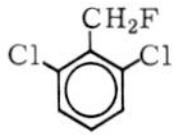 CH_2F, Cl, Cl	Canet [146]	T	G, I
$C_7H_5Cl_3$	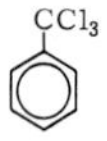 CCl_3	Canet [138]	T	S
C_7H_5FO	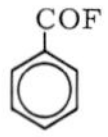 COF	Burnell [124]	T	S, G, I
	 CHO, F	Emsley [425] Wong [1103]	Sm, ^1H, ^2H, ^{19}F (C_6H_4FCDO) T	S, G, I S, G, I
$C_7H_5FO_2$	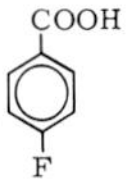 COOH, F	MacLean [779]	L, 2T, ^1H, ^{19}F	S, G, ΔJ

Formula	Structure	Author and ref.	Methods	Results
$C_7H_5F_3$	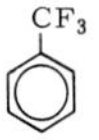	Diehl [297]	T, ^1H, ^{19}F	G, I, ΔJ_{FF}
C_7H_5N		Canet [138] Goldstein [507] Schaumburg [957] Veracini [1056]	T L T, ^1H, ^{13}C ($C_6{}^{13}CH_5N$) 3T, ^1H	S S, G G, S G
	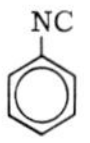	Canet [138]	T	S
C_7H_5NO		Canet [138]	T	S
$C_7H_5NO_2$	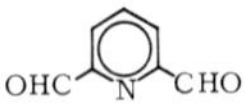	Veracini [1063]	T	G, I
$C_7H_5NO_3$		Wong [1103]	T	S, G, I
$C_7H_5NO_4$	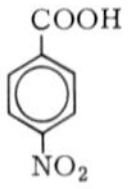	Reeves [907]	2L	S
C_7H_5NS	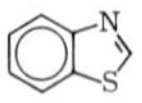	Diehl [319]	T	G
		Canet [138]	T	S
$C_7H_5N_3$	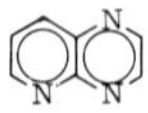	Diehl [319]	T	G
$C_7H_5O_4Re$	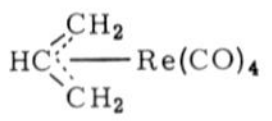 $HC\overset{CH_2}{\underset{CH_2}{\lessgtr}}Re(CO)_4$	Emsley [408]	T	G, I
C_7H_6		Günther [521]	T	G, S
$C_7H_6Cl_2$	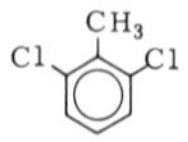	Diehl [282]	T	G, I
	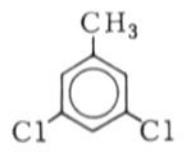	Diehl [283]	T	G, I
$C_7H_6Cl_2O$	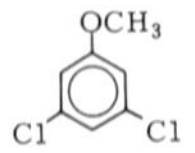	Emsley [420]	T	S, G, I

Formula	Structure	Author and ref.	Methods	Results
$C_7H_6FeO_3$	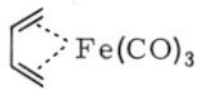 Fe(CO)$_3$	Diehl [322]	T	G
	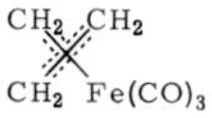 CH$_2$ CH$_2$ / CH$_2$ Fe(CO)$_3$	Buckingham [113]	T, ^{1}H (^{13}CO)	G, I
C_7H_6O	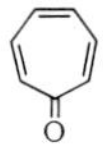 (O)	Emsley [399] Veracini [1054]	T T	G G
	 CHO	Diehl [291]	T	G, S, I
$C_7H_6O_2$	 COOH	MacLean [779]	L, 2T	S, G, ΔJ
	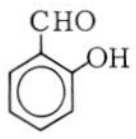 CHO OH	Diehl [286]	T	G, S, I
	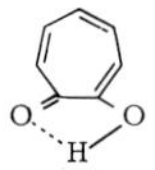 O...O H	Emsley [404]	3T	G, I
C_7H_7Br	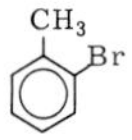 CH$_3$ Br	Diehl [289]	T	G, S, I
C_7H_7Cl	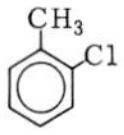 CH$_3$ Cl	Diehl [288]	T	G, S, I
C_7H_7I	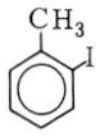 CH$_3$ I	Diehl [289]	T	G, S, I
C_7H_8	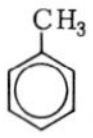 CH$_3$	Diehl [285]	T	G, S, I
	(cyclopentadiene)	Diehl [294] Emsley [415]	T T	G, S G, I
C_7H_8O	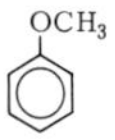 OCH$_3$	Diehl [324]	T, ^{1}H-{^{2}H}, ^{2}H, ^{2}H-{^{1}H} ($C_6H_5OCD_3$)	G, I
	 CH$_3$ OH	Goldstein [508]	L	G, I

Formula	Structure	Author and ref.	Methods	Results
$C_7H_9Cl_2NPt$	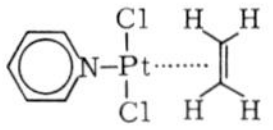	Drago [373]	L	G
$C_7H_{14}O_2$	$n\text{-}C_6H_{13}COOH$	Reeves [911]	L, ^2H ($n\text{-}C_5H_{11}CD_2COOH$)	S
C_7H_{16}	$n\text{-}C_7H_{16}$	Reeves [911]	L, ^2H (deut.)	S
$C_7H_{16}O$	$n\text{-}C_7H_{15}OH$	Reeves [911]	L, ^2H (deut.)	S
$C_8H_4Br_2S_2$	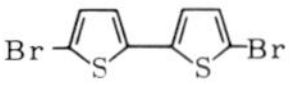	Veracini [1059]	T	I
$C_8H_4Cl_2S_2$	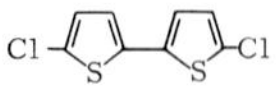	Veracini [1059]	T	I
$C_8H_4N_2$	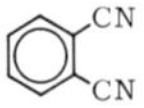	de Lange [250] Diehl [270]	T T	G G, S
	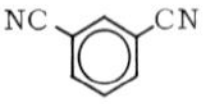	de Lange [250]	T	G
		de Lange [250]	T	G
$C_8H_4N_2O_4S_2$	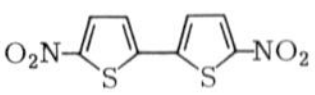	Veracini [1059]	T	I
$C_8H_5MnO_3$	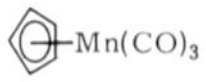	Buckingham [112] Dailey [223] Khetrapal [573] Saupe [948]	T, ^1H (^{13}CO) T, ^1H (^{13}C-Sat) T T, ^1H (^{13}C-Sat)	G, S, I G, $\Delta\sigma_H$ S, $\Delta\sigma_H$ G, I
C_8H_6	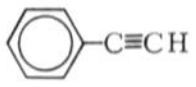	Dereppe [261] Diehl [281] Fung [461]	T T T ($C_6H_5C{\equiv}CD$)	G G QCC
$C_8H_6Br_2O$		Emsley [420]	T	S, G, I
$C_8H_6Cl_4$	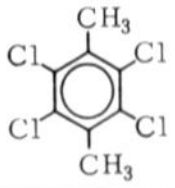	Canet [137] Saupe [926]	T T	G, I S
$C_8H_6N_2$	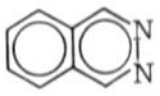	Saupe [943]	T	G, S, J
	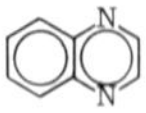	Khetrapal [575]	T	G, S, J

Formula	Structure	Author and ref.	Methods	Results
	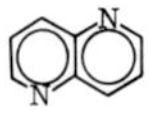	Gilson [490]	T	S, G
	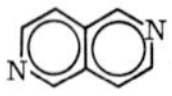	Gilson [490] Veracini [1067]	T T	S, G G
$C_8H_6N_2O$	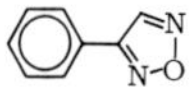	Veracini [1071]	T	G, I
$C_8H_6N_2S$	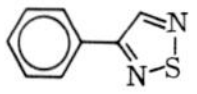	Veracini [1071]	T	G, I
$C_8H_6N_2Se$	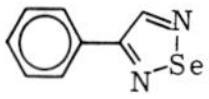	Veracini [1071]	T	G, I
$C_8H_6N_4$	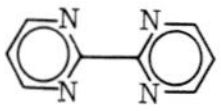	Courtieu [207]	T	G, I
$C_8H_6O_2$	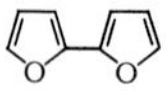	Veracini [1070]	T	G, I
$C_8H_6O_3W$	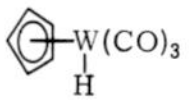 —W(CO)₃	Buckingham [111] McIvor [756]	T (^{13}CO) T	G, I G, I
C_8H_6S	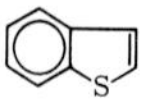	Saupe [945]	T	S, G
$C_8H_6S_2$	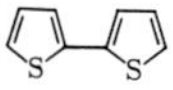	Khetrapal [581] Veracini [1062]	T T	G, I G, I
	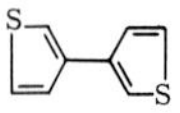	Veracini [1068]	T	G, I
C_8H_8		Meiboom [811]	5T, ^1H-$\{^2H\}$ $(C_8H_6D_2)$	G, I
$C_8H_8Cl_2$	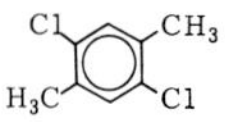	Canet [139]	T	G, I
C_8H_8O	 COCH₃	Diehl [324] Emsley [418]	T T, ^1H-$\{^2H\}$, ^2H $(C_6H_5COCD_3)$	G, I S, G, I
$C_8H_8O_2$	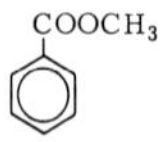 COOCH₃	Diehl [324]	T	G, I
C_8H_{10}	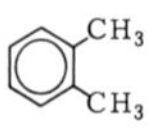 CH₃ CH₃	Diehl [296]	T	G, I

Formula	Structure	Author and ref.	Methods	Results
	CH_3 benzene ring CH_3	Canet [142]	T	G, S, I
		Phillips [12:299]	L	S
$C_8H_{16}O_2$	$n\text{-}C_7H_{15}COOH$	Reeves [911]	L, ^{2}H $(n\text{-}C_6H_{13}CD_2COOH)$	S
C_8H_{18}	$n\text{-}C_8H_{18}$	Reeves [911]	L, ^{2}H (deut.)	S
$C_8H_{18}O$	$n\text{-}C_8H_{17}OH$	Reeves [911]	L, ^{2}H (deut.)	S
$C_8H_{20}P_4$	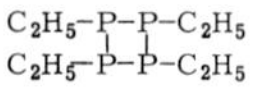 $C_2H_5\text{-}P\text{-}P\text{-}C_2H_5$ $C_2H_5\text{-}P\text{-}P\text{-}C_2H_5$	Albrand [5]	T, ^{31}P	$\Delta\sigma_P$, G, I, J
$C_9H_6CrO_3$	benzene–$Cr(CO)_3$	Diehl [287]	T	G, I, $\Delta\sigma_H$
		Diehl [327]	T, ^{1}H (^{13}C-Sat)	G, J
$C_9H_6O_2$	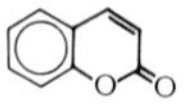 coumarin structure	Segre [983]	T	G, S
$C_9H_7MnO_3$	CH_3–cyclopentadienyl–$Mn(CO)_3$	Saupe [941]	T	G, S, I
C_9H_7N	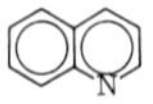 quinoline structure	Diehl [319]	T	G
C_9H_8	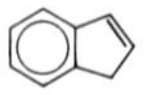 indene structure	Diehl [314]	T	G, S
$C_9H_{18}O_2$	$n\text{-}C_8H_{17}COOH$	Reeves [911]	L, ^{2}H $(n\text{-}C_7H_{15}CD_2COOH)$	S
$C_9H_{20}O$	$n\text{-}C_9H_{19}OH$	Reeves [911]	L, ^{2}H (deut.)	S
$C_{10}H_4Cl_2O_2$	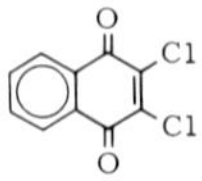 dichloronaphthoquinone structure	Dereppe [263]	T	S, G
$C_{10}H_6ClNO_2$	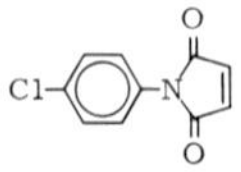 Cl–phenyl–N-maleimide structure	Saupe [950]	T	G, I
$C_{10}H_6O_2$	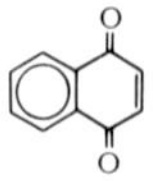 naphthoquinone structure	Dereppe [256, 258, 263]	T	G, S
$C_{10}H_7Cl$	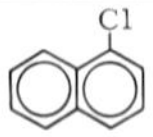 chloronaphthalene structure	Diehl [306]	T	G, S
$C_{10}H_7NO_2$	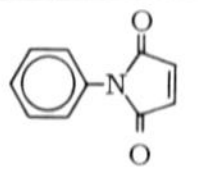 phenyl-N-maleimide structure	Diehl [325]	T	G, I

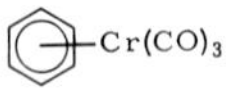

Formula	Structure	Author and ref.	Methods	Results
$C_{10}H_8$	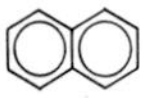	Dereppe [259]	T	G, S
$C_{10}H_8N_2$	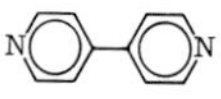	Emsley [412]	T	G, I
$C_{10}H_{10}$		Yannoni [1113]	T	S
$C_{10}H_{10}Hg$	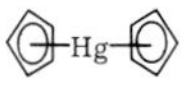	Emsley [403]	T	G, S
$C_{10}H_{16}$		Frey [451]	T	S, G
$C_{10}H_{20}O_2$	n-$C_9H_{19}COOH$	Reeves [911]	L, ^2H (n-$C_8H_{17}CD_2COOH$)	S
$C_{10}H_{22}$	n-$C_{10}H_{22}$	Reeves [911]	L, ^2H (deut.)	S
$C_{10}H_{22}O$	n-$C_{10}H_{21}OH$	Diehl [313] Reeves [911]	2T, ^2H (deut.) L, ^2H (deut.)	S, QCC S
$C_{11}H_6O_9Os_3$	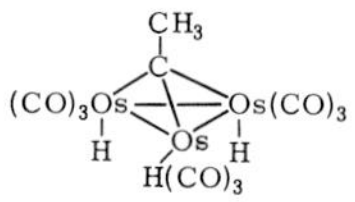 CH_3 $(CO)_3Os$—$Os(CO)_3$ H Os H $H(CO)_3$	Bailey [22] Buckingham [115]	T T	G
$C_{11}H_6O_9Ru_3$	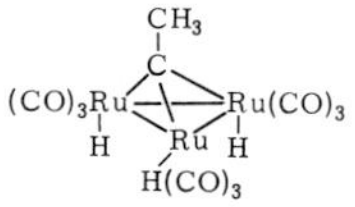 CH_3 $(CO)_3Ru$—$Ru(CO)_3$ H Ru H $H(CO)_3$	Buckingham [110, 115]	T	G
$C_{11}H_{22}O_2$	n-$C_{10}H_{21}COOH$	Reeves [911]	L, ^2H (n-$C_9H_{19}CD_2COOH$)	S
$C_{12}H_4N_4$	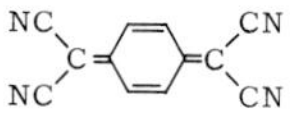 NC C=... CN NC CN	Burnell [123]	T, ^1H (^{13}C-Sat)	S, G
$C_{12}H_6Br_4$	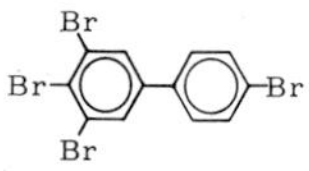 Br Br—...—Br Br	Lunazzi [720]	T	G, I
$C_{12}H_6Cl_4$	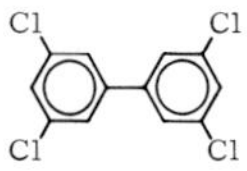 Cl Cl ... Cl Cl	Lunazzi [717]	T	G, I
$C_{12}H_8Cl_2$	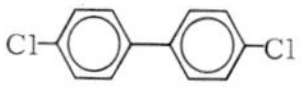 Cl—...—Cl	Diehl [300]	T	G, I
$C_{12}H_{24}O_2$	n-$C_{11}H_{23}COOH$	Reeves [911]	L, ^2H (n-$C_{10}H_{21}CD_2COOH$)	S
$C_{12}H_{28}P_4$	Me_2CH–P–P–$CHMe_2$ Me_2CH–P–P–$CHMe_2$	Albrand [5]	T, ^{31}P	$\Delta\sigma_P$, G, I, J

Formula	Structure	Author and ref.	Methods	Results
$C_{14}H_{28}O_2$	$n\text{-}C_{13}H_{27}COOH$	Reeves [911]	L, ^2H $(n\text{-}C_{12}H_{25}CD_2COOH)$	S
$C_{16}H_{32}O_2$	$n\text{-}C_{15}H_{31}COOH$	Reeves [911]	L, ^2H $(n\text{-}C_{14}H_{29}CD_2COOH)$	S
$C_{16}H_{36}P_4$	$Me_3C\text{-}P\text{-}P\text{-}CMe_3$ $Me_3C\text{-}P\text{-}P\text{-}CMe_3$	Albrand [3] Albrand [5]	T, ^1H, ^{31}P T, ^{31}P	G, $\Delta\sigma_P$ $\Delta\sigma_P$, G, I, J
$C_{24}H_{44}P_4$	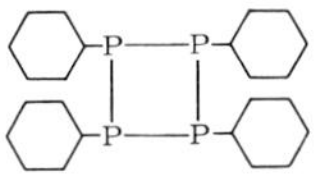	Albrand [5]	T, ^{31}P	$\Delta\sigma_P$, G, I, J
$C_{36}H_{30}P_6$	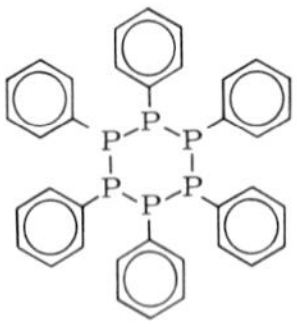	Albrand [2]	T, ^{31}P	G

10.4 ESR Investigations

10.4.1 Table of Mesophases

Thermotropic mesophases and synthetic lyophases (mark L) are listed together.

Formula	Structure	Author and ref.
$C_8H_{15}O_2^-$	$C_7H_{15}COO^-$ (L, $+$decanol)	Yamaoka [1109]
$C_8H_{16}O_2$	$C_7H_{15}COOH$ (L, $+$ Na octanoate)	Campbell [136]
$C_{10}H_{20}O_2$	$C_9H_{19}COOH$ (L, $+$decanol)	Setaka [984]
$C_{10}H_{22}O$	$C_{10}H_{21}OH$ (L, $+$ Na octanoate)	Yamaoka [1109]
	$C_{10}H_{21}OH$ (L, $+$decanoate)	Setaka [984]
$C_{11}H_{14}O_3$	 C_4H_9O—⟨⟩—COOH	Corvaja [199]
$C_{12}H_{16}O_3$	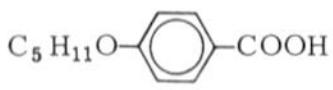 $C_5H_{11}O$—⟨⟩—COOH	Luckhurst [689]
$C_{12}H_{23}NaO_2$	$C_{11}H_{23}COONa$ (L)	Bikchantaev [47] Schara [956]
$C_{12}H_{25}O_3S^-$	$C_{12}H_{25}SO_3^-$ (L)	Oakes [848]

Formula	Structure	Author and ref.
$C_{12}H_{25}O_4S^-$	$C_{12}H_{25}OSO_3^-$ (L)	Cyr [211] Griffith [516, 517] Oakes [847, 849] Yamaoka [1109]
$C_{13}H_{18}O_3$	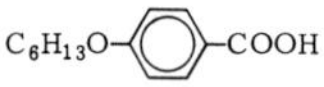 $C_6H_{13}O$—⟨◯⟩—COOH	Hudson [553]
$C_{14}H_{14}N_2O_3$	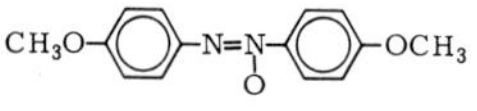 CH_3O—⟨◯⟩—N=N—⟨◯⟩—OCH_3 　　　　　　⏜O (PAA)	Corvaja [199, 200] Fryburg [459] Gelerinter [470] Glarum [492–494] Haustein [527–529, 531] Klein [589] Kothe [596, 597] Luckhurst [676, 678–680, 682, 686–689, 699, 701, 702, 707, 708, 712, 715] Pedulli [854] Schara [951]
$C_{14}H_{20}O_3$	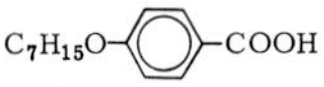 $C_7H_{15}O$—⟨◯⟩—COOH	Bikchantaev [45] Luckhurst [686]
$C_{15}H_{14}N_2O_3$	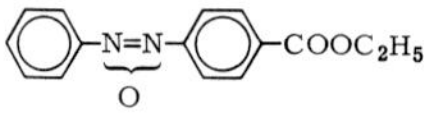 ⟨◯⟩—N=N—⟨◯⟩—$COOC_2H_5$ 　　　　⏜O	Luckhurst [696, 698]
$C_{15}H_{16}N_2O_2$	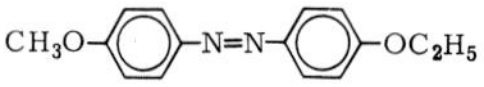 CH_3O—⟨◯⟩—N=N—⟨◯⟩—OC_2H_5	Bikchantaev [45]
	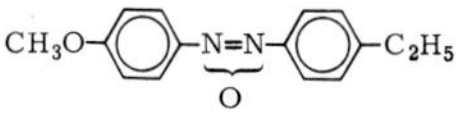 CH_3O—⟨◯⟩—N=N—⟨◯⟩—C_2H_5 　　　　　　⏜O	Eastman [393] Ero-Gecs [430] Freed [446, 447, 450] Kothe [598] Luckhurst [713]
$C_{15}H_{22}O_3$	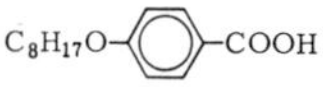 $C_8H_{17}O$—⟨◯⟩—COOH	Klein [589] Luckhurst [686]
$C_{16}H_{15}NO_3$	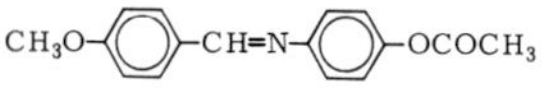 CH_3O—⟨◯⟩—CH=N—⟨◯⟩—$OCOCH_3$ (APAPA)	Luckhurst [678, 686]
$C_{16}H_{16}N_2O_2$	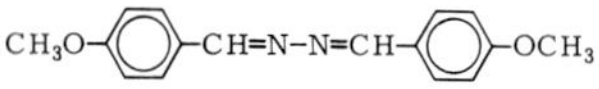 CH_3O—⟨◯⟩—CH=N–N=CH—⟨◯⟩—OCH_3	Klein [589] Luckhurst [683, 686]
$C_{16}H_{18}N_2O_3$	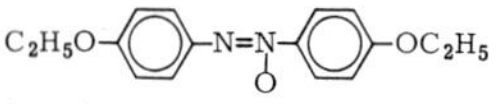 C_2H_5O—⟨◯⟩—N=N—⟨◯⟩—OC_2H_5 　　　　　　⏜O (PAP)	Kuznetsov [601] Luckhurst [701]

Formula	Structure	Author and ref.
$C_{16}H_{31}O_2^-$	$C_{15}H_{31}COO^-$ (L)	Brotherus [96] Kuznetsov [600, 602, 604, 605]
$C_{17}H_{20}N_2O_2$	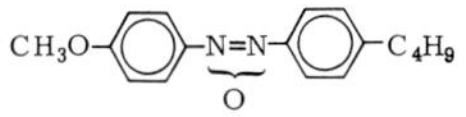 $CH_3O-\!\!\bigcirc\!\!-N=N-\!\!\bigcirc\!\!-C_4H_9$ $\overset{\mid}{O}$	Eastman [393] Ero-Gecs [430] Freed [446, 447, 450] Haustein [530, 532] Kothe [598] Luckhurst [695, 704, 707, 713]
$C_{18}H_{19}N$	 $C_5H_{11}-\!\!\bigcirc\!\!-\!\!\bigcirc\!\!-CN$	Luckhurst [713]
$C_{18}H_{21}NO$	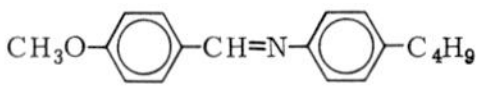 $CH_3O-\!\!\bigcirc\!\!-CH=N-\!\!\bigcirc\!\!-C_4H_9$ (MBBA)	Bales [24] Barbarin [25, 27] Bikchantaev [47, 48] Brog [95] Casini [149] Dong [345, 351] Ero-Gecs [430] Freed [446, 448] Heppke [545] Hoffman [548] Le Pesant [622] Luckhurst [688, 713] Pudzianowski [879] Shimoyama [988]
$C_{18}H_{21}NO_2$	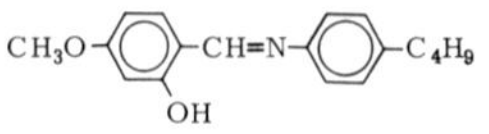 $CH_3O-\!\!\bigcirc\!\!-CH=N-\!\!\bigcirc\!\!-C_4H_9$ $\overset{\mid}{OH}$	Freed [446]
$C_{18}H_{22}N_2O$	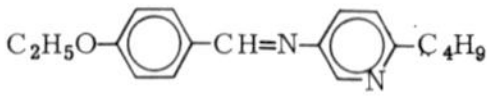 $C_2H_5O-\!\!\bigcirc\!\!-CH=N-\!\!\bigcirc\!\!-C_4H_9$ $\underset{N}{}$	Luckhurst [705]
$C_{19}H_{19}NO_3$	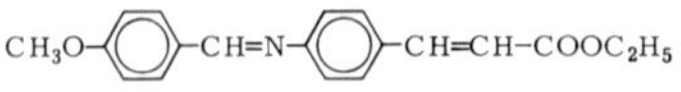 $CH_3O-\!\!\bigcirc\!\!-CH=N-\!\!\bigcirc\!\!-CH=CH-COOC_2H_5$	Klein [589] Ptak [877, 878]
$C_{19}H_{21}NO_2$	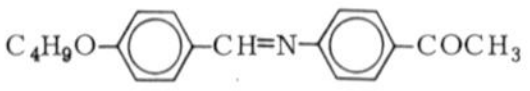 $C_4H_9O-\!\!\bigcirc\!\!-CH=N-\!\!\bigcirc\!\!-COCH_3$	Fryburg [454, 459] Luckhurst [697]
$C_{19}H_{21}NO_3$	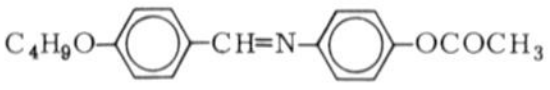 $C_4H_9O-\!\!\bigcirc\!\!-CH=N-\!\!\bigcirc\!\!-OCOCH_3$	Fryburg [458]
$C_{19}H_{22}N_2O_3$	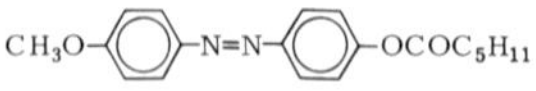 $CH_3O-\!\!\bigcirc\!\!-N=N-\!\!\bigcirc\!\!-OCOC_5H_{11}$	Dong [345, 351] Luckhurst [690]
$C_{19}H_{23}NO$	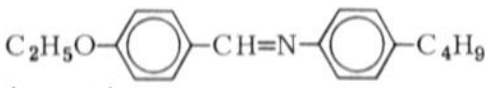 $C_2H_5O-\!\!\bigcirc\!\!-CH=N-\!\!\bigcirc\!\!-C_4H_9$ (EBBA)	Barbarin [26] Bikchantaev [45, 48]

Formula	Structure	Author and ref.
$C_{20}H_{17}NO$	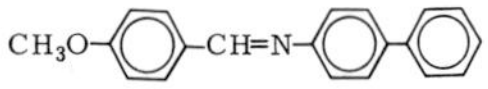 CH_3O—⟨⟩—$CH=N$—⟨⟩—⟨⟩	Luckhurst [686]
$C_{20}H_{17}N_3O$	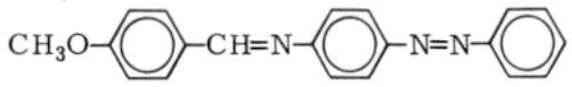 CH_3O—⟨⟩—$CH=N$—⟨⟩—$N=N$—⟨⟩	Blinc [69] Klein [589] Luckhurst [686]
$C_{20}H_{21}NO_3$	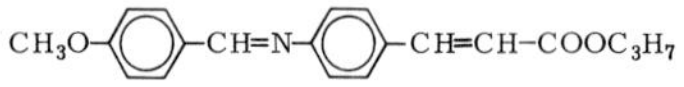 CH_3O—⟨⟩—$CH=N$—⟨⟩—$CH=CH-COOC_3H_7$	Ptak [877]
$C_{20}H_{22}O_6$	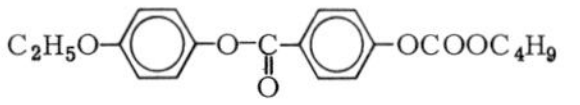 C_2H_5O—⟨⟩—$O-\underset{O}{\overset{}{C}}$—⟨⟩—$OCOOC_4H_9$	Eastman [393] Fackler [432, 433] Freed [446]
$C_{20}H_{24}N_2O_3$	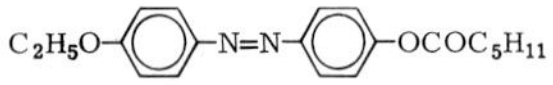 C_2H_5O—⟨⟩—$N=N$—⟨⟩—$OCOC_5H_{11}$	Freed [446] Klein [588, 589]
$C_{20}H_{26}N_2O_2$	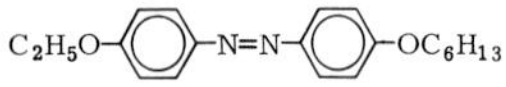 C_2H_5O—⟨⟩—$N=N$—⟨⟩—OC_6H_{13}	Freed [446]
$C_{20}H_{26}N_2O_3$	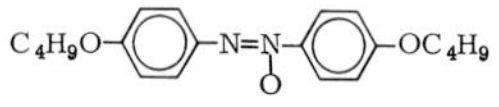 C_4H_9O—⟨⟩—$N=\underset{O}{N}$—⟨⟩—OC_4H_9	Luckhurst [710]
	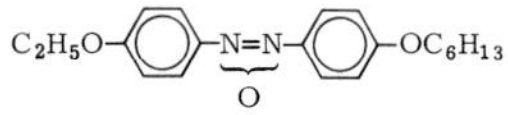 C_2H_5O—⟨⟩—$N=N$—⟨⟩—OC_6H_{13} (with O below)	Freed [446]
$C_{20}H_{37}O_7S^-$	$ROOC-CH_2-\underset{SO_3^-}{CH}-COOR$ $R:\ -CH_2-CH\overset{C_2H_5}{\underset{C_4H_9}{<}}$	Setaka [984]
$C_{21}H_{23}NO_3$	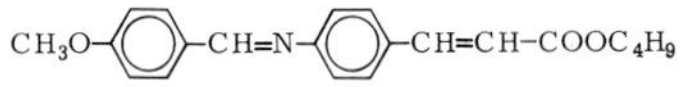 CH_3O—⟨⟩—$CH=N$—⟨⟩—$CH=CH-COOC_4H_9$	Ptak [877]
	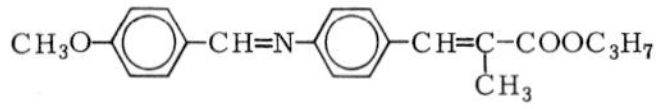 CH_3O—⟨⟩—$CH=N$—⟨⟩—$CH=\underset{CH_3}{C}-COOC_3H_7$	Diehl [278, 279] Dong [345, 347, 351] Luckhurst [690] Ptak [875]
$C_{21}H_{25}N$	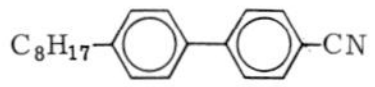 C_8H_{17}—⟨⟩—⟨⟩—CN	Luckhurst [709]
$C_{21}H_{25}NO$	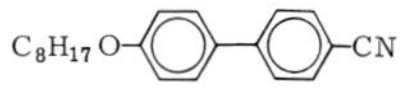 $C_8H_{17}O$—⟨⟩—⟨⟩—CN	Luckhurst [709]
$C_{22}H_{27}NO_3$	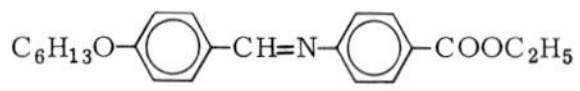 $C_6H_{13}O$—⟨⟩—$CH=N$—⟨⟩—$COOC_2H_5$	Fryburg [456]
$C_{23}H_{27}NO_3$	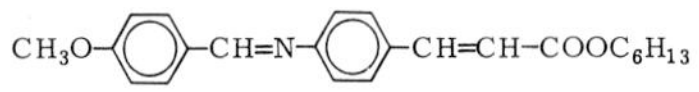 CH_3O—⟨⟩—$CH=N$—⟨⟩—$CH=CH-COOC_6H_{13}$	Ptak [877, 878]

Formula	Structure	Author and ref.
$C_{23}H_{31}NO$	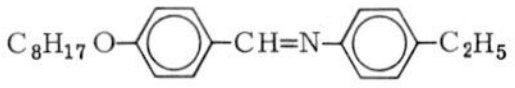 $C_8H_{17}O$—⟨◯⟩—$CH=N$—⟨◯⟩—C_2H_5	Fryburg [454, 458]
$C_{24}H_{31}NO_3$	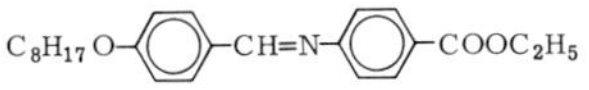 $C_8H_{17}O$—⟨◯⟩—$CH=N$—⟨◯⟩—$COOC_2H_5$	Fryburg [456]
$C_{24}H_{34}N_2O_3$	$C_6H_{13}O$—⟨◯⟩—$N=N$—⟨◯⟩—OC_6H_{13} $\overset{\|}{O}$	Klein [589] Luckhurst [686, 701]
$C_{25}H_{32}N_2O_3$	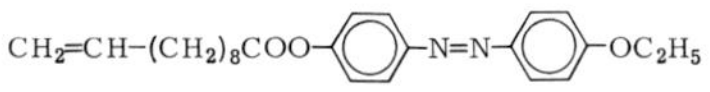 $CH_2=CH-(CH_2)_8COO$—⟨◯⟩—$N=N$—⟨◯⟩—OC_2H_5	Klein [589]
$C_{25}H_{35}NO$	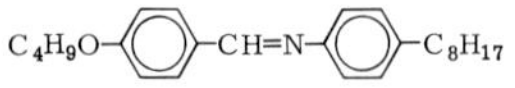 C_4H_9O—⟨◯⟩—$CH=N$—⟨◯⟩—C_8H_{17}	Schara [954, 955]
$C_{26}H_{38}N_2O_3$	$C_7H_{15}O$—⟨◯⟩—$N=N$—⟨◯⟩—OC_7H_{15} $\overset{\|}{O}$	Blinc [69] Fryburg [453] Gelerinter [471] Luckhurst [685, 700, 701] Marusic [794, 796] Ptak [876] Schara [952]
$C_{27}H_{45}Cl$	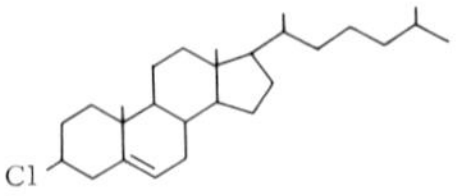 (steroid structure, Cl)	Luckhurst [702, 708] Sackmann [917, 919, 925]
$C_{28}H_{32}N_2$	 C_4H_9—⟨◯⟩—$N=CH$—⟨◯⟩—$CH=N$—⟨◯⟩—C_4H_9 (TBBA)	Luckhurst [694] Luz [722]
$C_{29}H_{35}N_3O$	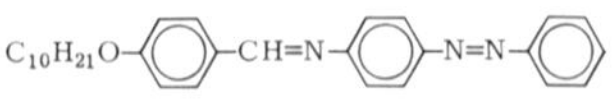 $C_{10}H_{21}O$—⟨◯⟩—$CH=N$—⟨◯⟩—$N=N$—⟨◯⟩	Blinc [69]
$C_{36}H_{47}ClN_2O_2$	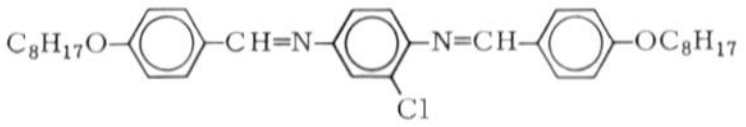 $C_8H_{17}O$—⟨◯⟩—$CH=N$—⟨◯⟩—$N=CH$—⟨◯⟩—OC_8H_{17} Cl	Freed [446] Fryburg [452, 455]
$C_{39}H_{68}O_2$	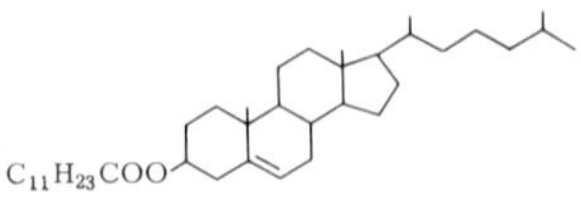 (steroid structure, $C_{11}H_{23}COO$)	Sackmann [925]
$C_{39}H_{70}O$	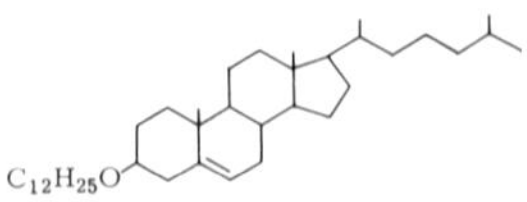 (steroid structure, $C_{12}H_{25}O$)	Sackmann [917, 919]
$C_{48}H_{82}N_2O_3$	$C_{18}H_{37}O$—⟨◯⟩—$N=N$—⟨◯⟩—$OC_{18}H_{37}$ $\overset{\|}{O}$	Dvolaitzky [387–391]

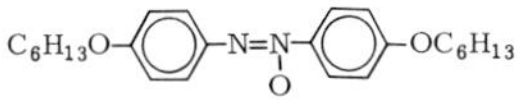

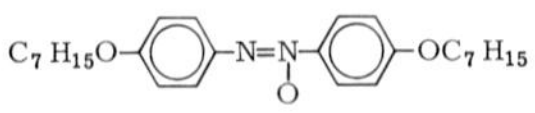

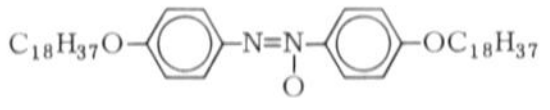

10.4.2 Table of Paramagnetic Solutes

Formula	Structure	Author and ref.
$C_6N_4^-$	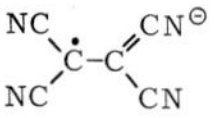	Haustein [528] Luckhurst [676]
$C_6H_4O_2^-$	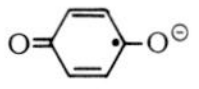	Haustein [529]
$C_6H_5N_2O_2$	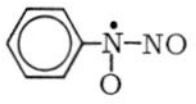	Freed [446]
$C_6H_{12}CuN_2S_4$	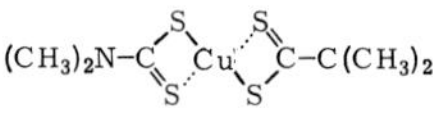	Bikchantaev [47]
$C_8H_{16}NO_2$	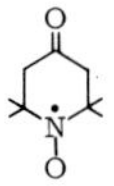	Freed [446, 448]
$C_8H_{18}NO$	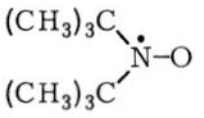	Bales [24] Freed [446]
$C_9H_{13}N_2O$	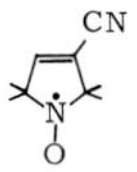	Setaka [984]
$C_9H_{14}NO_3$	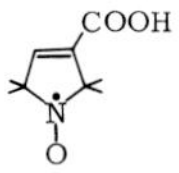	Kuznetsov [605] Setaka [984]
$C_9H_{15}N_2O_2$	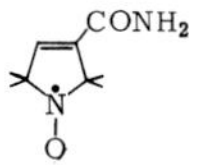	Corvaja [199] Fryburg [455] Haustein [532] Setaka [984]
$C_9H_{16}NO_2$	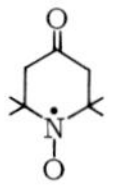	Barbarin [25] Brotherus [96] Freed [447]
$C_9H_{18}NO$	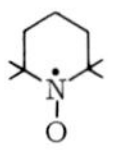	Oakes [848]
$C_9H_{18}NO_2$	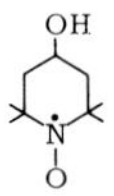	Casini [149] Fryburg [455] Haustein [532] Kuznetsov [602, 605] Ptak [878]
$C_{10}N_4O_2^-$	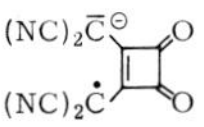	Corvaja [200]

Formula	Structure	Author and ref.
$C_{10}H_2CuF_{12}O_4$	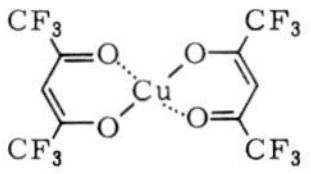	Fackler [433]
$C_{10}H_6O_2^-$	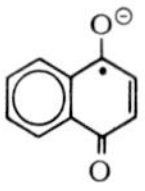	Haustein [529]
$C_{10}H_8$	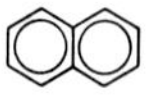 + h*ν*	Sackmann [925]
$C_{10}H_{14}CuO_4$	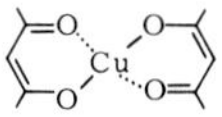	Bikchantaev [45, 48] Fackler [433]
$C_{10}H_{14}O_5V$	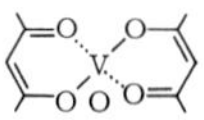	Bikchantaev [45, 47] Blinc [69] Brog [95] Diehl [278, 279] Dong [345, 347, 351] Eastman [393] Ero-Gecs [429, 430] Fackler [432, 433] Fryburg [452–455] Glarum [493] Heppke [545] Luckhurst [685–687, 690, 694, 699, 712] Marusic [796] Ptak [875] Schara [951, 952] Shimoyama [988]
$C_{10}H_{16}N_2^+$	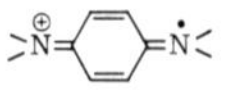	Luckhurst [689]
$C_{10}H_{20}CuN_2S_4$	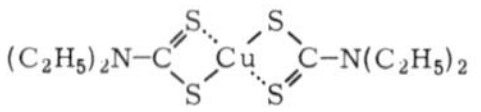 $(C_2H_5)_2N-C\overset{S}{\underset{S}{<}}\,\,Cu\,\,\overset{S}{\underset{S}{>}}C-N(C_2H_5)_2$	Bikchantaev [45] Fackler [432, 433] Nordio [843]
$C_{10}H_{20}CuN_2Se_4$	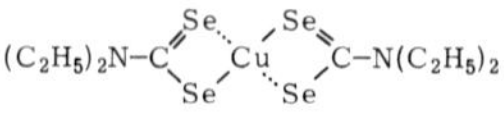 $(C_2H_5)_2N-C\overset{Se}{\underset{Se}{<}}\,\,Cu\,\,\overset{Se}{\underset{Se}{>}}C-N(C_2H_5)_2$	Nordio [843]
$C_{10}H_{20}N_2OS_4V$	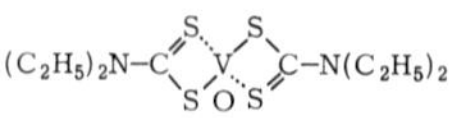 $(C_2H_5)_2N-C\overset{S}{\underset{S}{<}}\overset{}{\underset{O}{V}}\overset{S}{\underset{S}{>}}C-N(C_2H_5)_2$	Bikchantaev [45]
$C_{11}H_{18}F_3N_2O$	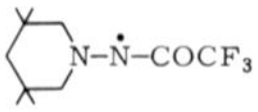	Luckhurst [689]
$C_{11}H_{21}N_2O_2$	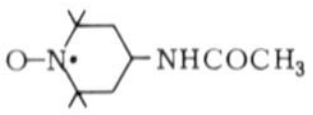	Haustein [532]

Formula	Structure	Author and ref.
$C_{11}H_{22}NO_2$	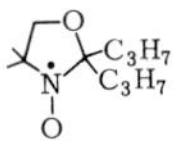	Ptak [878]
$C_{12}H_4N_4^-$	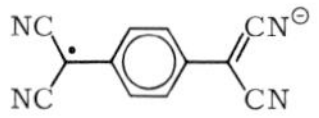	Corvaja [200] Haustein [531]
$C_{12}H_8NO$	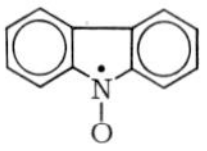	Luckhurst [682, 689]
$C_{12}H_{10}NO$	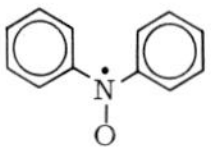	Luckhurst [689]
$C_{12}H_{18}CoN_2O_2$	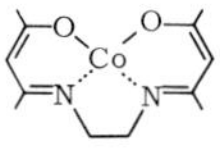	Hoffman [548]
$C_{13}H_9S_3$	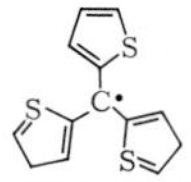	Pedulli [854]
$C_{13}H_{13}$		Glarum [492] Haustein [527, 528, 530] Luckhurst [680, 689]
$C_{13}H_{17}BrN_3O_4$	 O–N NHCOOCH$_2$CH$_2$NHCOCH$_2$Br	Freed [446]
$C_{13}H_{17}N_2O$	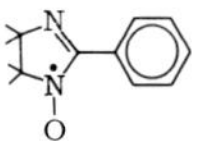	Luckhurst [688]
$C_{13}H_{17}N_2O_2$	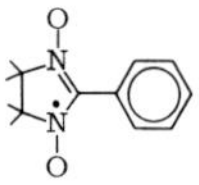	Luckhurst [688]
$C_{13}H_{19}$	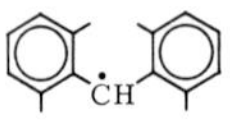	Luckhurst [689]
$C_{13}H_{20}N_2O_3V$	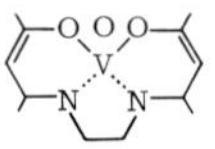	Fackler [433]
$C_{14}H_8O_2^-$	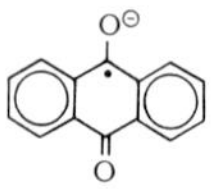	Haustein [529]
$C_{14}H_{10}$	 + hv	Sackmann [925]

Formula	Structure	Author and ref.
	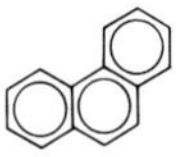 + hν	Sackmann [925]
$C_{14}H_{10}CuO_4$	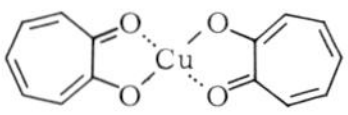	Fackler [433]
$C_{14}H_{13}O_2$	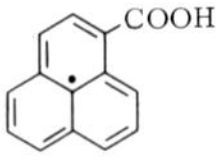	Luckhurst [689]
$C_{14}H_{22}CoN_2O_2$	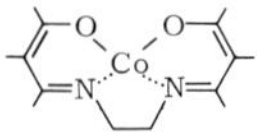	Hoffman [548]
$C_{14}H_{22}CuO_4$	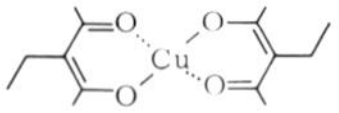	Fackler [432, 433]
$C_{14}H_{22}NO$	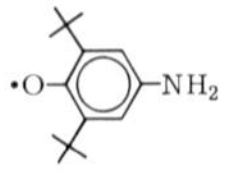	Luckhurst [689]
$C_{14}H_{24}N_2O_4$	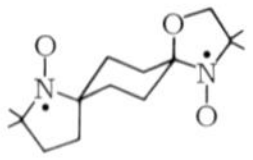	Luckhurst [710]
$C_{15}H_{18}N_2O$	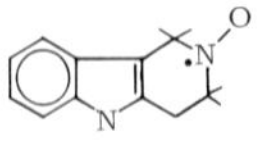	Griffith [516]
$C_{15}H_{19}N_2O$	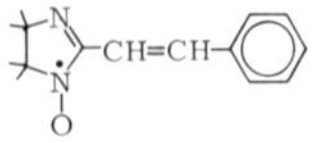	Luckhurst [688]
$C_{15}H_{19}N_2O_2$		Corvaja [199]
	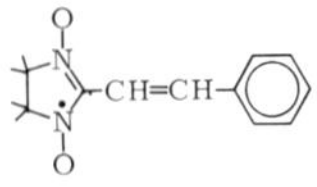	Luckhurst [688]
$C_{15}H_{20}N_5O_5$	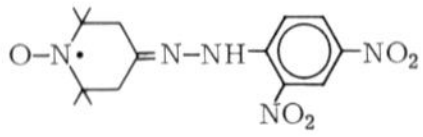	Barbarin [25] Griffith [516] Oakes [847, 848]
$C_{15}H_{21}CrO_6$	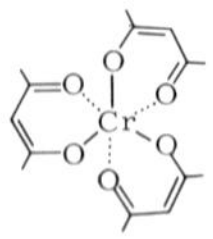	Bikchantaev [46]

Formula	Structure	Author and ref.
$C_{16}H_{10}$	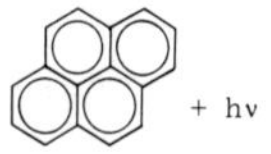 + hv	Sackmann [925]
$C_{16}H_{16}CuN_2O_2$	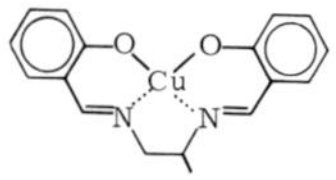	Fackler [433]
$C_{16}H_{24}NO_4S$	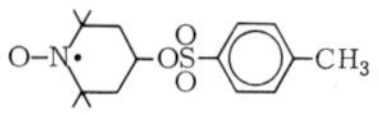	Fryburg [455, 458]
$C_{17}H_{16}N_2O_3V$	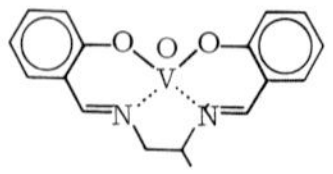	Fackler [433]
$C_{17}H_{25}N_2O_2$	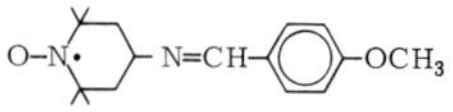	Pudzianowski [879]
$C_{17}H_{32}NO_3$	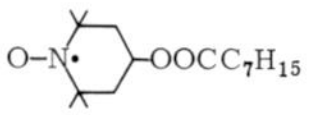	Griffith [517]
$C_{18}H_{12}$	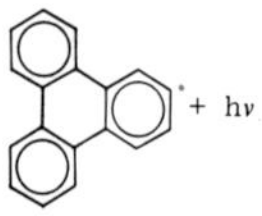 + hv	Sackmann [925]
$C_{18}H_{12}N_5O_6$	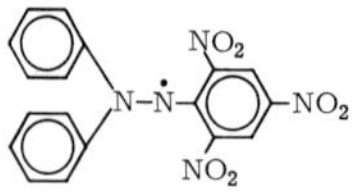	Luckhurst [676]
$C_{18}H_{18}CoN_2O_4$	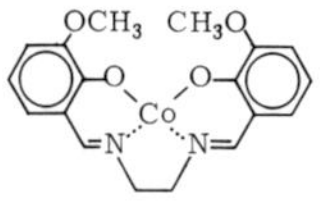	Hoffman [548]
$C_{18}H_{18}N_2O_3V$	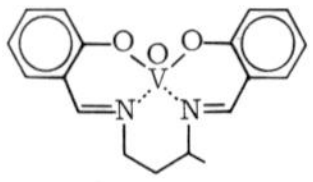	Fackler [433]
$C_{18}H_{19}N_2O$	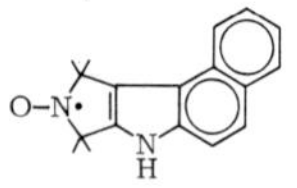	Kuznetsov [602, 605]
$C_{18}H_{23}N_2O_5$	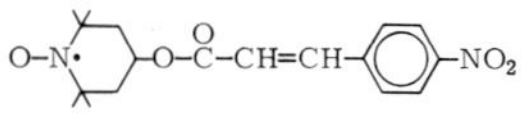	Fryburg [455]
$C_{18}H_{27}N_2O_3$	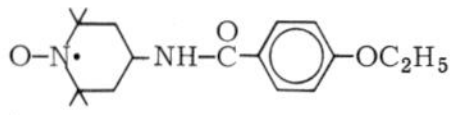	Barbarin [27]

Formula	Structure	Author and ref.
$C_{18}H_{29}O$	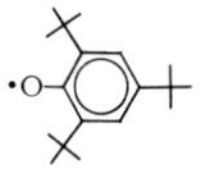	Haustein [530]
$C_{18}H_{35}N_2O$	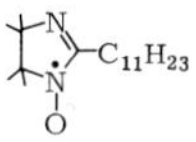 $-C_{11}H_{23}$	Luckhurst [688]
$C_{18}H_{35}N_2O_2$	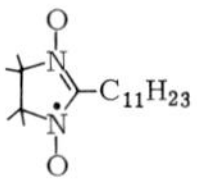 $-C_{11}H_{23}$	Luckhurst [688]
$C_{19}H_{15}$	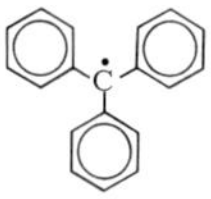	Haustein [527] Luckhurst [689]
$C_{19}H_{20}N_2O_3V$	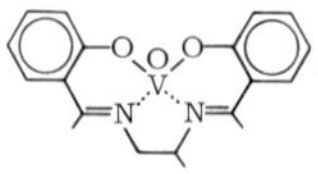	Fackler [433]
$C_{19}H_{34}N_2O_5$	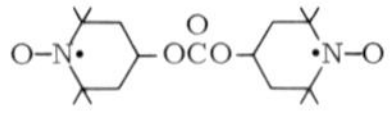 $O-N$ OCO $N-O$	Glarum [494]
$C_{19}H_{36}NO_3$	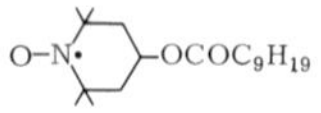 $O-N$ $OCOC_9H_{19}$	Luckhurst [695]
$C_{19}H_{36}N_4O_3$	 $O-N$ $NHCNH$ $N-O$	Cyr [211]
$C_{19}H_{38}NO_2$	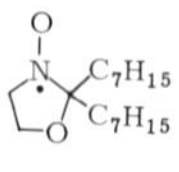 C_7H_{15} C_7H_{15}	Fryburg [456] Ptak [876, 878]
$C_{20}H_{18}CuO_4$	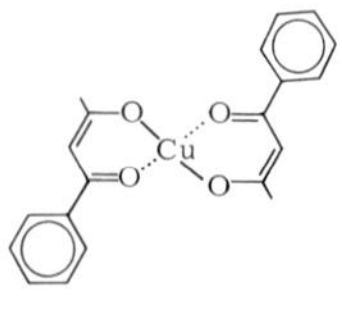	Fackler [433]
$C_{20}H_{18}O_5V$	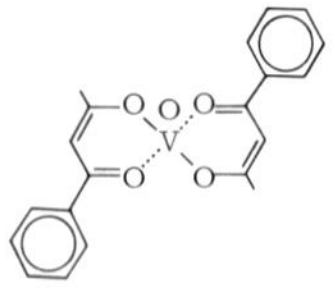	Glarum [493] Luckhurst [699]
$C_{20}H_{20}N_2O_3V$	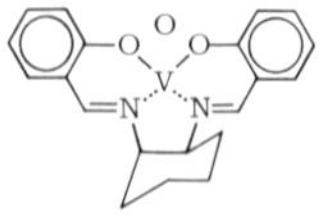	Fackler [433]

Formula	Structure	Author and ref.
$C_{20}H_{27}N_2O_6$	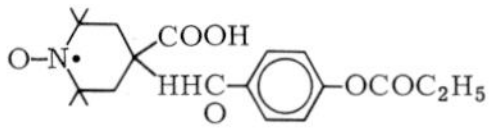	Klein [588]
$C_{20}H_{31}N_2O_3$	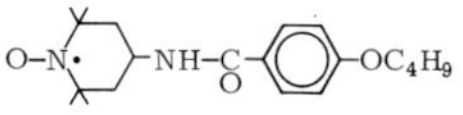	Barbarin [27]
$C_{20}H_{34}N_2O_6$	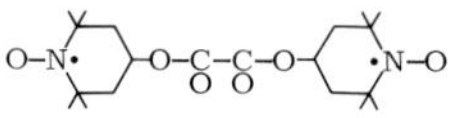	Glarum [494]
$C_{21}H_{36}N_2O_6$	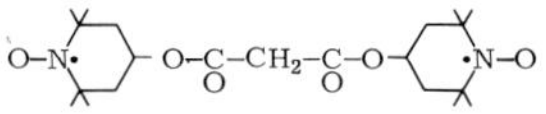	Glarum [494]
$C_{21}H_{40}NO_3$	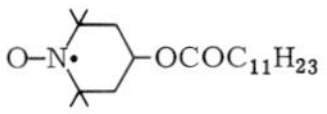	Griffith [517]
$C_{22}H_{22}CoN_2O_2$	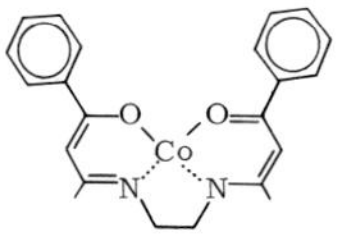	Hoffman [548]
$C_{22}H_{22}CuO_4$	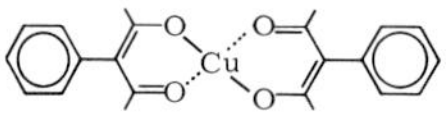	Fackler [433]
$C_{22}H_{22}O_5V$	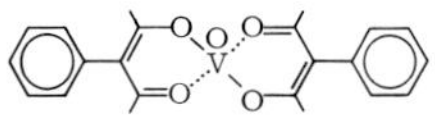	Glarum [493] Luckhurst [699]
$C_{22}H_{26}N_4O_2^+$	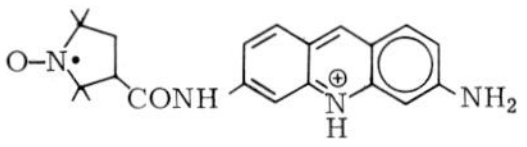	Yamaoka [1109]
$C_{22}H_{35}N_2O_3$	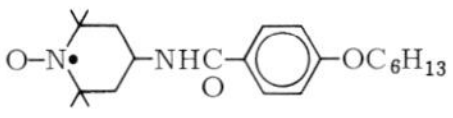	Barbarin [27]
$C_{22}H_{38}CuO_4$	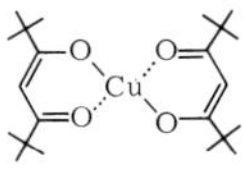	Fackler [433]
$C_{22}H_{38}N_2O_6$	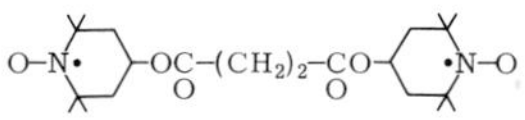	Glarum [494]
$C_{22}H_{42}NO_4$	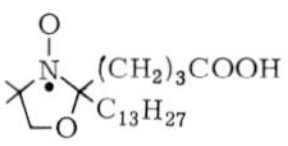	Schara [954, 956] Seteka [984]
	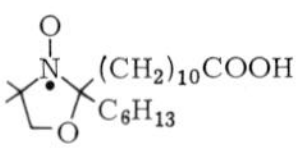	Seteka [984]
$C_{23}H_{38}NO_2$	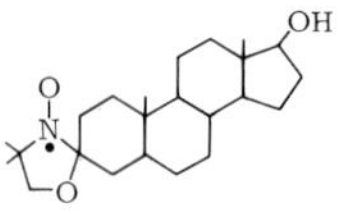	Luckhurst [696, 698, 700] Luz [722] Ptak [875–878]

Formula	Structure	Author and ref.
$C_{23}H_{40}N_2O_6$	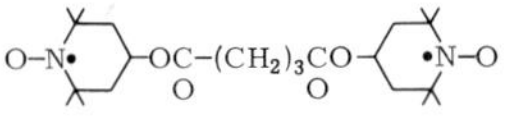 O–N•⟩–OC–(CH$_2$)$_3$CO–⟨•N–O	Luckhurst [678]
$C_{23}H_{48}N_2O^+$	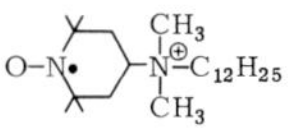 O–N•⟩–N$^⊕$(CH$_3$)$_2$C$_{12}$H$_{25}$	Oakes [848]
$C_{24}H_{12}$	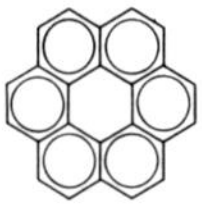 + hν	Sackmann [925]
$C_{24}H_{18}CuN_4S_2$	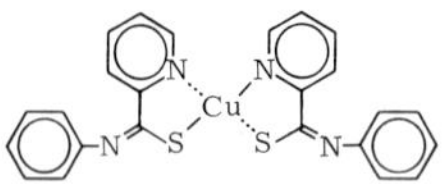	Fackler [433]
$C_{24}H_{29}N_2O_3$	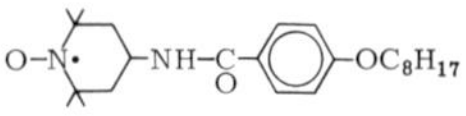 O–N•⟩–NH–C(O)–⟨⟩–OC$_8$H$_{17}$	Klein [589]
$C_{24}H_{32}N_4O_4$	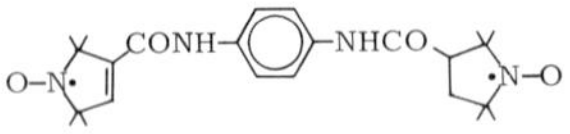 O–N•⟩–CONH–⟨⟩–NHCO–⟨•N–O	Corvaja [199]
$C_{24}H_{38}NO_4$	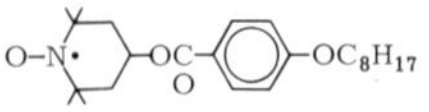 O–N•⟩–OC(O)–⟨⟩–OC$_8$H$_{17}$	Fryburg [455]
$C_{24}H_{39}N_2O_3$	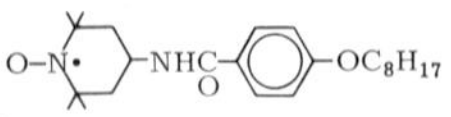 O–N•⟩–NHC(O)–⟨⟩–OC$_8$H$_{17}$	Barbarin [26, 27]
$C_{24}H_{42}N_2O_6$	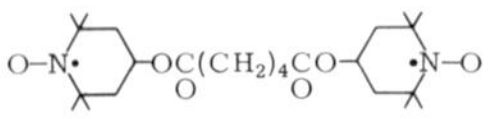 O–N•⟩–OC(CH$_2$)$_4$CO–⟨•N–O	Glarum [494]
$C_{25}H_{29}N_2O_5$	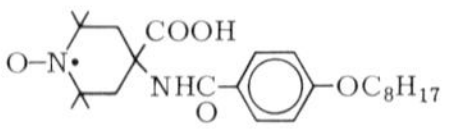 O–N•⟩(COOH)–NHC(O)–⟨⟩–OC$_8$H$_{17}$	Klein [588]
$C_{25}H_{30}N_3O_4$	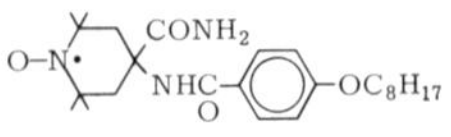 O–N•⟩(CONH$_2$)–NHC(O)–⟨⟩–OC$_8$H$_{17}$	Klein [588]
$C_{25}H_{49}N_2O_2$	 O–N•⟩–NHCOC$_{15}$H$_{31}$	Kuznetsov [602, 604]
$C_{26}H_{34}CuO_6$	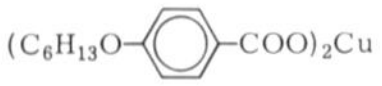 (C$_6$H$_{13}$O–⟨⟩–COO)$_2$Cu	Hudson [553]
$C_{26}H_{38}N_2O_6$	O–N•⟩–O–C(O)–⟨⟩–C(O)–O–⟨•N–O	*para*- Glarum [494] *para*- Luckhurst [678] *meta*- Glarum [494] *ortho*- Glarum [494]

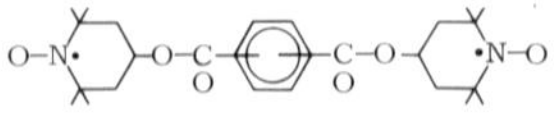

Formula	Structure	Author and ref.
$C_{26}H_{41}N_4O_3$	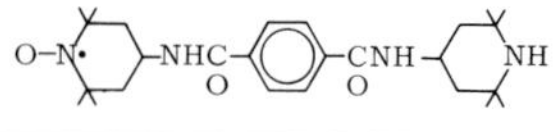	Klein [588]
$C_{26}H_{46}O_5V$	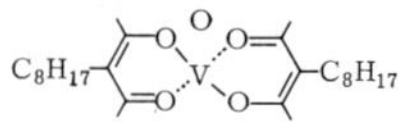	Glarum [493]
$C_{28}H_{40}NO_2$	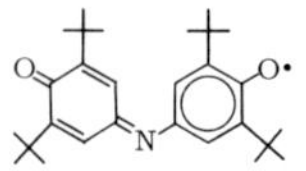	Luckhurst [689]
$C_{28}H_{50}N_2O_6$	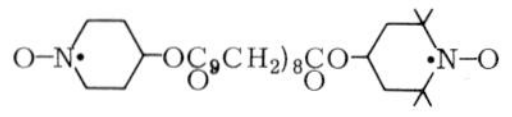	Glarum [494]
$C_{29}H_{41}O_2$	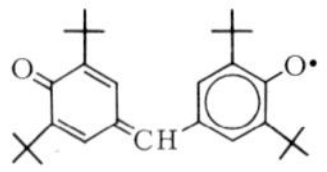	Luckhurst [679]
$C_{30}H_{22}O_5V$	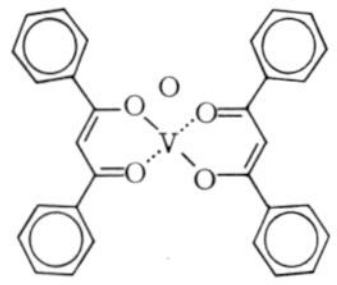	Glarum [493] Luckhurst [699]
$C_{30}H_{36}N_4O_4$	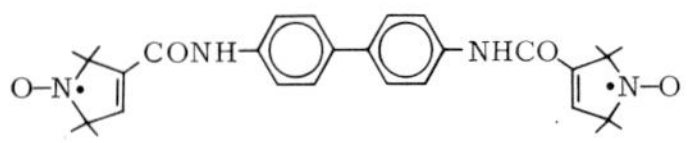	Corvaja [199]
$C_{31}H_{54}NO_2$	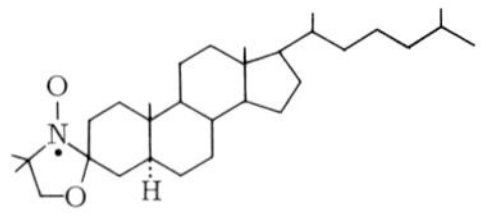	Freed [450] Fryburg [455, 458, 459] Gelerinter [471] Luckhurst [697, 701, 702, 704, 705, 708, 709, 713, 715] Ptak [878] Schara [954, 955]
	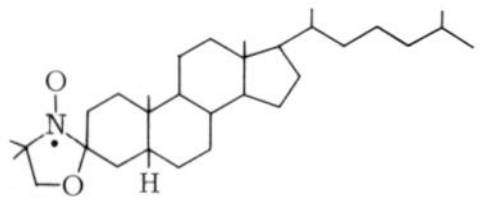	Luckhurst [701]
$C_{35}H_{25}$	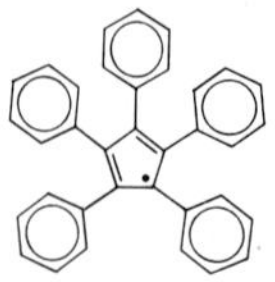	Haustein [527]
$C_{36}H_{40}N_4O_4$	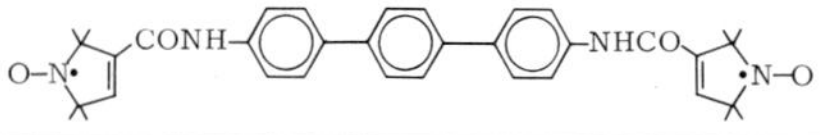	Corvaja [199] Nordio [838]
$C_{36}H_{54}N_3O_9$	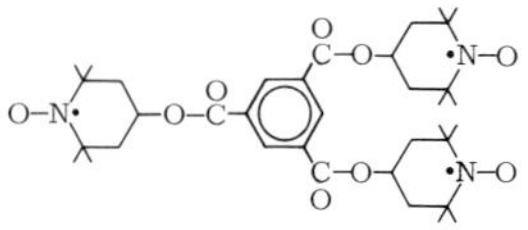	Luckhurst [683]

Formula	Structure	Author and ref.
$C_{46}H_{38}O_5V$		Glarum [493] Luckhurst [699]
$C_{48}H_{39}N_{12}$		Kothe [596–598]
$C_{50}H_{84}N_3O_4$	$(CH_2)_7CH_3$; $(CH_2)_7O$—⟨⟩—N=N—⟨⟩—$OC_{18}H_{37}$	Dvolaitzky [387]
	$(CH_2)_mCH_3$; $(CH_2)_nO$—⟨⟩—N=N—⟨⟩—$OC_{18}H_{37}$ $n+1 = 4,5,6,7,8$ $m+n = 13$	Dvolaitzky [388]
$C_{52}H_{88}N_3O_4$	$(CH_2)_mCH_3$; $(CH_2)_nO$—⟨⟩—N=N—⟨⟩—$OC_{18}H_{37}$ $n+1 = 12,14$ $m+n = 16$	Dvolaitzky [388]
	$n+1 = 4,5,6,7,8,12,16$ $m+n = 16$	Dvolaitzky [390]
$C_{55}H_{92}N_3O_3$	$C_{14}H_{29}$; O—N·—C—CH_2O—⟨⟩—N=N—⟨⟩—$OC_{18}H_{37}$	Dvolaitzky [389]
	Charge transfer complexes	Sackmann [917, 919]
Inorganics	$K_3[Fe(CN)_6]$	Kuznetsov [600]
	Mn^{2+}	Oakes [849]
	VO^{2+}	Campbell [136]

10.4.3 List of Spin Label Studies in Biological Systems

Lecithins (phosphatidylcholines). H. J. C. Berendsen [33], E. Gelerinter [472], M. A. Hemminga [541, 542], K. Inoue [555], G. Lindblom [640], H. M. McConnell [735, 736, 738, 739, 741], C. Mailer [785], D. Marsh [786, 787], E. Sackmann [920–922].

Phosphatidylethanolamines. G. Lindblom [640], H. M. McConnell [739].

Membranes. J. M. Boggs [85], D. A. Cadenhead [133], O. H. Griffith [519, 520], G. Lindblom [640], H. M. McConnell [728–738, 740, 742], J. D. Morrisett [824], E. Sackmann [924], U. Schummer [966], J. Seelig [967, 968, 970–972, 977], G. G. Smith [989].

Lyotropic Mesomorphism

11

11 Lyotropic Mesomorphism

11.1 Introduction

The difficulties encountered in differentiating thermotropic and lyotropic systems have already been mentioned in chapter 1. The solvent itself is not a reliable criterion, because solvent and solubilizate in three component systems can be of entirely different natures and gradually exchange roles with variations in composition. On the other hand the molecular dispersion and the molecular structure of the solute also do not give characteristic features in every case. Normally it is molecularly disperse in thermotropic mesophases and either micellar or lamellar in lyotropic mesophases, therefore the latter are closely related to colloids. This relationship was perspicuously discerned by Wo. Ostwald [279, see also 159, 364, 365a]. For this reason, the present chapter will begin with a few remarks concerning the formation of micelles. Such structures are illustrated in fig. 11.1.

Fig. 11.1. Schematic drawings showing isotropic solutions with a, molecular dispersion; b, spherical micelles; and c, cylindrical micelles.

The critical micelle concentration (c.m.c.) is an important parameter which is generally defined as the concentration where an abrupt change takes place in the derivative of some properties of the system (e. g., surface tension, turbidity, density) with respect to concentration. Fig. 11.2 illustrating the influence of the c.m.c. has been taken from W. Philippoff's highly recommended summary [290]. These schematic representations disregard the fact that an association which is not of micellar form takes place even below the c.m.c. However, the threshold concentration for this association, called the limiting association concentration (l.m.c.), is much less apparent.

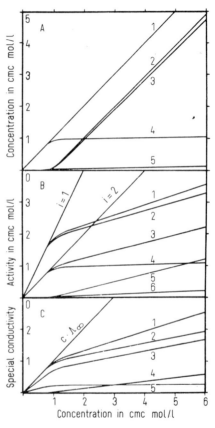

Fig. 11.2. Calculated micelle equilibrium and the concentration dependence of the osmotic activity and specific conductivity.
Fig. A. — 1. Total concentration. 2. Concentration involved in micelle formation. 3. Concentration of micelles. 4. Concentration of free molecules. 5. Concentration of counterions of the micelles.
Fig. B. — 1. Total osmotic activity. 2. Ionic activity. 3. Total counterion activity. 4. Micelle forming ion activity. 5. Activity of counterions from micelles. 6. Activity of the micelles. $i=1$ curve for undissociated molecules; $i=2$ curve for fully dissociated molecules.
Fig. C. — 1. Total conductivity. 2. Ionic conductivity. 3. Counterion conductivity. 4. Micelle conductivity. 5. Micelle forming ion conductivity. $c \cdot \Lambda_\infty$ curve for non-aggregating fully dissociated molecules (from [290]).

The models illustrated by various authors frequently differ significantly with respect to the configuration of the long paraffin chains [317]. In fact, it is astonishing to see the crowded packing of paraffin chains toward the center of the micelle portrayed by the models. X-ray and other studies indicate that the chains in the interior of the micelle approach a liquid rather than a crystalline state. Recently it has been possible by means of fluorescence-depolarization to study the motion of a probe molecule, perylene, in the interior of a micelle during its short excitation time (in ns). During this time the motions of the micelle itself can be neglected while the probe molecule can be shown to rotate within the micelle like in an isotropic solution. Mobile foreign molecules cause, due to their relatively rapid rotational motion, a depolarization of the incoming polarized light. Perrin's equation correlating the degree of depolarization with the anisotropy of polarizability in the case of spherical fluorescent particles has been extended by M. Shinitzky et al. to form-anisotropic bodies with corresponding rotational aberrations [322, 323]. The comparison of paraffin hydrocarbons, micellar systems, and high-viscosity glycerol/glycol mixtures led these investigators to conclude that the symbathically progressing dispersion of rotational velocity in these systems is subject to the same random distribution as that of chains. Thus, the anisotropy of the depolarization effects is a function only of the foreign molecule.

Like the long-chain aliphatic hydrocarbons, the soaps lose part of their fine structure in the IR spectrum upon melting [386], cf. p. 542 and chapter 6. X-ray scattering experiments have been interpreted in terms of quasi-liquid-crystal regions in long-chain aliphatic iodides [19].

Water is by far the most important solvent for micelle formation. Detergent action, solubilization and emulsifying action are closely interrelated phenomena. In water and waterlike media, the normal types of micelle are formed with the polar groups on the surface and the hydrocarbon chains directed toward the interior of the micelle (fig. 11.1). In contrast, nonpolar media contain micelles of reverse orientation with the polar groups in the core and the hydrocarbon chains directed outward. Either type of micelle can be transformed into the other by reversing the character of the solvent. Slightly bent or lamellar structures can be assumed for moderate polarities. Table 11.1 presents data on the formation of micelles of *p-t*-nonylphenoxypolyethoxyethanol (NPE$_9$, Igepal® CO-630) in various hydrophilic solvents [299]. The only common feature of these micelle-forming solvents is that the molecules have two or more potential hydrogen bonding centers and are therefore most likely to be capable of forming three-dimensional hydrogen bonded network structures. There is no correlation

Table 11.1. NPE$_9$ micelle formation in hydrophilic solvents (from [299]).

Solvent	Dielectric constant	Surface tension, dynes/cm	c.m.c. of NPE$_9$ at 27.5 °C, mol fraction	$-\Delta G_m = -RT \ln\{\text{c.m.c.}\}$, kJ/mol, at higher c.m.c. approximately only
Water	78.5	72	$5.6 \cdot 10^{-5}$	24.46
Glycerol	42.5	63	$8.7 \cdot 10^{-5}$	22.75
Ethylene glycol	37.7	48.9	$1.25 \cdot 10^{-2}$	10.94
2-Aminoethanol	—	50.3	$1.53 \cdot 10^{-2}$	10.43
Formamide	109.5	57.6	$1.57 \cdot 10^{-2}$	10.39
1,3-Propanediol	—	49.2	$6.3 \cdot 10^{-2}$	6.91
1,4-Butanediol	—	47.4	$1.60 \cdot 10^{-1}$	4.57
Ethylenediamine	14.2	42.9	$3.39 \cdot 10^{-1}$	2.72
1,3-Butanediol	—	39.1	$4.0 \cdot 10^{-1}$	2.26
2-Mercaptoethanol	—	45.7	$4.0 \cdot 10^{-1}$	2.26
Formic acid	58.5	38.1	$4.5 \cdot 10^{-1}$	2.01
1,2-Propanediol	—	38.0	$5.0 \cdot 10^{-1}$	1.72
1-Amino-2-propanol	—	36.5	$5.0 \cdot 10^{-1}$	1.72
Methanol	32.6	22.6	—	—
Ethanol	24.3	22.3	—	—
Toluene	2.4	28.5	—	—

to the bulk properties of the solvents such as dielectric constant, surface tension, or solubility parameters.

The formation of micelles is characteristic of amphiphilic substances, i.e., compounds that exhibit both hydrophilic and hydrophobic groups within the molecule. The classical examples are the soaps, but in recent decades these have been replaced by the synthetic washing agents, the detergents, in many applications. If all synthetic surface active compounds are included under the even more general term "tensides", then many new applications are added. In their work "Konstitution und Eigenschaften von Tensiden", H. Köbel and P. Kurzendörfer trace their development [177]. All these substances have in common the amphiphilic character which leads generally to the formation of micelles and mesophases.

If a solution of a nonionogenic tenside is heated to a certain temperature, then the cloud point phenomenon is observed due to a rapid increase in the size of the micelles. Rising temperature or increasing electrolyte concentration lead to a dehydration of the hydrophilic groups and, with the increase in micelle size, the c.m.c. is simultaneously reduced [5], fig. 11.3. The change in molar volume during the formation of micelles is always positive [15].

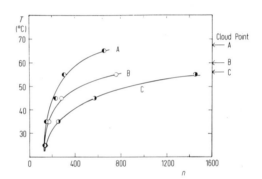

Fig. 11.3. Effect of temperature, T, on the aggregation number of poly-oxyethylene-8-lauryl ether, n, in the presence of Na_2SO_4: A, 0.5; B, 0.1; C, 0.18 mol/l (from [5]).

Nonequilibrium solutions containing an appreciable concentration of micelles can exist below the c.m.c. for a relatively long duration due to the time required for demicellization [14]. The literature on micellar systems reports classification [42a, 156, 166, 192, 193, 409], thermodynamics [38, 271, 373] and kinetics [27, 264] of micelle formation, electrochemical determinations of ionic activity [41, 67, 139, 172, 199, 200, 256, 257], electrophoretic behavior [377, 378], structure of the electric double layer [359, 375], membrane potentials [164], micellar effects on acidity functions [22, 52, 97], intrinsic viscosity and flexibility [357], vapor-pressure depression [379], estimation of the micellar weight from gel-filtration [376], estimation of micellar shape and dimensions from small-angle X-ray scattering [215], comparison of micelle formation in H_2O and in D_2O [270], micelle formation in the homologous series of the sodium n-alkylsulfonates [188], micelle formation of optically active [269] and zwitterionic amphiphiles [151], the formation of micellar emulsions (microemulsions) [1, 374], and counterion binding [189].

Summaries on lyotropic mesomorphism have been written by H. Zocher [440], L. Mandell [258], A. Skoulios [345]. P. Ekwall [68, 79, 94], A. S. C. Lawrence [202], P. A. Winsor [422, 426, 428, 429], and other authors [42, 45a, 130b, 186, 314a].

11.2 Mesophases in Soap Solutions and their Textures

Many early investigations of these phases were performed in order to elucidate the processes occurring on soap boiling. Thus aqueous solutions of soaps [23, 98–100, 180, 291–221, 224–226, 228, 255, 278, 288, 355, 410, 443] and ammonium oleate [1:217, 1:249, 191] became classical study objects. J. B. McBain counts as normal components of an aqueous soap system the following phases [224]: true lamellar crystals, crystalline curd fibers, anisotropic neat soap, anisotropic middle soap, and isotropic liquid. This prediction has proven true in many cases, and therefore neat and middle phases will be discussed here. Both are semitransparent and birefringent. In spite of a substantially higher water content, the middle phase is much stiffer than the neat phase (but exceeded here by the pseudoisotropic mesophases discussed below). Polarization microscopy is the most frequently used investigation method permitting a definite distinction of neat and middle phases [1:303, 56, 58, 305, 314].

The lamellar structure of the neat phase manifests itself in the formation of the homeotropic texture and terraced droplets which do not occur in the middle phase. There is a remarkable similarity between the neat and smectic-A phases. Both additionally exhibit focal-conic texture, oily streaks, pinwheel shaped crosses, and fine mosaic texture. Each uniformly oriented domain of the neat phase can be regarded as a uniaxial crystal with the optical axis normal to the planes of the lamellae, i.e. an analog to homeotropic texture. If, in addition, the C—C bonds are predominantly perpendicular to the layer planes, the birefringence is positive. The neat phase of branched-chain amphiphiles may show positive or negative birefringence. The middle phase exhibits angular, fanlike, or nongeometric textures. According to the two-dimensional hexagonal array of cylindrical micelles, the middle phase is optically uniaxial with the axis lying with the cylinders. This model indicates a negative birefringence. Lyotropic mesophases are separated from other phases by the geometric configurations of bâtonnets, rounded droplets, platelets, and isotropic polyhedra [154, 423].

Other observations of textures [30] and an expedient technique for determining anisotropic/isotropic phase boundaries [190a] are reported in the literature.

11.3 X-Ray Analysis

In the preceding text, some results of microscopic and X-ray investigation have already been contrasted. To avoid confusion it seems practical to first deal with the structures and to describe later the occurrence and properties of the phases. This need not be justified, since already in the early twenties G. Friedel [1:62, 5:69, 131] and J. W. McBain [222, 223, 227] had begun their X-ray studies of soap solutions. Other early works were reported by K. Hess et al. [152, 153, 175] and other authors [368]. The reader can also be referred to two detailed reviews [104, 218]. The X-ray diagrams of soap solutions do not display the normal patterns of liquids but show sharper interferences. An interference within the water ring is found to correspond to the lateral distance between the molecules and is independent of the chain length of the soap studied. A remarkably sharp interference is additionally observed pointing to a large layer spacing which varies with chain length and concentration of the soap solution. The data presented in fig. 11.4 applies to the lamellar mesophase (neat phase) of sodium oleate. H. Kießig has also explained the solubilization of benzene in this system [175].

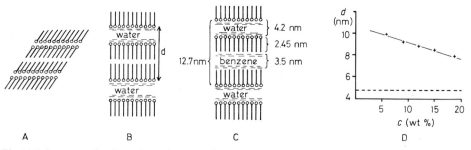

Fig. 11.4. Structure of sodium oleate (from [175]).
Scheme of molecular arrangement (A) in the solid state; (B) in the aqueous mesophase; (C) in the mesophase formed by a 9.12% aqueous solution containing 0.791 g benzene/g oleate. (D): dependence of the
layer distance, d, on the concentration, c, in aqueous solution. The dotted line indicates the theoretical
length of the double molecule.

In the forties, W. J. McBain et al. were almost the only investigators in this area.
This team included X-ray diffraction analysis of detergent solutions in their work [245, 247,
249–253, 405]. A few more papers stem from this period [146, 300] with the most important
being P. A. Winsor's interpretation of the previously published X-ray studies on the basis
of an intermicellar equilibrium [414]. More recent research has been undertaken by several
teams, especially V. Luzzati and A. Skoulios in Strasbourg [210–216, 325, 326, 328, 329,
334, 335, 337, 338, 340, 341, 343], P. Ekwall and K. Fontell in Stockholm [71–73, 76, 101–103],
in England by P. A. Winsor at Shell Research [414, 422, 425], and by the Procter and Gamble
team [8–11]. A. J. Mabis determined the neat phase of the sodium oleate/water system to
have Hermann's SS(RD) structure (cf. chapter 5), and the viscous neat phase, called "soluble
green", to have the SS(RP$_0$) structure [5:204]. V. Luzzati et al. confirmed the lamellar structure
of the neat phase and exposed the hexagonal structure of the middle phase. The data for
the system potassium palmitate/water are reproduced in fig. 11.5 [325].

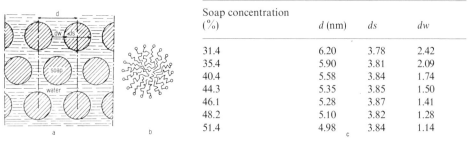

Soap concentration (%)	d (nm)	ds	dw
31.4	6.20	3.78	2.42
35.4	5.90	3.81	2.09
40.4	5.58	3.84	1.74
44.3	5.35	3.85	1.50
46.1	5.28	3.87	1.41
48.2	5.10	3.82	1.28
51.4	4.98	3.84	1.14

Fig. 11.5. Middle soap: (a) schematic structure, (b) cross-section of a cylinder, (c) table of structure
parameters, ds: diameter of the cylinders, dw: distance of the cylinders (from [325]).

In the layers as well as in the micelles of the mesophases, the arrangement of the
hydrocarbon chains corresponds to that in a liquid hydrocarbon of equivalent molecular
weight. This is indicated by the weak and relatively diffuse X-ray diffraction band at high
angles corresponding to a Bragg spacing of about 0.45 nm. Another diffuse band corresponding
to a distance of 0.32 nm is characteristic of all aqueous micellar solutions. It must not be
forgotten that the water in such systems cannot be treated as a continuous free medium but

rather as an associate of the structureal units despite the traditional notation in which only the solute positions are marked in models of lyophases [204], cf. chapter 12.

As mentioned with the neat phase, sharp diffraction bands are found at low angles relating to the layer distance (i.e. the periodicity of the structure, d, as shown in figs. 11.4B and 11.6). Several orders are observed giving rise to calculated Bragg spacings in the ratios $1:1/2:1/3:1/4$. A series of Bragg distances in the ratio $1:1/\sqrt{3}:1/\sqrt{4}:1/\sqrt{7}$ is characteristic of the middle phase indicating a two-dimensional hexagonal array of parallel cylindrical micelles.

B. Gilg, J. François, and A. Skoulios have thoroughly studied the structure parameter, S, the mean lateral packing area per ionized group [338, 343]. S for a cyclindrical structure is calculated from the molecular weight, M_a, the specific volume, $\bar{v}_a$, the concentration of the amphiphile, c_a, and the specific volume of the water, $\bar{v}_w$, according to the equation

$$S = \frac{4 M_a \bar{v}_a}{N d} \sqrt{\frac{\pi}{2\sqrt{3}} \left(1 + \frac{\bar{v}_w}{v_a} \frac{1 - c_a}{c_a} \right)}$$

where d is the distance between the axes of two adjacent cylinders. S for a lamellar structure is calculated according to the equation

$$S = \frac{2 M_a \bar{v}_a}{N d} \left(1 + \frac{\bar{v}_w}{\bar{v}_a} \frac{1 - c_a}{c_a} \right)$$

where d is the layer distance. Consider S_0 as applying to the solution without additive. If S is then observed in a three-component system in the presence of a water-soluble additive, then the ratio S/S_0 always changes in the same way with respect to the dielectric constant of the aqueous phase. This change is independent of the chemical nature of either the ionic amphiphile or added component, and it is also independent of the structure. The ratio S/S_0 rises significantly with decreasing dielectric constant but in aqueous solutions of nonionic amphiphiles this dependence does not exist. S itself, the specific surface of the polar groups, varies with the temperature, the nature of the polar groups, the structure of the system, and the water content of the mixture, but it is independent of the chain-length of the soap. From the values of the cylinder diameter, d_3, and the specific surface area S per polar end-group for the middle phase of different soaps at 100°C being

soap	KC$_{12}$	NaC$_{12}$	KC$_{14}$	NaC$_{14}$	KC$_{16}$	NaC$_{16}$	KC$_{18}$	NaC$_{18}$
d_3 (nm)	2.91	2.90	3.23	3.24	3.74	3.73	4.18	4.19
S (nm^2)	0.545	0.54	0.545	0.55	0.535	0.53	0.53	0.535

V. Luzzati, H. Mustacchi, and A. Skoulios conclude
(i) the diameter of the cylinders does not depend on the cation but only on the length of the hydrocarbon chain;
(ii) the diameter of the cylinder varies regularly with the length of the hydrocarbon chain;
(iii) the specific surface area per polar end-group is the same for all the soaps [326].

In a series of solutions of alkaline-earth soaps, the expansion of the hexagonal lattice obeys the equation

$$\left(\frac{d_c}{d_0} \right)^2 = 1 + \text{const.} \frac{1 - c}{c}$$

where d is the distance between the hexagonal axes at solvent concentrations c and 0, and the constant is the quotient of the densities of the soap and of the solvent [337].

Using the example of a lamellar mesophase, P. A. Winsor compares the idealized structure of the model with the actual existing arrangement [426], fig. 11.6. Other structural models of lyophases must also be evaluated analogously.

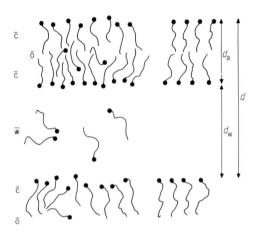

Fig. 11.6. Schematic representation of the structure of the neat phase in a binary amphiphile/water system. On the left is the situation in principle: $\bar{C}$ layer contains most of the amphiphile, but both amphiphile and water are distributed throughout the $\bar{C}$, $\bar{O}$, and $\bar{W}$ regions so as to maintain their respective activities as uniform throughout. On the right is the working-approximation usually employed in X-ray studies: $\bar{C}$ layer contains only and entirely all of the amphiphile present (from [426]).

The lamellar mesophases present some special problems whose connections have not yet been clearly understood. A Skoulios et al. have postulated the existence of another lamellar structure in aqueous potassium soaps by interpretation of X-ray data as indicated in fig. 11.7. The soap molecules in this structure lie with chains stretched alongside one another in a hexagonal arrangement, and the polar groups point alternately out of both sides of the lamellae towards the interstitial water layers.

The Swedish study group also encountered a neat type mesophase in the system sodium caprylate/1,8-octanediol/water showing a layer thickness of 1.35 to 1.47 nm which differs from

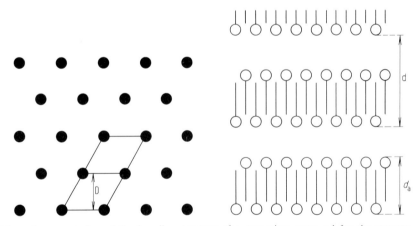

Fig. 11.7. Schematic presentation of the lamellar structure of a potassium soap gel for the stearate, $d_a = 2.52$ nm (from [335]).

that of the bilayer neat phase and can best be correlated with a monolayer structure [71, 81, see also 31a].

K. Fontell compared the X-ray diffraction data obtained by various research teams and found a "gap" in the middle of the lamellar mesophase region [103], fig. 11.8. This was then confirmed by his own measurements. Possibly there exists a range where the molecular arrangement inside the amphiphilic and aqueous layers undergoes a reorganization. The nature of this, however, remains unknown.

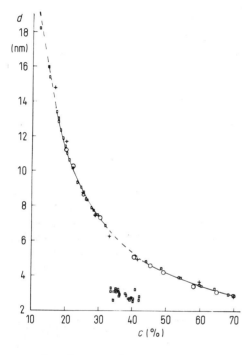

Fig. 11.8. The fundamental repeat distance, d, vs. the Aerosol[®] OT content, c, in the system Aerosol[®] OT/water at 20°C. X-ray diffraction data for the lamellar mesophase (from [103]).

Lamellar mesophases can behave in different ways toward water [75, 217]:

a) By intercalation of water to an almost indefinite amount, the thickness of the water layer varies enormously (from 0.5 nm to more than 25 nm).

b) The water layer can expand only to a limited extent within the range of 0.5 to 2.5 nm and thus water is taken up only to a saturation limit, and any excess forms a separate phase without additional change.

c) The intercalation of water is again limited, but excess water causes a phase change.

Soaps and tensides form mesophases of the third type, the neat phase. Biological materials provide examples of the other two types (cf. chapter 12). The water-insoluble fatty alcohols provide an example of the water intercalation into the bimolecular layer lattice of the solid state [201].

R. D. Vold [396] and J. W. McBain et al. [238] relate the existence of a much larger number of lyophases than those already mentioned. This does not include the thermotropic mesophases of anhydrous soaps that can incorporate only a small amount of water resulting in a lowering of transition temperatures [397]. Addition of larger amounts of water destroys the original phases and leads to the formation of new phases which may have no counterpart in the anhydrous soaps. Like soaps, water-insoluble fatty alcohols can incorporate water in their bimolecular layer lattice. The following outline, which uses P. Ekwall's notation, gives an idea of the multiplicity of lyophases [79].

1. Mesophases with lamellar structure
a) Neat phase, mesophase D.
b) Mucous woven type, mesophase B.
c) Single-layered lamellar type, mesophase D_S.

2. Mesophases with particle structure
a) Normal two-dimensional tetragonal type, mesophase C.
b) Reversed two-dimensional tetragonal type, mesophase K.
c) Normal two-dimensional rectangular type, mesophase R.
d) Middle phase, normal two-dimensional hexagonal type, mesophase E.
e) Reversed two-dimensional hexagonal type, mesophase F.
f) Complex two-dimensional hexagonal type, mesophase H_c.
g) Normal face-centered cubic type, mesophase I_{f1}.
h) Reversed face-centered cubic type, mesophase I_{f2}.
i) Complex face-centered cubic type, mesophase I_{fc}.
j) Normal body-centered cubic type, mesophase I_{b1}.
k) Complex body-centered cubic type, mesophase I_{bc}.

Nevertheless, the possibilities for the construction of mesomorphous structures from 1. lamellae, bands, and discs; 2. cylinders; and 3. spheres are not yet exhausted [216].

The structures determined to date by X-ray diffraction cannot be discussed here in individual detail, and only the models are presented. As indicated in the outline, the mucous woven phase is closely related to the neat phase (fig. 11.9). Examples of the single-layered type have already been given (p. 520).

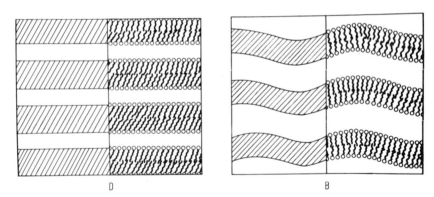

D   B

Fig. 11.9. Schematic drawings of lamellar mesomorphous structures. D: Coherent double layers of amphiphile molecules and ions separated by water layers. Presumed structure for the neat phase type with tilted molecules. B: Coherent double layers of amphiphile molecules and ions separated by water layers. Presumed structure for the mucous woven type with vertical molecules (from [79]).

The three types of hexagonal mesophase are compared in fig. 11.10 [79, 161, 326].

The distances, d, between the axes of the cylinders of the complex hexagonal type (H_c) are

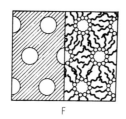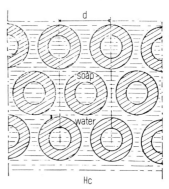

Fig. 11.10. Schematic drawings of two-dimensional hexagonal mesomorphous structures. E: Rod-like particles with hydrocarbon core in water environment; rods in hexagonal array. Proposed structure of the middle phase type. F: Rod-like particles with water core in hydrocarbon environment; rods in hexagonal array. Proposed structure for the reversed two-dimensional hexagonal type (from [79]). H_c: Rod-like particles with water core in water environment (from [326]).

in the soap	NaC_{14}	NaC_{16}	NaC_{18}	KC_{16}	KC_{18}
d (nm)	9.3	10.8	12.4	10.5	11.4

The long-range order of these mesophases is as perfect as that of a crystal in spite of the short-range disorder of the liquid-like paraffin chains [326].

The models proposed by P. Ekwall [79] and V. Luzzati and A. Skoulios [210–212, 341] to describe the tetragonal and rectangular type are shown in fig. 11.11.

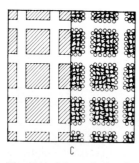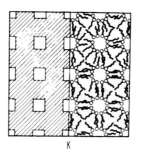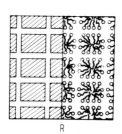

Fig. 11.11. Schematic drawings of tetragonal and orthorhombic mesomorphous structures (from [79]). C: Rod-like particles with hydrocarbon core in aqueous environment; rods with predominantly quadratic cross-section in tetragonal array. Proposed structure for the normal two-dimensional tetragonal type. K: Rod-like particles with water core in hydrocarbon environment; rods with predominantly quadratic cross-section in tetragonal array. Proposed structure for the reversed two-dimensional tetragonal type. R: Rod-like particles with hydrocarbon core in aqueous environment; rods with rectangular cross-section in an orthorhombic array. Proposed structure for the normal two-dimensional rectangular type.

Optically isotropic mesophases occur with several types of structural units (normal, reversed, and complex spherical micelles) and in numerous systems [76, 214, 329]. The former doubt as to whether the cubic mesophases were to be considered an independent type of

Fig. 11.12. Schematic drawings of cubic mesomorphous structures. I_1: Particles with hydrocarbon core in aqueous environment. Proposed structure for the normal face-centred cubic type. I_2: Particles with water core in hydrocarbon environment. Proposed structure for the reversed face-centred cubic type (from [79]).

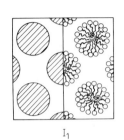 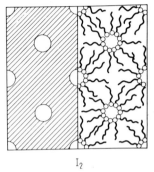

I_1 I_2

mesophases no longer exists [420]. Body-centered and face-centered lattices are mentioned in fig. 11.12 and in the literature [102].

Recently it has been possible to clarify and explain the morphological relationship between the cubic mesophases and the plastic crystals [1:98, 140, 427]. The addition of plastic crystals to the mesophase system, which can be completely reconciled with Lehmann's original model, completes the spectrum of phases from the linear-nematic via the planar-smectic as far as three-dimensional structures under the influence of lattice forces.

Kettle wax, a product obtained on soap boiling was proven by X-ray studies not to be a single phase, but it is a mixture of curd, neat, and lye [334]. Aqueous systems of α,ω-bifunctional soaps have stirred little interest. Experience to date indicate they can form only one type of mesophase, the lamellar [132]. Neutron scattering experiments in the system ammonium perfluorooctanoate/water revelaed a diffusion coefficient of interlamellar water which depends on the layer distance and the orientation of the sample which was adjusted by shear [148].

11.4 Electron Microscopy

W. Stoeckenius [358] and many more recent investigators have been able to confirm by electron microscopy the lyophase structures determined by X-ray analysis. The periodicity of these structures could be resolved thanks to the large lattice spacings of lyophases, fig. 11.13. The main problems arise from the preparation techniques. Electron micrographs of dried droplets of neat and middle mesophases indicate a stepped growth pattern [8]. In both cases the step-height measurements agree with the interlayer spacings determined by small angle X-ray scattering and are consistent with the length of two surface active molecules. The edges of the steps are scalloped in the neat phase and angular in the middle phase.

In order to isolate spurious data resulting from preparation technique, S. Eins selected the system potassium oleate/water as study object, because the structural parameters of the mesophase are well known and thus suitable for comparison work. In addition, the double bond eases fixation with OsO_4, and the chain length is comparable to that of lipids in biological membranes [55]. He investigated the following techniques:

a) fixation in closed capsule by OsO_4-vapor,
b) dehydration in graded concentrations by use of acetone or similar agents,
c) embedding in methacrylate, or similar media, and
d) ultra-thin sectioning by a microtome.

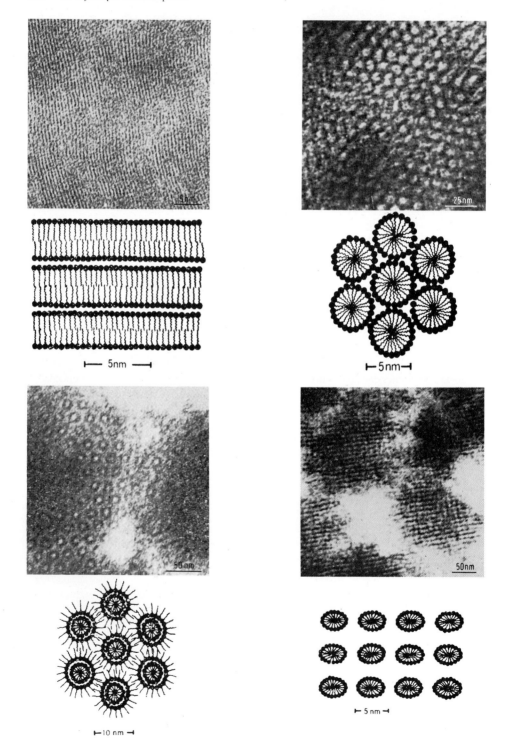

Deviations were found among the samples prepared in various ways, but these were deemed tolerable. It was also difficult to ascribe any deviation to a particular preparation step. A further paper discusses the possibilities and limitations of various preparation methods, and we quote directly from the conclusion, "The conventional surface replication technique can only be applied to those systems which give anhydrous mesophases, while freeze replication must be restricted to systems which do not crystallize at low temperatures. Negative staining and polymerization are not universally applicable and ... the osmium tetroxide fixation and thin sectioning technique appears to be the most versatile currently available for examining thin sections" [35]. Fig. 11.13 shows picture segments of electron micrographs from the work of R. R. Balmbra, D. A. B. Bucknall, and J. S. Clunie [11]. Other structure determinations [104c] and other aspects of the subject are treated in the literature [9, 54]. J. F. Goodman and J. S. Clunie have written a recent summary [37].

In chapter 12 we return to the most important current application of electron microscopy, the investigation of biological materials.

11.5 Composition of Lyotropic Mesophases

11.5.1 Binary Systems with Soaps

R. D. Vold and J. W. McBain have made the most comprehensive studies of aqueous soap solutions [240, 244], figs. 11.14 and 11.15.

The existence of liquid-crystalline solutions in certain areas of concentration is a common feature of aqueous soap solutions. Also common is the definite temperature at which a good water-solubility appears (usually about 20–30%), although the solubility is small just a few degrees below this point. This temperature, the socalled Krafft point, is between 45 and 55°C for common commercial sodium soaps. This is also the lowest temperature at which the middle phase can exist [244].

The effect of the soap molecule chain length on the phase diagram (see fig. 11.15) involves at least two opposing factors in which the longer chain soaps are less soluble at low temperatures and low soap concentrations while at higher temperatures and higher soap concentrations the shorter chain soaps are the less soluble [240]. With increasing water demand to hydrate K^+, Na^+, and Li^+ ions, the minimum water content required for the existence of phases with reversed type micelles increases in the isotropic and the mesomorphous phase [74]. The phase diagrams of lithium soaps otherwise show no special features [399, 407] just as the technical and mixed sodium potassium soaps behave qualitatively like single pure soaps with water [242].

The electric conductivity of soap solutions has been repeatedly investigated, but the results were confusing and difficult to evaluate. In some cases, time-dependent changes were observed [46, 105, 106, 145a, 149, 276a, 346, 348, 404]. J. François made the most definite statement when he recognized that the activation energies of the electric conductivity of cylindrical type soap/water mesophases are close to those for the corresponding ordinary electrolytic salt solutions. He had therefore shown that a classical conduction mechanism is involved [106].

◀ Fig. 11.13. Electron micrographs of thin sections OsO$_4$-fixed mesophases of the system potassium-oleate/water, and models of the investigated phases: neat, middle, complex hexagonal, and rectangular (from [11]).

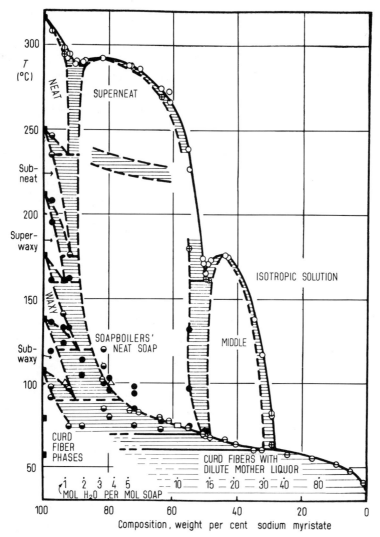

Fig. 11.14. Phase diagram of the system sodium myristate/water (from [240]).

Phase transitions in soap solutions are reported to be readily observable by dielectric measurements [206].

R. D. Vold discusses in detail the application of differential calorimetry to binary systems such as sodium oleate/water [402], but this method is still less important for the investigation of lyotropic mesophases than for thermotropic ones [356a, 361]. Viscosity [229] and hydrodynamic properties [371] have not been very closely examined.

Vapor-pressure studies can be successfully used in the examination of phase equilibria [4, 394, 395]. Fig. 11.16 shows that the ranges of existence for various mesophases can be seen more or less distinctly in the vapor pressure curve. Like water, many other solvents form lyophases with soaps [239]. The liquid-crystal phase of potassium oleate, existing between

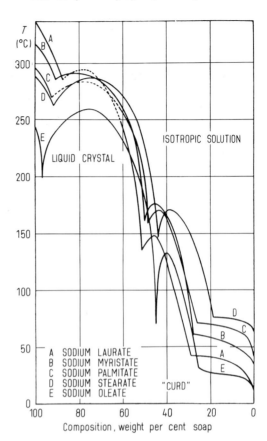

Fig. 11.15. Phase boundaries in various soap/water systems (from [240]; as to sodium stearate, see [40]).

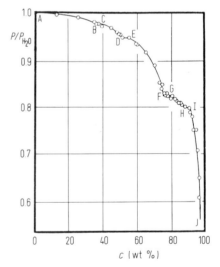

Fig. 11.16. Relative vapor pressure, P/P_{H_2O}, of aqueous sodium laurate solutions at 90°C vs. the concentration of sodium laurate, c. The composition ranges (given in % by weight of sodium laurate) of the various phases are:

isotropic solution   0.0–35.8 (A–B)
middle soap   38.0–52.0 (C–D)
neat soap   55.4–75.0 (E–F)
waxy soap   77.0–88.0 (G–H)
curd fiber phase   91.3–100 (I–J)
(from [395]).

258.5°C and 324.2°C, is extended by addition of oleic acid to the two-component system and there it begins to appear at a temperature as low as 107°C [233]. In the sodium palmitate/palmitic acid system, the depression goes down to a minimum at 154°C [232]. The mesophase is a neat phase in both cases. In the potassium laurate/lauric acid system, the neat phase reaches no lower than 240°C [230]. The subneat phases of pure sodium soaps similarly can incorporate considerable quantities of free fatty acids [339], (see p. 538). The literature also describes the soap/alcohol system [342].

Several mesophases appear in binary soap/hydrocarbon systems. Two exist within a wide range: one described as white and waxy, and the other golden or orange [248]. T. M. Doscher and R. D. Vold [400, 404] report a middle phase (fig. 11.17), but this classification is in doubt since A. Skoulios found a lamellar structure in the analogous zones of similar systems [328]. Hexagonal or rectangular structures appear at higher soap concentrations.

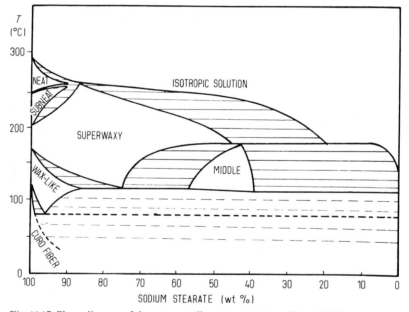

Fig. 11.17. Phase diagram of the system sodium stearate/cetane (from [403]).

11.5.2 Binary Systems with other Amphiphiles

Compounds with soaplike character, i.e., those that combine distinctly hydrophilic and lyophilic solubilities, can be of ionic or nonionic nature. Among the ionic amphiphiles, sulfonates are the most important. These are followed by sulfates and the socalled invert soaps which are salts of long-chain amines or other nitrogen bases. Nonionic amphiphiles usually derive their hydrophilic nature from the polyoxyethyl group.

Since no new structures among the mesomorphous phases appear (other than those found in soap solutions), it is sufficient here to discuss only a few special aspects. While the phase diagrams of long-chain alkali sulfonate/water systems are quite like those of soap solutions, this similarity does not extend to the free acids. Mesophases and definite hydrates occur only with free sulfonic acids as shown in fig. 11.18 [398].

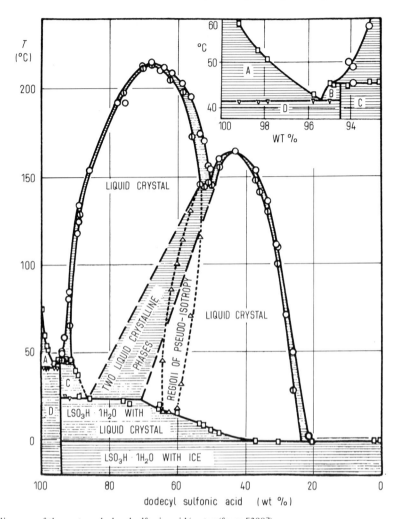

Fig. 11.18. Phase diagram of the system dodecylsulfonic acid/water (from [398]).

Lyophases are also found in sodium alkylbenzene sulfonates [372] and Aerosol® OT [89, 421, 424, 425] which is the sodium salt of di-(2-ethylhexyl)sulfosuccinic acid ester. The neat phase of the system Aerosol®-OT/water is remarkable due to a concentration-dependent change in the sign of the birefringence that occurs at the same time as a significant dispersion of the birefringence (figs. 11.19 and 11.20).

Like sulfonates, sulfates form lyophases in many systems [254a]. A. S. C. Lawrence investigated mostly ternary mixtures containing sodium dodecylsulfonate and water [196–198]. In the two-phase equilibrium comprising an isotropic and a mesomorphous solution of sodium dodecyl sulfate in water, sodium iodide distributes itself such that the iodide-ion concentration in the isotropic solution is greater than in the liquid crystal, where water is not so readily available for hydration as in the isotropic solution [171]. Lyophases are also presumed to exist

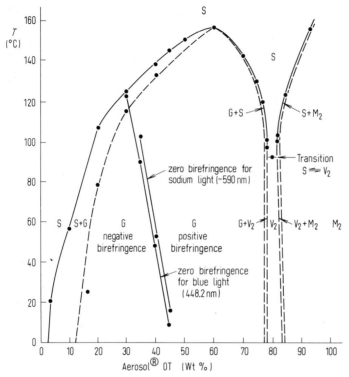

Fig. 11.19. Phase diagram of the system Aerosol®-OT/water. The dotted lines indicate tentative boundaries. S, mobile isotropic phase; G, neat phase; M_2, inverse middle phase; V_2, reverse viscous isotropic phase (from [424]).

in sulfite, sulfate and cotton pulps [277] and in the system tall oil/water [302] and hydroxypropyl cellulose/water [411].

   As components of lyophases, nitrogen compounds can function in ionic form (as salts of long-chain amines [91, 157, 184a, 297] or long-chain carboxylic acids [365]), as well as in a nonionic (free) form [298], where it still remains to be proven that one of the phases is nematic, as reported. Of the remaining data on amines and their salts [6, 21, 114, 246, 303], the existence of the neat phase in propellant compositions should be noted, e.g., in 1,1-dichloro-1,2,2,2-tetrafluoroethane (Freon® 114), 1-octanoic acid, and aqueous 1-aminooctane [117]. Two cubic mesophases occurring in the system dodecyltrimethylammonium chloride/water are quite remarkable in several respects [10], fig. 11.21. One of the two phases is formed only by decyl-, dodecyl- and tetradecyltrimethylammonium chlorides, but not from the hexadecyl- and octadecyltrimethylammonium homologs or by the corresponding bromides. The most probable space groups are P43n or Pm3n. Both cubic phases exist at room temperature, are optically isotropic, highly viscous, and exhibit high resolution NMR spectra.

   No special phenomena are found in the phase diagram, density measurements, or X-ray diffraction patterns of mesophases containing the zwitterionic compound N,N,N-trimethylamino-dodecanimide: $C_{11}H_{23}CON\overset{\ominus}{N}\overset{\oplus}{(CH_3)_3}$ dissolved in water or deuteriumoxide [32]. Mesophases are also described in the system monooctadecyl phosphate/water [136] and with other phosphated

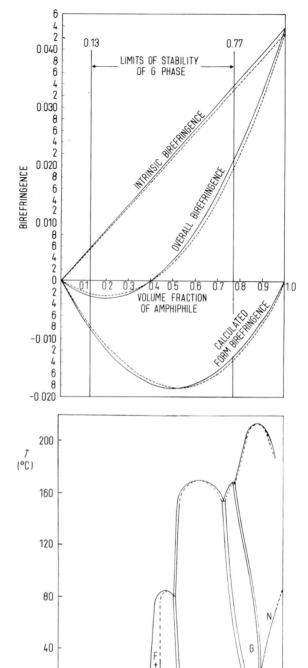

Fig. 11.20. Birefringence at 20°C of the lamellar (G) phase vs. volume fraction in the system Aerosol®-OT/ water (from [424]).

Fig. 11.21. Phase diagram (temperature, T, vs. composition, γ) of the system dodecyltrimethylammonium chloride (II)/water (I). F, fluid isotropic phase; C and C′, cubic phases; M, middle phase; N, neat phase; ——— = experimentally determined boundary; ––––– = interpolated boundary (from [10]).

surfactants [142]. D. H. Chen and D. G. Hall observed in a ternary system the neat⇌middle phase-transition at constant water content [29].

The nonionic amphiphiles have a smaller tendency to form liquid-crystalline solutions than do the ionic ones [138, 272, 324]. Mesophases have been described for hexaethyleneglycol *n*-decyl ether and related compounds [187, 272, 388], polysorbate-80 [276], octylglucoside [24], and some amides [387]. Neat and middle phases, plus one additional mesophase, were found in the system dimethyldodecylamine oxide/water [209]. Mesophases with glycerides are described in chapter 12, see also [254]. Bile acid salts are not amphiphilic and form no mesophases with water [241]. H. Zocher observed soaplike mesophases among naphthenates and xanthogenates [438]. Only sparse information is available to compare the tendency of fluorocarbon and hydrocarbon surfactants to form liquid-crystalline solutions [369, 370]. The examples known show different types and regions of existence of mesophases due to the low mutual solubility of fluorocarbon and hydrocarbon chains.

P. A. Winsor's basic paper on solubilization contains data for many phase systems [412].

11.5.3 Ternary and Multicomponent Systems

The vast variety of lyophases emerges when multicomponent systems are considered. In a binary system it is possible to define the range of existence in one diagram using the two parameters of composition and temperature. Thus, the clear isotropic solution of all soap/water systems occupies a concentration range that widens with increasing temperature until complete miscibility is attained at the true melting point of the soap. In ternary systems, a diagram can only reflect the composition. We begin the discussion with a consideration of the general system soap/salt/water [235, 236, 243] diagrammed in fig. 11.22. Two separate liquid phases are in mutual equilibrium in the isotropic "bay" region:
1. The lye (Unterlauge or dünne Lauge) containing much salt and little soap.
2. The nigre (Leimseife or Leim) containing a moderate concentration of soap.

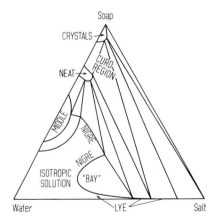

Fig. 11.22. Schematic equilibrium diagram for a system soap/salt/water near 100°C (from [235]).

The neat and middle phases are both birefringent, clear or translucent, and mutually immiscible. While hot, the neat phase includes the soapboiler's fitted and settled soap (geschliffene oder glatte Kernseife). The crystalline soap phases consist of lamellar crystals or curd fibers.

The least soluble phase of sodium palmitate is actually made as soluble as sodium laurate by the addition of sodium laurate forming the quaternary system palmitate/laurate/chloride/water [237].

The Stockholm school carried out exemplary research on the system sodium caprylate/decanol/water [59, 60, 62, 65, 66, 69, 70, 78, 101, 259, 260], fig. 11.23. This literature contains detailed descriptions of the individual phases, but here it is necessary to restrict the discussion to a few selected aspects of the system.

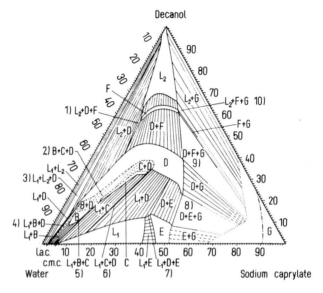

Fig. 11.23. Phase-equilibrium diagram of the system sodium caprylate/decanol/water at 20°C, concentrations in weightpercent: L_1, homogeneous isotropic aqueous solutions; L_2, homogeneous isotropic decanol solutions; B, C, D, E, F, homogeneous mesomorphous phases (structures, see p. 521); G, not further investigated (in part, neat soap containing decanol); 1–10, ternary ranges (from [65]).

The mesophases of this system may be characterized as follows:

Lamellar structures: phase D (neat phase) and phase B (mucous woven type).

Normal micellar structures: phase E, hexagonal (middle phase) and phase C, tetragonal.

Reversed micellar structure: phase F, hexagonal.

Further study of this system revealed a close correlation between the structure and the rheological behavior of the lyophases depending on:

the geometry of the mesophase,

the surface characteristics of the aggregates, and

water bonding [78, 93].

This behavior may be used to characterize the mesophases since analogous mesophases in different systems exhibit similar rheological behavior (fig. 11.24).

Aqueous phases, being in equilibrium with each other, should have the same water activity (expressed by the water vapor pressure) but differ in their composition as shown in fig. 11.25 ([93]).

Substitution of K for Na in the caprylate/decanol/water system changes the range of existence of mesophases to lower water content due to the smaller hydration of the K^+ ion. Ca^{2+} ions have the opposite effect [83, 84, 260].

Figs. 11.26a and 11.26c illustrate the occurrence of cubic structures: phases I (normal) and J (complex). Fig. 11.26a also illustrates a rectangular structure (phase R), and fig. 11.26b illustrates a reversed two-dimensional tetragonal structure (phase K).

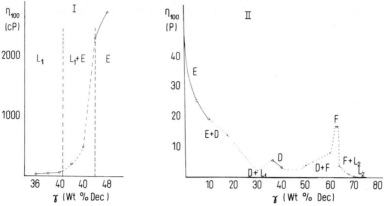

Fig. 11.24. Change in apparent viscosity, η_{100}, vs. composition, γ, I) at the transition from solution L_1 to mesophase E; II) at the transition from mesophase E (via mesophases D and F) to solution L_2 in the system sodium caprylate/decanol/water (from [78]).

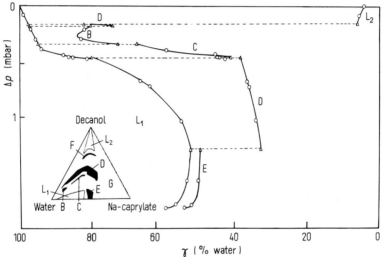

Fig. 11.25. Equilibrium water vapor pressure between the aqueous solution L_1 and the decanolic solution L_2 and the mesophases B, C, D, and E, respectively, in the system sodium caprylate/decanol/water at 25 °C (from [93]).

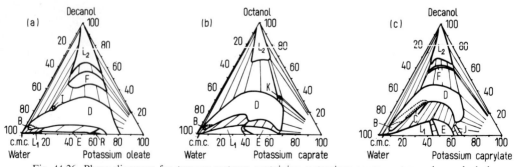

Fig. 11.26. Phase diagrams for ternary systems containing potassium soaps, water and an alcohol: (a) Potassium oleate/decanol/water at 20 °C, (b) Potassium caprate/octanol/water at 20 °C, (c) Potassium caprylate/decanol/water at 20 °C (from [79]).

Two types of mesophases, with normal and reversed micellar particles, are recognized in hexagonal, tetragonal, and cubic structures.

In the normal case, the interior of the micelle is hydrophobic while the micelle itself is in an hydrophilic matrix. In the reversed case, the interior of the micelle is the hydrophilic portion while the matrix is hydrophobic. Fig. 11.27 indicates the occurrence of normal and reversed type mesophases in the phase diagrams.

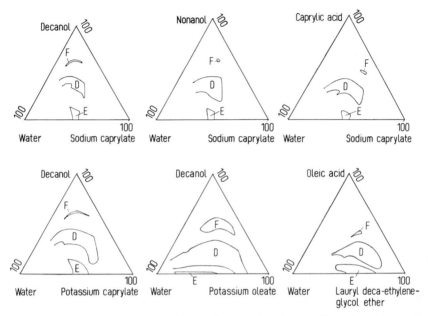

Fig. 11.27. Phase diagrams showing the location of the regions E and F with two-dimensional hexagonal mesophases in various ternary systems. D: Lamellar structure (neat phase); E: Normal two-dimensional hexagonal structure (middle phase); F: Reversed two-dimensional hexagonal structure (from [72]).

Fig. 11.28. Schematic diagram showing the occurrence of mesophases with normal micelles (signature ∼) and reversed micelles (signature ×) in relation to the neat phase and the two isotropic solutions L_1 and L_2.

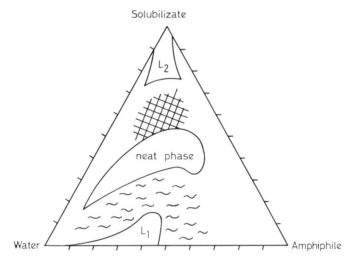

The key position of the lamellar neat phase became apparent in many other investigations by P. Ekwall et al. [72, 73, 76, 79, 88, 92] from which the phase diagram in fig. 11.28 has been abstracted.

The results can be summarized as follows: if phases with normal and reversed structures both occur in the same system, then their ranges of existence lie on each side of a lamellar-meso-phase region with the normal type on the side with greater amount of water. The complex structures occur on both sides of this region. All transitions from a normal to a reversed mesomorphous structure observed up to now take place via the lamellar mesophase. The lamellar stage is also an intermediate in most transitions from normal to reversed micellar structures. Fig. 11.29 summarizes the transitions observed by P. Ekwall.

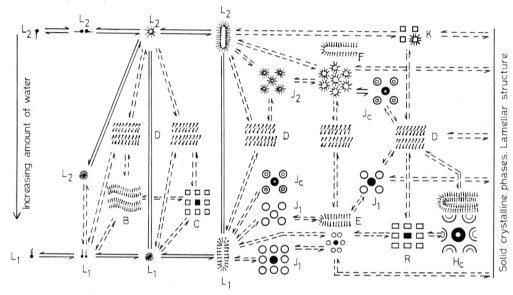

Fig. 11.29. Schematic presentation of transitions in micellar and lyotropic mesomorphous solutions; ⟹ transition of association stage; ⟶ phase transitions. Terminology, see p. 521; in the iso-tropic solutions, L_1 and L_2, simple molecules, small aggregates, spherical and cylindrical micelles are indicated (from [79]).

Reversed structures can exist only below a certain specific water content which appears to be determined by the maximum capacity of the hydrophilic groups for binding water. A minimum water content (sufficient for the hydration of the alkali ions) is required by the neat phase and the phases B, C, and I_1 of the normal type to enable their formation in aqueous soap solutions. Lyophases lacking an amphiphile and water have not been detected.

The neat phase displays the greatest tolerance of variations in both the relative amount of amphiphile and water content. Examples are known of both extremes and of intermediate stages.

In some systems, the normal hexagonal phase can exist over quite large ranges of concentration of water and amphiphile. We refer to the literature for data on the existence ranges of other mesophases as well as on the density of the amphiphile parts [79]. In binary amphiphile/water systems, a transition from a normal to a reversed type mesophase has not been observed, but this is normal in many ternary systems.

Due to their optical activity and unusual compositions, lyotropic liquid crystals containing cholesterol or cholesteryl esters hold possibilities of great interest [26, 31, 44, 61, 203, 267, 268, 318]. Other phase diagrams of ternary systems are described in the literature [39, 77, 82, 89, 90, 254a, 272a, 289] and in the following section.

11.6 Solubilization

The core of the micelle and the matrix surrounding the micelle can be media of very different polarity, but they together form one uniform phase. Due to the micellar structure, such phases dissolve water as well as hydrocarbons and distribute the solute in a definite equilibrium ratio between core and matrix. When the solute shows a preference for the core of the micelle, the phenomenon is called solubilization, and is also known as intramicellar solubility. P. A. Winsor [412, 413, 415, 417, 419] and P. Ekwall [57, 63, 64, 79] have been particularly active in studying the potentials of these phenomena. An older summary also exists [176]. In the cited literature, many ternary systems containing water/amphiphile/solubilizate or hydrocarbon/amphiphile/solubilizate are described. Mesophases normally occur in these systems. The solubilization properties are closely correlated to the formation of lyophases. Hydrocarbon molecules can in extreme cases be incorporated into the hydrocarbon core of the micelles. Conversely, reversed type micelles can solubilize water in oil.

Amphiphilic molecules can be solubilized in such a way that they form mixed micelles with the micelle forming substance. In this case, a slight solubilization effect is observed even below the c.m.c., beginning at the limiting association concentration. This effect is closely related to Neuberg's hydrotropic phenomenon encountered in solutions of such amphiphilic substances that do not form micelles at all. The solubilization capacity of a colloid solution depends on the total amount of micellar substance and the type of micelles. The solubilization of a hydrocarbon thus increases linearly with the amount of micellar substance present but only within a range where the form of the micelles does not change (near c.m.c.). The maximum capacity of the micelle to dissolve a foreign substance is attained when, in the distribution equilibrium, a saturation of the intermicellar solution with solubilizate is reached so that any excess separates in more or less pure form. In most cases, however, mesophases are formed continuing the solubilization. This happens before the solubilization capacity of the micelles is reached.

The neat soap with the greatest solubilization capacity can incorporate 40 to 50% of some additives without significant structural change, but other properties may undergo modification. To a more moderate extent, this also holds for the middle soap which can incorporate up to 27% of foreign substances.

P. Ekwall has systematically investigated the mesophases occurring with solubilization in aqueous sodium caprylate [79]. Fig. 11.30 shows that such a solution can solubilize only small amounts of *p*-xylene or other completely lyophilic substances forming an aqueous phase, L_1, and two mesophases: the middle phase, E, and the optically isotropic cubic phase I.

A weakly amphiphilic substance having a weakly hydrophilic group, e.g., capryl aldehyde, can be solubilized to a significantly greater extent (fig. 11.31). The only mesophases observed

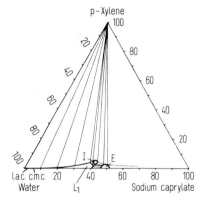

Fig. 11.30. Phase diagram of the ternary system sodium caprylate/p-xylene/water at 20°C (from [79]).

are the neat phase, D, and middle phase, E. An example of a moderately amphiphilic substance, n-decanol, has already been considered (fig. 11.23). A multiplicity of mesophases and two areas of isotropic solutions (L_1 with normal and L_2 with reversed micelles) give the phase diagrams a completely different character similar to that found with carboxylic acids (fig. 11.32).

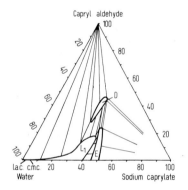

Fig. 11.31. Phase diagram of the ternary system sodium caprylate/capryl aldehyde/water at 20°C (from [79]).

Several mesophases occur in soap/fatty acid/water systems [80, 85–87, 108, 231, 234]. In the system sodium caprylate/caprylic acid/water containing mesophases B, C, D, E, and F (Ekwall's nomenclature, see p. 521) the most remarkable feature is the high solubility

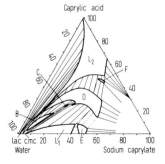

Fig. 11.32. Phase diagram of the ternary system sodium caprylate/caprylic acid/water at 20°C (from [79]).

of the soap in the anhydrous fatty acid. This is attributed to the formation of the molecular compound (Na caprylate)·2(caprylic acid) [80].

In general, the phase diagrams of ternary ionic amphiphile/alcohol/water systems show common features independent of whether the ionic association colloid is sodium caprylate, sodium octylsulfate, sodium octylsulfonate, octylammonium chloride, or octyltrimethylammonium bromide [79]. The introduction of a nonionic amphiphile does not change the situation in principle. This, however is not true if one considers that the temperature dependence of the solubilization capacity changes very little with ionic amphiphiles but drastically with nonionics [129]. The transition from normal to reversed type aggregates occurs in a narrow phase inversion temperature range. An especially rich palett of lyophases is found in the ternary system lauryl decaethyleneglycol ether/oleic acid/water (fig. 11.33).

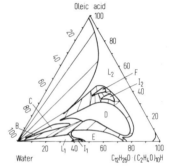

Fig. 11.33. Phase diagram of the ternary system lauryl deca-ethyleneglycol ether/oleic acid/water at 20°C (from [79]).

As indicated by the phase diagram of the ternary system sodium caprylate/butanol/water, shortening the chain and raising the temperature reduces the number of phases (fig. 11.34).

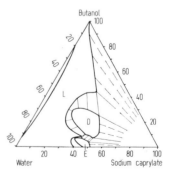

Fig. 11.34. Phase diagram of the ternary system sodium caprylate/butanol/water at 20°C (from [79]).

Many biological materials (cf. chapter 12) belong to a special class of amphiphiles that are only slightly soluble in water but which incorporate water with a simultaneous swelling.

The literature reports other solubilization data [162, 194, 261, 281, 304, 344, 401] and the partition of an additional component [126].

11.7 Other Properties of Lyotropic Mesophases

Lyophases having nematic properties seem to be exceedingly rare. Only the systems sodium decyl sulfate/decanol/sodium sulfate/water, potassium laurate/decanol/potassium chloride/deuterium oxide, and a few systems mentioned in chapter 10 come to mind [207]. As genuine nematic is one mesophase described occurring in the system disodium chromoglycate (I)/water [147].

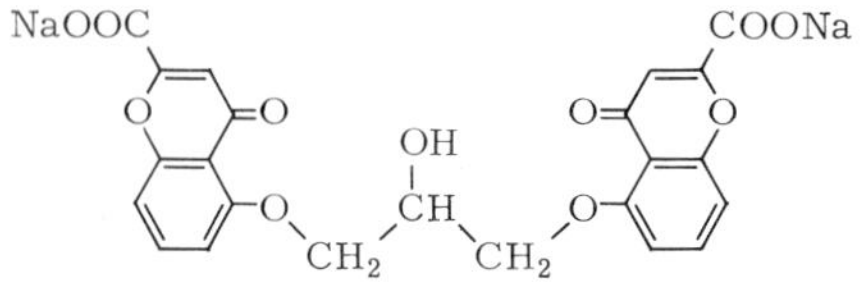

(I)

However, there is some uncertainty about stating this firmly because many lyophases, although long known, have not yet been thoroughly investigated using the latest methods. Nematic-like textures have additionally been reported that could also be due to deformed lamellar phases [147].

Of the earlier observations of nematic lyophases, only that of Salvarsan by H. Zocher can be considered certain [432, 437]. The most convincing arguments are that the typical twist textures are induced by orientation at the boundary layers (Mauguin texture), and the cholesteric behavior is induced by addition of cane sugar or glucose. The spatial element is optically negative in the smallest droplets, while it is optically positive in the homeotropic layer.

The ordered coagulation of micellar solutions by foreign ions leads to an arrangement very similar to that of liquid crystals, but the correlations between ionotropy and lyotropic mesomorphism have not been further pursued [366, 367]. Gels with laminated structure are formed, if convection is excluded, during Ca^{2+} ion diffusion into sodium alginate solution [380].

Ultrasonic studies of lyophases require measurements over a wide range of composition, temperature, and frequency. In the *n*-octylamine/water system, an anomalously large value was found for sound absorption at the mesomorphous-isotropic transition, while the discontinuity of the velocity showed nothing unusual [53, 275], fig. 11.35. Measurements of density and thermal expansion performed in the same system demonstrate the pretransitional behavior [275], fig. 11.36.

Although P. A. Winsor found as early as 1954 an electro-optical turbidity effect in the liquid-crystalline solutions of some amphiphilic salts, no further attention has been given to these phenomena [416]. They are specific of the liquid-crystalline state of colloidal electrolytes. The excitation and decay times, on the order of several minutes, promise no possibility of a technical application [418]. This explains why only one recent paper is concerned with this effect [362].

Further investigations are necessary to state general features of electric conductivity in lyophases [104a, 208, 430, cf. p. 525] and the observation of ferroelectric properties attributed to an intermediate lyotropic state should arouse interest [206, 206a]. The literature also reports volta potential studies of liquid-crystalline solutions [18].

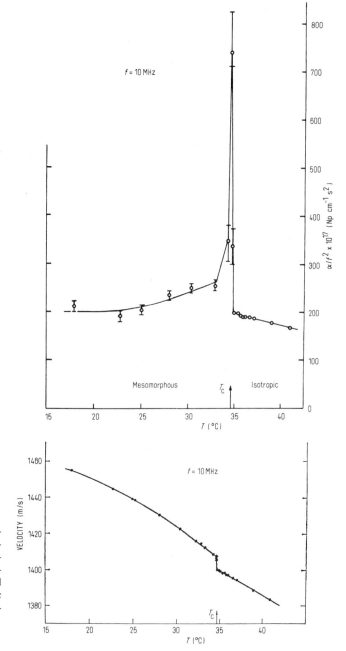

Fig. 11.35. Temperature dependence of ultrasonic characteristics of the system *n*-octylamine/water at the critical concentration of 0.854 mol fraction of water: a, ultrasonic absorption coefficient; b, ultrasonic velocity (from [53]).

L. W. Reeves et al. found lyophases having positive and negative diamagnetic anisotropy [299a]. There is a profound dependence on the electrolyte concentration which may convert the positive to a negative diamagnetic anisotropy with the addition of less than 1% of electrolyte.

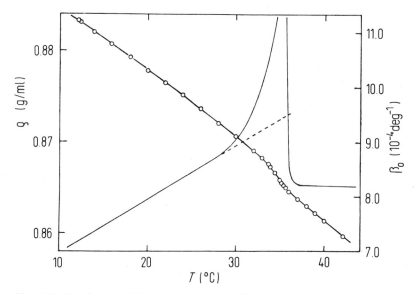

Fig. 11.36. Density, ρ, and thermal expansion coefficient, β_0, vs. temperature, T, in the *n*-octylamine/ water system at 0.85 mol fraction of water (from [275]).

At first J. S. Clunie et al. could find no measurable discontinuities in the heat of mixing and in the vapor pressure at the transition of the middle phase to the isotropic liquid [34]. However, the same authors were later able to demonstrate that the transition from a lyophase to a randomly ordered isotropic solution is accompanied by an increase in volume and in enthalpy. However the changes are extremely small, being two orders of magnitude lower than those for the melting of organic crystals of comparable molecular structure and one order of magnitude lower than those associated with the melting of thermotropic mesophases [36, 356a].

Among the attempts trying to elucidate the role of water in lyophases, near-IR spectroscopic studies have brought valuable information. Thus the spectra of water in the 900 to 1300 nm range differ markedly for different states of water as shown in fig. 11.37.

The band at 977 nm has been assigned to the combination $2v_1 + v_3$ and the bands at 1160 and 1200 nm to the combination of $v_1 + v_2 + v_3$ with v_1 (symmetric stretching), v_2 (bending), and v_3 (asymmetric stretching) being the fundamental vibrations. Absorptions at

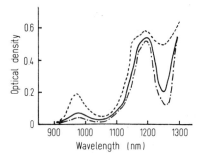

Fig. 11.37. Spectra of pure water (– – – –) and water in liquid crystals containing 60 % (———) or 80 % (–·–·–·–) of polyoxyethylene-nonylphenylether (from [3]).

1 160, 1 200, and 1 250 nm are assigned to water species having 0, 1, and 2 hydrogen bondings, respectively. As the spectra shown in fig. 11.37 suggest, the mole fractions of nonbonded and single-bonded water molecules (bands at 977 and 1 160 nm, respectively) decrease in the lamellar liquid crystal when compared with pure water [3]. Near-IR spectroscopic observations of mesomorphous soap solutions have been interpreted in terms of a two-group model of water. One kind of water in the closest proximity to the structural elements is assumed to contain the most part of counterions, while the remaining water comprises little or no counterions [107]. Raman studies are indicative of conformational changes occurring within the lamellar mesophase [95a]. Lamellar mesophases can be used as orientational matrix for linear dichroism measurements of solutes [104b].

Theoretical considerations on the formation of lyophases have only concerned partial problems [43, 45, 145a, 266, 286, 287, 360, 431]. Evaluations of conformational energies and intermicellar energies have been reported [266]. V. A. Parsegian has tried to strike an energy balance of the transition between neat and middle phases [287, see also 391a, 391b]. R. J. Prime and O. E. Hileman, Jr. found indications of an energy barrier to mesophase transformations in lyotropic liquid crystals [295]. From hydrodynamic considerations and a comparison with smectics A, F. Brochard and P. G. de Gennes expect a second sound mode and a slip mode to be observable in lamellar mesophases [43]. W. Helfrich's calculations suggest that a slight "internal twist" within the bilayer might couple to bilayer curvature so that corrugation may lower the total elastic energy of lyotropic layered structures [150]. Three models have been examined to describe the asymmetric diffusion of small molecules [17]. The rheological behavior can be complicated by thixotropy or dilatancy besides plasticity [143, 363, 365a]. A lyotropic surface layer on a semipermeable membrane was observed in a stagnation flow field [95]. Also noteworthy are the investigations on the fluid birefringence [185], NMR spectroscopy in lyophases as is discussed in chapter 10 [28], and a review of dynamic phenomena [265].

11.8 Mesomorphous Solutions of Synthetic Polymers and Copolymers

11.8.1 Polymers

The liquid-crystalline character of many polymer solutions is not surprising since many polymers are paracrystalline and closely related to thermotropic liquid crystals (cf. chapter 13). However, the soaps remain the prime examples of the tendency toward thermotropic and lyotropic mesomorphism, and numerous other examples of tensides can be added.

Russian chemists have been especially active in the investigation of liquid-crystalline polymer solutions. They have studied poly-(2-methyl-5-vinylpyridine) [389], poly-(p-benzamide):

(NH—⟨○⟩—CO)$_n$ [13, 165, 170, 173, 174, 178, 183, 205, 282, 285, 285a, 291–294, 294a, 296, 319], poly-(p-hexyloxybenzoyloxy)phenylmethacryate) [390], poly-(hexadecylacrylate) [321], poly-(p-(p-nonyloxy)benzamidostyrene) [306], poly-(p-phenyleneterephthalamide) [141, 184, 205, 354a, 356b, 393], and other polymers [53a, 391, see also 273].

The reader interested in these problems can find wealth of information in the literature [41a, 41b] and in K. Solc's book "Order in polymer solutions" [356]. Solutions of poly-(hexadecyl-acrylate) in alcohols are smectic and nematic in paraffins. The low concentration of polymers (0.3 to 0.35%) necessary for the formation of mesophases is remarkable [321].

Anisotropic solutions of poly-(*p*-benzamide) in dimethylacetamide are of the nematic type [294] and consist of spherical rod-like aggregates as indicated by small-angle scattering of polarized light [174]. The structure formation during precipitation of poly-(*p*-benzamide) from anisotropic solutions deserves special interest [13, 173, 284], since the textile strength of poly-(*p*-benzamide) fibers is highest when they are prepared from anisotropic solutions [165, 282]. Russian authors report many other data on the anisotropic solutions of poly-(*p*-benz-amide), e. g., action of magnetic fields [178, 292, 294], IR-dichroism [291, 292], X-ray diffraction [292], domain formation [291], sedimentation by ultracentrifuging [296], viscosity and molecular weight distribution [170, 183, 283, 285], phase diagrams [285, 285a], flow orientation [294a], and surface tension [293]. Cholesteric polypeptide solutions are described in section 12.1.

11.8.2 Copolymers

The leading studies of the lyophases of block copolymers were by the French teams in Strasbourg [307–309, 327, 330–333, 336, 347, 349–353] and in Orléans [47–51, 137]. As early as 1961 A. Skoulios et al. proposed a structural model for the mesomorphous gel of a styrene/ethylene oxide copolymer [330], fig. 11.38. The structural parameters of these meso-phases as well as illustrations of the textures have been published [331, 336]. A. Skoulios himself gives a detailed review [354].

Fig. 11.38. Mesomorphous gels of copolymers. I, cylindrical structure observed in nitromethane; II, lamellar structure observed in butyl phthalate. In all cases, the styrene sequences (A) are amorphous. The ethylene oxide sequences (B) are crystalline in the lamellar but amorphous in the cylindrical structure (circles-solvents, S). Typical sizes are 8.8 nm cylinder diameter and 10.5 nm layer thickness (from [330]).

Mesomorphous gels also occur in ternary systems, e. g., the copolymer styrene/ethylene oxide/ethylbenzene/water [327]. C. Sadron has synthesized organized polymers in which the liquid-crystalline structure was pseudomorphically fixed by the polymerization of mesomorphous solutions of copolymers. A monomer was used as solvent in these syntheses [308]. A. Douy et al. determined the structure of the block mesophases in an A–B block copolymer in a preferential solvent of one block using small-angle X-ray scattering [48]. The characteristic parameters of the lamellar, hexagonal, and centered cubic mesophases are given in figs. 11.39, 11.40, and 11.41, respectively.

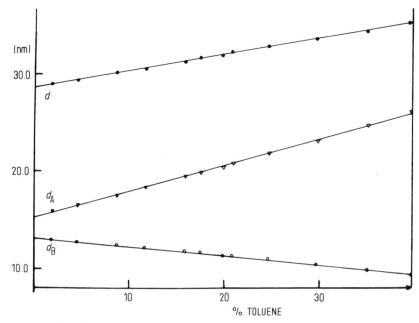

Fig. 11.39. Parameters of the lamellar mesophase vs. concentration of the preferential solvent. d: inter-layer spacing, d_A: thickness of the soluble layer, d_B: thickness of the insoluble layer. Solute: polystyrene-polyisoprene copolymer with the total mass 40 500 and 57% of weight polystyrene (from [48]).

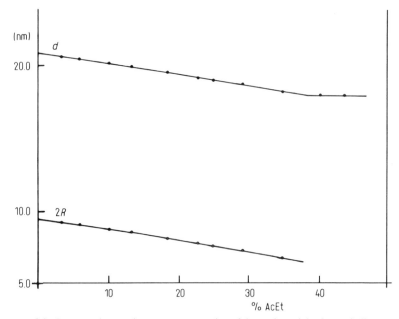

Fig. 11.40. Parameters of the hexagonal mesophase vs. concentration of the preferential solvent. d: distance between the axes of two neighboring cylinders, $2R$: diameter of the cylinders. Solute: polyisoprene-poly-2-vinyl-pyridine copolymer with the total mass 21 100 and 79% of weight polyisoprene (from [48]).

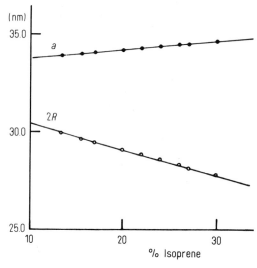

Fig. 11.41. Parameters of the centered cubic mesophase vs. concentration of the preferential solvent. *a*: side of the centered cubic cell, 2*R*: diameter of the spheres. Solute: polystyrene-polybutadiene copolymer with the total mass 117000 and 61% of weight polystyrene (from [48]).

The results of these investigations lead to the conclusion that the type of structure is determined mainly by the ratio of the volumes of the phase containing the insoluble sequence and the volume of the phase formed by the solution [310]. A. Douy and B. Gallot developed the following technique to compare the results of small angle X-ray scattering and electron microscopy on these mesophases [49]. First, by using small angle X-ray scattering, they ascertain the structure of the mesomorphous gel prepared with the monomer as a preferential solvent. Then the solvent is completely polymerized by UV-radiation. As shown by a subsequent X-ray examination, this leaves the structure intact with but minor changes in the parameters. Now the material is ready to be cut with an ultramicrotome and treated with osmium tetroxide vapor. It is probable through this method that an equilibrium state between solvent and solute is fixed, but no reproducible preparations of copolymers could be obtained by solvent evaporation. Table 11.2 contrasts the results.

Table 11.2. Structural parameters of A–B polystyrene/polybutadiene block copolymers as found experimentally by small angle X-ray scattering (XR) and by electron microscopy (EM). % PB = percentage of polybutadiene (from [49]).

% PB	XR (nm)	EM (nm)	Parameter
30.5	38.0	36.0	Distance between adjacent cylinder axes in the
71.7	66.3	71.0	hexagonal structure
30.5	20.5	18.0	Cylinder diameter
71.7	46.0	40.0	
39.8	36.5	37.5	Interlayer spacing in the lamellar structure
50.2	42.4	44.0	
39.8	24.7	24.0	Thickness of the (soluble) polystyrene layer
50.2	25.3	24.0	
39.8	11.8	13.5	Thickness of the (insoluble) polybutadiene layer
50.2	17.1	20.0	

Block copolymers of polystyrene and polybutadiene exhibit liquid-crystalline structures in preferential solvents of polystyrene for solvent concentrations smaller than 45%. A. Douy and B. R. Gallot have determined the following structures [49]:

Hexagonal structure if the copolymer contains less than 35% polybutadiene.

Lamellar structure if the copolymer contains between 35 and 60% polybutadiene.

Hexagonal structure if the copolymer contains more than 60% polybutadiene.

The two hexagonal structures are of entirely different natures. The cylinders are filled with the insoluble block in the case of copolymers containing less than 35% polybutadiene, but the cylinders are filled with solution in the copolymers containing more than 60% polybutadiene. This can be seen clearly in the electron micrographs since, due to its double bonds, only the polybutadiene is fixed by osmium tetroxide. Furthermore, it can be impressively visualized how a copolymer dissolved in styrene proceeds from dilute solutions with micellar structure through aggregates of various forms and sizes upon increasing the solute concentration to finally reach the liquid-crystalline order.

The same group has enlarged our knowledge by other details, e.g., the variation of the parameters of the lamellar structure with temperature, and solvent concentration [51]. They also produced the phase diagrams of the copolymer systems polystyrene/polybutadiene/toluene, polystyrene/polyethylene oxide/diethylphthalate [137], and polystyrene/polyethylene oxide/nitromethane. From these, the following generalization can be abstracted:

"Block copolymers possessing two amorphous blocks exhibit only one mesophase whose structure is of the lamellar type when lengths of the two blocks are comparable. Block copolymers with an amorphous block and a crystallizable block exhibit two mesophases; both with a lamellar structure but differing by the state of the crystallizable block. If the nature of the blocks determines the number of the mesophases, then the nature of the solvent only determines the domain of stability of the mesophases: the lamellar structure with one block in the crystallized state disappears at lower solvent concentrations and at lower temperature if the solvent used is a solvent of the crystallizable block than if it is a solvent of the amorphous block" [51].

In the normal type of mesophase comprising a copolymer and a preferential solvent, the insoluble sequences form the structural elements: lamellae, cylinders, or spherical units. These are embedded in a matrix formed by the solvent and the dissolved sequences. Recently reversed structures have also been found that show the solution inside the cylinders and the undissolved chains between the cylinders [353].

Furthermore, mesophases of the following copolymers are known in a "semidissolved" state: styrene/4-vinylpyridine block copolymer [349], styrene/2-vinylpyridine copolymer [352], 2-vinylpyridine/4-vinylpyridine copolymer [350], and methyl methacrylate/hexyl methacrylate copolymer [351]. Mesophases with lamellar or cylindrical structure, depending on the degree of neutralization, exist in the copolymer system maleic acid/hexadecylvinyl ether/water [347]. Copolymers comprising three sequences, e.g., polyethylene-oxide/polypropylene-oxide/polyethylene-oxide, are as capable of forming mesophases as those comprising only two sequences. A. Skoulios et al. found two lamellar structures (fig. 11.42) and a phase with spherical structure, probably face-centered cubic [332, 333].

The copolymer of the three block polybutadiene/polystyrene/polybutadiene forms solutions with lamellar and hexagonal structures [50]. Here might also be mentioned the monodisperse latexes that macroscopically indicate a two-dimensional perodicity of the array when the interference is studied in visible light as monolayer film [181]. Other studies on copolymer solutions are reported in the literature [179].

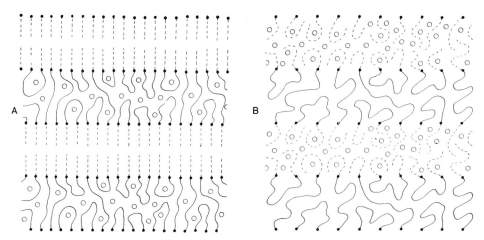

Fig. 11.42. Structures proposed for the two lamellar arrangements in the copolymer sequences of poly-ethylene oxide and polypropylene oxide: A, polyethylene-oxide sequences in the crystalline state and polypropylene-oxide sequences dissolved; B, both sequences unordered with polyethylene oxide dissolved and polypropylene oxide undissolved (from [332]).

11.9 Mesomorphous Solutions of Miscellaneous Types of Substances

Lyotropic mesomorphism is a domain of aliphatic compounds, but it is not limited to these substances. Many examples comprising aromatic compounds cannot be fitted into the system amphiphile/solvent/solubilizate. However, these mesophases have not been reexamined using modern methods, so their structures remain completely unknown. Only their existence has been stated using optical methods.

H. Sandquist recognized the mesomorphous character of aqueous 10-bromopenanthrene-3-(or 6-)sulfonic acid solutions [1:247, 311–313].

H. Zocher observed the birefringence of a solution of benzopurpurine upon addition of sodium chloride as well as the anisotropy of other gels and sols under the influence of electrolytes, electric fields, and mechanical deformation [433, 435, 436]. In the broader class of paracrystals H. Zocher's "tactosols" must be included. "Tactosols" separate from many colloidal solutions in such a way that the nonspherical particles spontaneously associate into zones in which they are parallel to each other [439, 442]. H. Zocher also prepared the way to the color-switching technique being of technical interest today by his investigations of the dichroism and fluorescence polarization of dyestuffs dissolved in mesophases [441]. There is no direct connection between mesomorphous systems and the observation of dichroism and birefringence in Ag/AgCl layers, the socalled photochloride [434].

To continue enumerating the lyophases of unknown structure, we should mention naphthylamine disulfonic acid derivatives [7], tartrazine [133, 134], methylene blue toluylene (neutral) red [135], 1,1'-diethyl-Ψ-cyanine chloride (concerning which the authorities are not in agreement) [167, 168, 315, 316, 320], K-methyl orange [20], naphthol yellow S [169], thiazine dyes [262, 263], 2-ethoxy-6,9-diaminoacridine lactate [274], and solutions of glyceric acid/Cd-phosphate [392]. Since many dyestuffs are encountered in this list, it has been supposed that the salting out of a dye from an aqueous solution begins with an aggregation that is retained as liquid-crystalline phase at low concentrations [16]. The structural parameters of all such mesophases remain unknown. Furthermore, there has not yet been a comment on the publications of I. A. Trapeznikov concerning the mesophases of the hydrates of higher aliphatic compounds [381–384]. N-1-naphthyl-N'-phenylenecarboxyalkyl ureas of the formula

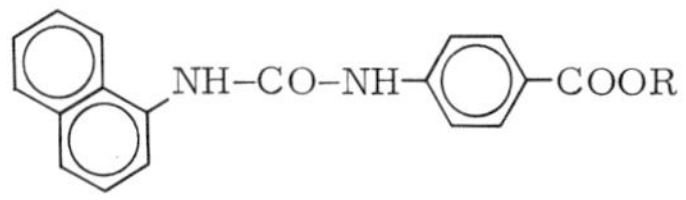

NH–CO–NH–⟨◯⟩–COOR

have been suggested as agents for protecting the skin from light. They exhibit the typical characteristics of lyophases in heptane solutions [158].

11.10 Technical Importance of Lyotropic Mesophases

Soaps, detergents, and emulsifiers commonly tend to form liquid-crystalline solutions. These phases play no part in washing processes per se, but they are of importance in the manufacture of these washing agents, e.g., in soap boiling. Occasionally mesomorphous solutions are marketed, e.g. as automobile or textile soaps, shampoos, and previously as lubricating greases [406, 408]. Heterogeneous distribution systems such as foams, emulsions, and suspensions may be much more stable in the mesomorphous state than they are in normal liquids.

11.10.1 Foams and "Black Foam Films"

Most lamellae drawn directly from a soap solution are too thick to exhibit interference colors, and must first be allowed to thin by gravity to a thickness between 0.1 and a few μm. Under favorable conditions the film will drain further and become so thin that reflection from the front side and the back side are very nearly in counterphase and the film looks black. R. Hooke (1672) and I. Newton have already observed black holes in soap bubbles often sharply delineated from the colored part. The principal forces leading to such thin films have been given much consideration, i.e. electrostatic repulsion and van der Waals attraction [160, 163, 280]. As seen from the energy profile shown in fig. 11.43, these films do not break spontaneously.

The thickness of socalled second black films is close to 4.3 nm and certainly varies over at most a 1 nm range as a function of ionic strength and ionic composition. All observations indicate that these "second black films" possess a well-defined molecular organization that shows some structural similarities to liquid crystals [125, 160].

"First black films" being formed spontaneously from colored films can vary considerably in thickness between 8 and more than 20 nm. By careful vaporization these can be converted to the above-mentioned second black films. Both films are in equilibrium when a certain critical concentration (analogous to the flocculation value for rapid flocculation) is attained [163]. Data on the equilibrium thickness and the disjoining pressure as a function of the film thickness have been published in the literature [33]. The equilibrium thickness of black foam films and the interlayer spacings of the neat phase rise symbathically as the salt concentration increases, and they fall together as the ionic strength is increased even further [12]. Cholesterol black lipid membranes lead toward biological problems [2].

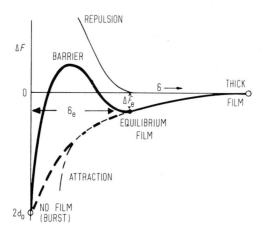

Fig. 11.43. Changes in the partial specific surface free energy ΔF with film thickness δ. A (meta) stable film forms if a sufficient minimum results from a combination of attractive and repulsive potentials (from [160]).

While lyotropic mesomorphism, black films, and the stability of foams are certainly closely related, detailed predictions are still speculative [118, 301]. Enhanced stability was observed only when a liquid-crystalline solution was combined with the aqueous isotropic phase [122, 130a]. Alcohols of medium chain length can act both as foam stabilizers and foam breakers [122]. Aerosol foams were also studied [124]. A recommended introduction to the basic problem of stability in droplets, bubbles, and lamellae is K. L. Wolf's booklet "Tropfen, Blasen und Lamellen" [1:399].

11.10.2 Emulsions

A. S. C. Lawrence [195] and S. Friberg [109–113, 115, 116, 119, 120, 123, 127, 128] have many times proven the stabilizing influence of mesophases on emulsions. S. Friberg has also written a detailed review [130]. Emulsions in the three-phase zone $L_1 + L_2 + C$ tend to separate much less than those in the two-phase zone $L_1 + L_2$ [110], fig. 11.44. The stability and the rheological behavior of emulsions change abruptly when a mesophase is formed [112,

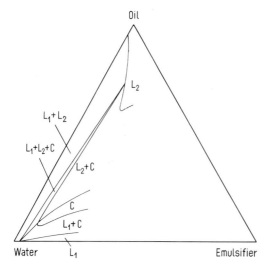

Fig. 11.44. Phase diagram of a system water/oil/emulsifier: L_1, water solution of emulsifier; L_2, oil solution of emulsifier; C, mesophase (from [110]).

113], but this does not appear to depend on the means by which the mesophase is created [115]. In the system 1-aminooctane/1-octanoic acid/p-xylene/water, it is not the amine-acid compound stabilizing the emulsions, but rather the mesophase [116]. It can be shown microscopically that the formation of liquid-crystalline layers on the surface of the emulsified droplets is responsible for the stabilization. An estimation of the reduction of the interaction energy between two emulsified drops covered by a liquid crystal layer shows that the thickness of the covering liquid crystal layer is less important than the ratio of the Hamaker interaction constants [119, 155].

Other studies on emulsion stability in relation to lyotropic mesomorphism are reported in the literature [3a, 25, 96, 144, 145, 182, 190, 385].

11.10.3 Suspensions

A dispersion stabilized by liquid crystals is very sensitive to salts, as the following experiment of S. Friberg shows [121]: A suspension of 10 g/l kaolin containing 4.5% by weight of sodium caprylate settles within a few minutes. Addition of 1% caprylic acid produces a suspension that is stable for months. It finally settles as large flocs which can mechanically be redispersed. This stable suspension contains a phase B liquid crystal (the "lamellar mucous woven" phase in Ekwall's nomenclature). This disappears upon addition of 1 mol/l NaCl solution, and the suspension immediately begins to precipitate.

Liquid Crystals in Living Systems

12

12 Liquid Crystals in Living Systems

All living things exhibiting optical anisotropy in their organic tissues contain either lyophases or pseudomorphs of previously lyotropic states. This entire chapter has been devoted to them because of their special importance. Since the first observations of R. Virchow [1:380] and C. Mettenheimer [1:275, 266] there has been a steady growth in the interest of biochemists in liquid crystals. Of the earlier investigators, some of whom have already been mentioned in chapter 1, the following should be noted here: O. Lehmann [1:136, 1:151, 1:158, 1:159, 1:162, 1:165–1:174, 1:178–1:184, 1:187, 1:188, 1:190, 1:191, 1:197, 1:199, 1:219, 1:230, 1:232, 1:251, 1:253, 1:257, 1:258], G. Quincke [326, 327], E. Sommerfeldt [385], S. S. Chalatov [58, 59], O. Rosenheim [343], F. O. Schmitt [368], F. Rinne [332–336], A. L. von Muralt and J. T. Edsall [448, 449]. Several summaries have been published [10, 15, 37, 42, 73, 77, 77a, 122, 131, 196, 197a, 215, 238, 276, 383a, 400, 404, 415, 438] and W. J. Schmidt's classical review gives the best survey of older investigations with numerous citations of liquid crystalline phases [367a]. An introduction must be omitted here, but the texts on biochemistry that give an introduction are legion. A few are cited here [231, 253, 254, 287, 401]. The most recent publication is the book "Liquid Crystals and Biological Structures" by G. H. Brown and J. J. Wolken (Academic Press: New York). 1979. 200 pp.

12.1 Polypeptide Model Substances

One of the most thoroughly investigated polypeptides is poly-γ-benzyl-L-glutamate (PBLG), which has often served as a synthetic model substance for lyophases of biological materials. This polypeptide has the following structure

$$\left[\begin{matrix} O=C \\ \dot{H}C-(CH_2)_2-COOCH_2C_6H_5 \\ HN \end{matrix} \right]_n$$

Lowering the molecular weight widens the liquid crystal-isotropic miscibility gap in fig. 12.1 shifting it to higher polymer concentrations, and it simultaneously lowers the liquid-crystal critical solution temperature [272, 273].

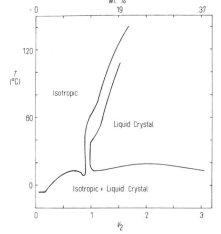

Fig. 12.1. Phase diagram (temperature, T, vs. volume fraction, V_2) of the system PBLG (molecular weight 310,000)/dimethylformamide (from [272]).

Behavior similar to that of PBLG is exhibited by the polypeptides poly-γ-methyl-L-glutamate, poly-γ-ethyl-L-glutamate, poly-β-benzyl-L-aspartate, ε-carbobenzoyloxy-L-lysine, and related compounds [7:244, 47, 187–189, 194a, 278, 286, 290, 293, 342, 363, 416, 417, 427]. Detailed reviews [193, 360] and brief reviews [432] have appeared. These compounds display in solution numerous mesophases with cholesteric, hexagonal, complex hexagonal, quadratic, and lamellar structures [245, 320]. Thus, in the PBLG/pyridine system there is a cholesteric phase and a complex hexagonal one [245].

Of all of the various mesophases, the cholesteric deserves the greatest attention. The PBLG molecules exist as an α-helix in this phase as well as in the solid state [156, 199, 308]. The cylindrical molecules are about 1.8 nm in diameter and several hundred nm long. Their dipole moment is about 1930 D in *trans*-1,1-dichloroethylene or 1410 D in dioxane [156]. Two successive residues that are projected into the helical axis repeat at intervals of approximately 0.15 nm [199]. The side chains, which project from the helix in a regular pattern, provide the lateral adherence. The elementary cell of a 70–80% PBLG solution in dimethylformamide is remarkably symmetrical having the parameters $a=b=3.003$ nm, $c=6.414$ nm, and $\alpha=\beta=\gamma=90°$ [386]. Other data is given in the literature [386, 387, 389]. Supermolecular PBLG structures, including low-molecular-weight PBLG, have been investigated in very different solvent types [19, 118, 320, 392]. High-molecular-weight PBLG has also been synthesized [30], and several authors have studied the viscosity of PBLG solutions [101, 169, 170, 188, 273, 319, 364, 364d].

Observations of the helix-coil transformation in dilute PBLG solutions show that it is a priori impossible to decide whether the helical or randomly coiled configuration is stable at high temperature [205, see also 364a, 364d]. The Moffitt parameter can be obtained from ORD measurements and it can be used to show that the high temperature form of PBLG (with positive $[\alpha]_D$) corresponds to the helical conformation of the polypeptide (figs. 12.2 and 12.3).

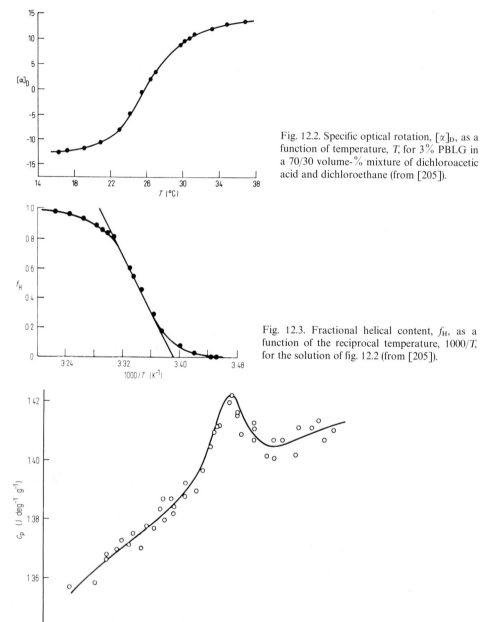

Fig. 12.2. Specific optical rotation, $[\alpha]_D$, as a function of temperature, T, for 3% PBLG in a 70/30 volume-% mixture of dichloroacetic acid and dichloroethane (from [205]).

Fig. 12.3. Fractional helical content, f_H, as a function of the reciprocal temperature, $1000/T$, for the solution of fig. 12.2 (from [205]).

Fig. 12.4. Specific heat, c_p, vs. temperature, T, measured in a solution containing 20.02 mg PBLG per gram of solution; solvent as in fig. 12.2 (from [205]).

The heat of transformation in the same system was also measured, (fig. 12.4) and the behavior of the deuterated compound considered in the paper [205].

The cholesteric solutions of PBLG show a pattern of symbathic lines with a periodicity, s, of about $10\,\mu m$. These lines can have an approximately parallel or spiral arrangement, but only one direction of spiral is present. The texture also shows a high degree of light scattering over a limited range of wavelengths, reaching a maximum at the wavelength where the optical rotation changes sign [338]. Periods of 2 to $100\,\mu m$ were observed microscopically. This has been found to be independent of the shape of the vessel containing the solution, the thickness of the observed solution, and the wavelengths of the incident light [342]. This texture is the lyotropic analog to the fingerprint texture of thermotropic cholesterics. In addition to the fingerprint texture observed perpendicular to the helical axis, textureless zones appear parallel to it corresponding to the Grandjean texture. Microscopic observations on PBLG solutions involving a longer time have also been published [220] as well as studies of the influence of temperature and pressure [260]. Other morphological observations are reported [136].

X-ray diffraction data indicates a structure in which the twist is superimposed on a hexagonal array of rods (in the α-helix conformation) that deviate only slightly from parallel orientation [137, 340]*. X-ray studies deduced a remarkably high degree of order [364b].

The optical rotation of PBLG solutions obeys the theoretical predictions of the de Vries theory as shown in fig. 12.5 [339–342, cf. p. 301].

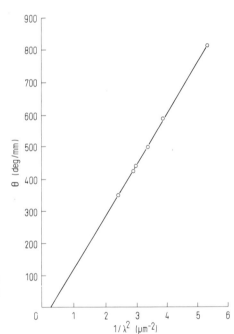

Fig. 12.5. Optical rotation, θ, vs. the wavelength (plotted as $1/\lambda^2$) for PBLG in chloroform, 18 g/100 g, showing the periodicity $s = 25\,\mu m$ (from [341]).

---

* This means that there are two helical axes at right angles, that of the polymer molecules and that of the supermolecular structure. The pitch and the considerations always refer to the latter in this case.

Actually, PBLG forms not only solutions with either twist, e.g., negative in dioxane and positive in methylene chloride, but it also forms an inactive (compensated) nematic solution in mixtures of proper proportions [47, 341], fig. 12.6.

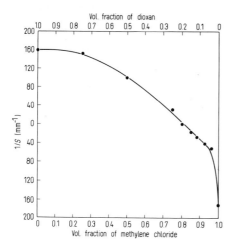

Fig. 12.6. Periodicity, $s = p/2$, vs. solvent composition for PBLG in mixtures of methylene chloride and dioxane (from [341]).

Another inactive, purely nematic solution is obtained by mixing equal parts of PBLG and PBDG. The measured birefringence is 0.026 while that calculated from the de Vries equation is 0.025. From the de Vries equation, it is seen that the optical rotation, θ, becomes infinite at $\lambda^2/p^2n^2 = 1$ and also changes sign in this area. Colored light is reflected in obedience to the Bragg equation. Solutions of poly-γ-ethyl-L-glutamate in ethyl acetate have their pitch near the wavelength of visible light and behave analogously to thermotropic cholesterics. For perpendicular incidence, θ becomes infinite at $\lambda = 532$ nm [397]. Incident light of approximately this wavelength is split up into two circularly polarized components of which one is reflected while the other is transmitted. The dispersion of the optical rotation can be described using the two-term equation proposed by W. Moffitt [31, 102, 103, 279]. This implies opposite twist for the PBLG and the poly-β-benzyl-L-aspartate helices. The competing influences of the opposing helices have been investigated, especially in copolymers [38, 39].

Copolymers with non-polypeptide sequences were also synthesized [311]. These copolymers also indicate interesting parallels with thermotropic systems, and experiments on fluorescence polarization are being given increased attention [148].

Theoretical calculations and some experimental results show that the optical rotation has a considerable influence on the light-scattering patterns of cholesteric PBLG solutions [316, 373]. Quasielastic-light-scattering spectroscopy revealed two new purely dissipative modes which are probably due to the theoretically predicted twist and viscous splay modes [104, 106b, 108, 110, see also 407b]. The IR dichroism was described in some of the earliest papers [6, 74, see also 192, 447e]. CD spectra were measured in the visible and UV range with aromatic solutes [190, 426]. No definite explanation can be offered to account for the distinct difference in LCICD observed with anthracene in a lyotropic (PBLG/dioxane) and a thermotropic (cholesteryl chloride/cholesteryl nonanoate) cholesteric mesophase [351, see also 290] and section 7.2.

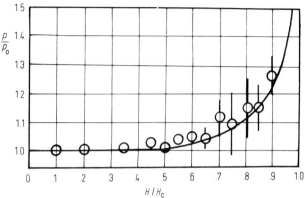

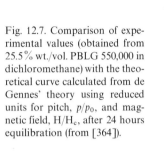

Fig. 12.7. Comparison of experimental values (obtained from 25.5% wt./vol. PBLG 550,000 in dichloromethane) with the theoretical curve calculated from de Gennes' theory using reduced units for pitch, p/p_0, and magnetic field, H/H_c, after 24 hours equilibration (from [364]).

Many papers are concerned with the behavior in magnetic and electric fields [16, 105, 106, 107, 109, 111, 139, 183, 186, 187, 189, 200, 299, 363, 364, 364a, 364d, 403, 414, 416, 417, 419, 420, 433].

Cholesteric PBLG solutions as well as thermotropic cholesterics can be transformed into the nematic state (fig. 12.7), and the twist elastic constant can thus be determined [364d]. The axes of the molecular helices orient themselves parallel to the field and the supermolecular helix is untwisted.

It is difficult to abstract common features from the various observations in magnetic and electric fields, but it can be stated that "the extreme differences in the hydrodynamic shape of the constituent molecules in the polypeptide liquid crystals (length to breadth 100) as compared to that of thermotropic systems (with this ratio about 2 to 3) must be responsible for the striking differences in the response times for these two classes of liquid crystals" [364]. PBLG molecules are oriented by an electric field even as weak as 84 V/cm. The presence of molecular clusters is indicated by the high dipole moment and by the light-scattering patterns [186, see also 191]. The IR dichroitic ratio and the birefringence of PBLG indicate that complete order is not attained in an electric field, although the very small current in liquid-crystal solutions seems to have no effect on order [182].

Thorough investigations of the Kerr effect in PBLG solutions revealed different concentration dependencies between highly dielectric solvents (e. g., ethylene chloride) and in low dielectric solvents (e. g., benzene, dioxane) [319, 428, 429, 431]. This is explained by assuming two different modes of aggregation, one linear and the other antiparallel [318], fig. 12.8. As soon as more than 12% PBLG is dissolved in methylene chloride, the solvent NMR spectrum singlet splits into a doublet because the liquid-crystalline matrix restricts the rotation of the solvent molecules such that the direct dipole-dipole interaction between the solvent molecules is no longer compensated to zero. The splitting of the doublet increases with concentration and also, at the onset, with the length of time the magnetic field has been applied. This continues until a nematic phase is formed. Due to the high viscosity, it could be shown that the doublet splitting in the parallel-equilibrium state is twice as great as when there is a 90° angle between the field direction and the helical axes [354]. The degree of orientation for nonspherical molecules in thermotropics is on the order of 0.1, and that found in PBLG solutions is one or two orders of magnitude smaller. Therefore, the dipolar proton and quadrupolar [14]NMR splittings can be seen even in nitromethane, acetonitrile, and nitrobenzene-1,3,5-D$_3$

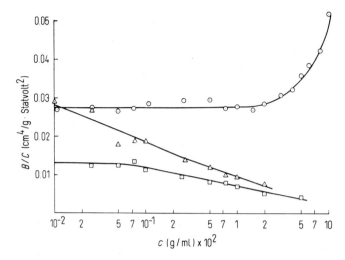

Fig. 12.8. Specific Kerr constant, B/C, vs. concentration, $\log c$, for PBLG 67,000 in ⊙ ethylene dichloride, ☐ dioxane, and △ benzene (from [318]).

[144]. Numerous other publications have also studied NMR spectroscopy in solutions of PBLG [126, 127, 139, 140, 145, 147, 184, 299, 359, 361, 362, 407a, 447d, 466a].

The literature contains further data on flow birefringence [428, 429, 431], surface films of PBLG [237], molecular interaction [364c], and thermodynamic and dynamic properties [274].

[1]HMR-spectra and X-ray studies in the cholesteric solution show no discontinuity on passing to the solid state to suggest that the local structure in each state must be quite similar [354]. Various authors have observed that the cholesteric structure is retained in solid films cast from PBLG solutions [352, 353, 357], especially when the material is plasticized with 35 volume-percent 3,3'-dimethylbiphenyl [135]. This method applied to PBLG solutions under a strong magnetic field yields nematic films which are uniaxially oriented and have a slightly distorted α-helix [355, 356, 358]. The small-angle light scattering of such films has been investigated [455]. Solid liquid-crystal films are also known from poly-γ-methyl-L-glutamate [406, see also 195a, 425] and can also be used to study induced CD. Liquid membranes composed of cholesteric solutions of poly-γ-methyl-D-glutamate have shown identical permeation coefficients for the D-, L-, and DL-isomers of mandelic acid [280].

12.2 Cholesteric Pseudomorphosis in Crustacean and Insect Cuticles

An uninterrupted connection of the following text with the previous description of polypeptide model compounds would require the synthesis of a polymeric "solid liquid crystal" film reflecting visible light, but this seems to be merely a question of time (cf. H. Ringsdorf et al., [9:147h, 9:147i]).

If a collection of beetles is examined with a polarizer absorbing left-circularly polarized light, the colors of many species disappear and the insects then appear black. A. A. Michelson in his 1911 work on the metallic colouring of birds and insects first called attention to the reflection of circularly polarized light shown by the beetle Plusiotis resplendens in ordinary daylight [268]. His conclusion that a helical aggregation of probably molecular ultramicroscopic particles must be responsible for this behavior has been confirmed. In 1924 P. Gaubert recognized the connection with nematic-cholesteric phases [1:74]. As with optically active mesophases, the cuticle color does not disappear at any position when the specimen is rotated and observed under crossed Nicols. P. Gaubert correctly characterized the reflected light as left-cirularly polarized, and he also mentioned that right-circularly polarized light is occasionally, but very rarely, found.

    Distinct progress was signaled by J. P. Mathieu and N. Ferragi in their development of a technique enabling them to study the light transmission through very thin preparations [263]. On approaching the wavelength where the sense of rotation changes sign, an insect preparation, as typical of any cholesteric material, reflects light being circularly polarized in the direction opposite to that of the transmitted light. Until very recently, little new work was performed to expand this area of knowledge.

    While those insect colors due to scaled platelets disappear or change abruptly upon immersion, beetles showing cholesteric reflection colors become even more vivid when immersed in a medium having an index of refraction close to that of chitin, which is 1.5. The color-producing layer is probably created in the late chrysalis stage from a liquid-crystalline glandular secretion which hardens quickly into a pseudomorph on the surface. The reflecting layer, having a helical structure of molecules of unknown chemical constitution lies in the outer $5-20\,\mu m$ of the exocuticle. According to Bragg equation the reflection color is an interference color due to diffraction from molecular layers having equal orientation within the helical structure. Since the distance between the scattering planes can be determined by electron microscopy, and the index of refraction ($n=1.525$ perpendicular to the surface) is known, it is possible to compare the values of λ calculated from the Bragg equation with the experimentally determined value λ_0 obtained from the dispersion curve. A. C. Neville and S. Caveney found a sufficiently good agreement between the two values to give the theory plausible support, but the discrepancies are large enough to require an explanation [288]. In addition, one must consider the simplification introduced by inserting constant, averaged values for pitch and index of refraction in the equation since in natural samples these parameters vary with layer depth, e.g., $535<\lambda_0<586$ or $440<\lambda_0<507$ nm. Mixed colors such as bronze are created by increasing the gradient of the layer. The left-circular polarization of reflected light can be considered normal but not universal. The dispersion curve for short-wave light shows negative rotatory power, while it is positive for longer wavelengths, fig. 12.9. Only one beetle, Plusiotis optima, is contrary.

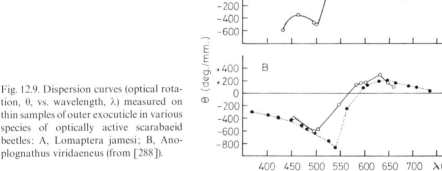

Fig. 12.9. Dispersion curves (optical rotation, θ, vs. wavelength, λ) measured on thin samples of outer exocuticle in various species of optically active scarabaeid beetles: A, Lomaptera jamesi; B, Anoplognathus viridaeneus (from [288]).

Natural samples support the predictions of the extended de Vries theory which states that there should exist zero rotation at two wavelengths (fig. 12.9). The agreement of the measured dispersion curves with those derived from the theories of de Vries and of Chandrasekhar and Rao (fig. 12.10) may be considered additional proof of the cholesteric helical structure in the exocuticle of beetles.

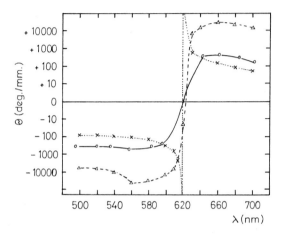

Fig. 12.10. Dispersion curve (optical rotation, θ, vs. wavelength, λ) measured on Potosia speciosissima outer exocuticle (o————o) compared with the theoretical curves calculated according to Chandrasekhar and Rao: △———△ and according to de Vries: ×······× (from [288]).

In Lomaptera, the interference color and thus the pitch of the helix is concurrent with the bilateral body symmetry, but the beetle becomes asymmetric with regard to the sense of rotation of the helix which is the same on the left and the right side of the animal. This suggests that the sense of rotation is determined by a self-orientation that is independent of the cell and universally dependent only on the chirality of the molecules.

S. Caveney found a new variant of cuticle reflectivity in Plusiotis resplendens [57]. The reflecting layer of this ruteline scarab beetle has an anticlockwise helicoidal architecture and interferes with only the left circularly polarized component of incident light of visible wavelengths. Below this first reflecting layer there is an untwisted layer of 1.8 μm thickness which functions as a perfect halfwave plate at 590 nm. Transmitted right-circularly polarized light, which is transformed into left-circularly polarized light in the halfwave layer, is then reflected from a second helicoidal layer, passed through the halfwave layer again to be retransformed into right-circularly polarized light, passed again through the outer reflecting layer, and is thus reflected as well. This means that the reflectivity of Plusiotis resplendens is greater than in species that do not possess a halfwave plate. In addition, the cuticle in any Anoplognathus species contains uric acid. This increases the birefringence of the system by a factor of five and causes the cuticle to have twenty times greater optical rotatory power compared with reflecting layers lacking this component [57].

Using electron microscopy Y. Bouligand succeeded in the visualization of a twisted arrangement in the cuticle of the beetle Cetonia which reflects circularly polarized light [1:17]. Plusiotis gloriosa has been analogously investigated [296]. A helical structure also arises by self-assembly of the protein in gland lumen of praying mantids (Sphodromatins tenuidentata, Miomantis monacha), and this has also been proven by electron microscopy [289].

The paracrystallinity exhibited by collagen, the chief protein of hide, tendon, cartilage, bone, and teeth, has a periodicity of 64 nm [1]. Collagen is regarded as a biological analog to smectics A showing a layer thickness which is determined by the amino acid sequence [180a]. It may transform to a chiral smectic C.

Y. Bouligand describes great variety of twisted biological materials and details their arrangement and dislocations using electron and optical microscopy [35, 36].

12.3 *Binary Aqueous Mesophases of Lipids*

As an introduction, we present the structures of the substances under consideration here. While these do not appear dissolved in pure form in human or animal tissues, they compose many complex systems, e.g., secretions or membranes whose state may be considered liquid-crystalline.

Monoglycerides	$RCOOCH_2$ $CHOH$ CH_2OH
Phospholipids Lysolecithins	$RCOOCH_2$ $CHOH$ $CH_2OPO_2^{\ominus}OCH_2CH_2\overset{\oplus}{N}(CH_3)_3$
Lecithins (Phosphatidyl- cholines)	$RCOOCH_2$ $R'COOCH$ $CH_2OPO_2^{\ominus}OCH_2CH_2\overset{\oplus}{N}(CH_3)_3$
Phosphatidyl- ethanolamines	$RCOOCH_2$ $R'COOCH$ $CH_2OPO_2^{\ominus}OCH_2CH_2\overset{\oplus}{N}H_3$
Phosphatidyl- serines	$RCOOCH_2$ $R'COOCH \quad \overset{\oplus}{N}H_3$ $CH_2OPO_2^{\ominus}OCH_2CH-COOH$
Sphingomyelins	$CH_3(CH_2)_{12}CH=CH-CH-CH-CH_2OPO_2^{\ominus}OCH_2CH_2\overset{\oplus}{N}(CH_3)_3$ $\qquad\qquad\qquad\quad OH \quad NHCOR$
Cerebrosides	CH_2OH $\quad OH$ $O-CH-CH-CH=CH-(CH_2)_{12}CH_3$ $\quad NHCOR$

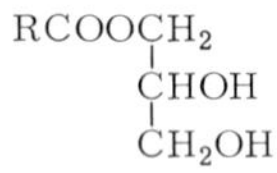

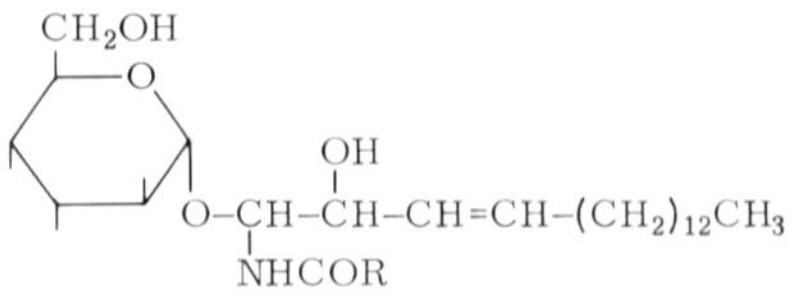

Speaking of lipids in the narrower sense, we can regard the naturally occurring amphiphiles not belonging to a single class of compounds as is reflected by the given structures. The extent to which lipids interact with water is crucial to the organization, structure, and function of the living cell.

As a result of X-ray analysis, F. Reiss-Husson recognized the following mesomorphous lipid structures which are all known from the lyotropic systems discussed in chapter 11 [331, see also 67, 75, 245, 246]:

a) lamellar structure in the systems
water/glycerine-1-decanoate (25°C),
water/glycerine-1-dodecanoate (45°C),
water/sphingomyelin (45°C),
water/cerebrosides (72°C),
water/phosphatidylethanolamine (20°C)
water/egg lecithin (5 to 35°C), and
anhydrous lecithin at high temperatures;
b) hexagonal structure in the system water/lysolecithin (35°C);
c) reversed hexagonal structure in the system water/phosphatidylethanolamine (55°C);
d) cubic structure in anhydrous lecithin at high temperature.

The reader is referred to reviews [69, 228, 229] and to R. C. Waldbillig's dissertation describing the mesophases of the phospholipid/water systems [450].

Cerebrosides form also aqueous liquid crystals and myelin forms [269, 371].

12.3.1 Monoglycerides

Di- and triglycerides do not exhibit mesomorphism alone or with water, but appreciable amounts can be solubilized in monoglyceride/water mesophases [226, 370], and mesophases also appear in binary and ternary systems containing cholesteryl esters and triglycerides [240a]. The 1-monoglycerides can dissolve up to 50% water and form lamellar, hexagonal, or cubic mesophases. The neat phase has been found in all systems and is the only mesophase at lower chain lengths, e.g., monolaurine. With longer chain lengths, viscous isotropic and then finally middle phases appear, as shown in fig. 12.11 [242]. The phase diagrams for systems of unsaturated compounds resemble those for systems of saturated compounds, but corresponding phase regions occur at lower temperatures.

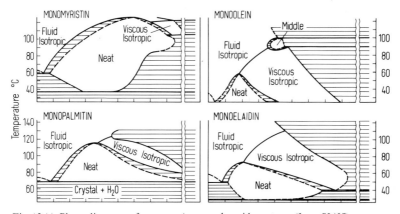

Fig. 12.11. Phase diagrams of aqueous 1-monoglyceride systems (from [242]).

The literature reports microscopic observations [367], IR investigations [252], and thermodynamic data [78]. Bis-monoglycerides of the structure CH₂OH—CHOH—CH₂O—CO—(CH₂)ₙ—COOCH₂—CHOH—CH₂OH form a neat phase with water [227].

12.3.2 Phospholipids

One normally finds a saturated fatty-acid radical in the 3-position (e. g., stearyl, palmitoyl, myristoyl) and an unsaturated one in the 2-position (e. g., oleyl). Phospholipids exhibit thermotropic and lyotropic mesomorphism. Highly unsaturated phospholipids such as egg-yolk lecithin swell to myelin forms in water, and this occurs even at room temperature. Completely saturated phospholipids, however, require higher temperature to do so [64]. D. M. Small describes most precisely the system of egg-lecithin/water [378, see also 151], fig. 12.12.

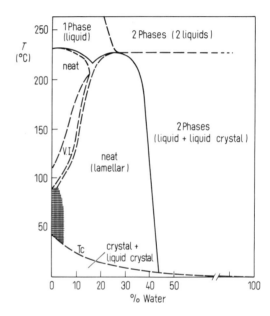

Fig. 12.12. Swelling of pure egg-lecithin. in water: the binary phase diagram as a function of temperature. V. I. is the viscous isotropic phase (face-centered cubic). Tc is the ill-defined boundary of the crystal/mesophase transition. The hatched area between 0 and 5% water from 45 to 90°C is the poorly-defined zone in which the lamellar liquid-crystalline phase may coexist with another mesophase (from [378]).

The long X-ray spacing of the lamellar phase in the system lecithin/water increases gradually with increasing water content to 6.5 nm at 45% water, after which it remains constant (fig. 12.13).

From measurements of the diffusion coefficient and electric conductivity, R. T. Roberts and G. P. Jones conclude that lecithin in solutions up to 5% forms spherical micelles that change into cylindrical micelles at a second critical micelle concentration (5–10%), as does potassium oleate [337]. Dynamic phenomena in lipid mesophases are reviewed in the literature [11:265].

Pure egg-lecithin at 40°C is in a form that is at least partially crystalline, and it transforms into a waxlike phase at about 80°C. It exists in a face-centered cubic lattice between 88 and 109°C, and it exists as a lamellar neat phase above this and up to the melting point at 231°C. These mesophases also exist in mixtures with water in the ranges indicated in the phase diagram [378], fig. 12.12. The lamellar mesophase comprising lecithin and water can incorporate a foreign amphiphile [132]. The transition temperature T_1 approaches a limiting

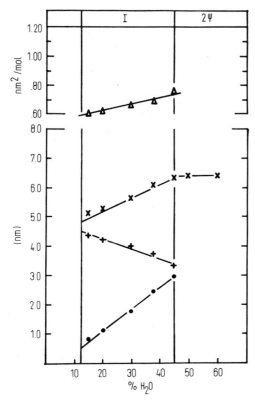

Fig. 12.13. Dimensions of the structural elements of lamellar phase of the lecithin/water system. △, average surface area per molecule; ×, the repeat distance; +, the calculated thickness of the lipid layer; ●, the calculated thickness of the water layer; I, one homogeneous phase, type lamellar; 2ψ, two phases (from [377]).

value T_1^*, the minimum temperature required for the water to penetrate between the layers of lipid molecules [65]. This is shown even more distinctly than fig. 12.12 indicates by the system 1,2-dipalmitoyl-L-phosphatidylcholine/water (fig. 12.14).

A similar situation occurs with the soaps where the ability to disperse in water occurs only above a temperature, referred to as the Krafft temperature, which is nearly constant over a wide range of concentration (cf. p. 525). The lamellar mesophase cannot take up more than ~40% water (figs. 12.12 and 12.14).

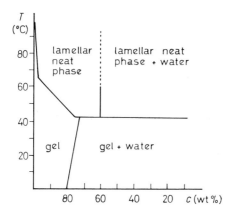

Fig. 12.14. Phase diagram (temperature, T, vs. concentration, c) of the system 1,2-dipalmitoyl-L-phosphatidylcholine/water (from [65]).

An abrupt increase in volume ($\Delta V/V = 1.4\%$) occurs in the system dipalmitoylphosphatidyl-choline/water at 44 °C, the transition temperature from the crystalline to the liquid-crystalline phase [422]. This value is about one order of magnitude smaller than a typical value for the melting dilatation of hydrocarbons.

Lysolecithin, which contains only one hydrocarbon chain (not two) for each polar group in the molecule, behaves unusually by forming only a hexagonal mesophase with water, but it does not form a lamellar one [67].

The question of the conformation and orientation of the "head groups" in the mesophases of phospholipids is still unanswered.

Measurements of the relative intensity of the methylene deformation vibration band at 1470 cm^{-1} gives an insight into the hydrocarbon chain fluidity as has been shown in the ternary system lecithin/phosphatidylserine/water [50–52]. In the temperature range of 110 to 120 °C, the remaining fine structure in the IR spectrum of anhydrous DL-α-dipalmitoylethanol-amine disappears and the spectrum resembles that of a liquid rather than a solid, but the substance does not melt until about 194–195 °C [60]. This "melting" occurs at lower temperatures for lipids having unsaturated chains than for those with saturated ones.

NMR spectroscopy, X-ray diffraction, and DTA indicate the (thermotropic) formation of a liquid-crystalline phase many tens of degrees below the published capillary melting points [61, 62]. Fig. 12.15 compares the results of various investigation methods [63].

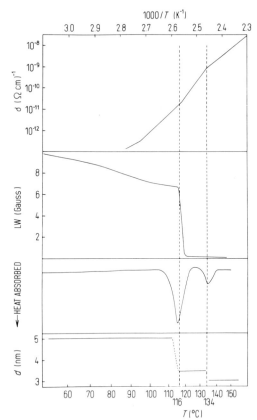

Fig. 12.15. Discontinuities in the curves of d.c. conductivity, σ, NMR line-width, LW, DTA measurements, and X-ray analysis: long spacing, *d*, for 1,2-dimyristoyl-DL-phos-phatidylethanolamine vs. temperature (from [63]).

Attention is also called to the sharp heat absorption in the DTA curve for lecithin at the gel/mesophase transition and the heat absorption preceeding this transition [174, see also 264], fig. 12.16.

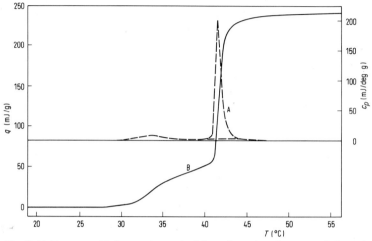

Fig. 12.16. Excess specific heat, c_p (curve A, right ordinates), and excess enthalpy, q (curve B, left ordinates), vs. temperature, T, during the gel/mesophase transition of dipalmitoyl-L-α-lecithin in aqueous suspension at 3.88 mg/ml lipid concentration (from [174]).

Other calorimetric studies, including phospholipid-mixtures are reported in the literature [20a, 32, 137a, 445].

Electrolytes have little influence on the thickness of the bilayer in aqueous lecithin mesophases. With LiCl, NaCl, Na_2SO_4, KCl, and CsCl, the bilayer thickness is less than with pure water [150], fig. 12.17. The maximum reduction in bilayer thickness with these

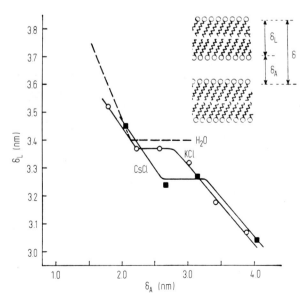

Fig. 12.17. Bilayer thickness, δ_L, vs. distance between the bilayers, δ_A: $\circ$ = KCl solutions; $\blacksquare$ = CsCl solutions of 1-octadec-9-enyl-2-hexadecylglycerophosphocholine (from [150]).

electrolytes is about 10% and occurs with mesophases of a high content of KCl and CsCl solutions. The bilayer thickness with HCl solutions is about 5% greater than with pure water, and with CaCl₂ solutions the bilayer thickness are about the same as with pure water.

An estimation of ion binding constants in aqueous dimyristoyl lecithin mesophases could not find evidence of binding with NaCl and KCl [153].

The lamellar phase and the type II hexagonal phase can definitely be recognized in the system phosphatidylethanolamine/water [202].

S. Eins has used electron microscopy (fixation with OsO_4, dehydration with acetone, and embedding in Epon®) to determine the range of existence of the lamellar phase in the system phosphatidylserine/water. He thereby excluded the formation of a hexagonal mesophase [115]. This method was also applied to phospholipids of microorganisms [116], for the detection of defect structures in lamellar phospholipid phases [213b] and the hexagonal structure of phospholipids [11:358].

Substantial differences were found in the electrophoretic mobility of phospholipids in liquid-crystalline phases (fig. 12.18).

Fig. 12.18. Electrophoretic mobility, Δl, vs. p_H for phospholipid mesophase particles. The bulk phase contained 145 mMol NaCl and the p_H was adjusted with HCl or NaOH. ● = lecithin (PC), ○ = phosphatidylethanolamine (PE), △ = 10% phosphatidic acid in PC, □ = phosphatidylserine (PS), ■ = 20% PS in PC (from [302]).

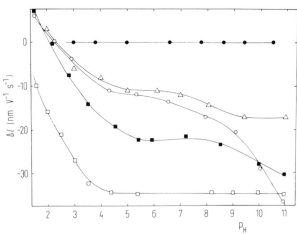

A calculation of free energy in the lamellar lecithin/water bilayer system has been suggested [82, see also 275]. The CD spectrum of aqueous sphingomyelin shows a negative peak at 210 nm which is probably due to the amide chromophor, and it also shows a positive peak at 290 nm. The rotational strengths of both transitions decrease suddenly in certain temperature ranges [210]. Other special studies concern the preparation of mesomorphous phospholipid "monocrystals" [321], defects and textures [213], osmotic properties [18], rheological properties [106a], and partition coefficients [172].

12.4 Classification of Biological Lipids

D. M. Small, in his basic work, proposed the following classification of lipids in the broader sense based upon their interaction with water [379]:

1. Nonpolar lipids
Substances insoluble in water and not forming monomolecular layers on the water surface, e.g., carotene, squalene. As a class, rare in living organisms.

2. Polar lipids

a) Insoluble, nonswelling lipids. These substances are insoluble and do not swell in water, but they possess a surface solubility as shown by the formation of stable monolayers upon water. Compounds in this class comprise the majority of lipids in higher animals, e.g., di- and triglycerides, long-chain protonated fatty acids, waxes, sterol esters (and many other sterols), long-chain alcohols, phytols, retinols, and vitamins A, E, and K. Due to their insolubility in water, these lipids can form aqueous mesophases only in the presence of other components.

b) Insoluble swelling amphiphiles. Although these substances are virtually insoluble in water, they swell to form certain well-defined liquid-crystalline phases. Biological substances of this type have already been discussed (cf. pages 563–569), e.g., monoglycerides, phosphatidylethanolamines, lecithins, phosphatidic acid, phosphatidylserine, sphingomyelin, phosphatidylinositol, cerebrosides, and cardiolipins.

c) Soluble amphiphiles. As salts of long-chain carboxylic acids, e.g., bile acids, these play an important biological role. While no mesophases are formed by bile acid salts with water alone [11:241], they do so in many three- and other multicomponent systems. All soaps and those detergents discussed in chapter 11 could be correctly assigned here.

The discussion on the prebiological membrane formation also has to consider the formation of amphiphilic lyomesogens that was observed during the UV irradiation of C_{12}–C_{14} petroleum hydrocarbons in an aqueous medium similar to sea water [371a].

12.5 Phase Characteristics of Lipids in Systems with Three or More Components

The cholesterol-containing lipid complexes of certain normal human and animal tissues, such as the ovaries and the ardrenal cortex, exist as mesophases at body temperature [395–398, 405a, 435]. Under crossed Nicols, the corpus luteum exhibits spherulitic droplets having axial crosses ("stars"). Chemically, these aqueous mesophases contain mainly free and esterified cholesterol, triglycerides, phospholipids, ketosteroid hormones, and unknown steroids [397]. Ternary systems of phospholipids with water and a nonionic substance (such as cholesterol or a cholesteryl ester) have been the subjects of many investigations [88, 89, 123, 180, 211, 233, 380, 447a]. Free cholesterol is much more soluble (up to one molecule per molecule of lecithin) in the lamellar aqueous lecithin phase than its long chain polyunsaturated esters. This is shown in fig. 12.19 [380, 381].

The incorporation of cholesterol into the lamellar lecithin mesophase causes an increase in lipid layer thickness and a corresponding reduction in the surface area of lecithin [239], fig. 12.20.

Varying amounts of cholesteryl linolenate are incorporated into the thermotropic mesophases of lecithin [382]. The literature reports data on the interaction between lecithin and cholesterol [100, 237a, 344, 345, 447], the distribution of cholesterol in phospholipid bilayers [446], calorimetric data of phospholipid/cholesterol/water systems [214], the exchange of fatty acids between phospholipids and cholesteryl esters [255, see also 198a], and the chemical constitution of tumor lipids [463]. The affinity of cholesterol for the neutral phospholipids (as deduced from DSC experiments) decreases in the order sphingomyelin > phosphatidylcholine > phosphatidylethanolamine [93a].

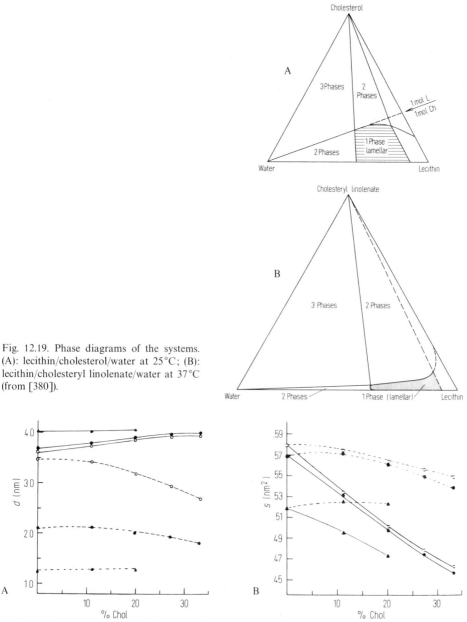

Fig. 12.19. Phase diagrams of the systems. (A): lecithin/cholesterol/water at 25°C; (B): lecithin/cholesteryl linolenate/water at 37°C (from [380]).

Fig. 12.20. Parameters of the lamellar mesophases related to a fixed lecithin to water ratio within one curve (▲, 9:1; ●, 3:1; ○, 3:2) as a function of the dry weight percentage of cholesterol (from [239]). (A): the thickness of the lipid layer (continuous lines), and the thickness of the water layer (broken lines); (B): the mean molecular surface area (continuous lines), and the surface area of lecithin (broken lines).

Besides the wide-spread cholesteryl esters, small quantities of cholesteryl ethers also appear as physiological substances in rare cases [138, 146].

12.5.1 Bile Analog Systems

In fresh lithogenic specimens and in stored specimens of normal and abnormal human bile, there exist birefringent droplets with a diameter of about five µm. Their features indicate a liquid-crystalline state with a lamellar structure [48, 176, 177, 435]. The addition of cholesterol to the ternary system water/lecithin/sodium cholate markedly affects the ranges of existence of mesophases [375], fig. 12.21. The phase diagram and the textures of the water/lecithin/bile-salt system [376] and of the water/lecithin/cholesterol/bile-salt system [176] are discussed in detail in the literature; see fig. 12.22.

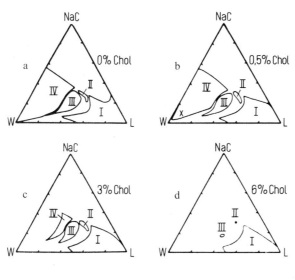

Fig. 12.21. Phase diagrams of the quaternary system water/lecithin/ sodium cholate/cholesterol (W/L/ NaC/Chol). Phase I is analogous to the neat soap, phase II is probably made up of dodecahedrally deformed spherical micelles packed in a face-centered cubic lattice, phase III is analogous to the middle soap, and phase IV is an isotropic micellar solution (from [375]).

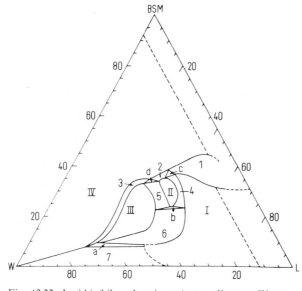

I zone of the neat phase.
II zone of the cubic phase.
III zone of the middle phase.
IV zone of the isotropic solution.
1 zone of separation of neat phase and isotropic solution.
2 zone of separation of cubic phase and isotropic solution.
3 zone of separation of middle phase and isotropic solution.
4 zone of separation of neat phase and cubic phase.
5 zone of separation of cubic phase and middle phase.
6 zone of separation of neat phase and middle phase.
7 zone of separation of neat phase and isotropic solution.
a b c d, zones of separation of three phases whose composition for each zone are indicated by the apices of the triangle (from [376]).

Fig. 12.22. Lecithin/bile salt mixture/water diagram. W, water. L, lecithin. BSM, bile salt mixture.

An average normal human gall bladder bile contains about 92% water, 3.5% bile salt, 3% lecithin, 0.5% cholesterol, and 1% other components such as bile pigments (mostly in the form of water-soluble bilirubin diglucuronide), protein, and inorganic anions and cations. If one neglects the minor changes that these other components may produce, one can compare bile with a system comprising 93% water, 3.5% sodium cholate, 3% lecithin, and 0.5% cholesterol. This falls within the domain of the isotropic phase IV at the point marked X in fig. 12.21b.

From the diagrams, it is apparent that a change in the composition can lead to the formation of cholesterol crystals or a mesophase (phases I to III in fig. 12.21). Only within the confines of phase IV bile is a normal micellar solution; in pathological cases, however, its composition can vary widely.

An intravenous infusion of lecithin increased the cholesterol synthesis from acetate in the rat liver [198]. The plasma lipoprotein was found to contain a 1:1 molar lecithin-free cholesterol smectic phase in the low-density flotation region of s series of patients with obstructed biliary tracts [325].

Concerning this, we also mention the mesomorphous phases in cholesterol/estradiol mixtures [384] and in mixtures of deoxycholic acid with nonphysiological substances such as pyrene and other π-systems [56, 56a].

12.5.2 Atherosclerotic Lesions

The atheromatous deposits in atherosclerotic arteries is in a liquid-crystalline state exhibiting spherulites with axial crosses, secondary rings, and beginning as optically positive [240, 298, 397, 398]. X-ray analysis revealed a structural spacing of about 3.5 nm [121a]. Cholesteryl esters are the main component of these deposits that, in spite of their great importance for living man, have not yet been sufficiently researched. A great accumulation of lipids is the most striking feature of atherosclerotic lesions which can be imitated in animals and in vitro from cholesterol, lecithin, and water. Thus the cell membrane function is disturbed, presumably by impairment of membrane enzyme activity [305], cf. section 12.7. The changes in plasma induced by an atherogenic diet have been repeatedly studied [29, 218]. Local cholesterol esterification is said to be important in maintaining the high levels of cholesterol esters found in atherosclerotic lesions [79, 324]. Cholesterol is much more rapidly esterified with palmityl-CoA using cell-free homogenates from rabbit aorta when the aorta is atherosclerotic rather than normal [25, 162].

In an in vitro study, acetate-1-^{14}C was incorporated into cholesterol esters of atherosclerotic intima from cholesterol-fed rabbits [282]. In normal serum, 80% of the radioactivity incorporated into neutral cell lipids was recovered in cholesterol, but only 7% in cholesteryl esters and 5% in triglycerides. In lipemic serum, only 3% was recovered in cholesterol, but 65% was found in cholesteryl esters and 25% in triglycerides [17]. In cells that were brought in vitro from a low lipid serum to a high lipid serum state, the content of all major classes of lipids was increased. The growth in lipemic serum was especially favorable for the formation of cholesteryl esters relative to cholesterol [17]. These lipids were mainly formed from the serum of the medium.

The composition of the cholesteryl esters and the phospholipids has been investigated in normal intima, normal media, and in fatty streak lesions of aortas [90, 173]. Even the damage to rabbit arteries caused by serum and the more intense damage due to plasma following the feeding of cholesterol can be imitated in models [383b]. The literature provides details and reviews [3:291, 91, 113, 125, 149, 222, 313, 388, 439]. O. Stein et al. have found that aortic smooth muscle cells in culture may serve as a good model to study the role of the lysosomal systems in atherogenesis [394a].

12.6 Myelin

Mesomorphous lipid droplets and tube-like structures, the socalled myelin forms, were discovered in nerve tissue very early. These exhibit axial crosses and can be either optically positive or negative. Chemically, myelin is composed of lipoproteins on which ionic substances such as fatty acids or phosphatides are still bound, and nonionic compounds such as glycerides and cholesterol [53, 97, 134, 267] and it can be separated into different fractions [444]. Studies using 4^{14}C-cholesterol show that most of the cholesterol of the myelin in the brain and spinal cord is synthesized in situ, while it is derived from the blood in peripheral nerves [405]. Electron micrographs and X-ray studies have shown that a smectic layer lattice is formed by intercalation of water, and it resembles a lamellar stacking of membranes in paracrystalline order [181, 181a, 215a]. The irregular and bizarre forms with a double contour, known as myelin forms and first described in 1854 by R. Virchow [1:380] have been found in many other biological materials besides the first-investigated medulla of nerves [11:314a]. J. G. Adami and L. Aschoff enumerate examples of intracellular myelin bodies that appear physiologically, pathologically, post mortem, and also examples of myelin and myelin-like substances in secretions [2]. Myelin forms were often obtained artificially, e.g. on saponification [391].

Vesicles can separate under certain conditions from the injured myelin sheath of nerve fibers containing marrow. These vesicles have a size distribution that exhibits two maxima. The average diameter of the smaller ones is 0.43 µm, as shown by counting of electron micrographs of cut-samples. The average diameter of the larger is about 3.2 µm. G. Albrecht-Bühler has worked out an ultramicroscopic method for their investigation by using light optics applied directly to the native vesicles in an aqueous solution without a preparation step [3]. To this end, the two reflections are observed arising on them in the same way as they macroscopically form on transparent hollow globules that are illuminated perpendicularly to the direction of observation and have a diameter exceeding 1.4 µm. Since the distance of the two reflections depends on the difference of refraction index between the inner and the outer space, this method offers the possibility of studying diffusion phenomena through the vesicle membranes. The two points of light are regarded as a real and a virtual image of the light source, and it is reasonable because the points shine with the color of the light source and lie along a straight line. W. Helfrich et al. have used phase contrast microscopy to discern the phase transitions of some lecithin membranes forming large vesicles [168d]. Such vesicles may show circular or elongated domains due to lateral phase separation [350a, see also 401a]. When the myelin sheath swells in distilled water, the outer myelin layers separate from one another. The first-loosened ones are transforming into vesicles enclosed by a myelin monolayer while the inner layers are still closely packed. Multiconcentric vesicles can be constricted from the loops of more than one layer, and these are probably the predecessors of the larger vesicles. Thus the two maxima of the micelle size-distribution curve can be explained. The cited dissertation also reports other interesting details [3].

Pathological changes of myelin were observed in demyelinating diseases [452] and encephalomyelitis [453]. Quantitative analysis of electron micrographs have shown that two inhibitors of sterol biosynthesis, triparanol and AY 9944, influence the peripheral nerve myelination in two ways: by retarding the "triggering" of myelination in unmyelinated axons, and by decreasing the rate of that myelination already in progress [329]. Two lamellar phases occurring in the water/lipid/protein system have been studied by X-ray analysis and are closely related to biological membranes, which may be conceived as two-dimensional mesophases comprising the same components [247, 248].

12.7 The Liquid-Crystalline State in Biological Membranes

All cells build a membrane to shield the intracellular space from the extracellular in order to maintain their biological reaction media more or less constant. Even organelles within cells are enclosed by membranes, often for this same reason. Biological membranes comprise multifaceted systems that perform a whole range of important functions: as selectively permeable boundaries; as organized substrates for the adsorption of ribosomes, enzymes, antigens, and other substances; and as a material basis in the information process with the acceptance, transport, processing, and storage of information. These varied tasks can be performed only by an organized system that is at the same time flexible in its organization. The vis vitalis of an organism depends on the existence of an order and the ability of the order to adapt. These two characteristics are found simultaneously only in the mesophase. Actual crystallization within the cell cytoplasm generally leads to cell degeneration or death.

Electron micrographs of the cell membranes of Euglena gracilis indicate a thickness of about 9 nm. Lamellar bilayers have also been found in all organelles of this cell, e.g., in the internal membranes of the endoplasmic reticulum, the Golgi apparatus, and the mitochondria (fig. 11.23). The chloroplasts show a lamellar double membrane about 25 nm thick comprising single membranes each 5–10 nm thick [461]. This might obviously be a mesomorphous state that includes pigment, lipoprotein, and lipid, while the aqueous protein, salt, and enzyme solutions are assigned to the space outside the double membrane.

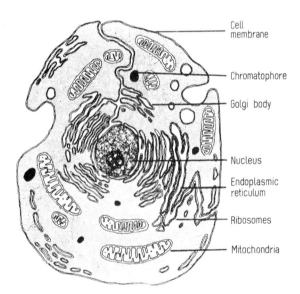

Cell membrane

Chromatophore

Golgi body

Nucleus

Endoplasmic reticulum

Ribosomes

Mitochondria

Fig. 12.23. Schematic illustration of an animal cell (from [461]).

Moreover, the energy transfer can be optimized by a regular arrangement of chlorophyll and carotene molecules forming a lamellar system with a large surface accessible to enzymes and other molecules. The possibility of energy transfer through mesophase systems has been considered [46].

The cited paper provides some additional interesting information on Euglena gracilis [461]. The chloroplast lipids are mainly composed of mono- and digalactoglycerides with an exceptionally high proportion of polyunsaturated acids. During growth in the dark, the Euglena cells lack not only chloroplast structures but also a large part of the α-linolenate content (0.6%). The α-linolenate rises to 7.8% when the cells are exposed to light for 22 hours, and 85% of this is found in the chloroplasts. This reversible response of the cell to light suggests a correlation between the α-linolenate content and the ability to photosynthesize and photophosphorylate, but γ-linolenic acid, arachidonic acid, and C_{20}-polyenoic acids appear to be important for the dark processes of respiration and oxidative phosphorylation. Above 33°C a chloroplast-free Euglena mutant appears lacking α-linolenic acid. In addition to the chlorophyll of the chloroplasts, all biological photoreceptors contain carotinoids that are probably organized in a monomolecular layer on the external surface of the lipid portion.

Some reviews [43, 70–72, 197, 410, 460] and books [117, 133, 203, 232, 261] provide further information on structure and properties of biological membranes.

12.7.1 Constituents of Membranes

Water, comprising 30 to 50% of the membrane volume, is the major component for whose importance the membrane models (to be discussed below) can least explain [129, 460].

Lipids in the broader sense (e. g., phospholipids, cholesterol, plant-membrane glycolipids) are constituents of all biological membranes. Phospholipids are believed to occur in all membranes. The lipid content is especially high in the myelin component of the brain, while is it very low in muscle tissue. Comparative data on phospholipids in the tissues of various animals can be taken from the literature [67, 143].

Proteins take part as structural or transport proteins, enzymes, and glycoproteins. ORD and CD investigations indicate that the membrane proteins possess a globular structure and a large amount of the α-helical component. This agrees with their predominantly hydrophobic character.

Ribonucleic acids have been occasionally detected in bacterial membranes. The membrane components are so closely bound that the membrane can often be isolated in its entirety. However, this binding is not covalent since even the influence of solvents or detergents can destroy the membrane structure. This explains the experience that phospholipids cannot be extracted from animal tissue with diethyl ether without previous treatment using a polar solvent such as ethanol, since the substances must first be freed from their anchoring matrix.

Phospholipids in the membranes of most animals are more than 50% choline derivatives. This is perhaps favored because the choline phosphoglycerides have a net zero charge at physiological pH. The particular charge distribution setup for the phosphoglycerides in a membrane may be important for determining the specific interactions with proteins. Among the particularly well investigated membrane phospholipids of Escherichia coli are found *cis*-9,10-methylenehexadecanoic acid and *cis*-11,12-methyleneoctadecanoic acid. They are originated by methylation of the unsaturated phospholipid fatty acids with S-adenyl methionine [85]. Plasma membranes form squid retinal axon are especially rich in long-chain polyunsaturated fatty acids with the phosphatidylethanolamine comprising about 20% of the 22:6 (number of carbon atoms:number of double bonds) fatty acid [467].

The cholesterol content varies sharply with the type of membrane. Thus, the molar ratio cholesterol/phospholipid is 0.76 for the plasma membrane, 0.24 for the smooth endoplasmic reticulum, 0.12 for the outer membrane and microsomes, and 0.06 for the rough endoplasmic reticulum. The inner mitochondrial membrane appears to contain no cholesterol [81] (which includes additional data on the chemical composition of the lipids in these membranes). The cell-surface membrane (plasma membrane) often has an appreciably higher cholesterol content than cytoplasmic membranes, and the lipid hydrocarbon chains have a significantly lower level of unsaturation [128]. As the similarity to liquid crystals suggests, only free sterols with a flat configuration function in the permeability of plant membranes [160]. Additional information on cholesterol [96, 201] and β-sitosterol [160a] in membranes and on protein conformation [434] can be found in the original papers.

12.7.2 Solubilization and Reconstitution

A pronounced dissociation of the components has been observed when a membrane is treated with solutions of various detergents, e.g., sodium dodecylsulfate, deoxycholate, or Triton® X 100. The dissociation is not reversed by removal of the detergent [121, 277]. Various detergents, e.g. those mentioned above, differ in both the quantitative and selective progress of solubilization, and therefore the use of several detergents has been recommended. This sensitivity is the reason for the uncertainty factor in many membrane studies. Even water can have a destructive influence when bound to the hydrophilic polar groups of the phospholipid beyond the capacity of the internal lattice [399]. The literature reports the peptidolysis of proteins in plasma membranes [175].

S. Razin considers the possibility of restoring membranes from solubilized membrane constituents [330]. The gentler nonionic detergents and sodium deoxycholate seem to enable many solubilized membrane proteins and lipoprotein complexes to retain their original conformations, so that their binding to membrane lipids upon reconstitution may be expected to be the same as that in the native membrane. In this way, it has proven possible to restore the activity of several multienzyme systems in reconstituted membranes. The reconstitution experiments indicate that the divalent cations, especially Mg^{2+}, are needed to overcome the electrostatic repulsion due to excess negative charges on membrane proteins and lipids at physiological p_H, and this enables the proteins and lipids to approach close enough for hydrophobic bond formation. The main interest is in the analysis and reconstitution of membrane components that carry out special functions, e.g., complex electron transport, oxidative phosphorylation, photophosphorylation, and active ion transport. The reconstitution of membranes performing specific functions will be of great importance for immunological studies. S. C. Kinsky was able to imitate the immunological damage to membranes, and he thus stimulated further work on the molecular basis of antibody-complement interaction [212]. Trilamellar structures have also been observed in reconstitution experiments.

Structure and Membrane Models

W. Groß presents a comparison of the various membrane models [158]; the most important are illustrated in fig. 12.24.

Agreement has not yet been reached as to how the Danielli-Dawson model must be modified, and it is not now clear whether one or more membrane structures exist. The lipid

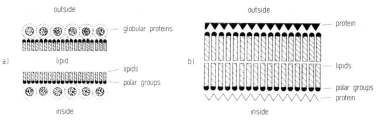

Fig. 12.24. Membrane models of Danielli and Dawson (a), and Robertson's "unit membrane" (b) (from [158]).

bilayer and the globular proteins can be considered definite partial structures, but to date it is not yet certain whether these bilayer membranes correspond to the native biomembrane or if they are already denatured products.

E. J. Ambrose suggests a cytoskeleton determining the shape of mammalian cells. It consists of the plasma membrane and a highly ordered assembly of microfilaments that lie parallel beneath the plasma membrane [7–11]. This arrangement remarkably resembles a smectic state, and thus the structure of the cell surface complex might be that indicated in fig. 12.25.

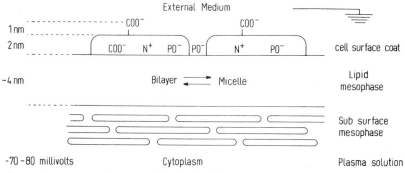

Fig. 12.25. Structure of the cell-surface complex (from [8]).

The outer cell surface coat consists of glycoproteins, proteins, and mucopolysaccharide molecules. Stereoscan electron micrographs indicate a complex topology of the cell surface [9], and the membrane is subject to constant changes in form [8]. In cancer tissues, a reduction in molecular order is symptomatic and manifests itself in reduced cell-cell adhesiveness and a disturbance in the liquid-crystalline order of the subsurface microfilaments [11].

Cell electrophoresis can be used to study cell surface structures. This requires dissection of the tissue under investigation into individual cells whose electric mobility can be measured in an electrophoresis chamber under a microscope [7]. In solutions of physiological ionic strength, this method is sensitive to the ionic distribution at the surface and penetrating to a depth of about 1 nm. The charge carriers take influence up to 3 nm below the surface in solutions of lower ionic strength. Disclinations in the surface coat of unicellular organisms have been described [283].

The lipids in a membrane can be considered as having their own dynamics, and these include transversal molecular rotations, lateral diffusion processes, and cation-induced organizational modifications. W. Kreutz discusses p_H-dependent phospholipid-bilayer conformation changes that are influenced by ligand bonding [217], fig. 12.26.

Fig. 12.26. Ca^{2+}-ion and pH-dependent conformation changes in phosphatic acid bilayers (from [217]).

Membrane proteins can either be loosely associated on the membrane, or they can be tightly anchored within the membrane as an integral component of it, in which case they are termed "structural proteins". These are distinguished from other proteins by their high volume and their tendency to form two-dimensional lattices [174a]. The properties of the lipid bilayer are distinctly affected in the vicinity of these proteins [369a]. W. Kreutz exhaustively describes the best known two-dimensional lattice protein in the photosynthesis membrane based on X-ray diffraction and electron-microscope studies [217]. Fig. 12.27 illustrates the known structural details of this membrane.

Chlorophylls a and b and phytol are immiscible with monogalactosyl diglyceride, which is the principal component of the chloroplast lipids and comprises some two-thirds of the

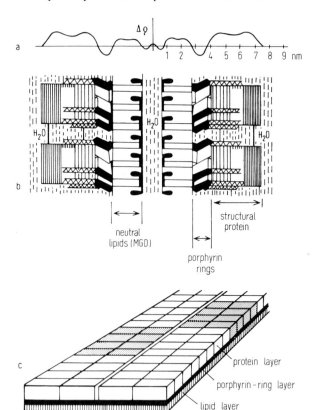

Fig. 12.27. Structure of the photosynthesis membrane in higher plants: a) electron density (ρ) distribution across a section of the vesicle in vivo; b) interpretation of this distribution; and c) projection view of a single asymmetric membrane (from [217]).

total lipid. The greater part of the chlorophyll remains associated with the protein when the membrane is destroyed with detergents. The chlorophyll molecules are probably anchored in the protein grid via the phytol chain such that the porphyrin rings protrude from one side of the protein layer. In the chlorophyll/protein complex, the lamellae are about 1.2 nm thicker than those of pure protein layers. The orientation of dye molecules in photosynthetic membranes was studied by using measurements of the linear electrochromism [330a].

Most other membranes have a much higher proportion of phospholipids than the photosynthetic membranes. They can vary considerably in chemical composition, but not in their rather constant content of about $30 \pm 5\%$ lecithin (phosphatidylcholine). It is hypothetically considered as having universal importance for protein-lipid coupling. Fig. 12.28 shows a structure assumed from X-ray diffraction intensities obtained with myelin membranes. According to this concept, the myelin membrane possesses $C=C$ bond planes in the lipid layer and in the position of the lecithin which extends into the protein lattice.

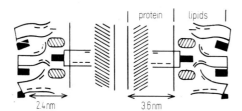

Fig. 12.28. Structure of myelin membrane comprising two inner protein layers and two outer lipid layers. The coupling is via lecithins and sterines. A water space must be assumed between the protein layers and between the lipid layers (from [217]).

A similar structure is found in the outer segments of retina rods in which not a single membrane was observed, but it consists of a mirror-symmetrical arrangement of two asymmetrical membranes. The "unit membrane" is considered to be the product of a collapse of the membrane system that arranges proteins and lipids so that it is an example of a generalized form of a denaturated membrane rather than a native form. Many more individual objects must be studied before any concepts concerning membranes can be considered valid. There exist papers suggesting structures for mitochondrial and chloroplast membranes [163], photosynthetic bacteria [291], and others [124, 436, 437, 454]. Mitochondrial membranes having two hydrocarbon chains of different lengths exhibit special characteristics [68].

Electron microscopy has contributed a great portion of our knowledge of the structure of biological membranes [223, 301, 310, 346, 413, 465, 468]. It should be added that H. P. Zingsheim has been very critical of electron microscopy as a reliable method for membrane studies, and questions whether the presently attainable cooling rates are high enough to avoid membrane splitting [468]. A complement to electron microscopy is suggested in photoelectron microscopy as a new means of mapping organic and biological surfaces [157].

A new method, torsional briad analysis, promises to gain insight on the molecular interactions between membrane constituents [462]. Some additional X-ray [76, 128, 178, 249] and Raman studies [447a, 447b, 447c] on membranes must also be mentioned here. In these, there have been differences observed between one membrane and another, their lipid extracts, and as well as between each membrane and its corresponding lipid extract when compared at an appropriate hydration level [128].

Lipids having a conical molecular shape lead to an equilibrium curvature of the lipid bilayer membrane [159]. The elastic energy of such an arrangement has been calculated as has been the change of membrane elastic energy when one molecule changes its position from the internal monolayer to the external [314]. In this paper one also finds an estimation of the flexoelectric coefficient and experimental evidence for the existence of curvature induced polarization in bilayer membranes. Curvature is also assumed to arise from nonequilibrium lipid distribution between the monolayers of a lipid bilayer membrane [166, see also 44a]. W. Helfrich et al. have measured a value of $2.3 \cdot 10^{-12}$ erg for the curvature-elastic modul of an egg lecithin bilayer at room temperature [168a, see also 213a].

For research on monomolecular layers, the reader is referred to the literature [54, 55, 94, 164, 265, 285, 297, 300, 302, 350b, 451]. Chapter 10 itemizes the very important NMR- and ESR-spectroscopic studies that provide valuable information on the state of the fatty acid chains [235, 465].

The simplification is certainly being carried too far when the model of a lipid bilayer with most globular proteins partly or completely immersed within the layer and some attached to its surface is considered to be a valid generalization applicable to the structure of all biomembranes.

12.7.4 Properties and Membrane Functions

Biological membranes are extremely vulnerable as the discussion of their structure has shown. Most of the studies to be discussed now have therefore employed lipid bilayer membranes as models of biological membranes. A paper on lipoidic structures can be considered to link the lipid bilayer membranes and the "black foam films" [34], cf. p. 549.

In membranes, an optimal fluidity of hydrocarbon chains is required to allow diffusion and material-exchange processes. Experiments on model substances are concerned with the influence of chainlength, the degree of unsaturation at the crystal/mesophase transition [67, 69], fig. 12.29, for a review, see [264a], and the effect of proteins [306]. Fig. 12.30 shows the remarkable difference in the room-temperature IR spectra of a crystalline and a liquid-crystalline phospholipid.

In biological objects, the temperature of the crystalline to liquid-crystalline transition, T_c, must be lower than the physiological temperature. This may explain why the unsaturated lecithins are so common in natural systems once the possibility of lowering T_c by shortening the chains to less than ten carbon atoms has been eliminated [66]. Temperature intervals of 6 to 7 °C including T_c have been measured using dyestuff indicators, and the values obtained agree within a few degrees with those found by calorimetry [421, 424]. The transition temperatures observed in membranes in all cases correspond to those found for the extracted lipids in aqueous mesophases. The membrane can thus in a certain sense be considered a two-dimensional liquid crystal [393, 394]. The literature discusses the analogy between membranes and lamellar mesophases [309].

The movement and arrangement of hydrocarbon chains in membranes are considered to be of substantial importance for diffusion through the membranes. For this reason, an attempt was made to calculate the degree of order and to determine it from specific-heat data [257]. S. Marcelja [258] and R. M. J. Cotterill [83a] have presented theoretical treatments. For small molecules such as water, there is a possibility of universal diffusion through lipid membranes which is passed by the "diffusion" of conformational disturbances such as kinks. A coefficient of 10^{-5} cm²/s has been calculated for this "kink diffusion" [29a, 423]. In the presence of NaI or CsCl, bilayer lipid membranes can possess three times the permeability coefficient of water (25 µm/s) as in the presence of NaCl (8.4 µm/s) [411]. For the diffusion of nonelectrolytes through liposome membranes, an approximately linear correlation has been

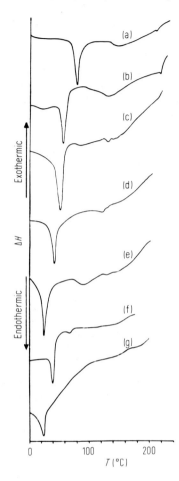

Fig. 12.29. DTA curves of 1,2-diacyl-L-phosphatidylcholine monohydrates (α_1 form). (a) distearoyl; (b) dipalmitoyl; (c) dimyristoyl; (d) dilauroyl; (e) dicapryl; (f) 1-stearoyl-2-oleoyl; (g) egg yolk (from [65]).

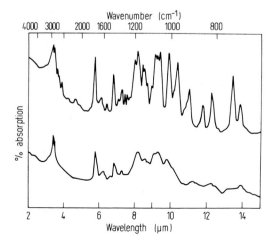

Fig. 12.30. IR spectra of (top) dielaidoyl and (bottom) dioleoylphosphatidylethanolamine at room temperature (from [67]).

found between $\log(P_{rel})/(K_{oil})$ and the molecular weight, where (P_{rel}) is the relative permeability and (K_{oil}) is the olive-oil/water partition coefficient [80], for the glucose permeation, see [29a].

Phosphatidylserine will interact with Ca^{2+} or Mg^{2+} ions, and it preferentially adsorbs K^+ rather than Na^+ from aqueous solutions [67]. Phosphatidic acid, which is purported to be synthesized from ATP and diglyceride on the inner surface of the membrane, is seen as the carrier molecule for Na^+ ions. Certain polar groups can also take part in the ion transport, leading to a lower Na^+ and a higher K^+ concentration in the cell than in the surrounding fluid. No ion-transport theory has as yet found general acceptance, and the discussion is still open concerning the view that the phospholipid merely provides a useful matrix for organizing the enzyme systems involved in the transport process. The literature reports other studies on osmotic behavior [408, 440]. In these considerations studies of the lateral compressibility deserve increased interest [315]. Concerning the state and transport of inorganic ions in membranes, F. W. Cope states, "The findings that ions in cells occupy discrete binding sites (or energy levels) like electrons in a semiconductor, and that a logarithmic conduction law probably describes ion conduction across the cell surface as is true for electron conduction across a semiconductor surface, suggest that the cell, like a semiconductor, may possess a degree of crystallinity. It is then reasonable to expect that concepts derived from studies of liquid crystals may prove applicable to cellular phenomena" [83].

The literature deals at length with the carrier model [174b], e.g., for glucose transport across the membrane of the human red blood cell [234] and RNA transport across the nuclear membrane [464]. A decrease in the permeability (of both hydrophilic and amphiphilic nonelectrolytes) is caused by cholesterol loading of erythrocyte membranes. It also implies the active and passive components of Na^+ efflux [219].

Fig. 12.31 illustrates the crystalline/liquid-crystalline phase transition observed in dipalmitoyllecithin bilayer membranes using 8-anilino-1-naphthalene sulfonate as fluorescence probe.

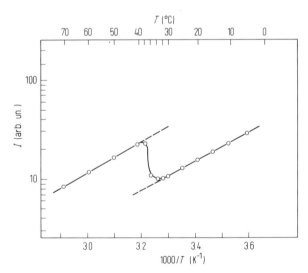

Fig. 12.31. Relative fluorescence intensity, I, (excitation: 370, emission 485 nm) of ANS (see text) in a lipid bilayer membrane as a function of temperature (from [347]).

Since the excimer formation in lipid bilayer membranes is a diffusion-controlled process, the lateral diffusion coefficient, $D_{diff.}$, can be obtained from the second order rate constant of the excimer formation, k_a, which is determined from the ratio of excimer to monomer fluorescence quantum yields, Φ'/Φ [347, 349] (other methods are also briefly mentioned here). This bimolecular process is characterized as follows

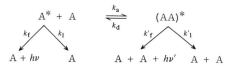

$$A^* + A \underset{k_d}{\overset{k_a}{\rightleftharpoons}} (AA)^*$$

k_d, the dissociation constant, is small compared to the transition rate, $1/\tau'_0$, so that the fluorescence intensity ratio, I'/I, does not depend on k_d at the temperatures being considered. k_f and $k_{f'}$ are the transition probabilities of the radiation emitting decay of the excited monomer and the excimer, respectively. They are molecular characteristics, and thus the ratio $k_f/k_{f'}$ can be assumed to be independent on temperature and the nature of the solvent. In this case the value $k_f/k_{f'} \approx 0.1$ is used for pyrene. The radiationless transition probabilities are k_1 and $k_{1'}$. In a separate experiment the proportionality coefficient, $\varkappa$, in the equation $I'/I = \varkappa \Phi'/\Phi$ was measured to be 0.8. Other studies could show that every collision between A and A* is effective in excimer formation. The collision rate, $v_{col.}$, and the concentration, c (defined as the number of molecules per Å^2), are therefore simply related by $v_{col.} = k_a c$. However, in the final equation

$$D_{\text{diff.}} = \frac{\lambda k_a}{4 d_c} = \frac{I'}{\varkappa I} \frac{k_f}{k_{f'}} \frac{\lambda}{4 d_c \tau'_0 c}$$

the two parameters, λ and d_c, the length of one diffusional jump, and the critical interaction distance for the onset of the excimer formation, respectively, can only be estimated. The lateral mobility of pyrene incorporated in the hydrophobic part of phospholipid bilayer membranes does not depend on the hydrocarbon chain length for membranes formed by distearyllecithin, dipalmitoyllecithin, dimyristoyllecithin, and dilauryllecithin [348].

Using pyrenedecanoic acid as fluorescence probe, E. Sackmann et al. studied lipid bilayer membranes containing different molar ratios of dipalmitoyl phosphatidic acid and dipalmitoyllecithin [350]. They found a remarkable lateral phase separation induced by Ca^{2+} ions and by polylysine indicating that the local distribution of lipids in biological membranes will be greatly affected by the proteins. Such cooperative and simultaneous changes comprising the proteins and the lipid matrix to form domains may play a role in switching membrane permeability and memory functions.

In the living cell, an electric potential arises from the constant ion transport across the permeability barrier against the concentration gradient [8]. Recently the noise in the current through nerve cell membranes has been related to the normal modes of vibration in liquid crystals [241, see also 314]. G. Boheim studied models that purported to imitate the function of excitability [33]. Under the surface of and between two aqueous solutions, one in a glass chamber and the other in a Teflon® chamber, a membrane solution is painted using a marten hair pencil across a one-millimeter hole in the Teflon®. After a variable time, a bilayer membrane forms which is discerible by the growing black zones (cf. p. 549). Certain additives can dramatically modify the permeability and selectivity of the pure lipid membranes. Neutral macrocyclic antibiotics induce pore-formation by a mechanism different from that of polyene antibiotics [33, 206]. G. Boheim developed a functional mechanism to explain

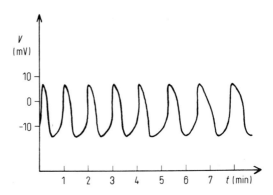

Fig. 12.32. Periodic behavior of the system alamethicine/protamine. The membrane current is 1 μA cm^{-2}. The membrane solution is 20 mg phosphatidylcholine and 4 mg phosphatidylserine in 1 ml decane and 0.5 ml squalene. The chamber solutions are 0.1 and 0.001 N KCl in the glass – Teflon® chamber, and, 10^{-7} g/ml alamethicine and $5 \cdot 10^{-7}$ g/ml salmine 1 in the glass chamber (from [33]).

the observations shown in fig. 12.32 by correlating the changing cation-anion selectivity with the conformational changes of a single type of pore [33, see also 407].

In general, electrochemical investigations promise to yield still more information on membrane processes [221, 365]. An electric field, for example, reduces the phase difference between ordinary and extraordinary rays in nerve-fiber membranes and in "excitable" bimolecular membranes. However, this electro-optical effect is subject to wide deviations due to the membrane itself and added charge carriers [26, 209]. The dc-electric resistance of membranes comprising acidic or neutral phospholipids is about 10^7 to $10^8 \, \Omega \, cm^2$ concurrent on films and vesicles [303], figs. 12.33 and 12.34. Other studies on vesicles are reported in the literature [402a]. Electrodes coated with a liquid crystal membrane may be profitable in electrochemical reactions [2a, 2b].

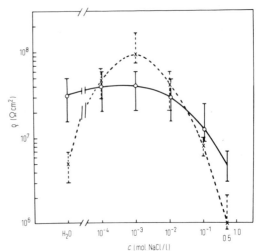

Fig. 12.33. Electric resistance, ρ, of phosphatidylcholine (o) and phosphatidylserine (×) bilayer membranes in aqueous solution containing NaCl and 0.2 mMol tris-Cl at pH 7.4 (from [303]).

Experiments with vesicles indicate a significantly higher relative diffusion rate for $^{36}Cl^-$ compared to that for $^{22}Na^+$ ions, the ratio being about 40:1, and also similar rates for $^{42}K^+$ and $^{22}Na^+$ [303].

The influence of voltage and frequency on the capacitance of bilayer membranes has also been studied [407, 459]. Light induces a photo voltage between two aqueous solutions

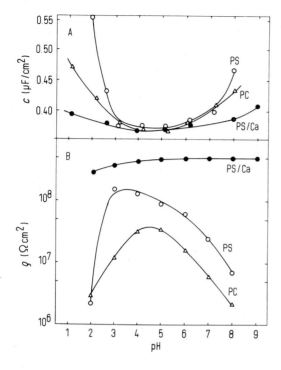

Fig. 12.34. Electric resistance, ρ, and capacitance, c, of phosphatidylcholine (PC) and phosphatidylserine (PS) membranes in aqueous solution containing 100 mmol NaCl at various p_H values. Addition of 1 mmol CaCl$_2$ does not affect the PC curve, but it transforms the PS membrane into a very stable state (PS/Ca) (from [303]).

separated by certain lipid bilayer membranes containing chlorophyll or carotinoid pigments [412]. A series of investigations, from which table 12.1 has been taken, concerns the influence of local anesthetics on membrane functions [171, 275, 304].

Table 12.1. Comparison of relative potencies of local anesthetics on nerve and phospholipid membranes (from [304]).

	Procaine	Lidocaine	Cocaine	Tetracaine	Dibucaine
Minimum blocking concentration (mmol/l)	4.6	—	2.6	0.01	0.005
Relative blocking potency	1.0	—	1.8	460	920
Relative concentration in lipid monolayer	1.0	—	1.1	0.4	0.6
Concentration (mmol/l) for 5 mV reduction in zeta potential	5.0	—	0.8	0.1	0.02
Relative anesthetic potency	1.0	3.8	—	36.5	53.8
Relative inhibition of Ca^{2+} binding	1.0	3.5	—	8.7	10.1
Minimum concentration (mmol/l) for Ca^{2+} blocking	13.5	5.5	3.5	0.3	0.1
Relative potency for Ca^{2+} blocking	1.0	2.5	3.9	45	135

12.8 Nucleic Acids and Viruses

A number of in vitro studies has stated the mesomorphous character of nucleic acid and virus preparations and indicating the presence of this state in the living organism although this has still to be widely proved by in vivo studies.

Nematic-cholesteric phases similar to those of PBLG solutions have been found frequently in DNA and RNA solutions. Thus, DNA solutions exhibit a striped pattern with a periodicity of about 1 µm that decreases with increasing concentration. The analogy goes farther in the case of transfer-RNA which shows birefringent and isotropic phases in equilibrium solutions and a texture completely analogous to that of PBLG: "spheres" of equidistant concentric or double-spiral bands [20, 41, 49, 155, 204, 208, 430]. Small angle X-ray scattering by solutions of transport-RNA indicate an ordered arrangement of molecules that can be represented by an elliptic cylinder or by a triaxial ellipsoid with an anisometry of 2.5 or 3 [93]. Y. Bouligand has visualized a twisted fibrous structure in dinoflagellate chromosomes [35]. In the electron micrograph of a longitudinal section the DNA filaments appear in a periodicity of 100 to 150 nm (this is the half pitch) where they are cut parallel and alternating with bands cut at right angle. Transverse or slightly oblique sections revealed a series of bow-shaped patterns. Y. Bouligand also describes a variety of textures of DNA observed in chromosomes or certain inclusions [36]. Measurements of the dichroism led M. L. Sipski and T. E. Wagner to ascribe the DNA quaternary ordering of equine sperm chromosomal fibers to a cholesteric state [374a]. A liquid-crystalline structure is likely to exist in concentrated solutions of the Na salt of DNA [194b].

Several optical methods, e.g., CD, ORD [154], and absorption spectroscopy, indicate that in addition to the Watson-Crick double helix, biologically important polynucleotides in solution assume a second type of organized structure for which a single-chain stacked-base arrangement has been suggested [40]. For dichroitic studies on cylindrical biological molecules a theory has been presented relating the orientation of the transition moment in a chromophore to linearly polarized incident light at an arbitrary angle [141].

F. Crick has developed a model of chromosomal DNA by assuming that fibrous DNA alone codes for proteins, while globular DNA, which is the major component, takes over control functions using its unpaired regions as recognition sites [84].

M. Delbrück offers an explanation of the transduction of a chemical signal from the environment into the DNA transcription [92]. J. Lapointe and D. A. Marvin isolated flexible deoxyribonucleoprotein rods measuring about 6 by 900 nm from the fd-type filamentous bacterial viruses [225]. These largely unknown and very complicated substances form cholesteric gels exhibiting an iridescent Bragg reflection. Other studies of aqueous solutions of nucleoproteins are reported in the literature [243, 244, 307].

J. D. Bernal and K. Fankuchen suggest classifying viruses on the basis of their X-ray patterns [27]. They first observed a new kind of liquid crystal having a regular hexagonal arrangement in cross-section [28]. The distances between the particles, which indicate homogeneous distribution, depend on the p_H and the concentration of the virus and the salt. The elongated shape of particles in virus preparations, e.g., approximately cylindrical with a diameter of about 15 and a length of about 150 nm, is not a biological necessity. Spherical particles have also been observed in a body-centered cubic close-packing [28]. X-ray diffraction patterns reveal that the virus particles consist of smaller subunits in a regular arrangement which remains unchanged upon drying, and they can thus contain little water. From the Onsager theory, A. Ishihara derived for the prolate spheroid particle of tobacco mosaic virus a critical volume concentration, C_r, that marks the beginning of the anisotropic phase:

$$C_r = 3.4 (a/b)$$

where $2a$ is the minor axis and $2b$ is the major axis of the ellipsoid [195]. Similar results are obtained from considerations on other particle shape, e.g., oblate spheroid or circular cylinder. I. Langmuir discussed the role of attractive and repulsive forces in the formation of tactoids of tobacco mosaic virus [224]. Very intensive studies on tobacco mosaic virus have more recently been presented by A. C. H. Durham [112], while others have examined other virus core proteins [262]. The dichroism measured on tobacco mosaic virus can in a theoretical model be described as a function of dipole moment, excess polarizability, field strength, and the angle between the dipole moment and the transition moment of the absorption band [4]. A theoretical treatment and experimental verification were also given for light scattering [328]. The literature reports many additional investigations of this virus [45, 230, 256, 294, 295, 458]. A liquid-crystalline texture is exhibited not only by the tobacco mosaic virus proper but also by the protein in the sap obtained from virus-infected plants [22]. Lyophases are also found in the solutions of other plant viruses [24] such as cucumber virus [23]. Narcissus mosaic virus solutions behave optically like cholesteric phases [457].

12.9 Blood

While the importance of blood viscosity's influence on general circulation has been stressed in many papers, only a few works concern the viscosity of the fluid within the red blood cells. This is also of great importance since the red cells can change their shape in many ways, and the internal viscosity as well as the nature of the cell membrane both help determine the total viscosity.

L. Dintenfass postulated a type of liquid-crystal state within the red cells because of the low viscosity and its dependence on shear [98]. He also supposed a mesophase to be involved in artificial thrombi, whose viscosities in some diseases (hypertension, renal failure, unsuccessful kidney transplantation, or arterosclerosis) were always found higher (even manyfold) than in normal man [99]. Only in the case of hemophilics does the viscosity of thrombi distinctly correlate to the fibrinogen concentration. Concentrated fibrinogen solutions show nematic properties under shear [216, see also 11:95]. The low density lipoproteins of human plasma exhibit a reversible transition in a broad temperature range between 20 and 40°C, and it has been described as a cooperative liquid-crystalline to liquid phase change involving the cholesteryl ester enrichment in a separate region of the lipoprotein [383, see also 383c]. This behavior is important in the studies on the etiology of atherosclerosis (see section 12.5.2) and the degenerative hepatic cholesterogenesis induced by carcinogenic mycotoxins [250]. J. W. McBain and E. Jameson [251] project horse serum globulin to be in a liquid-crystalline state that has also been found in the plasma of humans having hyperlipaemia [48] and in duckling serums [250].

Sickle-cell anemia hemoglobin exhibits a liquid-crystalline texture in Ringer solution [312]. When the concentration of sickle-cell hemoglobin is raised above 12 g/100 ml (see fig. 12.35 B), the viscosity increases much more rapidly than that of normal adult hemoglobin, until at 16 g/100 ml a completely rigid, birefringent liquid crystalline phase is formed [5].

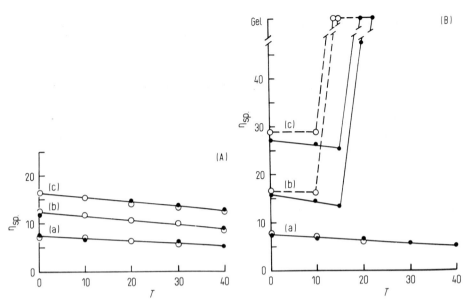

Fig. 12.35. Specific viscosity, η_{sp}, of sickle-cell oxyhemoglobin (A) and sickle-cell hemoglobin (B) vs. temperature, T. Curve (a), 12.0 g/100 ml; (b) 14.6 g/100 ml; (c) 16.4 g/100 ml (from [5]).

A critical examination has proved the liquid-crystalline (probably nematic) character of concentrated solutions of deoxygenated sickle-cell hemoglobin [194]. Normal adult hemoglobin and oxyhemoglobin curves are similar to those for sickle-cell oxyhemoglobin (fig. 12.35).

Living red blood cells usually possess a biconcave-discoid shape. Such shapes have been reproduced by computer calculations on the basis of curvature elasticity. Using lipid bilayer spheres as a model, W. Helfrich has studied these forms and their deformations theoretically [165–168, 168b, 168c, see also 11:150, 236]. Addition of chemical agents in the outside medium creates spontaneous curvatures forming the socalled echinocytes and stomatocytes. Another reason to be discussed in these shape transformation is a nonequilibrium lipid distribution between the monolayers of the membrane. Concerning the cell membrane of erythrocytes, we refer the reader to the literature [152, 322] and the treatment on biological membranes in this chapter. Comprehensive information is found in the cited book [402]. The literature reports on low-angle scattering diagrams of normal, swollen, and shrunken human erythrocytes [21] and on the frequency of the flicker phenomenon due to the rhythmic changes in shape of human erythrocytes [44].

12.10 Muscle Tissue

Many plant and animal cells contain fibers in their cytoplasm: microfilaments (mainly of 6 and 10 nm diameter) and microtubules, i.e. hollow filaments that are 27 nm in diameter, extremely uniform and highly rigid.

After removing the plasma membrane by ion etching, E. J. Ambrose could visualize the 6 nm microfilaments just beneath the plasma membrane [10]. They were found in many types of mammalian cells in the form of highly oriented sheets or cylindrical bundles lying parallel to the surface. In their chemical constitution the 6 nm microfilaments are described as almost identical to the actin filaments of smooth muscle. The arrangement of filaments shows increasing order comparing sub-surface forms, smooth muscle, and striated muscle. Actin- and myosin filaments associated in the actinomyosin complex perform the contraction by conversion of chemical energy stored in ATP into mechanical work, e.g. by sliding of actin and myosin molecules relative to each other. A living cell requires a constant supply of energy to maintain its shape even in a steady state, otherwise it will die rapidly. The models assume a three-dimensional smectic order in striated muscle which is reduced in smooth muscle by a staggering of fibers along their axes. In their function the striated muscle allows especially rapid and highly coordinated contractions (insect flight muscle), while the staggering of the elements in smooth muscle offers the greater possibilities of maintaining tension. In a similar manner the sub-surface microfilaments play an important biological role providing high and active mobility to the cell surface. Ion etching experiments have indicated that in cancer cells the order of the sub-surface microfilaments is severly disturbed leading to high irregularity in shape and movement [10, 11]. Degeneration was also detected in the plasma membrane.

Microtubules were discerned as the rigid constituents of the cytoskeleton. On the other hand microtubules are also employed in movements which are, however, slow and occur within the cytoplasm. Microtubules form the main structure of the spindle during mitosis; the centromeres attached to chromosome fibers carry the chromosomes to the daughter poles along the microtubules. Certain drugs were shown to specifically affect either the microfilaments or the microtubules.

Electron microscopy revealed three paracrystal types of the muscle protein actin F. Two types form a flat net structure with a rhombic unit cell of the same size and shape differing only in the amount of material in the unit cell. The third type of paracrystals are a side-by-side aggregate of F-actin filaments [466]. Analogous studies of actinomyosin showed a parallel alignment of thin and thick filaments with crossbindings between them [87]. Proteins that can be isolated from vertebrate skeletal muscle (i.e., actin, tropomyosin, and troponin complex) can be combined to form paracrystals showing a cross-striation that resembles the natural cross-striation in the thin filament assembly of the myofibril [161], (including an extensive bibliography). It can be concluded from electron micrographs that the cross-striation can be attributed to the location of part or all of the troponin complex at sites spaced at regular intervals along the filaments. G. F. Elliot has studied the factors underlying the interfilament distance in striated muscle, the liquid-crystalline structure of the muscle fibers proper, and its significance in the contractile mechanism [119, 120].

E. W. April has been concerned with the influence of ionic strength [12]. The long-range forces in a muscle filament lattice have been calculated [270]. Water filling the interfilament space is suggested to be in an ordered state thus providing a static support for the protein lattice [142, see also 129]. Finally, we mention some other papers mainly dealing with the paracrystalline properties of F-actin [77b, 207, 292, 441–443].

12.11 Other Biological Objects

In addition to the examples discussed in detail above, many other papers concern the liquid-crystalline state in biological objects. Mesomorphous sterols have just recently been discovered in plants, appearing to be wide-spread [13, 14]. We also mention two older works on the dielectric [374] and thermodynamic properties [372] of mesomorphous protein solutions (see also [418]), and a lyophase separating from a solution of 2% folic acid in N,N-dimethyl-acetamide [114].

Paracrystalline characteristics of biological materials have been studied frequently, e.g. collagen [179, 323, 369, 456], glycogen, other polysaccharides [95, 259], enzyme preparations [271, 284], other materials [86, 130, 163a, 317, 331a, 366, 390], related phenomena in polyuronic acids [409], and in fossilized organic material [281]. Y. Bouligand has published a tentative list of twisted fibrous materials [35].

12.12 Final Remark

The problem of defining lyotropic mesomorphism has often been the reason for avoiding this term; this frequently means, especially in the case of biochemical phenomena, that a liquid crystal is investigated and even described without being explicitly mentioned in the report. This is the case in many works on the properties of membranes and the structurally influenced optical activity. At various points it has been attempted to show the correlation between mesomorphism, paracrystallinity, and biological objects, however, the biochemical aspects must be sharply curtailed in this discussion.

Liquid-Crystalline Order in Polymers

13

13 Liquid-Crystalline Order in Polymers

13.1 Introduction

Even before H. Staudinger has finally recognized the true nature of macromolecular compounds, D. Vorländer had considered the possible mesomorphism of "infinitely long molecules" and concluded that it must exist [2:799]. In order to treat the properties that characterize the intermediate state between the two extreme cases of crystalline and amorphous order, the general term paracrystals has been coined. This chapter therefore expands some concepts introduced in chapter 5, p. 227, because X-ray analysis is the most important tool for the structure investigation of polymers. NMR spectroscopy [34], electron microscopy [7, 81], selected-area small-angle electron diffraction [86], depolarized light intensity measurements [3], and many other methods have also been applied. An introduction to paracrystallinity would be superfluous here because R. Hosemann has already written an exceptional treatise on this theme [29] and also comments on misunderstandings [5:154]. Fig. 13.1 introduces some liquid-crystalline polymer structures.

For a more detailed study of general questions, the following literature citations are given concerning polymeric single crystals [18], polymer crystallization [19], theory of phase transitions [8], and synthetic fibers [55]. The works of J. D. Bernal [6] and K. Thinius [77] are also very enlightening. J. D. Bernal describes the configurations and close interactions of macromolecules in solutions, liquid crystals, and the solid state. K. Thinius' summary of this subject is dedicated to the memory of D. Vorländer. The literature also contains

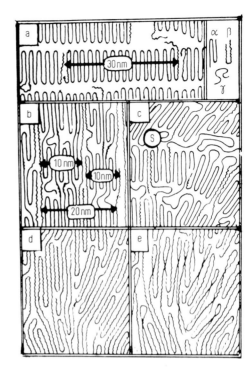

Fig. 13.1. Schematic drawing of the most important paracrystalline arrangements from the main-chain conformations α,β, and γ: a, layered bundle of single crystals; b, ultrafibrils in a nematic arrangement; c, lateral lamellae in smectic arrangement and amorphous zones in the transition range; d, melt of rigid chains, now also nematic from chain to chain; e, as d, but with twistable chain-molecules (from [25]).

further review articles [9:24b, 9:157b, 14, 38, 39, 43, 44, 60], a monograph entitled "Structure and Properties of Oriented Polymers" [83], the proceedings of a symposium [9d], and one book directly dealing with this theme [60b].

Fig. 13.2. Nematic "spaghetti" order (from [74]).

As a general remark to the following detailed description, it must be placed in front that the model of a random coil of a macromolecule in dilute solution is no longer valid for the condensed state. For melts, glasses, and crystals, a nematic "spaghetti" order (fig. 13.2) may come closer to reality than the concept of a molecular blanket since the intermolecular forces favor a parallel orientation and a reduction in the number of chain-crossings [61, 62, 66, 74]. Even in the socalled amorphous polymers there still exist traces of a nematic order and a beady structure [64].

A largely unregarded paper of P. H. Hermans and P. Platzek in 1939 deserves special mention [20]. In this work on macromolecular filaments an orientation parameter, f, has been introduced which is related to the mean angle of orientation, α_m, in a fashion similar to Maier-Saupe:

$$f = 1 - \tfrac{3}{2}\sin^2\alpha_m$$

where $f = \Gamma/\Gamma_0$ with Γ being the actual birefringence and Γ_0 the actual birefringence at ideal orientation (completely parallel particles). These considerations were based on a highly symmetrical model of a filament, and the physical and mathematical situation are clearly distinct from Maier-Saupe calculations (see section 3.4).

For the X-ray analysis of polymers, the theory of paracrystals is better applied in the form of the variance method rather than the integral-breadth method [49]. Since the real paracrystals are composed of crystallites that differ from each other in size, order, and orientation, the primitive translations of an ideal lattice are replaced by vectors which vary both in magnitude and direction from cell to cell. For this reason, an attempt was made to separate the size and distortion effects using Fourier transform analysis of the radial diffraction profiles [51].

The formation of mesophases can be understood such that a nematic order results whenever all parts of the molecules are compatible with one another, but a mesophase will be smectic whenever the molecules contain two laterally locking parts. In the latter case, the two parts segregate with the creation of an interface and arrange in lamellae, cylinders, or spheres. Amphiphiles of high molecular weight, like block copolymers, behave as smectogens in this respect although no characteristic optical textures have been observed so far. M. Szwarc and C. Schuerch discuss the principles of mesomorphism in block polymers [75]. C. Sadron has synthesized organized copolymers of styrene and isoprene in solution, and he also obtained them in solid form after evaporation of the solvent [67].

With regard to their mesomorphous properties, polymers can be classified as follows
a) polymers without a typical mesogenic group,
b) polymers with a typical mesogenic group in the main chain,
c) polymers with a typical mesogenic group in the side chain directly attached to the main chain,
d) polymers with a typical mesogenic group in the side chain and a flexible spacer inserted between the main chain and the rigid mesogenic group.

13.2 Polymers without a Typical Mesogenic Group

Polymers of this type exhibit only a distant analogy to the liquid crystals. Nevertheless, we will mention some details because of their great technical importance.

The paracrystalline properties of polyethylene have been thoroughly studied, especially by R. Hosemann [5:149, 5:152–5:154, 21, 23–31, 34, 35] and J. Petermann [61, 62].

For the very complicated methodology of these investigations, we refer the reader to the original papers. Here only the results can be briefly sketched. The X-ray determination of particle size showed that only lattice sections with a thickness of about 12 nm and a length of between 5 and 100 nm scatter coherently, but the socalled single crystals are rhombic platelets of approximately 10 nm thickness and with a lateral extension of 1 000 nm or more. The "single crystals" were revealed to be a lattice of mosaic blocks [37a]. Within these blocks the chains are regularly folded back forming the true single crystals (fig. 13.3 and 13.4 right). Only a few chains cross (usually strongly bent) to the next mosaic block, so that there is an amorphous zone in between, and the material shows the properties of a greasy powder.

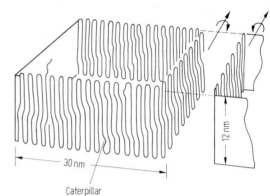

Fig. 13.3. Paracrystalline mosaic block in a polyethylene "single-crystal" showing twist boundaries indicated by the screw axes and "caterpillars" at the places of enhanced gauche conformation (from [26]).

Polyethylene extruded from the melt through nozzles or a product that is later cold-stretched exhibits the other extreme with few backfoldings and having a small fraction of the paracrystalline domains in the direction perpendicular to that of the chains (fig. 13.4 left).

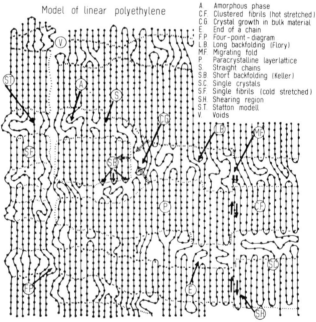

Model of linear polyethylene

A.	Amorphous phase
C.F.	Clustered fibrils (hot stretched)
C.G.	Crystal growth in bulk material
E.	End of a chain
F.P.	Four-point-diagram
L.B.	Long backfolding (Flory)
M.F.	Migrating fold
P.	Paracrystalline layerlattice
S.	Straight chains
S.B.	Short backfolding (Keller)
S.C.	Single crystals
S.F.	Single fibrils (cold stretched)
S.H.	Shearing region
S.T.	Statton modell
V.	Voids

Fig. 13.4. Model of linear high polymer. At the left is cold-stretched, single ultrafibril; at the right is annealed bulk material, and at the bottom right is a single crystal (from [23, 24]).

During tempering, the ultrafibrils arrange themselves in layered structures and "stick" together, but many chains still cross creating a material that is suitably tearproof (fig. 13.4 center). The fibrils shown structurally at the far left of fig. 13.4 undergo the following history upon tempering. Fig. 13.5a illustrates the superstructure of cold-stretched polyethylene in a section parallel to the direction of the chains with the paracrystalline areas indicated by black and the amorphous zones by white.

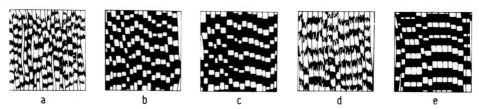

a   b   c   d   e

Fig. 13.5. Structural changes of microparacrystallites and their mutual arrangement in paracrystalline superlattices, schematically sketched for polyethylene during tempering (from [29]).

Fig. 13.5b shows the effect of annealing to 100°C for 500 hours, and fig. 13.5c shows the sample after five minutes at 120°C. In both cases, the paracrystals become more sharply demarcated, and in the latter case they also grow longer with a simultaneous disappearance of amorphous domains. After ten minutes at 120°C, the frontal areas of the paracrystals suddenly become less sharply defined (fig. 13.5d). This state then passes upon further annealing (500 hours at 120°C) into an arrangement with larger paracrystals (fig. 13.5e). While the mosaic blocks enlarge, the paracrystalline g-values remain almost constant, e.g., 1.5% in the (110) plane and 1.8% in the (200) plane [26]. R. Hosemann presents the data in table 13.1 for technical polyethylene.

Table 13.1. Structures and superstructures of bundles of polyethylene single crystals (from [27]).

Lattice	Hostalen® G (Hoechst)	Lupolen® 6001 H (BASF)
Microlattice		
d (nm)	0.41	0.41
g_{11} (%)	1.88 ± 0.07	2.14 ± 0.06
Macrolattice		
L_1 (nm)	29.9 ± 2	32.2 ± 1.5
L_1/L_1 (%)	≈ 50	≈ 50
d_3 (nm)	12.0	12.2
g_{33} (%)	< 10	< 10

The bulk polyethylene sometimes offers more fibrillar and sometimes more layer-like structure that changes over from a nematic-type to a smectic-type order with many gradual intermediate stages. Here we must refrain from more structural description since these may be found in the literature; especially the changes taking place in stretching, annealing, and grafting processes are discussed in detail [5:155, 28, 30, 35, 37, 68, 82]. Of particular interest is R. Hosemann's comparison of a synthetic polymer with a biopolymer using the examples of polyethylene and rat-tail collagen [32]. Paraffins (up to about $C_{50}H_{102}$), which are often compared with polyethylene, can crystallize without chain folding forming large stacks of lamellae that build up the paracrystalline superlattice of the different solid phases [33].

Similarly, the liquid-crystalline properties of polypropylene, first described by G. Natta et al., have been the focus of much attention [58]. Smectic polypropylene differs distinctly from crystalline isotactic polypropylene and from the amorphous preparation by density and X-ray diffraction, but the IR spectra of all three phases are very similar. Only high resolution IR spectroscopy can find differences in band position of the smectic and crystalline modifications of isotactic polypropylene, isotactic *trans*-1,4-polybutadiene-1,3, and isotactic *trans*-1,4-polyhexadiene-1,3, but the intensities are sometimes markedly different [15]. Drawing of isotactic polypropylene films at low rates and temperatures (<40 to 50 °C) produces oriented mesomorphous structures [2, 45]. The appearance of a smectic form is due to the deformation of united chain folds [2]. Full crystallization is found when the X-ray pattern of the drawn sample is taken again after aging three months [45]. Differential scanning calorimetry investigations revealed a series of liquid-crystal transitions in isotactic polypropylene and polyethylene [1, 10, esp. 73]. In an exceptional study, which should stimulate further research, P. A. A. Smit has compared these transitions with the "smectic" transitions in sodium soaps (table 13.2).

The mesomorphous character of polymers does not require a helical structure of the macromolecules, because this is lacking in atactic poly-*p*-acylstyrenes having 10-, 12-, and

Table 13.2. Transition temperatures (°C) of linear polyolefins and anhydrous sodium soaps (from [73]).

Transition	isotactic poly-propylene	lin. poly-ethylene	Sodium laurate (12)		Sodium myristate (14)		Sodium palmitate (16)		Sodium stearate (18)		Transition
			Calor.	lit.	Calor.	lit.	Calor.	lit.	Calor.	lit.	
T_{g_1}	−56	n. d.	—	—	—	—	—	—	—	—	
T_{g_2}	+47	n. d.	—	—	—	—	—	—	—	—	
T_{cc_1}	100	n. d.	—	—	80	80	—	—	89	90	curd-curd
T_{cc_2}	131	n. d.	98	100	106	107	114	117	114	117	curd-subwaxy
T_m	162	134	130	141	133	141	135	135	134	132	subwaxy-waxy
T (s. c.)	169	138	187	182	—	176	—	172	—	167	waxy-superwaxy
T_{ll_2}	200 } d 207	168	—	—	204 } d 215	—	—	—	—	—	superwaxy-subneat
T_{ll_3}	217 }	203	220	220	220 }	217	209	208	208	205	subneat
T_{ll_4}	234	230	—	—	233	—	237	—	238	—	subneat
T_{ll_5}	256	261	—	244	—	245	—	253	—	257	neat
T_{ll_6}	283	291	—	—	247	—	242	295	280	288	neat-isotropic
T_{ll_7}	308 (?)	—	324	336	—	310	—	—	—	—	isotropic

( ) = number of carbon atoms in hydrocarbon chain of soap
s. c. = shear crystallization
d = double transition
n. d. = not determined

14-carbon atoms in the acyl chains. The same is true for atactic poly-*p*-hexadecylstyrene and isotactic poly-α-olefins of 10- and 12-carbons (fig. 13.6).

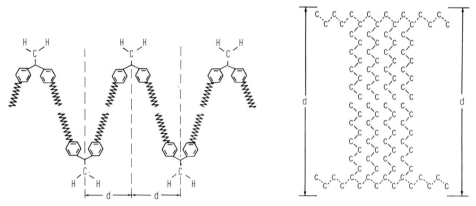

Fig. 13.6. Polymer structures: left, end view of *p*-substituted styrene polymers; right, isotactic poly-*n*-decene-1 (from [13]).

V. N. Tsvetkov et al. confirm that vinyl polymers with long side-chains can form a liquid-crystalline order [79]. Such long side chain polymers considered in the solid, fluid, and gel states reveal a tendency to ordering on account of strong intra- and intermolecular interactions of side methylene chains or other groups [63]. Without going into details here, we cite works on the paracrystallinity of polyvinyl alcohol [54], polyacrylonitrile [22], polyacrylonitrile-copolymer fibers [57], cellulose diacetate films [5], polyvinylchloride [52, 56, 76, 80], poly-*p*-xylene [59], poly(phenylethylisocyanide) [46, 47], polydiethylsiloxane [4], *m*- and *p*-poly(bis chlorophenoxy)-phosphazene and other polyphosphazenes [16, 16a, 16b], poly(N$^\alpha$-acyl-N$^\varepsilon$-methacryloyl-L-lysine) [2:664], linear polyamides [17, 19a], poly(ethylene terephthalate) [53], polyester fibers [12a], and amphipathic copolymers [72]. No details can be given on refs. [9, 70]. X-ray studies conclude that the axial order is highest in ramie cellulose fiber, lowest in tufcel polynosic viscose fiber, and of intermediate state in jute ligno-cellulose fiber [50]. R. Hosemann et al. have determined the mean lateral crystallite size to be 4.3 nm in tufcel and 5.0 nm in fortisan [36]. Chemical treatment of native cotton also manifests itself in paracrystalline lattice distortions [69], the domains of different order differ in their reactivity [71c].

13.3 Polymers with a Typical Mesogenic Group in the Main Chain

Polymers of this type have not yet been studied systematically. Thermotropic mesomorphism was found with the polycondensation polymers of the following structures

$$-\left[OC-\bigcirc-COO-\bigcirc-COOCH_2CH_2O\right]_x- \qquad [40, 87]$$

$$-\left[O-\bigcirc-C(CH_3)=N-N=C(CH_3)-\bigcirc-OOC(CH_2)_nCO\right]_x- \qquad [65]$$

P. G. de Gennes discusses the elastic constants of nematic polymers [15b] and the mechanical deformation of such polymers that might induce an i/n transition [15a]. The polymer might thus resemble an elastomer. S. P. Papkov considers the influence of mesomorphism on the orientation of fibers formed by polymers with rigid chains [60a].

13.4 Polymers with a Typical Mesogenic Group in the Side Chain Directly Attached to the Main Chain

H. Ringsdorf, E. Perplies, and J. H. Wendorff have demonstrated that a mesomorphous polymer (e. g. of smectic order) can result from the polymerization of a mesomorphous monomer having different order (e. g. nematic N-(p-acryloyloxy)benzylidene-p'-ethoxyaniline) [2:551, 9:145–9:147,9:147a, 84, 85], see section 9.2.2. A few other examples of such polymers are known, but usually polymerizations carried out in mesophases only yield a frozen liquid-crystalline structure which is irreversibly lost above the glass transition temperature. It was not possible to decide in all publications which type of mesomorphism (real thermotropic or confined to the glassy state) has been obtained, and the literature is therefore cited altogether [2:713, 9:23, 9:24, 9:24a, 9:74, 9:106–9:109, 9:115, 9a, 9b, 9c, 12, 48, 78]. Flow and electric birefringence of these polymers have been studied [66a, 79a]. With a view to the synthesis of mesomorphous polymers, this structural principle is superseded by the following one.

13.5 Polymers with a Typical Mesogenic Group in the Side Chain and a Flexible Spacer Inserted between the Main Chain and the Rigid Mesogenic Group

This principle is the most productive in the synthesis of mesomorphous polymers. The motions of the polymer main chain and those of the anisotropically oriented side chains can be decoupled by the intercalation of the long flexible spacer. H. Ringsdorf et al. obtained nematic and smectic polymers from radical polymerization of p-R-phenyl-p-(ω-methacryloyloxy)-alkoxybenzoates [9:147h], e. g.

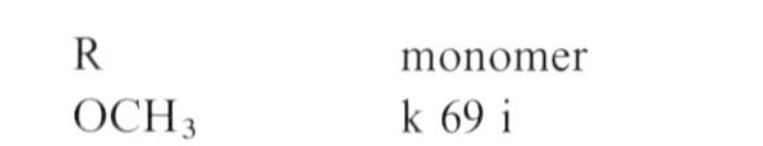

n	R	monomer	polymer
2	OCH_3	k 69 i	g 101 n 121 i
2	OC_3H_7	k 67 i	g 120 s 129 i
6	OC_6H_{13}	k 47 n 53 i	g 60 s 115 i
6	C_6H_5	k 64 s 68 n 92 i	g 130 s 164 n 184 i

g designates a glassy state.

Other mesomorphous polymers are derived from the following monomers:

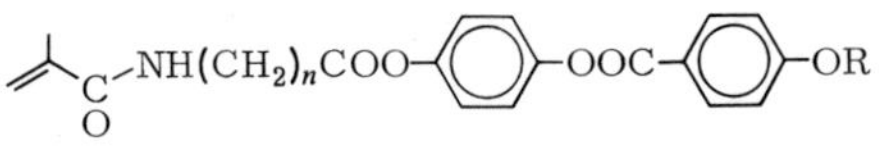

[9:157a]

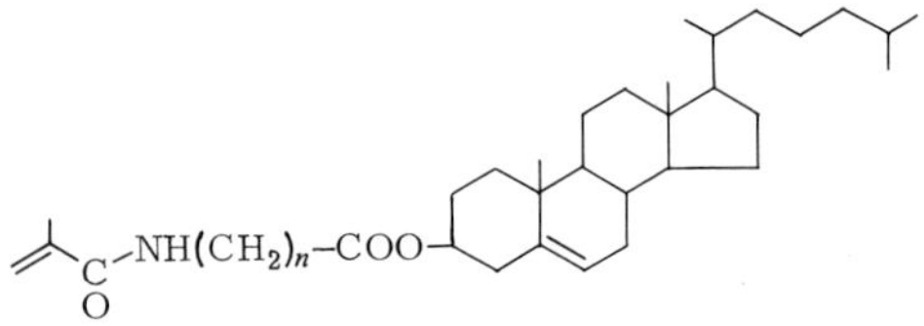

[11, 41, 42, 71, 71a, 71b]

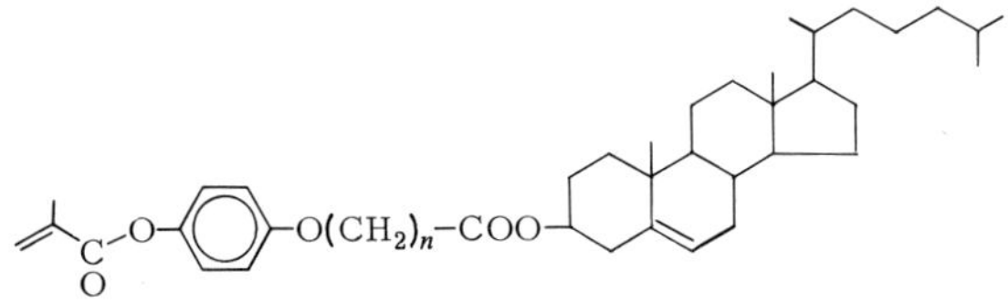

[9:147i]

The cited cholesteryl esters form smectic and cholesteric polymers but only weak color reflection has been observed as has also been with non-sterol cholesteric copolymers [16c].

Technical Applications

14

14 Technical Applications

In his pioneering work more than fifty years ago, D. Vorländer had already considered the idea of technical applications of liquid crystals without, however, finding a possibility [1:389]. Only in the last two decades has a renewed and intensive study been begun to make technological use of the unique properties of liquid crystals. Two events have prepared the way for this development, which in many respects is still in its infancy: the discovery of the dynamic scattering mode [1287] and the synthesis of MBBA, a room temperature nematic [2:392]. Meanwhile, more than 500 publications on the chemical and physical properties of this compound and its mixtures have appeared, and other nematics with low solidification points have also been synthesized, see p. 59. Cholesterics have proven useful as temperature indicators, especially in medicine and the nondestructive testing of materials. In addition, they can be used with an absorber for the recording of wave fields, e.g., microwaves, IR, UV, and ultrasonic waves. Nematics are more important for imaging systems, although image storage can be better realized with cholesterics. The possibilities of technical application have been discussed in a series of review articles [129, 174, 226, 227, 269, 299, 317, 330, 401, 465, 475, 576, 602a, 614, 658, 678, 898, 905, 1059, 1100, 1176], and some books [523, 524, 741].

The literature including the patents is already so volumous and is increasing so rapidly that even a monograph only dealing with the technical applications of liquid crystals could hardly do this in an exhaustive manner. On the other hand, the cited books provide a good overview so that this theme could be sharply curtailed in this presentation.

14.1 Embodiments

One often works with unsealed liquid crystals in thermography when long-term stability is not important, but otherwise a sealing process is necessary. The simplest method consists of enclosing the liquid crystal between cemented glass plates and, for electro-optical applications, carrying an electrically conductive coating [368, 369]. The literature recommends materials for sealing liquid crystals, and such sandwich cells are in general use [28, 48, 83, 157, 213, 424, 425, 505a, 822, 1050, 1082, 1099, 1328, 1328a]. Contact with organic material (e. g., enclosure within foils, dispersion in plastics, microencapsulation) always brings with it the danger that soluble or diffusible substances can alter the liquid crystal material. Several techniques, mostly the subjects of patents, have been developed for thermography with cholesterics.

Cholesterics can be sandwiched between two plastic films (a transparent cover and a black substrate) or can be protected by similar procedures [106, 217, 236, 624, 750, 899, 1295]. The influence of polymer surfaces on the texture has been studied [260, see also p. 11]. Textile fabrics coated with cholesterics are described in the literature [849, 855, 882], as are cholesteric filled hollow fibers (e.g. of nylon 6) than can reversibly change color with temperature [46]. Yarns with temperature sensitive colors can be obtained in the following way [715]: A polyester film is gravure coated with a black ink, gravure coated with a mixture of 30% aqueous polyacrylamide and 30% aqueous slurry of a microencapsulated cholesteric material, and top coated with a solution of maleic acid-vinylacetate-vinylchloride graft copolymer. The film is slit to 0.4 mm and wound around a 120-dernier yarn.

Thermosetting resins (e.g., acrylic resin [76, 77, 828, 876, 1020a] or melamine resin [883], polyvinyl alcohol [179, 180, 700] and other emulsion media [93, 238, 291, 621, 877, 919a, 1020a, 1168, 1192] have been suggested for polymer matrix dispersions. This technique can be applied to the manufacture of molded articles containing cholesterics [883] or adjustable window panes that can be made either opaque or transparent by temperature control [76].

The microencapsulation process seems especially well suited to protect liquid crystals [45, 152, 178, 183, 184, 480, 542, 623, 643, 647, 648, 887, 1236]. We have selected the following procedure as a typical example of these processes [152]: 1.25 g of acid-extracted pigskin gelatin and 1.25 g of gum arabic are stirred with 125 g of distilled water at 55°C to yield a solution of about pH 4.5. When the solution was formed, the pH is adjusted to 6.0 by the drop-by-drop addition of 20% by weight aqueous sodium hydroxide solution. Then the cholesteric mixture consisting of 2 g cholesteryl chloride, 8 g cholesteryl nonanoate and 1 g MBBA is added and emulsified to an average droplet size of 5 to 30 μm. By the dropwise addition of 14% by weight aqueous acetic acid solution the pH is slowly reduced to finally reach about 5, whereby the polymeric material is phased out to deposit on the liquid crystal droplets. The system is then chilled to 10°C under continuous agitation, and 0.6 ml of a 25% by weight aqueous solution of pentanedial, a chemical hardening agent for the gelatin, is added. After 12 hours stirring while slowly returning to room temperature, the capsule walls are firm and sufficiently hardened to be sieved through a wire mesh sieve. If a silver halide is incorporated in a similar procedure, a photosensitive liquid crystal microcapsule material can be obtained [642, 644, 645, see also 25]. W. Sliwka has presented a comprehensive summary of encapsulation techniques [1055].

Possibilities for the further processing of microencapsulated materials as temperature-indicating devices include ink compositions [476, 483], synthetic paints [884, 885], sprays [466], film-forming pastes [900], and sheets [277, 377, 481, 656, 843, 1253]. A suggestion for a medical application forsees a 50 µm thick support sheet (e. g., of Mylar®) coated with a 0.2 mm thick heat-conducting layer and a 0.2 mm liquid crystal layer. The heat-conducting layer provides good heat-transfer to the liquid crystal layer with a minimum of lateral heat dissipation. The heat-sensitive layer is a dispersion of gelatin-encapsulated liquid crystals in latex [1194]. Incorporation within an elastic membrane has the advantage of protecting the microencapsulated particles from internal comprehensive stress [893]. The response time of various commercially available microencapsulated cholesteric liquid crystal films and coating materials have been measured and recorded [892]. Electro-optical display devices containing microencapsulated cholesterics are claimed [181, 182, 185, 186, 468]. For the electro-optical storage mode, it is important that the scattering state, after removing the electric field, is retained far longer than it would with the same substance not microencapsulated [181]. Matrix-dispersed cholesteric droplets behave similarly [182, 185, 186]. The preparation of vitrified nematic and cholesteric liquid-crystal films [171, 672] and of decorative iridescent coatings have been described [218, 697–699].

14.2 Optical Devices

Apart from the action of electric fields and temperature dependence, only a few of the possible applications of liquid crystals are based solely on their optical properties. The selective reflectivity characteristics of cholesterics can be applied to optical filters [4–10]. Thus if one encloses a cholesteric substance, such as a mixture of MBBA and cholesteryl oleyl carbonate, between two crossed linear polarizers, then the assembly is transparent to wavelengths near λ_0, while other wavelengths are extinguished [5]. This method can be used to narrow the bandwidth of an optical signal [9]. With the aid of a filter consisting of a half-wave plate between two cholesteric layers having the same λ_0 and direction of optical rotation, one can obtain a complete reflection of this wavelength from white light [8]. In an assembly of two layers having the same λ_0 but opposite sense of rotation, this wavelength can be removed from white light [6]. Further tuning of optical properties is offered by the addition of dyestuffs [1317–1319], by the angular dependence of cholesteric reflection [7], and its dependence on temperature, pressure, shear, and electric or magnetic fields [10]. Disturbances due to temperature dependence are unconcerned with suitable mixtures, e.g., of nematics and cholesterics [518].

Based on the experiences of H. Zocher [1357], J. F. Dreyer developed light-polarization films containing a dichroitic dye that are oriented in the nematic state [239–243, 245–247]. These were usually made by evaporation of dichroitic dye solutions that were oriented in the lyotropic nematic state. A variant uses long-chain liquid crystal polymer molecules [710, 1086]. The literature reports a tunable, internally distributed-feedback dye laser that employs a mixture of a strongly fluorescent dye, 7-diethyl-4-methylcoumarin and a cholesteric liquid crystal [357]. Other dye lasers have been studied in smectic, nematic, and isotropic phases [655, 1174].

14.3 Thermography

Cholesteric liquid crystals have special advantages as temperature indicators in cases where temperature differences are to be visualized. In order to observe the cholesteric reflection colors without disturbance, it is advisable to inhibit reflections from the object under investigation. This is accomplished by first applying a thin black film which is followed by uniform thin layer of liquid crystal. During color comparisons, lighting and observation should be made at constant angles whenever possible. A hysteresis of the thermogram is limited to rare cases. For the cholesteric mixtures used to date, $T_{(blue)} > T_{(green)} > T_{(red)}$ is the usual relationship. These mixtures are not limited to cholesterics, but they can also be obtained from nematics and cholesterics (cf. section 7.1.6), in which cases their colors can be largely independent of temperature. There can even be a reversed temperature relationship observed, $T_{(red)} > T_{(blue)}$. Numerous cholesteric liquid crystal compositions have been recommended for thermographic applications [138, 219, 228, 270, 354, 355, 396, 477, 483a, 622, 631, 646, 957, 985, 1019, 1042, 1094a, 1175, 1239, 1294, 1328b, including a lyotropic system in ref. 693a].

Since the temperature range in which a cholesteric reflection color appears is generally in the neighborhood of a phase-transition, a significant improvement can be obtained by an addition of a crystallization inhibitor such as cholesteryl erucyl carbonate [354], cholesteryl *p*-nonylphenyl carbonate [355], or a steroid derivative of isostearylcarbonate [131, 132]. Cholesteric liquid-crystal preparations must be protected from ambient light during storage because their degradation is mainly caused by UV radiation [902]. In order to increase stability, the addition of antioxidants [176, 176a, 622, 983, 984, 1355] and other stabilizers [176, 236, 1355] has been suggested.

Experiments on a irreversible temperature-indicator based on cholesterics have not yet led to technologically applicable results. This is due to the fact that reversibility of color/temperature change is the rule except when a phase transition occurs, i.e., the clearing point is exceeded [216, 287]. For this, the dicholesteryl esters of α,ω-dicarboxylic acids are purported to be especially applicable [294].

Substances having a color range of 3–4 °C permit visual recognition of 0.2 °C temperature differences by means of color change. At the expense of the indicator range, which reduces to 0.1 °C, the temperature resolution can even reach 0.007 °C [686]. Under optimal conditions, a one μm wide warmed zone can be recognized microscopically as a color stripe of equal width. A width of 20 nm is reported to be the spatial limit of resolution being necessary for the formation of the color reflection by the socalled Grandjean texture. The temporal limit of resolution is given as 30 ms (33 Hz), thus there is a possibility of visual observations of speed [686]. A liquid crystal on the surface of an object does not indicate the exact temperature if there is a large temperature difference between the object and its environment. However, this need not necessarily be a significant detriment [901]. The temperature can thus be controlled on thermostatic objects as on those having a defined temperature gradient [352, 1356]. A warning light or flashing device employing a thermochromic lens filled with a thin layer of a cholesteric liquid crystal has been claimed in a patent [497].

Instead of the visible reflection color, the optical rotatory power of cholesteric liquid-crystal films can be used as a criterion in the measurement of temperature [1056], cf. chapter 7. The literature reports thermography using cholesterics irradiated with monochromatic light [347], a thermometer [513, 1013], a fiberoptic temperature probe [505], and a brief notice for a method of measuring relative porosity [214]. Cholesterics can furthermore be used to make large-area display screens that are thermally addressed [53, 286, 423, 469, 506, 1307]. Imagewise heating above the clearing point followed by cooling generates colorless image areas on colored backgrounds of a cholesteric liquid-crystal layer [405, 738, 841, 1063]. Thermally induced texture transformations in a s_A phase are suggested for storage and display of high-resolution graphic images with an IR laser-addressed light valve [520]. The critical properties of a thermally addressed smectic liquid crystal storage display (e.g., laser writing speed, contrast, and electric erase) have also been studied [1166]. Dopants which are soluble in the isotropic phase and insoluble in the smectic phase have enhanced the contrast and the range of laser writing speeds. The application of electrothermo-optical effects to display has been summarized [972a].

There are many reports containing more or less general information on thermography with cholesterics [19, 109, 177, 259, 297, 366, 386, 397, 600, 695, 739, 906, 1075, 1255, 1260, 1352, 1354]. Compared to the cholesterics, variable-tilt s_C phases [1170, see also 222] and nematics [1080] play only a very minor role in thermography. The three principle areas of application are mentioned below as follows: imaging of skin temperature, nondestructive testing of materials, and use in radiation- and wave-detectors.

14.3.1 Medical Applications

Cholesteric liquid crystals can serve as diagnostic aids in cases where a disease is associated with a temperature change near the skin surface. Thus, eight cancer lesions (malignant melanoma, chondrosarcoma, bladder carcinoma, renal carcinoma, breast carcinoma, carcinoma vulva, bronchogenic carcinoma, malignant Schwanneroma) showed a 0.9 to 3.3°C higher temperature than normal surrounding skin [1010]. "This means that the skin overlying a tumor represents marginal tumor area and that cooler areas of the overlying skin herald impeding tumor break down; areas of central tumor necrosis are 0.5°C cooler than the surroundings. The thermographically measured tumor size in some cases may considerably exceed the pulpable tumor mass. Veins draining larger, well vascularized tumors were up to 0.9°C warmer than the surrounding skin. Benign tumors such as lipomas and fibromas were either isothermic or cooler than the surroundings." Other results, e.g. obtained in human breast cancer or rat tumors, advise caution about making generalizations, and U. Moller and J. Bojsen comment upon experience as follows [759]: "It has been suggested that it was mainly the benign tumors in which the temperature of the overlying skin was colder than that of the surrounding skin area, but the fact that the skin above malignant tumors does not always have a temperature higher than the surrounding skin areas decreases the usefulness of thermography in tumor diagnosis". Thus no definite conclusion about the reliability is possible today. False-negative results were obtained by liquid-crystal thermography in five of 64 cases of breast cancer [122]. For the detection of breast cancer see also [220, 337, 373a, 641a, 647a, 666, 682, 1010, 1195].

Acute cellulitis following radiotherapy or caused by a superficial skin burn has also been studied. The erythema lesions were warmer than the surrounding skin by 0.6 and 1.4°C, respectively. Cold stress augmented the temperature gradients from 0.6 to 3.0°C in the patient with radiation dermatitis, while the patient with a burn showed no increased gradient. A patient with Wegener's granulomatosis had cyanotic areas about 1.7°C cooler than the surroundings in a highly irregular temperature pattern [1010]. Liquid-crystal skin thermography has also been used to study peripheral vascular disorders and the action of drugs [26, 123, 163a, 693b, 1094a]. Papaverin increased heat dissipation due to vasodilatation while methoxamine decreased heat dissipation due to peripheral vasoconstriction [1172]. The temperature patterns of skin over the site of injection of histamine or epinephrine or after UV-irradiation have been described [295]. In blood vessel surgery (e.g., direct arterial operation for Claudicatio intermittens), the degree of success can be measured by the rise of skin temperature above the anterior distal shin artery or dorsal foot artery [1085], Placental localization [706, 897] and allergy skin-test reactions [630] may be mentioned as further applications of liquid-crystal thermography. The literature reports still other biological and clinical applications of cholesterics [44, 231b, 280, 301, 335, 336, 344, 350, 383, 470, 640, 641, 672a, 672b, 714, 919, 924, 1046, 1057a, 1170a]. One work considers the application of cholesterics for blood pressure measurements [504]. There is also a suggestion for using the electro-optical properties of nematics in the measurement of visual faculty [1241].

14.3.2 Nondestructive Testing

Cholesterics can be used to measure the surface temperature of a test object that is heated either directly (e.g., by electric current or mechanical vibration) or indirectly (e.g., by radiation). A material flaw near the surface can be detected by an increased temperature, e.g. due to a decreased heat conductivity in a composite structure. Special apparati are usually not necessary for such tests, but they have been developed for special cases [1267, 1272]. Surface and subsurface flaws can be detected in: metals, e.g. Lueder lines in Al and Al alloys [1297, 1299], welded metals, e.g. cracks, voids or leaks in a maraging steel weld or a welded pressure vessel [1298], metal adhesive bonds [664, 1296, 1297], bonded structures, e.g. honeycomb sandwich structures [108, 134, 1296], and other composite structures [133, 135, 988a]. They can also be found in rivets in structures where they have to be insulated from dissimilar metals [664], electric components and electronic circuits [398, 664], e.g., resistors [398], resistance heating of windshields [1020, 1299], transistors [398], transducers [703], semiconductor devices [333], and circuit board interconnections having excessive resistance [1296]; switching phenomena in Au, B, or Si films [231, 281], effects of annealing and laser irradiation on amorphous films of As, Te, and Ge [571], and coalescer elements, e.g. voids, split seams, end-cap leaks and cracks, material imperfections and epoxy-filled voids in fuel filters [912–914].

The nucleation of bubbles in a liquid boiling on thin metallic plates takes place on sites which can be visualized by coating with a layer of a cholesteric liquid crystal [920]. Other studies are concerned with the localization of plastic deformation during fatigue tests of metals [1276, see also 379a, 1009] and the convective heat transfer [198]. General information on the use of cholesterics in nondestructive testing can be extracted from the literature [136, 238a, 298, 433, 628, 694, 696, 701, 758, 895, 946].

14.3.3 Electromagnetic Radiation Detectors

In these "thermal image transformers", the heating pattern of a suitable absorber (e.g., for microwaves, IR, UV, or X-rays) is visualized by cholesterics. The term "thermal radiography using liquid crystals" stems from this. Technical details on the construction and operation of such detectors can be found in the patents [283, 284, 502, 1003, 1274]. These usually are membrane-type displays comprising one thin layer of carbon black and another of cholesteric liquid crystal supported on a thin foil; these are frequently applied for the visualization of IR radiation [22, 130, 432, 438, 486, 1073, 1083, 1181, 1252]. Such detectors have become an important aid in the observation of IR laser modes, e.g. the two-dimensional interference pattern of a gas plasma jet can be visualized on a cholesteric liquid crystal screen [573–575]. Full contrast of a detector membrane (3 nm chromium, 3 μm Hostaphan® foil and 7 μm cholesteric layer) was found at an energy density of $4 \, \text{mJ/cm}^2$ [575]. In the optimization of a 337 μm HCN gas laser, a liquid crystal screen was used to demonstrate the characteristics of several modes of the cavity [676]; the patterns of a 3.39 μm helium-neon laser have also been observed [860]. Other emission parameters of continuous and pulsed IR lasers have been studied [601, 1059a, 1182a, also using nematics 602].

A thermal imaging device that transduces IR radiation by means of an absorbing membrane coated with cholesteryl oleyl carbonate offers the possibility of night vision [267]. In devices employing such beam addressing, a cholesteric layer can be addressed thermally by means of an X-Y deflected intensity-modulated IR laser beam to create an image that remains stored until erased electrically [27, 743, 970–972, 986, 1069b, see also 736 and p. 607]. Information written with an IR absorptive substance, e.g. on a document, can be thermographically reproduced by using cholesterics [339]. The literature also reports an application of cholesterics in photographic products [24], and still more imaging devices using the thermo-optical effects of liquid crystals [419, 969].

We have less extensive experience with the detection of microwaves using cholesterics [82, 130, 351, 356, 951, 1182]. The microwave properties and characteristics of high-dielectric, low loss resonators have been studied [1012]. Thermal mapping was performed on the surface of a body exposed to a super-high-frequency load [81]. A microwave detection apparatus to detect concealed weapons in particular has been claimed [916], but the same purpose can be served by merely displaying the hologram (without reconstruction) by means of liquid crystals [620].

Dissolving radiation sensitive compounds in cholesterics, e.g. iodine compounds being decomposed by X-rays [289, 293] or UV light [3, 404, 844, 1346], offers the possibility of indicating radiation with liquid crystal dosimeters or producing color images corresponding to the areas of different total exposure (see also [672]). Information concerning the radiation stability of liquid crystals is also given in the literature [20a, 577a, 949, 949a, cf. p. 250]. The use of liquid crystals as a medium for the registration of elementary particles has been suggested [373, 501, cf. p. 250]. In order to cover a larger temperature range with one encapsulated commercial material, lowering the temperature of colored reflection by exposure to γ-radiation has been suggested. This shift was found to be linear with the dose, and the temperature response remains stable after irradiation [20, 282, 659].

The use of cholesterics for aerodynamic studies has occasionally been mentioned [595, 1017, 1353, 1355a]. With regard to the measurement of local skin friction in aerodynamic testing, a calibration of shear versus wavelength together with static calibration (effects of temperature and pressure) has been performed [595]. A display device utilizing stress-deformation of a cholesteric material has been developed [1018].

14.3.4 *Sound Wave and Pressure Detectors*

The intensity distribution in an ultrasound field can be registered by using cholesterics on a thin membrane that becomes warmer in the areas of absorption. The literature describes acoustic images or holograms being visualized by the color response of cholesterics [128, 197, 234, 384, 449, 490, 491, 516, 545, 546, 1273]. The patterns of interfering coherent acoustic fields retain sufficient phase information for image reconstruction. Fig. 14.1 shows the application of a liquid crystal detector in acoustic holography.

P. Greguss developed another acoustical-to-optical conversion device by utilizing the change in the birefringence of a nematic liquid-crystal layer exposed to an acoustic field [388, see also 448a, 546a]. Acoustic modulation of light by a nematic liquid crystal and a piezoelectric driver has been reported [99, 248, 249, 437, 1007, 1008]. Compared to the nematic phase, the s_A phase of CBOOA shows an acoustic light modulation only at very high frequencies [84].

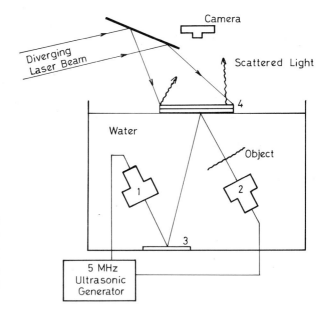

Fig. 14.1. Holographic set up for optical mapping of acoustic holograms by means of liquid crystals. Key: 1, reference transduces; 2, illuminating transduces; 3, acoustic mirror; 4, liquid crystal layer on a vinyl sheet (from [490]).

Ultrasound waves can also be indicated by means of a nematic liquid-crystal layer, which turns from a clear state to a light-scattering state when the intensity of the ultrasonic field exceeds a threshold value [579]. The sensitivity of this detector is enhanced by an appropriate electric field applied to the cell [516].

As noted in chapter 7, the reflection color of cholesterics depends on the pressure as well as temperature, but to date this dependence has hardly been used in pressure-measurement technology [747]. Similarly, pressure-sensitive devices using nematics have as yet found no practical application [244, 633].

14.4 Electro-Optical Display Devices

The reawakened interest in nematics during the last few years has been mainly directed by the technical applications of some long-known phenomena in electric fields that are suitable for altering within wide of ranges the light-scattering, birefringence, or the absorption of a liquid-crystal layer. This makes possible the manufacture of liquid-crystal display devices that can be controlled by low voltages, the influence of the liquid crystal itself, by utilizing boundary effects, and the influence of suitable additives. Furthermore, technical applications of cholesterics are indicated which depend on the many changes that can take place in the helical structure under the influence of electric fields. The effects that are the basis for such display devices are the Frederiks deformation, the hydrodynamic deformation, the cholesteric/nematic transition, and variants of these effects. Since these were discussed in chapter 4 in detail, the present notice is limited to a short summary of the technologically important aspects. Review articles and introductory texts to this field have already been published in many languages [15, 100, 104, 105, 125, 143, 147, 149, 160, 161, 262–265, 279, 313, 329, 365b, 367, 370, 371, 385, 392a, 395, 402, 442, 446, 462, 464, 464a, 467, 472, 515, 581, 583, 584, 613, 618, 626, 627, 629, 636,

638, 639, 667, 675, 688a, 705a, 713, 740, 761, 894, 895a, 896, 993, 995, 997, 998, 1001, 1014, 1049, 1065, 1066, 1081, 1087, 1089–1091, 1116, 1120, 1190, 1257, 1258, 1258a, 1263, 1268, 1284, 1285].

Note also the reviews cited on p. 604. We especially recommend M. Schiekel's article [995] as an introduction and the monographs for a more comprehensive study [523, 524, 741, 1177, 1286].

A comparison of liquid crystal displays with other "Nonemissive Electrooptical Displays" can best be drawn from the proceedings of the BBC symposium on the cited theme [605].

Among the uses of liquid-crystal displays, time-indicating devices are most numerous [105a, 118, 189, 194, 348, 349, 400, 471, 547a, 607, 685, 752, 935, 936, 1265]. Space limitations prohibit any discussion here of the state of the art regarding other possible uses [115, 116, 172, 464b, 840, 1266, 1289].

The abundance of patents on nematic liquid-crystal compositions for electro-optical display devices has already become so overwhelming that we can give only an enumeration here [13, 17, 29–35, 37–43, 43a, 60, 66, 95, 107, 112–114, 139, 156–158, 164, 173, 193, 200–202, 205, 208, 209, 215, 224, 230, 231a, 232, 233, 250–252, 255, 256, 272–276, 277a, 292, 306, 307, 319, 322–328, 338a, 358, 359, 361, 362, 409, 410, 413, 414, 426, 427, 430, 434, 466a, 466b, 467a, 478, 479, 483b, 483c, 483d, 484, 484a, 489, 492–495, 495a, 507, 521, 525, 527–540, 540a, 553–569, 572, 585, 586, 589, 593, 594, 612, 615, 635, 636a, 649–651, 653, 653a, 655a, 681, 688, 717–729, 737, 744–746, 753–756, 753a, 762, 765–776, 779–814, 814a, 816–818, 820, 821, 826, 827, 829, 832–835, 838a, 839, 839a, 839b, 850, 851, 858, 859, 859a, 861, 862, 864, 873, 878–880, 888, 891a, 907, 922, 922a, 925, 939, 941, 943, 959, 963–965, 974–976, 976b, 976c, 981, 982, 982a, 991, 992, 1021–1041, 1038a, 1045, 1047, 1057, 1062, 1076, 1094–1098, 1098a, 1100a, 1102–1111, 1117–1119, 1123–1161, 1169, 1185, 1191, 1193, 1196, 1212, 1215–1218, 1221, 1225, 1234, 1235, 1237, 1243, 1247, 1259, 1268a, 1269–1271, 1292, 1293, 1324–1327, 1331–1337, 1339, 1342–1345, 1345a, 1345b].

We also refer to the reference list on low temperature mesomorphs given on p. 59. In this literature which mainly describes the preparation and properties of the pure mesogens occasionally mixtures for technical uses are mentioned. Many other nematic liquid crystal compositions are described together with the propositions of additives itemized on p. 613.

MBBA plays the principal part in these mixtures where dielectrically negative compositions are desired. However, since dielectrically positive ones begin to gain in importance, there is also a high level of interest in mixtures with nitriles bearing the cyano group at the 4 or 4′ position of the benzene ring. A mixture of smectics and nematics proposed for electro-optical display devices is mentioned as a curiosity [202].

Some nematic compositions cannot be regarded as lyotropic in the sense defined in chapter 1 and 11 [507, 508, 1156]. A paper compares the properties of those liquid crystals that are commercially available for display devices [206].

The nature and function of doping agents for nematic liquid crystal compositions may be classified as follows:

a) cholesterics which give a memory effect
b) dyes which induce a color effect
c) additives for a uniform alignment at zero field
d) additives which lower the threshold voltage
e) additives which lower the electric resistance

f) additives which lower the response times

g) additives which act as stabilizers and increase operational life time.

Salts of quaternary nitrogen bases are mentioned most often as additives [35, 69, 92, 107b, 139, 140, 430, 448, 482, 510, 512, 587, 588, 590–593, 591a, 591b, 729, 732b, 745, 753c, 762, 780–783, 818, 819, 824, 836, 838b, 853, 960–962, 975, 976a, 1068, 1185–1189, 1347, 1348].

Other suggestions include:

amines [175, 176b, 191, 448, 587, 591a, 838b, 1348]

alkanolamines [191, 732c, 790]

amine-borontrifluoride complexes [68]

aminoxide-type nonionic surfactants [2, 175, 176b]

aryldiazonium salts [71]

urea and thiourea compounds [31, 139, 1347]

Versamid [409–411]

alcohols [191, 271, 428, 450, 792]

phenolic compounds [32, 173, 175, 176b, 324, 587, 728, 1062, 1132, 1187, 1189]

quinoid compounds [32, 175, 660, 668, 720, 861–863, 866, 868, 888]

alkyl alkoxybenzoates [50]

alkoxyphenylalkanoates [52]

alkoxyphenyl alkyl ketones [51]

esters of polyalcohols [775, 776, 778, 784, 788, 789, 791]

polyacrylates [716]

carboxylic acids [67, 70, 191, 570, 745, 777, 838b, 981, 1054, 1057]

sulfonic acids and derivatives [511, 512, 779, 958, 1045]

phosphites and phosphates [785–787]

fluoro compounds [49]

nitriles [191, 591b, 680a, 720a, 974, 982, 1054, 1293]

silane derivatives [170]

4-phenylurazole [889]

2,5-disubstituted 1,3,4-thiadiazoles [72]

other heterocyclic compounds [890]

crown ether complex salts [431]

metalorganic compounds [412, 680a, 751]

chromium complexes [730, 732a, 732d]

redox dopants which are more easily oxidized or reduzed than the liquid crystal [80, 680, 705b]

and other organic compounds [191, 450, 512, 591, 652, 753b, 817, 838c, 868, 960, 961, 1067a, 1330a, 1348a].

Alignment additives are usually amphiphilic compounds having one long aliphatic chain (see also p. 11).

Concerning electro-optical applications of cholesterics, the most frequently mentioned are the socalled memory liquid-crystal compositions. K. Tsukamoto et al. propose such mixtures not containing steroids [11, 159, 192, 210, 374, 375, 418, 420, 422b, 435, 459, 569, 764, 837, 856, 867, 934, 952–955, 987, 1197–1211, 1213, 1214, 1219, 1220, 1222–1224, 1226, 1227, 1229, 1304]. For example, a mixture of MBBA and cholesteryl oleyl carbonate is suggested because of its broad optical bandwidth combined with a weak temperature-dependence [518].

The liquid crystal is usually sandwiched between two glass plates bearing transparent electrodes on either side (fig. 14.2).

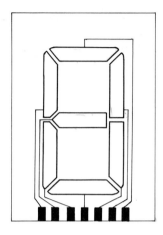 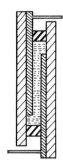

Fig. 14.2. Liquid crystal cell with a seven-segment front electrode.

Indium oxide and tin dioxide are the standard materials for sputtering transparent conductive metal oxides [16, 101, 117, 119, 346, 749, 870, 871, 1047, 1240, 1329, see also 541, 619, 1088]. The electrode material helps determine the uniformity of the liquid crystal layer orientation [379, 403, 874], and thus various methods of surface treatment [21, 36, 163, 211, 255a, 268, 368, 369, 382, 838], ion etching and depletion [372, 680b, 1069a] have been suggested. Conversely, the electric field effects allow the visualization of defects on glass surfaces [465a]. The life of liquid-crystal cells can be increased by a protective coating on the electrodes [124, 749b, 944], e.g., SiO which is widely used to align nematic liquid crystal layers [151, 207, 830, 926, 933, 1071, 1350, cf. p. 616]. A different alignment at both electrodes can be induced by a different surface treatment [225, 731, 732]. A liquid-crystal cell with side-by-side electrodes where only one of the flat substrates is provided with electrodes has been described [154, 940, see also 167, 168]. The literature also describes the measurement of the thickness of the liquid crystal layer [549, 604], the preparation of reflector plates [1330], and antireflection-conductive plates [341]. Other papers mainly deal with the methods for hermetically sealing liquid crystal cells (see p. 605), pressure effects in those cells [1288], and the methods for characterizing displays [78, 85a].

Details on the construction, manufacture, and performance of liquid-crystal display devices can be obtained from the original papers, mostly patents [1, 14, 23, 59, 85, 144, 148, 235, 257, 266, 290, 313a, 314, 321, 340, 342, 389, 429, 436, 451, 452, 458, 499, 503, 507, 547, 571a, 596–598, 606, 608–610, 615, 679, 711, 733, 760, 823, 823a, 847, 886, 904, 908, 911, 915, 921, 922, 927, 937, 938, 942, 973, 1002, 1011, 1016, 1061, 1067, 1069, 1112–1114, 1162, 1228, 1242, 1248, 1254, 1264a, 1305, 1308, 1309, 1314, 1338, 1340] and the monographs [523, 524, 741, 1177, 1286].

Many of these works are also concerned with the addressing of liquid-crystal display devices [120]. Appropriate circuits for driving the displays can markedly improve their operation, e.g., increasing the contrast, reducing the response time, or suppressing crosstalk.

The literature has a full length discussion of the techniques of discrimination in liquid-crystal matrix displays [363, 365a].

A nonlinear resistant material can be provided to suppress crosstalk [380, see also 103a], or a ferroelectric ceramic layer that additionally stores the polarization state can be utilized [381, see also 331, 1120a]. The additional nonlinear electronic components must be connected to each of the individual matrix elements. The potential advantages of liquid-crystal displays are limited by the necessity of an individual drive for each resolution element.

For matrix addressing, it is important that the voltage threshold of the dynamic scattering mode, or of the planar to focal-conic transition in cholesterics, can be increased by a superimposed voltage of sufficiently high frequency [1279–1281]. If in a matrix, the voltage $V_1 + V_2$ is applied to one row and the voltage $-V_1 + V_2$ to the opposing row, the selected picture element of the matrix will scatter with an intensity corresponding to a voltage of $2V_1$ because the suppressive V_2 voltage does not operate here, although it does so in neighboring areas. Frequency-coincidence matrix addressing of liquid-crystal displays is also described [1077, 1078]. For parallel operation, there is a characteristic necessity of having a separate electric contact for each display segment. For a seven-segment plate for the display of n identical figures, this would be $7n+1$ connections; this number is reduced by $n+7$ for multiplex addressing. In order to make an XY addressing system possible, identical segments are connected to each other (matrix rows), but each figure has a separate back electrode (matrix columns). In multiplex operation, this arrangement is addressed by cyclic impulse signals whose amplitude is limited by the cross talk. Publications compare the multiplexing behavior and techniques [485, 603, 948b, 996, 1282], and they discuss the advantages of nematics having different signs of dielectric anisotropy at different frequencies (positive at low and negative at high frequencies) for two-frequency addressing [141, 948a]. Circuits for driving liquid-crystal display devices have been described in many papers [18, 102, 119, 338, 364, 439, 453, 454, 456, 457, 460, 474, 670, 683, 702, 709, 825, 846, 947, 1167, 1264].

As detailed earlier (p. 607 and 609), many methods of beam addressing of the picture elements have been tried, especially with photoconductor liquid-crystal display devices (see section 14.4.6). Electron-beam addressing in a cathode-ray tube can be applied to imaging systems that transform a cholesteric liquid crystal from the Grandjean texture to the homeotropic nematic texture [1303, see also 440].

Three variations of the Frederiks deformation of nematics are suitable for display devices, namely:
a) the DAP mode, **D**eformation of **A**ligned **P**hases; also called electrically controlled birefringence, electrically tunable optical birefringence, or electro-optical index modulation
b) the Color-switching mode in the socalled guest-host-type device
c) the Schadt-Helfrich mode in the socalled twisted-nematic device or liquid-crystal twist device.

14.4.1 The DAP Mode

The first patent for the application of nematics in a sandwich cell in which the birefringence can be varied in a small slit between two metal electrodes was contemporary with Frederiks' pioneering studies [677]. The literature also describes another early trial for light modulation by this method [707]. The great advantage of this method is the possibility of generating all colors in a single cell and to alter them electrically, as explained in section 4.4.3. The possible phase and amplitude modulations of light are switched at low voltages and the influence of the incident light angle has been studied experimentally and theoretically [1044c]. Although technical applications are not yet attained by the current state of the art, some possibilities appear in outlines [54, 57, 86, 94, 169, 278, 279a, 378, 441, 447, 517, 691, 693, 872, 948, 950, 977, 980, 994, 1044, 1122, 1238].

Light deflectors and waveguides utilizing the electro-optical effect of index modulation in nematics have been suggested [312, 345, 632, 663, 1274a]. Electro-optical light modulation is still possible in the isotropic phase up to about one degree (C) above the n/i transition temperature when a marked cluster formation induces a high Kerr constant [111].

14.4.2 The Color-Switching Mode

G. Heilmeier was the first to use a pleochroitic dyestuff dissolved in a nematic solvent to control the absorption by orientation with an electric field [455]. To reproduce all of the colors of the visible spectrum, three liquid-crystal cells are sufficient when arranged in tandem; each cell is composed of a solution of a pleochroitic dye in a nematic solvent whose alignment can be electrically controlled [155]. A voltage-controlled color film of the guest-host-type can be provided for a monochromatic cathode-ray tube display device [221]. Furthermore, this area has not yet been intensively studied with respect to its technical applications. Papers and patents which also propose many pleochroitic dyes must be mentioned [107a, 153, 190, 196, 315, 429a, 657, 810, 814, 933a, 1101, 1231, 1232, 1290], and the spectroscopic uses were discussed in chapter 6. We should also mention a new display device with a pleochroitic dye in a cholesteric liquid crystal which exhibits strong absorption in the Grandjean texture and weak absorption if an electric field is applied to unwind the cholesteric material to the homeotropic nematic state [1163, see also 193b]. The electrically switched guest-host alignment of a pleochroitic dye in a cholesteric liquid crystal also permits the operation of reflective liquid-crystal display devices [1275]. Another suggestion introduces a quarter-wave retarder between a dichroitic nematic display cell and a diffuse metallic reflector [193a].

14.4.3 The Schadt-Helfrich Mode

When compared with the other display devices employing liquid crystals, the most important today is the twisted nematic type, but this must be limited to small alphanumeric displays at present because other systems have not yet been developed to a state of production readiness. Large-area liquid crystal displays exist at prototype stage [669, 881]. As detailed in section 4.4.4, TN-devices (i.e. twisted nematic devices) basically require dielectrically positive nematics, and the mixtures suggested for this purpose have been included in the literature summary on page 612. Suitable substances exhibit low threshold voltages (e.g., 0.9 V), quick responses, and yield a high contrast ratio (e.g., 100:1 at 1.3 V) [110, see also 509, 854, 930, 988].

Disarrangement in Schadt-Helfrich cells can be due to areas of reverse twist that are separated by twist disclinations and which give patches of varying contrast [103, 1070, 1249]. These areas of opposite twist can also be recognized by observation of the Williams domains [928]. In TN displays, another type of patch can arise because the electric field induces both clockwise and counterclockwise reorientation about an axis normal to both the field and the director [929]. Either type of patch can be eliminated by the addition of a surface-alignment agent combined with a cholesteric substance [752a, 929]. Advantages are also obtained by off-ninety-degrees twisted nematic layers [617] which are produced by an oblique evaporation of the electrode or protective material layer [212, 237, 253, 254, 372b, 422a, 584a, 752a, 831, 933]. Twisted nematic layers containing optically active additives have also been studied [13a, 428a, 929, 932, 1178]. A twist angle of the nematic molecules equal to 45° was found to be optimal for the voltage-controlled color formation by means of twisted structures [1044b].

The change in molecular orientation with time when an electric field is applied to a twisted nematic layer has been calculated both neglecting [97] and considering [98, 1245, 1246] the effect of backflow. The literature includes suggestions for two-frequency addressing to reduce turnoff time [74, 165, 343, see also 228a] and to attain a memory effect [229], and it covers transmission characteristics [87, 96, 394a, 749a, 1244], addressing requirements and possibilities [304, 372a, 376a, 584a], see also p. 614 and 615 and miscellaneous data [47, 88, 302–304, 343a, 684, 692, 875, 990a, 1044a, 1093, 1121, 1283, 1341].

TN cells can be combined with a birefringent film or a pleochroitic filter to display colors in transmission [989, see also 931, 1015, 1043]. A new reflective mode, TN display scheme has been described that uses only one sheet-type polarizing filter [89, 989, 990]. In this device the reflector and the second polarizer of the conventional TN display are replaced by a combination of a quarter-wave plate and a cholesteric liquid crystal layer (fig. 14.3).

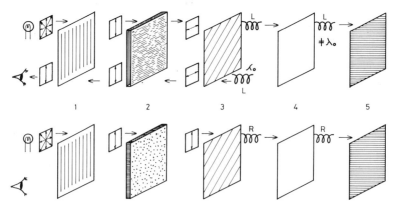

Fig. 14.3. Exploded view of a reflection display device. At each element the polarization state is indicated (L and R refer to left- and right circularly polarized light) in the field-off (upper) and field-on state (lower scheme). λ_0 and $\neq\lambda_0$ refer to light falling inside and outside of the cholesteric reflection band, respectively. Key: 1, polarizer; 2, TN cell; 3, quarter wave plate with the slow axis at a 45° angle with the transmitted E-field direction of the polarizer; 4, lefthanded cholesteric layer; 5, absorber (from [990]).

This arrangement is equivalent to a combination consisting of a linear polarizer, a color filter, and a diffuse reflector. The display scheme shows either black characters against an unusually brilliant iridescent background or the converse situation of iridescent characters against a black field.

14.4.4 *The Dynamic Scattering Mode*

The first generation of liquid-crystal display devices was due to the works of R. Williams [1287] and G. Heilmeier [461] in this area. The description in section 4.4.5 is expanded here only by a few details that are important for technological applications [261]. For instance, the electric properties of liquid-crystal cells are equivalent to those of the circuits shown in fig. 14.4.

R_1 and C_1 can be calculated from the cell dimensions, the specific resistance, and the dielectric constants. R_3 stands for the resistance of the electrodes and leads; R_2 and

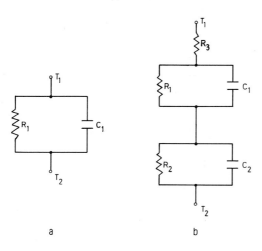

a   b

Fig. 14.4. Equivalent circuits of a liquid-crystal cell: a, simplified; and b, general version (from [671]).

C_2 can then become important when an additional layer (e. g., of Al_2O_3) is present on the boundary surface between the liquid crystal and the electrode, which can be of Al, for example. As fig. 14.5 shows, the optical properties of a dynamic-scattering cell the simplest type, without polarization foils, exhibit a strong angular dependence.

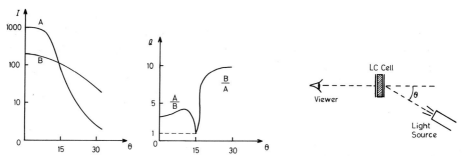

Fig. 14.5. Relative brightness, I, and contrast, Q, of a liquid-crystal transmission cell. A and B denote the clear and scattering cell, respectively. θ is the angle of illumination as shown at right hand (from [671]).

In a transmission cell, the sign of the contrast changes at an illumination angle of 15°. The literature describes contrast enhancement by spatial filtering [162]. Both the contrast ratio and the display-cell current strongly depend on the liquid-crystal material, the thickness and temperature of the cell, the applied voltage, and the polarization [484b]. Fig. 14.6 illustrates the characteristic times of the dynamic behavior when it is excited by a rectangular voltage pulse where τ_1 is the delay time, τ_2 is the rise time, and τ_3 is the decay time, see also [12, 199, 201, 708]. In the case of MBBA, a slower response of the homeotropic layer was found when compared to the homogeneous alignment [204].

The angular distribution of the intensity of light scattered by a nematic liquid crystal in the dynamic scattering mode has been measured [199, 500, see also 1005]. Continuous d. c. operation of more than one year of a dynamic scattering cell has been reported [865]; this was achieved by an addition of the hydroquinone-benzoquinone complex at the expense of increased current density (for dopants, see p. 613). Use of the dynamic-scattering mode permits the design of the following: a mirror for reflected images of variable brightness [146],

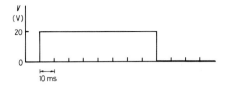

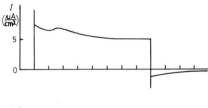

Fig. 14.6. Transient response (current density, I, and contrast ratio, Q) and applied voltage, V, for a 6 μm thick N-014 cell at 25°C. τ_1, delay time; τ_2, rise time; τ_3, decay time (from [203]).

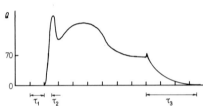

nonselective filters and lenses with adjustable transparency [522, 634, 668, 896a, 1316, see also 1251] a chamber with two liquid-crystal cell windows [187], a new type of optical processor [488], and an application for the nondestructive testing of p-n and Schottky junctions [599, see also p. 609].

Table 14.1 summarizes some of the properties of liquid-crystal display modes used in the discussed display devices.

Table 14.1. Comparison of liquid-crystal display modes (from [399]).

	Dynamic Scattering	DAP	Twisted Nematic
Cost	Lowest	Higher	Highest
Response time	Function of voltage and cell construction		
Power	Highest	Low	Low
Operating voltage	Highest (15 V or more)	Low (1–50 V)	Lowest (1–50 V)
Expected cell life	Lowest	High	High
Washout	Yes	No	No
Angular viewing range	Highest	Lowest	Medium (~30°)
Colors	No (except for filters, etc.)	Yes	Yes
Contrast	Good	Better	Best

14.4.5 *The Cholesteric Storage Mode*

In the case of positive dielectric anisotropy, the nematic homeotropic state and a distinct hysteresis effect can be observed in cholesterics, as fig. 14.7 shows [390, 391].

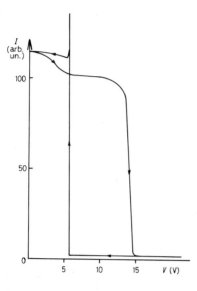

Fig. 14.7. Light intensity, I, vs. voltage, V, transmitted through a cholesteric liquid-crystal cell between crossed polarizers. One full cycle of the hysteresis curve: 3 min (from [391]).

Fig. 14.8 shows the angular dependence of light scattering in the three states of a cholesteric mixture containing 7.5% cholesteryl oleyl carbonate in a composition of equivalent quantities of the Schiff bases $CH_3C_6H_4CHNC_6H_4OCOR$ where R is CH_3, C_3H_7, and C_4H_9 [748, see also 571b].

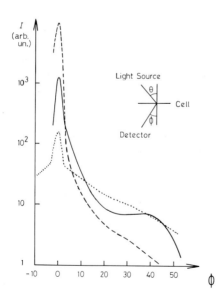

Fig. 14.8. Light scattering of a cholesteric mixture (transmitted intensity, I, vs. detector angle, ϕ, for $\theta = 0$) in the Grandjean texture (-----), in the focal conic texture (———), and in the dynamic scattering state (········) (from [748]).

Analogous to the way in which an undisturbed focal-conic texture forms when the cholesteric phase is usually obtained from the isotropic melt, the same texture forms after the removal of an electric field that had been used to untwist a cholesteric helix. Electro-optical storage effects are known in cholesterics under varied conditions. In all cases, the transparent planar (Grandjean) texture is transformed into a focal-conic or a fingerprint texture that remains for a relatively long time and exhibits moderately strong light scattering. With negative as well as with positive dielectric anisotropy, technologically useful storage effects are possible [1262, see also 1048].

Details on storage effects in cholesterics may be found in the literature [360, 407, 415, 578, 684a, 815a, 869, 1230, 1302, 1310, 1311, 1322, 1323, see also 365, 463, 836, 1313]. Concerning a xerographic technique, an electrostatic latent image can be visualized in an erasable and reusable imaging device comprising a cholesteric liquid crystal layer [406, 416, 417].

14.4.6 Liquid-Crystal Cells with Photoconductors

Several drafts have been suggested for optic-electric-optic transformation systems containing a photoconductor and a liquid crystal. These devices allow simultaneous addressing of picture elements, dependent on the photoproperties of the sensor. A photoconductor sensitive to UV (e.g., ZnS) can be used to record an image which can be viewed or displayed with visible light [498, 1261]. Fig. 14.9 illustrates one such device, while fig. 14.10 shows a version of a light-valve comprising a photoconductor, light-absorbing layer, dielectric mirror, liquid-crystal layer, and transparent electrodes in a multilayer structure [91, 310, 311]. Multilayer dielectric mirrors have been described [257a].

Research in this field usually attempts the modulation of a high-intensity light beam with a comparatively weak light source for application in television projection displays (fig. 14.11).

Basically, all electro-optical effects in liquid crystals can be photo-addressed in this fashion, e.g. the dynamic scattering mode [55, 73], the DAP mode [56, 393, 394, 499a, 1250], the color-switching mode [58], the Schadt-Helfrich mode [422, 1084], and the various storage modes of cholesterics [309, 421, 945, 1306]. The high resistance of the photoconductors in

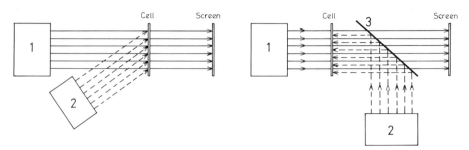

Fig. 14.9. Photo-addressing of a transmission-type projection display. Key: 1, display light source; 2, imaging light source; 3, partial mirror from [498]).

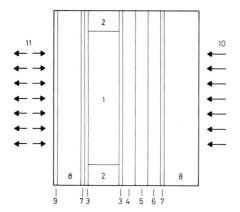

Fig. 14.10. Setup of a reflection-type liquid crystal light valve. Key: 1, liquid crystal; 2, spacer; 3, SiO_2 insulating layer; 4, dielectric mirror; 5, CdTe layer; 6, CdS photoconductor; 7, transparent conductive electrode; 8, glass; 9, antireflection coating; 10, writing light; 11, projection light (from [91]).

the dark ensures that the voltage is across the photoconductor, thus leaving the liquid-crystal film unaffected. However, if an image inscribed with violet or ultraviolet light is focused on the photoconductor, then the resistance is lowered following the illumination, and a portion of the voltage is applied to the liquid-crystal layer, inducing an electro-optical effect.

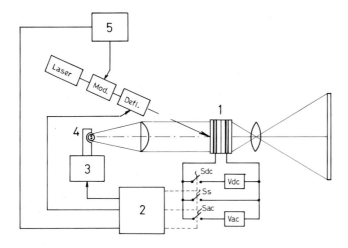

Fig. 14.11. Principle of an optically scanned liquid crystal display apparatus. Key: 1, photoconductor liquid crystal cell; 2, timing control unit; 3, power supply; 4, flash tube; 5, video signal source (from [376]).

The literature reports other details of imaging systems and image converters; some works use the cholesteric/nematic transition [64, 90, 126, 127, 195, 235a, 308, 353, 394b, 473, 690, 704, 742, 815, 845, 857, 943a, 1060, 1072, 1074, 1079, 1092, 1173, 1184, 1251a, 1256, 1300, 1312, 1315, 1320, 1321, 1349]. The cholesteric/nematic transition has also been studied in matrix displays [869a].

14.4.7 Special Electro-Optical Devices

The literature suggests different modes of operation for multiple liquid-crystal cells, e.g., in arrays, tandem, cascades, etc. [734, 735, 848, 1233, 1301]. Since the transmission of a cholesteric layer can be controlled electrically, the cascaded combination of three color filters offers good control across the entire visible spectrum [978]. Considering the various combinations of different types of cells, many possibilities suggest themselves. Thus, a color display device composed two cells, one DAP and one TN, has been developed [979]. The application of color-sensitive polarizers has been suggested [238c]. A cell combining the dynamic-scattering mode [543, 544] or the Schadt-Helfrich mode [302–304] with electroluminescence can be operated at any level of illumination, including absolute darkness.

Fig. 14.12 shows an arrangement having special optical properties and which is called a cholophor because it combines a cholesteric liquid-crystal layer and a photoluminescent phosphor [519]. The literature also describes an electroluminescent dispersion containing a nematic [763], fluorescent [665, 1350a] and luminescent liquid crystal compositions [141a], and a large brightness intensification attained in fluorescence-activated liquid crystal displays [88a, 91a, 392]. No visible cathodoluminescence was found with liquid crystals themselves [223].

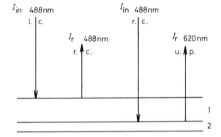

Fig. 14.12. Structure and optical properties of a cholophor. 1, cholesteric liquid crystal; 2, phosphor. Wavelengths and polarization states for incident and reflected light are given (from [519]).

An appropriate voltage applied to a nematic induces a periodically recurring distribution of the nematic director which may act as a thick phase grating of electrically tunable spatial frequency. The possibilities of application of this variable grating mode to color projection displays is discussed in the literature [548, see also p. 196]. A contribution at half the expected spatial frequency observed in the diffraction pattern is attributed to nonorthogonal transversals of the thick phase grating [550].

The field-induced, optically biaxial state of nematics has additionally been employed in electro-optical systems [408]. An electro-optical device operating with a nematic liquid crystal layer between two glass prisms of appropriate refractive index provides a method for controlling the transmission or total reflection by applying an electric field to the nematic layer [551]. The literature describes other nematic liquid crystal prism cells [980a] and a retrodirective corner reflector of tetrahedral configuration with at least one liquid crystal cell [903]. Electrochromic displays must be mentioned as one of the alternatives to liquid-crystal displays [137, 305, 611, 891, 910, 999, 1000, 1291]. Other patents claim an arrangement for the second-harmonic generation of light in nematics [316], see also section 7.9 or electrets containing cholesterics [332].

Due to their high viscosities, smectics exhibit storage effects, these being an exception in nematics [552]. s_A phases which can be switched from a transparent homeotropic texture to a scattering focal-conic texture are suitable [444, see also 443] and p. 607. A smectic light valve has been proposed which utilizes thermal writing with an XY deflected IR laser beam and electric erasure of the stored picture [445a]. A striped texture of a s_A phase can grow from a nematic phase and be stable under certain conditions [188]. Smectics have also been used for hard copy producing of printed circuit board artwork [464c]. A summary of possible applications suggested for smectics is given in the literature [445]. The few substances proposed are deceptive, because a large amount of compounds exhibiting smectic mesomorphism at low temperatures can be taken from the literature [387].

14.5 Holography and Interferometry

Cholesterics, as well as nematics, are valuable aids in holography. Fig. 14.13 illustrates a device using a 200 µm thick layer of a cholesteric liquid crystal as the holographic recording medium [1051].

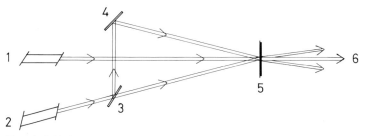

Fig. 14.13. Holographic setup. Key: 1, He-Ne laser; 2, CO_2 laser; 3, beam splitter; 4, mirror; 5, liquid crystal layer; 6, diffracted He-Ne laser beams (from [1051]).

In this device, the two IR beams that impinge on the liquid crystal layer cause slight temperature differences between the regions of constructive and destructive interference, thus varying optical density for the reconstructing He-Ne laser [1051]. Moreover, cholesterics were used as recording medium for IR holography [966, see also 1053].

In microwave holography, cholesterics are applied on a thin metallized plastic film to record the interference pattern, e. g. of any type of antenna [61–63, 487]; for objects permeable to microwaves, a nondestructive internal examination by recording interference pattern is suggested [63, 620].

In a photoconductor liquid crystal cell, the Frederiks deformation of a nematic layer can serve as a reversible and nearly real-time recorder for phase type holograms [705, 909]. The grating which was recorded showed up to 500 cycles/mm, but higher resolution could be possible [909]. A page composer may use the electrically tunable birefringence of a nematic liquid crystal to generate an optical pattern for holography; such a device acts as a spatial wave polarization modulator and is shown in fig. 14.14 [661, 662].

Concerning a page composer, G. W. Taylor compares the properties of a nematic liquid crystal in the dynamic scattering mode with those of other materials [1164, 1165]. Displays utilizing dynamic scattering in a room temperature nematic liquid crystal were found suitable as input planes both for the correlation with and recording of matched filters [687].

Nematics in the dynamic scattering state reduce the coherence of a laser beam and reveal speckle-free projection screens [75, 580], see also p. 197; the time-averaged degree of coherence of laser beams can be controlled in this way [757]. Screens for image projections have been described [842, 1351] including a rear projection screen which is essentially free of sparkle, and it allows a brightness adjustment by variation of the voltage applied to the liquid crystal in the dynamic scattering state [150].

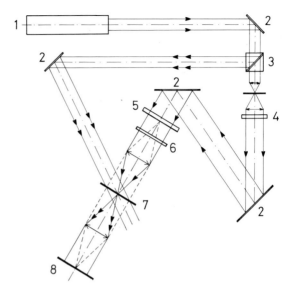

Fig. 14.14. Fourier hologram optical bench. Key: 1, Ar-laser and shutter; 2, mirror; 3, beam splitter; 4, half wave plate; 5, page composer; 6, analyser; 7, Fourier transform hologram plane; 8, photodiode array (from [661]).

The light transmitted through a Schadt-Helfrich cell is sufficiently coherent to allow holographic recording; such cells are therefore proposed as a light valve in the page composer that is used for writing into a holographic memory [142]. Interferometry and holography with liquid crystals have been studied further [238b, 320, 616, 625, 654, 712, 1006, 1184, 1251]. Liquid crystals have also been proposed for the real-time statistical analysis of random patterns, e.g., the continuous monitoring of flow velocities, etc. [65].

14.6 *Miscellaneous Possibilities of Application*

The pronounced voltage threshold of electro-optical effects in liquid crystals permits their use in voltage indicators. For this purpose, the liquid crystal is sandwiched between two electrodes that have a large difference in surface resistivity [1277, 1278]. Liquid crystal voltmeters using the dynamic scattering [577, 1064] and the DAP mode [1115], e.g., for a still or motion-picture camera [121], and a flying spot scanner have been suggested [1064]. Other voltage-indicating devices contain cholesterics that are actuated with an accompanying visible color change [288, 334, 956].

A variety of electric components can be nondestructively tested with the aid of the electro-optical effects. For example, the dynamic scattering mode can be applied to the testing of insulating layers of all kinds on metallic or semiconducting substrates [673, 1171], dielectric films [1183], and thin-film $Al\text{-}Al_2O_3\text{-}Al$ capacitors [674]. Special low-resistance nematic materials can be used for detection of the reverse characteristics of *p-n* junctions and the detection of sensitivity of photodiodes [923, 1171]. In addition, integrated circuit operation and large-scale integration can be examined [166, 258, 345a].

Cholesterics can delineate discharge figures on an insulating surface with high resolving power ($\approx 1\,\mu m$); such surface discharge-figures in the liquid crystal method and the dust figure method have been compared [1058].

One patent claims the generation of electric energy by means of a photovoltaic effect in a nematic liquid crystal [526, see also 4:473].

Properties of liquid crystals changing in a visible or easily measured manner upon addition of small quantities of solute molecules permit the liquid crystals to be used in analytical detectors, utilizing for example, the change in cholesteric reflection color [285, 296, 300, 917, 1179, 1180] or the change in birefringence of nematics [852, 917]. Organic vapors can thus be nonselectively detected in the ppm range [215a, 386, 917, 918]. An apparatus has been developed for measuring halothane concentrations in anesthesia [496, 496a].

Chapter 11 has already discussed the application of lyophases, and here we only recall the stabilization of suspensions, emulsions, and foams [318, 637, 648a, 968, ref. 967] describes a sunscreen cream. The spinning in the liquid crystalline state was also discussed in that chapter; such fibers obtained from a lyophase can have moduli several times greater than those spun from similar isotropic solutions [145, 582], see also p. 544. Speaking of fibers, we also mention pitch fibers having a high mesophase content which can be converted by heat treatment into carbon fibers with a high Youngs modulus of elasticity and tensile strength [79, 689, 1004, 1052]. Among other materials, salts having liquid-crystalline melts have been suggested as additives for poly-α-olefins [514].

History, Reviews, and Books

15

15 History, Reviews, and Books

15.1 History

In the first two decades after its discovery, the "theoretical possibility of the existence of liquid crystals" — as B. Weinberg's discussion [145] is entitled — has repeatedly been denied. W. Nernst [108, 109], T. Rotarski [119, 120], G. Tammann [133–135], G. Wulf [147–149], and others [41, 145] mainly stated thermodynamic arguments against their existence. However, these authors offered an emulsion theory that was not recognized due to its inability to explain the crystal-optical observations. From today's view, this criticism might be associated with the pretransition phenomena and the second-order phase transitions discussed in the preceding chapters.

An essay on the history of liquid crystals will hardly be necessary here; the reader can be referred to a comprehensive treatment by H. Kelker [77]. Remarks on the history are found in the introductory text of the various chapters, especially chapter 1, and there is a special report on the liquid crystal research in Poland [103].

15.2 Reviews

The reviews dealing with liquid crystals are legion; those concerning a special theme, however, have already been mentioned in the preceding chapters. The key-word "review" in the subject index gives a compilation. Here, we only list the articles reporting a more or less comprehensive theme.

Brief and popular reviews:

in English and American
[2, 3, 6, 13–15, 18, 20–22, 24, 26, 29, 51–54, 56, 58, 62, 66, 76, 91–93, 95, 96, 104, 113–115, 125, 138, 140, 142, 146, 150]

in German
[10, 11, 49, 50, 61, 63, 65, 73, 74, 78, 81, 88, 98, 101, 107, 112, 126, 127, 136, 139, 143]

in French
[12, 28, 46, 57, 64]

in Russian
[30–32, 34–36, 38, 39, 55, 60, 72, 75]

in Japanese
[4, 67, 82–86, 89, 90, 105, 106, 110, 111, 121, 129, 130, 132, 137]

in other languages
[1, 7, 8, 37, 40, 42, 44, 45, 59, 68–71, 79, 80, 94, 102, 116, 118, 124, 131, 144]

Detailed reviews:

in English and American [1:21, 5, 9, 16, 17, 19, 23, 25, 43, 47, 48, 97, 122, 128]
in German [1:116, 1:262, 1:321, 2:870, 123, 141]
in French [1:62, 1:65, 27, 99, 117]
in other languages [1:30, 33, 87, 100].

15.3 Books and Monographs

Here we itemize all books and monographs in chronological order.

O. Lehmann. Flüssige Kristalle, sowie Plastizität von Kristallen im Allgemeinen, molekulare Umlagerungen und Aggregatzustandsänderungen (Engelmann: Leipzig). 1904. 264 pp.

R. Schenck. Kristallinische Flüssigkeiten und flüssige Kristalle (Engelmann: Leipzig). 1905. 159 pp.

O. Lehmann. Flüssige Kristalle und die Theorien des Lebens (Barth: Leipzig). 1. Aufl. 1906, 55 pp. 2. Aufl. 1908, 70 pp.

D. Vorländer. Kristallinisch-flüssige Substanzen (Enke: Stuttgart). 1908. 82 pp.

O. Lehmann. Das Krystallisationsmikroskop und die damit gemachten Entdeckungen, insbesondere die flüssigen Kristalle (Vieweg: Braunschweig). 1910. 112 pp.

O. Lehmann. Die neue Welt der flüssigen Kristalle und deren Bedeutung für Physik, Chemie, Technik und Biologie (Akad. Verlagsges.: Leipzig). 1911. 388 pp.

O. Lehmann. Die Lehre von den flüssigen Kristallen und ihre Beziehung zu den Problemen der Biologie (Bergmann: Wiesbaden). 1920. 254 pp.

O. Lehmann. Flüssige Kristalle und ihr scheinbares Leben (Voss: Leipzig). 1921. 72 pp.

D. Vorländer. Chemische Kristallographie der Flüssigkeiten (Akad. Verlagsges.: Leipzig). 1924. 90 pp.

R. Brauns. Flüssige Kristalle und Lebewesen (Schweizerbart: Stuttgart). 1931. 111 pp.

C. Weygand. Chemische Morphologie der Flüssigkeiten und Kristalle. Hand- und Jahrbuch der chemischen Physik, Bd. 2, Abschn. 3C (Akad. Verlagsges.: Leipzig). 1941. 192 pp.

G. W. Gray. Molecular structure and the properties of liquid crystals (Academic: New York). 1962. 314 pp.

I. G. Chistyakov. Liquid crystals (Nauka: Moscow). 1966. 127 pp.

P. Diehl, C. L. Khetrapal. NMR studies of molecules oriented in the nematic phase of liquid crystals. NMR basic principles and progress, Vol. 1 (Springer: Berlin). 1969. 95 pp.

T. Kallard. Liquid crystals and their applications (Optosonic Press: New York). 1970. 219 pp.

I. V. Sushkin, Edt. Structure and properties of liquid crystals, No. 2 (Proc. of the D. A. Furamanov Ivanovo State Pedagogical Institute, Vol. 77 (Ivan. Gos. Ped. Inst.: Ivanovo). 1970. 120 pp.

G. H. Brown, J. W. Doane, V. D. Neff. Structure and physical properties of liquid crystals (Butterworth: London). 1971. 86 pp.

T. Kallard. Liquid crystal devices (Optosonic Press: New York). 1973. 365 pp.

P. G. de Gennes. The physics of liquid crystals (Clarendon: Oxford). 1974. 333 pp.

D. Demus, H. Demus, H. Zaschke. Flüssige Kristalle in Tabellen (Deutscher Verlag Grundstoffind.: Leipzig). 1. Aufl. 1974, 356 pp. 2. Aufl. 1976, 356 pp.

G. W. Gray, P. A. Winsor, Edts. Ellis Horwood series in physical chemistry: Liquid crystals and plastic crystals (Halsted: New York). Vol. 1, 1974, 383 pp. Vol. 2, 1974, 314 pp.

G. H. Brown, Edt. Advances in liquid crystals (Academic: New York). Vol. 1, 1975, 320 pp. Vol. 2, 1976, 308 pp. Vol. 3, 1978, 278 pp.

C. L. Khetrapal, A. C. Kunwar, A. S. Tracey, P. Diehl. NMR studies in lyotropic liquid crystals. NMR basic principles and progress, Vol. 9 (Springer: Berlin). 1975. 85 pp.

Eastman Kodak Co. Liquid crystal bibliography (Kodak publication No. JJ-193: Rochester, N. Y.). 1975. Microfilm bibliography.

G. Meier, E. Sackmann, J. G. Grabmaier. Applications of liquid crystals (Springer: Berlin). 1975. 160 pp.

E. B. Priestley, P. J. Wojtowicz, P. Sheng, Edts. Introduction to liquid crystals (Plenum: New York). 1975. 356 pp.

E. L. Williams. Liquid crystals for electronic devices (NDC: Park Ridge). 1975. 264 pp.

L. Vistin, I. G. Chistyakov. New in life, science, and technology. Physics series, No. 8. Liquid crystals (Znanie: Moscow). 1975. 63 pp.

A. Adamczyk, Z. Strugalski. Liquid crystals (WNT: Warsaw). 1976. 204 pp.

M. Tobias. International handbook of liquid-crystal displays 1975–76 (Ovum: London). 1976. 181 pp.

I. G. Chistyakov. Liquid crystals (Ivanov. Gos. Univ.: Ivanovo). 1976. 158 pp.

S. Chandrasekhar. Liquid crystals (Cambridge Univ. Press: Cambridge). 1977. 342 pp.

M. Kléman. Points, lignes, parois dans les fluides anisotropes et les solides cristallins (Editions de Physique: Orsay). 1977. 181 pp.

S. P. Papkov, V. G. Kulichikhin. Liquid-crystalline state of Polymers (Khimiya: Moscow). 1977. 240 pp.

D. Demus, L. Richter. Textures of liquid crystals (Verlag Chemie: Weinheim). 1978. 228 pp.

L. Liebert, Edt. Solid state physics, supplement 14. Liquid crystals (Academic: New York). 1978. 328 pp.

L. M. Blinov. Electro- and magnetooptics of liquid crystals (Nauka: Moscow). 1978. 384 pp.

M. F. Smith. Liquid crystals (A bibliography with abstracts). Report for 1964 – July, 1978 (NTIS: Springfield). 1978. 296 pp.

A. P. Kapustin. Experimental study of liquid crystals (Nauka: Moscow). 1978. 368 pp.

G. H. Brown, J. J. Wolken. Liquid crystals and biological structures (Academic: New York). 1979. 200 pp.

15.4 Symposia

Only the most important symposia and the proceedings can be mentioned here.

Symposium on liquid crystals and anisotropic melts, 58th Meeting of the Faraday Society, London 24th and 25th April 1933. Trans. Faraday Soc. *29*, 881–1085 (1933)

Symposium on configuration and interactions of macromolecules and liquid crystals, Leeds 15th to 17th April 1958. Discussions of the Faraday Society *25* (1958) 235 pp.

Symposium on steric effects in conjugated systems, Hull 15th to 17th July 1958 (Butterworth: London). 1958. 181 pp.

Symposium on liquid crystals, London 13th and 14th December 1971. Symposia of the Faraday Society Nr. 5 (1971). 186 pp.

1st International liquid crystal conference, Kent 16th to 20th August 1965. G. H. Brown, G. J. Dienes, M. M. Labes, Edts. (Gordon and Breach: New York). 1967. 486 pp.

2nd International liquid crystal conference, Kent 12th to 16th August 1968. Molecular Crystals and Liquid Crystals *7* (1968), 486 pp. and *8* (1969), 487 pp.

3rd International liquid crystal conference, Berlin 24th to 28th August 1970. G. H. Brown, M. M. Labes, Edts. (Gordon and Breach: London). 1972. 1141 pp.

4th International liquid crystal conference, Kent 21st to 25th August 1972. Molecular Crystals and Liquid Crystals *21*, 1–274 (1973).

5th International liquid crystal conference, Stockholm 17th to 21st June 1974. Journal de Physique Colloque C-1 *36* (1975). 421 pp.

6th International liquid crystal conference, Kent 23rd to 27th August 1976. Molecular Crystals and Liquid Crystals Vols. *37* (1976), *38, 40, 42* (1977)

7th International liquid crystal conference, Bordeaux 1st to 5th July 1978. Journal de Physique Colloque C-3, *40* (1979). 536 pp.

5th International liquid crystal conference, Stockholm 17th to 21st June 1974. Lyotropic liquid crystals and the structure of biomembranes. S. Friberg, Edt. Advances in Chemistry Series 152 (ACS: Washington). 1976. 156 pp.

Symposium on ordered fluids and liquid crystals, 150th Meeting of the ACS, Atlantic City 14th and 15th Sept. 1965. R. F. Gould, Edt. Advances in Chemistry Series, No. 63 (ACS: Washington). 1967. 332 pp.

Symposium on liquid crystals and ordered fluids, 158th Meeting of the ACS, New York 10th to 12th Sept. 1969. J. F. Johnson, R. S. Porter, Edts. (Plenum: New York). 1970. 494 pp.

Symposium on liquid crystals and ordered fluids, Chicago 27th to 31st August 1973. J. F. Johnson, R. S. Porter, Edts. (Plenum: New York). 1974. 783 pp.

Symposium on liquid crystals and ordered fluids. Chicago August 30th to September 4th 1976. J. F. Johnson, R. S. Porter, Edts. (Plenum: New York). 1978. 559 pp.

1st Gordon research conference on the chemistry and physics of liquid crystals, Santa Barbara 17th to 22nd January 1972

2nd Gordon research conference on the chemistry and physics of liquid crystals, Santa Barbara 14th to 18th January 1974

3rd Gordon research conference on the chemistry and physics of liquid crystals, Santa Barbara 5th to 9th January 1976

Symposium on liquid crystals, Meeting of the ACS, Washington 15th and 16th September 1971

ACS Symposium Series, Vol. 74: Mesomorphic Order in Polymers and Polymerization in Liquid-Crystalline Media. Chicago, 29th August to 2nd September 1977. A. Blumstein, Edt. (ACS: Washington). 1978. 264 pp.

Colloque sur les cristaux liquides, Montpellier 5th to 7th June 1969. Journal de Physique, Colloque C4, *30*, No. 11–12 (1969). 126 pp.

Colloque sur les cristaux liquides, Pont-a-mousson, 27th June to 3rd July 1971

Symposium on thermotropic smectics and their applications, Les Arcs, France, 15th to 18th December 1975. Journal de Physique *37*, C-3 (1976). 164 pp.

Symposium on physics and applications of smectic and lyotropic liquid crystals, Madonna di Campiglio, 9th to 13th January 1978

Liquid Crystals. Proceedings of the International Conference, Held at the Raman Research Institute, Bangalore, 3–8 December 1973. S. Chandrasekhar, Edt. (Ind. Acad. Sci.: Bangalore). 1975. 568 pp.

Symposium on non-emissive electro-optic displays, BBC Forschungszentrum Baden/ Schweiz 29th and 30th September 1975

Symposium über Übergänge zwischen Ordnung und Unordnung in festen und flüssigen Phasen, Darmstadt 28th and 29th December 1938. Z. Elektrochem. *45*, (1939)

Symposium über physikalisch-chemische Aspekte flüssiger Kristalle, Königstein/Ts. 20th to 22nd March 1974. Ber. Bunsenges. Phys. Chem. *78*, (9), 818–965 (1974)

A national annual meeting is held since 1971 at the Institut für Angewandte Festkörperphysik der Fraunhofergesellschaft, Freiburg i. Br.

1st Symposium on liquid crystals, Ivanovo 17th to 19th November 1970. Acad. Sci. USSR *99* (1972). 336 pp.

2nd Symposium on liquid crystals, Ivanovo 27th to 29th June 1972. I. G. Chistyakov et al., Edts. (Ivanov. Gos. Pedagog. Inst.: Ivanovo). 1973. 326 pp.

1. Flüssigkristall-Konferenz sozialistischer Länder, Halle, Januar 1976. H. Sackmann, Herausg. (Wiss. Beiträge Martin Luther Univ. Halle-Wittenberg: Halle). 1976. 113 pp.

2nd Liquid crystal conference of socialist countries, Sunny Beach, Bulgaria, 27th to 30th September 1977.

The first symposium on liquid crystals, still being prepared, has never taken place, and only a synopsis of lectures and discussions has been published in Z. Krist. *79*, 1–347 (1931).

References*

Chapter 1

[1] P. Adamski, A. Dylik-Gromiec, S. Klimczyk, M. Wojciechowski, Mol. Cryst. Liq. Cryst. 35, 171 (1976), C.A. 85, 114877b (1976); Study of isothermal crystallization in cholesteryl nonanoate by optical methods. [1a] A. Adamszyk, Mol. Cryst. Liq. Cryst. 42, 81 (1977); Stable inhomogeneities of mixtures and permanent liquid-crystalline diffraction gratings. [2] S. L. Arora, T. R. Taylor, J. L. Fergason, Liq. Cryst. Ord. Fluids 1970, 321, C.A. 78, 152452x (1973); Polymorphism of smectics A. [3] E. M. Barrall, II, R. S. Porter, J. F. Johnson, Mol. Cryst. 3, 103 (1967), C.A. 68, 7279k (1968); Polymorphism of cholesteryl esters: differential thermal and microscopic measurements on cholesteryl myristate. [4] E. M. Barrall, II, M. A. Sweeney, Mol. Cryst. 5, 257 (1969), C.A. 70, 119047f (1969); Depolarized light intensity and optical microscopy of some mesogens. [5] M. J. Vogel, E. M. Barrall, II, C. P. Mignosa, Mol. Cryst. Liq. Cryst. 15, 49 (1971), C.A. 75, 144736r (1971); Polymorphism of cholesteryl palmitate. [6] G. Baur, V. Wittwer, Phys. Lett. A 56, 142 (1976), Determination of tilt angles at surfaces of substrates in liquid-crystal cells. [7] P. Bennema, G. H. Gilmer, North-Holland Ser. Cryst. Growth 1, 263 (1973), C.A. 84, 67921d (1976); Kinetics of crystal growth. [8] D. W. Berreman, Phys. Rev. Lett. 28, 1683 (1972), C.A. 77, 67216h (1972); Solid surface shape and the alignment of an adjacent nematic. [9] D. W. Berreman, Mol. Cryst. Liq. Cryst. 23, 215 (1973), C.A. 80, 88151k (1974); Alignment of liquid crystals by grooved surfaces. [10] J. Billard, R. Cerne, Mol. Cryst. 2, 27 (1967), C.A. 66, 32489v (1967); Recherche d'une phase nématique à la température ambiante.
[11] M. Leclercq, J. Billard, J. Jacques, C. R. Acad. Sci., Paris, Sér. C 266, 654 (1968), C.A. 69, 51443d (1968); Composés cholestériques par dédoublement de racémiques nématiques. [12] J. P.

---

* The titles have been abridged using the symbols of phases described on pp. 6 and 7 and the symbols of substances introduced in section 1.7.

Meunier, J. Billard, Mol. Cryst. Liq. Cryst. *7*, 421 (1969), C.A. *71*, 85804d (1969); Faciès d'une stase smectique. [13] M. Leclercq, J. Billard, J. Jacques, Mol. Cryst. Liq. Cryst. *8*, 367 (1969), C.A. *71*, 85223p (1969); Séparation de racémiques nématiques en deux antipodes cholésteriques. [14] J. Billard, Bull. Soc. Fr. Mineral. Cristallogr. *95*, 206 (1972), C.A. *77*, 131745j (1972); Les phases mésomorphes et leur identification. [15] W. Z. Urbach, J. Billard, C. R. Acad. Sci., Sér. B, *274*, 1287 (1972), C.A. *77*, 106427z (1972); Sur les phases smectiques C. [16] J. M. Blakely, Introduction to the properties of crystal surfaces (Pergamon: Elmsford). 1973. 272 pp., C.A. *80*, 75348e (1974). [17] Y. Bouligand, J. Phys. (Paris) C-4, *30*, 90 (1969); Sur l'existence de "pseudomorphoses cholésteriques" chez divers organismes vivants. [18] Y. Bouligand, J. Phys. (Paris) *33*, 715 (1972), C.A. *78*, 34877s (1973); Les champs polygonaux dans les cholestériques. [19] Y. Bouligand, J. Phys. (Paris) *34*, 603 (1973), C.A. *79*, 84266h (1973); Les plages a éventails dans les cholestériques. [20] Y. Bouligand, J. Phys. (Paris), C-1, *36*, 173 (1975); Presentation of a film entitled: "Textures of nematic and cholesteric liquid crystals".

[21] G. H. Brown, W. G. Shaw, Chem. Rev. *57*, 1049 (1957), C.A. *52*, 5071d (1958); The mesomorphic state. Liquid crystals. [22] G. H. Brown, Advances in liquid crystals (Academic: New York). Vol. 1: 1975. 320 pp., Vol. 2: 1976. 308 pp., Vol. 3: 1978. 278 pp., C.A. *84*, 11196d (1976) and *86*, 163908s (1977). [23] V. K. Burima, Visn. L'vov. Univ., Ser. Fiz., No. 7, 78 (1972), C.A. *79*, 24419a (1973); Structure of cholesteryl succinate liquid crystals. [24] V. K. Burima, Zh. Fis. Khim. *48*, 777 (1974), C.A. *81*, 42522z (1974); Texture of some representatives of a homologous series of cholesteryl esters of dicarboxylic acids studied in a polarization microscope. Electron-microscopic textures. [25] C. Burri, Das Polarisationsmikroskop (Birkhäuser: Basel). 1950. C.A. *44*, 7596d (1950). [25a] S. Chandrasekhar, Liquid crystals (Cambridge Univ. Press: Cambridge). 1977. 342 pp., C.A. *87*, 192395y (1977). [26] P. Chatelain, Compt. rend. *213*, 875 (1941), C.A. *37*, 3987⁵ (1943); Sur l'orientation des cristaux liquides par les surfaces frottées; étude expérimentale. [27] P. Chatelain, Compt. rend. *214*, 32 (1942), C.A. *37*, 2238⁹ (1943); Sur l'orientation des cristaux liquides par les surfaces frottées; étude théoretique et conséquences. [28] P. Chatelain, Bull. soc. franç. minéral. *66*, 105 (1943), C.A. *39*, 4784⁵ (1945); Sur l'orientation des cristaux liquides par les surfaces frottées. [29] P. Chatelain, M. Brunet, J. Cano, Service du Film de Recherche, Paris 96, boulevard Raspail. Nr. 22246-44 548 95-25; Propriétés optiques des cristaux liquides des types nématiques et cholestériques. [30] V. A. Usol'tseva, I. G. Chistyakov, Uspekhi Khim. *32*, 1124 (1963), C.A. *60*, 78a (1964); Chemical peculiarities, structure, and properties of liquid crystals.

[31] I. G. Chistyakov, Kristallografiya *8*, 123 (1963), C.A. *58*, 12033c (1963); The growth of smectic domains. [32] I. G. Chistyakov, E. A. Kosterin, Rost Kristallov, Akad. Nauk SSSR, Inst. Kristallogr. *4*, 68 (1964), C.A. *61*, 15450e (1964); Vitrified liquid crystal films. [33] I. G. Chistyakov, Rost Kristallov, Akad. Nauk SSSR, Inst. Kristallogr. *4*, 74 (1964), C.A. *61*, 15450c (1964); Growth of single-crystal nematic films. [34] I. G. Chistyakov, L. A. Gusakova, Kristallografiya *14*, 153 (1969), C.A. *70*, 100700s (1969); Textures of cholesterics. [35] I. G. Chistyakov, L. S. Shabishev, R. I. Yarenov, L. A. Gusakova, Mol. Cryst. Liq. Cryst. *7*, 279 (1969), C.A. *71*, 95875s (1969); Smectic polymorphism. [36] L. K. Vistin, I. G. Chistyakov, New in Life, Science, and Technology. Physics Series, No. 8: Liquid Crystals. (Znanie: Moscow). 1975. 63 pp., C.A. *84*, 82970n (1976). [37] P. E. Cladis, M. Kléman, P. Pieranski, C. R. Acad. Sci., Sér. B, *273*, 275 (1971), C.A. *75*, 134141d (1971); Sur une nouvelle méthode de décoration de la phase mésomorphe du MBBA. [38] L. T. Creagh, A. R. Kmetz, Mol. Cryst. Liq. Cryst. *24*, 59 (1973), C.A. *81*, 142298z (1974); Mechanism of surface alignment in nematics. [39] W. A. Crossland, J. H. Morrissy, B. Needham, J. Phys. D *9*, 2001 (1976), C.A. *85*, 134830j (1976); Tilt angle measurements of nematic cyano-biphenyls aligned by obliquely evaporated films. [40] E. Dubois-Violette, P. G. de Gennes, J. Colloid Interface Sci. *57*, 403 (1976), C.A. *86*, 36551v (1977); Effects of long range van der Waals forces on the anchoring of a nematic at an interface.

[41] A. Kolbe, D. Demus, Z. Naturforsch. A *23*, 1237 (1968), C.A. *69*, 105488t (1968); H-D-Isotopeneffekte an kristallin flüssigen Alkoxybenzoesäuren. [42] D. Demus, Wiss. Z. Martin-Luther-Univ., Halle-Wittenberg, Math.-Naturwiss. Reihe *21*, 41 (1972), C.A. *77*, 157363h (1972); Über die Texturen der s_B Modifikationen. [43] D. Demus, H. Demus, H. Zaschke, Flüssige Kristalle in Tabellen (Deut. Verlag Grundstoffind.: Leipzig). 2nd ed. 1976. 356 pp., C.A. *87*, 192396z (1977). [44] D. Demus, Krist. Tech. *10*, 933 (1975), C.A. *83*, 124487x (1975); Schlieren textures in smectics. [44a] D. Demus, L. Richter, Textures of Liquid Crystals (Verlag Chemie: Weinheim). 1978. 228 pp. [44b] K. U. Deniz, T. K. Bhattacharya, C. Manohar, Proc. Nucl. Phys. Solid State Phys. Symp. *18C*, 299 (1975), C.A. *87*, 93790s (1977); Effects of thin metal films on the alignment properties of MBBA. [45] A. de Vries, Mol. Cryst. Liq. Cryst. *10*, 31 (1970), C.A. *73*, 49655p (1970); Evidence for the existence of more than one type of nematic phase. [46] J. F. Dreyer, J. Phys. (Paris), C-4, *30*, 114 (1969), C.A. *72*,

116188h (1970); Light polarization from films of lyotropic nematics. [47] J. F. Dreyer, 3rd Int. Liq. Cryst. Conf. Berlin, Aug. 1970, S 1.1; Epitaxy of nematics. [48] J. C. Dubois, M. Gazard, A. Zann, Appl. Phys. Lett. *24*, 297 (1974), C.A. *80*, 137836y (1974); Plasma-polymerized films as orientating layers for liquid crystals. [49] J. C. Dubois, M. Gazard, A. Zann, J. Appl. Phys. *47*, 1270 (1976), C.A. *85*, 12489j (1976); Liquid-crystal orientation induced by polymeric surfaces. [50] Eastman Kodak Co., Department 454, Rochester, N. Y. Liquid Crystal Bibliography. No. JJ-193 (1975).

[51] R. D. Ennulat, Mol. Cryst. *3*, 405 (1968), C.A. *69*, 39548j (1968); Mesophase identification. [52] M. Françon, R. Genty, F. Taboury, Compt. rend. *230*, 2082 (1950), C.A. *44*, 10425e (1950); Étude des couches monomoléculaires par contraste de phase. [53] V. Frederiks, Acta Physicochim. URSS *3*, 741 (1935), C.A. *30*, 4063⁷ (1936); Moderne Strukturvorstellungen über anisotrope Flüssigkeiten und ihre Grundlagen. [54] G. Friedel, Bull. soc. franç. min. *30*, 69 (1907), C.A. *1*, 1833⁹ (1907); Observations relatent aux cristaux fluides. [55] G. Friedel, F. Grandjean, Compt. rend. *151*, 327, 442 (1910), Bull. soc. franç. min. *33*, 192 (1910), C.A. *5*, 1219² (1911); Les liquides anisotropes de Lehmann. [56] G. Friedel, F. Grandjean, Compt. rend. *151*, 762 (1910), C.A. *5*, 1219³ (1911); Les liquides à coniques focales. [57] G. Friedel, F. Grandjean, Compt. rend. *151*, 988 (1910), C.A. *5*, 1219⁶ (1911); Liquides anisotropes. [58] G. Friedel, F. Grandjean, Compt. rend. *152*, 322 (1911), C.A. *5*, 1545¹ (1911); Structure des liquides à coniques focales. [59] G. Friedel, J. chim. phys. *11*, 478 (1913), C.A. *8*, 278⁵ (1914); Examen critical de la théorie de Curie-Wulff sur les formes cristallines. Application aux liquides anisotropes. [60] G. Friedel, L. Royer, Compt. rend. *173*, 1320 (1921), C.A. *16*, 1527² (1922); Mélanges de liquides anisotropes et l'identité des liquides stratifiés de Grandjean avec les liquides du type PAP.

[61] G. Friedel, L. Royer, Compt. rend. *174*, 1523, 1607 (1922), C.A. *16*, 3243 (1922); Liquides à plans équidistants de Grandjean. [62] G. Friedel, Ann. Physique *18*, 273 (1922), C.A. *17*, 3267⁷ (1923); Les état mésomorphes de la matière. [63] G. Friedel, Compt. rend. *176*, 475 (1923), C.A. *17*, 3633⁷ (1923); Corps cholestériques. [64] G. Friedel, Colloid Chemistry, edited by J. Alexander, Vol. I, p. 102, The Chemical Cataloque Company, Inc. New York 1926; Mesomorphic states of matter. [65] G. Friedel, E. Friedel, Z. Krist. *79*, 1 (1931), C.A. *26*, 2629⁴ (1932); Les propriétés physiques des states mésomorphes en général et leur importance comme principe de classification. [66] G. Friedel, E. Friedel, J. Phys. et Radium *2*, 133 (1931), Texture à coniques focales dans les corps mésomorphes. [67] R. Fürth, K. Sitte, Ann. Physik *30*, 388 (1937), C.A. *32*, 15⁴ (1938); Begründung der Schwarmtheorie der flüssigen Kristalle. [68] R. Fürth, K. Sitte, Ann. Physik. *31*, 579 (1938), C.A. *32*, 4848⁹ (1938); Bemerkungen zu der vorstehenden Arbeit v. H. Zocher: „Über die Kontinuumtheorie und die Schwarmtheorie der nematischen Phasen". [69] Y. Furuhata, K. Toriyama, Appl. Phys. Lett. *23*, 361 (1973), C.A. *80*, 31585a (1974); New liquid-crystal method for revealing ferroelectric domains. [69a] E. K. Galanov, G. K. Kostyuk, M. V. Mukhina, I. D. Kostrov, R. I. Mel'nik, Zh. Strukt. Khim. *17*, 698 (1976), C.A. *86*, 10862x (1977); Phase transitions in some cholesterics. [70] L. Gattermann, A. Ritschke, Ber. *23*, 1738 (1890), Über Azoxyphenoläther.

[71] L. Gattermann, Liebigs Ann. Chem. *347*, 347 (1906); Synthesen aromatischer Aldehyde. [72] L. Gattermann, Liebigs Ann. Chem. *357*, 313 (1908); C.A. *2*, 820 (1908); Synthesen aromatischer Aldehyde. [73] P. Gaubert, Bull. soc. franç. min. *32*, 62 (1909); Cristaux liquides de quelques composés nouveaux de la cholestérine et sur ceux de l'ergostérine. [74] P. Gaubert, Compt. rend. *179*, 1148 (1924); Polarisation circulaire de la lumière réfléchie par les insectes. [75] P. Gaubert, Compt. rend. *200*, 304 (1935), Anisotropie des liquides autour des bulles gazeuses. [76] P. Gaubert, Compt. rend. *205*, 997 (1937), C.A. *32*, 406⁶ (1938); Cristaux liquides obtenus par sublimation. [77] P. Gaubert, Compt. rend. *206*, 62 (1938), C.A. *32*, 1995³ (1938); Formation et orientation de cristaux mous maclés dans les gouttes liquides anisotropes de PAA. [78] P. Gaubert, Compt. rend. *206*, 1030 (1938), C.A. *32*, 4034⁶ (1938); Rôle des matières étrangères dans la structure des liquides crystallins. [79] A. K. Ghatak, L. S. Kothari, Introduction to lattice dynamics (Addison-Wesley: Reading). 1972. 234 pp., C.A. *78*, 49369b (1973). [80] F. Grandjean, Compt. rend. *163*, 394 (1916); L'orientation des liquides anisotropes sur les cristaux.

[81] F. Grandjean, Bull. soc. franç. min. *39*, 164 (1916), C.A. *11*, 1939⁶ (1917); L'orientation des liquides anisotropes sur les cristaux. [82] F. Grandjean, Bull. soc. franç. min. *40*, 69 (1917), C.A. *13*, 809³ (1919); L'orientation des liquides anisotropes au contact des cristaux. [83] F. Grandjean, Compt. rend. *164*, 105 (1917), C.A. *11*, 750⁴ (1917); L'orientation des liquides anisotropes sur les clivages des cristaux. [84] F. Grandjean, Compt. rend. *164*, 431 (1917), C.A. *11*, 2425⁶ (1917); La visibilité, au-dessus de la température de fusion isotrope, des plages de contact entre les liquides anisotropes et les cristaux. [85] F. Grandjean, Compt. rend. *164*, 636 (1917), C.A. *11*, 2064⁶ (1917); Orientation

des sels de cholestérine et des oléates liquides anisotropes sur les cristaux. [86] F. Grandjean, Compt. rend. *166*, 165 (1917), C.A. *12*, 1141[1] (1918); Structure en gradins dans certains liquides anisotropes. [87] F. Grandjean, Compt. rend. *172*, 71 (1921), C.A. *15*, 1448[2] (1921); Existence de plans differenciés équidistants normaux à l'axe optique dans les liquides anisotropes. [88] W. G. Gray, Nature *172*, 1137 (1953), C.A. *48*, 8597h (1954); Heating instrument for the accurate determination of mesomorphic and polymorphic transition temperatures. [89] G. W. Gray, A. Ibbotson, Nature *176*, 1160 (1955), C.A. *50*, 6104b (1956); Projection instrument to aid in the determination of mesomorphic and polymorphic transition temperatures. [90] G. W. Gray, Molecular structure and the properties of liquid crystals (Academic: London). 1962. 314 pp., C.A. *57*, 1658g (1962).

[91] G. W. Gray, Mol. Cryst. *2*, 189 (1966), C.A. *66*, 69738b (1967); Nomenclature of liquid crystals. [92] G. W. Gray, P. A. Winsor, Liq. Cryst. Plast. Cryst. *1*, 1 (1974), C.A. *83*, 19074n (1975); Liquid crystals and plastic crystals. Introduction. [93] G. W. Gray, P. A. Winsor, Editors Ellis Horwood Series in Physical Chemistry: Liquid Crystals and Plastic Crystals. (Halsted: New York) Vol. 1, 383 pp. (1974); Vol. 2, 314 pp. (1974); C.A. *82*, 163659d; *83*, 200578v (1975). [94] D. Coates, G. W. Gray, J. Chem. Soc., Chem. Commun. *1974*, 101, C.A. *81*, 7182s (1974); Classification of the smectic phase of cholesteryl n-alkanoates. [95] G. W. Gray, J. E. Lydon, Nature *252*, 221 (1974), C.A. *82*, 148647e (1975); New type of smectic mesophase? [96] G. W. Gray, Int. Ser. Monogr. Sci. Solid State *6*, 601 (1975), C.A. *83*, 69174r (1975); Liquid crystals. [97] D. Coates, G. W. Gray, Microscope *24*, 117 (1976), C.A. *85*, 39340r (1976); Structures and microscopic textures of smectics. [98] G. W. Gray, P. A. Winsor, Adv. Chem. Ser. *152*, 1 (1976), C.A. *85*, 152085z (1976); Generic relations between nonamphiphilic and amphiphilic mesophases of the "fused" type. Relation of cubic mesophases ("plastic crystals") formed by nonamphiphilic globular molecules to cubic mesophases of the amphiphilic series. [99] H. Hakemi, M. M. Labes, J. Chem. Phys. *63*, 3708 (1975), C.A. *84*, 24660a (1976); Self-diffusion coefficients of a nematic via an optical method. [100] I. Haller, Appl. Phys. Lett. *24*, 349 (1974), C.A. *81*, 7091m (1974); Alignment and wetting properties of nematics.

[101] J. T. Hartman, Diss. Drexel Univ., Philadelphia, Pa. 1973, 152 pp. Avail. Univ. Microfilms, Ann Arbor, Mich., Order No. 73-25, 523, C.A. *80*, 20612f (1974); Fluctuation phenomena in solid-state photoconductors and liquid crystals. [102] N. H. Hartshorne, M. H. Roberts, J. chem. Soc. *1951*, 1097, C.A. *45*, 8859i (1951); Investigation of the linear rate of transformation of monoclinic into rhombic sulphur. [103] N. H. Hartshorne, A. Stuart, Crystals and the Polarizing Microscope, 4th ed. (Elsevier: New York). *1970*. 614 pp., C.A. *74*, 131561j (1971). [104] N. H. Hartshorne, Microscope *19*, 424 (1971); Study of a liquid crystal system. [105] N. H. Hartshorne, Liq. Cryst. Plast. Cryst. *2*, 24 (1974), C.A. *83*, 185576x (1975); Optical properties of liquid crystals. [106] N. H. Hartshorne, Microscope *23*, 177 (1975), C.A. *83*, 156019n (1975); Hot-wire stage and its applications. [107] G. Heppke, F. Oestreicher, Z. Naturforsch. A *32*, 899 (1977), C.A. *87*, 144295u (1977); Bestimmung des Helixdrehsinnes cholesterinischer Phasen mit der Grandjean-Cano-Methode. [108] N. Isaert, B. Soulestin, J. Malthete, Mol. Cryst. Liq. Cryst. *37*, 321 (1976), C.A. *86*, 113912t (1977); Détermination des sens de torsion absolus de phases cholestériques et s_C chirales. [109] S. A. Jabarin, Diss. Univ. Massachusetts, Amherst, Mass. 1971, Avail. Univ. Microfilms, Ann. Arbor, Mich. Order No. 72-3750, C.A. *76*, 127261k (1972); Light scattering and microscopic investigations of mesophase transitions of cholesteryl-myristate. [110] S. A. Jabarin, R. S. Stein, J. Phys. Chem. *77*, 399 (1973), C.A. *78*, 89485t (1973); Light scattering and microscopic investigations of mesophase transition of cholesteryl myristate. I. Morphology of the cholesteric phase.

[111] S. A. Jabarin, R. S. Stein, J. Phys. Chem. *77*, 409 (1973), C.A. *78*, 89481p (1973); II. Kinetics of spherulite formation (cf. [110]). [112] J. L. Janning, Appl. Phys. Lett. *21*, 173 (1972), C.A. *77*, 106347y (1972); Thin-film surface orientation for liquid crystals. [113] F. J. Kahn, Appl. Phys. Lett. *22*, 386 (1973), C.A. *78*, 152482g (1973); Orientation of liquid crystals by surface coupling agents. [114] J. F. Kahn, G. N. Taylor, H. Schonhorn, Proc. IEEE *61*, 823 (1973), C.A. *79*, 97733h (1973); Surface-produced alignment of liquid crystals. [115] W. Kast, Physik. Z. *29*, 293 (1928), C.A. *22*, 2866[9] (1928); Grenzflächenwirkungen in anisotropen Flüssigkeiten. [116] W. Kast, Landolt-Börnstein: 6. Aufl. (Springer: Berlin). Bd. II, Teil 2a, S. 266 (1960); Zahlenwerte und Funktionen aus Physik, Chemie, Astronomie, Geophysik und Technik. [117] H. Kelker, E. v. Schivizhoffen, Advan. Chromatogr. *6*, 247 (1968), C.A. *69*, 69882f (1968); Liquid crystals in gas chromatography. [118] H. Kelker, Mol. Cryst. Liq. Cryst. *15*, 347 (1972), C.A. *76*, 64590u (1972); Konzentrationsbedingte Grandjean-Stufen cholesterinischer Phasen. [119] H. Kelker, R. Hatz, Ber. Bunsenges. Phys. Chem. *78*, 819 (1974), C.A. *82*, 10060b (1975); Morphologie und Phasenübergänge flüssig-kristalliner Systeme. [120] A. I. Kitaigorods-

kii, Molecular crystals and molecules (Academic: New York). 1973. 572 pp. Translated from Russ., C.A. *80*, 125791s (1974).

[**121**] A. Koehler, H. Dunken, R. Korch, Z. Chem. *16*, 199 (1976), C.A. *85*, 70893x (1976); Einfluß der Substratoberfläche auf Fließtexturen nematischer Flüssigkristallschichten. [**122**] A. Koehler, H. H. Dunger, E. Dietel, Z. Phys. Chem. (Leipzig) *257*, 1049 (1976), C.A. *86*, 24628n (1977); Einfluß der Substratoberfläche auf den i/n Phasenübergang. [**123**] L. Kofler, A. Kofler, Thermo-Mikro-Methoden zur Kennzeichnung organischer Stoffe und Stoffgemische (Verlag Chemie: Weinheim). 1954. 608 pp., C.A. *48*, 10046d (1954). [**123a**] H. Kozawaguchi, M. Wada, Oyo Butsuri *45*, 771 (1976), C.A. *85*, 200758v (1976); Determination of the chirality of cholesterics by observing a circularly polarized light. [**124**] F. A. Kroeger, The chemistry of imperfect crystals (North Holland: Amsterdam). 2nd ed. 1974. Vol. 1, 328 pp., C.A. *81*, 128090u (1974), Vol. 2, 1000 pp., C.A. *83*, 156245h (1975), Vol. 3, 306 pp., C.A. *83*, 156244g (1975). [**125**] F. Kroupa, L. Lejcek, Cesk. Cas. Fyz. *26*, 128 (1976), C.A. *85*, 85583v (1976); Disclinations — defects in solid and liquid crystals. [**126**] H. Krüger, W. Greubel, 3. Arbeitstagung Flüssigkristalle, Inst. Angew. Festkörperphysik Freiburg, 11. 5. 1973; Einfluß von Oberflächen auf die Orientierung von Flüssigkristallschichten. [**126a**] T. P. Kudryashova, V. V. Polyukhov, V. V. Tuturina, Sintez Vysokomolekul. Produktov na Osnove Sapropelitov i Kremniiorgan. Soedin. *1976*, (Ch. 2), 16, C.A. *87*, 75827e (1977); Thermooptical behavior of nematocholesteric systems. [**127**] K. S. Kunihisa, S. Hagiwara, Bull. Chem. Soc. Jpn. *49*, 1204 (1976), C.A. *85*, 54887t (1976); Study of cholesteryl formate and cholesteryl octanoate by means of thermal analytic microscopy with a thermoelement. [**128**] O. Lehmann, Diss. Straßburg 1877; Physikalische Isomerie. [**128a**] O. Lehmann, Molekularphysik mit besonderer Berücksichtigung mikroskopischer Untersuchungen und Anleitung zu solchen. 2. Band (W. Engelmann: Leipzig) 1889. 697 pp. [**129**] O. Lehmann, Z. physik. Chem. *4*, 462 (1889); Fließende Kristalle. [**130**] O. Lehmann, Z. physik. Chem. *5*, 427 (1890); Struktur krystallinischer Flüssigkeiten.

[**131**] O. Lehmann, Ann. Physik *40*, 401 (1890); Tropfbarflüssige Kristalle. [**132**] O. Lehmann, Ann. Physik *40*, 525 (1890); Krystallinische Flüssigkeiten. [**133**] O. Lehmann, Ann. Physik *41*, 525 (1890); Krystallinische Flüssigkeiten. [**134**] O. Lehmann, Z. Kristallogr. *18*, 457 (1890); Definition des Begriffes „Krystall". [**135**] O. Lehmann, Z. Kristallogr. *18*, 464 (1890); Einige Fälle von Allotropie. [**136**] O. Lehmann, Ann. Physik *56*, 771 (1895); Contaktbewegung und Myelinformen. [**137**] O. Lehmann, Z. physik. Chem. *18*, 91 (1895); Zusammenfließen und Ausheilen fließend-weicher Kristalle. [**138**] O. Lehmann, Ann. Physik *2*, 649 (1900); Struktur, System und magnetisches Verhalten flüssiger Kristalle und deren Mischbarkeit mit festen. [**139**] O. Lehmann, Verh. deut. phys. Ges. *16*, 72 (1900); Struktur, System und magnetisches Verhalten flüssiger Krystalle. [**140**] O. Lehmann, Verh. Nat. Ver. Karlsruhe *13*, 619, 630 (1900); Flüssige Krystalle.

[**141**] O. Lehmann, Ann. Physik *5*, 236 (1901); Flüssige Kristalle, Entgegnungen auf die Bemerkungen des Herrn G. Tammann. [**142**] O. Lehmann, Ann. Physik *8*, 908 (1902); Über künstlichen Dichroismus bei flüssigen Kristallen und Herrn Tammanns Ansicht. [**143**] O. Lehmann, Ann. Physik *9*, 727 (1902); Berichtigung zu ibid. *8*, 914 (1902). [**144**] O. Lehmann, Ann. Physik. *12*, 311 (1903); Plastische, fließende und flüssige Kristalle; erzwungene und spontane Homöotropie derselben. [**145**] O. Lehmann, Flüssige Kristalle, sowie Plastizität von Kristallen im Allgemeinen, molekulare Umlagerungen und Aggregatzustandsänderungen. W. Engelmann, Leipzig 1904, 264 Seiten. [**146**] O. Lehmann, Ann. Physik *16*, 160 (1905); Flüssige Misch- und Schichtkristalle. [**147**] O. Lehmann, Ann. Physik *17*, 728 (1905); Gleichgewichtsform fester und flüssiger Kristalle. [**148**] O. Lehmann, Ann. Physik *18*, 796 (1905); Näherungsweise Bestimmung der Doppelbrechung fester und flüssiger Kristalle. [**149**] O. Lehmann, Ann. Physik *18*, 808 (1905); Drehung der Polarisationsebene und der Absorptionsrichtung bei flüssigen Kristallen. [**150**] O. Lehmann, Z. Elektrochemie *11*, 955 (1905); Demonstration der flüssigen Kristalle bei der Deutschen Bunsengesellschaft.

[**151**] O. Lehmann, Chem. Ztg. *30*, 1 (1906); Scheinbar lebende weiche Kristalle. [**152**] O. Lehmann, Ann. Physik *19*, 22 (1906); Fließend-kristallinische Trichiten, deren Kraftwirkungen und Bewegungserscheinungen. [**153**] O. Lehmann, Ann. Physik *19*, 1 (1906); Homöotropie und Zwillingsbildung bei fließend-weichen Kristallen. [**154**] O. Lehmann, Ann. Physik *20*, 63 (1906); Struktur der scheinbar lebenden Kristalle. [**155**] O. Lehmann, Ann. Physik *20*, 77 (1906); Kontinuität der Aggregatzustände und die flüssigen Kristalle. [**156**] O. Lehmann, Ann. Physik *21*, 181 (1906), C.A. *1*, 280 (1907); Erweiterung des Existenzbereiches flüssiger Kristalle durch Beimischungen. [**157**] O. Lehmann, Ann. Physik *21*, 381 (1906); Molekulare Drehmomente bei enantiotroper Umwandlung. [**158**] O. Lehmann, Archiv f. Entwicklungsmechanik der Organismen, XXI. Band, 3. Heft (1906); Fließende Kristalle und Organismen. [**159**] O. Lehmann, Verh. Nat. Ver. Karlsruhe *19*, 107 (1906); Bedeutung der flüssigen

und scheinbar lebenden Kristalle für die Theorie der Molekularkräfte. **[160]** O. Lehmann, Verh. deut. phys. Ges. *8*, No. 16 (1906); Bemerkungen zu K. Fuchs, die Gestaltungskraft fließender Kristalle. **[161]** O. Lehmann, Verh. deut. phys. Ges. *8*, 143 (1906); Gestaltungskraft fließender Kristalle. **[162]** O. Lehmann, Verh. deut. phys. Ges. *8*, 528 (1906); Flüssige und scheinbar lebende Kristalle. **[163]** O. Lehmann, Z. physik. Chem. *56*, 750 (1906); Stoffe mit drei flüssigen Zuständen, einem isotrop- und zwei kristallinisch-flüssigen. **[164]** O. Lehmann, Physik. Z. 7, 578 (1906); Farberscheinungen bei fließenden Kristallen. **[165]** O. Lehmann, Flüssige Kristalle und die Theorien des Lebens (Barth: Leipzig). *1906*, 55 Seiten. **[166]** O. Lehmann, Umschau *10*, 1906, No. 17; Scheinbar lebende fließende Kristalle. **[167]** O. Lehmann, Verh. Ges. d. Naturf. u. Ärzte *78*, 139 (1906); Flüssige und scheinbar lebende Kristalle. **[168]** O. Lehmann, Jahresber. physik. Verein Frankfurt/M. *1906/7*, 68; Flüssige Kristalle, ihre Entdeckung, Bedeutung und Ähnlichkeit mit Lebewesen. **[169]** O. Lehmann, Illustrierte Zeitung *128*, 806 v. 9. 5. 1907; Flüssige Kristalle und deren scheinbares Leben. **[170]** O. Lehmann, Scheinbar lebende Kristalle, Anleitung zur Demonstration ihrer Eigenschaften, sowie ihrer Beziehungen zu anderen flüssigen und zu den festen Kristallen in Form eines Dreigesprächs (Schreiber: Esslingen). 1907, 68 Seiten.

[171] O. Lehmann, Himmel und Erde *19*, 434 (1907); Flüssige Kristalle und scheinbare Übergänge zu den niedrigsten Lebewesen. **[172]** O. Lehmann, „Deutsche Revue", September 1907; Gibt es lebende flüssige Kristalle? **[173]** O. Lehmann, Natur und Schule *6*, 111 (1907); Scheinbar lebende flüssige Kristalle. **[174]** O. Lehmann, Kosmos, Handweiser für Naturfreunde, Bd. IV, Heft 1/2 (1907); Flüssige Kristalle und ihre Analogien zu den niedrigsten Lebewesen. **[175]** O. Lehmann, Physik. Z. *8*, 42 (1907), C.A. *1*, 681[6] (1907); Flüssige Kristalle. **[176]** O. Lehmann, Physik. Z. *8*, 386 (1907); Flüssige Kristalle und mechanische Technologie. **[177]** O. Lehmann, Ann. Physik *22*, 469 (1907), C.A. *1*, 1355 (1907); Die van der Waalssche Formel und die Kontinuität der Aggregatzustände, Erwiderung an K. Fuchs. **[178]** O. Lehmann, Wissensch. Beilage z. Germania Nr. 36 v. 5. 9. 1907; Flüssige Kristalle und Leben, Antwort an Herrn Kathariner. **[179]** O. Lehmann, Vierteljahrsb. Wien. Vereins Förd. phys. chem. Unterr. *12*, 239 (1907); Flüssige und scheinbar lebende Kristalle. **[180]** O. Lehmann, Flüssige Kristalle und die Theorien des Lebens, 2. Auflage (Barth: Leipzig). *1908*, 70 Seiten.

[181] O. Lehmann, Aus der Natur *4*, 7 (1908); Flüssige und scheinbar lebende fest-flüssige Kristalle. **[182]** O. Lehmann, Archiv für Entwicklungsmechanik der Organismen, Bd. XXVI, Heft 3 (1908); Scheinbar lebende Krystalle und Myelinformen. **[183]** O. Lehmann, Biol. Centralblatt, *28*, 481 (1908); Scheinbar lebende Kristalle, Pseudopodien, Cilien und Muskeln. **[184]** O. Lehmann, Rivista di Scienza "Scientia" 4, NVIII (1908); Scheinbar lebende fließende Kristalle, künstliche Zellen und Muskeln. **[185]** O. Lehmann, Ann. Physik *25*, 852 (1908), C.A. *2*, 2481[3] (1908); Geschichte der flüssigen Kristalle. **[186]** O. Lehmann, Ber. *41*, 3774 (1908), C.A. *3*, 506[2] (1909); Bemerkungen zu den Abhandlungen von D. Vorländer und A. Prins über flüssige Kristalle. **[187]** O. Lehmann, Flüssige Krystalle und die Theorien des Lebens (Barth: Leipzig). 1908, C.A. *3*, 20[8] (1909). **[188]** O. Lehmann, Scientia (Bologna) *4*, 283 (1908); Scheinbar lebende fließende Kristalle. **[189]** O. Lehmann, Z. VDI *52*, 387 (1908); Flüssige Kristalle und mechanische Technologie. **[190]** O. Lehmann, Verh. deut. phys. Ges. *10*, 321 (1908) identisch mit Ber. physik. Ges. *6*, 321 (1908), C.A. *2*, 2481[3] (1908); Flüssige Kristalle, Myelinformen und Muskelkraft.

[191] O. Lehmann, Ber. physik. Ges. *6*, 406 (1908), C.A. *2*, 3182[8] (1908); Künstliche Zellen mit flüssig-kristallinischen Wänden. **[192]** O. Lehmann, Ann. Physik *27*, 1099 (1908), C.A. *3*, 1239[7] (1909); Bemerkungen zu Fr. Reinitzers Mitteilung über die Geschichte der flüssigen Kristalle. **[193]** O. Lehmann, Mathem.-naturwiss. Blätter, Berlin 7, 1 (1910); Flüssige Kristalle. **[194]** O. Lehmann, Arch. sci. phys. nat. *28*, 205 (1909), C.A. *4*, 10[1] (1910); Cristaux liquides et modèles moléculaires. **[195]** O. Lehmann, Rev. gén. chim. *12*, 184 (1909), C.A. *3*, 2400[2] (1909); Flüssige Kristalle. **[196]** O. Lehmann, J. physique 7, 713 (1909), C.A. *4*, 9[6] (1910); Cristaux liquides. **[197]** O. Lehmann, „Deutsche Revue", Januar 1909; Künstliche Zellen und Muskeln aus fließenden Kristallen. **[198]** O. Lehmann, Physik. Z. *10*, 553 (1909), C.A. *4*, 10[2] (1910); Demonstrationen und Modelle zur Lehre von den flüssigen Kristallen. **[199]** O. Lehmann, Flüssige Kristalle, Myelinformen und Muskelkraft. (Isaria Verlag: München) 1910, 43 Seiten; C.A. *4*, 1421 (1910). **[200]** O. Lehmann, Physik. Z. *11*, 44 (1910), C.A. *4*, 1256[4] (1910); Selbstreinigung flüssiger Kristalle.

[201] O. Lehmann, Physik. Z. *11*, 575 (1910), C.A. *4*, 2594[4] (1910); Pseudoisotropie und Schiller-farben bei flüssigen Kristallen. **[202]** O. Lehmann, Bull. soc. franç. min. *33*, 300 (1910), Bull. assoc. chim. sucr. dist. *28*, 389, C.A. *5*, 1382[5] (1911); Flüssige Kristalle. Antwort an G. Friedel und F. Grandjean. **[203]** O. Lehmann, Das Krystallisationsmikroskop und die damit gemachten Entdeckungen, insbesondere die flüssigen Kristalle (Vieweg: Braunschweig). *1910*, 112 Seiten, C.A. 5, 26[4] (1911). **[204]** O. Lehmann,

Z. physik. Chem. *71*, 355 (1910), C.A. *4*, 1256[8] (1910); Flüssige Kristalle und Avogadros Gesetz. [**205**] O. Lehmann, Z. physik. Chem. *73*, 598 (1910), C.A. *4*, 2763[6] (1910); Dimorphie sowie Mischkristalle bei flüssig-kristallinischen Stoffen und Phasenlehre. [**206**] O. Lehmann, Umschau *14*, 950 (1910); Die Selbstreinigung flüssiger Kristalle. [**207**] O. Lehmann, Ber. physik. Ges. *1911*, 338, C.A. *5*, 3747[7] (1911); Konische Strukturstörungen bei flüssigen Pseudokristallen. [**208**] O. Lehmann, Ber. physik. Ges. *1911*, 945, C.A. *6*, 437[3] (1912); Kristallinische und amorphe Flüssigkeiten. [**209**] O. Lehmann, Physik. Z. *12*, 540 (1911), C.A. *5*, 3747[7] (1911); Struktur und Optik großer Kristalltropfen. [**210**] O. Lehmann, Physik. Z. *12*, 1032 (1911), C.A. *6*, 437[3] (1912); Kristalline und amorphe Flüssigkeiten.

[**211**] O. Lehmann, Die neue Welt der flüssigen Kristalle und deren Bedeutung für Physik, Chemie, Technik und Biologie (Akad. Verlagsges.: Leipzig). *1911*, 388 Seiten, C.A. *5*, 2217[9] (1911). [**212**] O. Lehmann, Ann. Physik *35*, 193 (1911), C.A. *5*, 3747[7] (1911); Molekularstruktur und Optik flüssiger Kristalle. [**213**] O. Lehmann, Verh. deut. phys. Ges. *13*, 945 (1911); Kristallinische und amorphe Flüssigkeiten. [**214**] O. Lehmann, Arch. sci. phys. nat. *32*, 5 (1911), C.A. *5*, 3747[7] (1911); La structure des grands cristaux liquides. [**215**] O. Lehmann, Physik. Z. *13*, 550 (1912), C.A. *7*, 12[2] (1913); Magnetische Analyse flüssiger Kristalle. [**216**] O. Lehmann, Ann. Physik *39*, 80 (1912), C.A. *7*, 566[3] (1913); Einfluß von Wirbelbewegung auf die Struktur flüssiger Kristalle. [**217**] O. Lehmann, Z. Kryst. Min. *52*, 592 (1912/13), C.A. *8*, 38 (1914); Flüssige Kristalle des Ammoniumoleats. [**218**] O. Lehmann, Physik. Z. *14*, 1129 (1913), C.A. *8*, 851[3] (1914); Quellung flüssiger Kristalle. [**219**] O. Lehmann, Prometheus *25*, 2 (1913); Flüssige Kristalle, Moleküle und Lebewesen. [**220**] O. Lehmann, Ber. physik. Ges. *1913*, 413, C.A. *7*, 3882[2] (1913); Erforschung der Molekularkräfte durch Störung des molekularen Gleichgewichtes bei flüssigen Kristallen.

[**221**] O. Lehmann, Sitzb. Heidelberger Akad. Wissenschaften (A), I. 1911, 22. Abh., II. 1912, 13. Abh., III. 1913, 13. Abh., C.A. *8*, 1689[8] (1914); Untersuchungen über flüssige Kristalle. [**222**] O. Lehmann, Verh. Nat. Ver. Karlsruhe *25*, 164 (1913); Flüssige Kristalle, Molekularstruktur und Molekularkräfte. [**223**] O. Lehmann, Verh. deut. physik. Ges. *15*, 953 (1913), C.A. *8*, 851[3] (1914); Quellung flüssiger Kristalle. [**224**] O. Lehmann, Compt. rend. *158*, 389 (1914), C.A. *8*, 1530[4] (1914); Un brusque changement de la forme des cristaux liquides, causé par une transformation moléculaire. [**225**] O. Lehmann, Compt. rend. *158*, 1100 (1914), C.A. *8*, 2286[7] (1914); Effets de succion observés dans les cristaux liquides en voie de bourgonnement. [**226**] O. Lehmann, Physik. Z. *15*, 617 (1914), C.A. *8*, 2978[1] (1914); Optische Anisotropie der flüssigen Kristalle. [**227**] O. Lehmann, Verh. deut. physik. Ges. *16*, 443 (1914), C.A. *8*, 3140[8] (1914); Zentralkraft- und Richtkraftoberflächenspannung bei flüssigen Kristallen. [**228**] O. Lehmann, Ann. Physik *44*, 112 (1914), C.A. *8*, 2288[5] (1914); Plötzliche Gestaltsänderung flüssiger Kristalle infolge Änderung der molekularen Richtkraft aus Anlaß polymorpher Umwandlung. [**229**] O. Lehmann, Ann. Physik *44*, 969 (1914), C.A. *8*, 3140[6] (1914); Saugkraft quellbarer myelinartiger flüssiger Kristalle. [**230**] O. Lehmann, Biochem. Z. *63*, 74 (1914), C.A. *8*, 2403[2] (1914); Flüssige Krystalle und Biologie.

[**231**] O. Lehmann, Int. Z. Metallog. *6*, 217 (1914), C.A. *9*, 779 (1915); Spontane und erzwungene Homöotropie. [**232**] O. Lehmann, Kolloid-Z. *15*, 65 (1914), C.A. *9*, 464 (1915); Flüssige Kristalle und Kolloide. [**233**] O. Lehmann, Scientific American Supplement No. 2039, 80 (1915); Optical anisotropy of liquid crystals. [**234**] O. Lehmann, Elster-Geitel Festschrift *1915*, 381, Beiblätter Ann. Physik *40*, 407 (1916), C.A. *11*, 1071[6] (1917); Ausscheidungen, Niederschläge und flüssige Kristalle. [**235**] O. Lehmann, Ann. Physik *47*, 832 (1915), C.A. *10*, 411[9] (1916); Lösung und Ausscheidung von Stoffen, insbesondere flüssiger Kristalle. [**236**] O. Lehmann, Ann. Physik *48*, 177 (1915), C.A. *10*, 838[3] (1916); Erzeugung von Bewegung durch molekulare Richtkraft bei flüssigen Kristallen. [**237**] O. Lehmann, Ann. Physik *48*, 725 (1915), C.A. *10*, 838[5] (1916); Struktur schleimig-kristallinischer Flüssigkeiten. [**238**] O. Lehmann, Physik. Z. *17*, 241 (1916), C.A. *11*, 907[1] (1917); Ölige Streifen schleimig und tropfbarflüssiger Kristalle. [**239**] O. Lehmann, Ann. Physik *50*, 555 (1916), C.A. *11*, 750 (1917); Plastizität fester Kristalle und erzwungene Homöotropie I. und II. Art. [**240**] O. Lehmann, Ann. Physik *51*, 353 (1916), C.A. *11*, 2425[8] (1917); Störung der Struktur homogen tropfbar-flüssiger Kristalle durch Verdrillung.

[**241**] O. Lehmann, Ann. Physik *52*, 445 (1917), C.A. *12*, 552[5] (1918); Struktur inhomogener tropfbar-flüssig-kristallinischer Schichten (Spurlinien, Fäden und Höfe). [**242**] O. Lehmann, Ann. Physik *52*, 527 (1917), C.A. *12*, 552[5] (1918); Störung der Struktur tropfbar-flüssiger Kristalle durch Beimischungen. [**243**] O. Lehmann, Ann. Physik *52*, 541 (1917), C.A. *12*, 552[5] (1918); Fortschreitende Strukturwellen (scheinbare Rotationen) bei flüssigen Kristallen. [**244**] O. Lehmann, Ann. Physik *52*, 736 (1917), C.A. *12*, 552[7] (1918); Tropfen und Säulen kristallinischer Flüssigkeiten mit verdrehter Struktur. [**245**] O. Lehmann, Physik. Z. *19*, 73 (1918), C.A. *13*, 1040 (1919); Hauptsätze der Lehre von den flüssigen

Kristallen. [246] O. Lehmann, Physik. Z. *19*, 88. (1918), C.A. *13*, 1040 (1919); Hauptsätze der Lehre von den flüssigen Kristallen. [247] O. Lehmann, Ann. Physik *55*, 81 (1918), C.A. *12*, 2269[1] (1918); Flüssige Kristalle der 10-Bromphenanthren-3- oder -6-sulfosäurehydrate. [248] O. Lehmann, Ann. Physik *56*, 321 (1918), C.A. *13*, 3047[5] (1919); Hydrodynamik schleimig-kristallinischer Flüssigkeiten. [249] O. Lehmann, Ann. Physik *57*, 244 (1918), C.A. *13*, 2800[1] (1919); Ionenwanderung in flüssigen Kristallen von Ammoniumoleathydrat. [250] O. Lehmann, Ber. physik. Ges. *20*, 63 (1918), C.A. *13*, 1274[5] (1919); Bestimmung der Elastizitätsgrenze.

[251] O. Lehmann, Ergebn. Physiol. *16*, 256 (1918), C.A. *13*, 3198[2] (1919); Die Lehre von den flüssigen Krystallen und ihre Beziehung zu den Problemen der Biologie. [252] O. Lehmann, Ann. Physik *58*, 631 (1919), C.A. *14*, 2886[4] (1920); Beziehung zwischen mechanischer und chemischer Verdrehung der Struktur flüssiger Kristalle. [253] O. Lehmann, Die Lehre von den flüssigen Kristallen und ihre Beziehung zu den Problemen der Biologie (Bergmann: Wiesbaden). 1917. 254 pp., C.A. *14*, 1689[1] (1920). [254] O. Lehmann, Z. anorg. Allgem. Chem. *113*, 253 (1920), C.A. *15*, 2763[3] (1921); Molekulare Richtkraft flüssiger Kristalle. [255] O. Lehmann, Z. Physik *2*, 127 (1920), C.A. *15*, 1429[7] (1921); Molekularkräfte flüssiger Kristalle und ihre Beziehung zu bekannten Kräften. [256] O. Lehmann, Ann. Physik *61*, 501 (1920), C.A. *14*, 2290[6] (1920); Strukturverdrehung bei schleimig-flüssigen Kristallen. [257] O. Lehmann, Flüssige Kristalle und ihr scheinbares Leben (Voss: Leipzig). 1921, 72 Seiten, C.A. *16*, 1699[6] (1922). [258] O. Lehmann, Flüssige Kristalle und ihr scheinbares Leben (Forschungsergebnisse, dargestellt in einem Kinofilm) (Barth: Leipzig) 1921. [259] O. Lehmann, Z. Metallkunde *13*, 57, 81, 113 (1921), C.A. *16*, 371[3] (1922); Erhaltung, Änderung und Wiederherstellung der Struktur flüssiger und weicher fester Kristalle beim Fließen derselben. [260] O. Lehmann, Ann. Physik *66*, 323 (1921), C.A. *16*, 3778[9] (1922); Struktur tropfbar flüssiger Mischkristalle.

[261] O. Lehmann, Z. physik. Chem. *102*, 91 (1922), C.A. *16*, 4118[8] (1922); Aggregatzustände und flüssige Kristalle. [262] O. Lehmann, in E. Abderhalben, Handbuch der biologischen Arbeitsmethoden, Ab. III, Teil A, Heft 2, Berlin–Wien 1922, 123, C.A. *17*, 1489[3] (1923); Methoden zur Darstellung und Untersuchung flüssiger Kristalle. [263] A. M. Levelut, J. Phys. (Paris) *36*, 1029 (1975), C.A. *83*, 200346t (1975); Croissance épitaxique de phases smectiques ordonnées sur du mica. [263a] A. Loesche, Wiss. Z., Karl-Marx-Univ., Leipzig, Math.-Naturwiss. Reihe *25*, 601 (1976), C.A. *86*, 198053m (1977); Experimentelle Untersuchungen zur Phasenstruktur von Flüssigkristallen. [264] S. Matsumoto, M. Kawamoto, N. Kaneko, Appl. Phys. Lett. *27*, 268 (1975), C.A. *83*, 140075s (1975); Surface-induced molecular orientation of liquid crystals by carboxylatochromium complexes. [265] S. Matsumoto, D. Nakagawa, N. Kaneko, K. Mizunoya, Appl. Phys. Lett. *29*, 67 (1976), C.A. *85*, 85772f (1976); Surface-produced parallel alignment of nematics by polynuclear dicarboxylatochromium complexes. [266] C. Mauguin, Compt. rend. *151*, 886 (1910), C.A. *5*, 1219[5] (1911); Cristaux liquides en lumière convergente. [267] C. Mauguin, Compt. rend. *151*, 1141 (1910), C.A. *5*, 1219[7] (1911); Liquides biréfringents à structure hélicoidale. [268] C. Mauguin, Bull. soc. franç. min. *34*, 71 (1911), Les cristaux liquides de Lehmann. [269] C. Mauguin, Physik. Z. *12*, 1011 (1911), C.A. *6*, 437[2] (1912); O. Lehmanns flüssige Kristalle. [270] C. Mauguin, Compt. rend. *154*, 1359 (1912), C.A. *7*, 1646[3] (1913); L'agitation interne des cristaux liquides.

[271] C. Mauguin, Comp. rend. *156*, 1246 (1913), C.A. *7*, 3060[3] (1913); Orientation des cristaux liquides par les lames de mica. [272] P. Kassubek, G. Meier, Mol. Cryst. Liq. Cryst. *8*, 305 (1969), C.A. *71*, 85576f (1969); Optical studies on Grandjean planes in cholesterics. [273] H. Gruler, G. Meier, Mol. Cryst. Liq. Cryst. *12*, 289 (1971), C.A. *75*, 54881e (1971); Correlation between electrical properties and optical behaviour of nematics. [273a] D. Melzer, F. R. N. Nabarro, Philos. Mag. *35*, 901, 907 (1977), C.A. *87*, 151499s, 151500k (1977); Optical studies of a nematic with circumferential surface orientation in a capillary. Cols and noeuds in a nematic with a homeotropic cylindrical boundary. [274] W. G. Merritt, G. D. Cole, W. W. Walker, Mol. Cryst. Liq. Cryst. *15*, 105 (1971), C.A. *76*, 19002t (1972); Polymorphism and mesomorphism of four esters of cholesterol. [275] C. Mettenheimer, Korr. Blatt des Vereins f. gemeinschaftl. Arbeit zur Förderung wiss. Heilkunde *24*, 331 (1857); Mikroskopische Beobachtungen mit polarisiertem Licht. [276] R. B. Meyer, Solid State Commun. *12*, 585 (1973), C.A. *78*, 151923q (1973); Interaction between a disclination in a nematic and a rubbed surface. [276a] P. Möckel, Naturwissenschaften *64*, 40 (1977), C.A. *86*, 81998g (1977); Strukturbildung bei der Diffusion flüssig-kristalliner Phasen. [276b] J. H. Morrissy, W. A. Crossland, B. Needham, J. Phys. D *10*, L175 (1977), C.A. *87*, 125593y (1977); Tilt angle measurements of nematic cyano-biphenyl eutectic mixtures. [277] J. Nehring, M. A. Osman, Z. Naturforsch. A *31*, 786 (1976), C.A. *85*, 102647p (1976); Liquid crystalline phases in N-(4-*n*-alkylbenzylidene)-4'-*n*-alkylanilines. [278] M. E. Neubert, P. Norton, D.

L. Fishel, Mol. Cryst. Liq. Cryst. *31*, 253 (1975), C.A. *84*, 107501a (1976); Using the Mettler FP-2 hot stage at below-room temperatures. [**279**] A. K. Niessen, A. den Ouden, Philips Res. Rep. *29*, 119 (1974), C.A. *81*, 142295w (1974); Conoscopic observations on some smectics. [**280**] G. Nomarski, J. de Phys. et de Rad. *16*, 9 (1955), Microinterféromètre différentiel à ondes polarisées. [**280a**] I. C. Petrea, An. Univ. Bucuresti, Stiint. Nat. *25*, 35 (1976), C.A. *87*, 160032m (1977); Structural-textural interrelation in smectophase.

[**281**] F. Pockels, Lehrbuch der Kristalloptik (Teubner: Leipzig). 1906. 519 pp. (Johnson Reprint: New York). [**282**] E. A. Poe, Doubleday & Co., Inc. Garden City, New York 1966, 706; The narrative of A. Gordon Pym of Nantucket. Complete stories and poems of Edgar Allan Poe. [**283**] L. Pohl, R. Steinsträsser, Z. Naturforsch. B *26*, 26 (1971), C.A. *74*, 92320h (1971); Existenz nematischer Sekundär-strukturen. [**284**] L. Pohl, 6. Arbeitstagung Flüssigkristalle, Inst. Angew. Festkörperphysik, Freiburg, 2. 4. 1976; Verfahren zur Herstellung stabiler homogener oder verdrillter Flüssigkristallzellen. [**285**] Y. M. Popovskii, G. P. Silenko, Sb. Dokl. Vses. Nauch. Konf. Zhidk Krist. Simp. Ikh Prakt. Primen., 2nd, *1972*, 226, C.A. *82*, 24594r (1975); Liquid crystal state of the interfacial phase of nitrobenzene formed on a glass surface. [**286**] G. Porte, J. Phys. (Paris) *37*, 1245 (1976), C.A. *85*, 169978r (1976); Tilted alignment of MBBA induced by short-chain surfactants. [**287**] M. Pospisil, Chem. Listy *68*, 194 (1974), C.A. *80*, 125597h (1974); Heated stand for microscopic analysis of liquid crystals. [**288**] M. J. Press, A. S. Arrott, Phys. Rev. Lett. *33*, 403 (1974), C.A. *81*, 127914d (1974); Theory and experiments on configurations with cylindrical symmetry in liquid-crystal droplets. [**289**] F. P. Price, A. K. Fritzsche, J. Phys. Chem. *77*, 396 (1973), C.A. *78*, 89397r (1973); Kinetics of spherulite growth in cholesteryl esters. [**290**] F. P. Price, C. S. Bak, Mol. Cryst. Liq. Cryst. *29*, 225 (1975), C.A. *83*, 106524c (1975); Orienting effects of substrates on cholesteryl esters.

[**291**] E. B. Priestley, P. J. Wojtowicz, P. Sheng, Edts. Introduction to Liquid Crystals. (Plenum: New York). 1975, 356 pp., C.A. *84*, 82971p (1976). [**292**] J. E. Proust, L. Ter-Minassian-Saraga, E. Guyon, Solid State Commun. *11*, 1227 (1972), C.A. *78*, 21131u (1973); Orientation of a nematic by suitable boundary surfaces. [**293**] J. E. Proust, L. Ter-Minassian-Saraga, C. R. Acad. Sci., Sér. C *276*, 1731 (1973), C.A. *79*, 97225u (1973); Effet sur l'énergie d'adhésion, de la structure d'un support solide et de l'orientation à son interface d'un cristal liquide nématique. [**294**] J. E. Proust, L. Ter-Minassian-Saraga, C. R. Acad. Sci., Sér. C *279*, 615 (1974), C.A. *83*, 36008m (1975); Ancrage physicochimique d'un cristal liquide nématique aux interfaces solide/liquide et énergie libre d'adhésion. Hystérésis de mouillage en phase nématique et isotrope. [**295**] J. E. Proust, L. Ter-Minassian-Saraga, J. Phys. (Paris) C-1, *36*, 77 (1975), C.A. *83*, 36028t (1975); Orientation d'un cristal liquide par les surfaces et énergie libre d'adhésion. [**296**] E. Perez, J. E. Proust, C. R. Acad. Sci., Sér. C, *282*, 559 (1976), C.A. *85*, 54851b (1976); Interactions moléculaires dans un cristal liquide nématique. [**297**] J. E. Proust, L. Ter Minassian-Saraga, Colloid Polym. Sci. *254*, 492 (1976), C.A. *85*, 102608b (1976); Structure, free energy of adhesion and disjoining pressure in a solid-nematic thermotrope system. [**297a**] J. E. Proust, E. Perez, J. Phys. (Paris), Lett. *38*, 91 (1977), C.A. *86*, 131189g (1977); Films minces smectiques symétriques et asymétriques. [**297b**] E. Perez, J. E. Proust, L. Ter-Minassian-Saraga, Mol. Cryst. Liq. Cryst. *42*, 167 (1977); Interfacial origin of liquid crystal anchorage. [**298**] Publikationen zur Flüssigkristallforschung, Titelverzeichnis der Martin-Luther-Universität Halle-Wittenberg, Sektion Chemie. [**299**] R. Rath, Theoretische Grundlagen der allgemeinen Kristalldiagnose im durchfallenden Licht (Springer: Berlin). 1969. 133 pp. [**300**] F. Reinitzer, Monatshefte *9*, 421 (1888); Zur Kenntnis des Cholesterins.

[**301**] F. Reinitzer, Ann. Physik *27*, 213 (1908), C.A. *3*, 1239[6] (1909); Zur Geschichte der flüssigen Kristalle. [**302**] F. Rinne, M. Berek, Anleitung zu optischen Untersuchungen mit dem Polarisationsmikroskop (Jänecke: Leipzig). 1934. 279 pp., C.A. *29*, 1706[3] (1935). [**303**] F. B. Rosevear, J. Am. Oil Chem. Soc. *31*, 628 (1954), C.A. *49*, 2100b (1955); Microscopy of the neat and middle phases of soaps and synthetic detergents. [**304**] G. Ryschenkow, J. Phys. (Paris) *36*, 243 (1975), C.A. *82*, 178498z (1975); Méthode d'étude des défauts de surface des lames nématiques orientées par un support directionnel. [**305**] H. Arnold, H. Sackmann, Z. Phys. Chem. (Leipzig) *213*, 137 (1960), C.A. *54*, 9413i (1960); Polymorphie und Mischbarkeit bei kristallinen Flüssigkeiten. [**306**] H. Arnold, H. Sackmann, Z. Phys. Chem. (Leipzig) *213*, 145 (1960), C.A. *54*, 9414a (1960); Mischbarkeit in binären Systemen aus Dialkoxy-azoxybenzolen. [**307**] H. Arnold, H. Sackmann, Z. Phys. Chem. (Leipzig) *213*, 262 (1960), C.A. *54*, 12705e (1960); Mischbarkeit zwischen Azoxybenzoesäure-diäthylester und PAP. [**308**] H. Arnold, H. Sackmann, Z. Elektrochem. *63*, 1171 (1959), C.A. *54*, 6283c (1960); Mischbarkeit in binären Systemen mit mehreren smektischen Phasen. [**309**] D. Demus, H. Sackmann, Z. Phys. Chem. (Leipzig) *222*, 127 (1963), C.A. *59*, 5883d (1963); Umwandlungsvorgänge von 67 Verbindungen mit smektischen Phasen.

[**310**] H. Arnold, D. Demus, H. Sackmann, Z. Phys. Chem. (Leipzig) *222*, 15 (1963), C.A. *58*, 10813b (1963); Texturen kristallin-flüssiger Phasen. [**311**] H. Sackmann, D. Demus, Z. Phys. Chem. (Leipzig) *222*, 143 (1963), C.A. *59*, 5883d (1963); Mischbarkeitsbeziehungen in binären Systemen, deren beide Komponenten entweder über s$_B$ oder s$_C$ Phasen verfügen. [**312**] H. Sackmann, D. Demus, Z. Phys. Chem. (Leipzig) *224*, 177 (1963), C.A. *60*, 9970a (1964); Mischbarkeitsbeziehungen in binären Systemen, deren beide Komponenten über s$_A$ Phasen verfügen. [**313**] H. Sackmann, D. Demus, Z. Phys. Chem. (Leipzig) *227*, 1 (1964), C.A. *62*, 4663h (1965); Ausbildung von „Mischungslücken" in binären Systemen, deren Komponenten über kristallin-flüssige Phasen verschiedener Kennzeichnung verfügen. [**314**] H. Sackmann, Chem. Phys. Appl. Surface Active Subst., Proc. Int. Congr., 4th 1964 (Pub. 1967), 2, 721, C.A. *72*, 16432q (1970); Die Natur kristallin-flüssiger Stoffzustände demonstriert am Verhalten ihrer Grenzflächen in binären Systemen. [**315**] H. Sackmann, D. Demus, Z. Phys. Chem. (Leipzig) *230*, 285 (1965), C.A. *64*, 10456f (1966); Kristallin-flüssige Zwischenphasen. [**316**] H. Sackmann, D. Demus, Mol. Cryst. 2, 81 (1966), C.A. *66*, 32721q (1967); Polymorphism of liquid crystals. [**317**] G. Pelzl, D. Demus, H. Sackmann, Z. Phys. Chem. (Leipzig) *238*, 22 (1968), C.A. *69*, 91022x (1968); Kristallin-flüssige Phasen in der homologen Reihe der *p,p'*-Azoxyzimtsäure-di-*n*-alkylester und ihr Mischbarkeitsverhalten mit *p,p'*-Azoxy-benzal-bis-(*m*-toluidin). [**318**] D. Demus, H. Sackmann, Z. Phys. Chem. (Leipzig) *238*, 215 (1968), C.A. *70*, 62002f (1969); Neues über s$_B$ Phasen. [**319**] D. Demus, H. Sackmann, G. Kunicke, G. Pelzl, R. Salffner, Z. Naturforsch. A *23*, 76 (1968), C.A. *69*, 71355e (1968); Polymorphie bei kristallin-flüssigen Substanzen: Smektische Trimorphie. [**320**] D. Demus, G. Kunicke, J. Neelsen, H. Sackmann, Z. Naturforsch. A *23*, 84 (1968), C.A. *69*, 54663m (1968); Polymorphie der kristallin-flüssigen Modifikationen in der homologen Reihe der 4'*n*-Alkoxy-3'-nitrodiphenyl-4-carbonsäuren.

[**321**] H. Sackmann, D. Demus, Fortschr. chem. Forsch. *12*, 349 (1969), C.A. *71*, 129502k (1969); Eigenschaften und Strukturen thermotroper kristallin flüssiger Zustände. [**322**] E. Baum, D. Demus, H. Sackmann, Wiss. Z. Martin-Luther-Univ. Halle-Wittenberg, Math.-Naturwiss. Reihe *19*, 37 (1970), C.A. *75*, 91655e (1971); Kennzeichnung der neat-Modifikationen der Salze aliphatischer Carbonsäuren durch ihre Texturen und Mischbarkeitsbeziehungen. [**323**] D. Demus, H. Sackmann, K. Seibert, Wiss. Z. Martin-Luther-Univ. Halle-Wittenberg, Math.-Naturwiss. Reihe *19*, 47 (1970), C.A. *75*, 123351w (1971); Kristallin-flüssige Zustände von Salzen aromatischer Carbonsäuren. [**324**] G. Pelzl, H. Sackmann, Mol. Cryst. Liq. Cryst. *15*, 75 (1971), C.A. *75*, 156422y (1971); Birefringence of smectic modifications of the homologous thallium soaps. [**325**] D. Demus, S. Diele, M. Klapperstück, V. Link, H. Zaschke, Mol. Cryst. Liq. Cryst. *15*, 161 (1971), C.A. *76*, 18891v (1972); Smectic tetramorphous substance. [**326**] D. Demus, M. Klapperstück, R. Rurainski, D. Marzotko, Z. Phys. Chem. (Leipzig), *246*, 385 (1971), C.A. *75*, 68626n (1971); Eine neue Variante von Polymorphie im kristallin-flüssigen Zustand: Eine Substanz mit den Modifikationen smektisch B und nematisch. [**327**] G. Pelzl, H. Sackmann, Symp. Faraday Soc. *1971*, No. 5, 68, C.A. *79*, 24183u (1973); Birefringence and polymorphism of liquid crystals. [**328**] D. Demus, K. H. Koelz, H. Sackmann, Z. Phys. Chem. (Leipzig) *249*, 217 (1972), C.A. *76*, 132579z (1972); Polymorphie der kristallin-flüssigen Modifikationen in der homologen Reihe der 2,5-Bis-(4-*n*-alkylphenyl)-pyrazine. [**329**] H. Sackmann, D. Demus, Mol. Cryst. Liq. Cryst. *21*, 239 (1973), C.A. *79*, 108875j (1973); Polymorphism in liquid crystals. [**330**] D. Demus, H. Sackmann, R. Salffner, Wiss. Z. Martin-Luther-Univ., Halle-Wittenberg, Math.-Naturwiss. Reihe *22*, 143 (1973), C.A. *79*, 130099f (1973); Polymorphie der kristallin-flüssigen Modifikationen in der homologen Reihe der 2,5-Bis-(4-*n*-alkoxyphenyl)-pyrazine.

[**331**] D. Demus, K. H. Koelz, H. Sackmann, Z. Phys. Chem. (Leipzig) *252*, 93 (1973), C.A. *78*, 152429v (1973); Polymorphie der kristallin-flüssigen p-Terphenyl-4,4''-dicarbonsäure-di-*n*-alkylester; s$_E$-Modifikationen. [**332**] D. Demus, G. Kunicke, G. Pelzl, B. Röhlig, H. Sackmann, R. Salffner, Z. Phys. Chem. (Leipzig) *254*, 373 (1973), C.A. *80*, 137909z (1974); Neue Substanzen mit kristallin-flüssigen Modifikationen vom Typ s$_A$, s$_B$ oder s$_C$. [**333**] D. Demus, G. Kunicke, J. Neelsen, H. Sackmann, Z. Phys. Chem. (Leipzig) *255*, 71 (1974), C.A. *81*, 127894x (1974); Mischbarkeitsbeziehungen der kristallin-flüssigen Modifikationen der 4-*n*-Alkoxydiphenyl-4-carbonsäuren und ihrer 3'-substituierten Derivate. [**334**] H. Sackmann, Pure Appl. Chem. *38*, 505 (1974), C.A. *83*, 49095n (1975); Thermodynamic aspects of polymorphism in liquid crystals. [**335**] A. Biering, D. Demus, G. W. Gray, H. Sackmann, Mol. Cryst. Liq. Cryst. *28*, 275 (1974), C.A. *83*, 200450x (1975); Classification of the liquid crystalline modifications in some homologous series. [**336**] L. Richter, D. Demus, H. Sackmann, J. Phys. (Paris), C-3, *37*, 41 (1976), C.A. *85*, 102618e (1976); Relations between smectics B, E, and G. [**337**] H. Sackmann, Wiss. Beitr. Martin-Luther-Univ., Halle-Wittenberg *20*, 85 (1976), C.A. *86*, 63595u (1977); Polymorphie

flüssiger Kristalle. [**338**] N. M. Sakevich, Uch. Zap., Ivanov, Gos. Pedagog. Inst. No. *99*, 83 (1972), C.A. *78*, 152431q (1973); Kinetics of formation of liquid crystal seeds and nuclei. [**339**] A. Saupe, Liq. Cryst. Plast. Cryst. *1*, 18 (1974), C.A. *83*, 19075p (1975); Smectic, nematic, and cholesteric mesophases formed by nonamphiphilic compounds. [**340**] R. Schenck, Habilitationsschrift, Marburg 1897; Untersuchungen über die kristallinischen Flüssigkeiten.

[**341**] R. Schenck, Z. physik. Chem. *25*, 337 (1898); Untersuchungen über die kristallinischen Flüssigkeiten I. [**342**] R. Schenck, Z. physik. Chem. *27*, 167 (1898); Untersuchungen über die kristallinischen Flüssigkeiten II. [**343**] R. Schenck, Z. physik. Chem. *28*, 280 (1899); Untersuchungen über die kristallinischen Flüssigkeiten III. [**344**] R. Schenck, F. Schneider, Z. physik. Chem. *29*, 546 (1899); Untersuchungen über die kristallinischen Flüssigkeiten IV. [**345**] R. Schenck, Physik. Z. *1*, 409 und 425 (1900); Ergebnisse der bisherigen Untersuchungen über die flüssigen Kristalle. [**346**] R. Schenck, Z. physik. Chem. *32*, 564 (1900); Erwiderung an Herrn Bakhuis-Roozeboom. [**347**] R. Schenck, Ann. Physik *9*, 1053 (1902); Die Natur der flüssigen Kristalle. [**348**] R. Schenck, E. Eichwald, Ber. *36*, 3873 (1903); Flüssige Krystalle. [**349**] R. Schenck, Kristallinische Flüssigkeiten und flüssige Kristalle (Engelmann: Leipzig). 1905. 159 pp. [**350**] G. E. W. Schulze, T. Kunz, Mol. Cryst. Liq. Cryst. *36*, 223 (1976), C.A. *86*, 63880h (1977); Mechanical twinning of smectic mosaic textures.

[**351**] I. Singstad, Fra Fys. Verden *37*, 88 (1975), C.A. *84*, 114234u (1976); The orientation of liquid crystals. [**352**] I. Singstad, Fra Fys. Verden *38*, 21 (1976), C.A. *85*, 134509m (1976); The orientation of liquid crystals. II. [**353**] E. R. Smith, B. W. Ninham, Physica *66*, 111 (1973), C.A. *79*, 46538r (1973); Response of nematics to Van der Waals forces. [**354**] E. Sommerfeldt, Physik. Z. *9*, 234 (1908), C.A. *2*, 2327[9] (1908); Untersuchung flüssiger Kristalle im konvergenten polarisierten Licht. [**354a**] G. J. Sprokel, Mol. Cryst. Liq. Cryst. *42*, 233 (1977); Molecular order induced by cell walls. Experimental results. [**355**] M. Steers, M. Kléman, C. Williams, J. Phys. (Paris), Lett. *35*, L21 (1974), C.A. *80*, 113516t (1974); Observations au microscope polarisant de la phase smectique du diéthyl 4,4'-azoxydibenzoate. [**356**] M. Brehm, H. Finkelmann, H. Stegemeyer, Ber. Bunsenges. Phys. Chem. *78*, 883 (1974), C.A. *82*, 10182t (1975); Orientierung cholesterischer Mesophasen an mit Lecithin behandelten Oberflächen. [**357**] G. W. Stewart, R. M. Morrow, Phys. Rev. *30*, 232 (1927), C.A. *21*, 3549 (1927); X-ray diffraction in liquids: primary normal alcohols. [**358**] G. W. Stewart, Phys. Rev. *38*, 931 (1931), C.A. *26*, 372[3] (1932); X-ray study of the magnetic character of liquid crystalline PAA and a comparison with the isotropic liquid. [**359**] G. W. Stewart, Trans. Faraday Soc. *29*, 982 (1933), C.A. *28*, 1239[2] (1934); Alternations in the nature of a fluid from a gaseous to liquid crystalline conditions as shown by X-rays. [**360**] H. R. Letner, G. W. Stewart, Phys. Rev. *45*, 332 (1934), C.A. *30*, 7443[9] (1936); Comparison by X-ray diffraction of PAA in liquid and liquid-crystalline phases.

[**361**] G. W. Stewart, J. Chem. Physics *4*, 231 (1936), C.A. *30*, 3292[2] (1936); X-ray diffraction intensity of the two liquid phases of PAA. [**362**] G. W. Stewart, Phys. Rev. *69*, 51 (1946), C.A. *41*, 3683d (1947); Heat-conduction effects with liquid crystals and suspended particles. [**363**] R. E. Stoiber, S. A. Morse, Microscopic identification of crystals (Ronald: New York). 1972. 278 pp., C.A. *76*, 159655h (1972). [**364**] H. Stolzenberg, Z. physik. Chem. *77*, 73 (1911), C.A. *5*, 2586[7] (1911); Schmelzpunktsbestimmung kristallinisch-flüssiger Körper. [**365**] F. Stumpf, Physik. Z. *11*, 780 (1910), C.A. *4*, 3161[3] (1910); Optische Untersuchung einer optisch aktiven flüssig-kristallinischen Substanz. [**366**] F. Stumpf, Diss. Göttingen 1911; Optische Beobachtungen an einer flüssig-kristallinischen aktiven Substanz. [**367**] F. Stumpf, Ann. Physik *37*, 351 (1912), C.A. *6*, 2204[4] (1912); Optische Beobachtungen an einer flüssig-kristallinischen aktiven Substanz. [**368**] F. Stumpf, Jahrb. Radioakt. Elektronik *15*, 1 (1918), C.A. *12*, 2488[6] (1918); Doppelbrechung und optische Aktivität flüssig-kristalliner Substanzen. [**369**] W. Stürmer, Photographie und Forschung, Zeiss Ikon, Stuttgart 7, 233 (1957); Ein einfacher Mikroheiztisch für Temperaturen bis über 400°C. [**370**] H. Suter, GIT Fachz. Lab. *18*, 1251 (1974), C.A. *82*, 178684g (1975); Thermomikroskopische Untersuchungen von flüssigen Kristallen.

[**371**] T. R. Taylor, J. L. Fergason, S. L. Arora, Phys. Rev. Lett. *24*, 359 (1970), C.A. *72*, 94275e (1970); Biaxial liquid crystals. [**372**] T. R. Taylor, S. L. Arora, J. L. Fergason, Phys. Rev. Lett. *25*, 722 (1970), C.A. *73*, 114021v (1970); Temperature-dependent tilt angle in a s_C phase. [**373**] J. Timmermans, J. chim. phys. *35*, 331 (1938); C.A. *33*, 1564[7] (1939); Un nouvel état mésomorphe les cristaux organiques plastiques. [**374**] H. Tropper, Ann. Physik *30*, 371 (1937), C.A. *32*, 15[5] (1938); Schwankungserscheinungen an flüssigen Kristallen. [**375**] T. Uchida, H. Watanabe, M. Wada, Jap. J. Appl. Phys. *11*, 1559 (1972), C.A. *77*, 169882q (1972); Molecular arrangement of nematics. [**375a**] T. Uchida, C. Shishido, M. Wada, Electr. Commun. Jpn. *58*, 132 (1976); Molecular alignment of nematics on silane treated glass surfaces. [**376**] W. Urbach, M. Boix, E. Guyon, Appl. Phys. Lett.

25, 479 (1974), C.A. *81*, 178196d (1974); Alignment of nematics and smectics on evaporated films. [377] J. van der Veen, H. B. Haanstra, J. Phys. (Paris), Lett. *37*, 43 (1976), C.A. *84*, 158235z (1976); Fine structure in a smectic stepped drop. [378] G. van Iterson, Jr., Trans. Faraday Soc. *29*, 915 (1933), C.A. *28*, 1238^4 (1934); Simple arrangement to demonstrate liquid crystals. [379] L. Verbit, Mol. Cryst. Liq. Cryst. *15*, 89 (1971), C.A. *75*, 155783e (1971); Line notation for thermotropic liquid crystal phase transitions. [380] R. Virchow, Virchows Archiv pathol. Anatomie und Physiologie *6*, 562 (1854); Über das ausgebreitete Vorkommen einer dem Nervenmark analogen Substanz in den thierischen Geweben.

[381] W. Voigt, Lehrbuch der Kristallphysik (Teubner: Leipzig). 1910. 978 pp., C.A. *5*, 827 (1911) (Johnson Reprint: New York). 1966. [382] W. Voigt, Ber. physik. Ges. *14*, 649 (1912), C.A. *6*, 3049^3 (1912); Optische Anomalie gewisser flüssiger Kristalle im konvergenten polarisierten Licht. [383] P. P. von Veimarn, J. Russ. Phys. Chem. Soc. *41*, 28 (1909), C.A. *3*, 1714^2 (1909); Ultramicroscopic investigation of liquid crystals. Preliminary communication. [384] D. Vorländer, H. Hauswaldt, Abh. d. Kaiserl. Leop. Carol. Dtsch. Akad. d. Naturf. *90*, 107 (1909); Systembestimmung und Achsenbilder flüssiger Kristalle. [385] D. Vorländer, M. E. Huth, Z. physik. Chem. *83*, 424 (1913), C.A. *7*, 3073^5 (1913); Doppelbrechung pleochroitischer flüssiger Kristalle. [386] D. Vorländer, M. E. Huth, Z. physik. Chem. *83*, 723 (1913), C.A. *7*, 3566^3 (1913); Interferenzerscheinungen bei pleochroitischen flüssigen Kristallen im konvergenten polarisierten Licht. [387] D. Vorländer, F. Janecke, Z. physik. Chem. *85*, 691 (1913), C.A. *8*, 608^1 (1914); Vergleich flüssiger Kristalle von racemischen und optisch aktiven Amylestern. [388] D. Vorländer, F. Janecke, Z. physik. Chem. *85*, 697 (1913), C.A. *8*, 602^7 (1914); Entstehung zirkularpolarisierender flüssiger Kristalle aus optisch inaktiven liquokristallinen Substanzen durch Beimischungen. [389] D. Vorländer, Chemische Kristallographie der Flüssigkeiten (Akad. Verlagsges.: Leipzig). 1924. 90 pp., C.A. *19*, 954^9 (1925). [390] D. Vorländer, J. Fischer, Ber. *63*B, 2506 (1930), C.A. *25*, 243^4 (1931); Einachsige Aufrichtung der kristallinen Flüssigkeiten.

[391] F. Wallerant, Compt. rend. *143*, 605 (1906); Cristaux liquides de propionate de cholestéryle. [392] F. Wallerant, Compt. rend. *148*, 1291 (1909), C.A. *3*, 1955^2 (1909); Liquides cristallisés biaxes. [393] C. Weygand, Chemische Morphologie der Flüssigkeiten und Kristalle. Hand- und Jahrbuch der Chemischen Physik Bd. 2, Abschn. 3C (Akad. Verlagsges.: Leipzig). 1941. 192 pp., C.A. *37*, 1078^6 (1943). [394] C. Weygand, Daniel Vorländer, Nekrolog Ber. *76*A, 41 (1943). [395] E. T. Wherry, Am. Mineral. *9*, 45 (1924), C.A. *18*, 1408^3 (1924); At the surface of a crystal. [396] A. Wigand, Ernst Dorn, Nekrolog Phys. Z. *17*, 297 (1916). [397] C. Williams, V. Vitek, M. Kléman, Solid State Commun. *12*, 581 (1973), C.A. *78*, 152464c (1973); Surface disclination lines in MBBA. [398] B. Williamson Irvin, J. Colloid Interface Sci. *23*, 221 (1967), C.A. *66*, 88925b (1967); Gel formation and spherulite production in the mesomorphic melts of cholesteryl esters. [399] K. L. Wolf, Tropfen, Blasen und Lamellen — oder von den Formen flüssiger Körper (Springer: Berlin). 1968. 83 pp. [400] U. Wolff, W. Greubel, H. Krueger, Mol. Cryst. Liq. Cryst. *23*, 187 (1973), C.A. *80*, 88148q (1974); Homogeneous alignment of liquid crystal layers.

[401] M. Yamashita, Y. Amemiya, Jpn. J. Appl. Phys. *15*, 2087 (1976), C.A. *86*, 10864z (1977); Effect of substrate surface on alignment of liquid crystal molecules. [401a] A. K. Zarzouk, et al., Solid State Electron. *19*, 133 (1976); Polarity dependent oxide defects located using liquid crystals. [402] Ya. B. Zel'dovich, Zh. Eksp. Teor. Fiz. *67*, 2357 (1974), C.A. *82*, 163353z (1975); Pseudoscalar liquid crystals. [403] H. Zocher, K. Coper, Z. physik. Chem. *132*, 295 (1928), C.A. *22*, 1913 (1928); Erzeugung der Anisotropie von Oberflächen. [404] H. Zocher, Z. Krist. *79*, 122 (1931), C.A. *26*, 2368^6 (1932); Optik der Mesophasen. [405] H. Zocher, Mol. Cryst. Liq. Cryst. *7*, 177 (1969), C.A. *71*, 74370h (1969); Nematics and smectics of higher order. [406] H. Zocher, Persönliche Mitteilung. [407] Mol. Cryst. *2*, 189 (1966); Discussion of nomenclature of liquid crystals.

Chapter 2

[1] M. A. Abert-Mellah, A. Zann, J. C. Dubois, Ger. Offen. 2,618,609 (30. 4. 1975), C.A. *86*, 139641d (1977); Organic compounds with strongly negative dielectric anisotropy. [1a] S. O. Adeosun, S. J. Sime, Thermochim. Acta *17*, 351 (1976), C.A. *86*, 22603b (1977); Quantitative DTA study of

melting and mesophase formation in lead(II) carboxylates. [**1b**] S. O. Adeosun, W. J. Sime, S. J. Sime, Thermochim. Acta *19*, 275 (1977), C.A. *87*, 29817d (1977); Quantitative DTA study of mesophase formation in the systems lead(II)dodecanoate/lead(II) oxide and lead(II)dodecanoate/hendecane. [**1c**] P. Adomenas, J. Daugvila, G. Denys, V. Ceponyte, U.S.S.R. 562,547 (21. 4. 1975), C.A. *87*, 109457b (1977); Nematics with negative dielectric anisotropy. [**2**] Agence Nationale de Valorisation de la Recherche. Fr. Demande 2,243,925 (14. 9. 1973), C.A. *83*, 131376p (1975); Mesomorphic derivatives of dihydrophenanthrene. [**3**] G. I. Agren, D. E. Martire, J. Chem. Phys. *61*, 3959 (1974), C.A. *82*, 50024z (1975); End-chain flexibility and the n/i transition. Lattice model of hard particies with rigid, rodlike, central cores, and semiflexible pendant segments. [**4**] N. A. Akmanova, N. R. Khairullin, Y. V. Svetkin, Zh. Obshch. Khim. *46*, 704 (1976), C.A. *84*, 163766c (1976); Association and mesomorphism of amidonitrones. [**5**] A. Amerio, Nuovo Cim. *2*, 281 (1902); Sui cristalli liquidi del Lehmann. [**6**] A. G. Amit, H. Hope, Acta Chem. Scand. *20*, 835 (1966), C.A. *65*, 8121b (1966); Crystal and molecular structure of *trans-p,p'*-dibromoazobenzene. [**7**] T. Aoyagi, K. Toriyama, F. Hori, B. Kato, N. Arima, Japan. Kokai 73 40,746 (4. 10. 1971), C.A. *79*, 104956p (1973); *p*-Alkoxyphenyl *p*-alkoxybenzoates for use as liquid crystals. [**8**] T. Aoyagi, K. Toriyama, F. Hori, N. Arima, B. Kato, Japan. Kokai 73 75,535 (7. 1. 1972), C.A. *80*, 59706q (1974); *p*-Alkoxyphenyl *p*-alkylbenzoate esters for use as liquid crystals. [**9**] A. Apel, Diss. Halle 1932; Die Richtung der Kohlenstoff-Valenzen in Benzolabkömmlingen; see also [818]. [**10**] T. Araki, S. Kanbe, Japan. Kokai 76,109,291 (24. 3. 1975), C.A. *86*, 36734g (1977); Liquid crystal compound and composite.

[**11**] N. Bravo, J. W. Doane, S. L. Arora, J. L. Fergason, J. Chem. Phys. *50*, 1398 (1969), C.A. *70*, 72688y (1969); NMR study of molecular configuration and order in a fluorinated liquid-crystalline Schiff base. [**12**] S. L. Arora, T. R. Taylor, J. L. Fergason, A. Saupe, J. Am. Chem. Soc. *91*, 3671 (1969), C.A. *71*, 43219m (1969); Liquid-crystal polymorphism in bis-(4-*n*-alkoxybenzal)-1,4-phenylenediamines. [**13**] S. L. Arora, T. R. Taylor, J. L. Fergason, Liq. Cryst. Ord. Fluids *1970*, 321 (1969), C.A. *78*, 152452x (1973); Polymorphism of smectics with s$_A$ morphology. [**14**] S. L. Arora, J. L. Fergason, A. Saupe, Mol. Cryst. Liq. Cryst. *10*, 243 (1970), C.A. *73*, 70883w (1970); Two liquid crystal phases with nematic morphology in latterally substituted phenylenediamine derivatives. [**15**] S. L. Arora, T. R. Taylor, J. L. Fergason, J. Org. Chem. *35*, 1705 (1970), C.A. *73*, 8242h (1970); Mesomorphism of alkoxybenzylideneaminoacetophenones. [**16**] S. L. Arora, J. L. Fergason, T. R. Taylor, J. Org. Chem. *35*, 4055 (1970), C.A. *74*, 16738k (1971); Molecular structure and liquid crystallinity. Phenylene-bis-(alkoxybenzoates). [**17**] S. L. Arora, J. L. Fergason, Symp. Faraday Soc. *1971*, No. 5, 97, C.A. *79*, 24379n (1973); Effects of molecular geometry on the formation of smectics. [**18**] S. L. Arora, U.S. 3,965,029 (4. 2. 1974), C.A. *85*, 123556y (1976); Liquid crystal materials. [**19**] A. M. Atallah, H. J. Nicholas, Mol. Cryst. Liq. Cryst. *17*, 1 (1971), C.A. *77*, 34729z (1972); 31-Norcycloartanol fatty acid esters: Cholesterics from a triterpene of plant origin. [**20**] A. M. Atallah, H. J. Nicholas, Mol. Cryst. Liq. Cryst. *18*, 321 (1972), C.A. *77*, 164972j (1972); Mesomorphism of fatty acid esters of lophenol, a cholesterol biosynthetic intermediate.

[**21**] A. M. Atallah, H. J. Nicholas, Mol. Cryst. Liq. Cryst. *18*, 339 (1972), C.A. *78*, 4369k (1973); Influence of the position of ring unsaturation in steroids and triterpenes on the type and formation of mesophases. I. Influence of the Δ^8 double bond. [**22**] A. M. Atallah, H. J. Nicholas, Mol. Cryst. Liq. Cryst. *19*, 217 (1973), C.A. *78*, 84642g (1973); II. Influence of the Δ^4-double bond [cf. 21]. [**23**] A. M. Atallah, H. J. Nicholas, Mol. Cryst. Liq. Cryst. *24*, 213 (1973), C.A. *82*, 178628s (1975); Mesomorphism of pollinastanol fatty acid esters. [**24**] M. V. Averina, I. A. Kondrat'ev, I. M. Rozenman, Khim. Tverd. Topl. *1974*, 70, C.A. *81*, 155570g (1974); Change in the shape of mesophase spherical particles during heat treatment. [**25**] S. Baentsch, Diss. Halle 1931; Kristallin-flüssige Harze und Lacke. [**26**] J. A. W. Barnard, J. E. Lydon, Mol. Cryst. Liq. Cryst. *26*, 285 (1974), C.A. *82*, 66568r (1975); Crystallographic examination of 14 straight chain alkyl esters of cholesterol. [**27**] E. M. Barrall, II, J. F. Johnson, R. S. Porter, Mol. Cryst. Liq. Cryst. *8*, 27 (1969), C.A. *71*, 85260y (1969); DSC of aromatic, difunctional, unsaturated and substituted acid esters of cholesterol. [**28**] E. M. Barrall, II, Anal. Calorimetry, Proc. Symp., 2nd *1970*, 121, C.A. *75*, 11708m (1971); Effects of substituent chains on the mesomorphism of the Schiff bases of *p*-aminocinnamic acid esters. [**29**] E. M. Barrall, II, K. E. Bredfeldt, M. J. Vogel, Mol. Cryst. Liq. Cryst. *18*, 195 (1972), C.A. *77*, 144642y (1972); Effect of substituent location on the thermodynamic properties of cholesteryl halobenzoates and halocinnamates. [**30**] E. M. Barrall, II, K. E. Bredfeldt, M. J. Vogel, IBM-Research RJ 1080 (1972); Mesomorphism of two N-(*p*-alkoxybenzylidene)-*p-n*-butylanilines.

[**31**] O. Bastiansen, Acta Chem. Scand. *3*, 408 (1949), C.A. *44*, 6212a (1950); Molecular structure of biphenyl and some of its derivatives. [**32**] O. Bastiansen, Acta Chem. Scand. *4*, 926 (1950), C.A. *45*, 4105i (1951); Molecular structure of biphenyl and some of its derivatives. II. [**33**] O. Bastiansen, L. Smedvik, Acta Chem. Scand. *8*, 1593 (1954), C.A. *49*, 9999i (1955); Electron diffraction studies on fluoroderivatives of biphenyl. [**34**] D. C. Batesky, U.S. 3,772,209 (21. 1. 1972), C.A. *80*, 31400m (1974); Liquid crystal compositions. [**35**] F. Bättenhausen, Diss. Halle 1925; Einfluß der molekularen Gestalt organischer Verbindungen auf ihre flüssig-kristallinen Eigenschaften. [**36**] G. H. Beaven, D. M. Hall, J. Chem. Soc. *1956*, 4637, C.A. *51*, 2392i (1957); Relation between configuration and conjugation in biphenyl derivatives. VII. Halobiphenyls. [**37**] D. P. Benton, P. G. Howe, R. Farnand, J. E. Puddington, Can. J. Chem. *33*, 1798 (1955), C.A. *50*, 11688i (1956); Mesomorphism of anhydrous soaps. Densities of alkali metal stearates. [**38**] D. Berchet, A. Hochapfel, R. Viovy, C. R. Acad. Sci., Sér. C, *270*, 1065 (1970), C.A. *73*, 29136s (1970); Variation de la viscosité de la phase nématique en fonction de la température. [**39**] K. Bergt, Diss. Halle 1925; Einfluß der *p*-Substituenten auf die flüssig-kristallinen Eigenschaften der Oxyazobenzolester. [**40**] J. D. Bernal, D. Crowfoot, Trans. Faraday Soc. *29*, 1032 (1933), C.A. *28*, 1240¹ (1934); Crystalline phases of some substances studied as liquid crystals.

[**41**] J. Bernstein, I. Izak, J. Chem. Soc., Perkin Trans. 2, *1976*, 429, C.A. *84*, 150266q (1976); Crystal and molecular structure of the stable form of N-(*p*-chlorobenzylidene)-*p*-chloraniline. [**42**] J. P. Berthault, P. Keller, Bull. Soc. Chim. Fr. *1976*, 135, C.A. *86*, 42920f (1977); Dérivés mésomorphes de l'acide α-cyanocinnamique. [**43**] J. Berthou, J. Elguero, R. Jacquier, C. Marzin, C. Rérat, C. R. Acad. Sci., Sér. C 265, 513 (1967), C.A. *68*, 16841g (1968); Structure de la dibromo-4,4′-cinnamaldazine. [**44**] J. Berthou, C. Marzin, B. Rérat, C. Rérat, Y. Uesu, C. R. Acad. Sci. Sér. C *270*, 918 (1970), C.A. *73*, 8224d (1970); Structure de la dibromo-4,4′-diméthyl-α,α′-cinnamaldazine. [**45**] V. Bertleff, Diss. Halle 1908; Morphotropie und Isomorphie kristallinisch-flüssiger Substanzen. [**46**] J. Billard, J. P. Meunier, C. R. Acad. Sci., Sér. C, *266*, 937 (1968), C.A. *69*, 13596y (1968); Faciès d'une phase smectique. [**47**] J. Canceill, C. Gros, J. Billard, J. Jacques, Liq. Cryst., Proc. Int. Conf., Bangalore *1973*, 397, C.A. *84*, 114471u (1976); New series of thermotropic liquid crystals with low temperature s_A phase. [**48**] J. Canceill, J. Jacques, J. Billard, Chem. Ind. (London) *1974*, 615, C.A. *82*, 4054m (1975); New low temperature smectics-A of 9,10-dihydrophenanthrene. [**49**] J. Billard, J. C. Dubois, A. Zann, J. Phys., (Paris), C-1, 36, 355 (1975), C.A. *83*, 113423x (1975); Benzylidène anilines *p,p′* disubstituées. Nouvelles séries de mésomorphes a anisotropie diélectrique positive. [**50**] J. Canceill, J. Gabard, J. Jacques, J. Billard, Bull. Soc. Chim. Fr. *1975*, 2066, C.A. *85*, 158810d (1976); Matériaux smectiques dérivés du fluorène. [**50a**] A. Béguin, J. Billard, J. C. Dubois, Nguyen Huu Tinh, A. Zann, J. Phys. (Paris) C-3, *40*, 15 (1979); Discotic mesophases potentialities.

[**51**] N. Bircan, Diss. Leipzig 1940; Anormale Wasserstoffaufnahme der Maleinsäure bei der katalytischen Hydrierung (1). Neue Klasse kristallinflüssiger Substanzen (2). [**51a**] A. Bloom, L. K. Hung, U.S. 4,032,470 (22. 12. 1975), C.A. *87*, 76425j (1977); Electro-optic device. [**52**] A. Bogojawlenski, N. Winogradow, Z. physik. Chem. *60*, 433 (1907), C.A. *2*, 744⁴ (1908); Verhalten von Schmelz- und Klärungskurven der flüssigen Kristalle und ihrer Mischungen. [**53**] A. Bogojawlensky, N. Winogradow, Z. physik. Chem. *64*, 229 (1908), C.A. *3*, 8⁵ (1909); Verhalten von Schmelz- und Klärungskurven flüssiger Kristalle und ihrer Mischungen. [**54**] B. M. Bogoslovskii, Akad. Nauk S.S.S.R., Inst. Org. Khim., Sintezy Org. Soedinenii, Sbornik 2, 5, C.A. *48*, 621f (1954); 4,4′-Azoxydianisole. [**55**] W. Böhme, Diss. Halle 1924; Flüssig-kristalline Eigenschaften einiger Benzidinverbindungen. [**56**] A. Boller, H. Scherrer, Ger. Offen. 2,306,738 (23. 2. 1972), C.A. *79*, 146264s (1973); Schiff base liquid crystals. [**57**] A. Boller, H. Scherrer, Ger. Offen. 2,306,739 (23. 2. 1972), C.A. *79*, 146267v (1973); Low-melting liquid crystalline 4-cyanophenyl benzoates. [**58**] A. Boller, H. Scherrer, Ger. Offen. 2,407,818 (2. 3. 1973), C.A. *82*, 86229j (1975); Liquid crystal aromatic isonitriles. [**59**] A. Boller, H. Scherrer, Ger. Offen. 2,447,098 (3. 10. 1973), C.A. *83*, 205983h (1975); Liquid crystalline esters. [**60**] A. Boller, H. Scherrer, Ger. Offen. 2,447,099 (3. 10. 1973), C.A. *84*, 10937j (1976); Schiff base liquid crystals.

[**61**] A. Boller, M. Cereghetti, H. Scherrer, Ger. Offen. 2,547,737 (25. 10. 1974), C.A. *85*, 46742b (1976); Phenylpyrimidines. [**62**] A. Boller, H. Scherrer, Ger. Offen. 2,454,570 (19. 11. 1973), C.A. *83*, 178606d (1975); Liquid crystalline biphenyls. [**63**] A. Boller, H. Scherrer, Ger. Offen. 2,459,374 (17. 12. 1973), C.A. *83*, 178608f (1975); Liquid crystalline compounds. [**63a**] A. Boller, M. Cereghetti, H. Scherrer, Ger. Offen. 2,641,724 (19. 9. 1975), C.A. *87*, 85035j (1977); Pyrimidine derivatives. [**63b**] A. Boller, M. Cereghetti, M. Schadt, H. Scherrer, Mol. Cryst. Liq. Cryst. *42*, 215 (1977); Mesomorphism of phenylpyrimidines. [**64**] B. M. Bolomin, L. K. Tarygina, R. V. Poponova, D. E. Ostromogol'skii, Zh. Org. Khim. *11*, 782 (1975), C.A. *83*, 9353p (1975); Mesomorphism of sterically hindered azoxy

compounds. [**64a**] B. M. Bolotin, N. B. Etingen, R. P. Lastovskii, L. S. Zeryukina, R. U. Safina, Zh. Org. Khim. *13*, 375 (1977), C.A. *87*, 67643b (1977); Effect of acoplanarity on the mesomorphism of azomethines. [**65**] A. Bömer, K. Winter, Z. Unt. Nahr. Genus. *4*, 865 (1901); Ester des Cholesterins und Phytosterins. [**66**] E. G. Boonstra, Acta Cryst. *16*, 816 (1963), C.A. *59*, 10838b (1963); Crystal and molecular structure of 4,4'-dinitrodiphenyl. [**67**] H. Boysen, Diss. Halle 1925; Acylverbindungen des Benzidin. [**68**] H. Brandt, Diss. Halle 1922; Abkömmlinge des 1,4-Diphenylbutans und des 1,5-Diphenylpentans. [**69**] G. Bredig, G. v. Schukowsky, Ber. *37*, 3419 (1904); Prüfung der Natur der flüssigen Krystalle mittels elektrischer Kataphorese. [**70**] L. W. Breed, E. Murrill, Inorg. Chem. *10*, 641 (1971), C.A. *74*, 116894b (1971); Solid-solid phase transitions determined by DSC. III. Organosilicon compounds.

[**71**] J. Bremer, Diss. Halle 1924; 1. Hochsiedende Braunkohlenteeröle. 2. Abkömmlinge des Benzidins, o-Tolidins und Dianisidins. [**72**] M. Bresse, Fr. Demande 2,243,023 (6. 9. 1973), C.A. *83*, 171287s (1975); Nematic liquid crystals. [**73**] C. J. Brown, Acta Cryst. *7*, 97 (1954), C.A. *48*, 4924f (1954); Crystal structure of 4,4'-dimethyldibenzyl. [**74**] C. J. Brown, Acta Cryst. *21*, 146 (1966), C.A. *65*, 6432f (1966); Refinement of the crystal structure of azobenzene. [**75**] C. J. Brown, Acta Cryst. *21*, 153 (1966), C.A. *65*, 6432g (1966); Crystal structure of *p*-azotoluene. [**76**] W. G. Shaw, G. H. Brown, J. Am. Chem. Soc. *81*, 2532 (1959), C.A. *53*, 21776e (1959); Mesomorphic 4,4'-di-(*n*-alkoxy)benzalazines. [**77**] G. H. Brown, W. G. Shaw, J. Org. Chem. *24*, 132 (1959), C.A. *54*, 5219h (1960); Phototropy of *p-n*-nonoxybenzalphenylhydrazone and *p-n*-decycloxybenzalphenylhydrazone. [**78**] R. F. Bryan, J. Chem. Soc. *1960*, 2517, C.A. *54*, 21921b (1960); X-ray study of the *p*-(*n*-alkoxy)benzoic acids. [**79**] R. F. Bryan, H. H. Mills, J. C. Speakman, J. Chem. Soc. *1963*, 4350, C.A. *59*, 9411g (1963); Crystal structure of ammonium hydrogen dicinnamate. [**80**] R. F. Bryan, J. Chem. Soc. B *1967*, 1311, C.A. *68*, 34028j (1968); Crystal structure of anisic acid.

[**81**] R. F. Bryan, L. Fallon, III, J. Chem. Soc., Perkin Trans. 2, *1975*, 1175, C.A. *83*, 200530y (1975); Crystal structure of p-butoxybenzoic acid [cf. 78]. [**82**] R. F. Bryan, D. P. Freyberg, J. Chem. Soc., Perkin Trans. 2, *1975*, 1835, C.A. *84*, 82846b (1976); Crystal structures of α-*trans*- and p-methoxycinnamic acids and their relation to thermal mesomorphism. [**83**] C. Bühner, Diss. Marburg 1906; Zur Kenntnis der kristallinischen Flüssigkeiten. [**84**] H. B. Buergi, J. D. Dunitz, Helv. Chim. Acta *53*, 1747 (1970), C.A. *73*, 134882e (1970); Crystal and molecular structures of N-benzylideneaniline, N-benzylidene-4-carboxyaniline and N-(*p*-methylbenzylidene)-*p*-nitroaniline. [**84a**] B. J. Bulkin, R. K. Rose, A. Santoro, Mol. Cryst. Liq. Cryst. *43*, 53 (1977); Possibly mesomorphic Palladium chelate. [**85**] V. K. Burima, Y. I. Dutchak, Visnik l'vov. un-tu. Ser. fiz. *1974*, 101, C.A. *83*, 79474c (1975); Structural studies of the cholesteric ester of adipic acid. [**86**] H. Bürki, W. Nowacki, Z. Krist. *108*, 206 (1956), C.A. *51*, 3228c (1957); Kristallstruktur von linksdrehendem ($=\alpha$) 7-Bromcholesterylbromid, -chlorid und -methyläther (isotype Reihe). [**87**] N. Campbell, A. Henderson, D. Taylor, Mikrochemie ver. Mikrochim. Acta *38*, 376 (1951), C.A. *46*, 2867a (1952); Polymorphism and liquid crystal formation of some azo- and azoxy-compounds. [**88**] R. Cano, C. R. Acad. Sci. *251*, 1139 (1960), C.A. *55*, 14361g (1961); Pouvoir rotatoire des corps cholestériques. [**89**] C. H. Carlisle, D. Crowfoot, Proc. Roy. Soc. A *184*, 64 (1945), C.A. *40*, 262[8] (1946); Crystal structure of cholesteryl iodide. [**90**] J. A. Castellano, J. E. Goldmacher, L. A. Barton, J. S. Kane, J. Org. Chem. *33*, 3501 (1968), C.A. *69*, 67013z (1968); Effects of terminal group substitution on the mesomorphism of benzylideneanilines.

[**91**] J. A. Castellano, M. T. McCaffrey, J. E. Goldmacher, Mol. Cryst. Liq. Cryst. *12*, 345 (1971), C.A. *75*, 41754r (1971); Nematic *p*-alkylcarbonato-*p*-alkoxyphenyl benzoates. [**92**] M. T. McCaffrey, J. A. Castellano, Mol. Cryst. Liq. Cryst. *18*, 209 (1972), C.A. *77*, 157326y (1972); Mesomorphism of homologous *p*-alkoxy-*p'*-acyloxyazoxybenzenes. [**93**] J. A. Castellano, C. S. Oh, M. T. McCaffrey, Mol. Cryst. Liq. Cryst. *27*, 417 (1974), C.A. *82*, 178570s (1975); Mesomorphism of optically active aromatic Schiff's bases. [**94**] J. A. Castellano, J. E. Goldmacher, Ger. Offen. 1,618,827 (9. 6. 1966); Nematische Flüssigkristall-Zusammensetzung. [**94a**] F. Cavatorta, M. P. Fontana, N. Kirov, Mol. Cryst. Liq. Cryst. *34*, 241 (1977), C.A. *87*, 76623x (1977); Raman and calorimetric evidences for the existence of three solid modifications in EBBA. [**95**] R. A. Champa, Mol. Cryst. Liq. Cryst. *16*, 175 (1972), C.A. *76*, 77781g (1972); Low-temperature liquid crystal exhibiting smectic morphology. [**96**] R. A. Champa, Liq. Cryst. Ord. Fluids 2, 507 (1974), C.A. *87*, 14137h (1977); Heterocyclic liquid crystals and some air force applications of mesomorphic compounds. [**97**] R. A. Champa, Mol. Cryst. Liq. Cryst. *19*, 233 (1973), C.A. *78*, 102956u (1973); Low melting liquid crystalline heterocyclic anils. [**98**] N. V. Madhusudana, S. Chandrasekhar, Liq. Cryst., Proc. Int. Conf., Bangalore, *1973*, 57, C.A. *84*, 114463t (1976); Role of permanent dipoles in nematic order. [**99**] S. Chandrasekhar, B. K. Sadashiva,

K. A. Suresh, Pramana *9*, 471 (1977); Liquid crystals of disc-like molecules. [**100**] S. A. Chawdhury, A. Hargreaves, S. H. Rizvi, Acta Cryst. B *24*, 1633 (1968), C.A. *70*, 32432d (1969); Crystal and molecular structure of 4,4'-diamino-3,3'-dichlorobiphenyl.
[**101**] I. G. Chistyakov, V. A. Usol'tseva, Izv. Vysshikh Uchebn. Zavedenii, Khim. i Khim. Tekhnol. *5*, 585 (1962), C.A. *58*, 13239d (1963); Liquid-crystalline cholesterol compounds. [**102**] I. G. Chistyakov, V. A. Usol'tseva, M. D. Nasyrova, L. I. Ershova, Izv. Vysshikh Uchebn. Zavedenii, Khim. i Khim. Tekhnol. *6*, 257 (1963), C.A. *59*, 9014a (1963); Liquid-crystalline cholesteryl caprylate and cholesteryl decanoate. [**103**] I. G. Chistyakov, V. A. Usol'tseva, M. D. Nasyrova, Izv. Vysshikh Uchebn. Zavedenii, Khim. i Khim. Tekhnol. *6*, 434 (1963), C.A. *59*, 14679b (1963); Liquid-crystalline *p*-nonyloxybenzylidene-*p*-toluidine. [**104**] G. G. Maidachenko, I. G. Chistyakov, Zh. Obshch. Khim *37*, 1730 (1967), C.A. *68*, 34045n (1968); Liquid-crystalline stigmasterol esters. [**105**] L. A. Gusakova, B. P. Smirnov, I. G. Chistyakov, Uch. Zap. Ivanov. Gos. Pedagog. Inst. *62*, 98 (1967), C.A. *71*, 49629s (1969); Synthesis of certain liquid crystals. [**106**] B. P. Smirnov, I. G. Chistyakov, Izv. Vyssh. Ucheb. Zaved., Khim. Khim. Tekhnol. *13*, 217 (1970), C.A. *73*, 49660m (1970); Liquid crystals derived from *p*-substituted benzaldehydes. [**107**] G. G. Maidachenko, I. G. Chistyakov, Uch. Zap. Ivanov. Gos. Pedagog. Inst. *77*, 61 (1970), C.A. *76*, 18886x (1972); Liquid-crystalline stigmasterol esters. [**108**] I. G. Chistyakov, Uch. Zap. Ivanov. Gos. Pedagog. Inst. No. *99*, 43 (1972), C.A. *78*, 152441t (1973); Structure of ethyl *p*-(anisalamino)cinnamate liquid crystals. [**109**] J. Y. C. Chu, J. Chem. Soc., Chem. Commun. *1974*, 374, C.A. *81*, 105805s (1974); Mesomorphism of 5α-cholest-8(14)en-3β-yl alkanoates. [**110**] J. Y. C. Chu, J. Phys. Chem. *79*, 119 (1975), C.A. *82*, 86491p (1975); Thermal transitions of 5α-cholest-8(14)-en-3β-yl alkanoates.
[**111**] R. E. D. Clark, Chem. & Ind. *49*, 396 (1930), C.A. *24*, 3409⁹ (1930); Meso-states and stereochemistry. [**112**] H. S. Cole, J. R. Sowa, Mol. Cryst. Liq. Cryst. *30*, 149 (1975), C.A. *83*, 130976x (1975); Low-melting nematic N,N'-bis(*p*-alkylbenzylidene)-2-chloro-1,4-phenylenediamines. [**113**] M. Cotrait, D. Sy, M. Ptak, Acta Cryst. *31B*, 1869 (1975), C.A. *83*, 124456m (1975); Structure cristalline d'un composé nématogène: le (méthoxybenzylidèneamino-4')-4α-methyl cinnamate de propyle. [**114**] M. Cotrait, C. Destrade, H. Gasparoux, Acta Cryst. *31B*, 2704 (1975), C.A. *84*, 24754j (1976); Structure cristalline d'un composé smectogène: le (méthyl-2-butyl)-2-(oxo-1-pentyl)-7-dihydro-9,10-phénanthrène. [**114a**] M. Cotrait, P. Marsau, Acta Cryst. *32B*, 2993 (1976), C.A. *86*, 24667z (1977); Structure cristalline d'un mésogène: 2-butyl-7-(oxo-1-pentyl)-9,10-dihydrophenanthrène. [**115**] M. Cotrait, C. Destrade, H. Gasparoux, Mol. Cryst. Liq. Cryst. *39*, 159 (1977), C.A. *87*, 32316v (1977); Arrangement moléculaire à l'état solide de deux *p*-méthoxy-*p*'-alkyltolanes. [**116**] A. Couttet, J. C. Dubois, A. Zann, Fr. Demande 2,213,803 (29. 2. 1972), C.A. *82*, 105332f (1975); Nematics with great stability. [**117**] R. J. Cox, Mol. Cryst. Liq. Cryst. *19*, 111 (1972), C.A. *78*, 71179m (1973); Liquid crystalline methyl-substituted stilbenes. [**118**] R. J. Cox, N. J. Clecak, Mol. Cryst. Liq. Cryst. *37*, 241 (1976), C.A. *86*, 171055b (1977); 4-cyano-4'-alkyltolanes, a new series of liquid crystals. [**119**] R. J. Cox, N. J. Clecak, Mol. Cryst. Liq. Cryst. *37*, 263 (1976), C.A. *86*, 171056c (1977); 4-alkyl-4'-cyanostilbenes, a new series of liquid crystals. [**120**] R. J. Cox, R. C. Gaskill, J. F. Johnson, N. J. Clecak, Thermochim. Acta *18*, 37 (1977), C.A. *86*, 63871f (1977); Thermal properties of 4-alkyl-4'-cyanotolanes.
[**121**] B. M. Craven, G. T. DeTitta, J. Chem. Soc., Perkin Trans. 2, *1976*, 814, C.A. *85*, 46932p (1976); Cholesteryl myristate. Structures of the crystalline solid and mesophases. [**122**] T. R. Criswell, B. H. Klanderman, D. C. Batesky, Mol. Cryst. Liq. Cryst. *22*, 211 (1973), C.A. *79*, 146081e (1973); Alkyl carbonato terminally substituted anils. [**123**] W. Cwikiewicz, M. Jawdosiuk, Pr. Nauk. Inst. Metrol. Elektr. Politech. Wroclaw. *7*, 29 (1975), C.A. *86*, 88431a (1977); Preparation methods for liquid crystals. [**124**] J. Daugvila, A. Tubalyte, G. Denys, P. Adomenas, Zh. Obshch. Khim. *46*, 2125 (1976), C.A. *85*, 185036n (1976); Liquid crystals of α-cyanostilbene type. [**125**] E. Däumer, Diss. Halle 1912; Kristallin-flüssige Cholesterinverbindungen. [**126**] J. S. Dave, J. M. Lohar, Proc. Natl. Acad. Sci. India Sect. A *29*, 35 (1960), C.A. *55*, 12992h (1961); Liquid crystallinity in Schiff's bases. [**127**] J. S. Dave, P. R. Patel, Indian J. Chem. *2*, 164 (1964), C.A. *61*, 5038c (1964); Liquid crystallinity in homologous series of Schiff's bases. [**128**] J. S. Dave, P. R. Patel, Mol. Cryst. *2*, 103 (1967), C.A. *66*, 41296g (1967); Influence of molecular structure on liquid crystalline properties and phase transitions in these structures. I. [**129**] J. S. Dave, P. R. Patel, Mol. Cryst. *2*, 115 (1967), C.A. *66*, 41297h (1967); Part II (cf. [**128**]). [**130**] J. S. Dave, R. A. Vora, Liq. Cryst. Ord. Fluids *1970*, 477, C.A. *79*, 5501e (1973); Mesomorphism of *p-n*-alkoxybenzoates of cholesterol.
[**131**] J. S. Dave, G. Kurian, A. P. Prajapati, R. A. Vora, Mol. Cryst. Liq. Cryst. *14*, 307 (1971), C.A. *75*, 144761v (1971); Mesomorphism of N,N'-bis-(4-alkoxy-1-naphthylidene)-benzidines. [**132**]

J. S. Dave, R. A. Vora, Mol. Cryst. Liq. Cryst. *14*, 319 (1971), C.A. *75*, 144721g (1971); Mesomorphism of *trans-p*-alkoxy cinnamates of cholesterol. [**133**] J. S. Dave, G. Kurian, A. P. Prajapati, R. A. Vora, Curr. Sci. *41*, 415 (1972), C.A. *77*, 80742w (1972); Mesomorphism in a homologous series of naphthylidene Schiff bases. [**134**] J. S. Dave, G. Kurian, A. P. Prajapati, R. A. Vora, Indian J. Chem. *10*, 754 (1972), C.A. *78*, 102949u (1973); Mesomorphism of Schiff bases in a homologous series comprising a naphthalene moiety. [**135**] J. S. Dave, R. A. Vora, Indian J. Chem. *11*, 19 (1973), C.A. *78*, 159976e (1973); Mesomorphism of substituted benzoates of cholesterol. [**136**] J. S. Dave, G. Kurian, Indian J. Chem. *11*, 833 (1973), C.A. *79*, 150357s (1973); Mesomorphism in a homologous series of cholesteryl esters containing a naphthalene nucleus. [**137**] J. S. Dave, G. Kurian, Curr. Sci. *42*, 200 (1973), C.A. *78*, 147215t (1973); Mesomorphism in a homologous series of naphthylidene Schiff's base compounds of cholesterol. [**138**] J. S. Dave, B. C. Joshi, G. Kurian, Curr. Sci. *42*, 349 (1973), C.A. *79*, 78442w (1973); Mesomorphism of naphthalene derivatives. [**139**] J. S. Dave, G. Kurian, Liq. Cryst., Proc. Int. Conf., Bangalore, *1973*, 427, C.A. *84*, 114473w (1976); Mesomorphic cholesteryl 6-alkoxy-2-naphthoates. [**140**] J. S. Dave, A. P. Prajapati, Liq. Cryst., Proc. Int. Conf., Bangalore, *1973*, 435, C.A. *84*, 114474x (1976); Mesomorphic N,N'-di(4-*n*-alkoxy-1-naphthylidene)*p*-azoanilines and N(4-*n*-alkoxy-1-naphthylidene)4'-aminoazobenzenes.

[**141**] J. S. Dave, R. A. Vora, Liq. Cryst., Proc. Int. Conf., Bangalore, *1973*, 447, C.A. *84*, 114475y (1976); Mesomorphic *p*-(*p'-n*-alkoxybenzoyloxy)toluenes. [**142**] J. S. Dave, G. Kurian, Mol. Cryst. Liq. Cryst. *24*, 347 (1973), C.A. *82*, 178634r (1975); Mesomorphic *p*-(*n*-alkoxybenzylidene)-*p*-amino-benzoates of cholesterol. [**143**] J. S. Dave, R. A. Vora, Mol. Cryst. Liq. Cryst. *28*, 269 (1974), C.A. *82*, 178709u (1975); Mesomorphic biphenyl 4-*trans-p-n*-alkoxycinnamates. [**144**] J. S. Dave, G. Kurian, J. Phys. (Paris), C-1, *36*, 403 (1975), C.A. *83*, 113873u (1975); Low melting mesomorphic *p*-(*p'-n*-alkoxyben-zoyloxy)acetophenones and *p*-(*p'-n*-alkoxybenzoyloxy)benzaldehydes. [**145**] J. S. Dave, A. P. Prajapati, Curr. Sci. *45*, 95 (1976), C.A. *85*, 20410s (1976); Mesomorphism of Schiff bases comprising naphthalene moiety. [**145a**] J. S. Dave, G. Kurian, N. R. Patel, Curr. Sci. *46*, 300 (1977), C. A. *86*, 198244z (1977); Mesomorphic 4-*n*-alkoxy-1-naphthylidene-*p-n*-butoxyanilines. [**145b**] J. S. Dave, G. Kurian, Mol. Cryst. Liq. Cryst. *42*, 175 (1977), Mesomorphism of *p*(*p'-n*-alkoxybenzoyloxy)benzylidene-*p''*-anisidines and -toluidines. [**145c**] G. Kurian, J. S. Dave, Mol. Cryst. Liq. Cryst. *42*, 193 (1977), Mesomorphism of 4-*n*-alkoxy-1-naphthylidene-*p*-aminobenzoates of cholesterol. [**146**] H. Dehne, P. Wolff, H. G. Fuchs, Z. Chem. *9*, 423 (1969); C.A. *72*, 43059e (1970); Kristallin-flüssige *p*-Terphenyl-4,4''-derivate. [**147**] J. Dejace, Bull. Soc. fr. Minéral Cristallogr. *92*, 141 (1969), C.A. *71*, 25494t (1969); Structure cristalline du paraterphényle. [**148**] W. H. de Jeu, J. van der Veen, Philips Res. Rep. *27*, 172 (1972), C.A. *77*, 131773s (1972); Relation between molecular structure and liquid-crystalline behavior. [**149**] W. H. de Jeu, J. Van der Veen, W. J. A. Goossens, Solid State Commun. *12*, 405 (1973), C.A. *78*, 146917m (1973); Dependence of the clearing temperature on alkyl chain length in nematic homologous series. [**149a**] W. H. de Jeu, J. Van der Veen, Mol. Cryst. Liq. Cryst. *40*, 1 (1977), C.A. *87*, 125406q (1977); Molecular structure and nematic behavior. [**149b**] W. H. de Jeu, J. Phys. (Paris) *38*, 1265 (1977), C.A. *87*, 175839d (1977); Molecular structure and the occurrence of s$_A$ and s$_C$ phases. [**150**] A. C. de Kock, Z. physik. Chem. *48*, 129 (1904); Bildung und Umwandlung von fließenden Mischkristallen.

[**151**] D. Demus, Z. Chem. *15*, 1 (1975), C.A. *83*, 19060e (1975); Eigenschaften, Theorien und Molekülbau flüssiger Kristalle. [**152**] D. Demus, L. Richter, C. E. Rürup, H. Sackmann, H. Schubert, J. Phys. (Paris), C-1, *36*, 349 (1975), C.A. *83*, 27801n (1975); Mesomorphism of 4,4'-disubstituted biphenyls. [**153**] F. Kuschel, D. Demus, Z. Chem. *15*, 350 (1975), C.A. *83*, 171130k (1975); Kristallin-flüssige 4-Nitro-zimtsäureester. [**154**] H. J. Deutscher, F. Kuschel, S. König, H. Kresse, D. Pfeiffer, A. Wiegeleben, J. Wulf, D. Demus, Z. Chem. *17*, 64 (1977), C.A. *86*, 148993x (1977); Flüssig-kristalline *trans*-4-*n*-Alkylcyclo-hexancarbonsäure 4-cyanophenylester. [**155**] H. J. Deutscher, W. Weißflog, D. Demus, G. Pelzl, Ger. Offen. 2,123,175 (15. 6. 1970), C.A. *77*, 67372f (1972); Nematische Flüssigkristalle. [**156**] D. Demus, H. Schubert, K. Hanemann, Fr. Demande 2,133,788 (19. 4. 1971), C.A. *78*, 159226k (1973); Nematic liquid crystals. [**157**] D. Demus, H. Kresse, H. Schubert, W. Weißflog, A. Wiegeleben, Ger. Offen. 2,348,193 (5. 10. 1972), C.A. *81*, 7205b (1974); Nematische Flüssigkristalle. [**158**] D. Demus, F. Kuschel, P. Moeckel, W. Weißflog, H. Zaschke, Ger. Offen. 2,544,577 (30. 10. 1974), C.A. *85*, 54650k (1976); Nematische Flüssigkristalle. [**159**] A. Denat, B. Gosse, P. Gosse, J. Chim. Phys. Physicochim. Biol. *70*, 319 (1973), C.A. *78*, 158546c (1973); MBBA: Stabilité chimique et conductibilités ioniques. [**160**] A. Denat, B. Gosse, J. P. Gosse, Chem. Phys. Lett. *18*, 235 (1973), C.A. *78*, 110326d (1973); Chemical and electrochemical stability of MBBA. [**160a**] C. Destrade, H. Gasparoux, F. Guillon, Mol. Cryst.

Liq. Cryst. *40*, 163 (1977), C.A. *87*, 125603b (1977); Molecular correlations in isotropic phases. A previsional evaluation of the mesogenic character.
[**161**] M. J. S. Dewar, J. P. Schroeder, J. Org. Chem. *30*, 2296 (1965), C.A. *63*, 5550a (1965); *p*-Alkoxy- and *p*-carbalkoxybenzoates of diphenols. New series of liquid crystals. [**162**] M. J. S. Dewar, R. S. Goldberg, J. Org. Chem. *35*, 2711 (1970), C.A. *73*, 66215n (1970); Effects of central and terminal groups on nematic mesophase stability. [**163**] M. J. S. Dewar, R. S. Goldberg, J. Am. Chem. Soc. *92*, 1582 (1970), C.A. *72*, 104981y (1970); Role of *p*-phenylene groups in nematics. [**164**] M. J. S. Dewar, A. Griffin, R. M. Riddle, Liq. Cryst. Ord. Fluids *2*, 733 (1974), C.A. *86*, 181002w (1977); Effect of structure on the stability of nematics. [**165**] M. J. S. Dewar, U.S.N.T.I.S., AD/A Rep. 1974, No. 003698/8GA, 30 pp., C.A. *83*, 77860b (1975); Prediction of properties and behavior of materials. [**166**] M. J. S. Dewar, R. M. Riddle, J. Am. Chem. Soc. *97*, 6658 (1975), C.A. *83*, 211550f (1975); Factors influencing the stabilities of nematics. [**167**] M. J. S. Dewar, A. C. Griffin, J. Am. Chem. Soc. *97*, 6662 (1975), C.A. *83*, 211551g (1975); Thermodynamic study of the role of the central group on the stability of nematics. [**168**] M. J. S. Dewar, A. C. Griffin, J. Chem. Soc., Perkin Trans. 2, *1976*, 710, C.A. *85*, 54910v (1976); Ester linkage in the nematic phase. Thermodynamic study. [**169**] M. J. S. Dewar, A. C. Griffin, J. Chem. Soc., Perkin Trans. 2, *1976*, 713, C.A. *85*, 54911w (1976); Terminal groups in nematics and cholesterics. Thermodynamic study. [**170**] H. J. Dietrich, E. L. Steiger, Mol. Cryst. Liq. Cryst. *16*, 263 (1972), C.A. *76*, 132624k (1972); Mesomorphism of compounds of low thermal phase stability.
[**171**] H. J. Dietrich, E. L. Steiger, U.S. 3,743,681 (18. 5. 1971), C.A. *79*, 84476b (1973); Preparation of liquid crystals. [**172**] H. J. Dietrich, E. L. Steiger, U.S. 3,751,467 (19. 5. 1971), C.A. *79*, 91768b (1973); Liquid-crystal compounds. [**173**] H. J. Dietrich, E. L. Steiger, U.S. 3,769,327 (4. 6. 1971), C.A. *79*, 146243j (1973); *p*-[[*p*-(Alkenyloxy)benzylidene]amino]phenylacetates. [**174**] H. J. Dietrich, E. L. Steiger, U.S. 4,005,064 (18. 5. 1971), C.A. *86*, 149142f (1977); Mesomorphic 4-methoxyalkoxy-4'-alkyl azoxybenzenes. [**175**] G. D. Dixon, L. C. Scala, Mol. Cryst. Liq. Cryst. *10*, 327 (1970), C.A. *73*, 77476z (1970); Thermal decomposition of cholesteryl oleyl carbonate. [**176**] D. Dolphin, Z. Muljiani, J. Cheng, R. B. Meyer, J. Chem. Phys. *58*, 413 (1973), C.A. *78*, 76840e (1973); Low-temperature chiral nematics derived from β-methylbutylaniline. [**177**] E. Donat, O. Stierstadt, Ann. Physik *17*, 897 (1933), C.A. *27*, 5602 (1933); Flüssige Metalleinkristalle. [**178**] J. B. Donnet, J. Lahaye, A. Voet, G. Prado, Carbon *12*, 212 (1974), C.A. *81*, 79856n (1974); Are carbon-blacks formed from liquid-crystalline systems? [**178a**] J. Doucet, A. M. Levelut, M. Lambert, Acta Cryst. B *33*, 1710 (1977), C.A. *87*, 32372k (1977); Structure cristalline du TBBA à température ambiante. [**179**] J. C. Dubois, A. Couttet, A. Zann, Ger. Offen. 2,309,501 (29. 2. 1972), C.A. *79*, 136837s (1973); Nematics derived from tolan. [**180**] J. C. Dubois, A. Zann, A. Couttet, 3rd ACS Symp. Ord. Fluids Liq. Cryst., Chicago, Aug. 1973, Abstr. No. 123; *pp'*-disubstituted benzylideneanilines — new series of liquid crystals with positive dielectric anisotropy.
[**181**] J. C. Dubois, A. Zann, A. Couttet, Mol. Cryst. Liq. Cryst. *27*, 187 (1974), C.A. *82*, 79031k (1975); Tolanes nématiques. [**182**] J. C. Dubois, A. Zann, J. Phys. (Paris), C-3, *37*, 35 (1976), C.A. *86*, 4698g (1977); New mesomorphic 4,4'-disubstituted biphenyls. [**183**] J. C. Dubois, A. Zann, Inf. Chim. *157*, 275 (1976), C.A. *86*, 42581w (1977); Cristaux liquides dérivés du *p*-biphényle. [**183a**] J. C. Dubois, A. Zann, H. T. Nguyen, C. R. Acad. Sci., Sér. C *284*, 137 (1977), C.A. *86*, 170636e (1977); Stilbènes mésomorphes. [**183b**] J. C. Dubois, A. Zann, A. Béguin, Mol. Cryst. Liq. Cryst. *42*, 139 (1977), Mesomorphism of *m*-cyano or *m*-halogeno-benzoate derivatives. [**183c**] Tinh Nguyen Huu, J. C. Dubois, J. Malthete, C. Destrade, C. R. Acad. Sci., Sér. C *286*, 463 (1978), Synthèse de l'hexaalcoxy-triphénylène, de nouvelles mésophases. [**184**] J. H. Dumbleton, T. R. Lomer, Acta Cryst. *19*, 301 (1965), C.A. *63*, 10792g (1965); Crystal structure of potassium palmitate (Form B). [**184a**] Y. I. Dutchak, Z. M. Mikityuk, M. I. Fedyshin, Kristallografiya *21*, 1218 (1976), C.A. *87*, 125555n (1977); Large periods in cholesteryl acetate. [**185**] M. Dvolaitzky, J. Billard, F. Poldy, C. R. Acad. Sci., Sér. C, *279*, 533 (1974), C.A. *82*, 178742z (1975); Radicaux libres mésogènes. [**186**] M. Dvolaitzky, J. Billard, F. Poldy, Tetrahedron *32*, 1835 (1976), C.A. *86*, 29760j (1977); Smectic E, C and A free radicals. [**187**] C. Eaborn, N. H. Hartshorne, J. Chem. Soc. *1955*, 549, C.A. *49*, 7948f (1955); Mesomorphism of diisobutylsilanediol. [**188**] E. Eichwald, Diss. Marburg 1905; Untersuchungen über flüssige Kristalle. [**189**] R. Eidenschink, D. Erdmann, J. Krause, L. Pohl, Angew. Chem. *89*, 103 (1977), C.A. *86*, 89349s (1977); Substituierte Phenylcyclohexane — neue Klasse flüssiger Kristalle. [**190**] R. Eidenschink, J. Krause, 7. Arbeitstagung Flüssigkristalle, Inst. Angew. Festkörperphysik Freiburg, 4. 3. 1977, Substituierte Phenylcyclohexane — neues Bauprinzip nematischer Flüssigkristalle.

[191] L. Pohl, R. Eidenschink, G. Krause, D. Erdmann, Phys. Lett. A *60*, 421 (1977), C.A. *86*, 181007b (1977); Physical properties of nematic phenylcyclohexanes, new class of low melting liquid crystals with positive dielectric anisotropy. [191a] R. Eidenschink, D. Erdmann, J. Krause, L. Pohl, Angew. Chem. *90*, 133 (1978), Substituierte Bicyclohexyle — neue Klasse nematischer Flüssigkristalle. [192] Z. M. Elashvili, G. S. Chilaya, S. N. Aronishidze, M. I. Brodzeli, K. G. Dzhaparidze, Soobshch. Akad. Nauk Gruz. SSR *81*, 105 (1976), C.A. *85*, 32556g (1976); Mesomorphism of chiral *p*-alkoxybenzyli-dene(*p'*-isopentylcarboxy)anilines. [193] M. A. El-Bayoumi, M. El-Aasser, F. Abdel-Halim, J. Am. Chem. Soc. *93*, 586 (1971), C.A. *74*, 75761k (1971); Electronic spectra and structures of Schiff bases. I. Benzanils. [194] W. Elser, Mol. Cryst. *2*, 1 (1966), C.A. *66*, 32968a (1967); Mesomorphism of cholesteryl-*n*-alkylcar-bonates. [195] W. Elser, Mol. Cryst. Liq. Cryst. *8*, 219 (1969), C.A. *71*, 113165s (1969); Mesomorphism of sulfur containing steroid derivatives. [196] W. Elser, R. D. Ennulat, J. Phys. Chem. *74*, 1545 (1970), C.A. *72*, 125846y (1970); Mesomorphism of cholesteryl-S-alkylthiocarbonates. [197] W. Elser, J. L. W. Pohlmann, P. R. Boyd, Mol. Cryst. Liq. Cryst. *11*, 279 (1970), C.A. *74*, 54096g (1971); S-Cholesteryl alkanethioates. [198] W. Elser, J. L. W. Pohlmann, P. R. Boyd, Mol. Cryst. Liq. Cryst. *13*, 255 (1971), C.A. *75*, 64098u (1971); Structure dependence of cholesterics. Minor changes within the 17 β-side chain of cholesterol. [199] W. Elser, J. L. W. Pohlmann, P. R. Boyd, Mol. Cryst. Liq. Cryst. *15*, 175 (1971), C.A. *76*, 141161j (1972); Odd-even effect in steroidal ω-phenylalkanoates. [200] W. Elser, J. L. W. Pohlmann, P. R. Boyd, Mol. Cryst. Liq. Cryst. *20*, 77 (1973), C.A. *78*, 102976a (1973); Cholesteryl n-alkyl carbonates.

[201] W. Elser, J. L. W. Pohlmann, P. R. Boyd, Mol. Cryst. Liq. Cryst. *27*, 325 (1974), C.A. *82*, 148583f (1975); Mesomorphism of S-cholesteryl ω-phenylalkanethioates. [202] W. Elser, R. D. Ennulat, J. L. W. Pohlmann, Mol. Cryst. Liq. Cryst. *27*, 375 (1974), C.A. *82*, 132200t (1975); Mesomorphism of S-cholesteryl alkyl thiocarbonates. [203] W. Elser, J. L. W. Pohlmann, P. R. Boyd, U.S. NTIS, AD Rep. 1975, AD-AO23526, 53 pp., C.A. *85*, 152090x (1976); Chemistry of mesomorphic compounds. [204] W. Elser, J. L. W. Pohlmann, P. R. Boyd, Mol. Cryst. Liq. Cryst. *36*, 279 (1976), C.A. *86*, 81975x (1977); High-purity cholesteryl oleyl carbonate. [205] W. W. Emerson, Nature *178*, 1248 (1956), C.A. *51*, 5647i (1957); Liquid crystals of montmorillonite. [206] R. D. Ennulat, Mol. Cryst. Liq. Cryst. *8*, 247 (1969), C.A. *71*, 85807g (1969); Mesomorphism of homologous series. I. [207] R. D. Ennulat, A. J. Brown, Mol. Cryst. Liq. Cryst. *12*, 367 (1971), C.A. *75*, 54888n (1971); II. Odd-even effect (cf. [206]). [208] H. Esaki, M. Fukai, Y. Moriyama, K. Mori, Japan. Kokai 76 59,077 (20. 11. 1974), C.A. *86*, 148842x (1977); Nematic liquid crystal compositions for field-effect display devices. [209] K. Eulner, Diss. Halle 1930; Versuche mit p-Azobenzaldehyd und p-Azoxybenzaldehyd. [210] J. L. Fergason, N. N. Goldberg, R. J. Nadalin, Mol. Cryst. *1*, 309 (1966), C.A. *66*, 14678b (1967); Cholesterics. Chemical significance.

[211] H. Filss, Diss. Halle 1926; Einfluß der Stellungsisomerie auf die flüssig-kristallinen Eigen-schaften von Benzoesäureestern. [212] E. Fischer, Diss. Halle 1922; Zur Kenntnis des Diphenylbiphenyls und des symmetrischen Triphenylbenzols. [213] D. L. Fishel, P. R. Patel, Mol. Cryst. Liq. Cryst. *17*, 139 (1972), C.A. *77*, 53461x (1972); Transition temperatures and mesophase identifications for some anils. [214] J. B. Flannery, Jr., W. Haas, J. Phys. Chem. *74*, 3611 (1970), C.A. *73*, 114018z (1970); Low-temperature mesomorphism in terminally substituted benzylideneanilines. [215] J. E. Fletcher, Engineering *106*, 382 (1918), C.A. *13*, 2630[9] (1919); Liquid crystals. [216] H. W. Foote, International Critical Tables Bd. 1, 314 (McGraw-Hill: New York). 1926; Liquid crystals. [217] B. C. Freasier, Diss. Louisiana State Univ., Baton Rouge, La., 1973, 54 pp. Avail. Univ. Microfilms, Ann Arbor, Mich., Order No. 74-7222, C.A. *81*, 7081h (1974); Equilibrium states of a dimer model with angular forces. [218] R. Freymann, Bull. Belg. Phys. Soc. *1967*, 295, C.A. *69*, 6541y (1968); Intermolecular activity in liquids and solids. IR, NMR, dielectric absorption. [219] P. Friar, Dipl.-Arbeit, Univ. Frank-furt/M, 1974; Synthese und Eigenschaften speziell substituierter Thiobenzanilide im Hinblick auf ihre mögliche Verwendung als flüssige Kristalle. [220] K. Roberts, S. Friberg, Kolloid-Z. Z. Polym. *230*, 357 (1969), C.A. *71*, 42649q (1969); Phase transitions of adsorbed carboxylic acids on zinc oxide and of zinc soaps. IR and X-ray diffraction investigations.

[221] G. Friedel, Compt. rend. *180*, 892 (1925), C.A. *19*, 1646[3] (1925); éthyle anisal-p-aminocinna-mate. [222] E. Froelich, Diss. Halle 1910; Einfluß der Halogene und des CN-Radikals auf die kristallinisch-flüssigen Eigenschaften aromatischer Substanzen. [223] Y. Fujii, S. Mori, N. Matsumura, Japan. Kokai 74,109,328 (27. 2. 1973), C.A. *82*, 139789a (1975); Azomethines having enantiomorphoric dimorphism nematic state. [224] Y. Fujii, N. Matsumura, Japan. Kokai 76 75,050 (19. 12. 1974), C.A. *85*, 159722p (1976); 4-Alkyl (or alkoxy)-4'-cyanobiphenyls. [225] H. Fürst, H. J. Dietz, J. prakt. Chem. *4*, 147

(1956), C.A. *51*, 12087b (1957); Langkettige Pyridinderivate und ihre tertiären Immoniumsalze. [226] S. Furuyama, Y. Murakami, K. Morimoto, Japan. Kokai 76,115,435 (1. 4. 1975), C.A. *86*, 89421j (1977); Aromatic esters. [226a] S. Furuyama, Y. Murakami, K. Morimoto, Japan. Kokai 76,136,643 (19. 5. 1975), C.A. *86*, 171109x (1977); Aromatic ester derivatives. [226b] S. Furuyama, Y. Murakami, K. Morimoto, Japan. Kokai 77 73,836 (12. 12. 1975), C.A. *87*, 184225c (1977); Aromatic ester derivatives. [227] R. Gabler, Diss. Leipzig 1939; Einfluß des Molekülbaus auf die kristallin-flüssigen Eigenschaften von Kohlenstoffverbindungen. [228] A. Gahren, Diss. Halle 1908; Entstehung kristallinischer Flüssigkeiten durch Mischung von Substanzen. [229] J. L. Galigné, J. Feilgueirettes, Acta Cryst. B *24*, 1523 (1968), C.A. *70*, 23917p (1969); Structure cristalline de l'anisaldehyd-azine. [230] J. L. Galigné, J. Phys. (Paris), C-4, *30*, 4 (1969), C.A. *72*, 94270z (1970); Structures cristallines de produits donnant une phase nématique a la fusion.

[231] J. L. Galigné, J. Feilgueirettes, C. R. Acad. Sci., Sér. C *268*, 938 (1969), C.A. *70*, 109842x (1969); Structure cristalline du 4,4′-azodiphénétole. [232] J. L. Galigné, Acta Cryst. B *26*, 1977 (1970), C.A. *74*, 46758j (1971); Structure cristalline du 4,4-azodiphénétole. [233] P. Gaubert, Compt. rend. *147*, 498 (1908), C.A. *3*, 139[8] (1909); Les cristaux liquides des éthers-sels de l'ergostérine. [234] P. Gaubert, Compt. rend. *149*, 608 (1909), C.A. *4*, 138 (1910); Les cristaux liquides des combinaisons de la cholestérine et de l'ergostérine avec l'urée. [235] P. Gaubert, Compt. rend. *156*, 149 (1913), C.A. *7*, 1484[3] (1913); Quelques composés de la cholestérine donnant des cristaux liquides. [236] P. Gaubert, Bull. soc. franc. min. *32*, 62 (1909), Cristaux liquides de quelques composés de la cholestérine et ceux de l'ergostérine. [237] P. Gaubert, Compt. rend. *163*, 392 (1916), C.A. *11*, 229[7] (1917); Les liquides cristallins obtenus par évaporation d'un solution. [238] P. Gaubert, Compt. rend. *168*, 277 (1919), C.A. *13*, 925[9] (1919); Les cristaux liquides de l'acide agaricique. [239] P. Gaubert, Compt. rend. *174*, 1115 (1922), C.A. *16*, 4118[9] (1922); Les cristaux liquides de phosphate de calcium. [240] D. M. Gavrilovic, U.S. 3,925,238 (28. 6. 1974), C.A. *84*, 105222m (1976); Liquid crystal electrooptic devices. [240a] D. M. Gavrilovic, U.S. 4,013,582 (17. 6. 1976), C.A. *86*, 198026e (1977); Liquid crystal compounds and electrooptic devices incorporating them.

[241] H. W. Gibson, J. M. Pochan, J. Phys. Chem. *77*, 837 (1973), C.A. *78*, 129056n (1973); Effect of cholesteryl alkanoate structure on liquid crystal transition thermodynamics. Pure and in binary mixtures. [242] H. W. Gibson, Mol. Cryst. Liq. Cryst. *27*, 43 (1974), C.A. *83*, 10585x (1975); New cholesteryl esters of chiral alkanoic acids. [243] K. Gieseler, Diss. Halle 1927; *p*-Methoxy-5-phenyl-pentadienal-(1) und p-Methoxy-cinnamylidenessigsäure. [244] H. K. Gilliam, M. P. Whittaker, Bienn. Conf. Carbon, Ext. Abstr. Program, 11th *1973*, 211, C.A. *83*, 47182q (1975); Mesophase graphite. [244a] A. M. Giroud, U. T. Müller-Westerhoff, Mol. Cryst. Liq. Cryst. *41*, 11 (1977), C.A. *87*, 109755x (1977); Mesomorphic transition metal complexes. [245] T. I. Gnilomedova, V. P. Sevost'yanov, A. V. Bondarenko, Zh. Prikl. Khim. *49*, 1337 (1976), C.A. *85*, 85777m (1976); Nematic alkoxyphenyl esters of benzoic acids. [246] R. S. Goldberg, Diss. Univ. of Texas, Austin, 1969, 129 pp. Avail. Univ. Microfilms, Ann Arbor, Mich. Order Nr. 69-15,814, C.A. *72*, 125238b (1970); Molecular structure and mesophase stability. [247] J. Goldmacher, L. A. Barton, J. Org. Chem. *32*, 476 (1967), C.A. *67*, 81869y (1967); Liquid crystals. I. Fluorinated anils. [248] J. E. Goldmacher, M. T. McCaffrey, Liq. Cryst. Ord. Fluids *1970*, 375, C.A. *78*, 152439y (1973); Nematic benzylidene anils containing a terminal alcohol group. [249] J. F. Goldmacher, M. T. McCaffrey, Ger. Offen. 2,026,280 (2. 6. 1969), C.A. *74*, 99644a (1971); *p*-(*p*-Alkoxybenzylidenamino)-phenyl-carboxylates. [250] H. Goodarzi, G. Hermon, M. Iley, H. Marsh, Fuel *54*, 105 (1975), C.A. *83*, 63206f (1975); Effect of preoxidation of vitrinites upon coking properties.

[251] A. P. Gorlov, U.S.S.R. 498,316 (11. 2. 1974), C.A. *84*, 114706z (1976); Sulfur-containing derivatives of cholesteryl chlorocarbonate. [252] A. P. Gorlov, Yu. I. Rysakova, V. V. Zezina, U.S.S.R. 516,684 (11. 2. 1974), C.A. *85*, 102805p (1976); Metal-containing derivatives of cholesteryl chlorocarbonate. [253] P. J. Graham, U.S. 2,849,469 (8. 6. 1953), C.A. *53*, 4298e (1959); Formyl substituted bis-(cyclopentadienyl)-iron compounds. [254] G. W. Gray, B. Jones, Nature *167*, 83 (1951), C.A. *45*, 7993d (1951), Mesomorphism of alkoxynaphthoic acids. [255] G. W. Gray, B. Jones, Nature *170*, 451 (1952), C.A. *47*, 9092f (1953); Mesomorphism of *p-n*-alkoxybenzoic and 6-*n*-alkoxy-2-naphthoic acids. [256] G. W. Gray, B. Jones, J. Chem. Soc. *1953*, 4179, C.A. *48*, 6218b (1954); Mesomorphic transition points of *p-n*-alkoxybenzoic acids. [257] G. W. Gray, B. Jones, J. Chem. Soc. *1954*, 678, C.A. *49*, 1679f (1955); 4- and 5-*n*-Alkoxy-1-naphthoic and 6- and 7-*n*-alkoxy-2-naphthoic acids. [258] G. W. Gray, B. Jones, J. Chem. Soc. *1954*, 683, C.A. *49*, 1676i (1955); Mesomorphic *n*-alkoxynaphthoic acids. [259]

G. W. Gray, B. Jones, J. Chem. Soc. *1954*, 1467, C.A. *49*, 4577d (1955); Mesomorphic *trans-p-n*-alkoxycinnamic acids. [**260**] G. W. Gray, B. Jones, J. Chem. Soc. *1954*, 2556, C.A. *49*, 8868c (1955); Effect of halogen substitution on the mesomorphism of the 4-alkoxybenzoic acids.

[**261**] G. W. Gray, B. Jones, J. Chem. Soc. *1955*, 236, C.A. *50*, 917c (1956); Effect of substitution on the mesomorphism of the 6-*n*-alkoxy-2-naphthoic acids. [**262**] G. W. Gray, J. B. Hartley, B. Jones, J. Chem. Soc. *1955*, 1412, C.A. *50*, 3336e (1956); Mesomorphism of 4'-*n*-alkoxydiphenyl-4-carboxylic acids and their simple alkyl esters. [**263**] G. W. Gray, J. B. Hartley, A. Ibbotson, J. Chem. Soc. *1955*, 2686, C.A. *50*, 7092h (1956); Friedel-Crafts reaction on 2-methoxyfluorene and the preparation of 7-methoxyfluorenone-2-carboxylic acid. [**264**] G. W. Gray, J. B. Hartley, A. Ibbotson, B. Jones, J. Chem. Soc. *1955*, 4359, C.A. *50*, 10061i (1956); Mono- and dianils of the benzene, biphenyl, fluorene, and fluorenone series. [**265**] G. W. Gray, B. Jones, F. Marson, J. Chem. Soc. *1956*, 1417, C.A. *51*, 1087i (1957); Effect of halogen substitution on the mesomorphism of *trans-p-n*-alkoxycinnamic acids. [**266**] G. W. Gray, J. Chem. Soc. *1956*, 3733, C.A. *51*, 3259c (1957); Mesomorphism of the fatty esters of cholesterol. [**267**] G. W. Gray, B. Jones, F. Marson, J. Chem. Soc. *1957*, 393, C.A. *51*, 8701a (1957); Effect of 3'-substituents on the mesomorphism of 4'-*n*-alkoxydiphenyl-4-carboxylic acids and their alkyl esters. [**268**] G. W. Gray, A. Ibbotson, J. Chem. Soc. *1957*, 3228, C.A. *51*, 16379g (1957); Mesomorphism of 7-n-alkoxy-fluorene- and -fluorenone-2-carboxylic acids and their *n*-propyl esters. [**269**] G. W. Gray, J. Chem. Soc. *1958*, 552, C.A. *52*, 10990i (1958); 2-*p-n*-Alkoxybenzylideneaminophenanthrenes and evidence for the planarity of the biphenyl ring system in the mesomorphic state. [**270**] G. W. Gray, Steric Effects Conjugated Systems, Proc. Symposium, Hull *1958*, 160, C.A. *53*, 14049a (1959); Steric effects on the mesomorphism of mono- and dibenzylideneaminobiphenyls.

[**271**] G. W. Gray, B. M. Worrall, J. Chem. Soc. *1959*, 1545, C.A. *53*, 17962h (1959); Mesomorphic transition temperatures of 3'-substituted 4'-*n*-octyloxydiphenyl-4-carboxylic acids. [**272**] P. Culling, G. W. Gray, D. Lewis, J. Chem. Soc. *1960*, 2699, C.A. *54*, 23557i (1960); Mesomorphism and polymorphism in simple derivatives of *p*-terphenyl. [**273**] D. J. Byron, G. W. Gray, A. Ibbotson, B. M. Worrall, J. Chem. Soc. *1963*, 2246, C.A. *58*, 12398 (1963); Mesomorphism of substituted 4-*p-n*-alkoxybenzylidene-aminobiphenyls. [**274**] S. J. Branch, D. J. Byron, G. W. Gray, A. Ibbotson, B. M. Worrall, J. Chem. Soc. *1964*, 3279, C.A. *61*, 11876d (1964); Mesomorphism of substituted 4,4'-bis-(*p-n*-alkoxybenzylidene-amino)-biphenyls. [**275**] D. J. Byron, G. W. Gray, B. M. Worrall, J. Chem. Soc. *1965*, 3706, C.A. *63*, 4140c (1965); Mesomorphic di-, tri-, and tetrasubstituted 4,4'-bis-(*p-n*-alkoxy-benzylideneamino)-biphenyls. [**276**] G. W. Gray, Mol. Cryst. *1*, 333 (1966), C.A. *65*, 14535g (1966); Influence of molecular structure on liquid crystalline properties. [**277**] G. W. Gray, Mol. Cryst. Liq. Cryst. *7*, 127 (1969), C.A. *71*, 85318y (1969); Advances in synthesis and the role of molecular geometry in liquid crystallinity. [**278**] G. W. Gray, K. J. Harrison, Mol. Cryst. Liq. Cryst. *13*, 37 (1971), C.A. *75*, 54884h (1971); Mesomorphic alkyl, aryl and arylalkyl 4-(*p*-substituted benzylideneamino)cinnamates and α-methylcinnamates. [**279**] G. W. Gray, K. J. Harrison, Symp. Faraday Soc. *1971*, No. 5, 54, C.A. *79*, 24373f (1973); Effects of molecular structural change on liquid crystalline properties. [**280**] G. W. Gray, 2. Arbeitstagung Flüssigkristalle, Inst. Angew. Festkörperphysik Freiburg, 21. 4. 1972; Liquid crystals for display devices.

[**281**] G. W. Gray, K. J. Harrison, J. A. Nash, Electron, Lett. *9*, 130 (1973), C.A. *78*, 152926e (1973); New family of nematics for displays. [**282**] G. W. Gray, K. J. Harrison, J. A. Nash, E. P. Raynes, Electron. Lett. *9*, 616 (1973), C.A. *80*, 102264h (1974); New cholesterics for displays. [**283**] G. W. Gray, K. J. Harrison, J. A. Nash, Liq. Cryst., Proc. Int. Conf., Bangalore, *1973*, 381, C.A. *84*, 114470t (1976); Recent developments concerning biphenyl mesogens and structurally related compounds. [**284**] G. W. Gray, Mol. Cryst. Liq. Cryst. *21*, 161 (1973), C.A. *79*, 98011h (1973); Synthetic chemistry related to liquid crystals. [**285**] D. Coates, K. J. Harrison, G. W. Gray, Mol. Cryst. Liq. Cryst. *22*, 99 (1973), C.A. *79*, 109053b (1973); Mesophase transformations for certain Schiff's base esters. [**286**] G. W. Gray, K. J. Harrison, J. A. Nash, J. Chem. Soc., Chem. Commun. *1974*, 431, C.A. *81*, 160479p (1974); Wide range nematic mixtures incorporating 4''-*n*-alkyl-4-cyano-*p*-terphenyls. [**287**] J. A. Nash, G. W. Gray, Mol. Cryst. Liq. Cryst. *25*, 299 (1974), C.A. *81*, 55350s (1974); Heterocyclic mesogens. [**288**] G. W. Gray, K. J. Harrison, J. A. Nash, J. Constant, D. S. Hulme, J. Kirton, E. P. Raynes, Liq. Cryst. Ord. Fluids *2*, 617 (1974), C.A. *87*, 60722t (1977); Stable, low melting nematogens of positive dielectric anisotropy for display devices. [**289**] G. W. Gray, Liq. Cryst. Plast. Cryst. *1*, 103 (1974), C.A. *83*, 19082p (1975); Influence of molecular structure on the liquid crystals formed by single component nonamphiphilic systems. [**290**] G. W. Gray, J. Phys. (Paris) C-1, *36*, 337 (1975), C.A. *83*, 113422w (1975); Some new mesogens.

[291] D. Coates, G. W. Gray, J. Phys. (Paris) C-1, *36*, 365 (1975), C.A. *83*, 106484q (1975); Novel smectic polymorphism in homologous series of mesogens. [292] D. Coates, G. W. Gray, Mol. Cryst. Liq. Cryst. *31*, 275 (1975), C.A. *84*, 120645v (1976); Mesomorphism of 4',4''-disubstituted phenyl biphenyl-4-carboxylates. [293] D. Coates, G. W. Gray, J. Chem. Soc., Chem. Commun. *1975*, 514, C.A. *83*, 113890x (1975); Effect of light on the liquid crystal transition temperatures of 4-(4-pentylphenyl)-vinyl cyanide. [294] G. W. Gray, D. G. McDonnell, Electron. Lett. *11*, 556 (1975), C.A. *84*, 172102k (1976); New low-melting cholesterogens for electrooptical displays and surface thermography. [295] J. W. Goodby, G. W. Gray, J. Phys. (Paris) C-3, *37*, 17 (1976), C.A. *86*, 81950k (1977); Molecular structure and smectic polymorphism. [296] G. W. Gray, A. Mosley, J. Chem. Soc., Perkin Trans. 2, *1976*, 97, C.A. *84*, 104890r (1976); Trends in the n/i transition temperatures for the homologous series of 4-n-alkoxy- and 4-n-alkyl-4'-cyanobiphenyls. [297] D. Coates, G. W. Gray, J. Chem. Soc., Perkin Trans. 2, *1976*, 300, C.A. *84*, 114541s (1976); Effect of the terminal trifluoromethyl group on nematic liquid crystal thermal stability. [298] D. Coates, G. W. Gray, J. Chem. Soc., Perkin Trans. 2, *1976*, 863, C.A. *85*, 70942n (1976); Mesomorphism of 4'-substituted 4-(β-p-substituted arylethyl)biphenyls. [299] D. Coates, G. W. Gray, Mol. Cryst. Liq. Cryst. *34*, 1 (1976), C.A. *85*, 85812u (1976); A chiral (tilted) s_B phase. [300] G. W. Gray, J. W. Goodby, Mol. Cryst. Liq. Cryst. *37*, 157 (1976), C.A. *86*, 188575u (1977); Effects of small changes in molecular framework on the incidence of smectics-C and other smectics in esters.

[301] G. W. Gray, D. G. McDonnell, Mol. Cryst. Liq. Cryst. *37*, 189 (1976), C.A. *86*, 155116s (1977); Mesomorphism of chiral alkylcyanobiphenyls (and -p-terphenyls) and of some related chiral compounds derived from biphenyl. [302] G. W. Gray, A. Mosley, Mol. Cryst. Liq. Cryst. *37*, 213 (1976), C.A. *86*, 170720c (1977); Mesomorphic transition temperatures for the homologous series of 4-n-alkyl-4'-cyanotolanes and other related compounds. [303] D. Coates, G. W. Gray, Mol. Cryst. Liq. Cryst. *37*, 249 (1976), C.A. *86*, 155052t (1977); Mesomorphism of cyanosubstituted aryl esters. [304] G. W. Gray, Adv. Liq. Cryst. *2*, 1 (1976), C.A. *87*, 4805k (1977); Molecular geometry and the properties of nonamphiphilic liquid crystals. [305] D. Coates, G. W. Gray, Ger. Offen. 2,617,593 (22. 4. 1975), C.A. *86*, 72247j (1977); Biphenyl derivatives. [305a] J. W. Goodby, G. W. Gray, D. G. McDonnell, Mol. Cryst. Liq. Cryst. *34*, 183 (1977), C.A. *87*, 46804q (1977); Dipole moments and the s_C phase. [305b] D. Coates, G. W. Gray, Mol. Cryst. Liq. Cryst. *39*, 361 (1977), Liquid-crystalline 4',4''-disubstituted phenyl biphenyl-4-carboxylates. [305c] G. W. Gray, A. Mosley, Mol. Cryst. Liq. Cryst. *41*, 75 (1977), Transition temperatures of some deuterated liquid crystals. [305d] G. W. Gray, D. G. McDonnell, Ger. Offen. 2,639,838 (3. 9. 1975), C.A. *87*, 39142u (1977); Optically active cyanodiphenyl compounds and liquid crystal materials and devices. [306] A. C. Griffin, III, Diss. Univ. Texas, Austin. 1975. 113 pp. Avail. Xerox Univ. Microfilms, Ann Arbor, Mich., Order No. 75-16,678, C.A. *83*, 178032p (1975); Structure-stability relations in the mesophase. Thermodynamic study. [307] A. C. Griffin, Mol. Cryst. Liq. Cryst. *34*, 111 (1976), C.A. *86*, 36555z (1977); Dependence of n/i transition temperatures on the anisotropy of polarizability of the C-X bond for terminal substituents. [308] A. C. Griffin, S. F. Thames, M. S. Bonner, Mol. Cryst. Liq. Cryst. *34*, 135 (1977), C.A. *86*, 149025v (1977); Two series of novel methylene-bridged liquid crystals. [309] A. C. Griffin, Mol. Cryst. Liq. Cryst. *34*, 141 (1977), C.A. *86*, 149026w (1977); Factors governing smectic polymorphism. [310] D. Gross, B. Böttcher, Z. Naturforsch. B *25*, 1099 (1970), C.A. *74*, 23063q (1971); Eigenschaften von β-Alkoxypropionsäureestern des Cholesterins.

[311] D. Gross, Z. Naturforsch. B *27*, 472 (1972), C.A. *77*, 88782e (1972); Dicholesterinester von Dicarbonsäuren. [312] T. Gunjima, Y. Masuda, Japan. Kokai 74 55,579 (3. 10. 1972), C.A. *81*, 144292s (1974); Electrooptical display device containing nematics. [313] T. Gunjima, Y. Masuda, Japan. Kokai 74 86,278 (23. 12. 1972), C.A. *82*, 49919p (1975); Liquid crystal electrooptical device. [314] E. Günther, Diss. Halle 1922; Kondensationsprodukte von aromatischen Aldehyden mit p,p'-Diaminobenzylanilin und mit p,p'-Diaminobenzanilid. [315] J. L. Haberfeld, E. C. Hsu, J. F. Johnson, Mol. Cryst. Liq. Cryst. *24*, 1 (1973), C.A. *81*, 142310x (1974); Liquid crystal purification by zone refining. [316] U. Haberland, Diss. Halle 1924; Mikrobestimmung von Schmelz- und Übergangspunkten polymorpher Substanzen. [317] T. Haga, H. Fukutani, H. Nagasaka, T. Ohya, Japan. Kokai 74 79,983 (11. 11. 1972), C.A. *82*, 49914h (1975); Liquid crystal display devices. [318] H. Hakemi, Diss. Temple Univ., Philadelphia. 1976. 262 pp. Avail. Xerox Univ. Microfilms, Ann Arbor, Mich., Order No. 76-12,000, C.A. *85*, 70916g (1976); Branched chain esters of cholesterol and an optical method for determining diffusion coefficients of liquid crystals. [319] I. Haller, R. J. Cox, Liq. Cryst. Ord. Fluids *1970*, 393, C.A. *78*, 152451w (1973); Effect of end-chain polarity on the mesophase stability

of substituted Schiff-bases. [**320**] I. Haller, H. A. Huggins, H. R. Lilienthal, T. R. McGuire, J. Phys. Chem. *77*, 950 (1973), C.A. *78*, 141285u (1973); Order-related properties of some nematics. [**320a**] Z. Halmos, K. Seybold, T. Meisel, Therm. Anal., Proc. Int. Conf., 4th, *2*, 429 (1974), C.A. *87*, 93718z (1977); Investigation of fatty acid salts of thallium(I) by electrical thermal analysis.

[**321**] P. Hansen, Diss. Halle 1907; Flüssig-kristallinische Substanzen. [**322**] A. Hargreaves, S. H. Rizvi, Acta Cryst. *15*, 365 (1962), C.A. *57*, 1656f (1962); Crystal and molecular structure of biphenyl. [**323**] F. L. Hedberg, H. Rosenberg, J. Therm. Anal. *6*, 571 (1974), C.A. *82*, 42827y (1975); Thermal and oxidative stability of polychlorinated metallocenes. [**324**] W. Helfrich, C. S. Oh, Mol. Cryst. Liq. Cryst. *14*, 289 (1971), C.A. *75*, 133989z (1971); Optically active smectic liquid crystal. [**325**] H. Hempel, Diss. Halle 1926; Reduktionsprodukte der Pyridiniumsalze. [**326**] W. Hennicke, Diss. Halle 1924; p,p'-Biphenyldicarbonsäure. [**327**] F. Hentze, Diss. Halle 1925; Darstellung fester und flüssiger pleochroitischer optisch aktiver Kristalle. [**327a**] G. Heppke, J. Martens, K. Präfcke, H. Simon, Angew. Chem. *89*, 328 (1977), C.A. *87*, 134264u (1977); Flüssigkristalline Selenolester. [**328**] C. L. Hillemann, G. R. Van Hecke, J. Phys. Chem. *80*, 944 (1976), C.A. *85*, 10895q (1976); Homologous *trans*-4-ethoxy-4'-cycloalkanecarbonyloxyazobenzenes. Calorimetry. [**329**] S. M. Hiscock, D. A. Swann, J. H. Turnbull, J. Chem. Soc. D *1970*, 1310, C.A. *74*, 13330k (1971); Axial and equatorial selenols: 3α- and 3β-selenyl derivatives of cholestane and androstanone. [**330**] A. Hochapfel, D. Berchet, R. Perron, J. Petit, Mol. Cryst. Liq. Cryst. *13*, 165 (1971), C.A. *75*, 41755s (1971); Polymorphisme d'un composé nématique dérivé de l'acide cinnamique.

[**331**] A. Hochapfel, D. Lecoin, R. Viovy, Mol. Cryst. Liq. Cryst. *37*, 109 (1976), C.A. *86*, 171042v (1977); Syntheses de nouveaux composés mésomorphes de la série des *p*-alkoxyphenylazoxy-phenylesters. [**332**] H. Hoffmann, Diss. Halle 1923; Derivate des Biphenyls. [**333**] S. Hoffmann, W. Brandt, Z. Chem. *15*, 306 (1975), C.A. *83*, 211508y (1975); Ferne Membrankomponenten-Analoge. II. Mesogene Östronderivate. [**334**] H. Honda, H. Kimura, Y. Sanada, S. Sugawara, T. Furuta, Carbon *8*, 181 (1970), C.A. *73*, 27384x (1970); Optical mesophase texture and X-ray diffraction pattern of the early-stage carbonization of pitches. [**335**] H. Hope, D. Victor, Acta Cryst. B *25*, 1849 (1969), C.A. *71*, 95952q (1969), Structure of *trans-p,p'*-dichloroazobenzene. [**336**] P. Horbach, Diss. Halle 1922; Derivate der *p*-Oxybenzoesäure. [**337**] F. Hori, B. Kato, N. Arima, Japan. Kokai 73 75,484 (13. 1. 1972), C.A. *80*, 102358s (1974); Nematic liquid crystal for electrooptical display devices. [**338**] F. Hori, Japan. Kokai 74,127,884 (13. 4. 1973), C.A. *82*, 163085p (1975); Nematics with positive dielectric anisotropy. [**339**] F. Hori, B. Kato, N. Arima, Japan. Kokai 75 52,039 (13. 9. 1973), C.A. *83*, 200594x (1975); p-Alkoxybenzylidene-p'-alkylanilines for use as nematics. [**340**] R. Hosemann, G. Willmann, B. Roessler, Phys. Rev. A *6*, 2243 (1972), C.A. *78*, 21316h (1973); Paracrystalline structure of molten metals.

[**341**] K. Lemm, R. Hosemann, Stekloobrazn. Sostoyanie *147*, 177 (1971), C.A. *78*, 8903c (1973); Calculation of liquid structure by a paracrystal-distortion method. [**342**] Y. Y. Hsu, Diss. Kent State Univ., Kent, Ohio. Avail. Univ. Microfilms, Ann Arbor, Mich., Order No. 73-6625, C.A. *78*, 159116z (1973); Syntheses and characterization of selectively deuterated mesogens. Syntheses of novel Schiff bases and the investigation of binary mesomorphic systems. [**343**] Y. Y. Hsu, D. Dolphin, Liq. Cryst. Ord. Fluids *2*, 461 (1974), C.A. *86*,180994j (1977); Mesomorphism of optically active anils: 4-*n*-alkoxyben-zylidene-4'-methylalkylanilines. [**344**] E. C. Hsu, L. K. Lim, R. B. Blumstein, A. Blumstein, Mol. Cryst. Liq. Cryst. *33*, 35 (1976), C.A. *85*, 123463r (1976); Mesomorphism of vinyl compounds and their saturated analogs. [**344a**] Y. Y. Hsu, Mol. Cryst. Liq. Cryst. *42*, 263 (1977); Synthesis of optically active (+)-4-*n*-alkyl-phenyl-4'-(4''-methylalkylbenzoyloxy)benzoates and effect on LC display properties. [**345**] J. E. Hulme, Diss. Halle 1907; Krystallinisch-flüssige Substanzen. [**346**] M. E. Huth, Diss. Halle 1909; 1. Der Einfluß ungleicher *p*-Substituenten auf den kristallin-flüssigen Zustand aromatischer Verbindungen, 2. Pleo-chroismus und Zirkularpolarisation flüssiger Kristalle. [**347**] K. J. Hüttinger, Chem.-Ing.-Tech. *43*, 1145 (1971), C.A. *76*, 35679z (1972); Bildung graphitischer Kohlenstoffe durch Flüssigphasenpyrolyse. [**348**] K. J. Hüttinger, Ber. Deut. Keram. Ges. *48*, 216 (1971), C.A. *75*, 8306y (1971); Intermediäre, kristallinflüssige Phasen beim thermischen Abbau von Aromatengemischen zu graphitischem Kohlenstoff. [**349**] K. J. Hüttinger, Erdöl, Kohle, Erdgas, Petrochem. Brennstoff-Chem. *26*, 21 (1973), C.A. *79*, 21424n (1973); Rolle der unlöslichen Bestandteile bei den Mesophasen-Transformationen von Pechen. [**350**] S. Ignasiak, M. J. Rafuse, Mol. Cryst. Liq. Cryst. *30*, 125 (1975), C.A. *83*, 146621p (1975); Cyanoaryl alkylaryl compounds. Preparation and liquid crystal properties.

[**351**] M. Inagaki, M. Ishihara, S. Naka, Carbon *14*, 88 (1976), C.A. *85*, 169740g (1976); Crack formation in the separated mesophase spherules. [**351a**] M. Inagaki, M. Ishihara, S. Naka, High Temp. — High Pressures *8*, 279 (1976), C.A. *87*, 25745f (1977); Mesophase formation during carbonization

under pressure. [**352**] T. Inukai, H. Sato, S. Sugimori, T. Ishibe, Ger. Offen. 2,545,121 (11. 10. 1974), C.A. *85*, 32663q (1976); p-Cyanophenyl-4-alkyl-4'-biphenyl carboxylates. [**353**] T. Inukai, H. Sato, T. Ishibe, Japan. Kokai 76 78,793 (28. 12. 1974), C.A. *86*, 36365n (1977); Liquid crystal composite. [**354**] T. Inukai, H. Sato, H. Inoue, Japan. Kokai 76 87,181 (29. 1. 1975), C.A. *86*, 81729v (1977); Liquid crystal composite containing optically active compound. [**354a**] T. Inukai, T. Ishibe, H. Sato, S. Sugimori, Japan. Kokai 77 68,155 (12. 4. 1975), C.A. *87*, 152280a (1977); 4'-Alkyl-4-biphenylcarbonitriles. [**355**] T. Ishikawa, J. Tanaka, Japan. Kokai 75 87,472 (10. 12. 1973), C.A. *84*, 124023p (1976); Needle-shaped coke from waste plastics. [**356**] J. Itakura, S. Kojima, A. Onishi, Japan. Kokai 74,135,936 (12. 5. 1973), C.A. *82*, 155775y (1975); N-(p-Alkoxybenzylidene)-p-butylanilines. [**357**] J. Jacques, Ger. Offen. 2,226,376 (2. 6. 1971), C.A. *78*, 71651j (1973); Tolans as liquid crystals. [**358**] J. Jacques, Fr. Addn. 2,161,285 (2. 6. 1971), C.A. *80*, 3243u (1974); Tolans. [**359**] J. Jacques, J. Billard, J. Malthete, J. Gabard, Fr. Demande 2,234,261 (22. 6. 1973), C.A. *83*, 96714y (1975); Substituted tolanes. [**360**] F. M. Jaeger, Rec. trav. chim. *25*, 334 (1906), C.A. *1*, 327[9] (1907); Les éthers-sels des acides gras avec la cholestérine et la phytostérine, et les phases liquides anisotropes des dérivés de la cholestérine.

[**361**] F. M. Jaeger, Proc. Kon. Nederl. Akad. Wetenschapen *9*, 78 (1906); Fatty esters of cholesterol and phytosterol, and the anisotropous liquid phases of the cholesterol-derivatives. [**362**] F. M. Jaeger, Rec. trav. chim. *26*, 311 (1907), C.A. *1*, 3060 (1907); Les éthers-sels des acides gras avec les deux phytostérines de la graisse calabar, et les dérivés analogues de la cholestérine, qui possèdent trois phases liquid stabiles. [**363**] F. Janecke, Diss. Halle 1910; Kondensationsprodukte von Aldehyden mit aromatischen Aminen. [**364**] T. Jinnai, K. Totani, Ger. Offen. 2,519,659 (19. 12. 1974), C.A. *85*, 159721n (1976); Nematic liquid crystal compounds. [**365**] T. Jinnai, K. Iwasaki, G. Matsumoto, Japan. Kokai 76,146,435 (12. 6. 1975), C.A. *87*, 52953d (1977); 4'-n-Alkoxyphenyl 4-cyanocinnamates. [**366**] A. E. Bradfield, B. Jones, J. Chem. Soc. *1929*, 2660, C.A. *24*, 1776[8] (1930); Two apparent cases of liquid-crystal formation. [**367**] B. Jones, J. Chem. Soc. *1935*, 1874, C.A. *30*, 1775[7] (1936); Apparent cases of liquid-crystal formation in p-alkoxybenzoic acids. [**368**] G. M. Bennett, B. Jones, J. Chem. Soc. *1939*, 420, C.A. *33*, 4220[3] (1939); Mesomorphism of some alkoxybenzoic and p-alkoxycinnamic acids. [**369**] F. B. Jones, Jr., J. J. Ratto, J. Chem. Soc., Chem. Commun. *1973*, 841, C.A. *80*, 70491m (1974); Mesomorphism of alkyl and alkoxycinnamic acid esters. [**370**] F. B. Jones, Jr., J. J. Ratto, Liq. Cryst. Ord. Fluids *2*, 723 (1974), C.A. *86*, 181001v (1977); Mesomorphism of *trans* cinnamic acid esters.

[**371**] F. B. Jones, Jr., J. J. Ratto, J. Phys. (Paris) C-1, *36*, 413 (1975), C.A. *83*, 27820t (1975); p-Cyano substituted cinnamic acid esters. [**372**] F. B. Jones, Jr., J. J. Ratto, U.S. 3,926,834 (1. 10. 1973), C.A. *84*, 98160j (1976); Low melting point disubstituted-p,p'-phenyl cinnamate liquid crystals. [**373**] M. Kakuta, N. Tuchiya, M. Kooriki, Tanso *85*, 55 (1976), C.A. *85*, 110670f (1976); Relations between characteristics of petroleum residues and physical properties of petroleum cokes derived from them. [**374**] S. Kambe, Japan. Kokai 75,111,042 (13. 2. 1974), C.A. *84*, 73944e (1976); p-Alkylphenyl p-cyanobenzoates. [**375**] S. Kambe, Japan. Kokai 75,111,043 (15. 2. 1974), C.A. *84*, 73943d (1976); p-Cyanophenyl p-alkylbenzoates. [**376**] S. Kambe, Japan. Kokai 75,137,963 (23. 4. 1974), C.A. *84*, 58986b (1976); Biphenyls. [**377**] S. Kanbe, Y. Yamazaki, T. Suzuki, Japan. Kokai 74 78,683 (6. 12. 1972), C.A. *82*, 24385y (1975); Liquid crystal display devices employing mesomorphic p-cyano-phenyl benzoates. [**377a**] S. Kanbe, Japan. Kokai 76,139,581 (29. 5. 1975), C.A. *86*, 180746e (1977); Liquid crystal compositions. [**377b**] S. Kanbe, Japan. Kokai 76,142,485 (4. 6. 1975), C.A. *87*, 93566y (1977); Nematic liquid crystal compositions for display devices. [**377c**] S. Kanbe, Japan. Kokai 77 05,731 (3. 7. 1975), C.A. *87*, 76416g (1977); Liquid crystal compounds and their mixtures. [**378**] L. A. Karamysheva, E. I. Kovshev, V. V. Titov, Zh. Org. Khim. *12*, 1508 (1976), C.A. *85*, 123517m (1976); Mesomorphism of esters of 4,4'-dihydroxybiphenyl and its analogs. [**379**] L. A. Karamysheva, E. I. Kovshev, V. V. Titov, Zh. Org. Khim. *12*, 2628 (1976), C.A. *86*, 63874j (1977); Mesomorphism of 2,6-disubstituted naphthalenes. [**380**] L. A. Karamysheva, E. I. Kovshev, M. I. Barnik, Mol. Cryst. Liq. Cryst. *37*, 29 (1976), C.A. *86*, 131381y (1977); Mesomorphism and dielectric properties of phenyl 4-alkylbiphenyl-4'-carboxylates and phenyl 4-(4-alkylphenyl)cyclohexanecarboxylates. [**381**] K. Karg, Diss. Halle 1934; Mikroskopische Ermittlung der Übergangspunkte kristallin-flüssiger polymorpher Substanzen. [**382**] J. I. Kaplan, J. Chem. Phys. *57*, 3015 (1972), C.A. *77*, 131724b (1972); Dependence of n/i transition temperature on alkyl chain length in homologous series. [**383**] W. Kast, Naturwissenschaften *25*, 234 (1937), C.A. *31*, 5236[2] (1937); Bedingungen für das Auftreten einer anisotrop-flüssigen Phase. [**384**] W. Kast, Angew. Chem. *67*, 592 (1955), C.A. *50*, 1398h (1956); Molekelstruktur der Verbindungen mit kristallin-flüssigen Schmelzen. [**385**] W. Kasten, Diss. Halle

1909; Kristallinisch-flüssige Derivate der α-alkylierten Zimtsäuren. [386] Y. Katagiri, Y. Miyata, O. Nagasaki, Ger. Offen. 2,121,487 (2. 5. 1970), C.A. *76*, 77918g (1972); Aromatic nematics. [387] Y. Katagiri, Y. Miyata, Ger. Offen. 2,261,548 (16. 12. 1971), C.A. *79*, 58816t (1973); Nematic liquid crystals. [388] Y. Katagiri, Japan. Kokai 74 55,577 (3. 10. 1972), C.A. *81*, 180005j (1974); Electrooptical display device. [389] Y. Katagiri, Japan. Kokai 74 55,578 (3. 10. 1972), C.A. *81*, 144287u (1974); Electrooptical display device containing nematics. [390] G. Kelbg, Sitzungsber. Plenums Kl. Akad. Wiss. DDR, *1972*, 5, C.A. *81*, 17238d (1974); Recent results of fluid physics and their meaning for natural science and technology.

[391] H. Kelker, B. Scheurle, J. Phys. (Paris), C-4, *30*, 104 (1969), C.A. *72*, 104453c (1970); Synthèse de nouvelles substances donnant des phases mésomorphes. [392] H. Kelker, B. Scheurle, Angew. Chem. *81*, 903 (1969), C.A. *72*, 36734t (1970); Nematische Phase mit besonders niedrigem Erstarrungspunkt. [393] H. Kelker, B. Scheurle, Mol. Cryst. Liq. Cryst. *7*, 381 (1969), C.A. *71*, 85812e (1969); Liquid-crystalline behavior and constitutional analysis of 1,5-dimethoxynaphthalenedicarboxylic acid derivatives. [394] B. Scheurle, H. Kelker, Ger. Offen. 1,928,242 (3. 6. 1969), C.A. *74*, 42128f (1971); Azomethine mit enantiotroper nematischer Phase. [395] H. Kelker, B. Scheurle, R. Hatz, W. Bartsch, Angew. Chem. *82*, 984 (1970), C.A. *74*, 52865h (1971); Flüssig-kristalline Phasen mit besonders niedrigem Erstarrungspunkt. [396] P. Keller, L. Liebert, L. Strzelecki, J. Phys. (Paris), C-3, *37*, 27 (1976), C.A. *85*, 191996s (1976); Synthèse de cristaux liquides ferroélectriques. [397] P. Keller, S. Juge, L. Liebert, L. Strzelecki, C. R. Acad. Sci., Sér. C, *282*, 639 (1976), C. A. *85*, 158811e (1976); Mésomorphisme de *p*-alkoxybenzylidène-*p'*-aminocinnamates de R(−)chloro-2 propyle. [398] J. O. Kessler, J. E. Lydon, Liq. Cryst. Ord. Fluids *2*, 331 (1974), C.A. *86*, 180988k (1977); Structure and thermal conductivity of supercooled MBBA. [399] P. H. Keyes, W. B. Daniels, Bull. Am. Phys. Soc. *20*, 886 (1975), Dominant role of repulsive forces in nematic ordering and the influence of flexible end chains. [400] Y. B. Kim, M. Seno, Mol. Cryst. Liq. Cryst. *36*, 293 (1976), C.A. *86*, 139199r (1977); Thermodynamic and mesomorphic properties of phenylthiobenzoate derivatives.

[401] A. M. King, W. E. Garner, J. Chem. Soc. *1936*, 1368; C.A. *30*, 7939⁹ (1936); Melting points of long-chain carbon compounds. [402] H. Kimura, Y. Sanada, S. Sugawara, T. Furuta, H. Honda, H. Sugimura, M. Kumagai, Nenryo Kyokai-shi *49*, 752 (1970), C.A. *74*, 55995m (1971); Formation of anisotropic texture from various carbonaceous materials in the early stage of carbonization. [403] B. H. Klanderman, T. R. Criswell, J. Am. Chem. Soc. *97*, 1585 (1975), C.A. *82*, 163386n (1975); Novel stable cholesterics. [404] T. J. Klingen, J. H. Kindsvater, Mol. Cryst. Liq. Cryst. *26*, 365 (1974), C.A. *82*, 50055k (1975); Plastic crystallinity in a series of organosubstituted carboranes. [405] L. E. Knaak, H. M. Rosenberg, M. P. Servé, Mol. Cryst. Liq. Cryst. *17*, 171 (1972), C.A. *77*, 74669m (1972); Estimation of n/i transition temperatures. [406] F. F. Knapp, H. J. Nicholas, J. Org. Chem. *33*, 3995 (1968), C.A. *69*, 106943z (1968); 24ξ-Methyl-9,19-cyclolanostan-3β-yl palmitate. A new liquid crystal. [407] F. F. Knapp, H. J. Nicholas, J. P. Schroeder, J. Org. Chem. *34*, 3328 (1969), C.A. *72*, 12907a (1970); Cholesteric 9,19-cyclopropane triterpene fatty acid esters. [408] F. F. Knapp, H. J. Nicholas, Liq. Cryst. Ord. Fluids *1970*, 147, C.A. *79*, 10809r (1973); Structural studies of the cholesteric mesophase. [409] F. F. Knapp, H. J. Nicholas, Mol. Cryst. Liq. Cryst. *6*, 319 (1970), C.A. *73*, 10721g (1970); Cycloartenyl palmitate: a naturally occurring ester that forms a cholesteric mesophase. [410] F. F. Knapp, H. J. Nicholas, Mol. Cryst. Liq. Cryst. *10*, 173 (1970), C.A. *73*, 59680g (1970); Mesomorphism of ergosteryl-fatty acid esters.

[411] H. J. Nicholas, F. F. Knapp, Jr., U.S. 3,686,235 (14. 10. 1969), C.A. *78*, 4377m (1973); Triterpene liquid crystals. [412] H. J. Nicholas, F. F. Knapp, Jr., U.S. 3,852,311 (11. 5. 1970), C.A. *82*, 105339p (1975); Mesomorphic ergosteryl esters. [413] G. A. Knight, B. D. Shaw, J. Chem. Soc. *1938*, 682, C.A. *32*, 5835⁷ (1938); Long-chain alkylpyridines and their derivatives. New liquid crystals. [414] W. Knudsen, Diss. Halle 1924; Darstellung langer Ketten mit Hilfe des *p*-Aminobenzaldehyds. [415] H. Kobayashi, M. Yoshino, Y. Takahara, Japan. Kokai 73 67,248 (20. 12. 1971), C.A. *80*, 3273d (1974); *p*-(*p*-Alkoxybenzylideneamino)phenyl alkyl carbonates. [416] S. Kojima, A. Onishi, H. Tanaka, J. Itakura, Japan. Kokai 76 65,728 (3. 12. 1974), C.A. *85*, 94058e (1976); N-(*p*-Alkoxybenzylidene)-*p-n*-alkyl-anilines. [416a] S. Kojima, A. Onishi, H. Tanaka, J. Itakura, Japan. Kokai 77 46,039 (7. 10. 1975), C.A. *87*, 84699s (1977); Cyanoaniline derivatives. [417] H. Kölbel, D. Klamann, P. Kurzendörfer, Liebigs Ann. *632*, 16 (1960), C.A. *55*, 425e (1961); Natrium-N-methyl-N-acyl-sulfanilate als kristallin-flüssig aufschmelzende, grenzflächenaktive Verbindungen. [417a] T. Kompolthy, G. Bencz, Hung. Teljes 13,132 (4. 10. 1974), C.A. *87*, 152223j (1977); Ditriazolobenzenes. [418] I. I. Konstantinov, Y. B. Amerik, B. A. Krentsel, E. V. Polunin, Mol. Cryst. Liq. Cryst. *29*, 1 (1974), C.A. *82*, 163360z (1975); Mesomorphic

p-acylphenyl esters of *p*-*n*-alkoxybenzoic acids. [419] V. Kosanke, Diss. Halle 1924; *p*-Aminophenyl-Ester der *p*-Aminobenzoesäure, Bernsteinsäure, Adipinsäure und Glutarsäure. [420] M. Kozutsumi, Y. Miyazawa, J. Matsumura, M. Gonda, Japan. Kokai 74 33,945 (7. 10. 1970), C.A. *82*, 139625u (1975); Schiff base.

[421] M. Kotsutsumi, Y. Miyazawa, J. Matsumura, M. Gonda, Japan. Kokai 74 45,855 (10. 4. 1970), C.A. *83*, 9458b (1975); Schiff's bases. [421a] J. Krause, L. Pohl, Ger. Offen. 2,603,293 (29. 1. 1976), C.A. *87*, 151863f (1977); Liquid crystals and dielectrics. [422] G. Kreiß, Diss. Halle 1925; Polymorphe Umwandlungen und der doppelbrechende weiße Phosphor. [423] W. R. Krigbaum, Y. Chatani, P. G. Barber, Acta Cryst. B *26*, 97 (1970), C.A. *72*, 94244u (1970); Crystal structure of PAA. [424] W. R. Krigbaum, T. Taga, Mol. Cryst. Liq. Cryst. *28*, 85 (1974), C.A. *82*, 178785r (1975); Crystal structure of the high temperature polymorph of ethyl p-azoxybenzoate. [425] D. Krishnamurti, K. S. Krishnamurthy, R. Shashidhar, Mol. Cryst. Liq. Cryst. *8*, 339 (1969), C.A. *71*, 85514j (1969); Thermal, optical, X-ray, IR, and NMR studies on the α-phase of some saturated aliphatic esters. [426] R. Kühnemann, Diss. Halle 1922; Einfluß der Kettenlänge alkylierter aromatischer Amine auf den kristallinisch-flüssigen Zustand. [427] C. Kuhrmann, Diss. Halle 1927; Darstellung kristalliner Harze, Lacke und Gläser. [428] A. Kuksis, J. M. R. Beveridge, J. Org. Chem. *25*, 1209 (1960), C.A. *55*, 5574i (1961); Preparation and physical properties of plant steryl esters (β- and γ-sitosterol, stigmasterol and ergosterol). [428a] K. S. Kunihisa, M. Gotoh, Mol. Cryst. Liq. Cryst. *42*, 97 (1977); Phase transitions and solid polymorphs of cholesteryl acetate. [429] M. V. Kurik, V. A. Shayuk, Fiz. Tverd. Tela *16*, 2734 (1974), C.A. *81*, 160398m (1974); Existence of a liquid-crystal phase for stilbene. [430] S. Kusabayashi, Kyoritsu Kagaku Raiburari *1*, 66 (1973), C.A. *83*, 50842k (1975); Chemistry of liquid crystals.

[431] M. M. Kusakov, M. V. Shishkina, V. L. Khodzhaeva, Dokl. Akad. Nauk SSSR *186*, 366 (1969), C.A. *71*, 54995n (1969); Mesomorphic *p*-(*n*-alkoxy)-benzoic acids. [432] K. Kuwahara, S. Akiyama, T. Akiyama, Japan. Kokai 76 70,746 (12. 12. 1974), C.A. *85*, 123643z (1976); The (+)-*p*-(*p*-2-methylbutylbenzylideneamino)benzonitrile. [433] M. M. Labes, U.S. 3,827,780 (14. 10. 1971), C.A. *82*, 132165k (1975); Stable anil-type nematics. [434] M. M. Labes, U.S. 3,853,785 (14. 10. 1971), C.A. *83*, 19038d (1975); Stable liquid crystal mixtures including anil-type nematics. [435] H. Hakemi, M. M. Labes, J. Chem. Phys. *58*, 1318 (1973), C.A. *78*, 124808z (1973); Helical twisting power of a diasteromeric pair of branched-chain cholesteryl esters. [436] H. Hirata, S. N. Waxman, I. Teucher, M. M. Labes, Mol. Cryst. Liq. Cryst. *20*, 343 (1973), C.A. *79*, 77570z (1973); Properties of a homologous series of *o*-hydroxy substituted anils and some binary mixtures. [437] V. T. Lazareva, V. V. Titov, E. I. Kovshev, Zh. Obshch. Khim. *45*, 244 (1975), C.A. *82*, 178608k (1975); Mesomorphic 4-substituted benzonitrile derivatives. [438] V. T. Lazareva, V. V. Titov, K. V. Roitman, Zh. Org. Khim. *12*, 149 (1976), C.A. *84*, 105163t (1976); Mesomorphic *p*-substituted *p'*-cyanoazobenzenes and *p'*-cyanoazoxybenzenes. [439] M. D. Lebedeva, Izv. Vyssh. Ucheb. Zaved., Khim. Khim. Tekhnol. *14*, 1014 (1971), C.A. *75*, 144792f (1971); Physicochemical analysis of cholesteryl caprylate, caprate, and pelargonate. [440] M. D. Lebedeva, Izv. Vyssh. Ucheb. Zaved., Khim. Khim. Tekhnol. *14*, 1335 (1971), C.A. *76*, 18945r (1972); Physicochemical analysis of cholesteryl laurate, myristate, palmitate, and stearate.

[441] M. Leclercq, J. Billard, J. Jacques, Mol. Cryst. Liq. Cryst. *10*, 429 (1970), C.A. *73*, 81743b (1970); Substances mésomorphes. Arylidène amino-cinnamates substitués. [442] D. Lecoin, A. Hochapfel, R. Viovy, Mol. Cryst. Liq. Cryst. *31*, 233 (1975), C.A. *84*, 89750z (1976); Composés mésomorphes de la série des *p*-alkoxyphénylazo-*p'*-phénylesters. [443] L. B. Leder, Chem. Phys. Lett. *6*, 285 (1970), C.A. *73*, 124604b (1970); Mesomorphic cholesteryl-2-(2-ethoxyethoxy)-ethylcarbonate and related compounds. [444] L. B. Leder, J. Chem. Phys. *54*, 4671 (1971), C.A. *75*, 11707k (1971); Mesomorphic steryl chlorides. [445] L. B. Leder, J. Chem. Phys. *58*, 1118 (1973), C.A. *78*, 97882k (1973); Right-rotatory cholesterics. Derivatives of $\Delta^{8(14)}$-cholestanol. [446] L. B. Leder, U.S. 3,888,892 (24. 10. 1972), C.A. *83*, 106739b (1975); Liquid crystals. [447] L. B. Leder, U.S. 3,907,406 (24. 10. 1972), C.A. *83*, 186784a (1975); Liquid crystals. [448] D. P. Lesser, A. de Vries, J. W. Reed, G. H. Brown, Acta Cryst. B *31*, 653 (1975), C.A. *82*, 178850h (1975); Crystal structure of the nematogen 2,2'-dibromo-4,4'-bis-(*p*-methoxybenzylideneamino)biphenyl. [449] A. W. Levine, Introd. Liq. Cryst. *1974*, 15, C.A. *84*, 104556m (1976); Structure-property relationships in thermotropic organic liquid crystals. [450] E. L. V. Lewis, T. R. Lomer, Acta Cryst. B *25*, 702 (1969), C.A. *70*, 119105y (1969); Refinement of the crystal structure of potassium caprate (form A).

[451] I. C. Lewis, L. S. Singer, Carbon *10*, 336 (1972); Chemistry of the mesophase transformation of acenaphthylene. [451a] I. C. Lewis, E. R. McHenry, L. S. Singer, U.S. 4,017,327 (11. 12. 1973), C.A. *87*, 8088b (1977); Mesophase pitch. [451b] I. C. Lewis, R. Didchenko, Carbon '76, Int. Carbon

Conf., Prepr., 2nd *1976*, 385, C.A. *87*, 55491n (1977); Molecular weight constitution of mesophase pitch. [**451c**] I. C. Lewis, R. Didchenko, Carbon *14*, 302 (1976), Molecular weight constitution of mesophase pitch. [**452**] C. H. Lochmüller, R. W. Souter, J. Phys. Chem. *77*, 3016 (1973), C.A. *80*, 15167k (1974); Thermotropic mesomorphism in chiral carbonyl bis(amino acid esters). [**453**] W. Lützkendorf, Diss. Halle 1922; Zur Kenntnis der Sulfanilsäure, des Dehydrothiotoluidins und der Azoxy-bzw. Azo-Verbindungen des Benzaldehyds und des Benzylalkohols. [**454**] V. Luzzati, A. Tardieu, T. Gulik-Krzywicki, Nature *217*, 1028 (1968), C.A. *69*, 39546g (1968); Polymorphism of lipids. [**454a**] E. R. McHenry, Ger. Offen. 2,462,369 (11. 12. 1973), C.A. *87*, 55652r (1977); Pitch containing a mesophase. [**455**] G. G. Maidachenko, Izv. Vyssh. Ucheb. Zaved., Khim. Tekhnol. *14*, 1115 (1971), C.A. *75*, 141048g (1971); Mesomorphic cholesteryl oleyl carbonate. [**456**] G. G. Maidachenko, N. B. Makarov, Uch. Zap., Ivanov. Gos. Pedagog. Inst. No. 99, 205 (1972), C.A. *78*, 135803v (1973); Mesomorphic N-(p-alkoxybenzylidene)-p-acyloxyanilines. [**457**] G. G. Maidachenko, L. A. Gusakova, Sb. Dokl. Vses. Nauch. Konf. Zhidk. Krist. Simp. Ikh Prakt. Primen., 2nd, *1972*, 165, C.A. *81*, 127816y (1974); Mesomorphism of homologous p-alkoxybenzylidene-p'-toluidines. [**458**] G. G. Maidachenko, B. N. Makarov, Sb. Dokl. Vses. Nauch. Konf. Zhidk. Krist. Simp. Ikh Prakt. Primen., 2nd, *1972*, 172, C.A. *81*, 127923f (1974); Mesomorphism of homologous p-alkyloxybenzylidene-p'-aminobenzonitriles. [**459**] G. G. Maidachenko, B. P. Smirnov, Uch. zap. Ivanov. un-t, *1974*, 106, C.A. *83*, 131279j (1975); Mesomorphic derivatives of amyloxycinnamaldehyde. [**460**] B. N. Makarov, G. G. Maidachenko, Uch. zap. Ivanov. un-t, *1974*, 110, C.A. *83*, 156012e (1975); Low-temperature liquid-crystal azo-azoxy compounds.

[**461**] W. Maier, G. Baumgartner, Z. Naturforsch. *7a*, 172 (1952), C.A. *46*, 10729e (1952); Elektrische Dipolmomente der homologen 4,4'-Di-n-alkoxyazoxybenzole. [**462**] W. Maier, A. Saupe, Z. Naturforsch. *12a*, 668 (1957), C.A. *52*, 4277f (1958); Klärpunkt und Anisotropie der molekularen Polarisierbarkeit flüssiger Kristalle. [**463**] K. Markau, W. Maier, Ber. *95*, 889 (1962), C.A. *57*, 3279i (1962); Nichtaromatische flüssige Kristalle. [**463a**] M. Makabe, H. Itoh, K. Ouchi, Carbon *14*, 365 (1976), C.A. *87*, 154629v (1977); Mesophase formation of pitch under reduced pressure. [**464**] H. Mallison, Diss. Göttingen 1908; Kondensation von Cyclohexanon und seinen Homologen mit Aldehyden. [**465**] J. Malthête, M. Leclercq, J. Gabard, J. Billard, J. Jacques, C. R. Acad. Sci., Sér. C, *273*, 265 (1971), C.A. *71*, 133979w (1971); Tolanes nématiques. [**466**] J. Malthête, M. Leclercq, M. Dvolaitzky, J. Gabard, J. Billard, V. Pontikis, J. Jacques, Mol. Cryst. Liq. Cryst. *23*, 233 (1973), C.A. *80*, 95393u (1974); Tolanes nématiques. [**467**] J. Malthête, J. Billard, J. Jacques, Bull. Soc. Chim. Fr. *1974*, 1199, C.A. *81*, 831i7q (1974); Nouveaux esters stéroides présentant des propriétés cholestériques. [**468**] J. Malthête, J. Billard, J. Jacques, C. R. Acad. Sci., Sér. C, *281*, 333 (1975), C.A. *84*, 58362v (1976); Tetrabenzoates de naphtalène-tétrol mésomorphes. [**469**] J. Malthête, J. Billard, J. Canceill, J. Gabard, J. Jacques, J. Phys. (Paris), C-3, *37*, 1 (1976), C.A. *86*, 29247x (1977); Structure moléculaire et propriétés smectiques. [**470**] J. Malthête, J. Billard, Mol. Cryst. Liq. Cryst. *34*, 117 (1976), C.A. *87*, 6149y (1977); Mesomorphic derivatives of ferrocene. [**470a**] J. Malthête, J. Billard, J. Jacques, Mol. Cryst. Liq. Cryst. *41*, 15 (1977), C.A. *87*, 109756y (1977); Géométrie moléculaire, facteurs polaires et propriétés mésomorphes.

[**471**] S. Marcelja, Solid State Commun. *13*, 759 (1973), C.A. *79*, 150363r (1973); End-chain ordering in nematics. [**472**] S. Marcelja, J. Chem. Phys. *60*, 3599 (1974), C.A. *81*, 42530a (1974); Chain ordering in liquid crystals. Even-odd effect. [**473**] H. Marsh, Carbon *11*, 254 (1973), C.A. *79*, 109034w (1973); Are carbon blacks formed from liquid-crystals? [**474**] H. Marsh, J. M. Foster, G. Hermon, M. Iley, Carbon *11*, 424 (1973), C.A. *80*, 70090e (1974); Enhanced graphitization of fluorene and carbazole. Mesophase formation. [**475**] H. Marsh, Fuel *52*, 205 (1973), C.A. *79*, 94581w (1973); Significance of the mesophase during carbonization of coking coals. [**476**] H. Marsh, J. M. Foster, G. Hermon, M. Iley, Fuel *52*, 234 (1973), C.A. *80*, 50272n (1974); Cocarbonization of aromatic and organic dye compounds, and influence of inerts. [**477**] H. Marsh, J. M. Foster, G. Hermon, M. Iley, J. N. Melvin, Fuel *52*, 243 (1973), C.A. *80*, 50273p (1974); Cocarbonization of aromatic and heterocyclic compounds containing oxygen, nitrogen, and sulfur. [**478**] H. Marsh, F. Dachille, M. Iley, P. L. Walker Jr., P. W. Whang, Fuel *52*, 253 (1973), C.A. *80*, 50274q (1974); Carbonization of coal-tar pitches and coals of increasing rank. [**479**] H. Marsh, G. Hermon, C. Cornford, Fuel *53*, 168 (1974), C.A. *81*, 124044j (1974); Carbonization and mesophase. Importance of eutectic zones formed from nematics. [**480**] H. Marsh, Prepr., Div. Pet. Chem., Am. Chem. Soc. *20*, 432 (1975), C.A. *86*, 123758m (1977); Graphitizability and mesophase. Factors influencing growth and structural characteristics.

[**481**] H. Marsh, C. Cornford, ACS Symp. Ser. *21*, 266 (1976), C.A. *84*, 182224r (1976); Mesophase: the precursor to graphitizable carbon. [**481a**] H. Marsh, Carbon *14*, 298 (1976), Importance of mesophase to graphitization process. [**482**] Y. Matsufuji, Japan. Kokai 75,104,780 (25. 1. 1974), C.A. *84*, 114226t

(1976); Liquid crystal compositions for TN display devices. [**483**] N. Matsumura, Japan. Kokai 72 39,070 (26. 4. 1971), C.A. *78*, 63656z (1973); Nematic liquid crystals. [**483a**] M. Matsuo, I. Tsunoda, Japan. Kokai 77 13,484 (24. 7. 1975), C.A. *87*, 60789v (1977); Storage-type liquid crystal for memory-type display devices. [**484**] E. Mauerhoff, Diss. Halle 1922; Diphenylbenzole. [**485**] C. Maze, H. G. Hughes, Liq. Cryst. Ord. Fluids *2*, 613 (1974), C.A. *86*, 180998p (1977); Mesomorphic N-(*p*-azidobenzylidene)-anilines. [**485a**] T. Meisel, K. Seybold, Z. Halmos, J. Roth, C. Melykuti, J. Therm. Anal. *10*, 419 (1976), C.A. *87*, 67649h (1977); Thermal behavior of thallium(I) fatty acid salts. [**486**] O. Meißner, Diss. Halle 1922; Kristallinisch-flüssige Derivate der β-Methylzimtsäure. [**487**] O. Meye, Diss. Halle 1908; Diphenyloxypropiophenon und kristallinisch-flüssige Nitroverbindungen. [**488**] F. Meyer, K. Dahlem, Liebigs Ann. *326*, 331 (1903); Azo- und Azoxybenzoesäureester. [**489**] S. Minezaki, H. Kumazaki, T. Nakaya, M. Imoto, Mem. Fac. Eng., Osaka City Univ. *14*, 183 (1973), C.A. *82*, 66502q (1975); Preparation and properties of liquid crystals. [**490**] V. I. Minkin, Y. A. Zhadanov, E. A. Medyantzeva, Y. A. Ostroumov, Tetrahedron *23*, 3651 (1967), C.A. *67*, 73044k (1967); Acoplanarity of aromatic azomethines.

[**491**] T. Mitote, Japan. Kokai 74,101,347 (9. 2. 1973), C.A. *83*, 9473c (1975); *p*-Alkoxy-benzal-*p*-butylanilines for use as nematics. [**491a**] T. Mitote, Y. Fujii, Japan. Kokai 76,146,445 (10. 6. 1975), C.A. *87*, 5649z (1977); Liquid crystal esters and their compositions. [**492**] Y. Miyazawa, M. Ozutsumi, M. Gonda, Japan. Kokai 72 25,151 (23. 3. 1971), C.A. *78*, 15766p (1973); N-(*p*-Cyanobenzylidene)-aniline derivatives. [**493**] Y. Mizuguchi, Japan. Kokai 75,159,480 (13. 6. 1974), C.A. *85*, 27398c (1976); Nematic liquid crystal compositions for display devices. [**494**] A. Mlodzieyovskii, Z. Kryst. Min. *52*, 1 (1913), C.A. *7*, 2713⁶ (1913); Fließende Krystalle des Ammoniumoleats. [**495**] S. Mori, N. Matsumura, Oyo Butsuri *42*, 60 (1973), C.A. *79*, 24396r (1973); Mesomorphic range and life time of Schiff bases. [**496**] A. Moriyama, M. Fukai, H. Tasuta, H. Takahashi, H. Egaki, Japan. Kokai 75 93,883 (25. 12. 1973), C.A. *84*, 24455n (1976); Nematic liquid crystal compositions for electrooptical display devices. [**497**] M. V. Mukhina, V. G. Atabekyan, A. S. Sopova, Zh. Org. Khim. *12*, 2263 (1976), C.A. *86*, 81956s (1977); Structure and mesomorphism of sterol esters. [**498**] A. Müller, Ber. *54*, 1481 (1921), C.A. *15*, 3826¹ (1921); Neuer Fall von Anisotropie im Schmelzpunkt. [**499**] S. Münch, Diss. Marburg 1903; p-Dioxystilben. [**500**] Y. Murakami, K. Morimoto, Japan. Kokai 75,160,225 (17. 6. 1974), C.A. *85*, 20822c (1976); Azoxybenzene derivative.

[**501**] Y. Murakami, K. Morimoto, Japan. Kokai 75,160,226 (17. 6. 1974), C.A. *85*, 20823d (1976); Azoxybenzene derivative. [**502**] Y. Murakami, K. Morimoto, Japan. Kokai 75,160,227 (17. 6. 1974), C.A. *85*, 20820a (1976); Azoxybenzene derivative. [**503**] Y. Murakami, K. Morimoto, Japan. Kokai 75,160,228 (17. 6. 1974), C.A. *85*, 20821b (1976); Azoxybenzene derivative. [**504**] Y. Murakami, S. Furuyama, K. Morimoto, Japan. Kokai 75,160,232 (17. 6. 1974), C.A. *85*, 20851m (1976); Benzalaniline derivative. [**505**] Y. Murakami, S. Furuyama, K. Morimoto, Japan. Kokai 75,160,237 (17. 6. 1974), C.A. *85*, 20895d (1976); Benzalcyanoaniline derivative. [**506**] Y. Murakami, K. Morimoto, Japan. Kokai 76 01,425 (18. 6. 1974), C.A. *85*, 20839p (1976); Azoxybenzene derivatives. [**507**] Y. Murakami, K. Morimoto, Japan. Kokai 76 01,426 (18. 6. 1974), C.A. *85*, 20838n (1976); Azoxybenzene derivatives. [**508**] Y. Murakami, K. Morimoto, Japan. Kokai 76 01,427 (18. 6. 1974), C.A. *85*, 20841h (1976); Azobenzene derivatives. [**509**] Y. Murakami, K. Morimoto, Japan. Kokai 76 01,428 (18. 6. 1974), C.A. *85*, 20840g (1976); Azobenzene derivatives. [**510**] Y. Murakami, K. Morimoto, Japan. Kokai 76 11,731 (18. 7. 1974), C.A. *85*, 46226m (1976); Azobenzene derivatives.

[**511**] Y. Murakami, K. Morimoto, Japan. Kokai 76 11,732 (19. 7. 1974), C.A. *84*, 164501z (1976); Azoxybenzene derivatives. [**512**] Y. Murakami, K. Morimoto, Japan. Kokai 76 13,738 (23. 7. 1974), C.A. *84*, 179858v (1976); Azobenzene derivative. [**513**] Y. Murakami, K. Morimoto, Japan. Kokai 76 13,739 (23. 7. 1974), C.A. *84*, 179857u (1976); Azoxybenzene derivatives. [**514**] Y. Murakami, K. Morimoto, Japan. Kokai 76 13,740 (23. 7. 1974), C.A. *84*, 179850m (1976); Benzalaniline derivative. [**515**] Y. Murakami, K. Morimoto, Japan. Kokai 76 13,741 (23. 7. 1974), C.A. *84*, 179856t (1976); Azobenzene derivatives. [**516**] Y. Murakami, K. Morimoto, Japan. Kokai 76 13,742 (23. 7. 1974), C.A. *84*, 179855s (1976); Azoxybenzene derivatives. [**517**] Y. Murakami, K. Morimoto, Japan. Kokai 76 80,831 (27. 12. 1974), C.A. *86*, 55151f (1977); Azoxybenzene derivatives. [**518**] Y. Murakami, K. Morimoto, Japan. Kokai 76 80,832 (27. 12. 1974), C.A. *86*, 72189s (1977); Azobenzene derivatives. [**519**] Y. Murakami, S. Furuyama, K. Morimoto, Japan. Kokai 76,118,734 (10. 4. 1975), C.A. *86*, 120963p (1977); Acrylic acid derivative. [**520**] Y. Murakami, K. Morimoto, Japan. Kokai 76,118,735 (11. 4. 1975), C.A. *86*, 106161u (1977); Vinyl phenyl ethers.

[521] K. Murase, Japan. Kokai 73 14,641 (5. 7. 1971), C.A. *78*, 124275y (1973); Mesomorphic Schiff bases. [522] K. Murase, Chem. Lett. *1972*, 471, C.A. *77*, 47817b (1972); Mesomorphism of laterally substituted benzylideneanilines. [523] K. Murase, Bull. Chem. Soc. Jap. *45*, 1772 (1972), C.A. *77*, 61402u (1972); Mesomorphism of terminally substituted benzylideneanilines. [524] K. Murase, Denki Tsushin Kenkyujo Kenkyu Jitsuyoka Hokoku *21*, 1159 (1972), C.A. *77*, 151572d (1972); Mesomorphism of terminally substituted benzylideneanilines. [525] K. Murase, Rev. Elec. Commun. Lab. *20*, 1125 (1972), C.A. *78*, 116248b (1973); Mesomorphism of terminally substituted benzylideneanilines. [526] K. Murase, H. Watanabe, Bull. Chem. Soc. Jap. *46*, 3142 (1973), C.A. *80*, 7904h (1974); Mesomorphism of substituted azo- and azoxybenzenes. [527] K. Murase, T. Wachi, Denki Tsushin Kenkyujo Kenkyu Jitsuyoka Hokoku *23*, 2855 (1974), C.A. *83*, 171101b (1975); Mesomorphism of substituted azo- and azoxybenzenes. [527a] M. M. Murza, G. P. Tataurov, L. I. Popov, Y. V. Svetkin, Zh. Org. Khim. *13*, 1046 (1977), C.A. *87*, 134286c (1977); Mesomorphic fluorine-containing aromatic azomethines. [528] C. W. Nash, Chem. Eng. Mining Rev. *12*, 363 (1920), C.A. *14*, 3393³ (1920); Liquid crystals, their role in the solidification of metals. [529] W. Naucke, Diss. Halle 1922; I. *p*-Aminobenzyl- und *p*-Aminoben-zoyl-*p*-aminobenzoesäureester. II. *p,p′*-Diamidobenzil. [530] N. A. Nedostup, V. V. Gal'tsev, Zh. Fiz. Khim. *51*, 214 (1977), C.A. *86*, 131374y (1977); Relation between the molecular structure of liquid crystals and their thermodynamic properties.

[531] V. D. Neff, M. K. H. Chang, D. L. Fishel, Mol. Cryst. Liq. Cryst. *17*, 369 (1972), C.A. *77*, 131742f (1972); Anomalous melting behavior of nematogenic azobenzene derivatives. [532] B. Nekras-sov, Z. Physik. Chem. *128*, 203 (1927), C.A. *22*, 56 (1928); Homologe Reihen und *cis-trans*-Formen. [533] N. N. Neronova, Zh. Strukt. Khim. *9*, 147 (1968), C.A. *68*, 117819u (1968); Crystal structure of decafluorobiphenyl. [534] M. E. Neubert, L. T. Carlino, R. D'Sidocky, D. L. Fishel, Liq. Cryst. Ord. Fluids *2*, 293 (1974); Mesomorphism of asymmetrically 4,4′-disubstituted phenyl benzoates. [534a] M. E. Neubert, L. J. Maurer, Mol. Cryst. Liq. Cryst. *43*, 313 (1977), Mesomorphism of homologous terephthal-bis-*n*-alkylanilines. [535] B. S. Neumann, K. G. Sansom, Ger. Offen. 2,028,652 (10. 6. 1969), C.A. *74*, 77902f (1971); Synthetic silicate with a smectic clay structure. [535a] D. T. Nguyen, T. C. Pham, Tap Chi Hoa Hoc *14*, 13 (1976), C.A. *87*, 68043t (1977); Mesomorphic N-furfurylidenaniline derivatives. [536] M. Nishida, S. Miyagishi, T. Tarutani, M. Okano, Yukagaku *24*, 687 (1975), C.A. *83*, 200486p (1975); Liquid crystal of long chain alkylamines. [537] C. S. Oh, Mol. Cryst. Liq. Cryst. *19*, 95 (1972), C.A. *78*, 63488w (1973); Effect of heterocyclic nitrogen on the mesomorphism of 4-alkoxybenzy-lidene-2′-alkoxy-5′-aminopyridines. [537a] C. S. Oh, E. F. Pasierb, U.S. 3,996,260 (17. 10. 1972), C.A. *86*, 171115w (1977); Alkoxybenzylidene-aminobenzonitriles. [538] S. Oi, Y. Yamada, H. Honda, Tanso *76*, 23 (1974), C.A. *82*, 61641z (1975); Formation of carbonaceous mesophase spherules having a new molecular arrangement from coal-tar pitch containing mesophase. [539] S. Oi, Y. Yamada, H. Honda, Tanso *85*, 47 (1976), C.A. *85*, 110693r (1976); Formation of carbonaceous mesophase in pitch. Effect of ferrocene on nucleation and growth processes of mesophase. [540] A. S. Orakhovats, T. Z. Radeva, S. L. Spasov, Dokl. Bolg. Akad. Nauk *26*, 663 (1973), C.A. *79*, 146131w (1973); Nematic nitrile-containing Schiff bases with positive dielectric anisotropy.

[541] M. A. Osman, Z. Naturforsch. B *31*, 801 (1976), C.A. *85*, 108361n (1976); Low melting mesomorphic N-(4-*n*-alkylbenzylidene)-4′-*n*-alkylanilines. [541a] M. A. Osman, Ger. Offen. 2,557,267 (19. 12. 1975), C.A. *87*, 134473m (1977); Liquid crystal Schiff bases. [542] H. Ottensmeyer, Diss. Halle 1924; Einfluß des Schwefels auf den kristallin-flüssigen Zustand aromatischer Verbindungen. [542a] M. Ozeki, H. Sato, K. Morita, T. Yamagi, H. Takatsu, Y. Arai, K. Kimura, Japan. Kokai 77 23,040 (14. 8. 1975), C.A. *87*, 84697q (1977); 4-Acyloxy-4′-cyanobiphenyls. [543] I. H. Page, H. Rudy, Biochem. Z. *220*, 304 (1930), C.A. *24*, 4050 (1930); Fettsäureester des Cholesterins. [544] C. M. Paleos, T. M. Laronge, M. M. Labes, Chem. Commun. *1968*, 1115, C.A. *69*, 106053j (1968); Liquid crystal monomers: N-(*p*-alkoxybenzylidene)-*p*-aminostyrenes. [545] L. Pastorelli, G. Chiavari, A. Arcelli, Ann. Chim. (Rome) *63*, 195 (1973), C.A. *84*, 114435k (1976); Mesomorphic naphthalene derivatives. [546] P. R. Patel, J. Indian Chem. Soc. *50*, 514 (1973), C.A. *80*, 75202c (1974); Mesomorphism of two homologous series of anils. [547] A. I. Pavlyuchenko, E. I. Kovshev, V. V. Titov, G. Purvaneckas, Zh. Org. Khim. *12*, 375 (1976), C.A. *84*, 150294x (1976); Synthesis of *p*-alkylbenzaldehydes and *p*-alkylcinnamic acids. [548] A. I. Pavlyuchenko, N. I. Smirnova, E. I. Kovshev, V. V. Titov, G. Purvaneckas, Zh. Org. Khim. *12*, 1054 (1976), C.A. *85*, 77830h (1976); Synthesis of *p*-alkyl-*p′*-cyanobiphenyls. [549] A. I. Pavlyuchenko, N. I. Smirnova, E. I. Kovshev, V. V. Titov, Zh. Org. Khim. *12*, 1511 (1976), C.A. *85*, 159597b (1976); Methyl-substituted *p*-alkyl- and *p*-alkoxycinnamic acids and their cyanophenyl esters. [550] A. I. Pavlyuchenko, N. I. Smirnova, V. V. Titov, E. I. Kovshev, K. M. Djumaev, Mol.

Cryst. Liq. Cryst. *37*, 35 (1976), C.A. *87*, 4816q (1977); Nematics containing pyridine and benzazole rings.

[551] E. Perplies, Diss. Mainz 1975; Flüssigkristalline Monomere und Polymere. [552] Philips' Gloeilampenfabrieken Fr. Demande 2,135,153 (3. 4. 1971), C.A. *79*, 42135j (1973); Azoxy compound liquid crystals. [553] Philips' Gloeilampenfabrieken Neth. Appl. 72 11,383 (18. 8. 1972), C.A. *81*, 13297z (1974); Liquid crystal compounds. [554] A. Pines, D. J. Ruben, S. Allison, Phys. Rev. Lett. *33*, 1002 (1974); Molecular ordering and even-odd effect in a homologous series of nematics. [555] D. A. Pink, J. Chem. Phys. *63*, 2533 (1975), C.A. *83*, 171119p (1975); Even-odd effect in liquid crystals. Simple model. [556] K. Pinter, KFKI (Rep.), *1974*, 74–85, 24 pp., C.A. *83*, 69155k (1975); Chemistry of cholesterics. [557] J. M. Pochan, H. W. Gibson, J. Am. Chem. Soc. *94*, 5573 (1972), C.A. *77*, 93828d (1972); Crystal nucleation studies in supercooled mesophases of cholesteryl derivatives. [558] J. M. Pochan, J. E. Kuder, J. Y. C. Chu, D. Wychick, D. F. Hinman, Can. J. Chem. *53*, 578 (1975), C.A. *82*, 163849r (1975); Electric dipole moments of a series of $\Delta^{8(14)}$ cholestenyl esters. [559] J. L. W. Pohlmann, Mol. Cryst. *2*, 15 (1966), C.A. *66*, 32720p (1967); Mesomorphism of stigmasteryl carbonates. [560] J. L. W. Pohlmann, Mol. Cryst. Liq. Cryst. *8*, 417 (1969), C.A. *71*, 102113x (1969); Structure dependence of cholesterics.

[561] J. L. W. Pohlmann, W. Elser, Mol. Cryst. Liq. Cryst. *8*, 427 (1969), C.A. *71*, 113167u (1969); Synthesis of pure mesogens. [562] J. L. W. Pohlmann, W. Elser, P. R. Boyd, Mol. Cryst. Liq. Cryst. *13*, 243 (1971), C.A. *75*, 64097t (1971); Structure dependence of cholesterics. [563] J. L. W. Pohlmann, W. Elser, P. R. Boyd, Mol. Cryst. Liq. Cryst. *13*, 271 (1971), C.A. *75*, 64099v (1971); 17β-Alkyl substituted androst-5-en-3β-ols. [564] J. L. W. Pohlmann, W. Elser, P. R. Boyd, Mol. Cryst. Liquid Cryst. *20*, 87 (1973), C.A. *78*, 97880h (1973); Cholesteryl ω-phenylalkanoates. [565] J. L. W. Pohlmann, W. Elser, P. R. Boyd, Mol. Cryst. Liq. Cryst. *26*, 59 (1974), C.A. *81*, 112239p (1974); Mesomorphic 5α-cholestan-3β-yl ω-phenylalkanoates. [566] J. C. Poirier, U.S. Nat. Tech. Inform. Serv., AD Rep. *1973*, No. 759200, 270873, C.A. *79*, 97905r (1973); Synthesis of novel organic and organometallic compounds potentially exhibiting mesophases. [567] J. Politis, Liebigs Ann. *255*, 293 (1889); Anisaldehyd und Bernsteinsäure. [568] G. J. Davis, R. S. Porter, Mol. Cryst. Liq. Cryst. *6*, 377 (1970), C.A. *72*, 104494s (1970); Some solubility characteristics of cholesteryl esters. [569] G. J. Davis, R. S. Porter, E. M. Barrall II, Mol. Cryst. Liq. Cryst. *10*, 1 (1970), C.A. *73*, 49189q (1970); Thermal transitions in some cholesteryl esters of saturated aliphatic acids. [570] G. J. Davis, R. S. Porter, J. W. Steiner, D. M. Small, Mol. Cryst. Liq. Cryst. *10*, 331 (1969), C.A. *73*, 70653w (1970); Thermal transitions of cholesteryl esters of C_{18}-aliphatic acids.

[571] G. J. Davis, R. S. Porter, E. M. Barrall II, Mol. Cryst. Liq. Cryst. *11*, 319 (1970), C.A. *74*, 80492c (1971); Intercomparison of temperatures and heats of transition for cholesteryl esters. [572] B. D. Powell, J. E. Puddington, Can. J. Chem. *31*, 828 (1953), C.A. *48*, 1772i 1954); Flow properties and mesomorphism of anhydrous sodium stearate above 200°. [573] R. J. Price, J. L. White, Carbon *10*, 343 (1972); Development of mesophase microstructures during pyrolysis of selected coker feedstocks. [574] E. B. Priestley, Introd. Liq. Cryst. *1974*, 1, C.A. *84*, 82631c (1976); Liquid crystal mesophases. [575] A. Prins, Z. physik. Chem. *67*, 689 (1909); C.A. *4*, 704⁶ (1910); Flüssige Mischkristalle in binären Systemen. [576] A. Prins, Z. physik. Chem. *75*, 681 (1911), C.A. *5*, 1218⁹ (1911); Mischkristalle bei flüssigen Kristallen und die Phasenlehre. Entgegnung an Herrn O. Lehmann. [577] M. J. Rafuse, R. A. Soref, Mol. Cryst. Liq. Cryst. *18*, 95 (1972), C.A. *77*, 119263t (1972); Carbonate Schiff base nematics. Synthesis and electrooptic properties. [578] N. V. V. Raghavan, L. A. Paddock, Org. Prep. Proced. Int. *7*, 311 (1975), C.A. *84*, 150265p (1976); Improved synthesis and purification of MBBA. [579] W. Rasche, Diss. Halle 1921; Verbindungen von Diamino-azo-azoxy- und -hydrazobenzol mit aromatischen Aldehyden. [580] J. Rault, L. Liebert, L. Strzelecki, Bull. Soc. Chim. Fr. *1975*, 1175, C.A. *84*, 158203n (1976); Mésogènes "dimérisés".

[581] G. B. Ravich, G. G. Tsurinov, Dokl. Akad. Nauk S.S.S.R. *38*, 331 (1943), C.A. *37*, 6514³ (1943); Polymorphism of trilaurin. [582] C. Reichardt, Diss. Halle 1909; Einfluß der Bromierung auf den kristallinisch-flüssigen Zustand. [583] K. Reinknecht, Diss. Halle 1924; *p,p'*-Dioxydiphenylmethan und *p,p'*-Dioxydiphenyl. [584] H. Rembges, F. Kröhnke, I. Vogt, Ber. *103*, 3427 (1970), C.A. *74*, 12726g (1971); Reaktion von Nitro-benzylpyridiniumsalzen mit Phenylhydroxylamin. [585] D. O. Rester, C. R. Rowe, Carbon *12*, 218 (1974), C.A. *82*, 78970d (1975); Effect of gas-bubble formation and percolation on the carbonaceous mesophase from acenaphthylene. [586] R. M. Reynolds, C. Maze, E. Oppenheim, Mol. Cryst. Liq. Cryst. *36*, 41 (1976), C.A. *86*, 49316x (1977); Nematic *p,p'*-disubstituted phenyl thiolbenzoates. [587] R. M. Riddle, Diss. Univ. Texas, Austin, Tex. 1971, 147 pp. Avail. Univ. Microfilms,

Ann Arbor, Mich., Order No. 72-15,820, C.A. *77*, 113667f (1972); Mesophase stability and molecular structure. [**588**] F. Rinne, Z. Krist. *83*, 227 (1932), C.A. *26*, 5469 (1932); Parakristallines und kristallines Anisal-α-benzolazo-α₁-naphthylamin. [**589**] A. Rising, Ber. *37*, 43 (1904); Methyl- und Äthyläther des p-Oxyphenylhydroxylamins und daraus dargestellter Azoxyverbindungen. [**590**] O. Ritter, Diss. Halle 1928; p-Äthoxy- und Methoxyamidoazobenzol.

[**591**] K. Ritvay, KFKI (Rep.), *1974*, 74-25, 36 pp., C.A. *81*, 104864y (1974); Trends and results in the preparation of nematics. [**592**] J. M. Robertson, I. Woodward, Proc. Roy. Soc. A *162*, 568 (1937), C.A. *32*, 1539³ (1938); X-ray analysis of stilbene. Detailed structure. [**593**] J. M. Robertson, I. Woodward, Proc. Roy. Soc. A *164*, 436 (1938), C.A. *32*, 3683⁷ (1938); X-ray analysis of tolane and the triple bond. [**594**] R. C. Robinder, J. C. Poirier, J. Am. Chem. Soc. *90*, 4760 (1968), C.A. *69*, 90988e (1968); Monotropic crystalline phases of PAA from the nematic melt. [**595**] L. A. Roche, Diss. American Univ., Washington. 1975. 118 pp. Avail. Xerox Univ. Microfilms, Ann Arbor, Mich., Order No. 76-8940, C.A. *85*, 15831a (1976); Intermolecular forces of mesophases. [**596**] E. Rolle, Diss. Halle 1921; Derivate des Phenylbenzyläthers und des 1,3-Diphenylpropans. [**597**] R. K. Rose, Diss. City Univ. New York, New York. 1976. 303 pp. Avail. Xerox Univ. Microfilms, Ann Arbor, Mich., Order No. 76-10,644, C.A. *85*, 63153c (1976); Potentially mesomorphic organopalladium compounds. [**598**] H. M. Rosenberg, R. A. Champa, Mol. Cryst. Liq. Cryst. *11*, 191 (1970), C.A. *74*, 16530m (1971); Thermal nematic stability of Schiff's bases upon reversal of terminal substituents. [**599**] H. M. Rosenberg, M. P. Serve, J. Chem. Eng. Data *16*, 496 (1971), C.A. *75*, 151480f (1971); Mesomorphic N-(p-alkoxybenzylidene)-p'-acyloxyanilines. [**600**] D. L. Ross, D. M. Gavrilovic, U.S. 3,925,237 (28. 6. 1974), C.A. *84*, 105223n (1976); Liquid crystal electrooptic devices.

[**601**] T. Rotarski, Ber. *41*, 1994 (1908), C.A. *2*, 2799¹ (1908); Übersehene Angaben betreffs flüssiger Krystalle. [**602**] A. Roviello, A. Sirigu, Mol. Cryst. Liq. Cryst. *33*, 19 (1976), C.A. *84*, 187796p (1976); Solid and mesomorphic phases of aliphatic esters of 4,4'-dihydroxy-α,α'-dimethylbenzalazine. [**603**] A. Roviello, A. Sirigu, Mol. Cryst. Liq. Cryst. *35*, 155 (1976), C.A. *85*, 114992k (1976); Solid and mesomorphic phases of aliphatic esters of 4,4'-dihydroxybenzalazine. [**604**] E. I. Ryumtsev, T. A. Rotinyan, A. P. Kovshik, Yu. Yu. Daugvila, G. I. Donis, V. N. Tsvetkov, Opt. Spektrosk. *41*, 65 (1976), C.A. *85*, 142438c (1976); Value and direction of dipole moments in molecules of nematics containing nitrile groups. [**605**] B. K. Sadashiva, G. S. R. S. Rao, Curr. Sci. *44*, 222 (1975), C.A. *82*, 170294j (1975); Mesomorphic biphenylylbenzoates. [**606**] B. K. Sadashiva, Mol. Cryst. Liq. Cryst. *35*, 205 (1976), C.A. *86*, 88936u (1977); Mesomorphism of α-methylcinnamic acids and their esters. [**607**] B. K. Sadashiva, G. S. R. S. Rao, Mol. Cryst. Liq. Cryst. *38*, 345 (1977), C.A. *87*, 102050e (1977); Convenient method for the preparation of 4-n-alkyl-4''-cyano-p-terphenyl. [**608**] F. D. Saeva, R. L. Schank, U.S. 3,819,531 (24. 2. 1972), C.A. *81*, 96987y (1974); Liquid crystal compositions. [**609**] S. Sakagami, A. Takase, M. Nakamizo, Mol. Cryst. Liq. Cryst. *36*, 261 (1976), C.A. *86*, 81974w (1977); Liquid crystalline compounds exhibiting two s_B modifications. [**609a**] S. Sakagami, M. Nakamizo, Bull. Chem. Soc. Jpn. *50*, 1009 (1977), C.A. *87*, 125658y (1977); Mesomorphism of 2-(4-n-alkoxybenzyli-deneamino)anthracenes. [**610**] Y. Sanada, Sekiyu Gakkai Shi *15*, 182 (1972), C.A. *76*, 156323n (1972); Carbonization of heavy oils.

[**611**] Y. Sanada, T. Furuta, H. Kimura, H. Honda, Carbon *10*, 644 (1972), C.A. *78*, 6381 (1973), Development of optical anisotropic mesophases from naphtha-tar pitch in magnetic field. [**612**] Y. Sanada, T. Furuta, H. Kimura, H. Honda, Fuel *52*, 143 (1973), C.A. *79*, 7755q (1973); Mesophase formation from various carbonaceous materials in early stages of carbonization. [**613**] Y. Sanada, T. Furuta, J. Kumai, H. Kimura, Sekiyu Gakkai Shi *16*, 409 (1973), C.A. *79*, 81274y (1973); Carbonization of heavy oils. 3. Changes of volatile matter, stacking of lamellae, and heat of combustion during mesophase formation. [**614**] Y. Sanada, T. Furuta, J. Kumai, H. Kimura, Sekiyu Gakkai Shi *16*, 746 (1973), C.A. *80*, 29049k (1974); 4. Changes of benzene insolubles during mesophase formation (cf. [613]). [**615**] Y. Sanada, T. Furuta, J. Kumai, H. Kimura, Sekiyu Gakkai Shi *16*, 902 (1973), C.A. *80*, 61836a (1974); 5. Properties of mesophase isolated from the matrix by solvent extraction (cf. [613]). [**616**] J. W. Sanders, Diss. Kent State Univ., Kent, Ohio, 1969, 174 pp. Avail. Univ. Microfilms, Ann Arbor, Mich., Order No. 70-11,356, C.A. *74*, 112305n (1971); Sterol benzoate liquid crystals. Correlations of structure and transition constants. [**617**] Y. Satomi, Japan. Kokai 73 68,560 (2. 12. 1971), C.A. *80*, 48246p (1974); Cholesteryl ferulate derivatives. [**618**] K. Sawatari, T. Mukai, K. Tsukinuki, S. Kamenosono, Japan. Kokai 74 80,059 (8. 12. 1972), C.A. *82*, 112202s (1975); Cholesterol derivatives. [**619**] L. C. Scala, G. D. Dixon, Mol. Cryst. Liq. Cryst. *7*, 443 (1969), C.A. *71*, 85813f (1969); Long

term stability of cholesterics. I. [**620**] L. C. Scala, G. D. Dixon, Mol. Cryst. Liq. Cryst. *10*, 411 (1970), C.A. *73*, 81744c (1970); Part II (cf. [**619**]).

[**621**] H. Schade, Diss. Halle 1933; Kenntnis der *p*-Azo- und *p*-Azoxybenzoesäuren. [**622**] E. Schaefer, Diss. Halle 1911; Dichte, Reibung und Kapillarität kristallinischer Flüssigkeiten. [**623**] E. Scheil, H. Baach, Z. Metallkunde *50*, 386 (1959), C.A. *56*, 10993g (1962); Dampfdruck von geschmolzenen Cd/Sb-Legierungen. [**624**] H. Scherrer, A. Boller, U.S. 3,952,046 (14. 11. 1974), C.A. *85*, 94086n (1976); Liquid crystalline biphenyls. [**625**] K. Schoenemann, Diss. Halle 1922; α-Ungesättigte Phenolketone. [**626**] S. A. Haut, D. C. Schroeder, J. P. Schroeder, J. Org. Chem. *37*, 1425 (1972), C.A. *77*, 10731t (1972); Unsymmetrical *p*-phenylene di-*p*-*n*-alkoxybenzoates. [**627**] J. P. Schroeder, D. W. Bristol, J. Org. Chem. *38*, 3160 (1973), C.A. *79*, 97891h (1973); Effects of terminal substituents on the nematic mesomorphism of p-phenylene dibenzoates. [**628**] D. W. Bristol, J. P. Schroeder, J. Org. Chem. *39*, 3138 (1974), C.A. *81*, 160392e (1974); Molecular structural effects on the mesomorphism of phenylene esters. [**629**] D. C. Schroeder, J. P. Schroeder, J. Am. Chem. Soc. *96*, 4347 (1974), C.A. *81*, 135305w (1974); p-Phenylene di-*p*-amino- and di-*p*-hydroxybenzoate. [**630**] D. C. Schroeder, J. P. Schroeder, J. Org. Chem. *41*, 2566 (1976), C.A. *85*, 62499h (1976); Mesomorphic p-phenylene dibenzoates with terminal hydroxy and amino groups.

[**631**] D. C. Schroeder, J. P. Schroeder, Mol. Cryst. Liq. Cryst. *34*, 43 (1976), C.A. *85*, 152108j (1976); Smectic mesomorphism of *p*-*n*-alkoxyaniline hydrochlorides. [**632**] E. Schroedter, Diss. Halle 1925; Untersuchungen der Struktur aromatischer Verbindungen. [**633**] E. Schröter, Diss. Halle 1927; Substanzen mit bei niedriger Temperatur beständigen flüssig-kristallinen Phasen. [**634**] G. Schroeter, Ber. *41*, 5 (1908), C.A. *2*, 1136 (1908); β-Alkylzimtsäuren. [**635**] R. Schroeter, Diss. Halle 1923; Abkömmlinge der *p*-Aminobenzoesäure. [**636**] H. Schubert, F. Trefflich, Z. Chem. *4*, 228 (1964), C.A. *61*, 7014 (1964); Kristallin-flüssige Hydroxypyrazine. [**637**] H. Schubert, H. J. Lorenz, R. Hoffmann, F. Franke, Z. Chem. *6*, 337 (1966); Neue kristallin-flüssige *p*-Terphenylderivate. [**638**] H. Schubert, R. Koch, C. Weinbrecher, Z. Chem. *6*, 467 (1966), C.A. *66*, 54977e (1967); Kristallin-flüssige trans-Dibenzoyläthylene. [**639**] H. Schubert, I. Eissfeldt, R. Lange, F. Trefflich, Mitt. Chem. Ges. DDR *13*, 135 (1966); Kristallin-flüssige Hydroxypyrazine. [**640**] H. Schubert, I. Eissfeldt, R. Lange, F. Trefflich, J. Prakt. Chem. *33*, 265 (1966), C.A. *66*, 46399c (1967); Kristallin-flüssige *n*-Alkyl- und *n*-Alkoxyderivate des 3-Hydroxy-2,5-diphenyl-pyrazins.

[**641**] H. Schubert, R. Hacker, K. Kindermann, J. Prakt. Chem. *37*, 12 (1968), C.A. *68*, 39588b (1968); Kristallin-flüssige *n*-Alkyl- und *n*-Alkoxyderivate des 2,5-Diphenyl-pyrazins. [**642**] H. Schubert, H. Zaschke, J. Prakt. Chem. *312*, 494 (1970), C.A. *73*, 98890v (1970); Kristallin-flüssige *n*-Alkyl- und *n*-Alkoxyderivate des 2,5-Diphenyl-pyrimidins. [**643**] H. Schubert, Wiss. Z. Martin-Luther-Univ. Halle-Wittenberg, Math. Naturwiss. Reihe *19*, 1 (1970), C.A. *75*, 98467d (1971); Kristallin-flüssige Heterocyclen. [**644**] H. Schubert, R. Dehne, V. Uhlig, Z. Chem. *12*, 219 (1972), C.A. *77*, 164084w (1972); Kristallin-flüssige *trans*-4-*n*-Alkyl-cyclohexan-1-carbonsäuren. [**645**] H. Schubert, H. Dehne, Z. Chem. *12*, 241 (1972), C.A. *77*, 151553y (1972); Kristallin-flüssige *p*-Oligophenyle. [**646**] H. Zaschke, H. Schubert, J. Prakt. Chem. *315*, 1113 (1973), C.A. *80*, 82864y (1974); Kristallin-flüssige 2,5-Diphenylpyrimidine mit nur einer Flügelgruppe. [**647**] D. Heydenhauss, G. Jaenecke, H. Schubert, Z. Chem. *13*, 295 (1973), C.A. *80*, 3704v (1974); Cholesterinester von ω-Ferrocenylalkansäuren. [**648**] S. Hoffmann, W. Brandt, H. Schubert, Z. Chem. *15*, 59 (1975), C.A. *83*, 10563p (1975); Ferne Membrankomponenten-Analoge. Östradiol-3,17-bis-(*p*-(*n*-alkoxy)benzoate). [**649**] H. Zaschke, H. Debacq, H. Schubert, Z. Chem. *15*, 100 (1975), C.A. *83*, 58364b (1975); Kristallin-flüssige Azobenzole. [**650**] H. Schubert, I. Sagitdinov, Y. V. Svetkin, Z. Chem. *15*, 222 (1975), C.A. *83*, 106547n (1975); Kristallin-flüssige 2,5-Diarylthiophene.

[**651**] H. Schubert, W. Schulze, H. J. Deutscher, V. Uhlig, R. Kuppe, J. Phys. (Paris), C-1, *36*, 379 (1975), C.A. *83*, 78266z (1975); Liquid crystals with hydroaromatic and hydroheterocyclic structures. [**652**] I. Herrmann, H. Schubert, B. Grossmann, L. Richter, L. Vogel, D. Demus, Z. Phys. Chem. (Leipzig) *257*, 563 (1976), C.A. *85*, 85794q (1976); Mesomorphie von 2-(4'-*n*-alkyloxy-biphenylyl-(4))-chinoxalinen und 2-(4'-*n*-alkyl-biphenylyl-(4))-chinoxalinen. [**653**] W. Weissflog, H. Schubert, J. Prakt. Chem. *318*, 785 (1976), C.A. *86*, 10835r (1977); Kristallin-flüssige Aldoximester. [**654**] H. Schubert, Wiss. Beitr. Martin-Luther-Univ. Halle-Wittenberg *20*, 5 (1976), C.A. *86*, 63592r (1977); Azolstruktur und Mesomorphie. [**655**] H. Zaschke, H. Schubert, F. Kuschel, F. Dinger, D. Demus, Ger. (East) 95,892 (22. 12. 1971), C.A. *80*, 32446t (1974); Nematic liquid crystals. [**656**] J. Hausschild, H. Kresse, H. Schubert, D. Demus, Ger. (East) 117,014 (15. 1. 1975), C.A. *84*, 187561h (1976); Nematic liquid crystals. [**656a**] H. Zaschke, K. Nitsche, H. Schubert, J. Prakt. Chem. *319*, 475 (1977), C.A. *87*, 152140e (1977); Kristallin-flüssige 3,6-Diphenyl-1,2,4-triazine. [**656b**] W. Weissflog, H. Schubert, S. König, D. Demus, L. Vogel,

J. Prakt. Chem. *319*, 507 (1977), C.A. *87*, 125574t (1977); Kristallin-flüssige Ketoximester. [**656c**] H. Zaschke, S. Arndt, V. Wagner, H. Schubert, Z. Chem. *17*, 293 (1977), C.A. *87*, 184456d (1977); Niedrigschmelzende kristallinflüssige Heterocyclen: 2-n-Alkylthio-5-(4-subst.-phenyl)-pyrimidine. [**656d**] K. Hanemann, H. Schubert, D. Demus, G. Pelzl, Ger. (East) 115,283 (30. 10. 1974), C.A. *87*, 76418j (1977); Nematic liquid crystal compositions. [**657**] H. Schuster, Diss. Halle 1933; Die isomeren *p*-Aminoazoxybenzole (cf. [826]). [**658**] A. Schweig, Angew. Chem. *84*, 1074 (1972); Eigenschaften von Molekülen in elektrischen Feldern. [**659**] V. P. Sevost'yanov, T. I. Gnilomedova, Zh. Org. Khim. *12*, 2400 (1976), C.A. *86*, 89307b (1977); Mesomorphic benzeneazo- and *p*-cyanobenzeneazophenyl esters of *p'*-alkoxybenzoic acids. [**660**] V. P. Sevost'yanov, Y. M. Zharov, T. I. Gnilomedova, Zh. Org. Khim. *12*, 2562 (1976), C.A. *86*, 139567j (1977); Synthesis of new electrically conducting additives for nematics.

[**661**] E. Shapiro, S. Ohki, J. Colloid Interface Sci. *47*, 38 (1974), C.A. *80*, 132577e (1974); Interaction energy between hydrocarbon chains. [**662**] V. A. Shaposhnikova, Z. I. Kurteeva, Sb. nauch. tr. NII elektrod. prom-sti, *1974*, 73, C.A. *83*, 181851k (1975); Change in composition of an isotropic medium during mesophase transformations in coal pitch. [**663**] J. L. Sheumaker, Diss. Univ. Iowa, Iowa City, 1972, 137 pp. Avail. Univ. Microfilms, Ann Arbor, Mich., Order No. 73-689, C.A. *78*, 120638w (1973); Physical properties of cholesterol and cholesteryl esters. [**664**] V. P. Shibaev, R. V. Tal'roze, F. I. Karakhanova, A. V. Kharitonov, N. A. Plate, Dokl. Akad. Nauk SSSR *225*, 632 (1975), C.A. *84*, 74746d (1976); Liquid-crystal state in monomer and polymer N$^\alpha$-acyl derivatives of N$^\varepsilon$-methacryloyl-L-lysine. [**665**] K. Shibayama, H. Ono, Japan. Kokai 72 31,887 (11. 3. 1971), C.A. *79*, 10912u (1973); Terephthalate derivative liquid crystal compositions. [**666**] M. Shiraishi, Kogai Shigen Kenkyusho Hokoku *6*, 1 (1975), C.A. *83*, 88942f (1975); X-ray study on the graphitization of carbon. [**667**] M. Shiraishi, Y. Sanada, Nippon Kagaku Kaishi *1976*, 153, C.A. *84*, 108224z (1976); Layer stacking of mesophase in early stages of carbonization. [**668**] H. N. Shrivastava, J. C. Speakman, J. Chem. Soc. *1961*, 1151, C.A. *55*, 12991f (1961); Crystal structure of rubidium hydrogen di-*o*-nitrobenzoate and potassium hydrogen di-*p*-nitrobenzoate. [**668a**] L. S. Singer, U.S. 3,991,170 (27. 4. 1973), C.A. *87*, 8533t (1977); Producing orientation in mesophase pitch by rotational motion relative to a magnetic field and carbonization of the oriented mesophase. [**669**] J. M. Skinner, G. M. D. Stewart, J. C. Speakman, J. Chem. Soc. *1954*, 180, C.A. *48*, 4923g (1954); Crystal structure of potassium hydrogen dibenzoate. [**670**] A. Skoulios, V. Luzzati, Nature *183*, 1310 (1959), C.A. *54*, 8115g (1960); Structure of anhydrous sodium soaps at high temperatures.

[**671**] P. Spegt, A. Skoulios, C. R. Acad. Sci. *251*, 2199 (1960), C.A. *55*, 12992g (1961); Structure des phases mésomorphes du stéarate de calcium anhydre. [**672**] A. Skoulios, B. Gallot, C. R. Acad. Sci. *252*, 142 (1961), C.A. *55*, 14009b (1961); Structure des phases mésomorphes du stéarate de potassium anhydre. [**673**] P. Spegt, A. Skoulios, C. R. Acad. Sci. *254*, 4316 (1962), C.A. *57*, 6723e (1962); Structure des phases mésomorphes du stéarate de magnésium. [**674**] P. A. Spegt, A. E. Skoulios, Acta Cryst. *16*, 301 (1963), C.A. *59*, 1125a (1963); Structure des savons de cadmium à température élevée. [**675**] P. A. Spegt, A. E. Skoulios, Acta Cryst. *17*, 198 (1964), C.A. *60*, 6290e (1964); Structure des savons de calcium à température ordinaire et à température élevée. [**676**] B. Gallot, A. Skoulios, Kolloid-Z. Z. Polym. *213*, 143 (1966), C.A. *67*, 30246q (1967); Structure des savons de cesium et comparaison des monosavons alcalins. [**677**] B. E. North, G. G. Shipley, D. M. Small, Biochim. Biophys. Acta *424*, 376 (1976), C.A. *84*, 158225w (1976); Thermal transitions and structural properties of 5α-cholestan-3β-ol esters of aliphatic acids. [**678**] D. L. Smare, Acta Cryst. *1*, 150 (1948), C.A. *42*, 8045g (1948); Crystal structure of 2,2'-dichlorobenzidine. [**679**] B. P. Smirnov, Probl. Poluch. Poluprod. Prom. Org. Sin., Akad. Nauk SSSR, Otd. Obshch. Tekh. Khim. *1967*, 120, C.A. *69*, 100780x (1968); Mesomorphic *p*-isoalkoxybenzaldehyde derivatives. [**680**] Z. G. Gardlund, R. J. Curtis, G. W. Smith, J. Chem. Soc., Chem. Commun. *1973*, 202, C.A. *78*, 159092p (1973); Low-temperature mesomorphism in ring-substituted benzylideneanilines.

[**681**] G. W. Smith, Z. G. Gardlund, R. J. Curtis, Mol. Cryst. Liq. Cryst. *19*, 327 (1973), C.A. *78*, 89539p (1973); Phase transitions in mesomorphic benzylidene anilines. [**682**] Z. G. Gardlund, R. J. Curtis, G. W. Smith, Liq. Cryst. Ord. Fluids *2*, 541 (1974), C.A. *87*, 22124s (1977); Influence of molecular structural changes on the mesomorphism of benzylideneanilines. [**683**] H. Sorkin, A. Denny, RCA Rev. *34*, 308 (1973), C.A. *79*, 140199c (1973); Equilibrium properties of Schiff-base liquid-crystal mixtures. [**684**] G. Stainsby, R. Fernand, T. E. Puddington, Can. J. Chem. *29*, 838 (1951), C.A. *46*, 3300i (1952); Preparation of anhydrous sodium stearate. [**685**] E. L. Steiger, H. J. Dietrich, Mol. Cryst. Liq. Cryst. *16*, 279 (1972), C.A. *76*, 146201g (1972); Mesomorphic Schiff's bases derived from *p*-acetoxy-, *p*-*n*-butyl-, and (*p*-*n*-butoxy-anilines. [**686**] E. L. Steiger, H. J. Dietrich, U.S. 3,742,067 (8.

1. 1971), C.A. *79*, 78381a (1973); Alkylstilbene liquid crystals. [687] E. L. Steiger, H. J. Dietrich, U.S. 3,769,314 (14. 7. 1971), C.A. *80*, 36856z (1974); N-(*p-n*-Alkoxycarbonyloxybenzylidene)-*p'*-butoxy-aniline. [688] E. L. Steiger, H. J. Dietrich, U.S. 3,784,513 (8. 8. 1971), C.A. *80*, 82428j (1974); N-[*p*-(Alkanoyloxy)benzylidene]-*p'*-butylaniline. [689] E. L. Steiger, H. J. Dietrich, U.S. 3,799,971 (23. 8. 1971), C.A. *80*, 133068b (1974); N-(*p-n*-Alkanoyloxybenzylidene)-*p'*-aminophenyl acetate. [690] E. L. Steiger, H. J. Dietrich, U.S. 4,000,124 (18. 5. 1971), C.A. *86*, 120973s (1977); Liquid-crystal *p*-(phenylazo)phenol derivatives.

[691] M. Steineck, Diss. Halle 1924; Darstellung von Abkömmlingen des sym. Triphenylbenzols als Beitrag zur Frage: Bilden sternförmige Moleküle flüssige Kristalle? [692] R. Steinsträsser, L. Pohl, Z. Naturforsch. B *26*, 87 (1971), C.A. *74*, 147571y (1971); Tiefschmelzende nematische Eutektika. [693] R. Steinsträsser, L. Pohl, Z. Naturforsch. B *26*, 577 (1971), C.A. *75*, 76300s (1971); Niedrigschmelzende *p*-Alkyl-*p'*-alkoxy- und *p*-Alkyl-*p'*-acyloxy-azobenzole. [694] R. Steinsträsser, L. Pohl, Tetrahedron Lett. *22*, 1921 (1971), C.A. *75*, 63282n (1971); Niedrig schmelzende *p*-Alkyl-*p'*-alkoxy- und *p*-Alkyl-*p'*-acyloxyazoxybenzole. [695] R. Steinsträsser, Z. Naturforsch. B *27*, 774 (1972), C.A. *77*, 131760k (1972); Nematische *p,p'*-disubstituierte Benzoesäurephenylester und niedrig schmelzende eutektische Gemische. [696] R. Steinsträsser, Angew. Chem. *84*, 636 (1972), C.A. *77*, 131723a (1972); Nematische Flüssigkeiten mit positiver dielektrischer Anisotropie. [697] R. Steinsträsser, L. Pohl, B. Hampel, Ger. Offen. 1,951,092 (10. 10. 1969), C.A. *75*, 5501x (1971); Nematische Substanzen. [698] R. Steinsträsser, Ger. Offen. 2,014,989 (28. 3. 1970), C.A. *76*, 14080g (1972); Nematische Substanzen. [699] R. Steinsträsser, Ger. Offen. 2,065,465 (28. 3. 1970), C.A. *79*, 146172k (1973); Nematische Azobenzol-Derivate. [700] R. Steinsträsser, Ger. Offen. 2,139,628 (7. 8. 1971), C.A. *78*, 124301d (1973); Nematische Phenylbenzoat-Derivate.

[701] R. Steinsträsser, Ger. Offen. 2,240,864 (19. 8. 1972), C.A. *81*, 25404v (1974); Nematische Ester und elektrooptische Eigenschaften. [702] R. Steinsträsser, F. del Pino, Ger. Offen. 2,450,088 (22. 10. 1974), C.A. *85*, 46219m (1976); Biphenylester. [703] R. Steinsträsser, F. del Pino, Ger. Offen. 2,535,046 (6. 8. 1975), C.A. *86*, 189552w (1977); Biphenylester. [704] H. Stenschke, Solid State Commun. *10*, 653 (1972), C.A. *76*, 146001s (1972); Reduction of the n/i transition temperature due to thermal fluctuations of the conformation of long molecules. [705] R. N. Steppel, L. E. Knaak, R. E. Rondeau, H. M. Rosenberg, Mol. Cryst. Liq. Cryst. *21*, 275 (1973), C.A. *79*, 109044z (1973); Transiton point studies of 4(4')-alkoxyformyloxy-4'(4)-alkylazoxybenzenes. [706] H. Stoltzenberg, Diss. Halle 1911; Optische Aktivität und kristallinisch-flüssiger Zustand. [707] H. Stoltzenberg, M. E. Huth, Z. physik. Chem. *71*, 641 (1910), C.A. *4*, 1417³ (1910); Kristallinisch-flüssige Phasen bei den Monohalogeniden des Thalliums und Silbers. [708] R. Störmer, Ber. *44*, 637 (1911), C.A. *5*, 1783 (1911); Umlagerung der stabilen Stereoisomeren in labile durch UV-Licht. [709] R. Störmer, F. Wodarg, Ber. *61*, 2323 (1928), C.A. *23*, 1396³ (1929); Zur Frage der flüssigen Krystalle bei stereoisomeren Zimtsäuren. [710] E. L. Strebel, Ger. Offen. 2,017,727 (17. 4. 1969), C.A. *74*, 42155n (1971); Thermotropic nematic N-benzylideneanilines.

[711] L. Strzelecki, L. Liebert, Bull. Soc. Chim. Fr. Pt. 2, *1973*, 605, C.A. *79*, 66839t (1973); Monomères mésomorphes. Polymérisation de la *p*-acryloyloxybenzylidène *p*-carboxyaniline. [712] L. Strzelecki, L. Liebert, D. Vacogne, Bull. Soc. Chim. Fr. Pt. 2, *1974*, 2849, C.A. *83*, 192309r (1975); Monomères mésomorphes. Le mésomorphisme des *p*-acryloyl (ou méthacryloyl) oxybenzylidène *p*-alkyl (ou alkyloxy) anilines. [713] L. Strzelecki, L. Liebert, P. Keller, Bull. Soc. Chim. Fr. Pt. 2, *1975*, 2750, C.A. *84*, 165233a (1976); Série homologue des *p*-acryloyloxybenzylidène *p*-aminoalkylcinnamates. [714] L. Strzelecki, P. Keller, L. Liebert, H. Guyot, C. R. Acad. Sci., Sér. C, *283*, 219 (1976), C.A. *86*, 71053f (1977); Synthèse de nouveaux monomères mésomorphes, optiquement actifs. [714a] S. Sugimori, T. Inukai, H. Sato, Japan. Kokai 77 17,438 (30. 7. 1975), C.A. *87*, 84690g (1977); Methylhydroquinone 1,4-bis(4'-substituted benzoates). [714b] S. Sugimori, T. Inukai, H. Sato, Japan. Kokai 77 17,441 (30. 7. 1975), C.A. *87*, 84696p (1977); 4-Cyanophenyl 4-substituted benzoates. [715] C. Sultze, Diss. Halle 1908; Dielektrizitätskonstante flüssiger Kristalle. [716] H. H. Sutherland, T. G. Hoy, Acta Cryst. B *24*, 1207 (1968), C.A. *69*, 90992b (1968); Crystal structure of 4-acetyl-2'-chlorobiphenyl. [717] Suwa Seikosha, Brit. 1,334,961 (29. 12. 1970), C.A. *80*, 53791s (1974); Nematic liquid crystal. [718] T. Suzuki, Japan. Kokai 74 28,588 (13. 7. 1972), C.A. *81*, 97823d (1974); Liquid crystal compositions for TN display devices. [719] T. Suzuki, Japan. Kokai 75,108,230 (7. 2. 1974), C.A. *84*, 37556w (1976); Liquid crystals having azoxy linkage. [720] T. Suzuki, Japan. Kokai 75,111,031 (13. 2. 1974), C.A. *84*, 73884k (1976); *p*-Alkyl-*p'*-alkoxyazoxybenzene liquid crystals.

[721] K. Tabei, E. Saitou, Bull. Chem. Soc. Jap. *42*, 1440 (1969), C.A. *71*, 44203p (1969); NMR and IR spectra of aromatic azomethines. [722] W. Täglich, Diss. Halle 1923; Abkömmlinge der *p*-Oxyben-

zoesäure. [**723**] H. Tanaka, K. Maruyama, E. Yasuda, S. Kimura, Tanso *86*, 107 (1976); C.A. *86*, 92867q (1977); Effect of carbon felt on mesophase formation of petroleum pitch. [**723a**] K. Tanaka, K. Kuwahara, S. Akiyama, T. Akiyama, Y. Ueno, Japan. Kokai 77 71,452 (8. 12. 1975), C.A. *87*, 167769r (1977); 4-Alkyl-4'-cyanostilbenes. [**724**] Y. Tanaka, H. Shiozaki, Y. Shimura, A. Okada, Japan. Kokai 73 04,457 (1. 6. 1971), C.A. *78*, 136543x (1973); Cholesteryl α,β-unsaturated esters. [**725**] L. Tanguy, M. Leclercq, J. Billard, J. Jacques, Bull. Soc. Chim. Fr. Pt. 2, *1974*, 640, C.A. *81*, 160469k (1974); Dérivés mésomorphes du *trans* stilbène et du *trans* hexahydrochrysène. [**726**] R. F. Tarvin, F. W. Neetzow, J. Chem. Soc., Chem. Commun. *1973*, 396, C.A. *79*, 71010v (1973); Unsymmetrical liquid crystalline esters. [**727**] H. Tatsuta, M. Fukai, K. Asai, H. Takahashi, Ger. Offen. 2,352,151 (20. 10. 1972), C.A. *81*, 56700t (1974); Display device with liquid crystals. [**728**] H. Tatsuta, M. Fukai, H. Takahashi, Ger. Offen. 2,535,125 (6. 8. 1974), C.A. *85*, 54643k (1976); Liquid crystal composition. [**729**] I. Teucher, C. M. Paleos, M. M. Labes, Mol. Cryst. Liq. Cryst. *11*, 187 (1970), C.A. *74*, 16575e (1971); Properties of structurally stabilized anil-type nematics. [**730**] C. Thinius, Diss. Halle 1928; Bestimmung der Umwandlungspunkte polymorpher kristalliner Flüssigkeiten mit der Roberts-Austenschen Differentialmethode.

[**731**] V. G. Tishchenko, R. M. Cherkashina, Kholestericheskie Zhidk. Krist. *1976*, 26, C.A. *87*, 175704f (1977); Molecular structure and mesomorphism of steroid derivatives. [**732**] V. V. Titov, E. I. Kovshev, A. I. Pavluchenko, V. T. Lazareva, M. F. Grebenkin, J. Phys. (Paris), C-1, *36*, 387 (1975); Synthesis and properties of nematics exhibiting a positive dielectric anisotropy. [**733**] V. V. Titov, E. I. Kovshev, L. A. Karamysheva, G. Purvaneckas, V. Dauksas, Khim. Geterotsikl. Soedin. *1975*, 1364, C.A. *84*, 30837e (1976); Reaction of *p*-alkylbenzaldehydes with *p*-aminobenzonitrile. [**734**] J. Toussaint, Bull. Soc. Chim. Belg. *54*, 319 (1945), C.A. *41*, 2297h (1947); Structures moléculaires de quelques composés aromatiques sulfurés (4,4'-dibromdiphenylsulfid). [**735**] J. Toussaint, Bull. Soc. Roy. Sci. Liège *15*, 86 (1946), C.A. *42*, 7128e (1948); Structure cristalline de l'oxyde de phenyle *p*-dibrome. [**736**] J. Trotter, Acta Cryst. *13*, 732 (1960), C.A. *55*, 54i (1961); Crystal structure of 1-naphthoic acid. [**737**] J. Trotter, Acta Cryst. *14*, 101 (1961), C.A. *55*, 8992i (1961); Crystal structure of 2-naphthoic acid. [**738**] J. Trotter, Acta Cryst. *14*, 1135 (1961), C.A. *56*, 4189c (1962); Crystal and molecular structure of biphenyl. [**739**] K. Tsuji, M. Sorei, S. Seki, Bull. Chem. Soc. Jap. *44*, 1452 (1971), C.A. *75*, 54876g (1971); Glassy liquid crystal. Nonequilibrium state of cholesteryl hydrogen phthalate. [**740**] K. Tsukamoto, T. Otsuka, K. Morimoto, Y. Murakami, Japan. Kokai 75,158,580 (11. 6. 1974), C.A. *85*, 170116q (1976); Liquid crystals.

[**741**] H. Tsutsui, K. Fukuda, S. Oi, Y. Yamada, H. Honda, Tanso *82*, 87, 93 (1975), C.A. *84*, 7385x, 7386y (1976); Influence of coexistent materials on the formation of carbonaceous mesophase in pitch. 1. Behavior of mesocarbon microbeads (type-C) in pitch. 2. Correlation between mesocarbon microbeads added and mesophase formed in pitch. [**742**] L. S. Tyan, A. S. Fialkov, Y. V. Aleksandrov, V. I. Perepechenykh, Dokl. Akad. Nauk SSSR *225*, 384 (1975), C.A. *84*, 52728x (1976); Switching effect in pitch mesophase. [**743**] J. N. Andrews, A. R. Ubbelohde, Proc. Roy. Soc. A *228*, 435 (1955), C.A. *49*, 14408i (1955); Melting and crystal structure. [**744**] A. R. Ubbelohde, H. J. Michels, J. J. Duruz, Nature *228*, 50 (1970), C.A. *73*, 134914s (1970); Liquid crystals in molten salt systems. [**745**] A. R. Ubbelohde, Nature *244*, 487 (1973), C.A. *79*, 108877m (1973); Organic ionic melts. [**746**] A. R. Ubbelohde, Rev. Int. Hautes Temp. Refract. *13*, 5 (1976), C.A. *86*, 4338q (1977); Organic ionic melts. [**747**] M. Ueda, F. Hori, Japan. Kokai 73 62,733 (8. 12. 1971), C.A. *79*, 146223c (1973); 2-Chloro-4-alkoxyphenyl 4-alkoxybenzoate liquid crystals. [**748**] United Kingdom Secretary of State for Defence, London, Ger. Offen. 2,356,085 (9. 11. 1972), C.A. *81*, 96988z (1974); Liquid crystal materials. [**749**] J. Van der Veen, A. H. Grobben, Mol. Cryst. Liq. Cryst. *15*, 239 (1971), C.A. *76*, 77726t (1972); Conformation of aromatic Schiff bases in connection with mesomorphism. [**750**] J. Van der Veen, W. H. de Jeu, A. H. Grobben, J. Boven, Mol. Cryst. Liq. Cryst. *17*, 291 (1972), C.A. *77*, 67322q (1972); Low-melting mesomorphic *p,p'*-di-*n*-alkylazoxy- and azobenzenes.

[**751**] J. Van der Veen, T. C. J. Hegge, Angew. Chem. *86*, 378 (1974), C.A. *81*, 63267k (1974); Niedrigschmelzende, flüssigkristalline *o*-Hydroxyazo- und *o*-Hydroxyazoxybenzole. [**752**] J. Van der Veen, W. H. de Jeu, Mol. Cryst. Liq. Cryst. *27*, 251 (1974), C.A. *82*, 178606h (1975); Liquid crystalline bridge substituted Schiff bases. [**753**] J. Van der Veen, J. Phys. (Paris), C-1, *36*, 375 (1975), C.A. *83*, 69394n (1975); Influence of terminal substituents upon the n/i transition temperature. [**754**] J. Van der Veen, J. Phys. (Paris), C-3, *37*, 13 (1976), C.A. *86*, 43367t (1977); Liquid crystalline isothiocyanates. [**754a**] G. V. Vani, K. Vijayan, Mol. Cryst. Liq. Cryst. *42*, 249 (1977); Crystal and molecular structure of CBOOA. [**755**] J. P. Van Meter, Org. Chem. Bull. *45*, 1 (1973), C.A. *78*, 89379m (1973); Chemistry

of liquid crystals. [**756**] J. P. Van Meter, B. H. Klanderman, J. Am. Chem. Soc. *95*, 626 (1973), C.A. *78*, 76872s (1973); Low-melting mesomorphic phenyl 4-benzoyloxybenzoates. [**757**] J. P. Van Meter, B. H. Klanderman, Mol. Cryst. Liq. Cryst. *22*, 271 (1973), C.A. *79*, 140470j (1973); Mesomorphism of phenyl benzoates. [**758**] J. P. Van Meter, B. H. Klanderman, Mol. Cryst. Liq. Cryst. *22*, 285 (1973), C.A. *79*, 145478c (1973); Mesomorphism of phenyl 4-benzoyloxybenzoates. [**759**] J. P. Van Meter, Chem. Double-Bonded Funct. Groups *1*, 93 (1977), C.A. *87*, 38266a (1977); Liquid crystals with X=Y groups. [**760**] J. P. Van Meter, A. K. Seidel, J. Org. Chem. *40*, 2998 (1975), C.A. *83*, 171086a (1975); Low-melting nematic phenyl 4-benzoyloxybenzoates.

[**761**] J. P. Van Meter, B. H. Klanderman, U.S. 3,971,824 (26. 4. 1972), C.A. *86*, 72220v (1977); *p*-Benzoyloxybenzoate. [**762**] P. van Romburgh, Proc. Kon. Nederl. Akad. Wetenschapen *9*, 9 (1901); Over het gekristalliseerde bestanddeel van de aetherische olie van Kaempferia Galanga L. [**763**] W. Vaupel, Diss. Halle 1911; Dielektrizitätskonstante flüssiger Kristalle. [**764**] L. Verbit, R. L. Tuggey, 3rd Int. Liq. Cryst. Conf., Berlin, Aug. 1970 M 4.1; Computer program for liquid crystal searching. [**765**] L. Verbit, R. L. Tuggey, Mol. Cryst. Liq. Cryst. *17*, 49 (1972), C.A. *77*, 74700q (1972); Mesomorphism of acetylene derivatives. [**766**] L. Verbit, R. L. Tuggey, Liq. Cryst. Ord. Fluids *2*, 307 (1974), C.A. *86*, 180987j (1977); Effect of certain central groups on the mesomorphism of dicarboxylic esters. [**767**] L. Verbit, G. A. Lorenzo, Mol. Cryst. Liq. Cryst. *30*, 87 (1975), C.A. *83*, 171084y (1975); Mesomorphism of urethanes. [**768**] L. Verbit, R. L. Tuggey, A. R. Pinhas, Mol. Cryst. Liq. Cryst. *30*, 201 (1975), C.A. *83*, 130668y (1975); Mesomorphism of dicarboxylic esters. [**769**] L. Verbit, T. R. Halbert, Mol. Cryst. Liq. Cryst. *30*, 209 (1975), C.A. *83*, 130669z (1975); Thermotropic behavior of ferrocene-carboxalde-hyde and some ferrocene Schiff's base derivatives. [**770**] K. Vijayan, G. V. Vani, Liq. Cryst., Proc. Int. Conf., Bangalore, *1973*, 75, C.A. *84*, 114665k (1976); Crystal structure of *n-p*-methoxybenzylidene-*p*-phenylazoaniline.

[**771**] R. Viterbo, M. Mastursi, G. C. Perri, S. African 72 05,587 (14. 8. 1972), C.A. *80*, 14642z (1974); Mannich bases of cyclopentanones and cyclopent-2-enones. [**772**] R. D. Vold, M. J. Vold, J. Am. Chem. Soc. *61*, 808 (1939), C.A. *33*, 9108[7] (1939); Successive phases in the transformation of anhydrous sodium palmitate from crystal to liquid. [**773**] R. D. Vold, F. B. Rosevear, R. H. Ferguson, Oil Soap *16*, 48 (1939), C.A. *33*, 3619[6] (1939); New allotropic forms of anhydrous sodium palmitate. [**774**] M. J. Vold, M. Macomber, R. D. Vold, J. Am. Chem. Soc. *63*, 168 (1941), C.A. *35*, 1658[2] (1941); Stable phase occurring between true crystals and true liquids for single pure anhydrous soaps. [**775**] R. D. Vold, J. Am. Chem. Soc. *63*, 2915 (1941), C.A. *36*, 323[8] (1942); Anhydrous sodium soaps: Heats of transition and phase classification. [**776**] R. D. Vold, M. J. Vold, J. Phys. Chem. *49*, 32 (1945), C.A. *39*, 1588[6] (1945); Thermal transitions of the alkali palmitates. [**777**] R. D. Vold, J. Phys. Chem. *49*, 315 (1945), C.A. *40*, 225[1] (1946); Polymorphism and transitions of anhydrous and hydrous sodium stearate. [**778**] R. D. Vold, J. D. Grandine, M. J. Vold, J. Colloid Sci. *3*, 339 (1948), C.A. *43*, 29c (1949); Polymorphic transformations of calcium stearate and calcium stearate monohydrate. [**779**] M. J. Vold, H. Funakoshi, R. D. Vold, J. Phys. Chem. *80*, 1753 (1976), C.A. *85*, 85750x (1976); Polymorphism of lithium palmitate. [**780**] R. A. Vora, Curr. Sci. *45*, 538 (1976), C.A. *86*, 4632f (1977); Effect of substitution and its location on the mesomorphism of cholesterylbenzoate and cinnamate.

[**781**] R. A. Vora, J. Maharaja Sayajirao Univ. Baroda *24*, 53 (1976), C.A. *86*, 131420k (1977); Effect of molecular structure on cholesteric and nematic properties. [**782**] D. Vorländer, Ber. *39*, 803 (1906); Krystallinisch-flüssige Substanzen. [**783**] D. Vorländer, Z. physik. Chem. *57*, 357 (1907), C.A. *1*, 266[6] (1907); Neue Erscheinungen beim Schmelzen und Kristallisieren. [**784**] D. Vorländer, Ber. *40*, 1415 (1907), C.A. *1*, 1813 (1907); Substanzen mit mehreren festen und mehreren flüssigen Phasen. [**785**] D. Vorländer, A. Gahren, Ber. *40*, 1966 (1907), C.A. *1*, 2114[1] (1907); Latent kristallin-flüssige Substanzen. [**786**] D. Vorländer, Ber. *40*, 1970 (1907), C.A. *1*, 2114[4] (1907); Einfluß der molekularen Gestalt auf den kristallinisch-flüssigen Zustand. [**787**] D. Vorländer, Ber. *40*, 4527 (1907), C.A. *2*, 743[7] (1908); Polymorphie von Flüssigkeiten. [**788**] D. Vorländer, Sammlung chem. und chem.-techn. Vorträge *12*, 321 (1907); Kristallinisch-flüssige Substanzen. [**789**] D. Vorländer, Ber. *41*, 2033 (1908), C.A. *2*, 2798[7] (1908); Durchsichtig klare, krystallinische Flüssigkeiten. Achsenbilder flüssiger Krystalle. [**790**] D. Vorländer, Z. physik. Chem. *61*, 166 (1908), C.A. *2*, 743[5] (1908); Systembestimmung und Achsenbilder flüssiger Kristalle.

[**791**] D. Vorländer, Ber. *43*, 3120 (1910), C.A. *5*, 506[9] (1911); Verhalten der Salze organischer Säuren beim Schmelzen. [**792**] D. Vorländer, Z. physik. Chem. *85*, 701 (1913), C.A. *8*, 1225[9] (1914); Kolloide Lösungen von Farbstoffen und von Kolophonium in flüssigen Kristallen. [**793**] D. Vorländer, Physik. Z. *15*, 141 (1914), C.A. *8*, 1538[2] (1914); Optische Anisotropie flüssiger Kristalle. [**794**] D.

Vorländer, Z. physik. Chem. *93*, 516 (1919), C.A. *13*, 2789[5] (1919); Flüssige Kristalle. [**795**] D. Vorländer, I. Ernst, Z. physik. Chem. *93*, 521 (1919), C.A. *13*, 2796 (1919); Rhythmisches Erstarren. [**796**] D. Vorländer, Physik. Z. *21*, 590 (1920), C.A. *15*, 1429[6] (1921); Ableitung der molekularen Gestalt aus dem kristallinisch-flüssigen Zustand. [**797**] D. Vorländer, Ber. *54*, 2261 (1921), C.A. *16*, 188[5] (1922); Krystallinisch-flüssige Eigenschaften α-ungesättigter Ketone. [**798**] D. Vorländer, Z. angew. Chem. *35*, 249 (1922), C.A. *16*, 2800[2] (1922); Ableitung der molekularen Gestalt aus dem kristallinisch-flüssigen Zustand. [**799**] D. Vorländer, Z. physik. Chem. *105*, 211 (1923), C.A. *18*, 1072[3] (1924); Erforschung der molekularen Gestalt mit Hilfe der kristallinischen Flüssigkeiten. [**800**] D. Vorländer, Ber. *58*, 118 (1925), C.A. *19*, 1413 (1925); Säure, Salz, Ester, Addukt.

[**801**] D. Vorländer, E. Fischer, K. Kunze, Ber. *58*, 1284 (1925), C.A. *19*, 2940 (1925); 5-Phenylpenta-dienal-(1) und 7-Phenylheptatrienal-(1). [**802**] D. Vorländer, Ber. *58*, 1893 (1925), C.A. *20*, 584 (1926); Die Lehre von den innermolekularen Gegensätzen und die Lenkung der Substituenten im Benzol (II). [**803**] D. Vorländer, U. Haberland, Ber. *58*, 2652 (1925), C.A. *20*, 1168 (1926); Mikrobestimmung von Schmelz- und Übergangspunkten. [**804**] D. Vorländer, W. Zeh, H. Enderlein, Ber. *60*, 849 (1927), C.A. *21*, 1972 (1927); Symmetrisches Bisbenzolazoäthylen. [**805**] D. Vorländer, Z. physik. Chem. *126*, 449 (1927), C.A. *24*, 2026[3] (1930); Die Natur der Kohlenstoffketten in flüssigen Kristallen. [**806**] D. Vorländer, W. Selke, Z. physik. Chem. *129*, 435 (1927), C.A. *24*, 4198[1] (1930); Einachsige Aufrichtung von festen weichen Kristallmassen und von kristallinen Flüssigkeiten. [**807**] D. Vorländer, Naturwissen-schaften *16*, 759 (1928), C.A. *23*, 765[7] (1929); Absorptionsspektren kristalliner Flüssigkeiten. [**808**] D. Vorländer, E. Daehn, Ber. *62*, 541 (1929), C.A. *23*, 3687 (1929); 5-Phenylpentadienal-(1) und 7-Phenylhep-tatrienal-(1). [**809**] D. Vorländer, E. Daehn, Ber. *62*, 545 (1929), C.A. *23*, 3688 (1929); 7-Phenylheptatrien-säure. [**810**] D. Vorländer, Ber. *62*, 2824 (1929), C.A. *24*, 1358[2] (1930); Richtung der Kohlenstoffvalenzen in Methanabkömmlingen.

[**811**] D. Vorländer, Ber. *62*, 2831 (1929), C.A. *24*, 1358[9] (1930); Richtung der Kohlenstoffvalenzen in Benzolabkömmlingen. [**812**] D. Vorländer, K. Gieseler, J. prakt. Chem. *121*, 237 (1929), C.A. *23*, 3911 (1929); *p*-Methoxyzimtaldehyd und *p*-Methoxy-5-phenylpentadienal-1. [**813**] D. Vorländer, K. Gieseler, J. prakt. Chem. *121*, 247 (1929), C.A. *23*, 3912 (1929); *p*-Methoxycinnamylidenessigsäure. [**814**] D. Vorländer, Z. angew. Chem. *43*, 13 (1930), C.A. *24*, 4944 (1930); Amorphe und kristalline Harze und Lacke. [**815**] D. Vorländer, Physik. Z. *31*, 428 (1930), C.A. *24*, 4433[2] (1930); Optisch leere, kristalline Flüssigkeiten und verschiedene Arten der kristallinen Flüssigkeiten. [**816**] D. Vorländer, Z. Krist. *79*, 61 (1931), C.A. *26*, 2357[8] (1932); Chemie der kristallinen Flüssigkeiten. [**817**] D. Vorländer, Z. Krist. *79*, 274 (1931); Kristalline Flüssigkeiten. [**818**] D. Vorländer, A. Apel, Ber. *65*, 1101 (1932), C.A. *26*, 5083[4] (1932); Richtung der Kohlenstoffvalenzen in Benzolabkömmlingen II. [**819**] D. Vorländer, Ber. *66*, 1666 (1933), C.A. *28*, 470[7] (1934); *p*-Azoxybiphenyl und *p*-Azobiphenyl. [**820**] D. Vorländer, Trans. Faraday Soc. *29*, 899 (1933), C.A. *28*, 1238[3] (1934); Systems with mixed dimensions.

[**821**] D. Vorländer, Trans. Faraday Soc. *29*, 902 (1933), C.A. *28*, 1241[1] (1934); Supracrystallinity of *p*-azoxybenzoic acid. [**822**] D. Vorländer, Trans. Faraday Soc. *29*, 907 (1933), C.A. *28*, 1205[3] (1934); Liquocrystalline resins and lacquers. [**823**] D. Vorländer, Trans. Faraday Soc. *29*, 910 (1933), C.A. *28*, 1242[4] (1934); Cyclo-pentanone and cyclo-hexanone. [**824**] D. Vorländer, Trans. Faraday Soc. *29*, 913 (1933), C.A. *28*, 1238[3] (1934); Polymorphism of liquid crystals. A demonstration. [**825**] D. Vorländer, Naturwissenschaften *21*, 781 (1933), C.A. *28*, 2235[1] (1934); Kristalle und Molekeln als mischdimensionale Systeme. [**826**] D. Vorländer, H. Schuster, J. prakt. Chem. *140*, 193 (1934), C.A. *28*, 6428 (1934); Isomere *p*-Aminoazoxybenzole. [**827**] D. Vorländer, A. Fröhlich, Ber. *67*, 1556 (1934), C.A. *29*, 1400[2] (1935); Isomerie der krystallin-flüssigen *p*-Phenetolazoxy-benzoesäureester. [**828**] D. Vorländer, Ber. *68*, 453 (1935), C.A. *29*, 3671[4] (1935); *p*-Phenylzimtsäure. [**829**] D. Vorländer, Naturwissenschaften *24*, 113 (1936), C.A. *30*, 4374[6] (1936); Suprakristalline organische Verbindungen. [**830**] D. Vorländer, Ber. *70*, 1202 (1937), C.A. *31*, 6524[8] (1937); Polymorphie der kristallinen Flüssigkeiten.

[**831**] D. Vorländer, Ber. *70*, 2096 (1937), C.A. *32*, 3355[7] (1938); Krystallin-flüssige Kombinationen von *p*-Azozimtsäureestern mit *p*-Azophenolderivaten; zur Kenntnis der Assoziations-Vorgänge. [**832**] D. Vorländer, R. Wilke, H. Hempel, U. Haberland, J. Fischer, Z. Krist. *97*, 485 (1937), C.A. *32*, 2011[8] (1938); Kristallin-flüssige und feste Formen des Anisal-*p*-amino-zimtsäure-äthylesters. [**833**] D. Vorländer, R. Wilke, U. Haberland, K. Thinius, H. Hempel, J. Fischer, Ber. *71*, 501 (1938), C.A. *32*, 4847[4] (1938); Polymorphie der krystallin-flüssigen Aryliden-*p*-amino-zimtsäureester; zur Kenntnis der Assoziationsvor-gänge. [**834**] D. Vorländer, Ber. *71*, 1688 (1938), C.A. *33*, 30[7] (1939); Schmelzpunktserniedrigung durch gemischte Struktur der Molekeln von kristallin-festen und kristallin-flüssigen Substanzen. [**835**] F. Wallerant, Compt. rend. *143*, 694 (1906), C.A. *1*, 320[6] (1907); Cristaux liquides d'oléate d'ammonium.

[836] R. Feldtkeller, R. Walter, Z. Krist. 60, 349 (1924), C.A. 19, 1072[4] (1925); Strukturbeobachtungen an der zweiten kristallinflüssigen Phase des Anisal-p-aminozimtsäureäthylesters. [837] R. Walter, Ber. 58, 2303 (1925), C.A. 20, 528[6] (1926); Kenntnis der flüssigen Mischkristalle. [838] R. Walter, Ber. 59, 962 (1926), C.A. 20, 2817[3] (1926); Krystallinisch-flüssige Thallosalze organischer Säuren. [839] L. W. Wasum, Diss. Halle 1928; Optisch aktive pleochroitische kristallin-flüssige Substanzen. [839a] H. Watanabe, D. Nakagawa, M. Miyamura, M. Hashimoto, Japan. Kokai 76,136,630 (23. 5. 1975), C.A. 87, 5617n (1977); Liquid crystal materials. [840] T. Watanabe, S. Kanbe, Japan. Kokai 75 92,872 (19. 12. 1973), C.A. 84, 128796r (1976); Benzoyloxychloroazobenzene liquid crystal compositions for display devices.

[841] F. H. Weber, Diss. Halle 1914; p-Substitutionsprodukte des Diphenyläthers, Diphenylthioäthers und Diphenylamins. [842] W. Weissflog, H. Heberer, K. Mohr, H. Zaschke, H. Kresse, S. Koenig, D. Demus, Ger. (East) 116,732 (5. 12. 1975), C.A. 84, 187562j (1976); Nematic liquid crystals. [843] W. Weißwange, Diss. Halle 1925; I. Versuche mit Trichlormethylsulfonsäurechlorid. II. Neue Abkömmlinge des Dehydrothio-p-toluidins. [844] J. H. Wendorff, F. P. Price, Mol. Cryst. Liq. Cryst. 22, 85 (1973), C.A. 79, 109066h (1973); X-ray diffraction study of crystalline cholesteryl myristate and stearate. [845] C. Weygand, R. Gabler, Ber. 71, 2399 (1938), C.A. 33, 2008[9] (1939); Thermische Persistenz von flüssigen Kristallen. [846] C. Weygand, W. Lanzendorf, J. prakt. Chem. 151, 204 (1938), C.A. 33, 562[6] (1939); Stereoisomere, homologe Dibenzoyläthylene und zugehörige Dibenzoyläthane. [847] C. Weygand, R. Gabler, J. prakt. Chem. 151, 215 (1938), C.A. 33, 153[3] (1939); Auftreten von mehr als zwei polymorphen, kristallin-flüssigen Phasen bei Azomethinen. [848] C. Weygand, W. Lanzendorf, J. prakt. Chem. 151, 221 (1938), C.A. 33, 153[4] (1939); Flüssige Kristalle: Homologe p,p'-Diphenyl-pyridazine. [849] C. Weygand, R. Gabler, Z. physik. Chem. B 44, 69 (1939), C.A. 33, 9068[6] (1939); Verzögerung von Umordnungen zwischen gewöhnlichen und kristallinen Flüssigkeiten sowie zwischen deren Typen untereinander. [850] C. Weygand, R. Gabler, Naturwissenschaften 27, 28 (1939), C.A. 33, 4099[8] (1939); Einfluß des Molekülbaus auf das Vorkommen von kristallinen Flüssigkeiten.

[851] C. Weygand, R. Gabler, J. prakt. Chem. 155, 332 (1940), C.A. 35, 1775[9] (1941); Höhere Homologe der Azophenol-, Azoxyphenol- und Azomethinphenoläther. [852] C. Weygand, R. Gabler, Z. physik. Chem. B 46, 270 (1940), C.A. 35, 947[2] (1941); Die einfachsten flüssigen Kristalle; chemische Morphologie der Flüssigkeiten. [853] C. Weygand, R. Gabler, Z. physik. Chem. B 48, 148 (1941), C.A. 35, 3869[2] (1941); Deutung der Klär- und Umwandlungspunkts-Regelmäßigkeiten in homologen Reihen von flüssigen Kristallen. [854] C. Weygand, R. Gabler, J. Hoffmann, Z. physik. Chem. B 50, 124 (1941), C.A. 36, 2459[8] (1942); Kristallin-flüssige aliphatische Monocarbonsäuren. [855] C. Weygand, R. Gabler, N. Bircan, J. prakt. Chem. 158, 266 (1941), C.A. 36, 1023[5] (1942); Flüssige Kristalle mit neuartigen Flügelgruppen vom Typus RO(CH₂)ₙO—. [856] C. Weygand, Naturwissenschaften 31, 571 (1943), C.A. 40, 4930[9] (1946); Molekulare Ordnung in kristallinen Flüssigkeiten. [857] C. Weygand, Z. physik. Chem. 53B, 75 (1943), C.A. 37, 4607[5] (1943); Formbeständige, isolierte Flüssigkristalle. [858] C. Weygand, Ber. A 76, 41 (1943); D. Vorländer (Nachruf und Bibliographie). [859] J. L. White, G. L. Guthrie, J. O. Gardner, Carbon 5, 517 (1967), C.A. 68, 31877z (1968); Mesophase microstructures in carbonized coal-tar pitch. [860] J. L. White, R. J. Price, Bienn. Conf. Carbon, Ext. Abstr. Program, 11th, 1973, 209, (CONF-730601), C.A. 83, 100507c (1975); Mesophase behavior during coke formation.

[861] J. L. White, R. J. Price, Carbon 12, 321 (1974), C.A. 82, 33092c (1975); Mesophase microstructures during the pyrolysis of selected coker feedstocks. [862] J. L. White, U.S. Nat. Tech. Inform. Serv., AD Rep., No. 777814/5GA, 1974, 119 pp., C.A. 81, 83041k (1974); Microstructure in graphitizable carbons. [863] J. L. White, ACS Symp. Ser. 21, 282 (1976), C.A. 84, 182225s (1976); Mesophase mechanism in the formation of the microstructure of petroleum coke. [864] J. L. White, J. E. Zimmer, Surf. Defect Prop. Solids 5, 16 (1976), C.A. 86, 123950t (1977); Disclination structures in carbonaceous mesophase and graphite. [865] C. Wiegand, E. Merkel, Z. Naturforsch. 3b, 313 (1948), C.A. 43, 6025g (1949); Flüssige Kristalle und ebener Molekelbau. [866] C. Wiegand, Z. Naturforsch. 4b, 249 (1949), C.A. 44, 2309f (1950); Beziehungen zwischen dem räumlichen Bau und der Bildung flüssiger Kristalle bei Cholesterin-Derivaten. [867] C. Wiegand, Angew. Chem. 63, 127 (1951); Flüssige Kristalle und Molekelbau. [868] C. Wiegand, Z. Naturforsch. 6b, 240 (1951); Dianisalbenzidine. [869] C. Wiegand, Z. Naturforsch. 9b, 516 (1954), C.A. 50, 13543e (1956); Dianisalnaphthylen-diamine. [870] C. Wiegand, in Houben-Weyl-Müller: Methoden der Organischen Chemie Bd. 3/1, 681 (1955) (Thieme: Stuttgart); Charakterisierung und Untersuchungsmethoden flüssiger Kristalle.

[871] C. Wiegand, Z. Naturforsch. 12b, 512 (1957), C.A. 52, 3446a (1958); Konstitutionseinfluß auf Schmelzpunkt und Klärpunkt flüssiger Kristalle. [872] R. Wilke, Diss. Halle 1909; Polymorphie

der kristallinischen Flüssigkeiten und kristallinische Harze. [**873**] H. Wille, Diss. Halle 1927; Symmetrisches Triphenylbenzol. [**873a**] J. M. Wilson, R. Harden, J. Phillips, Mol. Cryst. Liq. Cryst. *34*, 237 (1977), C.A. *87*, 76622w (1977); Plastic crystal state of ferrocenecarboxaldehyde. [**874**] F. Wodarg, Diss. Rostock 1925; Zur Frage der flüssigen Kristalle bei stereoisomeren Zimtsäuren. [**875**] F. Würstlin, Z. Krist. *88*, 185 (1934), C.A. *28*, 5735⁵ (1934); Strukturbestimmung von PAA. [**876**] Y. Yamada, H. Honda, Tanso *73*, 51 (1973), C.A. *79*, 109816c (1973); Carbonaceous mesophase spherules with new molecular arrangement. [**877**] Y. Yamada, H. Honda et al., Bienn. Conf. Carbon, Ext. Abstr. Program, 11th, *1973*, 213, 215, 217, 219, C.A. *83*, 45633p, 82404k, 82405m, 63167u (1975); Characteristics of meso-carbon microbeads separated from pitches. Characteristics of heat-treated meso-carbon microbeads. High-density isotropic carbon solids made of mesocarbon microbeads. Carbonaceous mesophase spherules having a new molecular arrangement. [**878**] Y. Yamazaki, Japan. Kokai 73 20,787 (21. 7. 1971), C.A. *79*, 36030e (1973); Nematic azobenzene derivatives. [**879**] Y. Yamazaki. Kokai 73 36,128 (17. 9. 1971), C.A. *79*, 42134h (1973); Nematic azobenzene derivatives. [**880**] D. W. Young, P. Tollin, H. H. Sutherland, Acta Cryst. B *24*, 161 (1968), C.A. *68*, 82140m (1968); Crystal structure of 4-acetyl-2′-fluorobiphenyl.

[**881**] W. R. Young, I. Haller, L. Williams, Liq. Cryst. Ord. Fluids *1970*, 383, C.A. *78*, 152450v (1973); Mesomorphic heterocyclic analogs of benzylidene-4-amino-4′-methoxybiphenyl. [**882**] W. R. Young, Mol. Cryst. Liq. Cryst. *10*, 237 (1970), C.A. *73*, 66176a (1970); Nitrones: a novel class of liquid crystals (first description by D. Vorländer, 784). [**883**] W. R. Young, A. Aviram, R. J. Cox, Angew. Chem. *83*, 399 (1971), C.A. *75*, 88223g (1971); Nichtplanare trans-Stilbene mit nematischer Phase bei Raumtemperatur. [**884**] W. R. Young, I. Haller, D. C. Green, Mol. Cryst. Liq. Cryst. *13*, 305 (1971), C.A. *75*, 123297h (1971); Mesogens containing some group IV elements. [**885**] W. R. Young, I. Haller, A. Aviram, Mol. Cryst. Liq. Cryst. *13*, 357 (1971), C.A. *75*, 123296g (1971); Preparation and thermodynamics of some homologous nitrones. [**886**] W. R. Young, I. Haller, A. Aviram, Mol. Cryst. Liq. Cryst. *15*, 311 (1972), C.A. *76*, 64629p (1972); Mesomorphism in the 4,4′-dialkoxy-*trans*-stilbenes. [**887**] W. R. Young, A. Aviram, R. Cox, J. Am. Chem. Soc. *94*, 3976 (1972), C.A. *77*, 47974a (1972); Stilbene derivatives. New class of room temperature nematics. [**888**] W. R. Young, I. Haller, D. C. Green, J. Org. Chem. *37*, 3707 (1972), C.A. *78*, 29107m (1973); Mesomorphism of ring-methylated phenyl benzoyloxybenzoates. [**889**] W. R. Young, D. C. Green, Mol. Cryst. Liq. Cryst. *26*, 7 (1974), C.A. *81*, 127902y (1974); Low-melting nematic phenyl 3-methyl-benzoyloxybenzoates. [**890**] R. J. Cox, W. R. Young, A. Aviram, Ger. Offen. 2,165,700 (31. 12. 1970), C.A. *77*, 114084a (1972); Nematic, thermotropic stilbenes for electrooptical devices.

[**891**] D. C. Green, W. R. Young, Ger. Offen. 2,352,368 (24. 10. 1972), C.A. *81*, 128105c (1974); Nematic liquid crystals. [**892**] J. G. J. Ypma, G. Vertogen, J. Phys. (Paris) *37*, 1331 (1976), C.A. *86*, 10855x (1977); Effect of permanent dipoles on the n/i transition. [**893**] H. Zaschke, J. Prakt. Chem. *317*, 617 (1975), C.A. *83*, 193218x (1975); Niedrigschmelzende kristallinflüssige 5-*n*-Alkyl-2-(4-*n*-alkoxyphenyl)-pyrimidine. [**894**] H. Zaschke, R. Stolle, Z. Chem. *15*, 441 (1975), C.A. *84*, 90108c (1976); Niedrigschmelzende kristallinflüssige 5-*n*-Alkyl-2-[4-*n*-alkanoyloxyphenyl]pyrimidine. [**894a**] H. Zaschke, Z. Chem. *17*, 63 (1977), C.A. *87*, 5897d (1977); Flüssig-kristalline 2-Cyanopyrimidine. [**895**] W. Zeh, Diss. Halle 1926; I. Methon als Aldehydreagenz. II. Kondensationsprodukte dicarbonylhaltiger Körper. [**896**] J. E. Zimmer, J. L. White, Mol. Cryst. Liq. Cryst. *38*, 177 (1977), C.A. *87*, 70715g (1977); Disclinations in the carbonaceous mesophases. [**897**] B. M. Zuev, A. P. Filippova, N. A. Palikhov, Izv. Akad. Nauk SSSR, Ser. Khim. *1975*, 1884, C.A. *84*, 58848h (1976); bis(*p*-Carballyloxyphenyl) esters of aliphatic and aromatic dicarboxylic acids.

Chapter 3

[**1**] J. Adams, W. Haas, Mol. Cryst. Liq. Cryst. *16*, 33 (1972), C.A. *76*, 77756c (1972); Relation between pitch change and stimulus in cholesterics. [**2**] E. L. Aero, A. N. Bulygin, Fiz. Tverd. Tela *13*, 1701 (1971), C.A. *75*, 102384u (1971); Linear mechanics of liquid-crystalline media. [**3**] E. L. Aero, A. N. Bulygin, Zh. Tekh. Fiz. *42*, 880 (1972), C.A. *77*, 25103p (1972); Viscosity and thermal conductivity of liquid crystals in a magnetic field. [**4**] E. L. Aero, A. N. Bulygin, 1. Wiss. Konferenz über flüssige Kristalle, 17.–19. 11. 1970, Iwanowo, Sammlung der Vorträge (1972), Nr. 20; Mechanik der flüssigen

Kristalle als Teil einer Kontinuum-Theorie mit kollektiven Rotationsfreiheitsgraden. [5] E. L. Aero, A. N. Bulygin, Gidromekhanika *1973*, 106, C.A. *83*, 51163b (1975); Hydromechanics of liquid crystals. [6] E. L. Aero, Fiz. Tverd. Tela *16*, 1245 (1974), C.A. *81*, 7087q (1974); Dispersion of the velocity and absorption of sound in liquid crystals. [7] T. Akahane, T. Tako, Jpn. J. Appl. Phys. *15*, 1559 (1976), C.A. *85*, 134806f (1976); Molecular alignment of bubble domains in cholesteric-nematic mixtures. [8] T. Akahane, T. Tako, Mol. Cryst. Liq. Cryst. *38*, 251 (1977), C.A. *86*, 181029k (1977); Molecular alignment of bubble domains in large pitch cholesterics. [9] R. Alben, Mol. Cryst. Liq. Cryst. *13*, 193 (1971), C.A. *75*, 68621g (1971); Pretransition effects in nematics. Model calculations. [10] C.-S. Shih, R. Alben, J. Chem. Phys. *57*, 3055 (1972), C.A. *77*, 157371j (1972); Lattice model for biaxial liquid crystals.

[11] R. Alben, J. R. McColl, C.-S. Shih, Solid State Commun. *11*, 1081 (1972), C.A. *77*, 169843c (1972); Characterization of order in nematics. [12] R. Alben, Liq. Cryst. Ord. Fluids *2*, 81 (1974); Possible phase diagrams for mixtures of positive and negative nematics. [13] Yu. S. Alekhin, V. S. Samsonov, Nauch. Tr. Vses. Zaoch. Mashinostr. In-t, *47*, 31 (1975), C.A. *86*, 10841q (1977); Characteristics of method for studying the acoustical properties of liquid crystals in a longitudinal magnetic field at frequencies above 0.4 GHz. [14] Yu. S. Alekhin, Nauch. Tr. Vses. Zaoch. Mashinostr. In-t, *47*, 35 (1975), C.A. *85*, 200779c (1976); Study of the acoustical properties of liquid crystals at high frequencies. [15] Yu. S. Alekhin, A. S. Lagunov, S. V. Pasechnik, Yu. V. Reztsov, Akust. Zh. *23*, 342 (1977), C.A. *87*, 122942a (1977); Temperature dependence of relaxation parameters in nematics. [16] S. I. Anisimov, I. E. Dzyaloshinskii, Zh. Eksp. Teor. Fiz. *63*, 1460 (1972), C.A. *78*, 102933j (1973); New type of disclinations in liquid crystals and the stability of various types of disclinations. [16a] I. M. Aref'ev, N. B. Lezhnev, A. A. Shamov, Izv. Akad. Nauk Turkm. SSR, Ser. Fiz.-Tekh., Khim. Geol. Nauk *1977*, 28, C.A. *87*, 173081b (1977); Propagation of ultrasound in the phase transition region of EBBA. [17] T. Ariman, M. A. Turk, N. D. Sylvester, Int. J. Eng. Sci. *12*, 273 (1974); Applications of microcontinuum fluid mechanics. [18] T. Asada, Y. Maruhashi, S. Onogi, J. Phys. (Paris), C-1, *36*, 299 (1975), C.A. *83*, 106523b (1975); Preliminary experiments on dynamic mechanical properties of cholesteryl myristate. [19] A. Askar, Lett. Appl. Eng. Sci. *2*, 265 (1974), C.A. *83*, 19436g (1975); Nematics viewed as a micropolar fluid. Numerical values of material coefficients for PAA. [20] A. Askar, Int. J. Eng. Sci. *14*, 259 (1976), C.A. *84*, 129017z (1976); Stability of nematics under a temperature gradient. Calculations for PAA.

[21] A. Askar, A. S. Cakmak, Proc.-Int. Congr. Rheol., 7th, *1976*, 602, C.A. *86*, 64029z (1977); Stability of nematics under a temperature gradient. [22] E. W. Aslaksen, Phys. kondens. Mat. *14*, 80 (1971); Continuum mechanics of nematics. [23] J. C. Bacri, C. R. Acad. Sci., Sér. B *270*, 1589 (1970), C.A. *73*, 81748g (1970); Mesures de l'absorption et de la dispersion des ondes ultrasonores longitudinales à 1 GHz dans trois corps possédant une stase cholestérique. [24] J. C. Bacri, J. Phys. (Paris), Lett. *35*, 141 (1974), C.A. *81*, 142327h (1974); Mesure de quelques coefficients de viscosité dans la phase nématique. [25] J. C. Bacri, J. Phys. (Paris) *35*, 601 (1974), C.A. *81*, 83152x (1974); Effets d'un champ magnétique sur l'attention et la vitesse des ondes ultrasonores dans un nématique. [26] J. C. Bacri, J. Phys. (Paris), Lett. *36*, 177 (1975), C.A. *83*, 85021u (1975); Anisotropie de l'atténuation ultrasonore au-dessus d'une transition s_A/n. Détermination d'un temps de relaxation. [27] J. C. Bacri, J. Phys. (Paris), Lett. *36*, 259 (1975), Atténuation et vitesse ultrasonores critique au-dessus d'une transition s_A/c faiblement du premier ordre. [28] J. C. Bacri, J. Phys. (Paris), C-1, *36*, 123 (1975), C.A. *83*, 106522a (1975); Divergence de la constante de Frank k_{33} et de l'atténuation ultrasonore $\alpha_\parallel$ au-dessus d'une transition s_A/n. [29] J. C. Bacri, J. Phys. (Paris), C-3, *37*, 119 (1976), C.A. *85*, 134771r (1976); Mise en évidence du second son par une méthode acoustique dans la phase smectique a d'un cristal liquide. [30] H. Unal, J. C. Bacri, J. Phys. (Paris), Lett. *38*, 111 (1977), C.A. *86*, 131417q (1977); Ondes transverses dans les smectiques B_A, B_C et E_A.

[31] H. Baessler, P. A. G. Malya, W. R. Nes, M. M. Labes, Mol. Cryst. Liq. Cryst. *6*, 329 (1970), C.A. *72*, 104982z (1970); The absence of helical inversion in single component cholesterics. [32] H. Baessler, M. M. Labes, J. Chem. Phys. *52*, 631 (1970), C.A. *72*, 67184t (1970); Helical twisting power of steroidal solutes in cholesterics. [33] I. Teucher, H. Baessler, M. M. Labes, Nature, Phys. Sci. *229*, 25 (1971), C.A. *75*, 41753q (1971); Diffusion through nematics. [34] B. Bahadur, J. Prakash, K. Tripathi, S. Chandra, Acustica *33*, 217 (1975), C.A. *83*, 171073u (1975); Specific volume of nematic N-(*p*-hexyloxybenzylidene)-*p*-toluidine. [35] B. Bahadur, Acustica *33*, 277 (1975), C.A. *84*, 24631s (1976); Ultrasonic studies of some nematics. [36] B. Bahadur, Acustica *34*, 86 (1975), Ultrasonic velocity studies in two cholesterics, cholesteryl benzoate and cholesteryl heptylate. [37] B. Bahadur, J. Phys.

C 9, 11 (1976), C.A. *84*, 114458v (1976); Ultrasonic velocity and absorption of nematic HBT. [**38**] P. J. Barratt, Mol. Cryst. Liq. Cryst. *24*, 223 (1973), C.A. *82*, 178629t (1975); Features of a continuum description of disclination lines in nematics. [**39**] P. J. Barratt, Q. J. Mech. Appl. Math. *27*, Pt. 4, 505 (1974); A continuum model for disclination lines in nematics. [**39a**] P. J. Barratt, D. M. Sloan, J. Phys. A *9*, 1987 (1976); Thermal instabilities in nematics. [**40**] R. Bartolino, F. Scudieri, D. Sette, A. Slivinski, J. Phys. (Paris), C-1, *36*, 121 (1975), C.A. *83*, 19457q (1975); Ultrasonic absorption and order of the s_A/n transition. [**40a**] R. Bartolino, G. Durand, Mol. Cryst. Liq. Cryst. *40*, 117 (1977), C.A. *87*, 109771z (1977); Dislocation effects on the viscoelastic properties of a smectic A.

[**41**] S. H. Bastow, F. P. Bowden, Proc. Roy. Soc. A *151*, 220 (1935), C.A. *29*, 7143[9] (1935); Viscous flow of liquid films; range of action of surface forces. [**42**] L. Bata, A. Buka, G. Molnar, Mol. Cryst. Liq. Cryst. *38*, 155 (1977), C.A. *87*, 32191a (1977); Rotary motion of molecules about their short axis by dielectric and splay viscosity measurements. [**43**] A. V. Belousov, Nauch. Tr. Vses. Zaoch. Mashinostr. In-t *47*, 52 (1975), C.A. *85*, 200777a (1976); Effect of a magnetic field on the absorption coefficient of ultrasound in nematic *p-n*-heptyloxybenzoic acid. [**44**] A. M. Anikin, A. V. Belousov, A. S. Lagunov, Nauch. Tr. Vses. Zaoch. Mashinostr. In-t *47*, 82 (1975), C.A. *85*, 200781x (1976); Acoustical properties of smectics. [**44a**] A. M. Anikin, A. V. Belousov, A. S. Lagunov, Akust. Zh. *23*, 459 (1977), C.A. *87*, 125581t (1977); Effect of a magnetic field on the acoustic properties of nematics. [**45**] D. W. Berreman, J. Chem. Phys. *62*, 776 (1975), C.A. *83*, 19395t (1975); Elastic continuum theory cutoffs and order in nematics and solids. [**46**] D. W. Berreman, J. Chem. Phys. *63*, 1041 (1975), C.A. *83*, 106559t (1975); Elastic continuum theory cutoffs and order in nematics and solids. Reply to comments. [**47**] M. Bertolotti, F. Scudieri, A. Ferrarri, D. Apostol, Appl. Opt. *15*, 2468 (1976), C.A. *85*, 169045j (1976); Spatial coherence properties of light scattered by disclinations in a liquid crystal. [**47a**] S. Bhattacharya, C. J. Umrigar, J. B. Ketterson, Mol. Cryst. Liq. Cryst. *40*, 79 (1977), C.A. *87*, 109770y (1977); Anisotropic ultrasound propagation in a smectic C. [**48**] E. C. Bingham, G. F. White, J. Am. Chem. Soc. *33*, 1257 (1911), C.A. *5*, 3191[3] (1911); The viscosity and fluidity of emulsions, crystalline liquids and colloidal solutions. [**48a**] H. Birecki, J. D. Litster, Mol. Cryst. Liq. Cryst. *42*, 33 (1977), Director bend mode behavior near a n/s_A transition. [**49**] Y. Björnstahl, O. Snellman, Kolloid-Z. *86*, 223 (1939), C.A. *33*, 3212[9] (1939); Einfaches Couette-Viskosimeter mit Thermostateinrichtung sowie einige Versuche, mittels dessen die Einwirkung eines elektrischen Feldes auf die Viskosität anisotroper Flüssigkeiten zu untersuchen. [**50**] S. Blaha, Phys. Rev. Lett. *36*, 874 (1976), C.A. *84*, 169880g (1976); Quantization rules for point singularities in superfluid helium-3 and liquid crystals.

[**51**] R. Blinc, S. Lugomer, B. Zeks, Liq. Cryst., Proc. Int. Conf., Bangalore, *1973*, 277, C.A. *84*, 114466w (1976); Soft mode dynamics in nematics. [**52**] R. Blinc, S. Lugomer, B. Zeks, Phys. Rev. A *9*, 2214 (1974), Soft-mode dynamics in nematics. [**53**] N. Boccara, R. Mejdani, L. de Seze, Mol. Cryst. Liq. Cryst. *35*, 91 (1976), C.A. *85*, 102596w (1976); Lattice gas of axially symmetric ellipsoids. [**54**] M. Born, Sitzb. kgl. preuß. Akad. Wiss. *1916*, 614, C.A. *11*, 2063 (1917); Theorie der flüssigen Kristalle und des elektrischen Kerr Effekts in Flüssigkeiten. [**55**] E. Bose, Physik. Z. *8*, 347 (1907), C.A. *1*, 1942[3] (1907); Physikalische Eigenschaften von Emulsionen, insbesondere deren Beziehung zu den kristallinischen Flüssigkeiten. [**56**] E. Bose, Physik. Z. *8*, 513 (1907), C.A. *1*, 2662[2] (1907); Für und wider die Emulsionsnatur der kristallinischen Flüssigkeiten. [**57**] E. Bose, F. Conrat, Physik. Z. *9*, 169 (1908), C.A. *2*, 1916[2] (1908); Viskositätsanomalien beim Klärungspunkte sogenannter kristallinischer Flüssigkeiten. [**58**] E. Bose, Physik. Z. *9*, 707 (1908), C.A. *3*, 139[9] (1909); Viskositätsanomalien von Emulsionen und von anisotropen Flüssigkeiten. [**59**] E. Bose, Physik. Z. *9*, 708 (1908), C.A. *3*, 140[1] (1909); Theorie der anisotropen Flüssigkeiten. [**60**] E. Bose, Physik. Z. *10*, 32 (1909), C.A. *3*, 737[3] (1909); Viskositätsanomalien anisotroper Flüssigkeiten im hydraulischen Strömungszustand.

[**61**] E. Bose, Physik. Z. *10*, 230 (1909), C.A. *3*, 1609[2] (1909); Theorie der anisotropen Flüssigkeiten. [**62**] E. Bose, Physik. Z. *12*, 60 (1911), C.A. *5*, 1219[1] (1911); Experimentalbeitrag zur Schwarmtheorie der anisotropen Flüssigkeiten. [**63**] Y. Bouligand, M. Kléman, J. Phys. (Paris) *31*, 1041 (1970), C.A. *74*, 92315k (1971); Paires de disinclinaisons helicoidales dans les cholestériques. [**64**] Y. Bouligand, J. Phys. (Paris) *33*, 525 (1972); Les arrangements focaux dans les smectiques: Rappels et considérations théoriques. [**65**] Y. Bouligand, J. Microsc. (Paris) *17*, 145 (1973), C.A. *79*, 130075v (1973); Chevrons et quadrilatères dans les plages a éventaile des cholestériques. [**66**] Y. Bouligand, J. Phys. (Paris) *34*, 1011 (1973), C.A. *80*, 75216k (1974); Texture a plans et la morphogenèse des principales textures dans les cholestériques. [**67**] Y. Bouligand, J. Phys. (Paris) *35*, 215 (1974), C.A. *80*, 137816s (1974); Noyaux, fils et rubans de Moebius dans les nématiques et les cholestériques peu torsadés. [**68**] C.

Williams, Y. Bouligand, J. Phys. (Paris) *35*, 589 (1974), C.A. *81*, 69554y (1974); Fils et disinclinaisons dans un nématique en tube capillaire. [**69**] Y. Bouligand, J. Phys. (Paris) *35*, 959 (1974), C.A. *82*, 105294v (1975); Dislocations coins et signification des cloisons de Grandjean-Cano dans les cholestériques. [**70**] W. Bragg, Trans. Faraday Soc. *29*, 1056 (1933); Focal conic structures.

[**71**] W. H. Bragg, Nature *133*, 445 (1934), C.A. *28*, 2964^3 (1934); Liquid crystals. [**72**] S. E. Bresler, P. F. Pokhil, Bull. acad. sci. U.R.S.S., Classe sci. math. nat. Sér. chim. *1937*, 413, C.A. *31*, 7305^6 (1937); Structure of surface layers of liquids and films. [**73**] F. Brochard, J. Phys. (Paris) *32*, 685 (1971), C.A. *76*, 63292m (1972); Mesures impédométriques des viscosités dans les cholestériques. [**74**] F. Brochard, J. Phys. (Paris) *34*, 411 (1973), C.A. *79*, 71018d (1973); Dynamique des fluctuations près d'une transition s$_A$/n du 2e ordre. [**75**] F. Brochard, Phys. Lett. A *49*, 315 (1974); Damping of second sound for smectics A. [**76**] F. Jaehnig, F. Brochard, J. Phys. (Paris) *35*, 301 (1974), C.A. *80*, 137849e (1974); Critical elastic constants and viscosities above a s$_A$/n transition of second order. [**77**] F. Brochard, J. Phys. (Paris), C-3, *37*, 85 (1976), C.A. *85*, 85785n (1976); Relaxation time of the cybotactic groups at the s$_A$/n transition. [**78**] Y. S. Lee, S. L. Golub, G. H. Brown, J. Phys. Chem. *76*, 2409 (1972), C.A. *77*, 106349a (1972); Ultrasonic shear wave study of the mechanical properties of a nematic. [**79**] W. F. Brown Jr., S. Shtrikman, H. Thomas, Mol. Cryst. Liq. Cryst. *22*, 375 (1973), C.A. *79*, 149521j (1973); Stability of one-dimensional director structures in ferromagnets and liquid crystals. [**80**] M. Brunet-Germain, C. R. Acad. Sci., Sér. B *274*, 1036 (1972), C.A. *77*, 40112j (1972); Interprétation des discontinuités de Grandjean dans une structure cholestérique à l'aide de la théorie de De Vries.

[**81**] A. V. Bulgadaev, Poverkhn. Sily Tonkikh Plenkakh Ustoich. Kolloidov, Sb. Dokl. Konf., 5th, *1972*, 56, C.A. *83*, 19440d (1975); Shear elasticity of liquid crystal films in contact with solids. [**82**] S. Bulygin, V. Ostroumov, Kolloid. Zhur. *10*, 1 (1948), C.A. *42*, 8677c (1948); Streaming potentials of an anisotropic liquid. [**83**] N. T. Bykova, Primen. Ul'traakust. Issled. Veshchestva *1971*, No. 25, 256, C.A. *78*, 129324y (1973); Effect of shear and bulk viscosity on the absorption of ultrasound in liquid-crystal phase-transition regions. [**84**] D. Cabib, L. Benguigui, J. Phys. (Paris) *38*, 419 (1977), C.A. *86*, 149035y (1977); The s$_C$ phase. [**85**] S. Nagai, A. Peters, S. Candau, Rev. Phys. Appl. *12*, 21 (1977), C.A. *86*, 62876z (1977); Acousto-optical effects in a nematic. [**86**] R. Cano, Bull. Soc. Fr. Mineral. Cristallogr. *91*, 20 (1968), C.A. *69*, 22837e (1968); Interprétation des discontinuités de Grandjean. [**87**] C. R. Carrigan, E. Guyon, J. Phys. (Paris), Lett. *36*, 415 (1975), Convection driven by centrifugal buoyancy in nematics. [**88**] C. A. Castro, A. Hikata, C. Elbaum, Mol. Cryst. Liq. Cryst. *25*, 167 (1974), C.A. *83*, 124422x (1975); Ultrasonic second harmonics in a nematic. [**89**] C. A. Castro, A. Hikata, C. Elbaum, Mol. Cryst. Liq. Cryst. *34*, 165 (1977), C.A. *86*, 198229y (1977); Nonlinear effects in ultrasonic propagation in a nematic. [**90**] S. Chakravarti, C. W. Woo, Phys. Rev. A *11*, 713 (1975), Short-range correlations in two-dimensional liquid crystals. I.

[**91**] S. Chakravarti, C. W. Woo, Phys. Rev. A *12*, 245 (1975), II. Smectic and crystalline phases (cf. [**90**]). [**92**] S. Chandrasekhar, Mol. Cryst. *2*, 71 (1967), C.A. *66*, 41315n (1967); Surface tension of liquid crystals. [**93**] S. Chandrasekhar, N. V. Madhusudana, J. Phys. (Paris), C-4, *30*, 24 (1969), C.A. *72*, 94269f (1970); Orientational order in PAA, PAP, and their mixtures. [**94**] S. Chandrasekhar, N. V. Madhusudana, Mol. Cryst. Liq. Cryst. *10*, 151 (1970), C.A. *73*, 49656q (1970); Statistical theory of orientational order in nematics. [**95**] S. Chandrasekhar, N. V. Madhusudana, Acta. Cryst. A *26*, 153 (1970), C.A. *72*, 94268e (1970); Interferometric study of liquid crystalline surfaces. [**96**] N. V. Madhusudana, R. Shashidhar, S. Chandrasekhar, Mol. Cryst. Liq. Cryst. *13*, 61 (1971), C.A. *75*, 54887m (1971); Orientational order in nematic anisalazine. [**97**] S. Chandrasekhar, N. V. Madhusudana, K. Shubha, Symp. Faraday Soc. *1971*, No. 5, 26, C.A. *79*, 24394p (1973); Relation between elasticity and orientational order in nematics. [**98**] S. Chandrasekhar, N. V. Madhusudana, Acta. Cryst. A *27*, 303 (1971), C.A. *75*, 92254k (1971); Molecular statistical theory of nematics. [**99**] S. Chandrasekhar, N. V. Madhusudana, K. Shubha, Acta Cryst. A *28*, 28 (1972), C.A. *76*, 38452f (1972); Molecular statistical theory of nematics. Relation between elasticity and orientational order. [**100**] S. Chandrasekhar, N. V. Madhusudana, Mol. Cryst. Liq. Cryst. *17*, 37 (1972), C.A. *77*, 10704m (1972); Molecular theory of nematics.

[**101**] S. Chandrasekhar, N. V. Madhusudana, Mol. Cryst. Liq. Cryst. *24*, 179 (1973), C.A. *81*, 142302w (1974); Nematic order in PAA and its dependence on pressure, volume and temperature. [**102**] C. A. Croxton, S. Chandrasekhar, Liq. Cryst., Proc. Int. Conf., Bangalore, *1973*, 237, C.A. *84*, 112044h (1976); Statistical thermodynamics of the nematic liquid crystal free surface. [**103**] N. V. Madhusudana, K. L. Savithramma, S. Chandrasekhar, Pramana *8*, 22 (1977), C.A. *86*, 99280b (1977);

Short range orientational order in nematics. [104] R. Chang, F. B. Jones, Jr., J. J. Ratto, Mol. Cryst. Liq. Cryst. *33*, 13 (1976), C.A. *85*, 93204n (1976); Molecular order and an odd-even effect in an homologous series of nematic Schiff bases. [105] R. Cano, P. Châtelain, Compt. rend. *253*, 2081 (1961); Les plans de Grandjean existent-ils dans le monocristal cholestérique? [106] J. Cheng, Bull. Am. Phys. Soc. *17*, 350 (1972); Pretransitional phenomena in the isotropic phase of cholesterics. [107] V. G. Chigrinov, S. A. Pikin, Kristallografiya *21*, 589 (1976), C.A. *85*, 54903v (1976); Structure of the nematic mesophase flow in a cylindrical capillary. [108] I. G. Chistyakov, Kristallografiya *7*, 764 (1962), C.A. *58*, 6274e (1963); Various properties in liquid crystal preferred orientations of cholesterol and nematics. [109] N. I. Kalinnikova, P. P. Pugachevich, I. G. Chistyakov, U. V. Sushkin, Uch. Zap. Ivanov. Gos. Pedagog. Inst. *62*, 176 (1967), C.A. *70*, 40956t (1969); Surface tension of melts of liquid crystals. [110] N. I. Kalinnikova, I. G. Chistyakov, P. P. Pugachevich, Tr. Kafedry Teor. Eksp. Fiz. Kaliningrad. Univ. *1969*, 25, C.A. *75*, 10501b (1971); Surface tension of liquid crystals.

[111] I. G. Chistyakov, S. K. Sukharev, Kristallografiya *16*, 1052 (1971), C.A. *76*, 51126x (1972); Effect of flux on the orientation of nematic PAA. [112] I. G. Chistyakov, Uch. Zap., Ivanov. Gos. Pedagog. Inst. *1972*, No. 99, 19, C.A. *78*, 152443v (1973); Temperature dependence of the degree of orientation in nematic PAA. [113] I. G. Chistyakov, S. K. Sucharev, 1. Wiss. Konferenz über flüssige Kristalle, 17.–19. 11. 1970, Iwanowo, Sammlung der Vorträge (1972), Nr. 16; Einfluß der Strömung auf die Orientierung des nematischen PAA. [114] V. M. Chaikovskii, I. G. Chistyakov, 1. Wiss. Konferenz über flüssige Kristalle, 17.–19. 11. 1970, Iwanowo, Sammlung der Vorträge (1972), Nr. 39; Möglichkeiten zur Bestimmung des mittleren Neigungswinkels der Moleküle und des Ordnungsgrades in flüssigen Kristallen. [115] I. M. Aref'ev, V. N. Biryukov, V. A. Gladkii, S. V. Krivokhizha, I. L. Fabelinskii, I. G. Chistyakov, Zh. Eksp. Teor. Fiz. *63*, 1729 (1972), C.A. *78*, 49232b (1973); Propagation of hypersound and ultrasound in the isotropic phase of a nematic in the phase transition region. [116] I. G. Chistyakov, S. K. Sukharev, Kristallografiya *17*, 1264 (1972), C.A. *78*, 63508c (1973); Structure of nematic *p*-anisal-aminoazobenzene. [117] T. W. Chou, J. Appl. Phys. *42*, 4931 (1971); Elastic behavior of disclinations in nonhomogeneous media. [118] K. S. Chu, B. L. Richards, D. S. Moroi, W. M. Franklin, Liq. Cryst. Ord. Fluids *2*, 73 (1974); Rotational diffusion in the nematic phase. [119] K. S. Chu, D. S. Moroi, J. Phys. (Paris), C-1, *36*, 99 (1975); Self-diffusion in nematics. [120] K. S. Chu, N. K. Ailawadi, D. S. Moroi, Mol. Cryst. Liq. Cryst. *38*, 45 (1977), C.A. *87*, 14456m (1977); Diffusion of impurities in s_A phases.

[121] P. L. Chung, Rev. Sci. Instrum. *44*, 1669 (1973), C.A. *80*, 49633z (1974); Linear oscillatory viscometer. [122] D. Churchill, L. W. Bailey, Mol. Cryst. Liq. Cryst. *7*, 285 (1969), C.A. *71*, 74349h (1969); Surface tension of cholesterics: cholesteryl myristate. [123] P. E. Cladis, M. Kléman, Mol. Cryst. Liq. Cryst. *16*, 1 (1972), C.A. *76*, 77754a (1972); Cholesteric domain texture. [124] P. E. Cladis, Phys. Rev. Lett. *28*, 1629 (1972), C.A. *77*, 53403e (1972); New method for measuring the twist elastic constant K_{22}/χ_a and the shear viscosity γ_1/χ_a for nematics. [125] P. E. Cladis, M. Kléman, J. Phys. (Paris) *33*, 591 (1972), C.A. *77*, 157343b (1972); Nonsingular disclinations of strength $S = +1$ in nematics. [126] P. E. Cladis, Phys. Rev. Lett. *31*, 1200 (1973), C.A. *80*, 7900d (1974); Bend elastic constant near a s_A/n transition. [127] P. E. Cladis, Phil. Mag. *29*, 641 (1974), C.A. *81*, 83151w (1974); n and s_A phases of CBOOA in tubes. [128] P. E. Cladis, Phys. Rev. Lett. A *48*, 179 (1974), C.A. *81*, 69543u (1974); Impurity dependence of the critical exponent of the nematic bend elastic constant, $K_3(T)$, near a s_A transition. [129] P. E. Cladis, S. Torza, Phys. Rev. Lett. *35*, 1283 (1975), C.A. *84*, 11093t (1976); Stability of nematics in Couette flow. [130] P. E. Cladis, J. Phys. (Paris) C-3, *37*, 137 (1976), C.A. *85*, 85791m (1976); Beading of smectic disclination lines. [130a] P. E. Cladis, S. Torza, Colloid Interface Sci., 50th, *4*, 487 (1976), C.A. *87*, 61013z (1977); Flow instabilities in Couette flow in nematics. [130b] A. E. White, P. E. Cladis, S. Torza, Mol. Cryst. Liq. Cryst. *43*, 13 (1977), Liquid crystals in flow. I. Conventional viscosimetry and density measurements.

[131] N. A. Clark, P. S. Pershan, Phys. Rev. Lett. *30*, 3 (1973), C.A. *78*, 49204u (1973); Light scattering by deformation of the plane texture of smectics and cholesterics. [132] N. A. Clark, Phys. Lett. A *46*, 171 (1973), C.A. *80*, 113490e (1974); Continuity of flow alignment parameters across the n/i transition. [133] M. G. Clark, Mol. Phys. *31*, 1287 (1976), C.A. *85*, 199067k (1976); Algebraic derivation of the free-energy of a distorted nematic. [134] R. L. Coldwell, T. P. Henry, C. W. Woo, Phys. Rev. A *10*, 897 (1974); Free energy of a system of spherocylinders serving as a simple model for liquid crystals. [135] M. A. Cotter, Diss. Georgetown Univ., Washington, D. C. 1969, 279 pp. Avail. Univ. Microfilms, Ann. Arbor, Mich. Order No. 70-4636, C.A. *74*, 7559a (1971); Statistical mechanics of rodlike particles. Application to the n/i transition in liquid crystals. [136] M. A. Cotter,

Phys. Rev. A *10*, 625 (1974); Hard-rod fluid-scaled particle theory revisited. [137] M. A. Cotter, Mol. Cryst. Liq. Cryst. *35*, 33 (1976), C.A. *85*, 102655q (1976); Lattice models for thermotropic liquid crystals. [138] M. A. Cotter, J. Chem. Phys. *66*, 1098 (1977), C.A. *86*, 99281c (1977); Hard spherocylinders in an anisotropic mean field: a simple model for a nematic. [138a] M. A. Cotter, J. Chem. Phys. *66*, 4710 (1977), C.A. *87*, 32203f (1977); Generalized van der Waals theory of nematics: an alternative formulation. [139] M. A. Cotter, Mol. Cryst. Liq. Cryst. *39*, 173 (1977), C.A. *87*, 31255n (1977); Consistency of mean field theories of nematics. [140] P. K. Currie, Mol. Cryst. Liq. Cryst. *19*, 249 (1973), C.A. *78*, 89524e (1973); Propagating plane disinclination surfaces in nematics.

[141] P. K. Currie, Solid State Commun. *12*, 31 (1973), Implications of Parodi's relation for waves in nematics. [142] P. K. Currie, Rheol. Acta *12*, 165 (1973), C.A. *79*, 109029y (1973); Orientation of liquid crystals by temperature gradients. [143] P. K. Currie, J. Acoust. Soc. *56*, 765 (1974); Decay of weak waves in liquid crystals. [144] P. K. Currie, Mol. Cryst. Liq. Cryst. *28*, 335 (1974), C.A. *82*, 178711p (1975); Parodi's relation as a stability condition for nematics. [145] P. K. Currie, Rheol. Acta *14*, 688 (1975), Approximate solutions for Poiseuille flow of liquid crystals. [145a] G. P. Mac Sithigh, P. K. Currie, J. Phys. D *10*, 1471 (1977), C.A. *87*, 93761h (1977); Apparent viscosity during simple shearing flow of nematics. [146] L. E. Davis, J. Chambers, Electron Lett. *7*, 287 (1971); Optical scattering in a nematic induced by acoustic surface waves. [147] L. E. Davis, IEEE Trans. Son. Ultrason. *19*, 390 (1972); Liquid crystals in ultrasonic, electric and optical fields. [148] L. Davison, Physics of Fluids *10*, 2333 (1967), Linear theory of mechanical equilibrium of nematics. [149] L. Davison, Phys. Rev. *180*, 232 (1969), C.A. *70*, 91048f (1969); Linear theory of heat conduction and dissipation in nematics. [150] L. Davison, D. E. Amos, Phys. Rev. *183*, 288 (1969), C.A. *71*, 43032v (1969); Dissipation in liquid crystals.

[151] J. H. de Boer, Nederland. Tijdschr. Natuurk. *15*, 205 (1949), C.A. *44*, 2824g (1950); Orientation by van der Waals forces. [152] P. Debye, Der feste Körper *1938*, 42, C.A. *32*, 8865² (1938); Quasikristalline Struktur von Flüssigkeiten. [153] P. Debye, Z. Elektrochem. *45*, 174 (1939), C.A. *33*, 3649¹ (1939); Quasikristalline Struktur von Flüssigkeiten. [154] P. G. de Gennes, Compt. Rend. Acad. Sci., Sér. B *266*, 15 (1968); Fluctuation d'orientation et diffusion Rayleigh dans un cristal nématique. [155] P. G. de Gennes, C. R. Acad. Sci., Sér. B *266*, 571 (1968); Structure des cloisons de Grandjean-Cano. [156] P. G. de Gennes, Phys. Lett. A *30*, 454 (1969); Phenomenology of short-range-order effects in the i phase of nematogens. [157] P. G. de Gennes, J. Phys. (Paris) C-4, *30*, 65 (1969), C.A. *72*, 94274d (1970); Conjectures sur l'état smectique. [158] P. G. de Gennes, Mol. Cryst. Liq. Cryst. *7*, 325 (1969), C.A. *71*, 85343c (1969); Long range order and thermal fluctuations in liquid crystals. [159] P. G. de Gennes, Mol. Cryst. Liq. Cryst. *12*, 193 (1971), C.A. *75*, 41393x (1971); Short-range order effects in the isotropic phase of nematics and cholesterics. [160] P. G. de Gennes, J. Phys. (Paris) C-5a, *32*, 3 (1971), C.A. *77*, 10703k (1972); Cristaux liquides nématiques.

[161] P. G. de Gennes, Symp. Faraday Soc. No. 5, *1971*, 16, Possible experiments on two-dimensional nematics. [162] P. G. de Gennes, G. Sarma, Phys. Lett. A *38*, 219 (1972); Tentative model for the s_B phase. [163] P. G. de Gennes, Phys. Lett. A *41*, 479 (1972); Mecanochromatic effect in nematics. [164] P. G. de Gennes, Solid State Commun. *10*, 753 (1972); Analogy between superconductors and smectics A. [165] P. G. de Gennes, C. R. Acad. Sci., Sér. B *275*, 319 (1972); Types de singularités permises dans une phase ordonnée. [166] P. G. de Gennes, C. R. Acad. Sci., Sér. B *275*, 549 (1972); Structure du coeur des coniques focales dans les smectiques A. [167] P. G. de Gennes, C. R. Acad. Sci., Sér. B *275*, 939 (1972); Dislocations coin dans un smectique A. [168] P. G. de Gennes, Mol. Cryst. Liq. Cryst. *21*, 49 (1973), C.A. *79*, 97895n (1973); Polymorphism of smectics. [169] P. G. de Gennes, Proc. Int. Sch. Phys. "Enrico Fermi" *59*, 691 (1973), C.A. *86*, 63580k (1977); Short guide to the physics of mesophases. [170] P. G. de Gennes, Collect. Prop. Phys. Syst., Proc. Nobel Symp. 24th, *1973*, 228, C.A. *81*, 83047s (1974); Recent results in the physics of liquid crystals.

[171] P. G. de Gennes, Int. Conf. on Liquid Crystals, Raman Res. Inst., Bangalore, December 3–8, 1973, Abstr. No. 1; Recent progress in the physics of smectics A. [172] P. G. de Gennes, Liq. Cryst. Plast. Cryst. *1*, 67 (1974), C.A. *83*, 19079t (1975); Microstructures of liquid crystals. Principles of the continuum theory. [173] P. G. de Gennes, Liq. Cryst. Plast. Cryst. *1*, 92 (1974), C.A. *83*, 19081n (1975); Microstructures of liquid crystals. Dynamical effects. [174] P. G. de Gennes, Phys. Fluids *17*, 1645 (1974); Viscous flow in smectics A. [175] P. G. de Gennes, Recherche *5*, 1022 (1974), C.A. *82*, 160338n (1975); Fluctuations géantes et phénomènes critiques. [176] P. G. de Gennes (Oxford Univ. Press: Fair Lawn, N. J.), 1974, 333 pp., C.A. *81*, 112350t (1974); International Series of Monographs on Physics: The Physics of Liquid Crystals. [177] E. Dubois-Violette, P. G. de Gennes, J. Phys. (Paris)

C-1, *36*, 293 (1975); Convection and permeation in cholesterics. [**178**] P. G. de Gennes, P. Pincus, J. Phys. (Paris) *37*, 1359 (1976), C.A. *86*, 24610a (1977); Instabilities under mechanical tension in a smectic cylinder. [**179**] P. G. de Gennes, Mol. Cryst. Liq. Cryst. *34*, 91 (1976), C.A. *86*, 47383t (1977); Effect of shear flows on critical fluctuations in fluids. [**180**] W. H. de Jeu, 7. Arbeitstagung Flüssigkristalle, Inst. Angew. Festkörperphysik, Freiburg, 4. 3. 1977; The elastic constants of nematics.

[**181**] M. Delaye, R. Ribotta, G. Durand, Phys. Lett. A *44*, 139 (1973), C.A. *79*, 58708j (1973); Buckling instability of the layers in a smectic A. [**182**] M. Delaye, J. Phys. (Paris) C-3, *37*, 99 (1976), C.A. *85*, 85787q (1976); Damping of the twist director fluctuations above a s_A/n transition. [**183**] A. I. Derzhanski, I. Bivas, A. G. Petrov, Dokl. Bolg. Akad. Nauk *28*, 1327 (1975); One-dimensional approach to spontaneous deformations in nematics. [**183a**] J. P. Dias, J. Mecanique *15*, 698 (1976); Equation system in relation to two-dimensional evolution equations of nematic liquid crystal. [**184**] F. Dickenschied, Diss. Halle *1908*, Untersuchungen über Dichte, Reibung, Kapillarität kristalliner Flüssig-keiten. [**185**] D. S. Dimitrov, I. B. Ivanov, Dokl. Bolg. Akad. Nauk *28*, 1513 (1975); Hydrodynamics of thin films of nematics. [**185a**] J. L. Dion, A. D. Jacob, Appl. Phys. Lett. *31*, 490 (1977), C.A. *87*, 192277m (1977); A new hypothesis on ultrasonic interaction with a nematic. [**185b**] J. L. Dion, C. R. Acad. Sci., Sér. B *284*, 219 (1977), C.A. *87*, 93732z (1977); Nouvel effet des ultrasons sur l'orientation d'un cristal liquide. [**186**] S. G. Dmitriev, Zh. Eksp. Teor. Fiz. *67*, 1949 (1974); Instability in smectic A. [**187**] H. D. Doerfler, W. Kerscher, H. Sackmann, Z. Phys. Chem. (Leipzig) *251*, 314 (1972), C.A. *78*, 140675j (1973); Oberflächenfilme von Azoxy-α-methylzimtsäureestern auf wäßrigem Substrat. [**188**] R. Dreher, Z. Naturforsch. A *29*, 125 (1974), C.A. *80*, 149690j (1974); High precision method for determining twist elastic constants of nematics. I. Theory. [**189**] Y. A. Dreizin, A. M. Dykhne, Zh. Eksp. Teor. Fiz. *61*, 2140 (1971), C.A. *76*, 91231j (1972); Planar problems in elasticity theory of nematics. [**190**] V. P. Dremina, B. B. Kudryavtsev, Primenenie Ul'traakustiki k Issledovan. Veshchestva Sbornik *1955*, No. 2, 25, C.A. *52*, 17856h (1958); Dispersion of ultrasound in organic substances.

[**191**] J. F. Dreyer, Liq. Cryst. Ord. Fluids *1970*, 311, C.A. *78*, 165408n (1973); Alignment of molecules in the nematic state. [**192**] C. Caroli, E. Dubois-Violette, Solid State Commun. *7*, 799 (1969), C.A. *71*, 74228t (1969); Energy of a disclination line in a cholesteric. [**193**] E. Dubois-Violette, C. R. Acad. Sci., Sér. B *273*, 923 (1971), C.A. *76*, 77698k (1972); Instabilités hydrodynamiques d'un nématique soumis à un gradient thermique. [**194**] E. Dubois-Violette, J. Phys. (Paris) *34*, 107 (1973), C.A. *78*, 129274g (1973); Hydrodynamic instabilities of cholesterics under a thermal gradient. [**195**] P. Pieranski, E. Dubois-Violette, E. Guyon, Phys. Rev. Lett. *30*, 736 (1973), C.A. *78*, 152481f (1973); Heat convection in liquid crystals heated from above. [**196**] E. Dubois-Violette, Solid State Commun. *14*, 767 (1974); Determination of thermal instabilities thresholds for homeotropic and planar nematics. [**197**] E. Dubois-Violette, E. Guyon, P. Pieranski, Mol. Cryst. Liq. Cryst. *26*, 193 (1974), C.A. *82*, 77826z (1975); Heat convection in a nematic. [**197a**] P. Manneville, E. Dubois-Violette, J. Phys. (Paris) *37*, 285 (1976); Shear-flow instability in nematics. Theory of steady simple shear flows. [**197b**] P. Manneville, E. Dubois-Violette, J. Phys. (Paris) *37*, 1115 (1976); Steady Poiseuille flow in nematics. Theory of uniform instability. [**198**] N. T. Dunwoody, J. Appl. Math. Phys. *26*, 105 (1975); Balance laws for liquid crystal mixtures. [**199**] C. Dupin, Applications de Géométrie et de Mécanique, Bachelier, Paris, 1822, p. 200. [**200**] J. F. Dyro, P. D. Edmonds, Mol. Cryst. Liq. Cryst. *25*, 175 (1974), C.A. *83*, 140059q (1974); Ultrasonic absorption and dispersion in cholesteryl esters.

[**201**] J. F. Dyro, P. D. Edmonds, Mol. Cryst. Liq. Cryst. *29*, 263 (1975), C.A. *83*, 51166e (1975); Dynamic viscoelastic properties of cholesterics. [**202**] D. Eden, C. W. Garland, R. C. Willamson, J. Chem. Phys. *58*, 1861 (1973), C.A. *78*, 116254a (1973); Ultrasonic investigation of the n/i transition in MBBA. [**203**] P. D. Edmonds, D. A. Orr, Mol. Cryst. *2*, 135 (1967), C.A. *66*, 41314m (1967); Ultrasonic absorption and dispersion at phase transitions in liquid crystals. [**204**] G. Eilenberger, Plenarvortr. Physikertag., 36th, *1972*, 419, C.A. *77*, 25602a (1972); Ordnungsvorgänge in flüssigen Kri-stallen. [**204a**] S. Engelsberg, Phys. Lett. A *62*, 223 (1977), C.A. *87*, 141513c (1977); Anisotropy of the electric susceptibility in superfluid helium-3 A. [**205**] J. L. Ericksen, J. Polymer Sci. *47*, 327 (1960); Vorticity effect in anisotropic fluids. [**206**] J. L. Ericksen, Trans. Soc. Rheol. *4*, 29 (1960), C.A. *55*, 19379i (1961); Theory of anisotropic fluids. [**207**] J. L. Ericksen, Kolloid-Z. *173*, 117 (1960); Transversely isotropic fluids. [**208**] J. L. Ericksen, Arch. Rat'l Mech. Anal. *4*, 231 (1960); Anisotropic fluids. [**209**] J. L. Ericksen, Trans. Soc. Rheol. *5*, 23 (1961), C.A. *56*, 942f (1962); Conservation laws for liquid crystals. [**210**] J. L. Ericksen, Arch. Rat'l Mech. Anal. *8*, 1 (1961); Poiseuille flow of certain anisotropic fluids.

[211] J. L. Ericksen, Arch. Rat'l Mech. Anal. *9*, 371 (1962); Hydrostatic theory of liquid crystals. [212] J. L. Ericksen, Arch. Rat'l Mech. Anal. *10*, 189 (1962); Nilpotent energies in liquid crystal theory. [213] J. L. Ericksen, Int'l J. Engr. Sci. *1*, 157 (1963); Singular surfaces in anisotropic fluids. [214] J. L. Ericksen, Phys. Fluids *9*, 1205 (1966); Inequalities in liquid crystal theory. [215] J. L. Ericksen, Q. J. Mech. Appl. Math. *19*, 455 (1966); Instability in Couette flow of anisotropic fluids. [216] J. L. Ericksen, Trans. Soc. Rheol. *11*, 5 (1967); General solutions in the hydrostatic theory of liquid crystals. [217] J. L. Ericksen, J. Fluid Mech. *27*, 59 (1967), C.A. *69*, 100783a (1968); Twisting of liquid crystals. [218] J. L. Ericksen, Appl. Mech. Rev. *20*, 1029 (1967); Continuum theory of liquid crystals. [219] J. L. Ericksen, J. Acoust. Soc. Am. *44*, 444 (1968); Propagation of weak waves in nematics. [220] J. L. Ericksen, Q. J. Mech. Appl. Math. *21*, 463 (1968); Twist waves in liquid crystals.

[221] J. L. Ericksen, Q. Appl. Math. *25*, 474 (1968); Twisting of partially oriented liquid crystals. [222] J. L. Ericksen, Trans. Soc. Rheol. *13*, 9 (1969), C.A. *71*, 7326k (1969); Boundary-layer effect in viscometry of liquid crystals. [223] J. L. Ericksen, Mol. Cryst. Liq. Cryst. 7, 153 (1969), C.A. *71*, 85329c (1969); Continuum theory of nematics. [224] J. L. Ericksen, Liq. Cryst. Ord. Fluids *1970*, 181, C.A. *78*, 165396g (1973); Singular solutions in liquid crystal theory. [225] J. L. Ericksen, Q. J. Mech. Appl. Math. *27*, Pt. 2, 213 (1974); Liquid crystals and Cosserat surfaces. [226] J. L. Ericksen, Adv. Liq. Cryst. *2*, 233 (1976), C.A. *86*, 180805y (1977); Equilibrium theory of liquid crystals. [226a] J. L. Ericksen, Q. J. Mech. Appl. Math. *29*, 203 (1976); Equations of motion for liquid crystals. [226b] J. L. Ericksen, AMD *22*, 47 (1977), C.A. *87*, 192137r (1977); The mechanics of nematics. [227] A. C. Eringen, J. D. Lee, U. S. Clearinghouse Fed. Sci. Techn. Inform., AD 1970, No. 717396, 31 pp., C.A. *75*, 52995w (1971); Wave propagation in nematics. [228] J. D. Lee, A. C. Eringen, J. Chem. Phys. *54*, 5027 (1971), C.A. *75*, 26200p (1971); Wave propagation in nematics. [229] J. D. Lee, A. C. Eringen, J. Chem. Phys. *55*, 4504 (1971), C.A. *75*, 144731k (1971); Alignment of nematics. [230] J. D. Lee, A. C. Eringen, J. Chem. Phys. *55*, 4509 (1971), C.A. *75*, 144729r (1971); Boundary effects of orientation of nematics.

[231] J. D. Lee, A. C. Eringen, J. Chem. Phys. *58*, 4203 (1973), C.A. *79*, 10803j (1973); Continuum theory of smectics. [232] J. D. Lee, A. C. Eringen, Liq. Cryst. Ord. Fluids *2*, 315 (1974); Relations of two continuum theories of liquid crystals. [233] A. C. Eringen, J. D. Lee, Liq. Cryst. Ord. Fluids *2*, 383 (1974); Continuum theory of cholesterics. [234] M. N. L. Narasimhan, A. C. Eringen, Mol. Cryst. Liq. Cryst. *29*, 57 (1974), C.A. *82*, 178672b (1975); Orientational effects in heat-conducting nematics. [235] J. D. Lee, A. C. Eringen, J. Chem. Phys. *63*, 1321 (1975), C.A. *83*, 106590w (1975); Alignment of nematics. Reply to comments. [236] M. N. L. Narasimhan, A. C. Eringen, Int. J. Eng. Sci. *13*, 233 (1975), C.A. *83*, 33856a (1975); Thermomicropolar theory of heat-conducting nematics. [236a] T. E. Faber, Proc. R. Soc. London, A *353*, 247, 261, 277 (1977), C.A. *86*, 163783x, 163784y, 163785z (1977); Continuum theory of disorder in nematics. I. II. Intermolecular correlations. III. Nuclear magnetic relaxation. [237] C. P. Fan, J. M. Stephen, Phys. Rev. Lett. *25*, 500 (1970), C.A. *73*, 81750b (1970); i/n transition in liquid crystals. [238] C. P. Fan, Phys. Lett. A *34*, 335 (1971), C.A. *75*, 25583d (1971); Disclination lines in liquid crystals. [239] A. Ferguson, S. J. Kennedy, Phil. Mag. *26*, 41 (1938), C.A. *32*, 7793[7] (1938); Surface tension of liquid crystals. [240] B. A. Finlayson, Liq. Cryst. Ord. Fluids *2*, 211 (1974), C.A. *86*, 180982d (1977); Numerical computations for the flow of liquid crystals.

[241] W. L. Kuhn, B. A. Finlayson, Mol. Cryst. Liq. Cryst. *39*, 101 (1977); Average orientation in a three-dimensional layer of nematic liquid crystal. [242] J. Wahl, F. Fischer, Optics Commun. *5*, 341 (1972); New optical method for studying the viscoelastic behaviour of nematics. [243] J. Wahl, F. Fischer, Mol. Cryst. Liq. Cryst. *22*, 359 (1973), C.A. *79*, 140469r (1973); Elastic and viscosity constants of nematics from a new optical method. [244] T. Waltermann, F. Fischer, Z. Naturforsch. A *30*, 519 (1975), C.A. *83*, 36744y (1975); Einfluß magnetischer Felder auf die Torsionsscherung von MBBA. [245] K. Hiltrop, F. Fischer, Z. Naturforsch. A *31*, 800 (1976), C.A. *85*, 102648q (1976); Radial Poiseuille flow of a homeotropic nematic layer. [245a] F. Fischer, Z. Naturforsch. A *32*, 429 (1977), C.A. *86*, 198245a (1977); Periodic ripples in a free s_A surface skin. [246] J. A. Fisher, Diss. Univ. of Minnesota, Minneapolis 1969, 125 pp. Avail. Univ. Microfilms, Ann Arbor, Mich., Order No. 69-16,409, C.A. *72*, 125236z (1970); Transport phenomena in liquid crystals. [247] W. Flügge, Tensor Analysis and Continuum Mechanics (Springer: New York). 1972. 207 pp. [248] W. G. F. Ford, J. Chem. Phys. *56*, 6270 (1972), C.A. *77*, 54205d (1972); Temperature dependence of PAA in crystal and nematic phases. [249] D. Forster, T. C. Lubensky, P. C. Martin, J. Swift, P. S. Pershan, Phys. Rev. Lett. *26*, 1016 (1971), C.A. *75*, 11711g (1971); Hydrodynamics of liquid crystals. [250] D. Forster, Phys. Rev. Lett. *32*, 1161 (1974), C.A. *81*, 17823r (1974); Microscopic theory of flow alignment in nematics.

[251] D. Forster, Ann. Physics *84*, 505 (1974); Hydrodynamics and correlation functions in ordered systems — nematics. [252] D. Forster, Adv. Chem. Phys. *31*, 231 (1975), C.A. *84*, 52184s (1976); Theory of liquid crystals. [253] F. C. Frank, Physik. Z. *39*, 530 (1938), C.A. *32*, 6921¹ (1938); Quasi-kristalline und kristalline Flüssigkeiten. [254] F. C. Frank, Discussions Faraday Soc. *25*, 19 (1958), C.A. *54*, 4093i (1960); Theory of liquid crystals. [255] W. Franklin, Mol. Cryst. Liq. Cryst. *14*, 227 (1971), C.A. *75*, 133188n (1971); Diffusion theory in liquid crystals. [256] W. Franklin, Phys. Lett. A *48*, 247 (1974), C.A. *81*, 142285t (1974); Effects of anisotropy, order parameter, and viscosity on translational self-diffusion in nematics. [257] W. Franklin, Phys. Rev. A *11*, 2156 (1975), C.A. *83*, 69425y (1975); Theory of translational diffusion in nematics. [257a] E. Moritz, W. Franklin, Phys. Rev. A *14*, 2334 (1976); Nonlinearities in nematic stress tensor. [257b] W. Franklin, Mol. Cryst. Liq. Cryst. *40*, 91 (1977), C.A. *87*, 125599e (1977); Relations between rotational diffusion and frequency spectra of nematics. [258] J. J. C. Picot, A. G. Fredrickson, Ind. Eng. Chem., Fundam. 7, 84 (1968), C.A. *68*, 54159f (1968); Interfacial and electrical effects on thermal conductivity of nematics. [259] J. Fisher, A. G. Fredrickson, Mol. Cryst. Liq. Cryst. *6*, 255 (1969), C.A. *72*, 59916t (1970); Transport processes in anisotropic fluids. II. Coupling of momentum and energy transport in a nematic. [260] J. Fisher, A. G. Fredrickson, Mol. Cryst. Liq. Cryst. *8*, 267 (1969), C.A. *71*, 95134z (1969); Interfacial effects on the viscosity of a nematic.

[261] C. K. Yun, A. G. Fredrickson, Mol. Cryst. Liq. Cryst. *12*, 73 (1970), C.A. *74*, 92316m (1971); Anisotropic mass diffusion in liquid crystals. [262] C. K. Yun, J. J. C. Picot, A. G. Fredrickson, J. Appl. Phys. *42*, 4764 (1971), C.A. *76*, 50922s (1972); Thermal conduction near an interface in a nematic subjected to a magnetic field. [263] C. K. Yun, A. G. Fredrickson, J. Chem. Phys. *57*, 4313 (1972), C.A. *78*, 20322v (1973); Anisotropic transport processes in highly polarizable fluids. [264] C. K. Yun, A. G. Fredrickson, Mol. Cryst. Liq. Cryst. *24*, 69 (1973), C.A. *81*, 176405x (1974); Polarization of liquids. [265] M. J. Freiser, Phys. Rev. Lett. *24*, 1041 (1970), C.A. *73*, 8048z (1970); Ordered states of a nematic. [266] M. J. Freiser, Mol. Cryst. Liq. Cryst. *14*, 165 (1971), C.A. *75*, 123340s (1971); Successive transitions in a nematic. [267] Ya. I. Frenkel, Kinetic theory of liquids (Moskau 1945), (Oxford Univ. Press: New York). 1947. 544 pp., C.A. *41*, 2315a (1947); Kinetische Theorie der Flüssigkeiten (Deut. Verlag Wissensch.: Berlin). 1957. 510 pp., C.A. *52*, 9744c (1958). [268] M. Kléman, J. Friedel, J. Phys. (Paris), C-4, *30*, 43 (1969), C.A. *72*, 94273c (1970); Lignes de dislocation dans les cholestériques. [269] J. Friedel, M. Kléman, Nat. Bur. Stand (U.S.), Spec. Publ. No. *317*, 607 (1970), C.A. *75*, 92055w (1971); Application of dislocation theory to liquid crystals. [270] J. Friedel, Phys. 50 Years Later, Gen. Assem. Int. Union Pure Appl. Phys., 14th, *1972*, 193, C.A. *80*, 20217t (1974); Dislocations in solids and liquids.

[271] I. Gabrielli, L. Verdini, Nuovo cimento 2, 526 (1955), C.A. *50*, 6122b (1956); Velocity of propagation and coefficient of absorption of ultrasound in mesophases. [272] C. Gaehwiller, Phys. Lett. A *36*, 311 (1971), C.A. *75*, 144064v (1971); Viscosity coefficients of MBBA. [273] C. Gaehwiller, Phys. Rev. Lett. *28*, 1554 (1972), C.A. *77*, 67172r (1972); Temperature dependence of flow alignment in nematics. [274] C. Gaehwiller, Mol. Cryst. Liq. Cryst. *20*, 301 (1973), C.A. *79*, 58657s (1973); Direct determination of the five independent viscosity coefficients of nematics. [275] W. M. Gelbart, B. A. Baron, J. Chem. Phys. *66*, 207 (1977), C.A. *86*, 81991z (1977); Generalized van der Waals theory of the i/n transition. [275a] W. M. Gelbart, A. Gelbart, Mol. Phys. *33*, 1387 (1977), C.A. *87*, 141384m (1977); Effective one-body potentials for orientationally anisotropic fluids. [276] C. J. Gerritsma, W. J. A. Goossens, A. K. Niessen, Phys. Lett. A *34*, 354 (1971), C.A. *75*, 26541a (1971); Helical twist in a cholesteric Grandjean-Cano pattern. [277] C. J. Gerritsma, P. van Zanten, Phys. Lett. A *37*, 47 (1971), C.A. *76*, 38366f (1972); Periodic perturbations in the cholesteric plane texture. [278] J. A. Geurst, Phys. Lett. A *34*, 283 (1971), C.A. *75*, 11716n (1971); Continuum theory and focal conic texture for smectics. [279] J. A. Geurst, Phys. Lett. A *36*, 63 (1971); Generalised vorticity in the theory of liquid crystals. [280] J. A. Geurst, Phys. Lett. A *37*, 279 (1971), C.A. *76*, 77896y (1972); Continuum theory of smectics A.

[281] J. A. Geurst, Philips Res. Rep. *30*, 171 (1975); Continuum theory for smectics A. [282] S. K. Ghosh, Mol. Cryst. Liq. Cryst. *37*, 9 (1976), C.A. *86*, 113904s (1977); An empirical form for the orientation-ordering intermolecular potential in PAA. [283] R. Gibrat, Compt. rend. *185*, 1491 (1927), C.A. *22*, 1509² (1928); Structure à focales des corps smectiques. [284] R. Gibrat, Compt. rend. *188*, 183 (1929), C.A. *23*, 4607⁸ (1929); Variation avec la direction de la constante capillaire des corps smectiques. [285] A. Gierer, K. Wirtz, Z. Naturforsch. A 5, 270 (1950), C.A. *44*, 8186c (1950); Theorie der Ultraschallabsorption von Flüssigkeiten. [285a] P. I. Golubnichii, K. F. Olzoev, A. D. Filonenko,

Ukr. Fiz. Zh. *22*, 1039 (1977), C.A. *87*, 58672h (1977); Plasma-acoustic cavitation phenomena initiated by a laser pulse in liquid crystals. [286] W. J. A. Goossens, Phys. Lett. A *31*, 413 (1970), C.A. *73*, 49657r (1970); Molecular theory of the cholesteric phase. [287] W. J. A. Goossens, Mol. Cryst. Liq. Cryst. *12*, 237 (1971), C.A. *75*, 41757u (1971); Molecular theory of the cholesteric phase and of the twisting power of optically active molecules in a nematic. [288] F. K. Gorskii, N. M. Sakevich, Kristallogra-fiya *12*, 674 (1967), C.A. *68*, 16379v (1968); Determination of interphase energy in the solid-liquid crystalline phase boundary of PAA and cholesteryl caprinate. [289] F. K. Gorskii, M. E. Mikhlin, Krist. Fazovye Prevrashch. *1971*, 5, C.A. *76*, 18913d (1972); Change in viscosity in the isotropic-liquid/liquid-crystal transition point. [290] F. K. Gorskii, M. E. Mikhlin, Sb. Dokl. Vses. Nauch. Konf. Zhidk. Krist. Simp. Ikh Prakt. Primen., 2nd, *1972*, 99, C.A. *81*, 112805v (1974); Effect of a magnetic field on the viscosity of cholesteryl acetate.

[291] F. K. Gorskii, M. E. Mikhlin, A. I. Kubarko, Kinet. Mekh. Krist. *1973*, 176, C.A. *81*, 7056d (1974); Structural viscosity of liquid crystals. [292] F. Grandjean, Compt. rend. *164*, 280 (1917), C.A. *11*, 1350^5 (1917); Application de la théorie du magnétisme aux liquides anisotropes. [293] M. F. Grandjean, Bull. soc. franç. min. *42*, 42 (1919), C.A. *14*, 1773^9 (1920); Propriétés optiques de certaines structures de liquides anisotropes. [294] A. E. Green, R. S. Rivlin, Arch. Rat'l Mech. Anal. *17*, 113 (1964), C.A. *67*, 47813r (1967); Multipolar continuum mechanics. [295] A. E. Green, P. M. Naghdi, R. S. Rivlin, Int. J. Engng. Sci. *2*, 611 (1965); Directors and multipolar displacements in continuum mechanics. [296] H. Gruler, Z. Naturforsch. A *28*, 474 (1973), C.A. *79*, 31289m (1973); Elastic constants of nematics. Splay (k_{11}) and bend (k_{33}) elastic constants of 4,4'-dialkoxyazoxybenzene for n: 1 to 6. [297] H. Gruler, J. Chem. Phys. *61*, 5408 (1974), C.A. *83*, 19405w (1975); Elastic properties of the nematic phase influenced by molecular properties. [298] H. Gruler, Z. Naturforsch. A *30*, 230 (1975), C.A. *82*, 178636t (1975); Elastic constants of a nematic. [299] E. Guyon, P. Pieranski, F. Brochard, C. R. Acad. Sci., Sér. B *273*, 486 (1971), C.A. *75*, 155496p (1971); Étude de la conductivité thermique d'un film nématique sous champ magnétique. [300] E. Guyon, P. Pieranski, C. R. Acad. Sci., Sér. B *274*, 656 (1972), C.A. *76*, 146023a (1972); Étude expérimentale de la convection dans un film de cristal liquide nématique.

[301] E. Guyon, C. D. Mitescu, Thin Solid Films *12*, 355 (1972), C.A. *77*, 157356h (1972); Comparative study of size effects in solid and liquid films. [302] E. Guyon, P. Pieranski, J. Phys. (Paris), C-1, *36*, 203 (1975); Poiseuille flow instabilities in nematics. [302a] D. Horn, E. Guyon, P. Pieranski, Rev. Phys. Appl. Mathem. *11*, 139 (1976); Thermal convection in tilted nematic cells. [303] I. H. Ibrahim, W. Haase, Z. Naturforsch. A *31*, 1644 (1976), C.A. *86*, 63867j (1977); Order parameter temperature dependence of nematics related to magnetic and optical anisotropies. [304] I. Haller, J. Chem. Phys. *57*, 1400 (1972), C.A. *77*, 93943n (1972); Elastic constants of nematic MBBA. [305] B. I. Halperin, T. C. Lubensky, Solid State Commun. *14*, 997 (1974); Analogy between smectics A and superconductors. [306] F. Hardouin, M. F. Achard, H. Gasparoux, Solid State Commun. *14*, 453 (1974), C.A. *80*, 149647a (1974); Divergence du coefficient de viscosité γ_1 mesure en phase n au voisinage d'une transition s_A/n. [307] F. Hardouin, M. F. Achard, G. Sigaud, H. Gasparoux, Phys. Lett. A *49*, 25 (1974), C.A. *81*, 142267p (1974); Twist viscosity coefficient γ_1 divergence near the n/s_A transition. [307a] T. Hatakeyama, Y. Kagawa, J. Chem. Phys. *65*, 4128 (1976), C.A. *86*, 36754p (1977); Effects of electric fields on the ultrasonic attenuation in liquid crystals. [308] C. F. Hayes, J. Acoust. Soc. Am. *60*, 1227 (1976), C.A. *86*, 36552w (1977); Rayleigh disk in an aligned nematic. [308a] C. Sripaipan, C. F. Hayes, G. T. Fang, Phys. Rev. A *15*, 1297 (1977), C.A. *86*, 162955t (1977); Ultrasonically-induced optical effect in a nematic. [309] W. Helfrich, J. Chem. Phys. *50*, 100 (1969), C.A. *70*, 62167p (1969); Molecular theory of flow alignment of nematics. [310] W. Helfrich, Phys. Rev. Lett. *23*, 372 (1969), C.A. *71*, 94995u (1969); Capillary flow of cholesterics and smectics.

[311] W. Helfrich, Liq. Cryst. Ord. Fluids *1970*, 405, C.A. *78*, 152446y (1973); Capillary viscosimetry of cholesterics. [312] W. Helfrich, J. Chem. Phys. *53*, 2267 (1970), C.A. *73*, 92499c (1970); Torques in sheared nematics: simple model in terms of the theory of dense fluids. [313] W. Helfrich, J. Chem. Phys. *56*, 3187 (1972), C.A. *76*, 118439x (1972); Flow alignment of weakly ordered nematics. [314] W. Helfrich, Phys. Rev. Lett. *29*, 1583 (1972), C.A. *78*, 49182k (1973); Orienting action of sound on nematics. [315] W. Helfrich, Helv. Phys. Acta *45*, 35 (1972); Flow alignment of nematics. [316] W. Helfrich, Mol. Cryst. Liq. Cryst. *26*, 1 (1974), C.A. *81*, 112219g (1974); Inherent bounds to the elasticity and flexoelectricity of liquid crystals. [317] W. Helfrich, Ber. Bunsenges. Phys. Chem. *78*, 886 (1974), C.A. *82*, 10063e (1975); Elastizität und Hydrodynamik flüssiger Kristalle. [318] C. Heppke, F. Schneider, Z. Naturforsch. A *29*, 1356 (1974), C.A. *81*, 178165t (1974); Dynamik des Zerfalls von Inversionswänden

in einem nematischen flüssigen Kristall. [**319**] R. O. Herzog, H. Kudar, Trans. Faraday Soc. *29*, 1006 (1933), C.A. *28*, 1239^5 (1934); Viscosity of liquid crystals. [**320**] S. Hess, Z. Naturforsch. A *30*, 728 (1975); Irreversible thermodynamics of nonequilibrium alignment phenomena in molecular liquids and in liquid crystals. I. Derivation of nonlinear constitutive laws, relaxation of the alignment, phase transition. [**321**] S. Hess, Z. Naturforsch. A *30*, 1224 (1975); II. Viscous flow and flow alignment in the isotropic (stable and metastable) and nematic phases (cf. [320]). [**322**] S. Hess, Z. Naturforsch. A *31*, 1034 (1976), C.A. *85*, 169975n (1976); Fokker-Planck equation approach to flow alignment in liquid crystals. [**323**] S. Hess, Z. Naturforsch. A *31*, 1507 (1976), C.A. *86*, 81970s (1977); Pre- and post-transitional behavior of the flow alignment and flow-induced phase transition in liquid crystals. [**324**] S. Holmström, S. T. Lagerwall, Mol. Cryst. Liq. Cryst. *38*, 141 (1977); Shear flow in the presence of electric fields. [**325**] W. A. Hoyer, Phys. Rev. *94*, 812 (1954), C.A. *49*, 9977c (1955); The compressibility of liquid cholesterol benzoate at the isotropic-anisotropic transition. [**326**] W. A. Hoyer, A. W. Nolle, J. Chem. Phys. *24*, 803 (1956), C.A. *50*, 10472b (1956); Behavior of liquid crystals near the isotropic-anisotropic transition. [**327**] H. W. Huang, Phys. Rev. Lett. *26*, 1525 (1971), C.A. *75*, 54890g (1971); Hydrodynamics of nematics. [**328**] H. W. Huang, Bull. Am. Phys. Soc. *16*, 523 (1971); New macroscopic theory of nematics. [**329**] W. Huang, T. Mura, J. Appl. Phys. *43*, 239 (1972); Elastic energy of an elliptical-edge disclination. [**330**] H. Imura, K. Okano, Jap. J. Appl. Phys. *11*, 1440 (1972), C.A. *78*, 20300m (1973); Temperature dependence of the viscosity coefficients of liquid crystals.

[**331**] H. Imura, K. Okano, J. Fac. Eng., Univ. Tokyo, Ser. B *31*, 757 (1972), C.A. *79*, 46514e (1973); Fluctuation and relaxation in liquid crystals with special emphasis on pretransition phenomena just above the clearing point of nematics. [**332**] H. Imura, K. Okano, Oyo Butsuri *41*, 1177 (1972), C.A. *78*, 165174h (1973); Fluctuation and relaxation phenomena in liquid crystals. [**333**] H. Imura, K. Okano, Chem. Phys. Lett. *19*, 387 (1973), C.A. *78*, 165451w (1973); Theory of anomalous ultrasonic absorption and dispersion of nematics just above the clearing point. [**334**] H. Imura, K. Okano, Phys. Lett. A *42*, 403 (1973), C.A. *78*, 89547q (1973); Friction coefficient for a moving disclination in a nematic. [**335**] H. Imura, K. Okano, Phys. Lett. A *42*, 405 (1973), C.A. *78*, 89550k (1973); Interaction between disclinations in nematics. [**336**] Y. Ishida, Seisan-Kenkyu, *24*, 477 (1972), C.A. *78*, 63212b (1973); Disclination, a crystal lattice defect recently drawing attention. [**337**] F. M. Jaeger, Z. anorg. allgem. Chem. *101*, 1 (1917), C.A. *12*, 875 (1912); Temperaturabhängigkeit der molekularen freien Oberflächenenergie von Flüssigkeiten im Temperaturbereich von −80 bis +1650°C. [**338**] F. Jähnig, H. Schmidt, Symp. Faraday Soc. *1971*, No. 5, 9, C.A. *79*, 24182t (1973); Continuum theory of thermotropic systems. Hydrodynamics of liquid crystals. [**339**] F. Jähnig, H. Schmidt, Ann. Phys. (N. Y.) *71*, 129 (1972), C.A. *76*, 159484b (1972); Hydrodynamics of liquid crystals. [**340**] F. Jähnig, Z. Phys. *258*, 199 (1973), C.A. *79*, 24404s (1973); Dispersion and absorption of sound in nematics.

[**341**] F. Jähnig, Chem. Phys. Lett. *23*, 262 (1973), C.A. *80*, 124869m (1974); Interpretation of ultrasonic relaxation phenomena in nematics. [**342**] F. Jähnig, Liq. Cryst., Proc. Int. Conf., Bangalore, *1973*, 31, C.A. *84*, 114462s (1976); Temperature dependence of the viscosities of a nematic. [**343**] F. Jähnig, J. Phys. (Paris) *36*, 315 (1975), C.A. *82*, 178697p (1975); Critical damping of first and second sound at a s_A/n transition. [**344**] H. J. F. Jansen, G. Vertogen, J. G. J. Ypma, Mol. Cryst. Liq. Cryst. *38*, 87 (1977), C.A. *86*, 181024e (1977); Monte Carlo calculation of the n/i transition. [**345**] J. T. Jenkins, J. Fluid Mech. *45*, 465 (1971); Cholesteric energies. [**346**] J. T. Jenkins, Mol. Cryst. Liq. Cryst. *18*, 309 (1972), C.A. *77*, 144951y (1972); Material coefficient in cholesterics. [**347**] J. T. Jenkins, P. J. Barratt, Q. J. Mech. Appl. Math. *27*, Pt. 1, 111 (1974); Interfacial effects in the static theory of nematics. [**348**] J. T. Jenkins, Mol. Cryst. Liq. Cryst. *27*, 105 (1974), C.A. *82*, 79033n (1975); Propagating plane disclination surfaces in nematics. [**349**] M. Kahlweit, W. Ostner. Chem. Phys. Lett. *18*, 589 (1973), C.A. *78*, 140670d (1973); Estimation of the interfacial tension between the i and n states of a liquid crystal. [**349a**] V. G. Kamenskii, E. I. Kats, Zh. Eksp. Teor. Fiz. *71*, 2168 (1976); Van der Waals forces in liquid crystals with a large dielectric anisotropy. [**350**] J. I. Kaplan, E. Drauglis, Chem. Phys. Lett. *9*, 645 (1971), C.A. *75*, 101538k (1971); Statistical theory of the nematic mesophase.

[**351**] G. E. Zvereva, A. P. Kapustin, Primenenie Ul'traakustiki k Issled. Veshchestva *15*, 69 (1961), C.A. *59*, 1115g (1963); Behavior of nematic PAP in ultrasonic and electric fields. [**352**] A. P. Kapustin, L. M. Dmitriev, Kristallografiya *7*, 332 (1962), C.A. *57*, 9314e (1962); Effect of ultrasonic vibration on the domain structure of liquid crystals. [**353**] G. E. Zvereva, A. P. Kapustin, Akust. Zh. *10*, 122 (1964), C.A. *61*, 4983a (1964); Measurement of ultra-acoustical parameters in liquid-crystalline cholesteryl caprate. [**354**] G. E. Zvereva, A. P. Kapustin, Primenenie Ul'traakustiki k Issled. Veshchestva

20, 87 (1964), C.A. *64*, 16663e (1966); Absorption of ultrasound in mesomorphic *p',p*-nonyloxybenzaltoluidine. [**355**] A. P. Kapustin, G. E. Zvereva, Kristallografiya *10*, 723 (1965), C.A. *64*, 2818e (1966); Study of phase transition in polymesomorphic liquid crystals by ultrasonic methods. [**356**] A. P. Kapustin, N. T. Bykova, Kristallografiya *11*, 330 (1966), C.A. *65*, 3109a (1966); Phase transition in liquid crystals. [**357**] A. P. Kapustin, Izv. Vyssh. Ucheb. Zaved., Fiz. *10*, 55 (1967), C.A. *68*, 108160v (1968); Properties of liquid crystals. [**358**] N. T. Bykova, A. P. Kapustin, Primen. Ul'traakust. Issled. Veshchestva *22*, 149 (1967), C.A. *68*, 43418p (1968); Ultraacoustical studies of the properties of *p-n*-heptyloxybenzoic acid. [**359**] A. P. Kapustin, L. I. Mart'yanova, Primen. Ul'traakust. Issled. Veshchestva *23*, 213 (1967), C.A. *70*, 118332h (1969); Ultrasound velocity and absorption in a homologous series of dialkoxybenzenes as a function of alkyl chain length. [**360**] A. P. Kapustin, N. T. Bykova, Kristallografiya *13*, 345 (1968), C.A. *69*, 39407n (1968); Phase transitions in polymesomorphic liquid crystals.

[**361**] A. P. Kapustin, L. I. Mart'yanova, Kristallografiya *14*, 480 (1969), C.A. *71*, 117283r (1969); Phase transformations in liquid crystals of the homologs of dialkoxyazoxybenzene. [**362**] A. P. Kapustin, Kristallografiya *14*, 943 (1969), C.A. *72*, 59245y (1970); Elastic vibrations of nematics. [**363**] A. P. Kapustin, L. I. Mart'yanova, Kolloid. Zh. *32*, 60 (1970), C.A. *72*, 125321y (1970); Relation between mesophases and colloidal systems. [**364**] A. P. Kapustin, L. I. Mart'yanova, Kristallografiya *16*, 648 (1971), C.A. *75*, 53390g (1971); Temperature dependence of ultrasound velocity in liquid crystals in the region of phase transitions. [**365**] A. P. Kapustin, Tr. Akust. Inst. *14*, 78 (1971), C.A. *76*, 159480x (1972); Absorptions and dispersion of ultrasound in liquid crystals. [**366**] A. P. Kapustin, Z. K. Kuvatov, A. N. Trofimov, Izv. Vyssh. Ucheb. Zaved., Fiz. *14*, 150 (1971), C.A. *75*, 54879k (1971); Photoelastic phenomena in liquid crystals. [**367**] A. P. Kapustin, A. S. Lagunov, A. Rakhimov, Zh. Fiz. Khim. *46*, 1320 (1972), C.A. *77*, 67184w (1972); Acoustic relaxation of cholesterics. [**368**] A. P. Kapustin, L. I. Mart'yanova, Uch. Zap., Ivanov. Gos. Pedagog. Inst. *1972*, No. 99, 71, C.A. *78*, 152279w (1973); Ultrasound spectroscopy and its application to liquid crystals. [**369**] L. F. Lependin, A. P. Kapustin, V. A. Poyarkova, Uch. Zap., Ivanov. Gos. Pedagog. Inst. *1972*, No. 99, 77, C.A. *78*, 129322w (1973); Measuring rheological properties of liquid crystals by ultrasound methods. [**370**] A. P. Kapustin, Sb. Dokl. Vses. Nauch. Konf. Zhidk. Krist. Simp. Ikh Prakt. Primen., 2nd, *1972*, 137, C.A. *81*, 127623h (1974); Experimental studies of liquid crystals by acoustic method.

[**371**] A. P. Kapustin, O. A. Kapustina, Krist. Tech. *8*, 237 (1973), C.A. *79*, 84267j (1973); Properties of liquid crystals in the acoustic field. [**372**] A. P. Kapustin, A. S. Lagunov, A. Rakhimov, Int. Conf. on Liquid Crystals, Raman Res. Inst., Bangalore, December 3–8, 1973, Abstr. No. 7; Investigation of acoustic relaxation in liquid crystal esters of cholesterol. [**373**] A. V. Belousov, A. P. Kapustin, A. S. Lagunov, Akust. Zh. *19*, 905 (1973), C.A. *80*, 64650w (1974); Effect of a transverse magnetic field on acoustic relaxation in nematics. [**374**] A. V. Belousov, A. S. Lagunov, A. P. Kapustin, Zh. Fiz. Khim. *47*, 1564 (1973), C.A. *79*, 97929b (1973); Effect of a transverse magnetic field on the acoustic properties of nematics. [**375**] A. V. Belousov, A. P. Kapustin, A. S. Lagunov, Sov. Phys. Acoust. *19*, 579 (1974); Effect of a transverse magnetic field on acoustic relaxation in nematics. [**376**] A. Rakhimov, A. S. Lagunov, A. P. Kapustin, Zh. Fiz. Khim. *50*, 68 (1976), C.A. *84*, 140954y (1976); Study of acoustical relaxation in the anisotropic phase of cholesteryl formate and cholesteryl acetate. [**377**] O. A. Kapustina, Y. G. Statnikov, Zh. Eksp. Teor. Fiz. *64*, 226 (1973), C.A. *78*, 89543k (1973); Effect of ultrasonic waves on liquid crystals. [**378**] O. A. Kapustina, Akust. Zh. *20*, 1 (1974), C.A. *80*, 149459r (1974); Properties of liquid crystals in an acoustic field. [**379**] O. A. Kapustina, Y. G. Statnikov, Akust. Zh. *20*, 248 (1974), C.A. *81*, 42528f (1974); Orientation action of an acoustic wave on nematics. [**380**] O. A. Kapustina, Akust. Zh. *20*, 482 (1974); Acoustical properties and potential application of liquid crystals.

[**381**] O. A. Kapustina, Y. G. Statnikov, Zh. Eksp. Teor. Fiz. *66*, 635 (1974), C.A. *80*, 113478g (1974); Rheological properties of liquid crystals studied by acoustic currents. [**382**] O. A. Kapustina, V. N. Lupanov, Zh. Eksp. Teor. Fiz. *71*, 2324 (1976), C.A. *86*, 80988s (1977); Acousto-optical properties of a nematic layer with homogeneous orientation. [**382a**] P. P. Karat, N. V. Madhusudana, Mol. Cryst. Liq. Cryst. *40*, 239 (1977), C.A. *87*, 125609h (1977); Elasticity and orientational order in some 4'-*n*-alkyl-4-cyanobiphenyls. [**383**] C. G. Kartha, A. R. K. L. Padmini, J. Phys. Soc. Jap. *28*, 470 (1970), C.A. *72*, 83761j (1970); Viscous behavior of cholesterics. [**384**] C. G. Kartha, A. R. K. L. Padmini, G. S. Sastry, J. Phys. Soc. Jap. *31*, 617 (1971), C.A. *75*, 155780b (1971); Ultrasonic absorption in cholesterics. [**385**] C. G. Kartha, A. R. K. L. Padmini, J. Phys. Soc. Jap. *31*, 904 (1971), C.A. *75*, 121647y (1971); Ultrasonic and viscous behavior of polymesomorphic liquid crystal. [**386**] C. G. Kartha, A. R. K. L. Padmini, Indian J. Pure Appl. Phys. *9*, 725 (1971), C.A. *76*, 7521y (1972); Ultrasonic

velocity studies in cholesterics. [387] C. G. Kartha, J. F. Otia, A. R. K. L. Padmini, Indian J. Pure Appl. Phys. *12*, 648 (1974), C.A. *82*, 79008h (1975); Ultrasonic studies in mixed liquid crystals. [388] C. G. Kartha, A. R. K. L. Padmini, Mol. Cryst. Liq. Cryst. *29*, 243 (1975), C.A. *83*, 19463p (1975); Ultrasonic behavior of mixed liquid crystals near phase transitions. [389] P. P. Karat, N. V. Madhusudana, Mol. Cryst. Liq. Cryst. *36*, 51 (1976); Elastic and optical properties of 4'-*n*-alkyl-4-cyanobiphenyls. [390] W. Kast, W. Maier, Physica *4*, 957 (1937), C.A. *32*, 3224^9 (1938); Wechselwirkung der Moleküle in den anisotropen Flüssigkeiten.

[391] G. Becherer, W. Kast, Ann. Physik *41*, 355 (1942), C.A. *36*, 6390^5 (1942); Messungen der Viskosität kristalliner Flüssigkeiten nach der Helmholtzschen Methode. [392] Y. Kawamura, Y. Maeda, K. Okano, S. Iwayanagi, Jap. J. Appl. Phys. *12*, 1510 (1973), C.A. *79*, 150364s (1973); Anomalous ultrasonic absorption and dispersion of nematics near the clearing point. [393] Y. Kawamura, S. Iwayanagi, Sci. Pap. Inst. Phys. Chem. Res. (Jpn.) *70*, 38 (1976), C.A. *85*, 83540e (1976); Complex shear viscosity in the isotropic phase of MBBA. [394] Y. Kawamura, S. Iwayanagi, Mol. Cryst. Liq. Cryst. *38*, 239 (1977), C.A. *86*, 181028j (1977); Complex shear viscosity of liquid crystals in the isotropic phase of nematics. [395] P. N. Keating, Mol. Cryst. Liq. Cryst. *8*, 315 (1969), C.A. *71*, 85806f (1969); Theory of the cholesteric mesophase. [396] K. A. Kemp, Diss. Univ. Rhode Island, Kingston. 1974. 198 pp. Avail. Xerox Univ. Microfilms, Ann Arbor, Mich., Order No. 74-26,690, C.A. *82*, 103559m (1975); Ultrasonic propagation in nematics. [397] L. W. Kessler, S. P. Sawyer, Appl. Phys. Lett. *17*, 440 (1970), C.A. *74*, 45820m (1971); Ultrasonic stimulation of optical scattering in nematics. [398] A. G. Khachaturyan, J. Phys. Chem. Solids *36*, 1055 (1975), C.A. *83*, 156070x (1975); Development of helical cholesteric structure in a nematic due to the dipole-dipole interaction. [399] D. G. Kim, A. N. Rep'eva, Trans. Leningrad Ind. Inst. No. 7, Sect. Phys. Math. No. *4*, 3 (1937), C.A. *33*, 5718^1 (1939); Structure of smectic mesophase. [400] M. G. Kim, S. Park, M. Cooper, S. V. Letcher, Mol. Cryst. Liq. Cryst. *36*, 143 (1976), C.A. *86*, 49318z (1977); Shear viscosity measurements in CBOOA.

[401] H. Kimura, J. Phys. Soc. Jap. *30*, 1273 (1971), C.A. *75*, 42484q (1971); Theory of optical activity in nematics. [402] H. Kimura, Bussei Kenkyu *23*, B14 (1974), C.A. *83*, 69154j (1975); Molecular theory of nematics. [403] H. Kimura, J. Phys. Soc. Jap. *36*, 1280 (1974), C.A. *81*, 69539x (1974); Nematic ordering of rod-like molecules interacting via anisotropic dispersion forces as well as rigid-body repulsions. [404] H. Kimura, Phys. Lett. A *47*, 173 (1974), C.A. *80*, 137837z (1974); Theory of excluded volume effects in nematics. [405] H. Kimura, J. Phys. Soc. Jap. *37*, 1204 (1974); Orientational order and phase transition in Maier-Saupe model of liquid crystals. [406] U. D. Kini, G. S. Ranganath, Pramana *4*, 19 (1975), C.A. *83*, 88978x (1975); Effect of an axial magnetic field on the Poiseuille flow of a nematic. [407] U. D. Kini, G. S. Ranganath, S. Chandrasekhar, Pramana *5*, 101 (1975), C.A. *83*, 156074b (1975); Flow along the helical axis of cholesterics. [408] U. D. Kini, Pramana *7*, 378 (1976), C.A. *86*, 81984z (1977); Effect of magnetic fields and boundary conditions on the shear flow of nematics. [409] U. D. Kini, G. S. Ranganath, Mol. Cryst. Liq. Cryst. *38*, 311 (1977), C.A. *86*, 181031e (1977); Singularities in nematics. Effect of elastic constant variations. [410] E. J. Klein, A. P. Margozzi, Rev. Sci. Instrum. *41*, 238 (1970), C.A. *72*, 94091s (1970); Apparatus for the calibration of shear sensitive liquid crystals.

[411] M. Kléman, Phil. Mag. *27*, 1057 (1973), C.A. *79*, 10808q (1973); Defect densities in directional media, mainly liquid crystals. [412] M. Kléman, Bull. Soc. Fr. Mineral. Cristallogr. *95*, 215 (1972), C.A. *77*, 144694s (1972); Défauts dans les cristaux liquides. [413] M. Kléman, C. Williams, J. Phys. (Paris) Lett. *35*, 39 (1974); Interaction between parallel edge dislocation lines in a smectic A. [414] M. Kléman, J. Phys. (Paris) *35*, 595 (1974), C.A. *81*, 69551v (1974); Linear theory of dislocations in a smectic A. [415] M. Kléman, Liq. Cryst. Plast. Cryst. *1*, 76 (1974), C.A. *83*, 19080m (1975); Microstructures of liquid crystals. [416] M. Kléman, Adv. Liq. Cryst. *1*, 267 (1975), C.A. *84*, 37349f (1976); Defects in liquid crystals. [417] V. Vitek, M. Kléman, J. Phys. (Paris) *36*, 59 (1975), Surface disclinations in nematics. [418] M. Kléman, O. Parodi, J. Phys. (Paris) *36*, 671 (1975); Covariant elasticity for smectics A. [419] G. Ryschenkow, M. Kléman, J. Chem. Phys. *64*, 404 (1976), C.A. *84*, 82789k (1976); Surface defects and structural transitions in very low anchoring energy nematic thin films. [420] M. Kléman, G. Ryschenkow, J. Chem. Phys. *64*, 413 (1976), C.A. *84*, 82790d (1976); Two-dimensional nematics.

[421] M. Kléman, Philos. Mag. *34*, 79 (1976), C.A. *85*, 134880a (1976); Dislocations vis et surfaces minima dans les smectiques A. [421a] M. Kléman, J. Phys. (Paris) Lett. *37*, 93 (1976); Remarks concerning an analogy between dislocations in smectics and electric-current lines. [421b] M. Kléman, (Les Editions de Physique: Orsay). 1977. Vol. 1: 182 pp. Vol. 2: 139 pp. Points, lignes, parois dans

les fluides anisotropes et les solides cristallins. [**422**] N. V. Klyagina, G. E. Nevskaya, Akust. Zh. *20*, 916 (1974), C.A. *82*, 163369j (1975); Effect of a constant electric field on absorption of ultrasound in nematic PAA. [**423**] K. Kobayashi, Oyo Butsuri *40*, 532 (1971), C.A. *76*, 64394h (1972); Statistical mechanical theories of liquid crystal. [**424**] K. K. Kobayashi, W. M. Franklin, D. S. Moroi, Phys. Rev. A *7*, 1781 (1973), C.A. *78*, 165457c (1973); Molecular theory of collective modes in nematics. [**425**] K. Kobayashi, Kyoritsu Kagaku Raiburari *1*, 23 (1973), C.A. *83*, 35773v (1975); Physics of liquid crystals. [**426**] K. K. Kobayashi, W. M. Franklin, D. S. Moroi, Int. Conf. on Liquid Crystals, Raman Res. Inst., Bangalore, December 3–8, 1973, Abstr. No. 45; Molecular theory of collective modes in liquid crystals. [**427**] F. Kohler, The liquid state. (Verlag Chemie: Weinheim). 1972. 256 pp., C.A. *78*, 48265c (1973). [**428**] S. K. Kor, S. K. Pandey, Proc. Nucl. Phys. Solid State Phys., Symp. *17C*, 291 (1974), C.A. *84*, 52355y (1976); Ultrasonic propagation constants at the phase transition in two cholesterics. [**429**] S. K. Kor, S. K. Pandey, J. Chem. Phys. *64*, 1333 (1976), C.A. *84*, 112002t (1976); Ultrasonic investigation of cholesterics. [**430**] N. I. Koshkin, Primen. Ul'traakust. Issled. Veshchestva *22*, 29 (1967), C.A. *69*, 22207z (1968); Acoustic method for studying the molecular interaction in organic liquids near the melting point. [**430a**] N. I. Koshkin, A. A. Tabidze, S. V. Chumakova, Sb. Dokl. I Vses. Simpoz. po Akustooptich Spektroskopii. *1976*, 85, C.A. *86*, 198209s (1977); Acoustic spectroscopy of nematics on low-frequency shift waves.

[**431**] E. N. Kozhevnikov, I. A. Chaban, Akust. Zh. *21*, 421 (1975), C.A. *84*, 50986f (1976); Propagation of sound near the i/n transition. [**432**] T. J. Krieger, H. M. James, J. Chem. Phys. *22*, 796 (1954), C.A. *48*, 11867f (1954); Successive orientational transitions in crystals. [**433**] S. Krishnaswamy, R. Shashidhar, Liq. Cryst., Proc. Int. Conf., Bangalore, *1973*, 247, C.A. *84*, 112045j (1976); Experimental determination of the surface tension of nematics. [**434**] S. Krishnaswamy, R. Shashidhar, Mol. Cryst. Liq. Cryst. *35*, 253 (1976), C.A. *85*, 198673t (1976); Experimental studies of the surface tension of nematics. [**435**] S. Krishnaswamy, R. Shashidhar, Mol. Cryst. Liq. Cryst. *38*, 353 (1977), C.A. *86*, 177815w (1977); Surface tension in CBOOA. [**436**] B. Kronberg, D. F. R. Gilson, D. Patterson, J. Chem. Soc., Faraday Trans. 2, *72*, 1673 (1976), C.A. *86*, 36419h (1977); Effect of solute size and shape on orientational order in liquid crystals. [**437**] E. Kröner, K. H. Anthony, Ann. Rev. Mater. *5*, 43 (1975); Dislocations and disclinations in material structures: the basic topological concepts. [**438**] F. Krüger, Physik. Z. *14*, 651 (1913), C.A. *7*, 3439³ (1913); Viskosität der anisotropen Flüssigkeiten. [**439**] A. H. Krummacher, Diss. Halle 1929; Messung der Zähigkeit von kristallinischen Flüssigkeiten. [**440**] J. Kushick, B. J. Berne, J. Chem. Phys. *64*, 1362 (1976), C.A. *84*, 155920h (1976); Computer simulation of anisotropic molecular fluids.

[**441**] E. Kuss, Chem.-Ing.-Techn. *48*, 909 (1976); Viskositäts-Druckverhalten von flüssigen Kristallen. [**441a**] L. Lam, Z. Phys. B *27*, 349 (1977), C.A. *87*, 125578c (1977); Constraints, dissipation functions and cholesterics. [**442**] M. A. Bouchiat, D. Langevin-Cruchon, Phys. Lett. A *34*, 331 (1971), C.A. *75*, 10211g (1971); Molecular order at the free surface of a nematic from light reflectivity measurements. [**443**] D. Langevin, M. A. Bouchiat, J. Phys. (Paris) *33*, 77 (1972), C.A. *80*, 20953t (1974); Light scattering from the free surface of a nematic. [**444**] D. Langevin, M. A. Bouchiat, J. Phys. (Paris) *33*, 101 (1972), C.A. *77*, 39446q (1972); Spectre des fluctuations thermiques a la surface libre d'un cristal liquide nématique. [**445**] D. Langevin, J. Phys. (Paris) *33*, 249 (1972), C.A. *77*, 80612d (1972); Analyse spectrale de la lumière diffusée par la surface libre d'un cristal liquide nématique. Mesure de la tension superficielle et des coefficients de viscosité. [**446**] D. Langevin, M. A. Bouchiat, Mol. Cryst. Liq. Cryst. *22*, 317 (1973), C.A. *79*, 150360n (1973); Molecular order and surface tension for the n/i interface of MBBA deduced from light reflectivity and light scattering measurements. [**447**] D. Langevin, J. Phys. (Paris) *36*, 745 (1975), C.A. *83*, 89017v (1975); Diffusion de la lumière par la surface libre d'un cristal liquide au voisinage d'une transition n/s$_A$ du 2e ordre. [**448**] D. Langevin, J. Phys. (Paris) *37*, 901 (1976), C.A. *85*, 101454m (1976); Light scattering from the free surface near a second order n/s$_A$ transition. [**449**] G. Lasher, J. Chem. Phys. *53*, 4141 (1970), C.A. *74*, 7292h (1971); Nematic ordering of hard rods derived from a scaled particle treatment. [**450**] G. Lasher, Phys. Rev. A *5*, 1350 (1972), C.A. *76*, 104914m (1972); Monte-Carlo results for a discrete-lattice model of nematic ordering.

[**451**] P. A. Lebwohl, G. Lasher, Phys. Rev. A *6*, 426 (1972); Nematic-liquid-crystal order — a Monte Carlo calculation. [**452**] J. D. Y. Lee, Diss. Princeton Univ., Princeton, N. J. 1971, 138 pp. Avail. Univ. Microfilms, Ann Arbor, Mich., Order No. 72-2727, C.A. *76*, 118355s (1972); Mechanics of liquid crystals. [**453**] E. D. Lieberman, J. D. Lee, F. C. Moon, Appl. Phys. Lett. *18*, 280 (1971), C.A. *74*, 146623e (1971); Anisotropic ultrasonic wave propagation in a nematic placed in a magnetic field. [**453a**] M. A. Lee, C. W. Woo, Phys. Rev. A *16*, 750 (1977), C.A. *87*, 160138a (1977); Statistical-mechani-

cal calculations for nematics. [**454**] Y. S. Lee, Diss. Kent State Univ., Kent, Ohio *1971*, 171 pp. Avail. Univ. Microfilms, Ann Arbor, Mich., Order No. 72-9271, C.A. *77*, 10781j (1972); Ultrasonic shear wave study of the mechanical properties of a nematic. [**455**] F. Leenhouts, F. Van der Woude, A. J. Dekker, Phys. Lett. A *58*, 242 (1976), C.A. *85*, 166898d (1976); Twist elastic constant as a function of temperature for nematic MBBA. [**456**] L. Léger, Phys. Lett. A *44*, 535 (1973), C.A. *79*, 119285t (1973); Divergence of the bend elastic constant above a n/s_A quasi second order transition. [**457**] L. Léger, A. Martinet, J. Phys. (Paris) C-3, *37*, 89 (1976), C.A. *85*, 85786p (1976); Viscosity measurements in the n and s_A phases of a liquid crystal critical behavior. [**458**] H. Leipholz, Einführung in die Elastizitätstheorie (Braun: Karlsruhe). 1968. 312 pp. [**459**] H. N. W. Lekkerkerker, D. Carle, W. G. Laidlaw, J. Phys. (Paris) *37*, 1061 (1976), C.A. *85*, 152118n (1976); Hydrodynamic correlation functions in nematics. [**459a**] H. N. W. Lekkerkerker, J. Phys. (Paris), Lett. *38*, 277 (1977), C.A. *87*, 109758a (1977); Oscillatory convective instabilities in nematics. [**460**] F. M. Leslie, Q. J. Mech. Appl. Math. *19*, 347 (1966), C.A. *67*, 4267y (1967); Constitutive equations for anisotropic fluids.

[**461**] F. M. Leslie, Proc. Roy. Soc. A *307*, 359 (1968), C.A. *69*, 110911e (1968); Thermal effects in cholesterics. [**462**] F. M. Leslie, Mol. Cryst. Liq. Cryst. *7*, 407 (1969), C.A. *71*, 85814g (1969); Continuum theory of cholesterics. [**463**] F. M. Leslie, Symp. Faraday Soc. *1971*, No. 5, 33; Thermo-mechanical coupling in cholesterics. [**464**] F. M. Leslie, Liq. Cryst., Proc. Int. Conf., Bangalore, *1973*, 41, C.A. *84*, 114664j (1976); Distorted twisted orientation patterns in nematics. [**464a**] F. M. Leslie, J. Phys. D *9*, 925 (1976); Analysis of a flow instability in nematics. [**465**] S. V. Letcher, A. J. Barlow, Phys. Rev. Lett. *26*, 172 (1971), C.A. *74*, 68768q (1971); Dynamic shear properties of some smectics. [**466**] K. A. Kemp, S. V. Letcher, Phys. Rev. Lett. *27*, 1634 (1971), C.A. *76*, 51354v (1972); Ultrasonic determination of anisotropic shear and bulk viscosities in nematics. [**467**] K. A. Kemp, S. V. Letcher, J. Acoust. Soc. Am. *50*, 125 (1971); Ultrasonic propagation in oriented nematics. [**468**] K. A. Kemp, S. V. Letcher, IEEE Trans. Son. Ultrason. *19*, 408 (1972); Ultrasonic absorption and viscosity anisotropy in oriented nematics. [**469**] K. A. Kemp, S. V. Letcher, Liq. Cryst. Ord. Fluids *2*, 351 (1974), C.A. *86*, 180989m (1977); Bulk viscosities of MBBA from ultrasonic measurements. [**470**] S. P. Levitskii, A. T. Listrov, Prikl. Mat. *38*, 1031 (1974); Stability of flow of a liquid crystal layer on an inclined plane.

[**471**] Y. R. Lin-Liu, Y. M. Shih, C. W. Woo, H. T. Tan, Phys. Rev. A *14*, 445 (1976), C.A. *85*, 102649r (1976); Molecular model for cholesterics. [**472**] Y. R. Lin-Liu, Y. M. Shih, C. W. Woo, Phys. Lett. A *57*, 43 (1976), C.A. *85*, 70897b (1976); Relation between Frank constants and viscosity coefficients in an ellipsoid model for nematics. [**472a**] Y. R. Lin-Liu, Y. M. Shih, C. W. Woo, Phys. Rev. A *15*, 2550 (1977), C.A. *87*, 60982c (1977); Molecular theory of cholesterics. [**473**] L. N. Lisetskii, B. L. Timan, V. G. Tishchenko, Ukr. Fiz. Zh. (Russ. Ed.) *21*, 2064 (1976), C.A. *86*, 81966v (1977); Short-range order in the molecular-statistical theory of translation phase transition in liquid crystals. [**474**] L. N. Lisetskii, V. G. Tishchenko, Fiz. Tverd. Tela *19*, 280 (1977), C.A. *86*, 113885m (1977); Intermolecular interaction in cholesterics. [**474a**] L. N. Lisetskii, V. G. Tishchenko, Fiz. Tverd Tela *18*, 3674 (1976); Effect of dipole-dipole interaction on translational order at phase transitions in liquid crystals. [**475**] J. D. Litster, T. W. Stinson III, J. Appl. Phys. *41*, 996 (1970), C.A. *72*, 137307m (1970); Critical slowing of fluctuations in a nematic. [**476**] J. D. Litster, T. W. Stinson, N. A. Clark, Bull. Am. Phys. Soc. *16*, 519 (1971); Observation of coupling between orientational order fluctuations and hydrodynamic shear modes in a liquid crystal. [**477**] J. D. Litster, Crit. Phenomena Alloys, Magnets, Supercond., Battelle Inst. Mater. Sci. Colloq., 5th, *1970*, 393, C.A. *76*, 159474y (1972); Critical properties of ferromagnets and liquid crystals. [**478**] A. E. Lord, Jr., M. M. Labes, Phys. Rev. Lett. *25*, 570 (1970), C.A. *73*, 92400p (1970); Anisotropic ultrasonic properties of a nematic. [**479**] A. E. Lord, Jr., Mol. Cryst. Liq. Cryst. *18*, 313 (1972), C.A. *77*, 144861u (1972); Ultrasonic investigation of the c/n transition. [**480**] A. E. Lord, Jr., Phys. Rev. Lett. *29*, 1366 (1972), C.A. *79*, 58692z (1973); Anisotropic ultrasonic properties of a smectic.

[**481**] B. A. Lowry, M. R. Woodward, S. E. Monroe, Jr., G. C. Wetsel, Jr., Bull. Am. Phys. Soc. *17*, 350 (1972); Ultrasonic-wave attenuation in nematic EBBA. [**482**] T. C. Lubensky, Phys. Lett. A *33*, 202 (1970), C.A. *74*, 35841e (1971); Calculation of the elastic constant k_{11} for a nematic. [**483**] T. C. Lubensky, Phys. Rev. A *2*, 2497 (1970), C.A. *74*, 35844h (1971); Molecular description of nematics. [**484**] T. C. Lubensky, Phys. Rev. Lett. *29*, 206 (1972), C.A. *77*, 80615g (1972); Low-temperature phase of infinite cholesterics. [**485**] T. C. Lubensky, Phys. Rev. A *6*, 452 (1972); Hydrodynamics of cholesterics. [**486**] T. C. Lubensky, J. Phys. Chem. Solids *34*, 365 (1973), C.A. *78*, 140605m (1973); Spin model for cholesterics. [**487**] R. L. Humphries, P. G. James, G. R. Luckhurst, Symp. Faraday Soc. *1971*,

No. 5, 107, C.A. *79*, 24393n (1973); Molecular field treatment of liquid crystalline mixtures. **[488]** P. G. James, G. R. Luckhurst, Mol. Phys. *20*, 761 (1971); Representation of the anisotropic pseudopotential for nematogens. **[489]** R. L. Humphries, G. R. Luckhurst, Chem. Phys. Lett. *17*, 514 (1972), C.A. *78*, 97005b (1973); Molecular field theory of the nematic mesophase. **[490]** R. L. Humphries, P. G. James, G. R. Luckhurst, J. Chem. Soc. Faraday Trans. 2, *68*, 1031 (1972), C.A. *77*, 40018h (1972); Molecular field treatment of nematics.

[491] R. L. Humphries, G. R. Luckhurst, Mol. Cryst. Liq. Cryst. *26*, 269 (1974), C.A. *82*, 66456c (1975); Determination of the parameter in the pseudopotential for nematics. **[492]** G. R. Luckhurst, Spectrum *12*, 11 (1974), C.A. *82*, 42378c (1975); What makes a liquid crystal liquid? **[493]** G. R. Luckhurst, C. Zannoni, P. L. Nordio, U. Segre, Mol. Phys. *30*, 1345 (1975), C.A. *84*, 82870e (1976); Molecular field theory for uniaxial nematics formed by noncylindrically symmetric molecules. **[494]** G. R. Luckhurst, R. N. Yeates, Mol. Cryst. Liq. Cryst. *34*, 57 (1976), C.A. *85*, 152109k (1976); Negative order parameters for nematics? **[495]** R. L. Humphries, G. R. Luckhurst, Proc. Roy. Soc. A *352*, 41 (1976), C.A. *86*, 63857f (1977); Statistical theory of liquid crystalline mixtures: phase separation. **[495a]** J. Y. Denham, R. L. Humphries, G. R. Luckhurst, Mol. Cryst. Liq. Cryst. *41*, 67 (1977); Computer simulation studies of anisotropic systems. I. Linear lattice. **[495b]** G. R. Luckhurst, C. Zannoni, Nature *267*, 412 (1977), C.A. *87*, 175885r (1977); Why is the Maier-Saupe theory of nematics so successful? **[496]** J. E. Lydon, D. G. Robinson, Biochim. Biophys. Acta *260*, 298 (1972), C.A. *76*, 150258y (1972); Structures of cholesteryl ester mesophases revealed by freeze-fracturing. **[497]** J. R. McColl, C. S. Shih, Phys. Rev. Lett. *29*, 85 (1972), C.A. *77*, 67181t (1972); Temperature dependence of orientational order in a nematic at constant molar volume. **[498]** J. B. McCoy, L. S. Kowalczyk, Chem. Eng. Progr., Symp. Ser. *56*, 11 (1960), C.A. *60*, 11391g (1964); Thermal conductivity of some mesophases. **[499]** T. J. McKee, Diss. Yale Univ., New Haven, Conn. 1975. 126 pp. Avail. Xerox Univ. Microfilms, Ann Arbor, Mich., Order No. 76-13,212, C.A. *85*, 54898x (1976); Orientational order measurements in nematics and smectics. **[500]** W. L. McMillan, Phys. Rev. A *4*, 1238 (1971), C.A. *75*, 102335d (1971); Simple molecular model for smectics A.

[501] W. L. McMillan, Phys. Rev. A *8*, 1921 (1973), C.A. *80*, 7898j (1974); Simple molecular theory of the s$_C$ phase. **[502]** W. L. McMillan, Liq. Cryst. Ord. Fluids *2*, 141 (1974); Molecular order and molecular theories of liquid crystals. **[503]** D. McQueen, J. Phys. D *6*, 2273 (1973), C.A. *80*, 32367t (1974); Inelastic light scattering from the free surfaces of two nematics. **[504]** D. H. McQueen, V. K. Singhal, J. Phys. D *7*, 1983 (1974), C.A. *81*, 129152c (1974); Viscosities of nematics and smectics A by inelastic light scattering. **[505]** S. M. Ma, H. Eyring, Proc. Natl. Acad. Sci. U.S.A. *72*, 78 (1975), C.A. *83*, 51149b (1975); Significant structure theory applied to a mesophase system. **[505a]** V. B. Magalinskii, Izv. Vyssh. Uchebn. Zaved., Fiz. *20*, 114 (1977), C.A. *87*, 192291m (1977); Integral equation for the angular distribution function of anisotropic liquid taking into account short-range order in a quasichemical approximation. **[506]** W. Maier, A. Saupe, Z. Naturforsch. A *13*, 564 (1958), C.A. *53*, 3818i (1959); Einfache molekularstatistische Theorie der nematischen Phase, Dispersionswechselwirkung. **[507]** W. Maier, A. Saupe, Z. Naturforsch. A *14*, 882 (1959), C.A. *54*, 5200a (1960); Einfache molekularstatistische Theorie der nematischen Phase. **[508]** W. Maier, A. Saupe, Z. Naturforsch. A *15*, 287 (1960), C.A. *54*, 21912a (1960); Einfache molekularstatistische Theorie der nematischen Phase. **[509]** A. Saupe, W. Maier, Z. Naturforsch. A *16*, 816 (1961), C.A. *56*, 8114i (1962); Methoden zur Bestimmung des Ordnungsgrades nematischer Schichten. **[510]** W. Maier, Angew. Chem. *73*, 660 (1961); Struktur und Eigenschaften nematischer Phasen.

[511] H. Mailer, K. L. Likins, T. R. Taylor, J. L. Fergason, Appl. Phys. Lett. *18*, 105 (1971), C.A. *74*, 131560h (1971); Effect of ultrasound on a nematic. **[512]** T. Malkin, Trans. Faraday Soc. *29*, 977 (1933), C.A. *28*, 1239^1 (1934); Rotating molecules and the liquid crystalline state. **[513]** D. Marsh, J. Pochan, P. Erhardt, J. Chem. Phys. *58*, 5795 (1973), C.A. *79*, 46527m (1973); Mechanism of shear induced structural changes in liquid crystals. Cholesteric-polymer solutions. **[514]** P. Martinoty, Diss. Univ. Straßburg 1970; Quelques études sur les cristaux liquides par des methodes dynamiques. **[515]** P. Martinoty, S. Candau, C. R. Acad. Sci., Sér. B *271*, 107 (1970), C.A. *73*, 113819z (1970); Relaxation ultrasonore dans un cristal liquide nématique. **[516]** P. Martinoty, S. Candau, Mol. Cryst. Liq. Cryst. *14*, 243 (1971), C.A. *75*, 133266m (1971); Determination of viscosity coefficient of a nematic using shear waves reflectance technique. **[517]** P. Martinoty, S. Candau, F. Debeauvais, Phys. Rev. Lett. *27*, 1123 (1971), C.A. *75*, 144801h (1971); Dynamic properties near the n/i transition. **[518]** P. Martinoty, S. Candau, Phys. Rev. Lett. *28*, 1361 (1972), C.A. *77*, 10716s (1972); Ultrasonic impedometric studies in cholesterics. **[519]** P. Martinoty, S. Candau, J. Phys. (Paris), C-6, *33*, 81 (1972), C.A. *79*,

46523g (1973); Mesure de l'impédance mécanique de cisaillement des mésophases nématique et cholestérique aux fréquences ultrasonores. [**520**] S. Candau, P. Martinoty, F. Debeauvais, C. R. Acad. Sci., Sér. A *277*, 769 (1973), C.A. *80*, 20326r (1974); Écoulements non newtoniens de cholestériques.

[**521**] S. Candau, P. Martinoty, R. Zana, J. Phys. (Paris), Lett. *36*, 13 (1975), C.A. *82*, 116471b (1975); Ultrasonic investigation of rotational isomerism on mesophases. [**522**] Y. Thiriet, P. Martinoty, J. Phys. (Paris), Lett. *36*, 125 (1975), C.A. *83*, 51196q (1975); Ultrasonic impedometric studies in s_B N-(*p*-butyloxybenzylidene)-*p-n*-octylaniline. [**523**] S. Nagai, P. Martinoty, S. Candau, R. Zana, Mol. Cryst. Liq. Cryst. *31*, 243 (1975), C.A. *84*, 114421e (1976); Intramolecular ultrasonic relaxation of nematics far below the transition temperature. [**524**] F. Kiry, P. Martinoty, J. Phys. (Paris) C-3, *37*, 113 (1976), C.A. *85*, 85789s (1976); Ultrasonic absorption and pretransitional phenomena near a second order n/s_A transition. [**525**] S. Nagai, P. Martinoty, S. Candau, J. Phys. (Paris) *37*, 769 (1976), C.A. *85*, 70919k (1976); Ultrasonic investigation of nematics in the i and n phases. [**526**] F. Kiry, P. Martinoty, J. Phys. (Paris) *38*, 153 (1977), C.A. *86*, 82004s (1977); Ultrasonic investigation of anisotropic viscosities in a nematic. [**527**] P. Martinoty, F. Kiry, S. Nagai, S. Candau, F. Debeauvais, J. Phys. (Paris) *38*, 159 (1977), C.A. *86*, 82005t (1977); Viscosity coefficients in the i phase of a nematic. [**527a**] S. Candau, P. Martinoty, Proc. Int. Sch. Phys. "Enrico Fermi" *63*, 165 (1974), C.A. *87*, 31985a (1977); Absorption des ondes ultrasonores longitudinales et transversales dans les cristaux liquides. [**527b**] F. Kiry, P. Martinoty, J. Phys. (Paris), Lett. *38*, 389 (1977), C.A. *87*, 192296s (1977); Shear wave attenuation in smectics A. [**528**] A. F. Martins, Phys. Lett. A *38*, 211 (1972), C.A. *76*, 104990h (1972); Simple model for self-diffusion in the i phase of nematics. [**529**] A. F. Martins, Port. Phys. *9*, 1 (1974), C.A. *81*, 112212z (1974); Molecular approach to the hydrodynamic viscosities of nematics. [**529a**] A. F. Martins, A. C. Diogo, Port. Phys. *9*, 129 (1975), C.A. *86*, 177509z (1977); Simple molecular statistical interpretation of the nematic viscosity γ_1. [**530**] M. A. Cotter, D. E. Martire, Mol. Cryst. Liq. Cryst. *7*, 295 (1969), C.A. *71*, 85488d (1969); Quasi-chemical lattice treatment of rod-like molecules. Application to the n/i transition.

[**531**] M. A. Cotter, D. E. Martire, J. Chem. Phys. *52*, 1902 (1970), C.A. *72*, 83168w (1970); Statistical mechanics of rodlike particles. I. Scaled particle treatment of a fluid of perfectly aligned rigid cylinders. [**532**] M. A. Cotter, D. E. Martire, J. Chem. Phys. *52*, 1909 (1970), C.A. *72*, 83169x (1970); II. Scaled particle investigation of the aligned → isotropic transition in a fluid of rigid spherocylinders (cf. [531]). [**533**] M. A. Cotter, D. E. Martire, J. Chem. Phys. *53*, 4500 (1970), C.A. *74*, 25744m (1971); III. Fluid of rigid spherocylinders with restricted orientational freedom (cf. [531]). [**534**] H. T. Peterson, D. E. Martire, M. A. Cotter, J. Chem. Phys. *61*, 3547 (1974), C.A. *82*, 24591n (1975); Lattice model for a binary mixture of hard rods of different lengths. Application to solute induced n/i transitions. [**535**] G. I. Ågren, D. E. Martire, J. Phys. (Paris), C-1, *36*, 141 (1975); Lattice model for a binary mixture of hard rods and hard cubes. Application to solute induced n/i transitions. [**536**] L. I. Mart'yanova, Sb. Dokl. Vses. Nauch. Konf. Zhidk. Krist. Simp. Ikh Prakt. Primen., 2nd, *1972*, 150, C.A. *81*, 178202c (1974); Shear and dilatational viscosities in liquid crystals. [**536a**] L. I. Mart'yanova, S. V. Chumakova, Pis'ma Zh. Eksp. Teor. Fiz. *25*, 266 (1977), C.A. *86*, 198198n (1977); Viscoelastic relaxation of nematics in the i/n transition region. [**537**] S. Meiboom, R. C. Hewitt, Phys. Rev. Lett. *30*, 261 (1973), C.A. *78*, 89528j (1973); Measurements of the rotational viscosity coefficient and the shear-alignment angle in nematics. [**538**] S. Meiboom, R. C. Hewitt, Phys. Rev. Lett. *34*, 1146 (1975), C.A. *83*, 36068f (1975); Rotational viscosity of a smectic C. [**538a**] S. Meiboom, R. C. Hewitt, Phys. Rev. A *15*, 2444 (1977), C.A. *87*, 60979g (1977); Rotational viscosity in smectic TBBA. [**539**] T. J. Scheffer, H. Gruler, G. Meier, Solid State Commun. *11*, 253 (1972), C.A. *77*, 93954s (1972); Periodic free-surface disclination at the n/s_C transition. [**540**] H. Gruler, T. J. Scheffer, G. Meier, Z. Naturforsch. A *27*, 966 (1972), C.A. *77*, 106377h (1972); Elastic constants of nematics. Theory of the normal deformation.

[**541**] H. Gruler, G. Meier, Mol. Cryst. Liq. Cryst. *23*, 261 (1973), C.A. *80*, 88152m (1974); Elastic constants of the nematic homologous 4,4'-di(n-alkoxy)azoxybenzenes. [**541a**] H. Meirovitch, Chem. Phys. *21*, 251 (1977), C.A. *86*, 198214q (1977); Computer simulation of a lattice model of nematic liquid crystal. [**542**] L. Melamed, J. Opt. Soc. Am. *61*, 1551 (1971); Stress-optic effects in a liquid crystal velocimeter. [**543**] R. B. Meyer, Mol. Cryst. Liq. Cryst. *16*, 355 (1972), C.A. *76*, 132593x (1972); Point disclinations at a n/i interface. [**544**] R. B. Meyer, Phil. Mag. *27*, 405 (1973), C.A. *78*, 89522c (1973); Existence of even indexed disclinations in nematics. [**545**] L. Cheung, R. B. Meyer, H. Gruler, Phys. Rev. Lett. *31*, 349 (1973), C.A. *79*, 84283m (1973); Measurements of nematic elastic constants near a second order n/s_A phase change. [**546**] N. A. Clark, R. B. Meyer, Appl. Phys. Lett. *22*, 493

(1973), C.A. *79*, 24426a (1973); Strain-induced instability of monodomain smectics A and cholesterics. [547] L. Cheung, R. B. Meyer, Phys. Lett. A *43*, 261 (1973), C.A. *78*, 165439y (1973); Pretransitional anomaly in the bend elastic constant for a n/s_A transition. [548] L. Cheung, R. B. Meyer, Bull. Am. Phys. Soc. *17*, 349 (1972); Theory for optical and dielectric effects in cooperatively aligned nematic under electric field. [549] R. J. Meyer, W. L. McMillan, Phys. Rev. A *9*, 899 (1974), C.A. *80*, 113271j (1974); Simple molecular theory of the s_C, s_B, and s_H phases. [550] R. J. Meyer, Phys. Rev. A *12*, 1066 (1975), C.A. *83*, 200473g (1975); Molecular theory of the s_E, s_H, and s_{VI} phases. [550a] R. B. Meyer, T. C. Lubensky, Phys. Rev. A *14*, 2307 (1976); Mean-field theory of n/s_A transition. [550b] C. S. Rosenblatt, R. Pindak, N. A. Clark, R. B. Meyer, J. Phys. (Paris) *38*, 1105 (1977), C.A. *87*, 160147c (1977); The parabolic focal conic: a new s_A defect.

[551] M. Miesowicz, Nature *136*, 261 (1935), C.A. *29*, 7136[1] (1935); Influence of a magnetic field on the viscosity of PAA. [552] M. Miesowicz, Bull intern. acad. polon. sci., Classe sci. math. nat. *1936A*, 228, C.A. *31*, 3354[6] (1937); Effect of a magnetic field on the viscosity of anisotropic liquids. [553] M. Miesowicz, Nature *158*, 27 (1946), C.A. *40*, 5614[4] (1946); Three coefficients of viscosity of anisotropic liquids. [553a] R. I. Mints, E. V. Krivopishina, Pis'ma Zh. Tekh. Fiz. *3*, 485 (1977), C.A. *87*, 61015b (1977); Morphology of disclinations in a cholesteric. [554] K. Miyakawa, J. Phys. Jap. *39*, 628 (1975), Fluctuations in nematics under thermal gradient. [555] K. Miyano, J. B. Ketterson, Phys. Rev. Lett. *31*, 1047 (1973), C.A. *79*, 150367v (1973); Anisotropic ultrasound propagation in a smectic A. [556] K. Miyano, J. B. Ketterson, Phys. Rev. A *12*, 615 (1975), C.A. *83*, 156050r (1975); Ultrasonic study of liquid crystals. [557] K. Miyano, Diss. Northwest, Univ., Evanston, Ill. 1975. 126 pp. Avail. Xerox Univ. Microfilms, Ann Arbor, Mich., Order No. 75-29,711, C.A. *84*, 114456t (1976); Ultrasonic study of liquid crystals. [558] K. Miyano, Y. R. Shen, Appl. Phys. Lett. *28*, 473 (1976), C.A. *84*, 172365y (1976); Domain pattern excited by surface acoustic waves in a nematic film. [558a] K. Miyano, Y. R. Shen, Phys. Rev. A *15*, 2471 (1977), C.A. *87*, 60980a (1977); Excitation of stripe domain patterns by propagating acoustic waves in an oriented nematic film. [559] S. N. Mochalin, Uch. Zap., Ivanov Gos. Pedagog. Inst. *1972*, No. 99, 321, C.A. *78*, 151866y (1973); Liquid crystal surface tension. [560] S. E. Monroe, Jr., G. C. Wetsel, Jr., M. R. Woodard, B. A. Lowry, J. Chem. Phys. *63*, 5139 (1975), C.A. *84*, 52443a (1976); Ultrasonic investigation of viscosity coefficients in nematic EBBA.

[561] S. E. Monroe, Jr., Diss. South. Methodist Univ., Dallas, Tex. 1976. 181 pp. Avail. Xerox Univ. Microfilms, Ann Arbor, Mich., Order No. 76-29,372, C.A. *86*, 63877n (1977); Investigation of the ultrasonic attenuation and viscosity coefficients of nematic EBBA and measurements of the attenuation of transverse ultrasonic waves in calcium fluoride using the method of Bragg scattering. [561a] M. A. D. Moura, T. C. Lubensky, Y. Imry, A. Rony, Phys. Rev. B *13*, 2176 (1976); Coupling to anisotropic elastic media. Magnetic and liquid-crystal phase transitions. [562] M. E. Mullen, B. Leuthi, M. J. Stephen, Phys. Rev. Lett. *28*, 799 (1972), C.A. *76*, 132585y (1972); Sound velocity in a nematic. [563] T. Mura, Arch. Mech. (Warszawa) *24*, 449 (1972); Semi-microscopic plastic distortion and disclinations. [564] J. Murakami, J. Phys. Soc. Jpn. *42*, 210 (1977), C.A. *86*, 127753d (1977); Molecular theory of surface tension for liquid crystal. [565] I. Muscutariu, S. Bhattacharya, J. B. Ketterson, Phys. Rev. Lett. *35*, 1584 (1975), C.A. *84*, 37462n (1976); Acoustic Brillouin-zone effect in a cholesteric. [566] F. R. N. Nabarro, J. Phys. (Paris) *33*, 1089 (1972); Singular lines and singular points of ferromagnetic spin systems and of nematics. [567] S. Nagai, K. Iizuka, Keiryo Kenkyusho Hokoku *23*, 163 (1974), C.A. *82*, 178681d (1975); Effect of ultrasound on nematic liquid crystal. [568] S. Nagai, K. Iizuka, Jpn. J. Appl. Phys. *13*, 189 (1974); Effect of ultrasound to nematic liquid crystal. [569] S. Nagai, K. Iizuka, Jpn. J. Appl. Phys. *14*, 567 (1975), C.A. *82*, 178704p (1975); Measurement of the dynamic viscosity of MBBA by a torsional quartz oscillator. [570] S. Nagai, K. Iizuka, Bull. Natl. Res. Lab. Metrol. Tokyo *31*, 17 (1975), C.A. *84*, 187777h (1976); Measurement of the dynamic viscosity of MBBA by a torsional quartz oscillator.

[571] J. F. Nagle, A. Yanagawa, J. Stecki, Mol. Cryst. Liq. Cryst. *37*, 127 (1976), C.A. *86*, 113910r (1977); Model calculations for nematic ordering. [572] M. Nakamizo, S. Sakagami, A. Takase, Chem. Lett. *1972*, 71, C.A. *76*, 132583w (1972); Molecular arrangement in the nematic droplet. [572a] M. N. L. Narasimhan, Proc.-Annu. Meet. Soc. Eng. Sci., 12th *1975*, 61, C.A. *86*, 198241w (1977); Continuum theory of heat-conducting smectics. [573] G. G. Natale, D. E. Commins, Phys. Rev. Lett. *28*, 1439 (1972), C.A. *77*, 24966k (1972); Temperature dependence of anisotropic-ultrasonic propagation in a nematic. [574] J. Nehring, Phys. Rev. A *7*, 1737 (1973), C.A. *78*, 165453y (1973); Calculation of the structure and energy of nematic threads. [575] V. B. Nemtsov, Nauch. Tr. Vses. Zaoch. Mashinostr.

In-t *47*, 42 (1975), C.A. *85*, 200778b (1976); Statistical theory of the relaxation of the orientation order in nematics. [575a] V. B. Nemtsov, Physica A *86*, 513 (1977), C.A. *86*, 198219v (1977); Statistical hydrodynamics of cholesterics. [576] M. W. Neufeld, Physik. Z. *14*, 646 (1913), C.A. *7*, 3075[6] (1913); Einfluß eines Magnetfeldes auf die Ausflußgeschwindigkeit anisotroper Flüssigkeiten aus Kapillaren. [577] P. J. Sell, A. W. Neumann, Kolloid-Z. Z. Polym. *222*, 160 (1968), C.A. *69*, 69897q (1968); Temperaturabhängigkeit der Oberflächenspannung von Cholesterylstearat. [578] A. W. Neumann, P. J. Sell, Z. Phys. Chem. *65*, 19 (1969), C.A. *71*, 84811s (1969); Oberflächen- und Volumeneigenschaften von Cholesterylestern homologer Fettsäuren. II. Temperaturabhängigkeit der Oberflächenspannung. [579] A. W. Neumann, L. J. Klementowski, R. W. Springer, J. Colloid Interface Sci. *41*, 538 (1972), C.A. *78*, 62549e (1973); Surface and bulk properties of cholesteryl stearate. [580] A. W. Neumann, R. W. Springer, R. T. Bruce, Mol. Cryst. Liq. Cryst. *27*, 23 (1974), C.A. *82*, 118332f (1975); Surface and bulk properties of PAA.

[581] G. E. Nevskaya, Sb. Dokl. Vses. Nauch. Konf. Zhidk. Krist. Simp. Ikh Prakt. Primen., 2nd, *1972*, 144, C.A. *81*, 112191s (1974); Effect of an electric field on ultrasound absorption in PAA. [582] G. E. Nevskaya, M. A. Shorokhova, Akust. Zh. *21*, 127 (1975), C.A. *83*, 51515t (1975); Effect oa an a.c. electric field on the absorption of ultrasound in PAA. [582a] G. E. Nevskaya, Kholestericheskie Zhidk. Krist. *1976*, 81, C.A. *87*, 175707j (1977); Acoustical properties of cholesterics. [583] A. M. North, R. A. Pethrick, Transfer Storage Energy Mol. *4*, 441 (1969), C.A. *83*, 35780v (1975); Acoustic studies of liquid crystals and molecular solids. [584] T. F. North, W. G. B. Britton, R. W. B. Stephens, Ultrason. Int. Conf. Proc. *1975*, 120, C.A. *84*, 129399a (1976); Interaction of surface waves with liquid crystals. [585] L. Novakovic, G. C. Shukla, Fizika *4*, 29 (1972), C.A. *76*, 132588b (1972); Quantum theory of liquid crystals. [585a] V. F. Nozdrev, A. S. Lagunov, S. V. Pasechnik, Sb. Dokl. I Vses. Simpoz. po Akustooptich. Spektroskopii. *1976*, 98, C.A. *86*, 181017e (1977); Orientational order and acoustic properties of liquid crystals. [586] L. S. Ornstein, Z. Krist. *79*, 90 (1931), C.A. *26*, 2357[7] (1932); Experimentelle und theoretische Begründung der Schwarmbildung in kristallinen Flüssigkeiten. [587] L. S. Ornstein, W. Kast, Trans. Faraday Soc. *29*, 931 (1933), C.A. *28*, 1238[7] (1934); New arguments for the swarm theory of liquid crystals. [588] L. S. Ornstein, Kolloid-Z. *69*, 137 (1934), C.A. *29*, 1301[8] (1935); Schwarmtheorie der flüssigen Kristalle. [589] L. S. Ornstein, Proc. Acad. Sci. Amsterdam *37*, 318 (1934), C.A. *28*, 6040[5] (1934); Swarm theory of liquid crystals. [590] L. S. Ornstein, Proc. Acad. Sci. Amsterdam *41*, 1046 (1938), C.A. *33*, 8465[2] (1939); Theory of liquid crystals.

[591] L. S. Ornstein, Nederland. Tijdschr. Natuurkunde *7*, 373 (1940), C.A. *35*, 6495[8] (1941); Liquid crystals. [592] Orsay Liquid Crystal Group, Phys. Rev. Lett. *22*, 1361 (1969), Errate *23*, 208 (1969); Quasielastic Rayleigh scattering in nematics. [593] Groupe d'étude des Cristaux Liquides, J. Chem. Phys. *51*, 816 (1969), C.A. *71*, 64296z (1969); Dynamics of fluctuations in nematics. [594] Orsay Liquid Crystal Group, J. Phys. (Paris), C-4, *30*, 38 (1969), C.A. *72*, 94272b (1970); Lignes doubles de désinclinaison dans les cristaux liquides cholestériques a grands pas. [595] Orsay Liquid Crystal Group, Mol. Cryst. Liq. Cryst. *13*, 187 (1971), C.A. *75*, 25686q (1971); Viscosity measurements by quasi elastic light scattering in PAA. [596] Orsay Liquid Crystal Group, Solid State Commun. *9*, 653 (1971); Simplified elastic theory for smectics C. [597] Orsay Liquid Crystal Group, J. Phys., Paris, C-1, *36*, 305 (1975); Flow properties of smectics A. [598] C. W. Oseen, Stockholm: Almqvist & Wikesell, Berlin: R. Friedländer 1920 und 1922, 39 pp., C.A. *17*, 922[5] (1923); Kinetische Theorie der kristallinischen Flüssigkeiten. [599] C. W. Oseen, Arkiv. Mat. Astron. Fysik *18*, No. 4, 25 pp. (1923), C.A. *18*, 605[7] (1924); Theorie der anisotropen Flüssigkeiten. [600] C. W. Oseen, Arkiv. Mat. Astron. Fysik *18*, No. 8, 23 pp. (1923), C.A. *18*, 1072[5] (1924); Theory of anisotropic liquids. III.

[601] C. W. Oseen, Arkiv. Mat. Astron. Fysik *18*, No. 13, 36 pp. (1924), C.A. *18*, 1608[1] (1924); Theory of anisotropic liquids. IV. Propagation of light through an anisotropic liquid. [602] C. W. Oseen, Arkiv. Mat. Astron. Fysik *18*, No. 15, 1 (1924), C.A. *19*, 427[6] (1925); Theory of anisotropic liquids. V. Explanation of the focal conics liquids discovered by Friedel and Grandjean. [603] C. W. Oseen, Arkiv. Mat. Astron. Fysik *19*A, No. 5, 1 (1925), C.A. *19*, 1646[1] (1925); Theorie der anisotropen Flüssigkeiten. VI. [604] C. W. Oseen, Arkiv. Mat. Astron. Fysik *19*A, No. 9A, 1, C.A. *19*, 1646[1] (1925); Theory of anisotropic liquids. VII. [605] C. W. Oseen, Arkiv. Mat. Astron. Fysik *21*A, No. 11, 1 (1928), C.A. *23*, 4118[7] (1929); Theory of anisotropic liquids. VIII. Plane structures. [606] C. W. Oseen, Arkiv. Mat. Astron. Fysik *21*A, No. 11, 14 (1928), C.A. *23*, 4118[7] (1929); Theory of anisotropic liquids. IX. Propagation of light through a helicoidal twisted structure parallel to the axis. [607] C. W. Oseen, Arkiv. Mat. Astron. Fysik *21*A, 17 pp., C.A. *24*, 3409[8] (1930); Theory of anisotropic liquids. X. [608] C. W. Oseen, Arkiv. Mat. Astron. Fysik *21*A, No. 25, 10 pp. (1929), C.A. *24*, 764[4]

(1930); Theory of anisotropic liquids. XI. Reflection laws of an obliquely truncated, helicoidal twisted structure. [609] C. W. Oseen (Borntraeger: Berlin). 1929. 87 pp., C.A. *23*, 4401[6] (1929); Die anisotropen Flüssigkeiten (Tatsachen und Theorien). [610] C. W. Oseen, Fortschritte Chem., Physik physik. Chem. *20*B, Heft 2 (1929), C.A. *23*, 2085[8] (1929); Die anisotropen Flüssigkeiten, Tatsachen und Theorien. [611] C. W. Oseen, Arkiv. Mat. Astron. Fysik *22*A, No. 12, 20 pp. (1930), C.A. *25*, 10[6] (1931); Theory of anisotropic liquids. XII. Relations between molecular structure and density fluctuations. [612] C. W. Oseen, Arkiv. Mat. Astron. Fysik *22*A, No. 17, 12 pp. (1931), C.A. *25*, 2919[9] (1931); Theory of anisotropic liquids. XIII. Optical activity of twisted structures. [613] C. W. Oseen, Arkiv Mat. Astron. Fysik *22*A, No. 18, 19 pp. (1931), C.A. *25*, 2920[2] (1931); Theory of anisotropic liquids. XIV. Possible structures of nematics. [614] C. W. Oseen, Arkiv. Mat. Astron. Fysik *22*A, 1 (1931), C.A. *25*, 4453[9] (1931); Theory of anisotropic liquids. XV. Geometric optics of nematics. [615] C. W. Oseen, Z. Krist. *79*, 173 (1931), C.A. *26*, 5805[2] (1932); Probleme für die Theorie der anisotropen Flüssigkeiten. [616] C. W. Oseen, Arkiv. Mat. Astron. Fysik *23*A, No. 3, 1 (1932), C.A. *26*, 3159[5] (1932); Theory of anisotropic liquids. XVI. Thermodynamic theory of the motion of an anisotropic liquid. [617] C. W. Oseen, Arkiv. Mat. Astron. Physik *23*A, No. 17, 27 pp. (1933), C.A. *27*, 4145[8] (1933); Theory of anisotropic liquids. XVII. Doubly rotating structures and their optics. [618] C. W. Oseen, Arkiv. Mat. Astron. Fysik *23*A, No. 24, 1 (1933), C.A. *27*, 4974[3] (1933); Theory of anisotropic liquids. XVIII. Structures of nematic cholesterol substances. [619] C. W. Oseen, Arkiv. Mat. Astron. Fysik *23*A, No. 25, 1 (1933); Theory of anisotropic liquids. XIX. Dependence of anisotropy upon temperature in the presence of external forces. [620] C. W. Oseen, Arkiv. Mat. Astron. Fysik *24*A, No. 14, 18 pp. (1933), C.A. *28*, 2963[2] (1934); Theory of anisotropic liquids. XX. Simplest cases of the symmetry of free energy of molecules and their orientation.

[621] C. W. Oseen, Trans. Faraday Soc. *29*, 883 (1933), C.A. *27*, 5601[4] (1933); Theory of liquid crystals. [622] C. W. Oseen, Arkiv. Mat. Astron. Fysik *24*A, No. 19, 11 pp. (1934), C.A. *28*, 3633[7] (1934); Theory of anisotropic liquids. XXI. Molecular forces that produce liquid crystals. [623] C. W. Oseen, Arkiv. Mat. Astron. Fysik *26*A, No. 5, 10 pp. (1937), C.A. *32*, 2799[6] (1938); Theory of anisotropic liquids. [624] Wo. Ostwald, Trans. Faraday Soc. *29*, 1002 (1933), C.A. *28*, 1239[5] (1934); Anomalous viscosity in mesomorphic melts. [625] Wo. Ostwald, H. Malss, Kolloid-Z. *63*, 192 (1933), C.A. *27*, 3378 (1933); Viskositätsanomalien sich entmischender Systeme. Strukturviskosität mesomorpher Schmelzen. [626] J. R. Otia, A. R. K. L. Padmini, Indian J. Pure Appl. Phys. *11*, 190 (1973), C.A. *79*, 58695c (1973); Ultrasonic velocity and other parameters in a nematic. [627] J. R. Otia, A. R. K. L. Padmini, Mol. Cryst. Liq. Cryst. *36*, 25 (1976), C.A. *86*, 49315w (1977); Behavior of ultrasonic velocity in cholesterics. [628] M. Papoular, J. Phys. (Paris) *30*, 406 (1969); Structure et dispersion des ondes de surface d'un liquide némEatique. [629] M. Papoular, A. Rapini, Solid State Commun. *7*, 1639 (1969), C.A. *72*, 26171t (1970); Surface waves in nematics. [630] M. Papoular, Phys. Lett. *30*, 5 (1969); Behavior of viscosity at the n/i transition.

[631] M. Papoular, A. Rapini, J. Phys. (Paris), C-1, *31*, 27 (1970), C.A. *73*, 81753e (1970); Ondes de surface dans un cristal liquide némEatique. [632] M. Papoular, Phys. Lett. A *31*, 65 (1970), C.A. *72*, 136667s (1970); Structural relaxation and sound propagation in cholesterics. [633] M. Papoular, Phys. Lett. A *38*, 173 (1972); Magnetic stabilization of nematic turbulence. [634] O. Parodi, J. Phys. (Paris) *31*, 581 (1970), C.A. *73*, 124606d (1970); Stress tensor for a nematic. [635] J. D. Parsons, C. F. Hayes, Phys. Rev. A *10*, 2341 (1974), C.A. *82*, 66531y (1975); Surface waves in liquid crystals. [636] J. D. Parsons, Mol. Cryst. Liq. Cryst. *31*, 79 (1975), C.A. *84*, 11091r (1976); Ordering and distortion at the n/i fluid interface. [637] J. D. Parsons, Solid State Commun. *17*, 935 (1975); Surface waves in cholesterics in a magnetic field. [638] J. D. Parsons, J. Phys. (Paris) *36*, 1363 (1975); Fluctuations near the convective instability in a cholesteric. [639] J. D. Parsons, J. Phys. (Paris) *37*, 1187 (1976), C.A. *85*, 167047n (1976); Molecular theory of surface tension in nematics. [640] M. N. Patharkar, V. S. V. Rajan, J. J. C. Picot, Mol. Cryst. Liq. Cryst. *15*, 225 (1971), C.A. *76*, 38140c (1972); Interfacial and temperature gradient effects on thermal conductivity of a liquid crystal.

[641] M. N. Patharkar, Diss. Univ. New Brunswick 1971, Avail. Nat. Libr. Canada, Ottawa, Ont., C.A. *77*, 131514h (1972); Heat conduction in nematics. [642] E. Perez, J. E. Proust, J. Phys. (Paris), Lett. *38*, 117 (1977), C.A. *86*, 148996a (1977); Orientation of a smectic having different interfaces and structures induced by them. [643] A. Perrier, Arch. sci. phys. nat. *6*, 233 (1924), C.A. *18*, 3529[9] (1924); Transparence and anisotropic diffusion of oriented liquid crystals. [644] P. C. Martin, P. S. Pershan, J. Swift, Phys. Rev. Lett. *25*, 844 (1970), C.A. *73*, 113839f (1970); New elastic-hydrodynamic theory of liquid crystals. [645] Y. Liao, N. A. Clark, P. S. Pershan, Phys. Rev. Lett. *30*, 639 (1973),

C.A. *78*, 141309e (1973); Brillouin scattering from smectics. [**646**] P. S. Pershan, J. Appl. Phys. *45*, 1590 (1974), C.A. *80*, 149686n (1974); Dislocations effects in smectics A. [**647**] P. S. Pershan, J. Prost, J. Appl. Phys. *46*, 2343 (1975), C.A. *83*, 88971q (1975); Dislocations and impurity effects in smectics A. [**648**] S. Peter, H. Peters, Z. physik. Chem. *3*, 103 (1955), C.A. *49*, 5058b (1955); Strukturviskosität des kristallin-flüssigen PAA. [**649**] S. Peter, Angew. Chem. *75*, 194 (1963), C.A. *58*, 8420a (1963); Fortschritte und Probleme der Rheologie. [**650**] P. Petrescu, Opt. Commun. *19*, 284 (1976), C.A. *86*, 63841w (1977); Hexagonal arrays in nematics. [**650a**] G. P. Petrova, A. B. Aleinikov, Sb. Dokl. I Vses. Simpoz. po Akustooptich. Spektroskopii *1976*, 242, C.A. *86*, 181016d (1977); Hyperacoustic properties of some liquid crystals near the i/n transition.

[**651**] H. Pick, Z. physik. Chem. *77*, 577 (1911), C.A. *5*, 3747⁹ (1911); Innere Reibung kristallinisch-flüssiger Gemische von PAA und PAP. [**652**] P. Pieranski, E. Guyon, Solid State Commun. *13*, 435 (1973), C.A. *79*, 108177h (1973); Shear-flow-induced transition in nematics. [**653**] P. Pieranski, E. Guyon, Phys. Lett. A *49*, 237 (1974); Transverse effects in nematic flows. [**654**] P. Pieranski, E. Guyon, P. Keller, J. Phys. (Paris) *36*, 1005 (1975), C.A. *83*, 200865m (1975); Shear flow induced polarization in ferroelectric smectics C. [**655**] P. Pieranski, E. Guyon, S. A. Pikin, J. Phys. (Paris), C-1, *37*, 3 (1976), C.A. *84*, 187783g (1976); Nouvelles instabilités de cisaillement dans les nématiques. [**656**] P. Pieranski, E. Guyon, Phys. Rev. Lett. *32*, 924 (1974), C.A. *80*, 149646z (1974); Two shear-flow regimes in nematic *p-n*-hexyloxybenzylidene-*p'*-aminobenzonitrile. [**657**] P. Pieranski, E. Guyon, Phys. Rev. A *9*, 404 (1974), C.A. *80*, 85139b (1974); Instability of certain shear flows in nematics. [**658**] P. Pieranski, E. Guyon, Commun. Phys. *1*, 45 (1976), C.A. *85*, 39507a (1976); Shear flow instabilities in nematic CBOOA. [**658a**] I. Janossi, P. Pieranski, E. Guyon, J. Phys. (Paris) *37*, 1105 (1976); Poiseuille flow in nematics. Experimental studies of instabilities. [**659**] R. Pindak, Bull. Am. Phys. Soc. *20*, 502 (1975); Elastic constant measurements near a n/s$_A$ transition. [**660**] J. M. Pochan, P. F. Erhardt, Phys. Rev. Lett. *27*, 790 (1971), C.A. *75*, 122354n (1971); Shear-induced texture changes in cholesteric mixtures.

[**661**] J. M. Pochan, D. G. Marsh, J. Chem. Phys. *57*, 1193 (1972), C.A. *77*, 80604c (1972); Mechanism of shear-induced structural changes in cholesteric mixtures. [**662**] P. F. Erhardt, J. M. Pochan, W. C. Richards, J. Chem. Phys. *57*, 3596 (1972), C.A. *78*, 34829c (1973); Normal stress effects in cholesterics. [**663**] J. M. Pochan, D. G. Marsh, J. Chem. Phys. *57*, 5154 (1972), C.A. *78*, 49244g (1973); Effect of thickness on the layer structure of Grandjean texture in a sheared cholesteric. [**664**] J. Pochan, P. Erhardt, W. C. Richards, Liq. Cryst. Ord. Fluids *2*, 449 (1974), C.A. *86*, 180993h (1977); Temperature dependence and rheological behavior of the shear-induced Grandjean to focal conic transition in the cholesteric mesophase. [**665**] Y. Poggi, J. Robert, J. Borel, Mol. Cryst. Liq. Cryst. *29*, 311 (1975), C.A. *82*, 175405f (1975); Relations between liquid crystal order parameter and macroscopic physical coefficients—experimental proof. [**666**] Y. Poggi, J. C. Filippini, R. Aleonard, Phys. Lett. A *57*, 53 (1976), C.A. *85*, 70898c (1976); Free energy as a function of the order parameter in nematics. [**666a**] G. Porte, J. Phys. (Paris) *38*, 509 (1977), C.A. *86*, 198231t (1977); Surface disclination lines observed in nematics when the surfaces induce homogeneously tilted alignment. [**667**] P. Pollmann, Z. Naturforsch. A *27*, 719 (1972), C.A. *77*, 40171c (1972); Fließverhalten kompensierter cholesterischer Mischungen. [**668**] R. S. Porter, J. F. Johnson, J. Phys. Chem. *66*, 1826 (1962), C.A. *58*, 41e (1963); Order and flow of liquid crystals. Orientation of nematics. [**669**] R. S. Porter, J. F. Johnson, J. Appl. Phys. *34*, 51 (1963), C.A. *58*, 10747b (1963); Order and flow of liquid crystals: the nematic mesophase. [**670**] R. S. Porter, J. F. Johnson, J. Appl. Phys. *34*, 55 (1963), C.A. *58*, 1920f (1963); Order and flow of liquid crystals: the cholesteric mesophase.

[**671**] R. S. Porter, E. M. Barrall, II, J. F. Johnson, J. Chem. Phys. *45*, 1452 (1966), C.A. *65*, 12874h (1966); Flow characteristics of mesophases. [**672**] R. S. Porter, J. F. Johnson, F. Eirich, editor (J. Wiley: New York). 317 (1968); Rheology of liquid crystals. A chapter in "Rheology. IV." [**673**] A. Torgalkar, R. S. Porter, E. M. Barrall, II, J. F. Johnson, J. Chem. Phys. *48*, 3897 (1968), C.A. *69*, 39416q (1968); Interpretation of mesophase transitions. [**674**] K. Sakamoto, R. S. Porter, J. F. Johnson, Mol. Cryst. Liq. Cryst. *8*, 443 (1969), C.A. *71*, 95046x (1969); Viscosity of mesophases formed by cholesteryl myristate. [**675**] R. S. Porter, C. Griffen, J. F. Johnson, Mol. Cryst. Liq. Cryst. *25*, 131 (1974), C.A. *83*, 140058p (1975); Viscosity of mesophase blends. Cholesteryl acetate-myristate pair. [**676**] E. M. Friedman, R. S. Porter, Mol. Cryst. Liq. Cryst. *31*, 47 (1975), C.A. *83*, 211544g (1975); Nematic-like flow behavior in a nonsteroid cholesteric. [**676a**] E. Praestgaard, J. Rotne, Chem. Phys. *24*, 125 (1977), C.A. *87*, 144305x (1977); Nematic and crystalline ordering in the lattice model of hard rods with a mean field interaction. [**677**] M. J. Press, A. S. Arrott, J. Phys. (Paris), C-1, *36*, 177 (1975); Elastic energies and director fields in liquid crystal droplets. I. Cylindrical symmetry. [**678**]

M. J. Press, A. S. Arrott, J. Phys. (Paris) *37*, 387 (1976), C.A. *84*, 172363w (1976); Static strain waves in cholesterics. Homeotropic boundary conditions. [**679**] M. J. Press, A. S. Arrott, Mol. Cryst. Liq. Cryst. *37*, 81 (1976), C.A. *86*, 113907v (1977); Static strain waves in cholesterics. Response to magnetic and electric fields. [**680**] F. P. Price, C. S. Bak, Liq. Cryst. Ord. Fluids *2*, 411 (1974), C.A. *86*, 179736v (1977); Cylindrically symmetric textures in mesophases of cholesteryl esters.

[**681**] R. G. Priest, Phys. Rev. Lett. *26*, 423 (1971), C.A. *74*, 92313h (1971); Comments on the lattice model of liquid crystals. [**682**] R. G. Priest, Mol. Cryst. Liq. Cryst. *17*, 129 (1972), C.A. *77*, 53460w (1972); Calculation of the elastic constants of a nematic. [**683**] R. G. Priest, Diss. Univ. Pennsylvania, Philadelphia, Pa. 1972, 153 pp., C.A. *78*, 102994e (1973); Development of a model for nematics. [**684**] R. G. Priest, Phys. Rev. A *8*, 3191 (1973), C.A. *80*, 53171q (1974); Treatment of the steric model of nematics based on the empirical equation of state. [**685**] R. G. Priest, T. C. Lubensky, Phys. Rev. A *9*, 893 (1974), C.A. *80*, 113518v (1974); Biaxial model of cholesterics. [**686**] R. G. Priest, J. Phys. (Paris) *36*, 437 (1975); Simple model for the s_A/s_C transition. [**687**] R. G. Priest, Solid State Commun. *17*, 519 (1975); Comment on the rank two spherical harmonic model for nematics. [**688**] R. G. Priest, J. Chem. Phys. *65*, 408 (1976), C.A. *85*, 134768v (1976); Molecular statistical model of smectics A and smectics C. [**689**] R. G. Priest, Mol. Cryst. Liq. Cryst. *37*, 101 (1976), C.A. *86*, 113908w (1977); Generalized rank two tensor statistical model of the s_A state. [**690**] E. B. Priestley, Introd. Liq. Cryst. *1974*, 71, C.A. *84*, 97899v (1976); Nematic order: the long range orientational distribution function.

[**691**] J. Prost, H. Gasparoux, Phys. Lett. A *36*, 245 (1971), C.A. *75*, 144115n (1971); Determination of twist viscosity coefficient in nematics. [**692**] J. Prost, Solid State Commun. *11*, 183 (1972); Comments on thermomechanical coupling in cholesterics. [**693**] J. Prost, H. Gasparoux, Mol. Cryst. Liq. Cryst. *22*, 25 (1973), C.A. *79*, 109050y (1973); Dynamics of molecular orientation, during the c/n transition. [**694**] J. Prost, G. Sigaud, B. Regaya, J. Phys. (Paris), Lett. *37*, 341 (1976), C.A. *86*, 63859h (1977); Thermal dependence of the twist viscosity in nematics. [**695**] J. E. Proust, E. Perez, L. Ter-Minassian-Saraga, Colloid Polym. Sci. *254*, 672 (1976), C.A. *85*, 114984j (1976); Thin films of nematic liquid crystal on liquid substrate. Bulk and surface structures, surface and line tensions. [**696**] F. Pusnik, Obz. Mat. Fiz. *22*, 4 (1975), C.A. *84*, 52179u (1976); Microscopic theory of liquid crystals. [**697**] R. Pynn, Solid State Commun. *14*, 29 (1974), C.A. *80*, 74450v (1974); Theory of static correlations in a fluid of linear molecules. [**698**] G. H. Quincke, Wied. Ann. Physik *62*, 1 (1897); Klebrigkeit isolierender Flüssigkeiten im constanten elektrischen Felde. [**699**] J. C. Raich, R. D. Etters, L. Flax, Chem. Phys. Lett. *6*, 491 (1970), C.A. *73*, 124374b (1970); Orientational transition in liquid crystals. [**700**] J. C. Raich, R. D. Etters, J. Chem. Phys. *55*, 3901 (1971); Soft modes in molecular crystals.

[**701**] V. S. V. Rajan, J. J. C. Picot, Mol. Cryst. Liq. Cryst. *17*, 109 (1972), C.A. *77*, 53006c (1972); Interfacial effect on the heat conduction in a nematic. [**702**] V. S. V. Rajan, M. N. Patharkar, J. J. C. Picot, Mol. Cryst. Liq. Cryst. *18*, 279 (1972), C.A. *77*, 144586h (1972); Anisotropic thermal diffusion in nematic APAPA. [**703**] V. S. V. Rajan, J. J. C. Picot, Mol. Cryst. Liq. Cryst. *20*, 55 (1973), C.A. *78*, 102703j (1973); Thermal transport phenomena in nematics. [**704**] V. S. V. Rajan, J. J. C. Picot, Mol. Cryst. Liq. Cryst. *20*, 69 (1973), C.A. *78*, 102750x (1973); Non-Fourier heat conduction in a nematic. [**705**] V. S. V. Rajan, J. J. C. Picot, Rev. Sci. Instrum. *44*, 238 (1973), C.A. *78*, 89299k (1973); Simple technique for determining the sign of thermal conductivity anisotropy in nematics. [**706**] V. S. V. Rajan, J. J. C. Picot, Mol. Cryst. Liq. Cryst. *22*, 185 (1973), C.A. *79*, 108870d (1973); Discontinuity in the properties of PAA at the n/i transition. [**707**] V. S. V. Rajan, J. J. C. Picot, Liq. Cryst., Proc. Int. Conr., Bangalore, *1973*, 305, C.A. *84*, 112575g (1976); Thermal instability in a nematic layer. [**708**] V. S. V. Rajan, Acta Phys. Pol. A, *46*, 407 (1974), C.A. *82*, 50058p (1975); Application of the continuum theory to the thermoconvective instability in nematics. [**709**] G. S. Ranganath, Mol. Cryst. Liq. Cryst. *34*, 71 (1976), C.A. *85*, 200782y (1976); Dislocations and disclinations. [**709a**] G. S. Ranganath, Mol. Cryst. Liq. Cryst. *40*, 143 (1977), C.A. *87*, 125601z (1977); Interaction between a cavity and a singularity in nematics. [**710**] S. N. Rao, K. S. Rao, Acustica *35*, 276 (1976), C.A. *86*, 8694a (1977); Ultrasonic absorption in the isotropic region of cholesteryl propionate and valerate.

[**711**] A. Rapini, C. R. Acad. Sci., Sér. B *275*, 701 (1972); Pouvoir rotatoire acoustique d'un verre cholésterique. [**712**] A. Rapini, Can. J. Phys. *53*, 968 (1975); Ondes de surface dans les cristaux liquides smectiques et cholestériques. [**712a**] A. Rapini, J. Phys. (Paris) Lett. *37*, 49 (1976); Permeation boundary layer in a s_A by reflection of transverse ultrasonic waves. [**712b**] M. S. Rapport, Diss. Univ. Maryland, College Park, Md. 1976. 132 pp. Avail. Univ. Microfilms Int., Order No. 77-10,418,

C.A. *87*, 76600n (1977); Statistical theories of the nematic mesophase. [**713**] J. Rault, C. R. Acad. Sci., Sér. B *272*, 1275 (1971), C.A. *75*, 68627p (1971); Méthode nouvelle d'étude de l'orientation moléculaire à la surface d'un cholestérique. [**714**] J. Rault, J. Phys. (Paris) *33*, 383 (1972), C.A. *77*, 119294d (1972); Irrégularités sur les lignes de dislocation dans les nématiques et dans les cholestériques. [**715**] J. Rault, Philos. Mag. *28*, 11 (1973), C.A. *79*, 70931j (1973); Dislocations χ dans les cholestériques. I. Propriétés. [**716**] J. Rault, Philos. Mag. *30*, 621 (1974), C.A. *82*, 66451x (1975); II. Modèles (cf. [715]). [**717**] J. Rault, C. R. Acad. Sci., Sér. B, *280*, 417 (1975); Interprétation des «stries huileuses» dans les cristaux liquides. [**718**] S. K. Ray, J. Indian Chem. Soc. *13*, 194 (1936), C.A. *30*, 5850[1] (1936); Parachor and chemical constitution. V. Structure of liquid crystals. [**719**] Yu. V. Reztsov, V. S. Samsonov, Nauch. Tr. Vses. Zaoch. Mashinostr. In-t *47*, 58 (1975), C.A. *86*, 10840p (1977); Magnetoacoustical properties of the nematic hydroquinone ester of *n*-heptyloxybenzoic acid and *p*-anisalaminoazobenzene. [**720**] Yu. V. Reztsov, Nauch. Tr. Vses. Zaoch. Mashinostr. In-t *47*, 66 (1975), C.A. *85*, 200780w (1976); Mechanism of the absorption of ultrasound in nematic anisaldazine.

[**721**] R. Ribotta, C. R. Acad. Sci., Sér. B *279*, 295 (1974), C.A. *82*, 78996s (1975); Mesure de la longueur de pénétration dans un smectique A au voisinage d'une transition s$_A$/n du 2e ordre. [**722**] R. Ribotta, J. Phys. (Paris) C-3, *37*, 149 (1976), C.A. *85*, 85792n (1976); Thermomechanical instability of layers in a smectic A. [**723**] R. Ribotta, G. Durand, J. Phys. (Paris) *38*, 179 (1977), C.A. *86*, 82006u (1977); Mechanical instabilities of smectics A under dilative or compressive stresses. [**724**] C. Robinson, Surface Phenomena Chem. Biol. *1958*, 133 (J. F. Danielli, K. G. A. Pankhurst and A. C. Riddiford, editors. Pergamon Press), C.A. *53*, 5780b (1959); Surface structures in liquid crystals. [**725**] F. Rondelez, Solid State Commun. *14*, 815 (1974), C.A. *81*, 30684r (1974); Measurements of diffusion anisotropies in liquid crystals by use of dyes. [**725a**] W. Urbach, F. Rondelez, P. Pieranski, F. Rothen, J. Phys. (Paris) *38*, 1275 (1977), C.A. *87*, 192262c (1977); Marangoni effect in nematics. [**726**] L. K. Runnels, C. Colvin, J. Chem. Phys. *53*, 4219 (1970), C.A. *74*, 7560u (1971); Nature of the rigid-rod mesophase. [**727**] L. K. Runnels, C. Colvin, Mol. Cryst. Liq. Cryst. *12*, 299 (1971), C.A. *75*, 10279k (1971); Fluid phases of highly asymmetric molecules: plate-shaped molecules. [**728**] C. Rymarz, Biul. Wojsk. Akad. Tech. *25*, 105 (1976), C.A. *85*, 102466d (1976); Phenomenological models of a liquid crystal. [**729**] S. Sakagami, A. Takase, M. Nakamizo, H. Kakiyama, Mol. Cryst. Liq. Cryst. *19*, 303 (1973), C.A. *78*, 89523d (1973); Microscopic study on the molecular arrangements in s$_A$, s$_B$, and s$_C$ modifications. [**730**] A. Saupe, Z. Naturforsch. A *15*, 810 (1960), C.A. *55*, 3137b (1961); Temperaturabhängigkeit und Größe der Deformationskonstanten nematischer Flüssigkeiten.

[**731**] A. Saupe, Z. Naturforsch. A *15*, 815 (1960), C.A. *55*, 3136i (1961); Biegungselastizität von nematischem PAA. [**732**] A. Saupe, Mol. Cryst. Liq. Cryst. *7*, 59 (1969), C.A. *71*, 85815h (1969); Molecular structure and physical properties of thermotropic liquid crystals. [**733**] J. Nehring, A. Saupe, J. Chem. Phys. *54*, 337 (1971), C.A. *74*, 25819q (1971); Elastic theory of uniaxial liquid crystals. [**734**] J. Nehring, A. Saupe, J. Chem. Phys. *56*, 5527 (1972), C.A. *77*, 25767h (1972); Calculation of the elastic constants of nematics. [**735**] J. Nehring, A. Saupe, J. Chem. Soc., Faraday Trans. (2), *68*, 1 (1972), C.A. *76*, 38351x (1972); Schlieren texture in nematics and smectics. [**736**] A. Saupe, Mol. Cryst. Liq. Cryst. *21*, 211 (1973), C.A. *79*, 108874h (1973); Disclinations and properties of the directorfield in nematics and cholesterics. [**727**] A. Saupe, Ber. Bunsenges. Phys. Chem. *78*, 846 (1974), C.A. *82*, 10061c (1975); Statistical theories of nematics. [**738**] P. J. Photinos, A. Saupe, Phys. Rev. A *13*, 1926 (1976), C.A. *85*, 54865j (1976); Mean-field study of uniaxial smectics with polarized layers. [**738a**] D. Johnson, A. Saupe, Phys. Rev. A *15*, 2079 (1977), C.A. *87*, 32231p (1977); Undulation instabilities in smectics C. [**739**] R. Schachenmeier, Ann. Physik *46*, 569 (1915), C.A. *9*, 2023[3] (1915); Anomalie des Reibungskoeffizienten bei anisotropen Flüssigkeiten. [**740**] T. J. Scheffer, Phys. Rev. A *5*, 1327 (1972), C.A. *76*, 104915n (1972); Structures and energies of Grandjean-Cano liquid-crystal disclinations. [**740a**] H. Schröder, J. Chem. Phys. *67*, 16 (1977), C.A. *87*, 125571q (1977); Molecular statistical theory for inhomogeneous nematics with boundary conditions.

[**741**] T. D. Schultz, Mol. Cryst. Liq. Cryst. *14*, 147 (1971), C.A. *75*, 123339y (1971); Validity of the Maier-Saupe theory of the nematic transition. [**742**] W. M. Schwartz, H. W. Moseley, J. Phys. & Colloid Chem. *51*, 826 (1947), C.A. *41*, 4985h (1947); Surface tension of liquid crystals. [**743**] F. Scudieri, Appl. Phys. Lett. *29*, 398 (1976), C.A. *85*, 152137t (1976); High-frequency shear instability in nematics. [**744**] F. Scudieri, M. Bertolotti, S. Melone, G. Albertini, J. Appl. Phys. *47*, 3781 (1976), C.A. *85*, 169959k (1976); Acoustohydrodynamic instability in nematics. [**745**] F. Scudieri, A. Ferrari, D. Apostol, Rev. Roum. Phys. *21*, 677 (1976), C.A. *86*, 113903r (1977); Hydrodynamic instabilities in nematics by ultrasound. [**745a**] V. K. Semenchenko, B. Chotchaev, Zh. Fiz. Khim. *51*, 718 (1977),

C.A. *86*, 178497f (1977); Statistical theory of nematics. [**746**] M. Shahinpoor, Q. J. Mech. Appl. Math. *28*, 223 (1975); Finite twist waves in liquid crystals. [**747**] M. Shahinpoor, J. Chem. Phys. *63*, 1319 (1975), C.A. *83*, 106589c (1975); Alignment of nematics. [**748**] M. Shahinpoor, Mol. Cryst. Liq. Cryst. *37*, 121 (1976), C.A. *86*, 113909x (1977); Effect of material nonlinearity on the acceleration twist waves in liquid crystals. [**748a**] M. Shahinpoor, Rheol. Acta *15*, 99 (1976); Stress tensor in nematics. [**748b**] M. Shahinpoor, Rheol. Acta *15*, 215 (1977); Lehmann's effect. [**749**] S. A. Shaya, J. Yu, J. Chem. Phys. *63*, 221 (1975), C.A. *83*, 124375j (1975); Order fluctuations dynamics of nematic binary system. MBBA/biphenyl. [**749a**] L. Shen, H. K. Sim, Y. M. Shih, C. W. Woo, Mol. Cryst. Liq. Cryst. *39*, 229 (1977), C.A. *87*, 109766b (1977); Systematic solution of the mean field equation for liquid crystals. [**750**] P. Sheng, Introd. Liq. Cryst. *1974*, 59; Hard rod model of the n/i transition.

[**751**] P. Sheng, Introd. Liq. Cryst. *1974*, 103, C.A. *84*, 97901q (1976); Elastic continuum theory of liquid crystals. [**752**] P. Sheng, E. B. Priestley, P. J. Wojtowicz, J. Chem. Phys. *63*, 1040 (1975), C.A. *83*, 106558s (1975); Elastic continuum theory cutoffs and order in nematics and solids. [**753**] P. Sheng, P. J. Wojtowicz, Phys. Rev. A *14*, 1883 (1976), C.A. *86*, 24626k (1977); Constant-coupling theory of nematics. [**754**] P. Sheng, Solid State Commun. *18*, 1165 (1976); Effect of director fluctuations on the nematic distribution function. [**755**] C. H. Shih, Diss. Yale Univ., New Haven, Conn. 1971, 109 pp. Avail. Univ. Microfilms, Ann Arbor, Mich., Order No. 71-31,010, C.A. *76*, 90363s (1972); Theory of orientation correlations in nematics. [**756**] Y. M. Shih, Y. R. Lin-Liu, C. W. Woo, Phys. Rev. A *14*, 1895 (1976), C.A. *86*, 36529u (1977); Theoretical analysis of i/n transition properties. [**757**] Y. M. Shih, H. M. Huang, C. W. Woo, Mol. Cryst. Liq. Cryst. *34*, 7 (1976), C.A. *85*, 102637k (1976); Pretransitional effects in nematogens. [**758**] J. Sivardiere, J. Phys. (Paris) *37*, 1267 (1976), C.A. *86*, 11301a (1977); Double spin one-half lattice gas model. [**759**] A. Smekal, Ann. Physik *83*, 1202 (1927), C.A. *22*, 3074³ (1928); Größenordnung der ideal gebauten Gitterbereiche in Realkristallen. [**760**] A. S. Sonin, A. N. Chuvyrov, A. S. Sobachkin, V. L. Ovchinnikov, Fiz. Tverd. Tela *18*, 3099 (1976), C.A. *86*, 63821q (1977); Disclination annihilation with an m = +2 Frank index in nematics.

[**761**] R. S. Speer, G. C. Wetsel, Jr., A. L. Lowry, M. R. Woodard, Bull. Am. Phys. Soc. *17*, 59 (1972); Effects of a magnetic field on the propagation of ultrasonic waves in nematics. [**762**] R. S. Speer, Diss. South. Methodist Univ., Dallas, Tex. 1974. 234 pp. Avail. Xerox Univ. Microfilms, Ann Arbor, Mich., Order No. 75-15,279, C.A. *83*, 171100a (1975); Measurement of ultrasonic attenuation in dielectric solids using the method of Bragg scattering and ultrasonic attenuation in nematics in a magnetic field. [**763**] A. M. J. Spruijt, Solid State Commun. *13*, 1919 (1973), C.A. *80*, 53160k (1974); Twist-disclination line in planar-oriented samples of liquid crystals. [**764**] J. Stecki, Bull. Acad. Pol. Sci., Ser. Sci. Chim. *23*, 995 (1975), C.A. *84*, 128974x (1976); Nematic ordering in the lattice model of rigid rods. I. Athermal lattice gas in the Huggins-Miller-Guggenheim approximation. [**764a**] J. Stecki, J. Chem. Phys. *67*, 948 (1977), C.A. *87*, 109762x (1977); Nematic ordering in the lattice model with attractive isotropic interaction. [**765**] M. J. Stephen, Phys. Rev. A *2*, 1558 (1970), C.A. *73*, 134915t (1970); Hydrodynamics of liquid crystals. [**766**] G. W. Stewart, D. O. Holland, Proc. Iowa Acad. Sci. *43*, 267 (1936), CA. *32*, 4033⁹ (1938); Orientation of liquid crystals of PAA occurring with temperature gradient and no convection. [**767**] D. O. Holland, G. W. Stewart, Phys. Rev. *51*, 62 (1937), C.A. *32*, 5268⁴ (1938); Orientation of liquid crystals by heat conduction. [**768**] R. C. Davis, G. W. Stewart, Proc. Iowa Acad. Sci. *45*, 189 (1938), C.A. *33*, 7637¹ (1939); Swarm structure of the liquid crystal PAA. [**769**] G. W. Stewart, D. O. Holland, L. M. Reynolds, Phys. Rev. *58*, 174 (1940), C.A. *34*, 6499⁹ (1940); Orientation of liquid crystals by heat conduction. [**770**] S. Bereskin, G. W. Stewart, Proc. Iowa Acad. Sci. *48*, 305 (1941), C.A. *36*, 3722⁸ (1942); Interesting case where the heat flow decreases the hat conductivity of a fluid.

[**771**] G. W. Stewart, Phys. Rev. *69*, 51 (1945), C.A. *41*, 3683d (1947); Heat conduction effects with liquid crystals and suspended particles. [**772**] T. W. Stinson, J. D. Litster, N. A. Clark, J. Phys. (Paris) *33*, 69 (1972), C.A. *80*, 20340r (1974); Static and dynamic behavior near the order-disorder transition of nematics. [**773**] T. W. Stinson, J. D. Litster, Phys. Rev. Lett. *30*, 688 (1973), C.A. *78*, 141273p (1973); Correlation range of fluctuations of short-range order in the i phase of a liquid crystal. [**774**] J. P. Straley, Phys. Rev. A *4*, 675 (1971), C.A. *75*, 68625m (1971); Liquid crystals in two dimensions. [**775**] J. P. Straley, J. Chem. Phys. *57*, 3694 (1972), C.A. *77*, 157350b (1972); Zwanzig model for liquid crystals. [**776**] J. P. Straley, Mol. Cryst. Liq. Cryst. *24*, 7 (1973), C.A. *81*, 142319g (1974); Third virial coefficient for the gas of long rods. [**777**] J. P. Straley, Phys. Rev. A *8*, 2181 (1973), C.A. *80*, 7899k (1974); Frank elastic constants of the hard-rod liquid crystal. Comments. [**778**] J. P. Straley, Phys. Rev. A *14*, 1835 (1976), C.A. *86*, 24625j (1977); Theory of piezoelectricity in nematics, and of the

cholesteric ordering. [**779**] H. A. Stuart, Z. Elektrochem. *45*, 180 (1939); Diskussionsbemerkung zu P. Debye. [**780**] V. I. Sugakov, Fiz. Tverd. Tela *17*, 3381 (1975); Boundary conditions in theory of liquid crystals.

[**781**] C. C. Sung, Chem. Phys. Lett. *10*, 35 (1971), C.A. *75*, 114635q (1971); NMR in liquid crystals. [**782**] J. T. Sullivan, Diss. Univ. of Minnesota, Minneapolis, 1966, 113 pp. Avail. Univ. Microfilms (Ann Arbor, Mich.), Order No. 67-7792, C.A. *67*, 111614f (1967); Transport phenomena in liquid crystals. [**783**] J. W. Summerford, J. R. Boyd, B. A. Lowry, J. Appl. Phys. *46*, 970 (1975), C.A. *82*, 178657a (1975); Angular and temperature dependence of viscosity coefficients in a plane normal to the direction of flow in MBBA. [**784**] I. V. Sushkin, Uch. Zap. Ivanov. Gos. Pedagog. Inst. *77*, 65 (1970), C.A. *76*, 28644f (1972); Consequences of the molecular theory of nematics. [**785**] T. Svedberg, Kolloid. Z. *22*, 68 (1918), C.A. *12*, 2148^8 (1918); Diffusion in anisotropen Flüssigkeiten. [**786**] A. A. Tabidze, N. I. Koshkin, Nauch. Tr. Kursk. Gos. Ped. In-T *40*, 267 (1974), C.A. *84*, 52451b (1976); Use of an ultrasonic impedance method in the study of liquid crystals. [**787**] B. Tamamushi, Bull. Chem. Soc. Jap. *9*, 475 (1934), C.A. *29*, 2812^3 (1935); Zweidimensionale Zustandsgleichung und der Bau von Grenzflächenschichten. [**788**] B. Tamamushi, Chem. Phys. Chem. Anwendungstech. Grenzflächenaktiven Stoffe, Ber. Int. Kongr., 6th, *2*, Teil 2, 431 (1972), C.A. *82*, 10211b (1975); Surface tension anomalies of liquid crystals. [**789**] B. Tamamushi, Biorheology *10*, 239 (1973), C.A. *79*, 83718v (1973); Flow properties of smectics. [**790**] B. Tamamushi, Rheol. Acta *13*, 247 (1974), C.A. *81*, 83127t (1974); Flow properties of smectics. [**790a**] B. Tamamushi, Colloid Interface Sci., 50th, *5*, 453 (1976), C.A. *87*, 58809h (1977); Surface state of thermotropic liquid crystals.

[**791**] H. T. Tan, Y. M. Shih, Bull. Am. Phys. Soc. *20*, 25 (1975); Mean-field calculation on the cholesteric phase. [**792**] G. Toulouse, J. Phys. (Paris), Lett. *38*, 67 (1977), C.A. *86*, 99276e (1977); For biaxial nematics. [**793**] M. Trautz, E. Fröschel, Ann. Physik *22*, 223 (1935), C.A. *29*, 3565^9 (1935); Änderung der inneren Reibung von paramagnetischen Gasen im Magnetfeld. [**793a**] S. Trimper, Phys. Lett. A *60*, 314 (1977), C.A. *86*, 163795c (1977); Static and dynamic behavior in nematics. [**794**] H. C. Tseng, D. L. Silver, B. A. Finlayson, Phys. Fluids *15*, 1213 (1972), C.A. *77*, 66419q (1972); Application of the continuum theory to nematics. [**795**] H.-C. Tseng, Diss. Univ. Washington, Seattle, Wash. 1975. 147 pp. Avail. Xerox Univ. Microfilms, Ann Arbor, Mich., Order No. 76-17,663, C.A. *85*, 134825m (1976); Viscosity measurements for nematic PAA. [**796**] V. N. Tsvetkov, G. M. Mikhailov, Acta Physico-chim. U.R.S.S. *8*, 77 (1938), C.A. *33*, 3645^2 (1939); Einfluß eines Magnetfelds auf die Viskosität von nematischem PAA. [**797**] G. M. Mikhailov, V. N. Tsvetkov, J. Exptl. Theoret. Phys. (U.S.S.R.) *9*, 597 (1939), Acta Physicochim. *10*, 775 (1939), C.A. *33*, 8081^7 (1939); Wirkung eines magnetischen und elektrischen Feldes auf die Strömungsgeschwindigkeit von nematischem PAA in einem Kapillar-rohr. G. M. Mikhailov, V. N. Tsvetkov, Acta Physicochim. U.R.S.S. *10*, 415 (1939), C.A. *33*, 6106^3 (1939); Einfluß eines elektrischen Feldes auf die Strömungsgeschwindigkeit von nematischem PAA in einem Kapillarrohr. [**798**] V. N. Tsvetkov, Acta Physicochim. U.R.S.S. *11*, 97 (1939), C.A. *34*, 3554^5 (1940); Größe und Form von Molekülschwärmen in anisotropen Flüssigkeiten. [**799**] V. Marinin, V. Tsvetkov, Acta Physicochim. U.R.S.S. *11*, 837 (1939), J. Exptl. Theoret. Phys. (U.S.S.R.) *9*, 1388, C.A. *34*, 4625^2 (1940); Dielektrizitätskonstante einer strömenden anisotropen Flüssigkeit. [**800**] V. N. Tsvetkov, V. Marinin, Acta Physicochim. U.R.S.S. *13*, 219, J. Exptl. Theoret. Phys. (U.S.S.R.) *10*, 929 (1940), C.A. *35*, 1679^6 (1941); Strömungsorientierung anisotroper Flüssigkeiten.

[**801**] V. N. Tsvetkov, Akad. Nauk S.S.S.R., Otdel. Tekh. Nauk, Inst. Mashinovedeniya, Sovesh-chanie Vyazkosti Zhidkostei i Kolloid. Rastvorov (Conf. on Viscosity of Liquids and Colloidal Solns.) *1*, 47 (1941), C.A. *40*, 3033^3 (1946); Effect of the magnetic and electric fields on the viscosity and molecular orientation of anisotropic liquids. [**802**] V. N. Tsvetkov, Acta Physicochim. U.R.S.S. *16*, 132 (1942), C.A. *37*, 2238^7 (1943); Molekulare Ordnung in anisotropen Flüssigkeiten. [**803**] V. N. Tsvetkov, S. P. Krozer, Zhur. Tekh. Phys. *28*, 1444 (1958), C.A. *53*, 20998i (1959); Pretransition phenomena in PAA near the transition point. [**804**] A. L. Tsykalo, A. D. Bagmet, Zh. Fiz. Khim. *50*, 751 (1976), C.A. *84*, 187794m (1976); Detection of nematic mesophase during modeling of a liquid crystal by the method of molecular dynamics. [**805**] A. L. Tsykalo, A. D. Bagmet, Kristallografiya *21*, 1101 (1976), C.A. *86*, 81962r (1977); Study of the dynamics of particles, self-diffusion, and structure of nematics using molecular-dynamic "machine experiments". [**806**] R. Turner, Philos. Mag. *30*, 13 (1974), C.A. *82*, 10270v (1975); Twist walls in nematics. [**807**] R. Turner, T. E. Faber, Phys. Lett. A *49*, 423 (1974), C.A. *82*, 66496r (1975); Relaxation of twist in nematics. [**808**] R. Turner, Philos. Mag. *31*, 719 (1975), C.A. *83*, 36148g (1975); Numerical computation of the director field in a twist wall. [**809**] A. R. Ubbelohde, H. J. Michels, J. J. Duruz, J. Phys. E *5*, 283 (1972), C.A. *76*, 104100t (1972); Relaxation

dynamometer for studying molten salt mesophases. [**810**] H. J. Michels, A. R. Ubbelohde, Proc. Roy Soc. A *338*, 447 (1974), C.A. *81*, 55277y (1974); Horizon formation by molten organic salts as a test of their fluidity.

[**811**] V. I. Uchastkin, N. V. Solov'eva, E. Y. Tagunov, L. K. Vistin, I. E. Pitovranova, Sb. Dokl. Vses. Nauch. Konf. Zhidk. Krist. Simp. Ikh Prakt. Primen., 2nd, *1972*, 160, C.A. *82*, 24595s (1975); Study of liquid crystals using elastic surface waves. [**812**] B. W. van der Meer, G. Vertogen, A. J. Dekker, J. G. J. Ypma, J. Chem. Phys. *65*, 3935 (1976), C.A. *86*, 49305t (1977); Molecular-statistical theory of the temperature-dependent pitch in cholesterics. [**813**] B. W. van der Meer, G. Vertogen, Phys. Lett. A *59*, 279 (1976), C.A. *86*, 99246v (1977); Molecular biaxiality and temperature dependence of the cholesteric pitch. [**814**] D. C. van Eck, W. Westera, Mol. Cryst. Liq. Cryst. *38*, 319 (1977), C.A. *86*, 181032f (1977); Visco-elastic properties of some nematics. [**815**] J. Vieillard-Baron, J. Phys. (Paris), C-4, *30*, 22 (1969); Les transitions de phase du système d'ellipses dures classiques. [**816**] J. Vieillard-Baron, J. Chem. Phys. *56*, 4729 (1972), C.A. *77*, 10732u (1972); Phase transitions of the classical hard-ellipse system. [**817**] J. Vieillard-Baron, Mol. Phys. *28*, 809 (1974), C.A. *82*, 90154y (1975); Equation of state of a system of hard spherocylinders. [**818**] R. Vilanove, E. Guyon, C. Mitescu, P. Pieranski, J. Phys. (Paris) *35*, 153 (1974), C.A. *80*, 137648p (1974); Mesure de la conductivité thermique et détermination de l'orientation des molécules a l'interface n/i de MBBA. [**818a**] G. E. Volovik, V. P. Mineev, Zh. Eksp. Teor. Fiz. *72*, 2256 (1977), C.A. *87*, 73619q (1977); Study of singularities in superfluid helium-3 and liquid crystals by homotopic topology methods. [**819**] P. A. Vuillermot, M. V. Romerio, J. Phys. C *6*, 2922 (1973), C.A. *79*, 150341g (1973); Exact solution of the Maier-Saupe model for a nematic on a one-dimensional lattice. [**820**] M. Wadati, A. Isihara, Mol. Cryst. Liq. Cryst. *17*, 95 (1972), C.A. *77*, 53459c (1972); Theory of liquid crystals.

[**821**] F. E. Wargocki, Diss. Drexel Univ., Philadelphia, Pa. 1973. 156 pp. Avail. Univ. Microfilms, Ann Arbor, Mich., Order No. 73-25,520, C.A. *80*, 75113z (1974); Ordering mechanisms in nematics. [**822**] M. Watabe, Y. Odaka, Kogakuin Daigaku Kenkyu Hokoku *1971*, 1, C.A. *77*, 105816p (1972); Liquid crystals 1. Measurement of surface tension. [**822a**] P. Weiss, Compt. rend. *143*, 1136 (1906); La variation du ferromagnétisme avec la température. [**823**] G. C. Wetsel, Jr., R. S. Speer, B. A. Lowry, M. R. Woodard, J. Appl. Phys. *43*, 1495 (1972), C.A. *76*, 131804g (1972); Effects of magnetic field on attenuation of ultrasonic waves in a nematic. [**824**] C. Williams, P. E. Cladis, Solid State Commun. *10*, 357 (1972), C.A. *76*, 118427s (1972); Measurement of the elastic constants of twist and bend for nematic MBBA. [**825**] C. Williams, P. Pieranski, P. E. Cladis, Phys. Rev. Lett. *29*, 90 (1972), C.A. *77*, 67180s (1972); Nonsingular S = +1 screw disclination lines in nematics. [**826**] C. E. Williams, P. E. Cladis, M. Kléman, Mol. Cryst. Liq. Cryst. *21*, 355 (1973), C.A. *79*, 109021q (1973); Screw disclinations in nematics with cylindrical symmetry. [**827**] C. E. Williams, M. Kléman, J. Phys. (Paris), Lett. *35*, 33 (1974), C.A. *80*, 125577b (1974); Observation of edge dislocation lines in a smectic A. [**828**] C. E. Williams, Philos. Mag. *32*, 313 (1975), C.A. *83*, 171046n (1975); Helical disclination lines in smectics A. [**829**] C. E. Williams, M. Kléman, J. Phys. (Paris), C-1, *36*, 315 (1975); Dislocations, grain boundaries and focal conics in smectics A. [**830**] C. E. Williams, M. Kléman, Philos. Mag. *33*, 213 (1976), C.A. *84*, 128904z (1976); Association des lignes de dislocations vis et des coniques focales dans les smectiques A.

[**831**] R. Williams, Mol. Cryst. Liq. Cryst. *35*, 349 (1976), C.A. *85*, 198681u (1976); Measurement of the interfacial tension between n and i phases of MBBA. [**832**] P. J. Wojtowicz, Introd. Liq. Cryst. *1974*, 31, C.A. *84*, 82632d (1976); Introduction to the molecular theory of nematics. [**833**] P. J. Wojtowicz, Introd. Liq. Cryst. *1974*, 45, C.A. *84*, 82633e (1976); Generalized mean field theory of nematics. [**834**] P. J. Wojtowicz, Introd. Liq. Cryst. *1974*, 83, C.A. *84*, 97900p (1976); Introduction to the molecular theory of smectics A. [**835**] H. Workman, M. Fixman, J. Chem. Phys. *58*, 5024 (1973), C.A. *79*, 24413u (1973); Integral equation approach to liquid crystals. [**836**] A. Wulf, A. G. de Rocco, Liq. Cryst. Ord. Fluids *1970*, 227, C.A. *78*, 152418r (1973); Can a model system of rod-like particles exhibit both a fluid-fluid and a fluid-solid phase transition? [**837**] A. Wulf, Diss. Univ. Maryland, College Park, Md. 1970, 87 pp. Avail. Univ. Microfilms, Ann Arbor, Mich., Order No. 71-4102, C.A. *75*, 54549c (1971); Statistical mechanical models for systems of long linear molecules: application to the i/n transition. [**838**] A. Wulf, A. G. de Rocco, J. Chem. Phys. *55*, 12 (1971), C.A. *75*, 54892j (1971); Statistical mechanics for long semiflexible molecules. Model for the nematic mesophase. [**839**] A. Wulf, J. Chem. Phys. *55*, 4512 (1971), C.A. *75*, 144730j (1971); Distribution function theory of nematics. [**840**] A. Wulf, Acta Cient. Venezolana *22*, 38 (1971); Distribution function method applied to nematics.

[841] A. Wulf, Phys. Rev. A *8*, 2017 (1973), C.A. *80*, 7897h (1974); Fluctuations in the low temperature phase of a model cholesteric liquid crystal. [842] A. Wulf, J. Chem. Phys. *59*, 6596 (1973), C.A. *80*, 75165t (1974); Biaxial order in cholesterics. Phenomenological argument. [843] A. Wulf, J. Chem. Phys. *59*, 1487 (1973), C.A. *79*, 119296x (1973); Cholesteric twist in a model system. [844] A. Wulf, Phys. Rev. A *11*, 365 (1975), C.A. *82*, 79014g (1975); Steric model for the s_C phase. [845] A. Wulf, J. Chem. Phys. *64*, 104 (1976), C.A. *84*, 82788j (1976); Difficulties with the Maier-Saupe theory. [845a] A. Wulf, J. Chem. Phys. *67*, 2245 (1977), C.A. *87*, 160156e (1977); Short-range correlations and the effective orientational energy in liquid crystals. [846] T. Yamada, E. Fukuda, Jap. J. Appl. Phys. *12*, 68 (1973), C.A. *78*, 75951e (1973); Non-Newtonian viscosity of liquid-crystalline cholesteryl oleyl carbonate. [847] E. Yamazaki, J. Chem. Soc. (Japan) *43*, 134 (1922), C.A. *16*, 2245^9 (1922); Theory of formation of liquid crystals or crystalline liquids and their properties elements of crystals and nucleus of crystals. [848] C. C. Yang, H. S. D. Shih, Bull. Am. Phys. Soc. *17*, 59 (1972); Short range order studies in the pretransitional region of an isotropic cholesteric. [849] T. Yasunaga, Kanwa Gensho No Kagaku *1973*, 277, C.A. *84*, 169765y (1976); Ultrasonic relaxation. [850] K. M. Yin, Bull. Am. Phys. Soc. *20*, 886 (1975); Dielectric short-range order in liquid crystals.

[851] J. G. J. Ypma, G. Vertogen, Solid State Commun. *18*, 475 (1976), C.A. *84*, 129043e (1976); Short range order in nematics. [852] J. G. J. Ypma, G. Vertogen, H. T. Koster, Mol. Cryst. Liq. Cryst. *37*, 57 (1976), C.A. *86*, 113906u (1977); Molecular statistical calculation of pretransitional effects in nematics. [852a] J. G. J. Ypma, G. Vertogen, Phys. Lett. A *60*, 212 (1977), C.A. *86*, 198197m (1977); Equation of state for the Maier-Saupe model of nematics. [852b] J. G. J. Ypma, G. Vertogen, Phys. Lett. A *61*, 45 (1977), C.A. *86*, 198234w (1977); Equation of state for nematics. [852c] J. G. J. Ypma, G. Vertogen, Phys. Lett. A *61*, 125 (1977), C.A. *87*, 14471n (1977); Incorporation of short-range orientational order in the equation of state for nematics. [853] C. K. Yun, Diss. Univ. Minnesota, Minneapolis, Minn. 1970, 367 pp. Avail. Univ. Microfilms, Ann Arbor, Mich., Order No. 71-18,844, C.A. *76*, 64716q (1972); Transport phenomena in liquid crystals. [854] C. K. Yun, Phys. Lett. A *43*, 369 (1973), C.A. *79*, 24415w (1973); Friction coefficient of nematics. [855] C. K. Yun, Phys. Lett. A *45*, 119 (1973), C.A. *79*, 140462h (1973); Inertial coefficient of liquid crystals. Proposal for its measurements. [856] C. K. Yun, Mol. Cryst. Liq. Cryst. *35*, 181 (1976), C.A. *85*, 114993m (1976); Activation energy of a friction coefficient of liquid crystals. [857] H. Zocher, V. Birstein, Z. physik. Chem. A *141*, 413 (1929), C.A. *23*, 4387^7 (1929); Wesen der Mesophasen. [858] H. Zocher, V. Birstein, Z. physik. Chem. A *142*, 113 (1929), C.A. *23*, 4607^9 (1929); Gegenseitige Lagerung der Raumelemente einer Mesophase. [859] H. Zocher, Kolloid-Z. *75*, 161 (1936), C.A. *30*, 5088^3 (1936); Anwendbarkeit der Theorie der thermischen Schwankungen auf das Gebiet der Mesophasen. [860] H. Zocher, Ann. Physik. *31*, 570 (1938), C.A. *32*, 4848^8 (1938); Kontinuumtheorie und die Schwarmtheorie der nematischen Phasen.

[861] H. Zocher, G. Ungar, Z. Physik *110*, 529 (1938), C.A. *33*, 446^9 (1939); Struktur undeformierter und einfacher deformierter Gebiete in nematischen Schichten größerer Dicke. [862] H. Zocher, Liq. Cryst. Plast. Cryst. *1*, 64 (1974), C.A. *83*, 19078s (1975); Microstructure of liquid crystals. Swarm theory and the theory of the elastic continuum of liquid crystals. [863] V. Zolina, Acta Physicochim. U.R.S.S. *4*. 85 (1936), C.A. *31*, 3355^2 (1937); Elastische Schwingungen einer anisotropen Flüssigkeit. [864] G. E. Zvereva, Akust. Zh. *11*, 251 (1965), C.A. *63*, 12346g (1965); Absorption of ultrasound in cholesteryl caprate liquid crystals.

Chapter 4

[1] R. Abegg, W. Seitz, Z. physik. Chem. *29*, 491 (1899); Dielektrisches Verhalten einer krystallinischen Flüssigkeit. [2] J. J. Wysocki, J. Adams, W. Haas, Phys. Rev. Lett. *20*, 1024 (1968); Electric-field-induced phase change in cholesterics. [3] J. J. Wysocki, J. Adams, W. E. L. Haas, J. Appl. Phys. *40*, 3865 (1969), C.A. *71*, 106420t (1969); Electroviscosity of a cholesteric mixture. [4] J. J. Wysocki, J. Adams, W. E. L. Haas, Mol. Cryst. Liq. Cryst. *8*, 471 (1969), C.A. *71*, 85810c (1969); Electric-field induced phase change in cholesterics. [5] W. E. L. Haas, J. E. Adams, J. J. Wysocki, Appl. Opt., Suppl. *1969*, No. 3, 196, C.A. *74*, 92317n (1971); Imagewise deformation and color change of liquid crystals in electric fields. [6] W. Haas, J. E. Adams, J. B. Flannery, Phys. Rev. Lett. *24*, 577 (1970), C.A. *72*, 104983a (1970); A.c.-field-induced Grandjean plane texture in mixtures of room-temperature

nematics and cholesterics. [7] J. J. Wysocki, J. E. Adams, D. J. Olechna, Liq. Cryst. Ord. Fluids *1970*, 419, C.A. *78*, 152437w (1973); Kinetic study of the electric field-induced c/n transition in liquid crystals films. I. Relaxation to the cholesteric state. [8] W. Haas, J. E. Adams, J. B. Flannery, Phys. Rev. Lett. *25*, 1326 (1970), C.A. *74*, 16808h (1971); New electrooptic effect in a room-temperature nematic. [9] W. Haas, J. Adams, J. Electrochem. Soc. *118*, 1372 (1971), C.A. *75*, 113463v (1971); Electric field effects on the system oleyl cholesteryl carbonate/cholesteryl chloride. [10] W. Haas, J. Adams, G. Dir, Chem. Phys. Lett. *14*, 95 (1972), C.A. *77*, 40808r (1972); Optical storage effects in liquid crystals.

[11] G. Dir, J. Adams, W. Haas, Liq. Cryst. Ord. Fluids *2*, 429 (1974), C.A. *87*, 32498f (1977); Conductivity differences in the cholesteric textures. [12] G. A. Dir, J. Adams, W. Haas, Mol. Cryst. Liq. Cryst. *25*, 19 (1974), C.A. *83*, 140057n (1975); Dynamics of texture transitions in cholesteric-nematic mixtures. [13] A. Adamczyk, Pr. Nauk. Inst. Metrol. Elektr. Politech. Wroclaw. *7*, 15 (1975), C.A. *86*, 62628v (1977); Ionic polarization effect in liquid crystals. [14] A. Adamczyk, Pomiary, Autom., Kontrola *22*, 313 (1976), C.A. *86*, 98205u (1977); Static liquid-crystalline diffraction gratings. [14a] A. Adamczyk, A. Szymanski, Bull. Acad. Pol. Sci., Ser. Sci., Math., Astron. Phys. *20*, 955 (1972), C.A. *78*, 76817c (1973); Interface potential maxima in liquid crystals. I. Cholesteric structure. [15] P. Adamski, A. Dylik-Gromiec, M. Wojciechowski, Zesz. Nauk. Politech. Lodz., Fiz. *2*, 77 (1973), C.A. *83*, 36380b (1975); Dielectric properties of liquid-crystalline cholesteryl nonanoate. [16] A. Adamski, A. Dylik-Gromiec, M. Wojciechowski, R. Peeva, Dokl. Bolg. Akad. Nauk *27*, 39 (1974), C.A. *80*, 149977h (1974); Dielectric properties of cholesteryl decanoate. [17] P. Adamski, A. Dylik-Gromiec, M. Wojciechowski, Kristallografiya *19*, 1234 (1974), C.A. *82*, 66533a (1975); Temperature dependence of the dielectric constants of cholesterics. [18] P. Adamski, A. Dylik-Gromiec, M. Wojciechowski, Zesz. Nauk. Politech. Lodz., Fiz. *3*, 37 (1976), C.A. *86*, 99557x (1977); Dielectric anisotropy of nematics. [19] P. Adamski, A. Dylik-Gromiec, M. Wojciechowski, Zesz. Nauk. Politech. Lodz., Fiz. *3*, 45 (1976), C.A. *86*, 99127g (1977); Studies of isothermal crystallization of cholesteryl pelargonate by measurement of its static dielectric constant. [20] V. K. Agarwal, A. H. Price, J. Chem. Soc., Faraday Trans. 2, *70*, 188 (1974), C.A. *80*, 149982f (1974); Dielectric measurements of MBBA in the frequency range 1 kHz to 120 MHz.

[21] T. Akahane, M. Nakao, T. Tako, Jpn. J. Appl. Phys. *16*, 241 (1977), C.A. *86*, 113129t (1977); Structure of striped domains in c/n mixtures. [22] S. Akahoshi, K. Miyakawa, A. Takase, Jpn. J. Appl. Phys. *15*, 1839 (1976), C.A. *85*, 185009f (1976); Frequency measurement of the oscillatory motion of the Williams domain in the nematic liquid crystal. [22a] S. Akahoshi, K. Miyakawa, J. Phys. Soc. Jpn. *42*, 1997 (1977), C.A. *87*, 46955q (1977); Non-linear analysis of the Williams domain mode in nematics. [23] P. M. Alt, M. J. Freiser, J. Appl. Phys. *45*, 3237 (1974), C.A. *81*, 112222c (1974); Texturing. New effect in the dynamic scattering regime of LC cells. [24] S. M. Arakelvan, V. B. Pakhalov, A. S. Chirkin, Kvantovaya Elektron. *2*, 1205 (1975), C.A. *83*, 186013y (1975); Measurement of agitation time in liquid crystals using laser interferometry. [25] S. N. Aronishidze, M. I. Brodzeli, S. P. Ivchenko, M. N. Kushnirenko, M. D. Mkhatvrishvili, K. G. Tevdorashvili, G. S. Chilaya, Fiz. Tverd. Tela *17*, 555 (1975), C.A. *82*, 178648y (1975); Determination of elastic constants of a nematic. [25a] S. N. Aronishidze, M. I. Brodzeli, S. P. Ivchenko, G. S. Chilaya, Deposited Doc. *1974*, VINITI 137-75, 9 pp., C.A. *86*, 197068h (1977); Anisotropy of the dynamic scattering mode of polarized light in nematics. [26] A. Ashford, J. Constant, J. Kirton, E. P. Raynes, Electron. Lett. *9*, 118 (1973), C.A. *78*, 129710w (1973); Electro-optic performance of a new room temperature nematic. [27] E. W. Aslaksen, B. Ineichen, J. Appl. Phys. *42*, 882 (1971), C.A. *74*, 92319q (1971); Two-dimensional order in a nematic near threshold for dynamic scattering. [28] E. W. Aslaksen, Mol. Cryst. Liq. Cryst. *15*, 121 (1971), C.A. *76*, 64638r (1972); Steady state of a nematic above the treshold for dynamic scattering. [29] E. W. Aslaksen, J. Appl. Phys. *43*, 776 (1972), C.A. *76*, 104910g (1972); Electrohydrodynamic instability in nematics. [30] G. Assouline, E. Leiba, J. Phys. (Paris), C-4, *30*, 109 (1969), C.A. *72*, 126740w (1970); Diffusion optique induite par le champ électrique dans les structures némátiques.

[31] G. Assouline, A. Dmitrieff, M. Hareng, E. Leiba, J. Appl. Phys. *42*, 2567 (1971), C.A. *75*, 42710k (1971); Scattering of light by nematics. [32] G. Assouline, M. Hareng, E. Leiba, IEEE Trans. Electron Devices *18*, 959 (1971), C.A. *75*, 144943f (1971); Electric measurements in nematics. [33] M. Hareng, E. Leiba, G. Assouline, Mol. Cryst. Liq. Cryst. *17*, 361 (1972), C.A. *77*, 67968e (1972); Effet du champ électrique sur la biréfringence de cristaux liquides némátiques. [34] M. Hareng, G. Assouline, E. Leiba, Appl. Opt. *11*, 2920 (1972); La biréfringence électriquement contrôlée dans les cristaux liquides némátiques. [35] P. Atten, J. Mecanique *14*, 461 (1975); Electrohydrodynamic stability

of liquids with weak conductivity. [**36**] M. N. Avadhanlu, C. R. K. Murty, Liq. Cryst., Proc. Int. Conf., Bangalore *1973*, 289, C.A. *84*, 114468y (1976); Anomalous alignment and domain formation in nematic ethoxyphenylazophenyl hexanoate. [**37**] A. Axmann, Z. Naturforsch. A *21*, 290 (1966), C.A. *65*, 1526f (1966); Dielektrische Untersuchungen an kristallinflüssigen Azophenoläthern im Mikrowellenbereich. [**38**] A. Axmann, Z. Naturforsch. A *21*, 615 (1966), C.A. *65*, 9872a (1966); Dielektrische Untersuchungen an kristallinflüssigen 4,4'-Di-*n*-alkoxyazoxybenzolen im Mikrowellenbereich. [**39**] H. Weise, A. Axmann, Z. Naturforsch. A *21*, 1316 (1966), C.A. *65*, 17831b (1966); Messung der Hauptdielektrizitätskonstanten von homogen geordnetem nematischem PAA bei Frequenzen zwischen 10 MHz und 50 MHz. [**40**] A. Axmann, Mol. Cryst. *3*, 471 (1968), C.A. *69*, 47111b (1968); Berechnung der molekularen Hauptpolarisierbarkeiten von 4,4'-Di-*n*-alkoxyazoxybenzolen aus den Haupt-DK der nematischen Schichten.

[**41**] A. S. Babaev, G. S. Filatov, N. S. Saidov, V. I. Stafeev, Izv. Vyssh. Uchebn. Zaved. Radiofiz. *18*, 604 (1975), C.A. *84*, 24854s (1976); Current-voltage characteristics of molecular cholesteryl myristate films between metallic electrodes. [**42**] A. S. Babaev, A. I. Bogdanov, Dokl. Akad. Nauk Tadzh. SSR *19*, 17 (1976), C.A. *85*, 115239q (1976); Electric conductivity of cholesteryl stearate. [**43**] A. S. Babaev, A. I. Bogdanov, Dokl. Akad. Nauk Tadzh. SSR *19*, 25 (1976), C.A. *86*, 131691z (1977); Study of dynamic current-voltage characteristics of EBBA. [**43a**] A. S. Babaev, A. I. Bogdanov, V. I. Stafeev, Elektrokhimiya *13*, 1235 (1977), C.A. *87*, 144286s (1977); Electric conductivity and viscosity of cholesteryl esters. [**44**] H. Baessler, M. M. Labes, Phys. Rev. Lett. *21*, 1791 (1968), C.A. *70*, 41513h (1969); Relation between electric field strength and helix pitch in induced c/n transitions. [**45**] H. Baessler, M. M. Labes, J. Chem. Phys. *51*, 1846 (1969), C.A. *71*, 96044g (1969); Electric field effects on the dielectric properties and molecular arrangements of cholesterics. [**46**] H. Baessler, T. M. Laronge, M. M. Labes, J. Chem. Phys. *51*, 3213 (1969), C.A. *72*, 7445u (1970); Electric field effects on the optical rotatory power of a compensated cholesteric. [**47**] T. M. Laronge, H. Baessler, M. M. Labes, J. Chem. Phys. *51*, 4186 (1969), C.A. *72*, 36658w (1970); Magnetic-field effects on a compensated cholesteric. [**48**] H. Baessler, M. M. Labes, J. Chem. Phys. *51*, 5397 (1969), C.A. *72*, 71611x (1970); A.c. field induced c/n transitions. [**49**] H. Baessler, R. B. Beard, M. M. Labes, J. Chem. Phys. *52*, 2292 (1970), C.A. *72*, 94374m (1970); Dipole relaxation in a liquid crystal. [**50**] S. Bahadur, S. Chandra, Proc. Nucl. Phys. Solid State Phys. Symp. (16) *1972*, C, 225, C.A. *78*, 152766c (1973); Dielectric behavior of cholesteric cholesteryl laurate.

[**51**] B. Bahadur, R. A. Yadav, S. Chandra, Curr. Sci. *42*, 51 (1973), C.A. *78*, 110079a (1973); Magnetic susceptibility of MBBA. [**52**] B. Bahadur, S. Chandra, N. K. Sanyal, Phys. Status Solidi A *35*, 387 (1976), C.A. *85*, 27541u (1976); Phase transition studies of the liquid crystal N-(*p*-octyloxybenzylidene)-*p*-toluidine. [**52a**] B. Bahadur, J. Chem. Phys. *67*, 3272 (1977), C.A. *87*, 192850t (1977); Magnetic susceptibility of the liquid crystal N-(*p*-hexyloxybenzylidene)-*p*-toluidine. [**53**] P. Balog, D. L. Johnson, S. H. Christensen, Bull. Am. Phys. Soc. *19*, 173 (1974); Intermediate state of type-1 smectics A. [**54**] J. Baran, J. Kedzierski, Z. Raszewski, J. Zmija, Biul. Wojsk. Akad. Tech. *25*, 37 (1976), C.A. *86*, 64240m (1977); Measurement of the anisotropy of dielectric permittivity of liquid crystals. [**55**] M. I. Barnik, L. M. Blinov, M. F. Grebenkin, S. A. Pikin, V. G. Chigrinov, Zh. Eksp. Teor. Fiz. *69*, 1080 (1975), C.A. *83*, 211519c (1975); Electrohydrodynamic instability in nematics. [**56**] M. I. Barnik, L. M. Blinov, M. F. Grebenkin, S. A. Pikin, V. G. Chigrinov, Phys. Lett. A *51*, 175 (1975), C.A. *83*, 19449p (1975); Experimental verification of the theory of electrohydrodynamic instability in nematics. [**57**] M. I. Barnik, L. M. Blinov, M. F. Grebenkin, A. N. Trufanov, Mol. Cryst. Liq. Cryst. *37*, 47 (1976), C.A. *86*, 131832w (1977); Dielectric regime of electrohydrodynamic instability in nematics. [**58**] M. I. Barnik, L. M. Blinov, S. A. Pikin, A. N. Trufanov, Zh. Eksp. Teor. Fiz. *72*, 756 (1977), C.A. *86*, 131385c (1977); Instability mechanism in the n and i phases of liquid crystals with positive dielectric anisotropy. [**59**] S. Barret, F. Gaspard, R. Herino, F. Mondon, J. Appl. Phys. *47*, 2375 (1976), C.A. *86*, 10054k (1977); Dynamic scattering in nematics under dc conditions. Basic electrochemical analysis. [**60**] E. L. Bartashevskii, V. M. Dolgov, V. A. Krasovskii, Sb. Dokl. Vses. Nauch. Konf. Zhidk. Krist. Simp. Ikh Prakt. Primen., 2nd, *1972*, 75, C.A. *81*, 112181p (1974); Complex-permittivity anisotropy of MBBA at superhigh frequencies.

[**61**] E. L. Bartashevskii, V. M. Dolgov, N. V. Palii, Elektrodinamika i fiz. sVCH *1973*, 144, C.A. *83*, 20135q (1975); Dielectric constant of liquid crystals at superhigh frequencies. [**62**] G. Barzilai, P. Maltese, C. M. Ottavi, Electrotecn. *61*, 679 (1974); Theoretical interpretation of dynamic deformation effect in nematics. [**62a**] G. Barzilai, P. Maltese, C. M. Ottavi, Alta Freq. *45*, 53 (1976); Theoretical interpretation of dynamic deformation effect in nematics. [**63**] G. G. Basistov, G. Z. Blyum, Yu. G.

Nikiforov, Tr., Vses. Nauchno-Issled. Inst. Khim. Reakt. Osobo Chist. Khim. Veshchestv *36*, 55 (1974), C.A. *85*, 169953d (1976); Purification of MBBA by electrodialysis. [63a] G. G. Basistov, G. Z. Blyum, Yu. G. Nikiforov, M. K. Tuganova, Tr. VNII Khim. Reaktivov i Osobo Chistykh Khim. Veshchestv *1975*, 145, C.A. *87*, 32194d (1977); Electrodialysis purification of liquid crystals. [64] L. Bata, A. Buka, I. Janossy, Kozp. Fiz. Kut. Intez. *1973*, KFKI-73-32, 8 pp., C.A. *79*, 109012n (1973); Electric field induced instabilities in nematics. [65] L. Bata, A. Buka, I. Janossy, Solid State Commun. *15*, 647 (1974), C.A. *81*, 112185t (1974); Reorientation of nematic films by alternating and static fields. [66] L. Bata, C. Haranadh, G. Molnar, KFKI (Rep.), KFKI-74-55, 21 pp., C.A. *81*, 142787h (1974); Dielectric dispersion of a room-temperature nematic mixture with positive anisotropy. [67] L. Bata, G. Molnar, Chem. Phys. Lett. *33*, 535 (1975), C.A. *83*, 140478a (1975); Dielectric dispersion measurements in a nematic mixture. [68] L. Bata, A. Buka, G. Molnar, Hung. Acad. Sci., Cent. Res. Inst. Phys., KFKI 1976, KFKI-76-70, 9 pp., C.A. *86*, 88494y (1977); Rotary motion of molecules about their short axis by dielectric and splay viscosity measurements. [69] E. Bauer, Compt. rend. *182*, 1541 (1926), C.A. *20*, 2923[7] (1926); La structure électrique des molécules, particulièrement des corps mésomorphes. [70] M. S. Beevers, Mol. Cryst. Liq. Cryst. *31*, 333 (1975), C.A. *84*, 120666c (1976); Electro-optical Kerr effect in solutions of MBBA. [70a] M. S. Beevers, G. Williams, J. Chem. Soc., Faraday Trans. 2, *72*, 2171 (1976), C.A. *86*, 105666a (1977); Electrooptical Kerr effect in solutions of benzylidene aniline and its derivatives.

[71] B. I. Bel'tsov, V. M. Chaikovskii, Uch. zap. Jvanov. un-t *1974*, 25, C.A. *83*, 124408x (1975); Degree of orientation S and function D(α) for MBBA in relation to a magnetic field. [71a] S. V. Belyaev, V. G. Rumyantsev, V. V. Belyaev, Zh. Eksp. Teor. Fiz. *73*, 644 (1977), C.A. *87*, 174928v (1977); Optical and electrooptical properties of confocal cholesteric textures. [72] S. I. Ben-Abraham, Phys. Rev. A *14*, 1251 (1976), C.A. *85*, 169956g (1976); Nonlinear effects in liquid crystals. [73] D. Beneshevich, Acta Physicochim. U.R.S.S. *4*, 607 (1936), C.A. *31*, 5640[4] (1937); Volume of an anisotropic liquid in a magnetic field. [74] J. G. Berberian, Rev. Sci. Instrum. *46*, 107 (1975), C.A. *82*, 148704w (1975); Method for measuring very small conductances. [75] P. D. Berezin, I. N. Kompanets, V. V. Nikitin, S. A. Pikin, Zh. Eksp. teor. Fiz. *64*, 599 (1973), C.A. *78*, 116594m (1973); Orienting effect of an electric field on nematics. [76] D. Berg, Natl. Acad. Sci. Natl. Res. Council Publ. No. *1356*, 23 (1965), C.A. *65*, 4767h (1966); Electrical and physical properties of cholesterics. [77] A. L. Berman, Diss. Kent State Univ., Kent, Ohio 1975. 107 pp. Avail. Xerox Univ. Microfilms, Ann Arbor, Mich., Order No. 76-14,354, C.A. *85*, 85764e (1976); Optical studies of electric field effects in nematics that have some smectic like ordering. [78] A. L. Berman, E. Gelerinter, Chem. Phys. Lett. *30*, 143 (1975), C.A. *82*, 105302w (1975); New technique for measuring the critical magnetic field of c/n mixtures. [79] M. Bertolotti, B. Daino, F. Scudieri, D. Sette, Mol. Cryst. Liq. Cryst. *15*, 133 (1971), C.A. *76*, 19058r (1972); Spatial distribution of light scattered by PAA in applied electric field. [80] M. Bertolotti, B. Daino, P. Di Porto, F. Scudieri, D. Sette, J. Phys. A *4*, L97 (1971), C.A. *75*, 113473y (1971); Light scattering by electrohydrodynamic fluctuations in nematics.

[81] M. Bertolotti, B. Daino, P. Di Porto, F. Scudieri, D. Sette, J. Phys. (Paris) *33*, 63 (1972), C.A. *80*, 20952s (1974); Light scattering by fluctuations induced by an electric field in nematic APAPA. [82] M. Bertolotti, F. Scudieri, D. Sette, R. Bartalino, J. Appl. Phys. *43*, 3914 (1972); Electric and optic behavior of aligned samples of nematics. [83] M. Bertolotti, S. Lagomarsino, E. Scudieri, D. Sette, J. Phys. C *6*, L177 (1973), C.A. *79*, 10938g (1973); Velocity measurements of electrohydrodynamic flow in a nematic. [84] M. Zulauf, M. Bertolotti, F. Scudieri, J. Appl. Phys. *46*, 5152 (1975), C.A. *84*, 114420b (1976); Photon-correlation measurements on light scattered by a nematic under a dc electric field. [85] B. V. Bhide, R. D. Bhide, Rasayanam *1*, 121 (1938), C.A. *33*, 1564[5] (1939); Dielectric properties of PAA and cholesteryl benzoate. [86] V. G. Bhide, S. Chandra, S. C. Jain, R. K. Medhekar, J. Appl. Phys. *47*, 120 (1976), C.A. *84*, 98028x (1976); Structure and properties of bubble domains in c/n mixtures. [86a] V. G. Bhide, S. C. Jain, S. Chandra, J. Appl. Phys. *48*, 3349 (1977), C.A. *87*, 125591w (1977); Structure and properties of bubble domains in c/n mixtures. [87] S. Bini, R. Capelletti, Electrets, Charge Storage Transp. Dielectr. (Int. Conf.), 2nd, *1972*, 66, C.A. *83*, 156325j (1975); Conductivity and thermocurrents in liquid crystals. [87a] T. Bischofberger, R. Yu, Y. R. Shen, Mol. Cryst. Liq. Cryst. *43*, 287 (1977); Measurements of dc Kerr constants for a homologous series of nematics. [88] Y. Björnståhl, Ann. Physik *56*, 161 (1918), C.A. *13*, 3047[5] (1919); Untersuchungen über anisotrope Flüssigkeiten. [89] Y. Björnståhl, Physics *6*, 257 (1935), C.A. *29*, 7143[5] (1935); Effect of an electric field on the viscosity of aeolotropic liquids. [90] Y. Björnståhl, Z. physik. Chem. A *175*, 17 (1935), C.A. *30*, 2062[5] (1936); Extinktion von mesomorphen Flüssigkeiten im Magnetfeld.

[91] R. Blinc, Phys. Stat. Sol. B *70*, K29 (1975); Soft mode dynamics of ferroelectric smectics C. [92] R. Blinc, Wiss. Beitr. Martin-Luther-Univ. Halle-Wittenberg *1976*/20, 75, C.A. *86*, 82406z (1977); Ferroelectric liquid crystals. [93] R. Blinc, Ferroelectrics *14*, 603 (1976), C.A. *86*, 131933e (1977); Soft mode dynamics in ferroelectric liquid crystals. [94] L. Blinov, Usp. Fiz. Nauk *114*, 67 (1974), C.A. *81*, 177954n (1974); Electrooptical effects in liquid crystals. [95] S. V. Belyaev, L. M. Blinov, Zh. Eksp. Teor. Fiz. 70, 184 (1976), C.A. *84*, 129002r (1976); Instability of the planar texture of cholesterics in an electric field. [95a] L. M. Blinov, S. V. Belyaev, Kholestericheskie Zhidk. Krist. *1976*, 69, C.A. *87*, 175706h (1977); Electro- and magnetooptical effects in cholesterics. [96] V. I. Bobrov, Sb. Dokl. Vses. Nauch. Konf. Zhidk. Krist. Simp. Ikh Prakt. Primen., 2nd, *1972*, 112, C.A. *82*, 49431y (1975); Relation between the electric properties and optical behavior of a nematic. [97] V. I. Bobrov, Sb. Dokl. Vses. Nauch. Konf. Zhidk. Krist. Simp. Ikh Prakt. Primen., 2nd, *1972*, 118, C.A. *81*, 112180n (1974); Electric conductivity of 4,4′-bis(heptyloxy)azoxy-benzene. [98] V. I. Bobrov, Uch. zap. Ivanov. un-t *1974*, 75, C.A. *83*, 156392d (1975); Electric conductivity of cholesteryl myristate. [99] L. I. Boguslavskii, Y. B. Amerik, M. I. Gugeshashvili, Elektrokhimiya 9, 662 (1973), C.A. *79*, 72869u (1973); Phase transition in a liquid crystal studied by Volta potential measurements. [100] P. H. Bolomey, Mol. Cryst. Liq. Cryst. *29*, 103 (1974), C.A. *82*, 163364d (1975); Comportement d'un nématique sous l'action du champ électrique.

[101] P. H. Bolomey, Mol. Cryst. Liq. Cryst. *31*, 145 (1975), C.A. *83*, 211548m (1975); Déformation de l'alignement moléculaire et formation des chevrons dans un nématique en régime diélectrique. [102] P. H. Bolomey, C. Dimitropoulos, Mol. Cryst. Liq. Cryst. *36*, 75 (1976), C.A. *86*, 49317y (1977); Formation d'un réseau bidimensionnel dans un nématique soumis à un champ électrique alternatif. [103] B. M. Bolotin, M. I. Barnik, R. U. Safina, L. S. Zeryukina, R. S. Shishova, Zh. Vses. Khim. O-va. *21*, 594 (1976), C.A. *86*, 24883s (1977); Dielectric anisotropy of *p*-(*p*-fluorobenzoyloxy)benzoates. [104] C. J. F. Böttcher, Theory of electric polarization. (Elsevier: New York). Vol. 1. Dielectrics in static fields. 2nd ed. 1973. 377 pp., C.A. *81*, 112747c (1974). [105] M. A. Bouchiat, D. Langevin-Cruchon, Phys. Lett. A *34*, 331 (1971), C.A. *75*, 10211g (1971); Molecular order at the free surface of a nematic from light reflectivity measurements. [106] P. J. Bouma, Arch. neerland. sci. *IIIA*, 219 (1933), C.A. *28*, 1902⁶ (1934); The dynamics of liquid crystals. [107] G. Brière, F. Gaspard, R. Herino, Chem. Phys. Lett. 9, 285 (1971), C.A. *75*, 81383a (1971); Ionic residual conduction in the isotropic phase of a nematic. [108] G. Brière, R. Herino, F. Modon, Mol. Cryst. Liq. Cryst. *19*, 157 (1972), C.A. *78*, 51721d (1973); Correlation between chemical and electrochemical reactivity and dc conduction in n and i MBBA. [109] F. Brochard, P. Pieranski, E. Guyon, Phys. Rev. Lett. *28*, 1681 (1972), C.A. *77*, 80677d (1972); Dynamics of the orientation of a nematic film in a variable magnetic field. [110] F. Brochard, J. Phys. (Paris) *33*, 607 (1972), C.A. *77*, 157364j (1972); Mouvements de parois dans une lame mince nématique.

[111] P. Pieranski, F. Brochard, E. Guyon, J. Phys. (Paris) *33*, 681 (1972), C.A. *78*, 34638q (1973); Static and dynamic behavior of a nematic in a magnetic field. [112] F. Brochard, Mol. Cryst. Liq. Cryst. *23*, 51 (1973), C.A. *80*, 7106z (1974); Backflow effects in nematics. [113] F. Brochard, J. Phys. (Paris), Lett. *35*, 19 (1974), C.A. *80*, 126127s (1974); Distorsion d'un film nématique sous champ magnétique tournant. [114] E. Brochard, L. Léger, R. B. Meyer, J. Phys. (Paris), C-1, *36*, 209 (1975); Freedericksz transition of a homeotropic nematic in rotating magnetic fields. [115] A. A. Bronnikova, Uch. Zap., Ivanov. Gos. Pedagog. Inst. *1972*, No. 99, 112, C.A. *78*, 102988f (1973); Behavior of *p*-(*n*-heptyloxy)benzoic acid in an electric field. [116] A. A. Bronnikova, E. A. Kirsanov, Sb. Dokl. Vses. Nauch. Konf. Zhidk. Krist. Simp. Ikh Prakt. Primen., 2nd, *1972*, 133, C.A. *81*, 128317y (1974); Behavior of a nematic in dielectric state. [117] A. D. Buckingham, Proc. Phys. Soc. (London) *70*B, 753 (1957), C.A. *52*, 12484i (1958); Influence of a strong magnetic field on the dielectric constant of a diamagnetic fluid. [118] A. S. Cakmak, A. Askar, Proc.-Int. Congr. Rheol., 7th *1976*, 600, C.A. *86*, 64028y (1977); Electrohydrodynamic stability of nematics. Calculations for PAA. [119] H. Callen, U.S. Nat. Tech. Inform. Serv., AD Rep. No. 757435 (1972), 12 pp., C.A. *79*, 24955x (1973); Magnetic bubble materials. [120] S. Candau, R. Le Roy, F. Debeauvais, Mol. Cryst. Liq. Cryst. *23*, 283 (1973), C.A. *80*, 88153n (1974); Magnetic field effects in nematic and cholesteric droplets suspended in an isotropic liquid.

[121] E. F. Carr, R. D. Spence, Phys. Rev. *90*, 339 (1953), C.A. *48*, 7944g (1954); Influence of a magnetic field on the microwave dielectric constant of a liquid crystal. [122] E. F. Carr, R. D. Spence, J. Chem. Phys. *22*, 1481 (1954), C.A. *49*, 19e (1955); Influence of a magnetic field on the microwave dielectric constant of a liquid crystal. [123] E. F. Carr, J. Chem. Phys. *26*, 420 (1957),

C.A. *51*, 8496a (1957); Microwave dielectric measurements in liquid crystals. [**124**] E. F. Carr, J. Chem. Phys. *30*, 600 (1959), C.A. *53*, 12772f (1959); Microwave dielectric measurements in the liquid crystal ethyl *p*-azoxybenzoate. [**125**] E. F. Carr, J. Chem. Phys. *32*, 620 (1960), C.A. *54*, 14841b (1960); Microwave dielectric measurements in a polymesomorphic liquid crystal. [**126**] E. F. Carr, J. Chem. Phys. *37*, 104 (1962), C.A. *57*, 15933g (1962); Dielectric loss in the liquid crystal PAA. [**127**] E. F. Carr, J. Chem. Phys. *38*, 1536 (1963), C.A. *58*, 10822b (1963); Influence of electric and magnetic fields on the dielectric constant and loss of the liquid crystal anisaldazine. [**128**] E. F. Carr, J. Chem. Phys. *39*, 1979 (1963), C.A. *59*, 10854a (1963); Influence of an electric field on the dielectric loss of the liquid crystal PAA. [**129**] E. F. Carr, J. Chem. Phys. *42*, 738 (1965), C.A. *62*, 5969a (1965); Influence of electric and magnetic fields on the microwave dielectric constant of a liquid crystal with a positive dielectric anisotropy. [**130**] E. F. Carr, J. Chem. Phys. *43*, 3905 (1965), C.A. *64*, 1366h (1966); Influence of electric and magnetic fields on the molecular alignment in the liquid crystal anisal-*p*-aminoazobenzene.

[**131**] E. F. Carr, Advan. Chem. Soc. *63*, 76 (1967), C.A. *67*, 57939a (1967); Ordering in liquid crystals owing to electric and magnetic fields. [**132**] R. P. Twitchell, E. F. Carr, J. Chem. Phys. *46*, 2765 (1967), C.A. *67*, 26907j (1967); Influence of electric fields on the molecular alignment in the liquid-crystal PAA. [**133**] E. F. Carr, Mol. Cryst. Liq. Cryst. *7*, 253 (1969), C.A. *71*, 85578h (1969); Influence of electric fields on the molecular alignment in the liquid crystal APAPA. [**134**] E. F. Carr, J. H. Parker, D. P. McLemore, Liq. Cryst. Ord. Fluids *1970*, 201, C.A. *78*, 152448a (1973); Effect of electric fields on mixtures of nematics and cholesterics. [**135**] E. F. Carr, Phys. Rev. Lett. *24*, 807 (1970), C.A. *72*, 125880e (1970); Anomalous alignment in a smectic owing to an electric field. [**136**] E. F. Carr, Mol. Cryst. Liq. Cryst. *13*, 27 (1971), C.A. *75*, 54886k (1971); Ordering in the smectic phase owing to electric fields. [**137**] J. H. Parker, E. F. Carr, J. Chem. Phys. *55*, 1846 (1971), C.A. *75*, 92255m (1971); Anomalous alignment and domains in a nematic. [**138**] D. P. McLemore, E. F. Carr, B. Am. Phys. Soc. *16*, 186 (1971); Ordering effects of electric fields near the critical frequency in the liquid crystal PAA. [**139**] E. F. Carr, C. R. K. Murty, Mol. Cryst. Liq. Cryst. *18*, 369 (1972), C.A. *77*, 144884d (1972); Angular-dependent studies in liquid crystals involving permeability and dielectric anisotropies. [**140**] D. P. McLemore, E. F. Carr, J. Chem. Phys. *57*, 3245 (1972), C.A. *77*, 145108r (1972); Anomalous effects owing to electric fields in nematic 4,4'-di-*n*-heptyloxyazoxybenzene.

[**141**] L. S. Chou, E. F. Carr, Phys. Rev. A *7*, 1639 (1973), C.A. *78*, 165456b (1973); Molecular alignment owing to electric and high magnetic fields in the liquid crystal *p-n*-nonyloxybenzoic acid. [**142**] E. F. Carr, L. S. Chou, J. Appl. Phys. *44*, 3365 (1973), C.A. *79*, 84274j (1973); Molecular alignment and conductivity anisotropy in a nematic. [**143**] W. T. Flint, E. F. Carr, Mol. Cryst. Liq. Cryst. *22*, 1 (1973), C.A. *79*, 109051z (1973); Anomalous alignment and dynamic scattering in liquid crystals exhibiting a positive dielectric anisotropy. [**144**] E. F. Carr, Liq. Cryst., Proc. Int. Conf., Bangalore *1973*, 263, C.A. *84*, 158398e (1976); Electrical properties and ordering in the *p-n*-alkoxybenzoic acids. [**145**] L. S. Chou, E. F. Carr, Liq. Cryst. Ord. Fluids *2*, 39 (1974), C.A. *86*, 181196n (1977); Electric field effects in n and s *p-n*-nonyloxybenzoic acid. [**146**] E. F. Carr, W. T. Flint, J. H. Parker, Phys. Rev. A *11*, 1732 (1975), C.A. *83*, 89044b (1975); Effects of dielectric and conductivity anisotropies on molecular alignment in a liquid crystal. [**147**] E. F. Carr. Phys. Rev. A *12*, 327 (1975), C.A. *83*, 124642u (1975); Conductivity anisotropy in *p-n*-nonyloxybenzoic acid. Comments. [**148**] E. J. Sinclair, E. F. Carr, Mol. Cryst. Liq. Cryst. *35*, 143 (1976), C.A. *85*, 102657s (1976); Flow patterns in bulk samples of a nematic due to electric fields. [**149**] E. J. Sinclair, E. F. Carr, Mol. Cryst. Liq. Cryst. *37*, 303 (1976), C.A. *86*, 131833x (1977); Flow patterns and molecular alignment in a nematic due to electric fields. [**150**] E. F. Carr, Mol. Cryst. Liq. Cryst. *34*, 159 (1977), C.A. *86*, 198228x (1977); Domains due to magnetic fields in a nematic. [**150a**] E. F. Carr, P. H. Ackroyd, J. K. Newell, Mol. Cryst. Liq. Cryst. *43*, 93 (1977); Dynamic scattering and material flow in a nematic.

[**151**] T. O. Carroll, J. Appl. Phys. *43*, 767 (1972), C.A. *76*, 104913k (1972); Liquid-crystal diffraction grating. [**152**] T. O. Carroll, J. Appl. Phys. *43*, 1342 (1972), C.A. *76*, 132601a (1972); Dependence of conduction-induced alignment of nematics upon voltage above threshold. [**153**] J. A. Castellano, M. T. McCaffrey, Liq. Cryst. Ord. Fluids *1970*, 293, C.A. *79*, 24399u (1973); Electrooptic effects in *p*-alkoxybenzylidene-*p'*-aminoalkylphenones and related compounds. [**154**] S. Chandra, B. Bahadur, Curr. Sci. *41*, 806 (1972), C.A. *78*, 29064v (1973); Dipole moments of some cholesteryl esters. [**155**] S. Chandra, B. Bahadur, J. Chim. Phys. Physicochim. Biol. *70*, 605 (1973), C.A. *79*, 42744g (1973); Odd-even effect in dipole moments of liquid crystalline cholesteryl esters. [**156**] N. V. Madhusudana, P. P. Karat, S. Chandrasekhar, Curr. Sci. *42*, 147 (1973), C.A. *78*, 123498m (1973); Electrohydrodynamic distortion patterns in nematics. [**157**] N. V. Madhusudana, S. Chandrasekhar, Solid State Commun.

13, 377 (1973), C.A. *79*, 109098v (1973); Short range order in the i phase of nematogens. [**158**] N. V. Madhusudana, P. P. Karat, S. Chandrasekhar, Liq. Cryst., Proc. Int. Conf., Bangalore *1973*, 225, C.A. *84*, 114464u (1976); Experimental determination of the twist elastic constant of nematics. [**159**] N. V. Madhusudana, S. Chandrasekhar, Pramana *1*, 12 (1973), C.A. *79*, 140459n (1973); Magnetic and electric birefringence in the i phase of nematogens. [**160**] N. V. Madhusudana, S. Chandrasekhar, Liq. Cryst. Ord. Fluids *2*, 657 (1974), C.A. *86*, 179739y (1977); Kerr effect in isotropic PAA.

[**161**] R. Chang, J. Appl. Phys. *44*, 1885 (1973), C.A. *78*, 141301w (1973); Secondary dynamic scattering in negative nematic films. [**162**] R. Chang, Mol. Cryst. Liq. Cryst. *20*, 267 (1973), C.A. *79*, 58660n (1973); Microstructural observations of MBBA liquid crystal films. [**163**] T. S. Chang, P. E. Greene, E. E. Loebner, Liq. Cryst. Ord. Fluids *2*, 115 (1974), C.A. *86*, 180980b (1977); Kinetics of field alignment and elastic relaxation in TN liquid crystals. [**164**] R. Chang, Liq. Cryst. Ord. Fluids *2*, 367 (1974), C.A. *86*, 181342g (1977); Anisotropic electric conductivity of MBBA containing alkyl ammonium halides. [**165**] R. Chang, J. M. Richardson, Mol. Cryst. Liq. Cryst. *28*, 189 (1974), C.A. *83*, 16624f (1975); Anisotropic electric conductivity of MBBA containing tetrabutylammonium tetraphenyl-boride. [**166**] D. J. Channin, Appl. Phys. Lett. *22*, 365 (1973), C.A. *78*, 153179a (1973); Optical waveguide modulation using nematic liquid crystal. [**167**] D. J. Channin, Introd. Liq. Cryst. *1974*, 281, C.A. *84*, 113572r (1976); Liquid-crystal optical waveguides. [**168**] V. G. Chigrinov, M. F. Grebenkin, Kristallo-grafiya *20*, 1240 (1975), C.A. *84*, 98006p (1976); Determination of elasticity constants k_{11} and k_{33} and viscosity factor γ_1 of nematics from oriented electrooptical effects. [**168a**] V. G. Chigrinov, V. Belyaev, Kristallografiya *22*, 603 (1977), C.A. *87*, 144261e (1977); Time characteristics of orientational electrooptical effects in nematics. [**169**] G. S. Chilaya, V. T. Lazareva, L. M. Blinov, Kristallografiya *18*, 203 (1973), C.A. *78*, 116275h (1973); Dynamic scattering with memory in liquid crystal mixtures. [**170**] G. S. Chilaya, S. N. Aronishidze, Z. M. Elashvili, M. N. Kushnirenko, M. I. Brodzeli, Soobshch. Akad. Nauk Gruz. SSR, *84*, 81 (1976), C.A. *86*, 62867x (1977); Study of the parameters of the racemic and optically active forms of a liquid crystal. [**170a**] V. N. Chirkov, D. F. Aliev, A. K. Zeinally, Kristallografiya *22*, 660 (1977), C.A. *87*, 192251y (1977); Behavior of a nematic in a.c. electric fields. [**170b**] V. N. Chirkov, D. F. Aliev, A. K. Zeinally, Zh. Eksp. Teor. Fiz. *73*, 761 (1977), C.A. *87*, 175830u (1977); Electrotropic c/n transition and the memory effect based on it.

[**171**] I. G. Chistyakov, I. V. Sushkin, V. M. Chaikovskii, E. A. Kosterin, Uch. Zap. Ivanov. Gos. Pedagog. Inst. *77*, 13 (1970), C.A. *76*, 29158n (1972); Magnetic and electric properties of liquid crystals. [**172**] I. G. Chistyakov, S. K. Sukharev, Uch. Zap., Ivanov. Gos. Pedagog. Inst. *1972*, No. 99, 108, C.A. *78*, 152428u (1973); Effect of current on orientation of nematic PAA. [**173**] I. G. Chistyakov, L. K. Vistin, Kristallografiya *18*, 873 (1973), C.A. *79*, 130083w (1973); Phase transitions in liquid crystals induced by an electric field. [**174**] I. G. Chistyakov, L. K. Vistin, Kristallografiya *19*, 195 (1974), C.A. *80*, 100821p (1974); Domains in liquid crystals. [**175**] L. K. Vistin, I. G. Chistyakov, R. I. Zharenov, S. S. Yakovenko, Kristallografiya *21*, 173 (1976), C.A. *84*, 129015x (1976); Change in the domain pattern of nematics in electric fields. [**176**] L. K. Vistin, I. G. Chistyakov, R. I. Zharenov, Kristallografiya *21*, 853 (1976), C.A. *85*, 133061x (1976); Effect of an electric field on the orientation of the optical indicatrix of a liquid crystal. [**177**] I. G. Chistyakov, B. N. Makarov, L. K. Vistin, S. P. Chumakova, Dokl. Akad. Nauk SSSR *229*, 1350 (1976), C.A. *86*, 24605c (1977); Preferred orientation and phase transformations in liquid crystals caused by electric fields and by pressure. [**177a**] B. K. Vainshtein, I. G. Chistyakov, L. K. Vistin, R. I. Zharenov, Bulg. J. Phys. *3*, 292 (1976), C.A. *87*, 46825x (1977); Effect of electric fields on the structure and properties of liquid crystals. [**177b**] L. Vistin, I. G. Chistyakov, Dokl. Akad. Nauk SSSR *234*, 1063 (1977), C.A. *87*, 125565r (1977); "Frozen" space charge in liquid crystals. [**177c**] L. K. Vistin, I. G. Chistyakov, S. P. Chumakova, V. V. Parkhomenko, Pis'ma Zh. Eksp. Teor. Fiz. *25*, 201 (1977), C.A. *87*, 14454j (1977); Temperature dependence of threshold voltages in nematics. [**178**] L. S. T. Chou, Diss. Univ. Maine, Orono, Maine 1973. 89 pp. Avail. Univ. Microfilms, Ann Arbor, Mich., Order No. 73-32,321, C.A. *81*, 30658k (1974); Molecular alignment, conductivity anisotropy, and dielectric properties of the liquid crystals p-nonyloxybenzoic acid and PAA. [**179**] A. N. Chuvyrov, A. N. Trofimov, Kristallografiya *17*, 1205 (1972), C.A. *78*, 49224a (1973); Orientation oscillations domain structures of liquid crystals. Mechanism of formation of hexagonal domain structures in constant electric fields. [**180**] A. N. Chuvyrov, Zh. Eksp. Teor. Fiz. *63*, 958 (1972), C.A. *77*, 157539v (1972); Experimental study of the frequency of domain structure oriented oscillations in liquid crystals.

[**181**] A. N. Chuvyrov, Z. K. Kuvatov, Kristallografiya *18*, 344 (1973), C.A. *78*, 165399k (1973); Effect of domain rotation in smectics. [**182**] A. N. Chuvyrov, Kristallografiya *19*, 297 (1974), C.A.

81, 42516a (1974); Cylindrical domains of liquid crystals. [**183**] A. N. Chuvyrov, Fiz. Tverd. Tela *16*, 321 (1974), C.A. *80*, 150906r (1974); Electrooptical effect of nematics. [**184**] A. N. Chuvyrov, A. S. Sonin, A. D. Zakirova, Fiz. Tverd. Tela *18*, 3084 (1976), C.A. *86*, 98203s (1977); Transverse electrooptical effect in nematics with negative dielectric anisotropy. [**184a**] A. N. Chuvyrov, A. N. Lachinov, V. S. Avzyanov, A. S. Sonin, Fiz. Tverd. Tela *19*, 1191 (1977), C.A. *87*, 45892e (1977); Study of a nematic liquid crystal/semiconductor interface. [**185**] K. S. Cole, R. H. Cole, J. Chem. Phys. *9*, 341 (1941), C.A. *35*, 3495[7] (1941); Dispersion and absorption in dielectrics. I. A.c. characteristics. [**186**] H. J. Coles, B. R. Jennings, Mol. Phys. *31*, 571 (1976), C.A. *84*, 179123b (1976); Static and optical Kerr effect in MBBA. [**187**] H. J. Coles, B. R. Jennings, Mol. Phys. *31*, 1225 (1976), C.A. *85*, 113931c (1976); Electric birefringence of solutions of nematics. [**188**] H. J. Coles, B. R. Jennings, Electro-Opt./Laser Int. UK, Conf. Proc. *1976*, 109, C.A. *86*, 98739w (1977); Laser and electric field Kerr effect of liquids and liquid crystals. [**189**] G. H. Conners, K. B. Paxton, J. Appl. Phys. *43*, 2959 (1972), C.A. *77*, 40114m (1972); Oscillation modes of nematics. [**190**] J. Constant, E. P. Raynes, Electron. Lett. *9*, 561 (1973), C.A. *80*, 64746g (1974); c/n Phase-change effect in mixtures containing the liquid crystal PCB.

[**191**] P. G. Cummins, D. A. Dunmur, N. E. Jessup, Liq. Cryst. Ord. Fluids *2*, 341 (1974), C.A. *86*, 181391x (1977); Dielectric properties of nematic MBBA in the presence of electric and magnetic fields. [**192**] P. G. Cummins, D. A. Dunmur, D. A. Laidler, Mol. Cryst. Liq. Cryst. *30*, 109 (1975); Dielectric properties of nematic 4,4-*n*-pentylcyanobiphenyl. [**193**] A. L. Dalisa, R. J. Seymour, Proc. IEEE *61*, 981 (1973), C.A. *79*, 110010s (1973); Convolution scattering for ferroelectric ceramics and other display media. [**194**] M. Davies, R. Moutran, A. H. Price, M. S. Beevers, G. Williams, J. Chem. Soc., Faraday Trans. 2, *72*, 1447 (1976), C.A. *85*, 152497k (1976); Dielectric and optical studies of nematic 4,4-*n*-heptyl-cyanobiphenyl. [**195**] G. W. Day, O. L. Gaddy, Proc. IEEE *56*, 1113 (1968); Electric field induced optical rotation in cholesterics. [**195a**] F. Dazai, T. Uchida, M. Wada, Mol. Cryst. Liq. Cryst. *34*, 197 (1977), C.A. *87*, 46063x (1977); Electro-optic effects of smectics. [**196**] P. G. de Gennes, Solid State Commun. *6*, 163 (1968), C.A. *69*, 5850m (1968); Calcul de la distorsion d'une structure cholestérique par un champ magnétique. [**197**] J. Friedel, P. G. de Gennes, C. R. Acad. Sci., Sér. B *268*, 257 (1969), C.A. *70*, 91478w (1969); Boucles de disinclination dans les cristaux liquides. [**198**] P. G. de Gennes, Solid State Commun. *8*, 213 (1970), C.A. *72*, 84274q (1970); Structures en domaines dans un nématique sous champ magnétique. [**199**] P. G. de Gennes, Comments Solid State Phys. *3*, 35 (1970), C.A. *74*, 131559q (1971); Electrohydrodynamic effects in nematics. D.c. fields. [**200**] F. Brochard, P. G. de Gennes, J. Phys. (Paris) *31*, 691 (1970), C.A. *73*, 124983z (1970); Theory of magnetic suspensions in liquid crystals.

[**201**] G. Durand, P. G. de Gennes, M. Veyssie, Semin. Chim. Etat Solide *1970*, No. 5, 1, C.A. *77*, 67384m (1972); Effets de champ électrique dans les cristaux liquides. [**202**] E. Dubois-Violette, P. G. de Gennes, O. Parodi, J. Phys. (Paris) *32*, 305 (1971), C.A. *75*, 81386d (1971); Hydrodynamic instabilities of nematics under a.c. electric fields. [**203**] P. G. de Gennes, J. Phys. (Paris) *32*, 789 (1971); Mouvements de parois dans un nématique sous champ tournant. [**204**] E. Dubois-Violette, P. G. de Gennes, J. Phys. (Paris), Lett. *36*, 255 (1975); Local Frederiks transitions near a solid/nematic interface. [**205**] P. G. de Gennes, C. R. Acad. Sci., Sér. B, *280*, 9 (1975); Effet d'un écoulement sur les domaines de Williams. [**206**] W. H. de Jeu, C. J. Gerritsma, A. M. van Boxtel, Phys. Lett. A *34*, 203 (1971), C.A. *75*, 11715m (1971); Electrohydrodynamic instabilities in nematics. [**207**] W. H. de Jeu, Phys. Lett. A *37*, 365 (1971), C.A. *76*, 77967x (1972); Instabilities of nematics in pulsating electric fields. [**208**] W. H. de Jeu, C. J. Gerritsma, T. W. Lathouwers, Chem. Phys. Lett. *14*, 503 (1972), C.A. *77*, 67475s (1972); Instabilities in electric fields of nematics with positive dielectric anisotropy. Domains, loop domains, and reorientation. [**209**] W. H. de Jeu, C. J. Gerritsma, J. Chem. Phys. *56*, 4752 (1972), C.A. *77*, 10945r (1972); Electrohydrodynamic instabilities in some nematic azoxy compounds with dielectric anisotropies of different sign. [**210**] W. H. de Jeu, C. J. Gerritsma, P. van Zanten, W. J. A. Goossens, Phys. Lett. A *39*, 355 (1972), C.A. *77*, 67428d (1972); Relaxation of the dielectric constant and electrohydrodynamic instabilities in a liquid crystal.

[**211**] W. H. de Jeu, J. van der Veen, Phys. Lett. A *44*, 277 (1973), C.A. *79*, 71017c (1973); Instabilities in electric fields of a nematic with large negative dielectric anisotropy. [**212**] W. H. de Jeu, T. W. Lathouwers, Mol. Cryst. Liq. Cryst. *26*, 225 (1974), C.A. *82*, 79334m (1975); Nematic phenyl benzoates in electric fields. I. Static and dynamics properties of the dielectric permittivity. [**213**] W. H. de Jeu, T. W. Lathouwers, Mol. Cryst. Liq. Cryst. *26*, 235 (1974), C.A. *82*, 50336c (1975); II. Instabilities around the frequency of dielectric isotropy (cf. [212]). [**214**] W. H. de Jeu, T. W. Lathouwers,

P. Bordewijk, Phys. Rev. Lett. *32*, 40 (1974), C.A. *80*, 100974r (1974); Dielectric properties of n and s_A *p,p'*-di-*n*-heptylazoxybenzene. [**215**] W. H. de Jeu, T. W. Lathouwers, Z. Naturforsch. A *29*, 905 (1974), C.A. *81*, 83157c (1974); Dielectric constants and molecular structure of nematics. I. Terminally substituted azobenzenes and azoxybenzenes. [**216**] W. H. de Jeu, T. W. Lathouwers, Chem. Phys. Lett. *28*, 239 (1974), C.A. *82*, 10358e (1975); Dielectric properties of some nematics for dynamic scattering displays. [**217**] W. H. de Jeu, W. J. A. Goossens, P. Bordewijk, J. Chem. Phys. *61*, 1985 (1974), C.A. *81*, 142784e (1974); Influence of smectic order on the static dielectric permittivity of liquid crystals. [**218**] W. H. de Jeu, T. W. Lathouwers, Z. Naturforsch. A *30*, 79 (1975), C.A. *82*, 105397f (1975); II. Variation of the bridging group (cf. [215]). [**219**] W. H. de Jeu, W. A. P. Claassen, A. M. J. Spruijt, Mol. Cryst. Liq. Cryst. *37*, 269 (1976), C.A. *86*, 113911s (1977); Determination of the elastic constants of nematics. [**220**] W. H. de Jeu, Advances in liquid crystal research. Dielectric permittivity of liquid crystals.

[**221**] J. M. Delrien, J. Chem. Phys. *60*, 1081 (1974), C.A. *80*, 125575s (1974); Comparison between square, triangular, or one-dimensional lattice of distortions in smectics A and cholesterics for superposed strain and magnetic field of any directions. [**222**] A. Denat, B. Gosse, J. P. Gosse, J. Chim. Phys. Physicochim. Biol. *70*, 327 (1973), C.A. *78*, 167756y (1973); Comportement électrochimique du MBBA et relations avec le comportement sous champ électrique. [**223**] A. Denat, B. Gosse, Chem. Phys. Lett. *22*, 91 (1973), C.A. *80*, 33234j (1974); Electrochemical behavior of nematic tolanes. [**224**] A. I. Derzhanski, A. G. Petrov, Dokl. Bolg. Akad. Nauk *24*, 569 (1971), C.A. *75*, 113679v (1971); Dielectric properties of a nematic with ellipsoidal molecules. [**225**] A. Derzhanski, A. G. Petrov, Phys. Lett. A *34*, 427 (1971), C.A. *75*, 41874e (1971); Possible relation between the dielectric and the piezoelectric properties of nematics. [**226**] A. Derzhanski, A. G. Petrov, Phys. Lett. A *36*, 307 (1971), C.A. *75*, 144893q (1971); Inverse currents and contact behavior of some nematics. [**227**] A. Derzhanski, A. G. Petrov, Phys. Lett. A *36*, 483 (1971), C.A. *76*, 38767n (1972); Molecular-statistical approach to the piezoelectric properties of nematics. [**228**] A. I. Derzhanski, L. Grigorov, B. Tenchov, Dokl. Bolg. Akad. Nauk *25*, 163 (1972), C.A. *77*, 25964v (1972); A.c. conductivity of liquid crystals in electric and magnetic fields. [**229**] A. I. Derzhanski, A. G. Petrov, Dokl. Bolg. Akad. Nauk *25*, 167 (1972), C.A. *77*, 26069u (1972); Molecular-statistical approach to the piezoelectric properties of nematics. One-dimensional model. [**230**] A. Derzhanski, A. Petrov, K. Khinov, B. Markovski, Bulg. J. Phys. *1*, 165 (1974), C.A. *83*, 20307x (1975); Piezoelectric deformations of nematics in nonhomogeneous d.c. electric fields.

[**231**] A. I. Derzhanski, M. D. Mitov, C. P. Khinov, Dokl. Bolg. Akad. Nauk *27*, 453 (1974); Piezoelectric deformation of liquid crystal layer in a slightly nonhomogeneous electric field. [**232**] K. P. Khinov, A. I. Derzhanski, K. S. Avramova, Dokl. Bolg. Akad. Nauk *27*, 1339 (1974), C.A. *82*, 105298z (1975); Orienting action of an a.c. electric field on a nematic layer. [**233**] A. I. Derzhanski, M. D. Mitov, Dokl. Bolg. Akad. Nauk *28*, 1331 (1975); Statistical characteristics of a deformed liquid-crystal layer in a slightly nonhomogeneous dc electric field. [**234**] A. Derzhanski, Kh. Khinov, Phys. Lett. A *56*, 465 (1976), C.A. *85*, 70901y (1976); Influence of second order elasticity on the piezoelectric bending of homeotropic nematic layers. [**234a**] A. Derzhanski, Kh. Khinov, J. Phys. (Paris) *38*, 1013 (1977), C.A. *87*, 125636q (1977); Polar flexoelectric deformations and second order elasticity in nematics. [**234b**] A. Petrov, A. Derzhanski, Mol. Cryst. Liq. Cryst. *41*, 41 (1977), C.A. *87*, 175861e (1977); Flexoelectricity and surface polarization. [**234c**] A. Derzhanski, Kh. Khinov, Phys. Lett. A *62*, 36 (1977), C.A. *87*, 109697a (1977); Method for measuring e_{1z}/e_{3x} in nematics. [**235**] H. J. Deuling, Mol. Cryst. Liq. Cryst. *19*, 123 (1972), C.A. *78*, 63535j (1973); Deformation of nematics in an electric field. [**236**] H. J. Deuling, Mol. Cryst. Liq. Cryst. *26*, 281 (1974), C.A. *82*, 50418f (1975); Piezoelectric effect in nematic layers. [**237**] H. J. Deuling, Mol. Cryst. Liq. Cryst. *27*, 81 (1974), C.A. *82*, 92102d (1975); Deformation pattern of TN liquid crystal layers in an electric field. [**238**] H. J. Deuling, Solid State Commun. *14*, 1073 (1974); Method of measure flexo-electric coefficients of nematics. [**239**] H. J. Deuling, E. Guyon, P. Pieranski, Solid State Commun. *15*, 277 (1974); Deformation of nematic layers in crossed electric and magnetic fields. [**240**] H. J. Deuling, M. Gabay, E. Guyon, P. Pieranski, J. Phys. (Paris) *36*, 689 (1975), C.A. *83*, 89015t (1975); Freedericksz transition of nematics in an oblique magnetic field.

[**241**] H. J. Deuling, A. Buka, I. Janossy, J. Phys. (Paris) *37*, 965 (1976), C.A. *85*, 85802r (1976); Two Freedericksz transitions in crossed electric and magnetic fields. [**242**] C. Deutsch, P. N. Keating, J. Appl. Phys. *40*, 4049 (1969), C.A. *71*, 107470w (1969); Scattering of coherent light from nematics in the dynamic scattering mode. [**243**] M. de Zwart, T. W. Lathouwers, Phys. Lett. A *55*, 41 (1975),

C.A. *84*, 114479c (1976); Electric field-induced pitch contraction in the planar cholesteric texture of a liquid crystal with a large negative dielectric anisotropy. [**244**] M. de Zwart, 6. Arbeitstagung Flüssigkristalle, Inst. Angew. Festkörperphysik Freiburg, 2. 4. 1976; Influence of negative dielectric anisotropy on the electric field-induced distortion mode in the planar cholesteric texture. [**245**] M. de Zwart, 7. Arbeitstagung Flüssigkristalle, Inst. Angew. Festkörperphysik Freiburg, 4. 3. 1977; Effect of pitch variation on the electrohydrodynamic deformation in the cholesteric planar texture. [**246**] D. D'Humières, L. Léger, J. Phys. (Paris), C-1, *36*, 113 (1975); Critical behaviour above a n/s$_A$ transition. [**246a**] J. P. Dias, C. R. Acad. Sci., Sér. A *282*, 71 (1976); Equations d'évolution d'un nématique incompressible soumis à l'action d'un champ magnétique homogène. [**247**] S. G. Dmitriev, Zh. Eksp. Teor. Fiz. *61*, 2049 (1971), C.A. *76*, 38763h (1972); Piezoelectric domains in liquid crystals. [**248**] R. Dreher, Solid State Commun. *13*, 1571 (1973), C.A. *80*, 42252r (1974); Distortion of a cholesteric structure by a magnetic field. [**249**] C. Druon, J. M. Wacrenier, J. Phys. (Paris) *38*, 47 (1977), C.A. *86*, 64245s (1977); Propriétés diélectriques a large bande de fréquences du 4-cyano-4'-octylbiphenyl en phases s, n et i. [**250**] E. Dubois-Violette, O. Parodi, J. Phys. (Paris), C-4, *30*, 57 (1969), C.A. *72*, 104137c (1970); Émulsions nématiques. Effets de champ magnétique et effets piézoélectriques.

[**251**] E. Dubois-Violette, J. Phys. (Paris) *33*, 95 (1972); Theory of instabilities of nematics under a.c. electric fields: Special effects near the cut off frequency. [**252**] D. A. Dunmur, Chem. Phys. Lett. *10*, 49 (1971), C.A. *75*, 113481z (1971); Local field effects in oriented nematics. [**253**] G. Durand, L. Léger, F. Rondelez, M. Veyssie, Phys. Rev. Lett. *22*, 227 (1969), C.A. *70*, 100491z (1969); Magnetically induced c/n transition in liquid crystals. [**254**] G. Durand, M. Veyssie, F. Rondelez, L. Léger, C. R. Acad. Sci., Sér. B *270*, 97 (1970), C.A. *72*, 94327y (1970); Effet électrohydrodynamique dans un nématique. [**255**] D. Diquet, F. Rondelez, G. Durand, C. R. Acad. Sci., Sér. B *271*, 954 (1970), C.A. *74*, 58077z (1971); Anisotropie de la constante diélectrique et de la conductivité du MBBA en phase nématique. [**256**] F. Rondelez, D. Diguets, G. Durand, Mol. Cryst. Liq. Cryst. *15*, 183 (1971), C.A. *72*, 19128p (1972); Dielectric and resistivity measurements on nematic MBBA. [**257**] J. L. Martinand, G. Durand, Solid State Commun. *10*, 815 (1972), C.A. *77*, 107237z (1972); Electric field quenching of thermal fluctuations of orientation in a nematic. [**258**] G. Durand, C. R. Acad. Sci., Sér. B *275*, 629 (1972); Sur la portée des ondulations de couches dans les smectiques A. [**258a**] D. Dvorjetski, Y. Silberberg, E. Wiener-Avnear, Mol. Cryst. Liq. Cryst. *42*, 273 (1977), C.A. *87*, 175874m (1977); Temperature dependence of the electrohydrodynamic instability in nematics. [**259**] L. Ebert, H. v. Hartel, Physik. Z. *28*, 786 (1927), C.A. *22*, 3076 (1928); Dipolmoment und Anisotropie von Flüssigkeiten. [**260**] G. Elliot, J. G. Gibson, Nature *205*, 995 (1965); Domain structures in liquid crystals induced by electric fields.

[**261**] G. Elliott, D. Harvey, M. G. Williams, Electron. Lett. *9*, 399 (1973), C.A. *79*, 119555f (1973); Dielectric anisotropy of new liquid-crystal mixtures and its effect on dynamic scattering. [**262**] E. C. Ellis, Diss. Duke Univ., Durham, N. C. 1975. 184 pp. Avail. Xerox Univ. Microfilms, Ann Arbor, Mich., Order No. 76-8736, C.A. *84*, 187842a (1976); Electrohydrodynamic instabilities in thin homogeneous liquid crystalline materials. [**263**] J. L. Ericksen, Arch. Ration. Mech. Anal. *23*, 266 (1966), C.A. *67*, 58672v (1967); Magnetohydrodynamic effects in liquid crystals. [**264**] J. L. Ericksen, Z. Angew. Math. Phys. *20*, 383 (1969); Twisting of liquid crystals by magnetic fields. [**265**] J. Errera, Physik. Z. *29*, 426 (1928), C.A. *22*, 3573 (1928); Das elektrische Moment des PAA. [**266**] S. Faetti, L. Fronzoni, P. A. Rolla, G. Stoppini, Lett. Nuovo Cimento Soc. Ital. Fis. *17*, 475 (1976), C.A. *86*, 63827w (1977); Behavior of nematic suspended films submitted to electric fields. [**267**] C.-P. Fan, M. J. Stephen, Phys. Rev. Lett. *25*, 500 (1970), C.A. *73*, 81750b (1970); i/n Transition in liquid crystals. [**268**] C.-P. Fan, Mol. Cryst. Liq. Cryst. *13*, 9 (1971), C.A. *75*, 54877h (1971); Piezoelectric effect in liquid crystals. [**269**] N. J. Felici, R. Tobazeon, Proc. IEEE *60*, 241 (1972); Comments on "simplified electrohydrodynamic treatment of threshold effects in nematics" (reply of G. H. Heilmeier). [**270**] N. Felici, B. Gosse, J. P. Gosse, Conduct. Breakdown Dielectr. Liq., Proc. Int. Conf., 5th, *1975*, 23, C.A. *85*, 151738c (1976); Electric conduction in nematics.

[**271**] J. C. Filippini, C. R. Acad. Sci., Sér. B *275*, 349 (1972), C.A. *77*, 158118u (1972); Biréfringence électrique du MBBA en phase isotrope. [**272**] J. C. Filippini, Y. Poggi, J. Phys. (Paris) Lett. *35*, 99 (1974), C.A. *81*, 142331e (1974); Kerr effect in the i phase of nematogens. [**273**] J. C. Filippini, Y. Poggi, Phys. Lett. A *49*, 291 (1974), C.A. *82*, 9454b (1975); Action of impurities on the Kerr effect of MBBA. [**274**] J. C. Filippini, Y. Poggi, C. R. Acad. Sci., Sér. B *279*, 605 (1974), C.A. *82*, 162023c (1975); Biréfringence magnétique et phénomènes prétransitionnels dans les nématiques. [**275**] J. C. Filippini, J. C. Lacroix, C. R. Acad. Sci., Sér. B *280*, 305 (1975), C.A. *83*, 18800j (1975); Cellule de Kerr à champ électrique

tournant. [**276**] J. C. Filippini, Y. Poggi, J. Phys. (Paris), C-1, *36*, 137 (1975); Kerr effect and pretransitional phenomena in nematics. [**277**] J. C. Filippini, Y. Poggi, G. Maret, Colloq. Int. C.N.R.S. *242* (Phys. Champs Magn. Intenses), 67 (1975), C.A. *85*, 185812n (1976); Biréfringence magnétique dans la phase i des nématiques. Effets prétransitionnels. [**278**] J. C. Filippini, Y. Poggi, IEE Conf. Publ. *129*, 307 (1975), C.A. *86*, 10129p (1977); Problems arising from the use of liquid dielectrics in Kerr cells. Application to nitrobenzene and liquid crystals in the i phase. [**279**] J. C. Filippini, Y. Poggi, J. Phys. D *8*, 152 (1975), C.A. *83*, 154880n (1975); Nematogens. New class of materials for Kerr cells. [**280**] J. C. Filippini, Y. Poggi, J. Phys. (Paris), Lett. *37*, 17 (1976), C.A. *84*, 82798n (1976); Domaine de température des phénomènes prétransitionnels dans la phase isotrope des nématiques.

[**281**] F. Fischer, I. Wahl, T. Waltermann, Ber. Bunsenges. Phys. Chem. *78*, 891 (1974), C.A. *82*, 10184v (1975); Torsionsscherung nematischer Schichten in elektrischen und magnetischen Feldern. [**282**] F. Fischer, Z. Naturforsch. A *31*, 41 (1976); Critical pitch in thin cholesteric films with homeotropic boundaries. [**282a**] F. Fischer, Z. Naturforsch. A *31*, 302 (1976); Critical splay-bend deformation of nematic films. [**283**] P. J. Flanders, S. Shtrikman, Amer. Inst. Phys. Conf. Proc. 1972, No. 10, Pt. 2, 1515, C.A. *79*, 59302j (1973); Torque measurements on liquid crystals in a rotating magnetic field. [**284**] P. J. Flanders, Mol. Cryst. Liq. Cryst. *29*, 19 (1974), C.A. *82*, 163362b (1975); Twist viscosity in nematics. [**285**] P. J. Flanders, Appl. Phys. Lett. *28*, 571 (1976), C.A. *85*, 12499n (1976); Torque curves and rotational hysteresis in a smectic C. [**285a**] W. H. Flint, Jr., Diss. Univ. Maine, Portland, Maine. 1976. 90 pp. Avail. Univ. Microfilms Int., Order No. 77-13,732, C.A. *87*, 109699g (1977); Anomalous alignment and electrohydrodynamic effects in nematics. [**286**] G. Foex, L. Royer, Compt. rend. *180*, 1912 (1925), C.A. *20*, 1024^3 (1926); Diamagnétisme des substances nématiques. [**287**] G. Foex, Compt. rend. *184*, 147 (1927); Diamagnétisme des substances mésomorphes; orientation des corps smectiques par le champ magnétique. [**288**] G. Foex, Compt. rend. *187*, 822 (1928), C.A. *23*, 1795^5 (1929); Cristallisation des substances mésomorphes dans le champ magnétique. Obtention d'un solide à molécules orientées. [**289**] G. Foex, J. phys. radium *10*, 421 (1929), C.A. *24*, 2344^1 (1930); Magnetic properties of mesomorphic substances. [**290**] G. Foex, Trans. Faraday Soc. *29*, 958 (1933), C.A. *28*, 1238^9 (1934); Magnetic properties of mesomorphic substances. Analogies with ferromagnetics.

[**291**] V. Frederiks, A. Repiewa, Z. Physik *42*, 532 (1927), C.A. *21*, 2207^4 (1927); Theoretisches und Experimentelles zur Frage nach der Natur anisotroper Flüssigkeiten. [**292**] V. Frederiks, V. Zolina, Trans. Am. Electrochem. Soc. *55*, 85 (1929), C.A. *23*, 2662^8 (1929); Use of a magnetic field in the measurement of the forces tending to orient an anisotropic liquid in a thin homogeneous layer. [**293**] V. K. Frederiks, V. Zolina, J. Russ. Phys.-Chem. Soc., Phys. Pt. *62*, 457 (1930), C.A. *27*, 3368^3 (1933); Use of a magnetic field for measuring orientation forces in thin uniform layers of anisotropic liquids. [**294**] V. Frederiks, V. Zolina, Z. Krist. *79*, 255 (1931), C.A. *26*, 2356^9 (1932); Doppelbrechung dünner anisotrop-flüssiger Schichten im Magnetfeld und diese Schicht orientierende Kräfte. [**295**] V. Frederiks, V. Zolina, Trans. Faraday Soc. *29*, 919 (1933), C.A. *28*, 1238^5 (1934); Forces causing the orientation of an anisotropic liquid. [**296**] V. Frederiks, V. Tsvetkov, Physik. Z. Sowjetunion *6*, 490 (1934), C.A. *29*, 2040^8 (1935); Orientierung der anisotropen Flüssigkeiten in dünnen Schichten und Messung einiger ihrer elastischen Eigenschaften charakterisierenden Konstanten. [**297**] V. Frederiks, G. Mikhailov, D. Beneshevich, Compt. rend. acad. sci. U.R.S.S. *2*, 208 (1935), C.A. *29*, 7142^6 (1935); Conductivity of an anisotropic liquid. [**298**] V. Frederiks, G. Mikhailov, D. Beneshevich, Compt. rend. acad. sci. U.R.S.S. *2*, 469 (1935), C.A. *29*, 7138^8 (1935); Dielectric loss in anisotropic liquids. [**299**] V. Frederiks, V. Tsvetkov, Compt. rend. acad. sci. U.R.S.S. *2*, 528 (1935), C.A. *29*, 7728^3 (1935); Orienting effect of electric fields on the molecules of anisotropic liquids. [**300**] V. Frederiks, V. Tsvetkov, Compt. rend. acad. sci. U.R.S.S. *2*, 548 (1934), C.A. *28*, 6040^1 (1934); Orientation of molecules in thin layers of anisotropic liquids and measurement of two of their elastic constants.

[**301**] V. Frederiks, V. Tsvetkov, Acta Physicochim. U.R.S.S. *3*, 879 (1935); Bewegung anisotroper Flüssigkeiten im elektrischen Feld. [**302**] V. Frederiks, V. Tsvetkov, Acta Physicochim. U.R.S.S. *3*, 895 (1935); Orientierung anisotroper Flüssigkeiten im elektrischen Feld. [**303**] V. Frederiks, V. Tsvetkov, Compt. rend. acad. sci. U.R.S.S. *4*, 131 (1935), C.A. *30*, 2443^9 (1936); Orienting effect of electric fields on anisotropic liquids. [**304**] V. Frederiks, A. Repieva, Acta Physicochim. U.R.S.S. *4*, 91 (1936), C.A. *31*, 3354^9 (1937); Einwirkung der elektrischen Feldes auf die smektische Mesophase. [**305**] V. Frederiks, G. Michailov, Acta Physicochim. U.R.S.S. *5*, 451 (1936), C.A. *31*, 6070^7 (1937); Behavior of an anisotropic liquid in d.c. and a.c. electric fields. [**306**] O. Fuchs, K. L. Wolf, Hand- und Jahrbuch der chemischen Physik, Bd. VI, 1B, 237 (Akad. Verlagsges.: Leipzig) 1935; Dielektrische Polarisation, C.A. *29*, 1714^8 (1935). [**307**] B. L. Funt, S. G. Mason, Can. J. Chem. *29*, 848 (1951), C.A. *47*, 10928c (1953); Influence

of shear on the dielectric constant of liquids. [**308**] Y. Galerne, G. Durand, M. Veyssié, V. Pontikis, Phys. Lett. A 38, 449 (1972), C.A. *76*, 146303s (1972); Electrohydrodynamic instability in a nematic. Effect of an additional stabilizing a.c. electric field on the spatial period of chevrons. [**309**] Y. Galerne, G. Durand, M. Veyssié, Phys. Rev. A *6*, 484 (1972); Spatial period of bend oscillations in the dielectric electrohydrodynamic instability of a nematic. [**310**] Y. Galerne, C. R. Acad. Sci., Sér. B, *278*, 347 (1974), C.A. *81*, 7312j (1974); Relaxation diélectrique d'un cristal liquide à la transition n/s.

[**311**] F. Gaspard, R. Herino, F. Mondon, Mol. Cryst. Liq. Cryst. *24*, 145 (1973), C.A. *81*, 142303x (1974); Low field conduction of nematics studied by means of electrodialysis. [**312**] F. Gaspard, R. Herino, F. Mondon, Chem. Phys. Lett. *25*, 449 (1974), C.A. *81*, 7222e (1974); Electrohydrodynamic instabilities in d.c. fields of a nematic with negative dielectric anisotropy. [**313**] F. Gaspard, R. Herino, Appl. Phys. Lett. *24*, 452 (1974), C.A. *80*, 152050f (1974); Effect of charge-transfer acceptors on dynamic scattering in a nematic. Comments. [**314**] H. Gasparoux, B. Regaya, J. Prost, C. R. Acad. Sci., Sér. B *272*, 1168 (1971), C.A. *75*, 55428z (1971); Propriétés magnétiques de substances nématiques. Étude du PAA et du MBBA. [**315**] J. Prost, H. Gasparoux, C. R. Acad. Sci., Sér. B *273*, 335 (1971), C.A. *76*, 8174z (1972); Modification, sous l'action d'un champ magnétique, de l'orientation des molécules au sein d'une goutte de substance nématique MBBA. [**316**] H. Gasparoux, J. Prost, J. Phys. (Paris) *32*, 953 (1971), C.A. *76*, 133428m (1972); Détermination directe de l'anisotropie magnétique de cristaux nématiques. [**317**] B. Regaya, H. Gasparoux, J. Prost, Rev. Phys. Appl. *7*, 83 (1972), C.A. *77*, 169857k (1972); Application des mesures de susceptibilité magnétique a l'étude de la transition c/n. [**318**] H. Gasparoux, G. Sigaud, J. Prost, Mol. Cryst. Liq. Cryst. *22*, 189 (1973), C.A. *79*, 140900z (1973); Magnetic properties of cholesterics. [**319**] H. Gasparoux, F. Hardouin, M. F. Achard, Liq. Cryst., Proc. Int. Conf., Bangalore *1973*, 215, C.A. *84*, 115193s (1976); Magnetic properties of smectics. [**320**] H. Gasparoux, F. Hardouin, M. F. Achard, G. Sigaud, J. Phys. (Paris), C-1, *36*, 107 (1975), C.A. *83*, 88962n (1975); Compared action of a rotating magnetic field on smectics A and nematics. Application to the study of the s_A/n transition.

[**321**] F. Hardouin, H. Gasparoux, P. Delhaes, J. Phys. (Paris), C-1, *36*, 127 (1975), C.A. *83*, 19458r (1975); Calorimetric and magnetic studies of phase transition in liquid crystals. [**322**] M. F. Achard, H. Gasparoux, F. Hardouin, G. Sigaud, J. Phys. (Paris), C-3, *37*, 107 (1976), C.A. *85*, 85788r (1976); Advances in the investigation of the n/s_A transition: magnetic study of mixtures of mesogens. [**323**] A. L. Berman, E. Gelerinter, A. De Vries, Mol. Cryst. Liq. Cryst. *33*, 55 (1976), C.A. *85*, 134752k (1976); Optical studies of electric field effects in nematics that have some smectic ordering. [**323a**] R. C. Weir, E. Gelerinter, Mol. Cryst. Liq. Cryst. *40*, 199 (1977), C.A. *87*, 125605d (1977); Comparative optical studies of electrohydrodynamic instabilities in selected nematics with some smectic ordering. [**324**] C. J. Gerritsma, P. van Zanten, Mol. Cryst. Liq. Cryst. *15*, 257 (1971), C.A. *76*, 38451e (1972); Electric-field-induced texture transformation and pitch contraction in a cholesteric. [**325**] C. J. Gerritsma, W. H. de Jeu, P. van Zanten, Phys. Lett. A *36*, 389 (1971), C.A. *76*, 51925g (1972); Distortion of a TN by a magnetic field. [**326**] C. J. Gerritsma, P. van Zanten, Phys. Lett. A *42*, 127 (1972), C.A. *78*, 49478m (1973); Dependence of the electric-field-induced c/n transition on the dielectric anisotropy. [**327**] C. J. Gerritsma, P. van Zanten, Phys. Lett. A *42*, 329 (1972), C.A. *78*, 64718h (1973); Explanation of the observed field-induced blue shift of an imperfectly aligned planar cholesteric texture. [**328**] C. J. Gerritsma, J. A. Geurst, A. M. J. Spruijt, Phys. Lett. A *43*, 356 (1973), C.A. *79*, 59288j (1973); Magnetic-field-induced motion of disclinations in a TN layer. [**329**] C. J. Gerritsma, P. van Zanten, Liq. Cryst. Ord. Fluids *2*, 437 (1974), C.A. *86*, 197164m (1977); Electro-optical properties of imperfectly ordered planar cholesteric layers. [**330**] C. J. Gerritsma, C. Z. van Doorn, P. van Zanten, Phys. Lett. A *48*, 263 (1974); Transient effects in electrically controlled light transmission of a TN layer.

[**331**] J. A. Geurst, W. J. A. Goossens, Phys. Lett. A *41*, 369 (1972), C.A. *77*, 157352d (1972); Theory of electrically induced hydrodynamic instabilities in smectics. [**332**] J. A. Geurst, A. M. J. Spruijt, C. J. Gerritsma, J. Phys. (Paris) *36*, 653 (1975), C.A. *83*, 89013r (1975); Dynamics of S = 1/2 disclinations in TN. [**333**] M. Giurgea, T. Serban, L. Matei, T. Beica, L. Nasta, A. Lupu, Phys. Lett. A *57*, 336 (1976), C.A. *85*, 102645m (1976); Dependence of Williams domains width on threshold voltage. [**334**] G. L. Gladstone, Phys. Lett. A *37*, 325 (1971), C.A. *76*, 64828c (1972); Nonlinear behavior of the electrohydrodynamic instability threshold in nematics. [**335**] C. H. Gooch, H. A. Tarry, J. Phys. D *5*, L25 (1972), C.A. *76*, 159458w (1972); Dynamic scattering in the homeotropic and homogeneous textures of a nematic. [**336**] W. J. A. Goossens, Phys. Lett. A *40*, 95 (1972); Conduction regime in a nematic with negative dielectric anisotropy at frequencies above the dielectric relaxation frequency. [**337**] C. V. S. S. V. Gopalakrishna, C. Haranadh, C. R. K. Murty, Trans. Faraday Soc. *63*, 1953

(1967), C.A. *67*, 82304x (1967); Dipole moments of cholesteryl compounds. [**338**] C. V. S. S. V. Gopala-krishna, M. N. Avadhanulu, V. V. S. Sarma, C. R. K. Murty, Indian J. Pure Appl. Phys. *5*, 579 (1967), C.A. *68*, 108995w (1968); Dielectric properties of cholesterics. [**339**] C. V. S. S. V. Gopalakrishna, C. Haranadh, C. Murty, K. Radha, Indian J. Pure Appl. Phys. *6*, 375 (1968), C.A. *70*, 7364a (1969); Temperature variation of static dielectric constant in cholesteryl esters. [**340**] C. V. S. S. V. Gopalakrishna, M. N. Avadhanulu, C. Murty, K. Radha, Indian J. Pure Appl. Phys. *6*, 713 (1968), C.A. *70*, 51988f (1969); Influence of magnetic and electric fields on the dielectric constant of liquid-crystalline anisylidene-*p*-aminoazobenzene.

[**341**] C. V. S. S. V. Gopalakrishna, M. M. Avadhanulu, C. Murty, K. Radha, Indian J. Pure Appl. Phys. *8*, 178 (1970), C.A. *73*, 71008b (1970); Influence of electric and magnetic fields on the dielectric constant of cholesterics. [**342**] L. S. Gorbatenko, V. S. Dronov, Sb. Dokl. Vses. Nauch. Konf. Zhidk. Krist. Simp. Ikh Prakt. Primen., 2nd, *1972*, 91, C.A. *81*, 112138e (1974); Influence of external effects on phase transitions in liquid crystals. [**343**] F. K. Gorskii, N. M. Sakevich, Zh. Fiz. Khim. *45*, 255 (1971), C.A. *74*, 117082x (1971); Change in the energy of electric conductivity activation of liquid crystals during i/a transition. [**344**] F. K. Gorskii, M. L. Bashun, Sb. Dokl. Vses. Nauch. Konf. Zhidk. Krist. Simp. Ikh Prakt. Primen., 2nd, *1972*, 103, C.A. *81*, 128190b (1974); Temperature dependences of the electric conductivities of liquid crystals. [**345**] M. Goscianski, L. Léger, J. Phys. (Paris), C-1, *36*, 231 (1975); Electrohydrodynamic instabilities above a n/s$_A$ or s$_C$ transition. [**346**] M. Goscianski, L. Léger, A. Mircea-Roussel, J. Phys. (Paris), Lett. *36*, 313 (1975), C.A. *84*, 52446d (1976); Field induced transitions in smectics A. [**347**] M. Goscianski, Philips Res. Rep. *30*, 37 (1975); Electrohydro-dynamic instabilities in a nematic above a s$_A$ or s$_C$/n transition. [**348**] B. Gosse, J. P. Gosse, J. Appl. Electrochem. *6*, 515 (1976), C.A. *86*, 35735w (1977); Degradation of liquid crystal devices under d.c. excitation and their electrochemistry. [**349**] J. Grabmaier, W. Greubel, 2. Arbeitstagung Flüssigkristalle, Inst. Angew. Festkörperphysik Freiburg, 21. 4. 1972; Texturübergänge in cholesterinischen Flüssigkri-stallmischungen im elektrischen Feld. [**350**] F. Grandjean, Compt. rend. *167*, 494 (1918), C.A. *13*, 90³ (1919); Franges d'interférence développées par le frottement et l'électricité dans certains liquides anisotropes.

[**351**] M. F. Grebenkin, G. S. Chilaya, V. T. Lazareva, K. V. Roitman, L. M. Blinov, V. V. Titov, Kristallografiya *18*, 429 (1973), C.A. *78*, 166765p (1973); Dynamic scattering of light in liquid crystalline azo compounds. [**352**] M. F. Grebenkin, G. S. Chilaya, V. T. Lazareva, K. V. Roitman, L. M. Blinov, V. V. Titov, Sb. Dokl. Vses. Nauch. Konf. Zhidk. Krist. Simp. Ikh Prakt. Primen., 2nd, *1972*, 184, C.A. *82*, 49422w (1975); Contrast and time characteristics of the dynamic scattering in liquid crystalline azo and azoxy compounds. [**353**] M. F. Grebenkin, V. A. Seliverstov, L. M. Blinov, V. G. Chigrinov, Kristallografiya *20*, 984 (1975), C.A. *84*, 37288k (1976); Properties of nematics with positive dielectric anisotropy. [**354**] W. Greubel, U. Wolff, Appl. Phys. Lett. *19*, 213 (1971), C.A. *75*, 144944g (1971); Electrically controllable domains in nematics. [**355**] W. Greubel, H. Krüger, U. Wolff, 2. Arbeitstagung Flüssigkristalle, Inst. Angew. Festkörperphysik Freiburg, 21. 4. 1972; Neuere Farbeffekte mit nematischen Flüssigkristallen. [**356**] W. Greubel, U. Wolff, H. Krüger, Mol. Cryst. Liq. Cryst. *24*, 103 (1973), C.A. *81*, 142299a (1974); Electric field induced texture changes in certain n/c mixtures. [**357**] M. J. Gross, J. E. Porter, Nature *212*, 1343 (1966), C.A. *66*, 41531e (1967); Electrically induced convection in dielectric liquids. [**358**] M. Grubic, J. Voss, J. Phys. E *9*, 3 (1976), C.A. *84*, 68104b (1976); Simple 2.5 kV pulse generator for liquid crystal relaxation measurements. [**359**] H. Gruler, 1. Arbeitstagung Flüssigkristalle, Inst. Angew. Festkörperphysik Freiburg, 23. 4. 1971; Verhalten nematischer Flüssigkristalle im elektrischen Feld. [**360**] H. Gruler, Diss. Freiburg 1972; Deformation nematischer Flüssigkristalle im elektrischen und magnetischen Feld.

[**361**] H. Gruler, Mol. Cryst. Liq. Cryst. *27*, 31 (1974), C.A. *83*, 106518d (1975); Williams domains and dielectric alignment in a nematic. [**362**] H. Gruler, L. Cheung, J. Appl. Phys. *46*, 5097 (1975), C.A. *84*, 98283b (1976); Dielectric alignment in an electrically conducting nematic. [**363**] R. Guettich, Proc.-Int. Symp. Techn. Comm. Photon-Detect., Int. Meas. Confed., 6th, *1974*, 321, C.A. *83*, 124773n (1975); Pyroelectric detectors made of noncrystalline pyroelectrics. [**364**] G. Le Guillanton, M. Cariou, L. Lebouc, Bull. Soc. Chim. Fr. 2, *1974*, 2980, C.A. *83*, 42583m (1975); Réduction électrochimique de propène nitriles substitués. 2. Alkyl(aryl)oxy-3-phényl-2 propène nitriles z et e(1). [**364a**] K. Guminski, L. Stobinski, T. Zyczkowska, Rocz. Chem. *50*, 1947 (1976), C.A. *86*, 82319y (1977); Electric conductivity of polycrystalline N-dodecylpyridinium and quinolinium halides. [**365**] R. W. Gurtler, W. Casey, Mol. Cryst. Liq. Cryst. *35*, 275 (1976), C.A. *86*, 24609g (1977); Surface tilt distribution of homogeneously aligned liquid crystals. [**366**] E. Guyon, P. Pieranski, Physica *73*, 184 (1974), C.A. *81*, 55114t (1974); Convective instabilities in nematics. [**367**] E. Guyon, P. Pieranski, C. R. Acad. Sci., Sér. B *280*, 187

(1975); Instabilités de Williams en présence d'un écoulement: expériences. [**368**] W. E. L. Haas, J. E. Adams, Appl. Phys. Lett. *25*, 263 (1974), C.A. *81*, 112215c (1974); Electrically variable diffraction in spherulitic liquid crystals. [**369**] W. Haase, D. Pötzsch, Mol. Cryst. Liq. Cryst. *38*, 77 (1977), C.A. *86*, 179761z (1977); Light transmission experiments with nematics showing positive and negative dielectric anisotropy. [**370**] E. G. Hanson, G. K. L. Wong, Y. R. Shen, IEEE J. Q. Electron. *10*, 697 (1974); Third-order nonlinearity in liquid crystals.

[**371**] E. G. Hanson, G. K. L. Wong, Y. R. Shen, Dig. Tech. Pap. — Int. Quantum Electron. Conf., 8th, *1974*, 21, C.A. *85*, 84878h (1976); Third-order nonlinearity in liquid crystals. [**372**] E. G. Hanson, Y. R. Shen, G. K. L. Wong, Phys. Rev. A *14*, 1281 (1976), C.A. *85*, 151161j (1976); Optical-field-induced refractive indexes and orientational relaxation times in a homologous series of isotropic nematogens. [**373**] F. Hardouin, M. F. Achard, H. Gasparoux, C. R. Acad. Sci., Sér. C *277*, 551 (1973), C.A. *80*, 64706u (1974); Évolution des propriétés magnétiques d'une phase mésomorphe lors d'une transition s_A/n. [**373a**] F. Hardouin, M. F. Achard, G. Sigaud, H. Gasparoux, Mol. Cryst. Liq. Cryst. *39*, 241 (1977), C.A. *87*, 125590v (1977); Magnetic anisotropy and polymorphism in liquid crystals. [**374**] M. Hareng, S. Le Berre, L. Thirant, Appl. Phys. Lett. *25*, 683 (1974), C.A. *82*, 50335b (1975); Electric field effects on biphenyl smectics A. [**375**] M. Hareng, S. Le Berre, J. J. Metzger, Appl. Phys. Lett. *27*, 575 (1975), C.A. *84*, 11115b (1976); Planar-to-homeotropic structure transition under electric field in a smectic A. [**376**] W. J. Harper, Mol. Cryst. *1*, 325 (1966), C.A. *65*, 14537h (1966); Voltage effects in cholesterics. [**377**] C. F. Hayes, Mol. Cryst. Liq. Cryst. *36*, 245 (1976), C.A. *86*, 63881j (1977); Magnetic platelets in a nematic. [**378**] G. H. Heilmeier, J. Chem. Phys. *44*, 644 (1966), C.A. *64*, 7464h (1966); Transient behavior of domains in liquid crystals. [**379**] G. H. Heilmeier, P. M. Heyman, Phys. Rev. Lett. *18*, 583 (1967), C.A. *67*, 68417e (1967); Transient current measurements in liquid crystals and related systems. [**380**] G. H. Heilmeier, Advan. Chem. Ser. No. *63*, 68 (1967), C.A. *67*, 15906g (1967); Cooperative effects in butyl *p*-anisylidene-*p'*-aminocinnamate.

[**381**] G. H. Heilmeier, L. A. Zanoni, L. A. Barton, Appl. Phys. Lett. *13*, 46 (1968), C.A. *69*, 91023y (1968); Dynamic scattering in nematics. [**382**] G. H. Heilmeier, L. A. Zanoni, Appl. Phys. Lett. *13*, 91 (1968), C.A. *69*, 111067w (1968); Guest-host interactions in nematics. A new electrooptic effect. [**383**] G. H. Heilmeier, J. E. Goldmacher, Appl. Phys. Lett. *13*, 132 (1968), C.A. *70*, 41973h (1969); Electric-field-controlled reflective optical storage effect in mixed liquid crystals. [**384**] G. Heilmeier, L. Zanoni, L. Barton, Proc. IEEE *56*, 1162 (1968); Dynamic scattering. A new electro-optic effect in certain classes of nematics. [**385**] G. H. Heilmeier, J. E. Goldmacher, J. Chem. Phys. *51*, 1258 (1969), C.A. *71*, 75051s (1969); Electric-field-induced c/n phase change. [**386**] G. H. Heilmeier, J. A. Castellano, L. A. Zanoni, Mol. Cryst. Liq. Cryst. *8*, 293 (1969), C.A. *71*, 85930s (1969); Guest-host interactions in nematics. [**387**] G. H. Heilmeier, L. A. Zanoni, J. E. Goldmacher, Liq. Cryst. Ord. Fluids *1970*, 215, C.A. *78*, 165431q (1973); Electric field induced structural changes in a mixed liquid crystal. [**388**] G. H. Heilmeier, W. Helfrich, Appl. Phys. Lett. *16*, 155 (1970), C.A. *72*, 126081g (1970); Orientational oscillations in nematics. [**389**] G. H. Heilmeier, L. A. Zanoni, L. A. Barton, IEEE Trans. Electron Devices *17*, 22 (1970), C.A. *72*, 71926d (1970); Dynamic scattering mode in nematics. [**390**] G. H. Heilmeier, Proc. IEEE *59*, 422 (1971), C.A. *74*, 147615r (1971); Simplified electrohydrodynamic treatment of threshold effects of nematics.

[**391**] W. Helfrich, Phys. Rev. Lett. *21*, 1518 (1968), C.A. *70*, 51821w (1969); Alignment-inversion walls in nematics in the presence of a magnetic field. [**392**] W. Helfrich, J. Chem. Phys. *51*, 4092 (1969), C.A. *72*, 16482f (1970), Erratum: J. Chem. Phys. *52*, 4318 (1970), C.A. *73*, 8018q (1970); Conduction-induced alignment of nematics. Basic model and stability considerations. [**393**] W. Helfrich, J. Chem. Phys. *51*, 2755 (1969), C.A. *71*, 129670p (1969); Orientation pattern of domains in nematic PAA. [**394**] W. Helfrich, Appl. Phys. Lett. *17*, 531 (1970), C.A. *74*, 58089e (1971); Deformation of cholesterics with low threshold voltage. [**395**] W. Helfrich, Phys. Rev. Lett. *24*, 201 (1970), C.A. *72*, 83654b (1970); Effect of electric fields on the temperatures of phase transitions of liquid crystals. [**396**] W. Helfrich, J. Chem. Phys. *55*, 839 (1971), C.A. *75*, 68623j (1971); Electrohydrodynamic and dielectric instabilities of cholesterics. [**397**] W. Helfrich, Phys. Lett. A *35*, 393 (1971), C.A. *75*, 123826e (1971); Simple method to observe the piezoelectricity of liquid crystals. [**398**] W. Helfrich, M. Schadt, Phys. Rev. Lett. *27*, 561 (1971), C.A. *75*, 113522p (1971); Birefringence of nematogens caused by electric conduction. [**399**] W. Helfrich, Z. Naturforsch. A *26*, 833 (1971), C.A. *75*, 55089q (1971); Strength of piezoelectricity in liquid crystals. [**400**] W. Helfrich, Mol. Cryst. Liq. Cryst. *21*, 187 (1973), C.A. *79*, 109124a (1973); Electric alignment of liquid crystals.

[**401**] D. Schmidt, M. Schadt, W. Helfrich, Z. Naturforsch. A *27*, 277 (1972), C.A. *76*, 132995g (1972); Liquid-crystalline piezoelectricity; the bending mode of MBBA. [**402**] W. Helfrich, Appl. Phys. Lett. *24*, 451 (1974); Polarity-dependent electrooptic effect of nematics. [**403**] H. J. Deuling, W. Helfrich, Appl. Phys. Lett. *25*, 129 (1974); Hysteresis in the deformation of nematic layers with homeotropic orientation. [**404**] G. Heppke, F. Schneider, Ber. Bunsenges. Phys. Chem. *75*, 1231 (1971), C.A. *76*, 28392x (1972); Elektrische Leitfähigkeit in nematischen Lösungen. [**405**] G. Heppke, F. Schneider, Chem. Phys. Lett. *13*, 548 (1972), C.A. *77*, 10875t (1972); Field-induced surface charges in nematic electrolyte systems. [**406**] G. Heppke, F. Schneider, Z. Naturforsch. A *27*, 976 (1972), C.A. *77*, 106335t (1972); Geschwindigkeit der Ausrichtung nematischer Flüssigkeiten durch Magnetfelder. [**407**] G. Heppke, F. Schneider, Ber. Bunsenges. Phys. Chem. *76*, 1107 (1972); Zeitabhängigkeit der Orientierung nematischer Phasen in Magnetfeldern. [**408**] G. Heppke, F. Schneider, Ber. Bunsenges. Phys. Chem. *76*, 1107 (1972); Anisotropie der elektrischen Leitfähigkeit nematischer Elektrolytlösungen. [**409**] G. Heppke, F. Schneider, Z. Naturforsch. A *28*, 497 (1973), C.A. *79*, 35703q (1973); Einfluß der Probengeometrie auf die elektrischen Eigenschaften nematischer Elektrolytlösungen. [**410**] G. Heppke, F. Schneider, Z. Naturforsch. A *28*, 994 (1974), C.A. *79*, 130650d (1973); Bestimmung der magnetischen Suszeptibilitätsanisotropie und der Rotationsviskosität nematischer Flüssigkeiten.

[**411**] G. Heppke, F. Schneider, Z. Naturforsch. A *28*, 1044 (1973), C.A. *79*, 130726h (1973); Inversionswände in einer nematischen Flüssigkeit. [**412**] G. Heppke, F. Schneider, Ber. Bunsenges. Phys. Chem. *78*, 903 (1974), C.A. *82*, 10268a (1975); Dynamik von magnetisch erzeugten Deformationen in nematischen Flüssigkristallen. [**413**] G. Heppke, F. Schneider, Ber. Bunsenges. Phys. Chem. *78*, 910 (1974), C.A. *82*, 10213d (1975); Leitfähigkeitsuntersuchungen an smektischem 3-[N-(*p*-Äthoxybenzyliden)amino]-6-*n*-butylpyridin. [**414**] G. Heppke, F. Schneider, Ber. Bunsenges. Phys. Chem. *78*, 981 (1974), C.A. *82*, 66535c (1975); Magnetfeldinduzierte Leitfähigkeitsänderungen in einer s_C Phase. [**415**] G. Heppke, F. Schneider, Z. Naturforsch. A *29*, 310 (1974), C.A. *81*, 69535t (1974); Metastabile Deformationen in nematischen Flüssigkristallen. [**416**] G. Heppke, W. E. M. Benavent, F. Schneider, Z. Naturforsch. A *29*, 728 (1974), C.A. *81*, 69720z (1974); Elektrische Leitfähigkeitsuntersuchungen an einem smektischen Flüssigkristall mit mehreren Mesophasen. [**417**] G. Heppke, F. Schneider, Z. Naturforsch. A *30*, 316 (1975), C.A. *83*, 16629m (1975); Elektrische Leitfähigkeit homologer nematischer 4,4-Di-*n*-alkyl-oxy-azoxybenzole. [**418**] M. Greulich, G. Heppke, F. Schneider, Z. Naturforsch. A *30*, 515 (1975), C.A. *83*, 51173e (1975); Bestimmung elastischer Konstanten von MBBA und EBBA mit der elektrischen Leitfähigkeitsmethode. [**419**] G. Heppke, F. Schneider, Z. Naturforsch. A *30*, 1640 (1975), C.A. *84*, 82781b (1976); Untersuchung von Phasenumwandlungen flüssiger Kristalle durch elektrische Leitfähigkeitsmessungen. [**420**] G. Heppke, F. Schneider, 6. Arbeitstagung Flüssigkristalle, Inst. Angew. Festkörperphysik Freiburg, 2. 4. 1976; Dynamische Streuung in einer nematischen Phase mit positiver und negativer Leitfähigkeitsanisotropie.

[**421**] G. Heppke, F. Schneider, Z. Naturforsch. A *31*, 611 (1976), C.A. *85*, 69081z (1976); Anisotropie der elektrischen Leitfähigkeit verschiedener Elektrolyte in einem nematischen Flüssigkristall. [**422**] G. Heppke, F. Schneider, A. Sterzl, Z. Naturforsch. A *31*, 1700 (1976), C.A. *86*, 64182u (1977); Elektrische Leitfähigkeit homologer N-(4-*n*-Alkoxybenzylidene)-4'-*n*-butylaniline in n und s Phasen. [**423**] P. R. Herczfeld, J. T. Hartman, Mol. Cryst. Liq. Cryst. *18*, 157 (1972), C.A. *77*, 119519f (1972); Noise measurements in nematics. [**424**] H. Hervet, J. P. Hurault, F. Rondelez, Phys. Rev. A *8*, 3055 (1973), C.A. *80*, 53165r (1974); Static one-dimensional distortions in cholesterics. [**425**] Hidemasa, S. Oi, K. Fukada, Tanso *75*, 133 (1973), C.A. *82*, 10166r (1975); Behavior of the mesophase spherules having new molecular arrangements in a magnetic field. [**426**] N. Hijikuro, Prog. Theor. Phys. *54*, 592 (1975), C.A. *83*, 186613u (1975); Reductive perturbation approach to electrohydrodynamic instabilities and dissipative structures in liquid crystals. [**426a**] H. P. Hinov, J. Phys. (Paris), Lett. *38*, 215 (1977), C.A. *87*, 60984e (1977); Influence of second-order elasticity on local Frederiks transitions near a nematic liquid crystal-anisotropic crystal interface. [**426b**] K. Hirakawa, S. Kai, Mol. Cryst. Liq. Cryst. *40*, 261 (1977), C.A. *87*, 125611c (1977); Analogy between hydrodynamic instabilities in nematic and classical fluid. [**427**] S. Hisamitsu, K. Yoshino, Y. Inuishi, Technol. Rep. Osaka Univ. *22*, 201 (1972), C.A. *77*, 144917s (1972); Electrical and optical properties of a nematic. [**428**] W. W. Holloway, Jr., M. J. Rafuse, J. Appl. Phys. *42*, 5395 (1971), C.A. *76*, 28661j (1972); Dynamic scattering by a nematic where the applied electric field is normal to the incident light. [**429**] R. M. Hornreich, S. Shtrikman, Solid State Commun. *17*, 1141 (1975); Magnetic properties of a multi-domain model for smectics C. [**429a**] R. M. Hornreich, S. Shtrikman, Phys. Lett. A *63*, 39 (1977), C.A. *87*, 192288r (1977); Critical behavior of the s_A/s_C transition in a magnetic field. [**429b**] R. M. Hornreich, Phys. Rev. A *15*, 1767

(1977), C.A. *86*, 198225u (1977); Instability threshold of a nematic in a biased rotating magnetic field. [430] W. Hottinger, A. R. Kmetz, Rev. Sci. Instrum. *47*, 390 (1976); Swept-amplitude square-wave generator for liquid crystal measurements.

[431] C. Hu, J. R. Whinnery, N. M. Amer, IEEE J. Quantum Electron. *10*, 218 (1974), C.A. *80*, 126613x (1974); Optical deflection in thin-film nematic liquid crystal waveguides. [432] C. M. Hu, J. R. Whinnery, IEEE J. Quantum Electron. *10*, 556 (1974), C.A. *81*, 70947k (1974); Field-realigned nematic-liquid-crystal optical waveguides. [433] C. C. Huang, R. S. Pindak, P. J. Flanders, J. T. Ho, Phys. Rev. Lett. *33*, 400 (1974), C.A. *81*, 127911a (1974); Dynamics of Freedericksz deformation near a n/s$_A$ transition. [434] J. P. Hulin, Appl. Phys. Lett. *21*, 455 (1972), C.A. *78*, 21872m (1973); Parametric study of the optical storage effect in mixed liquid crystals. [435] J. Hurault, Amer. Inst. Phys. Conf. Prov. *1972*, No. 10, Pt. 2, 1459, C.A. *79*, 10619d (1973); Magnetic properties of liquid crystals. [436] J. P. Hurault, J. Chem. Phys. *59*, 2068 (1973), C.A. *79*, 130151s (1973); Static distortions of a cholesteric planar structure induced by magnetic or a.c. electric fields. [437] S. Ikeno, M. Yokoyama, H. Mikawa, Mol. Cryst. Liq. Cryst. *36*, 89 (1976), C.A. *86*, 49520j (1977); Dielectric behavior of nematics under dc electric field. [438] I. P. Il'chishin, E. A. Tikhonov, M. T. Shpak, A. A. Doroshkin, Pis'ma Zh. Eksp. Teor. Fiz. *24*, 336 (1976), C.A. *86*, 36150p (1977); Generation of induced radiation by organic dyes in a nematic. [439] I. V. Ioffe, Zh. Tekh. Fiz. *44*, 1619 (1974); Theory of behavior of liquid crystals in external fields. [440] I. V. Ioffe, B. I. Lembrikov, Fiz. Tverd. Tela *16*, 3536 (1974); Theory of behavior of cholesterics in magnetic field.

[441] I. V. Ioffe, B. I. Lembrikov, Fiz. Tverd. Tela *17*, 1451 (1975), C.A. *83*, 51204r (1975); Theory of domain formation in nematic PAA in an electric field. [441a] Y. Ishii, T. Uchida, M. Wada, Oyo Butsuri *46*, 155 (1977), C.A. *87*, 144071t (1977); Effects of dielectric anisotropy on DSM display devices. [442] E. Jakeman, E. P. Raynes, Phys. Lett. A *39*, 69 (1972), C.A. *76*, 159693u (1972); Electrooptic response times in liquid crystals. [443] E. Jakeman, P. N. Pusey, J. M. Vaughan, Opt. Commun. *17*, 305 (1976), C.A. *85*, 70183r (1976); Intensity fluctuation light-scattering spectroscopy using a conventional light source. [444] J. A. Janik, S. Wrobel, J. M. Janik, A. Migdal, S. Urban, Faraday Symp. Chem. Soc. *1972*, No. 3, 48, C.A. *79*, 46750d (1973); Rotational diffusion in PAA. [445] J. A. Janik, J. M. Janik, Nguyen Thi Thoa, K. Rosciszewski, S. Wrobel, Phys. Status Solidi A *18*, K143 (1973), C.A. *79*, 119306a (1973); Estimation of rotational correlation times for PAA and MBBA molecules by various methods. [446] J. A. Janik, J. M. Janik, K. Otnes, K. Rosciszewski, S. Wrobel, Liq. Cryst., Proc. Int. Conf., Bangalore, *1973*, 253, C.A. *84*, 114465v (1976); Estimation of rotational correlation times for PAA and MBBA by the dielectric relaxation and the neutron quasi-elastic scattering methods. [447] S. Wrobel, J. A. Janik, J. Moscicki, S. Urban, Acta Phys. Pol. A *48*, 215 (1975), C.A. *83*, 156441u (1975); Dielectric relaxation in the i, n, and k phases of PAA in the radio and microwave frequency range. [447a] J. K. Moscicki, X. P. Nguyen, S. Urban, S. Wrobel, M. Rachwalska, J. A. Janik, Mol. Cryst. Liq. Cryst. *40*, 177 (1977), C.A. *87*, 125604c (1977); Calorimetric and dielectric investigations of MBBA and HAB. [448] B. R. Jennings, H. Watanabe, J. Appl. Phys. *47*, 4709 (1976), C.A. *85*, 185038q (1976); Voltage-frequency thresholds for MBBA viewed in transverse electric fields. [449] M. Jezewski, Ann. Physik *75*, 108 (1924), C.A. *18*, 3502^3 (1924); Anisotropie der flüssigen Kristalle bezüglich ihrer Dielektrizitätskonstante und ihrer elektrischen Leitfähigkeit. [450] M. Jezewski, J. phys. radium *5*, 59 (1924), C.A. *18*, 2102^6 (1924); Influence du champ magnétique sur la constante diélectrique de cristaux liquides.

[451] M. Jezewski, Z. Physik *40*, 153 (1927), C.A. *21*, 3538^3 (1927); Elektrische Anisotropie der kristallinischen Flüssigkeiten. [452] M. Jezewski, Z. Physik *51*, 159 (1928), C.A. *23*, 25^8 (1929); Einfluß des elektrostatischen Feldes auf die Dielektrizitätskonstante der Körper in nematischer Phase. [453] M. Jezewski, Z. Physik *52*, 268 (1928), C.A. *23*, 1031^7 (1929); Dielektrische Anisotropie der nematischen Flüssigkeiten im magnetischen Felde. [454] M. Jezewski, Z. Physik *52*, 878 (1928), C.A. *23*, 1793^4 (1929); Dielektrische Eigenschaften der nematischen Flüssigkeiten im gleichzeitigen magneto- und elektrostatischen Feld. [455] M. Miesowicz, M. Jezewski, Phys. Z. *36*, 107 (1935); Über den thermischen, vom magnetischen Felde hervorgerufenen Effekt in anisotropen Flüssigkeiten und den Einfluß des elektrischen Feldes auf denselben. [456] A. R. Johnston, J. Appl. Phys. *44*, 2971 (1973), C.A. *79*, 109382h (1973); Kerr response of nematics. [457] D. Jones, L. Creagh, S. Lu, Appl. Phys. Lett. *16*, 61 (1970), C.A. *72*, 104978c (1970); Dynamic scattering in a room-temperature nematic. [458] W. Jung, K. N. Kang, S. K. Min, Y. B. Chae, Q. W. Choi, Proc. Conf. (Int.) Solid State Devices, 5th, *1973*, 136, C.A. *81*, 127892v (1974); Electrooptical rotation of vertically aligned nematic MBBA layers. [459] Y. Kagawa, T. Hatakeyama, Y. Tanaka, J. Sound Vibr. *41*, 1 (1975); Vibro-optical and

vibro-dielectric effects in a nematic layer. [**460**] Y. Kagawa, T. Hatakeyama, Appl. Phys. Lett. *29*, 71 (1976), C.A. *85*, 86085w (1976); Piezoelectric effect in a cholesteric layer subjected to shear vibration. [**461**] F. J. Kahn, Phys. Rev. Lett. *24*, 209 (1970), C.A. *72*, 71647p (1970); Electric-field-induced color changes and pitch dilation in cholesterics. [**462**] F. J. Kahn, Appl. Phys. Lett. *20*, 199 (1972), C.A. *76*, 132837g (1972); Electric-field-induced orientational deformation of nematics. Tunable bire-fringence. [**463**] F. J. Kahn, Mol. Cryst. Liq. Cryst. *38*, 109 (1977), C.A. *87*, 14193y (1977); Capacitive analysis of TN displays. [**464**] S. Kai, N. Yoshitsune, K. Hirakawa, J. Phys. Soc. Jpn. *38*, 1789 (1975), C.A. *83*, 124383k (1975); Measurement of the flow velocity in a nematic. [**465**] S. Kai, M. Araoka, K. Hirakawa, J. Phys. Soc. Jpn. *39*, 849 (1975), C.A. *83*, 211528e (1975); Fluctuation noise of flow in a nematic. [**466**] S. Kai, K. Yamaguchi, K. Hirakawa, Jpn. J. Appl. Phys. *14*, 1385 (1975); New pattern in a nematic. [**467**] S. Kai, K. Yamaguchi, K. Hirakawa, Jpn. J. Appl. Phys. *14*, 1653 (1975), C.A. *84*, 11233p (1976); Flow figures in nematic MBBA. [**468**] S. Kai, N. Yoshitsune, K. Hirakawa, J. Phys. Soc. Jpn. *40*, 267 (1976), C.A. *84*, 114433h (1976); Instability of the flow in nematic MBBA. [**469**] S. Kai, K. Hirakawa, J. Phys. Soc. Jpn. *40*, 301 (1976), C.A. *84*, 128994d (1976); Analogy between instabilities in i and n fluids. [**470**] S. Kai, M. Araoka, H. Yamazaki, K. Hirakawa, J. Phys. Soc. Jpn. *40*, 305 (1976), C.A. *84*, 128995e (1976); Fluctuation of the transmitted-intensity of light in the nematic liquid crystal.

[**471**] S. Kai, K. Hirakawa, Solid State Commun. *18*, 1573 (1976), C.A. *85*, 55134g (1976); Phase diagram of dissipative structures in the nematic liquid crystal under a.c. field. [**472**] S. Kai, K. Hirakawa, Solid State Commun. *18*, 1579 (1976), C.A. *85*, 54878r (1976); Anomalies near the electrohy-drodynamic instability points in nematic MBBA. [**472a**] S. Kai, M. Araoka, H. Yamazaki, K. Hirakawa, Mem. Fac. Eng., Kyushu Univ. *36*, 243 (1976), C.A. *87*, 14486w (1977); Electrohydrodynamic instabilities in nematic MBBA. I. Characteristic behaviors of autocorrelations and power spectra in successive transitions from two-dimensional-laminar flow to turbulent flow. [**472b**] T. Kai, S. Kai, K. Hirakawa, J. Phys. Soc. Jpn. *43*, 717 (1977), C.A. *87*, 125642p (1977); Current density in the electrohydrodynamic instability in a nematic. [**472c**] S. Kai, K. Hirakawa, Mem. Fac. Eng., Kyushu Univ. *36*, 269 (1977), C.A. *87*, 60963x (1977); II. Flow patterns and phase diagrams (cf. [472a]). [**473**] H. Kamei, Y. Katayama, T. Ozawa, Jap. J. Appl. Phys. *11*, 1385 (1972), C.A. *77*, 144873z (1972); Photovoltaic effect in the nematic liquid crystal. [**474**] L. T. Kantardzhyan, A. I. Oganisyan, T. B. Rudnenko, Izv. Akad. Nauk Arm. SSR, Fiz. *9*, 236 (1974), C.A. *81*, 113176c (1974); Spectrophotometric investigation of the dependence of light scattering from a liquid crystal on the voltage applied. [**475**] G. E. Zvereva, A. P. Kapustin, Primenenie Ul'traakustiki k Issled. Veshchestva *1961*, No. 15, 69, C.A. *59*, 1115f (1963); Behavior of nematic PAP in ultrasonic and electric fields. [**476**] A. P. Kapustin, L. S. Larionova, Kristallografiya *9*, 297 (1964), C.A. *61*, 57b (1964); Behavior of anisotropic liquids in an electric field. [**477**] A. P. Kapustin, L. K. Vistin, Kristallografiya *10*, 118 (1965), C.A. *62*, 11261b (1965); Ferroelectric properties of liquid crystals. [**478**] L. K. Vistin, A. P. Kapustin, Primen. Ultraakust. Issled. Veshchestva, Moscow No. *21*, 153 (1965), C.A. *66*, 59771t (1967); Behavior of anisotropic liquids in an electric field. [**479**] L. K. Vistin, A. P. Kapustin, Opt. Spektrosk. *24*, 650 (1968), C.A. *69*, 31212z (1968); Electrooptical effect in liquid crystals. [**480**] L. K. Vistin, A. P. Kapustin, Kristallografiya *13*, 349 (1968), C.A. *69*, 39549k (1968); Domains in smectics.

[**481**] L. K. Vistin, A. P. Kapustin, Kristallografiya *14*, 740 (1969), C.A. *71*, 106455h (1969); Domain structure of liquid crystals. [**482**] A. P. Kapustin, S. P. Chumakova, Kristallografiya *15*, 1091 (1970), C.A. *74*, 35891w (1971); Electrooptic phenomena in thin layers of liquid crystals. [**483**] A. P. Kapustin, Z. K. Kuvatov, L. S. Mamaeva, A. N. Trofimov, Izv. Vyssh. Ucheb. Zaved., Fiz. *13*, 22 (1970), C.A. *74*, 92312g (1971); Effect of a varying electric field on the transparency of a thin layer of smectic *p*-(*n*-octyloxy)-benzoic acid. [**484**] A. P. Kapustin, A. N. Trofimov, A. N. Chuvyrev, Kristallografiya *16*, 833 (1971), C.A. *75*, 123311h (1971); Periodic orientation oscillations of liquid-crystal domains in a constant electric field. [**485**] A. P. Kapustin, L. S. Mamajev, A. N. Trofimov, 1. Wiss. Konf. über flüssige Kristalle, 17.–19. 11. 1970, Ivanovo; Sammlung der Vorträge (1972), Nr. 7; Einfluß eines elektrischen Feldes auf die Transmission einer Mischung von APAPA und Cholesterylnonanoat. [**486**] A. P. Kapustin, C. H. Kuvatov, A. N. Trofimov, 1. Wiss. Konf. über flüssige Kristalle, 17.–19. 11. 1970, Ivanovo, Sammlung der Vorträge (1972), Nr. 8; Elastische Deformationen in flüssigen Kristallen. [**487**] A. P. Kapustin, A. N. Trofimov, A. N. Chuvyrev, Kristallografiya *17*, 194 (1972), C.A. *76*, 118742r (1972); Periodic orientational oscillations of the domain structure of liquid crystals in variable electric fields. [**488**] A. P. Kapustin, Z. K. Kuvatov, A. N. Trofimov, Kristallografiya *18*, 416 (1973), C.A. *78*, 165759c (1973); Electret effect in PAA. [**489**] A. P. Kapustin, Z. K. Kuvatov, A. N. Trofimov,

Kristallografiya *18*, 647 (1973), C.A. *79*, 71198n (1973); Thermodielectric effect during a liquid crystal-solid crystal phase transition. [**490**] Z. K. Kuvatov, A. P. Kapustin, A. N. Trofimov, Pis'ma Zh. Eksp. Teor. Fiz. *19*, 89 (1974), C.A. *80*, 101255u (1974); Change in the anisotropy sign of the electric conductivity of a nematic as a function of temperature.

[**491**] A. P. Kapustin, Z. K. Kuvatov, A. N. Trofimov, D. Demus, Kristallografiya *20*, 449 (1975), C.A. *83*, 20144s (1975); Dielectric properties of liquid crystalline 4-(4-*n*-heptyloxyazoxybenzal)-butylaniline. [**492**] P. P. Karat, N. V. Madhusudana, Liq. Cryst., Proc. Int. Conf., Bangalore *1973*, 285, C.A. *84*, 114467x (1976); New types of electrohydrodynamic flow patterns in nematics. [**492a**] P. P. Karat, N. V. Madhusudana, Mol. Cryst. Liq. Cryst. *40*, 171 (1977), C.A. *87*, 144075x (1977); Verification of Leslie's expression for the threshold field of a cell. [**492b**] P. P. Karat, N. V. Madhusudana, Mol. Cryst. Liq. Cryst. *42*, 57 (1977); Study of the dielectric relaxation in nematics using the Frederiks transition technique. [**493**] R. A. Kashnow, H. S. Cole, J. Appl. Phys. *42*, 2134 (1971), C.A. *75*, 11709n (1971); Electrohydrodynamic instabilities in a high-purity nematic. [**494**] R. A. Kashnow, J. E. Bigelow, H. S. Cole, C. R. Stein, Appl. Phys. Lett. *23*, 290 (1973), C.A. *79*, 130056q (1973); Transient observations of field-induced n/c relaxation. [**495**] R. A. Kashnow, H. S. Cole, Mol. Cryst. Liq. Cryst. *23*, 329 (1973), C.A. *80*, 88465r (1974); Electric effects in MBBA/PEBAB mixtures. [**496**] R. A. Kashnow, J. E. Bigelow, H. S. Cole, C. R. Stein, Liq. Cryst. Ord. Fluids *2*, 483 (1974), C.A. *86*, 181392y (1977); Comments on the relaxation process in the n/c transition. [**497**] R. A. Kashnow, Phys. Lett. A *48*, 163 (1974), C.A. *81*, 83328j (1974); Dielectric observations of nematic instabilities. [**498**] W. Kast, Ann. Physik *73*, 145 (1923), C.A. *18*, 934² (1924); Anisotropie der flüssigen Kristalle bezüglich ihrer Dielektrizitätskonstanten und ihrer elektrischen Leitfähigkeit. [**499**] W. Kast, Z. Physik *37*, 233 (1926), C.A. *21*, 1734¹ (1927); Bemerkung zur Arbeit von G. Szivessy, Zur Bornschen Dipoltheorie der anisotropen Flüssigkeiten. [**500**] W. Kast, Z. Physik *39*, 490 (1926), C.A. *21*, 2827³ (1927); Bemerkung zur Erwiderung des Herrn G. Szivessy auf meine Kritik seiner Arbeit: Zur Bornschen Dipoltheorie der anisotropen Flüssigkeiten.

[**501**] W. Kast, Z. Physik *42*, 81 (1927), C.A. *21*, 1920⁴ (1927); Dielektrische Untersuchungen an der anisotropen Schmelze von PAA. [**502**] W. Kast, Z. Physik *42*, 91 (1927), C.A. *21*, 1909³ (1927); Dritte Bemerkung zu der Arbeit des Herrn G. Szivessy: Zur Bornschen Dipoltheorie der anisotropen Flüssigkeiten. [**503**] W. Kast, Ann. Physik *83*, 391 (1927), C.A. *22*, 907⁴ (1928); Dielektrische Untersuchungen an der anisotropen Schmelze von PAA. [**504**] W. Kast, Z. Krist. *79*, 146 (1931), C.A. *26*, 2095¹ (1932); Dielektrizitätskonstanten mesomorpher Substanzen. [**505**] W. Kast, Z. Krist. *79*, 161 (1931), C.A. *26*, 2093² (1932); Magnetische Suszeptibilität mesomorpher Substanzen. [**506**] W. Kast, P. J. Bouma, Z. Physik *87*, 753 (1933), C.A. *28*, 2963³ (1934); Messungen des dielektrischen Verlustes von kristallin-flüssigem PAA. [**507**] W. Kast, L. S. Ornstein, Z. Physik *87*, 763 (1933), C.A. *28*, 2963⁴ (1934); Registrierungen der Lichtdurchlässigkeit der anisotropen Schmelze des PAA im Magnetfeld. Beitrag zur Schwarmtheorie der flüssigen Kristalle. [**508**] W. Kast, Physik. Z. *36*, 869 (1935), C.A. *30*, 1622⁶ (1936); Feldstärke- und Frequenzabhängigkeit der Dielektrizitätskonstanten anisotroper Flüssig-keiten. [**509**] W. Kast, Z. techn. Physik *16*, 475 (1935); Feldstärke- und Frequenzabhängigkeit der Dielektrizitätskonstante anisotroper Flüssigkeiten. [**510**] K. Kato, Japan. Kokai 75 53,282 (12. 9. 1973), C.A. *83*, 200260k (1975); Liquid crystal compositions for electrooptical display devices.

[**511**] E. I. Kats, Zh. Eksp. Teor. Fiz. *63*, 329 (1972), C.A. *77*, 106354y (1972); Effect of periodic perturbations on the cholesteric mesophase. [**512**] E. I. Kats, Zh. Eksp. Teor. Fiz. *70*, 1394 (1976), C.A. *85*, 27542v (1976); Effect of van der Waals forces on the orientation of a nematic film. [**513**] E. I. Kats, Zh. Eksp. Teor. Fiz. *70*, 1941 (1976), C.A. *85*, 70909g (1976); Surface oscillations in liquid crystals. [**514**] M. Kawachi, O. Kogure, Y. Kato, Jpn. J. Appl. Phys. *13*, 1457 (1974), C.A. *81*, 160444y (1974); Bubble domain texture of a liquid crystal. [**515**] M. Kawachi, O. Kogure, S. Yoshii, Y. Kato, Jpn. J. Appl. Phys. *14*, 1063 (1975), C.A. *83*, 106574u (1975); Field-induced n/c relaxation in a small angle wedge. [**516**] M. Kawachi, O. Kogure, Jpn. J. Appl. Phys. *15*, 1557 (1976), C.A. *85*, 134805e (1976); Movable bubble domains of large pitch cholesterics. [**516a**] M. Kawachi, O. Kogure, Jpn. J. Appl. Phys. *16*, 1673 (1977), C.A. *87*, 160146b (1977); Hysteresis behavior of texture in the field-induced n/c relaxation. [**517**] W. Kaye, Mol. Cryst. Liq. Cryst. *37*, 137 (1976), C.A. *86*, 163575f (1977); Interpretation and nomenclature for the transmittance vs. voltage curves for LCDs. [**518**] W. Kaye, Mol. Cryst. Liq. Cryst. *37*, 147 (1976), C.A. *86*, 180655z (1977); Molecular orientation in TN cells. [**519**] B. Kerlleñevich, A. Coche, J. Appl. Phys. *42*, 5313 (1971), C.A. *76*, 28639h (1972); Effects of the addition of cholesterics on nematic liquid crystal properties. [**520**] B. Kerlleñevich, A. Coche, Mol. Cryst. Liq. Cryst. *24*, 113 (1973), C.A. *81*, 142305z (1974); Relaxation of light scattering in nematics and in n/c mixtures.

[**521**] B. Kerlleñevich, A. Coche, Liq. Cryst. Ord. Fluids *2*, 705 (1974), C.A. *86*, 181000u (1977); Erasure of textures stored in n/c mixtures. [**522**] B. Kerlleñevich, A. Coche, Electron. Lett. *11*, 421 (1975), C.A. *83*, 186600n (1975); Instabilities in some nematics of positive dielectric anisotropy. [**523**] J. O. Kessler, M. Longley-Cook, W. O. Rasmussen, Mol. Cryst. Liq. Cryst. *8*, 327 (1969), C.A. *71*, 85908r (1969); Low frequency electric properties of nematic PAA. [**524**] J. O. Kessler, Liq. Cryst. Ord. Fluids *1970*, 361, C.A. *78*, 152973t (1973); Magnetic alignment of nematics. [**525**] H. Kiemle, 1. Arbeitstagung Flüssigkristalle, Inst. Angew. Festkörperphysik Freiburg, 23. 4. 1971; Dynamische Streuung von kohärentem Licht und Möglichkeiten über technische Anwendungen. [**526**] C. Kikuchi, Phys. Rev. *93*, 934 (1954), C.A. *49*, 9976g (1955); Dielectric absorption of liquid crystalline biaxial molecules. [**527**] E. A. Kirsanov, Sb. Dokl. Vses. Nauch. Konf. Zhidk. Krist. Simp. Ikh Prakt. Primen., 2nd, *1972*, 122, C.A. *81*, 142294v (1974); Diffraction of light on the domain structure of a liquid crystal. [**528**] E. A. Kirsanov, Uch. zap. Ivanov. gos. ped. in-t *1974*, 115, C.A. *83*, 20136r (1975); Electrohydrodynamic instability in nematics with different dielectric anisotropy. [**529**] E. A. Kirsanov, Uch. zap. Ivanov. un-t *1974*, 36, 44, 52, 58; Electrohydrodynamic instability in nematics, I. Complex domain systems, C.A. *83*, 156059a (1975); II. Changes in the domain system on increasing the potential beyond the threshold. Special forms of the domains, C.A. *83*, 156060u (1975); III. Processes of formation and destruction of domain systems in nematics with negative dielectric anisotropy, C.A. *83*, 156061v (1975); IV. Dependence of the character of the instability on the dielectric anisotropy, C.A. *83*, 156062w (1975). [**530**] J. Kirton, E. P. Raynes, Endeavour *32*, 71 (1973), C.A. *79*, 47244d (1973); Electrooptic effects in liquid crystals.

[**531**] Z. I. Kir'yashkina, V. F. Nazvanov, F. Y. Filipchenko, G. A. Lebedina, V. A. Elistratov, Pis'ma Zh. Tekh. Fiz. *1*, 1044 (1975), C.A. *84*, 52833c (1976); Photoconductor-liquid crystal contact. [**532**] G. G. Klimenko, S. A. Nesvedov, V. M. Shoshin, Sb. Dokl. Vses. Nauch. Konf. Zhidk. Krist. Simp. Ikh Prakt. Primen., 2nd, *1972*, 268, C.A. *81*, 97724x (1974); Possibility of constructing matrix display devices using the dynamic scattering effect in nematics. [**533**] R. T. Klingbiel, D. J. Genova, T. R. Criswell, J. P. Van Meter, J. Am. Chem. Soc. *96*, 7651 (1974), C.A. *82*, 24866f (1975); Comparison of the dielectric behavior of several Schiff-base and phenyl benzoate liquid crystals. [**534**] R. T. Klingbiel, D. J. Genova, H. K. Buecher, Mol. Cryst. Liq. Cryst. *27*, 1 (1974), C.A. *82*, 148685r (1975); Temperature dependence of the dielectric and conductivity anisotropies of liquid crystals. [**535**] L. M. Klyukin, I. D. Samodurova, A. S. Sonin, Fiz. Tverd. Tela *17*, 1173 (1975), C.A. *83*, 36055z (1975); Temperature dependence of characteristics of dynamic scattering in nematics. [**536**] H. W. Ko, Diss. Drexel Univ., Philadelphia, 1973, 208 pp. Avail. Univ. Microfilms, Ann Arbor, Mich., Order No. 73-22,831, C.A. *79*, 150931z (1973); Magnetically induced birefringence in nematics. [**537**] K. Kobayashi, Bussei Kenkyu *24*, C3 (1975), C.A. *85*, 85592x (1976); Thermodynamic and hydrodynamic instabilities in liquid crystals. [**538**] H. Koelmans, A. M. van Boxtel, Phys. Lett. A *32*, 32 (1970), C.A. *73*, 103104y (1970); Electrohydrodynamic flow in nematics. [**539**] H. Koelmans, A. M. van Boxtel, Mol. Cryst. Liq. Cryst. *12*, 185 (1971), C.A. *75*, 26544d (1971); Electrohydrodynamic flow in nematics. [**540**] O. Kogure, K. Murase, Jap. J. Appl. Phys. *9*, 1280 (1970), C.A. *74*, 35882u (1971); Electric field effect of nematic ethyl *p*-(anisylidene-amino)cinnamate.

[**541**] C. Konak, Cesk. Cas. Fyz. *25*, 521 (1975), C.A. *84*, 37450g (1976); Ferroelectric liquid crystals. [**541a**] A. A. Kovalev, L. V. Syt'ko, I. P. Mazur, G. L. Nekrasov, V. A. Grozhik, Kristallografiya *22*, 586 (1977), C.A. *87*, 109683x (1977); Stimulated rearrangement of domain structure in nematics. [**542**] N. L. Kramarenko, I. V. Kurnosov, Y. V. Naboikin, Phys. Status Solidi A *33*, 773 (1976), C.A. *84*, 171617v (1976); Light scattering by liquid crystals in an electric field. [**543**] H. Kresse, D. Demus, C. Krinzer, Z. Phys. Chem. *256*, 7 (1975), C.A. *83*, 156428v (1975); Messungen der Dielektrizitäts-konstante und des dielektrischen Verlustes in nematischem 4-*n*-Pentyloxybenzoesäure-4-*n*-octyloxyphenyl-ester. [**544**] H. Kresse, P. Schmidt, D. Demus, Phys. Status Solidi A *32*, 315 (1975), C.A. *84*, 37677m (1976); Dielectric behavior of the nematic 5-*n*-hexyl-2-(4-n-alkyloxyphenyl)-pyrimidines. [**545**] H. Kresse, P. Schmidt, Z. Phys. Chem. (Leipzig) *256*, 987 (1975), C.A. *84*, 120878y (1976); Dielektrische Untersuchungen am nematischen 2,6-Bis-(4-*n*-butyloxybenzal)-cyclohexanon. [**546**] H. Kresse, K. H. Lücke, H. J. Deut-scher, Z. Chem. *16*, 55 (1976), C.A. *84*, 143416k (1976); Dielektrisches Verhalten nematischer 4-*n*-Alkoxy-benzoesäuren. [**546a**] H. Kresse, D. Demus, S. König, Phys. Status Solid A *41*, K67 (1977), C.A. *87*, 47020t (1977); Dielectric relaxation of nematics and smectics. [**546b**] H. Kresse. C. Selbmann, W. Weissflog, Z. Chem. *17*, 137 (1977), C.A. *87*, 67681n (1977); Kristallin-flüssige 4-*n*-Butoxycarbonyloxy-benzoesäure-4'-*n*-alkoxyphenylester. [**546c**] H. Kresse, K. H. Lücke, P. Schmidt, D. Demus, Z. Phys. Chem. (Leipzig) *258*, 785 (1977), C.A. *87*, 125975z (1977); Dielektrische Anisotropie nematischer 4-*n*-Alkoxy-

benzoesäuren. [**547**] T. A. Kruglova, Sb. Dokl. Vses. Nauch. Konf. Zhidk. Krist. Simp. Ikh Prakt. Primen., 2nd, *1972*, 108, C.A. *82*, 49419a (1975); Transparency of MBBA in sound-frequency electrical fields. [**548**] T. A. Kruglova, E. A. Kirsanov, Uch. zap. Ivanov. un-t *1974*, 65, C.A. *83*, 140051f (1975); Effect of an electric field of various frequencies on the orientation of MBBA molecules. [**549**] T. Krupkowski, W. Vieth, Bull. Acad. Pol. Sci., Ser. Sci. Chim. *22*, 823 (1974), C.A. *82*, 50027c (1975); Influence of an electric field on the dielectric permittivity of nematic PAA. [**550**] W. L. Kuhn, B. A. Finlayson, Mol. Cryst. Liq. Cryst. *36*, 307 (1976), C.A. *86*, 81976y (1977); Orientation of nematics in thick layers.

[**551**] A. N. Kuznetsov, V. A. Livshits, S. G. Cheskis, Sb. Dokl. Vses. Nauch. Konf. Zhidk. Krist. Simp. Ikh Prakt. Primen., 2nd, *1972*, 85, C.A. *81*, 112182q (1974); Theory of the anisotropy of the permittivity of nematics. [**552**] A. N. Kuznetsov, V. A. Livshits, S. G. Cheskis, Kristallografiya *20*, 231 (1975), C.A. *83*, 20143r (1975); Theory of anisotropy of the dielectric constant of nematics. [**553**] A. N. Kuznetsov, T. P. Kulagina, Zh. Eksp. Teor. Fiz. *68*, 1501 (1975), C.A. *83*, 69412s (1975); Theory of the magnetohydrodynamic effect in nematics. [**554**] S. Kusabayashi, M. M. Labes, Mol. Cryst. Liq. Cryst. *7*, 395 (1969), C.A. *71*, 75396h (1969); Conductivity in liquid crystals. [**555**] A. I. Baise, I. Teucher, M. M. Labes, Appl. Phys. Lett. *21*, 142 (1972), C.A. *77*, 107033e (1972); Effect of charge-transfer acceptors on dynamic scattering in a nematic. [**556**] N. Oron, M. M. Labes, Appl. Phys. Lett. *21*, 243 (1972), C.A. *77*, 144929x (1972); The a.c.-d.c. technique for rapid conical-helical perturbation in cholesterics. [**557**] A. I. Baise, M. M. Labes, J. Chem. Phys. *59*, 551 (1973), C.A. *79*, 84272g (1973); Dynamic scattering in a liquid crystal having positive dielectric anisotropy. [**558**] N. Oron, L. J. Yu, M. M. Labes, Mol. Cryst. Liq. Cryst. *21*, 333 (1973), C.A. *79*, 120035t (1973); Rapid conical-helical perturbation in doped cholesterics. [**559**] L. J. Yu, M. M. Labes, Mol. Cryst. Liq. Cryst. *28*, 423 (1974), C.A. *82*, 178712q (1975); Nonuniform distortions during electric field induced unwinding of a cholesteric. [**560**] M. M. Labes, A. I. Baise, U.S.N.T.I.S., AD/A Rep., No. 004201/OGA, 74 pp. (1974), C.A. *83*, 89006r (1975); Mechanism of dynamic scattering in nematics suitable for display applications.

[**561**] A. I. Baise, M. M. Labes, Appl. Phys. Lett. *24*, 298 (1974), C.A. *80*, 138048e (1974); Effect of dielectric anisotropy on twisted nematics. [**562**] L. J. Yu, H. Lee, C. S. Bak, M. M. Labes, Phys. Rev. Lett. *36*, 388 (1976), C.A. *84*, 129398z (1976); Observation of pyroelectricity in chiral smectics C and H. [**563**] J. W. Park, M. M. Labes, J. Appl. Phys. *48*, 22 (1977), C.A. *86*, 82800y (1977); Dielectric, elastic, and electro-optic properties of a liquid crystalline molecular complex. [**564**] G. Labrunie, J. Robert, Centre d'Études Nucléaires de Grenoble, Note technique LETI/ME n° 849; Transient behavior of the electrically controlled birefringence in a nematic. [**565**] G. Labrunie, J. Robert, J. Appl. Phys. *44*, 4869 (1973), C.A. *80*, 31453f (1974); Transient behavior of the electrically controlled birefringence in a nematic. [**566**] J. C. Lacroix, R. Tobazeon, Appl. Phys. Lett. *20*, 251 (1972), C.A. *76*, 132730s (1972); Behavior of a deionized liquid crystal subjected to unipolar injection in n and i phase. [**567**] J. C. Lacroix, R. Tobazeon, C. R. Acad. Sci., Sér. B *278*, 623 (1974), C.A. *81*, 30868d (1974); Mesure de mobilités ioniques dans un nématique. [**568**] D. Langevin, Phys. Lett. A *56*, 61 (1976), C.A. *84*, 143191h (1976); Periodic structure at the free surface of smectics A. [**569**] D. Langevin, J. Phys. (Paris) *37*, 737 (1976), C.A. *85*, 54900s (1976); Spectre des fluctuations thermiques a la surface libre d'un smectique A en présence d'un champ magnétique horizontal. [**570**] D. Langevin, J. Phys. (Paris) *37*, 755 (1976), C.A. *85*, 54901t (1976); Structure induite par un champ magnétique a la surface libre des smectiques A.

[**571**] J. D. Lee, A. C. Eringen, U.S. Clearinghouse Fed. Sci. Tech. Inform., AD *1970*, No. 721107, 25 pp., C.A. *75*, 81385c (1971); Alignment of nematics. [**572**] L. Léger, Solid State Commun. *10*, 697 (1972), C.A. *76*, 145985d (1972); Wall motions in nematics. [**573**] L. Léger, Solid State Commun. *11*, 1499 (1972), C.A. *78*, 49860e (1973); Static and dynamic behavior of walls in nematics above a Frederiks transition. [**574**] L. Léger, Mol. Cryst. Liq. Cryst. *24*, 33 (1973), C.A. *81*, 142308c (1974); Walls in nematic. [**575**] P. le Roy, F. Debeauvais, S. Candau, C. R. Acad. Sci., Sér. B *274*, 419 (1972), C.A. *76*, 146027e (1972); Effets de champ magnétique sur la structure de gouttelettes de liquides nématiques en émulsion dans un liquide isotrope. [**576**] F. M. Leslie, Mol. Cryst. Liq. Cryst. *12*, 57 (1970), C.A. *74*, 92318p (1971); Distortion of twisted orientation patterns in liquid crystals by magnetic fields. [**577**] F. M. Leslie, Mol. Cryst. Liq. Cryst. *37*, 335 (1976), C.A. *86*, 113913u (1977); Magnetohydrodynamic instabilities in nematics. [**578**] M. Levy, Compt. rend. *227*, 278 (1948), C.A. *43*, 453d (1949); Étude de l'oléate de cholestéryle à l'état mésomorphe. Action du champ magnétique. [**579**] K. Lichtenecker, Physik. Z. *27*, 115 (1926), C.A. *20*, 3124^5 (1926); Dielektrizitätskonstante natürlicher und künstlicher

Mischkörper. [580] H. S. Lim, J. D. Margerum, J. Electrochem. Soc. *123*, 837 (1976), C.A. *85*, 85755c (1976); Dopant effects on d.c. dynamic scattering in a liquid crystal: microscopic pattern studies. [581] M. J. Little, H. S. Lim, J. D. Margerum, Mol. Cryst. Liq. Cryst. *38*, 207 (1977), C.A. *86*, 197897c (1977); Alignment effects on the dynamic scattering characteristics of an ester liquid crystal. [581a] H. S. Lim, J. D. Margerum, A. Graube, J. Electrochem. Soc. *124*, 1389 (1977), C.A. *87*, 174763n (1977); Electrochemical properties of dopants and the d.c. dynamic scattering of a nematic. [582] A. Lomax, R. Hirasawa, A. J. Bard, J. Electrochem. Soc. *119*, 1679 (1972), C.A. *78*, 37189e (1973); Electrochemistry of MBBA. Role of electrode reactions in dynamic scattering. [583] M. Longley-Cook, J. O. Kessler, Mol. Cryst. Liq. Cryst. *12*, 315 (1971), C.A. *75*, 11249n (1971); Heat transport in liquid crystals. [584] S. Lu, D. Jones, Appl. Phys. Lett. *16*, 484 (1970), C.A. *73*, 60150r (1970); Electric field distribution associated with dynamic scattering in nematics. [585] S. Lu, D. Jones, J. Appl. Phys. *42*, 2138 (1971), C.A. *75*, 27756m (1971); Light diffraction phenomena in an ac-excited nematic. [586] M. Luban, D. Mukamel, S. Shtrikman, Phys. Rev. A *10*, 360 (1974), C.A. *81*, 69557b (1974); Transition from the cholesteric storage mode to the nematic phase in critical restricted geometries. [587] T. C. Lubensky, Mol. Cryst. Liq. Cryst. *23*, 99 (1973), C.A. *79*, 150990t (1973); Hydrodynamics of cholesterics in an external magnetic field. [588] G. R. Luckhurst, C. Zannoni, Proc. Roy. Soc. A *343*, 389 (1975), C.A. *83*, 69840e (1975); Theory of dielectric relaxation in anisotropic systems. [589] G. R. Luckhurst, R. N. Yeates, Chem. Phys. Lett. *38*, 551 (1976), C.A. *84*, 172355v (1976); Interpretation of dielectric relaxation times for uniaxial liquid crystals. [590] D. P. McLemore, Diss. Univ. Maine, Orono. 1973, 103 pp. Avail. Univ. Microfilms, Ann Arbor, Mich., Order No. 73-32,330, C.A. *81*, 30657j (1974); Anomalous effects owing to electric fields in nematic 4,4'-di-*n*-heptyloxyazoxybenzene. [591] H. Mada, S. Kobayashi, Mol. Cryst. Liq. Cryst. *33*, 47 (1976), C.A. *84*, 186808g (1976); Wavelength and voltage dependences of refractive indices of nematics. [592] W. Maier, Ann. Phys. *33*, 210 (1938), C.A. *33*, 908⁴ (1939); Feldstärkeabhängigkeit der Dielektrizitätskonstante des PAA. [593] W. Maier, Physik. Z. *45*, 285 (1944), C.A. *40*, 6926¹ (1946); Änderung der Dielektrizitätskonstante der pl-Phase von PAA durch magnetische Felder. [594] W. Maier, Z. Naturforsch. A *2*, 458 (1947), C.A. *42*, 2831c (1948); Dielektrische Anisotropie geordneter kristalliner Flüssigkeiten vom Typ des PAA. [595] W. Maier, G. Barth, H. E. Wiehl, Z. Elektrochem. *58*, 674 (1954), C.A. *49*, 5053e (1955); Messungen der Hauptdielektrizitätskonstanten magnetisch geordneter kristallin-flüssiger Phasen. [596] W. Maier, G. Meier, Z. physik. Chem. *13*, 251 (1957), C.A. *52*, 2484a (1958); Dielektrische Suszeptibilitäten der normal-flüssigen und der kristallin-flüssigen Phasen zweier Azobenzolderivate. [597] W. Maier, G. Meier, Z. Naturforsch. A *16*, 262 (1961), C.A. *55*, 19403c (1961); Einfache Theorie der dielektrischen Eigenschaften von homogen orientierten nematischen Phasen. [598] W. Maier, G. Meier, Z. Naturforsch. A *16*, 470 (1961), C.A. *56*, 4204a (1962); Hauptdielektrizitätskonstanten von homogen geordnetem nematischem PAA. [599] W. Maier, G. Meier, Z. Naturforsch. A *16*, 1200 (1961), C.A. *56*, 13653h (1962); Anisotrope DK-Dispersion im Radiofrequenzgebiet bei homogen geordneten kristallinen Flüssigkeiten. [600] W. Maier, G. Meier, Z. Elektrochem. *65*, 556 (1961); Dielektrische Untersuchungen an den kristallin-flüssigen Phasen einiger Azophenoläther. Bestimmung der molekularen Hauptpolarisierbarkeiten. [601] H. Mailer, Diss. Ohio State Univ., Columbus. 1973, 212 pp. Avail. Univ. Microfilms, Ann Arbor, Mich., Order No. 73-26,862, C.A. *80*, 75196d (1974); Effects of electric, magnetic, and acoustic fields on the optical and NMR properties of nematic MBBA. [602] G. Malet, J. Marignan, O. Parodi, J. Phys. (Paris), Lett. *36*, 317 (1975), C.A. *84*, 68098c (1976); Dynamical analysis of magnetic field effects on a cholesteric Cano wedge. [603] G. Malet, J. Marignan, O. Parodi, J. Phys. (Paris) *37*, 865 (1976), C.A. *85*, 86400v (1976); Dynamics of the first single Grandjean-Cano line in cholesterics under weak applied magnetic fields. [604] B. Malraison, P. Pieranski, E. Guyon, J. Phys. (Paris), Lett. *35*, 9 (1974), C.A. *80*, 114157p (1974); Distorsion d'un film nématique dans un champ magnétique presque parallèle a l'axe optique. [604a] P. Maltese, C. M. Ottavi, Alta Freq. *44*, 727 (1975); Experimental study of properties of contact between semiconductor and liquid crystal layers. [605] R. Malvano, P. Rava, Atti Accad. Sci. Torino, Cl. Sci. Fis., Mat. Nat. *110*, 225 (1976), C.A. *86*, 105628q (1977); Optical study of the second dynamic scattering regime in MBBA. [606] J. D. Margerum, H. S. Lim, P. O. Braatz, A. M. Lackner, Mol. Cryst. Liq. Cryst. *38*, 219 (1977), C.A. *86*, 181369w (1977); Effect of dopants on the conductivity anisotropy and ac dynamic scattering of liquid crystals. [607] J. Marignan, G. Malet, O. Parodi, J. Phys. (Paris) *37*, 365 (1976), C.A. *84*, 172961q (1976); Dynamics of the director alignment in cholesterics under applied magnetic field. [607a] B. Markowska, A. Szymanski, Mater. Sci. *2*, 33 (1976), C.A. *87*, 13619e (1977); Electrooptical properties of cholesterics I. Textural transitions. [608] G. A. Mann, R. D. Spence, Phys. Rev. *92*, 844 (1953); Rotation of microwave polarization in liquid crystals in a transverse magnetic field. [609] G. A. Mann, R. D. Spence, J. Appl. Phys. *25*, 271 (1954), C.A. *48*, 5586a (1954); Rotation of microwave polarization in liquid crystals. [610] A. M. Marks, Appl. Opt. *8*, 1397 (1969), C.A. *71*, 96539x (1969); Electrooptical characteristics of dipole suspensions.

[611] P. Martinot-Lagarde, J. Phys. (Paris), C-3, *37*, 129 (1976), C.A. *85*, 114978k (1976); Observation of ferroelectric monodomains in chiral smectics C. [612] P. Martinot-Lagarde, J. Phys. (Paris), Lett. *38*, 17 (1977), C.A. *86*, 82361f (1977); Direct electric measurement of the permanent polarization of a ferroelectric chiral smectic C. [613] C. H. Massen, J. A. Poulis, R. D. Spence, Bull. Soc. Belge Phys. No. 6, 395 (1965), C.A. *64*, 18654e (1966); Field dependence of the magnetic susceptibility of liquid crystals. [614] C. H. Massen, J. A. Poulis, R. D. Spence, Advan. Chem. Ser. No. *63*, 72 (1967), C.A. *67*, 27319f (1967); Field dependence of the magnetic susceptibility of nematic PAA. [615] S. Matsumoto, M. Kawamoto, T. Tsukada, Chem. Lett. *1973*, 837, C.A. *79*, 109010k (1973); Conductivity effects on cut-off frequency of electrohydrodynamic instability in nematics. [616] S. Matsumoto, M. Kawamoto, T. Tsukada, Jpn. J. Appl. Phys. *14*, 965 (1975), C.A. *83*, 186813j (1975); Temperature effect on cut-off frequency of electrohydrodynamic instability in nematics. [617] S. Matsumoto, K. Mizunoya, H. Ikeya, Oyo Butsuri *45*, 763 (1976), C.A. *85*, 169043g (1976); Effects of initial molecular alignment on the electrooptical properties of dynamic scattering liquid crystal devices. [618] Y. F. Matyushin, A. M. Kovnatskii, Zh. Tekh. Fiz. *45*, 661 (1975); Electrohydrodynamic effect in piezoelectric nematic. [619] C. Mauguin, Compt. rend. *152*, 1680 (1911), C.A. *5*, 3187³ (1911); Orientation des cristaux liquides par le champ magnétique. [620] P. Maurel, A. H. Price, J. Chem. Soc., Faraday Trans. 2, *69*, 1486 (1973), C.A. *80*, 53452g (1974); Dipole moment of MBBA.

[621] C. Maze, D. Johnson, Mol. Cryst. Liq. Cryst. *33*, 213 (1976), C.A. *85*, 102653n (1976); Determination of nematic liquid crystal elastic and dielectric constants from birefringence experiments. [622] G. Meier, A. Saupe, Mol. Cryst. *1*, 515 (1966), C.A. *66*, 6351h (1967); Dielectric relaxation in nematics. [623] A. J. Martin, G. Meier, A. Saupe, Symp. Faraday Soc. *1971*, No. 5, 119, C.A. *79*, 24388q (1973); Extended Debye theory for dielectric relaxations in nematics. [624] H. Gruler, G. Meier, Mol. Cryst. Liq. Cryst. *16*, 299 (1972), C.A. *76*, 132594a (1972); Electric field-induced deformations in oriented nematics. [625] H. Gruler, T. J. Scheffer, G. Meier, Z. Naturforsch. A *27*, 966 (1972), C.A. *77*, 106377h (1972); Elastic constants of nematics. I. Theory of the normal deformation. [626] G. Meier, 3. Arbeitstagung Flüssigkristalle, Inst. Angew. Festkörperphysik Freiburg, 11. 5. 1973; Schwellwerte elektrooptischer Feldeffekte in nematischen Flüssigkristallen. [627] H. Gruler, G. Meier, Mol. Cryst. Liq. Cryst. *12*, 289 (1971), C.A. *75*, 54881e (1971); Correlation between the electrical properties and optical behavior of nematics. [628] G. Meier, Dielectr. Relat. Mol. Processes *2*, 183 (1973), C.A. *83*, 36276x (1975); Dielectric properties of liquid crystals. [629] G. Baur, A. Stieb, G. Meier, Liq. Cryst. Ord. Fluids *2*, 645 (1974), C.A. *86*, 180999q (1977); Electric field induced deformation in nematic phenyl benzoates. [630] A. Stieb, G. Baur, G. Meier, Ber. Bunsenges. Phys. Chem. *78*, 899 (1974), C.A. *82*, 10237q (1975); Inversionswände in dielektrisch deformierten Flüssigkristallschichten.

[631] G. Meier, Ber. Bunsenges. Phys. Chem. *78*, 905 (1974), C.A. *82*, 10064f (1975); Elektrische und magnetische Feldeffekte in flüssigen Kristallen. [632] A. Stieb, G. Baur, G. Meier, J. Phys. (Paris), C-1, *36*, 185 (1975); Alignment inversion walls in nematic layers deformed by an electric field. [633] G. Baur, A. Stieb, G. Meier, Appl. Phys. *6*, 309 (1975), C.A. *82*, 163399u (1975); Freedericksz deformation in nematics with frequency dependent dielectric constant. [634] L. Melamed, D. Rubin, Appl. Phys. Lett. *16*, 149 (1970), C.A. *72*, 115683d (1970); Electric field hysteresis effects in cholesterics. [635] R. Meyer, Appl. Phys. Lett. *12*, 281 (1968), C.A. *69*, 111068x (1968); Effects of electric and magnetic fields on the structure of cholesterics. [636] R. B. Meyer, Appl. Phys. Lett. *14*, 208 (1969), C.A. *71*, 16578w (1969); Distortion of a cholesteric structure by a magnetic field. [637] R. B. Meyer, Phys. Rev. Lett. *22*, 918 (1969), C.A. *71*, 7687d (1969); Piezoelectric effects in liquid crystals. [638] S. C. Chou L. Cheung, R. B. Meyer, Solid State Commun. *11*, 977 (1972), C.A. *77*, 170648z (1972); Effects of a magnetic field on the optical transmission in cholesterics. [639] R. B. Meyer, P. S. Pershan, Solid State Commun. *13*, 989 (1973), C.A. *79*, 150374v (1973); Surface polarity induced domains in liquid crystals. [640] H. Gruler, L. K. Cheung, R. B. Meyer, 3rd ACS Symp. Ord. Fluids Liquid Cryst., Chicago, Aug. 1973, Abstr. No. 25; Strong evidence for flexo-electric effect in nematic liquid crystal.

[641] R. B. Meyer, L. Liebert, L. Strzelecki, P. Keller, J. Phys. (Paris), Lett. *36*, 69 (1975), C.A. *82*, 179333d (1975); Ferroelectric liquid crystals. [641a] R. B. Meyer, Mol. Cryst. Liq. Cryst. *40*, 33 (1977), C.A. *87*, 125806v (1977); Ferroelectric liquid crystals; a review. [641b] S. Garoff, R.

B. Meyer, Phys. Rev. Lett. *38*, 848 (1977), C.A. *86*, 163792z (1977); Electroclinic effect at the A-C phase change in a chiral smectic. [**642**] D. Meyerhofer, Introd. Liq. Cryst. *1974*, 129, C.A. *84*, 97902r (1976); Electrohydrodynamic instabilities in nematics. [**643**] D. Meyerhofer, J. Appl. Phys. *46*, 5084 (1975), C.A. *84*, 98282a (1976); Elastic and dielectric constants in mixtures of nematics. [**644**] D. Meyerhofer, Phys. Lett. A *51*, 407 (1975), C.A. *83*, 69401n (1975); Field induced distortions of a liquid crystal with various surface alignments. [**645**] D. Meyerhofer, Mol. Cryst. Liq. Cryst. *34*, 13 (1976), C.A. *85*, 85813v (1976); Distortion of liquid crystals in the twisted field effect configuration. [**645a**] A. Michelson, D. Cabib, J. Phys. (Paris), Lett. *38*, 321 (1977), C.A. *87*, 144301t (1977); Electric field induced tricritical point in chiral polarized liquid crystals. [**645b**] H. Miike, T. Kohno, K. Koga, Y. Ebina, J. Phys. Soc. Jpn. *42*, 1419 (1977), C.A. *87*, 14192x (1977); Dynamic analysis of the growing process to DSM in nematic MBBA. [**645c**] H. Miike, T. Kohno, K. Koga, Y. Ebina, J. Phys. Soc. Jpn. *43*, 727 (1977), C.A. *87*, 125643q (1977); Dissipative structures in n/c (MBBA/CN) mixture. [**646**] V. I. Minkin, O. A. Osipov, Yu. A. Zhdanov, Dipole moments in organic chemistry (Plenum: New York). 1970. 288 pp. Translated from Russ., C.A. *73*, 76580s (1970). [**647**] A. Mircea-Roussel, F. Rondelez, J. Chem. Phys. *63*, 2311 (1975), C.A. *83*, 171118n (1975); Pretransitional behavior of conductivity and dielectric properties above a s_C/n transition in 4-4'-di(n-alkyloxy)azoxybenzenes. [**648**] A. Mircea-Roussel, L. Léger, F. Rondelez, W. H. de Jeu, J. Phys. (Paris), C-1, *36*, 93 (1975), C.A. *83*, 20014z (1975); Measurements of transport properties in nematics and smectics. [**649**] K. Miyakawa, S. Akahoshi, A. Takase, J. Phys. Soc. Jpn. *40*, 1785 (1976), C.A. *85*, 54895u (1976); Spatial fluctuations of Williams domain mode in nematics. [**650**] K. Miyakawa, J. Phys. Soc. Jpn. *42*, 18 (1977), C.A. *86*, 130259c (1977); Quasi-elastic light scattering near the instability point in nematics. I. Conduction regime. [**650a**] K. Miyakawa, S. Akahoshi, Solid State Commun. *22*, 647 (1977), C.A. *87*, 76616x (1977); Three dimensional convective instability in a nematic under an a.c. field.

[**651**] H. Miyazawa, Kotai Butsuri *8*, 633 (1973), C.A. *80*, 101598h (1974); Magnetic properties of liquid crystals. [**652**] M. Morita, S. Imamura, K. Yatabe, J. Phys. Soc. Jpn. *37*, 1710 (1974), C.A. *82*, 92277q (1975); Magnetic state of liquid crystal MBBA which is analogous to a ferromagnetic anisotropy. [**653**] M. Morita, S. Imamura, K. Yatabe, Jpn. J. Appl. Phys. *14*, 315 (1975), C.A. *82*, 163400n (1975); Guest-host effect in nematic MBBA. [**654**] E. Moritz, W. Franklin, Bull. Am. Phys. Soc. *20*, 855 (1975); Non-linear hydrodynamic instabilities of nematics under electromagnetic fields. [**655**] E. Moritz, Diss. Kent State Univ., Kent, Ohio. 1976. 80 pp. Avail. Xerox Univ. Microfilms, Ann Arbor, Mich., Order No. 77-3837, C.A. *86*, 131403g (1977); A class of nonlinear electrohydrodynamic effects in nematics. [**655a**] E. Moritz, W. Franklin, Mol. Cryst. Liq. Cryst. *40*, 229 (1977), C.A. *87*, 125608g (1977); Nonlinear electrohydrodynamic effects in nematics. [**656**] D. S. Moroi, Mol. Cryst. Liq. Cryst. *18*, 327 (1972), C.A. *77*, 145171f (1972); Equivalent permittivity tensor for parallel anisotropic homogeneous stratified media. [**657**] J. K. Moscicki, Solid State Commun. *20*, 481 (1976), C.A. *85*, 200994u (1976); Dielectric properties of the metastable and stable solid phase modifications of MBBA. [**658**] C. Motoc, P. Sterian, E. Sofron, Rev. Roum. Phys. *21*, 683 (1976), C.A. *86*, 113484e (1977); Morphological changes in nematics and their implications in laser light modulation. [**658a**] C. Motoc, I. Cuculescu, M. Honciuc, I. Baciu, Rev. Roum. Phys. *22*, 283 (1977), C.A. *87*, 159302t (1977); Transmission of polarized light in a cholesteric mixture in the presence of a d.c. electric field. [**659**] L. N. Mulay, I. L. Mulay, Anal. Chem *48*, 314R (1976), C.A. *84*, 173289p (1976); Magnetic susceptibility instrumentation and applications. [**660**] J. H. Muller, Z. Naturforsch. A *20*, 849 (1965), C.A. *63*, 10795h (1965); Electric field effects in cholesterics.

[**661**] J. H. Muller, U.S. Dep. Comm. AD 634627 (1966), C.A. *66*, 119631j (1967); Influence of electromagnetic fields on liquid crystals. [**662**] J. H. Muller, Mol. Cryst. *2*, 167 (1967), C.A. *66*, 32723s (1967); Effects of electric fields on cholesteryl nonanoate liquid crystals. [**663**] V. Naggiar, Compt. rend. *200*, 903 (1935), C.A. *29*, 3208[4] (1935); Production des fils et des tourbillons dans les liquides nématiques. [**664**] V. Naggiar, Compt. rend. *208*, 1916 (1939), C.A. *33*, 6105[8] (1939); Orientation dans le champ magnétique d'une goutte de liquides anisotrope suspendue. [**665**] I. Nakada, F. Hori, J. Phys. Soc. Jap. *33*, 1726 (1972), C.A. *78*, 103257d (1973); Electric conduction of PAP. [**666**] M. Nakagaki, Y. Naito, Yakugaku Zasshi *93*, 634 (1973), C.A. *79*, 45721h (1973); Physicochemical studies on the phase transition of the mixture of cholesteryl propionate and cholesterol. [**667**] M. Nakagaki, Y. Naito, Yakugaku Zasshi *95*, 390 (1975), C.A. *83*, 33629d (1975); Dielectric properties of the mixtures of cholesterol and cetyl alcohol. [**668**] M. Nakagaki, Y. Naito, Yakugaku Zasshi *96*, 791 (1976), C.A. *85*, 68224m (1976); Dielectric properties of the mixture of cholesteryl nonanoate and cholesteryl chloride. [**668a**] T. Nakagomi, S. Hasegawa, K. Toriyama, Japan. Kokai 77 59,081 (12. 11. 1975), C.A. *87*,

125397n (1977); Purification of liquid crystals. [**669**] F. Nakano, K. Toriyama, M. Sato, T. Muroi, Proc. Conf. (Int.) Solid State Devices, 5th, *1973*, 141, C.A. *81*, 127891u (1974); Light scattering in n/c mixtures at a low electric field. [**670**] J. Nakauchi, M. Yokoyama, K. Kato, K. Okamoto, H. Mikawa, S. Kusabayashi, Chem. Lett. *1973*, 313, C.A. *78*, 116236w (1973); Electrooptical properties of high-purity nematics under vacuum conditions.

[**671**] J. Nakauchi, M. Yokoyama, H. Sawa, K. Okamoto, H. Mikawa, S. Kusabayashi, Bull. Chem. Soc. Jap. *46*, 3321 (1973), C.A. *80*, 53164q (1974); Impurity effect on domain formation in a nematic. [**672**] J. Nehring, 2. Arbeitstagung Flüssigkristalle, Inst. Angew. Festkörperphysik Freiburg, 21. 4. 1972; Die dynamische Streuung. [**673**] J. Nehring, M. S. Petty, Phys. Lett. A *40*, 307 (1972), C.A. *77*, 119246q (1972); Formation of threads in the dynamic scattering mode of nematics. [**674**] J. Nehring, A. R. Kmetz, T. J. Scheffer, J. Appl. Phys. *47*, 850 (1976); Weak-boundary-coupling effects in LC displays. [**674a**] A. Nomura, S. Ogawa, Jpn. J. Appl. Phys. *16*, 639 (1977), C.A. *87*, 32187d (1977); Effect of field application on the flow speed of some dye-mixed nematics at a constant pressure. [**675**] P. L. Nordio, G. Rigatti, U. Segre, Mol. Phys. *25*, 129 (1973), C.A. *78*, 77058t (1973); Dielectric relaxation theory in nematics. [**676**] G. Agostini, P. L. Nordio, G. Rigatti, U. Segre, Atti Accad. Naz. Lincei, Mem., Cl. Sci. Fis., Mat. Nat., Sez. 2a, *13*, 20 (1975), C.A. *85*, 70728x (1976); Relaxation phenomena in liquid crystals. [**677**] C. Obayashi, J. Phys. Soc. Jpn. *38*, 1787 (1975), C.A. *83*, 140342b (1975); Enhancement of thermally stimulated current in PAA polarized by magnetic field. [**678**] Y. Odaka, Kogakuin Diagaku Kenkyu Hokoku *37*, 1 (1974), C.A. *83*, 186955g (1975); Liquid crystals. Measurements of dielectric constant. [**678a**] F. Ogawa, C. Tani, F. Saito, Electron. Lett. *12*, 70 (1976); New electro-optical effect. Optical activity of electric field-induced TN liquid crystal. [**679**] T. Ohta, Y. Takasaki, Bussei *14*, 452 (1973), C.A. *79*, 98028u (1973); Mechanism of the electric conductance in nematics. [**680**] M. Ohtsu, T. Akahane, T. Takao, Jap. J. Appl. Phys. *13*, 621 (1974), C.A. *81*, 18660x (1974); Birefringence of n-type nematic due to electrically induced deformations of vertical alignment.

[**681**] T. Ohtsuka, M. Tsukamoto, Jap. J. Appl. Phys. *10*, 1046 (1971), C.A. *75*, 123474p (1971); Electrooptical properties of nematic films. [**682**] T. Ohtsuka, M. Tsukamoto, Jap. J. Appl. Phys. *12*, 22 (1973), C.A. *78*, 76811w (1973); The a.c. electric-field-induced c/n transition in mixed liquid crystal films. [**683**] H. Onnagawa, T. Ootake, K. Miyashita, Oyo Butsuri *40*, 510 (1971), C.A. *76*, 7657x (1972); Electrooptic effect in cells of MBBA. [**684**] H. Onnagawa, K. Miyashita, Jpn. J. Appl. Phys. *13*, 1741 (1974), C.A. *82*, 37523s (1975); Abrupt change of molecular alignments in nematic liquid crystal cells by a rotating magnetic field. [**685**] W. J. H. Moll, L. S. Ornstein, Verslag Akad. Wetenschappen Amsterdam *25*, 682 (1917), C.A. *11*, 2425³ (1917); Contribution to the study of liquid crystals. [**686**] W. J. H. Moll, L. S. Ornstein, Verslag Akad. Wetenschappen Amsterdam *25*, 1112 (1917), C.A. *11*, 3170² (1917); Liquid crystals. Influence of temperature on extinction. Influence of a magnetic field. [**687**] W. J. H. Moll, L. S. Ornstein, Proc. Acad. Sci. Amsterdam *20*, 210 (1917), C.A. *12*, 444⁶ (1918); Liquid crystals. Influence of temperature on extinction. Influence of a magnetic field. [**688**] W. J. H. Moll, L. S. Ornstein, Verslag Akad. Wetenschappen Amsterdam *26*, 1442 (1918), C.A. *12*, 2477¹ (1918); Liquid crystals. Thermal effect of a magnetic field. [**689**] W. H. J. Moll, L. S. Ornstein, Proc. Acad. Sci. Amsterdam *21*, 254 and 259 (1919), C.A. *13*, 1172⁸ (1919); Liquid crystals. Melting and congelation phenomena with PAA. Thermal effect of the magnetic field. [**690**] L. S. Ornstein, Ann. Physik *74*, 445 (1924), C.A. *18*, 2827⁹ (1924); Anisotropie der flüssigen Kristalle bezüglich ihrer Dielektrizitätskonstante und ihrer Leitfähigkeit.

[**691**] L. S. Ornstein, Physik. Z. *29*, 668 (1928), C.A. *23*, 556⁴ (1929); Zur Frage der flüssigen Kristalle. [**692**] L. S. Ornstein, W. Kast, P. J. Bouma, Proc. Acad. Sci. Amsterdam *35*, 1209 (1932), C.A. *27*, 2073² (1933); Kristallin-flüssiger Charakter von Dipolflüssigkeiten am Schmelzpunkt. [**693**] W. de Braaf, L. S. Ornstein, Kolloid-Beihefte *44*, 427 (1936), C.A. *31*, 1269⁶ (1937); Lichtzerstreuung des nematischen PAA. [**694**] L. S. Ornstein, W. de Braaf, Proc. Acad. Sci. Amsterdam *42*, 105 (1939), C.A. *33*, 8464⁶ (1939); Scattering of light by anisotropic liquids. [**695**] Orsay Liquid Crystal Group, Phys. Lett. A *28*, 687 (1969), C.A. *70*, 100507j (1969); Existence and magnetic instability of double disclination lines in cholesterics. [**696**] Orsay Liquid Crystal Group, Phys. Rev. Lett. *25*, 1642 (1970), C.A. *74*, 46497y (1971); Hydrodynamic instabilities in nematics under a.c. electric fields. [**697**] Orsay Liquid Crystal Group, Liq. Cryst. Ord. Fluids *1970*, 447, C.A. *78*, 152467f (1973); Investigations in nematics and cholesterics. [**698**] Orsay Liquid Crystal Group, Mol. Cryst. Liq. Cryst. *12*, 251 (1971), C.A. *75*, 26543c (1971); A.c. and d.c. regimes of the electrohydrodynamic instabilities in nematics. [**699**] Orsay Liquid Crystal Group, Phys. Lett. A *39*, 181 (1972), C.A. *77*, 40286u (1972); Transition

between conduction and dielectric regimes of the electrohydrodynamic instabilities in a nematic. [**700**] B. I. Ostrovskii, A. Z. Rabinovich, A. S. Sonin, B. A. Strukov, N. I. Chernova, Pis'ma Zh. Eksp. Teor. Fiz. *25*, 80 (1977), C.A. *86*, 99708x (1977); Ferroelectric properties of a smectic.

[**701**] S. Pan, C. H. Wang, Liq. Cryst., Proc. Int. Conf., Bangalore *1973*, 299, C.A. *84*, 114469z (1976); Relaxation study of the electric field induced hydrodynamic turbulence in nematics by Raman scattering. [**702**] S. Pan, C. H. Wang, U.S. Nat. Tech. Inform. Serv., AD Rep. 1973, No. 770595/7GA, 18 pp., C.A. *80*, 113669v (1974); Relaxation study of the electric field-induced hydrodynamic turbulence in nematics by Raman scattering. [**702a**] J. H. Parker, Jr., Diss. Univ. of Maine, Orono, 1971, 117 pp. Avail. Univ. Microfilms, Ann Arbor, Mich., Order No. 72-15649, C.A. *77*, 106855n (1972); Anomalous alignment and domains in a nematic. [**703**] D. S. Parmar, A. K. Jalaluddin, Chem. Phys. Lett. *25*, 417 (1974), C.A. *81*, 7221d (1974); Preelectrohydrodynamic relaxation phenomenon in nematics. [**704**] J. P. Parneix, A. Chapoton, E. Constant, J. Phys. (Paris), *36*, 1143 (1975), C.A. *84*, 24933s (1976); Propriétés diélectriques du *p*-méthoxyphénylazoxy-*p'*-butylbenzène en phase n et i. [**705**] O. Parodi, Solid State Commun. *11*, 1503 (1972), C.A. *78*, 49855g (1973); Possible magnetic transition in smectics A. [**706**] J. D. Parsons, C. F. Hayes, Phys. Rev. A *9*, 2652 (1974), C.A. *81*, 43069u (1974); Fluctuations of a cholesteric in a static magnetic field. [**707**] P. A. Penz, Phys. Rev. Lett. *24*, 1405 (1970), C.A. *73*, 49505q (1970); Voltage-induced vorticity and optical focusing in liquid crystals. [**708**] P. A. Penz, Bull. Am. Phys. Soc. Ser. II, *15*, 59 (1970); Voltage dependent conductivity of nematics. [**709**] P. A. Penz, Mol. Cryst. Liq. Cryst. *15*, 141 (1971), C.A. *76*, 19094z (1972); Order parameter distribution for the electrohydrodynamic mode of a nematic. [**710**] P. A. Penz, G. W. Ford, Appl. Phys. Lett. *20*, 415 (1972), C.A. *77*, 80744y (1972); Electrohydrodynamic solutions for nematics.

[**711**] P. A. Penz, G. W. Ford, Phys. Rev. A *6*, 414 (1972); Electromagnetic hydrodynamics of liquid crystals. [**712**] P. A. Penz, G. W. Ford, Phys. Rev. A *6*, 1676 (1972), C.A. *77*, 119290z (1972); Electrohydrodynamic solutions for the homeotropic nematic geometry. [**713**] A. P. Penz, Mol. Cryst. Liq. Cryst. *23*, 1 (1973), C.A. *79*, 150539c (1973); Electrohydrodynamic solutions for nematics with positive dielectric anisotropy. [**714**] P. A. Penz, Phys. Rev. A *10*, 1300 (1974), C.A. *82*, 24658q (1975); Electrohydrodynamic wavelengths and response rates for a nematic. [**715**] P. A. Penz, Phys. Rev. A *12*, 1585 (1975), C.A. *83*, 211537g (1975); Propagating electrohydrodynamic mode in a nematic. [**716**] R. Pepinsky, A. J. Adams, G. W. Christoph, J. C. Nelander, Bull. Am. Phys. Soc. *19*, 1097 (1974); Moll-Ornstein effect in nematics. [**717**] P. Petrescu, M. Giurgea, Phys. Lett. A *59*, 41 (1976), C.A. *86*, 36537v (1977); New type of domain structure in nematics. [**718**] A. G. Petrov, Dokl. Bolg. Akad. Nauk *24*, 573 (1971), C.A. *76*, 51477n (1972); Molecular parameters and dielectric anisotropy of nematic PAA. [**719**] P. Pieranski, F. Brochard, E. Guyon, J. Phys. (Paris) *34*, 35 (1973), C.A. *78*, 141289y (1973); Static and dynamic behavior of a nematic in a magnetic field. Dynamics. [**720**] P. Pieranski, E. Guyon, P. Keller, J. Phys. (Paris), C-3, *37*, 133 (1976); Shear flow induced polarization in ferroelectric smectics C.

[**721**] M. Cohen, P. Pieranski, E. Guyon, C. D. Mitescu, Mol. Cryst. Liq. Cryst. *38*, 97 (1977), C.A. *86*, 181025f (1977); New type of instability in nematic CBOOA. [**721a**] L. Petit, P. Pieranski, E. Guyon, C. R. Acad. Sci., Sér. B *284*, 535 (1977), C.A. *87*, 94174f (1977); Mesure de polarisation spontanée dans les smectiques chiraux. [**722**] S. A. Pikin, Zh. Eksp. Teor. Fiz. *60*, 1185 (1971), C.A. *75*, 40568w (1971); Steady state flow of a nematic in an external electric field. [**723**] S. A. Pikin, Zh. Eksp. Teor. Fiz. *61*, 2133 (1971), C.A. *76*, 38624p (1972); High-frequency electrohydrodynamic effect in liquid crystals. [**724**] S. A. Pikin, Zh. Eksp. Teor. Fiz. *63*, 1115 (1972), C.A. *77*, 157524m (1972); Turbulent flow of liquid crystals in an electric field. [**725**] S. A. Pikin, A. A. Shtol'berg, Kristallografiya *18*, 445 (1973), C.A. *79*, 58688c (1973); Theory of the electrohydrodynamic effect in liquid crystals. [**726**] S. A. Pikin, V. L. Indenbom, Kristallografiya *20*, 1127 (1975), C.A. *84*, 82773a (1976); New type of electrohydrodynamic instability in a liquid crystal. [**727**] S. A. Pikin, G. Ryschenkow, W. Urbach, J. Phys. (Paris), *37*, 241 (1976), C.A. *84*, 158234y (1976); New type of electrohydrodynamic instability in tilted nematic layers. [**728**] S. A. Pikin, Priroda (Moscow) *1976*, 31, C.A. *85*, 114861s (1976); Liquid (crystal) ferroelectrics. [**729**] S. A. Pikin, V. G. Chigrinov, V. L. Indenbom, Mol. Cryst. Liq. Cryst. *37*, 313 (1976), C.A. *86*, 131653p (1977); New types of instabilities in liquid crystals with tilted orientation. [**730**] M. A. Piliavin, R. M. Hornreich, Mol. Cryst. Liq. Cryst. *35*, 185 (1976), C.A. *86*, 10843s (1977); Instability thresholds of nematics and smectics C in elliptically polarized rotating magnetic fields.

[**731**] P. A. Pincus, J. Appl. Phys. *41*, 974 (1970), C.A. *73*, 8646t (1970); Magnetic properties of liquid crystals. [**732**] Y. G. Pliner, Tr. Leningrad. Tekhnol. Inst. Tsellyul.-Bum. Prom. *25*, 136

(1970), C.A. *75*, 25540n (1971); Structure of liquid crystals, viscous solutions, and boundary phases induced by electric fields. [**733**] Yu. I. Plotnikov, A. V. Rakov, Yu. P. Smirnov, Mikroelektronika (Akad. Nauk SSSR) *5*, 378 (1976), C.A. *85*, 135286y (1976); Conductivity of liquid crystal cells under dynamic scattering conditions. [**734**] J. M. Pochan, P. F. Erhardt, W. C. Richards, Mol. Cryst. Liq. Cryst. *24*, 89 (1973), C.A. *81*, 142300u (1974); Mechanism of shear induced texture changes in cholesterics. Electric field effects. [**735**] J. M. Pochan, H. W. Gibson, J. Phys. Chem. *78*, 1740 (1974), C.A. *81*, 105799t (1974); Electric field effects on a uniaxial smectic phase of mixtures of cholesteryl alkanoates. [**736**] Y. Poggi, R. Aleonard, C. R. Acad. Sci., Sér. B *276*, 643 (1973), C.A. *79*, 97890g (1973); Mesure de l'anisotropie diamagnétique d'un nématique orienté par un champ magnétique. [**737**] Y. Poggi, J. Robert, R. Aleonard, C. R. Acad. Sci., Sér. B *277*, 123 (1973), C.A. *79*, 119290r (1973); Détermination du paramètre d'ordre dans les nématiques par une mesure d'anisotropie diamagnétique. Comparaison de ces résultats à ceux obtenus par la mesure des constantes élastiques du corps étudié. [**738**] Y. Poggi, R. Aleonard, J. Robert, Phys. Lett. A *54*, 393 (1975), C.A. *84*, 11110w (1976); Transition of an oriented nematic to a glass with a nematic order. Determination of the order parameter. [**739**] Y. Poggi, P. Atten, R. Aleonard, Phys. Rev. A *14*, 466 (1976), C.A. *85*, 102650j (1976); Application of the Landau approximation to nematics. [**739a**] Y. Poggi, J. C. Filippini, Phys. Rev. Lett. *39*, 150 (1977), C.A. *87*, 77580t (1977); Magnetic-field dependence of the order parameter in a nematic single crystal. [**740**] J. M. Pollack, J. B. Flannery, Liq. Cryst. Ord. Fluids *2*, 557 (1974), C.A. *87*, 31911y (1977); Domain formation in homogeneous nematics. [**740a**] V. N. Pokrovskii, Zh. Eksp. Teor. Fiz. *71*, 1880 (1976); Theory of relaxation processes in molecular liquids and liquid crystals.

[**741**] M. J. Press, A. S. Arrott, AIP Conf. Proc. *24*, 252 (1974), C.A. *83*, 125109n (1975); Magnetic field dependence of molecular orientations in liquid crystal droplets. [**742**] J. Prost, R. Canet, C. R. Acad. Sci., Sér. B *274*, 54 (1972), C.A. *76*, 104883a (1972); Structure périodique induite dans une phase nématique par un champ magnétique tournant. [**743**] J. Prost, P. S. Pershan, J. Appl. Phys. *47*, 2298 (1976), C.A. *85*, 134759t (1976); Flexoelectricity in nematics and smectics A. [**744**] P. N. Pusey, E. Jakeman, J. Phys. A: Math. Gen. *8*, 392 (1975), C.A. *82*, 147340n (1975); Non-Gaussian fluctuations in electromagnetic radiation scattered by a random phase screen. II. Application to dynamic scattering in a liquid crystal. [**745**] W. S. Quon, E. Wiener-Avnear, Solid State Commun. *15*, 1761 (1974), C.A. *82*, 132207a (1975); Normal mode oscillations of the Williams domains in thin films of nematics. [**746**] W. S. Quon, E. Wiener-Avnear, Appl. Phys. Lett. *24*, 529 (1974), C.A. *81*, 19032f (1974); Transient laser phenomena in nematics subject to a.c. electric fields. [**747**] N. V. Raghavan, M. S. Vijaya, T. S. Chang, J. Chem. Phys. *63*, 5493 (1975), C.A. *84*, 58069e (1976); Conjugative and inductive effects of cyano group on the electrooptic properties in substituted cyanophenyl benzoate and phenyl cyanobenzoate esters. [**748**] N. V. S. Rao, P. R. Kishore, T. F. S. Raj, M. N. Avadhanlu, C. R. K. Murty, Z. Naturforsch. A *31*, 283 (1976), C.A. *84*, 158428q (1976); Electric and magnetic field effects in MBBA. [**749**] N. V. S. Rao, P. R. Kishore, T. F. S. Raj, M. N. Avadhanlu, C. R. K. Murty, Mol. Cryst. Liq. Cryst. *36*, 65 (1976), C.A. *86*, 63831t (1977); Influence of electric and magnetic fields on the molecular alignment in EBBA. [**750**] A. Rapini, M. Papoular, P. Pincus, C. R. Acad. Sci., Sér. B *267*, 1230 (1968); Distorsion de structure d'un film nématique sous champ magnétique.

[**751**] A. Rapini, M. Papoular, J. Phys. (Paris), C-4, *30*, 54 (1969); Distorsion d'une lamelle nématique sous champ magnétique conditions d'ancrage aux parois. [**752**] A. Rapini, J. Phys. (Paris) *33*, 237 (1972); Instabilités magnétiques d'un smectique C. [**753**] A. Rapini, L. Léger, A. Martinet, J. Phys. (Paris), C-1, *36*, 189 (1975); Umbilics: Static and dynamic properties. [**754**] B. R. Ratna, M. S. Vijaya, R. Shashidhar, B. K. Sadashiva, Liq. Cryst., Proc. Int. Conf., Bangalore *1973*, 69, C.A. *84*, 114742h (1976); Experimental studies of short range order in nematogens of strong positive dielectric anisotropy. [**755**] B. R. Ratna, R. Shashidhar, Pramana *6*, 278 (1976), C.A. *85*, 159164h (1976); Dielectric properties of nematic 4'-alkyl-4-cyanobiphenyls. [**755a**] B. R. Ratna, R. Shashidhar, Mol. Cryst. Liq. Cryst. *42*, 113 (1977); Dielectric studies on liquid crystals of strong positive dielectric anisotropy. [**755b**] B. R. Ratna, R. Shashidhar, Mol. Cryst. Liq. Cryst. *42*, 185 (1977); Dielectric dispersion in 4'-*n*-alkyl-4-cyanobiphenyls. [**756**] J. Rault, P. E. Cladis, J. P. Burger, Phys. Lett. A *32*, 199 (1970), C.A. *73*, 71343g (1970); Ferronematics. [**757**] I. Rault, P. E. Cladis, Mol. Cryst. Liq. Cryst. *15*, 1 (1971), C.A. *75*, 144710c (1971); Cholesteric texture near T_c and in the presence of a magnetic field. [**758**] J. Rault, Mol. Cryst. Liq. Cryst. *16*, 143 (1972), C.A. *76*, 77778m (1972); Création de lignes de dislocation dans un cholésterique par l'application d'un champ magnétique. [**759**] J. Rault, Liq. Cryst. Ord. Fluids *2*, 677 (1974); Periodic distortions in cholesterics. [**760**] P. Richmond, L. R. White,

Mol. Cryst. Liq. Cryst. *27*, 217 (1974), C.A. *82*, 148585h (1975); Influence of van der Waals forces on the alignment of nematics in magnetic fields.

[**761**] R. A. Rigopoulos, H. M. Zenginoglou, Mol. Cryst. Liq. Cryst. *35*, 307 (1976), C.A. *85*, 200783z (1976); Electrohydrodynamic instability limits of nematics. [**762**] J. Robert, G. Labrunie, Centre d'Études Nucléaires de Grenoble, 1972, Note Technique LETI/ME n° 817; Variation de l'orientation collective des molécules d'un cristal liquide nématique soumises à l'effet du champ électrique en fonction du temps. [**763**] J. Robert, G. Labrunie, Centre d'Études Nucléaires de Grenoble Note Technique LETI/ME n° 841; Application des propriétés optiques des cristaux liquides a la déflection de la lumière. [**764**] J. Robert, G. Labrunie, J. Borel, Mol. Cryst. Liq. Cryst. *23*, 197 (1973), C.A. *80*, 88372h (1974); Static and transient electric field effect on homeotropic thin nematic layers. [**765**] F. Gharadjedaghi, J. Robert, Rev. Phys. Appl. *10*, 69 (1975); Optical and electro-optical behavior of a liquid crystal helicoidal structure. [**766**] F. Gharadjedaghi, J. Robert, Rev. Phys. Appl. *10*, 281 (1975); Étude mathémati-que de la réponse optique d'une structure nématique en hélice distordue par un champ normal a son axe. [**767**] R. C. Robinder, Diss. Duke Univ., Durham, N. C. 1973, 243 pp. Avail. Univ. Microfilms, Ann Arbor, Mich., Order No. 73-19,503, C.A. *80*, 8063v (1974); Low-frequency electrical and electrooptical studies on nonoriented thick films of four liquid crystals. [**768**] F. Rondelez, Diss. Univ. de Paris, Faculté des Sciences d'Orsay, 1970; Contribution à l'étude des instabilités électrohydrodynamiques dans les cristaux liquides nématiques. [**769**] F. Rondelez, H. Arnould, C. R. Acad. Sci., Sér. B *273*, 549 (1971), C.A. *76*, 19065r (1972); Déformations de la texture planaire d'un cholestérique à grand pas sous l'action d'un champ électrique. [**770**] F. Rondelez, J. P. Hulin, Solid State Commun. *10*, 1009 (1972), C.A. *77*, 40660m (1972); Distortions of a planar cholesteric induced by a magnetic field.

[**771**] F. Rondelez, Solid State Commun. *11*, 1675 (1972), C.A. *78*, 63738c (1973); Conductance measurements above a s_C/n transition. [**772**] F. Rondelez, H. Arnould, C. J. Gerritsma, Phys. Rev. Lett. *28*, 735 (1972), C.A. *77*, 157526p (1972); Electrohydrodynamic effects in cholesterics under a.c. electric fields. [**773**] H. Arnould-Netillard, F. Rondelez, Mol. Cryst. Liq. Cryst. *26*, 11 (1974), C.A. *81*, 127849m (1974); Electrohydrodynamic instabilities in cholesterics with negative dielectric anisotropy. [**774**] F. Rondelez, A. Mircea-Roussel, Mol. Cryst. Liq. Cryst. *28*, 173 (1974), C.A. *82*, 163837k (1975); Dielectric relaxation in the radio frequency range for nematic MBBA. [**775**] F. Rondelez, Philips Res. Rep., Suppl. 1974, 157 pp., C.A. *82*, 10181s (1975); Effets de champ dans les nématiques et cholestériques. [**776**] P. I. Rose, Mol. Cryst. Liq. Cryst. *26*, 75 (1974), C.A. *81*, 127850e (1974); Determination of the diamagnetic anisotropy of nematic MBBA. [**777**] L. Royer, J. phys. radium *5*, 208 (1924), C.A. *18*, 3309[7] (1924); États mésomorphes et la biréfringence magnétique. [**778**] V. G. Rumyantsev, P. D. Berezin, L. M. Blinov, I. N. Kompanets, Kristallografiya *18*, 1104 (1973), C.A. *80*, 7873k (1974); Orientation order and molecular parameters of liquid crystals with positive dielectric anisotropy. [**779**] E. I. Ryumtsev, Akust. Zh. *20*, 484 (1974); Relaxation effects in liquid crystals in electric fields. [**780**] E. Sackmann, S. Meiboom, L. C. Snyder, A. E. Meixner, R. E. Dietz, J. Am. Chem. Soc. *90*, 3567 (1968), C.A. *69*, 31153f (1968); Structure of the liquid crystalline state of cholesterol derivatives.

[**781**] H. U. Rega, E. Sackmann, Ber. Bunsenges. Phys. Chem. *78*, 915 (1974), C.A. *82*, 10283b (1975); Kinetics of texture changes in electric field-induced c/n transitions. [**782**] S. Sakagami, A. Takase, M. Nakamizo, Bull. Chem. Soc. Jap. *46*, 3573 (1973), C.A. *80*, 75337a (1974); Deformation of the molecular orientation of homeotropic nematics in an electric field. [**783**] S. Sakagami, A. Takase, M. Nakamizo, Chem. Phys. Lett. *38*, 547 (1976), C.A. *84*, 186969k (1976); Electro-optic effects in smectics C. [**784**] S. Sakagami, A. Takase, M. Nakamizo, Chem. Lett. *1975*, 95, C.A. *82*, 178658b (1975); Anomalous electrooptic effect in nematic 4-pentyloxybenzylidene-4-heptylaniline. [**785**] A. Sakamoto, K. Yoshino, U. Kubo, Y. Inuishi, Jpn. J. Appl. Phys. *15*, 545 (1976), C.A. *84*, 187781e (1976); Effects of the magnetic field on the s_A/n transition temperature. [**786**] A. Sakamoto, K. Yoshino, U. Kubo, Y. Inuishi, Jpn. J. Appl. Phys. *15*, 745 (1976), C.A. *84*, 172364x (1976); Temperature dependence of critical magnetic field on Freedericksz transition near s_A/n transition. [**786a**] A. Sakamoto, U. Kubo, K. Yoshino, Y. Inuishi, Mol. Cryst. Liq. Cryst. *43*, 249 (1977); Magnetic field dependence of laser light scattering through nematic CBOOA. [**786b**] A. Sakamoto, K. Yoshino, U. Kubo, Y. Inuishi, Technol. Rep. Osaka Univ. *27*, 129 (1977), C.A. *86*, 180325s (1977); Magnetic field dependence of laser light scattering through nematic CBOOA. [**787**] I. D. Samodurova, A. S. Sonin, Uch. Zap., Ivanov. Gos. Pedagog. Inst. *1972*, No. 99, 89, C.A. *78*, 141304z (1973); Domain structure of PAA. [**788**] I. D. Samodurova, A. S. Sonin, Fiz. Tverd. Tela *15*, 2359 (1973), C.A. *79*, 119288w (1973); Domain structure of nematic PAA studied by light diffraction. [**789**] I. D. Samodurova, Deposited Publ. VINITI 5464-73, 11 pp. (1972), C.A. *85*, 27529w (1976); Study of the domain structure of nematic PAA using light diffraction.

[**790**] I. D. Samodurova, Deposited Publ. VINITI 5463-73, 13 pp. (1972), C.A. *85*, 26888a (1976); Study of the angular distribution of the intensity of light scattered by liquid crystals under dynamic scattering conditions.
[**791**] I. D. Samodurova, A. S. Sonin, A. B. Uspenskii, Opt. Spektrosk. *36*, 1165 (1974), C.A. *81*, 112203x (1974); Angular distribution of the intensity of light scattered by liquid crystals under conditions of dynamic scattering. [**792**] I. D. Samodurova, A. S. Sonin, Fiz. Tverd. Tela *16*, 255 (1974), C.A. *80*, 113515s (1974); Dynamic scattering and concentration of scattering centers in nematic PAA. [**793**] A. T. Sarkisyan, A. E. Dingchyan, Izv. Akad. Nauk Arm. SSR, Fiz. *11*, 146 (1976), C.A. *86*, 11009m (1977); Electroconductivity of a liquid crystal of MBBA and 2,2,6,6-tetramethyl-4-oxypiperidin-1-oxyl. [**794**] A. T. Sarkisyan, V. K. Mirzoyan, A. E. Dingchyan, S. S. Arakelyan, Izv. Akad. Nauk Arm. SSR, Fiz. *11*, 399 (1976), C.A. *86*, 130220h (1977); Effect of paramagnetic centers on dynamic scattering of light by MBBA and EBBA. [**794a**] A. T. Sarkisyan, A. E. Dingchyan, S. S. Arakelyan, L. L. Zaraelyan, Izv. Akad. Nauk Arm. SSR. Fiz. *11*, 476 (1976), C.A. *87*, 14457n (1977); Effect of some additives on the domain structure parameters of MBBA. [**795**] G. Sarma, Solid State Commun. *10*, 1049 (1972); Non linear response of two dimensional nematics to an applied field. [**796**] H. Sasabe, K. Ooizumi, Jap. J. Appl. Phys. *11*, 1750 (1972), C.A. *78*, 63768n (1973); Anomalous dielectric absorption in MBBA at the n/i transition. [**797**] M. Sato, K. Hirakawa, J. Phys. Soc. Jpn. *42*, 433 (1977), C.A. *86*, 131384b (1977); Observation of linear and nonlinear relaxations of fluid dynamical motion in a liquid crystal. [**798**] S. Sato, M. Wada, Jap. J. Appl. Phys. *10*, 1106 (1971), C.A. *75*, 123407u (1971); c/n Transitions in compensated liquid crystals. [**799**] S. Sato, M. Wada, Jap. J. Appl. Phys. *11*, 1566 (1972), C.A. *77*, 170108y (1972); Molecular orientation effects in compensated liquid crystals. [**800**] S. Sato, M. Wada, Proc. Conf. Solid State Devices, 4th, *1972*, 276, C.A. *79*, 140675e (1973); Optical and electrical properties of liquid crystal mixtures.
[**801**] P. Sengupta, A. Saupe, Phys. Rev. A *9*, 2698 (1974), C.A. *81*, 55274v (1974); Bend-shear-mode instabilities of nematics in a.c. fields. [**802**] M. Schadt, W. Helfrich, Appl. Phys. Lett. *18*, 127 (1971), C.A. *75*, 11713j (1971); Voltage-dependent optical activity of a twisted nematic. [**803**] M. Schadt, J. Chem. Phys. *56*, 1494 (1972), C.A. *76*, 64939q (1972); Dielectric properties of some nematics with strong positive dielectric anisotropy. [**804**] M. Schadt, W. Helfrich, Mol. Cryst. Liq. Cryst. *17*, 355 (1972), C.A. *77*, 67385n (1972); Kerr effect in the i phase of some nematogens. [**805**] A. Boller, H. Scherrer, M. Schadt, P. Wild, Proc. IEEE *60*, 1002 (1972), C.A. *77*, 131928w (1972); Low electrooptic threshold in new liquid crystals. [**806**] M. Schadt, C. Von Planta, J. Chem. Phys. *63*, 4379 (1975), C.A. *84*, 68261a (1976); Conductivity relaxation in positive dielectric liquid crystals. [**807**] M. Schadt, Phys. Lett. A *57*, 442 (1976), C.A. *85*, 151737b (1976); Solute-induced transmission changes in liquid crystal twist cells. [**808**] M. Schadt, 7. Arbeitstagung Flüssigkristalle, Inst. Angew. Festkörperphysik Freiburg, 4. 3. 1977; Static and dynamic Kerr effect in the i phase of strongly positive dielectric liquid crystals. [**808a**] M. Schadt, J. Chem. Phys. *67*, 210 (1977), C.A. *87*, 93768r (1977); Kerr effect and orientation relaxation of pretransitional domains and individual molecules in positive dielectric liquid crystals. [**809**] T. J. Scheffer, Phys. Rev. Lett. *28*, 593 (1972), C.A. *76*, 104891b (1972); Electric and magnetic field investigations of the periodic gridlike deformation of a cholesteric. [**810**] T. J. Scheffer, 3. Arbeitstagung Flüssigkristalle, Inst. Angew. Festkörperphysik Freiburg, 11. 5. 1973; Influence of electric fields on cholesterics.
[**811**] A. F. Schenz, V. D. Neff, T. W. Schenz, Mol. Cryst. Liq. Cryst. *23*, 59 (1973), C.A. *79*, 150840u (1973); Faraday rotation in nematics. [**812**] A. F. Schenz, Diss. Kent State Univ., Kent, Ohio. 1974, 115 pp. Avail. Univ. Microfilms, Ann Arbor, Mich., Order No. 74-19,402, C.A. *82*, 9476k (1975); Optical and Faraday studies of liquid crystals. [**813**] M. F. Schiekel, K. Fahrenschon, Appl. Phys. Lett. *19*, 391 (1971), C.A. *76*, 51400g (1972); Deformation of nematics with vertical orientation in electric fields. [**814**] M. Schiekel, K. Fahrenschon, 2. Arbeitstagung Flüssigkristalle, Inst. Angew. Festkörperphysik Freiburg, 21. 4. 1972; Deformation aufgerichteter Phasen nematischer Flüssigkristalle als Basis für elektronisch durchstimmbare optische Vielpunkte-Interferenzfilter. [**815**] M. Schiekel, K. Fahrenschon, H. Gruler, 3. Arbeitstagung Flüssigkristalle, Inst. Angew. Festkörperphysik Freiburg, 11. 5. 1973; Ansprechzeiten und Multiplexverhalten von nematischen Flüssigkristallen. [**816**] M. Schiekel, Int. Elektron. Rundschau *27*, 184, 203 (1973); Verhalten nematischer Flüssigkristalle im elektrischen Feld. [**817**] K. Fahrenschon, H. Gruler, M. Schiekel, 4. Arbeitstagung Flüssigkristalle, Inst. Angew. Festkörperphysik Freiburg, 26. 4. 1974; Einfluß der elektrischen Leitfähigkeit auf die dielektrische Deformation von nematischen Flüssigkristallen. [**818**] H. Gruler, M. Schiekel, K. Fahrenschon, Wiss. Ber. AEG-Telefunken *48*, 133 (1975), C.A. *84*, 10552e (1976); Dielektrische Ausrichtung elektrisch leitender

nematischer Flüssigkristalle. [**819**] K. Fahrenschon, H. Gruler, M. F. Schiekel, Appl. Phys. *11*, 67 (1976), C.A. *85*, 152095c (1976); Deformation of a pretilted nematic layer in an electric field. [**819a**] A. Schiraldi, G. Chiodelli, J. Phys. E *10*, 596 (1977), C.A. *87*, 109657s (1977); Determination of phase transitions through a.c. conductance. [**820**] F. Schneider, Z. Naturforsch. A *28*, 1660 (1973), C.A. *80*, 64597j (1974); Bestimmung von Elastizitätskoeffizienten einer nematischen Flüssigkeit durch elektrische Leitfähigkeitsmessung.

[**821**] F. Schneider, Ber. Bunsenges. Phys. Chem. *81*, 66 (1977), C.A. *86*, 99268d (1977); Elastisches Verhalten und Orientierung induzierter cholesterischer Flüssigkristalle im Magnetfeld. [**822**] F. Scudieri, N. Bertolotti, J. Opt. Soc. Am. *64*, 776 (1974), C.A. *81*, 55928m (1974); Irradiance statistics of light scattered by electrohydrodynamic fluctuations in MBBA. [**823**] F. Scudieri, M. Bertolotti, R. Bartolino, Appl. Opt. *13*, 181 (1974); Light scattered by a liquid crystal: new quasi-thermal source. [**824**] P. Sengupta, Diss. Kent State Univ., Kent, Ohio. 1973. 103 pp. Avail. Univ. Microfilms, Ann Arbor, Mich., Order No. 74-15,086, C.A. *81*, 96555f (1974); Theoretical studies of the electrohydrodynamic instabilities in nematics. [**825**] D. G. Shaw, Diss. Northwest. Univ. Evanston, Ill., 1971, 90 pp. Avail. Univ. Microfilms, Ann Arbor, Mich., Order No. 71-30,947, C.A. *76*, 91517g (1972); Electrical properties of cholesterics. [**826**] D. G. Shaw, J. W. Kauffman, J. Chem. Phys. *54*, 2424 (1971), C.A. *74*, 104363q (1971); Electric conductivity in cholesterics. [**827**] D. G. Shaw, J. W. Kauffman, Phys. Status Solidi A *4*, 467 (1971), C.A. *74*, 117086b (1971); Current-voltage measurements on cholesterics. [**828**] D. G. Shaw, J. W. Kauffman, Phys. Status Solidi A *12*, 637 (1972), C.A. *77*, 119553n (1972); Dielectric properties of cholesterics. [**829**] J. P. Sheridan, J. M. Schnur, T. G. Giallorenzi, Appl. Phys. Lett. *22*, 560 (1973), C.A. *79*, 25149z (1973); Electrooptic switching in low-loss liquid-crystal waveguides. [**830**] J. P. Sheridan, T. G. Giallorenzi, J. Appl. Phys. *45*, 5160 (1974), C.A. *82*, 79504s (1975); Electrooptically induced deflection in liquid-crystal waveguides.

[**831**] S. Shtrikman, E. P. Wohlfahrt, Y. Wand, Phys. Lett. A *37*, 369 (1971), C.A. *76*, 78500b (1972); Magnetic field dependence of the capacity of a TN cell. [**832**] R. D. Shulvas-Sorokina, M. V. Posnova, Physik. Z. Sowjetunion *8*, 319 (1935), C.A. *30*, 1277^9 (1936); Struktur anisotroper Flüssigkeiten. [**833**] I. V. Shuvalov, Uch. Zap., Ivanov. Gos. Pedagog. Inst. *1972*, No. 99, 95, C.A. *78*, 89808a (1973); Temperature dependence of the electric conductivity of cholesterics. [**834**] J. Sicart, J. Phys. (Paris), Lett. *37*, 25 (1976), C.A. *84*, 82799p (1976); Method of measuring the anchoring energy of a nematic. Application to anchoring on an untreated plate. [**835**] G. Siguad, H. Gasparoux, J. Chim. Phys. Physicochim. Biol. *70*, 699 (1973), C.A. *79*, 97982p (1973); Étude comparée des propriétés magnétiques du MBBA présentant des temperatures de transition n/i différentes. [**835a**] P. Simova, M. Petrov, N. Kirov, Mol. Cryst. Liq. Cryst. *42*, 295 (1977), C.A. *87*, 192275j (1977); Coherent light diffraction in a smectic C. [**836**] A. Smilgevicius, R. Zilenas, A. Sarpis, R. Baltrusaitis, Uch. Zap., Ivanov. Gos. Pedagog. Inst. *1972*, No. 99, 117, C.A. *78*, 103195g (1973); Electrooptical characteristics of a nematic thin film. [**837**] A. Smilgevicius, V. Kazeliene, R. Baltrusaitis, P. Adomenas, Liet. Fiz. Rinkinys *12*, 869 (1972), C.A. *79*, 37102s (1973); Electrooptical properties of liquid crystals. [**838**] A. Smilgevicius, R. Baltrusaitis, V. Zurauskiene, V. Kazeliene, Sb. Dokl. Vses. Nauch. Konf. Zhidk. Krist. Simp. Ikh Prakt. Primen., 2nd, *1972*, 275, C.A. *81*, 97722v (1974); Electrooptical effects in mixtures of nematics and cholesterics. [**839**] R. E. Michel, G. W. Smith, J. Appl. Phys. *45*, 3234 (1974), C.A. *81*, 142816s (1974); Dependence of birefringence threshold voltage on dielectric anisotropy in a nematic. [**840**] I. W. Smith, Y. Galerne, S. T. Lagerwall, E. Dubois-Violette, G. Durand, J. Phys. (Paris), C-1, *36*, 237 (1975); Dynamics of electrohydrodynamic instabilities in nematics.

[**841**] C. P. Smyth, Dielectric Behavior and Structure. (McGraw-Hill: New York). 1955. 441 pp., C.A. *49*, 12106d (1955). [**842**] S. Sokerov, N. Aneva, P. Stefanov, A. Petrov, S. Stoilov, V. Encheva, Izv. Khim. *9*, 304 (1976), C.A. *86*, 113880f (1977); Electrooptical study of the ZLI-207 nematic. [**843**] R. A. Soref, Appl. Phys. Lett. *22*, 165 (1973), C.A. *78*, 103865a (1973); Transverse field effects in nematics. [**844**] R. A. Soref, M. J. Rafuse, J. Appl. Phys. *43*, 2029 (1972); Electrically controlled birefringence of thin nematic films. [**845**] R. A. Soref, J. Appl. Phys. *45*, 5466 (1974), C.A. *82*, 105151w (1975); Field effects in nematics obtained with interdigital electrodes. [**846**] B. Specht, Diss. Halle 1908; Dielektrizitätskonstante flüssiger Kristalle. [**846a**] G. Spott, W. Thiel, D. Demus, Wiss. Z. Martin-Luther-Univ., Halle-Wittenberg, Math.-Naturwiss. Reihe *26*, 111 (1977), C.A. *87*, 109675w (1977); Messung der Verteilung des elektrischen Potentials in Flüssigkristallen. [**847**] G. J. Sprokel, J. Electrochem. Soc. *119*, 241C (1972); Effect of electrode space charge on the measurement of conductivity and permittivity of nematics. [**848**] G. J. Sprokel, Mol. Cryst. Liq. Cryst. *22*, 249 (1973), C.A. *79*, 140636t (1973); Resistivity, permittivity and the electrode space charge of nematics. [**849**] G. J. Sprokel, Mol. Cryst. Liq. Cryst. *26*, 45 (1974),

C.A. *81*, 112240g (1974); Conductivity, permittivity, and the electrode spacecharge of nematics. [**850**] G. J. Sprokel, Mol. Cryst. Liq. Cryst. *29*, 231 (1975), C.A. *83*, 36384f (1975); Anisotropy changes in the c/n transition.

[**851**] M. Steers, A. Mircea-Roussel, J. Phys. (Paris), C-3, *37*, 145 (1976), C.A. *86*, 49162u (1977); Novel electro-optic storage mode in smectics A. [**851a**] P. Stepanek, B. Sedlacek, Mol. Cryst. Liq. Cryst. *43*, 197 (1977); Intensity and anisotropy of the dynamic light scattering in nematics. [**852**] T. W. Stinson, III, J. D. Litster, Phys. Rev. Lett. *25*, 503 (1970), C.A. *73*, 81751c (1970); Pretransitional phenomena in the i phase of a nematic. [**853**] S. S. Sukiasyan, S. A. Akopyan, A. T. Sarkisyan, S. Z. Petrosyan, Izv. Akad. Nauk Arm. SSR, Fiz. *11*, 72 (1976), C.A. *85*, 135354u (1976); Temperature and frequency dependence of permittivity and the angle of dielectric losses of cholesteryl capronate and cholesteryl acetate. [**854**] M. Sukigara, O. Nagasaki, K. Honda, Bull. Chem. Soc. Jap. *45*, 959 (1972), C.A. *76*, 132731t (1972); Orientation patterns in nematics. [**855**] A. Sussman, Mol. Cryst. Liq. Cryst. *14*, 183 (1971), C.A. *75*, 113294r (1971); Ionic equilibrium and ionic conductance in the system tetraisopentylammonium nitrate/PAA. [**856**] D. Meyerhofer, A. Sussman, Appl. Phys. Lett. *20*, 337 (1972); Electrohydrodynamic instabilities in nematics in low-frequency fields. [**857**] A. Sussman, Appl. Phys. Lett. *21*, 126 (1972), C.A. *77*, 119293c (1972); Dynamic scattering life in nematic APAPA as influenced by current density. [**858**] A. Sussman, Appl. Phys. Lett. *21*, 269 (1972), C.A. *77*, 143878t (1972); Secondary hydrodynamic structure in dynamic scattering. [**859**] D. A. Sussman, 3rd ACS Symp. Ord. Fluids Liquid Cryst., Chicago, Aug. 1973, Abstr. No. 31; Secondary hydrodynamic structure in dynamic scattering. II. Field-flow-induced off-state birefringence. [**860**] A. Sussman, Introd. Liq. Cryst. *1974*, 319, C.A. *84*, 81456n (1976); Electrochemistry in nematic solvents.

[**861**] A. Sussmann, Appl. Phys. Lett. *29*, 633 (1976); Electrohydrodynamic instabilities in nematics of positive dielectric anisotropy. [**862**] T. Svedberg, Ark. Mat. Astron. Fysik. *9*, No. 9 and 21 (1913/14), C.A. *8*, 1372⁵ and 2097⁶ (1914); Elektrisches Leitvermögen anisotroper Flüssigkeiten in magnetischen und elektrischen Feldern. [**863**] T. Svedberg, Ann. Physik. *44*, 1121 (1914), C.A. *9*, 11² (1915); Elektrizitäts-Leitung in anisotropen Flüssigkeiten. [**864**] T. Svedberg, Kolloid-Z. *16*, 103 (1915), C.A. *9*, 2830⁴ (1915); Struktur kristalliner Flüssigkeiten. [**865**] T. Svedberg, Jahrb. Radioakt. Elektronik *12*, 129 (1915), C.A. *10*, 2550 (1916); Methode zur Ermittlung der geometrischen Dissymmetrie der Moleküle. [**866**] T. Svedberg, Ann. Physik *49*, 437 (1916), C.A. *11*, 1078⁹ (1917); Elektrizitäts-Leitung in anisotropen Flüssigkeiten. [**867**] G. Szivessy, Z. Physik *34*, 474 (1925), C.A. *20*, 1752⁷ (1926); Bornsche Dipoltheorie anisotroper Flüssigkeiten. [**868**] G. Szivessy, Z. Physik *38*, 159 (1926), C.A. *21*, 2827² (1927); Bornsche Dipoltheorie anisotroper Flüssigkeiten. Erwiderung auf Kritik von W. Kast. [**869**] A. Szymanski, 3rd Int. Liq. Cryst. Conf., Berlin, 1970, S. 7.12; Pyroelectric effects in liquid crystals. [**870**] T. Tachibana, T. Takamatsu, E. Fukada, Chem. Lett. *1973*, 907, C.A. *79*, 109059h (1973); Thermally stimulated current in smectics.

[**871**] A. Takase, S. Sakagami, M. Nakamizo, Jpn. J. Appl. Phys. *12*, 1255 (1973), C.A. *79*, 109824d (1973); Light diffraction in a nematic with positive dielectric anisotropy. [**872**] A. Takase, S. Sakagami, M. Nakamizo, Jpn. J. Appl. Phys. *14*, 228 (1975), C.A. *83*, 51277s (1975); Light diffraction and light scattering in nematics with a positive dielectric anisotropy. [**873**] C. Tani, Appl. Phys. Lett. *19*, 241 (1971), C.A. *76*, 7682b (1972); Novel electrooptical storage effect in a smectic. [**874**] D. T. Teaney, A. Migliori, J. Appl. Phys. *41*, 998 (1970), C.A. *72*, 137308n (1970); Current- and magnetic-field-induced order and disorder in ordered nematics. [**875**] I. Teucher, M. M. Labes, J. Chem. Phys. *54*, 4130 (1971), C.A. *75*, 11710f (1971); Magnetic field effects on the dynamic scattering threshold in a nematic. [**876**] V. V. Titov, Y. N. Gerulaitis, N. T. Lazareva, E. I. Balabanov, A. I. Vasil'ev, Y. M. Bunakov, M. F. Grebenkin, K. V. Roitman, Sb. Dokl. Vses. Nauch. Konf. Zhidk Krist. Simp. Ikh Prakt. Primen., 2nd, *1972*, 178, C.A. *81*, 142663q (1974); Synthesis and purification of liquid crystal-azo and azoxy compounds and of Schiff bases for electrooptical apparatus. [**877**] K. Toda, S. Nagaura, Mem. Fac. Eng., Osaka City Univ. *16*, 97 (1975), C.A. *86*, 11044u (1977); Electric conductivity in nematics. [**878**] A. V. Tolmachev, A. S. Sonin, Fiz. Tverd. Tela *17*, 3096 (1975); Optical properties of a cholesteric layer with account of dielectric boundaries. [**879**] N. A. Tolstoi, J. Exptl. Theoret. Phys. *17*, 724 (1947), C.A. *42*, 5322g (1948); Long-range fluctuations in the i phase of PAA. [**880**] N. A. Tolstoi, Zhur. Eksptl. Teoret. Fiz. *19*, 319 (1949), C.A. *45*, 9954g (1951); Kerr effect in dilute solutions of PAA and PAP.

[**881**] K. Toriyama, M. Koga, S. Nomura, Jap. J. Appl. Phys. *8*, 498 (1969), C.A. *71*, 8359k (1969); Temperature dependence of transmitted light and electrooptic effect by PAA. [**882**] K. Toriyama, T. Aoyagi, S. Nomura, Jap. J. Appl. Phys. *9*, 584 (1970), C.A. *73*, 49662p (1970); Mixed liquid crystals

with new electro-optic effect. [**883**] K. Toriyama, Jap. J. Appl. Phys. *9*, 1190 (1970), C.A. *73*, 114060g (1970); Optical transient behavior of nematics in an electric field. [**884**] H. Tsuchiya, K. Nakamura, Mol. Cryst. Liq. Cryst. *29*, 89 (1974), C.A. *82*, 163363c (1975); Effects of magnetic field on domain formation in nematic MBBA. [**885**] V. Tsvetkov, Acta Physicochim. U.R.S.S. *6*, 865 (1937), C.A. *32*, 4026[8] (1938); Einfluß magnetischer und elektrischer Felder auf anisotrop-flüssige Mischungen. [**886**] V. Tsvetkov, Acta Physicochim. U.R.S.S. *6*, 885 (1937), C.A. *32*, 4027[2] (1938); Ursache der Bewegung der anisotropen Flüssigkeiten im elektrischen Felde. [**887**] V. N. Tsvetkov, J. Exptl. Theoret. Phys. U.S.S.R. *8*, 855 (1938), C.A. *33*, 5293[1] (1939); Dispersion of light in anisotropic liquids. [**888**] V. N. Tsvetkov, Acta Physicochim. U.R.S.S. *10*, 555 (1939), C.A. *33*, 6672[9] (1939); Bewegung anisotroper Flüssigkeiten in einem rotierenden Magnetfeld. [**889**] V. N. Tsvetkov, Acta Physicochim. U.R.S.S. *11*, 537, J. Exptl. Theoret. Phys. U.S.S.R. *9*, 947 (1939), C.A. *34*, 3148[2] (1940); Optische Eigenschaften anisotrop-flüssiger Schichten in einem rotierenden Magnetfeld. [**890**] V. Tsvetkov, A. Sosnovskii, Acta Physicochim. U.R.S.S. *18*, 358, J. Exptl. Theoret. Phys. U.S.S.R. *13*, 353 (1943), C.A. *39*, 848[7] (1945); Diamagnetic anisotropy of crystalline liquids.

[**891**] V. N. Tsvetkov, J. Exptl. Theoret. Phys. U.S.S.R. *14*, 35 (1944), Acta Physicochim. U.R.S.S. *19*, 86 (1944), C.A. *39*, 1336[8] (1945); Magnetic and dynamic double refraction in the i phase of mesogens. [**892**] V. N. Tsvetkov, V. Marinin, Zhur. Eksptl. Teoret. Fiz. *18*, 641 (1948), C.A. *43*, 3675b (1949); Dipole moments of molecules of liquid crystals and electric birefringence of their solutions. [**893**] V. N. Tsvetkov, E. I. Ryumtsev, Dokl. Akad. Nauk SSSR *176*, 382 (1967), C.A. *68*, 55044b (1968); Orientation fluctuations in the i phase of mesogens and their electrooptical properties. [**894**] V. N. Tsvetkov, E. I. Ryumtsev, Kristallografiya *13*, 290 (1968), C.A. *69*, 5911g (1968); Pretransition phenomena and electrooptical properties of liquid crystals. [**895**] V. N. Tsvetkov, E. I. Ryumtsev, Tepl. Dvizhenie Mol. Mezhmol. Vzaimodeistvie Zhidk. Rastvorakh *1969*, 278, C.A. *74*, 58182e (1971); Pretransition phenomenon and electrooptical properties of liquid crystals. [**896**] E. I. Ryumtsev, V. N. Tsvetkov, Opt. Spektrosk. *26*, 607 (1969), C.A. *71*, 34401n (1969); Electrooptical properties of the i phase of mesogens. [**897**] V. N. Tsvetkov, E. I. Ryumtsev, I. P. Kolomiets, Dokl. Akad. Nauk SSSR *189*, 1310 (1969), C.A. *72*, 94271a (1970); Orientation of anisotropic-liquid anisalaminoazobenzene in electric and magnetic fields. [**898**] V. N. Tsvetkov, Kristallografiya *14*, 681 (1969), C.A. *71*, 106218h (1969); Hindered rotational motion of molecules, and dielectric anisotropy of nematics. [**899**] V. N. Tsvetkov, Vestn. Leningrad. Univ., Fiz., Khim. *1970*, 26, C.A. *73*, 29830p (1970); Theory of the dielectric anisotropy of nematics. [**900**] V. N. Tsvetkov, Kolloid. Zh. *33*, 154 (1971), C.A. *74*, 104254e (1971); Theory of dielectric anisotropy of nematics.

[**901**] V. N. Tsvetkov, E. I. Ryumtsev, I. P. Kolomiets, A. P. Kovshik, Dokl. Akad. Nauk SSSR *203*, 1122 (1972), C.A. *77*, 26025b (1972); Dielectric properties and anisotropy of molecular rotation in nematic anisalaminobenzene. [**902**] E. I. Ryumtsev, M. V. Mukhina, V. N. Tsvetkov, Dokl. Akad. Nauk. SSSR *204*, 397 (1972), C.A. *77*, 106488v (1972); Electrooptical properties of an i phase of cholesterics. [**903**] V. N. Tsvetkov, I. P. Kolomiets, E. I. Ryumtsev, F. M. Aliev, Dokl. Akad. Nauk SSSR *209*, 1074 (1973), C.A. *79*, 150939h (1973); Rotating magnetic field as a method for determining the diamagnetic anisotropy of nematics. [**904** V. N. Tsvetkov, E. I. Ryumtsev, I. P. Kolomiets, A. P. Kovshik, Dokl. Akad. Nauk SSSR *211*, 821 (1973), C.A. *79*, 130348m (1973); Macroscopic equivalence and difference in the molecular mechanisms of the orienting action of electric and magnetic fields on nematics. [**905**] V. A. Tsvetkov, N. A. Morozov, M. I. Elinson, Mikroelektronika (Akad. Nauk SSSR) *3*, 160 (1974), C.A. *81*, 18698r (1974); Dynamic properties of the twisted structure of a liquid crystal. [**906**] V. A. Tsvetkov, N. A. Morozov, L. K. Vistin, Kristallografiya *19*, 305 (1974), C.A. *81*, 18669g (1974); Optical activity and electrooptical properties of liquid crystals with twisted structure. [**907**] V. N. Tsvetkov, E. I. Ryumtsev, A. P. Kovshik, G. I. Denis, J. Daugvila, Dokl. Akad. Nauk SSSR *216*, 1105 (1974), C.A. *81*, 113080s (1974); Kerr effect in the i phase of nematics containing a cyano group in the molecules. [**908**] V. N. Tsvetkov, E. I. Ryumtsev, A. P. Kovshik, I. P. Kolomiets, Kristallografiya *20*, 865 (1975), C.A. *83*, 186951c (1975); Radio frequency dispersion of the dielectric constant perpendicular component in liquid-crystal α-cyanostilbenes. [**909**] V. N. Tsvetkov, A. P. Kovshik, E. I. Ryumtsev, I. P. Kolomiets, M. A. Makar'ev, Y. Y. Daugvila, Dokl. Akad. Nauk SSSR *222*, 1393 (1975), C.A. *83*, 156421n (1975); Dielectric relaxation in liquid-crystalline α-cyanostilbene. [**909a**] A. S. Lagunov, V. S. Samsonov, V. A. Tsvetkov, Sb. Dokl. I Vses. Simpoz. po Akustooptich. Spektroskopii *1976*, 103, C.A. *86*, 182021p (1977); Magnetoacoustic phenomena in solutions of nematics. [**910**] J. S. van der Lingen, Verh. deut. physik. Ges. *15*, 913 (1913), C.A. *8*, 602[5] (1914); Molekularer Bau flüssiger Kristalle.

[911] C. Z. van Doorn, Phys. Lett. A *42*, 537 (1973), C.A. *78*, 116940w (1973); Magnetic threshold for the alignment of a twisted nematic. [912] C. Z. van Doorn, 4. Arbeitstagung Flüssigkristalle, Inst. Angew. Festkörperphysik Freiburg, 26. 4. 1974; Transient behavior of a TN layer in switched electric or magnetic fields. [913] J. P. van Meter, R. T. Klingbiel, D. J. Genova, Solid State Commun. *16*, 315 (1975), C.A. *83*, 20096c (1975); Dielectric properties of several mesomorphic esters. [914] A. van Wyk, Ann. Physik *3*, 879 (1929), C.A. *24*, 1776⁹ (1930); Orientierende Einflüsse von Magnetfeld, Wand und gegenseitige Wechselwirkung auf die Schwärme des nematischen PAA. [915] J. C. Varney, L. E. Davis, Electron. Lett. *10*, 331 (1974); Structural oscillation in a c/n mixture. [916] W. E. Vaughan, G. P. Johar, Digest of literature on dielectrics, Vol. 37 (Nat. Acad. Sci.: Washington). 1975. 740 pp., C.A. *85*, 135898z (1976). [917] V. G. Veselago, Y. V. Korobkin, Y. S. Leonov, Pis'ma Zh. Eksp. Teor. Fiz. *17*, 552 (1973), C.A. *79*, 47317e (1973); Effect of a magnetic field on light scattering in nematics. [918] G. Vieth, Diss. Halle 1910; Magnetische Drehung der Polarisationsebene in flüssigen Kristallen. [919] G. Vieth, Physik. Z. *11*, 526 (1910), C.A. *4*, 2232¹ (1910); Magnetische Drehung der Polarisationsebene in flüssigen Kristallen. [920] G. Vieth, Physik. Z. *12*, 546 (1911), C.A. *5*, 3747⁸ (1911); Einfluß eines magnetischen Feldes auf flüssige Kristalle. [920a] J. L. Viovy, M. Laurent, L. Soulie, R. Viovy, J. Phys. (Paris) *38*, 877 (1977), C.A. *87*, 77051w (1977); Dielectric properties of compounds of the *p*-alkoxy-phenylazo-*p'*-phenyl and *p*-alkoxyphenylazoxy-*p'*-phenyl esters series.

[921] L. K. Vistin, Kristallografiya *15*, 594 (1970), C.A. *73*, 102851c (1970); Electrostructural effect and optical properties of a specific class of liquid crystals and their binary mixtures. [922] L. K. Vistin, Dokl. Akad. Nauk SSSR *194*, 1318 (1970), C.A. *74*, 35842f (1971); New electrostructural phenomenon in nematics. [923] L. K. Vistin, Kristallografiya *17*, 842 (1972), C.A. *77*, 145152a (1972); Dielectric hysteresis in liquid single crystals. [924] L. K. Vistin, S. P. Chumakova, Sb. Dokl. Vses. Nauch. Konf. Zhidk. Krist. Simp. Ikh Prakt. Primen., 2nd, *1972*, 129, C.A. *81*, 128176b (1974); Threshold characteristics of the formation of domains in smectic *p-n*-heptyloxybenzoic acid. [925] W. Voigt, Ann. Physik *50*, 222 (1916), C.A. *11*, 1069⁹ (1917); Bemerkungen zu den Svedbergschen Beobachtungen über Elektrizitätsleitung in anisotropen Flüssigkeiten. [926] M. Voinov, J. S. Dunnett, J. Electrochem. Soc. *120*, 922 (1973), C.A. *79*, 152264h (1973); Electrochemistry of nematics. [927] H. von Wartenberg, Physik. Z. *12*, 837 (1911), C.A. *6*, 173⁹ (1912); Zur Kenntnis der kristallinen Flüssigkeiten. [928] T. Wako, K. Nakamura, Mol. Cryst. Liq. Cryst. *19*, 141 (1972), C.A. *78*, 63490r (1973); Voltage effect on cholesterics. [929] F. E. Wargocki, A. E. Lord, Jr., J. Appl. Phys. *44*, 531 (1973), C.A. *78*, 63515c (1973); Variation of d.c. domain threshold in a nematic under continual scattering. [930] H. Watanabe, B. R. Jennings, J. Chem. Soc., Faraday Trans. 2, *72*, 1730 (1976), C.A. *86*, 36590g (1977); Transverse observations of Williams patterns in nematic MBBA.

[931] E. Wiener-Avnear, Appl. Phys. Lett. *29*, 635 (1976); Laser speckle studies of the electric field suppression of the fluctuations in nematics. [932] P. J. Wild, 2. Arbeitstagung Flüssigkristalle, Inst. Angew. Festkörperphysik Freiburg, 21. 4. 1972; Elektronische Beeinflussung der elektro-hydro-dynamischen Schwellenspannung in Flüssigkristallen. [933] R. Williams, J. Chem. Phys. *39*, 384 (1963), C.A. *59*, 3389d (1963); Domains in liquid crystals. [934] R. Williams, Nature *199*, 273 (1963), C.A. *59*, 8228h (1963); Liquid crystals in an electric field. [935] R. Williams, G. Heilmeier, J. Chem. Phys. *44*, 638 (1966), C.A. *64*, 7482b (1966); Possible ferroelectric effects in liquid crystals and related liquids. [936] R. Williams, Advan. Chem. Ser. No. *63*, 61 (1967), C.A. *67*, 5986u (1967); Interfaces in nematics. [937] R. Williams, Phys. Rev. Lett. *21*, 342 (1968), C.A. *69*, 62282b (1968); Optical rotatory effect in nematic PAA. [938] R. Williams, J. Chem. Phys. *50*, 1324 (1969), C.A. *70*, 72175d (1969); Optical-rotatory power and linear electrooptic effect in nematic PAA. [939] R. Williams, J. Chem. Phys. *56*, 147 (1972), C.A. *76*, 38350w (1972); Liquid-crystal domains in a longitudinal electric field. [940] D. Meyerhofer, A. Sussman, R. Williams, J. Appl. Phys. *43*, 3685 (1972), C.A. *77*, 106485s (1972); Electrooptic and hydrodynamic properties of nematic films with free surfaces.

[941] R. Williams, Liq. Cryst. Plast. Cryst. *2*, 110 (1974), C.A. *83*, 200277w (1975); Electric properties of liquid crystals. Nonamphiphilic systems. [942] H. E. Wirth, W. W. Wellman, J. Phys. Chem. *60*, 921 (1956), C.A. *50*, 16224a (1956); Phase transitions in sodium palmitate by dielectric-constant measurements. [943] P. J. Wojtowicz, P. Sheng, Phys. Lett. A *48*, 235 (1974); Critical point in magnetic field temperature phase diagram of nematics. [944] G. K. L. Wong, Y. R. Shen, Int. Conf. Liq. Cryst., Bangalore, *1973*, Abstr. No. 22; Optical Kerr effect and related phenomena in the i phase of a nematic. [945] J. J. Wright, J. F. Dawson, Phys. Lett. A *43*, 145 (1973), C.A. *78*, 141417p (1973); Electric field induced domains in twisted nematics. [946] J. J. Wysocki, Mol. Cryst. Liq. Cryst. *14*, 71 (1971), C.A. *75*, 134054c (1971); Continued kinetic study of the c/n transition in a liquid crystal film. [947]

T. Yamaguchi, J. Chem. Phys. *64*, 1555 (1976), C.A. *84*, 113660t (1976); Statistical properties of scattered field from nematics in dynamic scattering mode. [**947a**] R. Yamamoto, S. Ishihara, S. Hayakawa, K. Morimoto, Phys. Lett. A *60*, 414 (1977), C.A. *87*, 22130r (1977); Even-odd effect in the Kerr effect for nematic homologous series. [**948**] T. Yanagisawa, H. Matsumoto, K. Yahagi, Jpn. J. Appl. Phys. *16*, 45 (1977), C.A. *86*, 82285j (1977); Transient electric current in MBBA. [**949**] S. Yano, T. Kasatori, M. Kuwahara, K. Aoki, J. Chem. Phys. *57*, 571 (1972), C.A. *77*, 80929n (1972); Dielectric anisotropy of nematic PAA produced by magnetic field. [**950**] S. Yano, T. Kasatori, M. Kuwahara, K. Aoki, Jpn. J. Appl. Phys. *14*, 1149 (1975), C.A. *83*, 140486b (1975); Dielectric anisotropy in nematic PAA. [**950a**] S. Yano, Y. Hayashi, M. Kuwahara, K. Aoki, Jpn. J. Appl. Phys. *16*, 649 (1977), C.A. *86*, 181387a (1977); Dielectric relaxation in binary nematic mixtures of MBBA.

[**951**] K. M. Yin, J. Chem. Phys. *60*, 4621 (1974), C.A. *81*, 55491p (1974); Dielectric anisotropy and rotational orders of PAA and PAP. [**952**] K. Yoshino, K. Yamashiro, Y. Inuishi, Technol. Rep. Osaka Univ. *24*, 521 (1974), C.A. *83*, 89298n (1975); Electric conductivity in cholesterics. [**953**] K. Yoshino, K. Yamashiro, Y. Tabuchi, Y. Inuishi, Technol. Rep. Osaka Univ. *24*, 537 (1974), C.A. *83*, 89299p (1975); Carrier transport in nematics and smectics. [**954**] K. Yoshino, K. Yamashiro, Y. Inuishi, Jpn. J. Appl. Phys. *13*, 1471 (1974), C.A. *81*, 160867p (1974); Electric conductivity in cholesterics. [**955**] K. Yoshino, K. Yamashiro, Y. Inuishi, Jpn. J. Appl. Phys. *14*, 216 (1975), C.A. *82*, 179173b (1975); Carrier transport and electrooptic effects in a smectic. [**956**] K. Yoshino, K. Yamashiro, Y. Tabuchi, Y. Inuishi, IEE Conf. Publ. *129*, 295 (1975), C.A. *86*, 11057a (1977); Electric conduction mechanisms and electrooptical effects in liquid crystals. [**957**] K. Yoshino, K. Yamashiro, Y. Inuishi, Conduct. Breakdown Dielectr. Liq., Proc. Int. Conf., 5th, *1975*, 19, C.A. *85*, 135295a (1976); Carrier mobility in liquid crystals. [**958**] K. Yoshino, N. Tanaka, Y. Inuishi, Jpn. J. Appl. Phys. *15*, 735 (1976), C.A. *84*, 188106a (1976); Anomalous carrier mobility in a smectic. [**958a**] K. Yoshino, N. Tanaka, K. Kaneto, Y. Inuishi, Technol. Rep. Osaka Univ. *26*, 445 (1976); C.A. *86*, 149187z (1977); Anisotropy of carrier mobility and electrooptical effects in smectics. [**958b**] K. Yoshino, T. Uemoto, Y. Inuishi, Jpn. J. Appl. Phys. *16*, 571 (1977), C.A. *86*, 181457y (1977); Ferroelectric behavior of smectic liquid crystal. [**959**] C. K. Yun, A. G. Fredrickson, Liq. Cryst. Ord. Fluids *1970*, 239, C.A. *78*, 165461z (1973); Heat generation in nematics subject to magnetic fields. [**960**] J. Zadoc-Kahn, Compt. rend. *190*, 672 (1930), C.A. *24*, 2924 (1930); Biréfringence magnétique du PAA à des températures supérieures au point de disparition de l'état mésomorphe.

[**961**] H. M. Zenginoglou, R. A. Rigopoulos, I. A. Kosmopoulos, Mol. Cryst. Liq. Cryst. *39*, 27 (1977); Electrohydrodynamic instability limits of nematics under the action of sinusoidal electric fields. [**961a**] H. M. Zenginoglou, I. A. Kosmopoulos, Mol. Cryst. Liq. Cryst. *43*, 265 (1977); Ability of homogeneously aligned nematics with positive dielectric anisotropy to exhibit Williams domains as a threshold effect. [**962**] H. Zocher, Physik. Z. *28*, 790 (1927), C.A. *22*, 3076[7] (1928); Einwirkung magnetischer, elektrischer und mechanischer Kräfte auf Mesophasen. [**963**] H. Zocher, Z. physik. Chem. A *142*, 186 (1929), C.A. *23*, 4860[7] (1929); Beeinflussung der Mesophasen durch das elektrische und magnetische Feld. [**964**] H. Zocher, Trans. Faraday Soc. *29*, 945 (1933), C.A. *28*, 1238[8] (1934); Effect of a magnetic field upon the nematic state. [**965**] M. Zulauf, M. Bertolotti, F. Scudieri, J. Phys. (Paris), C-1, *36*, 265 (1975); Statistical properties of dynamic scattering in MBBA.

Chapter 5

[**1**] H. J. Ache, Angew. Chem. *84*, 234 (1972), C.A. *76*, 147302c (1972); Chemie des Positrons und Positroniums. [**1a**] V. O. Aimiuwu, Diss. Kent State Univ., Kent, Ohio. 1976. 81 pp. Avail. Xerox Univ. Microfilms, Ann Arbor, Mich., Order No. 76-25,359, C.A. *86*, 63086d (1977); Fe-57 Moessbauer study of four ferrocene derivatives in a s_B glass. [**1b**] Z. B. Alfassi, L. Feldman, A. P. Kushelvesky, Radiat. Eff. *32*, 67 (1977), C.A. *87*, 13241a (1977); Gamma rays dosimetry by solutions of cholesterics. [**1c**] J. Als-Nielsen, R. J. Birgeneau, M. Kaplan, J. D. Litster, C. R. Safinya, Phys. Rev. Lett. *39*, 352 (1977), C.A. *87*, 109735r (1977); High-resolution x-ray study of a second-order n/s_A transition. [**2**] B. A. Asherov, B. M. Ginzburg, S. Y. Frenkel, Vysokomol. Soedin., Ser. A, *18*, 1748 (1976), C.A. *85*, 152124m (1976); Small angle scattering of x-rays from ideal "supramolecular" paracrystals. [**3**]

L. L. Ban, P. C. Vegvari, W. M. Hess, Norelco Rep. *20*, 1 (1973), C.A. *82*, 66447a (1975); Application of optical diffraction techniques and electron optical considerations to the study of paracrystalline distortions in carbon blacks. [4] L. Bata, I. Vizi, KFKI (Rep.) 1973, KFKI-73-60, 12 pp., C.A. *80*, 88137k (1974); Rotational diffusion of PAA molecules in the n and i states. [5] L. Bata, I. Vizi, S. Kugler, KFKI (Rep.) 1974, KFKI-74-75, 12 pp., C.A. *82*, 24705c (1975); Interpretation of the quasielastic neutron scattering on PAA by rotational diffusion models. [6] L. Bata, I. Vizi, S. Kugler, Solid State Commun. *18*, 55 (1976), C.A. *84*, 114437n (1976); Rotational diffusion motion of PAA molecules in liquid crystal state. [7] L. Bata, I. Vizi, Phys. Lett. A *56*, 92 (1976), C.A. *84*, 143203p (1976); Temperature dependence of the circular random walk relaxation time in nematic. [7a] L. Bata, I. Tütto, Hung. Acad. Sci., Cent. Res. Inst. Phys., KFKI 1976, 76–69, 9 pp., C.A. *86*, 63888s (1977); Investigation of the circular random walk motion in nematics. [7b] L. Bata, I. Tütto, Mol. Cryst. Liq. Cryst. *38*, 163 (1977), C.A. *87*, 29179x (1977); Investigation of the circular random walk motion in nematic. [8] V. A. Belyakov, V. P. Orlov, Phys. Lett. A *42*, 3 (1972), C.A. *78*, 35713x (1973); Cerenkov structure radiation in cholesterics. [9] J. D. Bernal, Z. Krist. *112*, 4 (1959), C.A. *54*, 7278e (1960); Order and disorder and their expression in diffraction. [9a] V. G. Bhide, M. C. Kandpal, S. Chandra, Solid State Commun. *23*, 459 (1977), C.A. *87*, 159620v (1977); Moessbauer studies of a Fe-57 bearing solute in a supercooled monotropic smectic-B. [10] H. Bjerrum-Moeller, T. Riste, Phys. Rev. Lett. *34*, 996 (1975), C.A. *83*, 19474t (1975); Neutron-scattering study of transitions to convection and turbulence in nematic PAA.

[11] R. Blinc, V. Dimic, Phys. Lett. A *31*, 531 (1970), C.A. *73*, 70705q (1970); Neutron scattering study of self-diffusion in liquid crystals. [12] R. Blinc, V. Dimic, J. Pirs, M. Vilfan, I. Zupancic, Mol. Cryst. Liq. Cryst. *14*, 97 (1971), C.A. *75*, 134052a (1971); Self-diffusion in liquid crystals. [13] R. Blinc, V. Dimic, J. Pirs, M. Vilfan, I. Zupancic, Inst. Jozef Stefan, IJS Rep. *1970*, R-592, 18 pp., C.A. *77*, 105738q (1972); Self-diffusion in liquid crystals. [14] V. Dimic, L. Barbic, R. Blinc, Phys. Stat. Sol. B *54*, 121 (1972), C.A. *78*, 21132v (1973); Study of molecular motions in nematic MBBA by cold neutron scattering. [14a] R. Braemer, W. Ruland, Makromol. Chem. *177*, 3601 (1976), C.A. *86*, 55811w (1977); Limitations of the paracrystalline model of disorder. [14b] V. L. Broude, N. Kroo, G. Pepy, L. Rosta, Hung. Acad. Sci., Cent. Res. Inst. Phys., KFKI 1977, 77-25, 5 pp., C.A. *87*, 14473q (1977); Neutron scattering by collective excitations in PAA. [15] L. W. Gulrich, G. H. Brown, Mol. Cryst. *3*, 493 (1968), C.A. *69*, 39547h (1968); X-ray diffraction study of nematic and liquid N-(*p*-methoxybenzyli-dene)-*p*-cyanoaniline. [15a] G. H. Brown, J. Colloid Interface Sci. *58*, 534 (1977), C.A. *86*, 163791y (1977); Structures and properties of liquid crystals. [15b] O. Bruemmer, U. Marx, G. Dlubek, V. Andreichev, Zentralinst. Kernforsch., Rossendorf Dresden, 1973, ZfK-262, 171, C.A. *86*, 131370u (1977); Messungen der Lebensdauer von Positronen in Flüssigkristallen mit und ohne elektrische Feldeffekte. [16] E. Buchwald, Ann. Physik *10*, 558 (1931), C.A. *25*, 5620 (1931); Theorie der Röntgeninterferenzen in PAA. [17] A. Caille, C. R. Acad. Sci., Sér. B *274*, 891 (1972), C.A. *76*, 159488f (1972); Diffusion des rayons X dans les smectiques. [18] C. J. Carlile, K. Krebs, EURATOM (Rep.), EUR *5260*, 216 (1973), C.A. *83*, 51202p (1975); Quasielastic neutron scattering and orientational motions in liquid crystals. [19] C. J. Carlile, K. Krebs, Mol. Cryst. Liq. Cryst. *29*, 43 (1974), C.A. *82*, 178671a (1975); Quasielastic neutron scattering and orientational motions in liquid crystals. [20] V. M. Chaikovskii, Uch. Zap. Ivanov. Gos. Pedagog. Inst. *77*, 90 (1970), C.A. *76*, 28646h (1972); Structure of liquid crystals in magnetic fields.

[21] V. M. Chaikovskii, Sb. Dokl. Vses. Nauch. Konf. Zhidk. Krist. Simp. Ikh Prakt. Primen., 2nd, *1972*, 27, C.A. *81*, 112234h (1974); Structure of smectics A in a magnetic field. [21a] K. A. Suresh, S. Chandrasekhar, Mol. Cryst. Liq. Cryst. *40*, 133 (1977), C.A. *87*, 125600y (1977); Optical and X-ray studies on the twisted smectics C and twisted nematics. Evidence for a skew-cybotactic type of cholesteric structure. [22] I. G. Chistyakov, B. K. Vainshtein, Kristallografiya *8*, 570 (1963), C.A. *59*, 12261h (1963); Structure of 1-phenylazo-4-(*p*-methoxybenzylideneamino)-naphthalene in the frozen liquid crystalline state. [23] I. G. Chistyakov, Kristallografiya *8*, 859 (1963), C.A. *61*, 112f (1964); Structure of N-(*p*-nonyloxybenzylidene)-*p*-toluidine and cholesteryl caprate in liquid crystal state. [24] I. G. Chistyakov, Zh. Strukt. Khim. *5*, 550 (1964), C.A. *61*, 13976g (1964); Structure of some liquid crystals in the supercooled state. [25] B. K. Vainsthein, I. G. Chistyakov, Rost Kristallov, Akad. Nauk SSSR, Inst. Kristallogr. *5*, 163 (1965), C.A. *65*, 8121e (1966); Molecular structure and growth of liquid crystals. [26] B. K. Vainshtein, I. G. Chistyakov, E. A. Kosterin, V. M. Chaikovskii, Dokl. Akad. Nauk SSSR *174*, 341 (1967), C.A. *67*, 85848p (1967); X-ray diffraction study of nematics in electric and magnetic fields with the aid of distribution functions. [27] B. K. Vainsthein, I. G. Chistyakov,

Uch. Zap. Ivanov. Gos. Pedagog. Inst. *62*, 5 (1967), C.A. *72*, 60204r (1970); Structure of liquid crystals. [**28**] I. G. Chistyakov, V. M. Chaikovskii, Kristallografiya *12*, 883 (1967), C.A. *68*, 43993x (1968); Structure of nematic PAA in magnetic fields. [**29**] V. M. Chaikovskii, I. G. Chistyakov, Kristallografiya *13*, 158 (1968), C.A. *68*, 90647x (1968); Apparatus for x-ray diffraction studies of liquid crystals in magnetic fields. [**30**] E. A. Kosterin, I. G. Chistyakov, Kristallografiya *13*, 295 (1968), C.A. *69*, 5992j (1968); Structure of nematic PAA in constant electric fields.

[**31**] L. A. Gusakova, I. G. Chistyakov, Kristallografiya *13*, 545 (1968), C.A. *69*, 55217f (1968); Structure and properties of liquid-crystalline mixtures of 1-[(p-isoalkoxybenzylidene)amino]naphthalene-4-azobenzenes. [**32**] E. A. Kosterin, I. G. Chistyakov, Kristallografiya *14*, 321 (1969), C.A. *71*, 16709q (1969); X-ray diffraction study of the structure of PAA in alternating electric fields. [**33**] I. G. Chistyakov, V. M. Chaikovskii, Mol. Cryst. Liq. Cryst. *7*, 269 (1969), C.A. *71*, 106392k (1969); Structure of p-azoxybenzenes in magnetic fields. [**34**] B. K. Vainshtein, I. G. Chistyakov, E. A. Kosterin, V. M. Chaikovskii, Mol. Cryst. Liq. Cryst. *8*, 457 (1969), C.A. *71*, 85809j (1969); Structure of nematic PAA in electric and magnetic fields. [**35**] V. M. Chaikovskii, I. G. Chistyakov, Uch. Zap. Ivanov. Gos. Pedagog. Inst. *77*, 68 (1970), C.A. *76*, 28643e (1972); Apparatus for x-ray diffraction studies of liquid crystals in magnetic fields. [**36**] B. K. Vainshtein, E. A. Kosterin, I. G. Chistyakov, Uch. Zap. Ivanov. Gos. Pedagog. Inst. *77*, 75 (1970), C.A. *76*, 38554r (1972); Structure of PAA in constant electric fields. [**37**] B. K. Vainshtein, I. G. Chistyakov, E. A. Kosterin, V. M. Chaikovskii, A. D. Inozemtseva, Kristallografiya *16*, 717 (1971), C.A. *75*, 123341t (1971); Optical synthesis of the cylindrical function of interatomic distances. [**38**] B. K. Vainshtein, E. A. Kosterin, I. G. Chistyakov, Dokl. Akad. Nauk SSSR *199*, 323 (1971), C.A. *75*, 144743r (1971); Cylindrical function of interatomic distances for liquid crystals. [**39**] I. G. Chistyakov, Vestn. Akad. Nauk SSSR *1972*, 28, C.A. *78*, 8852k (1973); Structure of liquid crystals. [**40**] I. G. Chistyakov, L. S. Shabishev, 1. Wiss. Konferenz über flüssige Kristalle, 17.–19. 11. 1970, Iwanowo, Sammlung der Vorträge (1972), Nr. 5; Struktur der flüssige Kristalle von Anisyliden-p-aminozimtsäureäthylester.

[**41**] B. K. Vainshtein, E. A. Kosterin, I. G. Chistyakov, Uch. Zap., Ivanov. Gos. Pedagog. Inst. *1972*, No. 99, 5, C.A. *78*, 152432r (1973); Study of liquid crystal structure by use of two-dimensional interatomic distances. [**42**] L. A. Gusakova, I. G. Chistyakov, Uch. Zap., Ivanov. Gos. Pedagog. Inst. *1972*, No. 99, 26, C.A. *78*, 129286n (1973); Structure of nematic mixtures of some binary liquid-crystal systems. [**43**] I. G. Chistyakov, L. S. Shabyshev, Uch. Zap., Ivanov. Gos. Pedagog. Inst. *1972*, No. 99, 50, C.A. *78*, 147161x (1973); Structure of p-nonyloxybenzoic acid in a constant electric field. [**44**] R. I. Zharenov, I. G. Chistyakov, Uch. Zap., Ivanov. Gos. Pedagog. Inst. *1972*, No. 99, 216, C.A. *78*, 116280f (1973); Structure of p-azoxybenzenes in static electric fields. [**45**] V. M. Chaikovskii, I. G. Chistyakov, Uch. Zap., Ivanov. Gos. Pedagog. Inst. *1972*, No. 99, 298, C.A. *78*, 152430p (1973); Approximate evaluation of the average angle of molecular inclination and degree of orientation in liquid crystals. [**46**] V. M. Chaikovskii, I. G. Chistyakov, Uch. Zap., Ivanov. Gos. Pedagog. Inst. *1972*, No. 99, 301, C.A. *78*, 152442u (1973); Structure changes in smectics and nematics in a magnetic field with a rise in temperature. [**47**] V. M. Chaikovskii, I. G. Chistyakov, Sb. Dokl. Vses. Nauch. Konf. Zhidk. Krist. Simp. Ikh Prakt. Primen., 2nd, *1972*, 18, C.A. *81*, 127899c (1974); Aggregation of molecules in smectic and nematic layers of liquid crystals of the homologous p,p'-dialkoxyazoxybenzenes in magnetic fields. [**48**] I. G. Chistyakov, A. I. Aleksandrov, L. S. Shabyshev, V. I. Gerasimov, Sb. Dokl. Vses. Nauch. Konf. Zhidk. Krist. Simp. Ikh Prakt. Primen., 2nd, *1972*, 36, C.A. *82*, 24593q (1975); X-ray studies of transition phenomena in polyphase liquid crystals. [**49**] V. I. Gol'danskii, O. P. Kevdin, N. K. Kivrina, V. Ya. Rochev, E. F. Makarov, R. A. Stukan, I. G. Chistyakov, L. S. Shabyshev, Sb. Dokl. Vses. Nauch. Konf. Zhidk. Krist. Simp. Ikh Prakt. Primen., 2nd, *1972*, 56, C.A. *82*, 10161k (1975); Study of the systems diacetylferrocene + 4,4'-di-n-heptyl(octyl)oxyazoxybenzene and ferrocene + cholesteryl myristate by Moessbauer. [**50**] B. K. Vainshtein, I. G. Chistyakov, A. D. Inozemtseva, Kristallografiya *17*, 484 (1972), C.A. *77*, 67196b (1972); Modeling possibilities for studying the structure of mesophases.

[**51**] B. K. Vainsthein, I. G. Chistyakov, Acta Cryst. A *28*, S134 (1972); Structure of liquid crystals. [**52**] I. G. Chistyakov, V. M. Chaikovskii, Kristallografiya *18*, 293 (1973), C.A. *78*, 165427t (1973); Structure of liquid crystals of homologous p-alkoxyazoxybenzenes. [**53**] V. I. Gol'danskii, O. P. Kevdin, N. K. Kivrina, E. F. Makarov, V. Y. Rochev, R. A. Stukan, I. G. Chistyakov, L. S. Shabyshev, Dokl. Akad. Nauk SSSR *209*, 1139 (1973), C.A. *79*, 36838z (1973); Effect of external magnetic fields on the Mössbauer spectra of cholesterics. [**54**] V. I. Gol'danskii, O. P. Kevdin, N. K. Kivrina, E. F. Makarov, V. Ya. Rochev, R. A. Stukan, I. G. Chistyakov, Int. Conf. Liq. Cryst.,

Bangalore, *1973*, Abstr. No. 11; Mössbauer investigations of liquid crystals. [**55**] B. K. Vainshtein, I. G. Chistyakov, Liq. Cryst., Proc. Int. Conf., Bangalore, *1973*, 79, C.A. *84*, 114246z (1976); Structure of liquid crystals. [**56**] V. M. Chaikovskii, I. G. Chistyakov, Uch. zap. Ivanov. un-t, *1974*, 5, C.A. *83*, 124407w (1975); X-ray diffraction study of the effect of a magnetic field on the structure of liquid crystals. [**57**] I. G. Chistyakov, Adv. Liq. Cryst. *1*, 143 (1975), C.A. *84*, 24478x (1976); Ordering and structure of liquid crystals. [**58**] I. G. Chistyakov, A. V. Mirenskii, V. D. Belilovskii, R. I. Zharenov, Prib. Tekh. Eksp. *1975*, 231, C.A. *85*, 7632w (1976); X-ray temperature chamber for studying liquid crystals in electric fields. [**59**] I. G. Chistyakov, A. D. Inozemtseva, R. I. Zharenov, Kristallografiya *21*, 564 (1976), C.A. *85*, 102685z (1976); Characteristics of the structure of p'-nonyloxybenzoic acid using a cylindrically symmetric function of interatomic distances. [**60**] D. B. V. Chung, Diss. Kent State Univ., Kent, 1974, 130 pp. Avail. Xerox Univ. Microfilms, Ann Arbor, Mich., Order No. 75-11,987, C.A. *83*, 124398u (1975); X-ray study of the crystal structure and the s_E structure of dipropyl-p-terphenyl-4,4''-carboxylate.

[**61**] G. D. Cole, Diss. Univ. of Alabama, Univ. Microfilms (Ann Arbor, Mich.), Order No. 64-9117, 60 pp., C.A. *62*, 2439d (1965); Positron lifetimes in liquid crystals. [**62**] G. D. Cole, W. W. Walker, J. Chem. Phys. *39*, 850 (1963), C.A. *59*, 10842c (1963); Positronium decay in cholesteryl acetate. [**63**] G. D. Cole, W. W. Walker, J. Chem. Phys. *42*, 1692 (1965), C.A. *62*, 9820g (1965); Positron annihilation in liquid crystals. [**64**] G. D. Cole, W. G. Merritt, W. W. Walker, J. Chem. Phys. *49*, 1980 (1968), C.A. *69*, 102062p (1968); Positron lifetimes in cholesteryl propionate. [**65**] H. M. Conrad, H. H. Stiller, R. Stockmeyer, Phys. Rev. Lett. *36*, 264 (1976), C.A. *84*, 98027w (1976); Dispersion of collective excitations in a nematic. [**65a**] H. Conrad, H. Stiller, R. Stockmeyer, Phys. Rev. Lett. *38*, 575 (1977), C.A. *87*, 109674v (1977); Propagating or nonpropagating modes: Reply to the comment by Pelizarri and Postol. [**65b**] H. M. Conrad, H. H. Stiller, C. G. B. Frischkorn, G. Shirane, Solid State Commun. *23*, 571 (1977), C.A. *87*, 160164f (1977); s_A Order parameter measurements and critical scattering near the s_A/n transition in CBOOA. [**66**] M. J. Costello, III, Diss. Duke Univ., Durham, N. C. 1971, 182 pp., Avail. Univ. Microfilms, Ann Arbor, Mich., Order No. 72-16,982, C.A. *77*, 131813e (1972); X-ray diffraction studies of the smectic mesophase. [**67**] B. Cvikl, M. Copic, V. Dimic, J. W. Doane, W. Franklin, J. Phys. (Paris) *36*, 441 (1975); Molecular translation-rotational coupling contribution to neutron incident line broadening in nematics. [**68**] B. Cvikl, Phys. Lett. A *57*, 239 (1976), C.A. *85*, 85805u (1976); Contribution of the correlations of the director fluctuations to the neutron diffraction in nematic. [**69**] M. de Broglie, E. Friedel, Compt. rend. *176*, 738 (1923), C.A. *17*, 3832^8 (1923); Diffraction des rayons X par les corps smectiques. [**70**] P. P. Debye, H. Menke, Physik. Z. *31*, 797 (1930), C.A. *24*, 5606 (1930); Bestimmung der inneren Struktur von Flüssigkeiten mit Röntgenstrahlen.

[**71**] P. G. de Gennes, C. R. Acad. Sci., Sér. B *274*, 142 (1972), C.A. *76*, 118402e (1972); Diffusion des rayons X par les fluides nématiques. [**72**] P. G. de Gennes, J. Phys. (Paris) *36*, 603 (1975), C.A. *83*, 123713f (1975); Sur une éventuelle application de l'effet Mössbauer ou des neutrons a l'étude des interfaces fluides et des smectiques. [**72a**] W. H. de Jeu, J. A. de Poorter, Phys. Lett. A *61*, 114 (1977), C.A. *87*, 14470m (1977); X-ray diffraction of the smectics of N-(p-n-heptyloxybenzylidene)-p'-n-pentylaniline. [**73**] P. Delord, J. Phys. (Paris), C-4, *30*, 14 (1969), C.A. *72*, 104980x (1970); Diffusion des rayons X par des préparations nématiques orientées. [**74**] P. Delord, G. Malet, C. R. Acad. Sci., Sér. B *270*, 1107 (1970), C.A. *73*, 70885y (1970); Effet de diffraction sur une distribution cylindrique d'atomes; application à la diffusion X d'une phase nématique de PAA. [**75**] C. Cabos, G. Malet, P. Delord, C. R. Acad. Sci., Sér. B *273*, 199 (1971), C.A. *75*, 145725y (1971); Effet de la double diffusion des rayons X sur l'intensité diffractée par une phase nématique de PAP. [**76**] C. Cabos, P. Delord, C. R. Acad. Sci., Sér. B *275*, 387 (1972), C. A. *78*, 21153c (1973); Diffusion des rayons X par une phase n orientée de PAP. Distributions cylindriques d'atomes. [**77**] P. Delord, G. Malet, Mol. Cryst. Liq. Cryst. *27*, 231 (1974), C.A. *82*, 92103e (1975); Diffusion des rayons X par une phase nématique orientée. I. Calcul et validité des fonctions de distributions cylindriques d'atomes et d'axes moléculaires. [**78**] P. Delord, G. Malet, Mol. Cryst. Liq. Cryst. *28*, 223 (1974), C.A. *82*, 178708t (1975); Diffusion des rayons X par une phase nématique orientée. L'ordre à courte distance dans le PAA. [**79**] K. U. Deniz, U. R. K. Rao, Phys. Lett. A *59*, 208 (1976), C.A. *86*, 36566d (1977); DSC and X-ray diffraction studies of HxBPA. [**79a**] K. U. Deniz, U. R. K. Rao, A. S. Paranjpe, P. S. Parvathanathan, Proc. Nucl. Phys. Solid State Phys. Symp. *18C*, 296 (1975), C.A. *87*, 93789y (1977); Calorimetric, X-ray and Raman scattering investigations of the crystalline and liquid-crystalline phases of HBPA. [**79b**] K. U. Deniz, U. R. K. Rao, A. I. Mehta, A. S. Paranjpe, P. S. Parvathanathan, Mol. Cryst. Liq. Cryst.

42, 127 (1977); Thermal and structural studies of HxBPA. [**80**] M. Descamps, G. Coulon, Solid State Commun. *20*, 379 (1976), C.A. *85*, 200775y (1976); Correlations in a s_B plane.

[**81**] R. E. Detjen, Diss. Kent State Univ., Kent., 1973, 143 pp. Avail. Univ. Microfilms, Ann Arbor, Mich., Order No. 74-7306, C.A. *81*, 7080g (1974); Moessbauer investigation of the lattice dynamics of smectic. [**82**] A. de Vries, Mol. Cryst. Liq. Cryst. *10*, 219 (1970), C.A. *73*, 49654n (1970); X-ray photographic studies of liquid crystals. I. Cybotactic nematic phase. [**83**] A. de Vries, Mol. Cryst. Liq. Cryst. *11*, 361 (1970), C.A. *74*, 80808k (1971); II. Apparent molecular length and thickness in three phases of ethyl *p*-[(*p*-ethoxybenzylidene)-amino]benzoate (cf. [82]). [**84**] A. de Vries, D. L. Fishel, Mol. Cryst. Liq. Cryst. *16*, 311 (1972), C.A. *76*, 159611r (1972); III. Structure determination of smectic N-(4-butoxybenzal)-4-ethylaniline (cf. [82]). [**85**] A. de Vries, Acta Cryst. A *28*, 659 (1972); Comments on cylindrical functions, with special reference to X-ray studies of liquid crystals. [**86**] A. de Vries, J. Chem. Phys. *56*, 4489 (1972), C.A. *76*, 146063p (1972); Calculation of the molecular cylindrical distribution function and the order parameter from X-ray diffraction data of liquid crystals. [**87**] A. de Vries, Mol. Cryst. Liq. Cryst. *20*, 119 (1973), C.A. *78*, 129320u (1973); IV. s_A, n, and i phase of some 4-alkoxybenzal-4'-ethylanilines (cf. [82]). [**88**] A. de Vries, Liq. Cryst., Proc. Int. Conf., Bangalore, *1973*, 93, C.A. *84*, 114666m (1976); V. Classification of thermotropic liquid crystals and discussion of intermolecular distances (cf. [82]). [**89**] A. de Vries, Mol. Cryst. Liq. Cryst. *24*, 337 (1973), C.A. *82*, 178633q (1975); New classification system for thermotropic smectics. [**90**] A. de Vries, Chem. Phys. Lett. *28*, 252 (1974), C.A. *82*, 24655m (1975); Different kinds of s_B phases.

[**91**] A. de Vries, J. Chem. Phys. *61*, 2367 (1974), C.A. *81*, 142272m (1974); Structure of the s_H phase. [**92**] A. de Vries, J. Phys. (Paris), C-1, *36*, 2 (1975), C.A. *83*, 50846q (1975); Experimental investigations of the structure of thermotropic liquid crystals. [**93**] A. J. Dianoux, F. Volino, H. Hervet, Mol. Phys. *30*, 1181 (1975), C.A. *84*, 23289z (1976); Incoherent scattering law for neutron quasielastic scattering in liquid crystals. [**94**] A. J. Dianoux, F. Volino, A. Heidemann, H. Hervet, J. Phys. (Paris), Lett. *36*, 275 (1975), C.A. *84*, 11100t (1976); Neutron quasi-elastic scattering study of translational motions in the s_A, s_C, and s_H phases of TBBA. [**94a**] A. J. Dianoux, A. Heidemann, F. Volino, H. Hervet, Mol. Phys. *32*, 1521 (1976), C.A. *86*, 181042j (1977); Self-diffusion and undulation modes in a smectic A. High-resolution neutron-scattering study. [**94b**] A. J. Dianoux, H. Hervet, F. Volino, J. Phys. (Paris) *38*, 809 (1977), C.A. *87*, 61008b (1977); Orientational order in tilted smectics: high resolution neutron quasi-elastic scattering study. [**95**] S. Diele, Phys. Status Solidi A *25*, K183 (1974), C.A. *82*, 66705h (1975); X-ray studies of dipropyl-*p*-terphenyl-4,4''-carboxylate in the s_E modification. [**96**] V. Dimic, M. Osredkar, Mol. Cryst. Liq. Cryst. *19*, 189 (1973), C.A. *78*, 89514b (1973); Low frequency mode in liquid crystals. [**97**] J. Doucet, M. Lambert, A. M. Levelut, J. Phys. (Paris), C-5, *32*, 247 (1971), C.A. *77*, 10693g (1972); Structure des smectiques B. [**98**] J. Doucet, A. M. Levelut, M. Lambert, Mol. Cryst. Liq. Cryst. *24*, 317 (1973), C.A. *82*, 178632p (1975); Long and short range order in the crystalline and s_B phases of TBBA. [**99**] J. Doucet, A. M. Levelut, M. Lambert, Phys. Rev. Lett. *32*, 301 (1974), C.A. *81*, 127864n (1974); Polymorphism of mesomorphic TBBA. [**100**] J. Doucet, A. M. Levelut, M. Lambert, L. Liebert, L. Strzelecki, J. Phys. (Paris), C-1, *36*, 13 (1975), C.A. *83*, 36025q (1975); Nature de la phase s_E.

[**101**] J. Doucet, A. M. Levelut, M. Lambert, Acta Cryst. *31A*, S160 (1975); Relations between solid and ordered smectic phases. [**101a**] J. Doucet, A. M. Levelut, J. Phys. (Paris) *38*, 1163 (1977), C.A. *87*, 160148d (1977); X-ray study of the ordered smectics in some benzylideneanilines. [**102**] Ya. I. Dutchak, V. K. Burima, Visn. L'viv. Derzh. Univ., Ser. Fiz. *1973*, No. 8, 106, C.A. *81*, 142334h (1974); Device for x-ray diffraction study of liquid crystals. [**103**] Ya. I. Dutchak, V. K. Burima, Kristallografiya *18*, 1078 (1973), C.A. *80*, 7876a (1974); X-ray diffraction study of some representatives of a homologous series of cholesteryl esters of dicarboxylic acids. [**104**] S. Ergun, J. Bayer, W. van Buren, J. Appl. Phys. *38*, 3540 (1967), C.A. *67*, 112045h (1967); Normalization and absorption correction of arbitrary X-ray scattering intensities of paracrystals. [**104a**] S. N. Evstaf'ev, G. S. Yur'ev, V. V. Polyukhov, V. V. Tuturina, Gidrodinamika i Yavleniya Perenosa v Dvukhfaz. Dispers. Sistemakh *1976*, 179, C.A. *87*, 192289s (1977); X-ray diffraction analysis of cholesterics. [**105**] J. Falgueirettes, Compt. rend. *241*, 71 (1955), C.A. *49*, 15329g (1955); Diffusion des rayons X par un monocristal liquide nématique. [**106**] J. Falgueirettes, Compt. rend. *241*, 225 (1955), C.A. *49*, 15329f (1955); Répartition des molécules dans le monocristal liquide, deduit des mesures de diffusion des rayons X. [**107**] J. Falgueirettes, Bull. soc. franç. minéral. et crist. *82*, 171 (1959), C.A. *55*, 12992i (1961); Diffusion des rayons X par un monocristal liquide nématique. [**108**] P. Delord, J. Falgueirettes, Compt. rend. *260*, 2468 (1965), C.A. *62*, 15529d (1965); Fonctions de réparation des molécules dans un monocristal liquide

orienté de PAP à différentes températures. [**109**] P. Delord, J. Falgueirettes, C. R. Acad. Sci., Sér. C *267*, 1177 (1968), C.A. *70*, 41521j (1969); Diffusion des rayons X par un monocristal liquide de PAA. Distributions linéaires d'atomes. [**110**] P. Delord, J. Falgueirettes, C. R. Acad. Sci., Sér. C *267*, 1437 (1968), C.A. *70*, 51823y (1969); Diffractions des rayons X par un monocristal liquide de PAA. Distribution cylindrique d'atomes et d'axes de molécules.

[**111**] P. Delord, J. Falgueirettes, C. R. Acad. Sci., Sér. C *267*, 1528 (1968), C.A. *70*, 62168q (1969); Structure du PAA nématique. [**112**] J. Falgueirettes, P. Delord, Liq. Cryst. Plast. Cryst. *2*, 62 (1974), C.A. *83*, 186433k (1975); X-ray diffraction by liquid crystals. Nonamphiphilic systems. [**112a**] M. I. Fedishin, Visnik L'viv. Un-tu. Ser. Fiz. *1977*, 95, C.A. *87*, 144369w (1977); Study of the structure of cholesteryl caprate by the method of low-angle X-ray diffraction. [**113**] E. Friedel, Compt. rend. *180*, 269 (1925), C.A. *19*, 1072[3] (1925); Corps smectiques et rayons X. [**114**] G. Friedel, Compt. rend. *182*, 425 (1926), C.A. *20*, 1736[8] (1926); Acides gras et corps smectiques. [**115**] K. Fuchs, Physik. Z. *8*, 417 (1907), C.A. *1*, 2661[6] (1907); Flüssige Kristalle, Bemerkung über eine Veröffentlichung von O. Lehmann. [**116**] V. G. Gol'danskii, O. P. Kevdin, N. K. Kivrina, E. F. Makarov, V. Y. Rochev, R. A. Stukan, Zh. Eksp. Teor. Fiz. *63*, 2323 (1972), C.A. *78*, 64824q (1973); Determination of the dynamic and structural characteristics of liquid crystals on the basis of Moessbauer spectra. [**117**] V. I. Gol'danskii, O. P. Kevdin, N. K. Kivrina, E. F. Makarov, V. Y. Rochev, R. A. Stukan, Sb. Dokl. Vses. Nauch. Konf. Zhidk. Krist. Simp. Ikh Prakt. Primen, 2nd, *1972*, 63, C.A. *81*, 127882s (1974); Determination of displacement anisotropy and the angle of inclination of layers in smectics from Moessbauer spectra. [**118**] V. I. Gol'danskii, O. P. Kevdin, N. K. Kivrina, V. Y. Rochev, R. A. Stukan, I. G. Chistyakov, L. S. Shabishev, Mol. Cryst. Liq. Cryst. *24*, 239 (1973), C.A. *82*, 131670j (1975); Moessbauer investigations of liquid crystals. [**119**] V. I. Gol'danskii, O. P. Kevdin, N. K. Kivrina, E. F. Makarov, V. A. Rocev, Wiss. Beitr. Martin-Luther-Univ. Halle-Wittenberg *1976*, 99, C.A. *86*, 130055h (1977); Untersuchung thermotroper flüssiger Kristalle mit der Methode der γ-Resonanz-Mößbauer-Spektroskopie. [**120**] L. S. Gorbatenko, V. S. Dronov, Izv. Vyssh. Ubech. Zaved., Fiz. *14*, 129 (1971), C.A. *75*, 102239a (1971); Detection of ions in a supercooled isotropic melt of liquid crystals.

[**121**] S. Goshen, D. Mukamel, S. Shtrikman, Mol. Cryst. Liq. Cryst. *31*, 171 (1975), C.A. *83*, 211549n (1975); Classification of the possible symmetry groups of liquid crystals. [**122**] C. C. Gravatt, G. W. Brady, Mol. Cryst. Liq. Cryst. *7*, 355 (1969), C.A. *71*, 85805e (1969); Small angle X-ray studies of liquid-crystal phase transitions. I. PAA. [**123**] C. C. Gravatt, G. W. Brady, Liq. Cryst. Ord. Fluids *1970*, 455, C.A. *78*, 152436v (1973); II. Surface, impurity and electric field effects (cf. [122]). [**124**] G. W. Brady, C. C. Gravatt, J. Appl. Cryst. *4*, 424 (1971); Small angle X-ray scattering studies of liquid crystal phase-transition. [**124a**] A. C. Griffin, J. F. Johnson, J. Am. Chem. Soc. *99*, 4859 (1977), C.A. *87*, 76624y (1977); 4-Nitrophenyl 4'-decyloxybenzoate: a liquid crystal with novel mesomorphic properties. [**125**] H. U. Gruber, H. Krebs, Z. Anorg. Allgem. Chem. *369*, 184 (1969), C.A. *72*, 6991g (1970); Nahordnung des geschmolzenen Antimons. [**126**] L. W. Gulrich, Jr., Diss. Kent State Univ., Kent, 1967, 151 pp. Avail. Univ. Microfilms, Ann Arbor, Mich. Order No. 68-6210, C.A. *69*, 55216e (1968); X-ray diffraction studies of nematic *p*-methoxybenzylidene-*p'*-cyanoaniline. [**127**] C. Hermann, Z. Krist. *79*, 186 (1931), C.A. *26*, 2359[7] (1932); Symmetriegruppen der amorphen und mesomorphen Phasen. [**128**] E. Alexander, K. Herrmann, Z. Krist. *69*, 285 (1928), C.A. *23*, 2335 (1929); Theorie der flüssigen Kristalle. [**129**] P. W. Glamann, K. Herrmann, A. H. Krummacher, Z. Krist. *74*, 73 (1930), C.A. *24*, 4438[4] (1930); Röntgenuntersuchungen an kristallin-flüssigen Substanzen. I. PAA. [**130**] K. Herrmann, A. H. Krummacher, Z. Krist. *79*, 134 (1931), C.A. *26*, 2357[7] (1932); II. Phenetolazoxy-benzoesäureallylester (cf. [129]).

[**131**] K. Herrmann, A. H. Krummacher, Z. Physik *70*, 758 (1931), C.A. *25*, 5814[4] (1931); Röntgenuntersuchungen an kristallin-flüssigen Substanzen. [**132**] K. Herrmann, A. H. Krummacher, K. May, Z. Physik *73*, 419 (1931), C.A. *26*, 2356[4] (1932); Verhalten flüssiger Kristalle im elektrischen Feld (röntgenographische und optische Untersuchungen). [**133**] K. Herrmann, Ergebn. techn. Röntgenkunde *2*, 23 (1931); Röntgenbilder flüssiger Kristalle mit magnetischen und elektrischen Feldern. [**134**] K. Herrmann, A. H. Krummacher, Z. Krist. *81*, 317 (1932), C.A. *26*, 5804[9] (1932); Röntgenuntersuchungen an kristallin-flüssigen Substanzen. [**135**] K. Herrmann, Trans. Faraday Soc. *29*, 972 (1933), C.A. *28*, 1239[1] (1934); Inclination of molecules in crystalline fluids. [**136**] K. Herrmann, Z. Krist. *92*, 49 (1935), C.A. *30*, 2444[3] (1936); Röntgenuntersuchungen an kristallin-flüssigen Substanzen. [**137**] H. Hervet, F. Volino, A. J. Dianoux, R. E. Lechner, J. Phys. (Paris), Lett. *35*, 151 (1974), C.A. *82*, 16176r (1975); Uniaxial rotational diffusion in the s_B phase of TBBA observed by quasielastic neutron scattering.

[138] H. Hervet, F. Volino, A. J. Dianoux, R. E. Lechner, Phys. Rev. Lett. *34*, 451 (1975), C.A. *83*, 8972c (1975); Nature of the molecular alignment in a s_H phase. [139] H. Hervet, S. Lagomarsino, F. Rustichelli, F. Volino, Solid State Commun. *17*, 1533 (1975), C.A. *84*, 98036y (1976); Direct measurement of tilt angle in smectics by neutron diffraction. [140] H. Hervet, S. Lagomarsino, F. Rustichelli, F. Volino, Acta Cryst. A *31*, S160 (1975); Orientational order in smectic TBBA by neutron diffraction. [141] H. Hervet, S. Lagomarsino, F. Rustichelli, F. Volino, Acta Cryst. A *32*, 166 (1976), C.A. *84*, 52416u (1976); Neutron diffraction from a s_A monodomain. [142] H. Hervet, S. Lagomarsino, F. Rustichelli, F. Volino, J. Phys. (Paris), C-3, *37*, 127 (1976); Direct measurement of tilt angle in smectics by neutron diffraction. [142a] H. Hervet, A. J. Dianoux, R. E. Lechner, F. Volino, J. Phys. (Paris) *37*, 587 (1976), C.A. *85*, 68418c (1976); Neutron scattering study of methyl group rotation in solid PAA. [143] O. A. Hoffman, Rev. Sci. Instrum. *19*, 277 (1948), C.A. *42*, 4800h (1948); Apparatus for taking x-ray diffraction pictures of volatile systems at high temperatures. [144] R. Hosemann, Z. Physik *128*, 465 (1950), C.A. *45*, 4131g (1951); Der ideale Parakristall und die von ihm gestreute kohärente Röntgenstrahlung. [145] R. Hosemann, Acta Cryst. *4*, 520 (1951), C.A. *46*, 7395a (1952); Parakristalline Feinstruktur natürlicher und synthetischer Eiweiße. Visuelles Näherungsverfahren zur Bestimmung der Schwankungstensoren von Gitterzellen. [146] R. Hosemann, S. N. Bagchi, Acta Cryst. *5*, 612 (1952), C.A. *46*, 10749b (1952); Interference theory of ideal paracrystals. [147] R. Hosemann, Naturwissenschaften *41*, 440 (1954), C.A. *49*, 10698i (1955); Parakristalline Strukturen. [148] R. Hosemann, K. Lemm, Phys. Non-Crystalline Solids Proc. Intern. Conf., Delft, Neth. *1964*, 85, C.A. *64*, 1758f (1966); Parakristallinität und dreidimensionale Analyse der radialen Dichteverteilung in geschmolzenen Metallen. [149] R. Hosemann, K. Lemm, W. Wilke, Mol. Cryst. *2*, 333 (1967), C.A. *67*, 85847n (1967); Paracrystal as a model for liquid crystals. [150] R. Hosemann, B. Mueller, Mol. Cryst. Liq. Cryst. *10*, 273 (1970), C.A. *73*, 70884x (1970); Paracrystalline lattices in mesophases and real structures.

[151] W. Vogel, R. Hosemann, Acta Cryst. A *26*, 272 (1970), C.A. *73*, 8128a (1970); Evaluation of paracrystalline distortions from line broadening. [152] R. Hosemann, J. Loboda-Cackovic, H. Cackovic, Ber. Bunsenges. Phys. Chem. *77*, 1044 (1973), C.A. *81*, 26050p (1974); New type of liquid crystal. [153] W. Vogel, J. Haase, R. Hosemann, Z. Naturforsch. A *29*, 1152 (1974), C.A. *82*, 31661p (1975); Linienprofilanalyse von Röntgen-Weitwinkelreflexen mittels Fourier-Transformation zur Bestimmung von Mikrospannungen und parakristallinen Störungen. [154] R. Hosemann, Makromol. Chem., Suppl. 1, 559 (1975), C.A. *83*, 115204a (1975); Microparacrystallites and paracrystalline superstructures. [155] R. Hosemann, J. Loboda-Cackovic, J. Polym. Sci., Polym. Symp. *53*, 159 (1975), C.A. *84*, 106244g (1976); Role of microparacrystallites in polymer network. [156] B. Steffen, R. Hosemann, Ber. Bunsenges. Phys. Chem. *80*, 710 (1976), C.A. *85*, 152216t (1976); Paracrystalline structure of molten lead. [157] E. Hückel, Diss. Göttingen 1921; Physik. Z. *22*, 561 (1921), C.A. *16*, 873[8] (1922); Zerstreuung von Röntgenstrahlen durch anisotrope Flüssigkeiten. [158] A. D. Inozemtseva, Kristallografiya *17*, 656 (1972), C.A. *77*, 119265v (1972); Structure of liquid crystals studied by x-ray diffraction and optical methods. [159] A. D. Inozemtseva, Uch. Zap., Ivanov. Gos. Pedagog. Inst. No. 99, 33 (1972), C.A. *78*, 165405j (1973); Use of optical modeling for liquid-crystal structure studies. [160] A. D. Inozemtseva, L. S. Shabyshev, Izv. Vyssh. Ubech. Zaved., Fiz. *16*, 28 (1973), C.A. *79*, 10744r (1973); Use of optical analogy for determining the structure of liquid crystals. I. Structural characteristics of smectics.

[161] A. D. Inozemtseva, L. S. Shabyshev, Izv. Vyssh. Ucheb. Zaved., Fiz. *16*, 130 (1973), C.A. *79*, 109017t (1973); II. Nematics (cf. [160]). [162] A. D. Inozemtseva, Uch. Zap.-Ivanov. Gos. Univ. *128*, 29 (1974), C.A. *85*, 54857h (1976); Possibilities of optical modeling for studying order perturbations. [163] J. A. Janik, S. Kraśnicki, A. Murasik, Acta Phys. Polon. *17*, 483 (1958), C.A. *53*, 14676e (1959); Influence of polarization of liquid crystal molecules on the scattering of slow neutrons. [164] J. A. Janik, J. M. Janik, Acta Phys. Polon. *25*, 845 (1964), C.A. *63*, 6415b (1965); Total neutron scattering cross-section study of molecular motions in PAA. [165] J. A. Janik, J. M. Janik, K. Otnes, T. Riste, Mol. Cryst. Liq. Cryst. *15*, 189 (1971), C.A. *76*, 38454h (1972); Anisotropy of self-diffusion in a liquid crystal studied by neutron quasielastic scattering. [166] J. A. Janik, J. M. Janik, K. Otnes, K. Rosciszewski, Physica *77*, 514 (1974), C.A. *82*, 124559b (1975); Stochastic proton jumps in nematic MBBA studied by neutron scattering. [167] J. A. Janik, Wiss. Beitr. Martin-Luther-Univ. Halle-Wittenberg *1976*, 43, C.A. *86*, 63594t (1977); Neutron scattering experiments with liquid crystals. [168] J. I. Kaplan, M. L. Glasser, Mol. Cryst. Liq. Cryst. *11*, 103 (1970), C.A. *73*, 135741v (1970); Moessbauer effect in the smectic mesophase. [169] W. Kast, Ann. Physik *83*, 418 (1927), C.A. *22*, 912[4] (1928); Röntgenuntersuchungen an festem kristallinischem und anisotropflüssigem PAA. [170] W. Kast, Z. Physik. *71*, 39 (1931), C.A. *25*, 5807[1] (1931); Anisotrope Flüssigkeiten im elektrischen Feld.

[**171**] W. Kast, Z. Physik *76*, 19 (1932), C.A. *26*, 5805[1] (1932); Anisotrope Flüssigkeiten im elektrischen Feld. [**172**] W. Kast, Naturwissenschaften *21*, 737 (1933), C.A. *28*, 697[5] (1934); Vergleich der Röntgenbilder der kristallin-flüssigen und der normal-flüssigen Phase derselben Substanz. [**173**] W. Kast, Ann. Physik *19*, 571 (1934), C.A. *28*, 3957[5] (1934); Vergleichende röntgenographische und optische Untersuchungen an der n und i Schmelze des PAA. [**174**] E. I. Kats, Zh. Eksp. Teor. Fiz. *61*, 1686 (1971), C.A. *76*, 7604c (1972); Cherenkov radiation in cholesterics. [**175**] J. R. Katz, Naturwissenschaften *16*, 758 (1928), C.A. *23*, 765[6] (1929); Weitgehende Übereinstimmung im Röntgenspektrum der flüssig-kristallinischen und der flüssigen Phase derselben Substanz. [**176**] J. R. Katz, Scienca Rondo No. *9/10*, 5 (1949), C.A. *45*, 9959h (1951); Rigidity and configuration of molecules in the liquid state according to the Röntgen spectrograms. [**177**] M. A. Khan, D. J. Morantz, 3rd Int. Liq. Cryst. Conf., Berlin, 1970, S. 2.10; X-ray optical examination of binary cholesteric systems. [**178**] M. V. King, M. Young, J. Appl. Cryst. *6* (Pt. 4), 289 (1973), C.A. *79*, 84286q (1973); Specimen cell for studying liquid crystals in X-ray diffraction and by polarized light. [**179**] E. A. Kosterin, Kristallografiya *14*, 524 (1969), C.A. *71*, 34090k (1969); Effect of temperature on the texture of nematic PAA oriented by an electric field. [**180**] E. A. Kosterin, Kristallografiya *17*, 639 (1972), C.A. *77*, 80652s (1972); Calculation of the parameters of molecular disorder and optical modeling of structures of mesomorphic ethyl *p*-anisalaminocinnamate.

[**181**] W. R. Krigbaum, J. C. Poirier, M. J. Costello, Mol. Cryst. Liq. Cryst. *20*, 133 (1973), C.A. *78*, 129323x (1973); X-ray diffraction study of three smectics A. [**182**] V. G. Kulkarni, N. K. Dave, Proc. Nucl. Phys. Solid State Phys. Symp. *16C*, 218 (1973), C.A. *82*, 163378m (1975); Positron annihilation in liquid and plastic crystals. [**183**] A. J. Leadbetter, F. P. Temme, A. Heidemann, W. S. Howells, Chem. Phys. Lett. *34*, 363 (1975), C.A. *83*, 156066a (1975); Self-diffusion tensor for two nematics from incoherent quasielastic neutron scattering at low momentum transfer. [**184**] A. J. Leadbetter, R. M. Richardson, C. N. Colling, J. Phys. (Paris), C-1, *36*, 37 (1975), C.A. *83*, 69392k (1975); Structure of a number of nematogens. [**185**] A. J. Leadbetter, R. M. Richardson, C. J. Carlile, J. Phys. (Paris), C-3, *37*, 65 (1976), C.A. *85*, 85783k (1976); Nature of the s_E phase. [**186**] A. J. Leadbetter, R. M. Richardson, B. A. Dasannacharya, W. S. Howells, Chem. Phys. Lett. *39*, 501 (1976), C.A. *85*, 39475p (1976); Incoherent neutron quasi-elastic scattering studies of the anisotropic self-diffusion in n and s_A phases of ethyl 4(4'-acetoxybenzylidene)aminocinnamate. [**186a**] B. A. Dasannacharya, A. J. Leadbetter, R. Richardson, A. Heidemann, W. Howell, Proc. Nucl. Phys. Solid State Phys. Symp. *18C*, 338 (1975), C.A. *87*, 93791t (1977); Temperature dependence of diffusion constants in liquid crystal EABAC. [**187**] K. Lemm, Mol. Cryst. Liq. Cryst. *10*, 259 (1970), C.A. *73*, 81752d (1970); Calculation of liquid structures by paracrystalline distortions. [**188**] A. M. Levelut, M. Lambert, C. R. Acad. Sci., Sér. A *272*, 1018 (1971), C.A. *75*, 41751n (1971); Structure des cristaux liquides s_B. [**189**] A. M. Levelut, M. Lambert, Acta Cryst. A *28*, S130 (1972); Linear local order in molecular crystals and liquid crystals. [**190**] A. M. Levelut, J. Doucet, M. Lambert, J. Phys. (Paris) *35*, 773 (1974), C.A. *81*, 160548k (1974); Étude par diffusion de rayons X de la nature des phases s_B et de la transition k/s_B.

[**191**] M. Lambert, A. M. Levelut, NATO Adv. Study Inst. Ser., Ser. E, *1*, 375 (1974), C.A. *84*, 24643x (1976); X-ray studies of the smectic structure and the k/s transition. [**192**] A. M. Levelut, J. Phys. (Paris), C-3, *37*, 51 (1976), C.A. *85*, 134769w (1976); Étude de l'ordre local lié a la rotation des molécules dans la phase s_B. [**193**] V. Luzzati, P. A. Spegt, Nature *215*, 701 (1967), C.A. *69*, 31098s (1968); Polymorphism of lipids. [**194**] V. Luzzati, A. Tardieu, T. Gulik-Krzywicki, Nature *217*, 1028 (1968), C.A. *69*, 39546g (1968); Polymorphism of lipids. [**195**] J. E. Lydon, 3rd Int. Liq. Cryst. Conf., Berlin, 1970, S. 4.3; X-ray study of mesomorphic cholesteryl stearate. [**196**] J. E. Lydon, C. J. Coakley, J. Phys. (Paris), C-1, *36*, 45 (1975), C.A. *83*, 106597d (1975); Structural study of the smectics of two biphenyl compounds and an X-ray investigation of the miscibility criterion. [**197**] J. E. Lydon, J. O. Kessler, J. Phys. (Paris), C-1, *36*, 153 (1975), C.A. *83*, 19459s (1975); Phase transitions observed on warming fast-quenched MBBA. [**197a**] C. J. Coakley, J. E. Lydon, J. Phys. E *10*, 296 (1977), C.A. *86*, 131597y (1977); Temperature-scanning x-ray diffraction camera. [**198**] A. De Bretteville, Jr., J. W. McBain, J. Chem. Phys. *11*, 426 (1943), C.A. *37*, 6177[5] (1943); X-ray-diffraction investigation of sodium stearate from room temperature to the melting point. [**199**] W. L. McMillan, Phys. Rev. A *6*, 936 (1972), C.A. *77*, 119296f (1972); X-ray scattering from liquid crystals. I. Cholesteryl nonanoate and myristate. [**200**] W. L. McMillan, Phys. Rev. A *7*, 1419 (1973), C.A. *78*, 165421m (1973); Measurement of s_A-phase order-parameter fluctuations near a second-order s_A/n transition.

[**201**] W. L. McMillan, Phys. Rev. A *7*, 1673 (1973), C.A. *78*, 165454z (1973); Measurement of s_A-phase order-parameter fluctuations in nematic *p-n*-octyloxybenzylidene-*p'*-toluidine. [**202**] W. L.

McMillan, Phys. Rev. A *8*, 328 (1973), C.A. *79*, 84264f (1973); Measurement of smectic-phase order-parameter fluctuations in nematic heptyloxyazoxybenzene. [203] J. D. McNutt, W. W. Kinnison, M. D. Searcey, Phys. Rev. B *5*, 826 (1972), C.A. *76*, 63241u (1972); Effect of intermolecular forces on the annihilation of positrons in a nematic. [204] A. J. Mabis, Acta Cryst. *15*, 1152 (1962), C.A. *58*, 2920a (1963); Structure of mesophases. [205] W. G. Merritt, Diss. Univ. Alabama, 1971, 121 pp., Avail. Univ. Microfilms, Ann Arbor, Mich., Order No. 71-20,119, C.A. *76*, 51179s (1972); Polymorphic and mesomorphic behavior of some organic compounds. [205a] R. J. Meyer, Phys. Rev. A *13*, 1613 (1976); Molecular order in smectic E. [206] V. M. Mikhailov, Uch. Zap., Ivanov. Gos. Pedagog. Inst. *1972*, No. 99, 105, C.A. *78*, 129284k (1973); Small-angle X-ray diffraction study of ethyl-*p*-(anisalamino)-cinnamate in a constant electric field. [206a] Z. M. Mikityuk, Visnik L'viv. Un-tu. Ser. Fiz. *1977*, 98, C.A. *87*, 144298x (1977); Method of low-angle X-ray diffraction for the study of some cholesteryl problems. [207] A. Müller, Trans. Faraday Soc. *29*, 990 (1933), C.A. *28*, 1239³ (1934); Arrangement of chain molecules in liquid *n*-paraffins. [208] E. L. Mueller, W. W. Walker, Phys. Lett. A *44*, 320 (1973), C.A. *79*, 97938d (1973); Evidence for a phase transition in N-(*p*-butyloxybenzal)-*p*-ethylaniline. [209] D. E. Nagle, J. W. Doane, R. Madey, A. Saupe, Mol. Cryst. Liq. Cryst. *26*, 71 (1974), C.A. *81*, 127851f (1974); Effects of the passage of ionizing particles through a liquid crystal. [210] J. B. Nicholas, H. J. Ache, J. Chem. Phys. *57*, 1597 (1972), C.A. *77*, 93023n (1972); Phase and temperature dependence of positron annihilation in liquid crystals.

[211] N. Niimura, Phys. Lett. A *48*, 375 (1974), C.A. *81*, 127942m (1974); Neutron diffraction from a nematic. [212] N. Niimura, M. Muto, Nucl. Instrum. Methods *126*, 87 (1975), C.A. *83*, 124332t (1975); Analysis of transient phenomena in crystalline and liquid structures using neutron time-of-flight techniques. [213] N. Niimura, Mol. Cryst. Liq. Cryst. *31*, 123 (1975), C.A. *83*, 211547k (1975); Neutron diffraction from n and i phase of PAA. [214] H. Nordsieck, F. B. Rosevear, R. H. Ferguson, J. Chem. Phys. *16*, 175 (1948), C.A. *42*, 3236e (1948); X-ray study of the stepwise melting of anhydrous sodium palmitate. [215] A. Olivei, Acta Cryst. A *29*, 692 (1973), C.A. *80*, 7920k (1974); Cold-neutron incoherent scattering by homogeneously oriented nematics. [216] A. Olivei, Acta Cryst. A *32*, 983 (1976), C.A. *85*, 185032h (1976); Neutron small-angle scattering by dislocations in homogeneously oriented nematics. [217] O. Parodi, J. Phys. (Paris), Lett. *37*, 143 (1976), C.A. *85*, 70918j (1976); Permeation and self-diffusion in smectics A. [217a] C. A. Pelizzari, T. A. Postol, Phys. Rev. Lett. *38*, 573 (1977), C.A. *86*, 131415n (1977); Comment on "Dispersion of collective excitations in a nematic". [218] M. J. Potasek, E. Muenck, J. L. Groves, P. G. Debrunner, Chem. Phys. Lett. *15*, 55 (1972), C.A. *77*, 107371p (1972); Observation of alignment in a quenched liquid crystal with the Moessbauer effect. [219] M. J. Potasek, Diss. Univ. Illinois, Urbana, 1974, 50 pp. Avail. Xerox Univ. Microfilms, Ann Arbor, Mich., Order No. 75-404, C.A. *82*, 131985r (1975); I-129 Moessbauer spectroscopy. Probing the molecular orientation in liquid crystals. [220] M. J. Potasek, P. G. Debrunner, G. DePasquali, Phys. Rev. A *13*, 1605 (1976), C.A. *84*, 172358y (1976); I-129 Moessbauer spectroscopy. Probing the molecular orientation in liquid crystals.

[221] R. Pynn, K. Otnes, T. Riste, Solid State Commun. *11*, 1365 (1972), C.A. *78*, 34841a (1973); Coherent neutron scattering by a nematic. [222] R. Pynn, J. Phys. Chem. Solids *34*, 735 (1973), C.A. *78*, 129290j (1973); Calculations of neutron diffraction patterns obtained with a nematic. [223] T. Riste, R. Pynn, Solid State Commun. *12*, 409 (1973), C.A. *78*, 129249c (1973); Pretransitional effects associated with the melting of a nematogen. [224] R. Pynn, Acta Cryst. A *31*, 323 (1975), C.A. *83*, 51183h (1975); X-ray and neutron diffraction by nematics. [225] M. Kohli, K. Otnes, R. Pynn, T. Riste, Z. Phys. B, *24*, 147 (1976), C.A. *85*, 39480m (1976); Investigation of nematic order by coherent neutron scattering. [226] C. V. Raman, K. R. Ramanathan, Proc. Indian Assoc. Cultivation of Science *8*, 127 (1923), C.A. *18*, 2839⁵ (1924); Diffraction of X-rays in liquids, fluid crystals and amorphous solids. [227] T. Riste, Phys. Norveg. *8*, 64 (1975); Neutron scattering study of orientational order and fluctuations at n/i transition of PAA. [227a] T. Riste, Proc. Conf. Neutron Scattering *1*, 379 (1976), C.A. *87*, 125412p (1977); Liquid crystals and neutron scattering. [227b] M. Roder, K. Pinter, K. Ritvay, Izotoptechnika *19*, 326 (1976), C.A. *86*, 130997s (1977); Radiation stability of liquid crystals. [228] K. Rosciszewski, Inst. Nucl. Phys., Cracow, Rep. *1971*, INP No. 776/PS, 27 pp., C.A. *76*, 158498x (1972); Incoherent cross section for neutron quasi-elastic scattering in a liquid crystal. [229] K. Rosciszewski, Acta Phys. Pol. A *41*, 549 (1972), C.A. *77*, 40169h (1972); Incoherent cross-section for neutron quasielastic scattering in a liquid crystal. [230] K. Rosciszewski, Physica *75*, 268 (1974), C.A. *81*, 176358j (1974); Incoherent cross section for neutron quasielastic scattering in liquids, liquid crystals, and molecular crystals.

[231] J. Voss, U. Wuerz, E. Sackmann, Ber. Bunsenges. Phys. Chem. *78*, 874 (1974), C.A. *82*, 10269b (1975); X-ray studies of c/s pretransitions in mixtures of cholesteryl chloride and cholesteryl nonanoate. [232] S. Diele, P. Brand, H. Sackmann, Mol. Cryst. Liq. Cryst. *16*, 105 (1972), C.A. *76*, 77753z (1972); X-ray diffraction and polymorphism of smectics: I. A-, B-, and C-modifications. [233] S. Diele, P. Brand, H. Sackmann, Mol. Cryst. Liq. Cryst. *17*, 163 (1972), C.A. *77*, 53458b (1972); II. D- and E-modifications (cf. [232]). [233a] E. Saito, J. Belloni, Rev. Sci. Instrum. *47*, 629 (1976); All-silica cell for pulse-radiolysis studies of liquid crystals with low energy (600 keV) electrons. [234] W. G. Shaw, Diss. Univ. of Cincinnati, Cincinnati, 1957, 140 pp. Univ. Microfilms (Ann Arbor, Mich.), Publ. No. 23243, C.A. *52*, 1720h (1958); X-ray diffraction studies of mesomorphic anisaldazine and 4,4'-dialkoxybenzalazine homologs (Radial distribution analysis of anisaldazine and synthesis of substituted benzalazines). [235] S. K. Sinha, K. U. Deniz, G. Venkataraman, B. A. Dasannacharya, A. S. Paranjape, P. S. Parvathanathan, Proc. Nucl. Phys. Solid State Phys. Symp. *16C*, 197 (1973), C.A. *82*, 163375h (1975); Dynamics of the molecule of 4-*n*-hexyloxybenzylidene-4'-propylaniline in its mesophases. [236] A. Skoulios, V. Luzzati, Acta Cryst. *14*, 278 (1961), C.A. *55*, 11030h (1961); La structure des colloïdes d'association. III. Description des phases mésomorphes des savons de sodium purs, rencontrées au-dessus de 100°C. [237] B. Gallot, A. E. Skoulios, Acta Cryst. *15*, 826 (1962), C.A. *57*, 16777c (1962); VI. Polymorphisme des groupes polaires dans des phases mésomorphes des savons alcalins purs (cf. [236]). [238] P. A. Spegt, A. E. Skoulios, Acta Cryst. *21*, 892 (1966), C.A. *66*, 14747y (1967); Structure des savons de Sr en fonction de la température. [239] B. Gallot, A. Skoulios, Kolloid-Z. Z. Polym. *209*, 164 (1966), C.A. *65*, 10811c (1966); Structure des savons alcalins I. Généralités et savons de Li. [240] B. Gallot, A. Skoulios, Kolloid-Z. Z. Polym. *210*, 143 (1966), C.A. *65*, 13965h (1966); II. Savons de K (cf. [239]).

[241] B. Gallot, A. Skoulios, Mol. Cryst. *1*, 263 (1966), C.A. *65*, 14556d (1966); Structure des savons de Rb. [242] B. Gallot, A. Skoulios, Kolloid-Z. Z. Polym. *222*, 51 (1968), C.A. *68*, 88409c (1968); IV. Disavons de Li, K, Rb, Cs, Na (cf. [239]). [243] D. Guillon, A. Mathis, A. Skoulios, J. Phys. (Paris) *36*, 695 (1975), C.A. *83*, 89016u (1975); Étude par diffraction des rayons X aux petits angles du polymorphisme smectique du di-(*p-n*-octadécyloxybenzylidèneamino)-4,4'-diphényle. [244] D. Guillon, A. Skoulios, J. Phys. (Paris), C-3, *37*, 83 (1976), C.A. *85*, 134770q (1976); Étude du polymorphisme smectique par dilatométrie et diffractométrie X. [244a] D. Guillon, A. Skoulios, Mol. Cryst. Liq. Cryst. *38*, 32 (1977), C.A. *86*, 181023d (1977); Polymorphisme smectique. III. Aire moléculaire et angle d'inclinaison des molécules dans la série des 4,4'-di(*p,n*-alcoxybenzylidènamino)biphényles. [244b] D. Guillon, A. Skoulios, Mol. Cryst. Liq. Cryst. *39*, 183 (1977), C.A. *87*, 109765a (1977); Diffraction des rayons X de la série des 4,4'-di(*p-n*-alcoxybenzylidèneamino)biphényles. [245] G. W. Stewart, H. R. Letner, Proc. Iowa Acad. Sci. *42*, 153 (1935), C.A. *30*, 7943[5] (1936); Liquid-crystalline and isotropic states with special reference to PAA. [246] H. T. Tan, Phys. Lett. A *48*, 309 (1974), C.A. *81*, 126930u (1974); X-ray scattering from s$_A$ phase. [247] A. Tardieu, J. Billard, J. Phys. (Paris), C-3, *37*, 79 (1976), C.A. *85*, 85784m (1976); Structure of the s$_D$ modification. [247a] F. P. Temme, Chem. Phys. Lett. *48*, 518 (1977), C.A. *87*, 93751e (1977); Neutron scattering: form of the elastic incoherent structure factor for "monocrystal" aligned nematogen with specific Q to n orientation. [248] H. Terauchi, K. Nakatsu, S. Kusabayashi, Jap. J. Appl. Phys. *11*, 763 (1972), C.A. *77*, 40131q (1972); Small angle X-ray scattering near n/i transition temperature. [249] H. Terauchi, T. Takeuchi, S. Kusabayashi, Jap. J. Appl. Phys. *11*, 1862 (1972), C.A. *78*, 63523d (1973); X-ray scattering from a smectic. [250] T. Takeuchi, H. Terauchi, K. Nakatsu, S. Kusabayashi, Jap. J. Appl. Phys. *12*, 1639 (1973), C.A. *79*, 150378z (1973); X-ray halo scattering from liquid crystals.

[251] H. Terauchi, T. Takeuchi, K. Nakatsu, N. Maruyama, Jpn. J. Appl. Phys. *13*, 1203 (1974), C.A. *81*, 142263j (1974); X-ray study of phase transitions in liquid crystals. [252] H. Terauchi, R. Ohnishi, J. Phys. Soc. Jpn. *40*, 915 (1976), C.A. *84*, 158232w (1976); Fluctuation of smectic phase order parameter in nematic phase. [253] J. Toepler, B. Alefeld, T. Springer, Mol. Cryst. Liq. Cryst. *26*, 297 (1974), C.A. *82*, 116457b (1975); Quasielastic neutron scattering to determine self-diffusion constants in nematic PAA. [254] J. J. Duruz, A. R. Ubbelohde, Proc. Roy. Soc. A *330*, 1 (1972), C.A. *77*, 131741e (1972); Structure of organic ionic melt mesophases. [255] D. L. Uhrich, J. M. Wilson, W. A. Resch, Phys. Rev. Lett. *24*, 355 (1970), C.A. *72*, 84736s (1970); Moessbauer investigation of the smectic state. [256] J. M. Wilson, D. L. Uhrich, Mol. Cryst. Liq. Cryst. *13*, 85 (1971), C.A. *75*, 54875f (1971); Theory of Mössbauer spectral asymmetry of quadrupole split lines in liquid crystals. [257] R. E. Detjen, D. L. Uhrich, C. F. Sheley, Phys. Lett. A *42*, 522 (1973), C.A. *78*, 102957v (1973); Moessbauer comparison of the recoil-free fraction of a supercooled smectic with its solid state. [258]

D. L. Uhrich, Y. Y. Hsu, D. L. Fishel, J. M. Wilson, Mol. Cryst. Liq. Cryst. *20*, 349 (1973), C.A. *79*, 98946s (1973); Moessbauer study of a Sn-119 bearing solute in an ordered smectic at 77 K. [**259**] D. L. Uhrich, R. E. Detjen, J. M. Wilson, Moessbauer Eff. Methodol. *8*, 175 (1973), C.A. *81*, 43746n (1974); Use of liquid crystals in Moessbauer studies and the use of the Moessbauer effect in liquid crystal studies. [**260**] D. L. Uhrich, J. Stroh, R. D'Sidocky, D. L. Fishel, Chem. Phys. Lett. *24*, 539 (1974), C.A. *80*, 137835x (1974); Moessbauer measurement of some lattice properties of a smectic H. [**261**] J. M. Wilson, D. L. Uhrich, Mol. Cryst. Liq. Cryst. *25*, 113 (1974), C.A. *83*, 139513b (1975); Reinterpretation of the Fe-57 Moessbauer effect of 1,1′-diacetylferrocene in 4,4′-bis(heptyloxy)az-oxybenzene. [**262**] D. L. Uhrich, V. O. Aimiuwu, P. I. Ktorides, W. J. LaPrice, Phys. Rev. A *12*, 211 (1975), C.A. *83*, 123731k (1975); s_B glass (at 77 K) as seen by the Moessbauer effect of tin-bearing solute molecules. [**263**] V. O. Aimiuwu, D. L. Uhrich, Bull. Am. Phys. Soc. *20*, 887 (1975); Fe-57 Moessbauer studies in a s_B glass. [**264**] P. I. Ktorides, D. L. Uhrich, R. M. D'Sidocky, D. L. Fishel, Bull. Am. Phys. Soc. *20*, 886 (1975); Sn-119 Moessbauer study of the smectic glass phase of p-(11-trimethyl-tin)-undecyloxybenzylidene-p′-n-butylaniline. [**264a**] R. E. Detjen, D. L. Uhrich, Moessbauer Eff. Metho-dol. *9*, 113 (1974), C.A. *86*, 149043z (1977); Moessbauer observation of anisotropic diffusion near the glass transition of a smectic H. [**264b**] P. I. Ktorides, D. L. Uhrich, Mol. Cryst. Liq. Cryst. *40*, 285 (1977), C.A. *87*, 125061y (1977); Orientation dependence of an unresolved Moessbauer quadrupole doublet in an aligned liquid crystalline glass. [**264c**] V. O. Aimiuwu, D. L. Uhrich, Mol. Cryst. Liq. Cryst. *43*, 295 (1977); ^{57}Fe Mössbauer study of four ferrocene derivatives in a s_B glass. [**265**] J. S. van der Lingen, J. Franklin Inst. *192*, 511 (1921), C.A. *16*, 19³ (1922); X-ray and IR investigations of the molecular structure of liquid crystals. [**266**] J. S. van der Lingen, J. Franklin Inst. *191*, 651 (1921), C.A. *15*, 2570⁶ (1921); Anisotropic liquids. [**267**] F. Volino, A. J. Dianoux, R. E. Lechner, H. Hervet, J. Phys. (Paris), C-1, *36*, 83 (1975), C.A. *83*, 78265y (1975); End chain motion in the solid phase of TBBA. [**268**] F. Volino, A. J. Dianoux, H. Hervet, Solid State Commun. *18*, 453 (1976), C.A. *84*, 172336q (1976); s_H/s_{VI} pretransitional effect by orientational ordering in TBBA. [**269**] F. Volino, A. J. Dianoux, H. Hervet, J. Phys. (Paris), C-3, *37*, 55 (1976), C.A. *85*, 102619f (1976); Neutron quasi-elastic scattering study of rotational motions in the s, s_H, and s_{VI} phases of TBBA. [**269a**] F. Volino, A. J. Dianoux, H. Hervet, Mol. Cryst. Liq. Cryst. *38*, 125 (1977), C.A. *86*, 181026g (1977); Incoherent neutron quasi-elastic scattering as a tool to study molecular ordering in liquid crystals. [**270**] W. W. Walker, E. L. Mueller, III, Appl. Phys. *3*, 155 (1974), C.A. *81*, 7097t (1974); Positron lifetimes in five phases of 4-butoxybenzal-4′-ethylaniline.

[**271**] J. H. Wendorff, F. P. Price, ACS Symp. New York *1971*, Abstr. Nr. 179; X-ray diffraction studies on mesogens. [**272**] J. H. Wendorff, F. P. Price, Mol. Cryst. Liq. Cryst. *24*, 129 (1973), C.A. *81*, 142307b (1974); Structure of mesophases of cholesteryl esters. [**273**] J. H. Wendorff, F. P. Price, Mol. Cryst. Liq. Cryst. *25*, 71 (1974), C.A. *83*, 124548t (1975); X-ray diffraction studies of the solid phases of cholesteryl acetate. [**274**] A. N. Weselov, W. W. Andrushkvich, 1. Wiss. Konferenz über flüssige Kristalle, 17.–19. 11. 1970, Iwanowo, Sammlung der Vorträge (1972) Nr. 14; Automatischer Temperaturregler für Röntgenkammern. [**275**] J. Wesolowski, M. Szuszkiewicz, S. Szuszkiewicz, Acta Phys. Polon. *29*, 97 (1966), C.A. *67*, 38655p (1967); Effect of ordering on the angular distribution of annihilation quanta from PAA. [**276**] J. M. Wilson, Diss. Kent State Univ., Kent, 1972, 109 pp. Avail. Univ. Microfilms, Ann Arbor, Mich. Ord. No. 73-13,313, C.A. *79*, 130100z (1973); Moessbauer spectroscopy in smectics. [**277**] R. I. Zharenov, Sb. Dokl. Vses. Nauch. Konf. Zhidk. Krist. Simp. Ikh Prakt. Primen., 2nd, *1972*, 68, C.A. *81*, 127939r (1974); X-ray analysis of the structure of binary mixtures of two first homologs of p-alkoxybenzal p′-butylanilines.

Chapter 6

[**1**] N. I. Afanas'eva, V. M. Burlakov, G. N. Zhizhin, Pis'ma Zh. Eksp. Teor. Fiz. *23*, 506 (1976); JETP Lett. *23*, 461 (1976), C.A. *85*, 54879s (1976); Phonon spectra in the neighborhood of a k/n transition and ordering parameters. [**1a**] A. A. Adkhamov, I. M. Aref'ev, B. S. Umarov, Dokl. Akad. Nauk Tadzh. SSR *20*, 22 (1977), C.A. *87*, 46048w (1977); Spectra of Mandelshtain-Brillouin scattering in the i phase of MBBA in the transition region. [**1b**] N. I. Afanas'eva, G. N. Zhizhin,

B. M. Zuev, V. I. Kovalenko, L. N. Konnova, Fiz. Tverd. Tela *19*, 1137 (1977), C.A. *87*, 13700z (1977); Low-frequency spectra of liquid crystals of terephthalates. [**1c**] N. I. Afanas'eva, E. I. Balabanov, A. V. Bobrov, V. M. Burlakov, A. I. Vasil'ev, G. N. Zhizhin, Fiz. Tverd. Tela *19*, 1932 (1977), C.A. *87*, 124840q (1977); Low-frequency spectra of butoxybenzylindeneaminobenzonitrile in crystal and liquid-crystal states. [**1d**] S. A. Akopvan, S. M. Arakelyan, L. E. Arushanyan, Yu. S. Chilingaryan, Kvantovaya Elektron *4*, 1387 (1977), C.A. *87*, 143896d (1977); Spatial coherency of laser radiation passing through a liquid crystal near the i transition. [**1e**] A. B. Aleinikov, G. P. Petrova, Opt. Spektrosk. *43*, 267 (1977), C.A. *87*, 143473p (1977); Spectra of the molecular scattering of light in the i phase of nematics. [**2**] G. R. Alms, T. D. Gierke, W. H. Flygare, J. Chem. Phys. *61*, 4083 (1974), C.A. *82*, 49425z (1975); Depolarized Rayleigh scattering in liquids. Density and temperature dependence of the orientational pair correlations in liquid composed of anisotropic molecules. [**3**] N. M. Amer, Y. R. Shen, H. Rosen, Phys. Rev. Lett. *24*, 718 (1970), C.A. *72*, 116479d (1970); Raman study of PAA at the phase transitions. [**4**] N. M. Amer, Y. R. Shen, J. Chem. Phys. *56*, 2654 (1972), C.A. *76*, 105900r (1972); Raman scattering from nematic azoxybenzenes. [**5**] N. M. Amer, Y. R. Shen, Solid State Commun. *12*, 263 (1973), C.A. *78*, 116283j (1973); Low-frequency Raman mode in smectics near the phase transitions. [**6**] N. M. Amer, Y. S. Lin, Y. R. Shen, Solid State Commun. *16*, 1157 (1975), C.A. *83*, 51217x (1975); Temperature dependence of Rayleigh-wing scattering from nematic MBBA. [**7**] A. Azima, C. W. Brown, S. S. Mitra, Spectrochim. Acta A *31*, 1475 (1975), C.A. *83*, 178082e (1975); Temperature dependence of the IR spectra of liquid crystalline *p*-heptoxybenzoic acid and *p*-octoxybenzoic acid. [**7a**] A. Azima, Diss. Univ. Rhode Island, Kingston, R. I. 1976, 123 pp. Avail. Xerox Univ. Microfilms, Ann Arbor, Mich., Order No. 77-7623, C.A. *87*, 14462k (1977); Investigation of thermotropic liquid crystals by IR and Raman spectroscopy. [**8**] D. A. Balzarini, Phys. Rev. Lett. *25*, 914 (1970), C.A. *73*, 134731e (1970); Temperature dependence of birefringence in liquid crystals. [**9**] J. W. Baran, J. Kedzierski, Z. Raszewski, J. Zmija, Biul. Wojsk. Akad. Tech. *25*, 147 (1976), C.A. *86*, 80985p (1977); Determination of the anisotropic dispersion of the refractive index of liquid crystals by the interference wedge method. [**10**] D. Barbero, R. Malvano, M. Omini, Mol. Cryst. Liq. Cryst. *39*, 69 (1977), C.A. *87*, 31287z (1977); Refractive indexes of nematics.

[**11**] C. Bästlein, Diss. Halle 1912; Untersuchungen über Brechungskoeffizienten flüssiger Kristalle. [**12**] G. Baur, Proc. Int. Sch. Phys. "Enrico Fermi" *59*, 751 (1973), C.A. *86*, 88631r (1977); Fluorescence in liquid crystals. [**13**] D. P. Benton, P. G. Howe, I. E. Puddington, Can. J. Chem. *33*, 1384 (1955), C.A. *50*, 16271c (1956); Mesomorphism of anhydrous soaps. I. Light transmission by alkali metal stearates. [**14**] V. I. Berezin, Y. I. Nedranets, V. P. Sevest'yanov, Izv. Vyssh. Uchebn. Zaved., Fiz. *19*, 158 (1976), C.A. *86*, 71640v (1977); Electronic absorption spectra, stoichiometric composition, and stability constants of complexes with charge transfer in liquid crystals. [**14a**] V. I. Berezin, N. V. Bogachev, Yu. I. Nedranets, V. P. Sevost'yanov, Zh. Fiz. Khim. *51*, 1814 (1977), C.A. *87*, 133368u (1977); Charge-transfer complexes of azoxybenzenes. [**15**] D. Berger, J. P. Heger, R. Mercier, Helv. Phys. Acta *47*, 426 (1974), C.A. *82*, 50026b (1975); Mesure du paramètre d'ordre dans une phase nématique par diffusion Rayleigh. [**16**] J. Billard, M. Delhaye, J. C. Merlin, G. Vergoten, C. R. Acad. Sci., Sér. B *273*, 1105 (1971), C.A. *76*, 147010f (1972); Spectres Raman de basses fréquences du MBBA. [**17**] V. G. Rumyantsev, L. M. Blinov, V. A. Kizel, Sb. Dokl. Vses. Nauch. Konf. Zhidk. Krist. Simp. Ikh Prakt. Primen., 2nd, *1972*, 191, C.A. *82*, 49347a (1975); Spectroscopy of the molecules of dyes in liquid crystal matrixes. [**18**] V. G. Rumyantsev, L. M. Blinov, V. A. Kizel, Kristallografiya *18*, 1101 (1973), C.A. *80*, 20375f (1974); Determination of the degree of order of liquid crystals according to the dichroism of dyes dissolved in them. [**19**] L. M. Blinov, V. A. Kizel, V. G. Rumyantsev, V. V. Titov, J. Phys. (Paris) C-1, *36*, 69 (1975), C.A. *83*, 36027s (1975); Study of nematics by means of the guest dye dichroism. [**20**] L. M. Blinov, G. G. Dyadyusha, F. A. Mikhailenko, I. L. Mushkalo, V. G. Rumyantsev, Dokl. Akad. Nauk SSSR *220*, 860 (1975), C.A. *83*, 50175v (1975); Polarization of absorption bands of biscyanine dye solutions in liquid crystals.

[**21**] L. M. Blinov, V. A. Kizel, V. G. Rumyantsev, V. V. Titov, Kristallografiya *20*, 1245 (1975), C.A. *84*, 82774b (1976); Structure of nematics studied by optical methods. [**22**] V. G. Rumyantsev, L. M. Blinov, J. Freimanis, J. Dregeris, Zh. Strukt. Khim. *16*, 222 (1975), C.A. *83*, 96155s (1975); Structure of auto complexes and nature of charge transfer studied from the dichroism of their solutions in liquid crystals. [**22a**] A. Bloom, P. L. K. Hung, Mol. Cryst. Liq. Cryst. *40*, 213 (1977), C.A. *87*, 125606e (1977); Effect of dye structure on order parameter in a nematic host. [**22b**] A. Bloom, P. L. K. Hung, D. Meyerhofer, Mol. Cryst. Liq. Cryst. *41*, 1 (1977), C.A. *87*, 125580s (1977); Effect of host on pleochroic dye order parameter. [**23**] Y. S. Bobovich, N. M. Belyaevskaya, Zh. Prikl.

Spektrosk. *8*, 1018 (1968), C.A. *69*, 91629a (1968); Raman spectra of N-anisylidene-4-(phenylazo)-1-naphthylamine in the crystalline and vitrified liquid-crystalline states. [24] Y. S. Bobovich, N. M. Belyaevskaya, Zh. Prikl. Spektrosk. *10*, 679 (1969), C.A. *71*, 26480x (1969); Spectroscopic indications of the crystallization of viscous films of vitreous N-(methoxybenzylidene)-α-(phenylazo)-α'-naphthylamine on stretching. [25] W. J. Borer, S. S. Mitra, C. W. Brown, Phys. Rev. Lett. *27*, 379 (1971), C.A. *75*, 113429p (1971); Crystal to liquid-crystal transition studied by Raman scattering. [26] M. Born, F. Stumpf, Sitzb. kgl. preuss. Akad. Wissenschaften *1916*, 1043, C.A. *11*, 2986[8] (1917); Anisotrope Flüssigkeiten II. Temperaturabhängigkeit der Brechungsindices senkrecht zur optischen Achse. [27] C. C. Bott, T. Kurucsev, J. Chem. Soc. Faraday Trans. 2, *71*, 749 (1975), C.A. *82*, 169621g (1975); Determination of transition moment directions by means of dichroitic spectra in stretched polymer films. I. Orientation of solutes. [27a] G. W. Bradberry, J. M. Vaughan, J. Phys. C *9*, 3905 (1976), C.A. *86*, 131425r (1977); Brillouin scattering in s_A and i phases. [27b] G. W. Bradberry, J. M. Vaughan, Phys. Lett. A *62*, 225 (1977), C.A. *87*, 125596b (1977); Measurement of the interlayer elastic constant in a liquid crystal close to the s_A/n transition. [28] V. D. Neff, W. L. Gulrich, G. H. Brown, Mol. Cryst. *1*, 225 (1966), C.A. *65*, 8105b (1966); Determination of the degree of orientation in thin films of nematics from IR dichroic measurements in a homogeneous electric field. [29] M. Brunet-Germain, Mol. Cryst. Liq. Cryst. *11*, 289 (1970), C.A. *74*, 58760s (1971); Indices des mélanges de PAA et de PAP dans l'état nématique. Interprétation des résultats à l'aide de la théorie de Maier et Saupe. [30] M. Brunet-Germain, C. R. Acad. Sci., Sér. B *271*, 1075 (1970), C.A. *74*, 69598w (1971); Indices du MBBA.

[31] M. Brunet, J. C. Martin, C. R. Acad. Sci., Sér. B *278*, 283 (1974), C.A. *80*, 125579d (1974); Indices du propoxy-4-heptyl-4'-tolane. [32] M. Brunet, C. Cabos, J. Sicart, C. R. Acad. Sci., Sér. B *281*, 109 (1975), C.A. *83*, 185622j (1975); Indices d'un mélange équimoléculaire de deux tolanes, nématique à l'ambiante: propoxy-4-heptyl-4'-tolane et méthoxy-4-pentyl-4'-tolane. Interprétation à l'aide de la théorie de Vuks. [33] B. J. Bulkin, D. Grunbaum, A. V. Santoro, J. Chem. Phys. *51*, 1602 (1969), C.A. *71*, 96484a (1969); Changes in the IR spectrum at the k/n transition. [34] B. J. Bulkin, D. Grunbaum, Liq. Cryst. Ord. Fluids *1970*, 303, C.A. *78*, 152447z (1973); IR spectroscopic measurements on the k/n transition. [35] B. J. Bulkin, F. T. Prochaska, J. Chem. Phys. *54*, 635 (1971), C.A. *74*, 47777b (1971); Raman spectrum of PAA in k, n, and i phases, 10–100 cm^{-1} region. [36] B. J. Bulkin, J. O. Lephardt, K. Krishnan, Mol. Cryst. Liq. Cryst. *19*, 295 (1973), C.A. *78*, 90473u (1973); Relative intensities in Raman spectra of cholesteric and nematic solutions. [37] B. J. Bulkin, W. B. Lok, J. Phys. Chem. *77*, 326 (1973), C.A. *78*, 90677p (1973); Far-IR study of intermolecular modes in PAA and MBBA. [38] B. J. Bulkin, T. Kennelly, W. Bong Lok, Liq. Cryst. Ord. Fluids *2*, 85 (1974), C.A. *86*, 180978g (1977); IR spectroscopic measurements of order in nematics and nematic solutions. [39] B. J. Bulkin, D. Grunbaum, T. Kennelly, W. B. Lok, Liq. Cryst., Proc. Int. Conf., Bangalore, *1973*, 155, C.A. *84*, 113831z (1976); IR and Raman spectra of nematics and nematogenic crystals. [40] D. Grunbaum, B. J. Bulkin, J. Phys. Chem. *79*, 821 (1975), C.A. *82*, 177657p (1975); Calculation of IR and Raman active lattice vibration frequencies of PAA.

[41] B. J. Bulkin, Adv. Liq. Cryst. *2*, 199 (1976), C.A. *86*, 179633j (1977); Vibrational spectroscopy of liquid crystals. [42] B. J. Bulkin, U.S.NTIS, AD Rep. *1976*, AD-A029404, 96 pp., C.A. *86*, 63869m (1977); Interaction of small molecules with liquid crystals; a spectroscopic study. [42a] B. J. Bulkin, Prog. Anal. Chem. *6*, 1 (1973), C.A. *86*, 197014n (1977); Raman spectroscopy for the study of nematics. [43] L. M. Cameron, Mol. Cryst. Liq. Cryst. *7*, 235 (1969), C.A. *71*, 85516m (1969); Depolarization of light scattered from liquid crystals. [44] S. J. Candau, Ann. Phys. (Paris) *4*, 21 (1969), C.A. *72*, 126737a (1970); Diffusion inélastique de la lumière par les liquides. [45] G. P. Ceasar, H. B. Gray, J. Am. Chem. Soc. *91*, 191 (1969), C.A. *70*, 62597d (1969); Polarized electronic spectroscopy of molecules oriented by a nematic. [46] G. P. Ceasar, R. A. Levenson, H. B. Gray, J. Am. Chem. Soc. *91*, 772 (1969), C.A. *70*, 91962z (1969); Polarized IR spectroscopy of molecules oriented in a nematic. Application to decacarbonyldiamanganese (0) and decacarbonyldiruthenium (0). [47] E. D. Cehelnik, R. B. Cundall, C. J. Timmons, R. M. Bowley, Proc. Roy. Soc. A *335*, 387 (1973), C.A. *79*, 151462j (1973); Spectroscopic studies of trans-1,6-diphenyl-1,3,5-hexatriene in ordered liquid crystal solutions. [48] E. D. Cehelnik, R. B. Cundall, J. R. Lockwood, T. F. Palmer, J. Chem. Soc. Faraday Trans. 2, *70*, 244 (1974), C.A. *80*, 144963r (1974); Time dependent fluorescence polarization studies using isotropic and liquid crystal media. [49] E. D. Cehelnik, K. D. Mielenz, R. B. Cundall, J. Res. Natl. Bur. Stand., Sect. A, *80*, 15 (1976), C.A. *85*, 70282x (1976); Polarization of fluorescence of ordered systems with application to ordered liquid crystals. [50] S. Chandrasekhar, D. Krishnamurti, Nature *212*, 746 (1966), C.A. *66*, 24005u (1967); Vibrational spectrum of liquid crystalline methyl stearate.

[51] S. Chandrasekhar, D. Krishnamurti, Phys. Lett. *23*, 459 (1966), C.A. *66*, 50024u (1967); Birefringence of nematics. [52] S. Chandrasekhar, D. Krishnamurti, N. V. Madhusudana, Mol. Cryst. Liq. Cryst. *8*, 45 (1969), C.A. *71*, 85564a (1969); Theory of birefringence of nematics. [53] S. Chandrasekhar, N. V. Madhusudana, Appl. Spectrosc. Rev. *6*, 189 (1972), C.A. *78*, 9511k (1973); Spectroscopy of liquid crystals. [54] R. Chang, Mater. Res. Bull. *7*, 267 (1972), C.A. *76*, 160254h (1972); Application of polarimetry and interferometry to liquid crystal-film research. [55] R. Chang, F. B. Jones, G. D. Simpson, Bull. Am. Phys. Soc. *19*, 1145 (1974); Temperature and wave-number dependence of birefringence of oriented nematic films. [56] R. Chang, Mol. Cryst. Liq. Cryst. *28*, 1 (1974), C.A. *82*, 162028k (1975); Anisotropic refractive indexes of aligned MBBA films. [57] R. Chang, Mol. Cryst. Liq. Cryst. *30*, 155 (1975), C.A. *83*, 106562p (1975); Orientational order in MBBA from optical anisotropy measurements. [58] R. Chang, Mol. Cryst. Liq. Cryst. *34*, 65 (1976); The anisotropic refractive indices of MBBA. [59] P. Châtelain, Compt. rend. *200*, 412 (1935), C.A. *29*, 2039[8] (1935); Mesure des indices du PAP à l'état de liquide anisotrope. [60] P. Châtelain, Compt. rend. *203*, 1169 (1936), C.A. *31*, 589[4] (1937); Étude du PAA sous les états solide, liquide anisotrope et liquide isotrope.

[61] P. Châtelain, Compt. rend. *204*, 1352 (1937), C.A. *31*, 4866[2] (1937); La biréfringence des cristaux liquides est-elle indépendante de l'action des parois ou de l'action du champ magnétique? [62] P. Châtelain, Compt. rend. *218*, 652 (1944), C.A. *40*, 2707[6] (1946); La diffusion de la lumière par les cristaux liquides. [63] P. Châtelain, Compt. rend. *222*, 229 (1946), C.A. *40*, 3662[5] (1946); La diffusion, par les cristaux liquides, de la lumière polarisée. [64] P. Châtelain, Compt. rend. *224*, 130 (1947), C.A. *41*, 4019e (1947); La lumière diffusée vers l'arrière par les cristaux liquides. [65] P. Châtelain, Acta Cryst. *1*, 315 (1948), C.A. *43*, 3257i (1949); La diffusion, par les cristaux liquides du type nématique, de la lumière polarisée. [66] P. Châtelain, Compt. rend. *227*, 136 (1948), C.A. *42*, 8564a (1948); L'interprétation, par les fluctuations d'orientation, de l'état de polarisation de la lumière diffusée par les cristaux liquides. [67] O. Pellet, P. Châtelain, Bull. soc. franç. minéral. *73*, 154 (1950), C.A. *45*, 919c (1951); Indices des cristaux liquides: mesure par la méthode du prisme et étude théorique. [68] P. Châtelain, Acta Cryst. *4*, 453 (1951), C.A. *45*, 9969b (1951); Étude théorique de la diffusion de la lumière par un fluide présentant un seul axe d'isotropie: application aux cristaux liquides nématiques. [69] P. Châtelain, Bull. soc. franç. minéral. *77*, 353 (1954), C.A. *49*, 7913c (1955); La lumière diffusée par les cristaux liquides nématiques. [70] P. Châtelain, Bull. soc. franç. minéral. *78*, 262 (1955), C.A. *49*, 15368e (1955); Determination du facteur d'orientation du monocristal liquide nématique a partir des valeurs des indices.

[71] P. Châtelain, M. Germain, Compt. rend. *259*, 127 (1964), C.A. *61*, 10162d (1964); Indices des mélanges de PAA et de PAP dans l'état nématique. [72] P. Châtelain, J. Phys. (Paris), C-4, *30*, 3 (1969); Détermination de la structure des monocristaux liquides nématiques par l'étude des propriétés optiques. [73] B. Chu, C. S. Bak, F. L. Lin, Phys. Rev. Lett. *28*, 1111 (1972), C.A. *76*, 146067t (1972); Coherence length in the i phase of a roomtemperature nematic. [74] B. Chu, C. S. Bak, F. L. Lin, J. Chem. Phys. *56*, 3717 (1972), C.A. *76*, 132768k (1972); Coherence lenght in isotropic liquid MBBA. [75] K. C. Chu, W. L. McMillan, Phys. Rev. A *11*, 1059 (1975), C.A. *82*, 163335v (1975); Static and dynamic behavior near a second-order s_A/n transition by light scattering. [76] E. Gulari, B. Chu, J. Chem. Phys. *62*, 798 (1975), C.A. *83*, 19241q (1975); Short-range order fluctuations in isotropic liquid MBBA. [77] P. E. Cladis, A. E. White, Appl. Phys. *47*, 1256 (1976), C.A. *84*, 187419t (1976); A polarizing oriented smectic beam splitter. [78] N. A. Clark, Y. Liao, J. Chem. Phys. *63*, 4133 (1975), C.A. *84*, 23869p (1976); Inelastic light scattering from isotropic MBBA. [79] M. Copic, B. B. Lavrencic, J. Phys. (Paris) C-1, *36*, 89 (1975), C.A. *83*, 17916q (1975); Brillouin scattering in liquid crystals. [80] E. Courtens, G. Koren, Phys. Rev. Lett. *35*, 1711 (1975), C.A. *84*, 52433x (1976); Measurement of coherence-lenght anisotropy in the i phase of nematics. [80a] E. Courtens, J. Chem. Phys. *66*, 3995 (1977), C.A. *86*, 198227w (1977); Coherence-length anisotropy above the n/i transition in a series of benzylidene anilines.

[81] B. Cvikl, D. Moroi, W. Franklin, Mol. Cryst. Liq. Cryst. *12*, 267 (1971), C.A. *75*, 42490q (1971); Form birefringence of smectics. [82] A. Davidsson, B. Norden, Chem. Scr. *8*, 95 (1975), C.A. *83*, 185614h (1975); Correction of quantitative linear dichroism. [83] P. G. de Gennes, J. Phys. (Paris) Lett. *35*, 217 (1974); Light-scattering from random disclinations in a nematic. [84] M. Delaye, P. Keller, Phys. Rev. Lett. *37*, 1065 (1976), C.A. *85*, 185031g (1976); Critical angular fluctuations of molecules above a second-order s_A/s_C transition. [85] C. Destrade, H. Gasparoux, J. Phys. (Paris), Lett. *36*, 105 (1975), C.A. *82*, 178699r (1975); Étude de la structure des chaînes dans les différentes phases, k, n, i du MBBA. [86] C. Destrade, F. Guillon, H. Gasparoux, Mol. Cryst. Liq. Cryst. *36*, 115 (1976),

C.A. *86*, 139171a (1977); Application de la diffusion Raman à la détermination de l'ordre de chaîne alkyl dans deux séries homologues de composés némátogénes. [87] V. K. Dolganov, Fiz. Tverd. Tela *18*, 1786 (1976), C.A. *85*, 84966k (1976); Polarized luminescence of a nematic. [88] E. Dorn, W. Lohmann, Ann. Physik *29*, 533 (1908), C.A. *3*, 2525[4] (1909); Bestimmung der optischen Konstanten flüssiger Kristalle. [89] E. Dorn, Physik. Z. *11*, 777 (1910), C.A. *4*, 3039[4] (1910); Optik flüssiger Kristalle. [90] H. Dunken, N. Trzebowski, Vorträge Originalfassung Intern. Kongr. Grenzflächenaktive Stoffe, 3, Cologne, *1*, 289 (1960), C.A. *57*, 9367a (1962); IR-spektroskopische Untersuchungen an Alkalistearaten.

[91] G. Durand, L. Léger, F. Rondelez, M. Veyssié, Phys. Rev. Lett. *22*, 1361 (1969), C.A. *71*, 43220e (1969); Quasi-elastic Rayleigh scattering in nematics. [92] M. Lefevre, J. L. Martinand, G. Durand, M. Veyssié, C. R. Acad. Sci., Sér. B *273*, 403 (1971), C.A. *76*, 65688u (1972); Biréfringence de deux textures uniformes du di-(4-n-décyloxybenzal)-2-chloro-1,4-phénylènediamine en phase s$_C$. [93] Y. Galerne, J. L. Martinand, G. Durand, M. Veyssié, Phys. Rev. Lett. *29*, 562 (1972), C.A. *77*, 120402u (1972); Quasielectric Rayleigh scattering in a smectic C. [94] R. Ribotta, G. Durand, J. D. Litster, Solid State Commun. *12*, 27 (1973), C.A. *78*, 90659f (1973); Rayleigh scattering induced by static bends of layers in a smectic A. [95] M. Delaye, R. Ribotta, G. Durand, Phys. Rev. Lett. *31*, 443 (1973), C.A. *79*, 97903p (1973); Rayleigh scattering at a second-order n/s$_A$ transition. [96] G. Durand, Liq. Cryst., Proc. Int. Conf., Bangalore, *1973*, 23, C.A. *84*, 114245y (1976); Rayleigh scattering of light in smectics A. [97] R. Ribotta, D. Salin, G. Durand, Phys. Rev. Lett. *32*, 6 (1974), C.A. *80*, 101010k (1974); Quasielastic Rayleigh scattering in a smectic A. [98] D. Salin, I. W. Smith, G. Durand, J. Phys. (Paris), Lett. *35*, 165 (1974), C.A. *81*, 142324e (1974); Dynamics of angular fluctuations in a liquid crystal near a second-order n/s$_A$ transition. [99] D. Dvorjeski, V. Volterra, E. Wiener-Avnear, Phys. Rev. A *12*, 681 (1975), C.A. *83*, 140073q (1975); Raman study on several smectic phases in TBBA. [100] I. E. Dzyaloshinskii, S. G. Dmitriev, E. I. Katz, JETP Lett. *19*, 305 (1974); Influence of long-range van der Waals forces on the scattering of light in liquid crystals.

[101] I. E. Dzyaloshinskii, S. G. Dmitriev, E. I. Kats, Zh. Eksp. Teor. Fiz. *68*, 2335 (1975); Van der Waals forces and scattering of light in liquid crystals. [101a] I. B. Dzyaloshinskii, Theory Light Scattering Solids, Proc. Sov.-Am. Symp., 1st *1975*, 79, C.A. *87*, 124789e (1977); Van der Waals forces and light scattering in liquid crystals. [102] M. Evans, M. Davies, I. Larkin, J. Chem. Soc. Faraday Trans. 2, *69*, 1011 (1973), C.A. *79*, 71862z (1973); Molecular motion and molecular interaction in the n and i phases of a liquid crystal. [102a] G. J. Evans, M. Evans, J. Chem. Soc. Faraday Trans. 2, *73*, 285 (1977), C.A. *86*, 197108w (1977); High and low frequency torsional absorptions in nematic K21. [103] F. Falgueirettes, Compt. rend. *234*, 2619 (1952), C.A. *46*, 8922b (1952); Propriétés optiques de l'acide p-butyloxybenzoique à l'état de liquide nématique et de liquide isotrope. [104] H. G. Fellner, Diss. Kent State Univ., Kent, Ohio, 1973, 140 pp. Avail. Univ. Microfilms, Ann Arbor, Mich., Order No. 73-32,340, C.A. *81*, 31326n (1974); Light scattering from liquid crystals. [105] H. Fellner, W. Franklin, S. Christensen, Phys. Rev. A *11*, 1440 (1975), C.A. *83*, 35052j (1975); Quasielastic light scattering from nematic MBBA. [106] J. R. Fernandes, S. Venugopalan, Mol. Cryst. Liq. Cryst. *35*, 113 (1976), C.A. *85*, 114100t (1976); IR spectroscopic study of orientational order and phase transformations in liquid crystalline CBOOA. [107] J. Fischer, Diss. Halle 1932; Lichtabsorption kristallin-flüssiger Substanzen. [108] J. Fischer, Z. physik. Chem. A *160*, 101 (1932), C.A. *26*, 3971[9] (1932); Lichtabsorption kristallin-flüssiger Substanzen. [109] W. H. Flygare, T. D. Gierke, Annu. Rev. Mater. Sci. *4*, 255 (1974), C.A. *81*, 129059c (1974); Light scattering in noncrystalline solids and liquid crystals. [110] C. Flytzanis, Y. R. Shen, Phys. Rev. Lett. *33*, 14 (1974), C.A. *81*, 55243j (1974); Molecular theory of orientational fluctuations and optical Kerr effect in the i phase of a liquid crystal.

[111] M. P. Fontana, S. Bini, Phys. Rev. A *14*, 1555 (1976), C.A. *85*, 184394r (1976); Temperature dependence of Raman scattering from monocrystalline TBBA. Low-frequency spectra. [111a] D. Fracko-wiak, D. Bauman, H. Manikowski, T. Martynski, Biophys. Chem. *6*, 369 (1977), C.A. *87*, 81412g (1977); Spectral properties of chlorophyll a in liquid crystal. [112] M. J. Freiser, R. J. Joenk, Phys. Lett. A *24*, 683 (1967), C.A. *68*, 73540q (1968); Enhancement of self-trapping by cooperative phenomena; application to liquid crystals. [113] R. Freymann, R. Servant, Ann. phys. *20*, 131 (1945), C.A. *40*, 1394[4] (1946); L'effet Raman du PAA. [114] M. Schadt, P. Rihak, H. H. Guenthard, Z. Naturforsch. A *31*, 1098 (1976), C.A. *85*, 169976p (1976); Hydrocarbon chain ordering in liquid crystals investigated by means of IR attenuated total reflection spectroscopy. [114a] Y. Galerne, S. T. Lagerwall, I. W. Smith, Opt. Commun. *19*, 147 (1976); Biaxialité critique d'un smectique C. [115] P. Gaubert, Compt. rend. *153*, 573 (1911), C.A. *6*, 822[4] (1912); Indices de réfraction des cristaux liquides. [116] P. Gaubert, Compt. rend. *153*, 1158 (1911), C.A. *6*, 822[5] (1912); Indices de réfraction des cristaux

liquides mixtes. [**117**] P. Gaubert, Compt. rend. *154*, 995 (1912), C.A. *6*, 2025[7] (1912); Polarisation circulaire des cristaux liquides. [**118**] P. Gaubert, Bull. soc. franç. min. *36*, 174 (1913), C.A. *8*, 1224[3] (1914); Indices de réfraction des cristaux liquides. [**119**] P. Gaubert, Compt. rend. *167*, 1073 (1918), C.A. *13*, 926[6] (1919); Coloration artificiel des cristaux liquides. [**120**] P. Gaubert, Compt. rend. *176*, 907 (1923), C.A. *17*, 2109[4] (1923); Cristaux liquides de l'anisal-*p*-amido-azotoluol.

[**121**] T. G. Giallorenzi, J. P. Sheridan, J. Appl. Phys. *46*, 1271 (1975), C.A. *82*, 178693j (1975); Light scattering from nematic liquid crystal waveguides. [**121a**] T. G. Giallorenzi, J. A. Weiss, J. P. Sheridan, J. Appl. Phys. *47*, 1820 (1976); Light scattering from smectic liquid-crystal waveguides. [**122**] T. D. Gierke, Diss. Univ. Illinois, Urbana. 1974. 176 pp. Avail. Xerox Univ. Microfilms, Ann Arbor, Mich., Order No. 75-11,575, C.A. *83*, 103529x (1975); Depolarized Rayleigh light scattering studies of orientational pair correlations in a nematic, the coupled translational-rotational diffusion of rodlike macromolecules and the method of atom dipoles. [**123**] T. D. Gierke, W. H. Flygare, J. Chem. Phys. *61*, 2231 (1974), C.A. *81*, 161497e (1974); Depolarized Rayleigh scattering in liquids. Molecular reorientation and orientation pair correlations in nematic MBBA. [**124**] F. Grandjean, Compt. rend. *168*, 91 (1919), C.A. *13*, 1040[6] (1919); Calcul des rayons extraordinaires pour certaines structures de liquides anisotropes. [**125**] F. Grandjean, Compt. rend. *168*, 408 (1919), C.A. *13*, 1040[6] (1919); Calcul des rayons extraordinaires pour certaines structures de liquides anisotropes. [**126**] G. W. Gray, A. Mosley, Mol. Cryst. Liq. Cryst. *35*, 71 (1976), C.A. *85*, 114099z (1976); Raman spectra of 4-cyano-4'-pentylbiphenyl and 4-cyano-4'-pentyl-d$_{11}$-biphenyl. [**126a**] Z. A. Grigoryan, G. G. Petrosyan, V. K. Mirzoyan, A. Kh. Pochikyan, Arm. Khim. Zh. *29*, 916 (1976), C.A. *86*, 130348f (1977); Effect of *p*-xylene on the temperature dependence of MBBA IR spectra. [**126b**] Z. A. Grigoryan, S. G. Kazaryan, O. V. Avakyan, S. A. Kazaryan, A. Kh. Pochikyan, Arm. Khim. Zh. *29*, 925 (1976), C.A. *86*, 130349g (1977); Dependence between IR spectra and liquid crystal properties in a series of Schiff bases and hippuric acid derivatives. [**127**] A. Gruger, N. Lecalve, F. Romain, J. Mol. Struct. *21*, 97 (1974); Spectres de vibration du PAA dans les phases solide et nématique. [**128**] H. Wedel, W. Haase, Ber. Bunsenges. Phys. Chem. *79*, 1169 (1975); Messung der Polarisation von Elektronenübergängen mehrkerniger Kupfer(II)-Komplexe unter Anwendung nematischer Flüssigkristalle als Orientierungsmatrizen. [**129**] H. Wedel, W. Haase, Ber. Bunsenges. Phys. Chem. *80*, 1342 (1976), C.A. *86*, 81145b (1977); Nematische Mesophasen als Orientierungsmatrizen in der optischen Absorptionsspektroskopie. [**130**] W. Haase, H. Wedel, Mol. Cryst. Liq. Cryst. *38*, 61 (1977), C.A. *87*, 31347u (1977); Application of nematics as anisotropic solvents in the optical absorption spectroscopy for the determination of the polarization of electronic transitions of organic molecules and Cu(II)-complexes.

[**131**] I. Haller, J. D. Litster, Phys. Rev. Lett. *25*, 1550 (1970), C.A. *74*, 35840d (1971); Temperature dependence of normal modes in a nematic. [**132**] I. Haller, J. D. Litster, Mol. Cryst. Liq. Cryst. *12*, 277 (1971), C.A. *75*, 27701q (1971); Light scattering spectrum of a nematic. [**133**] I. Haller, H. A. Huggins, M. J. Freiser, Mol. Cryst. Liq. Cryst. *16*, 53 (1972), C.A. *76*, 105729s (1972); Measurement of indices of refraction of nematics. [**134**] I. Haller, Prog. Solid State Chem. *10*, Pt. 2, 103 (1975), C.A. *83*, 186632z (1975); Thermodynamic and static properties of liquid crystals. [**135**] T. S. Hansen, Z. Naturforsch. A *24*, 866 (1969), C.A. *71*, 43852n (1969); Liquid crystal as a solvent in IR spectroscopy. [**136**] E. G. Hanson, Y. R. Shen, Mol. Cryst. Liq. Cryst. *36*, 193 (1976), C.A. *86*, 105716s (1977); Refractive indices and optical anisotropy of homologous liquid crystals. [**136a**] E. G. Hanson, Report *1976*, LBL-6022, 36 pp., C.A. *87*, 191883u (1977); Nonlinear optical effects in liquid crystals. [**136b**] E. G. Hanson, Y. R. Shen, G. K. L. Wong, Appl. Phys. *14*, 65 (1977), C.A. *87*, 109151x (1977); Experimental study of self-focusing in a liquid crystal. [**136c**] T. Harada, P. P. Crooker, Mol. Cryst. Liq. Cryst. *42*, 283 (1977), C.A. *87*, 191355y (1977); Brillouin scattering in the i phase of liquid crystals. [**137**] W. Harz, Diss. Halle 1917; Optische Eigenschaften des Äthoxybenzalamino-α-methylzimtsäureäthylesters. [**137a**] T. Hashimoto, K. Yamaguchi, H. Kawai, Polym. J. *9*, 405 (1977), C.A. *87*, 168523t (1977); Light scattering from polymer films having optically anisotropic rodlike texture. II. Principle to estimate its size. [**138**] J. P. Heger, R. Mercier, Helv. Phys. Acta *45*, 886 (1972), C.A. *78*, 35876c (1973); Effet Raman dans un liquide anisotrope. [**139**] J. P. Heger, J. Phys. (Paris), Lett. *36*, 209 (1975), C.A. *83*, 124403s (1975); Raman scattering. Measurements of depolarization ratios and order parameters in an oriented nematic. [**140**] A. Hochapfel, J. A. Hiver, R. Viovy, C. R. Acad. Sci. Ser. C *272*, 1265 (1971), C.A. *75*, 42602b (1971); Orientation de la chlorophylle a dans un nématique.

[**141**] A. Hochapfel, J. A. Hiver, R. Viovy, Eur. Biophys. Congr. Proc., 1st *4*, 155 (1971), C.A. *76*, 109420g (1972); Orientation de la chlorophylle dans un nématique. [**142**] A. Hochapfel, D. Lecoin, R. Viovy, C. R. Acad. Sci., Sér. C *276*, 221 (1973), C.A. *78*, 96991h (1973); Orientation de colorants

dans un nématique. [**143**] G. Holzwarth, I. Chabay, N. A. W. Holzwarth, J. Chem. Phys. *58*, 4816 (1973), C.A. *79*, 36642f (1973); IRCD and LD of liquid crystals. [**144**] C. Hu, J. R. Whinnery, J. Opt. Soc. Am. *64*, 1424 (1974); Losses of a nematic liquid-crystal optical waveguide. [**145**] C. Hu, J. R. Whinnery, Dig. Tech. Pap. — Top. Meet. Integr. Opt. 1974, TuA6, 4 pp., C.A. *82*, 131980k (1975); Nematic liquid crystal optical waveguides. [**146**] C. C. Huang, R. S. Pindak, J. T. Ho, J. Phys. (Paris), Lett. *35*, 185 (1974), C.A. *81*, 161485z (1974); Birefringence and order parameter in nematic CBOOA. [**147**] N. Isaert, J. Billard, Mol. Cryst. Liq. Cryst. *38*, 1 (1977), C.A. *86*, 179760y (1977); Propriétés optiques des phases s$_C$. [**148**] H. Itoh, H. Nakatsuka, M. Matsuoka, J. Phys. Soc. Jap. *34*, 841 (1973), C.A. *78*, 152461z (1973); Frequency shifts in Raman spectra of liquid crystals at the phase transitions. [**149**] E. Sciesinska, J. Sciesinski, J. Twardowski, J. A. Janik, Inst. Nucl. Phys., Cracow, Rep. 1973, INP-No. 847/PS, 22 pp., C.A. *80*, 138830d (1974); Absorption spectra of MBBA in the 80–400 cm^{-1} range at temperatures between -200 and $+70°$. [**150**] E. Sciesinska, J. Sciesinski, J. Twardowski, J. A. Janik, Mol. Cryst. Liq. Cryst. *27*, 125 (1974), C.A. *82*, 162090z (1975); Absorption spectra of MBBA in the 80–400 cm^{-1} range at temperatures between -200 and $+70°$.

[**151**] J. A. Janik, J. M. Janik, J. Mayer, E. Sciesinska, J. Sciesinski, J. Twardowski, T. Waluga, W. Witko, J. Phys. (Paris) C-1, *36*, 159 (1975), C.A. *83*, 96124f (1975); Calorimetric and IR study of the phase situation in solid MBBA. [**152**] S. Jen, N. A. Clark, P. S. Pershan, E. B. Priestley, Phys. Rev. Lett. *31*, 1552 (1973), C.A. *80*, 42506b (1974); Raman scattering from a nematic. Orientational statistics. [**152a**] S. Jen, N. A. Clark, P. S. Pershan, E. B. Priestley, J. Chem. Phys. *66*, 4635 (1977), C.A. *87*, 32202e (1977); Polarized Raman scattering studies of orientational order in uniaxial liquid crystals. [**153**] M. A. Jeppesen, W. T. Hughes, Amer. J. Phys. *38*, 199 (1970), C.A. *72*, 128527f (1970); Liquid crystals and Newton's rings. [**154**] A. P. Kapustin, L. S. Gorbatenko, Izv. Vyssh. Uchebn. Zaved., Fiz. *18*, 158 (1975), C.A. *83*, 77968t (1975); IR absorption spectra of homologous *p,p*-dialkoxyazoxy-benzenes during phase transitions. [**155**] W. Kast, L. S. Ornstein, Z. Physik *87*, 763 (1934), C.A. *28*, 2963^4 (1934); Registrierungen der Lichtdurchlässigkeit der anisotropen Schmelze von PAA im Magnet-feld. Zur Schwarmtheorie der flüssigen Kristalle. [**156**] W. Kast, Ann. Physik *33*, 185 (1938), C.A. *33*, 908^7 (1938); Doppelbrechung der anisotropen Flüssigkeiten. [**157**] H. Kelker, Ber. Bunsenges. Phys. Chem. *75*, 1128 (1971); Untersuchungen des IR-Dichroismus in flüssig-kristallinen Schichten ohne Verwen-dung von polarisiertem Licht. [**158**] H. Kelker, R. Hatz, G. Wirzing, Z. Analyt. Chem. *267*, 161 (1973), C.A. *80*, 54033h (1974); Untersuchung des IR-Dichroismus in flüssig-kristallinen Schichten ohne Verwen-dung von polarisiertem Licht. [**159**] S. R. F. Kess, Diss. Kent State Univ., Kent, Ohio. 1973. 77 pp. Avail. Univ. Microfilms, Ann Arbor, Mich., Order No. 74-7318. C.A. *81*, 7847n (1974); Spectral distribution of light scattered by a nematic. [**160**] V. L. Khodzhaeva, Izv. Akad. Nauk SSSR, Ser. Khim. *1969*, 2409, C.A. *72*, 60878p (1970); IR absorption spectra study of the liquid-crystalline state of *p-(n*-alkoxy)benzoic acids.

[**161**] V. L. Khodzhaeva, M. V. Shishkina, I. I. Konstantinov, Mol. Cryst. Liq. Cryst. *31*, 21 (1975), C.A. *84*, 3994x (1976); *p*-Acylphenyl esters of *p-n*-alkoxybenzoic acids. II. Spectral features of the mesomorphic state. [**162**] N. Kirov, P. Simova, Phys. Lett. A *37*, 51 (1971), C.A. *76*, 7582u (1972); Potential barriers of the PAA molecules in k, n and i phases. [**163**] N. Kirov, P. Simova, Spectrochim. Acta, A *29*, 55 (1973), C.A. *78*, 83491v (1973); IR absorption spectra of liquid crystals. Determination of the preorientation potential barriers of the molecules in k, n and i phases. [**164**] N. Kirov, Izv. Fiz. Inst. ANEB, Bulg. Akad. Nauk. *24*, 133 (1973), C.A. *81*, 43541s (1974); IR spectra of benzylideneaniline, EBBA, and *p*-aminophenyl acetate in the k phase. [**165**] N. Kirov, P. Simova, Mol. Cryst. Liq. Cryst. *30*, 59 (1975), C.A. *83*, 105704z (1975); Integral intensity temperature dependence of IR bands in k, n, and i phases. [**166**] N. Kirov, P. Simova, M. Sabeva, Mol. Cryst. Liq. Cryst. *33*, 189 (1976), C.A. *85*, 114408z (1976); Depolarization of light scattered in nematics. [**167**] V. I. Klenin, V. A. Kolchanov, Issled. V Obl. Elektrokhimii I Fiz.-Khimii Polimerov *1975*, 62, C.A. *84*, 52408t (1976); Determination of the microstructure of heterogeneous liquid crystal substances by turbidity spectra. [**168**] K. Koller, K. Lorenzen, G. M. Schwab, Z. Physik. Chem. (Frankfurt) *44*, 101 (1965), C.A. *63*, 1368a (1965); Raman-Spektroskopie einer quasikristallinen Flüssigkeit *p-n*-Butyloxybenzoesäure. [**169**] V. K. Kondratev, M. M. Fartzdinov, A. N. Chuvyrov, Fiz. Tverd. Tela *17*, 795 (1975); Photo-elastic effect in nematics. [**170**] I. I. Konstantinov, V. L. Khodzhaeva, M. V. Shishkina, Y. B. Amerik, J. Phys. (Paris), C-1, *36*, 55 (1975), C.A. *83*, 106521z (1975); Determination of the degree of order of mesomorphic *p-n*-alkoxybenzoic acids by IR dichroism method.

[**171**] F. Kopp, U. P. Fringeli, K. Muehlethaler, H. H. Guenthard, Z. Naturforsch. C *30*, 711 (1975), C.A. *84*, 104821u (1976); Spontaneous rearrangement in Langmuir-Blodgett layers of tripalmitin

studied by means of ATR IR spectroscopy and electron microscopy. [172] I. N. Kozlov, A. M. Sarzhevskii, Dokl. Akad. Nauk Beloruss. SSR *16*, 893 (1972), C.A. *78*, 21837d (1973); Calculation of secondary reabsorption during a measurement of fluorescence polarization in solid isotropic and anisotropic media. [173] D. Krishnamurti, H. S. Subramhanyam, Mol. Cryst. Liq. Cryst. *14*, 209 (1971), C.A. *75*, 133963m (1971); Scattering of light by nematic PAA. [174] D. Krishnamurti, H. S. Subramhanyam, Mol. Cryst. Liq. Cryst. *31*, 153 (1975), C.A. *84*, 11092s (1976); Polarization field and molecular order in smectics. [175] W. Kuczynski, B. Stryla, Mol. Cryst. Liq. Cryst. *31*, 267 (1975), C.A. *84*, 81912h (1976); Interference method for the determination of refractive indices and birefringence of liquid crystals. [175a] W. Kuczynski, P. Pieranski, K. Wojciechowski, B. Stryla, Mol. Cryst. Liq. Cryst. *34*, 203 (1977), C.A. *87*, 75806x (1977); Methods of optical birefringence determination in liquid crystals from interference measurements. [176] M. M. Kusakov, V. L. Khodzhaeva, M. V. Shishkina, I. I. Konstantinov, Kristallografiya *14*, 485 (1969), C.A. *71*, 43794v (1969); Study of liquid crystals of p-alkoxybenzoic acids by IR dichroism. [177] G. Labrunie, M. Bresse, C. R. Acad. Sci., Sér. A *276*, 647 (1973), C.A. *79*, 25120h (1973); Indices du p-méthoxy-p′-pentyldiphénylacétylène nématique. [178] J. R. Lalanne, Phys. Lett. A *51*, 74 (1975), C.A. *82*, 162710b (1975); Picosecond investigation of optical field induced birefringence of the i phase of a nematogen. [179] J. R. Lalanne, R. Lefevre, J. Chim. Phys. Phys.-Chim. Biol. *73*, 337 (1976), C.A. *85*, 70892w (1976); Réalisation d'un laser a modes bloqués en phase et application a la mesure des temps de relaxation orientationelle de molécules nématogènes en phase i. [180] J. R. Lalanne, B. Martin, B. Pouligny, S. Kielich, Opt. Commun. *19*, 440 (1976), C.A. *86*, 80989t (1977); Fast picosecond reorientation in the i phase of nematogens. [180a] J. R. Lalanne, B. Martin, B. Pouligny, S. Kielich, Mol. Cryst. Liq. Cryst. *42*, 153 (1977); Direct observation of ps reorientation of molecules in the i phases of nematogens.

[181] M. Lamotte, J. Chim. Phys. Phys.-Chim. Biol. *72*, 803 (1975), C.A. *83*, 147992x (1975); Interprétation des paramètres d'ordre dans les cas de carbures non polaires orientés dans le polyéthylène étiré. [182] D. Langevin, Solid State Commun. *14*, 435 (1974), C.A. *80*, 126257j (1974); Anisotropy of the turbidity of a nematic. [183] D. Langevin, M. A. Bouchiat, J. Phys. (Paris) C-1, *36*, 197 (1975); Anisotropy of the turbidity of an oriented nematic. [184] R. D. Larrabee, RCA Rev. *34*, 329 (1973), C.A. *79*, 136099c (1973); Fluorescence switching by means of liquid crystals. [185] M. Laurent, R. Journeaux, Mol. Cryst. Liq. Cryst. *36*, 171 (1976), C.A. *86*, 48694g (1977); Mesure de l'anisotropie des indices de réfraction des cristaux liquides. Interprétation des différences observées lors de l'utilisation des diverses méthodes. [186] B. Lavrencic, S. Lugomer, Advan. Raman Spectrosc. *1*, 215 (1972), C.A. *79*, 151185w (1973); Raman depolarization ratio in smectics and nematics. [187] M. Leclercq, F. Wallart, J. Raman Spectr. *1*, 587 (1973); Description of a zero-dispersion fore-monochromator for low frequency Raman spectroscopy. [188] A. Ledoux, Bull. soc. franç. min. *40*, 119 (1917), C.A. *13*, 926 (1919); Détermination des indices de réfraction principaux de substances anisotropes par l'observation du retard de lames minces en lumière parallèle oblique. [189] E. Lehmann, Diss. Halle 1910; Doppelbrechung flüssiger Kristalle. [190] R. A. Levenson, Diss. Columbia Univ., New York, N. Y. 1970, 195 pp. Avail. Univ. Microfilms, Ann Arbor, Mich. Order No. 71-17,518, C.A. *76*, 6834r (1972); Liquid crystal spectroscopy. Electronic structure of compounds containing metal-metal bonds.

[191] R. A. Levenson, H. B. Gray, G. P. Ceasar, J. Amer. Chem. Soc. *92*, 3653 (1970), C.A. *73*, 50290x (1970); Electronic and vibrational spectroscopy in a nematic solvent. Band polarizations for binuclear metal carbonyls. [192] S. Lugomer, Fizika (Zagreb) *6*, 1 (1974), C.A. *81*, 128033c (1974); IR correlation function analysis of benzene ring rotations in liquid crystals. [193] S. Lugomer, B. Lavrencic, Solid State Commun. *15*, 177 (1974), C.A. *81*, 97150g (1974); Raman spectroscopy and elastic properties of liquid crystals. [194] S. Lugomer, Mol. Cryst. Liq. Cryst. *29*, 141 (1974), C.A. *82*, 163365e (1975); Benzene ring rotations in liquid crystals. [195] A. S. L'vova, M. M. Sushchinskii, Optika i Spektroskopyia, Akad. Nauk SSSR, Otd. Fiz.-Mat. Nauk, Sb. Statei *2*, 266 (1963), C.A. *60*, 118b (1964); IR absorption spectra of liquid crystals. [196] A. S. L'vova, L. M. Sabirov, I. M. Aref'ev, M. M. Sushchinskii, Opt. Spektrosk. *24*, 613 (1968), C.A. *69*, 23369r (1968); The 100-cm^{-1} absorption band of k, n, and i PAP. [197] D. McQueen, K. Edgren, L. Einarson, J. Phys. D *6*, 885 (1973), C.A. *79*, 24397s (1973); Inelastic light scattering from nematic PAP. [198] W. Maier, A. Saupe, Z. physik. Chem. (Frankfurt) *6*, 327 (1956), C.A. *50*, 7534e (1956); Untersuchungen mit linear polarisiertem UV an k, n und i PAA. I. Experimenteller Teil. [199] W. Maier, A. Saupe, A. Englert, Z. physik. Chem. (Frankfurt) *10*, 273 (1957), C.A. *51*, 10235d (1957); Die im nahen UV gelegenen Elektronenübergänge des Azoxybenzols und einiger seiner p-Derivate. [200] W. Maier, G. Englert,

Z. physik. Chem. (Frankfurt) *12*, 123 (1957), C.A. *51*, 13576d (1957); IR-Untersuchungen an flüssigen Kristallen.
[201] W. Maier, G. Englert, Z. Elektrochem. *62*, 1020 (1958), C.A. *53*, 2788g (1959); IR-Spektren einiger 4,4′-Derivate des Azobenzols und Azoxybenzols. [202] W. Maier, G. Englert, Z. physik. Chem. (Frankfurt) *19*, 168 (1959), C.A. *54*, 1074a (1960); IR-Untersuchungen an flüssigen Kristallen. [203] W. Maier, G. Englert, Z. Elektrochem. *64*, 689 (1960), C.A. *54*, 21911i (1960); Ordnungsgradbestimmungen an flüssig-kristallinen Schichten durch Messung des IR-Dichroismus. [204] W. Maier, K. Markau, Z. physik. Chem. (Frankfurt) *28*, 190 (1961), C.A. *55*, 21709i (1961); Bestimmung des Ordnungsgrades von 3 aliphatischen Diensäuren durch Messung des IR-Dichroismus. [205] W. Maier, Landolt-Börnstein: 6. Aufl. (Springer: Berlin) Bd. II, Teil 6, S. 607 (1959) und Teil 8, S. 4 (1962); Refraktometrische Messungen. [206] W. H. Martin, Trans. Roy. Soc. Can., Sec. III *19*, 36 (1925), C.A. *20*, 329[1] (1926); Scattering of light by anisotropic liquids. [207] C. Mauguin, Bull. Soc. chim. Belg. *36*, 172 (1927), C.A. *21*, 2597 (1927); Mesure des deux indices de réfraction d'un cristal liquide dans tout son domaine d'existence. [208] G. Baur, A. Stieb, G. Meier, J. Appl. Phys. *44*, 1905 (1973), C.A. *78*, 166445j (1973); Quenching of fluorescence at the n/i transition. [209] G. Baur, A. Stieb, G. Meier, Mol. Cryst. Liq. Cryst. *22*, 261 (1973), C.A. *79*, 151244q (1973); Polarized fluorescence of dyes oriented in room temperature nematics. [210] E. B. Priestley, P. S. Pershan, R. B. Meyer, D. H. Dolphin, Vijnana Parishad Anusandhan Patrika *14*, 93 (1971), C.A. *78*, 165400d (1973); Raman scattering from nematics. Determination of the degree of ordering. [210a] K. Miyano, Phys. Lett. A *63*, 37 (1977), C.A. *87*, 192287q (1977); Order parameters of a nematic measured by Raman scattering. [210b] E. Moritz, Mol. Cryst. Liq. Cryst. *41*, 63 (1977); Evanescence in liquid crystals.
[211] T. P. Myasnikova, L. S. Gorbatenko, R. Y. Evseeva, Ural. Konf. Spektrosk., 7th, *1971*, No. 2, 28, C.A. *77*, 158238h (1972); IR spectra of liquid crystals near the phase transition. [211a] M. Nakagake, Y. Naito, Yakugaku Zasshi *97*, 330 (1977), C.A. *86*, 149044a (1977); Phase transition by addition of cetyl alcohol and anthracene on a compensated nematic. [211b] K. A. Narinyan, Z. A. Grigoryan, A. Kh. Pochikyan, Arm. Khim. Zh. *29*, 921 (1976), C.A. *86*, 170201j (1977); Temperature dependence of MBBA and EBBA IR spectra. [212] V. D. Neff, Liq. Cryst. Plast. Cryst. *2*, 231 (1974), C.A. *83*, 199532s (1975); IR, Raman, VIS, and UV spectroscopy of liquid crystals. [213] J. J. Nemec, Diss. Univ. Houston, Houston, Tex., 1970, 94 pp. Avail. Univ. Microfilms, Ann Arbor, Mich., Order No. 71-15,467, C.A. *76*, 28679w (1972); Light scattering study of correlations in molecular orientation of a nematic. [214] L. Oberlaender, Diss. Halle 1914; Brechungskoeffizienten flüssiger Kristalle bei höheren Temperaturen. [215] K. Z. Ogorodnik, Fiz. Tverd. Tela *17*, 2781 (1975), C.A. *83*, 211514x (1975); Basis for a discrete-statistical representation of the nematic phase. [216] Y. Ohnishi, Jap. J. Appl. Phys. *12*, 1079 (1973), C.A. *79*, 109923k (1973); IR spectra of a nematic in dynamic scattering mode. [217] Orsay Liquid Crystal Group, J. Phys. (Paris), C-4, *30*, 71 (1969); Étude, par diffusion quasi-elastique de la lumière, de l'amortissement des fluctuations thermiques d'orientation moléculaire dans un nématique. [218] Orsay Liquid Crystal Group, Liq. Cryst. Ord. Fluids 1970, 195, C.A. *78*, 165409p (1973); Theory of light scattering by nematics. [219] Y. H. Pao, H. L. Frisch, R. Bersohn, Int. J. Quantum Chem., Symp. 1967, 829, C.A. *69*, 111507q (1968); Double quantum light scattering and orientational correlations in liquids. [220] G. D. Patterson, J. Chem. Phys. *63*, 4032 (1975), C.A. *83*, 210854w (1975); Rayleigh scattering in a dense medium. [220a] G. D. Patterson, A. P. Kennedy, J. P. Latham, Macromolecules *10*, 667 (1977), C.A. *87*, 46812r (1977); Temperature dependence of orientation correlations in *n*-alkane liquids.
[221] R. D. Peacock, B. Samori, J. Chem. Soc. Faraday Trans. 2, *72*, 1909 (1975), C.A. *84*, 3953h (1976); Polarisations of the lowest-energy transitions of pyridine 1-oxide and 4-methylpyridine 1-oxide. [222] G. Pelzl, Z. Phys. Chem. (Leipzig) *255*, 602 (1974), C.A. *82*, 23850j (1975); Brechungsindices binärer smektischer Mischungen vom Typ A in dem System 4,4-Azoxy-α-methylzimtsäure-di-*n*-butyl- und -hexylester. [223] G. Pelzl, D. Demus, Z. Phys. Chem. (Leipzig) *256*, 305 (1975), C.A. *83*, 170222e (1975); Doppelbrechung niedrig schmelzender nematischer Substanzen. [223a] I. Penchev, I. Dozov, Phys. Lett. A *60*, 34 (1977), C.A. *86*, 113916x (1977); Determining the S_2 and S_4 order parameters through fluorescent peasurments. [223b] P. S. Pershan, Theory Light Scattering Solids, Proc. Sov.-Am. Symp., 1st, *1*, 61 (1975), C.A. *87*, 160140v (1977); Measurement of liquid crystal orientational statistics by Raman scattering. [224] G. P. Petrova, R. M. Bashirova, E. I. Koshel'nik, Opt. Spektrosk. *39*, 399 (1975), C.A. *83*, 185609k (1975); Spectra of Brillouin light scattering in the i phase of a liquid crystal. [225] Y. Poggi, G. Labrunie, J. Robert, C. R. Acad. Sci., Sér. B, *277*, 561 (1973), C.A. *80*, 64690j (1974); Détermination du paramètre d'ordre dans les nématiques par des mesures d'anisotropie

magnétique. Proportionnalité entre paramètre d'ordre et anisotropie d'indice. [226] K. R. Popov, L. V. Smirnov, Opt. Spektrosk. *30*, 178 (1971), C.A. *74*, 105198b (1971); Polarization of electron transitions in stilbene and azobenzene molecules. [227] N. R. Posledovich, V sb., Spektroskopiya i Lyuminestsentsiya. *1975*, 71, C.A. *85*, 70222c (1976); IR spectroscopic study of a "new" property of the liquid-crystal state of alkoxybenzoic acids. [228] J. S. Prasad, H. S. Subramhanyam, Mol. Cryst. Liq. Cryst. *33*, 77 (1976), C.A. *85*, 11751b (1976); Refractive indices and molecular order in nematic 4,4'-bis(pentyloxy)azoxybenzene. [229] E. B. Priestley, P. S. Pershan, Mol. Cryst. Liq. Cryst. *23*, 369 (1973), C.A. *80*, 88155q (1974); Investigation of nematic ordering using Raman scattering. [230] S. Procopiu, Ann. sci. univ. Jassy I, *29*, 1 (1943), C.A. *42*, 2151c (1948); Longitudinal depolarization of light by liquid crystals and soft crystals in relation to temperature.

[231] S. Procopiu, Kolloid-Z. *109*, 90 (1944), C.A. *41*, 2964c (1947); Longitudinale Depolarisation des Lichtes durch kristalline Flüssigkeiten und durch die weichen Kristalle in Abhängigkeit von der Temperatur. [232] D. V. G. L. N. Rao, D. K. Agrawal, Phys. Lett. A *37*, 383 (1971), C.A. *76*, 92398f (1972); Stimulated Raman scattering in a nematic. [233] D. V. G. L. N. Rao, D. K. Agrawal, Bull. Am. Phys. Soc. *17*, 67 (1972); Stimulated Raman scattering in a nematic. [234] I. M. Asher, D. V. G. L. N. Rao, Bull. Am. Phys. Soc. *17*, 575 (1972); Nonlinear optical effects in the i phase of a nematic. [235] D. V. G. L. N. Rao, S. Jayaraman, Appl. Phys. Lett. *23*, 539 (1973), C.A. *80*, 42722u (1974); Self-focusing of laser light in the i phase of a nematic. [236] D. V. G. L. N. Rao, S. Jayaraman, Phys. Rev. A *10*, 2457 (1974), C.A. *82*, 118058w (1975); Pretransitional behavior of self-focusing in nematics. [237] S. Jayaraman, D. V. G. L. N. Rao, Bull. Am. Phys. Soc. *19*, 1087 (1974); Stimulated light scattering in a liquid crystal. [238] S. Jayaraman, D. V. G. L. N. Rao, Bull. Am. Phys. Soc. *19*, 53 (1974); Self-focusing of laser light in a liquid crystal. [239] R. Ribotta, Proc. Int. Conf. Light Scattering Solids, 3rd, *1975*, 713, C.A. *85*, 151040u (1976); Rayleigh scattering of light in smectics. [240] R. Ribotta, Phys. Lett. A *56*, 130 (1976), C.A. *84*, 158246d (1976); Observation of undulations of layers in smectics B and H.

[241] R. Riwlin, Proc. Acad. Sci. Amsterdam *23*, 807 (1921), C.A. *15*, 3588[6] (1921); Liquid crystals. V. Photographic absorption and extinction measurements. [242] R. Riwlin, Arch. néerland sci. IIIA, *7*, 95 (1922), C.A. *18*, 2999[1] (1924); Dispersion of light in liquid crystals. [243] R. Riwlin, Diss. Utrecht 1923; Lichtzerstreuung in flüssigen Kristallen. [244] F. Rondelez, H. Birecki, R. Schaetzing, J. D. Litster, J. Phys. (Paris) C-3, *37*, 122 (1976); Diffusion de lumière par les fluctuations du directeur dans la phase s_A du CBOOA. [245] H. Birecki, R. Schaetzing, F. Rondelez, J. D. Litster, Proc. Int. Conf. Light Scattering Solids, 3rd, *1975*, 707, C.A. *85*, 169017b (1976); Light scattering study of director fluctuations in the s_A phase of CBOOA. [246] H. Birecki, R. Schaetzing, F. Rondelez, J. D. Litster, Phys. Rev. Lett. *36*, 1376 (1976), C.A. *85*, 54873k (1976); Light-scattering study of a s_A phase near the s_A/n transition. [247] A. Rosencwaig, Bull. Am. Phys. Soc. *20*, 494 (1975); Studies of transition metal oxides, liquid crystals and polymers by means of photoacoustic spectroscopy. [248] L. Royer, Compt. rend. *178*, 1066 (1924), C.A. *18*, 2834[2] (1924); Les états mésomorphes et la biréfringence magnétique. [249] E. Sackmann, J. Am. Chem. Soc. *90*, 3569 (1968), C.A. *69*, 55958y (1968); Polarization of optical transitions of dye molecules oriented in an ordered glass matrix. [250] E. Sackmann, Chem. Phys. Lett. *3*, 253 (1969), C.A. *71*, 25495u (1969); Electric-field-induced orientation of liquid crystals and optical absorption experiments.

[251] E. Sackmann, D. Rehm, Chem. Phys. Lett. *4*, 537 (1970), C.A. *72*, 116433j (1970); Fluorescence polarization measurements on molecules oriented in liquid crystals. [252] H. Beens, H. Moehwald, D. Rehm, E. Sackmann, A. Weller, Chem. Phys. Lett. *8*, 341 (1971), C.A. *75*, 27840j (1971); Fluorescence polarization studies of heteroexcimers oriented in liquid crystals. [253] E. Sackmann, H. Moehwald, Chem. Phys. Lett. *12*, 467 (1972), C.A. *76*, 104905j (1972); Relation between the principal polarizabilities of a molecule and its average orientation in nematics. [254] E. Sackmann, Habilitationsschrift Göttingen; Flüssige Kristalle als anisotrope Lösungsmittel und ihre Anwendung in der optischen Spektroskopie. Neue Methode zur Bestimmung der Polarisation optischer Vorgänge. [255] E. Sackmann, H. Moehwald, J. Chem. Phys. *58*, 5407 (1973), C.A. *79*, 47523u (1973); Optical polarization measurements in liquid crystals. [256] E. Sackmann, P. Krebs, H. U. Rega, J. Voss, H. Moehwald, Mol. Cryst. Liq. Cryst. *24*, 283 (1973), C.A. *82*, 178630m (1975); Optical studies of the anisotropic solute-solvent interaction in liquid crystals. [257] F. D. Saeva, G. R. Olin, Mol. Cryst. Liq. Cryst. *35*, 319 (1976), C.A. *85*, 200189k (1976); Utilization of the c/n transition for the evaluation of polarization spectra with unpolarized light. [258] S. Sakagami, A. Takase, M. Nakamizo, H. Kakiyama, Bull. Chem. Soc. Jap. *46*, 2062 (1973), C.A. *79*, 84291n (1973); New method of determining the molecular arrangement in liquid crystals

by the use of the polarization of the fluorescence. [259] A. Takase, S. Sakagami, M. Nakamizo, Mol. Cryst. Liq. Cryst. *22*, 67 (1973), C.A. *79*, 109049e (1973); Light scattering study on orientations in smectics. [260] S. Sakagami, A. Takase, M. Nakamizo, H. Kakiyama, Liq. Cryst. Ord. Fluids *2*, 125 (1974), C.A. *86*, 180981c (1977); Studies on the molecular arrangement in liquid crystals by polarization of fluorescence.

[261] A. Sakamoto, K. Yoshino, U. Kubo, Y. Inuishi, Technol. Rep. Osaka Univ. *23*, 485 (1973), C.A. *81*, 56223q (1974); Raman and Rayleigh scattering in liquid crystals. [262] A. Sakamoto, K. Yoshino, U. Kubo, Y. Inuishi, Jpn. J. Appl. Phys. *13*, 359 (1974), C.A. *80*, 114312k (1974); Raman and Rayleigh scattering in liquid crystal. [263] A. Sakamoto, K. Yoshino, U. Kubo, Y. Inuishi, Jpn. J. Appl. Phys. *13*, 1285 (1974), C.A. *81*, 97249w (1974); Raman scattering in ethyl *p*-[(*p*-methoxybenzyli-dene)amino]cinnamate. [264] A. Sakamoto, K. Yoshino, U. Kubo, Y. Inuishi, Jpn. J. Appl. Phys. *13*, 1691 (1974), C.A. *82*, 24629f (1975); Raman scattering in nematics and smectics. [265] A. Sakamoto, U. Kubo, K. Yoshino, Y. Inuishi, Kinki Daigaku Rikogakubu Kenkyu Hokoku *10*, 103 (1975), C.A. *83*, 200482j (1975); Raman scattering in nematics. [266] I. D. Samodurova, A. S. Sonin, Opt. Spektrosk. *38*, 980 (1975), C.A. *83*, 88970p (1975); Distribution of scales of turbulence in nematics found by the method of low-angle light scattering. [267] A. Santoro, M. Esposito, J. Therm. Anal. *6*, 101 (1974), C.A. *80*, 100999c (1974); Rapid detection and identification of liquid crystal states by light scattering. [268] A. Saupe, Z. Naturforsch. A *18*, 336 (1963), C.A. *59*, 128d (1963); UV-Untersuchungen an Stilben, Benzalanilin, Azobenzol, Azoxybenzol und kristallin-flüssigen Alkoxyderivaten. [269] A. Saupe, Mol. Cryst. Liq. Cryst. *16*, 87 (1972), C.A. *76*, 105643j (1972); UV-, IR-, and NMR spectroscopy on liquid crystals. [270] J. M. Schnur, M. Hass, W. L. Adair, Phys. Lett. A *41*, 326 (1972), C.A. *77*, 158246j (1972); Raman spectra of various phases of PAA.

[271] J. M. Schnur, Phys. Rev. Lett. *29*, 1141 (1972), C.A. *77*, 170789w (1972); Raman spectral evidence for conformational changes in the liquid-crystal homologous series of the alkoxyazoxybenzenes. [272] J. M. Schnur, Mol. Cryst. Liq. Cryst. *23*, 155 (1973), C.A. *79*, 151153j (1973); Raman spectra of the liquid crystal homologous series of the alkoxyazoxybenzenes. [273] J. M. Schnur, J. P. Sheridan, M. Fontana, Liq. Cryst., Proc. Int. Conf., Bangalore, *1973*, 175, C.A. *84*, 114407c (1976); Raman spectral and thermodynamic studies on the liquid crystal TBBA. [274] J. M. Schnur, M. Fontana, J. Phys. (Paris), Lett. *35*, 53 (1974), C.A. *80*, 138864t (1974); Raman spectra of smectic and nematic TBBA. [275] H. Schulze, W. Burkersrode, Ann. Phys. (Leipzig) *34*, 30 (1977), C.A. *86*, 180970y (1977); Probleme der optischen Anisotropie nematischer Flüssigkristalle. [276] J. F. Scott, Vib. Spectra Struct. *5*, 67 (1976), C.A. *86*, 64019w (1977); Vibrational Raman spectroscopy as a probe of solid state structure and structural phase transitions. [277] L. G. Shaltiko, A. A. Shepelewskij, S. J. Frenkel, 1. Wiss. Konferenz über flüssige Kristalle, 17.–19. 11. 1970, Iwanowo, Sammlung der Vorträge (1972), Nr. 19; Kleinwinkelstreuung von Licht in flüssigen Kristallen. [278] T. Y. Shen, S. S. Mitra, C. W. Brown, J. Raman Spectros. *3*, 109 (1975), C.A. *83*, 124337y (1975); Raman scattering study of the liquid crystal *p*-(*p*-ethoxyphenylazo)phenyl undecylenate. [279] K. Shibata, M. Kutsukake, H. Takahashi, K. Higasi, Bull. Chem. Soc. Jpn. *49*, 406 (1976), C.A. *84*, 128362w (1976); IR and Raman spectra of k, n, and i phases of PAA. [280] P. Simova, N. Kirov, Advan. Mol. Relaxation Processes *5*, 233 (1973), C.A. *80*, 36433j (1974); IR absorption spectra of liquid crystals. Preorientation potential barriers of the molecules from the homologous series of PAA in k, n, and i phases.

[281] P. Simova, I. T. Savatinova, Dokl. Bolg. Akad. Nauk *27*, 173 (1974), C.A. *81*, 31364y (1974); Low-frequency Raman study of APAPA. [282] P. Simova, N. Kirov, Spectrosc. Lett. *7*, 55 (1974), C.A. *81*, 42533d (1974); Anisotropy of rotational diffusion in n and i phases. [283] P. Simova, N. Kirov, Spectrosc. Lett. *8*, 561 (1975), C.A. *84*, 52427y (1976); IR spectra of liquid crystals. Rotational diffusion in smectic and cholesteric. [284] K. J. Mainusch, U. Müller, P. Pollmann, H. Stegemeyer, Z. Naturforsch. A *27*, 1677 (1972), C.A. *78*, 70944v (1973); Absorption and fluorescence measurements in compensated cholesterics. I. Orientation of chromophores in liquid crystal solvents. [285] K. J. Mainusch, P. Pollmann, H. Stegemeyer, Z. Naturforsch. A *28*, 1476 (1973), C.A. *80*, 47023b (1974); II. Elektronenübergänge mit unterschiedlicher Polarisationsrichtung (cf. [284]). [286] T. R. Steger Jr., J. D. Litster, W. R. Young, Liq. Cryst. Ord. Fluids *2*, 33 (1974), C.A. *86*, 180976e (1977); Pretransitional behavior in the i phase of homologous compounds showing n and s_C type order. [287] T. R. Steger Jr., J. D. Litster, Liq. Cryst. Ord. Fluids *2*, 671 (1974), C.A. *86*, 195351w (1977); Brillouin scattering in the i phase of MBBA. [288] R. S. Stein, Mol. Cryst. Liq. Cryst. *6*, 125 (1969), C.A. *71*, 129552b (1969); Specification of order in mesophases. [289] H. S. Subramhanyam, D. Krishnamurti, Mol. Cryst. Liq. Cryst. *22*, 239 (1973), C.A. *79*, 150359u (1973); Polarization field and molecular order in nematics.

[290] H. S. Subramhanyam, C. S. Prabha, D. Krishnamurti, Mol. Cryst. Liq. Cryst. *28*, 201 (1974), C.A. *83*, 18007f (1975); Optical anisotropy of nematics.

[291] H. S. Subramhanyam, J. S. Prasad, Mol. Cryst. Liq. Cryst. *37*, 23 (1976), C.A. *86*, 113905t (1977); Refractive indices, densities, polarizabilities and molecular order in nematic MBBA. [292] H. S. Subramhanyam, Acta Cienc. Indica *2*, 187 (1976), C.A. *86*, 99241q (1977); Orientational order in nematic mixtures. [292a] H. Takahashi, M. Kutsukake, I. Tajima, Proc. Int. Conf. Raman Spectrosc., 5th *1976*, 546, C.A. *87*, 192266g (1977); Raman studies on the conformational changes in liquid crystals at phase transitions. [293] A. Takase, S. Sakagami, M. Nakamizo, Chem. Lett. *1975*, 797, C.A. *83*, 139336w (1975); Raman spectrum of liquid-crystalline TBBA. [293a] A. Takase, S. Sakagami, M. Naka-mizo, Jpn. J. Appl. Phys. *16*, 549 (1977), C.A. *86*, 198203k (1977); Deformation vibrations of the alkyl chains for the homologous series of TBAA in the smectic B. [294] Y. Tanizaki, H. Inoue, T. Hoshi, J. Shiraishi, Z. Phys. Chem. (Frankfurt) *74*, 45 (1971), C.A. *75*, 103182p (1971); Localized and delocalized electronic transitions in diphenylacetylene, stilbene, and diphenylbutadiene. [295] R. Taschek, D. Williams, Phys. Rev. *52*, 852 (1937), C.A. *33*, 5278[8] (1939); IR absorption spectrum of a liquid crystal. [296] R. Taschek, D. Williams, J. Chem. Phys. *6*, 546 (1938), C.A. *32*, 8269[4] (1938); IR study of liquid crystals. [297] K. Toda, S. Nagaura, M. Sukigara, K. Honda, Denki Kagaku Oyobi Kogyo Butsuri Kagaku *43*, 403 (1975), C.A. *83*, 211504u (1975); Determination of apparent order parameters of the orientation of nematics by the guest-host effect. [297a] A. N. Trofimov, Z. K. Kuvatov, L. S. Mamaeva, F. M. Khusnullin, Kristallografiya *22*, 204 (1977), C.A. *86*, 161595p (1977); Calculation of the principal polarizabilities of molecules using the refractive indices of homologous nematic *p*-alkoxybenzoic acids. [298] N. Trzebowski, E. Langholf, Z. Chem. *7*, 245 (1967), C.A. *67*, 69011e (1967); IR-spektroskopische Unterscheidung der verschiedenen Formen von Natriumseifen bei Raumtemperatur. [299] N. Trzebowski, E. Langholf, Z. Chem. *7*, 282 (1967), C.A. *67*, 86220q (1967); Veränderungen in den IR-Spektren bei den kristallinen und flüssig-kristallinen Phasenübergängen von Natriumstearat. [300] M. Y. Tsenter, Y. S. Bobovich, N. M. Belyaevskaya, Opt. Spektrosk. *29*, 53 (1970), C.A. *74*, 8154b (1971); Manifestations of nematic order in the Raman spectrum of vitrified α-phenylazo-N-anisal-α'-naphthylamine liquid crystals.

[301] V. N. Tsvetkov, Acta Physicochim. U.R.S.S. *9*, 111, 130 (1938), C.A. *33*, 7169[5] (1939); Lichtstreuung anisotroper Flüssigkeiten. I. Untersuchungen mit transmittiertem Licht. II. Depolarisation des Streulichts. [302] V. N. Tsvetkov, E. I. Ryumtsev, I. P. Kolomiets, A. P. Kovshik, N. L. Gantseva, Opt. Spektrosk. *35*, 880 (1973), C.A. *80*, 54022d (1974); Anisotropy of molar refraction of mesomorphic *p,p'*-dialkoxyazoxybenzenes. [303] A. P. Kovshik, Y. I. Denite, E. I. Ryumtsev, V. N. Tsvetkov, Kristallo-grafiya *20*, 861 (1975), C.A. *84*, 51605t (1976); Optical anisotropy of liquid-crystal 4-cyanophenyl esters of 4'-*n*-alkyloxybenzoic and 4'-*n*-alkylbenzoic acids. [304] K. Vacek, E. Vavrinec, V. Cerna, O. Vinduskova, J. Naus, P. Lokai, Stud. Biophys. *44*, 155 (1974), C.A. *82*, 104953x (1975); Fluorescence of chlorophyll a in different polymer matrixes after conventional and laser excitation. [305] J. M. Vaughan, Phys. Lett. A *58*, 325 (1976), C.A. *85*, 184209j (1976); Brillouin scattering in the n and i phases of a liquid crystal. [306] S. Venugopalan, Liq. Cryst., Proc. Int. Conf., Bangalore, *1973*, 167, C.A. *84*, 113832a (1976); Far-IR absorption spectrum of PAA. [307] S. Venugopalan, J. R. Fernandes, G. V. Vani, Mol. Cryst. Liq. Cryst. *31*, 29 (1975), C.A. *83*, 211012g (1975); Far-IR and Raman spectra of solid CBOOA. [307a] S. Venugopalan, J. R. Fernandes, V. Surendranath, Mol. Cryst. Liq. Cryst. *40*, 149 (1977), C.A. *87*, 125602a (1977); Far-IR absorption in highly ordered smectic TBBA. [308] G. Vergoten, Advan. Raman Spectrosc. *1*, 219 (1972), C.A. *79*, 151204b (1973); Raman studies of liquid crystals. [309] G. Vergoten, R. Demol, G. Fleury, Trav. Soc. Pharm. Montpellier *33*, 321 (1973), C.A. *80*, 125590a (1974); Étude par effet Raman des mésophases. [310] G. Vergoten, G. Fleury, Mol. Cryst. Liq. Cryst. *30*, 213 (1975), C.A. *83*, 123557h (1975); Raman spectra of MBBA in the k, n, and i phases.

[311] G. Vergoten, G. Fleury, Mol. Spectrosc. Dense Phases, Proc. Eur. Congr. Mol. Spectrosc., 12th, *1975*, 315, C.A. *86*, 36545w (1977); Mesophase transitions as studied by means of the Raman scattering. [312] G. Vergoten, G. Fleury, J. Mol. Struct. *30*, 347 (1976), C.A. *84*, 120672b (1976); Normal coordinate analysis of benzaniline: basis structure for some liquid crystals. [313] G. Vergoten, G. Fleury, R. N. Jones, A. Nadeau, Mol. Cryst. Liq. Cryst. *36*, 327 (1976), C.A. *86*, 81163f (1977); IR and far-IR spectra of the MBBA and EBBA liquid crystals. [314] M. Veyssié, Ber. Bunsenges. Phys. Chem. *76*, 279 (1972); Fluctuations in nematics. Static and dynamic properties: Study by inelastic light scattering. [315] L. K. Vistin, S. S. Yakovenko, Kristallografiya *21*, 571 (1976), C.A. *85*, 102609c (1976); Use of an IR spectroscopic method for determining the orientation of molecules in liquid-crystal

layers which strongly scatter light. [316] L. V. Volod'ko, N. R. Posledovich, A. I. Serafimovich, Zh. Prikl. Spektrosk. *17*, 542 (1972), C.A. *77*, 171134r (1972); Uniform temperature cuvette for optical studies in an electric field. [317] L. V. Volod'ko, N. R. Posledovich, Zh. Prikl. Spektrosk. *21*, 115 (1974), C.A. *81*, 113255c (1974); Changes in the vibrational spectra of *p*-(benzalamino)-α-methylcinnamates during phase transitions. [318] V. Volterra, E. Weiner-Avnear, Opt. Commun. *12*, 194 (1974), C.A. *82*, 24135s (1975); Continuous wave thermal lens effect in thin nematic layer. [319] D. Vorländer, M. E. Huth, Z. physik. Chem. *75*, 641 (1911), C.A. *5*, 1357[1] (1911); Charakter der Doppelbrechung flüssiger Kristalle. [320] C. H. Wang, A. L. Leu, J. Am. Chem. Soc. *94*, 8605 (1972), C.A. *78*, 34830w (1973); Raman depolarization ratio and short-range order in liquid crystals.

[321] T. Watanabe, M. Sukigara, K. Honda, Seisan-Kenkyu *26*, 188 (1974), C.A. *81*, 113023a (1974); Nematic liquid crystal as an anisotropic solvent in spectroscopic measurements. [322] T. Watanabe, M. Sukigara, K. Honda, K. Toda, S. Nagaura, Mol. Cryst. Liq. Cryst. *31*, 285 (1975); C.A. *84*, 82115n (1976); Étude d'absorption des composés organiques dissous dans un nématique orienté. [323] O. Wiener, S. Ber. Sächs. Ges. Wissensch. *61*, 113 (1909); Theorie der Stäbchendoppelbrechung. [324] G. K. L. Wong, Y. R. Shen, Phys. Rev. Lett. *30*, 895 (1973), C.A. *78*, 166759q (1973); Optical-field-induced ordering in the i phase of a nematogen. [325] G. K. L. Wong, Y. R. Shen, Phys. Rev. A *10*, 1277 (1974), C.A. *82*, 9452z (1975); Pretransitional behavior of laser field induced molecular alignment in isotropic nematogens. [326] G. K. L. Wong, Y. R. Shen, Phys. Rev. Lett. *32*, 527 (1974), C.A. *80*, 114651v (1974); Transient self-focusing in the i phase of a nematogen. [327] G. K. L. Wong, Y. R. Shen, Dig. Tech. Pap.–Int. Quantum Electron. Conf., 8th, *1974*, 28, C.A. *85*, 85285z (1976); Quantitative experimental study of transient self-focusing. [328] K. S. Yun, Diss. Univ. of Cincinnati, Cincinnati, Ohio, 1962, 66 pp. Avail. Univ. Microfilms (Ann Arbor, Mich.), Order No. 61-5238, C.A. *57*, 5317h (1962); Theoretical study of light scattering in the mesomorphic state (application to the nematic structure). [329] A. S. Zhdanova, L. F. Morozova, G. V. Peregudov, M. M. Sushchinskii, Opt. Spektrosk. *26*, 209 (1969), C.A. *71*, 8234r (1969); Raman effect method for studying liquid crystals. [330] A. S. Zhdanova, V. S. Gorelik, M. M. Sushchinskii, Opt. Spektrosk. *31*, 903 (1971), C.A. *76*, 105804n (1972); Raman effect in liquid crystals studied using an argon laser.

[331] H. Zocher, Naturwiss. *13*, 1015 (1925), C.A. *20*, 699 (1926); Optische Anisotropie selektiv absorbierender Stoffe und über mechanische Erzeugung von Anisotropie. [332] V. M. Zolotarev, N. M. Belyaevskaya, Y. S. Bobovich, Opt. Spektrosk. *28*, 195 (1970), C.A. *72*, 126940m (1970); Perturbed total internal reflection spectra of liquid-crystals.

Chapter 7

[1] W. Haas, J. Adams, J. Wysocki, Mol. Cryst. Liq. Cryst. *7*, 371 (1969), C.A. *71*, 86316h (1969); Interaction between UV radiation and cholesterics. [2] J. Adams, W. E. L. Haas, J. J. Wysocki, Mol. Cryst. Liq. Cryst. *8*, 9 (1969), C.A. *71*, 86175m (1969); Light scattering properties of cholesteric films. [3] J. Adams, W. E. L. Haas, J. Wysocki, Phys. Rev. Lett. *22*, 92 (1969), C.A. *70*, 51772f (1969); Dependence of pitch on composition in cholesterics. [4] J. E. Adams, W. E. L. Haas, J. Wysocki, J. Chem. Phys. *50*, 2458 (1969), C.A. *70*, 110232e (1969); Optical properties of cholesteric films. [5] J. Adams, W. E. L. Haas, Mol. Cryst. Liq. Cryst. *11*, 229 (1970), C.A. *74*, 46761e (1971); Dispersive reflection in cholesterics. [6] J. Adams, L. Leder, Chem. Phys. Lett. *6*, 90 (1970), C.A. *73*, 92912g (1970); Effective rotary power of cholesterol. [7] J. E. Adams, W. Haas, J. J. Wysocki, Liq. Cryst. Ord. Fluids *1970*, 463, C.A. *79*, 24444e (1973); Effective rotary power of the fatty esters of cholesterol. [8] J. E. Adams, W. E. L. Haas, Mol. Cryst. Liq. Cryst. *15*, 27 (1971), C.A. *75*, 144737s (1971); Characterization of molecular role in pitch determination in liquid crystal mixtures. [9] J. Adams, G. Dir, W. Haas, Liq. Cryst. Ord. Fluids *2*, 421 (1974), C.A. *86*, 180992g (1977); Induced rotary power in ternary mixtures of liquid crystals. [10] J. Adams, W. Haas, Mol. Cryst. Liq. Cryst. *30*, 1 (1975), C.A. *84*, 11079t (1976); Pitch dependence on composition in mixtures of liquid crystals.

[11] J. E. Adams, K. F. Nelson, Mol. Cryst. Liq. Cryst. *31*, 319 (1975), C.A. *84*, 98026v (1976); Pitch dependence on strain in cholesteric films. [12] P. Adamski, A. Dylik-Gromiec, Zesz. Nauk. Politech. Lodz., Fiz. *2*, 51 (1973), C.A. *83*, 51159e (1975); Optical properties of cholesterics. [13] P.

Adamski, A. Dylik-Gromiec, Mol. Cryst. Liq. Cryst. 25, 273 (1974), C.A. 81, 55265t (1974); Determination of the cholesteric temperature range by birefringence measurements in an Abbe refractometer. [14] P. Adamski, A. Dylik-Gromiec, Mol. Cryst. Liq. Cryst. 25, 281 (1974), C.A. 81, 55259u (1974); Temperature range and optical properties of COC, cholesteryl nonanoate, and their mixtures. [15] P. Adamski, A. Dylik-Gromiec, Mol. Cryst. Liq. Cryst. 35, 337 (1976), C.A. 86, 10844t (1977); Determination of order of molecular arrangement and polarizability of cholesteryl nonanoate. [15a] P. Adamski, Z. Mastelarz, K. Robakowski, Zesz. Nauk. Politech. Lodz., Fiz. 3, 55 (1976), C.A. 86, 162971v (1977); Temperature hysteresis of light transmission through cholesterics. [16] M. Aihara, H. Inaba, Opt. Commun. 3, 77 (1971), C.A. 75, 27723y (1971); Optical wave propagation in cholesterics with finite thickness. [17] M. Aihara, H. Inaba, Rep. Res. Inst. Elec. Commun., Tohoku Univ. 22, 89 (1971), C.A. 76, 78648f (1972); Analytical study on selective reflection, CD and ORD of cholesterics with finite thickness. [18] M. Aihara, H. Inaba, Oyo Butsuri 41, 338 (1972), C.A. 77, 145689f (1972); Optical properties of cholesterics. I. [19] M. Aihara, H. Inaba, Oyo Butsuri 41, 345 (1972), C.A. 77, 145691a (1972); Optical properties of cholesterics. II. [19a] S. A. Akopyan, S. M. Arakelyan, R. V. Kochikyan, S. Ts. Nersisyan, Yu. S. Chilingaryan, Kvantovaya Elektron 4, 1441 (1977), C.A. 87, 175478k (1977); Efficiency of nonlinear radiation conversion in THG in a cholesteric. [20] R. Alben, Mol. Cryst. Liq. Cryst. 20, 231 (1973), C.A. 79, 58659u (1973); Theory of the change in cholesteric pitch near c/s transitions.

[21] T. Asada, Kyoto Daigaku Nippon Kagakuseni Kenkyusho Koenshu 31, 1 (1974), C.A. 82, 118283r (1975); Formation process of liquid crystal structure. [21a] A. Azima, C. W. Brown, S. S. Mitra, J. Mol. Struct. 36, 219 (1977), C.A. 86, 98463b (1977); Raman spectra of magnetic field induced c/n transition in doped p-octoxybenzoic acid. [22] R. M. A. Azzam, N. M. Bashara, J. Opt. Soc. Am. 62, 1252 (1972), C.A. 77, 158061v (1972); Simplified approach to the propagation of polarized light in anisotropic media. Application to liquid crystals. [23] R. M. A. Azzam, B. E. Merrill, N. M. Bashara, Appl. Opt. 12, 764 (1973); Trajectories describing the evolution of polarized light in homogeneous anisotropic media and liquid crystals. [24] B. E. Merrill, R. M. A. Azzam, N. M. Bashara, J. Opt. Soc. Am. 64, 731 (1974); Numerical solution for the evolution of the ellipse of polarization in inhomogeneous anisotropic media. [25] H. Baessler, M. M. Labes, Mol. Cryst. Liq. Cryst. 6, 419 (1970), C.A. 72, 105414c (1970); Determination of the pitch of a cholesteric by IR transmission measurements. [26] C. S. Bak, M. M. Labes, J. Chem. Phys. 62, 3066 (1975), C.A. 83, 51170b (1975); Pitch-concentration relations in multicomponent liquid crystal mixtures. [27] C. S. Bak, M. M. Labes, J. Chem. Phys. 63, 805 (1975), C.A. 83, 106556q (1975); Analysis of pitch-concentration dependences in some binary and ternary liquid crystal mixtures. [28] L. D. Barron, Nature 238, 17 (1972), C.A. 77, 81401w (1972); Parity and optical activity. [29] V. A. Belyakov, V. E. Dmitrienko, Fiz. Tverd. Tela 17, 491 (1975); Theory of optical properties of cholesterics in an external field. [30] V. A. Belyakov, V. E. Dmitrienko, Acta Cryst. A 31, S161 (1975); Diffraction theory of optical properties of cholesterics. [30a] V. A. Belyakov, V. E. Dmitrienko, Fiz. Tverd. Tela 18, 2880 (1976); Optics of absorbing cholesterics. [30b] V. A. Belyakov, V. E. Dmitrienko, V. P. Orlov, Kholestericheskie Zhidk. Krist. 1976, 35, C.A. 87, 174875a (1977); Theory of the optical properties of cholesterics. [30c] V. E. Dmitrienko, V. A. Belyakov, Zh. Eksp. Teor. Fiz. 73, 681 (1977), C.A. 87, 143480p (1977); Theory of the optical properties of imperfect cholesterics.

[31] D. W. Berreman, T. J. Scheffer, Phys. Rev. Lett. 25, 577, 902 (1970), C.A. 73, 92401q (1970); Bragg reflection of light from single-domain cholesteric films. [32] D. W. Berreman, T. J. Scheffer, Mol. Cryst. Liq. Cryst. 11, 395 (1970), C.A. 74, 69870d (1971); Reflection and transmission by single-domain cholesteric films. [33] D. W. Berreman, T. J. Scheffer, J. Opt. Soc. Am. 61, 679 (1971); Optical effects of boundary proximity in cholesteric films. [34] T. J. Scheffer, D. W. Berreman, Bull. Am. Phys. Soc. 16, 147 (1971); Reflectance spectrum of a single-domain room-temperature cholesteric film. [35] D. W. Berreman, J. Opt. Soc. Am. 62, 502 (1972), C.A. 76, 133585k (1972); Optics in stratified and anisotropic media. 4 × 4-Matrix formulation. [36] D. W. Berreman, T. J. Scheffer, Phys. Rev. A 5, 1397 (1972), C.A. 76, 105921y (1972); Order versus temperature in cholesterics from reflectance spectra. [37] D. W. Berreman, Mol. Cryst. Liq. Cryst. 22, 175 (1973), C.A. 79, 109047c (1973); s* phase. Unique optical properties. [37a] J. E. Bigelow, R. A. Kashnow, Appl. Opt. 16, 2090 (1977); Poincaré sphere analysis of liquid crystal optics. [38] J. Billard, Compt. rend. 261, 939 (1965), C.A. 63, 15659h (1965); Étude expérimentale des ondes électromagnétiques planes qui se propagent dans une lame nématique heliocoidale. [39] J. Billard, Mol. Cryst. 3, 227 (1967), C.A. 68, 54264m (1968); Electromagnetic waves, not satisfying the Airy conditions, can propagate in heliocoidal nematic films.

[40] J. P. Penot, J. Jacques, J. Billard, Tetrahedron Lett. *37*, 4013 (1968), C.A. *69*, 95796a (1968); Pour reduire la part de hasard dans la recherche des dedoublements spontanes. II. Un micro-diagnostic de l'activite optique et son application.

[41] J. Billard, C. R. Acad. Sci., Sér. B *274*, 333 (1972), C.A. *76*, 159467y (1972); Microdiagnostic rapide du sens de torsion de phases mésomorphes. [42] G. Joly, J. Billard, C. R. Acad. Sci., Sér. B *275*, 485 (1972); Propagation d'ondes électromagnétiques dans les piles de Reusch. [43] M. Goscianski, J. Billard, C. R. Acad. Sci., Sér. B *278*, 279 (1974), C.A. *80*, 126241z (1974); Méthode microscopique de polarimétrie en solvant nématique. [43a] J. P. Berthault, J. Billard, J. Jacques, C. R. Acad. Sci., Sér. C *284*, 155 (1977), C.A. *86*, 170724g (1977); Influence de la structure des solutés chiraux sur l'hélicité induite dans un solvant nématique. [44] B. Boettcher, Materialprüfung *13*, 11 (1971), C.A. *74*, 132507h (1971); Optische Eigenschaften cholesterinischer Flüssigkeiten bei schrägem Lichteinfall. [45] B. Boettcher, G. Graber, Mol. Cryst. Liq. Cryst. *14*, 1 (1971), C.A. *75*, 123332r (1971); Doppelrechnung an dünnen Schichten cholesterinischer Flüssigkeiten. [46] B. Böttcher, Diss. Techn. Univ. Berlin 1972; Optische Eigenschaften cholesterinischer Flüssigkeiten. [47] B. Böttcher, Chem.-Ztg. *96*, 214 (1972), C.A. *77*, 40811m (1972); Temperaturabhängigkeit der optischen Eigenschaften cholesterinischer Flüssigkeiten. [48] B. Böttcher, BAM-Berichte Nr. 41 (1976); Optische Eigenschaften cholesterinischer Flüssigkeiten. [49] M. Born, Ann. Physik *55*, 177 (1918), C.A. *13*, 1560³ (1919); Elektronentheorie des natürlichen optischen Drehvermögens isotroper und anisotroper Flüssigkeiten. [50] S. A. Brazovskii, S. G. Dmitriev, Zh. Eksp. Teor. Fiz. *69*, 979 (1975), C.A. *83*, 200469k (1975); Phase transitions in cholesterics.

[51] M. Brunet-Germain, J. Phys. (Paris), C-4, *30*, 28 (1969); Propriétés optiques de mélanges n/c. [52] M. Brunet-Germain, Acta Cryst. A *26*, 595 (1970), C.A. *74*, 25885h (1971); Mesures du pouvoir rotatoire spécifique et du pas de mélanges de benzoate de cholestérol et de PAA. Comparaison avec les résultats de la theorie de Mauguin-de Vries. [53] M. Brunet, J. Phys. (Paris), C-1, *36*, 321 (1975); Optical properties of a s$_C^*$. [54] S. Bualek, Diss. Dortmund 1975; IRRD induziert-cholesterischer Phasen: Aussagemöglichkeiten über die absolute Konfiguration chiraler Moleküle. [55] O. N. Bubel, N. R. Posledovich, V. M. Astaf'ev. Vestn. Belorus. un-ta ser. 1, *1974*, 32, C.A. *83*, 123635g (1975); DTA and IR spectra of new cholesterics. [56] A. D. Buckingham, G. P. Ceasar, M. B. Dunn, Chem. Phys. Lett. *3*, 540 (1969), C.A. *71*, 105952f (1969); Addition of optically active compounds to nematics. [57] B. J. Bulkin, K. Krishnan, J. Am. Chem. Soc. *93*, 5998 (1971), C.A. *76*, 39581j (1972); Raman spectra of crystal, cholesteric, and isotropic cholesteryl esters, 2800–3100-cm^{-1} region. [58] B. J. Bulkin, Liq. Cryst., Proc. Int. Conf., Bangalore, 1973, Abstr. No. 17; IR optical activity-probe of order in nematics. [59] R. Cano, Compt. rend. *251*, 1139 (1960), C.A. *55*, 14361g (1961); Pouvoir rotatoire des corps cholestériques. [60] R. Cano, Bull. Soc. Fr. Mineral. Cristallogr. *90*, 333 (1967), C.A. *68*, 54252f (1968); Pouvoir rotatoire des cholestériques.

[61] R. Cano, J. Phys. (Paris), C-4, *30*, 28 (1969); Propriétés optiques des cholestériques. [62] J. C. Martin, R. Cano, C. R. Acad. Sci., Sér. B *278*, 219 (1974), C.A. *80*, 114274z (1974); Étude des vibrations se propageant dans un cholestérique au voisinage de la bande d'inversion. Pouvoir rotatoire apparent. [63] P. L. Carroll (Ovum: London). 1973. 280 pp. Cholesterics: their technology and applications. [64] F. Castano, An. Fis. *65*, 55 (1969), C.A. *71*, 13273g (1969); Structural mechanism of the transduction of the mesomorphic aggregate between deoxycholic acid and π-systems. [65] I. Chabay, Chem. Phys. Lett. *17*, 283 (1972), C.A. *78*, 49939n (1973); Absorptive and scattering CD of cholesterics in the IR. [66] I. Chabay, G. Holzwarth, Appl. Opt. *14*, 454 (1975), C.A. *82*, 131996v (1975); IRCD and a LD spectrophotometer. [67] S. Chandrasekhar, K. N. S. Rao, Acta Cryst. A *24*, 445 (1968), C.A. *69*, 31605e (1968); Optical rotatory power of liquid crystals. [68] S. Chandrasekhar, J. S. Prasad, Phys. Solid State, *1969*, 77, C.A. *76*, 38493v (1972); Diffraction of light by cholesterics. Incident beam normal to the optical axis. [69] S. Chandrasekhar, J. S. Prasad, Mol. Cryst. Liq. Cryst. *14*, 115 (1971), C.A. *75*, 123290a (1971); Theory of rotatory dispersion of cholesterics. [70] G. S. Ranganath, S. Chandrasekhar, U. D. Kini, K. A. Suresh, S. Ramaseshan, Chem. Phys. Lett. *19*, 556 (1973), C.A. *79*, 11504z (1973); Optical properties of mixtures of right- and left-handed cholesterics.

[71] R. Nityananda, U. D. Kini, S. Chandrasekhar, K. A. Suresh, Liq. Cryst., Proc. Int. Conf., Bangalore, *1973*, 325, C.A. *84*, 128166k (1976); Anomalous transmission (Borrmann effect) in absorbing cholesterics. [72] S. Chandrasekhar, G. S. Ranganath, K. A. Suresh, Liq. Cryst., Proc. Int. Conf., Bangalore, *1973*, 341; Dynamical theory of reflexion from cholesterics. [73] S. Chandrasekhar, G. S. Ranganath, U. D. Kini, K. A. Suresh, Mol. Cryst. Liq. Cryst. *24*, 201 (1973), C.A. *82*, 131242w (1975); Theory of the optical properties of nonabsorbing compensated cholesterics. [74] S. Chandrasekhar, G. S. Ranganath, Mol. Cryst. Liq. Cryst. *25*, 195 (1974), C.A. *81*, 56009z (1974); Spectral width of

reflection from a cholesteric. [**75**] S. Chandrasekhar, B. R. Ratna, Mol. Cryst. Liq. Cryst. *35*, 109 (1976), C.A. *85*, 114991j (1976); Pressure dependence of the pitch of COC. [**76**] R. Chang, Mol. Cryst. Liq. Cryst. *12*, 105 (1971), C.A. *75*, 42796t (1971); Low-frequency Raman scattering of cholesteryl propionate nonanoate and palmitate. [**77**] R. Cano, P. Châtelain, Compt. rend. *253*, 1815 (1961), C.A. *57*, 69c (1962); Variations de l'équidistance des plans de Grandjean avec le titre des mélanges de *p*-cyanobenzalaminocinnamate d'amyle actif et inactif. [**78**] R. Cano, P. Châtelain, Compt. rend. *259*, 352 (1964), C.A. *61*, 13967e (1964); Vérification d'une théorie sur le pouvoir rotatoire des corps cholestériques. [**79**] P. Châtelain, M. Brunet-Germain, C. R. Acad. Sci., Sér. C *268*, 205 (1969), C.A. *70*, 72177f (1969); Mesures du pouvoir rotatoire de mélanges de PAA et de benzoate de cholestérol, et vérification de la théorie de De Vries. [**80**] P. Châtelain, J. C. Martin, C. R. Acad. Sci., Sér. C *268*, 758 (1969), C.A. *71*, 107165a (1969); Separation des deux vibrations circulaires se propageant dans un cholestérique.

[**81**] P. Châtelain, J. C. Martin, C. R. Acad. Sci., Sér. C *268*, 898 (1969), C.A. *72*, 26027a (1970); Calcul théorique des indices pour les vibrations circulaires se propageant dans un cholestérique et comparaison avec l'expérience. [**82**] P. Châtelain, M. Brunet-Germain, C. R. Acad. Sci., Sér. C *268*, 1016 (1969), C.A. *71*, 7590s (1969); Nouveaux types de discontinuités de Grandjean-Cano dans une structure cholestérique et interprétation hypothétique. [**82a**] S. H. Chen, L. S. Chou, Mol. Cryst. Liq. Cryst. *40*, 223 (1977), C.A. *87*, 125607f (1977); Light reflection in cholesteric mixtures. [**83**] I. G. Chistyakov, Kristallografiya *8*, 79 (1963), C.A. *58*, 12033a (1963); Transparency of liquid crystals. [**84**] V. N. Aleksandrov, I. G. Chistyakov, Kristallografiya *14*, 520 (1969), C.A. *71*, 49079n (1969); Optical activity of cholesteryl propionate-PAA mixtures. [**85**] V. N. Aleksandrov, I. G. Chistyakov, Mol. Cryst. Liq. Cryst. *8*, 19 (1969), C.A. *71*, 102114y (1969); Optical activity of the mixtures of PAA and cholesteryl propionate. [**86**] A. I. Aleksandrov, I. G. Chistyakov, Uch. Zap. Ivanov. Gos. Pedagog. Inst. *77*, 117 (1970), C.A. *76*, 18890u (1972); Apparatus for studying the temperature dependence of the color of cholesterics. [**86a**] I. G. Chistyakov, I. I. Gorina, M. Yu. Rubtsova, Kristallografiya *22*, 149 (1977), C.A. *86*, 179714m (1977); Optical properties of multicomponent cholesterics. [**87**] D. F. Ciliberti, G. D. Dixon, L. C. Scala, Mol. Cryst. Liq. Cryst. *20*, 27 (1973), C.A. *78*, 102973x (1973); Shear effects on cholesterics. [**88**] G. H. Conners, J. Opt. Soc. Am. *58*, 875 (1968), C.A. *69*, 41262e (1968); Electromagnetic wave propagation in cholesterics. [**88a**] I. Cuculescu, C. Motoc, M. Honciuc, O. Savin, Stud. Cercet. Fiz. *29*, 19 (1977), C.A. *87*, 60965z (1977); Determination of the birefringence of cholesteric mixtures with the Abbe refractometer. [**89**] D. Demus, G. Wartenberg, Liq. Cryst., Proc. Int. Conf., Bangalore, *1973*, 363, C.A. *84*, 113657x (1976); Selective reflection of light in cholesteryl esters. [**90**] Y. V. Denisov, V. A. Kizel, V. V. Mnev, E. P. Sukhenko, V. G. Tishchenko, Pis'ma Zh. Eksp. Teor. Fiz. *22*, 242 (1975), C.A. *83*, 210838u (1975); Gyrotropism at vibrational transitions.

[**91**] Y. V. Denisov, V. A. Kizel, E. P. Sukhenko, Zh. Eksp. Teor. Fiz. *71*, 679 (1976), C.A. *85*, 152125n (1976); Study of the ordering of cholesterics on basis of their optical parameters. [**92**] Y. V. Denisov, V. A. Kizel, E. P. Sukhenko, V. G. Tishchenko, Kristallografiya *21*, 991 (1976), C.A. *86*, 63814q (1977); Gyrotropy of cholesterics. I. Dependence on the layer thickness. [**92a**] Y. V. Denisov, V. A. Kizel, E. P. Sukhenko, Kristallografiya *22*, 339 (1977), C.A. *87*, 31339t (1977); II. Region of electronic absorption band (cf. [**92**]). [**93**] H. de Vries, Acta Cryst. *4*, 219 (1951), C.A. *45*, 8834i (1951); Rotatory power and other optical properties of certain liquid crystals. [**94**] G. D. Dixon, L. C. Scala, Mol. Cryst. Liq. Cryst. *10*, 317 (1970), C.A. *73*, 70698q (1970); Thermal hysteresis in cholesteric color responses. [**95**] R. Dreher, Dokumentationszentrum der Bundeswehr, 53 Bonn, Fr.-Ebert-Allee 34, April 1972; Das selektive optische Reflexionsvermögen cholesterinischer Flüssigkristalle. [**96**] R. Dreher, Solid State Commun. *12*, 519 (1973), C.A. *78*, 153264z (1973); Reflection properties of distorted cholesterics. [**97**] R. Dreher, H. Schomburg, Chem. Phys. Lett. *25*, 527 (1974), C.A. *81*, 18716v (1974); Prolongation of fluorescence decay time by structural changes of the environment of the emitting molecule. [**98**] R. J. Dudley, S. F. Mason, R. D. Peacock, J. Chem. Soc., Chem. Commun. *1972*, 1084, C.A. *77*, 163718u (1972); IR vibrational CD. [**99**] R. J. Dudley, S. F. Mason, R. D. Peacock, J. Chem. Soc., Faraday Trans. 2, *71*, 997 (1975), C.A. *83*, 68224v (1975); Electronic and vibrational LD and CD of nematics and cholesterics. [**100**] G. Durand, D. V. L. G. N. Rao, Bull. Am. Phys. Soc. *12*, 1054 (1967); Brillouin scattering of light in liquid crystals.

[**101**] G. Durand, C. R. Acad. Sci., Sér. B *264*, 1251 (1967), C.A. *67*, 103491y (1967); Diffraction de la lumière et inversion du pas de la mésophase cholestérique. [**102**] G. Durand, C. H. Lee, C. R. Acad. Sci., Sér. B *264*, 1397 (1967), C.A. *67*, 112505h (1967); Origine de la génération d'harmonique lumineux dans un cristal liquide. [**103**] G. Durand, C. H. Lee, Mol. Cryst. *5*, 171 (1968), C.A. *70*,

91481s (1969); Origin of SHG of light in liquid crystals. [104] G. Durand, D. V. G. L. N. Rao, Phys. Lett. A *27*, 455 (1968), C.A. *70*, 15515r (1969); Brillouin scattering of light in a liquid crystal. [105] C. Elachi, C. Yeh, J. Opt. Soc. Am. *63*, 840 (1973), C.A. *79*, 47287v (1973); Stop bands for optical wave propagation in cholesterics. [106] W. Elser, R. D. Ennulat, Adv. Liq. Cryst. *2*, 73 (1976), C.A. *86*, 198041f (1977); Selective reflection of cholesterics. [107] R. D. Ennulat, Mol. Cryst. Liq. Cryst. *13*, 337 (1971), C.A. *75*, 135449x (1971); Selective light reflection by plane textures. [108] R. D. Ennulat, L. E. Garn, J. D. White, Mol. Cryst. Liq. Cryst. *26*, 245 (1974), C.A. *82*, 66203t (1975); Temperature sensitivity of the selective reflection by cholesterics and its possible limitations. [109] M. Evans, R. Moutran, A. H. Price, J. Chem. Soc., Faraday Trans. 2, *71*, 1854 (1975), C.A. *84*, 4265x (1976); Dielectric properties, refractive index, and far-IR spectrum of COC. [110] H. Falk, O. Hofer, H. Lehner, Monatsh. Chem. *105*, 169 (1974), C.A. *81*, 104161s (1974); LCICD einiger Pyrrolmethenderivate in cholesterischer Mesophase.

[111] C. P. Fan, L. Kramer, J. M. Stephen, Phys. Rev. A *2*, 2482 (1970), C.A. *74*, 35839k (1971); Fluctuations and light scattering in cholesterics. [112] J. L. Fergasen, Mol. Cryst. *1*, 293 (1966), C.A. *66*, 14677a (1967); Cholesteric structure. I. Optical properties. [113] H. Finkelmann, Diss. Paderborn 1975; Einfluß von Molekülstruktur und Temperatur auf die Ganghöhe cholesterischer Mischphasen. [114] H. Franke, Diss. Halle 1912; Natürliche Drehung der Polarisationsebene in flüssigen Kristallen. [115] I. Freund, P. M. Rentzepis, Phys. Rev. Lett. *18*, 393 (1967), C.A. *66*, 120705t (1967); SHG in liquid crystals. [116] G. Friedel, Compt. rend. *176*, 475 (1923); C.A. *17*, 3633 (1923); Les corps cholestériques. [117] E. K. Galanov, R. I. Mel'nik, M. V. Mukhina, Opt. Spektrosk. *40*, 1006 (1976), C.A. *86*, 35845g (1977); Azimuthal dependence of the polarization characteristics of cholesterics. [118] E. K. Galanov, Opt. Spektrosk. *41*, 440 (1976), C.A. *86*, 23732e (1977); Theory of optical properties of cholesterics. [119] P. Gaubert, Compt. rend. *157*, 1446 (1913), C.A. *8*, 1370[6] (1914); Cristaux liquides mixtes. [120] P. Gaubert, Compt. rend. *164*, 405 (1917), C.A. *11*, 1586[9] (1917); Pouvoir rotatoire des cristaux liquides.

[121] P. Gaubert, Bull. soc. franç. min. *40*, 5 (1917), C.A. *12*, 2548[1] (1918); Polymorphisme de quelques substances (cristaux liquides et sphérolites à enroulement helicoïdal). [122] P. Gaubert, Compt. rend. *177*, 698 (1923), C.A. *18*, 1072[4] (1924); Sur les plans de Grandjean. [123] P. Gaubert, Compt. rend. *207*, 1052 (1938), C.A. *33*, 1191[2] (1939); Anneaux mobiles dans les gouttes anisotropes de PAA contenant une petit quantité de phloridzine. [124] P. Gaubert, Compt. rend. *208*, 43 (1939), C.A. *33*, 1564[4] (1939); Anneaux mobiles produits dans les gouttes anisotropes de la phase nématique par l'addition d'une substance possédant le pouvoir rotatoire. [125] E. V. Generalova, Z. P. Dobronevskaya, T. N. Ignatovich, E. L. Kitaeva, B. I. Ostrovskii, A. S. Sonin, N. B. Titova, N. A. Tukmanova, Kristallografiya *20*, 1253 (1975), C.A. *84*, 68092w (1976); Phase diagram and optical properties of stable cholesterics. [126] H. J. Gerritsen, R. T. Yamaguchi, Amer. J. Phys. *39*, 920 (1971), C.A. *75*, 81911w (1971); Microwave analog of optical rotation in cholesterics. [127] H. W. Gibson, J. M. Pochan, D. D. Hinman, Liq. Cryst. Ord. Fluids *2*, 593 (1974), C.A. *86*, 180997n (1977); Effect of cholesteryl alkanoate structure on the pitch of the cholesteric mesophase. [128] F. Giesel, Physik. Z. *11*, 192 (1910); Polarisationserscheinungen an flüssigen Kristallen der Cholesterinester. [129] L. S. Goldberg, J. M. Schnur, Appl. Phys. Lett. *14*, 306 (1969), C.A. *71*, 26468z (1969); Optical SHG and THG in cholesteryl nonanoate liquid crystal. [130] L. S. Goldberg, J. M. Schnur, Radio Electron. Eng. *39*, 279 (1970); C.A. *73*, 71680q (1970); Optical harmonic generation in liquid crystals.

[131] C. H. Gooch, H. A. Tarry, Electron. Lett. *10*, 2 (1974), C.A. *80*, 76050p (1974); Optical characteristics of twisted nematic films. [132] C. H. Gooch, H. A. Tarry, J. Phys. D *8*, 1575 (1975); Optical properties of twisted nematics with twist angles less than 90°. [132a] L. S. Gorbatenko, Deposited Doc. 1975, VINITI 564, 11 pp., C.A. *86*, 179900u (1977); IR spectra of twisting and fan-shaped vibrations of methylene groups in cholesterics at phase transitions. [132b] L. S. Gorbatenko, T. P. Myasnikova, R. Ya. Evseeva, V. N. Lyubimenko, Deposited Doc. 1975, VINITI 565, 7 pp., C.A. *86*, 179901v (1977); IR spectra of cholesteryl acetate and cholesteryl propionate at phase transitions. [133] G. Gottarelli, B. Samori, S. Marzocchi, C. Stremmenos, Tetrahedron Lett. *1975*, 1981, C.A. *83*, 113197b (1975); Induction of a cholesteric mesophase in a nematic by optically active alcohols. Possible method for the correlation of configurations. [134] G. Gottarelli, B. Samori, C. Stremmenos, Chem. Phys. Lett. *40*, 308 (1976), C.A. *85*, 108034h (1976); Model for the induction of a cholesteric mesophase in nematic MBBA by the addition of optically active 1-phenylethanol. [135] D. Coates, G. W. Gray, Phys. Lett. A *45*, 115 (1973), C.A. *79*, 140463j (1973); Optical studies of the i/c transition. Blue phase. [136] D. Coates, G. W. Gray, Mol. Cryst. Liq. Cryst. *24*, 163 (1973), C.A. *81*, 142277s (1974); Synthesis

of a cholesterogen with hydrogen-deuterium asymmetry. [**137**] D. Coates, G. W. Gray, Phys. Lett. A *51*, 335 (1975), C.A. *83*, 51123p (1975); Correlation of optical features of i/c transitions. [**137a**] G. W. Gray, D. G. McDonnell, Mol. Cryst. Liq. Cryst. *34*, 211 (1977), C.A. *87*, 46819y (1977); Relationship between helical twist sense, absolute configuration and molecular structure for non-sterol cholesterics. [**138**] H. Hanson, A. J. Dekker, F. van der Woude, J. Chem. Phys. *62*, 1941 (1975), C.A. *82*, 178666c (1975); Analysis of the pitch in binary cholesteric mixtures. [**138a**] H. Hanson, A. J. Dekker, F. van der Woude, Mol. Cryst. Liq. Cryst. *42*, 15 (1977); Composition and temperature dependence of the pitch in cholesteric binary mixtures. [**139**] T. Harada, P. P. Crooker, Phys. Rev. Lett. *34*, 1259 (1975), C.A. *83*, 34945r (1975); Light scattering through the i/c transition of a cholesteric. [**140**] T. Harada, P. Crooker, Mol. Cryst. Liq. Cryst. *30*, 79 (1975), C.A. *84*, 11080m (1976); Temperature dependence of the pitch of CEEC.

[**141**] T. Harada, P. Crooker, Bull. Am. Phys. Soc. *20*, 886 (1975); Brillouin scattering in i phases of liquid crystals. [**141a**] T. B. Harvey, III, Mol. Cryst. Liq. Cryst. *34*, 225 (1977), C.A. *87*, 76620u (1977); Boundary induced c/n transition. [**142**] G. Holzwarth, N. A. W. Holzwarth, J. Opt. Soc. Am. *63*, 324 (1973), C.A. *78*, 141935f (1973); CD and ORD near absorption bands of cholesterics. [**143**] C. C. Huang, R. S. Pindak, J. T. Ho, Phys. Lett. A *47*, 263 (1974), C.A. *80*, 149669j (1974); Effect of impurity on the c/s$_A$ transition. [**144**] A. P. Kapustin, Kristallografiya *12*, 516 (1967), C.A. *67*, 85846m (1967); Effect of temperature on light scattering by cholesterics. [**145**] E. I. Kats, Zh. Eksp. Teor. Fiz. *59*, 1854 (1970), C.A. *74*, 69792e (1971); Optical properties of cholesterics. [**145a**] E. I. Kats, Zh. Eksp. Teor. Fiz. *73*, 212 (1977), C.A. *87*, 90929h (1977); Influence of nonlocal effects on Van der Waals interaction. [**146**] A. G. Khachaturyan, Phys. Lett. A *51*, 103 (1975), C.A. *82*, 163384k (1975); Helical domain structure of liquid ferroelectrics. [**146a**] G. M. Kharkova, V. M. Khachaturyan, Aerofiz. Issled. *5*, 38 (1975), C.A. *87*, 124758u (1977); Relation of selective light scattering to temperatures for binary systems of cholesteryl esters. [**147**] V. A. Kizel, Yu. I. Krasilov, V. I. Burkov, Usp. Fiz. Nauk *114*, 295 (1974), C.A. *82*, 66363v (1975); Experimental studies of crystal gyrotropy. [**147a**] V. A. Kizel, S. I. Kudashev, Zh. Eksp. Teor. Fiz. *72*, 2180 (1977), C.A. *87*, 61004x (1977); Ordering mechanism in cholesterics. [**148**] L. Kopf, J. Opt. Soc. Am. *58*, 269 (1968), C.A. *69*, 6392a (1968); Refractive indexes of liquid crystal cholesteryl 2-(2-ethoxyethoxy)ethyl carbonate. [**149**] E. H. Korte, B. Schrader, Messtechnik *81*, 371 (1973), C.A. *81*, 8280j (1974); Polarimeter für den IR-Spektralbereich. [**150**] E. H. Korte, S. Bualek, B. Schrader, Ber. Bunsenges. Phys. Chem. *78*, 876 (1974), C.A. *82*, 10164p (1975); IRRD induziert cholesterischer Lösungen.

[**151**] E. H. Korte, B. Schrader, Appl. Spectrosc. *29*, 389 (1975), C.A. *83*, 170757h (1975); Measurement of the IRRD of cholesterics. [**152**] E. H. Korte, J. Chem. Phys. *66*, 99 (1977), C.A. *86*, 81990y (1977); IRRD on binary cholesteric mixtures. [**152a**] E. H. Korte, S. Bualek, Colloq. Spectrosc. Int., (Proc.) 18th, *2*, 1975, 561, C.A. *87*, 151183j (1977); Assertions concerning the absolute configuration of chiral molecules through the rotation dispersion-induced cholesteric solutions. [**153**] K. Kosai, T. Higashino, T. Minowa, Shikizai Kyokaishi *48*, 729 (1975), C.A. *84*, 82801h (1976); Cholesterics. II. Effect on the coloration by mixing of cholesteryl carboxylic esters. [**154**] H. Kozawaguchi, M. Wada, Jpn. J. Appl. Phys. *14*, 651 (1975), C.A. *83*, 36076g (1975); Helical twisting power in cholesterics. I. Experimental results. [**155**] H. Kozawaguchi, M. Wada, Jpn. J. Appl. Phys. *14*, 657 (1975), C.A. *83*, 36077h (1975); II. Theoretical treatment (cf. [**154**]). [**156**] H. Kozawaguchi, M. Wada, Oyo Butsuri *45*, 543 (1976), C.A. *85*, 169925w (1976); Composition dependence of the helical pitch in n/c mixtures. [**156a**] H. Kozawaguchi, M. Wada, Oyo Butsuri *46*, 158 (1977), C.A. *87*, 109751t (1977); Composition dependence of helical pitch in n/c mixtures. [**156b**] H. Kozawaguchi, M. Wada, Oyo Butsuri *46*, 161 (1977), C.A. *87*, 109752u (1977); Composition dependence of helical pitch in ternary cholesteric mixtures. [**157**] N. L. Kramarenko, I. V. Kurnosov, Y. V. Naboikin, Phys. Status Solidi A *25*, 329 (1974), C.A. *81*, 142281p (1974); Structure investigation of liquid crystals by light scattering. [**158**] N. L. Kramarenko, I. V. Kurnosov, Y. V. Naboikin, V. G. Tishchenko, Ukr. Fiz. Zh. (Russ. Ed.) *20*, 84 (1975), C.A. *82*, 178610e (1975); Optical properties of cholesterics. [**159**] W. Kreide, Physik. Z. *14*, 979 (1913), C.A. *8*, 292^9 (1914); Untersuchungen über die Brechungskoeffizienten flüssiger Kristalle. [**160**] H. G. Kuball, T. Karstens, Angew. Chem. *87*, 200 (1975), C.A. *83*, 34909g (1975); Optische Aktivität orientierter Moleküle.

[**161**] H. Hakemi, M. M. Labes, J. Chem. Phys. *61*, 4020 (1974), C.A. *82*, 50056m (1975); Optical method for studying anisotropic diffusion in liquid crystals. [**162**] J. W. Park, M. M. Labes, Mol. Cryst. Liq. Cryst. *31*, 355 (1975); Helical twisting power of α-phenethylamine in nematics. [**162a**] M. M. Labes, U.S. 4.011.046 (12. 11. 1975), C.A. *87*, 35455a (1977); Liquid crystal quantitative analysis

method for optically active compounds. [163] L. B. Leder, J. Chem. Phys. *55*, 2649 (1971), C.A. *75*, 113504j (1971); Rotatory sense and pitch of cholesterics. [164] L. B. Leder, D. Olechna, Opt. Commun. *3*, 295 (1971), C.A. *75*, 114372b (1971); Halfwidth of the cholesteric reflection band. [165] J. D. Lee, H. Liebowitz, U.S.NTIS, AD-A Rep. *1975*, No. 008577, 35 pp., C.A. *83*, 171105f (1975); Change of optical activity in cholesterics. [166] B. Lenk, Diss. Halle 1913; Anormale ORD flüssig-kristallinischer Cholesterylderivate. [166a] L. N. Lisetskii, V. G. Tishchenko, Kholestericheskie Zhidk. Krist. *1976*, 14, C.A. *87*, 175703e (1977); Problems of the theory of cholesterics. [167] N. P. Lobko, Vestsi Akad. Navuk B. SSR, Ser. Fiz.-Mat. Navuk *1976*, 86, C.A. *84*, 186782u (1976); Toward the problem concerning the temperature dependence of the optical activity of a cholesteric. [168] D. S. Mahler, P. H. Keyes, W. B. Daniels, Phys. Rev. Lett. *36*, 491 (1976), C.A. *84*, 128176p (1976); Light-scattering study of the pretransitional phenomena in the i phase of COC. [169] W. Mahler, M. Panar, J. Am. Chem. Soc. *94*, 7195 (1972), C.A. *77*, 164971h (1972); Cholesteric solids. [170] A. S. Marathay, J. Opt. Soc. Am. *61*, 1363 (1971), C.A. *75*, 145551p (1971); Matrix-operator description of the propagation of polarized light through cholesterics.

[171] J. C. Martin, J. Phys. (Paris), C-4, *30*, 29 (1969), C.A. *72*, 94071k (1970); Vibrations circulaires dans les cholestériques. [172] J. C. Martin, R. Cano, Nouv. Rev. Opt. *7*, 265 (1976), C.A. *85*, 151165p (1976); Light propagation in cholesterics in a domain including inversion range. [173] S. F. Mason, R. D. Peacock, Chem. Phys. Lett. *21*, 406 (1973), C.A. *79*, 145526s (1973); Location of the ^1L_b transition of anthracene by dichroism techniques. [174] S. Masubuchi, T. Akahane, K. Nakao, T. Tako, Mol. Cryst. Liq. Cryst. *35*, 135 (1976), C.A. *85*, 113958s (1976); Investigation of the temperature dependence of the pitch of a cholesteric. [175] S. Masubuchi, T. Akahane, K. Nakao, T. Tako, Mol. Cryst. Liq. Cryst. *38*, 265 (1977), C.A. *86*, 181030d (1977); Temperature and composition dependence of the pitch of cholesteryl chloride/cholesteryl nonanoate mixtures. [176] J. P. Mathieu, Bull. soc. franç. minéral. *61*, 174 (1938), C.A. *33*, 2008^8 (1939); Propriétés optiques fondamentales des substances cholestériques. [177] K. Matsumura, S. Iwayanagi, Oyo Butsuri *43*, 126 (1974), C.A. *81*, 43368r (1974); Determination of helical pitch and optical rotatory power of a mixed cholesteric by the Cano wedge arrangement. [178] S. Mazkedian, S. Melone, F. Rustichelli, J. Phys. (Paris), C-1, *36*, 283 (1975); Light diffraction by cholesterics with a pitch gradient. [179] S. Mazkedian, S. Melone, F. Rustichelli, Acta Cryst. A *31*, S160 (1975), CD and ORD in cholesterics with a pitch gradient. [180] S. Mazkedian, S. Melone, F. Rustichelli, J. Phys. (Paris) *37*, 731 (1976), C.A. *85*, 84859c (1976); CD and ORD in cholesterics with a pitch gradient.

[181] S. Mazkedian, S. Melone, F. Rustichelli, Phys. Rev. A *14*, 1190 (1976), C.A. *85*, 152136s (1976); Light Pendelloesung fringes in Bragg diffraction by cholesterics. [182] R. Dreher, G. Meier, A. Saupe, Mol. Cryst. Liq. Cryst. *13*, 17 (1971), C.A. *75*, 54882f (1971); Selective reflection by cholesterics. [183] R. Dreher, G. Meier, Phys. Rev. A *8*, 1616 (1973), C.A. *79*, 130061n (1973); Optical properties of cholesterics. [184] L. Melamed, D. Rubin, Appl. Opt. *10*, 1103 (1971), C.A. *75*, 12564y (1971); Optical properties of mixtures of cholesterics. [185] J. Cheng, R. B. Meyer, Phys. Rev. Lett. *29*, 1240 (1972), C.A. *77*, 170680d (1972); Pretransitional optical rotation in the i phase of cholesterics. [186] J. Cheng, R. B. Meyer, Phys. Rev. A *9*, 2744 (1974), C.A. *81*, 43365n (1974); Pretransitional optical rotation in the i phase of cholesterics. [187] C. Mioskowski, J. Bourguignon, S. Candau, G. Solladie, Chem. Phys. Lett. *38*, 456 (1976), C.A. *85*, 134100w (1976); Photochemically induced c/n transition in liquid crystals. [188] M. Morita, K. Murata, T. Okamura, M. Seki, T. Kase, Y. Yokoyama, Seikei Daigaku Kogakubu Kogaku Hokoku *1971*, 935, C.A. *76*, 51089n (1972); Phase transition and the IR spectrum of cholesteryl nonanoate liquid crystals. [189] T. P. Myasnikova, L. S. Corbatenko, Sb. Dokl. Vses. Nauch. Konf. Zhidk. Krist. Simp. Ikh Prakt. Primen., 2nd, *1972*, 206, C.A. *81*, 129415r (1974); Temperature effect on the parameters of the IR spectral absorption bands of cholesterics. [190] T. Nakagiri, Phys. Lett. A *36*, 427 (1971), C.A. *75*, 155805n (1971); Helical twisting power in mixtures of nematics and cholesterics.

[191] T. Nakagiri, H. Kodama, K. K. Kobayashi, Phys. Rev. Lett. *27*, 564 (1971), C.A. *75*, 114368e (1971); Helical twisting power in mixtures of nematics and cholesterics. [192] R. Nityananda, Mol. Cryst. Liq. Cryst. *21*, 315 (1973), C.A. *79*, 109023s (1973); Theory of light propagation in cholesterics. [193] R. Nityananda, U. D. Kini, Liq. Cryst., Proc. Int. Conf. Bangalore, *1973*, 311; Theory of reflexion and transmission by plane parallel cholesteric films. [194] W. A. Nordland, J. Appl. Phys. *39*, 5033 (1968), C.A. *70*, 7596c (1969); Preliminary investigation of Brillouin scattering in a liquid crystal. [195] T. J. Novak, E. J. Poziomek, R. A. Mackay, Mol. Cryst. Liq. Cryst. *20*, 203 (1973), C.A. *79*, 58658t (1973); Effect of bulk impurities on the transparency of cholesteryl nonanoate. [196] T. J. Novak,

R. A. Mackay, E. J. Poziomek, Mol. Cryst. Liq. Cryst. 20, 213 (1973), C.A. 79, 71992s (1973); Fluorescence of pyrene and phenanthrene in cholesteryl nonanoate as a function of temperature. [197] T. Ohtsuka, Oyo Butsuri 45, 498 (1976), C.A. 85, 169721b (1976); Helical structure of cholesterics. [198] N. Oron, J. L. Yu, M. M. Labes, Appl. Phys. Lett. 23, 217 (1973), C.A. 80, 32151t (1974); Angular dependence of optical scattering in mixed nematic-cholesterics. [199] N. Oron, K. Ko, L. J. Yu, M. M. Labes, Liq. Cryst. Ord. Fluids 2, 403 (1974), C.A. 86, 180991f (1977); Chirality in mixed nematics and cholesterics. [200] J. C. Martin, O. Parodi, J. Phys. (Paris), C-1, 36, 273 (1975); Light refraction by a cholesteric prism.

[201] O. Parodi, J. Phys. (Paris), C-1, 36, 325 (1975); Light propagation along the helical axis in s$_C^*$. [202] J. D. Parsons, C. F. Hayes, Solid State Commun. 15, 429 (1974); Brillouin zone effect in cholesterics. [203] J. D. Parsons, C. F. Hayes, Mol. Cryst. Liq. Cryst. 29, 295 (1975), C.A. 83, 17923q (1975); Fluctuations and light scattering in a compressible cholesteric. [203a] G. Pelzl, Z. Chem. 17, 264 (1977), C.A. 87, 175000y (1977); Doppelbrechung von cholesterinischem 4-Ethoxy-4'-[(4-methoxyphenyl)-α-methylpropionyl]-azobenzol. [204] P. Pieranski, E. Guyon, P. Keller, L. Liebert, W. Kuczynski, P. Pieranski, Mol. Cryst. Liq. Cryst. 38, 275 (1977), C.A. 86, 179762a (1977); Optical study of s$_C^*$ under shear. [205] P. A. Pincus, C. R. Acad. Sci., Sér. B 267, 1290 (1968), C.A. 70, 91869z (1969); Diffusion Rayleigh dans un cholestérique. [206] R. S. Pindak, C. C. Huang, J. T. Ho, Phys. Rev. Lett. 32, 43 (1974), C.A. 80, 100966q (1974); Divergence of cholesteric pitch near a s$_A$ transition. [207] R. S. Pindak, C. C. Huang, J. T. Ho, Solid State Commun. 14, 821 (1974), C.A. 81, 30660e (1974); Intrinsic pitch of cholesteryl nonanoate. [208] R. Pindak, J. T. Ho, Phys. Lett. A 59, 277 (1976), C.A. 86, 63854c (1977); Cholesteric pitch of cholesteryl decanoate near the s$_A$ transition. [209] J. M. Pochan, D. D. Hinman, J. Phys. Chem. 78, 1206 (1974), C. A. 81, 30652d (1974); Theory of molecular association in c/n mixtures. [210] P. Pollmann, J. Phys. E 7, 490 (1974), C.A. 81, 19132p (1974); Apparatus for the measurement of light reflection of cholesterics at high pressures.

[211] E. J. Poziomek, T. J. Novak, R. A. Mackay, Mol. Cryst. Liq. Cryst. 15, 283 (1972), C.A. 76, 65931t (1972); Transparency characteristics of several cholesteryl esters. [212] J. S. Prasad, M. S. Madhava, Mol. Cryst. Liq. Cryst. 22, 165 (1973), C.A. 79, 109803w (1973); ORD of cholesterics. [213] J. S. Prasad, Liq. Cryst. Ord. Fluids 2, 607 (1974), C.A. 86, 179738x (1977); Theories of optical reflection from cholesteric films. [214] J. S. Prasad, Opt. Commun. 12, 389 (1974), C.A. 82, 78589m (1975); Optical reflection from cholesteric films. [215] J. S. Prasad, J. Phys. (Paris), C-1, 36, 289 (1975), C.A. 83, 17917r (1975); CD in liquid crystals. [216] J. S. Prasad, Opt. Commun. 16, 190 (1976), C.A. 84, 128183p (1976); Theories of form optical rotation. Existence of two zeros in the ORD curve. [217] E. B. Priestley, Introd. Liq. Cryst. 1974, 203, C.A. 84, 82635g (1976); Optical properties of cholesterics and chiral nematics. [218] H. Rabin, P. P. Bey, Phys. Rev. 156, 1010 (1967), C.A. 67, 27417m (1967); Phase matching in harmonic generation employing ORD. [218a] B. Rajagopalan, Diss. Univ. Minnesota, Minneapolis, 1976, 132 pp. Avail. Xerox Univ. Microfilms, Ann Arbor, Mich., Order No. 77-6994, C.A. 87, 14460h (1977); Quasielastic laser light scattering study of c/i transition. [219] G. S. Ranganath, K. A. Suresh, S. R. Rajagopalan, U. D. Kini, Liq. Cryst., Proc. Int. Conf. Bangalore, 1973, 353, C.A. 84, 128167m (1976); CD in absorbing mixtures of right- and left-handed cholesterics. [220] G. S. Ranganath, Opt. Commun. 16, 369 (1976), C.A. 84, 128178r (1976); ORD in cholesterics.

[221] D. V. G. L. N. Rao, Phys. Lett. A 32, 533 (1970), C.A. 74, 7899m (1971); Stimulated Brillouin scattering in a liquid crystal. [222] J. Rault, Solid State Commun. 9, 1965 (1971), C.A. 76, 118466d (1972); Lignes de dislocation helicoidales dans les cholestériques. [223] J. Rault, Mol. Cryst. Liq. Cryst. 26, 349 (1974), C.A. 82, 50025a (1975); Interpretation of "comma" in cholesterics. [224] R. Rettig, G. Pelzl, D. Demus, J. Prakt. Chem. 318, 450 (1976), C.A. 85, 53920m (1976); Doppelbrechung von cholesterinischen Cholesteryl-n-alkylcarbonaten und Cholesterylchlorid. [225] J. Robert, F. Gharadjedaghi, C. R. Acad. Sci., Sér. B 278, 73 (1974), C.A. 80, 89015f (1974); Rotation du plan de polarisation de la lumière dans une structure nématique en hélice. [226] H. Rosen, Y. R. Shen, Mol. Cryst. Liq. Cryst. 18, 285 (1972), C.A. 77, 145693c (1972); Brillouin scattering in a cholesteric near the c/i transition. [227] L. Royer, Compt. rend. 174, 1182 (1922), C.A. 16, 2633[1] (1922); Inversion du pouvoir rotatoire dans les liquides anisotropes. [228] L. Royer, Compt. rend. 180, 148 (1925), C.A. 19, 2595[1] (1925); Pouvoir rotatoire des corps cholestériques. [229] E. I. Ryumtsev, A. N. Cherkasov, V. G. Tishchenko, M. M. Fetisova, Kristallografiya 18, 1049 (1973), C.A. 80, 7879d (1974); Pretransition phenomena and optical activity of the i phase of cholesterics. [230] E. Sackmann, J. Am. Chem. Soc. 93, 7088 (1971), C.A. 76, 38356c (1972); Photochemically induced reversible color changes in cholesterics.

[**231**] E. Sackmann, J. Voss, Chem. Phys. Lett. *14*, 528 (1972), C.A. *77*, 67941r (1972); CD of helically arranged molecules in cholesterics. [**232**] J. Voss, E. Sackmann, Z. Naturforsch. A *28*, 1496 (1973), C.A. *80*, 31279d (1974); Solute and temperature induced pitch changes and pretransitional effects in cholesterics. [**233**] G. Pelzl, H. Sackmann, Z. Phys. Chem. (Leipzig) *254*, 354 (1973), C.A. *80*, 144932e (1974); Doppelbrechung der cholesterinischen und smektischen homologen *n*-Fettsäurechole-sterylester. [**234**] F. D. Saeva, J. J. Wysocki, J. Am. Chem. Soc. *93*, 5928 (1971), C.A. *76*, 8354h (1972); Cholesteric LCICD. [**235**] F. D. Saeva, Mol. Cryst. Liq. Cryst. *18*, 375 (1972), C.A. *77*, 145699j (1972); Method for determining the existence and chirality of cholesterics. [**236**] F. D. Saeva, J. Am. Chem. Soc. *94*, 5135 (1972), C.A. *77*, 67953w (1972); Cholesteric LCICD of achiral solutes. [**237**] F. D. Saeva, Mol. Cryst. Liq. Cryst. *23*, 171 (1973), C.A. *79*, 150344k (1973); Relation between cholesterics and nematics. [**238**] F. D. Saeva, P. E. Sharpe, G. R. Olin, J. Am. Chem. Soc. *95*, 7656 (1973), C.A. *80*, 8566m (1974); Cholesteric LCICD. Mechanistic aspects. [**239**] F. D. Saeva, P. E. Sharpe, G. R. Olin, J. Am. Chem. Soc. *95*, 7660 (1973), C.A. *80*, 8568p (1974); Cholesteric LCICD. Behavior of benzene and some of its mono- and disubstituted derivatives. [**240**] F. D. Saeva, Liq. Cryst. Ord. Fluids *2*, 581 (1974), C.A. *86*, 179737w (1977); Mechanistic aspects of the cholesteric LCICD phenomenon.

[**241**] F. D. Saeva, Pure Appl. Chem. *38*, 25 (1974), C.A. *82*, 178292c (1975); Optical properties of anisotropically ordered solutes in cholesterics. [**242**] F. D. Saeva, Mol. Cryst. Liq. Cryst. *31*, 327 (1975), C.A. *84*, 81913j (1976); Analysis of the CD behavior of some disubstituted benzenes in cholesterics. [**243**] F. D. Saeva, G. R. Olin, J. Am. Chem. Soc. *98*, 2709 (1976), C.A. *85*, 11767m (1976); Extrinsic CD in twisted nematics. [**244**] K. Sakamoto, R. Yoshida, M. Hatano, Chem. Lett. *1976*, 1401, C.A. *86*, 63833v (1977); Liquid crystals of N-acylamino acid. I. CD in liquid crystals of N-lauroyl-L-glutamic acid and aromatic solvents. [**245**] T. Sarada, Diss. American Univ., Washington, D.C. 1972, 184 pp. Avail. Univ. Microfilms, Ann Arbor, Mich., Order No. 73-16,619, C.A. *80*, 4129y (1974); Effect of linear polymers on the optical properties of cholesterics. [**246**] S. Sato, Kagaku No Ryoiki *27*, 983 (1973), C.A. *80*, 101770h (1974); Cholesteric LCICD. [**247**] B. Schrader, E. H. Korte, Angew. Chem. *84*, 218 (1972), C.A. *77*, 81404z (1972); IRRD. [**248**] J. W. Shelton, Y. R. Shen, Phys. Rev. Lett. *25*, 23 (1970), C.A. *73*, 60074u (1970); Phase matched THG in cholesterics. [**249**] J. W. Shelton, Y. R. Shen, Phys. Rev. Lett. *26*, 538 (1971), C.A. *74*, 104255f (1971); Umklapp optical THG in cholesterics. [**250**] J. W. Shelton, Y. R. Shen, Phys. Rev. A *5*, 1867 (1972), C.A. *76*, 119702q (1972); Phase-matched normal and Umklapp THG processes in cholesterics.

[**251**] Y. R. Shen, Phys. Opto-Electron. Mater., Proc. Symp. *1970*, 17, C.A. *77*, 26692s (1972); Nonlinear optical effects. [**251a**] Y. R. Shen, Proc. Int. Sch. Phys. "Enrico Fermi" *64*, 201 (1975), C.A. *87*, 46072z (1977); Nonlinear optics in a one-dimensional periodic medium. [**251b**] Y. R. Shen, Proc. Int. Sch. Phys. "Enrico Fermi" *64*, 210 (1975), C.A. *87*, 60005m (1977); Nonlinear optical study of pretransitional behavior in liquid crystals. [**252**] S. Shtrikman, M. Tur, J. Opt. Soc. Am. *64*, 1178 (1974); Optical properties of distorted cholesterics. [**252a**] P. E. Sokol, J. T. Ho, Appl. Phys. Lett. *31*, 487 (1977), C.A. *87*, 191356z (1977); Optical rotatory power near the c/s_A transition. [**253**] A. S. Sonin, A. V. Tolmachov, V. G. Tishchenko, V. G. Rak, Zh. Eksp. Teor. Fiz. *68*, 1951 (1975), C.A. *83*, 171077y (1975); Optical activity of plane texture of cholesteryl esters. [**254**] A. S. Sonin, A. V. Tolmachev, V. G. Tishchenko, Kristallografiya *21*, 1164 (1976), C.A. *86*, 80980h (1977); CD and structural features of the planar cholesteric texture. [**255**] H. Stegemeyer, K. J. Mainusch, Chem. Phys. Lett. *6*, 5 (1970), C.A. *73*, 82233x (1970); Optical rotatory power of liquid crystal mixtures. [**256**] H. Stegemeyer, K. J. Mainusch, E. Steigner, Chem. Phys. Lett. *8*, 425 (1971), C.A. *75*, 42492r (1971); Optical rotatory power of mixtures of a nematogen and nonmesomorphic chiralic compound. [**257**] H. Stegemeyer, K. J. Mainusch, Naturwissenschaften *58*, 599 (1971), C.A. *76*, 92265k (1972); Induzierung von optischer Aktivität und CD in nematischen Phasen durch chirale Moleküle. [**258**] H. Stegemeyer, U. Müller, Naturwissenschaften *58*, 621 (1971), C.A. *76*, 65699y (1972); Gleichzeitige Existenz zweier kristallinflüssiger Phasen mit Helix-Struktur. [**259**] K. J. Mainusch, H. Stegemeyer, Z. phys. Chem. (Frankfurt) *77*, 210 (1972), C.A. *77*, 81427j (1972); Induktion eines Cotton-Effekts durch den Einfluß einer cholesterinischen Lösungsphase auf achirale Moleküle. [**260**] U. Müller, H. Stegemeyer, 2. Arbeitstagung Flüssigkristalle, Inst. Angew. Festkörperphysik Freiburg, 21.4. 1972; CD eines binären Systems von Cholesterinestern entgegengesetzter Helixstruktur.

[**261**] H. Stegemeyer, Angew. Chem. *84*, 720 (1972), C.A. *82*, 23857s (1975); Optische Eigenschaften cholesterinischer Flüssigkristalle. [**262**] H. Stegemeyer, K. J. Mainusch, Chem. Phys. Lett. *16*, 38 (1972), C.A. *77*, 158095j (1972); Induction of optical activity in a nematic by 1-menthol. [**263**] W. U. Müller, H. Stegemeyer, Ber. Bunsenges. Phys. Chem. *77*, 20 (1973), C.A. *78*, 116284k (1973); Birefringence

of compensated cholesterics. [**264**] K. J. Mainusch, P. Pollmann, H. Stegemeyer, Naturwissenschaften *60*, 48 (1973), C.A. *78*, 110072t (1973); Circularpolarisation der Fluoreszenz achiraler Moleküle in cholesterischen Mesophasen. [**265**] H. Finkelmann, H. Stegemeyer, Z. Naturforsch. A *28*, 799 (1973), C.A. *79*, 65369c (1973); Helixinversion in einem binären n/c Mischsystem. [**266**] H. Finkelmann, H. Stegemeyer, Z. Naturforsch. A *28*, 1046 (1973), C.A. *79*, 145546y (1973); Doppelte Helixinversion in einem binären System cholesterischer Komponenten mit Rechtshelix. [**267**] P. Pollmann, H. Stegemeyer, Chem. Phys. Lett. *20*, 87 (1973), C.A. *79*, 35892a (1973); Pressure dependence of the helical structure of cholesterics. [**268**] H. Stegemeyer, H. Finkelmann, Chem. Phys. Lett. *23*, 227 (1973), C.A. *80*, 54027j (1974); Treatment of cholesteric mixtures by means of the Goossens theory. [**269**] P. Pollmann, H. Stegemeyer, Ber. Bunsenges. Phys. Chem. *78*, 843 (1974), C.A. *82*, 10199d (1975); Einfluß allseitigen Druckes auf die Struktur cholesterischer Mesophasen. [**270**] H. Stegemeyer, Ber. Bunsenges. Phys. Chem. *78*, 860 (1974), C.A. *82*, 56910g (1975); Helixstruktur und optische Aktivität in Flüssigkristallen.

[**271**] H. Finkelmann, H. Stegemeyer, Ber. Bunsenges. Phys. Chem. *78*, 869 (1974), C.A. *82*, 10275a (1975); Beschreibung cholesterischer Mischsysteme mit einer erweiterten Goossens-Theorie. [**272**] K. J. Mainusch, H. Stegemeyer, Ber. Bunsenges. Phys. Chem. *78*, 927 (1974), C.A. *82*, 36971z (1975); Einfluß der Molekülgeometrie auf die Fluoreszenz-Circularpolarisation in cholesterischen Phasen. [**273**] H. Stegemeyer, W. U. Müller, K. J. Mainusch, H. Finkelmann, Arch. Eisenhüttenwes. *46*, 609 (1975), C.A. *83*, 211357y (1975); Selektive Lichtreflexion an cholesterischen Flüssigkristallen. [**274**] H. Stegemeyer, H. Finkelmann, Naturwissenschaften *62*, 436 (1975), C.A. *83*, 186631y (1975); Temperature dependence of helical pitch of induced cholesterics. [**274a**] P. Pollmann, K. J. Mainusch, H. Stegemeyer, Z. Phys. Chem. (Frankfurt) *103*, 295 (1976), C.A. *86*, 170227x (1977); Circularpolarisation der Fluoreszenz achiraler Moleküle in cholesterischen Flüssigkristallen. [**275**] R. S. Stein, M. B. Rhodes, R. S. Porter, J. Colloid Interface Sci. *27*, 336 (1968), C.A. *69*, 46929n (1968); Light scattering by liquid crystals. [**276**] M. B. Rhodes, R. S. Stein, W. Chu, R. S. Porter, U. S. Clearinghouse Fed. Sci. Tech. Inform., AD *1968*, AD-679186, 44 pp., C.A. *71*, 7954p (1969); Light scattering studies of orientation correlations in cholesteryl esters. [**277**] M. B. Rhodes, R. S. Porter, W. Chu, R. S. Stein, Mol. Cryst. Liq. Cryst. *10*, 295 (1970), C.A. *73*, 60867e (1970); Light scattering studies of orientation correlations in cholesteryl esters. [**277a**] T. H. Sterling, C. F. Hayes, Mol. Cryst. Liq. Cryst. *43*, 279 (1977); Optical rotatory power of a cholesteric for obliquely incident light using multiple scaling. [**278**] S. V. Subramanyam, Appl. Opt. *10*, 317 (1971), C.A. *74*, 105227k (1971); Optical reflection from cholesteric films. [**279**] K. A. Suresh, Mol. Cryst. Liq. Cryst. *35*, 267 (1976), C.A. *86*, 23878g (1977); Experimental study of the anomalous transmission (Borrmann effect) in absorbing cholesterics. [**280**] T. Tako, T. Akahane, S. Masubuchi, Jap. J. Appl. Phys. *14*, 425 (1975); Measurement of the pitch and refractive indices of cholesterics using selective reflections and total reflections.

[**281**] D. Taupin, J. Phys. (Paris), C-4, *30*, 32 (1969), C.A. *72*, 126942p (1970); Étude quantitative théorique des réflexions sélectives de la lumière par les cholestériques parfaits. [**282**] I. Teucher, K. Ko, M. M. Labes, J. Chem. Phys. *56*, 3308 (1972), C.A. *76*, 118340h (1972); Birefringence and ORD of a compensated cholesteric. [**283**] K. Ko, I. Teucher, M. M. Labes, Mol. Cryst. Liq. Cryst. *22*, 203 (1973), C.A. *80*, 3700r (1974); Helical twisting power of steroidal solutes in cholesterics. Carbonate esters of cholesterol and dicholesteryl compounds. [**284**] A. V. Tolmachev, A. S. Sonin, Kristallografiya *21*, 794 (1976), C.A. *85*, 151119b (1976); Optical properties of a cholesteric layer. [**284a**] A. V. Tolmachev, A. N. Chuvyrov, V. P. Mikheev, Kholestericheskie Zhidk. Krist. *1976*, 50, C.A. *87*, 174876b (1977); Methods for studying the optical constants of cholesterics. [**284b**] A. V. Tolmachev, V. G. Tishchenko, L. N. Lisetskii, Fiz. Tverd. Tela *19*, 1886 (1977), C.A. *87*, 125567t (1977); Evaluation of orientation order in cholesterics. [**285**] Y. Tomkiewicz, A. Weinreb, Chem. Phys. Lett. *3*, 229 (1969), C.A. *71*, 34832d (1969); Decay time of pyrene in a liquid crystal. [**286**] E. E. Topchiashvili, Z. M. Elashvili, G. S. Chilaya, M. D. Museridze, Z. G. Dzotsenidze, Soobshch. Akad. Nauk Gruz. SSR *83*, 101 (1976), C.A. *85*, 200752p (1976); Induction of chiral structure in nematics. [**287**] M. Tur, Mol. Cryst. Liq. Cryst. *29*, 345 (1975), C.A. *83*, 35033d (1975); Reflection at the boundary between glass and cholesterics. [**288**] M. Tsukamoto, T. Ohtsuka, K. Morimoto, Y. Murakami, Jpn. J. Appl. Phys. *14*, 1307 (1975), C.A. *83*, 178127y (1975); Pitch and sense of helix in mixtures of optically active azo or azoxy compounds and nematics. [**289**] G. Vergoten, G. Fleury, Mol. Cryst. Liq. Cryst. *30*, 223 (1975), C.A. *83*, 123558j (1975); Very low frequency Raman spectra of some polycrystalline cholesteryl esters. [**290**] J. Voss, B. Voss, Z. Naturforsch. A *31*, 1661 (1976), C.A. *86*, 63868k (1977); SEM studies of cholesterics.

[**291**] C. H. Wang, Y. Y. Huang, U.S.N.T.I.S., AD Rep. *1974*, No. 783543/2GA, 19 pp., C.A. *82*, 131219u (1975); Krishnan effect in cholesteryl myristate. [**292**] C. H. Wang, Y. Y. Huang, U.S.N.T.I.S.,

AD Rep. *1974*, No. 78599/6GA, 13 pp., C.A. *82*, 160425p (1975); Brillouin scattering of light in a cholesteric. [**293**] C. H. Wang, Y. Y. Huang, J. Chem. Phys. *62*, 3834 (1975), C.A. *83*, 34920d (1975); Brillouin scattering of light in a cholesteric. [**294**] P. F. Waters, T. Sarada, Mol. Cryst. Liq. Cryst. *25*, 1 (1974), C.A. *83*, 139158q (1975); Measurement of the bulk refractive indexes of cholesterics. [**295**] J. J. White, III, Appl. Phys. *5*, 57 (1974), C.A. *82*, 24588s (1975); Estimation of confidence intervals for the nonlinear parameters of a least-squares fit. Application to cholesteryl nonanoate data. [**296**] D. G. Willey, D. E. Martire, Mol. Cryst. Liq. Cryst. *18*, 55 (1972), C.A. *77*, 66963n (1972); Effect of organic solutes on the selective reflection of visible light by cholesteric mixtures. [**297**] J. J. Wright, J. F. Dawson, J. Opt. Soc. Am. *64*, 250 (1974), C.A. *80*, 126242a (1974); Transmission of polarized light as a function of sample thickness for cholesteric films. [**298**] A. Wulf, J. Chem. Phys. *60*, 3994 (1974), C.A. *81*, 55377f (1974); Helical pitch in mixtures of cholesterics. [**299**] C. C. Yang, Phys. Rev. Lett. *28*, 955 (1972), C.A. *76*, 159417g (1972); Light-scattering study of the dynamical behavior of ordering just above the phase transition to a cholesteric. [**300**] G. M. Zharkova, V. M. Khachaturyan, Kholestericheskie Zhidk. Krist. *1976*, 4, C.A. *87*, 175702d (1977); Cholesteric liquid crystals.

[**301**] G. M. Zharkova, Kholestericheskie Zhidk. Krist. *1976*, 56, C.A. *87*, 175705g (1977); Effect of external actions on the pitch of cholesteric spirals.

Chapter 8

[**1**] M. F. Achard, F. Hardouin, G. Sigaud, H. Gasparoux, J. Chem. Phys. *65*, 1387 (1976), C.A. *85*, 134803c (1976); Orientational order and enthalpic measurements on binary mixtures at the n/s$_A$ transition: Comparison with McMillan's models. [**2**] K. Adachi, O. Haida, T. Matsuo, M. Sorai, H. Suga, S. Seki, Acta Cryst. A *28*, S129 (1972); Glassy state of liquid, liquid crystal and of crystal. [**3**] J. Adams, Phase Transitions, Proc. Conf. Phase Transitions Their Appl. Mater. Sci. *1973*, 61, C.A. *81*, 112025r (1974); Phase transitions in liquid crystals. [**3a**] P. Adomenas, V. A. Grozhik, Vestsi Akad. Navuk BSSR, Ser. Khim Navuk *1977*, 39, C.A. *87*, 29178w (1977); Temperature dependence of the density of homologous *p-n*-butyl-*p'-n*-alkoxyazobenzene liquid crystals. [**4**] R. S. Alben, Mol. Cryst. Liq. Cryst. *10*, 21 (1970), C.A. *73*, 49188p (1970); Interpretation of thermodynamic derivatives near the nematic phase transition in PAA. [**5**] R. Alben, Solid State Commun. *13*, 1783 (1973), C.A. *80*, 53161m (1974); n/s Transitions in mixtures. Liquid crystal tricritical point. [**6**] R. Alben, J. Chem. Phys. *59*, 4299 (1973), C.A. *80*, 53170p (1974); Liquid crystal phase transitions in mixtures of rodlike and platelike molecules. [**7**] J. T. S. Andrews, W. E. Bacon, J. Chem. Thermodyn. *6*, 515 (1974), C.A. *81*, 69357m (1974); Adiabatic calorimetry of bis(*p*-methoxyphenyl)*trans*-cyclohexane-1,4-dicarboxylate. [**7a**] M. A. Anisimov, S. R. Garber, V. S. Esipov, V. M. Mamnitskii, G. I. Ovodov, L. A. Smolenko, E. L. Sorkin, Zh. Eksp. Teor. Fiz. *72*, 1983 (1977), C.A. *87*, 30014w (1977); Anomaly in the specific heat and the nature of the i/n transition. [**7b**] O. P. Anisimova, G. A. Vardanyan, Dokl. Akad. Nauk Arm.-SSR *64*, 224 (1977), C.A. *87*, 175817v (19777); Theory of coupled scalar fields. [**8**] K. V. Arkhangel'skii, Mekh. Kinet. Krist. *1969*, 142, C.A. *76*, 18505x (1972); Mesophase transitions and isomeric effect in the supercritical immicibility region of binary systems. [**9**] K. V. Arkhangel'skii, Kinet. Mekh. Krist. *1973*, 29, C.A. *81*, 17843x (1974); Relation of polymorphous transformations and phase transitions. [**10**] H. Arnold, Z. Phys. Chem. (Leipzig) *225*, 45 (1964), C.A. *61*, 90g (1964); Adiabatisches Präzisionskalorimeter zur Bestimmung der Wärmekapazitäten und latenten Wärmen kondensierter Stoffe.

[**11**] H. Arnold, Z. Phys. Chem. (Leipzig) *226*, 146 (1964), C.A. *61*, 11395a (1964); Kalorimetrische Messungen an 12 homologen Dialkoxyazoxybenzolen. [**12**] H. Arnold, Z. Chem. *4*, 211 (1964), C.A. *61*, 10105b (1964); Kalorische Daten und molekulare Wechselwirkung in der Reihe der Dialkoxy-azoxybenzole. [**13**] H. Arnold, P. Roediger, Z. Phys. Chem. (Leipzig) *231*, 407 (1966), C.A. *65*, 1476g (1966); Kalorimetrische Messungen an Azoxybenzoesäure-diäthylester und Äthoxybenzalaminozimtsäure-äthylester. [**14**] H. Arnold, Mol. Cryst. *2*, 63 (1967)-C.A. *66*, 41229n (1967); Heat capacity and enthalpy of transition of aromatic liquid crystals. [**15**] H. Arnold, E. B. El-Jazairi, H. König, Z. Phys. Chem. (Leipzig) *234*, 401 (1967), C.A. *68*, 2509a (1968); Kalorimetrie kristallin-flüssiger Azoxymethylzimtsäure-dialkylester. [**16**] H. Arnold, P. Roediger, Z. Phys. Chem. (Leipzig) *239*, 283 (1968), C.A. *70*, 100369r

(1969); Kalorimetrie kristallin-flüssiger Cholesterylester. [17] H. Arnold, J. Jacobs, O. Sonntag, Z. Phys. Chem. (Leipzig) *240*, 177 (1969), C.A. *71*, 16512v (1969); Kalorimetrie und Umwandlungen smekti- scher Phasen. [18] H. Arnold, D. Demus, H. J. Koch, A. Nelles, H. Sackmann, Z. Phys. Chem. (Leipzig) *240*, 185 (1969), C.A. *71*, 7159h (1969); Umwandlungswärmen und Systematik kristalliner Flüssigkeiten. [19] K. Auwers, Z. Phys. Chem. *32*, 39 (1900); Vermischte kryoskopische Beobachtungen. [20] K. Auwers, Z. Phys. Chem. *42*, 629 (1903); Kryoskopische Notizen.

[21] B. Bahadur, Z. Naturforsch. A *30*, 1093 (1975), C.A. *83*, 124419b (1975); Specific volume studies on some nematics. [22] B. Bahadur, Phys. Lett. A *55*, 133 (1975); First order phase-change during i/n and n/s$_A$ transitions for NPOB. [23] B. Bahadur, S. Chandra, J. Phys. C *9*, 5 (1976), C.A. *84*, 114457u (1976); Specific volume studies of nematic EBBA. [24] B. Bahadur, Mol. Cryst. Liq. Cryst. *35*, 83 (1976), C.A. *85*, 102656r (1976); Specific volume studies on PAA. [25] B. Bahadur, J. Chim. Phys. Phys.-Chim. Biol. *73*, 255 (1976), C.A. *85*, 54670s (1976); Review on the specific volume of liquid crystals. [26] B. Bahadur, S. Chandra, J. Phys. Soc. Jpn. *41*, 237 (1976), C.A. *85*, 85811t (1976); Wada constant of some nematics. [27] E. M. Barrall, II, R. S. Porter, J. F. Johnson, J. Phys. Chem. *68*, 2810 (1964), C.A. *61*, 13956d (1964); Heats of transition for nematics. [28] E. M. Barrall, II, R. S. Porter, J. F. Johnson, J. Phys. Chem. *70*, 385 (1966), C.A. *64*, 10470a (1966); Temperatures of liquid crystal transitions in cholesteryl esters by DTA. [29] E. M. Barrall, II, R. S. Porter, J. F. Johnson, J. Phys. Chem. *71*, 895 (1967); C.A. *66*, 80121r (1967); Specific heats of nematics, smectics, and cholesterics by DSC. [30] E. M. Barrall, II, R. S. Porter, J. F. Johnson, J. Phys. Chem. *71*, 1224 (1967), C.A. *66*, 108977a (1967); Heats of transition for cholesteryl esters by DSC.

[31] E. M. Barrall, II, R. S. Porter, J. F. Johnson, Mol. Cryst. *3*, 299 (1968), C.A. *68*, 99428j (1968); Heats and temperatures of transition of aromatic mesogens. [32] E. M. Barrall, II, M. J. Vogel, IBM-Research RJ 603 (1969); Effect of purity on the thermodynamic properties of cholesteryl heptadecanoate. [33] E. M. Barrall, II, J. F. Johnson, R. S. Porter, Thermal Analysis, Vol. I, 555 (1969), (Academic: New York); Homologous series of aliphatic esters of cholesterol: thermodynamic properties. [34] R. S. Porter, E. M. Barrall, II, J. F. Johnson, Thermal Analysis, Vol. I, 597 (1969), (Academic: New York); Mesophase transition thermodynamics for homologous series. [35] J. F. Johnson, R. S. Porter, E. M. Barrall, II, Mol. Cryst. Liq. Cryst. *8*, 1 (1969), C.A. *71*, 85236v (1969); Thermodynamics of mesophase transitions from calorimetric measurements. [36] R. S. Porter, E. M. Barrall, II, J. F. Johnson, Accounts Chem. Res. *2*, 53 (1969), C.A. *70*, 91452h (1969); Thermodynamic order in mesophases. [37] M. J. Vogel, E. M. Barrall, II, C. P. Mignosa, Liq. Cryst. Ord. Fluids *1970*, 333, C.A. *79*, 35765m (1973); Effect of solvent type on the thermodynamic properties of normal aliphatic cholesteryl esters. [38] E. M. Barrall, II, M. J. Vogel, Thermochimica Acta *1*, 127 (1970), C.A. *73*, 49401c (1970); Effect of purity on the thermodynamic properties of cholesteryl heptadecanoate. [39] W. R. Young, E. M. Barrall, II, A. Aviram, Anal. Calorimetry, Proc. 2nd Symp. *1970*, 113, C.A. *74*, 147570x (1971); Scanning calorimetry of mesophase transitions: Marker's acid. [40] E. M. Barrall, II, J. F. Johnson, Liq. Cryst. Plast. Cryst. *2*, 254 (1974), C.A. *83*, 210000q (1975); Thermal properties of liquid crystals.

[41] L. Bata, V. L. Broude, V. G. Fedotov, N. Kroo, L. Rosta, J. Szabon, L. M. Umgarov, I. Vizi, Hung. Acad. Sci., Cent. Res. Inst. Phys. KFKI *1976*, KFKI-76-42, 16 pp., C.A. *85*, 184985j (1976); Solid state polymorphism of PAA. [42] J. Baturic-Rubcic, D. Durek, J. Phys. E *6*, 995 (1973), C.A. *79*, 129858w (1973); Accurate latent heat measurements of small liquid samples. [42a] J. Baturic-Rub- cic, D. Durek, A. Rubcic, J. Phys. E *10*, 373 (1977), C.A. *86*, 177661t (1977); Heat capacity measurements on samples of low thermal diffusivity. [43] E. Bauer, J. Bernamont, J. phys. radium *7*, 19 (1936), C.A. *30*, 5088[1] (1936); La dilatation du PAP et la nature du changement de phase n/i. [44] J. T. Bendler, Diss. Yale Univ., New Haven, Conn. 1974, 107 pp. Avail. Xerox Univ. Microfilms, Ann Arbor, Mich., Order No. 75-11,276. C.A. *83*, 124399v (1975); Theory of pretransitional effects in the i phase of liquid crystals. [45] J. Bendler, Mol. Cryst. Liq. Cryst. *38*, 19 (1977), C.A. *86*, 198212n (1977); Compressibility and thermal expansion anomalies in the i phase of mesogens. [46] R. A. Bernheim, T. A. Shuhler, J. Phys. Chem. *76*, 925 (1972), C.A. *76*, 117997r (1972); Phase diagrams of liquid crystal solvents used in NMR-studies. [47] J. Billard, C. R. Acad. Sci., Sér. B *280*, 573 (1975), C.A. *83*, 69419z (1975); Existence des phases s$_F$ et s$_G$. [48] N. Boccara, Mol. Cryst. Liq. Cryst. *32*, 1 (1976), C.A. *84*, 184992p (1976); Quelques aspects de la théorie des transitions du deuxième ordre. [49] J. D. Boyd, C. H. Wang, U.S. Nat. Tech. Inform. Serv., AD Rep. *1973*, No. 766803/1, 6 pp., C.A. *80*, 53143g (1974); Effect of hydrostatic pressure on the phase transition temperatures of a nematic. [50] J. D. Boyd, C. H. Wang, J. Chem. Phys. *60*, 1185 (1974), C.A. *80*, 125574y (1974); Effect of hydrostatic pressure on the phase transition temperatures of a nematic.

[51] W. L. Bragg, E. J. Williams, Proc. Roy. Soc. A *145*, 699 (1934); *151*, 540 (1935), C.A. *28*, 5794[9] (1934); *30*, 662[7] (1936); Effect of thermal agitation on atomic arrangement in alloys. [52] S. Chandrasekhar, R. Shashidhar, N. Tara, Mol. Cryst. Liq. Cryst. *10*, 337 (1970), C.A. *73*, 81745d (1970); Theory of melting of molecular crystals. I. Liquid crystalline phase. [53] S. Chandrasekhar, R. Shashidhar, N. Tara, Mol. Cryst. Liq. Cryst. *12*, 245 (1971), C.A. *75*, 25730z (1971); II. Solid-solid and melting transitions (cf. [52]). [54] S. Chandrasekhar, R. Shashidhar, Indian J. Pure Appl. Phys. *9*, 975 (1971), C.A. *76*, 104879d (1972); Theory of melting of molecular crystals. Nematic liquid crystalline phase. [55] S. Chandrasekhar, R. Shashidhar, Mol. Cryst. Liq. Cryst. *16*, 21 (1972), C.A. *76*, 77755b (1972); III. Effect of short-range orientational order on liquid crystalline transitions (cf. [52]). [56] S. Chandrasekhar, S. Ramaseshan, A. S. Reshamwala, B. K. Sadashiva, R. Shashidhar, V. Surendranath, Liq. Cryst., Proc. Int. Conf., Bangalore, *1973*, 117, C.A. *84*, 114406b (1976); Pressure induced mesomorphism. [57] R. Shashidhar, S. Chandrasekhar, J. Phys. (Paris), C-1, *36*, 49 (1975), C.A. *83*, 69393m (1975); Pressure studies on liquid crystals. [58] R. Shashidhar, S. Ramaseshan, S. Chandrasekhar, Curr. Sci. *45*, 1 (1976), C.A. *84*, 114485b (1976); Optical high pressure cell for liquid crystals. [59] R. Chang, Solid State Commun. *14*, 403 (1974), C.A. *81*, 29840g (1974); Pretransition and critical phenomena in nematic MBBA. [60] R. Chang, J. C. Gysbers, J. Phys. (Paris), C-1, *36*, 147 (1975); Evaluation of volume-temperature relationships in nematic MBBA near its n/i transition by means of computer minimization methods.

[61] R. Chang, Chem. Phys. Lett. *32*, 493 (1975), C.A. *83*, 36093k (1975); Thermodynamics of nematic mixtures. Regular solution approximation. [62] J. H. Chen, T. C. Lubensky, Phys. Rev. A *14*, 1202 (1976), C.A. *85*, 169955f (1976); Landau-Ginzburg mean-field theory for the n/s_C and n/s_A transitions. [62a] H. Chihara, N. Nakamura, Mol. Cryst. Liq. Cryst. *41*, 21 (1977), C.A. *87*, 109757z (1977); Dynamical nature of the phase transition and melting of plastic crystals. [63] J. C. H. Chin, Diss. Kent State Univ., Kent. 1974, 168 pp. Avail. Xerox Univ. Microfilms, Ann Arbor, Mich., Order No. 75-11,986, C.A. *83*, 137938v (1975); Effect of compressibility on the thermodynamic properties of liquid crystals. [64] J. C. Chin, V. D. Neff, Mol. Cryst. Liq. Cryst. *31*, 69 (1975), C.A. *83*, 211546j (1975); Effect of compressibility on the thermodynamic properties of the n/i transition. [65] I. G. Chistyakov, V. A. Usol'tseva, Izv. Vysshikh Uchebn. Zavedenii, Khim. i Khim. Tekhnol. *6*, 436 (1963), C.A. *59*, 14635f (1963); Investigation of liquid crystals: PAA and PAP. [66] V. A. Usol'tseva, I. G. Chistyakov, Uch. Zap. Ivanov. Gos. Pedagog. Inst. *77*, 101 (1970), C.A. *76*, 28645g (1972); Systems with liquid-crystal state. [67] I. G. Chistyakov, L. A. Gusakova, G. G. Maidachenko, Izv. Vyssh. Ucheb. Zaved., Khim. Khim. Tekhnol. *14*, 1433 (1971), C.A. *76*, 18985d (1972); Low-temperature nematic mixtures. [68] L. A. Gusakova, I. G. Chistyakov, 1. Wiss. Konf. über flüssige Kristalle, 17.–19. 11. 1970, Ivanovo. Sammlung der Vorträge Nr. 3; Struktur nematischer, binärer flüssig-kristalliner Systeme. [69] B. K. Vainshtein, I. G. Chistyakov, G. G. Maidachenko, L. A. Gusakova, V. D. Belilovskii, V. M. Chaikovskii, L. K. Vistin, S. P. Chumakova, Dokl. Akad. Nauk SSSR *220*, 1349 (1975), C.A. *82*, 163338y (1975); Formation of a smectic in mixtures of nematics. [69a] B. K. Vainshtein, I. G. Chistyakov, B. N. Makarov, G. G. Maidachenko, L. A. Gusakova, V. M. Chaikovskii, L. A. Mineev, V. D. Belilovskii, L. Vistins, S. P. Chumakova, Kristallografiya *22*, 592 (1977), C.A. *87*, 32234s (1977); Formation of a smectic in mixtures of nematics. [70] P. E. Cladis, J. Rault, J. P. Burger, Mol. Cryst. Liq. Cryst. *13*, 1 (1971), C.A. *75*, 54885j (1971); Binary mixtures of rod-like molecules with MBBA. [71] P. E. Cladis, Phys. Rev. Lett. *35*, 48 (1975), C.A. *83*, 89003n (1975); New liquid-crystal phase diagram. [71a] P. E. Cladis, R. K. Bogardus, W. B. Daniels, G. N. Taylor, Phys. Rev. Lett. *39*, 720 (1977), C.A. *87*, 144309b (1977); High-pressure investigation of the reentrant n/bilayer-s_A transition. [72] N. A. Clark, Phys. Rev. A *14*, 1551 (1976), C.A. *85*, 185037p (1976); Pretransitional mechanical effects in a smectic A. [73] F. Conrat, Physik. Z. *10*, 202 (1909), C.A. *3*, 1609[5] (1909); Verhalten der Dichte des Anisaldazins beim Klärpunkt. [74] S. J. Dave, M. J. S. Dewar, J. Chem. Soc. *1954*, 4616, C.A. *49*, 4355e (1955); Mixed liquid crystals. [75] J. S. Dave, M. J. S. Dewar, J. Chem. Soc. *1955*, 4305, C.A. *50*, 5384h (1956); Effect of structure on the transition temperatures of mixed liquid crystals. [76] J. S. Dave, J. M. Lohar, Chem. & Ind. *1959*, 597, C.A. *53*, 19503i (1959); Formation of mixed liquid crystals in mixtures of Schiff's bases. I. [77] J. S. Dave, J. M. Lohar, Chem. & Ind. *1960*, 494, C.A. *54*, 22426d (1960); II. Explanation (cf. [76]). [78] J. S. Dave, J. M. Lohar, Proc. Natl. Acad. Sci., India, Sect. A *32*, 105 (1962), C.A. *57*, 10618i (1962); Mixed mesomorphism. [79] J. S. Dave, J. M. Lohar, Indian J. Chem. *4*, 386 (1966), C.A. *66*, 32719v (1967); Mesomorphism of binary systems containing *p*-methoxycinnamic acid, PAA, and Schiff bases. [80] J. S. Dave, P. R. Patel, K. L. Vasanth, Indian J. Chem. *4*, 505 (1966), C.A. *67*, 15867v (1967); Influence of molecular

structure on liquid crystalline properties and phase transitions in mixed liquid crystals in s phase. [81] J. S. Dave, J. M. Lohar, J. Chem. Soc. A *1967*, 1473, C.A. *67*, 94830d (1967); Mixed liquid crystals: additive effect of terminal polar groups in Schiff bases. [82] J. S. Dave, P. R. Patel, K. L. Vasanth, Mol. Cryst. Liq. Cryst. *8*, 93 (1969), C.A. *71*, 85803c (1969); Mixed mesomorphism in binary systems forming s/n phases. [83] J. S. Dave, K. L. Vasanth, Indian J. Chem. *7*, 498 (1969), C.A. *71*, 25193u (1969); Mixed mesomorphism in mixtures of Schiff bases. [84] J. S. Dave, P. R. Patel, J. Indian Chem. Soc. *47*, 815 (1970), C.A. *74*, 46759k (1971); Mixed polymesomorphism. [85] J. S. Dave, K. L. Vasanth, Liq. Cryst., Proc. Int. Conf., Bangalore, *1973*, 415, C.A. *84*, 114472v (1976); Influence of molecular structure on mixed mesomorphism in binary systems. [86] J. S. Dave, R. A. Vora, Liq. Cryst. Plast. Cryst. *1*, 153 (1974), C.A. *83*, 19083q (1975); Influence of molecular structure and composition of the liquid crystals formed by mixtures of nonamphiphilic compounds. [87] P. G. de Gennes, C. R. Acad. Sci., Sér. B *274*, 758 (1972); Sur la transition s_A/s_C. [88] W. H. de Jeu, Solid State Commun. *13*, 1521 (1973), C.A. *80*, 41446v (1974); First- and second-order n/s_A transitions in the series of di-*n*-alkyl azoxybenzenes. [89] T. Deleanu, I. V. Ionescu, V. Nahorniak, D. A. Isacescu, Rev. Roum. Chim. *17*, 103 (1972), C.A. *76*, 158538k (1972); Deduction d'une classification des transitions de phase. [90] B. Deloche, B. Cabane, D. Jerome, Mol. Cryst. Liq. Cryst. *15*, 197 (1971), C.A. *76*, 38453g (1972); Effect of pressure on the mesomorphic transitions in PAA.

[91] D. Demus, Z. Naturforsch. A *22*, 285 (1967), C.A. *68*, 16844t (1968); Bei niedrigen Temperaturen stabile nematische Flüssigkeiten. [92] D. Demus, R. Rurainski, Mol. Cryst. Liq. Cryst. *16*, 171 (1972), C.A. *76*, 77782h (1972); Anomalous densities in liquid crystalline bis(4-*n*-alkoxybenzal)-1,4-phenylenedi-amines. [93] D. Demus, R. Rurainski, Z. Phys. Chem. (Leipzig) *253*, 53 (1973), C.A. *79*, 140471k (1973); Dichtemessungen an smektischen Flüssigkristallen. [94] D. Demus, H. Koenig, D. Marzotko, R. Rurainski, Mol. Cryst. Liq. Cryst. *23*, 207 (1973), C.A. *80*, 88149r (1974); Smectic dimorphism of hexadecyloxy- and octadecyloxyazoxybenzene. [95] D. Marzotko, D. Demus, Liq. Cryst., Proc. Int. Conf., Bangalore, *1973*, 189, C.A. *84*, 112525r (1976); Calorimetric investigation of liquid crystals. [96] D. Demus, C. Fietkau, R. Schubert, H. Kehlen, Mol. Cryst. Liq. Cryst. *25*, 215 (1974), C.A. *81*, 69149v (1974); Calculation and experimental verification of eutectic systems with nematics. [97] A. de Vries, J. Phys. (Paris) Lett. *35*, 139 (1974), C.A. *81*, 142321b (1974); Relation between miscibility and structure for s_A, s_B, and s_C phases. [98] A. de Vries, J. Phys. (Paris) Lett. *35*, 157 (1974), C.A. *81*, 142323d (1974); Different types of s_H phases. [98a] A. de Vries, Mol. Cryst. Liq. Cryst. *41*, 27 (1977), C.A. *87*, 175860d (1977); Experimental evidence concerning two different kinds of s_C/s_A transitions. [99] D. Djurek, J. Baturic-Rubcic, K. Franulovic, Phys. Rev. Lett. *33*, 1126 (1974), C.A. *81*, 178175w (1974); Specific heat critical exponents near the n/s_A transition. [100] Y. Do, M. S. Jhon, T. Ree, Taehan Hwahak Hoechi *20*, 118 (1976), C.A. *85*, 52591t (1976); Theoretical prediction of the thermodynamic properties of nematic PAA.

[101] M. Domon, Thèse, Lille, 1973; Contribution a l'étude des diagrammes de phase des mélanges mésomorphes. [102] M. Domon, J. Billard, Liq. Cryst., Proc. Int. Conf., Bangalore, *1973*, 131, C.A. *84*, 112401x (1976); Prediction of phase diagrams for liquid crystalline mixtures. [103] E. Donth, Z. Phys. *207*, 342 (1967), C.A. *68*, 117513q (1968); Phasenumwandlung 2. Art als kritische Erscheinung eines inneren Phasengleichgewichtes. [103a] F. Dowell, Diss. Georgetown Univ., Washington, D.C. 1977. 244 pp. Avail. Univ. Microfilms Int., Order No. 77-16,847, C.A. *87*, 144313y (1977); Lattice model study of the effect of flexibility on the n/i transition in pure liquid crystals and in mixtures of chain solutes in liquid crystal solvents. [104] G. Dupont, O. Lozac'h, Compt. rend. *221*, 751 (1945), C.A. *40*, 4585[8] (1946); Cryoscopie dans le PAA. [105] D. Durek, J. Baturic-Rubcic, S. Marcelja, J. W. Doane, Phys. Lett. A *43*, 273 (1973), C.A. *78*, 152223y (1973); n/s_A Transition entropies in a homologous series. [106] M. Dvolaitzky, R. K. Bogardus, Jr., W. B. Daniels, J. Billard, C. R. Acad. Sci., Sér. C *284*, 5 (1977), C.A. *86*, 127488w (1977); Diagramme de phase du 4,4'-di-*n*-octadécyloxyazoxybenzène. [107] P. Ehrenfest, Proc. Acad. Sci. Amsterdam *36*, 153 (1933), C.A. *27*, 5235 (1933); Phase changes in the ordinary and extended sense classified according to the corresponding singularities of the thermodynamic potential [108] R. D. Ennulat, Anal. Calorimetry, Proc. 155th Amer. Chem. Soc. Symp. *1968*, 219, C.A. *70*, 91368k (1969); Thermal analysis of mesophases. [109] A. Eucken, E. Bartholomé, Nachr. Ges. Wiss. Göttingen, Math.-phys. Klasse, Fachgr. II, *2*, 51 (1936), C.A. *31*, 1690[5] (1937); Thermische Hysterese der Methanumwandlung bei 20,4 K. [110] H. L. Finke, M. E. Gross, G. Waddington, H. M. Huffman, J. Am. Chem. Soc. *76*, 333 (1954), C.A. *48*, 5634b (1954); Low-temperature thermal data for the nine normal paraffin hydrocarbons from octane to hexadecane. [111] D. L. Fishel, Y. Y. Hsu, J. Chem. Soc. D *1971*, 1557, C.A. *76*, 64576u (1972); Extended

range ambient temperature nematic in binary systems of N-(4-alkoxybenzylidene)-4-alkylanilines. [112] J. R. Flick, A. S. Marshall, S. E. B. Petrie, Liq. Cryst. Ord. Fluids *2*, 97 (1974), C.A. *86*, 180979h (1977); Changes in thermodynamic and optical properties associated with mesomorphic transitions. [113] C. Flick, J. W. Doane, Phys. Lett. A *47*, 331 (1974), C.A. *81*, 7064e (1974); Approach to a second order n/s$_A$ transition using mixtures. [113a] M. Fodor, L. Hodany, K. Pinter, K. Ritvay, Therm. Anal., Proc. Int. Conf., 4th, *1974*, 2, 417, C.A. *87*, 46823v (1977); Thermoanalytical investigation of new liquid crystals. [114] F. C. Frank, K. Wirtz, Naturwissenschaften *26*, 687 (1938), C.A. *33*, 2007[7] (1939); Ordnung und Umwandlungen in kondensierten Phasen. [114a] F. C. Frank, J. Phys. (Paris) Lett. *38*, 207 (1977), C.A. *87*, 46792j (1977); Thermodynamics of smectic mixtures. [115] S. G. Frank, B. G. Byrd, J. Pharm. Sci. *61*, 1762 (1972), C.A. *78*, 20152q (1973); Phase transitions in binary mixtures of cholesteryl esters. [116] S. G. Frank, J. Pharm. Sci. *63*, 795 (1974), C.A. *81*, 34809h (1974); DSC of mixtures of cholesteryl myristate and cholesteryl palmitate. [117] P. D. Garn, J. Am. Chem. Soc. *91*, 5382 (1969), C.A. *71*, 95815x (1969); Pressure-induced shifts of mesomorphic phase transition temperatures. [118] P. D. Garn, R. J. Richardson, Therm. Anal., Proc. Int. Conf. *3*, 123 (1971), C.A. *78*, 129314v (1973); Pressure dependence of phase transitions in bis(4'-*n*-alkoxybenzal)-1,4-phenylene-diamines. [119] P. Gaubert, Compt. rend. *202*, 141 (1936), C.A. *30*, 1625[3] (1936); Cristaux liquides de quelques composés de la cholestérine et leur surfusion cristalline. [120] S. K. Ghosh, S. Amadesi, Phys. Lett. A *59*, 282 (1976), C.A. *86*, 63855d (1977); Temperature dependence of the specific volume in nematic PAA.

[121] H. W. Gibson, J. Phys. Chem. *80*, 1310 (1976), C.A. *85*, 27547a (1976); Effect of structure on the mesomorphism of cholesteryl alkanoates. Effect of configuration in chiral alkanoates. [122] S. Goshen, D. Mukamel, S. Shtrikman, Solid State Commun. *9*, 649 (1971), C.A. *75*, 41756t (1971); Application of the Landau theory to liquids-liquid crystals transitions. [123] D. J. Byron, G. W. Gray, Chem. & Ind. *1959*, 1021, C.A. *54*, 1960c (1960); Comments on a reported case of liquid crystallinity in mixtures of two Schiff's bases. [124] H. Gruler, F. Jones, J. Phys. (Paris), C-1, *36*, 53 (1975), C.A. *83*, 33866d (1975); Thermal expansion and specific heat of a nematic. [125] E. Gulari, B. Chu, J. Chem. Phys. *62*, 795 (1975), C.A. *83*, 19353c (1975); Density of MBBA about the i/n phase transition. [126] N. Gurusamy, K. L. Vasanth, Technology (Coimbatore, India) *24*, 57 (1976), C.A. *85*, 167383u (1976); Binary systems of *p*-methoxycinnamic acid and substituted benzoic acids. [127] L. A. Gusakova, V. A. Nikitin, Uch. Zap. Ivanov. Gos. Pedagog. Inst. *77*, 103 (1970), C.A. *76*, 104447m (1972); Miscibility of a system of cholesteryl acetate and propionate. [128] R. Haase, Thermodynamik der Mischphasen (Springer: Berlin). 1956. 610 pp., C.A. *50*, 13590f (1956). [129] J. L. Haberfeld, Diss. Univ. Connecticut, Storrs, Conn. 1975. 215 pp. Avail. Xerox Univ. Microfilms, Ann Arbor, Mich., Order No. 75-16,514, C.A. *83*, 164773u (1975); Specialized applications of thermal analysis. Characterization of crosslinked polyethylenes. Multicomponent liquid crystal systems. [130] L. E. Hajdo, A. C. Eringen, A. E. Lord, Jr., Lett. Appl. Eng. Sci. *2*, 367 (1974), C.A. *83*, 19359j (1975); Density change at the clearing point of smectic diethyl 4,4'-azoxydibenzoate.

[131] L. E. Hajdo, A. C. Eringen, J. Giancola, A. E. Lord, Jr., Lett. Appl. Eng. Sci. *3*, 61 (1975), C.A. *83*, 69402p (1975); Thermal expansion coefficients of cholesterics. [132] B. I. Halperin, T. C. Lubensky, S. K. Ma, Phys. Rev. Lett. *32*, 292 (1974); First order phase transitions in superconductors and smectics A. [132a] C. Hanawa, T. Shirakawa, T. Tokuda, Chem. Lett. *1977*, 1223, C.A. *87*, 192282j (1977); Pressure-volume-temperature relations in mesomorphic *p*-pentoxybenzilidene-*p*-*n*-butylaniline. [133] F. Hardouin, H. Gasparoux, P. Delhaes, C. R. Acad. Sci., Sér. B *278*, 811 (1974), C.A. *81*, 55278z (1974); Étude de la nature de la transition s$_A$/n par mesure de la chaleur spécifique. [133a] G. Sigaud, M. F. Achard, F. Hardouin, H. Gasparoux, Chem. Phys. Lett. *48*, 122 (1977), C.A. *87*, 60991e (1977); Effect of non-mesomorphic plate-like solutes on thermal behavior of liquid crystalline solvents. [133b] F. Hardouin, G. Sigaud, M. F. Achard, H. Gasparoux, Solid State Commun. *22*, 343 (1977), C.A. *87*, 29776q (1977); Experimental tricritical point n/s$_A$ in a two components-system. [133c] G. Sigaud, F. Hardouin, M. F. Achard, Solid State Commun. *23*, 35 (1977), C.A. *87*, 93774q (1977); Experimental system for a n/s$_A$/s$_C$ Lifshitz's point. [134] W. Helfrich, Phys. Lett. A *58*, 457 (1976), C.A. *86*, 49302q (1977); Defect model of the k/s$_B$ transition. [134a] G. Heppke, E. J. Richter, Z. Naturforsch. A *33*, 185 (1978); Induktion smektischer Phasen in binären Mischungen nematischer Flüssigkristalle. [135] A. J. Herbert, Trans. Faraday Soc. *63*, 555 (1967), C.A. *66*, 109216p (1967); Transition temperatures and transition energies of *p*-*n*-alkoxy benzoic acids. [136] W. Herz, Z. anorg. allgem. Chem. *161*, 228 (1927), C.A. *21*, 2083[5] (1927); Klärpunkte kristallinischer Flüssigkeiten. [137] N. Hijikuro, K. Miyakawa, H. Mori, Phys. Lett. A *45*, 257 (1973), C.A. *79*, 150338m (1973); n/k

Transitions in liquid crystals. [138] N. Hijikuro, K. Miyakawa, H. Mori, J. Phys. Soc. Jpn. *37*, 928 (1974), C.A. *81*, 160410j (1974); Theory of phase transitions in nematics. [139] C. L. Hillemann, G. R. van Hecke, S. R. Peak, J. B. Winther, M. A. Rudat, D. A. Kalman, M. L. White, J. Phys. Chem. *79*, 1566 (1975), C.A. *83*, 96192b (1975); Calorimetric studies of homologous *trans*-4-ethoxy-4'-*n*-alkanoyl-oxyazobenzenes. [140] K. Hirakawa, S. Kai, J. Phys. Soc. Jap. *37*, 1472 (1974), C.A. *82*, 103960k (1975); Anomalous specific heat of MBBA near n/i transition.

[141] J. Homer, A. R. Dudley, J. Chem. Soc., Chem. Commun. *1972*, 926, C.A. *77*, 131725c (1972); Phase diagrams of binary mixtures of liquid crystals. [142] E. C. H. Hsu, J. F. Johnson, Mol. Cryst. Liq. Cryst. *20*, 177 (1973), C.A. *78*, 128974y (1973); Phase diagrams of binary nematic systems. [143] E. C. H. Hsu, Diss. Univ. Connecticut, Storrs, Conn. 1974, 216 pp. Avail. Univ. Microfilms, Ann Arbor, Mich. Order No. 73-28,517, C.A. *80*, 74966t (1974); Thermodynamic properties of binary mesophase systems. [144] E. C. H. Hsu, J. F. Johnson, Mol. Cryst. Liq. Cryst. *25*, 145 (1974), C.A. *83*, 153302p (1975); Phase diagrams of binary nematic systems. [145] E. C. H. Hsu, J. F. Johnson, Mol. Cryst. Liq. Cryst. *27*, 95 (1974), C.A. *82*, 90746t (1975); Prediction of eutectic temperatures, compositions, and phase diagrams for binary mesophase systems. [146] E. C. H. Hsu, J. L. Haberfeld, J. F. Johnson, E. M. Barrall, II, Mol. Cryst. Liq. Cryst. *27*, 269 (1974), C.A. *82*, 105249j (1975); Thermal properties of binary mixtures of liquid crystals. [147] B. A. Hubermann, D. M. Lublin, S. Doniach, Solid State Commun. *17*, 485 (1975); Theory of melting in liquid crystals. [148] G. A. Hulett, Z. physik. Chem. *28*, 629 (1899); Der stetige Übergang fest-flüssig. [149] D. S. Hulme, P. E. Raynes, K. J. Harrison, J. Chem. Soc., Chem. Commun. *1974*, 98, C.A. *81*, 7049d (1974); Eutectic mixtures of nematic 4'-substituted 4-cyanobiphenyls. [150] M. Ikeda, T. Hatakeyama, Mol. Cryst. Liq. Cryst. *33*, 201 (1976), C.A. *85*, 102652m (1976); Thermal studies on the phase transitions of *p-n*-octadecyloxyben-zoic acid.

[151] M. Ikeda, T. Hatakeyama, Mol. Cryst. Liq. Cryst. *39*, 109 (1977); Thermal properties of *p-n*-octadecyloxybenzoic acid. [152] H. Imura, K. Okano, Chem. Phys. Lett. *17*, 111 (1972), C.A. *78*, 34851d (1973); Theory of the anomalous heat capacity and thermal expansion of nematics above the clearing point. [153] V. L. Indenbom, S. A. Pikin, E. B. Loginov, Kristallografiya *21*, 1093 (1976), C.A. *86*, 63837z (1977); Phase transitions and ferroelectric structures in liquid crystals. [154] K. U. Ingold, I. E. Puddington, J. Inst. Petrol. *44*, 41 (1958), C.A. *52*, 6774i (1958); Mesomorphism of sodium and lithium soaps prepared from oxidized paraffin wax. [155] A. Isihara, M. Wadati, J. Chem. Phys. *55*, 4678 (1971), C.A. *76*, 7439c (1972); Phase transition in two-dimensional liquid crystals. Comments. [156] A. V. Ivashchenko, V. V. Titov, E. I. Kovshev, Mol. Cryst. Liq. Cryst. *33*, 195 (1976), C.A. *85*, 102651k (1976); Applicability of Schröder-Van Laar equations to liquid crystal mixtures. [157] D. L. Johnson, C. Maze, E. Oppenheim, R. Reynolds, Phys. Rev. Lett. *34*, 1143 (1975), C.A. *83*, 19478x (1975); Evidence for a s_A/n tricritical point. Binary mixtures. [157a] D. Johnson, D. Allender, R. DeHoff, C. Maze, E. Oppenheim, R. Reynolds, Phys. Rev. B *16*, 470 (1977), C.A. *87*, 109737t (1977); n/s_A/s_C Polycritical point: experimental evidence and a Landau theory. [158] J. F. Johnson, G. W. Miller, Thermochimica Acta *1*, 373 (1970), C.A. *73*, 92259z (1970); Thermal analyses of mesophases. [159] E. Justi, M. von Laue, Sitzber. preuss. Akad. Wiss. Phys.-math. Klasse *1934*, 237, C.A. *28*, 6615[6] (1934); Neuartige Phasenumwandlung bei einheitlichen Stoffen. [160] E. Justi, M. von Laue, Physik. Z. *35*, 945 (1934), C.A. *29*, 2430[8] (1935); Phasengleichgewichte dritter Art.

[161] S. Kai, K. Hirakawa, Kyushu Daigaku Kogaku Shuho *46*, 650 (1973), C.A. *80*, 94775h (1974); Specific heat anomaly of nematic APAPA. [162] S. Kai, K. Hirakawa, Kyushu Daigaku Kogaku Shuho *47*, 109 (1974), C.A. *81*, 111986t (1974); Anomalous specific heats at the phase transition of MBBA. [163] S. Kai, M. Araoka, K. Hirakawa, Kyushu Daigaku Kogaku Shuho *47*, 479 (1974), C.A. *82*, 78997t (1975); Thermal properties of nematic MBBA. Quasi-second-order phase transition theory and magnetic field dependence of the specific heat. [164] T. A. Kalinina, I. A. Kleinman, A. Z. Rabinovich, M. B. Roitberg, Prib. Tek. Eksp. *1973*, 235, C.A. *79*, 140466n (1973); Pyroelectric thermography for indicating phase transitions in liquid crystals. [165] J. O. Kessler, E. P. Raynes, Phys. Lett. A *50*, 335 (1974); Glassy liquid crystals. Observation of a quenched twisted nematic. [166] P. H. Keyes, H. T. Weston, W. B. Daniels, Phys. Rev. Lett. *31*, 628 (1973), C.A. *79*, 109033v (1973); Tricritical behavior in a liquid crystal system. [167] P. H. Keyes, H. T. Weston, W. J. Lin, W. B. Daniels, J. Chem. Phys. *63*, 5006 (1975), C.A. *84*, 37466s (1976); Liquid crystal phase diagrams. Seven thermotropic materials. [167a] B. B. Khanukaev, Y. A. Avakyan, N. S. Khanukaeva, A. K. Pochikyan, Arm. Khim. Zh. *30*, 208 (1977), C.A. *87*, 60724v (1977); Calorimetry of several cholesterics. [167b] Y. B. Kim, K. Ogino, Phys. Lett. A *61*, 40 (1977), C.A. *86*, 198233v (1977); Studies on n/i transitions.

[168] W. Klement, Jr., L. H. Cohen, Mol. Cryst. Liq. Cryst. *27*, 359 (1974), C.A. *83*, 51110g (1975); Determination to 2 kbar of the n/i and melting transitions in PAA. [169] K. K. Kobayashi, Phys. Lett. A *31*, 125 (1970); Theory of translational and orientational melting with application to liquid crystals. [170] K. Kobayashi, Bussei Kenkyu *15*, C63 (1970), C.A. *80*, 53032v (1974); Orientational relaxation and the transformation of lattices.

[171] K. K. Kobayashi, Mol. Cryst. Liq. Cryst. *13*, 137 (1971), C.A. *75*, 41758v (1971); Theory of translational and orientational melting with application to liquid crystals. [172] E. Kordes, Z. physik. Chem. A *152*, 161 (1931), C.A. *25*, 2043 (1931); Phasengleichgewichte in binären Systemen mit kontinuierlichen Mischkristallreihen. [173] I. A. Kotze, Diss. Univ. of Virginia, Charlottesville, Va. 1968, 174 pp. Avail. Univ. Microfilms, Ann Arbor, Mich., Order No. 70-4805, C.A. *74*, 7324v (1971); Melting and structure of liquids. [174] K. Kreutzer, W. Kast, Naturwissenschaften *25*, 233 (1937), C.A. *31*, 5257[8] (1937); Kalorimetrische Messungen beim Übergang der anisotrop-flüssigen Phase in die isotrope. [175] C. Kreutzer, Ann. Physik *33*, 192 (1938), C.A. *33*, 918[6] (1939); Kalorimetrische Messungen beim Übergang von der anisotropen zur isotropen flüssigen Phase. [176] B. Kronberg, D. Patterson, J. Chem. Soc., Faraday Trans. 2, *72*, 1686 (1976), C.A. *85*, 182944w (1976); Application of the Flory-Huggins theory to n/i equilibriums. [177] K. S. Kunihisa, T. Shinoda, Bull. Chem. Soc. Jpn. *48*, 3506 (1975), C.A. *84*, 68118j (1976); Studies of phase transitions in cholesteryl myristate by means of simultaneous measurements of polarizing microscopy and thermal analyses. [178] K. S. Kunihisa, S. Hagiwara, Bull. Chem. Soc. Jpn. *49*, 2658 (1976), C.A. *86*, 10847w (1977); Apparatus for thermal analytic microscopy and its application to cholesteryl nonanoate. [179] M. V. Kurik, V. A. Shayuk, Fiz. Tverd. Tela *17*, 2320 (1975), C.A. *83*, 186602q (1975); Mechanism of melting of molecular crystals. [179a] M. V. Kurik, Fiz. Tverd. Tela *19*, 1849 (1977), C.A. *87*, 93720u (1977); Dependence of phase transition temperatures on crystal thickness in stilbene and PAA. [180] M. Feyz, E. Kuss, Ber. Bunsenges. Phys. Chem. *78*, 834 (1974), C.A. *82*, 57137x (1975); Druckabhängigkeit der Schmelz- und Klärtemperatur von 4,4'-disubstituierten Azo- und Azoxybenzolen und einiger Schiffscher Basen.

[181] K. Lakatos, J. Chem. Phys. *55*, 4679 (1971), C.A. *76*, 7440w (1972); Phase transitions in two-dimensional liquid crystals. Reply to comments. [182] L. Landau, Physik. Z. Sowjetunion *8*, 113 (1935), C.A. *30*, 1645[3] (1936); Theory of specific heat anomalies. [183] L. Landau, Physik. Z. Sowjetunion *11*, 26 (1937), J. Exptl. Theoret. Phys. USSR *7*, 19 (1937), C.A. *32*, 415[4] (1938); Theorie der Phasenübergänge. I. [184] L. Landau, Physik. Z. Sowjetunion *11*, 545 (1937), C.A. *31*, 6075[9] (1937); Theorie der Phasenübergänge. II. [185] L. D. Landau, E. M. Lifschitz, Lehrbuch der Theoretischen Physik, Band V: Statistische Physik (Akademie Verlag: Berlin) 1966. [185a] A. J. Leadbetter, J. L. A. Durrant, M. Rugman, Mol. Cryst. Liq. Cryst. *34*, 231 (1977), C.A. *87*, 76621v (1977); Density of 4-*n*-octyl-4'-cyanobiphenyl. [186] M. Leclerq, J. Billard, J. Jacques, C. R. Acad. Sci., Sér. C *264*, 1789 (1967), C.A. *67*, 94582z (1967); Microanalyse enthalpique différentielle appliquée aux substances mésomorphes. [187] F. T. Lee, H. T. Tan, Y. M. Shih, C. W. Woo, Phys. Rev. Lett. *31*, 1117 (1973), C.A. *79*, 150353n (1973); Phase diagram for liquid crystals. [188] F. T. Lee, H. T. Tan, C. W. Woo, Phys. Lett. A *48*, 68 (1974), C.A. *81*, 55252m (1974); s_A/k Transition in the mean field approximation. [188a] K. C. Lee, Proc. Coll. Nat. Sci., Sect. 2 (Seoul Natl. Univ.) *1*, 69 (1976), C.A. *87*, 175867m (1977); Eutectic behavior of a binary mixture of liquid crystals. [189] J. E. Lennard-Jones, A. F. Devonshire, Proc. Roy. Soc. A *169*, 317; *170*, 464 (1939), C.A. *33*, 4098[8], 9071[8] (1939); Critical and cooperative phenomena. III. Theory of melting and structure of liquids. IV. Theory of disorder in solids and liquids and the process of melting. [190] J. O. Lephardt, B. J. Bulkin, Mol. Cryst. Liq. Cryst. *24*, 187 (1973), C.A. *81*, 142311y (1974); Anomalous spherulitic crystallization from a cholesteric. [190a] A. W. Levine, K. D. Tomeczek, Mol. Cryst. Liq. Cryst. *43*, 183 (1977); Enthalpy of fusion of nematogens. [190b] L. Liebert, W. B. Daniels, J. Phys. (Paris), Lett. *38*, 333 (1977), C.A. *87*, 144310v (1977); Comportement sous pression de smectiques A bicouches. [190c] L. Liebert, W. B. Daniels, J. Billard, Mol. Cryst. Liq. Cryst. *41*, 57 (1977); Effets de la pression sur des tolanes nématiques.

[191] W. J. Lin, P. H. Keyes, W. B. Daniels, Phys. Lett. A *49*, 453 (1974), C.A. *82*, 24706d (1975); High pressure studies of the phase transitions in CBOOA. [192] J. D. Litster, Dyn. Aspects Crit. Phenomena, (Proc.) Conf. *1970*, 152, C.A. *80*, 53190v (1974); Critical points and almost critical points. [193] J. M. Lohar, D. S. Shah, Mol. Cryst. Liq. Cryst. *28*, 293 (1974), C.A. *83*, 19366j (1975); Mixed mesomorphism. Determination of latent transition temperatures by extrapolation. [194] J. M. Lohar, J. Phys. (Paris), C-1, *36*, 393 (1975), C.A. *83*, 88964q (1975); Exhibition of nematic mesophase in binary mixtures of Schiff's bases. [195] J. M. Lohar, J. Phys. (Paris), C-1, *36*, 399 (1975), C.A.

83, 106485r (1975); Mixed liquid crystal formation in mixture of *p*-methoxy and *p*-ethoxy benzoic acids. [**196**] J. M. Lohar, G. H. Patel, Current Sci. *44*, 887 (1975), C.A. *84*, 98012n (1976); Studies in mixed liquid crystals. Reliability of extrapolation method. [**197**] A. E. Lord, Jr., F. E. Wargocki, L. E. Hajdo, A. C. Eringen, Lett. Appl. Eng. Sci. *3*, 125 (1975), C.A. *83*, 88976v (1975); Density changes at the clearing points of nematics cholesterics and smectics. [**198**] T. C. Lubensky, J. Phys. (Paris), C-1, *36*, 151 (1975); Latent heat of the c/s_A transition. [**199**] R. L. Humphries, G. R. Luckhurst, Chem. Phys. Lett. *23*, 567 (1973), C.A. *80*, 88122b (1974); Statistical theory of liquid crystalline mixtures. Components of different size. [**200**] C. G. Lyons, E. K. Rideal, Proc. Cambridge Phil. Soc. *26*, 419 (1930), C.A. *24*, 5203[4] (1930); Phase diagrams for unimolecular films.

[**201**] E. F. Lype, Phys. Rev. *69*, 652 (1946), C.A. *40*, 5326[1] (1946); Thermodynamic equilibria of higher order. [**202**] P. V. E. McClintock, Nature *253*, 590 (1975); Textured superfluids. [**203**] D. W. McClure, J. Chem. Phys. *49*, 1830 (1968), C.A. *69*, 100766x (1968); Nature of the rotational phase transition in paraffin crystals. [**204**] W. L. McMillan, Phys. Rev. A *9*, 1720 (1974); Time-dependent Landau theory for s_A/n transition. [**205**] W. L. McMillan, J. Phys. (Paris), C-1, *36*, 103 (1975); Phase transitions in liquid crystals. [**205a**] K. C. Chu, W. L. McMillan, Phys. Rev. A *15*, 1181 (1977), C.A. *86*, 163789d (1977); Unified Landau theory for the nematics, smectics A, and smectics C. [**206**] D. McQueen, K. A. Edgren, S. A. Rydman, J. Phys. D *7*, 935 (1974), C.A. *80*, 149691k (1974); High-special n/k transition in PAP. [**206a**] R. G. Makitra, Y. M. Tsikanchuk, Zh. Obshch. Khim. *47*, 1123 (1977), C.A. *87*, 93738f (1977); Latent liquid-crystal properties of hexachloroethane. [**207**] H. Martin, F. H. Müller, Kolloid-Z. *187*, 107 (1963), C.A. *59*, 77a (1963); Umwandlungswärme von PAA beim n/i Übergang. [**208**] D. E. Martire, Mol. Cryst. Liq. Cryst. *28*, 63 (1974), C.A. *83*, 27312d (1975); Question of molecular flexibility in nematogens. [**209**] D. E. Martire, G. A. Oweimreen, G. I. Agren, S. G. Ryan, H. T. Peterson, J. Chem. Phys. *64*, 1456 (1976), C.A. *84*, 114481x (1976); Effect of quasispherical solutes on the n/i transition. [**210**] T. Matsumoto, Kogakuin Daigaku Kenkyu Hokoku *33*, 1 (1973), C.A. *79*, 129897h (1973); Specific heat of liquid crystals.

[**211**] T. Matsumoto, Kogakuin Daigaku Kenkyu Hokoku *34*, 1 (1973), C.A. *80*, 149676j (1974); Measurement of volume expansion of liquid crystals. [**212**] T. Matsumoto, Kogakuin Daigaku Kenkyu Hokoku *36*, 1 (1974), C.A. *83*, 171116k (1975); Different phase transition in nematics and cholesterics. [**213**] J. Mayer, T. Waluga, J. A. Janik, Phys. Lett. A *41*, 102 (1972), C.A. *78*, 29086d (1973); Evidence of the existence of two solid phases of MBBA in specific heat measurements. [**214**] L. Michel, J. Phys. (Paris), C-7, *36*, 41 (1975), C.A. *85*, 37332r (1976); Les brisures spontanées de symétrie en physique. [**215**] A. Michelson, Phys. Lett. A *60*, 29 (1977), C.A. *86*, 113915w (1977); Effect of chiral solute on the s_A/s_C transition of second order. [**215a**] A. Michelson, D. Cabib, L. Benguigui, J. Phys. (Paris) *38*, 961 (1977), C.A. *87*, 125635p (1977); Symmetry changes and dipole orderings in the s_A/s_C transitions of second order. [**215b**] A. Michelson, L. Benguigui, D. Cabib, Phys. Rev. A *16*, 394 (1977), C.A. *87*, 109731m (1977); Phenomenological theory of the polarized helicoidal smectic C. [**215c**] A. Michelson, Phys. Rev. Lett. *39*, 464 (1977), C.A. *87*, 125592x (1977); Physical realization of a Lifshitz point in liquid crystals. [**216**] K. Miyakawa, N. Hijikuro, H. Mori, Phys. Lett. A *48*, 133 (1974), C.A. *81*, 55254p (1974); n/s Phase transitions. [**217**] K. Miyakawa, N. Hijikuro, H. Mori, J. Phys. Soc. Jap. *36*, 944 (1974), C.A. *80*, 149692m (1974); n/s Phase transitions. [**218**] S. N. Mochalin, P. P. Pugachevich, Uch. Zap., Ivanov. Gos. Pedagog. Inst. *99*, 200 (1972), C.A. *78*, 152444w (1973); Relation of cholesteric-mesophase density to temperature. [**219**] S. N. Mochalin, Uch. zap. Ivanov. un-t *1974*, 86, C.A. *83*, 152689h (1975); Temperature dependence of the density of PAA, cholesteryl propionate, and their mixtures. [**219a**] V. A. Molochko, O. P. Chernova, G. M. Kurdyumov, G. I. Karpushkina, Izv. Vyssh. Uchebn. Zaved., Khim. Khim. Tekhnol. *20*, 1088 (1977), C.A. *87*, 175832w (1977); Polymorphism of mesomorphic aromatic esters. [**220**] V. A. Molochko, O. P. Chernova, G. M. Kurdyumov, Izv. Vyssh. Uchebn. Zaved., Khim. Khim. Tekhnol. *19*, 1459 (1976), C.A. *86*, 34805g (1977); Phase equilibriums in binary systems of aromatic esters with a nematic mesophase.

[**221**] V. A. Molochko, O. P. Chernova, G. M. Kurdyumov, Zh. Prikl. Khim. (Leningrad) *50*, 45 (1977), C.A. *86*, 96758c (1977); Prediction of coordinates of eutectics in liquid crystals. [**222**] U. Müller, Diss. Berlin 1974; Phasenumwandlungen und Strukturen cholesterischer Flüssigkristalle. [**223**] M. Nakagaki, Y. Naito, Yakugaku Zasshi *92*, 1225 (1972), C.A. *78*, 8849q (1973); Phase transition of cholesteryl esters and variation of dielectric properties. [**224**] N. A. Nedostup, V. V. Gal'tsev, Zh. Fiz. Khim. *50*, 2949 (1976), C.A. *86*, 81961q (1977); Change in the compressibility of cholesteryl stearate. [**225**] A. W. Neumann, L. J. Klementowski, J. Therm. Anal. *6*, 67 (1974), C.A. *80*, 100879p (1974); Effect of zone refining and recrystallization on organic materials exhibiting allotropic transitions. [**226**]

F. C. Nix, W. Shockley, Rev. Mod. Phys. *10*, 1 (1938), C.A. *32*, 4927[4] (1938); Order-disorder transitions in alloys. [**226a**] K. Z. Ogorodnik, Mol. Cryst. Liq. Cryst. *42*, 53 (1977); Crystallization and melting of metastable solid crystalline EBBA. [**226b**] C. S. Oh, Mol. Cryst. Liq. Cryst. *42*, 1 (1977); Induced smectic mesomorphism by incompatible nematogens. [**227**] R. A. Orwoll, R. H. Rhyne, Jr., S. D. Christesen, S. N. Young, J. Phys. Chem. *81*, 181 (1977), C.A. *86*, 71654c (1977); Volume changes of mixing for the system *p,p'*-di-*n*-hexyloxyazoxybenzene + xylene. [**228**] W. Ostner, S. K. Chan, M. Kahlweit, Ber. Bunsenges. Phys. Chem. *77*, 1122 (1973), C.A. *80*, 88087 (1974); i/n Transformation of PAA. [**229**] B. I. Ostrovskii, S. A. Taraskin, B. A. Strukov, A. S. Sonin, Zh. Eksp. Teor. Fiz. *71*, 692 (1976), C.A. *85*, 134832m (1976); Temperature dependence of the specific heat of MBBA during the n/i transition. [**230**] P. Pacor, H. L. Spier, J. Am. Oil Chem. Soc. *45*, 338 (1968), C.A. *69*, 11629n (1968); Thermal analysis and calorimetry of some fatty acid sodium soaps.

[**231**] P. Papon, J. P. Le Pesant, Chem. Phys. Lett. *12*, 331 (1971), C.A. *76*, 77738y (1972); Statistical model for transitions in nematics. [**232**] M. Papoular, Solid State Commun. *7*, 1691 (1969), C.A. *72*, 71603w (1970); Liquid crystal phase diagrams. [**233**] M. Papoular, J. P. Laheurte, Solid State Commun. *12*, 71 (1973), C.A. *78*, 88744q (1973); Possible tricritical situations in liquids: an analogy. [**234**] J. W. Park, C. S. Bak, M. M. Labes, J. Am. Chem. Soc. *97*, 4398 (1975), C.A. *83*, 106580t (1975); Effects of molecular complexing on the properties of binary nematic mixtures. [**235**] J. W. Park, M. M. Labes, Mol. Cryst. Liq. Cryst. *34*, 147 (1977), C.A. *86*, 149027x (1977); Broadening of the nematic temperature range by a non-mesogenic solute in a nematic. [**236**] H. T. Peterson, Diss. Univ. Georgetown, Washington, D.C. 1972, 170 pp. Avail. Univ. Microfilms, Ann Arbor, Mich., Order No. 73-11,800, C.A. *79*, 46259a (1973); Solute-induced n/i transitions as studied through solution thermodynamics and lattice statistical mechanics. [**237**] R. S. Pindak, Diss. Univ. Pennsylvania, Philadelphia, Pa. 1975, 187 pp. Avail. Xerox Univ. Microfilms, Ann Arbor, Mich., Order No. 75-24,115, C.A. *84*, 52357a (1976); Experimental study of pretransitional effects near n/s_A transitions. [**238**] J. M. Pochan, H. W. Gibson, J. Am. Chem. Soc. *93*, 1279 (1971), C.A. *74*, 117083y (1971); Nucleation studies of supercooled cholesterics. [**239**] Y. Poggi, P. Atten, J. C. Filippini, Mol. Cryst. Liq. Cryst. *37*, 1 (1976), C.A. *86*, 131380x (1977); Validity of the Landau approximation in the n phase. [**240**] P. Pollmann, Ber. Bunsenges. Phys. Chem. *78*, 374 (1974), C.A. *81*, 69726f (1974); Existenzbereich cholesterischer Mesophasen bei hohen Drücken. Die Phasenübergänge s/c und c/i. [**240a**] P. Pollmann, G. Scherer, Chem. Phys. Lett. *47*, 286 (1977), C.A. *87*, 14479w (1977); Pressure induced change of order of c/s_A transition. [**240b**] P. Pollmann, G. Scherer, Mol. Cryst. Liq. Cryst. *34*, 189 (1977), C.A. *87*, 46805r (1977); Pressure dependence of the cholesteric pitch below and above a tricritical point. Cholesteryl myristate and COC near the c/s_A transition.

[**241**] J. A. Pople, F. E. Karasz, J. Phys. and Chem. Solids *18*, 28 (1961); *20*, 294 (1961), C.A. *55*, 18227a (1961); *56*, 2962g (1962); Theory of fusion of molecular crystals. I. Effects of hindered rotation. II. Phase diagrams and relations with solid-state transitions. [**242**] A. V. Galanti, R. S. Porter, J. Phys. Chem. *76*, 3089 (1972), C.A. *77*, 157354f (1972); Thermal transitions and phase relations for binary mixtures of cholesteryl esters. [**243**] C. W. Griffen, R. S. Porter, Mol. Cryst. Liq. Cryst. *21*, 77 (1973), C.A. *79*, 119282q (1973); Phase studies on binary systems of cholesteryl esters. I. Two aliphatic ester pairs. [**244**] R. J. Krzewki, R. S. Porter, Mol. Cryst. Liq. Cryst. *21*, 99 (1973), C.A. *79*, 119281p (1973); II. Three C_{18} ester pairs (cf. [**243**]). [**245**] R. J. Krzewki, R. S. Porter, A. M. Atallah, H. J. Nicholas, Mol. Cryst. Liq. Cryst. *29*, 127 (1974), C.A. *82*, 178673c (1975); Calorimetric study of liquid crystalline behavior for some 9,19-cyclopropane tetracyclic triterpene palmitates. [**246**] R. S. Porter, Mol. Cryst. Liq. Cryst. *33*, 227 (1976), C.A. *85*, 114989 (1976); Relationships between mesophase temperature range and transition entropies for steroids. [**247**] M. J. Press, A. S. Arrott, Phys. Rev. A *8*, 1459 (1973), C.A. *79*, 130072s (1973); Expansion coefficient of MBBA through the liquid-crystal phase transition. [**248**] F. P. Price, J. H. Wendorff, J. Phys. Chem. *75*, 2839 (1971), C.A. *75*, 113530q (1971); Transformation kinetics and pretransition effects in cholesteryl myristate. [**249**] F. P. Price, J. H. Wendorff, J. Phys. Chem. *75*, 2849 (1971), C.A. *75*, 113529m (1971); Transformation kinetics and properties of cholesteryl acetate. [**250**] F. P. Price, J. H. Wendorff, J. Phys. Chem. *76*, 276 (1972), C.A. *76*, 72707c (1972); Transformation kinetics and textural changes in cholesteryl nonanoate.

[**251**] F. P. Price, J. H. Wendorff, J. Phys. Chem. *76*, 2605 (1972), C.A. *77*, 131763p (1972); Transformation behavior and pretransition effects in PAA. [**252**] D. Armitage, F. P. Price, J. Phys. (Paris), C-1, *36*, 133 (1975), C.A. *83*, 86051r (1975); Calorimetry of liquid crystal phase transitions. [**253**] D. Armitage, F. P. Price, Bull. Am. Phys. Soc. *20*, 886 (1975); Pretransition density behavior. [**254**] D. Armitage, F. P. Price, J. Appl. Phys. *47*, 2735 (1976), C.A. *85*, 54883p (1976); Precision

recording dilatometer with results for cholesteryl nonanoate. [**255**] D. Armitage, F. P. Price, Chem. Phys. Lett. *44*, 305 (1976), C.A. *86*, 63851z (1977); Size and surface effects on phase transitions. [**256**] D. Armitage, F. P. Price, Mol. Cryst. Liq. Cryst. *38*, 229 (1977), C.A. *86*, 181027h (1977); Volumetric behavior of liquid crystal CBOOA. [**256a**] D. Armitage, F. P. Price, J. Chem. Phys. *66*, 3414 (1977), C.A. *87*, 144317c (1977); Volumetric pretransition in cholesterics. [**256b**] D. Armitage, F. P. Price, Phys. Rev. A *15*, 2069 (1977), C.A. *87*, 46790g (1977); Volumetric study of liquid-crystal heptyloxyazoxybenzene. [**256c**] D. Armitage, F. P. Price, Phys. Rev. A *15*, 2496 (1977), C.A. *87*, 60981b (1977); Volumetric study of the n/i pretransition region. [**257**] R. G. Priest, Phys. Lett. A *47*, 475 (1974), C.A. *81*, 42149h (1974); Tricritical point for nematics. [**258**] V. V. Prisedskii, V. V. Klimov, Probl. issled. svoistv segnetoelektrikov. *1974*, 144, C.A. *83*, 156011d (1975); Phase transitions in CBOOA at high pressures. [**259**] Proceedings of the 14th Conference on Chemistry at the University of Brussels, May 1969. (Interscience: London). 1971. 256 pp. [**260**] P. Pruzan, L. Ter Minassian, Conf. Int. Thermodyn. Chim., 4th, *8*, 47 (1975), C.A. *84*, 127405a (1976); Application des methodes piezo-thermiques a la mesure directe des coefficients de dilatation et chaleurs de transformation de cristaux organiques sous hautes pressions.

[**261**] N. A. Pushkin, I. V. Grebenshchikov, Z. physik. Chem. *124*, 270 (1926); C.A. *21*, 843[5] (1927); Einfluß des Druckes auf die Kristallisationstemperatur von PAA und α-Naphthylamin. [**262**] R. Pynn, J. Chem. Phys. *60*, 4579 (1974), C.A. *81*, 55262q (1974); Density and temperature dependence of the i/n transition. [**263**] W. Pyzuk, W. Vieth, Bull. Acad. Pol. Sci., Ser. Sci. Chim. *24*, 677 (1976), C.A. *86*, 63875k (1977); Determination of latent isotropic melting parameters of N-(4′-methoxybenzylidene)-4-methoxyaniline. [**264**] V. S. V. Rajan, J. J. C. Picot, Mater. Res. Bull. *9*, 311 (1974), C.A. *81*, 55023n (1974); Heat transport in nematic MBBA. [**265**] J. Rault, Philos. Mag. *34*, 753 (1976), C.A. *86*, 99164s (1977); Nucleation of the focal conic texture in lamellar liquid crystals. [**266**] F. I. G. Rawlins, Trans. Faraday Soc. *29*, 993 (1933), C.A. *28*, 1239[3] (1934); Anisotropic melts: a study in change of state. [**266a**] A. S. Reshamwala, R. Shashidhar, J. Phys. E *10*, 180 (1977), C.A. *86*, 163767v (1977); Coaxial DTA cell for the study of liquid crystalline transitions at elevated pressures. [**267**] R. Ribotta, R. B. Meyer, G. Durand, J. Phys. (Paris) Lett. *35*, 161 (1974), C.A. *81*, 142325f (1974); Compression induced s_A/s_C transition. [**268**] J. Robberecht, Bull. soc. chim. Belg. *47*, 597 (1938), C.A. *33*, 2008[4] (1939); Recherches piézométriques. V. Contribution à l'étude sous pression des liquides anisotropes. [**269**] H. W. B. Roozeboom, Z. physik. Chem. *30*, 413 (1899); Umwandlungspunkte bei Mischkristallen. [**270**] K. Rosciszewski, Postepy Fiz. *24*, 393 (1972), C.A. *80*, 53037a (1974); Theory of liquid crystal phase and phase transitions related to them.

[**271**] J. Roth, T. Meisel, K. Seybold, Z. Halmos, J. Therm. Anal. *10*, 223 (1976), C.A. *86*, 36547y (1977); Investigation of the thermal behavior of fatty acid sodium salts. [**272**] L. K. Runnels, J. Chem. Phys. *60*, 4086 (1974), C.A. *81*, 55240f (1974); Two-dimensional liquid crystals again. [**273**] G. S. Sachan, K. L. Vasanth, Technology (Coimbatore, India) *24*, 30 (1976), C.A. *85*, 167382t (1976); Graphical method for calculation of eutectic points in binary systems. [**274**] N. M. Sakevich, Izv. Vyssh. Ucheb. Zaved., Fiz. *10*, 52 (1967), C.A. *68*, 108677u (1968); Determination of the heat of reaction during the transition from liquid-crystalline state of PAA and cholesteryl caprate into isotropic liquid and solid phases. [**275**] N. M. Sakevich, Izv. Vyssh. Ucheb. Zaved., Fiz. *10*, 56 (1967), C.A. *69*, 22905a (1968); Determination of the work of formation and the critical size of centers of PAA and cholesteryl caprinate liquid crystals. [**276**] N. M. Sakevich, Zh. Fiz. Khim. *42*, 2930 (1968), C.A. *70*, 51616h (1969); Temperature dependence of the specific volume of some liquid crystals in the region of an isotropic-liquid phase/liquid crystalline phase transition. [**277**] N. M. Sakevich, Kristallografiya *16*, 650 (1971), C.A. *75*, 68624k (1971); Temperature dependence of the linear rate of growth of nematic PAA and PAP. [**278**] A. V. Santoro, G. I. Spielholtz, Anal. Chim. Acta *42*, 537 (1968), C.A. *70*, 7227h (1969); Heats of transitions and kinetic study of liquid crystals by DTA. [**279**] H. Sasabe, K. Ooizumi, Jap. J. Appl. Phys. *11*, 1751 (1972), C.A. *78*, 76768n (1973); Pressure dependence of the k/n transition temperature in MBBA. [**280**] S. Sato, M. Wada, Jap. J. Appl. Phys. *11*, 1752 (1972), C.A. *78*, 63517e (1973); c/n Transitions in mixtures of cholesteryl chloride and nematics.

[**281**] F. Schneider, Diss. Marburg 1899; Zur Kenntnis der kristallinischen Flüssigkeiten. [**282**] H. Schroeder, Ber. Bunsenges. Phys. Chem. *78*, 855 (1974), C.A. *82*, 10236p (1975); Mikroskopische Theorie zur Erklärung des Phasendiagramms flüssiger Kristalle. [**283**] J. P. Schroeder, D. C. Schroeder, J. Org. Chem. *33*, 591 (1968), C.A. *68*, 63487p (1968); Stable smectic mixtures of 4,4′-di-*n*-hexyloxyazoxybenzene and *p*-nitro-substituted aromatic compounds. [**284**] W. Schröer, Ber. Bunsenges. Phys. Chem. *78*, 626 (1974), C.A. *81*, 119702t (1974); Nahordnung in Flüssigkeiten. Statistisch-thermodynamisches Flüssigkeitsmodell auf der Basis von Kontaktwahrscheinlichkeiten. [**285**] P. J. Sell, A. W. Neumann,

Z. Phys. Chem. (Frankfurt) *65*, 13 (1969), C.A. *71*, 84810r (1969); DTA und Polarisationsmikroskopie von Cholesterinestern homologer Fettsäuren. [**286**] V. K. Semenchenko, Dokl. Akad. Nauk Belorus. SSR *3*, 445 (1959), C.A. *54*, 23693c (1960); Thermodynamics of high polymers and liquid crystals. [**287**] V. K. Semenchenko, Primeneneie Ul'traakustiki k Issled. Veshchestva *1962*, No. 16, 101, C.A. *58*, 13190e (1963); Thermodynamics of mesophases and properties of polymers and liquid crystals. [**288**] V. K. Semenchenko, M. M. Martynyuk, Kolloidn. Zh. *24*, 611 (1962), C.A. *58*, 9240d (1963); Comparison of conclusions derived from mesophase thermodynamics with experimental data. [**289**] V. K. Semenchenko, V. N. Kuznetsova, Kolloid. Zh. *30*, 279 (1968), C.A. *69*, 80792x (1968); Thermodynamics of liquid crystals. [**290**] V. K. Semenchenko, Khim. Svyaz Krist. *1969*, 213, C.A. *72*, 48261z (1970); Phase equilibria and transitions in low-temperature zones.

[**291**] V. Y. Baskakov, V. K. Semenchenko, Pis'ma Zh. Eksp. Teor. Fiz. *17*, 580 (1973), C.A. *79*, 84265g (1973); Cholesteryl valerate under pressure. [**292**] V. Y. Baskakov, V. K. Semenchenko, V. M. Byankin, Zh. Eksp. Teor. Fiz. *66*, 792 (1974), C.A. *80*, 113479h (1974); Possibility of a critical transition in PAA. [**293**] V. M. Byankin, V. K. Semenchenko, V. Y. Baskakov, Zh. Fiz. Khim. *48*, 1250 (1974), C.A. *81*, 112230d (1974); Change in the volume of cholesteryl caproate under pressure. [**294**] V. Y. Baskakov, V. K. Semenchenko, N. A. Nedostup, Kristallografiya *19*, 185 (1974), C.A. *80*, 100971n (1974); Pressure-volume-temperature diagram of PAA to 400 atm. [**295**] V. K. Semenchenko, V. M. Byankin, V. Y. Baskakov, Zh. Fiz. Khim. *48*, 2353 (1974), C.A. *82*, 66529d (1975); Effect of pressure on phase transitions in cholesteryl propionate. [**296**] V. K. Semenchenko, V. Y. Baskakov, N. A. Nedostup, Zh. Fiz. Khim. *48*, 2444 (1974), C.A. *82*, 163310h (1975); Thermodynamic stability of homologs of the dialkoxyazoxybenzene series. [**297**] V. K. Semenchenko, V. M. Byankin, V. Y. Baskakov, Kristallografiya *20*, 187 (1975), C.A. *83*, 19388t (1975); Effect of pressure on phase transitions in cholesteryl myristate and cholesteryl pelargonate. [**298**] V. K. Semenchenko, N. A. Nedostup, V. Y. Baskakov, Zh. Fiz. Khim. *49*, 1543 (1975), C.A. *83*, 140043e (1975); Liquid crystals studied under pressure. [**299**] V. K. Semenchenko, N. A. Nedostup, V. Y. Baskakov, Zh. Fiz. Khim. *49*, 1547 (1975), C.A. *83*, 140044f (1975); Volume effects in cholesteryl butyrate studied under pressure. [**300**] V. Y. Baskakov, V. K. Semenchenko, V. M. Byankin, Zh. Fiz. Khim. *50*, 200 (1976), C.A. *84*, 128972v (1976); Effect of pressures on phase transitions in dioctyloxyazoxybenzene. [**300a**] N. A. Nedostup, V. K. Semenchenko, Zh. Fiz. Khim. *51*, 1632 (1977), C.A. *87*, 175825w (1977); Thermodynamic stability of mesophases at different temperatures and pressures. [**300b**] N. A. Nedostup, V. K. Semenchenko, Zh. Fiz. Khim. *51*, 1628 (1977), C.A. *87*, 175824v (1977); Analytical representation of the results of studying mesophases at different temperatures and pressures. [**300c**] N. K. Semendyaeva, G. I. Karpushkina, A. A. Kotylar, V. N. Shoshin, Proc. Eur. Symp. Therm. Anal., 1st *1976*, 188, C.A. *87*, 60969d (1977); DTA investigation of binary systems forming liquid crystals.

[**301**] L. S. Shabyshev, Uch. Zap. Ivanov. Gos. Pedagog, Inst. *77*, 108 (1970), C.A. *76*, 91238s (1972); DTA of liquid crystals. [**302**] L. S. Shabyshev, B. N. Makarov, A. I. Aleksandrov, Uch. Zap. — Ivanov. Gos. Univ. *128*, 113 (1974), C.A. *85*, 110408b (1976); Apparatus for zone purification of liquid crystals. [**302a**] R. Shashidhar, Mol. Cryst. Liq. Cryst. *43*, 71 (1977); High pressure studies on mesomorphic and polymesomorphic transitions. [**303**] S. A. Shaya, Diss. Univ. Wisconsin, Madison, Wis. 1974, 261 pp. Avail. Xerox Univ. Microfilms, Ann Arbor, Mich., Order No. 74-18,956, C.A. *82*, 77678c (1975); Equilibrium and dynamic properties in the nematic MBBA/biphenyl system. [**304**] S. A. Shaya, H. Yu, J. Phys. (Paris) C-1, *36*, 59 (1975), C.A. *83*, 36026r (1975); Nematic order of a binary system: MBBA/biphenyl. [**305**] Y. R. Shen, Report 1975, LBL-4178, 16 pp., C.A. *85*, 185018h (1976); Nonlinear optical study of pretransitional behavior in liquid crystals. [**306**] P. Sheng, E. B. Priestley, Introd. Liq. Cryst. 1974, 143, C.A. *84*, 82634f (1976); Landau-de Gennes theory of liquid crystal phase transitions. [**307**] P. Sheng, Phys. Rev. Lett. *37*, 1059 (1976), C.A. *85*, 185030f (1976); Phase transition in surface-aligned nematic films. [**308**] T. Shinoda, Y. Maeda, H. Enokido, J. Chem. Thermodyn. *6*, 921 (1974), C.A. *82*, 30803z (1975); Thermodynamic properties of MBBA from 2 K to its i phase. [**309**] Y. Shiwa, Prog. Theor. Phys. *55*, 629 (1976), C.A. *84*, 128975y (1976); Dynamic renormalization-group theory above a s_A/n transition. [**310**] D. Guillon, A. Skoulios, C. R. Acad. Sci., Sér. C *278*, 389 (1974), C.A. *81*, 30649h (1974); Étude dilatométrique du di-(p-n-octadécyloxybenzylidèneamino)-4,4'-diphényle.

[**311**] D. Guillon, A. Skoulios, J. Phys. (Paris) *37*, 797 (1976), C.A. *85*, 70920d (1976); Smectic polymorphism and melting processes of molecules in the case of 4,4'-di(p,n-alkoxybenzylideneamino)biphenyls. [**312**] D. Guillon, A. Skoulios, J. Phys. (Paris) *38*, 79 (1977), C.A. *86*, 63879q (1977); Polymorphisme smectique du TBBA. [**313**] D. Guillon, A. Skoulios, Mol. Cryst. Liq. Cryst. *39*, 139 (1977), C.A. *87*,

32201d (1977); Polymorphisme smectique. Étude dilatométrique de la série des 4,4'-di(*p,n*-alkoxybenzyli-dèneamino)biphényles. [**314**] A. R. Sluzas, Diss. Univ. Wisconsin, Madison, Wis. 1975. 168 pp. Avail. Xerox Univ. Microfilms, Ann Arbor, Mich., Order No. 76-6108., C.A. *85*, 54899y (1976); Twist diffusivity and thermal properties of a binary nematic-nonmesomorph system: MBBA/benzene. [**314a**] J. E. Smaar-dyk, Diss. Univ. Illinois, Urbana, Ill. 1977, 111 pp. Avail. Univ. Microfilms Int., Order No. 77-15,018, C.A. *87*, 93797z (1977); Heat capacity studies of phase transitions in a series of homologous liquid crystals using the ac heat capacity technique. [**314b**] B. E. North, D. M. Small, J. Phys. Chem. *81*, 723 (1977), C.A. *86*, 180954w (1977); Thermal and structural properties of the cholestanyl myristate-choles-teryl myristate and cholestanyl myristate-cholesteryl oleate binary systems. [**315**] G. W. Smith, Z. G. Gardlund, J. Chem. Phys. *59*, 3214 (1973), C.A. *80*, 7902f (1974); Liquid crystals in a doubly homologous series of benzylideneanilines. Textures and DSC. [**316**] G. W. Smith, Z. G. Garlund, R. J. Curtis, Liq. Cryst. Ord. Fluids *2*, 573 (1974), C.A. *86*, 180996m (1977); Phase diagram of mixed mesomorphic benzylideneanilines, MBBA/EBBA. [**317**] G. W. Smith, Mol. Cryst. Liq. Cryst. *30*, 101 (1975), C.A. *83*, 106505x (1975); Influence of a metastable solid phase on eutectic formation of a binary nematic. [**318**] G. W. Smith, Adv. Liq. Cryst. *1*, 189 (1975), C.A. *84*, 37348e (1976); Plastic crystals, liquid crystals, and the melting phenomenon. Importance of order. [**319**] G. W. Smith, Mol. Cryst. Liq. Cryst. *34*, 87 (1976), C.A. *86*, 9312m (1977); Thermal parameters for crystal/mesophase and crystal/crystal transitions of some benzylideneanilines. [**320**] H. M. Smith, W. H. McClelland, J. Am. Chem. Soc. *26*, 1446 (1904), Molecular depression constant of PAA.

[**321**] M. Sorai, S. Seki, Bull. Chem. Soc. Jap. *44*, 2887 (1971), C.A. *75*, 155804n (1971); Glassy liquid crystal of nematic N-(*o*-hydroxy-*p*-methoxybenzylidene)-*p*-butylaniline. [**322**] M. Sorai, R. Naka-mura, S. Seki, Liq. Cryst., Proc. Int. Conf., Bangalore, *1973*, 503, C.A. *84*, 114476z (1976); Thermal studies of benzylideneaniline liquid crystals. [**323**] M. Sorai, S. Seki, Mol. Cryst. Liq. Cryst. *23*, 299 (1973), C.A. *80*, 137641f (1974); Heat capacity of N-(*o*-hydroxy-*p*-methoxybenzylidene)-*p*-butylaniline. Glassy nematic. [**324**] M. Sorai, T. Nakamura, S. Seki, Bull. Chem. Soc. Jpn. *47*, 2192 (1974), C.A. *81*, 177913y (1974); Heat capacity of nematogenic EBBA between 14 and 375 K. [**325**] M. E. Spaght, S. B. Thomas, G. S. Parks, J. Phys. Chem. *36*, 882 (1932), C.A. *26*, 2641[7] (1932); Heat-capacity data on organic compounds obtained with a radiation calorimeter. [**326**] W. Spratte, G. M. Schneider, Ber. Bunsenges. Phys. Chem. *80*, 886 (1976), C.A. *85*, 169957h (1976); DTA under high pressure. Phase transitions of some liquid crystals up to 3 kbar. [**327**] H. E. Stanley, Cooperative phenomena near phase transitions. Bibliography with selected readings (MIT Press: Cambridge, Mass.). 1973. 308 pp., C.A. *78*, 115516a (1973). [**328**] L. A. K. Staveley, Quart. Rev. *3*, 65 (1949), C.A. *43*, 6481f (1949); Transitions in solids and liquids. [**329**] W. U. Müller, H. Stegemeyer, Chem. Phys. Lett. *27*, 130 (1974), C.A. *81*, 142266n (1974); Order of c/s_A transition in cholesteric mixtures. [**330**] W. U. Müller, H. Stegemeyer, Ber. Bunsenges. Phys. Chem. *78*, 880 (1974), C.A. *82*, 10285d (1975); Ordnung der Phasenumwandlung c/s_A in cholesterischen Mischungen.

[**331**] H. Stegemeyer, W. U. Müller, Naturwissenschaften *63*, 388 (1976), C.A. *85*, 105660y (1976); Effect of cholesterol on the c/s_A transition. [**332**] S. M. Stishov, V. A. Ivanov, V. N. Kachinskii, Pis'ma Zh. Eksp. Teor. Fiz. *24*, 329 (1976), C.A. *86*, 61323e (1977); Thermodynamics of the n/i transition in PAA at high pressures. [**333**] H. Suga, S. Seki, J. Non-Cryst. Solids *16*, 171 (1974), C.A. *82*, 48268p (1975); Thermodynamic investigation on glassy states of pure simple compounds. [**333a**] J. Swift, Phys. Rev. A *14*, 2274 (1976); Fluctuations near n/s_C transition. [**334**] Z. Szabo, KFKI (Kozp. Fiz. Kut. Intez) (Rep.) 1974, 74, C.A. *81*, 30653e (1974); Phase diagram of liquid-crystal mixtures. [**335**] Z. Szabo, I. Kosa-Somogyi, Mol. Cryst. Liq. Cryst. *31*, 161 (1975), C.A. *83*, 209892a (1975); Phase diagrams of low-melting nematogenic esters. [**336**] J. Szabon, L. Bata, K. Pinter, KFKI (Rep.) *1974*, 74-82, 14 pp., C.A. *82*, 118321b (1975); Thermal properties of disubstituted liquid crystalline phenyl benzoates. [**337**] G. Tammann, Aggregatzustände (Leopold Voss: Leipzig). 1922. 294 pp., C.A. *16*, 2066 (1922). [**338**] D. Ter Haar, Collected Papers of L. D. Landau (Pergamon: Oxford). 1965. 836 pp. [**338a**] L. Ter Minassian, P. Pruzan, J. Chem. Thermodyn. *9*, 375 (1977), C.A. *87*, 12482t (1977); High-pressure expansivity of materials determined by piezo-thermal analysis. [**339**] N. A. Tikhomirova, L. K. Vistin, V. N. Nosov, Kristallografiya *17*, 1000 (1972), C.A. *78*, 8834f (1973); Effect of pressure on phase transitions in nematics. [**340**] N. A. Tikhomirova, A. V. Ginzberg, Kristallografiya *22*, 155 (1977), C.A. *86*, 113889r (1977); P-T Diagrams of cholesterics to 5000 kg/cm[2].

[**341**] S. Torza, P. E. Cladis, Phys. Rev. Lett. *32*, 1406 (1974), C.A. *81*, 42503u (1974); Volumetric study of the n/s_A transition of CBOOA. [**342**] T. Toshiaki, O. Hirokuni, F. Eiichi, Rika Gaku Kenkyusho Hokoku *47*, 116 (1971), C.A. *76*, 51098q (1972); Electret thermal analysis of liquid crystals. [**342a**]

S. Trimper, Phys. Status Solidi B *82*, K75 (1977), C.A. *87*, 144277q (1977); Phase transition in ferroelectric liquid crystals. [**343**] A. R. Ubbelohde, Pure Appl. Chem. *2*, 251 (1961), C.A. *57*, 75c (1962); Prefreezing and premelting. [**344**] A. R. Ubbelohde, Z. Physik. Chem. (Frankfurt) *37*, 183 (1963), C.A. *59*, 10777a (1963); Premonitory phenomena in phase transitions of solids. [**345**] E. McLaughlin, M. A. Shakespeare, A. R. Ubbelohde, Trans. Faraday Soc. *60*, 25 (1964), C.A. *60*, 8732d (1964); Pre-freezing phenomena in relation to liquid-crystal formation. [**346**] A. R. Ubbelohde, Angew. Chem. *77*, 614 (1965), C.A. *63*, 9146d (1965); Schmelzvorgang und Kristallstruktur. [**347**] A. R. Ubbelohde, Melting and Crystal Structure (Clarendon: Oxford). 1965. 325 pp., C.A. *64*, 16750e (1966). [**348**] J. van der Veen, W. H. de Jeu, M. W. M. Wanninkhof, C. A. M. Tienhoven, J. Phys. Chem. *77*, 2153 (1973), C.A. *79*, 108790c (1973); Transition entropies and mesomorphism of *p*-disubstituted azoxybenzenes. [**349**] G. R. van Hecke, J. Chem. Educ. *53*, 161 (1976), C.A. *84*, 163613a (1976); Thermotropic liquid crystals. Use of chemical potential-temperature phase diagrams. [**350**] J. P. van Meter, S. E. B. Petrie, ACS Symp. New York 1971, Abstr. No. 177; Atypical heats of transition of smectics.

[**351**] K. L. Vasanth, Technology (Coimbatore, India) *23*, 9 (1975), C.A. *84*, 37453k (1976); Nematic mixed liquid crystals in binary systems. [**352**] W. Vieth, W. Pyzuk, Bull. Acad. Pol. Sci., Ser. Sci. Chim. *24*, 671 (1976), C.A. *86*, 105711m (1977); Isotropic melting point and transition heat in the mixture of two 4,4'-di-*n*-alkoxyazoxybenzenes. [**353**] P. B. Vigman, V. M. Filev, Zh. Eksp. Teor. Fiz. *69*, 1466 (1975); c/s Phase transition. [**354**] L. K. Vistin, V. I. Uchastkin, Zh. Eksp. Teor. Fiz. *70*, 1798 (1976), C.A. *85*, 70907e (1976); Peculiarity of the n/i transition. [**355**] M. J. Vold, J. Am. Chem. Soc. *63*, 160 (1941), C.A. *35*, 1657[7] (1941); Liquid crystalline, waxy and crystalline phases in binary mixtures of pure anhydrous soaps. [**356**] U. Wannagat, R. Braun, L. Gerschler, H. J. Wismar, J. Organomet. Chem. *26*, 321 (1971), C.A. *74*, 112118a (1971); Chemie der Si—N-Verbindungen. Dodeca-methylcyclotetrasilazan. [**357**] U. Wannagat, D. Schmid, Chem.-Ztg. *97*, 448 (1973), C.A. *79*, 119060r (1973); Tris(silyl)amine: eine Stoffklasse mit ungewöhnlich hohen kryoskopischen Konstanten. [**357a**] P. F. Waters, L. R. Farmer, Colloid Interface Sci., (Proc. Int. Conf.) 50th, *5*, 97 (1976), C.A. *87*, 61012y (1977); First-order transition in the i phase of cholesteryl stearate. [**358**] H. T. Weston, P. H. Keyes, W. B. Daniels, Bull. Am. Phys. Soc. *19*, 284 (1974); High-pressure investigations of liquid crystals. [**359**] W. Woycicki, J. Stecki, Bull. Acad. Pol. Sci., Ser. Sci. Chim. *22*, 241 (1974), C.A. *81*, 6850w (1974); Enthalpy of mixing and solid-liquid equilibrium in mixtures of MBBA and some benzenes and naphthenes. [**360**] S. Yano, J. Yasue, K. Aoki, J. Phys. Chem. *80*, 88 (1976), C.A. *84*, 52367d (1976); Polymorphism of cholesteryl acrylate.

[**361**] E. D. Yorke, A. G. de Rocco, J. Chem. Phys. *59*, 92 (1973), C.A. *79*, 84271f (1973); Can two-dimensional mesophases solidify? [**362**] U. Yoshida, Mem. Coll. Sci. Kyoto Imp. Univ. A *24*, 1 (1942), C.A. *44*, 2816g (1950); Relaxation of the cybotactic structure of amorphous solids and the mechanism of formation of crystal nuclei. [**363**] U. Yoshida, Mem. Coll. Sci. Kyoto Imp. Univ. A *24*, 135 (1944), C.A. *44*, 892d (1950); Structural relaxation of amorphous solids and the cybotactic structure of supercooled liquids. [**364**] J. G. J. Ypma, G. Vertogen, J. Phys. (Paris) *37*, 557 (1976), C.A. *85*, 27534u (1976); Molecule statistical treatment of pretransitional effects in nematics. [**365**] A. C. Zawisza, J. Stecki, Solid State Commun. *19*, 1173 (1976), C.A. *85*, 169951b (1976); Compressibility of MBBA. [**366**] B. Zeks, Ferroelectrics *12*, 91 (1976), C.A. *86*, 24950m (1977); Random ferroelectrics. [**367**] G. M. Zharkova, V. M. Khachaturyan, Sb. Dokl. Vses. Nauch. Konf. Zhidk. Krist. Simp. Ikh Prakt. Primen, 2nd, *1972*, 197, C.A. *82*, 37491e (1975); DTA study of cholesterics. [**368**] G. M. Zharkova, V. M. Khachaturyan, Acrofiz. Issled. *1972*, 41, C.A. *80*, 31293d (1974); Thermal-analysis studies of liquid crystals. [**369**] G. M. Zharkova, V. M. Khachaturyan, Izv. Sib. Otd. Akad. Nauk SSSR, Ser. Tekh. Nauk *1976*, 102, C.A. *85*, 134781u (1976); Thermographic studies of binary liquid-crystal systems. [**370**] I. P. Zhuk, A. A. Sitnov, Vestsi Akad. Navuk BSSR, Ser. Fiz.-Energ. Navuk *1975*, 115, C.A. *83*, 186570c (1975); Thermodynamic characteristics and mesomorphism.

[**371**] I. P. Zhuk, A. A. Sitnov, Jnzh.-Fiz. Zh. *28*, 1025 (1975), C.A. *83*, 153432f (1975); Calorimetric studies of liquid crystals.

Chapter 9

[**1**] A. G. Altenau, R. E. Kramer, D. J. McAdoo, C. Merrit, Jr., J. Gas Chromatogr. *4*, 96 (1966), C.A. *64*, 18459b (1966); Behavior of stationary phases at cryogenic temperatures. [**2**] Y. B.

Amerik, B. A. Krentsel, I. I. Konstantinov, Dokl. Akad. Nauk SSSR *165*, 1097 (1965), C.A. *64*, 9825e (1966); Polymerization of vinyl oleate in the liquid-crystal state. [3] Y. B. Amerik, I. I. Konstantinov, B. A. Krentsel, J. Polym. Sci. Part C, No. 23, 231 (1966), C.A. *69*, 77776q (1968); Polymerization of *p*-(methacryloyloxy)benzoic acid in mesomorphic and in liquid states. [4] Y. B. Amerik, B. A. Krentsel, J. Polym. Sci. Part C, No. 16, 1383 (1967), C.A. *67*, 44128k (1967); Polymerization of certain vinyl monomers in liquid crystals. [5] Y. B. Amerik, I. I. Konstantinov, B. A. Krentsel, E. M. Malakhaev, Vysokomol. Soedin. Ser. A *9*, 2591 (1967), C.A. *68*, 50145a (1968); Polymerization of *p*-methacryloyloxyben-zoic acid in the liquid crystalline state. [6] L. G. Shaltyko, A. A. Shepelevskii, I. I. Konstantinov, Y. B. Amerik, S. Y. Frenkel, Vysokomol. Soedin. Ser. B *11*, 824 (1969), C.A. *72*, 55988v (1970); Structural transformations during the polymerization of organized monomers. [7] B. A. Krentsel, Y. B. Amerik, Vysokomol. Soedin. Ser. A *13*, 1358 (1971), C.A. *75*, 88999q (1971); Radical polymerization in anisotropic media in conditions of weak intermolecular interaction. [8] A. A. Baturin, Y. B. Amerik, B. A. Krentsel, Mol. Cryst. Liq. Cryst. *16*, 117 (1972), C.A. *76*, 127543d (1972); Liquid crystalline state copolymerization. [9] A. A. Baturin, Y. B. Amerik, B. A. Krentsel, V. N. Tsvetkov, I. N. Shtennikova, E. I. Ryumtsev, Dokl. Akad. Nauk SSSR *202*, 586 (1972), C.A. *76*, 141350v (1972); Reactivity and lifetime of macroradicals during the polymerization of *p*-[*p*-(cetyloxy)benzoyloxy]phenyl methacrylate. [10] Y. B. Amerik, B. A. Krentsel, Usp. Khim. Fiz. Polim. *1973*, 97, C.A. *80*, 108878h (1974); Role of anisotropic states in polymerization processes.

[11] I. I. Konstantinoff, Y. B. Amerik, L. Vogel, D. Demus, Wiss. Z. Martin-Luther-Univ., Halle-Wittenberg, Math.-Naturwiss. Reihe *22*, 37 (1973), C.A. *82*, 73524b (1975); Polymerisation und Vorumwandlungserscheinungen kristalliner Monomerer mit anisotropen Eigenschaften. [12] V. A. Averin, V. Y. Shnol, A. P. Domarev, U.S.S.R. 501,350 (27. 7. 1973), C.A. *85*, 56255j (1976); GC separation of mixtures of substances. [13] A. Aviram, J. Polym. Sci., Polym. Lett. Ed. *14*, 757 (1976), C.A. *86*, 30232v (1977); Crosslinking of lyophases in magnetic fields. [14] G. Aviv, J. Sagiv, A. Yogev, Mol. Cryst. Liq. Cryst. *36*, 349 (1976), C.A. *86*, 120406 (1977); Photodimerization of tetraphenylbutatriene in polyethylene and nematic matrices. [15] W. E. Bacon, Liq. Cryst., Proc. Int. Conf., Bangalore, *1973*, 455, C.A. *84*, 134719j (1976); Influence of liquid crystalline solvents in chemical reactions. [16] W. E. Bacon, J. Phys. (Paris), C-1, *36*, 409 (1975), C.A. *83*, 59439s (1975); Polymerization of phenylacetylene in n, c, and i solvents. [16a] I. M. Barclay, J. A. V. Butler, Trans. Faraday Soc. *34*, 1445 (1938), C.A. *33*, 2395[6] (1939); The entropy of solution. [17] W. E. Barnett, W. H. Sohn, J. Chem. Soc. D *1971*, 1002, C.A. *75*, 140067a (1971); Xanthate pyrolysis-enhanced olefin production in n solvents. [18] E. M. Barrall, II, R. S. Porter, J. F. Johnson, J. Chromatog. *21*, 392 (1966), C.A. *65*, 1421g (1966); GC using cholesteryl ester liquid phases. [19] P. J. Taylor, R. A. Culp, C. H. Lochmueller, L. B. Rogers, E. M. Barrall, II, Separ. Sci. *6*, 841 (1971), C.A. *75*, 133326f (1971); Effect of an electric field on GC retentions by liquid crystals. [19a] R. P. Bell, Trans. Faraday Soc. *33*, 496 (1937), C.A. *31*, 4874[6] (1937); Relations between the energy and entropy of solution and their significance. [20] A. Blumstein, N. Kitagawa, R. Blumstein, Mol. Cryst. Liq. Cryst. *12*, 215 (1971), C.A. *75*, 49623t (1971); Polymerization of *p*-(methacryloyloxy)benzoic acid within liquid crystalline media.

[21] A. Blumstein, R. Blumstein, G. J. Murphy, C. Wilson, J. Billard, Liq. Cryst. Ord. Fluids *2*, 277 (1974), C.A. *87*, 6429q (1977); Polymerization of *p*-methacryloyloxybenzoic acid in mesomorphic *n*-alkoxybenzoic acids. [22] A. Blumstein, J. Billard, P. Blumstein, Mol. Cryst. Liq. Cryst. *25*, 83 (1974), C.A. *83*, 184238b (1975); Polymerization of *p*-methacryloyloxybenzoic acid in liquid crystalline solvents. Phase diagrams of model systems. [23] A. Blumstein, R. B. Blumstein, S. B. Clough, E. C. Hsu, Macromolecules *8*, 73 (1975), C.A. *82*, 125723n (1975); Oriented polymer growth in thermotropic mesophases. [24] S. B. Clough, A. Blumstein, E. C. Hsu, Macromolecules *9*, 123 (1976), C.A. *84*, 106123s (1976); Structure and thermal expansion of polymers with mesomorphic ordering. [24a] E. C. Hsu, A. Blumstein, J. Polym. Sci., Polym. Lett. Ed. *15*, 129 (1977), C.A. *86*, 156052e (1977); Polymerization of nematic N-(*p*-cyanobenzylidene)-*p*-aminostyrene. [24b] A. Blumstein, Midl. Macromol. Monogr. *3*, 133 (1977), C.A. *87*, 184983y (1977); Polymerization of thermotropic mesophases and formation of organized polymers. [24c] J. F. Bocquet, C. Pommier, J. Chromatogr. *117*, 315 (1976), C.A. *84*, 127439q (1976); Cristaux liquides comme phases stationnaires en GLC. Thermodynamique des solutions de décane dans le 4,4′-dihexyloxyazoxybenzène. [25] W. E. Bacon, G. H. Brown, Mol. Cryst. Liq. Cryst. *6*, 155 (1969), C.A. *72*, 42551x (1970); Liquid crystal solvents as reaction media for the Claisen rearrange-ment. [26] W. E. Bacon, G. H. Brown, Mol. Cryst. Liq. Cryst. *12*, 229 (1971), C.A. *75*, 48167k (1971); n Solvents as media for the Claisen rearrangement. [27] W. E. Bacon, G. H. Brown, ACS Symposium

New York 1971, Abstr. Nr. 208; Thermal isomerization of 2,4,6-trimethoxy-*s*-triazine in liquid crystalline solvent. [**28**] D. G. Willey, G. H. Brown, J. Phys. Chem. *76*, 99 (1972); C.A. *76*, 77420p (1972); Thermodynamic study of solutions with liquid crystal solvents by GLC. [**29**] C. A. Bunton, L. Robinson, J. Org. Chem. *34*, 773 (1969), C.A. *71*, 12203x (1969); Micellar effects upon the reaction of *p*-nitrophenyl diphenyl phosphate with hydroxide and fluoride ions. [**30**] C. A. Bunton, L. Robinson, J. Org. Chem. *34*, 780 (1969), C.A. *71*, 12204y (1969); Electrolyte and micellar effects upon the reaction of 2,4-dinitrofluorobenzene with hydroxide ion.

[**31**] C. Camargo, Rev. Dep. Quim., Univ. Nac. Colomb. *1969*, No. 2, 12, C.A. *74*, 103358e (1971); Liquid crystals as stationary phases in GLC. [**32**] G. Chiavari, Atti Accad. Naz. Lincei, Cl. Sci. Fis., Mat. Natur., Rend., Ser. 8, *51*, 531 (1971), C.A. *77*, 83287n (1972); GC on 2,6-naphthylene bis(p-alkoxybenzoate) as the liquid-crystal phase. [**33**] G. Chiavari, A. Arcelli, A. M. Di Pietra, Atti Accad. Naz. Lincei, Cl. Sci. Fis., Mat. Natur., Rend., Ser. 8, *52*, 381 (1972), C.A. *78*, 43108u (1973); Synthesis, heats and temperatures of transitions and behavior as GC solvents of isomeric mesogenic esters of dihydroxynaphthalene. [**34**] G. Chiavari, L. Pastorelli, Chromatographia *7*, 30 (1974), C.A. *80*, 100545b (1974); GC separation of isomeric 1- and 2-substituted naphthalenes on liquid crystals as stationary phases. [**35**] L. C. L. Chow, Diss. Georgetown Univ., Washington, D.C. 1970, 216 pp., Avail. Univ. Microfilms, Ann Arbor, Mich., Order No. 70-21,280, C.A. *75*, 11091e (1971); Thermodynamic investigation of dilute solutions in two nematics. [**36**] R. R. Claeys, H. Freund, J. Gas Chromatogr. *6*, 421 (1968), C.A. *69*, 80488c (1968); GC behavior of the stationary phase near the freezing point. [**37**] L. E. Cook, R. C. Spangelo, Anal. Chem. *46*, 122 (1974), C.A. *80*, 59615j (1974); Separation of monosubstituted phenol isomers using liquid crystals. [**38**] M. Croucher, B. Kronberg, D. F. R. Gilson, J. Chem. Soc., Chem. Commun. *1975*, 686, C.A. *83*, 171115j (1975); Effects of orientational order in solution thermodynamics. Mixtures containing a liquid crystal. [**39**] P. G. de Gennes, Phys. Lett. A *28*, 725 (1969), C.A. *71*, 13636c (1969); Possibilités offertes par la réticulation de polymères en présence d'un cristal liquide. [**40**] A. C. de Visser, J. Feyen, K. de Groot, A. Bantjes, J. Polym. Sci. B *8*, 805 (1970), C.A. *74*, 64465r (1971); Bulk polymerization of cholesteryl acrylate.

[**41**] A. C. de Visser, K. de Groot, J. Feyen, A. Bantjes, J. Polym. Sci. A-1, *9*, 1893 (1971), C.A. *75*, 152135j (1971); Thermal bulk polymerization of cholesteryl acrylate. [**42**] A. C. de Visser, K. de Groot, F. Feyen, A. Bantjes, J. Polym. Sci. B *10*, 851 (1972), C.A. *78*, 44028e (1973); Liquid crystalline acrylates and their polymers. [**43**] A. C. de Visser, J. W. A. van den Berg, A. Bantjes, Makromol. Chem. *176*, 495 (1975), C.A. *82*, 171548g (1975); Bulk and solution polymerization of 5β-cholestan-3-yl methacrylate. [**44**] M. J. S. Dewar, J. P. Schroeder, J. Am. Chem. Soc. *86*, 5235 (1964), C.A. *62*, 1111g (1965); Use of nematics and smectics in GLC. [**45**] M. J. S. Dewar, J. P. Schroeder, J. Org. Chem. *30*, 3485 (1965), C.A. *63*, 14749a (1965); Liquid crystals as stationary phases in GLC. [**46**] M. J. S. Dewar, J. P. Schroeder, D. C. Schroeder, J. Org. Chem. *32*, 1692 (1967), C.A. *67*, 15376j (1967); Molecular order in the nematics of 4,4′-bi-*n*-hexyloxyazoxybenzene and its mixtures with PAA. [**47**] M. J. S. Dewar, B. D. Nahlovsky, J. Am. Chem. Soc. *96*, 460 (1974), C.A. *80*, 81773f (1974); Claisen rearrangement of cinnamyl phenyl ether in i and n solvents and in a clathrate. [**48**] J. Fendler, E. Fendler, Catalysis in micellar and macromolecular systems. (Academic: New York) 1975. 552 pp., C.A. *83*, 103988w (1975). [**49**] W. Fiddler, R. C. Doerr, J. Chromatogr. *21*, 481 (1966), C.A. *64*, 16598c (1966); Low temperature discontinuity of retention times with Carbowax 20M. [**50**] P. J. Flory, J. Chem. Phys. *9*, 660 (1941); *10*, 51 (1942), C.A. *35*, 6174^6 (1941); *36*, 1229^4 (1952); Thermodynamics of high-polymer solutions.

[**51**] P. J. Flory, J. Chem. Phys. *13*, 453 (1945), C.A. *40*, 512^6 (1946); Thermodynamics of dilute solutions of high polymers. [**52**] S. Frenkel, J. Polym. Sci., Polym. Symp. *44*, 49 (1974), C.A. *82*, 86751y (1975); Nature of polymeric liquid crystals. [**53**] Y. Frenkel, J. Chem. Phys. *7*, 538 (1939), C.A. *33*, 6674^5 (1939); General theory of heterophase fluctuations and pretransition phenomena. [**54**] S. I. Ahmad, S. Friberg, J. Am. Chem. Soc. *94*, 5196 (1972), C.A. *77*, 74690m (1972); Catalysis in micellar and liquid-crystalline phases. I. System water/hexadecyltrimethylammonium bromide/hexanol. [**55**] S. Friberg, S. I. Ahmad, Liq. Cryst. Ord. Fluids *2*, 515 (1974), C.A. *87*, 5006 (1977); Catalysis in micellar and liquid crystalline phases. [**55a**] R. K. Gabitova, M. S. Vigdergauz, Zh. Fiz. Khim. *51*, 1205 (1977), C.A. *87*, 73777q (1977); Effect of a solid support on the selectivity of a liquid-crystal stationary phase. [**56**] E. Grushka, J. F. Solsky, Anal. Chem. *45*, 1836 (1973), C.A. *79*, 97243y (1973); PAA liquid crystal as a stationary phase for capillary column GC. [**57**] E. Grushka, J. F. Solsky, J. Chromatogr. *99*, 135 (1974), C.A. *82*, 22150u (1975); Behavior of *p*-(*p*-ethoxyphenylazo)phenyl undecy-

lenate liquid crystal as a chromatographic liquid phase coated on glass beads. [**58**] E. Grushka, J. F. Solsky, J. Chromatogr. *112*, 145 (1975), C.A. *84*, 22530j (1976); Solid support surface effects on the chromatographic behavior of a liquid crystal stationary phase. [**59**] E. A. Guggenheim, Thermodynamics, an Advanced Treatment for Chemists and Physicists. 5th ed. (North-Holland: Amsterdam). 1967. 390 pp., C.A. *67*, 94615n (1967). [**60**] O. Smidsroed, J. E. Guillet, Macromolecules *2*, 272 (1969), C.A. *71*, 39564c (1969); Study of polymer-solute interactions by GC.

[**61**] A. Lavoie, J. E. Guillet, Macromolecules *2*, 443 (1969), C.A. *71*, 92110c (1969); Estimation of glass transition temperatures from GC studies on polymers. [**62**] J. E. Guillet, A. N. Stein, Macromolecules *3*, 102 (1970), C.A. *72*, 101184e (1970); Study of crystallinity in polymers by the use of "molecular probes". [**63**] J. E. Guillet, J. Macromol. Sci., Chem. *4*, 1669 (1970), C.A. *73*, 77626y (1970); Molecular probes in the study of polymer structure. [**64**] D. Patterson, Y. B. Tewari, H. P. Schreiber, J. E. Guillet, Macromolecules *4*, 356 (1971); Application of GLC to the thermodynamics of solution. [**65**] B. T. Guran, L. B. Rogers, J. Gas Chromatogr. *3*, 269 (1965), C.A. *64*, 1326d (1966); Influence of a crystal phase transition of thallium(I) nitrate on GC behavior of organic adsorbates. [**66**] B. T. Guran, L. B. Rogers, J. Gas Chromatogr. *5*, 574 (1967), C.A. *69*, 13173h (1968); Effects of thermochromic transitions of Cu_2HgI_4 and Ag_2HgI_4 on the adsorption behavior of C_8-hydrocarbons. [**67**] M. Hall, D. N. B. Mallen, J. Chromatogr. *118*, 268 (1976), C.A. *84*, 155757k (1976); GC separation of isomers of Benoxaprofen using liquid crystals. [**67a**] M. Hall, D. N. B. Mallen, J. Chromatogr. Sci. *14*, 451 (1976), C.A. *85*, 172555f (1976); Separation of benefin and trifluralin by GC using a liquid crystal column. [**68**] G. Hardy, N. Fedorova, G. Kovacs, J. Boros-Gyevi, J. Polym. Sci. C No. 16, 2675 (1967), C.A. *68*, 30149b (1968); Radiation polymerization behavior in the supercooled liquid and mesophase. [**69**] G. Hardy, K. Nyitrai, Magy. Kem. Foly. *76*, 182 (1970), C.A. *73*, 26015r (1970); Polymerization of cholesteryl acrylate and its two-component systems. [**70**] G. Hardy, K. Nyitrai, F. Cser, Kinet. Mech. Polyreactions, Int. Symp. Macromol. Chem., Prepr. *4*, 121 (1969), C.A. *75*, 64344w (1971); Polymerization of smectic cholesteryl acrylate.

[**71**] G. Hardy, F. Cser, A. Kallo, K. Nyitrai, G. Bodor, B. Lengyel, Magy. Kem. Foly. *76*, 176 (1970), C.A. *73*, 26013p (1970); Structural studies related to the polymerization of cholesteryl acrylate. [**72**] G. Hardy, F. Cser, N. Fedorova, M. Batky, Magy. Kem. Foly. *82*, 191 (1976), C.A. *85*, 47122m (1976); Polymerization in liquid crystals. I. Vinyl oleate. [**73**] K. Nyitrai, F. Cser, M. Lengyel, E. Seyfried, G. Hardy, Magy. Kem. Foly. *82*, 195 (1976), C.A. *85*, 47123n (1976); II. Mesophase copolymerization of *p*-methyl-*p'*-acryloyloxyazoxybenzene and cholesteryl vinyl succinate (cf. [**72**]). [**74**] F. Cser, K. Nyitrai, E. Seyfried, G. Hardy, Magy. Kem. Foly. *82*, 207 (1976), C.A. *85*, 63570m (1976); III. Structural studies on azoxyacrylate polymers (cf. [**72**]). [**75**] K. Nyitrai, F. Cser, Bui Duc Ngoc, G. Hardy, Magy. Kem. Foly. *82*, 210 (1976), C.A. *85*, 47125q (1976); IV. Polymerization studies on cholesteryl vinyl succinate (cf. [**72**]). [**76**] W. Helfrich, Z. Naturforsch. A *28*, 1968 (1973), C.A. *80*, 125145j (1974); Possible chromatographic effect of liquid-crystalline permeation. [**77**] J. H. Hildebrand, Solubility. American Chemical Society Monograph (Chem. Catalog Co.: New York). 1924. 206 pp., C.A. *18*, 1609 (1924). [**78**] J. H. Hildebrand, R. L. Scott, Regular Solutions. (Prentice-Hall: New York). 1962. 180 pp., C.A. *57*, 6694a (1962). [**79**] M. L. Huggins, J. Chem. Phys. *9*, 440 (1941), C.A. *35*, 3875[3] (1941); J. Phys. Chem. *46*, 151 (1942), C.A. *36*, 2197[8] (1942); Solutions of long-chain compounds. [**80**] M. L. Huggins, Ann. N. Y. Acad. Sci. *43*, 1 (1942), C.A. *36*, 3725[3] (1942); Thermodynamic properties of solutions of long-chain compounds.

[**81**] J. Illy, Magy. Kem. Lapja *29*, 216 (1974), C.A. *81*, 176531k (1974); Liquid crystals in GC. [**81a**] B. S. Iyengar, Diss. Univ. Lowell, Lowell, Mass. 1976, 134 pp. Avail. Univ. Microfilms Int. Order No. 77-13,360, C.A. *87*, 101842c (1977); Influence of anisotropic solvents on the stereochemistry of model biomimetic reactions. [**82**] G. M. Janini, K. Johnston, W. L. Zielinski, Jr., Anal. Chem. *47*, 670 (1975), C.A. *82*, 132598d (1975); Use of a nematic for GLC separation of polyaromatic hydrocarbons. [**83**] G. M. Janini, G. M. Muschik, W. L. Zielinski, Jr., Anal. Chem. *48*, 809 (1976), C.A. *84*, 189077d (1976); N,N'-Bis[*p*-butoxybenzylidene]-αα'-bi-*p*-toluidine: thermally stable liquid crystal for unique GLC separations of polycyclic aromatic hydrocarbons. [**83a**] G. M. Janini, G. M. Muschik, J. A. Schroer, W. L. Zielinski, Jr., Anal. Chem. *48*, 1879 (1976), C.A. *86*, 84285h (1977); GLC evaluation and GC/MS application of new high-temperature liquid crystal stationary phases for polycyclic aromatic hydrocarbon separations. [**84**] G. M. Janini, B. Shaikh, W. L. Zielinski, Jr., J. Chromatogr. *132*, 136 (1977); GLC analysis of benzo(a)pyrene in cigarette smoke on a nematic. [**85**] A. A. Jeknavorian, E. F. Barry, J. Chromatogr. *101*, 299 (1974), C.A. *82*, 103610w (1975); Thermodynamic study of a liquid crystal as a liquid phase in GLC. I. Nematic liquid phase. [**86**] A. A. Jeknavorian, P. Barret, A. C. Watterson,

E. F. Barry, J. Chromatogr. *107*, 317 (1975), C.A. *83*, 36044v (1975); II. Cholesteric liquid crystal (cf. [85]). [**87**] M. Jernejcie, L. Premru, J. Chromatogr. *28*, 409 (1967), C.A. *67*, 67911z (1967); Behavior of 7,8-benzoquinoline as the stationary phase in GLC in the vicinity of the melting point. [**88**] H. Kelker, Ber. Bunsenges. Physik. Chem. *67*, 698 (1963), C.A. *59*, 12144d (1963); Optisch anisotrope Schmelzen als stationäre Phase in der GLC. [**89**] H. Kelker, Z. Anal. Chem. *198*, 254 (1963), C.A. *60*, 3550d (1964); Flüssigkristalle als stationäre Phasen in der GLC. [**90**] H. Kelker, Abh. Deut. Akad. Wiss. Berlin, Kl. Chem., Geol. Biol. *1966*, 49, C.A. *67*, 76516b (1967); Vergleich der Löseeigenschaften von i, n und c Schmelzen.

[**91**] H. Kelker, H. Winterscheidt, Z. Anal. Chem. *220*, 1 (1966), C.A.*65*, 11401d (1966); Cholesteric mesophase and its mixtures with PAP as solvents in GLC. [**92**] H. Kelker, B. Scheurle, H. Winderscheidt, Anal. Chim. Acta *38*, 17 (1967), C.A. *67*, 29107j (1967); Flüssigkristalle bei Temperaturen unterhalb 100°C und oberhalb 200°C als stationäre Phasen in der GC. [**93**] H. Kelker, A. Verhelst, J. Chromatogr. Sci. *7*, 79 (1969), C.A. *70*, 91335x (1969); Pretransformation behavior of nematics in relation to retention volumes. [**94**] H. Kelker, B. Scheurle, J. Sabel, J. Jainz, H. Winterscheidt, Mol. Cryst. Liq. Cryst. *12*, 113 (1971), C.A. *75*, 29740a (1971); Nematische Phasen in der GC. [**95**] H. Kelker, J. Chromatogr. *112*, 165 (1975), C.A. *83*, 209889e (1975); Partition coefficients at the clearing point. [**96**] D. N. Kirk, P. M. Shaw, J. Chem. Soc. C *1971*, 3979, C.A. *76*, 34464p (1972); GC of steroids. Steroid derivatives as stationary phases. [**97**] T. J. Klingen, J. R. Wright, Mol. Cryst. Liq. Cryst. *13*, 173 (1971), C.A. *75*, 49620q (1971); Radiolytically induced polymer formation in alkenyl carboranes: phase effects. [**98**] N. Kostev, D. Shopov, Dokl. Bolg. Akad. Nauk *21*, 889 (1968), C.A. *70*, 32083r (1969); Determination of stability constants of an olefin/liquid crystal system by means of GLC. [**99**] G. Kraus, K. Seifert, H. Zaschke, H. Schubert, Z. Chem. *11*, 22 (1971), C.A. *75*, 44658y (1971); GC-Isomerentrennung an Flüssigkristallen. [**100**] G. Kraus, K. Seifert, H. Schubert, Z. Chem. *11*, 428 (1971), C.A. *76*, 63625x (1972); Flüssigkristalle als stationäre Phasen in der Kapillar-GC.

[**101**] G. Kraus, K. Seifert, H. Schubert, J. Chromatogr. *100*, 101 (1974), C.A. *82*, 139008v (1975); Löseverhalten kristallin-flüssiger Phasen in der Kapillar-GC. [**102**] J. W. Park, M. M. Labes, Mol. Cryst. Liq. Cryst. *34*, 25 (1976), C.A. *85*, 158984p (1976); Reaction kinetics determined by pitch changes in liquid crystals. [**103**] S. H. Langer, J. H. Purnell, J. Phys. Chem. *67*, 263 (1963), C.A. *58*, 5058a (1963); GLC study of the thermodynamics of solution of aromatic compounds. [**104**] D. Lecoin, A. Hochapfel, R. Viovy, J. Chim. Phys. Phys.-Chim. Biol. *72*, 1029 (1975), C.A. *84*, 44743h (1976); Comportement de la phase n dans des réactions de polymérisation. [**105**] J. O. Lephardt, B. J. Bulkin, Anal. Chem. *45*, 706 (1973), C.A. *78*, 143536a (1973); On-the-fly GC-IR spectrometry using a cholesteric-effluent interface. [**106**] L. Liebert, L. Strzelecki, C. R. Acad. Sci., Sér. C, *276*, 647 (1973), C.A. *79*, 19193t (1973); Copolymérisation des monomères mésomorphes dans la phase n en présence d'un champ magnétique. [**107**] L. Strzelecki, L. Liebert, Bull. Soc. Chim. Fr., Pt. 2, *1973*, 597, C.A. *78*, 136738q (1973); Synthése et polymérisation de monomères mésomorphes. [**108**] L. Liebert, L. Strzelecki, Bull. Soc. Chim. Fr., Pt. 2, *1973*, 603, C.A. *78*, 136739r (1973); Nouveau type de polymères à structure s, n et c. [**109**] L. Liebert, L. Strzelecki, D. Vacogne, Bull. Soc. Chim. Fr., Pt. 2, *1975*, 2073, C.A. *85*, 6101k (1976); Polymérisation de monomères mésomorphes initiée par rayonnement UV. [**110**] S. A. Liebmann, D. H. Ahlstrom, C. R. Foltz, J. Chromatogr. *67*, 153 (1972), C.A. *77*, 48831p (1972); Inverse GLC for detection of order-disorder in poly(vinyl chloride).

[**111**] G. C. Lin, Diss. Georgetown Univ., Washington, D.C. 1975. 162 pp. Avail. Xerox Univ. Microfilms, Ann Arbor, Mich., Order No. 76-11,661, C.A. *85*, 54872j (1976); Thermotropic liquid crystals: a cell model for the s_A/n transition and a thermodynamic investigation of dilute solutions in nematic and isotropic MBBA. [**112**] C. H. Lochmüller, R. W. Souter, J. Chromatogr. *87*, 243 (1973), C.A. *80*, 60164t (1974); Direct GC resolution of enantiomers on optically active mesophases. I. Smectic carbonylbis(D-leucine isopropyl ester). [**113**] C. H. Lochmüller, R. W. Souter, J. Chromatogr. *88*, 41 (1974), C.A. *80*, 60174w (1974); II. Effects of stationary phase structure on selectivity (cf. [112]). [**114**] C. H. Lochmüller, R. W. Souter, J. Chromatogr. *113*, 283 (1975), C.A. *84*, 98907h (1976); Chromatographic resolution of enantiomers. Selective review. [**115**] H. J. Lorkowski, F. Reuther, Plaste Kautsch. *23*, 81 (1976), C.A. *84*, 136089c (1976); Anisotrope Polymerisation. Methode zur Herstellung einheitlich orientierter Polymerer. [**116**] J. Herz, F. Husson, V. Luzzati, Compt. rend. *252*, 3462 (1961), C.A. *55*, 26606h (1961); Polymérisation et réticulation d'un monomère-savon, operées en phase mésomorphe. [**117**] J. Herz, F. Reiss Husson, P. Rempp, V. Luzzati, J. Polymer. Sci. Pt. C, *4*, 1275 (1964), C.A. *60*, 8141d (1964); Exemples de polymérisation en phase mésomorphe. [**118**] P. F. McCrea, J. H. Purnell, Nature *219*, 261 (1968), C.A. *69*, 61762w (1968); Temperature-independent retention in GC. [**119**]

G. G. Maidachenko, R. V. Vigalok, M. S. Vigdergauz, U.S.S.R. 455,276 (24. 7. 1972), C.A. *83*, 53009e (1975); Quantitative analysis of aromatic hydrocarbon mixtures. [**120**] D. E. Martire, P. A. Blasco, P. F. Carone, L. C. Chow, H. Vicini, J. Phys. Chem. *72*, 3489 (1968), C.A. *69*, 110575y (1968); Thermodynamics of solutions with liquid-crystal solvents. I. GLC study of cholesteryl myristate.

[**121**] D. E. Martire, L. Z. Pollara, Advan. Chromatogr. *1*, 335 (1965), C.A. *65*, 17727e (1966); Interactions of the solute with the liquid phase in GLC. [**122**] L. C. Chow, D. E. Martire, J. Phys. Chem. *73*, 1127 (1969), C.A. *71*, 7130s (1969); II. Surface effects with nematics (cf. [120]). [**123**] W. L. Zielinski, Jr., D. H. Freeman, D. E. Martire, L. C. Chow, Anal. Chem. *42*, 176 (1970), C.A. *72*, 66502h (1970); GC and thermodynamics of divinylbenzene separations on 4,4'-bis(hexyloxy)azoxybenzene liquid crystal. [**124**] L. C. Chow, D. E. Martire, Mol. Cryst. Liq. Cryst. *14*, 293 (1971), C.A. *75*, 122954h (1971); IV. GLC determination of the degree of order in a nematic (cf. [120]). [**125**] J. M. Schnur, D. E. Martire, Anal. Chem. *43*, 1201 (1971), C.A. *75*, 81036q (1971); V. Surface effects with cholesterics (cf. [120]). [**126**] D. E. Martire, L. C. Chow, J. Phys. Chem. *75*, 2005 (1971), C.A. *75*, 54174b (1971); III. Molecular interpretation of solubility in nematics (cf. [120]). [**127**] H. T. Peterson, D. E. Martire, W. Lindner, J. Phys. Chem. *76*, 596 (1972), C.A. *76*, 90871f (1972); Activity coefficient of *n*-heptane in 4,4'-dihexyloxyazoxybenzene liquid crystal. [**128**] H. T. Peterson, D. E. Martire, Mol. Cryst. Liq. Cryst. *25*, 89 (1974), C.A. *83*, 153493b (1975); VIII. Solute induced n/i transitions (cf. [120]). [**129**] J. M. Schnur, D. E. Martire, Mol. Cryst. Liq. Cryst. *26*, 213 (1974), C.A. *82*, 50022x (1975); VII. Mixture of cholesteryl chloride and cholesteryl myristate (cf. [120]). [**129a**] D. E. Martire, U. S. NTIS, AD Rep. *1977*, AD-A037105, 6 pp., C.A. *87*, 46623e (1977); Thermodynamics and statistical mechanics of liquid crystals and their solutions. [**130**] R. K. Mar'yakhin, V. A. Ezrets, R. V. Vigalok, M. S. Vigdergauz, Usp. Gaz Khomatogr. No. *3*, 43 (1973), C.A. *81*, 32798e (1974); Effect of various factors on the structural selectivity of azoxy ether liquid crystals.

[**131**] S. Minezaki, T. Nakaya, M. Imoto, Makromol. Chem. *175*, 3017 (1974), C.A. *82*, 43801x (1975); Synthesis and polymerization of cholesteryl 11-methacryloyloxyundecanoate. [**132**] P. Molina, A. Soler, J. Cambronero, An. Quim. *71*, 629 (1975), C.A. *83*, 178114s (1975); Behavior of (±)-γ-lactones in GLC over chiral phases. [**133**] J. V. Mortimer, P. L. Gent, Nature *197*, 789 (1963), C.A. *58*, 10706e (1963); Modified Bentone 34 for the GC separation of aromatic hydrocarbons. [**134**] K. S. Murthy, E. G. Rippie, J. Pharm. Sci. *59*, 459 (1970), C.A. *72*, 136352k (1970); Hydrolysis of procaine and its quaternary derivatives within lyophases. [**134a**] T. Nakaya, Kobunshi *26*, 709 (1977), C.A. *87*, 184977z (1977); Liquid crystalline state polymerization. [**135**] A. Ono, Nippon Kagaku Kaishi *1972*, 811, C.A. *77*, 5056x (1972); Separation behavior of alkylbenzenes studied by GLC with liquid crystals as stationary phases. [**136**] M. Pailer, V. Hlozek, J. Chromatogr. *128*, 163 (1976), C.A. *86*, 83273r (1977); Nematic liquid crystal for the GC separation of aza-heterocyclic compounds. [**137**] C. M. Paleos, M. M. Labes, Mol. Cryst. Liq. Cryst. *11*, 385 (1970), C.A. *74*, 64503b (1971); Polymerization of a nematic monomer. [**138**] C. M. Paleos, Diss. Drexel Inst. Technol., Philadelphia, Pa. 1970, 178 pp. Avail. Univ. Microfilms, Ann Arbor, Mich., Order No. 70-22,267, C.A. *75*, 75564u (1971); Reactions in the liquid crystalline phase. [**138a**] J. W. Park, Diss. Temple Univ., Philadelphia, Pa. 1977. 203 pp. Avail. Univ. Microfilms Int., Order No. 77-13,581, C.A. *87*, 93764m (1977); Reactions and interactions in liquid crystals. [**139**] W. Parr, P. Y. Howard, J. Chromatogr. *71*, 193 (1972), C.A. *77*, 152537h (1972); Molecular interaction in a unique solvent-solute system. [**140**] Dr. Petrowitz (BAM, Berlin), Privatmitteilung. [**140a**] W. H. Pirkle, P. L. Rinaldi, J. Am. Chem. Soc. *99*, 3510 (1977), C.A. *87*, 52637d (1977); Asymmetric synthesis in liquid crystals: independence of stereochemistry on the handedness of the cholesteric.

[**141**] P. J. Porcaro, P. Shubiak, J. Chromatogr. Sci. *9*, 690 (1971), C.A. *79*, 100253j (1973); Liquid crystals as substrates in the GLC of aroma chemicals. [**142**] I. Prigogine, Molecular Theory of Solutions. (M. Nijhoff: The Hague). 1957. 450 pp., C.A. *51*, 10218c (1957). [**143**] H. Purnell, Gas Chromatography. (Wiley: New York). 1962. 441 pp., C.A. *58*, 3891g (1963). [**144**] A. B. Richmond, J. Chromatogr. Sci. *9*, 571 (1971), C.A. *75*, 140378c (1971); Liquid crystals for the separation of position isomers of distributed benzenes. [**145**] E. Perplies, H. Ringsdorf, J. H. Wendorff, Makromol. Chem. *175*, 553 (1974), C.A. *81*, 106065n (1974); Polyreaktionen in orientierten Systemen. 3. Polymerisation ungesättigter flüssig-kristalliner Benzylidenaniline. [**146**] E. Perplies, H. Ringsdorf, J. H. Wendorff, Ber. Bunsenges. Phys. Chem. *78*, 921 (1974), C.A. *83*, 147846c (1975); 5. Flüssig-kristalline Polyacryl- und methacryl-Schiffsche Basen (cf. [145]). [**147**] E. Perplies, H. Ringsdorf, J. H. Wendorff, J. Polym. Sci., Polym. Lett. Ed. *13*, 243 (1975), C.A. *83*, 79673s (1975); 8. Polymerization of a liquid crystalline monomer under the influence of a high magnetic field (cf. [145]). [**147a**] E. Perplies, H. Ringsdorf,

J. H. Wendorff, Midl. Macromol. Monogr. *3*, 149 (1977), C.A. *87*, 185186c (1977); Monomeric and polymeric Schiff bases exhibiting mesomorphism. [**147b**] R. Ackermann, O. Inacker, H. Ringsdorf, Kolloid-Z. Z. Polym. *249*, 1118 (1971), C.A. *77*, 20181c (1972); 1. Polymerisation von Acryl- und Methacryl-verbindungen in monomolekularen Schichten (cf. [145]). [**147c**] I. Mielke, H. Ringsdorf, Makromol. Chem. *153*, 307 (1972), C.A. *77*, 5854f (1972); 2. Polymerisation von 4-Vinylpyridinium-Salzen in micellaren Lösungen. [**147d**] R. Ackermann, D. Naegele, H. Ringsdorf, Makromol. Chem. *175*, 699 (1974), C.A. *83*, 59338h (1975); 4. Fotoreaktionen von Fumar- und Maleinsäurederivaten in Multischichten (cf. [145]). [**147e**] V. Martin, H. Ringsdorf, H. Ritter, W. Sutter, Makromol. Chem. *176*, 2029 (1975), C.A. *83*, 131994p (1975); 6. Mizellare Assoziationen in 4-Vinylpyridiniumsalzlösungen (cf. [145]). [**147f**] B. Tieke, G. Wegner, D. Naegele, H. Ringsdorf, Angew. Chem. *88*, 805 (1976), C.A. *86*, 30105f (1977); Polymerisation von Tricosa-10,11-diin-1-säure in Multischichten. [**147g**] D. Naegele, H. Ringsdorf, J. Polym. Sci., Polym. Chem. Ed. *15*, 2821 (1977); Polymerization of octadecyl methacrylate in monolayers at the gas/water interface. [**147h**] H. Finkelmann, H. Ringsdorf, J. H. Wendorff, Makromol. Chem. *179*, 273 (1978); Model considerations and examples of enantiotropic liquid crystalline polymers. [**147i**] H. Finkelmann, H. Ringsdorf, W. Siol, J. H. Wendorff, Makromol. Chem. *179*, 829 (1978); Synthesis of cholesteric polymers. [**148**] E. G. Rippie, H. G. Ibrahim, Thermochim. Acta *11*, 125 (1975), C.A. *82*, 116980s (1975); Nematic-isotropic solution thermodynamics in bis(*p*-methoxyphenyl)-*trans*-cyclohexane-1,4-dicarboxylate. [**148a**] H. G. Ibrahim, E. G. Rippie, J. Pharm. Sci. *65*, 1639 (1976), C.A. *86*, 105358b (1977); Microenvironmental kinetic effects within lyotropic smectic biophase model: conformational restrictions in Fischer indole cyclization. [**149**] G. L. Roberts, S. J. Hawkes, J. Chromatogr. Sci. *11*, 16 (1973), C.A. *79*, 115950h (1973); Definition of the activity coefficient for polymeric liquid phases. Regression equations for retention data of homologous series. [**150**] A. E. Russell, D. R. Cooper, Biochemistry *10*, 3890 (1971), C.A. *76*, 26411d (1972); Comparison of lyotropic and chromatographic effects of polar organic solvents on collagen and cellulose.

[**151**] H. Saeki, K. Iimura, M. Takeda, Polym. J. *3*, 414 (1972), C.A. *77*, 127114k (1972); Polymerization of mesomorphic cholesteryl methacrylate. [**152**] F. D. Saeva, P. E. Sharpe, G. R. Olin, J. Am. Chem. Soc. *97*, 204 (1975), C.A. *82*, 97777n (1975); Asymmetric synthesis in a cholesteric solvent. [**152a**] K. Sanui, H. Imamura, N. Ogata, Kobunshi Ronbunshu *34*, 249 (1977), C.A. *86*, 171985e (1977); Radical polymerization of methacrylates in liquid crystalline solvents. [**153**] J. M. Schnur, Diss. Georgetown Univ., Washington, D.C. 1972, 162 pp. Avail. Univ. Microfilms. Ann Arbor, Mich., Order No. 72-16,039, C.A. *77*, 106184t (1972); Thermodynamic investigation of dilute solutions in two cholesterics. [**153a**] B. Schnuriger, J. Bourdon, J. Chim. Phys. Phys.-Chim. Biol. *73*, 795 (1976), C.A. *86*, 88850m (1977); Photoisomérisation dans les milieux cholésteriques. Influence sur les propriétés optiques du milieu. [**154**] J. P. Schroeder, D. C. Schroeder, M. Katsikas, Liq. Cryst. Ord. Fluids *1970*, 169; Nematic mixtures as stationary liquid phases in GLC. [**155**] M. A. Andrews, D. C. Schroeder, J. P. Schroeder, J. Chromatogr. *71*, 233 (1972), C.A. *77*, 151575g (1972); 4,4'-Di-*n*-alkoxyazoxybenzenes and their mixtures as stationary liquid phases in GLC. [**156**] J. P. Schröder, Liq. Cryst. Plast. Cryst. *1*, 356 (1974), C.A. *83*, 19087u (1975); Liquid crystals in GLC. [**157**] V. P. Shibaev, Y. S. Freidzon, N. A. Plate, U.S.S.R. 525,709 (10. 3. 1975), C.A. *85*, 193309u (1976); Polymers by radical polymerization of methacrylic compounds containing cholesterol in the side chain. [**157a**] V. P. Shibaev, V. M. Moisenko, N. Yu. Lukin, Dokl. Akad. Nauk *237*, 401 (1977); Phase transitions in liquid-crystalline comb-like polymers. [**157b**] V. P. Shibaev, N. A. Plate, Vysokomol. Soedin., Ser. A *19*, 923 (1977), C.A. *87*, 6384w (1977); Liquid-crystalline polymers. [**158**] G. Finaz, A. Skoulios, C. Sadron, Compt. rend. *253*, 265 (1961), C.A. *56*, 7485h (1962); Obtention de polymères organisés par polymérisation de solutions concentrées d'un copolymère séquencé styrolène/oxyde d'éthylène. [**159**] C. Sadron, G. Finaz, A. Skoulios, Fr. 1,295,524 (27. 4. 1961), C.A. *57*, 16876e (1962); Polymers from mesophases. [**160**] J. Herz, F. Husson, V. Luzzati, A. Skoulios, Fr. 1,295,525 (27. 4. 1961), C.A. *57*, 16876d (1962); Polymers with organized structure from soap-containing mesophases.

[**161**] C. Sadron, G. Finaz, A. Skoulios, Fr. Addn. 82,070 (18. 4. 1962), Addn. to Fr. 1,295,524, C.A. *60*, 13342a (1964); Polymers from mesophases. [**162**] J. F. Solsky, E. Grushka, J. Phys. Chem. *78*, 275 (1974), C.A. *80*, 87631e (1974); Vapor pressure measurements of PAA. [**163**] T. Svedberg, Kolloid-Z. *18*, 54, 101 (1916); *20*, 73 (1917); *21*, 19 (1917), C.A. *10*, 2429^2, 2826^8 (1916); *11*, 3145^7 (1917); *12*, 552^2 (1918); Chemische Reaktionen in anisotropen Flüssigkeiten. I–IV. [**164**] Y. Tanaka, S. Kabaya, Y. Shimura, A. Okada, Y. Kurihara, Y. Sakakibara, J. Polym. Sci. B *10*, 261 (1972), C.A. *77*, 20064s (1972); Bulk polymerization of mesomorphic cholesteryl methacrylate. [**165**] Y. Tanaka, Yuki Gosei Kagaku Kyokai Shi *34*, 2 (1976), C.A. *85*, 20184w (1976); Chemical reactions in the liquid

crystal state. [**165a**] Y. Tanaka, M. Hitotsuyanagi, Y. Shimura, A. Okada, H. Sakuraba, T. Sakata, Makromol. Chem. *177*, 3035 (1976), C.A. *86*, 30125n (1977); Bulk polymerization of mesomorphic 4-(2-vinyloxyethoxy)benzoic acid. [**166**] P. Tancrede, M. D. Croucher, D. Patterson, Polym. Prepr., Am. Chem. Soc., Div. Polym. Chem. *15*, 251 (1974), C.A. *84*, 180657k (1976); Effects of orientational order in solution thermodynamics. [**167**] P. J. Taylor, A. O. Ntukogu, S. S. Metcalf, L. B. Rogers, Separ. Sci. *8*, 245 (1973), C.A. *78*, 140718a (1973); Effect of an electric field on the GC properties of cholesterics as stationary phases. [**168**] S. I. Torgova, E. I. Kovshev, V. V. Titov, Zh. Org. Khim. *12*, 1593 (1976), C.A. *85*, 142385h (1976); Thermal isomerization of *cis*-stilbene to *trans*-stilbene in an anisotropic solvent. [**169**] W. J. Toth, A. V. Tobolsky, J. Polym. Sci. B *8*, 289 (1970), C.A. *73*, 35807c (1970); Polymerization of liquid crystal monomers. [**170**] W. J. Toth, Diss. Princeton Univ., Princeton, N. J. 1972, 197 pp. Avail. Univ. Microfilms, Ann Arbor, Mich., Order No. 72-24,706, C.A. *78*, 30314h (1973); Polymeric liquid crystals. [**170a**] T. Tsutsui, T. Tanaka, J. Polym. Sci., Polym. Lett. Ed. *15*, 475 (1977), C.A. *87*, 85304w (1977); Polymerization of N,N-dimethylacrylamide in c mesophase.

[**171**] L. Verbit, T. R. Halbert, R. B. Patterson, J. Org. Chem. *40*, 1649 (1975), C.A. *83*, 27392e (1975); Asymmetric decarboxylation of ethylphenylmalonic acid in a cholesteric solvent. [**172**] Z. P. Vetrova, D. A. Vyakhirev, N. T. Karabanov, G. G. Maidatsenko, Y. I. Yashin, Chromatographia *8*, 643 (1975), C.A. *84*, 38207v (1976); GC on monolayers of liquid crystals. [**172a**] Z. P. Vetrova, N. T. Karabanov, J. A. Jashin, Chromatographia *10*, 341 (1977), C.A. *87*, 107089j (1977); Sorbents for liquid chromatography based on thin films of liquid crystals supported on Silochrom. [**173**] M. S. Vigdergauz, R. V. Vigalok, Mater. Nauch. Konf., Inst. Org. Fiz. Khim., Akad. Nauk SSSR *1969*, 166, C.A. *78*, 131804y (1973); GC on solid organic crystals. [**174**] M. S. Vigdergauz, R. V. Vigalok, Neftekhimiya *11*, 141 (1971), C.A. *74*, 134721x (1971); Chromatographic analysis on columns with liquid-crystal fixed phases. [**175**] R. V. Vigalok, M. S. Vigdergauz, Ivz. Akad. Nauk SSSR, Ser. Khim. *1972*, 718, C.A. *77*, 93131w (1972); Chromatographic study of binary mixtures based on liquid crystals. [**176**] R. V. Vigalok, N. A. Palikhov, M. S. Vigdergauz, Kristallografiya *17*, 837 (1972), C.A. *78*, 21172h (1973); Chromatographic and thermooptical study of phase transitions in liquid crystals. [**177**] R. V. Vigalok, N. A. Palikhov, G. A. Seichasova, G. G. Maidachenko, M. S. Vigdergauz, Usp. Gazov. Khromatogr. *3*, 34 (1973), C.A. *83*, 48736k (1975); Chromatographic properties of liquid crystal *p,p'*-methoxyethoxyazoxybenzene. [**178**] R. V. Vigalok, G. G. Maidachenko, G. A. Seichasova, R. K. Nasybullina, T. R. Bankovskaya, N. A. Palikhov, V sb., Uspekhi Gaz. Khromatografii, Kazan *1975*, 115, C.A. *85*, 169944b (1976); Comparative study of chromatographic and thermooptical properties of liquid crystals. [**179**] R. V. Vigalok, R. K. Gabitova, N. P. Anoshina, N. A. Palikhov, G. G. Maidachenko, M. S. Vigdergauz, Zh. Anal. Khim. *31*, 644 (1976), C.A. *85*, 103489a (1976); Liquid-crystal adsorbents for GC of aromatic isomers. [**180**] S. Wasik, S. Chesler, J. Chromatogr. *122*, 451 (1976), C.A. *85*, 57594f (1976); Nematic liquid crystal for the GLC separation of naphthalene homologs.

[**181**] S. P. Wasik, J. Chromatogr. Sci. *14*, 516 (1976), C.A. *86*, 49296r (1977); Determination of transition temperatures on sodium stearate using GC. [**182**] R. B. Westerberg, F. J. van Lenten, L. B. Rogers, Sep. Sci. *10*, 593 (1975), C.A. *84*, 22519n (1976); Chromatographic behavior of a cholesteryl myristate stationary phase in an electric field. [**183**] D. G. Willey, Diss. Kent State Univ., Kent, Ohio. 1970. 111 pp. Avail. Univ. Microfilms, Ann Arbor, Mich., Order No. 71-10,879, C.A. *76*, 28461u (1972); Thermodynamics of dilute solutions using selected liquid crystalline solvents. [**183a**] Z. Witkiewicz, Wiad. Chem. *31*, 19 (1977), C.A. *86*, 150081s (1977); Liquid crystals as stationary phases in GC. [**184**] W. L. Zielinski, Jr., K. Johnston, G. M. Muschik, Anal. Chem. *48*, 907 (1976), C.A. *84*, 176071x (1976); Nematic liquid crystal for GLC separation of steroid epimers.

Chapter 10

[**1**] H. Akutsu, Y. Kyogoku, Tampakushitsu Kakusan Koso, Bessatsu *1974*, 93, C.A. *81*, 87277q (1974); Biomembrane studies by NMR. [**2**] J. P. Albrand, A. Cogne, J. B. Robert, Chem. Phys. Lett. *42*, 498 (1976), C.A. *85*, 184497b (1976); NMR spectral analysis of partially oriented hexaphenylcyclohexaphosphine. Relation between ^{1}J(PP) NMR coupling and stereochemistry. [**3**] J. P. Albrand, A. Cogne, D. Gagnaire, J. B. Robert, Mol. Phys. *31*, 1021 (1976), C.A. *85*, 122777j (1976); Spectres ^{31}P RMN

de deux cyclotétraphosphines en phase nématique. Géometrie du cycle, couplage [1]J(P—P), anisotropie du déplacement chimique du P. [**4**] J. P. Albrand, A. Cogne, J. B. Robert, Chem. Phys. Lett. *48*, 524 (1977), C.A. *87*, 76076w (1977); Structure, ^{31}P and ^{13}C chemical shift anisotropy of trimethylphosphine, trimethylphosphine oxide, trimethylphosphine sulfide and trimethylphosphine selenide as determined by liquid crystal NMR spectroscopy. [**5**] J. P. Albrand, A. Cogne, J. B. Robert, J. Am. Chem. Soc. *100*, 2600 (1978), C.A. *89*, 107044g (1978); Cyclotetraphosphanes: geometry, [1]J(PP) NMR coupling constants, and P chemical shift anisotropy. NMR study in liquid crystals. [**6**] S. A. Allison, Report *1975*, LBL-4506, 133 pp., C.A. *87*, 32227s (1977); [13]CMR studies of liquid crystals. [**7**] J. M. Anderson, A. C. F. Lee, J. Magn. Reson. *3*, 427 (1970), C.A. *74*, 59131z (1971); Calculation of exchange-broadened NMR lineshapes. I. Relaxation times and direct coupling. [**8**] J. M. Anderson, A. C. F. Lee, J. Magn. Reson. *4*, 160 (1971), C.A. *74*, 148844b (1971); NMR spectra of N,N-dimethylacetamide-d$_3$ in a lyotropic solvent. [**9**] J. M. Anderson, J. Magn. Reson. *4*, 231 (1971), C.A. *75*, 56420c (1971); Inertia and ordering in anisotropic solvents. Substituted benzenes. [**10**] A. C. Balazs, J. M. Anderson, J. Magn. Reson. *20*, 177 (1975), C.A. *84*, 113890t (1976); NMR relaxation in an anisotropic environment.

[**11**] V. A. Andreev, N. I. Lebovka, Y. A. Marazuev, G. Y. Shimanskaya, Zh. Eksp. Teor. Fiz. *72*, 1926 (1977), C.A. *87*, 46270n (1977); Angular dependence of the NMR spectra of cholesterics. [**12**] M. N. Avadhanlu, J. S. M. Sarma, C. R. K. Murty, Ber. Bunsenges. Phys. Chem. *77*, 275 (1973), C.A. *79*, 31071j (1973); [1]HMR in nematics. 4-[(4-Ethoxybenzylidene)amino]azobenzene, N-(4-acetoxybenzylidene)-*p*-phenetidine, and N-(4-butoxybenzylidene)-*p*-phenetidine. [**13**] M. N. Avadhanlu, C. R. K. Murty, Mol. Cryst. Liq. Cryst. *20*, 221 (1973), C.A. *79*, 85295d (1973); [1]HMR in nematics. Acetoxybenzal-*p*-anisidine, acetoxybenzal-*p*-aminoazobenzene, and anisal-*p*-aminoazobenzene. [**14**] M. N. Avadhanlu, N. V. S. Rao, T. F. S. Raj, A. S. N. Rao, C. R. K. Murty, Magn. Reson. Relat. Phenom., Proc. Congr. Ampère, 18th, *1*, 211 (1974), C.A. *83*, 170586b (1975); [1]HMR in nematic MBBA, EBBA, and BBPP. [**15**] F. S. Axel, Biophys. Struct. Mech. *2*, 181 (1976), C.A. *86*, 116899e (1977); Biophysics with nitroxyl radicals. [**16**] M. Zaucer, A. Azman, Z. Naturforsch. A *27*, 1534 (1972), C.A. *78*, 57179e (1973); NMR study of 3,5-dichloropyridine in a nematic. [**17**] M. Zaucer, A. Azman, J. Magn. Reson. *11*, 105 (1973), C.A. *79*, 98927m (1973); [1]HMR and [2]HMR spectra of oriented acetone and DMSO. [**18**] D. Pumpernik, A. Azman, J. Mol. Struct. *22*, 463 (1974), C.A. *81*, 168763x (1974); NMR spectra of oriented 2,2,2-trifluoroethanol. [**19**] D. Pumpernik, A. Azman, Magn. Reson. Relat. Phenom., Proc. Congr. Ampère 18th *1*, 207 (1974), C.A. *83*, 170584z (1975); Dipolar relaxation of molecules oriented in a nematic. [**20**] D. Pumpernik, A. Azman, Z. Naturforsch. A *29*, 527 (1974), C.A. *81*, 12674h (1974); NMR spectrum of oriented 2-chloroethanol.

[**21**] D. Pumpernik, A. Azman, Adv. Mol. Relaxation Interact. Processes *10*, 291 (1977), C.A. *87*, 125017p (1977); Longitudinal NMR relaxation of methyl iodide dissolved in the nematic phase Merck V. [**22**] J. P. Yesinowski, D. Bailey, J. Organometal. Chem. *65*, C27 (1974), C.A. *80*, 70182m (1974); Structure of H$_3$Os$_3$(CO)$_9$CCH$_3$ by nematic-phase [1]HMR and X-ray powder photography. [**23**] D. Bailey, J. P. Yesinowski, J. Chem. Soc., Dalton Trans. *1975*, 498, C.A. *83*, 35244y (1975); Fluxional motion in bis(ethane-1,2-dithiolato)nickel(IV) from nematic-phase [1]HMR. [**24**] B. L. Bales, J. A. Swenson, R. N. Schwartz, Mol. Cryst. Liq. Cryst. *28*, 143 (1974), C.A. *82*, 131691s (1975); ESR studies of Heisenberg spin exchange in a nematic. [**25**] F. Barbarin, J. P. Germain, C. Vialle, C. R. Acad. Sci., Sér. B *278*, 595 (1974), C.A. *81*, 17731j (1974); Paramètre d'ordre de sondes nitroxydes dissoutes dans nématiques. [**26**] F. Barbarin, M. Gauriat, J. P. Germain, Magn. Reson. Relat. Phenom., Proc. Congr. Ampère, 19th *1976*, 205, C.A. *86*, 197386k (1977); ESR relaxation of a probe dissolved in nematic EBBA. [**27**] F. Barbarin, J. P. Germain, C. Fabre, Mol. Cryst. Liq. Cryst. *39*, 217 (1977), C.A. *87*, 159615x (1977); Étude RSE de la diffusion rotationelle et du paramètre d'ordre de sondes nitroxydes dissoutes dans le MBBA. [**28**] M. Barfield, Chem. Phys. Lett. *4*, 518 (1970), C.A. *72*, 105710w (1970); Anisotropy of the indirect nuclear spin-spin coupling constant. [**29**] E. M. Barrall, II, J. C. Guffy, Advan. Chem. Ser. No. *63*, 1 (1967), C.A. *67*, 4008q (1967); Polymorphism of tristearin. [**30**] J. G. Batchelor, Diss. Yale Univ., New Haven. 1973. 261 pp. Avail. Univ. Microfilms, Ann Arbor, Mich., Order No. 74-12,481., C.A. *81*, 101512w (1974); Application of [13]CMR chemical shift measurements to the conformational analysis of fatty acyl chains in lipid membranes.

[**31**] J. J. de Vries, H. J. C. Berendsen, Nature *221*, 1139 (1969), C.A. *70*, 110362x (1969); NMR measurements on a macroscopically ordered smectic. [**32**] E. T. Samulski, H. J. C. Berendsen, J. Chem. Phys. *56*, 3920 (1972), C.A. *76*, 133873c (1972); [1]HMR, [2]HMR, and [14]NMR of dimethylformamide in nematic polypeptide liquid crystals. [**33**] M. A. Hemminga, H. J. C. Berendsen, J. Magn. Reson. *8*, 133 (1972), C.A. *77*, 132975w (1972); NMR in ordered lecithin-cholesterol multilayers. [**34**] C. Dijkema,

H. J. C. Berendsen, J. Magn. Reson. *14*, 251 (1974), C.A. *81*, 90687d (1974); ¹HMR in membrane model systems. Octylammonium chloride and potassium oleate in the oriented lamellar phase. [**35**] L. D. Bergel'son, L. I. Barsukov, N. I. Dubrovina, V. F. Bystrov, Dokl. Akad. Nauk SSSR *194*, 708 (1970), C.A. *74*, 9716y (1971); Differentiation of the interior and exterior surfaces of phospholipid membranes by NMR spectroscopy. [**36**] V. F. Bystrov, N. I. Dubrovina, L. I. Barsukov, L. D. Bergel'son, Chem. Phys. Lipids *6*, 343 (1971), C.A. *76*, 22203q (1972); NMR differentiation of the internal and external phospholipid membrane surfaces using paramagnetic Mn²⁺ and Eu³⁺ ions. [**37**] L. I. Barsukov, Y. E. Shapiro, A. V. Viktorov, V. I. Volkova, V. F. Bystrov, L. D. Bergel'son, Biochem. Biophys. Res. Commun. *60*, 196 (1974), C.A. *81*, 164920y (1974); Study of intervesicular phospholipid exchange by NMR. [**38**] C. A. Berglund, Diss. Univ. Illinois, Urbana. 1975. 110 pp. Avail. Xerox Univ. Microfilms, Ann Arbor, Mich., Order No. 75-24,259, C.A. *84*, 37463p (1976); NMR relaxation in liquid crystals and in alkali hexafluoroniobates and hexafluorotantalates. [**39**] R. A. Bernheim, B. J. Lavery, J. Am. Chem. Soc. *89*, 1279 (1967), C.A. *66*, 89963z (1967); Absolute signs of indirect nuclear spin-spin coupling constants. [**40**] R. A. Bernheim, T. R. Krugh, J. Am. Chem. Soc. *89*, 6784 (1967), C.A. *68*, 44599k (1968); Chemical shift anisotropies from NMR studies in liquid crystals.

[**41**] R. A. Bernheim, B. J. Lavery, J. Colloid Interface Sci. *26*, 291 (1968), C.A. *68*, 110075q (1968); NMR of molecules oriented by liquid crystals. [**42**] T. R. Krugh, R. A. Bernheim, J. Am. Chem. Soc. *91*, 2385 (1969), C.A. *71*, 17405z (1969); Anisotropic nuclear spin-spin coupling in methyl fluoride. [**43**] R. A. Bernheim, D. J. Hoy, T. R. Krugh, B. J. Lavery, J. Chem. Phys. *50*, 1350 (1969), C.A. *70*, 72690t (1969); ¹HMR and ¹⁹FMR of partially oriented fluoromethanes. [**44**] T. R. Krugh, R. A. Bernheim, J. Chem. Phys. *52*, 4942 (1970), C.A. *73*, 20202c (1970); Anisotropies and absolute signs of the indirect spin-spin coupling constants in ¹³CH₃F. [**45**] I. G. Bikchantaev, I. V. Ovchinnikov, Izv. Akad. Nauk SSSR, Ser. Khim. *1975*, 1178, C.A. *83*, 139459p (1975); ESR study of the orientation of copper and vanadyl acetylacetonate and diethyldithiocarbamate complexes in liquid crystal matrixes. [**46**] I. G. Bikchantaev, I. V. Ovchinnikov, Fiz. Tverd. Tela *18*, 1479 (1976), C.A. *85*, 101788y (1976); Orientation of chromium(III) acetylacetonate in a vitrified liquid crystal. [**47**] I. V. Ovchinnikov, I. G. Bikchantaev, N. E. Domracheva, Fiz. Tverd. Tela *18*, 3573 (1976), C.A. *86*, 81312d (1977); Features of ESR spectra in a partially ordered matrix. [**48**] I. G. Bikchantaev, V. N. Konstantinov, I. V. Ovchinnikov, Zh. Strukt. Khim. *17*, 821 (1976), C.A. *86*, 98553f (1977); Angular dependence of ESR spectra of copper complexes in an oriented vitrified liquid crystal. [**49**] N. J. M. Birdsall, A. G. Lee, Y. K. Levine, J. C. Metcalfe, Biochim. Biophys. Acta *241*, 693 (1971), C.A. *75*, 137021a (1971); ¹⁹FMR of monofluoro-stearic acids in lecithin vesicles. [**50**] J. C. Metcalfe, N. J. M. Birdsall, J. Feeney, A. G. Lee, Y. K. Levine, P. Partington, Nature *233*, 199 (1971), C.A. *76*, 1026u (1972); ¹³CMR spectra of lecithin vesicles and erythrocyte membranes.

[**51**] J. D. Robinson, N. J. M. Birdsall, A. G. Lee, J. C. Metcalfe, Biochemistry *11*, 2903 (1972), C.A. *77*, 71708n (1972); ¹³CMR and ¹HMR relaxation measurements of the liquids of sarcoplasmic reticulum membranes. [**52**] J. C. Metcalfe, N. J. M. Birdsall, A. G. Lee, FEBS Lett. *21*, 335 (1972), C.A. *77*, 2201e (1972); ¹³CMR spectra of Acholeplasma membranes containing ¹³C-labeled phospholipids. [**53**] Y. K. Levine, P. Partington, G. C. K. Roberts, N. J. M. Birdsall, A. G. Lee, J. C. Metcalfe, FEBS Lett. *23*, 203 (1972), C.A. *77*, 84765d (1972); ¹³CMR relaxation times and models for chain motion in lecithin vesicles. [**54**] N. J. M. Birdsall, J. Feeney, A. G. Lee, Y. K. Levine, J. C. Metcalfe, J. Chem. Soc., Perkin Trans. 2, *1972*, 1441, C.A. *77*, 87644z (1972); Dipalmitoyllecithin. Assignment of the ¹HMR and ¹³CMR spectra, and conformational studies. [**55**] J. C. Metcalfe, N. J. M. Birdsall, A. G. Lee, Ann. N. Y. Acad. Sci. *222*, 460 (1973), C.A. *81*, 680u (1974); NMR studies of lipids in bilayers and membranes. [**56**] A. G. Lee, N. J. M. Birdsall, J. C. Metcalfe, Biochemistry *12*, 1650 (1973), C.A. *78*, 144535m (1973); Measurement of fast lateral diffusion of lipids in vesicles and in biological membranes by ¹HMR. [**57**] Y. K. Levine, A. G. Lee, N. J. M. Birdsall, J. C. Metcalfe, J. D. Robinson, Biochim. Biophys. Acta *291*, 592 (1973), C.A. *78*, 107307t (1973); Interaction of paramagnetic ions and spin labels with lecithin bilayers. [**58**] A. G. Lee, N. J. M. Birdsall, J. C. Metcalfe, Chem. Brit. *9*, 116 (1973), C.A. *79*, 1388a (1973); NMR studies of biological membranes. [**59**] R. Blinc, D. E. O'Reilly, E. M. Peterson, G. Lahajnar, I. Levstek, Solid State Commun. *6*, 839 (1968), C.A. *70*, 82822b (1969); ¹H spin-lattice relaxation study of the liquid crystal transition in *p*-anisaldehyde azine. [**60**] R. Blinc, D. L. Hogenboom, D. E. O'Reilly, E. M. Peterson, Phys. Rev. Lett. *23*, 969 (1969), C.A. *72*, 6396s (1970); Spin relaxation and self-diffusion in liquid crystals.

[**61**] R. Blinc, K. Easwaran, J. Pirs, M. Vilfan, I. Zupancic, Phys. Rev. Lett. *25*, 1327 (1970), C.A. *74*, 17863j (1971); Self-diffusion and molecular order in lyophases. [**62**] R. Blinc, Inst. Jozef

Stefan, IJS-Rep., R-595, 42 pp. (1971), C.A. *76*, 65574d (1972); NMR and NQR studies of critical effects in structural phase transitions. [**63**] R. Blinc, V. Dimic, J. Pirs, M. Vilfan, I. Zupancic, Mol. Cryst. Liq. Cryst. *14*, 97 (1971), C.A. *75*, 134052a (1971); Self-diffusion in liquid crystals. [**64**] M. Vilfan, R. Blinc, J. W. Doane, Solid State Commun. *11*, 1073 (1972), C.A. *77*, 170949y (1972); Mechanisms for spin-lattice relaxation in nematics. [**65**] R. Blinc, J. Pirs, I. Zupancic, Phys. Rev. Lett. *30*, 546 (1973), C.A. *78*, 129315w (1973); Measurement of self-diffusion in liquid crystals by a multiple-pulse NMR method. [**66**] R. Blinc, Proc. Int. Sch. Phys. "Enrico Fermi" *59*, 165 (1973), C.A. *86*, 63570g (1977); Double resonance of rare nuclei. [**67**] R. Blinc, Proc. Int. Sch. Phys. "Enrico Fermi" *59*, 693 (1973), C.A. *86*, 63872g (1977); Order parameter fluctuations, spin relaxation and self-diffusion in liquid crystals. [**68**] I. Zupancic, J. Pirs, M. Luzar, R. Blinc, Pulsed Nucl. Magn. Reson. Spin Dyn. Solids, Proc. Spec. Colloq. Ampère, 1st, *1973*, 148, C.A. *81*, 68883t (1974); Measurement of self-diffusion in nematic MBBA by a multiple-pulse NMR method. [**69**] I. Zupancic, M. Vilfan, M. Sentjurc, M. Schara, F. Pusnik, J. Pirs, R. Blinc, Liq. Cryst. Ord. Fluids *2*, 525 (1974), C.A. *86*, 180995k (1977); Liquid crystal dynamics as studied by NMR and ESR. [**70**] R. Blinc, M. Burger, M. Luzar, J. Pirs, I. Zupancic, S. Zumer, Phys. Rev. Lett. *33*, 1192 (1974), C.A. *82*, 24602s (1975); Anisotropy of self-diffusion in smectics A and C.

[**71**] I. Zupancic, J. Pirs, M. Luzar, R. Blinc, J. W. Doane, Solid State Commun. *15*, 227 (1974), C.A. *81*, 96969u (1974); Anisotropy of the self-diffusion tensor in nematic MBBA. [**72**] R. Blinc, M. Luzar, M. Vilfan, M. Burgar, J. Chem. Phys. *63*, 3445 (1975), C.A. *83*, 199941z (1975); ^{1}H spin-lattice relaxation in smectic TBBA. [**73**] R. Blinc, M. Vilfan, V. Rutar, Solid State Commun. *17*, 171 (1975), C.A. *83*, 105913s (1975); Nature of spin-lattice relaxation in nematic MBBA. [**74**] R. Blinc, M. Luzar, M. Mali, R. Osredkar, J. Seliger, M. Vilfan, J. Phys. (Paris), C-3, *37*, 73 (1976), C.A. *85*, 85157j (1976); Frequency dispersion of the proton Zeeman spin-lattice relaxation in smectic TBBA. [**75**] R. Blinc, NMR: Basic Princ. Prog. *13*, 97 (1976), C.A. *86*, 62798a (1977); Spin-lattice relaxation in nematics via the modulation of the intramolecular dipolar interactions by order fluctuations. [**76**] I. Zupancic, V. Zagar, M. Rozmarin, I. Levstik, F. Kogovsek, R. Blinc, Solid State Commun. *18*, 1591 (1976), C.A. *85*, 54270m (1976); Angular dependence of the ^{1}H spin-spin and spin-lattice relaxation in nematic MBBA-EBBA mixtures. [**77**] R. E. Block, G. P. Maxwell, G. L. Irvin, J. L. Hudson, D. L. Prudhomme, Miami Winter Symp. *8*, 253 (1974), C.A. *81*, 149866s (1974); NMR studies of membranes of normal and cancer cells. [**78**] M. Bloom, Pulsed Nucl. Magn. Reson. Spin Dyn. Solids, Proc. Spec. Colloq. Ampère, 1st, *1973*, 80, C.A. *81*, 65527u (1974); Nuclear spin dynamics in soap solutions and related systems. [**79**] N. Boden, Chemical Society, Specialist Periodical Reports, Nucl. Magn. Reson. *1*, 115 (1972), C.A. *78*, 64315z (1973); Nuclear spin relaxation. [**80**] N. Boden, Chemical Society, Specialist Periodical Reports, Nucl. Magn. Reson. *2*, 112 (1973), C.A. *79*, 98658z (1973); Nuclear spin relaxation.

[**81**] N. Boden, Y. K. Levine, D. Lightowlers, R. T. Squires, Chem. Phys. Lett. *31*, 511 (1975), C.A. *82*, 177811j (1975); NMR dipolar echoes in liquid crystals. [**82**] N. Boden, Y. K. Levine, D. Lightowlers, R. T. Squires, Chem. Phys. Lett. *34*, 63 (1975), C.A. *83*, 140009y (1975); Internal molecular disorder in thermotropic nematics and smectics studied by NMR SPDE experiments. [**83**] N. Boden, P. Jackson, Y. K. Levine, A. J. I. Ward, Biochim. Biophys. Acta *419*, 395 (1976), C.A. *84*, 70666e (1976); Intramolecular disorder and its relation to mesophase structure in lipid/water mixtures. [**84**] N. Boden, P. Jackson, Y. K. Levine, A. J. I. Ward, Chem. Phys. Lett. *37*, 100 (1976), C.A. *84*, 98015r (1976); Intramolecular disorder and its relation to mesophase structures in lyophases. [**85**] J. M. Boggs, J. C. Hsia, Can. J. Biochem. *51*, 1451 (1973), C.A. *80*, 44979g (1974); Structural characteristics of hydrated glycerol- and sphingolipids. Spin label study. [**86**] T. T. Bopp, N. S. Balakrishnan, Mol. Cryst. Liq. Cryst. *41*, 47 (1977), C.A. *87*, 175376a (1977); ^{1}HMR spectrum of N,N-dimethylformamide in a nematic solvent. [**87**] P. M. Borodin, Y. A. Ignat'ev, Vestn. Leningrad. Univ., Fiz., Khim. *46*, 157 (1972), C.A. *77*, 54521d (1972); ^{1}HMR spectra of dichloromethane and 1,4-dibromo-2-butyne molecules in a nematic solvent. [**88**] P. M. Borodin, Y. A. Ignat'ev, Yad. Magn. Rezon. *5*, 3 (1974), C.A. *83*, 105473y (1975); NMR spectroscopy of molecules dissolved in liquid crystals. [**89**] Y. A. Ignat'ev, B. V. Semakov, P. M. Borodin, Zh. Strukt. Khim. *15*, 210 (1974), C.A. *80*, 144996d (1974); Structure and properties of acetaldehyde molecules in a nematic solvent according to high-resolution NMR spectra. [**90**] Y. A. Ignat'ev, B. V. Semakov, P. M. Borodin, Teor. Eksp. Khim. *11*, 103 (1975), C.A. *82*, 138746x (1975); Acetone in a nematic solvent studied by high-resolution NMR spectra.

[**91**] P. M. Borodin, Y. V. Molchanov, I. P. Kolomiets, Kristallografiya *22*, 658 (1977), C.A. *87*, 108996q (1977); Premelting of substances possessing liquid-crystalline states. [**92**] P. R. Bossard,

V. J. Cicerone, J. S. Karra, J. Appl. Phys. *47*, 988 (1976); Free-induction NMR line-shapes of various liquid-sample configurations and diffusion constant of a liquid crystal by spin echo. [**93**] L. F. Williams, A. A. Bothner-By, J. Magn. Reson. *11*, 314 (1973), C.A. *79*, 125407a (1973); ¹HMR spectrum of enriched propene-2-¹³C oriented in a nematic. [**94**] V. Breternitz, H. Kresse, Phys. Lett. A *54*, 148 (1975), C.A. *83*, 200459g (1975); Anisotropy of diffusion in nematic alkyloxybenzoic acids. [**95**] E. Drauglis, K. C. Brog, N. F. Hartmann, W. H. Jones, Jr., C. E. Moeller, US Nat. Tech. Inform. Serv., AD Rep. No. 734151, 25 pp. (1971), C.A. *76*, 117812h (1972); Thin film rheology of boundary lubricating surface films. 2. [**96**] J. Brotherus, P. Tormala, Colloid Polym. Sci. *252*, 543 (1974), C.A. *82*, 77389j (1975); Motion of a spin probe in hexagonal and lamellar potassium palmitate. [**97**] A. D. Buckingham, J. A. Pople, Trans. Faraday Soc. *59*, 2421 (1963), C.A. *60*, 2466a (1964); High-resolution NMR spectra in electric fields. [**98**] A. D. Buckingham, E. E. Burnell, J. Am. Chem. Soc. *89*, 3341 (1967), C.A. *67*, 59260w (1967); Chemical shift anisotropies from NMR studies of oriented molecules. [**99**] A. D. Buckingham, E. E. Burnell, C. A. de Lange, J. Chem. Soc., Chem. Commun. *1968*, 1408, C.A. *70*, 24522t (1969); NMR spectra of hydrogen in a nematic. [**100**] A. D. Buckingham, E. E. Burnell, C. A. de Lange, J. Am. Chem. Soc. *90*, 2972 (1968), C.A. *69*, 14677n (1968); Determination of NMR shielding anisotropies of solutes in liquid-crystal solvents.

[**101**] A. D. Buckingham, E. E. Burnell, C. A. de Lange, Mol. Phys. *14*, 105 (1968), C.A. *69*, 31861k (1968); NMR studies of 3,3,3-trifluoropropyne dissolved in nematics. [**102**] A. D. Buckingham, E. E. Burnell, C. A. de Lange, Mol. Phys. *15*, 285 (1968), C.A. *70*, 15798k (1969); NMR studies of dimethylacetylene and perfluorodimethylacetylene in nematic solvents. [**103**] A. D. Buckingham, E. E. Burnell, C. A. de Lange, Mol. Phys. *16*, 191 (1969), C.A. *71*, 34900z (1969); NMR study of ethyl fluoride dissolved in a nematic. [**104**] A. D. Buckingham, E. E. Burnell, C. A. de Lange, Mol. Phys. *16*, 299 (1969), C.A. *71*, 34897d (1969); NMR studies of 1,1-difluoroethylene in nematic solvents. [**105**] A. D. Buckingham, E. E. Burnell, C. A. de Lange, Mol. Phys. *16*, 521 (1969), C.A. *72*, 17090p (1970); NMR studies of 1,4-cyclohexadiene oriented in a nematic solvent. [**106**] A. D. Buckingham, E. E. Burnell, C. A. de Lange, Mol. Phys. *17*, 205 (1969), C.A. *71*, 101093d (1969); NMR study of spiropentane in a nematic solvent. [**107**] A. D. Buckingham, I. Love, J. Magn. Reson. *2*, 338 (1970), C.A. *74*, 105332r (1971); Theory of the anisotropy of nuclear spin coupling. [**108**] A. D. Buckingham, M. B. Dunn, Mol. Phys. *19*, 721 (1970), C.A. *74*, 47845x (1971); NMR studies of cis-1,2-difluoroethylene and vinyl fluoride in a nematic solvent. [**109**] A. D. Buckingham, E. E. Burnell, C. A. de Lange, J. Chem. Phys. *54*, 3242 (1971), C.A. *74*, 147569d (1971); Measurement of the anisotropic shielding of protons in a nematic. Comments. [**110**] A. D. Buckingham, J. P. Yesinowski, A. J. Canty, A. J. Rest, J. Am. Chem. Soc. *95*, 2732 (1973), C.A. *78*, 148059g (1973); Structural study of a ruthenium hydride cluster by nematic-phase ¹HMR.

[**111**] D. Bailey, A. D. Buckingham, M. C. McIvor, A. J. Rest, J. Organometal. Chem. *61*, 311 (1973); C.A. *80*, 36426j (1974); Nematic phase ¹HMR spectrum of π-cyclopentadienyltricarbonyl-tungsten hydride. [**112**] D. Bailey, A. D. Buckingham, M. C. McIvor, A. J. Rest, Mol. Phys. *25*, 479 (1973), C.A. *78*, 130296x (1973); NMR study of ¹³C monoxide-enriched π-cyclopentadienylmanganese tricarbonyl in a nematic solvent. [**113**] A. D. Buckingham, A. J. Rest, J. P. Yesinowski, Mol. Phys. *25*, 1457 (1973), C.A. *79*, 91172c (1973); ¹HMR of tris(methylene)methaneiron tricarbonyl in a nematic. [**114**] D. Bailey, A. D. Buckingham, A. J. Rest, Mol. Phys. *26*, 233 (1973), C.A. *79*, 136053h (1973); Evidence for intramolecular rotation in π-cyclobutadienyliron tricarbonyl from nematic-phase NMR. [**115**] A. D. Buckingham, Pure Appl. Chem. *40*, 1 (1974), C.A. *83*, 35284m (1975); Molecular structure determination by NMR spectroscopy. [**116**] D. Bailey, A. D. Buckingham, F. Fujiwara, L. W. Reeves, J. Magn. Reson. *18*, 344 (1975), C.A. *83*, 88174g (1975); High-resolution NMR spectra of tetrahedral molecules and ions in anisotropic environments. [**117**] J. Bulthuis, NMR in liquid crystalline solvents; complications in determining molecular geometries. (Internal. Scholarly Book Services: Portland, Ore.). 1974. 104 pp., C.A. *82*, 148359n (1975). [**118**] J. Bulthuis, J. Mol. Struct. *38*, 149 (1977), C.A. *87*, 52653f (1977); NMR study of the conformation of glutaric anhydride oriented in a nematic solvent. [**119**] E. E. Burnell, M. A. J. Sweeney, Can. J. Chem. *52*, 3565 (1974), C.A. *82*, 72253a (1975); ¹HMR study of 1,4-dibromobenzene-¹³C partially oriented in nematics. [**120**] I. M. Armitage, E. E. Burnell, M. B. Dunn, L. D. Hall, R. B. Malcolm, J. Magn. Reson. *13*, 167 (1974), C.A. *80*, 139135t (1974); Effect of lanthanide shift reagents on the NMR spectra of molecules partially oriented in a nematic. Pyridine.

[**121**] E. E. Burnell, J. R. Council, S. E. Ulrich, Chem. Phys. Lett. *31*, 395 (1975), C.A. *83*, 18483h (1975); NMR spectra of methyl (carbon-13) fluoride partially oriented in nematics. Reinvestigation

and explanation of anomalous results. [**122**] E. E. Burnell, M. A. J. Sweeney, T. C. Wong, Chem. Phys. Lett. *39*, 489 (1976), C.A. *85*, 93296u (1976); ^{1}HMR study of furan-^{13}C partially oriented in a nematic with vibrational averaging. [**123**] T. C. Wong, E. E. Burnell, L. Weiler, Chem. Phys. Lett. *42*, 272 (1976), C.A. *85*, 169364n (1976); NMR study of tetrathiofulvalene and tetracyanoquinodimethane: relative proton geometries and order parameters using nematic solvents; C—H spin-spin coupling constants for TTF. [**124**] T. C. Wong, E. E. Burnell, J. Magn. Reson. *22*, 227 (1976), C.A. *86*, 29070j (1977); NMR study of benzoyl fluoride partially oriented in nematics: molecular structure, barrier to rotation, and order matrix. [**125**] T. C. Wong, E. E. Burnell, J. Mol. Struct. *33*, 217 (1976), C.A. *85*, 192094h (1976); Vibrationally averaged structure of partially oriented *p*-benzoquinone-^{13}C by ^{1}HMR. [**126**] T. C. Wong, E. E. Burnell, L. Weiler, Chem. Phys. Lett. *50*, 243 (1977), C.A. *88*, 6186d (1978); Molecular structure of tetrathiofulvalente-^{13}C from ^{1}HMR using nematic solvents. [**127**] M. Bloom, E. E. Burnell, S. B. W. Roeder, M. I. Valic, J. Chem. Phys. *66*, 3012 (1977); C.A. *86*, 198200g (1977); NMR line shapes in lyophases and related systems. [**128**] M. Weger, B. Cabane, J. Phys. (Paris), C-4, *30*, 72 (1969), C.A. *73*, 135785n (1970); Relaxation nucléaire dans un nématique. [**129**] B. Cabane, W. G. Clark, Phys. Rev. Lett. *25*, 91 (1970), C.A. *73*, 61064c (1970); Effects of order and fluctuations on the ^{14}NMR in a liquid crystal. [**130**] B. Cabane, Advan. Mol. Relaxation Processes *3*, 341 (1972), C.A. *78*, 64855a (1973); Nuclear relaxation in liquid crystals.

[**131**] B. Deloche, B. Cabane, Mol. Cryst. Liq. Cryst. *19*, 25 (1972), C.A. *78*, 15337z (1973); Coupling of hydrogen bonding to orientational fluctuation modes in the liquid crystal *p*-hexyloxybenzoic acid. [**132**] B. Cabane, W. G. Clark, Solid State Commun. *13*, 129 (1973), C.A. *79*, 97930v (1973); Orientational order in the vicinity of a second order s_A/n transition. [**133**] D. A. Cadenhead, F. Mueller-Landau, Protides Biol. Fluids, Proc. Colloq. *21*, 175 (1973), C.A. *81*, 147228m (1974); Impurity effects of spin-label probes. Evaluation using monomolecular film techniques. [**134**] A. Caillé, Solid State Commun. *10*, 571 (1972), C.A. *76*, 147063a (1972); Calculations of the nuclear relaxation time in a two-dimensional quasi-nematic system. [**135**] R. F. Campbell, Diss. Univ. Colorado, Boulder. 1975. 192 pp. Avail. Xerox Univ. Microfilms, Ann Arbor, Mich., Order No. 76-11,558; Vanadyl ion as an ESR probe of micelle-liquid crystal systems. [**136**] R. F. Campbell, M. W. Hanna, J. Phys. Chem. *80*, 1892 (1976), C.A. *85*, 99704x (1976); Vanadyl ion as an ESR probe of micelle-liquid crystal systems. [**137**] D. Canet, P. Granger, C. R. Acad. Sci., Sér. C *272*, 1345 (1971), C.A. *75*, 27952x (1971); Spectre RMN du tétrachloro-*p*-xylène dissous dans un nématique. [**138**] E. Haloui, D. Canet, C. R. Acad. Sci., Sér. C *275*, 447 (1972), C.A. *77*, 158435v (1972); Spectres RMN d'une série de benzènes monosubstitués partiellement orientés dans un nématique. [**139**] D. Canet, R. Price, J. Magn. Reson. *9*, 35 (1973), C.A. *78*, 64867f (1973); ^{1}HMR of 2,5-dichloro-*p*-xylene dissolved in a nematic. [**140**] D. Canet, E. Haloui, H. Nery, J. Magn. Reson. *10*, 121 (1973), C.A. *79*, 59777t (1973); ^{1}HMR spectrum of chlorobenzene partially oriented in a nematic. Use of the INDOR technique in a liquid crystal solution.

[**141**] E. Haloui, D. Canet, Chem. Phys. Lett. *26*, 261 (1974), C.A. *81*, 49162r (1974); FT ^{13}CMR and ^{1}HMR including ^{13}C-satellites of oriented cyanopropyne. [**142**] D. Canet, J. Barriol, Mol. Phys. *27*, 1705 (1974), C.A. *81*, 119779y (1974); Structure of *p*-xylene from its ^{1}HMR spectrum in the nematic phase. [**143**] E. Haloui, D. Canet, Org. Magn. Reson. *6*, 537 (1974), C.A. *82*, 154638u (1975); Structure du propyne par RMN en phase nématique. Utilisation des satellites du ^{13}C. [**144**] E. Haloui, D. Canet, J. Barriol, J. Chim. Phys. Phys.-Chim. Biol. *72*, 1097 (1975), C.A. *84*, 134950c (1976); Orientation des solutés en phase nématique. Étude par RMN et interprétation des résultats rélatifs a une série de benzènes monosubstitués et leur homologues p-disubstitués. [**145**] E. Haloui, D. Canet, J. Mol. Struct. *24*, 85 (1975), C.A. *82*, 49596f (1975); Spectres RM^{1}H et structures de l'oxyde et du sulfure d'éthylène en phase nématique. Utilisation des ^{13}C-satellites. [**146**] E. Gout, C. Beguin, D. Canet, J. P. Marchal, Org. Magn. Reson. *7*, 633 (1975), C.A. *84*, 120685h (1976); Détermination de la forme la plus stable de l'*o*-dichlorofluorure de benzyle par RMN en phase nématique. [**147**] J. P. Marchal, D. Canet, J. Chem. Phys. *66*, 2566 (1977), C.A. *86*, 163242v (1977); Measurements of ^{15}N-, ^{1}H-, and ^{13}C-^{1}H direct couplings in natural abundance from ^{1}HMR spectra of *s*-triazine in the nematic phase. [**148**] E. F. Carr, E. A. Hoar, W. T. MacDonald, J. Chem. Phys. *48*, 2822 (1968); Influence of electric fields on the NMR spectra of nematic APAPA. [**149**] R. Casini, S. Faetti, M. Martinelli, S. Santucci, M. Giordano, J. Magn. Reson. *26*, 201 (1977), C.A. *87*, 46277v (1977); Nematic mesophase orientation phenomena by ESR. [**150**] A. L. Segre, S. Castellano, J. Magn. Reson. *7*, 5 (1972), C.A. *77*, 27125w (1972); ^{1}HMR spectrum of 1,3-butadiene oriented in a nematic.

[**151**] G. P. Ceasar, Diss. Columbia Univ., New York. 1967. 144 pp. Avail. Univ. Microfilms, Ann Arbor, Mich., Order No. 68-8568, C.A. *69*, 82122c (1968); NMR spectroscopy in liquid crystals.

[152] S. I. Chan, G. W. Feigenson, C. H. A. Seiter, Nature *231*, 110 (1971), C.A. *75*, 56323y (1971); Nuclear relaxation studies of lecithin bilayers. [153] S. I. Chan, C. H. A. Seiter, G. W. Feigenson, Biochem. Biophys. Res. Commun. *46*, 1488 (1972), C.A. *76*, 137228u (1972); Anisotropic and restricted molecular motion in lecithin bilayers. [154] S. I. Chan, M. P. Sheetz, C. H. A. Seiter, G. W. Feigenson, M. C. Hsu, A. Lau, A. Yau, Ann. N. Y. Acad. Sci. *222*, 499 (1973), C.A. *81*, 59379f (1974); NMR studies of the structure of model membrane systems. Effect of surface curvature. [155] M. C. Hsu, S. I. Chan, Biochemistry *12*, 3872 (1973), C.A. *79*, 133724s (1973); NMR studies of the interaction of valinomycin with unsonicated lecithin bilayers. [156] C. H. A. Seiter, S. I. Chan, J. Am. Chem. Soc. *95*, 7541 (1973), C.A. *80*, 14132h (1974); Molecular motion in lipid bilayers. NMR line width study. [157] G. W. Feigenson, S. I. Chan, J. Am. Chem. Soc. *96*, 1312 (1974), C.A. *80*, 105118z (1974); NMR relaxation behavior of lecithin multilayers. [158] D. Chapman, S. A. Penkett, Nature *211*, 1304 (1966), C.A. *65*, 18906e (1966); NMR studies of the interaction of phospholipids with cholesterol. [159] D. Chapman, N. J. Salsbury, Trans. Faraday Soc. *62*, 2607 (1966), C.A. *65*, 14011a (1966); [1]HMR studies of molecular motion in 2,3-diacyl-DL-phosphatidylethanolamines. [160] D. Chapman, V. B. Kamat, J. de Gier, S. A. Penkett, Nature *213*, 74 (1967), C.A. *66*, 73918c (1967); NMR studies of erythrocyte membranes.

[161] N. J. Salsbury, D. Chapman, Biochim. Biophys. Acta *163*, 314 (1968), C.A. *70*, 8870z (1969); NMR studies of diacyl L-phosphatidylcholines. [162] D. Chapman, V. B. Kamat, J. de Gier, S. A. Penkett, J. Mol. Biol. *31*, 101 (1968), C.A. *68*, 65621b (1968); NMR studies of erythrocyte membranes. [163] T. J. Jenkinson, V. B. Kamat, D. Chapman, Biochim. Biophys. Acta *183*, 427 (1969), C.A. *71*, 87557z (1969); [1]HMR and IR spectroscopy of myelin. [164] Z. Veksli, N. J. Salsbury, D. Chapman, Biochim. Biophys. Acta *183*, 434 (1969), C.A. *71*, 98353f (1969); NMR studies of molecular motion in pure lecithin/water systems. [165] M. C. Phillips, V. B. Kamat, D. Chapman, Chem. Phys. Lipids *4*, 409 (1970), C.A. *74*, 9550q (1971); Interaction of cholesterol with the sterol free lipids of plasma membranes. [166] H. Hauser, E. G. Finer, D. Chapman, J. Mol. Biol. *53*, 419 (1970), C.A. *74*, 38381r (1971); NMR studies of the polypeptide alamethicin and its interaction with phospholipids. [167] N. J. Salsbury, D. Chapman, G. P. Jones, Trans. Faraday Soc. *66*, 1554 (1970), C.A. *73*, 82399f (1970); Hindered molecular rotation in 1,2-dipalmitoyl-L-phosphatidylcholine monohydrate by NMR spin-lattice relaxation in the rotating frame. [168] E. Oldfield, D. Chapman, Biochem. Biophys. Res. Commun. *43*, 610 (1971), C.A. *75*, 15165e (1971); Effects of cholesterol and cholesterol derivatives on hydrocarbon chain mobility in lipids. [169] E. Oldfield, D. Chapman, Biochem. Biophys. Res. Commun. *43*, 949 (1971), C.A. *75*, 30188h (1971); ^{13}C FT NMR of lecithins. [170] J. T. Daycock, A. Darke, D. Chapman, Chem. Phys. Lipids *6*, 205 (1971), C.A. *75*, 44916f (1971); Nuclear relaxation measurements of lecithin/water systems.

[171] E. Oldfield, J. Marsden, D. Chapman, Chem. Phys. Lipids *7*, 1 (1971), C.A. *76*, 11348q (1972); [1]HMR relaxation study of mobility in lipid/water systems. [172] E. Oldfield, D. Chapman, W. Derbyshire, FEBS Lett. *16*, 102 (1971), C.A. *75*, 115636c (1971); [2]HMR. Novel approach to the study of hydrocarbon chain mobility in membrane systems. [173] E. Oldfield, D. Chapman, W. Derbyshire, Chem. Phys. Lipids *9*, 69 (1972), C.A. *77*, 45437d (1972); Lipid mobility in Acholeplasma membranes using [2]HMR. [174] E. Oldfield, D. Chapman, FEBS Lett. *21*, 303 (1972), C.A. *77*, 2177b (1972); Molecular dynamics of cerebroside-cholesterol and sphingomyelin-cholesterol interactions. Implications for myelin membrane structure. [175] D. Chapman, E. Oldfield, D. Doskocilova, B. Schneider, FEBS Lett. *25*, 261 (1972), C.A. *78*, 144888d (1973); NMR of gel and liquid crystalline phospholipids spinning at the magic angle. [176] M. Martin-Lomas, D. Chapman, Chem. Phys. Lipids *10*, 152 (1973), C.A. *79*, 32244e (1973); Structural studies on glycolipids. 220 MHz [1]HMR spectra of acetylated galactocerebrosides. [177] J. Bermejo, A. Barbadillo, F. Tato, D. Chapman, FEBS Lett. *52*, 69 (1975), C.A. *83*, 37516n (1975); NMR studies on the interaction of antidepressants with lipid model membranes. [178] J. Charvolin, P. Rigny, J. Phys. (Paris), C-4, *30*, 76 (1969), C.A. *72*, 127027f (1970); RM[2]H dans les phases mésomorphes du système laurate de potassium/eau lourde. [179] J. Charvolin, P. Rigny, J. Magn. Reson. *4*, 40 (1971), C.A. *74*, 148930b (1971); Pulsed NMR in dynamically heterogeneous systems. [180] P. Rigny, J. Charvolin, J. Phys. (Paris), C-5a, *32*, 243 (1971), C.A. *77*, 41093r (1972); Relaxation des protons paraffiniques dans une phase smectique laurate de potassium/eau. [181] J. Charvolin, P. Rigny, Mol. Cryst. Liq. Cryst. *15*, 211 (1971), C.A. *76*, 52644b (1972); NMR study of molecular motions in the mesophases of potassium laurate/water-d$_2$ system. [182] J. Charvolin, P. Rigny, Nature, New Biol. *237*, 127 (1972), C.A. *77*, 58056d (1972); Transverse nuclear relaxation in lecithin bilayers. [183] J. Charvolin, P. Rigny, Chem. Phys. Lett. *18*, 515 (1973), C.A. *78*, 140673g (1973); ^{2}H relaxation study of deuterium oxide motions in a lyophase. [184] J. Charvolin,

P. Manneville, B. Deloche, Chem. Phys. Lett. *23*, 345 (1973), C.A. *81*, 39372v (1974); NMR of perdeuterated potassium laurate in oriented soap/water multilayers. [**185**] J. Charvolin, P. Rigny, J. Chem. Phys. *58*, 3999 (1973), C.A. *78*, 166712u (1973); ^{1}H relaxation study of paraffinic chain motions in a lyophase. [**186**] B. Mely, J. Charvolin, P. Keller, Chem. Phys. Lipids *15*, 161 (1975), C.A. *84*, 88977s (1976); Disorder of lipid chains as a function of their lateral packing in lyophases. [**187**] B. Deloche, J. Charvolin, L. Liebert, L. Strzelecki, J. Phys. (Paris), C-1, *36*, 21 (1975), C.A. *83*, 19456p (1975); ^{2}HMR study of molecular order in TBBA. [**188**] J. Charvolin, B. Mely, ACS Symp. Ser. *34*, 48 (1976), C.A. *86*, 9028y (1977); ^{2}HMR study of soap/water interfaces. [**189**] J. Charvolin, B. Deloche, J. Phys. (Paris), C-3, *37*, 69 (1976), C.A. *86*, 81951m (1977); Intramolecular rotations in TBBA. [**190**] B. Deloche, J. Charvolin, J. Phys. (Paris) *37*, 1497 (1976), C.A. *86*, 63850y (1977); ^{2}HMR comparison of mesogen molecules with octyl and octyloxy end-chains.

[**191**] B. Mely, J. Charvolin, Chem. Phys. Lipids *19*, 43 (1977), C.A. *87*, 83982s (1977); ^{2}HMR comparison of lyophases with ordered and disordered paraffinic chains. [**192**] D. M. Chen, Diss. Univ. Waterloo, Waterloo, Ont. 1975, C.A. *85*, 134822h (1976); Studies of oriented lyotropic nematics by NMR methods. [**193**] D. M. Chen, J. D. Glickson, J. Magn. Reson. *28*, 9 (1977), C.A. *88*, 30026x (1978); ^{14}NMR and quadrupole splitting of ammonium ions in a lyophase. [**194**] C. J. Clemett, J. Chem. Soc. A, *1970*, 2251, C.A. *73*, 82389c (1970); ^{1}H spin-lattice relaxation times in nonionic micellar solutions. [**195**] B. Clin, C. R. Acad. Sci., Sér. C *280*, 73 (1975), C.A. *82*, 162503m (1975); Spectres haute résolution de composés mésomorphes en phase orientée. [**196**] M. Cocivera, J. Chem. Phys. *47*, 3061 (1967), C.A. *68*, 7910j (1968); Analysis of ^{1}HMR of *s*-trioxane dissolved in a nematic solvent. [**197**] K. C. Cole, Diss. McGill Univ., Montreal, Que. 1974. Avail. Natl. Libr. Canada, Ottawa, Ont., C.A. *82*, 97272u (1975); Structural studies of small-ring compounds by nematic phase NMR spectroscopy. [**198**] P. J. Collings, Diss. Yale Univ., New Haven. 1976. 137 pp. Avail. Univ. Microfilms Int., Order No. 77-10,866, C.A. *87*, 76045k (1977); NMR spectroscopy in cholesterics. [**199**] C. Corvaja, G. Giacometti, K. D. Kopple, Ziauddin, J. Am. Chem. Soc. *92*, 3919 (1970), C.A. *73*, 55268f (1970); ESR studies of nitroxide radicals and biradicals in nematic solvents. [**200**] C. Corvaja, G. Farnai, B. Luneli, J. Chem. Soc., Faraday Trans. 2, *71*, 1293 (1975), C.A. *83*, 130988c (1975); ESR spectra and orientation of 3,4-bis(dicyanomethylene)cyclobutane-1,2-dione and tetracyanoquinodimethane radical anions in liquid crystals.

[**201**] J. Courtieu, Y. Gounelle, J. Jullien, Bull. Soc. Chim. Fr. *1969*, 4184, C.A. *72*, 71885q (1970); RMN en phase nématique: variation du facteur d'orientation avec la structure du soluté. [**202**] J. Courtieu, Y. Gounelle, Bull. Soc. Chim. Fr. *1970*, 2951, C.A. *74*, 3148t (1971); RMN en phase nématique: étude du groupement éthylène-cétal. [**203**] J. Courtieu, Y. Gounelle, Org. Magn. Reson. *3*, 533 (1971), C.A. *76*, 39924y (1972); Étude RMN de l'acétone en solution dans un nématique. [**204**] J. Courtieu, Y. Gounelle, Mol. Phys. *28*, 161 (1974), C.A. *82*, 57404g (1975); RMN en phase nématique: étude de la géométrie du 2,5-dihydrofuranne. [**205**] J. Courtieu, Y. Gounelle, Org. Magn. Reson. *6*, 11 (1974), C.A. *81*, 62621j (1974); RMN en phase nématique: exploration de la barrière d'empêchement à la libre rotation de la monofluoroacétone. [**206**] J. Courtieu, Y. Gounelle, P. Gonord, S. K. Kan, Org. Magn. Reson. *6*, 151 (1974), C.A. *81*, 62646w (1974); RMN en phase nématique: exploration de la barrière d'empêchement à la libre rotation pour le fluorure d'acroyle et l'acroléine. [**207**] J. Courtieu, Y. Gounelle, C. Duret, P. Gonord, S. K. Kan, Org. Magn. Reson. *6*, 622 (1974), C.A. *83*, 42483d (1975); Étude structurale de la bipyrimidine par RMN en phase nématique. [**208**] J. Courtieu, P. E. Fagerness, D. M. Grant, J. Chem. Phys. *65*, 1202 (1976), C.A. *85*, 114296m (1976); Elimination of degeneracies in spin systems of magnetically equivalent nuclei by nematic solvents and spin tickling. [**209**] J. M. Courtieu, C. L. Mayne, D. M. Grant, J. Chem. Phys. *66*, 2669 (1977), C.A. *86*, 148436t (1977); Nuclear relaxation in coupled spin systems dissolved in a nematic — the AX and A$_2$ spin systems. [**210**] D. Cutler, Mol. Cryst. Liq. Cryst. *8*, 85 (1969), C.A. *71*, 86440u (1969); Spin lattice relaxation time measurements on cholesterol derivatives.

[**211**] S. Ohnishi, T. J. R. Cyr, H. Fukushima, Bull. Chem. Soc. Jap. *43*, 673 (1970), C.A. *72*, 136752r (1970); Biradical spin-labeled micelles. [**212**] K. I. Dahlqvist, A. B. Hornfeldt, Chem. Scr. *1*, 125 (1971), C.A. *76*, 66018n (1972); NMR spectrum of oriented selenophene in a lyophase. [**213**] C. S. Yannoni, G. P. Ceasar, B. P. Dailey, J. Am. Chem. Soc. *89*, 2833 (1967), C.A. *67*, 38125r (1967); NMR spectrum of oriented (cyclobutadiene)iron tricarbonyl. [**214**] G. P. Ceasar, C. S. Yannoni, B. P. Dailey, J. Chem. Phys. *50*, 373 (1969), C.A. *70*, 62762d (1969); Chemical-shift anisotropy in liquid-crystal solvents. I. Experimental results for the methyl halides. [**215**] G. P. Ceasar, B. P. Dailey, J. Chem. Phys. *50*, 4200 (1969), C.A. *71*, 34902b (1969); II. Theoretical calculations for the methyl halides (cf.

[214]). [**216**] W. J. Caspary, F. Millett, M. Reichbach, B. P. Dailey, J. Chem. Phys. *51*, 623 (1969), C.A. *71*, 65850n (1969); NMR determination of ^{2}H quadrupole coupling constants in nematic solutions. [**217**] D. N. Silverman, B. P. Dailey, J. Chem. Phys. *51*, 655 (1969), C.A. *71*, 65784u (1969); III. NMR spectra of nematic solutions of ethane and 1,1,1-trifluoroethane (cf. [214]). [**218**] D. N. Silverman, B. P. Dailey, J. Chem. Phys. *51*, 1679 (1969), C.A. *71*, 96680m (1969); NMR spectra of nematic solutions of methyl isocyanide and methyl cyanide. [**219**] J. Lindon, B. P. Dailey, Mol. Phys. *19*, 285 (1970), C.A. *73*, 104002a (1970); ^{1}H shielding anisotropy in benzene. [**220**] C. S. Yannoni, B. P. Dailey, G. P. Ceasar, J. Chem. Phys. *54*, 4020 (1971), C.A. *74*, 132768u (1971); IV. Results for fluorine compounds (cf. [214]).

[**221**] F. Millett, B. P. Dailey, J. Chem. Phys. *54*, 5434 (1971), C.A. *75*, 56142p (1971); Anisotropy of the ^{13}C chemical shift in hydrogen cyanide. [**222**] J. Lindon, B. P. Dailey, Mol. Phys. *20*, 937 (1971), C.A. *75*, 56319b (1971); Orientation studies by NMR spectroscopy using a lyotropic solvent. [**223**] J. C. Lindon, B. P. Dailey, Mol. Phys. *22*, 465 (1971), C.A. *76*, 39825s (1972); NMR spectrum of cyclopentadienylmanganese tricarbonyl partially oriented in a nematic solvent. [**224**] F. S. Millett, B. P. Dailey, J. Chem. Phys. *56*, 3249 (1972), C.A. *76*, 119588g (1972); NMR determination of ^{2}H quadrupole coupling constants in nematic solutions. [**225**] P. K. Bhattacharyya, B. P. Dailey, J. Chem. Phys. *59*, 3737 (1973), C. A. *80*, 47337g (1974); ^{13}C chemical shift anisotropy and the H—C—H bond angle of methanol-^{13}C in a nematic solvent. [**226**] P. K. Bhattacharyya, B. P. Dailey, J. Chem. Phys. *59*, 5820 (1973), C.A. *80*, 76404g (1974); ^{15}NMR shielding anisotropies in nitrous (^{15}N, ^{15}N) oxide. [**227**] P. K. Bhattacharyya, B. P. Dailey, J. Magn. Reson. *12*, 36 (1973), C.A. *79*, 151315p (1973); ^{13}C and ^{1}H chemical-shift anisotropies in chloroform. [**228**] N. Zumbulyadis, B. P. Dailey, Mol. Phys. *26*, 777 (1973), C.A. *79*, 131079m (1973); ^{19}FMR and ^{31}PMR spectra of phosphorus trifluoride in nematic solution. [**229**] P. K. Bhattacharyya, B. P. Dailey, Mol. Phys. *26*, 1379 (1973), C.A. *80*, 107456g (1974); Anisotropies in ^{13}C chemical shift tensor, H—C—H bond angles and the signs of the coupling constants in iodomethane-^{13}C, bromomethane-^{13}C, and chloromethane-^{13}C. [**230**] B. R. Appleman, B. P. Dailey, Adv. Magn. Reson. *7*, 231 (1974), C.A. *81*, 179131r (1974); Magnetic shielding and susceptibility anisotropies.

[**231**] N. Zumbulyadis, B. P. Dailey, Chem. Phys. Lett. *26*, 273 (1974), C.A. *81*, 56261a (1974); ^{31}P chemical shielding anisotropies and the sign of J_{PP} in P_4S_3 from NMR studies in a nematic solvent. [**232**] N. Zumbulyadis, B. P. Dailey, J. Chem. Phys. *60*, 4223 (1974), C.A. *81*, 56291k (1974); Determination of the ^{2}H quadrupole coupling constant in phosphine-d_3 from studies in a nematic solvent. [**233**] N. Zumbulyadis, B. P. Dailey, J. Magn. Reson. *13*, 189 (1974), C.A. *80*, 150791z (1974); ^{31}PMR spectra of partially oriented cyclic halophosphazenes. [**234**] P. K. Bhattacharyya, B. P. Dailey, J. Magn. Reson. *13*, 317 (1974), C.A. *81*, 104166x (1974); Structural parameters and the ^{13}C and ^{19}F magnetic shielding anisotropies in ^{13}C labeled methyl fluoride from nematic phase studies. [**235**] B. R. Appleman, B. P. Dailey, J. Magn. Reson. *16*, 265 (1974), C.A. *82*, 57028n (1975); ^{19}F shielding anisotropy in $CFCl_3$. [**236**] N. Zumbulyadis, B. P. Dailey, Mol. Phys. *27*, 633 (1974), C.A. *81*, 43696w (1974); ^{1}HMR and ^{31}PMR spectra of phosphine in isotropic and nematic solvents. [**237**] P. K. Bhattacharyya, B. P. Dailey, Mol. Phys. *28*, 209 (1974), C.A. *81*, 179607a (1974); ^{19}F and ^{31}P magnetic shielding anisotropies in phosphoryl fluoride. [**238**] P. K. Bhattacharyya, B. P. Dailey, Chem. Phys. Lett. *32*, 305 (1975), C.A. *83*, 35313v (1975); ^{13}C magnetic shielding anisotropy in acetonitrile-1-^{13}C obtained from NMR studies in a liquid crystal solution. [**239**] P. K. Bhattacharyya, B. P. Dailey, J. Chem. Phys. *63*, 1336 (1975), C.A. *83*, 139514c (1975); ^{2}H quadrupole coupling constants in methyl-d_3 fluoride and 1,3,5-trifluorobenzene-d_3. [**240**] A. J. Montana, B. P. Dailey, Mol. Phys. *30*, 1521 (1975), C.A. *84*, 89023q (1976); ^{1}H chemical shielding anisotropies of acetylene and fluoroform, as measured in a smectic solvent.

[**241**] A. J. Montana, N. Zumbulyadis, B. P. Dailey, J. Chem. Phys. *65*, 4756 (1976), C.A. *86*, 48938q (1977); ^{19}F and ^{31}P magnetic shielding anisotropies and the F—P—F bond angle of thiophosphoryl fluoride in a smectic solvent. [**242**] A. J. Montana, B. P. Dailey, J. Magn. Reson. *21*, 25 (1976), C.A. *84*, 128451z (1976); ^{1}H chemical shielding anisotropies of the methyl halides as measured in smectic solvents. [**243**] A. J. Montana, B. P. Dailey, J. Magn. Reson. *22*, 117 (1976), C.A. *85*, 54237f (1976); NMR studies of spherical molecules dissolved in smectic solvents. [**244**] B. R. Appleman, B. P. Dailey, J. Magn. Reson. *22*, 375 (1976), C.A. *85*, 142156j (1976); ^{13}C shielding tensors of CCl_3X systems from studies of liquid-crystal solutions. [**245**] A. J. Montana, N. Zumbulyadis, B. P. Dailey, J. Am. Chem. Soc. *99*, 4290 (1977), C.A. *87*, 46314e (1977); Measurements of ^{31}P chemical shielding anisotropies and signs of spin-spin coupling constants of oriented $PX(CH_3)_3$ systems in smectic A

solution. [246] A. J. Montana, B. P. Dailey, J. Chem. Phys. 66, 989 (1977), C.A. 86, 113371r (1977); ^{19}F shielding anisotropies of the fluoromethanes obtained from NMR studies in a smectic solution. [247] A. J. Montana, B. R. Appleman, B. P. Dailey, J. Chem. Phys. 66, 1850 (1977), C.A. 86, 148414j (1977); ^{19}F FT-NMR studies of trifluoromethyl-containing systems in smectic solutions. [248] D. G. de Kowalewski, V. J. Kowalewski, J. Mol. Struct. 23, 203 (1974), C.A. 82, 97362y (1975); NMR study of 2,5-dihydrofuran partially oriented in a nematic. [249] E. E. Burnell, C. A. de Lange, Mol. Phys. 16, 95 (1969), C.A. 70, 110355x (1969); NMR studies of pyridazine and pyridine oriented in a nematic. [250] W. de Kieviet, C. A. de Lange, Chem. Phys. Lett. 22, 378 (1973), C.A. 80, 36410z (1974); NMR spectra of o-, m-, and p-dicyanobenzene in a nematic solvent.

[251] C. A. de Lange, Chem. Phys. Lett. 28, 526 (1974), C.A. 82, 9514w (1975); NMR study of trimethylacetic acid in a nematic. [252] J. Bulthuis, C. A. de Lange, J. Magn. Reson. 14, 13 (1974), C.A. 81, 17949m (1974); NMR study of phosphoryl fluoride oriented in a nematic. [253] C. A. de Lange, K. J. Peverelli, J. Magn. Reson. 16, 159 (1974), C.A. 82, 30880x (1975); NMR study of p-dioxene oriented in a nematic. [254] G. J. den Otter, C. A. de Lange, J. Bulthuis, J. Magn. Reson. 20, 67 (1975), C.A. 84, 97495k (1976); NMR investigation of partially oriented trifluoromethane and difluoromethane including ^{13}C satellites. [255] C. A. de Lange, J. Magn. Reson. 21, 37 (1976), C.A. 84, 163740 (1976); Ring-puckering vibrations of 1,3-dioxolane studied by nematic phase NMR. [256] J. M. Dereppe, J. Degelaen, M. van Meerssche, J. Chim. Phys. Physicochim. Biol. 67, 1875 (1970), C.A. 74, 59155k (1971); ^{1}HMR spectrum of 1,4-naphthoquinone in a nematic. [257] J. M. Dereppe, J. P. Morisse, M. van Meerssche, Org. Magn. Reson. 3, 583 (1971), C.A. 76, 39929d (1972); ^{1}HMR spectrum of thiophene dissolved in a nematic solvent. [258] J. M. Dereppe, J. Degelaen, M. van Meerssche, Org. Magn. Reson. 4, 551 (1972), C.A. 77, 151123h (1972); Analysis of the ^{1}HMR spectrum of 1,4-naphthoquinone in a nematic solvent. [259] J. M. Dereppe, J. Degelaen, M. van Meerssche, J. Mol. Struct. 17, 225 (1973), C.A. 79, 136066q (1973); Molecular structure of naphthalene as determined by NMR in a nematic. [260] J. Degelaen, E. Arte, J. M. Dereppe, J. Chem. Phys. 61, 5295 (1974), C.A. 82, 147923e (1975); ^{1}HMDR of ethylene oxide dissolved in a nematic solvent.

[261] J. M. Dereppe, E. Arte, M. van Meerssche, J. Cryst. Mol. Struct. 4, 193 (1974), C.A. 84, 82202p (1976); ^{1}HMR spectrum of phenylacetylene dissolved in a nematic. [262] E. Arte, J. Degelaen, J. M. Dereppe, J. Chem. Phys. 63, 3171 (1975), C.A. 83, 185946t (1975); Use of the GOE for the assignment of weak lines and double quantum transitions in the spectrum of oriented o-dichlorobenzene. [263] J. Degelaen, E. Arte, B. van Eerdewegh, M. van Meerssche, J. M. Dereppe, J. Mol. Struct. 36, 263 (1977), C.A. 86, 155109s (1977); Comparative NMR study in nematics of the bicyclic compounds: 1,4-naphthoquinone, 2,3-dichloro-1,4-naphthoquinone and naphthalene. [264] A. Derzhanski, S. Naidenova, L. Grigorov, A. Petrov, Magn. Reson. Relat. Phenomena, Proc. Congr. Ampère 16th, 1970, 841, C.A. 78, 36020f (1973); FT NMR spectroscopy in liquid crystals. [265] P. Diehl, C. L. Khetrapal, U. Lienhard, Can. J. Chem. 46, 2645 (1968), C.A. 69, 82116d (1968); NMR spectra of oriented 4-spin systems with C$_{2v}$ symmetry (AA'BB'). Spectrum of thiophene. [266] P. Diehl, C. L. Khetrapal, H. P. Kellerhals, Helv. Chim. Acta 51, 529 (1968), C.A. 68, 110001n (1968); ^{1}HMR-Spektrum von orientiertem Furan in nematischer Lösung. [267] P. Diehl, C. L. Khetrapal, Mol. Phys. 14, 283 (1968), C.A. 69, 14634w (1968); Temperature, concentration and spinning speed dependence in NMR spectra of oriented molecules. [268] P. Diehl, C. L. Khetrapal, Mol. Phys. 14, 327 (1968), C.A. 69, 31884v (1968); NMR spectra of pyrazine and p-benzoquinone oriented in a nematic. [269] P. Diehl, C. L. Khetrapal, U. Lienhard, Mol. Phys. 14, 465 (1968), C.A. 69, 56117k (1968); Use of the direct and the moment methods in the analysis of NMR spectra of oriented molecules. [270] P. Diehl, C. L. Khetrapal, Mol. Phys. 15, 201 (1968); NMR spectra of symmetrical o-disubstituted benzenes dissolved in the nematic phase.

[271] P. Diehl, C. L. Khetrapal, H. P. Kellerhals, Mol. Phys. 15, 333 (1968), C.A. 70, 24490f (1969); NMR spectrum of pyridine oriented in the nematic phase. [272] P. Diehl, C. L. Khetrapal, Mol. Phys. 15, 633 (1968), C.A. 70, 42697h (1969); NMR spectra of symmetrical m-disubstituted benzenes in a room temperature nematic. [273] P. Diehl, C. L. Khetrapal, Proc. Colloq. Ampère 15, 251 (1968), C.A. 72, 17085r (1970); Structure of benzofurazan oxide from ^{1}HMR in the nematic phase. [274] P. Diehl, C. L. Khetrapal, Can. J. Chem. 47, 1411 (1969), C.A. 70, 101482j (1969); Isotopic effects and determination of the quadrupole coupling constant from NMR spectra of monodeuterobenzene in a nematic. [275] P. Diehl, H. Kellerhals, J. Magn. Reson. 1, 196 (1969), C.A. 73, 30476j (1970); Uniqueness of analysis in NMR spectroscopy of oriented molecules. [276] P. Diehl, C. L. Khetrapal, J. Magn. Reson. 1, 524 (1969), C.A. 72, 95039t (1970); NMR spectra of partially deuterated acetonitrile dissolved in the nematic phase. [277] P. Diehl, C. L. Khetrapal, H. P. Kellerhals, U. Lienhard, W.

Niederberger, J. Magn. Reson. *1*, 527 (1969), C.A. *72*, 84684y (1970); Influence of an electric field on the NMR spectra of molecules oriented in the nematic phase. [**278**] C. F. Schwerdtfeger, P. Diehl, Mol. Phys. *17*, 417 (1969), C.A. *71*, 118217v (1969); ESR of vanadyl acetylacetonate dissolved in a room temperature liquid crystal. [**279**] P. Diehl, C. F. Schwerdtfeger, Mol. Phys. *17*, 423 (1969), C.A. *72*, 17140e (1970); ESR determination of the orientation distribution function of vanadyl acetylacetonate dissolved in a liquid crystal. [**280**] P. Diehl, C. L. Khetrapal, U. Lienhard, Org. Magn. Reson. *1*, 93 (1969), C.A. *71*, 60456d (1969); Deceptive simplicity in NMR spectra of oriented molecules.

[**281**] P. Diehl, C. L. Khetrapal, Org. Magn. Reson. *1*, 467 (1969), C.A. *72*, 95053t (1970); NMR spectrum of phenylacetylene oriented in the nematic phase. [**282**] P. Diehl, C. L. Khetrapal, W. Niederberger, P. Partington, J. Magn. Reson. *2*, 181 (1970), C.A. *74*, 105328u (1971); NMR spectrum of 2,6-dichlorotoluene partially oriented in the nematic phase. [**283**] P. Diehl, H. Kellerhals, W. Nieder- berger, J. Magn. Reson. *3*, 230 (1970), C.A. *74*, 31315p (1971); Influence of methyl group rotation on the NMR spectrum of 3,5-dichlorotoluene, partially oriented in a nematic. [**284**] C. L. Khetrapal, A. V. Patankar, P. Diehl, Org. Magn. Reson. *2*, 405 (1970), C.A. *74*, 22353x (1971); Molecular structure of pyrimidine by NMR in the nematic phase. [**285**] P. Diehl, H. Kellerhals, W. Niederberger, J. Magn. Reson. *4*, 352 (1971), C.A. *75*, 69334c (1971); Structure of toluene as determined by NMR of oriented molecules. [**286**] P. Diehl, P. M. Henrichs, J. Magn. Reson. *5*, 134 (1971), C.A. *75*, 103392g (1971); Structure of an intramolecular hydrogen bond from the NMR spectrum of salicylaldehyde, partially oriented in a nematic. [**287**] C. L. Khetrapal, A. C. Kunwar, C. R. Kanekar, P. Diehl, Mol. Cryst. Liq. Cryst. *12*, 179 (1971), C.A. *75*, 56376t (1971); NMR investigations on (benzene)chromium tricarbonyl oriented in a nematic. [**288**] P. Diehl, P. M. Henrichs, W. Niederberger, Mol. Phys. *20*, 139 (1971), C.A. *74*, 69962k (1971); Molecular structure and the barrier to methyl rotation in o-chlorotol- uene partially oriented in the nematic phase. [**289**] P. Diehl, P. M. Henrichs, W. Niederberger, J. Vogt, Mol. Phys. *21*, 377 (1971), C.A. *75*, 109739x (1971); Molecular structure and barrier to methyl rotation in o-bromotoluene and o-iodotoluene partially oriented in a nematic. [**290**] P. Diehl, C. L. Khetrapal, NMR-Basic Principles and Progress 1969, 1 (Springer: Berlin); NMR studies of molecules oriented in the nematic phase.

[**291**] P. Diehl, P. M. Henrichs, W. Niederberger, Org. Magn. Reson. *3*, 243 (1971), C.A. *75*, 56335d (1971); NMR spectrum of benzaldehyde partially oriented in the nematic phase. [**292**] P. Diehl, P. M. Henrichs, Org. Magn. Reson. *3*, 791 (1971), C.A. *76*, 98644e (1972); Molecular structure and unusual orientation of phenol in a liquid crystal solvent as determined by NMR spectroscopy. [**293**] G. Englert, P. Diehl, W. Niederberger, Z. Naturforsch. A *26*, 1829 (1971), C.A. *76*, 71588c (1972); [1]HMR of benzene and benzene-1-^{13}C in isotropic and nematic solution. [**294**] E. E. Burnell, P. Diehl, Can. J. Chem. *50*, 3566 (1972), C.A. *78*, 57210h (1973); NMR study of norbornadiene partially oriented in a nematic. [**295**] P. Diehl, P. M. Henrichs, Chemical Society, Specialist Periodical Reports, Nucl. Magn. Reson. *1*, 321 (1972), C.A. *78*, 64346k (1973); Oriented molecules. [**296**] E. E. Burnell, P. Diehl, Mol. Phys. *24*, 489 (1972), C.A. *78*, 3404f (1973); NMR study of the barrier to internal rotation, the molecular structure, and the indirect spin-spin coupling constants of o-xylene partially oriented in a nematic. [**297**] J. Degelaen, P. Diehl, W. Niederberger, Org. Magn. Reson. *4*, 721 (1972), C.A. *78*, 64914u (1973); NMR analysis of benzotrifluoride partially oriented in a nematic. [**298**] P. Diehl, Pure Appl. Chem. *32*, 111 (1972), C.A. *78*, 9884j (1973); Study of intramolecular motion by NMR of oriented molecules. [**299**] P. Diehl, W. Niederberger, J. Magn. Reson. *9*, 495 (1973), C.A. *78*, 135454g (1973); Vibrationally averaged molecular structure. Comparison of NMR data for oriented benzene with results from electron diffraction and Raman spectroscopy. [**300**] W. Niederberger, P. Diehl, L. Lunazzi, Mol. Phys. *26*, 571 (1973), C.A. *79*, 125433f (1973); Structure and conformation of 4,4′-dichlorobiphenyl determined by NMR of oriented molecules in nematic solvents.

[**301**] E. E. Burnell, P. Diehl, Org. Magn. Reson. *5*, 137 (1973), C.A. *79*, 17904b (1973); NMR study of cis-2-butene partially oriented in a nematic. [**302**] E. E. Burnell, P. Diehl, W. Niederberger, Can. J. Chem. *52*, 151 (1974), C.A. *80*, 107462f (1974); [1]HMR study of ^{13}C-labeled 1,4-dichlorobenzene partially oriented in nematics. [**303**] P. Diehl, S. Sykora, W. Niederberger, E. E. Burnell, J. Magn. Reson. *14*, 260 (1974), C.A. *81*, 77077k (1974); NMR spectra of oriented ^{13}C-labeled acetylene. Redetermina- tion of the shrinkage effects and analysis of the apparent temperature dependence of the molecular geometry. [**304**] P. Diehl, W. Niederberger, J. Magn. Reson. *15*, 391 (1974), C.A. *81*, 143933q (1974); ^{1}H decoupling in [2]HMR spectra of oriented molecules. [**305**] P. Diehl, W. Niederberger, Chemical Society, Specialist Periodical Reports, Nucl. Magn. Reson. *3*, 368 (1974), C.A. *82*, 9394g (1975); Oriented molecules. [**306**] P. Diehl, J. Vogt, Org. Magn. Reson. *6*, 33 (1974), C.A. *81*, 62617n (1974); NMR

spectrum of 1-chloronaphthalene orientated in the nematic phase. [**307**] P. Diehl, A. S. Tracey, Can. J. Chem. *53*, 2755 (1975), C.A. *84*, 19544m (1976); Use of ^{2}HMR in the determination of the structure and location of the anilinium ion in an aqueous anisotropic surfactant solution. [**308**] T. Bally, P. Diehl, E. Haselbach, A. S. Tracey, Helv. Chim. Acta *58*, 2398 (1975), C.A. *85*, 32328j (1976); Bond-rotational mobility of the guanidinium ion. [**309**] P. Diehl, S. Sykora, J. Vogt, J. Magn. Reson. *19*, 67 (1975), C.A. *83*, 170557t (1975); Automatic analysis of NMR spectra. Alternative approach. [**310**] P. Diehl, M. Reinhold, A. S. Tracey, J. Magn. Reson. *19*, 405 (1975), C.A. *84*, 10703e (1976); Determination of ^{2}H quadrupole coupling constants by high-resolution ^{2}HMR.

[**311**] P. Diehl, S. Sykora, E. Wullschleger, Mol. Phys. *29*, 305 (1975), C.A. *83*, 8955z (1975); Vibrationally averaged structures of ethylene, ethylene 1-^{13}C, and ethylene-1,2-^{13}C as determined by NMR of partially oriented molecules. [**312**] P. Diehl, A. S. Tracey, Mol. Phys. *30*, 1917 (1975), C.A. *84*, 82873h (1976); Order in pure nematics as determined by NMR of partially deuterated species. [**313**] P. Diehl, M. Reinhold, A. S. Tracey, E. Wullschleger, Mol. Phys. *30*, 1781 (1975), C.A. *84*, 89027u (1976); Interpretation of the anomalous results from a NMR study of methanol-^{13}C partially oriented in nematics. Implications for the determination of NMR parameters for small molecules. [**314**] P. Diehl, J. Vogt, Org. Magn. Reson. *7*, 81 (1975), C.A. *83*, 95834a (1975); ^{1}HMR spectrum of indene oriented in the nematic phase. [**315**] A. S. Tracey, P. Diehl, Can. J. Chem. *54*, 2283 (1976), C.A. *85*, 134786z (1976); ^{2}HMR study of a lamellar lyophase. [**316**] P. Diehl, Chemical Society, Specialist Periodical Reports, Nucl. Magn. Reson. *5*, 314 (1976), C.A. *85*, 113849g (1976); Oriented molecules. [**317**] P. Diehl, H. Boesiger, J. Vogt, R. Ader, J. Mol. Struct. *33*, 249 (1976), C.A. *86*, 4809u (1977); r_α-Structures of partially oriented thiophene and furan as determined from ^{1}HMR spectra including ^{13}C-satellites. [**318**] P. Diehl, H. Boesiger, A. S. Tracey, R. Ader, Org. Magn. Reson. *8*, 17 (1976), C.A. *84*, 163724n (1976); Structure of pyrazine determined from its ^{1}HMR spectrum, including ^{13}C-satellites. [**319**] P. Diehl, H. Zimmermann, Org. Magn. Reson. *8*, 155 (1976), C.A. *85*, 123204g (1976); ^{1}HMR spectra and structures of bicyclic compounds oriented in nematics. [**320**] P. Diehl, J. Vogt, Org. Magn. Reson. *8*, 638 (1976), C.A. *87*, 21730z (1977); Automatic analysis of NMR spectra; practical application to the spectra of oriented 2,4-dichlorobenzaldehyde and 2-chlorobenzaldehyde.

[**321**] P. Diehl, H. Boesiger, J. Mol. Struct. *42*, 103 (1977), C.A. *88*, 88699k (1978); r_α-Structures of partially oriented *o*-dichloro-, *o*-dibromo-, and *o*-diiodobenzene as determined from ^{1}HMR spectra with ^{13}C-satellites. [**322**] P. Diehl, A. C. Kunwar, H. Zimmermann, J. Organomet. Chem. *135*, 205 (1977), C.A. *87*, 133744p (1977); Structure of the proton skeleton in 1,3-butadieneiron tricarbonyl as determined from the NMR spectra of oriented molecules. [**323**] P. Diehl, H. Boesiger, Org. Magn. Reson. *9*, 98 (1977), C.A. *87*, 133291p (1977); r_α-Structure of partially oriented *m*-dichlorobenzene as determined from its ^{1}HMR spectrum including ^{13}C-satellites. [**324**] P. Diehl, H. Huber, A. C. Kunwar, M. Reinhold, Org. Magn. Reson. *9*, 374 (1977), C.A. *88*, 73947r (1978); Anisole, acetophenone, and methyl benzoate oriented in a nematic: structure and internal motion. [**325**] C. L. Khetrapal, P. Diehl, A. C. Kunwar, Org. Magn. Reson. *10*, 213 (1977), C.A. *89*, 145973s (1978); ^{1}HMR spectra of N-phenylmaleimide in i and n phases. [**326**] P. Diehl, A. C. Kunwar, H. Boesiger, J. Organomet. Chem. *145*, 303 (1978), C.A. *88*, 112946m (1978); NMR studies of dimethyl selenide and dimethyl telluride in a nematic. [**327**] P. Diehl, A. C. Kunwar, Org. Magn. Reson. *11*, 47 (1978), C.A. *89*, 41681d (1978); ^{1}HMR spectrum including ^{13}C-satellites of π-benzenechromium tricarbonyl in a nematic solvent. [**328**] J. W. Doane, J. J. Visintainer, Phys. Rev. Lett. *23*, 1421 (1969), C.A. *72*, 49421p (1970); ^{1}H spin-lattice relaxation in liquid crystals. [**329**] J. W. Doane, D. L. Johnson, Chem. Phys. Lett. *6*, 291 (1970), C.A. *73*, 114826m (1970); Spin-lattice relaxation in the nematic phase. [**330**] J. J. Visintainer, J. M. Doane, D. S. Moroi, Bull. Am. Phys. Soc. *16*, 317 (1971); Spin-lattice relaxation and order fluctuations in liquid crystals.

[**331**] J. W. Doane, D. S. Moroi, Chem. Phys. Lett. *11*, 339 (1971), C.A. *76*, 52558b (1972); Angular dependence of the spin-lattice relaxation time in liquid crystals. [**332**] J. J. Visintainer, J. W. Doane, D. L. Fishel, Mol. Cryst. Liq. Cryst. *13*, 69 (1971), C.A. *75*, 54880d (1971); Quadrupole and ^{1}H spin-lattice relaxation in the nematic phase. [**333**] J. A. Murphy, J. W. Doane, Mol. Cryst. Liq. Cryst. *13*, 93 (1971), C.A. *75*, 67701w (1971); NMR measurement of the diffusion anisotropy in a nematic. [**334**] J. W. Doane, R. S. Parker, Magn. Reson. Relat. Phenomena, Proc. Congr. Ampère, 17th, *1972*, 410, C.A. *80*, 137822r (1974); Measurement of the self-diffusion constant in smectics A. [**335**] J. W. Doane, R. S. Parker, B. Cvikl, D. L. Johnson, D. L. Fishel, Phys. Rev. Lett. *28*, 1694 (1972), C.A. *77*, 93983a (1972); Possible second-order n/s$_A$ transition. [**336**] J. A. Murphy, J. W. Doane, Y. Y. Hsu, D. L. Fishel, Mol. Cryst. Liq. Cryst. *22*, 133 (1973), C.A. *79*, 109045a (1973); Impurity

diffusion in smectics A and B. [**337**] R. A. Wise, D. H. Smith, J. W. Doane, Phys. Rev. A *7*, 1366 (1973), C.A. *78*, 152489q (1973); NMR in the smectic C. [**338**] J. A. Murphy, J. W. Doane, D. L. Fishel, Liq. Cryst. Ord. Fluids *2*, 63 (1974), C.A. *86*, 195349b (1977); Molecular diffusion in the n and s_C phase of 4,4'-di-*n*-heptyloxyazoxybenzene. [**339**] J. W. Doane, C. E. Tarr, M. A. Nickerson, Phys. Rev. Lett. *33*, 620 (1974), C.A. *81*, 128844t (1974); Nuclear spin-lattice relaxation in liquid crystals by fluctuations in the nematic director. [**340**] R. A. Wise, A. Olah, J. W. Doane, J. Phys. (Paris), C-1, *36*, 117 (1975), C.A. *83*, 88963p (1975); Measurements of γ_1 in nematic CBOOA and *p'*-butoxybenzylidene-*p*-heptylaniline by NMR.

[**341**] P. Ukleja, M. Neubert, J. W. Doane, Magn. Reson. Relat. Phenom., Proc. Congr. Ampère, 19th *1976*, 209, C.A. *87*, 38345a (1977); ^{2}H spin-lattice relaxation in the nematic phase. [**342**] P. Ukleja, J. Pirs, J. W. Doane, Phys. Rev. A *14*, 414 (1976), C.A. *85*, 101826j (1976); Theory for spin-lattice relaxation in nematics. [**343**] J. Pirs, P. Ukleja, J. W. Doane, Solid State Commun. *19*, 877 (1976), C.A. *85*, 114305p (1976); NMR in rapidly rotating samples of nematics. [**344**] P. J. Bos, J. Pirs, P. Ukleja, J. W. Doane, M. E. Neubert, Mol. Cryst. Liq. Cryst. *40*, 59 (1977), C.A. *87*, 125598d (1977); Molecular orientational order and NMR in the uniaxial and biaxial phases. [**345**] C. F. Schwerdtfeger, M. Marusic, A. Mackay, R. Y. Dong, Mol. Cryst. Liq. Cryst. *12*, 335 (1970), C.A. *75*, 54891h (1971); ESR study of the temperature dependence of molecular rotation in nematics. [**346**] R. Y. Dong, C. F. Schwerdtfeger, Solid State Commun. *8*, 707 (1970), C.A. *73*, 50428y (1970); Nuclear spin relaxation in a nematic. [**347**] R. Y. Dong, M. Marusic, C. F. Schwerdtfeger, Solid State Commun. *8*, 1577 (1970), C.A. *73*, 135739a (1970); Correlated NMR and ESR studies in a nematic. [**348**] R. Y. Dong, Chem. Phys. Lett. *9*, 600 (1971), C.A. *75*, 103404n (1971); Anisotropy of nuclear spin relaxation in a nematic. [**349**] R. Y. Dong, W. F. Forbes, M. M. Pintar, J. Chem. Phys. *55*, 145 (1971), C.A. *75*, 56383t (1971); ^{1}H spin relaxation in the mesogen PAA. [**350**] R. Y. Dong, M. M. Pintar, W. F. Forbes, J. Chem. Phys. *55*, 2449 (1971), C.A. *75*, 102342d (1971); ^{1}H spin relaxation study of the c/n transition.

[**351**] C. F. Schwerdtfeger, M. Marusic, A. Mackay, R. Y. Dong, Mol. Cryst. Liq. Cryst. *12*, 335 (1971), C.A. *75*, 26542b (1971); ESR study of the temperature dependence of molecular rotation in nematics. [**352**] R. Y. Dong, W. F. Forbes, M. M. Pintar, Solid State Commun. *9*, 151 (1971), C.A. *75*, 13130r (1971); Evidence for nonexponential time correlation functions in a liquid crystal. [**353**] R. Y. Dong, J. Magn. Reson. *7*, 60 (1972), C.A. *77*, 41096u (1972); ^{1}H spin-lattice relaxation in the nematic phase. [**354**] R. Y. Dong, W. F. Forbes, M. M. Pintar, Mol. Cryst. Liq. Cryst. *16*, 213 (1972), C.A. *76*, 133890f (1972); ^{1}H spin-lattice relaxation in nematic MBBA. [**355**] R. Y. Dong, M. Wiszniewska, E. Tomchuk, E. Bock, J. Chem. Phys. *59*, 6266 (1973), C.A. *80*, 76286v (1974); ^{1}H spin echo study of spin-spin relaxation in liquid crystals. [**356**] M. Wiszniewska, R. Y. Dong, E. Tomchuk, E. Bock, Can. J. Chem. *52*, 2294 (1974), C.A. *82*, 15860x (1975); ^{2}H relaxation study of the reorienting methyl groups in PAA. [**357**] R. Y. Dong, M. Wiszniewska, E. Tomchuk, E. Bock, Can. J. Phys. *52*, 766 (1974), C.A. *81*, 30688v (1974); ^{1}H spin relaxation study of order fluctuations above the n/i transition in MBBA. I. Applicability of BPP theory. [**358**] R. Y. Dong, M. Wiszniewska, E. Tomchuk, E. Bock, Can. J. Phys. *52*, 1331 (1974), C.A. *82*, 24636f (1975); ^{1}H spin-lattice relaxation study of order fluctuations above the s_A/n transition. [**359**] R. Y. Dong, M. Wiszniewska, E. Tomchuk, E. Bock, Chem. Phys. Lett. *25*, 299 (1974), C.A. *80*, 150755r (1974); ^{1}H spin relaxation in a smectic. [**360**] R. Y. Dong, E. Tomchuk, E. Bock, Magn. Reson. Relat. Phenom., Proc. Congr. Ampère, 18th, *1*, 205 (1974), C.A. *83*, 170583y (1975); Fine structure of the ^{1}HMR line in nematic MBBA by FFT.

[**361**] R. Y. Dong, M. Wiszniewska, E. Tomchuk, E. Bock, Mol. Cryst. Liq. Cryst. *27*, 259 (1974), C.A. *82*, 162451t (1975); ^{1}H spin relaxation study of the dipolar spin systems in nematics. [**362**] R. Y. Dong, E. Tomchuk, E. Bock, Mol. Cryst. Liq. Cryst. *29*, 117 (1974), C.A. *83*, 18420k (1975); ^{1}H spin-lattice relaxation near a weak first order s_A/n transition. [**363**] R. Y. Dong, M. Wiszniewska, E. Tomchuk, E. Bock, Solid State Commun. *14*, 691 (1974), C.A. *81*, 18970y (1974); ^{1}H spin echoes in nematics. [**364**] R. Y. Dong, E. Tomchuk, E. Bock, Can. J. Phys. *53*, 610 (1975), C.A. *83*, 18512s (1975); ^{1}H spin relaxation study of order fluctuations above the n/i transition in MBBA. II. Coherent and incoherent scattering. [**365**] J. J. Visintainer, E. Bock, R. Y. Dong, E. Tomchuk, Can. J. Phys. *53*, 1483 (1975), C.A. *83*, 170521b (1975); FFT study of the ^{1}HMR and ^{2}HMR line shapes of nematics. [**366**] R. Y. Dong, E. Tomchuk, J. J. Visintainer, E. Bock, Can. J. Phys. *54*, 1600 (1976), C.A. *85*, 101830f (1976); Order fluctuations in MBBA: ^{1}HMR study. [**367**] J. J. Visintainer, E. Bock, R. Y. Dong, E. Tomchuk, Can. J. Phys. *54*, 2282 (1976), C.A. *86*, 36093x (1977); ^{1}H FFT study of the aromatic and alkyl motions in the mesogen CBOOA: Orientational order fluctuations in the i phase.

[**368**] R. Y. Dong, E. Tomchuk, J. J. Visintainer, E. Bock, Mol. Cryst. Liq. Cryst. *33*, 101 (1976), C.A. *84*, 187212v (1976); ^{14}NMR study of order fluctuations in the i phase of mesogens. [**369**] J. J. Visintainer, R. Y. Dong, E. Bock, E. Tomchuk, D. E. Dewey, A. L. Kuo, C. G. Wade, J. Chem. Phys. *66*, 3343 (1977), C.A. *86*, 180224h (1977); ^{1}H and ^{2}H spin-lattice relaxation study of the partially deuterated nematogens PAA and MBBA. [**370**] R. Y. Dong, E. Tomchuk, C. G. Wade, J. J. Visintainer, E. Bock, J. Chem. Phys. *66*, 4121 (1977), C.A. *86*, 197437c (1977); ^{2}H line shape study of molecular order and conformation in the nematogens PAA and MBBA.

[**371**] R. Y. Dong, B. Nakka, E. Tomchuk, J. J. Visintainer, E. Bock, Mol. Cryst. Liq. Cryst. *38*, 53 (1977), C.A. *86*, 197359d (1977); Dipolar spin system in selectively deuterated nematogens. [**372**] R. Y. Dong, E. Tomchuk, E. Bock, Mol. Phys. *33*, 1503 (1977), C.A. *87*, 144195m (1977); NMR investigation of critical orientational order fluctuations below the n/i transition. [**373**] D. R. McMillin, R. S. Drago, Inorg. Chem. *13*, 546 (1974), C.A. *80*, 81568t (1974); NMR study of a platinum-ethylene complex in a liquid crystal solvent. [**374**] C. Schumann, H. Dreeskamp, K. Hildenbrand, J. Magn. Reson. *18*, 97 (1975), C.A. *83*, 68649n (1975); Anisotropy of indirect C—Hg and C—C nuclear spin coupling constants in dimethyl mercury. [**375**] Cl. Nicolau, H. Dreeskamp, D. Schulte-Frohlinde, FEBS Lett. *43*, 148 (1974), C.A. *81*, 116342q (1974); ^{13}CMR relaxation measurements of α-lecithin-peptide interaction in model membranes. [**376**] R. F. Grant, N. Hedgecock, B. A. Dunell, Can. J. Chem. *34*, 1514 (1956), C.A. *51*, 4136b (1957); NMR study of phase transitions in anhydrous sodium stearate. [**377**] R. F. Grant, B. A. Dunell, Can. J. Chem. *38*, 1951 (1960), C.A. *56*, 9600b (1962); ^{1}HMR absorption in anhydrous potassium stearate. [**378**] R. F. Grant, B. A. Dunell, Can. J. Chem. *38*, 2395 (1960), C.A. *55*, 10076h (1961); ^{1}HMR absorption in anhydrous sodium stearate. [**379**] R. F. Grant, B. A. Dunell, Can. J. Chem. *39*, 359 (1961), C.A. *56*, 13035g (1962); Phase transitions in anhydrous potassium soaps. [**380**] D. J. Shaw, B. A. Dunell, Trans. Faraday Soc. *58*, 132 (1962), C.A. *57*, 6771g (1962); ^{1}HMR study of phase transitions in rubidium and caesium stearates.

[**381**] W. R. Janzen, B. A. Dunell, Trans. Faraday Soc. *59*, 1260 (1963), C.A. *59*, 12324d (1963); ^{1}HMR absorption and molecular motion in anhydrous potassium caprylate and caproate. [**382**] M. R. Barr, B. A. Dunell, Can. J. Chem. *42*, 1098 (1964), C.A. *60*, 15321h (1964); ^{1}HMR absorption in high-temperature phases of anhydrous sodium stearate. [**383**] T. J. R. Cyr, W. R. Janzen, B. A. Dunell, Advan. Chem. Ser. No. *63*, 13 (1967), C.A. *67*, 12834w (1967); Effect of thermal history and impurities on phase transitions in long-chain fatty acid systems. [**384**] W. R. Janzen, B. A. Dunell, Can. J. Chem. *47*, 2722 (1969), C.A. *71*, 42813p (1969); Use of a saturation narrowed NMR spectrum to observe phase transitions in lithium stearate. [**385**] J. A. Ripmeester, B. A. Dunell, Can. J. Chem. *49*, 731 (1971), C.A. *74*, 105424x (1971); NMR studies of molecular motion in a number of stearates and oleates. [**386**] M. B. Dunn, Mol. Phys. *15*, 433 (1968), C.A. *70*, 62746b (1969); Chemical shift anisotropy in trifluoroacetic acid. [**387**] M. Dvolaitzky, F. Poldy, C. Taupin, Phys. Lett. A *45*, 454 (1973), C.A. *80*, 53195a (1974); Melting of the aliphatic chains in a k/s$_B$ transition. [**388**] F. Poldy, M. Dvolaitzky, C. Taupin, J. Phys. (Paris), C-1, *36*, 27 (1975), C.A. *83*, 8911g (1975); Spin label studies of chain organization in a smectic. [**389**] M. Dvolaitzky, C. Taupin, F. Poldy, Tetrahedron Lett. *1975*, 1469, C.A. *83*, 79042k (1975); Nitroxydes piperidiniques. Nouvelles sondes paramagnétiques. [**390**] F. Poldy, M. Dvolaitzky, C. Taupin, Chem. Phys. Lett. *42*, 449 (1976), C.A. *85*, 200766w (1976); Chain dynamics and residual order in the i phase of a smectogen.

[**391**] F. Poldy, M. Dvolaitzky, C. Taupin, J. Phys. (Paris), C-3, *37*, 77 (1976), C.A. *86*, 81952n (1977); ESR studies of smectics. [**392**] C. R. Dybowski, Diss. Univ. Texas, Austin. 1973, 200 pp. Avail. Univ. Microfilms, Ann Arbor, Mich., Order No. 73-26,002, C.A. *80*, 76321c (1974); NMR relaxation studies of liquid crystals. [**393**] M. P. Eastman, G. V. Bruno, J. O. Lawson, NASA Contract. Rep. 1977, NASA-CR-145201, 15 pp., C.A. *87*, 109057w (1977); ESR studies of the slow tumbling of vanadyl spin probes in nematics. [**394**] K. R. K. Easwaran, J. Magn. Reson. *9*, 190 (1973), C.A. *78*, 64818r (1973); NMR spectrum of nematic APAPA. [**395**] G. Gillberg, P. Ekwall, Acta Chem. Scand. *21*, 1630 (1967), C.A. *67*, 103223n (1967); Properties and structure of the decanolic solutions in the sodium caprylate/decanol/water system. NMR investigation of the solutions. [**396**] J. W. Emsley, J. C. Lindon, J. M. Tabony, T. H. Wilmshurst, J. Chem. Soc. D. *1971*, 1277, C.A. *76*, 24251c (1972); Simplification of the ^{1}HMR spectra of partially oriented molecules by partial deuteration and ^{2}H decoupling. [**397**] J. W. Emsley, J. C. Lindon, S. R. Salman, J. Chem. Soc., Faraday Trans. 2, *68*, 1343 (1972), C.A. *77*, 87305q (1972); ^{19}FMR spectrum of pentafluoropyridine partially oriented in a nematic. [**398**] J. W. Emsley, J. C. Lindon, G. R. Luckhurst, D. Shaw, Chem. Phys. Lett. *19*, 345 (1973), C.A. *78*, 166619u (1973); NMR studies of magnetohydrodynamic effects in the nematic phase. [**399**] J. W.

Emsley, J. C. Lindon, Mol. Phys. *25*, 641 (1973), C.A. *79*, 4502u (1973); Molecular structure of tropone from its ¹HMR spectrum in a nematic solvent. [**400**] J. W. Emsley, J. C. Lindon, J. Tabony, Mol. Phys. *26*, 1485 (1973), C.A. *80*, 107497w (1974); NMR spectra of ethanol and partially deuterated ethanols as solutes in a nematic.

[**401**] J. W. Emsley, J. C. Lindon, J. Tabony, Mol. Phys. *26*, 1499 (1973), C.A. *80*, 107496v (1974); Measurement of ²H quadrupole coupling constants of trideuteriomethyl groups by ¹H-{²H} double resonance studies on nematic solutions. [**402**] J. W. Emsley, J. C. Lindon, Chem. Phys. Lett. *26*, 361 (1974), C.A. *81*, 104196g (1974); Separation of isotropic from anisotropic interactions in the NMR spectra of solutes dissolved in nematics by using slow sample rotation. [**403**] I. R. Beattie, J. W. Emsley, R. M. Sabine, J. Chem. Soc., Faraday Trans. 2, *70*, 1356 (1974), C.A. *82*, 72264e (1975); ¹HMR spectra of partially oriented samples of cyclopentadienyl compounds. [**404**] J. W. Emsley, J. C. Lindon, D. S. Stephenson, Mol. Phys. *27*, 641 (1974), C.A. *81*, 36862u (1974); NMR spectrum of tropolone dissolved in nematic solvents. An exchanging spin system. [**405**] J. W. Emsley, J. C. Lindon, D. S. Stephenson, M. C. McIvor, Mol. Phys. *28*, 93 (1974), C.A. *82*, 56946y (1975); ¹HMR spectrum of cyclopentadiene dissolved in a nematic. [**406**] J. W. Emsley, J. C. Lindon, Mol. Phys. *28*, 1253 (1974), C.A. *82*, 124259d (1975); Analysis of partially oriented AA'XX' NMR spin systems with the aid of sample rotation. ¹⁹F spectrum of *cis*-difluoroethylene dissolved in a nematic. [**407**] J. W. Emsley, J. C. Lindon, Mol. Phys. *28*, 1373 (1974), C.A. *82*, 131175b (1975); Effect of vibrational averaging on the geometry of cyclobutadiene iron tricarbonyl derived from the ¹HMR spectrum of a nematic solution. [**408**] I. R. Beattie, J. W. Emsley, J. C. Lindon, R. M. Sabine, J. Chem. Soc., Dalton Trans. *1975*, 1264, C.A. *83*, 113514c (1975); Structural investigation of the η-allyl group in (η-allyl)tetracarbonylrhenium by nematic-phase NMR spectroscopy. [**409**] J. W. Emsley, J. C. Lindon, J. Tabony, J. Chem. Soc. Faraday Trans. 2, *71*, 579 (1975), C.A. *82*, 154663y (1975); Effect of vibrational averaging on the quadrupole coupling constant of deuterium in 4-[²H₁]pyridine determined from ¹H-{²H}-INDOR of a partially oriented sample. [**410**] J. W. Emsley, J. C. Lindon, J. Tabony, J. Chem. Soc., Faraday Trans. 2, *71*, 586 (1975), C.A. *82*, 155098e (1975); ¹H, ¹³C, and ¹H-{²H}-INDOR spectra of isotopically labeled acetaldehyde dissolved in a nematic.

[**411**] J. W. Emsley, J. C. Lindon, D. S. Stephenson, J. Chem. Soc., Perkin Trans. 2, *1975*, 1508, C.A. *84*, 30245k (1976); Structure of pentafluorobenzaldehyde determined from NMR spectra of nematic solutions. [**412**] J. W. Emsley, D. S. Stephenson, J. C. Lindon, L. Lunazzi, S. Pulga, J. Chem. Soc., Perkin Trans. 2, *1975*, 1541, C.A. *84*, 43167m (1976); Structure and conformation of 4,4'-bipyridyl by NMR spectroscopy of a nematic solution. [**413**] J. W. Emsley, J. C. Lindon, D. S. Stephenson, J. Chem. Soc., Perkin Trans. 2, *1975*, 1794, C.A. *84*, 89459m (1976); Structure of 2,3,5,6-tetrafluoroanisole determined from the analysis of a NMR spectrum of a nematic solution. [**414**] J. W. Emsley, J. Tabony, J. Magn. Reson. *17*, 233 (1975), C.A. *82*, 169614g (1975); Quadrupole coupling constants of deuterium in CH₂DCH₂Br and CH₃CD₂Br obtained from NMR in nematics and including vibrational averaging. [**415**] J. W. Emsley, J. C. Lindon, Mol. Phys. *29*, 531 (1975), C.A. *82*, 154979f (1975); ¹HMR spectrum of partially oriented norbornadiene. Molecular structure and the effect of vibrational averaging. [**416**] J. W. Emsley, J. C. Lindon, D. S. Stephenson, Mol. Phys. *30*, 1603 (1975), C.A. *84*, 89024r (1976); Evidence for an anisotropic contribution of ³J_{FF} coupling constant from the ¹⁹FMR spectrum of hexafluorocyclopropane dissolved in a nematic. [**417**] J. W. Emsley, J. C. Lindon, NMR spectroscopy using liquid crystal solvents. (Pergamon: Elmsford). 1975. 367 pp., C.A. *83*, 155568d (1975). [**418**] J. W. Emsley, J. C. Lindon, J. M. Street, G. E. Hawkes, J. Chem. Soc., Faraday Trans. 2, *72*, 1365 (1976), C.A. *86*, 4369a (1977); Conformation and reorientation of acetophenone in solution. ¹HMR and ²HMR study in a nematic solvent. [**419**] J. W. Emsley, J. C. Lindon, J. Chem. Soc., Faraday Trans. 2, *72*, 1436 (1976), C.A. *86*, 4701c (1977); Structures of 1,3,5-trichloro- and 1,3,5-trichlorotrifluorobenzene derived from NMR spectra of nematic solutions. [**420**] J. W. Emsley, J. C. Lindon, J. M. Street, J. Chem. Soc., Perkin Trans. 2, *1976*, 805, C.A. *85*, 77472t (1976); Conformations of 3,5-dichloroanisole and 3,5-dibromoacetophenone determined from ¹HMR spectra of nematic solutions.

[**421**] M. Duchene, J. W. Emsley, J. C. Lindon, D. S. Stephenson, S. R. Salman, J. Magn. Reson. *22*, 207 (1976), C.A. *86*, 4353r (1977); Structures of fluorinated pyridines from the analysis of NMR spectra of nematic solutions. [**422**] J. W. Emsley, E. W. Randall, J. Magn. Reson. *23*, 481 (1976), C.A. *86*, 88920j (1977); Effect of vibrational averaging on the molecular structure of pyrrole-¹⁵N derived from the NMR spectra of nematic solutions. [**423**] J. W. Emsley, J. C. Lindon, J. Tabony, Mol. Phys. *31*, 1617 (1976), C.A. *85*, 142180n (1976); Measurement of ²H quadrupole coupling constants of C²H₃ groups by ¹H-{²H}-NMR studies of nematic solutions. Erratum. [**424**] J. W. Emsley, J.

C. Lindon, G. R. Luckhurst, Mol. Phys. *32*, 1187 (1976), C.A. *86*, 105657y (1977); Molecular rotation in liquid crystals. Selective ^{2}H spin-lattice relaxation times in nematic 4-cyano-4'-d$_{17}$-n-octylbiphenyl. [**425**] M. Duchene, J. W. Emsley, J. C. Lindon, J. Overstall, D. S. Stephenson, S. R. Salman, Mol. Phys. *33*, 281 (1977), C.A. *87*, 52299v (1977); Molecular structure and reorientation of *p*-fluorobenzaldehyde dissolved in a smectic studied using NMR spectroscopy. [**426**] G. Englert, Z. Naturforsch. A *24*, 1074 (1969), C.A. *71*, 76009w (1969); ^{1}HMR study of dimethylmercury in a nematic. [**427**] G. Englert, Z. Naturforsch. A *27*, 715 (1972), C.A. *77*, 47544k (1972); ^{13}CMR chemical shift anisotropy in benzene-1-^{13}C. [**428**] G. Englert, Z. Naturforsch. A *27*, 1535 (1972), C.A. *78*, 57227u (1973); ^{13}CMR chemical shift anisotropy of acetylene. [**429**] M. Ero-Gecs, J. Menczel, KFKI (Rep.) *1974*, KFKI-74-13, 14 pp., C.A. *81*, 127888y (1974); ESR study of order in viscous nematics. [**430**] M. Ero-Gecs, J. Menczel, KFKI (Rep.) *1975*, KFKI-75-18, 13 pp., C.A. *83*, 124397t (1975); ESR study of the distribution function in nematics.

[**431**] R. Ewing, J. C. Lee, Phys. Rev. *94*, 1411 (1954), C.A. *49*, 10043d (1955); ^{1}HMR in compounds having several mesophases. [**432**] J. P. Fackler Jr., J. A. Smith, J. Am. Chem. Soc. *92*, 5787 (1970), C.A. *73*, 125581d (1970); ESR spectra of metal complexes oriented in nematic glasses. [**433**] J. P. Fackler, Jr., J. D. Levy, J. A. Smith, J. Am. Chem. Soc. *94*, 2436 (1972), C.A. *76*, 147081e (1972); ESR spectra of copper-(II) and oxovanadium(IV) complexes oriented in nematic glasses. [**434**] H. R. Falle, M. A. Whitehead, Can. J. Chem. *50*, 139 (1972), C.A. *76*, 112457e (1972); Theoretical anisotropic hyperfine tensors in neutral radicals. [**435**] M. E. Field, Diss. Univ. Maine, Portland. 1976. 118 pp. Avail. Univ. Microfilms Int., Order No. 77-13,731, C.A. *87*, 93796y (1977); Study of the molecular dynamics of cholesterics using pulsed NMR. [**436**] A. Darke, E. G. Finer, A. G. Flook, M. C. Phillips, FEBS Lett. *18*, 326 (1971), C.A. *76*, 82523d (1972); Complex and cluster formation in mixed lecithin-cholesterol bilayers. Cooperativity of motion in lipid systems. [**437**] E. G. Finer, A. G. Flook, H. Hauser, FEBS Lett. *18*, 331 (1971), C.A. *76*, 82521b (1972); Use of NMR spectra of sonicated phospholipid dispersions in studies on interactions with the bilayer. [**438**] E. G. Finer, A. G. Flook, H. Hauser, Biochim. Biophys. Acta *260*, 59 (1972), C.A. *76*, 109428r (1972); Nature and origin of the NMR spectrum of unsonicated and sonicated aqueous egg yolk lecithin dispersions. [**439**] E. G. Finer, Chem. Phys. Lipids *8*, 327 (1972), C.A. *77*, 71709p (1972); NMR studies of lipid/water systems. [**440**] A. Darke, E. G. Finer, A. G. Flook, M. C. Phillips, J. Mol. Biol. *63*, 265 (1972), C.A. *76*, 96063r (1972); NMR study of lecithin-cholesterol interactions.

[**441**] E. G. Finer, M. C. Phillips, Chem. Phys. Lipids *10*, 237 (1973), C.A. *79*, 14755t (1973); Factors affecting molecular packing in mixed lipid monolayers and bilayers. [**442**] E. G. Finer, J. Chem. Soc., Faraday Trans. 2, *69*, 1590 (1973), C.A. *80*, 34573f (1974); Interpretation of ^{2}HMR studies of the hydration of macromolecules. [**443**] E. G. Finer, A. Darke, Chem. Phys. Lipids *12*, 1 (1974), C.A. *81*, 22439z (1974); Phospholipid hydration studied by ^{2}HMR spectroscopy. [**444**] E. G. Finer, J. Magn. Reson. *13*, 76 (1974), C.A. *80*, 126490e (1974); Calculation of molecular motional correlation times from line widths in NMR spectra of aggregated systems. Effect of particle size on spectra of phospholipid dispersions. [**445**] C. F. Polnaszek, G. V. Bruno, J. H. Freed, J. Chem. Phys. *58*, 3185 (1973), C.A. *78*, 153504c (1973); ESR line shapes in the slow-motional region. Anisotropic liquids. [**446**] C. F. Polnaszek, J. H. Freed, J. Phys. Chem. *79*, 2283 (1975), C.A. *83*, 185961u (1975); ESR studies of anisotropic ordering, spin relaxation, and slow tumbling in liquid crystalline solvents. [**447**] J. S. Hwang, K. V. S. Rao, J. H. Freed, J. Phys. Chem. *80*, 1490 (1976), C.A. *85*, 54279w (1976); ESR study of pressure dependence of ordering and spin relaxation in a liquid-crystalline solvent. [**448**] K. V. S. Rao, J. S. Hwang, J. H. Freed, Phys. Rev. Lett. *37*, 515 (1976), C.A. *85*, 133621e (1976); Symmetry of orientational order fluctuations about the n/i transition: ESR study. [**449**] J. H. Freed, J. Chem. Phys. *66*, 4183 (1977), C.A. *87*, 31553h (1977); Stochastic-molecular theory of spin-relaxation for liquid crystals. [**450**] K. V. S. Rao, C. F. Polnaszek, J. H. Freed, J. Phys. Chem. *81*, 449 (1977), C.A. *86*, 130534p (1977); ESR studies of anisotropic ordering, spin relaxation, and slow tumbling in liquid crystalline solvents. 2.

[**451**] A. Frey, R. R. Ernst, Chem. Phys. Lett. *49*, 75 (1977), C.A. *87*, 109700a (1977); Deformation and orientation of solute molecules in nematics. [**452**] G. C. Fryburg, E. Gelerinter, J. Chem. Phys. *52*, 3378 (1970), C.A. *72*, 116634a (1970); ESR studies of a viscous nematic. I. [**453**] E. Gelerinter, G. C. Fryburg, Appl. Phys. Lett. *18*, 84 (1971), C.A. *74*, 93282j (1971); ESR study of a smectic C: new method for determining the tilt angle. [**454**] G. C. Fryburg, E. Gelerinter, D. L. Fishel, Mol. Cryst. Liq. Cryst. *16*, 39 (1972), C.A. *76*, 79029s (1972); ESR study of two smectics A. [**455**] W. E. Shutt, E. Gelerinter, G. C. Fryburg, C. F. Sheley, J. Chem. Phys. *59*, 143 (1973), C.A. *79*, 58686a

(1973); II. Evidence counter to a second-order phase change (cf. [452]). [456] G. C. Fryburg, E. Gelerinter, Mol. Cryst. Liq. Cryst. *22*, 77 (1973), C.A. *79*, 109048d (1973); ESR study of alignment induced by magnetic fields in two smectics A not exhibiting nematics. [457] J. I. Kaplan, E. Gelerinter, G. C. Fryburg, Mol. Cryst. Liq. Cryst. *23*, 69 (1973), C.A. *79*, 151291c (1973); Resonance spectra of a paramagnetic probe dissolved in a viscous medium. [458] A. Berman, E. Gelerinter, G. C. Fryburg, G. H. Brown, Liq. Cryst. Ord. Fluids *2*, 23 (1974), C.A. *86*, 180975d (1977); ESR investigation of the alignment of two smectics A. [459] E. Gelerinter, A. L. Bermann, G. C. Fryburg, S. L. Golub, Phys. Rev. A *9*, 2099 (1974), C.A. *81*, 18025u (1974); ESR observation of d.c. electric field effects in nematics. [460] A. M. Fuller, Jr., Diss. Univ. Maine, Orono. 1975. 118 pp. Avail. Xerox Univ. Microfilms, Ann Arbor, Mich., Order No. 76-51,118, C.A. *84*, 114455s (1976); Pulsed NMR study of the liquid crystal EMC.

[461] B. M. Fung, I. Y. Wei, J. Am. Chem. Soc. *92*, 1497 (1970), C.A. *72*, 105683q (1970); ^{1}HMR and ^{2}HMR of phenylsilane-d_3, phenylphosphine-d_2, and benzenethiol-d in liquid crystal solutions. [462] I. Y. Wei, B. M. Fung, J. Chem. Phys. *52*, 4917 (1970), C.A. *73*, 30458e (1970); ^{2}H quadrupole coupling constants in nitrobenzenes. [463] B. M. Fung, M. J. Gerace, J. Chem. Phys. *53*, 1171 (1970), C.A. *73*, 71860y (1970); ^{1}HMDR in a liquid-crystal solution. [464] M. J. Gerace, B. M. Fung, J. Chem. Phys. *53*, 2984 (1970), C.A. *73*, 114790v (1970); ^{14}NMR and quadrupole coupling constants in liquid crystalline PBLG solutions. [465] B. M. Fung, M. J. Gerace, L. S. Gerace, J. Phys. Chem. *74*, 83 (1970), C.A. *72*, 61150g (1970); NMR study of liquid crystalline solutions of PBLG in dichloromethane and 1,2-dichloroethane. [466] B. M. Fung, J. Magn. Reson. *15*, 170 (1974), C.A. *81*, 112208c (1974); Reorientation of a thermotropic liquid crystal in a magnetic field as studied by Fourier transform NMR. [467] I. J. Gazzard, N. Sheppard, Mol. Phys. *21*, 169 (1971), C.A. *75*, 69278n (1971); NMR spectra of oriented molecules. I. Internal rotation and conformational preference of 1,2,2,3-tetrachloropropane. [468] I. J. Gazzard, Mol. Phys. *21*, 175 (1971), C.A. *75*, 92783g (1971); II. Molecular geometry of ethylene oxide and ethylene sulfide (cf. [467]). [469] I. J. Gazzard, Mol. Phys. *25*, 469 (1973), C.A. *78*, 142228q (1973); IV. Molecular geometry of ethylenimine (cf. [467]). [470] A. L. Berman, E. Gelerinter, Solid State Commun. *16*, 61 (1975), C.A. *82*, 78653c (1975); ESR study of a nematic oriented by an electric field.

[471] E. Gelerinter, C. Flick, Chem. Phys. Lett. *44*, 300 (1976), C.A. *86*, 63108n (1977); Spin label study of a smectic C. [472] C. Flick, E. Gelerinter, R. Semer, Mol. Cryst. Liq. Cryst. *37*, 71 (1976), C.A. *87*, 123p (1977); Spin label study on nystadin and amphotericin B action on lipid planar multibilayers. [473] M. J. Gerace, Diss. Tufts Univ., Medford. 1970. 103 pp. Avail. Univ. Microfilms, Ann Arbor, Mich., Order No. 71-892, C.A. *76*, 39946g (1972); NMR studies of liquid crystalline solutions. [474] S. K. Ghosh, E. Tettamanti, P. L. Indovina, Phys. Rev. Lett. *29*, 638 (1972), C.A. *77*, 119308m (1972); Dynamical behavior of a nematogen just above the n/i transition from spin-lattice relaxation. [475] S. K. Ghosh, Solid State Commun. *11*, 1763 (1972), C.A. *78*, 64915v (1973); Order parameter of nematics from dipolar splittings of NMR spectra. [476] S. K. Ghosh, E. Tettamanti, Phys. Lett. A *43*, 361 (1973), C.A. *78*, 164316u (1973); Self-diffusion of nematogens in the i phase near the n/i transition. [477] E. Boilini, S. K. Ghosh, J. Appl. Phys. *46*, 78 (1975), C.A. *82*, 148593j (1975); Improved method for measuring the order parameter in nematics. [478] S. K. Ghosh, E. Tettamanti, P. L. Indovina, Z. Phys. B *24*, 227 (1976), C.A. *85*, 54267r (1976); Dynamic critical behavior in PAA from nuclear relaxation studies. [479] C. T. Yim, D. F. R. Gilson, Can. J. Chem. *46*, 2783 (1968), C.A. *69*, 91682n (1968); Analysis of the ^{1}HMR and ^{19}FMR spectra of 1,3,5-trifluorobenzene dissolved in a nematic. [480] C. T. Yim, D. F. R. Gilson, Can. J. Chem. *47*, 1057 (1969), C.A. *70*, 82825e (1969); NMR spectra of *o*-, *m*-, and *p*-difluorobenzene in a nematic solvent.

[481] C. T. Yim, D. F. R. Gilson, J. Am. Chem. Soc. *91*, 4360 (1969), C.A. *71*, 55367w (1969); Anisotropy of fluorine chemical shifts in substituted fluorobenzenes. [482] J. C. Robertson, C. T. Yim, D. F. R. Gilson, Can. J. Chem. *49*, 2345 (1971), C.A. *75*, 68622h (1971); Orientation of solute molecules in nematic solvents. [483] K. C. Cole, D. F. R. Gilson, J. Chem. Phys. *56*, 4363 (1972), C.A. *76*, 160543b (1972); Molecular structure of trimethylene oxide from its NMR spectrum in a nematic solvent. [484] K. C. Cole, D. F. R. Gilson, Can. J. Chem. *52*, 281 (1974), C.A. *80*, 95130f (1974); Molecular structure of 2,5-dihydrofuran from the analysis of the nematic liquid crystal NMR spectrum. [485] K. C. Cole, D. F. R. Gilson, J. Chem. Phy. *60*, 1191 (1974), C.A. *80*, 126481c (1974); Molecular structure of cyclobutane. Correction of the nematic phase NMR results for ring puckering motion. Comments. [486] K. C. Cole, D. F. R. Gilson, Can. J. Spectrosc. *20*, 61 (1975), C.A. *83*, 51365u (1975); Analysis of the NMR spectrum of partially oriented 2-cyclopentenone. Molecular structure

and orientation. [**487**] K. C. Cole, D. F. R. Gilson, J. Mol. Struct. *28*, 385 (1975), C.A. *83*, 163479r (1975); Molecular structures of ethylene and 1,1-dichlorocyclopropane. Comparison of averaged structures. [**488**] K. C. Cole, D. F. R. Gilson, Mol. Phys. *29*, 1749 (1975), C.A. *83*, 96270a (1975); Structures of trimethylene oxide and trimethylene sulfide by ^{1}HMR of the solutions in a nematic. [**489**] K. C. Cole, D. F. R. Gilson, Can. J. Chem. *54*, 657 (1976), C.A. *84*, 179136h (1976); Nematic phase NMR studies of the molecular structures of cyclobutanone and methylenecyclobutane. [**490**] Y. P. Lee, D. F. R. Gilson, Can. J. Chem. *54*, 2783 (1976), C.A. *86*, 15938m (1977); Nematic phase NMR studies of the structure and orientation of 1,5- and 2,6-naphthyridines.

[**491**] K. C. Cole, D. F. R. Gilson, Can. J. Chem. *54*, 2788 (1976), C.A. *86*, 54592p (1977); Nematic phase NMR spectra and vibrationally averaged structures of cyclopropyl chloride, bromide, and cyanide. [**492**] S. H. Glarum, J. H. Marshall, J. Chem. Phys. *44*, 2884 (1966), C.A. *64*, 16869a (1966); ESR of the perinaphthenyl radical in a liquid crystal. [**493**] S. H. Glarum, J. H. Marshall, J. Chem. Phys. *46*, 55 (1967), C.A. *66*, 42166b (1967); Paramagnetic relaxation in liquid-crystal solvents. [**494**] S. H. Glarum, J. H. Marshall, J. Chem. Phys. *47*, 1374 (1967), C.A. *67*, 86393y (1967); Spin exchange in nitroxide biradicals. [**495**] R. C. Long, Jr., J. H. Goldstein, J. Chem. Phys. *54*, 1563 (1971), C.A. *74*, 93262c (1971); NMR study of tetrafluoro-1,3-dithietane in an isotropic and a lyotropic phase. [**496**] R. C. Long, Jr., J. H. Goldstein, J. Mol. Spectrosc. *40*, 632 (1971), C.A. *76*, 29402n (1972); Orientation of p-dithiin in a lyophase. [**497**] R. C. Long, Jr., S. L. Baughcum, J. H. Goldstein, J. Magn. Reson. *7*, 253 (1972), C.A. *77*, 100317s (1972); NMR studies of furan and thiophene partially oriented in a lyophase. [**498**] R. C. Long, Jr., K. R. Long, J. H. Goldstein, Mol. Cryst. Liq. Cryst. *21*, 299 (1973), C.A. *79*, 109022r (1973); Orientation and structure of pyrazine, pyrimidine, and pyridazine in a lyophase. [**499**] R. C. Long, Jr., J. H. Goldstein, Mol. Cryst. Liq. Cryst. *23*, 137 (1973); NMR-studies of various solutes in a lyophase. [**500**] S. A. Spearman, J. H. Goldstein, J. Magn. Reson. *20*, 75 (1975), C.A. *84*, 97496m (1976); NMR studies of p-dioxene oriented in two lyophases.

[**501**] R. C. Long, Jr., J. H. Goldstein, Mol. Phys. *30*, 681 (1975), C.A. *83*, 205625t (1975); NMR of sodium methylphosphonate in a lyophase. Molecular structure and orientation. [**502**] S. A. Spearman, J. H. Goldstein, Spectrochim. Acta A *31*, 1565 (1975), C.A. *84*, 16404e (1976); NMR study of 4-pyrone oriented in two different lyophases. [**503**] S. A. Spearman, R. C. Long, Jr., J. H. Goldstein, J. Magn. Reson. *21*, 457 (1976), C.A. *85*, 142147g (1976); Structural and orientational studies of 4-methylpyridine oriented in a lyophase. [**504**] R. C. Long, Jr., J. H. Goldstein, J. Magn. Reson. *23*, 519 (1976), C.A. *86*, 43013z (1977); ^{2}HMR investigation of alkyl chain order in an oriented quaternary potassium laurate mesophase. [**505**] S. A. Spearman, J. H. Goldstein, J. Magn. Reson. *26*, 237 (1977), C.A. *87*, 83986w (1977); NMR studies of fluorobenzenes partially oriented in a lyophase. [**506**] S. A. Spearman, J. H. Goldstein, J. Mol. Struct. *36*, 243 (1977), C.A. *86*, 188612d (1977); NMR studies of p-substituted pyridine derivatives oriented in lyophases. [**507**] R. C. Long, Jr., J. H. Goldstein, Org. Magn. Reson. *9*, 148 (1977), C.A. *87*, 151201p (1977); NMR of oriented molecules in lyophases: pyridine, 4-cyanopyridine and benzonitrile. [**508**] R. C. Long, Jr., J. H. Goldstein, Org. Magn. Reson. *10*, 132 (1977), C.A. *89*, 145966s (1978); NMR of oriented molecules in lyophases: p-cresol. [**509**] J. M. Corkill, J. F. Goodman, J. A. Wyer, Trans. Faraday Soc. *65*, 9 (1969), C.A. *70*, 42663u (1969); NMR of aqueous solutions of alkyl(polyoxyethylene) glycol monoethers. [**510**] S. D. Goren, C. Korn, S. B. Marks, R. Potashnik, Chem. Phys. Lett. *24*, 249 (1974), C.A. *80*, 113485g (1974); ^{1}H spin-lattice relaxation in the k, n, and i phases of EBBA.

[**511**] S. D. Goren, S. B. Marks, R. Potashnik, Chem. Phys. Lett. *28*, 400 (1974), C.A. *82*, 9709p (1975); Effects of rotation on NMR linewidths in nematic solvents. [**512**] S. D. Goren, C. Korn, S. B. Marks, R. Potashnik, Magn. Reson. Relat. Phenom., Proc. Congr. Ampère, 18th, *1*, 209 (1974), C.A. *83*, 170585a (1975); ^{1}H spin lattice relaxation time in supercooled nematic. [**513**] S. B. Marks, R. Potashnik, S. D. Goren, J. Magn. Reson. *17*, 132 (1975), C.A. *82*, 147915d (1975); Oriented ^{1}HMR spectrum of methanol dissolved in a nematic solvent. [**514**] S. D. Goren, C. Korn, S. B. Marks, Phys. Rev. Lett. *34*, 1212 (1975), C.A. *83*, 19483v (1975); ^{1}HMR measurements of liquid crystals whose nematic range has been extended by supercooling. [**515**] M. Schwartz, P. E. Fagerness, C. H. Wang, D. M. Grant, J. Chem. Phys. *60*, 5066 (1974), C.A. *81*, 84102t (1974); Temperature dependent study of ^{13}C spin-lattice relaxation times in nematic PAA. [**516**] A. S. Waggoner, O. H. Griffith, C. R. Christensen, Proc. Nat. Acad. Sci. U.S. *57*, 1198 (1967), C.A. *68*, 73902j (1968); Magnetic resonance of nitroxide probes in micelle-containing solutions. [**517**] A. S. Waggoner, A. D. Keith, O. H. Griffith, J. Phys. Chem. *72*, 4129 (1968), C.A. *70*, 24536a (1969); ESR of solubilized long-chain nitroxides. [**518**] O. H. Griffith, A. S. Waggoner, Accounts Chem. Res. *2*, 17 (1969), C.A. *70*, 54117g (1969); Nitroxide

free radicals: spin labels for probing biomolecular structure. [519] A. S. Waggoner, T. J. Kingzett, S. Rottschaefer, O. H. Griffith, A. D. Keith, Chem. Phys. Lipids *3*, 245 (1969), C.A. *72*, 9329h (1970); Spin-labeled lipid for probing biological membranes. [520] O. H. Griffith, L. J. Libertini, G. B. Birrell, J. Phys. Chem. *75*, 3417 (1971), C.A. *75*, 145874w (1971); Role of lipid spin labels in membrane biophysics. [521] J. B. Pawliczek, H. Günther, J. Am. Chem. Soc. *93*, 2050 (1971); NMR spectra of oriented benzocyclopropene and 7,7-difluorobenzocyclopropene. [522] W. Herrig, H. Günther, J. Magn. Reson. *8*, 284 (1972), C.A. *77*, 170959b (1972); ¹HMR spectrum of partially oriented cyclobutene. [523] H. Günther, W. Herrig, J. B. Pawliczek, Z. Naturforsch. B *29*, 104 (1974), C.A. *81*, 37124s (1974); ¹HMR-Spektrum von partiell orientiertem Cyclopentadien. [524] C. J. Jameson, H. S. Gutowsky, J. Chem. Phys. *51*, 2790 (1969), C.A. *71*, 128958b (1969); Systematic trends in the coupling constants of directly bonded nuclei. [525] C. W. Haigh, S. Sykes, Chem. Phys. Lett. *19*, 571 (1973), C.A. *79*, 11491t (1973); Calculations of H-F and F-F NMR spin-spin coupling anisotropies for fluorinated ethenes and benzenes oriented in liquid crystal solvents. [526] R. K. Harris, V. J. Gazzard, Org. Magn. Reson. *3*, 495 (1971), C.A. *75*, 156758n (1971); Use of liquid crystal solvents for obtaining isotropic scalar nuclear spin-spin coupling constants. [527] K. Moebius, H. Haustein, M. Plato, Z. Naturforsch. A *23*, 1626 (1968), C.A. *70*, 42745x (1969); Hochauflösende ESR-Spektroskopie an organischen Radikalen in nematischen Flüssigkristallen. [528] H. Haustein, K. Moebius, K. P. Dinse, Z. Naturforsch. A *24*, 1764 (1969), C.A. *72*, 61193y (1970); Untersuchung des Einflusses von Leitsalz auf die ESR-Spektren neutraler und geladener Radikale in Flüssigkristallen. [529] H. Haustein, K. Moebius, K. P. Dinse, Z. Naturforsch. A *24*, 1768 (1969), C.A. *72*, 49505u (1970); ESR-Untersuchung an elektrolytisch erzeugten Semichinonen in Flüssigkristallen. [530] K. P. Dinse, R. Biehl, K. Moebius, H. Haustein, Chem. Phys. Lett. *12*, 399 (1971), C.A. *76*, 92715g (1972); ENDOR studies of organic radicals in liquid crystals.

[531] H. Haustein, K. P. Dinse, K. Moebius, Z. Naturforsch. A *26*, 1230 (1971), C.A. *75*, 135554c (1971); ESR-Untersuchung am Tetracyanochinodimethan-Anion in Flüssigkristallen. [532] K. P. Dinse, K. Moebius, M. Plato, R. Biehl, H. Haustein, Chem. Phys. Lett. *14*, 196 (1972), C.A. *77*, 40124q (1972); Observation of quadrupole splittings of organic radicals in solution by ENDOR in liquid crystals. [533] K. Hayamizu, O. Yamamoto, J. Chem. Phys. *51*, 1676 (1969), C.A. *71*, 96673m (1969); Measurement of the ¹H anisotropic shielding in a nematic. [534] K. Hayamizu, O. Yamamoto, J. Magn. Reson. *2*, 377 (1970), C.A. *74*, 105409w (1971); Determination of the ¹H chemical shift anisotropy by liquid crystals. [535] K. Hayamizu, O. Yamamoto, J. Chem. Phys. *54*, 3243 (1971), C.A. *74*, 147572z (1971); Measurement of the ¹H anisotropic shielding in a nematic. Reply to comments. [536] K. Hayamizu, O. Yamamoto, J. Magn. Reson. *5*, 94 (1971), C.A. *75*, 103370y (1971); Anisotropic ¹H chemical shieldings of methyl halides obtained from partially oriented molecules in a lyophase. [537] K. Hayamizu, O. Yamamoto, J. Chem. Phys. *66*, 1720 (1977), C.A. *86*, 170219w (1977); ¹H spin-lattice relaxation in selectively deuterated PAA. [538] H. Hayashi, H. Noda, S. Watanabe, K. Hirakawa, Kyushu Daigaku Kogaku Shuho *48*, 233 (1975), C.A. *86*, 149011n (1977); Gel-liquid crystal transition in a lyophase. ¹HMR of potassium palmitate. [539] H. Hayashi, K. Harada, M. Yamamoto, K. Hirakawa, Kyushu Daigaku Kogaku Shuho *49*, 297 (1976), C.A. *86*, 180969e (1977); Effect of potassium ion on the gel-liquid crystal transition in lyophases. ¹HMR of the potassium palmitate/water system. [540] R. J. Hayward, K. J. Packer, Mol. Phys. *26*, 1533 (1973), C.A. *80*, 107498x (1974); ¹H spin relaxation and self-diffusion in isotropic nematogens.

[541] M. A. Hemminga, Chem. Phys. *6*, 87 (1974), C.A. *82*, 37038n (1975); Angular dependent line widths of ESR spin probes in oriented smectics. Application to oriented lecithin-cholesterol multibilayers. [542] M. A. Hemminga, Chem. Phys. Lipids *14*, 151 (1975), C.A. *82*, 166241x (1975); ESR spin label study of structural and dynamical properties of the oriented lecithin-cholesterol multibilayers. [543] U. Henriksson, L. Ödberg, J. C. Eriksson, Mol. Cryst. Liq. Cryst. *30*, 73 (1975), C.A. *83*, 105850u (1975); Quadrupole splittings in ²HMR-spectra of the hexagonal phase in the system sodium octanoate-d₁₅/water/carbon tetrachloride. [544] U. Henriksson, T. Klason, L. Ödberg, J. C. Eriksson, Chem. Phys. Lett. *52*, 554 (1977), C.A. *88*, 68040k (1978); Solubilization of benzene and cyclohexane in aqueous solutions of hexadecyltrimethylammonium bromide: ²HMR study. [545] G. Heppke, F. Schneider, Ber. Bunsenges. Phys. Chem. *75*, 61 (1971), C.A. *74*, 93364n (1971); ESR-Untersuchungen des Ordnungsgrades von Vanadylacetylacetonat in nematischem und glasig erstarrtem MBBA. [546] D. F. Hillenbrand, Diss. Univ. Wisconsin. 1974, 139 pp. Avail Xerox Univ. Microfilms, Ann Arbor, Mich., Order No. 74-22,124, C.A. *82*, 154706q (1975); NMR studies of liquid crystal-nonliquid crystal mixtures, order, concentration, and temperature dependence in *p,p′-n*-hexyloxyazoxybenzene/monofluorobenzene and MBBA/biphenyl systems. [547] D. F. Hillenbrand, H. Yu, J. Chem. Phys. *67*, 957 (1977), C.A. *87*, 109764z (1977);

Solute orientational order in a nematic binary solution by NMR. [**548**] B. M. Hoffmann, F. Basolo, D. L. Diemente, J. Am. Chem. Soc. *95*, 6497 (1973), C.A. *79*, 140476r (1973); Spin-Hamiltonian axes of cobalt(II) Schiff base complexes from ESR in oriented nematic glass. [**549**] D. E. Holmes, Mod. Pharmacol. *1*, Pt. 2, 601 (1973), C.A. *81*, 45125h (1974); Drug induced changes in ESR spectra from spin labeled membranes. [**550**] B. Pedersen, J. Schaug, H. Hopf, Acta Chem. Scand. A *28*, 846 (1974), C.A. *82*, 85572d (1975); Interpretation of the ¹HMR spectrum of 1,2,4,5-hexatetraene oriented in a nematic solvent.

[**551**] D. L. Hoy, Diss. Pennsylvania, State Univ., University Park. 1970. 106 pp. Avail. Univ. Microfilms, Ann Arbor, Mich., Order No. 71-6319, C.A. *75*, 135548d (1971); NMR in liquid crystal solvents. Difluoroethylenes. [**552**] T. N. Huckerby, Tetrahedron Lett. *1971*, 3497, C.A. *76*, 3308m (1972); Nematic phase NMR spectra of furan- and thiophene-2,5-dialdehyde. New approach to conformational analysis. [**553**] A. Hudson, M. J. Kennedy, J. Inorg. Nucl. Chem. *32*, 2107 (1970), C.A. *73*, 93434q (1970); ESR of copper(II) dimers in a nematic solvent. [**554**] H. Inoue, T. Nakagawa, J. Phys. Chem. *70*, 1108 (1966), C.A. *64*, 16148g (1966); Shift of NMR signal caused by micelle formation. II. Micelle structure of mixed surfactants. [**555**] H. Utsumi, K. Inoue, S. Nojima, T. Kwan, Chem. Pharm. Bull. *24*, 1219 (1976), C.A. *85*, 73901j (1976); Motional state of spin-labeled stearates in lecithin-cholesterol liposomes and their incorporation capability. [**556**] J. Israelachvili, J. Sjosten, L. E. G. Eriksson, M. Ehrstrom, A. Graslund, A. Ehrenberg, Biochim. Biophys. Acta *339*, 164 (1974), C.A. *81*, 59700d (1974); Theoretical analysis of the molecular motion of spin labels in membranes. ESR spectra of labeled Bacilus subtilis membranes. [**557**] J. P. Jacobsen, H. K. Bildsoe, K. Schaumburg, J. Magn. Reson. *23*, 153 (1976), C.A. *85*, 114282d (1976); Application of density matrix formalism in NMR spectroscopy. II. The one-spin-1 case in anisotropic phase. [**558**] A. Jasinski, P. G. Morris, P. Mansfield, Magn. Reson. Relat. Phenom., Proc. Congr. Ampère, 19th *1976*, 185, C.A. *86*, 197383g (1977); ¹⁹F multipulse NMR study of the lamellar mesophase of amphiphilic liquid crystals. [**559**] T. Drakenberg, A. Johansson, S. Forsén, J. Phys. Chem. *74*, 4528 (1970), C.A. *74*, 36728k (1971); Magnetic susceptibility anisotropies in lyophases as studied by high-resolution ¹HMR. [**560**] N. O. Persson, A. Johansson, Acta Chem. Scand. *25*, 2118 (1971), C.A. *76*, 7527e (1972); Studies of lyophases containing ionic amphiphiles by means of ²HMR and ²³NaMR.

[**561**] A. Johansson, T. Drakenberg, Mol. Cryst. Liq. Cryst. *14*, 23 (1971), C.A. *75*, 123346y (1971); ¹HMR and ²HMR studies of lamellar lyophases. [**562**] C. S. Johnson, Jr., C. L. Watkins, J. Phys. Chem. *75*, 2452 (1971), C.A. *75*, 82180u (1971); Nuclear spin relaxation in nematics. [**563**] J. Jonas, Magn. Reson. Rev. *2*, 203 (1973), C.A. *80*, 53975e (1974); NMR relaxation in liquids and liquid crystals. [**564**] H. Kamei, Nippon Butsuri Gakkaishi *30*, 195 (1975), C.A. *83*, 36643q (1975); Nuclear spin relaxation in liquid crystals. [**565**] N. Kamezawa, J. Magn. Reson. *21*, 211 (1976), C.A. *84*, 187217a (1976); NMR spectra in the spiral mesophase consisting of nematic and optically active material. [**566**] N. Kamezawa, J. Magn. Reson. *27*, 325 (1977), C.A. *87*, 159643e (1977); Structure of spiral-like mesophases studied by high-resolution NMR of dissolved molecules. [**567**] J. I. Kaplan, Electron Spin Relaxation Liquids, Lect. NATO Advan. Study Inst. *1971*, 443, C.A. *78*, 50219j (1973); Two problems involving ESR in liquid crystals. [**568**] T. Yonezawa, I. Morishima, K. Deguchi, H. Kato, J. Chem. Phys. *51*, 5731 (1969), C.A. *72*, 72963a (1970); NMR spectra of ethylene derivatives dissolved in the two intermediate phases of the nematic solvent. [**569**] H. Nakatsuji, H. Kato, I. Morishima, T. Yonezawa, Chem. Phys. Lett. *4*, 607 (1970), C.A. *72*, 127056q (1970); Anisotropy of the indirect nuclear spin-spin coupling constant. I. [**570**] H. Nakatsuji, K. Hirao, H. Kato, T. Yonezawa, Chem. Phys. Lett. *6*, 541 (1970), C.A. *73*, 135779p (1970); II. Treatment by the finite perturbation method (cf. [**596**]).

[**571**] H. Nakatsuji, I. Morishima, H. Kato, T. Yonezawa, Bull. Chem. Soc. Jap. *44*, 2010 (1971), C.A. *75*, 124811h (1971); III. Problems in the structure determination of the molecule dissolved in a nematic solvent (cf. [**569**]). [**572**] A. D. Keith, M. Sharnoff, G. E. Cohn, Biochim. Biophys. Acta *300*, 379 (1973), C.A. *81*, 34626w (1974); Summary and evaluation of spin labels used as probes for biological membrane structure. [**573**] C. L. Khetrapal, A. C. Kunwar, C. R. Kanekar, Chem. Phys. Lett. *9*, 437 (1971), C.A. *75*, 56410z (1971); NMR investigations of π-cyclopentadienylmanganese tricarbonyl oriented in the nematic. [**574**] C. L. Khetrapal, A. C. Kunwar, J. Indian Chem. Soc. *48*, 649 (1971), C.A. *75*, 145953w (1971); NMR spectra of unsymmetrical *m*-disubstituted benzenes oriented in the nematic. Spectrum of *m*-chlorobromobenzene. [**575**] C. L. Khetrapal, A. C. Kunwar, Mol. Cryst. Liq. Cryst. *15*, 363 (1972), C.A. *76*, 58430a (1972); NMR spectrum of quinoxaline oriented in a nematic. [**576**] C. L. Khetrapal, A. V. Patankar, Mol. Cryst. Liq. Cryst. *15*, 367 (1972), C.A. *76*, 58431b (1972);

Structure of 2,1,3-benzothiadiazole from the NMR spectrum in the liquid crystalline phase. [577] C. L. Khetrapal, A. C. Kunwar, A. V. Patankar, Liq. Cryst., Proc. Int. Conf., Bangalore, *1973*, 471, C.A. *84*, 134760r (1976); NMR spectra of pyridazine, pyrimidine and pyrazine oriented in a lyophase. [578] C. L. Khetrapal, A. C. Kunwar, K. R. K. Easwaran, Liq. Cryst., Proc. Int. Conf., Bangalore, *1973*, 483, C.A. *84*, 135019t (1976); [1]HMR studies of N-methyl formamide oriented in a nematic. [579] C. L. Khetrapal, A. C. Kunwar, A. V. Patankar, J. Magn. Reson. *15*, 219 (1974), C.A. *82*, 3549q (1975); NMR spectra of pyridine and pyridine N-oxide oriented in a lyophase. [580] C. L. Khetrapal, A. C. Kunwar, J. Magn. Reson. *15*, 389 (1974), C.A. *81*, 129460b (1974); [1]HMR spectra of methyl alcohol in a nematic solvent with and without a lanthanide shift reagent.

[581] C. L. Khetrapal, A. C. Kunwar, Mol. Phys. *28*, 441 (1974), C.A. *82*, 85975n (1975); Conformation of 2,2'-bithiophene as determined from [1]HMR studies in a nematic solvent. [582] C. L. Khetrapal, A. C. Kunwar, A. V. Patankar, Org. Magn. Reson. *6*, 556 (1974), C.A. *82*, 154642r (1975); NMR spectrum of γ-hydroxypyridine oriented in a lyophase. [583] C. L. Khetrapal, A. C. Kunwar, A. S. Tracey, P. Diehl, NMR: Basic Princ. Prog. 1975, 9, 85 pp., C.A. *86*, 63589v (1977); NMR studies in lyophases. [584] S. Ramaprasad, H. P. Kellerhals, A. C. Kunwar, C. L. Khetrapal, Mol. Cryst. Liq. Cryst. *34*, 19 (1976), C.A. *86*, 43064s (1977); Amide planarity as studied by NMR of oriented molecules — the spectrum of N-methylacetamide. [585] C. L. Khetrapal, A. C. Kunwar, S. Ramaprasad, Mol. Cryst. Liq. Cryst. *34*, 123 (1976), C.A. *86*, 154842p (1977); NMR spectra of isotopically enriched N-methylformamide in isotropic and nematic media. [586] C. L. Khetrapal, A. C. Kunwar, A. V. Patankar, Mol. Cryst. Liq. Cryst. *34*, 219 (1977), C.A. *87*, 67509n (1977); [1]HMR spectrum of butadiene sulphone oriented in a lyotropic solvent. [587] A. Kumar, C. L. Khetrapal, J. Magn. Reson. *30*, 137 (1978), C.A. *89*, 82596y (1978); Two-dimensional NMR: application to oriented molecules. [588] G. Havach, P. Ferruti, D. Gill, M. P. Klein, J. Am. Chem. Soc. *91*, 7526 (1969), C.A. *72*, 66071s (1970); Nonunique ordering of solute molecules in nematic solvents. ESR observation of coexisting modes of solute alignment. [589] P. Ferruti, D. Gill, M. A. Harpold, M. P. Klein, J. Chem. Phys. *50*, 4545 (1969), C.A. *71*, 34927p (1969); ESR of spin-labeled nematogenlike probes dissolved in nematics. [590] A. F. Horwitz, W. J. Horsley, M. P. Klein, Proc. Nat. Acad. Sci. U.S. *69*, 590 (1972), C.A. *76*, 150259z (1972); NMR studies on membrane and model membrane systems. I. [1]HMR relaxation rates in sonicated lecithin dispersions.

[591] A. F. Horwitz, M. P. Klein, D. M. Michaelson, S. J. Kohler, Ann. N. Y. Acad. Sci. *222*, 468 (1973), C.A. *81*, 59378e (1974); V. Comparisons of aqueous dispersions of pure and mixed phospholipids (cf. [590]). [592] T. J. Klingen, J. R. Wright, Mol. Cryst. Liq. Cryst. *16*, 283 (1972), C.A. *76*, 146200f (1972); Further evidence for the plastic crystalline nature of 1-vinyl-*o*-carborane. [593] G. Klose, Wiss. Beitr. Martin-Luther-Univ. Halle-Wittenberg *1976* (20), 31, C.A. *86*, 96351q (1977); NMR-Untersuchungen an Phospholipid-Dispersionen. [594] J. G. Koch, Diss. Cornell Univ., Ithaca. 1972, 97 pp. Avail. Univ. Microfilms, Ann Arbor, Mich., Order No. 73-10,124, C.A. *79*, 17669d (1973); NMR studies of selected dimethyl compounds in a nematic solvent. [595] R. Köhler, Ann. Physik *6*, 241 (1960), C.A. *55*, 8046b (1961); Untersuchung der paramagnetischen Protonenresonanz in kristallinfestem Azoxyphenoldi-*p-n*-amyläther. [596] G. Kothe, E. Ohmes, Ber. Bunsenges. Phys. Chem. *78*, 924 (1974), C.A. *82*, 10216g (1975); Quartett-Radikale. Empfindliche Spinsonden zur Untersuchung molekularer Ordnungsstrukturen. [597] G. Kothe, A. Naujok, E. Ohmes, Mol. Phys. *32*, 1215 (1976), C.A. *86*, 148352n (1977); ESR of symmetrical three-spin systems in nematics. I. Fast-rotational spectra. [598] G. Kothe, Mol. Phys. *33*, 147 (1977), C.A. *87*, 31539h (1977); II. Slow-motional spectra (cf. [597]). [599] G. J. Krüger, R. Weiss, J. Phys. (Paris) *38*, 353 (1977), C.A. *86*, 127524e (1977); Self diffusion coefficients of TBBA. [600] A. N. Kuznetsov, V. A. Livshits, G. G. Malenkov, L. A. Mel'nik, B. G. Tenchov, Sb. Dokl. Vses. Nauch. Konf. Zhidk. Krist. Simp. Ikh Prakt. Primen., 2nd *1972*, 211, C.A. *82*, 37498n (1975); Paramagnetic-probe study of the potassium palmitate/water lyophase.

[601] A. N. Kuznetsov, V. A. Radtsig, Zh. Strukt. Khim. *13*, 802 (1972), C.A. *78*, 22230n (1973); Paramagnetic probe study of nematic PAP. [602] A. N. Kuznetsov, V. A. Livshits, Chem. Phys. Lett. *20*, 534 (1973), C.A. *79*, 85322k (1973); Spin probe ESR spectra in weakly anisotropic media. [603] A. N. Kuznetsov, V. A. Livshits, Zh. Fiz. Khim. *48*, 2995 (1974), C.A. *82*, 163453g (1975); Lyotropic liquid crystals studied by the spin probe method. I. Analysis of ESR spectra in microheterogeneous anisotropic medium. [604] A. N. Kuznetsov, V. A. Livshits, G. G. Malenkov, L. A. Mel'nik, V. I. Suskina, B. G. Tenchov, Zh. Fiz. Khim. *48*, 3000 (1974), C.A. *82*, 163314n (1975); II. Probe with a structure like that of potassium palmitate (cf. [603]). [605] A. N. Kuznetsov, V. A. Livshits, G. G. Malenkov, L. A. Mel'nik, B. G. Tenchov, Zh. Fiz. Khim. *48*, 3005 (1974), C.A. *82*, 163454h

(1975); III. Solubilizing probes (cf. [603]). **[606]** B. J. Lavery, Diss. Pennsylvania State Univ., University Park. 1967. 131 pp. Avail. Univ. Microfilms, Ann Arbor, Mich., Order No. 68-8718, C.A. *69*, 91652s (1968); NMR spectra of fluoromethane molecules in a liquid crystal solvent. **[607]** K. D. Lawson, T. J. Flautt, J. Phys. Chem. *69*, 3204 (1965), C.A. *64*, 7554b (1966); Measurement of the spin-lattice relaxation times of dimethyloctylamine oxide through the critical micelle concentration. **[608]** K. D. Lawson, T. J. Flautt, J. Phys. Chem. *69*, 4256 (1965), C.A. *64*, 4478a (1966); NMR absorption in anhydrous sodium soaps. **[609]** K. D. Lawson, T. J. Flautt, Mol. Cryst. *1*, 241 (1966), C.A. *65*, 8219b (1966); NMR studies of surfactant mesophases. **[610]** T. J. Flautt, K. D. Lawson, Advan. Chem. Soc. No. *63*, 26 (1967), C.A. *67*, 38103g (1967); Characterization of mesophases by NMR spectroscopy.

[611] K. D. Lawson, T. J. Flautt, J. Am. Chem. Soc. *89*, 5489 (1967), C.A. *67*, 111803s (1967); Magnetically oriented lyophases. **[612]** K. D. Lawson, T. J. Flautt, J. Phys. Chem. *72*, 2066 (1968), C.A. *69*, 31864p (1968); ^{1}HMR and ^{2}HMR studies of the dimethyldodecylamine oxide/deuterium oxide system. **[613]** P. J. Black, K. D. Lawson, T. J. Flautt, J. Chem. Phys. *50*, 542 (1969), C.A. *70*, 62742x (1969); ^{1}HMR spectrum of benzene oriented in a lyophase. **[614]** P. J. Black, K. D. Lawson, T. J. Flautt, Mol. Cryst. Liq. Cryst. *7*, 201 (1969), C.A. *71*, 65806c (1969); NMR spectra of molecules oriented in a lyophase. **[615]** J. R. Hansen, K. D. Lawson, Nature *225*, 542 (1970), C.A. *72*, 105239z (1970); Magnetic relaxation in ordered systems. **[616]** H. Lecar, G. Ehrenstein, I. Stillman, Biophys. J. *11*, 140 (1971), C.A. *74*, 94444a (1971); Detection of molecular motion in lyophilized myelin by NMR. **[617]** Y. S. Lee, Y. Y. Hsu, D. Dolphin, Liq. Cryst. Ord. Fluids *2*, 357 (1974), C.A. *86*, 180990e (1977); Order parameters and conformation of nematic MBBA by NMR studies of specifically deuterated derivatives. **[618]** R. Lenk, Chimia *28*, 51 (1974), C.A. *80*, 107431v (1974); Application d'une sonde radicalaire aux études des macromolécules, des cristaux plastiques et des cristaux liquides. **[619]** R. Lenk, Adv. Mol. Relaxation Processes *6*, 287 (1975), C.A. *82*, 172286g (1975); Diffusion and spin relaxation. **[620]** J. P. Le Pesant, P. Papon, 3rd Int. Liq. Cryst. Conf. Berlin, 1970, S 3.7; NMR study of the order parameter in nematic MBBA.

[621] J. P. Le Pesant, P. Papon, Mol. Cryst. Liq. Cryst. *24*, 305 (1973), C.A. *82*, 178631n (1975); Dynamic nuclear polarization study of MBBA with traces of nitroxide type molecules. **[622]** J. P. Le Pesant, P. Papon, Liq. Cryst. Ord. Fluids *2*, 663 (1974), C.A. *86*, 180198c (1977); Correlation time of the proton-electron interaction in MBBA with traces of nitroxyde molecules. **[623]** Y. K. Levine, Progr. Biophys. Mol. Biol. *24*, 1 (1972), C.A. *81*, 131607y (1974); Physical studies of membrane structure. **[624]** G. Lindblom, B. Lindman, L. Mandell, J. Colloid Interface Sci. *34*, 262 (1970), C.A. *73*, 134233u (1970); Study of counter-ion binding to reversed micelles by ^{81}BrMR quadrupole relaxation. **[625]** G. Lindblom, Acta Chem. Scand. *25*, 2767 (1971), C.A. *76*, 95834f (1972); Ion binding in liquid crystals studied by NMR. III. ^{23}Na quadrupolar effects in a model membrane system. **[626]** G. Lindblom, H. Wennerstrom, B. Lindman, Chem. Phys. Lett. *8*, 489 (1971), C.A. *75*, 117712h (1971); II. ^{35}Cl second-order quadrupole interactions in the lamellar mesophase (cf. [625]). **[627]** G. Lindblom, B. Lindman, Mol. Cryst. Liq. Cryst. *14*, 49 (1971), C.A. *75*, 123382g (1971); I. Cetyltrimethylammonium bromide/hexanol/water system (cf. [625]). **[628]** G. Lindblom, Acta. Chem. Scand. *26*, 1745 (1972), C.A. *77*, 95155f (1972); IV. ^{23}NaMR of macroscopically aligned lamellar mesophases (cf. [625]). **[629]** G. Lindblom, N. O. Persson, B. Lindman, Chem., Phys. Chem. Anwendungstech. Grenzflächenaktiven Stoffe, Ber. Int. Kongr., 6th, *2*, Teil 2, 925, 939 (1972), C.A. *82*, 77372y (1975), C.A. *82*, 77373z (1975); Nuclear quadrupole relaxation studies of counter-ion binding in some micellar and liquid crystalline solutions. NMR-nuclear quadrupole splittings in lyophases. **[630]** G. Lindblom, B. Lindman, L. Mandell, J. Colloid Interface Sci. *42*, 400 (1973), C.A. *78*, 88910r (1973); Effect of micellar shape and solubilization on counter ion binding studied by ^{81}BrMR.

[631] G. Lindblom, B. Lindman, J. Phys. Chem. *77*, 2531 (1973), C.A. *79*, 149709b (1973); Interaction between halide ions and amphiphilic organic cations in aqueous solutions studied by nuclear quadrupole relaxation. **[632]** G. Lindblom, B. Lindman, Mol. Cryst. Liq. Cryst. *22*, 45 (1973), C.A. *79*, 110045g (1973); V. Static quadrupolar effects for alkali nuclei (cf. [625]). **[633]** G. Lindblom, N. O. Persson, B. Lindman, G. Arvidson, Ber. Bunsenges. Phys. Chem. *78*, 955 (1974), C.A. *82*, 27512y (1975); Binding of water and sodium ions in model membrane systems studied by NMR. **[634]** N. O. Persson, G. Lindblom, B. Lindman, Chem. Phys. Lipids *12*, 261 (1974), C.A. *81*, 113328d (1974); ^{2}HMR and ^{23}NaMR studies of lecithin mesophases. **[635]** H. Wennerstrom, G. Lindblom, B. Lindman, Chem. Scr. *6*, 97 (1974), C.A. *81*, 129457f (1974); Theoretical aspects on the NMR of quadrupolar ionic nuclei in micellar solutions and amphiphilic liquid crystals. **[636]** H. Gustavsson, G. Lindblom, B. Lindman, N. O. Persson, H. Wennerstrom, Liq. Cryst. Ord. Fluids *2*, 161 (1974), C.A. *86*, 177937n

(1977); NMR studies of the interaction between sodium ions and anionic surfactants in amphiphile/water systems. [637] G. Lindblom, B. Lindman, G. J. T. Tiddy, Acta Chem. Scand. A 29, 876 (1975), C.A. 84, 82205s (1976); Counterion quadrupole splittings in lyophases. Determination of the sign of the order parameter. [638] J. Ulmius, H. Wennerstrom, G. Lindblom, G. Arvidson, Biochim. Biophys. Acta 389, 197 (1975), C.A. 83, 35339h (1975); ¹HMR bandshape studies of lamellar liquid crystals and gel phases containing lecithins and cholesterol. [639] G. Lindblom, H. Wennerstrom, B. Lindman, ACS Symp. Ser. 34, 372 (1976), C.A. 86, 10667n (1977); NMR quadrupole splitting method for studying ion binding in liquid crystals. [640] G. Lindblom, N. O. Persson, G. Arvidson, Adv. Chem. Ser. 152, 121 (1976), C.A. 85, 88883d (1976); Ion binding and water orientation in lipid model membrane systems studied by NMR.

[641] G. Lindblom, H. Wennerstrom, B. Lindman, J. Magn. Reson. 23, 177 (1976); C.A. 85, 114283e (1976); Multiple quantum transitions for spin-3/2 nuclei in the NMR spectra of lyophases. [642] J. Kowalewski, T. Lindblom, R. Vestin, T. Drakenberg, Mol. Phys. 31, 1669 (1976), C.A. 85, 158905p (1976); ²HMR of monodeuteroethene: isotropic and anisotropic phase spectra. [643] B. Lindman, P. Ekwall, Mol. Cryst. 5, 79 (1968), C.A. 70, 6968p (1969); ²³NaMR relaxation in different phases in the sodium caprylate/decanol/water system. [644] B. Lindman, I. Danielsson, J. Colloid Interface Sci. 39, 349 (1972), C.A. 77, 27050t (1972); ⁸⁵RbMR relaxation in aqueous soap solutions. [645] N. O. Persson, H. Wennerstrom, B. Lindman, Acta Chem. Scand. 27, 1667 (1973), C.A. 79, 129669k (1973); Counterion dependent water and amphiphile orientation and deuteron exchange in lyophases. [646] T. Drakenberg, B. Lindman, J. Colloid Interface Sci. 44, 184 (1973), C.A. 79, 72004h (1973); ¹³CMR of micellar solutions. [647] H. Wennerstrom, N. O. Persson, B. Lindman, J. Magn. Reson. 13, 348 (1974), C.A. 81, 18951t (1974); Double quantum transitions in ²HMR spectra of lyophases. [648] A. Johansson, B. Lindman, Liq. Cryst. Plast. Cryst. 2, 192 (1974), C.A. 83, 185577y (1975); NMR spectroscopy of liquid crystals. Amphiphilic systems. [649] T. Bull, B. Lindman, Mol. Cryst. Liq. Cryst. 28, 155 (1974), C.A. 82, 175563f (1975); Amphiphile diffusion in cubic lyophases. [650] H. Wennerstrom, N. O. Persson, B. Lindman, ACS Symp. Ser. 9, 253 (1975), C.A. 83, 16167j (1975); ²HMR studies on soap/water mesophases.

[651] N. O. Persson, B. Lindman, J. Phys. Chem. 79, 1410 (1975), C.A. 83, 104283z (1975); ²HMR in amphiphilic liquid crystals. Alkali ion dependent water and amphiphile orientation. [652] N. O. Persson, K. Fontell, B. Lindman, G. J. T. Tiddy, J. Colloid Interface Sci. 53, 461 (1975), C.A. 84, 65839j (1976); Mesophase structure studies by ²HMR. Observations for the sodium octanoate/decanol/water system. [653] N. O. Persson, B. Lindman, Mol. Cryst. Liq. Cryst. 38, 327 (1977), C.A. 86, 198213p (1977); ²H and ¹⁴N NMR studies of amphiphilic liquid crystals. Effect of solubilization, electrolyte and temperature on water orientation. [654] H. Lippmann, Ann. Physik [7], 1, 157 (1958), C.A. 52, 11500i (1958); Ordnungszustand einer im Magnetfeld orientierten Flüssigkristallprobe bei Rotation. [655] H. Lippmann, Ann. Physik [7], 2, 287 (1958), C.A. 53, 3901a (1959); Feinstruktur der ¹HMR in nematischem PAA. [656] A. Loesche, Ann. Phys. (Leipzig) 31, 182 (1974), C.A. 81, 113336e (1974); Struktur von NMR-Signalen in nematischen Phasen. [657] A. Loesche, S. Grande, Magn. Reson. Relat. Phenomena, Proc. Congr. Ampère, 18th, 1, 201 (1974), C.A. 83, 170582x (1975); ¹H chemical shifts in systems of oriented rotating molecules. [658] A. Loesche, Z. Phys. Chem. (Leipzig) 255, 1157 (1974), C.A. 83, 19385q (1975); NMR-Untersuchung der Anisotropie und des Orientierungsverhaltens nematischer Phasen. [659] A. Loesche, S. Grande, Magn. Reson. Relat. Phenom., Proc. Congr. Ampère, 19th 1976, 181, C.A. 86, 197382f (1977); Molecular dynamics in liquid crystals studied by NMR. [660] A. Loesche, S. Grande, P. M. Borodin, Y. V. Molchanov, Kristallografiya 21, 856 (1976), C.A. 85, 185772z (1976); Study of the reorientation of molecules of nematics in a constant magnetic field by the method of pulsed FT NMR spectroscopy.

[661] S. Grande, E. Heinze, A. Loesche, Phys. Lett. A 58, 102 (1976), C.A. 85, 133645r (1976); Nonexponential nuclear spin-lattice relaxation in the i phase of thermotropic mesogens. [662] A. Loesche, Wiss. Beitr. Martin-Luther-Univ. Halle-Wittenberg 1976 (20), 17, C.A. 86, 63593s (1977); Phasenstrukturen und -umwandlungen thermotroper Systeme aus dem Blickwinkel der NMR. [663] S. Limmer, A. Loesche, Z. Phys. Chem. (Leipzig) 257, 107 (1976), C.A. 84, 158191g (1976); Verhalten nematogener Substanzen am Phasenübergang k/n. [664] Y. Egozy, A. Loewenstein, B. L. Silver, Mol. Phys. 19, 177 (1970), C.A. 73, 71500f (1970); ²H relaxation in benzene-d₆ dissolved in a liquid crystal. [665] R. Ader, A. Loewenstein, Mol. Phys. 24, 455 (1972), C.A. 77, 145988c (1972); NMR spectra of deuterated methanes and 2,2-dimethylpropane in nematics. [666] R. Ader, A. Loewenstein, J. Am. Chem. Soc. 96, 5336 (1974), C.A. 81, 97400p (1974); ¹HMR and ²HMR studies of methylsilane and methylgermane dissolved

in a nematic. [**667**] R. Ader, A. Loewenstein, Mol. Phys. *27*, 1113 (1974), C.A. *81*, 77349a (1974); NMR study of methylsilane-d₃ dissolved in a nematic. [**668**] R. Ader, A. Loewenstein, Mol. Phys. *30*, 199 (1975), C.A. *83*, 155210f (1975); ¹HMR and ²HMR of methanes, silanes and GeH₃D dissolved in nematics. I. Experimental results. [**669**] A. Loewenstein, Chem. Phys. Lett. *38*, 543 (1976), C.A. *84*, 172354u (1976); Relaxation times, diffusion coefficients and high frequency NMR measurements of methane dissolved in MBBA. [**670**] A. Loewenstein, M. Brenman, R. Schwarzmann, J. Phys. Chem. *82*, 1744 (1978), C.A. *89*, 82554h (1978); NMR studies of cations and water in lyophases.

[**671**] R. C. Long, Jr., Diss. Emory Univ., Atlanta. 1972, 193 pp. Avail. Univ. Microfilms, Ann Arbor, Mich., Order No. 72-25,941, C.A. *78*, 9858d (1973); NMR study of molecules partially oriented in a lyophase. [**672**] R. C. Long, Jr., J. Magn. Reson. *12*, 216 (1973), C.A. *80*, 32268m (1974); New lyotropic nematic for NMR spectroscopy. [**673**] N. J. D. Lucas, Mol. Phys. *22*, 147 (1971), C.A. *75*, 145925p (1971); Influence of vibrations on molecular structure determinations from NMR in liquid crystals. I. Application to methyl fluoride. [**674**] N. J. D. Lucas, Mol. Phys. *22*, 233 (1971), C.A. *76*, 19848y (1972); II. Cyclopropane (cf. [673]). [**675**] N. J. D. Lucas, Mol. Phys. *23*, 825 (1972), C.A. *77*, 54216h (1972); Vibrationally averaged structure and vibrational expectation values. [**676**] A. Carrington, G. R. Luckhurst, Mol. Phys. *8*, 401 (1964), C.A. *62*, 7274f (1965); ESR spectra of free radicals dissolved in liquid crystals. [**677**] H. C. Longuet-Higgins, G. R. Luckhurst, Mol. Phys. *8*, 613 (1964), C.A. *62*, 15529e (1965); Composition of the nematic mesophase of PAA. [**678**] H. R. Falle, G. R. Luckhurst, H. Lemaire, Y. Marechal, A. Rassat, P. Rey, Mol. Phys. *11*, 49 (1966), C.A. *66*, 15353d (1967); ESR of ground state triplets in liquids crystal solutions. [**679**] G. R. Luckhurst, Mol. Phys. *11*, 205 (1966), C.A. *66*, 33419j (1967); Structure of Coppinger's radical and its orientation within a cluster of a liquid crystal. [**680**] H. R. Falle, G. R. Luckhurst, Mol. Phys. *11*, 299 (1966), C.A. *66*, 33407d (1967); ESR spectrum of perinaphthenyl dissolved in a liquid crystal.

[**681**] G. R. Luckhurst, Mol. Cryst. *2*, 363 (1967), C.A. *67*, 57924s (1967); Equivalence of the two theories of the nematic mesophase when applied to magnetic resonance experiments. [**682**] H. R. Falle, G. R. Luckhurst, Mol. Phys. *12*, 493 (1967), C.A. *68*, 64427f (1968); Linewidth variations in the ESR spectra of radicals dissolved in nematics. [**683**] G. R. Luckhurst, E. G. Rozantsev, Izv. Akad. Nauk SSSR, Ser. Khim. *1968*, 1708, C.A. *70*, 15872e (1969); ESR of a principal quartet state in liquid crystal solutions. [**684**] G. R. Luckhurst, Quart. Rev. *22*, 179 (1968), C.A. *69*, 23196g (1968); Liquid crystals as solvents in NMR. [**685**] P. D. Francis, G. R. Luckhurst, Chem. Phys. Lett. *3*, 213 (1969), C.A. *71*, 44277r (1969); ESR in the smectic. [**686**] D. H. Chen, P. G. James, G. R. Luckhurst, Mol. Cryst. Liq. Cryst. *8*, 71 (1969), C.A. *71*, 85808h (1969); Order in the nematic. [**687**] D. H. Chen, G. R. Luckhurst, Trans. Faraday Soc. *65*, 656 (1969), C.A. *70*, 71600b (1969); ESR study of the perturbation of the order in a nematic by a second component. [**688**] G. R. Luckhurst, F. Sundholm, Acta Chem. Scand. *24*, 3759 (1970), C.A. *74*, 80343e (1971); Solute alignment in the nematic. [**689**] H. R. Falle, G. R. Luckhurst, J. Magn. Reson. *3*, 161 (1970), C.A. *74*, 17882q (1971); ESR spectra of partially oriented radicals. [**690**] P. G. James, G. R. Luckhurst, Mol. Phys. *19*, 489 (1970), C.A. *73*, 124605c (1970); Anisotropic pseudopotential for nematics.

[**691**] G. R. Luckhurst, RIC Rev. *3*, 61 (1970), C.A. *73*, 92845n (1970); ESR in anisotropic solvents. [**692**] G. R. Luckhurst, Chem. Phys. Lett. *9*, 289 (1971), C.A. *75*, 121649a (1971); Alignment of cybotactic groups in a nematic by an electric field. [**693**] S. A. Brooks, G. R. Luckhurst, G. F. Pedulli, Chem. Phys. Lett. *11*, 159 (1971), C.A. *76*, 19860w (1972); Thermal fluctuations in the nematic. [**694**] G. R. Luckhurst, F. Sundholm, Mol. Phys. *21*, 349 (1971), C.A. *75*, 123516d (1971); Determination of a temperature-dependent tilt angle in a smectic C by ESR spectroscopy. [**695**] F. M. Leslie, G. R. Luckhurst, H. J. Smith, Chem. Phys. Lett. *13*, 368 (1972), C.A. *76*, 160161a (1972); Magnetohydrodynamic effects in the nematic. [**696**] G. R. Luckhurst, A. Sanson, Mol. Cryst. Liq. Cryst. *16*, 179 (1972), C.A. *76*, 91419b (1972); Molecular organization in smectic ethyl 4-azoxybenzoate. [**697**] G. R. Luckhurst, M. Setaka, Mol. Cryst. Liq. Cryst. *19*, 179 (1972), C.A. *78*, 76837j (1973); Molecular organization in nematic and smectic A N-(4-(butyloxy)benzylidene)-4'-acetoaniline. [**698**] G. R. Luckhurst, A. Sanson, Mol. Phys. *24*, 1297 (1972), C.A. *78*, 50224g (1973); Angular dependent linewidths for a spin probe dissolved in a liquid crystal. [**699**] R. L. Humphries, G. R. Luckhurst, J. Chem. Soc., Faraday Trans. 2, *69*, 1491 (1973), C.A. *80*, 53183v (1974); Anisotropic pseudopotential for a vanadyl chelate dissolved in a nematic. [**700**] G. R. Luckhurst, M. Ptak, A. Sanson, J. Chem. Soc., Faraday Trans. 2, *69*, 1752 (1973), C.A. *80*, 101020p (1974); Molecular organization within smectic C 4,4'-diheptyloxyazoxybenzene.

[**701**] G. R. Luckhurst, M. Setaka, Mol. Cryst. Liq. Cryst. *19*, 279 (1973), C.A. *78*, 83670c (1973); Solute-solvent interactions within the nematic. [**702**] G. R. Luckhurst, H. J. Smith, Mol. Cryst.

Liq. Cryst. *20*, 319 (1973), C.A. *79*, 58655q (1973); Structure of a cholesteric perturbed by a magnetic field. [**703**] G. R. Luckhurst, Mol. Cryst. Liq. Cryst. *21*, 125 (1973), C.A. *79*, 108882j (1973); Magnetic resonance studies of thermotropic liquid crystals. [**704**] G. R. Luckhurst, R. Poupko, Chem. Phys. Lett. *29*, 191 (1974), C.A. *82*, 78990k (1975); ESR study of the orientational order in a nematic. [**705**] G. R. Luckhurst, M. Setaka, C. Zannoni, Mol. Phys. *28*, 49 (1974), C.A. *81*, 161781t (1974); ESR investigation of molecular motion in the smectic. [**706**] G. R. Luckhurst, Liq. Cryst. Plast. Cryst. *2*, 144 (1974), C.A. *83*, 199531r (1975); NMR spectroscopy of liquid crystals. Nonamphiphilic systems. [**707**] S. G. Carr, G. R. Luckhurst, R. Poupko, H. J. Smith, Chem. Phys. *7*, 278 (1975), C.A. *83*, 51155a (1975); Director distribution in a spinning nematic subject to a static magnetic field. [**708**] G. R. Luckhurst, H. J. Smith, Mol. Phys. *29*, 317 (1975), C.A. *83*, 10584w (1975); Pairwise angular correlation in a cholesteric. [**709**] G. R. Luckhurst, R. Poupko, Mol. Phys. *29*, 1293 (1975), C.A. *83*, 36211x (1975); Molecular organization within nematic and smectic 4'-*n*-octyl-4-cyanobiphenyl and 4'-*n*-octyloxy-4-cyanobiphenyl. [**710**] G. R. Luckhurst, R. Poupko, C. Zannoni, Mol. Phys. *30*, 499 (1975), C.A. *83*, 177657j (1975); Spin relaxation for biradical spin probes in anisotropic environments.

[**711**] J. W. Emsley, J. C. Lindon, G. R. Luckhurst, Mol. Phys. *30*, 1913 (1975), C.A. *84*, 82872g (1976); Chain ordering in nematic 4-cyano-4'-*n*-pentylbiphenyl. NMR investigation. [**712**] S. A. Brooks, G. R. Luckhurst, G. F. Pedulli, J. Roberts, J. Chem. Soc., Faraday Trans. 2, *72*, 651 (1976), C.A. *84*, 158134r (1976); Molecular motion and orientational order in a nematic. ESR investigation. [**713**] G. R. Luckhurst, R. N. Yeates, J. Chem. Soc., Faraday Trans. 2, *72*, 996 (1976), C.A. *85*, 54915a (1976); Orientational order of a spin probe dissolved in nematics. ESR investigation. [**714**] G. R. Luckhurst, C. Zannoni, J. Magn. Reson. *23*, 275 (1976), C.A. *85*, 133609g (1976); Orientation-dependent spin relaxation. Triradical spin probes in anisotropic environments. [**715**] S. G. Carr, S. K. Khoo, G. R. Luckhurst, C. Zannoni, Mol. Cryst. Liq. Cryst. *35*, 7 (1976), C.A. *85*, 114990h (1976); Ordering matrix for the spin probe (3-spiro[2'-N-oxyl-3',3'-dimethyloxazolidine])-5α-cholestane, in nematic PAA. [**716**] C. A. Boicelli, A. Mangini, L. Lunazzi, M. Tiecco, J. Chem. Soc., Perkin Trans. 2, *1972*, 599, C.A. *76*, 125977z (1972); Liquid crystal ^{1}HMR spectrum of thieno(2,3-b)thiophene. [**717**] A. D'Annibale, L. Lunazzi, A. Boicelli, D. Macciantelli, J. Chem. Soc., Perkin Trans. 2, *1973*, 1396, C.A. *80*, 3007v (1974); Conformational problem of biphenyl in solution investigated by the liquid crystal NMR spectrum of 3,3',5,5'-tetrachlorobiphenyl. [**718**] A. D'Annibale, L. Lunazzi, G. Fronza, R. Mondelli, S. Bradamante, J. Chem. Soc., Perkin Trans. 2, *1973*, 1908, C.A. *80*, 47347k (1974); Conformational investigation of cyclobutanone and thietane in liquid crystalline solvents. [**719**] A. D'Annibale, L. Lunazzi, F. Fringuelli, A. Taticchi, Mol. Phys. *27*, 257 (1974), C.A. *80*, 144931d (1974); Isotropic and nematic phase NMR spectra of tellurophene. [**720**] L. Lunazzi, D. Macciantelli, Gazz. Chim. Ital. *105*, 657 (1975), C.A. *85*, 45854 (1976); Conformation of biphenyl in solution. NMR spectrum of 3,4,4',5-tetrabromobiphenyl in a nematic.

[**721**] L. Lunazzi, Determination Org. Struct. Phys. Methods *6*, 335 (1976), C.A. *87*, 21691n (1977); Molecular structures by NMR in liquid crystals. [**722**] E. Meirovtich, Z. Luz, Mol. Phys. *30*, 1589 (1975), C.A. *84*, 82871f (1976); Structure of smectic TBBA studied by ESR spectroscopy. [**723**] J. R. McColl, Phys. Lett. A *38*, 55 (1972), C.A. *76*, 77737x (1972); Effect of pressure on order in nematic PAA. [**724**] J. R. McColl, J. Chem. Phys. *62*, 1593 (1975), C.A. *82*, 178665b (1975); Orientational order in impure liquid crystals. [**725**] P. J. Collings, S. I. Goss, J. R. McColl, Phys. Rev. A *11*, 684 (1975), C.A. *82*, 178661x (1975); Methods to measure the orientational order in cholesterics. [**726**] R. J. McKee, J. R. McColl, Phys. Rev. Lett. *34*, 1076 (1975), C.A. *83*, 18537d (1975); Orientational order measurements near a possible n/s$_A$ tricritical point. [**727**] P. J. Collings, T. J. McKee, J. R. McColl, J. Chem. Phys. *65*, 3520 (1976), C.A. *86*, 10857z (1977); NMR spectroscopy in cholesterics. I. Orientational order parameter measurements. [**728**] W. L. Hubbell, H. M. McConnell, Proc. Nat. Acad. Sci. U.S. *61*, 12 (1968), C.A. *70*, 8771t (1969); Spin-label studies of the excitable membranes of nerve and muscle. [**729**] H. M. McConnell, J. Gen. Physiol. *54*, 277S (1969), C.A. *71*, 98891e (1969); Spin-labeled membranes. [**730**] W. L. Hubbell, H. M. McConnell, Proc. Nat. Acad. Sci. U.S. *63*, 16 (1969), C.A. *71*, 77875n (1969); Motion of steroid spin labels in membranes.

[**731**] W. L. Hubbell, H. M. McConnell, Proc. Nat. Acad. Sci. U.S. *64*, 20 (1969), C.A. *72*, 64010x (1970); Orientation and motion of amphiphilic spin labels in membranes. [**732**] S. Rottem, W. L. Hubbell, L. Hayflick, H. M. McConnell, Biochim. Biophys. Acta *219*, 104 (1970), C.A. *74*, 9545s (1971); Motion of fatty acid spin labels in the plasma membrane of Mycoplasma. [**733**] W. L. Hubbell, J. C. Metcalfe, S. M. Metcalfe, H. M. McConnell, Biochim. Biophys. Acta *219*, 415 (1970), C.A. *74*, 60966g (1971); Interaction of small molecules with spin-labeled erythrocyte membranes. [**734**]

H. M. McConnell, R. D. Kornberg, Biochemistry *10*, 1111 (1971), C.A. *74*, 107209e (1971); Inside-outside transitions of phospholipids in vesicle membranes. [**735**] W. L. Hubbell, H. M. McConnell, J. Am. Chem. Soc. *93*, 314 (1971); Molecular motion in spin-labeled phospholipids and membranes. [**736**] R. D. Kornberg, H. M. McConnell, Proc. Nat. Acad. Sci. U.S. *68*, 2564 (1971), C.A. *76*, 1063d (1972); Lateral diffusion of phospholipids in a vesicle membrane. [**737**] H. M. McConnell, K. L. Wright, B. G. McFarland, Biochem. Biophys. Res. Commun. *47*, 273 (1972), C.A. *77*, 44610t (1972); Fraction of the lipid in a biological membrane that is in a fluid state. Spin label assay. [**738**] P. Devaux, H. M. McConnell, J. Am. Chem. Soc. *94*, 4475 (1972), C.A. *77*, 58059g (1972); Lateral diffusion in spin-labeled phosphatidylcholine multilayers. [**739**] P. Devaux, H. M. McConnell, Ann. N. Y. Acad. Sci. *222*, 489 (1973), C.A. *81*, 59475j (1974); Equality of the rates of lateral diffusion of phosphatidylethanolamine and phosphatidylcholine spin labels in rabbit sarcoplasmic reticulum. [**740**] P. Devaux, C. J. Scandella, H. M. McConnell, J. Magn. Reson. *9*, 474 (1973), C.A. *79*, 50466h (1973); Spin-spin interactions between spin-labeled phospholipids incorporated into membranes.

[**741**] C. W. M. Grant, S. H. W. Wu, H. M. McConnell, Biochim. Biophys. Acta *363*, 151 (1974), C.A. *81*, 131878n (1974); Lateral phase separations in binary lipid mixtures. Correlation between spin label and freeze-fracture electron microscopic studies. [**742**] B. J. Gaffney, H. M. McConnell, Chem. Phys. Lett. *24*, 310 (1974), C.A. *81*, 4853a (1974); Effect of a magnetic field on phospholipid membranes. [**743**] M. P. McDonald, Arch. sci. (Geneva) *12*, Fasc. spec. 141 (1959), C.A. *54*, 5251i (1960); NMR in amphiphilic solutions. [**744**] A. S. C. Lawrence, M. P. McDonald, Mol. Cryst. *1*, 205 (1966), C.A. *65*, 17766f (1966); Lipid/water systems. Classical and NMR investigations. [**745**] B. Ellis, A. S. C. Lawrence, M. P. McDonald, W. E. Peel, Liq. Cryst. Ord. Fluids *1970*, 277, C.A. *78*, 152082b (1973); NMR in the mono-octanoin/D_2O system. [**746**] M. P. McDonald, W. E. Peel, Trans. Faraday Soc. *67*, 890 (1971), C.A. *74*, 118100p (1971); Lipid/water systems. ^{1}HMR in an ordered lyophase. [**747**] M. P. McDonald, W. E. Peel, Chem. Phys. Lipids *15*, 37 (1975), C.A. *84*, 89316n (1976); ^{1}HMR in ordered lyophases. [**748**] M. P. McDonald, W. E. Peel, J. Chem. Soc., Faraday Trans. 1, *72*, 2274 (1976), C.A. *86*, 47677s (1977); Solid and liquid crystalline phases in the sodium dodecyl sulfate/hexadecanoic acid/water system. [**749**] J. D. Kennedy, W. McFarlane, J. Chem. Soc., Chem. Commun. *1974*, 595, C.A. *81*, 179609c (1974); ^{199}Hg chemical shift anisotropy in methyl mercuric bromide. [**750**] J. D. Kennedy, W. McFarlane, Mol. Phys. *29*, 593 (1975), C.A. *82*, 154646v (1975); NMDR studies of ^{13}C and ^{15}N chemical shift anisotropies in methyl cyanide partially oriented in a nematic.

[**751**] J. D. Kennedy, W. McFarlane, J. Chem. Soc., Chem. Commun. *1976*, 666, C.A. *86*, 71096x (1977); ^{31}P chemical shift anisotropies in organophosphorus compounds. [**752**] J. D. Kennedy, W. McFarlane, J. Chem. Soc., Faraday Trans. 2, *72*, 1653 (1976), C.A. *86*, 15917d (1977); Studies of ^{199}Hg nuclear shielding anisotropies and their relation to isotropic chemical shifts. [**753**] J. D. Kennedy, W. McFarlane, B. Wrackmeyer, J. Mol. Struct. *30*, 125 (1976), C.A. *84*, 95879q (1976); Geometry of triethynyl phosphine partially oriented in a nematic. [**754**] I. J. Colquhoun, W. McFarlane, J. Chem. Soc., Faraday Trans. 2, *73*, 722 (1977), C.A. *87*, 101722p (1977); NMDR studies of isotopically labeled bis(dichlorophosphino)methylamine partially oriented in a nematic. [**755**] J. Dalton, J. D. Kennedy, W. McFarlane, J. R. Wedge, Mol. Phys. *34*, 215 (1977), C.A. *88*, 21516e (1978); NMR studies of dimethyl cadmium partially oriented in nematic EBBA. [**756**] M. C. McIvor, J. Organometal. Chem. *27*, C59 (1971), C.A. *74*, 99121c (1971); NMR of π-cyclopentadienyltungsten hydride in a nematic: further evidence for pseudorotation and pseudosymmetry. [**757**] L. A. McLachlan, D. F. S. Natusch, R. H. Newman, J. Magn. Reson. *4*, 358 (1971), C.A. *75*, 69297t (1971); Spin-lattice relaxation in ordered system. [**758**] L. A. McLachlan, D. F. S. Natusch, R. H. Newman, J. Magn. Reson. *10*, 34 (1973), C.A. *78*, 166629x (1973); Magnetic field gradient effect in the ^{2}HMR spectrum of a lyophase. [**759**] M. Ayres, K. A. McLauchlan, J. Wilkinson, J. Chem. Soc. D *1969*, 858, C.A. *71*, 80558s (1969); Structure of methylmercuric chloride. [**760**] G. E. Chapman, E. M. Long, K. A. McLauchlan, Mol. Phys. *17*, 189 (1969), C.A. *71*, 84673y (1969); Determination of absolute values of the orientation parameters of partially oriented solute molecules.

[**761**] J. Bulthuis, J. Gerritsen, C. W. Hilbers, C. MacLean, Rec. Trav. Chim. Pays-Bas *87*, 417 (1968), C.A. *69*, 40020n (1968); NMR spectra of substituted benzenes with D_{2h} symmetry in a nematic solvent. [**762**] W. Bovee, C. W. Hilbers, C. MacLean, Mol. Phys. *17*, 75 (1969), C.A. *71*, 86404k (1969); NMR spectrum of ethene in a nematic solvent. [**763**] J. Bulthuis, C. MacLean, J. Magn. Reson. *4*, 148 (1971), C.A. *74*, 148837b (1971); NMR in liquid crystalline solvents. Vibrational corrections on nuclear dipole-dipole interactions in the methyl halides. [**764**] J. Gerritsen, C. MacLean,

J. Magn. Reson. *5*, 44 (1971), C.A. *75*, 103377f (1971); NMR spectra of partially oriented 1,1-difluoroethene. Anisotropy of J_{FF}. [**765**] J. Gerritsen, C. MacLean, Mol. Cryst. Liq. Cryst. *12*, 97 (1971), C.A. *75*, 13145z (1971); NMR spectra of 1,2-difluorobenzene and 1,1-difluoroethene in nematic solvents. Anisotropy of indirect fluorine couplings and molecular geometry. [**766**] J. Gerritsen, C. MacLean, Spectrochim. Acta A *27*, 1495 (1971), C.A. *75*, 82215j (1971); NMR spectra of 1,2-difluorobenzene in nematic solvents. Anisotropy of indirect fluorine couplings and molecular geometry. [**767**] J. Gerritsen, G. Koopmans, H. S. Rollema, C. MacLean, J. Magn. Reson. *8*, 20 (1972), C.A. *77*, 120480t (1972); NMR spectra of partially oriented tetrafluorobenzenes. Anisotropies of indirect nuclear spin-spin couplings. [**768**] J. Bulthuis, C. W. Hilbers, C. MacLean, Med. Tech. Publ. Co. Int. Rev. Sci.: Phys. Chem., Ser. One, *4*, 201 (1972), C.A. *78*, 166220g (1973); NMR and ESR in liquid crystals. [**769**] J. Gerritsen, C. MacLean, Rec. Trav. Chim. Pays-Bas *91*, 1393 (1972), C.A. *78*, 64886m (1973); NMR spectra of partially oriented mono- and hexafluorobenzene. [**770**] G. J. den Otter, C. MacLean, Chem. Phys. Lett. *20*, 306 (1973), C.A. *79*, 65330h (1973); ^{19}F chemical shift anisotropy in 1,1-difluoroethene.

[**771**] J. Bulthuis, J. van den Berg, C. MacLean, J. Mol. Struct. *16*, 11 (1973), C.A. *78*, 135516d (1973); NMR spectra of 1,2-difluoroethane in nematic solvents. [**772**] G. J. den Otter, J. Gerritsen, C. MacLean, J. Mol. Struct. *16*, 379 (1973), C.A. *79*, 52450x (1973); NMR spectra of *m*-difluorobenzene in nematic solvents. Molecular geometry and anisotropy of the indirect F-F coupling. [**773**] G. J. den Otter, C. MacLean, Chem. Phys. *3*, 119 (1974), C.A. *80*, 139088e (1974); NMR spectra of *trans*-1,2-difluoroethene partially oriented in a lyotropic and a thermotropic liquid crystal. Signs and anisotropies of the indirect couplings. [**774**] G. J. den Otter, C. MacLean, C. W. Haigh, S. Sykes, J. Chem. Soc., Chem. Commun. *1974*, 24, C.A. *80*, 107504w (1974); Theoretical and experimental determination of the indirect F-F coupling anisotropy in *trans*-difluoroethene. [**775**] G. J. den Otter, W. Heijser, C. MacLean, J. Magn. Reson. *13*, 11 (1974), C.A. *80*, 132250t (1974); NMR spectra of *p*-difluorobenzene in a lyotropic nematic solvent. Molecular geometry and anisotropy of the indirect F-F coupling. [**776**] G. J. den Otter, J. G. van Dalen, C. MacLean, Chem. Phys. Lett. *33*, 463 (1975), C.A. *83*, 105894m (1975); NMR study of trifluoroethene partially oriented in nematic solvents. [**777**] G. J. den Otter, C. MacLean, J. Magn. Reson. *20*, 11 (1975), C.A. *84*, 97491f (1976); ^{13}C-satellites in the ^{19}FMR spectra of *cis*- and *trans*-1,2-difluoroethene partially oriented in nematic solvents. [**778**] G. J. den Otter, C. MacLean, J. Mol. Struct. *31*, 47 (1976), NMR spectra of 1,2,3,5-tetrafluorobenzene, pentafluorobenzene and hexafluorobenzene in nematic solvents. Molecular geometries and anisotropies of indirect F-F couplings. [**779**] G. J. den Otter, C. MacLean, J. Mol. Struct. *31*, 57 (1976), C.A. *85*, 4672y (1976); NMR investigation of benzoic acids partially oriented in nematic solvents. [**780**] G. J. den Otter, J. Bulthuis, C. A. de Lange, C. MacLean, Chem. Phys. Lett. *45*, 603 (1977), C.A. *86*, 130578f (1977); ^{19}F shielding anisotropies of 2,4,6-trifluoronitrobenzene from nematic phase NMR.

[**781**] J. A. Magnuson, D. S. Shelton, N. S. Magnuson, Biochem. Biophys. Res. Commun. *39*, 279 (1970), C.A. *73*, 31495b (1970); NMR study of sodium ion interaction with erythrocyte membranes. [**782**] J. A. Magnuson, N. S. Magnuson, Ann. N. Y. Acad. Sci. *1973*, 297, C.A. *79*, 14552z (1973); NMR studies of sodium and potassium in various biological tissues. [**783**] N. S. Magnuson, J. A. Magnuson, Biophys. J. *13*, 1117 (1973), C.A. *80*, 12410y (1974); ^{23}Na ion interaction with bacterial surfaces. NMR invisible signals. [**784**] T. M. Bray, J. A. Magnuson, J. R. Carlson, J. Biol. Chem. *249*, 914 (1974), C.A. *80*, 92374c (1974); NMR studies of lecithin-skatole interaction. [**785**] C. Mailer, C. P. S. Taylor, S. Schreier-Muccillo, I. C. P. Smith, Arch. Biochem. Biophys. *163*, 671 (1974), C.A. *82*, 12689n (1975); Influence of cholesterol on molecular motion in egg lecithin bilayers. Variable-frequency ESR study of a cholestane spin probe. [**786**] D. Marsh, I. C. P. Smith, Biochim. Biophys. Acta *298*, 133 (1973), C.A. *78*, 120653x (1973); Interacting spin label study of the fluidizing and condensing effects of cholesterol on lecithin bilayers. [**787**] D. Marsh, Can. J. Biochem. *52*, 631 (1974), C.A. *81*, 165127g (1974); Intermolecular steric interactions and lipid chain fluidity in phospholipid multibilayers. Spin label study. [**788**] T. H. Martin, Diss. Tufts Univ., Medford. 1974. 117 pp. Avail. Xerox Univ. Microfilms, Ann Arbor, Mich., Order No. 75-21,224, C.A. *83*, 202983x (1975); Magnetic relaxation of lyophases. [**789**] A. F. Martins, C. R. Acad. Sci., Sér. B *268*, 1731 (1969), C.A. *71*, 75994b (1969); Étude des mésophases de l'oléate de vinyle par relaxation RMN. [**790**] A. F. Martins, A. Rousseau, J. Phys. (Paris), C-5a, *32*, 249 (1971), C.A. *77*, 10705n (1972); Dynamique des effets d'ordre local dans la phase i des cristaux liquides.

[**791**] A. F. Martins, Mol. Cryst. Liq. Cryst. *14*, 85 (1971), C.A. *75*, 123406t (1971); Thermal fluctuations and ^{1}H spin-lattice relaxation in nematics. [**792**] A. F. Martins, Phys. Rev. Lett. *28*, 289 (1972), C.A. *76*, 79036s (1972); Interpretation of the nuclear spin-lattice relaxation-time behavior in

nematics. [**793**] A. F. Martins, Port. Phys. *8*, 166 (1972), C.A. *78*, 141272n (1973); Dynamique moléculaire dans les phases i et n des cristaux liquides. [**794**] P. Palffy, M. Marusic, D. Balzarini, C. F. Schwerdtfeger, Bull. Am. Phys. Soc. *16*, 845 (1971); Comparison of optical and ESR measurements of a nematic. [**795**] M. Marusic, Diss. Univ. British Columbia, Vancouver, B. C. 1973, Avail. Natl. Libr. Canada, Ottawa, Ont., C.A. *81*, 96972q (1974); Studies of nematics using the ESR technique. [**796**] M. Marusic, C. F. Schwerdtfeger, Mol. Cryst. Liq. Cryst. *28*, 131 (1974), C.A. *82*, 131690r (1975); ESR study of a nematic near the i/n transition. [**797**] R. M. C. Matthews, Diss. Univ. Texas, Austin. 1975. 147 pp. Avail. Xerox Univ. Microfilms, Ann Arbor, Mich., Order No. 75-24,914, C.A. *84*, 27213t (1976); NMR relaxation in biological liquid crystals. [**798**] L. C. Snyder, S. Meiboom, J. Chem. Phys. *44*, 4057 (1966), C.A. *65*, 6545c (1966); NMR of tetrahedral molecules in a nematic solvent. [**799**] S. Meiboom, L. C. Snyder, J. Am. Chem. Soc. *89*, 1038 (1967); Structure of cyclopropane and cyclobutane from ^{1}HMR in a nematic solvent. [**800**] E. Sackmann, S. Meiboom, L. C. Snyder, J. Am. Chem. Soc. *89*, 5981 (1967), C.A. *68*, 81974z (1968); Relation of nematics to cholesterics.

[**801**] L. C. Snyder, S. Meiboom, J. Chem. Phys. *47*, 1480 (1967), C.A. *67*, 86375u (1967); Molecular structure of cyclopropane from its ^{1}HMR in a nematic solvent. [**802**] E. Sackmann, S. Meiboom, L. C. Snyder, J. Am. Chem. Soc. *90*, 2183 (1968), C.A. *68*, 110045e (1968); NMR spectra of enantiomers in optically active liquid crystals. [**803**] S. Meiboom, L. C. Snyder, Science *162*, 1337 (1968), C.A. *70*, 52524p (1969); NMR in liquid crystals. [**804**] K. Wuethrich, S. Meiboom, L. C. Snyder, J. Chem. Phys. *52*, 230 (1970), C.A. *72*, 49400f (1970); NMR spectroscopy of bicyclobutane. [**805**] S. Meiboom, L. C. Snyder, J. Chem. Phys. *52*, 3857 (1970), C.A. *72*, 127079z (1970); Molecular structure of cyclobutane from its ^{1}HMR in a nematic solvent. [**806**] S. Meiboom, L. C. Snyder, Accounts Chem. Res. *4*, 81 (1971), C.A. *74*, 105329v (1971); NMR spectra in liquid crystals and molecular structure. [**807**] S. Meiboom, R. C. Hewitt, L. C. Snyder, Pure Appl. Chem. *32*, 251 (1972), C.A. *78*, 9882g (1973); Developments in NMR in liquid crystalline solvents. [**808**] R. C. Hewitt, S. Meiboom, L. C. Snyder, J. Chem. Phys. *58*, 5089 (1973), C.A. *79*, 36853a (1973); ^{1}HMR in nematic solvents. Use of ^{2}H decoupling. [**809**] L. C. Snyder, S. Meiboom, J. Chem. Phys. *58*, 5096 (1973), C.A. *79*, 25390w (1973); Theory of ^{1}HMR with ^{2}H decoupling in nematic solvents. [**810**] Z. Luz, S. Meiboom, J. Chem. Phys. *59*, 275 (1973), C.A. *79*, 72059e (1973); NMR studies of smectics.

[**811**] Z. Luz, S. Meiboom, J. Chem. Phys. *59*, 1077 (1973), C.A. *79*, 114905k (1973); Structure and bond shift kinetics of cyclooctatetraene studied by NMR in nematic solvents. [**812**] S. Meiboom, Z. Luz, Mol. Cryst. Liq. Cryst. *22*, 143 (1973), C.A. *79*, 109046b (1973); NMR study of smectics. [**813**] Z. Luz, R. C. Hewitt, S. Meiboom, J. Chem. Phys. *61*, 1758 (1974), C.A. *81*, 143867w (1974); ^{2}HMR study of a smectic. [**814**] J. C. Metcalfe, Colloq. Ges. Biol. Chem. *22*, 201 (1971), C.A. *77*, 15661e (1972); NMR studies of membranes and lipids. [**815**] J. C. Metcalfe, Chem. Phys. Lipids *8*, 333 (1972), C.A. *77*, 71710g (1972); ^{13}CMR studies of lipids in bilayers and membranes. [**816**] F. S. Millett, Diss. Columbia Univ., New York. 1970. 57 pp. Avail. Univ. Microfilms, Ann Arbor, Mich., Order No. 71-6227, C.A. *75*, 135558g (1971); NMR determination of ^{2}H quadrupole coupling constants in nematic solutions. [**817**] F. Millett, P. A. Hargrave, M. A. Raftery, Biochemistry *12*, 3591 (1973), C.A. *79*, 122842c (1973); Natural abundance ^{13}CMR spectra of the lipid in intact bovine retinal rod outer segment membranes. [**818**] K. Moebius, K. P. Dinse, Chimia *26*, 461 (1972), C.A. *77*, 158022h (1972); ENDOR of organic radicals in solution. [**819**] S. Mohanty, Mol. Phys. *25*, 1173 (1973), C.A. *79*, 47601t (1973); ^{1}HMR shielding anisotropy in acetylene. [**820**] H. Monoi, Biophys. J. *14*, 645 (1974), C.A. *81*, 167134f (1974); NMR of tissue ^{23}Na. I. ^{23}Na signal and Na$^+$ ion activity in homogenate.

[**821**] A. J. Montana, Diss. Columbia Univ., New York. 1976. 145 pp. Avail. Xerox Univ. Microfilms, Ann Arbor, Mich., Order No. 77-6654, C.A. *87*, 13862d (1977); NMR spectroscopy in smectic solvents. [**822**] I. Morishima, A. Mizuno, T. Yonezawa, Chem. Phys. Lett. *7*, 633 (1970), C.A. *74*, 93308x (1971); ^{13}CMR studies of oriented molecules in liquid crystal solvents. Anisotropy of ^{13}C shielding constants in methyl-^{13}C iodide and methyl-^{13}C cyanide. [**823**] I. Morishima, A. Mizuno, T. Yonezawa, J. Am. Chem. Soc. *93*, 1520 (1971), C.A. *74*, 132769v (1971); Anomalous features of NMR spectra of methyl and methylene halides oriented in a nematic solvent. Unusual ordering due to specific solute-solvent interaction. [**824**] J. D. Morrisett, H. J. Pownall, R. T. Plumlee, L. C. Smith, Z. E. Zehner, M. Esfahani, S. J. Wakil, J. Biol. Chem. *250*, 6969 (1975), C.A. *83*, 174331n (1975); Multiple thermotropic phase transitions in escherichia coli membranes and membrane lipids. [**825**] S. B. Christman, H. A. Moses, P. S. Cohen, E. P. Smith, J. J. Fink, J. Chem. Phys. *53*, 456 (1970), C.A. *73*, 71847z (1970); ^{1}HMR line in nematic APAPA. [**826**] J. J. Fink, H. A. Moses, P. S. Cohen, J. Chem. Phys. *56*, 6198 (1972), C.A. *77*, 41103u (1972); ^{1}HMR lines in liquid crystalline 4,4'-dipropoxyaz-

oxybenzene, 4,4′-dibutoxyazoxybenzene and diethyl 4,4′-azodibenzoate. [827] S. E. Gerofsky, H. A. Moses, J. J. Fink, A. Grover, J. Ross, J. Chem. Phys. *65*, 3526 (1976), C.A. *86*, 10435k (1977); ¹HMR lines in liquid crystalline *p*-[(*p*-methoxybenzylidene)amino]benzonitrile, *p*-[(*p*-ethoxybenzylidene)amino]benzonitrile, and *p*-[(*p*-ethoxybenzylidene)amino]phenyl acetate. [828] J. A. Murphy, Diss. Kent State Univ., Kent. 1973, 144 pp. Avail. Univ. Microfilms, Ann Arbor, Mich., Order No. 74-7328, C.A. *81*, 7079p (1974); NMR pulsed gradient studies of diffusion in liquid crystals. [829] J. I. Musher, J. Chem. Phys. *46*, 1537 (1967), C.A. *66*, 70736f (1967); Equivalence of nuclear spins. [830] J. I. Musher, J. Chem. Phys. *47*, 5460 (1967), C.A. *68*, 100450u (1968); Magnetic equivalence of nuclear spins.

[831] M. S. Gopinathan, P. T. Narasimhan, J. Magn. Reson. *6*, 147 (1972), C.A. *76*, 106141f (1972); Oriented ¹HMR spectra of acetone and dimethyl sulfoxide and molecular orbital studies of ⁴J_{HH}. [832] T. E. Needham, Diss. Univ. Illinois, Urbana. 1972, 190 pp. Avail. Univ. Microfilms, Ann Arbor, Mich., Order No. 72-19,892, C.A. *77*, 158471d (1972); Structure studies by NMR in liquid crystal solvents. [833] M. A. Nickerson, Diss. Univ. Maine, Portland. 1976. 101 pp. Avail. Xerox Univ. Microfilms, Ann Arbor, Mich., Order No. 77-8329, C.A. *87*, 13867j (1977); Spin-lattice relaxation by fluctuations in the nematic director. [834] W. Niederberger, Y. Tricot, J. Magn. Reson. *28*, 313 (1977), C.A. *88*, 46705d (1978); Water orientation in lyophases: ¹⁷OMR and ²HMR study. [835] W. Woelfel, F. Noack, M. Stohrer, Z. Naturforsch. A *30*, 437 (1975), C.A. *83*, 51172d (1975); Frequency dependence of ¹H spin relaxation in nematic PAA. [836] V. Graf, F. Noack, M. Stohrer, Z. Naturforsch. A *32*, 61 (1977), C.A. *86*, 98589x (1977); NMR investigation of order fluctuations, self-diffusion and rotational motions in nematic MBBA by t_1-relaxation spectroscopy. [837] W. R. Runyan, A. W. Nolle, J. Chem. Phys. *27*, 1081 (1957), C.A. *52*, 4328i (1958); NMR study of liquid crystal transitions. [838] P. L. Nordio, Chem. Phys. Lett. *6*, 250 (1970), C.A. *73*, 114844r (1970); Lineshape effects of anisotropic diffusion. [839] P. L. Nordio, P. Busolin, J. Chem. Phys. *55*, 5485 (1971), C.A. *76*, 8626y (1972); ESR line shapes in partially oriented systems. [840] P. L. Nordio, G. Rigatti, U. Segre, J. Chem. Phys. *56*, 2117 (1972), C.A. *76*, 90235b (1972); Spin relaxation of nematic solvents.

[841] P. L. Nordio, G. Rigatti, U. Segre, Chem. Phys. Lett. *19*, 295 (1973), C.A. *78*, 152763z (1973); Magnetic and dielectric relaxation in PAA. Effect of the reorienting methoxy groups. [842] G. Giacometti, P. L. Nordio, G. Rigatti, U. Segre, J. Chem. Soc., Faraday Trans. 2, *69*, 1815 (1973), C.A. *80*, 102060p (1974); Theory of NMR paramagnetic shifts in liquid crystalline solvents. [843] G. Agostini, P. L. Nordio, L. Pasimeni, U. Segre, J. Chem. Soc., Faraday Trans. 2, *70*, 621 (1974), C.A. *81*, 70702b (1974); ESR study of copper(II) bis(di-selenocarbamate) in liquid crystals. [844] P. L. Nordio, U. Segre, Gazz. Chim. Ital. *106*, 431 (1976), C.A. *85*, 101813c (1976); Nuclear spin relaxation in liquid crystals. [845] P. L. Nordio, U. Segre, Mol. Cryst. Liq. Cryst. *36*, 255 (1976), C.A. *86*, 81328p (1977); Frequency dependence of ¹H spin-lattice relaxation in MBBA and PAA. [846] P. L. Nordio, U. Segre, J. Magn. Reson. *27*, 465 (1977), C.A. *87*, 175386d (1977); Magnetic relaxation from first-rank interactions. [847] J. Oakes, Nature *231*, 38 (1971), C.A. *75*, 56497h (1971); ESR studies of the solubilization of nitroxide spin probes by micellar solutions. [848] J. Oakes, J. Chem. Soc., Faraday Trans. 2, *68*, 1464 (1972), C.A. *77*, 93146e (1972); Magnetic resonance studies in aqueous systems. I. Solubilization of spin probes by micellar solution. [849] J. Oakes, J. Chem. Soc., Faraday Trans. 2, *69*, 1321 (1973), C.A. *80*, 2779m (1974); III. ESR and NMR relaxation study of interactions between manganese ions and micelles (cf. [848]). [850] K. Orrell, V. Sik, J. Chem. Soc., Faraday Trans. 2, *71*, 1360 (1975), C.A. *83*, 131047g (1975); Nematic phase NMR studies of disubstituted pyridines. Molecular geometry of 2,6-difluoropyridine.

[851] K. G. Orrell, V. Sik, J. Chem. Soc., Faraday Trans. 2, *72*, 941 (1976), C.A. *85*, 77366m (1976); NMR investigation of pyridine-4-aldehyde oriented in a nematic. Internal rotation and molecular reorientation. [852] R. D. Orwoll, Diss. Univ. California, San Diego. 1971, 125 pp. Avail. Univ. Microfilms, Ann Arbor, Mich., Order No. 72-1072, C.A. *76*, 91207f (1972); NMR studies of liquid-crystalline solutions of PBLG. [853] J. B. Pawliczek, Diss. Köln 1971; NMR-Untersuchungen von organischen Molekülen in kristallinflüssigen Lösungen. [854] G. F. Pedulli, C. Zannoni, A. Alberti, J. Magn. Reson. *10*, 372 (1973), C.A. *79*, 77623u (1973); Molecular deformations induced by liquid crystalline solvents. [855] W. D. Phillips, J. C. Rowell, L. R. Melby, J. Chem. Phys. *41*, 2551 (1964), C.A. *62*, 2384f (1965); Quadrupole splittings in the ²HMR spectra of liquid crystals. [856] J. C. Rowell, W. D. Phillips, L. R. Melby, M. Panar, J. Chem. Phys. *43*, 3442 (1965), C.A. *64*, 1501b (1966); NMR studies of liquid crystals. [857] Th. Pietrzak, J. Phys. Chem. *76*, 672 (1972), C.A. *76*, 106129h (1972); Model for the simulation of ESR spectra of nitroxide-type probe molecules in a nematic. [858] P. A. Pincus, J. Phys. (Paris), C-4, *30*, 8 (1969), C.A. *72*, 104823y (1970); NMR determination of the orientational

order in liquid crystals. **[859]** P. A. Pincus, Solid State Commun. *7*, 415 (1969), C.A. *70*, 92122n (1969); Nuclear relaxation in a nematic. **[860]** P. A. Pincus, J. Appl. Phys. *41*, 974 (1970), C.A. *73*, 8646t (1970); Magnetic properties of liquid crystals.

[861] A. Pines, J. J. Chang, J. Am. Chem. Soc. *96*, 5590 (1974), C.A. *81*, 113340b (1974); Effect of phase transitions on ^{13}CMR spectra in nematic PAA. **[862]** A. Pines, J. J. Chang, Phys. Rev. A *10*, 946 (1974), C.A. *81*, 129447c (1974); Study of the i/n/k transitions in a liquid crystal by ^{13}C-$\{^1H\}$ double resonance. **[863]** M. M. Pintar, Pulsed Magn. Opt. Resonance, Proc. Ampère Int. Summer Sch. *2*, 251 (1971), C.A. *77*, 132939n (1972); Strong collision relaxation in the rotating frame. **[864]** R. T. Thompson, R. R. Knispel, M. M. Pintar, Chem. Phys. Lett. *22*, 335 (1973), C.A. *80*, 44993g (1974); Study of the proton exchange in tissue water by spin relaxation in the rotating frame. **[865]** A. R. Sharp, W. F. Forbes, M. M. Pintar, J. Chem. Phys. *59*, 460 (1973), C.A. *79*, 59723x (1973); Frequency dependence of the ^{1}H spin relaxation of the dipolar energy in a liquid crystal. **[866]** R. T. Thompson, D. W. Kydon, M. M. Pintar, J. Chem. Phys. *61*, 4646 (1974), C.A. *82*, 162473b (1975); ^{1}H spin thermometry and molecular dynamics of the mesogen PAA in the solid phase. **[867]** R. R. Knispel, R. T. Thompson, M. M. Pintar, J. Magn. Reson. *14*, 44 (1974), C.A. *81*, 61792d (1974); Dispersion of ^{1}H spin-lattice relaxation in tissues. **[868]** R. G. C. McElroy, R. T. Thompson, M. M. Pintar, Phys. Rev. A *10*, 403 (1974), C.A. *81*, 84066j (1974); ^{1}H-spin thermometry at low fields in liquid crystals. **[869]** R. T. Thompson, D. W. Kydon, M. M. Pintar, Chem. Phys. Lett. *42*, 586 (1976), C.A. *85*, 184498c (1976); Search for ultra slow molecular motion in liquid crystals by spin thermometry. **[870]** R. T. Thompson, D. W. Kydon, M. M. Pintar, Mol. Cryst. Liq. Cryst. *39*, 123 (1977), C.A. *87*, 31564n (1977); Spin thermometric analysis of spin-lattice relaxation by the order director fluctuation in nematic MBBA.

[871] I. Pocsik, K. Tompa, J. Lasanda, S. Kugler, G. Naray-Szabo, Hung. Acad. Sci., Cent. Res. Inst. Phys., KFKI-*1977*-40-16 pp., C.A. *87*, 109759b (1977); Motion of the methyl group in PAA in the solid. **[872]** C. F. Polnaszek, Diss. Cornell Univ., Ithaca. 1976. 618 pp. Avail. Xerox Univ. Microfilms, Ann Arbor, Mich., Order No. 76-21,121, C.A. *85*, 169340b (1976); ESR study of rotational reorientation and spin relaxation in liquid crystal media. **[873]** J. S. Prasad, J. Chem. Phys. *65*, 941 (1976), C.A. *85*, 169321w (1976); Orientational order parameters and conformation of nematic EBBA. **[874]** J. S. Prasad, Mol. Cryst. Liq. Cryst. *35*, 345 (1976), C.A. *85*, 200332b (1976); Orientational order parameter in 4,4′-bis(pentyloxy)azoxybenzene. **[875]** D. Sy, A. Sanson, M. Ptak, Solid State Commun. *10*, 985 (1972), C.A. *77*, 41098w (1972); ESR measurements of the degree of order in the nematic phase. **[876]** A. Sanson, M. Ptak, J. Phys. (Paris), C-8, *34*, 5 (1973), C.A. *81*, 77082h (1974); Sondes paramagnétiques a l'étude par ESR de phases mésomorphes. **[877]** D. Sy, M. Ptak, J. Phys. (Paris) *35*, 517 (1974), C.A. *81*, 113347j (1974); ESR study of 4-methoxybenzylidene-4′-amino-*n*-alkyl cinnamates. **[878]** D. Sy, M. Ptak, Mol. Cryst. Liq. Cryst. *39*, 53 (1977), C.A. *87*, 31563m (1977); ESR investigation of spin probes. Not unique ordering in smectics. **[879]** A. T. Pudzianowski, A. E. Stillman, R. N. Schwartz, B. L. Bales, E. S. Lesin, Mol. Cryst. Liq. Cryst. *34*, 33 (1976), C.A. *85*, 134821g (1976); Nematogen-like nitroxide free radical to probe the structure and dynamics of nematics. **[880]** N. F. Ramsey, Phys. Rev. *78*, 699 (1950), C.A. *44*, 8221h (1950); Magnetic shielding of nuclei in molecules.

[881] N. F. Ramsey, Phys. Rev. *91*, 303 (1953), C.A. *47*, 10998d (1953); Electron coupled interactions between nuclear spins in molecules. **[882]** J. M. Briggs, E. J. Rahkamaa, E. W. Randall, J. Magn. Reson. *17*, 55 (1975), C.A. *82*, 147909e (1975); ^{1}HMR studies of pyrrole-^{15}N oriented in nematic solvents. **[883]** L. W. Reeves, J. A. Vanin, V. R. Vanin, An. Acad. Brasil. Cienc. *44*, 431 (1972), C.A. *79*, 129848t (1973); Location of dissolved formamide in a lyotropic nematic. **[884]** M. A. Raza, L. W. Reeves, Can. J. Chem. *50*, 2370 (1972), C.A. *77*, 113295h (1972); NMR study of the symmetry of the ^{1}H spins in ethylene sulfite in a nematic solvent. **[885]** D. M. Chen, L. W. Reeves, J. Am. Chem. Soc. *94*, 4384 (1972), C.A. *77*, 54642u (1972); ^{23}NaMR in lyotropic nematics and the implications for observation in living systems. **[886]** M. A. Raza, L. W. Reeves, J. Chem. Phys. *57*, 821 (1972), C.A. *77*, 68255g (1972); ^{1}HMR spectrum of ethylene trithiocarbonate in a thermotropic nematic solvent. **[887]** M. A. Raza, L. W. Reeves, J. Magn. Reson. *8*, 222 (1972), C.A. *77*, 170981c (1972); Ethylene carbonate and ethylene monothiocarbonate as solutes in a thermotropic nematic. **[888]** M. A. Raza, L. W. Reeves, Mol. Phys. *23*, 1007 (1972), C.A. *77*, 68260e (1972); ^{1}HMR spectrum of β-propiolactone in two thermotropic nematic solvents. **[889]** L. W. Reeves, A. S. Tracey, M. M. Tracey, J. Am. Chem. Soc. *95*, 3799 (1973), C.A. *79*, 35689q (1973); Experimental method for the determination of the structure of complex negative ions in solution. **[890]** S. A. Barton, M. A. Raza, L. W. Reeves, J. Magn. Reson.

9, 45 (1973), C.A. *78*, 64860y (1973); Study of ethylene carbonate and ethylene monothiocarbonate oriented in a lyotropic nematic.

[**891**] L. W. Reeves, J. M. Riveros, R. A. Spragg, J. A. Vanin, Mol. Phys. *25*, 9 (1973), C.A. *78*, 96759p (1973); Study of ^{15}N formamide oriented in a lyotropic nematic. [**892**] L. W. Reeves, J. Sanches de Cara, M. Suzuki, A. S. Tracey, Mol. Phys. *25*, 1481 (1973), C.A. *79*, 91467c (1973); Structure of positively charged complex ions in solution. [**893**] L. W. Reeves, M. Suzuki, A. S. Tracey, J. A. Vanin, Inorg. Chem. *13*, 999 (1974), C.A. *80*, 101008r (1974); Membrane processes. II. Participation of the dimethyltin ion in an electric double layer. [**894**] L. W. Reeves, A. S. Tracey, J. Am. Chem. Soc. *96*, 365 (1974), C.A. *80*, 74878r (1974); I. Ammonium, ammonium-d_4, ammonium-d_3, and methylammonium-d_3 ions in equilibrium with an oriented electric double layer (cf. [893]). [**895**] L. W. Reeves, A. S. Tracey, J. Am. Chem. Soc. *96*, 1198 (1974), C.A. *80*, 107804u (1974); Structure of the methylammonium ion in lyotropic nematic solution. [**896**] F. Fujiwara, L. W. Reeves, A. S. Tracey, L. A. Wilson, J. Am. Chem. Soc. *96*, 5249 (1974), C.A. *81*, 119743g (1974); III. ^{2}HMR as a tool in studies of the lipophilic region of membrane systems (cf. [893]). [**897**] F. Fujiwara, L. W. Reeves, A. S. Tracey, J. Am. Chem. Soc. *96*, 5250 (1974), C.A. *81*, 104195f (1974); V. Distortion of tetrahedral ions in the electric double layer of a model membrane (cf. [893]). [**898**] D. M. Chen, K. Radley, L. W. Reeves, J. Am. Chem. Soc. *96*, 5251 (1974), C.A. *81*, 176564y (1974); VI. Monatomic ions in the electric double layer (cf. [893]). [**899**] D. M. Chen, L. W. Reeves, A. S. Tracey, M. M. Tracey, J. Am. Chem. Soc. *96*, 5349 (1974), C.A. *81*, 127002e (1974); IV. Structure of the acetate ion and degree of orientation of the ionic head groups in the electric double layer (cf. [893]). [**900**] L. W. Reeves, A. S. Tracey, J. Am. Chem. Soc. *96*, 7176 (1974), C.A. *82*, 9723p (1975); NMR studies of the di-, tri-, and tetramethylammonium ions oriented in the middle soap phases.

[**901**] Y. Lee, L. W. Reeves, Can. J. Chem. *53*, 161 (1975), C.A. *82*, 147952p (1975); Structure of the dimethylthallium ion oriented in a lyotropic nematic. [**902**] K. Radley, L. W. Reeves, Can. J. Chem. *53*, 2998 (1975), C.A. *84*, 37533m (1976); Studies of ternary nematics by NMR. Alkali metal decyl sulfates/decanol/deuterium oxide. [**903**] L. W. Reeves, Int. Rev. Sci.: Phys. Chem., Ser. Two *4*, 139 (1975), C.A. *85*, 15445c (1976); Study of ionic processes at membranes by NMR methods. [**904**] L. W. Reeves, A. S. Tracey, J. Am. Chem. Soc. *97*, 5729 (1975), C.A. *83*, 192087s (1975); VII. Hydrocarbon chain motions and the effect of changing counterions (cf. [893]). [**905**] L. W. Reeves, F. Y. Fujiwara, M. Suzuki, ACS Symp. Ser. *34*, 55 (1976), C.A. *86*, 81955r (1977); Chemical aspects of the hydrophobic/hydrophilic interface by NMR methods. [**906**] L. R. Baldo, L. W. Reeves, J. A. Vanin, An. Acad. Bras. Cienc. *48*, 37 (1976), C.A. *87*, 32347f (1977); Structure of γ-carboxyl pyridinium ion oriented in a lyotropic nematic. [**907**] Y. Lee, L. W. Reeves, Can. J. Chem. *54*, 500 (1976), C.A. *84*, 187792j (1976); Studies of p-nitrobenzoic acid and the p-nitrobenzoate ion in the electric double layer of oriented cationic and anionic nematics. [**908**] L. W. Reeves, M. Suzuki, J. A. Vanin, Inorg. Chem. *15*, 1035 (1976), C.A. *84*, 163706h (1976); Structures of simple organometallic ions derived from NMR studies of the oriented species. [**909**] F. Y. Fujiwara, L. W. Reeves, J. Am. Chem. Soc. *98*, 6790 (1976), C.A. *85*, 192065z (1976); Degree of order of lamellar and hexagonal mesophases with and without cholesterol additions. [**910**] K. Radley, L. W. Reeves, A. S. Tracey, J. Phys. Chem. *80*, 174 (1976), C.A. *84*, 68123g (1976); Effect of counterion substitution on the type and nature of nematic lyophases from NMR studies.

[**911**] D. M. Chen, F. Y. Fujiwara, L. W. Reeves, Can. J. Chem. *55*, 2404 (1977), C.A. *87*, 125442y (1977); Degrees of order of solubilized alkanes, alcohols, carboxylates, and carboxylic acid in oriented lyophases. [**912**] R. T. Roberts, C. Chachaty, Chem. Phys. Lett. *22*, 348 (1973), C.A. *80*, 32242y (1974); ^{13}C relaxation measurements of molecular motion in micellar solutions. [**913**] R. T. Roberts, Nature *242*, 348 (1973), C.A. *78*, 140483v (1973); Measurement of the diffusion coefficient in the concentrated phases of the soap/water system by NMR. [**914**] P. I. Rose, Mol. Cryst. Liq. Cryst. *26*, 75 (1974), C.A. *81*, 127850e (1974); Determination of the diamagnetic anisotropy of nematic MBBA. [**915**] J. Russell, Org. Magn. Reson. *4*, 433 (1972), C.A. *77*, 81831m (1972); NMR spectra of 1,4-dioxin and 1,4-dithiin partially oriented in a nematic. [**916**] S. S. Sabri, Diss. Univ. California, Los Angeles. 1972, 225 pp. Avail. Univ. Microfilms, Ann Arbor, Mich., Order No. 73-6396, C.A. *78*, 166655c (1973); NMR spectra of cyclic compounds in nematic solvents. [**917**] E. Sackmann, P. Krebs, Chem. Phys. Lett. *4*, 65 (1969), C.A. *72*, 26907n (1970); ESR and optical studies of charge-transfer complexes oriented in liquid crystals. [**918**] E. Sackmann, J. Chem. Phys. *51*, 2984 (1969), C.A. *71*, 130530z (1969); Molecular structure of allene from its ^{1}HMR in a nematic solvent. [**919**] P. Krebs, E. Sackmann, J. Schwartz, Chem. Phys. Lett. *8*, 417 (1971), C.A. *75*, 28071w (1971); Triplet ESR spectra of charge-transfer complexes

in liquid and single crystals. [**920**] E. Sackmann, H. Traeuble, J. Am. Chem. Soc. *94*, 4482 (1972), C.A. *77*, 54567y (1972); Crystalline-liquid crystalline phase transition of lipid model membranes. I. Use of spin labels and optical probes as indicators of the phase transition.

[**921**] E. Sackmann, H. Traeuble, J. Am. Chem. Soc. *94*, 4492 (1972), C.A. *77*, 58049d (1972); II. Analysis of ESR spectra of steroid labels incorporated into lipid membranes (cf. [920]). [**922**] H. Traeuble, E. Sackmann, J. Am. Chem. Soc. *94*, 4499 (1972), C.A. *77*, 54565w (1972); III. Structure of a steroid-lecithin system below and above the lipid-phase transition (cf. [920]). [**923**] P. Krebs, E. Sackmann, Mol. Phys. *23*, 437 (1972), C.A. *77*, 27062y (1972); Orientation distribution function of aromatic molecules in frozen liquid crystals from their ESR spectra. [**924**] H. J. Galla, E. Sackmann, Biochim. Biophys. Acta *401*, 509 (1975), C.A. *83*, 143346s (1975); Chemically induced lipid phase separation in model membranes containing charged lipids. Spin label study. [**925**] P. Krebs, E. Sackmann, J. Magn. Reson. *22*, 359 (1976), C.A. *85*, 85113s (1976); Triplet ESR spectroscopy in ordered glasses. Spectroscopic application of liquid crystals. [**926**] A. Saupe, G. Englert, Phys. Rev. Lett. *11*, 462 (1963), C.A. *60*, 3641e (1964); High-resolution NMR spectra of orientated molecules. [**927**] A. Saupe, Z. Naturforsch. A *19*, 161 (1964), C.A. *60*, 15318f (1964); NMR in Flüssigkristallen und kristallinflüssigen Lösungen. I. [**928**] G. Englert, A. Saupe, Z. Naturforsch. A *19*, 172 (1964), C.A. *61*, 1409h (1964); Teil II (cf. [927]). [**929**] A. Saupe, Z. Naturforsch. A *20*, 572 (1965), C.A. *63*, 4130c (1965); ^{1}HMR-Spektrum von orientiertem Benzol in nematischer Lösung. [**930**] G. Englert, A. Saupe, Z. Naturforsch. A *20*, 1401 (1965), C.A. *64*, 9107c (1966); Hochaufgelöste ^{1}HMR-Spektren mit direkter magnetischer Dipol-Dipol-Wechselwirkung.

[**931**] G. Englert, A. Saupe, Mol. Cryst. *1*, 503 (1966), C.A. *66*, 15306r (1967); ^{1}HMR spectra of oriented molecules: acetylenic compounds, acetonitrile, and methanol. [**932**] A. Saupe, Mol. Cryst. *1*, 527 (1966), C.A. *66*, 15304p (1967); Average orientation of solute molecules in nematics by ^{1}HMR measurements and orientation dependent intermolecular forces. [**933**] A. Saupe, G. Englert, A. Povh, Advan. Chem. Ser. No. *63*, 51 (1967), C.A. *67*, 38101e (1967); Determination of bond angle of CH_3 groups by ^{1}HMR in nematic solutions. [**934**] A. Saupe, J. Nehring, J. Chem. Phys. *47*, 5459 (1967), C.A. *68*, 100449a (1968); Magnetic equivalence of nuclear spins in oriented molecules. [**935**] A. Saupe, Angew. Chem. *80*, 99 (1968), C.A. *68*, 81938r (1968); Neuere Ergebnisse auf dem Gebiet der Flüssigkristalle. [**936**] G. Englert, A. Saupe, J. P. Weber, Z. Naturforsch. A *23*, 152 (1968), C.A. *69*, 6965q (1968); ^{1}HMR-Spektren orientierter Moleküle: Acetylenverbindungen. [**937**] G. Englert, A. Saupe, Mol. Cryst. Liq. Cryst. *8*, 233 (1969), C.A. *71*, 86430r (1969); Molecular geometry of acetonitrile, determined by ^{1}HMR in nematic solutions. [**938**] J. Nehring, A. Saupe, Mol. Cryst. Liq. Cryst. *8*, 403 (1969), C.A. *71*, 85565b (1969); Orientation studies on fluorobenzenes in nematics by NMR. [**939**] J. Nehring, A. Saupe, J. Chem. Phys. *52*, 1307 (1970), C.A. *72*, 84705f (1970); Anisotropies of the ^{19}F chemical shifts in fluorobenzene compounds from NMR in liquid crystals. [**940**] H. Spiesecke, A. Saupe, Mol. Cryst. Liq. Cryst. *6*, 287 (1970), C.A. *72*, 105677r (1970); NMR spectra of oriented 1,1-difluoroethylene and tetrafluoroethylene.

[**941**] C. L. Khetrapal, A. Saupe, A. C. Kunwar, C. R. Kanekar, Mol. Phys. *22*, 1119 (1971), C.A. *76*, 106157r (1972); Structure of π-methylcyclopentadienylmanganese tricarbonyl as determined from ^{1}HMR studies in the nematic phase. [**942**] C. L. Khetrapal, A. C. Kunwar, A. Saupe, J. Magn. Reson. *7*, 18 (1972), C.A. *77*, 27148f (1972); ^{1}HMR spectra of butadiene sulfone in isotropic and liquid crystalline media. [**943**] C. L. Khetrapal, A. Saupe, A. C. Kunwar, Mol. Cryst. Liq. Cryst. *17*, 121 (1972), C.A. *77*, 60964s (1972); NMR spectrum of phthalazine oriented in the nematic phase. [**944**] C. L. Khetrapal, A. Saupe, J. Magn. Reson. *9*, 275 (1973), C.A. *78*, 110010w (1973); Studies on the structure of γ-picoline by NMR in a nematic solution. [**945**] C. L. Khetrapal, A. C. Kunwar, A. Saupe, Liq. Cryst., Proc. Int. Conf., Bangalore, *1973*, 495, C.A. *84*, 163720h (1976); NMR spectrum of benzo(b)thiophene oriented in the nematic phase. [**946**] C. L. Khetrapal, A. Saupe, Mol. Cryst. Liq. Cryst. *19*, 195 (1973), C.A. *78*, 90735f (1973); ^{1}HMR studies of methylmercuric halides in isotropic and nematic solutions. [**947**] C. L. Khetrapal, A. C. Kunwar, A. Saupe, Mol. Phys. *25*, 1405 (1973), C.A. *79*, 91205r (1973); Ring puckering vibrations in trimethylene oxide and trimethylene sulfide as studied by ^{1}HMR spectroscopy in a liquid crystal. [**948**] C. L. Khetrapal, A. C. Kunwar, A. Saupe, Mol. Cryst. Liq. Cryst. *35*, 215 (1976), C.A. *86*, 105204y (1977); NMR spectra of π-cyclopentadienyl manganese tricarbonyl in nematic and isotropic solvents. [**949**] K. Radley, A. Saupe, Mol. Phys. *32*, 1167 (1976), C.A. *86*, 105226g (1977); Measurement of ^{1}H-^{1}H dipolar coupling for the ammonium ion using high resolution NMR. [**950**] C. L. Khetrapal, A. C. Kunwar, A. Saupe, Mol. Cryst. Liq. Cryst. *40*, 193 (1977), C.A. *87*, 125060x (1977); Structure and conformation of N,p-chlorophenylmaleimide in a nematic solvent by ^{1}HMR.

[**951**] M. Schara, M. Sentjurc, Solid State Commun. *8*, 593 (1970), C.A. *73*, 8041s (1970); Electric field ordering of nematic PAA molecules studied by ESR. [**952**] M. Sentjurc, M. Schara, Mol. Cryst. Liq. Cryst. *12*, 133 (1971), C.A. *75*, 41760q (1971); Liquid crystal ordering in the magnetic and electric fields studied in 4,4′-di-*n*-heptyloxyazoxybenzene by ESR. [**953**] M. Schara, Proc. Int. Sch. Phys. "Enrico Fermi" *59*, 765 (1973), C.A. *86*, 62792u (1977); ESR of liquid crystals. [**954**] F. Pusnik, M. Schara, M. Sentjurc, J. Phys. (Paris) *36*, 665 (1975), C.A. *83*, 89014s (1975); ESR relaxation study of a liquid crystal. [**955**] F. Pusnik, M. Schara, Chem. Phys. Lett. *37*, 106 (1976), C.A. *84*, 67531b (1976); ESR study of the orientational order and molecular dynamics in the smectic A and B. [**956**] M. Schara, F. Pusnik, M. Sentjurc, Croat. Chem. Acta *48*, 147 (1976), C.A. *85*, 37507b (1976); ESR of the sodium laurate/water lyophase and micellar solution. [**957**] J. P. Jacobsen, K. Schaumburg, Mol. Phys. *28*, 1505 (1974), C.A. *82*, 155066t (1975); Determination of the structure of benzonitrile by NMR. Analysis of ^{1}HMR and ^{13}CMR spectra of benzonitrile in i and n phase. [**958**] J. P. Jacobsen, K. Schaumburg, J. Magn. Reson. *24*, 173 (1976), C.A. *86*, 24151b (1977); Spin-lattice relaxation time measurements of water-d_2 in a lyophase. [**959**] J. P. Jacobsen, K. Schaumburg, J. Magn. Reson. *28*, 1 (1977), C.A. *88*, 30025w (1978); ^{1}HMR and ^{2}HMR spectra of chlorobenzene and bromobenzene in liquid crystals. [**960**] J. P. Jacobsen, K. Schaumburg, J. Magn. Reson. *28*, 191 (1977), C.A. *88*, 49873t (1978); Determination of ^{2}H quadrupole coupling constants in methyl groups by high-resolution NMR spectroscopy.

[**961**] G. E. Schenck, Diss. Univ. California, Los Angeles. 1971. 226 pp. Avail. Univ. Microfilms, Ann Arbor, Mich., Order No. 72-11,897, C.A. *77*, 47537k (1972); Solvent effects in NMR and the magnetic anisotropy of hydrocarbons. [**962**] I. A. Zlochower, J. H. Schulman, J. Colloid Interface Sci. *24*, 115 (1967), C.A. *67*, 67895x (1967); Study of molecular interactions and mobility at liquid/liquid interfaces by NMR spectroscopy. [**963**] H. Schulze, W. Burkersrode, Exp. Tech. Phys. *23*, 369 (1975); C.A. *84*, 37451h (1976); Bestimmung der molekularen Polarisierbarkeiten von nematischen Flüssigkristallen. [**964**] R. Price, C. Schumann, Mol. Cryst. Liq. Cryst. *16*, 291 (1972), C.A. *76*, 133924v (1972); Natural abundance ^{13}CMR of oriented 1,3,5-trichlorobenzene. [**965**] C. Schumann, R. Price, Angew. Chem. *85*, 989 (1973), C.A. *80*, 81571p (1974); ^{1}HMR-Spektrum von ^{15}N-Pyridin in nematischer Phase. [**966**] U. Schummer, D. Hegner, G. H. Schnepel, H. H. Wellhoener, Biochim. Biophys. Acta *394*, 93 (1975), C.A. *83*, 39132h (1975); Thermotropic phase changes in peripheral nerve of frog and rat. Spin label study. [**967**] J. Seelig, J. Am. Chem. Soc. *92*, 3881 (1970), C.A. *73*, 49658s (1970); Spin label studies of oriented smectics (a model system for bilayer membranes). [**968**] J. Seelig, H. Limacher, P. Bader, J. Am. Chem. Soc. *94*, 6364 (1972), C.A. *77*, 119273w (1972); Molecular architecture of liquid crystalline bilayers. [**969**] J. Seelig, Experientia *29*, 509 (1973), C.A. *79*, 50424t (1973); Magnetic resonance methods in membrane research. [**970**] F. Axel, J. Seelig, J. Am. Chem. Soc. *95*, 7972 (1973), C.A. *80*, 14143n (1974); Cis double bonds in liquid crystalline bilayers.

[**971**] H. Schindler, J. Seelig, J. Chem. Phys. *59*, 1841 (1973), C.A. *79*, 131102p (1973); ESR spectra of spin labels in lipid bilayers. [**972**] H. Schindler, J. Seelig, Ber. Bunsenges. Phys. Chem. *78*, 941 (1974), C.A. *82*, 27511x (1975); ESR-Spektren von Spin-Labeln in Modell-Membranen. [**973**] W. Niederberger, J. Seelig, Ber. Bunsenges. Phys. Chem. *78*, 947 (1974), C.A. *82*, 9744w (1975); ^{2}HMR-Spektroskopie an spezifisch deuterierten Flüssigkristallen. [**974**] J. Seelig, A. Seelig, Biochem. Biophys. Res. Commun. *57*, 406 (1974), C.A. *80*, 142120q (1974); ^{2}HMR studies of phospholipid bilayers. [**975**] J. Seelig, W. Niederberger, Biochemistry *13*, 1585 (1974), C.A. *81*, 116310c (1974); Two pictures of a lipid bilayer. Comparison between ^{2}H-label and spin-label experiments. [**976**] J. Seelig, W. Niederberger, J. Am. Chem. Soc. *96*, 2069 (1974), C.A. *80*, 139049t (1974); ^{2}H-labeled lipids as structural probes in liquid crystalline bilayers. ^{2}HMR study. [**977**] J. Seelig, H. Limacher, Mol. Cryst. Liq. Cryst. *25*, 105 (1974), C.A. *83*, 139512a (1975); Lipid molecules in lyophases with cylindrical structure. Spin label study. [**978**] J. Seelig, H. U. Gally, Biochemistry *15*, 5199 (1976), C.A. *86*, 38998v (1977); Investigation of phosphatidylethanolamine bilayers by ^{2}HMR and ^{31}PMR. [**979**] H. U. Gally, A. Seelig, J. Seelig, Hoppe-Seyler's Z. Physiol. Chem. *357*, 1447 (1976), C.A. *86*, 1376q (1977); Cholesterol-induced rod-like motion of fatty acyl chains in lipid bilayers. ^{2}HMR study. [**980**] W. Niederberger, J. Seelig, J. Am. Chem. Soc. *98*, 3704 (1976), C.A. *85*, 42583c (1976); ^{31}P chemical shift anisotropy in unsonicated phospholipid bilayers.

[**981**] J. Seelig, Spin Labeling 1976, 373, C.A. *84*, 158066v (1976); Anisotropic motion in liquid crystalline structures. [**982**] H. J. Segall, Diss. Univ. Illinois, Urbana. 1973. 82 pp. Avail. Univ. Microfilms, Ann Arbor, Mich., Order No. 74-12,178, C.A. *81*, 117317x (1974); Lipid spin-labeling of Hymenolepis diminuta. [**983**] E. Cappelli, A. di Nola, A. L. Segre, Mol. Phys. *27*, 1385 (1974), C.A. *81*, 104108e (1974); NMR spectrum of coumarin partially oriented in a nematic. [**984**] M. Setaka, C. Lagercrantz,

J. Am. Chem. Soc. *97*, 6013 (1975), C.A. *83*, 200555k (1975); Orientation of nitroxide spin labels in the lamellar mesophases of Aerosol-OT/water and decanol/decanoate/water systems. [**985**] D. O. Shah, Ann. N. Y. Acad. Sci. *1973*, 125, C.A. *79*, 73495f (1973); High resolution NMR studies on the structure of water in microemulsions and liquid crystals. [**986**] V. A. Shcherbakov, L. L. Shcherbakova, B. V. Semakov, Zh. Strukt. Khim. *15*, 925 (1974), C.A. *82*, 30558y (1975); NMR spectra of acetone oriented in an inorganic mesophase. [**987**] B. Sheard, Nature *223*, 1057 (1969), C.A. *71*, 119740x (1969); Internal mobility in phospholipids. [**988**] Y. Shimoyama, M. Shiotani, J. Sohma, Jpn. J. Appl. Phys. *16*, 1437 (1977), C.A *87*, 125579y (1977); ESR study of molecular order and motion in a nematic. [**989**] G. G. Smith, M. J. Ruwart, A. Haug, FEBS Lett. *45*, 96 (1974), C.A. *81*, 147253r (1974); Lipid phase transitions in membrane vesicles from Thermoplasma acidophila. [**990**] L. C. Snyder, E. W. Anderson, J. Am. Chem. Soc. *86*, 5023 (1964), C.A. *62*, 7269c (1965); Analysis of the ^{1}HMR spectrum of benzene in a nematic.

[**991**] L. C. Snyder, E. W. Anderson, J. Chem. Phys. *42*, 3336 (1965), C.A. *63*, 10881g (1965); Analysis of the ^{19}FMR spectrum of hexafluorobenzene in a nematic. [**992**] L. C. Snyder, J. Chem. Phys. *43*, 4041 (1965), C.A. *64*, 1499g (1966); Analysis of NMR spectra of molecules in liquid-crystal solvents. [**993**] R. Ditchfield, L. C. Snyder, J. Chem. Phys. *56*, 5823 (1972), C.A. *77*, 27143a (1972); Anisotropy of nuclear spin-spin coupling in methylfluoride. [**994**] S. Sobajima, J. Phys. Soc. Jap. *23*, 1070 (1967), C.A. *68*, 44626s (1968); NMR studies on orientation of liquid crystals of PBLG in magnetic fields. [**995**] R. D. Spence, H. A. Moses, P. L. Jain, J. Chem. Phys. *21*, 380 (1953), C.A. *47*, 7316e (1953); ^{1}HMR line in liquid crystals. [**996**] R. D. Spence, H. S. Gutowsky, C. H. Holm, J. Chem. Phys. *21*, 1891 (1953), C.A. *48*, 1094c (1954); Hindered molecular rotation in liquid crystals. [**997**] P. L. Jain, H. A. Moses, J. C. Lee, R. D. Spence, Phys. Rev. *92*, 844 (1953); ^{1}HMR in liquid crystals. [**998**] P. L. Jain, J. C. Lee, R. D. Spence, J. Chem. Phys. *23*, 878 (1955), C.A. *49*, 12124c (1955); ^{1}HMR in liquid crystals — orientation effects. [**999**] J. H. Muller, R. D. Spence, J. Chem. Phys. *29*, 1195 (1958), C.A. *53*, 5789a (1959); ^{1}HMR observation of mechanical destruction of magnetic ordering in liquid crystals. [**1000**] H. Spiesecke, J. Bellion-Jourdan, Angew. Chem. *79*, 475 (1967), C.A. *67*, 38084b (1967); Niedrigschmelzende nematische Phasen zur Aufnahme von NMR-Spektren orientierter Moleküle.

[**1001**] H. Spiesecke, Z. Naturforsch. A *23*, 467 (1968), C.A. *69*, 82127h (1968); Durch Kernresonanz bestimmte Strukturdaten des ^{13}CH$_3$NC. [**1002**] H. Spiesecke, Liq. Cryst. Ord. Fluids *1970*, 123, C.A. *78*, 153492x (1973); ^{1}HMR spectra of acetylene and its ^{13}C isomers in nematic solutions. [**1003**] H. Spiesecke, Z. Naturforsch. A *25*, 650 (1970), C.A. *73*, 61049b (1970); ^{1}HMR Spektrum des PH$_3$ in einem nematischen Lösungsmittel. [**1004**] G. J. Krüger, H. Spiesecke, Ber. Bunsenges. Phys. Chem. *75*, 272 (1971), C.A. *75*, 13202r (1971); Nuclear relaxation and diffusion in nematic solutions. [**1005**] K. Krebs, G. J. Krüger, H. Spiesecke, EURATOM Rep., No. *5060*, 171 (1972), C.A. *79*, 150375w (1973); Static and dynamic structure in liquid crystals. [**1006**] G. J. Krüger, H. Spiesecke, Z. Naturforsch. A *28*, 964 (1973), C.A. *80*, 2911y (1974); Anisotropy of the diffusion coefficient in nematic solutions measured by NMR techniques. [**1007**] G. J. Krüger, H. Spiesecke, R. van Steenwinkel, J. Phys. (Paris), C-1, *36*, 91 (1975), C.A. *83*, 77876m (1975); NMR relaxation and diffusion measurements in *p*-dodecanoyl-oxybenzylidene-*p'*-aminoazobenzene. [**1008**] G. J. Krüger, H. Spiesecke, R. Weiss, Phys. Lett. A *51*, 295 (1975), C.A. *83*, 19473s (1975); Simple spin echo measurement of the anisotropy of the self-diffusion coefficient in a smectic A and B. [**1009**] G. J. Krüger, H. Spiesecke, R. van Steenwinkel, J. Phys. (Paris), C-3, *37*, 123 (1976), C.A. *85*, 85790k (1976); NMR relaxation and self-diffusion in liquid crystals showing smectic polymorphism. [**1010**] G. J. Krüger, H. Spiesecke, R. van Steenwinkel, F. Noack, Mol. Cryst. Liq. Cryst. *40*, 103 (1977), C.A. *87*, 125059d (1977); NMR relaxation and self diffusion in a series of *p*-alkanoyloxybenzylidene-*p'*-aminoazobenzenes.

[**1011**] J. M. Steim, O. J. Edner, F. G. Bargoot, Science *162*, 909 (1968), C.A. *70*, 17099g (1969); Structure of human serum lipoproteins: NMR supports a micellar model. [**1012**] S. Kaufmann, J. M. Steim, J. H. Gibbs, Nature *225*, 743 (1970), C.A. *73*, 319t (1970); Nuclear relaxation in phospholipids and biological membranes. [**1013**] P. F. Swinton, G. Gatti, Spectrosc. Lett. *3*, 259 (1970), C.A. *74*, 47875g (1971); Nematic phase ^{1}HMR study of 1,2-dihaloethanes. [**1014**] P. F. Swinton, G. Gatti, J. Magn. Reson. *8*, 293 (1972), C.A. *78*, 35960a (1973); NMR study of ethylene carbonate and ethylene thiocarbonate partially oriented in the nematic phase. [**1015**] P. F Swinton, J. Magn. Reson. *13*, 304 (1974), C.A. *81*, 12656d (1974); NMR study of cyclobutanone and trimethylene sulfide partially oriented in the nematic phase. [**1016**] P. F. Swinton, J. Mol. Struct. *22*, 221 (1974), C.A. *81*, 97430y (1974); ^{1}HMR study of 2,5-dihydrofuran partially oriented in the nematic phase. [**1017**] C. E. Tarr, M. A.

Nickerson, C. W. Smith, Appl. Phys. Lett. *17*, 318 (1970), C.A. *74*, 26516a (1971); Anisotropic nuclear spin-lattice relaxation in a nematic. **[1018]** C. E. Tarr, A. M. Fuller, M. A. Nickerson, Appl. Phys. Lett. *19*, 179 (1971), C.A. *75*, 145963z (1971); Nuclear spin-lattice relaxation in a liquid crystal ordered in a.c. electric fields. **[1019]** C. E. Tarr, A. M. Fuller, M. A. Nickerson, Bull. Am. Phys. Soc. *16*, 317 (1971); NMR studies of a nematic in the presence of a.c. electric fields. **[1020]** C. E. Tarr, A. M. Fuller, Bull. Am. Phys. Soc. *17*, 332 (1972); NMR studies of a liquid crystal in the solid.

[1021] C. E. Tarr, A. M. Fuller, Mol. Cryst. Liq. Cryst. *22*, 123 (1973), C.A. *79*, 109052a (1973); Pulsed NMR study of molecular motions and ordering in ethyl [*p*-(*p*-methoxybenzylidene)amino]-cinnamate. **[1022]** C. E. Tarr, R. M. Dennery, A. M. Fuller, Liq. Cryst. Ord. Fluids *2*, 53 (1974), C.A. *86*, 180977f (1977); Stability of molecular order in the s$_A$ phase. **[1023]** T. E. Kubaska, C. E. Tarr, T. B. Tripp, Mol. Cryst. Liq. Cryst. *29*, 155 (1974), C.A. *82*, 163366f (1975); Influence of magnetic and d.c. electric fields on the molecular alignment of a nematic. **[1024]** C. E. Tarr, M. E. Field, Mol. Cryst. Liq. Cryst. *30*, 143 (1975), C.A. *84*, 11081n (1976); Frequency dependence of rotating-frame nuclear spin-lattice relaxation in n/c mixtures. **[1025]** C. E. Tarr, M. E. Field, L. R. Whalley, K. R. Brownstein, Mol. Cryst. Liq. Cryst. *35*, 231 (1976), C.A. *85*, 200331a (1976); Pulsed NMR study of molecular reordering in a n/c mixture. **[1026]** C. E. Tarr, M. E. Field, L. R. Whalley, Mol. Cryst. Liq. Cryst. *37*, 353 (1976), C.A. *86*, 113390w (1977); NMR study of short range s$_A$ order in nematic CBOOA. **[1027]** K. R. Brownstein, C. E. Tarr, Chem. Phys. Lett. *49*, 80 (1977), C.A. *87*, 93475t (1977); Nucleation and growth model for reordering of a n/c mixture. **[1028]** C. E. Tarr, F. Vosman, L. R. Whalley, J. Chem. Phys. *67*, 868 (1977), C.A. *87*, 125047y (1977); Orientational dependence of spin-lattice relaxation in nematics. **[1029]** R. T. Thompson, Diss. Univ. Waterloo, Waterloo, Ont. 1975. Avail. Natl. Libr. Canada, Ottawa, Ont., C.A. *85*, 152110d (1976); Study of molecular dynamics in liquid crystals by nuclear spin thermometry. **[1030]** J. Clifford, J. Oakes, G. J. T. Tiddy, Spec. Discuss. Faraday Soc. *1*, 175 (1970), C.A. *75*, 113125m (1971); NMR studies of water in disperse systems.

[1031] G. J. T. Tiddy, Nature, Phys. Sci. *230*, 136 (1971), C.A. *75*, 13173g (1971); Variable frequency pulsed NMR study of lyophases. **[1032]** G. J. T. Tiddy, Symp. Faraday Soc. *1971*, No. 5, 150, C.A. *79*, 11734z (1973); Fluorocarbon surfactant/water mesophases. II. ^{19}F and ^{1}H-pulsed NMR study of alkyl chains in the lamellar phase of the ammonium perfluorooctanoate/water system. **[1033]** G. J. T. Tiddy, J. Chem. Soc., Faraday Trans. 1, *68*, 369 (1972), C.A. *76*, 79017m (1972); NMR relaxation times of the lamellar phases of the system sodium caprylate/decanol/water. **[1034]** G. J. T. Tiddy, J. Chem. Soc., Faraday Trans. 1, *68*, 653 (1972), C.A. *76*, 117961z (1972); III. ^{1}H and ^{2}H pulsed NMR study of water in the lamellar phase of ammonium perfluorooctanoate (cf. [1032]). **[1035]** G. J. T. Tiddy, J. Chem. Soc., Faraday Trans. 1, *68*, 670 (1972), C.A. *76*, 117959e (1972); IV. Distribution of alkyl chain mobilities in the lamellar phase of the sodium 12,12,13,13,14,14,14-heptafluoro-myristate/heavy water system by ^{19}F and ^{1}H spin echo NMR (cf. [1032]). **[1036]** G. J. T. Tiddy, J. B. Hayter, A. M. Hecht, J. W. White, Ber. Bunsenges. Phys. Chem. *78*, 961 (1974), C.A. *82*, 21949z (1975); NMR studies of water self-diffusion in the lamellar phase. **[1037]** C. A. Barker, D. Saul, G. J. T. Tiddy, B. A. Wheeler, E. Wilis, J. Chem. Soc., Faraday Trans. 1, *70*, 154 (1974), C.A. *80*, 122674v (1974); Phase structure, NMR and rheological properties of viscoelastic sodium dodecyl sulfate and trimethylammonium bromide mixtures. **[1038]** G. J. T. Tiddy, P. A. Wheeler, J. Phys. (Paris), C-1, *36*, 167 (1975), C.A. *83*, 16059a (1975); Kinetics of formation and breakdown of prehexagonal phase aggregates in fluid-isotropic amphiphile solutions. **[1039]** G. J. T. Tiddy, Nucl. Magn. Reson. *4*, 233 (1975), C.A. *83*, 185539n (1975); NMR of liquid crystals and micellar solutions. **[1040]** G. J. T. Tiddy, E. Everiss, ACS Symp. Ser. *34*, 78 (1976), C.A. *86*, 63813p (1977); Self-diffusion in lyophases measured by NMR.

[1041] E. Everiss, G. J. T. Tiddy, B. A. Wheeler, J. Chem. Soc., Faraday Trans. 1, *72*, 1747 (1976), C.A. *85*, 134865z (1976); Phase diagram and NMR study of the lyophases formed by lithium perfluoro octanoate and water. **[1042]** G. J. T. Tiddy, Chemical Society, Specialist Periodical Reports, Nucl. Magn. Reson. *6*, 207 (1977); Liquid crystals and micellar solutions. **[1043]** J. A. Topich, Diss. Case West. Reserve Univ. 1975, 124 pp. Avail. Xerox Univ. Microfilms, Ann Arbor, Mich., Order No. 75-19,253, C.A. *83*, 211077g (1975); Nematic phase ESR and electronic properties of rhombic cobalt(II) complexes. **[1044]** A. S. Tracey, Mol. Phys. *33*, 339 (1977), C.A. *87*, 84276b (1977); Orientations of molecules dissolved in lyotropic surfactant solutions. Monohalogenated benzenes. **[1045]** P. L. M. Ukleja, Diss. Kent State Univ., Kent. 1976. 118 pp. Avail. Univ. Microfilms Int., Order No. 77-12,445, C.A. *87*, 93138k (1977); Spin-lattice relaxation and director fluctuations in nematics. **[1046]** K. van Putte, J. Magn. Reson. *2*, 23 (1970), C.A. *74*, 93351f (1971); Spin-lattice relaxation in anhydrous sodium and lithium soaps between 23°

and 180°. [1047] K. van Putte, Trans. Faraday Soc. *66*, 523 (1970), C.A. *72*, 137948w (1970); Alternation in spin-lattice relaxation times and barrier height to methyl reorientation in sodium and lithium soaps. [1048] K. van Putte, G. J. N. Egmond, J. Magn. Reson. *4*, 236 (1971), C.A. *75*, 56418h (1971); Experimental check on the dominant role of reorienting methyl groups on the spin-lattice relaxation behavior in a C_{17} lithium soap. [1049] G. Vass, Stud. Biophys. *47*, 215 (1974); Physical analysis of nerve water properties by NMR spectroscopy. [1050] P. L. Barili, C. A. Veracini, Chem. Phys. Lett. *8*, 229 (1971), C.A. *74*, 93315x (1971); Temperature dependence of the geometry of *p*-dinitrobenzene in a nematic.

[1051] P. Bucci, P. F. Franchini, A. M. Serra, C. A. Veracini, Chem. Phys. Lett. *8*, 421 (1971), C.A. *75*, 13201q (1971); NMR spectrum of 3,3'-bisisoxazole in a nematic. [1052] P. Bucci, C. A. Veracini, M. Longeri, Chem. Phys. Lett. *15*, 396 (1972), C.A. *77*, 120456q (1972); Conformational studies of molecules partially oriented in nematics. I. Furan-2,5-dialdehyde. [1053] P. Bucci, C. A. Veracini, J. Chem. Phys. *56*, 1290 (1972), C.A. *76*, 52598q (1972); II. NMR spectra of 5,5'-bisisoxazole (cf. [1052]). [1054] C. A. Veracini, F. Pietra, J. Chem. Soc., Chem. Commun. *1972*, 1262, C.A. *78*, 57248b (1973); Molecular structure of tropone from its ^{1}HMR spectrum in a nematic solvent. [1055] L. Lunazzi, G. F. Pedulli, M. Tiecco, C. A. Veracini, J. Chem. Soc., Perkin Trans. 2, *1972*, 755, C.A. *77*, 19051d (1972); Conformational analysis. NMR investigation of 2,5-thiophenedicarboxaldehyde in liquid crystals. [1056] C. A. Veracini, P. Bucci, P. L. Barili, Mol. Phys. *23*, 59 (1972), C.A. *76*, 133923u (1972); NMR of substituted benzenes in the nematic phase. Geometric distortion in benzonitrile, *p*-nitrobenzonitrile, and *p*-dinitrobenzene. [1057] P. L. Barili, L. Lunazzi, C. A. Veracini, Mol. Phys. *24*, 673 (1972), C.A. *78*, 42679u (1973); Nematic-phase NMR investigation of rotational isomerism. I. Conformations of 2-furaldehyde. [1058] C. A. Veracini, M. Longeri, P. L. Barili, Chem. Phys. Lett. *19*, 592 (1973), C.A. *79*, 10740m (1973); NMR investigation of the geometry of a pyridine-bromine complex in the nematic phase. [1059] C. A. Veracini, D. Macciantelli, L. Lunazzi, J. Chem. Soc., Perkin Trans. 2, *1973*, 751, C.A. *78*, 147246d (1973); Conformational analysis of bithienyl derivatives. Liquid crystal NMR approach. [1060] L. Lunazzi, C. A. Veracini, J. Chem. Soc., Perkin Trans. 2, *1973*, 1739, C.A. *80*, 59335t (1974); II. Evidence for the SO-cis-conformation in thiophene-2-carboxaldehyde (cf. [1957]).

[1061] C. A. Veracini, M. Guidi, M. Longeri, A. M. Serra, Chem. Phys. Lett. *24*, 99 (1974), C.A. *80*, 132351b (1974); Molecular structure of cyclopentadiene from its NMR spectrum in a nematic solvent. [1062] P. Bucci, M. Longeri, C. A. Veracini, L. Lunazzi, J. Am. Chem. Soc. *96*, 1305 (1974), C.A. *80*, 107884v (1974); III. Conformational preferences and interconversion barrier of 2,2'-bithienyl (cf. [1057]). [1063] P. L. Barili, M. Longeri, C. A. Veracini, Mol. Phys. *28*, 1101 (1974), C.A. *82*, 97559t (1975); III. 2,6-pyridinedicarboxaldehyde (cf. [1052]). [1064] G. Conti, E. Matteoli, C. Petrongolo, C. A. Veracini, M. Longeri, J. Chem. Soc., Perkin Trans. 2, *1975*, 1673, C.A. *84*, 89456h (1976); Nematic phase NMR, ultrasonic relaxation, and theoretical ab initio investigation of internal rotation in pyridine-2-carbaldehyde. [1065] A. Amanzi, P. L. Barili, P. Chidichimo, C. A. Veracini, Chem. Phys. Lett. *44*, 110 (1976), C.A. *86*, 36109g (1977); ^{1}HMR spectrum of hexamethylenetetramine in oriented mesophases. [1066] C. A. Veracini, L. Bellitto, M. Longeri, P. L. Barili, Gazz. Chim. Ital. *106*, 467 (1976), C.A. *85*, 101815e (1976); NMR investigation of pyridine-iodine interaction in a nematic. [1067] R. Danieli, L. Lunazzi, C. A. Veracini, J. Chem. Soc., Perkin Trans. 2, *1976*, 19, C.A. *84*, 89469q (1976); Liquid crystal NMR investigation on the structure of a molecule formed by two fused six-membered rings. 2,7-Naphthyridine. [1068] L. Lunazzi, F. Salvetti, C. A. Veracini, J. Chem. Soc., Perkin Trans. 2, *1976*, 1796, C.A. *86*, 139262f (1977); IV. Conformational study on the rotamers of 3,3'-bithienyl (cf. [1057]). [1069] A. Amanzi, D. Silvestri, C. A. Veracini, P. L. Barili, Chem. Phys. Lett. *51*, 116 (1977), C.A. *88*, 5765e (1978); NMR study of thiophene-2-carboxaldehyde and thiophene-2,5-dicarboxaldehyde in a lyophase. Molecular structure and orientation. [1070] L. Bellitto, C. Petrongolo, C. A. Veracini, M. Bambagiotti, J. Chem. Soc., Perkin Trans. 2, *1977*, 314, C.A. *87*, 5259x (1977); IV. NMR and theoretical investigation of 2,2'-bifuryl (cf. [1052]).

[1071] C. A. Veracini, A. de Munno, V. Bertini, M. Longeri, G. Chidichimo, J. Chem. Soc., Perkin Trans. 2, *1977*, 561, C.A. *87*, 101455d (1977); V. 3-phenyl-1,2,5-oxa-, -thia-, and -selena-diazole (cf. [1052]). [1072] L. Lunazzi, C. Zannoni, C. A. Veracini, A. Zandanel, Mol. Phys. *34*, 223 (1977), C.A. *88*, 21947w (1978); V. Conformation of pyridine-3-aldehyde (cf. [1057]). [1073] G. Chidichimo, F. Lelj, P. L. Barili, C. A. Veracini, Chem. Phys. Lett. *55*, 519 (1978), C.A. *89*, 59381r (1978); r_z-Structure of 1,2,5-selenadiazole partially oriented in ordered mesophases as determined by ^{1}HMR spectra including ^{13}C- and ^{77}Se-satellites. [1074] M. Vilfan, S. Zumer, Magn. Reson. Relat. Phenom., Proc. Congr. Ampère, 19th *1976*, 189, C.A. *86*, 197384h (1977); Intermolecular spin-lattice relaxation due to translational diffusion in nematics. [1075] J. J. Visintainer, Diss. Kent State Univ., Kent. 1973, 127 pp. Avail.

Univ. Microfilms, Ann Arbor, Mich., Order No. 73-27,260, C.A. *80*, 75210d (1974); Spin-lattice relaxation in the nematic phase. [**1076**] R. L. Vold, S. O. Chan, J. Chem. Phys. *53*, 449 (1970), C.A. *73*, 70702m (1970); Modulated spin echo trains from liquid crystals. [**1077**] R. D. Orwoll, R. L. Vold, J. Am. Chem. Soc. *93*, 5335 (1971), C.A. *75*, 144747v (1971); Molecular order in liquid crystalline solutions of PBLG in dichloromethane. [**1078**] R. R. Vold, R. L. Vold, J. Chem. Phys. *66*, 4018 (1977), C.A. *86*, 197436b (1977); ²H relaxation of chloroform in a nematic. [**1079**] E. T. Samulski, C. R. Dybowski, C. G. Wade, Chem. Phys. Lett. *11*, 113 (1971), C.A. *76*, 19657k (1972); Brownian-motion contributions to relaxation in nematics. [**1080**] C. R. Dybowski, C. G. Wade, J. Chem. Phys. *55*, 1576 (1971), C.A. *75*, 92896w (1971); NMR relaxation in cholesterol and cholesterics.

[**1081**] C. R. Dybowski, B. A. Smith, C. G. Wade, J. Phys. Chem. *75*, 3834 (1971), C.A. *76*, 8639e (1972); NMR relaxation in a homologous series of nematics. [**1082**] C. R. Dybowski, E. T. Samulski, C. G. Wade, Bull. Am. Phys. Soc. *17*, 128 (1972); Theory of NMR relaxation in cholesterics and nematics. [**1083**] E. T. Samulski, C. R. Dybowski, C. G. Wade, Phys. Rev. Lett. *29*, 340 (1972), C.A. *77*, 95119x (1972); Intermolecular and intramolecular contributions to ¹H relaxation in liquid crystals I. [**1084**] E. T. Samulski, B. A. Smith, C. G. Wade, Chem. Phys. Lett. *20*, 167 (1973), C.A. *79*, 47606y (1973); NMR free induction decay and spin echoes in oriented model membrane bilayers. [**1085**] M. Chien, E. T. Samulski, C. G. Wade, Macromolecules *6*, 638 (1973), C.A. *79*, 126857j (1973); NMR relaxation study of PBLG side-chain mobility in helix-coil transition. [**1086**] E. T. Samulski, C. R. Dybowski, C. G. Wade, Mol. Cryst. Liq. Cryst. *22*, 309 (1973), C.A. *79*, 141279r (1973); Part II (cf. [**1083**]). [**1087**] M. Chien, B. A. Smith, E. T. Samulski, C. G. Wade, Liq. Cryst. Ord. Fluids *2*, 67 (1974), C.A. *86*, 195350v (1977); Diffusion in oriented lamellar phases by pulsed NMR. [**1088**] R. M. C. Matthews, C. G. Wade, J. Magn. Reson. *19*, 166 (1975), C.A. *83*, 189542f (1975); Spin-lattice relaxation in a homologous series of C₁₈ esters of cholesterol. [**1089**] R. D. Orwoll, C. G. Wade, B. M. Fung, J. Chem. Phys. *63*, 986 (1975); C.A. *83*, 106557r (1975); ²H relaxation in liquid crystals. Di-*n*-alkoxyazoxybenzenes, *n* = 1,7. [**1090**] B. M. Fung, C. G. Wade, R. D. Orwoll, J. Chem. Phys. *64*, 148 (1976), C.A. *84*, 67520x (1976); Frequency dependence of ¹H relaxation in liquid crystals.

[**1091**] S. B. W. Roeder, E. E. Burnell, A. L. Kuo, C. G. Wade, J. Chem. Phys. *64*, 1848 (1976), C.A. *84*, 114483z (1976); Determination of the lateral diffusion coefficient of potassium oleate in the lamellar phase. [**1092**] M. Mehring, R. G. Griffin, J. S. Waugh, J. Chem. Phys. *55*, 746 (1971), C.A. *75*, 69298u (1971); ¹⁹F shielding tensors from coherently narrowed NMR powder spectra. [**1093**] J. Urbina, J. S. Waugh, Ann. N. Y. Acad. Sci. *222*, 733 (1973), C.A. *81*, 74456x (1974); Application of ¹H-enhanced nuclear induction spectroscopy to the study of membranes. [**1094**] H. Lippmann, K. H. Weber, Ann. Physik *20*, 265 (1957), C.A. *51*, 17436d (1957); Linienformen der ¹HMR-Absorption in Flüssigkristallen. [**1095**] K. H. Weber, Z. Naturforsch. A *13*, 1098 (1958), C.A. *53*, 8815b (1959); Untersuchung der ¹HMR in kristallin-flüssigem Azoxyphenyl-di-*p*-*n*-propyläther. [**1096**] K. H. Weber, Ann. Physik [7] *3*, 1 (1959), C.A. *53*, 11998e (1959); Deutung der ¹HMR an geordneten kristallin-flüssigen Azoxyphenol-di-*p*-*n*-äthern. [**1097**] K. H. Weber, Ann. Physik [7] *3*, 125 (1959), C.A. *53*, 12836i (1959); NMR-Untersuchungen an Flüssigkristallen. Frage der Molekülrotation in smektischen Phasen. [**1098**] K. H. Weber, Z. Naturforsch. A *14*, 112 (1959), C.A. *53*, 10970d (1959); NMR an Flüssigkristallen. Frage des molekularen Ordnungszustandes in magnetisch geordneten smektischen Phasen. [**1099**] I. Y. Wei, C. S. Johnson, Jr., J. Magn. Reson. *23*, 259 (1976), C.A. *86*, 15891r (1977); NMR spectra of pseudotetrahedral molecules dissolved in a nematic mixture. [**1100**] H. Wennerstrom, Chem. Phys. Lett. *18*, 41 (1973), C.A. *78*, 90708z (1973); ¹HMR lineshapes in lamellar liquid crystals.

[**1101**] L. F. Williams, Jr., Diss. Carnegie-Mellon Univ., Pittsburgh. 1971, 88 pp. Avail. Univ. Microfilms, Ann Arbor, Mich., Order No. 72-17,888, C.A. *77*, 158473f (1972); NMR studies of nematic solutions of hexafluorobenzene, 3,3,3-trifluoropropene, and propene. [**1102**] R. A. Wise, Diss. Kent State Univ., Kent 1973, 113 pp. Avail. Univ. Microfilms, Ann Arbor, Mich., Order No. 73-27,262, C.A. *80*, 76344n (1974); NMR study of smectics C. [**1103**] T. C. Wong, Mol. Phys. *34*, 921 (1977), C.A. *88*, 135796b (1978); NMR studies of molecular structure, reorientation and order matrix of substituted benzaldehydes partially oriented in a nematic. [**1104**] T. C. Wong, A. J. Ashe, III, J. Mol. Struct. *48*, 219 (1978), C.A. *89*, 97526z (1978); NMR studies of the molecular structures of phosphabenzene and arsabenzene partially oriented in a nematic solvent. [**1105**] C. M. Woodman, Mol. Phys. *13*, 365 (1967), C.A. *68*, 110082q (1968); Equivalence in anisotropic NMR spectra: spectrum of ethyl iodide in the nematic phase. [**1106**] J. B. Wooten, A. L. Beyerlein, J. Jacobus, G. B. Savitsky, J. Chem. Phys. *65*, 2476 (1976), C.A. *85*, 169345g (1976); Determination of ²H quadrupole coupling constants from direct ¹³C-²H couplings in ²HMR spectra. [**1107**] A. Wulf, J. Chem. Phys. *63*, 1564 (1975), C.A.

83, 139516e (1975); Molecular models of the s_C phase and NMR experiments. **[1108]** A. Wulf, M. P. Vecchi, Mol. Cryst. Liq. Cryst. *36*, 165 (1976), C.A. *86*, 48926j (1977); Remarks on ¹HMR of di-heptyl-oxyazoxybenzene in the s_C phase. **[1109]** K. Yamaoka, S. Noji, Chem. Lett. *1976*, 1351, C.A. *86*, 39802p (1977); Spin-labeled metachromatic dyes. I. ESR and optical properties of spin-labeled proflavine in solution, liquid crystal, and stretched film. **[1110]** C. S. Yannoni, Diss. Columbia Univ. N. Y., Avail. Univ. Microfilms, Ann Arbor, Mich., Order No. 67-12,282, 78 pp., C.A. *68*, 55229r (1968); Anisotropic NMR interactions in liquid crystal solution.

[1111] C. S. Yannoni, J. Am. Chem. Soc. *91*, 4611 (1969), C.A. *71*, 85811d (1969); Oriented smectic solutions. **[1112]** C. S. Yannoni, J. Chem. Phys. *51*, 1682 (1969), C.A. *71*, 107411c (1969); Orientation of CH_3D in a nematic solvent. **[1113]** C. S. Yannoni, J. Am. Chem. Soc. *92*, 5237 (1970), C.A. *73*, 104067a (1970); NMR spectrum of oriented bullvalene. **[1114]** C. S. Yannoni, J. Chem. Phys. *52*, 2005 (1970), C.A. *72*, 95098m (1970); Chemical-shift anisotropy and nuclear quadrupole coupling constant of ¹⁴N in methyl isocyanide. **[1115]** C. S. Yannoni, IBM J. Res. Develop. *15*, 59 (1971), C.A. *74*, 93267h (1971); Measuring NMR shielding anisotropies in liquid-crystal solvents. **[1116]** N. Zumbulyadis, Diss. Columbia Univ., New York. 1974. 85 pp. Avail. Xerox Univ. Microfilms, Ann Arbor, Mich., Order No. 75-18,457, C.A. *83*, 170593b (1975); Determination of magnetic shielding anisotropies in phosphorus compounds from NMR studies in liquid crystals. **[1117]** S. Zumer, Phys. Rev. A *15*, 378 (1977), C.A. *86*, 99277f (1977); Nuclear spin-lattice relaxation due to layer sliding in smectics B.

Chapter 11

[1] A. W. Adamson, J. Colloid Interface Sci. *29*, 261 (1969), C.A. *70*, 61473y (1969); Model for micellar emulsions. **[2]** J. Ahmad, J. A. Mann, M. J. Povich, J. Colloid Interface Sci. *49*, 1 (1974), C.A. *81*, 159421g (1974); Mixed monolayers of cholesterol and hexadecyltrimethylammonium bromide. **[3]** M. Aizawa, S. Suzuki, Bull. Chem. Soc. Jap. *46*, 2634 (1973), C.A. *79*, 141101b (1973); Near-IR studies of the states of water in lyophases. **[3a]** M. A. Al-Mamun, Bangladesh J. Sci. Ind. Res. *12*, 22 (1977), C.A. *87*, 157462w (1977); Studies of the ternary phase behaviors of non-ionic emulsifier blends/oil/water systems. **[4]** L. K. Altunina, E. P. Sokolova, T. G. Churyusova, Vestn. Leningrad. Univ., Fiz., Khim. *1973*, 83, C.A. *80*, 41558h (1974); Vapor pressure in the potassium palmitate/water system. Thermodynamic properties of the soap/water system. **[5]** H. Arai, J. Colloid Interface Sci. *29*, 166 (1969), C.A. *70*, 61476b (1969); Effect of temperature on the micellar weight in the presence of sodium sulfate. **[6]** S. Arakawa, T. Kawaguchi, H. Kato, Yakugaku Zasshi *78*, 278 (1958), C.A. *52*, 10690g (1958); Mutual solubility and anisotropic gel of the hexylamine/water system. **[7]** I. E. Balaban, H. King, J. Chem. Soc. *1927*, 3068, C.A. *22*, 959⁵ (1928); Trypanocidal action and chemical constitution. VII. *s*-Carbamides and arylamides of naphthylaminedi- and trisulfonic acids with observations on the mesomorphism. **[8]** R. R. Balmbra, J. S. Clunie, J. F. Goodman, Proc. Roy. Soc. A *285*, 534 (1965), C.A. *62*, 15464b (1965); Structure of mesophases by electron microscopy. **[9]** R. R. Balmbra, J. S. Clunie, J. F. Goodman, Mol. Cryst. *3*, 281 (1967), C.A. *69*, 71357g (1968); Structure of neat mesophase. **[10]** R. R. Balmbra, J. S. Clunie, J. F. Goodman, Nature *222*, 1159 (1969), C.A. *71*, 54535n (1969); Cubic mesophases.

[11] R. R. Balmbra, D. A. B. Bucknall, J. S. Clunie, Mol. Cryst. Liq. Cryst. *11*, 173 (1970), C.A. *74*, 7329a (1971); Mesophase structure in the potassium oleate/water system. **[12]** R. R. Balmbra, J. S. Clunie, J. F. Goodman, B. T. Ingram, J. Colloid Interface Sci. *42*, 226 (1973), C.A. *78*, 76134c (1973); Similarity between black foam films and neat mesophase. **[13]** S. I. Banduryan, M. M. Iovleva, V. D. Kalmykova, S. P. Papkov, Khim. Volokna *1975*, 73, C.A. *84*, 17984f (1976); Structure formation during precipitation of poly-*p*-benzamide from solutions. **[14]** P. Becher, N. K. Clifton, J. Colloid Sci. *14*, 519 (1959), C.A. *54*, 1979h (1960); Nonionic surface-active agents. II. Time dependence of micellar breakdown. **[15]** L. Benjamin, J. Phys. Chem. *70*, 3790 (1966), C.A. *66*, 20233v (1967); Partial molal volume changes during micellization and solution of nonionic surfactants, and perfluorocarboxylates using a magnetic density balance. **[16]** R. A. Berg, B. A. Haxby, Mol. Cryst. Liq. Cryst. *12*, 93 (1970), C.A. *74*, 117081w (1971); Aqueous dye aggregates as liquid crystals. **[17]** M. Bishop, E. A. Dimarzio, Mol. Cryst. Liq.

Cryst. *28*, 311 (1974), C.A. *82*, 178710n (1975); Models in diffusion in lyophases. [18] L. I. Boguslavskii, I. I. Konstantinov, A. B. Imenitov, Elektrokhimiya *9*, 435 (1973), C.A. *79*, 26465m (1973); Volta-potential study of liquid crystal solutions in the heptane/water system. [19] G. W. Brady, C. Cohen-Addad, E. F. X. Lyden, J. Chem. Phys. *51*, 4309 (1969), C.A. *72*, 26145n (1970); Structure studies of solutions of large organic molecules. $C_9H_{19}I$ and $1,18-C_{18}H_{36}I_2$ in decalin. [20] F. Branner, Nord. Kemikermøde, Forh. *5*, 207 (1939), C.A. *38*, 2872^6 (1944); Phase range for liquid crystals in the system K-methyl orange/water.

[21] F. K. Broome, C. W. Hoerr, H. J. Harwood, J. Am. Chem. Soc. *73*, 3350 (1951), C.A. *45*, 8865c (1951); Binary systems of water with dodecylammonium chloride and its N-methyl derivatives. [22] C. A. Bunton, L. B. Robinson, J. Phys. Chem. *73*, 4237 (1969), C.A. *72*, 36463d (1970); Micellar effects on acidity functions. [23] C. R. Bury, R. D. J. Owens, Trans. Faraday Soc. *32*, 782 (1936), C.A. *30*, 5108^2 (1936); The system lauric acid/sodium hydroxide/water. [24] C. R. Bury, J. Browning, Trans. Faraday Soc. *49*, 209 (1953), C.A. *47*, 8464g (1953); Comparison of ionic and nonionic detergents. [25] J. O. P. Calveras, Invest. Inf. Text. Tensioactivos *18*, 319 (1975), C.A. *84*, 35659h (1976); Stability of emulsions as a function of the interphase. Superficial rheology and liquid crystal films. [26] E. L. Cataline, L. Worrell, S. F. Jeffries, S. A. Aronson, J. Am. Pharm. Assoc. *33*, 107 (1944), C.A. *38*, 2764^7 (1944); Water-in-oil emulsifying agents. II. Synthesis of some cholesteryl and cetyl esters. [27] S. K. Chan, U. Herrmann, W. Ostner, M. Kahlweit, Ber. Bunsen-Ges. Phys. Chem. *81*, 60, 396 (1977), C.A. *86*, 96487p, 177976z (1977); Kinetics of the formation of ionic micelles. I. Analysis of the amplitudes. II. Analysis of the time constants. [28] J. Charvolin, Adv. Chem. Ser. *152*, 101 (1976), C.A. *85*, 151855p (1976); Aspects of liquid crystal microdynamics. [29] D. H. Chen, D. G. Hall, Kolloid-Z. Z. Polym. *251*, 41 (1973), C.A. *78*, 140964c (1973); Phase diagrams of the systems sodium dodecyl sulfate octyltrimethylammonium bromide/water and sodium dodecyl sulfate/dodecyltrimethylammonium bromide/water. [30] I. G. Chistyakov, Kristallografiya *6*, 479 (1961), C.A. *57*, 6724 (1962); Textures of liquid crystals of potassium oleate.

[31] L. G. Chistyakov, V. A. Usol'tseva, Izv. Vysshikh Uchebn. Zavedenii, Khim. i Khim. Tekhnol. *5*, 589 (1962), C.A. *58*, 13239f (1963); Investigation of liquid-crystalline cholesterol/cetyl alcohol and cholesterol/glycerol systems. [31a] M. K. Minasyants, A. A. Shaginyan, I. G. Chistyakov, Izv. Akad. Nauk Arm. SSR, Fiz. *12*, 67 (1977), C.A. *87*, 32196f (1977); X-ray structural study of a sodium pentadecylsulfonate lyophase. [32] J. S. Clunie, J. M. Corkill, J. F. Goodman, Proc. Roy. Soc. A *285*, 520 (1965); C.A. *62*, 15461d (1965); Structure of lyophases. [33] J. S. Clunie, J. F. Goodman, P. C. Symons, Nature *216*, 1203 (1967), C.A. *68*, 81609j (1968); Solvation forces in soap films. [34] J. S. Clunie, J. M. Corkill, J. F. Goodman, P. C. Symons, J. R. Tate, Trans. Faraday Soc. *63*, 2839 (1967); C.A. *68*, 6848h (1968); Thermodynamics of nonionic surface-active agent/water systems. [35] D. A. B. Bucknall, J. S. Clunie, J. F. Goodman, Mol. Cryst. Liq. Cryst. *7*, 215 (1969), C.A. *71*, 74372k (1969); Electron microscopy of lyophases. [36] J. S. Clunie, J. F. Goodman, P. C. Symons, Trans. Faraday Soc. *65*, 287 (1969), C.A. *70*, 41295p (1969); Phase equilibria of dodecylhexakis(oxyethylene) glycol monoether in water. [37] J. F. Goodman, J. S. Clunie, Liq. Cryst. Plast. Cryst. *2*, 1 (1974), C.A. *83*, 186432j (1975); Electron microscopy of liquid crystals. [38] J. M. Corkill, J. F. Goodman, Advan. Colloid Interface Sci. *2*, 297 (1969), C.A. *70*, 99889b (1969); Interaction of nonionic surface-active agents with water. [39] H. Cortopassi, Jr., V. R. Paoli, L. W. Reeves, R. R. Romano, J. A. Vanin, An. Acad. Bras. Cienc. *46*, 203 (1974), C.A. *83*, 184289u (1975); Ternary phase diagram of sodium lauryl sulfate/octanol/water. [40] J. L. Curat, R. Perron, Riv. Ital. Sostanze Grasse *53*, 198 (1976), C.A. *85*, 167399d (1976); Soap-water systems. Transition of coagel to mesophases (sodium stearate): establishment of equilibriums to 90°C.

[41] J. Czapkiewicz, B. Sliwa, Rocz. Chem. *44*, 1565 (1970), C.A. *74*, 35338w (1971); Conductometric studies of micellization of long-chain alkyldimethylbenzylammonium chlorides. [41a] A. K. Dadivanyan, Polym. Prepr., Am. Chem. Soc., Div. Polym. Chem. *16*, 654 (1977), C.A. *86*, 172074u (1977); Short-range orientational order in polymer solutions and in swollen polymers. [41b] A. K. Dadivanyan, G. A. Airapetyan, R. S. Avoyan, A. K. Mushegyan, V. Yu. Agasaryan, Izv. Akad. Nauk Arm. SSR. Fiz. *12*, 288 (1977), C.A. *87*, 185150m (1977); Optical and spectroscopic studies of short-range orientational order in polymer systems. [42] I. Danielsson, Adv. Chem. Ser. *152*, 13 (1976), C.A. *85*, 114862t (1976); Lyotropic mesomorphism. Phase equilibriums and relation to micellar systems. [42a] I. Danielsson, J. B. Rosenholm, P. Stenius, S. Backlund, Prog. Colloid Polym. Sci. *61*, 1 (1976), C.A. *86*, 60818h (1977); Lyotropic mesomorphism and aggregation in surfactant systems. [43] F. Brochard, P. G. de Gennes, Liq. Cryst., Proc. Int. Conf., Bangalore, *1973*, 1, C.A. *84*, 114461r (1976); Hydrodynamic properties

of fluid lamellar phases of lipid/water. [44] D. G. Dervichian, Trans. Faraday Soc. B *42*, 180 (1946), C.A. *42*, 8577i (1948); Swelling and molecular organization in colloidal electrolytes. [45] D. G. Dervichian, Mol. Cryst. *2*, 55 (1967), C.A. *66*, 32510v (1967); Conditions governing the formation of lyophases by molecular association. [45a] D. G. Dervichian, Mol. Cryst. Liq. Cryst. *40*, 19 (1977), C.A. *87*, 147498y (1977); Control of lyophase, biological and medical implications. [46] T. M. Doscher, S. Davis, J. Phys. Colloid Chem. *55*, 53 (1951), C.A. *46*, 3831i (1952); Transitions in soap/oil systems by dielectric absorption. [47] A. Douy, J. Rossi, B. Gallot, C. R. Acad. Sci., Sér. C *267*, 1392 (1968), C.A. *70*, 29468q (1969); Copolymères organisés à sequences amorphes. Influence de la nature du solvant sur les paramètres structuraux des gels mésomorphes. [48] A. Douy, R. Mayer, J. Rossi, B. Gallot, Mol. Cryst. Liq. Cryst. *7*, 103 (1969), C.A. *71*, 92010v (1969); Structure of mesophases from amorphous block copolymers. [49] A. Douy, B. R. Gallot, Mol. Cryst. Liq. Cryst. *14*, 191 (1971), C.A. *76*, 15103d (1972); Study of liquid-crystalline structures of polystyrene-polybutadiene block copolymers by small angle X-ray scattering and electron microscopy. [50] A. Douy, B. Gallot, C. R. Acad. Sci. Sér. C *272*, 1478 (1971), C.A. *75*, 36870w (1971); Structures mésomorphes des copolymères triséquencés polybuta-diène-polystyrène-polybutadiène. Étude par diffraction des rayons X et microscopie électronique.

[51] M. Gervais, A. Douy, B. Gallot, Liq. Cryst. *3*, Pt. I, 539 (1973), C.A. *79*, 43000y (1973); Phase diagram of systems: Block copolymer-preferential solvent of one block. [52] C. du Rietz, Proc. 2nd Scand. Symp. Surface Activ., Stockholm *1964*, 21, 38, C.A. *66*, 13076s (1967); Chemisorption of collectors. [53] J. F. Dyro, P. D. Edmonds, Mol. Cryst. Liq. Cryst. *8*, 141 (1969), C.A. *71*, 95070a (1969); Ultrasonic absorption and dispersion at phase transition in liquid crystals: n-octylamine/water. [53a] S. G. Efimova, N. P. Okromchedlidze, A.V. Volokhina, M. M. Iovleva, Vysokomol. Soedin. B *19*, 67 (1977), C.A. *86*, 107153e (1977); Properties of concentrated solutions of poly(arylene-1,3,4-oxadiazoles). [54] S. Eins, Naturwissenschaften *53*, 551 (1966), C.A. *66*, 32358b (1967); Elektronenmikroskopische Darstellung einer Intermediärphase des flüssigkristallinen Systems Kaliumoleat/Wasser. [55] S. Eins, Mol. Cryst. Liq. Cryst. *11*, 119 (1970), C.A. *74*, 7529r (1971); Electron microscopy of mesomorphic structures of the system potassium oleate/water. [56] P. Ekwall, Kolloid-Z. *85*, 16 (1938), C.A. *33*, 2392¹ (1939); Konstitution der verdünnten Seifenlösungen. Kristalline Abscheidungen. [57] P. Ekwall, J. Colloid Sci. 1954 Suppl. 1, 66, C.A. *48*, 11151b (1954); Concentration limits in association colloid solutions. [58] P. Ekwall, M. Salonen, I. Krokfors, I. Danielsson, Acta Chem. Scand. *10*, 1146 (1956), C.A. *52*, 7816b (1958); Microscopic investigation of the interaction between liquid alcohols and sodium laurate solutions. [59] P. Ekwall, I. Danielsson, L. Mandell, Vorträge Intern. Kongr. Grenzflächenaktive Stoffe, *3*, 189, 193 (1960), C.A. *57*, 5325h, 5348b (1962); Conditions leading to the formation of mesophases in aqueous association colloid solutions. Phase equilibria in the sodium caprylate/decanol/water system. [60] P. Ekwall, I. Danielsson, L. Mandell, Kolloid-Z. *169*, 113 (1960), C.A. *54*, 21935i (1960); Assoziations- und Phasengleichgewichte bei der Einwirkung von Paraffinkettenalkoholen an wäßrigen Lösungen von Assoziationskolloiden.

[61] P. Ekwall, L. Mandell, Acta Chem. Scand. *15*, 1403 (1961), C.A. *57*, 1615f (1962); Occurrence of cholesterol in watercontaining liquid-crystalline form. III. Formation of cholesterol-containing meso-phases of the interaction of sodium caprylate solutions with solid cholesterol crystals. IV. p. 1404. Solubility of cholesterol in sodium caprylate solutions at 20°. V. p. 1407. Equilibria between isotropic caprylate solutions and cholesterol-containing mesophases. [62] K. Fontell, P. Ekwall, L. Danielsson, Acta Chem. Scand. *16*, 2294 (1962), C.A. *58*, 7455c (1963); Structures of three mesophases in the sodium caprylate/decanol/water system. [63] L. Mandell, P. Ekwall, Chem. Phys. Appl. Surface Active Subst., Proc. 4 Int. Congr. *2*, 659 (1964), C.A. *72*, 25619h (1970); Phase equilibria in three-component systems containing an association colloid. [64] P. Ekwall, Wiss. Z. Friedrich-Schiller-Univ. Jena, Math.-Naturwiss. Reihe *14*, 181 (1965), C.A. *66*, 32184s (1967); Struktur, Solubilisierung und Bildung von Mesophasen in wäßrigen Assoziationskolloidsystemen. [65] P. Ekwall, L. Mandell, Acta Chem. Scand. *21*, 1612 (1967), C.A. *67*, 103221k (1967); Properties and structure of the decanolic solutions in the sodium caprylate/decanol/water system. I. Region of existence of the isotropic decanolic solutions. [66] P. Ekwall, P. Solyom, Acta Chem. Scand. *21*, 1619 (1967), C.A. *67*, 103222m (1967); II. Density and viscosity of the solutions (cf. [65]). [67] P. Ekwall, P. Stenius, Acta Chem. Scand. *21*, 1767 (1967), C.A. *68*, 43522t (1968); Aqueous sodium caprylate solutions. Activities of caprylate anions, and counterion binding to the micelles. [68] P. Ekwall, Svensk. Kem. Tidskr. *79*, 605 (1967), C.A. *68*, 75006u (1968); Association and ordered states in systems of amphiphilic substances. [69] L. Mandell, K. Fontell, P. Ekwall, Advan. Chem. Ser. *63*, 89 (1967), C.A. *67*, 6149k (1967); Occurrence of different mesophases in ternary systems of amphiphilic substances and water. [70] L. Mandell, P. Ekwall, Acta Polytech. Scand., Chem. Met. Ser. *74*, Pt. 1 (1968), C.A. *69*, 13346s, 13347t, 13348u (1968); Three-component

system sodium caprylate/decanol/water. I. Phase equilibria at 20°. L. Mandell, K. Fontell, H. Lehtinen, P. Ekwall, Pt. 2; II. Densities of the various phases and the partial specific volumes of the components; K. Fontell, L. Mandell, H. Lehtinen, P. Ekwall, Pt. 3; III. Structure of the mesophases.

[71] P. Ekwall, L. Mandell, K. Fontell, Acta Chem. Scand. *22*, 365 (1968), C.A. *68*, 108581h (1968); Lamellar mesophase with single amphiphile layers. [72] P. Ekwall, L. Mandell, K. Fontell, Acta Chem. Scand. *22*, 373 (1968), C.A. *68*, 108461u (1968); Lyophases with normal and reversed two-dimensional hexagonal structure. [73] P. Ekwall, L. Mandell, K. Fontell, Acta Chem. Scand. *22*, 697 (1968), C.A. *69*, 22465g (1968); Lyophases with normal and reversed two-dimensional tetragonal structure. [74] P. Ekwall, L. Mandell, Acta Chem. Scand. *22*, 699 (1968), C.A. *69*, 13482h (1968); Minimum water content of a number of reversed micellar and mesomorphous structures. [75] P. Ekwall, L. Mandell, K. Fontell, Acta Chem. Scand. *22*, 1543 (1968), C.A. *69*, 99977h (1968); Two types of neat soap in ternary systems. [76] K. Fontell, L. Mandell, P. Ekwall, Acta Chem. Scand. *22*, 3209 (1968), C.A. *70*, 71614j (1969); Isotropic mesophases in systems containing amphiphilic compounds. [77] P. Ekwall, L. Mandell, K. Fontell, J. Colloid Interface Sci. *28*, 219 (1968), C.A. *69*, 99948z (1968); Ternary systems of sodium caprylate and water with ethylene glycol, glycerol, and tetraethylene glycol. [78] P. Solyom, P. Ekwall, Rheol. Acta *8*, 316 (1969), C.A. *72*, 25648s (1970); Rheologische Eigenschaften der Mesophasen im Natriumcaprylat/Dekanol/Wasser System. [79] P. Ekwall, L. Mandell, K. Fontell, Mol. Cryst. Liq. Cryst. *8*, 157 (1969), C.A. *71*, 95442y (1969); Solubilization in micelles and mesophases and the transition from normal to reversed structures. [80] P. Ekwall, J. Colloid Interface Sci. *29*, 16 (1969), C.A. *70*, 41052g (1969); Two types of micelle formation in organic solvents.

[81] P. Ekwall, L. Mandell, K. Fontell, J. Colloid Interface Sci. *29*, 542 (1969), C.A. *70*, 81408j (1969); Phase equilibriums in the ternary system sodium caprylate/1,8-octanediol/water. [82] P. Ekwall, L. Mandell, K. Fontell, J. Colloid Interface Sci. *29*, 639 (1969), C.A. *70*, 109547e (1969); Cetyltrimethylammonium bromide/hexanol/water system. [83] P. Ekwall, L. Mandell, K. Fontell, J. Colloid Interface Sci. *31*, 508 (1969), C.A. *72*, 25611z (1970); Ternary system of potassium soap, alcohol, and water. I. Phase equilibriums and phase structures. [84] P. Ekwall, L. Mandell, K. Fontell, J. Colloid Interface Sci. *31*, 530 (1969), C.A. *72*, 25190m (1970); II. Structure and composition of the mesophases (cf. [83]). [85] P. Ekwall, L. Mandell, Kolloid-Z. Z. Polym. *233*, 938 (1969), C.A. *71*, 114458b (1969); Solutions of alkali soaps and water in fatty acids. I. Region of existence of the solutions. [86] P. Ekwall, P. Solyom, Kolloid-Z. Z. Polym. *233*, 945 (1969), C.A. *71*, 114459c (1969); II. Density and viscosity (cf. [85]). [87] S. Friberg, L. Mandell, P. Ekwall, Kolloid-Z. Z. Polym. *233*, 955 (1969); C.A. *71*, 114460w (1969); III. IR and NMR investigations (cf. [85]). [88] P. Ekwall, L. Mandell, K. Fontell, Chim. Phys. Appl. Prat. Ag. Surface, C. R. Congr. 5th Int. Deterg. *2* (Pt. 2) 1059 (1969), C.A. *74*, 14402x (1971); Transition from normal to reversed micellar and mesomorphous structures. [89] P. Ekwall, L. Mandell, K. Fontell, J. Colloid Interface Sci. *33*, 215 (1970), C.A. *73*, 29255e (1970); Binary and ternary Aerosol OT systems. [90] P. Ekwall, L. Mandell, P. Solyom, J. Colloid Interface Sci. *35*, 266 (1971), C.A. *74*, 91789f (1971); Solution phase with reversed micelles in the cetyltrimethylammonium bromide/hexanol/water system.

[91] P. Ekwall, L. Mandell, P. Solyom, J. Colloid Interface Sci. *35*, 519 (1971), C.A. *75*, 10519p (1971); Aqueous cetyltrimethylammonium bromide solutions. [92] P. Ekwall, I. Danielsson, P. Stenius, MTP Int. Rev. Sci., Phys. Chem., Ser. 1, *7*, 97 (1972), C.A. *78*, 151870v (1973); Aggregation in surfactant systems. [93] P. Ekwall, Liq. Cryst. Ord. Fluids *2*, 177 (1974), C.A. *86*, 180799z (1977); Dependence of some properties of aqueous liquid crystals on their water content. [94] P. Ekwall, Adv. Liq. Cryst. *1*, 1 (1975), C.A. *84*, 37347d (1976); Composition, properties, and structures of liquid crystals in systems of amphiphilic compounds. [95] S. F. Emory, A. A. Kozinski, Mol. Cryst. Liq. Cryst. *31*, 305 (1975), C.A. *84*, 98025u (1976); Lyophases in stagnation flow: Fibrinogen. [95a] R. Faiman, I. Lundstrom, Proc. Int. Conf. Raman Spectrosc., 5th *1976*, 542, C.A. *87*, 191472j (1977); Raman spectroscopic study of the aerosol OT/water system. [96] J. W. Falco, R. D. Walker, Jr., D. O. Shah, AIChE J. *20*, 510 (1974), C.A. *81*, 29909m (1974); Effect of phase-volume ratio and phase-inversion on viscosity of microemulsions and liquid crystals. [97] M. E. Feinstein, H. L. Rosano, J. Phys. Chem. *73*, 601 (1969), C.A. *70*, 100030y (1969); Influence of micelles on titrations of aqueous sodium and potassium soap solutions. [98] R. H. Ferguson, A. S. Richardson, Ind. Eng. Chem. *24*, 1329 (1932), C.A. *27*, 3354 (1933); Middle soap. [99] R. H. Ferguson, Oil and Soap *9*, 5, 25 (1932), C.A. *26*, 1465 (1932); Phase phenomena in commercial soap systems. [100] R. H. Ferguson, Oil and Soap *14*, 115 (1937), C.A. *31*, 4519[7] (1937); Effect of glycerol on the equilibria of hydrated soap systems.

[101] K. Fontell, Nord. Symp. Graensefladekemi, Fortryk Foredrag, 3rd 1967, N, 5 pp., C.A. *74*, 35211z (1971); Internal structure of the mesophases in the sodium caprylate/decanol/water system. [102] K. Fontell, J. Colloid Interface Sci. *43*, 156 (1973), C.A. *78*, 141282r (1973); Structure of the liquid crystalline optical isotropic viscous phase occurring in aerosol OT systems. [103] K. Fontell, J. Colloid Interface Sci. *44*, 318 (1973), C.A. *79*, 84277n (1973); Structure of the lamellar liquid crystalline phase in the Aerosol OT/water system. [104] K. Fontell, Liq. Cryst. Plast. Cryst. *2*, 80 (1974), C.A. *83*, 186434m (1975); X-ray diffraction by liquid crystals. Amphiphilic systems. [104a] I. Lundstrom, K. Fontell, J. Colloid Interface Sci. *59*, 360 (1977), C.A. *86*, 198481z (1977); Lateral electric conductivity in Aerosol-OT/water systems. [104b] I. Jonas, K. Fontell, G. Lindblom, B. Norden, Spectrosc. Lett. *10*, 501 (1977), C.A. *87*, 133722e (1977); Orientation studies of probe molecules in lamellar liquid crystalline lipid systems by linear dichroism. [104c] A. Forge, J. E. Lydon, J. Colloid Interface Sci. *59*, 186 (1977), C.A. *86*, 198207q (1977); Study of the structure of the C phase in the sodium octanoate/decanol/water system using electron microscopy. [105] J. François, J. Phys. (Paris), C-4, *30*, 84 (1969), C.A. *72*, 104609h (1970); Conductivité électrique de système savon/eau présentant une structure mésomorphe cylindrique. [106] J. François, C. R. Acad. Sci., Sér. C *272*, 876 (1971), C.A. *74*, 130718d (1971); Effet de la température sur la conductivité de gels mésomorphes de système savon/eau. [107] J. François, Kolloid-Z. Z. Polym. *251*, 594 (1973), C.A. *80*, 19817v (1974); Étude par spectroscopie proche IR de la structure de l'eau dans les solutions micellaires et les gels mésomorphes du système de laurate de potassium/eau. [108] S. Friberg, L. Mandell, P. Ekwall, Acta Chem. Scand. *20*, 2632 (1966); Preliminary IR and NMR investigations on the alkali soap/carboxylic acid/water systems. [109] S. Friberg, L. Mandell, K. Fontell, Acta Chem. Scand. *23*, 1055 (1969), C.A. *71*, 42810k (1969); Mesophases in systems of water/nonionic emulsifier/hydrocarbon. [110] S. Friberg, L. Mandell, M. Larsson, J. Colloid Interface Sci. *29*, 155 (1969), C.A. *70*, 81177h (1969); Mesophases, a factor of importance for the properties of emulsions.

[111] S. Friberg, L. Mandell, J. Am. Oil Chem. Soc. *47*, 149 (1970), C.A. *73*, 18749t (1970); Phase equilibriums and their influence on the properties of emulsions. [112] S. Friberg, P. Solyom, Kolloid-Z. Z. Polym. *236*, 173 (1970), C.A. *73*, 7588p (1970); Influence of mesophases on the rheological behavior of emulsions. [113] S. Friberg, I. Wilton, Am. Perfum. Cosmet. *85*, 27 (1970), C.A. *74*, 91449v (1971); Liquid crystals — the formula for emulsions. [114] S. Friberg, G. Soderlund, Kolloid-Z. Z. Polym. *243*, 56 (1971), C.A. *75*, 11714k (1971); The system water/1-aminooctane/octanoic acid. I. Determinations of phase equilibria and structure investigations of the liquid crystalline phase by IR-spectroscopy and X-ray diffraction. [115] S. Friberg, L. Rydhag, Kolloid-Z. Z. Polym. *244*, 233 (1971), C.A. *75*, 67802e (1971); The system: water/*p*-xylene/1-aminooctane/octanoic acid. II. Stability of emulsions in different regions. [116] S. Friberg, Kolloid-Z. Z. Polym. *244*, 333 (1971), C.A. *75*, 68008n (1971); Molecular complexes in liquid crystalline structures as a stabilizer at liquid interfaces. [117] S. Friberg, L. Rydhag, G. Jederstrom, J. Pharm. Sci. *60*, 1883 (1971), C.A. *76*, 37410d (1972); Mesophases in aerosol formulations. I. Phase equilibria in propellant compositions. [118] S. Friberg, S. I. Ahmad, J. Colloid Interface Sci. *35*, 175 (1971), C.A. *74*, 91437q (1971); Liquid crystals and the foaming capacity of an amine dissolved in water and *p*-xylene. [119] S. Friberg, J. Colloid Interface Sci. *37*, 291 (1971), C.A. *75*, 133388c (1971); Mesophases in emulsions. [120] S. Friberg, J. Am. Oil Chem. Soc. *48*, 578 (1971), C.A. *76*, 50449t (1972); Microemulsions, hydrotropic solutions and emulsions, a question of phase equilibria.

[121] S. Friberg, K. Roberts, Nature, Phys. Sci. *238*, 77 (1972), C.A. *77*, 105869h (1972); Liquid crystals in quick clays. [122] H. Saito, S. Friberg, Liq. Cryst., Proc. Int. Conf., Bangalore, *1973*, 537, C.A. *84*, 112046k (1976); Lyophases and foam stability. [123] L. Rydhag, S. Friberg, Chem., Phys. Chem. Anwendungstechn. Grenzflächenaktiven Stoffe, Ber. Int. Kongr., 6th, *1972*, 2, Teil 2, 483, C.A. *82*, 35257j (1975); Phase equilibriums of water/hydrocarbons/nonionic emulsifiers. [124] G. Jederstrom, L. Rydhag, S. Friberg, J. Pharm. Sci. *62*, 1979 (1973), C.A. *80*, 40999r (1974); II. Influence of mesophases on foam stability (cf. [117]). [125] S. Friberg, S. E. Linden, H. Saito, Nature *251*, 494 (1974); Thin films from liquid crystals. [126] S. Friberg, L. Larsson, J. Pharm. Sci. *64*, 822 (1975); Partition of an organophosphorus compound, dichlorvos, between liquid and liquid crystalline phases. [127] S. Friberg, P. O. Jansson, E. Cederberg, J. Colloid Interface Sci. *55*, 614 (1976), C.A. *85*, 37464k (1976); Surfactant association structure and emulsion stability. [128] P. O. Jansson, S. Friberg, Mol. Cryst. Liq. Cryst. *34*, 75 (1976), C.A. *85*, 198433q (1976); Van der Waals potential in coalescing emulsion drops with liquid crystals. [129] S. Friberg, L. Rydhag, T. Doi, Adv. Chem. Ser. *152*, 28 (1976), C.A. *85*, 114863u (1976); Micellar and lyotropic liquid crystalline phases containing nonionic active substances.

[130] S. Friberg, K. Larsson, Adv. Liq. Cryst. 2, 173 (1976), C.A. 86, 180804x (1977); Liquid crystals and emulsions. [130a] S. Friberg, H. Saito, Foams, Proc. Symp. 1975, 33, C.A. 87, 173227d (1977); Foam stability and association of surfactants. [130b] S. Friberg, Mol. Cryst. Liq. Cryst. 40, 49 (1977), C.A. 87, 125407r (1977); Lyotropic liquid crystals.

[131] G. Friedel, Rev. gén. Sci. pures et appl. 36, 162 (1925); Corps mésomorphes, solutions de savon et rayons X. [132] B. R. Gallot, Mol. Cryst. Liq. Cryst. 13, 323 (1971), C.A. 75, 123347z (1971); Liquid-crystalline structures of binary systems α-ω soap/water. [133] P. Gaubert, Compt. rend. 197, 1436 (1933), C.A. 28, 1239^8 (1934); Cristaux liquides produits par évaporation ou refroidissement d'une solution aqueuse de tartrazine. [134] P. Gaubert, Compt. rend. 198, 951 (1934), C.A. 28, 2964^3 (1934); Cristaux liquides obtenus par évaporation rapide d'une solution aqueuse. [135] P. Gaubert, Compt. rend. 200, 679 (1935), C.A. 29, 2802^8 (1935); Liquides anisotropes. [136] N. L. Gershfeld, C. Y. C. Pak, J. Colloid Interface Sci. 23, 215 (1967), C.A. 66, 88884n (1967); Surface chemistry of monooctadecyl phosphate at the air/water interface. Molecular aggregation in monolayers. [137] M. Gervais, G. Jouan, B. Gallot, C. R. Acad. Sci., Sér. C 275, 1243 (1972), C.A. 78, 72755h (1973); Étude des facteurs qui déterminent le domaine d'existence des mésophases présentées par les systèmes copolymère polystyrène-polyoxyéthylènephtalate de diéthyle. [138] P. E. Giua, 3rd ACS Symp. Ord. Fluids Liquid Cryst., Chicago, Aug. 1973, Abstr. No. 48; Micellar solutions and liquid crystals of nonionic emulsifiers. [139] K. G. Götz, K. Heckmann, J. Colloid Sci. 13, 266 (1958), C.A. 52, 15193h (1958); Shape of soap micelles and other polyions as obtained from anisotropy of electric conductivity. [140] G. W. Gray, P. A. Winsor, Mol. Cryst. Liq. Cryst. 26, 305 (1974), C.A. 82, 66656t (1975); New concept of the nature of the viscous isotropic or cubic mesophases of amphiphilic systems. Liquid crystalline character of plastic crystals. Constitutional analogies between amphiphilic and nonamphiphilic mesophases.

[141] A. V. Gribanov, N. G. Bel'nikevich, A. I. Kol'tsov, S. Y. Frenkel, Vysokomol. Soedin. B 18, 440 (1976), C.A. 85, 78530x (1976); Magnetic field effect on sulfuric acid solutions of poly(p-phenylene-terephthalamide). [142] M. J. Groves, R. M. A. Mustafa, J. Chromatogr. 97, 297 (1974), C.A. 82, 77050s (1975); Note on the thin layer chromatography of mixed surfactant systems. [143] M. J. Groves, A. B. Ahmad, Rheol. Acta 15, 501 (1976), C.A. 86, 10854w (1977); Rheological properties of lyophases formed by phosphated polyoxyethylene surfactants, n-hexane and water. [144] M. J. Groves, D. A. de Galindez, Acta Pharm. Suec. 13, 353 (1976), C.A. 86, 34591j (1977); Rheological characterisation of self-emulsifying oil/surfactant systems. [145] M. J. Groves, D. A. de Galindez, Acta Pharm. Suec. 13, 361 (1976), C.A. 86, 34592k (1977); Self-emulsifying action of mixed surfactants in oil. [145a] K. Harada, H. Hayashi, K. Hirakawa, Kyushu Daigaku Kogaku Shuho 49, 513 (1976), C.A. 86, 51897f (1977); Explanation for the phase transition in lyophases. [146] W. D. Harkins, R. W. Mattoon, M. L. Corrin, J. Am. Chem. Soc. 68, 220 (1946), C.A. 40, 2715^4 (1946); Structure of soap micelles indicated by X-rays and the theory of molecular orientation. I. Aqueous solutions. [147] N. H. Hartshorne, G. D. Woodard, Mol. Cryst. Liq. Cryst. 23, 343 (1973), C.A. 80, 88154p (1974); Mesomorphism in the system disodium chromoglycate/water. [148] J. B. Hayter, A. M. Hecht, J. W. White, G. J. T. Tiddy, Faraday Discuss. Chem. Soc. 57, 130 (1974), C.A. 82, 175567k (1975); Dynamics of water and amphiphile molecules in lamellar liquid crystals. [149] K. Heckmann, Naturwissenschaften 40, 478 (1953), C.A. 48, 7988i (1954); Leitfähigkeitsanisotropie bei Seifenlösungen. [150] W. Helfrich, Z. Naturforsch. C 29, 692 (1974), C.A. 82, 66543d (1975); Ripples in tilted bilayers and lyotropic smectics.

[151] K. W. Herrmann, J. Colloid Interface Sci. 22, 352 (1966), C.A. 66, 67037y (1967); Micellar properties of zwitterionic surfactants. [152] K. Hess, J. Gundermann, Ber. 70 B, 1800 (1937), C.A. 31, 7724^4 (1937); Röntgenographische Untersuchungen an ruhenden und strömenden kolloidalen Lösungen (Nachweis der Orientierung von Kolloidteilchen beim Strömen durch Capillaren durch das Auftreten von Faserdiagrammen; Hydratation von Kolloidteilchen in der Lösung). [153] K. Hess, W. Philippoff, H. Kiessig, Kolloid-Z. 88, 40 (1939), C.A. 33, 8471^6 (1939); Viskositätsbestimmungen, Dichtemessungen und Röntgenuntersuchungen an Seifenlösungen. [154] R. Heusch, A. Stessel, H. Schwentke, Fette, Seifen, Anstrichm. 78, 359 (1976), C.A. 86, 45051j (1977); Physikalische Methoden als Hilfe für die Herstellung und Entwicklung von Tensiden. II. Mikroskopische Untersuchungen. [155] P. C. Hiemenz, J. Chem. Educ. 49, 164 (1972), C.A. 76, 125890r (1972); Role of van der Waals forces in surface and colloid chemistry. [156] O. L. Hoerer, Acad. Rep. Populare Romine, Studii Cercetari Chim. 11, 95 (1963), C.A. 60, 2362d (1964); Criteria for systematization of the micelles of association colloids. [157] C. W. Hoerr, A. W. Ralston, J. Am. Chem. Soc. 64, 2824 (1942), C.A. 37, 816^8 (1943); Behavior of various salts of dodecylamine in water, ethanol and benzene. [158] U.

Hoppe, Ber. des 7. IFSCC-Kongresses, 12.–15. 9. 1972, Hamburg, J. Soc. Cosmet. Chem. *24*, 317 (1973), C.A. *79*, 70090j (1973); Neue hautaffine Lichtschutzsubstanzen. [159] R. Hosemann, Prog. Colloid Polym. Sci. *60*, 213 (1976), C.A. *85*, 167140n (1976); Significance of the paracrystalline state for colloid science. [160] F. Huisman, K. J. Mysels, J. Phys. Chem. *73*, 489 (1969), C.A. *70*, 99995h (1969); Contact angle and the depth of the free-energy minimum in thin liquid films. Their measurement and interpretation. [161] F. Husson, Compt. rend. *253*, 2948 (1961), C.A. *57*, 9279b (1962); Structure des phases méso-morphes du mélange eau/linolénat de sodium. Étude par diffraction des rayons X. [162] A. J. Hyde, Nature *170*, 234 (1952); Emulsion and films. Symp. at Sheffield. [163] G. Ibbotson, M. N. Jones, Trans. Faraday Soc. *65*, 1146 (1969), C.A. *70*, 99904c (1969); Effects of ion valency on formation of second black soap films. [164] T. Ingram, M. N. Jones, Trans. Faraday Soc. *65*, 297 (1969), C.A. *70*, 41104a (1969); Membrane potential studies on surfactant solutions. [165] M. M. Iovleva, S. I. Banduryan, V. D. Kalmykova, S. P. Papkov, Mater. Vses. Konf. Elektron. Mikrosk., 9th, *1973*, 303, C.A. *85*, 178817q (1976); Structure formation in the precipitation of poly-*p*-benzamide. [166] A. Ishihara, J. Chem. Phys. *19*, 1142 (1951), C.A. *46*, 810b (1952); Theory of anisotropic colloidal solutions. [167] E. E. Jelley, Nature *138*, 1009 (1936), C.A. *31*, 1703[7] (1937); Spectral absorption and fluorescence of dyes in the molecular state. [168] E. E. Jelley, Nature *139*, 631 (1937), C.A. *31*, 4561[9] (1937); Molecular, nematic and crystal states of 1,1'-diethyl-ψ-cyanine chloride. [169] S. H. Jury, R. C. Ernst, J. Phys. Colloid Chem. *53*, 609 (1949), C.A. *43*, 6493d (1949); First approximate conditions for the formation of liquid crystals in solution. [170] V. D. Kalmykova, G. I. Kudryavtsev, S. P. Papkov, A. V. Volokhina, M. M. Iovleva, L. P. Mil'kova, V. G. Kulichikhin, S. I. Banduryan, Vysokomol. Soedin. B *13*, 707 (1971), C.A. *76*, 60243k (1972); Mesomorphic state of poly(*p*-benzamide) solutions.

[171] Y. Kasai, M. Nakagaki, Nippon Kagaku Zasshi *91*, 19 (1970), C.A. *72*, 115090h (1970); Mesophase in multi-phase equilibrium of sodium dodecyl sulfate/oil/water system. [172] E. Keh, C. Gavach, J. Guastalla, C. R. Acad. Sci., Sér. C *263*, 1488 (1966), C.A. *66*, 49661t (1967); Détermination potentiométrique et conductométrique des dimensions des micelles d'halogénures d'ammonium qua-ternaire à longue chaîne. [173] O. A. Khanchich, A. T. Serkov, A. V. Volokhina, V. D. Kalmykova, Vysokomol. Soedin. A *17*, 579 (1975), C.A. *83*, 28708z (1975); Structure formation in the precipitation of poly-*p*-benzamide from isotropic and anisotropic solutions. [174] O. A. Khanchich, A. T. Serkov, V. D. Kalmykova, A. V. Volokhina, Khim. Volokna *1975*, 68, C.A. *84*, 44812e (1976); Small-angle scattering of polarized light by poly-*p*-benzamide liquid crystals. [175] H. Kießig, Kolloid-Z. *96*, 252 (1941), C.A. *36*, 6395[9] (1942); Röntgenographische Untersuchung der Struktur von Seifenlösungen. [176] H. B. Klevens, Chem. Rev. *47*, 1 (1950), C.A. *44*, 9217b (1950); Solubilization. [177] H. Kölbel, P. Kurzendörfer, Fortschr. Chem. Forsch. *12*, 252 (1969), C.A. *71*, 114453w (1969); Konstitution und Eigenschaften von Tensiden. [178] A. I. Kol'tsov, N. G. Bel'nikevich, A. V. Gribanov, S. P. Papkov, S. Y. Frenkel, Vysokomol. Soedin. B *15*, 645 (1973), C.A. *80*, 83762a (1974); Orientation of polymer and solvent molecules in poly-*p*-benzamide solutions under the action of a magnetic field. [179] E. V. Korneeva, P. N. Lavrenko, I. N. Shtennikova, N. A. Mikhailova, A. E. Polotskii, A. A. Baturin, Y. B. Amerik, B. Krentsel, V. N. Tsvetkov, Vysokomol. Soedin. A *17*, 2582 (1975), C.A. *84*, 44837s (1976); Intrinsic viscosity, diffusion, and sedimentation of copolymers of crystallike molecules. [180] F. Krafft, H. Wigelow, Ber. *28*, 2566 (1895); Verhalten der fettsauren Alkalien und der Seifen in Gegenwart von Wasser.

[181] I. M. Krieger, F. M. O'Neill, J. Am. Chem. Soc. *90*, 3114 (1968), C.A. *69*, 46322j (1968); Diffraction of light by arrays of colloidal spheres. [182] N. Krog, Fette, Seifen, Anstrichm. *77*, 267 (1975), C.A. *83*, 121181g (1975); Structures of emulsifier/water mesophases related to emulsion stability. [183] V. G. Kulichikhin, A. Y. Malkin, S. P. Papkov, O. N. Korol'kova, V. D. Kalmykova, A. V. Volokhina, O. B. Semenov, Vysokomol. Soedin. A *16*, 169 (1974), C.A. *81*, 64131y (1974); Viscometric criteria of the transition of poly-*p*-benzamide solutions into the liquid-crystalline state. [184] V. G. Kulichikhin, N. V. Vasil'eva, L. D. Serova, V. A. Platonov, L. P. Mil'kova, I. N. Andreeva, A. V. Volokhina, G. I. Kudryavtsev, S. P. Papkov, Vysokomol. Soedin. A *18*, 590 (1976), C.A. *84*, 180778a (1976); Flow of anisotropic solutions of poly(*p*-phenyleneterephthalamide). [184a] H. Kunieda, Nippon Kagaku Kaishi *1977*, 151, C.A. *86*, 128194j (1977); Mechanism of dissolution of quaternary ammonium chlorides containing two long-chain alkyl groups in water. [185] S. Kuroiwa, H. Matsuda, K. Kitazawa, Kogyo Kagaku Zasshi *70*, 2103 (1967), C.A. *68*, 60770q (1968); Fluid double refraction of nonionic surfactants. [186] S. Kuroiwa, Hyomen *11*, 579 (1973), C.A. *80*, 64470n (1974); Structures and properties of liquid crystals formed in the highly concentrated aqueous solutions of surfactants. [187] S. Kuroiwa, H. Matsuda, H. Fujimatsu, A. Miyazawa, Nippon Kagaku Kaishi *1976*, 1040, C.A. *85*, 110384r (1976);

Structures and properties of liquid crystals formed in highly concentrated aqueous solutions of non-ionic surface active agents. [188] H. Lange, M. J. Schwuger, Kolloid-Z. Z. Polym. *223*, 145 (1968), C.A. *69*, 70023h (1968); Mizellenbildung und Krafft-Punkte in der homologen Reihe der Na-n-alkylsulfonate einschließlich der ungeradzahligen Glieder. [189] J. W. Larsen, L. J. Magid, J. Am. Chem. Soc. *96*, 5774 (1974), C.A. *81*, 127004g (1974); Calorimetric and counterion binding studies of the interactions between micelles and ions. Observation of lyotropic series. [190] K. Larsson, Chem. Phys. Lipids *14*, 233 (1975), C.A. *83*, 73653p (1975); Significance of crystalline hydrocarbon chains in aqueous dispersions and emulsions of lipids. [190a] R. G. Laughlin, J. Colloid Interface Sci. *55*, 239 (1976), C.A. *84*, 185527c (1976); Expedient technique for determining solubility phase boundaries in surfactant/water systems.

[191] A. S. C. Lawrence, Trans. Faraday Soc. *29*, 1008 (1933), C.A. *28*, 1239[6] (1934); Lyotropic mesomorphism. [192] A. S. C. Lawrence, Trans. Faraday Soc. *31*, 189 (1935), C.A. *29*, 3186[5] (1935); Soap micelles. [193] A. S. C. Lawrence, Trans. Faraday Soc. *34*, 660 (1938), C.A. *32*, 5278[4] (1938); Metal soaps and the gelation of their paraffin solutions. [194] A. J. Hyde, D. M. Langbridge, A. S. C. Lawrence, Disc. Faraday Soc. *18*, 239 (1954), C.A. *49*, 14430c (1955); Soap/water/amphiphile systems. [195] A. S. C. Lawrence, 1. Congr. mondial détergence et prods. tensio-actifs, Paris *1*, 31 (1954), C.A. *51*, 10928c (1957); Soap/water/amphiphile systems. [196] D. M. Langbridge, A. S. C. Lawrence, R. Stenson, J. Colloid Sci. *11*, 585 (1956), C.A. *51*, 4114d (1975); Effect of soap on upper and lower consolute temperatures in the nicotine/water and triethylamine/ether systems. [197] A. S. C. Lawrence, Disc. Faraday Soc. *25*, 51 (1958), C.A. *53*, 11946g (1959); Solubility in soap solutions. [198] B. J. Boffey, R. Collison, A. S. C. Lawrence, Trans. Faraday Soc. *55*, 654 (1959), C.A. *54*, 4131a (1960); Solubility of organic substances in aqueous soap solutions. VIII. Sodium dodecyl sulfate/water/caproic acid and the hexadecyltrimethylammonium bromide/water/caproic acid systems. IX. Sodium dodecyl sulfate/water/*n*-octylamine system. [199] J. T. Pearson, A. S. C. Lawrence, Trans. Faraday Soc. *63*, 488 (1967), C.A. *66*, 88924a (1967); Behavior of calcium ions in micellar sodium dodecyl sulfate solution containing solubilized polar organic compounds. [200] A. S. C. Lawrence, J. T. Pearson, Trans. Faraday Soc. *63*, 495 (1967), C.A. *66*, 88934d (1967); Electric properties of soap/water/amphiphile systems. Electric conductance and sodium ion activity of aqueous 2% sodium dodecyl sulfate with added homologous *n*-aliphatic C_{2-7} alcohols.

[201] A. S. C. Lawrence, M. A. Al-Mamun, M. P. McDonald, Trans. Faraday Soc. *63*, 2789 (1967), C.A. *68*, 6816w (1968); Effect of water on the polymorphism of long chain alcohols and acids. [202] A. S. C. Lawrence, Mol. Cryst. Liq. Cryst. *7*, 1 (1969), C.A. *71*, 95440w (1970); Lyotropic mesomorphism in lipid/water systems. [203] A. S. C. Lawrence, Liq. Cryst. Ord. Fluids, *1970*, 289, C.A. *79*, 24374g (1973); Mesomorphism in cholesterol/fatty alcohol systems. [204] K. D. Lawson, A. J. Mabis, T. J. Flautt, J. Phys. Chem. *72*, 2058 (1968), C.A. *69*, 30370n (1968); Mesophases. X-ray studies of the dimethyldodecylamine oxide/deuterium oxide system. [205] G. D. Litovchenko, V. D. Kalmykova, T. S. Sokolova, A. V. Volokhina, G. I. Kudryavtsev, S. P. Papkov, Zh. Prikl. Spektrosk. *24*, 1089 (1976), C.A. *85*, 63552g (1976); Spectroscopic study of the crystallization of poly(*p*-benzamide) and poly(*p*-phenyleneterephthalamide). [206] A. V. Lobastova, Yu. F. Deinega, Usp. Kolloidn. Khim. *1973*, 300, C.A. *81*, 171870k (1974); Ferroelectric properties of hydrocarbon solutions of soaps. [206a] A. V. Lobastova, Yu. F. Deinega, Usp. Kolloidn. Khim. *1976*, 300, C.A. *87*, 192572d (1977); Ferroelectric properties of hydrocarbon solutions of soaps. [207] R. C. Long, Jr., J. H. Goldstein, Liq. Cryst. Ord. Fluids *2*, 147 (1974), C.A. *86*, 180195z (1977); New nematic lyophase. [208] I. Lundstrom, K. Fontell, Chem. Phys. Lipids *15*, 1 (1975), C.A. *84*, 1533x (1976); Lateral conductivity of lamellar mesophases in lipid/water systems. [209] E. S. Lutton, J. Am. Oil Chem. Soc. *43*, 28 (1966), C.A. *64*, 9961d (1966); Phase behavior of the dimethyldodecylamine oxide/water systems. [210] V. Luzzati, H. Mustacchi, A. Skoulios, F. Husson, Acta Cryst. *13*, 660 (1960), C.A. *54*, 23557g (1960); La structure des colloides d'association. I. Les phases liquide-cristallines des systèmes amphiphile/eau.

[211] F. Husson, H. Mustacchi, V. Luzzati, Acta Cryst. *13*, 668 (1960), C.A. *54*, 23557h (1960); II. Description des phases liquide-cristallines. Amphiphiles anioniques, cationiques, non-ioniques (cf. [210]). [212] F. Husson, J. Herz, V. Luzzati, Compt. rend. *252*, 3290 (1961), C.A. *56*, 8115c (1962); Structure des phases mésomorphes d'un savon-monomère seul et un présence d'eau. [213] V. Luzzati, F. Husson, J. Cell Biol. *12*, 207 (1962), C.A. *57*, 9314f (1962); Structure of the mesophase of lipid/water systems [214] V. Luzzati, F. Reiss-Husson, Nature *210*, 1351 (1966), C.A. *65*, 9240b (1966); Structure of the cubic phase of lipid/water systems. [215] F. Reiss-Husson, V. Luzzati, J. Colloid Interface Sci. *21*, 534 (1966), C.A. *65*, 4113b (1966); Small angle X-ray scattering of the structure of soap and

detergent micelles. [**216**] V. Luzzati, T. Gulik-Krzywicki, A. Tardieu, J. Phys. (Paris), C-4, *30*, 74 (1969); Structure des phases présentes dans les systèmes lipide/eau. [**217**] T. Gulik-Krzywicki, A. Tardieu, V. Luzzati, Mol. Cryst. Liq. Cryst. *8*, 285 (1969), C.A. *71*, 95125x (1969); Smectic phase of lipid/water systems. Properties related to the nature of the lipid and to the presence of net electric charges. [**218**] V. Luzzati, A. Tardieu, Annu. Rev. Phys. Chem. *25*, 79 (1974), C.A. *82*, 69214h (1975); Lipid phases. Structure and structural transitions. [**219**] C. G. Lyons, E. K. Rideal, Proc. Cambridge Phil. Soc. *26*, 419 (1930), C.A. *24*, 5203 (1930); Phase diagrams for unimolecular films. [**220**] J. W. McBain, A. J. Burnett, J. Chem. Soc. *121*, 1320 (1922), C.A. *16*, 3781 (1922); Effect of an electrolyte on solutions of pure soap. Phase-rule equilibria in the system sodium laurate/sodium chloride/water.

[**221**] M. E. Laying, J. W. McBain, Kolloid-Z. *35*, 18 (1924), C.A. *18*, 3304 (1924); Gallerten im Gegensatz zu Gelen und Flocken. Seifen in trocknem Alkohol. [**222**] J. W. McBain, Nature *113*, 534 (1924), C.A. *18*, 2095[7] (1924); Liquid crystals, soap solutions and X-rays. [**223**] J. W. McBain, Nature *114*, 49 (1924), C.A. *18*, 2827[8] (1924); Liquid crystals, soap solutions and X-rays. [**224**] J. W. McBain, G. M. Langdon, J. Chem. Soc. *127*, 852 (1925), C.A. *19*, 2139 (1925); Equilibria underlying the soap-boiling processes. I. Pure sodium palmitate. [**225**] J. W. McBain, M. C. Field, J. Phys. Chem. *30*, 1545 (1926), C.A. *21*, 663 (1927); II. The system potassium laurate/potassium chloride/water (cf. [**224**]). [**226**] J. W. McBain, W. J. Elford, J. Chem. Soc. *1926*, 421, C.A. *20*, 1725 (1926); III. The system potassium oleate/potassium chloride/water (cf. [**224**]). [**227**] J. W. McBain, J. Chem. Educ. *6*, 2115 (1929), C.A. *24*, 539[8] (1930); Structure in amorphous and colloidal matter. [**228**] J. W. McBain, L. H. Lazarus, A. V. Pitter, Z. physik. Chem. A *147*, 87 (1930), C.A. *24*, 2944 (1930); Anwendung der Phasenregel auf das Seifensieden. Das System Natriumpalmitat/Wasser/Natriumchlorid. [**229**] J. W. McBain, O. O. Watts, J. Rheol. *3*, 437 (1932), C.A. *27*, 649[9] (1933); Structural properties of anisotropic solutions of soap as determined by a new centrifugal fallingball method. [**230**] J. W. McBain, M. C. Field, J. Phys. Chem. *37*, 675 (1933), C.A. *27*, 4994 (1933); Phase-rule equilibria of acid soaps. I.

[**231**] J. W. McBain, M. C. Field, J. Am. Chem. Soc. *55*, 4776 (1933), C.A. *28*, 5744[7] (1934); IV. The three-component systems potassium laurate/lauric acid/water (cf. [**230**]). [**232**] J. W. McBain, M. C. Field, J. Chem. Soc. *1933*, 920, C.A. *27*, 5624 (1933); II. Anhydrous acid sodium palmitates (cf. [**230**]). [**233**] J. W. McBain, A. Stewart, J. Chem. Soc. *1933*, 924, C.A. *27*, 5624 (1933); III. Anhydrous acid potassium oleate (cf. [**230**]). [**234**] J. W. McBain, A. Stewart, J. Chem. Soc. *1933*, 928, C.A. *27*, 5616 (1933); Conductivity in the three-component system oleic acid/potassium oleate/water. [**235**] J. W. McBain, R. D. Vold, M. J. Vold, J. Am. Chem. Soc. *60*, 1866 (1938), C.A. *32*, 7800[1] (1938); Phase-rule studies of soap. I. Applicability of the phase rule. [**236**] J. W. McBain, G. C. Brock, R. D. Vold, M. J. Vold, J. Am. Chem. Soc. *60*, 1870 (1938), C.A. *32*, 7800[2] (1938); II. The system sodium laurate/sodium chloride/water (cf. [**235**]). [**237**] J. W. McBain, R. D. Vold, W. T. Jameson, J. Am. Chem. Soc. *61*, 30 (1939), C.A. *33*, 2748[6] (1938); Phase-rule study of the mixed soap system sodium palmitate/sodium laurate/sodium chloride at 90°. [**238**] J. W. McBain, R. D. Vold, M. Frick, J. Phys. Chem. *44*, 1013 (1940), C.A. *35*, 5025[2] (1941); Phase-rule study of the system sodium stearate/water. [**239**] R. D. Vold, C. W. Leggett, J. W. McBain, J. Phys. Chem. *44*, 1058 (1940), C.A. *35*, 5025[3] (1941); Systems of sodium palmitate in organic liquids. [**240**] R. D. Vold, R. Reivere, J. W. McBain, J. Am. Chem. Soc. *63*, 1293 (1941), C.A. *35*, 4272[4] (1941); Phase-rule study of the system sodium myristate/water.

[**241**] R. D. Vold, J. W. McBain, J. Am. Chem. Soc. *63*, 1296 (1941), C.A. *35*, 4272[6] (1941); Solubility curve of sodium desoxycholate in water. [**242**] J. W. McBain, M. J. Vold, J. L. Porter, Ind. Eng. Chem. *33*, 1049 (1941), C.A. *35*, 5776[7] (1941); Phase study of commercial soap and water. [**243**] J. W. McBain, W. W. Lee, Ind. Eng. Chem. *35*, 917 (1943), C.A. *37*, 5608[6] (1943); Kettle wax in the ternary system soap/water/salt. [**244**] J. W. McBain, W. W. Lee, Oil and Soap *20*, 17 (1943), C.A. *37*, 1887[1] (1943); Vapor pressure data and phase diagrams of some concentrated soap/water systems above room temperature. [**245**] S. Ross, J. W. McBain, J. Am. Chem. Soc. *68*, 296 (1946), C.A. *40*, 2716[5] (1946); Diffraction of X-rays by aqueous solutions of hexanolamine oleate. [**246**] E. Gonick, J. W. McBain, J. Am. Chem. Soc. *68*, 683 (1946), C.A. *40*, 3669[8] (1946); Isotropic and anisotropic liquid phases in the system hexanolamine oleate/water. [**247**] W. McBain, S. S. Marsden, Jr., J. Chem. Phys. *15*, 211 (1947), C.A. *41*, 3344i (1947); Nonionic detergents as association colloids giving long X-ray spacing in aqueous solution. [**248**] G. H. Smith, J. W. McBain, J. Phys. Colloid Chem. *51*, 1189 (1947), C.A. *42*, 23a (1948); Phase behavior of sodium stearate in anhydrous organic solvents. [**249**] S. S. Marsden, Jr., J. W. McBain, J. Phys. Colloid Chem. *52*, 110 (1948), C.A. *42*, 4022f (1948); Aqueous systems of nonionic detergents as studied by X-ray diffraction. [**250**] S. S. Marsden, Jr.,

J. W. McBain, J. Chem. Phys. *16*, 633 (1948), C.A. *42*, 6597b (1948); Oriented X-ray diffraction patterns produced by hydrous liquid crystals.
[**251**] J. W. McBain, S. S. Marsden, Jr., Acta Cryst. *1*, 270 (1948), C.A. *43*, 3289e (1949); Structural types of aqueous systems of surface-active substances and their X-ray diffraction characteristics. [**252**] S. S. Marsden, Jr., J. W. McBain, J. Am. Chem. Soc. *70*, 1973 (1948), C.A. *42*, 8601a (1948); X-ray diffraction in aqueous systems of dodecyl sulfonic acid. [**253**] W. Philippoff, J. W. McBain, Nature *164*, 885 (1949), C.A. *44*, 2824d (1950); Expansion of the lamellar crystal lattice of aerosol OT on the addition of water. [**254**] M. P. McDonald, Advan. Chem. Soc. *63*, 125 (1967), C.A. *67*, 6235k (1967); Molecular association in mono- and dihydric alcohol and alcohol/water systems. [**254a**] M. P. McDonald, W. E. Peel, J. Chem. Soc., Faraday Trans. I *72*, 2274 (1976), C.A. *86*, 47677s (1977); Solid and lyotropic phases in the sodium dodecyl sulfate/hexadecanoic acid/water system. [**255**] K. MacLennan, J. Soc. Chem. Ind. *42*, 393T (1923), C.A. *18*, 337 (1924); Microscopic structure of soap. [**256**] W. U. Malik, A. K. Jain, J. Electroanal. Chem. *14*, 37 (1967), C.A. *66*, 77253e (1967); Electrometric determination of the cmc of soap solutions. [**257**] W. U. Malik, A. K. Jain, J. Electroanal. Chim. Interfacial Electrochem. *16*, 442 (1968), C.A. *68*, 41333h (1968); Determination of the cmc of cationic surfactants from counter-ion activity measurements. [**258**] L. Mandell, Finska Kemistsamfundets Medd. *72*, 49 (1963), C.A. *60*, 3517g (1964); Mesophases in association colloid systems. [**259**] L. Mandell, Proc. Scand. Symp. Surface Activ., 2nd, Stockholm *1964*, 185, C.A. *66*, 32350t (1967); Phase equilibria in aqueous three-component systems of amphiphilic substances. [**260**] L. Mandell, Nord. Symp. Graensefladekemi, Fortryk Foredrag, 3rd *1967*, O, 4 pp., C.A. *74*, 35208d (1971); Phase equilibria in ternary systems containing surface-active substances.
[**261**] Z. N. Markina, E. V. Rybakova, A. V. Chinnikova, Kolloid. Zh. *30*, 75 (1968), C.A. *69*, 69919y (1968); The change of intramicellar solubility of hydrocarbons depending on the concentration of sodium oleate aqueous solutions at different temperatures. [**262**] N. M. Melankholin, Optica i Spektroskopiya *15*, 781 (1963), C.A. *60*, 6354a (1964); Absorption spectra of liquid crystals of thiazine dyes. [**263**] N. M. Melankholin, Rost Kristallov, Akad. Nauk SSSR, Inst. Kristallogr. *4*, 61 (1964), C.A. *61*, 15449h (1964); Type of growth and properties of liquid crystals of thiazine dyes. [**264**] P. F. Mijnlieff, R. Ditmarsch, Nature *208*, 889 (1965), C.A. *64*, 5791b (1966); Rate of micelle formation of sodium alkyl sulfates in water. [**265**] C. A. Miller, Adv. Chem. Ser. *152*, 85 (1976), C.A. *85*, 85591w (1976); Dynamic phenomena in lyophases. Brief review. [**266**] R. K. Mishra, N. K. Ropen, R. S. Tyagi, Liq. Cryst. Ord. Fluids *2*, 743, 759 (1974), C.A. *86*, 198205n (1977); Quantum chemical evaluation of intermolecular forces in a compound producing lyophase. [**267**] A. Mlodziejowski, Z. Physik *20*, 317 (1923), C.A. *18*, 1218^5 (1924); Bildung von flüssigen Kristallen in den Gemischen von Cholesterin und Cetylalkohol. [**268**] A. Mlodziejowski, Z. physik. Chem. *135*, 129 (1928), C.A. *23*, 1793^5 (1929); Dissoziation der flüssigen Kristalle. [**269**] R. A. Moss, W. L. Sunshine, J. Org. Chem. *35*, 3581 (1970), C.A. *73*, 113307f (1970); cmc of optically active and racemic 2-octylammonium and 2-octyltrimethylammonium ions. [**270**] P. Mukerjee, P. Kapauan, H. G. Meyer, J. Phys. Chem. *70*, 783 (1966), C.A. *64*, 11905b (1966); Micelle formation and hydrophobic bonding in deuterium oxide.
[**271**] P. Mukerjee, Kolloid-Z. Z. Polym. *236*, 76 (1970), C.A. *72*, 136782a (1970); Odd-even alternation in the chain length variation of micellar properties. Evidence of some solid-like character of the micelle core. [**272**] B. A. Mulley, A. D. Metcalf, J. Colloid Sci. *19*, 501 (1964), C.A. *61*, 9685e (1964); Phase equilibria in binary and ternary systems containing nonionic surface-active agents. [**272a**] R. M. A. Mustafa, F. H. Al-Jawad, M. Al-Adhami, J. Fac. Med., Baghdad *18*, 84 (1976), C.A. *87*, 12263x (1977); Phase equilibrium diagram of sodium lauryl sulfate/amyl alcohol/water system. [**273**] Y. Nakagawa, T. Noma, H. Mera, Ger. Offen. 2,530,875 (10. 7. 1974), C.A. *85*, 34555e (1976); Anisotropic liquids from aromatic polyamides. [**274**] V. P. Nazin, Uch. Zap., Ivanov. Gos. Pedagog. Inst. No. 99, 189 (1972), C.A. *78*, 129308w (1973); Optical studies of a liquid crystal layer of 2-ethoxy-6,9-diaminoacridine lactate in the presence of different chemical substances. [**275**] S. Nishikawa, J. Colloid Interface Sci. *45*, 259 (1973), C.A. *80*, 7886d (1974); Ultrasonic absorption in aqueous solutions of *n*-octylamine. Structural relaxation in mesophase. [**276**] J. R. Nixon, B. P. S. Chawla, J. Pharm. Pharmacol. *21*, 79 (1969), C.A. *70*, 60781k (1969); Solubilization and rheology of the system ascorbic acid/water/polysorbate 80: temperature effects. [**276a**] H. Noda, M. Yoshimaru, H. Hayashi, K. Hirakawa, Kyushu Daigaku Kogaku Shuho *48*, 239 (1975), C.A. *86*, 163983n (1977); Electric conduction in an aqueous lyophase. Effect of alkali metal ions. [**277**] E. Oltus, L. Kunial, D. Eliasova, B. Alince, Celuloza Hirtie (Bucharest) *14*, 297 (1965), C.A. *64*, 5287q (1966); Properties of decrystallized pulps. [**278**] Wo. Ostwald, H. Erbring, Kolloidchem. Beihefte *31*, 291 (1930), C.A. *25*, 3550 (1931); Flüssig-flüssige Entmischung der Natronseifen

höherer Fettsäuren mit Natriumsulfat und die Beziehungen dieser Systeme zur Phasenregel. [**279**] Wo. Ostwald, Z. Krist. *79*, 222 (1931), C.A. *26*, 2361[9] (1932); Mesomorphe und kolloidale Systeme. [**280**] J. T. G. Overbeek, J. Phys. Chem. *64*, 1178 (1960), C.A. *55*, 6992f (1961); Black soap films. [**281**] S. P. Palit, V. A. Moghe, B. Biswas, Trans. Faraday Soc. *55*, 463 (1959), C.A. *54*, 1979h (1960); Solubilization of water by cationic detergents. [**282**] A. A. Panfilova, V. D. Kalmykova, V. A. Platonov, A. V. Volokhina, V. G. Kulichikhin, S. P. Papkov, Khim. Volokna *1975*, 10, C.A. *83*, 116668s (1975); Physicochemical and technological properties of poly-*p*-benzamide solutions. [**283**] S. P. Papkov, V. G. Kulichikhin, A. Y. Malkin, V. D. Kalmykova, A. V. Volokhina, L. I. Gudim, Vysokomol. Soedin. B *14*, 244 (1972), C.A. *77*, 48940y (1970); Behavior of poly(*p*-benzamide) solutions in the low shear stress region. [**284**] S. P. Papkov, S. I. Banduryan, M. M. Iovleva, Vysokomol. Soedin. B *15*, 370 (1973), C.A. *79*, 66948c (1973); Structural characteristics of precipitates obtained during the separation of poly(*p*-benzamide) from solutions. [**285**] S. P. Papkov, V. G. Kulichikhin, V. D. Kalmykova, A. Y. Malkin, J. Polym. Sci., Polym. Phys. Ed. *12*, 1753 (1974), C.A. *81*, 170128n (1974); Rheological properties of anisotropic poly(benzamide) solutions. [**285a**] M. M. Iovleva, S. P. Papkov, L. P. Mil'kova, V. D. Kalmykova, A. V. Volokhina, G. I. Kudryavtsev, Vysokomol. Soedin. B *18*, 830 (1976), C.A. *87*, 53694g (1977); Temperature-concentration limits of the liquid crystals of poly(*p*-benzamide). [**286**] O. Parodi, J. Phys. (Paris), Lett. *37*, 295 (1976), C.A. *85*, 200787d (1976); Lyophases as defects in binary mixtures. [**287**] V. A. Parsegian, Trans. Faraday Soc. *62*, 848 (1966), C.A. *64*, 13445h (1966); Theory of liquid-crystal phase transition in lipid/water systems. [**288**] L. Paul, Z. deut. Öl-Fett-Ind. *40*, 425, 455, 469, 503, 521, 550, 584 (1920), C.A. *14*, 3809[7] (1920); Flüssige Harzseifen-krystalle. [**289**] J. T. Pearson, J. M. Smith, J. Pharm. Pharmacol. *26*, Suppl., 123P (1974), C.A. *82*, 169552k (1975); Effect of hydrotropic salts on the stability of liquid crystals. [**290**] W. Philippoff, Disc. Faraday Soc. *11*, 96, 147 (1951), C.A. *46*, 8468f (1952); The micelle and swollen micelle. Soap micelles.

[**291**] V. A. Platonov, O. A. Khanchich, T. A. Belousova, V. G. Kulichikhin, Vysokomol. Soedin. B *17*, 726 (1975), C.A. *84*, 31668n (1976); Formation of domains in anisotropic solutions of poly-*p*-benz-amide induced by a mechanical field. [**292**] V. A. Platonov, G. D. Litovchenko, L. P. Mil'kova, M. V. Shablygin, V. G. Kulichikhin, S. P. Papkov, Khim. Volokna *1975*, 70, C.A. *84*, 17982d (1976); Action of a magnetic field on anisotropic solutions of poly-*p*-benzamide. [**293**] A. A. Panfilova, V. A. Platonov, V. G. Kulichikhin, V. D. Kalmykova, S. P. Papkov, Koll. Zh. *37*, 210 (1975), C.A. *82*, 140621w (1975); Change in the surface tension of poly-*p*-benzamide solutions in the transition to the liquid-crystal state. [**294**] V. A. Platonov, G. D. Litovchenko, T. A. Belousova, L. P. Mil'kova, M. V. Shablygin, V. G. Kulichikhin, S. P. Papkov, Vysokomol. Soedin. A *18*, 221 (1976), C.A. *84*, 136170x (1976); Action of a magnetic field on anisotropic solutions of poly-*p*-benzamide. [**294a**] V. G. Kulichikhin, V. A. Platonov, L. P. Braverman, T. A. Belousova, V. F. Polyakov, M. V. Shablygin, A. V. Volokhina, A. Ya. Malkin, S. P. Papkov, Vysokomol. Soedin. A *18*, 2656 (1976); C.A. *86*, 99247w (1977); Structural-orientational processes in poly(*p*-benzamide) lyophases. [**295**] R. J. Prime, O. E. Hileman, Jr., Mol. Cryst. Liq. Cryst. *28*, 355 (1974), C.A. *83*, 51115n (1975); Energy barrier to mesophase transformation in lyotropic systems. [**296**] G. E. Prozorova, A. V. Pavlov, M. M. Iovleva, R. V. Antipova, V. D. Kalmykova, S. P. Papkov, Vysokomol. Soedin. B *18*, 111 (1976), C.A. *84*, 151146n (1976); Study of the molecular-mass characteristics of poly-*p*-benzamide by ultracentrifuging. [**297**] A. W. Ralston, E. J. Hoffman, C. W. Hoerr, W. M. Selby, J. Am. Chem. Soc. *63*, 1598 (1941), C.A. *35*, 5017[3] (1941); Behavior of the hydrochlorides of dodecylamine and octadecylamine in water. [**298**] A. W. Ralston, C. W. Hoerr, E. J. Hoffman, J. Am. Chem. Soc. *64*, 1516 (1942), C.A. *36*, 5416 (1942); The system octylamine/dodecylamine and octadecylamine/water. [**299**] A. Ray, Nature *231*, 313 (1971), C.A. *75*, 50707e (1971); Solvophobic interactions and micelle formation in structure forming nonaqueous solvents. [**299a**] D. M. Chen, F. Y. Fujiwara, L. W. Reeves, Can. J. Chem. *55*, 2396 (1977), C.A. *87*, 125588a (1977); Studies of the behavior in magnetic fields of lyophases with respect to electrolyte additions. [**300**] D. F. Riley, G. Oster, Disc. Faraday Soc. *11*, 107, 151 (1951), C.A. *46*, 8464i (1952); Theoretical and experimental studies of X-ray and light scattering by colloidal and macromolecular systems.

[**301**] K. Roberts, C. Axberg, R. Osterlund, H. Saito, Nature *255*, 53 (1975), C.A. *83*, 33353j (1975); Liquid crystals as lamellar reservoirs reduce thinning by drainage. [**302**] K. Roberts, R. Osterlund, C. Axberg, Tappi *59*, 156 (1976), C.A. *85*, 34893p (1976); Liquid crystals in systems of rosin and fatty acids: implications for tall oil recovery. [**303**] D. C. Robins, I. L. Thamas, J. Colloid Interface Sci. *26*, 422 (1968), C.A. *69*, 13142x (1968); Effect of counterions on micellar properties of 2-(dodecyl-amino)ethanol salts. III. Solubilization studies. [**304**] R. E. Rosenfeld, S. G. Frank, J. Pharm. Sci. *62*, 1194 (1973), C.A. *79*, 70143d (1973); Mesomorphism of dioctyl sodium sulfosuccinate in hydrocarbon

solvents. [**305**] F. B. Rosevear, J. Soc. Cosmet. Chem. *19*, 581 (1968), C.A. *69*, 81201r (1968); Mesomorphism of surfactant compositions. [**306**] E. I. Ryumtsev, I. N. Shtennikova, N. V. Pogodina, G. F. Kolbina, I. I. Konstantinova, Y. B. Amerik, Vysokomol. Soedin. A *18*, 439 (1976), C.A. *84*, 136214q (1976); Intramolecular liquid crystal structure in solutions of poly(nonyloxybenzamidostyrene). [**307**] C. Sadron, Atti Congr. Intern. Materie Plastiche *14*, 178 (1962), C.A. *59*, 14122b (1963); Les chaînes moléculaires hétérosolubles: quelques propriétés. [**308**] C. Sadron, Pure Appl. Chem. *4*, 347 (1962), C.A. *57*, 8717g (1962); Les polymères organisés. [**309**] C. Sadron, Angew. Chem. *75*, 472 (1963), C.A. *59*, 6523c (1963); Ungleichmäßig lösliche Makromoleküle. Heterogele und Heteropolymere. [**310**] C. Sadron, B. Gallot, Makromol. Chem. *164*, 301 (1973), C.A. *78*, 98206e (1973); Heterophases in block copolymer/solvent systems in the liquid and in the solid state.

[**311**] H. Sandquist, Ber. *48*, 2054 (1915), C.A. *10*, 715[9] (1916); Eine anisotrope Wasserlösung. [**312**] H. Sandquist, Kolloid-Z. *19*, 113 (1916), C.A. *11*, 1350 (1917); Eine elektrolyt-kolloid-kristallinische Flüssigkeit. [**313**] H. Sandquist, Arkiv för Kemi, Min. och Geol. *6*, Nr. 9, 38 pp. (1916); Anisotropie, Viskosität und Leitvermögen von 10-Bromphenanthren-3-oder-6-sulfonsäure. [**314**] K. Sato, K. Tamura, M. Okada, Hiroshima Daigaku Suichikusan Gakubu Kiyo *15*, 93 (1976), C.A. *85*, 169965j (1976); Optical studies of lyophases in the potassium caprate/water system. [**314a**] A. Saupe, J. Colloid Interface Sci. *58*, 549 (1977), C.A. *86*, 163677r (1977); Textures, deformations, and structural order of liquid crystals. [**315**] G. Scheibe, L. Kandler, H. Ecker, Naturwissenschaften *25*, 75 (1937), C.A. *31*, 3785[7] (1937); Polymerisation und polymere Adsorption als Ursache neuartiger Absorptionsbanden von organischen Farbstoffen. [**316**] G. Scheibe, L. Kandler, Naturwissenschaften *26*, 412 (1938), C.A. *32*, 8879[5] (1938); Anisotropie organischer Farbstoffmoleküle. Nebenvalenzbindung als Energieüberträger. [**317**] E. Segerman, Proc. 2nd Scand. Symp. Surface Activ., Stockholm 1964, 157, C.A. *66*, 23142z (1967); Ordered chain models for liquid-crystalline and micellar systems. [**318**] A. G. Semakova, Nauch. Tr., Kursk. Gos. Pedagog. Inst. No. 7, 198 (1972), C.A. *80*, 31294e (1974); Liquid crystals in mixtures of cholesterol with glycerol. [**319**] M. V. Shablygin, T. A. Belousova, V. G. Kulichikhin, V. A. Platonov, V. D. Kalmykova, S. P. Papkov, Vysokomol. Soedin. A *18*, 942 (1976), C.A. *85*, 33574y (1976); Spectroscopic method for determining the liquid crystal phase fraction in anisotropic solutions of polymers. [**320**] S. E. Sheppard, Science *93*, 42 (1941), C.A. *35*, 1699[6] (1941); Structure of the mesophase of certain cyanine dyes.

[**321**] V. P. Shibaev, R. V. Tal'roze, B. S. Petrukhin, N. A. Plate, Vysokomol. Soedin. B *13*, 4 and A *13*, 493 (1971), C.A. *74*, 112556v (1971); Liquid-crystal state of gels of poly(hexadecyl acrylate). [**322**] M. Shinitzky, A. C. Dianoux, C. Gitler, G. Weber, Biochem. *10*, 2106 (1971); Microviscosity and order in the hydrocarbon region of micelles and membranes determined with fluorescent probes. I. Synthetic micelles. [**323**] M. Shinitzky, Isr. J. Chem. *12*, 879 (1974), C.A. *82*, 103590q (1975); Fluidity and order in the hydrocarbon/water interface of synthetic and biological micelles as determined by fluorescence polarization. [**324**] K. Shinoda, H. Saito, J. Colloid Interface Sci. *26*, 70 (1968), C.A. *68*, 51180v (1968); Effect of temperature on the phase equilibria and the types of dispersions of the ternary system water/cyclohexane/nonionic surfactant. [**325**] V. Luzzati, H. Mustacchi, A. Skoulios, Nature *180*, 600 (1957), C.A. *52*, 4288f (1958); Structure of the mesophases of the soap/water system. Middle soap and neat soap. [**326**] V. Luzzati, H. Mustacchi, A. Skoulios, Disc. Faraday Soc. *25*, 43 (1958), C.A. *53*, 10901i (1959); Structure of the mesophases of soap/water systems. [**327**] A. Skoulios, G. Finaz, J. Parrod, Compt. rend. *251*, 739 (1960), C.A. *55*, 6024f (1961); Obtention de gels mésomorphes dans les mélanges de copolymères séquencés styrolène-oxyde d'éthylène avec différents solvants. [**328**] A. Skoulios, Acta Cryst. *14*, 419 (1961), C.A. *55*, 14019h (1961); Les phases mésomorphes des systèmes stéarate de sodium/tétradécane et stéarate de sodium/cyclohexane entre 120 et 160°C. Note préliminaire. [**329**] P. Spegt, A. Skoulios, V. Luzzati, Acta Cryst. *14*, 866 (1961), C.A. *55*, 26612b (1961); Description des phases mésomorphes d'un système savon/eau/hydrocarbure. [**330**] A. Skoulios, G. Finaz, Compt. rend. *252*, 3467 (1961), C.A. *55*, 22995e (1961); Influence de la nature du solvant sur la structure des gels mésomorphes d'un copolymère séquencé styrolène/oxyde d'éthylène.

[**331**] A. Skoulios, G. Finaz, J. Chim. Phys. *59*, 473 (1962), C.A. *57*, 12701f (1962); Caractère amphipatique et phases mésomorphes des copolymères séquencés styrolène/oxyde d'éthylène. [**332**] G. Tsouladze, A. Skoulios, J. Chim. Phys. *60*, 626 (1963), C.A. *59*, 7713h (1963); Caractère amphipatique et phases mésomorphes des copolymères séquencés oxyde de propylène/oxyde d'éthylène. [**333**] A. Skoulios, G. Tsouladze, E. Franta, J. Polymer Sci. C *4*, 507 (1963), C.A. *60*, 10808g (1964); Configuration des chaînes macromoléculaires des copolymères séquencés en solution concentrée. Copolymères séquencés polystyrolène/polyoxyéthylène et polyoxypropylène/polyoxyéthylène en solution dans des solvants préfé-

rentiels de l'une ou autre des séquencés. [**334**] J. M. Vincent, A. Skoulios, J. Am. Oil Chem. Soc. *40*, 20 (1963), C.A. *58*, 8147c (1963); Kettle wax in the sodium palmitate/water electrolyte system at 90°C. [**335**] J. M. Vincent, A. Skoulios, C *258*, 1229 (1964), C.A. *60*, 12239e (1964); Structure des gels de savon de potassium à 24°C. [**336**] E. Franta, A. Skoulios, P. Rempp, H. Benoit, Makromol. Chem. *87*, 271 (1965), C.A. *63*, 18363h (1965); Structure lamellaire des gels des copolymères séquencés polystyrolène/polyoxyéthylène. [**337**] P. Spegt, A. Skoulios, J. Chim. Phys. *62*, 377 (1965), C.A. *63*, 3172e (1965); Structure à cylindres des phases colloidales des savons alcalino-terreux a température élevée. Syncristallisation de différents savons et gonflement par un solvant organique. [**338**] B. Gilg, J. François, A. Skoulios, Kolloid-Z. Z. Polym. *205*, 139 (1965), C.A. *64*, 7402a (1966); Interactions électriques dans les phases mésomorphes des système amphiphile/eau. Rôle de la polarité du milieu. [**339**] B. Gallot, A. Skoulios, Compt. rend. *260*, 3033 (1965), C.A. *62*, 14935c (1965); Polymorphisme des groupes polaires dans les phases mésomorphes des savons de sodium. [**340**] J. M. Vincent, A. Skoulios, Acta Cryst. *20*, 432 (1966), C.A. *64*, 10439g (1966); Gel et coagel. I. Identification. Localisation dans un diagramme de phases et détermination de la structure du gel dans le cas de stéarate de potassium.

[**341**] J. M. Vincent, A. Skoulios, Acta Cryst. *20*, 441 (1966), C.A. *64*, 10440a (1966); II. Étude comparative de quelques amphiphiles (cf. [340]). [**342**] J. M. Vincent, A. Skoulios, Acta Cryst. *20*, 447 (1966), C.A. *64*, 10440b (1966); III. Étude du gel dans le mélange équimoléculaire stéarate de potassium/*n*-octadécanol (cf. [340]). [**343**] B. Gallot, A. Skoulios, Kolloid-Z. Z. Polym. *208*, 37 (1966), C.A. *64*, 18459f (1966); Interactions électriques dans les phases mésomorphes des systèms amphiphile/eau. Rôle de la teneur en eau, de la longueur de la chaîne paraffinique, de la nature du cation et de la température. [**344**] J. François, B. Gilg, P. A. Spegt, A. E. Skoulios, J. Colloid Interface Sci. *21*, 293 (1966), C.A. *64*, 16683a (1966); Étude de la solubilisation par les gels aqueux d'amphiphiles. [**345**] A. Skoulios, Advan. Colloid Interface Sci. *1*, 79 (1967), C.A. *66*, 98773q (1967); Structure des solutions aqueuses concentrées de savon. [**346**] J. François, A. Skoulios, Kolloid-Z. Z. Polym. *219*, 144 (1967), C.A. *68*, 6692c (1968); Propriétés électriques des gels eau/amphiphile. I. Technique expérimentale et résultats préliminaires de mesures de conductivité. [**347**] A. Schmitt, R. Varoqui, A. Skoulios, C. R. Acad. Sci., Sér. C *268*, 1469 (1969), C.A. *71*, 30878a (1969); Obtention de phases mésomorphes dans les solutions aqueuses concentrées d'un polyélectrolyte amphipathique. [**348**] J. François, A. Skoulios, C. R. Acad. Sci., Sér. C, *269*, 61 (1969), C.A. *71*, 95291y (1969); Influence de la texture des gels mésomorphes des systèmes savon/eau sur leur conductivité électrique. [**349**] P. Grosius, Y. Gallot, A. Skoulios, Eur. Polym. J. *6*, 355 (1970), C.A. *72*, 122081j (1970); Synthèse et propriétés des copolymères séquencés styrène/vinylpyridine. Étude des structures mésomorphes obtenues en milieu solvant préférentiel. [**350**] P. Grosius, Y. Gallot, A. Skoulios, Makromol. Chem. *136*, 191 (1970), C.A. *73*, 77649h (1970); Copolymères séquencés polyvinyl-2-pyridine/polyvinyl-4-pyridine. Synthèse, caractérisation et étude des phases méso-morphes.

[**351**] H. Ailhaud, Y. Gallot, A. Skoulios, Kolloid-Z. Z. Polym. *248*, 889 (1971); C.A. *76*, 127554h (1972); Structure of mesophases observed with block copolymers of different alkyl methacrylates. [**352**] A. Skoulios, P. Helffer, Y. Gallot, J. Selb, Makromol. Chem. *148*, 305 (1971), C.A. *76*, 25729b (1972); Solubilization and chain conformation in a block copolymer system. [**353**] H. Ailhaud, Y. Gallot, A. Skoulios, Makromol. Chem. *151*, 1 (1972), C.A. *76*, 113782u (1972); Caractère amphipathique et phases mésomorphes des copolymères biséquencés de différents méthacrylates d'alcoyle. [**354**] A. Skoulios, Adv. Liq. Cryst. *1*, 169 (1975), C.A. *84*, 17749h (1976); Mesomorphism of block copolymers. [**354a**] M. Arpin, C. Strazielle, A. Skoulios, J. Phys. (Paris) *38*, 307 (1977), C.A. *86*, 121951v (1977); Étude de la phase n d'un poly (téréphtalamide de paraphénylène). [**355**] E. L. Smith, J. Phys. Chem. *36*, 2455 (1932), C.A. *26*, 5477 (1932); Solvent properties of soap solutions. The system sodium oleate/sodium chloride/water/ethylacetate. [**356**] K. Solc, Order in polymer solutions (Gordon and Breach: London). 1976. 320 pp., C.A. *84*, 180852v (1976). [**356a**] J. Stenius, J. B. Rosenholm, M. R. Hakala, Colloid Interface Sci., (Proc. Int. Conf.) 50th, *1976*, 2, 397, C.A. *87*, 93748j (1977); Enthalpies of lyophases. Lamellar and hexagonal phases in the system *n*-pentanol/sodium *n*-octanoate/water. [**356b**] A. E. Stepan'yan, E. P. Krasnov, N. V. Lukasheva, Yu. A. Tolkachev, Vysokomol. Soedin. A *19*, 628 (1977), C.A. *86*, 171976c (1977); Conformational and structural features of poly(*p*-phenyleneterephthalamide). [**357**] D. Stigter, J. Phys. Chem. *70*, 1323 (1966), C.A. *65*, 4700b (1966); Intrinsic viscosity and flexibility of rodlike detergent micelles. [**358**] W. Stoeckenius, J. Cell Biol. *12*, 221 (1962), C.A. *57*, 4135c (1962); Electron microscopical observations on liquid-crystals in lipid/water systems. [**359**] S. Stoilov, S. S. Duhin, Kolloid. Zh. *32*, 757 (1970), C.A. *74*, 6789p (1971); Theory of polarization of the double layer

and its effect on electrooptical and electrokinetic phenomena and on the dielectric constant of disperse systems. Steady-state polarizability of highly elongated colloidal particles oriented parallel to the field. [360] J. P. Straley, Mol. Cryst. Liq. Cryst. *22*, 333 (1973), C.A. *79*, 140468q (1973); Gas of long rods as a model for lyophases.

[361] F. H. Stross, S. T. Abrams, J. Am. Chem. Soc. *73*, 2825 (1951), C.A. *45*, 6915d (1951); Thermal analysis of the system sodium stearate/cetane. [362] T. Tachibana, M. Tanaka, Bull. Chem. Soc. Jap. *44*, 1166 (1971), C.A. *75*, 54878j (1971); Electro-optic effect in lyophases. [363] B. Tamamushi, M. Matsumoto, Liq. Cryst. Ord. Fluids *2*, 711 (1974), C.A. *86*, 177569u (1977); Rheological properties of thermotropic and lyotropic mesophases formed by ammonium laurate. [364] B. Tamamushi, Prog. Colloid Polym. Sci. *60*, 152 (1976), C.A. *85*, 167137s (1976); Relation between mesomorphic and colloidal systems. [365] B. Tamamushi, Y. Kodaira, M. Matsumura, Colloid Polym. Sci. *254*, 571 (1976), C.A. *85*, 85775j (1976); Thermotropic and lyotropic mesophases formed by ammonium dodecanoate. [365a] B. Tamamushi, Pure Appl. Chem. *48*, 441 (1976), C.A. *87*, 73753d (1977); Colloid and surface aspects of mesophases. [366] H. Thiele, Naturwissenschaften *34*, 123 (1947), C.A. *43*, 4539e (1949); Richtwirkung von Ionen auf anisotrope Kolloide. Ionotropie. [367] H. Thiele, Kolloid-Z. *115*, 167 (1949), C.A. *44*, 9214b (1950); Ionotropie. [368] P. A. Thiessen, R. Spychalski, Z. physik. Chem. A *156*, 435 (1931). C.A. *26*, 2908 (1932); Anordnung der Moleküle in Seifenmicellen. [369] G. J. T. Tiddy, J. Chem. Soc., Faraday Trans. 1, *68*, 608 (1972), C.A. *76*, 117960y (1972); Fluorocarbon surfactant/water mesophases. I. Study of the ammonium perfluorooctanoate/water system by optical microscopy and low angle X-ray diffraction. [370] G. J. T. Tiddy, B. A. Wheeler, J. Colloid Interface Sci. *47*, 59 (1974), C.A. *81*, 17948k (1974); V. Phase diagram and X-ray diffraction study of the ammonium perfluorooctanoate/octanol/water system (cf. [369]).

[371] F. Tokiwa, K. Ohki, J. Phys. Chem. *71*, 1343 (1967), C.A. *66*, 108645r (1967); Micellar properties of a series of sodium dodecylpolyoxyethylene sulfates from hydrodynamic data. [372] F. Tokiwa, I. Kokubo, J. Colloid Interface Sci. *24*, 223 (1967), C.A. *68*, 108400y (1968); Phase behavior of the ternary systems alkylbenzene sulfonate/octanol/water. [373] F. Tokiwa, K. Ohki, Kolloid-Z. Z. Polym. *223*, 138 (1968), C.A. *69*, 70056w (1968); Thermodynamic studies of the micellization and micellar solubilization of a nonionic-cationic surfactant. [374] F. Tokiwa, J. Colloid Interface Sci. *28*, 145 (1968), C.A. *69*, 80667k (1968); Solubilization behavior of mixed surfactant micelles in connection with their zeta potentials. [375] F. Tokiwa, K. Ohki, Bull. Chem. Soc. Jap. *41*, 2828 (1968), C.A. *70*, 91122a (1969); Structures of an electric double layer around a surfactant micelle. [376] F. Tokiwa, K. Ohki, I. Kokubo, Bull. Chem. Soc. Jap. *42*, 575 (1969), C.A. *70*, 89075u (1969); Gel filtration of mixtures of surfactants on Sephadex. II. Estimation of micellar molecular weights of mixed surfactants. [377] F. Tokiwa, K. Ohki, Bull. Chem. Soc. Jap. *42*, 1216 (1969), C.A. *71*, 31642f (1969); Electrophoretic behavior of micelles of a polyether sulfate type surfactant. [378] F. Tokiwa, K. Ohki, Kolloid-Z. Z. Polym. *230*, 251 (1969), C.A. *71*, 6832k (1969); Paper electrophoretic properties of surfactants in aqueous systems. [379] F. Tokiwa, N. Moriyama, H. Sugihara, Nippon Kagaku Zasshi *90*, 673 (1969), C.A. *71*, 105666r (1969); Colloidal properties of mixed solutions of anionic and cationic surfactants. [380] V. B. Tolstoguzov, Colloid Polym. Sci. *253*, 109 (1975), C.A. *82*, 145638s (1975); Anisotropic gels of a laminated structure.

[381] I. A. Trapeznikov, Acta Physicochim. URSS *19*, 553 (1944), C.A. *39*, 2920^1 (1945); Temperature dependence of monolayer pressure as a method of investigating hydrates of higher aliphatic compounds. [382] A. Trapeznikov, Acta Physicochim. URSS *20*, 589 (1945), C.A. *40*, 4595^5 (1946); New phase changes in condensed monolayers and bulk hydrates of the higher alcohols. II. [383] A. Trapeznikov, Compt. rend. acad. sci. URSS *47*, 275 (1945), Doklady Akad. Nauk SSSR *1945*, 277, C.A. *40*, 3959^9 (1946); Interaction of hydrated fatty acid crystals with electrolytes and their polymorphism and equilibrium with monolayers. [384] A. Trapeznikov, Compt. rend. acad. sci. URSS *47*, 417 (1945), Doklady Akad. Nauk SSSR *47*, 435 (1945); Equilibrium between two- and three-dimensional hydrates of the higher alcohols and new phase transformations. [385] J. P. Treguier, I. Lo, M. Seiller, F. Puisieux, Pharm. Acta Helv. *50*, 421 (1975), C.A. *84*, 169593r (1976); Emulsions et diagrammes eau/surfactif/huile. Étude d'un système eau/Brijs 92 et 96/huile de vaseline. Influence de l'hydrophilie du surfactif. [386] N. Trzebowski, H. Kasch, Freiberger Forschungsh. A *251*, 143 (1962), C.A. *59*, 7757c (1963); Anwendung der IR-Spektroskopie bei der Aufklärung der Struktur von Seifen. [387] Y. M. Tsikanchuk, R. G. Makitra, Y. I. Fedishin, Dopov. Akad. Nauk Ukr. RSR. B *1976*, 154, C.A. *84*, 163998e (1976); Optic anisotropy in binary systems of hexachloroethane with some amides. [388] T. Tsumori, N. Nishikido, Y. Moroi, R. Matuura, Mem. Fac. Sci., Kyushu Univ. C *9*, 57 (1974), C.A. *81*, 96622a (1974); Phase

equilibrium of nonionic surface active agent/water systems. [389] V. N. Tsvetkov, S. Y. Lyubina, V. E. Bychkova, I. A. Strelina, Vysokomol. Soedin. *8*, 846 (1966), C.A. *65*, 5544e (1966); Birefringence and viscosity of solutions of poly(2-methyl-5-vinylpyridine). [390] V. N. Tsvetkov, E. Ryumtsev, I. I. Konstantinov, Y. B. Amerik, B. A. Krentsel, Vysokomol. Soedin. A *14*, 67 (1972), C.A. *76*, 127584t (1972); Intra- and supramolecular liquid-crystalline order in solutions of polymer molecules with anisotropically interacting chain side groups.

[391] V. N. Tsvetkov, E. I. Ryumtsev, I. N. Shtennikova, I. L. Konstaninov, Y. S. Amerik, B. A. Krentsel, Vysokomol. Soedin. A *15*, 2270 (1973), C.A. *80*, 83800m (1974); Intramolecular liquid-crystal order and electrooptical properties of alkoxybenzoic acid polymer molecules. [391a] R. S. Tyagi, G. C. Shukla, R. K. Mishra, Stud. Biophys. *54*, 187 (1976), C.A. *85*, 29756s (1976); Stability of the middle phase of ionic lipids. [391b] R. S. Tyagi, G. G. Shukla, R. K. Mishra, Stud. Biophys. *55*, 123 (1976), C.A. *85*, 73875d (1976); Lyophase transition in ionic lipids. [392] G. van Iterson, Jr., Proc. Acad. Sci. Amsterdam *37*, 367 (1934), C.A. *29*, 17[5] (1935); Remarkable properties of a birefringent liquid. [393] N. V. Vasil'eva, V. A. Platonov, V. G. Kulichikhin, S. P. Papkov, S. G. Efimova, Khim. Volokna *1975*, 71, C.A. *84*, 17983e (1976); Viscosity properties of anisotropic solutions of poly(*p*-phenyleneterephthalamide). [394] R. D. Vold, R. H. Ferguson, J. Am. Chem. Soc. *60*, 2066 (1938), C.A. *32*, 8903[2] (1938); Phase study of the system sodium palmitate/sodium chloride/water. [395] R. D. Vold, M. J. Vold, J. Am. Chem. Soc. *61*, 37 (1939), C.A. *33*, 2392[6] (1939); Thermodynamic behavior of liquid crystalline solutions of sodium palmitate and sodium laurate in water at 90°. [396] R. D. Vold, J. Phys. Chem. *43*, 1213 (1939), C.A. *34*, 1537[4] (1940); Phase-rule behavior of concentrated aqueous systems of a typical colloidal electrolyte/sodium oleate. [397] R. D. Vold, Soap *16*, 31, 71, 73 (1940), C.A. *34*, 4933[3] (1940); Soap phases. [398] M. J. Vold, J. Am. Chem. Soc. *63*, 1427 (1941), C.A. *35*, 4271[6] (1941); Phase behavior of dodecylsulfonic acid and of its alkali salts with water. [399] M. J. Vold, J. Am. Chem. Soc. *65*, 465 (1943), C.A. *37*, 2646[9] (1943); Phase behavior of lithium palmitate with water and with lithium chloride and water. [400] T. M. Doscher, R. D. Vold, J. Colloid Sci. *1*, 299 (1946), C.A. *40*, 6958[9] (1946); Phase relations in the system sodium stearate/cetane.

[401] R. D. Vold, J. M. Philipson, J. Phys. Chem. *50*, 39 (1946), C.A. *40*, 2063[1] (1946); Behavior of sodium stearate with cetane and water. [402] R. D. Vold, J. Phys. Colloid Chem. *51*, 797 (1947), C.A. *41*, 4995f (1947); Phase boundaries in concentrated systems of sodium oleate and water. [403] T. M. Doscher, R. D. Vold, J. Phys. Colloid Chem. *52*, 97 (1948), C.A. *42*, 4021c (1948); Colloidal structures in binary soap systems. [404] R. D. Vold, M. J. Heldman, J. Phys. Colloid Chem. *52*, 148 (1948), C.A. *42*, 4021f (1948); Electric conductivity of crystalline and liquid-crystalline soap/water systems. [405] R. D. Vold, M. J. Vold, J. Phys. Colloid Chem. *52*, 1424 (1948), C.A. *43*, 4081d (1949); Properties of systems of calcium stearate and cetane as deduced from X-ray diffraction patterns. [406] T. M. Doscher, R. D. Vold, J. Am. Oil Chem. Soc. *26*, 515 (1949), C.A. *44*, 828c (1950); Stability of sodium stearate gels. [407] M. J. Vold, R. D. Vold, J. Colloid Sci. *5*, 1 (1950), C.A. *44*, 6714f (1950); Phase behavior of lithium stearate in cetane and in decahydronaphthalene. [408] M. J. Vold, The Inst. Spokesman *16*, 8 (1952), C.A. *47*, 946g (1963); X-ray diffraction studies of oriented soap structures in grease-like systems. [409] M. J. Vold, R. D. Vold, Colloid Chemistry; the Science of Large Molecules, Small Particles, and Surfaces (Reinhold: New York). 1964. 118 pp., C.A. *62*, 4654a (1965). [410] P. P. von Weimarn, Z. Chem. Ind. Kolloide *3*, 166 (1908), C.A. *3*, 394[9] (1909); Der kristallinisch-flüssige Zustand als allgemeine Eigenschaft der Materie.

[411] R. S. Werbowyj, D. G. Gray, Mol. Cryst. Liq. Cryst. *34*, 97 (1976), C.A. *86*, 56985t (1977); Liquid crystalline structure in aqueous hydroxypropyl cellulose solutions. [412] P. A. Winsor, Trans. Faraday Soc. *44*, 376–398, 451–471 (1948), C.A. *43*, 465g (1949); I. Hydrotropy, solubilization, and related emulsification processes. II. Solubilization of petroleum ether with sodium secondary C_{10}-C_{18} alkyl sulfates. III. Solubilization of miscellaneous organic liquids with sodium secondary C_{10}-C_{18} alkyl sulfates. IV. Solubilization with miscellaneous amphiphiles. V. Solubilization in ethylene glycol. VI. Effect of temperature on the solubilization of liquids. VII. Solubilization of solids; freezing-point-depression effects in solubilized systems. VIII. Effect of constitution on amphiphilic properties. [413] P. A. Winsor, Trans. Faraday Soc. *46*, 762 (1950), C.A. *45*, 1843f (1951); IX. Electric conductivity and the water dispersibility of solubilized systems (cf. [412]). [414] P. A. Winsor, J. Phys. Chem. *56*, 391 (1952), C.A. *46*, 7844i (1952); Interpretation of published X-ray measurements on detergent solutions on the basis of an intermicellar equilibrium. [415] P. A. Winsor, Solvent properties of amphiphilic compounds (Butterworths: London). 1954. 207 pp., C.A. *49*, 11389b (1955). [416] P. A. Winsor, Nature *173*, 81 (1954), C.A. *48*, 5600d (1954); Electric effect in liquid-crystalline solutions containing amphiphilic salts.

[**417**] P. A. Winsor, J. Colloid Sci. *10*, 88 (1955), C.A. *49*, 7329d (1955); Structure of viscoelastic solutions of colloidal electrolytes. [**418**] P. A. Winsor, J. Colloid Sci. *10*, 101 (1955), C.A. *49*, 7332e (1955); Electrooptical effects in liquid-crystalline solutions containing amphiphilic salts. [**419**] P. A. Winsor, Chem. & Ind. *1960*, 632, C.A. *54*, 21935i (1960); Solubilization with amphiphilic compounds. [**420**] C. A. Gilchrist, J. Rogers, G. Steel, E. G. Vaal, P. A. Winsor, J. Colloid Interface Sci. *25*, 409 (1967), C.A. *69*, 30637e (1968); Constitution of aqueous liquid crystalline solutions of amphiphiles.

[**421**] J. Rogers, P. A. Winsor, Nature *216*, 477 (1967), C.A. *68*, 16845u (1968); Optically positive, isotropic, and negative lamellar liquid crystalline solutions. [**422**] P. A. Winsor, Chem. Rev. *68*, 1 (1968), C.A. *69*, 30588q (1968); Binary and multicomponent solutions of amphiphilic compounds. Solubilization and the formation, structure, and theoretical significance of liquid crystalline solutions. [**423**] J. Rogers, P. A. Winsor, Chim. Phys. Appl. Prat. Ag. Surface, C. R. Congr. Int. Deterg. 5th, *1968*, 2 (Pt. 2) 933, C.A. *74*, 46760d (1971); Shapes and structures of particles of a liquid crystalline solution phase as they separate from a second phase. [**424**] J. Rogers, P. A. Winsor, J. Colloid Interface Sci. *30*, 247 (1969), C.A. *71*, 33693x (1969); Change in the optic sign of the lamellar phase (G) in the aerosol OT/water system with composition or temperature. [**425**] D. Park, J. Rogers, R. W. Toft, P. A. Winsor, J. Colloid Interface Sci. *32*, 81 (1970), C.A. *72*, 47890s (1970); Structure of micellar solutions of ionic amphiphiles: the lamellar phase. X-ray diffraction measurements with the aerosol OT/water system. [**426**] P. A. Winsor, Mol. Cryst. Liq. Cryst. *12*, 141 (1971), C.A. *75*, 26540z (1971); Liquid crystallinity in relation to composition and temperature in amphiphilic systems. [**427**] P. A. Winsor, Liq. Cryst. Plast. Cryst. *1*, 48 (1974), C.A. *83*, 19076q (1975); Nonamphiphilic cubic mesophases. Plastic crystals. [**428**] P. A. Winsor, Liq. Cryst. Plast. Cryst. *1*, 60 (1974), C.A. *83*, 19077r (1975); Mesophases formed by amphiphilic compounds. [**429**] P. A. Winsor, Liq. Cryst. Plast. Cryst. *1*, 199 (1974), C.A. *83*, 19085s (1975); Influence of composition and temperature on the formation of mesophases in amphiphilic systems. R-Theory of fused micellar phases. [**430**] P. A. Winsor, Liq. Cryst. Plast. Cryst. *2*, 122 (1974), C.A. *83*, 200621d (1975); Electric properties of liquid crystals. Electric conduction by amphiphilic mesophases.

[**431**] L. T. Zimmer, Bull. Math. Biophys. *17*, 51 (1955), C.A. *51*, 4094i (1957); Possible phase transitions in dilute colloidal solutions. [**432**] H. Freundlich, R. Stern, H. Zocher, Biochem. Z. *138*, 307 (1923), C.A. *18*, 125 (1924); Kolloidchemische Beobachtungen an Salvarsan und Neosalvarsan. [**433**] H. Zocher, Kolloid-Z. *37*, 336 (1925), C.A. *20*, 2106[4] (1926); Optische Methoden zur Untersuchung der Anisotropie in Kolloiden. [**434**] H. Zocher, K. Coper, Z. physik. Chem. *132*, 302 (1928), C.A. *22*, 1913 (1928); Die durch den Weigert-Effekt in Photochlorid erzeugte Anisotropie. [**435**] H. Zocher, H. W. Albu, Kolloid-Z. *46*, 27 (1928), C.A. *22*, 4312 (1928); Sol-Gel-Systeme mit anisotropen Teilchen. I. Dibenzoylcystin. [**436**] H. Zocher, H. W. Albu, Kolloid-Z. *46*, 33 (1928), C.A. *22*, 4313 (1928); II. Bariummalonat (cf. [435]). [**437**] H. Zocher, V. Birstein, Z. physik. Chem. A *142*, 126 (1929), C.A. *23*, 4608 (1929); Die wäßrige Mesophase des Salvarsans. [**438**] H. Zocher, V. Birstein, Z. physik. Chem. A *142*, 177 (1929), C.A. *23*, 4860[6] (1929); Weitere Fälle wäßriger Mesophasen. [**439**] H. Zocher, K. Jacobsohn, Kolloidchem. Beihefte *28*, 167 (1929), C.A. *23*, 2868 (1929); Taktosole. [**440**] H. Zocher, M. Jacobowitz, Kolloid-Beihefte *37*, 427 (1933), C.A. *27*, 4456[1] (1933); Zwischenaggregatzustände.

[**441**] H. Zocher, Trans. Faraday Soc. *35*, 34 (1939), C.A. *33*, 2414[6] (1939); Polarization of the fluorescence of dyestuffs dissolved in mesophases. [**442**] H. Zocher, C. Torok, Kolloid-Z. *170*, 140 (1960), C.A. *54*, 23609i (1960); Neues Taktosol: Aluminiumhydroxyd. [**443**] R. Zsigmondy, W. Bachman, Z. Chem. Ind. Kolloide *11*, 145 (1912), C.A. *7*, 1644 (1913); Ultramikroskopische Studien an Seifenlösungen und -gallerten.

Chapter 12

[**1**] J. L. Abernethy, J. Chem. Educ. *44*, 364 (1967), C.A. *67*, 78779p (1967); Lyotropy, with particular application to collagen. [**2**] J. G. Adami, L. Aschoff, Proc. Roy. Soc. B *78*, 359 (1906), C.A. *1*, 863[6] (1907); Myelins, myelin bodies and potential fluid crystals of the organism. [**2a**] M. Aizawa, S. Suzuki, M. Kubo, Biochim. Biophys. Acta *444*, 886 (1976), C.A. *86*, 12838t (1977); Electrolytic regeneration of NADH from NAD[+] with a liquid crystal membrane electrode. [**2b**] M. Aizawa, M.

Kubo, S. Suzuki, Denki Kagaku Oyobi Kogyo Butsuri Kagaku *44*, 285 (1976), C.A. *86*, 147648b (1977); Electrolytic reduction of coenzymes at a liquid crystal membrane electrode. [3] G. Albrecht-Bühler, Diss. München 1970, 78 pp. Ges. f. Strahlenforschung, München-Neuherberg, Ber. B 288, Febr. 1971, Ultramikroskopisches Bild und Größenverteilung von Myelinmizellen. [4] F. S. Allen, K. E. van Holde, Biopolymers *10*, 865 (1971), C.A. *75*, 380x (1971); Dichroism of tobacco mosaic virus in pulsed electric fields. [5] A. C. Allison, Biochem. J. *65*, 212 (1957), C.A. *51*, 6719i (1957); Properties of sickle-cell hemoglobin. [6] A. Elliott, E. J. Ambrose, Disc. Faraday Soc. 1950, No. 9, 246, C.A. *46*, 3591g (1952); Evidence of chain folding of polypeptides and proteins. [7] E. J. Ambrose, Proc. Can. Cancer Res. Conf. *7*, 247 (1966), C.A. *71*, 98226s (1969); Biochemical and biophysical properties of cell membranes. [8] E. J. Ambrose, J. S. Osborne, P. R. Stuart, Liq. Cryst. Ord. Fluids 1970, 83, C.A. *79*, 1756u (1973); Structure and properties of the cell surface complex. [9] E. J. Ambrose, Faraday Soc. Symp. *1971*, No. 5, 175, C.A. *79*, 75007x (1973); Liquid crystalline phenomena at the cell surface. [10] E. J. Ambrose, Liq. Cryst., Proc. Int. Conf., Bangalore, *1973*, 523, C.A. *84*, 100949u (1976); Liquid crystalline properties of micro-filaments.

[11] E. J. Ambrose, Adv. Chem. Ser. *152*, 142 (1976), C.A. *85*, 75909s (1976); Liquid crystals and cancer. [12] E. W. April, Nature *257*, 139 (1975); Liquid-crystalline characteristics of the thick filament lattice of striated muscle. [13] A. M. Atallah, H. J. Nicholas, Chem. Technol. *2*, 486 (1972), C.A. *77*, 157210f (1972); What role have liquid crystals in nature? [14] A. M. Atallah, H. J. Nicholas, Lipids *9*, 613 (1974); Function of steryl esters in plants. Hypotheses that liquid-crystalline properties of some steryl esters may be significant in plant sterol-metabolism. [15] H. Athenstaedt, Naturwissenschaften *49*, 433 (1962); Mesomorphe Ordnungszustände biologisch bedeutsamer Stoffe. [16] A. Aviram, J. Polym. Sci., Polym. Lett. Ed. *14*, 757 (1976), C.A. *86*, 30232v (1977); Crosslinking of lyophases in magnetic fields. [17] P. J. Bailey, D. Keller, Atherosclerosis *13*, 333 (1971), C.A. *75*, 107107r (1971); Deposition of lipids from serum into cells cultured in vitro. [18] A. D. Bangham, J. de Gier, G. D. Greville, Chem. Phys. Lipids *1*, 225 (1967), C.A. *67*, 70603f (1967); Osmotic properties and water permeability of phospholipid liquid crystals. [19] V. G. Baranov, T. I. Volkov, S. Y. Frenkel, Dokl. Akad. Nauk SSSR *162*, 836 (1965), C.A. *63*, 8508g (1965); Polarization diffractometric investigation of the formation of a supermolecular structure in a solution of a spiral polypeptide. [20] E. Barbu, M. Joly, J. chim. phys. *53*, 951 (1956), C.A. *51*, 7806g (1957); Comportement en solutions des particules d'acide nucléique. I. Méthodes d'étude et généralités sur les dimensions des particules. [20a] Y. Barenholz, J. Suurkuusk, D. Mountcastle, T. E. Thompson, R. L. Biltonen, Biochemistry *15*, 2441 (1976), C.A. *85*, 42558y (1976); Calorimetric study of the thermotropic behavior of aqueous dispersions of natural and synthetic sphingomyelins.

[21] J. B. Bateman, S. S. Hsu, J. P. Knudsen, K. L. Yudowitch, Arch. Biochem. Biophys. *45*, 411 (1953), C.A. *47*, 11435d (1953); Hemoglobin spacing in erythrocytes. [22] F. C. Bawden, N. W. Pirie, J. D. Bernal, I. Fankuchen, Nature *138*, 1051 (1936), C.A. *31*, 2256⁷ (1937); Liquid crystals from virus-infected plants. [23] F. C. Bawden, N. W. Pirie, Nature *139*, 546 (1937), C.A. *31*, 4367¹ (1937); Liquid crystals from cucumber viruses 3 and 4. [24] F. C. Bawden, Plant viruses and virus diseases, 3rd ed. (Chronica Botanica: Waltham, Mass.). 1950. 349 pp. C.A. *44*, 6924h (1950). [25] F. P. Bell, Artery *2*, 173 (1976), C.A. *85*, 31372g (1976); Esterification of isotopic cholesterol presented in a membrane-bound form to arterial homogenates. [26] G. N. Berestovskii, G. M. Frank, E. A. Liberman, V. Z. Lunevskii, V. D. Razhin, Biochim. Biophys. Acta *219*, 263 (1970), C.A. *74*, 60960a (1971); Electrooptical phenomena in bimolecular phospholipid membranes. [27] J. D. Bernal, I. Fankuchen, Nature *139*, 923 (1937), C.A. *31*, 5410³ (1937); Structure types of protein "crystals" from virus-infected plants. [28] J. D. Bernal, I. Fankuchen, J. Gen. Physiol. *25*, 111, 120, 147 (1941), C.A. *36*, 4153¹ (1942); X-ray and crystallographic studies of plant virus preparations. I. Introduction and preparation of specimens. II. Modes of aggregation of the virus particles. III. Structure of the particles. Biological implications. [29] V. Blaton, A. N. Howard, G. A. Gresham, D. Vandamme, H. Peeters, Atherosclerosis *11*, 497 (1970), C.A. *74*, 2144b (1971); Lipid changes in the plasma lipoproteins of baboons given an atherogenic diet. I. Changes in the lipids of total plasma and of α- and β-lipoproteins. [29a] M. C. Blok, L. L. M. van Deenen, J. de Gier, Biochim. Biophys. Acta *433*, 1 (1976), C.A. *84*, 160892e (1976); Effect of the gel to liquid crystal transition on the osmotic behavior of phosphatidylcholine liposomes. [30] E. R. Blout, R, H. Karlson, J. Am. Chem. Soc. *78*, 941 (1956), C.A. *50*, 12819g (1956); Synthesis of high molecular weight PBLG.

[31] R. H. Karlson, K. S. Norland, G. D. Fasman, E. R. Blout, J. Am. Chem. Soc. *82*, 2268 (1960), C.A. *55*, 19812d (1961); Helical sense of poly(β-benzyl L-aspartate). Synthesis and rotatory

dispersion of copolymers of β-benzyl L- and D-asparate with γ-benzyl L-glutamate. [**32**] A. Blume, T. Ackermann, FEBS Lett. *43*, 71 (1974), C.A. *83*, 85946z (1975); Calorimetric study of the lipid phase transitions in aqueous dispersion of phosphorylcholine-phosphorylethanolamine mixtures. [**33**] G. Boheim, Diss. Aachen 1972, 136 pp. Erregbarkeit schwarzer Lipid-Membranen. Erregungsvorgänge in dem durch Alamethicin und Protamin modifizierten Membranmodellsystem. [**34**] C. Botre, C. del Vecchio, A. Memoli, M. Mascini, Anal. Chem. *47*, 1393 (1975), C.A. *83*, 107621u (1975); Polymorphic transitions in lipoidic structures and their analytical implications. [**35**] Y. Bouligand, Tissue Cell *4*, 189 (1972), C.A. *77*, 161294d (1972); Twisted fibrous arrangements in biological materials and cholesterics. [**36**] Y. Bouligand, J. Phys. (Paris), C-1, *36*, 331 (1975); Defects and textures in cholesteric analogs given by biological systems. [**37**] Y. Bouligand, Recherche *7*, 474 (1976), C.A. *85*, 29614u (1976); Les analogues biologiques des cristaux liquides. [**38**] E. M. Bradbury, A. R. Downie, A. Elliott, W. E. Hanby, Nature *187*, 321 (1960), C.A. *55*, 644c (1961); Screw sense of the α-helix in poly-β-benzyl-L-aspartate. [**39**] E. M. Bradbury, A. R. Downie, A. Elliott, W. E. Hanby, Proc. Roy. Soc. A *259*, 110 (1960), C.A. *55*, 8492g (1961); Stability and screw sense of the α-helix in poly(β-benzyl-L-aspartate). [**40**] J. Brahms, K. E. van Holde, Advan. Chem. Ser. *63*, 253 (1967), C.A. *67*, 166s (1967); Structure and thermodynamic properties of single-strand helical polynucleotides.

[**41**] S. Bram, P. Tougard, Nature, New Biol. *239*, 128 (1972), C.A. *78*, 1185m (1973); Polymorphism of natural DNA. [**42**] R. Brauns, Flüssige Kristalle und Lebewesen (Schweizerbart: Stuttgart). 1931. 111 pp. C.A. *26*, 1505⁵ (1932). [**43**] S. E. Bresler, V. M. Bresler, Dokl. Akad. Nauk SSSR *214*, 936 (1974); Liquid crystalline structure of biological membranes. [**44**] F. Brochard, J. F. Lennon, J. Phys. (Paris) *36*, 1035 (1975); Frequency spectrum of the flicker phenomenon in erythrocytes. [**44a**] F. Brochard, P. G. de Gennes, P. Pfeuty, J. Phys. (Paris) *37*, 1099 (1976); Surface tension and deformations of membrane structures. Relation to two-dimensional phase transitions. [**45**] A. V. Bromberg, V. M. Luk'yanovich, V. V. Nemtsova, L. V. Radushkevich, K. V. Chmutov, Dokl. Akad. Nauk SSSR *87*, 81 (1952), C.A. *49*, 10703a (1955); Electron-microscopic study of colloidal objectives by means of development. [**46**] J. L. Fergason, G. H. Brown, J. Am. Oil Chem. Soc. *45*, 120 (1968), C.A. *68*, 101952w (1968); Liquid crystals and living systems. [**47**] G. H. Brown, R. J. Mishra, J. Agr. Food Chem. *19*, 645 (1971), C.A. *75*, 44876t (1971); Liquid crystals as they relate to the structure of proteins. [**48**] M. F. Olszewski, R. T. Holzbach, A. Saupe, G. H. Brown, Nature *242*, 336 (1973); Liquid crystals in human bile. [**49**] E. Buder, Stud. Biophys. *1970*, 21, 347, C.A. *75*, 29870t (1971); Vergleich der optischen Anisotropieeigenschaften der isolierten DNS mit denen des isolierten DNP und der DNS in Zellen. [**50**] B. J. Bulkin, N. Krishnamachari, J. Am. Chem. Soc. *94*, 1109 (1972), C.A. *76*, 106073k (1972); IR- and Raman spectra of phospholipid/water mixtures.

[**51**] B. J. Bulkin, N. Krishnamachari, Mol. Cryst. Liq. Cryst. *24*, 53 (1973), C.A. *81*, 143724x (1974); IR spectra of lecithin/phosphatidyl serine/water systems. [**52**] B. J. Bulkin, Appl. Spectr. *30*, 261 (1976), C.A. *85*, 38742m (1976); IR- and Raman spectroscopy of liquid crystals. [**53**] H. G. Bungenberg de Jong, A. de Bakker, D. Andriesse, Konikl. Ned. Akad. Wetenschap., Proc. B *58*, 238 (1955), C.A. *50*, 4256d (1956); Colloid chemistry of phosphatides. [**54**] D. A. Cadenhead, J. Chem. Educ. *49*, 152 (1972), C.A. *76*, 136927j (1972); Film balance studies of membrane lipids and related molecules. Applying surface chemistry to molecular biology. [**55**] D. A. Cadenhead, B. M. J. Kellner, M. C. Phillips, J. Colloid Interface Sci. *57*, 224 (1976), C.A. *85*, 155432w (1976); Miscibility of dipalmitoyl phosphatidylcholine and cholesterol in monolayers. [**56**] F. Castano, An. Fis. *65*, 45, 55 (1969), C.A. *71*, 13272f, 13273g (1969); Spectroscopic behavior of cholesterics. Structural mechanism of the transduction of the mesomorphic aggregate between deoxycholic acid and π systems. [**56a**] F. Castano, A. Cigarran, An. Quim. *73*, 613 (1977), C.A. *87*, 167053c (1977); Spectral behavior of π-systems in lyotropic deoxycholic acid. [**57**] S. Caveney, Proc. Roy. Soc. B *178*, 205 (1971), C.A. *75*, 60420v (1971); Cuticle reflectivity and optical activity in scarab beetles: role of uric acid. [**58**] S. S. Chalatov, Frankf. Z. Path. *13*, 189 (1913), C.A. *7*, 4002 (1913); Flüssige Kristalle im tierischen Organismus, deren Entstehungsbedingungen und Eigenschaften. [**59**] S. S. Chalatov, Beitr. path. Anat. *57*, 85 (1914), C.A. *8*, 167⁹ (1914); Über experimentelle Cholesterin-Lebercirrhose in Verbindung mit eigenen neuen Erhebungen über flüssige Kristalle des Organismus und über den Umbau der Leber. Ein Beitrag zur Frage der Anisotropen Verfettung der Organe. [**60**] P. Byrne, D. Chapman, Nature *202*, 987 (1964), C.A. *61*, 7270f (1964); Liquid crystalline nature of phospholipids.

[**61**] D. Chapman, P. Byrne, G. G. Shipley, Proc. Roy. Soc. A *290*, 115 (1966), C.A. *64*, 6470e (1966); Solid state and mesomorphism of 2,3-diacyl-DL-phosphatidyl-ethanolamines. [**62**] D. Chapman, Ann. N. Y. Acad. Sci. *137*, 745 (1966), C.A. *65*, 14009c (1966); Liquid crystals and cell membrane.

[63] R. B. Leslie, D. Chapman, C. J. Hart, Biochim. Biophys. Acta *135*, 797 (1967), C.A. *68*, 27012w (1968); dc Electric conductivity of membrane phospholipids. [64] D. Chapman, Advan. Chem. Ser. *63*, 157 (1967), C.A. *67*, 7883g (1967); Liquid crystalline nature of phospholipids. [65] D. Chapman, R. M. Williams, B. D. Ladbrooke, Chem. Phys. Lipids *1*, 445 (1967), C.A. *68*, 27011v (1968); Thermotropic and lyotropic mesomorphism of 1,2-diacylphosphatidylcholines. [66] M. C. Phillips, R. M. Williams, D. Chapman, Chem. Phys. Lipids *3*, 234 (1969), C.A. *72*, 26016w (1970); Hydrocarbon chain motions in lipid liquid crystals. [67] R. M. Williams, D. Chapman, Progr. Chem. Fats Other Lipids *11*, 1 (1970), C.A. *76*, 95799y (1972); Phospholipids, liquid crystals, and cell membranes. [68] D. Chapman, Wiss. Veröff. Deut. Ges. Ernähr. *22*, 36 (1971), C.A. *76*, 55547h (1972); Physical studies of membranes. [69] D. Chapman, Symp. Faraday Soc. *1971*, No. 5, 163, C.A. *79*, 74979d (1973); Mesomorphism of phospholipids and biological membranes. [70] D. Chapman, G. H. Dodd, Struct. Funct. Biol. Membranes *1971*, 13, C.A. *80*, 105451c (1974); Physicochemical probes of membrane structure.

[71] D. Chapman, Pestic. Sci. *4*, 839 (1973), C.A. *80*, 92056a (1974); Lipid dynamic in cell membranes. [72] D. Chapman, Liq. Cryst. Plast. Cryst. *1*, 288 (1974), C.A. *83*, 2499g (1975); Significance of liquid crystals in biology. Liquid crystals and cell membranes. [73] D. Chapman, Pure Appl. Chem. *38*, 59 (1974), C.A. *82*, 134084g (1975); Liquid crystals of biological importance. [74] Y. N. Chirgadze, E. P. Rashevskaya, Biofizika *14*, 608 (1969), C.A. *71*, 113400q (1969); Intensities of the characteristic vibrations of the peptide group in the IR-spectrum of PBLG in the helical conformation. [75] I. G. Chistyakov, V. A. Usol'tseva, Uch. Zap. Ivanov. Gos. Pedagog. Inst. *62*, 68 (1967), C.A. *71*, 75306d (1969); Molecular structure of liquid crystals (protagon, cerebron, sphingomyelins). [76] W. M. Michailov, S. A. Celesnjev, I. G. Chistyakov, V. A. Usol'tseva, 1. Wiss. Konf. über flüssige Kristalle, Ivanovo 1970. Sammlung der Vorträge Nr. 23 (1972). Erforschung der Membran von Erythrocyten und dazu gehörender Lipide mit der Methode der Röntgenstrahlbeugung. [77] I. G. Chistyakov, V. A. Usol'tseva, S. A. Seleznev, N. M. Maksimova, Usp. Sovrem. Biol. *82*, 89 (1976), C.A. *85*, 172824t (1976); Liquid crystals and their biological significance. [77a] I. G. Chistyakov, S. A. Seleznev, Priroda *1977*, 38, C.A. *87*, 179079d (1977); Biological role of lyophases. [77b] P. K. Chowarshi, F. A. Pepe, J. Cell Biol. *74*, 136 (1977), C.A. *87*, 18085n (1977); Light meromyosin paracrystal formation. [78] T. G. Churyusova, E. P. Sokolova, A. G. Morachevskii, Vestn. Leningr. Univ., Fiz., Khim. *1976*, 106, C.A. *84*, 170384e (1976); Thermodynamic properties of glycerol α-monoester/water systems. [79] R. W. St. Clair, T. B. Clarkson, H. B. Lofland, Circ. Res. *31*, 664 (1972), C.A. *78*, 14222w (1973); Effects of regression of atherosclerotic lesions on the content and esterification of cholesterol by cell-free preparations of pigeon aorta. [80] B. E. Cohen, A. D. Bangham, Nature *236*, 173 (1972), C.A. *77*, 15670g (1972); Diffusion of small non-electrolytes across liposome membranes.

[81] A. Colbeau, J. Nachbaur, P. M. Vignais, Biochim. Biophys. Acta *249*, 462 (1971), C.A. *76*, 31340j (1972); Enzymic characterization and lipid composition of rat liver subcellular membranes. [82] K. Colbow, B. L. Jones, Biochim. Biophys. Acta *345*, 91 (1974), C.A. *81*, 46625h (1974); Stability of the liquid-crystalline lamellar lecithin/water system. [83] F. W. Cope, Mol. Cryst. *2*, 45 (1966), C.A. *66*, 35940w (1967); Solid state mechanisms of ion transport in biological systems. [83a] R. M. J. Cotterill, Biochim. Biophys. Acta *433*, 264 (1976), C.A. *84*, 175558z (1976); Computer simulation of model lipid membrane dynamics. [84] F. Crick, Nature *234*, 25 (1971), C.A. *76*, 42813s (1972); General model for the chromosomes of higher organisms. [85] J. E. Cronan, Jr., P. R. Vagelos, Biochim. Biophys. Acta *265*, 25 (1972), C.A. *76*, 110057a (1972); Metabolism and function of the membrane phospholipids of escherichia coli. [86] O. Csuka, J. Sugar, I. Sajo, Magy. Onkol. *20*, 81 (1976), C.A. *85*, 153792q (1976); Mechanism of action of Vinca alkaloids. [87] C. L. Davey, A. E. Graafhuis, Experientia *31*, 441 (1975), C.A. *83*, 2829q (1975); Paracrystallization of actomyosin. [88] G. J. Davis, Diss. Univ. Massachusetts, Amherst, Mass. 1970, 132 pp. Avail. Univ. Microfilms, Ann Arbor, Mich., Order No. 70-15,301, C.A. *75*, 94695d (1971); Thermal, purity, and solubility properties of cholesteryl esters and their thermal behavior in lipid/water systems. [89] G. J. Davis, R. S. Porter, Nat. Bur. Stand., Spec. Publ. No. *338*, 99 (1973), C.A. *79*, 133817z (1973); Thermal studies on lipid/water systems by DSC with reference to atherosclerosis. [90] A. J. Day, M. L. Wahlqvist, Exp. Mol. Pathol. *13*, 199 (1970), C.A. *74*, 11368e (1971); Cholesterol ester and phospholipid composition of normal aortas and of atherosclerotic lesions in children.

[91] S. Dayton, S. Hashimoto, Exp. Mol. Pathol. *13*, 253 (1970), C.A. *73*, 118205f (1970); Cholesterol flux and metabolism in arterial tissue and in atheromata. [92] M. Delbrück, Angew. Chem. *84*, 1 (1972), Angew. Chem. Int. Ed. Engl. *11*, 1 (1972), C.A. *76*, 82211a (1972); Signalwandler: terra incognita der Molekularbiologie. [93] A. T. Dembo, N. I. Sosfenov, L. A. Feigin, Kristallografiya *11*, 581 (1966),

C.A. *65*, 12461b (1966); Study of transport ribonucleic acid by the SAXS. [**93a**] R. A. Demel, J. W. C. M. Jansen, P. W. M. van Dijck, L. L. M. van Deenen, Biochim. Biophys. Acta *465*, 1 (1977), C.A. *86*, 116385j (1977); Preferential interaction of cholesterol with different classes of phospholipids. [**94**] D. G. Dervichian, J. Chem. Phys. *7*, 931 (1939), C.A. *33*, 9086³ (1939); Change of phase and transformations of higher order in monolayers. [**95**] H. de Wulf, Biochem. Glycosidic Linkage, Proc. Symp. *1971*, 399, C.A. *78*, 94000k (1973); Paracrystalline glycogen. [**96**] F. D'Hollander, F. Chevallier, J. Lipid. Res. *13*, 733 (1972), C.A. *78*, 27236d (1973); Movement of cholesterol in vitro in rat blood and quantitation of exchange of free cholesterol between plasma and erythrocytes. [**97**] V. Diamare, Rend. accad. Sci. (Napoli) *33*, 132 (1927), C.A. *22*, 2758⁷ (1928); Myelins of oleates, soaps and lipoids. Liquid lipoid crystals. [**98**] L. Dintenfass, Mol. Cryst. Liq. Cryst. *8*, 101 (1969); Internal viscosity of the red cell and the structure of the red cell membrane. Considerations of the liquid crystalline structure of the red cell interior and membrane from rheological data. [**99**] L. Dintenfass, Mol. Cryst. Liq. Cryst. *20*, 239 (1973), C.A. *79*, 63978b (1973); Possible liquid crystalline structures in artificial thrombi formed at arterial shear rates: effect of disease, protein concentrations, and AB0 blood groups. [**100**] M. Dobiasova, J. Linhart, Lipids *5*, 445 (1970), C.A. *73*, 31646b (1970); Association of phospholipid-cholesterol micelles with rat heart mitochondria: stimulators and inhibitors.

[**101**] P. Doty, J. H. Bradbury, A. M. Holtzer, J. Am. Chem. Soc. *78*, 947 (1956), C.A. *50*, 7165c (1956); Molecular weight, configuration, and association of PBLG in various solvents. [**102**] J. Tsi Yang, P. Doty, J. Am. Chem. Soc. *79*, 761 (1957), C.A. *51*, 7449d (1957); ORD of polypeptides and proteins in relation to configuration. [**103**] A. R. Downie, A. Elliott, W. E. Hanby, B. R. Malcolm, Proc. Roy. Soc. A *242*, 325 (1957), C.A. *52*, 7820g (1958); Optical rotation and molecular configuration of synthetic polypeptides in dilute solution. [**104**] R. W. Duke, Diss. Univ. Louisville, Louisville, Ky. 1974, 178 pp. Avail. Xerox Univ. Microfilms, Ann Arbor, Mich., Order No. 75-5462, C.A. *83*, 10814w (1975); Viscoelastic properties of a cholesteric polypeptide by direct and quasielastic laser light scattering techniques. [**105**] R. W. Duke, D. B. DuPre, Macromolecules *7*, 374 (1974), C.A. *81*, 106176z (1974); Examination of the electrotropic c/n transition of liquid crystal PBLG by laser light beating spectroscopy. [**106**] R. W. Duke, D. B. DuPre, J. Chem. Phys. *60*, 2759 (1974), C.A. *81*, 7065f (1974); Twist elastic constant of a cholesteric polypeptide. [**106a**] R. W. Duke, L. L. Chapoy, Rheol. Acta *15*, 548 (1976), C.A. *86*, 87891p (1977); Rheology and structure of lecithin in concentrated solution and the liquid crystal. [**106b**] R. W. Duke, D. B. DuPre, Mol. Cryst. Liq. Cryst. *43*, 33 (1977); Quasielastic light scattering from orientational fluctuations in a cholesteric. [**107**] D. B. DuPre, J. R. Hammersmith, Liq. Cryst. Ord. Fluids *2*, 237 (1974), C.A. *86*, 180984f (1977); Optical properties of nematic PBLG. [**108**] D. B. DuPre, R. W. Duke, Phys. Rev. Lett. *33*, 67 (1974), C.A. *81*, 69540r (1974); Light scattering from orientation fluctuations in a cholesteric. Angular and field dependences of the light-beating spectrum. [**109**] D. B. DuPre, J. R. Hammersmith, Mol. Cryst. Liq. Cryst. *28*, 365 (1974), C.A. *83*, 43900z (1975); Optical properties of the magnetotropic nematic phase of PBLG. [**110**] D. B. DuPre, R. W. Duke, Fed. Proc. *33*, 1510 (1974); Viscoelastic properties of a cholesteric polypeptide by direct optical and laser light beating techniques.

[**111**] D. B. DuPre, R. W. Duke, J. Chem. Phys. *63*, 143 (1975), C.A. *83*, 140038g (1975); Temperature, concentration, and molecular weight dependence of the twist elastic constant of cholesteric PBLG. [**112**] A. C. H. Durham et al., J. Mol. Biol. *67*, 289, 307, 315 (1972), C.A. *77*, 57970k, 57971m, 57955j (1972); Structures and roles of the polymorphic forms of tobacco mosaic virus protein. Electron microscope observations of the larger polymers. Model for the association of A-protein into discs. [**113**] J. F. Dyro, P. D. Edmonds, J. G. Berberian, D. Silage, IEEE Trans. Son. Ultrason. *19*, 403 (1972); Viscoelastic properties of liquid crystalline biological materials in the vicinity of c/i phase transition. [**114**] T. Eckert, N. Doerr, I. Reimann, Kolloid-Z. Z. Polym. *239*, 613 (1970), C.A. *73*, 102373y (1970); Mesophasen und Gelbildung bei Folsäure. [**115**] S. Eins, Chem. Phys. Lipids *8*, 26 (1972), C.A. *76*, 82525f (1972); Electron microscopy of aqueous lipid mesophases. III. Phosphatidylserine/water system. [**116**] S. Eins, H. Kahl, Z. Allg. Mikrobiol. *12*, 175 (1972), C.A. *77*, 72403w (1972); II. Isolated lipids and membrane structure of Polystictus versicolor (cf. [115]). [**117**] G. Eisenman, Membranes, Vol. 3: Lipid bilayers and biological membranes: Dynamic properties. (Dekker: New York). 1975. 538 pp., C.A. *84*, 160910j (1976). [**118**] H. G. Elias, J. Gerber, Makromol. Chem. *112*, 122, 142 (1968), C.A. *68*, 87665w, 87666x (1968); Multimerisation: Assoziation und Aggregation. VI. Geschlossene Assoziation von PBLG in gemischten Lösungsmitteln. VII. PBLG in reinen Lösungsmitteln. [**119**] G. F. Elliott, E. M. Rome, Mol. Cryst. Liq. Cryst. *8*, 215 (1969); Liquid-crystalline aspects

of muscle fibers. [**120**] G. F. Elliott, Ann. N. Y. Acad. Sci. *1973*, 564, C.A. *79*, 14497k (1973); Liquid-crystalline and hydraulic aspects of muscle fibers.

[**121**] P. Emmelot, H. vaz Dias, Biochim. Biophys. Acta *203*, 172 (1970), C.A. *73*, 83755f (1970); Separation of membrane components produced by anionic detergents and maintained after the latter's removal. [**121a**] D. M. Engelman, G. M. Hillman, J. Clin. Invest. *58*, 997 (1976), C.A. *85*, 175200x (1976); Molecular organization of the cholesteryl ester droplets in the fatty streaks of human aorta. [**122**] A. Engström, J. B. Finean, Biological Ultrastructure (Academic: New York). 1958. 326 pp., C.A. *52*, 7398b (1958). [**123**] R. Faiman, K. Larsson, D. A. Long, J. Raman Spectros. *5*, 3 (1976), C.A. *85*, 173138j (1976); Raman spectroscopic study of the effect of hydrocarbon chain length and chain unsaturation on lecithin/cholesterol interaction. [**124**] W. K. Farr, J. Phys. Chem. *42*, 1113 (1938), C.A. *33*, 1012³ (1939); Microscopic structure of plant cell membranes in relation to the micellar hypothesis. [**125**] V. Felt, P. Benes, Enzymol. Biol. Clin. *11*, 511 (1970), C.A. *73*, 118334x (1970); Cholesterol esterase (esterifying and hydrolyzing) activity in blood serum, liver, kidney, and aorta in rabbit atherosclerosis. [**126**] R. W. Filas, L. E. Hajdo, A. C. Eringen, J. Chem. Phys. *61*, 3037 (1974), C.A. *82*, 43935u (1975); Reorientation of PBLG liquid crystals in a magnetic field. [**127**] R. W. Filas, H. Stefanou, J. Phys. Chem. *79*, 941 (1975), C.A. *83*, 10825a (1975); Magnetic orientation of poly(γ-methyl-D-glutamate) liquid crystals. [**128**] J. B. Finean, S. Knutton, A. R. Limbrick, R. Coleman, Mol. Cryst. Liq. Cryst. *7*, 347 (1969), C.A. *71*, 85617v (1969); X-ray diffraction study of the physical state of the lipid phase in biological membranes. [**129**] E. Forslind, Adv. Chem. Ser. *152*, 114 (1976), C.A. *85*, 73910m (1976); Biowater. [**130**] W. G. Forssmann, J. Metz, Cell Tissue Res. *171*, 467 (1976), C.A. *86*, 664v (1977); Exocrine pancreas under experimental conditions. I. Formation of paracrystalline inclusions in vivo.

[**131**] A. Frey-Wyssling, Schweiz. mineralog. petrog. Mitt. *28*, 403 (1948), C.A. *42*, 8047d (1948); Optische Anisotropie biologischer Objekte. [**132**] S. Friberg, L. Rydhag, J. Am. Oil Chem. Soc. *48*, 113 (1971), C.A. *74*, 100832u (1971); Solubilization of triglycerides by hydrotropic interaction: liquid crystals. [**133**] S. Friberg, R. F. Gould, Advances in Chemistry Series, No. 152: Lyotropic Liquid Crystals and the Structure of Biomembranes (ACS: Washington). 1976. 156 pp., C.A. *85*, 105704r (1976). [**134**] G. Friedel, Compt. rend. *185*, 330 (1927), C.A. *22*, 1368 (1928); Formes que prend la myéline au contact de l'eau. [**135**] E. Friedman, C. Anderson, R. J. Roe, A. V. Tobolsky, U. S. Nat. Tech. Inform. Serv., AD Rep. *1972*, No. 74986, 10 pp., C.A. *78*, 4845n (1973); Solid state morphology of some polypeptide films. [**136**] E. M. Friedman, Diss. Princeton Univ., Princeton, N. J. 1973. 155 pp. Avail. Univ. Microfilms, Ann Arbor, Mich., Order No. 74-9685, C.A. *81*, 78340c (1974); PBG morphology and diffraction by cholesteric solids. [**137**] E. M. Friedman, R. J. Roe, Mol. Cryst. Liq. Cryst. *28*, 437 (1974), C.A. *82*, 178713r (1975); Effect of a cholesteric texture on X-ray diffraction patterns from liquid crystalline polypeptides. [**137a**] H. Frischleder, S. Gleichmann, Stud. Biophys. *64*, 95 (1977), C.A. *87*, 46820s (1977); Microcalorimetric study of the effect of alkylammonium iodides on the thermotropic phase transition of lamellar dipalmitoyl lecithin/water dispersions. [**138**] H. T. Funasaki, Diss. Univ. Pittsburgh, Pittsburgh, Pa. 1970, 145 pp. Avail. Univ. Microfilms, Ann Arbor, Mich. Order No. 70-24,211, C.A. *75*, 147821a (1971); Isolation identification, and biosynthesis of cholesteryl ethers. [**139**] B. M. Fung, T. H. Martin, Liq. Cryst. Ord. Fluids *2*, 267 (1974), C.A. *86*, 180197b (1977); Magnetic relaxation of PBLG solutions in deuterochloroform. [**140**] B. M. Fung, T. H. Martin, J. Chem. Phys. *61*, 1698 (1974), C.A. *81*, 143923m (1974); Magnetic relaxation and quadrupole splitting for liquid crystalline solutions of PBLG in chloroform-d and dichloromethane-d₂.

[**141**] R. Gabler, J. Bearden, I. Bendet, Biophys. J. *11*, 302 (1971), C.A. *75*, 213v (1971); Anisotropic absorption of linearly polarized light by cylindrical molecules. [**142**] I. A. Gamalei, A. B. Kaulin, Tsitologiya *15*, 690 (1973), C.A. *80*, 13184q (1974); Microscopic viscosity of muscle fibers. IV. Anisotropic structure of intracellular water. [**143**] B. P. Garcia, G. A. Pozuelo, Hosp. Gen. *11*, 149 (1971), C.A. *75*, 104977p (1971); Biological membranes. I. Composition and structure. [**144**] M. J. Gerace, B. M. Fung, J. Chem. Phys. *53*, 2984 (1970), C.A. *73*, 114790v (1970); N¹⁴-NMR and quadrupole coupling constants in liquid crystalline PBLG solutions. [**145**] B. M. Fung, M. J. Gerace, L. S. Gerace, J. Phys. Chem. *74*, 83 (1970), C.A. *72*, 61150q (1970); NMR-study of liquid crystalline solutions of PBLG in dichloromethane and 1,2-dichloroethane. [**146**] J. R. Gilbertson, H. H. Garlich, R. A. Gelman, J. Lipid Res. *11*, 201 (1970), C.A. *73*, 33051c (1970); Evidence for the existence of cholesteryl alkyl-1-enyl ethers in bovine and porcine cardiac muscle. [**147**] D. Gill, M. P. Klein, G. Kotowycz, J. Am. Chem. Soc. *90*, 6870 (1968), C.A. *70*, 15832s (1969); Complete determination of the alignment of dichloromethane-dideuteriodichloromethane molecules in PBLG solution by the NMR of ³⁵Cl, ²H, and ¹H. [**148**]

T. J. Gill, III, G. S. Omenn, Advan. Chem. Ser. *63*, 189 (1967), C.A. *67*, 113x (1967); Studies on the structure of synthetic polypeptides in solution by polarization of fluorescence techniques. [**149**] F. N. Gil'miyarova, Z. K. Tenesheva, Ukr. Biokhim. Zh. *43*, 292 (1971), C.A. *75*, 107349w (1971); Level of oxidized and reduced forms of nicotinamide coenzyme and α-glycerophosphate in myocardium and aorta during experimental atherosclerosis. [**150**] M. H. Gottlieb, E. D. Eanes, Biophys. J. *12*, 1533 (1972), C.A. *78*, 25655r (1973); Influence of electrolytes on the thickness of the phospholipid bilayer of lamellar lecithin mesophases.

[**151**] M. H. Gottlieb, E. D. Eanes, Biophys. J. *14*, 335 (1974), C.A. *81*, 22441u (1974); Coexistence of rigid crystals and liquid crystals in lecithin/water mixtures. [**152**] M. H. Gottlieb, E. D. Eanes, Biochim. Biophys. Acta *373*, 519 (1974), C.A. *82*, 94523r (1975); Phase transitions in erythrocyte membranes and extracted membrane lipids. [**153**] M. H. Gottlieb, J. Colloid Interface Sci. *48*, 394 (1974), C.A. *81*, 116294a (1974); Binding of alkaline and alkaline earth cations in lecithin mesophases. [**154**] W. B. Gratzer, D. A. Cowburn, Nature *222*, 426 (1969), C.A. *71*, 18711h (1969); Optical activity of biopolymers. [**155**] W. B. Gratzer, M. Haynes, R. A. Garrett, Biochemistry *9*, 4410 (1970), C.A. *74*, 174n (1971); Structure of nucleic acid/poly base complexes. [**156**] M. Gregson, G. P. Jones, M. Davies, Trans. Faraday Soc. *67*, 1630 (1971), C.A. *75*, 68670x (1971); High field pulse measurements of the electric anisotropy of polarizability and the electric dipole moment of PBLG. [**157**] O. H. Griffith, G. H. Lesch, G. F. Rempfer, G. B. Birrell, C. A. Burke, D. W. Schlosser, M. H. Mallon, G. B. Lee, R. G. Stafford, P. C. Jost, T. B. Marriott, Proc. Nat. Acad. Sci. USA *69*, 561 (1972); Photoelectron microscopy: a new approach to mapping organic and biological surfaces. [**158**] W. Gross, Angew. Chem. *83*, 419 (1971), Angew. Chem. Int. Ed. Engl. *10*, 388 (1971), C.A. *75*, 71222h (1971); Biologische Membranen. [**159**] H. Gruler, Z. Naturforsch. C *30*, 608 (1975), C.A. *83*, 174338v (1975); Chemoelastic effect of membranes. [**160**] C. Grunwald, Plant Physiol. *48*, 653 (1971), C.A. *76*, 54776v (1972); Effects of free sterols, steryl ester, and steryl glycoside on membrane permeability. [**160a**] R. Grupe, G. Menzel, E. Preusser, H. Goering, Stud. Biophys. *62*, 233 (1977), C.A. *86*, 116386k (1977); Vergleich des Einflusses von Cholesterin und β-Sitosterin auf den Phasenübergang der steroidhaltigen Dipalmitoyllecithin-Single- und -Multischichtliposomen.

[**161**] J. Hanson, Proc. Roy. Soc. B *183*, 39 (1973), C.A. *78*, 93954n (1973); Evidence from electron microscope studies on actin paracrystals concerning the origin of the cross-striation in the thin filaments of vertebrate skeletal muscle. [**162**] S. Hashimoto, S. Dayton, R. B. Alfin-Slater, Life Sci. *12*, 1 (1973), C.A. *78*, 95763y (1973); Esterification of cholesterol by homogenates of atherosclerotic and normal aortas. [**163**] G. Hauska, Angew. Chem. *84*, 123 (1972), Angew. Chem. Int. Ed. Engl. *11*, 153 (1972), C.A. *82*, 39918y (1975); Vergleichende Untersuchungen zur Topographie der Membranen in Mitochondrien und Chloroplasten. [**163a**] H. Hayashi, K. Hasegawa, J. Electron Microscopy *25*, 117 (1976); Phase transition of protein liquid crystal, yolk platelet proteins from newt, triturus pyrrhogaster. [**164**] M. Hayashi, Tampakushitsu, Kakusan, Koso *15*, 1088 (1970), C.A. *74*, 9531j (1971); Monolayers and bilayers of complex lipids. [**165**] W. Helfrich, Z. Naturforsch. C *29*, 182 (1974); Deformation of lipid bilayer spheres by electric fields. [**166**] W. Helfrich, Z. Naturforsch. C *29*, 510 (1974), C.A. *82*, 1354y (1975); Blocked lipid exchange in bilayers and its possible influence on the shape of vesicles. [**167**] W. Helfrich, J. J. Deuling, J. Phys. (Paris), C-1, *36*, 327 (1975); Theoretical shapes of red blood cells. [**168**] W. Helfrich, Z. Naturforsch. C *30*, 841 (1975); Out-of-plane fluctuations of lipid bilayers. [**168a**] R. M. Servuss, W. Harbich, W. Helfrich, Biochim. Biophys. Acta *436*, 900 (1976), C.A. *85*, 105671c (1976); Measurement of the curvature-elastic modulus of egg lecithin bilayers. [**168b**] H. J. Deuling, W. Helfrich, Biophys. J. *16*, 861 (1976); Red blood-cell shapes as explained on basis of curvature elasticity. [**168c**] H. J. Deuling, W. Helfrich, J. Phys. (Paris) *37*, 1335 (1976), C.A. *86*, 1395v (1977); Curvature elasticity of fluid membranes: a catalog of vesicle shapes. [**168d**] W. Harbich, R. M. Servuss, W. Helfrich, Phys. Lett. A *57*, 294 (1976), C.A. *85*, 88870x (1976); Optical studies of lecithin-membrane melting. [**169**] J. Hermans, Jr., J. Colloid Sci. *17*, 638 (1962), C.A. *58*, 1546a (1963); Viscosity of concentrated solutions of rigid, rodlike molecules (PBLG) in *m*-cresol. [**170**] J. Hermans, Jr., Advan. Chem. Ser. *63*, 282 (1967), C.A. *68*, 3267g (1968); Order and structure in concentrated polymer solutions and gels.

[**171**] M. W. Hill, Biochim. Biophys. Acta *356*, 117 (1974), C.A. *82*, 10976e (1975); Effect of anesthetic-like molecules on the phase transition in smectics of dipalmitoyllecithin. I. Normal alcohol up to C = 9 and three inhalation anesthetics. [**172**] M. W. Hill, Biochem. Soc. Trans. *3*, 149 (1975), C.A. *83*, 37910m (1975); Partition coefficients of anesthetic-like molecules between water and smectics of dipalmitoyl phosphatidylcholine. [**173**] G. M. Hillman, D. M. Engelman, J. Clin. Invest. *58*, 1008

(1976), C.A. *85*, 175201y (1976); Compositional mapping of cholesteryl ester droplets in the fatty streaks of human aorta. [174] H. J. Hinz, J. M. Sturtevant, J. Biol. Chem. *247*, 6071 (1972), C.A. *78*, 1322d (1973); Calorimetric studies of dilute aqueous suspensions of bilayers formed from synthetic L-α-lecithins. [174a] M. Hochli, C. R. Hackenbrock, Proc. Natl. Acad. Sci. USA *73*, 1636 (1976), C.A. *85*, 15886x (1976); Fluidity in mitochondrial membranes: Thermotropic lateral translational motion of intramembrane particles. [174b] M. Höfer, Transport durch biologische Membranen. Das Konzept der Trägerkatalyse (Verlag Chemie: Weinheim). 1977. 128 pp. [175] D. F. Hoelzl Wallach, Biochim. Biophys. Acta *265*, 61 (1972); Dispositions of proteins in the plasma membranes of animal cells: analytical approaches using controlled peptidolysis and protein labels. [176] R. T. Holzbach, M. Marsh, Mol. Cryst. Liq. Cryst. *28*, 217 (1974), C.A. *82*, 136713k (1975); Transient liquid crystals in human bile analogs. [177] R. T. Holzbach, C. Corbusier, M. Marsh, Gastroent. *67*, A-22/799 (1974); Transient biliary liquid crystals and metastable supersaturation in prairie dogs during rapid dietary induction of cholesterol gallstones. [178] R. Hosemann, W. Kreutz, Naturwissenschaften *53*, 298 (1966), C.A. *65*, 7497a (1966); Tertiary structure of the protein layers of chloroplasts. [179] W. Dreissig, R. Hosemann, T. Nemetschek, Z. Naturforsch. C *29*, 516 (1972), C.A. *81*, 152629x (1974); Parakristallite in nativem Kollagen. [180] F. K. Hui, P. G. Barton, Biochim. Biophys. Acta *296*, 510 (1973), C.A. *78*, 107314t (1973); Mesomorphism of phospholipids with aliphatic alcohols and other nonionic substances. [180a] D. W. L. Hukins, J. Woodhead-Galloway, Mol. Cryst. Liq. Cryst. *41*, 33 (1977), C.A. *87*, 179371t (1977); Collagen fibrils as examples of s_A biological fibers.

[181] A. Hybl, Mol. Cryst. Liq. Cryst. *36*, 271 (1976), C.A. *86*, 67075x (1977); Distribution of thickness irregularities shows myelin is paracrystalline. [181a] A. Hybl, J. Appl. Cryst. *10*, Pt. 3, 141 (1977), C.A. *87*, 17700x (1977); Paracrystalline nature of myelin. [182] E. Iizuka, Biochim. Biophys. Acta *243*, 1 (1971), C.A. *75*, 64512z (1971); Electric orientation of liquid crystals of PBLG. [183] E. Iizuka, Y. Go, J. Phys. Soc. Jap. *31*, 1205 (1971), C.A. *75*, 144751s (1971); Experiments on magnetic orientation of liquid crystals of PBLG. [184] E. Iizuka, J. Phys. Soc. Jap. *34*, 1054 (1973), C.A. *78*, 148364c (1973); NMR measurements of liquid crystals of PBLG in static electric fields. [185] E. Iizuka, Polym. J. *4*, 401 (1973), C.A. *79*, 53962c (1973); Effects of magnetic fields on the structure of cholesterics of polypeptides. [186] E. Iizuka, T. Keira, A. Wada, Mol. Cryst. Liq. Cryst. *23*, 13 (1973), C.A. *80*, 27608t (1974); Light scattering by liquid crystals of PBLG in an electric field. [187] E. Iizuka, Polym. J. *5*, 62 (1973), C.A. *80*, 48492r (1974); Electromagnetic orientation of liquid crystals of poly(γ-ethyl glutamates). [188] E. Iizuka, Mol. Cryst. Liq. Cryst. *25*, 287 (1974), C.A. *81*, 106185b (1974); Flow properties of liquid crystals of polypeptides. [189] E. Iizuka, Mol. Cryst. Liq. Cryst. *27*, 161 (1974), C.A. *82*, 162008d (1975); Birefringence of liquid crystals of polypeptides in a magnetic field. [190] E. Iizuka, J. T. Yang, Mol. Cryst. Liq. Cryst. *29*, 27 (1974), C.A. *83*, 10794q (1975); Circular dichroism of the liquid crystals of PBLG under static electric fields.

[191] E. Iizuka, Seibutsu Butsuri *15*, 133 (1975), C.A. *83*, 171104e (1975); Electromagnetic orientation of liquid crystals of polypeptides. [192] E. Iizuka, Polym. J. *7*, 650 (1975), C.A. *84*, 31657h (1976); Orientation of the atomic groups of PBLG in liquid crystalline states. [193] E. Iizuka, Adv. Polym. Sci. *20*, 79 (1976), C.A. *85*, 78321e (1976); Properties of liquid crystals of polypeptides with stress on the electromagnetic orientation. [194] E. Iizuka, Mol. Cryst. Liq. Cryst. *42*, 67 (1977); Liquid crystals of deoxygenated sickle-cell hemoglobin. [194a] E. Iizuka, Y. Kondo, Y. Ukai, Polym. J. *9*, 135 (1977), C.A. *87*, 23853r (1977); Liquid crystals of sodium salt of poly(glutamic acid) in aqueous solution. [194b] E. Iizuka, Polym. J. *9*, 173 (1977), C.A. *87*, 34576d (1977); Liquid crystals of the sodium salt of deoxyribonucleic acid. [195] A. Ishihara, J. Chem. Phys. *19*, 1142 (1951), C.A. *46*, 810b (1952); Theory of anisotropic colloidal solutions. [195a] K. Ito, T. Kajiyama, M. Takayanagi, Polym. J. *9*, 355 (1977), C.A. *87*, 185105a (1977); Aggregated states of molecular chains in solid-state films of poly(γ-n-alkyl D-glutamate) cast from various solutions. [196] S. Iwayanagi, Y. Sugiura, Oyo Butsuri *40*, 539 (1971), C.A. *76*, 69184z (1972); Liquid crystals related to biological systems. [197] S. Iwayanagi, Y. Sugiura, Seibutsu Butsuri *15*, 164 (1975), C.A. *84*, 55359p (1976); Liquid crystals and membrane structures. [197a] S. Iwayanagi, Hyomen *15*, 255 (1977), C.A. *87*, 113202g (1977); Biological organisms and liquid crystals. [198] L. Jakoi, S. H. Quarfordt, J. Biol. Chem. *249*, 5840 (1974), C.A. *82*, 39134q (1975); Induction of hepatic cholesterol synthesis in the rat by lecithin mesophase infusions. [198a] M. J. Janiak, Diss. Grad. Sch., Boston Univ., Boston, Mass. 1977. 341 pp. Avail. Univ. Microfilms Int., Order No. 77-11,400, C.A. *87*, 97730v (1977); Structure of phosphatidylcholines and their interaction with cholesteryl esters. [199] B. R. Jennings, H. G. Jerrard, Nature *210*, 90 (1966), C.A. *64*, 19765c

(1966); Helical conformation of PBLG. [**200**] B. R. Jennings, E. D. Baily, Nature *228*, 1309 (1970), C.A. *74*, 54311y (1971); Use of transient optical rotation for biopolymer characterization. [**201**] P. Joos, Chem. Phys. Lipids *4*, 162 (1970), C.A. *73*, 21467e (1970); Cholesterol as liquifier in phospholipid membranes studied by surface viscosity measurements of mixed monolayers. [**202**] E. Junger, H. Reinauer, Biochim. Biophys. Acta *183*, 304 (1969), C.A. *71*, 75059a (1969); Liquid crystals of hydrated phosphatidylethanolamine. [**203**] H. R. Kaback et al., Molecular aspects of membrane phenomena (Intern. Symp. Battelle Res. Center, Seattle, Nov. 4–6, 1974). (Springer: New York). 1975. 338 pp. C.A. *85*, 2450a (1976). [**204**] H. Kacser, Science *124*, 151 (1956); Molecular organization of genetic material. [**205**] F. E. Karasz, J. M. O'Reilly, Advan. Chem. Ser. *63*, 180 (1967), C.A. *67*, 22215j (1967); Helix-coil transition in deuterated PBLG. [**206**] K. M. Kasumov, E. A. Liberman, Biofizika *19*, 71 (1974), C.A. *81*, 21515j (1974); Role of cholesterol in increasing conductance of bimolecular membranes with help of polyenic antibiotics. [**207**] M. Kawamura, K. Maruyama, J. Biochem. (Tokyo) *68*, 885 (1970), C.A. *74*, 71735g (1971); Polymorphism of F-actin. I. Three forms of paracrystals. [**208**] N. Kellermayer, K. Jobst, Exp. Cell Res. *63*, 204 (1970), C.A. *74*, 19464d (1971); Ion-dependent anisotropy of deoxyribonucleoprotein structures in tissue cultures. [**209**] R. D. Keynes, Nature *239*, 29 (1972), C.A. *77*, 148612m (1972); Excitable membranes. [**210**] R. S. Khare, R. K. Mishra, W. H. Falor, A. J. Hopfinger, Curr. Sci. *43*, 67 (1974), C.A. *80*, 107512x (1974); CD of sphingomyelin.

[**211**] R. S. Khare, C. R. Worthington, Mol. Cryst. Liq. Cryst. *38*, 195 (1977); X-ray diffraction study of sphingomyelin/cholesterol interaction in oriented bilayers. [**212**] S. C. Kinsky, Biochim. Biophys. Acta *265*, 1 (1972), C.A. *76*, 125068d (1972); Antibody-complement interaction with lipid model membranes. [**213**] M. Kléman, C. Colliex, M. Veyssié, Adv. Chem. Ser. *152*, 71 (1976), C.A. *85*, 88882c (1976); Recognition of defects in water/lecithin Lα phases. [**213a**] M. Kléman, Proc. Roy. Soc. A *347*, 387 (1976); Remarks on a possible elasticity of membranes and lamellar media. Disordered layers. [**213b**] M. Kléman, C. E. Williams, M. J. Costello, T. Gulik-Krzywicki, Philos. Mag. *35*, 33 (1977), C.A. *87*, 14493w (1977); Defect structures in lyotropic smectics revealed by freeze-fracture electron microscopy. [**214**] W. E. Klopfenstein, B. de Kruyft, A. J. Verkleij, R. A. Demel, L. L. M. van Deenen, Chem. Phys. Lipids *13*, 215 (1974), C.A. *82*, 69480s (1975); DSC on mixtures of lecithin, lysolecithin, and cholesterol. [**215**] K. K. Kobayashi, Kotai Butsuri *5*, 420 (1970), C.A. *74*, 38217s (1971); Liquid crystals in biological systems. [**215a**] N. N. Kolotilov, Ya. T. Terletskaya, E. P. Kozulina, Ya. V. Belik, Tezisy Dokl. — Vses. Soveshch. Org. Kristallokhim., 1st *1974*, 102, C.A. *87*, 64384g (1977); Liquid crystals of myelin and its physicochemical properties. [**216**] A. A. Kozinski, G. J. Kizior, S. G. Wax, AIChE J. *20*, 1104 (1974), C.A. *82*, 60852g (1975); Separations with protein liquid crystals. [**217**] W. Kreutz, Angew. Chem. *84*, 597 (1972), Angew. Chem. Int. Ed. Engl. *11*, 551 (1972), C.A. *77*, 136302x (1972); Strukturprinzipien in Bio-Membranen. [**218**] D. Kritchevsky, Am. J. Clin. Nutr. *23*, 1105 (1970), C.A. *73*, 74393j (1970); Role of cholesterol vehicle in experimental atherosclerosis. [**219**] J. Kroes, R. Ostwald, Biochim. Biophys. Acta *249*, 647 (1971), C.A. *76*, 32495a (1972); Erythrocyte membranes. Effect of increased cholesterol content on permeability. [**220**] K. Kubo, S. Hiraga, K. Ogino, Jap. J. Appl. Phys. *11*, 427 (1972), C.A. *76*, 154327t (1972); Two-phase solution of PBLG.

[**221**] M. Kubo, I. Karube, S. Suzuki, Biochem. Biophys. Res. Commun. *69*, 731 (1976), C.A. *84*, 161196t (1976); Electric field control of enzyme membrane activity. [**222**] B. Kuennert, H. Krug, Acta Histochem., Suppl. *1971*, 165, C.A. *75*, 149853m (1971); Histochromatographische Untersuchungen der Cholesterinesterfraktionen bei Aortensklerose. [**223**] R. H. Lange, Princ. Tech. Electron Microsc. *6*, 241 (1976), C.A. *85*, 16484b (1976); Tilting experiments in the electron microscope. [**224**] I. Langmuir, J. Chem. Phys. *6*, 873 (1938), C.A. *33*, 1196[6] (1939); Role of attractive and repulsive forces in the formation of tactoids, thixotropic gels, protein crystals and coacervates. [**225**] J. Lapointe, D. A. Marvin, Mol. Cryst. Liq. Cryst. *19*, 269 (1973), C.A. *78*, 81270y (1973); Filamentous bacterial viruses. VIII. Liquid crystals of flexible deoxyribonucleoprotein. [**226**] K. Larsson, Z. Phys. Chem. (Frankfurt) *56*, 173 (1967), C.A. *69*, 22470e (1968); Structure of mesophases and micelles in aqueous glyceride systems. [**227**] K. Larsson, L. Sjostrom, Ark. Kemi *30*, 1 (1968), C.A. *70*, 32461n (1969); Behavior of the di-1-monoglyceride of dodecanedioic acid in the solid state and in aqueous systems. [**228**] K. Larsson, I. Lundstrom, Adv. Chem. Ser. *152*, 43 (1976), C.A. *85*, 105455k (1976); Liquid crystals in biological model systems. [**229**] K. Larsson, Food Sci. (N. Y.) *5* (Food Emulsions), 39 (1976), C.A. *85*, 102452w (1976); Crystal and liquid crystal structures of lipids. [**230**] M. A. Lauffer, W. M. Stanley, J. Biol. Chem. *123*, 507 (1938), C.A. *32*, 5419 (1938); Stream double refraction of virus proteins.

[**231**] A. L. Lehniger, Biochemistry. 2nd ed. (Worth Publishers: New York). 1976. 1104 pp. [**232**] L. Leive, Microbiology series, vol. 1: Bacterial membranes and walls (Dekker: New York).

1973. 495 pp. C.A. *85*, 2445c (1976). [**233**] Y. K. Levine, M. H. F. Wilkins, Nature, New Biol. *230*, 69 (1971), C.A. *75*, 15164d (1971); Structure of oriented lipid bilayers. [**234**] W. R. Lieb, W. D. Stein, Biochim. Biophys. Acta *265*, 187 (1972), C.A. *77*, 30363c (1972); Carrier and non-carrier models for sugar transport in the human red blood cell. [**235**] H. Limacher, J. Seelig, Angew. Chem. *84*, 950 (1972), Angew. Chem. Int. Ed. Engl. *11*, 920 (1972); C.A. *78*, 29121m (1973); Faltung von Fettsäuren in flüssig-kristallinen Doppelschichten. [**236**] J. D. Litster, Phys. Lett. A *53*, 193 (1975), C.A. *83*, 74043b (1975); Stability of lipid bilayers and red blood cell membranes. [**237**] G. I. Loeb, J. Colloid Interface Sci. *26*, 236 (1968), C.A. *68*, 87682z (1968); Study of the conformation of surface films of PBLG by multiple internal reflection spectroscopy. [**237a**] C. R. Loomis, Diss. Grad. Sch., Boston Univ., Boston, Mass. 1977. 597 pp. Avail. Univ. Microfilms Int., Order No. 77-11,408, C.A. *87*, 97729b (1977); Physical interactions of cholesteryl esters, cholesterol and phosphatidylcholine. [**238**] G. R. Luckhurst, IQ *68*, 46 (1974), C.A. *84*, 100928m (1976); Liquid crystals in the life processes. [**239**] B. Lundberg, Acta Chem. Scand. B *28*, 673 (1974), C.A. *82*, 12667d (1975); X-ray and vapor pressure study on lecithin/cholesterol/water interactions. [**240**] B. Lundberg, Chem. Phys. Lipids *14*, 309 (1975), C.A. *83*, 176387j (1975); Similarity between mesomorphic droplets in atherosclerotic lesions and cholesteryl ester suspensions. [**240a**] B. Lundberg, Acta Chem. Scand. B *30*, 150 (1976), C.A. *84*, 148906e (1976); Thermal properties of systems containing cholesteryl esters and triglycerides.

[**241**] I. Lundstroem, D. MacQueen, J. Theor. Biol. *45*, 405 (1974), C.A. *81*, 86997n (1974); Proposed 1/f noise mechanism in nerve cell membranes. [**242**] E. S. Lutton, J. Am. Oil Chem. Soc. *42*, 1068 (1965), C.A. *64*, 6913h (1966); Phase behavior of aqueous systems of monoglycerides. [**243**] A. Nicolaieff, A. Mazen-Knobloch, R. Verdrely, V. Luzzati, Compt. rend. *248*, 2805 (1959), C.A. *53*, 18995g (1959); Structure des gels aqueux de nucléoprotéines: d'origines différents: étude par la diffusion central des rayons X. [**244**] V. Luzzati, A. Nicolaieff, J. Mol. Biol. *1*, 127 (1959), C.A. *54*, 3539e (1960); Étude par diffusion des rayons X aux petits angles des gels d'acide DRN et de nucléoprotéines. [**245**] V. Luzzati, F. Reiss-Husson, P. Saludjian, Ciba Found. Symp., Principles Biomol. Organ. *1966*, 69, C.A. *67*, 87680b (1967); Phase changes in organized lipid and polypeptide structures. [**246**] V. Luzzati, T. Gulik-Krzywicki, A. Tardieu, Nature *218*, 1031 (1968), C.A. *69*, 33045w (1968); Polymorphism of lecithins. [**247**] T. Gulik-Krzywicki, E. Shechter, V. Luzzati, M. Faure, Nature *223*, 1116 (1969), C.A. *71*, 119763g (1969); Interactions of proteins and lipids: structure and polymorphism of protein/lipid/water phases. [**248**] E. Shechter, T. Gulik-Krzywicki, V. Luzzati, J. Phys. (Paris), C-4, *30*, 89 (1969); Propriétés optiques rotatoires de quelques systèmes a structure smectique formes de protéines, lipides et eau. [**249**] A. Tardieu, J. L. Ranck, L. Mateu, D. M. Sadler, T. Gulik-Krzywicki, V. Luzzati, Liq. Cryst., Proc. Int. Conf., Bangalore, *1973*, 115; Lipid/water systems: structure and structural transitions. [**250**] J. Q. Lynd, F. T. Lynd, Tex. Rep. Biol. Med. *33*, 423 (1975), C.A. *85*, 105051a (1976); Aflatoxin effects with liquid crystalline cholesteryl esters of duckling serums.

[**251**] J. W. McBain, E. Jameson, Trans. Faraday Soc. *26*, 768 (1930), C.A. *25*, 1546 (1931); Phase rule equilibria of horse serum globulin. [**252**] M. P. McDonald, L. D. R. Wilford, Liq. Cryst. Ord. Fluids *2*, 225 (1974), C.A. *86*, 180983e (1977); Lipid/water systems. V. IR-spectra of mesophases. [**253**] R. W. McGilvery, Biochemistry (Saunders: Philadelphia). 1970. 769 pp. C.A. *73*, 105628w (1970). [**254**] H. R. Mahler, E. H. Cordes, Biological Chemistry. 2nd ed. (Harper and Row: New York). 1977. 979 pp. [**255**] Y. Majima, Keio J. Med. *21*, 1 (1972), C.A. *78*, 121895q (1973); Interchangeability of fatty acids between phospholipids and esterified cholesterol in the plasma. [**256**] P. Manigault, G. Segrétain, Compt. rend. *224*, 152 (1947), C.A. *41*, 2303i (1947); Biréfringence magnétique du virus mosaique du tabac. [**257**] S. Marcelja, Nature, New Biol. *241*, 451 (1973), C.A. *78*, 120684h (1973); Molecular model for phase transitions in biological membranes. [**258**] S. Marcelja, Biochim. Biophys. Acta *367*, 165 (1974), C.A. *82*, 30866x (1975); Chain ordering in liquid crystals. II. Structure of bilayer membranes. [**259**] R. H. Marchessault, F. F. Morehead, N. M. Walter, Nature *184*, Suppl. No. 9, 632 (1959), C.A. *54*, 13221f (1960); Liquid crystal systems from fibrillar polysaccharides. [**260**] G. M. Martin, D. H. Reneker, Bull. Am. Phys. Soc. *17*, 317 (1972); Rotation of sheets of PBLG-molecules in a liquid crystal.

[**261**] A. Martonosi, Enzymes of Biological Membranes (Wiley: Chichester). 1976. Vol. 1: Physical and Chemical Techniques. 272 pp. Vol. 2: Biosynthesis of Cell Components. 674 pp. Vol. 3: Membrane Transport. 476 pp. Vol. 4: Electron Transport Systems and Receptors. 448 pp. C.A. *86*, 102537y to 102540u (1977). [**262**] R. Marusyk, E. Norrby, H. Marusyk, J. Gen. Virol. *14*, 261 (1972), C.A. *77*, 2701t (1972); Relation of adenovirus-induced paracrystalline structures to the virus core protein(s). [**263**] J. P. Mathieu, N. Faraggi, Compt. rend. *205*, 1378 (1937); Étude de la lumière polarisée circulairement

réfléchie par certains coléoptères. [264] B. W. Maxfield, R. M. Clegg, L. Avery, E. L. Elson, Biophys. J. *15*, A104 (1975); Kinetic aspects of gel liquid crystal transition in phospholipids. [264a] D. L. Melchior, J. M. Steim, Annu. Rev. Biophys. Bioeng. *5*, 205 (1976), C.A. *85*, 73508m (1976); Thermotropic transitions in biomembranes. [265] D. R. Merker, B. F. Daubert, J. Am. Chem. Soc. *86*, 1009 (1964), C.A. *60*, 9944b (1964); Molecular structure in monolayers of saturated triglycerides on water as related to three-dimensional polymorphic forms. [266] C. Mettenheimer, Corr. Blatt des Vereins für gem. Arbeit zur Förderung der wiss. Heilkunde *31*, 467 (1858). [267] C. Mezei, J. Neurochem. *17*, 1163 (1970), C.A. *73*, 96646w (1970); Cholesterol esters and hydrolytic cholesterol esterase during Wallerian degeneration. [268] A. A. Michelson, Phil. Mag. *21*, 554 (1911); Metallic colouring of birds and insects. [269] V. M. Mikhailov, S. A. Seleznev, Izv. Vyssh. Ucheb. Zaved., Khim. Khim. Tekhnol. *16*, 1597 (1973), C.A. *80*, 42535k (1974); IR spectroscopic characteristics of the smectic mesophase of the cerebron/water system. [270] A. Miller, J. Woodhead-Galloway, Nature *229*, 470 (1971); Long range forces in muscle.

[271] R. E. Miller, E. Shelton, E. R. Stadtman, Arch. Biochem. Biophys. *163*, 155 (1974), C.A. *81*, 165406x (1974); Zinc-induced paracrystalline aggregation of glutamine synthetase. [272] W. G. Miller, E. L. Wee, J. Phys. Chem. *75*, 1446 (1971), C.A. *75*, 10928w (1971); Liquid crystal/isotropic phase equilibria in the system PBLG/dimethylformamide. [273] W. G. Miller, J. H. Rai, E. L. Wee, Liq. Cryst. Ord. Fluids *2*, 243 (1974), C.A. *86*, 180985g (1977); Liquid crystal/isotropic phase equilibria in stiff chain polymers. [274] W. G. Miller, C. C. Wu, E. I. Wee, G. L. Santee, J. H. Rai, K. G. Goebel, Pure Appl. Chem. *38*, 37 (1974), C.A. *82*, 178293d (1975); Thermodynamics and dynamics of polypeptide liquid crystals. [275] R. K. Mishra, N. K. Roper, R. S. Tyagi, Liq. Cryst., Int. Conf. Bangalore, *1973*, Abstr. No. 31; Energetics of association in a lyotropic liquid crystalline compound. [276] R. K. Mishra, Mol. Cryst. Liq. Cryst. *29*, 201 (1975), C.A. *82*, 165981b (1975); Occurrence, fluctuations, and significance of liquid crystallinity in living systems. [277] D. N. Misra, C. T. Ladoulis, L. W. Estes, T. J. Gill III, Liq. Cryst. Ord. Fluids *2*, 495 (1974), C.A. *86*, 184790n (1977); Effect of detergents on isolated rat lymphocyte plasma membranes. [278] N. Miyata, K. Tohyama, Y. Go, J. Phys. Soc. Jap. *33*, 1180 (1972), C.A. *78*, 35560v (1973); Magnetic alignment of cholesteric poly-γ-ethyl-L-glutamate. [279] W. Moffitt, J.-T. Yang, Proc. Natl. Acad. Sci. U. S. *42*, 596 (1956), C.A. *52*, 20316b (1958); ORD of simple polypeptides. [280] Y. Mohadger, A. V. Tobolsky, T. K. Kwei, Macromolecules *7*, 924 (1974), C.A. *82*, 17403z (1975); Permeation of isomers of mandelic acid through isotropic and cholesteric fluid polypeptide membranes.

[281] G. Mueller, Nature *235*, 90 (1972); Organic microspheres from the precambrian of SW Africa. [282] T. Murakami, Hiroshima Daigaku Igaku Zasshi *20*, 31 (1972), C.A. *78*, 28329y (1973); Lipid metabolism in the atherosclerotic lesion and normal intima of cholesterol-fed rabbits. [283] F. R. N. Nabarro, W. F. Harris, Nature *232*, 423 (1971); Presence and function of disclinations in surface coats of unicellular organisms. [284] A. Nagayama, S. Dales, Proc. Nat. Acad. Sci. U.S. *66*, 464 (1970), C.A. *73*, 85920e (1970); Rapid purification and the immunological specifity of mammalian microtubular paracrystals possessing an ATPase activity. [285] M. Nakagaki, N. Funasaki, Nippon Kagaku Kaishi *1972*, 2255, C.A. *78*, 62566h (1973); Water permeability through mixed monolayers of glycerides with cholesteryl esters. [286] A. Nakajima, T. Hayashi, Kobunshi Kagaku *24*, 230, 235 (1967), C.A. *67*, 82497n, 82498p (1967); Structure and conformation of synthetic polypeptides. I. Synthesis and ORD of poly(γ-methyl-D-glutamate). II. Molecular conformation of poly(α-methyl-L-glutamate) in various two-component solvents. [287] H. Netter, Theoretische Biochemie; physikalisch-chemische Grundlagen der Lebensvorgänge. (Springer: Berlin). 1959. 816 pp. C.A. *53*, 18142f (1959). [288] A. C. Neville, S. Caveney, Biol. Rev. Cambridge Phil. Soc. *44*, 531 (1969), C.A. *72*, 97704e (1970); Scarabaeid beetle exocuticle as an optical analog of cholesterics. [289] A. C. Neville, B. M. Luke, J. Cell Sci. *8*, 93 (1971), C.A. *75*, 31945q (1971); Biological system producing a self-assembling cholesteric protein liquid crystal. [290] H. Nomori, N. Tsuchihashi, S. Takagi, M. Hatano, Bull. Chem. Soc. Jpn. *48*, 2522 (1975), C.A. *84*, 17703p (1976); Induced CD of benzyl chromophores bound to helical polypeptides.

[291] J. Oelze, G. Drews, Biochim. Biophys. Acta *265*, 209 (1972), C.A. *77*, 31514q (1972); Membranes of photosynthetic bacteria. [292] I. Ohtsuki, T. Wakabayashi, J. Biochem. (Tokyo) *72*, 369 (1972), C.A. *77*, 136503p (1972); Optical diffraction studies on the structure of troponin-tropomyosin-actin paracrystals. [293] Y. Okamoto, R. Sakamoto, Nippon Kagaku Zasshi *90*, 669 (1969), C.A. *71*, 81822k (1969); Conformation of poly(γ-methyl-L-glutamate) in solution. [294] C. T. O'Konski, A. J. Haltner, J. Am. Chem. Soc. *79*, 5634 (1957), C.A. *52*, 1731b (1958); Study of electric polarization in polyelectrolyte solutions by means of electric birefringence. [295] L. Onsager, Phys. Rev. *62*, 558

(1942), C.A. *38*, 2860² (1944); Anisotropic solutions of colloids. [**296**] A. Pace, Jr., Science *176*, 678 (1972); Cholesteric liquid crystal-like structure of the cuticle of plusiotis gloriosa. [**297**] R. E. Pagano, N. L. Gershfeld, J. Colloid Interface Sci. *44*, 382 (1973), C.A. *79*, 83746c (1973); Phase separation in cholesterol/dipalmitoyl lecithin mixed films and the condensing effect. [**298**] J. Paillas, J. Tricoire, J. Quillard, A. Vieillefond, Ann. Anat. Pathol. *20*, 121 (1975), C.A. *83*, 161831a (1975); Cristaux liquides de cholestérol substratum physique des premières lésions de l'athérome. [**299**] M. Panar, W. D. Phillips, J. Am. Chem. Soc. *90*, 3880 (1968), C.A. *69*, 36552v (1968); Magnetic ordering of PBLG-solutions. [**300**] A. D. Bangham, D. Papahadjopoulos, Biochim. Biophys. Acta *126*, 181, 185 (1966), C.A. *65*, 20417b, c (1966); Biophysical properties of phospholipids. I. Interaction of phosphatidylserine monolayers with metal ions. II. Permeability of phosphatidylserine liquid crystals to univalent ions.

[**301**] D. Papahadjopoulos, N. Miller, J. C. Watkins, Biochim. Biophys. Acta *135*, 624, 639 (1967), C.A. *68*, 33580w, 33581x (1968); Phospholipid model membranes. I. Structural characteristics of hydrated liquid crystals. II. Permeability properties of hydrated liquid crystals. [**302**] D. Papahadjo-poulos, Biochim. Biophys. Acta *163*, 240 (1968), C.A. *69*, 83439y (1968); Surface properties of acidic phospholipids; interaction of monolayers and hydrated liquid crystals with uni- and bivalent metal ions. [**303**] D. Papahadjopoulos, S. Ohki, Liq. Cryst. Ord. Fluids *1970*, 13, C.A. *79*, 1708e (1973); Conditions of stability for liquid-crystalline phospholipid membranes. [**304**] D. Papahadjopoulos, Bio-chim. Biophys. Acta *265*, 169 (1972), C.A. *77*, 28685r (1972); Studies on the mechanism of action of local anesthetics with phospholipid model membranes. [**305**] D. Papahadjopoulos, J. Theor. Biol. *43*, 329 (1974), C.A. *80*, 143743g (1974); Cholesterol and cell membrane function. Hypothesis concerning the etiology of atherosclerosis. [**306**] D. Papahadjopoulos, M. Moscarello, E. H. Eylar, T. Isac, Biochim. Biophys. Acta *401*, 317 (1975), C.A. *83*, 143312c (1975); Effects of proteins on thermotropic phase transitions of phospholipid membranes. [**307**] V. D. Paponov, Mol. Biol. (Moscow) *10*, 544 (1976), C.A. *85*, 42529q (1976); Rheological behavior of deoxyribonucleoprotein systems of chromatin. I. DNP in high ionic strength solutions. [**308**] D. A. D. Parry, A. Elliott, Nature *206*, 616 (1965), C.A. *63*, 12441f (1965); X-ray diffraction patterns of liquid crystalline solutions of PBLG. [**309**] V. A. Parsegian, Membrane Models Form. Biol. Membranes, Proc. Meet. Int. Conf. Biol. Membranes *1967*, 303, C.A. *70*, 84226j (1969); Energetic model of ionic lipids in the liquid-crystal state. [**310**] D. F. Parsons, J. R. Subjeck, Biochim. Biophys. Acta *265*, 85 (1972), C.A. *76*, 138462c (1972); Morphology of the polysaccharide coat of mammalian cells.

[**311**] B. Perly, A. Douy, B. Gallot, C. R. Acad. Sci., Sér. C *279*, 1109 (1974), C.A. *82*, 125638p (1975); Synthèse de copolymères séquencés polybutadiène/PBLG et étude structurale de leurs mésophases. [**312**] M. F. Perutz, A. M. Liquori, S. Eirich, Nature *167*, 929 (1951); X-ray and solubility studies of the haemoglobin of sickle-cell anaemia patients. [**313**] M. Peterson, A. J. Day, R. K. Tume, E. Eisenberg, Exp. Mol. Pathol. *15*, 157 (1971), C.A. *75*, 138769n (1971); Ultrastructure, fatty acid content, and metabolic activity of foam cells and other fractions separated from rabbit atherosclerotic lesions. [**314**] A. G. Petrov, A. Derzhanski, J. Phys. (Paris), C-3, *37*, 155 (1976); Problems in the theory of elastic and flexoelectric effects in bilayer lipid membranes and biomembranes. [**315**] M. C. Phillips, D. E. Graham, H. Hauser, Nature *254*, 154 (1975), C.A. *83*, 54942r (1975); Lateral compressibility and penetration into phospholipid monolayers and bilayer membranes. [**316**] C. Picot, R. S. Stein, J. Polym. Sci. A-2, *8*, 1491 (1970), C.A. *73*, 88339h (1970); Effect of optical rotation on low-angle light-scattering patterns. [**317**] J. Popinigis, T. Wrzolkowa, Experientia *28*, 1292 (1972), C.A. *78*, 39742d (1973); Protamine induced configurational changes in paracrystalline structures from heat disintegrated mitochondria. [**318**] J. C. Powers, Jr., W. L. Peticolas, Advan. Chem. Ser. *63*, 217 (1967); C.A. *68*, 3296r (1968); Electric ordering in polypeptide solutions. Molecular aggregation studied by the Kerr effect. [**319**] J. C. Powers, Jr., Liq. Cryst. Ord. Fluids *1970*, 365, C.A. *79*, 42946z (1973); Aggregation of PBLG in mixed solvent systems. [**320**] J. C. Powers, Jr., W. L. Peticolas, Biopolymers *9*, 195 (1970), C.A. *72*, 101181b (1970); Aggregation of PBLG in mixed solvent systems.

[**321**] L. Powers, N. A. Clark, Proc. Natl. Acad. Sci. USA *72*, 840 (1975), C.A. *83*, 4189e (1975); Preparation of large monodomain phospholipid bilayer smectics. [**322**] K. S. Pozhilenkova, Y. A. Shorokhov, Strukt. Biosint. Prevrashch. Lipidov Org. Zhivotn. Chel., Kratk. Tezisy Dokl. Simp. *1972*, 124, C.A. *82*, 110757c (1975); Change in several properties of the cell membrane of erythrocytes during nutritional hypercholesterolemia. [**323**] P. L. Privalov, E. I. Tiktopulo, I. N. Serdyuk, Biofizika *14*, 972 (1969), C.A. *72*, 74782q (1970); Viscosity of tropocollagen solutions at small gradients of flow velocity. [**324**] J. W. Proudlock, A. J. Day, R. K. Tume, Atherosclerosis *18*, 451 (1973), C.A. *80*, 46357b (1974); Cholesterol-esterifying enzymes of foam cells isolated from atherosclerotic rabbit intima.

[325] S. H. Quarfordt, H. Oelschlaeger, W. R. Krigbaum, J. Clin. Inv. *51*, 1979 (1972); Liquid crystalline lipid in the plasma of humans with biliary obstruction. [326] G. Quincke, Wiedemanns Ann. Phys. *53*, 593 (1894); Freiwillige Bildung von hohlen Blasen, Schaum und Myelinformen der ölsauren Alkalien und verwandte Erscheinungen, besonders des Protoplasmas. [327] G. Quincke, Ber. physik. Ges. *6*, 615 (1908), C.A. *2*, 3183[1] (1908); Flüssige Kristalle, Myelin-Formen und künstliche Zellen mit flüssig-kristallinen Wänden. [328] J. C. Ravey, P. Mazeron, Polymer *16*, 329 (1975), C.A. *83*, 75072x (1975); Light scattering in an electric field. Variations of the $H_H(\theta)$ component around $\theta = 90°$. [329] F. A. Rawlins, B. G. Uzman, J. Cell Biol. *46*, 505 (1970), C.A. *73*, 108012b (1970); Retardation of peripheral nerve myelination in mice treated with inhibitors of cholesterol biosynthesis. Quantitative electron microscopic study. [330] S. Razin, Biochim. Biophys. Acta *265*, 241 (1972), C.A. *77*, 30362b (1972); Reconstitution of biological membranes. [330a] R. Reich, R. Scheerer, K. U. Sewe, H. T. Witt, Biochim. Biophys. Acta *449*, 285 (1976), C.A. *86*, 27757q (1977); Effect of electric fields on the absorption spectrum of dye molecules in lipid layers. V. Refined analysis of the field-indicating absorption changes in photosynthetic membranes by comparison with electrochromic measurements in vitro.

[331] F. Reiss-Husson, J. Mol. Biol. *25*, 363 (1967), C.A. *67*, 17954v (1967); Structure des phases liquide-cristallines de différents phospholipides, monoglycérides, sphingolipides, anhydres ou en présence d'eau. [331a] R. Riehl, Z. Naturforsch. C *32*, 305 (1977), C.A. *86*, 186251m (1977); Parakristalline Körper in jungen Oocyten der Schmerle Noemacheilus barbatulus (Teleostei, Cobitidae). [332] F. Rinne, Nature *126*, 279 (1930), C.A. *24*, 5773[8] (1930); Sperms as living liquid crystals. [333] F. Rinne, Naturwissenschaften *18*, 837 (1930), C.A. *25*, 975[8] (1931); Spermien als lebende Flüssigkristalle. [334] F. Rinne, Kolloid-Z. *56*, 71 (1931), C.A. *25*, 5604[3] (1931); Parakristalline Lebewesen. [335] F. Rinne, Kolloid-Z. *60*, 288 (1932), C.A. *27*, 4456[2] (1933); Wesen der Parakristalle und ihre Beteiligung an Zerebrosiden und Phosphatiden als plasmatischen Bestandteilen. [336] F. Rinne, Trans. Faraday Soc. *29*, 1016 (1933), C.A. *28*, 1239[7] (1934); Paracrystallinity. [337] R. T. Roberts, G. P. Jones, Mol. Cryst. Liq. Cryst. *17*, 281 (1972), C.A. *77*, 57705c (1972); Studies around the second cmc in lecithin. [338] C. Robinson, Trans. Faraday Soc. *52*, 571 (1956), C.A. *50*, 15175a (1956); Liquid crystals in solutions of a polypeptide. [339] C. Robinson, J. C. Ward, Nature *180*, 1183 (1957), C.A. *52*, 8663b (1958); Liquid crystals in polypeptides. [340] C. Robinson, J. C. Ward, R. B. Beevers, Disc. Faraday Soc. *25*, 29 (1958), C.A. *53*, 10882h (1959); Liquid crystals in polypeptide solutions.

[341] C. Robinson, Tetrahedron *13*, 219 (1961), C.A. *55*, 24585d (1961); Liquid crystals in polypeptide solutions. [342] C. Robinson, Mol. Cryst. *1*, 467 (1966); Cholesteric phase in polypeptide solutions and biological structures. [343] O. Rosenheim, Biochem. J. *8*, 110, 121 (1914), C.A. *8*, 2398, 2399 (1914); Galactosides of the brain. II. Preparation of phrenosin and kerasin by the pyridine method. III. Liquid crystals and the melting point of phrenosin. [344] J. E. Rothman, D. M. Engelman, Nature, New Biol. *237*, 42 (1972), C.A. *77*, 44598v (1972); Molecular mechanism for the interaction of phospholipid with cholesterol. [345] D. M. Engelman, J. E. Rothman, J. Biol. Chem. *247*, 3697 (1972), C.A. *77*, 30639x (1972); Planar organization of lecithin/cholesterol bilayers. [346] C. Ruska, H. Ruska, Naturwissenschaften *55*, 230 (1968), C.A. *70*, 62171k (1969); Molekulare Ordnung in Lecithinschichten. [347] E. Sackmann, Ber. Bunsenges. Phys. Chem. *78*, 929 (1974), C.A. *82*, 10062d (1975); Flüssig-kristalline Zustände in künstlichen und biologischen Membranen. [348] H. J. Galla, E. Sackmann, Ber. Bunsenges. Phys. Chem. *78*, 949 (1974), C.A. *82*, 57313b (1975); Lateral mobility of pyrene in model membranes of phospholipids with different chain lengths. [349] H. J. Galla, E. Sackmann, Biochim. Biophys. Acta *339*, 103 (1974), C.A. *80*, 117439t (1974); Lateral diffusion in the hydrophobic region of membranes: Use of pyrene excimers as optical probes. [350] H. J. Galla, E. Sackmann, J. Am. Chem. Soc. *97*, 4114 (1975), C.A. *83*, 74061f (1975); Chemically induced phase separation in mixed vesicles containing phosphatic acid. An optical study. [350a] C. Gebhardt, H. Gruler, E. Sackmann, Z. Naturforsch. C *32*, 581 (1977), C.A. *87*, 147545m (1977); Domain structure and local curvature in lipid bilayers and biological membranes. [350b] O. Albrecht, H. Gruler, E. Sackmann, J. Phys. (Paris) *39*, 301 (1978); Polymorphism of phospholipid monolayers.

[351] F. D. Saeva, G. R. Olin, J. Am. Chem. Soc. *95*, 7882 (1973), C.A. *80*, 8567n (1974); Cholesteric LCICD. VII. Achiral solutes in lyophases. [352] E. T. Samulski, A. V. Tobolsky, Nature *216*, 997 (1967), C.A. *68*, 30456z (1968); Solid liquid crystal films of PBLG. [353] E. T. Samulski, A. V. Tobolsky, U. S. Clearinghouse Fed. Sci. Tech. Inform., AD *1968*, AD-680461, 33 pp. C.A. *70*, 119181v (1969); Liquid crystal of PBLG in solution and in the solid state. [354] E. T. Samulski, A. V. Tobolsky, Mol. Cryst. Liq. Cryst. *7*, 433 (1969), C.A. *71*, 106419z (1969); Liquid crystal of PBLG in solution and in the solid state. [355] A. V. Tobolsky, E. T. Samulski, Pure Appl. Chem.

23, 145 (1970), C.A. *74*, 142330j (1971); Solid liquid-crystalline films of synthetic polypeptides. [**356**] E. T. Samulski, A. V. Tobolsky, Liq. Cryst. Ord. Fluids *1970*, 111, C.A. *79*, 35864t (1973); Cholesteric and nematic structures of PBLG. [**357**] A. V. Tobolsky, E. T. Samulski, Advan. Chem. Phys. *21*, 529 (1971), C.A. *74*, 142521x (1971); Solid liquid-crystalline films of synthetic polypeptides. [**358**] E. T. Samulski, A. V. Tobolsky, Biopolymers *10*, 1013 (1971), C.A. *75*, 81384b (1971); Distorted α-helix for PBLG in the nematic solid state. [**359**] W. A. Hines, E. T. Samulski, Macromolecules *6*, 793 (1973), C.A. *80*, 15344r (1974); NMR spin-lattice relaxation in the lyotropic polypeptide liquid crystal. [**360**] E. T. Samulski, A. V. Tobolsky, Liq. Cryst. Plast. Cryst. *1*, 175 (1974), C.A. *83*, 19084r (1975); Cholesterics formed by certain polypeptides with organic solvents.

[**361**] W. A. Hines, E. T. Samulski, Liq. Cryst. Ord. Fluids *2*, 257 (1974), C.A. *86*, 180196a (1977); NMR in polypeptide liquid crystals. [**362**] W. A. Hines, E. T. Samulski, J. Polym. Sci., Polym. Symp. *44*, 11 (1974), C.A. *82*, 98624x (1975); Supramolecular structural transitions in polypeptide solutions. NMR study. [**363**] C. G. Sridhar, W. A. Hines, E. T. Samulski, J. Chem. Phys. *61*, 947 (1974), C.A. *82*, 31647p (1975); Magnetic susceptibility, twist elastic constant, rotational viscosity coefficient, and PBDLG side-chain conformation. [**364**] C. G. Sridhar, W. A. Hines, E. T. Samulski, J. Phys. (Paris), C-1, *36*, 269 (1975); Polypeptide liquid crystals. Diamagnetic anisotropy, twist elastic constant and rotational viscosity coefficient. [**364a**] R. W. Duke, D. B. DuPre, W. A. Hines, E. T. Samulski, J. Am. Chem. Soc. *98*, 3094 (1976), C.A. *85*, 21821p (1976); PBLG helix-coil transition. Pretransition phenomena in the liquid crystal. [**364b**] N. S. Murthy, J. R. Knox, E. T. Samulski, J. Chem. Phys. *65*, 4835 (1976), C.A. *86*, 131347s (1977); Order parameter measurements in polypeptide liquid crystals. [**364c**] T. V. Samulski, E. T. Samulski, J. Chem. Phys. *67*, 824 (1977), C.A. *87*, 109709k (1977); Van der Waals-Lifshitz forces in lyotropic polypeptide liquid crystals. [**364d**] D. B. DuPre, R. W. Duke, W. A. Hines, E. T. Samulski, Mol. Cryst. Liq. Cryst. *40*, 247 (1977), C.A. *87*, 125610b (1977); Effect of trifluoroacetic acid on the viscoelastic properties of a polypeptide liquid crystal. [**365**] J. Sandblom, F. Orme, Membranes *1*, 125 (1972), C.A. *80*, 141922r (1974); Liquid membranes as electrodes and biological models. [**366**] R. L. Sani, R. A. Hardwick, Lett. Heat Mass Transfer *3*, 565 (1976), C.A. *86*, 102103d (1977); Observations of the shear orientation of a protein liquid crystal. [**367**] K. Sato, Y. Tamura, M. Okada, Hiroshima Daigaku Suichikusan Gakubu Kiyo *14*, 69 (1975), C.A. *83*, 200479p (1975); Optical studies of lyophases in lipids. I. Monoglyceride/water system. [**367a**] W. J. Schmidt, Die Bausteine des Tierkörpers in polarisiertem Lichte. (F. Cohen: Bonn). 1924. [**368**] F. O. Schmitt, W. H. Chambers, Proc. Soc. Exptl. Biol. Med. *23*, 134 (1925), C.A. *20*, 3714[5] (1926); Fluid crystals and meristematic growth. [**369**] F. O. Schmitt, J. Am. Leather Chem. Assoc. *46*, 538 (1951), C.A. *46*, 7354b (1952); Structural and chemical studies on collagen. [**369a**] H. Schroeder, J. Chem. Phys. *67*, 1617 (1977), C.A. *87*, 163052d (1977); Aggregation of proteins in membranes. Example of fluctuation-induced interactions in liquid crystals. [**370**] A. Seher, Chem. Phys. Lipids *8*, 134 (1972), C.A. *76*, 132567u (1972); Thermische Untersuchung von Lipiden.

[**371**] S. A. Seleznev, T. A. Nikolaeva, F. K. Polyakova, V. M. Mikhailov, Biofizika *19*, 761 (1974), C.A. *82*, 53055w (1975); Time-dependent instability of lipid/water systems. [**371a**] S. A. Seleznev, L. M. Fedorov, S. O. Kuzina, A. I. Mikhailov, Rev. Fr. Corps Gras *24*, 191 (1977), C.A. *87*, 129197v (1977); UV-synthesis of amphiphilic molecules from *n*-alkanes and its biological significance. [**372**] V. K. Semenchenko, Zh. Fiz. Khim. *36*, 15 (1961), C.A. *58*, 12011h (1963); Thermodynamics of protoplasm. [**373**] L. G. Shaltyko, A. A. Shepelevskii, S. Y. Frenkel, Uch. Zap., Ivanov. Gos. Pedagog. Inst. *1972*, No. 99, 124, C.A. *78*, 129301p (1973); Small-angle light scattering in liquid crystals. [**374**] T. M. Shaw, E. F. Jansen, H. Lineweaver, J. Chem. Phys. *12*, 439 (1944), C.A. *39*, 657[9] (1945); Dielectric properties of β-lactoglobulin in aqueous glycine solutions and in the liquid crystalline state. [**374a**] M. L. Sipski, T. E. Wagner, Biopolymers *16*, 573 (1977), C.A. *86*, 135094e (1977); Probing DNA quaternary ordering with circular dichroism spectroscopy: studies of equine sperm chromosomal fibers. [**375**] D. M. Small, M. Bourges, D. G. Dervichian, Nature *211*, 816 (1966), C.A. *65*, 15718b (1966); Ternary and quaternary aqueous systems containing bile salt, lecithin, and cholesterol. [**376**] D. M. Small, M. Bourges, D. G. Dervichian, Biochim. Biophys. Acta *125*, 563 (1966), C.A. *66*, 43824b (1967); Biophysics of lipidic associations. I. Ternary systems lecithin/bile salt/water. [**377**] D. M. Small, M. Bourges, Mol. Cryst. *1*, 541 (1966), C.A. *66*, 69268v (1967); Lyotropic paracrystalline phases obtained with ternary and quaternary systems of amphiphilic substances in water: studies on agreous systems of lecithin, bile salt, and cholesterol. [**378**] D. M. Small, J. Lipid Res. *8*, 551 (1967), C.A. *68*, 9511d (1968); Phase equilibria and structure of dry and hydrated egg lecithin. [**379**] D. M. Small, J. Am. Oil Chem. Soc. *45*, 108 (1968), C.A. *68*, 101938w (1968); Classification of biologic lipids based upon their interaction

in aqueous systems. [**380**] D. M. Small, C. Loomis, M. Janiak, G. G. Shipley, Liq. Cryst. Ord. Fluids *2*, 11 (1974), C.A. *87*, 1466q (1977); Mesomorphism of biologically important lipids, polyunsaturated cholesteryl esters, and phospholipids.

[**381**] M. J. Janiak, C. R. Loomis, G. G. Shipley, D. M. Small, J. Mol. Biol. *86*, 325 (1974), C.A. *81*, 177438x (1974); Ternary phase diagram of lecithin, cholesteryl linolenate, and water. Phase behavior and structure. [**382**] C. R. Loomis, M. J. Janiak, D. M. Small, G. G. Shipley, J. Mol. Biol. *86*, 309 (1974), C.A. *81*, 177437w (1974); Binary phase diagram of lecithin and cholesteryl linolenate. [**383**] R. J. Deckelbaum, G. G. Shipley, D. M. Small, R. S. Lees, P. K. George, Science *190*, 392 (1975), C.A. *83*, 202943j (1975); Thermal transitions in human plasma low-density lipoproteins. [**383a**] D. M. Small, J. Colloid Interface Sci. *58*, 581 (1977), C.A. *86*, 166469k (1977); Liquid crystals in living and dying systems. [**383b**] G. H. Rothblat, J. M. Rosen, W. Insull, Jr., A. O. Yau, D. M. Small, Exp. Mol. Pathol. *26*, 318 (1977), C.A. *87*, 3535k (1977); Production of cholesteryl ester-rich, anisotropic inclusions by mammalian cells in culture. [**383c**] D. Armitage, R. J. Deckelbaum, G. G. Shipley, D. M. Small, Mol. Cryst. Liq. Cryst. *42*, 203 (1977); Size and surface effects related to phase transitions in human plasma low density lipoprotein. [**384**] R. S. Snart, Nature *215*, 957 (1967), C.A. *67*, 108824u (1967); Mesomorphism in mixtures of cholesterol with steroid hormones. [**385**] E. Sommerfeldt, Physik. Z. *8*, 799 (1908), C.A. *2*, 622^1 (1908); Flüssige und scheinbar lebende Kristalle. [**386**] J. M. Squire, A. Elliott, Mol. Cryst. Liq. Cryst. *7*, 457 (1969), C.A. *71*, 85316w (1969); Mesophases of PBLG in solution. [**387**] J. M. Squire, A. Elliott, J. Mol. Biol. *65*, 291 (1972), C.A. *77*, 75599a (1972); Side-chain conformations in dry and liquid-crystalline racemic PBG. [**388**] R. W. St. Clair, Ann. N. Y. Acad. Sci. *275*, 228 (1976), C.A. *86*, 14746k (1977); Cholesteryl ester metabolism in atherosclerotic arterial tissue. [**389**] J. B. Stamatoff, Mol. Cryst. Liq. Cryst. *16*, 137 (1972), C.A. *76*, 127640h (1972); Electrically oriented X ray diffraction patterns of liquid crystalline solutions of PBLG. [**390**] D. Starling, R. G. Burns, J. Ultrastruct. Res. *51*, 261 (1975), C.A. *83*, 73878r (1975); Ultrastructure of tubulin paracrystals from sea urchin eggs, with determination of spacings by electron and optical diffraction.

[**391**] A. Steiger, Mikrokosmos *35*, 54 (1941), C.A. *37*, 3648^6 (1943); Mikrochemische Verseifung und die dabei entstehenden Myelinformen. [**392**] J. Steigman, A. S. Verdini, C. Montagner, L. Strasorier, J. Am. Chem. Soc. *91*, 1829 (1969), C.A. *71*, 3640k (1969); Protonation of an amide and of a diamide in dichloroacetic acid, and the behavior of PBLG in dichloroacetic acid and in some mixed solvents. [**393**] J. M. Steim, Instrum. News *19*, 12 (1968), C.A. *71*, 98921q (1969); DSC of biological systems. [**394**] J. M. Steim, Liquid Cryst. Ord. Fluids *1970*, 1, C.A. *79*, 1707d (1973); Thermal phase transitions in biomembranes. [**394a**] O. Stein, J. Vanderhoek, Y. Stein, Atherosclerosis *26*, 465 (1977), C.A. *87*, 20186q (1977); Cholesteryl ester accumulation in cultured aortic smooth muscle cells. Induction of cholesteryl ester retention by chloroquine and low density lipoprotein and its reversion by mixtures of high density apolipoprotein and sphingomyelin. [**395**] G. T. Stewart, Nature *192*, 624 (1961), C.A. *56*, 15997d (1962); Physicochemical properties of paracrystalline spherulites of biological origin. [**396**] G. T. Stewart, J. Pathol. Bacteriol. *81*, 385 (1961), C.A. *55*, 21340b (1961); Mesomorphic forms of lipid in the structure of normal and atheromatous tissues. [**397**] G. T. Stewart, Mol. Cryst. *1*, 563 (1966), C.A. *66*, 16477j (1967); Liquid crystals in biological systems. [**398**] G. T. Stewart, Advan. Chem. Ser. *63*, 141 (1967), C.A. *67*, 7882f (1967); Liquid crystals as ordered components of living substance. [**399**] G. T. Stewart, Mol. Cryst. Liq. Cryst. *7*, 75 (1969), C.A. *71*, 109181b (1969); Change of phase and change of state in biological systems. [**400**] G. T. Stewart, Liq. Cryst. Plast. Cryst. *1*, 308 (1974), C.A. *83*, 2500a (1975); Significance of liquid crystals in biology. Role of liquid crystals in life processes.

[**401**] L. Stryer, Biochemistry (Freeman: San Francisco). 1975. 877 pp. C.A. *82*, 134537g (1975). [**401a**] Y. Suezaki, Phys. Lett. A *56*, 238 (1976); Mechanical rigidities and geometrical fluctuations of lipid bilayer spheres. [**402**] D. M. Surgenor, The red blood cell, vol. 2. 2nd ed. (Academic: New York). 1975. 760 pp. C.A. *85*, 3368s (1976). [**402a**] J. Suurkuusk, B. R. Lentz, Y. Barenholz, R. L. Biltonen, T. E. Thompson, Biochemistry *15*, 1393 (1976), C.A. *84*, 160897k (1976); Calorimetric and fluorescent probe study of the gel/liquid crystal transition in small, single-labellar dipalmitoylphosphatidylcholine vesicles. [**403**] H. Suzuki, M. Honda, Seibutsu Butsuri *10*, 205 (1970), C.A. *75*, 98m (1971); Liquid crystals and receptors. [**404**] H. Suzuki, Kyoritsu Kagaku Raiburari *1*, 104 (1973), C.A. *83*, 38833a (1975); Liquid crystals in biological systems. [**405**] O. Svanberg, Acta Physiol. Scand. *80*, 45 (1970), C.A. *73*, 117692a (1970); Incorporation of cholesterol into the nervous system of the newly hatched chick. [**405a**] D. Szabo, Histochemistry *53*, 341 (1977), C.A. *87*, 181355j (1977); Zonal differences in temperature-dependent birefringence in rat adrenocortical lipids. [**406**] T. Tachibana, E. Oda, Bull. Chem. Soc. Jap. *46*, 2583 (1973), C.A. *79*, 126981v (1973); CD evidence for the cholesteric phase in

solid films of poly(γ-methyl glutamate). [**407**] S. Takashima, H. P. Schwan, Liq. Cryst. Ord. Fluids *2*, 199 (1974), C.A. *86*, 184855n (1977); Dielectric relaxation in lipid bilayer membranes. [**407a**] H. Tanaka, A. Nishioka, Kobunshi Ronbunshu *34*, 405 (1977), C.A. *87*, 53724s (1977); NMR study of the orientation behavior of PBDG liquid crystals in the magnetic field. [**407b**] H. Tanji, H. Hamaguchi, H. Matsuura, I. Harada, T. Shimanouchi, Appl. Spectrosc. *31*, 470 (1977), C.A. *87*, 191944q (1977); Convenient polarization scrambler for Raman spectroscopy. [**408**] C. Taupin, M. Dvolaitzky, C. Sauterey, Biochemistry *14*, 4771 (1975), C.A. *83*, 203015v (1975); Osmotic pressure-induced pores in phospholipid vesicles. [**409**] H. Thiele, G. Andersen, Kolloid-Z. *140*, 76 (1955), C.A. *49*, 10009e (1955); Ionotrope Gele von Polyuronsäuren. [**410**] D. Thines-Sempoux, Methodol. Dev. Biochem. *1974*, 4 (Subcell. Stud.) 157, C.A. *85*, 30213u (1976); Approaches to the comparative study of rat liver cytomembranes.

[**411**] H. T. Tien, H. P. Ring, J. Colloid Interface Sci. *27*, 702 (1968), C.A. *69*, 89993w (1968); Permeation of water through bilayer lipid membranes. [**412**] H. T. Tien, S. P. Verma, Nature *227*, 1232 (1970), C.A. *74*, 19374z (1971); Electronic processes in bilayer lipid membranes. [**413**] D. O. Tinker, L. Pinteric, Biochemistry *10*, 860 (1971), C.A. *74*, 94502t (1971); Identification of lamellar and hexagonal phases in negatively stained phospholipid/water systems. [**414**] K. Tohyama, Sen'i Gakkaishi *29*, P167 (1973), C.A. *79*, 119842x (1973); Liquid crystals. Magnetic properties. [**415**] L. Tokody, Természet-tudományi Közlöny *73*, 480 (1941), C.A. *38*, 4484^8 (1944); Living crystals. [**416**] K. Tohyama, J. Phys. C *7*, L 270 (1974), C.A. *82*, 43898j (1975); Pitch-independent critical field in cholesteric poly(γ-ethyl-L-gluta-mate). [**417**] K. Tohyama, N. Miyata, Mol. Cryst. Liq. Cryst. *29*, 35 (1974), C.A. *83*, 11008y (1975); Magnetic alignment and diamagnetic anisotropy of cholesteric poly(γ-ethyl-L-glutamate). [**418**] M. P. Tombs, Brit. 1,393,537 (6. 8. 1969), C.A. *83*, 77269r (1975); Protein products. [**419**] W. J. Toth, A. V. Tobolsky, J. Polym. Sci. Pt. B *8*, 531 (1970), C.A. *74*, 32103e (1971); Electric field orientation of concentrated solutions of PBLG. [**420**] Y. Toyoshima, N. Minami, M. Sukigara, Mol. Cryst. Liq. Cryst. *35*, 325 (1976), C.A. *85*, 200784a (1976); Effect of the electric field on the phase transition in solutions of rod-like particles. I. Optical observations.

[**421**] H. Träuble, Naturwissenschaften *58*, 277 (1971), C.A. *75*, 71226n (1971); Phasenumwand-lungen in Lipiden. Mögliche Schaltprozesse in biologischen Membranen. [**422**] H. Träuble, D. H. Haynes, Chem. Phys. Lipids *7*, 324 (1971), C.A. *76*, 82526g (1972); Volume change in lipid bilayer lamellae at the crystalline/liquid crystalline phase transition. [**423**] H. Träuble, J. Membrane Biol. *4*, 193 (1971), C.A. *76*, 11354p (1972); Movement of molecules across lipid membranes. Molecular theory. [**424**] T. Y. Tsong, Proc. Natl. Acad. Sci. USA *71*, 2684 (1974), C.A. *81*, 147220c (1974); Kinetics of the crystalline/liquid crystalline phase transition of dimyristoyl-L-α-lecithin bilayers. [**425**] N. Tsuchi-hashi, H. Nomori, M. Hatano, S. Mori, Chem. Lett. *1974*, 823, C.A. *81*, 112174p (1974); Induced CD of dyes buried in solid liquid crystal films of poly(γ-methyl D-glutamate). [**426**] N. Tsuchihashi, H. Nomori, M. Hatano, S. Mori, Bull. Chem. Soc. Jap. *48*, 29 (1975); CD in lyophases of polyglutamate solutions. [**427**] T. Tsutsui, T. Tanaka, Chem. Lett. *1976*, 1315, C.A. *86*, 96688e (1977); Cholesteric mesophase in the poly-L-glutamic acid/poly(ethylene oxide)/dimethylformamide system. [**428**] V. N. Tsvetkov, Y. V. Mitin, V. R. Glushenkova, A. E. Grishchenko, N. N. Boitsova, S. Y. Lyubina, Vysokomol. Soedin. *5*, 453 (1963), C.A. *59*, 1772d (1963); Electric and dynamic birefringence of PBLG solutions. [**429**] V. N. Tsvetkov, I. N. Shtennikova, S. V. Skazka, E. I. Ryumtsev, J. Polym. Sci. C *16*, 3205 (1965), C.A. *70*, 4584m (1969); Conformation and flexibility of polypeptides in solution. [**430**] V. N. Tsvetkov, L. N. Andreeva, V. I. Sisenko, Molekul. Biofiz., Akad. Nauk SSSR, Inst. Biol. Fiz., Sb. Statei *1965*, 110, C.A. *64*, 17911h (1966); Dependence of optical anisotropy of DNA molecules on the molecular weight.

[**431**] V. N. Tsvetkov, I. N. Shtennikova, E. I. Ryumtsev, G. F. Pirogova, Vysokomol. Soedin. A *9*, 1575 (1967), C.A. *67*, 73966n (1967); Dynamooptic and electrooptic behavior of polypeptide molecules in helix and coil conformations. [**432**] I. Uematsu, Y. Uematsu, Kobunshi *25*, 175 (1976), C.A. *84*, 165209 (1976); Polymeric liquid crystals. Physical properties. [**433**] R. Ullman, ACS Symp. New York 1971, Abstr. No. 216; Dynamic response of rodlike molecules in a steady electric field. [**434**] D. W. Urry, Biochim. Biophys. Acta *265*, 115 (1972), C.A. *76*, 136937n (1972); Protein conformation in biomembranes. Optical rotation and absorption of membrane suspensions. [**435**] V. A. Usol'tseva, 1. Wiss. Konf. über flüssige Kristalle, 17.–19. 11. 1970, Iwanovo, Samml. der Vortr. Nr. 21 (1972); Chemische Eigenschaften der Flüssigkristalle, die im Stoffwechsel und in der Pathologie eine Rolle spielen. [**436**] S. A. Celesnjev, W. M. Michailov, V. A. Usol'tseva, 1. Wiss. Konf. über flüssige Kristalle, 17.–19. 11. 1970, Iwanovo, Samml. der Vortr. Nr. 22 (1972); Struktur der Biomembranen. [**437**] G. Vanderkooi, D. E. Green, BioScience *21*, 409 (1971), C.A. *75*, 30246a (1971); New insights into biomembrane

structure. [**438**] A. van Hook, Biodynamica 6, 81 (1947), C.A. 41, 5562c (1947); Atypical solid/liquid transitions and their possible biological significance. [**439**] M. van Winkle, L. Levy, J. Exp. Med. 132, 858 (1970), C.A. 74, 11371a (1971); Reversibility of serum sickness cholesterol-induced atherosclerosis. [**440**] E. J. J. van Zoelen, E. C. M. van der Neut-Kok, J. de Gier, L. L. M. van Deenen, Biochim. Biophys. Acta 394, 463 (1975), C.A. 83, 75295x (1975); Osmotic behavior of Acholeplasma laidlawii B cells with membrane lipids in liquid-crystalline and gel state.

[**441**] A. A. Vazina, B. K. Lemazhikhin, G. M. Frank, Biofizika 10, 420 (1965), C.A. 63, 5925a (1965); Liquid crystal structure in nonoriented gels and solutions of F-actin. [**442**] A. A. Vazina, L. A. Zheleznaya, B. K. Lemazhikhin, G. M. Frank, Biofiz. Myshechn. Sokrashcheniya 1966, 192, C.A. 66, 112184g (1967); Study of the liquid-crystalline structure in solutions of F-actin. [**443**] A. A. Vazina, L. A. Zheleznaya, V. M. Shelestov, A. M. Matyushin, Sb. Dokl. Vses. Nauch. Konf. Zhidk. Krist. Simp. Ikh Prakt. Primen., 2nd, 1972, 42, C.A. 82, 81864c (1975); Liquid crystals of contractile proteins. [**444**] L. Venkov, A. Stoykova, Dokl. Bolg. Akad. Nauk 28, 837 (1975), C.A. 83, 145031j (1975); Isolation of spinal cord myelin fractions and certain ultrastructural and biochemical characteristics. [**445**] A. J. Verkleij, B. de Kruyff, P. H. J. T. Ververgaert, J. F. Tocanne, L. L. M. van Deenen, Biochim. Biophys. Acta 339, 432 (1974), C.A. 81, 116282v (1974); Influence of pH, calcium ions, and protein on the thermotropic behavior of the negatively charged phospholipid phosphatidylglycerol. [**446**] A. J. Verkleij, P. H. J. T. Ververgaert, B. de Kruyff, L. L. M. van Deenen, Biochim. Biophys. Acta 373, 495 (1974), C.A. 82, 69478x (1975); Distribution of cholesterol in bilayers of phosphatidylcholines as visualized by freeze fracturing. [**447**] S. P. Verman, D. F. H. Wallach, Biochim. Biophys. Acta 330, 122 (1973), C.A. 80, 44977e (1974); Effects of cholesterol on the IR dichroism of phosphatide multibilayers. [**447a**] S. P. Verma, D. F. H. Wallach, Biophys. Acta 426, 616 (1976), C.A. 84, 146512z (1976); Effect of melittin on thermotropic lipid state transitions in phosphatidylcholine liposomes. [**447b**] S. P. Verma, D. F. H. Wallach, Biochim. Biophys. Acta 436, 307 (1976), C.A. 85, 29781w (1976); Multiple thermotropic state transitions in erythrocyte membranes. Laser-Raman study of the carbon-hydrogen stretching and acoustical regions. [**447c**] S. P. Verma, D. F. H. Wallach, Proc. Natl. Acad. Sci. USA 73, 3558 (1976), C.A. 85, 188117a (1976); Erythrocyte membranes undergo cooperative, pH-sensitive state transitions in the physiological temperature range: Evidence from Raman spectroscopy. [**447d**] B. Z. Volchek, A. V. Gribanov, A. I. Kol'tsov, A. V. Purkina, G. P. Vlasov, L. A. Ovsyannikova, Vysokomol. Soedin. A 19, 321 (1977), C.A. 86, 121906j (1977); Effect of organic acids on the properties of liquid crystalline PBLG. [**447e**] B. Z. Volchek, A. V. Gribanov, A. I. Kol'tsov, A. V. Purkina, G. P. Vlasov, L. A. Ovsyannikova, Vysokomol. Soedin. A 19, 519 (1977), C.A. 86, 156129k (1977); IR-spectroscopic study of the orientation of liquid-crystal domains of PBLG. [**448**] A. L. von Muralt, J. T. Edsall, Trans. Faraday Soc. 26, 837 (1930), C.A. 26, 1632 (1932); Double refraction of myosin and its relation to the structure of the muscle fiber. [**449**] A. L. von Muralt, J. T. Edsall, J. Biol. Chem. 89, 315, 351 (1930), C.A. 25, 1846 (1931); Physical chemistry of muscle globulin. III. Anisotropy of myosin and the angle of isocline. IV. Anisotropy of myosin and double refraction of flow. [**450**] R. C. Waldbillig, Diss. Univ. Rochester, Rochester, N. Y. 1973. 102 pp. Avail. Univ. Microfilms, Ann Arbor, Mich., Order No. 73-26,556. C.A. 80, 56777c (1974); Liquid crystalline states of phospholipid/water systems.

[**451**] N. D. Weiner, A. Felmeister, J. Lipid Res. 11, 220 (1970), C.A. 73, 21470a (1970); Comparison of physical models used to explain condensation effects in lecithin/cholesterol mixed films. [**452**] M. Wender, H. Filipek-Wender, J. Stanislawska, Clin. Chim. Acta 54, 269 (1974), C.A. 81, 149802t (1974); Cholesteryl esters of the brain in demyelinating diseases. [**453**] M. Wender, Z. Adamczewska, J. Pankrac, A. Goncerzewicz, Neuropatol. Pol. 13, 209 (1975), C.A. 83, 129747s (1975); Myelin lipids in experimental allergic encephalomyelitis. [**454**] M. H. F. Wilkins, A. E. Blaurock, D. M. Engelman, Nature, New Biol. 230, 72 (1971), C.A. 75, 29921k (1971); Bilayer structure in membranes. [**455**] G. L. Wilkes, Mol. Cryst. Liq. Cryst. 18, 165 (1972), C.A. 77, 152746a (1972); Superstructure in polypeptide films as noted by small angle light scattering. [**456**] G. L. Wilkes, P. H. Wilkes, Biopolymers 13, 411 (1974), C.A. 81, 61517t (1974); Evidence for in vivo anisotropic rod structure in collagenous tissue as noted by small angle light scattering. [**457**] H. R. Wilson, P. Tollin, J. Ultrastruct. Res. 33, 550 (1970), C.A. 74, 121606p (1971); Narcissus mosaic virus liquid crystals. [**458**] C. Wippler, J. chim. phys. 53, 328 (1956), C.A. 51, 4096e (1957); Diffusion de la lumière par les solutions macromoléculaires. II. Étude expérimentale de l'effet d'un champ électrique sur les particules rigides. [**459**] D. Wobschall, J. Colloid Interface Sci. 40, 417 (1972), C.A. 77, 119545m (1972); Voltage dependence of bilayer membrane

capacitance. [460] P. J. Wojtowicz, Introd. Liq. Cryst. 1974, 333, C.A. *84*, 85723v (1976); Lyotropic liquid crystals and biological membranes: the crucial role of water.

[461] J. J. Wolken, J. Am. Oil Chem. Soc. *45*, 241 (1968); Cellular organelles and lipids. [462] W. M. Wong, P. H. Geil, Mol. Cryst. Liq. Cryst. *37*, 281 (1976), C.A. *86*, 116400k (1977); Dynamic mechanical spectroscopy of biomembrane systems. [463] R. Wood, R. D. Harlow, Arch. Biochem. Biophys. *141*, 183 (1970), C.A. *74*, 21338j (1971); Tumor lipids: structural analyses of the phospholipids. [464] F. Wunderlich, W. Batz, V. Speth, D. F. H. Wallach, J. Cell Biol. *61*, 633 (1974), C.A. *81*, 35394f (1974); Reversible, thermotropic alteration of nuclear membrane structure, and nucleocytoplasmic RNA transport in tetrahymena. [465] F. Wunderlich, A. Ronai, V. Speth, J. Seelig, A. Blume, Biochemistry *14*, 3730 (1975), C.A. *83*, 127871d (1975); Thermotropic lipid clustering in tetrahymena membranes. [466] K. Yamamoto, M. Yanagida, M. Kawamura, K. Maruyama, H. Noda, J. Mol. Biol. *91*, 463 (1975), C.A. *82*, 107838z (1975); Structure of paracrystals of F-actin. [466a] Y. Yokoyama, M. Arai, A. Nishioka, Polym. J. *9*, 161 (1977), C.A. *87*, 6566g (1977); NMR study of the time dependent dipolar splitting of methylene chloride protons in PBLG. [467] F. Zambrano, M. Cellino, M. Canessa-Fischer, J. Membrane Biol. *6*, 289 (1971), C.A. *76*, 55541b (1972); Molecular organization of nerve membranes. IV. Lipid composition of plasma membranes from squid retinal axons. [468] H. P. Zingsheim, Biochim. Biophys. Acta *265*, 339 (1972), C.A. *77*, 123473d (1972); Membrane structure and electron microscopy. Significance of physical problems and techniques (freeze etching).

Chapter 13

[1] H. Awaya, Nippon Kagaku Zasshi *83*, 865 (1962), C.A. *58*, 14120d (1963); Study of hexagonal modification of isotactic polypropylene by DTA. [2] H. Awaya, Kobunshi Kagaku *20*, 1 (1963), C.A. *61*, 1957b (1964); Crystallite orientation of isotactic polypropylene under low elongation. [3] E. M. Barrall II, J. F. Johnson, R. S. Porter, Appl. Polym. Symp. *1969*, No. 8, 191, C.A. *71*, 13376t (1969); Use of depolarized light intensity measurements in polymer characterization. [4] C. L. Beatty, J. M. Pochan, M. F. Froix, D. D. Hinman, Macromolecules *8*, 547 (1975); Liquid crystalline type order in polydiethylsiloxane. [5] N. G. Bel'nikevich, L. S. Bolotnikova, E. S. Edilyan, Y. V. Brestkin, S. Y. Frenkel, Vysokomol. Soedin. B *18*, 485 (1976), C.A. *85*, 124502q (1976); Spontaneous formation of the n phase in cellulose diacetate films. [6] J. D. Bernal, Disc. Faraday Soc. *25*, 7 (1958), C.A. *53*, 10869f (1959); Structure arrangements of macromolecules. [7] L. I. Bezruk, Y. S. Lipatov, J. Polym. Sci. C *38*, 337 (1972), C.A. *78*, 174406b (1973); Change of spherulitic structure in meltcrystallized polymer fibers and films during uniaxial stretching. [8] S. Blasenbrey, W. Pechhold, Ber. Bunsenges. Phys. Chem. *74*, 784 (1970), C.A. *73*, 134679u (1970); Theory der Phasenumwandlung in Polymeren. [9] S. B. Clough, A. Blumstein, R. Blumstein, E. Hsu, L. Patel, Bull. Am. Phys. Soc. *20*, 313 (1975), Mesomorphic ordering within polymers; X-ray studies. [9a] A. Blumstein, S. B. Clough, L. Patel, L. K. Lim, E. C. Hsu, R. B. Blumstein, Polym. Prepr., Am. Chem. Soc., Div. Polym. Chem. *16*, 241 (1975), C.A. *87*, 6508q (1977); Crystallinity and order in polymers with mesomorphic or potentially mesomorphic side groups. [9b] E. C. Hsu, S. B. Clough, A. Blumstein, J. Polym. Sci., Polym. Lett. Ed. *15*, 545 (1977), C.A. *87*, 152639z (1977); Liquid crystalline order in polymers with cholesteric side groups. [9c] A. Blumstein, Macromolecules *10*, 872 (1977), C.A. *87*, 102820f (1977); Mesomorphic order in polymers with side groups containing elements of mesogenic structure. [9d] A. Blumstein. ACS Symposium Series, Vol. 74: Mesomorphic Order in Polymers and Polymerization in Liquid Crystalline Media (Am. Chem. Soc.: Washington). 1978. 264 pp. C.A. *89*, 60234b (1978). [10] N. Bordelius, V. K. Semenchenko, Mekh. Polim. *7*, 724 (1971), C.A. *76*, 100251v (1972); Thermodynamic stability of crystalline polymers in a transition region.

[11] T. I. Borisova, L. L. Burshtein, T. P. Stepanova, Y. S. Freidzon, V. P. Shibaev, N. A. Plate, Vysokomol. Soedin. B *18*, 628 (1976), C.A. *85*, 143694v (1976); Dipole moment and intramolecular orientation order in macromolecules of cholesteryl ester of poly(N-methacryloyl-ω-aminolauric acid). [12] Y. Bouligand, P. E. Cladis, L. Liebert, L. Strzelecki, Mol. Cryst. Liq. Cryst. *25*, 233 (1974), C.A. *81*, 121267y (1974); Study of sections of polymerized liquid crystals. [12a] P. Bouriot, J. Jacquemart, M. Sotton, Bull. Sci. Inst. Text. Fr. *6*, 9 (1977), C.A. *86*, 172963h (1977); Caractérisation d'une phase

intermediaire paracristalline ou mésomorphe dans les fibres de polyester. [13] D. W. Brownawell, I-Ming Feng, J. Polymer Sci. *60*, S19 (1962), C.A. *57*, 11364g (1962); Mesomorphic character in polymers. [14] P. Calvert, Nature *259*, 175 (1976); Polymeric liquid crystals. [15] F. Ciampelli, M. Cambini, M. P. Lachi, J. Polym. Sci. C *7*, 213 (1964), C.A. *62*, 651f (1965); IR study of the crystallinity of polymers. [15a] P. G. de Gennes, C. R. Acad. Sci., Sér. B *281*, 101 (1975); Réflexions sur un type de polymères nématiques. [15b] P. G. de Gennes, Mol. Cryst. Liq. Cryst. *34*, 177 (1977), C.A. *87*, 46803p (1977); Polymeric liquid crystals: Frank elasticity and light scattering. [16] C. R. Desper, N. S. Schneider, Macromolecules *9*, 424 (1976), C.A. *85*, 63481h (1976); Mesomorphic structure at elevated temperature in m- and p-forms of poly[bis(chlorophenoxy)phosphazene]. [16a] C. R. Desper, N. S. Schneider, E. Higginbotham, J. Polym. Sci., Polym. Lett. Ed. *15*, 457 (1977), C.A. *87*, 118224a (1977); Mesomorphic structure in poly(organophosphazenes). [16b] M. N. Alexander, C. R. Desper, P. L. Sagalyn, N. S. Schneider, Macromolecules *10*, 721 (1977), C.A. *87*, 39997b (1977); NMR- and X-ray study of the mesomorphic transition in poly[bis(2,2,2-trifluoroethoxy)phosphazene]. [16c] H. Finkelmann, M. Portugall, M. Happ, H. Ringsdorf, 8. Arbeitstagung Flüssigkristalle, Inst. Angew. Festkörperphysik Freiburg, 6. 4. 1978; Polyreaktionen enantiotrop flüssig-kristalliner Systeme. [17] C. S. Fuller, W. O. Baker, N. R. Pape, J. Am. Chem. Soc. *62*, 3275 (1940), C.A. *35*, 952⁴ (1941); Crystalline behavior of linear polyamides. Effect of heat-treatment. [18] P. H. Geil, Polymer Reviews Vol. 5 Polymer Single Crystals (Interscience: New York). 1963. 560 pp. C.A. *60*, 3121a (1964). [19] P. H. Geil, J. Polym. Sci. C *20*, 109 (1967), C.A. *67*, 117292w (1967); Polymer crystallization. [19a] A. Sh. Goikhman, Vysokomol. Soedin. B *19*, 269 (1977), C.A. *87*, 6534v (1977); Equilibrium melting temperature of the α-form of polycaprolactam. [20] P. H. Hermans, P. Platzek, Kolloid-Z. *88*, 68 (1939), C.A. *33*, 7174⁴ (1939); Deformationsmechanismus und Feinstruktur der Hydratzellulose. Theoretische Beziehung zwischen Quellungsanisotropie und Eigendoppelbrechung orientierter Fäden.

[21] R. Hosemann, J. Appl. Phys. *34*, 25 (1963), C.A. *58*, 2506d (1963); Crystalline and paracrystalline order in high polymers. [22] P. H. Lindenmeyer, R. Hosemann, J. Appl. Phys. *34*, 42 (1963), C.A. *58*, 1972d (1963); Application of the theory of paracrystals to the crystal structure analysis of polyacrylonitrile. [23] R. Hosemann, F. J. Balta-Calleja, W. Wilke, Abhandl. Deut. Akad. Wiss. Berlin, Kl. Chemie, Geol. Biol. *1965*, 79, C.A. *65*, 15531a (1966); Parakristalline Fibrillarstruktur von in verschiedener Weise verstrecktem und getempertem Polyäthylen. [24] R. Hosemann, Pure Appl. Chem. *12*, 311 (1966), C.A. *66*, 46712z (1967); Parakristallinität in linearem und kristallinem Polyäthylen als allgemeines Konzept von Makromolekülen. [25] F. J. Balta-Calleja, R. Hosemann, W. Wilke, Makromol. Chem. *92*, 25 (1966), C.A. *64*, 19812f (1966); Struktur von rekristallisiertem und verstrecktem Polyäthylen als Modell für lineare Hochpolymere. [26] R. Hosemann, H. Cackovic, W. Wilke, Naturwissenschaften *54*, 278 (1967), C.A. *67*, 44212h (1967); Sogenannte Einkristalle in Hochpolymeren. [27] R. Hosemann, J. Polym. Sci. C *20*, 1 (1967), C.A. *67*, 117448b (1967); Molecular and supramolecular paracrystalline structure of linear synthetic high polymers. [28] H. Cackovic, R. Hosemann, W. Wilke, Kolloid-Z. Z. Polym. *234*, 1000 (1969), C.A. *71*, 125113m (1969); Mosaikstruktur und Kerngrenzen in Polyäthyleneinkristallen. [29] R. Hosemann, Chem.-Ing.-Tech. *42*, 1252, 1325 (1970), C.A. *74*, 7358j, 7319x (1971); Parakristalline Phasen. I. Entstehung, Kennzeichnung, Eigenschaften. II. Anwendungsbeispiele und praktische Bedeutung. [30] J. Loboda-Cackovic, R. Hosemann, H. Cackovic, J. Mater. Sci. *6*, 269 (1971), C.A. *75*, 36872y (1971); n/s Transition in linear polyethylene.

[31] R. Hosemann, Umschau *72*, 749 (1972), C.A. *78*, 49219c (1973); Parakristalle. [32] R. Hosemann, Endeavour *32*, 99 (1973), C.A. *79*, 88375r (1973); Paracrystals in biopolymers and synthetic polymers. [33] O. Phaovibul, H. Cackovic, J. Loboda-Cackovic, R. Hosemann, J. Polym. Sci., Polym. Phys. Ed. *11*, 2377 (1973), C.A. *80*, 133964r (1974); Phase transition and paracrystalline order in solution crystallized solid *n*-paraffins. [34] O. Phaovibul, J. Loboda-Cackovic, H. Cackovic, R. Hosemann, Makromolekulare Chem. *175*, 2991 (1974), C.A. *83*, 60100f (1975); Chain conformation in linear polyethylene and paraffins defined by four NMR-components. [35] R. Hosemann, J. Polym. Sci., Polym. Symp. *50*, 265 (1975), C.A. *84*, 31680k (1976); Paracrystalline state of synthetic polymers. [36] J. Haase, R. Hosemann, B. Renwanz, Cellul. Chem. Technol. *9*, 513 (1975), C.A. *85*, 34502k (1976); Axiale und laterale Ordnung in Regeneratcellulose-Fasern. Parakristallinität von Cellulose. [37] G. S. Y. Yeh, R. Hosemann, J. Loboda-Cackovic, H. Cackovic, Polymer *17*, 309 (1976), C.A. *85*, 78373y (1976); Annealing effects of polymers and their underlying molecular mechanisms. [37a] J. Haase, R. Hosemann, H. Cackovic, Polymer *18*, 743 (1977), C.A. *87*, 184972u (1977); Mosaic blocks in polymer crystals. Concept of paracrystallinity. [38] E. Iizuka, Kobunshi *21*, 463 (1972), C.A. *77*, 165085j (1972); Liquid crystals of polymers with emphasis on their orientation. [39] S. Iwayanagi, Kobunshi *19*, 667 (1970),

C.A. *73*, 110152r (1970); High polymers and liquid crystals. [**40**] W. J. Jackson Jr., H. F. Kuhfuss, J. Polym. Sci., Polym. Chem. Ed. *14*, 2043 (1976), C.A. *85*, 160935s (1976); Liquid crystal polymers. I. Preparation and properties of *p*-hydroxybenzoic acid copolyesters.

[**41**] H. Kamogawa, Japan Kokai 73 05,892 (7. 6. 1971), C.A. *78*, 111974a (1973); Cholesteryl esters polymers with liquid crystal property at room temperature. [**42**] H. Kamogawa, J. Polym. Sci. B *10*, 7 (1972), C.A. *76*, 113611m (1972); Preparation of vinyl polymer containing a long cholesteric ester side chain. [**43**] N. Kasai, Nippon Kessho Gakkaishi 11, 140 (1969), C.A. *72*, 93925e (1970); X-ray diffraction of paracrystals. [**44**] H. H. Kausch-Blecken v. Schmeling, Kolloid-Z. Z. Polym. *237*, 251 (1970), C.A. *73*, 15274y (1970); Zusammenhänge von makroskopischer und molekularer Anisotropie in Hochpolymeren. [**45**] L. G. Kazaryan, D. Ya. Tsvankin, L. Z. Rogovina, Vysokomolekul. Soedin., Karbotsepnye Vysokomolekul. Soedin., Sb. Statei *1963*, 267, C.A. *61*, 3259b (1964); X-ray study of the orientation of crystallites in the films of polypropylene. [**46**] E. A. Kiamco, Diss. Univ. Missouri, Kansas City, Mo. 1975. 226 pp. Avail. Xerox Univ. Microfilms, Ann Arbor, Mich., Order No. 76-11,492. C.A. *85*, 47231w (1976); Liquid crystalline properties of poly(phenylethyl isocyanides). [**47**] E. A. Kiamco, S. Y. Huang, E. W. Hellmuth, Bull. Am. Phys. Soc. *20*, 313 (1975); Structure of poly(α-phenylethyl isocyanide) in the solid and liquid crystalline state. [**48**] I. I. Konstantinov, Y. B. Amerik, A. A. Baturin, B. A. Krentsel, Khim. Volokna *1975*, 67, C.A. *84*, 74728z (1976); Possible realization of the liquid crystalline state in polymers. [**49**] A. K. Kulshreshtha, N. E. Dweltz, T. Radhakrishnan, J. Appl. Cryst. *4*, 116 (1971), C.A. *75*, 77396q (1971); Analysis of polymeric X-ray diffraction profiles. Theory and application of the variance method. [**50**] A. K. Kulshreshtha, N. E. Dweltz, T. Radhakrishnan, Indian J. Pure Appl. Phys. *9*, 986 (1971), C.A. *76*, 155812j (1972); X-ray studies on the paracrystalline structure of fibrous polymers.

[**51**] A. K. Kulshreshtha, N. E. Dweltz, Acta Cryst. A *27*, 670 (1971), C.A. *76*, 72906s (1972); Fourier coefficients of paracrystalline X-ray diffraction. [**52**] V. P. Lebedev, D. Y. Tsvankin, A. I. Kitalgorodskii, Vysokomol. Soedin. B *13*, 813 (1971), C.A. *76*, 86321j (1972); Packing density of poly-(vinyl chloride) in crystalline and mesomorphic states. [**53**] W. L. Lindner, Polymer *14*, 9 (1973), C.A. *78*, 73445u (1973); Characterization of the crystalline, intermediate, and amorphous phase in poly(ethylene terephthalate) fibers by X-ray diffraction. [**54**] C. H. MacGillavry, Rec. trav. chim. *69*, 509 (1950), C.A. *44*, 5188a (1950); Anisotropy in the so called amorphous part of polyvinyl alcohol. [**55**] B. Magel, J. Polym. Sci. C *12*, 119 (1966), C.A. *64*, 19856h (1966); Synthetic fibers and polymer science. [**56**] M. Mammi, V. Nardi, Nature *199*, 247 (1963), C.A. *59*, 8892f (1963); Mesomorphic and crystalline states in poly(vinyl chloride) by X-ray diffraction. [**57**] D. Marin, M. Grindea, Bul. Inst. Politeh. Iasi *18*, 155 (1972), C.A. *79*, 137850c (1973); Structural modification of polyacrylonitrile fibers observed roentgenographically following hydrothermal treatment at 100°. [**58**] G. Natta, G. Peraldo, P. Corradini, Atti accad, nazl. Lincei, Rend., Classe sci. fis., mat. e nat. *26*, 14 (1959), C.A. *53*, 21049h (1959); Smectic mesomorphic form of isotactic polypropylene. [**59**] W. D. Niegisch, J. Appl. Phys. *37*, 4041 (1966), C.A. *65*, 20232c (1966); Crystallography of poly-*p*-xylene. [**60**] S. P. Papkov, Vysokomol. Soedin. A *19*, 3 (1977), C.A. *86*, 73145m (1977); Liquid crystal state of linear polymers. [**60a**] S. P. Papkov, Khim. Volokna *1977*, 7, C.A. *87*, 40618k (1977); Orientation phenomena in liquid crystal polymer systems. [**60b**] S. P. Papkov, V. G. Kulichikhin. Liquid-Crystalline State of Polymers. (Khimiya: Moscow). 1977. 240 pp. C.A. *89*, 90433h (1978).

[**61**] J. Petermann, H. Gleiter, Phil. Mag. *28*, 271 (1973), C.A. *80*, 27546w (1974); Molecular structure of molten polyethylene films. [**62**] J. Petermann, H. Gleiter, J. Macromol. Sci., Phys. *B 11*, 359 (1975), C.A. *84*, 17976e (1976); Molecular structure of thin molten polyethylene films. [**63**] N. A. Platé, V. P. Shibaev, J. Polym. Sci., Macromol. Rev. *8*, 117 (1974), C.A. *83*, 79626d (1975); Comb-like Polymers. Structure and Properties. [**64**] R. E. Robertson, Annu. Rev. Mater. Sci. *5*, 73 (1975), C.A. *83*, 147749y (1975); Molecular organization of amorphous polymers. [**65**] A. Roviello, A. Sirigu, J. Polym. Sci., Polym. Lett. Ed. *13*, 455 (1975), C.A. *83*, 179801a (1975); Mesophasic structures in polymers. Mesophases of poly(alkanoates) of *p,p'*-dihydroxy-α,α'-dimethylbenzalazine. [**66**] L. K. Runnels, B. C. Freasier, 3rd ACS Symp. Ord. Fluids Liquid Cryst., Chicago, Aug. 1973, Abstr. No. 97; Attractive forces and the liquid crystalline transition. [**66a**] E. I. Ryumtsev, I. N. Shtennikova, N. A. Mikhailova, G. I. Okhrimenko, N. V. Pogodina, Yu. B. Amerik, A. A. Baturin, Vysokomol. Soedin. A *19*, 500 (1977), C.A. *86*, 156127h (1977); Electrooptics of crystallike molecules of random copolymers. [**67**] C. Sadron, Chim. Ind., Genie Chim. *96*, 507 (1966), C.A. *66*, 18954m (1967); Copolymères organisés. [**68**] W. Schmidt, W. Vogel, Colloid Polym. Sci. *253*, 898 (1975), C.A. *84*, 31637b (1976); Temperaturabhängigkeit der parakristallinen Störungen in linearem schmelzkristallisiertem Polyäthylen 6041D unterhalb

des Schmelzpunktes. [69] S. G. Shenouda, A. Viswanathan, J. Appl. Polym. Sci. *16*, 395 (1972), C.A. 76, 142188k (1972); Crystalline character of native and chemically treated Egyptian cottons II. Computation of variance of X-ray line profile and paracrystalline lattice distortions. [70] V. P. Shibaev, Z. S. Freidzon, N. A. Plate, V sb., XI Medeleevsk. S'ezk po Obshch. i Prikl. Khimii. Ref. Dokl. i Soobshch. *1974*, 164, C.A. *84*, 151163r (1976); Structure and properties of polymer liquid crystal systems.

[71] V. P. Shibaev, Y. S. Freidzon, N. A. Plate, Dokl. Akad. Nauk SSSR *227*, 1412 (1976), C.A. *85*, 33618r (1976); Liquid-crystal cholesterol-containing polymers. [71a] E. V. Anufrieva, V. D. Pautov, Ya. S. Freidzon, V. P. Shibaev, Vysokomol. Soedin. A *19*, 755 (1977), C.A. *87*, 6501g (1977); Intramolecular interactions in cholesterol-containing polymers. [71b] V. P. Shibaev, Ya. S. Freidzon, I. M. Agranovich, V. D. Pautov, E. V. Anufrieva, N. A. Plate, Dokl. Akad. Nauk SSSR *232*, 401 (1977), C.A. *86*, 113888q (1977); Role of intramolecular structure formation in the realization of the liquid-crystal state of cholesterol-containing polymers. [71c] A. M. Shishko, S. M. Volkovich, A. I. Skrigan, Khim. Drev. *1977*, 69, C.A. *87*, 103508x (1977); Kinetics of the acid hydrolysis of accessible regions of cellulose. [72] A. Skoulios, J. Phys. (Paris), C-4, *30*, 74 (1969); Polymères amphipathiques. [73] P. P. A. Smit, Kolloid-Z. Z. Polymere *250*, 27 (1972), C.A. *77*, 35074u (1972); DSC-studies on transitions in polyolefins between −100°C and +300°C. [74] H. A. Stuart, Ber. Bunsenges. Phys. Chem. *74*, 739 (1970), C.A. *73*, 134681p (1970); Ordnungszustände u. Umwandlungserscheinungen in Polymeren. [75] M. Szwarc, C. Schuerch, Polym. Biol. Syst., Ciba Found. Symp. *1972*, 7, C.A. *79*, 5627a (1973); Synthetic polymers, biopolymers and block polymers. [76] Y. G. Tarasenko, M. V. Venediktov, V. P. Dushchenko, B. S. Kolupaev, Sin. Fiz.-Khim, Polim, No. 9, 127 (1971), C.A. *77*, 89231m (1972); Mesophase thermodynamics of filled plasticized poly(vinyl chloride). [77] K. Thinius, Plaste Kaut. *18*, 408 (1971), C.A. *75*, 36705w (1971); Bedeutung des kristallin-flüssigen Zustands für die Chemie und Technologie der Hochpolymeren. [78] V. N. Tsvetkov, E. I. Ryumtsev, I. N. Shtennikova, E. V. Korneeva, B. A. Krentsel, Y. B. Amerik, Eur. Polym. J. *9*, 481 (1973), C.A. *79*, 79373m (1973); Intramolecular liquid-crystal order in polymers with chain side groups. [79] V. N. Tsvetkov, E. I. Ryumtsev, I. N. Shtennikova, E. V. Korneeva, G. I. Okhrimenko, N. A. Mikhailova, A. A. Baturin, Y. B. Amerik, B. A. Krentsel, Vysokomol. Soedin. A *15*, 2570 (1973), C.A. *80*, 96478n (1974); Intramolecular order in copolymers of crystallike molecules. [79a] V. N. Tsvetkov, I. N. Shtennikova, E. I. Ryumtsev, N. V. Pogodina, G. F. Kolbina, E. V. Korneeva, P. N. Lavrenko, O. V. Okatova, Yu. B. Amerik, A. A. Baturin, Vysokomol. Soedin. A *18*, 2016 (1976), C.A. *86*, 30178g (1977); Kinetic and equilibrium flexibility of molecules of poly[cetyl *p*-(methacryloyloxy)benzoate]. [80] N. N. Vasil'ev, A. P. Ochkivskii, V. A. Pakharenko, V. T. Burmistrov, Khim. Tekhnol. (Kiev) *1975*, 18, C.A. *84*, 106461a (1976); Effect of rolling conditions on the formation of intermediate structures of poly(vinyl chloride).

[81] A. Visokinskas, J. Dudonis, Nauch.-Issled. Tr., Liet. Nauch.-Issled, Inst. Tekst. Prom. *1971*, No. 1, 233, C.A. *78*, 5269q (1973); Electron-microscopic study of the structure of high-bulk polyamide fibers. [82] W. Vogel, Kolloid-Z. Z. Polym. *250*, 499 (1972), C.A. *77*, 165176q (1972); Bestimmung der parakristallinen Gitterzelle sowie der a-priori Abstandsstatistik in linearem Polyäthylen aus den Röntgenweitwinkelreflexen. [83] I. M. Ward, Structure and properties of oriented polymers (Halsted: New York). 1975. 500 pp., C.A. *83*, 164821h (1975). [84] J. H. Wendorff, E. Perplies, H. Ringsdorf, Strukt. Polym.-Syst., Vortr. Diskuss. Hauptversamml. Kolloid-Ges., 26th *1973*, 272, C.A. *86*, 30118n (1977); Strukturen von Polymeren aus mesomorphen Monomeren. [85] J. H. Wendorff, E. Perplies, H. Ringsdorf, Prog. Colloid Polym. Sci. *57*, 272 (1975), C.A. *84*, 5490d (1976); Strukturen in Polymeren aus mesomorphen Monomeren. [86] G. S. Y. Yeh, P. H. Geil, J. Macromol. Sci., Phys. *2*, 29 (1968), C.A. *68*, 78931c (1968); Morphology of cold-drawn polyethylene. [87] C. Wang, G. Yeh, Polymer *18*, 1085 (1977); DRDF study of liquid crystalline poly(ethylene terephthalate) and *p*-hydroxybenzoic acid copolymer.

Chapter 14

[1] C. P. Abbott III, J. M. Reilly, U. S. 3,718,842 (21. 4. 1972); Liquid crystal display mounting structure. [2] H. Abe, F. Nakano, M. Ohkubo, T. Kitamura, Japan. Kokai 75,102,584 (16. 1. 1974), C.A. *84*, 67893w (1976); Electrooptical display devices with improved service lifetime. [3] J. Adams,

W. Haas, J. Electrochem. Soc. *118*, 2026 (1971), C.A. *76*, 64623g (1972); Sensitivity of cholesteric films to UV exposure. **[4]** J. Adams, W. Haas, J. Dailey, J. Appl. Phys. *42*, 4096 (1971), C.A. *75*, 135721e (1971); Cholesteric films as optical filters. **[5]** J. E. Adams, W. E. L. Haas, U. S. 3,669,525 (6. 1. 1971); Liquid crystal color filter. **[6]** J. E. Adams, L. B. Leder, W. E. L. Haas, U. S. 3,679,290 (6. 1. 1971), C.A. *77*, 95289c (1972); Liquid-crystal optical filter system. **[7]** J. E. Adams, L. B. Leder, U. S. 3,697,152 (6. 1. 1971); Tuning method for plural layer liquid crystal filters. **[8]** J. E. Adams, J. L. Dailey, U. S. 3,711,181 (8. 3. 1971); Optical notch filter. **[9]** J. E. Adams, W. E. L. Haas, U. S. 3,720,456 (29. 4. 1971); Method for narrowing the bandwidth of an optical signal. **[10]** J. E. Adams, W. E. L. Haas, U. S. 3,726,684 (28. 5. 1971); Light modulation system.

[11] P. Adomenas, A. Vaitkevichius, A. Girdziushas, G. Dienys, A. Smilgevicius, V. Ceponite, U.S.S.R. 462,855 (26. 1. 1973), C.A. *83*, 106258u (1975); Electrooptical liquid crystal compositions. **[12]** S. Aftergut, H. S. Cole, Soc. Inform. Display, Symp. 6.–8. 6. 1972, San Francisco, 92; Effect of boundary conditions on the performance of nematic liquid crystal displays. **[13]** S. Aftergut, H. S. Cole Jr., Ger. Offen. 2,518,725 (2. 5. 1974), C.A. *84*, 67900w (1976); Liquid crystals with positive dielectric anisotropy. **[13a]** S. Aftergut, H. S. Cole, Jr., Appl. Phys. Lett. *30*, 363 (1977), C.A. *86*, 180966b (1977); Decay time of twist cells with liquid crystals of shortened pitch. **[14]** W. R. Aiken, U. S. 3,734,598 (30. 4. 1971); Liquid crystal display device having an inclined rear reflector. **[15]** T. Akamatsu, S. Hotta, K. Yoshinaga, Senryo To Yakuhin *20*, 327 (1975), C.A. *85*, 12257g (1976); Introduction to recording techniques. Chromogenic materials. **[16]** K. Akeyoshi, Y. Okajima, R. Ohishi, Japan. Kokai 75,154,792 (23. 5. 1974), C.A. *84*, 158789h (1976); Transparent electrode films deposited in the form of precise patterns. **[17]** C. J. Alder, E. P. Raynes, J. Phys. D *6*, L33 (1973), C.A. *78*, 152754x (1973); Room-temperature nematic mixtures with positive dielectric anisotropy. **[18]** C. J. Alder, I. A. Shanks, Electron Lett. *12*, 326 (1976), C.A. *85*, 169648h (1976); Addressing scheme for the multiplexing of complex liquid-crystal displays. **[19]** T. A. Aleshina, A. N. Kalinin, V. Pak, Tr., Sib. Gos. Nauch.-Issled. Inst. Metrol., No. 6, 151 (1970), C.A. *79*, 70999n (1973); Determination of color transitions of heat detectors made of cholesterics. **[20]** Z. B. Alfassi, L. Feldman, A. P. Kushelevsky, Int. J. Appl. Radiat. Isot. *27*, 722 (1976), C.A. *86*, 147536p (1977); Effect of γ rays on encapsulated liquid crystals: ELC dosimeter. **[20a]** Z. B. Alfassi, A. P. Kushelevsky, L. Feldman, Mol. Cryst. Luq. Cryst. *39*, 33 (1977), C.A. *87*, 31912z (1977); Effect of γ irradiation of solutions of cholesterics on the color transition temperature.

[21] F. V. Allan, P. Y. Hsieh, U. S. 3,728,008 (1. 12. 1971), C.A. *79*, 37144g (1973); Liquid crystal display. **[22]** P. J. Allen, U. S. 3,604,930 (5. 3. 1970); Method and apparatus for displaying visual images of IR beams. **[23]** P. M. Alt, P. Pleshko, IEEE Device *21*, 146 (1974); Scanning limitations of liquid-crystal displays. **[24]** G. Altman, U. S. 3,802,884 (6. 7. 1967), C.A. *81*, 56597q (1974); Photographic products for direct observation and optical projection and photographic processes for their production and use. **[25]** G. Altman, U. S. Reissue 28,515 (6. 7. 1967), C.A. *83*, 211264r (1975); Photographic products for direct observation and optical projection. **[26]** C. Ambrosi, C. Bourde, Gaz. Med. France *82*, 628 (1975); Nouveauté thérapeutique médicale dans les artériopathies des membres inférieurs: Tanakan. Essai clinique et étude par les cristaux liquides. **[27]** L. K. Anderson, Bell. Lab. Rec. *52*, 223 (1974), C.A. *81*, 113626z (1974); Projecting images with liquid crystals. **[28]** K. Aoki, S. Matsuyama, H. Zama, M. Kanazaki, M. Koyama, K. Sasaki, Ger. Offen. 2,446,658 (5. 10. 1973), C.A. *83*, 98721x (1975); Bonding components of a liquid crystal indicating device. **[29]** T. Aoyagi, S. Nomura, K. Toriyama, Japan. Kokai 73 41,983 (4. 10. 1971), C.A. *79*, 120501s (1973); Electrooptical organic materials. **[30]** Y. Arai, S. Kinoshita, K. Kimura, Japan. Kokai 72 26,390 (29. 3. 1971), C.A. *78*, 76972z (1973); Liquid crystal composition based on MBBA and *p*-alkoxybenzylidene-*p'*-butylaniline.

[31] Y. Arai, S. Kinoshita, K. Kimura, Japan. Kokai 72 27,187 (30. 3. 1971), C.A. *78*, 91032m (1973); Liquid crystalline compositions containing urea or thiourea compounds. **[32]** Y. Arai, S. Kinoshita, K. Kimura, Japan. Kokai 72 27,188 (3. 4. 1971), C.A. *78*, 63654x (1973); Liquid crystalline compositions containing phenols or quinones. **[33]** Y. Arai, S. Kinoshita, K. Kimura, Japan. Kokai 73 34,085 (3. 9. 1971), C.A. *79*, 60005w (1973); Liquid crystal compositions. **[34]** Y. Arai, T. Okamoto, S. Kinoshita, T. Kato, K. Kimura, Japan. 74 23,107 (29. 8. 1970), C.A. *82*, 24777c (1975); Liquid crystal composition. **[35]** Y. Arai, S. Kinoshita, K. Kimura, T. Wada, H. Yamamoto, F. Funada, Japan. Kokai 74 44,989 (5. 9. 1972), C.A. *81*, 113737m (1974); Liquid crystal compositions for electrooptical display devices. **[36]** Y. Arai, S. Kinoshita, K. Kimura, U. S. 3,910,682 (15. 7. 1972), C.A. *84*, 37336z (1976); Liquid crystal cells. **[37]** Y. Arai, K. Kimura, Y. Fujita, Japan. Kokai 76 10,185 (17. 7. 1974), C.A. *85*, 12395a (1976); Nematic liquid crystal composition for display devices. **[38]** Y. Arai, K. Kimura,

Y. Fujita, Japan. Kokai 76 10,186 (17. 7. 1974), C.A. *85*, 12396b (1976); Nematic liquid crystal composition for display devices. **[39]** Y. Arai, M. Tazume, Y. Fujita, H. Sato, Japan. Kokai 76 28,585 (4. 9. 1974), C.A. *85*, 102440r (1976); Nematic liquid crystal compositions for display devices. **[40]** Y. Arai, K. Kimura, Y. Fujita, H. Sato, Japan. Kokai 76 28,586 (4. 9. 1974), C.A. *85*, 134485a (1976); Liquid crystal compositions for display devices.

[41] Y. Arai, M. Tazume, H. Sato, Japan. Kokai 76 28,587 (4. 9. 1974), C.A. *85*, 102441s (1976); Nematic liquid crystal compositions for display devices. **[42]** Y. Arai, H. Sato, K. Kimura, M. Ozeki, Y. Fujita, M. Tazume, Japan. Kokai 76,125,681 (25. 4. 1975), C.A. *86*, 63557h (1977); Nematic liquid crystal composite. **[43]** Y. Arai, K. Kimura, H. Sato, M. Ohzeki, T. Yamaki, K. Morita, H. Takatsu, Y. Fujita, M. Tazume, Ger. Offen. 2,633,928 (31. 7. 1975), C.A. *86*, 113740k (1977); Nematic liquid crystal compositions. **[43a]** Y. Arai, K. Kimura, H. Sato, M. Ohzeki, T. Yamagi, K. Morita, H. Takatsu, Y. Fujita, M. Tazume, Japan. Kokai 77 16,488 (31. 7. 1975), C.A. *87*, 109449a (1977); Nematic liquid crystal compositions for field effect-type display devices. **[44]** F. Archer, Diss. Univ. Strasbourg Faculté de Médicine Nr. 53, 1969; Utilisation des cristaux liquides en thermographie médicale. **[45]** H. Arimoto, T. Kakishita, M. Takada, Japan. Kokai 73 44,177 (25. 8. 1971), C.A. *79*, 93136t (1973); Liquid crystal microcapsules. **[46]** H. Arimoto, T. Kakishita, M. Takada, Japan. Kokai 73 44,522 (11. 10. 1971), C.A. *79*, 116286h (1973); Iris composite fibers containing cholesterics. **[47]** S. L. Arora, J. L. Fergason, Liq. Cryst., Proc. Int. Conf., Bangalore *1973*, 553; TN liquid crystals and their applications in the field effect displays. **[48]** Asahi Glass Co., Ltd., Fr. Demande 2,175,727 (14. 3. 1972); C.A. *80*, 99480t (1974); Liquid crystal cells. **[49]** Asahi Glass Co., Ltd., Brit. 1,385,294 (25. 5. 1972), C.A. *83*, 50775r (1975); Liquid crystal cell. **[50]** K. Asai, A. Moriyama, H. Tatsuta, M. Fukai, Japan. Kokai 74,130,376 (20. 4. 1973), C.A. *82*, 178284b (1975); Nematic liquid crystal display device.

[51] K. Asai, A. Moriyama, H. Tatsuta, M. Fukai, Japan. Kokai 74,130,377 (20. 4. 1973), C.A. *82*, 178285c (1975); Homeotropic nematic liquid crystal display device. **[52]** K. Asai, A. Moriyama, H. Tatsuta, M. Fukai, Japan. Kokai 74,130,378 (20. 4. 1973), C.A. *82*, 163083m (1975); Liquid crystal display devices using homeotropic liquid crystal compositions. **[53]** J. A. Asars, Brit. 1,119,253 (29. 6. 1965); Display device. **[54]** G. Assouline, M. Hareng, E. Leiba, Electron. Lett. *7*, 699 (1971); Affichage bicolore à cristal liquide. **[55]** G. Assouline, M. Hareng, E. Leiba, Proc. IEEE *59*, 1355 (1971); Liquid crystal and photoconductor image converter. **[56]** G. Assouline, M. Hareng, E. Leiba, C. R. Acad. Sci., Sér. B *274*, 692 (1972), C.A. *77*, 41333u (1972); Transformateur d'image à couche photoconductrice associée à un nématique. **[57]** G. Assouline, M. Hareng, E. Leiba, M. Roncillat, Electron. Lett. *8*, 45 (1972); Visualisation par cristaux liquides à biréfringence électriquement controlable. **[58]** G. Assouline, E. Leiba, E. Spitz, U. S. 3,663,086 (2. 7. 1969); Optical information storing system. **[59]** G. J. Assouline, M. Hareng, E. Leiba, U. S. 3,718,381 (17. 3. 1970); Liquid crystal electro-optical modulators. **[60]** G. Assouline, M. Hareng, E. Leiba, Fr. Demande 2,213,100 (10. 10. 1972), C.A. *82*, 66737v (1975); Liquid crystal material for optical use.

[61] C. F. Augustine, W. E. Kock, Proc. IEEE *1969*, 354; Microwave holograms using liquid crystal display. **[62]** C. F. Augustine, C. Deutsch, D. Fritzler, E. Marom, Proc. IEEE *1969*, 1333; Microwave holography using liquid crystal area detectors. **[63]** C. F. Augustine, U. S. 3,693,084 (17. 6. 1969); Method and apparatus for detecting microwave fields. **[64]** M. Auvergne, F. Roddier, C. R. Acad. Sci., Sér. B *273*, 1088 (1971), C.A. *76*, 134147f (1972); Application des cristaux liquides au traitement optique de l'information. **[65]** M. Auvergne, F. Roddier, Nouv. Rev. Opt. *4*, 199 (1973), C.A. *80*, 32487g (1974); Application des cristaux liquides a l'analyse statistique en temps réel d'un éclairement aléatoire. **[66]** A. Aviram, I. Haller, R. D. Miller, W. R. Young, Ger. Offen. 2,338,281 (3. 8. 1972), C.A. *80*, 126004z (1974); Doping agent for controlled alteration of the electrical properties of nematics. **[67]** N. Ayata, Y. Kasugayama, Japan. Kokai 75 75,583 (9. 11. 1973), C.A. *83*, 170959a (1975); Nematic liquid crystal electrooptical display devices. **[68]** N. Ayata, Y. Kasugayama, Japan. Kokai 75 77,279 (13. 11. 1973), C.A. *84*, 82588u (1976); Nematic liquid crystal composition for optical display devices. **[69]** N. Ayata, Y. Kasugayama, Japan. Kokai 75 77,280 (13. 11. 1973), C.A. *84*, 172189u (1976); Nematic liquid crystal composition for display devices. **[70]** N. Ayata, Y. Kasugayama, Japan. Kokai 75 77,281 (13. 11. 1973), C.A. *84*, 172188t (1976); Nematic liquid crystal compositions for display devices.

[71] N. Ayata, Y. Kasugayama, Japan. Kokai 75 77,282 (13. 11. 1973), C.A. *84*, 82589v (1976); Nematic liquid crystal compositions for optical display devices. **[72]** N. Ayata, Y. Kasugayama, Japan. Kokai 75 92,279 (18. 12. 1973), C.A. *84*, 187548j (1976); Thiadiazole-containing nematic liquid crystal compositions for display devices. **[73]** A. Bagdanavicius, R. Baltrusaitis, V. Gaidelis, R. Zilenas, A.

Smilgevicius, Liet. Fiz. Rinkinys *13*, 261 (1973), C.A. *79*, 119659t (1973); Liquid crystal-photoconductor system. [**74**] C. S. Bak, K. Ko, M. M. Labes, J. Appl. Phys. *46*, 1 (1975), C.A. *82*, 148596n (1975); Fast decay in a TN induced by frequency switching. [**75**] C. E. Baker, U. S. 3,650,608 (23. 12. 1969); Method and apparatus for displaying coherent light images. [**76**] D. H. Baltzer, U. S. 3,620,889 (11. 6. 1968); Liquid crystal systems. [**77**] D. H. Baltzer, Ger. Offen. 1,929,256 (11. 6. 1968), C.A. *72*, 80039u (1970); Flüssiges Kristallsystem. [**78**] G. G. Barna, Rev. Sci. Instrum. *47*, 1258 (1976), Apparatus for optical characterization of displays. [**79**] J. B. Barr, Ger. Offen. 2,556,126 (24. 12. 1974), C.A. *85*, 162649a (1976); Carbonaceous, graphitizable fibers. [**80**] S. Barret, L. Cahen, F. Gaspard, R. Herino, F. Mondon, H. Seinera, G. Pierre, D. Serve, Ger. Offen. 2,514,629 (5. 4. 1974), C.A. *84*, 10951j (1976); Liquid crystal substances.

[**81**] E. L. Bartashevskii, V. M. Dolgov, Sb. Dokl. Vses. Nauch. Konf. Zhidk. Krist. Simp. Ikh Prakt. Primen., 2nd *1972*, 301, C.A. *82*, 24633c (1975); Thermal mapping of absorbing superhigh-frequency loads using cholesteric mixtures. [**82**] E. L. Bartashevskii, V. M. Dolgov, V. A. Krasovskii, Izv. Vyssh. Ucheb. Zaved., Radiofiz. *17*, 734 (1974), C.A. *81*, 93433d (1974); Visualization of distribution of an electromagnetic field in a rectangular waveguide using liquid-crystal thermoindicators. [**83**] J. J. Bartfai, Ger. Offen. 2,331,414 (23. 6. 1972), C.A. *80*, 111852a (1974); Layered glass structures. [**84**] R. Bartolino, M. Bertolotti, F. Scudieri, D. Sette, J. Appl. Phys. *46*, 1928 (1975), C.A. *83*, 51781b (1975); Ultrasonic modulation of light with s_A and n phases. [**85**] I. Barycka, J. Luczyna, Pr. Nauk. Inst. Metrol. Elektr. Politech. Wroclaw. *7*, 55 (1975), C.A. *86*, 81645q (1977); Laboratory experiments on the preparation of seven-segment displays filled with liquid-crystal of MBBA type. [**85a**] I. Barycka, Pr. Nauk. Inst. Metrol. Elektr. Politech. Wroclaw. *7*, 67 (1975), C.A. *86*, 197889b (1977); Use of a Specol photocolorimeter for contrast measurements of thin liquid-crystal films. [**86**] N. G. Basov, P. D. Berezin, L. M. Blinov, I. N. Kompanets, V. N. Morozov, V. V. Nikitin, Pis'ma Zh. Eksp. Teor. Fiz. *15*, 200 (1972), C.A. *77*, 27205x (1972); Phase modulation of coherent light using liquid crystals. [**87**] G. Baur, G. Meier, Phys. Lett. A *50*, 149 (1974), C.A. *82*, 117936u (1975); Angular dependence of transmitted light in deformed liquid crystal twist cells. [**88**] G. Baur, F. Windscheid, D. W. Berreman, Appl. Phys. *8*, 101 (1975), C.A. *83*, 185627q (1975); Optical properties of a TN cell. [**88a**] G. Baur, W. Greubel, Appl. Phys. Lett. *31*, 4 (1977), C.A. *87*, 76348m (1977); Fluorescence-activated liquid-crystal display. [**89**] BBC, Fr. Demande 2,274,059 (7. 6. 1974), C.A. *85*, 114588h (1976); Apparatus for producing and modulating monochromatic light. [**90**] T. D. Beard, J. Opt. Soc. Amer. *61*, 1559 (1971); Photoconductor light-gated liquid crystals used for optical data.

[**91**] T. D. Beard, W. P. Bleha, S. Y. Wong, Appl. Phys. Lett. *22*, 90 (1973), C.A. *78*, 77273j (1973); a.c. Liquid-crystal light valve. [**91a**] M. Bechtler, H. Krüger, Electronics *50*, No. 25, 113 (1977); Dim light is no turnoff for fluorescence-activated LCD. [**92**] Beckman Instruments, Inc., Neth. Appl. 74 06,165 (17. 5. 1973), C.A. *83*, 50749k (1975); Nematic liquid crystal composition. [**93**] W. J. Benton, J. R. Quigley, U. S. 3,872,050 (19. 7. 1972), C.A. *83*, 29420m (1975); Polyurethane liquid crystal dispersion system and devices. [**94**] P. D. Berezin, L. M. Blinov, I. N. Kompanets, V. V. Nikitin, Kvantovaya Elektron. *1973*, 127, C.A. *79*, 98050v (1973); Electrooptical switching in oriented films of liquid crystals. [**95**] D. W. Berreman, D. L. White, Fr. Demande 2,189,767 (23. 6. 1972), C.A. *81*, 97828j (1974); Liquid crystal devices. [**96**] D. W. Berreman, J. Opt. Soc. Amer. *63*, 1374 (1973), C.A. *80*, 21029h (1974); Optics in smoothly varying anisotropic planar structures. Application to liquid-crystal twist cells. [**97**] D. W. Berreman, Appl. Phys. Lett. *25*, 12 (1974), C.A. *81*, 55250j (1974); Dynamics of liquid-crystal twist cells. [**98**] D. W. Berreman, J. Appl. Phys. *46*, 3746 (1975); Liquid-crystal twist cell dynamics with backflow. [**99**] M. Bertolotti, S. Martellucci, F. Scudieri, D. Sette, Appl. Phys. Lett. *21*, 74 (1972), C.A. *77*, 95263q (1972); Acoustic modulation of light by nematics. [**100**] M. Bertolotti, F. Scudieri, Electrotecn. *61*, 682 (1974); Electrooptical applications of liquid crystals.

[**101**] F. Bescond, J. Pompei, Colloq. Int. Pulverisation Cathodique Ses Appl., C. R., 1st; *1973*, 169, C.A. *81*, 83309d (1974); Adaption des procédés de depot par pulvérisation cathodique réactive a l'élaboration des dispositifs d'affichage a cristaux liquides. [**102**] J. E. Bigelow, R. A. Kashnow, C. R. Stein, IEEE Trans. Electron Dev. ED *22*, 22 (1975); Contrast optimization in matrix-addressed liquid crystal displays. [**103**] J. E. Bigelow, R. A. Kashnow, IEEE Trans. Electron Dev. ED *22*, 730 (1975); Observation of a bistable TN liquid-crystal effect. [**103a**] A. M. Bilkenkis, Telecommun. Rad. Engng. *30*, 104 (1976); Liquid-crystal displays with auxiliary nonlinear elements. [**104**] J. Billard, Rev. Phys. Appl. *1*, 311 (1966); Modulateurs électro-optiques. [**105**] J. Billard, Nucleus (Paris) *9*, 21 (1968), C.A. *69*, 31162h (1968); Modulateurs électro-optiques. [**105a**] P. A. Birnie, Wirel. World *82*, 38 (1976), Digital wristwatch. Single IC design using a liquid-crystal display. [**106**] D. E. Bishop, US Clearinghouse

Fed. Sci. Tech. Inform., PB Rep. *1968*, PB-183838, 6 pp. C.A. *71*, 95870m (1969); Procedure for applying a protected layer of liquid crystals. [**107**] L. M. Blinov, M. F. Grebenkin, N. M. Zhelyabovskaya, V. T. Lazareva, V. V. Titov, V. M. Shoshin, U.S.S.R. 523,441 (2. 10. 1970), C.A. *85*, 200485d (1976); Mesomorphic material. [**107a**] A. Bloom, E. B. Priestley, IEEE Trans. Electron Dev. ED *24*, 823 (1977), C.A. *87*, 125330k (1977); Criteria for evaluating pleochroic dye liquid-crystal displays. [**107b**] A. Bloom, D. L. Ross, U. S. 4,033,905 (5. 11. 1975), C.A. *87*, 94401c (1977); Increasing the conductivity of electrically resistive organic materials. [**108**] B. Böttcher, D. Gross, E. Mundry, Materialprüfung *11*, 156 (1969), C.A. *71*, 73293y (1969); Anwendung cholesterinischer Phasen in der zerstörungsfreien Materialprüfung mit Wärmeflußverfahren. [**109**] B. Böttcher, Matér. Constr. (Paris) *4*, 241 (1971), C.A. *76*, 34478w (1972); Utilisation de liquides cholestériques comme indicateurs colorés de température. [**110**] A. Boller, H. Scherrer, M. Schadt, P. Wild, Proc. IEEE *60*, 1002 (1972), C.A. *77*, 131928w (1972); Low electrooptic threshold in new liquid crystals.

[**111**] A. Boller, H. Scherrer, U. S. 3,795,436 (21. 6. 1972), C.A. *80*, 114009s (1974); Nematogenic materials which exhibit the Kerr effect at isotropic temperatures. [**112**] A. Boller, H. Scherrer, Ger. Offen. 2,415,929 (4. 4. 1973), C.A. *82*, 24776b (1975); Liquid crystal compositions. [**113**] B. M. Bolotin, V. A. Molochko, L. S. Zeryukina, R. U. Safina, N. B. Etingen, L. N. Stolyarova, U.S.S.R. 515,737 (25. 6. 1974), C.A. *86*, 131622c (1977); Nematic liquid-crystal mixture containing butyl *p*-[*p*-(hexyloxy)-phenoxycarbonyl]phenyl carbonate. [**114**] B. M. Bolotin, L. S. Zeryukina, R. U. Safina, D. E. Ostromogol's-kii, U.S.S.R. 516,677 (1. 3. 1974), C.A. *85*, 108429r (1976); Liquid crystal aromatic esters. [**115**] U. Bonne, J. P. Cummings, U. S. Nat. Tech. Inform. Ser., AD Rep. *1972*, No. 751667, 181 pp., C.A. *78*, 102952q (1973); Properties and limitations of liquid crystals for aircraft displays. [**116**] U. Bonne, J. P. Cummings, IEEE Trans. Electron Dev. *20*, 962 (1973), C.A. *80*, 53661z (1974); Nematic liquid-crystal displays. Properties and limitations. [**117**] H. C. Bordon, Jr., U. S. 3,702,723 (23. 4. 1971); Segmented master character for electronic display apparatus. [**118**] H. Borden, J. Mingione, P. Nance, Electronics *45*, 66 (1972); P/MOS chip drives liquid crystal display for digital alarm clock. [**119**] J. Borel, J. Robert, U. S. 3,716,290 (18. 10. 1971); Liquid-crystal display device. [**120**] J. Borel, G. Labrunie, J. Robert, J. Phys. (Paris), C-1, *36*, 215 (1975); New applications of liquid crystals.

[**121**] K. Borowski, A. Kubitzek, U. S. 3,727,527 (11. 6. 1971); Photographic apparatus with liquid crystal voltage indicator. [**122**] G. Bothmann, U. v. d. Bussche, F. Kubli, G. Seybold, Dtsch. med. Wschr. *99*, 730 (1974); Die Plattenthermographie. Eine neue Methode in der Diagnostik des Mammakarzinoms. [**123**] C. Bourde, C. Ambrosi, Thérapie *17*, 1091 (1972); Données nouvelles sur l'action de l'Hydergine en pathologie artérielle des membres. Étude critique par l'artériographie et les cristaux liquides. [**124**] U. Braatz, K. Lehmann, H. Wawra, W. Schwella, Ger. (East) 117,781 (17. 3. 1975), C.A. *85*, 70718u (1976); Liquid crystal device. [**125**] B. Brauer, D. Nemus, H. Klose, Probleme der Festkörperelektronik Bd. 4, S. 11 (Verlag Technik: Ostberlin). 1972. Flüssige Kristalle in der Optoelektronik. [**126**] M. Braunstein, W. P. Bleha, Jr., U. S. 3,732,429 (12. 11. 1971), C.A. *79*, 12016d (1973); Liquid-crystal device. [**127**] M. Braunstein, W. P. Bleha, Jr., U. S. 3,811,180 (12. 11. 1971), C.A. *81*, 113749s (1974); Liquid crystal device. [**128**] B. B. Brenden, H. R. Curtin, Brit. 1,194,544 (3. 8. 1966); Liquid crystal detector. [**129**] G. H. Brown, Anal. Chem. *41*, 26A (1969), C.A. *72*, 6798z (1970); Liquid crystals and their applications in chemistry. [**130**] G. H. Brown, US Bureau of Radiological Health. Seminar Program. Selected Papers, No. 9, *1970*, 21 pp., C.A. *76*, 19069v (1972); Properties of liquid crystals and their application to the measurement of microwaves and IR radiation.

[**131**] G. T. Brown, Jr., D. B. Clark, D. E. Koopman, U. S. 3,920,574 (26. 8. 1974), C.A. *84*, 37337a (1976); Solidification retardation of liquid crystalline compositions with steroid derivatives of isostearyl carbonate. [**132**] G. T. Brown, Jr., D. B. Clark, D. E. Koopman, U. S. 3,998,860 (26. 8. 1974), C.A. *86*, 82209n (1977); Solidification retardation of liquid crystalline compositions with steroid derivatives of isostearyl carbonate. [**133**] S. P. Brown, Mater. Eval. *26*, 163 (1968), C.A. *70*, 98306x (1969); Cholesterics for nondestructive testing. [**134**] S. P. Brown, US Clearinghouse Fed. Sci. Tech. Inform., AD *1967*, AD-816482, 26 pp., C.A. *71*, 16743w (1969); Detection of flaws in metal-honeycomb structures by means of liquid crystals. [**135**] S. P. Brown, Appl. Polym. Symp. No. *19*, 463 (1972), C.A. *78*, 59172w (1973); Cholesterics for nondestructive bond inspection. [**136**] S. P. Brown, AGARDograph, AG-201-Vol. *1*, 449 (1975), C.A. *84*, 183429y (1976); Liquid crystal and neutron radiography methods. [**137**] J. Bruinink, 5. Arbeitstagung Flüssigkristalle, Inst. Angew. Festkörperphysik Freiburg, 25. 4. 1975; Elektrochrome Anzeigen mit Viologenen. [**138**] O. N. Bubel, E. P. Kaloshkin, V. M. Koleshko, L. S. Novikov, V. P. Stepanov, U.S.S.R. 463,693 (14. 9. 1973), C.A. *83*, 88720g (1975); High-temperature thermographic mixtures containing alkoxybenzoic acid liquid crystal. [**139**] H. K. Buecher,

Fr. Demande 2,168,413 (17. 1. 1972), C.A. *80*, 32453t (1974); Liquid crystal composition for electrooptic cells. [**140**] H. K. Buecher, T. R. Criswell, U. S. 3,963,638 (8. 1. 1974), C.A. *86*, 24477n (1977); Liquid crystal compositions for electrooptical display devices. [**141**] H. K. Buecher, R. T. Klingbiel, J. P. van Meter, Appl. Phys. Lett. *25*, 186 (1974), C.A. *81*, 143202a (1974); Frequency-addressed liquid crystal field effect. [**141a**] Bunker Ramo Corp., Brit. 1,471,946 (29. 8. 1973), C.A. *87*, 160003c (1977); Integrated liquid crystal luminophor display. [**142**] C. B. Burckhardt, M. Schadt, W. Helfrich, Appl. Opt. *10*, 2196 (1971); Holographic recording with an electrooptic liquid crystal cell. [**143**] S. Burman, Z. Haladewicz, J. Pyzik, E. Soczewinska, Pr. Nauk. Inst. Metrol. Elektr. Politech. Wroclaw. *7*, 37 (1975), C.A. *86*, 81558p (1977); Application of liquid crystals in display devices. [**144**] E. N. Buyalo, L. M. Klyukin, V. I. Smirnov, A. S. Sonin, B. M. Stepanov, Zh. Nauch. Prikl. Fotogr. Kinematogr. *18*, 117 (1973); C.A. *79*, 151588e (1973); Recording of halftone images on liquid crystals. [**145**] P. Calvert, Nature *260*, 391 (1976); Strong fibres. [**146**] S. Caplan, U. S. 3,614,210 (6. 11. 1969); Liquid crystal day/night mirror. [**147**] J. C. Caradot, Mesures *39*, 49 (1974), C.A. *81*, 7108x (1974); Cristaux liquides: prêts pour 1974. [**148**] P. Carosi, P. Maltese, C. M. Ottavi, IEEE Trans. Electron Dev. *22*, 801 (1975); A 24-digit multiplexed liquid-crystal display working in dynamic deformation mode. [**149**] P. Carroll, J. Kinson, New Scientist *63*, 132 (1974); A new twist to liquid crystals. [**150**] J. V. Cartmell, D. Churchill, U. S. 3,674,338 (5. 8. 1970), C.A. *77*, 82195u (1972); Rear projection screen employing liquid crystals.

[**151**] J. V. Cartmell, D. Churchill, D. E. Koopman, U. S. 3,700,306 (22. 9. 1971); Electrooptic shutter having a thin glass or silicon oxide layer between the electrodes and the liquid crystal. [**152**] J. V. Cartmell, D. Churchill, U. S. 3,720,623 (8. 2. 1971); Encapsulated liquid crystals. [**153**] J. A. Castellano, U. S. 3,597,044 (3. 9. 1968), C.A. *77*, 54808c (1972); Electro-optic modulator. [**154**] J. A. Castellano, R. N. Friel, U. S. 3,674,342 (29. 12. 1970); Liquid crystal display device including side-by-side electrodes on a common substrate. [**155**] J. A. Castellano, U. S. 3,703,329 (29. 12. 1969), C.A. *75*, 125067g (1971); Liquid crystal color display. [**156**] J. A. Castellano, M. T. McCaffrey, Ger. Offen. 2,121,085 (25. 5. 1970), C.A. *76*, 79278x (1972); Electrooptical cells containing liquid crystal compositions. [**157**] J. A. Castellano, J. E. Goldmacher, W. H. Helfrich, M. T. McCaffrey, L. A. Zanoni, U. S. Nat. Tech. Inform. Serv., AD Rep. *1970*, No. 871971, 56 pp., C.A. *76*, 159617x (1972); Field effect liquid crystal. [**158**] J. A. Castellano, Ferroelectrics *3*, 29 (1971), C.A. *76*, 159670j (1972); Mesomorphic materials for electrooptical application. [**159**] J. A. Castellano, R. N. Friel, M. T. McCaffrey, D. Meyerhofer, C. S. Oh, U. S. Nat. Tech. Inform. Serv., AD Rep. 1971, No. 760173, 61 pp., C.A. *79*, 98047z (1973); Liquid crystal systems for electrooptical storage effects. [**160**] J. A. Castellano, RCA Rev. *33*, 296 (1972), C.A. *77*, 53559k (1972); Liquid crystals for electrooptical application.

[**161**] J. A. Castellano, Optics Laser Technol. *7*, 259 (1975); Fundamentals of liquid crystals and other liquid display technologies. [**162**] H. J. Caulfield, R. A. Soref, Appl. Phys. Lett. *18*, 5 (1971), C.A. *74*, 68756j (1971); Optical contrast enhancement in liquid crystal devices by spatial filtering. [**163**] R. Chabicovsky, G. Kochmann, IEEE Trans. Electron Dev. ED *24*, 807 (1977), C.A. *87*, 144067w (1977); Liquid-crystal cells with special electrodes for the generation of uniform colors by optical birefringence. [**163a**] C. Chandler, Southern Med. J. *69*, 614 (1976); Liquid crystal thermography. Measure of therapeutic response to reserpine given intraarterially. [**164**] R. Chang, J. P. Dobbins, Ger. Offen. 2,155,720 (6. 11. 1970), C.A. *77*, 120694r (1972); Nematic liquid crystal light valve. [**165**] T. S. Chang, E. E. Loebner, Appl. Phys. Lett. *25*, 1 (1974), C.A. *81*, 56582f (1974); Crossover frequencies and turn-off time reduction scheme for TN displays. [**166**] D. J. Channin, IEEE Trans. Electron Dev. *21*, 650 (1974); Liquid crystal technique for observing integrated circuit operation. [**167**] D. J. Channin, IEEE Trans. Electron. Dev. *22*, 1064 (1975); New liquid crystal electrooptical device. [**168**] D. J. Channin, Appl. Phys. Lett. *26*, 603 (1975); Triode optical gate: A new liquid crystal electrooptic device. [**169**] D. J. Channin, D. E. Carlson, Appl. Phys. Lett. *28*, 300 (1976); Rapid turn-off in triode optical gate liquid crystal devices. [**170**] D. J. Channin, E. B. Priestley, U. S. 3,973,057 (7. 3. 1975), C.A. *86*, 36361h (1977); Liquid crystal display.

[**171**] I. G. Chistyakov, E. A. Kosterin, Rost Kristallov, Akad. Nauk SSSR, Inst. Kristallogr. *4*, 68 (1964), C.A. *61*, 15450e (1964); Vitrified liquid crystal films. [**172**] I. G. Chistyakov, L. K. Vistin, Sb. Dokl. Vses. Nauch. Konf. Zhidk. Krist. Simp. Ikh Prakt. Primen., 2nd, *1972*, 232, C.A. *81*, 129830x (1974); Possible technical applications of liquid crystals. [**173**] I. G. Chistyakov, L. K. Vistin, V. M. Ivanov, I. I. Gorina, U.S.S.R. 429,083 (13. 1. 1972), C.A. *81*, 160678c (1974); Normal layer liquid crystal. [**174**] I. G. Chistyakov, L. K. Vistin, L. T. Kantardzhyan, R. V. Khalatyan, Prom. Arm. *1973*, 20, C.A. *79*, 47799p (1973); Properties of liquid crystals and their practical use. [**175**] I. G.

Chistyakov, I. I. Gorina, L. K. Vistin, U.S.S.R. 463,692 (4. 7. 1973), C.A. *83*, 89152k (1975); Stabilization of nematics of azomethine series. **[176]** I. G. Chistyakov, I. I. Gorina, L. K. Vistin, O. N. Karpov, A. N. Kuznetsov, U.S.S.R. 463,694 (4. 7. 1973), C.A. *83*, 89153m (1975); Stabilization of cholesterics. **[176a]** I. G. Chistyakov, I. I. Gorina, M. Y. Rubtsova, Kristallografiya *22*, 334 (1977), C.A. *87*, 32188e (1977); Stabilization of cholesterics. **[176b]** I. G. Chistyakov, I. I. Gorina, L. Vistins, Kristallografiya *22*, 598 (1977), C.A. *87*, 32235t (1977); Method for growing a homeotropic liquid crystal. **[177]** N. Chretien, Cah. Therm. *1973*, No. 3, II/18, C.A. *79*, 129951w (1973); Utilisation des cristaux liquides pour la mesure des températures. **[178]** D. Churchill, J. V. Cartmell, R. E. Miller, Brit. 1,138,590 (17. 6. 1966), C.A. *70*, 72918y (1969); Visual display device. **[179]** D. Churchill, J. V. Cartmell, R. E. Miller, P. D. Bouffard, Fr. 1,572,257 (27. 2. 1967), C.A. *72*, 123331c (1970); Visual posting of temperature with liquid crystals. **[180]** D. Churchill, J. V. Cartmell, R. E. Miller, P. D. Bouffard, Brit. 1,161,039 (27. 2. 1967); Visual display device.

[181] D. Churchill, J. V. Cartmell, U. S. 3,578,844 (23. 2. 1968), C.A. *75*, 69556b (1971); Radiation-sensitive display device containing encapsulated cholesterics. **[182]** D. Churchill, J. V. Cartmell, U. S. 3,600,060 (23. 2. 1968), C.A. *77*, 95382c (1972); Display device containing minute droplets of cholesterics in a substantially continuous polymeric matrix. **[183]** D. Churchill, J. V. Cartmell, R. E. Miller, U. S. 3,697,297 (22. 10. 1970); Gelatin-gum arabic capsules containing cholesteric material and dispersions of the capsules. **[184]** D. Churchill, J. V. Cartmell, R. E. Miller, U. S. 3,732,119 (5. 11. 1968); Temperature sensitive visual display device. **[185]** D. Churchill, J. V. Cartmell, U. S. 3,734,597 (9. 4. 1971); Process for producing a color state in a display device. **[186]** D. Churchill, J. V. Cartmell, U. S. 3,816,786 (9. 4. 1971), C.A. *81*, 113729k (1974); Display device comprising droplets of cholesteric liquid crystal in a substantially continuous polymeric matrix. **[187]** I. Cindrich, U. S. 3,572,907 (28. 4. 1969); Optical cell for attenuating, scattering and polarizing electromagnetic radiation. **[188]** P. E. Cladis, S. Torza, J. Appl. Phys. *46*, 584 (1975); Growth of a smectic A from a bent nematic and the smectic light valve. **[189]** D. D. Clegg, Wirel. World *81*, 298 (1975); Solid-state digital wristwatch. CMOS circuity and liquid-crystal display give long battery life. I. **[190]** D. Coates, G. W. Gray, D. G. McDonnell, Ger. Offen. 2,627,215 (17. 6. 1975), C.A. *86*, 141609t (1977); Dyes for liquid crystal materials.

[191] H. S. Cole, Jr., U. S. 3,951,845 (23. 6. 1972), C.A. *85*, 102439x (1976); Nematic liquid crystal mixtures having homogeneous boundary conditions. **[192]** H. S. Cole, Jr., R. A. Kashnow, U. S. 3,984,343 (29. 9. 1972), C.A. *86*, 148846b (1977); Liquid crystal mixtures with selectively variable dielectric anisotropies. **[193]** H. S. Cole, Jr., U. S. 3,984,344 (30. 4. 1975), C.A. *86*, 63551b (1977); Positive dielectric anisotropy liquid crystal compositions. **[193a]** H. S. Cole, R. A. Kashnow, Appl. Phys. Lett. *30*, 619 (1977), C.A. *87*, 76345h (1977); New reflective dichroic liquid-crystal display device. **[193b]** H. S. Cole, Jr., S. Aftergut, Appl. Phys. Lett. *31*, 58 (1977), C.A. *87*, 75825c (1977); Dependence of absorption and optical contrast of a dichroic dye guest on the pitch of a chiral nematic host. **[194]** Compagnie des Montres Longines, Francillon S. A. Brit. 1,303,947 (25. 5. 1970); Optoelectrical display element. **[195]** G. H. Conners, P. B. Mauer, U. S. 3,592,527 (12. 11. 1969); Image display device. **[196]** J. Constant, J. Kirton, E. P. Raynes, I. A. Shanks, D. Coates, G. W. Gray, D. G. McDonnell, Electron. Lett. *12*, 514 (1976), C.A. *86*, 163567e (1977); Pleochroic dyes with high-order parameters for liquid-crystal displays. **[197]** B. D. Cook, R. E. Werchan, Ultrasonics *9*, 101 (1971), C.A. *75*, 134007q (1971); Mapping ultrasonic fields with cholesterics. **[198]** T. E. Cooper, R. J. Field, J. F. Meyer, J. Heat. Trans. *97*, 442 (1975); Liquid crystal thermography and its application to study of convective heat transfer. **[199]** L. S. Cosentino, IEEE Trans. Electron Dev. *18*, 1192 (1971); Transient scattering of light by pulsed liquid crystal cells. **[200]** A. Coutet, J. C. Dubois, A. Zann, Fr. Demande 2,274,355 (14. 6. 1974), C.A. *85*, 134249b (1976); Liquid crystal mixture having a positive dielectric anisotropy and a wide range of mesomorphism.

[201] L. T. Creagh, U. S. 3,655,270 (28. 8. 1970), C.A. *77*, 41384m (1972); Electrooptical display devices using nematic mixtures with very wide temperature ranges. **[202]** L. T. Creagh, D. Jones, S. Lu, U. S. 3,716,289 (31. 8. 1970); Electro-optical display devices using smectic/nematic liquid crystal mixtures. **[203]** L. T. Creagh, A. R. Kmetz, R. A. Reynolds, IEEE Trans. Electron Dev. ED-*18*, 672 (1971); Performance characteristics of nematic liquid crystal display devices. **[204]** L. T. Creagh, A. R. Kmetz, Soc. Inform. Display, Symp. San Francisco, 1972, 90; Performance advantages of liquid crystal displays with surfactant-produced homogeneous alignment. **[205]** L. T. Creagh, A. R. Kmetz, J. Electron. Mater. *1*, 350 (1972), C.A. *78*, 89760d (1973); Selective doping of MBBA to optimize display performance. **[206]** L. T. Creagh, Proc. IEEE *61*, 814 (1973), C.A. *79*, 98008n (1973); Nematic liquid crystal materials for displays. **[207]** W. A. Crossland, Appl. Phys. Lett. *26*, 598 (1975), C.A.

83, 50156q (1975); Birefringence in silicon monoxide films used for aligning liquid crystal layers. [**208**] Daini Seikosha, Hattori Tokeiten, Brit. 1,318,011 (2. 5. 1970); Nematic liquid crystal material. [**209**] Daini Seikosha, K. Hattori, Brit. 1,318,012 (17. 4. 1970), C.A. *79*, 58815s (1973); Nematic liquid crystal compounds and mixtures. [**210**] Daini Seikosha, Brit. 1,327,133 (28. 12. 1970), C.A. *80*, 8358v (1974); Memory liquid crystal.

[**211**] Daini Seikosha, Fr. Demande 2,175,180 (8. 2. 1972), C.A. *80*, 114052a (1974); Liquid crystal display device and method of treating the surface of a glass for an electrode for this apparatus. [**212**] Dai Nippon Toryo, Neth. Appl. 74 14,843 (29. 6. 1974), C.A. *85*, 169716d (1976); Liquid crystal device. [**213**] Dai Nippon Toryo, Neth. Appl. 75 05,016 (14. 1. 1975), C.A. *86*, 99101u (1977); Liquid crystal cell. [**214**] E. D. Dalby, Rev. Sci. Instrum. *42*, 1540 (1971); Liquid crystal device for relative porosity measurements. [**215**] Y. Y. Daugvilla, G. I. Denis, E. Z. Kogan, G. D. Loninov, U.S.S.R. 507,611 (30. 7. 1973), C.A. *85*, 134483y (1976); Liquid crystal composition for electrooptical display devices. [**215a**] D. J. David, E. E. Hardy, U. S. 4,040,749 (5. 2. 1975), C.A. *87*, 110986e (1977); Organic vapor detection with liquid crystals. [**216**] F. Davis, U. S. 3,576,761 (18. 3. 1969), C.A. *75*, 13447z (1971); Thermometric compositions comprising a mesomorphic substance, a cholesteryl halide, and an oil soluble dye selected from the group consisting of disazo, indulene, and nigrosine dyes. [**217**] F. Davis, U. S. 3,619,254 (18. 3. 1969), C.A. *76*, 73402m (1972); Thermometric articles. [**218**] F. Davis, Ger. Offen. 2,459,618 (17. 12. 1973), C.A. *83*, 149326g (1975); Printing inks from cholesterics on water basis and laminates prepared from them. [**219**] F. Davis, U. S. 3,975,288 (2. 1. 1975), C.A. *85*, 135043s (1976); High temperature noncrystallizing cholesteric liquid crystal compositions. [**220**] T. W. Davison, K. L. Ewing, J. Fergason, M. Chapman, A. Can, C. C. Voorhis, Cancer *29*, 1123 (1972); Detection of breast cancer by liquid crystal thermography.

[**221**] D. E. DeBlance, U. S. 3,643,021 (2. 1. 1970), C.A. *76*, 134197x (1972); Voltage-controlled color film recording system. [**222**] H. A. de Koster, U. S. 3,524,726 (4. 4. 1968); Smectographic display. [**223**] M. de Mets, Microsc. Acta *76*, 405 (1975), C.A. *83*, 27140w (1975); Relation between cathodolumines-cence and molecular structure of organic compounds. [**224**] D. Demus, H. Schubert, K. Hanemann, Ger. Offen. 2,215,287 (29. 3. 72); Benzylcyclohexanone und diese enthaltende kristallinflüssige Gemische. [**225**] D. Demus, F. Kuschel, Ger. (East) 117,013 (21. 1. 1975), C.A. *84*, 172044t (1976); Electrooptical cells. [**226**] D. Demus, F. Kuschel, Mitt. Chem. Ges. DDR *22*, 144 (1975); Eigenschaften und Anwendungen von flüssigen Kristallen. [**227**] D. Demus, Wiss. Beitr. Martin-Luther-Univ. Halle-Wittenberg *1976* (20), 63, C.A. *86*, 98926e (1977); Applications of liquid crystals. [**228**] D. Demus, G. Wartenberg, Kristall und Technik *11*, 1197 (1976), Cholesterinische Gemische für die Thermographie. [**228a**] D. Demus, F. Kuschel, Ger. (East) 119,510 (7. 4. 1975), C.A. *87*, 14285e (1977); Electrooptical display cells containing liquid crystals. [**229**] A. Derzhanski, N. Shonova, Dokl. Bolg. Akad. Nauk *28*, 315 (1975), C.A. *83*, 106529h (1975); Memory effect at a TN layer. [**230**] H. J. Deutscher, W. Weissflog, D. Demus, G. Pelzl, H. Schubert, Brit. 1,368,136 (8. 10. 1971), C.A. *82*, 72634a (1975); Nematic liquid crystals.

[**231**] S. K. Dey, G. D. Dick, J. Vac. Sci. Technol. *11*, 97 (1974), C.A. *80*, 138061d (1974); Observation of switching phenomena in discontinuous gold films by liquid-crystal technique. [**231a**] M. de Zwart, T. W. Lathouwers, Ger. Offen. 2,641,861 (25. 9. 1975), C.A. *87*, 134735y (1977); Nematic liquid crystal mixture of α-cyanostilbenes and their use in image-reproducing apparatus. [**231b**] P. M. Diaz, Anesthesiol. *44*, 443 (1976); Use of liquid-crystal thermography to evaluate sympathetic blocks. [**232**] H. J. Dietrich, E. L. Steiger, U. S. 3,742,054 (2. 7. 1971), C.A. *79*, 78367a (1973); Mesomorphic aromatics. [**233**] H. J. Dietrich, E. L. Steiger, U. S. 3,769,313 (30. 6. 1971), C.A. *80*, 36855y (1974); N-(*p*-*n*-Alkoxycarbonyloxybenzylidene)-*p*'-aminophenyl acetate liquid crystals. [**234**] J. L. Dion, M. Bader, Proc. Soc. Photo-Opt. Instrum. Eng. *38*, 43 (1973), C.A. *81*, 71033w (1974); Circular dichroism in liquid-crystals mixtures and search for a piezooptic effect associated with ultrasonic fields. [**235**] G. Dir, J. Adams, W. Haas, J. Stephany, SID J. *11*, 14 (1974), C.A. *83*, 155703u (1975); Contrast enhancement for liquid crystal valves. [**235a**] G. A. Dir, W. E. L. Haas, J. E. Adams, Appl. Phys. Lett. *30*, 309 (1977), C.A. *86*, 197892x (1977); Light sensitivity enhancement of image intensifiers. [**236**] G. D. Dixon, L. C. Scala, U. S. 3,656,909 (16. 2. 1970), C.A. *77*, 27333n (1972); Cholesteric liquid crystal stabilizers for detector elements. [**237**] G. D. Dixon, T. P. Brody, W. A. Hester, Appl. Phys. Lett. *24*, 47 (1974), C.A. *82*, 9964t (1975); Alignment mechanism in TN layers. [**238**] G. D. Dixon, J. F. Meier, Mol. Cryst. Liq. Cryst. *37*, 233 (1976), C.A. *86*, 122675v (1977); Liquid crystal-rubber dispersions. [**238a**] G. D. Dixon, Mater. Eval. *35*, 51 (1977), C.A. *87*, 153722h (1977); Cholesterics in nondestructive testing. [**238b**] A. Dmitrieff, J. Perrin, B. Richard, W. Urbach, Rev. Phys. Appl.

11, 523 (1976), Transducteur électro-optique a cristal liquide pour l'affichage d'hologrammes acoustiques. [**238c**] R. Doriguzzi, T. Scheffer, Swiss 582,894 (17. 3. 1975), C.A. *87*, 76415f (1977); Liquid crystal display apparatus for color reproduction of information. [**239**] J. F. Dreyer, U. S. 2,400,877 (21. 3. 1941), C.A. *40*, 4960^4 (1946); Optical device. [**240**] J. F. Dreyer, U. S. 2,524,286 (14. 5. 1946), C.A. *45*, 3534e (1951); Flexible noncrystalline self-contained polarizing films.

[**241**] J. F. Dreyer, U.S. 2,544,659 (14. 5. 1946); Dichroic light-polarizing sheet materials and the like and the formation and the use thereof. [**242**] J. F. Dreyer, J. Phys. & Colloid Chem. *52*, 808 (1948), C.A. *42*, 7600h (1948); Fixing of molecular orientation. [**243**] J. F. Dreyer, U.S. 3,592,526 (15. 7. 1969); Means for rotating the polarization plane of light and for converting polarized light to nonpolarized light. [**244**] J. F. Dreyer, U.S. 3,597,043 (2. 5. 1969); Nematic liquid crystal optical elements. [**245**] J. F. Dreyer, U.S. 3,658,616 (15. 10. 1964), C.A. *77*, 82244j (1972); Making light-polarizing patterns. [**246**] J. F. Dreyer, J. Phys. (Paris), C-4, *30*, 114 (1969), C.A. *72*, 116188h (1970); Light polarization from films of lyotropic nematics. [**247**] J. F. Dreyer, Liq. Cryst. Ord. Fluids *1970*, 311, C.A. *78*, 165408n (1973); Alignment of molecules in the nematic liquid crystal state. [**248**] J. F. Dreyer, IEEE Son. Ultrason. *21*, 80 (1974); Characteristics of an acousto-optic transducer of nematic liquid crystal type. [**249**] J. F. Dreyer, J. Acoust. Soc. *55*, 407 (1974); Introduction to a nematic liquid crystal acousto-optic conversion cell. [**250**] J. C. Dubois, A. Zann, J. C. Lavenu, Ger. Offen. 2,324,760 (19. 5. 1972), C.A. *80*, 47668r (1974); Nematic 4-alkylphenyl 4-methoxybenzoates.

[**251**] J. C. Dubois, A. Zann, A. Couttet, Fr. Demande 2,214,523 (18. 1. 1973), C.A. *82*, 163071f (1975); Mixture of nematics, with strong positive dielectric anisotropy, large range of mesomorphism, and low threshold voltage. [**252**] J. C. Dubois, Fr. Demande 2,252,132 (28. 11. 1973), C.A. *83*, 211651q (1975); Mixture of nematics with strong positive dielectric anisotropy based on benzoic esters. [**253**] J. C. Dubois, A. Zann, Fr. Demande 2,289,931 (31. 10. 1974), C.A. *86*, 98921z (1977); Alignment of liquid crystals and its use in TN display devices. [**254**] J. C. Dubois, A. Zann, Fr. Demande 2,294,462 (13. 12. 1974), C.A. *86*, 113739s (1977); Alignment of liquid crystals in display devices. [**255**] J. C. Dubois, F. Barre, Ger. Offen. 2,600,558 (10. 1. 1975), C.A. *85*, 134506h (1976); Mesomorphic materials for liquid crystal cells. [**255a**] J. C. Dubois, U.S. 4,038,441 (23. 12. 1973), C.A. *87*, 192128p (1977); Liquid crystal device with plane alignment. [**256**] D. Earley, P. Franklin, R. Harris, S. Stevens, U.S. Air Force Syst. Command, Air Force Mater. Lab., Tech. Rep., AFML 1974, AFML-TR-74-37, 63 pp., C.A. *82*, 8299t (1975); Research and development on characterization of electromagnetic materials. [**257**] H. D. Edmonds, U.S. 3,832,034 (6. 4. 1973), C.A. *81*, 179999e (1974); Liquid crystal display assembly. [**257a**] H. D. Edmonds, Brit. 1,457,318 (3. 5. 1973), C.A. *86*, 181645h (1977); Improvements relating to protective coatings for field effect transistors. [**258**] W. Eggimann, Der Elektroniker *1975*, Nr. 8, EL 29; Flüssigkristalle in integrierten Schaltungen. [**259**] S. Elberg, P. Mathonnet, Report *1975*, CEA-N-1855, C.A. *85*, 126277u (1976); Thermographie à l'aide des cholestériques. [**260**] G. Elefante, H. F. Mark, J. E. Mark, C.F.S.T.I. AD-805 577 (1966); Investigation of the influence of polymer surfaces on the textures of the cholesteric mesophase.

[**261**] G. Elliott, Liq. Cryst. Disp. Appl. Symp. *1972*, 20, C.A. *84*, 52036v (1976); Principles of operation of dynamic scattering displays. [**262**] G. Elliott, Chem. Brit. *9*, 213 (1973), C.A. *79*, 36043m (1973); Liquid crystals for electrooptical displays. [**263**] G. Elliott, GEC J. Sci. Technol. *42*, 51 (1975), C.A. *85*, 70610c (1976); Electro-optical displays using liquid crystals. [**264**] G. Elliott, Radio Electron. Eng. *46*, 281 (1976), C.A. *85*, 169654g (1976); Recent developments in liquid crystals and other new display techniques. [**265**] G. Elliott, Microelectronics *7* (3), 19 (1976); Present status of liquid-crystal displays. [**266**] B. M. Elson, Aviation W. *89*, 71 (1968); Experimental display devices use films of liquid crystals. [**267**] R. D. Ennulat, J. L. Fergason, Mol. Cryst. Liq. Cryst. *13*, 149 (1971), C.A. *75*, 28227b (1971); Thermal radiography utilizing liquid crystals. [**268**] F. Erbe, E. Roll, Ger. (East) 111,410 (30. 4. 1974), C.A. *84*, 82592r (1976); Etching thin layers of tin oxide, antimony-containing tin oxide, indium oxide, or zinc-containing indium oxide. [**269**] D. Erdmann, Inf. Chim. *144*, 127 (1975), C.A. *83*, 170965z (1975); Cristaux liquides et leur application. [**270**] G. Erdos, I. Janossy, K. Pinter, Hung. Acad. Sci., Cent. Res. Inst. Phys., KFKI 1975, KFKI-75-65, 7 pp., C.A. *84*, 158212q (1976); Investigation of the properties of cholesterics.

[**271**] H. Esaki, M. Fukai, A. Moriyama, H. Tatsuta, Japan. Kokai 75 35,077 (1. 8. 1973), C.A. *83*, 106257t (1975); Electro-optical element. [**272**] H. Esaki, M. Fukai, A. Moriyama, Y. Takezako, Japan. Kokai 75,124,880 (20. 3. 1974), C.A. *85*, 54631e (1976); Nematic liquid crystal compositions with positive dielectric anisotropy. [**273**] H. Esaki, M. Fukai, A. Moriyama, Y. Takezako, Japan. Kokai 75,146,578 (17. 5. 1974), C.A. *84*, 128802q (1976); Nematic liquid crystal compositions with

positive dielectric anisotropy for display devices. [**274**] H. Esaki, M. Fukai, A. Moriyama, Y. Takezako, Japan. Kokai 75,147,488 (17. 5. 1974), C.A. *84*, 187558n (1976); Nematic liquid crystal compositions for display devices. [**275**] H. Esaki, M. Fukai, A. Moriyama, Y. Takezako, Japan. Kokai 75,147,489 (17. 5. 1974), C.A. *84*, 187557m (1976); Nematic liquid crystal compositions for display devices. [**276**] H. Esaki, M. Fukai, A. Moriyama, Y. Takezako, Japan. Kokai 75,147,490 (17. 5. 1974), C.A. *84*, 187556k (1976); Nematic liquid crystal compositions for display devices. [**277**] S. P. A. Eurand, F. di Roberto, Fr. Demande 2,216,121 (5. 2. 1973), C.A. *83*, 35748r (1975); Support for thermographic microencapsulated liquid crystal composition. [**277a**] H. Ezaki, M. Fukai, H. Tatsuta, Japan. Kokai 77 20,985 (11. 8. 1975), C.A. *86*, 198027f (1977); Nematic liquid crystal compositions for field effect type display devices. [**278**] K. Fahrenschon, R. Unbehaun, M. Schiekel, Comptes Rendus des Journees d'Electronique 1972, 84, EPFL-Lausanne; Nematische Flüssigkristall-Anzeigen kleiner Leistung. [**279**] K. Fahrenschon, M. Schiekel, Feinwerktech. Micronic *73*, 46 (1973); Anzeigeelemente mit Flüssigkristallen. [**279a**] K. Fahrenschon, M. F. Schiekel, J. Electrochem. Soc. *124*, 953 (1977), C.A. *87*, 60728z (1977); Properties of pretilted liquid crystal structures. [**280**] N. H. Farhat, Appl. Opt. *8*, 2562 (1969), C.A. *72*, 73137w (1970); Time-dependent resolution capability of liquid crystals.

[**281**] C. Feldman, K. Moorjani, Thin Solid Films *5*, R1 (1970), C.A. *73*, 29939f (1970); Filament formation in amorphous films during switching. [**282**] L. Feldman, Z. B. Alfassi, A. P. Kushelevsky, Trans., Annu. Meet. — Isr. Nucl. Soc. *4*, 90 (1976), C.A. *86*, 147463n (1977); Gamma ray dosimetry by solutions of cholesteryl nonanoate. [**283**] J. L. Fergason, P. Vogl, M. Garbuny, U.S. 3,114,836 (4. 3. 1960); Thermal imaging devices utilizing a cholesteric liquid crystal phase material. [**284**] J. L. Fergason, A. E. Anderson, U.S. 3,401,262 (29. 6. 1965); Radiation sensitive display system utilizing a cholesteric liquid crystal phase material. [**285**] J. L. Fergason, U.S. 3,409,404 (13. 11. 1963), C.A. *65*, 17689c (1966); Analytical methods and devices employing cholesteric liquid crystalline materials. [**286**] J. L. Fergason, A. E. Anderson, U.S. 3,410,999 (29. 6. 1965); Display system utilizing a liquid crystalline material of the cholesteric phase. [**287**] J. L. Fergason, N. N. Goldberg, U.S. 3,529,156 (13. 6. 1967), C.A. *71*, 25917h (1969); Hysteretic cholesteric liquid crystalline compositions and recording devices utilizing such compositions. [**288**] J. L. Fergason, U.S. 3,627,408 (29. 6. 1965); Electric field device. [**289**] J. L. Fergason, N. N. Goldberg, U.S. 3,663,390 (24. 9. 1970); Method for changing color play range of liquid crystal materials. [**290**] J. L. Fergason, U.S. 3,731,986 (22. 4. 1971), C.A. *79*, 47092c (1973); Display devices utilizing liquid crystal light modulation.

[**291**] J. L. Fergason, U.S. 3,885,982 (18. 6. 1968), C.A. *83*, 117312h (1975); Liquid crystal coating. [**292**] J. L. Fergason, U.S. 3,960,749 (9. 2. 1971), C.A. *85*, 184869z (1976); Liquid crystal compositions. [**293**] J. L. Fergason, N. N. Goldberg, S. African 6904,800 (22. 8. 1968), C.A. *73*, 72021n (1970); Detection of x-rays. [**294**] J. L. Fergason, N. N. Goldberg, Fr. Demande 2,040,490 (30. 4. 1969), C.A. *75*, 139590j (1971); Composition for indicating that a predetermined limiting temperature has been reached. [**295**] J. T. Crissey, J. L. Fergason, J. M. Bettenhausen, J. Invest. Dermatol. *45*, 329 (1965), C.A. *64*, 996f (1966); Cutaneous thermography with liquid crystals. [**296**] J. L. Fergason, N. N. Goldberg, C. H. Jones, et al., AD-62-940, Defense Documentation Center, Defense Supply Agency (Aug. 1965); Detection of liquid crystal gases (reactive materials). [**297**] J. L. Fergason, Trans. N. Y. Acad. Sci. *29*, 26 (1966), C.A. *67*, 85225h (1967); Cholesteric structure. III. Thermal mapping. [**298**] J. L. Fergason, Appl. Opt. *7*, 1729 (1968); Liquid crystals in nondestructive testing. [**299**] J. L. Fergason, T. R. Taylor, T. B. Harsch, Electro-Technol. (New York) *85*, 41 (1970), C.A. *75*, 123095r (1971); Liquid crystals and their applications. [**300**] J. L. Fergason, US Nat. Tech. Inform. Serv., AD Rep. *1971*, No. 741898, 115 pp., C.A. *77*, 121883p (1972); Qualitative and quantitative analysis of enclosed atmospheres by liquid crystals.

[**301**] W. I. Filatov, A. N. Savina, M. W. Muchina, 1. Wiss. Konferenz über flüssige Kristalle, 17.–19. 11. 1970, Iwanowo, Sammlung der Vorträge (1972), Nr. 30. Erfahrungen bei der klinischen Verwendung cholesterinischer Flüssigkristalle für die farbige Thermotopographie. [**302**] A. G. Fischer, D. Herbst, K. Koger, J. Knüfer, W. D. Frobenius, 6. Arbeitstagung Flüssigkristalle, Inst. Angew. Festkörperphysik Freiburg, 2. 4. 1976; Hinterbeleuchtung von TNLC-Anzeigen mittels ZnS-Pulver-Elektrolumineszenz-Leuchtplatten. [**303**] A. G. Fischer, Thin Solid Films *36*, 469 (1976), C.A. *86*, 148767b (1977); Thin film applications in flat image display panels. [**304**] A. G. Fischer, Microelectronics *7* (4), 5 (1976); Flat TV panels with polycrystalline layers. [**305**] J. K. Fischer, D. M. von Bruening, H. Labhart, Appl. Opt. *15*, 2812 (1976), C.A. *85*, 200192f (1976); Light modulation by electrochromism. [**306**] D. L. Fishel, U.S.N.T.I.S., AD/A Rep. 1974, No. 000344/2GA, 46 pp., C.A. *83*, 113780m (1975); Liquid crystals for air force displays. [**307**] J. B. Flannery, W. E. L. Haas, Ger. Offen. 2,233,540

(13. 9. 1971), C.A. *79*, 85689k (1973); Liquid crystals for imaging use. [**308**] J. B. Flannery, Jr., W. E. L. Haas, J. E. Adams, Fr. Demande 2,214,524 (24. 1. 1973), C.A. *82*, 163070e (1975); Transformation of optically negative liquid crystal compositions into an optically positive state, and its use in the formation of images. [**309**] E. Forest, C. K. Keller, U.S. 3,804,618 (16. 6. 1967); C.A. *81*, 84430y (1974); Liquid-crystal imaging system. [**310**] L. M. Fraas, J. Grinberg, W. P. Bleha, A. D. Jacobson, J. Appl. Phys. *47*, 576 (1976), C.A. *84*, 115030m (1976); Novel charge-storage-diode structure for use with light-activated displays.

[**311**] L. M. Fraas, W. P. Bleha, J. Grinberg, A. D. Jacobson, J. Appl. Phys. *47*, 584 (1976), C.A. *84*, 158664p (1976); ac Photoresponse of a large-area imaging cadmium sulfide/cadmium telluride heterojunction. [**312**] A. F. Fray, D. Jones, Electron. Lett. *11*, 358 (1975); Large-angle beam deflector using liquid crystals. [**313**] W. G. Freer, Electr. Pow. *21*, 631 (1975); Liquid-crystal displays. [**313a**] W. G. Freer, Microelectron. Rel. *15*, 315 (1976); Current liquid-crystal display technology. [**314**] M. J. Freiser, I. Haller, U.S. 3,625,591 (10. 11. 1969); Liquid crystal display elements. [**315**] L. J. French, U.S. 3,440,620 (10. 1. 1966); Electro-optical memory. [**316**] I. Freund, P. M. Rentzepis, U.S. 3,364,433 (10. 3. 1967); Molecular swarm liquid crystal optical parametric devices. [**317**] S. Friberg, Kem. Tidskr. *83*, 26 (1971), C.A. *75*, 11275t (1971); Thermotropic liquid crystals. [**318**] S. E. Friberg, K. Roberts, Ger. Offen. 2,215,263 (30. 3. 1971), C.A. *78*, 8272w (1973); Stable suspensions. [**319**] Y. Fujii, Japan. Kokai 76,133,187 (15. 5. 1975), C.A. *86*, 148852a (1977); Liquid crystal composite. [**320**] S. Fujiki, A. Hakoyama, M. Suzuki, Electr. Commun. Jap. *57*, 106 (1974); Real-time holography using a liquid crystal display device.

[**321**] S. Fujimura, J. Fac. Eng., Univ. Tokyo, A *1972*, No. 10, 48, C.A. *79*, 47746u (1973); Optical correlator using liquid crystal. [**322**] M. Fukai, K. Asai, A. Moriyama, S. Nagata, K. Hattori, H. Tatta, Japan. Kokai 73 69,775 (22. 12. 1971), C.A. *80*, 32528w (1974); Electrooptical display devices containing benzylidene aniline derivative mixtures. [**323**] M. Fukai, H. Tatsuta, A. Moriyama, K. Asai, Japan. Kokai 74 29,190 (17. 7. 1972), C.A. *81*, 71118c (1974); Nematic liquid crystal compositions for display devices. [**324**] M. Fukai, K. Asai, A. Moriyama, S. Nagata, K. Hattori, Japan. Kokai 74 29,292 (17. 7. 1972), C.A. *81*, 71119d (1974); Nematic liquid-crystal compositions for electrooptical display devices. [**325**] M. Fukai, K. Asai, S. Nagata, H. Tatsuta, K. Mori, Ger. Offen. 2,316,864 (5. 4. 1972), C.A. *80*, 31399t (1974); Liquid crystal composition. [**326**] F. Funada, T. Wada, Japan. Kokai 75 78,579 (15. 11. 1973), C.A. *83*, 170960u (1975); Liquid crystal compositions for electrooptical display devices. [**327**] F. Funada, M. Matsuura, Japan. Kokai 76 14,183 (26. 7. 1974), C.A. *85*, 102431p (1976); Liquid crystal compositions for TN display devices. [**328**] F. Funada, H. Kuwagaki, Japan. Kokai 76 60,692 (25. 11. 1974), C.A. *86*, 99093t (1977); Liquid crystal composition for TN display devices. [**329**] F. Funada, T. Wada, S. Mito, J. Electron. Engng. *98*, 16 (1975); Liquid crystal display — getting better all the time. [**330**] Y. Furubata, K. Toriyama, S. Nomura, Kotai Butsuri *4*, 242,303 (1969), C.A. *72*, 125786d (1970); Liquid crystals and their applications.

[**331**] Y. Furuhata, K. Toriyama, U.S. 3,832,033 (17. 12. 1971), C.A. *83*, 19041z (1975); Regular ferroelectrics-liquid crystal composite optical element. [**332**] A. Furushita, I. Yoshitake, S. Fujiwara, Y. Watanabe, T. Mizohata, Japan. 75 21,679 (24. 9. 1970), C.A. *83*, 211905a (1975); Electret. [**333**] P. L. Garbarino, R. D. Sandison, J. Electrochem. Soc. *120*, 834 (1973), C.A. *79*, 59082n (1973); Nondestructive location of oxide breakdowns on MOSFET structures. [**334**] A. Garfein, W. Rindner, D. C. Rubin, U.S. 3,667,039 (17. 6. 1970); Electricity measurement devices employing liquid crystalline materials. [**335**] M. Gautherie, J. Phys. (Paris), C-4, *30*, 122 (1969); Application des cholestériques à la thermographie cutanée. [**336**] M. Gautherie, Y. Quenneville, C. Gros, Pathol.-Biol. *22*, 553 (1974), C.A. *82*, 70091x (1975); Thermographie cholestérique. Feuilles de cristaux liquides. Applications cliniques, pharmacologiques et physiologiques et confrontation avec la thermographie infrarouge. [**337**] M. Gautherie, Y. Quenneville, C. Rempp, C. Gros, J. Radiol. *56*, 316 (1975); Valeur informative comparée de la téléthermographie (IR) et de la thermographie de contact en sénologie. [**338**] N. Gauthier, Toute Electron. *404*, 50 (1975); Cristaux liquides: circuits de commande et applications. [**338a**] D. M. Gavrilovic, D. L. Ross, U.S. 4,029,594 (17. 6. 1976), C.A. *87*, 60831c (1977); Liquid crystal compounds and electro-optic devices incorporating them. [**339**] J. Gaynor, T. G. Anderson, U.S. 3,733,485 (18. 3. 1971); Exposure meter for thermal imaging devices. [**340**] R. M. Gelber, E. A. Small, Jr., U.S. 3,736,047 (13. 8. 1971); Liquid crystal display device with internal anti-reflection casting.

[**341**] R. M. Gelber, E. A. Small, Jr., U.S. 3,971,869 (13. 8. 1971), C.A. *85*, 184873w (1976); Liquid crystal display device and method. [**342**] C. J. Gerritsma, J. H. J. Lorteye, Proc. IEEE *61*, 829 (1973), C.A. *79*, 98440x (1973); Hybrid liquid-crystal display with a small number of interconnections.

[343] C. J. Gerritsma, J. J. M. J. de Klerk, P. van Zanten, Solid State Commun. *17*, 1077 (1975), C.A. *84*, 37509h (1976); Changes of twist in TN liquid-crystal layers by frequency switching of applied electric fields. [343a] F. Gharadjedaghi, J. Robert, Rev. Phys. Appl. *11*, 467 (1976); Comportement électro-optique d'une structure nématique en hélice. Application a l'affichage. [344] E. Giachero, Minerva Chirurgica *30*, 797 (1975); La termografia a cristalli liquidi colesterici su placca. [345] T. G. Giallorenze, J. M. Schnur, U.S. Pat. Appl. 610,488 (4. 9. 1975), C.A. *85*, 134047j (1976); Active liquid core fibers. [345a] F. Giannini, P. Maltese, R. Sorrentino, Alta Freq. *46*, 170 (1977), C.A. *87*, 160654r (1977); Liquid crystal technique for field detection in microwave integrated circuitry. [346] F. H. Gillery, Inform. Disp. *9*, 17 (1972), C.A. *76*, 105097j (1972); Transparent, conductive coatings of indium oxide. [347] W. M. Ginsburg, W. I. Smirnov, A. S. Sonyin, B. M. Stepanov, 1. Wiss. Konferenz über flüssige Kristalle, 17.–19. 11. 1970, Iwanowo, Sammlung der Vorträge (1972), Nr. 26. Thermotopographie mit flüssigen Kristallen bei Bestrahlung mit monochromatischem Licht. [348] P. Girard, U.S. 3,691,755 (21. 10. 1969); Clock with digital display. [349] P. Girard, U.S. 3,712,047 (11. 5. 1970); Time display device for timepieces. [350] P. O. Glantz, S. Friberg, Odontol. Revy *22*, 341 (1971), C.A. *77*, 45056d (1972); New method to measure tooth temperature.

[351] G. N. Glazkov, A. M. Zhmud, G. G. Molokov, Yu. K. Nepochatov, Sb. Dokl. Vses. Nauch. Konf. Zhidk. Krist. Simp. Ikh. Prakt. Primen., 2nd *1972*, 296, C.A. *81*, 143224j (1974); Use of liquid crystals in the development of micro-miniature superhigh-frequency devices. [352] P. S. Glazyrin, T. A. Ivanova, V. Pak, Meas. Techn. *18*, 417 (1975); Methods and apparatus for calibrating thin-film liquid-crystal heat detectors. [353] W. L. Goffe, U.S. 3,907,559 (27. 5. 1966), C.A. *83*, 211303c (1975); Imaging process employing friction charging in the presence of an electrically insulating developer liquid. [354] N. N. Goldberg, J. L. Fergason, U.S. 3,580,864 (30. 4. 1969), C.A. *74*, 16774u (1971); Cholesteric-phase liquid-crystal compositions stabilized against true-solid formation, using cholesteryl erucyl carbonate. [355] N. N. Goldberg, Ger. Offen. 2,019,864 (30. 4. 1969), C.A. *74*, 16773t (1971); Cholesteryl *p*-nonylphenyl carbonate as solidification inhibitor of cholesterics. [356] N. N. Goldberg, J. L. Fergason, U.S. 3,627,699 (30. 4. 1969), C.A. *74*, 81757e (1971); Liquid crystal cholesteric material and sensitizing agent composition and method for detecting electromagnetic radiation. [357] L. S. Goldberg, J. M. Schnur, U.S. 3,771,065 (9. 8. 1972), C.A. *80*, 21324a (1974); Turnable internal-feedback liquid crystal-dye laser. [358] J. E. Goldmacher, G. H. Heilmeier, U.S. 3,499,702 (5. 12. 1967), C.A. *75*, 13446y (1971); Nematic liquid crystal mixtures for use in a light valve. [359] J. E. Goldmacher, J. A. Castellano, U.S. 3,540,796 (31. 3. 1967); Electro-optical compositions and devices. [360] J. E. Goldmacher, G. H. Heilmeier, U.S. 3,703,331 (26. 11. 1971); Liquid crystal display element having storage.

[361] J. E. Goldmacher, M. T. McCaffrey, U.S. 3,925,236 (2. 6. 1969), C.A. *84*, 67902y (1976); Electrooptic compositions. [362] J. E. Goldmacher, J. A. Castellano, Brit. 1,170,486 (9. 6. 1966), C.A. *71*, 44383x (1969); Schiff's bases and electro-optical compositions and devices. [363] C. H. Gooch, J. J. Low, J. Phys. D *5*, 1218 (1972); Matrix-addressed liquid crystal displays. [364] C. H. Gooch, R. Bottomley, J. J. Low, H. A. Tarry, J. Phys. E *6*, 485 (1973), C.A. *79*, 24395q (1973); Use of liquid crystals to make programmable graticules in optical systems. [365] C. H. Gooch, R. C. Bottomley, J. H. Firkins, J. J. Low, H. A. Tarry, J. Phys. D *6*, 1664 (1973); Storage cathode ray tube with liquid-crystal display. [365a] C. H. Gooch, NATO Adv. Study Inst. Ser., Ser. B *10*, 87 (1973), C.A. *86*, 197895a (1977); Liquid crystal matrix displays. [365b] C. H. Gooch, Mess. Pruef. *1974*, 225, C.A. *86*, 180560q (1977); Flüssigkristalle für die Datenanzeige. [366] J. B. Goodenough, Phase Transitions, Proc. Conf. Phase Transitions Their Appl. Mater. Sci. *1973*, 1, C.A. *81*, 96809s (1974); Applications of phase transitions in materials science. [367] L. A. Goodman, J. Vac. Sci. Technol. *10*, 804 (1973), C.A. *79*, 119130p (1973); Liquid crystal displays. [368] L. A. Goodman, Introd. Liq. Cryst. *1974*, 219, C.A. *85*, 102228c (1976); Liquid-crystal display packaging and surface treatments. [369] L. A. Goodman, RCA Rev. *35*, 447 (1974), C.A. *82*, 78736g (1975); Liquid-crystal displays. Packaging and surface treatments. [370] L. A. Goodman, Introd. Liq. Cryst. 1974, 241, C.A. *84*, 128686e (1976); Liquid-crystal displays-electro-optic effects and addressing techniques.

[371] L. A. Goodman, IEEE Trans. Cons. Electron. CE *21*, 247 (1975); Passive liquid displays: liquid crystals, electrophoretics and electrochromics. [372] L. A. Goodman, D. E. Carlson, U.S. 3,896,016 (5. 7. 1974), C.A. *83*, 124111v (1975); Liquid crystal devices. [372a] L. A. Goodman, D. Meyerhofer, S. DiGiovanni, IEEE Trans. Electron Dev. *23*, 1176 (1976); Effect of surface orientation on operation of multiplexed TN devices. [372b] L. A. Goodman, J. T. McGinn, C. H. Anderson, F. DiGeronimo, IEEE Trans. Electron Dev. ED *24*, 795 (1977), C.A. *87*, 109747w (1977); Topography of obliquely

evaporated silicon oxide films and its effect on liquid-crystal orientation. [**373**] L. S. Gorbatenko, W. S. Dronov, 1. Wiss. Konferenz über flüssige Kristalle, 17.–19. 11. 1970, Iwanowo, Sammlung der Vorträge (1972), Nr. 32; Anwendungsmöglichkeiten flüssiger Kristalle für die Registrierung von Teilchenbahnen. [**373a**] W. Gordenne, J. Belge Rad. *59*, 71 (1976); Thermographie en plaques de cristaux liquides en pathologie mammaire. [**374**] I. I. Gorina, I. G. Chistyakov, L. K. Vistin, G. G. Maidachenko, U.S.S.R. 487,923 (4. 7. 1973), C.A. *84*, 24823f (1976); Stable liquid crystal mixture. [**375**] I. I. Gorina, I. G. Chistyakov, Uch. zap. Ivanov, un-t *1974*, 90, C.A. *83*, 124409y (1975); Preferred orientation transformations of nonstabilized and stabilized mixtures of cholesterics during aging. [**376**] I. Gorog, U.S. 3,723,651 (27. 12. 1971); Optically-scanned liquid-crystal projection display. [**376a**] M. Goscianski, J. Appl. Phys. *48*, 1426 (1977), C.A. *86*, 179746y (1977); Optical characteristics of twisted nematics: application to the improvement of the scanning capability in matrix displays. [**377**] K. Goto, Japan. Kokai 73 19,488 (20. 7. 1971), C.A. *79*, 85685f (1973); Cholesteric liquid crystal sheet. [**378**] J. Goto, H. Okuyama, T. Narusawa, Y. Isozaki, M. Fujimori, Japan. Kokai 75,121,177 (11. 3. 1974), C.A. *85*, 27380r (1976); Nematic liquid crystal compositions for electrically controlled birefringence display device. [**379**] Z. Y. Gotra, V. V. Parkhomenko, L. M. Smerklo, B. N. Yavorskii, Prib. Sist. Upr. *20*, 49 (1975), C.A. *84*, 123858c (1976); Homogeneous orientation of liquid crystals for display devices. [**379a**] V. G. Govorkov, I. G. Chistyakov, N. L. Sizova, I. I. Gorina, B. V. Petukhov, M. S. Akchurin, Dokl. Akad. Nauk SSSR *234*, 1067 (1977), C.A. *87*, 76602q (1977); Study of heat fields appearing during the plastic deformation of crystals. [**380**] J. Grabmaier, H. Krüger, U. Wolff, U.S. 3,730,607 (23. 7. 1970); Indicator screen with controlled voltage to matrix crosspoints thereof.

[**381**] J. G. Grabmaier, W. F. Greubel, H. H. Krueger, Mol. Cryst. Liq. Cryst. *15*, 95 (1971), C.A. *76*, 19100y (1972); Liquid crystal matrix displays using additional solid layers for suppression of parasite currents. [**382**] J. Grabmaier, W. Geubel, H. Krüger, U. Wolff, Ger. Offen. 2,256,317 (16. 11. 1972), C.A. *81*, 129929m (1974); Homogeneous orientation of liquid crystal molecules for display devices. [**383**] Y. Grall, J. Tricoire, C. R. Soc. Biol. *16*, 1309 (1967); Thermographie cutanée par cristaux liquides d'éthers de cholestérol. [**384**] G. Grau, A. Fontanel, Brit. 1,316,497 (11. 12. 1969); Methods and apparatus for visual representation, in a model, of a surveyed medium. [**385**] G. W. Gray, Liq. Cryst. Disp. Appl., Symp. *1972*, 61, C.A. *84*, 52037w (1976); Liquid crystalline materials for display devices. [**386**] G. W. Gray, Liq. Cryst. Plast. Cryst. *1*, 327 (1974), C.A. *83*, 19086t (1975); Cholesterics in temperature measurement and vapor detection. [**387**] G. W. Gray, A. Mosley, J. Chem. Soc., Chem. Commun. *1976*, 147, C.A. *84*, 172311c (1976); Liquid crystal mixture for use in smectic liquid crystal display devices. [**388**] P. Greguss, 3rd ACS Symp. Ord. Fluids Liquid Cryst., Chicago, Aug. 1973, Abstr. No. 113; Optical replica of an ultrasonic wavefront via nematics. [**389**] W. Greubel, U.S. 3,725,899 (29. 7. 1970); Data exhibiting screen device with a liquid-crystal layer, and method of manufacture. [**390**] W. Greubel, M. Geffcken, H. Krüger, E. Reichel, M. Tauer, K. H. Walter, 4. Arbeitstagung Flüssigkristalle, Inst. Angew. Festkörperphysik Freiburg, 26. 4. 1974; Speichereffekt in cholesterinischen Flüssigkristallen mit positiver DK-Anisotropie.

[**391**] W. Greubel, Appl. Phys. Lett. *25*, 5 (1974), C.A. *81*, 55248q (1974); Bistability behavior of texture in cholesterics in an electric field. [**392**] W. Greubel, G. Baur, 7. Arbeitstagung Flüssigkristalle, Inst. Angew. Festkörperphysik Freiburg, 4. 3. 1977; Passive Helligkeitsverstärkung von Flüssigkristallanzeigen. [**392a**] W. Greubel, H. Krüger, Bundesminist. Forsch. Technol., Forschungsber., Technol. Forsch. Entwickl., 1976, BMFT-FB T 76-69, 75 pp., C.A. *87*, 192052j (1977); Nematische Flüssigkristalle. [**393**] J. Grinberg, W. P. Bleha, A. D. Jacobson, A. M. Lackner, G. D. Meyer, L. J. Miller, J. D. Margerum, L. M. Fraas, D. D. Boswell, IEEE Trans. Electron Dev. ED *22*, 775 (1975); Photoactivated birefringent liquid-crystal light valve for color symbology display. [**394**] J. Grinberg, A. Jacobson, W. Bleha, L. Willer, L. Fraas, D. Boswell, G. Myer, Opt. Eng. *14*, 217 (1975); New real-time noncoherent to coherent light image converter. Hybrid field-effect liquid-crystal light valve. [**394a**] J. Grinberg, A. D. Jacobson, J. Opt. Soc. Am. *66*, 1003 (1976); Transmission characteristics of a TN layer. [**394b**] J. Grinberg, L. M. Fraas, W. P. Bleha, Jr., P. O. Braatz, U.S. 4,032,954 (1. 6. 1976), C.A. *87*, 93581z (1977); Silicon single crystal charge storage diode. [**395**] A. A. Groshev, V. B. Sergeev, Usp. Fiz. Nauk *107*, 503 (1972), C.A. *77*, 131783v (1972); Liquid crystals in reflection devices. [**396**] B. Grossman, E. Rokicka, A. Szymanski, Zesz. Nauk. Politech. Lodz., Fiz. *2*, 117 (1973), C.A. *83*, 88961m (1975); Liquid-crystal temperature indicators. [**397**] B. Grossman, A. Lipinski, A. Szymanski, Pr. Nauk. Inst. Metrol. Electr. Politech. Wroclaw *7*, 99 (1975), C.A. *86*, 111816r (1977); Temperature indicators employing liquid-crystal materials. [**398**] E. Grzejdziak, A. Rogowski, R. Szylhabel, A. Szymanski, J. Hejwowski, Elektronika *13*, 234 (1972), C.A. *78*, 5785e (1973); Liquid-crystal temperature indicators and their use in electronics.

[399] R. W. Gurtler, C. Maze, IEEE spectrum, Nov. 1972, 25, Liquid crystal displays. [400] R. Gurtler, C. Maze, Microtecnic 27, 353 (1973), C.A. 80, 89531w (1974); Nematics and their application to the electronic watch.

[401] N. Gurusamy, K. L. Vasanth, Technology (Coimbatore, India) 22, 92 (1974), C.A. 84, 10964r (1976); Technological applications of liquid crystals. [402] R. Guye, J. Suisse Horlog. 1973, 77, C.A. 84, 24334x (1976); Liquid-crystal displays systems. [403] E. Guyon, P. Pieranski, M. Boix, Lett. Appl. Eng. Sci. 1, 19 (1973), C.A. 79, 70995h (1973); Different boundary conditions of nematic films deposited on obliquely evaporated plates. [404] W. E. L. Haas, J. E. Adams, J. H. Becker, J. J. Wysocki, U.S. 3,655,971 (12. 8. 1969), C.A. 74, 133076d (1971); Imaging system. [405] W. E. L. Haas, J. E. Adams, J. B. Flannery, Jr., B. Mechlowitz, U.S. 3,666,947 (6. 1. 1971), C.A. 77, 133228s (1972); Liquid crystal imaging system having an undisturbed image on a disturbed background and having a radiation absorptive material dispersed throughout the liquid crystal. [406] W. E. L. Haas, J. E. Adams, U.S. 3,671,231 (16. 6. 1967), C.A. 77, 133229t (1972); Liquid crystal imaging system. [407] W. E. L. Haas, J. E. Adams, Jr., J. B. Flannery, Jr., U.S. 3,680,950 (15. 3. 1971), C.A. 77, 120786x (1972); Grandjean-state liquid crystalline imaging system. [408] W. E. L. Haas, J. E. Adams, J. B. Flannery, Jr., U.S. 3,687,515 (6. 1. 1971), C.A. 77, 146264a (1972); Electrooptic liquid crystal system with a polyamide resin additive. [409] W. E. L. Haas, J. E. Adams, B. Mechlowitz, U.S. 3,803,050 (20. 8. 1971), C.A. 81, 129922d (1974); Liquid crystal compositions and imaging systems. [410] W. E. L. Haas, J. E. Adams, B. Mechlowitz, U.S. 3,843,230 (20. 8. 1971), C.A. 82, 37351j (1975); Imaging system with an aligned liquid crystal image contrasted by a nonaligned liquid crystal background.

[411] W. E. L. Haas, J. E. Adams, B. Mechlowitz, U.S. 3,871,904 (20. 8. 1971), C.A. 83, 35762r (1975); Liquid crystalline film. [412] W. E. L. Haas, U.S. 3,894,793 (7. 10. 1974), C.A. 84, 10945k (1976); Liquid crystal imaging system using tributyltin oxide or tributyltin chloride. [413] W. E. L. Haas, B. Mechlowitz, J. B. Flannery, Jr., J. E. Adams, Ger. Offen. 2,235,385 (13. 9. 1971), C.A. 79, 85688j (1973); Liquid crystal compositions for imaging use. [414] W. E. L. Haas, J. B. Flannery, B. Mechlowitz, J. E. Adams, Ger. Offen. 2,235,387 (13. 9. 1971), C.A. 79, 85687h (1973); Liquid crystal compositions for imaging use. [415] W. E. L. Haas, J. E. Adams, Ger. Offen. 2,522,535 (28. 5. 1974), C.A. 85, 70714q (1976); Liquid crystal material for image production systems. [416] W. Haas, J. Adams, Appl. Opt. 7, 1203 (1968), C.A. 69, 32020x (1968); Electrophotographic imaging with cholesterics. [417] W. E. Haas, J. E. Adams, J. Electrochem. Soc. 118, 220C (1971); Erasable display panel using cholesterics. [418] W. Haas, J. Adams, G. Dir, Soc. Inform. Display, Symp. San Francisco 1972, 94; Optical storage in mixture of nematics and nonmesomorphic, optically-active compounds. [419] W. E. Haas, K. F. Nelson, J. E. Adams, G. A. Dir, J. Electrochem. Soc. 121, 1667 (1974); C.A. 82, 78757q (1975); UV Imaging with nematic chlorostilbenes. [420] W. E. L. Haas, J. E. Adams, Appl. Phys. Lett. 25, 535 (1974), C.A. 82, 24332d (1975); New optical storage mode in liquid crystals.

[421] W. E. L. Haas, G. A. Dir, Appl. Phys. Lett. 29, 325 (1976), C.A. 85, 134033b (1976); Simple real-time light valves. [422] W. E. L. Haas, G. A. Dir, J. E. Adams, I. P. Gates, Appl. Phys. Lett. 29, 631 (1976), Ultralow-voltage image intensifiers. [422a] W. E. L. Haas, J. E. Adams, J. M. Pollack, J. Appl. Phys. 47, 772 (1976), C.A. 84, 154518w (1976); Diffraction effects in liquid-crystal aligning layers. [422b] W. E. L. Haas, J. E. Adams, Jr., B. Richter, U.S. 4,005,032 (25. 9. 1974), C.A. 87, 31968x (1977); Liquid crystal composition having mixed cholesteric-nematic properties. [423] S. A. Hadjistavros, US Nat. Tech. Inform. Serv., AD Rept. 1971, No. 734442, 65 pp., C.A. 76, 160436u (1972); Colormatrix. Thermally controlled liquid crystal alphanumeric display. [424] T. Haga, H. Fukutani, H. Nagasaka, T. Ohya, Japan. Kokai 74 103,917 (1. 2. 1973), C.A. 82, 178216f (1975); Glass cell for liquid crystal display devices. [425] T. Haga, H. Fukutani, H. Nagasaka, T. Ohya, Japan. Kokai 74,104,909 (3. 2. 1973), C.A. 82, 148544u (1975); Glass cells for liquid crystal display devices. [426] T. Haga, H. Fukutani, H. Nagasaka, T. Ohya, Japan. Kokai 74,130,373 (20. 4. 1973), C.A. 82, 163081j (1975); Liquid crystal display devices. [427] T. Haga, H. Fukutani, H. Nagasaka, T. Ohya, Japan. Kokai 74,130,374 (21. 4. 1974), C.A. 82, 163082k (1975); Phenyl benzoate liquid crystal display devices. [428] T. Haga, H. Fukutani, H. Nagasaka, T. Ohya, Japan. Kokai 74,130,375 (21. 4. 1973), C.A. 82, 178283a (1975); Homeotropic nematic liquid crystal compositions. [428a] A. Hakusui, K. Narita, Japan. Kokai 77 02,884 (24. 6. 1975), C.A. 86, 198022a (1977); TN liquid crystal display devices. [429] A. Halasz, KFKI (Rep.) 1975, KFKI-75-22, 13 pp., C.A. 83, 139815b (1975); Electrooptical study of p,p'-disubstituted phenyl benzoates liquid crystal mixture. [429a] G. Hallas, F. Jones, Ger. Offen. 2,640,624 (9. 9. 1975), C.A. 86, 191325t (1977); Dichroic azo dyes. [430] I. Haller, H. A. Huggins,

U.S. 3,656,834 (9. 12. 1970), C.A. *77*, 27336r (1972); Additive for nematic liquid crystal materials for electrooptic display devices.

[**431**] I. Haller, W. R. Young, C. L. Gladstone, D. T. Teaney, Mol. Cryst. Liq. Cryst. *24*, 249 (1973), C.A. *82*, 179070r (1975); Crown ether complex salts as conductive dopants for nematics. [**432**] S. A. Hamadto, Electr. Engng. *46*, 20 (1974); Liquid crystal viewer for observation of IR. [**433**] B. Hampel, Z. Werkstofftech. *3*, 149 (1972), C.A. *77*, 64046s (1972); Anwendung flüssiger Kristalle in der Materialprüfung. [**434**] B. Hampel, Liq. Cryst. Disp. Appl., Symp. *1972*, 68, C.A. *84*, 52038x (1976); Commercial development of liquid crystals. [**435**] K. Hanemann, H. Schubert, D. Demus, U. Bargenda, Ger. (East) 115,829 (30. 10. 1974), C.A. *86*, 149141e (1977); Cholesteric liquid crystal mixtures. [**436**] T. F. Hanlon, U.S. 3,569,614 (10. 4. 1969); Liquid crystal color modulator for electronic imaging systems. [**437**] T. F. Hanlon, U.S. 3,700,805 (26. 8. 1971); Black-and-white image control by ultrasonic modulation of nematics. [**438**] J. R. Hansen, J. L. Fergason, A. Okaya, Appl. Opt. *3*, 987 (1964); Display of IR-laser patterns by a liquid crystal viewer. [**439**] J. R. Hansen, R. J. Schneeberger, Scientific Paper 67-9C2-LIQXR-P1, Proprietary Class 3, Westinghouse Res. Lab., Pittsburgh, 15. 11. 1967; Liquid crystal media for electron beam recording. [**440**] J. R. Hansen, R. J. Schneeberger, IEEE Trans. Electron Dev. ED *15*, 896 (1968); Liquid crystal media for electron beam recording.

[**441**] M. Hareng, G. Assouline, E. Leiba, Proc. IEEE *60*, 913 (1972); Liquid crystal matrix display by electrically controlled birefringence. [**442**] M. Hareng, Rev. Tech. Thomson-CFS *5*, 319 (1973), C.A. *80*, 31141c (1974); Cristaux liquides. Propriétés et applications. [**443**] M. Hareng, S. LeBerre, Electron. Lett. *11*, 73 (1975), Formation of synthetic images on a laser-beam-addressed smectic liquid crystal display. [**444**] M. Hareng, S. LeBerre, J. Phys. (Paris), C-3, *37*, 135 (1976), C.A. *85*, 184802x (1976); Thermal relaxation recording device on smectics. [**445**] M. Hareng, J. Phys. (Paris), C-3, *37*, 161 (1976); Table ronde: applications des smectiques. [**445a**] M. Hareng, S. Le Berre, R. Hehlen, Rev. Tech. Thomson-CSF *9*, 373 (1977), C.A. *87*, 175633g (1977); Valve optique à cristaux liquides smectiques. [**446**] R. Hariharan, K. A. P. Menon, IEEE Trans. Instrum. Meas. *25*, 164 (1976); Fluidic display using liquid crystals. [**447**] T. B. Harsch, U.S. 3,785,721 (15. 7. 1971); Display devices utilizing liquid crystal light modulation with varying colors. [**448**] S. Hasegawa, H. Fukushima, T. Ishibashi, K. Nakamura, K. Totani, S. Furuta, Japan. Kokai 76 52,987 (6. 11. 1974), C.A. *86*, 99089w (1977); Nematic liquid crystal compositions for dynamic scattering-type display devices. [**448a**] T. Hatakeyama, Y. Kagawa, J. Sound Vibr. *46*, 551 (1976); Acousto-optical and acousto-dielectric effects in a nematic. [**449**] J. F. Havlice, Electron. Lett. *5*, 477 (1969); Visualization of acoustic beams using liquid crystals. [**450**] S. Hayashi, S. Wada, Japan. Kokai 76 68,483 (10. 12. 1974), C.A. *85*, 134501c (1976); Nematic liquid crystal compositions for dynamic scattering mode display devices.

[**451**] C. L. Hedman, Jr., K. D. S. Myrenne, U.S. 3,674,341 (8. 12. 1970); Liquid crystal display device having improved optical contrast. [**452**] G. H. Heilmeier, L. A. Zanoni, U.S. 3,499,112 (31. 3. 1967); Electro-optical device. [**453**] G. H. Heilmeier, L. A. Zanoni, U.S. 3,503,673 (14. 9. 1967); Reduction of turn-on delay in liquid crystal cell. [**454**] G. H. Heilmeier, U.S. 3,519,330 (14. 9. 1967); Turnoff method and circuit for liquid crystal display element. [**455**] G. H. Heilmeier, U.S. 3,551,026 (26. 4. 1965), C.A. *74*, 59298j (1971); Control of optical properties of materials with liquid crystals. [**456**] G. H. Heilmeier, U.S. 3,575,491 (16. 10. 1968), C.A. *78*, 36154c (1973); Decreasing response time of liquid crystals. [**457**] G. H. Heilmeier, U.S. 3,575,493 (5. 8. 1969); Fast self-quenching of dynamic scattering in liquid crystal devices. [**458**] G. H. Heilmeier, L. A. Zanoni, U.S. 3,600,061 (21. 3. 1969); Electro-optic device having grooves in the support plates to confine a liquid crystal by means of surface tension. [**459**] G. H. Heilmeier, J. E. Goldmacher, U.S. 3,650,603 (5. 12. 1967), C.A. *75*, 26966t (1971); Liquid crystal light valve containing a mixture of nematic and cholesteric materials in which the light scattering effect is reduced when an electric field is applied. [**460**] G. H. Heilmeier, U.S. 3,655,269 (25. 1. 1971); Liquid crystal display assembly having independent contrast and speed of response controls.

[**461**] G. H. Heilmeier, Appliance Eng. *2*, 21 (1968); Dynamic scattering in liquid crystals. [**462**] G. H. Heilmeier, RCA Engineer *15*, 14 (1969); Liquid crystals — The first electronic method for controlling the reflection of light. [**463**] G. H. Heilmeier, J. E. Goldmacher, Proc. IEEE *57*, 34 (1969); New electric field controlled reflective optical storage effect in mixed liquid crystal systems. [**464**] G. H. Heilmeier, Sci. Amer. *222*, 100 (1970), C.A. *72*, 126096r (1970); Liquid-crystal display devices. [**464a**] G. H. Heilmeier, IEEE Trans. Electron Dev. ED *23*, 780 (1976); LC displays. Experiment in interdisciplinary research that worked. [**464b**] R. A. Heinz, R. J. Klaiber, R. C. Oehrle, J. Opt. Soc. Am. *66*, 381 (1976); Liquid-crystal artwork generator. [**464c**] R. A. Heinz, R. C. Oehrle, West.

Electr. Eng. *21*, 3 (1977), C.A. *87*, 93486x (1977); Rapid generation of complex images with a liquid crystal. [465] W. Helfrich, Europhys. News *7*, 1 (1976), C.A. *85*, 200487f (1976); Applications of liquid crystals. [465a] Hempel, W. Nowak, Silikattechnik *28*, 169 (1977), C.A. *87*, 156016s (1977); Wechselwirkung von Defekten in Glasoberflächen und Flüssigkristallschichten. [466] R. Hesse, G. Edler, H. Keller, Ger. Offen. 2,201,121 (11. 1. 1972), C.A. *80*, 5037d (1974); Liquid crystal-based laquers. [466a] S. Hibino, T. Miyazaki, Japan. Kokai 77 04,485 (30. 6. 1975), C.A. *87*, 60822a (1977); Nematic liquid crystal compositions for display devices. [466b] S. Hibino, T. Miyazaki, Japan. Kokai 77 06,381 (7. 7. 1975), C.A. *86*, 189465v (1977); Liquid crystal composite. [467] G. Hiler, J. Sobanski, Pr. Nauk. Inst. Metrol. Elektr. Politech. Wroclaw. *7*, 73 (1975), C.A. *86*, 98987a (1977); Design and technology of analog-digital liquid-crystal displays. [467a] M. Hirano, Ger. Offen. 2,705,805 (13. 2. 1976), C.A. *87*, 160018m (1977); Electrooptical display device. [468] T. L. Hodson, J. V. Cartmell, D. Churchill, J. W. Jones, U.S. 3,585,381 (14. 4. 1969), C.A. *75*, 82318v (1971); Encapsulated cholesteric liquid crystal display device. [469] T. L. Hodson, J. G. Whitaker, J. W. Jones, U.S. 3,617,374 (14. 4. 1969); Display device. [470] R. Hoehn, B. Binkert, Plast. Reconstr. Surg. *48*, 209 (1971); Cholesterics. New visual aid to study of flap circulation.

[471] S. R. Hofstein, U.S. 3,505,804 (23. 4. 1968); Solid state clock. [472] F. Hoff, B. Stadnik, Krist. Tech. *7*, 855 (1972), C.A. *79*, 46406w (1973); Nutzung von Einkristallen für die Informationsspeicherung und -verarbeitung. [473] J. Hoimes, US Nat. Tech. Inform. Serv., PB Rep. 1973, No. 771070/OGA, 25 pp., C.A. *81*, 55699n (1974); Photodiode model of the CdS liquid crystal electrooptic transducer. [474] E. Holle, Neye-Enatechnik, Quickborn, Broschüre 2380-7.72; Heutige Möglichkeiten und Zukunftsaussichten für RCA Liquid-Crystal-Anzeigesysteme, die mit COS/MOS-Treiberschaltungen betrieben werden. [475] K. Honda, M. Sukigara, Kyoritsu Kagaku Raiburari *1*, 140 (1973), C.A. *83*, 106123w (1975); Application of liquid crystals. [476] F. Hori, B. Kato, N. Arima, Japan. Kokai 73 49,502 (22. 10. 1971), C.A. *79*, 116435f (1973); Liquid crystal microcapsule ink compositions. [477] F. Hori, Japan. Kokai 74 37,884 (11. 8. 1972), C.A. *81*, 97839p (1974); Transition temperature adjustment of electrosensitive cholesterics. [478] F. Hori, B. Kato, N. Arima, Japan Kokai 74 86,279 (23. 12. 1972), C.A. *82*, 92074w (1975); Nematic liquid crystals for electrooptical display devices. [479] F. Hori, N. Arima, Japan. Kokai 74 95,880 (18. 1. 1973), C.A. *82*, 163021q (1975); Low-temperature-nematic liquid crystal compositions. [480] F. Hori, B. Kato, N. Arima, Japan. Kokai 74,101,281 (3. 2. 1973), C.A. *83*, 19048g (1975); Liquid crystals and their encapsulated forms for display devices.

[481] F. Hori, B. Kato, N. Arima, Japan. Kokai 74,108,187 (19. 2. 1973), C.A. *83*, 35756s (1975); Plastic sheet containing microencapsulated liquid crystals for thermochromic display devices. [482] F. Hori, B. Kato, N. Arima, Japan. Kokai 75 13,280 (9. 6. 1973), C.A. *83*, 50826h (1975); Nematic liquid crystal compositions for electrooptical display devices. [483] F. Hori, B. Kato, Japan. Kokai 75,101,110 (14. 1. 1974), C.A. *84*, 32809w (1976); Liquid crystal ink compositions. [483a] B. Hori, B. Kato, N. Arima, Japan. Kokai 76, 136,586 (22. 5. 1975), C.A. *86*, 180747f (1977); Liquid crystal composition for temperature sensor. [483b] B. Hori, B. Kato, N. Arima, Japan. Kokai 77 82,685 (29. 12. 1975), C.A. *87*, 160019n (1977); Liquid crystal composition. [483c] B. Hori, B. Kato, N. Arima, Japan. Kokai 77 82,686 (29. 12. 1975), C.A. *87*, 175689e (1977); Nematic liquid crystal composition. [483d] Y. Hori, M. Fukai, H. Tatsuta, Japan. Kokai 77 71,393 (11. 12. 1975), C.A. *87*, 144131n (1977); Liquid crystal compositions. [484] P. Y. Hsieh, U.S. 3,815,972 (9. 8. 1972), C.A. *81*, 84459q (1974); Low voltage liquid crystal display. [484a] P. Y. Hsieh, Y. S. Lee, J. E. Jensen, U.S. 4,000,084 (17. 12. 1973), C.A. *86*, 198015a (1977); Liquid crystal mixtures for electro-optical display devices. [484b] C. Hu, Proc. IEEE *64*, 1737 (1976); Field-dependent light scattering in nematics. [485] R. C. Huener, S. J. Niemiec, D. K. Morgan, U.S. 3,740,717 (16. 12. 1971); Liquid crystals display. [486] M. Hugenschmidt, K. Vollrath, C. R. Acad. Sci., Sér. B *274*, 1221 (1972), C.A. *77*, 81962e (1972); Étude de phénomènes transitoires par un laser TEA-CO_2 associé à un détecteur à cristal liquide. [487] K. Jizuka, L. G. Gregoris, Appl. Phys. Lett. *17*, 509 (1970); Application of microwave holography in the study of the field from a radiating source. [488] S. Inokuchi, Y. Morita, Y. Sakurai, Appl. Opt. *11*, 2223 (1972); Optical pattern processing utilizing nematics. [489] International Liquid Xtal, Brit. 1,390,522 (9. 2. 1971), C.A. *83*, 78925p (1975); Nematic phase liquid crystal composition for use in a light shutter. [490] M. J. Intlekofer, D. C. Auth, Appl. Phys. Lett. *20*, 151 (1972), C.A. *76*, 132596c (1972); Liquid-crystal holography of high-frequency acoustic waves.

[491] M. J. Intlekofer, D. C. Auth, M. E. Fourney, Imaging Tech. Test. Insp., Semin.-in-Depth, Proc. Soc. Photo-Opt. Instrum. Eng. *1972*, 83, C.A. *79*, 25650f (1973); Display of hf acoustic holograms utilizing liquid crystals. [492] T. Inukai, H. Sato, S. Sugimori, Japan. Kokai 75 57,085 (22. 9. 1973),

C.A. *83*, 186413d (1975); Nematic liquid crystal compositions for optical display devices. **[493]** T. Inukai, H. Sato, T. Ishibe, Japan. Kokai 76 37,880 (26. 9. 1974), C.A. *85*, 151839m (1976); Liquid crystal compositions for TN display devices. **[494]** T. Inukai, H. Sato, T. Ishibe, Japan. Kokai 76 37,881 (26. 9. 1974), C.A. *85*, 151840e (1976); Liquid crystal compositions for TN display devices. **[495]** T. Inukai, H. Sato, T. Ishibe, Japan. Kokai 76 37,882 (26. 9. 1974), C.A. *85*, 151841f (1976); Liquid crystal composition for TN display devices. **[495a]** T. Inukai, H. Sato, T. Ishibe, Ger. Offen. 2,541,312 (16. 9. 1975), C.A. *86*, 163915s (1977); Liquid crystal compositions. **[496]** J. E. Jacobs, U. S. Pat. Appl. 322,057 (20. 9. 1974), C.A. *83*, 108622g (1975); Liquid crystal gas analyzer. **[496a]** J. E. Jacobs, U.S. 3,927,977 (20. 9. 1974), C.A. *87*, 103640j (1977); Liquid crystal gas analyzer. **[497]** J. W. Jacobs, U.S. 3,648,280 (15. 5. 1970); Thermochromic light-flashing system. **[498]** A. D. Jacobson, Soc. Inform. Display, Symp. San Francisco 1972, 70; Photoactivated liquid crystal light valves. **[499]** A. Jacobson, J. Grinberg, W. Bleha, L. Miller, L. Fraas, G. Myer, D. Boswell, Inf. Display *12*, 17 (1975); Real-time optical data-processing device. **[499a]** A. Jacobson, J. Grinberg, W. Bleha, L. Miller, L. Fraas, G. Myer, D. Boswell, Ann. N. Y. Acad. Sci. *267*, 417 (1976), C.A. *87*, 60716u (1977); Real-time optical data processing device. **[500]** E. Jakeman, P. N. Pusey, Phys. Lett. A *44*, 456 (1973), C.A. *79*, 119297y (1973); Light scattering from electrohydrodynamic turbulence in liquid crystals.

[501] A. K. Jalaluddin, H. Husain, Proc. Nucl. Phys. Solid State Phys. Symp., 15th, 2, *1970*, 527, C.A. *75*, 146947j (1971); Use of liquid crystals as media for continuously sensitive chambers for the registration of elementary particles. **[502]** G. Jankowitz, U.S. 3,527,945 (24. 9. 1968); Mounting structure for a liquid crystal thermal imaging device. **[503]** J. L. Janning, Fr. Demande 2,179,841 (10. 4. 1972), C.A. *81*, 8441n (1974); Liquid crystal film. **[504]** M. J. Jeudy, J. J. Robillard, Opt. Laser Technol. *8*, 117 (1976), C.A. *85*, 133329r (1976); Light scattering by liquid crystals under pressure variations: Application to blood pressure measurements. **[505]** C. C. Johnson, C. H. Durney, J. L. Lords, T. C. Rozzell, G. K. Livingston, Ann. N. Y. Acad. Sci. *247*, 527 (1975); Fiberoptic liquid crystal probe for absorbed radio-frequency power and temperature measurement in tissue during irradiation. **[505a]** M. R. Johnson, P. A. Penz, IEEE Trans. Electron Dev. ED *24*, 805 (1977), Low-tilt-angle nematic alignment compatible with frit sealing. **[506]** R. N. Johnson, U.S. 3,628,268 (28. 5. 1970); Pure fluid display. **[507]** D. Jones, S. Lu, U.S. 3,690,745 (3. 3. 1970), C.A. *77*, 146263z (1972); Electrooptical devices using lyotropic nematics. **[508]** D. Jones, S. Lu, U.S. 3,965,030 (3. 3. 1970), C.A. *85*, 184871u (1976); Lyotropic nematics for use in electro-optical display devices. **[509]** D. Jones, S. Lu, Soc. Inform. Display, Symp. San Francisco 1972, 100; Field-effect liquid crystal display devices. **[510]** F. B. Jones, Jr., R. M. Govan, U.S. 3,904,797 (7. 5. 1973), C.A. *84*, 24456p (1976); Homeotropic alignment of liquid crystals in a display cell by baked on ionic surfactants.

[511] F. B. Jones, Jr., R. Chang, E. P. Parry, U.S. 3,920,576 (7. 5. 1973), C.A. *84*, 37338b (1976); Doping of nematic liquid crystal. **[512]** F. B. Jones, Jr., U.S. 3,950,264 (7. 5. 1973), C.A. *85*, 102438w (1976); Schiff-base liquid crystals doped to raise dynamic scattering cutoff frequency. **[513]** W. J. Jones, U.S. 3,440,882 (9. 9. 1966); Thermometer. **[514]** F. B. Joyner, G. O. Cash, Jr., Fr. 1,450,982 (21. 10. 1964), C.A. *67*, 22441e (1967); Poly-α-olefin compositions. **[515]** W. Jung, K. N. Kang, Jonja Konghak Hoe Chi *10*, 111 (1973), C.A. *85*, 39626p (1976); Display device applications of liquid crystals. **[516]** Y. Kagawa, T. Hatakeyama, Y. Tanaka, J. Sound Vibr. *36*, 407 (1974); Detection and visualization of ultrasonic fields and vibrations by means of liquid crystals. **[517]** F. J. Kahn, U.S. 3,694,053 (22. 6. 1971); Nematic liquid crystal device. **[518]** F. J. Kahn, Appl. Phys. Lett. *18*, 231 (1971), C.A. *75*, 12865x (1971); Cholesterics for optical applications. **[519]** F. J. Kahn, J. T. La Macchia, IEEE Trans. Electron Dev. ED *18*, 733 (1971); The cholophor: A passive polarization-switched liquid crystal screen for multicolor laser displays. **[520]** F. J. Kahn, Appl. Phys. Lett. *22*, 111 (1973), C.A. *78*, 77882g (1973); IR-laser-addressed thermooptic smectic liquid-crystal storage displays.

[521] T. Kakeda, Japan. Kokai 74,106,485 (14. 2. 1973), C.A. *82*, 132171j (1975); Nematic liquid crystal compositions having positive dielectic anisotropy. **[522]** V. L. Kalinovskii, I. D. Samodurova, A. S. Sonin, Prib. Tekh. Eksp. *1976*, 176, C.A. *84*, 187400e (1976); Attenuation of optical radiation using nematics. **[523]** T. Kallard, Liquid crystals and their applications (Optosonic Press: New York). 1970. 219 pp., C.A. *73*, 49664r (1970). **[524]** T. Kallard, State of the Art Review, Vol. 7: Liquid-crystal devices. (Optosonic Press: New York). 1973, 365 pp., C.A. *79*, 119453w (1973). **[525]** H. Kamei, Japan 74 23,108 (29. 9. 1970), C.A. *82*, 118254g (1975); Liquid crystal compositions. **[526]** H. Kamei, T. Ozawa, K. Nozaki, T. Morikawa, U.S. 3,972,733 (29. 1. 1971), C.A. *86*, 82616t (1977); Producing electrical energy by means of liquid crystal devices. **[527]** N. Kamezawa, T. Kitamura, M. Sato, Japan. Kokai 76 08,183 (12. 7. 1974), C.A. *85*, 134478a (1976); Nematic liquid crystal compositions for display devices.

[528] S. Kanbe, Japan. Kokai 74 38,887 (16. 8. 1972), C.A. *81*, 113732f (1974); Electrooptical display devices based on TN effect. [529] S. Kanbe, Japan. Kokai 74 59,089 (12. 10. 1972), C.A. *82*, 10033v (1975); Electrooptical display device based on liquid crystals. [530] S. Kanbe, Japan. Kokai 74 93,283 (10. 1. 1973), C.A. *82*, 92065u (1975); Nematic liquid crystal compositions for electrooptical display devices.

[531] S. Kanbe, Japan. Kokai 75 23,385 (6. 7. 1973), C.A. *83*, 69142d (1975); Liquid crystals for electrooptical display devices. [532] S. Kanbe, Japan. Kokai 75 70,287 (26. 10. 1973), C.A. *84*, 82590p (1976); Nematic liquid crystal compositions containing azoxybenzene derivatives for electrooptical display devices. [533] S. Kanbe, Japan. Kokai 75 88,039 (12. 12. 1973), C.A. *83*, 178593x (1975); Ester liquid crystals. [534] S. Kanbe, Japan. Kokai 75 93,884 (25. 12. 1973), C.A. *84*, 52157k (1976); Electrooptical display devices based on TN phenomena. [535] S. Kanbe, Japan. Kokai 76 73,986 (23. 12. 1974), C.A. *85*, 134504f (1976); Nematics for field effect display devices. [536] S. Kanbe, Japan. Kokai 76 74,989 (25. 12. 1974), C.A. *86*, 63543a (1977); Nematic liquid crystal composition. [537] S. Kanbe, Japan. Kokai 76 75,682 (26. 12. 1974), C.A. *85*, 185143v (1976); Nematic liquid crystal composite. [538] S. Kanbe, Japan. Kokai 76 87,481 (30. 1. 1975), C.A. *86*, 63548f (1977); Nematic liquid crystal composition. [539] S. Kanbe, Japan. Kokai 76 91,886 (14. 1. 1975), C.A. *85*, 184879c (1976); Nematic liquid crystal compositions. [540] S. Kanbe, Japan. Kokai 76,147,488 (12. 6. 1975), C.A. *86*, 131207w (1977); Nematic liquid crystal compositions for display devices. [540a] S. Kanbe, Japan. Kokai 76,144,392 (9. 6. 1975), C.A. *86*, 163653e (1977); Nematic liquid crystal compositions.

[541] J. Kane, H. Schweizer, U.S. 3,944,684 (24. 8. 1973), C.A. *85*, 55482u (1976); Depositing transparent, electrically, conductive tin-containing oxide coatings on a substrate. [542] M. Kano, T. Ninomiya, Y. Nishimura, Japan. Kokai 73 71,377 (28. 12. 1971), C.A. *80*, 15612b (1974); Microencapsulation. [543] L. T. Kantardzhyan, K. A. Mantashyan, M. A. Khumaryan, Izv. Akad. Nauk Arm. SSR, Fiz. *9*, 516 (1974), C.A. *83*, 124018v (1975); Contrast of a cell with electroluminescent liquid-crystal suspension. [544] L. T. Kantardzhyan, K. A. Mantashyan, Izv. Akad. Nauk Arm. SSSR, Fiz. *8*, 309 (1973), C.A. *80*, 101391k (1974); Liquid crystals as dielectric media in electroluminescent cells. [545] O. A. Kapustina, A. A. Talashev, Akust. Zh. *19*, 626 (1973); Visualization of ultrasonic surface-waves by means of liquid crystals. [546] O. A. Kapustina, Akust. Zh. *20*, 482 (1974); Acoustical properties and potential applications of liquid crystals. [546a] O. A. Kapustina, V. N. Lupanov, Akust. Zh. *23*, 390 (1977), C.A. *87*, 175577s (1977); Experimental study of an acoustooptical transducer using a liquid crystal. [547] I. A. Karaseva, V. I. Lebedev, T. P. Sokolova, M. G. Tomilin, Opt.-Mekh. Prom-st. *42*, 47 (1975), C.A. *84*, 52074f (1976); Production and quality control of liquid-crystal cells. [547a] K. Kashima, M. Hirano, Japan. Kokai 77 68,220 (2. 12. 1975), C.A. *87*, 171867p (1977); Treatment of watch glasses. [548] R. A. Kashnow, J. E. Bigelow, Soc. Inform. Display, Symp. San Francisco 1972, 96; Nematic liquid-crystal as a phase grating of electrically-variable spatial frequency. [549] R. A. Kashnow, Rev. Sci. Instrum. *43*, 1837 (1972), C.A. *78*, 34846f (1973); Thickness measurements of nematic layers. [550] R. A. Kashnow, J. E. Bigelow, Appl. Opt. *12*, 2302 (1973), C.A. *79*, 130839x (1973); Diffraction from a liquid crystal phase grating.

[551] R. A. Kashnow, C. R. Stein, Appl. Opt. *12*, 2309 (1973), C.A. *79*, 131240g (1973); Total-reflection liquid-crystal electrooptic device. [552] Y. Katagiri, Y. Miyata, Japan. Kokai 73 08,673 (11. 6. 1971), C.A. *79*, 10919b (1973); Nematic liquid crystals. [553] Y. Katagiri, Y. Miyata, Japan. Kokai 73 66,582 (16. 12. 1971), C.A. *81*, 112355y (1974); Nematic liquid crystals. [554] Y. Katagiri, Y. Miyata, Japan. Kokai 73 92, 284 (7. 3. 1972), C.A. *81*, 112356z (1974); Composition of nematic liquid crystals. [555] Y. Katagiri, Y. Miyata, Japan. Kokai 73 92,285 (8. 3. 1972), C.A. *81*, 112354x (1974); Composition of nematic liquid crystals. [556] Y. Katagiri, Y. Miyata, Japan. Kokai 73 94,694 (15. 3. 1972), C.A. *81*, 128107e (1974); Composition of nematic liquid crystal. [557] Y. Katagiri, Y. Miyata, Japan. Kokai 73 97,781 (27. 3. 1972), C.A. *80*, 88324u (1974); Nematic liquid crystal composite. [558] Y. Katagiri, Y. Miyata, Japan. Kokai 73 97,782 (27. 3. 1972), C.A. *80*, 126819u (1974); Nematic liquid crystal compositions. [559] Y. Katagiri, K. Hamano, Japan. Kokai 73,100,384 (31. 3. 1972), C.A. *80*, 101175t (1974); Nematic liquid crystal composites. [560] Y. Katagiri, K. Hamano, Japan. Kokai 73,100,385 (31. 3. 1972), C.A. *81*, 31896y (1974); Nematic liquid crystal compositions.

[561] Y. Katagiri, Y. Miyata, Japan. Kokai 73,102,784 (10. 4. 1972), C.A. *81*, 112357a (1974); Nematic liquid crystal composites. [562] Y. Katagiri, Y. Miyata, Japan. Kokai 74 16,688 (7. 6. 1972), C.A. *80*, 151235h (1974); Nematic liquid crystal compositions with positive dielectric anisotropy for display devices. [563] Y. Katagiri, T. Miyata, O. Nagasaki, Japan. 74 37,351 (18. 5. 1970), C.A. *82*, 178984e (1975); Nematic liquid crystal. [564] Y. Katagiri, T. Miyata, O. Nagasaki, Japan. 74 37,352

(18. 5. 1970), C.A. *82*, 178983d (1975); Nematic liquid crystal. [**565**] Y. Katagiri, T. Miyata, O. Nagasaki, Japan. 74 37,515 (31. 7. 1970), C.A. *82*, 178985f (1975); Nematic liquid crystal. [**566**] Y. Katagiri, T. Miyata, O. Nagasaki, Japan. 74 38,988. (31. 7. 1970), C.A. *82*, 178981b (1975); Nematic liquid crystal. [**567**] Y. Katagiri, Japan. Kokai 75 91,581 (14. 12. 1973), C.A. *84*, 82591q (1976); Nematic liquid crystal compositions for display devices. [**568**] Y. Katagiri, Japan. Kokai 75,137,885 (24. 4. 1974), C.A. *84*, 82595u (1976); Nematic liquid crystal compositions for optical display devices. [**569**] Y. Katagiri, Y. Miyata, M. Sendai, Ger. Offen. 2,164,797 (28. 12. 1970); Kristalline Flüssigkeit (für Elektrooptik). [**570**] S. Kato, T. Miyazaki, Japan. Kokai 76,139,582 (28. 5. 1975), C.A. *86*, 131202r (1977); Liquid crystal display apparatus.

[**571**] S. Kato, T. Uhida, H. Watanabe, M. Wada, Oyo Butsuri *44*, 156 (1975), C.A. *83*, 19418c (1975); Temperature measurement in amorphous films by using liquid crystals. [**571a**] M. Kaufmann, Brit. 1,478,327 (7. 9. 1973), C.A. *87*, 192096b (1977); Liquid-crystal cell and a method for its manufacture. [**571b**] M. Kawachi, K. Kato, O. Kogure, Jpn. J. Appl. Phys. *16*, 1263 (1977), C.A. *87*, 92797n (1977); Light scattering properties of a n/c mixture with positive dielectric anisotropy. [**572**] M. Kawamoto, S. Matsumoto, Japan. Kokai 73 30,696 (24. 8. 1971), C.A. *79*, 84477c (1973); Nematic liquid crystal based on methoxybenzylideneamine derivatives. [**573**] F. Keilmann, Appl. Opt. *9*, 1319 (1970), C.A. *73*, 50342r (1970); IR interferometry with a carbon dioxide laser source and liquid crystal detection. [**574**] F. Keilmann, Laser angew. Strahlentechnik *1970*, No. 4, 77; Two-dimensional interferograms at 10,6 μm for plasma diagnostics. [**575**] F. Keilmann, K. F. Renk, Appl. Phys. Lett. *18*, 452 (1971), C.A. *75*, 56587n (1971); Visual observation of submillimeter wave laser beams. [**576**] H. Kelker, R. Hatz, Chem.-Ing.-Tech. *45*, 1005 (1973), C.A. *79*, 108872f (1973); Technische Anwendungen flüssiger Kristalle. [**577**] B. Kerllenevich, E. Chapunov, A. Coche, Rev. Sci. Instrum. *42*, 1545 (1971); Liquid crystal divice for the visualization of electric fields. [**577a**] B. Kerllenevich, A. Coche, Electron. Lett. *13*, 261 (1977), C.A. *87*, 18278c (1977); Effects of γ-irradiation on some cholesterics. [**578**] C. W. Kessler, T. T. Trzaska, U.S. 3,637,291 (11. 2. 1970); Display device with inherent memory. [**579**] L. W. Kessler, S. P. Sawyer, U.S. 3,707,323 (6. 11. 1970); Liquid crystal devices and systems for ultrasonic imaging. [**580**] H. Kiemle, U. Wolff, Opt. Commun. *3*, 26 (1971); Application de cristaux liquides an holographie optique.

[**581**] S. Kikuchi, Senryo To Yakuhin *17*, 34 (1972), C.A. *77*, 27353u (1972); Nonsilver halide photosensitive materials. [**582**] T. Kikuchi, Kobunshi *25*, 182 (1976), C.A. *84*, 137025r (1976); Spinning in the liquid crystal state. [**583**] J. Kirton, Liq. Cryst. Disp. Appl., Symp. 1972, 1, C.A. *84*, 52035u (1976); Electrooptic effects in liquid crystals. [**584**] J. Kirton, R. W. Sarginson, Optoelectronics *6*, 349 (1974), C.A. *82*, 105341h (1975); New display techniques. [**584a**] M. Kitamura, S. Iida, Asahi Garasu Kenkyu Hokoku *26*, 97 (1976), C.A. *87*, 93483u (1977); Multiplexing characteristics of TN displays. [**585**] T. Kitamura, K. Toriyama, Japan. Kokai 75 72,882 (31. 10. 1973), C.A. *84*, 82585r (1976); Nematic liquid crystal compositions for optical display devices. [**586**] T. Kitamura, K. Toriyama, N. Kamezawa, F. Nakano, K. Murao, Japan. Kokai 75 72,883 (31. 10. 1973), C.A. *84*, 82584q (1976); Nematic liquid crystal compositions for optical display devices. [**587**] T. Kitamura, N. Kamezawa, F. Nakano, M. Sato, Japan. Kokai 75,160,183 (19. 6. 1974), C.A. *85*, 12343g (1976); Liquid crystal composition for display devices. [**588**] T. Kitamura, N. Kamezawa, K. Toriyama, F. Nakano, M. Sato, Japan. Kokai 76 02,686 (26. 6. 1974), C.A. *85*, 12392x (1976); Nematic liquid crystal compositions for display devices. [**589**] T. Kitamura, N. Kamegawa, M. Sato, Japan. Kokai 76 05,280 (3. 7. 1974), C.A. *85*, 102430n (1976); Liquid crystal compositions for display devices. [**590**] T. Kitamura, H. Yokokura, F. Nakano, Japan. Kokai 76 71,884 (20. 10. 1974), C.A. *86*, 131186p (1977); Nematic liquid crystal compositions for dynamic scattering type display devices.

[**591**] T. Kitamura, H. Yokokura, Japan. Kokai 76 93,790 (17. 2. 1975), C.A. *85*, 184876z (1976); Nematic liquid crystal composition for display devices. [**591a**] T. Kitamura, Japan. Kokai 77 21,286 (13. 8. 1975), C.A. *87*, 192124j (1977); Nematic liquid crystal compositions for dynamic scattering mode display devices. [**591b**] T. Kitamura, H. Yokokura, Japan. Kokai 77 21,287 (13. 8. 1975), C.A. *87*, 192123h (1977); Nematic liquid crystal compositions for dynamic scattering mode display devices. [**592**] B. H. Klanderman, R. T. Klingbiel, U.S. 3,960,748 (24. 10. 1972), C.A. *86*, 24476m (1977); Nematic liquid crystal compositions. [**593**] B. H. Klanderman, D. P. Maier, U.S. 3,960,752 (24. 10. 1972), C.A. *85*, 200587p (1976); Liquid crystal compositions. [**594**] B. H. Klanderman, T. R. Criswell, Ger. Offen. 2,333,534 (3. 7. 1972), C.A. *80*, 101176u (1974); Nematic liquid-crystalline mixtures of Schiff bases. [**595**] E. J. Klein, A. P. Margozzi, Israel J. Technol. *7*, 173 (1969), C.A. *70*, 79499w (1969); Measurement of skin friction by means of liquid crystals. [**596**] R. I. Klein, S. Caplan, U.S. 3,612,654 (27. 5. 1970);

Liquid crystal display device. [**597**] R. I. Klein, S. Caplan, U.S. 3,647,280 (6. 11. 1969); Liquid crystal display device. [**598**] R. I. Klein, S. Caplan, R. T. Hansen, U.S. 3,689,131 (29. 6. 1970); Liquid crystal display device. [**599**] H. Klose, G. Mueller, K. Thiessen, Phys. Status Solidi A *16*, K97 (1973), C.A. *79*, 36353n (1973); Liquid crystal technique for the observation of electric fields in p-n and Schottky junctions. [**600**] L. M. Klyukin, A. S. Sonin, B. M. Stepanov, Kvantovaya Elektron. *1974*, 1700, C.A. *82*, 66339s (1975); Liquid crystal composition for photothermography.

[**601**] L. M. Klyukin, A. S. Sonin, B. M. Stepanov, I. N. Shibaev, Kvantovaya Elektron. *2*, 61 (1975), C.A. *83*, 50520d (1975); Use of thermophotography on films of cholesterics for study of parameters of continuous IR laser emission. [**602**] L. M. Klyukin, I. D. Samodurova, A. S. Sonin, Kvantovaya Elektron. *2*, 427 (1975), C.A. *83*, 88291t (1975); Use of nematics to record longwave laser emission. [**602a**] L. M. Klyukin, Kholestericheskie Zhidk. Krist. *1976*, 88, C.A. *87*, 175708k (1977); Practical use of cholesterics. [**603**] A. R. Kmetz, Soc. Inform. Display, Symp. San Francisco 1972, 66; Experimental comparison of multiplexing techniques for liquid crystal displays. [**604**] A. R. Kmetz, Appl. Opt. *14*, 2339 (1975); Liquid crystal wedge compensator for display thickness measurement. [**605**] A. R. Kmetz, F. K. von Willisen, Nonemissive electrooptic displays (Plenum: New York). 1976. 360 pp. [**606**] A. R. Kmetz, 7. Arbeitstagung Flüssigkristalle, Inst. Angew. Festkörperphysik Freiburg, 4. 3. 1977, TN dual bargraph system. [**607**] R. Knauer, Siemens-Bauteile-Informationen *9*, 71 (1971); Digitaluhr mit Flüssigkristallanzeige. [**608**] R. Knauer, G. Stenning, Electr. Engng. *46*, 17 (1974); Illumination of liquid crystal displays. [**609**] P. M. Knoll, W. Pauls, 6. Arbeitstagung Flüssigkristalle, Inst. Angew. Festkörperphysik Freiburg, 2. 4. 1976; Messung des Leitfähigkeits- und Ausschaltzeitenprofils von 64 × 64-Matrix-Displays. [**610**] P. M. Knoll, R. Cremers, W. Pauls, 7. Arbeitstagung Flüssigkristalle, Inst. Angew. Festkörperphysik Freiburg, 4. 3. 1977; Einfluß von Fertigungsparametern auf den Chromatographieeffekt.

[**611**] M. Kobale, H. P. Lorenz, 5. Arbeitstagung Flüssigkristalle, Inst. Angew. Festkörperphysik Freiburg, 25. 4. 1975; Eigenschaften von elektrochromen Schichten auf WO_3-Basis in Abhängigkeit von den Herstellungsverfahren. [**612**] H. Kobayashi, M. Yoshino, Y. Takahara, Japan. Kokai 73 74,487 (10. 1. 1972), C.A. *80*, 32530r (1974); Nematic liquid crystal compositions based on benzylidenamine derivatives. [**613**] S. Kobayashi, Nippon Kessho Gakkaishi *11*, 155 (1969), C.A. *72*, 94284g (1970); Applications of liquid crystals to electronics. [**614**] S. Kobayashi, Keiso *14*, 47 (1971), C.A. *76*, 91025v (1972); Liquid crystals and their application in measuring devices. [**615**] S. Kobayashi, T. Shimojo, K. Kasano, I. Tsunda, Soc. Inform. Display, Symp. San Francisco *1972*, 68; Preparation of alphanumeric indicators with liquid crystals. [**616**] S. Kobayashi, Konbunshi *22*, 350 (1973), C.A. *79*, 72205z (1973); Interferometry and holography with liquid crystals. [**617**] S. Kobayashi, T. Shimomura, Liq. Cryst., Proc. Int. Conf., Bangalore *1973*, 545; Multi-color display devices with twisted nematics. [**618**] S. Kobayashi, H. Mada, Kotai Butsuri *11*, 41 (1976), C.A. *85*, 38749u (1976); Liquid crystal displays. [**619**] A. Kochi, T. Ohtake, K. Inoue, K. Teraishi, H. Takeshita, Japan. Kokai 75 33,487 (31. 7. 1973), C.A. *83*, 89764m (1975); Patterned transparent electrically conductive metal oxide coatings. [**620**] W. E. Kock, Proc. IEEE *60*, 1105 (1972); Real-time detection of metallic objects using liquid-crystal microwave holograms.

[**621**] A. Koff, Ger. Offen. 2,018,028 (17. 4. 1969), C.A. *74*, 105716u (1971); Stable emulsions containing a hydrophobic oily phase of a cholesteryl liquid crystal substance. [**622**] A. Koff, Ger. Offen. 2,059,789 (15. 12. 1969), C.A. *75*, 146260e (1971); Antioxidant-containing mesomorphic material for use in thermography and thermometry. [**623**] A. Koff, Ger. Offen. 2,227,720 (14. 6. 1971), C.A. *79*, 24519h (1973); Stabiles Liquid-Crystal-Material. [**624**] E. A. Kolenko, L. G. Lopatina, V. A. Borodulya, Czech. Conf. Electron. Vac. Phys., (Proc.), 5th, *1972*, 1, IIa–17, C.A. *83*, 36064b (1975); Production of thin layers of liquid crystals. [**625**] I. N. Kompanets, V. N. Morozov, V. V. Nikitin, L. M. Blinov, Kvantovaya Elektron. *1972*, No. 3, 79, C.A. *78*, 166993m (1973); Controlled liquid-crystal transparency for recording holograms. [**626**] I. N. Kompanets, P. D. Berezin, A. A. Vasil'ev, V. V. Nikitin, L. M. Blinov, Sb. Dokl. Vses. Nauch. Konf. Zhidk. Krist. Simp. Ikh. Prakt. Primen., 2nd *1972*, 256, C.A. *81*, 113558d (1974); Study of liquid crystals for purposes of electrooptics. [**627**] I. N. Kompanets, V. V. Nikitin, Mikroelektronika (Akad. Nauk SSSR) *3*, 441 (1974), C.A. *82*, 24784c (1975); Nematic crystals in optoelectronics. [**628**] W. U. Kopp, Prakt. Metallogr. *9*, 370 (1972), C.A. *77*, 144896j (1972); Flüssige Kristalle und ihre Anwendung in der Werkstoffprüfung. [**629**] E. Kornstein, N. A. Luce, Electron. Engng. *31*, 41 (1972); The promise of liquid crystals. [**630**] J. L. Korotzer, Ann. Allergy *30*, 473 (1972); Interpreting allergy skin test reactions with liquid crystal tape.

[**631**] K. Kosai, T. Higashino, Shikizai Kyokaishi *48*, 85 (1975), C.A. *83*, 211552h (1975); Cholesterics. Coloration of two and three components. [**632**] K. M. Kosanke, W. W. Kulcke, E. Max, Brit. 1,138,985 (6. 10. 1967); Multi-frequency light deflectors. [**633**] T. Kovattana, A. Rosengreen, US Nat. Tech. Inform. Serv., AD Rep. 1973, No. 769081/1GA, 128 pp., C.A. *81*, 38522g (1974); Intrusion-detection devices and systems for air base security. [**634**] S. Kratomi, U.S. 3,737,567 (28. 2. 1972); Stereoscopic apparatus having liquid crystal filter viewer. [**635**] J. Krause, 6. Arbeitstagung Flüssigkristalle, Inst. Angew. Festkörperphysik Freiburg, 2. 4. 1976; Für die dynamische Streuung geeignete Estermischung mit breitem Mesophasenbereich und verminderter Viskosität. [**636**] J. Krause, Funk-Technik *31*, 8 (1976); Flüssige Kristalle, Grundlagen und Anwendung in der Anzeigetechnik. [**636a**] J. Krause, R. Steinstraesser, L. Pohl, F. del Pino, G. Weber, Ger. Offen. 2,548,360 (29. 10. 1975), C.A. *87*, 46607c (1977); Liquid crystals with reduced viscosity. [**637**] N. Krog, Fette, Seifen, Anstr. *76*, 493 (1974); Strukturen flüssig-kristalliner Phasen wäßriger Emulgatormischungen und ihre Beziehung zur Emulsionsstabilität. [**638**] H. Krüger, K. H. Walter, 5. Arbeitstagung Flüssigkristalle, Inst. Angew. Festkörperphysik Freiburg, 25. 4. 1975; Technologie von Flüssigkristallanzeigen. [**639**] H. Krüger, W. Geffcken, Tech. Mitt. *68*, 109 (1975), C.A. *83*, 106125y (1975); Verwendung nematischer Flüssigkristalle in Anzeige-Einheiten. [**640**] A. I. Kubarko, Uch. Zap., Ivanov. Gos. Pedagog. Inst. No. *99*, 227 (1972), C.A. *79*, 5502f (1973); Colored thermographic mixtures for medicinal study.
devices [**641**] A. I. Kubarko, E. P. Demitshik, 1. Wiss. Konferenz über flüssige Kristalle, 17.–19. 11. 1970, Iwanowo, Sammlung der Vorträge (1972), Nr. 29; Die farbige Thermotopographie und ihre klinische Verwendung. [**641a**] E. Kubista, H. Kucera, Österr. Monatsschrift Allgemeinmed. *31*, 743 (1977); Einsatz der Plattenthermographie in der Diagnostik des Mammakarzinoms. [**642**] S. Kubo, H. Arai, B. Hori, B. Kato, Japan. 74·20,967 (11. 7. 1970), C.A. *82*, 118172d (1975); Photosensitive liquid crystal microcapsule material. [**643**] S. Kubo, H. Arai, Chiba Daigaku Kogakubu Kenkyu Hokoku *21*, 163 (1970), C.A. *77*, 10696k (1972); Suitable conditions for microencapsulation of liquid crystals. [**644**] S. Kubo, H. Arai, Nippon Shashin Gakkaishi *35*, 40 (1972), C.A. *76*, 160788k (1972); Photographic sensitive element having liquid crystal core. [**645**] S. Kubo, Am. Chem. Soc., Div. Org. Coat. Plast. Chem. Pap. *33*, 549 (1973), C.A. *82*, 178155k (1975); Liquid-crystal-containing light sensitive microcapsules. [**646**] S. Kubo, H. Arai, Japan. 74 46,473 (30. 6. 1970), C.A. *83*, 186397b (1975); Color enhancement of cholesteryl liquid crystal compositions. [**647**] S. Kubo, H. Arai, F. Hori, T. Kato, Japan. 75 11,344 (21. 5. 1970), C.A. *83*, 124566x (1975); Liquid crystal microcapsules. [**647a**] H. Kucera, E. Kubista, Wien. Klin. Wochenschrift *88*, 737 (1976); Möglichkeiten zur Erfassung des Mammakarzinoms durch Einsatz der Flüssigkristall-Thermographie in der gynäkologischen Praxis. [**648**] F. Kuhn-Weiss, Mater.-Prüf. *16*, 140 (1974); Mikroverkapselte Flüssigkristalle. [**648a**] S. Kumagai, S. Fukushima, Colloid Interface Sci., 50th *1976*, 4, 91, C.A. *87*, 73949x (1977); Structure change with temperature in the aqueous solution of an ethoxylated surfactant and the stability of carbon black dispersions. [**649**] K. Kuroda, I. Tsunoda, Japan. Kokai 76 17,188 (1. 8. 1974), C.A. *85*, 114842m (1976); Nematic liquid crystal compositions for field effect display devices. [**650**] K. Kuroda, I. Tsunoda, Japan. Kokai 76 17,189 (2. 8. 1974), C.A. *85*, 134335b (1976); Nematic liquid crystal compositions for display devices.
[**651**] K. Kuroda, T. Toida, I. Tsunoda, Japan. Kokai 76 57,683 (15. 11. 1974), C.A. *86*, 99091r (1977); Liquid crystal compositions for TN display devices. [**652**] K. Kuroda, T. Toida, I. Tsunoda, Japan. Kokai 76 57,684 (15. 11. 1974), C.A. *86*, 99092s (1977); Liquid crystals composition for TN display devices. [**653**] K. Kuroda, T. Toida, I. Tsunoda, Japan. Kokai 76 97,586 (25. 2. 1975), C.A. *85*, 200590j (1976); Nematic liquid crystal compositions. [**653a**] K. Kuroda, H. Kawamoto, Japan. Kokai 76,119,387 (12. 4. 1975), C.A. *86*, 163644c (1977); Nematic liquid crystal compositions for display devices. [**654**] S. Kurda, M. Kimura, K. Kubota, Mol. Cryst. Liq. Cryst. *33*, 235 (1976), C.A. *85*, 102654p (1976); Dynamical study of heat conduction in liquid crystals by high-speed optical holography. [**655**] S. Kuroda, K. Kubota, Appl. Phys. Lett. *29*, 737 (1976), C.A. *86*, 24270q (1977); Dye laser action in a liquid crystal. [**655a**] T. Kuroda, H. Kudo, Japan. Kokai 76,107,292 (18. 3. 1975), C.A. *87*, 60817c (1977); Nematic liquid crystal. [**656**] K. Kurosawa, Japan. Kokai 73 97,777 (27. 3. 1972), C.A. *80*, 89449a (1974); Temperature-sensitive coloring matter. [**657**] F. Kuschel, D. Demus, G. Pelzl, Ger. (East) 116,116 (13. 11. 1974), C.A. *84*, 172203u (1976); Liquid crystals for electrooptical cells. [**658**] F. Kuschel, D. Demus, Wiss. Fortschr. *23*, 273 (1973); Flüssige Kristalle. Anwendungen in Wissenschaft und Technik. [**659**] A. P. Kushelevsky, L. Feldman, Z. B. Alfassi, Mol. Cryst. Liq. Cryst. *35*, 353 (1976), C.A. *86*, 10845u (1977); Gamma rays modification of encapsulated liquid crystals temperature range. [**660**] M. M. Labes, U.S. 3,932,298 (19. 7. 1973), C.A. *85*, 39322m (1976); Nematics with charge-transfer acceptors as dopants.

[661] G. Labrunie, J. Robert, J. Borel, Centre d'Études Nucléaires de Grenoble, Note technique LETI/ME n° 928 (18. 5. 1973); Nematic liquid crystal 1024 bits page composer. [662] G. Labrunie, J. Robert, J. Borel, Appl. Opt. *13*, 1355 (1974), C.A. *81*, 43980j (1974); Nematic liquid crystal 1024 bits page composer. [663] G. Labrunie, S. Valette, Appl. Opt. *13*, 1802 (1974), C.A. *81*, 84276c (1974); Nematic liquid crystal digital light deflector. [664] La Marr Sabourin, NASA Accession No. N66-34137, Rept. No. NASA-TM-X-57823 (1966), 12 pp., C.A. *66*, 106583v (1967); Nondestructive testing of bonding structures with liquid crystals. [665] R. D. Larrabee, U.S. 3,960,753 (30. 5. 1974), C.A. *86*, 10642a (1977); Fluorescent liquid crystals. [666] V. Lattanzio, N. Macarini, Y. Quenneville, J. Radiol. *56*, 631 (1975); Possibilités et limites des feuilles cholésteriques en sénologie. [667] V. I. Lebedev, V. I. Mordasov, M. G. Tomilin, Opt.-Mikh. Promst. *41*, 60 (1974), C.A. *81*, 143527k (1974); Liquid crystals in optics. [668] V. I. Lebedev, V. I. Mordasov, M. G. Tomilin, Opt.-Mekh. Prom-st. *43*, 60 (1976), C.A. *85*, 39221c (1976); Electrooptical properties of liquid crystal cells with dynamic scattering effect. [669] S. Le Berre, Electron. et Microelectron. Ind. *201*, 33 (1975); Projection d'images sur grand écran a l'aide de cristaux liquides. [670] B. J. Lechner, U.S. 3,532,813 (25. 9. 1967); Display circuit including charging circuit and fast reset circuit.

[671] B. J. Lechner, F. J. Marlowe, E. O. Nester, J. Tults, Proc. IEEE *59*, 1566 (1971); Liquid crystal matrix displays. [672] L. B. Leder, U.S. 3,789,225 (24. 8. 1971), C.A. *81*, 31898a (1974); Glassy liquid crystals for image formation. [672a] M. H. Lee, M. S. Sadove, S. I. Kim, Am. J. Acupuncture *4*, 145 (1976); Liquid-crystal thermography in acupuncture therapy. [672b] F. Lelik, G. Kezy, et al., Magyar Allatorv. *31*, 699 (1976); Use of liquid crystals for dermal thermography in lameness of horses and cattle. [673] M. Lescinsky, J. Fridrich, Thin Solid Films *14*, S17 (1972), C.A. *78*, 129558c (1973); Investigation of thin film capacitor defects using nematics. [674] M. Lescinsky, J. Fridrich, Czech. Conf. Electron. Vac. Phys., (Proc.), 5th *1972*, 1, IIa–18, C.A. *83*, 69846m (1975); Investigation of thin film properties by application of nematics. [675] M. Lescinsky, Slaboproudy Obz. *34*, 490 (1973), C.A. *80*, 75368m (1974); Liquid-crystal displays. [676] J. P. Lesieur, M. C. Sexton, D. Veron, J. Phys. D *5*, 1212 (1972), C.A. *77*, 95211w (1972); Gain and visualization of the modes of a thermally stabilized hydrogen cyanide laser. [677] B. Levin, N. Levin, Brit. 441,274 (13. 7. and 3. 8. 1934); Improvements in or relating to light valves. [678] H. Liebig, K. Wagner, Chem.-Ztg. *95*, 733 (1971), C.A. *75*, 123094q (1971); Flüssige Kristalle. Ihre Anwendung und deren Problematik. [679] L. T. Lipton, M. A. Meyer, H. G. Dill, D. O. Massetti, Int. Electron. Dev. Meeting, Washington, 9.–11. 12. 1974, Abstr. 15.9; SOS liquid crystal TV display. [680] H. S. Lim, J. D. Margerum, Appl. Phys. Lett. *28*, 478 (1976), C.A. *85*, 170164d (1976); Improved dc dynamic scattering with redox dopants in ester liquid crystals. [680a] H. S. Lim, Ger. Offen. 2,656,252 (30. 12. 1975), C.A. *87*, 93584c (1977); Doped liquid crystal material. [680b] M. J. Little, H. L. Garvin, Y. S. Lee, Fr. Demande 2,308,675 (21. 4. 1975), C.A. *87*, 125391f (1977); Liquid crystal cell.

[681] M. Lodolini, Ger. Offen. 2,300,055 (3. 1. 1972), C.A. *79*, 99125k (1973); Nematic liquid crystal mixtures. [682] W. W. Logan, Inv. Radiol. *9*, 329 (1974); Improved cutaneous liquid crystal breast thermography. [683] J. H. J. Lorteije, NTZ *28*, 196 (1975); Survey of drive methods for gasdischarge, light emitting diode, liquid crystal and electrochromic displays. [684] J. F. Lowe, Design News *30* (14), 44 (1975); Easier to read liquid crystal display. [684a] M. Luban, S. Shtrikman, J. Isaacson, Phys. Rev. A *15*, 1211 (1977), C.A. *86*, 163790x (1977); Director field and lifetime of the cholesteric storage mode. [685] N. A. Luce, Electronics *45*, 93 (1972); C/MOS digital wristwatch features liquid crystal display. [686] G. V. Lukianoff, Mol. Cryst. Liq. Cryst. *8*, 389 (1969), C.A. *71*, 95662v (1969); Information content and resolution aspects in thermal mappings using cholesterics. [687] R. B. MacAnally, Appl. Phys. Lett. *18*, 54 (1971), C.A. *74*, 93440j (1971); Liquid crystal displays for matched filtering. [688] M. T. McCaffrey, J. A. Castellano, Fr. Demande 2,189,121 (15. 6. 1972), C.A. *81*, 71103u (1974); Electrooptical devices such as liquid crystals. [688a] J. McDermott, Electron. Design *25* (10), 42, 44, 46 (1977); Liquid crystals, lasers and microprocessors reflected in tomorrow's displays. [689] E. R. McHenry, U.S. 3,974,264 (11. 12. 1973), C.A. *85*, 162658c (1976); Carbon fibers from mesophase pitch. [690] Y. Machi, Japan. 75 10,272 (7. 10. 1969), C.A. *83*, 211352t (1975); Electrooptical display devices.

[691] H. Mada, S. Kobayashi, Rev. Phys. Appl. *10*, 147 (1975), C.A. *83*, 106162h (1975); Electrooptical properties of twisted nematics. Application to voltage controllable color formation. [692] H. Mada, S. Kobayashi, J. Appl. Phys. *47*, 2898 (1976), C.A. *85*, 102310y (1976); Light scattering from nematics during transient time. [693] H. Mada, H. Kamimura, S. Kobayashi, Jpn. J. Appl. Phys. *15*, 1845 (1976), C.A. *86*, 63478h (1977); Temperature stability of liquid crystal display operated in FE-TB mode.

[**693a**] J. Maeno, Ger. Offen. 2,704,776 (6. 2. 1976), C.A. *87*, 144000u (1977); Liquid crystal element and its use. [**693b**] G. C. Maggi, F. M. di Roberto, Drugs Pharm. Sci. *3*, 103 (1976), C.A. *87*, 129699k (1977); Encapsulated liquid crystal thermography in the diagnosis and monitoring of peripheral vascular disorders. [**694**] M. Magne, P. Pinard, P. Thome, N. Chretien, Colloq. Met. *12*, 241 (1968), C.A. *74*, 92314j (1971); Utilisation des propriétés spécifiques des cristaux liquides au contrôle non destructif. [**695**] M. Magne, P. Pinard, J. Phys. (Paris), C-4, *30*, 117 (1969), C.A. *72*, 102150c (1970); Mesure des températures superficielles à l'aide de cristaux liquides. [**696**] M. Magne, P. Pinard, P. Thome, N. Chretien, Bull. Inform. Sci. Tech. (Paris) *136*, 45 (1969), C.A. *71*, 42566k (1969); Application des propriétés des cristaux liquides au contrôle non destructif. [**697**] W. Mahler, M. Panar, U.S. 3,766,061 (13. 4. 1970), C.A. *80*, 97475w (1974); Decorative iridescent compositions. [**698**] W. Mahler, M. Panar, Ger. Offen. 2,245,924 (19. 9. 1972), C.A. *81*, 31903y (1974); Film-forming mixture with cholesteric properties. [**699**] W. Mahler, M. Panar, Brit. 1,349,050 (13. 9. 1972), C.A. *82*, 60088n (1975); Decorative iridescent coating compositions. [**700**] A. P. Makhotilo, S. V. Shevchuk, V. P. Tkachenko, V. G. Tishchenko, U.S.S.R. 531,835 (8. 12. 1974), C.A. *86*, 17741r (1977); Thermochromic paste.

[**701**] J. C. Manaranche, J. Phys. D *5*, 1120 (1972), C.A. *77*, 67222g (1972); Application des cristaux liquides au contrôle nondestructif. [**702**] R. A. Mao, U.S. 3,653,745 (11. 6. 1970); Circuits for driving loads such as liquid crystal displays. [**703**] R. D. Maple, US Nat. Tech. Inform. Serv., AD Rep. *1972*, No. 744211, 38 pp., C.A. *77*, 169848h (1972); Utilisation of temperature sensitive liquid crystals for thermal analysis and an application to transducer investigations. [**704**] J. D. Margerum, J. Nimoy, S. Y. Wong, Appl. Phys. Lett. *17*, 51 (1970), C.A. *73*, 135604c (1970); Reversible UV imaging with liquid crystals. [**705**] J. D. Margerum, T. D. Beard, W. P. Bleha, Jr., S. Y. Wong, Appl. Phys. Lett. *19*, 216 (1971), C.A. *76*, 20003a (1972); Transparent phase images in photoactivated liquid crystals. [**705a**] J. D. Margerum, L. J. Miller, J. Colloid Interface Sci. *58*, 559 (1977), C.A. *86*, 180561r (1977); Electro-optical applications of liquid crystals. [**705b**] J. D. Margerum, U.S. NTIS, AD Rep. 1976, AD-AO32820, 137 pp., C.A. *86*, 198436p (1977); Molecular basis for liquid crystal field effects. [**706**] R. C. Margolis, L. S. Shaffer, J. Amer. Osteop. Assoc. *73*, 103 (1974); Placental localization by liquid crystal thermography. [**707**] A. M. Marks, M. M. Marks, U.S. 3,167,607 (11. 1. 1960); Multi-element electro-optic crystal shutter. [**708**] F. J. Marlowe, U.S. 3,503,672 (14. 9. 1967); Reduction of turn-on delay in liquid crystal cell. [**709**] F. J. Marlowe, E. O. Nester, U.S. 3,654,606 (6. 11. 1969); Alternating voltage excitation of liquid crystal display matrix. [**710**] T. B. Marsch, Fr. Demande 2,261,549 (15. 2. 1974), C.A. *85*, 12251a (1976); Orientation of liquid crystals on a surface, and polarizers manufactured using these crystals.

[**711**] D. H. Mash, Brit. 1,304,268 (11. 12. 1970); Image display device. [**712**] L. Massig-Palhes, Automatisme *21*, 72 (1976); Traitement hybride des images. [**713**] Y. Masuda, K. Sawada, Y. Nakagawa, Denki Kagaku Oyobi Kogyo Butsuri Kagaku *42*, 549 (1974), C.A. *82*, 162891m (1975); Liquid crystal display. [**714**] S. R. Matlin, Arch. Pod. Med. Foot Surg. *1*, 235 (1974); Liquid crystal thermography. [**715**] Y. Matsuda, S. Aramoto, K. Iwata, Japan. Kokai 74 66,976 (28. 10. 1972), C.A. *81*, 171225d (1974); Yarns with temperature-sensitive colors. [**716**] Y. Matsufuji, Japan. Kokai 75,103,485 (21. 1. 1974), C.A. *85*, 55064j (1976); Molecular orientation promoting agents for liquid crystal display devices. [**717**] Y. Matsufuji, Japan. Kokai 75,106,886 (31. 1. 1974), C.A. *84*, 128801p (1976); Nematic liquid crystal compositions with positive dielectric anisotropy. [**718**] Y. Matsufuji, Japan. Kokai 75,106,887 (31. 1. 1974), C.A. *84*, 158049s (1976); Positive dielectric anisotropy nematic liquid crystal compositions. [**719**] Y. Matsufuji, Japan. Kokai 75,152,987 (30. 5. 1974), C.A. *85*, 12388a (1976); Liquid crystal composition. [**720**] F. Matsukawa, H. Arai, Japan. Kokai 76 72,984 (17. 12. 1974), C.A. *85*, 134505g (1976); Liquid crystal composite. [**720a**] F. Matsukawa, H. Arai, Japan. Kokai 76,132,187 (30. 4. 1975), C.A. *86*, 163651c (1977); Liquid crystal composite.

[**721**] G. Matsumoto, T. Ginnai, K. Iwasaki, Japan. Kokai 76,145,477 (5. 6. 1975), C.A. *86*, 131211t (1977); Nematic liquid crystal compositions for display devices. [**722**] S. Matsumoto, M. Kawamoto, Fr. 2,107,242 (4. 9. 1970), C.A. *78*, 22412y (1973); Nematic liquid crystals. [**723**] S. Matsumoto, M. Kawamoto, Fr. 2,107,464 (9. 9. 1970), C.A. *78*, 22411x (1973); Nematic liquid crystals containing *p*-[N-(*p*-methoxybenzylidene)-amino]-phenyl-2-ethylhexanoate. [**724**] S. Matsumoto, M. Kawamoto, Japan. Kokai 73 05,677 (4. 6. 1971), C.A. *79*, 10917z (1973); Nematic liquid crystals for electrooptic applications. [**725**] S. Matsumoto, M. Kawamoto, Japan. Kokai 73 05,678 (4. 6. 1971), C.A. *79*, 10900p (1973); Nematic liquid crystals for electrooptic applications. [**726**] S. Matsumoto, M. Kawamoto, Japan. Kokai 73 05,679 (4. 6. 1971), C.A. *79*, 10899v (1973); Nematic liquid crystals for electrooptic applications. [**727**] S. Matsumoto, M. Kawamoto, Japan. 73 32,279 (16. 12. 1969), C.A. *80*, 139577p (1974); Nematic

crystalline liquid material. [**728**] S. Matsumoto, M. Kawamoto, Japan. Kokai 74 03,887 (2. 5. 1972), C.A. *80*, 139567k (1974); Phenol derivatives for nematic liquid crystal stabilization. [**729**] S. Matsumoto, M. Kawamoto, Japan. Kokai 74 86,280 (23. 12. 1972), C.A. *82*, 105239f (1975); Liquid crystals for electrooptical display devices. [**730**] S. Matsumoto, M. Kawamoto, N. Kaneko, Oyo Butsuri *44*, 875 (1975), C.A. *83*, 171124m (1975); Molecular orientation of liquid crystals by chromium-complexes.

[**731**] S. Matsumoto, M. Kawamoto, K. Mizunoya, J. Appl. Phys. *47*, 3842 (1976), C.A. *85*, 169658m (1976); Field-induced deformation of hybrid-aligned nematics: new multicolor liquid crystal display. [**732**] S. Matsumoto, M. Kawamoto, K. Mizunoya, Oyo Butsuri *45*, 853 (1976), C.A. *86*, 148012b (1977); New hybrid-aligned nematic multicolor liquid crystal display. [**732a**] S. Matsumoto, N. Kaneko, U.S. 3,991,241 (13. 6. 1974), C.A. *86*, 180741z (1977); Liquid crystal device. [**732b**] S. Matsumoto, D. Nakagawa, H. Ikeya, N. Kaneko, Japan. Kokai 77 02,881 (25. 6. 1975), C.A. *87*, 60820y (1977); Stabilizers for liquid crystal compositions for display devices. [**732c**] S. Matsumoto, H. Ikeya, D. Nakagawa, N. Kaneko, Japan. Kokai 77 02,882 (25. 6. 1975), C.A. *86*, 198023b (1977); Stabilizer for nematic liquid crystal composition for display device. [**732d**] S. Matsumoto, D. Nakagawa, K. Mizunoya, N. Kaneko, Japan. Kokai 77 47,745 (14. 10. 1975), C.A. *87*, 160014g (1977); TN liquid crystal display devices. [**733**] Matsushita Electric Industrial Co., Fr. Demande 2,246,006 (20. 7. 1973), C.A. *84*, 143079c (1976); Electrooptical display cell using liquid crystal. [**734**] D. L. Matthies, U.S. 3,622,226 (19. 11. 1969); Liquid crystal cells in a linear array. [**735**] D. L. Matthies, U.S. 3,661,444 (2. 12. 1969); Compounded liquid crystal cells. [**736**] D. Maydan, Proc. IEEE *61*, 1007 (1973), C.A. *79*, 99129q (1973); IR laser addressing of media for recording and displaying of high-resolution graphic information. [**737**] R. C. Maze, E. P. Oppenheim, R. M. Reynolds, Ger. Offen. 2,522,795 (10. 6. 1974), C.A. *84*, 97878n (1976); Liquid crystal. [**738**] B. Mechlowitz, J. E. Adams, W. E. L. Haas, U.S. 3,666,948 (6. 1. 1971), C.A. *77*, 133236t (1972); Liquid crystal thermal imaging system having an undisturbed image on a disturbed background. [**739**] G. Meier, Tech. Mitt. *68*, 105 (1975), C.A. *83*, 69161j (1975); Anwendungen cholesterinischer Flüssigkristalle. [**740**] G. Meier, NTZ *28*, 183 (1975); Neuere Entwicklungen auf dem Gebiet der Anzeigetechnik.

[**741**] G. Meier, E. Sackmann, J. G. Grabmaier, Applications of Liquid Crystals (Springer: Berlin). 1975. 164 pp., C.A. *85*, 71077w (1976). [**742**] H. Meier, W. Albrecht, U. Tschirwitz, Photogr. Sci. Eng. *20*, 72 (1976), C.A. *85*, 134223p (1976); Activation of electrooptical effects in liquid crystals by organic photoconductors. [**743**] H. Melchior, F. J. Kahn, D. Maydan, D. B. Fraser, Appl. Phys. Lett. *21*, 392 (1972), C.A. *77*, 171144u (1972); Thermally addressed electrically erased high-resolution liquid-crystal light valves. [**744**] Merck Patent, Fr. Demande 2,167,950 (11. 1. 1972), C.A. *80*, 42215f (1974); Nematic phases. [**745**] Merck Patent, Brit. 1,376,115 (26. 2. 1972), C.A. *83*, 69118a (1975); Modified nematogenic compositions. [**746**] Merck Patent, Fr. Demande 2,227,052 (28. 4. 1973), C.A. *84*, 67886w (1976); Modified nematic mixtures having a positive dielectric anisotropy. [**747**] R. Merran, Fr. 2,105,524 (10. 9. 1970), C.A. *78*, 21354u (1973); Liquid crystal compositions sensitive to pressure variations. [**748**] D. Meyerhofer, E. F. Pasierb, Mol. Cryst. Liq. Cryst. *20*, 279 (1973), C.A. *79*, 71732g (1973); Light scattering characteristics in liquid crystal storage materials. [**749**] D. Meyerhofer, Appl. Phys. Lett. *29*, 691 (1976), C.A. *86*, 24634m (1977); New technique of aligning liquid crystals on surfaces. [**749a**] D. Meyerhofer, J. Appl. Phys. *48*, 1179 (1977), C.A. *87*, 75985e (1977); Optical transmission of liquid-crystal field-effect cells. [**749b**] L. J. Miller, U.S. 4,022,934 (21. 4. 1975), C.A. *87*, 14134e (1977); Means for inducing perpendicular alignment of a nematic on a coated substrate. [**750**] F. A. Mina, Fr. Demande 2,038,104 (1. 4. 1969), C.A. *75*, 99075m (1971); Thermographic strips employing a color change.

[**751**] C. W. Mitchell, Jr., U.S. 3,894,794 (7. 10. 1974), C.A. *84*, 10946m (1976); Liquid crystal imaging system using tributyltin oxide. [**752**] H. Mitsui, U.S. 3,668,861 (17. 11. 1970); Solid state electronic watch. [**752a**] A. Miyaji, M. Yamaguchi, A. Toda, H. Mada, S. Kobayashi, IEEE Trans. Electron Dev. *24*, 811 (1977); Control and elimination of disclinations in TN displays. [**753**] Y. Miyata, H. Oyama, Japan. Kokai 76,143,584 (5. 6. 1975), C.A. *86*, 131621b (1977); Liquid crystal composite. [**753a**] T. Miyazaki, S. Kato, Japan. Kokai 77 02,883 (25. 6. 1975), C.A. *87*, 76412c (1977); Liquid crystal compositions for display devices. [**753b**] T. Miyazaki, S. Kato, Japan. Kokai 77 03,586 (27. 6. 1975), C.A. *86*, 180751c (1977); Liquid crystal composition for dynamic scattering-type display devices. [**753c**] T. Miyazaki, S. Kato, Ger. Offen. 2,629,698 (3. 7. 1975), C.A. *87*, 93565x (1977); Liquid crystal compositions. [**754**] Y. Mizuguchi, Japan. Kokai 73 23,683 (1. 8. 1971), C.A. *81*, 128108f (1974); Low temperature nematics and their production. [**755**] Y. Mizuguchi, T. Takahashi, H. Shimizu, Japan. Kokai 75 51,478 (7. 9. 1973), C.A. *83*, 170955w (1975); Liquid crystals for TN display devices. [**756**]

Y. Mizuguchi, H. Shimizu, Japan. Kokai 76 38,288 (30. 9. 1974), C.A. *85*, 134484z (1976); Nematic liquid crystal compositions for field-effect display devices. [**757**] H. Mizuno, S. Tanaka, Opt. Commun. *3*, 320 (1971), C.A. *75*, 124852x (1971); Application of nematics to control the coherence of laser beams. [**758**] J. A. Mock, Materials Eng. *69*, 66 (1969); Liquid crystals track flaws in a colorful way. [**759**] U. Moller, J. Bojsen, Cancer Res. *35*, 3116 (1975); Temperature and blood flow measurements in and around 7,12-dimethylbenz(α)anthracene-induced tumors and Walker 256 carcinosarcomas in rats. [**760**] M. A. Monahan, H. Caspers, L. B. Stotts, N. McLandrich, U.S.N.T.I.S., AD Rep. 1973 No. 917229/7GA, 51 pp., C.A. *82*, 131993s (1975); Liquid crystal devices.

[**761**] F. Moravec, M. Lescinsky, F. Hoff, Slaboproudy Obz. *33*, 195 (1972), C.A. *77*, 67376k (1972); Liquid crystals in electronics. [**762**] K. Mori, M. Fukai, K. Asai, A. Moriyama, S. Nagata, Japan. Kokai 74 29,291 (17. 7. 1972), C.A. *81*, 71121y (1974); Liquid crystal compositions for electrooptical display devices. [**763**] M. Morikawa, Ger. 2,249,867 (11. 10. 1971), C.A. *84*, 24453k (1976); Electroluminescent dispersion. [**764**] K. Morimoto, T. Ohtsuka, Y. Murakami, M. Tsukamoto, M. Tsuchiya, Natl. Tech. Rep. (Matsushita Electr. Ind. Co., Osaka) *22*, 213 (1976), C.A. *85*, 169659n (1976); Chiral nematics and their application to display devices. [**765**] A. Moriyama, M. Fukai, K. Asai, S. Nagata, H. Tatsuta, Japan. Kokai 73 94,684 (13. 3. 1972), C.A. *81*, 128113d (1974); Liquid crystal composition. [**766**] A. Moriyama, M. Fukai, K. Asai, S. Nagata, H. Tatsuta, Japan. Kokai 73 94,685 (13. 3. 1972), C.A. *81*, 128112c (1974); Liquid crystal composition. [**767**] A. Moriyama, M. Fukai, K. Asai, S. Nagata, H. Tatsuta, Japan. Kokai 73 94,686 (13. 3. 1972), C.A. *81*, 128109g (1974); Liquid crystal composition. [**768**] A. Moriyama, M. Fukai, K. Asai, S. Nagata, H. Tatsuta, Japan. Kokai 73 94,687 (13. 3. 1972), C.A. *81*, 128116g (1974); Liquid crystal composition. [**769**] A. Moriyama, M. Fukai, K. Asai, S. Nagata, H. Tatsuta, Japan. Kokai 73 94,688 (13. 3. 1972), C.A. *81*, 128114e (1974); Liquid crystal composition. [**770**] A. Moriyama, M. Fukai, K. Asai, S. Nagata, H. Tatsuta, Japan. Kokai 73 94,689 (13. 3. 1972), C.A. *81*, 128115f (1974); Liquid crystal composition.

[**771**] A. Moriyama, M. Fukai, K. Asai, S. Nagata, H. Tatsuta, Japan. Kokai 73 94,690 (13. 3. 1972), C.A. *81*, 128106d (1974); Liquid crystal composition. [**772**] A. Moriyama, M. Fukai, K. Asai, S. Nagata, H. Tatsuta, Japan. Kokai 73 94,691 (13. 3. 1972), C.A. *81*, 128111b (1974); Liquid crystal composition. [**773**] A. Moriyama, M. Fukai, K. Asai, S. Nagata, H. Tatsuta, Japan. Kokai 73 94,692 (13. 3. 1972), C.A. *81*, 96989a (1974); Liquid crystal composition. [**774**] A. Moriyama, M. Fukai, K. Asai, S. Nagata, H. Tatsuta, Japan. Kokai 73 94,693 (13. 3. 1972), C.A. *81*, 128110a (1974); Liquid crystal composition. [**775**] A. Moriyama, M. Fukai, K. Asai, K. Mori, Japan. Kokai 74 74,187 (17. 11. 1972), C.A. *82*, 148549z (1975); Homeotropic liquid crystal display device. [**776**] A. Moriyama, M. Fukai, K. Asai, K. Mori, Japan. Kokai 74 74,188 (17. 11. 1972), C.A. *82*, 92075x (1975); Nematic liquid crystal display device. [**777**] A. Moriyama, M. Fukai, K. Asai, K. Mori, Japan. Kokai 74 74,189 (17. 11. 1972), C.A. *82*, 148547x (1975); Liquid crystal display device based on homeotropic liquid crystal compositions. [**778**] A. Moriyama, M. Fukai, K. Asai, K. Mori, Japan. Kokai 74 74,190 (17. 11. 1972), C.A. *82*, 148546w (1975); Liquid crystal display devices based on homeotropic liquid crystal compositions. [**779**] A. Moriyama, M. Fukai, K. Asai, K. Mori, Japan. Kokai 74 74,676 (20. 11. 1972), C.A. *82*, 37354n (1975); Homeotropic liquid crystal display devices. [**780**] A. Moriyama, M. Fukai, K. Asai, K. Mori, Japan. Kokai 74 74,677 (20. 11. 1972), C.A. *82*, 37356q (1975); Nematic liquid crystal compositions with improved homeotropy for display devices.

[**781**] A. Moriyama, M. Fukai, K. Asai, K. Mori, Japan. Kokai 74 74,678 (20. 11. 1972), C.A. *82*, 118266n (1975); Nematic liquid crystal compositions with good homeotropy for electrooptical display devices. [**782**] A. Moriyama, M. Fukai, K. Asai, K. Mori, Japan. Kokai 74 74,679 (20. 11. 1972), C.A. *82*, 49908j (1975); Nematic liquid crystal display device with improved homeotropy. [**783**] A. Moriyama, M. Fukai, K. Asai, K. Mori, Japan. Kokai 74 74,680 (20. 11. 1972), C.A. *82*, 118265m (1975); Nematic liquid compositions with good homeotropy for electrooptical display devices. [**784**] A. Moriyama, M. Fukai, K. Asai, K. Mori, Japan. Kokai 74 74,681 (20. 11. 1972), C.A. *82*, 78790v (1975); Homeotropic liquid crystal electrooptical display device. [**785**] A. Moriyama, M. Fukai, K. Asai, K. Mori, Japan. Kokai 74 74,682 (20. 11. 1972), C.A. *82*, 92077z (1975); Liquid crystal display device. [**786**] A. Moriyama, M. Fukai, K. Asai, K. Mori, Japan. Kokai 74 74,683 (20. 11. 1972), C.A. *82*, 105234a (1975); Homeotropic liquid crystal electrooptical display device. [**787**] A. Moriyama, M. Fukai, K. Asai, K. Mori, Japan. Kokai 74 74,684 (20. 11. 1972), C.A. *82*, 105240z (1975); Homeotropic liquid crystal electrooptical display devices. [**788**] A. Moriyama, M. Fukai, K. Asai, K. Mori, Japan. Kokai 74 74,685 (20. 11. 1972), C.A. *82*, 78786y (1975); Liquid crystal electrooptical display device. [**789**] A. Moriyama, M. Fukai, K. Asai, K. Mori, Japan. Kokai 74 74,686 (20. 11. 1972), C.A. *82*,

37345k (1975); Nematic liquid crystal compositions with improved homeotropy for display devices. [790] A. Moriyama, M. Fukai, K. Asai, K. Mori, Japan. Kokai 74 74,687 (20. 11. 1972), C.A. *82*, 49920g (1975); Liquid crystal display devices.

[791] A. Moriyama, M. Fukai, K. Asai, K. Mori, Japan. Kokai 74 75,471 (16. 11. 1972), C.A. *82*, 37352k (1975); Liquid crystal display device. [792] A. Moriyama, M. Fukai, H. Tatsuta, H. Esaki, Japan. Kokai 75 35,076 (1. 8. 1973), C.A. *83*, 211341p (1975); Nematic liquid crystal composition for display devices. [793] A. Moriyama, M. Fukai, H. Tatsuta, H. Takahashi, H. Esaki, Japan. Kokai 75 43,069 (20. 8. 1973), C.A. *83*, 106260p (1975); Nematic liquid crystal composition with positive dielectric anisotropy. [794] A. Moriyama, M. Fukai, H. Tatsuta, H. Takahashi, H. Esaki, Japan. Kokai 75 43,070 (20. 8. 1973), C.A. *83*, 155795a (1975); Electro-optical element. [795] A. Moriyama, M. Fukai, H. Tatsuta, H. Takahashi, H. Esaki, Japan. Kokai 75 43,071 (20. 8. 1973), C.A. *83*, 186410a (1975); Liquid crystal composition for electrooptical display devices. [796] A. Moriyama, M. Fukai, H. Tatsuta, H. Takahashi, H. Esaki, Japan. Kokai 75 43,072 (20. 8. 1973), C.A. *83*, 139896d (1975); Liquid crystal composition for optical display devices. [797] A. Moriyama, M. Fukai, H. Tatsuta, H. Takahashi, H. Esaki, Japan. Kokai 75 43,073 (20. 8. 1973), C.A. *83*, 155794z (1975); Liquid crystal composition for electrooptical display. [798] A. Moriyama, M. Fukai, H. Tatsuta, H. Takahashi, H. Esaki, Japan. Kokai 75 43,074 (20. 8. 1973), C.A. *83*, 186409g (1975); Nematic liquid crystal composition for optical display devices. [799] A. Moriyama, M. Fukai, H. Tatsuta, H. Takahashi, H. Esaki, Japan. Kokai 75 43,075 (20. 8. 1973), C.A. *83*, 69147j (1975); Liquid crystal composition with positive dielectric anisotropy. [800] A. Moriyama, M. Fukai, H. Tatsuta, H. Takahashi, H. Esaki, Japan. Kokai 75 43,076 (20. 8. 1973), C.A. *83*, 139898f (1975); Electrooptical element for optical display.

[801] A. Moriyama, M. Fukai, H. Tatsuta, H. Takahashi, H. Esaki, Japan. Kokai 75 43,077 (20. 8. 1973), C.A. *83*, 139899g (1975); Liquid crystal composition having positive dielectric anisotropy. [802] A. Moriyama, M. Fukai, H. Tatsuta, H. Takahashi, H. Esaki, Japan. Kokai 75 43,078 (20. 8. 1973), C.A. *83*, 155796b (1975); Nematic liquid crystal composition with positive dielectric anisotropy. [803] A. Moriyama, M. Fukai, H. Tatsuta, H. Takahashi, H. Esaki, Japan. Kokai 75 43,079 (20. 8. 1973), C.A. *83*, 139897e (1975); Electrooptical composition for display devices. [804] A. Moriyama, M. Fukai, H. Tatsuta, H. Takahashi, H. Esaki, Japan. Kokai 75 43,080 (20. 8. 1973) C.A. *83*, 106259v (1975); Electrooptical liquid crystal composition. [805] A. Moriyama, M. Fukai, H. Tatsuta, H. Takahashi, H. Esaki, Japan. Kokai 75 43,081 (20. 8. 1973), C.A. *83*, 69148k (1975); Liquid crystal compositions for optical display devices. [806] A. Moriyama, M. Fukai, H. Tatsuta, H. Takahashi, H. Esaki, Japan. Kokai 75 43,082 (20. 8. 1973), C.A. *83*, 155797c (1975); Electrooptical liquid crystal composition. [807] A. Moriyama, M. Fukai, H. Tatsuta, H. Takahashi, H. Esaki, Japan. Kokai 75 43,083 (20. 8. 1973), C.A. *83*, 88765a (1975); Electrooptical liquid crystal compositions. [808] A. Moriyama, M. Fukai, H. Tatsuta, H. Takahashi, H. Esaki, Japan. Kokai 75 43,084 (20. 8. 1973), C.A. *83*, 69149m (1975); Liquid crystal composition. [809] A. Moriyama, M. Fukai, H. Tatsuta, H. Takahashi, H. Esaki, Japan. Kokai 75 43,085 (20. 8. 1973), C.A. *83*, 88766b (1975); Electrooptical elements from liquid crystal compositions. [810] A. Moriyama, M. Fukai, H. Takahashi, Japan. Kokai 75 56,386 (19. 9. 1973), C.A. *84*, 67888y (1976); Liquid crystal electrooptical display device.

[811] A. Moriyama, M. Fukai, H. Tatsuta, H. Takahashi, H. Esaki, Japan. Kokai 75 92,873 (19. 12. 1973), C.A. *84*, 128799u (1976); Nematic liquid crystal compositions for electrooptical devices. [812] A. Moriyama, M. Fukai, H. Tatsuta, H. Takahashi, H. Esaki, Japan. Kokai 75 92,874 (19. 12. 1973), C.A. *84*, 128798t (1976); Nematic liquid crystal display device. [813] A. Moriyama, M. Fukai, H. Tatsuta, H. Takahashi, H. Esaki, Japan. Kokai 75 92,875 (19. 12. 1973), C.A. *84*, 128797s (1976); Nematic liquid crystal compositions for electrooptical devices. [814] A. Moriyama, M. Fukai, K. Asai, Ger. Offen. 2,418,364 (16. 4. 1973), C.A. *83*, 35742j (1975); Nematic liquid crystal compositions for electrooptical display devices. [814a] A. Moriyama, M. Fukai, K. Asai, S. Nagata, M. Hattori, K. Mori, Japan. 76 06,635 (19. 10. 1970), C.A. *86*, 198020y (1977); Nematic liquid crystal compositions for electrooptical display devices. [815] J. E. Morse, U.S. 3,722,998 (19. 10. 1970); Liquid crystal apparatus for reducing contrast. [815a] C. Motoc, I. Cuculescu, I. Baciu, M. Honciuc, Rev. Roum. Phys. *22*, 53 (1977), C.A. *87*, 109719p (1977); Storage effects in cholesteric mixtures. [816] K. Motoyoshi, T. Terada, Japan. Kokai 75 23,348 (5. 7. 1973), C.A. *83*, 69143e (1975); Liquid crystals for electrooptical devices. [817] K. Murao, K. Toriyama, F. Nakano, T. Muroi, M. Sato, Japan. Kokai 74 82,592 (18. 12. 1972), C.A. *82*, 49917m (1975); Liquid crystal composition for display devices. [818] K. Murao, K. Toriyama, N. Kamezawa, T. Kitamura, Japan. Kokai 74,110,582 (23. 2. 1973), C.A. *82*, 92063s (1975); Nematic liquid crystal compositions for display devices. [819] K. Murao, F. Nakano, K. Toriyama,

T. Kitamura, Japan. Kokai 75 83,271 (28. 11. 1973), C.A. *83*, 200269v (1975); Nematic liquid crystal electrooptical display devices. [820] H. Murase, Japan. 74 27,509 (20. 2. 1970), C.A. *82*, 178980a (1975); Liquid crystals having a low temperature operation range.

[821] K. Murase, Japan. Kokai 73 89,178 (1. 3. 1972), C.A. *80*, 89572k (1974); Nematic liquid crystal composition for electrooptical display devices. [822] R. Muto, S. Furuuchi, H. Ukihashi, K. Uchijima, H. Nishimura, Fr. Demande 2,188,188 (6. 6. 1972), C.A. *81*, 50731g (1974); Liquid crystal cell. [823] K. D. S. Myrenne, C. L. Hedman, Jr., U.S. 3,728,007 (27. 5. 1971); Reflective type liquid crystal display device having improved optical contrast. [823a] A. W. Nagy, G. B. Trapani, U.S. 4,025,688 (1. 8. 1974), C.A. *87*, 24441s (1977); Polarizer lamination. [824] F. Naito, T. Kuroda, K. Arita, A. Hakusui, Japan. Kokai 75 03,976 (16. 5. 1973), C.A. *83*, 69132a (1975); Liquid crystal display device. [825] K. Nakada, T. Ishibashi, K. Toriyama, IEEE Trans. Electron Dev. ED *22*, 725 (1975); Design of multiplexing liquid-crystal display for calculators. [826] T. Nakagomi, K. Toriyama, M. Kanazaki, Japan. Kokai 76 72,985 (23. 12. 1974), C.A. *85*, 134402w (1976); Liquid crystal compositions for TN display devices. [827] T. Nakagomi, K. Toriyama, M. Kanazaki, Japan. Kokai 76 72,986 (23. 12. 1974), C.A. *86*, 131187q (1977); Liquid crystal composition for TN display devices. [828] H. Nakamachi, A. Ouchida, M. Shintani, Y. Wada, Ger. Offen. 2,442,176 (7. 9. 1973), C.A. *83*, 29200q (1975); Temperature-sensitive polymer compositions. [829] K. Nakamura, M. Yaguchi, Japan. Kokai 75 47,890 (31. 8. 1973), C.A. *83*, 124108z (1975); Liquid crystal compositions with positive anisotropy. [830] K. Nakamura, T. Jinnai, K. Totani, S. Furuta, Japan. Kokai 76 20,786 (25. 7. 1974), C.A. *85*, 54645n (1976); TN liquid crystal display device.

[831] K. Nakamura, T. Jinnai, S. Furuta, K. Totani, Japan. Kokai 76 28,588 (5. 9. 1974), C.A. *85*, 114854s (1976); TN liquid crystal display device. [832] K. Nakamura, T. Jinnai. S. Furuta, K. Totani, Japan. Kokai 76 29,386 (5. 9. 1974), C.A. *86*, 131182j (1977); Nematic liquid crystal compositions for TN display devices. [833] K. Nakamura, T. Jinnai, S. Furuta, K. Totani, Japan. Kokai 76 48,785 (31. 8. 1974), C.A. *86*, 99087u (1977); Nematic liquid crystal composition for TN display devices. [834] K. Nakamura, T. Jinnai, K. Totani, S. Furuta, Japan. Kokai 76 59,783 (25. 7. 1974), C.A. *85*, 134492a (1976); Liquid crystal compositions for TN display devices. [835] K. Nakamura, K. Totani, S. Furuta, T. Jinnai, M. Yaguchi, M. Nichogi, Ger. Offen. 2,456,077 (31. 7. 1974), C.A. *85*, 114843n (1976); Nematic liquid crystal composition. [836] F. Nakano, K. Toriyama, M. Sato, Japan. Kokai 74,114,586 (7. 3. 1973), C.A. *82*, 178281y (1975); Nematic liquid crystals for optical display devices. [837] F. Nakano, K. Toriyama, N. Nagata, M. Sato, Ger. Offen. 2,408,711 (23. 2. 1973), C.A. *82*, 66356v (1975); Nematic liquid crystal composition. [838] F. Nakano, M. Sato, Jpn. J. Appl. Phys. *15*, 1937 (1976), C.A. *86*, 81639r (1977); Alignment deterioration of homeotropically aligned liquid crystal cells under d.c. electric field. I. Effect of liquid crystal composition and surface treatment. [838a] F. Nakano, K. Murao, Ger. Offen. 2,640,902 (12. 9. 1975), C.A. *87*, 31973v (1977); Liquid crystal composition and its use. [838b] F. Nakano, T. Kutamura, M. Sato, S. Fuji, M. Yaguchi, Japan. Kokai 77 23,582 (18. 8. 1975), C.A. *87*, 93574z (1977); Nematic liquid crystal compositions for matrix-type display devices. [838c] F. Nakano, H. Yokokura, K. Murao, Japan. Kokai 77 36,586 (19. 9. 1975), C.A. *87*, 93577c (1977); Field effect-type liquid crystal display devices. [839] T. Narusawa, H. Okuyama, J. Goto, Y. Isozaki, M. Fujimori, Ger. Offen. 2,511,884 (25. 3. 1974), C.A. *83*, 200595y (1975); Nematic liquid-crystal mixture. [839a] T. Narusawa, H. Okuyama, Y. Isozaki, M. Fujimori, Japan. Kokai 76,149,184 (14. 3. 1975), C.A. *87*, 14286f (1977); Liquid crystal composition for dynamic scattering-type display devices. [839b] T. Narusawa, H. Okuyama, Y. Isozaki, M. Fujimori, Japan. Kokai 77 27,741 (28. 8. 1975), C.A. *87*, 160015h (1977); High-purity Schiff base liquid crystals. [840] E. G. Nassimbene, IBM Technic. Disclosure Bull. *18*, 231 (1975); Touch keyboard using liquid crystal material.

[841] NCR Co., Brit. 1,302,482 (11. 2. 1970); Visual display device. [842] NCR Co., Brit. 1,316,213 (5. 8. 1970); Rear projection system. [843] NCR Co., Brit. 1,387,389 (12. 5. 1972), C.A. *83*, 117366d (1975); Detection of counterfeiting in security documents. [844] K. F. Nelson, W. E. Haas, J. E. Adams, J. Electrochem. Soc. *122*, 1564 (1975), C.A. *84*, 67747b (1976); Identification of the UV irradiation product in nematic chlorostilbene imaging processes. [845] K. F. Nelson, Photogr. Sci. Eng. *20*, 268 (1976), C.A. *86*, 36308w (1977); Model for a photoconductor-liquid crystal image storage panel. [846] E. O. Nester, B. J. Lechner, U.S. 3,575,492 (10. 7. 1969); Turnoff method and circuit for liquid crystal display element. [847] D. T. Ngo, U.S. 3,645,604 (10. 8. 1970); Liquid crystal display. [848] L. J. Nicastro, U.S. 3,588,225 (27. 1. 1970); Electro-optic devices for portraying closed images. [849] Y. Nishijima, K. Shimizu, Japan. Kokai 74 14,784 (7. 6. 1972), C.A. *82*, 126529r (1975);

Fixing encapsulated liquid crystals on textiles. [**850**] K. Nishimura, Y. Mizuguchi, T. Takahashi, H. Shimizu, Japan. Kokai 74 66,584˙ (1. 11. 1972), C.A. *82*, 118260f (1975); Liquid crystal compositions for optical display devices.

[**851**] S. Nomura, K. Toriyama, T. Aoyagi, Japan. Kokai 73 41,984 (4. 10. 1971), C.A. *79*, 131415t (1973); Electrooptical organic materials. [**852**] T. J. Novak, E. J. Poziomek, R. A. Mackay, Anal. Lett. *5*, 187 (1972), C.A. *77*, 13721a (1972); Use of anisotropic materials as chemical detectors. [**853**] W. Nozaki, K. Watanabe, Japan. Kokai 76 02,687 (26. 6. 1974), C.A. *85*, 12391w (1976); Liquid crystal compositions for low temperature display devices. [**854**] F. Ogawa, Japan. Kokai 76 22,678 (20. 8. 1974), C.A. *85*, 114849u (1976); TN liquid crystal display devices. [**855**] N. Oguchi, I. Ikami, A. Ohe, Japan. 74 35,114 (24. 12. 1970), C.A. *82*, 141518e (1975); High-pressure treatment of fibers having voids and/or cavities. [**856**] C. S. Oh, E. F. Pasierb, U.S. 3,792,915 (17. 10. 1972), C.A. *80*, 114873u (1974); Liquid crystal electrooptical devices. [**857**] C. S. Oh, E. F. Pasierb, U.S. 3,923,685 (17. 10. 1972), C.A. *84*, 172195t (1976); Liquid crystal compositions. [**858**] C. S. Oh, U.S. 3,975,286 (3. 9. 1974), C.A. *86*, 49223q (1977); Low voltage actuated field effect liquid crystal compositions. [**859**] C. S. Oh, P. Y. Hsieh, U.S. 3,981,558 (12. 11. 1973), C.A. *86*, 63549g (1977); Liquid crystal electrooptical display. [**859a**] C. S. Oh, U.S. 4,020,002 (28. 2. 1974), C.A. *87*, 46604z (1977); Non-Schiff base field effect liquid crystal composition. [**860**] M. Ohi, Y. Akimoto, T. Tako, Oyo Butsuri *41*, 363 (1972), C.A. *77*, 146087v (1972); Observation of helium-neon 3.39 μm laser patterns by means of liquid crystals.

[**861**] M. Ohizumi, T. Yamamoto, M. Abe, Japan. Kokai 74 34,487 (2. 8. 1972), C.A. *81*, 84452g (1974); Electrooptic liquid crystal composition. [**862**] M. Ohizumi, T. Yamamoto, M. Abe, Japan. Kokai 74 74,675 (20. 11. 1972), C.A. *82*, 10041w (1975); Liquid crystal compositions for display devices. [**863**] M. Ohizumi, T. Yamamoto, Japan. Kokai 74,120,882 (22. 3. 1973), C.A. *83*, 19047f (1975); Nematic liquid crystal compositions. [**864**] Y. Ohkubo, Y. Matsufugi, Japan. Kokai 75,114,388 (19. 2. 1974), C.A. *84*, 97869k (1976); Nematic liquid crystal compositions for display devices. [**865**] Y. Ohnishi, M. Ozutsumi, Appl. Phys. Lett. *24*, 213 (1974), C.A. *80*, 126602t (1974); Properties of nematics doped with hydroquinone and *p*-benzoquinone: Long-term dynamic scattering under d.c. excitation. [**866**] Y. Ohnishi, M. Ozutsumi, Japan. Kokai 74,133,284 (25. 4. 1973), C.A. *83*, 50837n (1975); Liquid crystal composition. [**867**] Y. Ohnishi, M. Ozutsumi, Y. Miyazawa, M. Gonda, Japan. Kokai 75 71,587 (30. 10. 1973), C.A. *84*, 82587t (1976); Cholesteric liquid crystal compositions with controlled orientation. [**868**] Y. Ohnishi, M. Ozutsumi, Y. Miyazawa, M. Gonda, U.S. 3,975,285 (30. 10. 1972), C.A. *85*, 134048k (1976); Liquid crystal composition. [**869**] T. Ohtsuka, M. Tsukamoto, M. Tsuchiya, Jap. J. Appl. Phys. *12*, 371 (1973), C.A. *78*, 166996q (1973); Liquid crystal matrix display. [**869a**] T. Ohtsuka, M. Tsukamoto, M. Tsuchiya, Terebijon *29*, 491 (1975), C.A. *87*, 125329s (1977); Liquid crystal matrix displays using the c/n transition. [**870**] R. Oishi, Y. Okajima, S. Noguchi, K. Akeyoshi, Brit. 1,399,961 (30. 4. 1973), C.A. *83*, 139858t (1975); Formation of a patterned transparent electroconductive film on a substrate.

[**871**] R. Oishi, Y. Okajyama, S. Noguchi, K. Akeyoshi, Fr. Demande 2,232,617 (16. 5. 1973), C.A. *83*, 148591c (1975); Deposition of a printed, transparent, electroconductive layer on a substrate of a liquid crystal display device. [**872**] H. Okuyama, T. Narusawa, Y. Miyashita, M. Fujimori, A. Usui, J. Goto, Japan. Kokai 75 05,282 (21. 5. 1973), C.A. *83*, 170931k (1975); Nematic liquid crystals for display devices. [**873**] H. Okuyama, T. Narusawa, M. Fujimori, Fujitsu Sci. Tech. J. *11*, 175 (1975), C.A. *83*, 106555p (1975); Liquid crystal mixtures with wide mesomorphic range and high contrast. [**874**] H. Onnagawa, K. Miyashita, Oyo Butsuri *42*, 133 (1973), C.A. *79*, 24418z (1973); Molecular orientation and wall effect of nematic liquid crystal cells. [**875**] H. Onnagawa, K. Miyashita, Jpn. J. Appl. Phys. *14*, 1061 (1975), C.A. *83*, 124697r (1975); Electric field dependence of the capacitance of a TN liquid crystal cell. [**876**] D. Ono, T. Sawa, N. Tokuyama, Japan. 74 43,269 (14. 12. 1970), C.A. *82*, 141259w (1975); Shaped articles of synthetic high polymer having selectively scattering properties. [**877**] D. Ono, T. Ito, T. Sawa, N. Tokuyama, Japan. 75 15,489 (29. 12. 1970), C.A. *83*, 186783z (1975); High polymer composition having selective scattering property. [**878**] H. Ono, E. Jidai, Japan. Kokai 74 36,586 (10. 8. 1972), C.A. *81*, 162126v (1974); Nematics for use at room temperature. [**879**] H. Ono, E. Jidai, Japan. Kokai 74 36,587 (10. 8. 1972), C.A. *81*, 162127w (1974); Nematics for use at room temperature. [**880**] H. Ono, Mitsubishi Denki Giho *46*, 923 (1972), C.A. *79*, 130084x (1973); Phase transition temperatures of mixed nematics and electrooptical effect.

[**881**] K. Ono, S. Fujii, S. Furuuchi, J. Electron. Engng. *107*, 34 (1975); Large-area liquid crystal matrix display advances to prototype stage. [**882**] M. Ono, T. Ito, T. Sawa, T. Kato, Japan.

74 01,676 (3. 12. 1970), C.A. *81*, 92920y (1974); Textile fabrics whose color changes reversibily with temperature. [**883**] M. Ono, T. Ito, T. Sawa, N. Tokuyama, Japan. 74 07,595 (18. 12. 1970), C.A. *81*, 121900z (1974); Melamine resin molded articles having ability of scattering light of specific wave-length. [**884**] M. Ono, T. Ito, T. Sawa, M. Tokuyama, Japan. 75 04,370 (18. 12. 1970), C.A. *83*, 133512d (1975); Synthetic paint having selective wave length-scattering property. [**885**] M. Ono, T. Ito, T. Sawa, N. Tokuyama, Japan. 75 04,371 (29. 12. 1970), C.A. *83*, 133513e (1975); Synthetic paint having selective wave length-scattering property. [**886**] W. Opitz, Elektriker *14*, 211 (1975); Ziffernanzeigen mit Flüssigkristallen. [**887**] Z. Orido, Y. Uchida, T. Sawa, N. Tokuyama, T. Kato, Japan. 74 18,914 (17. 12. 1970), C.A. *83*, 50815d (1975); Capsule having wave length selective scattering ability. [**888**] M. Ozutsumi, Y. Onishi, Y. Miyazawa, M. Gonda, Japan. Kokai 74 65,995 (30. 10. 1972), C.A. *82*, 37615y (1975); Liquid crystal compositions. [**889**] M. Ozutsumi, Y. Ohnishi, Y. Miyazawa, M. Gonda, Japan. Kokai 75 57,083 (22. 9. 1973), C.A. *83*, 186411b (1975); Liquid crystal compositions for dynamic-scattering mode display devices. [**890**] M. Ozutsumi, Y. Ohnishi, Y. Miyazawa, M. Gonda, Japan. Kokai 75 57,084 (22. 9. 1973), C.A. *83*, 186412c (1975); Liquid crystal compositions for display devices.

[**891**] V. Paquet, D. Krause, 7. Arbeitstagung Flüssigkristalle, Inst. Angew. Festkörperphysik Freiburg, 4. 3. 1977, Ladungsmechanismen in festen elektrochromen Systemen. [**891a**] J. W. Park, M. M. Labes, J. Appl. Phys. *48*, 22 (1977), Dielectric, elastic, and electro-optic properties of a liquid crystalline molecular complex. [**892**] R. Parker, Mol. Cryst. Liq. Cryst. *20*, 99 (1973), C.A. *78*, 129278m (1973); Transient surface temperatures response of liquid crystal films. [**893**] J. A. Patterson, E. D. Finkle, U.S. 3,852,092 (5. 6. 1972), C.A. *82*, 92072u (1975); Thermally responsive elastic membrane. [**894**] M. Pausch, Funkschau *47* (21), 73 (1975); Elektrooptische Effekte bei Flüssigkristallen. [**895**] M. Pazdur, Hutnik *37*, 205 (1970), C.A. *73*, 58691z (1970); Thermal nondestructive testing. [**895a**] Peking Chem. Works, Hua Hsueh Tung Pao *1977*, 188, C.A. *87*, 159940z (1977); Liquid crystals for display. [**896**] P. A. Penz, Phys. Teach. *13*, 199 (1975), C.A. *86*, 154662e (1977); Hydro-optic effects in liquid crystals. [**896a**] C. Perrot, Electron. Microelectron. Ind. *228*, 39 (1976), Cristaux liquides au service de la securite routiere. [**897**] E. N. Peterson, G. D. Dixon, M. A. Levine, Obstet. Gyn. *37*, 468 (1971); Placental localization by liquid crystal thermography. [**898**] S. E. B. Petrie, H. K. Bücher, R. T. Klingbiel, P. I. Rose, Eastman Organic Chemical Bulletin *45*, Nr. 2 (1973); Physical properties and applications of liquid crystals. [**899**] N. A. Pichikyan, A. S. Sonin, N. B. Titova, Kvantovaya Electron. *3*, 1614 (1976), C.A. *86*, 23753n (1977); Selective light scattering by pseudocapsulated films of cholesterics. [**900**] P. G. Pick, Fr. 2,051,423 (30. 6. 1969), C.A. *76*, 155069x (1972); Film-forming paste containing liquid crystals.

[**901**] P. G. Pick, J. Fabijanic, Mol. Cryst. Liq. Cryst. *15*, 371 (1972), C.A. *76*, 51154e (1972); Cholesterics. Ambient temperature effects. [**902**] P. G. Pick, J. Fabijanic, A. Stewart, Mol. Cryst. Liq. Cryst. *20*, 47 (1973), C.A. *78*, 102977b (1973); Effect of ambient light on liquid crystal tapes. [**903**] R. Pietsch, B. L. Lewis, U.S. 3,675,989 (26. 11. 1969); Liquid crystal optical cell with selected energy scattering. [**904**] V. A. Pilipovich, V. P. Kustov, A. V. Guk, P. I. Kolennikov, Vestsi Akad. Navuk B. SSR, Ser. Fiz.-Mat. Navuk *1976*, 110, C.A. *86*, 131036q (1977); Use of a mosaic liquid-crystal controlled transparency for optical information recording. [**905**] C. B. Pimentel, Rev. Quim. Ind. *40*, 18 (1971), C.A. *77*, 39924u (1972); Use of liquid crystals in technology. [**906**] P. Pinard, P. Leyral, Cah. Therm. *1973*, No. 3, II/3, C.A. *79*, 130081u (1973); Études fondamentales sur le comportement des cristaux liquides. [**907**] L. Pohl, R. Steinstraesser, Ger. Offen. 2,024,269 (19. 5. 1970), C.A. *76*, 72264f (1972); Low melting nematic mixtures. [**908**] L. Pohl, R. Steinsträsser, B. Hampel, 2. Arbeitstagung Flüssigkristalle, Inst. Angew. Festkörperphysik Freiburg, 21. 4. 1972; Erfahrungen bei der Herstellung von Anzeigeelementen mit nematischen Flüssigkristallen, insbesondere mit der nematischen Phase VA. [**909**] F. Poisson, Opt. Commun. *6*, 43 (1972), C.A. *78*, 22500a (1973); Nematic liquid crystal used as an instantaneous holographic medium. [**910**] J. J. Ponjeé, H. T. van Dam, 6. Arbeitstagung Flüssigkristalle, Inst. Angew. Festkörperphysik Freiburg, 2. 4. 1976; Einige interessante Eigenschaften von Viologenen im Zusammenhang mit elektrochromen Anzeigen.

[**911**] T. M. Ponomarienko, W. N. Sintsov, 1. Wiss. Konferenz über flüssige Kristalle, 17.–19. 11. 1970, Iwanowo, Sammlung der Vorträge (1972), Nr. 38; Schwellwert von Flüssigkristall-Displaysystemen. [**912**] A. P. Pontello, US Nat. Tech. Inform. Serv., AD Rep. *1971*, No. 886071, 26 pp., C.A. *76*, 156171m (1972); Nondestructive testing of coalescer elements on a continuous basis using liquid crystals. [**913**] A. P. Pontello, Am. Soc. Mech. Eng. *94*, 61 (1972); Nondestructive testing of coalescers (fuel (filters) on a continuous basis using liquid crystals. [**914**] A. P. Pontello, US Nat. Tech. Inform. Serv., AD Rep. 1973, No. 772099/8GA, 22 pp., C.A. *81*, 65933y (1974); Nondestructive testing of fuel

handling and filtration equipment using liquid crystals and thermography. [**915**] M. Pospisil, Slaboproudy Obz. *35*, 63 (1974), C.A. *81*, 19227y (1974); Contrast of a liquid crystal display. [**916**] R. G. Pothier, U.S. 3,713,156 (12. 10. 1970); Surface and subsurface detection device. [**917**] E. J. Poziomek, T. J. Novak, R. A. Mackay, Mol. Cryst. Liq. Cryst. *27*, 175 (1974), C.A. *82*, 78988r (1975); Use of liquid crystals as vapor detectors. [**918**] E. J. Poziomek, T. J. Novak, R. A. Mackay, Edgewood Arsenal Spec. Publ. (US Dep. Army) 1976, EO-SP-76001, C.A. *86*, 126168e (1977); Research in the use of liquid crystals in chemical detection. [**919**] J. Puhl, L. A. Golding, Med. Sci. Sports *6*, 67 (1974); Application of liquid crystal thermography to exercise thermoregulation. [**919a**] J. R. Quigley, W. J. Benton, Mol. Cryst. Liq. Cryst. *42*, 43 (1977); Optical properties of dispersed cholesterics. [**920**] T. Raad, J. E. Myers, AIChE J. *17*, 1260 (1971), C.A. *75*, 142257m (1971); Nucleation studies in pool boiling on thin plates using liquid crystals.

[**921**] M. I. Rackman, U.S. 3,716,658 (9. 4. 1970); Liquid-crystal television system. [**922**] M. J. Rafuse, U.S. 3,675,987 (29. 3. 1971), C.A. *77*, 120787y (1972); Liquid crystal compositions and devices. [**922a**] N. V. V. Raghavan, U.S. 4,018,507 (5. 9. 1974), C.A. *86*, 198031c (1977); Liquid crystal device and composition. [**923**] S. Raith, Microsc. Acta *77*, 230 (1975); Delineation of pn-junctions in silicon electronic devices by DAP-effect of nematics. [**924**] A. M. Raso, S. M. Raso, Minerva Medica *66*, 3985 (1975); Correlazioni tra angiografia e termografia a cristalli liquidi su placca nelle sindromi di Raynaud. [**925**] E. P. Raynes, Ger. Offen. 2,264,147 (29. 12. 1971), C.A. *79*, 99126m (1973); Liquid crystal materials for electrooptical devices. [**926**] E. P. Raynes, Fr. Demande 2,262,319 (21. 2. 1974), C.A. *85*, 114840j (1976); Liquid crystal devices. [**927**] E. P. Raynes, Fr. Demande 2,268,277 (1. 3. 1974), C.A. *85*, 151851j (1976); Liquid crystal device. [**928**] E. P. Raynes, Electron Lett. *9*, 101 (1973), C.A. *78*, 129709c (1973); TN liquid-crystal electro-optic devices with areas of reverse twist. [**929**] E. P. Raynes, Electron. Lett. *10*, 141 (1974), C.A. *81*, 31119x (1974); Improved contrast uniformity in TN liquid-crystal electrooptic display devices. [**930**] P. Raynes, Liq. Cryst. Disp. Appl., Symp. *1972*, 49, C.A. *84*, 36792h (1976); TN electrooptic effect.

[**931**] E. P. Raynes, I. A. Shanks, Electron. Lett. *10*, 114 (1974); Fast-switching TN electrooptical shutter and color filter. [**932**] E. P. Raynes, Rev. Phys. Appl. *10*, 117 (1975), C.A. *83*, 106168q (1975); Optically active additives in TN devices. [**933**] E. P. Raynes, D. K. Rowell, I. A. Shanks, Mol. Cryst. Liq. Cryst. *34*, 105 (1976), C.A. *86*, 131037r (1977); Liquid crystal surface alignment treatment giving controlled low angle tilt. [**933a**] E. P. Raynes, I. A. Shanks, Brit. 1,459,046 (15. 3. 1974), C.A. *87*, 14235p (1977); Liquid crystal devices. [**934**] RCA Corp., Brit. 1,246,847 (13. 12. 1967), C.A. *75*, 157045q (1971); Liquid crystal display element having storage. [**935**] RCA Corp., Brit. 1,263,277 (23. 4. 1968); Display apparatus. [**936**] RCA Corp., Brit. 1,263,278 (23. 4. 1968); Liquid crystal display apparatus. [**937**] RCA Corp., Brit. 1,304,554 (6. 11. 1969); Liquid crystal display device. [**938**] RCA Corp., Brit. 1,308,208 (7. 8. 1969); Display system. [**939**] RCA Corp., Brit. 1,308,237 (2. 6. 1969); Imines and their use in electro-optic compositions and devices. [**940**] RCA Corp., Brit. 1,315,239 (29. 12. 1970); Liquid crystal display device.

[**941**] RCA Corp., Neth. Appl. 74 15,507 (5. 12. 1973), C.A. *85*, 54633g (1976); Liquid crystal composition for electrooptical cells. [**942**] RCA Corp., Japan. 76 17,509 (28. 11. 1968), C.A. *85*, 184763k (1976); Electro-optical plate system using nematic liquid crystal. [**943**] Reichel, H. Krüger, 5. Arbeitstagung Flüssigkristalle, Inst. Angew. Festkörperphysik Freiburg (25. 4. 1975); Einfluß der Leitfähigkeitsdotierung auf die elektrooptischen Kenndaten dynamisch streuender Flüssigkristallanzeigen. [**943a**] P. G. Reif, A. D. Jacobson, W. P. Bleha, J. Grinberg, Proc. Soc. Photo-Opt. Instrum. Eng. *83*, 34 (1976), C.A. *87*, 60719x (1977); Hybrid liquid crystal light valve-image tube devices for optical data processing. [**944**] P. Rheinberger, E. Zollinger, Swiss 575,632 (21. 12. 1973), C.A. *85*, 102806q (1976); Resistant orientation layers for liquid crystals. [**945**] S. P. Richard, A. S. Marathay, J. Opt. Soc. Amer. *61*, 1559 (1971); Cholesterics for optical processing. [**946**] D. Rieck, A. Wagner, Proc. S. Dak. Acad. Sci. *50*, 287 (1971), C.A. *80*, 150491b (1974); Cholesteric liquid crystals. [**947**] J. Robert F. Gharadjedaghi, Centre d'Études Nucléaires de Grenoble, Note technique LETI/ME No. 818 (1972); Affichage monolithique d'cristaux liquides et transistors MOS. [**948**] J. Robert, Fr. Demande 2,186,294 (31. 3. 1972), C.A. *81*, 8442p (1974); Collective orientation of the molecules of a liquid crystal and cell containing them. [**948a**] J. Robert, B. Dargent, IEEE Trans. Electron Dev. *24*, 694 (1977), Multiplexing techniques for liquid crystal displays. [**948b**] P. F. Robusto, L. T. Lipton, IEEE Trans. Electron Dev. *23*, 1344 (1976), Multiplexing and contrast ratio optimization for matrix addressed LC displays. [**949**] M. Roder, K. Pinter, K. Ritvay, R. Schiller, Radiat. Res. *59*, 281 (1974); Radiation stability of liquid crystals. [**949a**] M. Roder, K. Pinter, L. Hodany, Radiochem. Radioanal. Lett. *27*, 321 (1976), C.A. *86*, 98948p

(1977); Radiation stability of cholesteryl benzoate. [**950**] D. Roszeitis, Diss. Aachen 1975; Transformation ebener Leuchtdichte-Bilder mit Hilfe elektrooptischer Masken (Flüssigkristall-Matrix).

[**951**] W. M. Rudakov, G. M. Martshuk, M. W. Muchina, 1. Wiss. Konferenz über flüssige Kristalle, 17.–19. 11. 1970, Iwanowo, Sammlung der Vorträge (1972), Nr. 34; Verwendung flüssiger Kristalle zur Sichtbarmachung elektromagnetischer Feldverteilungen in der Radiointerferometrie. [**952**] F. D. Saeva, J. J. Wysocki, Ger. Offen. 2,239,700 (22. 10. 1971), C.A. *79*, 11927q (1973); Optically useful liquid crystal compositions with a cholesteric mesophase. [**953**] F. D. Saeva, R. L. Schank, U.S. 3,917,481 (24. 2. 1972), C.A. *84*, 187553g (1976); Liquid crystal compositions between electrodes, one of which is a photoconductor. [**954**] F. D. Saeva, R. L. Schank, U.S. Reissue 28,806 (24. 2. 1972), C.A. *86*, 49221n (1977); Liquid crystal compositions. [**955**] F. D. Saeva, R. L. Schank, U.S. 3,931,041 (24. 2. 1972), C.A. *84*, 143081x (1976); Liquid crystal compositions. [**956**] M. Sagane, Japan. Kokai 73 84,787 (17. 2. 1972), C.A. *81*, 128697x (1974); Liquid-crystal composites to indicate electric field. [**957**] M. Sagane, Japan. Kokai 74 35,427 (7. 8. 1972), C.A. *81*, 137702x (1974); Liquid crystal coating compositions. [**958**] K. Sageshima, S. Furuta, Japan. Kokai 74,102,584 (6. 2. 1973), C.A. *82*, 163079q (1975); Liquid crystal composition for display devices. [**959**] Y. Saito, K. Kinugawa, K. Toriyama, H. Sakurada, Japan. Kokai 76 33,785 (17. 9. 1974), C.A. *85*, 134490y (1976); Nematic liquid crystal composite. [**960**] Y. Saito, K. Kinugawa, K. Toriyama, Japan. Kokai 76 39,584 (2. 10. 1974), C.A. *85*, 134367p (1976); Nematics for dynamic scattering-type display devices.

[**961**] Y. Saito, K. Kinukawa, K. Toriyama, Japan. Kokai 76 39,585 (2. 10. 1974), C.A. *85*, 134368q (1976); Nematic liquid crystal compositions for dynamic scattering-type display devices. [**962**] Y. Saito, K. Kinugawa, K. Toriyama, Japan. Kokai 76 42,090 (9. 10. 1974), C.A. *85*, 134487c (1976); Nematic liquid crystal compositions for matrix type display devices. [**963**] Y. Saito, K. Kinugawa, K. Toriyama, Japan. Kokai 76 68,481 (11. 12. 1974), C.A. *85*, 134500b (1976); Nematic liquid crystal compositions for display devices. [**964**] Y. Saito, K. Kinugawa, K. Toriyama, Japan. Kokai 76 68,482 (11. 12. 1974), C.A. *85*, 134499h (1976); Nematic liquid crystal compositions for TN display devices. [**965**] Y. Saito, K. Kinugawa, K. Toriyama, Japan. Kokai 76 97,585 (26. 2. 1975), C.A. *85*, 184882y (1976); Liquid crystal compositions. [**966**] T. Sakusabe, S. Kobayashi, Jap. J. Appl. Phys. *10*, 758 (1971), C.A. *75*, 55839j (1971); IR holography with liquid crystals. [**967**] R. L. Sampson, Brit. 1,439,244 (29. 9. 1972), C.A. *86*, 60411v (1977); Liquid crystalline compositions. [**968**] P. A. Sanders, J. Soc. Cosmet. Chem. *21*, 377 (1970), C.A. *73*, 29243z (1970); Stabilization of aerosol emulsions and foams. [**969**] A. Sasaki, K. Kurahashi, T. Takagi, J. Appl. Phys. *45*, 4356 (1974), C.A. *81*, 179776e (1974); Liquid crystal thermooptic effects and two new information display devices. [**970**] A. Sasaki, T. Morioka, T. Takagi, T. Ishibashi, IEEE Trans. Electron Dev. *22*, 805 (1975); Thermally addressed liquid-crystal display for dynamic figures.

[**971**] A. Sasaki, T. Morioka, T. Ishibashi, T. Takagi, Tech. Dig. — Int. Electron Devices Meet. *1975*, 401, C.A. *85*, 169653f (1976); Thermally addressed liquid-crystal display for dynamic and static figures. [**972**] A. Sasaki, T. Morioka, T. Ishibashi, T. Takagi, Proc. Conf. Solid State Dev. *7*, 121 (1975), C.A. *86*, 148768c (1977); Liquid-crystal electrothermo-optic effects and their application to display. [**972a**] A. Sasaki, T. Morioka, T. Ishibashi, T. Takagi, Jap. J. Appl. Phys. *15*, 121 (1976); Liquid-crystal electrothermo-optical effects and their application to displays. [**973**] K. Sasaki, R. Ito, Japan. Kokai 76 92,790 (14. 2. 1975), C.A. *85*, 184878b (1976); Liquid crystal electrooptical cells. [**974**] H. Sato, T. Inukai, K. Shoji, A. Takahashi, K. Furukawa, Japan. Kokai 74,123,991 (3. 4. 1973), C.A. *82*, 92070s (1975); Liquid crystal compositions for electrooptical display devices. [**975**] H. Sato, K. Morita, H. Takatsu, Japan. Kokai 76 13,733 (25. 7. 1974), C.A. *85*, 32615a (1976); Schiff base compositions. [**976**] H. Sato, M. Tazume, Y. Arai, Y. Fujita, Japan. Kokai 76 47,586 (22. 10. 1974), C.A. *85*, 134489e (1976); Nematic liquid crystal compositions for dynamic scattering mode display devices. [**976a**] H. Sato, M. Tazume, T. Yamaki, Y. Fujita, Y. Arai, Ger. Offen. 2,646,485 (15. 10. 1975), C.A. *87*, 93579e (1977); 8-Alkyl-1,8-diazabicyclo[5.4.0]undecenium benzoate-containing liquid crystal compositions. [**976b**] H. Sato, T. Inukai, S. Sugimori, Japan. Kokai 75,150,684 (25. 5. 1974), C.A. *84*, 172197v (1976); Nematic liquid crystal compositions for display devices. [**976c**] H. Sato, S. Sugimori, T. Inukai, Japan. Kokai 77 47,583 (15. 10. 1975), C.A. *87*, 144126q (1977); Nematic liquid crystal compositions with positive dielectric anisotropy. [**977**] S. Sato, M. Wada, Jap. J. Appl. Phys. *13*, 559 (1974), C.A. *81*, 56578j (1974); Frequency color display by nematics. [**978**] S. Sato, M. Wada, IEEE Trans. Electron Dev. *21*, 171 (1974), C.A. *80*, 151086k (1974); Liquid-crystal color light valve. [**979**] S. Sato, M. Wada, IEEE Trans. Electron Dev. *21*, 312 (1974), C.A. *81*, 31826a (1974); Liquid-crystal color display by DAP-TN double-layered structures. [**980**] S. Sato, M. Wada, Oyo Butsuri *44*, 263 (1975), C.A. *83*,

51511p (1975); Angular dependence of electrically induced birefringence in nematic liquid-crystal cells. [**980a**] S. Sato, A. Kikuchi, Oyo Butsuri *45*, 938 (1976), C.A. *86*, 179719s (1977); Light deflection by nematic liquid-crystal cells.

[**981**] K. Sawada, Y. Masuda, Japan. Kokai 74 45,886 (7. 9. 1972), C.A. *81*, 180006k (1974); Liquid crystal display devices. [**982**] K. Sawada, Y. Masuda, Japan. Kokai 74 95,881 (19. 1. 1973), C.A. *82*, 105237d (1975); Liquid crystal display compositions containing charge-transfer complexes. [**982a**] K. Sawada, T. Miyashita, T. Koizumi, Y. Masuda, Japan. Kokai 76,141,786 (3. 6. 1975), C.A. *86*, 198021z (1977); Time-division multiplex-drive liquid crystal display devices. [**983**] K. Sawatari, T. Mukai, K. Tsukinuki, O. Nishida, Japan. Kokai 75 01,082 (8. 5. 1973), C.A. *83*, 35757t (1975); Stabilization of cholesteric liquid crystal compositions. [**984**] K. Sawatari, T. Mukai, K. Tsukinuki, O. Nishida, Japan. Kokai 75 32,082 (25. 7. 1973), C.A. *83*, 106255r (1975); Stabilized cholesteric liquid crystal compositions. [**985**] K. Sawatari, T. Mukai, K. Tsukinuki, T. Seto, Japan. Kokai 75 80,283 (19. 11. 1973), C.A. *84*, 152673u (1976); Cholesteric liquid crystal compositions for temperature indicating devices. [**986**] A. C. Saxman, K. H. Leners, R. V. Wick, Optoelectron. Laser Technol., IEEE Reg. Six Conf. Rec. *1974*, 5, C.A. *85*, 102027m (1976); Use of liquid crystals in profiling IR laser beams. [**987**] L. C. Scala, G. D. Dixon, D. F. Ciliberti, US Clearinghouse Fed. Sci. Tech. Inform., AD *1970*, No. 716003, 69 pp., C.A. *75*, 56011v (1971); Basic studies of liquid crystals as related to electrooptical and other devices. [**988**] M. Schadt, F. Müller, J. Chem. Phys. *65*, 2224 (1976), C.A. *85*, 185011a (1976); Influence of solutes on material constants of liquid crystals and on electro-optical properties of TN displays. [**988a**] R. T. Schaum, US NTIS, AD Rep. *1976*, AD-AO32322, 64 pp., C.A. *87*, 25189c (1977); Development of a non-destructive inspection technique for advanced composite materials using cholesterics. [**989**] T. J. Scheffer, J. Appl. Phys. *44*, 4799 (1973), C.A. *80*, 31469r (1974); New multicolor liquid crystal displays that use a TN electrooptical cell. [**990**] T. J. Scheffer, J. Phys. D *8*, 1441 (1975), C.A. *83*, 155704v (1975); TN display with cholesteric reflector. [**990a**] T. J. Scheffer, J. Nehring, IEEE Trans. Electron Dev. *24*, 816 (1977), Optimum polarizer combinations for TN displays.

[**991**] B. Scheurle, H. Kelker, Ger. Offen. 2,009,528 (28. 2. 1970), C.A. *75*, 134254t (1971); Nematogenic mixtures. [**992**] B. Scheurle, H. Kelker, Ger. Offen. 2,032,566 (28. 2. 1970), C.A. *76*, 51370x (1972); Nematogenic caproyloxybenzoate mixtures. [**993**] M. Schiekel, Funkschau *1972*, Heft 1, 17; Anzeigevorrichtungen für die Datenverarbeitung und andere Zwecke. [**994**] M. F. Schiekel, K. Fahrenschon, Soc. Inform. Display, Symp. San Francisco 1972, 98; Multicolor matrix-displays based on the deformation of vertically-aligned nematics. [**995**] M. Schiekel, Funkschau *47*, Hefte 1–6 (1975); Thermotrope Flüssigkristalle. [**996**] M. Schiekel, K. Fahrenschon, H. Gruler, Appl. Phys. 7, 99 (1975); Transient times and multiplex behavior of nematics in the electric field. [**997**] M. Schiekel, NTZ *28*, 189 (1975), Moderne Anzeigetechniken. [**998**] M. Schiekel, Elektroniker *15*, EL1 (1976); Stand und Trend bei Anzeige-Elementen mit Flüssigkristallen. [**999**] O. Schirmer, V. Wittwer, 6. Arbeitstagung Flüssigkristalle, Inst. Angew. Festkörperphysik Freiburg, 2. 4. 1976; Ursache der elektrochromen Verfärbung in WO₃ und MoO₃. [**1000**] O. F. Schirmer, V. Wittwer, 7. Arbeitstagung Flüssigkristalle, Inst. Angew. Festkörperphysik Freiburg, 4. 3. 1977, Dependence of WO₃ electrochromic absorption on crystallinity.

[**1001**] W. Schmidt, Elektronikpraxis 6, 7, 12 (1971); Der Flüssigkristall — potentieller Konkurrent der GaAs-Diode? [**1002**] A. Schneider, 2. Arbeitstagung Flüssigkristalle, Inst. Angew. Festkörperphysik Freiburg, 21. 4. 1972; Fertigungsverfahren von Flüssigkristallanzeigen. [**1003**] P. U. Schulthess, Swiss 520,939 (27. 8. 1970), C.A. *77*, 106805w (1972); Detector apparatus with a flat cholesteric layer. [**1004**] D. A. Schulz, U.S. 3,919,376 (26. 12. 1972), C.A. *84*, 19051y (1976); High mesophase content pitch fibers. [**1005**] A. H. Schwartz, Rev. Sci. Instrum. *42*, 1528 (1971); Apparatus for recording light scattering pattern of liquid crystals. [**1006**] F. Scudieri, M. Bertolotti, A. Ferrari, D. Apostol, Opt. Commun. *15*, 54 (1975); Nematic liquid crystal pi-filter for difference holography. [**1007**] F. Scudieri, A. Ferrari, M. Bertolotti, D. Apostol, Opt. Commun. *15*, 57 (1975); Opto-acoustic modulator with a nematic. [**1008**] F. Scudieri, Opt. Commun. *18*, 84 (1976), C.A. *85*, 101437h (1976); Opto-acoustic modulation of light with liquid crystals. [**1009**] F. Scudieri, S. Verginelli, A. Ferrari, Appl. Phys. *11*, 103 (1976), C.A. *85*, 134798e (1976); Optical visualization of deformations with nematics. [**1010**] O. S. Selawry, H. S. Selawry, J. F. Holland, Mol. Cryst. *1*, 495 (1966), C.A. *65*, 20502d (1966); Use of cholesterics for thermographic measurement of skin temperature in man. [**1011**] W. B. Sergejev, A. A. Groshev, 1. Wiss. Konferenz über flüssige Kristalle, 17.–19. 11. 1970, Iwanowo, Sammlung der Vorträge (1972), Nr. 33; Probleme beim Aufbau von Flüssigkristall-Displaysystemen für radiotechnische Zwecke. [**1012**] J. C. Sethares, M. R. Stiglitz, Appl. Opt. *8*, 2560 (1969);

Visual observation of high dielectric resonator modes. [**1013**] H. Seto, M. Ueda, H. Segawa, U.S. 3,704,625 (28. 12. 1970); Thermometer using liquid crystal compositions. [**1014**] I. A. Shanks, Electron. Engng. *46*, 30 (1974); Liquid crystal materials and device developments. [**1015**] I. A. Shanks, Electron. Lett. *10*, 90 (1974); Electrooptical color effects by twisted nematics. [**1016**] I. A. Shanks, Electron. Power *21*, 301 (1975); Properties and prospects of liquid-crystal displays. [**1017**] G. M. Sharkova, A. W. Lokotko, 1. Wiss. Konferenz über flüssige Kristalle, 17.–19. 11. 1970, Iwanowo, Sammlung der Vorträge (1970), Nr. 35; Untersuchung einiger flüssiger Kristalle im Windkanal. [**1018**] E. N. Sharpless, F. Davis, U.S. 3,647,279 (27. 5. 1970); Color display devices. [**1019**] E. N. Sharpless, S. African 74 04,168 (2. 7. 1973), C.A. *84*, 80622b (1976); Thermometric compositions including inert additives. [**1020**] H. E. Shaw, Jr., U.S. 3,590,371 (31. 12. 1969); Method utilizing the color change with temperature of a material for detecting discontinuities in a conductor member embedded within a windshield. [**1020a**] S. V. Shevchuk, A. P. Makhotilo, V. G. Tishchenko, Kholestericheskie Zhidk. Krist. *1976*, 67, C.A. *87*, 175846d (1977); Thermoindicator films containing cholesterics.

[**1021**] K. Shibayama, H. Ono, Japan. Kokai 72 31,880 (11. 3. 1971), C.A. *79*, 10918a (1973); Benzylideneamino derivative liquid crystal compositions. [**1022**] K. Shibayama, H. Ono, Japan. Kokai 72 31,881 (11. 3. 1971), C.A. *79*, 10914w (1973); Carbonic ester liquid compositions. [**1023**] K. Shibayama, H. Ono, Japan. Kokai 72 31,882 (11. 3. 1971), C.A. *79*, 10915x (1973); Benzylideneamino derivative liquid crystal compositions. [**1024**] K. Shibayama, H. Ono, Japan. Kokai 72 31,883 (11. 3. 1971), C.A. *79*, 10911t (1973); Benzylideneamino derivative liquid crystal compositions. [**1025**] K. Shibayama, H. Ono, Japan. Kokai 72 31,884 (11. 3. 1971), C.A. *79*, 10910s (1973); Benzylideneamino derivative liquid crystal compositions. [**1026**] K. Shibayama, H. Ono, Japan. Kokai 72 31,885 (11. 3. 1971), C.A. *79*, 10913v (1973); Liquid crystal compositions based on substituted phenoxycarbonylphenyl carbonates. [**1027**] K. Shibayama, H. Ono, Japan. Kokai 72 31,886 (11. 3. 1971), C.A. *79*, 10916y (1973); Terephthalate derivative based liquid crystal compositions. [**1028**] K. Shibayama, H. Ono, Japan. Kokai 74 37,885 (14. 8. 1972), C.A. *81*, 97827h (1974); Nematic liquid crystal compositions. [**1029**] K. Shibayama, H. Ono, Japan. Kokai 74 37,886 (14. 8. 1972), C.A. *81*, 97826g (1974); Liquid crystal compositions for room temperature use. [**1030**] K. Shibayama, H. Ono, Japan. Kokai 74 37,887 (15. 8. 1972), C.A. *81*, 180007m (1974); Nematic liquid crystal compositions for room temperature use.

[**1031**] K. Shibayama, H. Ono, Japan. 74 38,430 (25. 3. 1970), C.A. *82*, 178982c (1975); Nematic liquid crystal composition. [**1032**] K. Shibayama, H. Ono, Japan. 74 45,063 (27. 3. 1970), C.A. *82*, 178987h (1975); Nematic liquid crystal composition. [**1033**] K. Shibayama, H. Ono, Japan. 74 45,149 (14. 4. 1970), C.A. *83*, 106737z (1975); Nematic liquid crystal. [**1034**] K. Shibayama, H. Ono, Japan. 75 05,999 (25. 3. 1970), C.A. *83*, 106738a (1975); Nematic liquid crystal composition. [**1035**] K. Shibayama, H. Ono, Japan. 75 10,554 (25. 3. 1970), C.A. *83*, 106741w (1975); Nematic liquid crystal composition. [**1036**] K. Shibayama, H. Ono, Japan. 75 10,555 (25. 3. 1970), C.A. *83*, 106740v (1975); Nematic liquid crystal composition. [**1037**] K. Shibayama, H. Ono, Japan. 76 19,433 (11. 3. 1971), C.A. *86*, 36366p (1977); Terephthalate derivative-containing liquid crystal compositions. [**1038**] K. Shibayama, H. Ono, Japan. Kokai 76 44,582 (11. 3. 1971), C.A. *86*, 131184m (1977); Nematic liquid crystal compositions for display devices. [**1038a**] K. Shibayama, H. Ono, Japan. 77 14,237 (25. 3. 1970), C.A. *87*, 125393h (1977); Nematic liquid crystal compositions for display devices. [**1039**] Y. Shibayama, H. Ono, Japan. 75 28,917 (14. 4. 1970), C.A. *84*, 82896x (1976); Nematic liquid crystal composition. [**1040**] Y. Shibayama, H. Ono, Japan. 75 28,918 (14. 4. 1970), C.A. *84*, 82987y (1976); Nematic liquid crystal composition.

[**1041**] T. Shibutani, T. Hamanaka, Japan. Kokai 72 21,386 (24. 2. 1971), C.A. *79*, 151689p (1973); Liquid crystal compositions for electrooptical display apparatus. [**1042**] T. Shimomura, N. Yoshikawa, T. Murakami, Japan. 74 46,235 (11. 3. 1970), C.A. *83*, 186398c (1975); Cholesteryl liquid crystal composition. [**1043**] T. Shimomura, H. Mada, K. Uehara, S. K. Min, S. Kobayashi, Jpn. J. Appl. Phys. *14*, 1093 (1975); Voltage controllable color formation with a TN liquid crystal cell. [**1044**] T. Shimomura, H. Mada, S. Kobayashi, Jpn. J. Appl. Phys. *15*, 1479 (1976), C.A. *85*, 114210d (1976); Electrooptical properties of a nematic liquid crystal cell for obliquely incident light. [**1044a**] T. Shimomura, H. Mada, S. Kobayashi, Jpn. J. Appl. Phys. *15*, 1815 (1976), C.A. *86*, 163564b (1977); Angular dependence of voltage controlled color formation with a TN liquid crystal cell. [**1044b**] T. Shimomura, H. Mada, S. Kobayashi, Jpn. J. Appl. Phys. *16*, 1431 (1977), C.A. *87*, 175632f (1977); Electro-optical color effect of an off-ninety degree TN liquid crystal cell. [**1044c**] T. Shimomura, H. Mada, S. Kobayashi, Terebijon *31*, 480 (1977), C.A. *87*, 159986u (1977); Electrooptical color effect of nematic liquid crystal cell to obliquely incident light. [**1045**] T. Shimura, F. Hori, Japan. Kokai 74 02,789 (28. 4. 1972), C.A. *80*, 139580j (1974); Liquid crystal compositions. [**1046**] M. Shlens, M. R. Stoltz, A. Benjamin, West. J.

Med. *122*, 367 (1975); Orthopedic applications of liquid crystal thermography. [**1047**] K. Shoji, H. Sato, A. Takahashi, Japan. Kokai *74* 52,785 (22. 9. 1972), C.A. *81*, 180011h (1974); Liquid crystal electrooptical devices. [**1048**] W. M. Shoshin, 1. Wiss. Konferenz über flüssige Kristalle, 17.–19. 11. 1970, Iwanowo, Sammlung der Vorträge (1972), Nr. 36; Verwendung flüssiger Kristalle in Informationsspeichern. [**1049**] V. M. Shoshin, Y. P. Bobylev, Televizionnye Metody Ustroistva Otobrazheniya, Inf. *1975*, 193, C.A. *86*, 49113d (1977); Possible use of liquid crystal for display devices. [**1050**] Siemens AG, Fr. Demande 2,300,350 (10. 2. 1975), C.A. *86*, 148853b (1977); Liquid crystal cell with a glass sealed closing.

[**1051**] W. A. Simpson, W. E. Deeds, Appl. Opt. *9*, 499 (1970), C.A. *72*, 94809g (1970); Real-time visual reconstruction of IR holograms. [**1052**] L. S. Singer, U.S. 3,919,387 (26. 12. 1972), C.A. *84*, 19050x (1976); High mesophase content pitch fibers. [**1053**] W. M. Sintsov, 1. Wiss. Konferenz über flüssige Kristalle, 17.–19. 11. 1970, Iwanowo, Sammlung der Vorträge (1972), Nr. 37; Analogie zwischen den optischen Eigenschaften von Raumhologrammen und Domänen in cholesterinischen Flüssigkristallen. [**1054**] D. W. Skelly, G. J. Sewell, U.S. 3,972,589 (23. 6. 1972), C.A. *86*, 36362j (1977); Nematic liquid crystal mixtures with stable homeotropic boundary conditions. [**1055**] W. Sliwka, Angew. Chem. *87*, 556 (1975), C.A. *84*, 31775v (1976); Microverkapselung. [**1056**] C. W. Smith, D. G. Gisser, M. Young, S. R. Powers, Jr., Appl. Phys. Lett. *24*, 453 (1974), C.A. *81*, 27610h (1974); Liquid-crystal optical activity for temperature sensing. [**1057**] G. W. Smith, D. B. Hayden, U.S. 3,848,966 (15. 10. 1973), C.A. *82*, 92059v (1975); Homeotropic alignment additive for liquid crystals. [**1057a**] P. Smoranc, S. Pirkl, Česk. Časopis Fis. *26*, 312 (1976); Contact thermography with liquid crystals in medicine. [**1058**] M. Sone, K. Toriyama, Y. Toriyama, Appl. Phys. Lett. *24*, 115 (1974), C.A. *81*, 160892t (1974); Liquid-crystal Lichtenberg figure. [**1059**] A. S. Sonin, B. M. Stepanov, Priroda *1974*, 14, C.A. *82*, 78794z (1975); Instruments based on liquid crystals. [**1059a**] A. S. Sonin, I. N. Shibaev, M. I. Epshtein, Kvantovaya Elektron. *4*, 531 (1977), C.A. *87*, 76130j (1977); Large-dynamic-range visualizer. [**1060**] M. S. Sonin, V. S. Nefedov, N. F. Cherenkova, I. N. Kompanets. Mikroelektronika (Akad. Nauk SSSR) *5*, 284 (1976), C.A. *85*, 114722x (1976); Liquid crystal cell controlled by an MIS-IC.

[**1061**] R. A. Soref, U.S. 3,675,988 (25. 11. 1969); Liquid crystal electro-optical measurement and display devices. [**1062**] R. A. Soref, M. J. Rafuse, Fr. Demande 2,152,757 (9. 9. 1971), C.A. *79*, 99123h (1973); Compositions for electrooptical devices. [**1063**] R. A. Soref, J. Appl. Phys. *41*, 3022 (1970), C.A. *73*, 49485h (1970); Thermo-optic effects in nematic-cholesteric mixtures. [**1064**] R. A. Soref, Appl. Opt. *9*, 1323 (1970); Electronically scanned analog liquid crystal displays. [**1065**] R. A. Soref, Phys. Opto-Electron. Mater., Proc. Symp. *1970*, 207, C.A. *77*, 25748c (1972); Liquid-crystal light-control experiments. [**1066**] R. A. Soref, Proc. Soc. Photo-Opt. Instrum. Eng. *38*, 23 (1973), C.A. *81*, 84317s (1974); Liquid crystals. [**1067**] R. A. Soref, Proc. IEEE *62*, 1710 (1974); Interdigital TN displays. [**1067a**] H. Sorkin, U.S. 4,003,844 (17. 11. 1975), C.A. *87*, 60819e (1977); Liquid crystal devices. [**1068**] G. J. Sprokel, U.S. 3,882,039 (21. 6. 1973), C.A. *83*, 51460w (1975); Additive for liquid crystal materials. [**1069**] G. J. Sprokel, Fr. Demande 2,255,665 (19. 12. 1973), C.A. *84*, 143080w (1976); Display device using liquid crystals. [**1069a**] G. J. Sprokel, R. M. Gibson, J. Electrochem. Soc. *124*, 557 (1977), C.A. *86*, 180963y (1977); Liquid crystal alignment produced by RF plasma deposited films. [**1069b**] G. J. Sprokel, U.S. 3,999,838 (5. 6. 1975), C.A. *86*, 197948v (1977); Beam-addressed liquid crystal cells. [**1070**] A. M. J. Spruijt, 5. Arbeitstagung Flüssigkristalle, Inst. Angew. Festkörperphysik Freiburg, 25. 4. 1975; Structure of a $S = \frac{1}{2}$ disclination.

[**1071**] A. M. J. Spruijt, 6. Arbeitstagung Flüssigkristalle, Inst. Angew. Festkörperphysik Freiburg, 2. 4. 1976; Surface coupling of the director in nematics. [**1072**] R. W. Stabr, H. G. Franke, U.S. 3,795,516 (13. 11. 1972), C.A. *80*, 126809r (1974); Barrier layer for liquid crystal imaging elements. [**1073**] K. Stahl, Optik *27*, 11 (1968), C.A. *69*, 6318f (1968); IR-Detektoren. [**1074**] S. Stanevicius, I. Grinis, Sb. Dokl. Vses. Nauch Konf. Zhidk. Krist. Simp. Ikh Prakt. Primen, 2nd *1972*, 290, C.A. *82*, 118164c (1975); Projecting an image with a liquid crystal converter onto a screen. [**1075**] H. Stegemeyer, VDI Ber. *198*, 29 (1973), C.A. *83*, 198606g (1975); Anwendung von cholesterischen, flüssigen Kristallen zur Darstellung von Temperaturfeldern. [**1076**] E. L. Steiger, H. J. Dietrich, U.S. 3,742,053 (11. 6. 1971), C.A. *79*, 78368b (1973); Mesomorphic aromatics. [**1077**] C. R. Stein, R. A. Kashnow, Appl. Phys. Lett. *19*, 343 (1971), C.A. *76*, 29032s (1972); Two-frequency coincidence addressing scheme for nematic-liquid-crystal displays. [**1078**] C. R. Stein, R. A. Kashnow, Soc. Inform. Display, Symp. San Francisco *1972*, 64; Recent advances in frequency coincidence matrix addressing of liquid crystal displays. [**1079**] R. J. Stein, U.S. 3,666,881 (28. 12. 1970); Electro-optical display device employing liquid crystals. [**1080**] C. E. Stephens, F. N. Sinnadurai, J. Phys. E *7*, 641 (1974), C.A. *81*, 123479t (1974); Surface temperature limit detector using nematics with an application to microcircuits.

[**1081**] E. Stepke, Electrooptical Systems Design, Febr. *1972*, 20; Liquid crystals: perspectives, prospects and products. [**1082**] H. A. Stern, U.S. 3,862,830 (18. 7. 1973), C.A. *82*, 128640u (1975); Vitreous enclosures for liquid crystal cells. [**1083**] H. J. Stocker, US Nat. Tech. Inform. Serv., AD Rep. *1972*, No. 752201, 25 pp., C.A. *78*, 130390y (1973); Liquid crystal displays for the evaluation of carbon dioxide lasers. [**1084**] L. B. Stotts, A. Sussman, M. A. Monahan, Appl. Opt. *13*, 1752 (1974); Photoactivated TN device. [**1085**] D. E. Strandness, Jr., Manuskript Univ. Washington, med. Fakultät; Kontrolle der Hauttemperatur mit Hilfe von Flüssigkristallen. Möglichkeiten der Anwendung bei direkten Arterienoperationen. [**1086**] E. L. Strebel, Ger. Offen. 2,627,180 (16. 6. 1975), C.A. *86*, 108174z (1977); Lyotropic nematic compositions and sheet materials, especially transparent tape. [**1087**] M. Sukie, K. Honda, Senryo To Yakuhin *17*, 335 (1972), C.A. *78*, 49388g (1973); Electrooptical effect of liquid crystals and its use. [**1088**] A. Sussman, U.S. 3,938,242 (27. 9. 1973), C.A. *85*, 102427s (1976); Liquid crystal devices. [**1089**] A. Sussman, IEEE Trans. Parts, Hybrids, Packag. *8*, 24 (1972), C.A. *78*, 89750a (1973); Electrooptic liquid-crystal devices. Principles, applications. [**1090**] A. Sussman, Liq. Cryst. Plast. Cryst. *1*, 338 (1974), C.A. *83*, 35583h (1975); Liquid crystals in display systems.

[**1091**] A. Sussman, Introd. Liq. Cryst. *1974*, 297, C.A. *84*, 97736q (1976); Electro-optic transfer function in nematics. [**1092**] R. C. Sutton, U.S. 3,795,517 (13. 11. 1972), C.A. *80*, 126808q (1974); Barrier layer for liquid crystal imaging elements. [**1093**] Suwa Seikosha, Fr. Demande 2,196,199 (16. 8. 1972), C.A. *81*, 71114y (1974); Orienting a liquid crystal. [**1094**] Suwa Seikosha, Brit. 1,406,363 (13. 7. 1972), C.A. *84*, 10915a (1976); Aryl cyanobenzoate nematic liquid crystal materials and electrooptical display devices. [**1094a**] F. K. Suzuki, T. W. Davison, U.S. 4,015,591 (17. 2. 1976), C.A. *87*, 29058g (1977); Cholesteric liquid crystalline phase material-dye composition and venapuncture method employing the composition. [**1095**] T. Suzuki, Y. Yamazaki, S. Kanbe, Japan. Kokai 74 78,684 (6. 12. 1974), C.A. *82*, 92078a (1975); Azoxybenzene and phenyl benzoate type liquid crystals for display devices. [**1096**] T. Suzuki, Japan. Kokai 75,122,478 (14. 3. 1974), C.A. *84*, 187549k (1976); Nematic liquid crystal compositions for electrooptical display devices. [**1097**] T. Suzuki, Japan. Kokai 76 08,182 (10. 7. 1974), C.A. *85*, 134479b (1976); Nematic liquid crystal composition for TN display device. [**1098**] T. Suzuki, T. Yamazaki, Japan. 76 20,475 (24. 12. 1970), C.A. *86*, 113736p (1977); Electrooptical display device. [**1098a**] T. Suzuki, Japan. 76 19,832 (15. 6. 1971), C.A. *87*, 14279f (1977); Electrooptical display devices. [**1099**] L. V. Syt'ko, E. K. Belousov, A. M. Sapozhkov, I. G. Chistyakov, L. K. Vistin, Sb. Dokl. Vses. Nauch. Konf. Zhidk. Krist. Simp. Ikh Prakt. Primen., 2nd, *1972*, 281, C.A. *81*, 97723w (1974); Liquid crystal display devices. [**1100**] A. Szymanski, Pomiary, Automat., Kontr. *16*, 145 (1970), C.A. *74*, 7562w (1971); Application of liquid crystals in instrumentation and industrial control systems. [**1100a**] M. Taguchi, Y. Katagiri, Japan. Kokai 77 17,379 (1. 8. 1975), C.A. *87*, 76417h (1977); Nematic liquid crystal compositions for display devices.

[**1101**] H. Takahashi, M. Fukai, H. Tatsuta, A. Moriyama, Japan. Kokai 75 73,887 (31. 10. 1973), C.A. *84*, 82586s (1976); Nematic liquid crystal composition containing an anthraquinone dye for optical display devices. [**1102**] H. Takahashi, H. Tatsuta, M. Fukai, Japan. Kokai 75,114,389 (19. 2. 1974), C.A. *84*, 97870d (1976); Nematic liquid crystal compositions for display devices. [**1103**] H. Takahashi, H. Tatsuta, M. Fukai, Japan. Kokai 75,133,182 (9. 4. 1974), C.A. *85*, 27388z (1976); Nematic liquid crystal compositions for electrooptical display devices. [**1104**] H. Takahashi, H. Tatsuta, M. Fukai, Japan. Kokai 75,133,183 (9. 4. 1974), C.A. *85*, 12379y (1976); Nematic liquid crystal compositions for electrooptical display devices. [**1105**] H. Takahashi, H. Tatsuta, M. Fukai, Japan. Kokai 75,133,184 (9. 4. 1974), C.A. *85*, 27389a (1976); Nematic liquid crystal compositions for electrooptical display devices. [**1106**] H. Takahashi, H. Tatsuta, M. Fukai, Japan. Kokai 75,133,185 (9. 4. 1974), C.A. *85*, 27391v (1976); Nematic liquid crystal compositions for electrooptical display devices. [**1107**] H. Takahashi, H. Tatsuta, M. Fukai, Japan. Kokai 75,133,186 (9. 4. 1974), C.A. *85*, 27390u (1976); Nematic liquid crystal compositions for electrooptical display devices. [**1108**] H. Takahashi, H. Tatsuta, M. Fukai, Japan. Kokai 75,133,187 (9. 4. 1974), C.A. *85*, 27393x (1976); Nematic liquid crystal compositions for electrooptical display devices. [**1109**] H. Takahashi, H. Tatsuta, M. Fukai, Japan. Kokai 75,133,188 (9. 4. 1974), C.A. *85*, 27392w (1976); Nematic liquid crystal compositions for electrooptical display devices. [**1110**] H. Takahashi, H. Tatsuta, M. Fukai, Japan. Kokai 75,133,189 (9. 4. 1974), C.A. *85*, 12381t (1976); Nematic liquid crystal compositions for electrooptical display devices.

[**1111**] H. Takahashi, H. Tatsuta, M. Fukai, Japan. Kokai 75,133,190 (9. 4. 1974), C.A. *85*, 12380s (1976); Nematic liquid crystal compositions for electrooptical display devices. [**1112**] I. W.

Takahashi, I. U. Yamada, Ger. Offen. 2,226,633 (31. 5. 1972); Flüssigkristall-Anzeigevorrichtung. [1113] H. Takata, K. Murase, O. Kogure, K. Kato, Denki Tsushin Kenkyujo Kenkyu Jitsuyoka Hokoku 22, 1413 (1973), C.A. 81, 8356p (1974); Liquid crystal matrix displays. [1114] H. Takata, O. Kogure, K. Murase, IEEE Trans. Electron Dev. 20, 990 (1973), C.A. 80, 64935t (1974); Matrix-addressed liquid-crystal display. [1115] T. Tako, S. Masubuchi, T. Akahane, T. Nakada, Mol. Cryst. Liq. Cryst. 38, 303 (1977), C.A. 86, 181594r (1977); New type of analog voltmeter using electrooptic effects in nematics. [1116] T. Tanaka, Zairyo 20, 54 (1971), C.A. 76, 51373a (1972); New materials for electronics. [1117] C. Tani, M. Kozutsumi, Y. Miyazawa, Japan. 74 23,106 (22. 7. 1970), C.A. 82, 24780y (1975); Liquid crystal composition. [1118] C. Tani, Y. Ohnishi, M. Kozutsumi, Y. Miyazawa, Japan. 74 23,109 (29. 9. 1970), C.A. 82, 24778d (1975); Liquid crystal composite. [1119] K. Tani, Japan. 74 27,510 (6. 4. 1970), C.A. 82, 79180h (1975); Liquid crystal composite. [1120] L. E. Tannas, Jr., Electron. Design 14, July 5, 1974, p. 76; Liquid crystal displays are great-but... [1120a] L. E. Tannas, P. K. York, Ferroelectrics 10, 19 (1976); LC ferroelectric matrix addressable display.

[1121] H. A. Tarry, Electron. Lett. 11, 339 (1975), C.A. 84, 97777d (1976); Effect of temperature on the transient response of TN films. [1122] H. Tarry, Electron. Lett. 11, 471 (1975); Electrically tunable narrowband optical filter. [1123] H. Tatsuta, M. Fukai, K. Asai, A. Moriyama, Japan. Kokai 74 27,494 (6. 7. 1972), C.A. 81, 71120x (1974); Nematic liquid crystal compositions for display devices. [1124] H. Tatsuta, Japan. Kokai 74 62,390 (17. 10. 1972), C.A. 82, 37357r (1975); Liquid crystal display device. [1125] H. Tatsuta, M. Fukai, K. Asai, H. Takahashi, Japan. Kokai 74 63,672 (20. 10. 1972), C.A. 82, 37346m (1975); Nematics based on phenyl esters of benzoic acid for display devices. [1126] H. Tatsuta, M. Fukai, K. Asai, H. Takahashi, Japan. Kokai 74 63,673 (20. 10. 1972), C.A. 82, 78787z (1975); Nematics based on phenyl esters of benzoic acid for electrooptical display devices. [1127] H. Tatsuta, M. Fukai, K. Asai, H. Takahashi, Japan. Kokai 74 63,677 (20. 10. 1972), C.A. 82, 78788a (1975); Phenyl benzoate-type nematics for liquid crystal electrooptical display devices. [1128] H. Tatsuta, M. Fukai, K. Asai, H. Takahashi, Japan. Kokai 74 63,678 (20. 10. 1972), C.A. 82, 92076y (1975); Nematic liquid crystal compositions based on phenyl esters of benzoic acid for electrooptical display devices. [1129] H. Tatsuta, M. Fukai, K. Asai, H. Takahashi, Japan. Kokai 74 63,679 (20. 10. 1972), C.A. 82, 10034w (1975); Nematic liquid crystal compositions based on phenyl esters of benzoic acid for electrooptical display devices. [1130] H. Tatsuta, M. Fukai, K. Asai, H. Takahashi, Japan. Kokai 74 63,680 (20. 10. 1972), C.A. 82, 10045a (1975); Nematic liquid crystal compositions based on phenyl esters of benzoic acid for electrooptical display devices.

[1131] H. Tatsuta, M. Fukai, K. Asai, H. Takahashi, Japan. Kokai 74 64,579 (25. 10. 1972), C.A. 83, 139895c (1975); Nematic liquid crystal compositions for electrooptical display devices based on benzoic acid phenyl esters. [1132] H. Tatsuta, M. Takai, K. Asai, H. Takahashi, Japan. Kokai 74 64,580 (25. 10. 1972), C.A. 82, 37358s (1975); Nematic liquid crystal compositions based on benzoic acid phenyl ester derivatives for display devices. [1133] H. Tatsuta, M. Fukai, K. Asai, H. Takahashi, Japan. Kokai 74 64,584 (25. 10. 1972), C.A. 82, 118272m (1975); Nematic liquid crystal compositions based on benzoic acid phenyl ester derivates for electrooptical display devices. [1134] H. Tatsuta, M. Fukai, K. Asai, H. Takahashi, Japan. Kokai 74 65,390 (25. 10. 1972), C.A. 81, 180009p (1974); Nematic liquid crystal display device. [1135] H. Tatsuta, M. Fukai, K. Asai, H. Takahashi, Japan. Kokai 74 65,479 (25. 10. 1972), C.A. 82, 105241a (1975); Nematic liquid crystal compositions for electrooptical display devices based on benzoic acid phenyl esters. [1136] H. Tatsuta, M. Fukai, H. Takahashi, Japan. Kokai 74 74,182 (17. 11. 1972), C.A. 82, 148548y (1975); Liquid crystal electrooptical display device. [1137] H. Tatsuta, M. Fukai, K. Asai, H. Takahashi, Japan. Kokai 74 74,185 (17. 11. 1972), C.A. 82, 163073h (1975); Liquid crystal electrooptical display device using phenyl benzoate-type nematic compositions. [1138] H. Tatsuta, M. Fukai, K. Asai, H. Takahashi, Japan. Kokai 74 74,186 (17. 11. 1972), C.A. 82, 148545v (1975); Electrooptic display devices using phenyl benzoate type nematic liquid crystal compositions. [1139] H. Tatsuta, M. Fukai, K. Asai, H. Takahashi, Japan. Kokai 74 74,191 (17. 11. 1972), C.A. 82, 163080h (1975); Liquid crystal display device based on phenyl-benzoate type nematics. [1140] H. Tatsuta, M. Fukai, K. Asai, H. Takahashi, Japan. Kokai 74 75,470 (25. 10. 1972), C.A. 82, 10035x (1975); Optical display devices employing phenyl benzoate liquid crystals.

[1141] H. Tatsuta, M. Fukai, K. Asai, H. Takahashi, Japan. Kokai 74 76,484 (28. 11. 1972), C.A. 82, 10043y (1975); Liquid crystal display devices based on phenyl benzoate esters. [1142] H. Tatsuta, M. Fukai, K. Asai, Japan. Kokai 74 76,786 (28. 11. 1972), C.A. 82, 10036y (1975); Optical devices based on phenyl benzoate type liquid crystal composition. [1143] H. Tatsuta, M. Fukai, K. Asai, A. Takahashi, Japan. Kokai 74 77,890 (28. 11. 1972), C.A. 82, 10031t (1975); Phenyl benzoate type liquid crystal display device. [**1144**] H. Tatsuta, M. Fukai, K. Asai, H. Takahashi, Japan. Kokai 74 77,891 (28. 11. 1972), C.A. *82*, 10037z (1975); Liquid crystal electrooptical display devices. [**1145**] H. Tatsuta, M. Fukai, K. Asai, H. Takahashi, Japan. Kokai 74 78,686 (6. 12. 1972), C.A. *82*, 49918n (1975); Liquid crystal display devices. [**1146**] H. Tatsuta, M. Fukai, K. Asai, H. Takahashi, Japan. Kokai 74,112,877 (1. 3. 1973), C.A. *82*, 163077n (1975); Liquid crystal display device employing phenyl benzoate type compositions. [**1147**] H. Tatsuta, M. Fukai, K. Asai, H. Takahashi, Japan. Kokai 74,112,882 (1. 3. 1973), C.A. *82*, 163078p (1975); Liquid crystal display devices using phenyl benzoate type compositions. [**1148**] H. Tatsuta, M. Fukai, K. Asai, H. Takahashi, Japan. Kokai 74,112,889 (1. 3. 1973), C.A. *82*, 178282z (1975); Liquid crystal compositions for display devices based on phenyl benzoate derivatives. [**1149**] H. Tatsuta, Japan. Kokai 74,117,374 (13. 3. 1973), C.A. *83*, 36254p (1975); Liquid crystal indicator. [**1150**] H. Tatsuta, Japan. Kokai 74,117,375 (13. 3. 1973), C.A. *83*, 155791w (1975); Liquid crystal compositions.

[**1151**] H. Tatsuta, M. Fukai, K. Asai, H. Takahashi, Japan. Kokai 74,117,376 (13. 3. 1973), C.A. *83*, 36256r (1975); Liquid crystal indicator. [**1152**] H. Tatsuta, M. Fukai, K. Asai, H. Takahashi, Japan. Kokai 74,117,378 (13. 3. 1973), C.A. *83*, 88761w (1975); Liquid crystal composition for display. [**1153**] H. Tatsuta, M. Fukai, K. Asai, H. Takahashi, Japan. Kokai 74,117,379 (13. 3. 1973), C.A. *83*, 36257s (1975); Liquid crystal indicator. [**1154**] H. Tatsuta, M. Fukai, K. Asai, H. Takahashi, Japan. Kokai 74,117,380 (13. 3. 1973), C.A. *83*, 36255q (1975); Liquid crystal indicator. [**1155**] H. Tatsuta, M. Fukai, K. Asai, H. Takahashi, Japan. Kokai 74,117,383 (13. 3. 1973), C.A. *83*, 36253n (1975); Liquid crystal indicator. [**1156**] H. Tatsuta, M. Fukai, K. Asai, H. Takahashi, Japan. Kokai 75 24,175 (6. 4. 1973), C.A. *83*, 170935q (1975); Liquid crystal display device based on phenyl benzoate-type nematics. [**1157**] H. Tatsuta, M. Fukai, H. Esaki, Japan. Kokai 76,126,983 (30. 4. 1975), C.A. *86*, 131201q (1977); Field-effect liquid crystal display device. [**1158**] H. Tatsuta, M. Fukai, H. Esaki, Japan. Kokai 76,145,478 (10. 6. 1975), C.A. *86*, 131210s (1977); Field-effect liquid crystal display devices. [**1159**] H. Tatsuta, M. Fukai, H. Esaki, Japan. Kokai 76,149,886 (18. 6. 1975), C.A. *86*, 131209y (1977); Nematic liquid crystal compositions for field-effect display devices. [**1160**] H. Tatsuta, M. Fukai, K. Asai, H. Takahashi, Ger. Offen. 2,352,664 (20. 10. 1972), C.A. *81*, 56706z (1974); Display device using liquid crystals.

[**1161**] H. Tatsuta, M. Fukai, K. Asai, H. Takahashi, Ger. Offen. 2,353,315 (25. 10. 1972), C.A. *81*, 44146d (1974); Display device using liquid crystals. [**1162**] G. W. Taylor, A. Miller, U.S. 3,623,795 (24. 4. 1970); Electro-optical system. [**1163**] G. W. Taylor, D. L. White, U.S. 3,833,287 (8. 3. 1973), C.A. *82*, 10047c (1975); Guest-host liquid crystal device. [**1164**] G. W. Taylor, W. F. Kosonocky, IEEE Trans. Sonics Ultrason. *19*, 81 (1972), C.A. *77*, 40398g (1972); Ferroelectric light valve arrays for optical memories. [**1165**] G. W. Taylor, W. F. Kosonocky, Ferroelectrics *3*, 81 (1972), C.A. *77*, 11002t (1972); Ferroelectric light valve arrays for optical memories. [**1166**] G. N. Taylor, F. J. Kahn, J. Appl. Phys. *45*, 4330 (1974), C.A. *82*, 24330b (1975); Materials aspects of thermally addressed smectic and cholesteric storage displays. [**1167**] G. W. Taylor, Appl. Opt. *14*, 1485 (1975); Concurrent display of multiple functions. [**1168**] L. J. Taylor, U.S. 3,935,337 (4. 2. 1971), C.A. *85*, 102425q (1976); Liquid crystal-containing polymeric film. [**1169**] L. F. Taylor, U.S. 3,970,579 (2. 8. 1971), C.A. *86*, 63545c (1977); Low melting liquid-crystal mixtures. [**1170**] T. R. Taylor, J. L. Fergason, U.S. 3,723,346 (24. 5. 1971), C.A. *79*, 32995a (1973); Temperature indicator using a s_C phase. [**1170a**] W. Teichmann, D. Demus, W. Misselwitz, N. Mienert, Z. Inn. Med. *32*, 307 (1977); Hauttemperaturmessung mit Hilfe kristalliner Flüssigkeiten.

[**1171**] K. Thiessen, Le Trong Tuyen, Phys. Status Solidi A *13*, 73 (1972), C.A. *77*, 145358x (1972); Application of nematics for the investigation of p-n junctions and insulating layers. [**1172**] E. B. Thompson, W. H. Taylor, N. R. Cohen, Arch. Int. Pharmacodyn. Ther. *191*, 49 (1971), C.A. *75*, 61645j (1971); Use of cholesteric thermographic procedure for screening of vasoactive drugs. [**1173**] Thompson-CSF, Brit. 1,251,790 (11. 6. 1969); Image converter. [**1174**] E. Tikhonov, M. Bertolotti, F. Scudieri, Appl. Phys. *11*, 357 (1976), C.A. *86*, 24234f (1977); Planar dye lasers in isotropic and anisotropic oriented media. [**1175**] V. G. Tishchenko, M. M. Fetisova, A. P. Makhotilo, L. N. Onishchenko, V. V. Babenko, U.S.S.R. 449,923 (8. 1. 1973), C.A. *83*, 88757z (1975); Thermochromic composition containing cholesterics. [**1176**] M. Tobias, New Scientist *56*, 651 (1972); The fluid state of liquid crystals. [**1177**] M. Tobias, International Handbook of Liquid Crystal Displays 1975–76 (Ovum: London). 1975. 181 pp., C.A. *83*, 106172m (1975). [**1178**] T. Toida, I. Tsunoda, Japan. Kokai 76 16,285 (31. 7. 1974), C.A. *85*, 54644m (1976); Liquid crystal composition for TN display devices. [**1179**] W. H. Toliver, C. G. Roach, R. W. Roundy, P. E. Hoffman, Aerosp. Med. *40*, 35 (1969), C.A. *71*, 7593v

(1969); Liquid crystals. [1180] W. H. Toliver, J. L. Fergason, E. Sharpless, P. E. Hoffman, Aerosp. Med. *41*, 18 (1970), C.A. *74*, 83846g (1971); Liquid crystal trace contaminant vapor detector with an electronic input.

[1181] A. V. Tolmachev, V. M. Kuz'michev, Pis'ma Zh. Eksp. Teor. Fiz. *14*, 220 (1971), C.A. *76*, 40121r (1972); Use of liquid crystals for visually representing IR radiation. [1182] A. V. Tolmachev, E. Y. Govorun, V. M. Kuz'michev, Zh. Eksp. Teor. Fiz. *63*, 583 (1972), C.A. *77*, 145884r (1972); Visualization of millimeter and submillimeter wavelength radiation by means of liquid crystals. [1182a] A. V. Tolmachev, V. G. Tishchenko, A. I. Nikitin, V. M. Kuz'michev, Impul'snaya Fotometriya *4*, 101 (1975), C.A. *87*, 46349v (1977); Visualization of laser pulsed radiation using cholesterics. [1183] V. I. Tomilin, Prib. Tekh. Eksp. *1976*, 199, C.A. *86*, 82338d (1977); Monitoring dielectric films using liquid crystals. [1184] Y. Torii, T. Kaneko, Rev. Electr. Commun. Lab. *23*, 1121 (1975); Optical pattern display device by use of liquid crystal. [1185] K. Toriyama, F. Nakano, M. Sato, H. Abe, M. Kanesaki, Japan. Kokai 74 37,883 (11. 8. 1972), C.A. *81*, 97840g (1974); Liquid crystal display devices. [1186] K. Toriyama, H. Abe, K. Murao, Japan. Kokai 75 38,686 (8. 8. 1973), C.A. *83*, 155792x (1975); Nematic liquid crystal compositions for display devices. [1187] K. Toriyama, H. Abe, F. Nakano, K. Murao, M. Sato, Japan. Kokai 75 39,686 (15. 8. 1973), C.A. *83*, 106256s (1975); Nematic liquid crystal compositions for display devices. [1188] K. Toriyama, T. Kitamura, M. Sato, T. Muroi, Japan. Kokai 75 53,281 (12. 9. 1973), C.A. *83*, 170953u (1975); Nematic liquid crystal compositions for display devices. [1189] K. Toriyama, H. Abe, F. Nakano, K. Murao, M. Sato, Japan. Kokai 76 42,091 (8. 10. 1974), C.A. *85*, 134488d (1976); Nematic liquid crystal compositions for matrix type display devices. [1190] K. Toriyama, S. Nomura, Oyo Butsuri *38*, 698 (1969); Liquid crystals and their applications to electronics.

[1191] K. Toriyama, T. Aoyagi, Oyo Butsuri *40*, 560 (1971), C.A. *76*, 64395j (1972); Liquid crystal materials for the display device. [1192] K. Totani, S. Fuji, Japan. Kokai 74,122,541 (30. 3. 1973), C.A. *82*, 148551u (1975); Cholesteric liquid crystal coating compositions for display devices. [1193] K. Totani, G. Matsumoto, K. Iwasaki, T. Jinnai, Ger. Offen. 2,524,004 (28. 12. 1974), C.A. *85*, 134502d (1976); Nematic liquid crystal preparations. [1194] J. Tricoire, Fr. 2,110,505 (20. 10. 1970), C.A. *78*, 78173p (1973); Thermographic plate using liquid crystals. [1195] J. Tricoire, J. Gyn. Obst. Biol. Repr. *4*, 123 (1975); Étude du cancer du sein par la thermographie clinique. Thermographie en plaque. [1196] J. Tsukamoto, Japan. Kokai 74 91,083 (8. 1. 1973), C.A. *82*, 178277b (1975); Mixed nematic liquid crystals for electrooptical display devices. [1197] K. Tsukamoto, T. Otsuka, K. Morimoto, Y. Murakami, Japan. Kokai 75,153,782 (31. 5. 1974), C.A. *85*, 54637m (1976); Mixed cholesteric liquid crystal. [1198] K. Tsukamoto, T. Otsuka, K. Morimoto, Y. Murakami, Japan. Kokai 75,153,783 (31. 5. 1974), C.A. *85*, 114835m (1976); Mixed cholesteric liquid crystal composition. [1199] K. Tsukamoto, T. Otsuka, K. Morimoto, Y. Murakami, Japan. Kokai 75,155,481 (7. 6. 1974), C.A. *85*, 39324p (1976); Mixed cholesteric liquid crystal. [1200] K. Tsukamoto, T. Otsuka, K. Morimoto, Y. Murakami, Japan. Kokai 75,155,482 (7. 6. 1974), C.A. *85*, 114832h (1976); Mixed cholesteric liquid crystal.

[1201] K. Tsukamoto, T. Otsuka, K. Morimoto, Y. Murakami, Japan. Kokai 75,155,483 (7. 6. 1974), C.A. *85*, 114833j (1976); Mixed cholesteric liquid crystal compositions for electrooptical display devices. [1202] K. Tsukamoto, T. Otsuka, K. Morimoto, Y. Murakami, Japan. Kokai 75,155,484 (7. 6. 1974), C.A. *85*, 114834k (1976); Mixed cholesteric liquid crystal. [1203] K. Tsukamoto, T. Otsuka, K. Morimoto, Y. Murakami, Japan. Kokai 75,155,485 (7. 6. 1974), C.A. *85*, 134474w (1976); Mixed cholesteric liquid crystal composition. [1204] K. Tsukamoto, T. Otsuka, K. Morimoto, Y. Murakami, Japan. Kokai 75,155,486 (7. 6. 1974), C.A. *85*, 134475x (1976); Mixed cholesteric liquid crystal composition for electrooptical display devices. [1205] K. Tsukamoto, T. Otsuka, K. Morimoto, Y. Murakami, Japan. Kokai 75,155,487 (7. 6. 1974), C.A. *85*, 39325q (1976); Mixed cholesteric liquid crystal. [1206] K. Tsukamoto, T. Otsuka, K. Morimoto, Y. Murakami, Japan. Kokai 75,155,488 (7. 6. 1974), C.A. *85*, 39326r (1976); Mixed cholesteric liquid crystal composition. [1207] K. Tsukamoto, T. Otsuka, K. Morimoto, Y. Murakami, Japan. Kokai 75,155,489 (7. 6. 1974), C.A. *85*, 70715r (1976); Mixed cholesteric liquid crystal composition. [1208] K. Tsukamoto, T. Otsuka, K. Morimoto, Y. Murakami, Japan. Kokai 75,157,272 (10. 6. 1974), C.A. *85*, 114841k (1976); Mixed cholesteric liquid crystal composition. [1209] K. Tsukamoto, T. Otsuka, K. Morimoto, Y. Murakami, Japan. Kokai 75,157,273 (10. 6. 1974), C.A. *85*, 134477z (1976); Mixed cholesteric liquid crystal. [1210] K. Tsukamoto, T. Otsuka, K. Morimoto, Y. Murakami, Japan. Kokai 75,158,578 (11. 6. 1974), C.A. *85*, 170114n (1976); Mixed cholesteric liquid crystals.

[1211] K. Tsukamoto, T. Otsuka, K. Morimoto, Y. Murakami, Japan. Kokai 75,158,579 (11. 6. 1974), C.A. *85*, 170115p (1976); Mixed cholesteric liquid crystals. [1212] K. Tsukamoto, T. Otsuka,

K. Morimoto, Y. Murakami, Japan. Kokai 75,158,581 (11. 6. 1974), C.A. *85*, 114836n (1976); Liquid crystal compositions. [**1213**] K. Tsukamoto, T. Otsuka, K. Morimoto, Y. Murakami, Japan. Kokai 75,158,582 (12. 6. 1974), C.A. *85*, 54640g (1976); Mixed cholesteric liquid crystal. [**1214**] K. Tsukamoto, T. Otsuka, K. Morimoto, Y. Murakami, Japan. Kokai 75,158,583 (12. 6. 1974), C.A. *85*, 12389b (1976); Mixed cholesteric liquid crystal. [**1215**] K. Tsukamoto, T. Otsuka, K. Morimoto, Y. Murakami, Japan. Kokai 75,158,584 (12. 6. 1974), C.A. *85*, 12390v (1976); Liquid crystal composition. [**1216**] K. Tsukamoto, T. Otsuka, K. Morimoto, Y. Murakami, Japan. Kokai 75,158,585 (12. 6. 1974), C.A. *85*, 39329u (1976); Liquid crystal composition. [**1217**] K. Tsukamoto, T. Otsuka, K. Morimoto, Y. Murakami, Japan. Kokai 75,158,586 (12. 6. 1974), C.A. *85*, 114837p (1976); Liquid crystal composition. [**1218**] K. Tsukamoto, T. Otsuka, K. Morimoto, Y. Murakami, Japan. Kokai 75,158,587 (12. 6. 1974), C.A. *85*, 134476y (1976); Liquid crystal composition for optical display. [**1219**] K. Tsukamoto, T. Otsuka, K. Morimoto, Y. Murakami, Japan. Kokai 76 33,786 (17. 9. 1974), C.A. *85*, 114856u (1976); Mixed cholesteric liquid crystal. [**1220**] K. Tsukamoto, T. Otsuka, K. Morimoto, Y. Murakami, Japan. Kokai 76 34,883 (18. 9. 1974), C.A. *85*, 134486b (1976); Mixed cholesteric liquid crystal compositions for X-Y matrix type display devices.

[**1221**] M. Tsukamoto, T. Otsuka, K. Morimoto, Y. Murakami, Japan. Kokai 76 22,680 (20. 8. 1974), C.A. *85*, 114848t (1976); Liquid crystal compositions for display devices. [**1222**] M. Tsukamoto, T. Otsuka, K. Morimoto, Y. Murakami, Japan. Kokai 76 22,681 (20. 8. 1974), C.A. *85*, 114847s (1976); Cholesteric liquid crystal compositions for display devices. [**1223**] M. Tsukamoto, T. Otsuka, K. Morimoto, Y. Murakami, Japan. Kokai 76 22,682 (20. 8. 1974), C.A. *85*, 134482x (1976); Cholesteric liquid crystal compositions for display devices. [**1224**] M. Tsukamoto, T. Otsuka, K. Morimoto, Y. Murakami, Japan. Kokai 76 22,685 (20. 8. 1974), C.A. *85*, 114846r (1976); Cholesteric liquid crystal compositions for display devices. [**1225**] M. Tsukamoto, T. Otsuka, Japan. Kokai 76 52,382 (2. 11. 1974), C.A. *86*, 113727m (1977); Liquid crystal compositions for TN display devices. [**1226**] M. Tsukamoto, T. Otsuka, K. Morimoto, Y. Murakami, Japan. Kokai 76 78,791 (30. 12. 1974), C.A. *85*, 151850h (1976); Mixed cholesteric liquid crystal and apparatus for forming it. [**1227**] M. Tsukamoto, T. Otsuka, K. Morimoto, Y. Murakami, Japan. Kokai 76 78,792 (30. 12. 1974), C.A. *85*, 151849q (1976); Mixed cholesteric liquid crystal composition. [**1228**] M. Tsukamoto, T. Otsuka, Jpn. J. Appl. Phys. *13*, 1665 (1974), C.A. *82*, 9966v (1975); Liquid crystal matrix display device with a new panel structure. [**1229**] I. Tsunoda, H. Tateishi, Japan. Kokai 76,112,491 (28. 3. 1975), C.A. *86*, 81733s (1977); Liquid crystal composition for electronic watch. [**1230**] T. Uchida, C. Shishido, M. Wada, Electron. Commun. Jap. *57*, 103 (1975); Effect of surface treatments on the optical storage phenomenon in c/n phase-transition liquid crystal cells.

[**1231**] T. Uchida, C. Shishido, H. Seki, M. Wada, Mol. Cryst. Liq. Cryst. *34*, 153 (1977), C.A. *87*, 14186y (1977); Dichroitic dyes for guest-host interactions in liquid crystals. [**1232**] T. Uchida, C. Shishido, H. Seki, M. Wada, Mol. Cryst. Liq. Cryst. *39*, 39 (1977), C.A. *87*, 32200c (1977); Guest-host interactions in liquid crystals. [**1233**] T. Uchida, C. Shishido, M. Wada, Mol. Cryst. Liq. Cryst. *39*, 127 (1977), C.A. *87*, 31913a (1977); Liquid crystal color display device with phase transition. [**1234**] M. Ueda, B. Hori, B. Kato, Japan. 73 32,080 (11. 9. 1969), C.A. *80*, 114880u (1974); Nematic liquid crystal composition. [**1235**] M. Ueda, F. Hori, B. Kato, Japan. 73 37,913 (11. 10. 1969), C.A. *80*, 125797y (1974); Nematic liquid crystal composition. [**1236**] M. Ueda, F. Hori, B. Kato, N. Arima, Japan. Kokai 73 83,903 (12. 2. 1972), C.A. *80*, 97526p (1974); Liquid crystal microcapsule ink. [**1237**] M. Ueda, B. Hori, H. Kato, Japan. 74 23,105 (22. 5. 1970), C.A. *82*, 24779e (1975); Binary nematic liquid crystal composition. [**1238**] K. Uehara, H. Mada, S. Kobayashi, IEEE Trans. Electron Dev. *22*, 804 (1975); Reduction of electro-optical response times of a field-effect liquid-crystal device: application to dynamic-driven real-time matrix display. [**1239**] S. Ueno, S. Kobayashi, Farumashia *7*, 96 (1971), C.A. *75*, 54329f (1971); Liquid crystals and color. [**1240**] T. Uleman, Brit. 1,306,912 (19. 1. 1970), C.A. *75*, 146140r (1971); Liquid crystal cells.

[**1241**] H.-G. Ulrich, Optik *44*, 313 (1976), Sehleistungsmessung mit Sehzeichen aus Flüssigkristallen. [**1242**] R. Unbehaun, AEG-Telefunken, D79 Ulm; Hinweise für die Anwendung von Flüssigkristallanzeigen. [**1243**] J. van der Veen, T. C. J. M. Hegge, Ger. Offen. 2,456,804 (13. 12. 1973), C.A. *83*, 186415f (1975); Liquid crystalline compounds and mixtures. [**1244**] C. Z. van Doorn, J. L. A. Heldens, Phys. Lett. A *47*, 135 (1974); Angular dependent optical transmission of TN layers. [**1245**] C. Z. van Doorn, J. Appl. Phys. *46*, 3738 (1975); Dynamic behavior of TN layers in switched fields. [**1246**] C. Z. van Doorn, J. Phys. (Paris), C-1, *36*, 261 (1975); Transient behavior of a TN layer in an electric field. [**1247**] J. P. van Meter, B. H. Klanderman, U.S. 3,915,883 (26. 4. 1972), C.A.

84, 114228v (1976); Liquid crystal compositions. [**1248**] J. A. van Raalte, Proc. IEEE *56*, 2146 (1968); Reflective liquid crystal television display. [**1249**] P. van Zanten, C. J. Gerritsma, 5. Arbeitstagung Flüssigkristalle, Inst. Angew. Festkörperphysik Freiburg, 25. 4. 1975; Switching properties of TN layers. [**1250**] A. A. Vasil'ev, I. N. Kompanets, V. V. Nikitin, Kvantovaya Elektron. *1973*, 130, C.A. *79*, 119639m (1973); Switching in a photosemiconductor-liquid-crystal oriented film structure.

[**1251**] A. A. Vasil'ev, I. N. Kompanets, V. V. Nikitin, Opt. Metody Obrab. Inf. *1974*, 111, C.A. *83*, 35634a (1975); Controlled transparency in holographic information-processing systems. [**1251a**] A. A. Vasil'ev, P. V. Vashurin, I. N. Kompanets, Kvantovaya Elektron. *4*, 1714 (1977), C.A. *87*, 191970v (1977); Tunable spatial filters in optical signal converters. [**1252**] D. Veron, C. R. Acad. Sci., Sér. B *274*, 1013 (1972), C.A. *77*, 41168u (1972); Observation des modes d'un laser continu à acide cyanhydrique à l'aide de cristaux liquides. [**1253**] Visioterm Applications S.A., Swiss 582,352 (14. 8. 1974), C.A. *86*, 142103k (1977); Optical temperature indicators from microencapsulated cholesterics. [**1254**] L. K. Vistin, I. E. Polyakova, N. P. Udalov, I. G. Chistyakov, Sb. Dokl. Vses. Nauch. Konf. Zhidk. Krist. Simp. Ikh Prakt. Primen., 2nd, *1972*, 250, C.A. *81*, 113523p (1974); Liquid crystals in optoelectronic modulators. [**1255**] J. H. J. Vogelzangs, Ned. Tijdschr. Natuurk. *38*, 18 (1972), C.A. *76*, 117635w (1972); Temperature measurement by color changes in liquid crystals. [**1256**] V. Volterra, E. Wiener-Avnear, Appl. Phys. *6*, 257 (1975), C.A. *82*, 178687k (1975); Laser-induced isotropic holes in nematics. [**1257**] M. Wada, Oyo Butsuri *44*, 642 (1975), C.A. *83*, 139744c (1975); Solid state display device. Liquid crystals. [**1258**] M. Wada, J. Electron. Engng. *99*, 20 (1975); Liquid crystal display research: A report on the Tohoku university symposium. [**1258a**] M. Wada, T. Uchida, Oyo Butsuri *45*, 1153 (1976), C.A. *87*, 31837d (1977); Liquid crystal color display devices. [**1259**] K. Wagner, H. Liebig, H. Kelker, Ger. Offen. 2,235,072 (17. 7. 1972), C.A. *80*, 113620x (1974); Nematogene Mischungen. [**1260**] M. A. Wall, U.K., At. Energy Res. Estab., Bibliogr., AERE-B-b 181 (1972), 10 pp., C.A. *78*, 161287z (1973); Liquid crystals for nondestructive testing.

[**1261**] K. H. Walter, 4. Arbeitstagung Flüssigkristalle, Inst. Angew. Festkörperphysik Freiburg, 26. 4. 1974; Bilderzeugung in Flüssigkristallen. [**1262**] K. H. Walter, H. Krueger, Ber. Bunsenges. Phys. Chem. *78*, 912 (1974); C.A. *82*, 132114t (1975); Speichereffekte in cholesterinischen Flüssigkeiten mit positiver DK-Anisotropie. [**1263**] K. H. Walter, NTZ *28*, 184 (1975); Allgemeine Aspekte der Bilderzeugung und -adressierung bei den verschiedenen Anzeigetechniken. [**1264**] K. H. Walter, M. Tauer, 7. Arbeitstagung Flüssigkristalle, Inst. Angew. Festkörperphysik Freiburg, 4. 3. 1977; Matrixadressierte Flüssigkristalldisplays. [**1264a**] K. H. Walter, W. Geffcken, W. Greubel, M. Kobale, H. H. Krüger, E. Reichel, C. Rotter, M. Tauer, Bundesminist. Forsch. Technol., Forschungsber., Technol. Forsch. Entwickl. *1976*, BMFT-FB T 76-71, 120 pp., C.A. *87*, 192393w (1977); Flüssigkristalle. [**1265**] R. S. Walton, U.S. 3,613,351 (13. 5. 1969); Wristwatch with liquid crystal display. [**1266**] X. S. Wang, Wuli *5*, 18, 31 (1976), C.A. *85*, 134096z (1976); Liquid crystal digital display of the chemical composition of molten steel. [**1267**] M. R. Wank, U.S. 3,569,709 (27. 3. 1968); Thermal imaging system utilizing liquid crystal material. [**1268**] P. Wasserman, Electron. Design 16, Aug. 2, 1974, p. 76; Which LCD is best? [**1268a**] H. Watanabe, D. Nakagawa, Japan. Kokai 76,144,391 (6. 6. 1975), C.A. *86*, 163654f (1977); Liquid crystal compositions. [**1269**] T. Watanabe, Japan. Kokai 74 38,888 (18. 8. 1972), C.A. *81*, 97825f (1974); Liquid crystal compositions for electrooptical display devices based on TN effect. [**1270**] T. Watanabe, Japan. Kokai 74 96,983 (22. 1. 1973), C.A. *82*, 118249j (1975); Nematic liquid crystal composition for TN display devices.

[**1271**] T. Watanabe, Japan. Kokai 76 81,794 (14. 1. 1975), C.A. *85*, 185144w (1976); Nematic liquid crystal composite. [**1272**] G. L. Waterman, W. E. Woodmansee, U.S. 3,439,525 (22. 4. 1969); Nondestructive testing method using liquid crystals. [**1273**] R. Werchan, B. D. Cook, J. Acoust. Soc. Amer. *49*, 120 (1971); Acoustical holography using liquid crystals. [**1274**] Westinghouse Electric Corp., Brit. 1,309,558 (30. 4. 1969); Method of detecting electromagnetic radiation and homogeneous compositions for use in performing the method. [**1274a**] J. R. Whinnery, C. Hu, Y. S. Kwon, IEEE J. Quantum Electron. QE *13*, 262 (1977), C.A. *86*, 180496y (1977); Liquid-crystal waveguides for integrated optics. [**1275**] D. L. White, G. N. Taylor, J. Appl. Phys. *45*, 4718 (1974), C.A. *82*, 24333e (1975); New absorptive mode reflective liquid-crystals display device. [**1276**] B. Wielke, S. Stanzl, Ultrasonics *14*, 227 (1976), C.A. *86*, 46366c (1977); Distribution of plastic deformation during 21 kHz fatigue of copper samples. [**1277**] P. Wild, Brit. 1,318,007 (26. 3. 1970); Arrangement for locally altering the brightness and/or color of an indicator medium. [**1278**] P. Wild, U.S. 3,705,310 (27. 5. 1970); Liquid crystal voltage display device having photoconductive means to enhance the contrast at the indicating region. [**1279**] P. J. Wild, J. Nehring, Appl. Phys. Lett. *19*, 335 (1971); Turn-on time reduction and contrast enhancement in matrix-addressed liquid-crystal light valves. **[1280]** P. J. Wild, J. Nehring, Brown Boveri Research Report KLR-71-28 (1971); Improved matrix-addressed liquid crystal display.

[1281] P. J. Wild, Soc. Inform. Display, Symp. San Francisco *1972*, 62; Matrix-addressed liquid crystal projection display. **[1282]** P. J. Wild, 4. Arbeitstagung Flüssigkristalle. Inst. Angew. Festkörperphysik Freiburg, 26. 4. 1974; Ansteuerungstechniken für Flüssigkristall-Anzeigen. **[1283]** P. J. Wild, G-I-T *18*, 781 (1974); Elektro-optische Anzeigen mit Flüssigkristallen. **[1284]** P. J. Wild, Microtecnic *28*, 229 (1974); Stand der Entwicklung von Flüssigkristall-Anzeigen. **[1285]** P. J. Wild, Elektroniker *15*, M4 (1976); Flüssigkristall-Anzeigen. **[1286]** E. L. Williams, Liquid crystals for electronic devices (Noyes Data Corp.: Park Ridge). 1975. 264 pp., C.A. *83*, 70234s (1975). **[1287]** R. Williams, U.S. 3,322,485 (9. 11. 1962), C.A. *68*, 34254e (1968); Electro-optical elements utilizing an organic nematic compound. **[1288]** R. Williams, Introd. Liq. Cryst. *1974*, 235, C.A. *85*, 184995n (1976); Pressure effects in sealed liquid-crystal cells. **[1289]** J. H. Williamson, Electron. Engng. *47*, 46 (1975); Engineering liquid crystal display devices. **[1290]** R. M. Wilson, E. J. Gardner, R. H. Squire, J. Chem. Educ. *50*, 94 (1973), C.A. *78*, 109883b (1973); Absorption of light by oriented molecules.

[1291] V. Wittwer, 5. Arbeitstagung Flüssigkristalle. Inst. Angew. Festkörperphysik Freiburg, 25. 5. 1975; Elektrochrome Displays. **[1292]** S. Y. Wong, U.S. 3,826,757 (18. 9. 1972), C.A. *81*, 96986x (1974); Room temperature nematics. **[1293]** S. Y. Wong, U.S. 3,838,059 (22. 2. 1972), C.A. *82*, 10038a (1975); Liquid crystal composition. **[1294]** W. E. Woodmansee, U.S. 3,441,513 (5. 8. 1966), C.A. *71*, 34959a (1969); Liquid crystal compositions. **[1295]** W. Woodmansee, U.S. 3,511,086 (23. 11. 1966), C.A. *73*, 16831q (1970); Nondestructive testing with liquid crystals. **[1296]** W. E. Woodmansee, Mater. Eval. *24*, 564, 571 (1966), C.A. *66*, 30318q (1967); Cholesterics and their application to thermal nondestructive testing. **[1297]** W. E. Woodmansee, H. L. Southworth, Proc. Int. Conf. Nondestruct. Test., 5th, *1967*, 81, C.A. *73*, 81746e (1970); Thermal nondestructive testing with cholesterics. **[1298]** W. E. Woodmansee, H. L. Southworth, Mater. Eval. *26*, 149 (1968), C.A. *69*, 78793e (1968); Detection of material discontinuities with liquid crystals. **[1299]** W. E. Woodmansee, Appl. Opt. *7*, 1721 (1968); Aerospace thermal mapping applications of liquid crystals. **[1300]** M. H. Wu, Diss. Univ. Pennsylvania, Philadelphia 1975. 406 pp. Avail. Xerox Univ. Microfilms, Ann Arbor, Mich., Order No. 76-3233, C.A. *85*, 114710s (1976); PVK/TNF-liquid crystal noncoherent-to-coherent image converter.

[1301] J. J. Wysocki, R. W. Madrid, U.S. 3,622,224 (20. 8. 1969); Liquid crystal alpha-numeric electrooptic imaging device. **[1302]** J. J. Wysocki, J. E. Adams, R. W. Madrid, U.S. 3,642,348 (20. 10. 1969), C.A. *75*, 54901m (1971); Imaging system. **[1303]** J. J. Wysocki, J. E. Adams, J. H. Becker, R. W. Madrid, W. E. L. Haas, U.S. 3,652,148 (5. 5. 1969); Imaging system. **[1304]** J. J. Wysocki, U.S. 3,697,150 (6. 1. 1971), C.A. *78*, 104489t (1973); Electro-optic system in which an electrophoretic-like or dipolar material is dispersed throughout a liquid crystal to reduce the turn-off time. **[1305]** J. J. Wysocki, J. E. Adams, R. W. Madrid, U.S. 3,704,056 (31. 8. 1971); Imaging system. **[1306]** J. J. Wysocki, J. E. Adams, J. H. Becker, R. W. Madrid, W. E. L. Haas, U.S. 3,707,322 (5. 8. 1971); Electrostatic latent imaging system using a c/n transition. **[1307]** J. J. Wysocki, J. E. Adams, J. H. Becker, R. W. Madrid, W. E. L. Haas, U.S. 3,711,713 (5. 8. 1971); Electrically controlled thermal imaging system using a c/n transition. **[1308]** J. J. Wysocki, J. E. Adams, J. H. Becker, R. W. Madrid, W. E. L. Haas, U.S. 3,718,380 (5. 8. 1971); Imaging system in which either a liquid crystalline material or an electrode is shaped in an image configuration. **[1309]** J. J. Wysocki, J. E. Adams, J. H. Becker, R. W. Madrid, W. E. L. Haas, U.S. 3,718,382 (5. 8. 1971); Liquid crystal imaging system in which an electrical field is created by an X-Y address system. **[1310]** J. H. Wysocki, J. H. Becker, G. A. Dir, Ger. Offen. 2,130,504 (19. 6. 1970), C.A. *81*, 56648g (1974); Abbildungsverfahren.

[1311] Xerox Corp., Brit. 1,410,989 (14. 12. 1971), C.A. *84*, 24420x (1976); Imaging system. **[1312]** Xerox Corp., Brit. 1,417,719 (29. 11. 1971), C.A. *84*, 128777k (1976); Imaging system. **[1313]** Xerox Corp., Neth. Appl. 74 04,858 (9. 4. 1973), C.A. *83*, 35761q (1975); Electrooptical image forming system using liquid crystals. **[1314]** Xerox Corp., Neth. Appl. 74 15,860 (20. 12. 1973), C.A. *84*, 158051m (1976); Image formation using liquid crystals. **[1315]** Xerox Corp., Neth. Appl. 74 15,861 (20. 12. 1973), C.A. *84*, 67896z (1976); Image formation using liquid crystals. **[1316]** Xerox Corp., Neth. Appl. 75 02,041 (29. 3. 1974), C.A. *84*, 37316t (1976); Liquid crystal filters for electrophotographic printers. **[1317]** Xerox Corp., Neth. Appl. 75 03,504 (29. 4. 1974), C.A. *84*, 37248x (1976); Optical filter with liquid crystals. **[1318]** Xerox Corp., Neth. Appl. 75 03,508 (29. 4. 1974), C.A. *84*, 52028u (1976); Optical filter with liquid crystals. **[1319]** Xerox Corp., Neth. Appl. 75 03,511 (29. 4. 1974), C.A. *84*, 37249y (1976); Optical filter with liquid crystals. **[1320]** Xerox Corp., Neth. Appl. 75 05,845 (28. 5. 1974), C.A. *85*, 39327s (1976); Optical display device with liquid crystal mixture.

[**1321**] Xerox Corp., Neth. Appl. 75 05,846 (28. 5. 1974), C.A. *85*, 39328t (1976); Electrooptical display device with liquid crystal mixtures. [**1322**] Xerox Corp., Neth. Appl. 75 09,819 (19. 8. 1974), C.A. *82*, 163084n (1975); Nematic liquid crystal display devices. [**1323**] Xerox Corp., Neth. Appl. 75 10,892 (16. 9. 1974), C.A. *85*, 200589r (1976); Liquid crystal image-forming process. [**1324**] M. Yaguchi, Japan. Kokai 74 97,783 (23. 1. 1973), C.A. *82*, 105228b (1975); Cloudiness-free liquid crystal electrooptical device. [**1325**] M. Yaguchi, K. Totani, T. Jinnai, M. Shirotsuka, Japan. Kokai 74 130,882 (24. 4. 1973), C.A. *82*, 163084n (1975); Nematic liquid crystal display devices. [**1326**] M. Yaguchi, T. Jinnai, M. Shirozuka, K. Totani, Japan. Kokai 75 22,786 (30. 6. 1973), C.A. *83*, 106249s (1975); Nematic liquid crystal compositions for display devices. [**1327**] M. Yaguchi, K. Nakamura, Japan. Kokai 75 47,889 (31. 8. 1973), C.A. *83*, 124107y (1975); Liquid crystal composite. [**1328**] M. Yaguchi, Ger. Offen. 2,365,226 (30. 12. 1972), C.A. *81*, 144298y (1974); Liquid crystal cells. [**1328a**] M. Yaguchi, U.S. 4,007,077 (27. 12. 1973), C.A. *87*, 31970s (1977); Liquid crystal cells. [**1328b**] M. Yamaguchi, Japan. Kokai 77 24,992 (22. 8. 1975), C.A. *87*, 109447y (1977); Cholesteric liquid crystal compositions. [**1329**] K. Yamamura, K. Kakizawa, Y. Yamazaki, K. Kubota, I. Nishimura, U.S. 3,748,017 (5. 6. 1970), C.A. *79*, 98502u (1973); Electrodes for a liquid crystal device. [**1330**] S. Yamamura, Y. Ichinose, Y. Kando, S. Okudaira, Japan. Kokai 76 91,834 (10. 2. 1975), C.A. *86*, 113732j (1977); Reflector plates for liquid crystal display devices. [**1330a**] A. Yamashita, Fr. Demande 2,293,024 (26. 11. 1974), C.A. *86*, 180755g (1977); Display devices using liquid crystals comprising a styrene-based dye material with color change.

[**1331**] Y. Yamashita, R. Kawashima, S. Ohno, Japan. Kokai 76 05,281 (3. 7. 1974), C.A. *85*, 151834f (1976); Nematic liquid crystal compositions for display devices. [**1332**] Y. Yamashita, H. Kawashima, S. Ono, Japan. Kokai 76,122,679 (19. 4. 1975), C.A. *86*, 63558j (1977); p-Type nematic liquid crystal composite. [**1333**] Y. Yamazaki, Japan. Kokai 74 34,488 (3. 8. 1972), C.A. *81*, 144290q (1974); Liquid crystal electrooptical display device. [**1334**] Y. Yamazaki, Japan. Kokai 74 59,087 (11. 10. 1972), C.A. *82*, 10029y (1975); Liquid crystal display devices. [**1335**] Y. Yamazaki, Japan. Kokai 74 59,088 (12. 10. 1972), C.A. *82*, 10032u (1975); Liquid crystal display device. [**1336**] Y. Yamazaki, Japan. Kokai 74 88,791 (26. 12. 1972), C.A. *82*, 118271k (1975); Liquid crystal compositions for TN display devices. [**1337**] Y. Yamazaki, Japan. Kokai 74,117,373 (13. 3. 1973), C.A. *83*, 36258t (1975); Liquid crystal indicator. [**1338**] Y. Yamazaki, T. Suzuki, Japan. 75 25,912 (16. 7. 1970), C.A. *85*, 39321k (1976); Electrooptical display device. [**1339**] Y. Yamazaki, T. Suzuki, Japan. 75 27,832 (11. 7. 1970), C.A. *85*, 54636k (1976); Liquid crystal composition for electrooptical display devices. [**1340**] Y. Yamazaki, Japan. Kokai 75 65,484 (16. 10. 1973), C.A. *84*, 143077a (1976); Liquid crystal display devices.

[**1341**] Y. Yamazaki, Japan. Kokai 75,110,987 (13. 2. 1974), C.A. *84*, 97868j (1976); TN display devices. [**1342**] Y. Yamazaki, Japan. Kokai 75,130,690 (4. 4. 1974), C.A. *85*, 27385w (1976); Nematic liquid crystal compositions for field-effect display devices. [**1343**] Y. Yamazaki, Japan. 76 17,508 (11. 6. 1971), C.A. *85*, 151847n (1976); Nematic liquid crystal display device. [**1344**] Y. Yamazaki, Japan. Kokai 76 23,487 (9. 4. 1974), C.A. *85*, 134481w (1976); Nematic liquid crystal compositions for display devices. [**1345**] Y. Yamazaki, Japan. 76 44,501 (8. 12. 1970), C.A. *86*, 99103w (1977); Nematic liquid crystal compositions for electrooptical display device. [**1345a**] Y. Yamazaki, T. Suzuki, Japan. 77 08,273 (11. 7. 1970), C.A. *87*, 125395k (1977); Nematic liquid crystal compositions for display devices. [**1345b**] Y. Yamazaki, T. Suzuki, Japan. 77 08,274 (29. 7. 1970), C.A. *87*, 125394j (1977); Nematic liquid crystal compositions for display devices. [**1346**] Y. Yano, T. Takahashi, S. Harada, Japan. 74 16,806 (23. 7. 1969), C.A. *82*, 105242b (1975); UV radiation-sensitive imaging material. [**1347**] H. Yokokura, T. Kitamura, F. Nakano, Japan. Kokai 76 47,587 (23. 10. 1974), C.A. *85*, 151844j (1976); Nematic liquid crystal compositions for display devices. [**1348**] H. Yokokura, T. Kitamura, F. Nakano, Japan. Kokai 76 49,184 (28. 10. 1974), C.A. *86*, 99085s (1977); Nematic liquid crystal compositions for display devices. [**1348a**] H. Yokokura, F. Nakano, T. Kitamura, Japan. Kokai 77 56,086 (5. 11. 1975), C.A. *87*, 125396m (1977); Nematic liquid crystal compositions. [**1349**] S. Yoshikawa, M. Horie, H. Takahashi, Fujitsu Sci. Tech. J. *12*, 57 (1976), C.A. *86*, 81646r (1977); Construction of liquid crystal light valve in reflection mode. [**1350**] R. A. Young, U.S. 3,966,305 (10. 10. 1974), C.A. *85*, 85499x (1976); Liquid crystal cell with improved alignment. [**1350a**] L. J. Yu, M. M. Labes, Appl. Phys. Lett. *31*, 719 (1977); Fluorescent liquid-crystal display utilizing an electric-field induced c/n transition.

[**1351**] L. A. Zanoni, U.S. 3,576,364 (20. 5. 1969); Color advertising display employing liquid crystal. [**1352**] R. I. Zharenov, Uch. zap. Ivanov. un-t *1974*, 124, C.A. *83*, 149568n (1975); Hermetically sealed liquid-crystal temperature-indicating units. [**1353**] G. M. Zharkova, A. P. Kapustin, Izv. Sib. Otd. Akad. Nauk SSSR, Ser. Tekh. Nauk *1970*, 65, C.A. *74*, 80809m (1971); Characteristics of liquid crystals for aerodynamic studies. [**1354**] G. M. Zharkova, Izv. Sib. Otd. Akad. Nauk SSSR, Ser. Tekh.

Nauk, *1973*, 130, C.A. *79*, 83579a (1973); Use of liquid crystals as heat indicators. [**1355**] G. M. Zharkova, V. M. Khachaturyan, Aerofiz. Issled. *3*, 169 (1974), C.A. *84*, 98004m (1976); Stabilization of cholesteric liquid-crystal films. [**1355a**] G. M. Zharkova, Aeromekhanika *1976*, 238, C.A. *87*, 32186c (1977); Use of liquid crystals in an aerodynamic experiment. [**1356**] R. S. Ziernicki, W. F. Leonard, Rev. Sci. Instrum. *43*, 479 (1972), C.A. *76*, 105052r (1972); Detection of nonuniform thermal gradients in galvanothermomagnetic measurements using liquid crystals. [**1357**] H. Zocher, U.S. 1,873,951 (10. 11. 1925); Polarizers of light and a method of preparation.

Chapter 15

[**1**] A. Adamczyk, M. Jawdosiuk, Pr. Nauk. Inst. Metrol. Elektr. Politech. Wroclaw *7*, 5 (1975), C.A. *86*, 63564h (1977); Liquid crystals. General characteristics and chemical structure. [**2**] Anonymous, Engineering *106*, 349 (1918), C.A. *13*, 277[1] (1919); Liquid crystals. [**3**] Anonymous, Chem. Eng. News *49*, 20 (1971), Liquid crystals draw intense interest. [**4**] T. Aoyagi, K. Toriyama, Yuki Gosei Kagaku Kyokai Shi *28*, 309 (1970), C.A. *72*, 137258w (1970); Liquid crystals and their applications. [**5**] H. Baessler, Festkörperprobleme *11*, 99 (1971), C.A. *76*, 64390d (1972); Liquid crystals. [**6**] A. I. Baise, Pentacol *9*, 24 (1970), C.A. *75*, 26196s (1971); Liquid crystals. [**7**] D. Barbero, G. Fis. *13*, 211 (1972), C.A. *79*, 24180r (1973); Liquid crystals. [**8**] L. Bata, Fiz. Sz. *26*, 88 (1976), C.A. *85*, 134514j (1976); Liquid crystals in physical and biological systems and in practical applications. [**9**] Battelle Memorial Inst., Columbus Lab. (NTIS: Springfield, Va.) 1971, 121 pp. C.A. *79*, 150495k (1973); Liquid crystals. [**10**] H. Benedy, Chem.-Anlagen Verfahren *1971*, 85, C.A. *76*, 18782k (1972); Flüssige Kristalle.

[**11**] B. Böttcher, D. Groß, Umschau *69*, 574 (1969), C.A. *71*, 117163n (1969); Kristalline Flüssigkeiten und plastische Kristalle. [**12**] J. Bourdon, Mes., Regul., Automat. *32*, 174, 180 (1967), C.A. *67*, 85628s (1967); Cristaux liquides. [**13**] L. F. Brochard, Contemp. Phys. *18*, 247 (1977), C.A. *87*, 117059p (1977); Nematics. Some easy demonstration experiments. [**14**] G. H. Brown, Chemistry *40*, 10 (1967), C.A. *69*, 46930f (1968); Liquid crystals. [**15**] G. H. Brown, Electrochem. Soc., Fall Meeting, Detroit 1969, Abstr. No. 154; Structure and properties of liquid crystals. [**16**] G. H. Brown, J. W. Doane, V. D. Neff, Crit. Rev. Solid State Sci. *1*, 303 (1970), C.A. *74*, 131315g (1971); Structural and physical properties of liquid crystals. [**17**] G. H. Brown, US Clearinghouse Fed. Sci. Tech. Inform., AD 1970, No. 872377, 96 pp. C.A. *75*, 41752p (1971); Physical properties of the cholesteric mesophase. [**18**] G. H. Brown, Amer. Sci. *60*, 64 (1972), C.A. *76*, 91017u (1972); Liquid crystals and their roles in inanimate and animate systems. [**19**] G. H. Brown, US Nat. Tech. Inform. Serv., AD Rep. 1972, No. 736822, 56 pp. C.A. *77*, 67232k (1972); Basic studies of liquid crystals as related to electro-optical and other devices. [**20**] G. H. Brown, J. Electron. Mater. *2*, 403 (1973), C.A. *79*, 84132m (1973); Properties and applications of liquid crystals.

[**21**] G. H. Brown, J. Opt. Soc. Amer. *63*, 1505 (1973), C.A. *80*, 53035y (1974); Structure, properties, and some applications of liquid crystals. [**22**] J. A. Castellano, G. H. Brown, Chem. Technol. *3*, 47, 229 (1973), C.A. *78*, 116069u, 165163d (1973); Thermotropic liquid crystals. [**23**] J. T. S. Andrews, W. E. Bacon, G. H. Brown, A. De Vries, J. W. Doane, US Nat. Tech. Inform. Serv., AD Rep. 1973, No. 770685/6GA, 161 pp. C.A. *80*, 113477f (1974); Physical and structural properties of liquid crystals and their potential as detectors. [**24**] G. H. Brown, J. W. Doane, Appl. Phys. *4*, 1 (1974), C.A. *81*, 69395x (1974); Liquid crystals and some of their applications. [**25**] S. Chandrasekhar, Rep. Prog. Phys. *39*, 613 (1976), C.A. *86*, 24493q (1977); Liquid crystals. [**26**] D. Chapman, Sci. J. *1*, 32 (1965), C.A. *65*, 2976f (1966); Liquid crystals. [**27**] P. Châtelain, Bull. soc. franç. min. *77*, 323 (1954), C.A. *49*, 7913c (1966); Cristaux liquides. [**28**] P. Châtelain, M. Brunet, J. Cano, Manuskript. Film des Lab. de Minéral. Crist. Faculté des Sci., Montpellier; Propriétés optiques des cristaux liquides des types nématiques et cholestériques. [**29**] T. C. Chaudhari, Chem. News *117*, 269 (1918), C.A. *12*, 2476[7] (1918); Studies in liquid crystals. [**30**] I. G. Chistyakov, Kristallografiya *5*, 962 (1960), C.A. *56*, 8114i (1962); Liquid crystals.

[**31**] I. G. Chistyakov, V. A. Usol'tseva, Liquid crystals and their value in medicine and biology. (Lectures for students). Ivanovo 1962, 24 pp. C.A. *58*, 6275a (1963). [**32**] V. A. Usol'tseva, I. G. Chistyakov, Uspekhi Khim. *32*, 1124 (1963), C.A. *60*, 78a (1964); Chemical peculiarities, structure and properties of liquid crystals. [**33**] I. G. Chistyakov, Usp. Fiz. Nauk *89*, 563 (1966), C.A. *65*, 19399g (1966); Liquid crystals. [**34**] I. G. Chistyakov, V. N. Aleksandrov, Uch. Zap. Ivanov. Gos. Pedagog. Inst. *77*, 34

(1970), C.A. *76*, 18778p (1972); Properties of cholesterol liquid crystals. [**35**] I. G. Chistyakov, L. K. Vistin, Priroda *1972*, 76, C.A. *76*, 145792p (1972); Symmetry, structure and properties of liquid crystals. [**36**] I. G. Chistyakov, Sb. Dokl. Vses. Nauch. Konf. Zhidk. Krist. Simp. Ikh Prakt. Primen., 2nd *1972*, 3, C.A. *81*, 127622g (1974); Structure of liquid crystals. [**37**] I. G. Chistyakov, L. K. Vistin, Cesk. Casopis Fys. *24*, 131 (1974); Symmetry. Structure and properties of liquid crystals. [**38**] I. G. Chistyakov, L. K. Vistin, Khim. Zhizn 1975, 42, C.A. *84*, 10955p (1976); Liquid crystals. [**39**] B. K. Vainshtein, I. G. Chistyakov, V. sb., Probl. Sovrem. Kristallogr. *1975*, 12, C.A. *85*, 70928n (1976); Symmetry, structure, and properties of liquid crystals. [**40**] H. C. Chu, Wuli *1*, 89 (1972), C.A. *79*, 58430n (1973); Liquid crystals.

[**41**] A. Coehn, Z. Elektrochem. *10*, 856 (1904); Flüssige Kristalle. [**42**] V. L. Dang, H. P. Tran, T. T. Nguyen, T. Y. Nguyen, Tap Chi Hoa Hoc *14*, 20 (1976), C.A. *87*, 144137u, 184090e (1977); Liquid crystals. Synthesis and application. [**43**] Defense Documentation Center (NTIS: Springfield). 1971. 141 pp. C.A. *76*, 132790m (1972); Liquid crystals (AD-733750). [**44**] A. J. Dekker, Ned. Tijdschr. Natuurkd. A *43*, 47 (1977), C.A. *87*, 144140q (1977); Liquid crystals. [**45**] A. Derzhanski, S. Naidenova, K. Avramova, E. Kubleva, K. Khinov, Priroda *22*, 19 (1973), C.A. *81*, 69399b (1974); Liquid crystals. [**46**] E. Dubois-Violette, O. Parodi, Sci. Progr. Decouverte *1971*, No. 3438, 36, C.A. *76*, 132411p (1972); Cristaux liquides. [**47**] D. B. DuPre, E. T. Samulski, A. V. Tobolsky, Polym. Sci. Mater. *1971*, 123, C.A. *76*, 77520w (1972); Mesomorphic state. Liquid and plastic crystals. [**48**] G. Durand, J. D. Litster, Annu. Rev. Mater. Sci. *3*, 269 (1973), C.A. *80*, 64469u (1974); Recent advances in liquid crystals. [**49**] T. Eckert, W. R. Cramer, Pharm. Unserer Zeit *1*, 116 (1972), C.A. *77*, 110605y (1972); Flüssig-kristalline Mesophasen. [**50**] G. Ekert, Schweiz. Wochschr. *50*, 305 (1912), C.A. *6*, 2564⁶ (1912); Flüssige Kristalle.

[**51**] C. M. Ellis, Sch. Sci. Rev. *51*, 80 (1969), C.A. *72*, 106925g (1970); Liquid crystals. [**52**] J. L. Fergason, Sci. Amer. *211*, 76 (1964); Liquid crystals. [**53**] J. L. Fergason, Amer. J. Phys. *38*, 425 (1970), C.A. *72*, 128528g (1970); Experiments with cholesterics. [**54**] E. H. Frei, S. Shtrikman, US Nat. Tech. Inform. Serv., AD Rep. 1973, No. 772760/5GA, 8 pp. C.A. *81*, 6902q (1974); Physical properties of liquid crystals. [**55**] G. T. Frumin, Khim. Zhizn 1976, 27, C.A. *86*, 36370k (1977); Chemical functions of liquid crystals. [**56**] W. E. Garner, Sci. Progr. *17*, 357 (1923), *20*, 31 (1925), C.A. *17*, 1355⁴ (1923), *19*, 2889² (1925); Recent advances in science. Physical chemistry. [**57**] H. Gasparoux, J. R. Lalanne, P. Lalanne, S. Fourcade, Bull. Union Physisiens *70*, 1097 (1976), C.A. *85*, 134520h (1976); Cristaux liquides et leur application. [**58**] H. Gasparoux, J. Prost, Annu. Rev. Phys. Chem. *27*, 175 (1976), C.A. *86*, 36383s (1977); Liquid crystals. [**59**] C. J. Gerritsma, Natuurkd. Voordr. *53*, 63 (1975), C.A. *85*, 134512g (1976); Liquid crystals. [**60**] F. K. Gorskii, Kinet. Mekh. Krist. *1973*, 181, C.A. *81*, 6912t (1974); Properties and crystallization processes of liquid crystals.

[**61**] J. G. Grabmaier, H. H. Krüger, VDI Z. *115*, 629 (1973), C.A. *79*, 58433r (1973); Flüssige Kristalle. Grundlagen und technische Anwendungen. [**62**] G. W. Gray, A. J. Leadbetter, Phys. Bull. *28*, 28 (1977), C.A. *86*, 131227c (1977); Liquid crystals — what makes a mesophase. [**63**] B. Hampel, Laser, Angew. Strahlentechn. *1971* (3), 53; Flüssige Kristalle. [**64**] M. Hareng, Ing. EPCI 76, 3 (1972), C.A. *81*, 112032r (1974); Cristaux liquides. [**65**] E. Herlinger, Z. physik. chem. Unterricht *45*, 5 (1932), C.A. *26*, 5460 (1932); Der mesomorphe Aggregatzustand. [**66**] T. C. Huard, J. Cazes, Anal. Instrum. *15*, 37 (1977), C.A. *87*, 76431h (1977); Physical properties and analysis of liquid crystals. [**67**] S. Iwayanagi, Nippon Butsuri Gakkaishi *28*, 713 (1973), C.A. *80*, 64474s (1974); Guide to experiments on liquid crystals. [**68**] M. Jamsek-Vilfan, Obz. Mat. Fiz. *18*, 65 (1971), C.A. *76*, 77508y (1972); Liquid crystals. [**69**] J. A. Janik, Postepy Fiz. *11*, 551 (1960), C.A. *55*, 11996a (1961); Liquid crystals. [**70**] M. Jawdosiuk, E. Czarnecka, J. K. Walejko, Chemik *29*, 356 (1976), C.A. *86*, 198035g (1977); Liquid crystals. Chemical characteristics and application.

[**71**] M. Jawdosiuk, E. Czarnecka, Wiad. Chem. *31*, 329 (1977), C.A. *87*, 46612a (1977); General characteristics and chemical structure of liquid crystals. [**72**] A. P. Kapustin, Primen. Ul'traakust. Issled Veshchestva *23*, 207 (1967), C.A. *71*, 74965f (1969); Properties of liquid crystals. [**73**] W. Kast, Physik. Z. *38*, 627 (1937), C.A. *31*, 8288¹ (1937); Anisotrope Flüssigkeiten. [**74**] W. Kast, Z. Elektrochem. *45*, 184 (1939), C.A. *33*, 3649² (1939); Anisotrope Flüssigkeiten. [**75**] E. I. Kats, Fiz. Kondensir. Sostoyaniya i Rasseyaniya Neitronov *1976*, 23, C.A. *87*, 144297w (1977); Liquid crystals. [**76**] D. Kaye, Electronic Design *19*, 76 (1970); Liquid crystals. [**77**] H. Kelker, Mol. Cryst. Liq. Cryst. *21*, 1 (1973), C.A. *79*, 97736m (1973); History of liquid crystals. [**78**] H. Kelker, Tech. Mitt. *68*, 100 (1975), C.A. *83*, 69160h (1975); Physik und Chemie der flüssigen Kristalle. [**79**] J. Y. Kim, Hwahak Kwa Kongop Ui Chinbo *13*, 106, 128 (1973), C.A. *79*, 119124q (1973); Liquid crystals. [**80**] J. Y. Kim, Jonja Konghak Hoe Chi *10*, 98 (1973), C.A. *84*, 187568r (1976); Chemical characteristics of liquid crystals and its applicability.

[81] M. Kobale, H. Krüger, Phys. Unserer Zeit *6*, 66 (1975), C.A. *83*, 106289e (1975); Flüssige Kristalle. [82] K. Kobayashi, Bussei *11*, 469 (1970), C.A. *74*, 46374f (1971); Basic liquid-crystal theory. [83] S. Kobayashi, Kagaku *41*, 442 (1971), C.A. *76*, 50938b (1972); Liquid crystal and the laser. [84] S. Kobayashi, Kagaku To Kogyo (Tokyo) *29*, 813 (1976), C.A. *86*, 148882k (1977); Structure of liquids. Liquid crystals. [85] S. Kobayashi, T. Shimomura, H. Mada, Denshi Shashin *15*, 15 (1976), C.A. *86*, 148860b (1977); Recent developments in liquid crystal research. [86] K. Kosai, Yukagaku *21*, 871 (1972), C.A. *78*, 76652v (1973); Liquid crystals. [87] E. I. Kovshev, L. M. Blinov, V. V. Titov, Usp. Khim. *46*, 753 (1977), C.A. *87*, 32000n (1977); Thermotropic liquid crystals and their application. [88] J. Krause, Funk-Tech. *31*, 8 (1976), C.A. *87*, 191986e (1977); Flüssige Kristalle. Grundlagen und Verwendung in der Anzeigetechnik. [89] S. Kusabayashi, R. Matsuyama, Bussei *10*, 668 (1969), C.A. *72*, 115431v (1970); Liquid crystals. [90] S. Kusabayashi, Kagaku To Kogyo (Tokyo) *22*, 1429 (1969), C.A. *72*, 71444v (1970); Liquid crystals as a new organic material.

[91] J. R. Lalanne, F. Hare, J. Chem. Educ. *53*, 793 (1976), C.A. *86*, 70907a (1977); Three liquid-crystal teaching experiments. [92] A. S. C. Lawrence, J. Roy. Microscop. Soc. *58*, 30 (1938), C.A. *32*, 6921[1] (1938); Liquid crystals and anisotropic solutions. [93] J. D. Litster, Photon Correl. Light Beating Spectrosc., NATO Adv. Study Inst. *1973*, 475, C.A. *81*, 160097n (1974); Liquid crystals. [94] C. C. Liu, Hua Hsueh Tung Pao *1977*, 39, 64, C.A. *87*, 76428n (1977); Progress in the research of liquid crystals. [95] G. R. Luckhurst, Cron. Chim. *36*, 3 (1972), C.A. *78*, 116067s (1973); Liquid crystals. [96] G. R. Luckhurst, Proc. R. Inst. G. B. *49*, 159 (1976), C.A. *87*, 60835g (1977); Liquid crystals: the fourth state of matter. [97] T. E. Faber, G. R. Luckhurst, Annu. Rep. Prog. Chem., Sect. A: Phys. Inorg. Chem. *72*, 31 (1975), C.A. *86*, 24492p (1977); Liquid crystals. [98] A. Mannschreck, Chem.-Ztg., Chem. App. *92*, 69 (1968), C.A. *69*, 71356f (1968); Kristalline Flüssigkeiten, ein vierter Aggregatzustand? [99] C. Mauguin, Traitée de Chimie Organique *1*, 81 (1934); Cristaux liquides. [100] G. Mayr, Nuovo cimento *5*, XXV (1928), C.A. *23*, 3138[1] (1929); Mesomorphic states.

[101] G. Meier, Physik. Blätter *27*, 110 (1971); Flüssige Kristalle. [102] J. W. A. Meijer, Tijdschr. Chem. Instrum. *6*, 5 (1973), C.A. *78*, 116053j (1973); Liquid crystals. [103] M. Miesowicz, Postepy Fiz. *26*, 129 (1975), C.A. *83*, 124116a (1975); 50 Years of investigation on liquid crystals in Poland. [104] R. Morgan, A. J. Tomkins, U. K., At. Energy Res. Estab., Bibliogr. 1969, AERE-Bib 169, 17 pp. C.A. *72*, 93917d (1970); Effects of ultrasonic waves and magnetic and electric fields on liquid crystals. [105] O. Nagasaki, M. Sukigara, K. Honda, Kagaku No Ryoiki *26*, A23 (1972), C.A. *77*, 93759g (1972); Molecular orientation of nematic liquid crystal. [106] I. Nakada, Seramikkusu *6*, 264 (1971), C.A. *75*, 155540y (1971); Properties of liquid crystals and their uses. [107] J. Nehring, Neue Zürcher Ztg. vom 3. 1. 1973. Flüssige Kristalle. Eigenschaften und Anwendungen. [108] W. Nernst, Z. Elektrochem. *12*, 431 (1906); Diskussionsbemerkung über flüssige Kristalle. [109] W. Nernst, Z. Elektrochem. *16*, 702 (1910), C.A. *5*, 2014[3] (1911); Theorie der anisotropen Flüssigkeiten. [110] A. Nomura, Bussei *10*, 385 (1969), C.A. *71*, 117164p (1969); Physical properties of liquid crystals and their applications.

[111] A. Nomura, Kinzoku *40*, 44 (1970), C.A. *75*, 41297u (1971); Advances in liquid-crystal research. [112] M. A. Osman, Chimia *31*, 253 (1977), C.A. *87*, 109481e (1977); Flüssige Kristalle und ihre Anwendungen. [113] S. E. B. Petrie, H. K. Buecher, R. T. Klingbiel, P. I. Rose, Org. Chem. Bull. *45*, 1 (1973), C.A. *79*, 35796x (1973); Physical properties and applications of liquid crystals. [114] W. J. Pope, Chem. and Ind. *42*, 809 (1923), C.A. *17*, 3633[7] (1923); Crystalline liquids. [115] E. B. Pristley, RCA Rev. *35*, 81 (1974), C.A. *81*, 42393h (1974); Liquid crystal mesophases. [116] F. Pusnik, Obz. Mat. Fiz. *19*, 53 (1972), C.A. *78*, 9589s (1973); Optical properties of liquid crystals. [117] A. Rapini, Prog. Solid State Chem. *8*, 337 (1973), C.A. *81*, 112018r (1974); Propriétés et applications des cristaux liquides. [118] L. W. Reeves, Cienc. Cult. (Sao Paulo) *25*, 403 (1972), C.A. *79*, 129948a (1973); Liquid crystals. [119] T. Rotarski, Ber. *36*, 3158 (1903); Die sogenannten flüssigen Kristalle. [120] T. Rotarski, J. prakt. Chem. *82*, 23 (1910), C.A. *4*, 2763[7] (1910); Molekular-mechanische Theorie der anisotropen Flüssigkeiten oder der sogenannten flüssigen Kristalle.

[121] S. Sakagami, Hyomen *12*, 364 (1974), C.A. *81*, 160102k (1974); Various smectics and their properties. [122] A. Saupe, Annu. Rev. Phys. Chem. *24*, 441 (1973), C.A. *80*, 75005r (1974); Liquid crystals. [123] R. Schenck, Jahrb. Radioakt. Elektronik *6*, 572 (1909), C.A. *4*, 1256[4] (1910); Neuere Untersuchungen der kristallinen Flüssigkeiten. [124] J. Schreiber, Chem. Listy *66*, 594 (1972), C.A. *77*, 74325w (1972); Chemical structure, properties, and application of organic liquid crystals. [125] G. W. Smith, Int. Sci. Technology *6*, 72 (1967); Liquid-like solids. [126] R. Steinsträßer, Chem.-Ztg. *95*, 661 (1971), C.A. *75*, 144530u (1971); Flüssige Kristalle. Grundlagen und chemische Strukturen.

[127] R. Steinsträßer, L. Pohl, Angew. Chem. *85*, 706 (1973), C.A. *79*, 140353y (1973); Chemie und Verwendung flüssiger Kristalle. [128] M. J. Stephen, J. P. Straley, Rev. Mod. Phys. *46*, 617 (1974), C.A. *82*, 92084z (1975); Physics of liquid crystals. [129] H. Suga, Kagaku To Kogyo (Tokyo) *29*, 805 (1976), C.A. *87*, 70257j (1977); Transitions to the fourth states of aggregation. [130] M. Sukie, Kasen Geppo *29*, 86 (1976), C.A. *85*, 54654q (1976); Liquid crystals.

[131] O. Svekus, Termeszet Vilaga *100*, 542 (1969), C.A. *72*, 104728w (1970); Crystallized liquids. [132] T. Tachibana, Kyoritsu Kagaku Raiburari *1*, 1 (1973), C.A. *83*, 35772u (1975); Discovery of liquid crystals. [133] G. Tammann, Ann. Physik *4*, 524 (1901), *8*, 103 (1902); Die sogenannten flüssigen Kristalle. [134] G. Tammann, Ann. Physik *19*, 421 (1906); Die Natur der flüssigen Kristalle. III. [135] G. Tammann, Z. anorg. Chem. *90*, 297 (1914), C.A. *9*, 1563[5] (1915); Über den molekularen Aufbau fester, isotroper und anisotroper binärer Mischungen. [136] K. Tomopulos, Chem.-Ztg. *84*, 356 (1960), C.A. *54*, 20391g (1960); Kristallin-flüssige Zustände: I Phänomene in Schmelzen und wäßrigen Phasen. [137] S. Toshima, T. Harada, Kagaku Kogyo *21*, 741 (1970), C.A. *73*, 59815e (1970); Liquid crystals. [138] V. N. Tsvetkov, Liquid crystals; Encyclopaedic directory of physics (Pergamon: New York), Vol. 2, 14 (1963). [139] K. Überreiter, Lehrb. Experimentalphys. *4*, 361 (1975), C.A. *87*, 125418v (1977); Flüssige Kristalle. [140] L. Verbit, J. Chem. Educ. *49*, 36 (1972), C.A. *76*, 71521a (1972); Liquid crystals. Synthesis and properties. Experiment for the integrated organic and physical laboratory.

[141] W. Voigt, Physik. Z. *17*, 76, 128, 152, 305 (1916), C.A. *11*, 2063[9] and 1350[7] (1917); Flüssige Kristalle und anisotrope Flüssigkeiten. [142] C. Von Planta, Microtecnic *26*, 359; *27*, 280 (1973), C.A. *79*, 150180d (1973); Liquid crystals. [143] G. Weber, Chem. Labor Betrieb *25*, 193 (1974); Flüssige Kristalle und ihre Anwendung. [144] K. Wieczffinski, Chem. Szk. *22*, 169 (1976), C.A. *86*, 88355d (1977); What are liquid crystals? [145] B. Weinberg, Physik. Z. *7*, 831 (1906), C.A. *1*, 266[5] (1907); Die theoretische Möglichkeit der Existenz von flüssigen Kristallen. [146] J. J. Wright, Amer. J. Phys. *41*, 270 (1973), C.A. *78*, 96659f (1973); Optics experiments with nematics. [147] G. Wulff, Z. Kryst. Min. *44*, 209 (1908), C.A. *2*, 3323[3] (1908); Die Natur krystallinischer Flüssigkeiten. [148] G. Wulff, Z. Kryst. Min. *46*, 261 (1909), C.A. *3*, 1855[6] (1909); Die Natur flüssiger und fließender Krystalle. [149] G. Wulff, Ann. Physik *35*, 182 (1911), C.A. *5*, 2996[8] (1911); Die sogenannten Kern- und Konvergenzpunkte der kristallin-flüssigen Phase von PAP. [150] H. Zocher, Mol. Cryst. Liq. Cryst. *7*, 165 (1969), C.A. *71*, 85330w (1969); Topics of liquid crystals yet to be discussed.

Subject Index

Van't Hoff equation 373
Vapor pressure 375, 526, 527, 534
Variable grating mode 196, 623
Vesicles 574, 585
Viruses 587, 588
Viscosity 142–144, 148, 150, 163, 164, 244, 266, 290, 352, 364, 534, 589
Viscous isotropic phase, see structure cubic
Vitrified films 560, 606
Voltage indicators 625
Vuks relationship 261

Wada constants 372
Water (in micellar and lyotropic phases) 514, 517, 518, 520, 542, 543, 576
Waveguides 191, 262, 616
Williams domains 193–198, 202, 203, 616

Xanthate pyrolysis 413
X-ray studies 222-241, 250–252, 295, 514–523, 544, 546, 557, 564, 575, 579, 580, 587, 596, 597, 600
Xylene isomers 386–388, 392, 396–398, 405, 409, 410, 537

Yarns 605

Legal Note

The publisher and the authors of this book gratefully acknowledge the courtesy of the following publishers and copyright holders in grating permission to reproduce the figures indicated below.
We also express our thanks to the authors of the material reproduced. Their names, the journals, and the titles of the sources may be found in the bibliographies.
Reference to this note supersedes the usual credit line in order to avoid its hundred fold repetition and it also indicates that authorization can only be obtained from the source cited.

Fig. 1.27: Les Editions de Physique, Orsay
Fig. 2.1: Gordon and Breach, London
Fig. 2.3: Verlag Helvetica Chimica Acta, Fribourg
Fig. 2.4: International Union of Crystallography, Chester
Fig. 2.5: International Union of Crystallography, Chester
Fig. 2.8: VEB Deutscher Verlag für Grundstoffindustrie, Leipzig
Fig. 2.9: Gordon and Breach, London
Fig. 2.10: Les Editions de Physique, Orsay
Fig. 2.12: Gordon and Breach, London
Fig. 2.13: Gordon and Breach, London
Fig. 2.14: Gordon and Breach, London
Fig. 2.15: Gordon and Breach, London
Fig. 3.2: Gordon and Breach, London
Fig. 3.3: Verlag der Zeitschrift für Naturforschung, Oberkochen
Fig. 3.5: The Chemical Society, London
Fig. 3.12: Verlag der Zeitschrift für Naturforschung, Oberkochen
Fig. 3.13: The Chemical Society, London
Fig. 3.15: Indian Academy of Sciences, Bangalore

Fig. 3.16: Les Editions de Physique, Orsay
Fig. 3.17: Brookhaven National Laboratory, Upton
Fig. 3.18: Gordon and Breach, London
Fig. 3.20: Plenum, New York
Fig. 3.21: Les Editions de Physique, Orsay
Fig. 3.22: Les Editions de Physique, Orsay
Fig. 3.23: American Institute of Physics, New York
Fig. 3.25: Gordon and Breach, London
Fig. 4.2:   Gauthier-Villars, Montreuil
Fig. 4.3:   Gauthier-Villars, Montreuil
Fig. 4.4:   Akademische Verlagsgesellschaft, Wiesbaden
Fig. 4.4:   North-Holland, Amsterdam
Fig. 4.6:   Les Editions de Physique, Orsay
Fig. 4.7:   Les Editions de Physique, Orsay
Fig. 4.8:   Plenum, New York
Fig. 4.10: Gordon and Breach, London
Fig. 4.11: Indian Academy of Sciences, Bangalore
Fig. 4.12: American Institute of Physics, New York
Fig. 4.13: Brookhaven National Laboratory, Upton
Fig. 4.14: Pergamon Press, Elmsford
Fig. 4.15: Pergamon Press, Elmsford
Fig. 4.16: North-Holland, Amsterdam
Fig. 4.17: Pergamon Press, Elmsford
Fig. 4.18: Les Editions de Physique, Orsay
Fig. 4.19: Les Editions de Physique, Orsay
Fig. 4.20: Gordon and Breach, London
Fig. 4.21: Gordon and Breach, London
Fig. 4.22: Gauthier-Villars, Montreuil
Fig. 4.23: Gordon and Breach, London
Fig. 4.24: American Institute of Physics, New York
Fig. 4.25: North-Holland, Amsterdam
Fig. 4.27: Verlag der Zeitschrift für Naturforschung, Oberkochen
Fig. 4.28: Gordon and Breach, London
Fig. 4.29: Gordon and Breach, London
Fig. 4.30: Gordon and Breach, London
Fig. 4.31: Gordon and Breach, London
Fig. 4.34: North-Holland, Amsterdam
Fig. 4.35: Institute of Electrical and Electronic Engineers, Piscataway
Fig. 4.36: American Institute of Physics, New York
Fig. 4.37: American Institute of Physics, New York
Fig. 4.38: Verlag der Zeitschrift für Naturforschung, Oberkochen
Fig. 4.39: Gordon and Breach, London
Fig. 4.40: Gordon and Breach, London
Fig. 4.41: Gordon and Breach, London
Fig. 4.42: Centre d'Etudes Nucleaires de Grenoble, Grenoble
Fig. 4.43: Firma AEG-Telefunken, Ulm
Fig. 4.46: Verlag der Zeitschrift für Naturforschung, Oberkochen
Fig. 4.47: Verlag der Zeitschrift für Naturforschung, Oberkochen
Fig. 4.48: The Electrochemical Society, Princeton
Fig. 4.49: Gordon and Breach, London
Fig. 4.50: Gauthier-Villars, Montreuil
Fig. 5.41: American Institute of Physics, New York
Fig. 4.52: Brookhaven National Laboratory, Upton
Fig. 4.54: North-Holland, Amsterdam
Fig. 4.55: Gordon and Breach, London
Fig. 4.56: Gordon and Breach, London

Fig. 4.59: American Institute of Physics, New York
Fig. 4.60: American Institute of Physics, New York
Fig. 4.61: American Institute of Physics, New York
Fig. 4.62: American Institute of Physics, New York
Fig. 4.63: Plenum, New York
Fig. 4.65: Brookhaven National Laboratory, Upton
Fig. 4.67: Physical Society of Japan, Tokyo
Fig. 4.68: Gordon and Breach, London
Fig. 4.69: North-Holland, Amsterdam
Fig. 4.70: Verlag der Zeitschrift für Naturforschung, Oberkochen
Fig. 4.71: American Institute of Physics, New York
Fig. 4.72: Les Editions de Physique, Orsay
Fig. 4.73: Les Editions de Physique, Orsay
Fig. 5.1: The Chemical Society, London
Fig. 5.2: The Chemical Society, London
Fig. 5.3: International Union of Crystallography, Chester
Fig. 5.4: Hüthig & Wepf Verlag, Basel
Fig. 5.5: Hüthig & Wepf Verlag, Basel
Fig. 5.6: Springer Verlag, Berlin
Fig. 5.7: Gordon and Breach, London
Fig. 5.8: Gordon and Breach, London
Fig. 5.9: Gauthier-Villars, Montreuil
Fig. 5.11: Gordon and Breach, London
Fig. 5.12: Gordon and Breach, London
Fig. 5.14: Brookhaven National Laboratory, Upton
Fig. 5.15: Brookhaven National Laboratory, Upton
Fig. 5.16: Gordon and Breach, London
Fig. 5.17: Gordon and Breach, London
Fig. 5.18: Gordon and Breach, London
Fig. 5.19: Gordon and Breach, London
Fig. 5.20: American Institute of Physics, New York
Fig. 5.21: Les Editions de Physique, Orsay
Fig. 5.23: Steinkopff Verlag, Darmstadt
Fig. 5.24: International Union of Crystallography, Chester
Fig. 5.25: Gordon and Breach, London
Fig. 5.26: Gordon and Breach, London
Fig. 5.27: Gordon and Breach, London
Fig. 5.28: Gordon and Breach, London
Fig. 5.29: Gordon and Breach, London
Fig. 5.30: American Institute of Physics, New York
Fig. 5.31: American Institute of Physics, New York
Fig. 5.32: Springer Verlag, Berlin
Fig. 5.33: Gordon and Breach, London
Fig. 5.34: North-Holland, Amsterdam
Fig. 6.1: Hirzel Verlag, Leipzig
Fig. 6.2: Gordon and Breach, London
Fig. 6.3: The Chemical Society, London
Fig. 6.4: The Chemical Society, London
Fig. 6.5: The Chemical Society, London
Fig. 6.6: Gauthier-Villars, Montreuil
Fig. 6.7: Gordon and Breach, London
Fig. 6.9: Gauthier-Villars, Montreuil
Fig. 6.10: Brookhaven National Laboratory, Upton
Fig. 6.11: Les Editions de Physique, Orsay
Fig. 6.12: Gordon and Breach, London
Fig. 6.13: Indian Academy of Sciences, Bangalore

Fig. 6.14: Gordon and Breach, London
Fig. 6.15: Gordon and Breach, London
Fig. 6.16: Gordon and Breach, London
Fig. 6.17: Plenum, New York
Fig. 6.18: American Institute of Physics, New York
Fig. 6.19: Akademische Verlagsgesellschaft, Wiesbaden
Fig. 6.20: Gordon and Breach, London
Fig. 6.21: The Royal Society, London
Fig. 6.22: American Chemical Society, Washington
Fig. 6.23: Gordon and Breach, London
Fig. 6.24: American Institute of Physics, New York
Fig. 6.25: American Institute of Physics, New York
Fig. 6.26: Gordon and Breach, London
Fig. 6.27: Gordon and Breach, London
Fig. 6.28: Gordon and Breach, London
Fig. 6.29: North-Holland, Amsterdam
Fig. 6.30: North-Holland, Amsterdam
Fig. 6.31: The Chemical Society of Japan, Tokyo
Fig. 6.32: Gordon and Breach, London
Fig. 6.33: Springer Verlag, Berlin
Fig. 6.34: American Chemical Society, Washington
Fig. 6.35: American Chemical Society, Washington
Fig. 6.36: Indian Academy of Sciences, Bangalore
Fig. 6.37: Macmillan, London
Fig. 6.38: Heyden & Son, London
Fig. 6.39: Heyden & Son, London
Fig. 6.40: Indian Academy of Sciences, Bangalore
Fig. 6.41: North-Holland, Amsterdam
Fig. 6.42: Les Editions de Physique, Orsay
Fig. 7.7: Gordon and Breach, London
Fig. 7.8: Gordon and Breach, London
Fig. 7.9: American Institute of Physics, New York
Fig. 7.10: Indian Academy of Sciences, Bangalore
Fig. 7.11: Gordon and Breach, London
Fig. 7.12: Gordon and Breach, London
Fig. 7.13: Dokumentationszentrum der Bundeswehr, Bonn
Fig. 7.14: Dokumentationszentrum der Bundeswehr, Bonn
Fig. 7.16: Brookhaven National Laboratory, Upton
Fig. 7.17: Brookhaven National Laboratory, Upton
Fig. 7.18: North-Holland, Amsterdam
Fig. 7.19: Gordon and Breach, London
Fig. 7.20: Gordon and Breach, London
Fig. 7.21: Gordon and Breach, London
Fig. 7.22: American Chemical Society, Washington
Fig. 7.23: Gordon and Breach, London
Fig. 7.24: Gordon and Breach, London
Fig. 7.25: Société Française de Minéralogie et de Cristallographie, Paris
Fig. 7.26: Gordon and Breach, London
Fig. 7.27: North-Holland, Amsterdam
Fig. 7.28: Gordon and Breach, London
Fig. 7.29: Plenum, New York
Fig. 7.31: Gauthier-Villars, Montreuil
Fig. 7.32: North-Holland, Amsterdam
Fig. 7.33: Akademische Verlagsgesellschaft, Wiesbaden
Fig. 7.34: American Chemical Society, Washington
Fig. 7.35: American Chemical Society, Washington

Fig. 7.36: Gordon and Breach, London
Fig. 7.37: North-Holland, Amsterdam
Fig. 7.38: Springer Verlag, Berlin
Fig. 7.41: American Chemical Society, Washington
Fig. 7.43: Akad. Verlagsges. Geest & Portig, Leipzig
Fig. 7.44: Akad. Verlagsges. Geest & Portig, Leipzig
Fig. 7.45: American Institute of Physics, New York
Fig. 7.46: Brookhaven National Laboratory, Upton
Fig. 7.47: American Institute of Physics, New York
Fig. 7.48: Springer Verlag, Berlin
Fig. 7.49: Gordon and Breach, London
Fig. 8.3: American Chemical Society, Washington
Fig. 8.7: Gordon and Breach, London
Fig. 8.8: VEB Deutscher Verlag für Grundstoffindustrie, Leipzig
Fig. 8.9: Akad. Verlagsges. Geest & Portig, Leipzig
Fig. 8.10: Akad. Verlagsges. Geest & Portig, Leipzig
Fig. 8.11: Gordon and Breach, London
Fig. 8.12: American Chemical Society, Washington
Fig. 8.13: Gordon and Breach, London
Fig. 8.14: Gordon and Breach, London
Fig. 8.15: Firma IBM, Yorktown Heights
Fig. 8.16: Gordon and Breach, London
Fig. 8.17: Gordon and Breach, London
Fig. 8.18: Pergamon Press, Elmsford
Fig. 8.19: The American Oil Chemists' Society, Champaign
Fig. 8.20: The American Oil Chemists' Society, Champaign
Fig. 8.22: The Chemical Society of Japan, Tokyo
Fig. 8.23: Gordon and Breach, London
Fig. 8.24: American Chemical Society, Washington
Fig. 8.25: American Chemical Society, Washington
Fig. 8.26: Gordon and Breach, London
Fig. 8.27: Akad. Verlagsges. Geest & Portig, Leipzig
Fig. 8.28: Akad. Verlagsges. Geest & Portig, Leipzig
Fig. 8.29: American Institute of Physics, New York
Fig. 8.30: Les Editions de Physique, Orsay
Fig. 8.31: Les Editions de Physique, Orsay
Fig. 8.32: Les Editions de Physique, Orsay
Fig. 8.33: The Chemical Society, London
Fig. 8.34: Gordon and Breach, London
Fig. 8.35: The Council of Scientific & Industrial Research, New Delhi
Fig. 8.36: The Council of Scientific & Industrial Research, New Delhi
Fig. 8.37: The Council of Scientific & Industrial Research, New Delhi
Fig. 8.38: Society of Chemical Industry, London
Fig. 8.39: Gordon and Breach, London
Fig. 8.40: Gordon and Breach, London
Fig. 8.41: American Chemical Society, Washington
Fig. 8.42: American Chemical Society, Washington
Fig. 8.43: The American Oil Chemists' Society, Champaign
Fig. 8.44: Akad. Verlagsges. Geest & Portig, Leipzig
Fig. 8.45: Gordon and Breach, London
Fig. 8.46: Akad. Verlagsges. Geest & Portig, Leipzig
Fig. 8.47: Akad. Verlagsges. Geest & Portig, Leipzig
Fig. 8.48: Akad. Verlagsges. Geest & Portig, Leipzig
Fig. 8.49: Verlag der Zeitschrift für Naturforschung, Oberkochen
Fig. 8.50: Verlag der Zeitschrift für Naturforschung, Oberkochen
Fig. 8.51: Akad. Verlagsges. Geest & Portig, Leipzig

Fig. 8.52:   Akad. Verlagsges. Geest & Portig, Leipzig
Fig. 8.53:   Les Editions de Physique, Orsay
Fig. 9.3:   American Chemical Society, Washington
Fig. 9.5:   Springer Verlag, Berlin
Fig. 9.7:   Gordon and Breach, London
Fig. 9.8:   Preston, Niles
Fig. 9.9:   Preston, Niles
Fig. 9.10:   Springer Verlag, Berlin
Fig. 9.11:   Elsevier, Amsterdam
Fig. 9.12:   Dekker, New York
Fig. 9.14:   The Chemical Society, London
Fig. 9.15:   Wiley & Sons, New York
Fig. 9.16:   Hüthig & Wepf Verlag, Basel
Fig. 9.17:   Wiley & Sons, New York
Fig. 9.18:   Wiley & Sons, New York
Fig. 9.19:   Gordon and Breach, London
Fig. 9.20:   American Chemical Society, Washington
Fig. 9.21:   Wiley & Sons, New York
Fig. 9.22:   Wiley & Sons, New York
Fig. 11.2:   The Chemical Society, London
Fig. 11.3:   Academic Press, New York
Fig. 11.5:   Macmillan, London
Fig. 11.6:   Gordon and Breach, London
Fig. 11.7:   Gauthier-Villars, Montreuil
Fig. 11.8:   Academic Press, New York
Fig. 11.9:   Gordon and Breach, London
Fig. 11.10:   Gordon and Breach, London
Fig. 11.11:   Gordon and Breach, London
Fig. 11.12:   Gordon and Breach, London
Fig. 11.13:   Gordon and Breach, London
Fig. 11.14:   American Chemical Society, Washington
Fig. 11.15:   American Chemical Society, Washington
Fig. 11.17:   American Chemical Society, Washington
Fig. 11.18:   American Chemical Society, Washington
Fig. 11.19:   Academic Press, New York
Fig. 11.20:   Academic Press, New York
Fig. 11.21:   Macmillan, London
Fig. 11.23:   Acta Chemica Scandinavica, Stockholm
Fig. 11.24:   Steinkopff Verlag, Darmstadt
Fig. 11.25:   Plenum, New York
Fig. 11.26:   Gordon and Breach, London
Fig. 11.27:   Acta Chemica Scandinavica, Stockholm
Fig. 11.29:   Gordon and Breach, London
Fig. 11.30:   Gordon and Breach, London
Fig. 11.31:   Gordon and Breach, London
Fig. 11.32:   Gordon and Breach, London
Fig. 11.33:   Gordon and Breach, London
Fig. 11.34:   Gordon and Breach, London
Fig. 11.35:   Gordon and Breach, London
Fig. 11.36:   Academic Press, New York
Fig. 11.37:   The Chemical Society of Japan, Tokyo
Fig. 11.38:   Gauthier-Villars, Montreuil
Fig. 11.39:   Gordon and Breach, London
Fig. 11.40:   Gordon and Breach, London
Fig. 11.41:   Gordon and Breach, London
Fig. 11.42:   Journal de Chimie Physique, Paris

Fig. 11.43: American Chemical Society, Washington
Fig. 11.44: Academic Press, New York
Fig. 12.1: American Chemical Society, Washington
Fig. 12.2: American Chemical Society, Washington
Fig. 12.3: American Chemical Society, Washington
Fig. 12.4: American Chemical Society, Washington
Fig. 12.5: Pergamon Press, Elmsford
Fig. 12.6: Pergamon Press, Elmsford
Fig. 12.7: Les Editions de Physique, Orsay
Fig. 12.8: American Chemical Society, Washington
Fig. 12.9: Cambridge Univ. Press, Cambridge
Fig. 12.10: Cambridge Univ. Press, Cambridge
Fig. 12.11: The American Oil Chemists' Society, Champaign
Fig. 12.12: Journal of Lipid Research, Bronx
Fig. 12.13: Gordon and Breach, London
Fig. 12.14: Elsevier, Amsterdam
Fig. 12.15: Elsevier, Amsterdam
Fig. 12.16: The American Society of Biological Chemists, Bethesda
Fig. 12.17: Rockefeller Univ. Press, New York
Fig. 12.18: Elsevier, Amsterdam
Fig. 12.19: Plenum, New York
Fig. 12.20: Acta Chemica Scandinavica, Stockholm
Fig. 12.21: Macmillan, London
Fig. 12.22: Elsevier, Amsterdam
Fig. 12.23: The American Oil Chemists' Society, Champaign
Fig. 12.25: Plenum, New York
Fig. 12.29: Elsevier, Amsterdam
Fig. 12.30: Pergamon Press, Elmsford
Fig. 12.33: Plenum, New York
Fig. 12.34: Plenum, New York
Fig. 12.35: The Biochemical Society, London
Fig. 13.1: Hüthig & Wepf Verlag, Basel
Fig. 13.3: Springer Verlag, Berlin
Fig. 13.4: IUPAC, Oxford
Fig. 13.5: Umschau Verlag, Frankfurt/M.
Fig. 13.6: Wiley & Sons, New York